Essentials of Musculoskeletal Imaging

Georges Y. El-Khoury, MD
Professor of Radiology and Orthopaedics
University of Iowa College of Medicine
Iowa City, Iowa

Essentials of Musculoskeletal Imaging

Associate Editors

D. Lee Bennett, MD
Assistant Professor of Radiology
University of Iowa College of Medicine
Iowa City, Iowa

Mark D. Stanley, MD
Associate Professor of Radiology
University of Iowa College of Medicine
Iowa City, Iowa

Figure Editors

John Haller, PhD
Photography Lab
University of Iowa College of Medicine
Iowa City, Iowa

James Olsen
Photography Lab
University of Iowa College of Medicine
Iowa City, Iowa

An Imprint of Elsevier Science
New York Edinburgh London Philadelphia

An Imprint of Elsevier Science

The Curtis Center
Independence Square West
Philadelphia, PA 19106

ESSENTIALS OF MUSCULOSKELETAL IMAGING ISBN 0–443–06575–6

Distributed in the United Kingdom by Churchill Livingstone, Robert Stevenson House, 1-3 Baxter's Place, Leith Walk, Edinburgh EH1 3AF, and by associated companies, branches, and representatives throughout the world.

Notice

Medicine is an ever-changing field. Standard safety precautions must be followed, but as new research and clinical experience broaden our knowledge, changes in treatment and drug therapy may become necessary or appropriate. Readers are advised to check the most current product information provided by the manufacturer of each drug to be administered to verify the recommended dose, the method and duration of administration, and contraindications. It is the responsibility of the treating physician, relying on experience and knowledge of the patient, to determine dosages and the best treatment for each individual patient. Neither the Publisher nor the Editor assumes any liability for any injury and/or damage to persons or property arising from this publication.

The Publisher

Library of Congress Cataloging-in-Publication Data

Essentials of musculoskeletal imaging / [edited by] Georges Y. El-Khoury.
p. ; cm.
ISBN 0-443-06575-6
1. Musculoskeletal system—Imaging. I. El-Khoury, Georges Y.
[DNLM: 1. Diagnostic Imaging. 2. Musculoskeletal System—pathology. WE 141 E78 2002]
RC925.7 .E84 2002
616.7′0754—dc21 2002019287

Publishing Director: Richard Lampert
Acquisitions Editor: Stephanie Smith Donley
Project Manager: Tina Rebane
Book Designer: Ellen B. Zanolle
Indexer: Angela Holt

GW/MVY

Printed in the United States of America.

Last digit is the print number: 9 8 7 6 5 4 3 2 1

I dedicate this book to my wife and children
who gave me the time and freedom
to indulge in this gigantic undertaking.

Contributors

Carol J. Ashman, MD
Riverside Radiological Associates, Columbus, Ohio

E. Michel Azouz, MD
Professor of Clinical Radiology, Pediatric Radiology Section, University of Miami, Miami, Florida

Lori L. Barr, MD
Austin Radiological Association, Austin, Texas

D. Lee Bennett, MD
Assistant Professor, Department of Radiology, The University of Iowa, Iowa City, Iowa

Thomas D. Berg, MD
Clinical Lecturer, Department of Radiology, The University of Iowa, Iowa City, Iowa

Carol A. Boles, MD
Assistant Professor, Department of Radiology, Wake Forest University, Winston-Salem, North Carolina

Nathalie J. Bureau, MD, FRCPC
Associate Professor of Radiology, Université de Montréal, CHUM, Hôpital Saint Luc, Montreal, Quebec, Canada

John J. Callaghan, MD
Lawrence and Marilyn Dorr Chair and Professor, Department of Orthopaedic Surgery, The University of Iowa, Iowa City, Iowa

Étienne Cardinal, MD, FRCPC
Associate Professor of Radiology and Director, Musculoskeletal Section, Université de Montréal, CHUM, Hôpital Saint Luc, Montréal, Québec, Canada

Robert H. Choplin, MD
Professor of Clinical Radiology, Indiana University, Indianapolis, Indiana

Joseph G. Craig, MD
Staff Radiologist, Henry Ford Hospital, Department of Diagnostic Radiology, Division of Musculoskeletal and Emergency Radiology, Detroit, Michigan

Shigeru Ehara, MD
Professor and Chair, Department of Radiology, Iwate Medical University School of Medicine, Iwate, Japan

Georges Y. El-Khoury, MD, FACR
Professor of Radiology and Orthopaedics and Director, Musculoskeletal Radiology, Department of Radiology, The University of Iowa, Iowa City, Iowa

Shella Farooki, MD
Riverside Radiological Associates, Columbus, Ohio

Thomas J. Gilbert, Jr., MD
Clinical Associate Professor of Radiology, Center for Diagnostic Imaging and The University of Minnesota Hospital and Clinics, Minneapolis, Minnesota

M. Patricia Harty, MD
Clinical Associate Professor of Radiology, Thomas Jefferson University, and Medical Imaging Department, A. I. DuPont Hospital for Children, Wilmington, Delaware

Simon C. Kao, MD
Professor, Department of Radiology, The University of Iowa, Iowa City, Iowa

Nabil J. Khoury, MD
Assistant Professor of Radiology, American University of Beirut Medical Center, Beirut, Lebanon

Mitchell Kline, MD
Department of Radiology, University of Louisville, Louisville, Kentucky

Jonathan A. Lee, BS
The Ohio State University Medical Center, Columbus, Ohio

Leon Lenchik, MD
Assistant Professor, Department of Radiology, Wake Forest University, Winston-Salem, North Carolina

John Ly, MB, BS
Consultant Radiologist, Canberra Imaging Group, Canberra, ACT, Australia

Elias R. Melhem, MD
Associate Professor of Radiology and Neurosurgery and Director of Neuroradiology, Hospital of the University of Pennsylvania, Philadelphia, Pennsylvania

Timothy E. Moore, MD
Professor of Radiology, Department of Radiology, University of Nebraska Medical Center, Omaha, Nebraska

Donald L. Renfrew, MD
Center for Diagnostic Imaging, Winter Park, Florida

Bernard Roger, MD
Service de Radiologie, Unité Osteo-Articulaire, Hôpital Pitie–Salpetrière, Paris, France

Mark D. Stanley, MD
Associate Professor, Department of Radiology, The University of Iowa, Iowa City, Iowa

Murali Sundaram, MB, BS, FRCR
Professor of Radiology, Mayo Medical School, Rochester, Minnesota

Craig W. Walker, MD
Professor and Chair, Department of Radiology, University of Nebraska Medical Center, Omaha, Nebraska

John B. Weigele, MD, PhD
Assistant Professor, Department of Radiology, Division of Neuroradiology, University of Pennsylvania Hospital, Philadelphia, Pennsylvania

Joseph S. Yu, MD
Associate Professor of Radiology and Chief of Musculoskeletal Radiology, The Ohio State University Medical Center, Columbus, Ohio

Preface

In the imaging subspecialties, as in other fields of medicine, there is very little time and too much to learn. In the past decade, radiology has witnessed tremendous technical advances. Musculoskeletal imaging has been one of the prime beneficiaries of these advances. The knowledge base in this field has literally multiplied. It is no longer sufficient to stop at learning the radiologic signs of normal and abnormal states or to master a few diagnostic and interventional procedures to guarantee success in radiology. For example, a skilled musculoskeletal radiologist is above all an expert anatomist; he or she is also someone who is well versed in musculoskeletal pathology, orthopedics, rheumatology, oncology, and metabolic disease, as well as congenital and developmental musculoskeletal disorders.

As an educator, I have been faced with the ever-increasing challenge of imparting huge amounts of information to our trainees and teaching them manual skills in a thorough but condensed fashion and to fit it all in the limited time designated for training. Practicing radiologists, orthopedists, and rheumatologists are also faced with the problem of coping with the information overload generated in musculoskeletal imaging.

It is my hope that this book will mitigate some of the difficulties we are all facing in learning about musculoskeletal imaging. This book covers "the essentials of musculoskeletal imaging." It is intended to be comprehensive, yet concise.

The text covers most of the classical topics in musculoskeletal imaging, but also included are sections on the major pediatric musculoskeletal disorders, spinal diseases, magnetic resonance imaging of the major joints, and musculoskeletal ultrasound. Lately, interventional musculoskeletal procedures have flourished, so a section dedicated to these procedures has been included.

This book is intended for senior radiology residents, musculoskeletal fellows, and practicing radiologists. It is also suitable for orthopedic surgeons, who over the years seem to have coped with and adapted to the new developments in musculoskeletal imaging.

Georges Y. El-Khoury, MD

Acknowledgments

I would like to acknowledge the help of my associate editors and the contributing authors who worked closely with me in creating this book. I also would like to acknowledge the expert help of Mr. James E. Olson and Dr. John W. Haller in producing outstanding images. I am very grateful to my secretary, Mrs. Mary McBride, who spent countless hours typing hundreds of pages for this book. Many of my friends around the world have contributed images to the book, and I am grateful for their help. Finally, I would like to thank my department at the University of Iowa for an environment that fosters scholarship and creativity.

Contents

SECTION I

NEOPLASM

CHAPTER 1

Logical Approach to the Evaluation of Solitary Bone Lesions

Georges Y. El-Khoury, MD, and Murali Sundaram, MD

OVERVIEW

The spectrum of lesions that constitute a solitary bone lesion is wide and varied. They include solitary benign and malignant bone tumors; infection; solitary metastasis; and tumor mimics, such as stress fractures, postradiation changes, and myositis ossificans. The role of the radiologist is to detect the lesion and provide either a specific diagnosis or a limited cohesive differential diagnosis. The radiologist also is required to stage the disease, guide the tissue sampling (biopsy), and monitor therapeutic response. Depending on the differential diagnosis, the management can take one of four paths:

1. Ignore the lesion (e.g., nonossifying fibroma, bone island).
2. Recommend a follow-up examination (e.g., enchondroma, fibrous dysplasia).
3. Perform a biopsy of the lesion without staging it (e.g., intracompartmental lesions, such as an aneurysmal bone cyst or Brodie's abscess).
4. Perform further staging for
   a. Aggressive benign tumors (e.g., giant cell tumor).
   b. Lesions suspicious for metastasis.
   c. Lesions suspected of being primary sarcomas of bone.

IMAGING

Radiography

The radiograph is the most reliable predictor of the histology of a bone lesion. It remains the mainstay in guiding subsequent management.

Radionuclide Bone Scanning (Nuclear Medicine)

Scintigraphy plays a crucial role in confirming or negating that a radiographically identified solitary bone lesion is truly solitary. Radionuclide bone scans can be falsely negative in the presence of plasmacytoma, multiple myeloma, eosinophilic granuloma, and aggressive osteolytic lesions. A negative or mildly positive bone scan in the presence of a sclerotic lesion favors a benign process (Fig. 1–1).

Computed Tomography

Computed tomography (CT) is the preferred examination for determining the presence or absence of a nidus when an osteoid osteoma is suspected. CT is ideal for differentiating myositis ossificans from a surface osteosarcoma. In the setting of an abnormal bone scan of the flat bones, CT is useful in showing a lesion that is not well seen by radiography. CT scanning assists in the characterization of lesions by depicting the tumor matrix better than other modalities do.

Magnetic Resonance Imaging

Magnetic resonance (MR) imaging is the examination of choice for staging all suspected sarcomas and aggressive benign bone lesions with extracompartmental extension. MR imaging should not be used for lesions that do not require biopsy based on radiographic features, such as enchondromas, fibrous dysplasia, nonossifying fibromas, and simple bone cysts. MR imaging occasionally can assist in arriving at a specific diagnosis. This is particularly true for the diagnosis of intraosseous lipoma, hemangioma, and ganglion cyst. MR imaging produces misleading findings with some benign lesions, such as osteoid osteoma, osteoblastoma, chondroblastoma, eosinophilic granuloma, and myositis ossificans (in its early stages). MR imaging is relatively insensitive for the detection and characterization of cortical destruction, tumor matrix, and periosteal reaction.

APPROACH TO DIAGNOSIS

When a bone lesion is identified, it is advisable to use a systematic approach in formulating a reasonable differential diagnosis or suggesting a specific diagnosis. The radiologist should start by questioning whether the lesion is truly a neoplasm or not. There are many normal variants (*touch me not* lesions) and benign processes that can mimic a neoplastic process. After recognizing that a lesion is neoplastic, going systematically through a checklist can help narrow the differential diagnosis. This checklist includes the age of the patient, the location of the lesion, the biologic activity (rate of growth) of the lesion, the type

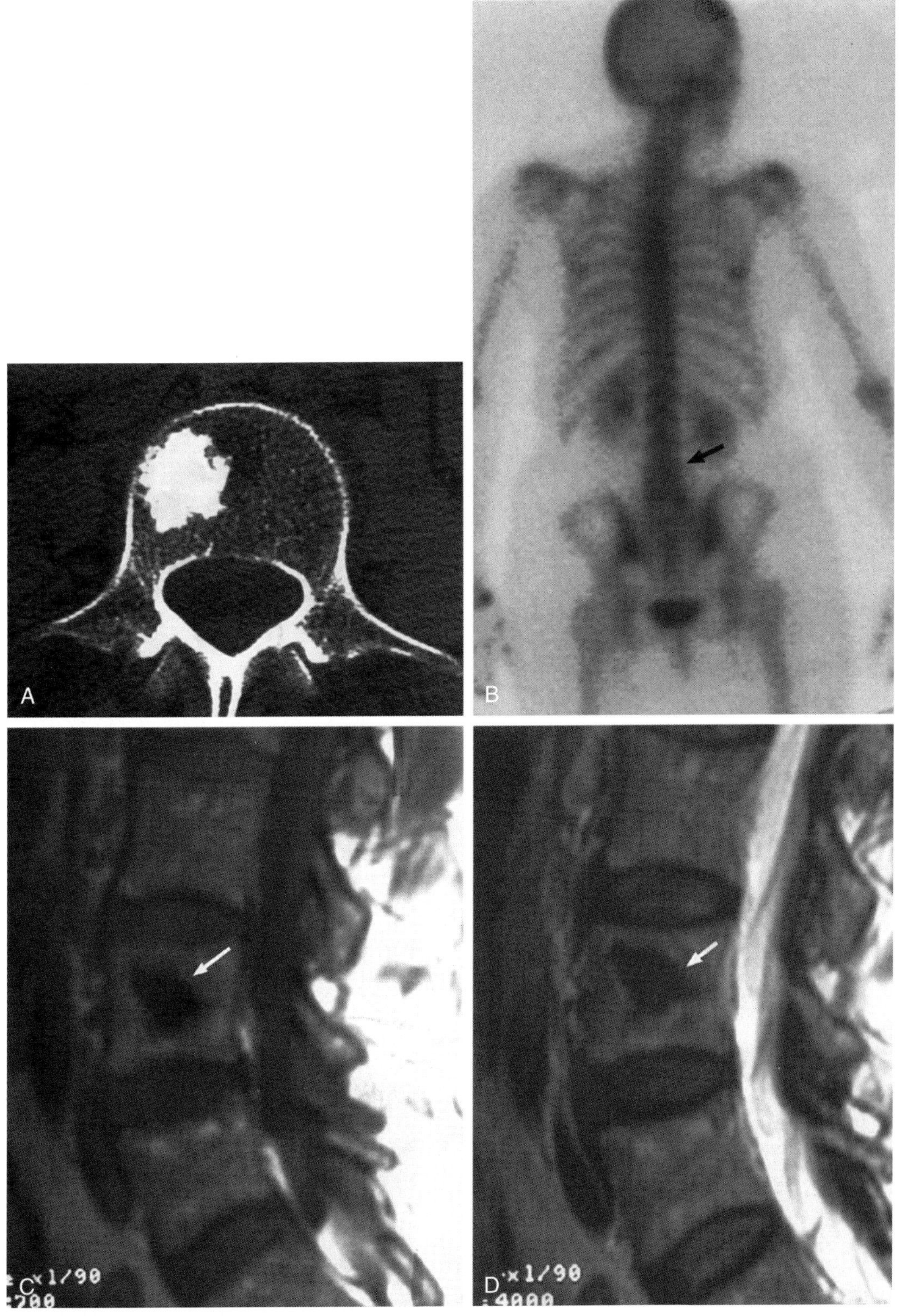

Figure 1–1. Minimal increased uptake in L4 on bone scan in a 62-year-old patient. *A,* Axial CT section shows a 2 × 2 cm irregular blastic density in the medullary space. Blastic metastasis was considered. *B,* Technetium 99m medronate bone scan shows minimally increased uptake in the L4 region *(arrow)*. *C* and *D,* T1- and T2-weighted MR images show an area of decreased signal intensity in L4 that corresponded to the lesion on CT *(arrows)*. The lesion was thought to represent a bone island. On a follow-up examination, the lesion was the same size.

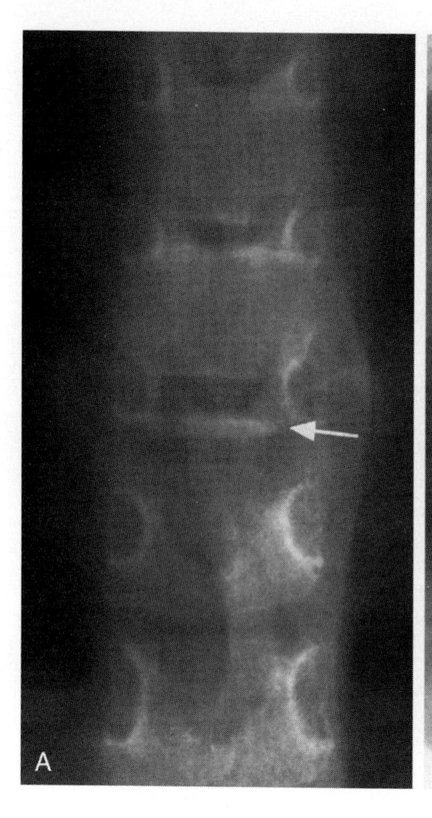

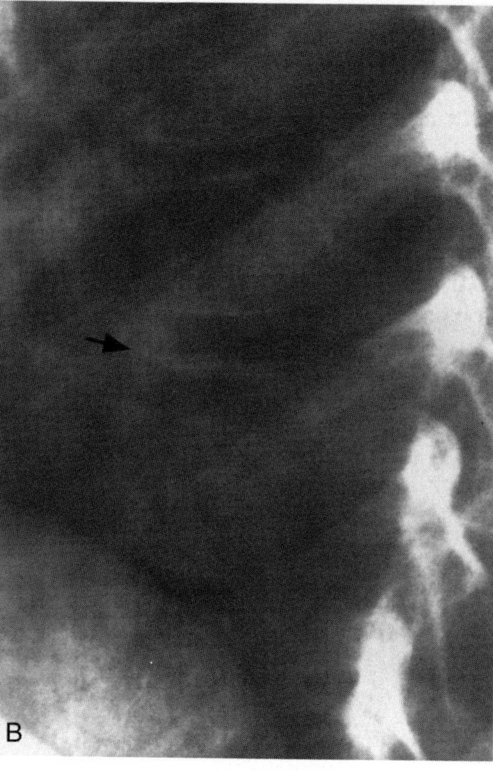

Figure 1–2. Vertebra plana in a 6-year-old child presenting with back pain. *A* and *B,* Anteroposterior and lateral views of the thoracic spine show vertebra plana *(arrows)* of the T8 vertebra associated with a paravertebral mass. At biopsy, the lesion proved to be an eosinophilic granuloma.

of matrix within the lesion, and the type of periosteal reaction.

Age

The age of the patient often is the most important clinical information used in the interpretation of bone tumors. Most primary bone tumors occur between the teens and 20s. A lytic lesion in a patient older than age 50 should raise the possibility of metastasis or multiple myeloma. A lytic lesion in a child should raise the possibility of Langerhans cell histiocytosis (eosinophilic granuloma). Sex and race occasionally may assist in arriving at a diagnosis. Giant cell tumors are more common in females, whereas osteoid osteomas and osteosarcomas are slightly more prevalent in males. Race is rarely important except in Ewing's sarcoma, which is rare in blacks.

Location of the Lesion

The location of the lesion provides a good indication as to what the lesion might be. The most frequent sites for metastasis or multiple myeloma are the spine, pelvis, and ribs. It is rare for metastasis to occur distal to the knee or elbow. Lesions in the clavicle or mandible are rarely metastatic. Primary tumors of bone, such as Ewing's sarcoma and chondrosarcoma, often occur in the pelvis. The ribs are a favored site for Ewing's sarcoma, chondrosarcoma, and fibrous dysplasia. Osteoblastoma has a predilection for occurring in the posterior elements of the vertebrae. A destructive lesion in the vertebral body of a child, causing collapse, is likely an eosinophilic granuloma (Fig. 1–2). An eccentric, lytic lesion in the subchondral bone at the knee is usually a giant cell tumor (Fig. 1–3). Lesions of the sternum almost always are malignant, and metastasis is the most common tumor of the sternum; the sources are the breast, lung, kidney, and thyroid. Primary malignant tumors do occur in the sternum with the most frequent, in descending order, being chondrosarcoma,

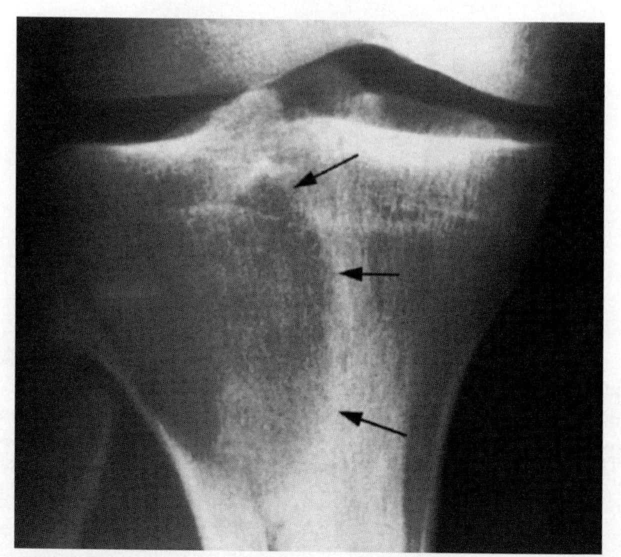

Figure 1–3. Typical location for giant cell tumor. Anteroposterior view of the knee with a lytic, eccentric lesion *(arrows)* reaching the subchondral bone.

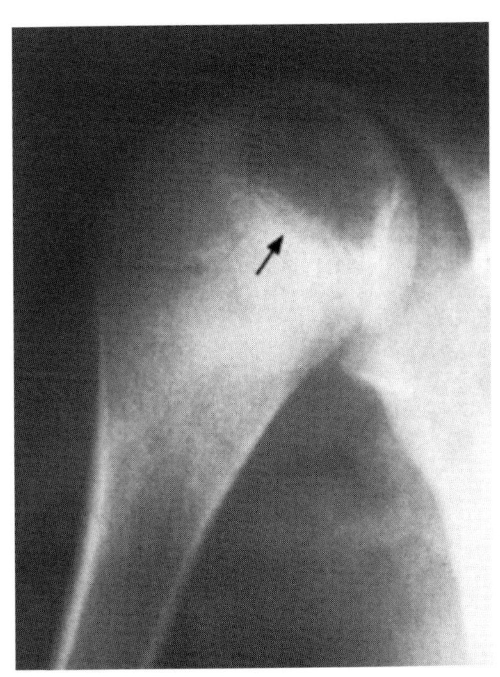

Figure 1–4. A lytic lesion in the epiphysis of a 14-year-old boy that proved to be a chondroblastoma at surgery. Anteroposterior view of the shoulder shows a well-circumscribed, lytic lesion in the proximal epiphysis of the humerus *(arrow)*. This appearance is typical for a chondroblastoma.

myeloma, lymphoma, and osteosarcoma. Most (90%) patellar tumors are benign, with the most common lesions being chondroblastoma followed by giant cell tumor. The radiographic features of benign versus malignant neoplasms in the patella are difficult to distinguish. A lytic lesion in the epiphysis of a skeletally immature individual is most likely a chondroblastoma (Fig. 1–4). It also is important to recognize if a lesion is central or eccentric. Giant cell tumor and nonossifying fibroma almost always are eccentric. Finally, it often is helpful to determine whether the lesion originated from the medullary space, cortex, or surface of the bone.

Biologic Activity of the Lesion

Radiographic evaluation of the rate of growth or biologic activity of a lesion can suggest whether the lesion is benign or malignant. Biologic activity can be assessed accurately based on the margin of the lesion (transitional zone) and pattern of bone destruction. A well-demarcated lesion with or without a sclerotic margin has a narrow transitional zone and is unlikely to be aggressive. A lesion with ill-defined margins has a wide zone of transition and is most likely aggressive. Lodwick and colleagues (1980a, b) assigned patterns of bone destruction into three grades: I, II, and III. Grade I is divided into three subgroups (A, B, and C). Grade I lesions have a predominantly geographic pattern of destruction. In grade IA, the lesion has a thick, well-defined sclerotic margin and a distinct separation between normal and abnormal tissue (Fig. 1–5). Such lesions are thought to have a narrow transitional zone. In most cases, such lesions are slow growing or stable, and they are considered benign. Grade IC lesions are usually expansile. They do not have a sclerotic margin, and the separation between the lesion and healthy tissue can be demarcated but not distinctly. Such lesions are thought to have a wide transitional zone (Fig. 1–6). Grade IB is between grade IA and grade IC in appearance. For the most part, grade IC lesions are considered aggressive. Grade II lesions have a moth-eaten pattern of bone destruction, and grade III lesions have a permeative pattern of bone destruction (Fig. 1–7). Grade II and grade III lesions have a wide transitional zone between normal tissue and tumor. When this scheme was tested for reliability, readers could not differentiate consistently between moth-eaten and permeative patterns of destruction, but this distinction in the clinical setting is not important because all lesions graded as IC or above are considered aggressive and are likely malignant.

Matrix

The matrix is the acellular substance produced by the mesenchymal cells in the tumor. Absence of a radiographically discernible matrix limits the ability of the radiologist to determine a specific diagnosis. Matrices produced by tumors include osteoid, chondroid, myxoid, and collagen. The presence of chondroid or osteoid matrices can be detected easily on radiographs. A chondroid matrix shows dotlike, ringlike, C-shaped, and popcorn-like calcifications; the presence of such a matrix signifies the cartilaginous nature of the lesion (Fig. 1–8). Osteoid matrix has

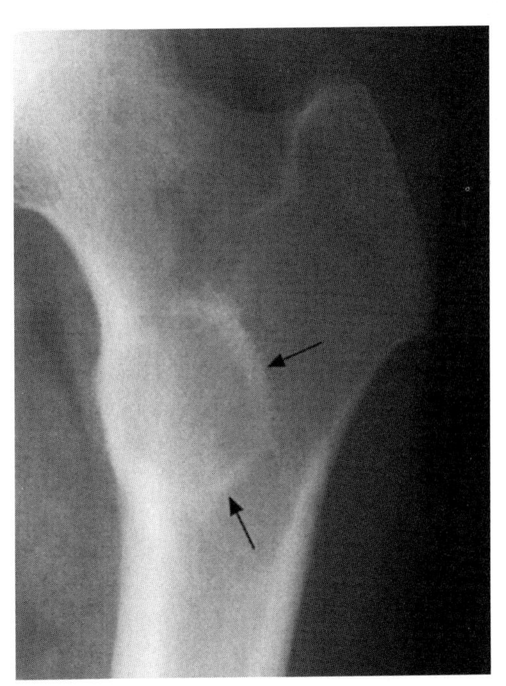

Figure 1–5. Grade IA geographic lesion in a patient with fibrous dysplasia. (Some authorities in the field have named this lesion liposclerosing myxofibrous tumor.) Anteroposterior view of the left hip shows a geographic lytic lesion (fibrous dysplasia) with a thick, well-defined, sclerotic margin *(arrows)* consistent with a grade IA lesion. This is also an example of a thick, sclerotic border (rind sign) that can be seen with fibrous dysplasia.

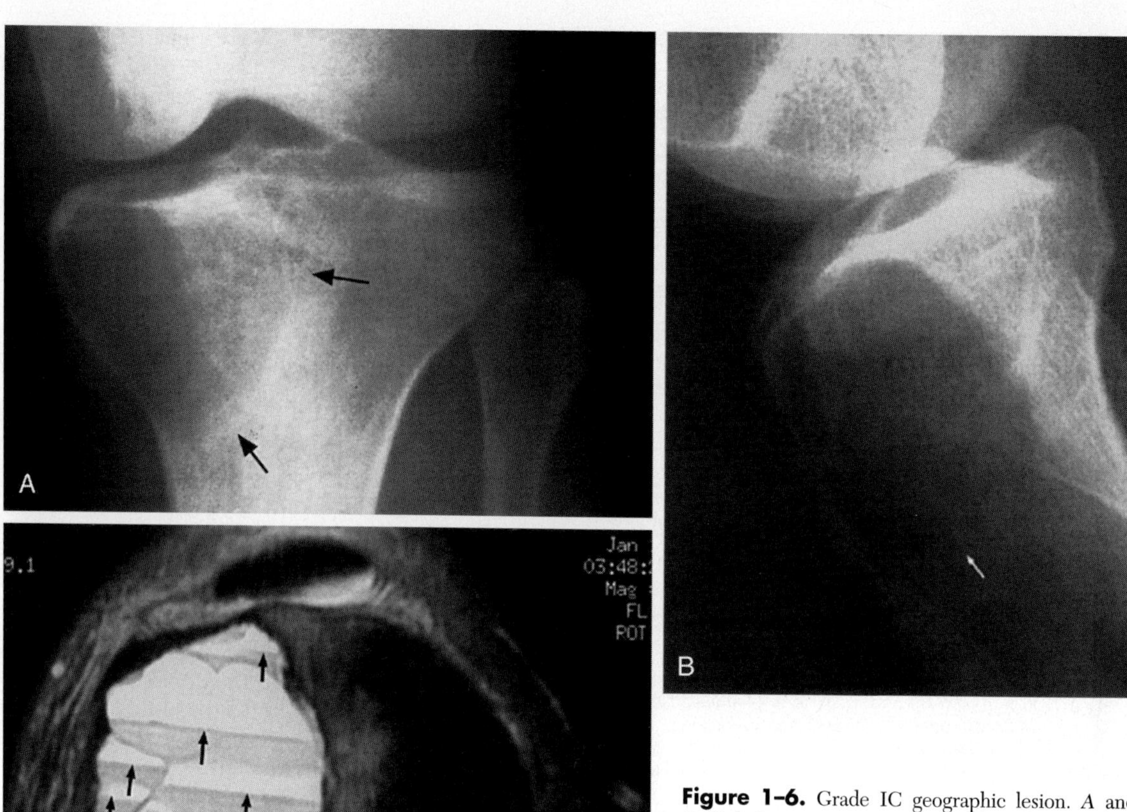

Figure 1–6. Grade IC geographic lesion. *A* and *B*, Right knee anteroposterior and lateral radiographs of an aneurysmal bone cyst show a grade IC lesion. The lesion is mildly expansile and does not have a sclerotic margin. Most of the margin is well seen *(black arrows)*; however, portions of the margin are not readily distinct *(white arrow)*. *C*, Axial fat-saturation FSE T2-weighted MR image of the right knee reveals multiple fluid-fluid levels *(arrows)* that can be seen with an aneurysmal bone cyst.

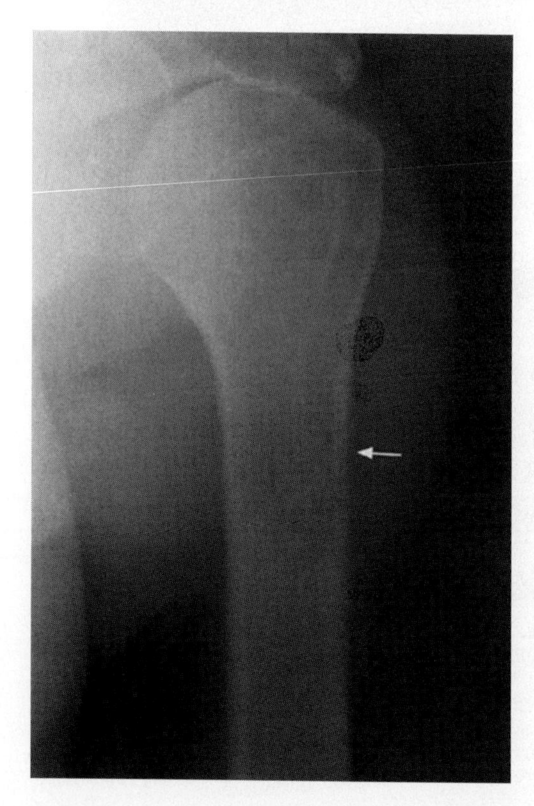

Figure 1–7. Grade III lytic lesion. Anteroposterior view of the shoulder shows a permeative pattern of bone destruction within the proximal shaft of the humerus *(arrow)*. Needle biopsy was performed and the lesion proved to be a metastatic deposit from a lung carcinoma.

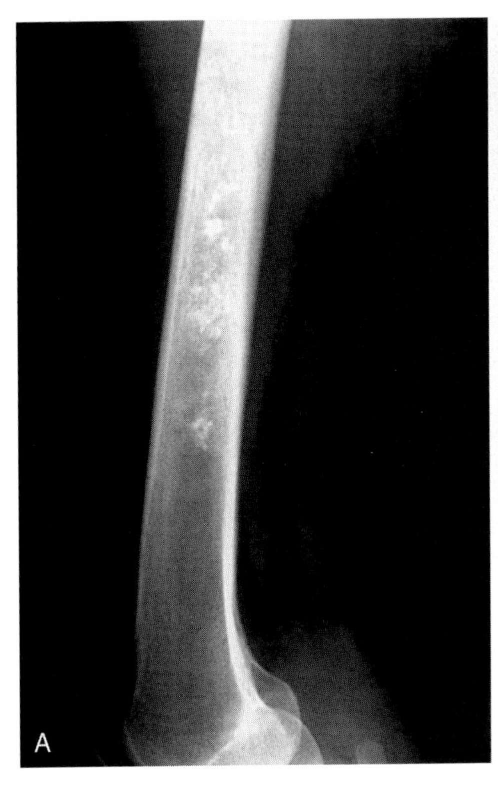

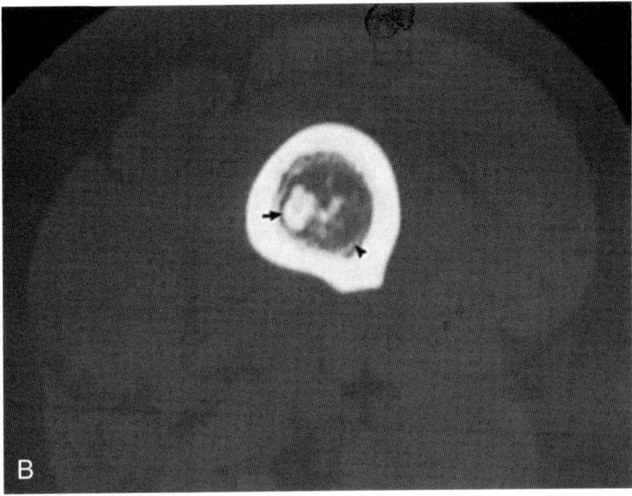

Figure 1–8. An enchondroma containing a chondroid matrix. *A*, Lateral radiograph of the left femur is shown. The medullary cavity contains a region of dot, ring, arc, and popcorn-like calcifications consistent with a chondroid matrix. *B*, CT scan (bone window) of the left midfemur also shows the typical findings of a chondroid matrix, popcorn-like *(arrow)* and arc-shaped *(arrowhead)* calcifications, which can be seen within the medullary cavity.

a confluent, cloudlike or cotton-like appearance, which can be faint or dense (Fig. 1–9). This type of matrix is found in lesions producing bone, such as osteoid osteoma, osteoblastoma, or osteosarcoma. Myositis ossificans and fracture callus also contain osteoid.

Periosteal Reaction

Current thinking pertaining to periosteal reactions is shaped by the classic work of Edeiken and colleagues (1966). Essentially, there are two types of periosteal reaction: solid (uninterrupted) and interrupted (Fig. 1–10). A solid, uninterrupted periosteal reaction almost always is associated with slow-growing benign lesions. An interrupted periosteal reaction signifies an aggressive lesion, and with few exceptions (e.g., fulminant infection or a fast-growing aneurysmal bone cyst), it almost always is associated with malignant tumors. Under these two types of periosteal reaction, there are several subtypes. Subtypes of solid periosteal reaction include thin straight, thin undulating, dense undulating, dense elliptical, and cloaking. Subtypes of interrupted periosteal reaction include lamellated or onion skin, sunburst with perpendicular spikes, and amorphous. The lamellated periosteal reaction produces Codman's triangle at the site where the bone lesion breaks through the newly formed periosteal layers (Fig. 1–10*B*). This appearance is not specific for malignant tumors in that it also can be seen with infection and aneurysmal bone cysts. The sunburst and amorphous periosteal reaction always are associated with malignant lesions.

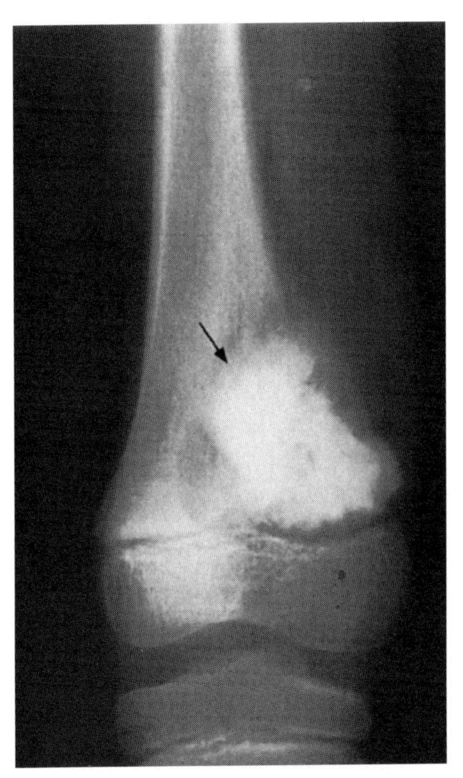

Figure 1–9. Osteoid matrix in a lesion in the distal metaphysis of the right femur in an 8-year-old child with multicentric osteosarcoma. Anteroposterior view of the right femur shows a destructive lytic and blastic lesion in the distal metaphysis of the right femur. The blastic component appears dense and cloudlike or cotton-like *(arrow)*, which is typical for an osteoid-forming lesion.

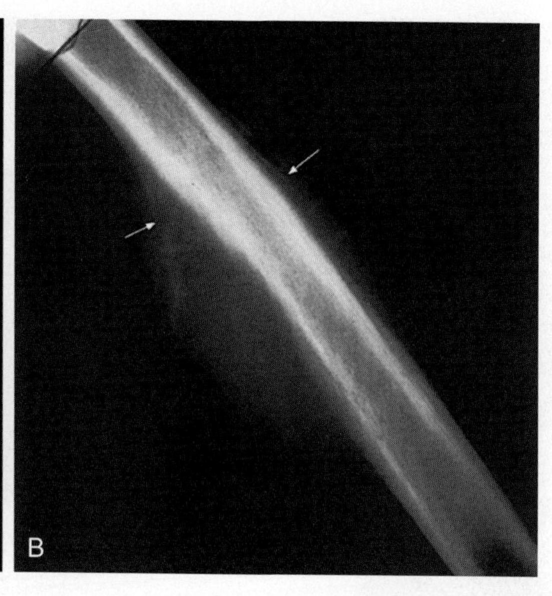

Figure 1–10. Solid periosteal reaction versus interrupted periosteal reaction. *A*, Solid, uninterrupted periosteal reaction in a 12-year-old child with eosinophilic granuloma. Anteroposterior view of the left humerus shows a lytic lesion in the shaft. The lesion is associated with a solid, uninterrupted periosteal reaction *(arrows)*. *B*, Classic Codman's triangle *(arrows)* with an interrupted periosteal reaction in a 17-year-old patient with Ewing's sarcoma.

CHARACTERISTIC RADIOGRAPHIC SIGNS

Some radiographic signs are often diagnostic of certain lesions. These are useful signs, and radiologists rely on them in suggesting a specific diagnosis.

Ivory Vertebra. An ivory vertebra suggests blastic metastasis (such as from prostate cancer), Hodgkin's disease, or Paget's disease (Fig. 1–11).

Picture Frame Pattern. A picture frame pattern in a vertebra suggests involvement of the vertebral body with Paget's disease (Fig. 1–12).

Corduroy Pattern. A corduroy pattern in a vertebral body suggests the presence of a hemangioma (Fig. 1–13).

Rind Sign. The rind sign (a dense, thick margin to a lesion) typically is seen with foci of fibrous dysplasia, especially in the femoral neck (see Fig. 1–5). A similar appearance is seen with an intraosseous lipoma of the femoral neck.

Ground-Glass Appearance. A ground-glass appearance typically is seen with fibrous dysplasia (Fig. 1–14).

Fallen Fragment Sign. A fallen fragment sign almost always indicates that a lesion is a simple bone cyst that has sustained a pathologic fracture (Fig. 1–15).

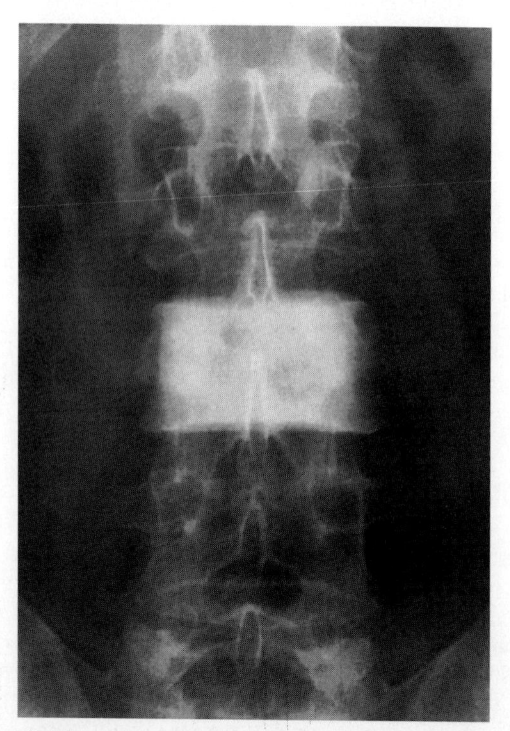

Figure 1–11. Ivory vertebra in a 65-year-old man with metastatic prostate cancer. Anteroposterior view of the lumbar spine shows a densely sclerotic L3 vertebral body.

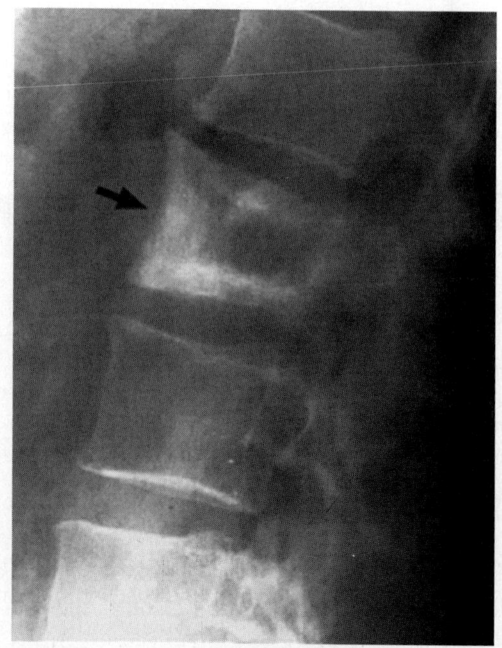

Figure 1–12. Picture frame vertebral body in Paget's disease. Lateral radiograph of the lumbar spine shows cortical thickening of a vertebral body *(arrow)* consistent with a classic picture frame vertebral body.

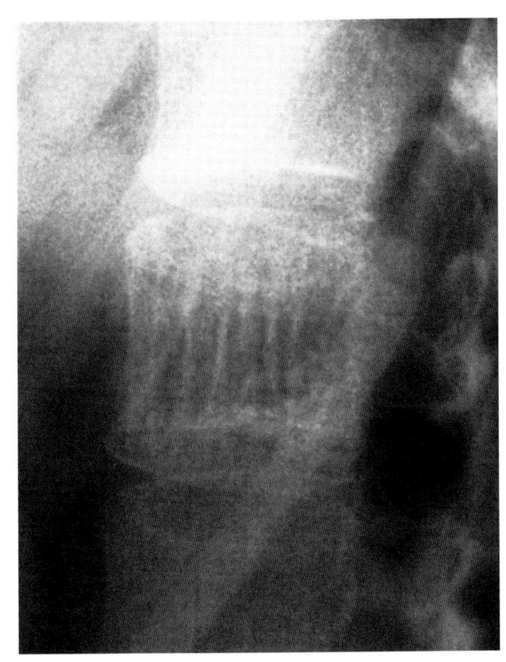

Figure 1–13. Corduroy pattern within a vertebral body. Lateral radiograph of the lumbar spine shows coarse thick trabeculae consistent with a vertebral body hemangioma.

Gas. Gas within a periarticular lucent lesion almost always indicates the presence of a degenerative cyst communicating with a joint (Fig. 1–16).

Fluid-Fluid Level. When a lesion containing a fluid-fluid level is seen on MR imaging, it is a nonspecific finding (see Fig. 1–6*C*). When a fluid-fluid level is present, the radiologist should think of an aneurysmal bone cyst, telangiectatic osteosarcoma, simple bone cyst with bleeding, giant cell tumor, fibrous dysplasia with cystic degeneration, and metastasis. When a lesion with a fluid-fluid level is seen on CT scan, it has a high likelihood of being an aneurysmal bone cyst (sensitivity, 87.5%; specificity, 99.7%).

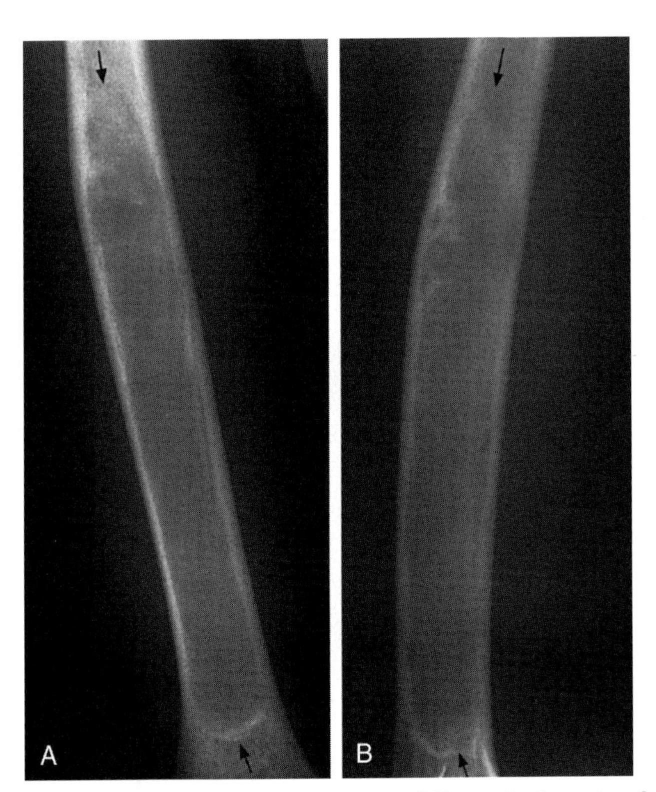

Figure 1–14. Ground-glass appearance of fibrous dysplasia. *A* and *B*, Anteroposterior and lateral radiographs of the distal right humerus show a case of fibrous dysplasia with a hazy or ground-glass appearance within the medullary cavity *(arrows)*.

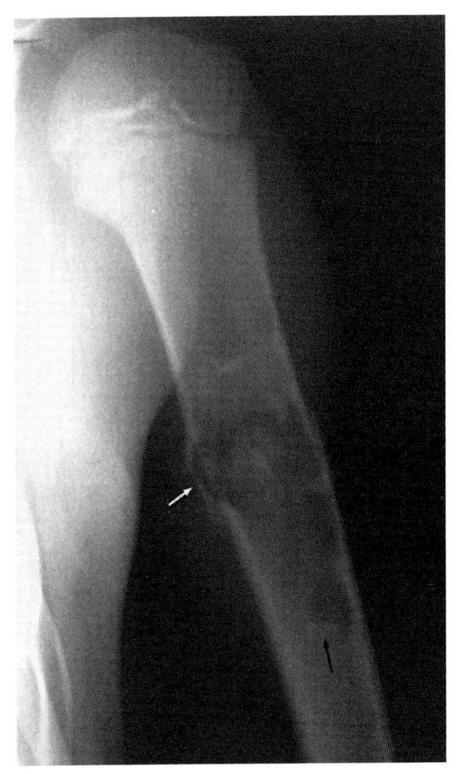

Figure 1–15. Fallen fragment sign. A simple bone cyst with a pathologic fracture *(white arrow)* and fallen fragment sign can be seen on this anteroposterior radiograph of the proximal humerus. The fallen fragment can be seen at the distal margin of the simple bone cyst *(black arrow)*.

STAGING

The approach to staging has been discussed previously. Local staging should precede a biopsy. Staging performed after a biopsy can be misleading because of edema and hemorrhage. All sarcomas should be studied for distant metastasis, including a chest CT scan to evaluate for the presence of pulmonary metastasis.

BIOPSY

Tissue sampling or biopsy is part of the staging process. The incision for the biopsy or needle track must be planned carefully to allow for eventual en bloc resection of the malignant neoplasm with the biopsy incision or needle track. Because a large tissue sample is required for thorough pathologic evaluation, most tumor surgeons prefer open biopsy for a primary malignant neoplasm. If chondrosarcoma is suspected, a percutaneous biopsy is contraindicated.

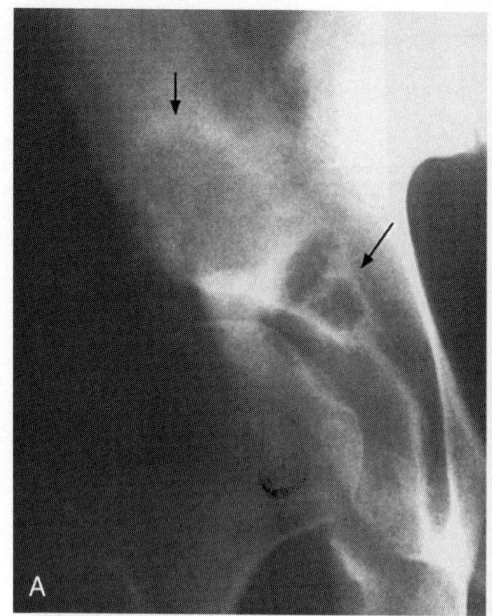

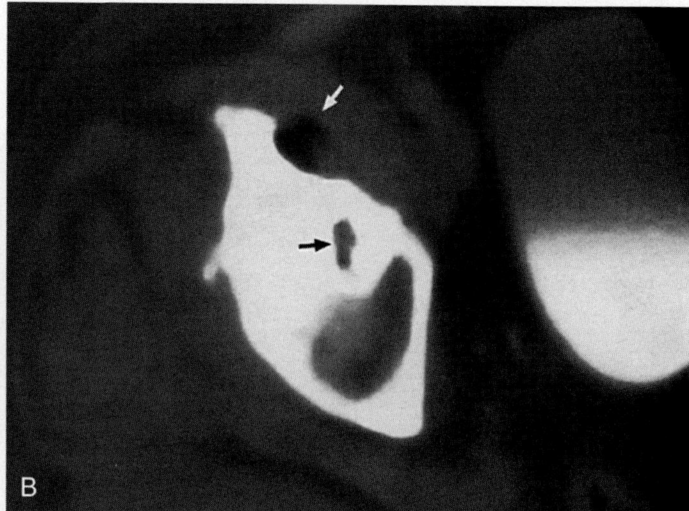

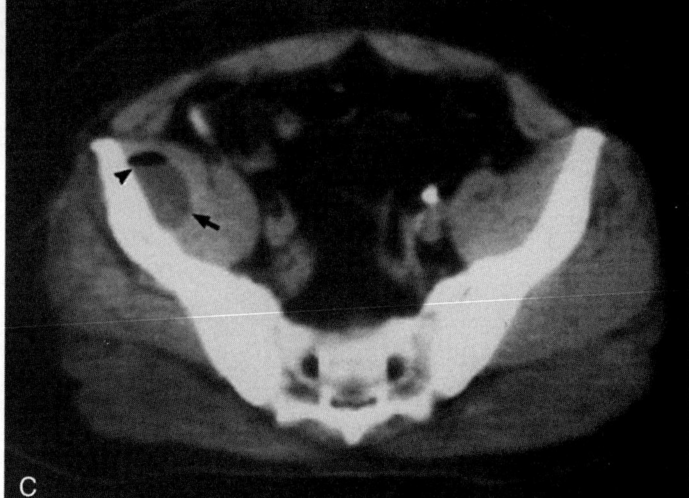

Figure 1–16. Periarticular gas containing lucent lesion proved at surgery to be a degenerative cyst. *A,* Anteroposterior view of the right hip shows mild acetabular dysplasia and a cystic lesion in the acetabular roof *(arrows). B,* CT section through the acetabular roof shows a defect in the acetabular roof *(black arrow)* and an air collection adjacent to the defect *(white arrow). C,* CT section through the iliac bone shows the cystic lesion *(arrow),* which contains an air-fluid level *(arrowhead).* On subsequent lower sections, this cyst communicates with the lytic lesion in the acetabular roof.

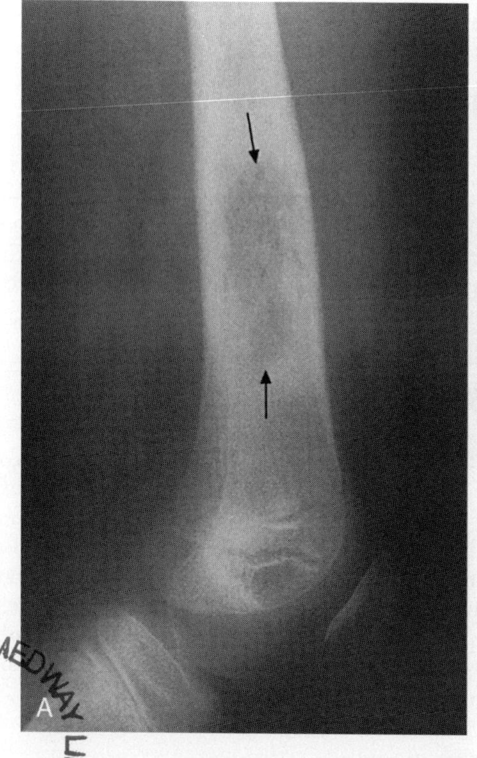

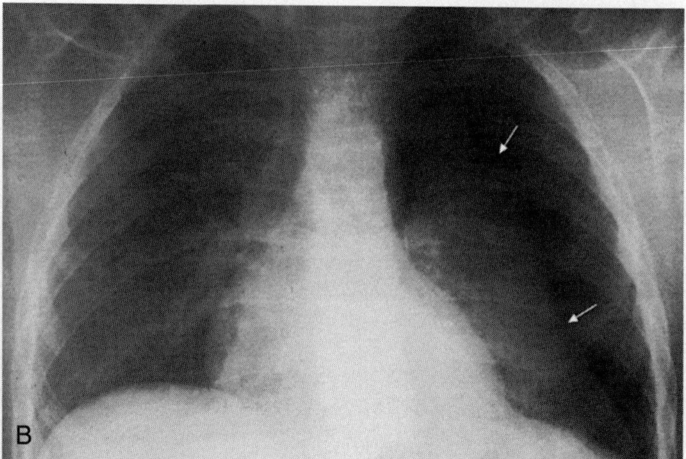

Figure 1–17. An 11-year-old child who has a known osteosarcoma of the distal left femur. *A,* Lateral view of the distal left femur shows a lytic lesion with ill-defined borders *(arrows)* with an associated soft tissue mass. Open biopsy revealed an osteosarcoma. *B,* Two months later, the patient presented with shortness of breath. An expiratory posteroanterior view of the chest showed a large pneumothorax on the left. This was the first sign that the patient had lung metastasis.

FOLLOW-UP

Monitoring therapeutic response is a crucial role for imaging. Local recurrence and distant metastasis always are possibilities in patients treated for malignant bone tumors. Frequently, there is metal hardware in the area of interest, making it difficult to study with CT or MR imaging. Radiography usually is used in the initial evaluation for local recurrence in the extremities. MR imaging and MR imaging with gadolinium are effective modalities in the follow-up of tumor recurrence. In patients with a previously resected sarcoma, follow-up chest radiographs frequently are used to look for pulmonary metastasis. If there is the slightest suspicion of a lesion on a follow-up chest radiograph, a chest CT scan is performed for further evaluation. A pneumothorax may be the first sign of pleural metastasis from an osteosarcoma (Fig. 1–17).

Recommended Readings

Brien EW, Mirra JM, Kerr R: Benign and malignant cartilage tumors of bone and joint: Their anatomic and theoretical basis with an emphasis on radiology, pathology and clinical biology. 1. The intramedullary cartilage tumors. Skeletal Radiol 26:325, 1997.

Davies AM, Cassar-Pullicino VN, Grimer RJ: The incidence and significance of fluid-fluid levels on computed tomography of osseous lesions. Br J Radiol 65:193, 1992.

Edeiken J, Hodes PJ, Caplan LH: New bone production and periosteal reaction. AJR Am J Roentgenol 97:708, 1966.

Enneking WF: Staging of musculoskeletal neoplasms. Musculoskeletal Tumor Society. Skeletal Radiol 13:183, 1985.

Gitelis S, Wilkins R, Conrad 2nd EU: Benign bone tumors. Instr Course Lect 45:425, 1996.

Grimer RJ, Sneath RS: Diagnosing malignant bone tumours [editorial]. J Bone Joint Surg Br 72:754, 1990.

Hayes CW, Conway WF, Sundaram M: Misleading aggressive MR imaging appearance of some benign musculoskeletal lesions. Radiographics 12:1119, 1992.

Lodwick GS, Wilson AJ, Farrell C, et al: Determining growth rates of focal lesions of bone from radiographs. Radiology 134:577, 1980a.

Lodwick GS, Wilson AJ, Farrell C, et al: Estimating rate of growth in bone lesions: Observer performance and error. Radiology 134:585, 1980b.

Murphy Jr WA: Imaging bone tumors in the 1990s. Cancer 67(4 Suppl):1169, 1991.

Peabody TD, Gibbs CP, Simon MA: Current concepts review: Evaluation and staging of musculoskeletal neoplasms. J Bone Joint Surg Am 80:1204, 1998.

Seeger LL, Yao L, Eckardt JJ: Surface lesions of bone. Radiology 206:17, 1998.

Sundaram M, McDonald DJ: The solitary tumor or tumor-like lesion of bone. Topics Magn Reson Imaging 1:17, 1989.

Sundaram M, McGuire MH: Computed tomography or magnetic resonance for evaluating the solitary tumor or tumor-like lesion of the bone? Skeletal Radiol 17:393, 1988.

CHAPTER 2

Benign Bone-Forming Tumors

Nabil J. Khoury, MD, Georges Y. El-Khoury, MD, and D. Lee Bennett, MD

BONE ISLAND (ENOSTOSIS)

Overview

A bone island is a common, solitary, small dense area of compact bone located within the medullary cavity (cancellous bone). It is formed of normal, mature lamellar compact bone. A bone island has no clinical significance except that some bone islands may be confused with blastic metastasis. It occurs in all age groups and has no sex predilection.

Skeletal Involvement

Any bone can be affected except the skull. Skeletal involvement includes the tubular and the flat bones. The overall prevalence of the lesion is 1% to 14% in the vertebral bodies, 1% in the pelvis, and 0.43% in the ribs. The proximal femur also is a frequent location. Bone islands are asymptomatic.

Imaging

Radiography

Radiography usually is diagnostic. The lesion appears as a round or ovoid, well-circumscribed dense area. The overall margin of a bone island is well demarcated with thorny or spiculated edges, which are seen best on close inspection (Fig. 2–1). The size of the lesion is variable. It usually is less than 1 cm in diameter but may reach 5 cm.

Computed Tomography

A computed tomography (CT) scan shows the same findings as a radiograph but to better advantage. This is particularly true in regard to visualizing the thorny margins of a bone island located in a complex anatomic site, such as the hindfoot, spine, or pelvis (see Fig. 1–1).

Magnetic Resonance Imaging

On magnetic resonance (MR) imaging, a bone island displays low signal intensity on all sequences. Neither MR imaging nor CT is required to confirm the radiographic finding of a bone island. A bone island may be detected, however, as an incidental finding on MR imaging studies.

Nuclear Medicine

The bone scan usually is normal. Some uptake may occur, however, in large lesions (>1.2 cm) or in growing lesions that are undergoing remodeling and osteoblastic activity (Fig. 2–2).

Natural History

A bone island can grow slowly or decrease in size and may even disappear.

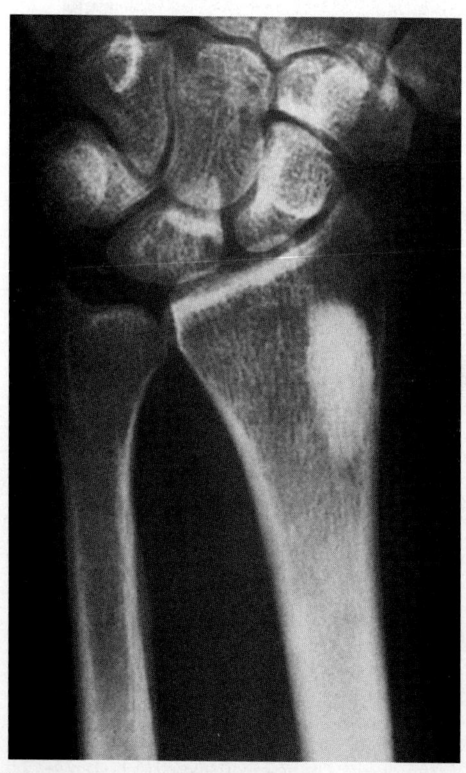

Figure 2–1. Bone island. Anteroposterior radiograph of the wrist shows an ovoid, dense mass in the distal radius. On close inspection, thorny or spiculated edges can be seen. This lesion has been stable over many years and is consistent in appearance with a bone island.

Differential Diagnosis

When a solitary bone island is found incidentally, it should be recognized for what it is, and no further imaging studies are necessary. If a bone island is large or multiple bone islands are present in a patient with a previous history of cancer, a bone scan would be helpful in ruling out blastic metastasis.

OSTEOMA

Overview

Osteomas are common in the skull and facial bones, but they are rare in the rest of the skeleton. Of all osteomas, 75% arise in the frontoethmoid region of the skull. They may arise secondary to an embryologic tissue maldevelopment or trauma. Osteomas may be solitary or multiple. When multiple osteomas are present, Gardner's syndrome should be considered (Fig. 2–3). They are mainly juxtacortical and rarely intramedullary. The juxtacortical lesions arise from the periosteum and are formed by dense, compact, mature lamellar bone similar to the cortex. Osteomas can be detected in a wide range of age groups (10 to 75 years); however, most are detected in the 30s to 40s. These lesions are usually asymptomatic, although occasionally they may present as a slow-growing mass or as chronic dull pain. The growth rate of osteomas is slow. Osteomas located in the frontoethmoid region can be the cause of sinusitis, headaches, exophthalmos, and diplopia.

Skeletal Involvement

Osteomas can be found in any bone. The craniofacial bones are particularly affected (Fig. 2–4). Rarely the lesion may be localized in the appendicular skeleton or the flat bones (vertebral bodies, scapulae, clavicles, and pelvis). In the long bones, osteomas may be either diaphyseal or metaphyseal in location.

Imaging

Radiography

Radiographs usually show a juxtacortical, homogeneous, circumscribed, round or lobular mass that is continuous with the underlying cortex. The lesion typically has a uniformly dense, sclerotic appearance described as an ivory-like mass (see Fig. 2–4). The margins are typically smooth. Intramedullary osteomas are similar to the juxtacortical lesions; however, they are intraosseous in location, abutting the endosteum. Osteomas are usually less than 2 cm in size, but occasionally they can reach 6.5 cm in diameter.

Computed Tomography

CT scans readily show the homogeneity and cortical location of the lesion (see Fig. 2–4).

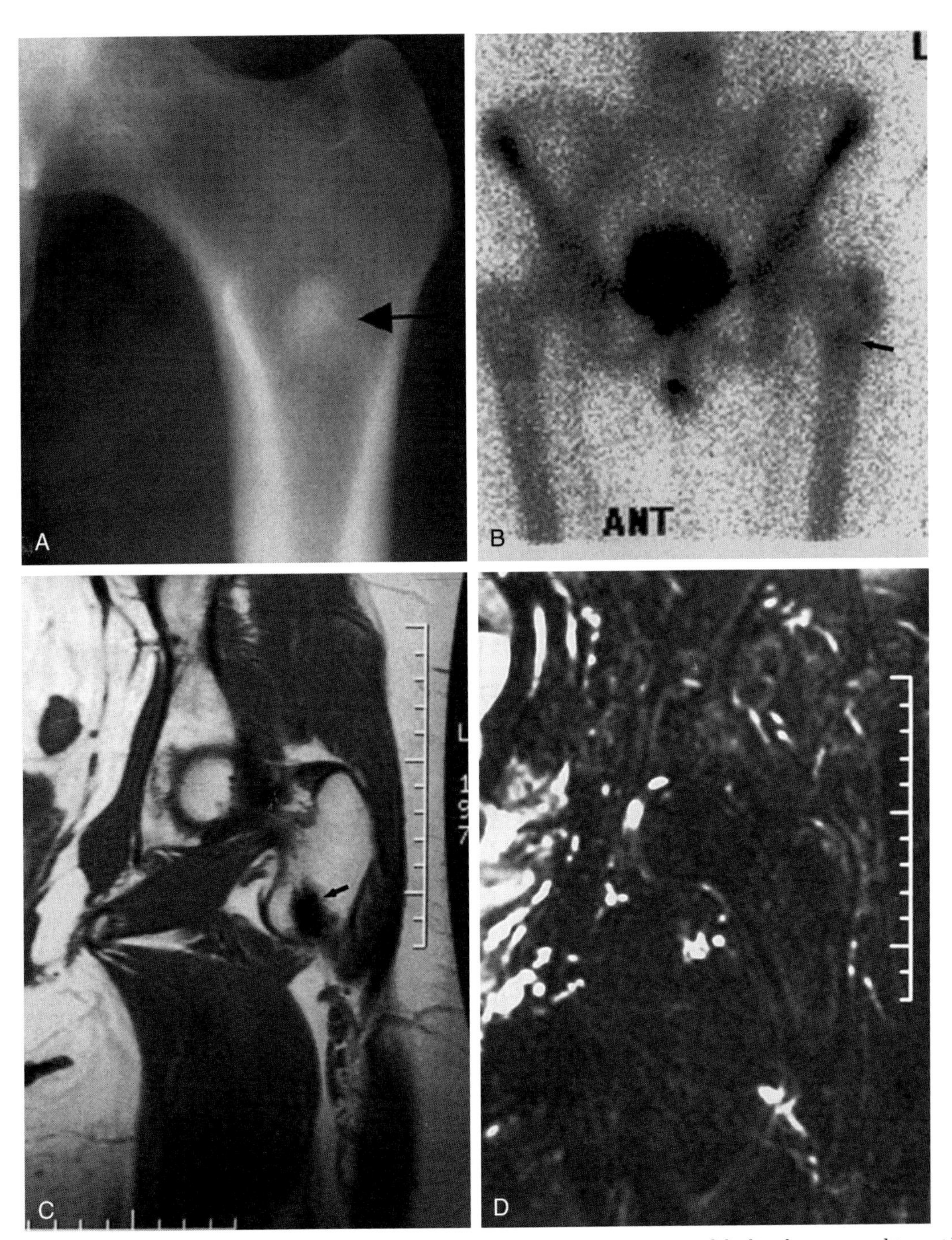

Figure 2–2. Radiographic, bone scan, and MR appearance of a bone island. *A*, Anteroposterior view of the hip shows a round-to-ovoid dense mass with spiculated margins near the intertrochanteric region of the femur *(arrow)*. *B*, Nuclear medicine bone scan shows mild uptake within the femoral lesion *(arrow)*, which can be seen with bone islands. *C* and *D*, T1-weighted *(C)* and fat-saturation FSE T2-weighted *(D)* MR images of the hips show the lesion is low in signal intensity *(arrow)* on all sequences, which would be consistent with a bone island.

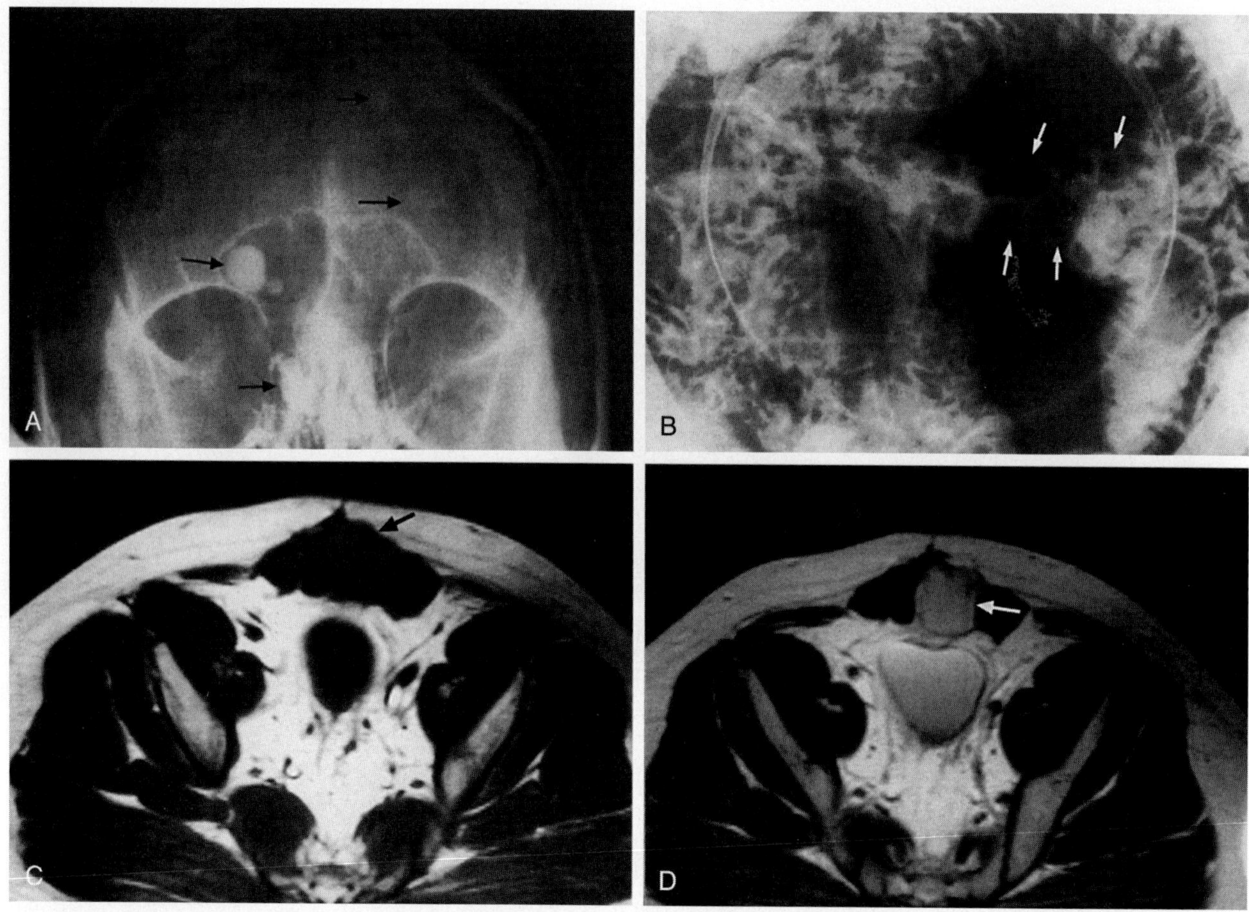

Figure 2–3. Gardner's syndrome. *A,* Multiple osteomas are present *(arrows),* including some within the sinuses. The lesions have a classic dense, sclerotic, ivory-like appearance. *B,* In the same patient, a spot film from a small bowel follow-through examination shows multiple polyps *(arrows).* Gardner's syndrome is associated with multiple adenomatous polyps. *C* and *D,* Axial T1-weighted *(C)* and axial postgadolinium T1-weighted *(D)* MR images show a desmoid tumor within the rectus abdominis musculature *(arrows).* Desmoid tumors also are associated with Gardner's syndrome.

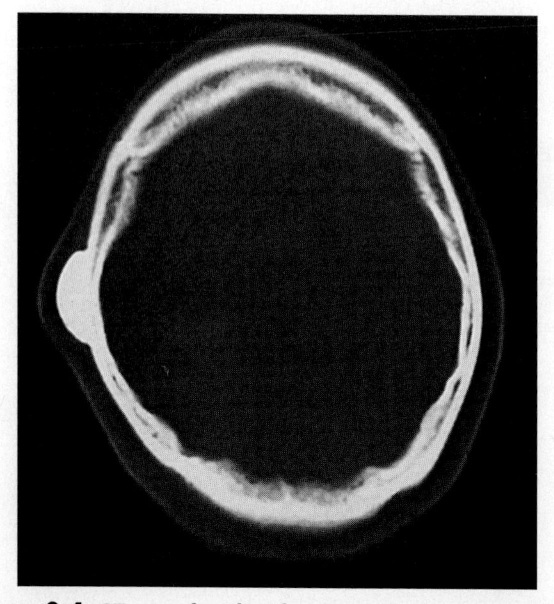

Figure 2–4. CT scan of a calvarial osteoma. There is a juxtacortical, uniformly sclerotic, circumscribed osteoma of the calvarium. It has an ivory-like appearance.

Magnetic Resonance Imaging

On MR imaging, osteomas have low signal intensity on all sequences.

Nuclear Medicine

Typically an osteoma shows no increase in uptake on a bone scan.

Complications

Frontoethmoidal osteomas can produce blockage of the frontonasal ducts, resulting in mucocele formation, sinusitis, and headaches. Rarely an osteoma can invade the orbit, causing exophthalmos.

Differential Diagnosis

Many lesions may mimic osteomas. This is particularly true in the appendicular skeleton, where osteomas are parosteal and rare. These lesions include a mature juxtacortical focus of myositis ossificans, parosteal osteosarcoma, and melorheostosis. Intramedullary osteomas should be differentiated from bone islands or healed nonossifying fibromas. Tuberous sclerosis should be considered if multiple osteomas are present in the metacarpals and metatarsals. A patient may have Gardner's syndrome if multiple osteomas are present in the calvaria, mandible, or tubular bones.

OSTEOID OSTEOMA

Overview

Osteoid osteoma is a fairly common benign lesion, representing about 10% to 12% of all benign bone tumors. It consists of a central core of vascular connective tissue and osteoid with surrounding reactive sclerotic bone. The lesion is uncalcified early, but when it matures, it develops osteoid calcifications that range from tiny speckles to dense bone within a highly vascularized connective tissue stroma. The lesion or nidus is well demarcated by the surrounding sclerosis. Most cases (90%) occur between the ages of 7 and 25 years. The lesion is rare in children younger than age 5 (3%) and adults older than age 50 (1% to 2%). It is more common in males, with a male-to-female ratio of 2:1 to 3:1. Typically, osteoid osteoma presents with pain that increases at night and is relieved by salicylates. Painless osteoid osteoma has been reported, but it is extremely rare. In the spine, patients present with painful scoliosis or torticollis. With intra-articular lesions, there may be signs and symptoms of synovitis.

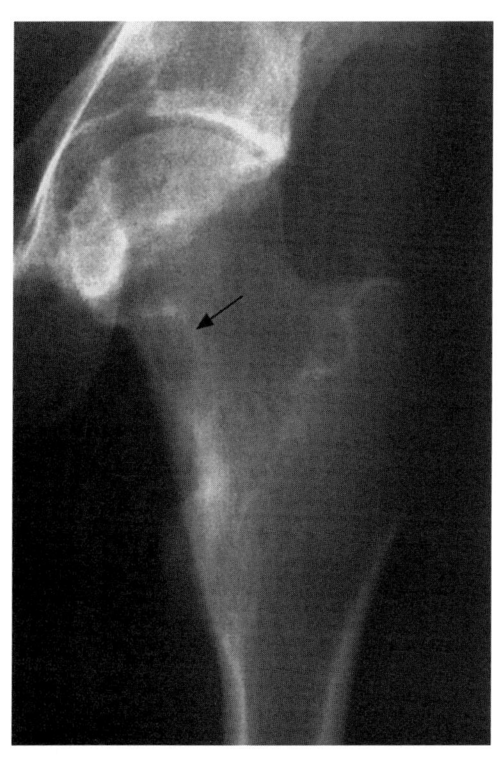

Figure 2–5. Osteoid osteoma of the femoral neck. The lucent nidus of this osteoid osteoma is located in the cortex of the femoral neck *(arrow)*. This portion of the femoral neck is intracapsular. There is buttressing of the femoral neck and surrounding reactive sclerosis.

Skeletal Involvement

Osteoid osteoma may occur in any bone. It is common in the lower extremity, with 40% of the lesions occurring in the femur. Two thirds of the femoral lesions are intertrochanteric or intracapsular (Figs. 2–5 and 2–6). The next most common locations are the tibial diaphysis and the humerus. Of osteoid osteomas, 10% to 18% occur in the spine, with a significant predilection for the posterior elements (Fig. 2–7). Vertebral body lesions are unusual. Osteoid osteoma is rare in the pelvis, craniofacial bones, clavicle, and sternum. Multifocal synchronous or metachronous lesions have been reported in a single site or at two different sites. In the long bones, osteoid osteoma is located most commonly in the diaphysis or metaphysis (Fig. 2–8). There are a few case reports of epiphyseal osteoid osteomas. The classic form of the tumor is the cortical osteoid osteoma (occurring within the cortex). It also can occur in the intramedullary (or cancellous) bone (Fig. 2–9). A subperiosteal location is the least common. Intracapsular lesions most commonly occur in the femoral neck. Other intracapsular sites include the elbow, foot, wrist, knee, and facet joints.

Imaging

Radiography

Findings on radiography vary depending on the location of the nidus. The size of the nidus is usually less than 1 to 2 cm. In the cortical location, the nidus appears as a round or ovoid lucency with a variable amount of calcification. It is surrounded by significant reactive fusiform sclerosis and cortical thickening, which may obscure the lucent nidus (see Fig. 2–8*A*). Cortical osteoid osteoma is seen mainly in the femur and tibia. In a medullary location, the nidus is similar in appearance to a cortical location (see Fig. 2–9); however, the reactive sclerosis is much less prominent. Intramedullary osteoid osteoma is seen primarily in the femoral neck, hand, foot, and neural arch. The subperiosteal osteoid osteoma initially presents as a soft tissue lesion abutting the cortex with almost no reactive sclerosis. At a later stage, it mimics an intracortical osteoid osteoma. Subperiosteal osteoid osteoma primarily occurs in the hand, foot, and talar neck (Fig. 2–10). In intra-articular osteoid osteomas, the nidus may be intramedullary or cortical in location, and the reactive endosteal new bone formation is minimal (see Fig. 2–5). Joint effusion, disuse osteopenia, and broadening of the femoral neck may occur. Spinal osteoid osteomas typically involve the posterior elements and often are accompanied by scoliosis. The nidus typically is located on the concave side of the curve at the apex of the curve.

Computed Tomography

A CT scan is the modality of choice for the diagnosis and localization of the nidus (see Figs. 2–6, 2–7, and 2–8*B*). CT

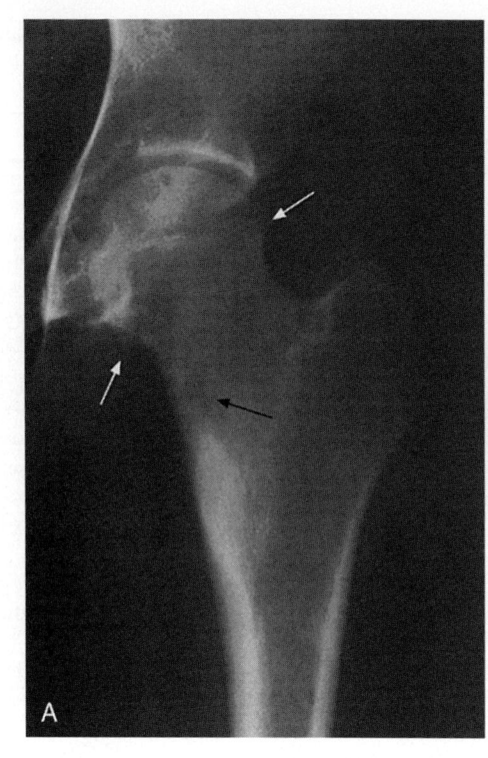

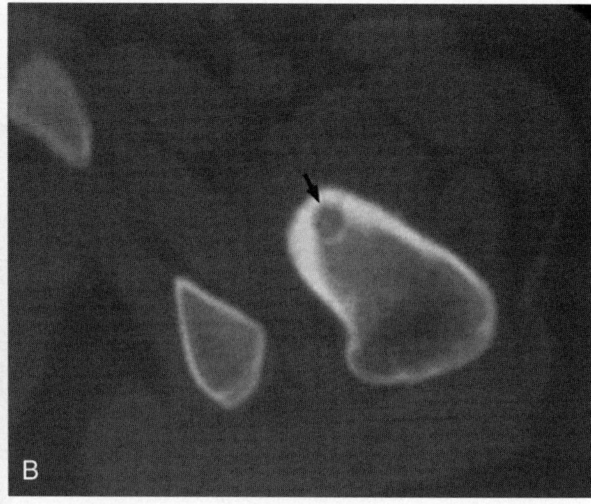

Figure 2–6. Radiograph and CT scan of an osteoid osteoma. *A,* Anteroposterior radiograph of the hip shows the lucent nidus in the femoral neck *(black arrow),* degenerative change of the hip joint *(white arrows),* and surrounding reactive sclerosis. *B,* CT scan shows the lucent nidus within the cortex of the femoral neck *(arrow),* surrounding reactive sclerosis, and calcification within the lucent nidus.

should be performed using thin collimation (≤2 mm). CT is commonly used to guide percutaneous radiofrequency ablation of an osteoid osteoma or percutaneous resection of an osteoid osteoma using a large-bore needle.

Magnetic Resonance Imaging

MR imaging shows the soft tissue and marrow changes (soft tissue mass affect and marrow edema) associated with an osteoid osteoma. These changes can be misleading because they appear aggressive and may resemble a malignant neoplasm (see Fig. 2–8*C, D*). The soft tissue and marrow abnormalities are less pronounced in patients receiving anti-inflammatory medications.

Nuclear Medicine

A bone scan may help in the early detection and localization of an osteoid osteoma (see Fig. 2–8*E*). It shows increased uptake in all three phases. Scintigraphy is a good modality for localizing the lesion when the nidus is in a complex anatomic location, such as the spine or foot.

Complications

Complications are related mainly to intra-articular lesions and include osteoarthritis and growth disturbance (see Fig. 2–6). Patients with long-standing synovitis, re-

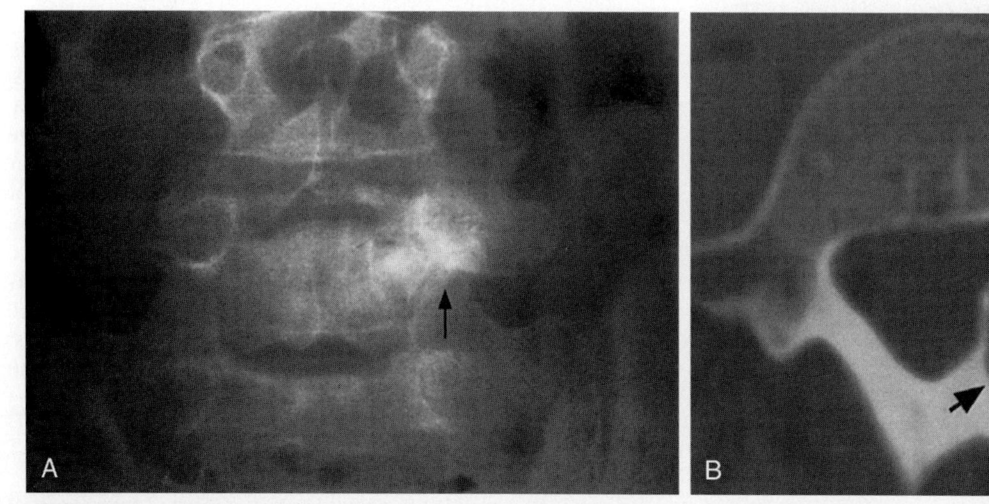

Figure 2–7. An osteoid osteoma of the lumbar spine. *A,* Anteroposterior radiograph shows reactive sclerosis in the left pedicle *(arrow)* and lamina of L5. *B,* CT scan of the osteoid osteoma at L5 shows the lucent nidus *(arrow),* bony matrix, and surrounding reactive sclerosis. The mildly expansile nature of the lesion causes mild narrowing of the ipsilateral lateral recess.

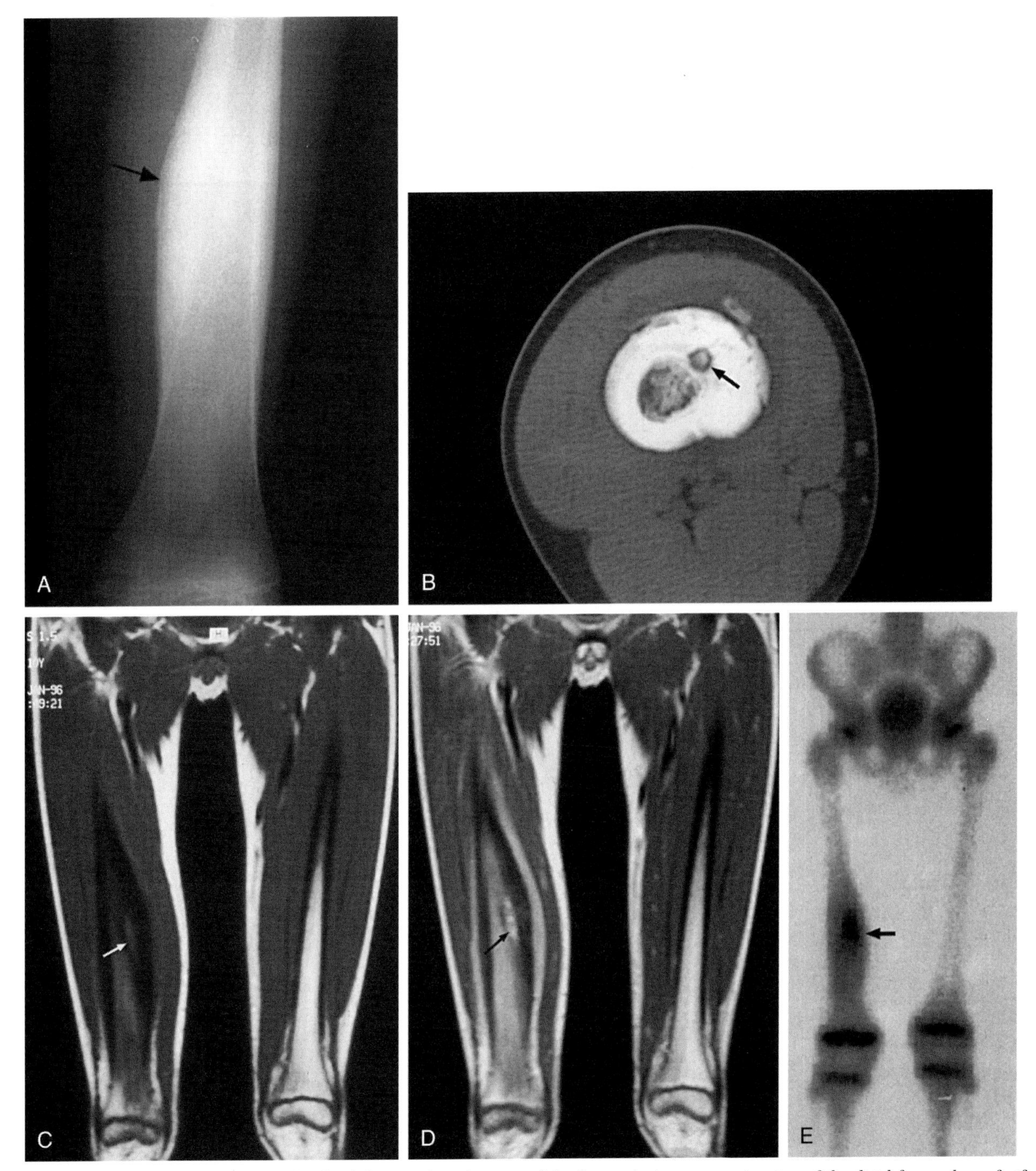

Figure 2–8. Osteoid osteoma involving the distal diametaphyseal region of the femur. *A,* Anteroposterior view of the distal femur shows fusiform sclerosis and cortical thickening *(arrow)* seen with osteoid osteoma. *B,* Axial CT scan shows the nidus *(arrow). C* and *D,* Coronal T1- *(C)* and T2-weighted *(D)* MR images of the femora. T1-weighted image shows extensive marrow edema in the involved femur. Both sequences show the nidus as well *(arrows). E,* Bone scan shows a double-density sign with more focal increased uptake at the location of the nidus *(arrow).*

sulting from an osteoid osteoma of the elbow, are at risk for developing a flexion contracture.

Differential Diagnosis

If the surrounding sclerotic reaction is seen and a nidus has not yet been identified, the differential diagnosis would include a stress fracture, osteoblastoma, Ewing's sarcoma, osteosarcoma, and infection. When the nidus is intra-articular, the clinical presentation and radiographic findings can be confused easily with an inflammatory arthritis.

OSTEOBLASTOMA

Overview

Osteoblastoma is a rare, benign neoplasm representing less than 3% of benign bone tumors. Osteoblastoma is histopathologically similar to osteoid osteoma. Most au-

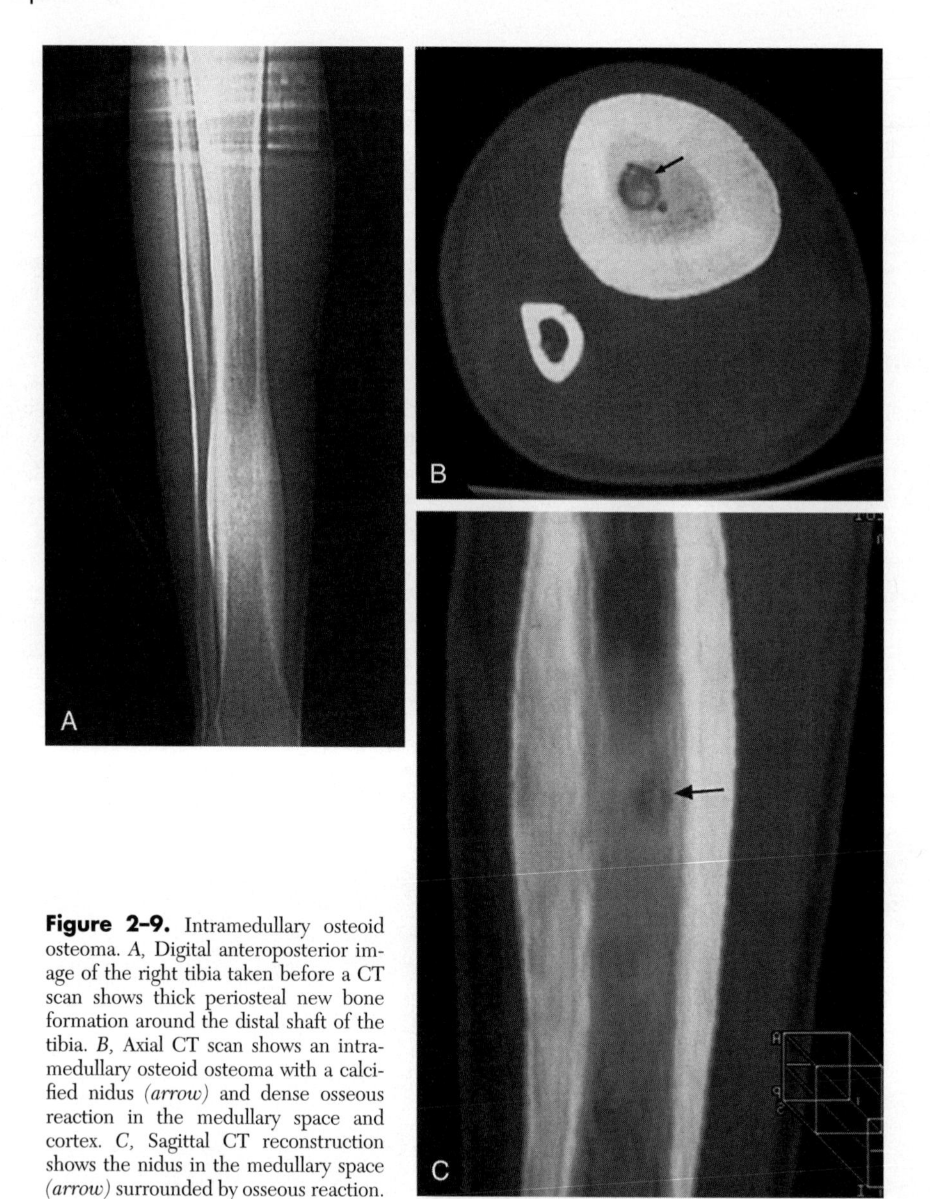

Figure 2–9. Intramedullary osteoid osteoma. *A,* Digital anteroposterior image of the right tibia taken before a CT scan shows thick periosteal new bone formation around the distal shaft of the tibia. *B,* Axial CT scan shows an intramedullary osteoid osteoma with a calcified nidus *(arrow)* and dense osseous reaction in the medullary space and cortex. *C,* Sagittal CT reconstruction shows the nidus in the medullary space *(arrow)* surrounded by osseous reaction.

thors differentiate osteoid osteoma from osteoblastoma on the basis of size, whereby a lesion larger than 1.5 to 2 cm is labeled an osteoblastoma. Another difference is based on the limited potential of osteoid osteomas to grow, whereas osteoblastomas can behave in an aggressive fashion. Osteoblastomas can occur between 3 and 78 years of age, with 90% of lesions occurring before age 30. The male-to-female ratio is similar to that of osteoid osteoma (2:1 to 3:1). Pain is a common presenting symptom (97%); it is usually mild, dull, persistent, and less localized than in osteoid osteoma. In spinal lesions, painful scoliosis as a result of muscle spasm is frequent but is less common than in osteoid osteoma.

Skeletal Involvement

The most common location of osteoblastoma is the axial skeleton, with about one third of osteoblastomas occurring in the spine (Fig. 2–11). In the spine, nearly all lesions are located within the posterior elements. The long bones also are frequent sites of involvement (30% to 35%), with the femur being the most common site. In the long bones, the diaphysis is involved in 75% of the cases. The foot and ankle are involved in about 12.5% of the cases, with the talus being a common location.

Imaging

Radiography

On plain radiographs, an osteoblastoma is usually a geographic, well-defined, lytic lesion and less frequently is osteoblastic. It has varying degrees of bone production and expansile behavior. Usually, there is an intact surrounding sclerotic rim. Cortical thinning with a secondary soft tissue mass sometimes is noted, however. Osteoblastomas have varying degrees of matrix mineralization: amorphous calcific, dense, solid, cloudy, flocculent, or ringlike (Fig. 2–12). When compared with osteoid osteomas, osteoblastomas are less sclerotic but more expansile. They have less surrounding bone reaction. In some instances, an osteoblastoma can show classic findings of osteoid osteoma except with a larger nidus. Most osteoblastomas grow much larger than osteoid osteomas, however, reaching several centimeters in size.

Computed Tomography

A CT scan is helpful in evaluating lesions located in anatomically complex areas, such as the spine. CT visualizes the sclerotic rim and the degree of matrix mineralization better than radiography. Matrix mineralization is seen in about 75% of cases (see Fig. 2–12).

Magnetic Resonance Imaging

MR imaging is nonspecific. Osteoblastomas have low signal intensity on all sequences when diffuse, calcified osteoid is present. The lesions also may appear isointense on T1-weighted images, with high signal intensity on T2-weighted images.

Complications

Osteoblastoma has a different clinical potential than osteoid osteoma. Osteoblastomas can be locally aggressive. Rarely, they can undergo malignant transformation into a low-grade or high-grade osteosarcoma (mainly in recurrent lesions). In the spine, an osteoblastoma may impinge on the spinal canal or neural foramina, resulting in spinal cord or nerve root impingement (Fig. 2–13). Rarely, osteoblastomas may produce toxic systemic symptoms, which possibly are due to prostaglandin production. The systemic toxic symptoms disappear with resection of the tumor.

Differential Diagnosis

Differential diagnosis includes osteosarcoma, osteoid osteoma, aneurysmal bone cyst, and eosinophilic granuloma.

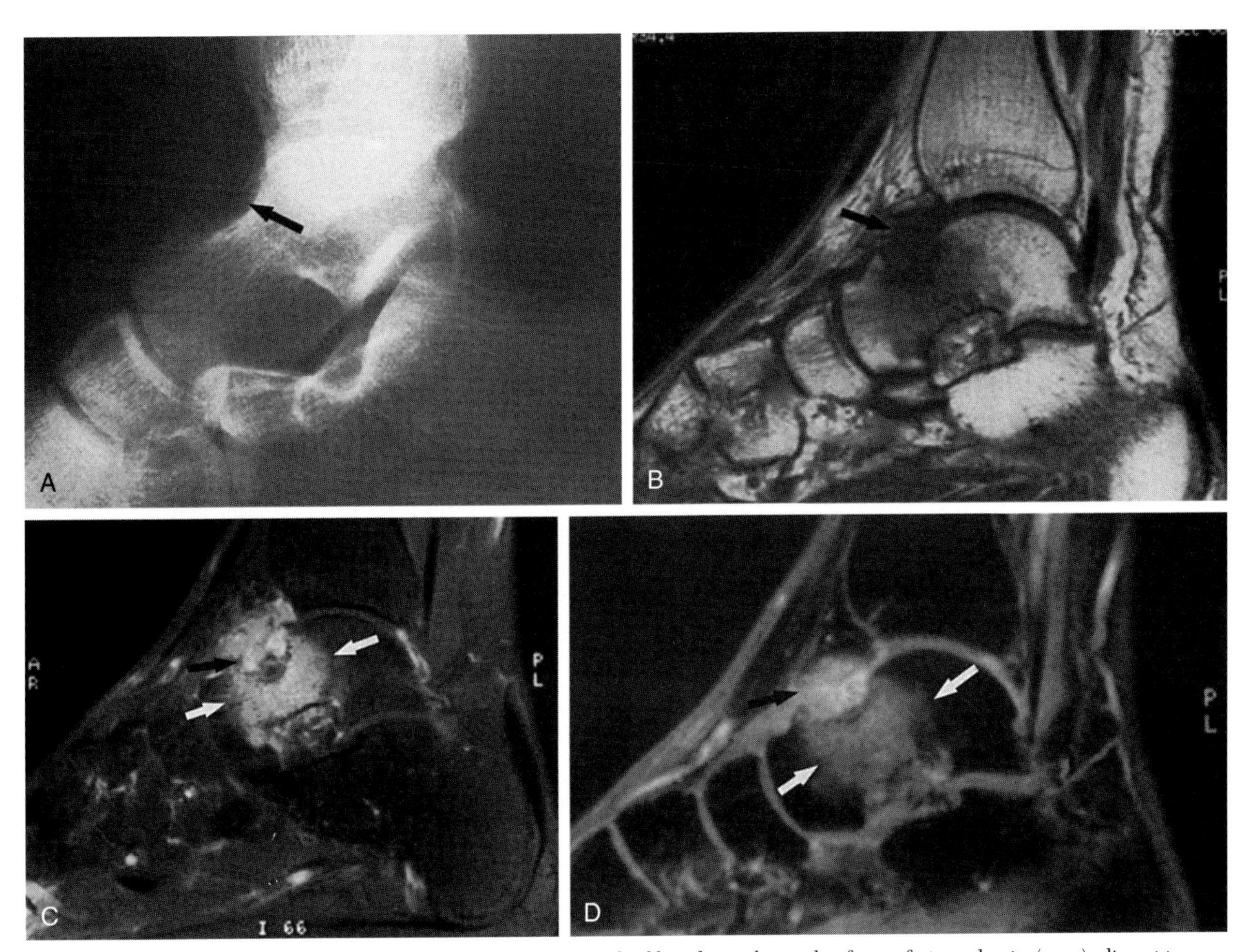

Figure 2–10. Subperiosteal osteoid osteoma of the talar neck. *A,* Lateral ankle radiograph reveals a faint soft tissue density (mass) adjacent to an area of cortical erosion in the talar neck *(arrow)*. *B,* Sagittal T1-weighted MR image shows low signal within the lesion *(arrow)*. There also is decreased signal in the marrow of the adjacent talus consistent with reactive marrow edema. *C,* Sagittal postgadolinium MR image shows enhancement within the lesion *(black arrow)* and in the adjacent marrow *(white arrows)*. *D,* A sagittal fat-saturation FSE T2-weighted MR image shows high signal within the lesion *(black arrow)*. There also is increased signal in the marrow of the adjacent talus consistent with reactive marrow edema *(white arrows)*.

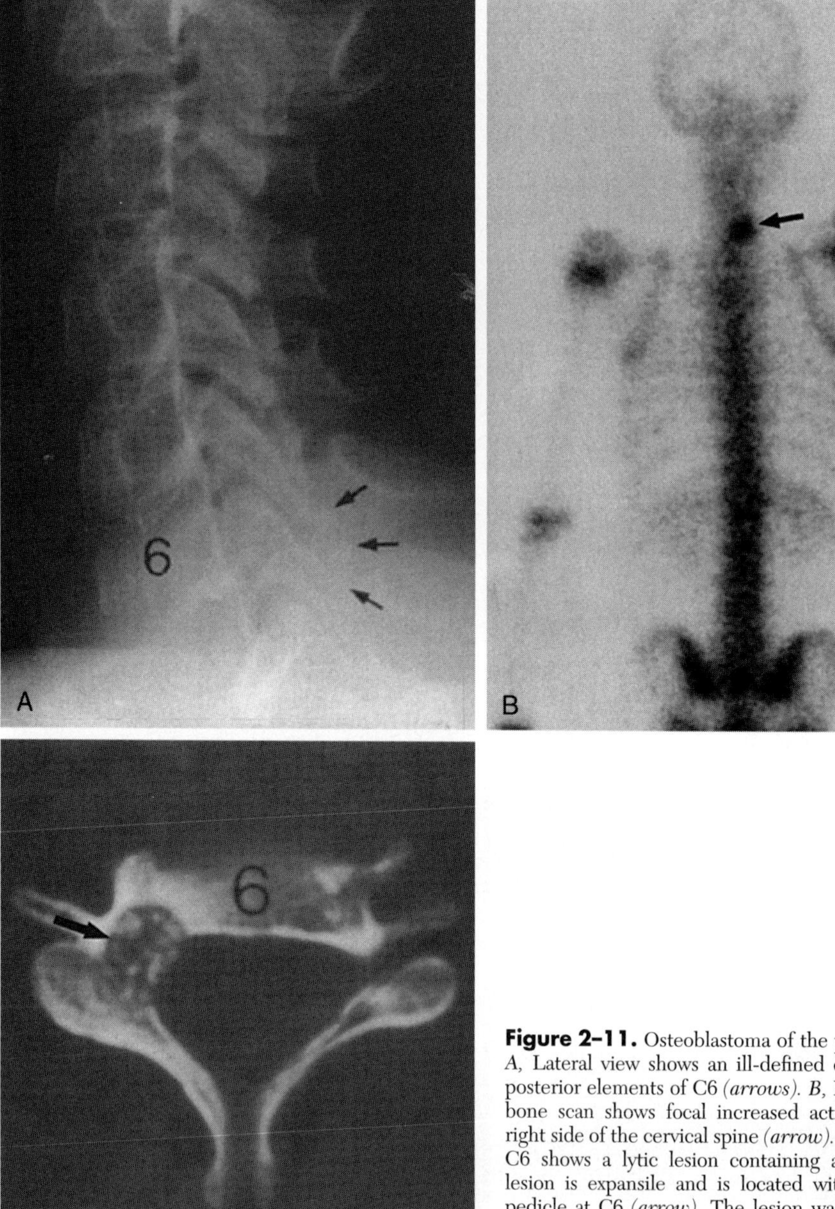

Figure 2–11. Osteoblastoma of the pedicle of C6. *A,* Lateral view shows an ill-defined density of the posterior elements of C6 *(arrows). B,* Posterior view bone scan shows focal increased activity over the right side of the cervical spine *(arrow). C,* CT scan at C6 shows a lytic lesion containing a matrix. The lesion is expansile and is located within the right pedicle at C6 *(arrow).* The lesion was found to be an osteoblastoma. (Case provided by Dr. Murali Sundaram, Rochester, MN.)

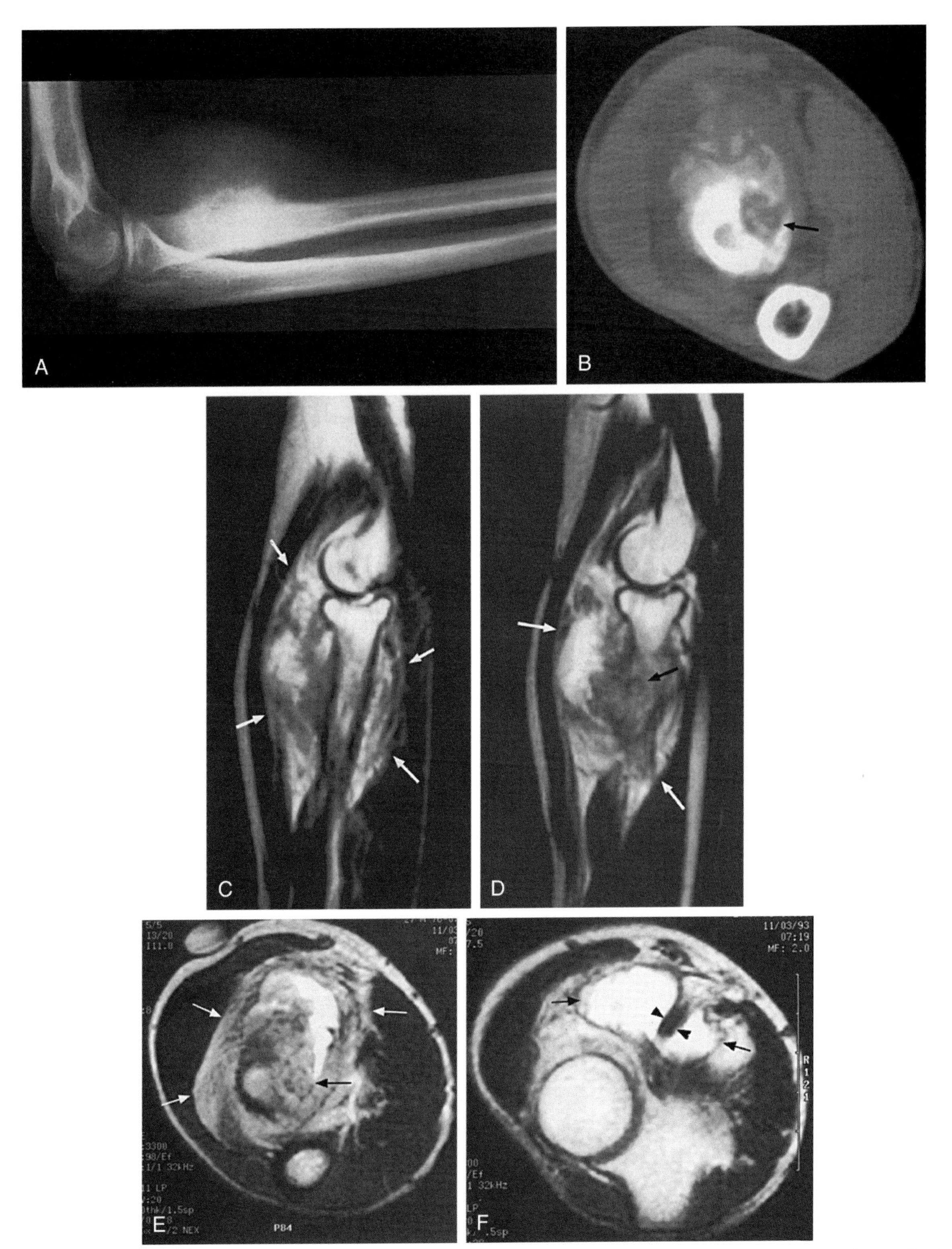

Figure 2–12. Osteoblastoma in the proximal metadiaphysis of the radius. The lesion is associated with extensive periosteal new bone and soft tissue reaction. *A,* Lateral view of the right elbow reveals an osseous reaction around the proximal radius. No definite nidus is identified. *B,* Axial CT section through the lesion shows a lucent nidus with punctate calcifications *(arrow)* surrounded by a dense osseous reaction. *C* and *D,* Sagittal T2-weighted MR images reveal extensive soft tissue edema around the lesion *(white arrows).* The nidus is not well shown on MR imaging *(black arrow). E* and *F,* Axial T2-weighted MR images show soft tissue edema *(white arrows)* and synovitis in the tendon sheath of the bicipital tendon *(black arrows).* The arrowheads are pointing to the bicipital tendon.

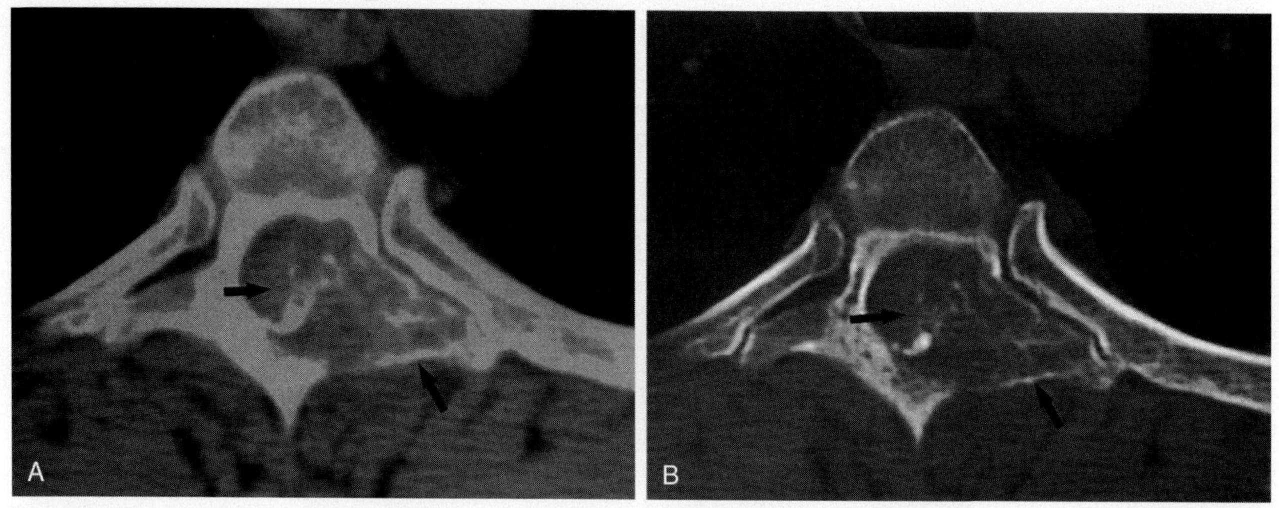

Figure 2–13. Osteoblastoma in the posterior elements of the thoracic spine. *A* and *B*, Axial CT scans (soft tissue *[A]* and bone windows *[B]*) show the tumor extending into the spinal canal and impinging on the spinal cord *(arrows)*.

Recommended Readings

Assoun J, Richardi G, Railhac JJ, et al: Osteoid osteoma: MR imaging versus CT. Radiology 191:217, 1994.

Bertoni F, Unni KK, Beabout JW, et al: Parosteal osteoma of bones other than of the skull and face. Cancer 75:2466, 1995.

Cerase A, Priolo F: Skeletal benign bone-forming lesions. Eur J Radiol 27(suppl 1):S91, 1998.

Crim JR, Mirra JM, Eckardt JJ, et al: Widespread inflammatory response to osteoblastoma: The flare phenomenon. Radiology 177:835, 1990.

Goldman AB, Schneider R, Pavlov H: Osteoid osteomas of the femoral neck: Report of four cases evaluated with isotopic bone scanning, CT, and MR imaging. Radiology 186:227, 1993.

Greenspan A: Benign bone-forming lesions: Osteoma, osteoid osteoma, and osteoblastoma. clinical, imaging, pathologic, and differential considerations. Skeletal Radiol 22:485, 1993.

Greenspan A, Klein MJ: Giant bone island. Skeletal Radiol 25:67, 1996.

Hall FM, Goldberg RP, Davies JAK, et al: Scintigraphic assessment of bone islands. Radiology 135:737, 1980.

Healey JH, Ghelman B: Osteoid osteoma and osteoblastoma: Current concepts and recent advances. Clin Orthop 204:76, 1986.

Houghton MJ, Heiner JP, De Smet AA: Osteoma of the innominate bone with intraosseous and parosteal involvement. Skeletal Radiol 24:455, 1995.

Kayser F, Resnick D, Haghighi P, et al: Evidence of the subperiosteal origin of osteoid osteomas in tubular bones: Analysis by CT and MR imaging. AJR Am J Roentgenol 170:609, 1998.

Kenan S, Abdelwahab IF, Klein MJ, et al: Lesions of juxtacortical origin (surface lesions of bone). Skeletal Radiol 22:337, 1993.

Kim SK, Barry WF: Bone islands. Radiology 99:77, 1968.

Lambiase RE, Levine SM, Terek RM, et al: Long bone surface osteomas: Imaging features that may help avoid unnecessary biopsies. AJR Am J Roentgenol 171:775, 1998.

McDermott MB, Kyriakos M, McEnery K: Painless osteoid osteoma of the rib in an adult: A case report and a review of the literature. Cancer 77:1442, 1996.

McGrath BE, Bush CH, Nelson TE, et al: Evaluation of suspected osteoid osteoma. Clin Orthop 327:247, 1996.

Nogues P, Marti-Bonmati L, Aparisi F, et al: MR imaging assessment of juxtacortical edema in osteoid osteoma in 28 patients. Eur Radiol 8:236, 1998.

Ozkal E, Erongun U, Cakir B, et al: CT and MR imaging of vertebral osteoblastoma: A report of two cases. Clin Imaging 20:37, 1996.

Onitsuka H: Roentgenologic aspects of bone islands. Radiology 123:607, 1977.

Peyser AB, Makley JT, Callewart CC, et al: Osteoma of the long bones and the spine: A study of eleven patients and a review of the literature. J Bone Joint Surg Am 78:1172, 1996.

Resnick D, Nemcek Jr AA, Haghighi P: Spinal enostosis (bone islands). Radiology 147:373, 1983.

Saifuddin A, White J, Sherazi Z, et al: Osteoid osteoma and osteoblastoma of the spine: Factors associated with the presence of scoliosis. Spine 23:47, 1998.

Seeger LL, Yao L, Eckardt JJ: Surface lesions of bone. Radiology 206:17, 1998.

Shaikh MI, Saifuddin A, Pringle J, et al: Spinal osteoblastoma: CT and MR imaging with pathological correlation. Skeletal Radiol 28:33, 1999.

Shankman S, Deasi P, Beltran J: Subperiosteal osteoid osteoma: Radiographic and pathologic manifestations. Skeletal Radiol 26:457, 1997.

Smith J: Giant bone islands. Radiology 107:35, 1973.

Sundaram M, Falbo S, McDonald D, et al: Surface osteomas of the appendicular skeleton. AJR Am J Roentgenol 167:1529, 1996.

CHAPTER 3

Primary Malignant Bone-Forming Tumors

Nabil J. Khoury, MD, Georges Y. El-Khoury, MD, and D. Lee Bennett, MD

OVERVIEW

Osteosarcoma is the second most common primary malignant bone lesion after multiple myeloma. It accounts for approximately 15% of all primary bone tumors. Osteosarcoma is more frequent in males (male-to-female ratio, 2:1) and is seen mostly in young patients (10 to 30 years old). Another smaller peak of occurrence is in patients between the fifth and seventh decades. Osteosarcoma can arise de novo in normal bone, but it also may develop in abnormal bone, such as in Paget's disease, fibrous dysplasia, and irradiated bone. Osteosarcomas arising from abnormal bone occur later in life and are referred to as *secondary osteosarcomas.* Osteosarcoma, in its conventional form, constitutes the majority of all osteosarcomas. It is typically unifocal and centrally located (intramedullary). Surface osteosarcoma presents in a variety of forms: parosteal, periosteal, high-grade surface, and intracortical. Osteosarcoma is uncommonly multifocal. Metastatic deposits typically are seen in the lungs, bones, and lymph nodes. Metastasis to the medullary space in the same long bone is called a *skip metastasis,* and it is reported to occur in 1% to 25% of conventional osteosarcomas. In the authors' experience, the incidence of skip metastasis is closer to the lower end of this range (<5%). Osteosarcoma is defined as a malignant mesenchymal tumor in which the tumor cells produce osteoid. Even if a small portion of the tumor produces osteoid, the tumor still is labeled an osteosarcoma. Histologically, osteosarcoma has been classified into osteoblastic, chondroblastic, and fibroblastic types. This classification is based on the predominant cell type; however, most lesions have a mixed content. The most common histologic type is osteoblastic osteosarcoma. Reparative processes, such as early myositis ossificans or a healing fracture, can resemble an osteosarcoma histologically.

CONVENTIONAL OSTEOSARCOMA

Skeletal Location

Conventional osteosarcomas usually affect the metaphysis of long bones, with two thirds of the lesions occurring around the knee. Extension into the diaphysis and the epiphysis may occur, even in patients with open growth plates. Initial involvement of the epiphysis is, however, extremely rare. Other skeletal locations include the humerus and flat bones (e.g., pelvis, scapula, and ribs).

Variants of Conventional Osteosarcoma

Two rare subtypes of conventional osteosarcoma are low-grade intraosseous osteosarcoma and small cell osteosarcoma.

Low-Grade Osteosarcoma

Low-grade intraosseous osteosarcoma is a well-differentiated tumor. Radiologically and pathologically, it simulates benign conditions, such as fibrous dysplasia and chondromyxoid fibroma. The lesion has a good prognosis.

Small Cell Osteosarcoma

Histologically, small cell osteosarcoma resembles Ewing's sarcoma except that it produces osteoid. Radiographically, small cell osteosarcoma appears as an aggressive lytic lesion with associated periosteal reaction and a soft tissue mass. The prognosis with this lesion is poor.

Imaging

Radiography

On radiographs, conventional osteosarcoma is a destructive lesion with a variable degree of mineralization leading to a mixture of sclerosis and permeative or moth-eaten lytic patterns. The lesion has a wide zone of transition with cortical destruction and an associated soft tissue mass in most cases. A cloudlike matrix is seen often within the soft tissue component (Fig. 3–1). There is an interrupted periosteal reaction with formation of Codman's triangle. A sunburst or spiculated periosteal reaction also can occur. Occasionally the lesion is either completely blastic or completely lytic.

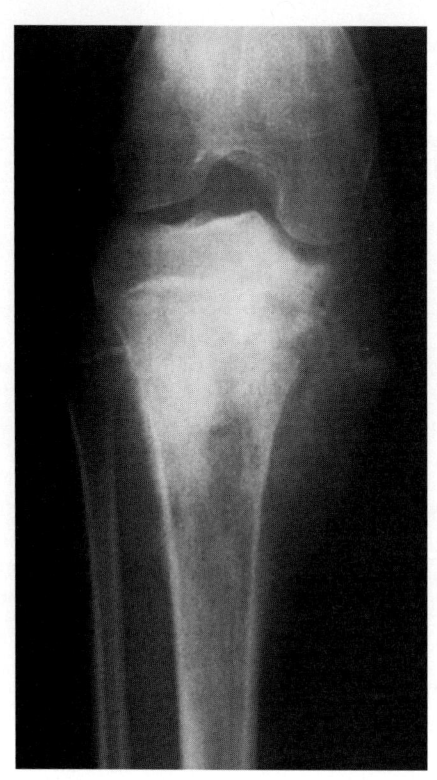

Figure 3–1. Osteosarcoma with invasion of the soft tissues in a 14-year-old girl. Anteroposterior view of the right knee shows a highly aggressive lesion with osteoid matrix. A large soft tissue component (containing osteoid matrix) is seen medially. The lesion crosses the epiphyseal plate into the epiphysis.

Computed Tomography

Computed tomography (CT) is helpful in evaluating lesions located in anatomically complex areas, such as the pelvis, chest wall, or scapula. CT also is essential in searching for metastasis to the lungs.

Magnetic Resonance Imaging

Magnetic resonance (MR) imaging is essential for staging osteosarcoma. It is the best modality to show the local extent of disease and to identify skip metastases (Fig. 3–2). Osteosarcoma is low in signal intensity on T1-weighted images and high in signal intensity on T2-weighted images. Areas showing low signal intensity on T1- and T2-weighted images represent calcifications within the tumor. MR imaging is effective in assessing the lesion's response to chemotherapy and in searching for local recurrence after surgery. Some authors have advocated the use of dynamic contrast-enhanced MR imaging, in which rapid enhancement (<2 minutes after injection) denotes viable tumor. Late enhancement denotes postoperative or postchemotherapy inflammatory response.

Nuclear Medicine

Bone scanning is effective in the detection of skeletal and extraskeletal metastatic lesions (Fig. 3–3).

Complications

Metastasis to the lungs is the most common complication. These lesions are commonly subpleural and may cause a pneumothorax (see Fig. 1–17). Treated patients presenting with a pneumothorax at follow-up are considered to have pulmonary metastasis until proved otherwise.

Differential Diagnosis

The differential diagnosis includes Ewing's sarcoma, fracture with callus formation, and metastasis.

SURFACE OSTEOSARCOMAS

Surface osteosarcomas include the following variants: parosteal, periosteal, intracortical, and high-grade surface osteosarcomas.

Parosteal Osteosarcoma. Parosteal osteosarcoma is the most common surface osteosarcoma. It represents 5% of all sarcomas. It is a low-grade tumor with a better prognosis than conventional osteosarcoma. The tumor is rich in fibrous stroma and may contain islands of cartilage. The lesion occurs in the second to fifth decades of life, and there is a female predominance. Clinically, it presents as a painless, slow-growing mass. It usually involves the metaphysis of a long tubular bone, especially the posterior aspect of the distal metaphysis of the femur (Fig. 3–4). The tibia, fibula, and humerus also can be common sites of involvement. Parosteal osteosarcoma appears as a broad-based, lobulated, dense protuberance that wraps around the host bone (see Fig. 3–4). On CT scan, a cleavage plane between the tumor and the host bone is seen in 60% of the cases except at the level of the stalk. Ossification of the lesion starts from the tumor base; the center of the lesion appears denser than the periphery. This feature is appreciated best on CT, and it helps in differentiating parosteal osteosarcoma from myositis ossificans, which has dense peripheral calcifications. In the adjacent long bone, the medullary space is invaded in about one third of the cases, as shown by MR imaging; however, this does not seem to influence the favorable prognosis of this lesion (Fig. 3–5). In about 10% of the cases, the tumor may become dedifferentiated into a high-grade osteosarcoma or malignant fibrous histiocytoma. Radiographically, this dedifferentiation is indicated by the development of an aggressive focal lucency within the tumor and an associated soft tissue component. Dedifferentiated parosteal osteosarcomas have a poor prognosis.

Periosteal Osteosarcoma. Periosteal osteosarcoma is an intermediate-grade lesion containing significant amounts of cartilaginous tissue. The prognosis for patients with periosteal osteosarcoma is better than for patients with conventional osteosarcoma but worse than for patients with parosteal osteosarcoma. It is typically a diaphyseal lesion, which is located most commonly in the proximal tibia, the femur, or the humerus. The lesion arises from the deep layers of the periosteum, and medullary invasion is rare. Radiography shows a fusiform soft tissue mass eroding the outer cortex with perpendicular bone spicules and Codman's triangle (Fig. 3–6). The underlying cortex may be thickened.

Intracortical Osteosarcoma. Intracortical osteosarcoma is the least common subtype of osteosarcoma (1%). It occurs in the teens or 20s, and it typically involves the shaft of the femur or tibia. Radiographically the lesion presents as a round or oval lucency confined to a thickened cortex. Medullary invasion is rare. Intracortical osteosarcoma should be differentiated from a fibrous cortical defect and Brodie's abscess.

High-Grade Surface Osteosarcoma. High-grade surface osteosarcomas have a prognosis and histology similar to conventional osteosarcomas. Seventy percent of these lesions occur in the teens or 20s, with males being affected more often than females. This lesion occurs in the long bones, mainly in the mid-diaphysis of the femur, tibia, or fibula. Radiographically the lesion resembles periosteal osteosarcoma; however, a high-grade surface osteosarcoma frequently invades the medullary canal.

MULTICENTRIC (MULTIFOCAL) OSTEOSARCOMA

Multicentric osteosarcoma is an uncommon disease encountered in childhood and adolescence (Fig. 3–7). The development of lesions can be metachronous or synchronous (osteosarcomatosis). Osteosarcomatosis may represent a rapid form of metastasis. In skeletally immature patients, osteosarcomatosis presents as bilateral, symmetrical, metaphyseal, dense lesions. Skeletally mature patients have asymmetrical lesions and fewer lesions, which have varying sizes. The prognosis in both forms is poor.

TELANGIECTATIC OSTEOSARCOMA

An osteosarcoma is defined as telangiectatic when more than 90% of the tumor has a telangiectatic pattern on histologic examination. The lesion typically is located in the metaphysis of long bones (femur, tibia, and humerus). Telangiectatic osteosarcoma is expansile, lytic, and has a wide zone of transition. Radiographically, osteoid matrix and periosteal new bone formation are not seen with telangiectatic osteosarcoma. On MR imaging, the presence of multiple fluid-fluid levels is shown in approximately 90% of cases (Fig. 3–8). The differential diagnosis for this lesion includes giant cell tumor and aneurysmal bone cyst.

EXTRASKELETAL OSTEOSARCOMA

Extraskeletal osteosarcoma is a highly malignant soft tissue tumor, and it usually is seen in patients in their 30s to 50s. Extraskeletal osteosarcoma accounts for about 1.2% of all soft tissue sarcomas. Most commonly, the lesion occurs in the lower extremities, upper extremities, breast, trunk, and retroperitoneum. Radiographically the mass is typically larger than 6 cm in diameter and has central calcifications or ossification; this differentiates extraskeletal osteosarcoma from myositis ossificans, which has calcifications at the periphery of the lesion. Synovial sarcoma contains amorphous calcifications in about 25% of cases, but it usually occurs in younger individuals. In soft tissue chondrosarcoma, the calcifications in the mass are ringlike or punctate. Liposarcoma can mimic extraskeletal osteosarcoma; however, the ossifications observed in lipo-

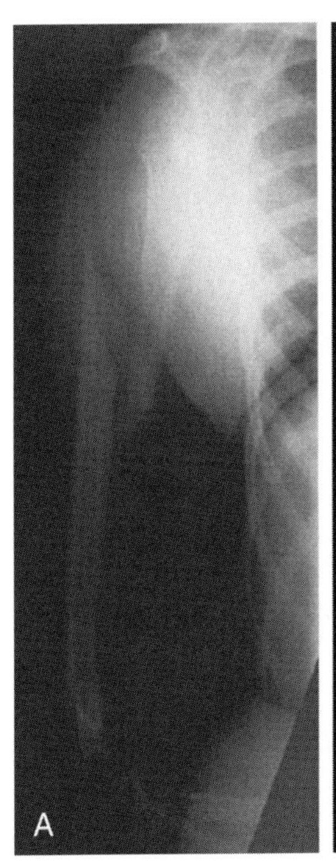

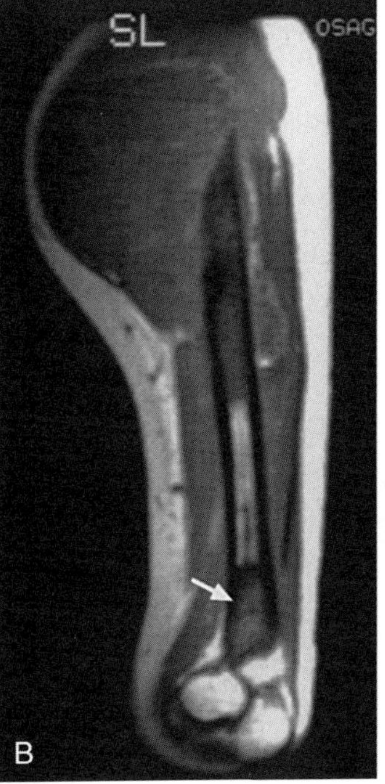

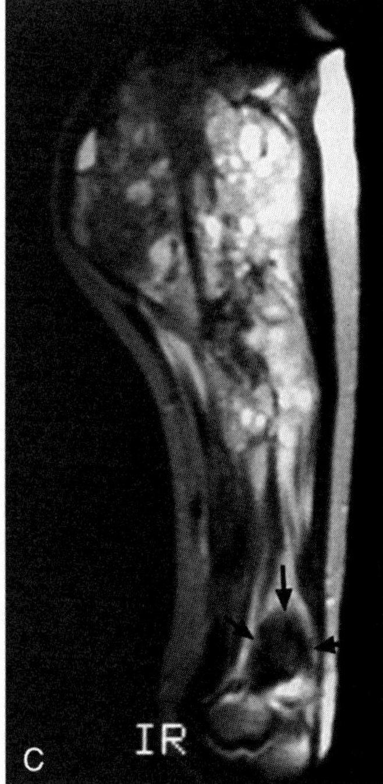

Figure 3–2. Osteosarcoma with a skip lesion. The patient is a 13-year-old girl who felt a snap in her right arm while playing. *A*, Lateral view of the right arm shows a pathologic fracture in the proximal shaft. This lesion appears moth-eaten. *B*, Sagittal T1-weighted MR image of the right humerus shows two lesions with normal marrow in between. The *arrow* identifies the more distal lesion. *C*, Sagittal T2-weighted MR image shows the lesions. The proximal lesion is mostly bright, indicating that it is primarily soft tissue. The distal lesion is dark, indicating it is blastic (*arrows*).

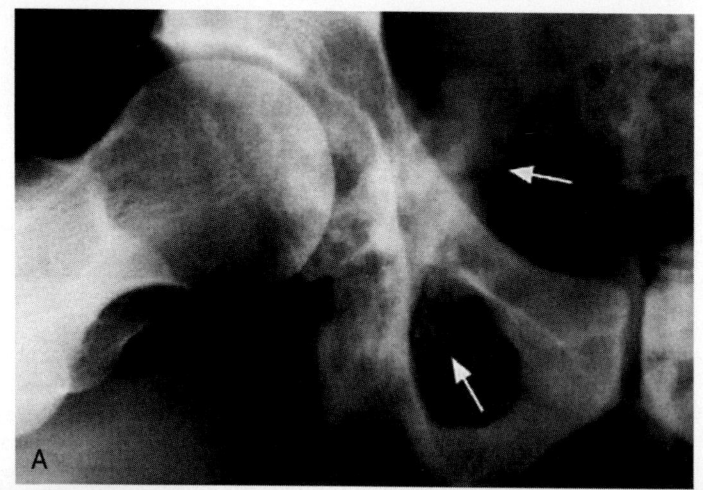

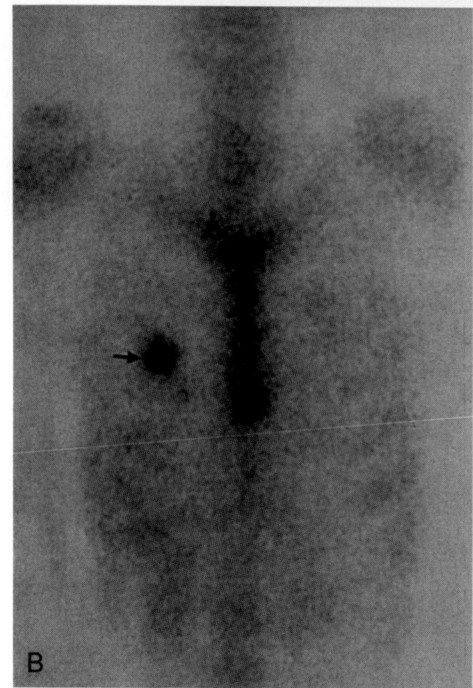

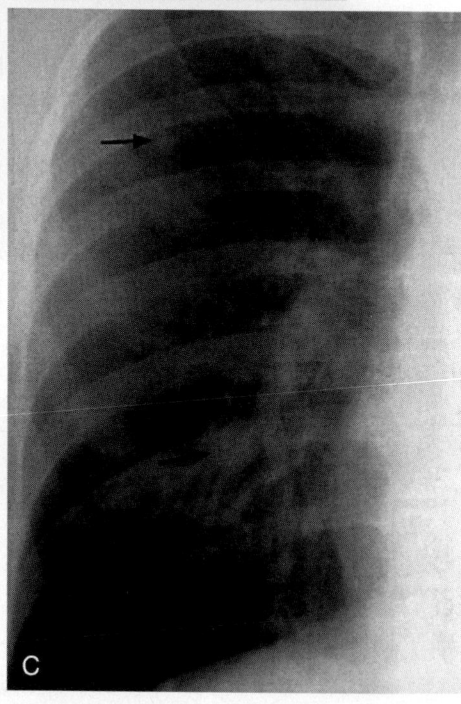

Figure 3–3. Osteosarcoma with metastasis shown by nuclear medicine bone scan. *A,* Frog-leg lateral view of the hip shows an osteosarcoma involving the acetabulum. Osteoid matrix can be seen extending into the adjacent soft tissues *(arrows)*. *B,* Bone scan reveals increased activity in the right thorax *(arrow)*, which was found to be metastatic osteosarcoma. *C,* Posteroanterior chest radiograph shows two nodules in the right lung *(arrows)*. These were foci of metastatic osteosarcoma.

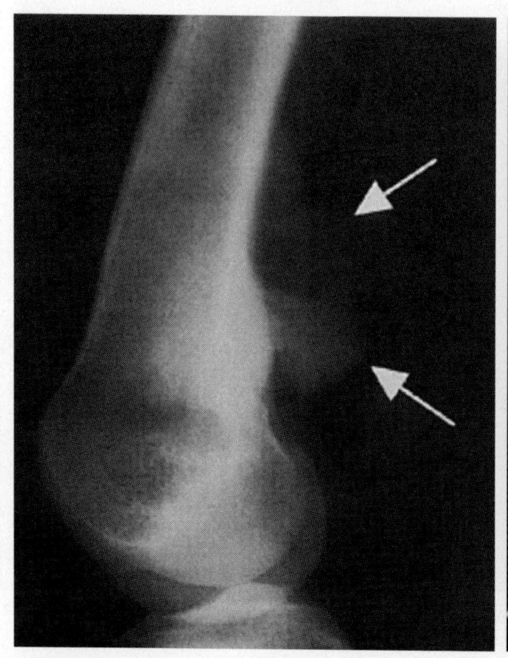

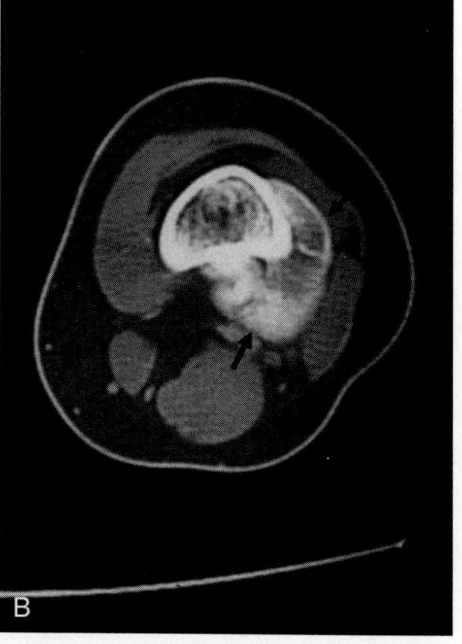

Figure 3–4. Parosteal osteosarcoma of the distal femur. *A,* Lateral radiograph of the knee shows an ossified, lobulated, protuberant mass posterior to the femur *(arrows)*. *B,* CT scan of the distal femur shows no evidence of invasion of the medullary space. The ossified mass can be seen to wrap partially around the distal femur *(arrows)*.

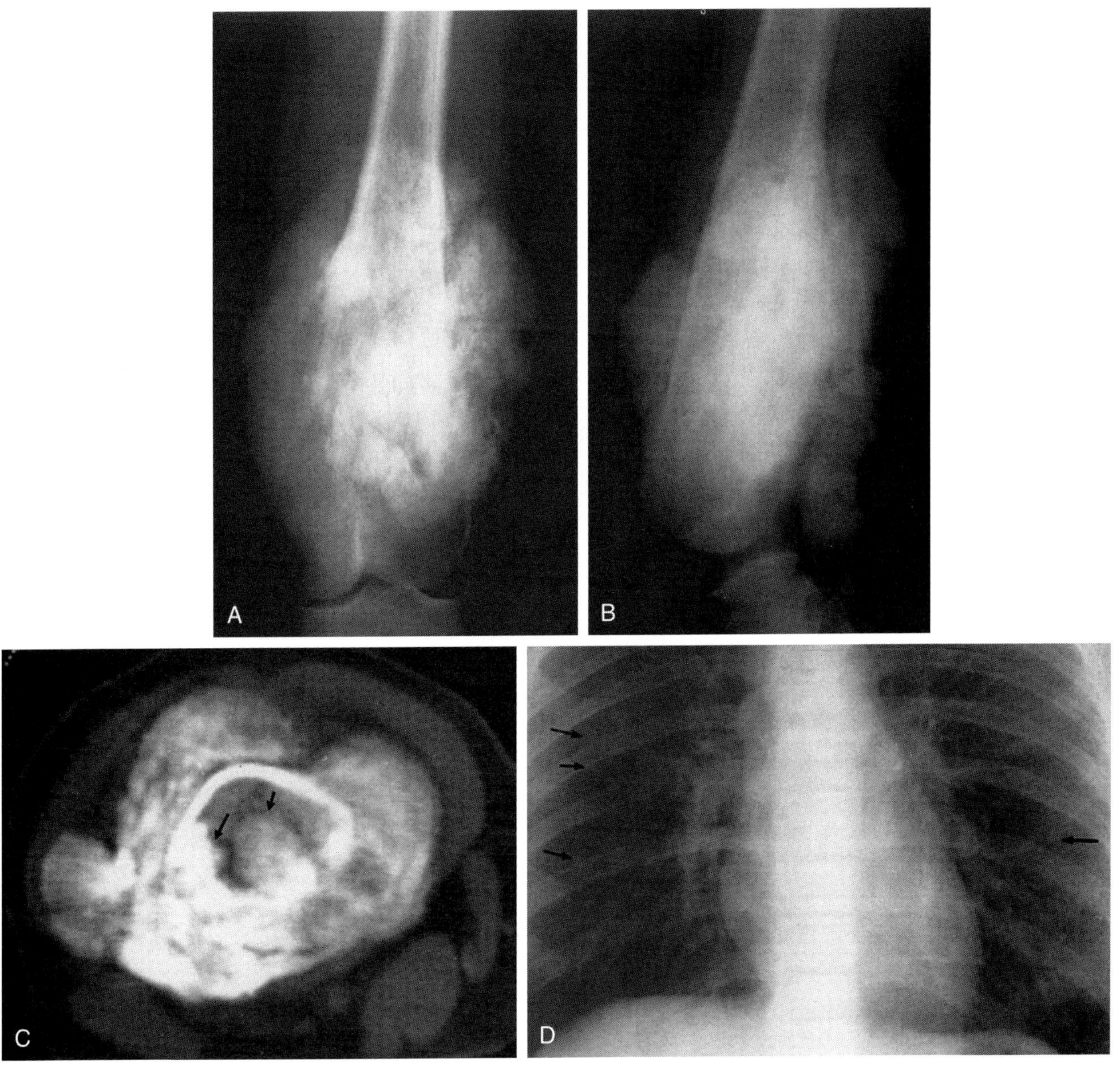

Figure 3–5. Large parosteal osteosarcoma of the distal femur with lung metastasis. *A* and *B*, Anteroposterior and lateral radiographs of the distal femur show a broad-based, lobulated, dense, protruding mass wrapping around the host bone (distal femur). Cloudlike osteoid matrix can be seen. *C*, CT scan shows the mass. Invasion of the medullary space is seen *(arrows)*, which occurs in about one third of cases of parosteal osteosarcoma. *D*, Posteroanterior view of the chest shows multiple nodules in both lungs *(arrows)*.

sarcoma are more organized than in osteosarcoma of the soft tissues.

POSTRADIATION OSTEOSARCOMA

Postradiation osteosarcomas have been noted in the radiated field 3 to 42 years after treatment. Characteristically, there is a history of orthovoltage (250 kVp) irradiation; however, there have been cases of radiation-induced sarcomas after megavoltage irradiation. The prognosis is poor. Radiographically the lesion can be lytic or blastic (Fig. 3–9) in appearance and usually is associated with changes of radiation osteitis (patchy areas of mixed density) in the surrounding bone.

OSTEOSARCOMA ASSOCIATED WITH PAGET'S DISEASE

Osteosarcoma occurs in about 2% of patients with Paget's disease; there is a special predilection for patients with long-standing polyostotic Paget's disease. Osteosarcoma accounts for about 60% to 80% of sarcomas arising in pagetoid bone. The pelvis and proximal long bones are common sites. Associated pathologic fractures are common (25%). The lesions usually are lytic and associated with a soft tissue mass (Fig. 3–10). Periosteal reaction is characteristically absent. In the peripheral skeleton, this lesion is studied best with MR imaging because peripheral Paget's disease typically has normal marrow signal on MR images, and Paget's disease associated with osteosarcoma appears infiltrated with tumor on T1-weighted images

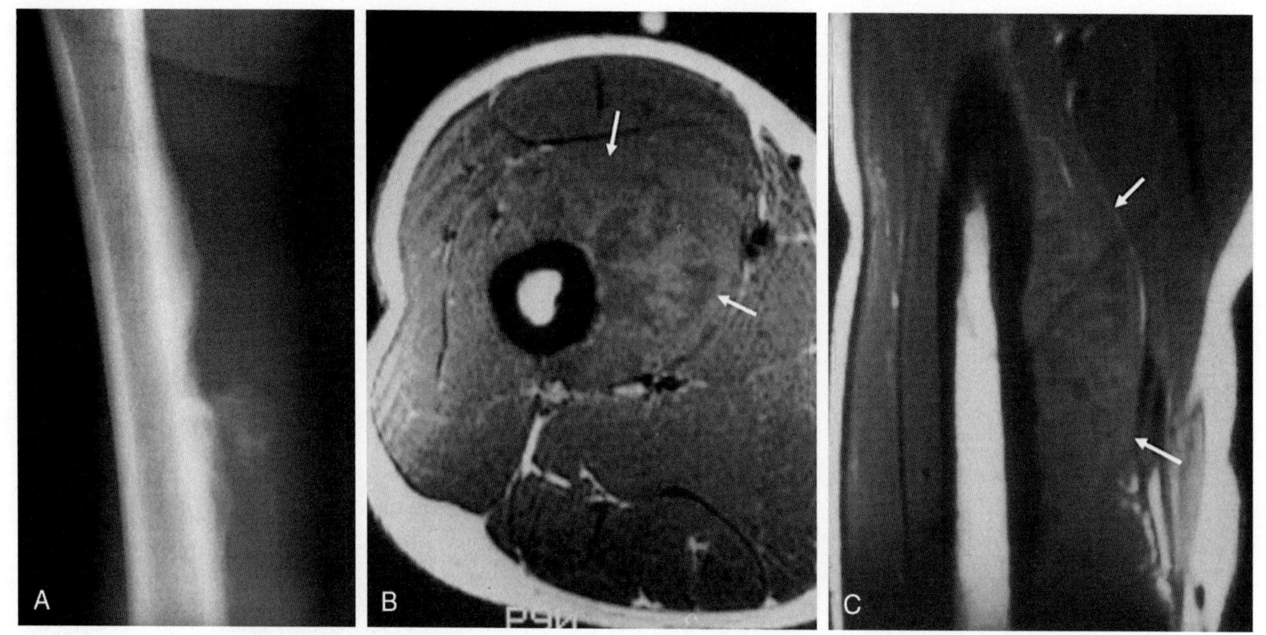

Figure 3–6. Periosteal osteosarcoma without invasion of the marrow space in an 18-year-old man. *A*, Anteroposterior view of the right femur shows an osteoid-forming lesion on the surface of the medial cortex. *B* and *C*, Axial and coronal T1-weighted MR images show the lesion *(arrows)* arising on the surface of the bone without invasion of the marrow space.

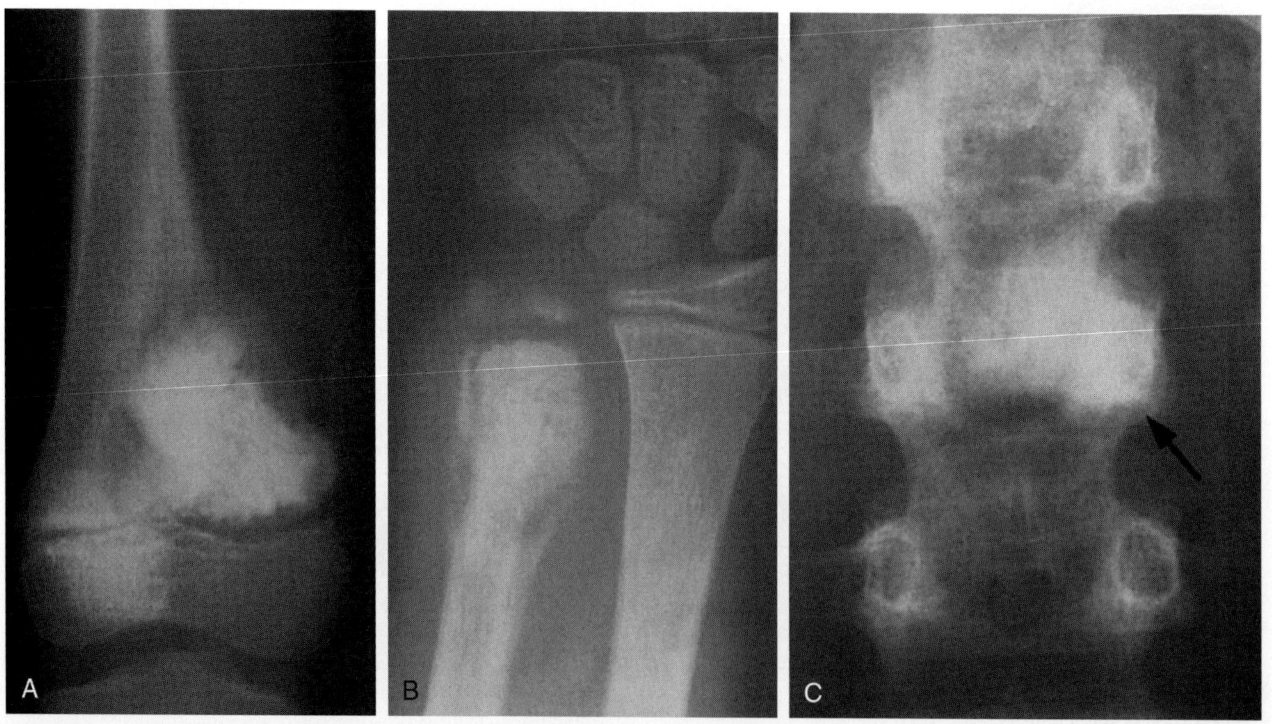

Figure 3–7. Multicentric osteosarcoma in an 8-year-old boy. The lesions were discovered within a few months of each other. *A*, Anteroposterior view of the right knee shows the initial lesion in the metaphysis of the distal femur. *B*, Posteroanterior view of the left wrist was obtained after pain and swelling developed on the ulnar side. A blastic, aggressive lesion involving the distal metaphysis and epiphysis of the ulna is shown. *C*, The patient also had back pain. Anteroposterior view of the spine shows a blastic lesion in the vertebra and left pedicle of L2 *(arrow)*.

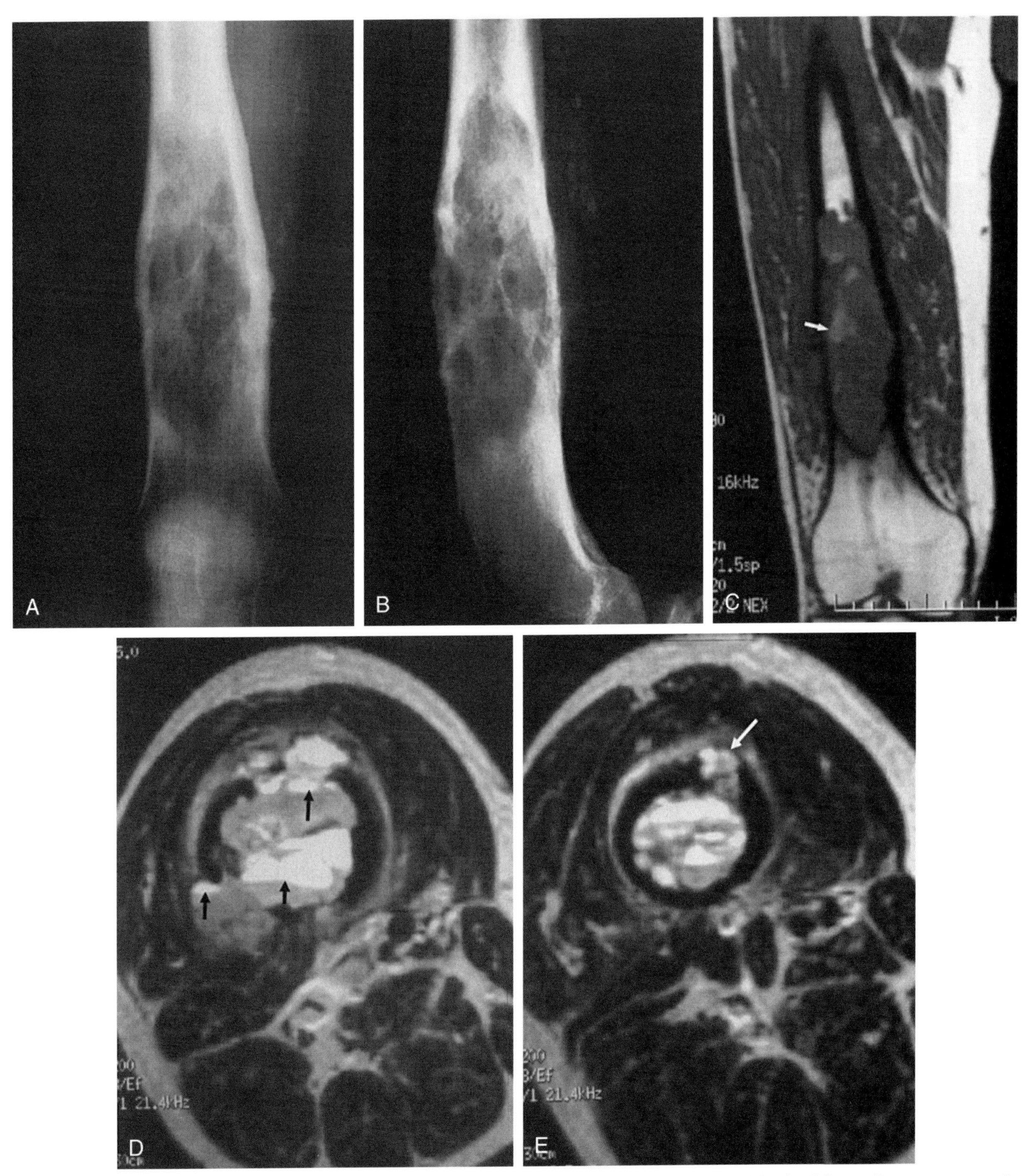

Figure 3–8. Telangiectatic osteosarcoma of the femur. *A* and *B*, Anteroposterior and lateral radiographs of the distal femur show an aggressive lesion in the diametaphyseal region. There is cortical destruction and extension into the surrounding soft tissues. *C*, Coronal T1-weighted MR image of the femur shows the mass within the distal femur. There are foci of mildly increased signal consistent with areas of intratumoral hemorrhage *(arrow)*. *D* and *E*, Axial fat-saturation FSE T2-weighted MR images of the distal femur show multiple fluid-fluid levels *(arrows)* that can be seen in telangiectatic osteosarcoma. Extension of the mass into the surrounding soft tissues also is visible.

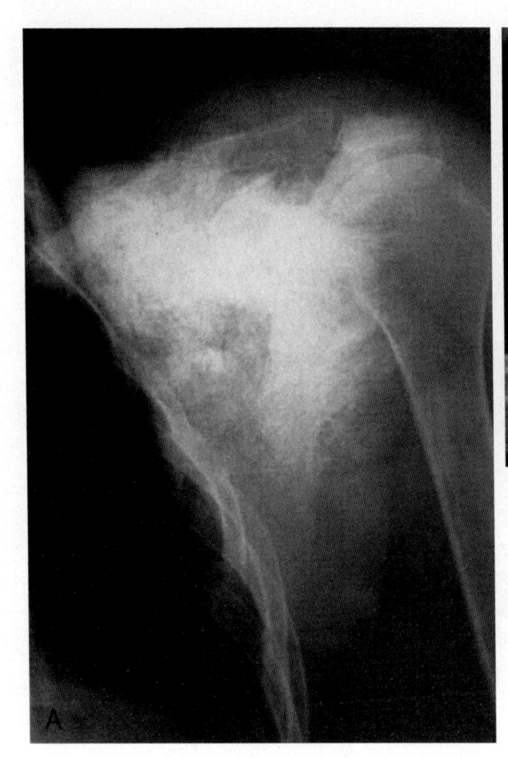

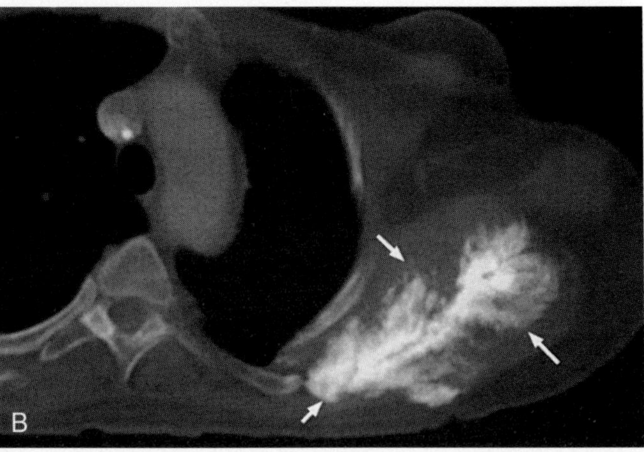

Figure 3–9. Postirradiation osteosarcoma in a 59-year-old woman who previously was irradiated for breast carcinoma. *A*, Anteroposterior view of the left shoulder shows an aggressive osteoid-forming lesion involving most of the left scapula. *B*, Axial CT scan through the scapula shows the lesion is an ossifying mass *(arrows)* extending from the scapula into the soft tissues.

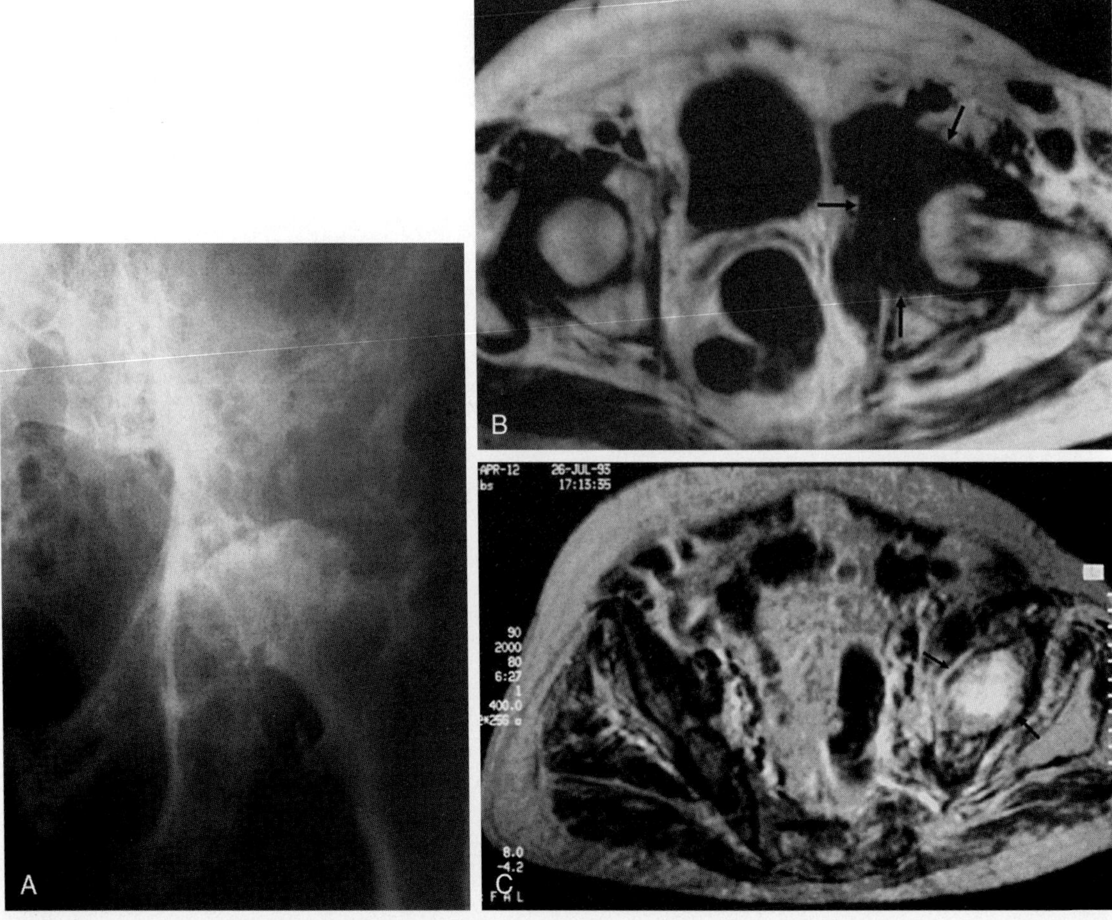

Figure 3–10. Paget's sarcoma in an 81-year-old man with a long-standing history of Paget's disease of the pelvis who presented with increasing pain in the left hip. *A*, Anteroposterior view of the left hip shows diffuse Paget's disease and a lytic lesion in the left acetabulum. *B*, Axial T1-weighted MR image through the left hip shows marrow replacement with a low signal intensity mass, which is expansile *(arrows)*. The mass is extending into the true pelvis. *C*, Axial T2-weighted MR image reveals a high signal intensity mass in the left acetabulum *(arrows)*.

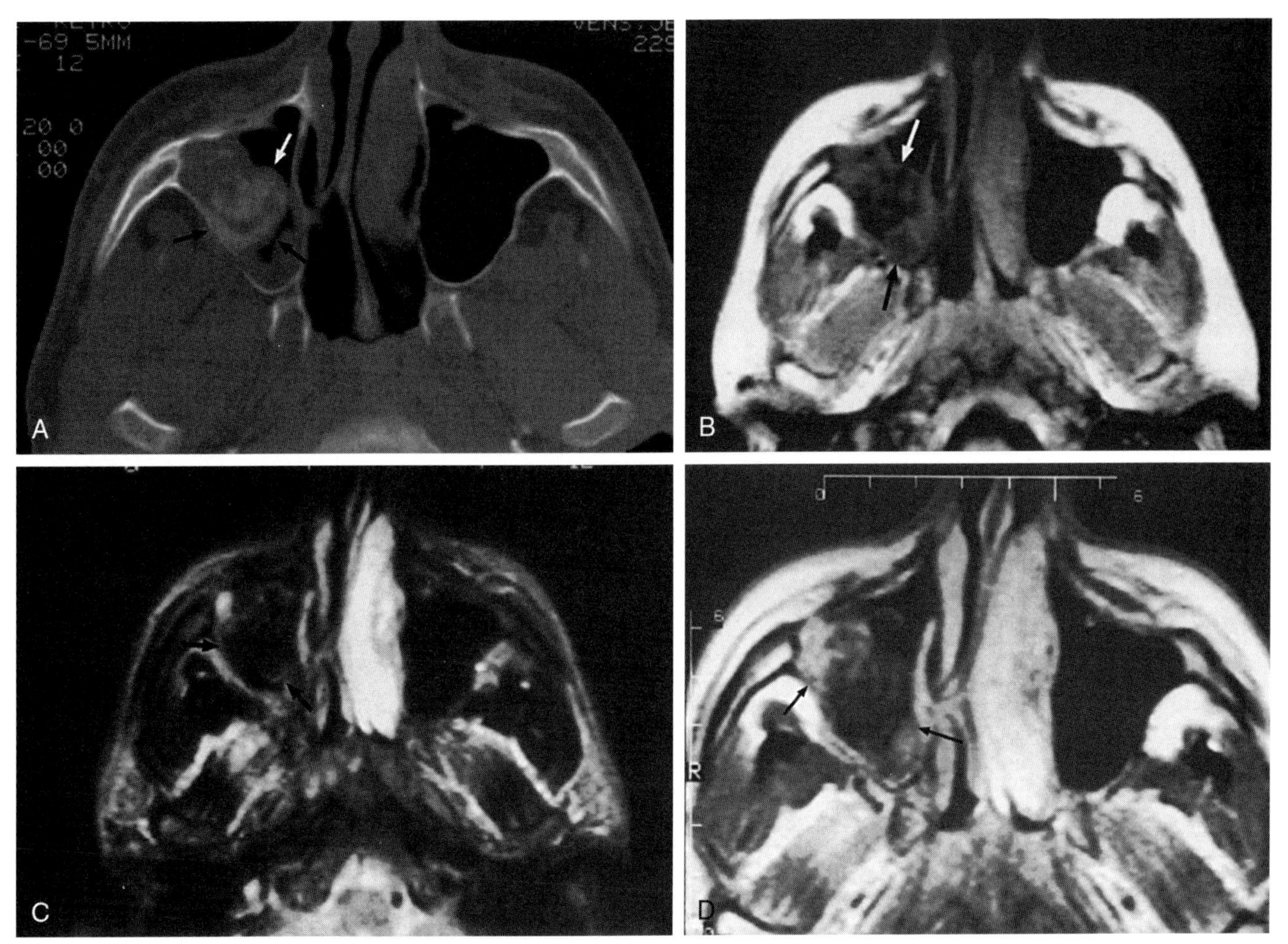

Figure 3–11. Gnathic osteosarcoma involving the right maxilla and antrum in a 32-year-old man. The lesion was resected surgically, and 11 years later the patient continues to be disease-free. *A,* Axial CT section through the maxillary sinus shows a densely mineralized mass in the right antrum *(arrows).* *B* and *C,* Axial T1- and T2-weighted MR images show the mass breaking through the anterior wall of the maxillary sinus. The *arrows* identify the mass. *D,* Axial postgadolinium T1-weighted MR image shows heterogeneous enhancement *(arrows)* and destruction of the anterior wall of the right maxillary sinus.

(see Fig. 3–10). In the spine, Paget's disease can have variable signal characteristics on MR imaging, which occasionally can resemble neoplasm.

GNATHIC OSTEOSARCOMA (OSTEOSARCOMA OF THE JAW)

Gnathic osteosarcoma is differentiated from conventional osteosarcoma because it occurs later in life (patients are 5 to 15 years older). The lesion usually is well differentiated and has a better prognosis than conventional osteosarcoma. Gnathic osteosarcoma is typically chondroblastic. It involves the mandible and maxilla equally. Radiographically the lesion can be lytic, blastic, or mixed (Fig. 3–11). It often is associated with a soft tissue mass.

Recommended Readings

Bloem JL, Taminiau AHM, Eulderink F, et al: Radiologic staging of primary bone sarcomas: MR imaging, scintigraphy, angiography, and CT correlated with pathologic examination. Radiology 169:805, 1988.

Griffith JF, Kumta SM, Chow LT, et al: Intracortical osteosarcoma. Skeletal Radiol 27:228, 1998.

Jelinek JS, Murphey MD, Kransdorf MJ, et al: Parosteal osteosarcoma: Value of MR imaging and CT in the prediction of histologic grade. Radiology 201:837, 1996.

Logan PM, Mitchell MJ, Munk PL: Imaging of variant osteosarcomas with an emphasis on CT and MR imaging. AJR Am J Roentgenol 171:1531, 1998.

Murphey MD, Robbin MR, McRae GA, et al: The many faces of osteosarcoma. Radiographics 17:1205, 1997.

Okada K, Unni KK, Swee RG, et al: High grade surface osteosarcoma: A clinicopathologic study of 46 cases. Cancer 85:1044, 1999.

Onikul E, Fletcher BD, Pharham DM, et al: Accuracy of MR imaging for estimating intraosseous extent of osteosarcoma. AJR Am J Roentgenol 167:1211, 1996.

Rosenberg ZS, Lev S, Schmahmann S, et al: Osteosarcoma: Subtle, rare, and misleading plain film features. AJR Am J Roentgenol 165:1209, 1995.

Schima W, Amann G, Stiglbauer R, et al: Preoperative staging of osteosarcoma: Efficacy of MR imaging in detecting joint involvement. AJR Am J Roentgenol 163:1171, 1994.

Seeger LL, Eckardt JJ, Bassett LW: Cross-sectional imaging in the evaluation of osteogenic sarcoma: MR imaging and CT. Semin Roentgenol 24:174, 1989.

Unni KK, Dahlin DC: Osteosarcoma: Pathology and classification. Semin Roentgenol 24:143, 1989.

Vanel D, Verstraete KL, Shapeero LG: Primary tumors of the musculoskeletal system. Radiol Clin North Am 35:213, 1997.

CHAPTER 4

Benign Cartilage-Forming Tumors

Nabil J. Khoury, MD, Georges Y. El-Khoury, MD, and D. Lee Bennett, MD

ENCHONDROMA

Solitary Enchondroma

Overview

Solitary enchondromas are a common medullary lesion, representing 3% to 17% of all primary bone tumors. They also are the most common lesion of the hand. Enchondromas may be solitary or multiple. When there are multiple lesions, the disease is called *multiple enchondromatosis* or *Ollier's disease.* When soft tissue hemangiomas are associated with multiple enchondromatosis, the disease is called *Maffucci's syndrome.* Enchondromas arise from cartilage rests displaced from the growth plate, which actively proliferate within regions where all other cartilaginous tissue has remodeled into bone. Enchondromas are variable in size. They usually are less than 5 cm in length but may reach 18 cm (Fig. 4–1). Lesions in the hands and feet usually are small. An enchondroma consists of small lobules of hyaline cartilage. Histologic examination reveals uniformly sized uninuclear cartilage cells with a few binuclear cells. The cells are benign in appearance. Mitosis is absent or rare. Most solitary enchondromas are diagnosed between the third and fourth decades of life. Enchondromas usually are asymptomatic, being discovered incidentally. Pain is present, however, if a pathologic fracture has occurred. Deformity, dysfunction, and reduced joint movement may be encountered, particularly in the digits.

Skeletal Involvement

Of solitary enchondromas, 50% occur in the tubular bones of the hands. They most commonly occur in the proximal phalanges, followed by the metacarpals and middle phalanges (Fig. 4–2). They rarely occur in the distal phalanges. Enchondromas also are common in the metatarsals. Of enchondromas, 25% involve the long bones, most commonly the femur, followed by the humerus, tibia, fibula, and ulna. Other less common sites include the carpal bones, sternum, patella, pelvis, and vertebrae. In the long bones, enchondromas are mainly metaphyseal or diaphyseal in location (see Fig. 4–1). In the short tubular bones, enchondromas occur in the diaphysis. An exophytic enchondroma of the long bones has been described and is termed *enchondroma protuberans.* This lesion resembles an osteochondroma, but the surgical treatment of this lesion is different.

Imaging

RADIOGRAPHY

Radiographs reveal a lytic, medullary lesion. The margins usually are well defined and lobulated; however, the margin may vary from being sclerotic to ill defined. Central chondroid calcifications are seen in about 95% of cases (see Fig. 4–1). These are either small flecks or popcorn-like in appearance. Enchondromas may be expansile, especially in the hands and feet. Frequently, endosteal scalloping is present. The surrounding cortex may be thinned, sclerotic, remodeled, or destroyed. These aggressive cortical features of benign enchondromas are seen most commonly in the hands and feet. If these features are noted in other locations, malignant degeneration should be considered. In some instances, enchondromas present as linear or tubular lucencies within the metaphysis (Fig. 4–3).

COMPUTED TOMOGRAPHY

Computed tomography (CT) findings are similar to the abnormalities seen on radiography. CT is superior to radiographs, however, in showing the presence of chondroid calcifications, endosteal scalloping, cortical destruction, and fractures (Fig. 4–4).

MAGNETIC RESONANCE IMAGING

On magnetic resonance (MR) imaging, an enchondroma appears as a well-defined lobular lesion that is hypointense to isointense to muscle on T1-weighted images. On T2-weighted and STIR (short tau inversion recovery) images, cartilage usually is hyperintense (Fig. 4–5; see also Fig. 4–1). Calcifications and septa within the lesion appear as foci of decreased signal intensity on all sequences. Speckled areas of increased signal may be seen on T1-weighted images as a result of remaining foci of normal yellow marrow within the cartilage lobules. Enchondromas do not display peritumoral edema or a soft tissue component on MR imaging, unless a fracture with hematoma develops.

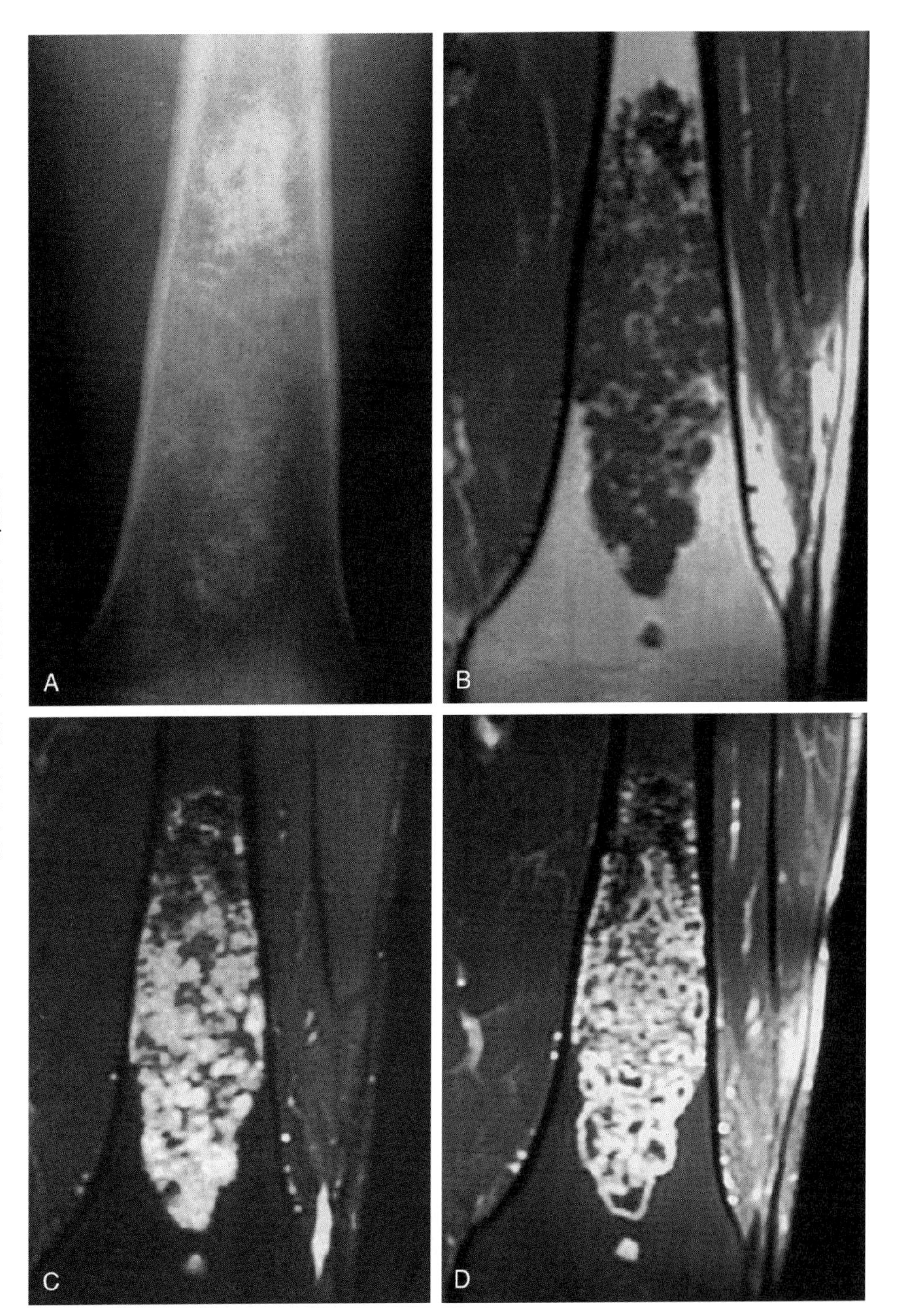

Figure 4–1. Large enchondroma. *A*, Anteroposterior radiograph of the distal femur shows a heavily calcified lesion occupying the medullary space of the left femur. The lesion is mildly expansile. The matrix calcifications, which are popcorn-like, are typical for a cartilaginous matrix. *B*, T1-weighted coronal MR image shows a lobulated low signal intensity lesion. *C*, STIR coronal MR image shows heterogeneous signal intensity reflecting nests of cartilage that appear bright, with areas of low signal intensity representing the calcified areas. *D*, T1-weighted fat-suppressed coronal MR image with gadolium enhancement reveals lobulated areas of increased signal intensity (enhancement) with focal areas of low signal intensity.

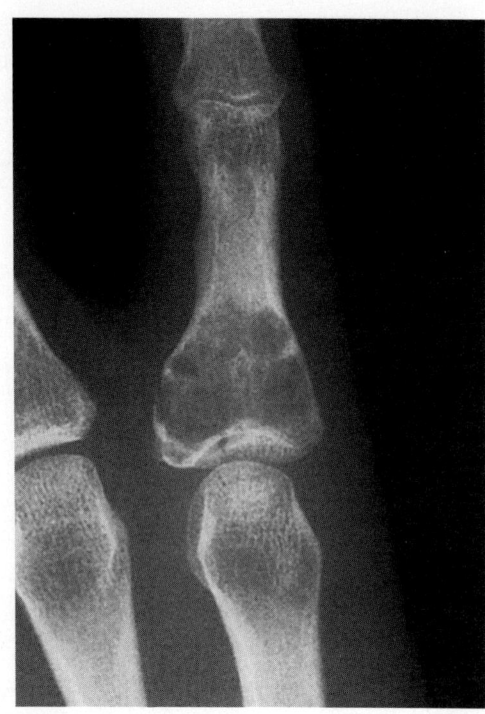

Figure 4–2. Enchondroma of the proximal phalanx. Anteroposterior view of the proximal phalanx of the right hand shows a lytic, expansile lesion centered within the medullary cavity. This is a classic location for an enchondroma.

Complications

The most common complication is malignant degeneration into a chondrosarcoma; however, this is extremely rare in the hands and feet. Occasionally an enchondroma may fracture (Fig. 4–6).

Differential Diagnosis

Low-Grade Chondrosarcoma. It may be difficult to differentiate an enchondroma from a low-grade chondrosarcoma. The degree of bone scalloping and the presence of pain are helpful in this differentiation.

Bone Infarct. A bone infarct can mimic an enchondroma. Bone infarcts have a serpentine calcified margin, however, rather than central calcifications, as seen in enchondromas.

Other. An enchondroma without radiographically detectable calcifications can resemble a simple bone cyst or fibrous dysplasia.

Multiple Enchondromatosis (Ollier's Disease)

Overview

Ollier's disease is a rare nonhereditary disease. It is characterized by persistent cartilaginous masses in the metaphyses and diaphyses. The limbs are involved asymmetrically with one side being either exclusively or predominantly affected. The histopathology of this disease is similar to that of a solitary enchondroma, but it shows more cellularity than in the solitary lesions. Enchondromatosis usually is diagnosed between 2 and 10 years of age. Clinically, Ollier's disease may present with palpable masses, extremity shortening, and deformity. There is an equal distribution between males and females.

Skeletal Involvement

The disease is predominantly unilateral, with one side of the body affected. It involves the tubular bones, most

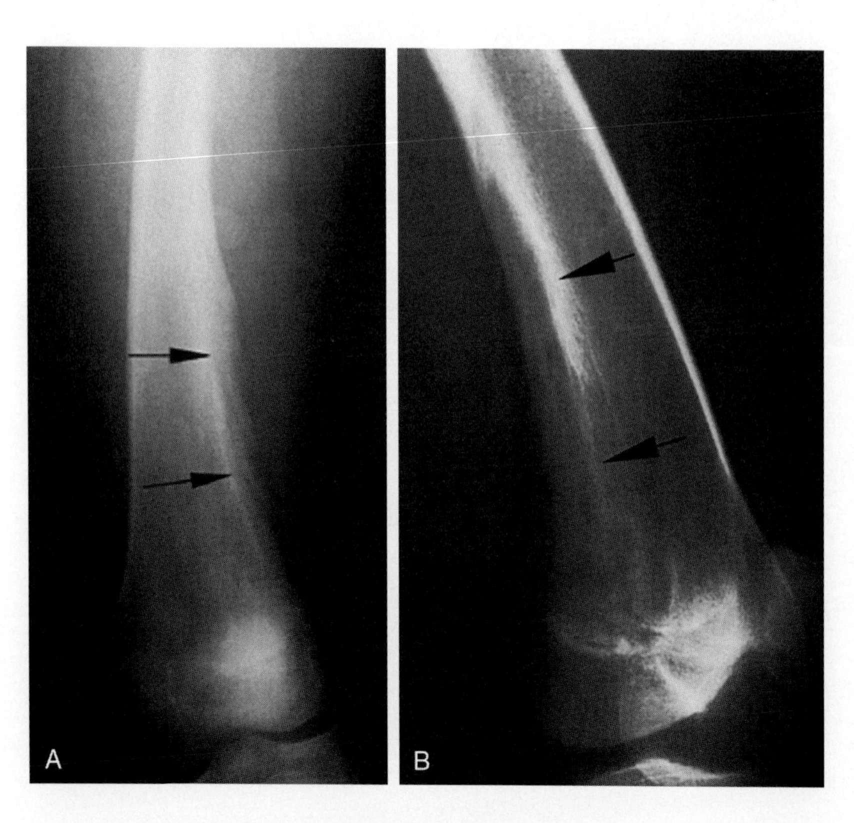

Figure 4–3. An enchondroma presenting as a linear, tubular lucency. *A* and *B*, Anteroposterior and lateral radiographs of the distal right femur show linear, tubular-like lucencies in the metaphyseal and diametaphyseal regions *(arrows)*. The lucencies are oriented more or less with the long axis of the bone.

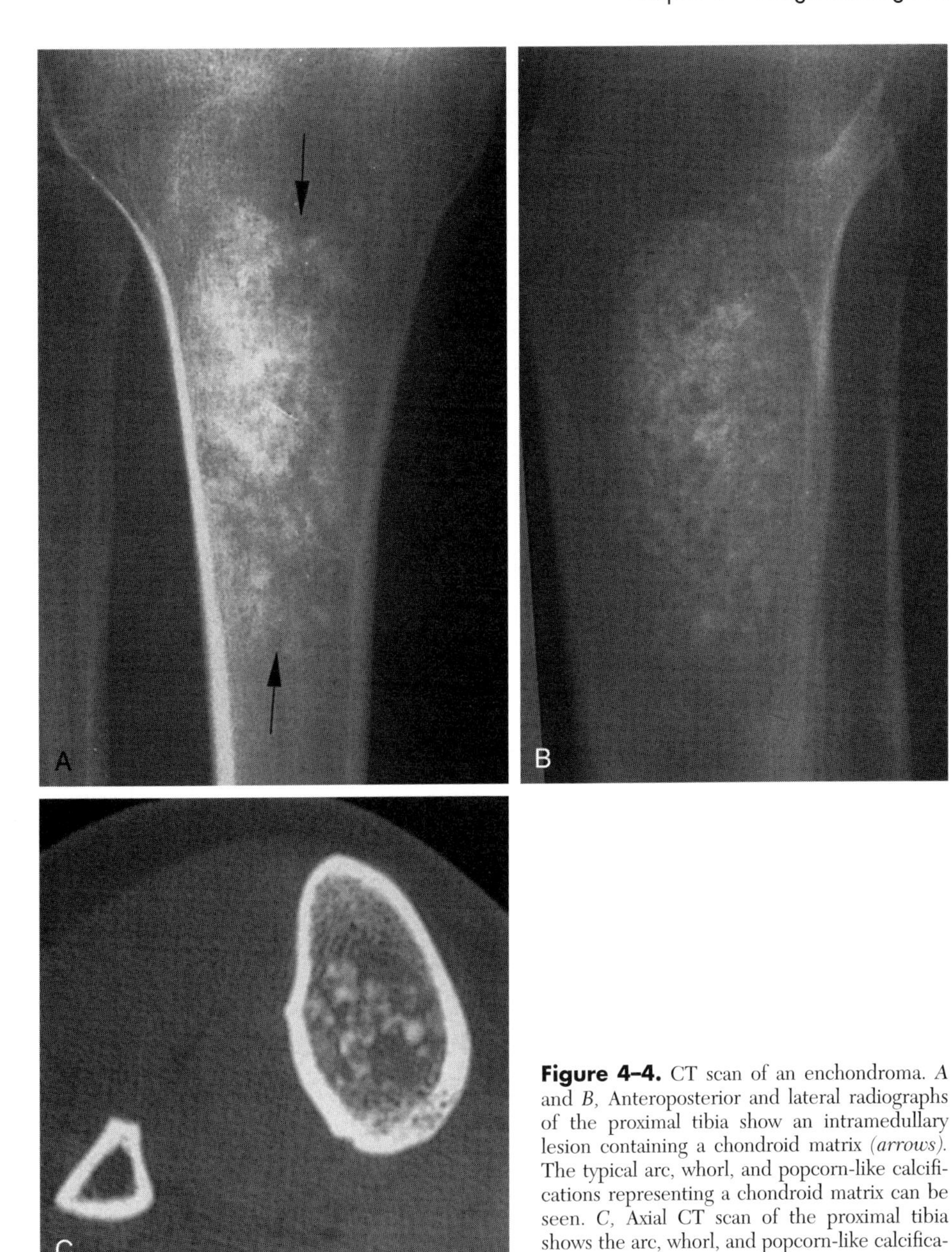

Figure 4–4. CT scan of an enchondroma. *A* and *B*, Anteroposterior and lateral radiographs of the proximal tibia show an intramedullary lesion containing a chondroid matrix *(arrows)*. The typical arc, whorl, and popcorn-like calcifications representing a chondroid matrix can be seen. *C*, Axial CT scan of the proximal tibia shows the arc, whorl, and popcorn-like calcifications indicative of a chondroid matrix.

frequently the femur, tibia, and bones in the hand (Fig. 4–7). To a lesser degree, lesions may be located in the flat bones, especially the pelvis (Fig. 4–8). The location within the affected bones is similar to that of a solitary enchondroma. Metaphyseal cartilaginous bands may extend into the shaft. The involved extremity almost always is shorter than the contralateral, normal limb. Pathologic fractures are common; they occur in 33% of patients, mainly affecting the femur.

Imaging

Radiographic findings are similar to solitary enchondromas. Ollier's disease displays prominent longitudinal, tubular lucencies in the metaphyses or diaphyses, with or without surrounding sclerosis (see Fig. 4–7). These correspond to the extension of cartilaginous bands. Similar lucencies also may be seen in the flat bones, particularly the iliac crest (see Fig. 4–8). Lesions may become large, causing cortical erosions, osseous deformities, and osseous shortening. Angular deformities of bone are most frequent in the femur and tibia. Calcifications are common, especially in the metaphyseal lesions.

Natural History and Complications

During growth, the lesions of Ollier's disease may regress. After growth cessation, the lesions remain stable. The incidence of malignant transformation into chondrosarcoma is higher than with a solitary enchondroma; it is reported to be 30%. Malignant transformation to other sarcomas is extremely rare; however, it has been reported. Malignant transformation almost always occurs during adulthood. Pathologic fractures may occur, particularly in childhood.

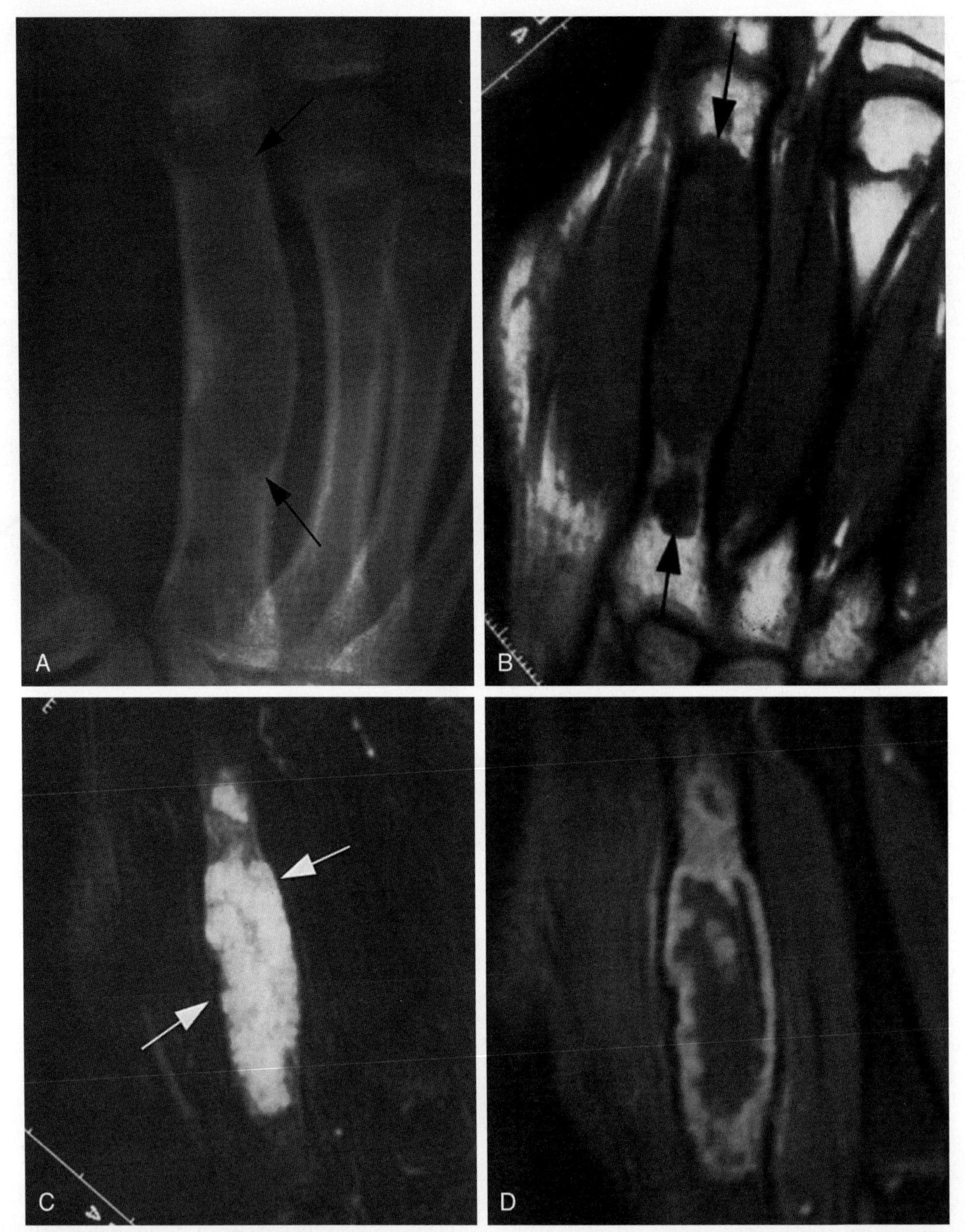

Figure 4–5. Enchondroma of the second metacarpal in a 15-year-old boy. *A,* Oblique view of the right hand shows a lytic, expansile lesion in the second metacarpal *(arrows)*. There is no identifiable matrix, but the location and appearance of the lesion are typical for an enchondroma. *B,* T1-weighted coronal MR image of the right hand shows a homogeneously low signal intensity lesion occupying most of the shaft of the second metacarpal *(arrows)*. *C,* Fat-suppressed T2-weighted coronal MR image shows the lesion to be fairly uniformly hyperintense *(arrows)*. The appearance is typical of cartilage. *D,* Postgadolinium fat-suppressed T1-weighted MR image shows predominantly peripheral enhancement of the lesion.

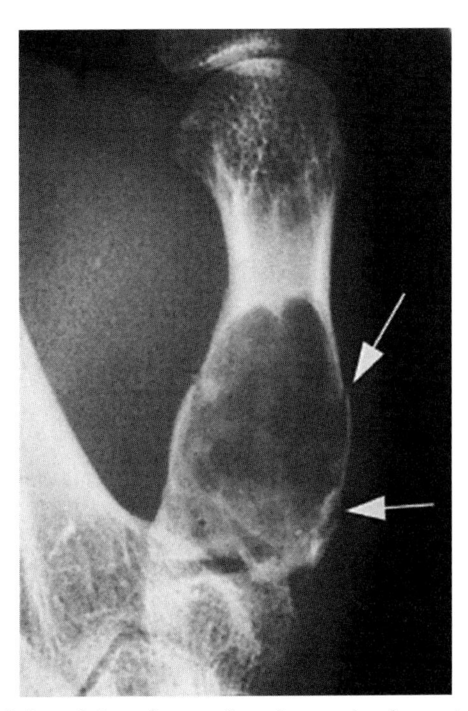

Figure 4–6. Pathologic fracture through an enchondroma. Anteroposterior radiograph of the first metacarpal shows an expansile, centrally located, lytic lesion in the proximal aspect of this bone. A pathologic fracture through the enchondroma is present *(arrows)*.

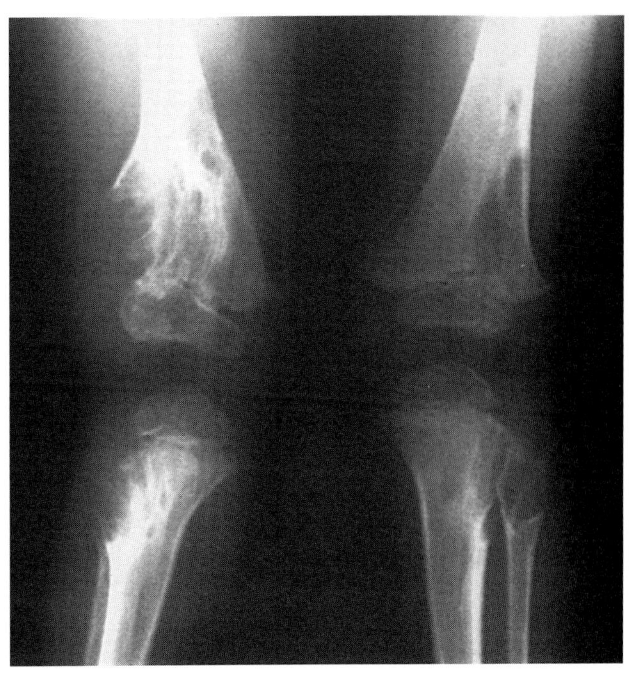

Figure 4–7. Ollier's disease involving the long bones. Anteroposterior radiograph of the knees shows large cartilage rests in the metaphyses of the distal femora, proximal tibiae, and proximal fibulae. A significant valgus deformity of the right knee is present.

Maffucci's Syndrome

Overview

Maffucci's syndrome is a rare, congenital, nonhereditary disease consisting of multiple enchondromas with multiple soft tissue hemangiomas. The hemangiomas appear as red-blue nodules, located in the subcutaneous tissues. Rarely, they are located on mucosal surfaces and within visceral organs. This syndrome usually is diagnosed by age 10.

Skeletal Involvement

Maffucci's syndrome is similar to Ollier's disease. Unilateral involvement occurs in 50% of cases.

Imaging

RADIOGRAPHY

Radiography shows findings similar to those of Ollier's disease. Also, phleboliths within the soft tissue hemangiomas can be seen, suggesting the proper diagnosis (Fig. 4–9). Hemihypertrophy sometimes is present, due to the hemangiomas.

MAGNETIC RESONANCE IMAGING

MR imaging can help detect malignant transformation by showing cortical destruction and soft tissue invasion on

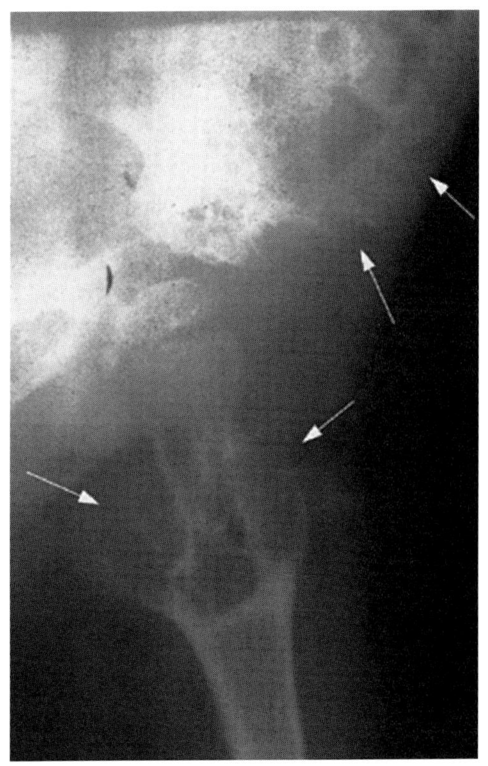

Figure 4–8. Ollier's disease. Anteroposterior radiograph of the hip shows extensive involvement of the proximal femur and iliac wing *(arrows)*.

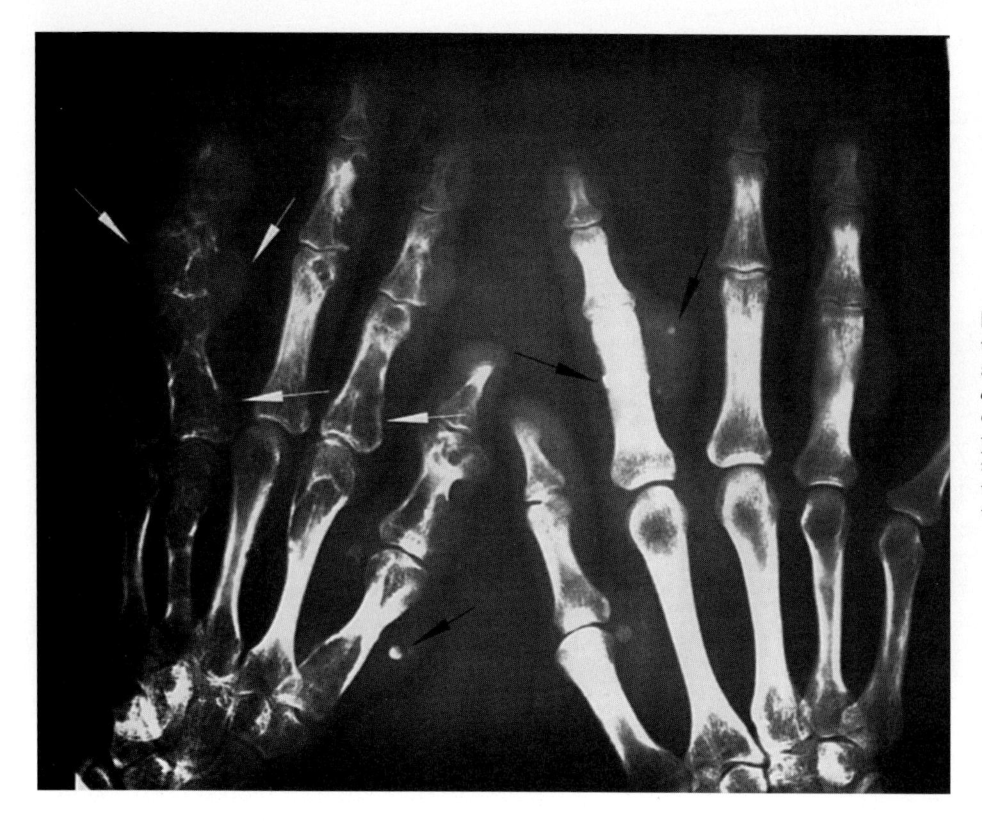

Figure 4–9. Maffucci's syndrome. Posteroanterior radiograph of the hands shows multiple enchondromas *(white arrows)*. Multiple soft tissue masses containing calcifications consistent with phleboliths also are present *(black arrows)*. This radiographic appearance of the soft tissues is consistent with multiple hemangiomas.

T2-weighted images. MR imaging also shows the extent and size of soft tissue hemangiomas.

Complications

Maffucci's syndrome has a high incidence of malignant degeneration of the enchondromas, usually occurring in patients older than age 40. Patients with Maffucci's syndrome also are prone to develop nonchondromatous malignancies, which include osteosarcoma, hepatobiliary carcinoma, pancreatic carcinoma, astrocytoma, glioma, and pituitary adenoma.

PERIOSTEAL (JUXTACORTICAL) CHONDROMA

Overview

Periosteal chondroma is a rare, solitary lesion representing 1% to 1.3% of all cartilaginous lesions. It is a surface variant of an enchondroma. It arises within the periosteum or adjacent connective tissue. It is usually small (<4 cm). The lesion is hypercellular and is formed by hyaline and myxoid cartilage. Most periosteal chondromas occur in children and young adults (<30 years old). The lesion is more frequent in males.

Skeletal Involvement

Most juxtacortical chondromas occur in the long tubular bones, with the proximal humerus being most commonly affected, followed by the femur and anterior tibial tubercle. The lesion also is seen occasionally in the tubular bones of the hands and feet. Juxtacortical chondromas usually are metaphyseal in location and frequently present at the insertion of tendons and ligaments.

Imaging

Radiography and Computed Tomography

Radiographs and CT scans usually reveal a juxtacortical low-density lesion with or without a sharp sclerotic margin (Fig. 4–10). Chondroid calcifications occur in 50% of the cases. Other findings may include endosteal sclerosis and a surrounding periosteal shell. The adjacent cortex may be intact, thickened, scalloped, or eroded.

Magnetic Resonance Imaging

On MR imaging, the lesions have intermediate signal intensity on T1-weighted sequences and increased signal intensity on T2-weighted sequences. If calcifications are present, they appear as low signal intensity foci on all sequences.

Differential Diagnosis

Periosteal chondrosarcomas can resemble periosteal chondromas.

Complications

Malignant degeneration is rare. The lesion is cured by curettage.

CHONDROBLASTOMA

Overview

Chondroblastoma is a rare, benign chondroid lesion accounting for about 1.4% of all benign bone tumors. It consists of polyhedral, moderately sized chondroblasts with interstitial connective tissue. Giant cells may be present as well as areas of hemorrhage and necrosis. The lesion usually is discovered before age 30, with most seen in the teens. The male-to-female ratio is approximately 1.8:1. Clinical manifestations include local pain and swelling. Joint symptoms with severe synovitis may be present.

Skeletal Involvement

Chondroblastomas usually occur in the epiphyses or apophyses of long tubular bones (Fig. 4–11). The femur is the most frequently affected bone, followed by the humerus, tibia, fibula, and radius. Bones of the hands and feet are affected in 10% of cases. In the long bones, the lesion typically is epiphyseal, extending to the articular surface or metaphysis.

Imaging

Radiography

Radiographs show a lytic, well-marginated, round or oval lesion with a thin, reactive, sclerotic rim (see Fig. 4–11). The metaphyseal extension may be less well marginated and without a rim. Minimal periosteal reaction may be present in the adjacent metaphysis. Aggressive features have been described in recurrent lesions. Amorphous calcifications consistent with a chondroid matrix are noted in 30% to 50% of cases (Fig. 4–12). The diagnosis often is made with confidence based on radiographs alone.

Computed Tomography

CT shows findings similar to those seen by radiographs but to better advantage. This is particularly true in visualizing the calcifications (see Fig. 4–12).

Magnetic Resonance Imaging

MR imaging features of chondroblastomas can be confusing (Fig. 4–13). Typically, there is extensive marrow and soft tissue edema, which can be confused for tumor infiltration. In most cases, MR imaging is not necessary.

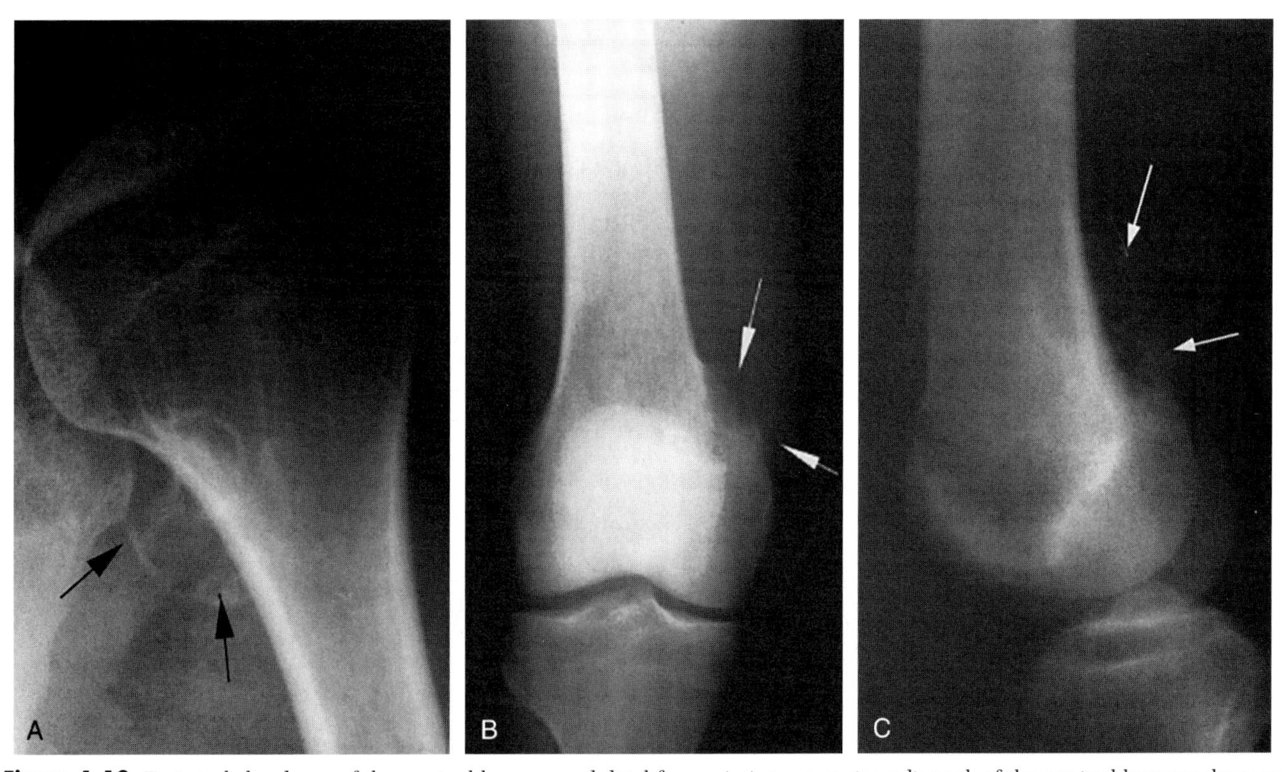

Figure 4–10. Periosteal chondroma of the proximal humerus and distal femur. *A*, Anteroposterior radiograph of the proximal humerus shows an exophytic lesion arising at the surface of the proximal humerus. The surface or periphery of the lesion has a thin sclerotic rim *(arrows)*. Periosteal chondroma of the distal femur from another patient. *B* and *C*, Anteroposterior and lateral radiographs of the distal femur reveal a juxtacortical mass *(arrows)*. Most of the margins are sharp, and calcifications consistent with a chondroid matrix can be seen. The calcifications are seen best on the lateral view.

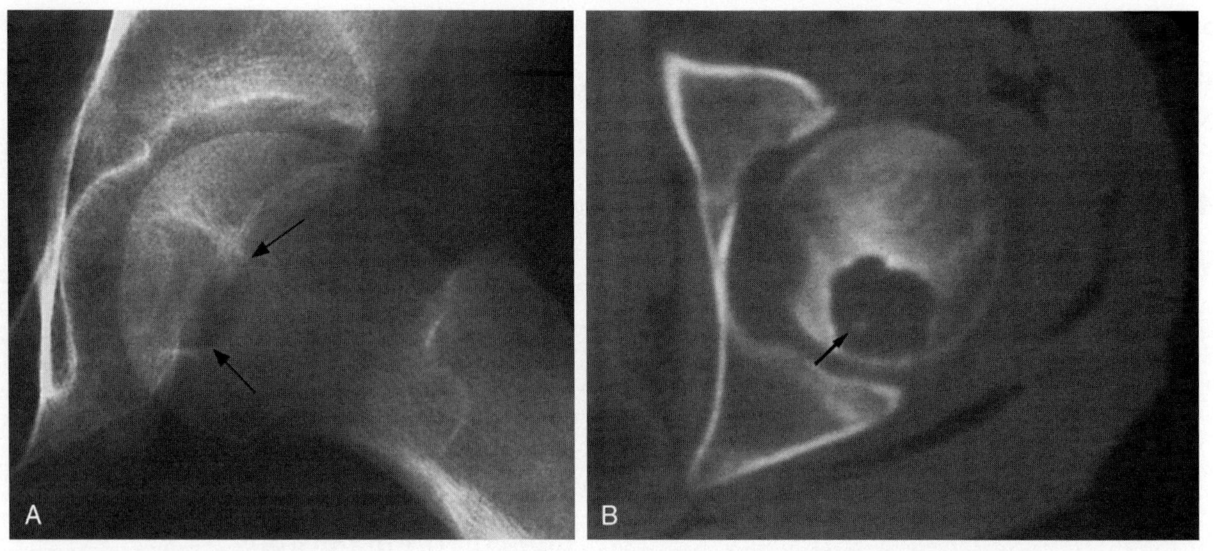

Figure 4–11. Chondroblastoma of the proximal femoral epiphysis in an 11-year-old girl. *A,* Frog-leg view of the left hip shows a well-demarcated epiphyseal lesion with a thin sclerotic margin *(arrows)*. There is no identifiable matrix within the lesion. *B,* Axial CT scan at the level of the lesion reveals a well-demarcated geographic lesion that contains a small calcification *(arrow)*.

Differential Diagnosis

Lesions that can resemble a chondroblastoma (on radiograph and on CT scan) include clear cell chondrosarcoma and a bone abscess. If the growth plate is closed, one may consider a giant cell tumor.

Complications

A few chondroblastomas behave aggressively, as indicated by involvement of the joint space. Reports of malignant chondroblastomas and distant metastases exist. Secondary aneurysmal bone cyst formation has been described. Because of epiphyseal involvement, growth disturbance may occur. Recurrence is a recognized complication because of incomplete curettage stemming from the intent of not damaging the growth plate during surgery.

CHONDROMYXOID FIBROMA

Overview

Chondromyxoid fibroma is a rare lesion; it is the least common benign cartilaginous lesion. It represents less than 1% of all primary bone tumors. It consists of stellate or spindle-shaped cells within chondroid and myxoid matrices. Chondromyxoid fibroma almost always is ar-

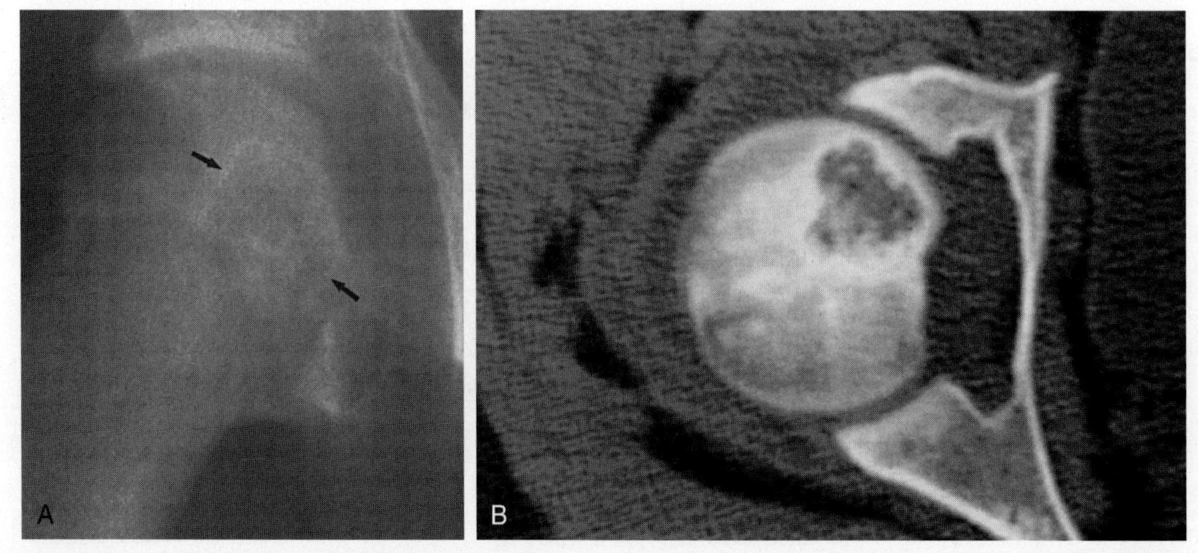

Figure 4–12. Chondroblastoma of the proximal femoral epiphysis. *A,* Anteroposterior radiograph of the hip shows a lytic lesion with a sclerotic margin *(arrows)*. The lesion is located in the proximal epiphysis of the femur. *B,* Axial CT scan at the level of the lesion shows popcorn-like calcifications within the lesion consistent with a chondroid matrix. A sclerotic rim and adjacent reactive sclerosis can be seen.

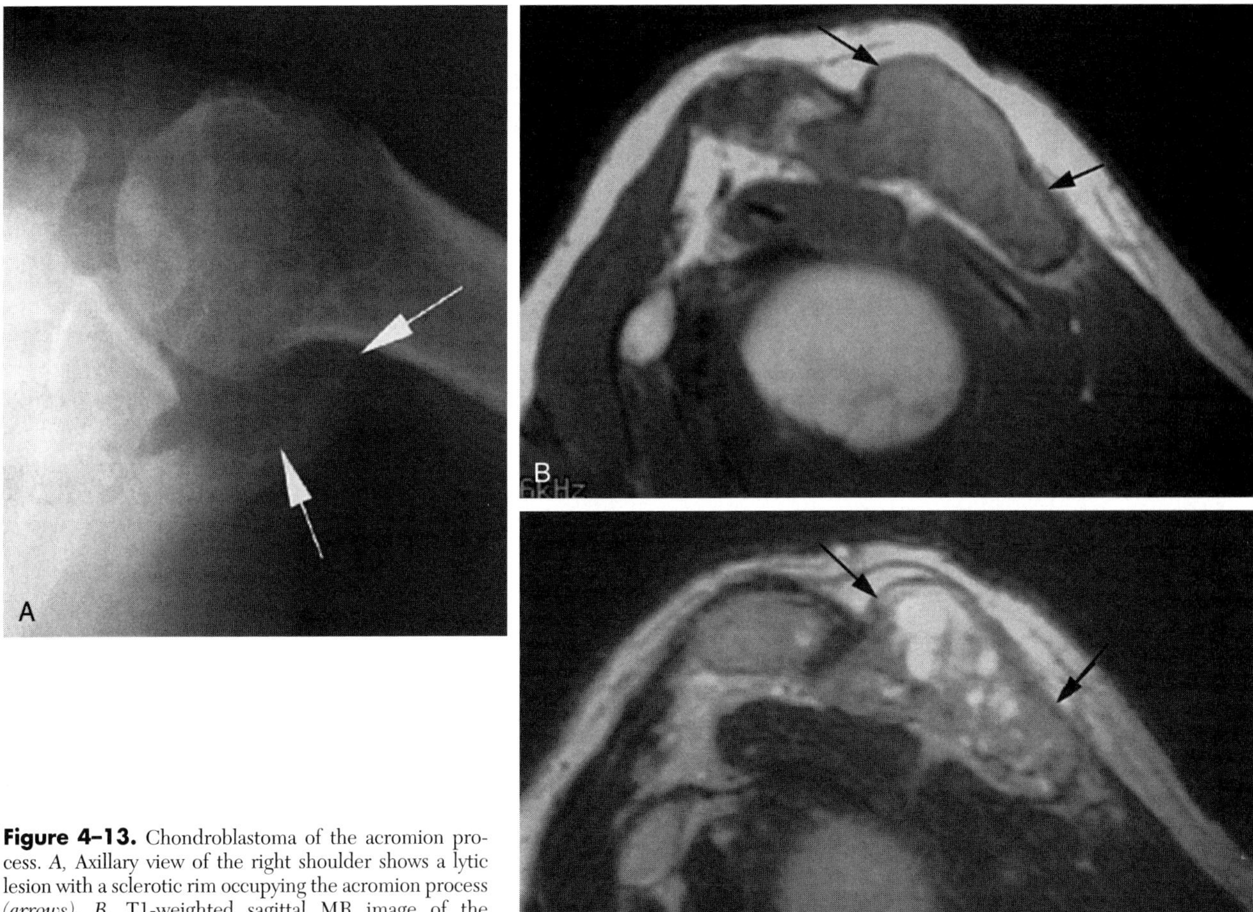

Figure 4–13. Chondroblastoma of the acromion process. *A,* Axillary view of the right shoulder shows a lytic lesion with a sclerotic rim occupying the acromion process *(arrows)*. *B,* T1-weighted sagittal MR image of the shoulder reveals a low signal intensity lesion of the acromion *(arrows)*. The lesion is almost isointense with muscle. *C,* T2-weighted sagittal MR image shows inhomogeneously increased signal intensity *(arrows)*.

ranged in lobules. Rarely, nuclear pleomorphism may be encountered. Chondromyxoid fibroma usually is diagnosed before age 30, with most discovered during the teens.

Skeletal Involvement

Most chondromyxoid fibromas occur in the long bones (58%), most frequently in the tibia, femur, and fibula. Of the lesions, 40% occur around the knee (Fig. 4–14). Other locations include bones of the feet (the second most common location, occurring in 17% to 24%), ilium, ribs, ulna, and radius. The lesion usually is medullary and eccentric (Fig. 4–15). In the tubular bones, it is primarily metaphyseal but may extend into the epiphysis or diaphysis. It also may be purely diaphyseal.

Imaging

Radiography and CT typically show an elongated, lucent, lobulated, sharply marginated lesion with a sclerotic rim (see Fig. 4–15). The cortex is expanded and thinned (see Fig. 4–15). Endosteal sclerosis is present with coarse trabeculations and septa. Less commonly, one or more of the following findings may be encountered: complete cortical erosion with soft tissue extension, stippled calcifications, or involvement of the full width of the bone when the lesion is in a short tubular bone. The mean size of a chondromyxoid fibroma is 3 cm, but it may reach 10 cm.

Differential Diagnosis

Lesions that can resemble a chondromyxoid fibroma radiographically include aneurysmal bone cyst, enchondroma, and nonossifying fibroma.

OSTEOCHONDROMA (OSTEOCARTILAGINOUS EXOSTOSIS)

Solitary Osteochondroma

Overview

An osteochondroma is a cartilage-covered osseous excrescence arising from a bone surface and communicating with the marrow space. It probably results from a

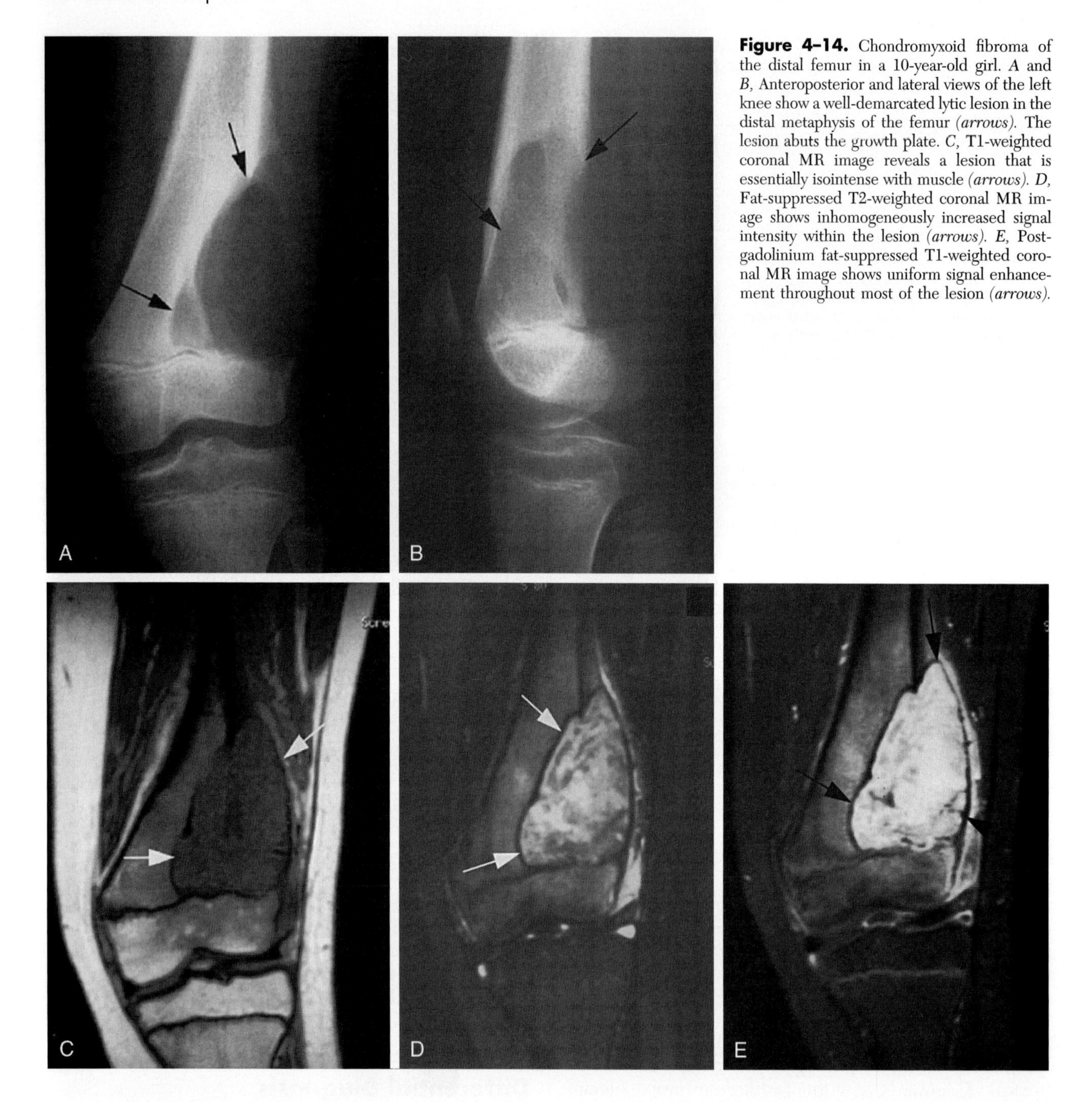

Figure 4–14. Chondromyxoid fibroma of the distal femur in a 10-year-old girl. *A* and *B*, Anteroposterior and lateral views of the left knee show a well-demarcated lytic lesion in the distal metaphysis of the femur *(arrows)*. The lesion abuts the growth plate. *C*, T1-weighted coronal MR image reveals a lesion that is essentially isointense with muscle *(arrows)*. *D*, Fat-suppressed T2-weighted coronal MR image shows inhomogeneously increased signal intensity within the lesion *(arrows)*. *E*, Post-gadolinium fat-suppressed T1-weighted coronal MR image shows uniform signal enhancement throughout most of the lesion *(arrows)*.

defect in the perichondrium covering the epiphyseal plate with separation of a piece of physeal cartilage that continues to grow. Osteochondromas may be induced by radiation therapy during childhood. Osteochondromas can be solitary or multiple. An epiphyseal location is designated by *dysplasia epiphysealis hemimelica* (DEH) or *Trevor's disease.* Osteochondromas grow from the cartilage cap and stop growing after skeletal maturity. The typical osteochondroma contains a spongiosa and cortex arising from the parent bone, and it is covered by a cap of mature hyaline cartilage. The cartilage cap usually measures less than 1 cm, and it may contain calcifications. The lesion may be pedunculated or sessile. Solitary osteochondromas constitute approximately 90% of all osteochondromas and 30% to 40% of all benign bone tumors. Most solitary osteochondromas are discovered before adulthood. Spontaneous regression of a solitary osteochondroma has been described, but it is rare. The most common presenting symptom of an osteochondroma is a painless, slow-growing mass. Frequently the lesion is discovered incidentally, however. Symptoms are more pronounced when complications occur.

Skeletal Involvement

Osteochondromas can arise from any bone preformed in cartilage. Most lesions occur in the long tubular bones. The most commonly involved bones are the femur, tibia, and humerus. Other locations include the small bones of the hands and feet (10%), innominate bone (5%), scapula

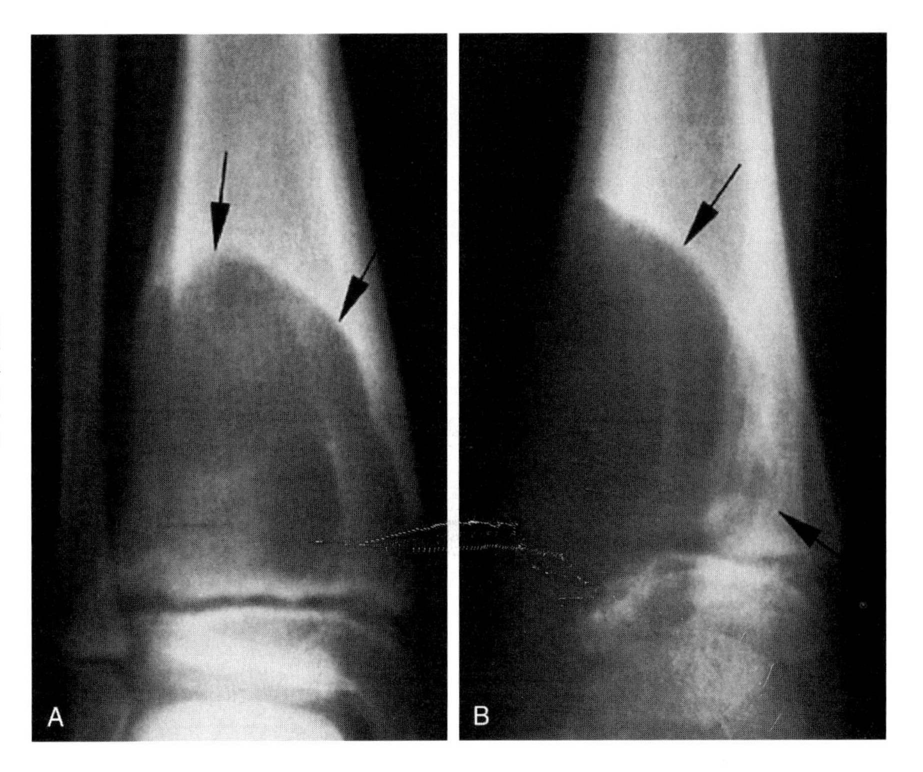

Figure 4–15. Chondromyxoid fibroma in the distal tibia. *A* and *B*, Anteroposterior and lateral radiographs of the distal tibia show a lucent, lobulated, sharply marginated lesion in the distal tibia *(arrows)*. The lesion is expansile and the anterior and lateral cortex are destroyed.

(4%), spine (1.3% to 4%), ribs, base of skull, and facial bones. In the long bones, the lesion typically is located within the metaphysis of the distal femur, proximal humerus, tibia, and fibula. The younger the patient, the closer the lesion is to the growth plate. In the spine, the posterior elements usually are involved. It is rare to have involvement of the vertebral body.

Imaging

RADIOGRAPHY

Radiographs are usually diagnostic. The classic radiographic appearance of an osteochondroma is an osseous protuberance arising from the surface of an underlying bone with continuity of its spongiosa with the marrow of the underlying parent bone. The lesion can be pedunculated or sessile (Figs. 4–16 and 4–17). It points away from the joint. Associated widening of the metaphysis may be present. In some cases, the cartilaginous cap contains stippled or dystrophic calcifications (Fig. 4–18).

COMPUTED TOMOGRAPHY

CT findings are similar to the abnormalities seen on radiographs. CT detects cap calcifications and cortical continuity of the base of the lesion with the parent bone better than radiography (see Fig. 4–18). CT also shows lesions of the spine and flat bones better than other modalities.

MAGNETIC RESONANCE IMAGING

MR imaging is particularly suited for evaluation of the cartilage cap thickness (Fig. 4–19). The cap shows increased signal intensity on T2-weighted images. MR imaging is the modality of choice for evaluating complications of osteochondromas, such as bursa formation, spinal canal impingement, malignant transformation (Fig. 4–20), mass effect on surrounding soft tissues (muscles, tendons, nerves, and vessels), and fractures.

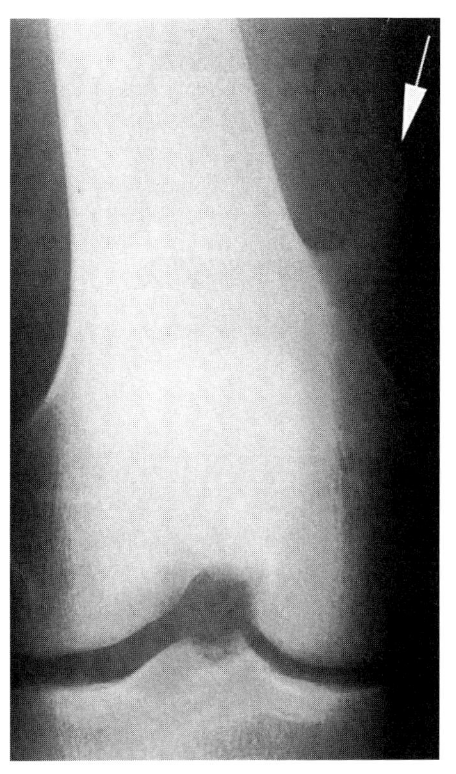

Figure 4–16. Pedunculated osteochondroma. Anteroposterior radiograph of the knee shows a pedunculated osteochondroma arising from the distal femur *(arrow)*. Note the continuity of the cortical and trabecular bone between the osteochondroma and the underlying femur.

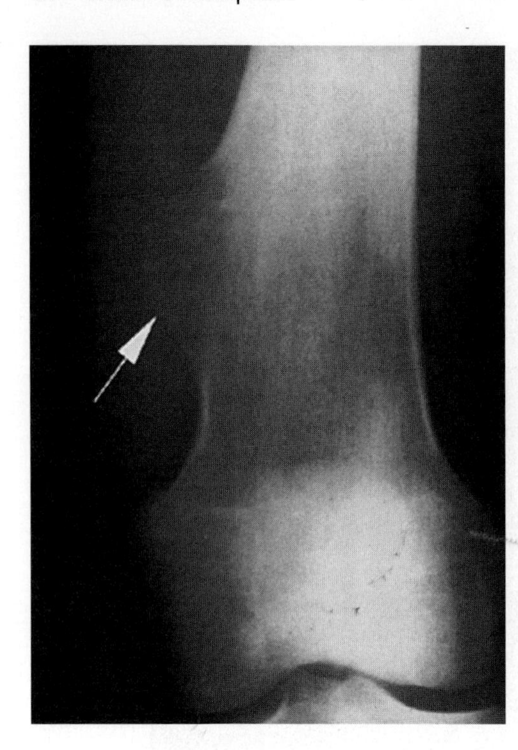

Figure 4–17. Sessile osteochondroma of the distal right femur. Anteroposterior radiograph of the distal femur shows a sessile osteochondroma *(arrow)*. The appearance is typical, and the differential diagnosis of such a lesion is limited.

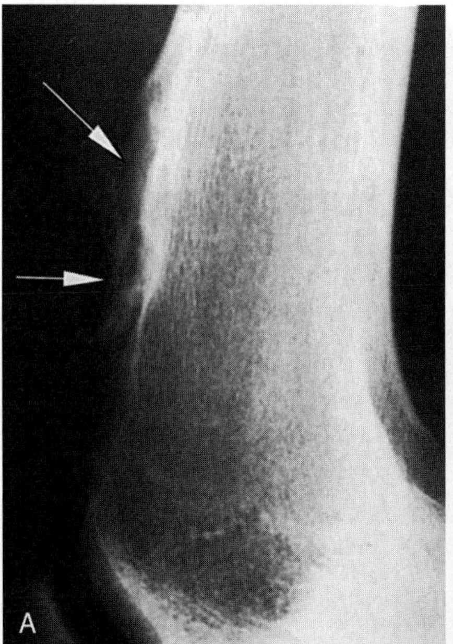

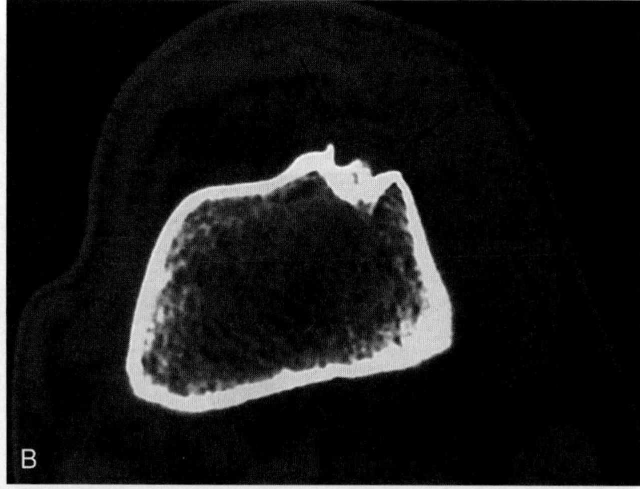

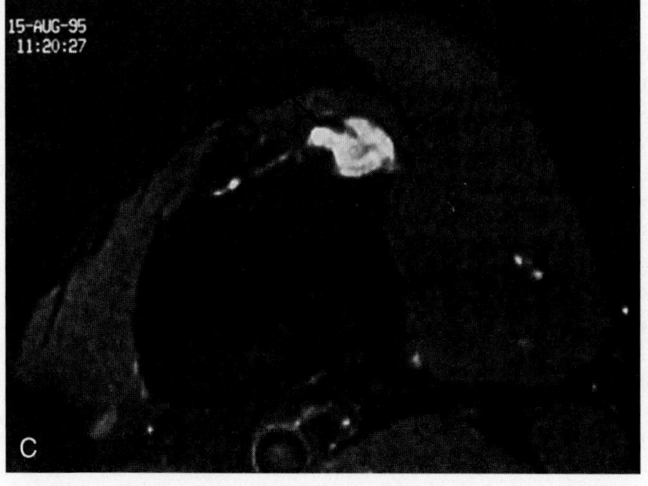

Figure 4–18. Radiograph, CT scan, and MR image of a sessile osteochondroma. *A,* Lateral radiograph of the knee shows a sessile osteochondroma along the anterior aspect of the distal femur *(arrows)*. *B,* Axial CT scan of the distal femur shows the osseous and the cartilaginous *(arrows)* portions of the lesion. A few calcifications are seen at the base of the groovelike area. *C,* Axial image from a STIR MR sequence shows the high signal within the cartilaginous portion of the osteochondroma *(arrows)*.

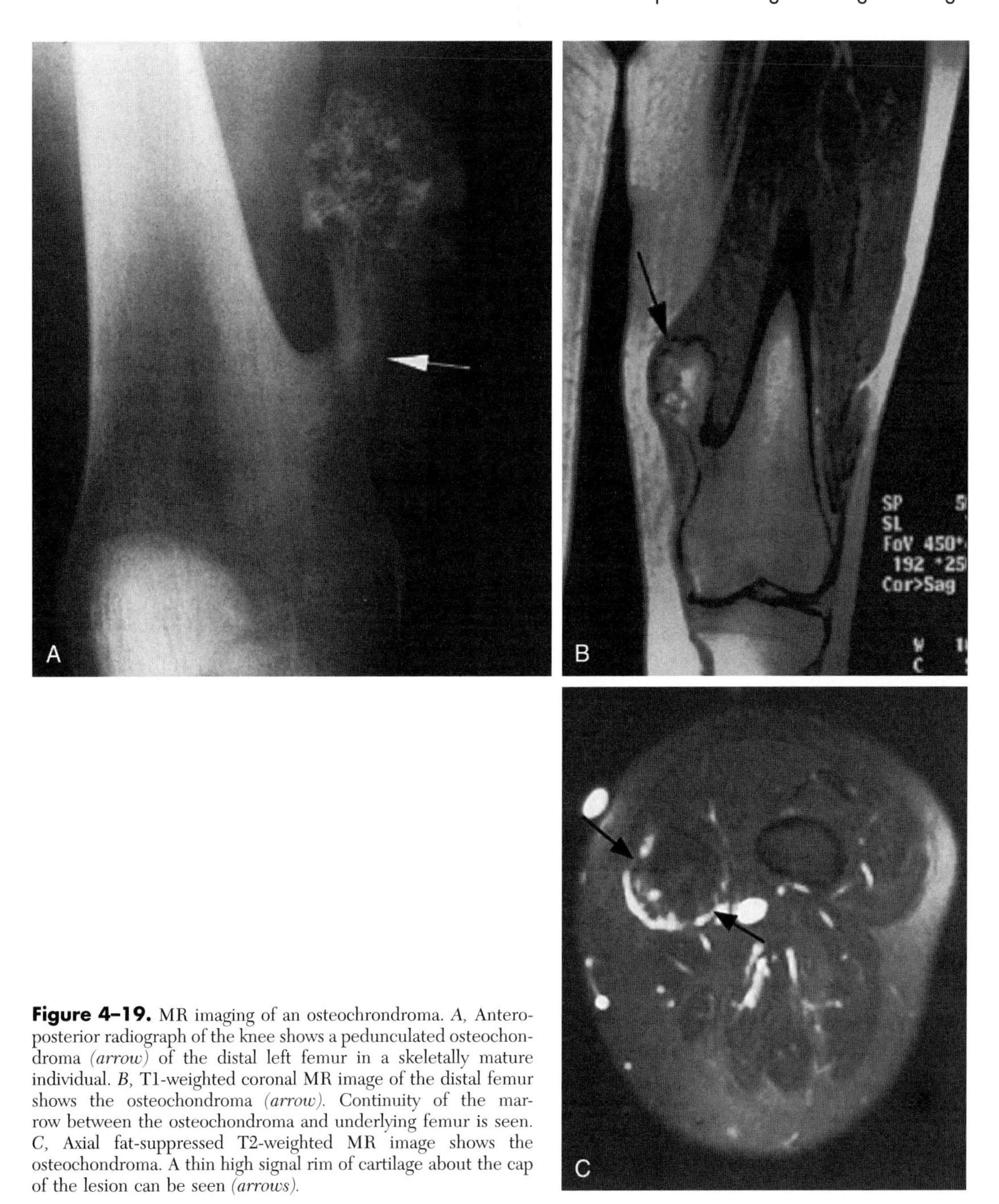

Figure 4–19. MR imaging of an osteochrondroma. *A,* Anteroposterior radiograph of the knee shows a pedunculated osteochondroma *(arrow)* of the distal left femur in a skeletally mature individual. *B,* T1-weighted coronal MR image of the distal femur shows the osteochondroma *(arrow).* Continuity of the marrow between the osteochondroma and underlying femur is seen. *C,* Axial fat-suppressed T2-weighted MR image shows the osteochondroma. A thin high signal rim of cartilage about the cap of the lesion can be seen *(arrows).*

ULTRASOUND

Ultrasound has been used to assess the cap thickness and to detect bursa formation around the lesion.

Complications

Fractures. A fracture through the stalk can occur.

Growth Disturbance. Deformity and growth disturbance of the bones and joints can occur.

Soft Tissue Complications. There can be mechanical irritation and compression of the surrounding tissues. Depending on the location, the following structures can be compressed, and secondary complications can occur.

Bursitis. A reactive bursa (Fig. 4–21) may form, which may bleed, become infected, or undergo chondrometaplastic change (i.e., formation of loose bodies).

Neurologic Abnormalities. Entrapment neuropathy, nerve palsy, spinal cord compression, or nerve root compression can occur. Secondary symptoms, resulting from skull base lesions, have been observed, including ataxic gait, headache, visual disturbances, and cranial nerve deficits.

Vascular Abnormalities. Popliteal artery pseudoaneurysm and deep vein thrombosis may be encountered, particularly in lesions around the knee.

Malignant Transformation. Malignant transformation is probably the most dreaded complication. It occurs in about 1% of cases. Malignant degeneration arises from the

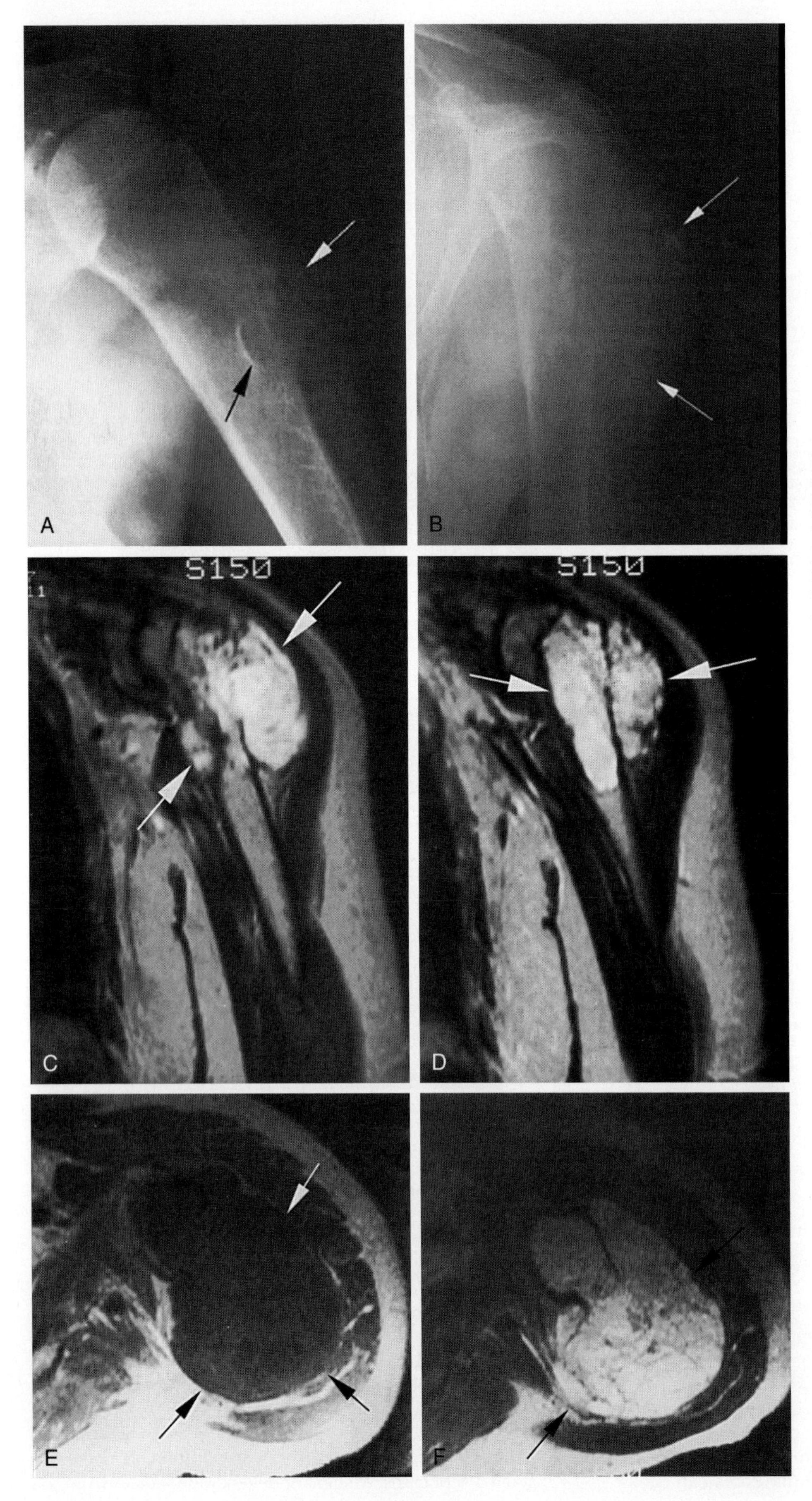

Figure 4–20. Malignant transformation of an osteochondroma. *A,* Anteroposterior radiograph of the proximal humerus in a 65-year-old woman shows a sessile osteochondroma of the proximal humerus *(arrows).* *B,* Anteroposterior radiograph of the proximal humerus of the same woman at age 70. There has been interval development of a soft tissue mass containing calcifications *(arrows)* consistent with a chondroid matrix. The mass is arising from the osteochondroma. There is osseous destruction at the previous location of the osteochondroma. This lesion was found to be a chondrosarcoma. *C* and *D,* Coronal T2-weighted MR images of the chondrosarcoma. The mass contains high signal consistent with a cartilaginous lesion. The *arrows* indicate involvement of the humerus and the soft tissues. *E* and *F,* Axial T1- and T2-weighted MR images of the mass *(arrows).* The mass has low signal intensity on T1-weighted images and high signal intensity on T2-weighted images, which would be consistent with a cartilaginous lesion.

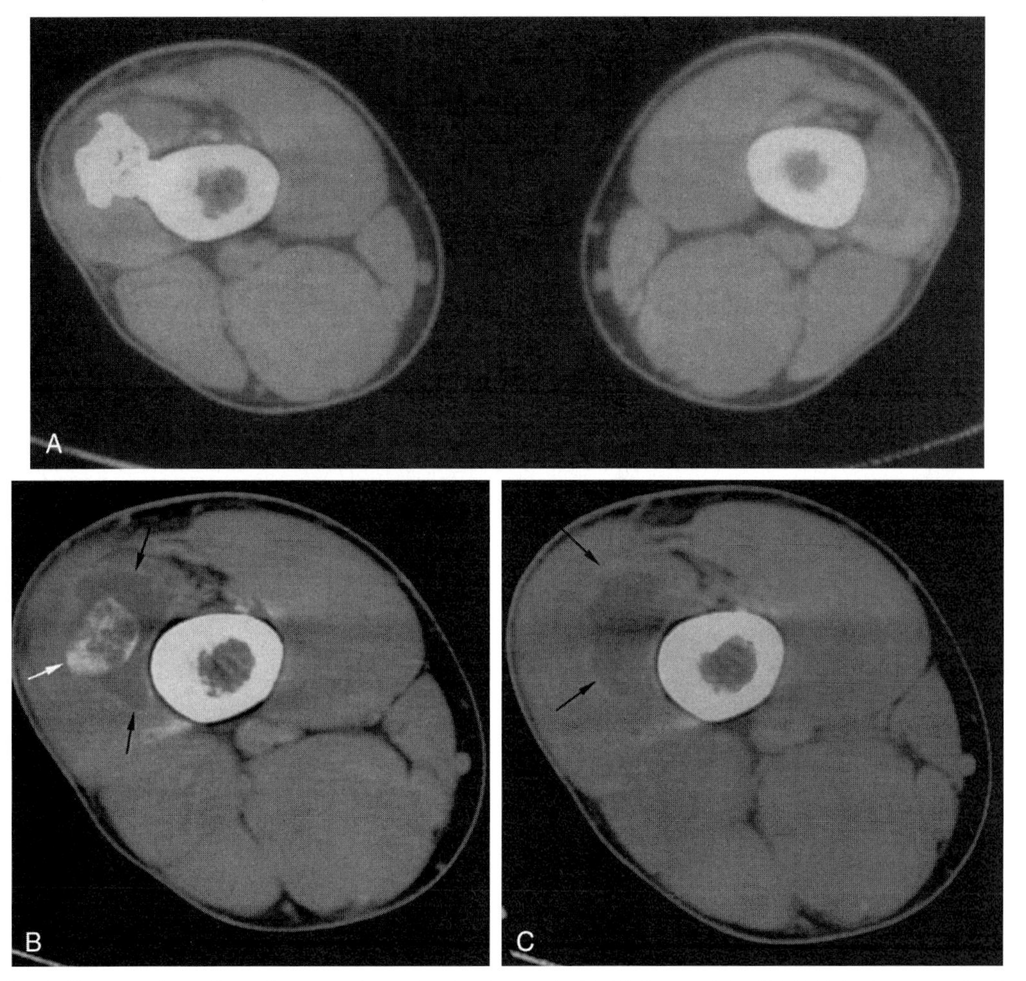

Figure 4–21. Surgically proven bursa formation around an osteochondroma. *A,* Axial CT scan of the distal thighs shows an osteochondroma arising from the right femur and projecting into the vastus lateralis muscle. *B* and *C,* Axial CT scans from around the tip of the osteochondroma reveal a low-density, fluid-containing structure *(black arrows)* surrounding the tip of the exostosis *(white arrow).* This was preoperatively diagnosed as bursa formation.

chondroid cells of the cartilaginous cap, leading to a secondary chondrosarcoma. Findings suggesting malignant transformation include a cartilaginous cap thickness greater than 1.5 to 2.0 cm, significant growth in the osteochondroma after skeletal maturity, and development of an associated soft tissue mass.

Radiation-Induced Osteochondroma

Radiation-induced osteochondroma arises in patients who received radiation therapy in childhood. Radiographically and histologically, radiation-induced osteochondromas are indistinguishable from spontaneous osteochondromas (Fig. 4–22). They may occur in any growing bone included in the radiation field. Libshitz and Cohen (1982) reported a 12% incidence of osteochondromas in irradiated children. The dose of radiation above which osteochondromas tend to develop exceeds 1300 rad.

Multiple Osteochondromas (Multiple Familial Exostoses)

Overview

An autosomal dominant disorder exists that consists of multiple, bilateral osteochondromas. In 10% of cases, however, no family history can be elicited. Multiple osteochondromas usually are discovered in childhood. The male-to-female ratio is about 1.5:1 to 2.0:1. The clinical manifestations are similar to those of solitary lesions. There are osseous and joint deformities, such as coxa valga and pseudo-Madelung deformity. Neurologic symptoms related to spinal osteochondromas have been reported.

Imaging

Imaging findings are similar to those seen in solitary osteochondromas (Fig. 4–23).

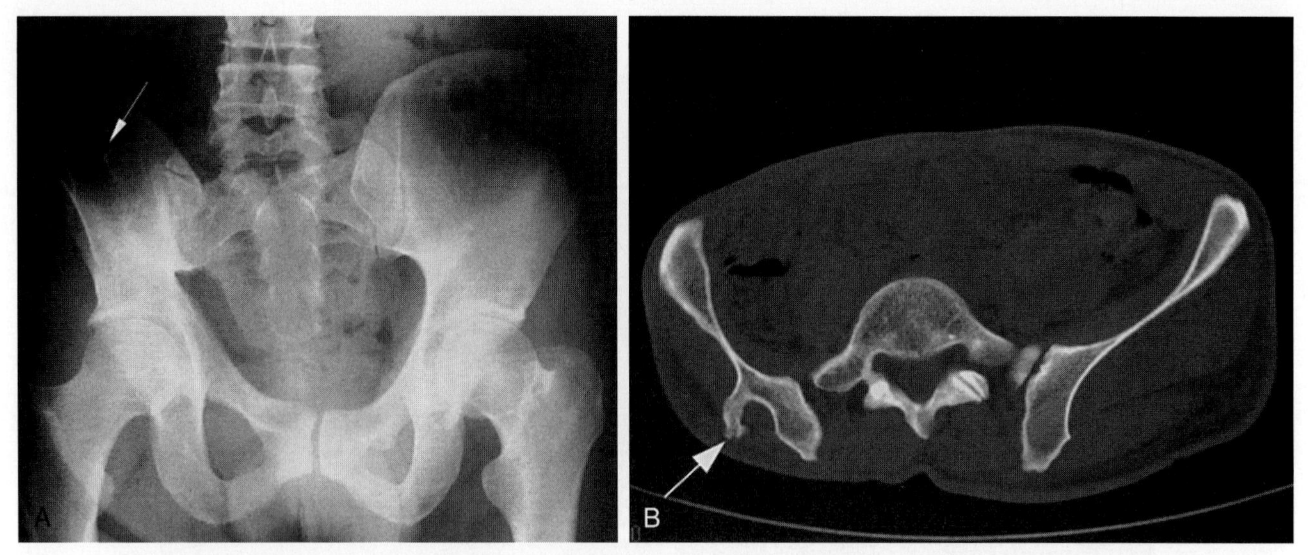

Figure 4–22. Radiation-induced osteochondroma. *A,* Anteroposterior view of the pelvis shows a small right hemipelvis with an associated osteochondroma of the right ilium *(arrow). B,* Axial CT scan shows the small right hemipelvis and the osteochondroma *(arrow)* to better advantage.

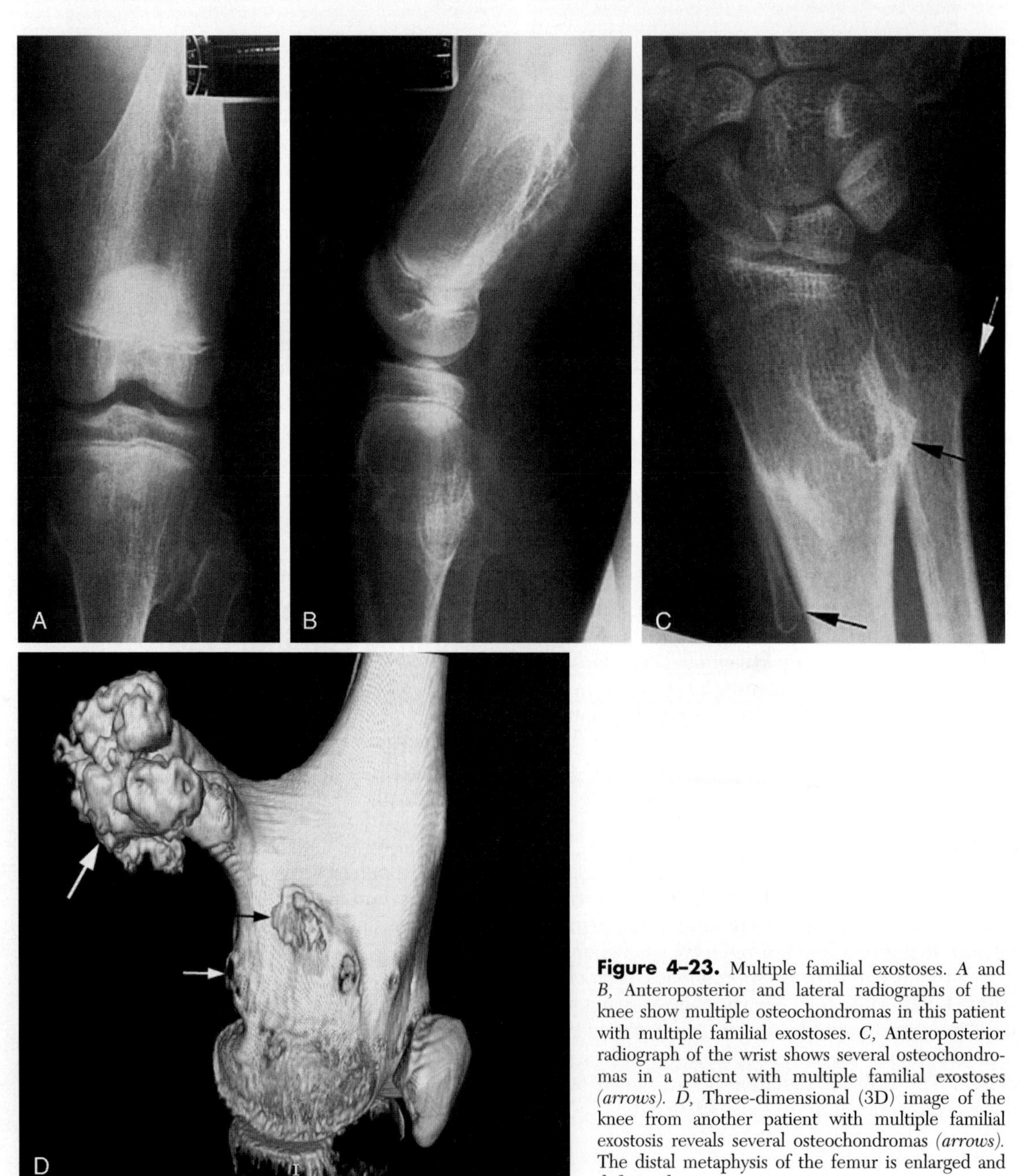

Figure 4–23. Multiple familial exostoses. *A* and *B,* Anteroposterior and lateral radiographs of the knee show multiple osteochondromas in this patient with multiple familial exostoses. *C,* Anteroposterior radiograph of the wrist shows several osteochondromas in a patient with multiple familial exostoses *(arrows). D,* Three-dimensional (3D) image of the knee from another patient with multiple familial exostosis reveals several osteochondromas *(arrows).* The distal metaphysis of the femur is enlarged and deformed.

Complications

Complications are similar to solitary lesions, but they are more frequent. In particular, malignant degeneration may increase by 20%.

Osteochondroma of the Epiphysis (Dysplasia Epiphysealis Hemimelica, or Trevor's Disease)

Overview

DEH is a rare, unilateral, asymmetrical disease consisting of an osteochondroma originating from the epiphysis. It usually is seen in children and young adults, diagnosed before the age of 15 years. Symptoms include swelling, pain, and deformity of the affected joint; lower extremity limb-length discrepancy; and gait abnormalities.

Location

Lesions typically are located in the distal femur, distal and proximal tibia, and talus (Fig. 4–24). There have been only sporadic reports of DEH in the upper extremity (Fig. 4–25). Within the affected bone, the lesion usually is noted in the distal epiphysis, involving one side of the ossification center (medial-to-lateral = 2:1). Rarely the entire epiphysis is involved.

Imaging

RADIOGRAPHY

On radiographs, there is an irregular osseous overgrowth from the epiphysis, with adjacent calcifications that coalesce to form an irregular lobular mass (see Fig. 4–25). Adjacent bone deformity and metaphyseal widening sometimes are present.

MAGNETIC RESONANCE IMAGING

MR imaging is effective in showing the cartilaginous component of DEH and the contour of the articular surface.

Complications

Growth disturbances (limb-length inequality) with either advanced bone age or arrest of a growth plate can be seen.

OSTEOCHONDROMA-LIKE LESIONS

Bizarre Parosteal Osteochondromatous Proliferation of Bone (Nora's Lesion)

Overview

Nora's lesion is a rare, benign disease consisting of proliferative heterotopic ossification. It may be due to trauma or a hamartomatous response of the periosteum. There is evidence to suggest that florid reactive periostitis, which until recently had been considered a distinct entity, if left untreated could progress to become bizarre parosteal osteochondromatous proliferation. Both of these lesions have been considered reactive lesions mimicking infection and tumor. These lesions have histologic and radiographic overlapping features. On histopathologic

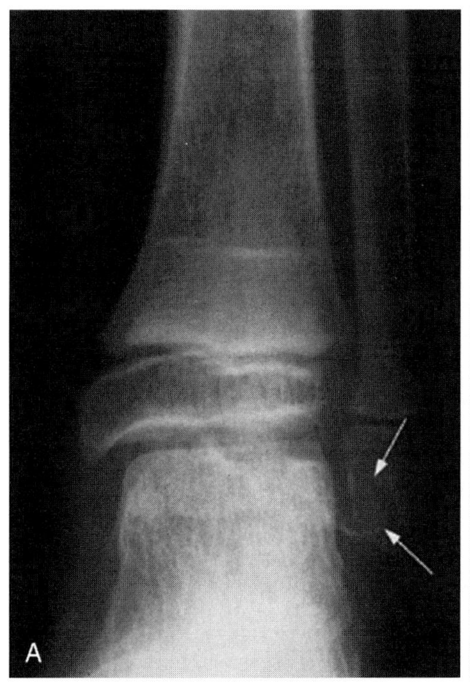

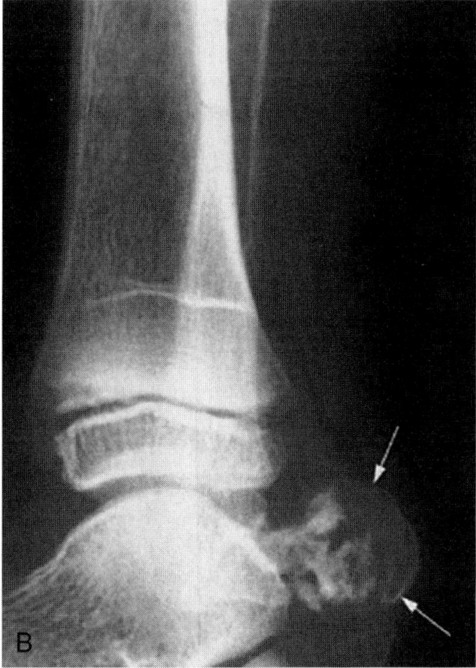

Figure 4–24. Dysplasia epiphysealis hemimelica (DEH) in the lower extremity. *A* and *B*, Anteroposterior and lateral radiographs of the ankle show an osteochondroma arising from the talus consistent with DEH *(arrows)*. The osteochondroma is causing an erosion on the distal fibula.

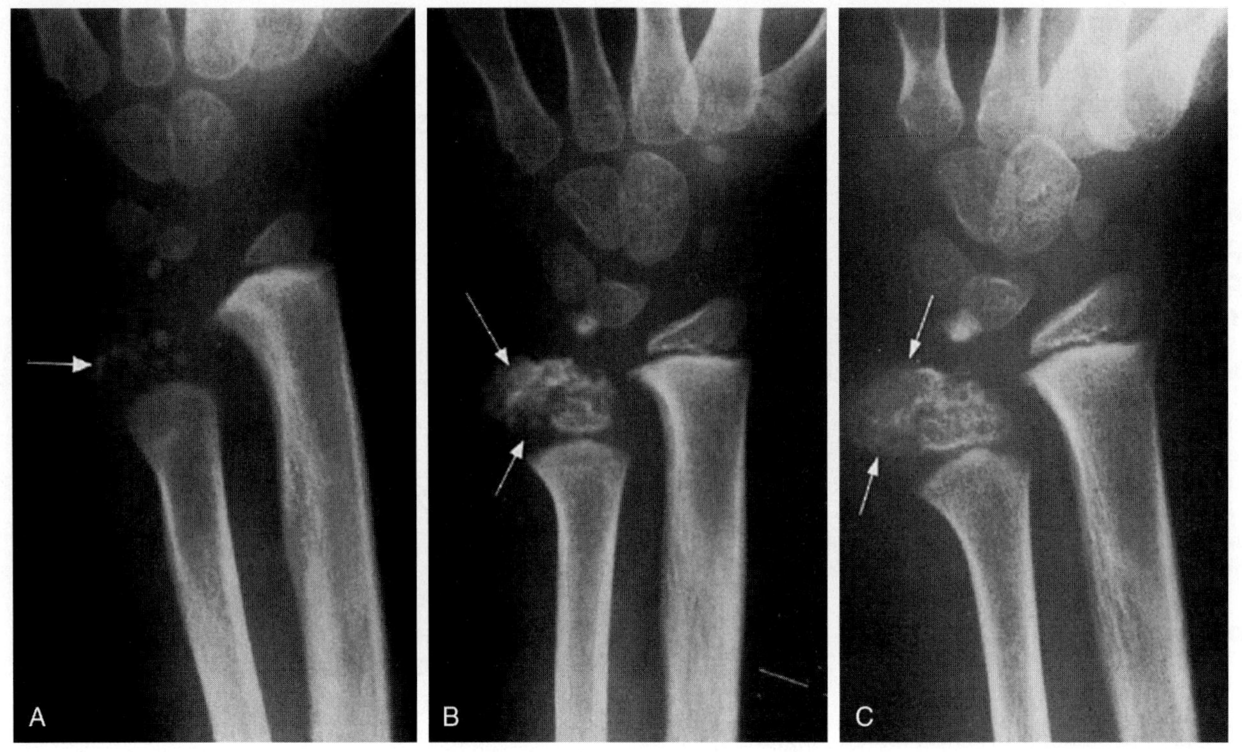

Figure 4–25. Dysplasia epiphysealis hemimelica (DEH) in the upper extremity. *A*, Anteroposterior radiograph of the wrist in a 3-year-old child with DEH. DEH is developing in the distal epiphysis of the ulna *(arrow)*. *B* and *C*, Anteroposterior radiographs of the wrist in the same patient at 5 and 6 years of age. Over time *(A–C)*, note the coalescence of adjacent calcifications to form an irregular lobular mass *(arrows)*.

examination, bizarre parosteal osteochondromatous proliferation of bone has irregular bony-cartilaginous interfaces with a mixture of fibroblasts and enlarged, bizarre, binucleate chondrocytes. Hypercellular cartilage showing maturation into trabecular bone with osteoblastic rimming also has been described. This lesion has been reported in patients between 8 and 74 years old with the peak in the 20s to 30s. Males and females are affected equally.

Skeletal Involvement

In 75% of cases, this lesion occurs in the tubular bones of the hands and feet. Less commonly, the long bones and skull can be involved.

Imaging

At an early stage of lesion development, radiographs show an irregularly ossified soft tissue mass abutting the bone with an adjacent periosteal reaction, which may be diagnosed as florid reactive periostitis (Fig. 4–26). At a later stage, the lesion appears irregular, pedunculated, and ossified. It has a broad base adjacent to an intact cortex. There is no continuity with the underlying marrow space of the affected bone (Fig. 4–27). The adjacent cortices may show some flaring. No periosteal reaction is seen in mature lesions. The size of the lesion may range from 0.4 to 3 cm.

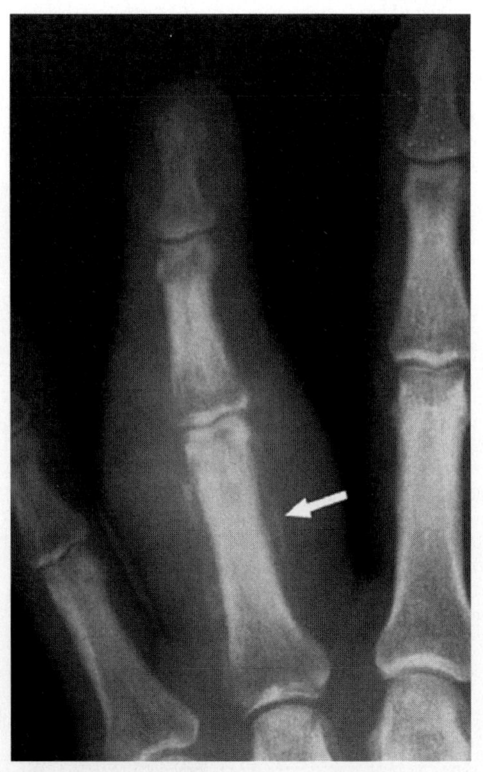

Figure 4–26. Florid reactive periostitis. Anteroposterior radiograph of the ring finger shows florid reactive periostitis *(arrow)*.

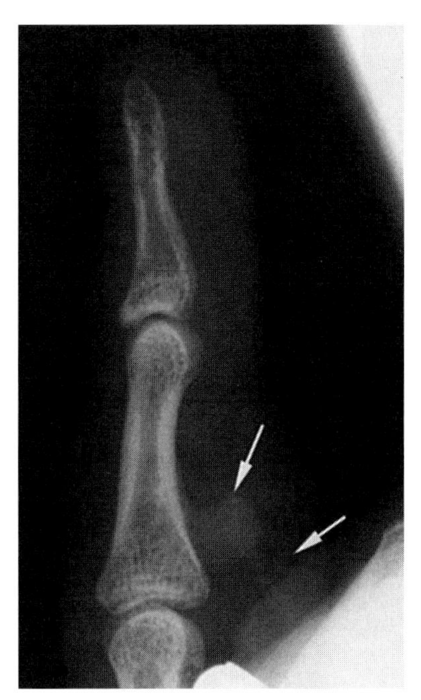

Figure 4–27. Nora's lesion. Lateral radiograph of the finger shows Nora's lesion. This is a mature lesion in that no adjacent periostitis is seen. The lesion is pedunculated and ossified *(arrows)*. The adjacent cortex of the phalanx is intact, and there is no continuity between the marrow space of the lesion and the adjacent bone.

Differential Diagnosis

The diagnosis of Nora's lesion may be difficult in the early stages of the disease. The differential diagnosis may include parosteal osteosarcoma. Fibro-osseous pseudotumor of the digit also may resemble bizarre parosteal osteochondromatous proliferation. Some authors believe that fibro-osseous pseudotumor of the digit represents a spectrum of the same disease process.

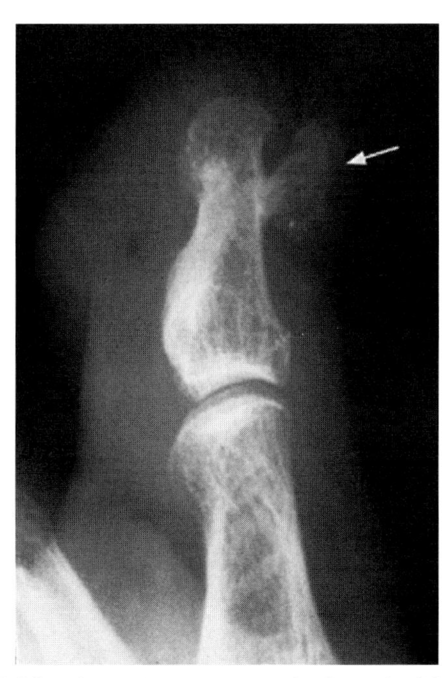

Figure 4–28. Subungual exostosis. Lateral radiograph of the great toe shows a bony protuberance based on the cortex of the distal phalanx *(arrow)*. There is no continuity between the marrow space of the lesion and the adjacent phalanx.

Complications

The lesion is benign and has no malignant potential. There is a 50% recurrence rate after local excision.

Subungual Exostosis

Overview

Subungual exostosis is an uncommon benign lesion. It is likely a reactive, non-neoplastic growth involving the distal phalanx of a finger or toe. It is located subjacent to the nail bed. Inciting factors include trauma and infection. A subungual exostosis begins as a growth of fibrous tissue and metaplastic cartilage with secondary enchondral ossification. It has a pedicle of mature bone with a fibrocartilaginous cap. Most (80%) of these lesions occur in the teens and 20s. The most common clinical manifestation of subungual exostosis is a painful spur at the distal phalanx of a digit. Swelling and ulceration of the nail bed may occur.

Skeletal Involvement

The great toe is the most common location for subungual exostosis (75%); however, any other digit may be affected. The exostosis usually occurs on the dorsal or dorsomedial aspect of the distal phalanx.

Imaging

On radiographs, the lesion appears as a bony protrusion based on the cortex of a distal phalanx (Fig. 4–28). There is no communication between the cortex or the central aspect of the lesion and that of the underlying bone. The size of the lesion ranges from 0.5 to 2 cm.

Complications

Subungual exostosis may have a rapid growth rate with secondary lifting of the nail and ulceration of the nail bed.

References and Recommended Readings

Albrecht S, Crutchfield JS, SeGall GK: On spinal osteochondromas. J Neurosurg 77:247, 1992.

Aoki J, Sone S, Fujioka F, et al: MR of enchondroma and chondrosarcoma: Rings and arcs of Gd-DTPA enhancement. J Comput Assist Tomogr 15:1011, 1991.

Azouz EM, Slomic AM, Marton D, et al: The variable manifestations of dysplasia epiphysealis hemimelica. Pediatr Radiol 15:44, 1985.

Beggs IG, Stoker DJ: Chondromyxoid fibroma of bone. Clin Radiol 33:671, 1982.

Bloem JL, Mulder JD: Chondroblastoma: A clinical and radiological study of 104 cases. Skeletal Radiol 14:1, 1985.

Bock GW, Reed MH: Forearm deformities in multiple cartilaginous exostoses. Skeletal Radiol 20:483, 1991.

Brien EW, Mirra JM, Luck Jr JV: Benign and malignant cartilage tumors of bone and joint: Their anatomic and theoretical basis with an emphasis on radiology, pathology and clinical biology: II. Juxtacortical cartilage tumors. Skeletal Radiol 28:1, 1999.
Carlson DH, Wilkinson RH: Variability of unilateral epiphyseal dysplasia (dysplasia epiphysealis hemimelica). Radiology 133:369, 1979.
Cohen EK, Kressel HY, Frank TS, et al: Hyaline cartilage-origin bone and soft-tissue neoplasms: MR appearance and histologic correlation. Radiology 167:477, 1988.
Connor JM, Horan FT, Beighton P: Dysplasia epiphysialis hemimelica: A clinical and genetic study. J Bone Joint Surg Br 65:350, 1983.
De Beuckeleer LH, De Schepper AM, Ramon F: Magnetic resonance imaging of cartilaginous tumors: Is it useful or necessary? Skeletal Radiol 25:137, 1996.
deSantos LA, Spjut HJ: Periosteal chondroma: A radiographic spectrum. Skeletal Radiol 6:15, 1981.
El-Khoury GY, Bassett GS: Symptomatic bursa formation with osteochondromas. AJR Am J Roentgenol 133:895, 1979.
Geirnaerdt MJ, Bloem JL, Eulderink F, et al: Cartilaginous tumors: Correlation of gadolinium-enhanced MR imaging and histopathologic findings. Radiology 186:813, 1993.
Geirnaerdt MJ, Hermans J, Bloem JL, et al: Usefulness of radiography differentiating enchondroma from central grade I chondrosarcoma. AJR Am J Roentgenol 169:1097, 1997.
Granter SR, Renshaw AA, Kozakewich HP, et al: The pericentromeric inversion, inv (6)(p25q13), is a novel diagnostic marker in chondromyxoid fibroma. Mod Pathol 11:1071, 1998.
Greenway G, Resnick D, Bookstein JJ: Popliteal pseudoaneurysm as a complication of an adjacent osteochondroma: Angiographic diagnosis. AJR Am J Roentgenol 132:294, 1979.
Hudson TM, Hawkins Jr IF: Radiological evaluation of chondroblastoma. Clin Radiol 139:1, 1981.
Janzen L, Logan PM, O'Connell JX, et al: Intramedullary chondroid tumors of bone: Correlation of abnormal peritumoral marrow and soft-tissue MR imaging signal with tumor type. Skeletal Radiol 26:100, 1997.
Karasick D, Schweitzer ME, Eschelman DJ: Symptomatic osteochondromas: Imaging features. AJR Am J Roentgenol 168:1507, 1997.
Kettelkamp DB, Campbell CJ, Bonfiglio M: Dysplasia epiphysialis hemimelica: A report of 15 cases and a review of the literature. J Bone Joint Surg Am 48:746, 1966.
Khosla A, Martin DS, Awwad EE: The solitary intraspinal vertebral osteochondroma: An unusual cause of compressive myelopathy: Features and literature review. Spine 24:77, 1999.
Landon GC, Johnson KA, Dahlin DC: Subungual exostosis. J Bone Joint Surg Am 61:256, 1979.
Lang IM, Azouz EM: MR imaging appearances of dysplasia epiphysealis hemimelica of the knee. Skeletal Radiol 26:226, 1997.
Lee JK, Yao L, Wirth CR: MR imaging of solitary osteochondromas: Report of eight cases. AJR Am J Roentgenol 149:557, 1987.
Lewis MM, Kenan S, Yabut SM, et al: Periosteal chondroma: A report of ten cases and review of the literature. Clin Orthop 256:185, 1990.
Libshitz HI, Cohen MA: Radiation-induced osteochondromas. Radiology 142:643, 1982.
Liu J, Hudkins PG, Swee RG, et al: Bone sarcomas associated with Ollier's disease. Cancer 59:1376, 1987.
Mehta M, White LM, Knapp T, et al: MR imaging of symptomatic osteochondromas with pathological correlation. Skeletal Radiol 27: 427, 1998.
Meneses MF, Unni KK, Swee RG: Bizarre parosteal osteochondromatous proliferation of bone (Nora's lesion). Am J Surg Pathol 17:691, 1993.
Miller-Breslow A, Dorfman HD: Dupuytren's (subungual) exostosis. Am J Surg Pathol 12:368, 1988.
Murphey MD, Flemming DJ, Boyea SR, et al: Enchondroma versus chondrosarcoma in the appendicular skeleton: Differentiating features. Radiographics 18:1213, 1998.
Nora FE, Dahlin DC, Beabout JW: Bizarre parosteal osteochondromatous proliferations of the hands and feet. Am J Surg Pathol 7:245, 1983.
Oxtoby JW, Davies AM: MR imaging characteristics of chondroblastoma. Clin Radiol 51:22, 1996.
Pazzaglia UE, Pedrotti L, Beluffi G, et al: Radiographic findings in hereditary multiple exostoses and a new theory of the pathogenesis of exostoses. Pediatr Radiol 20:594, 1990.
Quirini GE, Meyer JR, Herman M, et al: Osteochondroma of the thoracic spine: An unusual cause of spinal cord compression. AJNR Am J Neuroradiol 17:961, 1996.
Schmale GA, Conrad 3rd EU, Raskind WH: The natural history of hereditary multiple exostoses. J Bone Joint Surg Am 76:986, 1994.
Schuppers HA, van der Eijken JW: Chondroblastoma during the growing age. J Pediatr Orthop B 7:293, 1998.
Schutte HE, Vander Heul RO: Pseudomalignant, nonneoplastic osseous soft-tissue tumors of the hand and foot. Radiology 176:149, 1990.
Schwartz HS, Zimmerman NB, Simon MA, et al: The malignant potential of enchondromatosis. J Bone Joint Surg Am 69:269, 1987.
Seeger LL, Yao L, Eckardt JJ: Surface lesions of bone. Radiology 206:17, 1998.
Shapiro F: Ollier's disease: An assessment of angular deformity, shortening, and pathological fracture in twenty-one patients. J Bone Joint Surg Am 64:95, 1982.
Shapiro F, Simon S, Glimcher MJ: Hereditary multiple exostoses: Anthropometric, roentgenographic, and clinical aspects. J Bone Joint Surg Am 61:815, 1976.
Smith NC, Ellis AM, McCarthy S, et al: Bizarre parosteal osteochondromatous proliferation: A review of seven cases. Aust N Z J Surg 66:694, 1996.
Sundaram M, Wang L, Rotman M, et al: Florid reactive periostitis and bizarre parosteal osteochondromatous proliferation: Pre-biopsy imaging evolution, treatment and outcome. Skeletal Radiol 30:192, 2001.
Unger EC, Kessler HB, Kowalyshyn MJ, et al: MR imaging of Maffucci syndrome. AJR Am J Roentgenol 150:351, 1988.
Verstraete KL, De Deene Y, Roels H, et al: Benign and malignant musculoskeletal lesions: Dynamic contrast-enhanced MR imaging—parametric "first-pass" images depict tissue vascularization and perfusion. Radiology 192:835, 1994.
Weatherall PT, Maale GE, Mendelsohn DB, et al: Chondroblastoma: Classic and confusing appearance at MR imaging. Radiology 190:467, 1994.
Wilson AJ, Kyriakos M, Ackerman LV: Chondromyxoid fibroma: Radiographic appearance in 38 cases and in a review of the literature [published erratum appears in Radiology 1991;180:513]. Radiology 179:513, 1991.
Wu CT, Inwards CY, O'Laughlin S, et al: Chondromyxoid fibroma of bone: A clinicopathologic review of 278 cases. Hum Pathol 29:438, 1998.

CHAPTER 5

Primary Malignant Cartilage-Forming Tumors

Nabil J. Khoury, MD, Georges Y. El-Khoury, MD, and D. Lee Bennett, MD

OVERVIEW

Chondrosarcoma is the third most common primary malignant bone lesion, representing 10% to 15% of all malignant bone tumors. Chondrosarcomas may be primary (arising *de novo*) or secondary (arising from a preexisting enchondroma or osteochondroma). Primary chondrosarcomas most commonly arise within the medullary canal. This type of lesion is referred to as a *central* or *intramedullary chondrosarcoma.* Rarely a primary chondrosarcoma arises from the surface of the bone and is referred to as a *juxtacortical chondrosarcoma.* The age range is wide (9 to 86 years), with the mean age being 48 years. Only 4% to 16% of the tumors occur before 21 years of age. Generally, primary chondrosarcomas occur in adults and the elderly, whereas secondary chondrosarcomas affect younger individuals. The histologic diagnosis and grading of chondrosarcomas can be difficult. The tumor grade seems to correlate with the incidence of metastasis and the survival rate. Low-grade lesions tend to be slow growing and rarely metastasize (0% to 12%). High-grade chondrosarcomas metastasize early and often; they have a metastatic rate of 38% to 75%. Adequate surgery, with free margins, currently is the only effective treatment of chondrosarcomas.

Central (Conventional) Chondrosarcoma

Skeletal Involvement

Almost any bone can be involved. In adults, the pelvis is the most frequent location, followed by the femur, humerus, tibia, and ribs. In young patients, the most frequent location is the proximal end of the long bones, especially the femur and humerus. Within the long bones, the tumor mainly involves the metaphyses, followed by the diaphyses and the epiphyses. Epiphyseal lesions typically are of the clear cell variety. Other rare locations of tumor involvement include the spine, bones distal to the elbow and knee, mandible, sternum, and scapula.

Imaging

RADIOGRAPHY

Conventional chondrosarcoma appears as a lytic, lobulated, geographic, expansile lesion associated with either cortical thickening or deep endosteal scalloping (Fig. 5–1). Cortical break and aggressive periosteal reaction also may be seen. Eighty percent of lesions have chondroid matrix calcifications, which appear punctate, flocculent, amorphous, or as rings and arcs (see Fig. 5–1). Based on radiographs, it often is difficult to differentiate a benign enchondroma from a low-grade chondrosarcoma. A peripheral chondrosarcoma arising from the cartilage cap of a preexisting osteochondroma shows a rapid increase in size, with a destructive juxtacortical pattern and a large amount of chondroid calcification.

COMPUTED TOMOGRAPHY

Computed tomography (CT) is useful in assessing lesions that occur in bones that are difficult to image by radiographs, such as the pelvis. CT is the most sensitive modality for detecting chondroid calcifications (Fig. 5–2).

MAGNETIC RESONANCE IMAGING

On magnetic resonance (MR) imaging, chondrosarcomas appear lobulated, with intermediate to low signal intensity on T1-weighted images and high signal intensity on T2-weighted images. Punctate signal voids corresponding to calcifications may be seen. After intravenous gadolinium contrast injection, low-grade lesions show strong, homogeneous enhancement and enhancement of the margins and fibrovascular septa. In high-grade tumors, the enhancement may be inhomogeneous or homogeneous, with no septal enhancement. MR imaging can assess soft tissue invasion easily (see Fig. 5–1). In chondrosarcomas secondary to malignant degeneration of an osteochondroma, the cartilaginous cap thickness typically exceeds 2 to 3 cm (Fig. 5–3).

Differential Diagnosis

Central (conventional) chondrosarcomas should be differentiated from osteomyelitis, eosinophilic granuloma,

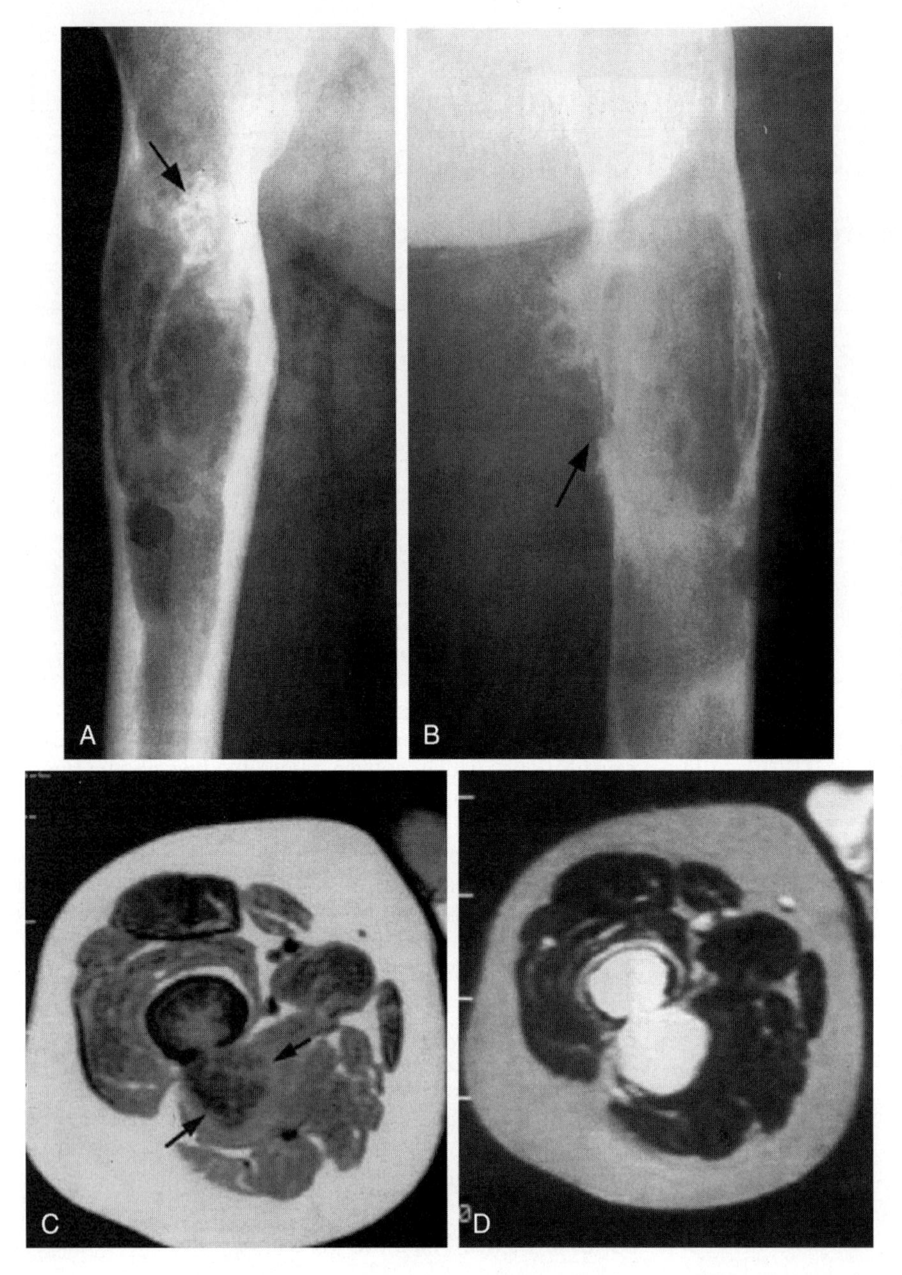

Figure 5–1. Central chondrosarcoma. *A* and *B*, Anteroposterior and lateral views of the proximal femur show a large expansile lesion arising within the medullary space. The tumor has broken through the cortex posteriorly (*arrow* in *B*). Chondral calcifications can be identified in the proximal portion of the tumor (*arrow* in *A*). *C* and *D*, Axial T1- and T2-weighted MR images show the lesion within the femur and in the soft tissues (*arrows*). Cartilage lesions are typically hyperintense on T2-weighted images.

fibrous dysplasia, osteosarcoma, and metastasis. A peripheral tumor can mimic an osteochondroma.

JUXTACORTICAL (PERIOSTEAL) CHONDROSARCOMA

Juxtacortical chondrosarcoma is a rare, low-grade surface lesion. It usually occurs after age 20 years (mean age, 35 years). Histologically, juxtacortical chondrosarcoma is characterized by the presence of lobules of hyaline cartilage with myxoid stroma, hypercellularity, anaplasia, and dystrophic calcifications. The tumor typically involves the metadiaphysis of a long tubular bone, in particular the proximal humerus and the distal femur. Radiographically, juxtacortical chondrosarcoma is usually a large tumor (>5 cm in diameter), round or fusiform, lobulated, and attached to the bone surface (Fig. 5–4). It is associated with cortical thickening, but no medullary extension is present. Codman's triangle may be noted. This lesion should be differentiated from a juxtacortical chondroma, its benign counterpart, and from periosteal osteosarcoma.

CLEAR CELL CHONDROSARCOMA

Clear cell chondrosarcomas represent about 2.2% of all chondrosarcomas. They are low-grade, slow-growing lesions occurring in the ends of long bones, but they may extend into the metaphyses. Osseous involvement typically occurs in the proximal femur, humerus, or tibia (Fig. 5–5). The age group involved is between the second and third decades. Radiographically the lesion appears lytic with a thin sclerotic margin. Chondroid calcifications

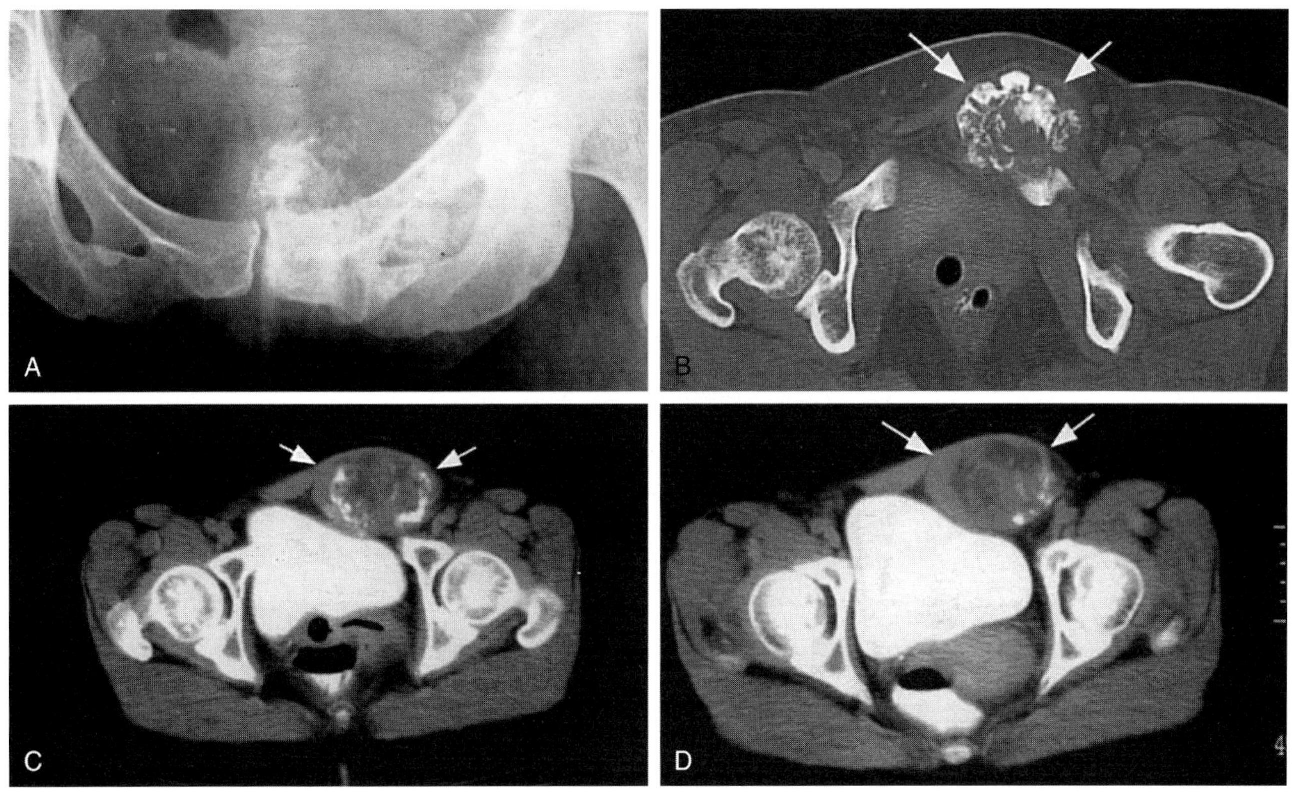

Figure 5–2. CT scan of a chondrosarcoma. *A*, Anteroposterior view of the symphysis pubis shows a mass at the region of the obturator ring and symphysis pubis. Calcifications are seen within the mass. *B*, Axial image from a pelvic CT scan (bone window) in the same patient shows calcifications consistent in appearance with a cartilage-containing lesion *(arrows)*. The mass is located just anterior to the left pubic rami. *C* and *D*, Axial CT scans (soft tissue window) of the mass show a prominent soft tissue component *(arrows)*. The location and size of the mass are appreciated better on CT scan.

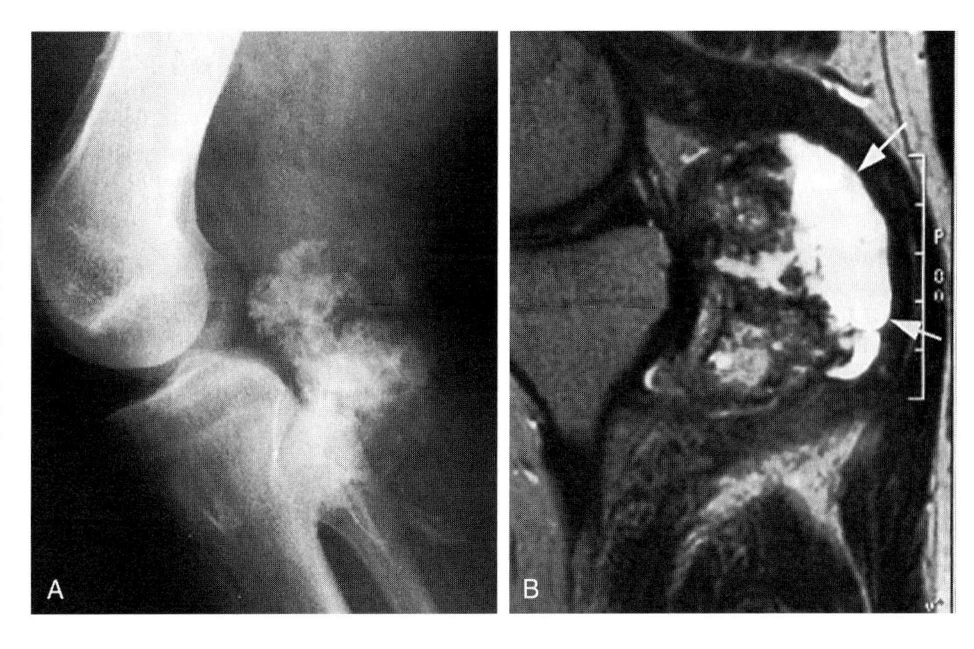

Figure 5–3. Osteochondroma with a thick cartilage cap proved to be a chondrosarcoma at biopsy. *A*, Lateral radiograph of the knee shows a large osteochondroma arising from the posterior aspect of the proximal tibia. Another small osteochondroma is seen arising from the proximal fibula. *B*, T2-weighted sagittal MR image shows a bright, thick cartilage cap measuring 2.5 cm at the thickest point *(arrows)*.

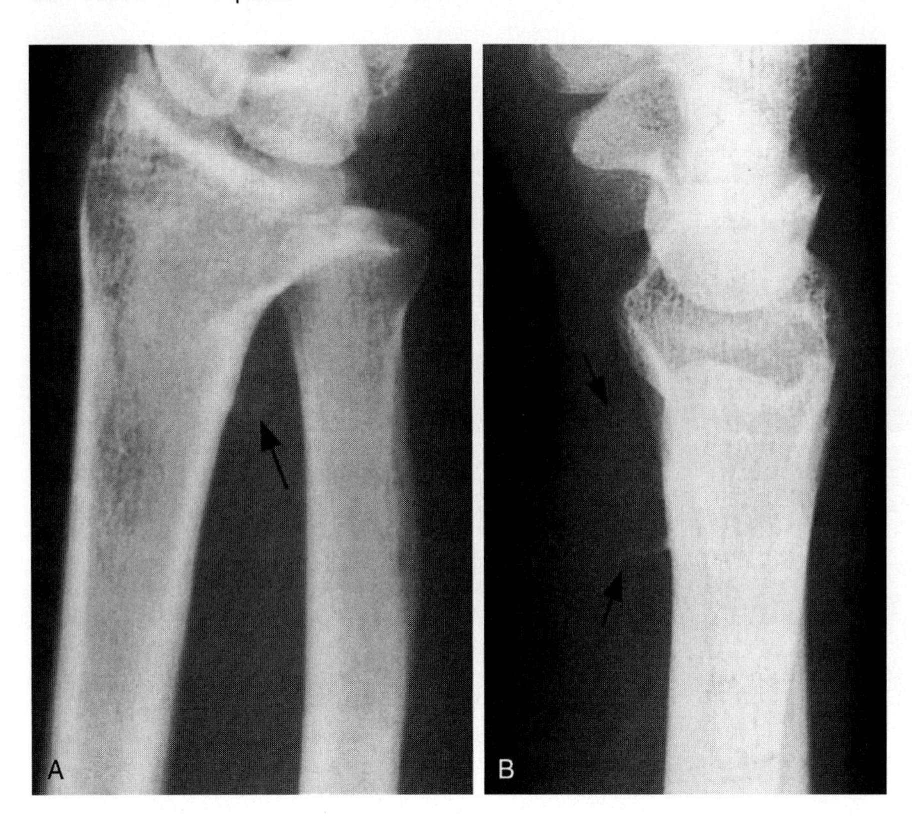

Figure 5–4. Periosteal chondrosarcoma. *A* and *B*, Oblique and lateral views of the distal forearm show a spiculated type of periosteal reaction *(arrows)* with a juxtacortical soft tissue mass. Biopsy revealed a chondrosarcoma.

sometimes are present. Radiographically the lesion closely resembles a chondroblastoma or giant cell tumor (see Fig. 5–5).

MESENCHYMAL CHONDROSARCOMA

Mesenchymal chondrosarcoma is a highly malignant tumor that frequently metastasizes to the lungs and other organs. Histologically, it contains alternating regions of cartilaginous tissue and undifferentiated small cells in varying amounts. Of mesenchymal chondrosarcomas, 60% to 80% occur in the teens and 20s. The lesion is more common in females. About 50% of cases develop in the soft tissues, including the meninges of the brain.

DEDIFFERENTIATED CHONDROSARCOMA

Dedifferentiated chondrosarcomas represent 10% of all chondrosarcomas. They usually occur after the age of 50 years, and they often present with a pathologic fracture. The lesion is highly aggressive, and metastasis is frequent. Histologically, dedifferentiated chondrosarcomas consist of two different components—a cartilaginous component and a noncartilaginous component—with an abrupt zone of transition between the two. The cartilaginous component consists of a low-grade chondrosarcoma juxtaposed to a high-grade osteosarcoma, malignant fibrous histiocytoma, or fibrosarcoma. The most common sites of involvement include the femur and humerus. On radiography, a central dedifferentiated chondrosarcoma may simulate a conventional central chondrosarcoma or a benign enchondroma adjacent to a destructive lytic lesion, which contains no calcifications. On MR imaging, two different regions of abnormal signal intensity may be present, reflecting the two different histologic components (Fig. 5–6).

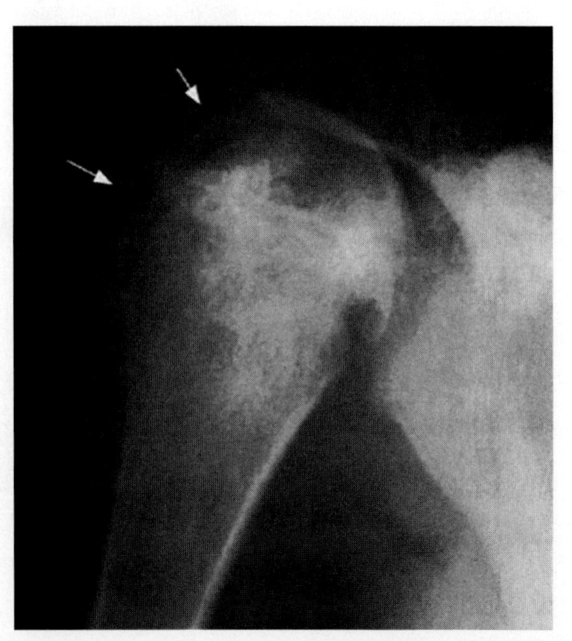

Figure 5–5. Clear cell chondrosarcoma in a 14-year-old child, which was diagnosed preoperatively as a chondroblastoma. Anteroposterior view of the shoulder shows a lytic lesion involving the lateral aspect of the humeral epiphysis in the region of the tumor *(arrows)*.

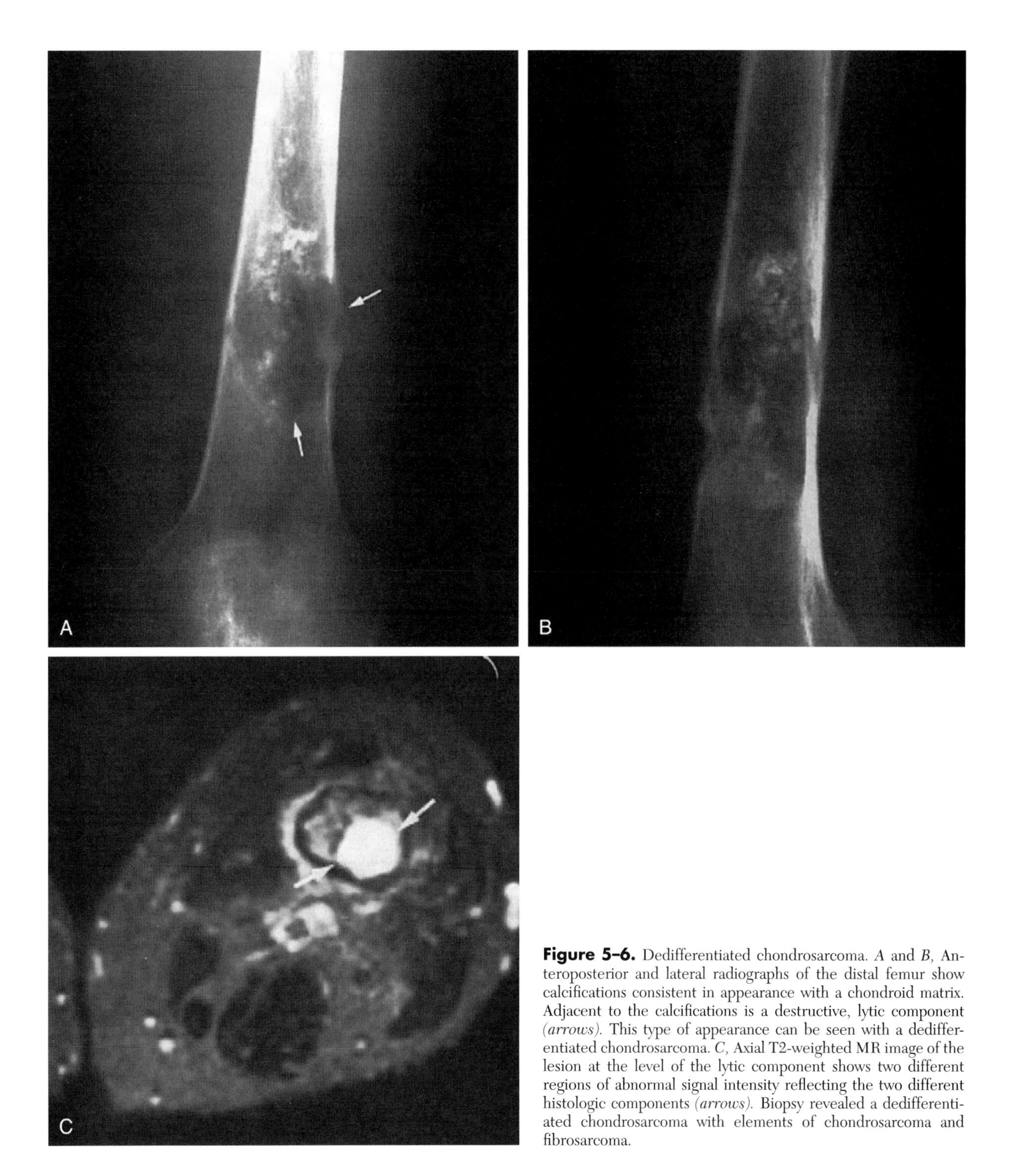

Figure 5–6. Dedifferentiated chondrosarcoma. *A* and *B*, Anteroposterior and lateral radiographs of the distal femur show calcifications consistent in appearance with a chondroid matrix. Adjacent to the calcifications is a destructive, lytic component *(arrows)*. This type of appearance can be seen with a dedifferentiated chondrosarcoma. *C*, Axial T2-weighted MR image of the lesion at the level of the lytic component shows two different regions of abnormal signal intensity reflecting the two different histologic components *(arrows)*. Biopsy revealed a dedifferentiated chondrosarcoma with elements of chondrosarcoma and fibrosarcoma.

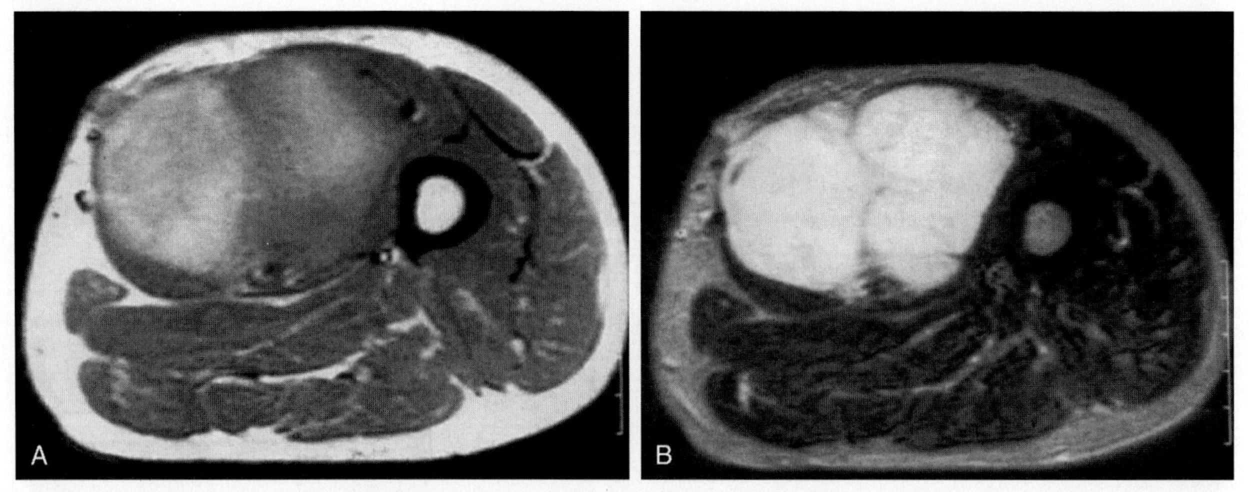

Figure 5–7. Myxoid chondrosarcoma in the adductor muscles of the thigh. *A* and *B*, Axial T1- and T2-weighted MR images show a large soft tissue mass in the proximal thigh. On the T1-weighted image *(A)*, the lesion is moderately intense, probably as a result of the high proteinaceous content of the lesion. The lesion is hyperintense on the T2-weighted image *(B)*.

EXTRASKELETAL (SOFT TISSUE) CHONDROSARCOMA

Extraskeletal chondrosarcomas are much less common than chondrosarcomas occurring in bone. The tumors are not attached to the bone, periosteum, or cartilage. About half of all mesenchymal chondrosarcomas arise in an extraskeletal location. Histologically, extraskeletal chondrosarcomas are divided into two types: myxoid and mesenchymal. The mesenchymal type presents with a well-defined soft tissue mass with speckled calcifications. The calcifications within the tumor can resemble soft tissue calcifications as a result of tumoral calcinosis, hyperparathyroidism, myositis ossificans, synovial sarcoma, or extraskeletal osteosarcoma. The myxoid type typically does not contain calcifications (Fig. 5–7). Its intense appearance on T2-weighted MR images closely resembles a soft tissue myxoma, but myxoid chondrosarcomas usually are much larger than myxomas.

Recommended Readings

Bjornsson J, McLeod RA, Unni KK, et al: Primary chondrosarcoma of long bones and limb girdles. Cancer 83:2105, 1998.

Brien EW, Mirra JM, Luck JV Jr: Benign and malignant cartilage tumors of bone and joint: Their anatomic and theoretical basis with an emphasis on radiology, pathology, and clinical biology: II. Juxtacortical cartilage tumors. Skeletal Radiol 28:1, 1999.

Eustace S, Baker N, Lan H, et al: MR imaging of dedifferentiated chondrosarcoma. Clin Imaging 21:170, 1997.

Geirnaerdt MJA, Bloem JL, Eulderink F, et al: Cartilaginous tumors: Correlation of gadolinium-enhanced MR imaging and histopathologic findings. Radiology 186:813, 1993.

Geirnaerdt MJA, Hermans J, Bloem JL, et al: Usefulness of radiography in differentiating enchondroma from central grade I chondrosarcoma. AJR Am J Roentgenol 169:1097, 1997.

Hudson TM, Springfield DS, Spanier SS, et al: Benign exostoses and exostotic chondrosarcomas: Evaluation of cartilage thickness by CT. Radiology 152:595, 1984.

Janzen L, Logan PM, O'Connell JX, et al: Intramedullary chondroid tumors of bone: Correlation of abnormal peritumoral marrow and soft-tissue MR imaging signal with tumor type. Skeletal Radiol 26:100, 1997.

Mercuri M, Picci P, Campanacci L, et al: Dedifferentiated chondrosarcoma. Skeletal Radiol 24:409, 1995.

Murphey MD, Flemming DJ, Boyea SR, et al: Enchondroma versus chondrosarcoma in the appendicular skeleton: Differentiating features. Radiographics 18:1213, 1998.

Present D, Bacchini P, Pignatti G, et al: Clear cell chondrosarcoma of bone: A report of 8 cases. Skeletal Radiol 20:187, 1991.

Varma DG, Ayala AG, Carrasco CH, et al: Chondrosarcoma: MR imaging with pathologic correlation. Radiographics 12:687, 1992.

CHAPTER 6

Benign Fibrous and Histiocytic Tumors

Nabil J. Khoury, MD, Georges Y. El-Khoury, MD, and D. Lee Bennett, MD

NONOSSIFYING FIBROMA AND FIBROUS CORTICAL DEFECT

Overview

Nonossifying fibroma and fibrous cortical defect represent the same histopathologic process. They are the most common benign lesions in the skeleton and are thought to represent ossification defects rather than true neoplasms. The differences between these two lesions are based on the size and age at which they develop. A fibrous cortical defect is usually smaller than 3 cm, and it usually develops before the age of 10 years. Nonossifying fibromas measure 4 to 15 cm in length, and they develop between the ages of 10 and 20 years. Histologically, these lesions consist of foam (or xanthoma) cells, giant cells, hemosiderin, and benign spindle cells arranged in a whorled or storiform configuration. The lesions are common in children; they are present in about one third of all children. Multiple lesions are seen in about half of patients. Nonossifying fibromas and fibrous cortical defects are asymptomatic unless fractures occur (Fig. 6–1), and they often are discovered incidentally. When the lesions are multiple, large, and bilateral, a more generalized mesodermal dysplasia, such as neurofibromatosis, should be suspected (Fig. 6–2). The incidence of neurofibromatosis in patients with multiple nonossifying fibromas is about 5%. Another interesting association is the one described by Campanacci and colleagues, in which multiple nonossifying fibromas are associated with café-au-lait spots, mental retardation, hypogonadism, cryptorchidism, megaureter, cardiovascular anomalies, and ocular anomalies; this is known as the *Jaffe-Campanacci syndrome.*

Skeletal Involvement

Nonossifying fibromas and fibrous cortical defects are discovered most commonly in the lower extremities, especially in the distal femur, proximal tibia, and fibula. They are seen occasionally in the humerus. They start in the cortex of the metaphysis of long bones. They almost always are eccentric, but when a narrow bone such as the fibula is involved, the entire width of the bone may become occupied by the lesion. As the bone continues to grow, the lesion migrates toward the diaphysis and assumes a more ovoid shape. The epiphyses are never involved. Both lesions have a self-limited course. Fibrous cortical defects usually regress by getting smaller and eventually disappear by the age of 10 years. When healing, nonossifying fibromas typically fill with bone and become densely sclerotic (Fig. 6–3).

Imaging

Radiography

The radiographic features of nonossifying fibromas and fibrous cortical defects are characteristic. In long bones, a nonossifying fibroma is an eccentric, ovoid, lytic lesion. It thins and expands the cortex (Fig. 6–4). The lesion has a multiloculated appearance, and its inner border is scalloped and sclerotic.

Magnetic Resonance Imaging

In most cases, magnetic resonance (MR) imaging is not indicated unless the lesion has atypical radiographic features. The lesion has low signal intensity on T1-weighted images and either low signal intensity (79%) or high signal intensity (21%) on T2-weighted images. The low signal intensity on T2-weighted images is a distinguishing feature related to the high collagen and hemosiderin content. The high signal intensity is due to the high content of foam (xanthoma) cells. After gadolinium injection, nonossifying fibromas show intense signal enhancement.

Nuclear Medicine

Radionuclide bone scans of nonossifying fibromas typically show only a mild increase in radiopharmaceutical uptake. This appearance helps to distinguish nonossifying fibromas from other benign or malignant abnormalities.

Complications

Nonossifying fibromas can be complicated by pathologic fractures; however, these fractures normally heal (see

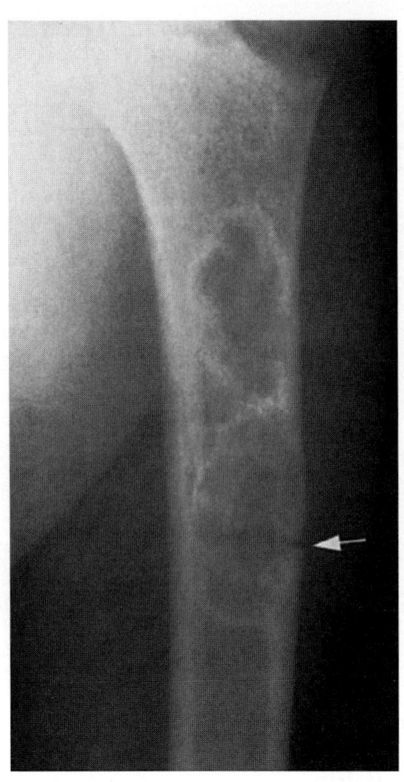

Figure 6–1. Pathologic fracture in a nonossifying fibroma in an 8-year-old boy. Anteroposterior view of the humerus reveals a well-corticated bubbly lesion in the proximal shaft. The lateral cortex is thinned, and a fracture line is seen *(arrow)*.

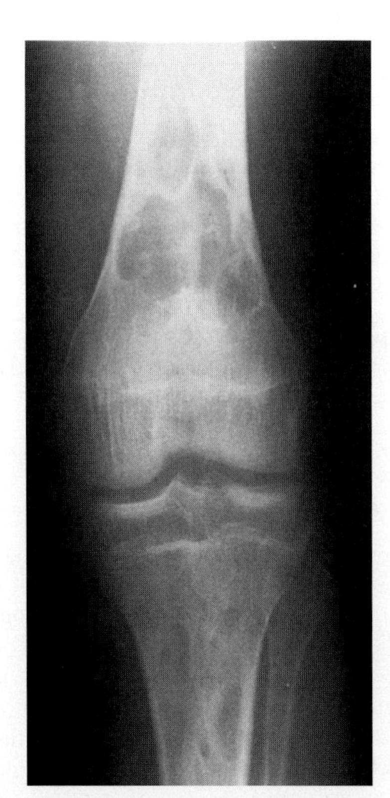

Figure 6–2. Large, multiple nonossifying fibromas in a patient with neurofibromatosis. Anteroposterior view of the left knee shows multiple, bubbly, lytic lesions with grade IA margins.

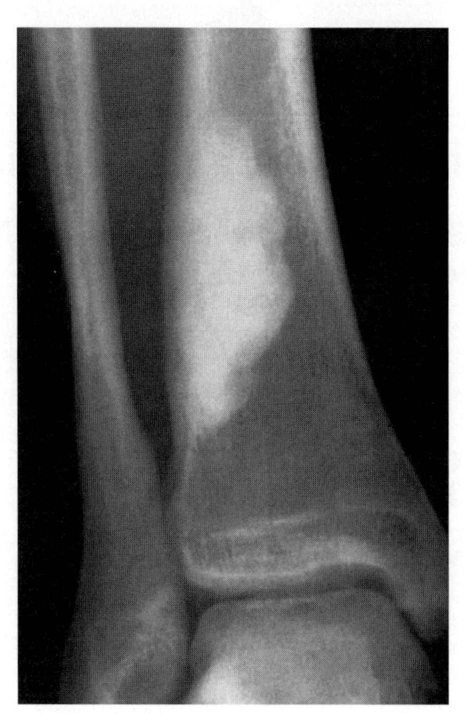

Figure 6–3. Healed nonossifying fibroma in the distal tibia. Anteroposterior view of the ankle shows the healing which sometimes takes the form of dense sclerotic bone replacing the fibrous tumor.

Fig. 6–1). Nonossifying fibromas do not undergo malignant transformation.

Differential Diagnosis

Nonossifying fibromas are easily diagnosable on radiographs. Other lesions should be considered when the radiographic features are not typical. These include chondromyxoid fibroma, fibrous dysplasia, and aneurysmal bone cyst.

FIBROUS DYSPLASIA

Overview

Fibrous dysplasia is not a true neoplasm but rather a developmental skeletal anomaly of unknown pathogenesis (Fig. 6–5). It is characterized by a tumor-like intramedullary proliferation of fibro-osseous tissue. The lesion usually presents before age 10. Males and females are affected equally. Fibrous dysplasia may involve a single bone (monostotic type) or multiple bones (polyostotic type). Monostotic fibrous dysplasia is about four times more common than the polyostotic type. Monostotic fibrous dysplasia often is discovered incidentally. Symptoms of polyostotic fibrous dysplasia (which include pain, swelling, and tenderness) may present before the age of 10 years. Polyostotic fibrous dysplasia is predominantly unilateral, symptomatic, and frequently associated with multiple

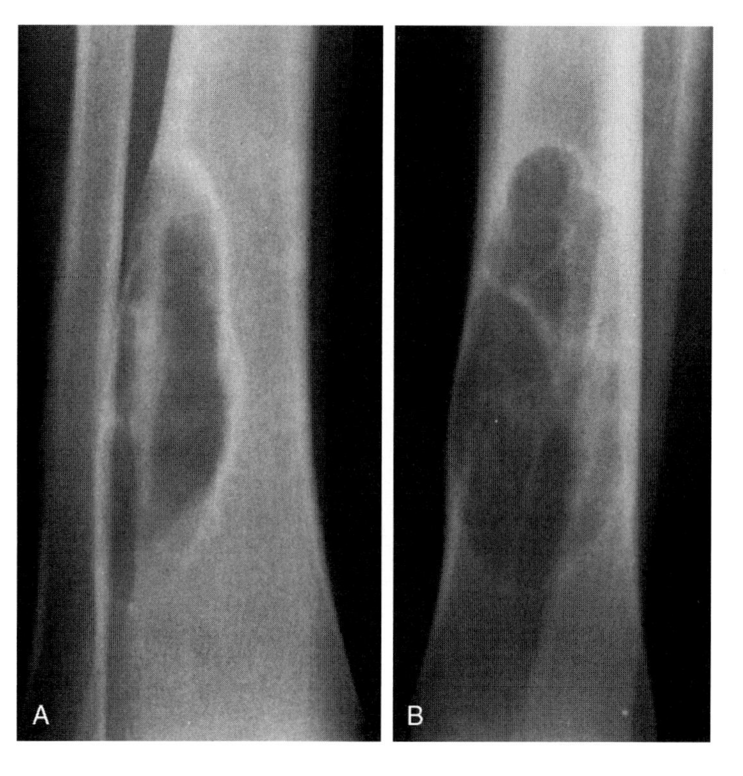

Figure 6–4. Typical appearance of a nonossifying fibroma. *A* and *B*, Anteroposterior and lateral radiographs of the distal tibia show a typical nonossifying fibroma. It is eccentric and ovoid (oriented with the long axis of the bone). It has a lytic bubbly appearance with a sclerotic (grade IA) margin. It also thins and expands the lateral cortex.

endocrine disorders, including McCune-Albright syndrome (polyostotic fibrous dysplasia, café-au-lait spots, and precocious puberty), hyperthyroidism, hyperparathyroidism, acromegaly, Cushing's syndrome, and diabetes mellitus. McCune-Albright syndrome can occur in either males or females; however, it is much more common in females. The association of fibrous dysplasia and intramuscular myxomas (Mazabraud's syndrome) (Fig. 6–6 *A* and *B*) also has been described. The myxomas tend to occur in the vicinity of the most severely affected bones and may be multiple. Histologically, fibrous dysplasia consists of whorled bundles of spindle cells with multiple interspersed trabeculae of varying sizes and shapes ("Chinese alphabet"). The trabeculae are composed of immature, woven (nonlamellar) bone produced by osseous metaplasia. Occasionally, islands of cartilage are seen. These islands of cartilage can calcify or form enchondral bone. The process of cartilage formation within fibrous dysplasia is referred to as *fibrocartilaginous dysplasia*.

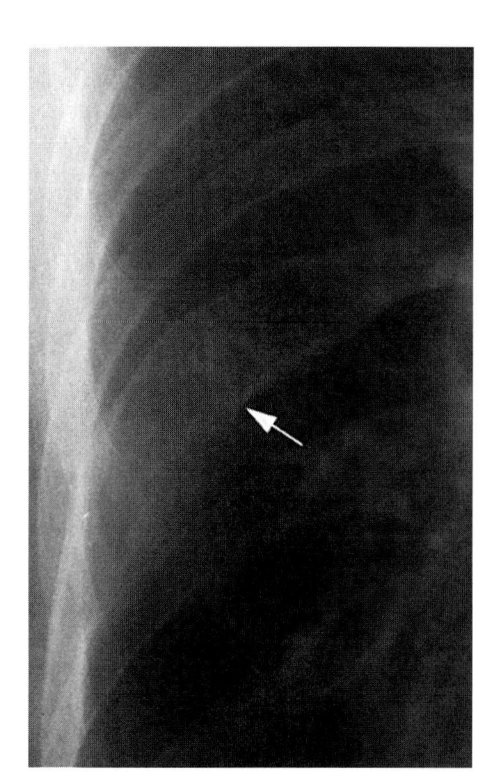

Figure 6–5. Fibrous dysplasia of the right eighth rib discovered incidentally in a 29-year-old asymptomatic man. Posteroanterior view of the chest shows an expansile lesion of the rib (*arrow*) with a ground-glass appearance.

Skeletal Involvement

Typical Skeletal Involvement

Typically the lesions in fibrous dysplasia arise within the marrow space; they expand the bone and cause deformity. Any bone may be involved; however, in the monostotic form, the most common sites are the ribs (see Fig. 6–5), proximal femora, and craniofacial bones. Lesions in the craniofacial bones closely resemble Paget's disease. Craniofacial disease in fibrous dysplasia is typically unilateral, whereas Paget's disease is usually bilateral and presents at a much later age. In the polyostotic form, the lesions may be present in the same extremity or on the same side of the body (Fig. 6–7). The most common locations include the femur, tibia, pelvis, and foot. When the pelvis is involved, the ipsilateral proximal femur almost always is affected. Spine involvement is rare and usually associated with the polyostotic form of fibrous dysplasia. The cervical

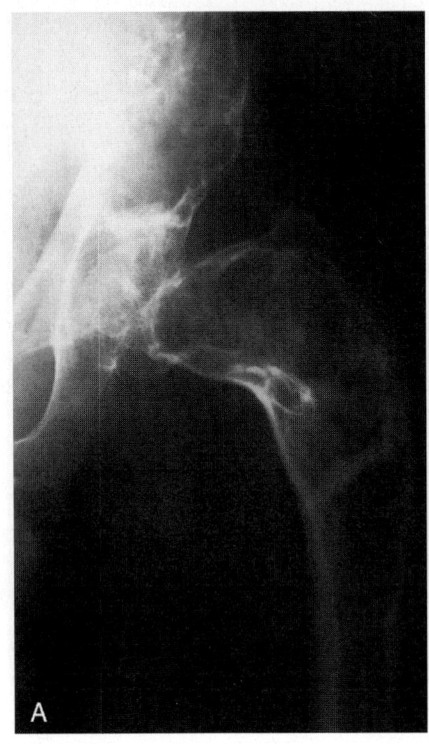

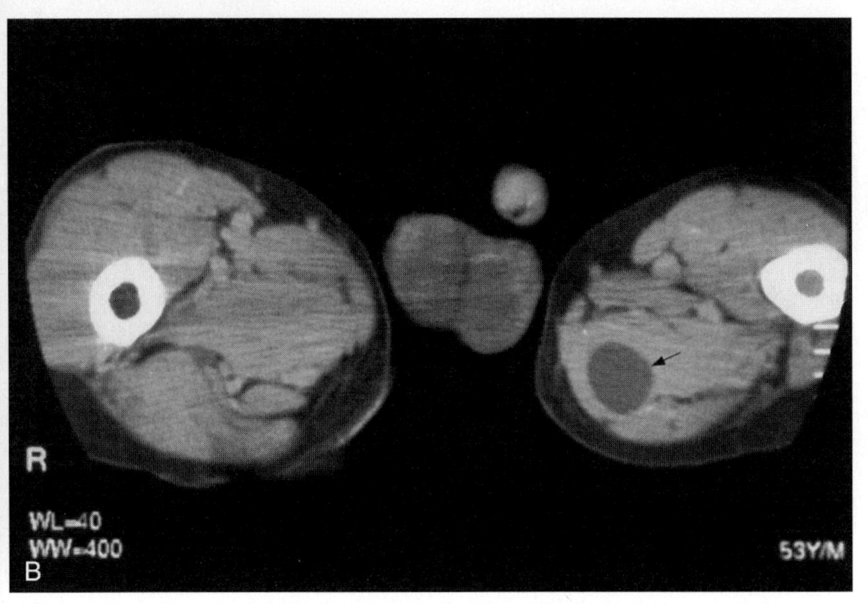

Figure 6–6. Mazabraud's syndrome in 42-year-old male. *A,* AP view of the left hip shows typical changes of fibrous dysplasia in the left ilium and proximal femur ("shepherd crook deformity"). *B,* Axial CT section through the proximal thigh shows an oval lesion *(arrow)* with low attenuation in the adductor magnus consistent with skeletal muscle myxoma.

spine and lumbar spine are reported to be involved more commonly. In the long bones, fibrous dysplasia is metaphyseal or diaphyseal, but it is not epiphyseal if the growth plate is still open. Pregnancy is known to exacerbate fibrous dysplasia lesions.

Variants of Skeletal Involvement

Two rare variants of fibrous dysplasia have been described.

Cherubism. This is a variant of fibrous dysplasia affecting the mandible. It is an autosomal dominant disease

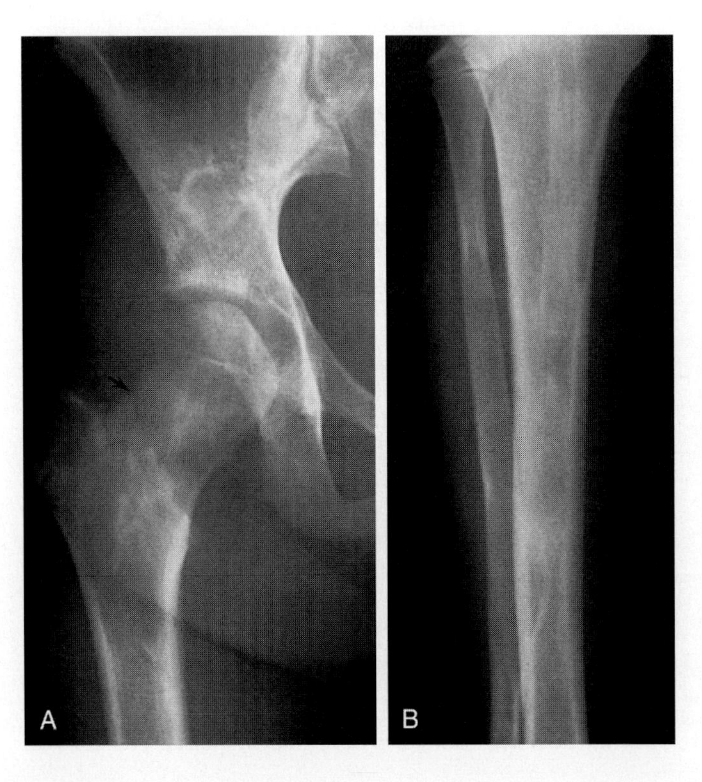

Figure 6–7. Polyostotic fibrous dysplasia. *A,* Anteroposterior view of the right hip shows ground-glass lesions in the ilium and proximal femur. A pathologic fracture is present in the femoral neck *(arrow). B,* Anteroposterior view of the right tibia and fibula in the same patient. Note the unilaterality of the lesions in this patient with polyostotic fibrous dysplasia.

with variable penetrance. The disease regresses after adolescence.

Fibrous Dysplasia Protuberans. This is a rare, exophytic variant of fibrous dysplasia, which can mimic an exostosis, osteoma, or surface osteosarcoma.

Imaging

Radiography

The radiographic appearance of fibrous dysplasia is characteristic. Typically the lesion is lucent and hazy with a ground-glass appearance (Fig. 6–8). The density of fibrous dysplasia depends on the degree of mineralization of the woven trabeculae. Rarely the lesion may appear cystic or may contain chondroid calcifications (Fig. 6–9). The margins of the lesion are well defined, and sometimes the border is thick and sclerotic. This thick border is termed the *rind sign* and often is seen in the femoral neck (see Fig. 1–5). Cortical scalloping and thinning also can be present. Bone deformity is assessed easily on radiographs, with a shepherd's crook deformity in the proximal femur being almost pathognomonic for fibrous dysplasia (Fig. 6–10). Three radiographic patterns of fibrous dysplasia have been described in the skull: sclerotic (Fig. 6–11), lytic (cystlike), and mixed. Malignant degeneration should be considered in cases with cortical destruction and an associated soft tissue mass. With malignant degeneration, areas that are initially mineralized may become lytic or develop periosteal reaction (Fig. 6–12).

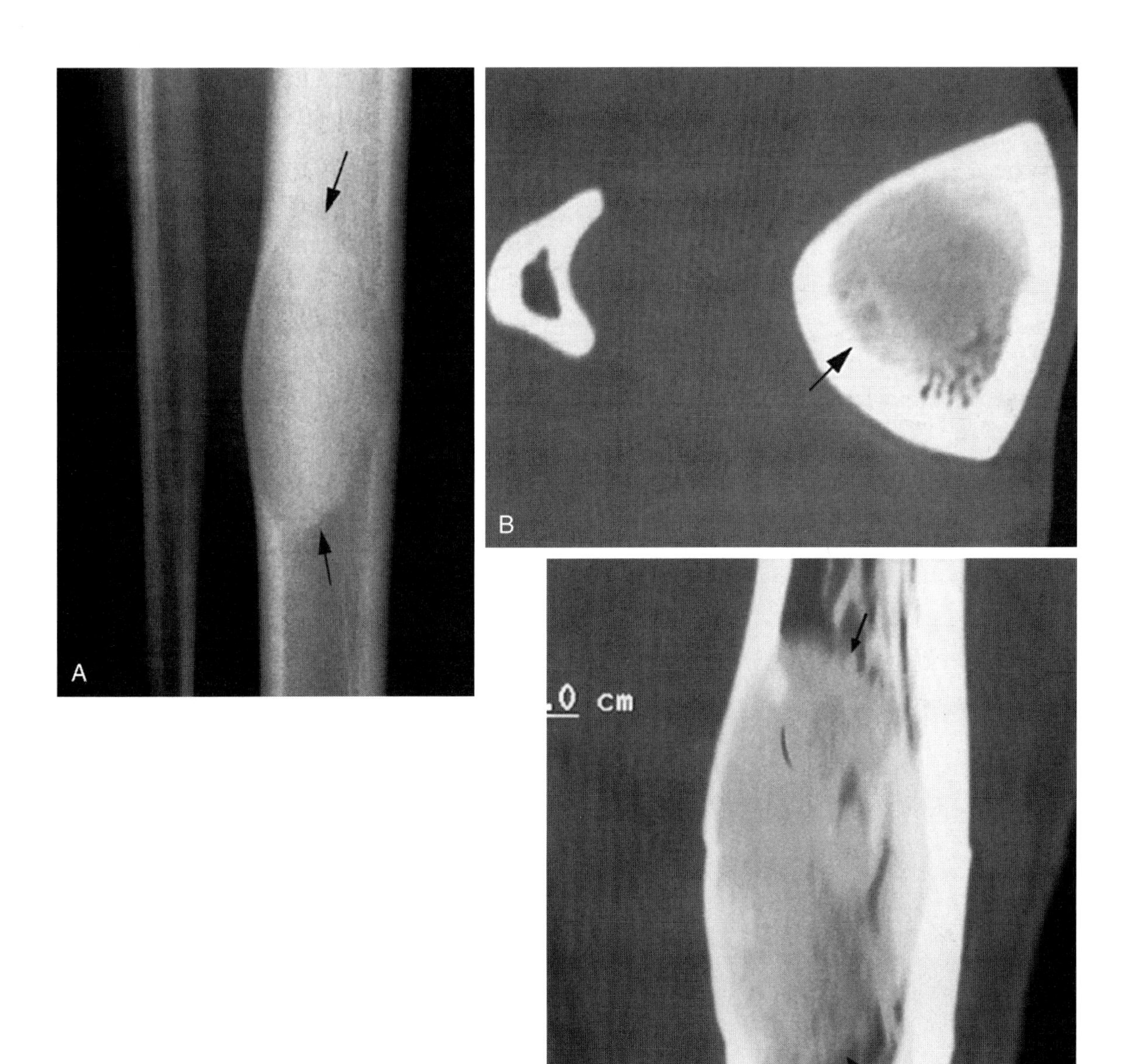

Figure 6–8. Typical ground-glass appearance of fibrous dysplasia in a 22-year-old man. *A*, Anteroposterior view of the leg shows an oval-shaped lesion in the midshaft of the tibia *(arrows)*. The lesion is moderately dense but not sclerotic. *B* and *C*, Axial and sagittal reconstructed CT images of the lesion *(arrows)* show the ground-glass appearance typical of fibrous dysplasia.

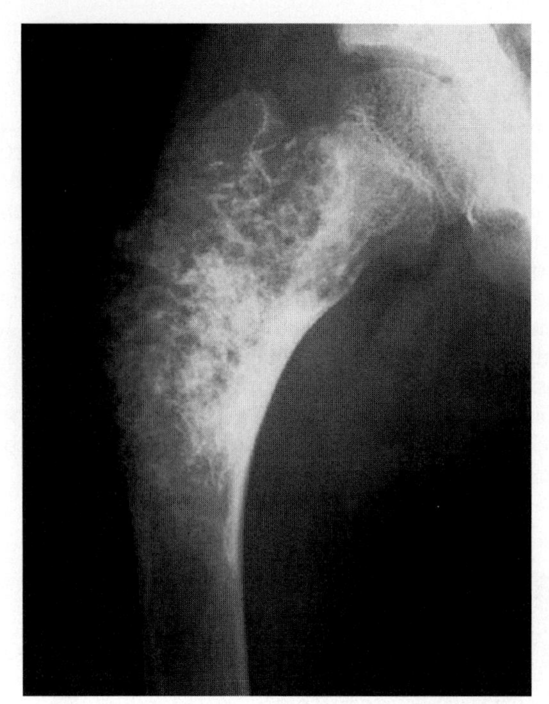

Figure 6–9. Fibrous dysplasia with chondroid calcifications. Anteroposterior view of the right hip shows arclike and whorl-like calcifications consistent with chondroid calcifications within fibrous dysplasia.

Computed Tomography

Computed tomography (CT) is useful in assessing the extent of fibrous dysplasia in anatomically complex areas, such as the spine and craniofacial bones.

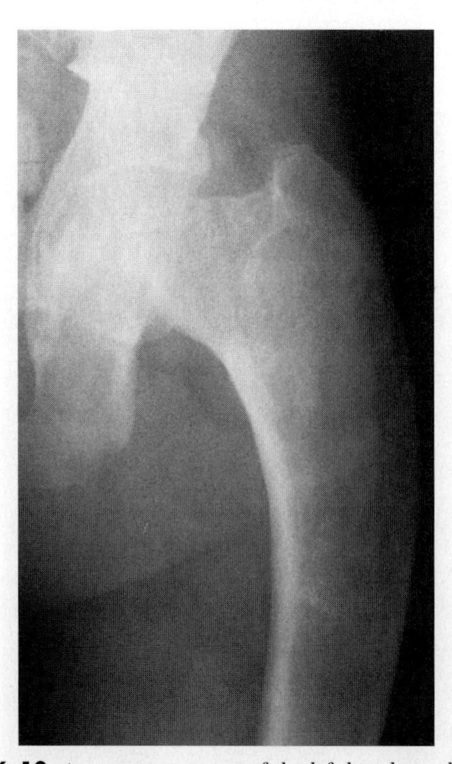

Figure 6–10. Anteroposterior view of the left hip shows the typical "shepherd's crook" deformity of the proximal femur in a patient with fibrous dysplasia.

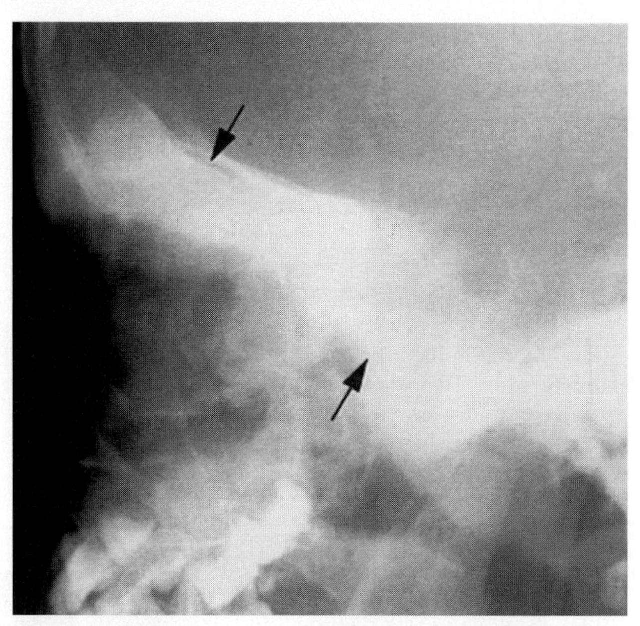

Figure 6–11. Fibrous dysplasia of the skull base in a 6-year-old child. Lateral view of the skull shows a diffusely thickened and dense skull base *(arrows).*

Magnetic Resonance Imaging

Fibrous dysplasia has low signal intensity on T1-weighted MR images. On T2-weighted MR images, 60% of the cases have increased signal intensity, and the remaining cases have either intermediate or low signal intensity. The degree of low signal intensity on T2-

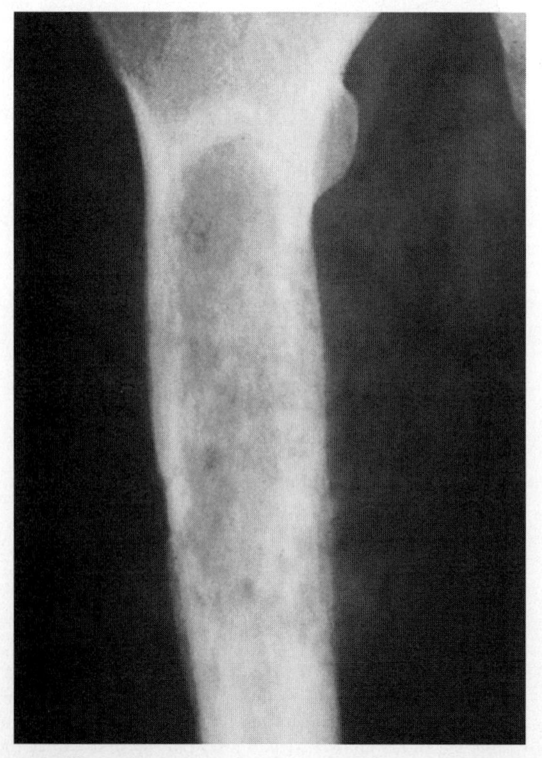

Figure 6–12. Malignant degeneration arising from fibrous dysplasia. The midportion of the lesion shows an aggressive periosteal reaction with osteoid matrix extending into the adjacent soft tissues. This is an example of osteosarcoma arising from fibrous dysplasia.

weighted images is directly proportional to the amount of osseous trabeculation within the lesion.

Nuclear Medicine

Radionuclide bone scans usually show increased radiopharmaceutical uptake in fibrous dysplasia.

Differential Diagnosis

The differential diagnosis includes simple bone cyst, giant cell tumor, nonossifying fibroma, enchondroma, eosinophilic granuloma, osteoblastoma (nonmineralized), and hemangioma.

Complications

Malignant degeneration is a rare complication; it is estimated to occur in about 0.5% of cases. Osteosarcoma, fibrosarcoma, and chondrosarcoma are the most common tumors occurring in fibrous dysplasia lesions (Fig. 6–12). About one third of patients with malignant degeneration have been irradiated previously.

BENIGN FIBROUS HISTIOCYTOMA

Overview

Benign fibrous histiocytoma is a rare, benign, fibrous lesion of bone that shares the same histologic features with nonossifying fibroma. It consists of a storiform pattern of fibrous tissue, histiocytes, and lipid-filled cells. Benign fibrous histiocytomas primarily occur in adults, whereas nonossifying fibromas occur in children and adolescents. Typically, benign fibrous histiocytomas are painful; however, nonossifying fibromas are asymptomatic and discovered incidentally or after a pathologic fracture.

Skeletal Involvement

The most commonly reported location for benign fibrous histiocytoma is the diaphysis or metaphysis of a long bone. Lesions starting in the metaphysis can extend into the epiphysis. Benign fibrous histiocytomas of bone also can occur in bones not commonly involved with nonossifying fibromas, such as the pelvis, ribs, clavicles, mandible, scapulae, and vertebrae.

Imaging

On radiographs and CT scans, benign fibrous histiocytomas are usually lytic, are slightly expansile, and have no identifiable matrix (Fig. 6–13). Typically, they are well defined and well marginated; however, they may show aggressive behavior and indistinct borders. Periosteal reaction is not present unless there is an associated pathologic fracture.

Differential Diagnosis

Benign fibrous histiocytomas may be distinguished from nonossifying fibromas by their atypical location (relative to nonossifying fibromas), occurrence in older patients, and

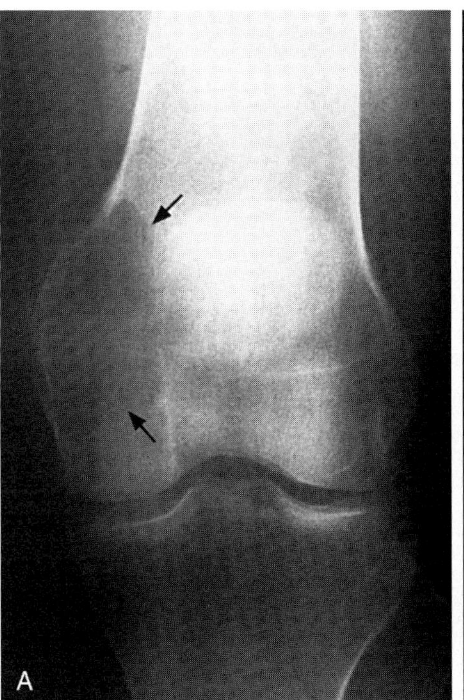

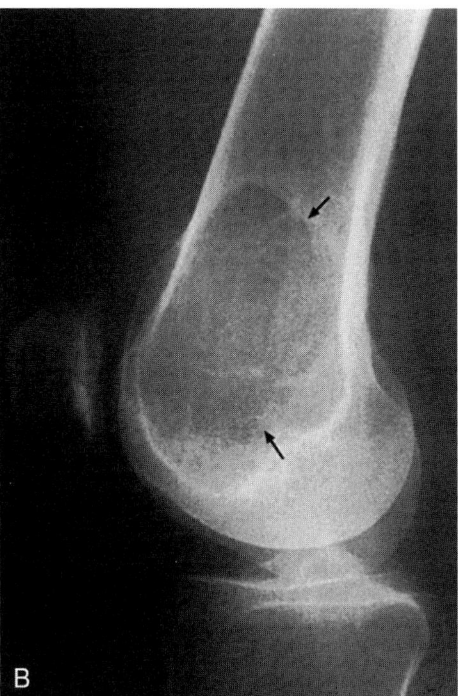

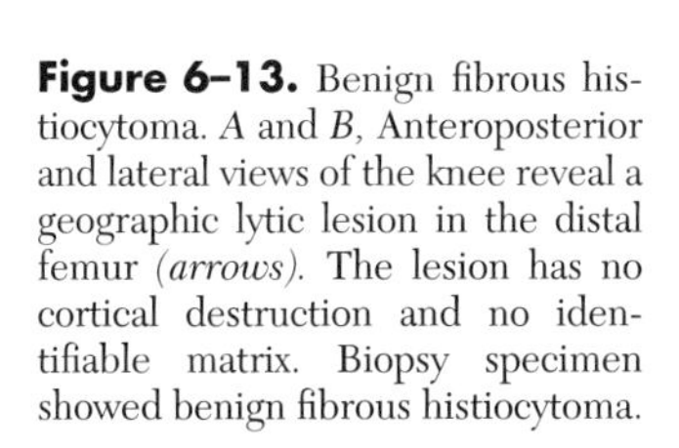
Figure 6–13. Benign fibrous histiocytoma. *A* and *B*, Anteroposterior and lateral views of the knee reveal a geographic lytic lesion in the distal femur *(arrows)*. The lesion has no cortical destruction and no identifiable matrix. Biopsy specimen showed benign fibrous histiocytoma.

atypical radiographic appearance. Also, benign fibrous histiocytomas can resemble radiographically giant cell tumors, aneurysmal bone cysts, fibrous dysplasia, and brown tumors.

PERIOSTEAL (JUXTACORTICAL) DESMOID

Overview

Periosteal (juxtacortical) desmoid is a common lesion that is known by different names, including cortical irregularities of the distal femur, cortical desmoid, avulsive cortical irregularity, and parosteal desmoid. Cortical desmoids are reported to occur in 11.5% of boys and 3.6% of girls between the ages of 3 and 17 years. The lesion is asymptomatic. It is bilateral in 35% of patients. The location of the lesion is typical; it involves the distal femur just proximal to the adductor tubercle. Most authors agree that it represents a chronic avulsive injury at the insertion of the adductor magnus muscle or origin of the medial head of the gastrocnemius. A similar avulsive lesion has been described in the proximal humerus at the insertion of the pectoralis major. This lesion is seen in young, high-caliber athletes, especially gymnasts.

Imaging

Radiography

Radiographs reveal a saucer-shaped cortical erosion in the typical location surrounded by sclerosis (Fig. 6–14). The lesion is shown best on oblique views with the knee in 20 to 40 degrees of external rotation.

Computed Tomography

CT scans show the cortical excavation, with a sclerotic margin, in the distal femoral metaphysis.

Magnetic Resonance Imaging

MR imaging findings can be confusing because of associated marrow edema raising the possibility of a malignant neoplasm.

Nuclear Medicine

A radionuclide bone scan can be helpful. Typically, bone scans show normal or only minimally increased uptake.

Complications

There are no complications reported with cortical desmoid except for the difficulty and confusion in differentiating a cortical desmoid from a malignant process. There is at least one report of an amputation in which the lesion was misdiagnosed as an osteosarcoma.

Differential Diagnosis

Osteosarcoma should be entertained as a differential diagnosis, but it can be ruled out easily when the lesion is

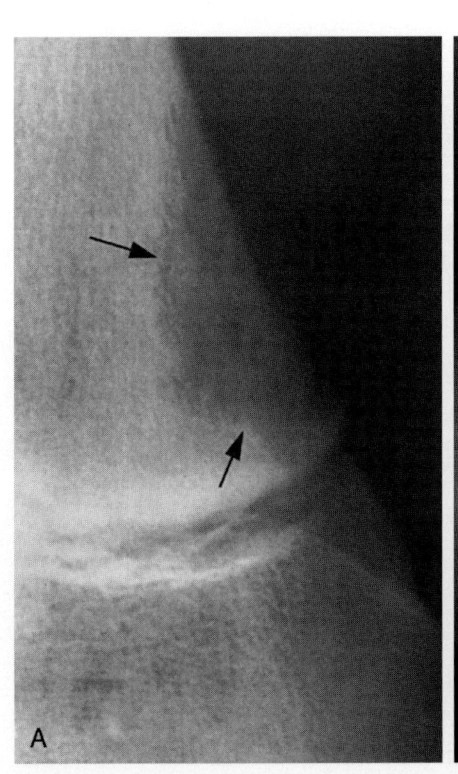

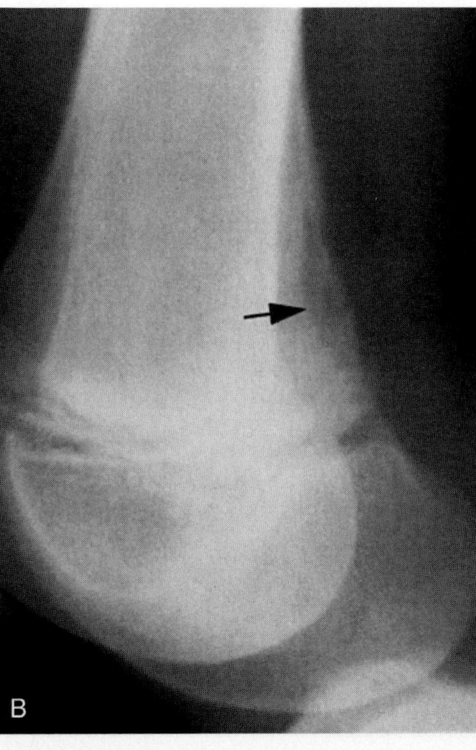

Figure 6–14. Juxtacortical desmoid. *A* and *B*, Oblique and lateral views of the knee show a saucer-shaped cortical erosion *(arrows)* at the posteromedial aspect of the femoral metaphysis.

bilateral or when the appearance on a CT or bone scan is characteristic.

DESMOPLASTIC FIBROMA OF BONE (INTRAOSSEOUS DESMOPLASTIC FIBROMA)

Overview

Desmoplastic fibroma of bone is a rare, benign, intraosseous lesion with fewer than 150 cases reported in the world literature. It is a locally aggressive tumor of fibrous origin representing the skeletal counterpart of soft tissue aggressive fibromatosis. Microscopically, there are spindle-shaped fibroblasts in a dense collagenous matrix. The histologic distinction between desmoplastic fibroma and grade I fibrosarcoma can be difficult. The lesion can occur at any age; it has been reported between the ages of 15 months and 75 years. It has no sex predilection.

Skeletal Involvement

Desmoplastic fibromas have been reported in almost all bones of the skeleton with equal frequency in flat and long bones.

Imaging

Desmoplastic fibroma of bone is difficult to diagnose preoperatively because of its rarity and lack of distinguishing imaging features.

Radiography

Radiographs reveal a lytic, geographic, expansile, trabeculated, honeycomb-like lesion with endosteal erosion (Fig. 6–15). The borders of the lesion are usually well defined with a narrow zone of transition and, rarely, sclerotic margins. Sometimes the lesion appears aggressive with cortical destruction and soft tissue extension. Periosteal reaction has not been described with this lesion.

Magnetic Resonance Imaging

MR imaging shows the extent of the lesion. Desmoplastic fibromas of bone have heterogeneous signal intensity. Areas of low signal intensity on T1- and T2-weighted sequences are seen because of the presence of dense, hypocellular, fibrous tissue rich in collagen.

Differential Diagnosis

A desmoplastic fibroma of bone should be differentiated from a nonossifying fibroma, giant cell tumor, aneurysmal bone cyst, unicameral bone cyst, fibrous dysplasia, fibrosarcoma, and malignant fibrous histiocytoma.

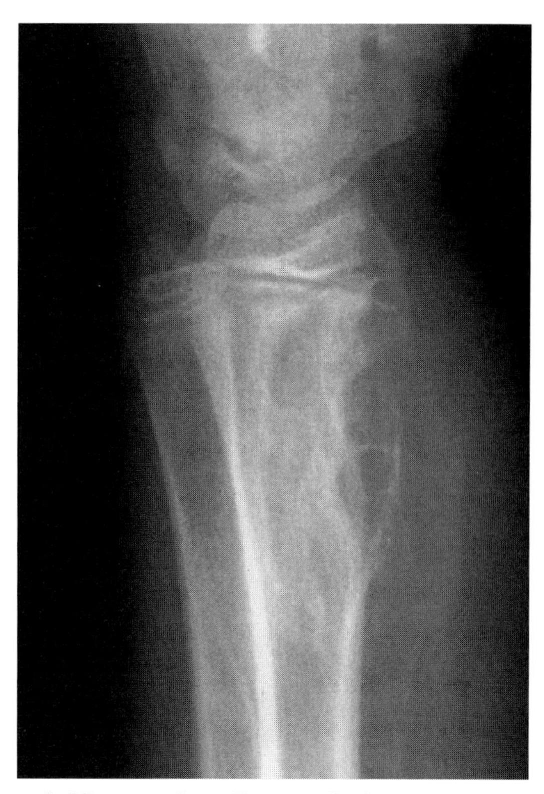

Figure 6–15. Desmoplastic fibroma in the forearm. Oblique view of the distal forearm shows an expansile, trabeculated, honeycomb-like lytic lesion with endosteal erosion. Biopsy specimen showed a desmoplastic fibroma.

OSTEOFIBROUS DYSPLASIA (OSSIFYING FIBROMA, CAMPANACCI LESION)

Overview

Osteofibrous dysplasia (ossifying fibroma, Campanacci lesion) is known by different names, but more recent publications have uniformly used the term *osteofibrous dysplasia*. The lesion has distinctive clinical and histologic features: It has a predilection to occur in the anterior cortex of the midtibia, it almost always is unilateral, and most cases occur before age 15 years. Greater than 60% of children affected are younger than 5 years old, and boys are affected slightly more commonly than girls. Occasionally, symptoms present shortly after birth. In these cases, the lesion is relatively large and may involve the entire width of the tibia. Typically, patients present with painless deformity of the leg, swelling, and anterior bowing of the involved lower leg. Histologically the lesion consists of fibrous tissue containing scattered osseous trabeculae rimmed by osteoblasts. This osteoblastic rimming distinguishes osteofibrous dysplasia from fibrous dysplasia. The histologic features of osteofibrous dysplasia are similar to those seen in ossifying fibroma of the mandible in adults. Some authors believe that osteofibrous dysplasia is related closely to adamantinoma because of the similarity in location, histology, and radiographic appearance.

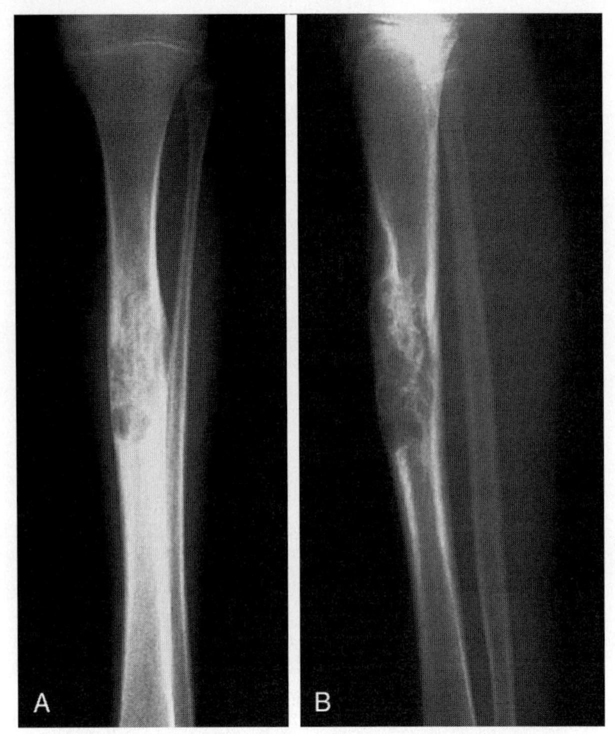

Figure 6–16. Osteofibrous dysplasia in a 13-year-old girl. *A* and *B*, Anteroposterior and lateral views of the tibia show typical intracortical osteofibrous dysplasia with invasion of the medullary space.

Skeletal Involvement

The lesion usually is seen in the tibia; however, the ipsilateral fibula may be affected in about 17% of cases. There are isolated reports in which the lesion is limited to the fibula. Bilateral osteofibrous dysplasia is extremely rare. The lesion typically is located anteriorly in the middle one third of the tibia (Fig. 6–16). In the fibula, it usually occurs in the distal diaphysis.

Imaging

Radiographic features include multiloculated intracortical lucencies involving the anterior diaphysis of the tibia (Fig. 6–17). The lytic areas are separated from the medullary space by a thick rim of sclerosis. The lytic appearance can assume the shape of a single focus or multiple satellites. The lesion can expand and coalesce until skeletal growth is complete, at which time it starts to regress. There is no soft tissue involvement and no periosteal reaction.

Complications

A pathologic fracture can occur resulting in pseudarthrosis. Local recurrence after surgery is frequent, especially in young children. Recurrence is unlikely if surgery is performed after the age of 15 years.

Differential Diagnosis

The differential diagnosis includes adamantinoma, fibrous dysplasia, nonossifying fibroma, and chondromyxoid fibroma. Adamantinomas can be difficult to differentiate radiographically from osteofibrous dysplasia; however, adamantinomas do not develop before the age of 15 years. Fibrous dysplasia can be differentiated by its intramedullary location and the more advanced patient age at the

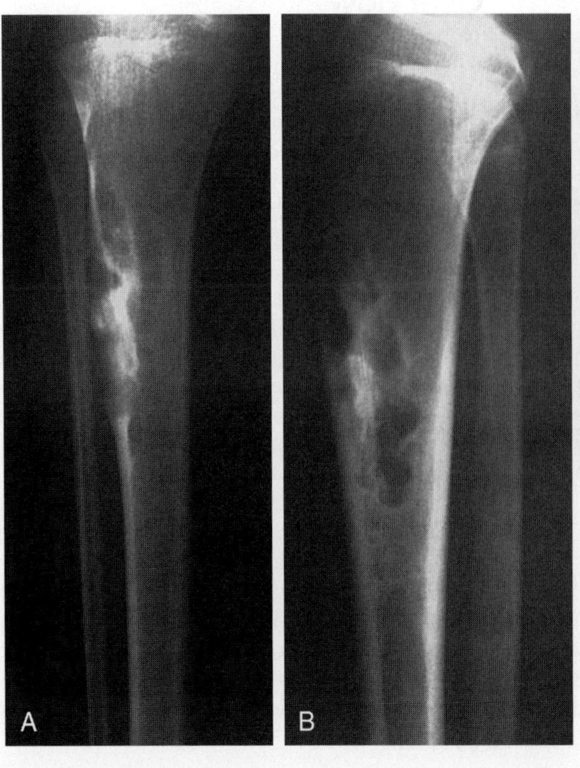

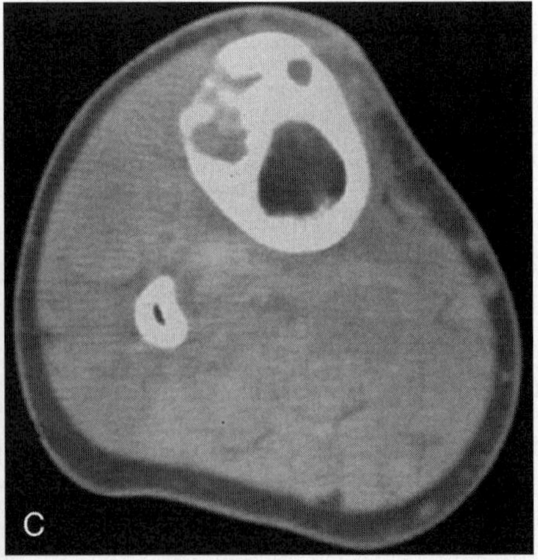

Figure 6–17. Osteofibrous dysplasia in a 16-year-old boy. *A* and *B*, Anteroposterior and lateral views of the leg show an intracortical bubbly lesion arising from the anterolateral cortex. *C*, Axial CT scan through the proximal tibia shows the intracortical origin of the lesion. The medullary space is not involved.

time of presentation. Chondromyxoid fibromas typically occur in the proximal metaphysis of the tibia rather than the middle one third of the tibia.

Recommended Readings

Arata MA, Peterson HA, Dahlin DC: Pathologic fractures through non-ossifying fibromas: Review of the Mayo Clinic experience. J Bone Joint Surg Am 63:980, 1981.

Bertoni F, Calderoni P, Bacchini P, et al: Desmoplastic fibroma of bone: A report of six cases. J Bone Joint Surg Br 66:265, 1984.

Bertoni F, Calderoni P, Bacchini P, et al: Benign fibrous histiocytoma of bone. J Bone Joint Surg Am 68:1225, 1986.

Bertoni F, Capanna R, Calderoni P, et al: Case report 223: Benign fibrous histiocytoma. Skeletal Radiol 9:215, 1983.

Blackwell JB, McCarthy SW, Xipell JM, et al: Osteofibrous dysplasia of the tibia and fibula. Pathology 20:227, 1988.

Bloem JL, van der Heul RO, Schuttevaer HM, et al: Fibrous dysplasia vs. adamantinoma of the tibia: Differentiation based on discriminant analysis of clinical and plain film findings. AJR Am J Roentgenol 156:1017, 1991.

Crim JR, Gold RH, Mirra JM, et al: Desmoplastic fibroma of bone: Radiographic analysis. Radiology 172:827, 1989.

Destouet JM, Kyriakos M, Gilula LA: Fibrous histiocytoma (fibroxanthoma) of a cervical vertebra: A report with a review of the literature. Skeletal Radiol 5:241, 1980.

Drolshagen LF, Reynolds WA, Marcus NW: Fibrocartilaginous dysplasia of bone. Radiology 156:32, 1985.

Dunham WK, Marcus NW, Enneking WF, et al: Developmental defects of the distal femoral metaphysis. J Bone Joint Surg Am 62:801, 1980.

Gebhardt MC, Campbell CJ, Schiller AL, et al: Desmoplastic fibroma of bone: A report of eight cases and review of the literature. J Bone Joint Surg Am 67:732, 1985.

Graudal N: Desmoplastic fibroma of bone: Case report and literature review. Acta Orthop Scand 55:215, 1984.

Greenspan A, Unmi KK: Case report 787: Desmoplastic fibroma. Skeletal Radiol 22:296, 1993.

Hamada T, Ito H, Araki Y, et al: Benign fibrous histiocytoma of the femur: Review of three cases. Skeletal Radiol 25:25, 1996.

Hyman AA, Heiser WJ, Kim SE, et al: An excavation of the distal femoral metaphysis: A magnetic resonance imaging study. J Bone Joint Surg Am 77:1897, 1995.

Jee WH, Choe BY, Kang HS, et al: Nonossifying fibroma: Characteristics at MR imaging with pathologic correlation. Radiology 209:197, 1998.

Jee WH, Choi KH, Choe BY, et al: Fibrous dysplasia: MR imaging characteristics with radiopathologic correlation. AJR Am J Roentgenol 167:1523, 1996.

Kransdorf MJ, Moser RP Jr, Gilkey FW: Fibrous dysplasia. Radiographics 10:519, 1990.

Kumar R, Madewell JE, Lindell MM, et al: Fibrous lesions of bones. Radiographics 10:237, 1990.

Marks KE, Bauer TW: Fibrous tumors of bone. Orthop Clin North Am 20:377, 1989.

Matsuno T: Benign fibrous histiocytoma involving the ends of long bone. Skeletal Radiol 19:561, 1990.

Moser RP Jr, Sweet DE, Hasemann DB, et al: Multiple skeletal fibroxanthomas: Radiologic-pathologic correlation of 72 cases. Skeletal Radiol 16:353, 1987.

Nguyen BD, Lugo-Olivieri CH, McCarthy EF, et al: Fibrous dysplasia with secondary aneurysmal bone cyst. Skeletal Radiol 25:88, 1996.

Pennes DR, Braunstein EM, Glazer GM: Computed tomography of cortical desmoid. Skeletal Radiol 12:40, 1984.

Posch TJ, Puckett ML: Marrow MR signal abnormality associated with bilateral avulsive cortical irregularities in a gymnast. Skeletal Radiol 27:511, 1998.

Savage PE, Stoker DJ: Fibrous dysplasia of the femoral neck. Skeletal Radiol 11:119, 1984.

Seeger LL, Yao L, Eckardt JJ: Surface lesions of bone. Radiology 206:17, 1998.

Steinmetz JC, Pilon VA, Lee JK: Jaffe-Campanacci syndrome. J Pediatr Orthop 8:602, 1988.

Stephenson RB, London MD, Hankin FM, et al: Fibrous dysplasia: An analysis of options for treatment. J Bone Joint Surg Am 69:400, 1987.

Suh JS, Cho JH, Shin KH, et al: MR appearance of distal femoral cortical irregularity (cortical desmoid). J Comput Assist Tomogr 20:328, 1996.

Sundaram M, McDonald DJ, Merenda G: Intramuscular myxoma: A rare but important association with fibrous dysplasia of bone. AJR Am J Roentgenol 153:107, 1989.

Sweet DE, Vinh TN, Devaney K: Cortical osteofibrous dysplasia of long bone and its relationship to adamantinoma: A clinicopathologic study of 30 cases. Am J Surg Pathol 16:282, 1992.

Taconis WK: Osteosarcoma in fibrous dysplasia. Skeletal Radiol 17:163, 1988.

Taconis WK, Schute HE, van der Heul RO: Desmoplastic fibroma of bone: A report of 18 cases. Skeletal Radiol 23:283, 1994.

Tehranzadeh J, Fung Y, Donohue M, et al: Computed tomography of Paget disease of the skull versus fibrous dysplasia. Skeletal Radiol 27:664, 1998.

Wang JW, Shih CH, Chen WJ: Osteofibrous dysplasia (ossifying fibroma of long bones): A report of four cases and review of the literature. Clin Orthop 278:235, 1992.

Young JW, Aisner SC, Levine AM, et al: Computed tomography of desmoid tumors of bone: Desmoplastic fibroma. Skeletal Radiol 17:333, 1988.

Yu JS, Lawrence S, Pathria M, et al: Desmoplastic fibroma of the calcaneus. Skeletal Radiol 24:451, 1995.

CHAPTER 7

Malignant Fibrous and Histiocytic Tumors

Nabil J. Khoury, MD, Georges Y. El-Khoury, MD, and D. Lee Bennett, MD

FIBROSARCOMA

Overview

Fibrosarcoma of bone is a malignant tumor of spindle-shaped cells that do not produce osteoid matrix. It is a rare lesion, representing about 3.3% of all primary malignant bone tumors, which is less than one sixth as common as osteosarcoma. About one fourth to one third of fibrosarcomas are secondary fibrosarcomas arising in preexisting skeletal lesions, such as Paget's disease or a bone infarct in previously irradiated bone. The age distribution is wide, usually occurring between the teens and 50s, with less than 5% occurring in individuals younger than age 15 years. A fibrosarcoma is characterized histologically by the presence of spindle-shaped cells and varying amounts of collagen. Until 1972, fibrosarcoma and malignant fibrous histiocytoma (MFH) of bone were lumped together, but MFH currently is distinguished from fibrosarcoma by the presence of giant cells and the arrangement of the cells and collagen fibers in a cartwheel or storiform pattern. Controversy still exists as to whether a significant difference exists in the biologic behavior of these two lesions.

Skeletal Involvement

Any bone may be involved, with the long tubular bones representing 70% of all locations. Most commonly, the distal femur and proximal tibia are affected. Fibrosarcomas are usually a central tumor (intramedullary), but they may be eccentric and even periosteal. They most commonly are metaphyseal or metadiaphyseal in location. Epiphyseal extension is uncommon.

Imaging

Radiography

The diagnosis of fibrosarcoma or MFH rarely is made on radiographs because these lesions have no distinguishing imaging features. The lesions are lytic with a moth-eaten or permeative pattern of destruction, and they have no visible matrix. Low-grade lesions have a geographic pattern of bone destruction with well-defined margins (Fig. 7–1). High-grade lesions have a permeative pattern of bone destruction without surrounding new bone formation. Older literature describes the presence of a sequestrum as a characteristic feature of a fibrosarcoma; however, this is not supported by more recent studies.

Magnetic Resonance Imaging

Magnetic resonance (MR) imaging is useful for defining the extent of the lesion. Fibrosarcomas appear low in signal intensity on T1-weighted MR images and inhomogeneously high in signal intensity on T2-weighted images. Low signal intensity areas may be present on T2-weighted images depending on the amount of fibrous tissue present.

Differential Diagnosis

The differential diagnosis includes malignant fibrous histiocytoma, lytic metastases, multiple myeloma, telangiectatic osteosarcoma, lymphoma, and desmoplastic fibroma.

MALIGNANT FIBROUS HISTIOCYTOMA

Overview

MFH is the most common soft tissue sarcoma in adults, but it rarely originates in bone. The lesion accounts for only about 5% of all malignant bone tumors. It occurs at any age between 6 and 80 years, with the peak prevalence in the 30s. There is a slight male predominance (1.5:1). Approximately 20% of patients present with a pathologic fracture. MFH is a pleomorphic sarcoma that contains fibroblasts and histiocytes in varying proportions. The spindle-shaped fibroblasts are arranged in a cartwheel (storiform) pattern. The principal histologic differential diagnosis is fibrosarcoma. The prognosis for MFH is poor; its rate of recurrence and rate of metastasis are high. About 20% of all intraosseous MFHs arise within abnormal bone. Examples of abnormal bone predisposed to MFH occurrence include irradiated bone, bone infarct, fibrous dysplasia, Paget's disease, enchondroma, chronic

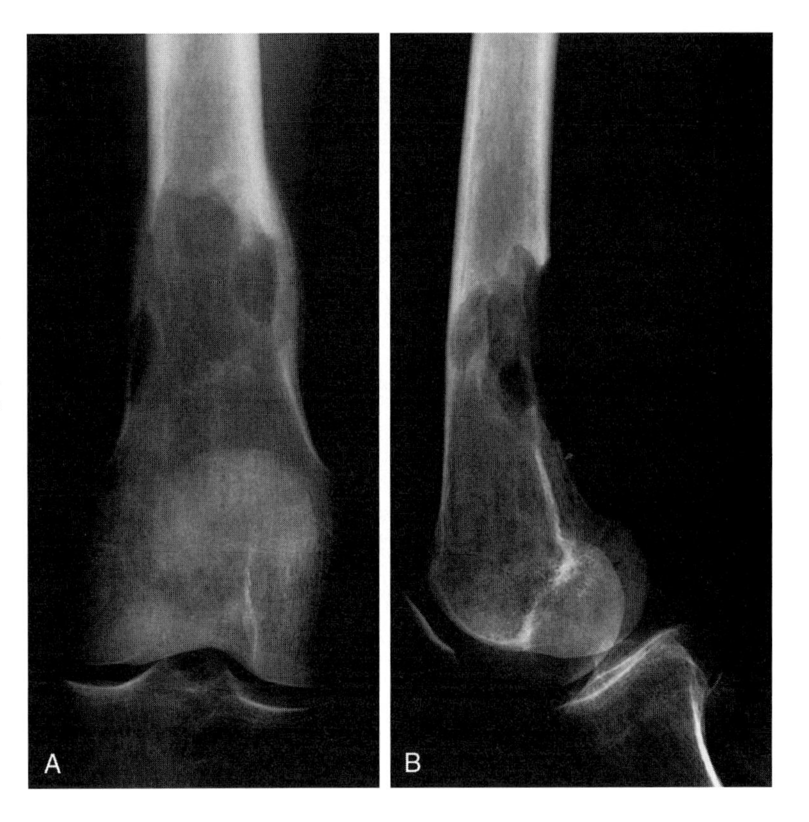

Figure 7–1. Fibrosarcoma of the distal femur. *A* and *B*, Anteroposterior and lateral views of the femur show a geographic destructive lesion in the distal shaft of the right femur. The lesion has a narrow zone of transition and no identifiable matrix. The posterior cortex is completely destroyed. (Courtesy of Ronald Swee, MD, Rochester, MN.)

osteomyelitis, metallic implants or shrapnel, and giant cell tumor. Irradiation is the most common cause of a secondary MFH. Multicentric MFHs have been described; they may occur as primary lesions or as metastatic foci from a primary soft tissue or primary osseous focus.

Skeletal Involvement

MFH occurs most commonly in the long bones of the lower extremity, primarily in the distal femur (45%) and proximal tibia (20%). Other locations where MFH may occur include the humerus, pelvis, spine, and ribs.

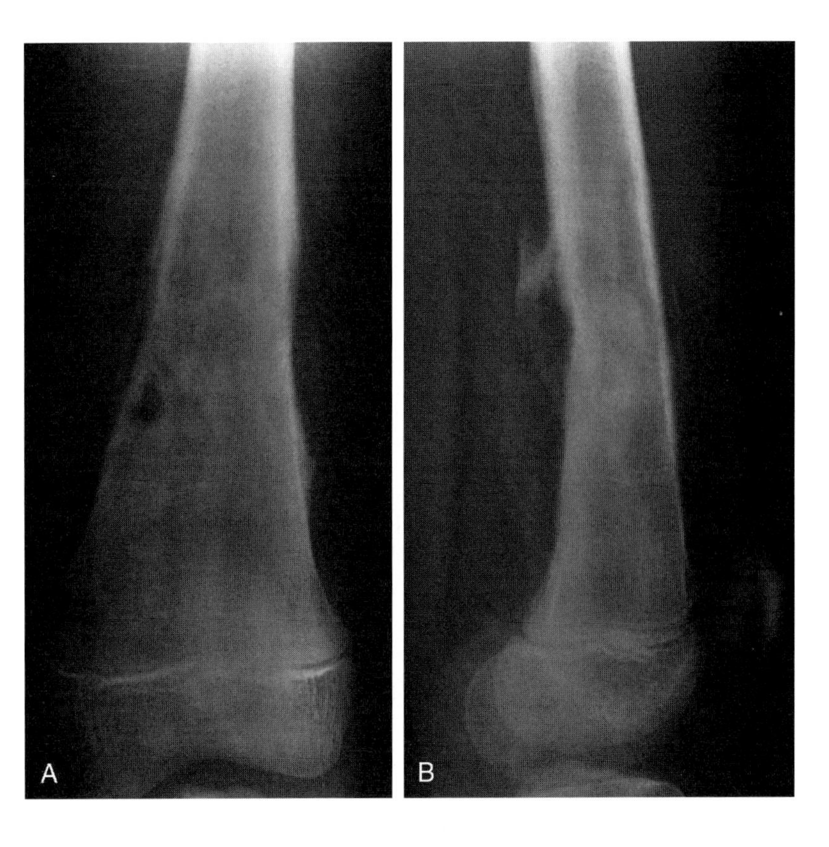

Figure 7–2. Malignant fibrous histiocytoma of the distal femur in a teenage boy. *A* and *B*, Anteroposterior and lateral views of the distal femur show a permeative destructive lesion with ill-defined margins. An interrupted periosteal reaction is associated with the lesion. (Courtesy of Ronald Swee, MD, Rochester, MN.)

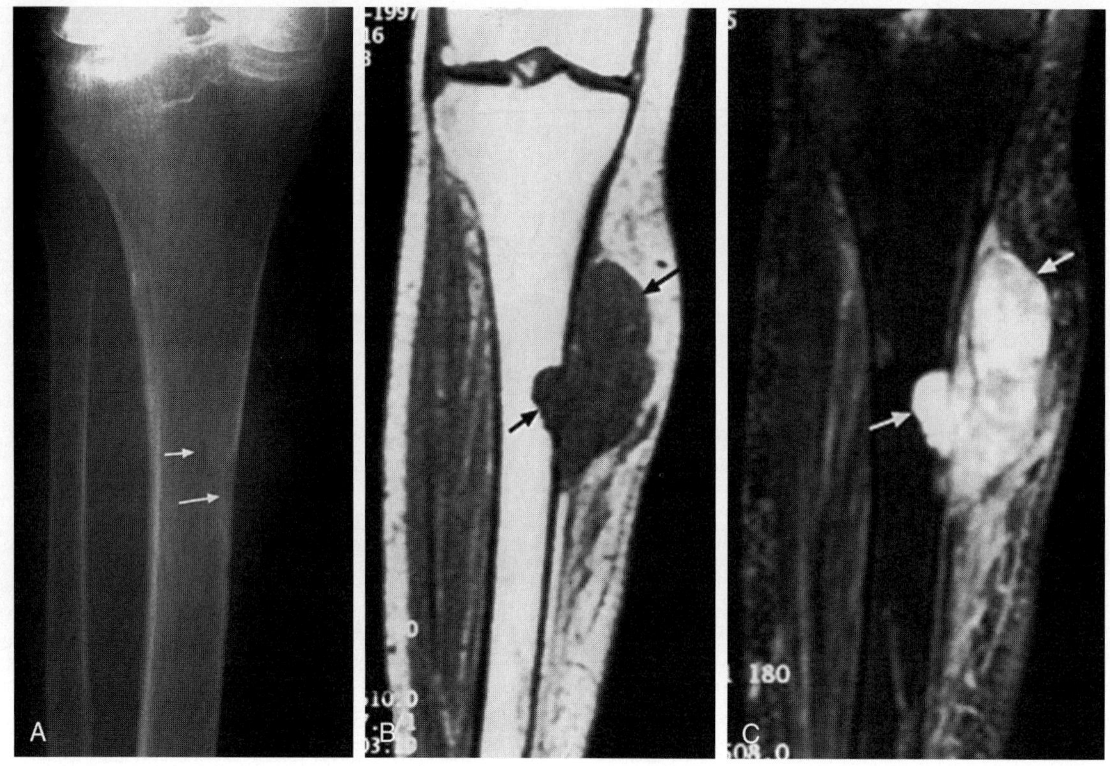

Figure 7–3. Malignant fibrous histiocytoma of the leg with involvement of the tibia. *A,* Anteroposterior view of the tibia shows a lytic lesion in the proximal tibia. The lesion has a motheaten appearance *(arrows).* A prominent soft tissue mass can be seen adjacent to the osseous lesion. *B* and *C,* Coronal T1-weighted and fat-saturation T2-weighted MR images of the tibia show a soft tissue mass with apparent extension into the adjacent bone *(arrows).* Biopsy specimen showed a malignant fibrous histiocytoma. It is rare for primary soft tissue sarcomas to invade bone.

Typically, when MFH occurs in a long bone, it is located centrally within the metaphysis.

Imaging

Radiography

Making a specific diagnosis of MFH on radiographs may be difficult. Radiography shows the lesion to be lytic and aggressive with a permeative or moth-eaten pattern of destruction (Fig. 7–2). Cortical destruction and soft tissue extension are frequent. Periosteal reaction is uncommon; however, it may be seen when a pathologic fracture is present. Calcifications and sequestra within the lesion can occur, but they are rare. Sometimes it is difficult to determine whether the tumor originated within the bone or adjacent soft tissues (Fig. 7–3), especially if the osseous and soft tissue components are equal in size.

Magnetic Resonance Imaging

On MR imaging, MFH has low signal intensity on T1-weighted images and inhomogeneous high signal intensity on T2-weighted images (see Fig. 7–3). It appears nodular and has peripheral enhancement after gadolinium injection. Some lesions may have fluid-fluid levels. MR imaging also shows the extent of soft tissue invasion.

Differential Diagnosis

The differential diagnosis includes metastases, plasmacytoma, lymphoma, osteolytic osteosarcoma, fibrosarcoma, and dedifferentiated chondrosarcoma.

Recommended Readings

Bertoni F, Cappana R, Calderoni P, et al: Primary central (medullary) fibrosarcoma of bone. Semin Diagn Pathol 1:185, 1984.

Cappana R, Bertoni F, Bacchini P, et al: Malignant fibrous histiocytoma of bone: The experience at the Rizzoli Institute: Report of 90 cases. Cancer 54:177, 1984.

Hudson TM, Stiles RG, Monson DK: Fibrous lesions of bone. Radiol Clin North Am 31:279, 1993.

Link TM, Haeussler MD, Poppek S, et al: Malignant fibrous histiocytoma of bone: Conventional x-ray and MR imaging features. Skeletal Radiol 27:552, 1998.

Marks KE, Bauer TW: Fibrous tumors of bone. Orthop Clin North Am 20:377, 1989.

Murphey MD, Gross TM, Rosenthal HG: From the archives of the AFIP: Musculoskeletal malignant fibrous histiocytoma: Radiologic-pathologic correlation. Radiographics 14:807, 1994.

Ros PR, Viamonte M Jr, Rywlin AM: Malignant fibrous histiocytoma: Mesenchymal tumor of ubiquitous origin. AJR Am J Roentgenol 142:753, 1984.

Taconis WK, Mulder JD: Fibrosarcoma and malignant fibrous histiocytoma of long bones: Radiographic features and grading. Skeletal Radiol 11:237, 1984.

CHAPTER 8

Benign Tumors Originating from Adipose Tissue

Nabil J. Khoury, MD, Georges Y. El-Khoury, MD, and D. Lee Bennett, MD

INTRAOSSEOUS LIPOMA

Overview

Intraosseous lipoma is a rare primary bone tumor. It is composed of mature fat lobules separated by fibrous septa. Intraosseous lipomas can occur in all age groups, but they are more frequent in adults. Males and females are affected equally. Some patients experience localized pain and soft tissue swelling; however, in about one third of patients, the lesion is discovered incidentally.

Skeletal Involvement

Intraosseous lipomas most commonly occur in the long tubular bones and the calcaneus (Fig. 8–1). In the

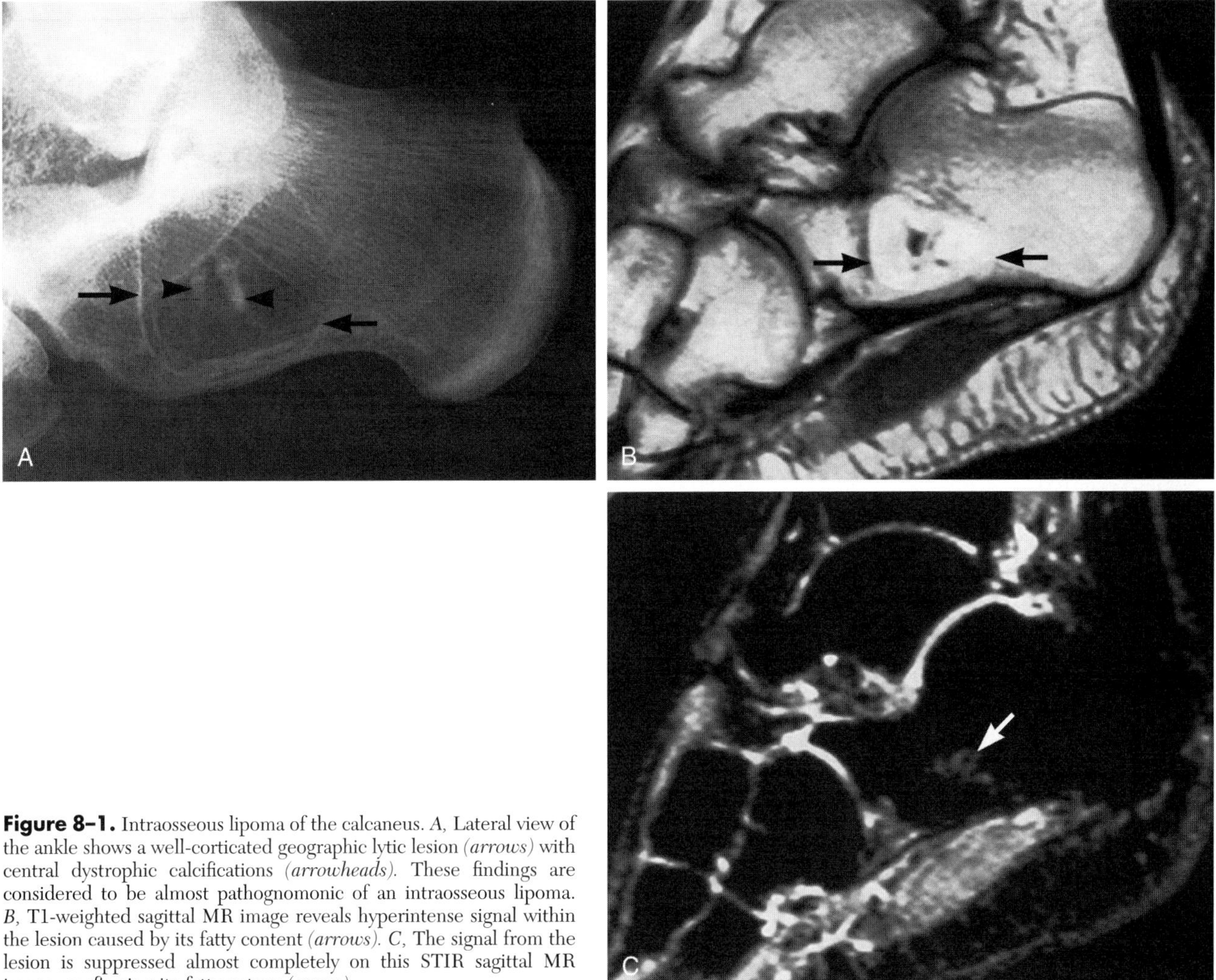

Figure 8–1. Intraosseous lipoma of the calcaneus. *A*, Lateral view of the ankle shows a well-corticated geographic lytic lesion *(arrows)* with central dystrophic calcifications *(arrowheads)*. These findings are considered to be almost pathognomonic of an intraosseous lipoma. *B*, T1-weighted sagittal MR image reveals hyperintense signal within the lesion caused by its fatty content *(arrows)*. *C*, The signal from the lesion is suppressed almost completely on this STIR sagittal MR image, confirming its fatty nature *(arrow)*.

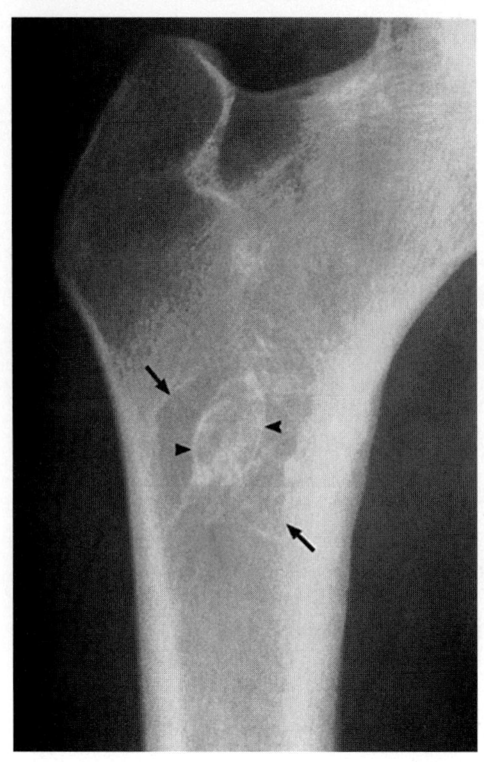

Figure 8–2. Anteroposterior view of the proximal femur shows an intraosseous lipoma *(arrows)* with central dystrophic calcifications *(arrowheads)*.

calcaneus, lipoma occurs in the triangular area between the major trabecular groups (see Fig. 8–1). In the long tubular bones, the femur, tibia, and fibula are the most frequently involved sites. In the femur, intraosseous lipomas are encountered in the neck and intertrochanteric region (Fig. 8–2). Other less common sites include the ribs, skull, pelvis, and spine.

Imaging

Radiography

The lack of specific radiographic features, the wide age range, and varying sites of occurrence make a specific diagnosis on radiographs difficult. When an intraosseous lipoma is located in the calcaneus or the proximal humerus, however, the radiographic features are almost pathognomonic. With fat involution, central dystrophic calcification and ossification develop. In the calcaneus, intraosseous lipomas occur in the same location as simple cysts, which is the main consideration in the differential diagnosis. Simple cysts do not have a central calcification, however, that is characteristic of an intraosseous lipoma. The tumor typically appears as a well-defined, nonaggressive, lytic lesion with thin sclerotic margins. Sometimes it is expansile and lobulated and has cortical thinning. Typically, there is no associated periosteal reaction. In the proximal femur, the margins may be thick and dense. Internal osseous ridges sometimes are seen.

Computed Tomography

A computed tomography (CT) scan can be diagnostic in showing the fat content of the lesion with Hounsfield units between −40 and −70.

Magnetic Resonance Imaging

Magnetic resonance (MR) imaging shows the tumor content, particularly the fat, which has high signal intensity on T1-weighted spin echo sequences (see Fig. 8–1). With tumor involution, MR imaging shows areas of necrosis and cystic degeneration, which appear as areas of low signal intensity on T1-weighted images and high signal intensity on T2-weighted images. When central calcifications are present, they appear as areas of low signal intensity on all pulse sequences (see Fig. 8–1).

Differential Diagnosis

On radiographs, intraosseous lipomas may simulate simple bone cysts, fibrous dysplasia (in particular, the femoral lesions), osteoblastomas, and chondromyxoid fibromas. An involuted lipoma should be differentiated from an enchondroma or bone infarct.

PAROSTEAL LIPOMA

Overview

Parosteal lipoma is a rare surface lesion arising from the tissues immediately superficial to the periosteum. The mean age of patients presenting with this lesion is approximately 50 years. The lesion may present as a painless soft tissue mass. Nerve entrapment with motor or sensory deficit has been reported. In decreasing order of frequency, involvement of the radial, sciatic, ulnar, and median nerves has been described.

Skeletal Involvement

The lesion most frequently occurs in the long bones of the upper and lower extremities. It is usually metaphyseal or diaphyseal in location. Rarely, it may occur in the scapula, clavicle, ribs, pelvis, skull, mandible, and short bones of the hands and feet.

Imaging

Radiography and Computed Tomography

Radiographic and CT findings are those of a well-defined, lucent juxtacortical lesion with adjacent reactive changes in the underlying cortex (Fig. 8–3*A*, *B*). These reactive changes primarily consist of bony excrescences resembling osteochondromas but without continuity with the cortex or communication with the marrow space (see

Fig. 8–3*B*). Other changes in the underlying cortex include thickening, a solid periosteal reaction, hyperostosis, pressure erosions, and bowing. The lipoma itself may contain areas of cartilage and bone.

Magnetic Resonance Imaging

On MR imaging, the lesion is isointense with fat (Fig. 8–3*C, D*). Fibrovascular tissue and hyaline cartilage present within the lesion appear as intermediate signal intensity areas on T1-weighted images and high signal intensity areas on T2-weighted images. Fibrous septa may be noted between the fat lobules and at the periphery of the lesion (see Fig. 8–3*C, D*). MR imaging can show sites of nerve impingement and muscle atrophy.

Recommended Readings

Barcelo M, Pathria MN, Abdul-Karim FW: Intraosseous lipoma: A clinicopathologic study of four cases. Arch Pathol Lab Med 116:947, 1992.

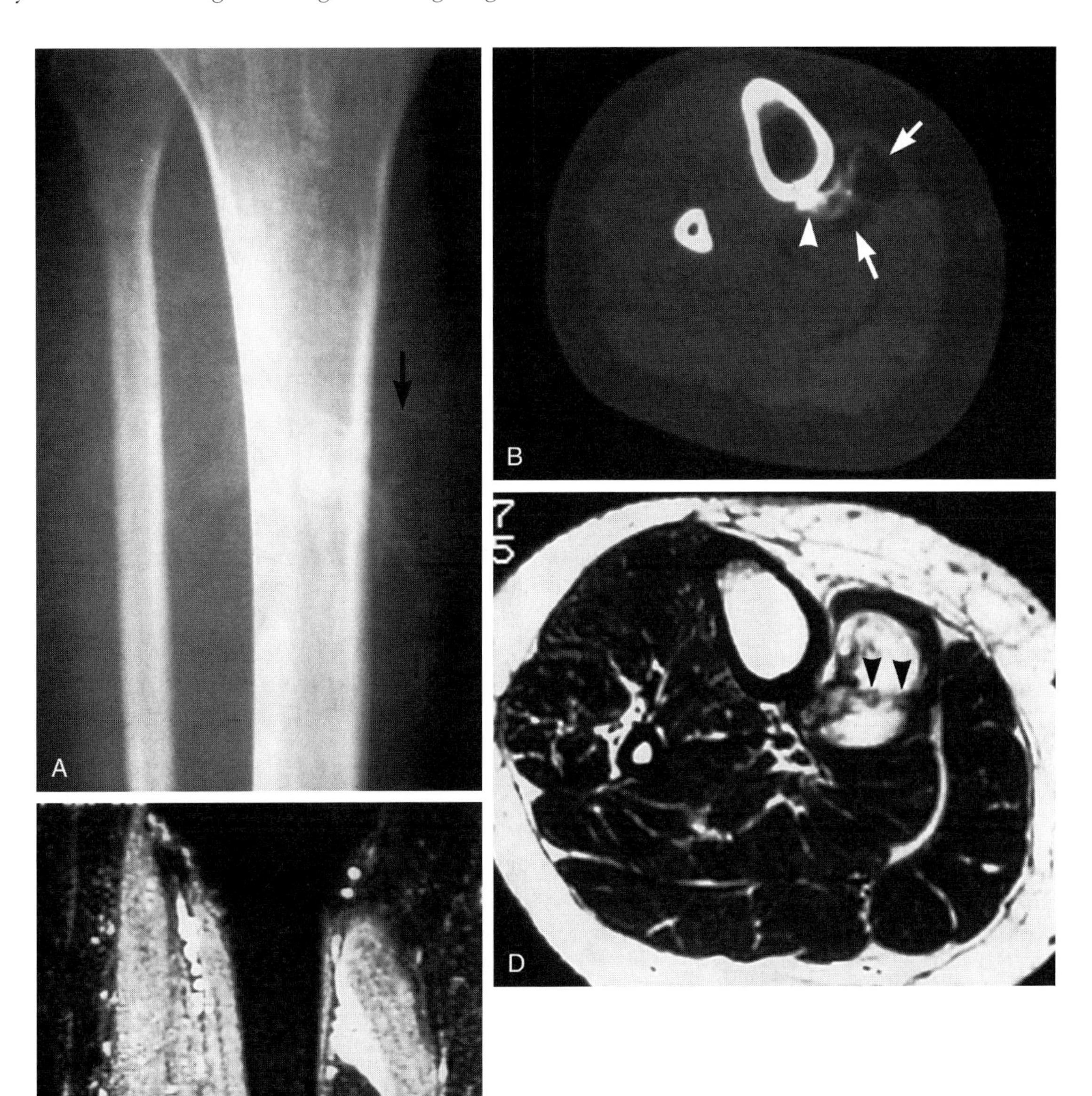

Figure 8–3. Parosteal lipoma. *A,* Anteroposterior radiograph of the lower leg shows a juxtacortical low-density mass adjacent to the diametaphyseal region of the tibia *(arrows).* There also is an osseous excrescence and bone within the lesion *(arrowhead). B,* Axial CT scan near the midportion of the lesion shows a well-defined, fat-containing mass *(arrows).* The osseous excrescence also is visible *(arrowhead). C* and *D,* Axial T1-weighted and coronal STIR MR images show the fatty content of the lesion. A septation, with isointense signal on the T1-weighted image *(arrowheads)* and hyperintense signal on the STIR image *(arrow),* also is seen. (Courtesy of the Armed Forces Institute of Pathology.)

Chow LT, Lee KC: Intraosseous lipoma: A clinicopathologic study of nine cases. Am J Surg Pathol 16:401, 1992.
Gonzalez JV, Stuck RM, Streit N: Intraosseous lipoma of the calcaneus: A clinicopathologic study of three cases. J Foot Ankle Surg 36:306, 1997.
Kenan S, Abdelwahab IF, Klein MJ, et al: Lesions of juxtacortical origin (surface lesions of bone). Skeletal Radiol 22:337, 1993.
Laorr A, Greenspan A: Parosteal lipoma with hyperostosis: Report of two pathologically proven cases evaluated by magnetic resonance imaging. Can Assoc Radiol J 44:285, 1993.
Lidor C, Lotem M, Hallel T: Parosteal lipoma of the proximal radius: A report of five cases. J Hand Surg Am 17:1095, 1992.
Milgram JW: Intraosseous lipomas: A clinicopathologic study of 66 cases. Clin Orthop 231:277, 1988.
Murphey MD, Johnson DL, Bhatia PS, et al: Parosteal lipoma: MR imaging characteristics. AJR Am J Roentgenol 162:105, 1994.
Ramos A, Castello J, Sartoris DJ, et al: Osseous lipoma: CT appearance. Radiology 157:615, 1985.
Williams CE, Close PJ, Meaney J, et al: Intraosseous lipomas. Clin Radiol 47:348, 1993.

CHAPTER 9

Malignant Tumors Originating from Adipose Tissue

Nabil J. Khoury, MD, Georges Y. El-Khoury, MD, and D. Lee Bennett, MD

LIPOSARCOMA

Overview

Liposarcomas are one of the most common malignant mesenchymal tumors of the soft tissues; however, liposarcomas arising in bone are extremely rare. The lesion occurs at all ages. Some authors expressed concern about the difficulty encountered when trying to ascertain whether a lesion actually had arisen from within the bone or if the bone had been invaded.

Skeletal Involvement

Intraosseous liposarcomas primarily occur in the long tubular bones, particularly the tibia. Usually, they are centrally located.

Imaging

Radiography

The radiographic appearance of intraosseous liposarcoma is nonspecific. It shows a lytic destructive appearance (Fig. 9–1). The borders of the lesion can be ill defined or well defined. Periosteal reaction is uncommon.

Magnetic Resonance Imaging

On magnetic resonance imaging, the lesion sometimes can show high signal intensity on T1-weighted images.

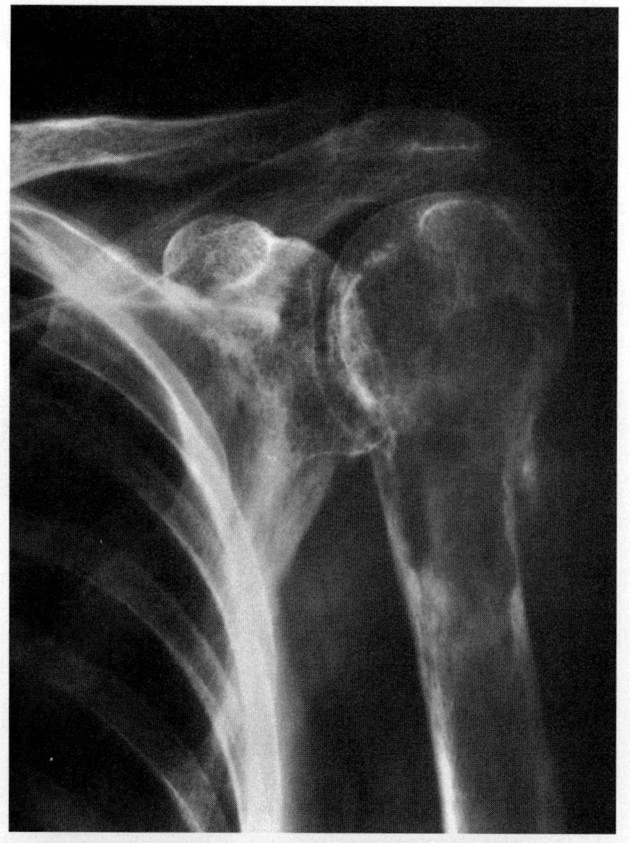

Figure 9–1. Liposarcoma of the proximal humerus. AP view of the left shoulder shows a lytic lesion with aggressive features in the proximal humerus. There is a pathologic fracture through the lesion. (Case provided by Dr. Mark Murphy, Washington, D.C.)

Recommended Readings

Kenan S, Klein M, Lewis MM: Juxtacortical liposarcoma: A case report and review of the literature. Clin Orthop 243:225, 1989.

Kenan S, Lewis MM, Abdelwahab IF, et al: Case report 652: Primary intraosseous low grade myxoid sarcoma of the scapula (myxoid liposarcoma). Skeletal Radiol 20:73, 1991.

Schwartz A, Shuster M, Becker SM, et al: Liposarcoma of bone: Report of a case and review of the literature. J Bone Joint Surg Am 52:171, 1970.

Torok G, Meller Y, Maor E: Primary liposarcoma of bone: Case report and review of the literature. Bull Hosp Jt Dis 43:28,a 1983.

CHAPTER 10

Benign Tumors of Vascular Origin

Nabil J. Khoury, MD, Georges Y. El-Khoury, MD, and D. Lee Bennett, MD

HEMANGIOMA AND HEMANGIOMATOSIS

Hemangioma

Overview

Vascular lesions of bone are fairly common neoplasms, with osseous hemangiomas being the most common. Large autopsy series revealed a prevalence of 10% to 14%. In one third of cases, hemangiomas are multiple. Hemangiomas usually are discovered incidentally in individuals in their 30s or 40s. They are rare in children. Histologically, hemangiomas consist of thin-walled vessels and sinuses lined by endothelium that are embedded in a connective tissue matrix and fat. Some authors have shown that vertebral hemangiomas with high fat content tend to be inactive, whereas lesions that have no fat are more aggressive and have the potential to grow and compress the spinal cord. Rarely, vertebral hemangiomas may cause neurologic symptoms as a result of cord compression or central stenosis (Fig. 10–1). Mechanisms causing cord compression include (1) vertebral body expansion resulting from the tumor involvement, (2) extension of the hemangioma into the epidural space (see Fig. 10–1), (3) fracture, and (4) hemorrhage. Pregnancy is a well-recognized condition during which vertebral hemangiomas may become symptomatic. Symptoms usually appear in the third trimester and are believed to be due to increased blood volume, elevated intra-abdominal pressure, and hormonal changes.

Skeletal Involvement

Hemangiomas are found most commonly in the spine (Fig. 10–2), and to a lesser extent, in the skull (Fig. 10–3). Many other locations have been described, but they are rare. In the spine, either a portion of or the entire vertebral body is involved with occasional extension into the posterior elements. The thoracic spine (followed by the lumbar spine) is the most common location for a vertebral hemangioma. Calvarial hemangiomas usually occur in the frontal and parietal bones, where they account for 20% of all hemangiomas; they are much more common in women. In the long bones, periosteal and intracortical hemangiomas have been described rarely.

Imaging

RADIOGRAPHY

On radiographs, medullary hemangiomas appear lucent with coarsened, thick trabeculae (see Fig. 10–2). In the skull and pelvis, the thick trabeculae may radiate from the center of the lesion, or they may be vertical as in the vertebral bodies, where the appearance is referred to as the *corduroy pattern* (see Fig. 10–2). Intracortical hemangiomas present as lytic lesions with multifocal serpiginous channels. The classic radiographic appearance of a periosteal hemangioma is a soft tissue mass with subjacent shallow, cup-shaped cortical erosions, which are surrounded by cortical thickening.

COMPUTED TOMOGRAPHY

In the spine, computed tomography (CT) shows to better advantage than radiographs the trabecular coarsening, which has the appearance of dense dots because the coarsened trabeculae are oriented perpendicularly to the axial CT scans (Fig. 10–4C). The fatty component can be appreciated by its low Hounsfield units.

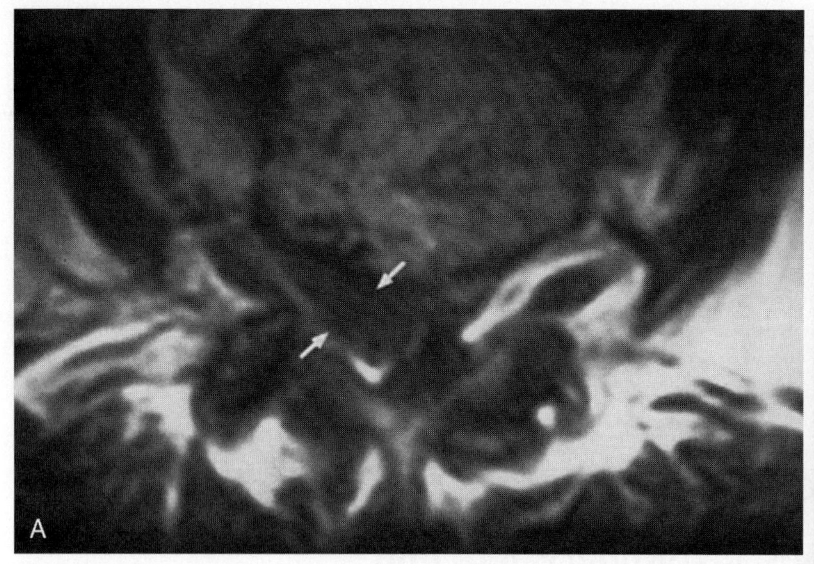

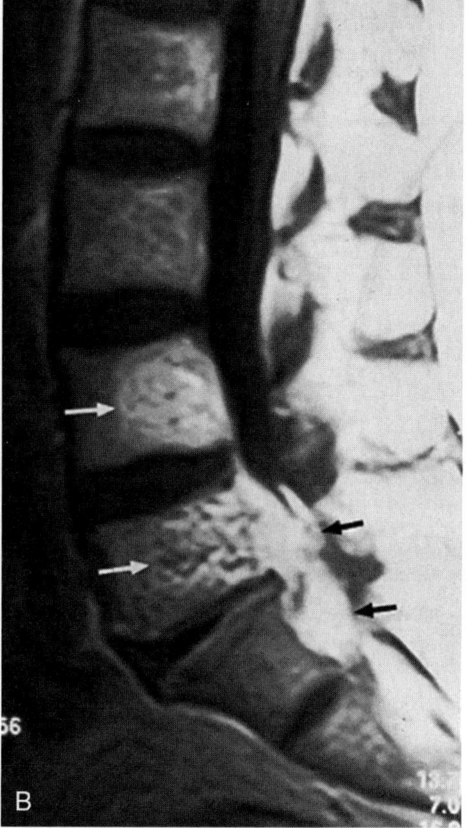

Figure 10–1. Vertebral hemangioma with epidural extension. *A,* Axial T1-weighted MR image of L5 shows a vertebral hemangioma with extension into the epidural space *(arrows).* The thecal sac is flattened by the epidural mass. *B,* Postgadolinium sagittal T1-weighted MR image shows increased signal in the lesion within L5 *(lower white arrow)* and in the epidural space *(black arrows).* Another hemangioma is seen in the vertebral body of L4 *(upper white arrow).*

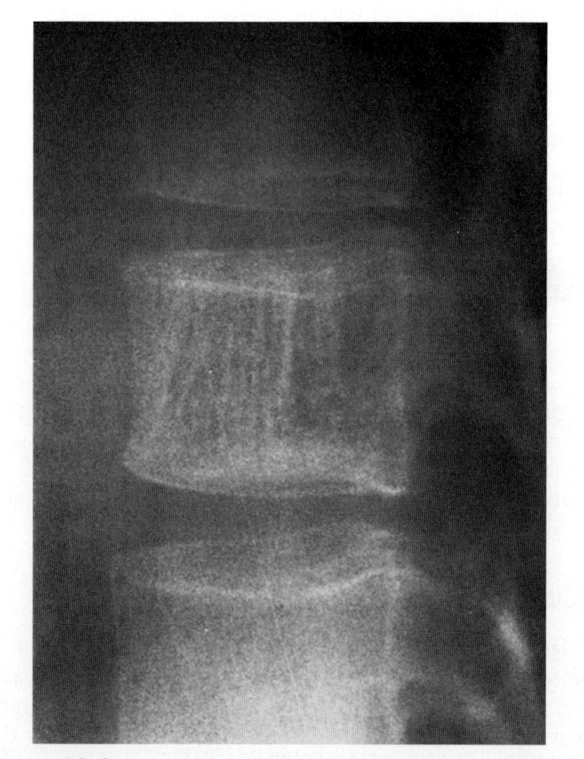

Figure 10–2. Lateral view of the lumbar spine shows the typical corduroy pattern of a hemangioma in the body of L2.

MAGNETIC RESONANCE IMAGING

Magnetic resonance (MR) imaging can be specific in lesions with a high fatty component, which results in high signal intensity on T1- and T2-weighted images (Fig. 10–5). Lesions with a low fat content have low signal intensity on T1-weighted images and high signal intensity on T2-weighted images (Fig. 10–4). This is especially true when the lesion extends into the epidural space or is extraosseous.

Differential Diagnosis

Vertebral body hemangiomas should be differentiated from plasmacytomas, giant cell tumors, eosinophilic granuloma, and Paget's disease.

Hemangiomatosis (Cystic Angiomatosis)

Overview

Hemangiomatosis is a rare condition of unknown cause. Lesions consist of hemangioma (vascular proliferation) mixed with lymphangioma. Abundant adipose tissue is found between trabeculae, mixed with angiomatous elements. Extraosseous components contain little or no fat. Visceral involvement is frequent; it occurs in 60% to 70% of patients. The natural history of the disease is variable

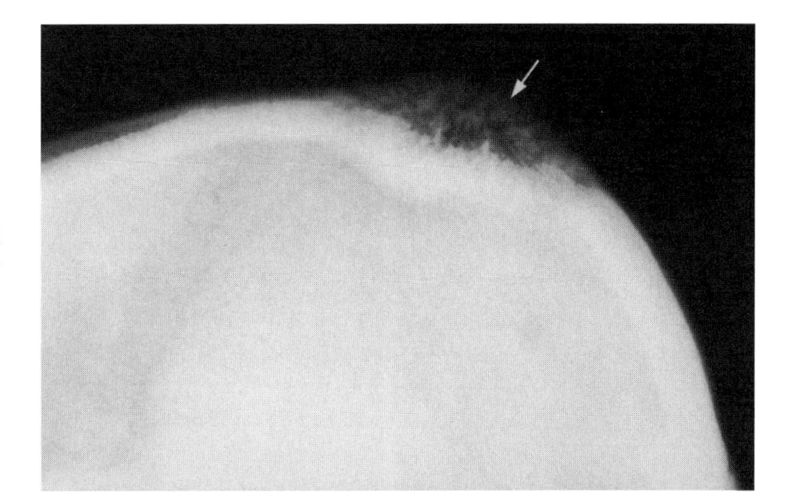

Figure 10–3. Lateral view of the skull shows the typical appearance of a calvarial hemangioma *(arrow)*. Thick trabeculae are radiating from the center of the lesion.

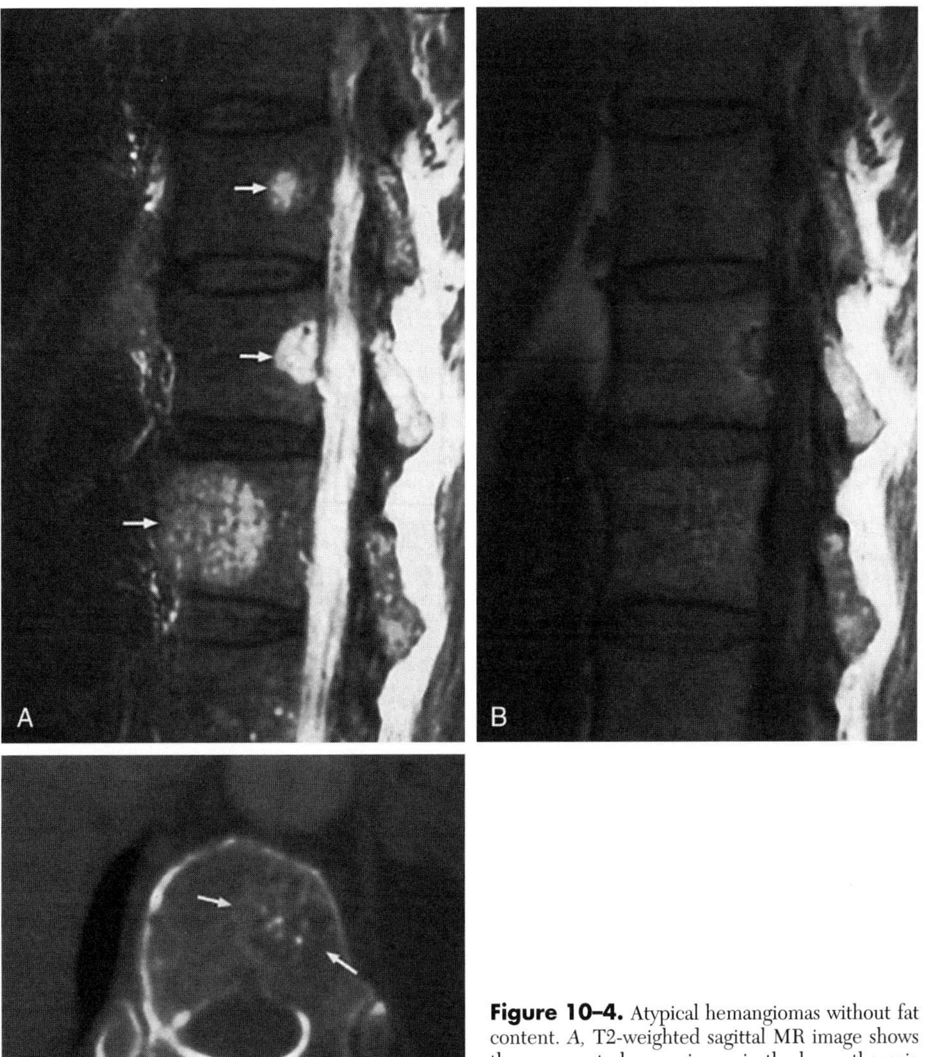

Figure 10–4. Atypical hemangiomas without fat content. *A,* T2-weighted sagittal MR image shows three separate hemangiomas in the lower thoracic spine *(arrows)*. *B,* T1-weighted sagittal MR image shows the atypical appearance in that the lesions are almost isointense with the rest of the vertebral marrow. *C,* Axial CT scan of one of the lesions shows the typical hemangioma appearance *(arrows)* with coarse primary trabeculae *(dense dot pattern)*.

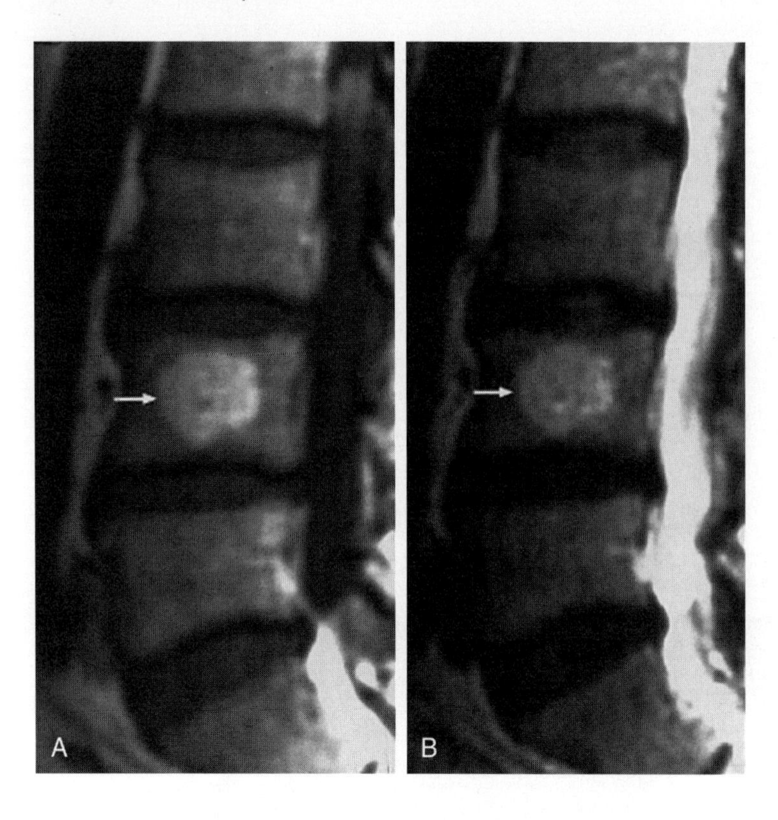

Figure 10–5. Classic MR appearance of a vertebral hemangioma. *A* and *B*, T1- and T2-weighted sagittal MR images show hyperintense signal in a hemangioma occupying the central portion of L4.

and primarily depends on the visceral component. The prognosis is poor if visceral involvement is present.

Skeletal Involvement

The spine, ribs, and pelvis are affected in nearly all patients. Other common sites include the skull, shoulder, and long bones.

Imaging

Radiography and CT show widespread cystic areas (Fig. 10–6). The lesions are round or oval, and they may or may not have well-demarcated, sclerotic edges. The lesions are usually medullary, and they can expand and scallop the cortex. Pathologic fractures are common. On MR imaging, cystic angiomatous lesions have variable signal intensity on T1- and T2-weighted images. The signal intensity depends on the fat content of the lesion and on the degree of reactive sclerosis.

Differential Diagnosis

The differential diagnosis includes polyostotic fibrous dysplasia, Langerhans cell histiocytosis, and metastasis.

LYMPHANGIOMA AND LYMPHANGIOMATOSIS (INTRAOSSEOUS)

Overview

Lymphangioma and intraosseous lymphangiomatosis is a rare condition that probably is due to a congenital malformation of the lymphatic system alone or combined with malformation of the vascular system. The disease can be benign or aggressive, causing serious morbidity, especially when associated with visceral involvement, such as the pleura, lung, and abdominal organs. Histologically, clinically, and radiographically, there is significant overlap between lymphangiomatosis, hemangiomatosis, and Gorham's disease (massive osteolysis). Patients with osse-

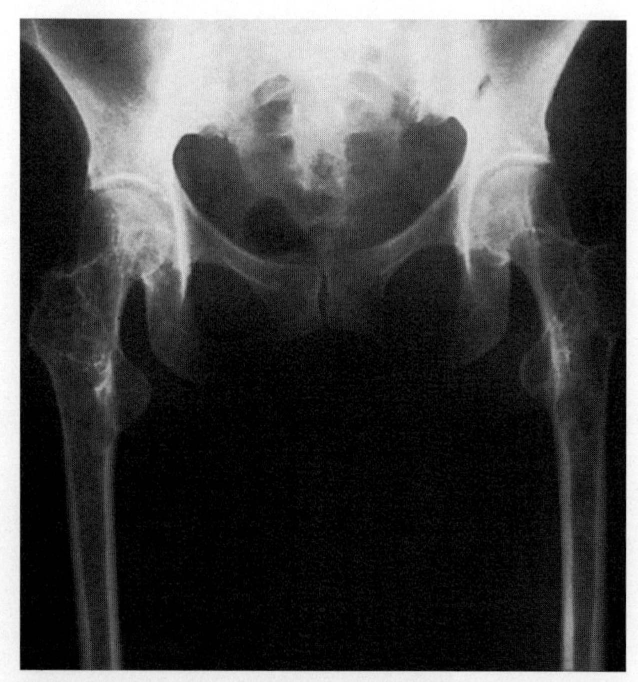

Figure 10–6. Cystic angiomatosis in a 40-year-old woman in whom lesions were discovered incidentally. Anteroposterior view of the pelvis reveals multiple well-defined cystic lesions in both proximal femora.

ous lymphangiomatosis usually present in childhood; patients with only bone involvement have a good prognosis. Visceral involvement affects 45% of patients with osseous lymphangiomas. Clinically, patients with visceral involvement commonly present with pleural effusion, pericardial effusion, or soft tissue lymphedema. At surgery, lymphangiomas either ooze clear milky fluid or are composed of empty cystic spaces.

Skeletal Involvement

Locations of involvement include the tibia, humerus, mandible, calvaria, ribs, and spine.

Imaging

Radiography

Cystic, multiloculated lesions expanding the cortex usually are seen. The margins of the lesions may be sclerotic and well defined, but not always.

Computed Tomography and Magnetic Resonance Imaging

Findings on CT and MR imaging are nonspecific. On CT, air has been described within the lesion after a biopsy or surgery.

GORHAM'S DISEASE (VANISHING OR DISAPPEARING BONE DISEASE)

Overview

Gorham's disease is a rare condition of unknown cause. The disease typically occurs in children and young adults. Males and females are affected equally. Often, osteolysis stabilizes after several years of localized bone destruction; however, it may progress and lead to death. The pathologic process is that of a locally aggressive angiomatosis.

Skeletal Involvement

Any bone can be affected, but the disease favors the shoulder girdle, hip girdle, mandible, and spine. The disease typically affects adjacent bones.

Imaging

Radiography

Radiographically the disease starts with patchy areas of osteoporosis, which progress to distinct lucencies (Fig. 10–7). With time, the lucencies enlarge and coalesce. The disease crosses joints; local invasion of the soft tissues with angiomatous masses occurs at later stages. As the bone resorbs (Fig. 10–8), it weakens and deformities develop, especially in the pelvis, rib cage, and spine. Pathologic fractures are common.

Magnetic Resonance Imaging

Only a few reports in the literature describe the MR imaging appearance of Gorham's disease. The signal intensities on T1- and T2-weighted MR images can be variable (see Fig. 10–7). In one report, the involved vertebrae appeared bright on T1-weighted images. Other reports described the lesions as being isointense or hypointense on T1-weighted images and hyperintense on T2-weighted images. This appearance was explained as representing a fibrotic reaction.

GLOMUS TUMOR

Overview

The glomus body is a specialized arteriovenous anastomosis, which controls thermoregulation. Glomus bodies are present in the dermis throughout the body; however, they are most abundant in the tips of the digits, particularly beneath the nails. Glomus tumors are benign tumors arising from glomus bodies, and they present in the teens through the 30s. Rarely, multiple lesions are present.

Musculoskeletal Involvement

Glomus tumors occur anywhere in the body; however, 75% occur in the hands, and 65% are subungual. Extraosseous glomus tumors are much more prevalent than intraosseous tumors.

Imaging

Radiography

Glomus tumors can produce pressure erosions on the adjacent bone. The erosion is smooth and well demarcated, typically involving the distal phalanx. This finding can be detected on radiographs in 14% to 60% of cases.

Magnetic Resonance Imaging

On MR imaging, the tumor has intermediate to low signal intensity on T1-weighted images and high signal intensity on T2-weighted images (Fig. 10–9).

Ultrasound

Ultrasound, using high-frequency transducers, has been shown to be effective for detecting glomus tumors in the fingers.

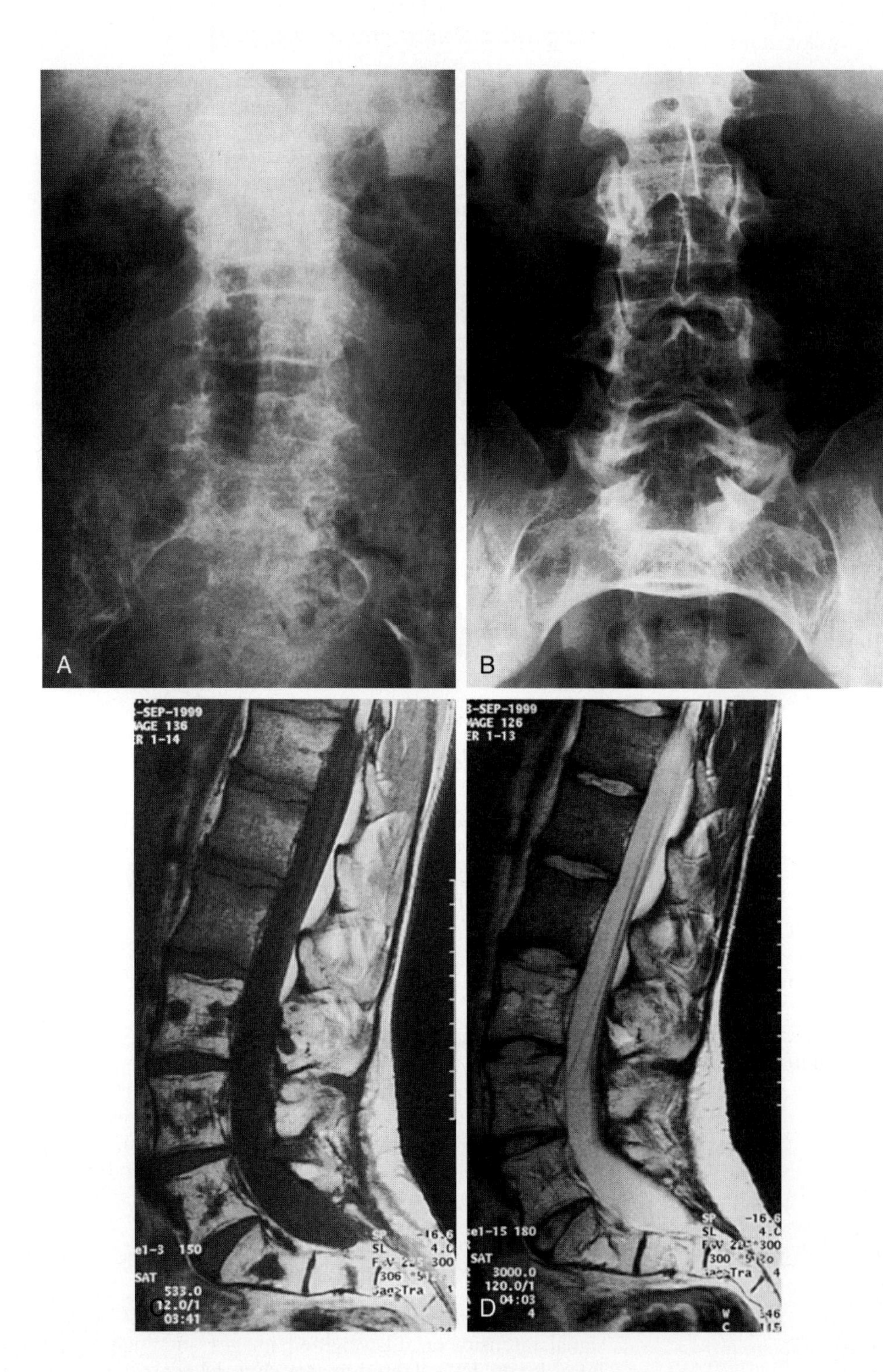

Figure 10–7. Gorham's disease affecting the lower lumbar spine and sacrum. *A,* Anteroposterior view of the lumbosacral spine taken at the age of 6 years shows profound osteopenia of the lower lumbar spine and sacrum. *B,* Anteroposterior view of the lumbosacral spine of the same patient at the age of 22 years shows osteopenia and coarse trabecular pattern of the lower three lumbar vertebrae and sacrum. *C* and *D,* Sagittal T1- and T2-weighted MR images reveal increased signal intensity involving the lower three lumbar vertebrae and sacrum.

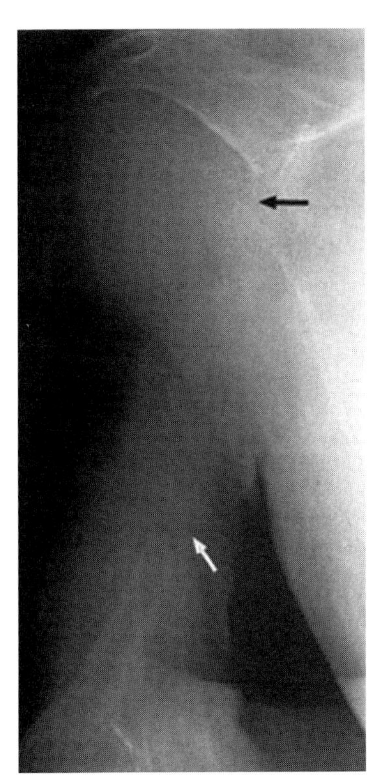

Figure 10–8. Advanced destruction resulting from Gorham's disease involving the right shoulder and humerus. The glenoid and proximal two thirds of the humerus are destroyed completely *(arrows)* on this anteroposterior view of the upper extremity.

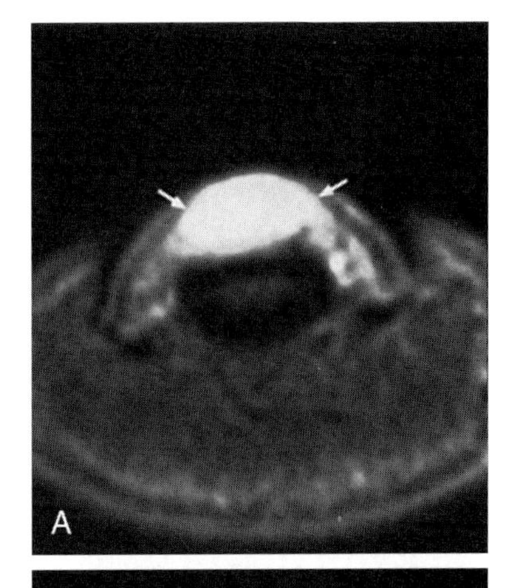

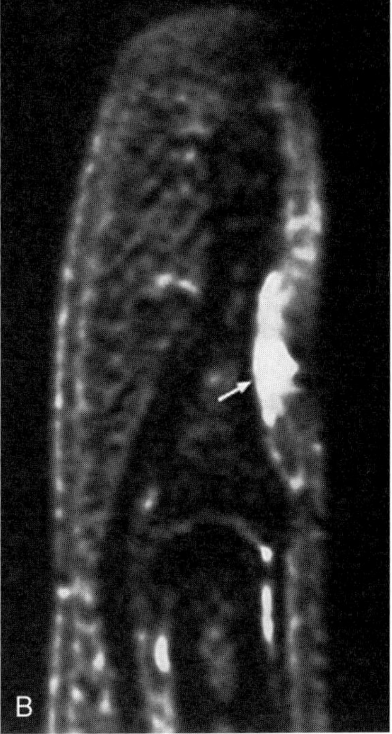

Figure 10–9. Glomus tumor. *A,* Postgadolinium axial fat-saturated T1-weighted MR image of the distal phalanx of the ring finger shows enhancement of a subungual lesion *(arrows).* The bone is not involved. *B,* Sagittal gradient echo MR image (with T2-weighting) shows the hyperintense lesion at the base of the nail *(arrow).*

Differential Diagnosis

The differential diagnosis includes osteomyelitis, sarcoid, epidermoid (inclusion) cyst, and keratoacanthoma.

Recommended Readings

Baker ND, Greenspan A, Neuwirth M: Symptomatic vertebral hemangiomas: A report of four cases [published erratum appears in Skeletal Radiol 1986; 15:686]. Skeletal Radiol 15:458, 1986.

Bjorkengren AG, Resnick D, Haghighi P, et al: Intraosseous glomus tumor: Report of a case and review of the literature. AJR Am J Roentgenol 147:739, 1986.

Dalrymple NC, Hayes J, Bessinger VJ, et al: MR imaging of multiple glomus tumors of the finger. Skeletal Radiol 26:664, 1997.

Devaney K, Vinh TN, Sweet DE: Skeletal-extraskeletal angiomatosis: A clinicopathological study of fourteen patients and nosologic considerations. J Bone Joint Surg Am 76:878, 1994.

Devaney K, Vinh TN, Sweet DE: Surface-based hemangiomas of bone: A review of 11 cases. Clin Orthop 300:233, 1994.

Fisher DR, Hinke ML: Musculoskeletal case of the day. AJR Am J Roentgenol 146:1087, 1986.

Fornage BD: Glomus tumors in the fingers: Diagnosis with US. Radiology 167:183, 1988.

Keenen TL, Buehler KC, Campbell JR: Solitary lymphangioma of the spine. Spine 20:102, 1995.

Kenan S, Abdelwahab IF, Klein MJ, et al: Hemangiomas of the long tubular bone. Clin Orthop 280:256, 1992.

Kenan S, Bonar S, Jones C, et al: Subperiosteal hemangiomas: A case report and review of the literature. Clin Orthop 232:279, 1988.

Laredo JD, Assouline E, Gelbert F, et al: Vertebral hemangiomas: Fat content as a sign of aggressiveness. Radiology 177:467, 1990.

Laredo JD, Reizine D, Bard M, et al: Vertebral hemangiomas: Radiologic evaluation. Radiology 161:183, 1986.

Lateur L, Simoens CJ, Gryspeerdt S, et al: Skeletal cystic angiomatosis. Skeletal Radiol 25:92, 1996.

Lomasney LM, Martinez S, Demos TC, et al: Multifocal vascular lesions of bone: Imaging characteristics. Skeletal Radiol 25:255, 1996.

Martinat P, Cotton A, Singer B, et al: Solitary cystic lymphangioma. Skeletal Radiol 24:556, 1995.

Mohan V, Gupta SK, Tuli SM, et al: Symptomatic vertebral haemangiomas. Clin Radiol 31:575, 1980.

Murphey MD, Fairbairn KJ, Parman LM, et al: From the archives of the AFIP: Musculoskeletal angiomatous lesions: Radiologic-pathologic correlation. Radiographics 15:893, 1995.

Ross JS, Masaryk TJ, Modic MT, et al: Vertebral hemangiomas: MR imaging. Radiology 165:165, 1987.

Rougraff BT, Deters ML, Ivancevich S: Surface-based hemangioma of bone: Three case studies and a review of the literature. Skeletal Radiol 27:182, 1998.

Rozmaryn LM, Sadler AH, Dorfman HD: Intraosseous glomus tumor in the ulna: A case report. Clin Orthop 220:126, 1987.

CHAPTER 11

Malignant Tumors of Vascular Origin

Nabil J. Khoury, MD, Georges Y. El-Khoury, MD, and D. Lee Bennett, MD

HEMANGIOENDOTHELIOMA (ANGIOSARCOMA)

Overview

Hemangioendothelioma is a rare lesion, representing less than 1% of all primary malignant bone tumors. It continues to pose a problem for radiologists because of its rarity and the confusion surrounding the nomenclature. The terms *hemangioendothelioma, hemangioendothelial sarcoma,* and *angiosarcoma* are used interchangeably in the literature. In more recent publications, the term *hemangioendothelioma* is preferred to describe lesions with this histology. The lesion can occur in multiple organs besides bone, including lung, liver, skin, heart, and soft tissues. It arises from the vascular endothelial cells and is composed of anastomotic channels lined by atypical endothelial cells. The tumor is of variable aggressiveness, and it may occur *de novo* or arise from several underlying processes. These underlying processes include osteomyelitis, osteonecrosis, Paget's disease, retained foreign bodies, or a hip prosthesis. Prolonged exposure to vinyl chloride may be a predisposing factor. The tumor occurs between the teens and 60s, with the highest prevalence in the 30s. It is slightly more frequent in males. The main presenting symptoms are local pain and swelling.

Skeletal Involvement

Hemangioendotheliomas most frequently occur in the long tubular bones, mainly the tibia, followed by the femur and the humerus (Fig. 11–1). Other locations include the pelvis, skull, vertebral bodies, and ribs. It is usually metadiaphyseal and rarely epiphyseal. The lesion can be either medullary or cortical in location. A characteristic feature of osseous hemangioendothelioma is the presence of synchronous or metachronous multicentric disease. This characteristic appearance occurs in 20% to 50% of cases. These multiple foci can occur in a single bone or in multiple bones of a single extremity.

Imaging

It is usually impossible to suggest the diagnosis of hemangioendothelioma on the basis of the radiographic appearance of a single lesion. Multifocality is helpful, however, in suggesting the diagnosis, especially when the lesions cluster in a single bone or anatomic region (Figs. 11–1 and 11–2). When multiple lytic lesions are seen involving the cortex of the tubular bones in the lower extremities, hemangioendothelioma should be a primary consideration.

Radiography

Individual lesions can vary in appearance. At one end of the spectrum, there is the well-defined lesion resembling a hemangioma; such lesions also are histologically well differentiated. On the other end of the spectrum, there is the aggressive, purely lytic, expansile, blowout lesion associated with a soft tissue mass. The margins are either

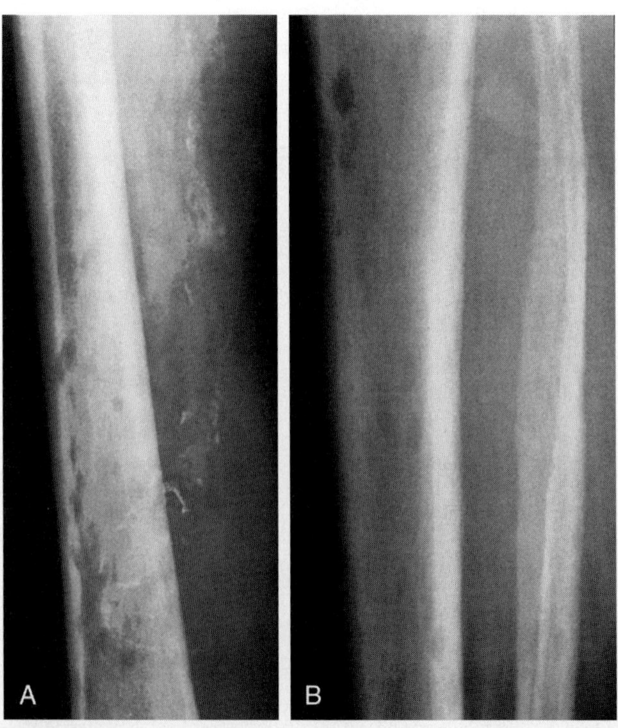

Figure 11–1. Hemangioendothelioma of the right femur and tibia. *A* and *B*, Anteroposterior views of the femur and lower leg show multifocal, cortical lucencies in the ipsilateral femur, tibia, and fibula. No other bones were involved. This is the characteristic appearance of multifocal hemangioendothelioma. This appearance occurs in 20% to 50% of the cases.

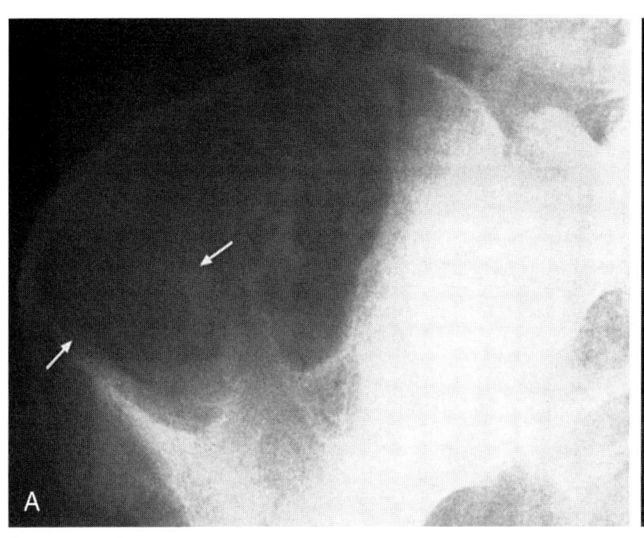

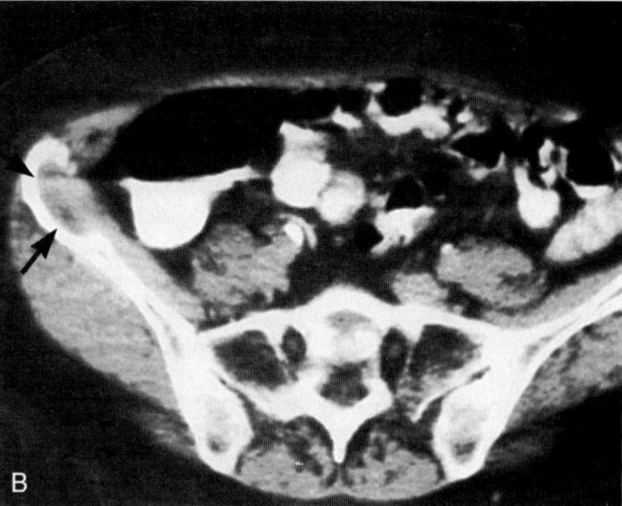

Figure 11–2. Hemangiopericytoma of the ilium. *A*, Anteroposterior view of the ilium shows a nonspecific, geographic, lytic lesion in the ilium *(arrows)*. *B*, Axial CT scan through the lesion *(arrows)* shows a lytic lesion with anterior cortical destruction. There is no associated soft tissue mass. Biopsy showed a hemangiopericytoma.

well or poorly defined, depending on the tumor grade. Peripheral sclerosis occasionally is present, and periosteal reaction is uncommon.

Magnetic Resonance Imaging

On magnetic resonance (MR) imaging, the lesion is heterogeneous. It is isointense on T1-weighted images and hyperintense on T2-weighted images. MR imaging is best at assessing cortical destruction and soft tissue involvement. MR imaging is the most sensitive modality for detecting multifocal lesions.

Differential Diagnosis

When a hemangioendothelioma is solitary (about 60% of cases), the differential diagnosis is broad and includes solitary metastasis, plasmacytoma, aneurysmal bone cyst, eosinophilic granuloma, and chronic infection. When the lesion is multifocal, its appearance is fairly distinctive.

HEMANGIOPERICYTOMA

Overview

Hemangiopericytoma is typically a lesion of the soft tissues, and it is extremely rare as a primary bone tumor. It represents only 0.1% of all malignant bone lesions. It arises from the pericytes surrounding the capillaries. The lesion usually occurs between the 30s and 60s. Men and women are affected equally. In contradistinction to hemangioendotheliomas, hemangiopericytomas are not multicentric. They can occur as a metastatic deposit secondary to a primary meningeal hemangiopericytoma. Some angioblastic meningiomas are difficult to separate histologically from a hemangiopericytoma. Some authors believe that all hemangiopericytomas of bone should be checked for a primary meningeal source. Hypophosphatemic osteomalacia can occur secondary to a hemangiopericytoma in the soft tissues or bones.

Skeletal Involvement

Most frequently, the lesion occurs in the pelvis and the proximal long bones (40%), mainly the humerus, femur, and tibia.

Imaging

Radiography

The radiographic appearance is nonspecific, and the diagnosis almost never is suggested based on the imaging findings (Fig. 11–2). Seventy percent of hemangiopericytomas are lytic, and 30% have variable degrees of sclerosis, honeycombing, and prominent trabeculae. Mild expansion may be seen. Margins can be ill or well defined. Sometimes, there is cortical destruction with soft tissue extension. Periosteal reaction is rare.

Magnetic Resonance Imaging

On MR imaging, the lesion has nonspecific signal characteristics. MR imaging may show serpentine vascular channels at the periphery of the mass or signal voids representing tubular vascular structures arranged radially in a spoke-wheel pattern within the mass. The vessels may have either high or low signal intensity depending on whether slow flow or high flow is present. Fluid-fluid levels are noted when intratumoral bleeding has occurred.

Recommended Readings

Abrahams TG, Bula W, Jones M: Epithelioid hemangioendothelioma of bone: A report of two cases and review of the literature. Skeletal Radiol 21:509, 1992.

Boutin RD, Spaeth HJ, Mangalik A, et al: Epithelioid hemangioendothelioma of bone. Skeletal Radiol 25:391, 1996.
Conway WF, Hayes CW: Miscellaneous lesions of bone. Radiol Clin North Am 31:339, 1993.
Ignacio EA, Palmer KM, Mathur SC, et al: Epithelioid hemangioendothelioma of the lower extremity. Radiographics 19:531, 1999.
Lomasney LM, Martinez S, Demos TC, et al: Multifocal vascular lesions of bone: Imaging characteristics. Skeletal Radiol 25:255, 1996.
Merine D, Fishman EK: Hemangioendothelioma of bone: CT findings. J Comput Assist Tomogr 13:1098, 1989.
Murphey MD, Fairbairn KJ, Parman LM, et al: From the archives of the AFIP: Musculoskeletal angiomatous lesions: Radiologic-pathologic correlation. Radiographics 15:893, 1995.
Tang JS, Gold RH, Mirra JM, et al: Hemangiopericytoma of bone. Cancer 62:848, 1988.

CHAPTER 12

Malignant Tumors Arising in the Bone Marrow and Lymphatic System

Nabil J. Khoury, MD, Georges Y. El-Khoury, MD, and D. Lee Bennett, MD

PLASMA CELL DISORDERS

Overview

The plasma cell is a specialized B lymphocyte, which produces antibodies in response to foreign antigens. The term *plasma cell dyscrasia* is used frequently in the literature; it refers to a group of plasma cell disorders with uncontrolled proliferation of plasma cells in the absence of a recognizable antigenic stimulus. These disorders include monoclonal gammopathy, Waldenström's macroglobulinemia, solitary plasmacytoma, multiple myeloma (MM), POEMS syndrome, and amyloidosis.

Monoclonal Gammopathy

Monoclonal gammopathies are a group of disorders affecting about 1% of the asymptomatic population. The incidence increases with age and reaches 3% of the population older than age 70 years. Monoclonal gammopathy is characterized by proliferation of a single clone of plasma cells, which produces a monoclonal protein (IgG or IgA globulin). Long-term follow-up has shown that 19% of patients with monoclonal gammopathy develop a hematologic neoplastic disease within 10 years. On magnetic resonance (MR) imaging of the spine, 19% of the patients have a spotty or focal abnormal marrow pattern.

Waldenström's Macroglobulinemia

Waldenström's macroglobulinemia is characterized by the presence of a homogeneous IgM. Clinically, Waldenström's macroglobulinemia manifests similar to a lymphoma with hepatosplenomegaly and lymphadenopathy; it can be relatively benign or progressive and malignant. Destructive bone changes and profound osteopenia, which are indistinguishable from multiple myeloma, have been noted in about 20% of patients. Symmetrical cystlike lesions in the supra-acetabular region also have been described. These lesions may be unique for this condition.

Solitary Plasmacytoma

Solitary plasmacytoma constitutes less than 10% of all plasma cell neoplasms. These tumors can be divided into extramedullary plasmacytomas and solitary plasmacytomas of bone (SPB).

Extramedullary Plasmacytoma

Extramedullary plasmacytomas are rare; most arise in the submucosa of the upper airway, particularly the nasal cavity. Extramedullary plasmacytoma has a better prognosis than MM. It also has a lower rate of conversion to MM than SPB. On computed tomography (CT) scan, the lesion is seen as a soft tissue mass impinging on the airway.

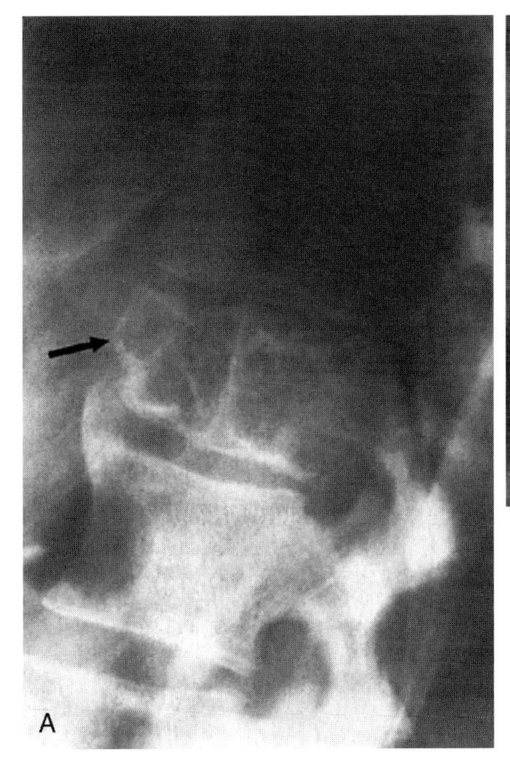

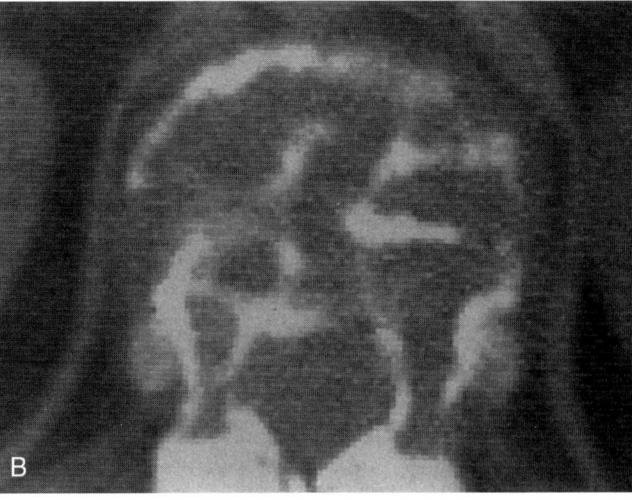

Figure 12–1. Solitary myeloma of T12. Bone marrow aspirate from the superior posterior iliac crest showed no plasmacytosis, and a bone survey was negative except for the lesion in T12. *A*, Lateral view of the spine shows a lytic lesion in T12 *(arrow)*. The vertebral body is moderately compressed. *B*, CT section shows what is known as the *minibrain* pattern of destruction, which is characteristic of myeloma.

Solitary Plasmacytoma of Bone (SPB)

SPB is thought to be MM in evolution because most cases of SPB eventually convert to MM. The mean age of patients with SPB is slightly younger than that of patients with MM. Most cases occur in the 50s. SPB usually occurs in the axial skeleton, pelvis, or ribs (Fig. 12–1). In osseous sites not involved by the lesion, the bone marrow shows no evidence of plasmacytosis on bone marrow biopsy. Radiographically, SPB manifests as a lytic, expansile lesion resembling metastasis, or it may present as a cystic, trabeculated lesion (Fig. 12–1A).

Multiple Myeloma

Overview

MM is the most common primary malignant tumor of bone. It almost always occurs after age 40, and its peak incidence is in the 60s. Most series report a male predominance. Almost all patients with MM have widespread plasmacytosis in the marrow detected by bone marrow aspiration. The prognosis in MM depends on the tumor mass and renal function. The most common presenting symptom is back pain, which can be progressive and disabling. The second most common symptom is easy fatigability resulting from severe anemia. Laboratory studies show an increased serum globulin fraction, typically IgG; anemia; and hypercalcemia.

Imaging

RADIOGRAPHY

Despite the advancements in imaging, radiography is still the most widely used method for assessing and following MM in the skeleton. The most common early radiographic manifestation of MM is diffuse osteopenia affecting the dorsal spine, lumbar spine, pelvis, proximal femur, proximal humerus, and ribs (Fig. 12–2A). This osteopenia is believed to be due to the production of an osteoclast-activating factor by the plasma cells. Vertebral compression is present in approximately two thirds of patients with MM (see Fig. 12–2A); in about half of these patients, vertebral compression is the presenting sign. The distribution of these fractures parallels vertebral compression fractures in postmenopausal osteoporosis involving T6 to L4. Lytic *punched-out* lesions without a sclerotic border can be seen. These lesions are seen best in the skull, but they can be present in other bones (see Fig. 12–2). The lesions of MM are intramedullary in location. They may be slightly expansile, producing endosteal scalloping and cortical destruction. Punched-out lesions in the vertebrae are difficult to see on radiographs; however, CT can help in detecting them (Fig. 12–2B). In the ribs, cortical destruction can be associated with an extrapleural soft tissue mass (Fig. 12–3). Occasionally, extensive bone destruction may reveal a moth-eaten or permeative pattern. In contradistinction to metastatic disease, MM lesions can be seen distal to the elbows and knees,

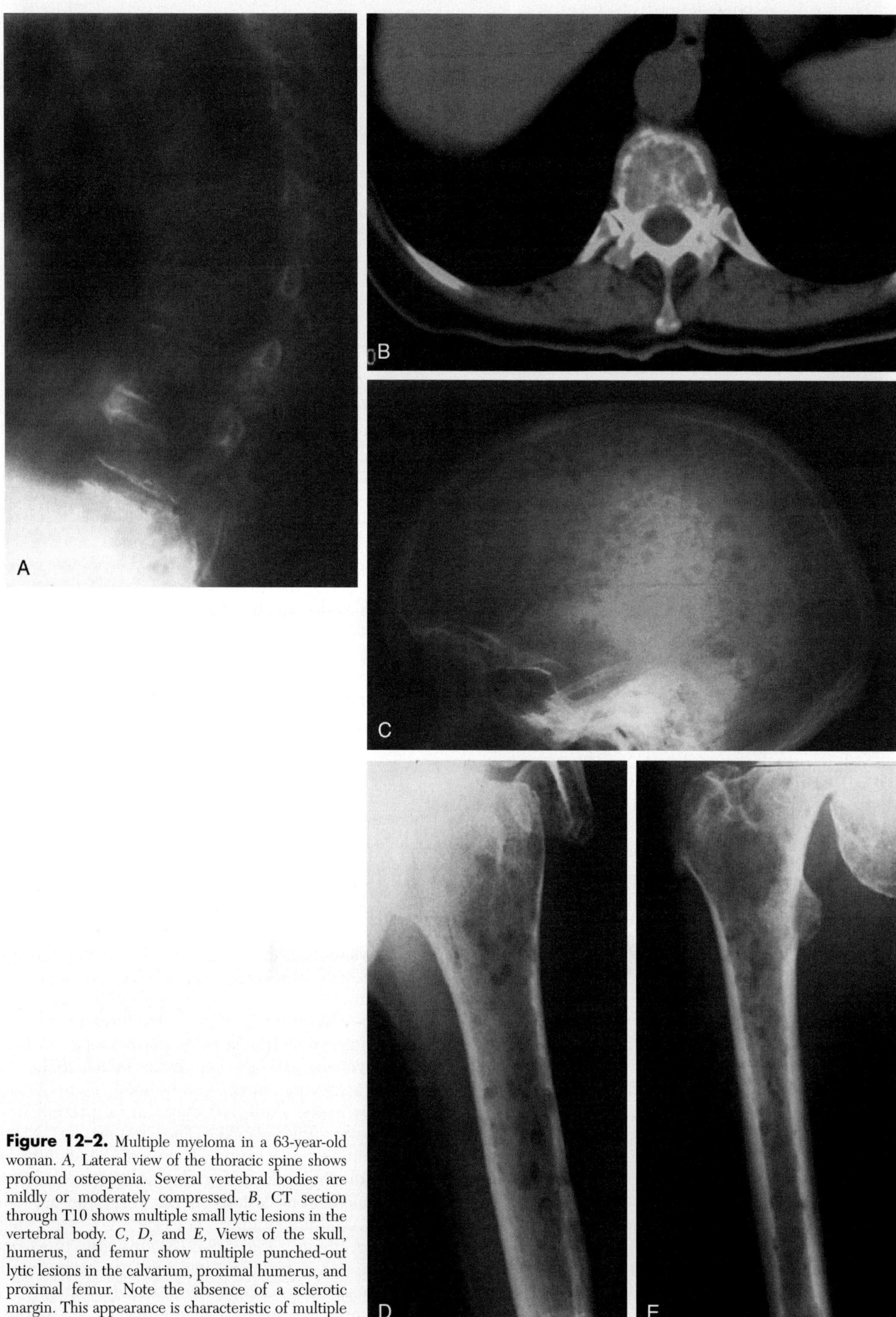

Figure 12–2. Multiple myeloma in a 63-year-old woman. *A,* Lateral view of the thoracic spine shows profound osteopenia. Several vertebral bodies are mildly or moderately compressed. *B,* CT section through T10 shows multiple small lytic lesions in the vertebral body. *C, D,* and *E,* Views of the skull, humerus, and femur show multiple punched-out lytic lesions in the calvarium, proximal humerus, and proximal femur. Note the absence of a sclerotic margin. This appearance is characteristic of multiple myeloma.

especially when the disease is advanced. Many MM patients do not have bone lesions at presentation and may not develop them until the tumor load reaches 1 × 10^{12} cells (about 1 kg).

A sclerotic variant of MM is seen in 1% to 2% of the patients with multiple myeloma (Fig. 12–4). Compared with the lytic type, the sclerotic variant occurs in younger individuals and has a better prognosis. About half of patients with sclerotic MM have a peripheral neuropathy. Rarely, sclerotic and lytic lesions may be present. The differential diagnosis in sclerotic MM should include blastic metastasis, myelofibrosis, and mastocytosis.

MAGNETIC RESONANCE IMAGING

The spine is the most common site for MM, and cord compression is a well-known complication of MM. MR imaging is helpful in patients with neurologic symptoms from cord compromise produced by vertebral compression fractures or epidural extension of the tumor (Fig. 12–5). At least four MR imaging patterns of spinal involvement have been described in MM, as follows. (1) A normal marrow pattern can be present. Most (67%) compression fractures in MM appear benign on MR imaging. In patients presenting with unexplained osteopenia and compression fractures, MM still should be suspected as a cause even when MR imaging shows a normal marrow pattern. (2) A mottled marrow pattern can be present. The mottled pattern also is known as the *salt-and-pepper* or variegated pattern (Fig. 12–6). (3) A mottled pattern with focal marrow involvement (Figs. 12–7 and 12–8). (4) Diffuse marrow involvement. The last two patterns have the worst prognosis, especially when the focal deposits are multiple (>10). Gadolinium-enhanced MR imaging has no definite role in the study of MM.

NUCLEAR MEDICINE

Bone scans add little to the management of MM patients. Bone scans are insensitive in detecting MM lesions, and they underestimate the extent of the disease. About 27% of all MM lesions can be missed by a bone scan.

POEMS Syndrome

POEMS syndrome is a rare multisystem disorder. The acronym *POEMS* represents *p*olyneuropathy, *o*rganomegaly (hepatosplenomegaly and lymphadenopathy), *e*ndocrinopathy (gynecomastia, impotence, amenorrhea, diabetes, and hypothyroidism), elevated *M* protein level, and *s*kin changes (skin thickening, hyperpigmentation, and hypertrichosis). Other significant findings include papilledema, peripheral edema, ascites, and clubbing of the fingers. In the skeleton, patients usually present with focal lesions in the axial skeleton that are typically sclerotic (see Fig. 12–4); however, occasionally the lesions may be mixed lytic and sclerotic or, rarely, purely lytic. With the mixed lesions, the rim of the lesion is sclerotic, and the center is lytic. The osteosclerotic pattern is the predominant pattern, however, in POEMS syndrome. Another prominent radiographic feature of POEMS syndrome is bony proliferation in the axial and peripheral skeleton resembling the enthesopathy (enthesitis) found in seronegative spondyloarthropathies.

Amyloidosis Associated with Multiple Myeloma

Amyloidosis occurs in approximately 10% to 15% of MM cases. Amyloid is more common with λ chain than with κ chain dyscrasias, but κ chain myeloma is more

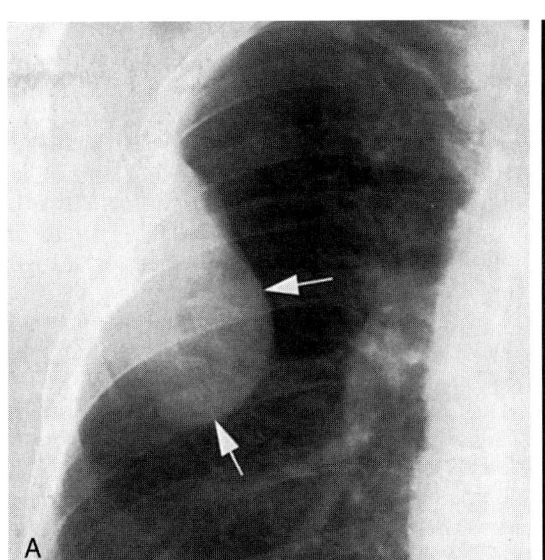

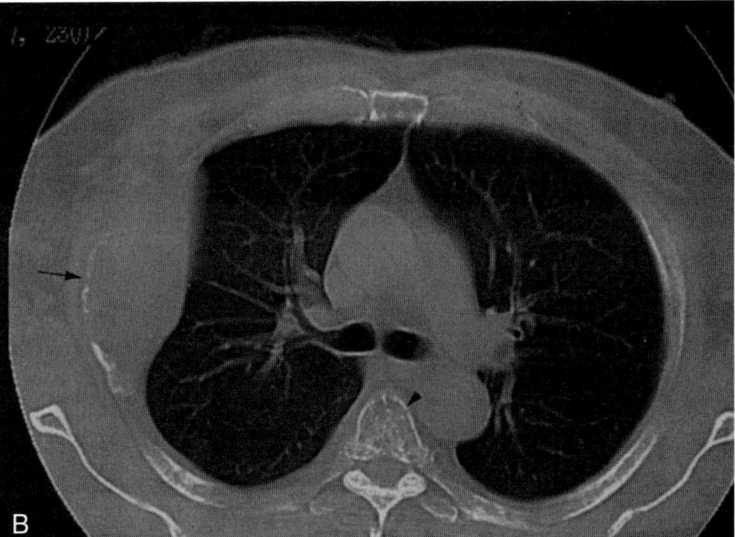

Figure 12–3. Large rib lesion resulting from multiple myeloma presenting as an extrapleural mass. *A*, Posteroanterior view of the chest shows an extrapleural mass *(arrows)*. *B*, CT scan shows the lesion arising at the fourth rib on the right *(arrow)*. The vertebrae and other ribs also are involved with multiple myeloma *(arrowhead)*.

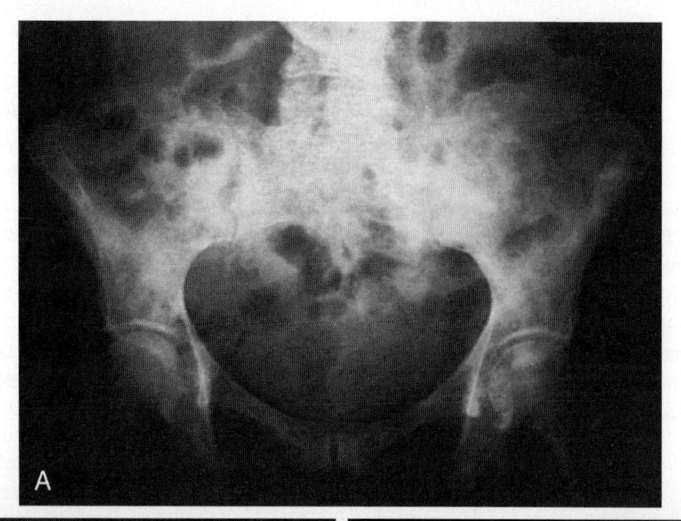

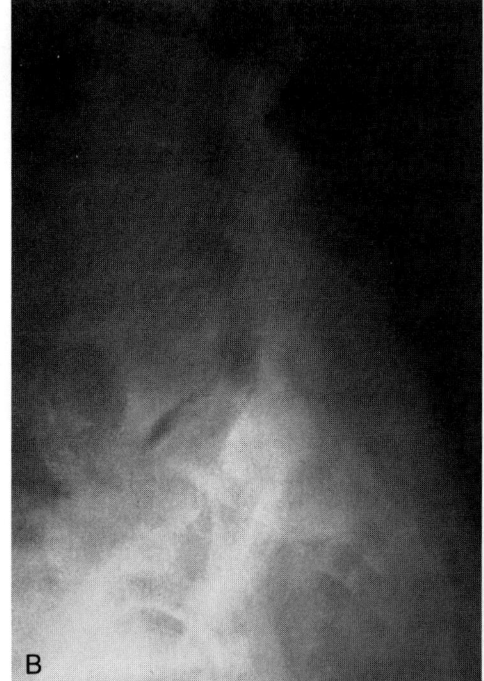

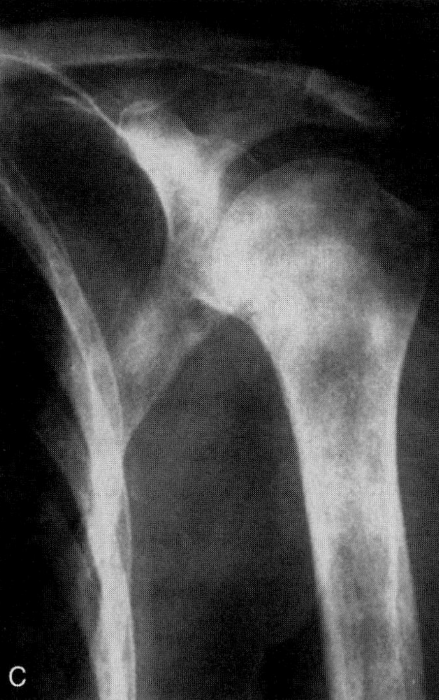

Figure 12–4. Sclerotic variant of multiple myeloma in an 83-year-old woman. *A,* Anteroposterior view of the pelvis. *B,* Lateral view of the lumbosacral spine. *C,* Anteroposterior view of the shoulder. All three sites show multiple sclerotic lesions resembling blastic metastasis.

common than λ chain myeloma. Amyloid deposits, which can calcify, are seen within the tumor, causing the lesion to resemble chondrosarcoma (Fig. 12–9).

LANGERHANS CELL HISTIOCYTOSIS

Overview

Langerhans cell histiocytosis (LCH) is a relatively uncommon disease representing about 1% of all bone tumors. The disease previously was referred to as *histiocytosis X.* It is characterized by inappropriate proliferation and infiltration of various tissues with a unique type of histiocyte called the *Langerhans cell.* The basic lesion seems to be a deficiency in the T-suppressor lymphocytes followed by uncontrolled proliferation of Langerhans cells. Normally, these cells are derived from $CD34^+$ stem cells in the bone marrow, from which they migrate through the bloodstream to the skin, lungs, thymus, and lymph nodes. A distinctive feature of Langerhans cells is the presence of Birbeck granules in their cytoplasm, which are identified by electron microscopy. LCH can occur at any age, but it has a predilection for children, with 75% of involved patients being younger than age 20. Lesions most commonly occur in the bone marrow, but they also can occur in the skin, lymph nodes, and lungs. Histologically the lesions consist of a mixture of mononuclear and inflammatory cells, mainly eosinophils and Langerhans cells. There are three forms of LCH: eosinophilic granuloma, Hand-Schüller-Christian disease, and Letterer-Siwe disease. These forms of LCH probably are variations of the same disease produced by progressively larger and more aggressive clones of Langerhans cells.

Forms of Langerhans Cell Histiocytosis

Eosinophilic Granuloma

Eosinophilic granuloma is the localized, chronic form of LCH. Clinically, it is the mildest and most common form of LCH. It accounts for approximately 70% of all LCH cases. Eosinophilic granuloma is usually, but not always, monostotic. It typically affects patients between the ages of 5 and 15 years. Some children presenting with localized disease can go on to develop additional lesions. Mirra and Gold (1989) described three phases in the development of eosinophilic granuloma lesions: incipient, midphase, and late phase. In the incipient phase, the lesion may appear radiographically aggressive, showing a permeative pattern, wide zone of transition, and lamellated periosteal reaction. Incipient lesions usually incite a severe soft tissue reaction and edema. Lesions in the late phase have a more benign appearance and a narrow zone of transition, resembling a bone abscess. Males are affected more commonly than females, with the most common sites in descending order of frequency being the skull, femur, jaw, ribs, pelvis, and spine. In adults, the disease occasionally occurs in bone, but lung involvement is more common. In eosinophilic granuloma, there is no extraskeletal involvement. With a single bone lesion, the prognosis is excellent, and spontaneous healing is commonly observed.

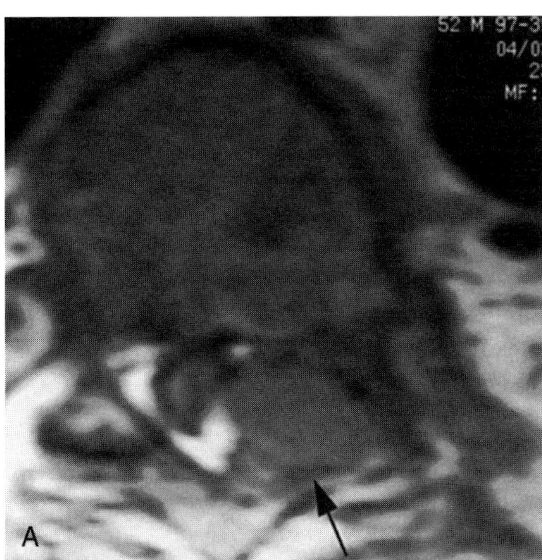

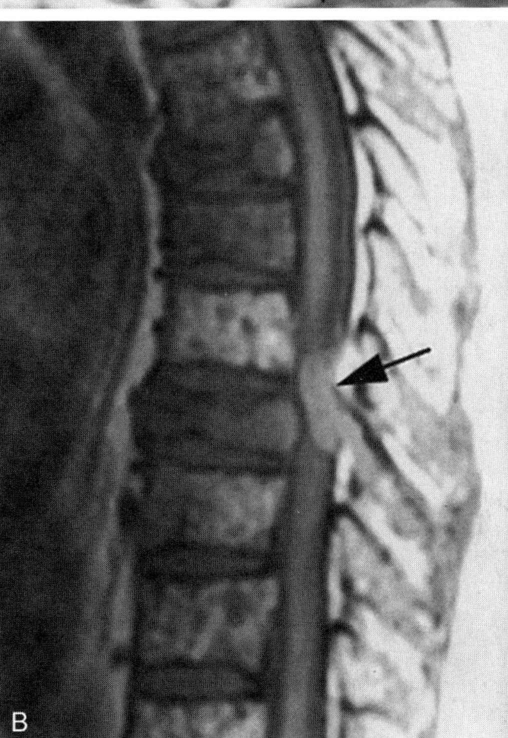

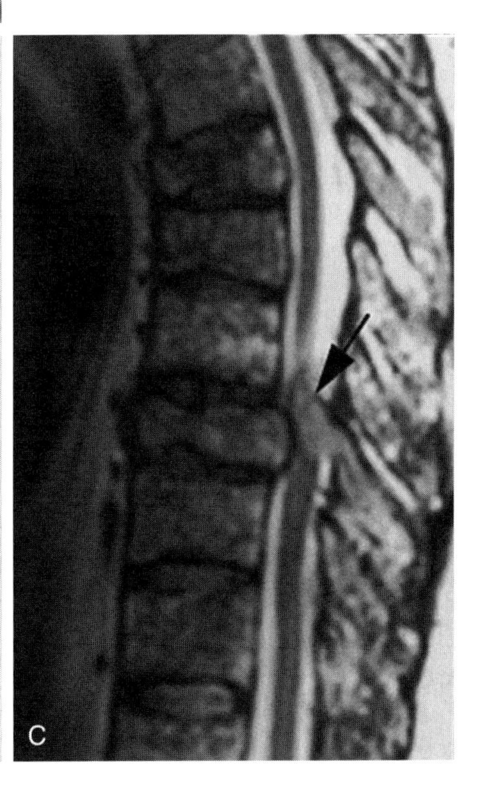

Figure 12–5. Multiple myeloma involving the entire spine and producing an epidural mass with compression of the cord. At least three vertebrae are collapsed. *A* and *B*, Axial and sagittal T1-weighted MR images show the epidural mass *(arrows)* compressing and displacing the cord. *C*, T2-weighted sagittal image shows the epidural mass *(arrow)*.

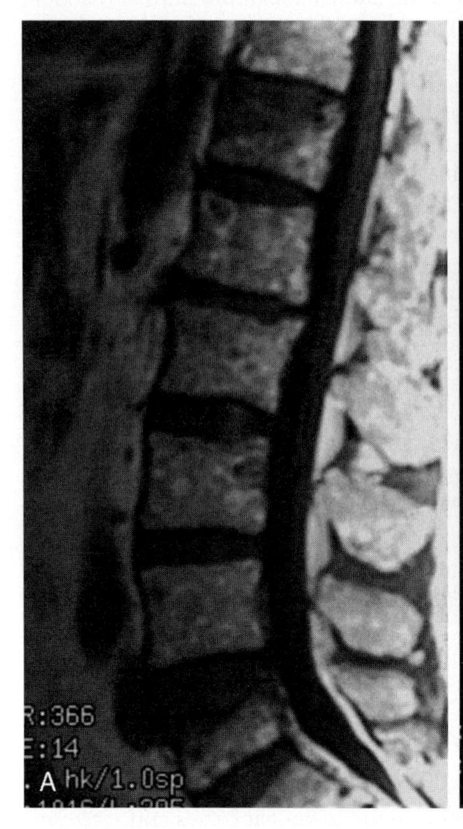

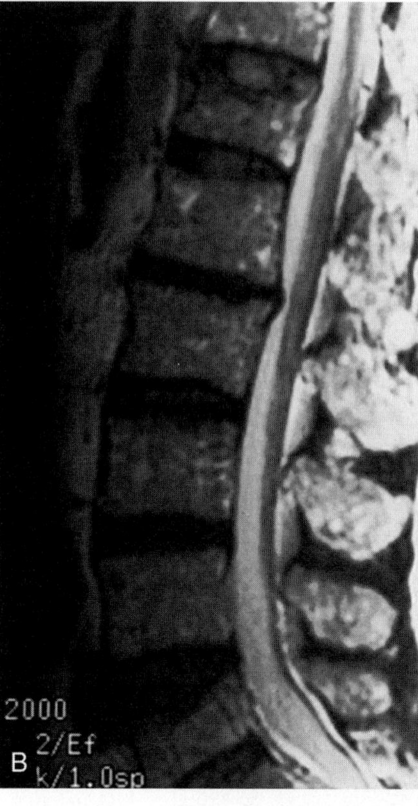

Figure 12–6. Mottled pattern in the spine resulting from multiple myeloma. *A,* T1-weighted sagittal MR image of the lumbar spine. *B,* T2-weighted sagittal image of the lumbar spine. Both sequences show a mottled pattern throughout the lumbar spine. T12 and L5 vertebral bodies are moderately compressed.

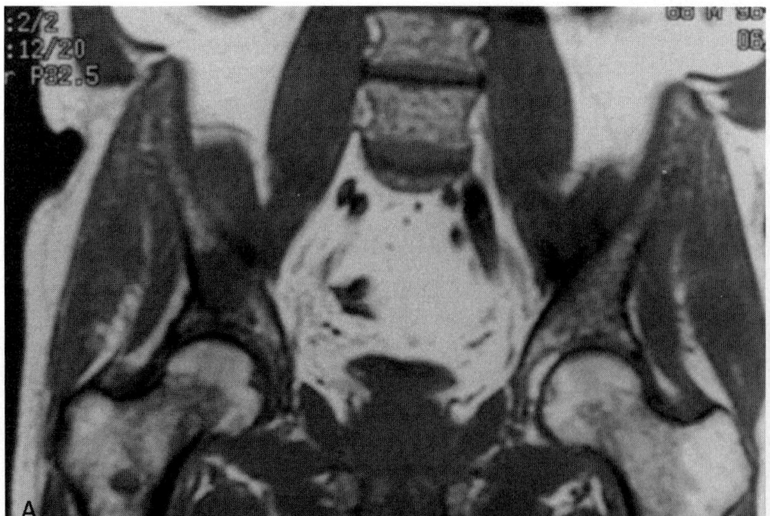

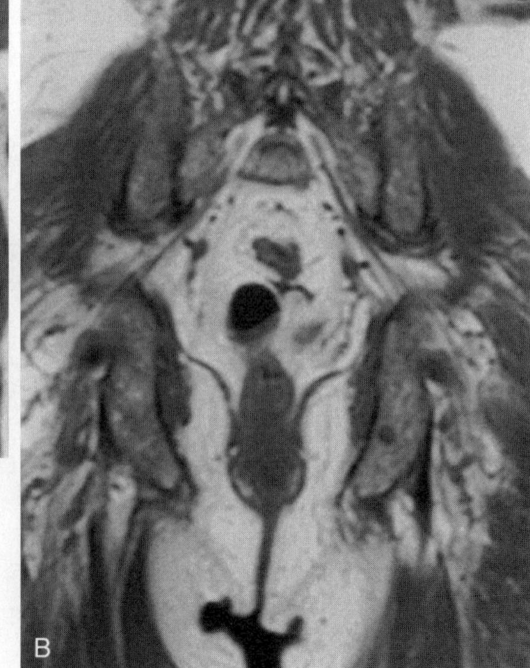

Figure 12–7. Mottled and focal marrow involvement with multiple myeloma in a 68-year-old man. *A* and *B,* T1-weighted coronal images of the pelvis show a mottled appearance and focal deposits in the left ischium, right acetabulum, and right femoral neck.

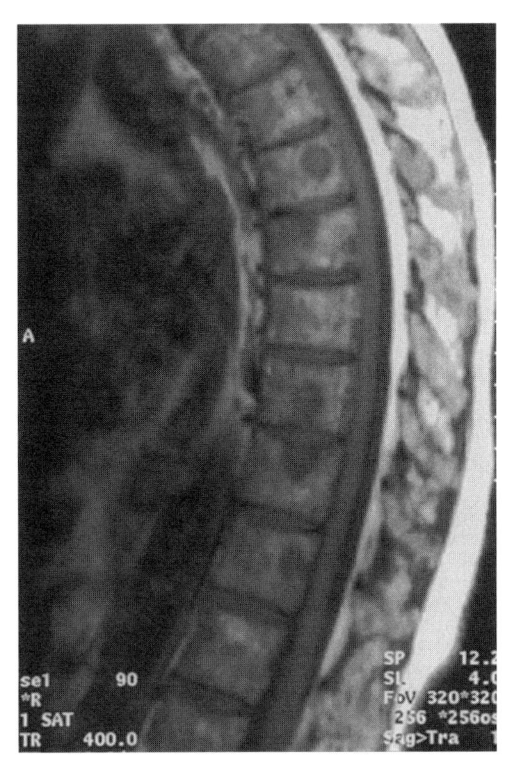

Figure 12–8. Predominant focal pattern on MR imaging in a 72-year-old man with multiple myeloma. Sagittal T1-weighted midline image of the thoracic spine shows several focal myeloma deposits in the vertebral bodies.

Hand-Schüller-Christian Disease

Hand-Schüller-Christian disease is the chronic form of LCH; it constitutes about 20% of cases. Skeletal and visceral sites are involved. More than two thirds of patients present before age 5 years. One of the most common extraskeletal sites of involvement in Hand-Schüller-Christian disease is the pituitary-thalamic axis. The classic triad of diabetes insipidus, exophthalmos, and lytic calvarial lesions is seen in 10% to 15% of patients. Hand-Schüller-Christian disease is fatal in about 15% of cases.

Letterer-Siwe Disease

Letterer-Siwe disease represents the acute fulminant or disseminated form of LCH. It constitutes about 10% of all LCH. Bone involvement is less common than in eosinophilic granuloma or Hand-Schüller-Christian disease; however, disseminated systemic infiltration is always present. It affects the liver, spleen, bone marrow, lymph nodes, and skin. Letterer-Siwe disease usually occurs in infants and children younger than 2 years, and it frequently is fatal.

Imaging

Radiography and Computed Tomography

LCH has a predilection for the flat bones and vertebral bodies (Figs. 12–10 and 12–11). The vertebral arch is involved only rarely. Calvarial lesions are common, and

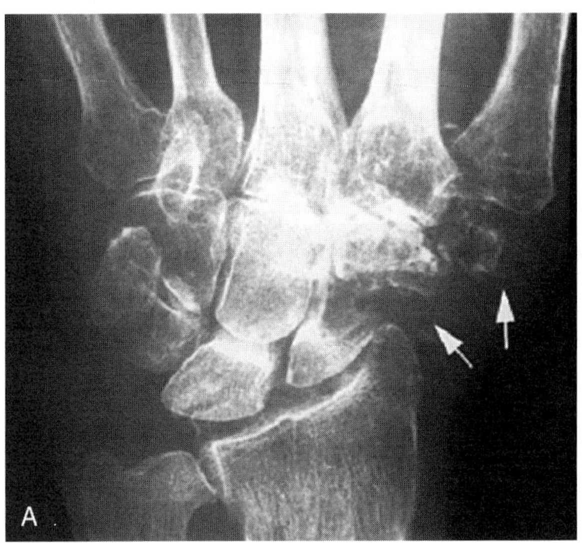

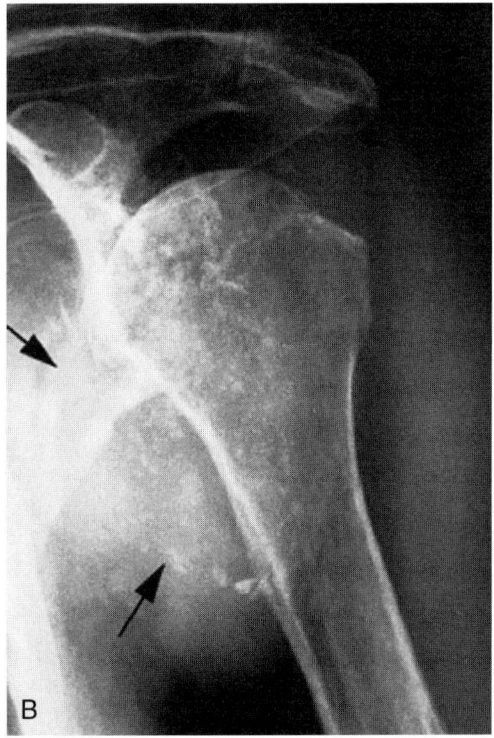

Figure 12–9. Amyloidosis of the joints in an 84-year-old woman with a long-standing history of gammopathy and Bence Jones protein in the urine. She recently developed multiple myeloma. She presented to the rheumatology clinic with multiple joint complaints. *A,* Posteroanterior view of the wrist shows almost complete destruction of the trapezium and distal pole of the scaphoid *(arrows)*. *B,* Anteroposterior view of the left shoulder shows a large calcified soft tissue mass projecting anterior and inferior to the shoulder joint *(arrows)*. This was shown to be amyloid deposit at biopsy.

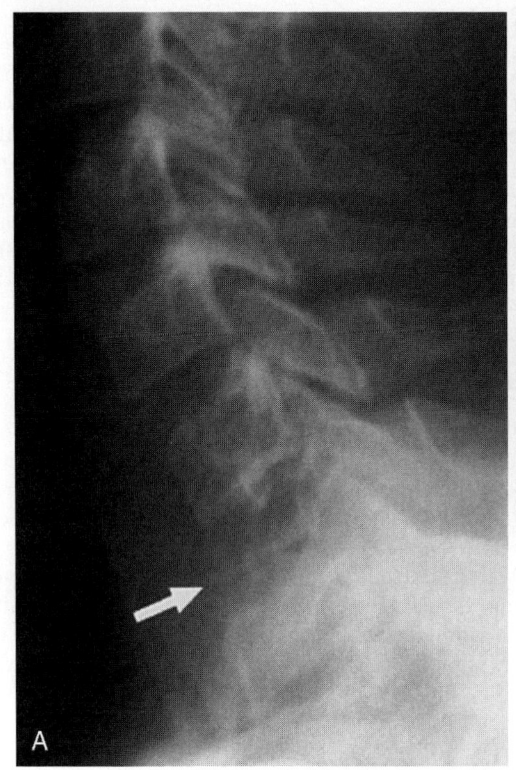

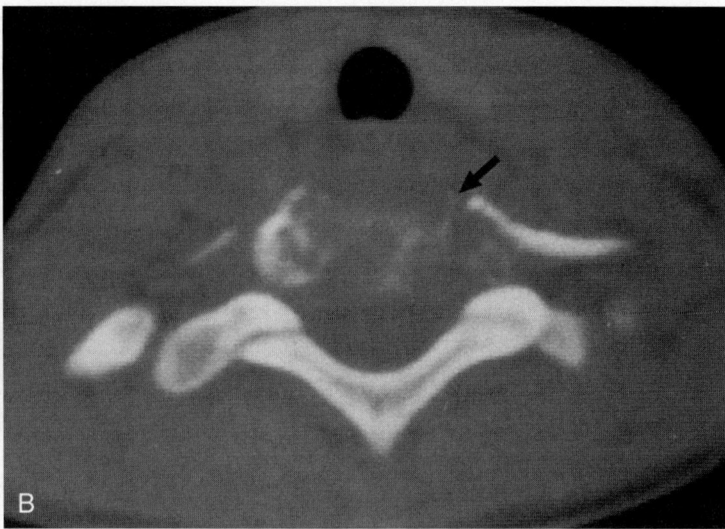

Figure 12–10. Eosinophilic granuloma involving C7 causing vertebra plana. *A*, Lateral view of the cervical spine shows complete flattening of C7 *(arrow)*. *B*, CT section through C7 shows almost complete destruction of the vertebral body *(arrow)*.

they tend to have a sharp border. Because of unequal destruction in flat bones, the edges of the lesions may appear beveled (Fig. 12–11*A*). Undestroyed bone within a lesion can be seen; this is referred to as a *button sequestrum*. In the mandible, bone destruction around a tooth gives the *floating tooth* appearance. In the spine, involved vertebral bodies show varying degrees of collapse, with some collapsing completely, resulting in vertebra plana (see Fig. 12–10). The most common site of involvement is the thoracic spine, followed by the lumbar, then the cervical spine. Epidural extension with cord compression is reported in about 15% of spinal lesions (MR imaging best shows this). Vertebra plana can occur at multiple levels. In the healing phase, a collapsed vertebra regains most of its height; this is believed to occur because the vertebral end plates are not involved with LCH. In the presence of a typical vertebra plana, some authorities advise against obtaining a vertebral biopsy specimen for diagnosis because obtaining a biopsy specimen potentially can damage the end plate and prevent future restoration of the vertebral height. Healing of vertebra plana also can take the form of fusion with an adjacent vertebral body across the intervertebral disk. In the long bones, the femur, humerus, and tibia are involved most often. Within a long bone, the diaphysis is the most common site of involvement, followed by the metaphysis and rarely the epiphysis (Figs. 12–12 and 12–13). The lesions are typically lytic with a geographic or moth-eaten pattern of bone destruction. Often there is endosteal scalloping (Fig. 12–13*B*). Thick single-layered or lamellated uninterrupted periosteal reaction often is present; this should be an indication that the lesion is benign. Ewing's sarcoma sometimes is difficult to exclude; however, the presence of an uninterrupted periosteal reaction and absence of a huge soft tissue mass in eosinophilic granuloma should help in differentiating the two conditions (see Fig. 12–13*B*). Rib lesions may be associated with an extrapleural soft tissue mass. CT is useful in mapping the extent of bone destruction, cortical erosion, and soft tissue involvement (see Fig. 12–12*B* and 12–14*B*).

Nuclear Medicine

Bone scintigraphy is less sensitive than radiography for the detection of old lesions (Fig. 12–14). A bone survey is currently the preferred imaging method to screen for multiple LHC lesions.

Magnetic Resonance Imaging

Because of the high sensitivity of MR imaging in the detection of inflammation and edema, images should be interpreted with caution, especially during the incipient phase of eosinophilic granuloma. The severe marrow and soft tissue inflammatory reaction may be misinterpreted as representing a malignant tumor. MR imaging is ideal for imaging soft tissue masses associated with eosinophilic granuloma, to help determine their origin and extent of the lesion (Fig. 12–15). MR imaging is especially effective in the evaluation of spinal lesions with epidural extension and cord compression.

Differential Diagnosis

In children, LCH can simulate osteomyelitis, Ewing's sarcoma, and lymphoma. In adults, lesions may simulate osteolytic metastasis and multiple myeloma.

ERDHEIM-CHESTER DISEASE

Overview

Erdheim-Chester disease (or lipoid granulomatosis) has been included as belonging to the spectrum of LCH based on histopathologic advances. This disease is characterized by the presence of histiocytes with irregular bone trabeculae, bone marrow fibrosis, and foci of lipid granulomas. It also is accompanied by xanthogranulomatous involvement of various visceral organs. The disease is seen in middle-aged individuals and in the elderly. Men and women are affected equally.

Skeletal Involvement

There is symmetrical involvement of the long tubular bones, predominantly in the metaphyses and diaphyses. The lesions are central (medullary). The axial skeleton usually is spared.

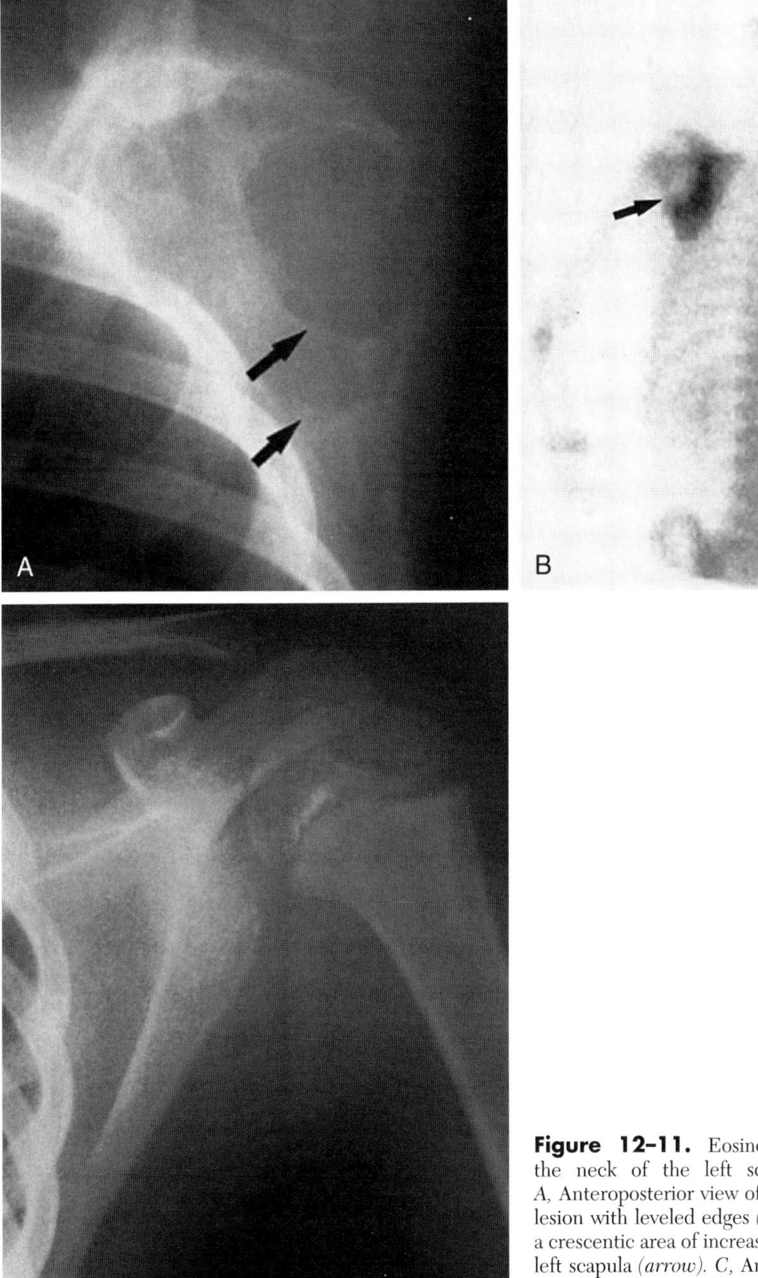

Figure 12–11. Eosinophilic granuloma involving the neck of the left scapula in a 3-year-old boy. *A,* Anteroposterior view of the left scapula shows a lytic lesion with leveled edges *(arrows)*. *B,* Bone scan shows a crescentic area of increased radionuclide uptake in the left scapula *(arrow)*. *C,* Anteroposterior view of the left scapula at the age of 6 years shows healing of the lytic lesion with some remodeling of the scapular neck.

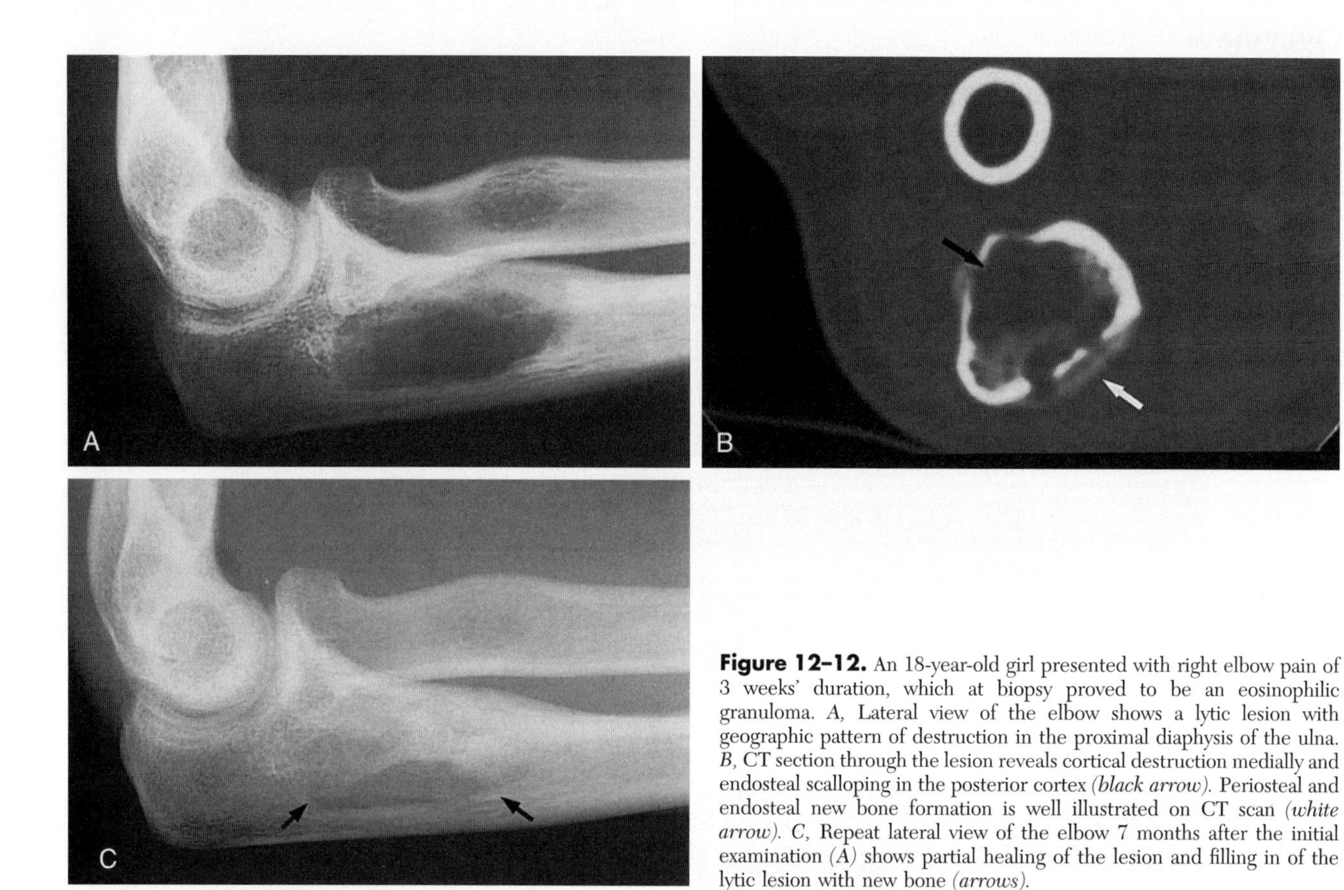

Figure 12–12. An 18-year-old girl presented with right elbow pain of 3 weeks' duration, which at biopsy proved to be an eosinophilic granuloma. *A,* Lateral view of the elbow shows a lytic lesion with geographic pattern of destruction in the proximal diaphysis of the ulna. *B,* CT section through the lesion reveals cortical destruction medially and endosteal scalloping in the posterior cortex *(black arrow).* Periosteal and endosteal new bone formation is well illustrated on CT scan *(white arrow). C,* Repeat lateral view of the elbow 7 months after the initial examination *(A)* shows partial healing of the lesion and filling in of the lytic lesion with new bone *(arrows).*

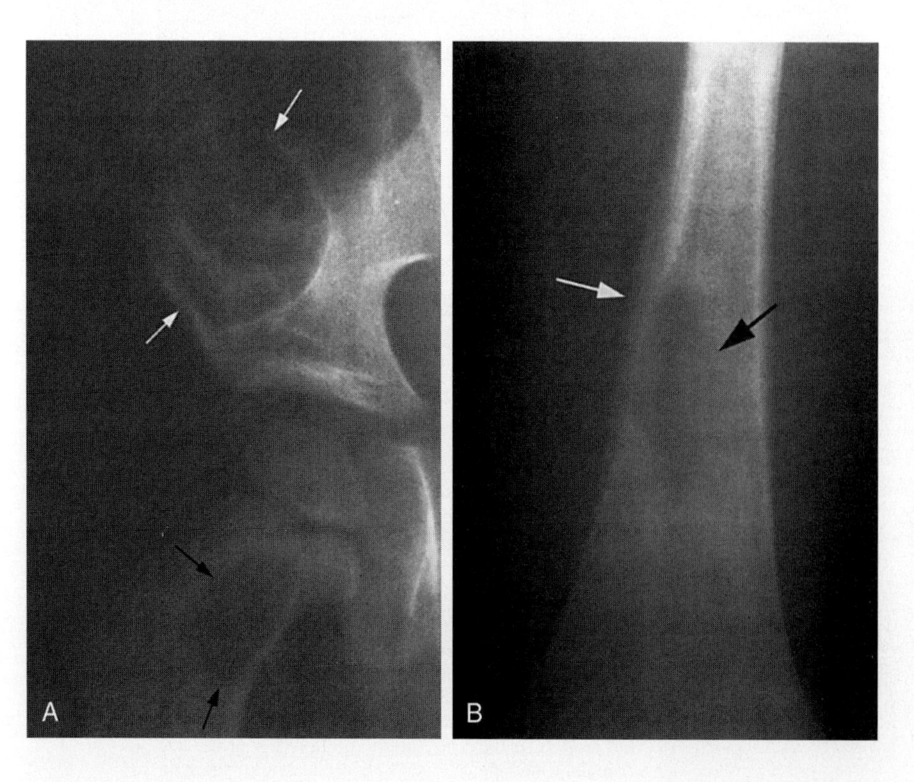

Figure 12–13. A 4-year-old boy with multiple skeletal lesions caused by Langerhans cell histiocytosis. *A,* Anteroposterior view of the right hip shows a well-circumscribed lytic lesion in the right ilium. The lesion has a thin sclerotic border *(white arrows).* Another lytic lesion is seen in the neck (metaphysis) of the right femur *(black arrows). B,* Anteroposterior view of the left femur shows an ill-defined lytic lesion *(black arrow),* which is eroding the medial cortex and has an associated solid uninterrupted periosteal reaction *(white arrow).* This lesion alone would be difficult to differentiate from a bone abscess.

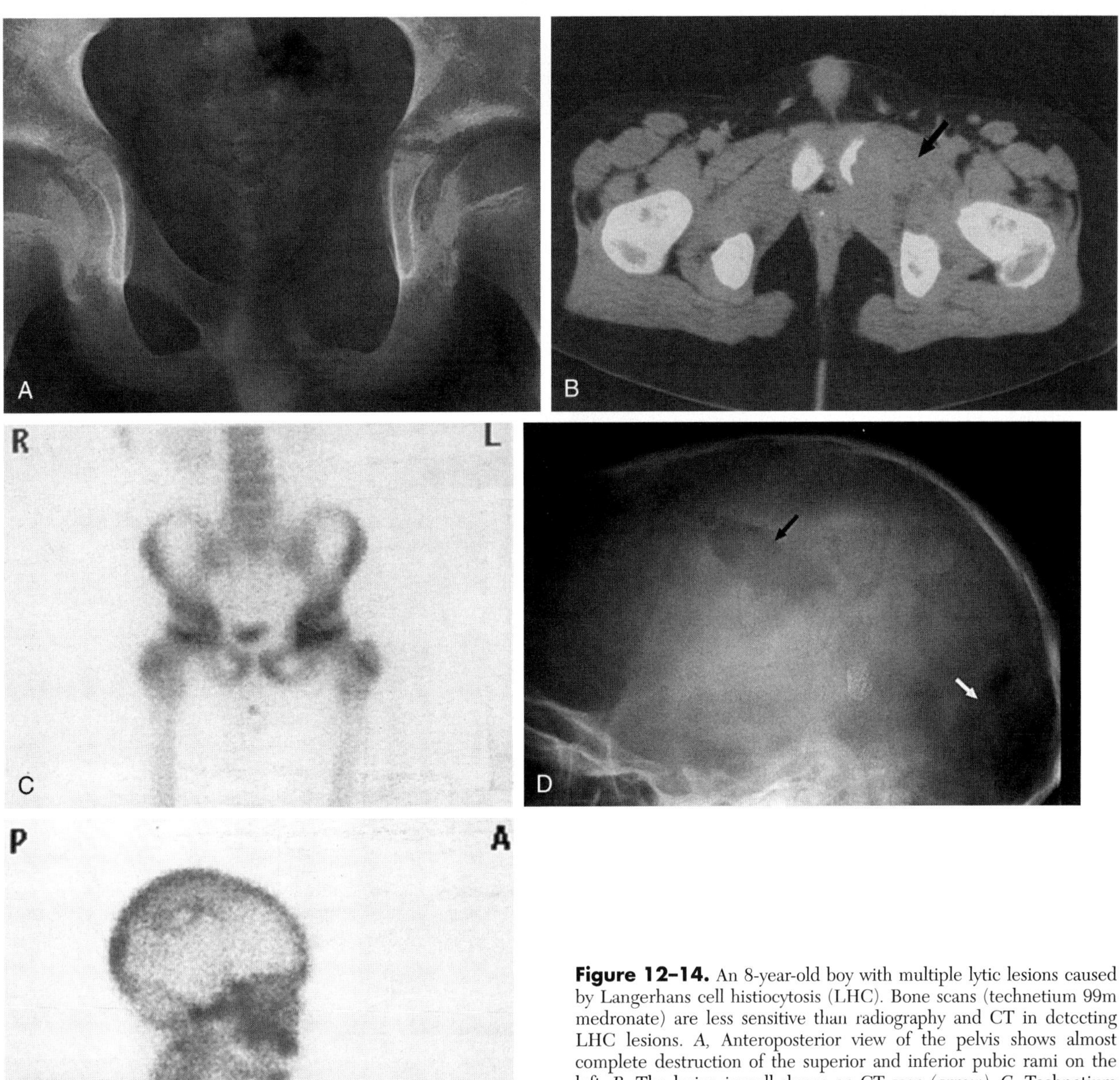

Figure 12–14. An 8-year-old boy with multiple lytic lesions caused by Langerhans cell histiocytosis (LHC). Bone scans (technetium 99m medronate) are less sensitive than radiography and CT in detecting LHC lesions. *A,* Anteroposterior view of the pelvis shows almost complete destruction of the superior and inferior pubic rami on the left. *B,* The lesion is well shown on CT scan *(arrow). C,* Technetium 99m bone scan performed at the same date as *A* and *B* is almost entirely normal. *D,* Lateral radiograph of the skull shows two large lytic lesions, one in the parietal region *(black arrow)* and another one in the occipital area *(white arrow). E,* Technetium 99m shows minimal increased uptake in the parietal region, but the occipital area is entirely normal.

Imaging

Radiography

On radiographs, lesions of Erdheim-Chester disease are focal, lytic, lobulated, and slightly expansile. They are associated with coarse, thickened trabeculae and diffuse sclerosis (Fig. 12–16*A*, *B*). There is associated mild periosteal reaction or endosteal scalloping.

Magnetic Resonance Imaging

On MR imaging, the lesions have low signal intensity on T1-weighted images and increased signal intensity on T2-weighted images (Fig. 12–16*C*, *D*).

Nuclear Medicine

Bone scintigraphy reveals bilateral, symmetrical high uptake, particularly around the knees.

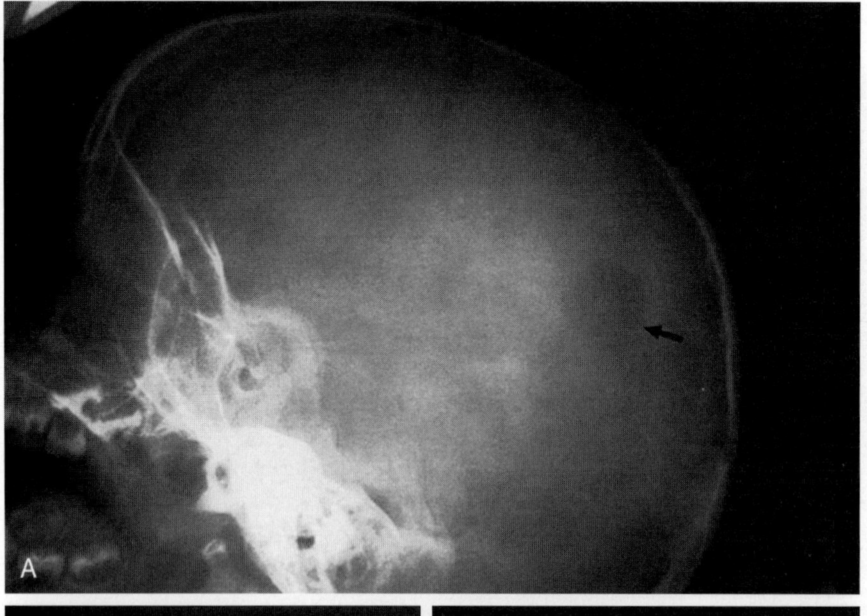

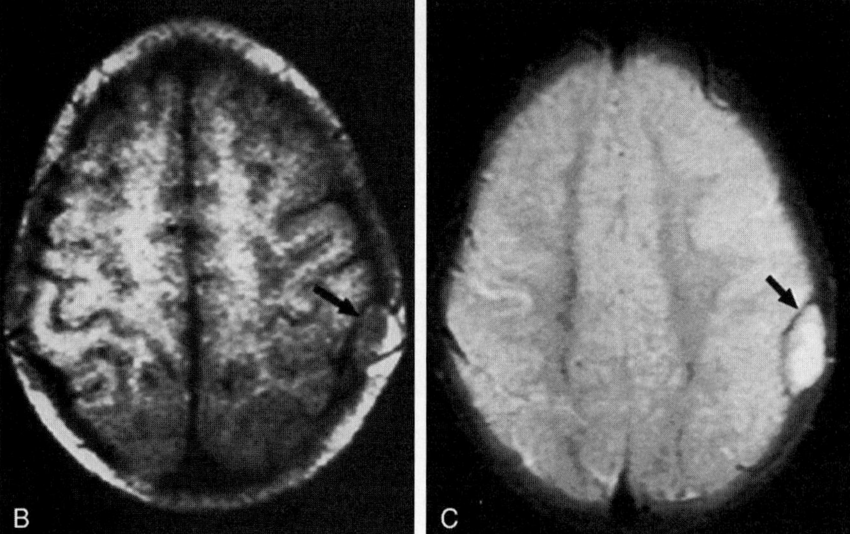

Figure 12–15. A 17-month-old boy presented with a soft tissue mass on the left side of the skull. *A,* Lateral radiograph reveals a lytic lesion in the parietal area measuring about 2.5 × 2.5 cm *(arrow),* which proved to be Langerhans' cell histiocytosis lesion. *B* and *C,* T1- and T2-weighted axial images show an oval lesion with its epicenter in the marrow space of the calvarium *(arrow).* The lesion is seen to compress the cortex of the brain.

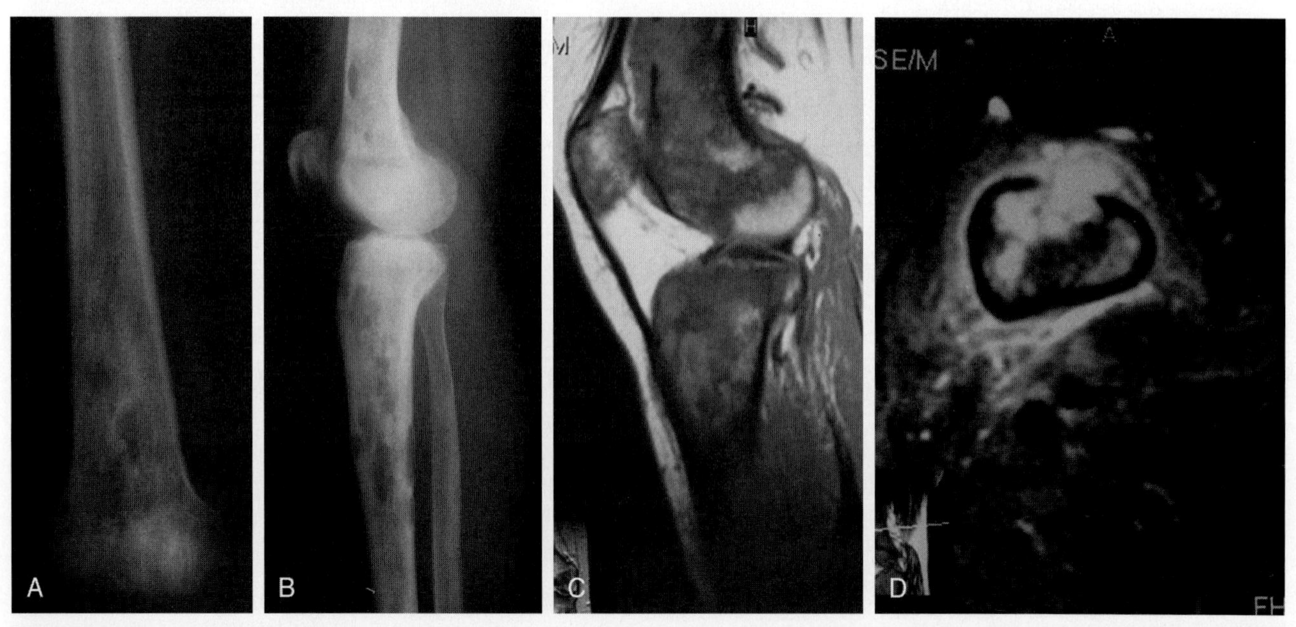

Figure 12–16. A 42-year-old woman with Erdheim-Chester disease. *A* and *B,* Anteroposterior view of the left femur and lateral view of the right knee show multiple lytic lesions within the medullary space of the long bones. The trabecular pattern is coarse, and the bones are fairly sclerotic. *C* and *D,* Lateral T1-weighted and axial STIR MR images of the right knee show extensive marrow infiltration. The anterior cortex of the distal femur is destroyed, and the tumor extends into the soft tissue.

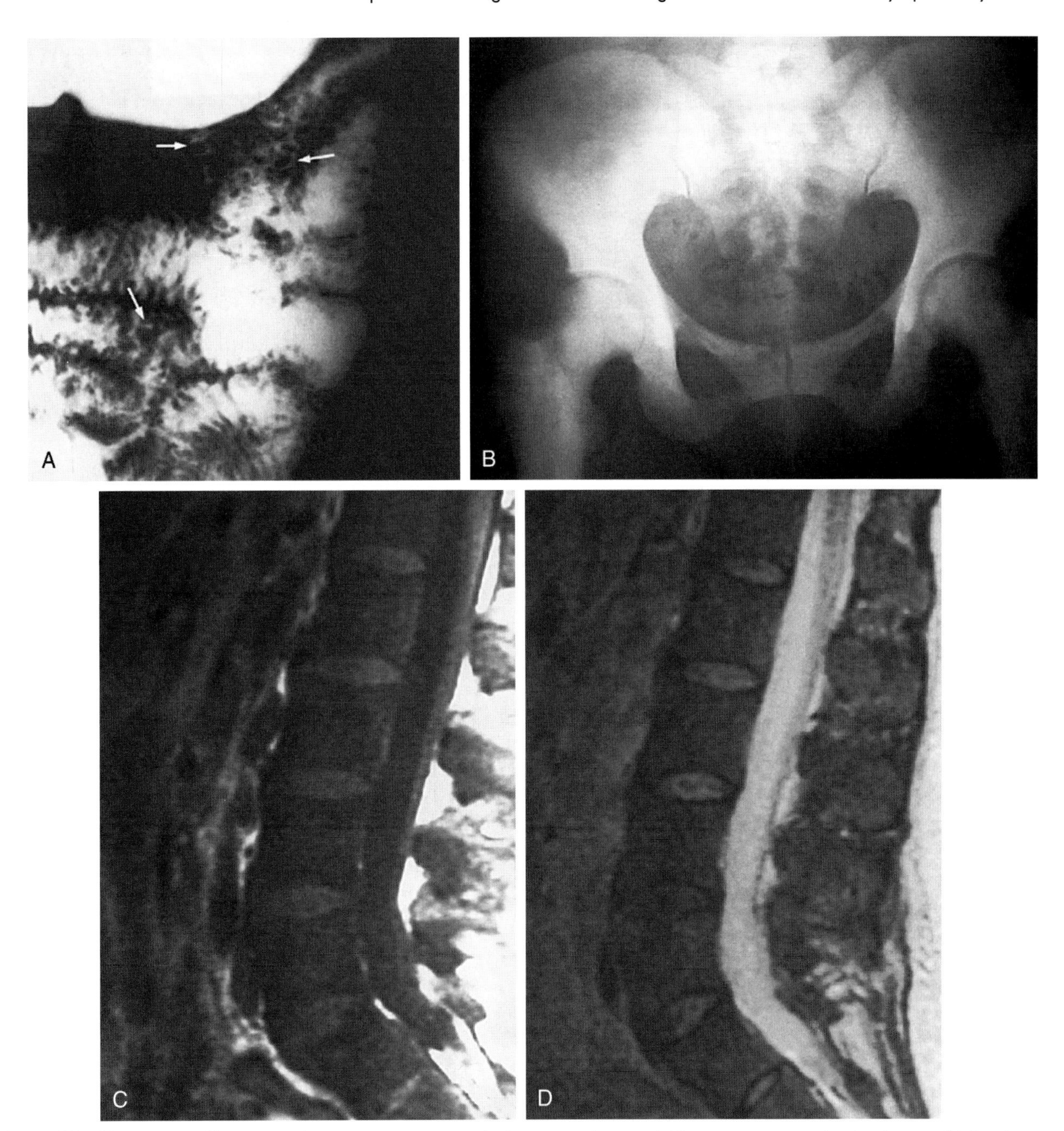

Figure 12–17. A 67-year-old woman with urticaria pigmentosa, which is rare at this age. *A,* Upper gastrointestinal series shows multiple submucosal nodules in the jejunum *(arrows). B,* Anteroposterior view of the pelvis reveals dense sclerosis of the pelvis and hips. *C* and *D,* T1- and T2-weighted sagittal images reveal low signal intensity throughout the spine on both sequences. Bone sclerosis is the cause of the low signal intensity. (Case provided by Dr. Craig Walker, University of Nebraska, Omaha.)

Differential Diagnosis

The differential diagnosis includes chronic osteomyelitis, Engelmann's disease, bone infarction, lymphoma, and metastases.

MASTOCYTOSIS

Overview

Mastocytosis is a rare disease characterized by the proliferation of mast cells in the skin and other organs, including the lymph nodes, spleen, liver, gastrointestinal tract, and bone marrow. Mastocytosis is divided into two clinical forms, cutaneous (urticaria pigmentosa) and systemic. The cutaneous form typically affects children, has a good prognosis, and resolves spontaneously by puberty. The systemic form is a chronic, progressive disease, and spontaneous regression does not occur. Systemic mastocytosis is a disease of middle-aged and older individuals of both genders. The mast cells secrete pharmacologically active substances, which play an important role in the clinical manifestations of this disease. These substances include histamine, heparin, prostaglandins, serotonin, and mucopolysaccharides.

Skeletal Involvement

Skeletal lesions affect 70% to 75% of patients with systemic mastocytosis. The most common sites of involvement are the axial skeleton, pelvis, and proximal ends of long bones.

Imaging

Three major patterns of radiographic involvement have been described in mastocytosis.

Diffuse Pattern. This is a pattern of diffuse skeletal involvement with a mixture of osteosclerosis and osteopenia, with the osteosclerotic component being more dominant (Fig. 12–17). Some patients present with only diffuse, profound osteopenia (Fig. 12–18), however. The profound osteopenia is believed to be due to the excessive secretion of heparin. The sclerotic lesions possibly may be secondary to fibrotic changes in response to excessive secretion of histamine. The fibrotic tissue consequently is converted to osteoid, which leads to the deposition of calcium salts resulting in sclerosis.

Circumscribed Pattern. This is a circumscribed pattern of involvement with multiple lytic or blastic lesions.

Solitary Pattern. This is a pattern of solitary blastic or lytic lesions. This pattern is rare, making the diagnosis of mastocytosis from a single lesion difficult. Cases of mastocytosis have been discovered incidentally on radiographs because some lesions are asymptomatic.

Differential Diagnosis

The diffuse sclerotic pattern resembles myelosclerosis (myelofibrosis) (Fig. 12–19), diffuse metastatic disease, Paget's disease, or fluorosis. The diffuse lytic pattern of mastocytosis closely resembles osteoporosis or multiple myeloma. The circumscribed form can be difficult to differentiate from metastasis. The solitary form resembles lytic or blastic metastasis, a plasmacytoma, or a primary neoplasm of bone.

EWING'S SARCOMA

Overview

Ewing's sarcoma is the fourth most common malignant tumor of bone and the second most common malignant bone tumor in children, after osteosarcoma. The origin of Ewing's sarcoma is uncertain, although there is some evidence suggesting a neuroectodermal origin. Ewing's sarcoma and primitive neuroectodermal tumors are related genetically, with Ewing's sarcoma representing the most undifferentiated form of the primitive neuroectodermal tumors. Ewing's sarcoma usually is encountered in individuals between the ages of 5 and 25 years. Males are affected more than females (2:1). Systemic symptoms, including fever, malaise, and weight loss, are seen commonly in Ewing's sarcoma patients. Metastasis typically involves the lungs and bones (Fig. 12–20). At presentation, 18% to 25% of patients have evidence of disseminated disease. Extraskeletal Ewing's sarcoma, which is rare, typically occurs in the soft tissues of the lower extremities, paravertebral area, or epidural space. These patients typically are older than patients with osseous Ewing's sarcoma (average age, 20 years) with males and females being affected equally. Another rare variant is periosteal Ewing's sarcoma, which occurs on the diaphyseal surface of long bones (Fig. 12–21). It presents as a subperiosteal mass with cortical thickening

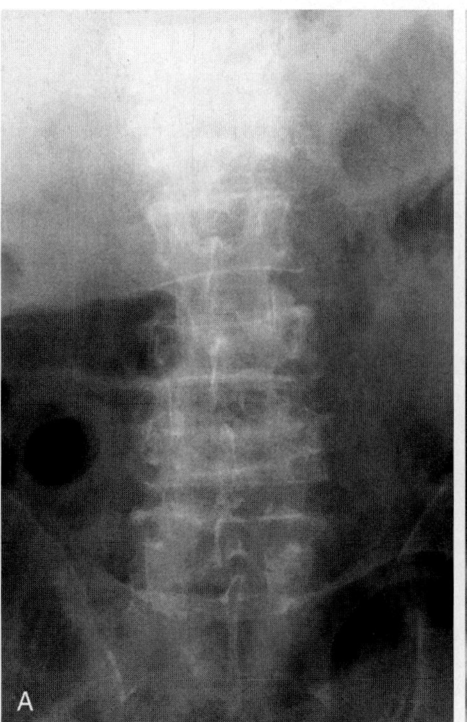

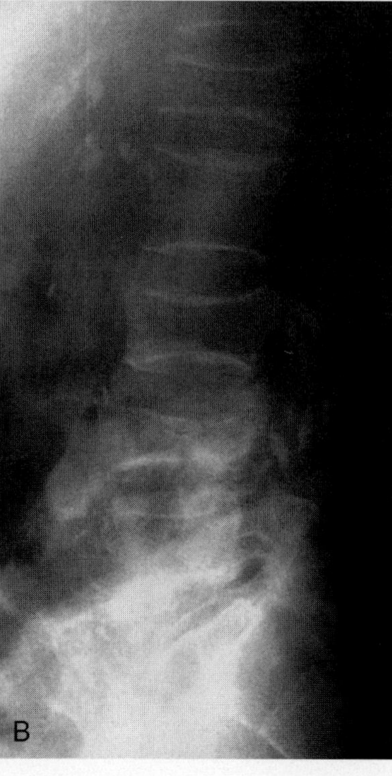

Figure 12–18. A 76-year-old man with systemic mastocytosis and osteopenia. *A* and *B*, Anteroposterior and lateral views of the lumbar spine show profound osteopenia with collapse of L1, L3, and L4.

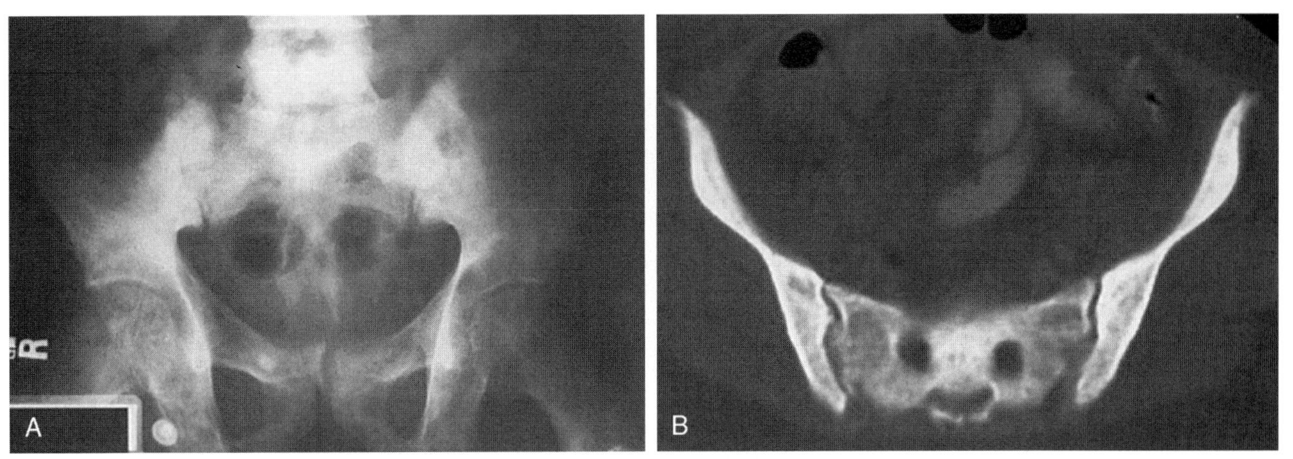

Figure 12–19. Radiographic and CT appearance of myelosclerosis (myelofibrosis). *A,* Anteroposterior view of the pelvis shows diffuse sclerosis and coarse trabecular pattern resembling mastocytosis and Paget's disease. *B,* CT scan of the pelvis reveals similar changes.

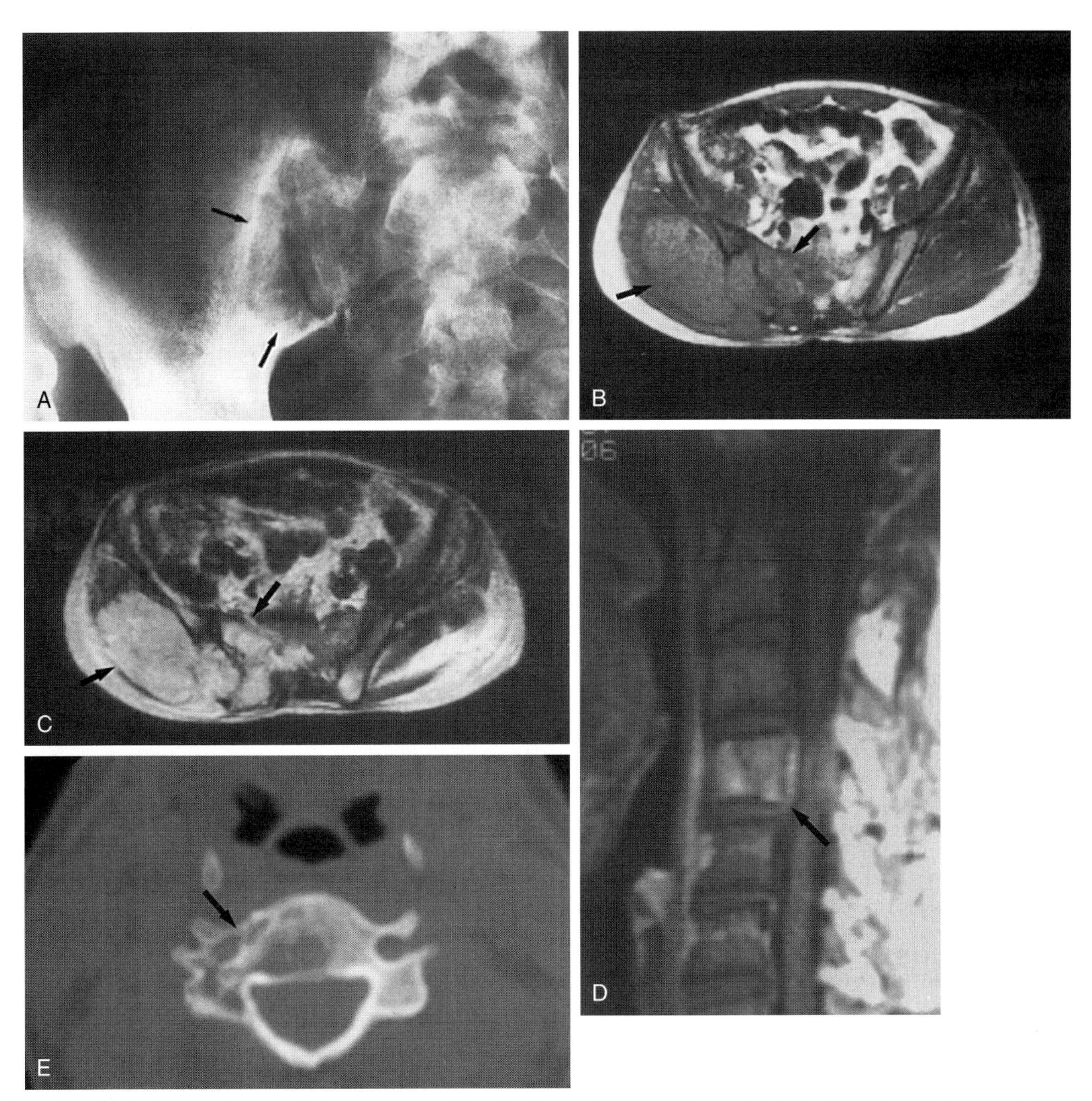

Figure 12–20. A 21-year-old man with Ewing's sarcoma of the right ilium and sacrum with metastasis to C4 5 years later. *A,* Anteroposterior view of the pelvis shows bone destruction on both sides of the sacroiliac joint *(arrows).* *B* and *C,* Axial T1- and T2-weighted MR images show the bony and soft tissue components of the lesion *(arrows).* The patient was treated with radiation and chemotherapy, and he was doing well. Five years later, he developed neck pain and neurologic symptoms. *D,* T1-weighted sagittal image of the cervical spine with gadolinium enhancement shows increased signal intensity in the body of C4 along with an epidural mass compressing the cord *(arrow).* *E,* CT section during a needle biopsy of C4 shows lytic destruction within the vertebral body and the lateral mass on the right *(arrow).*

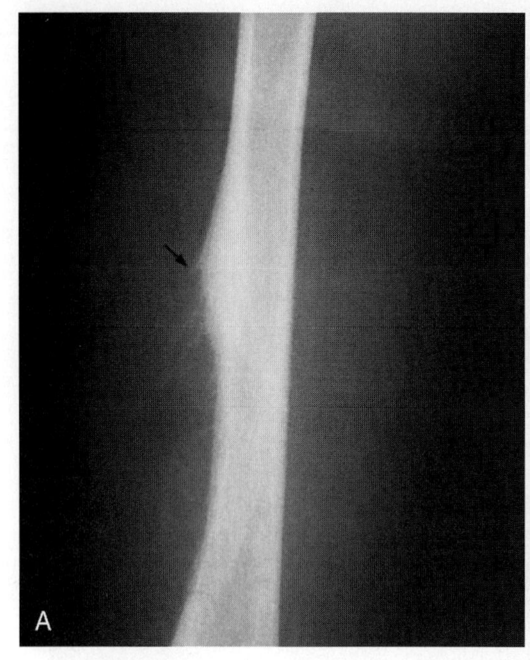

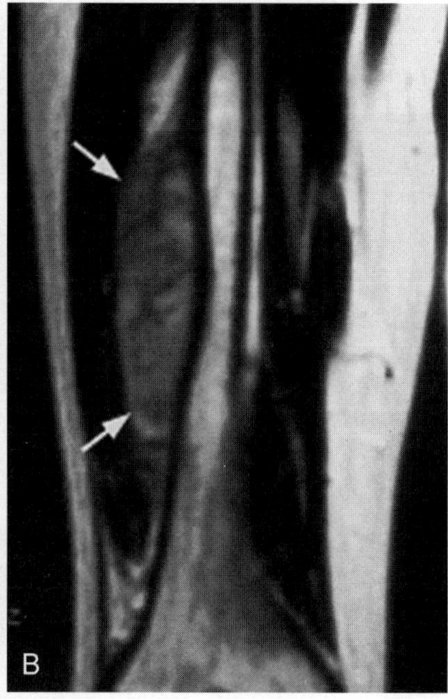

Figure 12–21. Periosteal Ewing's sarcoma in a 13-year-old boy. *A,* Anteroposterior view of the right femur shows a large surface lesion that is eroding the cortex (saucerization). Codman's triangles are on either end of the lesion *(arrow). B,* Sagittal T1-weighted image of the right femur shows the surface lesion *(arrows).* The marrow space is not invaded by the lesion.

or, occasionally, with a wide, shallow, cortical erosion (saucerization). There is an associated Codman's triangle, but no identifiable matrix within the mass. Periosteal Ewing's sarcoma does not invade the medullary space (Fig. 12–21*B*), and it usually has a more benign course than other types of Ewing's sarcoma.

In general, Ewing's sarcoma is grouped histologically under round, blue cell tumors. The tumor cells are closely packed, are poorly differentiated, have scant cytoplasm, and have indistinct cell membranes. Periodic acid–Schiff stain is positive for intracytoplasmic glycogen. Microscopically, Ewing's sarcoma must be distinguished from other malignant round, blue cell tumors, such as small cell osteosarcoma, mesenchymal chondrosarcoma, lymphoma, and neuroectodermal tumors of bone.

Skeletal Involvement

Ewing's sarcoma usually develops within the medullary cavity, and any bone can be involved. More than 50% of the lesions occur in the long bones, however, especially those of the lower extremity (femur, tibia, and fibula). Within the long bones, the tumor usually is located in the metadiaphysis, followed by the diaphysis and metaphysis. The flat bones also frequently are involved, particularly the ilium and ribs (Fig. 12–22). In the pelvic bones, Ewing's sarcoma is the most common primary malignant tumor in children (Fig. 12–23). In the pelvis, Ewing's sarcoma has a worse prognosis than in other locations. Calcaneal lesions also are reported to have a worse prognosis than lesions at other locations. About 7% to 10% of Ewing's sarcomas occur in ribs (see Fig. 12–22), making it the most common malignant lesion of the chest wall in childhood. Often an associated soft tissue mass or pleural effusion obscures rib lesions. In the spine, Ewing's sarcoma typically involves the vertebral body. Rarely, Ewing's sarcoma involves the hands and feet, being more common in the feet. Sclerosis is a prominent feature of Ewing's sarcoma occurring in the hands and feet.

Imaging

Radiography and Computed Tomography

Ewing's sarcoma has a protean array of plain film appearances. Most lesions of the long bones appear as permeative, destructive lesions of the diaphysis or metadiaphysis. The lesion often is associated with a lamellated or spiculated (sunburst) periosteal reaction and a large soft tissue mass. Saucerization, a shallow but wide erosion of the cortex, is described as being characteristic of Ewing's sarcoma (see Fig. 12–21). A diffusely sclerotic appearance is present in about 37% of cases (see Fig. 12–22). Radiographically, these sclerotic lesions are difficult to differentiate from osteosarcoma. Sclerotic Ewing's sarcoma tends to contain dead bone with occasional apposition of new bone on the dead bone. Mixed lytic and sclerotic lesions also have been described.

Magnetic Resonance Imaging

MR imaging is the most sensitive modality in detecting the tumor and in local staging of the tumor. The bone marrow and soft tissue extent are assessed well by MR imaging (see Figs. 12–20, 12–21, and 12–23). The lesions have low signal intensity on T1-weighted images and variable signal intensity on T2-weighted images but primarily high signal intensity. Skip lesions in the same bone are detected best by MR imaging.

Nuclear Medicine

A bone scan is a sensitive technique for the detection of Ewing's sarcoma (see Fig. 12–22) and for the detection of metastatic Ewing's sarcoma to bone.

Response to Therapy

On radiographic images, healed lesions are characterized by ossification of the soft tissue component and endosteal bone formation encroaching on the medullary cavity. MR imaging is the most useful imaging modality in assessing the response to chemotherapy. A good response is seen as a significant decrease in the tumor size and the size of the soft tissue component. On T1-weighted images, lesions responsive to treatment show a marked decrease in size or disappear completely. The reduction in tumor size also is appreciated on T2-weighted images. A decrease in signal intensity on T2-weighted images can be used as qualitative evidence of a positive chemotherapeutic response. After therapy, however, viable tumor cells have been found in shrunken tumors with either low or high signal intensity on T2-weighted images.

Differential Diagnosis

Radiographically the differential diagnosis includes osteomyelitis, osteosarcoma, eosinophilic granuloma, non-Hodgkin's lymphoma (NHL), and metastatic disease.

LYMPHOMAS

Generally, lymphomas have been difficult to understand because of the confusion with terminology, classification, staging, and the variety of ways these tumors involve bone. Lymphomas are the seventh leading cause of death in the United States. Lymphomas can be divided broadly into

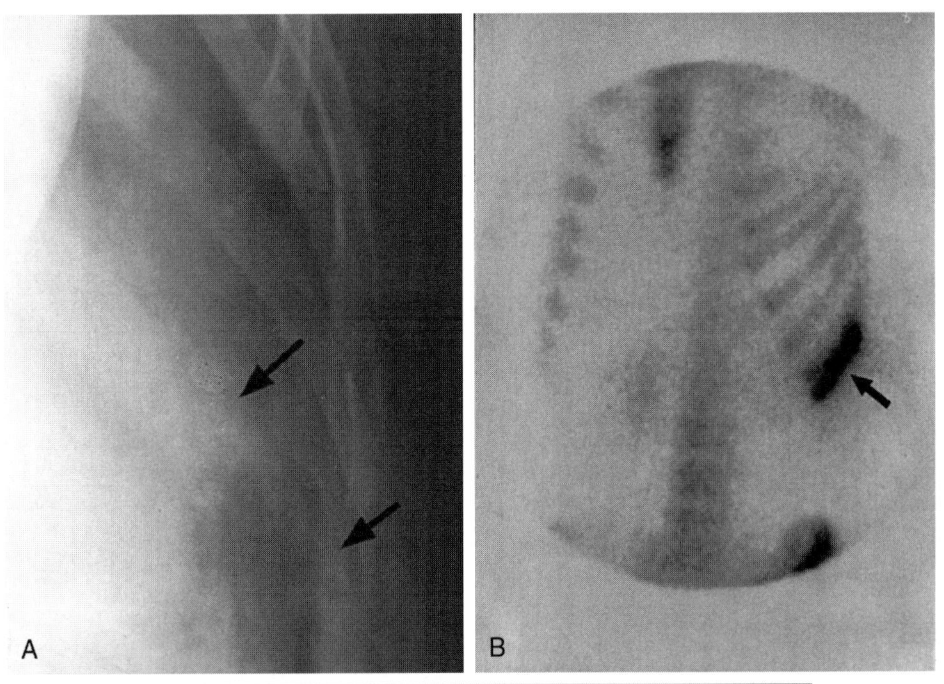

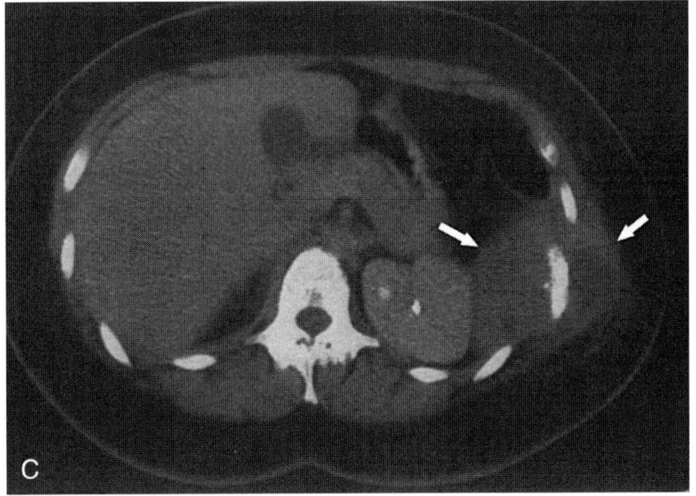

Figure 12–22. A 22-year-old woman with Ewing's sarcoma of the left 10th rib. *A*, Oblique view of the left chest wall shows blastic and lytic destruction of the anterior portion of the left 10th rib *(arrows)*. *B*, Technetium 99m MDP bone scan reveals intense uptake of the radionuclide *(arrow)*. *C*, CT scan through the lesion shows rib expansion. This expansion is associated with a large soft tissue mass *(arrows)*, possibly tumor necrosis just lateral to the involved rib.

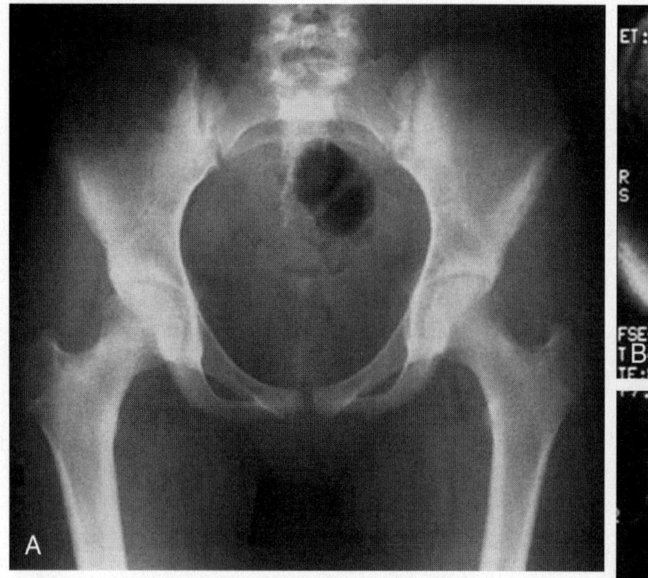

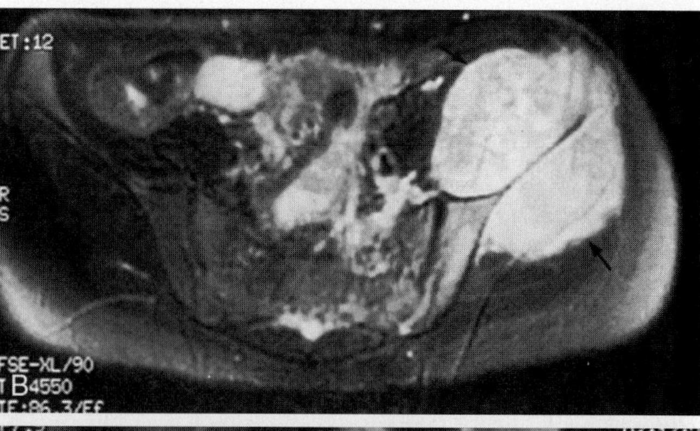

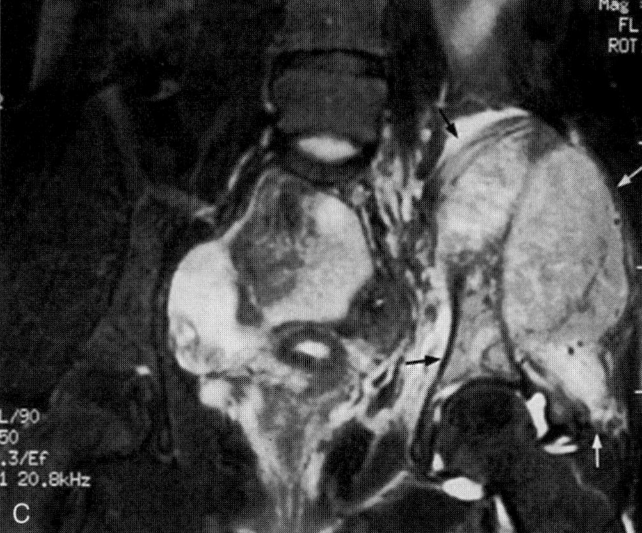

Figure 12–23. Ewing's sarcoma of the left ilium in a 15-year-old boy presenting with a growing mass on the left side of the pelvis. *A,* Anteroposterior view of the pelvis is almost entirely normal. *B* and *C,* Axial T2-weighted and coronal T2-weighted with fat suppression images show a huge soft tissue mass arising from the left ilium *(arrows).*

two groups: Hodgkin's disease (HD) and NHL, which is eight times more common than HD. They are the most common malignant tumor in adults between the ages of 20 and 40 years.

Hodgkin's Disease

Overview

HD is essentially a nodal disease that spreads in a predictable fashion through lymphatic channels to contiguous lymph node chains. Other means of spread include direct invasion from lymph nodes to adjacent structures or through hematogenous spread to the spleen, liver, lung, and bone marrow. Involvement of bone at one or more sites remote from the diseased lymph nodes is presumed to be due to hematogenous spread (stage IV). The neoplastic cell in HD is the Reed-Sternberg cell. With proper staging and treatment, cure can be achieved in greater than 70% of cases.

Skeletal Involvement

Axial involvement is more common (77%) (Fig. 12–24*A*, *B*) than appendicular involvement (23%) (Fig. 12–24*C*). The sites that are affected most frequently by HD are the spine, sternum, ribs, pelvis (Fig. 12–25), and proximal femur (Fig. 12–25*B;* see Fig. 12–24*C*). The presenting or primary site of HD is rarely bone. Most bone cases are multifocal, occurring during the advanced stages of the disease. There are three types of bone involvement in HD.

Primary Hodgkin's Disease of Bone. This presentation is extremely rare, consisting of less than 2% of all HD of bone. Patients present with a solitary bone lesion with no lymph node or visceral involvement.

Bone Involvement by Direct Invasion. This is a common type of involvement. It typically affects the lumbar spine (see Fig. 12–24), thoracic spine, and sternum.

Bone Marrow Infiltration with or Without Radiographic Evidence of Bone Involvement. This is the most common type of bone involvement in HD. Skeletal involvement by hematogenous dissemination occurs late in the disease, and it often is multifocal. Bone surveys at presentation typically are negative because HD initially is limited to lymph nodes. Most patients with disseminated disease (stage IV) have bone marrow involvement, but this is radiographically evident in only 15% to 25% of the cases because radiography underestimates the extent of marrow involvement. Currently the extent of marrow involvement is assessed best by MR imaging.

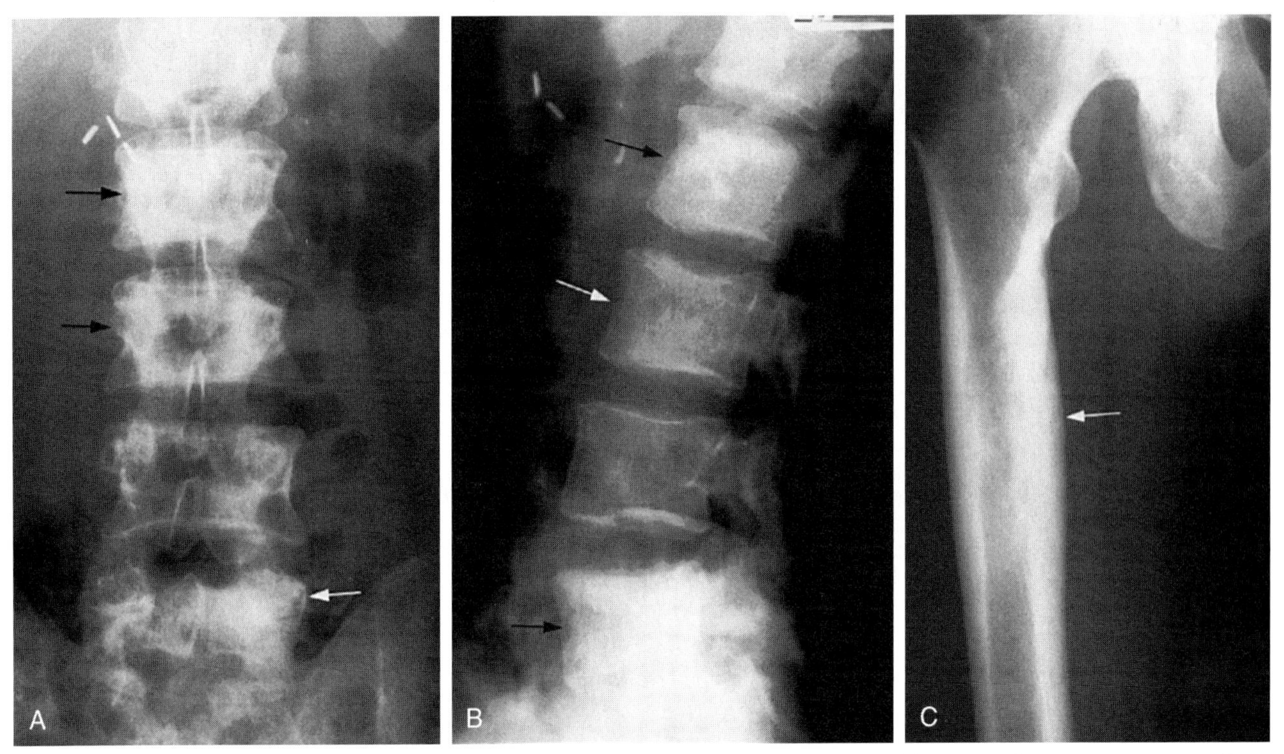

Figure 12–24. Stage IV Hodgkin's disease involving the spine, sacrum, and right femur in a 35-year-old man. *A* and *B*, Anteroposterior and lateral views of the lumbosacral spine show multiple blastic lesions in the vertebral bodies *(arrows)*. *C*, Anteroposterior view of the right femur shows an 8-cm-long blastic lesion involving the proximal shaft of the right femur *(arrow)*.

Imaging

Lesions can be single (33%) or multiple (66%); they range from small to extensive areas of permeative destruction with a wide zone of transition. Lamellated or sunburst periosteal reaction can be present. The lesions usually are lytic, but blastic lesions (see Fig. 12–25), such as the ivory vertebra (see Fig. 12–24), or mixed lesions have been described. Paravertebral

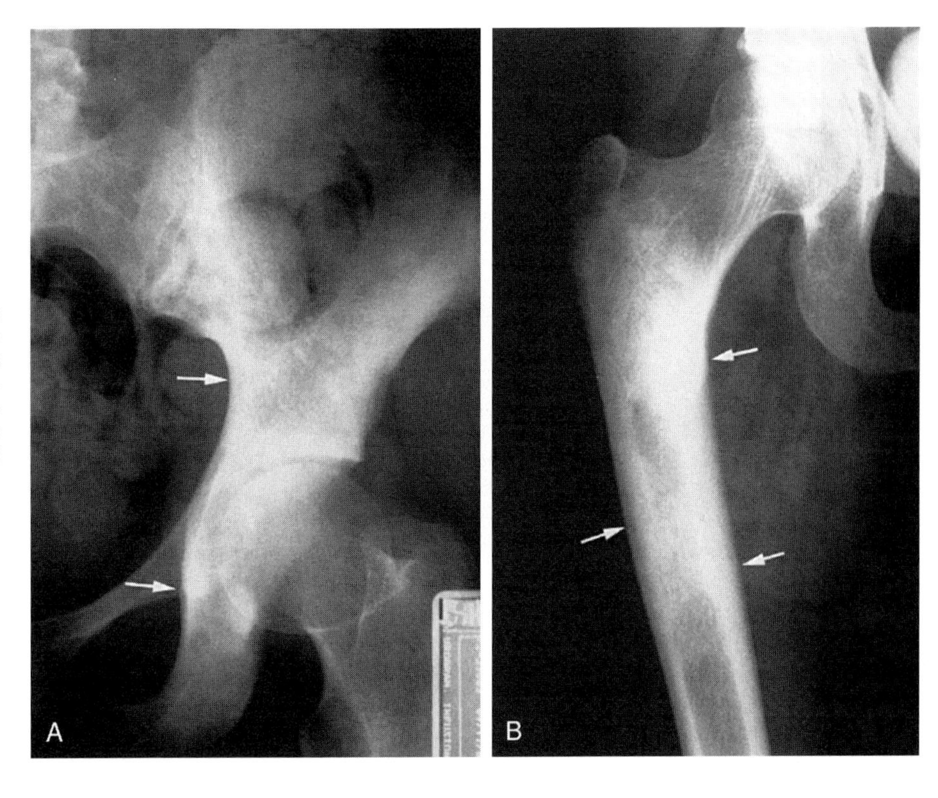

Figure 12–25. A 43-year-old woman with stage IV Hodgkin's disease. *A*, Anteroposterior view of the left hip shows densely sclerotic bone in the acetabulum and left iliac bone *(arrows)*. *B*, Anteroposterior view of the right femur shows similar changes in the proximal shaft *(arrows)*.

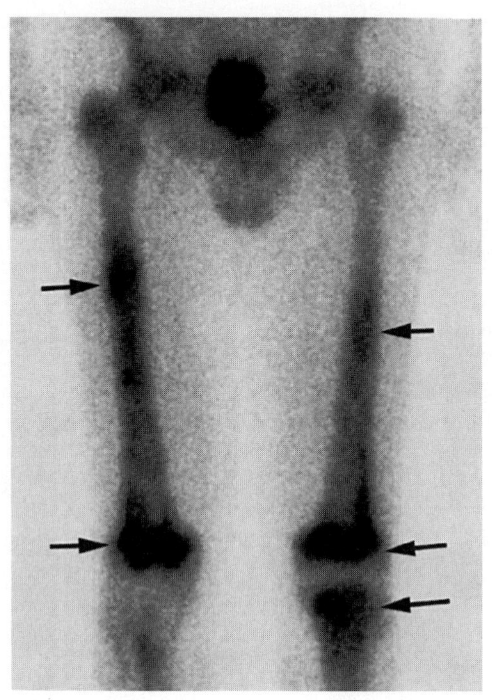

Figure 12–26. A 63-year-old man with B-cell lymphoma of Meckel's cave treated with radiation therapy who was thought to be in remission for 1 year, now presenting with pain in both thighs and shoulders. Technetium 99m MDP bone scan reveals increased radionuclide uptake in both femora and the left tibia *(arrows)*. The lesions are mainly in the diaphyses and metaphyses. This proved to be recurrent lymphoma at biopsy.

soft tissue masses, which are essentially affected lymph nodes, can be detected easily on CT. Large lymph nodes can erode into vertebral bodies, producing destruction of the anterior or lateral vertebral cortex.

Non-Hodgkin's Lymphoma

Overview

In contrast to HD, 5% to 25% of NHL is extranodal in origin, and 4% of all patients with NHL present with a skeletal lesion. The terminology for NHL of bone is confusing. At various times, NHL has been called *reticulum cell sarcoma, lymphosarcoma, large cell lymphoma,* and *histiocytic lymphoma.* Primary bone lymphomas account for less than 5% of all primary bone tumors. They have better prognosis than most malignant bone tumors. For systemic NHL there is currently a simple, widely used classification based on the immunochemical characteristics of the tumor: Low-grade type (40%); intermediate type (40%); high-grade type (5% to 10%). Most NHL are B-cell lymphomas. Patients with an impaired immune system, such as organ transplant patients or AIDS patients, have a significant increase in the incidence of NHL. Most of such cases are extranodal, and a small percentage arises in the skeletal system.

Skeletal Involvement

NHL most frequently occurs in the long bones of the lower extremities, followed by the pelvis. It arises in the metaphysis or diaphysis and rarely in the epiphysis (Figs. 12–26 and 12–27). Bone involvement in NHL is classified into three types, which has a significant bearing on the prognosis and treatment.

Primary Solitary Non-Hodgkin's Lymphoma of Bone. This type of NHL is always a single lesion with no evidence of disease elsewhere after a thorough diagnostic work-up (Figs. 12–28 and 12–29). It arises from a lymphomatous focus within the bone. Primary lymphoma of bone occurs in a younger age group (mean age, 49 years) and has a better prognosis than nodal or visceral

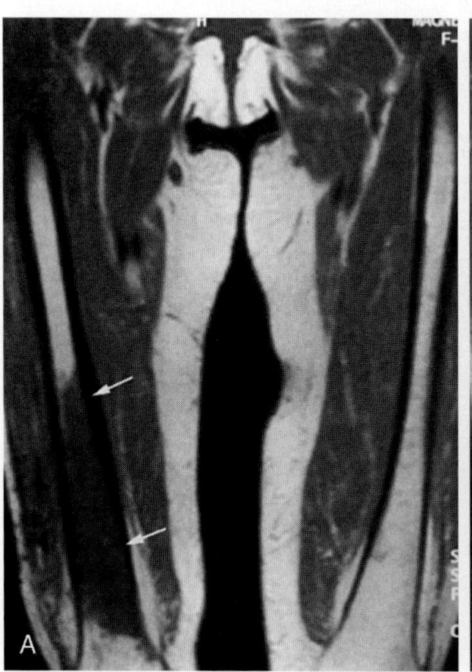

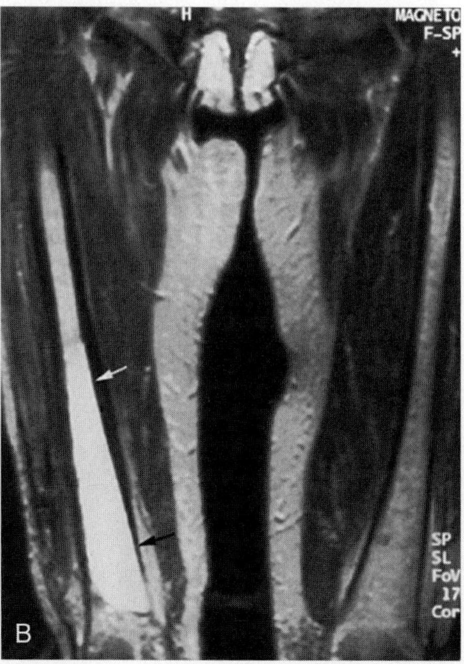

Figure 12–27. A 69-year-old woman with large cell lymphoma of the right femur. *A* and *B*, T1- and T2-weighted coronal images reveal a 12-cm lesion in the distal metadiaphysis of the right femur *(arrows)*, which is confined to the marrow space.

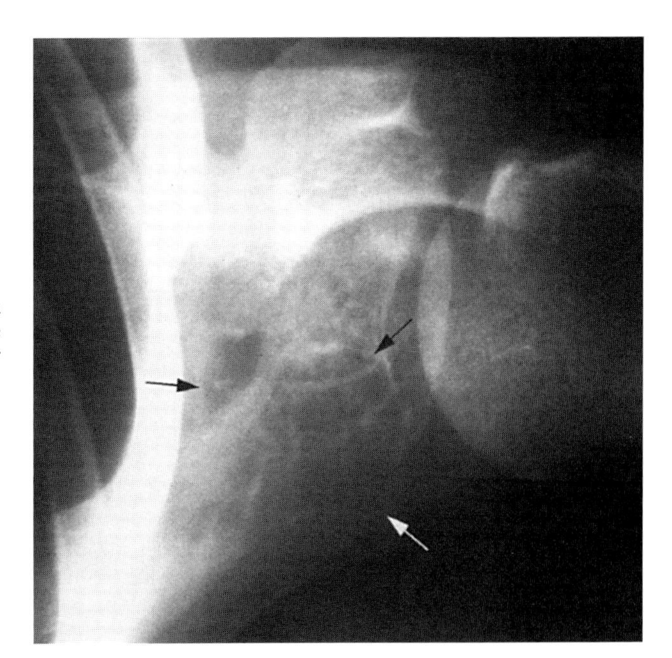

Figure 12–28. Primary solitary non-Hodgkin's lymphoma of bone in a 55-year-old man. Anteroposterior view of right shoulder shows a lytic expansile lesion involving the inferior portion of the glenoid and scapular neck *(arrows)*.

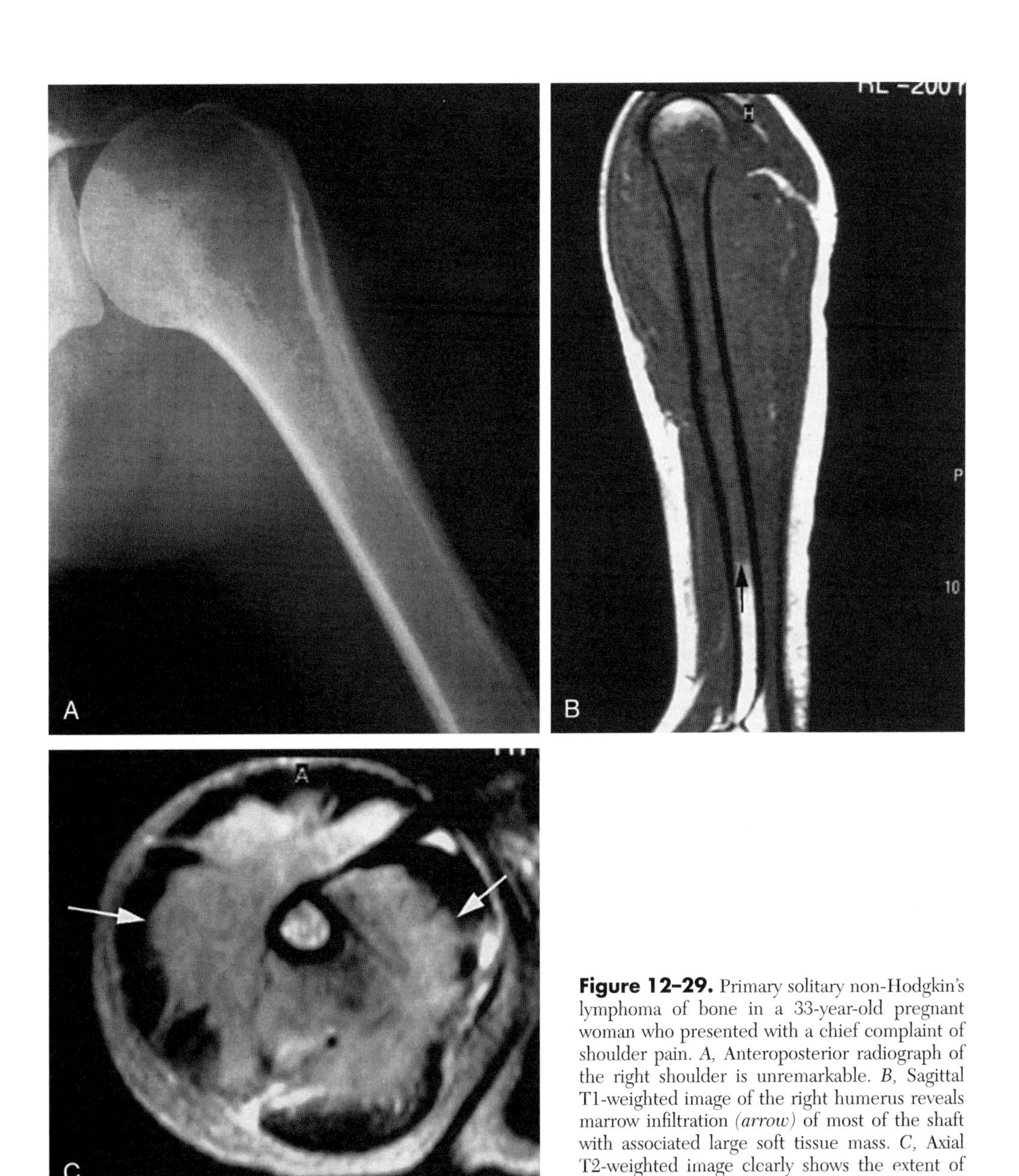

Figure 12–29. Primary solitary non-Hodgkin's lymphoma of bone in a 33-year-old pregnant woman who presented with a chief complaint of shoulder pain. *A*, Anteroposterior radiograph of the right shoulder is unremarkable. *B*, Sagittal T1-weighted image of the right humerus reveals marrow infiltration *(arrow)* of most of the shaft with associated large soft tissue mass. *C*, Axial T2-weighted image clearly shows the extent of the soft tissue mass *(arrows)*.

lymphomas. Of all malignant bone tumors, 3% to 5% are primary lymphomas of bone. Despite the aggressive appearance of this lesion, it has a relatively good prognosis compared with other primary malignant tumors of bone.

Primary Multifocal Non-Hodgkin's Lymphoma of Bone. This type of bone lymphoma has a peculiar distribution with a predilection for the skull and the bones around the knee. This unique distribution can be used to differentiate this entity from metastasis, which is uncommon around the knee and in the skull. With primary multifocal NHL of bone, distant nodal and visceral disease is absent, and the prognosis tends to be good. The 5-year survival for this type of NHL is about 42%.

Osseous Lesion Associated with Distant Disease. NHL lesions associated with distant nodal or visceral disease automatically make the disease process advanced (stage IV). Of patients with systemic NHL, 5% have bone involvement. The prognosis is poor, and the 5-year survival is approximately 22%.

Imaging

RADIOGRAPHY, COMPUTED TOMOGRAPHY, AND RADIONUCLIDE BONE SCAN

The appearance of NHL on radiographs and CT scans is nonspecific. Plain radiographs can occasionally be normal or near normal when the bone is positive. This appearance should be a clue that the lesion could be an NHL (Fig. 12–29). The lesions are predominantly lytic with a permeative pattern (Fig. 12–30). They typically occur in the metadiaphyses of long bones. In 43% of cases, the lesions are sclerotic. In approximately 16% of cases, the lesions are mixed (Fig. 12–31). Typically the lesions have a broad zone of transition, cortical destruction, and extension into the soft tissues (see Fig. 12–31). A sequestrum also may be present within the lesion.

MAGNETIC RESONANCE IMAGING

NHL is one of the rare lesions that occasionally can show low signal intensity on T1- and T2-weighted images; this is believed to be due to the extensive fibrosis within the tumor. Usually, NHL is characterized, however, by low signal intensity on T1-weighted images and variable signal intensity on T2-weighted images (see Figs. 12–27, 12–29, and 12–31). The suspicion for NHL should be raised when there is marrow involvement on MRI along with large soft tissue but minimal cortical destruction in a patient in the third or fourth decade of life (see Fig. 12–29).

Positron Emission Tomography Scanning

Currently positron emission tomography (PET) scanning ([^{18}F]-FDG) is used in the initial staging of NHL. Unsuspected disease is discovered in 25% of patients, and this impacts therapy.

Differential Diagnosis

The differential diagnosis includes lytic osteosarcoma, fibrosarcoma, multiple myeloma, and metastasis.

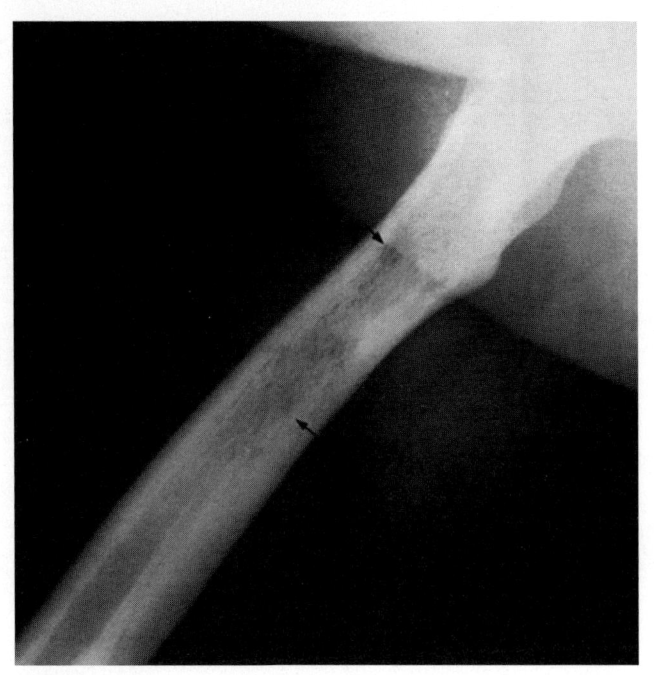

Figure 12–30. Non-Hodgkin's lymphoma of the left femur showing a permeative pattern *(arrows)*.

ADULT LEUKEMIAS

Chronic Lymphocytic Leukemia

In adults with leukemia, radiographically detectable bone lesions are not as common as in children with leukemia. The most common leukemia in adults is chronic lymphocytic leukemia. The proliferation and accumulation of neoplastic B lymphocytes in the blood, bone marrow, lymph nodes, and spleen characterize the disease. The disease typically affects the elderly, and survival varies between 1 and 10 years. Radiographically, bone involvement manifests as diffuse osteopenia. Extent of marrow involvement is well demonstrated by MRI (Fig. 12–32).

Chronic Myelogenous Leukemia

The second major leukemia affecting adults is chronic myelogenous leukemia, which may remain indolent; however, 25% of patients with chronic myelogenous leukemia convert to acute myelogenous leukemia each year, which is therapeutically difficult to control. As with most types of leukemia, diffuse marrow infiltration usually is present with myelogenous leukemia; this is shown on MR imaging as dark marrow on T1- and T2-weighted images (Fig. 12–33). Chronic myelogenous leukemia also can resemble multiple myeloma when it causes multiple, focal, lytic lesions in the axial skeleton.

A chloroma (granulocytic sarcoma) is an uncommon complication of acute and chronic myelogenous leukemia (Fig. 12–33*B*, *C*). Chloromas occur in 3% to 9% of patients with acute myelogenous leukemia. It represents a solid tumor consisting of myeloid precursors. Chloromas involv-

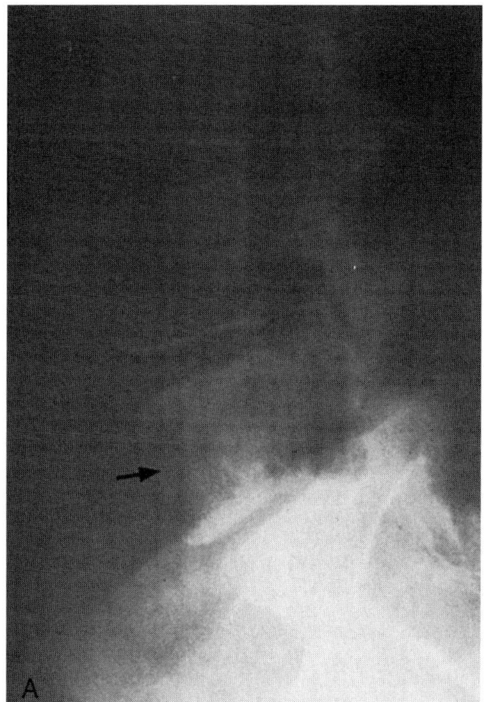

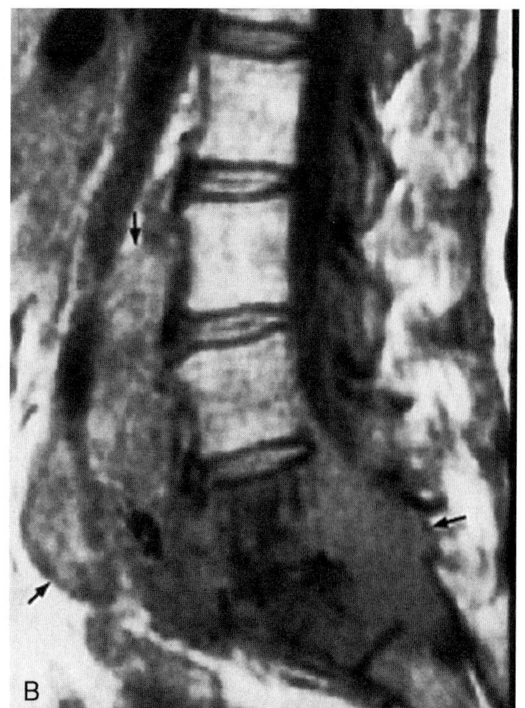

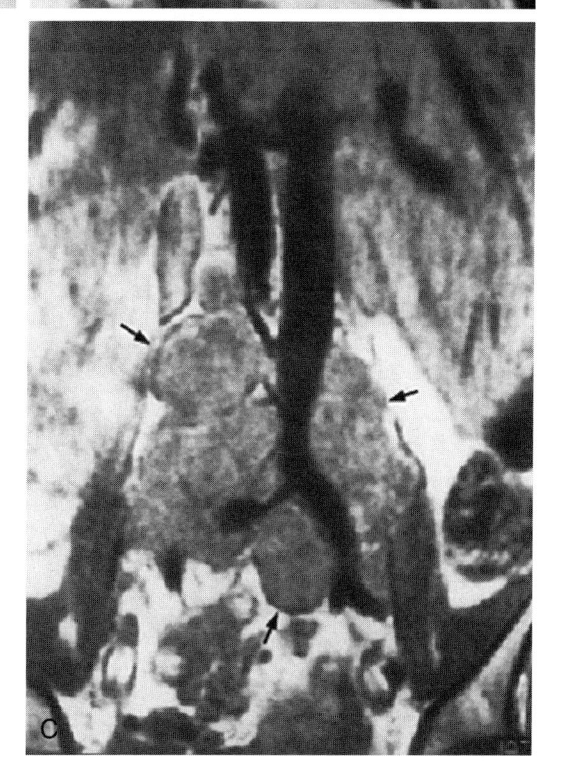

Figure 12–31. Non-Hodgkin's lymphoma of the spine. Mixed sclerotic and lytic pattern involving L5 and upper sacrum with extension into the para-aortic lymph nodes. *A,* Lateral view of the lumbosacral spine shows a lytic-blastic lesion of L5 *(arrow)*. *B,* T2-weighted sagittal MR image shows a large soft tissue mass *(arrows)* associated with the destructive lesion of L5. The lesion also involves S1 and the spinal canal. *C,* The soft tissue mass extends along the para-aortic lymph nodes *(arrows)*.

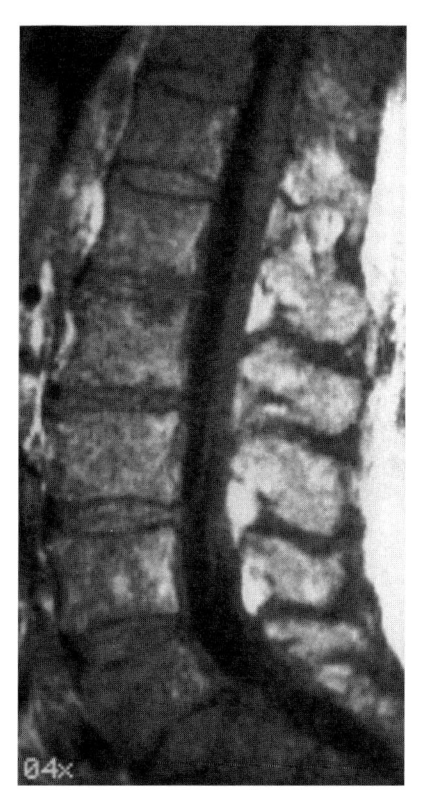

Figure 12–32. Sagittal T2-weighted MR image of the lumbar spine of an 82-year-old woman with chronic lymphocytic leukemia shows diffuse marrow infiltration throughout the spine.

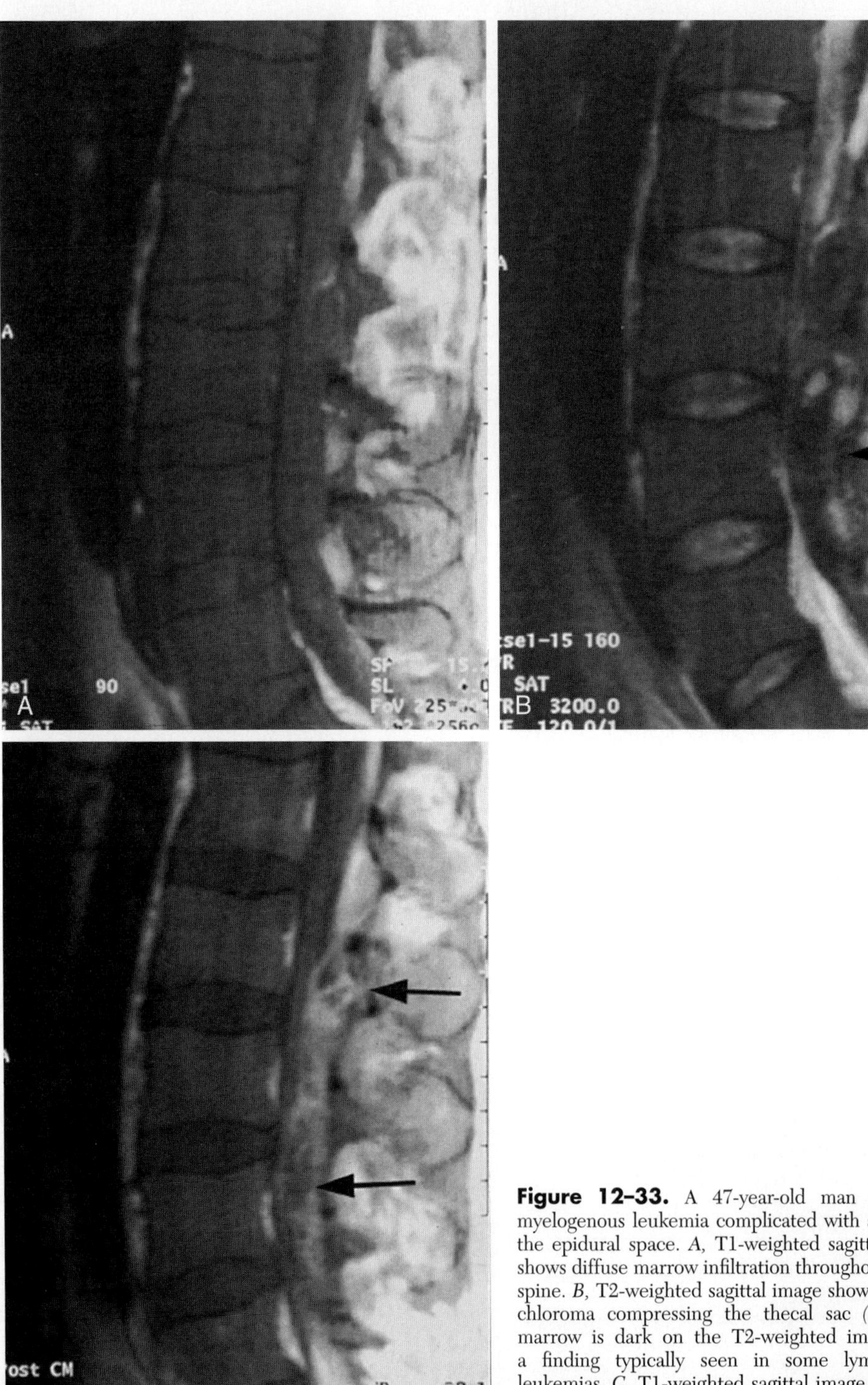

Figure 12–33. A 47-year-old man with chronic myelogenous leukemia complicated with a chloroma in the epidural space. *A,* T1-weighted sagittal MR image shows diffuse marrow infiltration throughout the lumbar spine. *B,* T2-weighted sagittal image shows the epidural chloroma compressing the thecal sac *(arrows).* The marrow is dark on the T2-weighted image, which is a finding typically seen in some lymphomas and leukemias. *C,* T1-weighted sagittal image after gadolinium enhancement shows the epidural chloroma well *(arrows).*

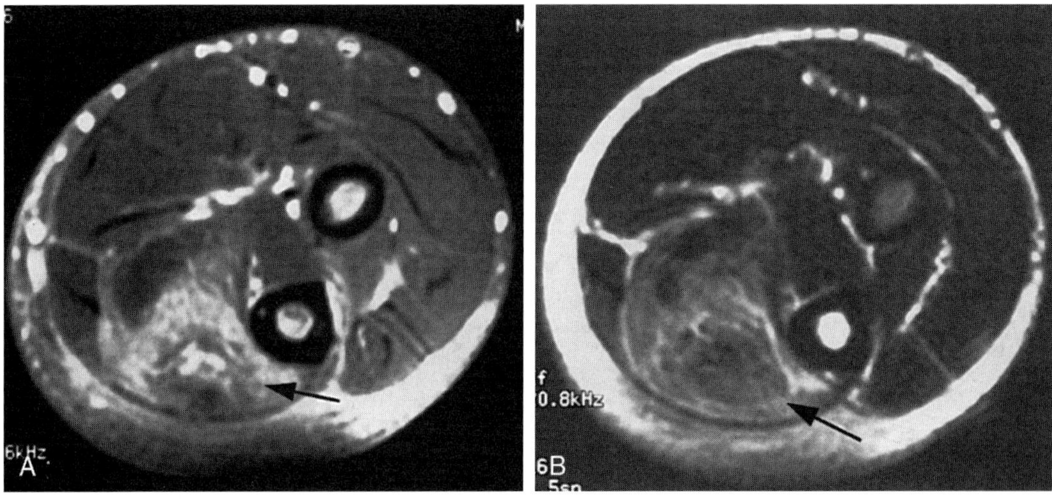

Figure 12–34. Chloroma in the flexor muscles of the forearm in a 57-year-old man with acute myelogenous leukemia. *A,* T2-weighted axial image shows mildly increased signals in the flexor digitorum profundus muscle group *(arrow). B,* T1-weighted axial image after gadolinium enhancement shows the chloroma *(arrow),* which subsequently was irradiated.

ing bone appear as expansile, lytic lesions with ill-defined margins. The lesion is associated with a soft tissue mass (Fig. 12–34). The sacrum, spine, sternum, and ribs are the most common sites for chloromas. Most of these tumors are found in association with bone or nervous tissue. Chloromas are named after their green color, which is attributed to the enzyme myeloperoxidase present in leukemia cells.

Recommended Readings

Aggarwal S, Goulatia RK, Sood A, et al: POEMS syndrome: A rare variety of plasma cell dyscrasia. AJR Am J Roentgenol 155:339, 1990.

Andrew SM, Freemont AJ: Skeletal mastocytosis. J Clin Pathol 46:1033, 1993.

Angervall L, Enzinger FM: Extraskeletal neoplasm resembling Ewing's sarcoma. Cancer 36:240, 1975.

Austin CB, Young JW, Park HJ, et al: Massive acroosteolysis in adult T-cell leukemia/lymphoma. Radiology 164:787, 1987.

Avila NA, Ling A, Metcalfe DD, et al: Mastocytosis: Magnetic resonance imaging patterns of marrow disease. Skeletal Radiol 27:119, 1998.

Bataille R, Chevalier J, Rossi M, et al: Bone scintigraphy in plasma-cell myeloma: A prospective study of 70 patients. Radiology 145:801, 1982.

Beltran J, Aparisi F, Bonmati LM, et al: Eosinophilic granuloma: MR imaging manifestations. Skeletal Radiol 22:157, 1992.

Bragg DG, Colby TV, Ward JH: New concepts in the non-Hodgkin lymphomas: Radiologic implications. Radiology 159:291, 1986.

Castellino RA: Hodgkin disease: Practical concepts for the diagnostic radiologist. Radiology 159:305, 1986.

David R, Oria RA, Kumar R, et al: Radiologic features of eosinophilic granuloma of bone. AJR Am J Roentgenol 153:1021, 1989.

de Gennes C, Kuntz D, de Vernejoul MC: Bone mastocytosis: A report of nine cases with a bone histomorphometric study. Clin Orthop 279:281, 1992.

Delsignore JL, Dvoretsky PM, Hicks DG, et al: Mastocytosis presenting as a skeletal disorder. Iowa Orthop J 16:126, 1996.

De Schepper AM, Ramon F, Van Marck E: MR imaging of eosinophilic granuloma: Report of 11 cases. Skeletal Radiol 22:163, 1993.

Eustace S, O'Regan R, Graham D, et al: Primary multifocal skeletal Hodgkin's disease confined to bone. Skeletal Radiol 24:61, 1995.

Geetha N, Jayaprakash M, Rekhanair A, et al: Plasma cell neoplasms in the young. Br J Radiol 72:1012, 1999.

Genez BM, Zirilli VL, Schlesinger AE, et al: Case report 487: Primary lymphoma of radius. Skeletal Radiol 17:306, 1988.

Grieser T, Minne HW: Systemic mastocytosis and skeletal lesions. Lancet 350:1103, 1997.

Hall FM, Gore SM: Osteosclerotic myeloma variants. Skeletal Radiol 17:101, 1988.

Haney K, Russell W, Raila FA, et al: MR imaging characteristics of systemic mastocytosis of the lumbosacral spine. Skeletal Radiol 25:171, 1996.

Hayes CW, Conway WF, Sundaram M: Misleading aggressive MR imaging appearance of some benign musculoskeletal lesions. Radiographics 12:1119, 1992.

Hermann G, Abdelwahab IF, Capozzi J, et al: Primary non-Hodgkin lymphoma of bone: Unusual manifestation of lymphoproliferative disease following liver transplantation. Skeletal Radiol 28:175, 1999.

Hermann G, Feldman F, Abdelwahab IF, et al: Skeletal manifestations of granulocytic sarcoma (chloroma). Skeletal Radiol 20:509, 1991.

Hermann G, Sherry H, Rabinowitz JG: Case report 151: Solitary plasmacytoma associated with peripheral neuropathy. Skeletal Radiol 6:217, 1981.

Herold CJ, Wittich GR, Schwarzinger I, et al: Skeletal involvement in hairy cell leukemia. Skeletal Radiol 17:171, 1988.

Hindman BW, Thomas RD, Young LW, et al: Langerhans cell histiocytosis: Unusual skeletal manifestations observed in thirty-four cases. Skeletal Radiol 27:177, 1998.

Holland J, Trenkner DA, Wasserman TH, et al: Plasmacytoma: Treatment results and conversion to myeloma. Cancer 69:1513, 1992.

Ishida T, Dorfman HD: Plasma cell myeloma in unusually young patients: A report of two cases and review of the literature. Skeletal Radiol 24:47, 1995.

Kaplan GR, Saifuddin A, Pringle JA, et al: Langerhans' cell histiocytosis of the spine: Use of MR imaging in guiding biopsy. Skeletal Radiol 27:673, 1998.

Klein MJ, Rudin BJ, Greenspan A, et al: Hodgkin disease presenting as a lesion in the wrist: A case report. J Bone Joint Surg Am 69:1246, 1987.

Laufer L, Benharroch D, Giryes H, et al: Pelvic granulocytic sarcoma. Skeletal Radiol 25:693, 1996.

Lavallee G, Lemarbre L, Bouchard R, et al: Ewing's sarcoma in adults. J Can Assoc Radiol 30:223, 1979.

Lecouvet FE, Vande Berg BC, Michaux L, et al: Early chronic lymphocytic leukemia: Prognostic value of quantitative bone marrow MR imaging findings and correlation with hematologic variables. Radiology 204:813, 1997.

Libshitz HI, Malthouse SR, Cunningham D, et al: Multiple myeloma: Appearance at MR imaging. Radiology 182:833, 1992.

Malloy PC, Fishman EK, Magid D: Lymphoma of bone, muscle, and skin: CT findings. AJR Am J Roentgenol 159:805, 1992.
Melhem RE, Saber TJ: Erosion of the medial cortex of the proximal humerus. Radiology 137:77, 1980.
Meyer JS, Harty MP, Mahboubi S, et al: Langerhans cell histiocytosis: Presentation and evolution of radiologic findings with clinical correlation. Radiographics 15:1135, 1995.
Mirra J, Gold R: Eosinophilic granuloma. In Mirra J (ed): Bone Tumors: Clinical, Radiologic and Pathologic Correlations. Philadelphia, Lea & Febiger, 1989, pp 1021–1039.
Moser Jr RP, Davis MJ, Gilkey FW, et al: Primary Ewing sarcoma of rib. Radiographics 10:899, 1990.
Mueller DL, Grant RM, Riding MD, et al: Cortical saucerization: An unusual imaging finding of Ewing sarcoma. AJR Am J Roentgenol 163:401, 1994.
Mulligan ME, Kransdorf MJ: Sequestra in primary lymphoma of bone: Prevalence and radiologic features. AJR Am J Roentgenol 160:1245, 1993.
Newcomer LN, Silverstein MB, Cadman EC, et al: Bone involvement in Hodgkin's disease. Cancer 49:338, 1982.
Reinus WR, Gilula LA, Shirley SK, et al: Radiographic appearance of Ewing sarcoma of the hands and feet: Report from the Intergroup Ewing Sarcoma Study. AJR Am J Roentgenol 144:331, 1985.
Reinus WR, Kyriakos M, Gilula LA: Plasma cell tumors with calcified amyloid deposition mistaken for chondrosarcoma. Radiology 189:505, 1993.
Resnick D, Greenway GD, Bardwick PA, et al: Plasma-cell dyscrasia with polyneuropathy, organomegaly, endocrinopathy, M-protein, and skin changes: The POEMS syndrome: Distinctive radiographic abnormalities. Radiology 140:17, 1981.
Resnick D, Greenway G, Genant H, et al: Erdheim-Chester disease. Radiology 142:289, 1982.
Roca M, Mota J, Giraldo P, et al: Systemic mastocytosis: MR imaging of bone marrow involvement. Eur Radiol 9:1094, 1999.
Rogalsky RJ, Black CB, Reed MH: Orthopaedic manifestations of leukemia in children. J Bone Joint Surg Am 68:494, 1986.
Schabel SI, Tyminski L, Holland RD, et al: The skeletal manifestations of chronic myelogenous leukemia. Skeletal Radiol 5:145, 1980.
Schreiman JS, McLeod RA, Kyle RA, et al: Multiple myeloma: Evaluation by CT. Radiology 154:483, 1985.
Sessa S, Sommelet D, Lascombes P, et al: Treatment of Langerhans-cell histiocytosis in children: Experience at the Children's Hospital of Nancy. J Bone Joint Surg Am 76:1513, 1994.
Shapeero LG, Vanel D, Sundaram M, et al: Periosteal Ewing sarcoma. Radiology 191:825, 1994.
Shirley SK, Gilula LA, Siegal GP, et al: Roentgenographic-pathologic correlation of diffuse sclerosis in Ewing sarcoma of bone. Skeletal Radiol 12:69, 1984.
Siegelman SS: Taking the X out of histiocytosis X. Radiology 204:322, 1997.
Stiglbauer R, Augustin I, Kramer J, et al: MR imaging in the diagnosis of primary lymphoma of bone: Correlation with histopathology. J Comput Assist Tomogr 16:248, 1992.
Sullivan WT, Solonick DM: Case report 414: Nodular sclerosing Hodgkin disease involving sternum and chest wall. Skeletal Radiol 16:166, 1987.
Van de Berg BC, Lecouvet FE, Michaux L, et al: Stage I multiple myeloma: Value of MR imaging of the bone marrow in the determination of prognosis. Radiology 201:243, 1996.
Van de Berg BC, Michaux L, Lecouvet FE, et al: Nonmyelomatous monoclonal gammopathy: Correlation of bone marrow MR images with laboratory findings and spontaneous clinical outcome. Radiology 202:247, 1997.
Vermess M, Pearson KD, Einstein AB, et al: Osseous manifestations of Waldenström's macroglobulinemia. Radiology 102:497, 1972.
White LM, Schweitzer ME, Khalili K, et al: MR imaging of primary lymphoma of bone: Variability of T2-weighted signal intensity. AJR Am J Roentgenol 170:1243, 1998.

CHAPTER 13

Benign Miscellaneous Tumors

Nabil J. Khoury, MD, Georges Y. El-Khoury, MD, and D. Lee Bennett, MD

GIANT CELL TUMOR

Overview

Giant cell tumor (GCT) is a fairly common bone tumor, representing about 5% of all primary bone tumors and 21% of benign bone tumors. Although benign, GCT can be locally aggressive and may metastasize. Histologically, GCT is characterized by uniformly distributed multinucleated giant cells in a stroma of mononuclear cells. Most GCTs (80%) present between the ages of 15 and 40 years. GCT rarely presents before physeal closure. The lesion is slightly more predominant in females. In the past, 30% of GCTs were thought to be malignant. The current thinking among bone pathologists is that primary or *de novo* malignant GCTs are rare. What were thought to be malignant GCTs most likely represented malignant fibrous histiocytomas or giant cell–rich variants of either fibrosarcoma or osteosarcoma. The older literature also reports a high rate of malignant transformation in patients who received radiation. When radical surgery is difficult or unfeasible, however, such as in the spine, modern radiotherapeutic techniques using megavoltage photons or proton beams have proved to be effective in controlling the tumor, and the risk of malignant transformation is low.

Skeletal Involvement

The lesions typically are located at the ends of long bones with most being around the knee, followed by the distal radius and the proximal humerus (Figs. 13–1 and 13–2). In the long bones, the lesion is subarticular. Rarely, GCTs occur in the metaphysis, and in such patients the physis is still open. The spine is the fourth leading location for GCT occurrence, with most lesions occurring in the sacrum (Fig. 13–3). GCT is the most common benign lesion of the sacrum, where it is usually eccentric and can cross the sacroiliac joint. In other parts of the spine, the lesion almost always involves the vertebral body (Fig. 13–4). GCTs rarely are multicentric. The lesions can present simultaneously or, more commonly, separated by months or years. It is not clear if these multiple lesions represent metastasis.

Imaging

Radiography

Radiographically, GCT has a distinctive appearance as an expansile lytic lesion without a matrix. The lesion blends imperceptibly with the healthy neighboring bone owing to lack of junctional sclerosis. In the long bones, GCT almost always is eccentric and extends to the subchondral bone (see Figs. 13–1 and 13–2). Occasionally the tumor appears mildly trabeculated (Fig. 13–5); it can appear aggressive, causing cortical destruction (see Figs. 13–1*A* and 13–2). Periosteal reaction is usually absent. Soft tissue recurrence after surgery typically presents as a soft tissue mass with or without rim calcifications (Fig. 13–6). The same is true for lung metastasis.

Computed Tomography

Computed tomography (CT) is useful in determining the integrity of the cortex (see Figs. 13–3*B* and 13–4*B*).

Magnetic Resonance Imaging

Magnetic resonance (MR) imaging can delineate the intramedullary and soft tissue extent of the lesion (see Fig. 13–4). MR imaging also can show the presence of fluid-fluid levels, which are a feature in other lesions, such as aneurysmal bone cyst, and telangiectatic osteosarcoma. The lesion has low-to-intermediate signal intensity on T1-weighted images and predominantly high signal intensity on T2-weighted images. After injection of gadolinium, the lesion usually shows heterogeneous signal enhancement (Fig. 13–7). Areas of low signal may be present on all sequences because of the presence of hemosiderin.

Complications

Older literature reports a high recurrence rate (50%). Modern surgical techniques using extended curettage, cryogenic agents, or heat associated with methyl methacrylate packing have resulted in recurrence rates of less than 10%. More radical surgery is reported to reduce the recurrence rate to almost 0%. Pathologic fractures attributed to cortical thinning may occur with GCTs. Benign

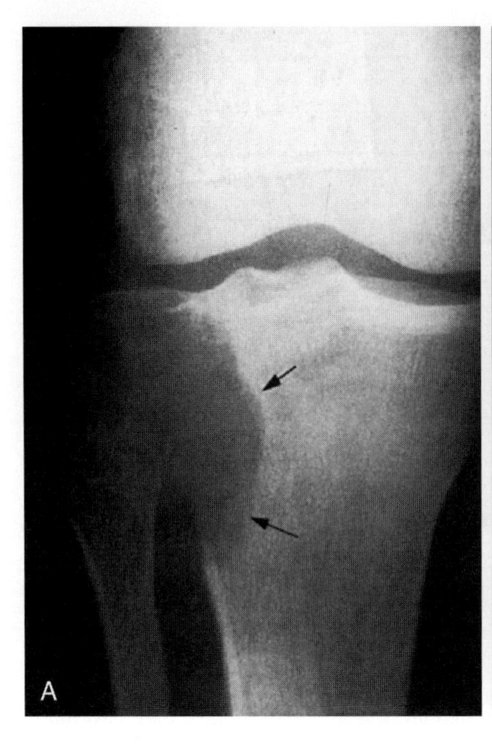

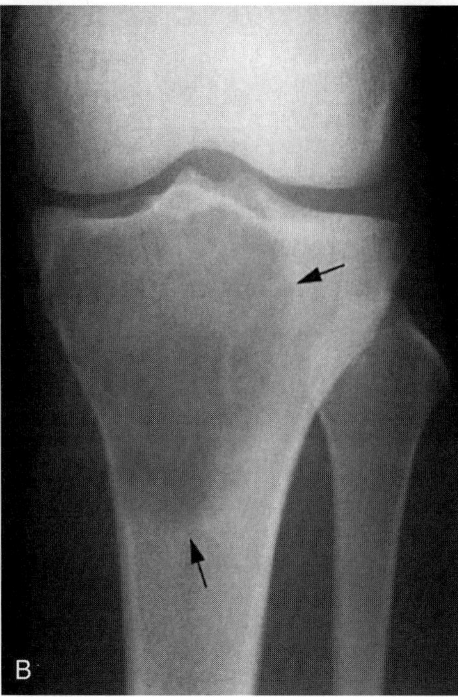

Figure 13–1. Typical giant cell tumors of the proximal tibia. *A,* Small giant cell tumor destroying the lateral tibial condyle. The lesion reaches the articular surface and has no cortical margin *(arrows).* The lateral cortex of the proximal tibia is destroyed. The lesion has no identifiable matrix. *B,* A large giant cell tumor from another patient *(arrows).* The radiographic features are typical.

metastasis, usually to the lung, is reported in 1% to 2% of patients. Most of these patients have a good prognosis with wedge resection. The paradoxical phenomenon of benign metastasis is not unique to GCT; it also is seen with leiomyoma of the uterus and chondroblastoma. Metastasis to mediastinal lymph nodes also has been described. The use of modern radiation therapy techniques has reduced significantly the risk of sarcomatous transformation.

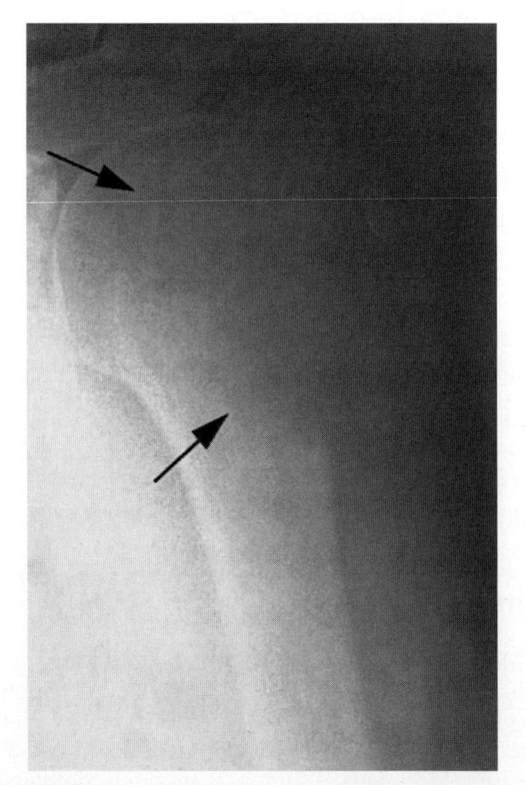

Figure 13–2. Giant cell tumor of the proximal humerus. The lesion reaches to the articular surface; it has no cortical margin (*arrows*) and no identifiable matrix.

Differential Diagnosis

The differential diagnosis of GCT is wide and includes aneurysmal bone cyst, chondroblastoma, telangiectatic osteosarcoma, malignant fibrous histiocytoma, and chondromyxoid fibroma. Brown tumors of hyperparathyroidism should be ruled out, especially when GCT is multicentric.

GIANT CELL REPARATIVE GRANULOMA

Overview

Giant cell reparative granuloma (GCRG) is an uncommon lesion with histologic and imaging features that overlap with other giant cell–producing lesions, such as GCT, aneurysmal bone cyst, and brown tumor. Originally the lesion was described by Jaffe as a lytic lesion of the mandible and maxilla. Histologically, similar lesions are seen in the hands (Figs. 13–8 and 13–9) and feet. It is unusual for GCRG to occur in other skeletal sites. GCRG can occur at any age up to the 60s, but the mean age is in the early to mid 20s. Histologically the lesion contains a fibrous stroma with spindle-shaped fibroblasts, varying amounts of collagen, hemorrhage, hemosiderin, giant cells, and mononuclear inflammatory cells. Newly formed bone or osteoid also is present.

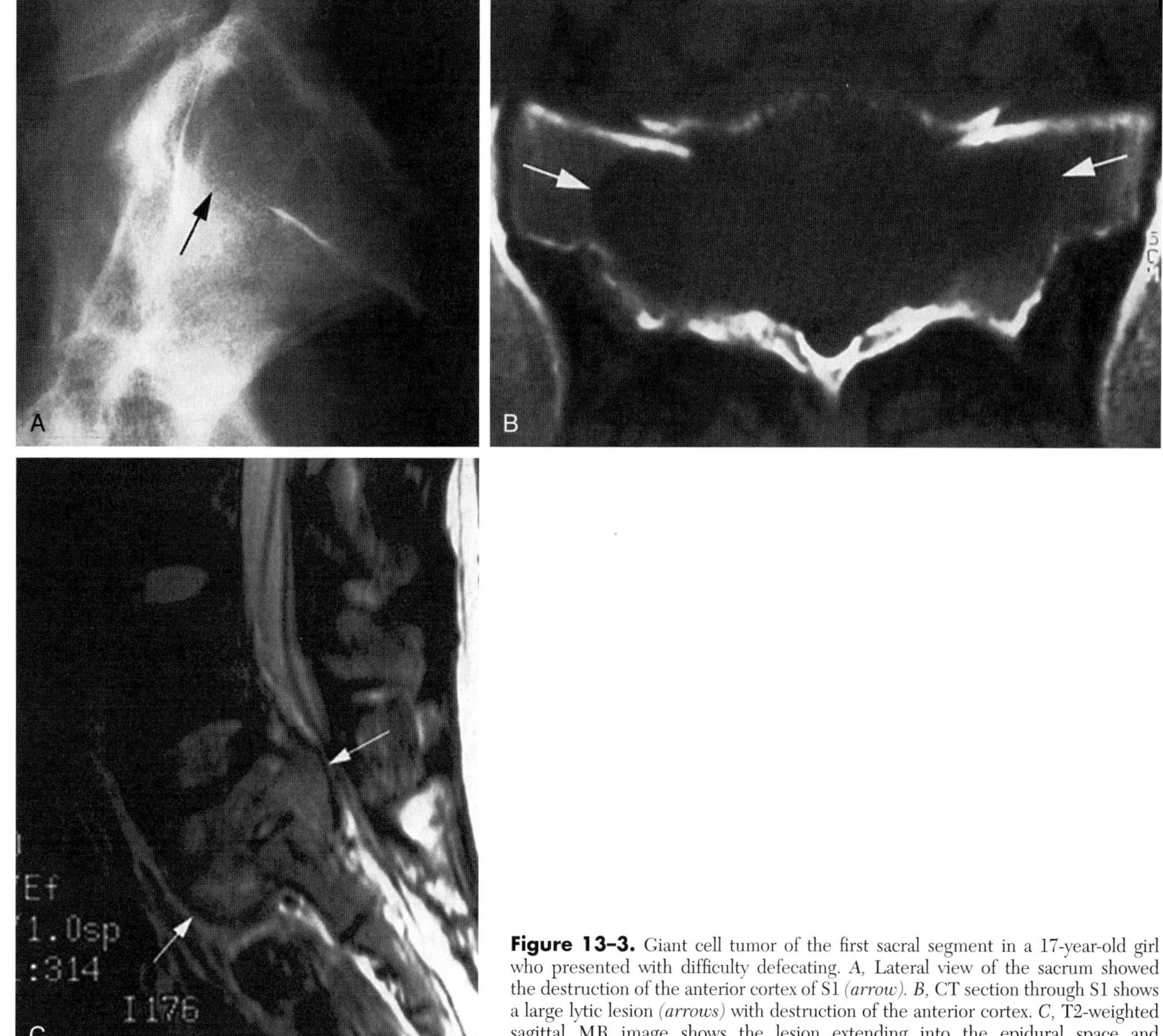

Figure 13–3. Giant cell tumor of the first sacral segment in a 17-year-old girl who presented with difficulty defecating. *A,* Lateral view of the sacrum showed the destruction of the anterior cortex of S1 *(arrow)*. *B,* CT section through S1 shows a large lytic lesion *(arrows)* with destruction of the anterior cortex. *C,* T2-weighted sagittal MR image shows the lesion extending into the epidural space and retroperitoneal soft tissues *(arrows)*.

Skeletal Involvement

In the hand, the lesions occur in the phalanges (see Fig. 13–9), whereas in the foot they typically occur in the metatarsals. Most lesions are metaphyseal or diaphyseal. Rarely a lesion may extend into the subarticular bone. In patients in whom the physis is still open, lesions occur in the metaphysis and do not cross the physis.

Imaging

GCRG should be suspected based on its location (hands, feet, mandible, and maxilla) and radiographic appearance. It presents as an expansile lytic lesion with thinning of the overlying cortex (see Figs. 13–8 and 13–9). It is rare for GCRG to extend into the soft tissues. The lesion does not contain matrix calcifications and usually is not associated with a periosteal reaction.

Differential Diagnosis

The differential diagnosis includes GCT, aneurysmal bone cyst, brown tumor, and enchondroma. GCT and aneurysmal bone cyst can resemble GCRG closely; however, GCTs usually seem more aggressive, and aneurysmal bone cysts rarely occur in the hands and feet. Enchondromas can be differentiated from GCRGs by the presence of characteristic cartilage matrix calcifications. Brown tumors can be multiple and are always associated with biochemical calcium and phosphorus abnormalities.

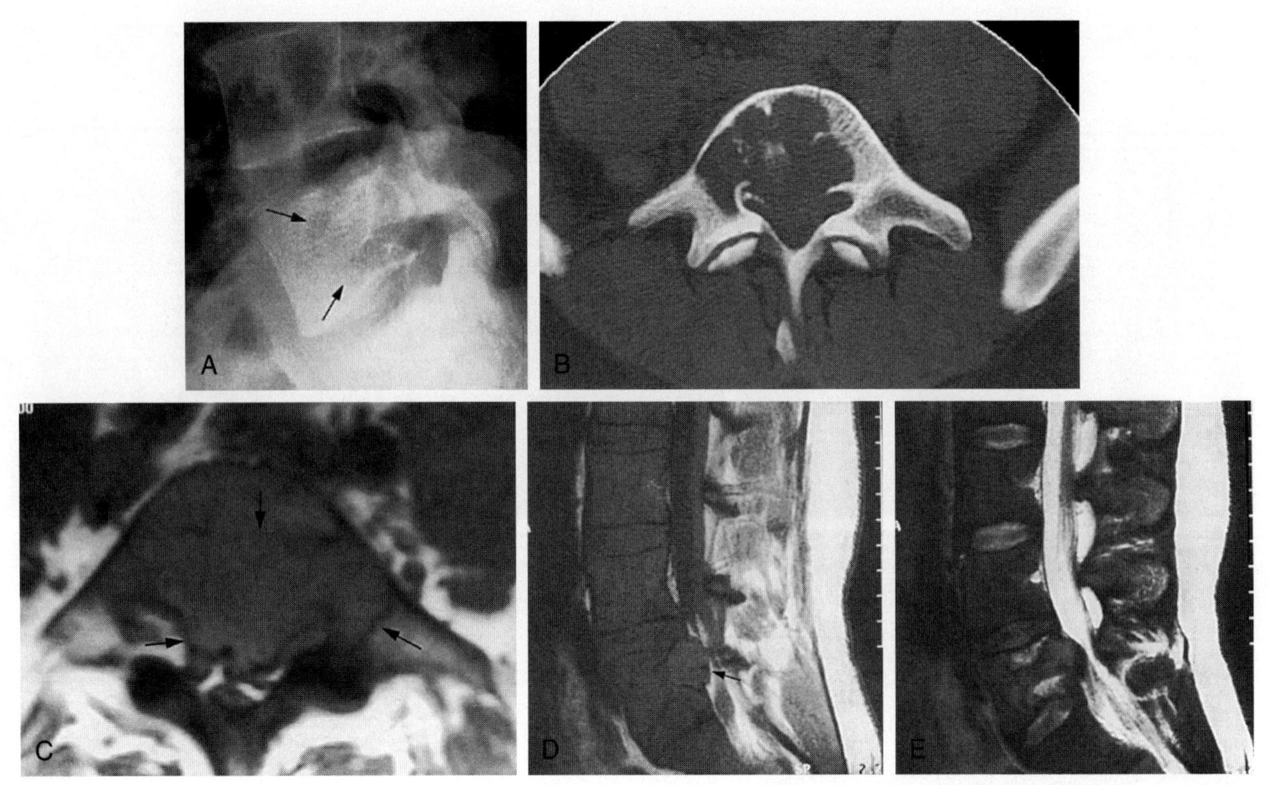

Figure 13–4. Giant cell tumor of L5 in a 22-year-old man. He presented with radicular symptoms and weakness in the lower extremities. *A,* Lateral view of the lumbar spine shows a lytic lesion in the body of L5 *(arrows). B,* CT scan shows the lytic lesion, which extends into the spinal canal. *C, D,* and *E,* T1-weighted axial and T1- and T2-weighted sagittal MR images show the extent of the lesion *(arrows)* and compression of the thecal sac.

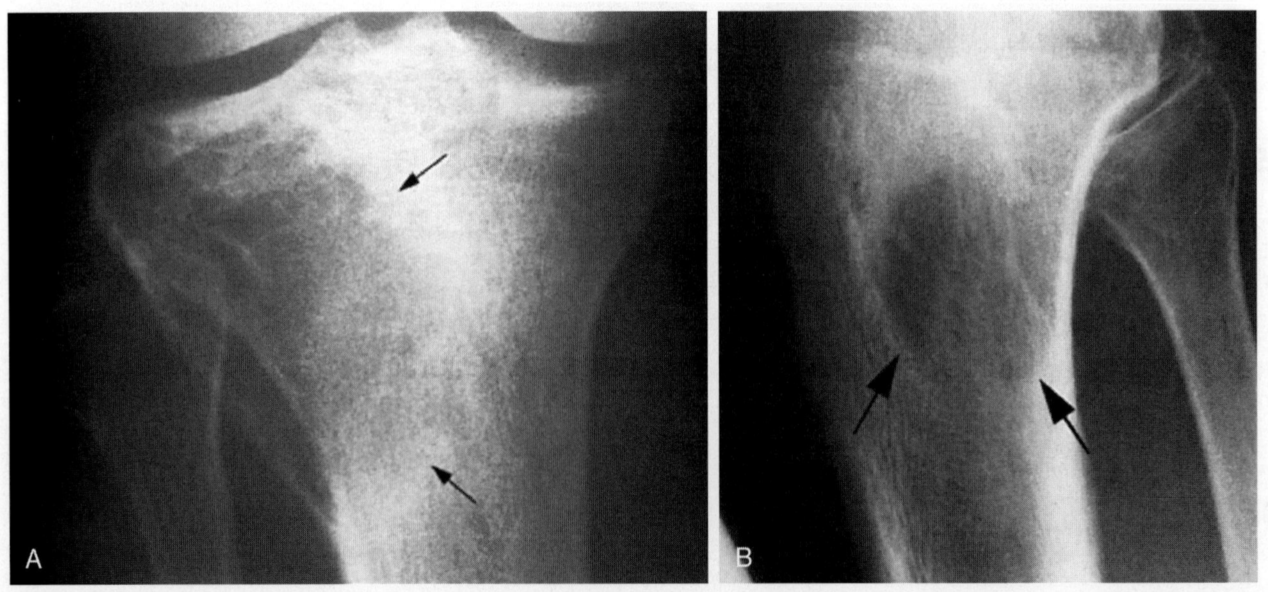

Figure 13–5. Mildly trabeculated giant cell tumor of the proximal tibia. *A* and *B,* Anteroposterior and lateral views of the right knee show a subchondral lytic lesion in the proximal tibia *(arrows).* Note the trabeculation in its superior portion.

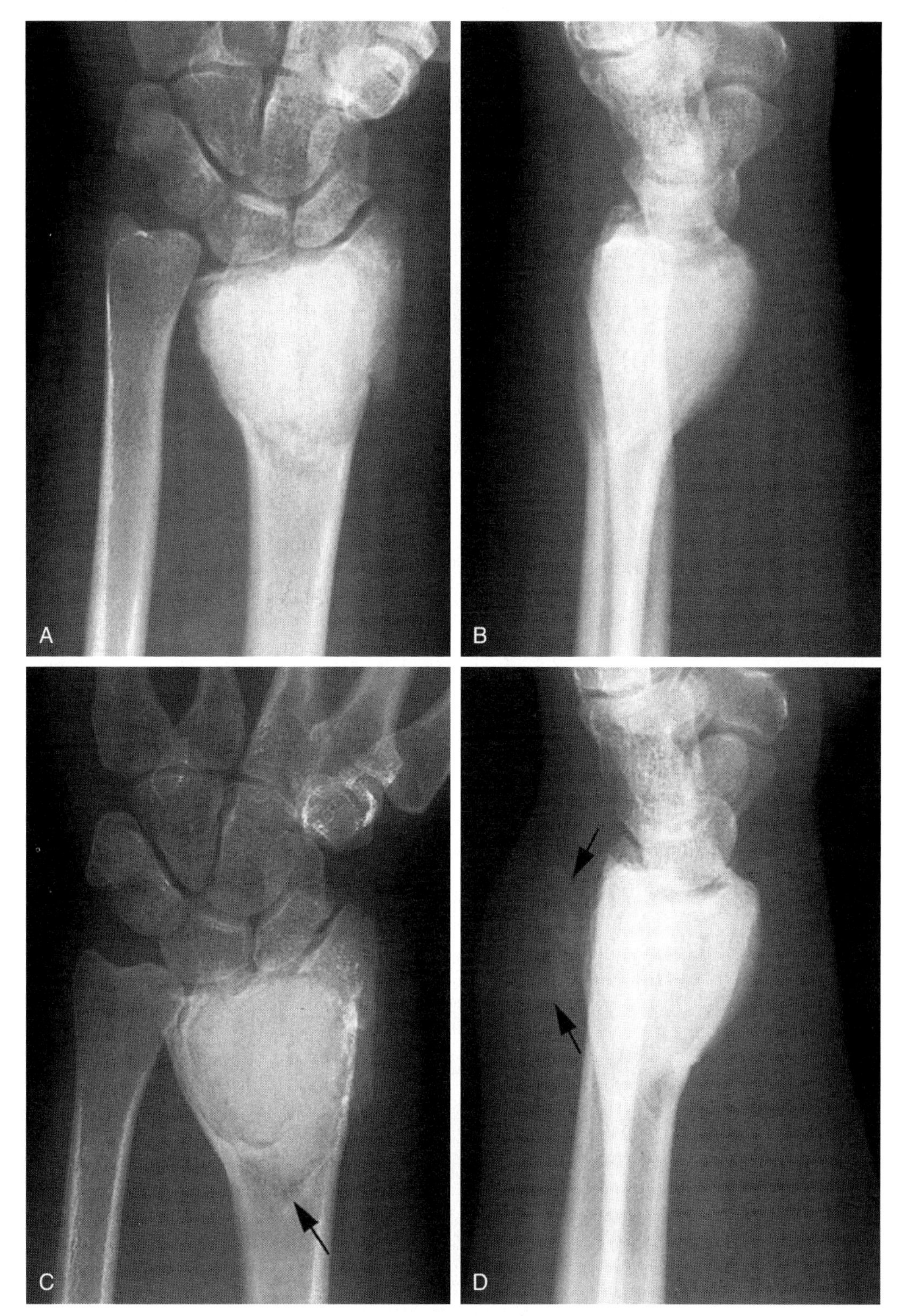

Figure 13–6. Soft tissue recurrence with rim calcifications. *A* and *B*, Giant cell tumor of the distal radius treated with curettage and packing with methyl methacrylate. No recurrence or complications were noted at this time. *C* and *D*, Posteroanterior and lateral views taken 8 months after *A* and *B*. There is a calcified mass (*arrows* in *D*) on the dorsum of the wrist and an area of bone destruction shown on the posteroanterior view (*arrow* in *C*).

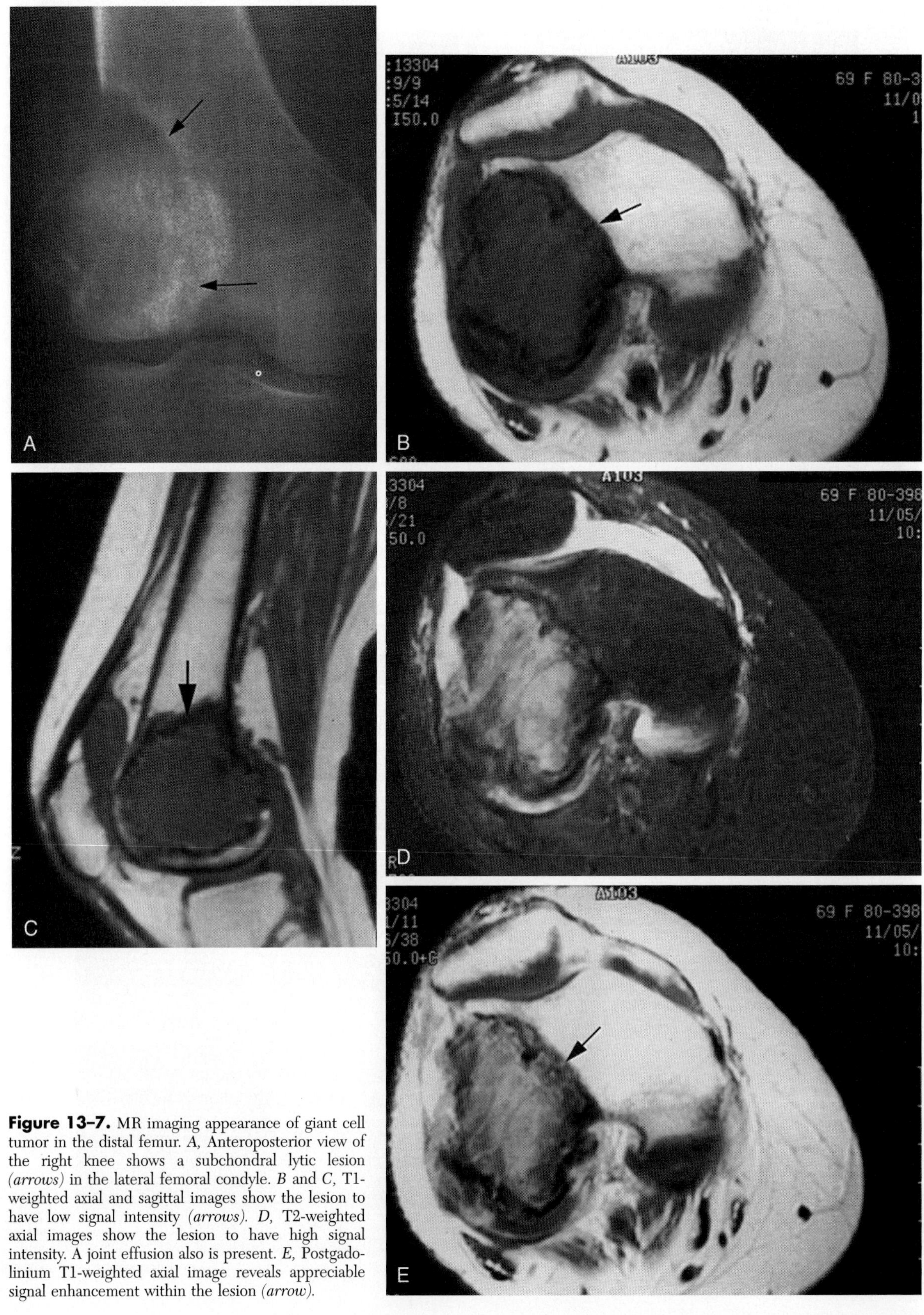

Figure 13–7. MR imaging appearance of giant cell tumor in the distal femur. *A,* Anteroposterior view of the right knee shows a subchondral lytic lesion *(arrows)* in the lateral femoral condyle. *B* and *C,* T1-weighted axial and sagittal images show the lesion to have low signal intensity *(arrows)*. *D,* T2-weighted axial images show the lesion to have high signal intensity. A joint effusion also is present. *E,* Postgadolinium T1-weighted axial image reveals appreciable signal enhancement within the lesion *(arrow)*.

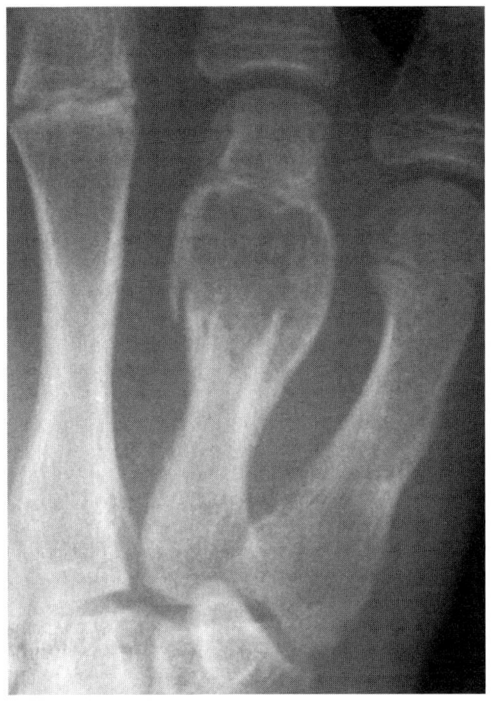

Figure 13–8. Giant cell reparative granuloma of the fourth metacarpal. The lesion is lytic and expansile. Periosteal reaction also is present.

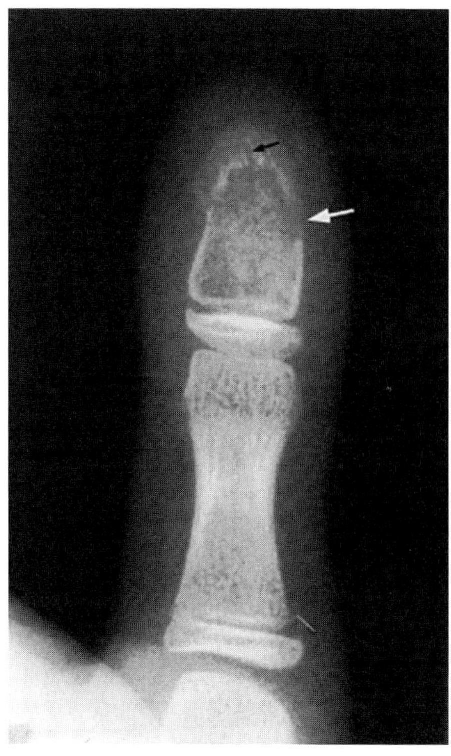

Figure 13–9. Giant cell reparative granuloma of the distal phalanx of the thumb. The lesion is lytic and expansile, occupying most of the distal phalanx *(white arrow)*. A pathologic fracture through the lesion also is present *(black arrow)*.

Recommended Readings

Bertoni F, Present D, Enneking WF: Giant-cell tumor of bone with pulmonary metastases. J Bone Joint Surg Am 67:890, 1985.

Campanacci M, Baldini N, Boriani S, et al: Giant-cell tumor of bone. J Bone Joint Surg Am 69:106, 1987.

Dahlin DC: Giant cell tumor of bone: Highlights of 407 cases. AJR Am J Roentgenol 144:955, 1985.

Herman SD, Mesgarzadeh M, Bonakdarpour A, et al: The role of magnetic resonance imaging in giant cell tumor of bone. Skeletal Radiol 16:635, 1987.

Hindman BW, Seeger LL, Stanley P, et al: Multicentric giant cell tumor: Report of five new cases. Skeletal Radiol 23:187, 1994.

Kaplan PA, Murphey M, Greenway G, et al: Fluid-fluid levels in giant cell tumors of bone: Report of two cases. J Comput Assist Tomogr 11:151, 1987.

Lee MJ, Sallomi DF, Munk PL, et al: Pictorial review: Giant cell tumours of bone. Clin Radiol 53:481, 1998.

Merkow RL, Bansal M, Inglis AE: Giant cell reparative granuloma in the hand: Report of three cases and review of the literature. J Hand Surg Am 10:733, 1985.

Picci P, Baldini N, Sudanese A, et al: Giant cell reparative granuloma and other giant cell lesions of the bones of the hands and feet. Skeletal Radiol 15:415, 1986.

Ratner V, Dorfman HD: Giant-cell reparative granuloma of the hand and foot bones. Clin Orthop 260:251, 1990.

Schajowicz F, Granato DB, McDonald DJ, et al: Clinical and radiological features of atypical giant cell tumours of bone. Br J Radiol 64:877, 1991.

Schutte HE, Taconis WK: Giant cell tumor in children and adolescents. Skeletal Radiol 22:173, 1993.

CHAPTER 14

Malignant Miscellaneous Tumors

Nabil J. Khoury, MD, Georges Y. El-Khoury, MD, and D. Lee Bennett, MD

ADAMANTINOMA

Overview

Adamantinoma is a rare, low-grade malignancy representing less than 0.5% of all malignant bone tumors. Advances in research support an epithelial origin for this tumor. Histologically, it consists of a fibrous stroma with epithelial cells of variable patterns. A history of trauma is frequently present, but it is unlikely that trauma is the primary cause of an adamantinoma. It has been shown that adamantinomas have a close relationship with osteofibrous dysplasia because of the similarity in location, histology, and radiologic appearance. An adamantinoma may present at any age, but it usually is encountered between the teens and 40s. There is a slight male predominance, and males seem to have a slightly more aggressive course than females. Symptoms include swelling with or without pain. The size of the tumor may reach 16 cm.

Skeletal Involvement

Adamantinomas typically are seen in the anterior aspect of the middle or distal third of the tibia (80% to 90% of cases). Almost all other long tubular bones may be affected. Multifocal lesions within the tibia or within the tibia and ipsilateral fibula have been described. The lesion typically is cortical and medullary in location.

Imaging

Radiography and Computed Tomography

On radiographs and computed tomography (CT) scans, adamantinomas appear as sharply outlined, expansile, and often bubbly lesions with a honeycomb appearance (Fig. 14–1). The lesion rarely appears as a single or multiloculated cystic mass (Fig. 14–2). The lesion occa-

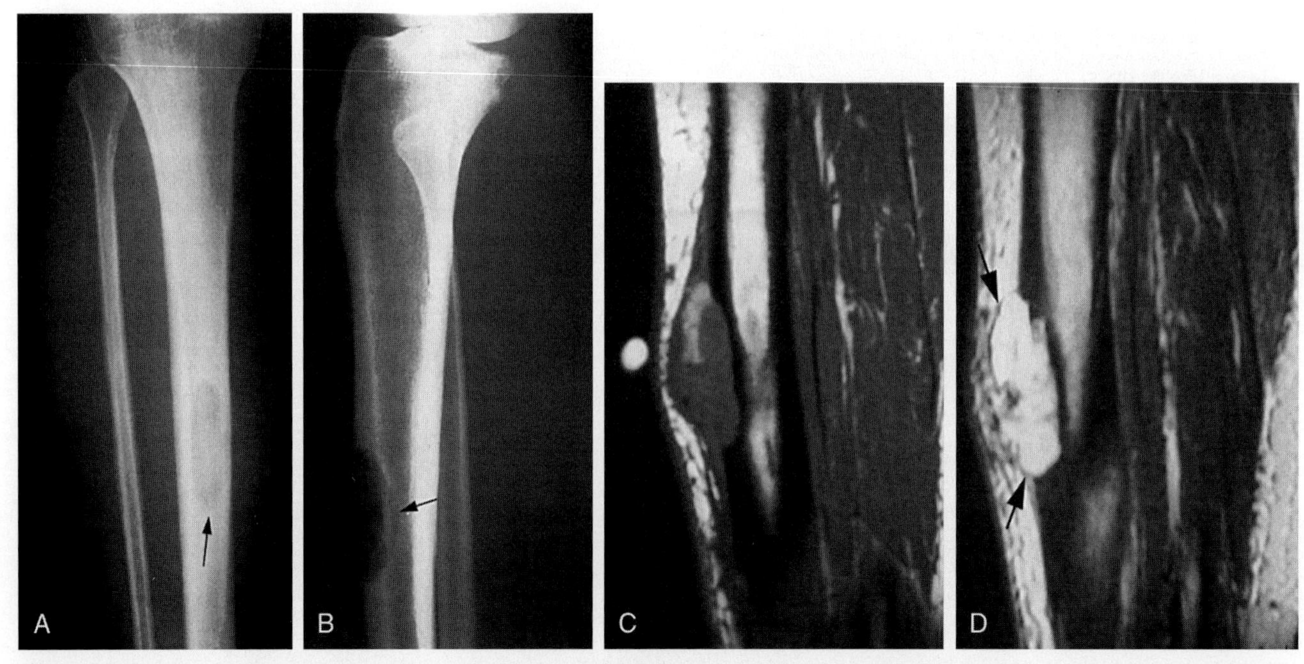

Figure 14–1. Adamantinoma in a 39-year-old woman who presented with an anterior leg mass. *A* and *B*, Anteroposterior and lateral views of the right tibia show a well-demarcated lytic lesion *(arrows)* destroying the anterior cortex. *C* and *D*, T1- and T2-weighted sagittal MR images show an exophytic lesion arising in the anterior cortex (*arrows* in *D*). The bright area within the lesion on the T1-weighted image most likely represents bleeding.

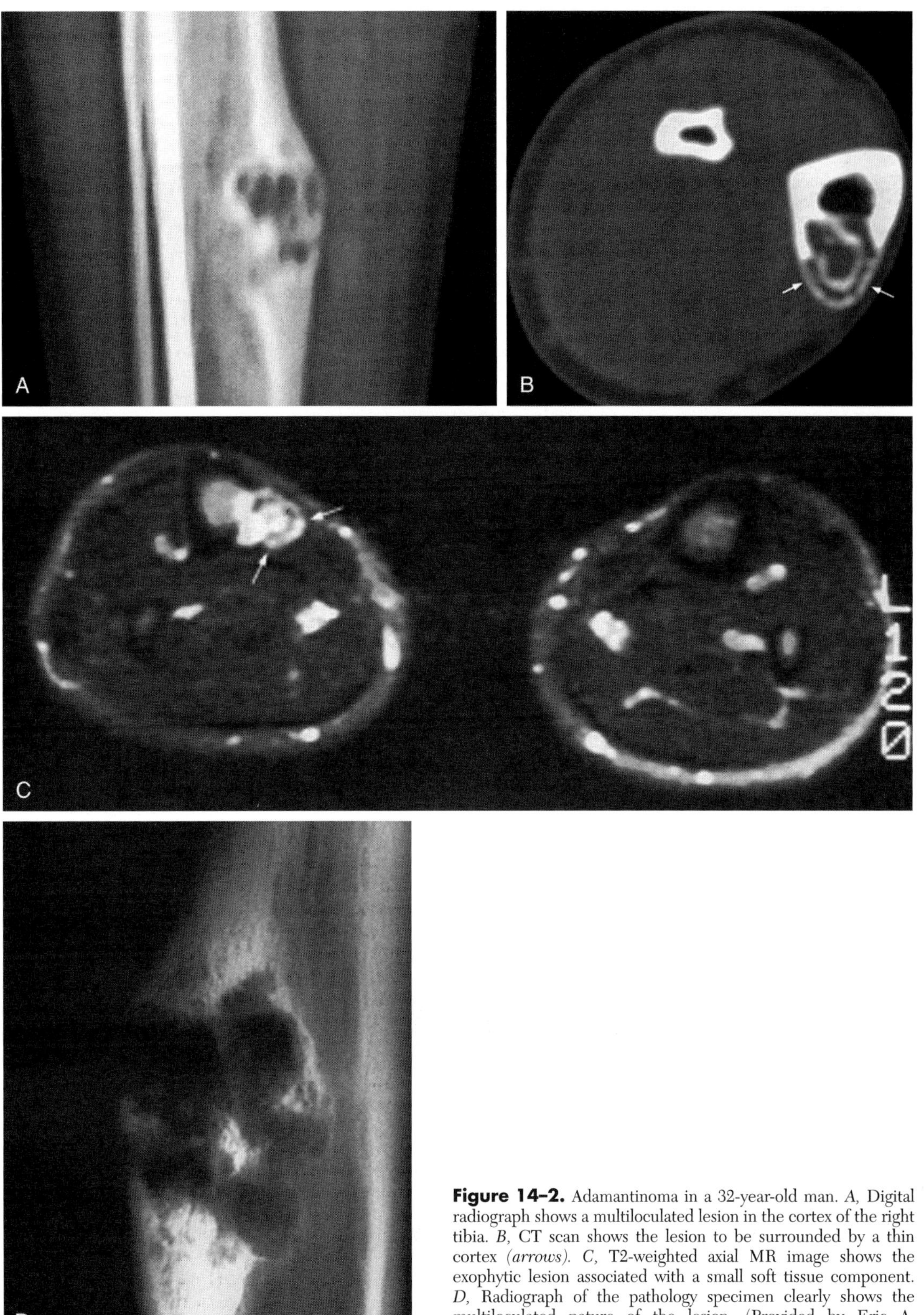

Figure 14–2. Adamantinoma in a 32-year-old man. *A,* Digital radiograph shows a multiloculated lesion in the cortex of the right tibia. *B,* CT scan shows the lesion to be surrounded by a thin cortex *(arrows).* *C,* T2-weighted axial MR image shows the exophytic lesion associated with a small soft tissue component. *D,* Radiograph of the pathology specimen clearly shows the multiloculated nature of the lesion. (Provided by Eric A. Brandser, MD, Cincinnati, OH.)

sionally is confined to the anterior cortex of the tibia and is surrounded by a rim of sclerotic bone (Fig. 14–2).

Magnetic Resonance Imaging

On magnetic resonance (MR) imaging, adamantinomas have intermediate-to-low signal intensity on T1-weighted images and heterogeneous high signal intensity on T2-weighted images (Figs. 14–1*C* and *D* and 14–2*B*). They may have areas consistent with fluid signal intensity. Occasionally, high signal intensity is seen on T1-weighted images because of bleeding (Fig. 14–1*C*). Low signal intensity on all sequences may be encountered when there is significant fibrous content and marginal sclerosis. MR imaging is useful for evaluating the medullary extent of the lesion and its soft tissue component (Fig. 14–1*C, D*).

Complications

Adamantinomas have a high local recurrence rate of about 30% after surgery. The rate of lung and lymph node metastases is 15% to 20%.

Differential Diagnosis

The main differential diagnosis includes osteofibrous dysplasia and fibrous dysplasia. Other lesions mimicking adamantinoma include aneurysmal bone cyst and chondromyxoid fibroma.

CHORDOMA

Overview

A chordoma is a relatively uncommon, slow-growing, malignant neoplasm that occurs almost exclusively in the axial skeleton. It represents about 1% to 4% of all primary malignant bone tumors. The tumor is thought to arise from notochordal remnants. Chordomas are the most common tumor affecting the sacrum. The tumor usually occurs between the 40s and 60s, but it also can occur in children and young adults. In the sacrococcygeal region, the male-to-female ratio is 2:1. In other areas, the male-to-female ratio is equal. The presenting symptom is usually pain, but neurologic symptoms also can occur depending on the spinal location.

Skeletal Involvement

The sacrum is the most common location for chordomas (Fig. 14–3). This location accounts for about 50% of all chordomas. The next most common location is the base of the skull (35%), in particular, the spheno-occipital region. The third most common location is the mobile spine (15%), where the cervical segment is the most commonly affected region. In the spine, chordomas usually affect the vertebral body with extension into the posterior elements. In the sacrum, the tumor typically arises from the more caudad segments, S3-S5, or even from the coccyx. This location helps in the definitive diagnosis because giant cell tumor, chondrosarcoma, and osteosarcoma are more common in the upper segments.

Imaging

Radiography

On plain radiographs, a chordoma is usually an ill-defined, lytic, destructive lesion with cortical expansion and disruption (see Fig. 14–3). The lesion almost always is central; this helps to differentiate it from eccentric lesions, such as a giant cell tumor. A chordoma also may be sclerotic or have a mixed lytic and sclerotic pattern. Amorphous or flocculent calcifications may be seen at

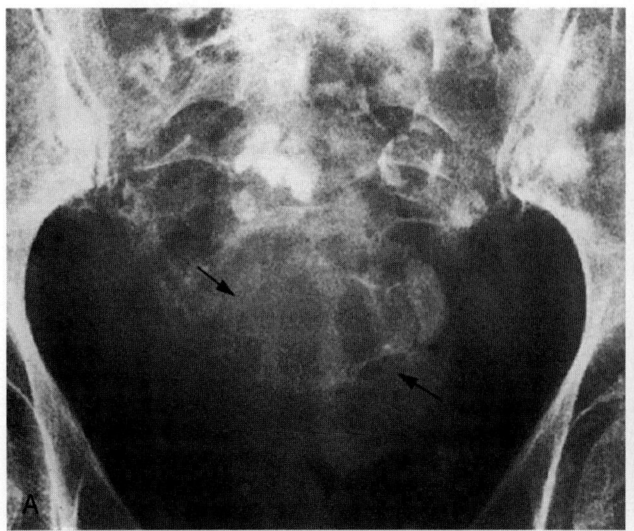

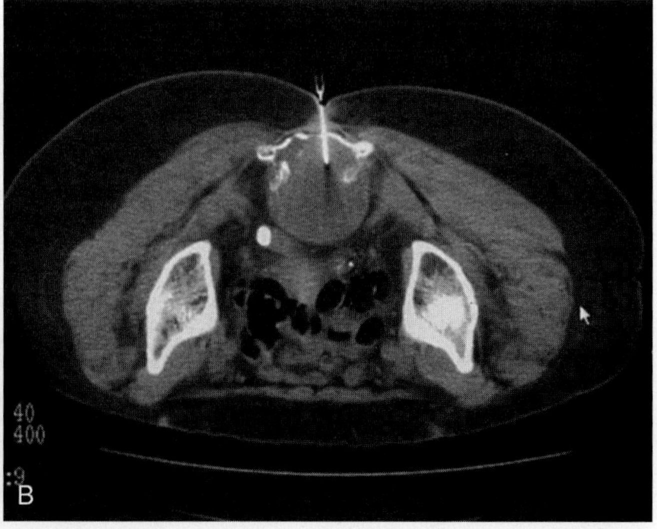

Figure 14–3. Sacral chordoma in an 81-year-old woman. *A*, Anteroposterior view of the pelvis shows a lytic lesion in the distal two sacral segments *(arrows)*. *B*, CT section obtained during a needle biopsy shows the extent of the mass.

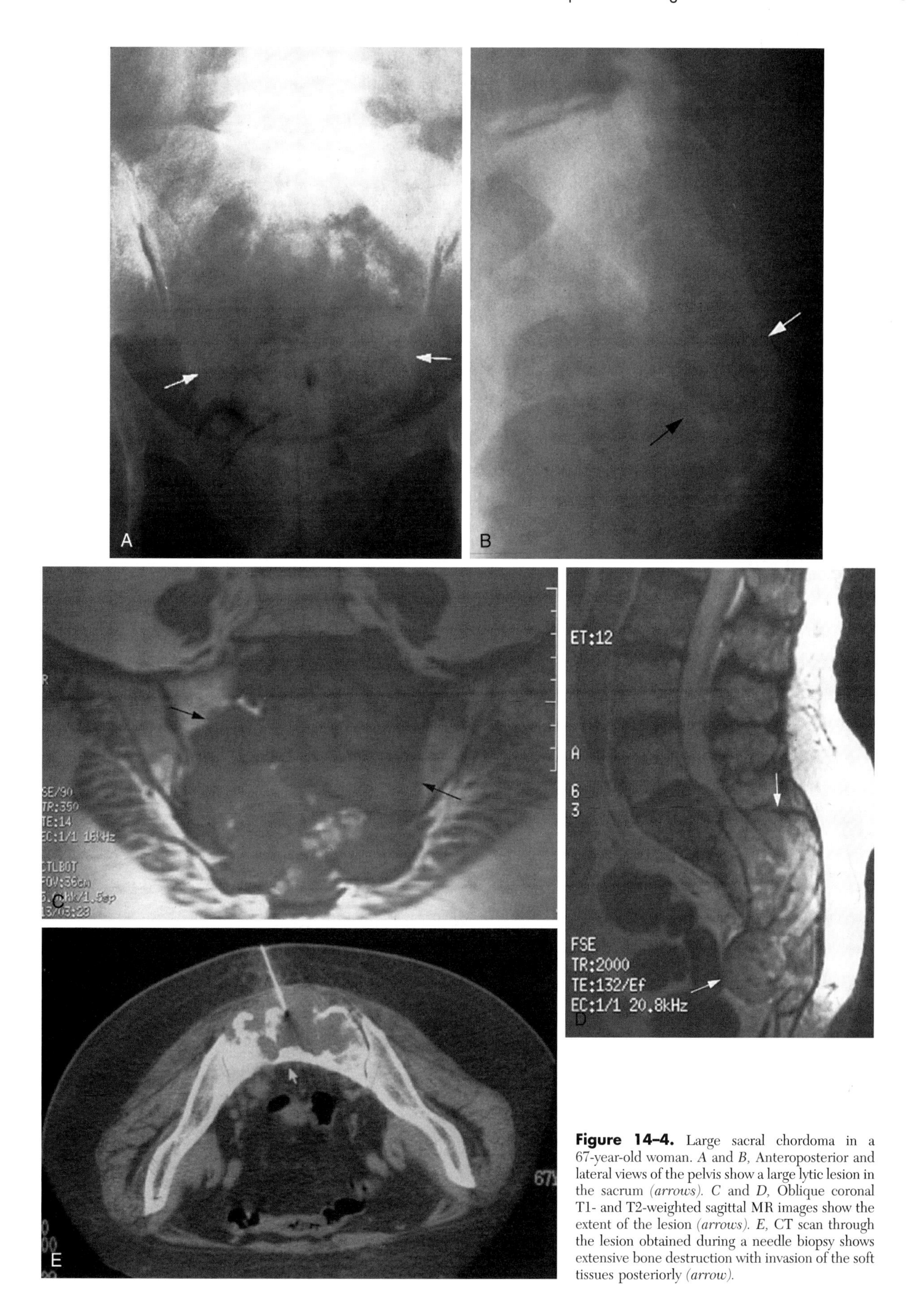

Figure 14–4. Large sacral chordoma in a 67-year-old woman. *A* and *B*, Anteroposterior and lateral views of the pelvis show a large lytic lesion in the sacrum *(arrows)*. *C* and *D*, Oblique coronal T1- and T2-weighted sagittal MR images show the extent of the lesion *(arrows)*. *E*, CT scan through the lesion obtained during a needle biopsy shows extensive bone destruction with invasion of the soft tissues posteriorly *(arrow)*.

the periphery or within the lesion. These calcifications probably represent sequestrated necrotic bone or calcified debris. In the spine, there is frequent involvement of two or more adjacent vertebrae with destruction of the intervening disk. Other findings in the spine include vertebral collapse and vertebral sclerosis (ivory vertebra). In the sacrum, chordomas usually extend anteriorly and rarely posteriorly.

Computed Tomography

The CT scan appearance is nonspecific. CT shows to better advantage, however, the extent of the lesion and surrounding soft tissue involvement (see Fig. 14–3). The calcifications also are seen better by CT.

Magnetic Resonance Imaging

MR imaging is the best modality for showing the extent of the tumor, particularly in the spine and soft tissues (Fig. 14–4). The signal intensity of the lesion is not specific. On T1-weighted images, chordomas either are isointense with muscle or have decreased signal intensity. On T2-weighted images, they have high signal intensity. In 70% of cases, there are radiating septa of low signal intensity seen within the lesion on T2-weighted images. Chondroid chordomas may have shorter T1 and T2 relaxation times.

Complications

Tumor recurrence is frequent. Distant metastases occur in 10% to 43% of patients, and metastatic sites include the liver, lungs, lymph nodes, bone, and skin. Metastasis does not seem to affect survival, however. Death mainly results from complications of paraplegia caused by the primary tumor.

Differential Diagnosis

The differential diagnosis is wide and includes plasmacytoma, multiple myeloma, metastasis, giant cell tumor, anterior meningocele, neurogenic tumor, and lymphoma. A newly described lesion called *benign notochordal rests* can resemble a chordoma. The treatment for these two conditions is very different and confusing them could have serious consequences. Notochordal rests do not destroy bone and do not increase in size over time. The contour of the vertebra is also maintained and on CT the trabecular pattern is thick. On MRI the signal intensity of the lesion is low on T1-weighted images and high on T2-weighted images.

Recommended Readings

Bloem JL, Van der Heul RO, Schuttevaer HM, et al: Fibrous dysplasia vs adamantinoma of the tibia: Differentiation based on discriminant analysis of clinical and plain film findings. AJR Am J Roentgenol 156:1017, 1991.

Conway WF, Hayes CW: Miscellaneous lesions of bone. Radiol Clin North Am 31:339, 1993.

DeBruine FT, Kroon HM: Spinal chordoma: Radiologic features in 14 cases. AJR Am J Roentgenol 150:861, 1988.

Garces P, Romano CC, Vellet AD, et al: Adamantinoma of the tibia: Plain-film, computed tomography and magnetic resonance imaging appearance. Can Assoc Radiol J 45:314, 1994.

Healey JH, Lane JM: Chordoma: A critical review of diagnosis and treatment. Orthop Clin North Am 20:417, 1989.

Judmaier W, Peer S, Krejzi T, et al: MR findings in tibial adamantinoma: A case report. Acta Radiol 39:276, 1998.

Mirra JM, Brien EW: Giant notochordal hamartoma of intraosseous origin: A newly reported benign entity to be distinguished from chordoma. Report of two cases. Skeletal Radiol 30:698, 2001.

Murphy JM, Wallis F, Toland J, et al: CT and MR imaging appearances of a thoracic chordoma. Eur Radiol 8:1677, 1998.

Rosenthal DI, Scott JA, Mankin HJ, et al: Sacrococcygeal chordoma: Magnetic resonance imaging and computed tomography. AJR Am J Roentgenol 145:143, 1985.

Smith J, Ludwig RL, Marcove RC: Sacrococcygeal chordoma: A clinicoradiological study of 60 patients. Skeletal Radiol 16:37, 1987.

Sze G, Uichanco LS III, Brant-Zawadzki MN, et al: Chordomas: MR imaging. Radiology 166:187, 1988.

Wippold FH II, Koeller KK, Smirniotopoulos JG: Clinical and imaging features of cervical chordoma. AJR Am J Roentgenol 172:1423, 1999.

Zehr RJ, Recht MP, Bauer TW: Adamantinoma. Skeletal Radiol 24:553, 1995.

CHAPTER 15

Cystic Lesions

Nabil J. Khoury, MD, Georges Y. El-Khoury, MD, and D. Lee Bennett, MD

SIMPLE (SOLITARY, UNICAMERAL) BONE CYST

Overview

A simple (unicameral or solitary) bone cyst is a common, benign lesion, representing approximately 5% of all bone tumors. It consists of a fluid-filled intramedullary cavity. The cause of this lesion is unknown. On pathologic examination, the lesion consists of a cavity lined by a fibrous membrane. The cavity contains clear fluid. Patients typically present before age 20. Males are affected more commonly than females (male-to-female ratio, 2:1). The typical simple bone cyst is often asymptomatic. Patients may present with pain, which frequently is due to a pathologic fracture (Fig. 15–1). Rarely a simple bone cyst may heal after a fracture (Fig. 15–2). Treatment traditionally consisted of curettage and grafting; however, more recently, cyst injection with corticosteroids has been used successfully. Simple bone cysts with loculations and multiple fibrous septa do not respond as well to corticosteroid therapy.

Skeletal Involvement

In the long bones, a simple bone cyst is essentially a central lesion. Before 20 years of age, simple bone cysts mainly occur in the long tubular bones (90% to 95%), particularly in the proximal humerus and proximal femur (see Figs. 15–1 and 15–2). After age 20, more than half of the lesions occur in the innominate bone and calcaneus (Fig. 15–3). In the long tubular bones, the lesion is most frequently metaphyseal in location. A diaphyseal location sometimes is observed because of skeletal growth with migration of the physis away from the lesion. Simple bone cysts have been reported to involve the epiphysis or apophysis, and in such cases, growth disturbances may occur.

Imaging

Radiography

Radiographs show a lucent, well-defined lesion with a narrow zone of transition. A simple bone cyst nearly always arises in the metaphysis, and when the humerus or femur is involved, the lesion virtually always arises in the proximal metaphysis. In the long bones, the lesion is central with its longitudinal axis parallel to the parent bone (see Figs. 15–1 and 15–2). The cyst is larger at the metaphyseal end and more narrow at the diaphysis. Sometimes, cysts may be multilocular. A pathognomonic finding in simple bone cysts is the *fallen fragment* sign, in which a free bone fragment is seen lying in the dependent portion of the cyst as a result of a pathologic fracture (Fig. 15–4).

Computed Tomography

Computed tomography (CT) scans show, to better advantage than radiographs, the extent of the cyst when it

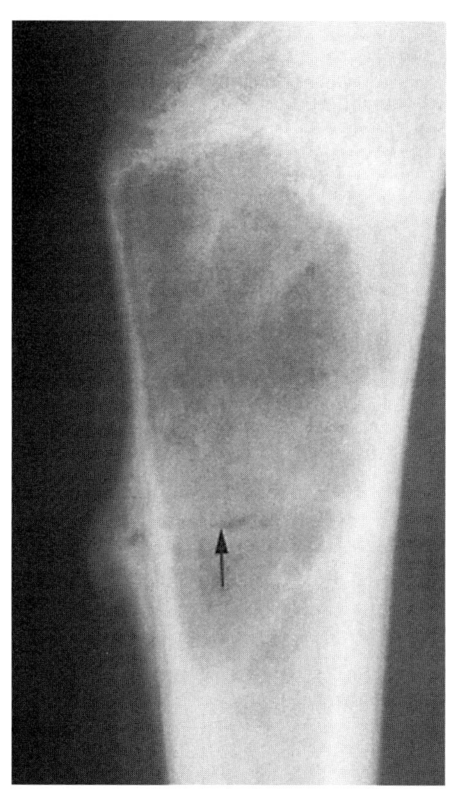

Figure 15–1. Simple bone cyst with a pathologic fracture. Anteroposterior view of the proximal femur shows a pathologic fracture through a simple bone cyst *(arrow)*. Periosteal reaction and callus formation can be seen.

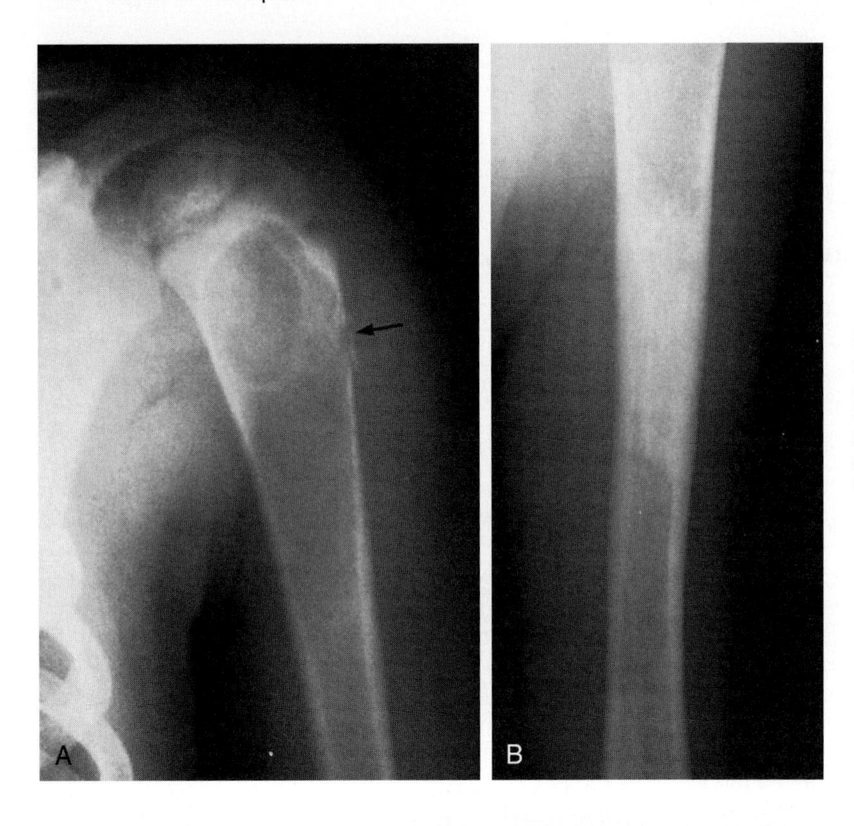

Figure 15–2. Simple bone cyst healing after a fracture. *A,* Anteroposterior view of the humerus in a 4-year-old child shows a pathologic fracture of a simple bone cyst *(arrow). B,* Radiograph of the same patient at age 10 shows progressive sclerosis (healing) of the simple bone cyst. There was no history of surgical intervention.

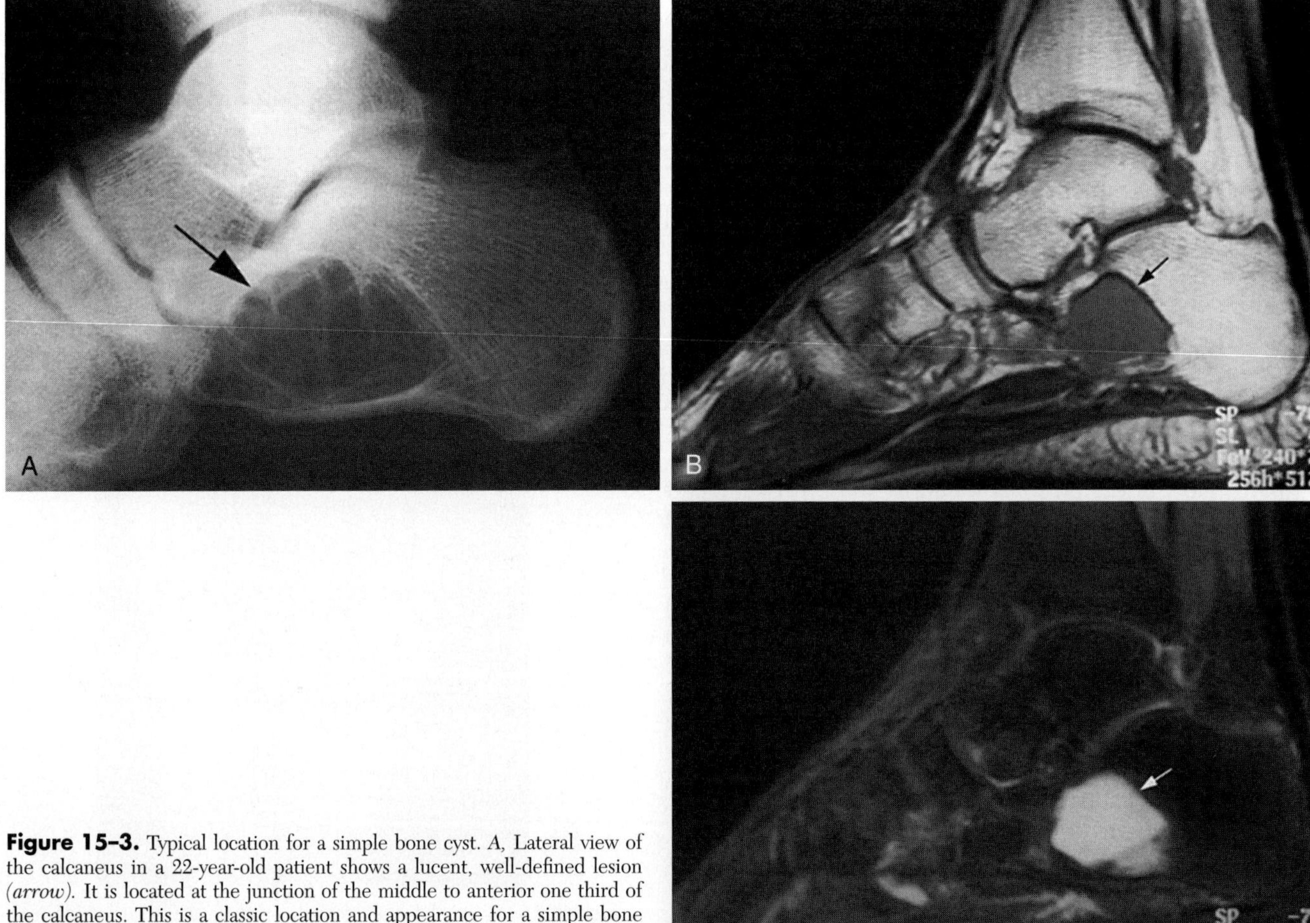

Figure 15–3. Typical location for a simple bone cyst. *A,* Lateral view of the calcaneus in a 22-year-old patient shows a lucent, well-defined lesion *(arrow).* It is located at the junction of the middle to anterior one third of the calcaneus. This is a classic location and appearance for a simple bone cyst. Pathologic evaluation revealed a simple bone cyst. *B* and *C,* Parasagittal T1-weighted and fat-saturated T2-weighted MR images show the typical fluid content of a simple bone cyst *(arrows).*

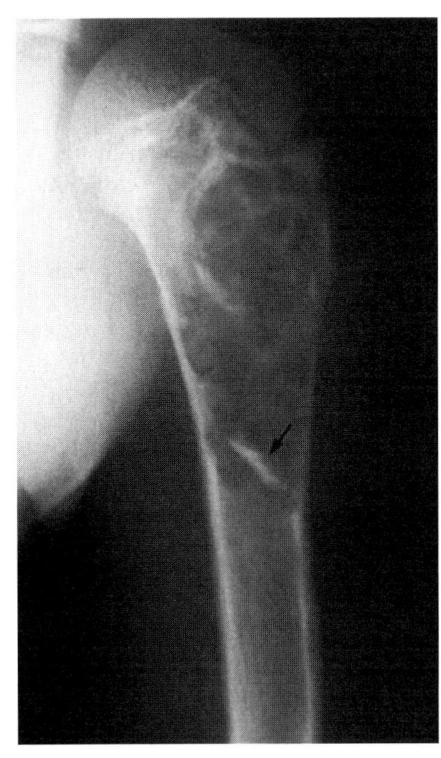

Figure 15–4. Fallen fragment sign. Upright anteroposterior view of the left shoulder shows a fragment of bone at the distal end of the lesion *(arrow)*. This sign indicates the cystic nature or fluid-filled cavity.

is located in an anatomically complex area, such as the pelvis or the spine. Fluid-fluid levels occasionally are encountered as a result of hemorrhage.

Magnetic Resonance Imaging

In most cases, magnetic resonance (MR) imaging is not required for the diagnosis of a simple bone cyst. If performed, MR imaging shows low signal intensity on T1-weighted images and high signal intensity on T2-weighted images, owing to the fluid content (see Fig. 15–3).

Complications

The most common complication of simple bone cysts is a pathologic fracture (see Figs. 15–1, 15–4, and 15–5). If the lesion involves the growth plate, growth disturbances (retardation or arrest) occur. Simple bone cysts have a high recurrence rate after various therapeutic modalities; recurrence is particularly common in children younger than age 10 (Fig. 15–5).

Differential Diagnosis

The differential diagnosis includes fibrous dysplasia, aneurysmal bone cyst (ABC), and eosinophilic granuloma.

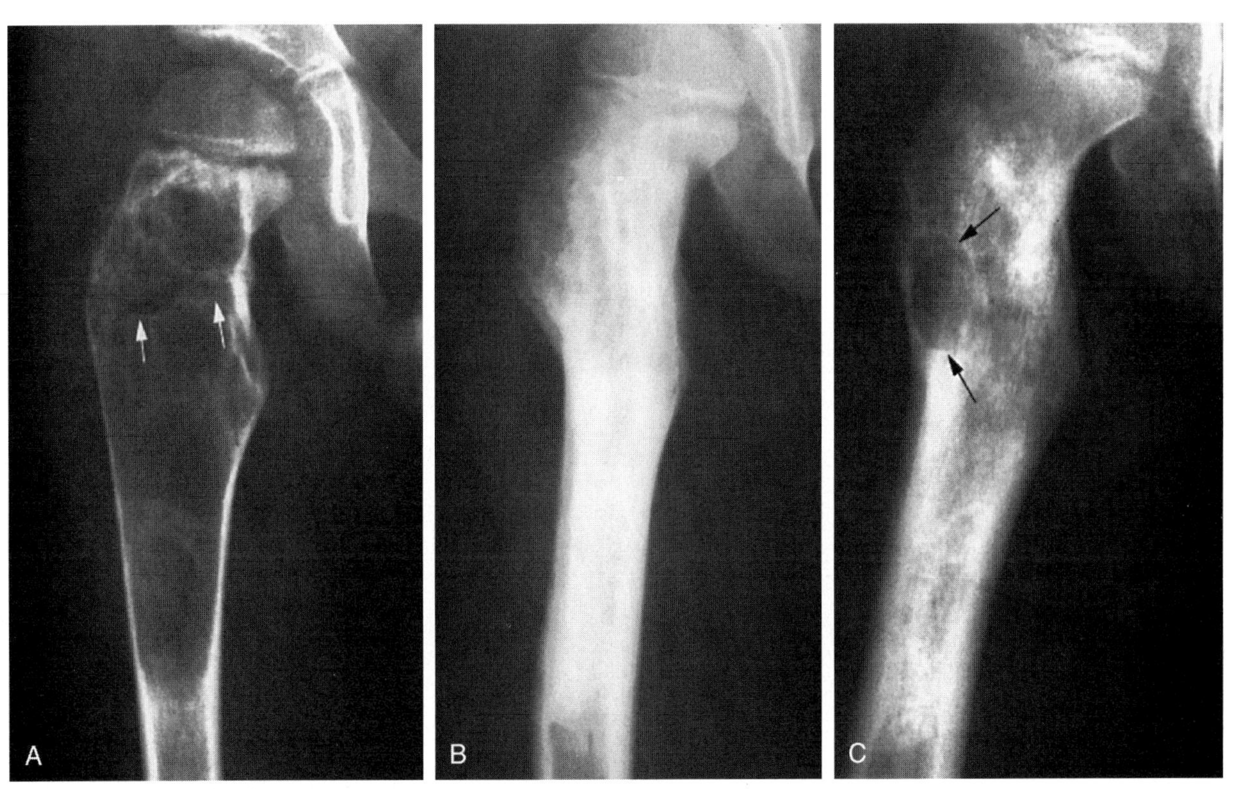

Figure 15–5. Example of a recurrent simple bone cyst. *A*, Anteroposterior view of the right hip in a 7-year-old child shows a pathologic fracture *(arrows)* of a simple bone cyst. *B*, Anteroposterior view of the right hip in the same patient shows the postoperative appearance several months after curettage and bone grafting. *C*, Anteroposterior view of the right hip 2 years later shows a new well-circumscribed, lytic lesion at the site of previous surgery *(arrows)*. This was a recurrent simple bone cyst.

In the calcaneus, intraosseous lipoma and normal trabecular sparsity can simulate a simple bone cyst.

ANEURYSMAL BONE CYST (ABC)

Overview

ABC is a benign lesion representing 1% of all bone tumors. There is much debate in the literature about its cause. Several cases of ABCs are reported to have developed at the site of a previous fracture. It generally is agreed that either ABC is a primary lesion, or it may have developed within a preexisting tumor. Tumors implicated in secondary ABC include giant cell tumor, chondroblastoma, osteoblastoma, chondromyxoid fibroma, fibrous dysplasia, benign and malignant fibrous histiocytoma, nonossifying fibroma, fibrosarcoma, osteosarcoma, eosinophilic granuloma, and metastatic carcinoma. Histologically, ABC consists of blood-filled spaces separated by connective tissue. It contains bone trabeculae, osteoid, and giant cells. A *solid* variant of ABC occurs in 5% of cases. Its histologic findings overlap with classic ABC; however, it lacks a cystic component. ABC has a good prognosis. It mainly occurs before age 20. It is extremely rare before age 5 years and after age 35. The solid variant has the same age distribution. Males and females are affected equally. The main presenting symptoms are pain and swelling. Patients with spinal lesions often present with neurologic deficit.

Skeletal Involvement

ABCs occur most frequently in the long tubular bones, in particular, the femur and tibia (Fig. 15–6). Flat bones commonly are affected with the pelvis being the most frequent site. The spine is involved in 11% of cases. Other locations are rare and include carpal and tarsal bones, short tubular bones, skull, mandible, maxilla, patella, and sacrum. The solid variant of ABC is more frequent in the axial skeleton, facial bones, and small tubular bones of the hands and feet. ABC is typically a medullary lesion; however, intracortical and subperiosteal locations have been reported. It is eccentric in the long tubular bones and central in the short tubular bones. Within long bones, ABC typically is metaphyseal or metadiaphyseal in location, and invasion of the growth plate can occur. In the spine, ABC always involves the posterior elements with frequent extension into the vertebral body or ribs. Extension into adjacent vertebrae through the posterior elements also can occur.

Imaging

Radiography

On radiography, ABC is typically a lytic, geographic, expansile lesion with thinning of the cortex and, sometimes, cortical loss (Figs. 15–7 and 15–8). Poorly defined

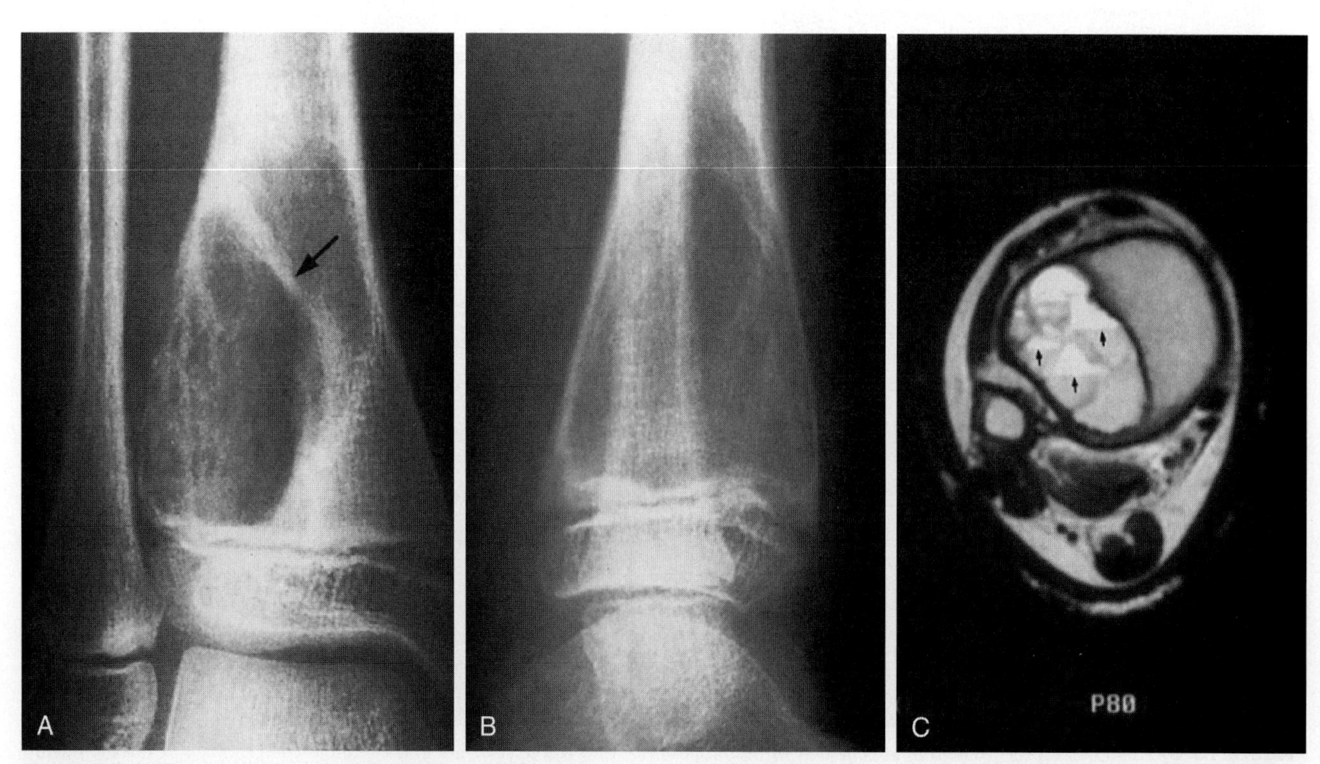

Figure 15–6. Aneurysmal bone cyst with multiple fluid-fluid levels on MR imaging. *A* and *B*, Anteroposterior and lateral views of the right ankle show an expansile lytic lesion in the distal metaphysis of the tibia *(arrow)*. The lesion is geographic with IC border. *C*, T2-weighted axial MR image through the lesion shows multiple fluid-fluid levels *(arrows)*.

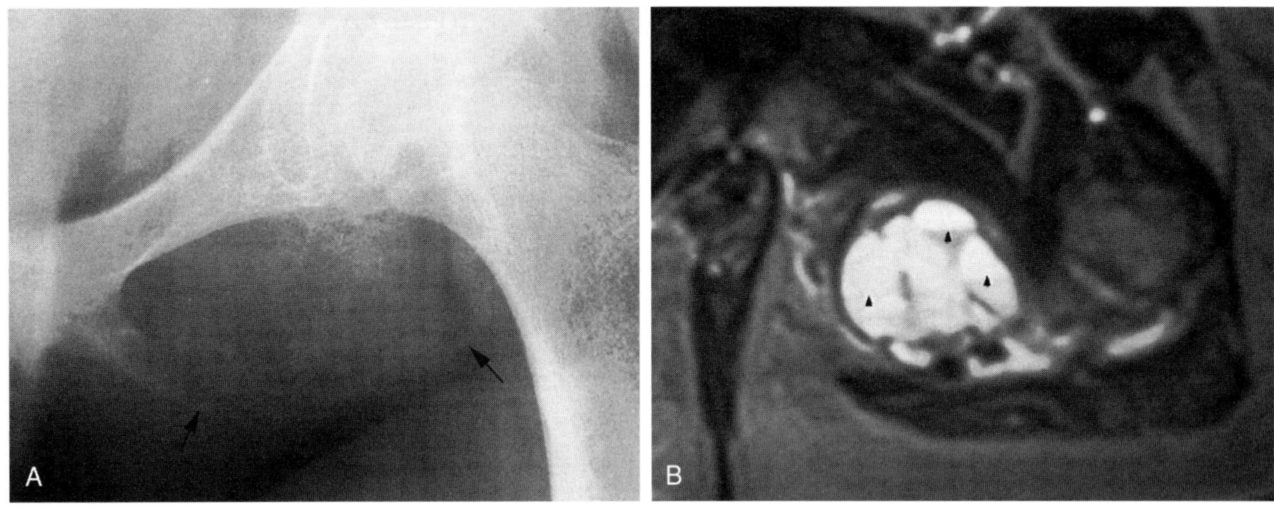

Figure 15–7. Aneurysmal bone cyst causing marked thinning of the cortex. *A*, Anteroposterior view of the left hip in a 12-year-old child shows a lytic lesion in the ischium. The lesion is expansile and is causing marked thinning of the cortex *(arrows)*. *B*, T2-weighted MR image of the lesion shows multiple fluid-fluid levels *(arrowheads)*.

margins are seen in 14% of cases. Internal trabeculae and septations are noted in about 42% of cases. Usually, there is no visible matrix. An interrupted periosteal reaction (Codman's triangle) is seen in about one fourth of cases mimicking an aggressive malignancy (Fig. 15–9). Classic and solid variants of ABC have similar radiographic findings.

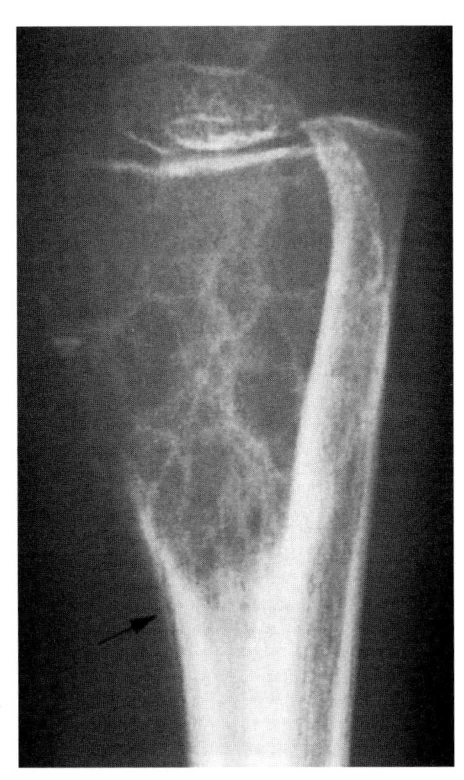

Figure 15–8. Expansile trabeculated aneurysmal bone cyst in the distal radius of a 6-year-old child. Small Codman's triangle is noted at the proximal end of the lesion *(arrow)*.

Computed Tomography

CT shows, to better advantage than radiography, the septations and the thinned, expanded cortical rim. It shows the extent of the lesion in difficult anatomic areas, such as the spine and pelvis. CT also can detect fluid-fluid levels (Fig. 15–10).

Magnetic Resonance Imaging

On MR imaging, ABC frequently has multiple fluid-fluid levels (see Figs. 15–6 and 15–7). This is a nonspecific finding that is encountered in many other tumors. ABC is the most common lesion to have this finding, however. MR imaging also shows the internal septations, which enhance after intravenous gadolinium injection.

Complications

ABCs can grow rapidly and exhibit locally aggressive behavior. Growth arrest or retardation may occur with involvement of the physis. Of patients with spinal lesions, 50% have a neurologic deficit. Postoperative recurrence is seen in 10% to 20% of cases, particularly in the long bones. Few cases have shown spontaneous regression after biopsy.

Differential Diagnosis

ABCs may mimic many osseous lesions, including giant cell tumor, unicameral bone cyst, nonossifying fibroma, enchondroma, fibrous dysplasia, and telangiectatic osteosarcoma. The solid variant may mimic giant cell tumor, fibroblastic osteosarcoma, and giant cell reparative granuloma.

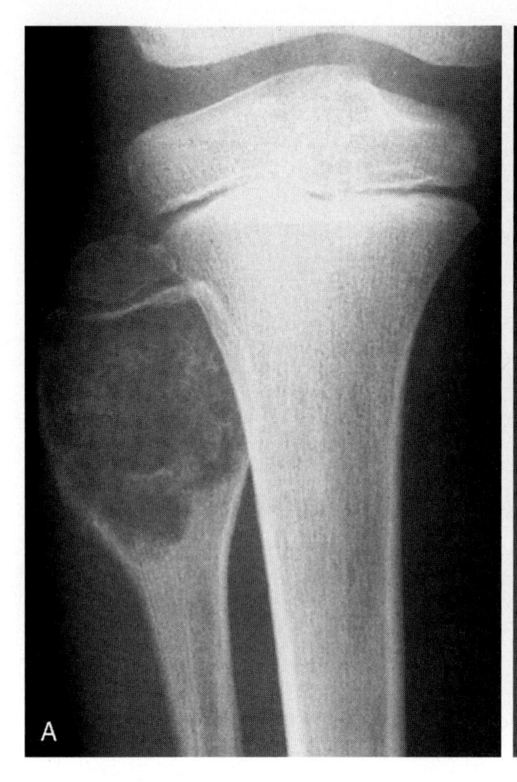

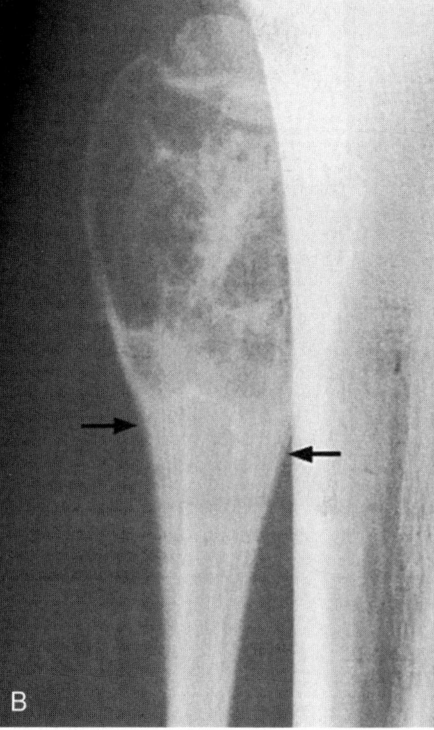

Figure 15–9. Aneurysmal bone cyst with periosteal reaction. *A* and *B*, Anteroposterior and lateral views of the proximal fibula in a 7-year-old child show an expansile, lytic lesion. A lamellated (onionskin) periosteal reaction is seen *(arrows)*. The periosteal reaction is also interrupted (Codman's triangle).

INTRAOSSEOUS GANGLION

Overview

Ganglion cysts have been divided into intraosseous ganglion cysts and soft tissue ganglion cysts. Soft tissue ganglion cysts are much more common, and they can be classified further into periosteal and intra-articular ganglion cysts. All of these lesions have the same histology. They are not true cysts because they do not have an epithelial lining but rather a fibrous lining. These cysts are septated and contain mucoid material. Most authorities believe that intraosseous ganglion cysts arise *de novo* within the bone with no radiographic evidence of

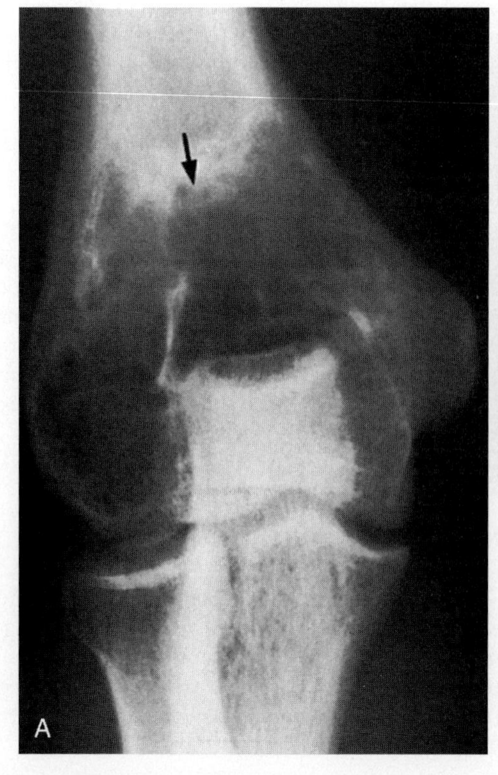

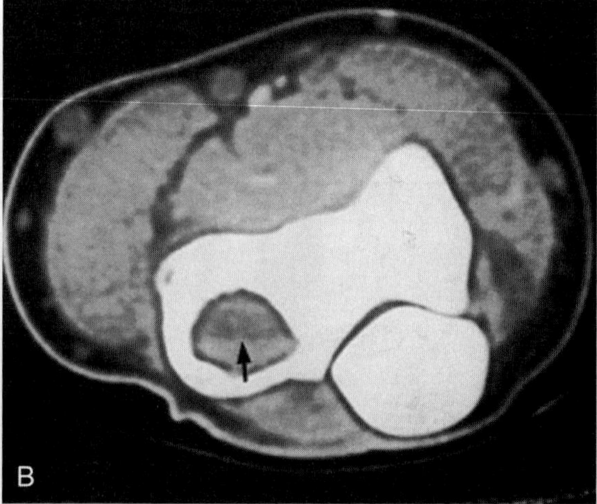

Figure 15–10. Fluid-fluid level in an aneurysmal bone cyst of the distal humerus shown by CT. *A*, Anteroposterior view of the right elbow shows a lytic trabeculated lesion in the distal humerus *(arrow)*. *B*, CT section through the lesion shows a fluid-fluid level *(arrow)*.

degenerative arthritis in the adjacent joint. The lesion is usually solitary; however, it can be multiple. It is frequently asymptomatic; however, mild, localized, progressive, or chronic pain and joint discomfort may occur. Ganglion cysts commonly are found in young and middle-aged adults.

Skeletal Involvement

The most common locations include the medial malleolus, femoral head, proximal tibia, and carpal bones. In the long bones, the lesion is typically epiphyseal and juxta-articular (Fig. 15–11). Isolated metaphyseal or diaphyseal lesions are rare. Metaphyseal lesions may extend into the subchondral area.

Imaging

Radiography

Plain radiographs show a well-defined, round or ovoid, lytic lesion with a surrounding thin rim of sclerosis (see Fig. 15–11). The lesion can penetrate the cortex, especially in the carpal bones. Ganglion cysts are often multiloculated. Their size may range from 2 mm to 10 cm.

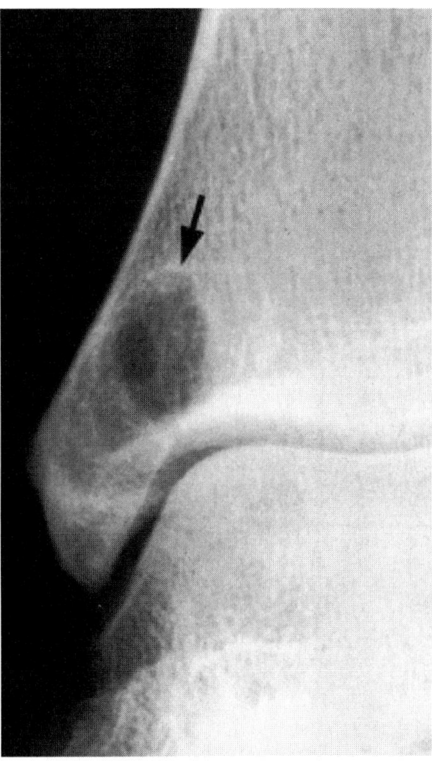

Figure 15–11. Classic appearance and location of an intraosseous ganglion cyst. Note the lobulated appearance with thin cortical margins of a lytic lesion in a subchondral location *(arrow)*.

Magnetic Resonance Imaging

MR imaging shows the typical finding of a cyst, showing low signal intensity on T1-weighted images and high signal intensity on T2-weighted images. Rarely, fluid-fluid levels have been described.

Differential Diagnosis

In the long bones, a ganglion cyst should be differentiated from a chondroblastoma, giant cell tumor, clear cell chondrosarcoma, ABC, and subchondral cyst of osteoarthritis. Osteoarthritic cysts are associated with degenerative joint changes, however, and the cyst often communicates with the joint.

PERIOSTEAL GANGLION

Overview

Periosteal ganglion is similar to an intraosseous ganglion; however, a periosteal ganglion arises within or beneath the periosteum. It usually presents as painless swelling. The mean age at presentation is 47 years.

Skeletal Involvement

The most frequent location is the medial aspect of the proximal tibia, under the insertion of the pes anserinus. Other locations include the tibial diaphysis, distal ulna, radius, femur, and medial malleolus.

Imaging

Radiography

Radiography shows cortical destruction and external scalloping of the bone adjacent to the periosteal ganglion cyst. Spicules of reactive periosteal new bone may radiate from the scalloped area.

Computed Tomography

CT shows the presence of a soft tissue mass and adjacent cortical changes.

Magnetic Resonance Imaging

On MR imaging, the periosteal ganglion has signal characteristics of a cyst. The lesion appears well defined and multilobulated.

Differential Diagnosis

The differential diagnosis includes periosteal chondroma and subperiosteal hematoma.

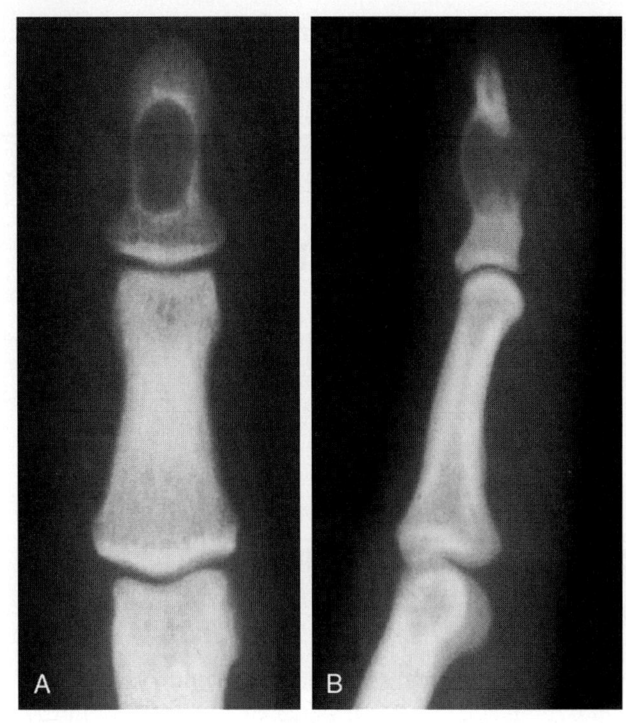

Figure 15–12. Epidermoid inclusion cyst in the finger. *A* and *B*, Anteroposterior and lateral views of the index finger show a classic epidermoid inclusion cyst. It is a sharply demarcated, unilocular, rounded, lytic lesion in a distal phalanx.

INTRAOSSEOUS EPIDERMOID CYST

Overview

Intraosseous epidermoid cyst is an uncommon bone lesion usually involving the fingers and skull. Frequently, there is a history of previous trauma, which resulted in implantation of epidermis into the involved bone. The cyst is lined by thick stratified squamous epithelium and contains laminated keratin. Clinically, lesions involving the fingers present with pain, gradual swelling of the digit, and inflammatory changes in the adjacent soft tissues and skin. Nail-plate deformity may be present. Epidermoid cysts occur more frequently in males (male-to-female ratio, 2:1). They are most common between the teens and 30s.

Skeletal Involvement

Most cysts are seen in the skull and phalanges of the hands, particularly of the thumb and middle finger (Fig. 15–12). More than 95% of the cysts in the digits occur in the tufts of the distal phalanges. Rare sites include the tibia, ulna, femur, and sternum.

Imaging

Radiography shows a sharply demarcated, unilocular, rounded, lytic lesion. Cortical expansion and thinning may be present.

Complications

Pathologic fracture is fairly common.

Differential Diagnosis

In the phalanges, a glomus tumor and an enchondroma can mimic an epidermoid cyst.

Recommended Readings

Beltran J, Simon DC, Levy M, et al: Aneurysmal bone cysts: MR imaging at 1.5 T. Radiology 158:689, 1986.

Bertoni F, Bacchini P, Capanna R, et al: Solid variant of aneurysmal bone cyst. Cancer 71:729, 1993.

Bollini G, Jouve JL, Cottalorda J, et al: Aneurysmal bone cyst in children: Analysis of twenty-seven patients. J Pediatr Orthop B 7:274, 1998.

Burr BA, Resnick D, Syklawer R, et al: Fluid-fluid levels in a unicameral bone cyst: CT and MR findings. J Comput Assist Tomogr 17:134, 1993.

Capanna R, Springfield DS, Biagini R, et al: Juxtaepiphyseal aneurysmal bone cyst. Skeletal Radiol 13:21, 1985.

Capanna R, Van-Horn J, Biagini R, et al: Aneurysmal bone cyst of the sacrum. Skeletal Radiol 18:109, 1989.

Capanna R, Van-Horn J, Ruggieri P, et al: Epiphyseal involvement in unicameral bone cysts. Skeletal Radiol 15:428, 1986.

De Kleuver M, Van der Heul RO, Veraart BE: Aneurysmal bone cyst of the spine: 31 cases and the importance of the surgical approach. J Pediatr Orthop B 7:286, 1998.

Freeby JA, Reinus WR, Wilson AJ: Quantitative analysis of the plain radiographic appearance of aneurysmal bone cysts. Invest Radiol 30:433, 1995.

Haims AH, Desai P, Present D, et al: Epiphyseal extension of a unicameral bone cyst. Skeletal Radiol 26:51, 1997.

Helwig U, Lang S, Baczynski M, et al: The intraosseous ganglion: A clinical-pathological report on 42 cases. Arch Orthop Trauma Surg 114:14, 1994.

Hinrichs RA: Epidermoid cyst of the terminal phalanx of the hand: Case report and brief review. JAMA 194:1253, 1965.

Kransdorf MJ, Sweet DE: Aneurysmal bone cyst: Concept, controversy, clinical presentation, and imaging. AJR Am J Roentgenol 164:573, 1995.

Lee JH, Reinus WR, Wilson AJ: Quantitative analysis of the plain radiographic appearance of unicameral bone cysts. Invest Radiol 34:28, 1999.

Lokiec F, Wientroub S: Simple bone cyst: Etiology, classification, pathology, and treatment modalities. J Pediatr Orthop B 7:262, 1998.

Magee TH, Rowedder AM, Degnan GG: Intraosseous ganglia of the wrist. Radiology 195:517, 1995.

Martinez V, Sissons HA: Aneurysmal bone cyst: A review of 123 cases including primary lesions and those secondary to other bone pathology. Cancer 61:2291, 1988.

Moore TE, King AR, Travis RC, et al: Post-traumatic cysts and cyst-like lesions of bone. Skeletal Radiol 18:93, 1989.

Moreau G, Letts M: Unicameral bone cyst of the calcaneus in children. J Pediatr Orthop 14:101, 1994.

Murff R, Ashry HR: Intraosseous ganglia of the foot. J Foot Ankle Surg 33:396, 1994.

Newland Z, Moore RM: Intraosseous ganglion of the ankle. J Foot Surg 25:241, 1986.

Oda Y, Tsuneyoshi M, Shinohara N: "Solid" variant of aneurysmal bone cyst (extragnathic giant cell reparative granuloma) in the axial skeleton and long bones: A study of its morphologic spectrum and distinction from allied giant cell lesions. Cancer 70:2642, 1992.

Okada K, Unoki E, Kubota H, et al: Periosteal ganglion: A report of three new cases including MR imaging findings and a review of the literature. Skeletal Radiol 25:153, 1996.

Pablos JM, Valdes JC, Gavilan F: Bilateral lunate intraosseous ganglia. Skeletal Radiol 27:708, 1998.

Pope TL, Fechner RE, Keats TE: Intra-osseous ganglion. Skeletal Radiol 18:185, 1989.
Rozbruch SR, Chang V, Bohne WH, et al: Ganglion cysts of the lower extremity: An analysis of 54 cases and review of the literature. Orthopedics 21:141, 1998.
Schoedel K, Shankman S, Desai P: Intracortical and subperiosteal aneurysmal bone cysts: A report of three cases. Skeletal Radiol 25:455, 1996.
Seeger LL, Yao L, Eckardt JJ: Surface lesions of bone. Radiology 206:17, 1998.
Stanton RP, Abdel-Mota'al MM: Growth arrest resulting from unicameral bone cyst. J Pediatr Orthop 18:198, 1998.
Struhl S, Edelson C, Pritzker H, et al: Solitary (unicameral) bone cyst: The fallen fragment sign revisited. Skeletal Radiol 18:261, 1989.
Tanaka H, Araki Y, Yamamoto H, et al: Intraosseous ganglion. Skeletal Radiol 24:155, 1995.
Trias A, Beauregard G: Epidermoid cyst of bone. Can J Surg 17:35,d 1974.
Vergel De Dios AM, Bond JR, Shives TC, et al: Aneurysmal bone cyst: A clinicopathologic study of 238 cases. Cancer 69:2921, 1992.

CHAPTER 16

Imaging of Metastatic Bone Disease

Murali Sundaram, MD

OVERVIEW

Metastasis to bone is the most common type of malignant bone tumor. Multiplicity is the usual presentation, although solitary lesions frequently are seen in patients with cancer of the kidney and thyroid. In decreasing order of frequency, the most common primary neoplasms with metastasis to bone are cancers of the breast, prostate, lung, kidney, and thyroid. In children, the primary malignancies that metastasize to bone are neuroblastoma, Ewing's sarcoma, and rhabdomyosarcoma. It is estimated that carcinoma of the prostate accounts for 60% of metastases to the bone in men, and carcinoma of the breast accounts for about 70% of metastases to the bone in women. Metastasis from prostate cancer usually occurs in men after their 40s, whereas metastatic breast carcinoma is encountered over a wider age range. Metastases from cancers of the lung, kidney, and thyroid usually are seen in middle-aged and older individuals.

PATHOPHYSIOLOGY OF SKELETAL METASTASIS

Overview

Mechanisms for tumor spread to bone are thought to be by direct invasion, retrograde venous flow, and systemic arterial circulation (after venous or lymphatic access); however, the predominant mechanism of metastatic spread is through the arterial circulation. Many theories have been proposed to explain why tumor cells implant and multiply in bone, but so far, the exact mechanism is not understood fully. The two most commonly cited theories are the *mechanical theory* and the *seed and soil theory*. The mechanical theory states that the vascular architecture of certain organs is conducive to implantation and growth of metastatic cells. Sites where vessels transition from large to small are sites where metastatic cells can become mechanically trapped.

As cancer cells circulate, they also interact with different antigens. The seed and soil theory implies that certain cancers preferentially interact with some antigens, increasing the probability of metastatic cells implanting at particular sites. When attached, cancer cells can migrate outside the sinusoidal wall and proceed to grow within the target tissue. The initial seeding of metastatic cells in bone occurs within the red marrow. Yellow marrow is hypovascular and is not a common site for metastasis. This fact explains the predominance of metastatic lesions within the axial skeleton, such as the spine, cranium, pelvis, proximal femora, and proximal humeri. Greater than 90% of metastatic lesions are found in this distribution. Metastasis distal to the elbows and knees is fairly uncommon. Most metastases to the wrists and hands are from the lung (Fig. 16–1). Subchondral metastasis (i.e., metastasis to the epiphysis of a long bone) usually does not occur because of the lack of red marrow at such sites (Fig. 16–2).

Vertebral Metastasis

The spine is the most common site for skeletal metastasis. Of patients who die of cancer, 40% have vertebral metastases. About half of patients with vertebral metastasis die within 6 months. Metastasis initially in-

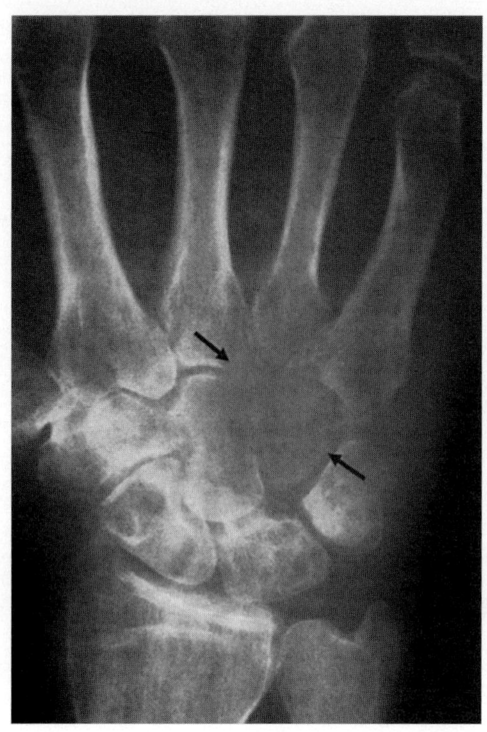

Figure 16–1. Metastasis to the carpal bones from carcinoma of the lung. Anteroposterior view of the wrist shows lytic destruction of the hamate and to a lesser extent the capitate bone *(arrows)*. Fine-needle aspiration revealed neoplastic cells compatible with carcinoma of the lung.

volves the vertebral body and not the pedicles; however, pedicle destruction is much easier to recognize on radiographs because of its high content of cortical bone (Fig. 16–3). The vertebral body, being predominantly cancellous bone, requires more extensive destruction before metastasis becomes visible on radiographs. Studies have shown that whenever radiographs reveal the presence of a destroyed pedicle, follow-up computed tomography (CT) or magnetic resonance (MR) imaging invariably shows that the metastatic deposit initially started in the vertebral body and subsequently extended into the pedicle (see Fig. 16–3). In the unusual case in which the vertebral arch (pedicles and laminae) harbors tumor deposits without vertebral body involvement, metastasis would be unlikely, and multiple myeloma should be considered. The most serious complication of vertebral metastasis is epidural invasion by the tumor, resulting in cord compression (Fig. 16–4). Cord compression most commonly occurs in the thoracic spine (70%) because of its length and narrow lumen; this is followed by the lumbar spine, then the cervical spine (see Fig. 16–4). Lymphoma is the only tumor that is reported to deposit primarily in the epidural space (i.e., lymphoma can metastasize to the epidural space without involvement of a vertebral body).

IMAGING

Radiography

Overview

Metastases usually are osteolytic, osteosclerotic (Fig. 16–5), or mixed; however, carcinoma of the breast can be lytic, blastic, or mixed (Fig. 16–6). Metastases from lung cancer usually are osteolytic (the rare bronchial carcinoid may produce blastic metastases), whereas metastases from prostate cancer almost always are blastic (Fig. 16–7). Purely lytic metastatic lesions may be seen from carcinomas of the thyroid and kidney, which have a propensity to

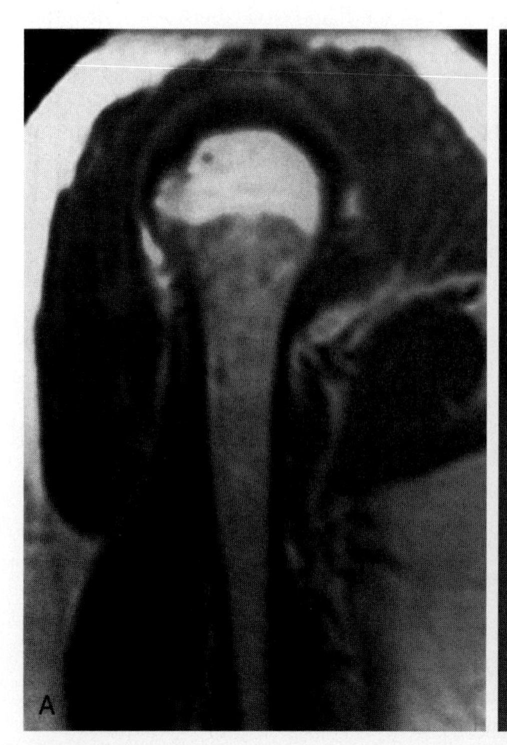

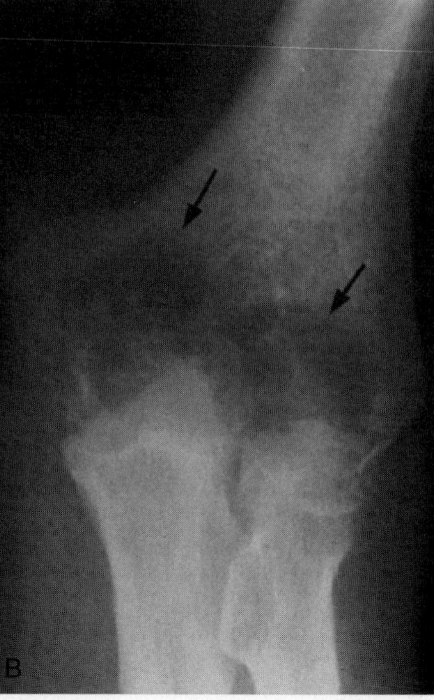

Figure 16–2. Red and yellow marrow distribution in long bones explaining the lack of metastasis in subchondral areas. *A,* Sagittal T1-weighted MR image of the humerus in a healthy 44-year-old woman shows yellow marrow occupying the epiphysis and red marrow within the diaphysis. *B,* Example of subchondral metastasis to the distal humerus in a patient known to have carcinoma of the urinary bladder. Anteroposterior view of the right elbow shows lytic destruction in the subchondral region of the distal humerus *(arrows)*, which on fine-needle aspiration revealed neoplastic cells.

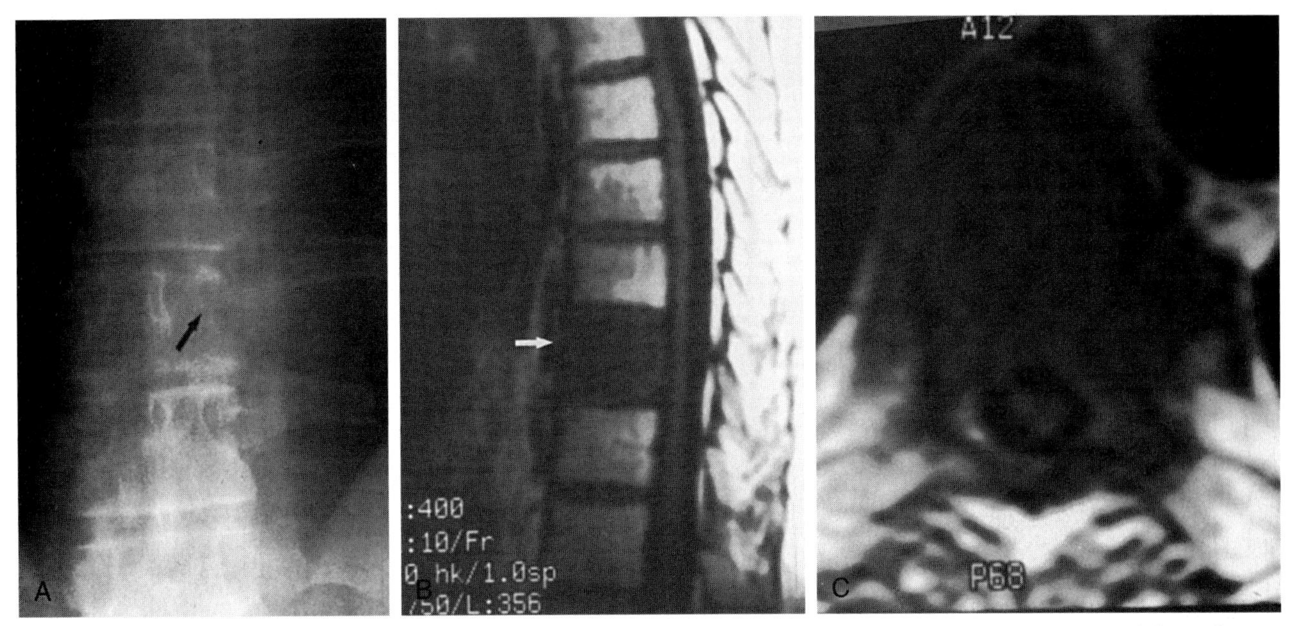

Figure 16–3. Absent pedicle resulting from metastasis with vertebral body involvement, which probably preceded the pedicle involvement. *A,* Anteroposterior view of the lower thoracic spine shows an absent left pedicle at T10 *(arrow),* suggesting metastasis. *B,* Sagittal T1-weighted MR image shows that the entire vertebral body is infiltrated with tumor *(arrow). C,* Axial T1-weighted MR image reveals that the vertebral body of T10 and the left pedicle are infiltrated with tumor.

blow out bone (Fig. 16–8). These patterns should serve as guidelines influenced by age and gender in suggesting the primary malignancy, but none of these patterns are specific. Lytic metastatic lesions may be geographic, moth-eaten, or permeative; they may have a narrow or wide zone of transition. Metastases may be intracompartmental (i.e., confined to bone), or they may breach the bone and produce an adjacent soft tissue mass (see Fig. 16–8). Profound periosteal reaction is an uncommon feature of metastatic lesions; however, prostatic metastases are known to produce exuberant periosteal reaction with a sunburst appearance mimicking osteosarcoma (Fig. 16–9). Certain tumors favor metastasis to particular sites. Lung carcinoma, more than any other primary source, tends to metastasize to cortical bone (Fig. 16–10). Gynecologic tumors tend to invade the pelvic brim. Pancoast tumors invade the chest wall directly, especially the first two ribs. Metastatic deposits from hypernephroma are typically vascular, making surgical intervention without prior embolization hazardous (Fig. 16–11). Some authors believe that metastases preferentially go to pagetic bone because it is more vascular (Fig. 16–12).

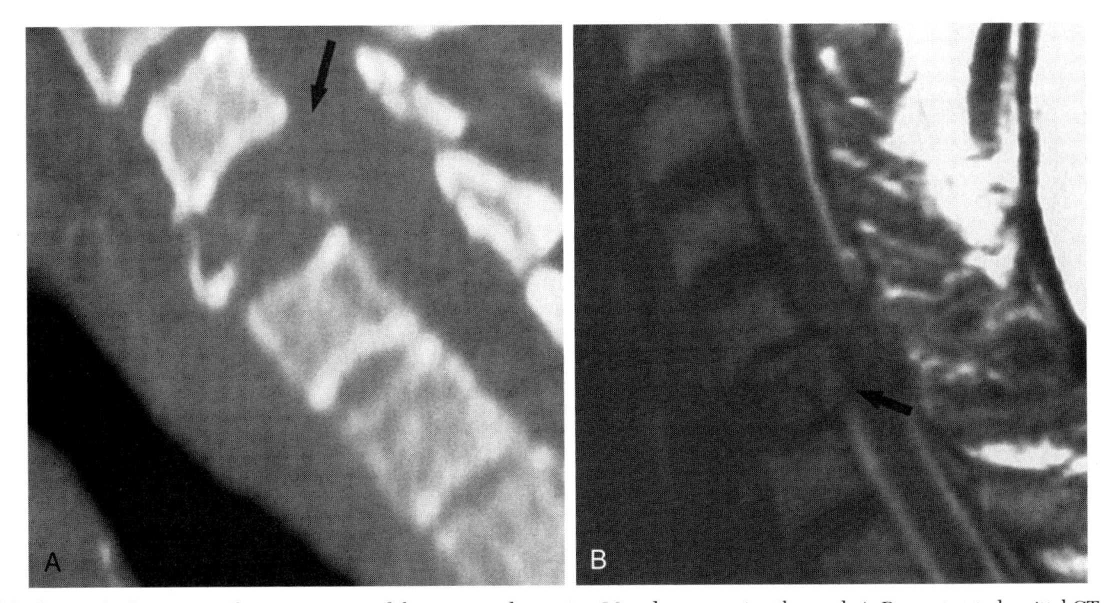

Figure 16–4. Vertebral metastasis from carcinoma of the prostate destroying C6 and compressing the cord. *A,* Reconstructed sagittal CT scan shows lytic destruction of the entire body of C6 *(arrow). B,* Sagittal T2-weighted MR image shows epidural extension of the tumor compressing the cord *(arrow).*

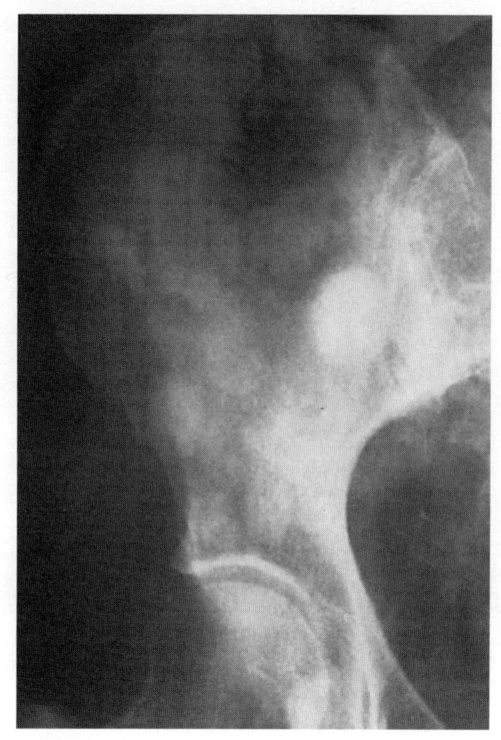

Figure 16–5. Blastic metastasis. Anteroposterior view of the right ilium shows multiple sclerotic lesions from metastatic carcinoma of the breast.

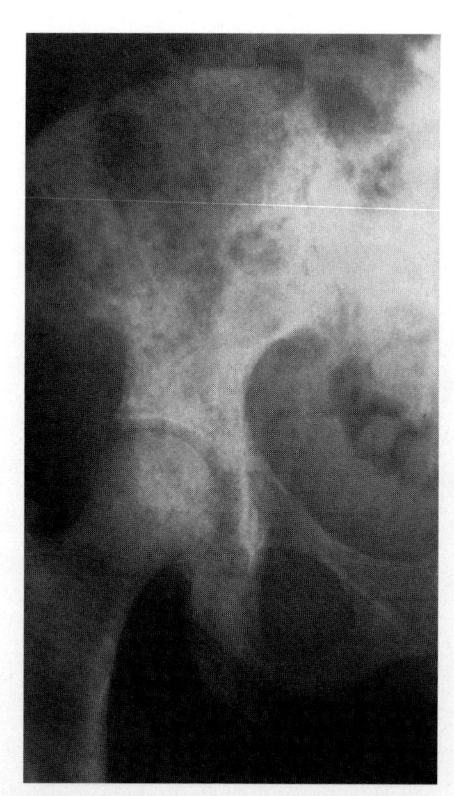

Figure 16–6. Mixed lytic and blastic metastasis from carcinoma of the breast. Anteroposterior view of the hip shows lytic and sclerotic areas in the iliac wing consistent with mixed lytic and blastic metastasis.

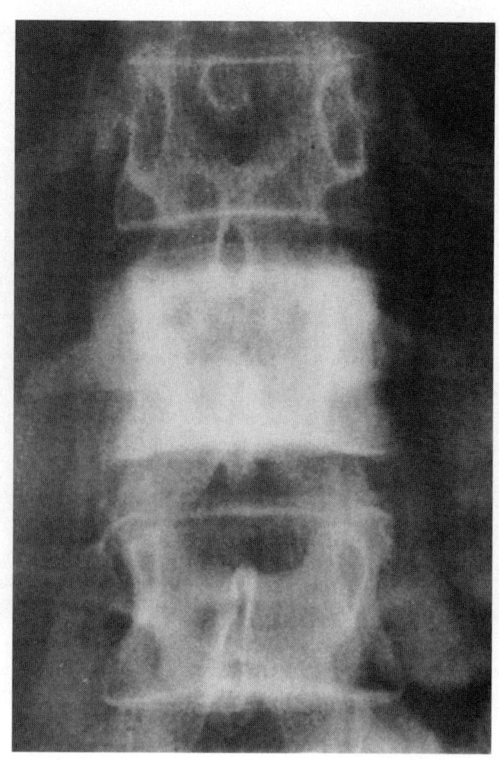

Figure 16–7. Blastic metastasis in a 62-year-old man with carcinoma of the prostate. Anteroposterior view of the lumbar spine reveals densely blastic L1 (ivory vertebra).

Vertebral Metastasis

In the vertebrae, metastases may be osteolytic, blastic, or mixed and may or may not be associated with vertebral collapse (Fig. 16–13). Metastatic carcinomas of the breast, lung, and prostate, in order of decreasing frequency, account for most cases of vertebral collapse. A helpful feature in the diagnosis of vertebral metastasis is the preservation of diskal height. Diskal height is a useful diagnostic aid in differentiating tumor from pyogenic

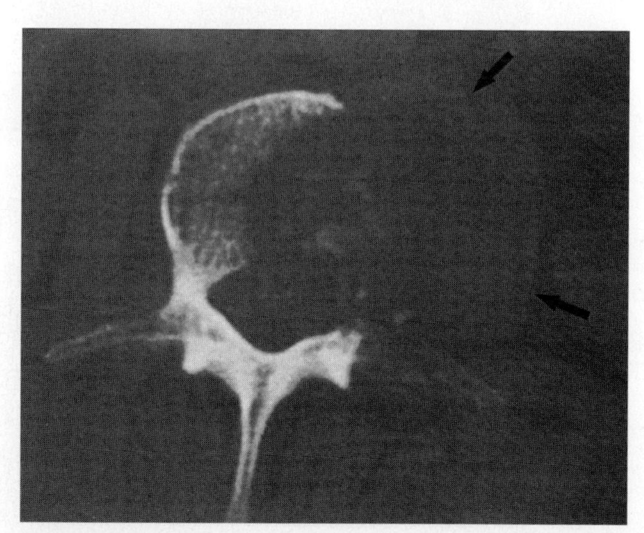

Figure 16–8. Blowout lytic metastasis from a hypernephroma to L2. Axial CT scan shows a large soft tissue component *(arrows)*. The vertebral body and pedicle on the left are involved.

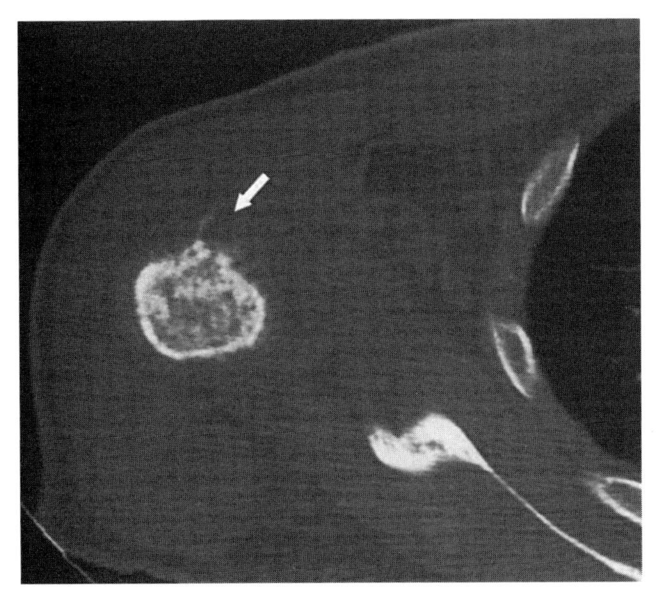

Figure 16–9. Sunburst appearance in prostate carcinoma metastasis to the right humerus. Axial CT scan through the proximal humerus shows the sunburst periosteal reaction resembling osteosarcoma *(arrow)*.

spondylodiskitis, which typically destroys the intervertebral disk.

Nuclear Medicine

Technetium bone scanning is a reliable imaging technique for determining the extent of metastatic disease. Radionuclide bone scanning can image the entire skeleton (see Fig. 16–13). There is a moderate rate of false-negative and false-positive bone scans, however. Solitary sites of increased uptake in patients with known malignancy can pose a diagnostic problem. Approximately 21% of patients with metastatic breast cancer present with a solitary lesion, which typically is located in the spine. Solitary rib metastasis in cancer patients is uncommon, and 90% of hot rib lesions are due to benign causes. In patients with prostate carcinoma, a bone scan usually is not indicated unless the prostate-specific antigen is greater than 20 ng/mL.

Computed Tomography

CT does not have a consistent role in the evaluation of metastatic lesions; however, it is used commonly to guide percutaneous needle biopsies. CT may be useful in imaging the pelvis if there is uncertainty about the presence of a lesion. CT is used in determining the amount of cortical destruction, especially in the femur, before prophylactic surgical fixation of a lytic metastatic lesion (Fig. 16–14). Blastic metastases do not require prophylactic fixation routinely because they do not weaken the bone.

Magnetic Resonance Imaging

MR imaging effectively confirms the presence or absence of metastatic lesions suspected on radiography or scintigraphy. Imaging of vertebral metastasis, especially in the presence of epidural extension, is accomplished best with MR imaging. MR imaging allows direct visualization of cord compression (see Fig. 16–4*B*). Multilevel cord compression occurs in 9% of patients with spinal metastases; this explains the rationale for performing total spine MR imaging. The usual spin echo MR imaging sequences cannot differentiate reliably acute osteopenic vertebral collapse from neoplastic collapse. Diffusion MR imaging has been advocated in differentiating these entities, but

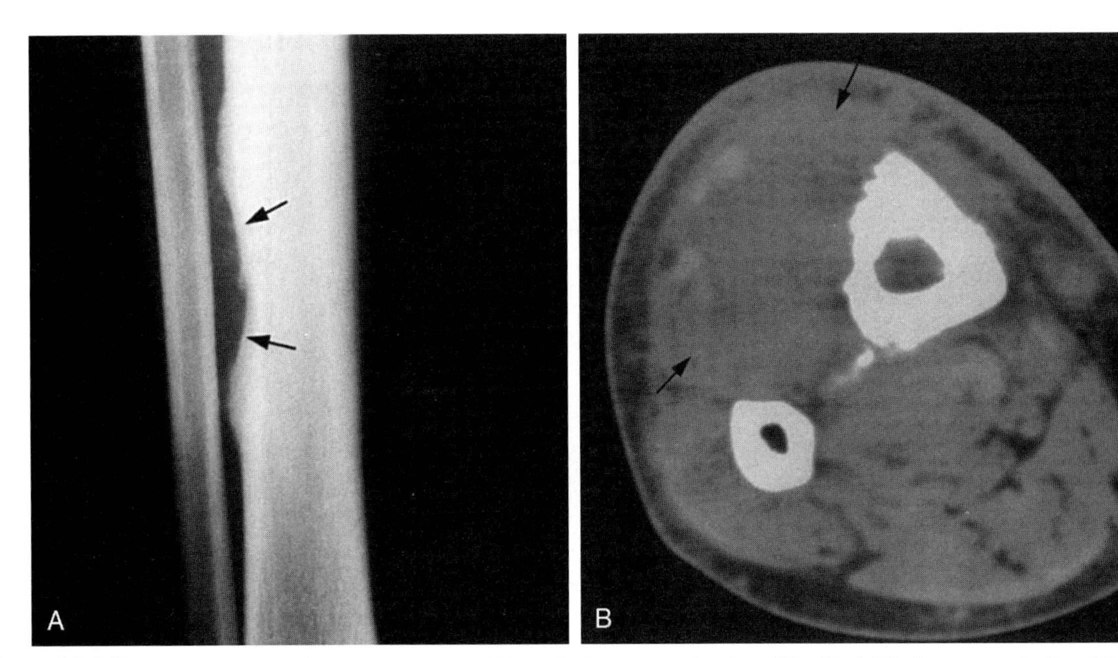

Figure 16–10. Lung carcinoma with metastasis to cortical bone. *A*, Anteroposterior view of the distal tibia shows saucerization of the cortex *(arrows)* with periosteal new bone. *B*, Axial CT scan at the level of the lesion shows cortical involvement and extension into the adjacent soft tissues *(arrows)*.

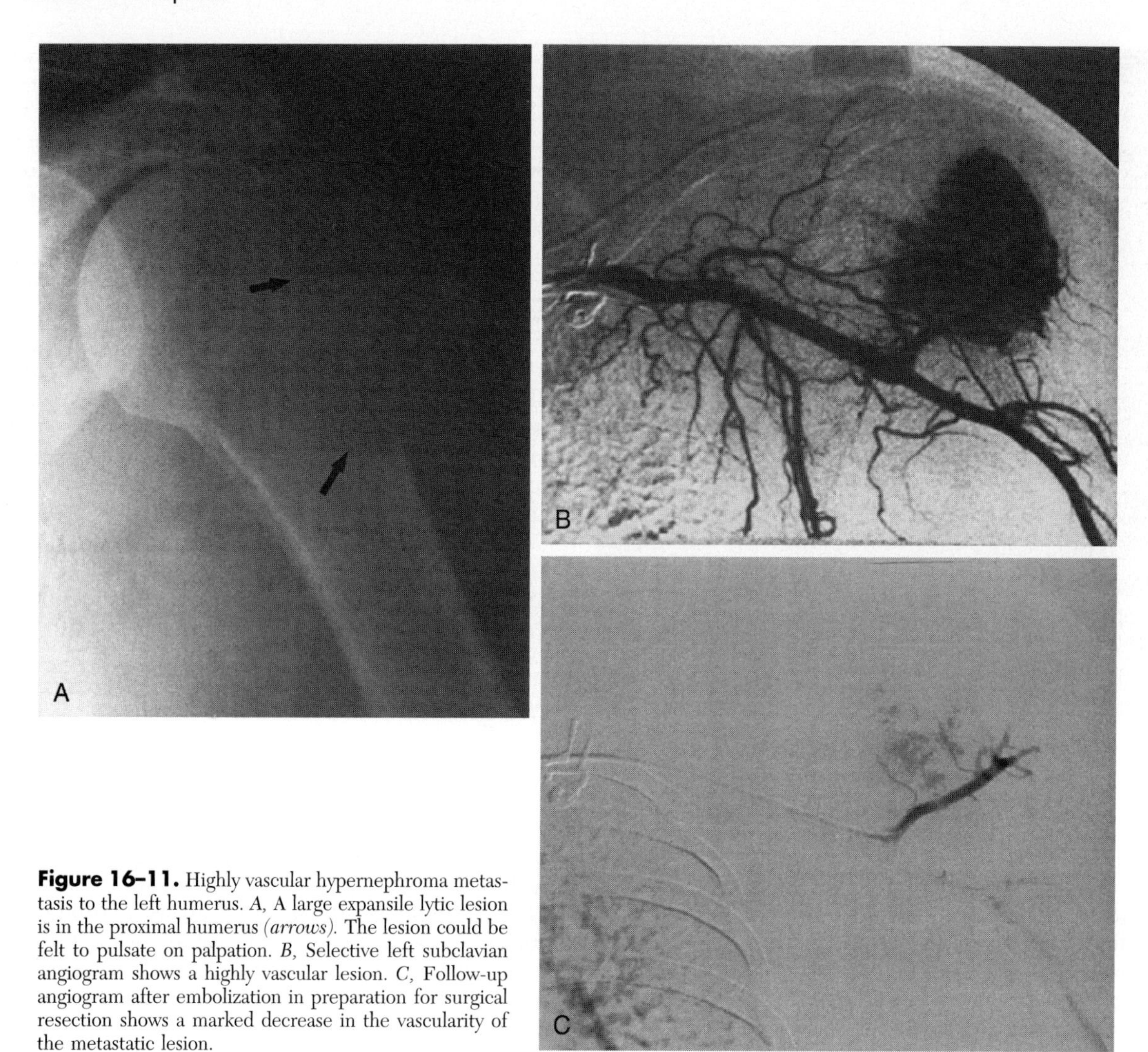

Figure 16–11. Highly vascular hypernephroma metastasis to the left humerus. *A,* A large expansile lytic lesion is in the proximal humerus *(arrows).* The lesion could be felt to pulsate on palpation. *B,* Selective left subclavian angiogram shows a highly vascular lesion. *C,* Follow-up angiogram after embolization in preparation for surgical resection shows a marked decrease in the vascularity of the metastatic lesion.

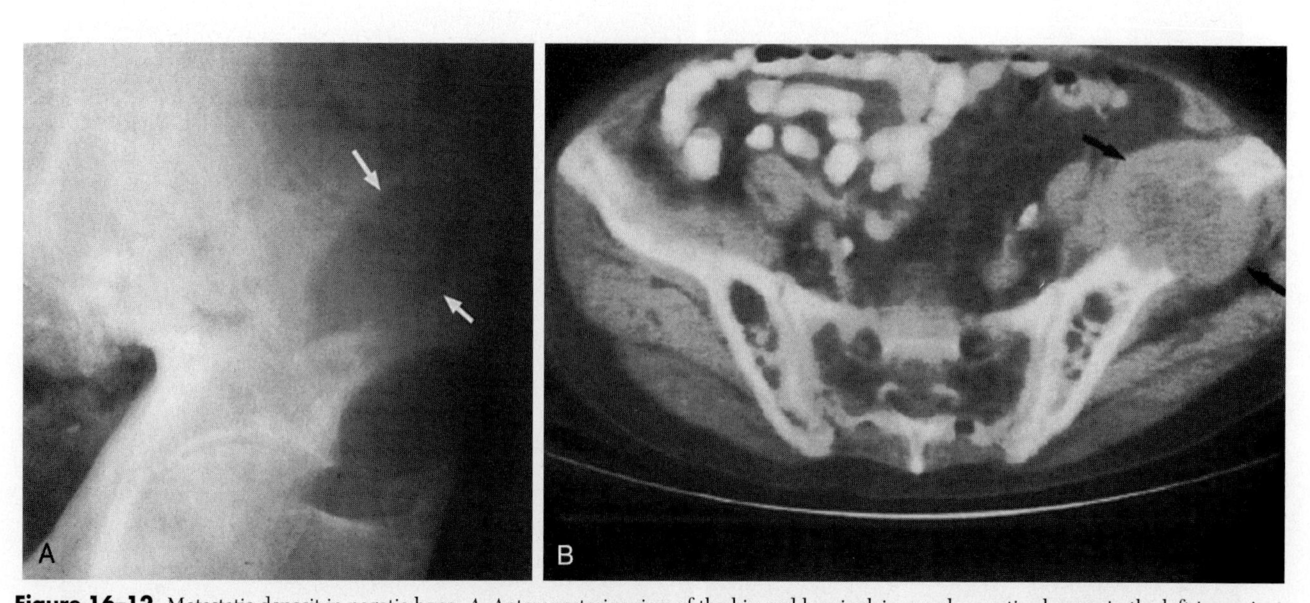

Figure 16–12. Metastatic deposit in pagetic bone. *A,* Anteroposterior view of the hip and hemipelvis reveals pagetic changes in the left innominate bone and a lytic lesion in the left ilium *(arrows).* *B,* CT scan through the pelvis reveals Paget's disease and a lytic expansile lesion in the left ilium *(arrows).* Needle aspiration from this lesion revealed metastatic carcinoma.

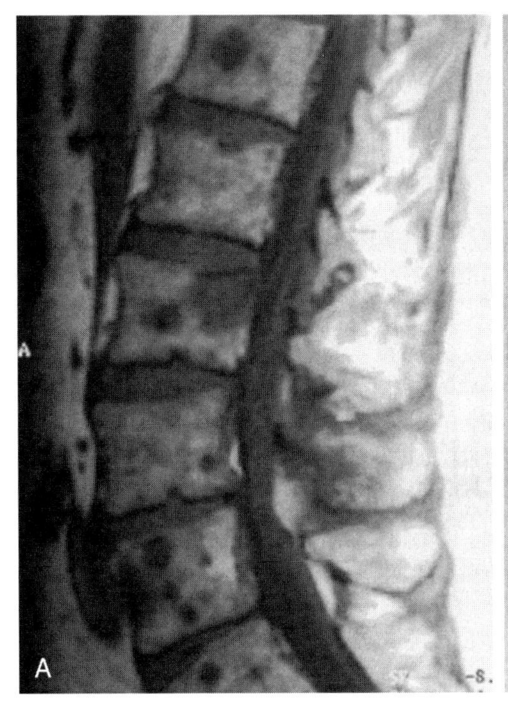

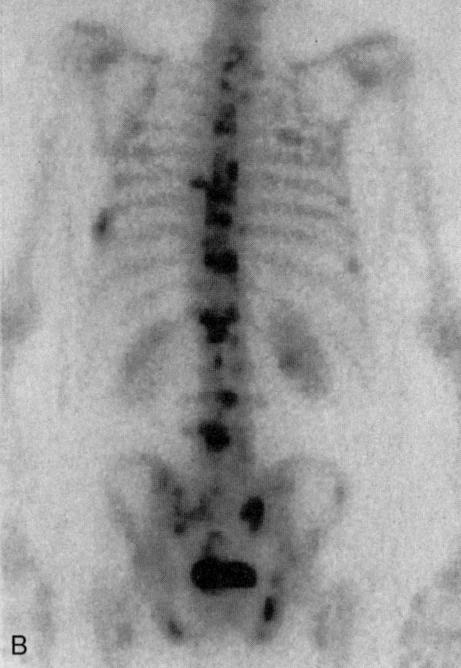

Figure 16–13. Multiple metastatic lesions in the spine from prostate carcinoma. *A,* T1-weighted sagittal MR image shows multiple, relatively small, metastatic deposits throughout the lumbosacral spine. *B,* Radionuclide bone scan reveals extensive metatases to the spine and pelvis.

the technique is not widely available at this time. There are, however, MR imaging signs that can be used to differentiate osteopenic from neoplastic vertebral collapse. It has been shown that a vertebra involved with metastasis does not lose height, or collapse, unless the entire body is infiltrated with neoplasm. The presence of fatty marrow on MR imaging within a collapsed vertebral body (i.e., marrow appearing hypointense on T1-weighted images in only a portion of the vertebral body) is a reliable sign of benign or osteopenic collapse (Fig. 16–15). Lack of pedicle involvement and absence of a soft tissue mass in association with the collapse are signs of benign or osteopenic collapse. Acute osteopenic collapse often shows a characteristic band of decreased signal intensity parallel to compressed vertebral end plates on T1-weighted images (Fig. 16–15*A*).

As MR imaging sequences continue to become faster, emerging evidence shows that whole-body fast short tau inversion recovery (STIR) MR imaging is feasible and that it can replace bone scintigraphy for the detection of metastatic bone disease (Fig. 16–16). Proponents of this technique indicate that whole-body MR imaging is more sensitive and more specific than bone scintigraphy. Whole-body MR imaging also is comparable in cost to

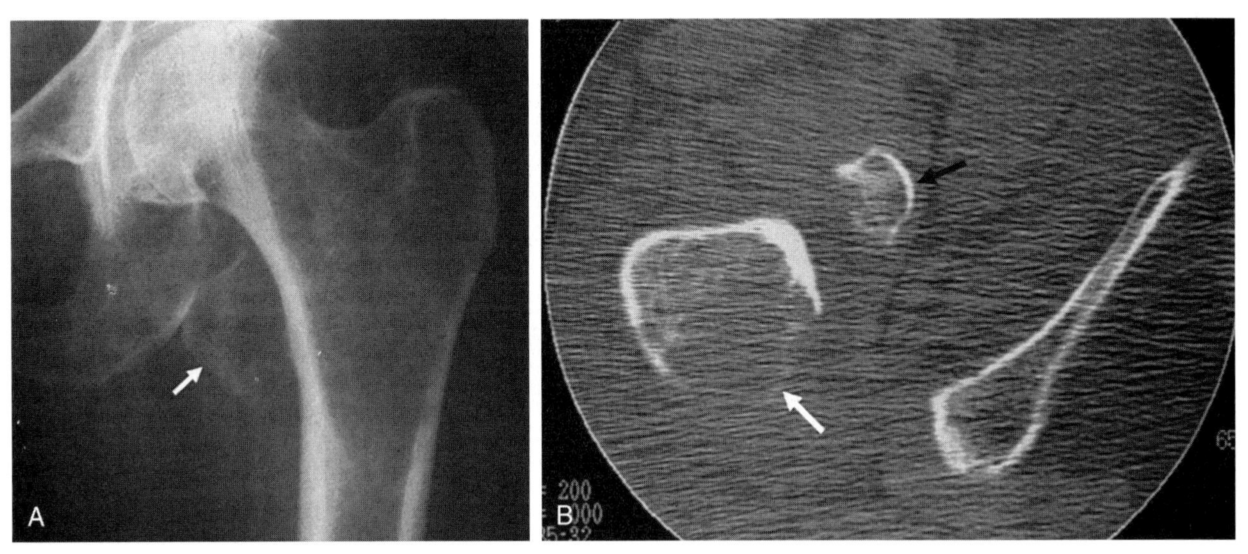

Figure 16–14. Avulsion of the lesser trochanter without a significant history of trauma. This finding in older patients always should be suspected as being the result of metastasis. *A,* Anteroposterior view of the left hip shows avulsion of the lesser trochanter *(arrow). B,* CT scan shows the extent of cortical destruction *(white arrow).* The avulsed lesser trochanter also is seen *(black arrow).*

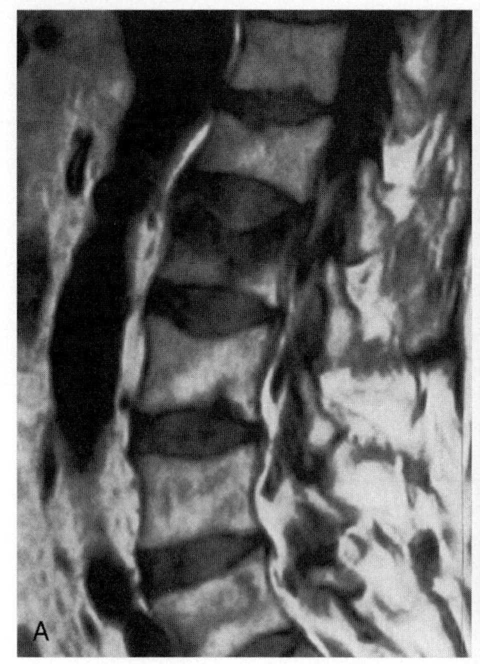

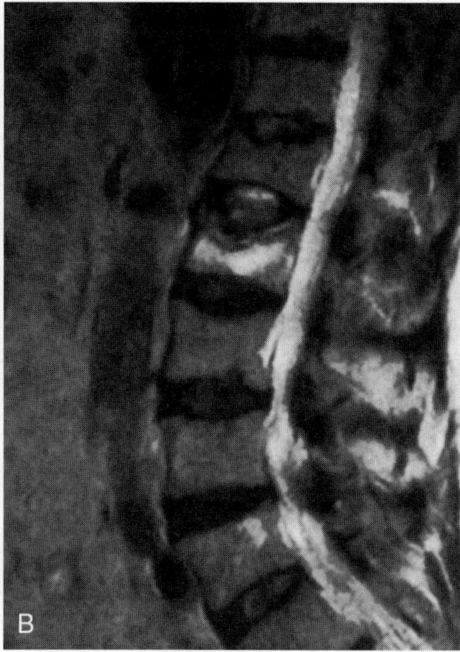

Figure 16–15. Osteopenic vertebral collapse of L1 and L2. *A*, T1-weighted sagittal MR image shows normal marrow signal intensity in L1 and hypointense marrow signal in the upper half of L2. Both vertebrae show partial collapse. *B*, T2-weighted sagittal MR image shows normal marrow signal intensity in L1 and increased marrow signal intensity in L2. These findings are typical of an old, healed compression fracture at L1 and an acute compression fracture at L2.

bone scintigraphy. There is no ionizing radiation involved with whole-body MR imaging, making this technique especially suited for pregnant patients with suspected bone metastasis. This technique is also used in the evaluation and follow-up of patients with multiple myeloma and breast cancer (Fig. 16–17).

Positron Emission Tomography Scanning

Positron emission tomography (PET) scanning is a rapidly developing technology with promising potential for staging and follow-up of patients with neoplasms. The

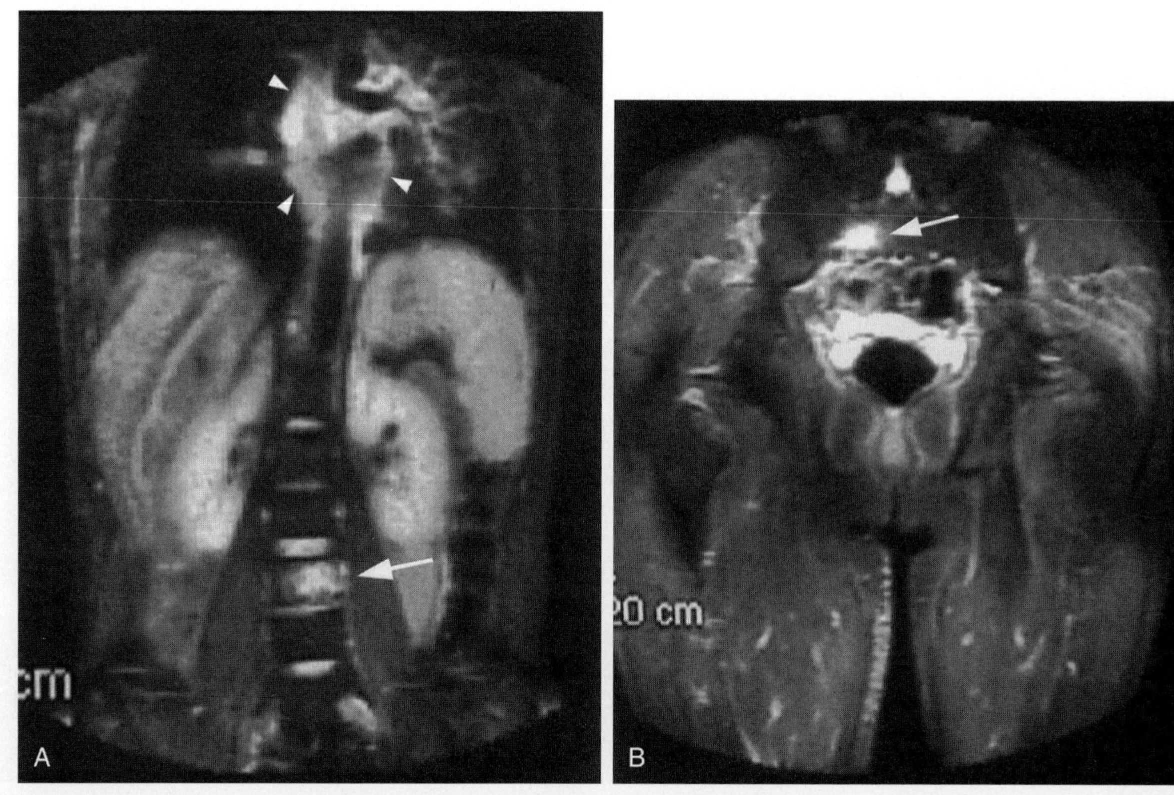

Figure 16–16. Total body MR using a STIR sequence in the coronal plane. The patient had a lung carcinoma. *A*, Coronal image of the chest and abdomen shows a metastatic lesion in L3 vertebral body *(arrow)*. The primary tumor *(arrowheads)* in the chest is also noted. *B*, Coronal image of the pelvis and proximal thighs shows another metastatic lesion in the sacrum *(arrow)*.

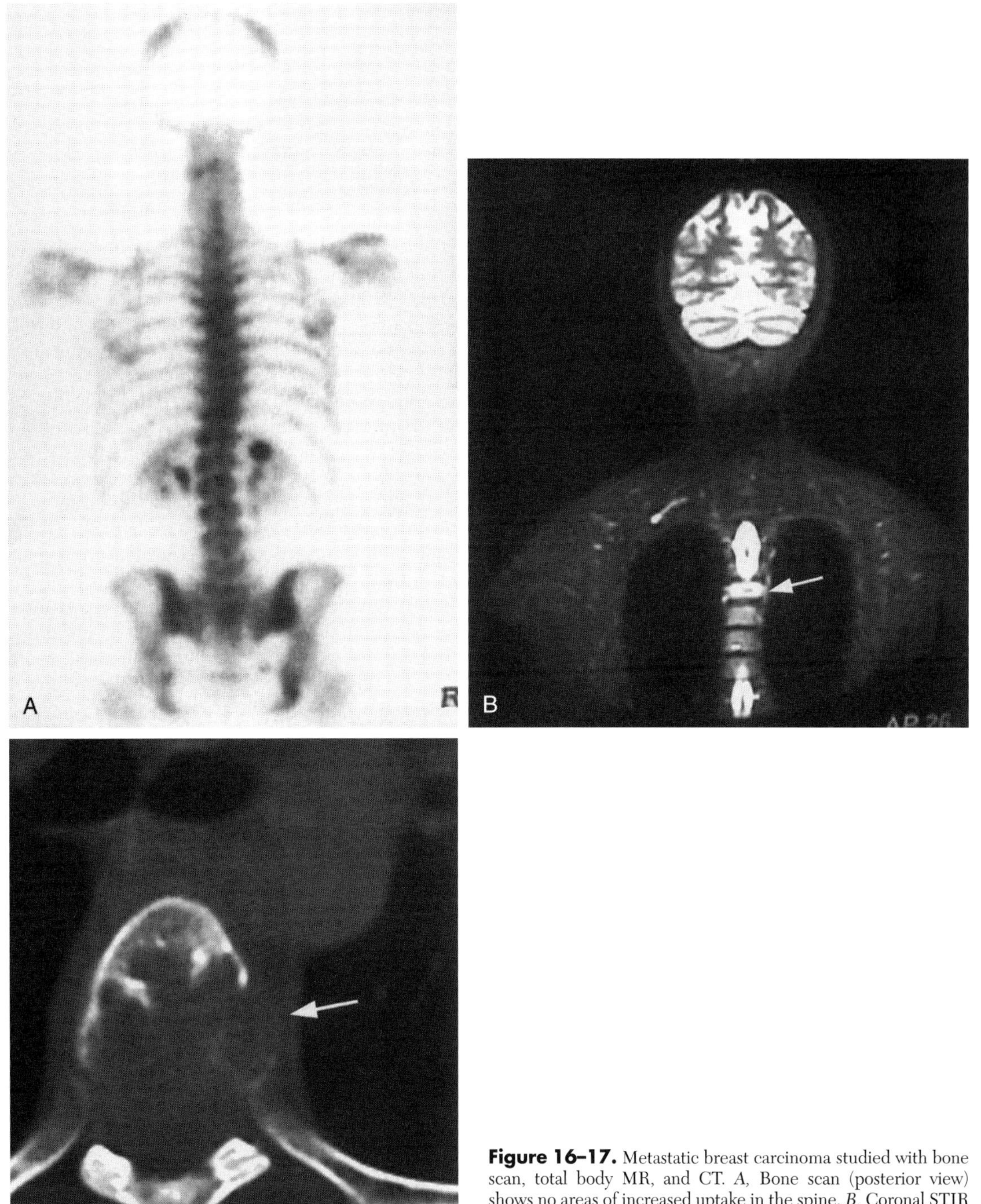

Figure 16–17. Metastatic breast carcinoma studied with bone scan, total body MR, and CT. *A,* Bone scan (posterior view) shows no areas of increased uptake in the spine. *B,* Coronal STIR image of the chest, neck, and head shows a metastatic lesion in the body of T6 *(arrow)*. *C,* On CT the destructive lesion in T6 vertebral body is confirmed *(arrow)*.

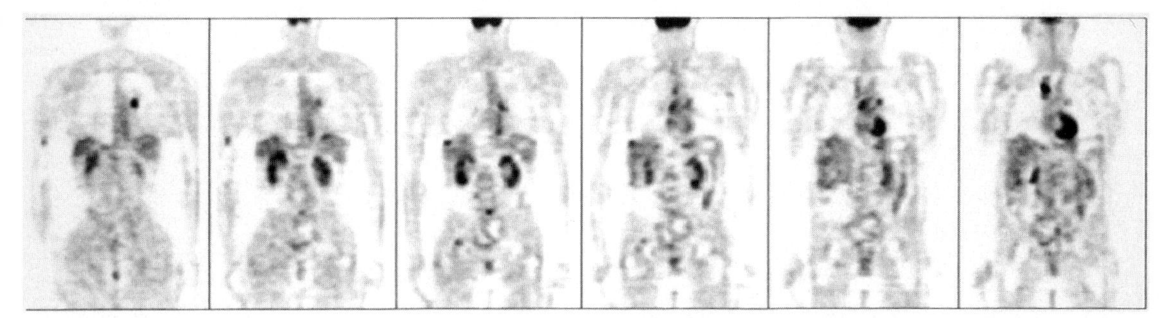

Figure 16–18. ^{18}FDG-PET scan for staging lung carcinoma. Six coronal tomographic sections starting in the back show the primary lesion on the first image in the left chest. The second and third images show bony metastases in the right humerus and L5 vertebra. The last two images show mediastinal metastasis.

most commonly used radiotracer is fluorodeoxyglucose (^{18}FDG), a glucose analogue that is actively metabolized by tumors (Fig. 16–18).

Currently PET is widely used in the diagnosis and staging of lung carcinoma. In 25% of patients with lung carcinoma, the management is changed based on the PET findings. PET is also used for the staging of head and neck tumors, lymphoma, melanoma, and esophageal carcinoma. In colon carcinoma PET is used to check for recurrent and metastatic disease. It is also used for the detection of metastatic breast carcinoma. Most authorities believe that it is only a matter of time before PET scanning is used to stage and follow-up bone and soft tissue sarcomas.

DIFFERENTIAL DIAGNOSIS

Osteopenic versus neoplastic vertebral collapse is a common vexing problem in daily practice (see previously for helpful MR imaging signs to differentiate these two conditions). Bone islands, osteopoikilosis, insufficiency fractures of the sacrum and pubis, and monostotic Paget's disease of the sacrum are some entities that can be mistaken for metastatic disease. Bone islands and osteopoikilosis can be ignored in healthy individuals. In patients with known prostate or breast cancer, bone scintigraphy may be required to differentiate bone islands or osteopoikilosis from blastic metastasis; benign lesions usually do not have increased radionuclide uptake; however, bone islands larger than 1.5 cm may show mild-to-moderate radionuclide uptake. Insufficiency fractures of the sacrum have a fairly characteristic appearance on CT. Their MR imaging appearance can be confusing, especially early, when they can resemble a neoplasm. Insufficiency fractures may be unilateral or bilateral and usually are seen in postmenopausal women or in women with a history of previous irradiation to the pelvis. On a radionuclide bone scan, the characteristic Honda sign (H) may be identified (Fig. 16–19). When the abnormality is unilateral, there may be less certainty about the diagnosis; however, this can be resolved by CT (see Fig. 16–19*B*). Monostotic Paget's disease of the sacrum is common. On radiographs, it might be identified as an ill-defined sclerotic area and on radionuclide bone scan as an area of intense radionuclide uptake. Under these circumstances, CT shows thickened cortices and coarse trabeculae secondary to Paget's disease. In the spine, one manifesta-

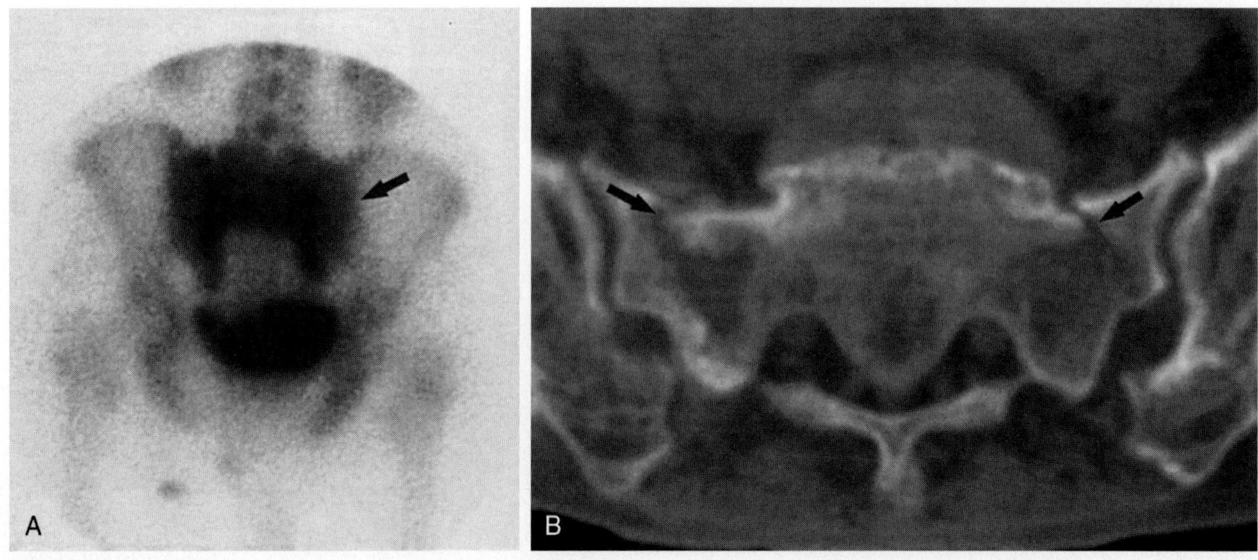

Figure 16–19. Insufficiency fracture of the sacrum. *A*, Radionuclide bone scan shows the characteristic Honda *(H)* sign in the sacrum *(arrow)*. *B*, Axial CT scan through the sacrum illustrates the fracture lines *(arrows)*.

tion of Paget's disease is the diffusely sclerotic vertebra—the so-called ivory vertebra. This appearance also may be seen in blastic metastasis and Hodgkin's disease. With Paget's disease, however, there is usually enlargement of the vertebra and no associated soft tissue mass.

Recommended Readings

Algra PR, Bloem JL, Tissing H, et al: Detection of vertebral metastases: Comparison between MR imaging and bone scintigraphy. Radiographics 11:219, 1991.

Baur A, Stabler A, Bruning R, et al: Diffusion-weighted MR imaging of bone marrow: Differentiation of benign versus pathologic compression fractures. Radiology 207:349, 1998.

Boxer DI, Todd CEC, Coleman R, et al: Bone secondaries in breast cancer: The solitary metastasis. J Nucl Med 30:1318, 1989.

Chybowski FM, Keller JJL, Bergstralh EJ, et al: Predicting radionuclide bone scan findings in patients with newly diagnosed, untreated prostate cancer: Prostate specific antigen is superior to all other clinical parameters. J Urol 145:313, 1991.

Eustace S, Tello R, DeCarvalho V, et al: A comparison of whole-body turbo STIR MR imaging and planar ^{99m}Tc-methylene diphosphonate scintigraphy in the examination of patients with suspected skeletal metastases. AJR Am J Roentgenol 169:1655, 1997.

Even-Sapir E, Martin RH, Barnes DC, et al: Role of SPECT in differentiating malignant from benign lesions in the lower thoracic and lumbar vertebrae. Radiology 187:193, 1993.

Holder LE: Clinical radionuclide bone imaging. Radiology 176:607, 1990.

Kunkler IH, Merrick MV, Rodger A: Bone scintigraphy in breast cancer: A nine-year follow-up. Clin Radiol 36:279, 1985.

Michel F, Soler M, Imhof E, et al: Initial staging of non-small cell lung cancer: Value of routine radioisotope bone scanning. Thorax 46:469, 1991.

Smoker WRK, Godersky JC, Knutson RK, et al: The role of MR imaging in evaluating metastatic spinal disease. AJR Am J Roentgenol 149:1241, 1987.

Tan SB, Kozak JA, Mawad ME: The limitations of magnetic resonance imaging in the diagnosis of pathologic vertebral fractures. Spine 16:919, 1991.

Tumeh SS, Beadle G, Kaplan WD: Clinical significance of solitary rib lesions in patients with extraskeletal malignancy. J Nucl Med 26:1140, 1985.

CHAPTER 17

Imaging of Soft Tissue Tumors

Murali Sundaram, MD

OVERVIEW

Magnetic resonance (MR) imaging has gained widespread acceptance as the modality of choice for imaging and follow-up of soft tissue tumors. For soft tissue masses, MR imaging can define the size; shape; location; and relationship to other structures, such as muscle compartments, fascial planes, bone, and neurovascular bundles. MR imaging also can provide information on hemorrhage, necrosis, edema, cystic degeneration, myxoid degeneration, and fibrosis. In general, MR imaging cannot predict the histologic diagnosis of soft tissue masses; however, several conditions can be diagnosed reliably based on location, morphology, and signal characteristics. Soft tissue masses can be categorized into three major groups: (1) non-neoplastic and reactive masses, (2) benign tumors, and (3) malignant tumors. Mesenchymal in origin, soft tissue neoplasms can arise from skeletal muscle, adipose tissue, vascular structures, connective tissue, and peripheral nerves.

IMAGING

Radiography

Radiographs, although of limited value, are an important first step in the imaging evaluation of soft tissue masses. Their value lies in (1) determining that a soft tissue mass closely apposed to bone has its origin in soft tissue and not bone and (2) showing calcification or ossification within a mass, such as hemangioma, synovial sarcoma, and myositis ossificans (Fig. 17–1). Large lipomas often can be identified on a radiograph (Fig. 17–2).

Ultrasound

Ultrasound is an attractive modality because of its widespread availability and relatively low cost. It can help identify aneurysms and differentiate cystic masses from solid tumors (Fig. 17–3). Ultrasound also can be used to

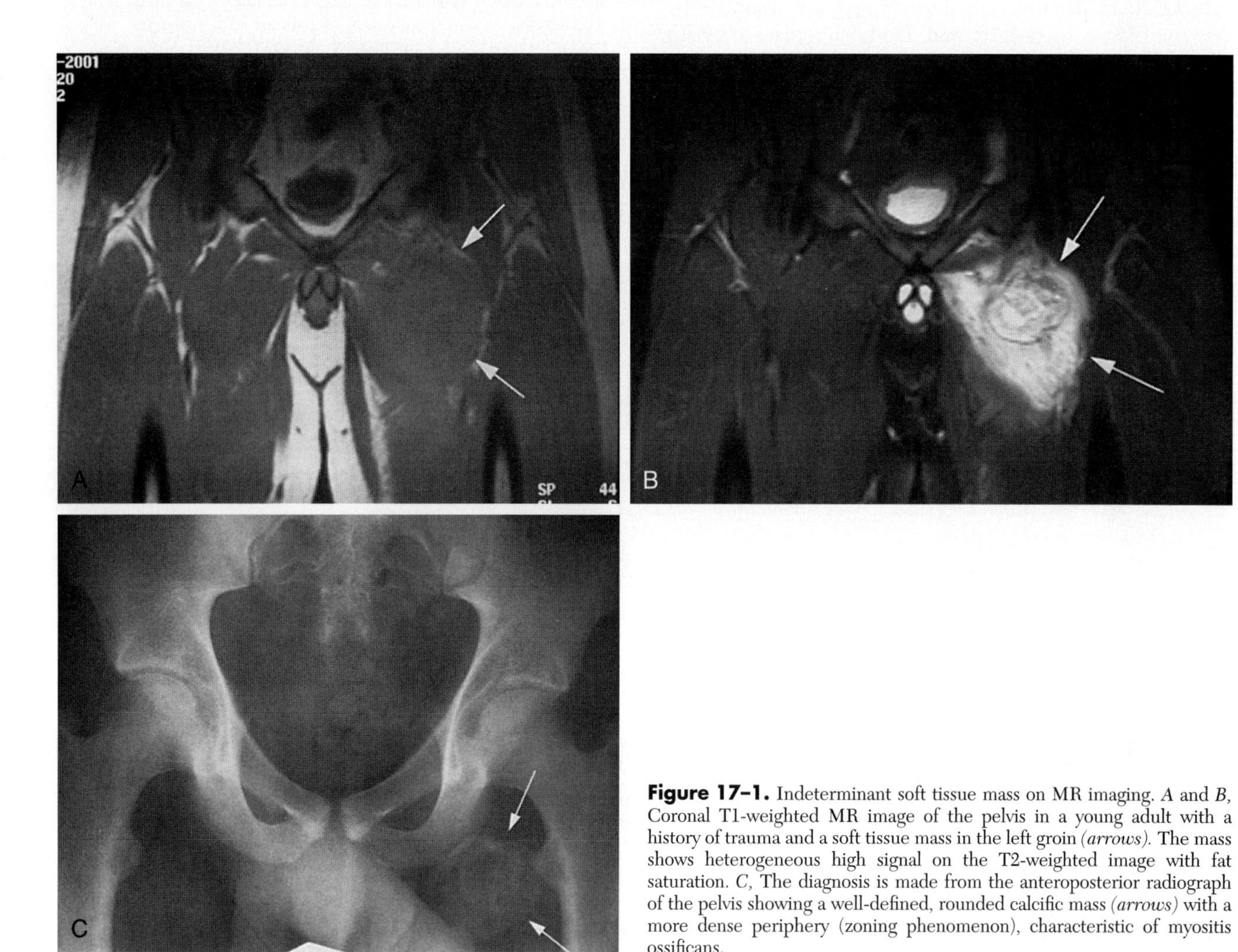

Figure 17–1. Indeterminant soft tissue mass on MR imaging. *A* and *B*, Coronal T1-weighted MR image of the pelvis in a young adult with a history of trauma and a soft tissue mass in the left groin *(arrows).* The mass shows heterogeneous high signal on the T2-weighted image with fat saturation. *C,* The diagnosis is made from the anteroposterior radiograph of the pelvis showing a well-defined, rounded calcific mass *(arrows)* with a more dense periphery (zoning phenomenon), characteristic of myositis ossificans.

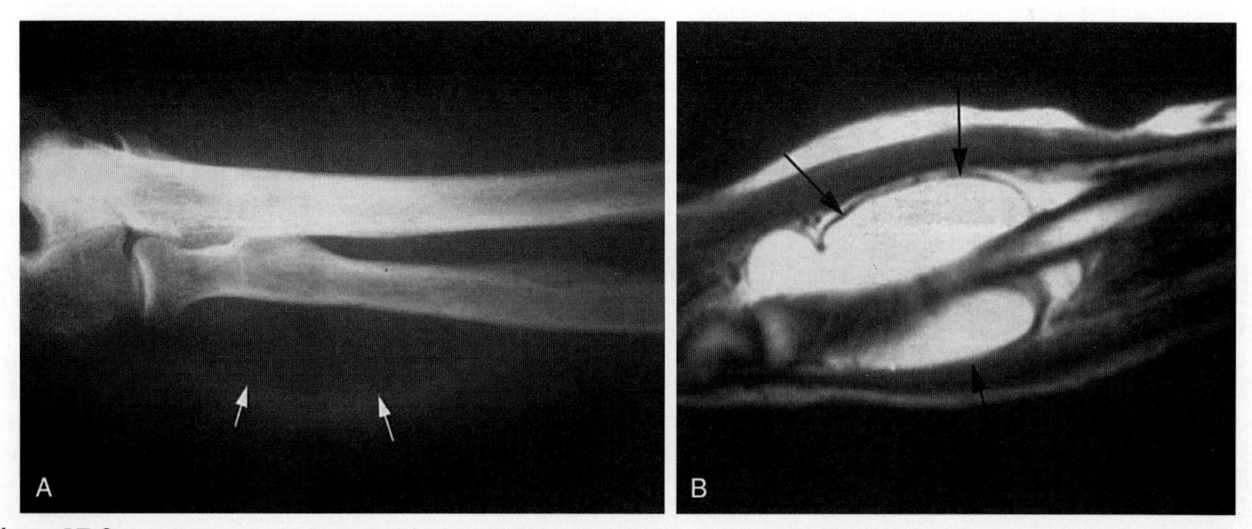

Figure 17–2. *A,* Anteroposterior radiograph of the forearm shows a large lucent area *(arrows)* adjacent to the radius, consistent with a fat-containing lesion. A lipoma is confirmed on this sagittal T1-weighted MR image. *B,* A large high signal region *(arrows)* envelops the proximal radius.

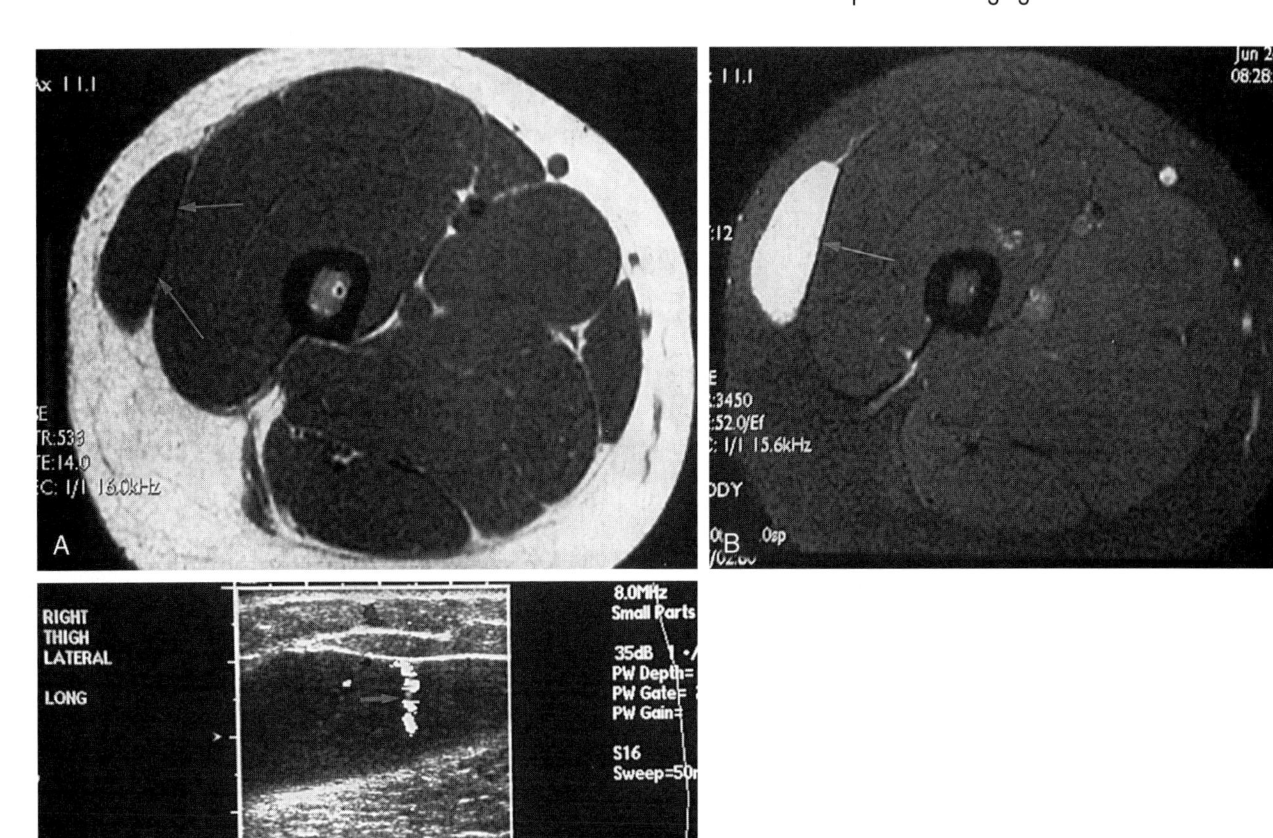

Figure 17–3. A pregnant patient with an asymptomatic palpable mass that developed in the thigh. *A* and *B*, Axial T1- and T2-weighted MR images show homogeneous signal within the mass, which could be cystic in nature. Gadolinium was not administered to the patient because it has not yet been approved for use during pregnancy. *C*, Ultrasound shows color Doppler *(arrow)* and pulse Doppler blood flow within the mass, indicating the mass is solid. The pathologic diagnosis was myxoid liposarcoma.

guide biopsy of solid soft tissue masses or aspiration of cystic lesions. Currently, ultrasound is not used as a primary imaging modality for detecting and staging soft tissue masses.

Computed Tomography

Computed tomography (CT) lacks the contrast resolution provided by MR imaging. Currently, CT has a limited role in the evaluation of soft tissue masses. Its major contribution is in evaluating the thorax and abdomen for staging of suspected sarcomas.

Positron Emission Tomography (PET)

Positron emission tomography is not used widely in the management of patients with soft tissue masses. It offers the potential for following patients with sarcomas when searching for recurrence or metastases.

Magnetic Resonance Imaging

MR imaging is the examination of choice for evaluating soft tissue lesions because it provides excellent discrimination between normal and abnormal tissues.

TUMOR LOCALIZATION AND STAGING BY MAGNETIC RESONANCE IMAGING

On T1-weighted spin echo sequences, most tumors have signal intensity equal to or lower than muscle. Discrimination between the tumor and surrounding fat is excellent. On T2-weighted sequences, most soft tissue tumors are hyperintense and are delineated clearly from the less intense muscle. Using T2-weighted spin echo and fast spin echo images, it often is difficult to distinguish the tumor from surrounding fat. Sequences using fat suppression are helpful in such situations. Useful MR imaging sequences include T1-weighted spin echo and T2-weighted (frequently with fat suppression) spin echo or fast spin echo sequences in the axial plane plus one other orthogonal plane. A short tau inversion recovery (STIR) sequence sometimes may be substituted for the fat-suppressed T2-weighted sequence. Gadolinium is not employed routinely. Gadolinium-enhanced studies can be useful, however, in detecting soft tissue abscess or in distinguishing a cystic lesion from a solid mass (Fig. 17–4). If gadolinium is used, fat saturation with the T1-weighted sequences often can define enhancing tissues better (Fig. 17–5). Gadolinium can be useful in tumor follow-up, detecting recurrent masses when suspicious high signal is present on T2-weighted images. MR imaging is useful in tumor staging because it provides good delineation of the anatomic compartments involved by the tumor and the tumor's size and its relationship to adjacent bone, fascial planes, and neurovascular bundles.

TISSUE CHARACTERIZATION BY MAGNETIC RESONANCE IMAGING

Diagnosis by Magnetic Resonance Imaging Criteria

Most soft tissue masses show low signal intensity on T1-weighted sequences and high signal intensity on T2-weighted sequences. Almost all sarcomas, unless there is hemorrhage within them, have these signal characteristics. A notable exception is well-differentiated liposarcoma, which has signal features similar to a lipoma. There are several benign entities that, based on a combination of signal features and location, permit a reasonably confident diagnosis, which in some instances may preclude biopsy or permit elective surgical removal. Entities that may be

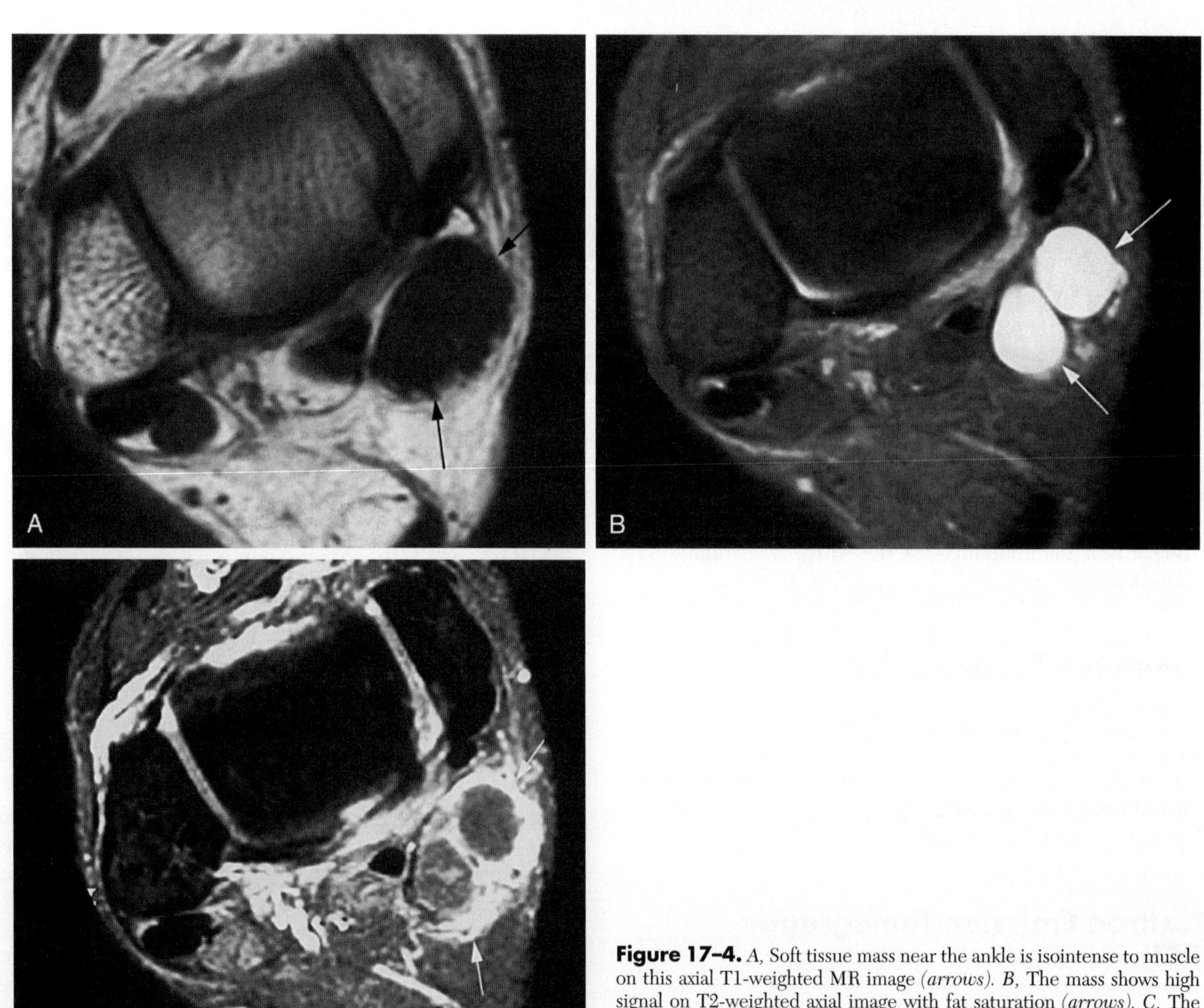

Figure 17–4. *A,* Soft tissue mass near the ankle is isointense to muscle on this axial T1-weighted MR image *(arrows)*. *B,* The mass shows high signal on T2-weighted axial image with fat saturation *(arrows)*. *C,* The cystic nature is shown on T1-weighted fat-saturation MR image after contrast administration, which shows peripheral *(arrows)* but no central enhancement in this ganglion cyst.

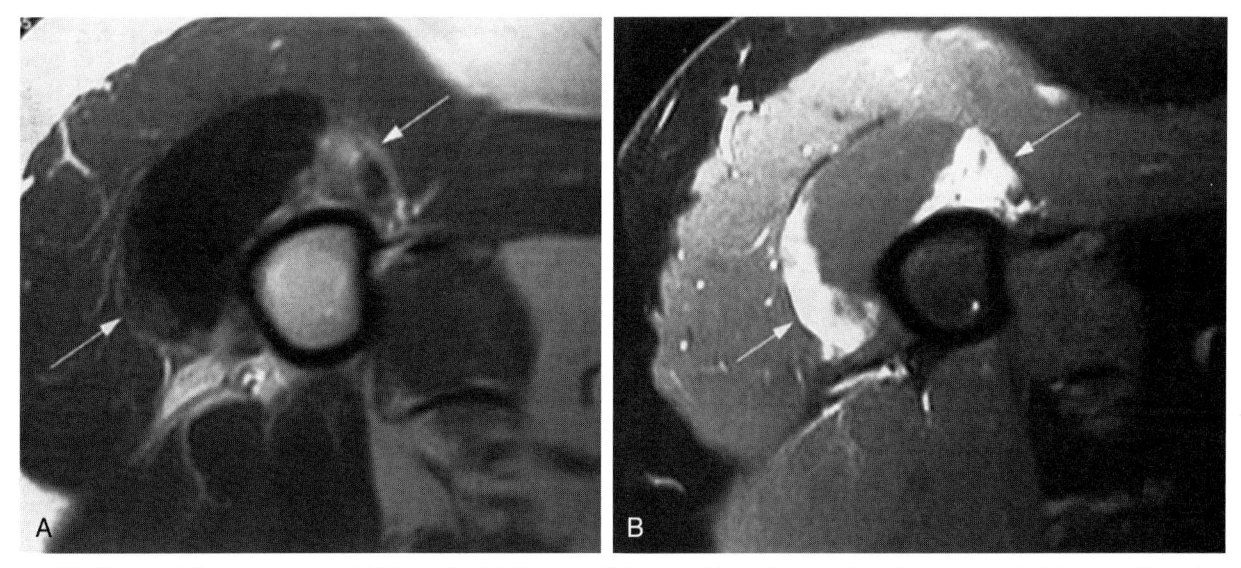

Figure 17–5. Myxoid liposarcoma. *A*, Axial T1-weighted MR image of the arm without fat saturation after contrast administration. The enhancing portion of the tumor *(arrows)* is difficult to distinguish from adjacent fat. *B*, The solid part of the tumor (high signal indicated by *arrows*) is seen much more easily on this axial T1-weighted postcontrast image with fat saturation.

Table 17–1. Characteristic Signs and Signal Patterns Useful for MR Imaging Diagnosis of Soft Tissue Masses

Masses with relatively high signal intensity on T1
- Lipoma
- Well-differentiated liposarcoma
- Primary melanotic melanoma
- Clear cell sarcoma
- Askin's tumor
- Alveolar soft parts sarcoma
- Hemangioma*
- Subacute hemorrhage

Clumps/streaks of high signal intensity within a low signal intensity mass on T1
- Hemangioma*
- Myxoid liposarcoma
- Infiltrative intramuscular lipoma
- Fibrolipomatous hamartoma

Lesions with low signal intensity on T2
- Fibromatosis
- Pigmented villonodular synovitis/giant cell tumor of tendon sheath
- Primary melanotic melanoma
- Granuloma annulare

Lesions with marked high signal intensity on T2
- Cysts
- Myxoma
- Hemangioma
- Necrotic/cystic malignant tumor

Fluid/fluid levels
- Synovial sarcoma
- Hemangioma
- Myositis ossificans
- Hemorrhage
- Lymphangioma

*Hemangiomas: heterogeneous, infiltrative; foci of high signal intensity on T1-weighted images (can be predominantly high signal intensity); progressively brighter on more T2-weighted images; arteriovenous malformations appear as large serpentine vessels.

diagnosed on the basis of MR imaging features are listed in Table 17–1. If definite diagnosis of a soft tissue mass on MR imaging cannot be made, it should be considered indeterminate, and a biopsy should be performed. An indeterminate lesion is considered a sarcoma until microscopically proved otherwise. Biopsy of a soft tissue mass always should be performed with close consultation of the referring surgeon to avoid tumor contamination of unaffected tissues and resultant unnecessary surgery.

Lesions with Predominantly High T1 Signal

Benign lipomatous tumors are a group of fat-containing neoplasms that generally have a characteristic signal pattern allowing accurate diagnosis.

Benign Lipoma

A benign lipoma is by far the most common mesenchymal neoplasm, outnumbering other benign and malignant tumors and representing the most commonly encountered soft tissue mass on MR imaging. The MR imaging diagnosis of lipoma can be made reliably and consistently because of its characteristic signal pattern paralleling that of subcutaneous fat (Fig. 17–6). Lipomas are well-encapsulated masses that may contain lobulations and fibrous septa. They may be simple or multiple, superficial or deep-seated, and they often constitute some of the largest and bulkiest tumors encountered. A well-differentiated liposarcoma may present with an MR imaging appearance similar to that of a lipoma; however,

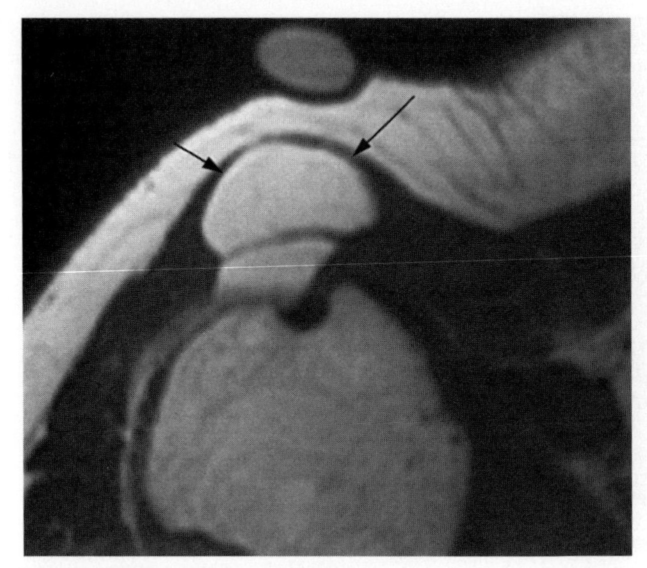

Figure 17–6. Axial T1-weighted MR image of the arm shows a well-defined high signal lesion *(arrows)* within the anterior deltoid muscle. The lesion is similar in signal to subcutaneous fat, consistent with an intramuscular lipoma.

tumor heterogeneity and multiple lobulations have been reported to be important clues to malignancy in a lipoid tumor (Fig. 17–7). In our experience, signal heterogeneity is the most useful sign of malignancy in a diffusely lipomatous tumor.

Lipoma Arborescens

Lipoma arborescens is a rare intra-articular lesion consisting of lipomatous proliferation of the synovium (see also Chapter 36). It occurs in the knee joint, especially the suprapatellar pouch. MR imaging findings include an intrasynovial frondlike mass paralleling fat in signal with an associated joint effusion (Fig. 17–8).

Macrodystrophia Lipomatosa

Macrodystrophia lipomatosa is a rare, localized gigantism characterized by an increase in all mesenchymal elements. It most frequently presents as enlargement of a digit or adjacent digits (Fig. 17–9). MR imaging reveals redundancy of fatty tissue and fibrous thickening of nerves.

Fibrolipomatous Hamartoma

Fibrolipomatous hamartoma is a rare, benign lesion that most commonly occurs in infants and children. In 80% of cases, it arises from the median nerve, but it also can occur in association with the ulnar nerve or in the dorsum of the foot. The MR imaging appearance is characteristic with longitudinally oriented cylindrical regions of signal void on all sequences, thought to represent epineural and perineural fibrosis separated by areas of high signal intensity corresponding to mature fat in the interfascicular connective tissue (Fig. 17–10). The differential diagnosis includes intraneuronal lipoma and intraneuronal hemangioma.

Lesions with Scattered Foci of High T1 Signal: Hemangiomas

Hemangiomas are a common soft tissue tumor, representing 7% of all benign tumors. They are the most common soft tissue tumor in infancy and childhood. Hemangiomas can be subcutaneous, intramuscular, or intrasynovial. The MR imaging appearance of an intramuscular hemangioma is characteristic but not pathognomonic. MR imaging findings consist of a poorly marginated, nonhomogeneous mass often with areas of increased signal on T1-weighting, which are believed to represent regions of fat or subacute hemorrhage

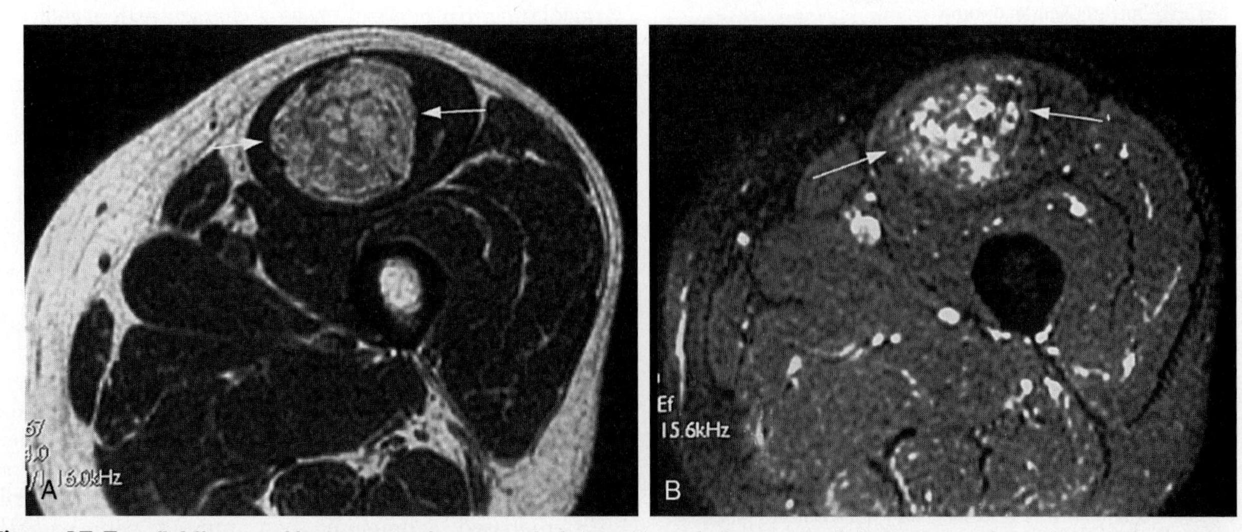

Figure 17–7. Well-differentiated liposarcoma in the anterior thigh *(arrows)*. *A,* The lesion is mostly high signal on this axial T1-weighted MR image. *B,* On axial T2-weighted fat-saturation image, there are regions of high signal within the lesion. The heterogeneity of this lesion helps to differentiate it from a simple lipoma.

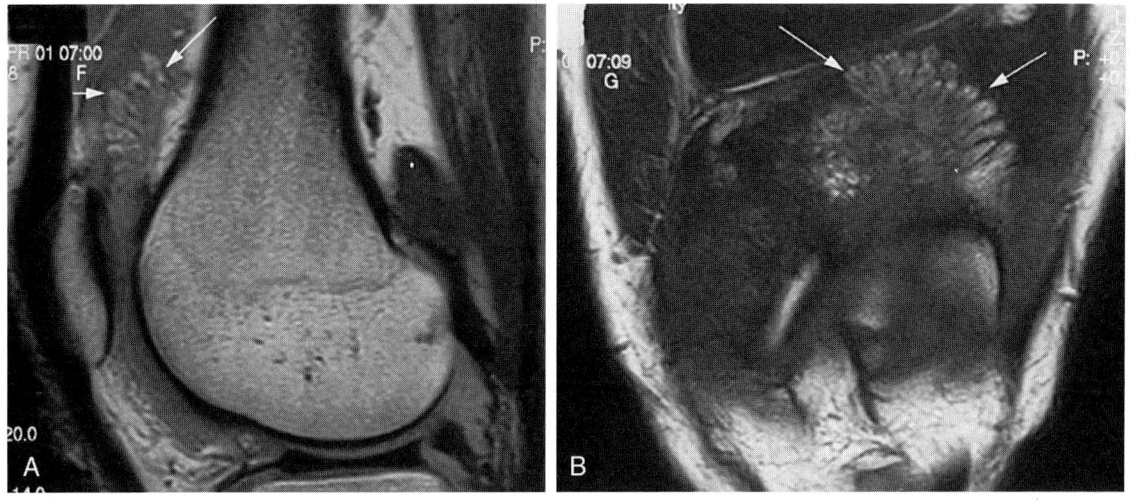

Figure 17–8. Lipoma arborescens. *A,* Sagittal proton-density MR image of the knee shows frondlike regions of high signal in the suprapatellar bursa representing the lipomatous proliferation of the synovium *(arrows). B,* Coronal T1-weighted image of the anterior knee shows the lipomatous tissue to better advantage against the darker joint fluid *(arrows).*

(Fig. 17–11). With T2-weighting, high signal lobulations with a central low signal intensity dot are characteristic. The cause of the low signal dot is uncertain but may represent higher velocity blood flow within vessels. Scattered serpentine areas of signal void sometimes can be seen, which represent vascular channels with high blood flow. Most hemangiomas, perhaps with the exception of the arteriovenous type, show high signal on gradient echo sequences, which are more sensitive to slow flow. Hemosiderin and phleboliths are shown as foci of signal void on all sequences. Hemangiomas typically show marked enhancement after gadolinium administration. The recognition and diagnosis of an intramuscular hemangioma by MR imaging may preclude biopsy.

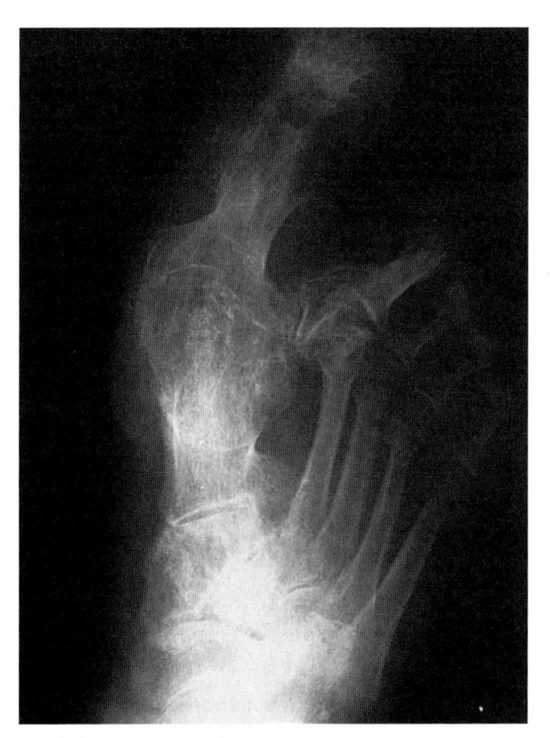

Figure 17–9. Macrodystrophia lipomatosa. Anteroposterior view of the foot shows enlargement and disfigurement of the first two rays.

Lesions with Low T2-Weighted Signal Intensity

Low signal on T2-weighting can be due to many things, such as paramagnetic effects of tumor content (hemosiderin), hypocellularity (low water content), high collagen content, fibrotic scar tissue, mineralized tissue, foreign body, air, and vessels with rapid blood flow. The following lesions have low cellularity and high collagen content.

Fibromatosis

Fibromatosis is a diverse group of soft tissue lesions characterized by proliferation of fibrous tissue and locally aggressive behavior with a tendency toward recurrence after resection. The MR imaging signal patterns of these lesions are highly variable. Fibromatosis can have low signal on T2, permitting diagnosis. High signal with T2-weighting in fibromatosis, similar to most other soft tissue tumors, is common, however (Fig. 17–12). Among soft tissue masses with low signal T2-weighting, fibromatosis is one of the more common in the appendicular skeleton. In rare instances, the lesions may be multiple within a single limb.

Nodular Fasciitis

Nodular fasciitis is a rapidly growing soft tissue mass that occurs in subcutaneous tissues or within muscle and presents in young adults. It can exhibit low signal on all pulse sequences, mimicking fibromatosis, or have high signal on T2-weighting, appearing similar to a sarcoma because of its rapid growth.

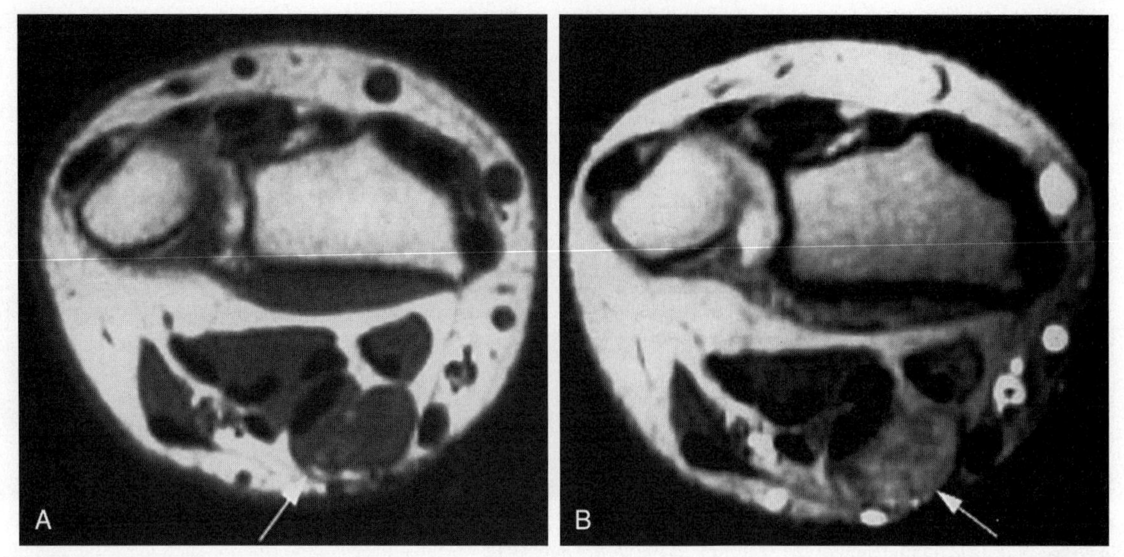

Figure 17–10. Fibrolipomatous hamartoma. *A,* Axial T1-weighted MR image of the distal forearm shows a round soft tissue mass in the region of the median nerve *(arrow).* Regions of high signal within the mass represent fat tissue. *B,* The mass heterogeneously enhances *(arrow)* on T1-weighted postcontrast MR image.

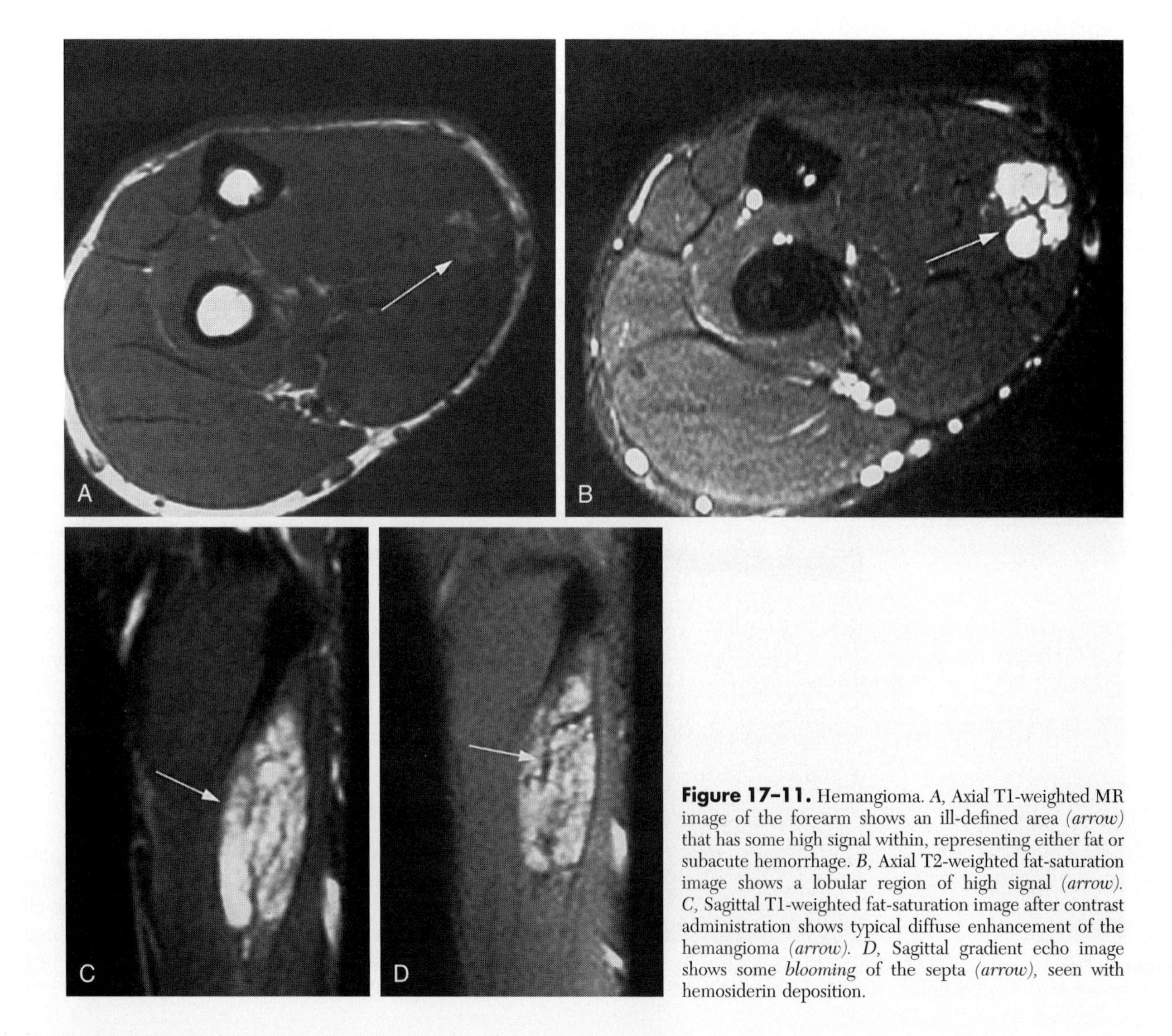

Figure 17–11. Hemangioma. *A,* Axial T1-weighted MR image of the forearm shows an ill-defined area *(arrow)* that has some high signal within, representing either fat or subacute hemorrhage. *B,* Axial T2-weighted fat-saturation image shows a lobular region of high signal *(arrow). C,* Sagittal T1-weighted fat-saturation image after contrast administration shows typical diffuse enhancement of the hemangioma *(arrow). D,* Sagittal gradient echo image shows some *blooming* of the septa *(arrow),* seen with hemosiderin deposition.

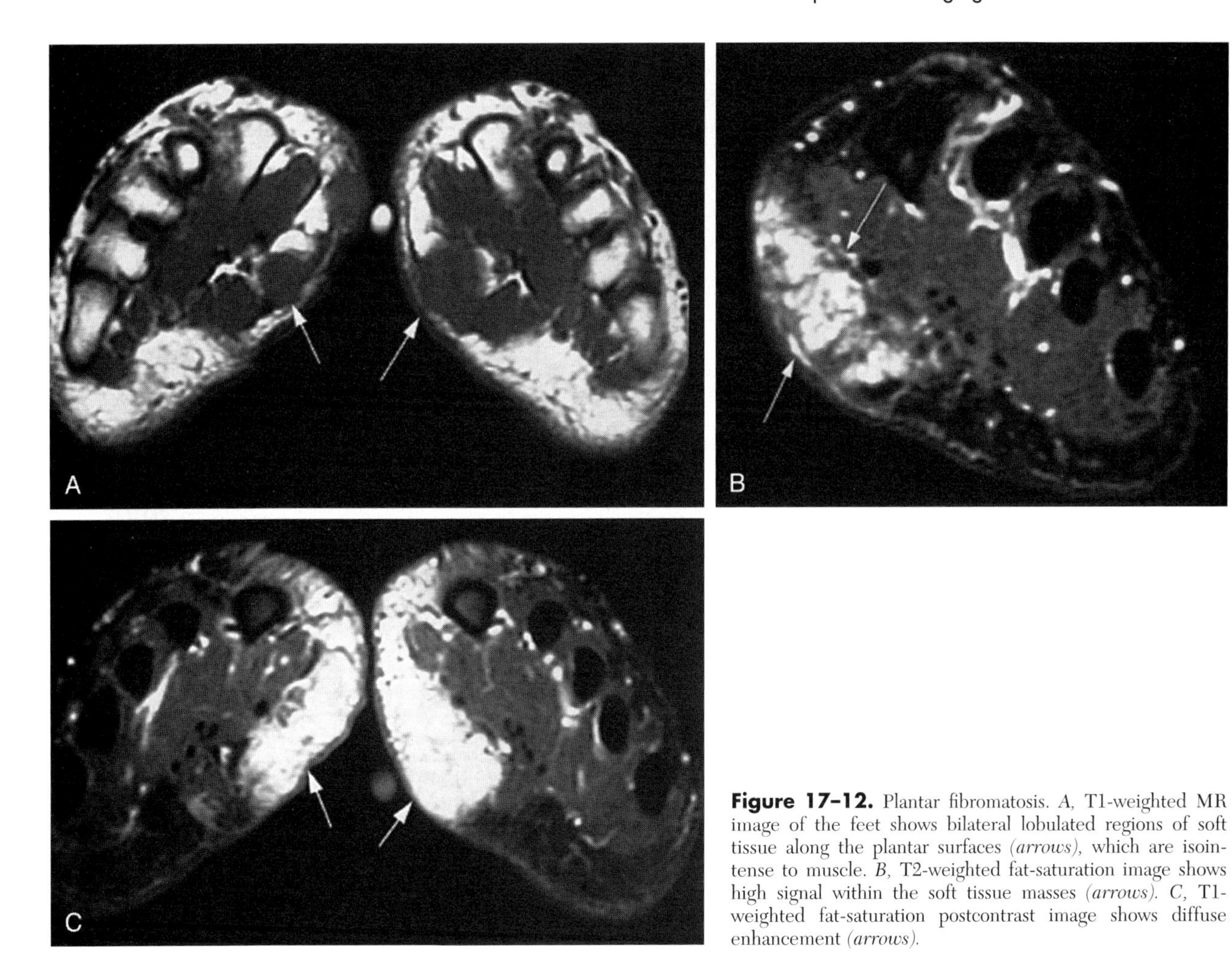

Figure 17–12. Plantar fibromatosis. *A*, T1-weighted MR image of the feet shows bilateral lobulated regions of soft tissue along the plantar surfaces *(arrows)*, which are isointense to muscle. *B*, T2-weighted fat-saturation image shows high signal within the soft tissue masses *(arrows)*. *C*, T1-weighted fat-saturation postcontrast image shows diffuse enhancement *(arrows)*.

Elastofibroma Dorsi

Elastofibroma dorsi is a slow-growing soft tissue pseudotumor in the subscapular posterolateral chest wall, believed to be due to friction between the scapular tip and the chest wall. A subscapular mass that is isointense to muscle, reflecting predominantly fibrous tissue, with interspersed areas of increased T1 signal corresponding to entrapped fat, is characteristic (Fig. 17–13). It may be bilateral. Because this lesion is removed usually only for cosmetic or symptomatic reasons, MR imaging diagnosis could eliminate an unnecessary biopsy.

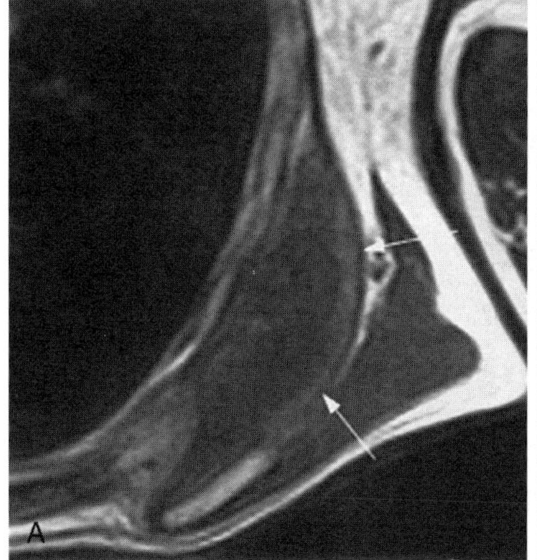

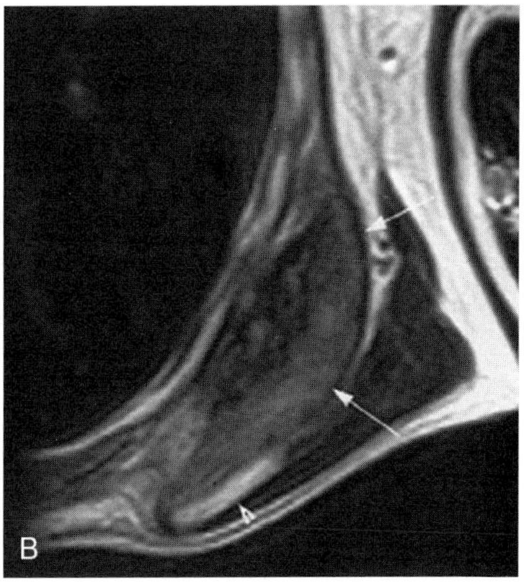

Figure 17–13. Elastofibroma. *A*, Axial T1-weighted MR image shows a low signal mass adjacent to the chest wall *(arrows)*, which also has some regions of high signal within, consistent with entrapped fat. *B*, The mass has some low signal regions on T2-weighting, reflecting the fibrous content *(arrows)*. The tip of the scapula is shown as a linear high signal structure *(arrowhead)*.

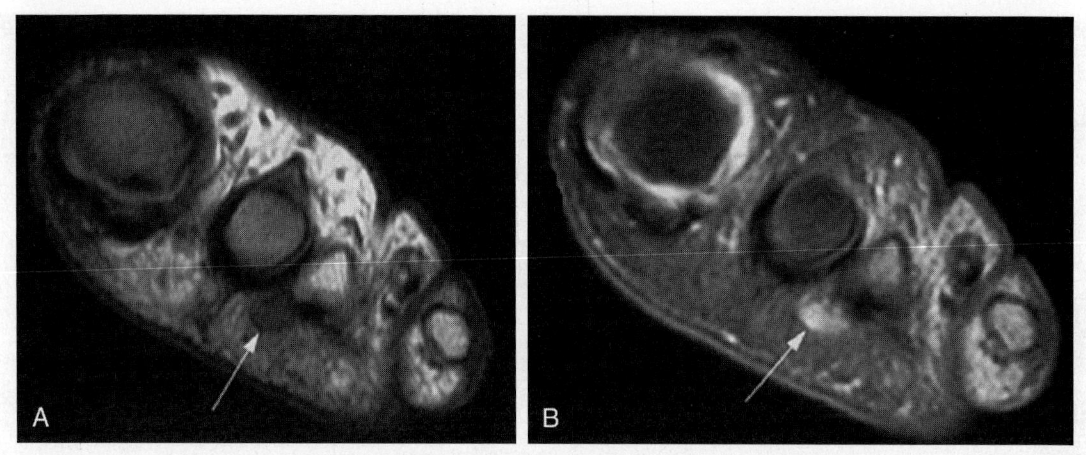

Figure 17–14. Morton's neuroma. *A,* T1-weighted MR image of the foot shows a small low signal mass between the heads of the second and third metatarsals *(arrows),* which enhances uniformly after contrast administration *(B).*

Morton's Neuroma

Morton's neuroma is a benign, fibrotic lesion consisting of perineural fibrosis and degeneration of the interdigital nerve in the interdigital space of the foot. It is found predominantly in females and usually is located between the heads of the third and fourth metatarsals. On MR imaging, the lesion may have low signal intensity on T1- and T2-weighted images and frequently may not be visualized at all. The tumor typically has moderate-to-marked enhancement after administration of gadolinium (Fig. 17–14).

Synovial Proliferative Lesions

Pigmented Villonodular Synovitis

Pigmented villonodular synovitis is a benign soft tissue lesion characterized by synovial proliferation and hemosiderin deposition (see also Chapter 35). It usually is seen in the knee and hip. The synovial space is distended with discrete masses of low signal intensity on T1- and T2-weighted images. These dark clumps are the key to the diagnosis and represent hemosiderin accumulation. Gradient echo sequences show characteristic *blooming* artifact of these hemosiderin deposits (Fig. 17–15).

Giant Cell Tumor of the Tendon Sheath

Pigmented villonodular synovitis, when extra-articular, is referred to as a *giant cell tumor of the tendon sheath* (see also Chapter 35). It typically arises in relation to small joints and is the second most common soft tissue mass of the hand after ganglion cyst. The tumor has a predominantly homogeneous low signal on T1-weighting and T2-weighted sequences (Figs. 17–16 and 17–17).

Cystic Lesions

A cyst usually can be diagnosed on MR imaging based on its ovoid shape, thin regular wall, lack of enhancement, and low signal T1-weighting and high signal T2-weighting, which reflect the composition of the fluid within the cyst. Cysts commonly present in the vicinity of a joint or tendon.

Popliteal (Baker's) Cyst

Baker's cysts are a common finding on MR imaging of the knee. The synovial-lined cysts are characteristic in location and occur by extension of joint fluid into the gastrocnemius-semimembranosus bursa (Fig. 17–18). A cystic-type signal pattern is shown on MR imaging, although debris and occasional septations can be seen.

Ganglion Cyst

Ganglion cysts are locular or multilocular, benign, cystic lesions containing variable amounts of mucinous fluid and are lined with a pseudosynovial capsule composed of flat spindle cells and a variable amount of collagen. On MR imaging, ganglion cysts often appear as a septated, ovoid, cystic mass in proximity to a joint or tendon sheath. Typically, ganglion cysts have high signal on T2-weighted images (Fig. 17–19).

Bursitis

Bursae are synovial-lined potential spaces that can fill with fluid because of inflammation, infection, or trauma, resulting in bursal enlargement. Common locations include the knee (pes anserine, prepatellar, infrapatellar), shoulder, and hip (greater trochanter, iliopsoas) (Fig. 17–20).

CHARACTERISTIC SHAPE, COURSE, AND LOCATION

Peripheral Nerve Sheath Tumors

Peripheral nerve sheath tumors are of two types: neurofibroma and schwannoma. They usually are elon-

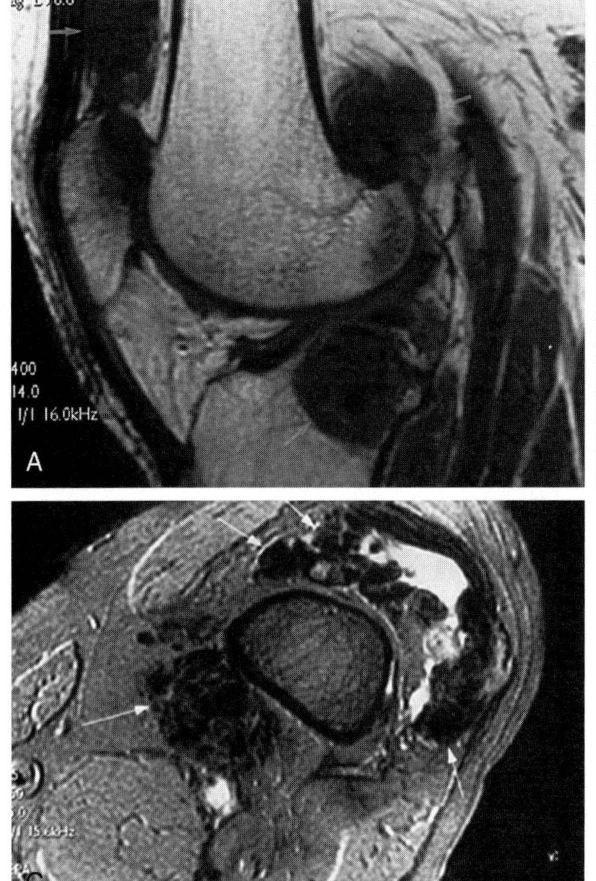

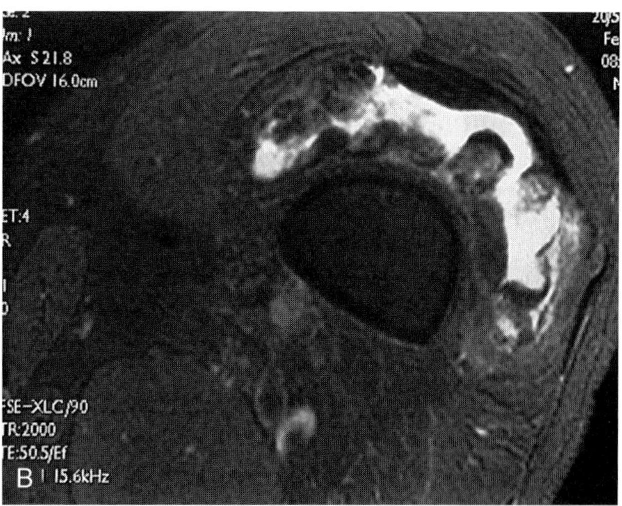

Figure 17–15. Diffuse pigmented villonodular synovitis. *A,* T1-weighted sagittal image of the knee shows masses about the knee *(arrows),* which are isointense to muscle. *B,* The masses within the suprapatellar bursa are low signal on axial T2-weighted fat-saturation image, outlined by the high signal joint effusion. *C,* On the axial gradient echo image, the masses show larger regions of signal void *(arrows),* which represent the *blooming* artifact caused by hemosiderin deposition.

gated along the long axis of the host nerve and are slow growing (Fig. 17–21). A feature described as a *target pattern* seen in schwannomas and neurofibromas is not identified in malignant nerve sheath tumors. On T2-weighted images, the target pattern refers to a central area of decreased signal intensity with a peripheral hyperintense rim (Fig. 17–22). This sign may aid in the diagnosis of benign nerve sheath tumors when associated with other features, such as typical location and shape. Peripheral nerve sheath tumors tend to show intense enhancement after contrast administration (Fig. 17–23).

Normal Muscles that Mimic Tumors

Anomalous or accessory muscles can mimic soft tissue tumors, and they can result in unnecessary surgery.

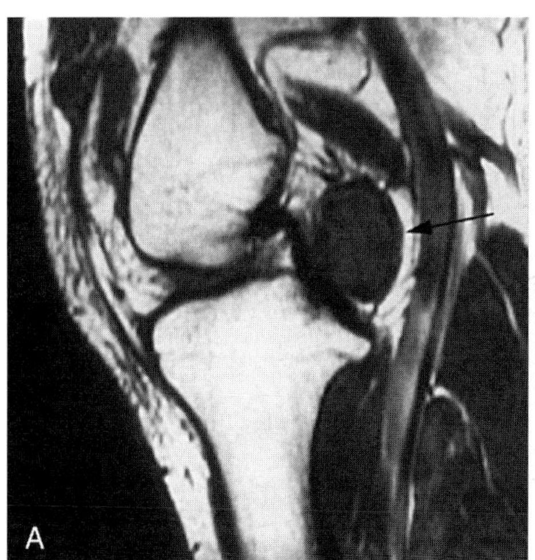

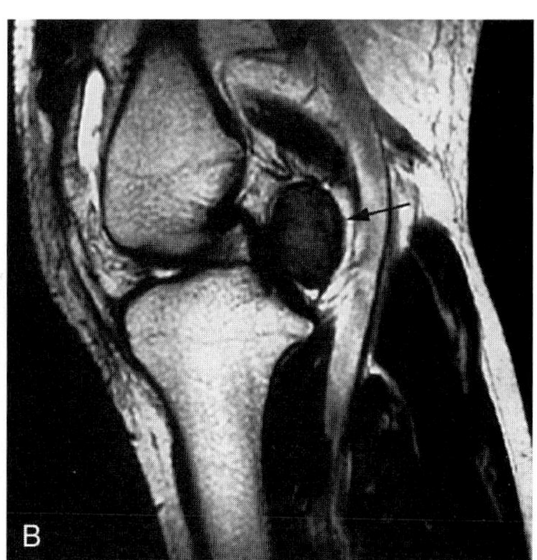

Figure 17–16. Giant cell tumor of the tendon sheath. *A* and *B,* Sagittal T1-weighted and T2-weighted MR images of the knee show a well-defined ovoid mass that is low signal, located posterior to the posterior cruciate ligament *(arrow).*

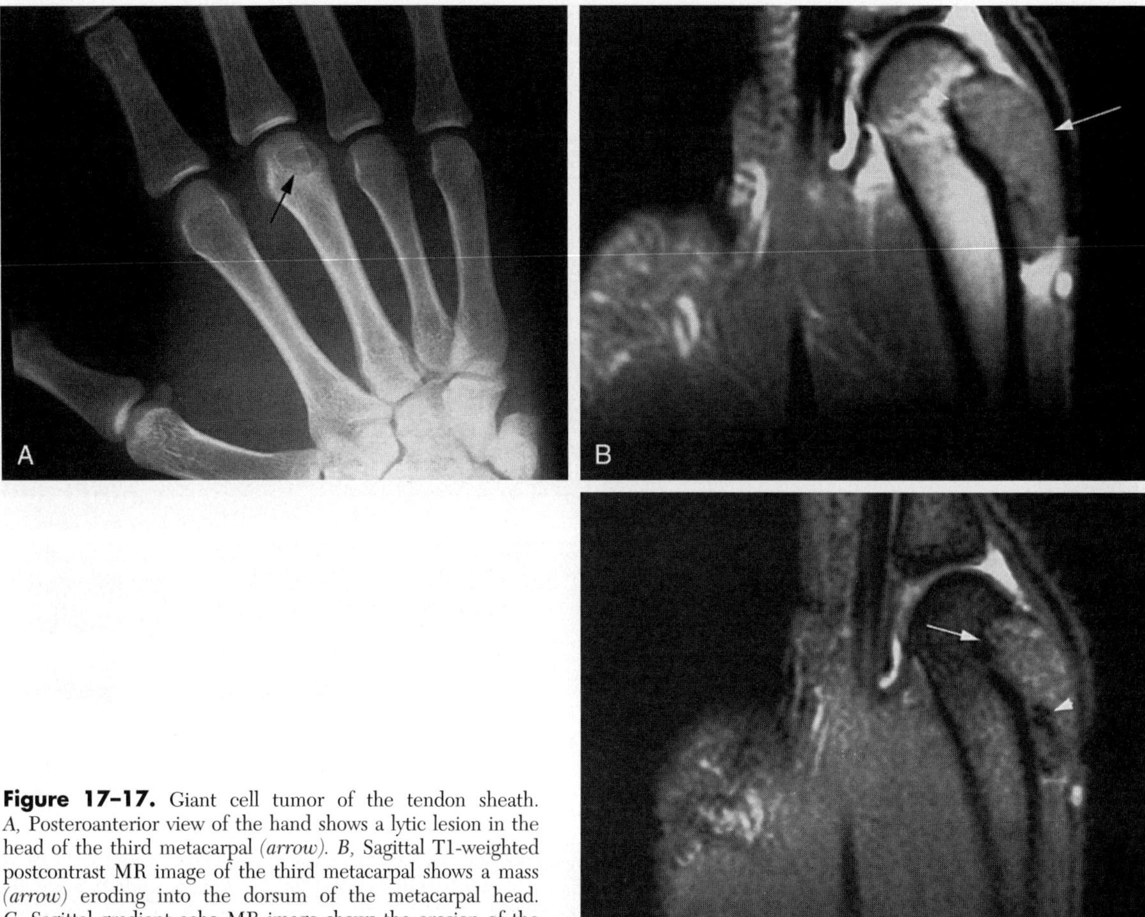

Figure 17–17. Giant cell tumor of the tendon sheath. *A*, Posteroanterior view of the hand shows a lytic lesion in the head of the third metacarpal *(arrow)*. *B*, Sagittal T1-weighted postcontrast MR image of the third metacarpal shows a mass *(arrow)* eroding into the dorsum of the metacarpal head. *C*, Sagittal gradient echo MR image shows the erosion of the metacarpal *(arrow)* and *blooming* of low signal within the mass consistent with hemosiderin deposition *(arrowhead)*.

Anomalous or accessory muscles occur most frequently in the upper extremity; however, their locations are less characteristic in the upper extremity. Reported anomalies include anomalous or accessory palmaris longus muscle, hypothenar muscle duplication, anomalous extensor indicis, and extensor digitorum brevis-manus muscle. In the lower extremity, almost all reported cases involve anatomic variations of the soleus muscle, which typically presents as a palpable mass between the medial malleolus and Achilles' tendon (Fig. 17–24). Normal muscle that is herniated through normal fascial planes can present as a palpable lump. The key to the diagnosis lies in the fact that these accessory muscles occur in predictable locations, and the MR imaging signal characteristics are identical to surrounding muscle.

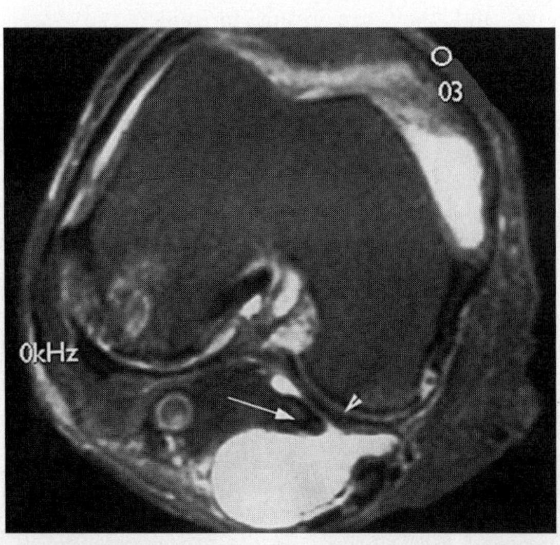

Figure 17–18. Axial T2-weighted MR image of the knee with fat saturation shows a high signal region in a characteristic location for Baker's (popliteal) cyst. The semimembranosus *(arrowhead)* and medial gastrocnemius *(arrow)* tendons are visible.

Cat-Scratch Disease

Cat-scratch disease is a common but often overlooked cause of unilateral lymphadenitis in young patients. On MR imaging, the abnormality is encountered most frequently around the elbow joint, where a mass with high signal on T2-weighting is found (Fig. 17–25). This mass consists of enlarged epitrochlear lymph nodes. A finding of a mass in this location merits careful questioning for a history of cat-scratch disease.

Summary

Most soft tissue tumors are cellular and have signal characteristics similar to water. In general, signal charac-

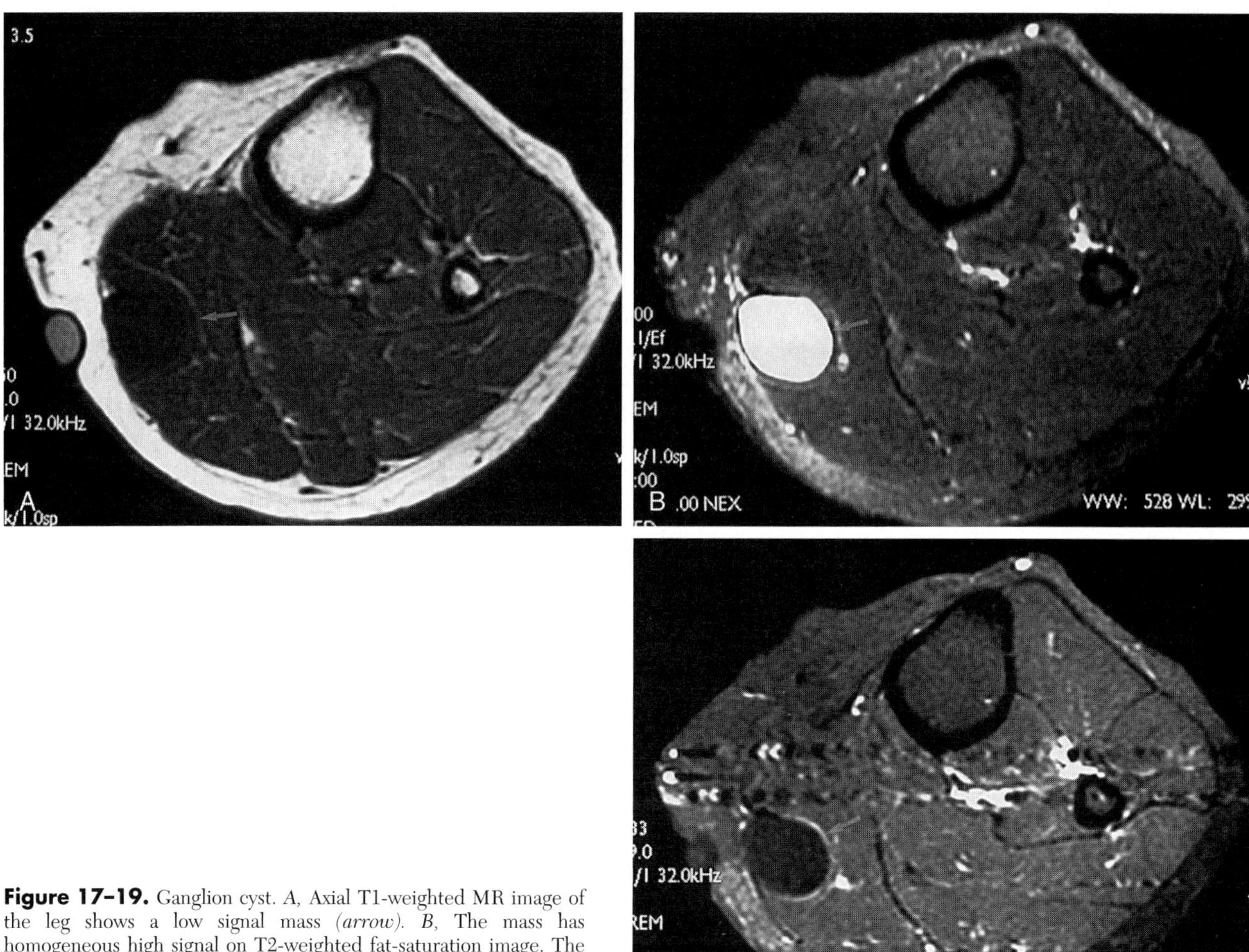

Figure 17–19. Ganglion cyst. *A,* Axial T1-weighted MR image of the leg shows a low signal mass *(arrow).* *B,* The mass has homogeneous high signal on T2-weighted fat-saturation image. The cystic nature of the mass is shown by faint peripheral enhancement *(arrow).* *C,* On T1-weighted fat-saturation postcontrast image. The central part of the cyst does not enhance *(arrow).*

teristics of soft tissue tumors overlap. In most soft tissue sarcomas, a specific preoperative diagnosis does not change the surgical management. An exception is lymphoma. Other exceptions occur when a tumor presents with specific MR imaging features unique to or suggestive of a benign soft tissue tumor. These features are determined by a certain tissue type (signal characteristics on MR imaging), characteristic shape, or characteristic location.

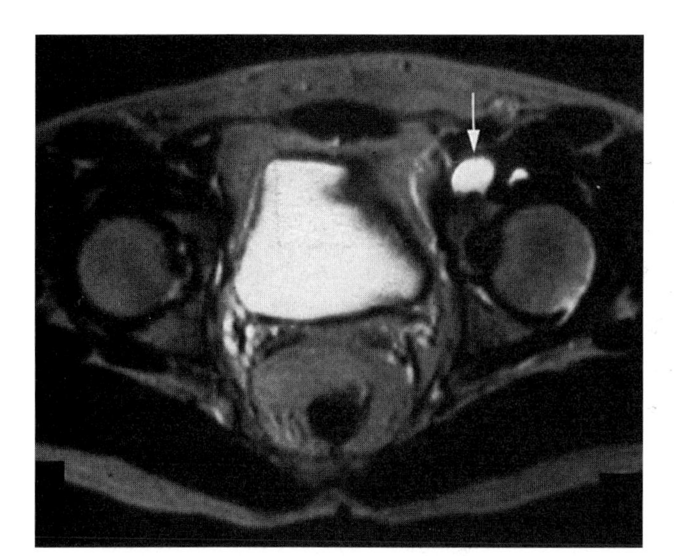

Figure 17–20. Iliopsoas bursa. Axial T2-weighted MR image of the pelvis shows a high signal mass *(arrow)* located medial to the iliopsoas muscle. The high signal represents fluid within the bursa.

Masses that have high signal on T1-weighted images contain fat, blood, or both (Fig. 17–26). Examples of soft tissue lesions with diffuse high signal on T1-weighted images are lipomas, lipoma-like (well-differentiated) liposarcomas, and subacute hematomas. On T2-weighted sequences, fat within a mass looks similar to subcutaneous fat, whereas subacute blood would be brighter than subcutaneous fat. These T2-weighted signal characteristics aid in distinguishing lipomatous masses from subacute hematomas. Masses containing foci of high signal on T1-weighted sequences include hemangioma, hemorrhage within a sarcoma (Fig. 17–27), myxoid liposarcoma, alveolar soft part sarcoma (Fig. 17–28), clear cell sarcoma, and dedifferentiated liposarcoma. Most soft tissue masses have high signal intensity on T2-weighted images; however, a subset of masses has low signal intensity on T2-weighted images. Common to these masses is the presence of considerable collagen, hypocellularity, or hemosiderin. Examples of such tumors include fibromatosis, pigmented villonodular synovitis and giant cell tumor of the tendon sheath, clear cell sarcoma, melanoma,

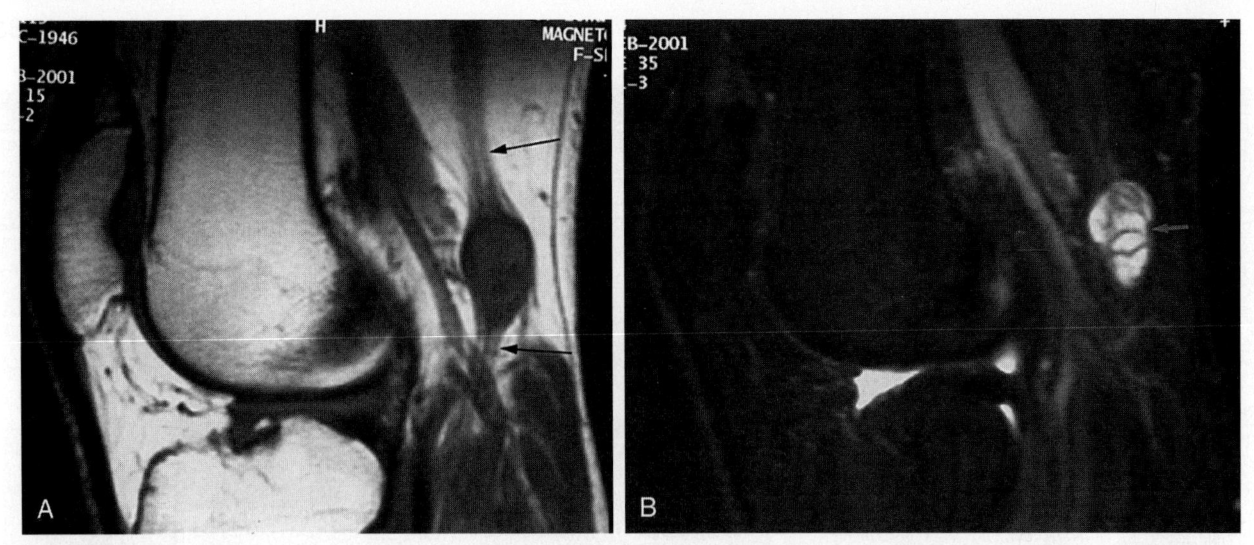

Figure 17–21. Peripheral nerve sheath tumor. *A,* Sagittal T1-weighted MR image of the knee shows a low signal mass elongated along the course of the peroneal nerve. The nerve fascicles are visible as parallel low signal lines *(arrows). B,* The mass *(arrow)* is predominantly high signal on the T2-weighted image with fat saturation.

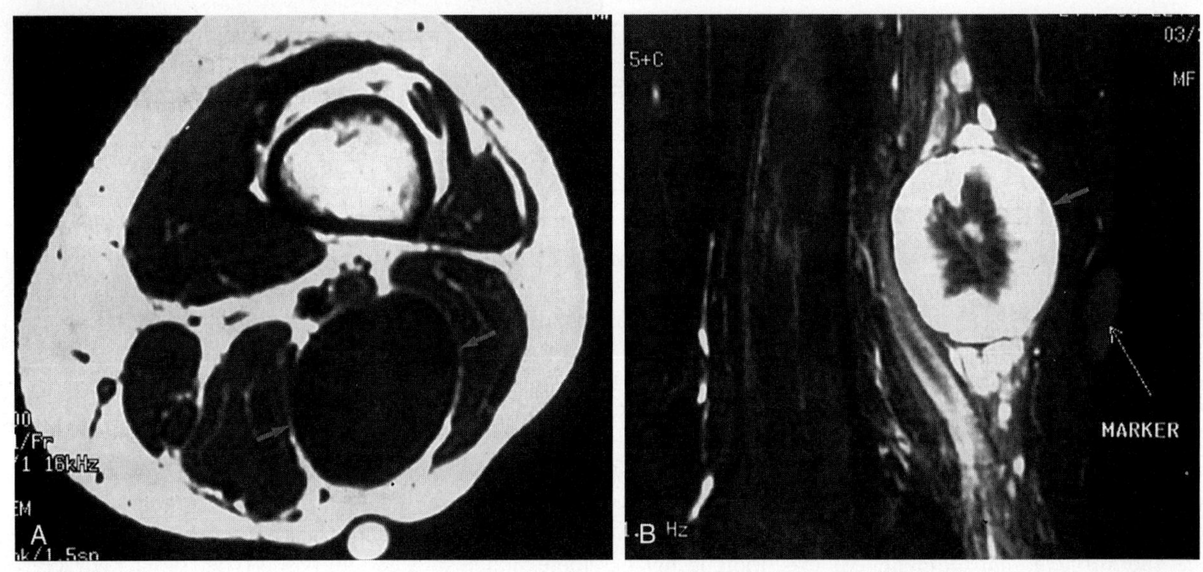

Figure 17–22. Target pattern in a peripheral nerve sheath tumor. *A,* Axial T1-weighted MR image of the knee shows a posterior rounded low signal mass *(arrows). B,* The mass shows high signal periphery with low signal center (target pattern *[arrow]*) on sagittal T2-weighted image with fat saturation. The low signal center is believed to represent fibrous tissue.

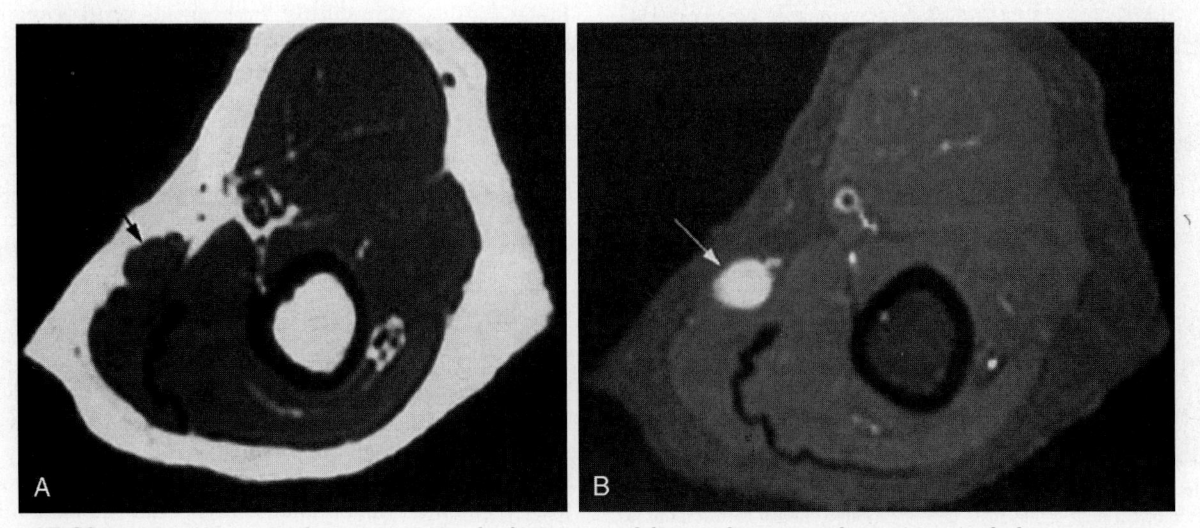

Figure 17–23. Peripheral nerve sheath tumor. *A,* T1-weighted MR image of the arm shows a round mass *(arrow),* which is isointense to muscle. *B,* The nerve sheath tumor enhances uniformly on T1-weighted fat-saturation image after contrast administration *(arrow).*

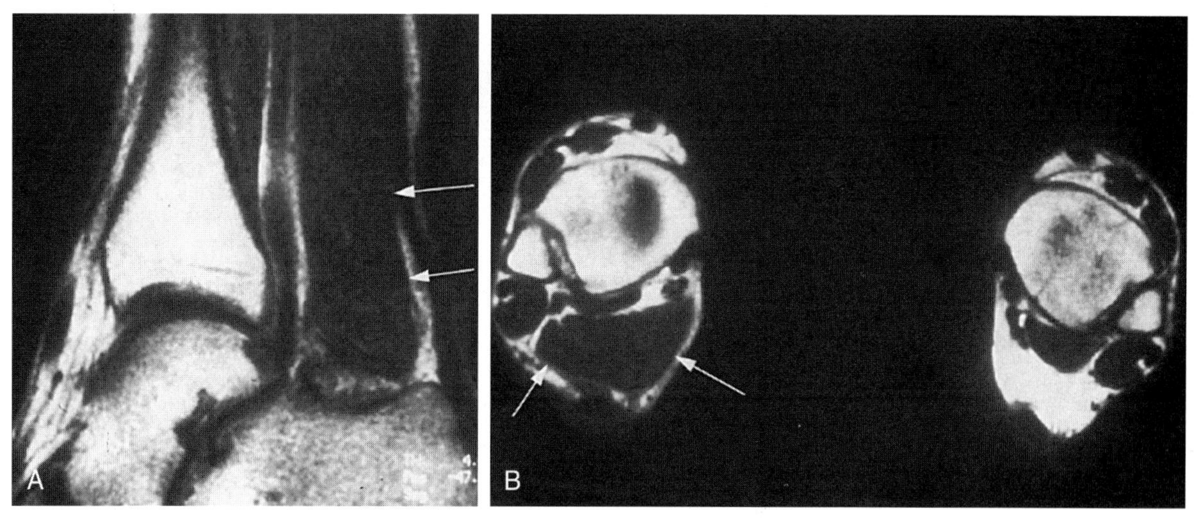

Figure 17–24. *A*, Sagittal T1-weighted MR image of the ankle shows an elongated mass *(arrows)* that is isointense to muscle in a characteristic location for an accessory soleus muscle. *B*, Axial T1-weighted image of both ankles shows the accessory soleus muscle on the right *(arrows)*.

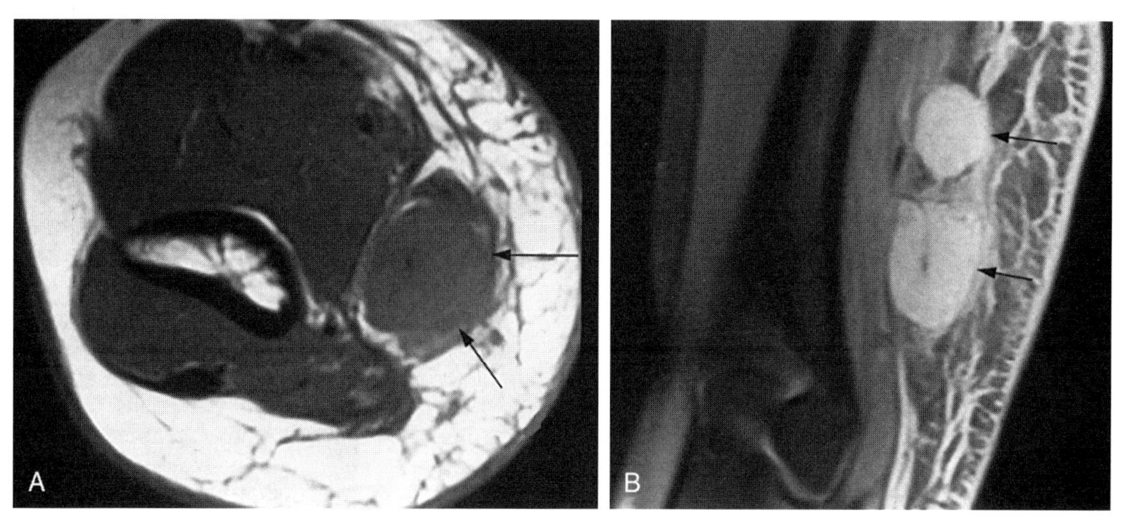

Figure 17–25. Cat-scratch disease. *A*, Axial T1-weighted MR image of the elbow shows a round mass that is slightly higher signal than muscle *(arrows)*. *B*, Sagittal fat-saturation proton-density image shows two high signal masses *(arrows)*; the more distal one has a low signal hilar region. These masses are inflamed lymph nodes. Edema (high signal) within the subcutaneous tissues also is present.

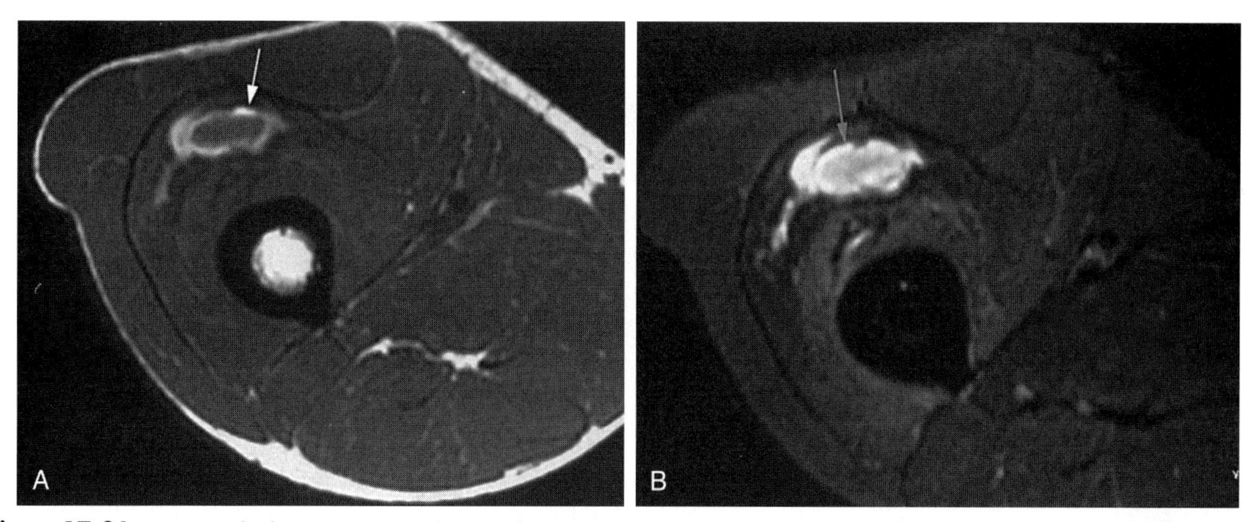

Figure 17–26. Intramuscular hematoma. *A*, Axial T1-weighted MR image of the thigh shows an oval mass that has a rim of high signal *(arrow)*. *B*, The high signal rim remains on T2-weighted image with fat saturation *(arrow)*, reflecting subacute hemorrhage. The center of the mass has signal characteristics of serous fluid.

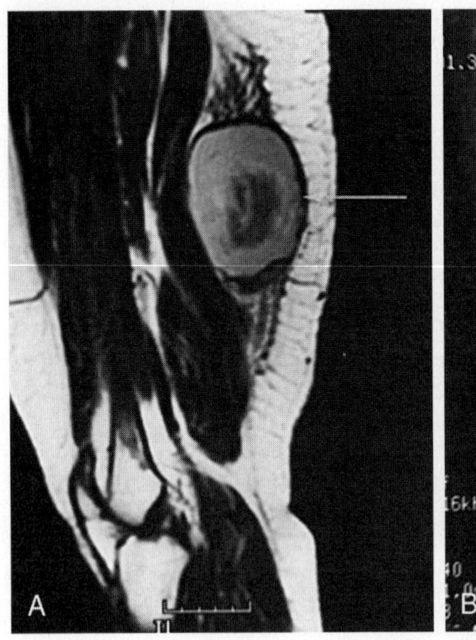

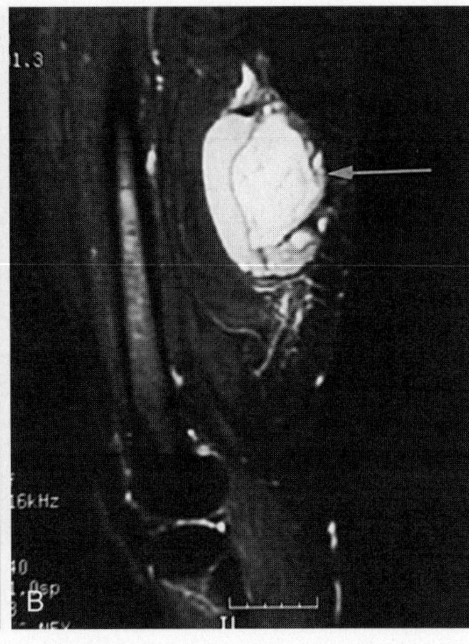

Figure 17–27. Hemorrhagic malignant fibrous histiocytoma. *A,* Sagittal T1-weighted MR image shows a 5- to 6-cm ovoid, well-marginated mass *(arrow)* located in the posteromedial aspect of the thigh. Almost 75% of the mass shows high signal with a central area of low signal. On T1-weighting, the high signal could represent fat or hemorrhage, and the low signal could represent an area of necrosis, myxoid tissue, or conventional tumor. *B,* Sagittal STIR image shows no suppression of signal in the tumor, and the entire tumor is bright *(arrow),* making it unlikely to represent a lipomatous neoplasm. The presumptive diagnosis was hemorrhagic sarcoma, probably malignant fibrous histiocytoma because it is statistically the most common sarcoma in soft tissues. This high-grade hemorrhagic sarcoma has an almost perfectly smooth margin. Smooth margins are a common feature of sarcomas and should not be relied on to distinguish benign from malignant.

granuloma annulare, and, rarely, malignant fibrous histiocytoma and fibrosarcoma. Soft tissue masses that do not show tumor-specific features on MR imaging should be considered indeterminate, and a biopsy should be performed.

DIFFERENTIATION OF BENIGN FROM MALIGNANT MASSES

Size

A tumor larger than 4 cm, located deep to the subcutaneous tissues, and having low signal T1 and high signal T2, should be considered malignant unless proved otherwise. Two large and deep benign tumors that can be diagnosed confidently because of their signal characteristics and morphology are lipomas and hemangiomas. One third of sarcomas may be smaller than 4 cm and superficial in location.

Tumor Margin

The tumor margin usually is not helpful in separating benign from malignant soft tissue masses. Synovial sarcomas are notorious for slow growth, smooth margins, and homogeneous signal intensity and can be mistaken for a benign lesion, such as ganglion cyst (Fig. 17–29). Gadolinium enhancement aids in differentiating a benign cyst from a synovial sarcoma. Benign lesions that can present with irregular or infiltrative margins include aggressive fibromatosis, post-traumatic hematoma, hemangioma, and pyomyositis.

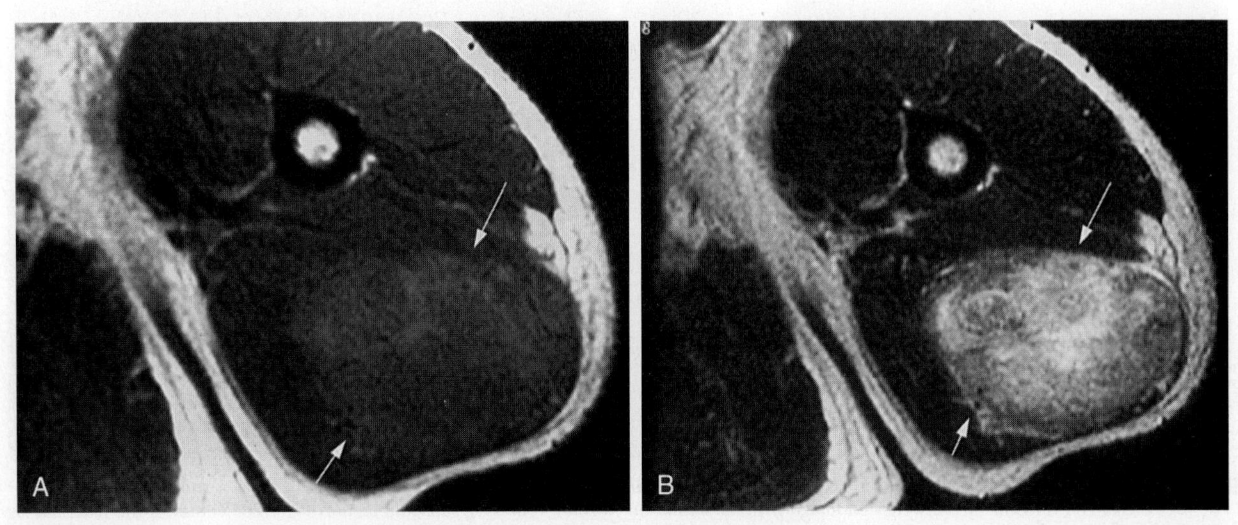

Figure 17–28. Alveolar soft part sarcoma. *A,* Axial T1-weighted MR image of the arm in a young adult shows a heterogeneous high signal (relative to muscle) mass *(arrows),* with heterogeneous high signal on the T2-weighted image *(B).*

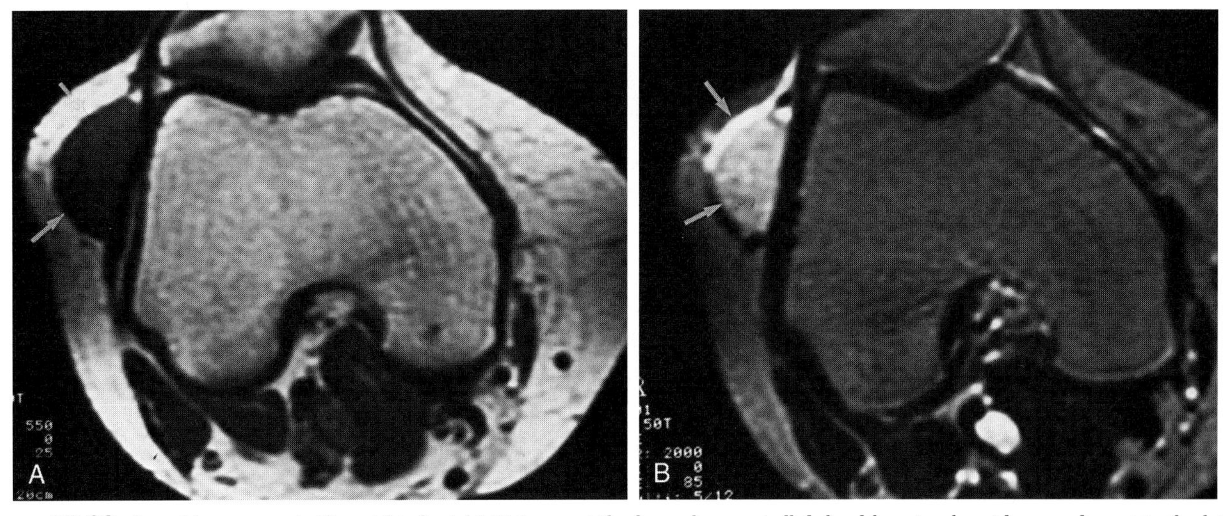

Figure 17–29. Synovial sarcoma. *A,* T1-weighted axial MR image at the knee shows a well-defined low signal ovoid mass adjacent to the lateral retinaculum *(arrows). B,* On T2-weighting, the mass *(arrows)* becomes homogeneously bright. There were no other findings, in particular, no accompanying joint effusion. This mass was, presumably because of its homogeneity, smooth margin, and location, interpreted as a cyst or ganglion, both of which are unlikely diagnoses without other accompanying signs. If a diagnosis of a ganglion or cyst is being considered in an atypical location, gadolinium enhancement or ultrasound should be considered to exclude a synovial sarcoma.

Location

Location aids in the diagnosis of a soft tissue mass primarily because of the predilection of certain lesions to occur in typical locations (Table 17–2). In the hand and wrist, there is a high prevalence of ganglion cysts, vascular tumors, and giant cell tumors of the tendon sheath. The foot is a common location for synovial sarcoma and clear cell sarcoma. The relationship of a tumor to other anatomic structures also can be helpful in arriving at a diagnosis. A mass located along the course of a large nerve is suggestive of a peripheral nerve sheath tumor. A mass associated with a joint space or tendon sheath raises suspicion for a ganglion cyst, pigmented villonodular synovitis, or giant cell tumor of the tendon sheath. Synovial sarcomas are most often extra-articular. A mass arising from the chest wall underneath the scapula, with low signal intensity on T1- and T2-weighted images, suggests an elastofibroma dorsi.

Signal Characteristics

Most soft tissue tumors, whether malignant or benign, are isointense or hypointense compared with skeletal muscle on T1-weighted images, and they show increased signal intensity on T2-weighted images. Commonly encountered sarcomas with low signal T1-weighting and high signal T2-weighting are malignant fibrous histiocytoma, synovial sarcoma, leiomyosarcoma, and malignant schwannoma. The most common malignant fatty tumor is myxoid liposarcoma, and the clue to the diagnosis is the presence of wisps of bright signal (representing fat) in a background of low signal (myxoid tissue) on T1-weighted images. On T2-weighted sequences, myxoid elements become bright, largely obscuring the fatty component (Fig. 17–30). Other MR imaging signal characteristics were discussed previously (see under Tissue Characterization by Magnetic Resonance Imaging).

Table 17–2. Lesions Characterized by Location on MR Imaging

- Foot
  - Ganglion cyst
  - Giant cell tumor of tendon sheath/pigmented villonodular synovitis
  - Hemangioma
  - Plantar fibromatosis
  - Morton's neuroma
  - Synovial sarcoma
- Hand and wrist
  - Ganglion cyst
  - Giant cell tumor of tendon sheath
  - Glomus tumor
  - Fibrolipomatous hamartoma
  - Macrodystrophica lipomatosa
- Retroperitoneum
  - Liposarcoma
  - Malignant fibrous histiocytoma
  - Lymphoma
- Posterior/lateral chest
  - Elastofibroma dorsi
- Mandible
  - Peripheral nerve sheath tumor
  - Rhabdomyosarcoma
- Origin within or from a vessel
  - Leiomyosarcoma
- Along the course of a major nerve trunk
  - Peripheral nerve sheath tumor
- Close to a joint or tendon
  - Ganglion cyst
  - Synovial sarcoma
  - Giant cell tumor of tendon sheath
  - Clear cell sarcoma
- Involving an entire single muscle
  - Calcific myonecrosis
  - Diabetic muscle infarct
- Contained within a joint
  - Pigmented villonodular synovitis
  - Synovial hemangioma
  - Lipoma arborescens
  - Synovial chondromatosis
  - Synovial hyperplasia
- Related to palmar/plantar aponeurosis
  - Fibromatosis

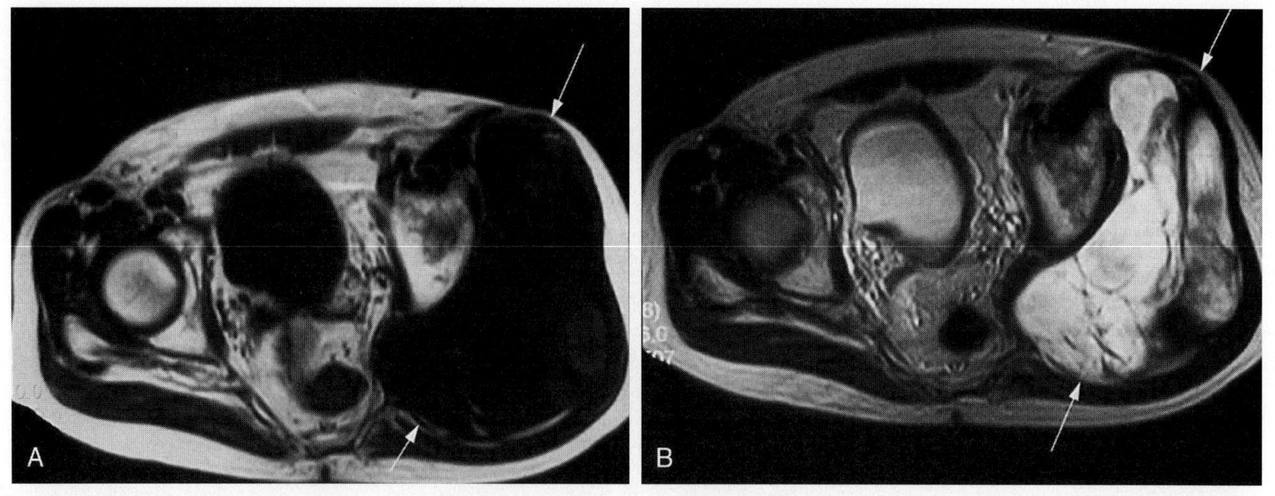

Figure 17–30. Myxoid liposarcoma. *A*, Axial T1-weighted MR image of the pelvis with a large, mostly low signal mass *(arrows)*. There are some faint regions of high signal within the mass representing fat. *B*, On the T2-weighted image, the mass is mostly high signal, consistent with its myxoid composition *(arrows)*. The regions of fat are now intermediate in signal intensity.

Homogeneity

Homogeneity of a soft tissue mass, as an isolated sign, is not specific enough to separate benign from malignant. Sarcomas can be homogeneous (Fig. 17–31), and benign masses can be heterogeneous. Peripheral nerve sheath tumors typically show the *target lesion* on T2-weighted images (see Fig. 17–22), where the center has a lower signal than the high signal periphery. Hemangiomas can have a similar finding.

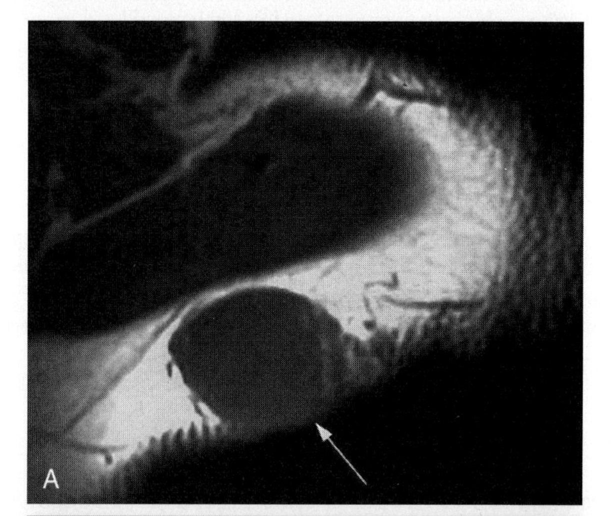

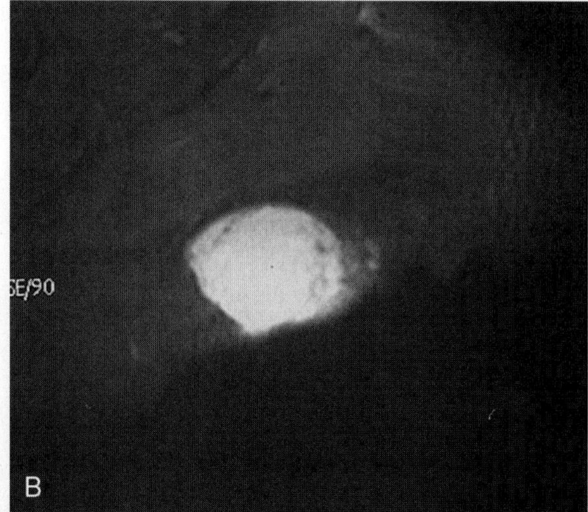

Figure 17–31. Dermatofibrosarcoma protuberans. *A*, Axial T1-weighted MR image in the supraclavicular region of the shoulder of a 41-year-old patient shows a fairly well-defined mass *(arrow)* that is mildly high signal relative to muscle. *B*, The mass has homogeneous high signal on T2-weighted image with fat saturation. The mass enhanced uniformly on postcontrast images (not shown), confirming the solid nature. This case shows that signal homogeneity cannot be relied on to distinguish benign from malignant lesions.

Involvement of Adjacent Structures

Involvement of adjacent structures is uncommon in benign and malignant conditions. It is rare for soft tissue sarcomas to show overt bone destruction or marrow invasion (Fig. 17–32). Also, major vascular invasion by soft tissue tumors is uncommon. Synovial sarcoma occasionally is reported to invade and destroy bone. Generally, when a mass engulfs a major artery or vein, one should consider that the mass might have originated from the vessel itself (i.e., an aneurysm if a major artery is involved and a leiomyosarcoma if a major vein is involved).

Hemorrhage and Surrounding Edema

Hemorrhage and surrounding edema are not helpful distinguishing features because they can be seen in benign and malignant soft tissue masses.

Necrosis

Newly discovered large lesions with extensive necrosis are likely to be sarcomas. T2-weighted signal intensities

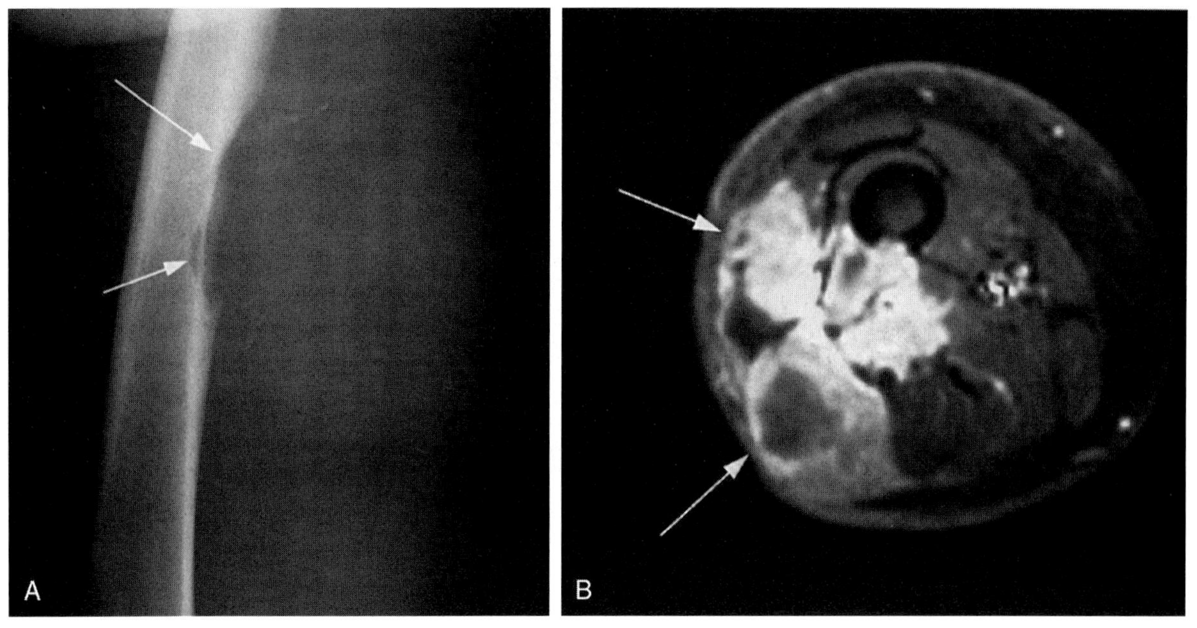

Figure 17–32. Malignant fibrous histiocytoma. *A,* Lateral view of the femur in a 70-year-old patient shows erosion of the posterior cortex *(arrows). B,* A large heterogeneous mass *(arrows)* is seen on postcontrast axial T1-weighted MR image with fat saturation. The posterior cortex of the femur is invaded by the tumor. Low signal regions within the mass represent necrosis.

from viable tumor, necrosis, edema, and hemorrhage can overlap. Gadolinium enhancement characterizes necrosis as nonenhancing areas.

Monitoring Therapeutic Response

Absence of recurrence confidently can be assumed when no increase in signal intensity is noted on T2-weighted images and when T1-weighted images after gadolinium injection reveal no tissue enhancement. The reverse is not true, however: Increased signal intensity on T2-weighted images or enhancing tissue on T1-weighted images with gadolinium could represent post-surgical changes or granulation tissue and recurrent tumor.

Recommended Readings

Kransdorf MJ, Murphey MD: Radiologic evaluation of soft tissue masses: A current perspective. AJR Am J Roentgenol 175:575, 2000.

Kransdorf MJ, Murphey MD, Smith SE: Imaging of soft tissue neoplasms in the adult: Benign tumors. Semin Musculoskeletal Radiol 3:21, 1999.

Kransdorf MJ, Murphey MD, Smith SE: Imaging of soft tissue neoplasms in the adult: Malignant tumors. Semin Musculoskeletal Radiol 3:39, 1999.

Sundaram M: MR imaging of soft tissue tumors: An overview. Semin Musculoskeletal Radiol 3:15, 1999.

Sundaram M, McGuire MH, Herbold DR, et al: High signal intensity soft tissue masses on T1-weighted pulsing sequences. Skeletal Radiol 16:30, 1987.

Sundaram M, McGuire MH, Schajowicz F: Soft tissue masses: Histologic basis for decreased signal (short T2) on T2-weighted MR images. AJR Am J Roentgenol 148:1247, 1987.

SECTION II

RHEUMATIC DISEASES

CHAPTER 18

Osteoarthritis

Georges Y. El-Khoury, MD, Mark D. Stanley, MD, and D. Lee Bennett, MD

OVERVIEW

Osteoarthritis (OA), also referred to as degenerative joint disease, is the most common rheumatic disorder and an important cause for disability in the aging population. OA is primarily a disease of hyaline cartilage accompanied by structural changes in the underlying bone. The exact etiology and pathogenesis of primary (or idiopathic) OA is not completely understood, but it is thought to be due to mechanical stresses that cause the chondrocytes to produce proteolytic enzymes that disrupt the cartilage matrix. The matrix consists of proteoglycans, collagen, and water. Collagen maintains the shape of cartilage, and proteoglycans, with their negatively charged macromolecules, are responsible for its elastic properties. Excess weight, repetitive trauma, or overuse of joints with certain occupations have been shown to cause OA. Unstable joints from ligamentous disruption predispose to OA. Once the osteoarthritic process starts, it progresses toward further joint space narrowing and joint damage. The disease, however, is nonlinear in its evolution, and it may progress at different rates in different patients.

Certain conditions seem to protect joints from developing OA. Studies have shown an inverse relation between osteoporosis and OA in women. Hemiplegia reduces the expression of OA in the hand on the paralyzed side.

JOINT INVOLVEMENT

Typically, the weight-bearing joints are involved, especially the knee and hip, with the knee being about twice as commonly involved as the hip (Fig. 18–1). Knee OA is more common in women, whereas OA of the hip is more common in men. In the hands, the distal interphalangeal, proximal interphalangeal, and first carpometacarpal joints are most frequently involved (Fig. 18–2), and in the feet, the first metatarsophalangeal joints are often affected (Fig. 18–3). The cervical and lumbar spines are frequently involved with OA as well, particularly the facet joints. Primary OA is rare in the wrist, elbow, shoulder, and ankle (Fig. 18–4). When OA is discovered in an unusual location, such as the metacarpophalangeal joints, talonavicular joint, and wrist, other underlying diseases such as calcium pyrophosphate dihydrate deposition disease (CPPD) or hemochromatosis should be suspected. Occasionally, the patellofemoral joint is preferentially involved with degenerative joint disease (OA) without significant involvement of the tibiofemoral joint; in such cases CPPD should be considered as the underlying cause of this presentation.

RADIOGRAPHIC ABNORMALITIES

The diagnosis of OA is usually made clinically and confirmed radiographically. Radiographic changes do not always correlate with the symptoms; however, patients with severe radiologic changes are more likely to be symptomatic. The radiographic abnormalities in OA are easily detectable on plain radiographs, and they are fairly diagnostic. Four radiographic signs constitute the major findings in OA.

Joint Space Narrowing

The joint space narrowing in OA is nonuniform, and it occurs in the weight-bearing portion of the joint. Articular cartilage thickness is evaluated indirectly on plain radiographs. In the knee, joint space narrowing typically occurs medially, and in the hip, it occurs superiorly (see Fig. 18–1). Specialized views have been advocated to improve the sensitivity of early cartilage loss. Because early cartilage loss in the knee has been shown to occur on the posterior aspect of the femoral condyles, some authors have advocated the use of a posteroanterior (PA) standing tunnel view in 30 degrees of flexion. A standing anteroposterior (AP) view of the knee is also effective in demonstrating articular cartilage loss (Fig. 18–5). Merchant's view is an excellent projection for the evaluation of OA of the patellofemoral joint (Fig. 18–6).

Osteophyte Formation

It is not clear why osteophytes develop, but it is generally accepted that the presence of osteophytes as the only abnormality does not necessarily signify the presence of OA. This is particularly true in the hip and knee where

Text continued on page 169

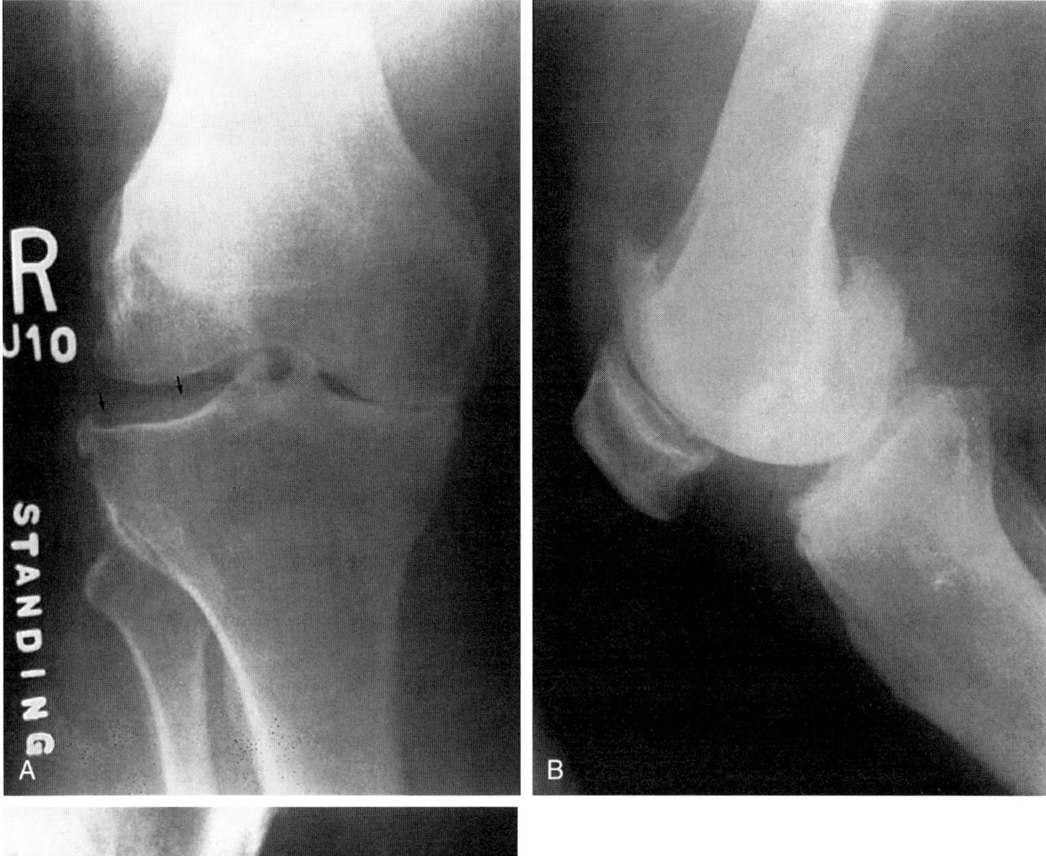

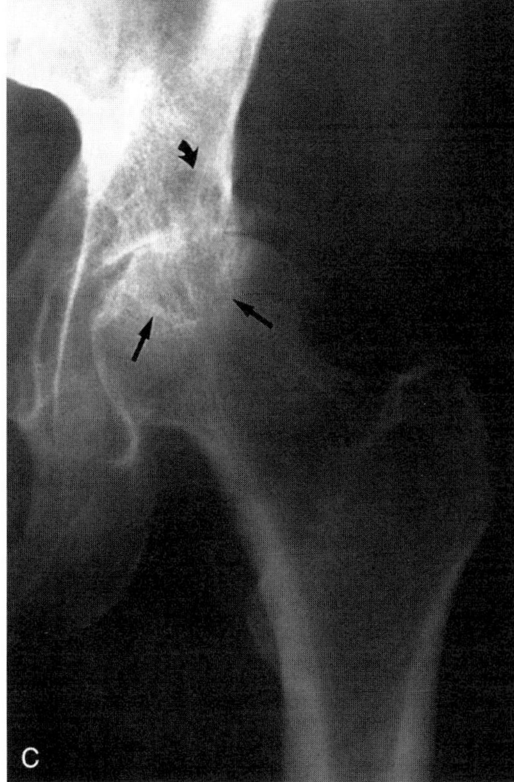

Figure 18–1. Osteoarthritis (OA) of the knee and the hip. *A*, Anteroposterior view of the right knee. There is marked narrowing of the joint space medially. A ridge of osteophytes is seen arising from the posterior rim of the lateral tibial plateau *(arrows)*. *B*, Lateral view of right knee shows OA that involves the patellofemoral joint as well as the tibiofemoral joint. There are large marginal osteophytes arising from the distal femur and patella. *C*, Anteroposterior view of the left hip. The joint space is markedly narrowed superiorly; there are degenerative cysts on both sides of the joint, which are referred to as kissing cysts *(arrows)*.

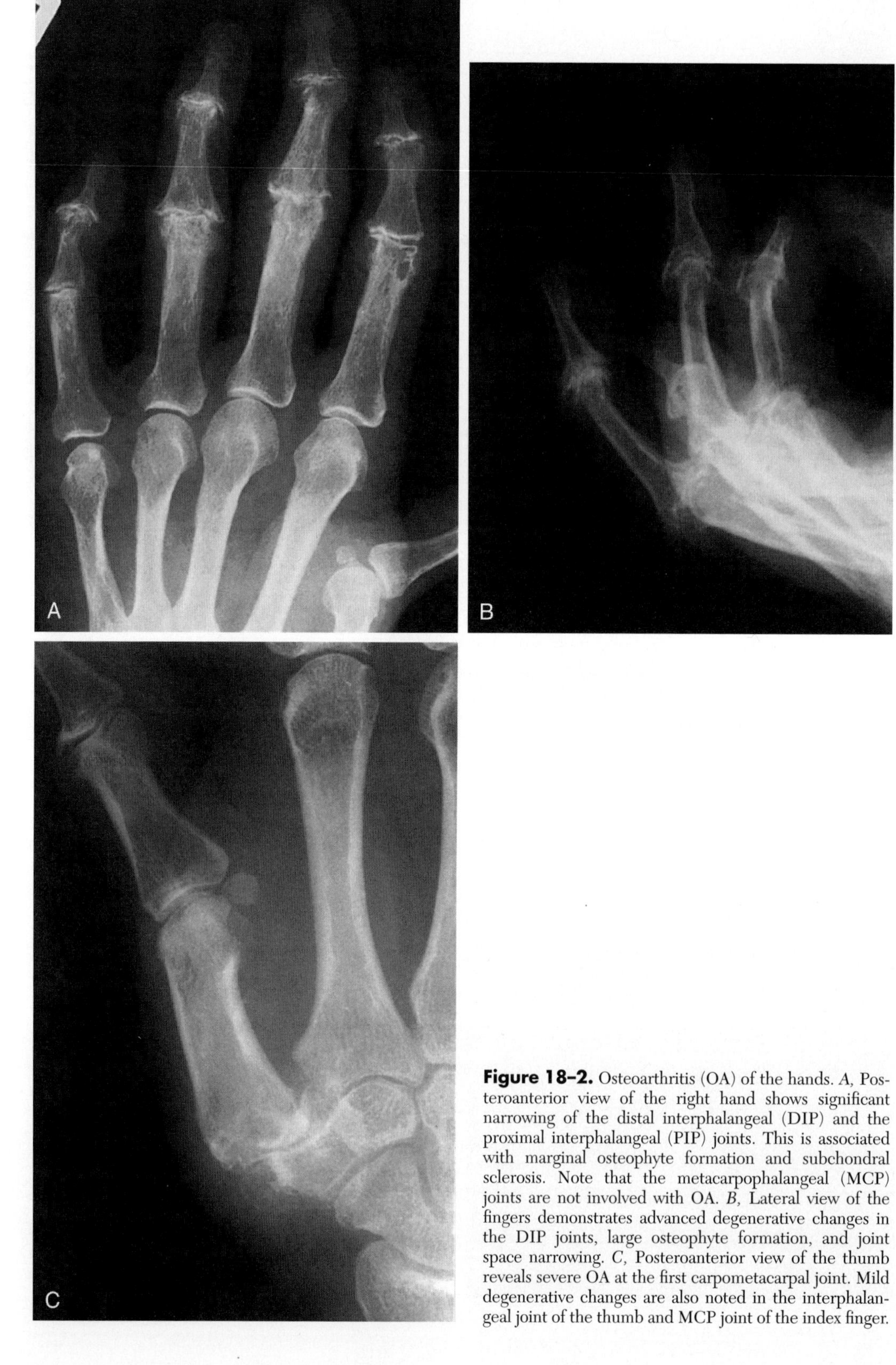

Figure 18–2. Osteoarthritis (OA) of the hands. *A*, Posteroanterior view of the right hand shows significant narrowing of the distal interphalangeal (DIP) and the proximal interphalangeal (PIP) joints. This is associated with marginal osteophyte formation and subchondral sclerosis. Note that the metacarpophalangeal (MCP) joints are not involved with OA. *B*, Lateral view of the fingers demonstrates advanced degenerative changes in the DIP joints, large osteophyte formation, and joint space narrowing. *C*, Posteroanterior view of the thumb reveals severe OA at the first carpometacarpal joint. Mild degenerative changes are also noted in the interphalangeal joint of the thumb and MCP joint of the index finger.

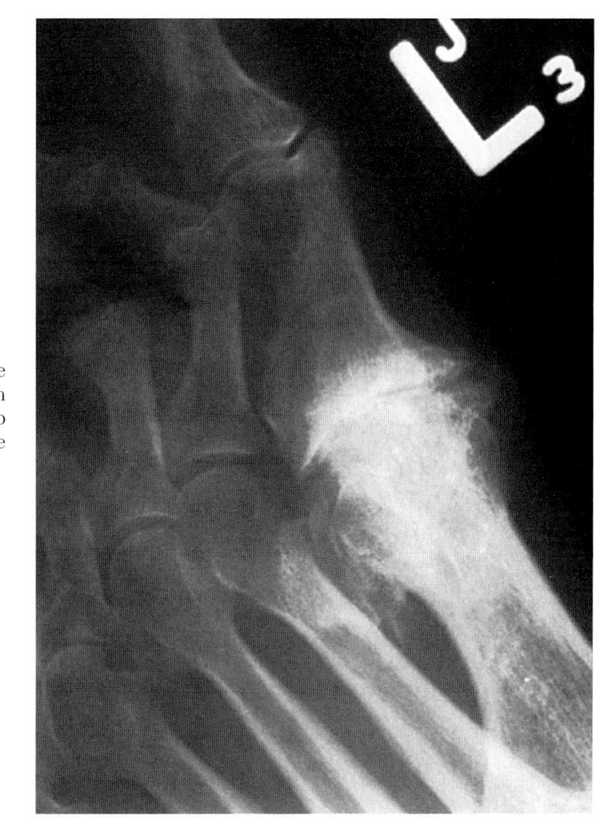

Figure 18–3. Osteoarthritis in the first metatarsophalangeal (MTP) joint. There is severe joint space narrowing and large osteophytes at the first MTP joint. Such changes often result in restricted motion of the great toe, a condition referred to as "hallux rigidus." There are also degenerative changes at the joints between the sesamoid bones and the first metatarsal head.

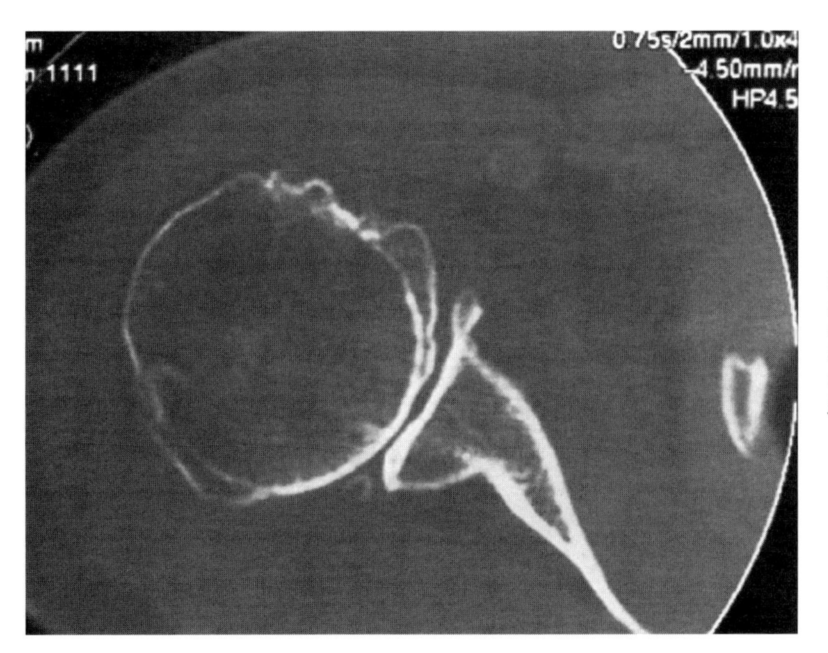

Figure 18–4. Primary osteoarthritis (OA) of the shoulder. A computed tomography scan of the right shoulder, performed prior to total shoulder replacement for end stage OA, demonstrates joint space narrowing and osteophyte formation. A loose body in the posterior joint space is also noted.

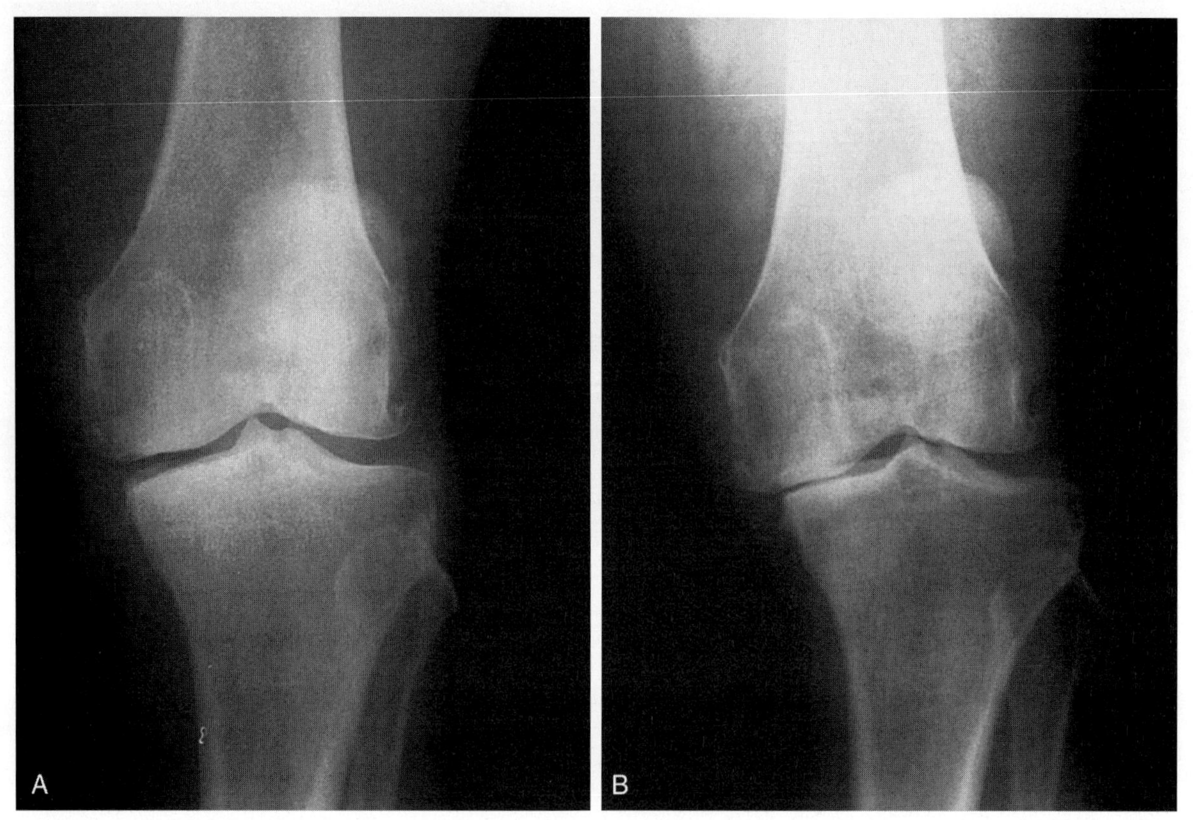

Figure 18–5. Standing anteroposterior (AP) view to demonstrate articular cartilage narrowing. *A*, Supine, non–weight bearing, anteroposterior view of the knee. *B*, Standing AP view of the same patient reveals severe loss of articular cartilage in the medial compartment.

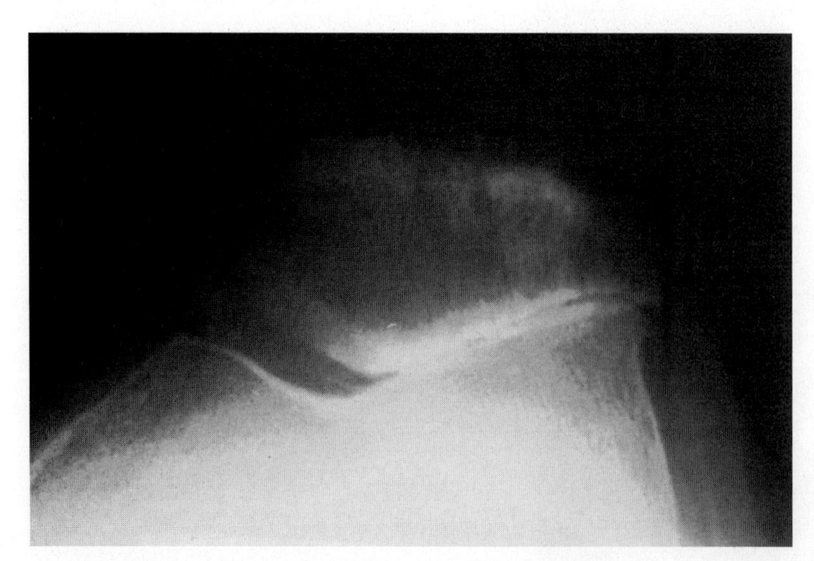

Figure 18–6. Patellofemoral joint osteoarthritis. Merchant's view demonstrates narrowing of the lateral joint space and marginal osteophyte formation.

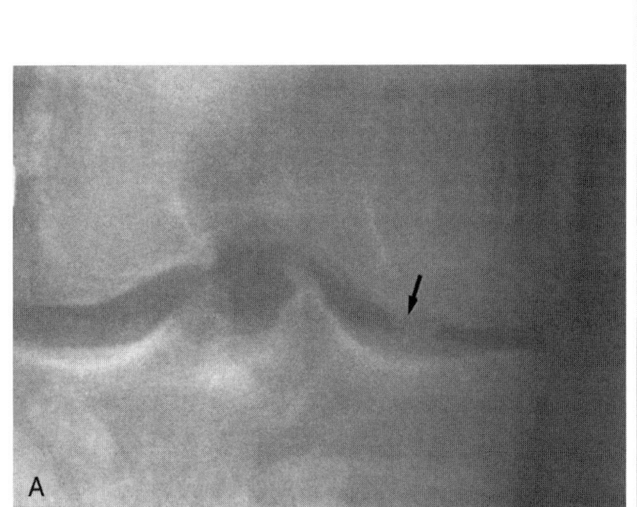

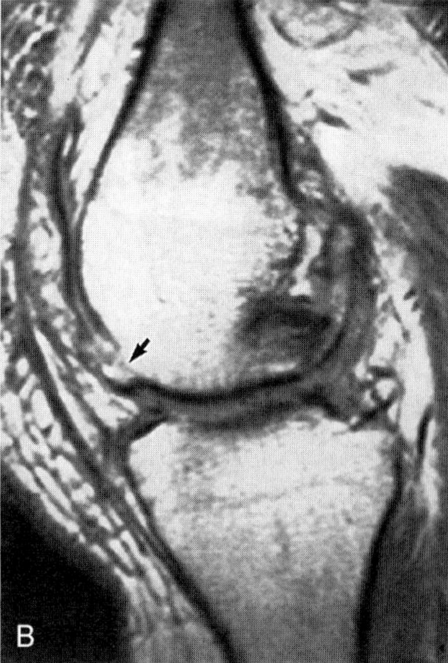

Figure 18–7. Central osteophyte in the medial femoral condyle. *A,* Anteroposterior view of the right knee showing a central osteophyte arising from the medial femoral condyle *(arrow). B,* Sagittal proton density image from another patient showing a central osteophyte arising from the medial femoral condyle *(arrow).*

small osteophytes may have greater association with aging than with OA. Marginal osteophytes develop at the periphery of the joint where articular cartilage is continuous with the synovial membrane and periosteum (see Figs. 18–1 and 18–4). In the knee, central osteophytes, which are button-like or flat excrescences, develop from the subchondral bone of the femur (Fig. 18–7). In the hip, peculiar osteophytes can develop along the medial aspect of the femoral head. These osteophytes displace the femoral head laterally. This appearance should not be confused with old slipped capital femoral epiphysis (Fig. 18–8). Similar osteophytes have been noted, on rare occasions, to develop on the medial surface of the humeral head (Figs. 18–8*B* and 18–8*C*). The sacroiliac (SI) joint, like other synovial joints, can be involved with OA. Osteophytes occasionally develop on the SI joint anteriorly. These osteophytes can resemble a blastic lesion on plain AP radiographs of the pelvis (Fig. 18–9). CT is helpful in demonstrating these osteophytes.

Subchondral Sclerosis

This is typically seen with moderate or advanced disease (see Fig. 18–8*A*). The term *eburnation,* which means ivory-like, is also used in this context. Sclerosis or eburnation are believed to result from trabecular compression with callus formation. Recent theories have implicated subchondral sclerosis in the pathogenesis of OA. Sclerotic subchondral bone is hard and therefore has less capacity to absorb stress during joint loading. Increased stresses damage the chondrocytes in the articular cartilage and lead to OA. It cannot be completely excluded that healthy cartilage protects subchondral bone from overload, thereby reducing subchondral sclerosis.

Cyst (Geode) Formation

Degenerative cysts, which are also referred to as geodes (see Fig. 18–1*C*), are commonly seen with OA. Degenerative cysts typically occur in the subchondral bone and are often multiple. Usually, they are a few millimeters in size (Fig. 18–10) but occasionally can reach several centimeters in diameter (see Fig. 18–10), such that they can expand the cortex or even extend beyond the cortex into the soft tissues. When two cysts occur across the joint from each other, they are called "kissing cysts" (see Fig. 18–1*C*). These degenerative cysts often communicate with the joint space and are seen to fill with air on computed tomography scans (Fig. 18–10*C*). There are two theories explaining the etiology of cysts in OA: (1) pressure-induced intrusion of synovial fluid into the subchondral bone; (2) osteonecrosis owing to mechanical stresses that precedes the entry of synovial fluid into the bone.

Bony Buttressing

Buttressing is defined as the deposition of new bone along the medial aspect of the femoral neck (Fig. 18–11). In severe OA, buttressing can also be seen along the lateral aspect of the femoral neck. The presence of buttressing is almost diagnostic of OA, although it has been reported in a small number of patients with rheumatoid arthritis and avascular necrosis. When buttressing occurs in arthritic diseases other than OA, the disease is usually advanced, and secondary OA is already a prominent feature of the arthritic process. Buttressing is reported to occur in approximately 50% of patients undergoing total hip arthroplasty. It is believed to be caused by altered hip mechanics.

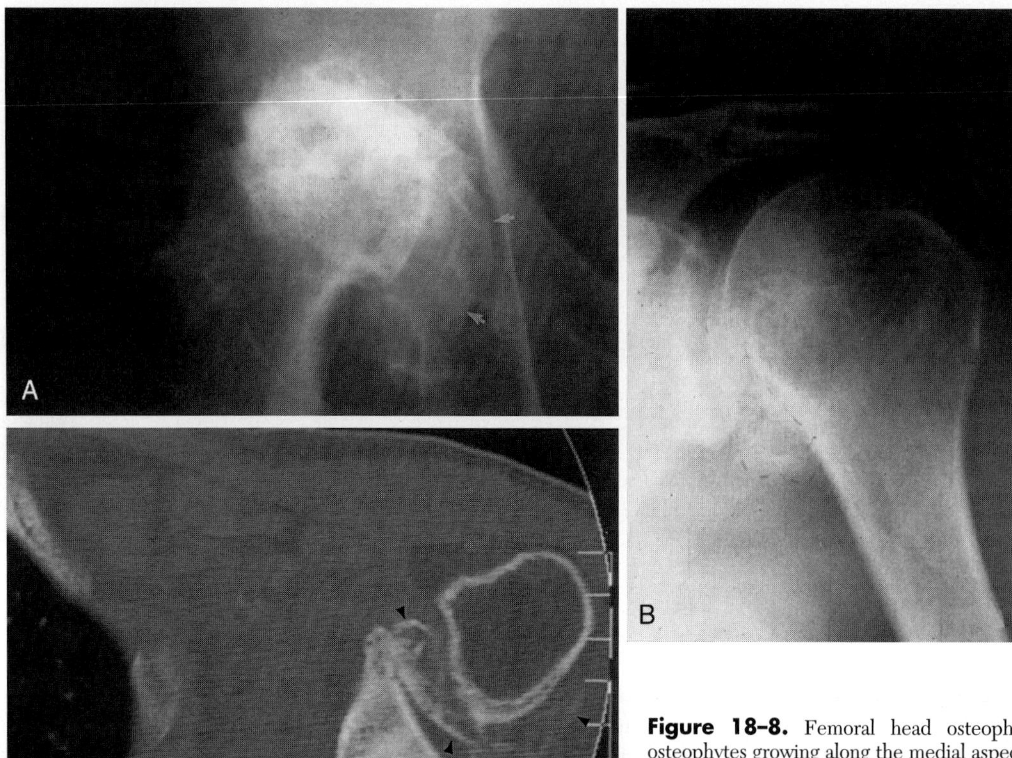

Figure 18–8. Femoral head osteophytes. *A,* Large osteophytes growing along the medial aspect of the femoral head *(arrows).* They should not be confused with old slipped capital femoral epiphysis. *B,* Similar osteophytes develop occasionally along the medial aspect of the humeral head. *C,* A computed tomography scan of the left shoulder demonstrating osteophytes arising from the medial and the inferior aspects of the humeral head *(arrowheads).*

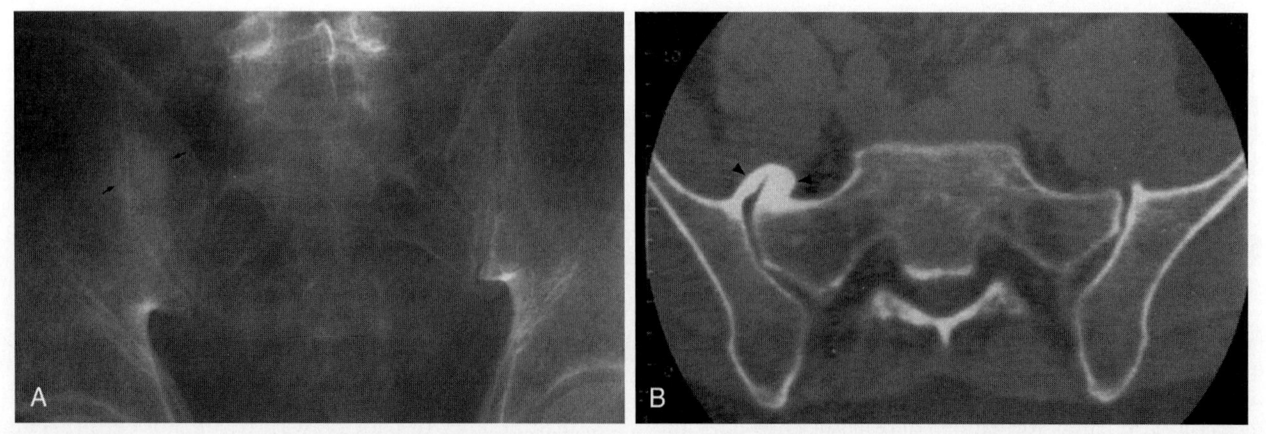

Figure 18–9. Sacroiliac joint anterior osteophyte in a 58-year-old male who was a football player in college. *A,* Anteroposterior view of the sacrum shows a rounded blastic lesion projecting over the right sacroiliac joint *(arrows). B,* A computed tomography scan shows that the blastic lesion represents a large osteophyte projecting anteriorly *(arrowheads).*

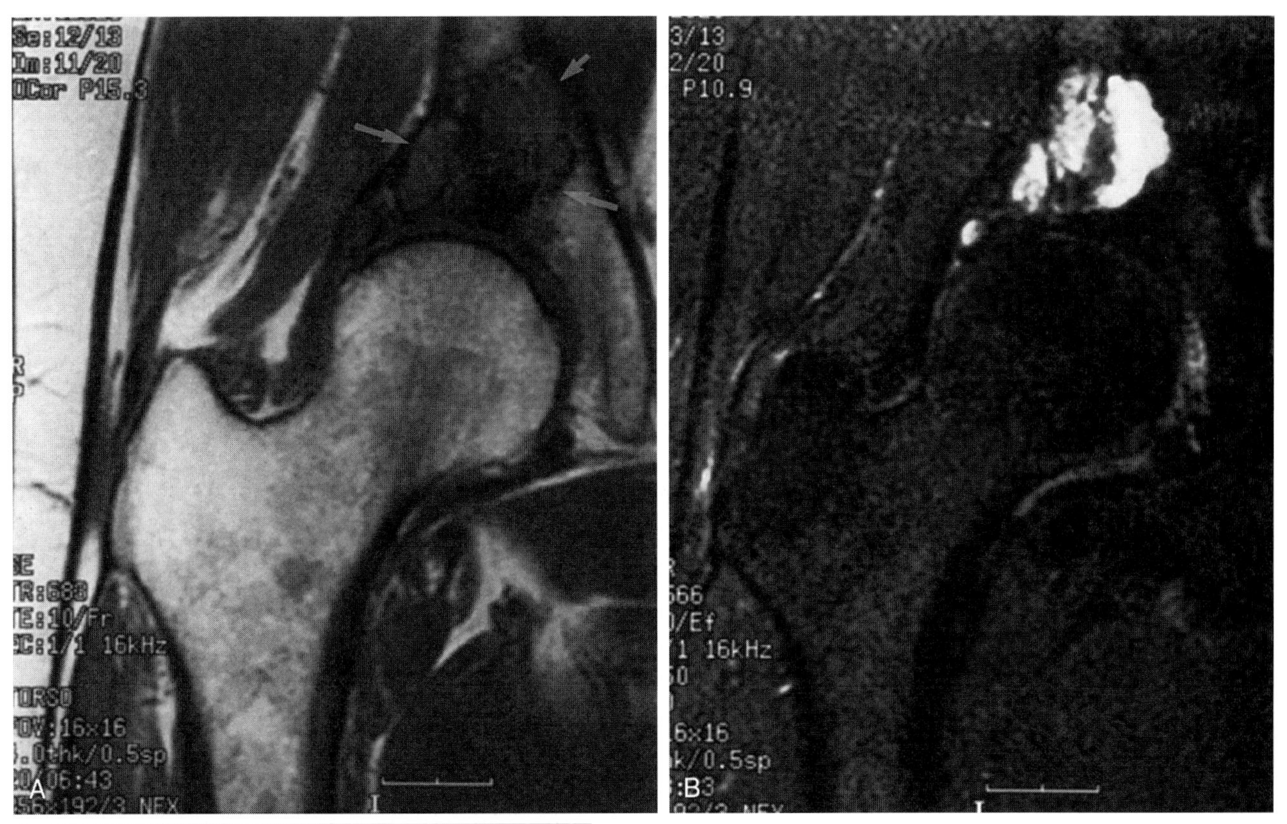

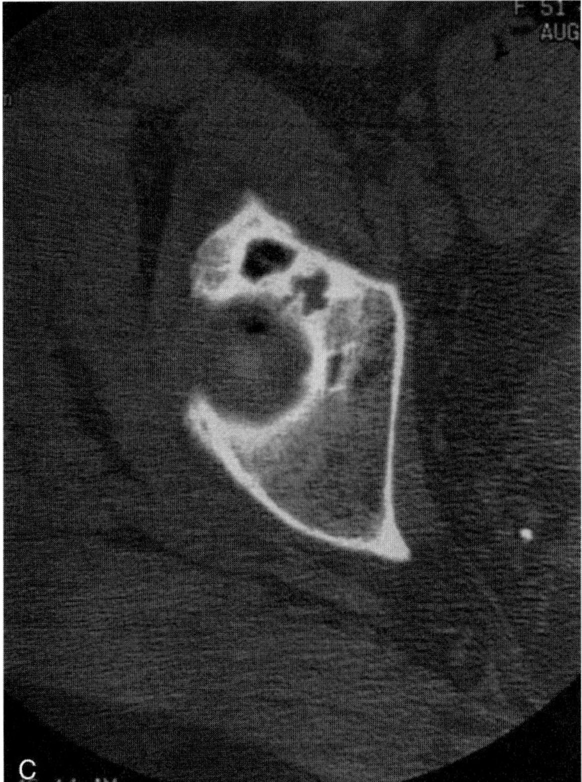

Figure 18–10. Degenerative cyst (geode). *A,* T1-weighted coronal image of the hip shows a lobulated low signal intensity lesion with septae in the roof of the right acetabulum *(arrows). B,* T2-weighted, fat-suppressed coronal image reveals the fluid containing the cystic nature of the lesion. *C,* A computed tomography scan section through a degenerative cyst of the right acetabulum demonstrates air within the cyst. The presence of air in a periarticular cystic structure is almost diagnostic of a degenerative cyst.

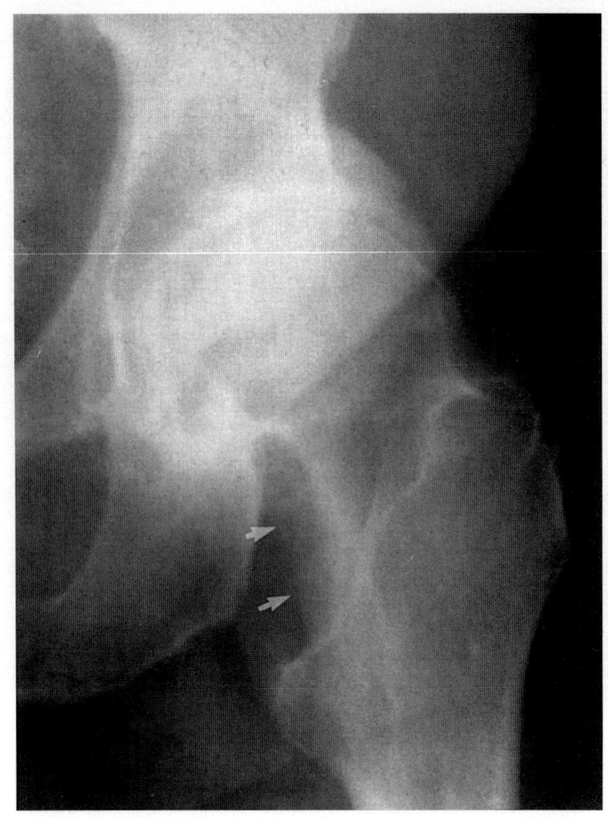

Figure 18–11. Buttressing. An anteroposterior view of the hip shows bone laid along the medial aspect of the femoral neck *(arrows)*. This is referred to as "buttressing," and it is characteristic of osteoarthritis.

Recommended Readings

Abrahim-Zadeh R, Yu JS, Resnick D: Central (interior) osteophytes of the distal femur: Imaging and pathologic findings. Invest Radiol 29:1001, 1994.

Adams JG, McAlindon T, Dimasi M, et al: Contribution of meniscal extrusion and cartilage loss to joint space narrowing in osteoarthritis. Clin Radiol 54:502, 1999.

Boegard T, Jonsson K: Radiography in osteoarthritis of the knee. Skeletal Radiol 28:605, 1999.

Buckland-Wright JC, MacFarlane DG, Lynch JA, et al: Quantitative microfocal radiographic assessment of progression in osteoarthritis of the hand. Arthritis Rheum 33:57, 1990.

Dixon T, Benjamin J, Lund P, et al: Femoral neck buttressing: A radiographic and histologic analysis. Skeletal Radiol 29:587, 2000.

Fife RS: Osteoarthritis. A. Epidemiology, pathology, and pathogenesis. In Klippel JH (ed): Primer on the rheumatic diseases, 11th ed. Atlanta, The Arthritis Foundation, 1997.

Fife RS, Brandt KD, Braunstein EM, et al: Relationship between arthroscopic evidence of cartilage damage and radiographic evidence of joint space narrowing in early osteoarthritis of the knee. Arthritis Rheum 34:377, 1991.

Glass TA, Dyer R, Fisher L, et al: Expansile subchondral bone cyst. AJR Am J Roentgenol 139:1210, 1982.

Hellmann DB, Helms CA, Genant HK: Chronic repetitive trauma: A cause of atypical degeneration joint disease. Skeletal Radiol 10:236, 1983.

Hochberg MC: Osteoarthritis. B. Clinical features and treatment. In Klippel JH (ed): Primer on the rheumatic diseases, 11th ed. Atlanta, The Arthritis Foundation, 1997.

Messieh SS, Fowler PJ, Munro T: Anteroposterior radiographs of the osteoarthritic knee. J Bone Joint Surg Br 72:639, 1990.

Resnick D: Patterns of migration of the femoral head in osteoarthritis of the hip. AJR Am J Roentgenol 124:62, 1975.

Segal R, Avrahami E, Lebdinski E, et al: The impact of hemiparalysis on the expression of osteoarthritis. Arthritis Rheum 41:2249, 1998.

CHAPTER 19

Osteoarthritis Variants

Georges Y. El-Khoury, MD, Mark D. Stanley, MD, and D. Lee Bennett, MD

SECONDARY OSTEOARTHRITIS

Secondary osteoarthritis (OA), which is fairly common, results from pre-existing joint disease that damages the articular cartilage. Significant joint trauma, intra-articular fractures (Figs. 19–1 and 19–2), septic arthritis, or inflammatory arthritis are common causes for secondary OA.

Some congenital and developmental joint conditions such as developmental dysplasia of the hip or Legg-Calvé-Perthes, slipped capital femoral epiphysis, and a variety of epiphyseal dysplasias can also lead to secondary OA.

Secondary OA is the final common pathway of all joint conditions that alter the mechanics of joints or damage the articular cartilage (see Figs. 19–1 and 19–2). Early in the disease, the radiographic findings vary depending on the underlying condition. In advanced secondary OA, the underlying condition is masked by the degenerative changes.

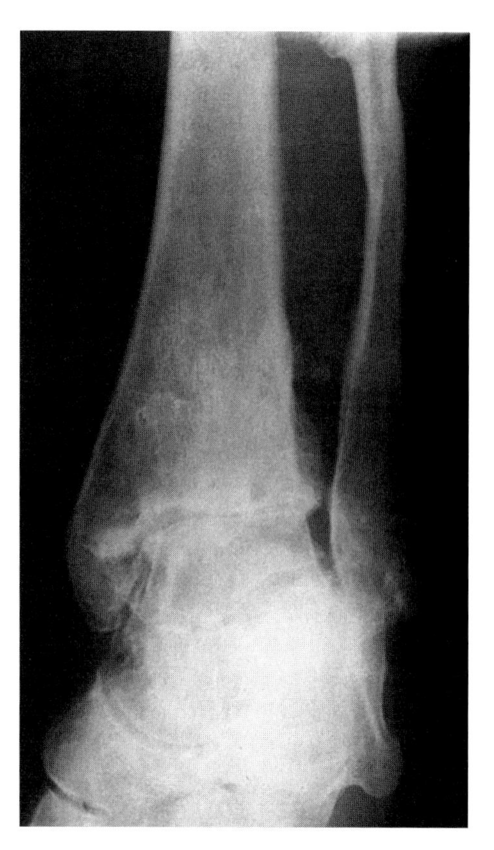

Figure 19–1. Secondary osteoarthritis of the ankle caused by previous fractures. An anteroposterior view of the ankle shows wide syndesmosis, narrowing of the ankle joint, osteophytes, subchondral cysts, and bony sclerosis. There is also evidence of an old healed fracture of the distal fibula.

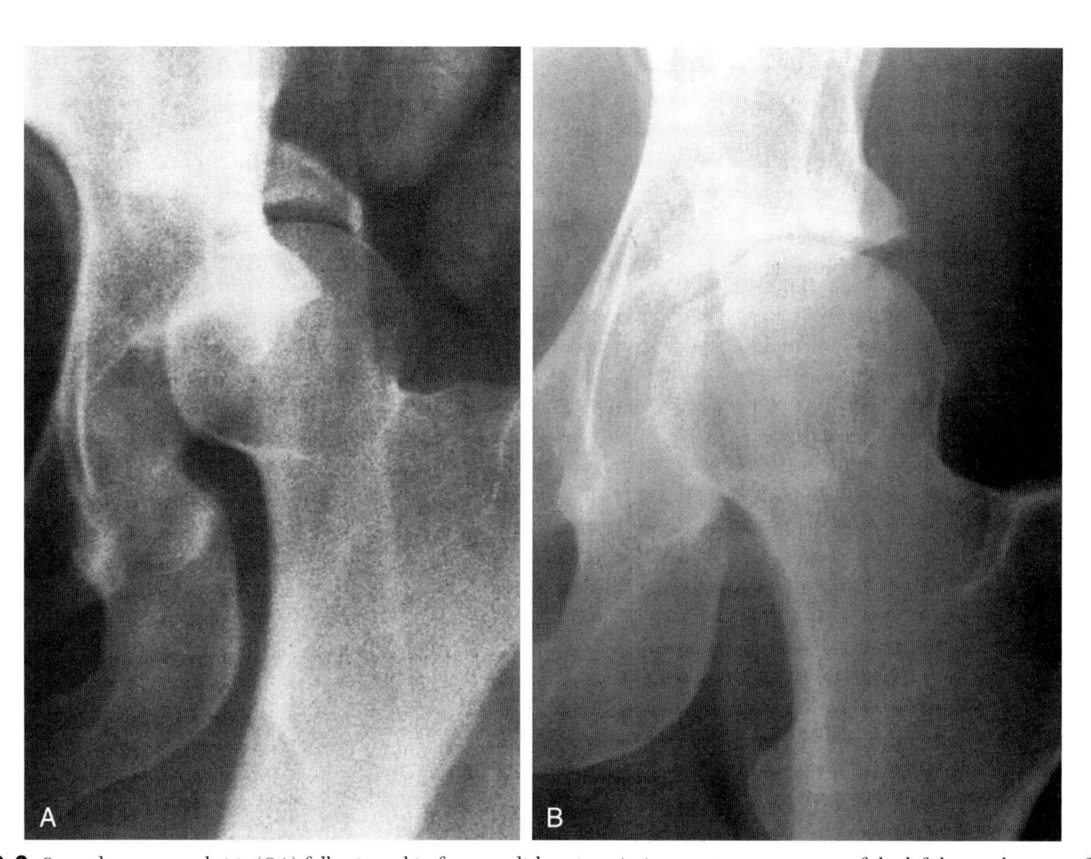

Figure 19–2. Secondary osteoarthritis (OA) following a hip fracture dislocation. *A*, An anteroposterior view of the left hip at the time of the injury. *B*, Early secondary OA 6 months after the initial injury.

Illustration continued on following page

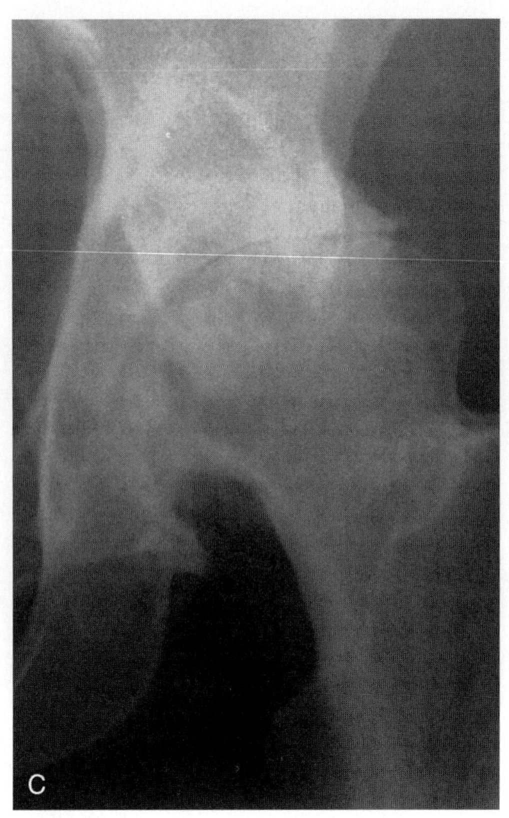

Figure 19–2 *Continued. C,* Advanced secondary OA 2½ years after the injury.

EROSIVE (INFLAMMATORY) OSTEOARTHRITIS

Erosive OA (EOA) typically affects middle-aged, postmenopausal women. It involves the distal interphalangeal (DIP) and first carpometacarpal joints. Occasionally, the proximal interphalangeal joints of the hand are also affected. Early in the disease, there is evidence of low-grade inflammation—thus the name inflammatory osteoarthritis.

Radiographically, there are subchondral erosions (Fig. 19–3) accompanied by sclerosis and marginal osteophytes. This results in wavy articular surfaces, also described as the "gull wing" deformity. Some affected joints eventually fuse (see Fig. 19–3). The radiographic changes in EOA can superficially resemble psoriatic arthritis, but the erosions in psoriasis are not subchondral; they occur at the corners of the distal phalanges, where the joint capsule inserts. Both erosive OA and psoriatic arthritis, however, show DIP joint fusion (see Fig. 19–3).

RAPIDLY DESTRUCTIVE OSTEOARTHRITIS OF THE HIP

Rapidly destructive OA (RDOA) of the hip is an uncommon form of osteoarthritis that involves the hip. It primarily affects middle-aged or elderly women. RDOA is characterized by rapid progression and significant joint destruction. The etiology for this rapid progression is not known. Radiographically, the hip changes can simulate septic arthritis, rheumatoid arthritis, or neuropathic arthropathy (Fig. 19–4).

METACARPOPHALANGEAL DEGENERATIVE ARTHROPATHY ASSOCIATED WITH MANUAL LABOR (MISSOURI METACARPAL SYNDROME)

Typically, the metacarpophalangeal (MCP) joints are not affected with OA. OA involving these joints is usually attributed to calcium pyrophosphate dihydrate deposition disease (CPPD) or hemochromatosis. Bilateral degenera-

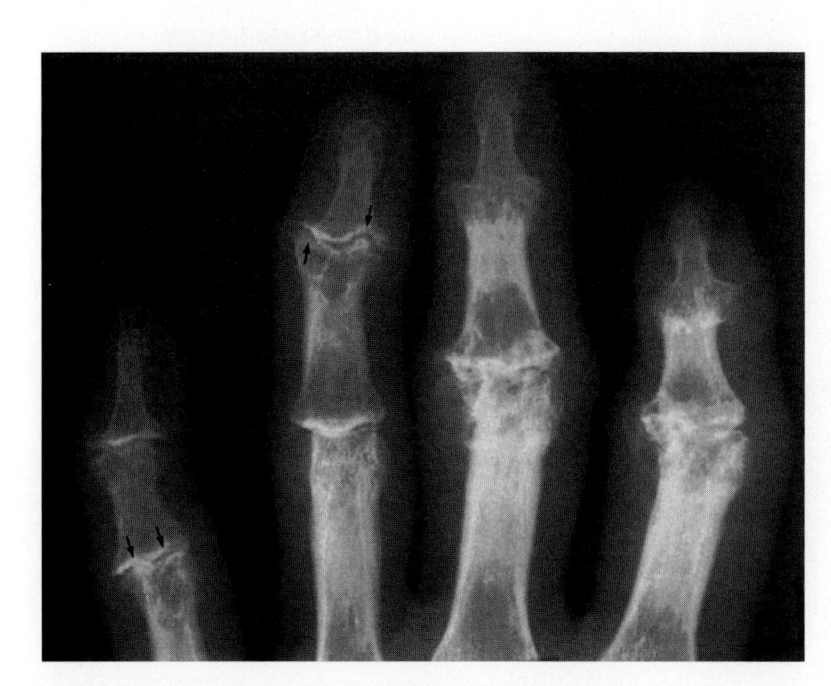

Figure 19–3. Erosive (inflammatory) osteoarthritis. Posteroanterior view of the hand shows subchondral erosions in the distal interphalangeal (DIP) joints of the fourth and proximal interphalangeal (PIP) joint of the fifth fingers *(arrows)*. The DIP joints of the second and third fingers are fused.

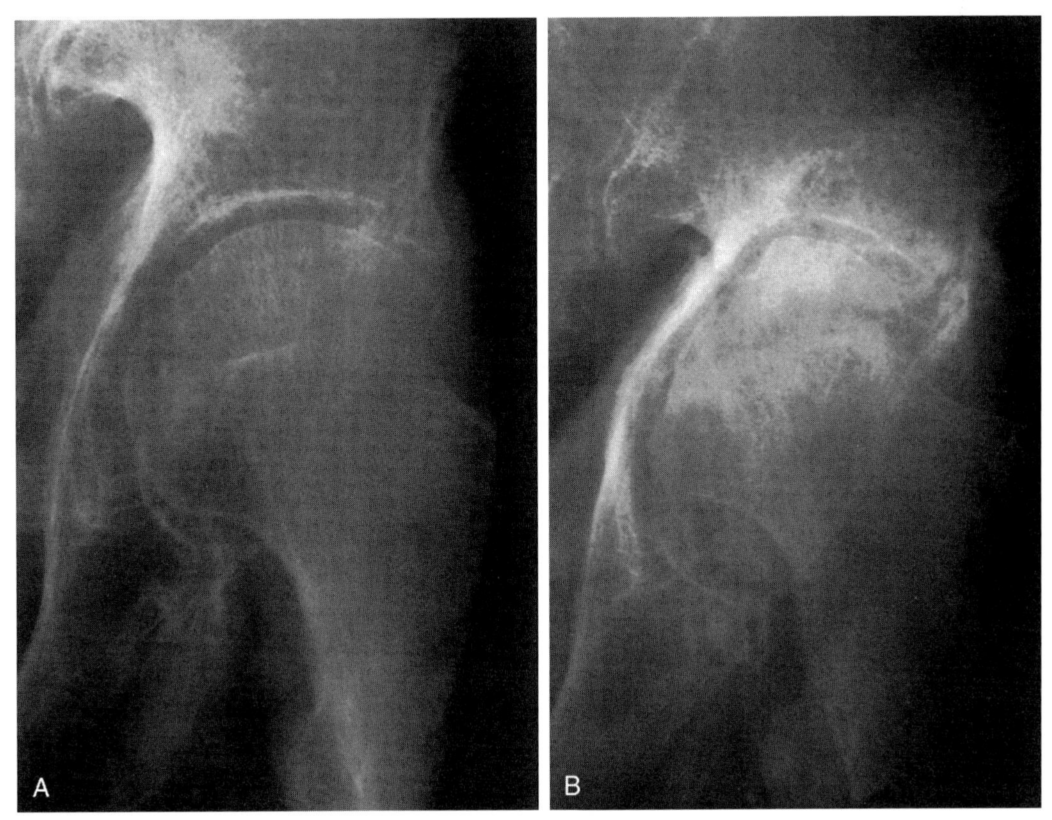

Figure 19–4. Rapidly destructive osteoarthritis. *A*, Anteroposterior (AP) radiograph of the left hip shows no significant abnormalities. *B*, Repeat AP radiograph taken 6 months later shows severe degenerative changes.

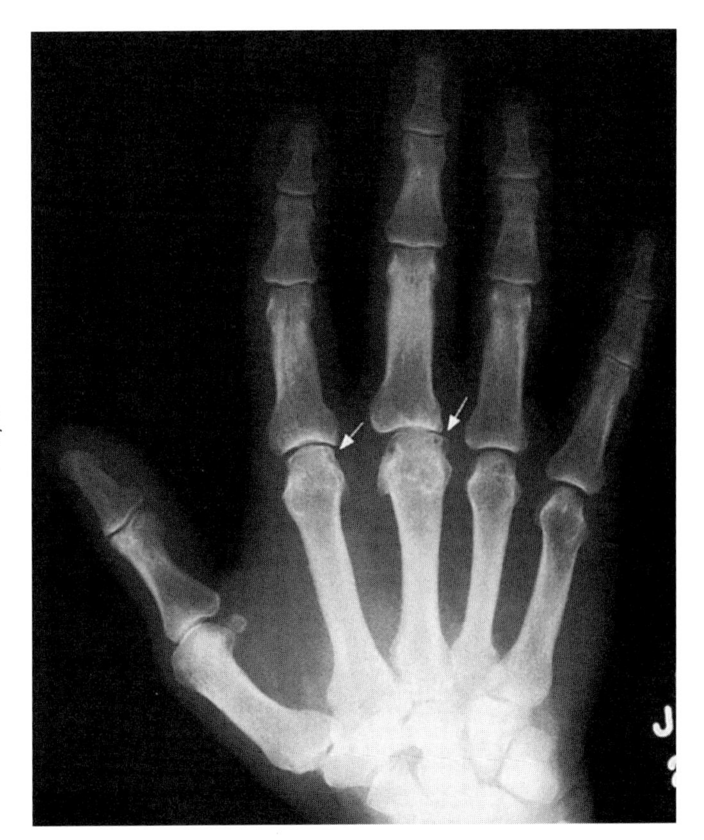

Figure 19–5. Metacarpophalangeal (MCP) osteoarthritis (OA) owing to manual labor in a 71-year-old farmer. Posteroanterior view of the left hand shows OA of the second and third MCP joints *(arrows)*.

tive arthropathy affecting mainly the second and third MCP joints is sometimes seen in patients who have performed manual labor most of their lives (Fig. 19–5).

Recommended Readings

Bock GW, Garcia A, Weisman MH, et al: Rapidly destructive hip disease: Clinical and imaging abnormalities. Radiology 186:461, 1993.

Cobby M, Cushnaghan J, Creamer P, et al: Erosive osteoarthritis: Is it a separate disease entity? Clin Radiol 42:258, 1990.

Greenway G, Resnick D, Weisman M, et al: Carpal involvement in inflammatory (erosive) osteoarthritis. Can Assoc Radiol J 30:95, 1979.

Martel W, Stuck KJ, Dworin AM, et al: Erosive osteoarthritis and psoriatic arthritis: A radiologic comparison in the hand, wrist, and foot. AJR Am J Roentgenol 134:125, 1980.

Rosenberg ZS, Shankman S, Steiner GC, et al: Rapid destructive osteoarthritis: Clinical, radiographic, and pathologic features. Radiology 182:213, 1992.

Williams WV, Cope R, Gaunt WD, et al: Metacarpophalangeal arthropathy associated with manual labor (Missouri Metacarpal Syndrome). Arthritis Rheum 30:1362, 1987.

CHAPTER 20

Rheumatoid Arthritis

Georges Y. El-Khoury, MD, Mark D. Stanley, MD, and D. Lee Bennett, MD

OVERVIEW

Rheumatoid arthritis (RA) is a systemic disease that affects approximately 1% of the population, with women being approximately 2.5 times more commonly affected than men. The disease can occur at any age, but the peak incidence is between the fourth and sixth decades. Clinical manifestations are most evident in the diarthrodial joints, although inflammatory changes in the small blood vessels can involve several organs, including the lungs (Fig. 20–1), pleura, pericardium, skin, and nerves. In the first few months of the disease, the diagnosis of RA can be difficult. Early characteristic findings include symmetric synovitis in the wrists and metacarpophalangeal (MCP) joints and serologically positive rheumatoid factor. The distinctive erosions may take several months to develop.

PATHOLOGY

The target tissue in RA is the synovium in diarthrodial joints, bursae, and tendon sheaths. Early synovial changes include increased cellularity of both types A and B synovial cell, lymphocytes (mainly T-cells), and synovial fibroblasts accompanied by capillary proliferation. This process results in granulation tissue proliferation, which is also known as pannus. Pannus is responsible for the bony erosions that are characteristic of RA. It directly invades the bone at joint margins or bare areas. The bare areas are defined as bony sites within the joint capsule that are not covered with cartilage. Areas covered with cartilage tend to be resistant to early erosion. Articular cartilage in RA is permeated with immune complexes, thus attracting neutrophils, which enzymatically degrade the cartilage by releasing collagenase and proteases. The enzymes destroy the proteoglycans, which hold water within the cartilage. Collagen within the articular cartilage is also very susceptible to enzymatic degradation.

Tenosynovitis is seen in most patients (Fig. 20–2). Inflammation of tendons with formation of inflammatory nodules and central necrosis can also be seen.

RADIOGRAPHIC CHANGES

Most synovial joints can be affected by RA, and the involvement is typically symmetrical. Unilateral disease is very rare, but it has been described in hemiplegic patients where the neurologic deficit seems to protect against RA. Wrist involvement is seen in virtually all RA patients. In the hand, the MCP and proximal interphalangeal (PIP) joints (Fig. 20–3) are typically involved, whereas the distal interphalangeal joints are relatively spared. Finger deformities and ulnar drift are common owing to ligamentous and tendinous disruption. The abnormalities that can be detected radiographically closely reflect the pathologic changes. These include the following.

Soft Tissue Swelling. Soft tissue swelling (Figs. 20–3 and 20–4) is due to joint effusion and synovial proliferation. In the PIP joints of the hand, the soft tissue swelling is typically fusiform (see Fig. 20–3). Soft tissue swelling can also develop as a result of synovial cysts or ruptured

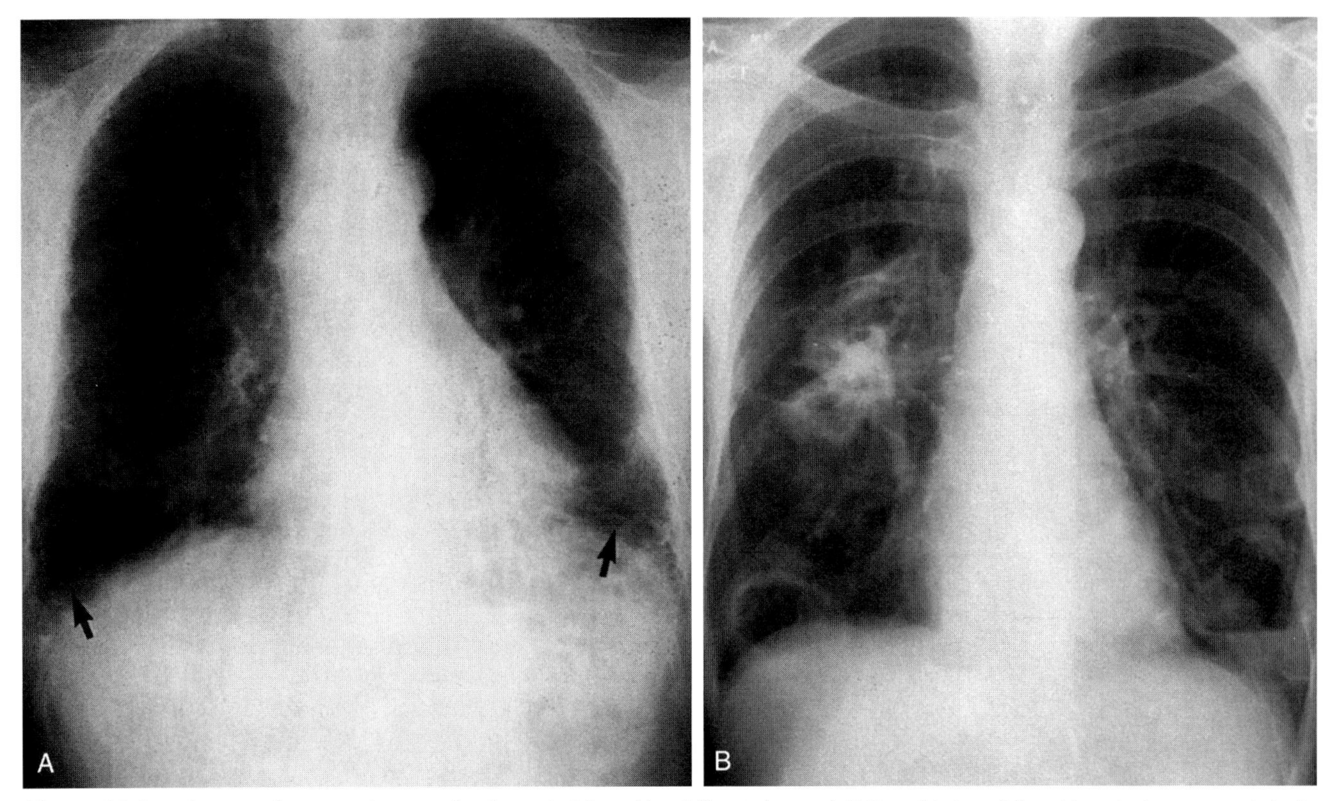

Figure 20–1. Pulmonary changes in rheumatoid arthritis. *A*, Bilateral basal fibrosis *(arrows)*. *B*, Necrobiotic nodules with cavitation are seen in both lungs.

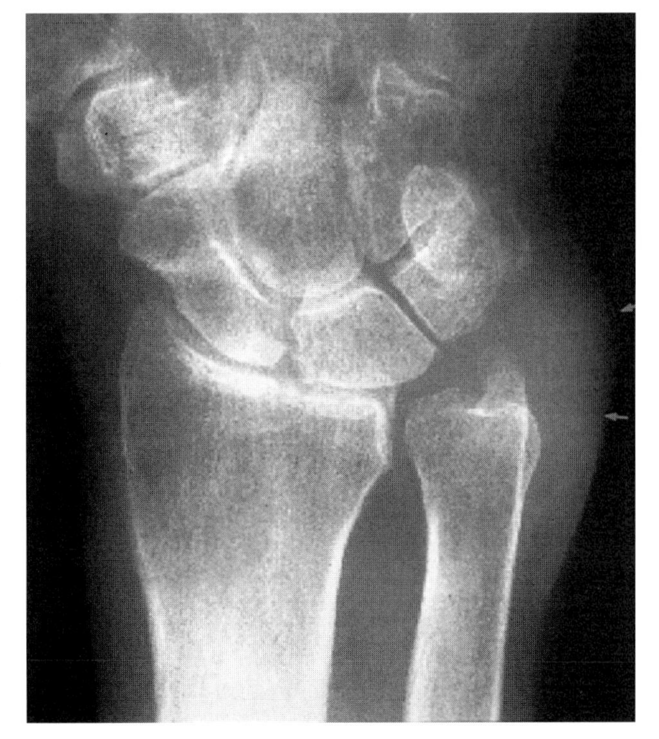

Figure 20–2. Tenosynovitis. Significant soft swelling of the tendon sheath of the extensor carpi ulnaris *(arrows)*.

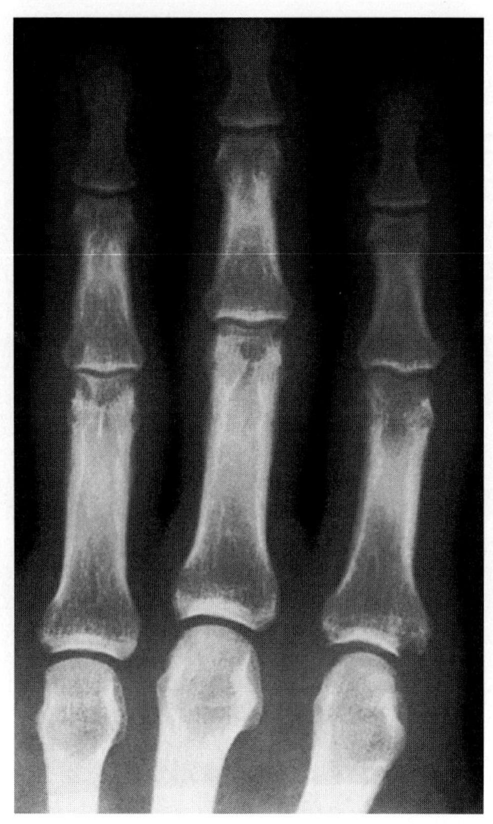

Figure 20–3. Metacarpophalangeal (MCP) and proximal interphalangeal (PIP) joint involvement with rheumatoid arthritis (RA). Note the erosions in the second MCP joint as well as the second, third, and fourth PIP joints. The fusiform soft tissue swelling around the PIP joints is typical of RA.

synovial cysts, which are common complications of RA. In the calf, inflamed or ruptured cysts can be clinically confused with thrombophlebitis (Fig. 20–5).

Periarticular Osteopenia. Periarticular osteopenia is seen early in the course of the disease and is followed by more generalized osteopenia in advanced disease. The periarticular osteopenia is due to synovial inflammation and hyperemia. The more generalized osteopenia that occurs later in the disease is due to disuse and bone atrophy.

Uniform Joint Space Narrowing. The cartilage is totally bathed with synovial fluid that is rich in inflammatory cells (neutrophils) (Fig. 20–6). Enzymes from the breakdown of these neutrophils degrade the articular cartilage, resulting in uniform joint space narrowing.

Bony Erosions Affecting the Bare Areas. During the active phase of the disease, the erosions do not provoke a bony reaction and, therefore, remain uncorticated (see Figs. 20–3 and 20–7). Sites in the hand and wrist where early marginal erosions are detectable include the radial side of the second and third metacarpal heads as well as the ulnar styloid process (see Fig. 20–7). These erosions tend to be symmetrical on both sides. Another form of erosion is called pressure or compressive erosions (Figs. 20–8 and 20–9). This develops late in the disease and is due to muscular forces compressing the ends of osteopenic bones together. Compressive erosions are occasionally seen in the proximal humerus medially, caused by pressure from the glenoid, and in the third through fifth ribs posteriorly, owing to pressure by the scapula.

Cyst or Pseudocyst Formation. The pathogenesis of cyst and pseudocyst formation (Fig. 20–10) is not determined; however, it seems that some cysts communicate with the joint whereas others do not. It is postulated that these cysts form as a result of pannus invasion into the bone. The cysts then become enclosed and lose their communication with the joint. A cystic form of RA has been described in which the dominant feature is intraosseous cyst formation, but with much less osteopenia and joint space narrowing (see Fig. 20–10A). Cystic RA is probably more common in males, and 50% of the patients are seronegative. Similar findings have been described in manual workers who are afflicted with RA. This type has been called "robust RA." It is not clear from the literature whether cystic RA and robust-type RA are the same disease.

Rheumatoid Nodules. Subcutaneous rheumatoid nodules are seen in 20% to 25% of patients with RA (Fig. 20–11). Rarely, rheumatoid nodules are seen without evidence of joint changes. This variant of RA has been called "rheumatoid nodulosis." The disease usually affects middle-aged men suffering from recurrent migratory arthritis. The nodules typically appear on the dorsal aspects of the hands, feet, and elbows.

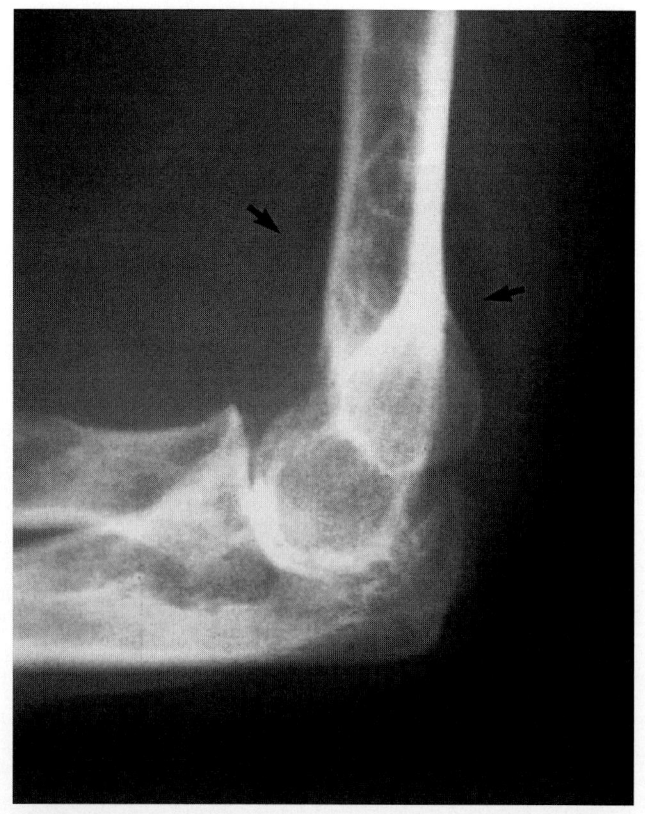

Figure 20–4. Joint effusion and synovial proliferation. Lateral view of the elbow shows elevation of the anterior and posterior fat pads in a patient with rheumatoid arthritis *(arrows)*. This finding had been present for several months, suggesting chronic effusion and synovial proliferation (pannus).

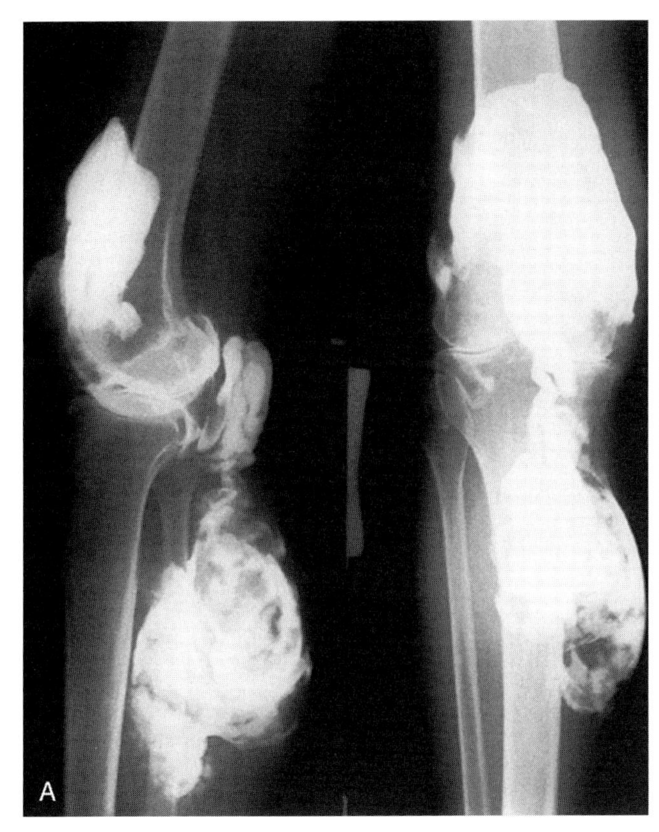

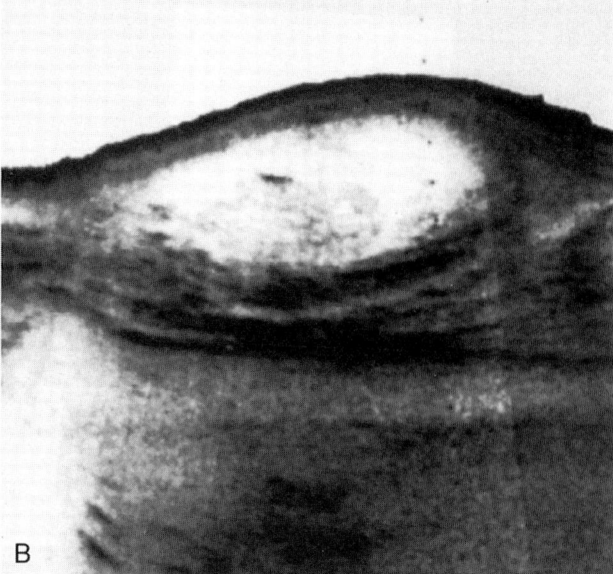

Figure 20–5. Ruptured synovial cyst in a patient with rheumatoid arthritis. Clinically, the patient was thought to have thrombophlebitis. *A*, Arthrogram shows that the cyst has ruptured into the calf muscles. *B*, Ultrasound shows the cyst and the debris within it.

Figure 20–6. Uniform joint space narrowing. Anteroposterior view of the knee shows uniform joint space narrowing and cystic changes.

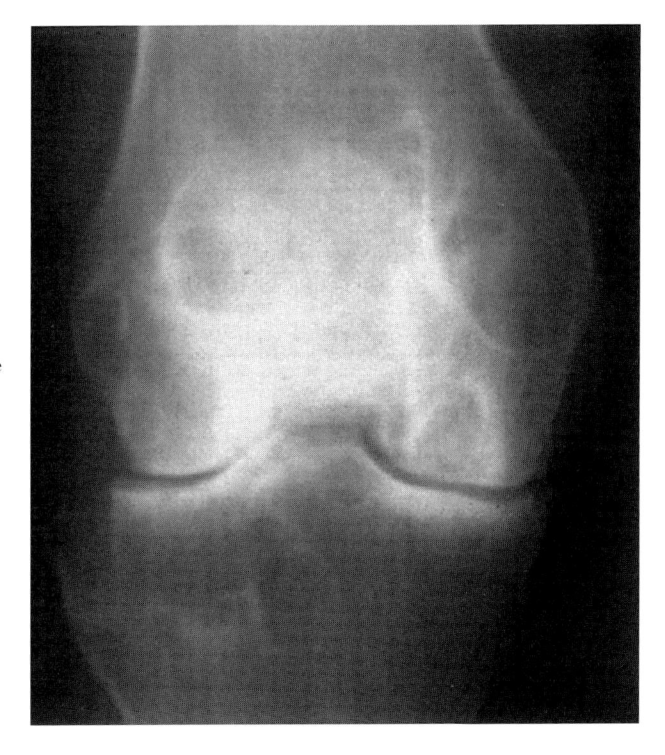

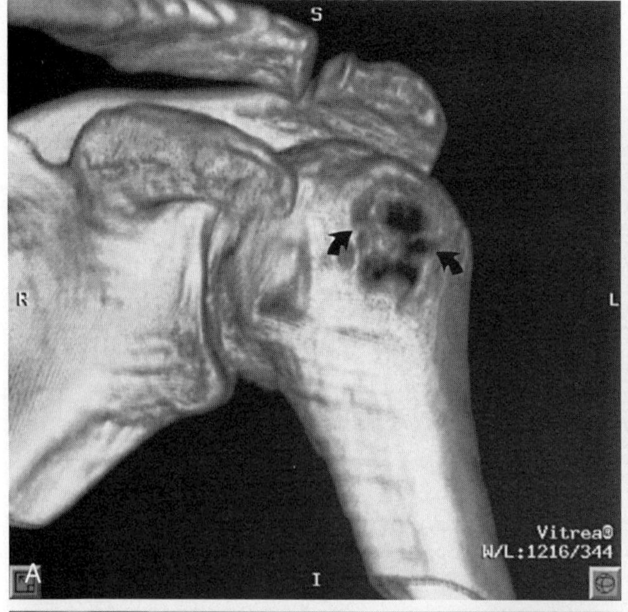

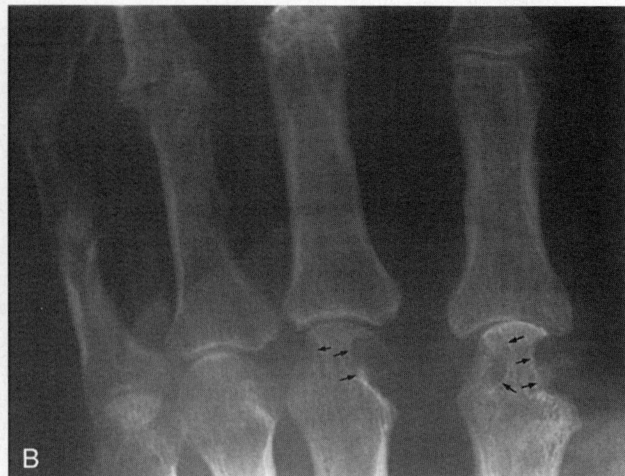

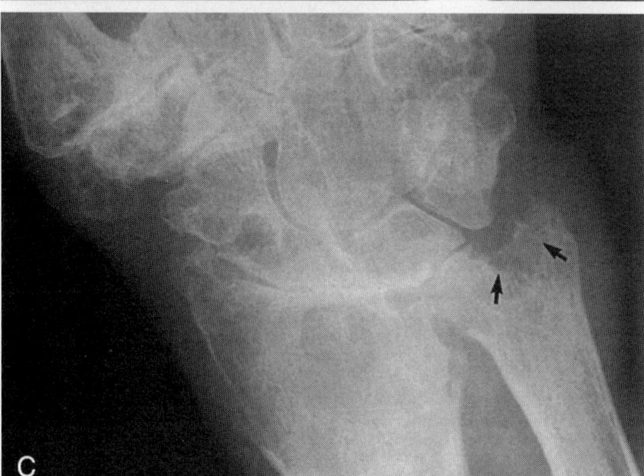

Figure 20–7. Erosions in rheumatoid arthritis (RA). *A,* Three-dimensional reconstruction of the left shoulder in a patient with RA shows a large erosion in the greater tuberosity *(arrows)*. *B,* Erosions in the second and third metacarpal heads *(arrows)*. *C,* Deep erosions affecting the ulnar styloid *(arrows)*. Cystic changes in the distal radius and carpal bones.

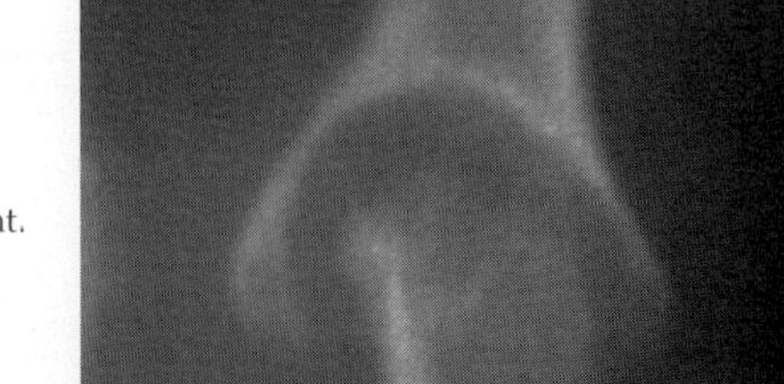

Figure 20–8. Pressure erosions at the second metacarpophalangeal joint.

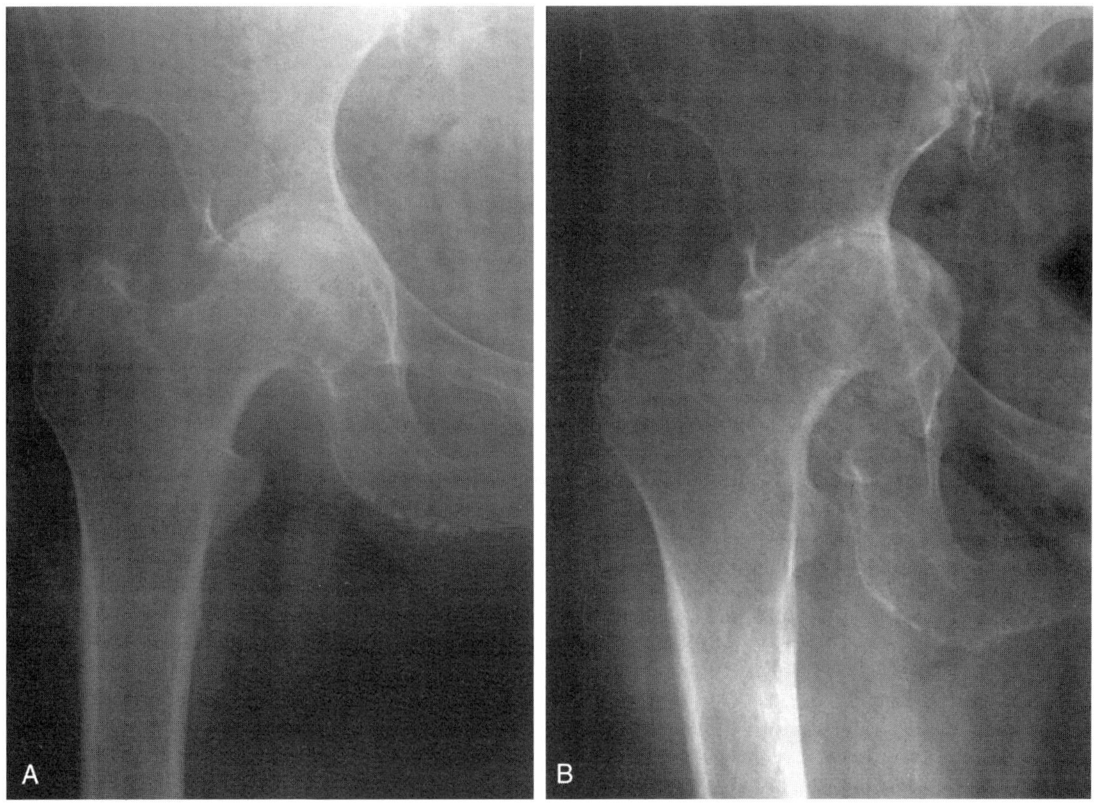

Figure 20–9. Progressive pressure erosions in the right hip. *A,* Anteroposterior view of the right hip shows osteopenia and uniform joint space narrowing. *B,* Repeat examination 1 year later shows extensive pressure erosions in the right acetabulum and femoral head, resulting in protrusion of the acetabulum.

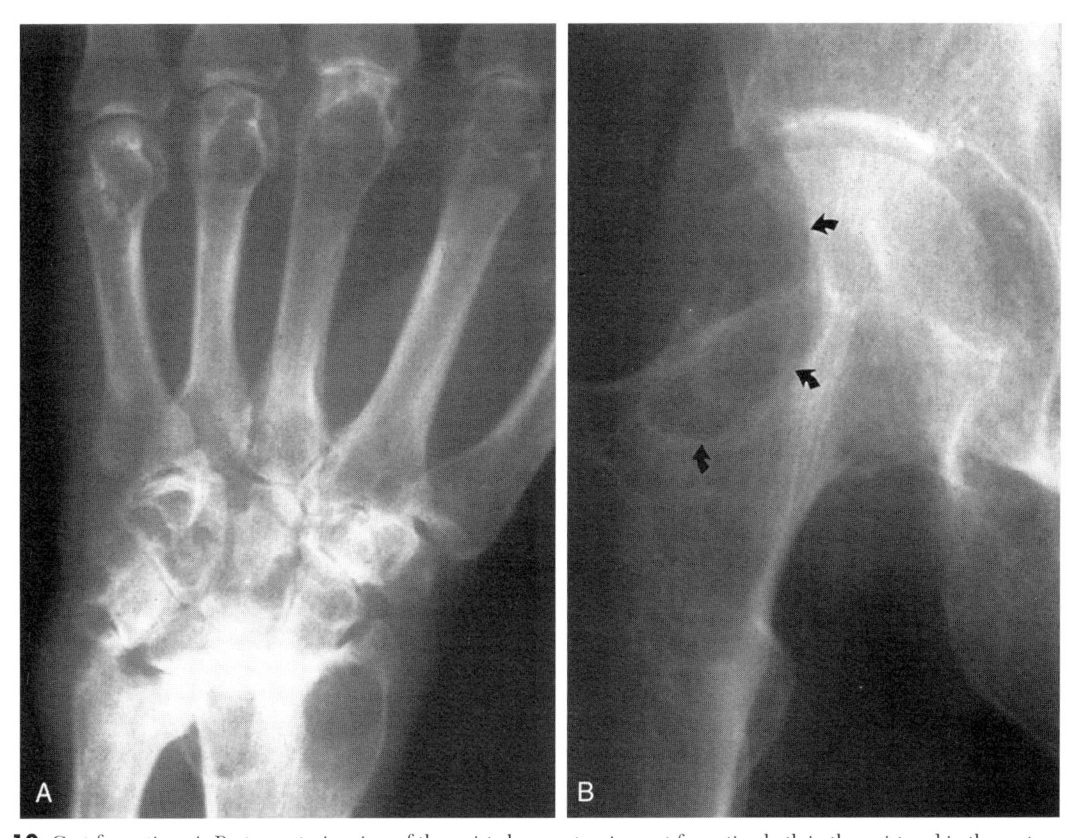

Figure 20–10. Cyst formation. *A,* Posteroanterior view of the wrist shows extensive cyst formation both in the wrist and in the metacarpophalangeal joints. *B,* Anteroposterior view of the hip shows a large cyst involving the lateral aspect of the femoral head and neck *(arrows).*

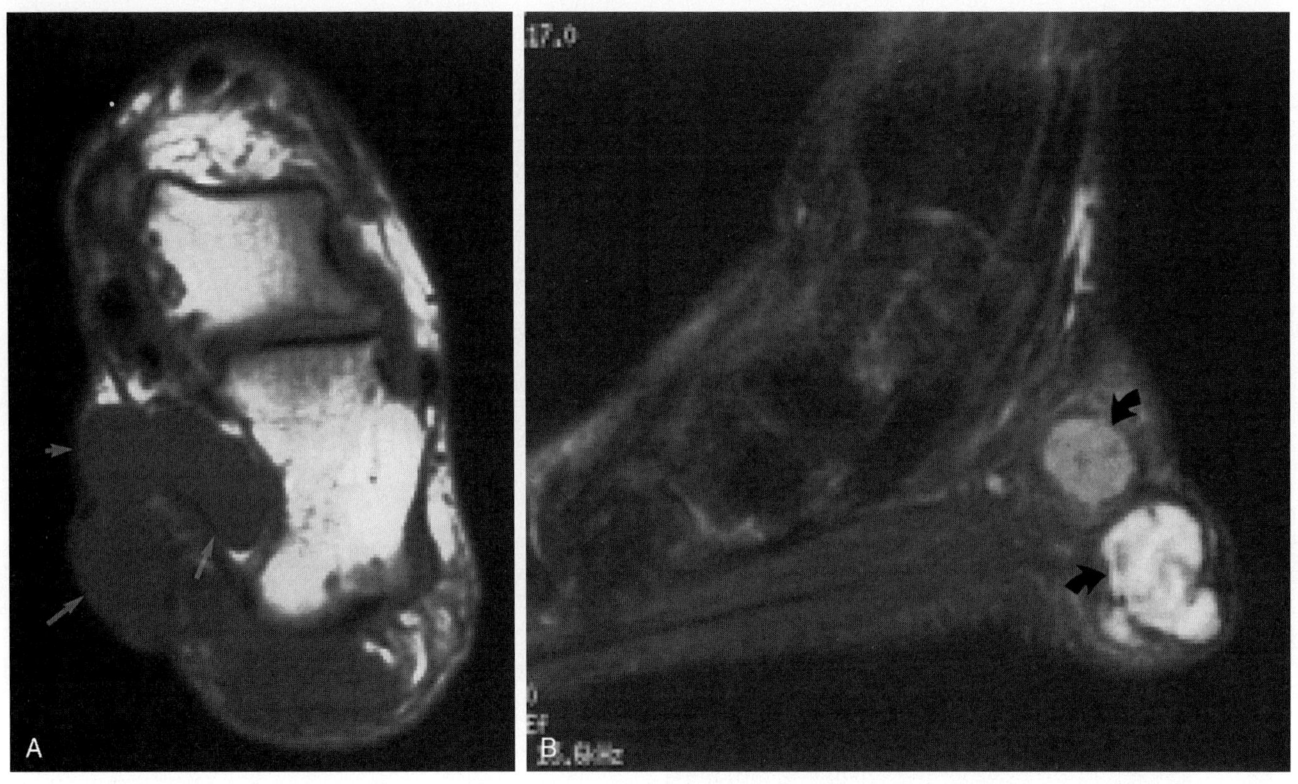

Figure 20–11. Rheumatoid nodules. *A*, Oblique axial T1-weighted MR image of the hindfoot shows subcutaneous nodules *(arrows)*. *B*, Sagittal T2-weighted images show the nodules *(arrows)*. The increase in signal intensity in the nodules is not uniform.

RHEUMATOID ARTHRITIS OF THE CERVICAL SPINE

This is a frequent problem affecting about 25% of patients with moderate to severe RA. The pathogenesis is similar to that seen in peripheral joints. Synovial inflammation and pannus formation result in ligament, cartilage, and bone destruction. Progressive instability of the upper cervical spine with C1-C2 subluxation followed by cranial settling may lead to compression of the medulla and upper cord causing severe neurologic deficit or even death. Cervical spine changes are typically more common in patients with severe peripheral arthritis. Radiographic changes include the following.

Odontoid Process (Dens) Erosions Resulting From Direct Pannus Invasion (Fig. 20–12). The extent of erosions can be evaluated by plain radiography, or by computed tomography. The dens can become markedly thinned, which makes it susceptible to fracture with resultant posterior subluxation of C1 on C2 (Fig. 20–13). Rarely, the entire dens becomes completely eroded by the pannus. The size of the pannus and its compressive effect on the upper cord or medulla are best evaluated by magnetic resonance imaging (Fig. 20–14).

Atlanto-Axial Subluxation (Fig. 20–15). Approximately 15% to 36% of rheumatoid patients have atlanto-axial instability, which is caused by transverse ligament disruption. On a lateral flexion view of the cervical spine, the atlanto-dental distance normally does not exceed 3 mm. As the atlanto-axial instability progresses, the antero-posterior diameter of the spinal canal, behind the dens, decreases. Follow-up of patients with severe atlanto-axial subluxation has shown that the deformity eventually develops into cranial settling, and as the cranial settling becomes established, the atlanto-axial subluxation diminishes.

Cranial Settling (Fig. 20–16). This is a well-known complication of advanced rheumatoid arthritis. It results from severe erosions and collapse of the lateral masses of C1. Erosions in the occipital condyles and superior articular facets of C2 are often present. This allows the skull to settle at a lower level on the cervical spine causing the odontoid process to project above the foramen magnum and compress the medulla. The exact incidence of cranial settling is not known but has been reported to occur in 5% to 8% of patients with rheumatoid arthritis. Other terms that have been used to describe cranial settling include upward migration, translocation, vertical subluxation of the odontoid, and basilar or pseudobasilar invagination. Chamberlain's and McGregor's lines can be used to assess cranial settling. Wackenheim's clivus baseline is also very useful in assessing cranial settling (see Fig. 20–16*A*).

Subaxial Subluxations (Fig. 20–17). These deformities occur between C2 and C7 and are reported in 7% to 15% of patients with RA. Subaxial subluxations, also called "stair stepping," are the result of apophyseal joint involvement with RA, ligamentous laxity, and discovertebral erosions (Figs. 20–17 and 20–18). The pathogenesis of discovertebral erosions and vertebral body destruction is

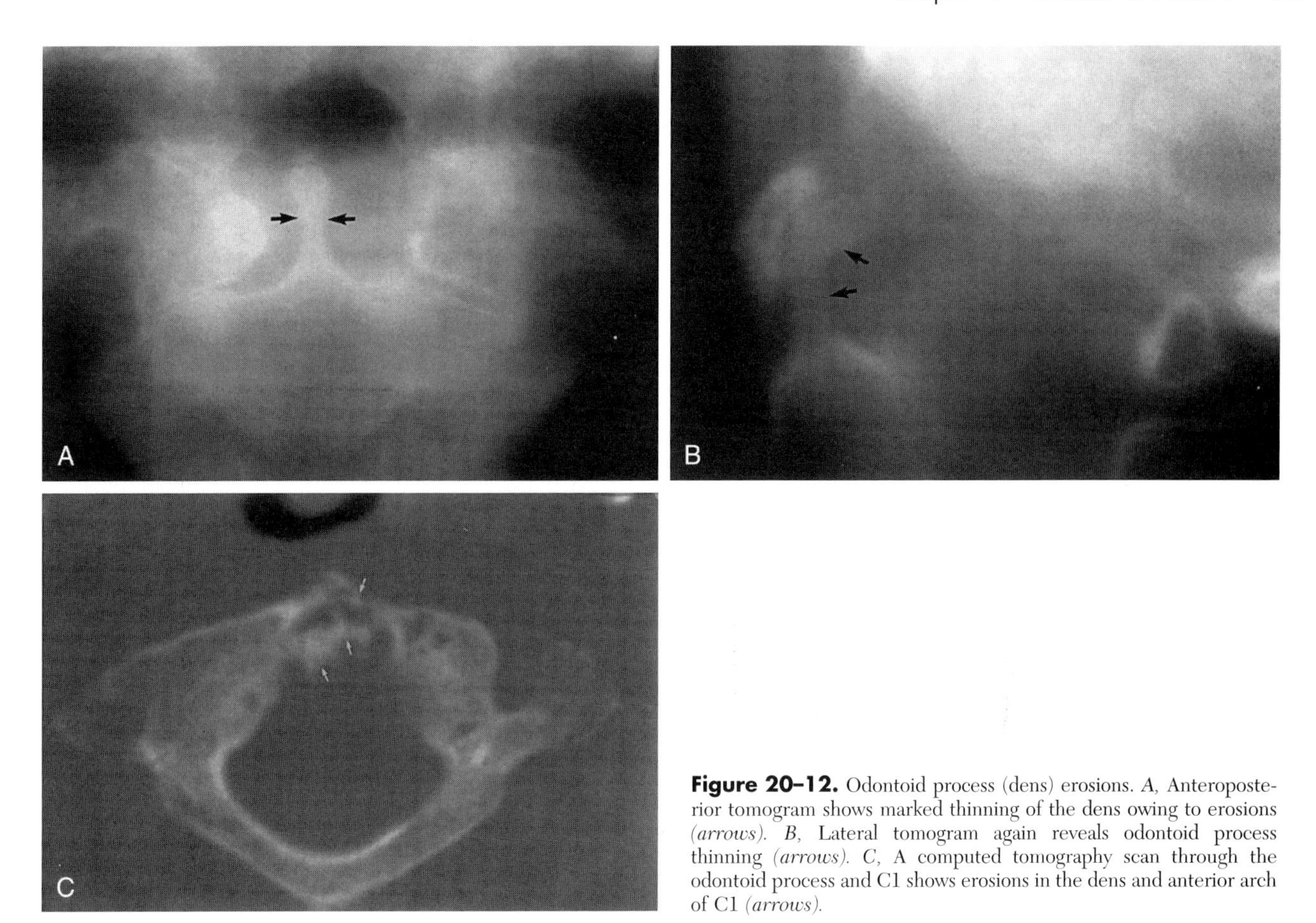

Figure 20–12. Odontoid process (dens) erosions. *A*, Anteroposterior tomogram shows marked thinning of the dens owing to erosions *(arrows)*. *B*, Lateral tomogram again reveals odontoid process thinning *(arrows)*. *C*, A computed tomography scan through the odontoid process and C1 shows erosions in the dens and anterior arch of C1 *(arrows)*.

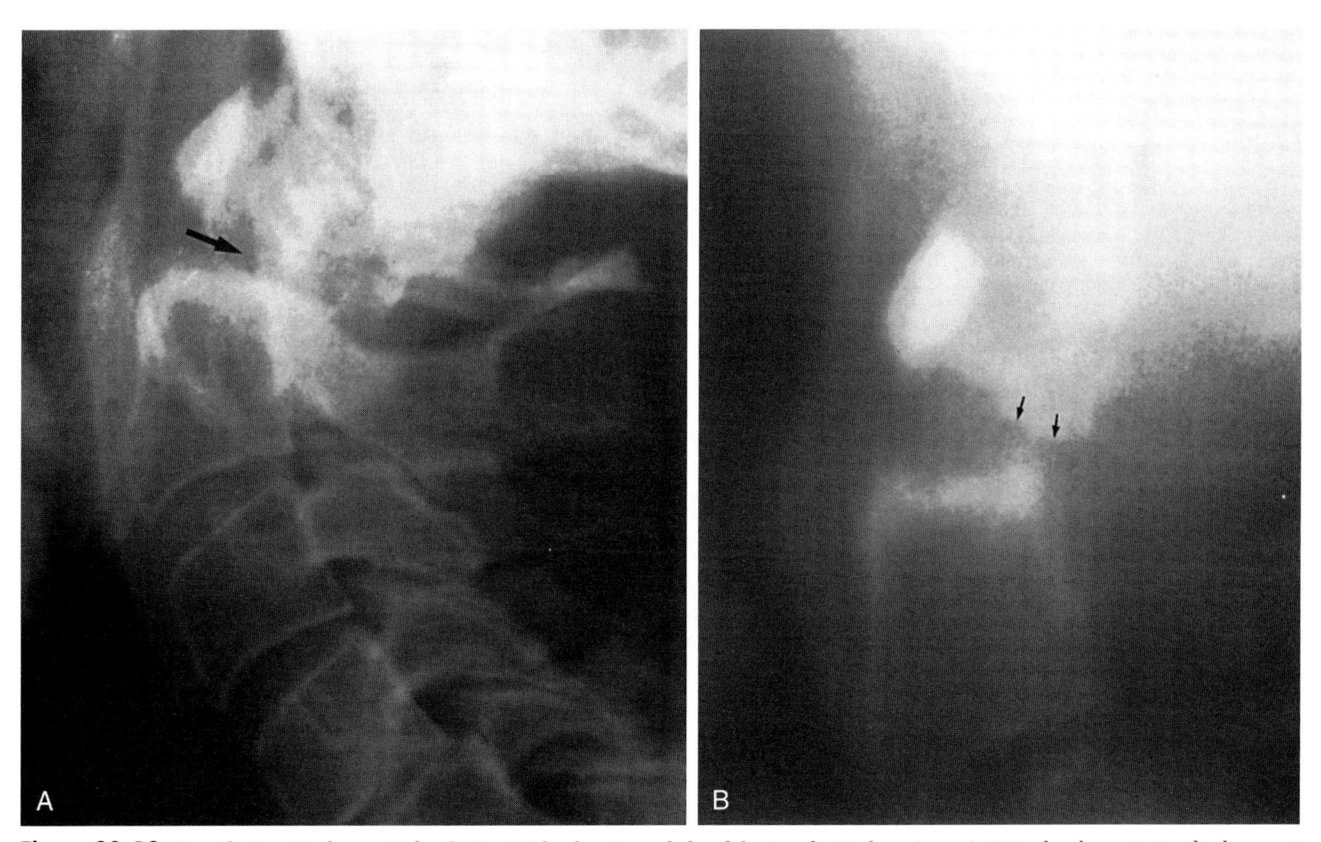

Figure 20–13. Dens fracture in rheumatoid arthritis. *A*, The dens is eroded and fractured at its base *(arrow)*. Note also the posterior displacement of C1 on C2. *B*, Lateral tomogram demonstrates the fracture *(arrows)* and posterior displacement of the dens.

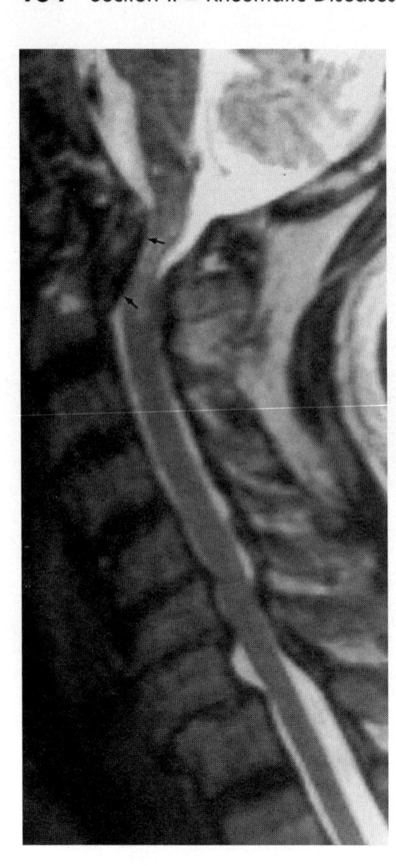

Figure 20–14. Pannus around the eroded dens. T2-weighted sagittal image shows a large, low signal intensity mass (pannus) around the dens and compressing the cord *(arrows).*

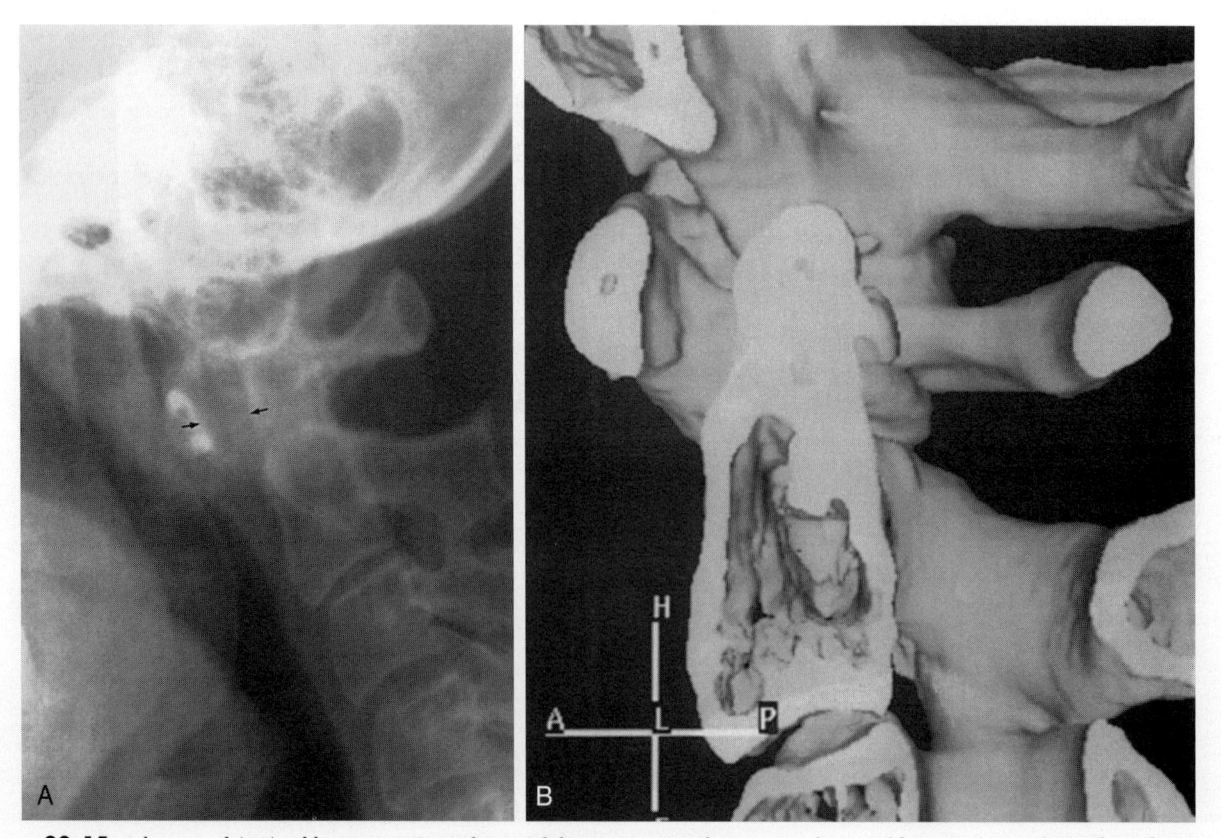

Figure 20–15. Atlanto-axial (AA) subluxation. *A,* Lateral view of the upper cervical spine reveals AA subluxation *(arrows). B,* Three-dimensional reconstruction from a computed tomography scan in another patient with AA subluxation.

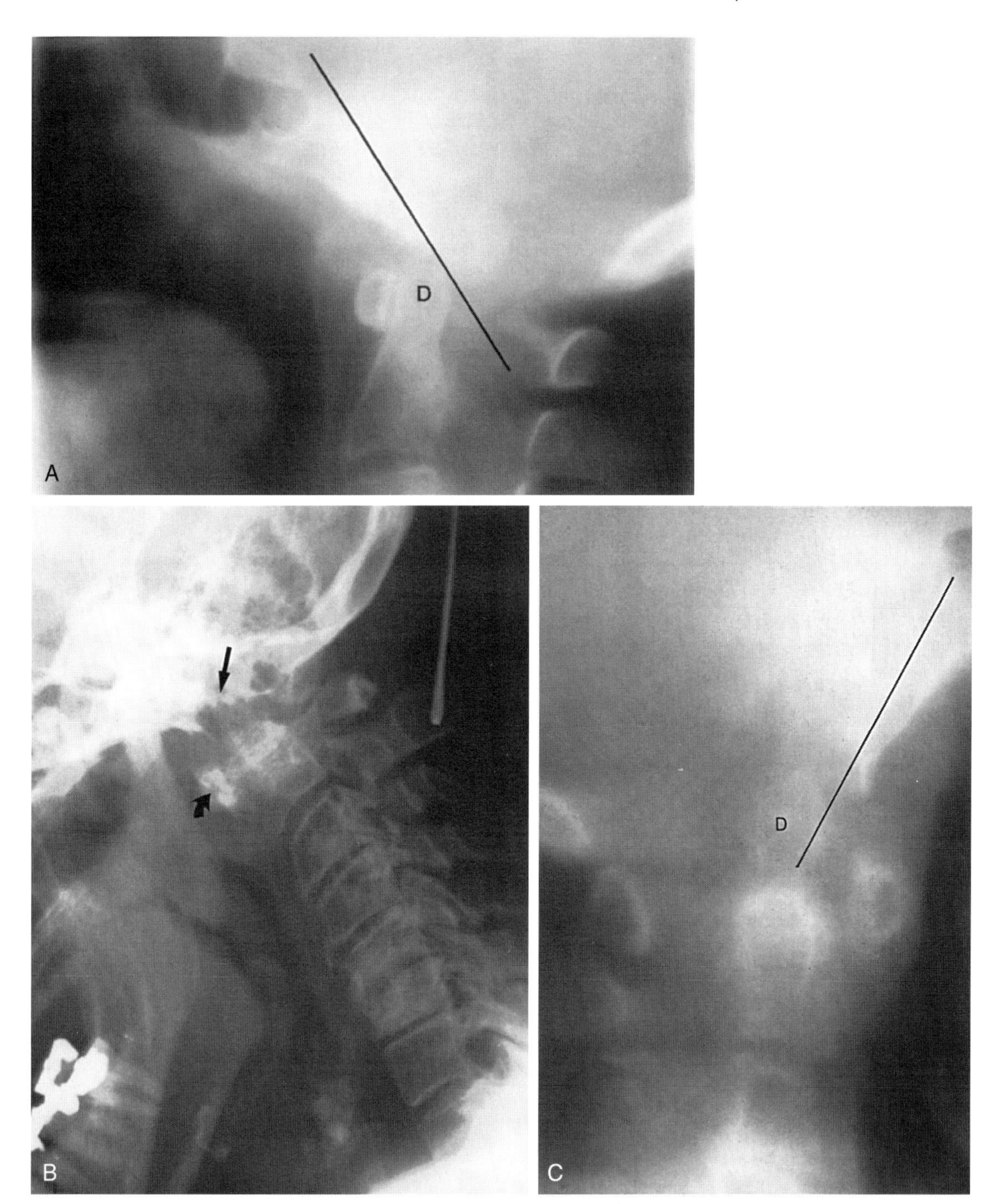

Figure 20–16. Cranial settling. *A,* Lateral tomogram showing the normal relationship between the clivus and the dens (D). Wackenheim's line is drawn along the surface of the clivus; its inferior continuation should be tangent to the tip of the dens. *B,* Lateral view of the cervical spine shows the typical changes of cranial settling where the dens *(straight arrow)* protrudes into the foramen magnum. Note that the tip of the dens lies above the tip of the clivus and the anterior arch of C1 *(curved arrow)* articulates with the body of C2. *C,* Lateral midline tomogram shows cranial settling. Wackenheim's line is seen intersecting the dens.

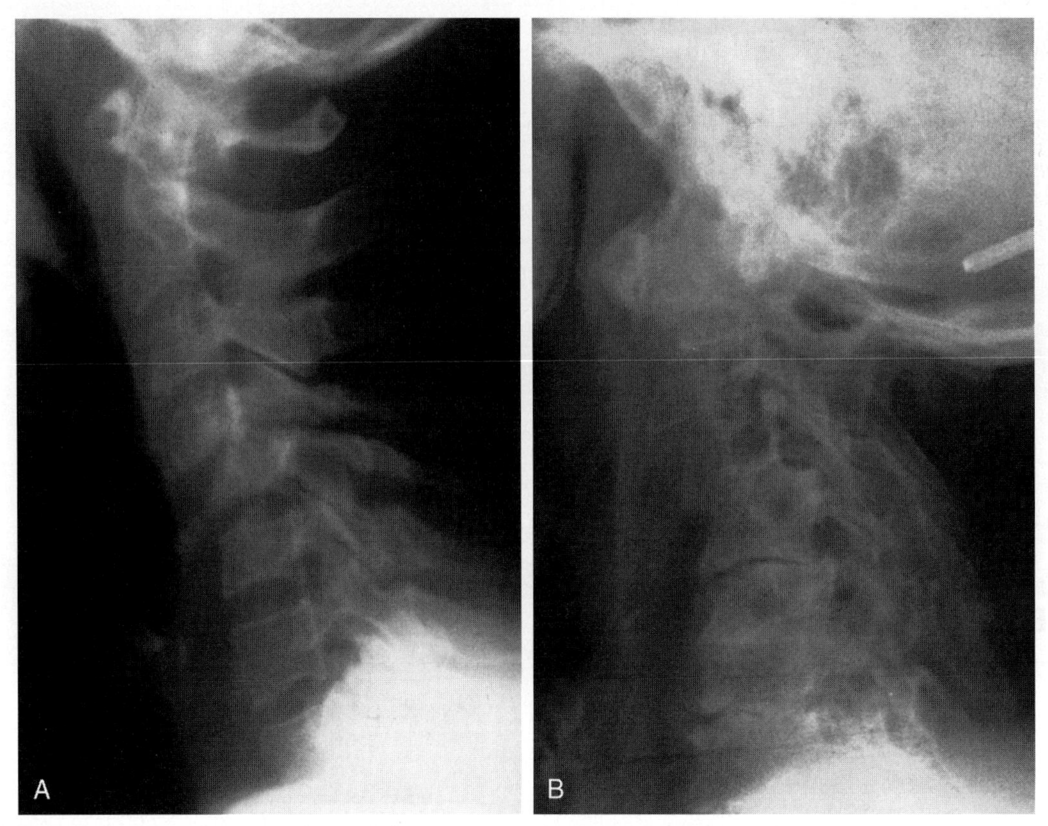

Figure 20–17. Subaxial subluxation. *A,* Lateral view of the cervical spine shows anterior translation of C4 on C5. Note the erosions at the vertebral endplates. *B,* Subaxial subluxation at three levels (C3-4, C4-5, and C5-6). There is also severe cranial settling.

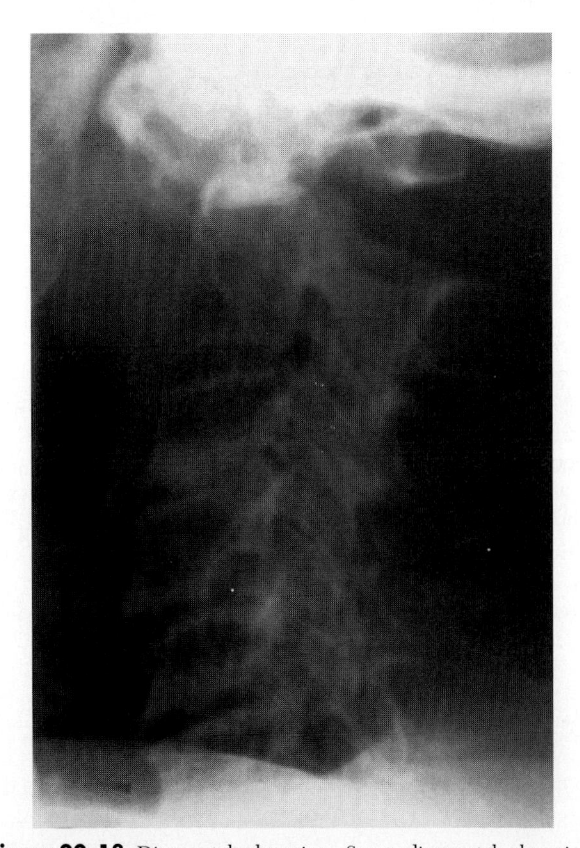

Figure 20–18. Discovertebral erosions. Severe discovertebral erosions at the C3-4, C4-5, and C5-6 levels. The patient also has atlanto-axial subluxation.

controversial. Some believe it is due to extension of the inflammatory process from the adjacent uncovertebral (joints of Luschka) joints, which are acquired synovial joints in adults. Others feel that the discovertebral destruction is mechanical. Excessive stresses on osteopenic bone resulting in bone destruction and instability (see Fig. 20–17).

Recommended Readings

Alpert M, Feldman F: The rib lesions of rheumatoid arthritis. Radiology 82:872, 1964.

Boden SD, Dodge LD, Bohlman HH, et al: Rheumatoid arthritis of the cervical spine. J Bone Joint Surg Am 75:1282, 1993.

Brower AC, NaPombejara C, Stechschulte DJ, et al: Rheumatoid nodulosis: Another cause of juxta-articular nodules. Radiology 125: 669, 1977.

El-Khoury GY, Wener MH, Menezes AH, et al: Cranial settling in rheumatoid arthritis. Radiology 137:637, 1980.

El-Noueam KI, Giuliano V, Schweitzer ME, et al: Rheumatoid nodules: MR/Pathological correlation. J Comput Assist Tomogr 21:796, 1997.

Goetz DD, Clark CR: Rheumatoid involvement of the cervical spine. Semin Orthop 4:147, 1989.

Gubler FM, Maas M, Dijkstra PF, et al: Cystic rheumatoid arthritis: Description of a nonerosive form. Radiology 177:829, 1990.

Levine RB, Sullivan KL: Rheumatoid arthritis: Skeletal manifestations observed on portable chest roentgenograms. Skeletal Radiol 13:295, 1985.

Martel W, Hayes JT, Duff IF: The pattern of bone erosion in the hand and wrist in rheumatoid arthritis. Radiology 84:204, 1965.

Monsees B, Destouet JM, Murphy WA, et al: Pressure erosions of bone in rheumatoid arthritis: A subject review. Radiology 155:53, 1985.

Park WM, O'Neill M, McCall IW: The radiology of rheumatoid involvement of the cervical spine. Skeletal Radiol 4:1, 1979.
Rana NA: Natural history of atlanto-axial subluxation in rheumatoid arthritis. Spine 14:1054, 1989.
Resnick D: Rheumatoid arthritis of the wrist: Why the ulnar styloid? Radiology 112:29, 1974.
Stiskal MA, Neuhold A, Szolar DH, et al: Rheumatoid arthritis of the craniocervical region by MR imaging: Detection and characterization. AJR Am J Roentgen 165:585, 1995.
Sugimoto H, Takeda A, Masuyama J-I, et al: Early-stage rheumatoid arthritis: Diagnostic accuracy of MR imaging. Radiology 198:185, 1996.
Wolfe BK, O'Keeffe D, Mitchell DM, et al: Rheumatoid arthritis of the cervical spine: Early and progressive radiographic features. Radiology 165:145, 1987.
Yaghmai I, Rooholamini SM, Faunce HF: Case reports. Unilateral rheumatoid arthritis: Protective effect of neurologic deficits. AJR Am J Roentgenol 128:299, 1977.

CHAPTER 21

Juvenile Rheumatoid Arthritis

Georges Y. El-Khoury, MD, Mark D. Stanley, MD, and D. Lee Bennett, MD

OVERVIEW

Juvenile rheumatoid arthritis (JRA) is the most common form of childhood arthritis. It is clinically and radiographically different from adult rheumatoid arthritis (RA). Generally, the radiographic features that distinguish JRA from adult RA include relatively late articular cartilage loss and bony erosions, growth disturbances, upper cervical spine ankylosis, and micrognathia. JRA begins before the age of 16 years, and the arthritis persists in one or more joints for at least 6 weeks before the diagnosis is made. JRA is divided into three distinct types: pauciarticular, polyarticular, and systemic.

Pauciarticular Disease

Pauciarticular disease affects about 30% to 40% of children with JRA. Four or fewer joints are involved. The disease is more common in females, and symptoms start in the large joints between the ages of 1 and 3 years.

Polyarticular Disease

Polyarticular disease affects about 40% to 50% of children with JRA. Five or more joints are involved, and typically the small joints are affected. Girls are twice as likely as boys to have this type of JRA.

Still's Disease

Systemic JRA (Still's disease) is characterized by daily high fever, presence of a distinctive rash, and arthritis. The disease occurs in 10% to 20% of children with JRA, and it affects males and females equally. Systemic manifestations include lymphadenopathy, hepatomegaly, pericarditis, pleuritis, anemia, and leukocytosis.

RADIOGRAPHIC FINDINGS

The peripheral joints most frequently affected are the knees, wrists, ankles, tarsal joints, and upper cervical spine. Radiographic abnormalities include periosteal reaction, articular changes, cervical spine changes, and growth disturbances.

Periosteal Reaction

Periosteal reaction is a rare and transient finding, which can be detected early after the onset of symptoms. It occurs along the shafts of the proximal phalanges, the metacarpals, and the metatarsals near affected joints. Subsequently, the periosteal new bone is incorporated with the cortex leading to a rectangular configuration of the metacarpals and phalanges. The periosteal reaction is thought to be due to the tenosynovitis of the flexor tendon sheaths.

Articular Changes

Articular changes depend on the age of onset. Early in life, loss of articular cartilage is generally a late finding (Fig. 21–1). Loss of articular cartilage followed by fusion is most frequently seen in the wrists, particularly affecting the carpometacarpal joints of the index and middle fingers

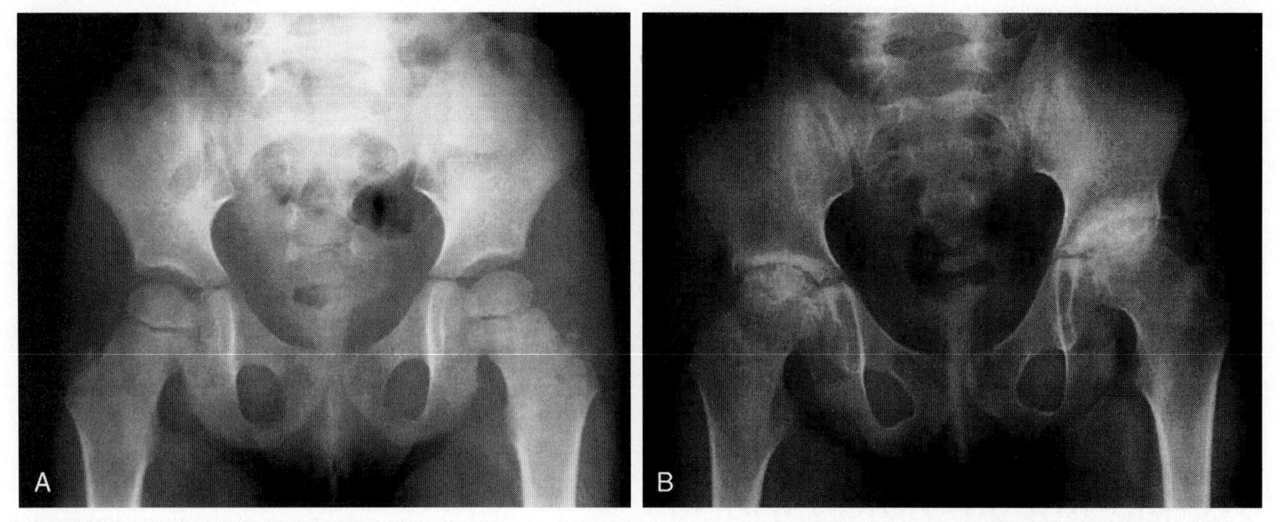

Figure 21–1. Loss of articular cartilage. *A,* Anteroposterior view of both hips in a 3-year-old child at approximately 4 months after onset of symptoms. The hip joints are normal at this time. *B,* Repeat examination of the same patient at 8 years of age reveals marked loss of the articular cartilage and narrowing of the joint space in both hips.

and the intercarpal joints (Fig. 21–2). Similar changes can occur in the tarsal bones. Occasionally, partial recovery of the joint space narrowing is detected radiographically in the small and large joints. Periarticular bony erosions are not as prominent as they are in patients with adult RA. When the disease starts before the age of 4 years, erosive changes and joint space narrowing are considerably delayed (Figs. 21–3 and 21–4). The erosions in JRA typically occur in the large joints. Adolescent patients with positive rheumatoid factor may show earlier erosive changes and a distribution that is similar to that found in adult RA.

Growth Disturbances

Growth disturbances are quite common in JRA. These include advanced skeletal maturation, enlargement of the

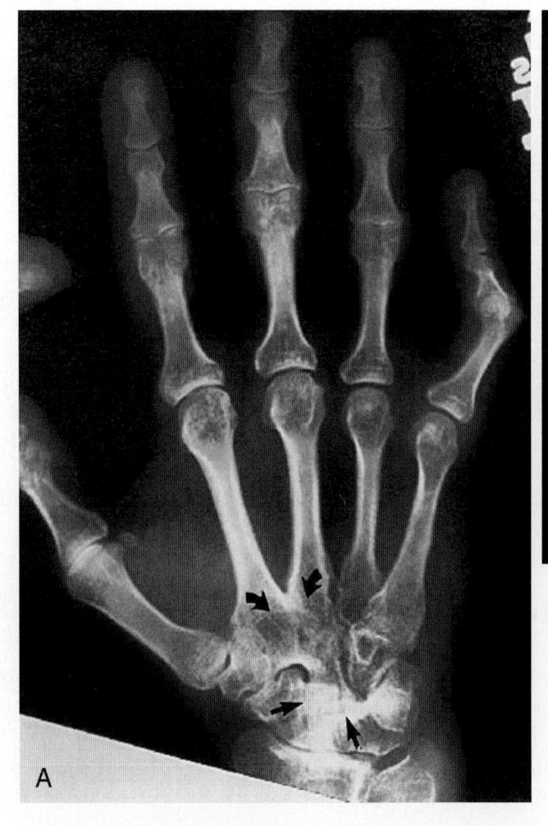

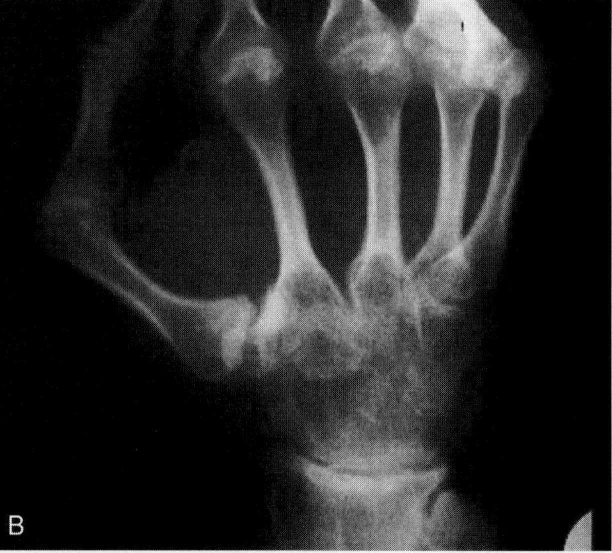

Figure 21–2. Wrist fusion in juvenile rheumatoid arthritis. *A,* Posteroanterior (PA) view of the wrist shows fusion at the second and third carpometacarpal joints *(curved arrows).* Some of the carpal bones *(arrows)* are also fused. *B,* PA view of the wrist shows complete fusion of the carpal bones. The second, third, and fourth carpometacarpal joints are also fused. The radiocarpal joint and first carpometacarpal joints are narrowed.

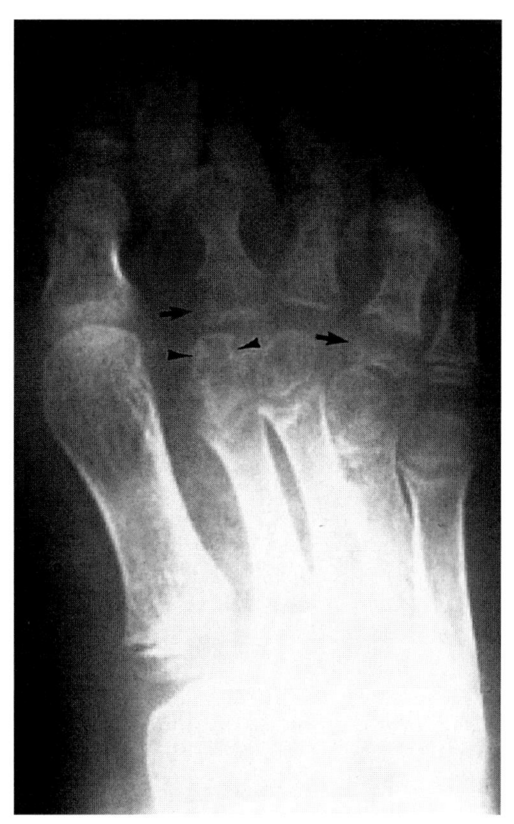

Figure 21–3. Bony erosions and epiphyseal overgrowth in a 5-year-old child with juvenile rheumatoid arthritis. Anteroposterior view of the left foot shows deep erosions in the head of the second metatarsal *(arrowheads)*. Note the overgrowth in the epiphyses of the proximal second and fourth phalanges *(arrows)*.

epiphyseal centers, accelerated longitudinal growth, shortening of small tubular bones, and generalized growth retardation (Fig. 21–5). When the disease develops before the age of 9 years, there is a tendency for the involved extremity to overgrow, but the overgrowth never exceeds 3 cm (Fig. 21–6). Rapid premature closure of the growth plates occurs when the disease develops after the age of 9 years. This typically leads to shortening of the affected extremity and occasionally to significant limb-length discrepancies. Early epiphyseal closure has been reported in 15% to 30% of patients; this is usually recognized in the distal radius, the metacarpals, and the metatarsals and results in shortening of these bones. The secondary epiphyseal centers show enlargement, osteopenia, and a coarse trabecular pattern (see Fig. 21–3). Similar epiphyseal changes are seen in hemophilia, pigmented villonodular synovitis, and synovial hemangioma. In the wrist and tarsal region, there is a tendency for the bones to overgrow with their corners developing sharper angles, which causes crowding (Fig. 21–7). Involvement of the temporomandibular joints (TMJ) usually results in mandibular underdevelopment (micrognathia), asymmetry, and deformity. This deformity is characterized by a concavity of the inferior border (antegonial notching) and exaggeration of the mandibular angle (Fig. 21–8). Unilateral TMJ involvement with JRA can result in asymmetry caused by underdevelopment of the jaw on the affected side.

Cervical Spine Changes

Cervical spine changes include:

1. C1-C2 subluxation caused by disruption of the transverse ligament (Fig. 21–9).
2. Erosions of the dens often affecting the anterior and posterior margins, leaving the apex intact, and producing what is known as the applecore odontoid.
3. Ankylosis of the apophyseal joints typically affecting the upper cervical spine (C2-C3 and C3-C4) (Fig. 21–10), although occasionally, most or all of the apophyseal joints become fused (Fig. 21–11).

When the apophyseal ankylosis occurs early in life, growth abnormalities in the vertebral bodies develop. There is a decrease in the anteroposterior and vertical dimensions of the vertebral bodies at the site of the fusion (see Fig. 21–11). The disc spaces at the fused levels show narrowing. Subaxial subluxations are not seen in children with JRA.

ADULT-ONSET STILL'S DISEASE

Adult-onset Still's disease (AOSD) is a systemic inflammatory disease dominated by constitutional manifestation and arthritis. The disease typically occurs in young adults, with most patients (76%) below the age of 35 years. The clinical features of AOSD are nonspecific, and the

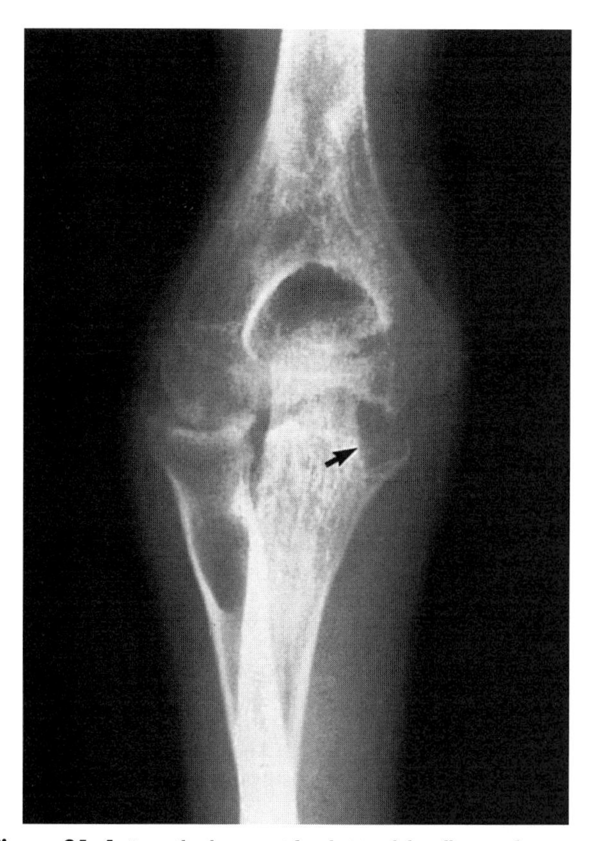

Figure 21–4. Juvenile rheumatoid arthritis of the elbow with erosions in a 9½-year-old child. Anteroposterior view of the elbow shows loss of the articular cartilage at the radiocapitellar joint. There are also large erosions on the medial aspect of the olecranon process *(arrow)*.

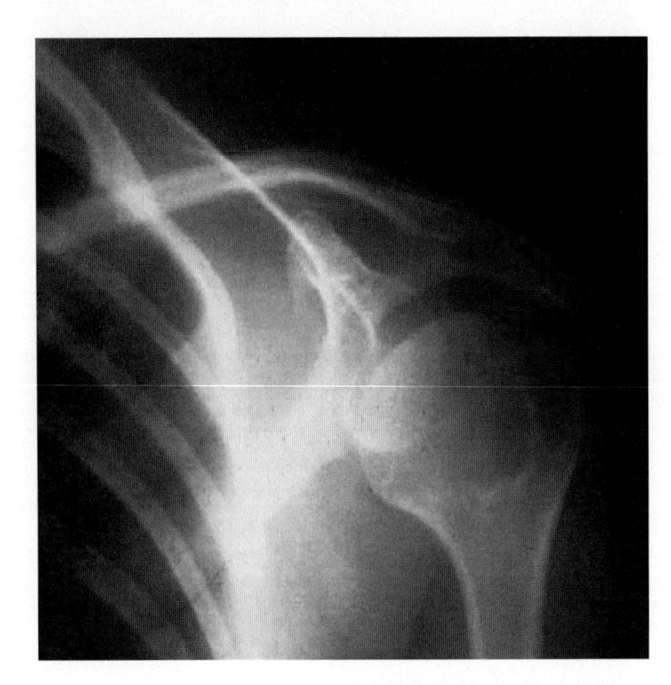

Figure 21–5. Epiphyseal enlargement in the left shoulder. This patient is a 26-year-old male who had juvenile rheumatoid arthritis in childhood. An anteroposterior view of the shoulder demonstrates previous epiphyseal enlargement of the humeral head.

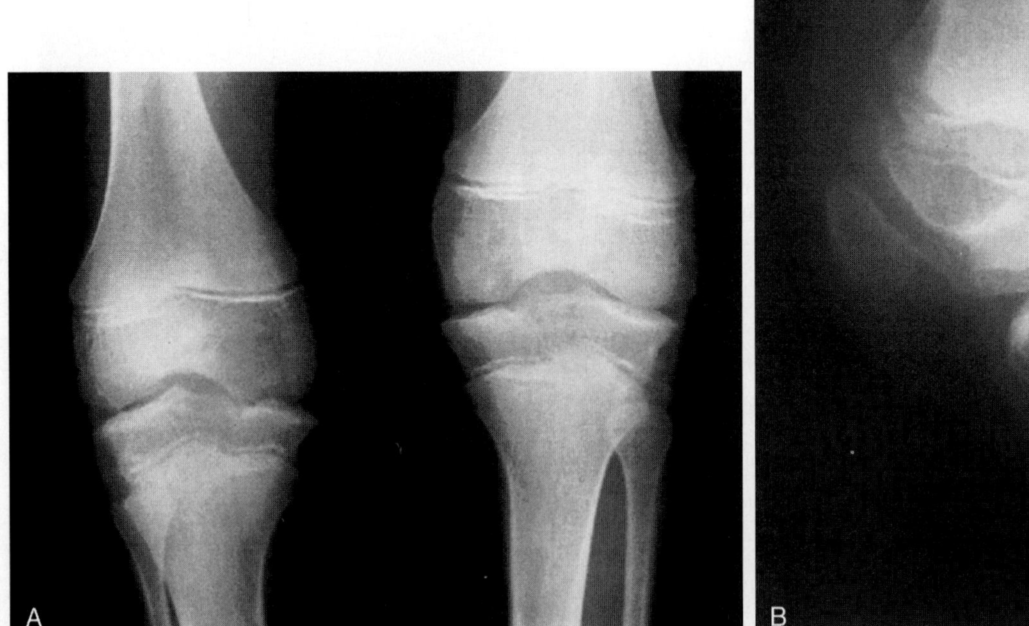

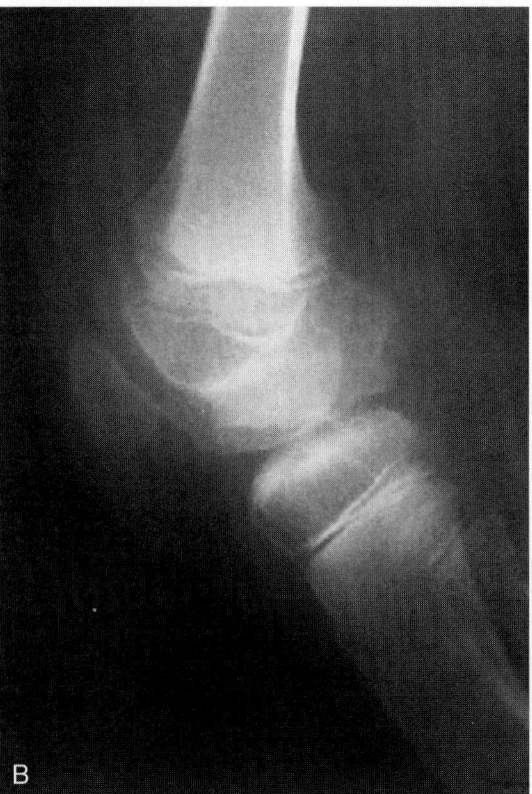

Figure 21–6. Epiphyseal enlargement, articular cartilage loss, and leg length discrepancy. *A,* Anteroposterior view of the knees shows typical changes of juvenile rheumatoid arthritis with epiphyseal enlargement, articular cartilage loss, and longer right femur. Note the sharp angles in the proximal epiphyses of the tibia bilaterally. *B,* Lateral view of the right knee, in the same patient, reveals marked enlargement of the distal femoral epiphysis and squaring of the inferior pole of the patella.

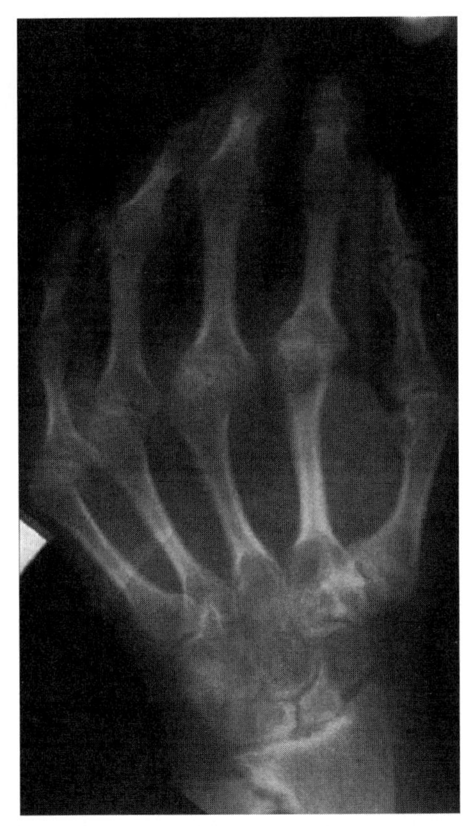

Figure 21-7. Crowding of the carpal bones owing to juvenile rheumatoid arthritis. Posteroanterior view of the right wrist shows the carpal bones with sharp angles, loss of the articular cartilage, and crowding. The radiocarpal joint and the metacarpophalangeal joints are all narrow.

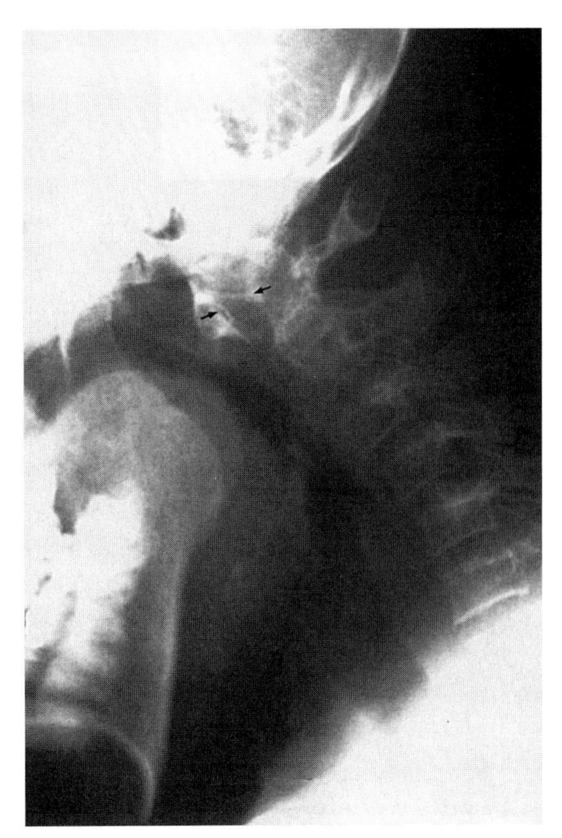

Figure 21-9. C1-C2 atlanto-axial subluxation. Lateral flexion view in a 9-year-old child with juvenile rheumatoid arthritis (JRA) shows C1-C2 subluxation *(arrows)*. Note also the characteristic JRA deformity in the mandible.

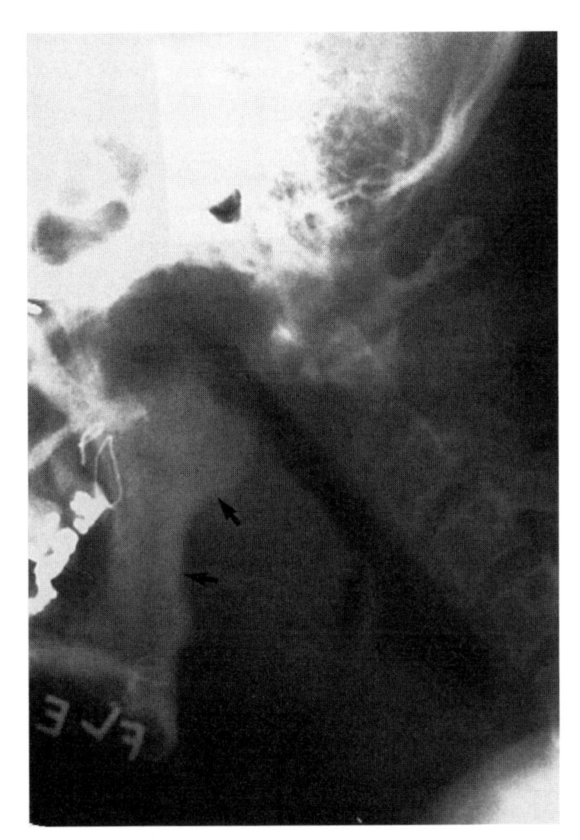

Figure 21-8. Mandibular deformity in juvenile rheumatoid arthritis (JRA). Lateral view of the upper cervical spine including the mandible shows the typical deformities of JRA. The mandible is short and underdeveloped (micrognathia). There is a large concavity on the inferior border of the body known as the antegonial notch *(arrows)*; the mandibular angle is more obtuse than normal.

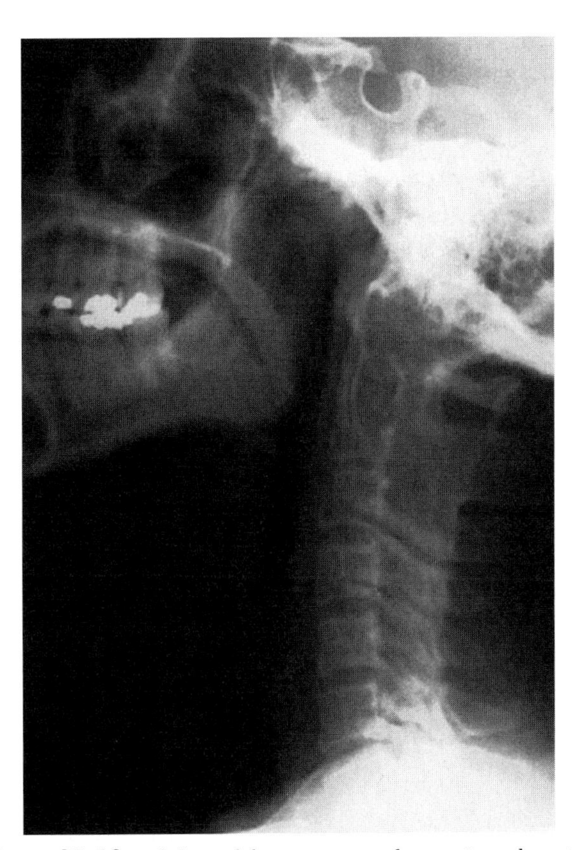

Figure 21-10. Ankylosis of the upper cervical spine. Lateral cervical spine shows the characteristic fusion of C1, C2, and C3 in a patient with juvenile rheumatoid arthritis.

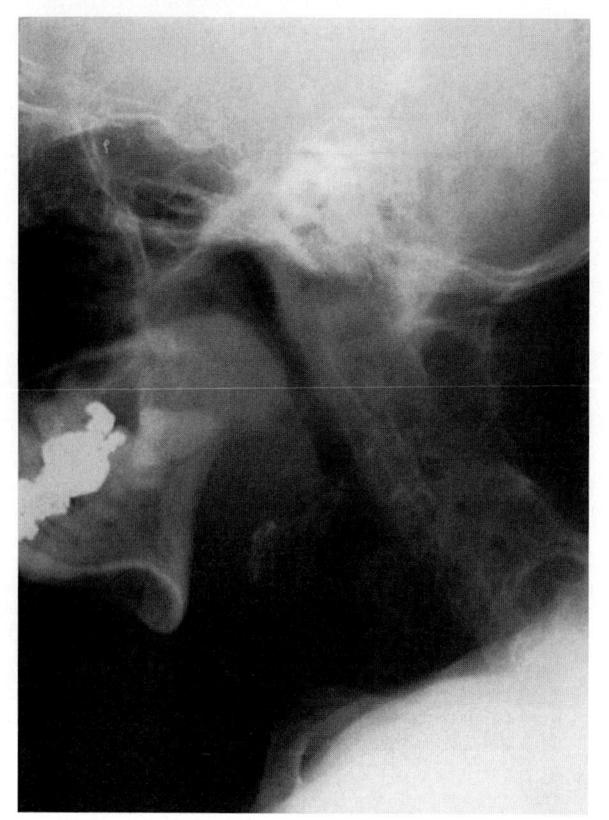

Figure 21–11. Fusion of the entire cervical spine in an adult who previously had juvenile rheumatoid arthritis (JRA). The vertebral bodies are underdeveloped. The entire cervical spine is fused. The mandible shows the typical deformities of JRA.

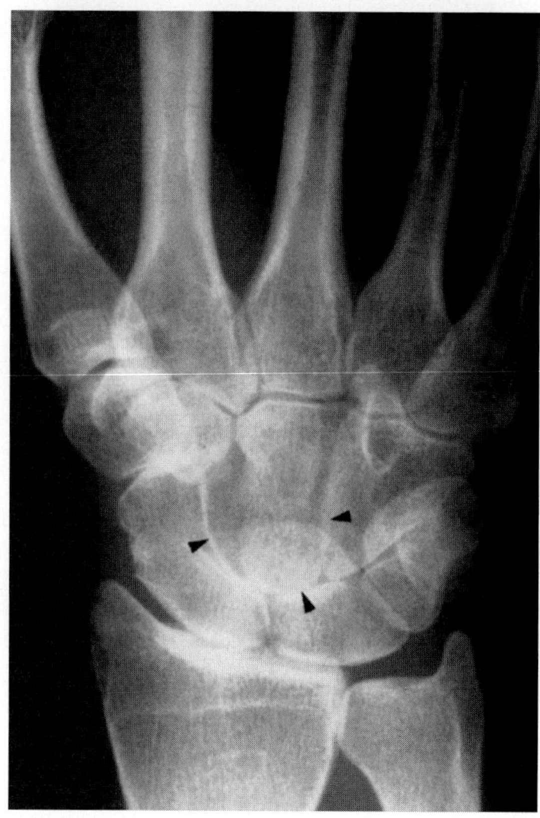

Figure 21–12. Adult-onset Still's disease (AOSD). Posteroanterior radiograph of the wrist shows arthritic changes in a patient with AOSD. Note that most of the change centers around the capitate *(arrowheads)*.

diagnosis is arrived at by exclusion. It often starts with a sore throat followed in a few days or weeks by fever, rash, arthralgias, myalgias, fatigue, anorexia, and rapid weight loss. Arthritis appears late and is found in more than 90% of patients with AOSD. Initially, the most commonly involved joints are knees, wrists, ankles, elbows, shoulders, proximal interphalangeal joints (PIPs), and cervical spine. As the arthritis becomes chronic (more than 6 months), the wrists are predominantly affected, and the tarsal joints, PIP joints, and cervical spine less commonly involved.

Radiographic Findings

Almost half of the patients develop pericapitate arthropathy and fusion (Fig. 21–12). In a few patients (19%), there is intertarsal and cervical facet joint fusion. Erosive arthritis is occasionally seen in patients with chronic joint symptoms.

Differential Diagnosis

Reiter's disease, dermatomyositis, and systemic febrile onset of RA can resemble AOSD.

Recommended Readings

Ansell BM, Kent PA: Radiological changes in juvenile chronic polyarthritis. Skeletal Radiol 1:129, 1977.
Cush JJ: Adult-onset Still's disease. Bull Rheum Dis 49:1, 2000.
Hensinger RN, DeVito PD, Ragsdale CG: Changes in the cervical spine in juvenile rheumatoid arthritis. J Bone Joint Surg Am 68:189, 1986.
Martel W, Holt JF, Cassidy JT: Roentgenologic manifestations of juvenile rheumatoid arthritis. AJR Am J Roentgenol 88:400, 1962.
Medsger T Jr, Christy WC: Carpal arthritis with ankylosis in late-onset Still's disease. Arthritis Rheum 19:232, 1976.
Simon S, Whiffen J, Shapiro F: Leg-length discrepancies in monoarticular and pauciarticular juvenile rheumatoid arthritis. J Bone Joint Surg Am 63:209, 1981.
Stabrun AE, Larheim TA, Hoyeraal HM, et al: Reduced mandibular dimensions and asymmetry in juvenile rheumatoid arthritis. Arthritis Rheum 31:602, 1988.

CHAPTER 22

Seronegative Spondyloarthropathies

Georges Y. El-Khoury, MD, and Shigeru Ehara, MD

OVERVIEW

The seronegative (rheumatoid factor negative) spondyloarthropathies are an interrelated group of multisystem inflammatory disorders that share a variety of clinical, radiologic, and genetic features. They include ankylosing spondylitis (AS), psoriatic arthritis (PA), reactive arthritis (Reiter's disease), arthritis associated with inflammatory bowel disease, SAPHO syndrome (sternoclavicular hyperostosis), and juvenile onset spondyloarthropathy. A small group of patients that is difficult to classify under one of the known clinical entities is referred to as having *undifferentiated spondyloarthropathy.* The spine, peripheral joints, and periarticular structures are affected by these diseases. Systemic manifestations are common, but they vary with each disease. There is also a strong association with the histocompatibility antigen HLA-B27; however, the presence of HLA-B27 is not always necessary for the development of any of these diseases.

PATHOLOGY

There are three basic pathologic processes present in seronegative spondyloarthropathies. These processes are responsible for the distinctive radiographic findings in these diseases.

Enthesitis

Enthesitis (or enthesopathy) is probably the most important process. It represents inflammation at the enthesis, which is the site of insertion of ligaments and tendons to bone. Pathologically, this lesion consists of granulation tissue that is gradually replaced by fibrocartilage and, later, by bone. Radiographically, it is characterized by shallow, short-lived erosions followed by fluffy bone formation and sclerosis.

Proliferation of Subchondral Granulation Tissue

Proliferation of subchondral granulation tissue with cartilage erosions is the dominant feature in sacroiliitis. The subchondral granulation tissue erodes the articular surface and is replaced with fibrocartilage, which eventually ossifies and results in joint fusion.

Synovitis

Synovitis with synovial hyperplasia, lymphoid infiltration, and pannus formation is found. The synovitis in spondyloarthropathies is different from the synovitis of rheumatoid arthritis in that it lacks proliferation of synovial villi, fibrin deposits, and ulceration.

ANKYLOSING SPONDYLITIS

Ankylosing spondylitis (AS) is a chronic inflammatory disease that primarily affects the axial skeleton. The disease tends to be familial, and its estimated prevalence in North American whites is 0.1% to 0.2%. Men are three times more affected than women. The disease tends to follow a more mild and benign course in women. Fewer than 20% of adult-onset AS cases progress to significant disability. The etiology of AS is not known, but the disease has strong association with HLA-B27—approximately 90% of patients are HLA-B27 positive.

Symptoms begin in late adolescence or early adulthood. Onset after the age of 40 years is very uncommon. Occasionally, AS occurs in association with psoriatic arthritis, Reiter's disease, ulcerative colitis, or Crohn's disease. In such cases it is called secondary AS. AS has a variable course with spontaneous remissions and exacerbations. In general, the disease has good outcome because it is mild and self-limited. Only patients with severe ankylosis of the hips and cervical spine are disabled.

Radiographic Changes in Ankylosing Spondylitis

Sacroiliitis

The earliest, most consistent and characteristic abnormalities in ankylosing spondylitis are seen in the sacroiliac (SI) joints. Virtually 100% of patients with AS have sacroiliitis; it is not possible to make the diagnosis of AS if there is not involvement of the SI joints. Sacroiliitis is typically bilateral and symmetric. The changes start on the iliac side and manifest with blurring of the subchondral plate, followed by erosions and sclerosis resembling postage-stamp serrations (Fig. 22–1). The sacral side of the joint, at least initially, appears protected from erosions because it is covered by thicker cartilage. Early in the disease, the joint becomes indistinct and, therefore, appears widened. Then, it starts to narrow, forms interosseous bridges, and eventually fuses. When the fusion is completed, the sclerosis disappears (see Fig. 22–1).

Sacroiliitis is often confused with osteitis condensans ilii, which typically occurs in women of childbearing age. It is characterized by dense sclerosis of the iliac side of the SI joints. There are no erosions, and the subchondral plate remains distinct (Fig. 22–2). Osteitis condensans ilii is very rarely unilateral, and it is rarely seen in men. The SI joint can be difficult to evaluate on straight anteroposterior views of the pelvis because of its almost horizontal orientation. A 30- to 45-degree cephalad angulated view can help display the SI joints. Computed tomography can be used to show the erosions better than radiographs, and magnetic resonance imaging (MRI) demonstrates the edema and inflammation surrounding the joints; however, these modalities are expensive and are rarely indicated.

Spondylitis

Spondylitis occurs in about 50% of patients with AS, and it typically starts at the thoracolumbar and lumbosacral junction. As the disease progresses, the midlumbar region,

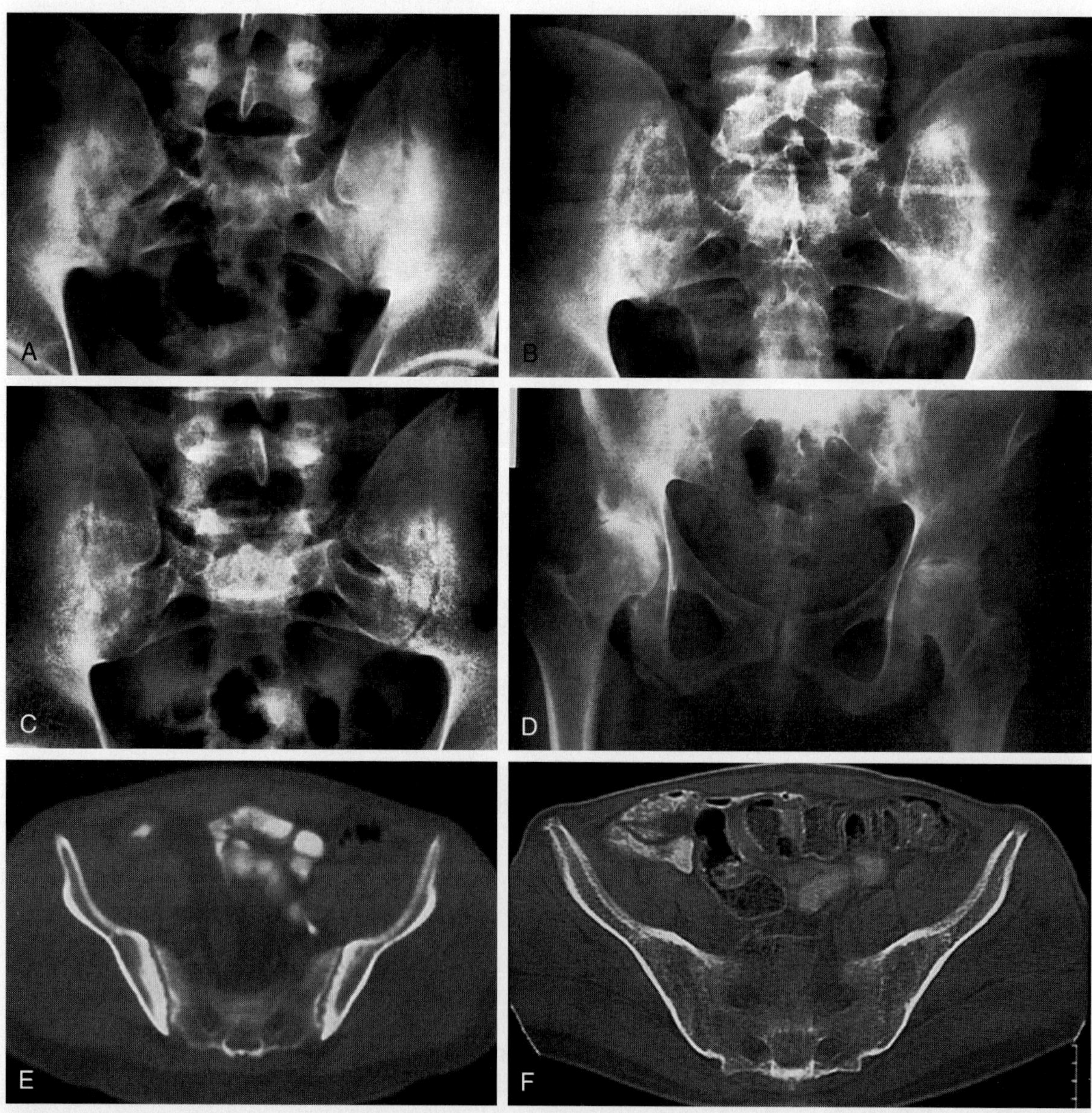

Figure 22–1. Sacroiliitis at different stages of activity in patients with ankylosing spondylitis (AS). *A,* Active sacroiliitis with large erosions and sclerosis mainly on the iliac side. The sacroiliac (SI) joint space is wide on the *right* because of the erosions. The articular surfaces are irregular bilaterally. *B,* Moderately advanced sacroiliitis. The SI joints are narrow, and the sclerosis is less pronounced than that in *A. C,* Late stage sacroiliitis with partial fusion of the right SI joint. Mild sclerosis at the SI joints is still seen. *D,* Complete fusion of the SI joints. The bony sclerosis has disappeared. Both hips are affected with inflammatory arthritis. *E,* A computed tomography scan section through the SI joints in a patient with Crohn's disease. Note the shallow erosions on the iliac side of the SI joints surrounded with sclerosis. *F,* Complete fusion of the SI joints in a 62-year-old man with AS. The sclerosis has totally disappeared.

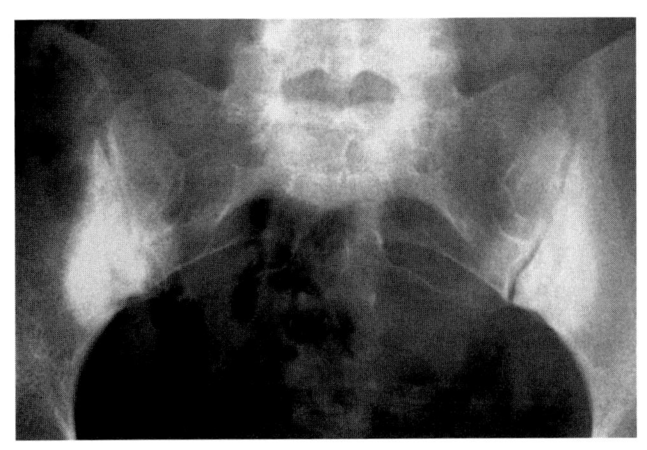

Figure 22–2. Osteitis condensans ilii in a 27-year-old woman. There is bony sclerosis on the iliac side of the sacroiliac (SI) joints bilaterally. The SI joints are distinct and do not show any erosions.

the upper thoracic spine, and the cervical spine become involved. Changes are rarely observed in the cervical or thoracic spine without lumbar involvement. Enthesitis in the spine manifests with small short-lived erosions at the insertion of the outer fibers of the annulus fibrosus to the vertebral bodies (Romanus lesions). These corner erosions are typically surrounded by bony sclerosis and are referred to as the *shiny corners* (Fig. 22–3). This is followed by ossification of the outer annular fibers and adjacent connective tissues to form what is known as marginal syndesmophytes (Fig. 22–4). The syndesmophytes of AS are different from the osteophytes seen in degenerative disk disease and paravertebral ossification (or nonmarginal syndesmophytes) associated with reactive arthritis (Reiter's disease) and PA. The syndesmophytes of AS extend vertically from one corner of the vertebral body to another corner across the disk. They are initially thin and delicate and are always symmetrical. When the disease starts in adolescence, syndesmophytes do not develop before the age of 20 years. The apophyseal joints and costovertebral joints undergo inflammatory changes and subchondral erosions, followed by ankylosis (Fig. 22–5). These processes at the disks and apophyseal joints can ultimately result in complete fusion of the spine, which is called the "bamboo spine" (Fig. 22–6). After spinal fusion is complete, the pain decreases and osteoporosis develops, making the spine vulnerable to fractures. The bridging vertebral osteophytes and calcified ligaments seen in diffuse idiopathic skeletal hyperostosis (DISH) may be confused with AS. The presence of sacroiliitis can be used to differentiate the two conditions, because it is only seen with AS. Vertebral body squaring is another characteristic feature of AS and is produced by the corner erosions and anterior apposition of periosteal new bone formation (Fig. 22–7). Squaring is easier to appreciate in the lumbar spine where vertebral bodies are normally concave anteriorly (Fig. 22–7*B*) as compared with thoracic vertebrae, which normally have straight anterior margins.

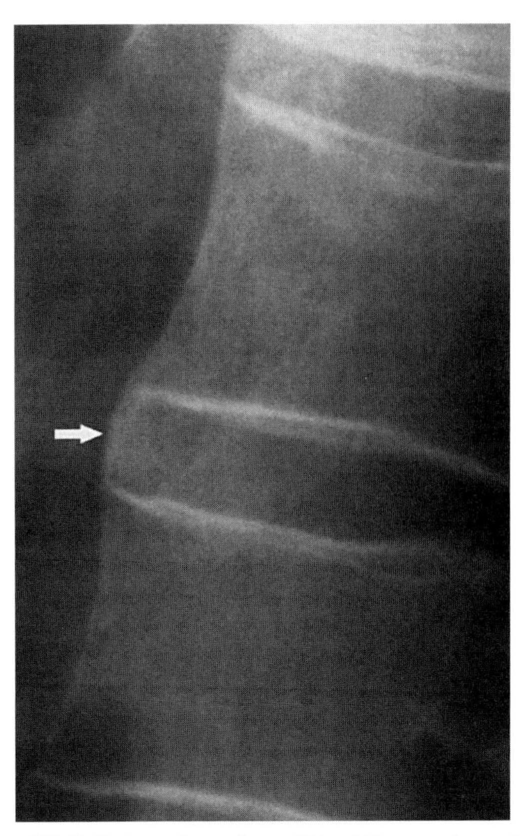

Figure 22–4. Early syndesmophytes. Thin, delicate syndesmophyte is seen bridging one corner of a vertebral body to another *(arrow)*.

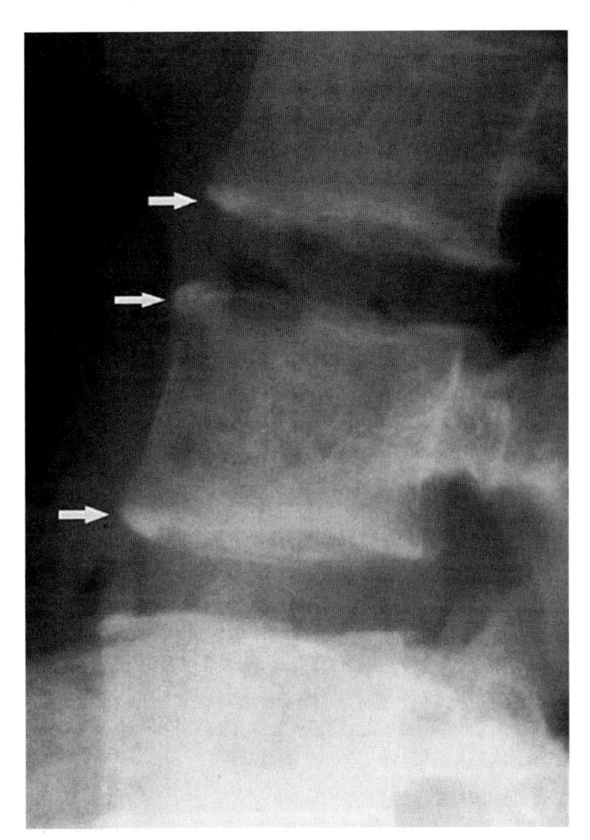

Figure 22–3. Shiny corners. Lateral view of the lumbar spine shows shallow erosions and bony sclerosis at the anterior corners of the vertebral bodies *(arrows)*. These sites are the insertions of the outer fibers of the annulus fibrosus.

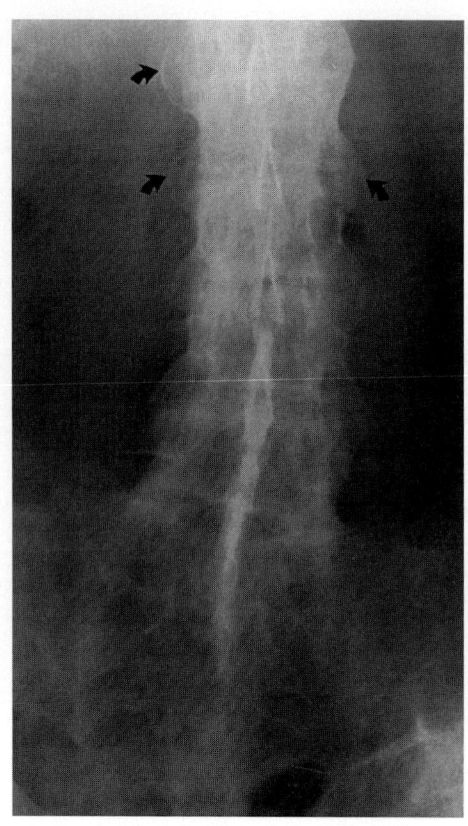

Figure 22–5. In advanced disease, the syndesmophytes become thicker and extend beyond the corners of the vertebral bodies *(arrows)*.

Spinal Fractures

The rigidity of the fused spine as well as the osteoporosis renders the spine brittle and, therefore, increases the risk for spinal fractures. When patients with long-standing AS and spinal fusion present with new back pain or neck pain, a stress fracture should be suspected. These fractures typically occur at the cervicothoracic (75%) and thoracolumbar junctions (Fig. 22–8). The mechanism of injury is often due to hyperextension, and the fracture always extends through all three columns, which by definition is an unstable injury. Such fractures may go undetected in the acute phase, and patients may eventually develop pseudarthrosis (Fig. 22–9). Because of the instability and high morbidity rate with such fractures, MRI is warranted for early detection and assessment of cord compromise. The incidence of cord injury with these fractures is reported to be approximately 69%. Commonly, the fracture traverses a fused intervertebral disk or, less often, it passes through a vertebral body that is adjacent to an endplate. Pseudarthrosis presents radiographically with diskovertebral erosions and bone destruction, an appearance resembling disk-space infection. These changes are referred to as *Andersson lesion* (see Fig. 22–9). The pseudarthrosis can also develop at one level or two noncontiguous disk levels without history of trauma. Some of these lesions are not due to trauma but rather due to increased stress at intervertebral disks that escaped fusion while other levels became fused. Atlanto-axial subluxation (AAS) is rarely seen in AS compared with rheumatoid

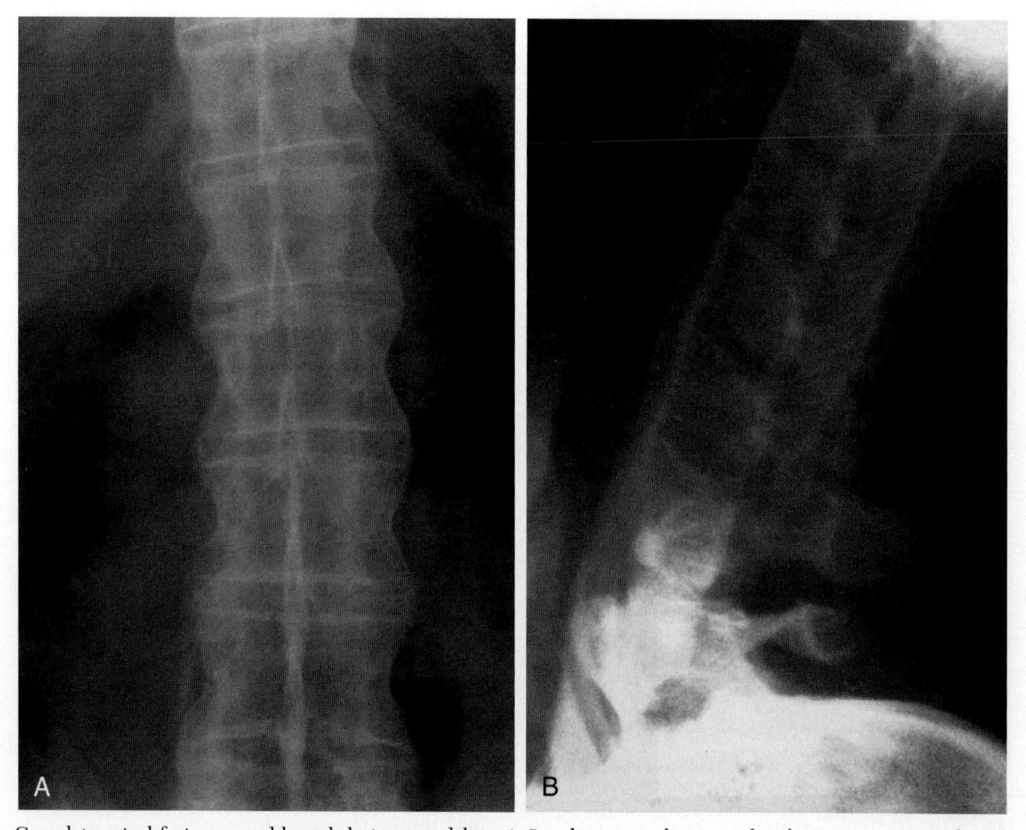

Figure 22–6. Complete spinal fusion caused by ankylosing spondylitis. *A*, Lumbar spine showing a bamboo spine. *B*, Fused cervical spine.

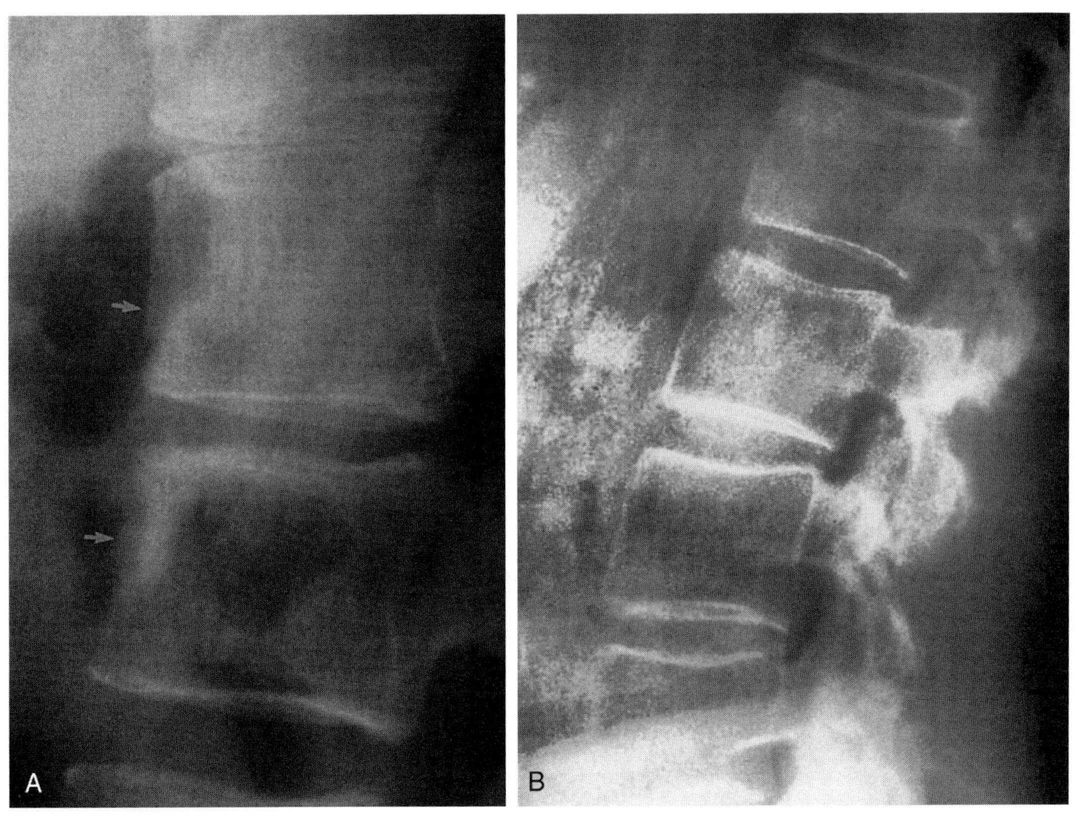

Figure 22–7. Squaring of the vertebral bodies in ankylosing spondylitis. *A,* Apposition of periosteal new bone filling the anterior concavity of the vertebra *(arrows)*. *B,* There is loss of the normal anterior concavity of the lumbar vertebrae.

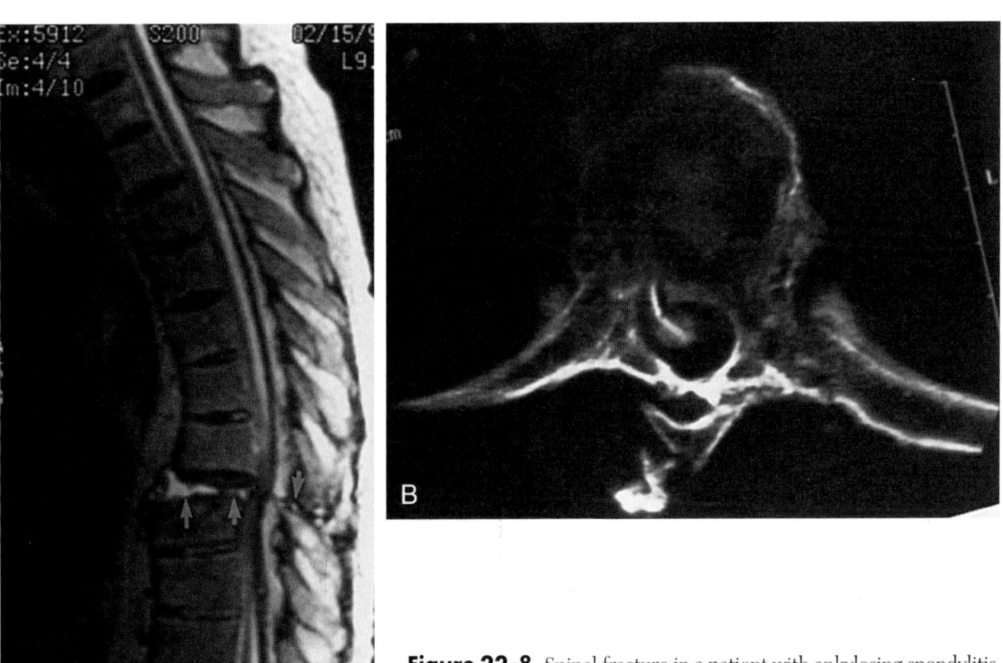

Figure 22–8. Spinal fracture in a patient with ankylosing spondylitis. *A,* T2-weighted mid-sagittal MR image shows a fracture through the T6-T7 ossified disk with extension into the posterior elements *(arrows)*. The patient became paraplegic after the injury. *B,* A computed tomography scan section through the fracture reveals a sharp fragment of bone within the spinal canal. As an incidental finding, the costovertebral joints are fused bilaterally.

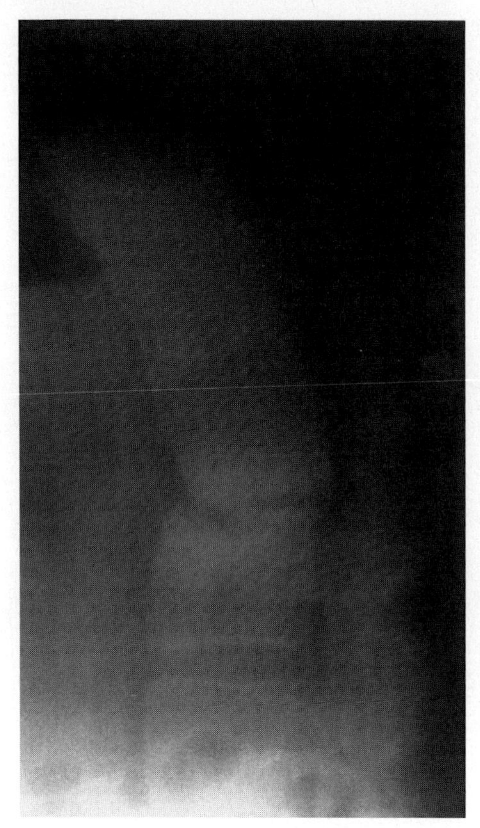

Figure 22–9. Pseudarthrosis at the thoracolumbar junction in a 45-year-old woman with long-standing ankylosing spondylitis. This abnormality is also referred to as the "Andersson lesion." It is often confused with disk-space infection.

arthritis. Typically, AAS occurs in the active or early phase of the disease.

Synovial Joints

The joints most commonly involved, in descending order of frequency, include the hips, shoulders, knees, and metatarsophalangeal joints. These joints are more commonly affected when the disease begins before the age of 21 years. With AS, osteopenia is typically absent in affected joints.

1. Hips are involved in up to 48% of patients, and the involvement is typically bilateral and symmetric. Sometimes the earliest symptoms may start in the hips. AS is characterized by symmetric concentric joint space narrowing, irregularity of the subchondral plate, axial migration of the femoral head, and collar-like osteophytes at the junction of the head with the neck (Fig. 22–10). Osteophytes develop also on the acetabular side. Protrusio acetabuli is present in approximately one third of the patients. A destructive form of hip arthritis occurring in older individuals has been described (Fig. 22–11). Eventual fusion of the hips occurs; this can be a major source of disability in patients with AS.
2. The shoulders are the second most frequently involved joints. There is concentric narrowing of the glenohumeral joint, and a large superolateral erosion develops in the region of the greater tuberosity (Fig. 22–12). Some authors believe this erosion is pathognomonic of AS. Fusion of the glenohumeral joint is unusual. Complete ankylosis of the acromioclavicular joint is, however, fairly frequent.
3. The knees are the third most frequently affected large joints. The radiographic findings include uniform joint-space narrowing and productive bony changes (enthesitis). Osteopenia and erosion are notably absent (Fig. 22–13).
4. Small joints of the hands and feet are rarely affected. Radiographic changes are differentiated from rheumatoid arthritis by the asymmetry of involvement, lack of demineralization, smaller and shallower erosions, and productive bony changes (enthesitis).

Enthesitis

Enthesitis at insertions of ligaments and tendons is fairly common with AS. In the early active phase of the disease, there are shallow erosions surrounded by bony sclerosis; as the disease becomes advanced, the involved areas show reactive bone formation known as *whiskering* (see Fig. 22–10). These changes are particularly evident at the ischial tuberosities, iliac crests, calcanei, femoral trochanters, and vertebral spinous processes. In women, osteitis pubis can be quite severe (Fig. 22–14), but pelvic whiskering is unusual.

Extraskeletal Manifestation

Major extraskeletal manifestations of AS have been observed: uveitis, aortitis, apical pulmonary fibrosis, and leptomeningeal diverticula. Uveitis is the most common, occurring in 25% to 30% of patients. Aortitis of the ascending aorta produces fibrosis and dilatation of the aortic ring, resulting in aortic valve insufficiency (Fig. 22–15*A*). Fibrosis of the subaortic area can also result in conduction abnormalities. Parenchymal lung disease is rare, usually asymptomatic, and occurs in about 1.3% of patients. Pulmonary changes typically are seen several years after joint manifestations. Slowly progressive fibrosis with cystic or bullous changes in the lung apices has been observed (Fig. 22–15*B*). Radiographically, these lesions resemble tuberculosis. The cysts can become colonized by aspergillus or bacteria, resulting in cough, dyspnea, and hemoptysis. Cauda equina syndrome has been described in some AS patients, and it is attributed to the development of leptomeningeal diverticula or sacculations in the lumbar region (Fig. 22–16). These abnormalities are best evaluated by MRI. Severe atrophy of the errector spinae muscles has also been reported in AS patients with complete spinal fusion (Fig. 22–17).

PSORIATIC ARTHRITIS (PA)

PA occurs in approximately 5% to 7% of patients with psoriasis. Nail involvement seems to identify patients who are likely to develop PA; nail changes are present in 80% of patients with PA but occur in only 30% of patients with cutaneous psoriasis alone. The male-to-female ratio is almost equal, and the peak age of onset of the arthritis is usually between 30 and 50 years. In a few patients (10%),

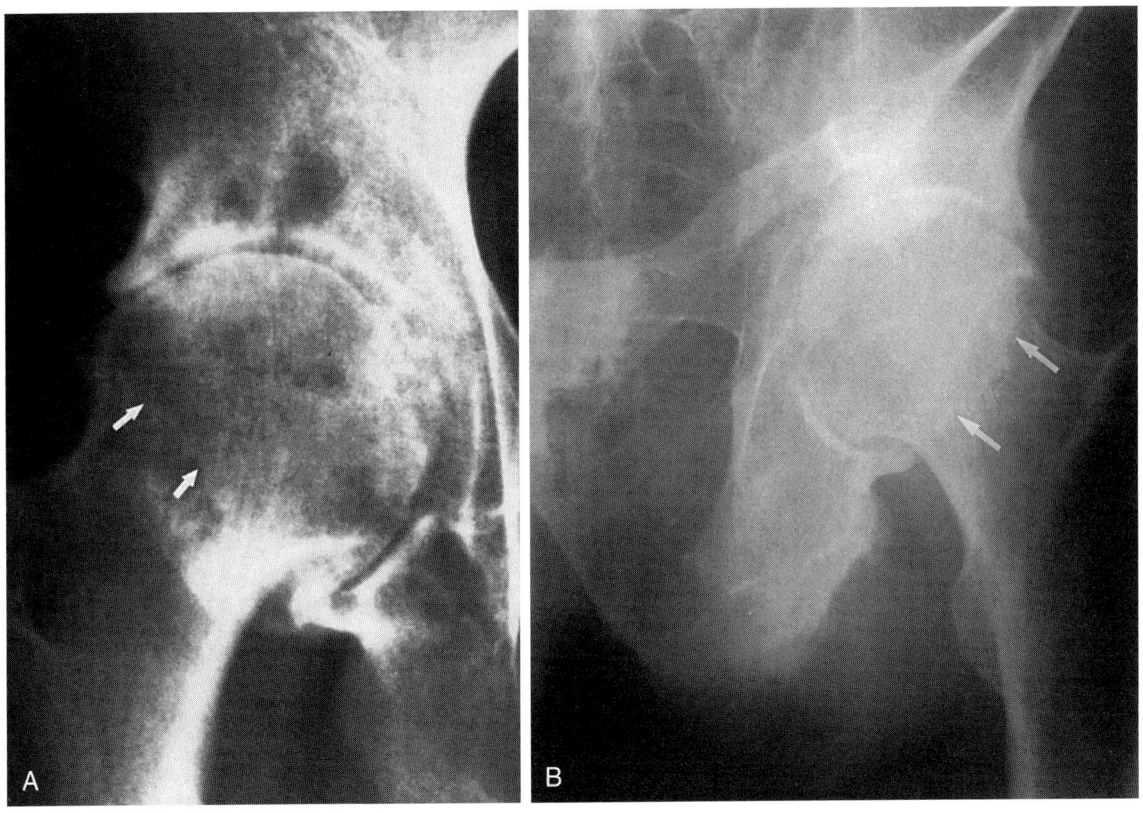

Figure 22–10. Moderately advanced changes of ankylosing spondylitis (AS) in the hips. *A*, Anteroposterior (AP) view of the right hip showing uniform joint space narrowing, osteopenia, and collar osteophytes *(arrows)*. *B*, AP view of the left hip shows typical changes of AS with collar osteophytes *(arrows)*. In addition, enthesitis in the ischium and fusion of the left sacroiliac joint can be visualized.

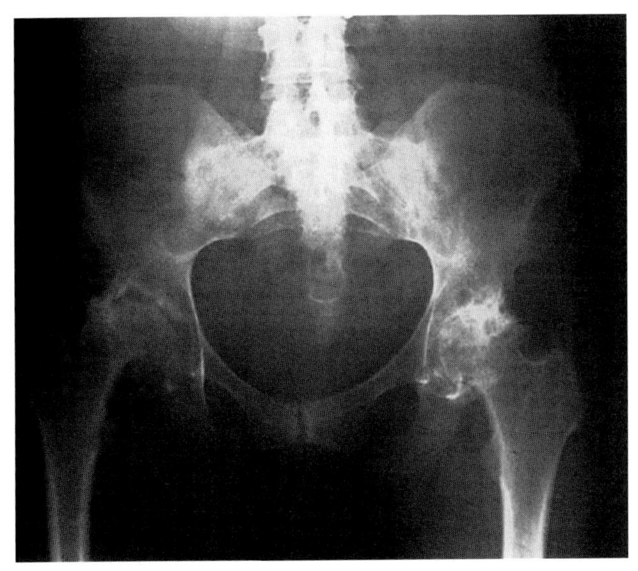

Figure 22–11. Anteroposterior radiograph of the pelvis shows bilateral hip involvement with ankylosing spondylitis. The changes in the right hip are more advanced and show erosions.

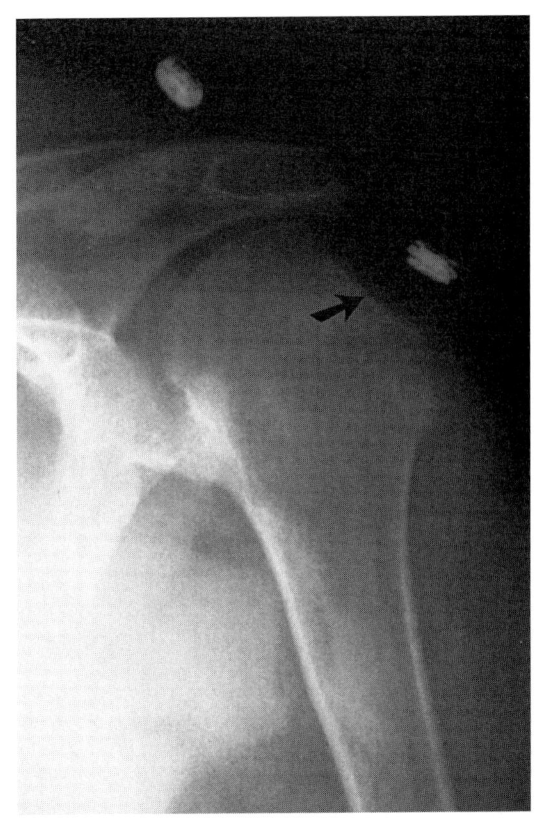

Figure 22–12. Shoulder involvement in ankylosing spondylitis. Anteroposterior view of the left shoulder shows narrowing of the glenohumeral joint and a large erosion in the region of the greater tuberosity *(arrow)*. This is sometimes called the "hatchet sign."

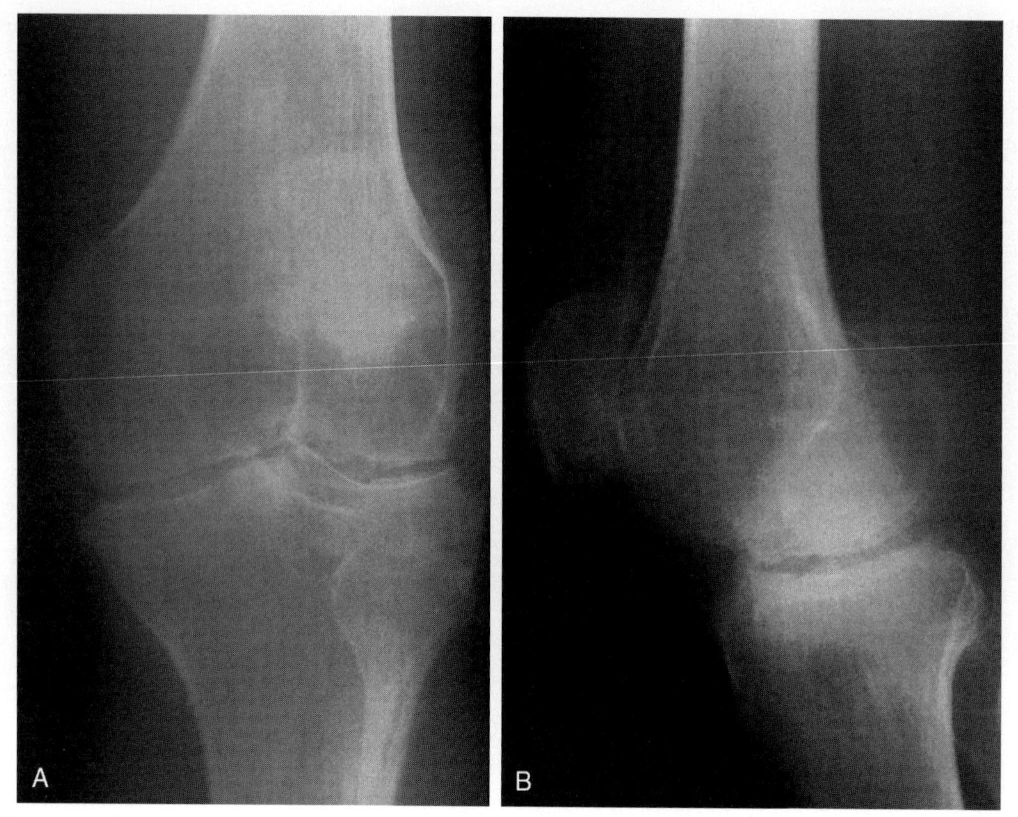

Figure 22–13. Knee involvement in a 62-year-old man with long-standing ankylosing spondylitis. *A*, Anteroposterior view. Note the uniform joint space narrowing, shallow erosions, productive bony changes, and lack of osteopenia. *B*, Lateral view of the knee in the same patient.

the arthritis appears before the skin changes. The disease is most commonly an asymmetric polyarticular disorder with predilection for the interphalangeal joints of the hands, especially the distal interphalangeal joints (DIPs), and the metatarsophalangeal and interphalangeal joints of the feet. The sacroiliac joints and spine are less frequently involved than are the hands and feet. PA rarely involves the hips and shoulders. Aortic insufficiency, uveitis, and upper lobe pulmonary fibrosis can occur in PA.

Radiographic Changes

Hands and Feet

The radiographic changes in the hands and feet are highly distinctive. In the hands, the typical pattern starts with an asymmetric erosive polyarthritis with predilection to the DIPs. Erosions in the fingers start in the bare areas and become surrounded by fluffy bone formation called

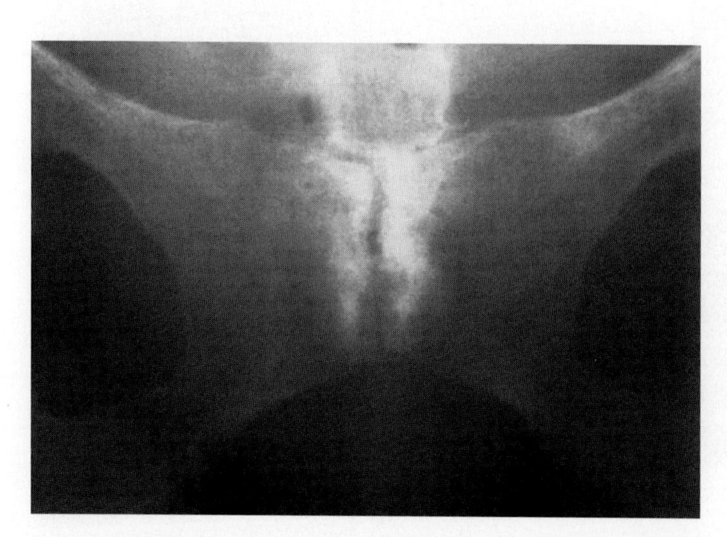

Figure 22–14. Posteroanterior view of the pelvis reveals severe osteitis pubis in a 35-year-old woman with ankylosing spondylitis.

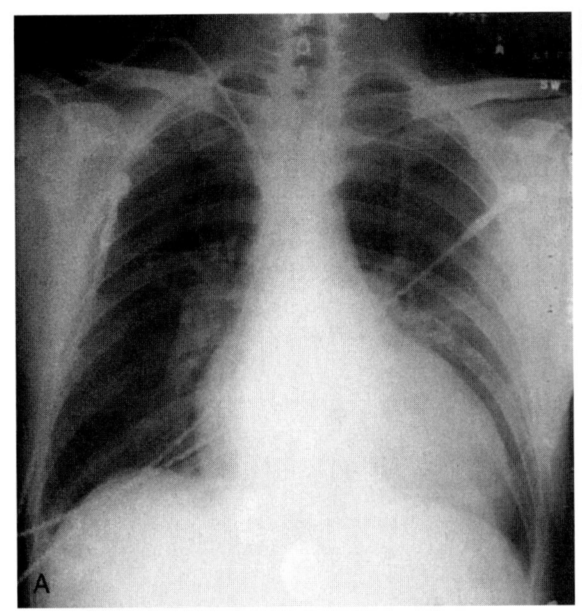

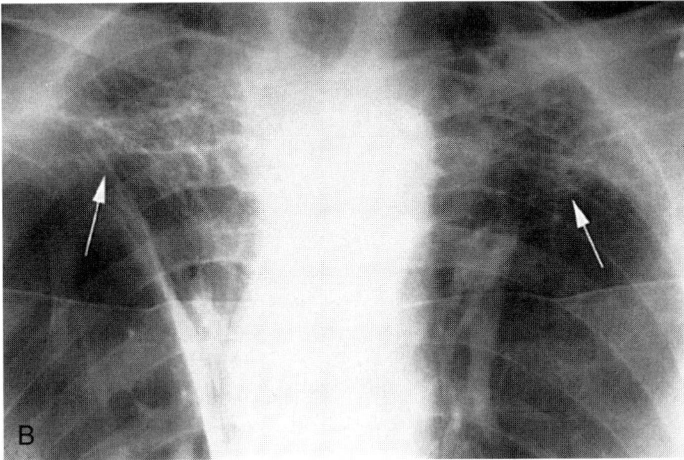

Figure 22–15. Cardiac and pulmonary changes in ankylosing spondylitis (AS). *A*, Aortic valve insufficiency. Posteroanterior view of the chest in a patient with long-standing AS shows cardiomegaly and left ventricular dilation owing to aortic regurgitation. *B*, Apical pulmonary fibrosis *(arrows)* in a 61-year-old man with a long-standing history of AS.

mouse ears (Fig. 22–18). The distal erosions can significantly enlarge and destroy entire joints, producing what is known as a *pencil-in-cup* deformity or arthritis mutilans (Fig. 22–19). These changes are considered by some to be pathognomonic of PA. Another manifestation includes interphalangeal joint space narrowing with or without ankylosis (Fig. 22–20). A tendency for ray distribution has also been described, whereby all the joints in one or two rays are destroyed while adjacent rays are relatively spared (Fig. 22–21). Resorption of the distal tufts (acrosteolysis) in the hands and feet can be seen. Erosive changes in the foot have a predilection for the interphalangeal joint of the great toe (Fig. 22–18*B*). In the distal phalanx of the great toe, tuft proliferation and sclerosis result in a shaggy dense distal phalanx known as the ivory phalanx (Fig. 22–22). In the absence of articular abnormalities, the ivory phalanx becomes an important sign of PA. Osteoporosis is notably absent in both PA and reactive arthritis.

Figure 22–16. Computed tomographic image of a lumbar vertebra showing thecal dilation and sacculations producing pressure erosions in the posterior body, laminae, and pedicles. Patient had long-standing ankylosing spondylitis with cauda equina syndrome.

Axial Disease

Axial disease can occur independent of peripheral arthritis, but usually develops several years after the peripheral arthritis. Sacroiliitis occurs in about one third of patients, frequently is asymptomatic and asymmetric, and may occur independent of spondylitis (Fig. 22–23). Fusion of the SI joints is relatively infrequent. Spondylitis may also occur without sacroiliitis. The cervical spine is rarely involved with atlanto-axial subluxation and subaxial sub-

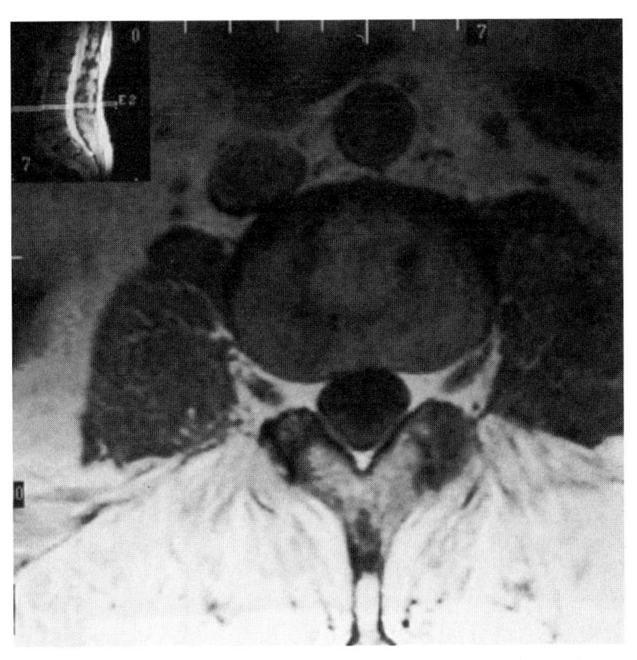

Figure 22–17. Erector spinae muscles in a patient with ankylosing spondylitis (AS) and complete spinal fusion. T1-weighted axial MR image through the L3-4 level shows severe atrophy and fatty replacement of the erector spinae muscles. This is in contradistinction to the psoas muscles, which are normal, suggesting that the hips are not fused or severely involved with AS.

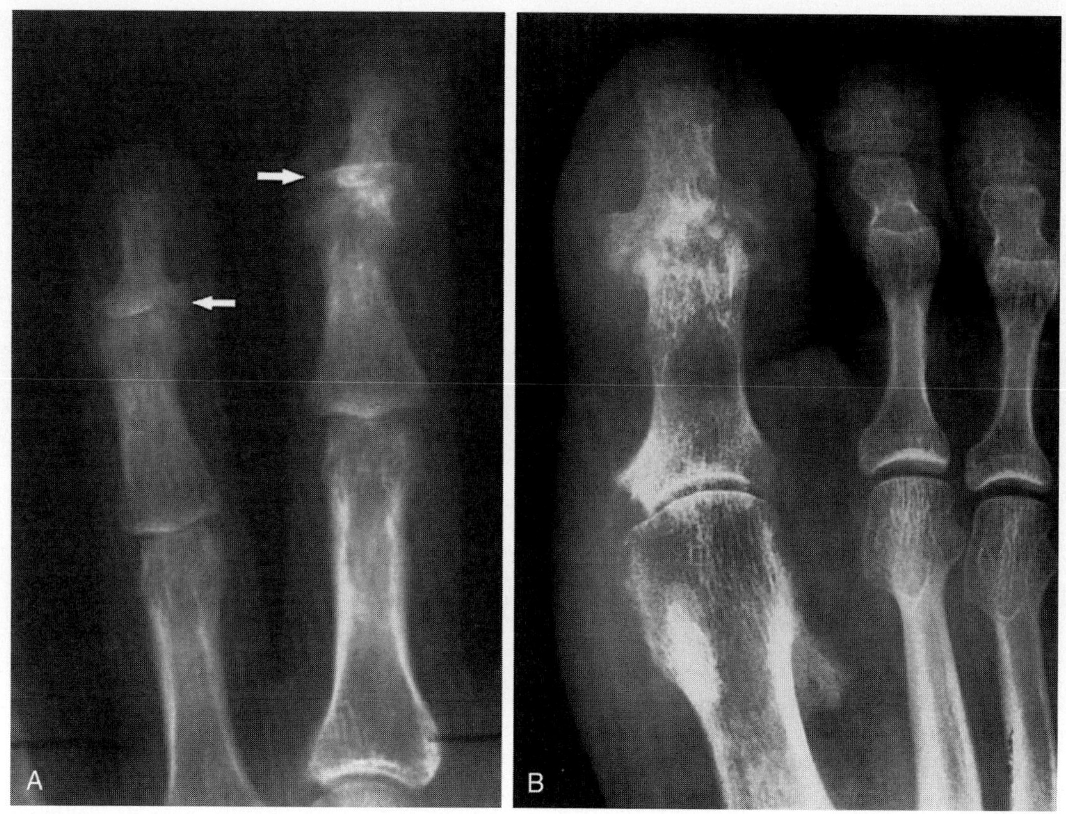

Figure 22–18. "Mouse ears" in psoriatic arthritis. *A,* Posteroanterior view of distal interphalangeal joints shows typical mouse ears *(arrows)* owing to enthesitis at the capsular attachments. Clinically, the patient had severe nail changes. *B,* "Mouse ears" at the base of the distal phalanx of the big toe. Extensive subchondral erosions at the interphalangeal joint. Metatarsophalangeal joint disease is also present. The lateral sesamoid of the great toe also shows enthesitis.

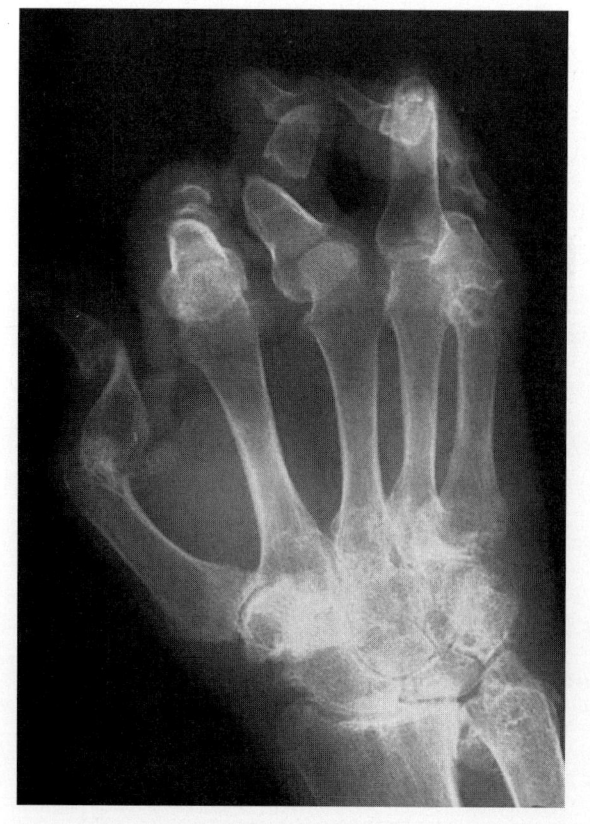

Figure 22–19. Arthritis mutilans in a 46-year-old man with long-standing psoriasis. Despite the severe and destructive disease, the patient had been consistently pain-free.

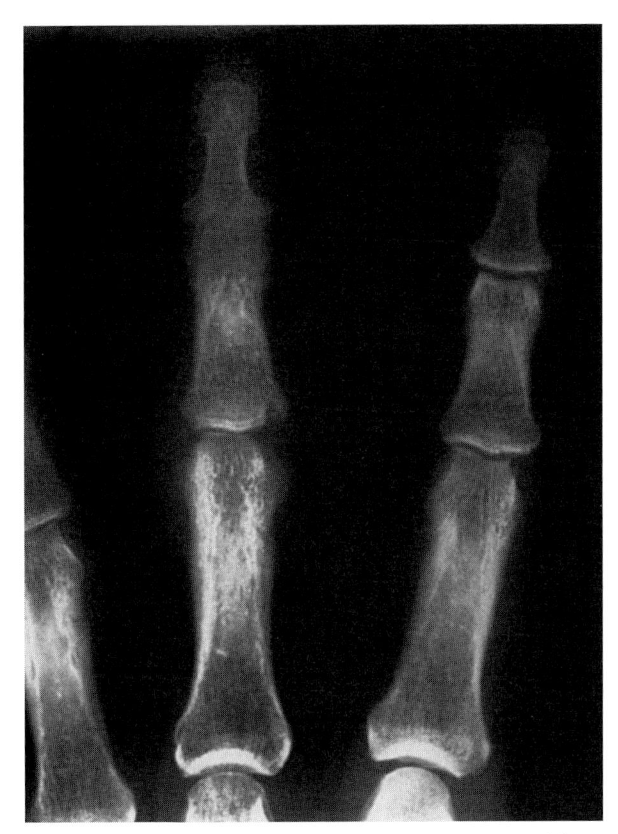

Figure 22–20. Ankylosis of the distal interphalangeal joint of the long finger in a patient with psoriatic arthritis. The proximal interphalangeal joint is involved with erosive and proliferative changes.

luxations. The spondylitis of psoriasis and reactive arthritis progresses in more random fashion than the spondylitis of AS. It is characterized by the presence of coarse, asymmetric paravertebral ossifications (nonmarginal syndesmophytes) at the lower thoracic and upper lumbar spine on alternate levels (Fig. 22–24). The lateral aspect of the vertebral bodies is particularly affected with paravertebral ossifications. Spine fusion, vertebral body squaring, and apophyseal joint involvement are rare. Another pattern of spondylitis that is identical to AS can occur in HLA-B27–positive patients with PA. Such patients are believed to have coincident or secondary AS.

Anterior Chest Wall

Manubriosternal and sternoclavicular joint involvement can be severe and includes soft-tissue swelling, erosions, sclerosis, and fusion (Fig. 22–25). These joints occasionally may be the only musculoskeletal sites affected in PA. It is easy to confuse these chest wall lesions radiographically with osteomyelitis or neoplasm. A related condition with similar findings in the anterior chest wall is the SAPHO syndrome (sternoclavicular hyperostosis), which some authors classify as a spondyloarthropathy.

Differential Diagnosis

Occasionally, erosive osteoarthritis (EOA) in the hands can be confused with PA. However, age of onset, female predominance in EOA, asymmetric disease with often single ray involvement in PA, and the typical subchondral erosions in erosive OA versus the erosions in the bare areas with PA should distinguish the two conditions. Reactive arthritis typically affects the lower extremities, especially the knees, ankles, and feet, but often spares the hands.

REACTIVE ARTHRITIS (REITER'S SYNDROME)

The terms "reactive arthritis" and "Reiter's syndrome" are used interchangeably, although the term reactive arthritis has essentially replaced the term Reiter's syndrome. Reactive arthritis is defined as a sterile inflammatory response to a previous infection remote from the affected joints. Reactive arthritis is typically detected following infections caused by chlamydia, yersinia, salmonella, and shigella. Reactive arthritis usually starts as an acute arthritis within about 1 month of infections involving the genitourinary or gastrointestinal tracts. Most endemic cases are venereally transmitted and usually affect young men (9:1), whereas arthritis following enteric infections affects men and women equally. Whites are affected more commonly than are blacks or other racial groups, who have lower frequency of the HLA-B27. In most cases, reactive arthritis is a self-limited disease, although in approximately

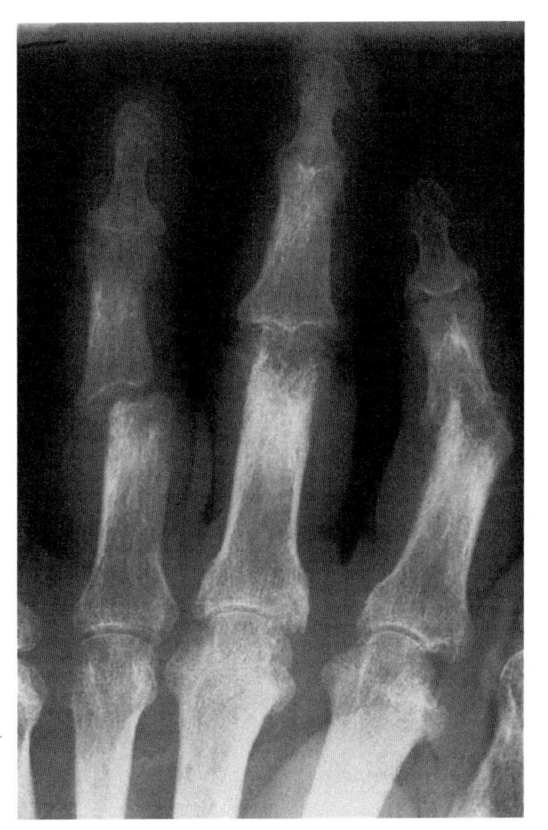

Figure 22–21. Ray distribution in psoriatic arthritis (PA). Some fingers in PA are severely affected whereas others can be mildly affected or completely disease-free.

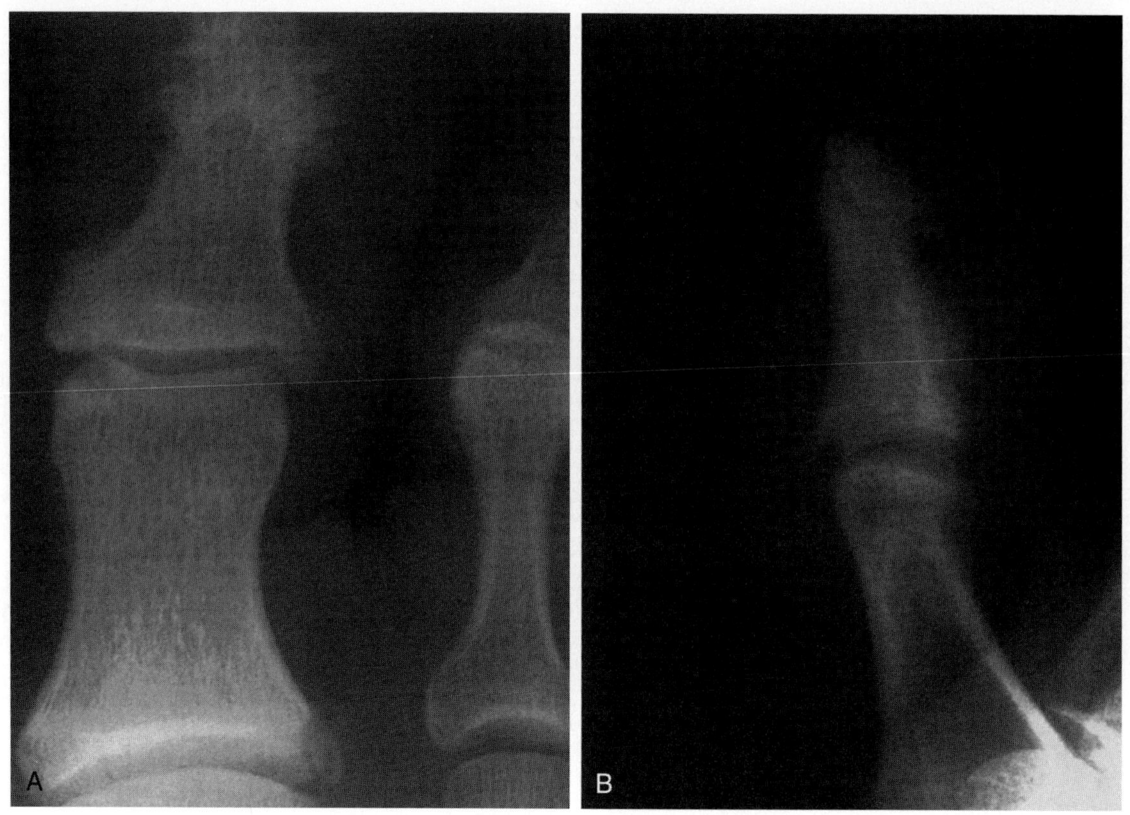

Figure 22–22. Involvement of the distal phalanx of the great toe with psoriatic arthritis. *A,* Anteroposterior view of the great toe in a patient with psoriasis and nail changes shows a shaggy and dense tuft. *B,* Lateral view in another patient in whom the entire distal phalanx is sclerotic shows productive bony changes.

15% of patients, the disease progresses to chronic destructive arthritis, enthesitis, and significant disability. Ten percent of patients develop AS. In 1% to 2% of patients with long-standing disease aortitis with aortic valve regurgitation can develop. There is no evidence that HIV causes reactive arthritis; however, patients who are HIV positive or have AIDS tend to develop more severe manifestations of reactive arthritis.

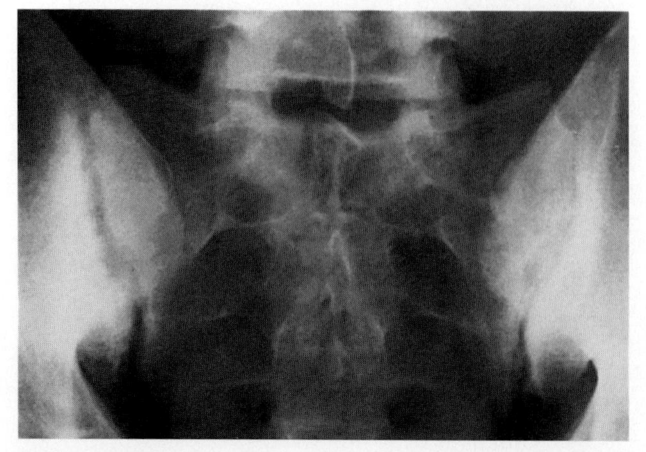

Figure 22–23. Asymmetric sacroiliitis in psoriatic arthritis. The right sacroiliac (SI) joint shows extensive erosive changes, but the left SI joint is relatively unaffected.

Radiographic Changes

Joint abnormalities become detectable radiographically several months after the disease has been active. Reactive arthritis has a predilection for the peripheral joints of the lower extremities, especially the knees, ankles, and feet. The most common site of involvement is the feet, particularly the calcaneus and metatarsophalangeal joints (Fig. 22–26). The enthesitis at the insertion of the Achilles' tendon and plantar fascia on the calcaneus is characteristic of reactive arthritis (Fig. 22–27). Shallow erosions develop at these sites followed by fluffy periosteal reaction and bony sclerosis. With time, inflammatory spurs develop on the plantar surface of the calcaneus. Inflammatory spurs are present in about 59% of patients. They can be differentiated from degenerative spurs by the absence of erosions and sclerosis, and by the continuity of the marrow space of the degenerative spur with the marrow space of the calcaneus. Similar changes in the calcaneus are occasionally seen with AS and PA, but the process is less aggressive. Subtle periosteal reaction, often fluffy, is seen at bony prominences such as the malleoli and the radial styloid. In the feet, there is a tendency for a destructive arthritis involving the small joints. These changes are identical to those seen in PA.

Sacroiliitis is reported to occur in 20% to 42% of patients with reactive arthritis. Early in the disease, it is often asymmetric or unilateral (Fig. 22–28). Complete SI joint fusion is much less frequent than in AS. Spondylitis

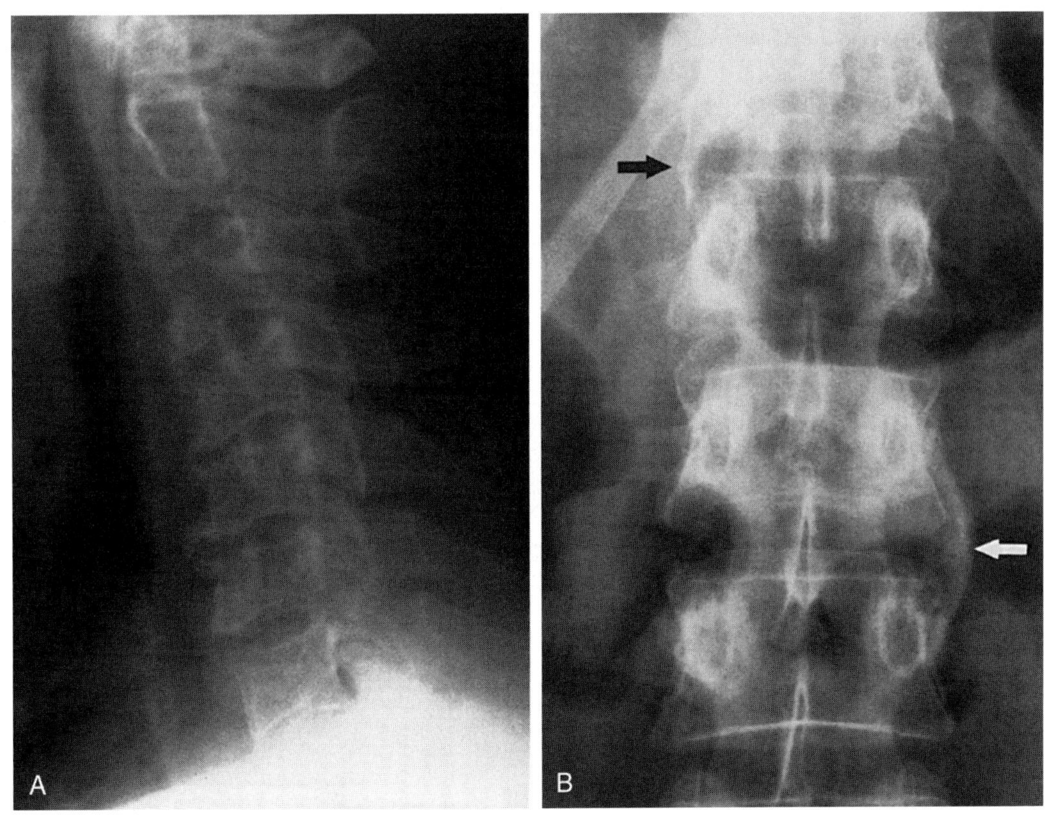

Figure 22–24. Spinal changes in psoriatic arthritis (PA). *A*, Lateral view of the cervical spine shows coarse syndesmophytes anteriorly. The C2, C3, and C4 vertebral bodies are fused. The apophyseal joints are not affected, which is typical in PA. *B*, Asymmetric paravertebral ossification with skip areas *(arrows)*. These findings can also be seen in Reiter's disease.

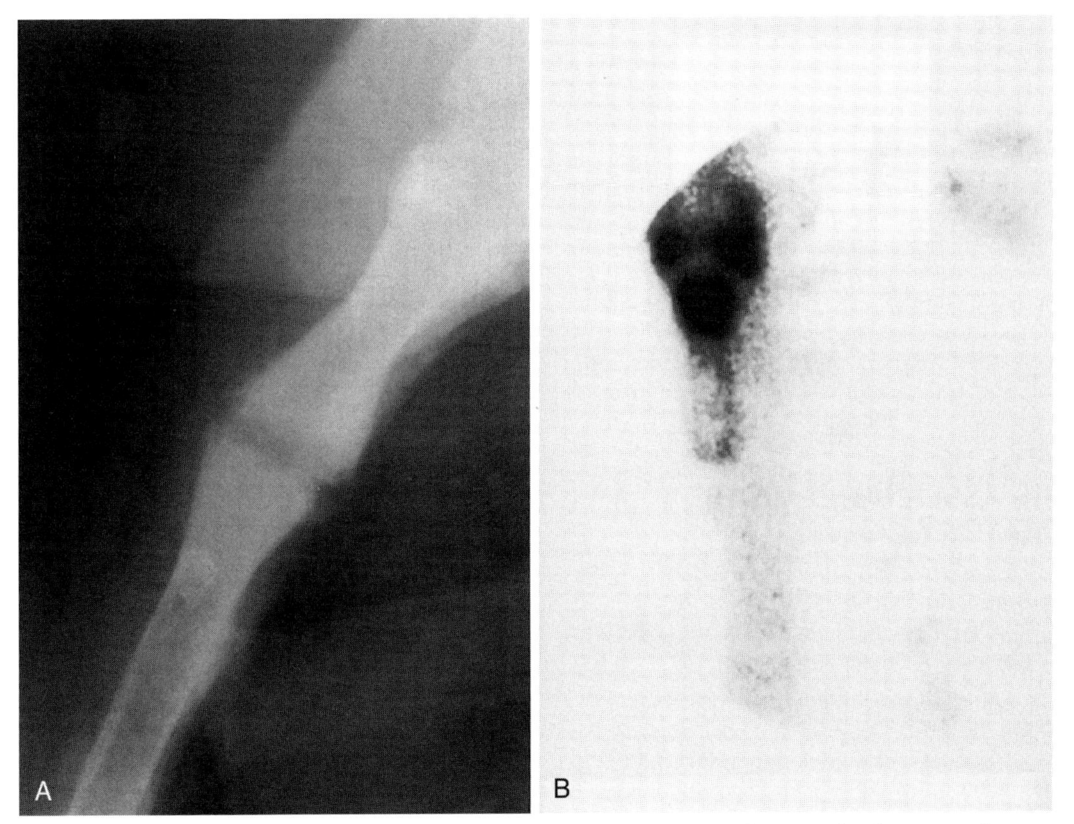

Figure 22–25. Manubriosternal and sternoclavicular involvement with hyperostosis, dense sclerosis, and soft-tissue swelling in a patient with psoriasis. The anterior chest wall changes closely resemble the changes in SAPHO (synovitis-acne-pustulosis-hyperostosis-osteitis syndrome). *A*, Lateral view of the upper sternum demonstrates hyperostosis and sclerosis in the manubrium and proximal body of the sternum. *B*, Bone scan of the area shows increased uptake in the clavicles and manubrium.

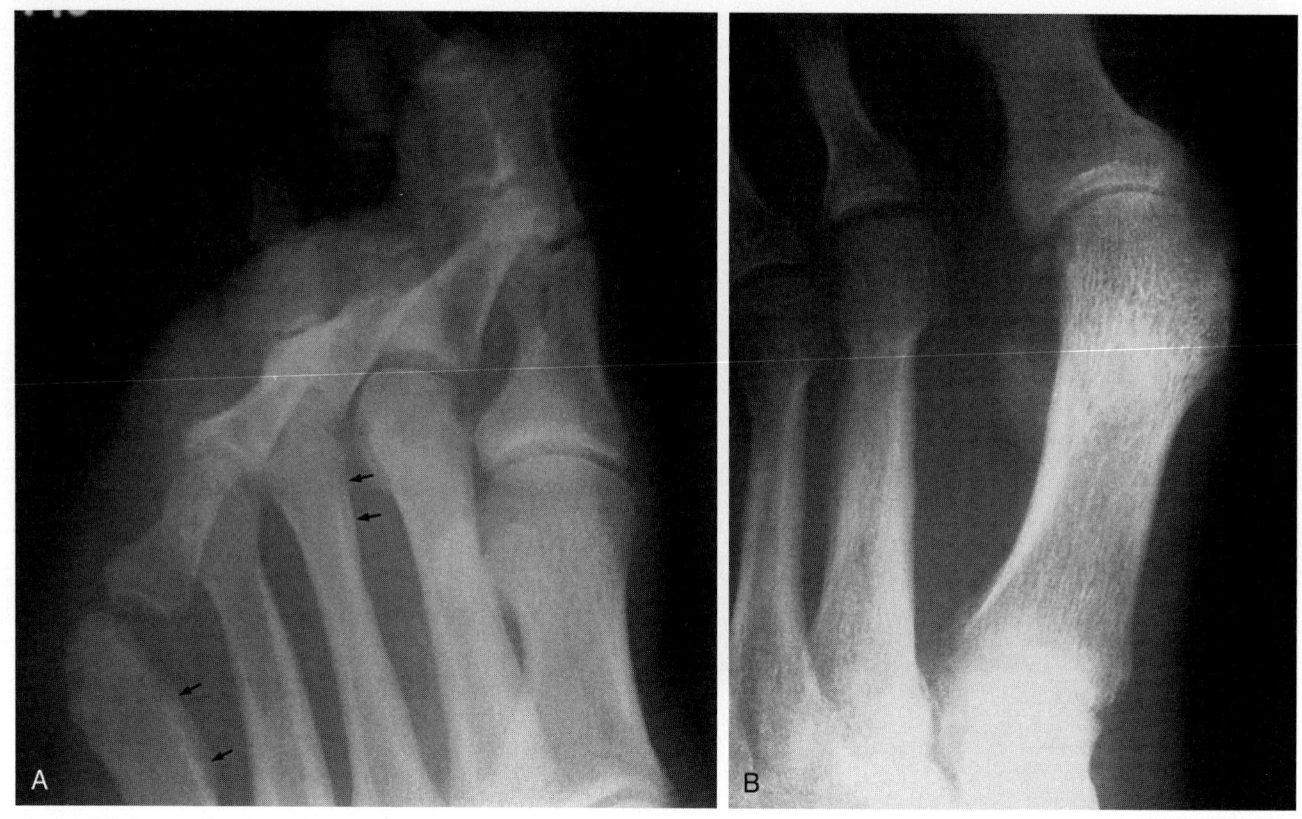

Figure 22–26. Involvement of the metatarsophalangeal (MP) joints with Reiter's disease. *A*, Oblique view of the right foot shows involvement of the second through the fifth MP joints. The fifth metatarsal head shows erosions. There is periostitis in the third and fifth metatarsals *(arrows)*. *B*, Anteroposterior view of the first MP joint from another patient with Reiter's disease shows inflammatory and erosive changes. Note the absence of osteopenia.

in reactive arthritis is identical to that seen in PA. The lumbar spine is involved in about one third of the patients, with changes characterized by paravertebral ossifications (nonmarginal syndesmophytes) with an asymmetric distribution. The most commonly affected segments of the spine are the lower thoracic and upper lumbar regions. The cervical spine is very rarely involved.

Differentiating reactive arthritis from PA on radiographs alone may be impossible. The scarcity of findings in the upper extremities, particularly in the DIP joints of the hand, may help in establishing the diagnosis of reactive arthritis.

ARTHRITIS ASSOCIATED WITH INFLAMMATORY BOWEL DISEASE

Essentially two types of joint problems affect patients with Crohn's disease and ulcerative colitis:

1. Peripheral migratory arthralgias or arthritis involving mainly the joints in the lower extremities. This occurs in 10% to 20% of patients, and it commonly involves the knees, ankles, elbows, and wrists. The arthritis is usually asymmetric, nondestructive, and frequently transient. The arthritic activity tends to parallel the flares in the underlying bowel disease. In ulcerative colitis, surgical removal of the diseased colon results in improvement of the arthritic condition.

2. Spondylitis and sacroiliitis affecting about 10% of patients with inflammatory bowel disease. It is frequently asymptomatic. The incidence is similar in both Crohn's disease and ulcerative colitis. The spondylitis is clinically and radiographically similar to AS (Fig. 22–29). The spondylitis in inflammatory bowel disease progresses independent of the bowel disease and a colectomy does not stop its progression.

SAPHO SYNDROME (STERNOCLAVICULAR HYPEROSTOSIS)

The acronym SAPHO was coined by Charmot and colleagues in 1987. It refers to synovitis, acne, palmoplantar pustulosis, hyperostosis, and osteitis. The most common site of the disease is the upper anterior chest wall. Extrathoracic involvement includes sacroiliitis, hyperostosis of the vertebral bodies, and sclerotic or lytic lesions with periosteal new bone formation involving the metaphysis and diaphysis of the long bones. SAPHO seems to be closely related to chronic relapsing multifocal osteomyelitis (CRMO) in children. Cultures are negative in both conditions.

Figure 22-27. Enthesitis (enthesopathy) at the insertion of the Achilles tendon and plantar fascia in patients with Reiter's disease. *A,* Bone scan in a patient with Reiter's disease. There is increased uptake in the calcaneal tuberosity at the insertion of the Achilles tendon and plantar fascia. *B,* Sagittal T1-weighted magnetic resonance image in a patient with Reiter's disease. Note the Achilles tendon thickening at its insertion. *C,* Lateral view of the calcaneus shows a large erosion at the insertion of the Achilles tendon *(arrows).* There is also a plantar inflammatory spur with eroded undersurface. *D,* A large plantar inflammatory spur associated with erosions at its base.

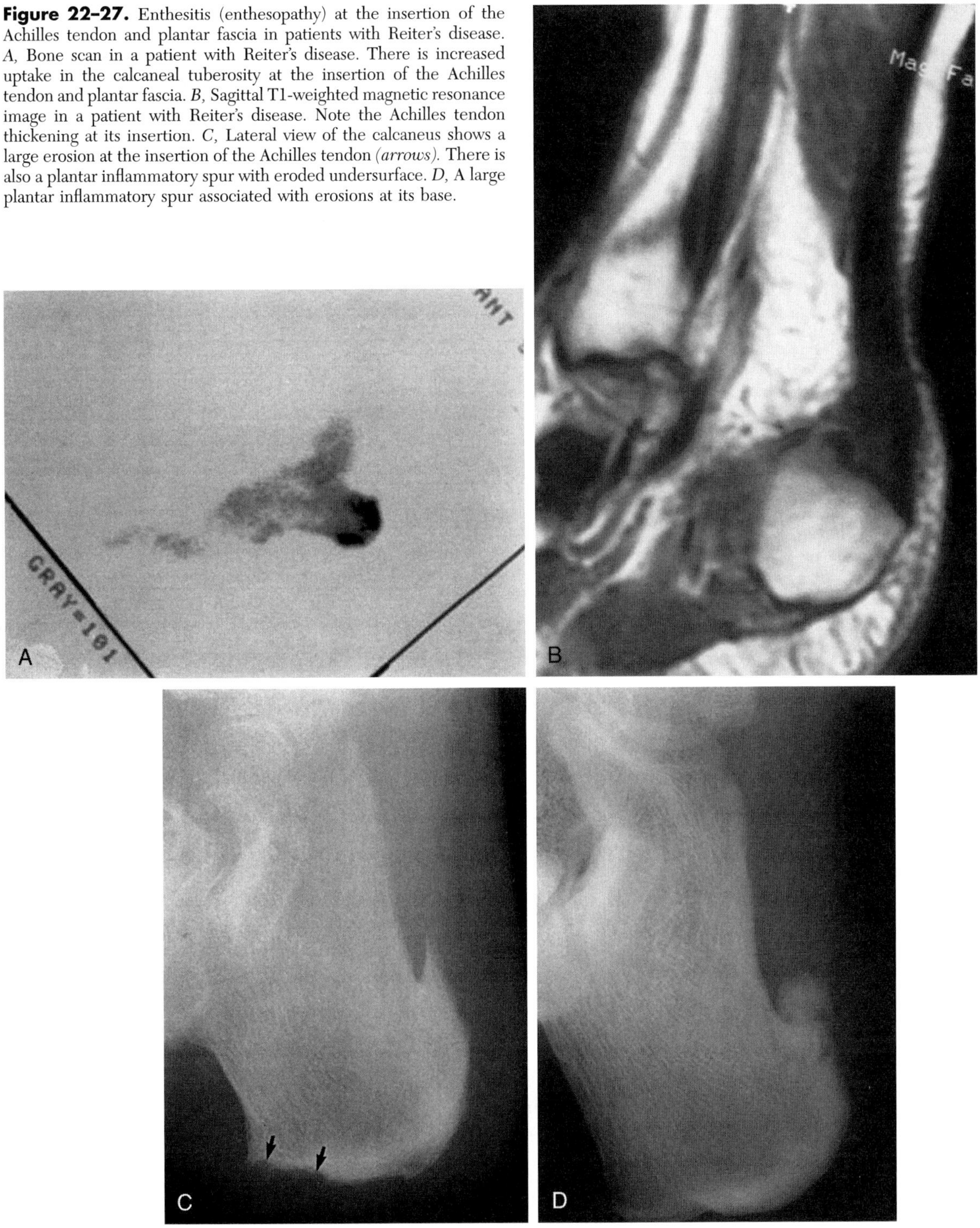

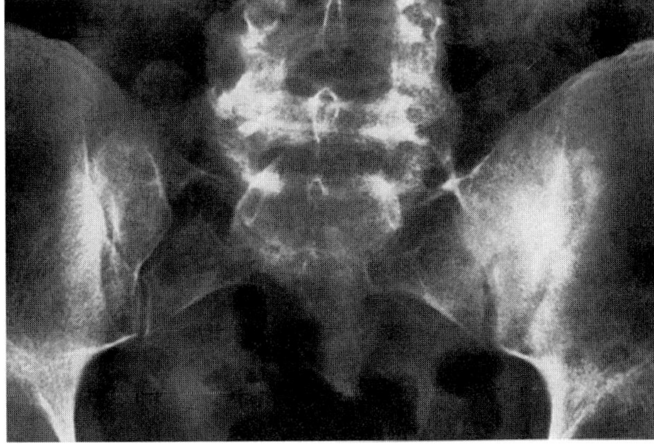

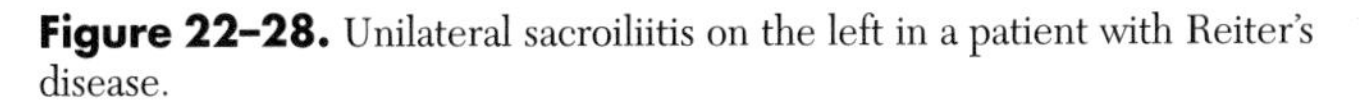

Figure 22-28. Unilateral sacroiliitis on the left in a patient with Reiter's disease.

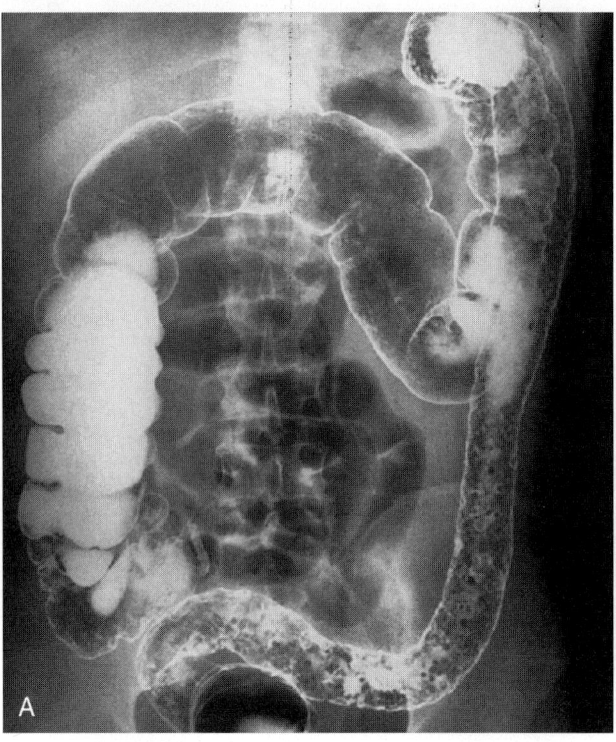

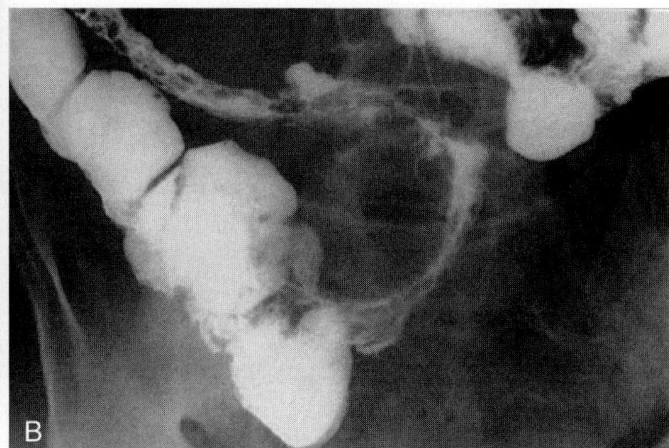

Figure 22–29. Sacroiliitis in inflammatory bowl disease. *A*, Patient with ulcerative colitis and sacroiliitis. *B*, Patient with Crohn's disease and sacroiliitis.

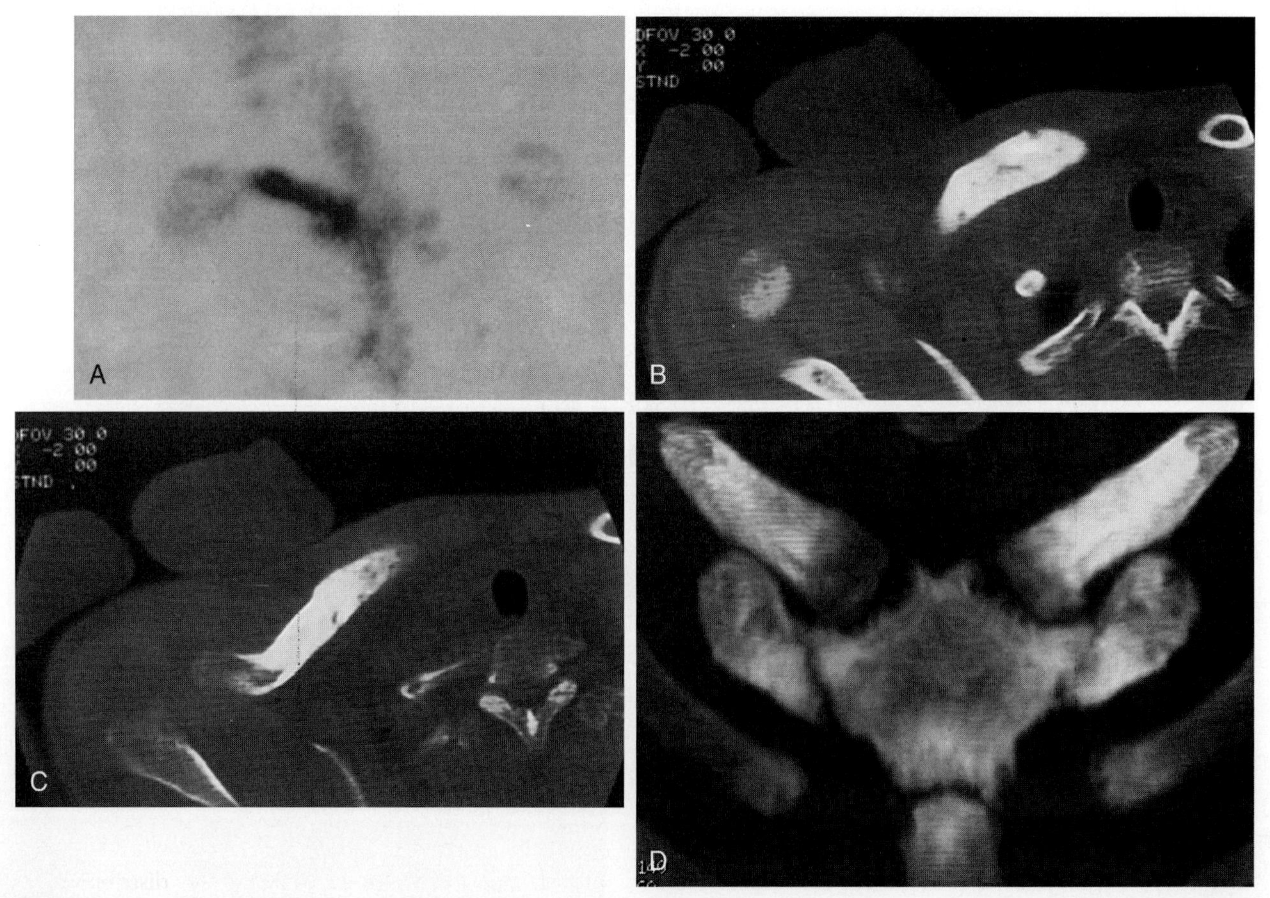

Figure 22–30. SAPHO (synovitis-acne-pustulosis-hyperostosis-osteitis syndrome) affecting the clavicle in a middle-aged man who was diagnosed initially with chronic osteomyelitis or neoplasm and was subjected to two bone biopsies of the right clavicle. *A*, Bone scan shows increased uptake involving the proximal two thirds of the clavicle. *B* and *C*, Computed tomography sections through the clavicle reveal hyperostosis and sclerosis of the clavicle. *D*, Oblique coronal reconstructed CT image of the anterior chest wall in another patient with SAPHO. There is involvement of the clavicles and the first costal cartilage bilaterally. The manubriosternal joint is irregular and sclerotic.

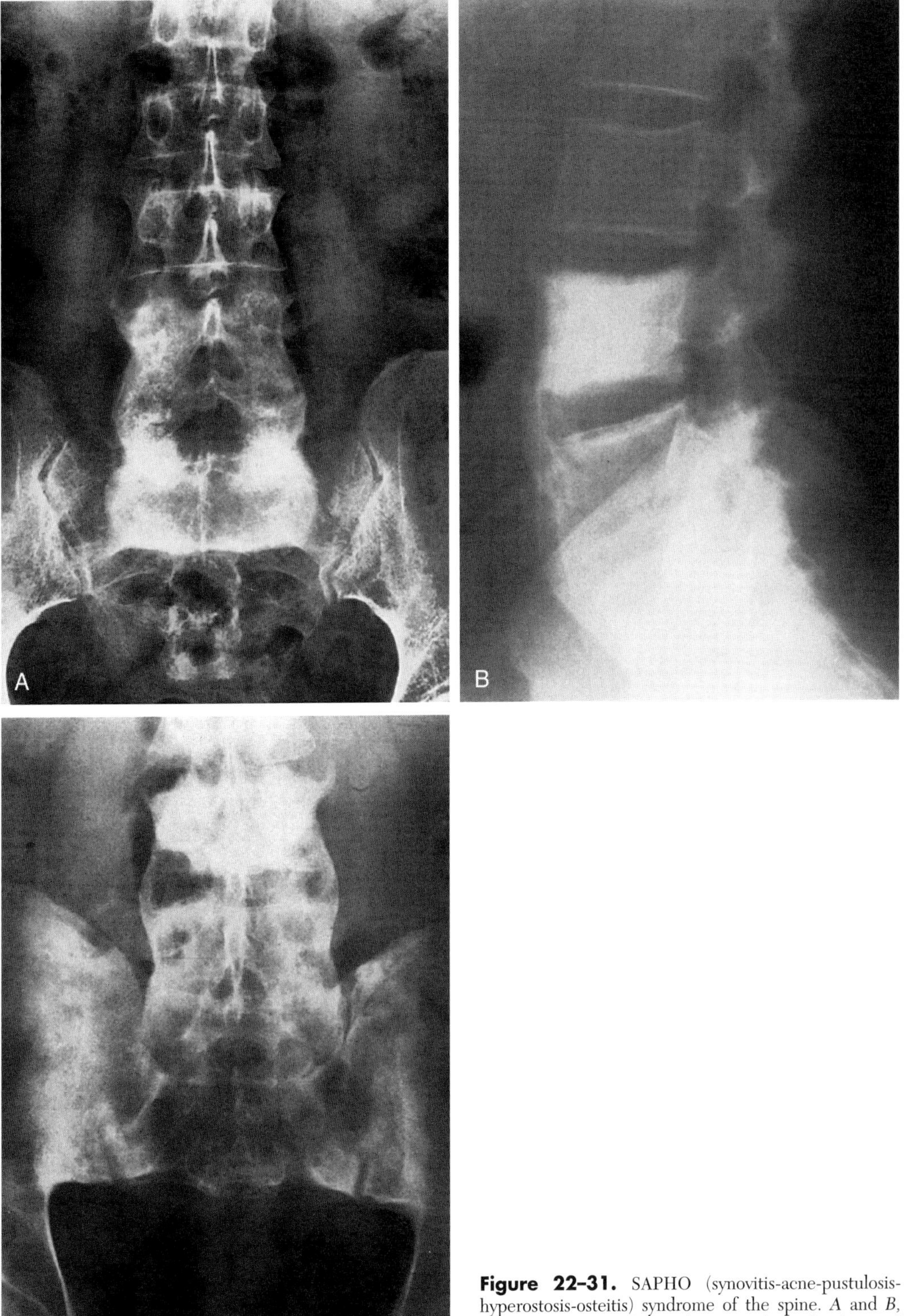

Figure 22–31. SAPHO (synovitis-acne-pustulosis-hyperostosis-osteitis) syndrome of the spine. *A* and *B*, Anteroposterior and lateral views of the lumbosacral spine reveal dense sclerosis and hyperostosis of L4 and L5. Syndesmophytes join the two vertebrae. The sacroiliac (SI) joints are not involved. *C*, Bilateral SI joint involvement in another patient with SAPHO.

Histologically, early bone changes are similar to osteomyelitis: acute inflammation, edema, and periosteal reaction. Late changes are similar to Paget's disease of the bone with sclerotic trabeculae and marrow fibrosis.

Plain radiography is usually adequate to detect and characterize these lesions, but bone scintigraphy is probably cost-effective to evaluate the distribution and activity of the inflammatory process. Computed tomography with multiplanar reconstruction is the best method to define lesions in the anterior chest wall (Fig. 22–30).

The changes in the anterior chest wall are characterized by periostitis and sclerosis of the sternum, clavicles, ribs,

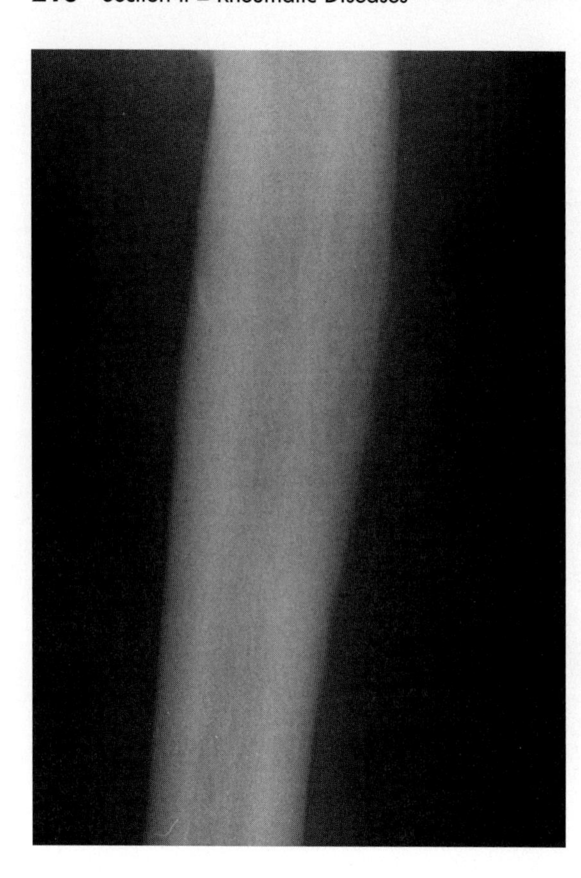

Figure 22–32. SAPHO (synovitis-acne-pustulosis-hyperostosis-osteitis) syndrome changes in the left femur. There is cortical thickening posteriorly and a mild periosteal reaction.

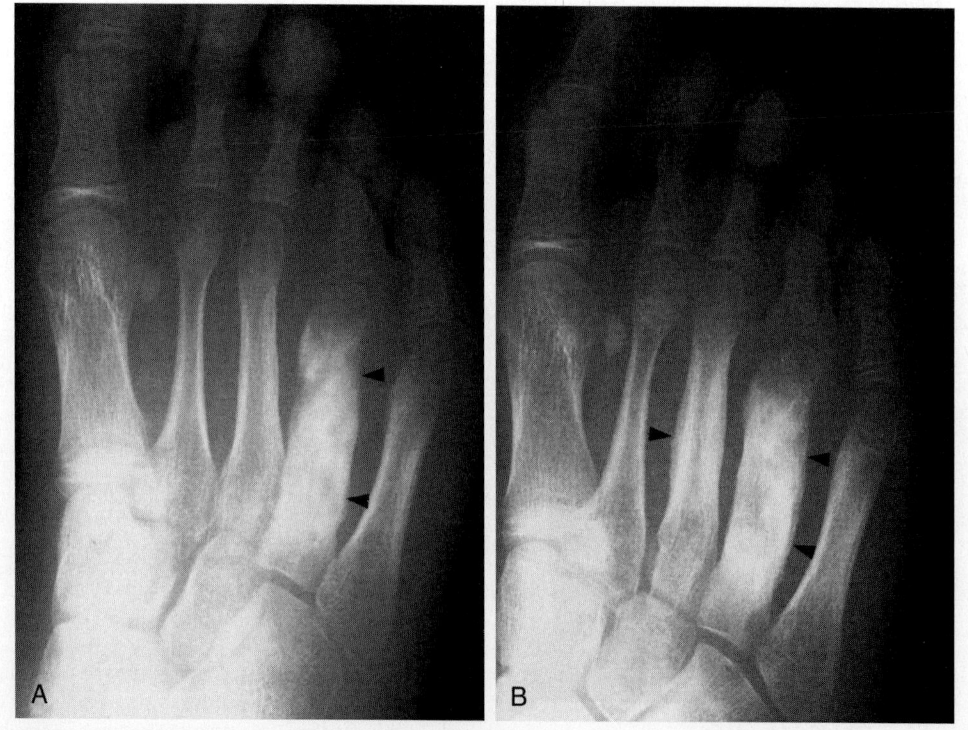

Figure 22–33. CRMO (chronic relapsing multifocal osteomyelitis) of the right foot in a 9-year-old girl presenting with foot pain and swelling. Biopsies of the bone and cultures did not yield any organisms. Anteroposterior views *(A, B* and *C)* of the left foot taken over 10 months reveal progressive osteitis involving the second through the fourth metatarsals *(arrows)*. (Cases contributed by Dr. James Tarter from St. Louis, Mo.)

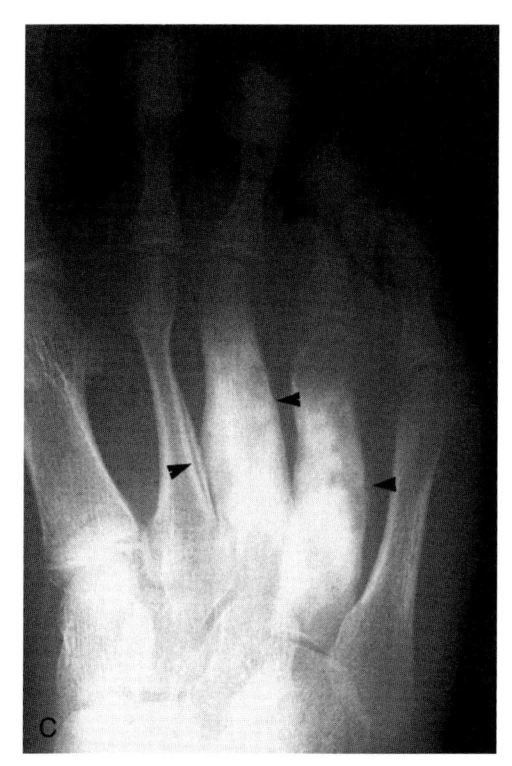

Figure 22–33 *Continued.* See legend on opposite page.

and costal cartilages. Hyperostosis and sclerosis around the sternoclavicular joint are the most common changes (see Fig. 22–30). There are also features of an erosive arthritis with subchondral erosion, sclerosis, and enthesopathy. Ossification and erosions at the ligamentous insertions, particularly at the clavicle, are frequently observed. The changes in the vertebrae are those of striking hyperostosis and spondylitis. Segmental and noncontiguous involvement is common, but contiguous involvement similar to AS is seen occasionally. Sclerosis and erosion subjacent to the end plate and at the anterior vertebral corner are reported to be most characteristic (Fig. 22–31).

Involvement of the sacroiliac joints is commonly bilateral. Erosions and subchondral sclerosis are the predominant features. Ankylosis is seen only in the advanced cases.

The changes in the long bones are those of nonspecific osteitis showing endosteal sclerosis and thick periosteal reaction (Fig. 22–32). In children with CRMO, osteitis (Fig. 22–33) as well as lytic permative lesions, can be seen, especially in the metaphyses.

Recommended Readings

Becker NJ, DeSmet AA, Cathcart-Rake W, et al: Psoriatic arthritis affecting the manubriosternal joint. Arthritis Rheum 29:1029, 1986.

Braunstein EM, Martel W, Moidel R: Ankylosing spondylitis in men and women: A clinical and radiographic comparison. Radiology 144:91, 1982.

Chigira M, Maehara S, Nagase M, et al: Sternocostoclavicular hyperostosis: A report of nineteen cases, with special reference to etiology and treatment. J Bone Joint Surg Am 68:103, 1986.

Dihlmann W: Current radiodiagnostic concept of ankylosing spondylitis. Skeletal Radiol 4:179, 1979.

Dwosh IL, Resnick D, Becker MA: Hip involvement in ankylosing spondylitis. Arthritis Rheum 19:683, 1976.

Emery RJH, Ho EKW, Leong JCY: The shoulder girdle in ankylosing spondylitis. J Bone Joint Surg Am 73:1526, 1991.

Fang D, Leong JCY, Ho EKW, et al: Spinal pseudarthrosis in ankylosing spondylitis. J Bone Joint Surg Br 70:443, 1988.

Fox MW, Onofrio BM, Kilgore JE: Neurological complications of ankylosing spondylitis. J Neurosurg 78:871, 1993.

Goldberg AL, Keaton NL, Rothfus WE, et al: Ankylosing spondylitis complicated by trauma: MR findings correlated with plain radiographs and CT. Skeletal Radiol 22:333, 1993.

Jurik AG, Helmig O, Graudal H: Skeletal disease, arthroosteitis, in adult patients with pustulosis palmoplantaris. Scand J Rheum 70(Suppl):1, 1988.

Kasperczyk A, Freyschmidt J: Pustulotic arthroosteitis: Spectrum of bone lesions with palmoplantar pustulosis. Radiology 191:207, 1994.

Khan MA: C: Seronegative spondyloarthropathies. Ankylosing spondylitis. In Klippel JH (ed): Primer on the rheumatic diseases, 11th ed. Atlanta, Arthritis Foundation, 1997.

Martel W, Braunstein EM, Borlaza G, et al: Radiologic features of Reiter disease. Radiology 132:1, 1979.

Martel W, Stuck KJ, Dworin AM, et al: Erosive osteoarthritis and psoriatic arthritis: A radiologic comparison in the hand, wrist, and foot. AJR Am J Roentgenol 134:125, 1980.

Reith JD, Bauer TW, Schils JP: Osseous manifestations of SAPHO (synovitis, acne, pustulosis, hyperostosis, osteitis) syndrome. Am J Surg Path 20:1368, 1996.

Resnick D, Dwosh IL, Goergen TG, et al: Clinical and radiographic abnormalities in ankylosing spondylitis: A comparison of men and women. Radiology 119:293, 1976.

Sonozaki H, Mitsui H, Miyanaga Y, et al: Clinical features of 53 cases with pustulotic arthro-osteitis. Ann Rheum Dis 40:547, 1981.

Sparling MJ, Bartleson JD, McLeod RA, et al: Magnetic resonance imaging of arachnoid diverticula associated with cauda equina syndrome in ankylosing spondylitis. J Rheumatol 16:1335, 1989.

Taurog JD: Seronegative spondyloarthropathies. A: Epidemiology, pathology, and pathogenesis. In Klippel JH (ed): Primer on the rheumatic diseases, 11th ed. Atlanta, Arthritis Foundation, 1997.

Vuillemin-Bodaghi V, Parlier-Cuau C, Cotton A, et al: MR imaging of vertebral involvement in SAPHO syndrome (RSNA 1998). Radiology 209:500, 1998.

CHAPTER 23

Scleroderma and CRST Syndrome

Georges Y. El-Khoury, MD, Mark D. Stanley, MD, and D. Lee Bennett, MD

Scleroderma (progressive systemic sclerosis [PSS]) is a systemic connective tissue disease of unknown etiology. It involves the skin, heart, lungs, gastrointestional tract, kidneys, muscles, and joints. Recent animal studies suggest that increased mast cell activity and production of growth factors may be responsible for the endothelial proliferation and excessive collagen synthesis and fibrosis that are found in scleroderma. In most organs, fibrosis is the critical damaging factor in scleroderma. Esophageal dysfunction is very common in scleroderma, manifesting with hypomotility, reflux, and stricturing (Fig. 23–1*A*). Patients with small bowel hypomotility (Fig. 23–1*B*) present clinically with abdominal pain, distension, vomiting, and, eventually, malabsorption. Interstitial lung disease and fibrosing alveolitis are also common in scleroderma (Fig. 23–1*C*).

A variant of scleroderma, known as the CRST syndrome, is characterized by calcinosis, Raynaud's phenomenon, sclerodactyly, and telangiectasia. Patients with CRST syndrome have more extensive calcinosis (Fig. 23–2) than patients with scleroderma do, but their prognosis is better than those with scleroderma.

RADIOGRAPHIC CHANGES

Radiographic changes of scleroderma (musculoskeletal) include:

1. Flexion contractures of the hands.
2. Soft tissue atrophy in the distal fingers (Fig. 23–3).
3. Calcific deposits in the hands, especially on the palmar aspects of the fingers distally (Figs. 23–3 and 23–4*A*).

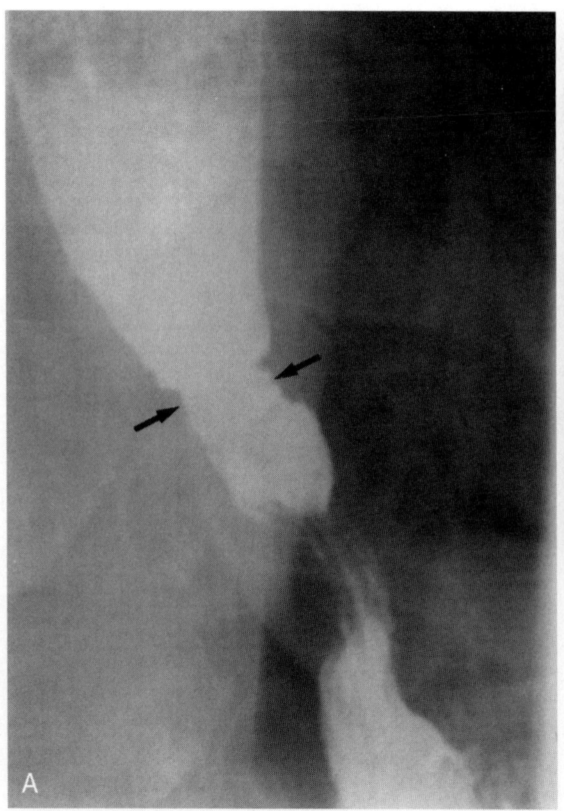

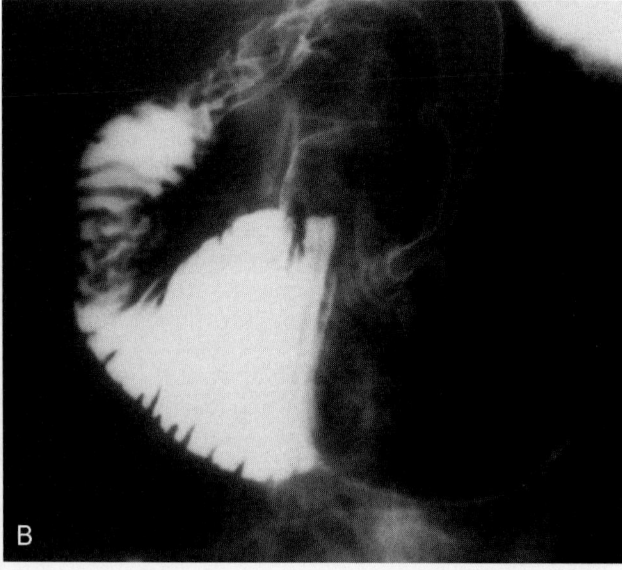

Figure 23–1. Changes in the esophagus, duodenum, and lung owing to scleroderma. *A*, Distal esophageal stricture *(arrows)* and proximal dilation owing to fibrosis in the muscularis causing decreased motility and reflux. *B*, Moderately dilated third part of the duodenum owing to fibrosis in the muscularis causing hypomotility.

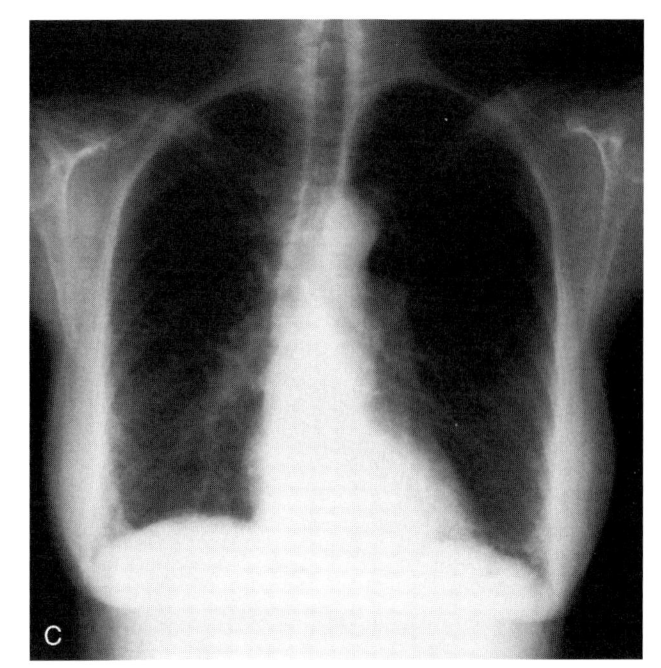

Figure 23–1 *Continued. C,* Fibrotic changes in both lung bases are nonspecific findings but are often seen in scleroderma.

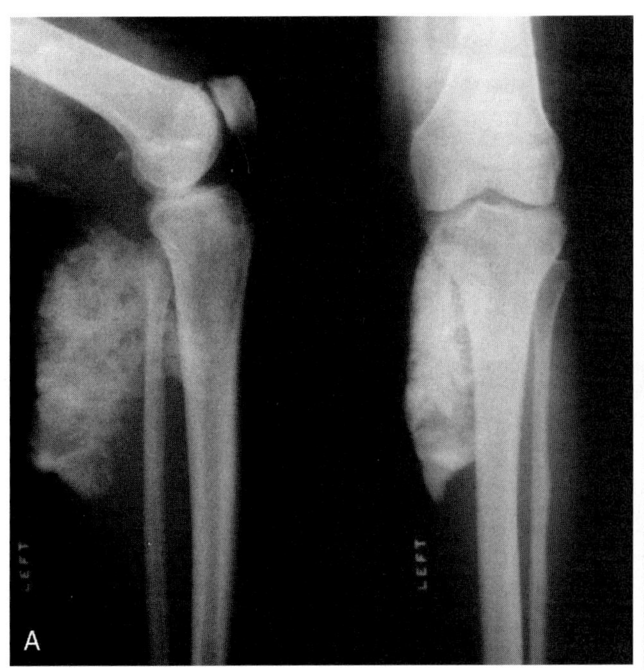

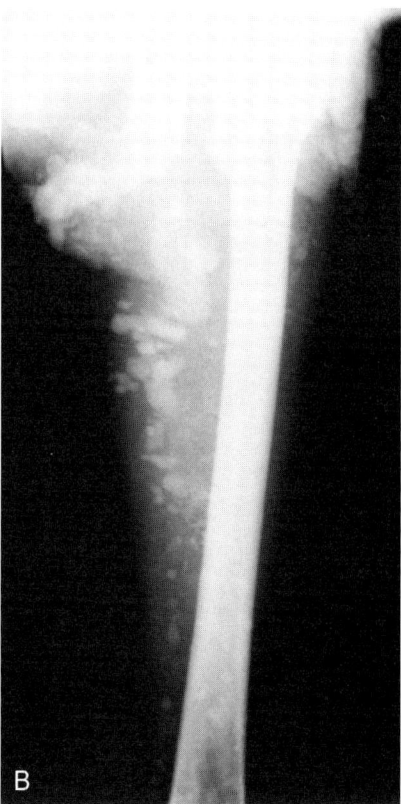

Figure 23–2. Calcinosis, Raynaud's phenomenon, sclerodactyly, and telangiectasia (CRST) syndrome. *A,* Anteroposterior (AP) and lateral radiographs of the leg show calcinosis in calf muscles. *B,* AP radiograph of the left thigh demonstrates extensive calcinosis in the soft tissue of the thigh and around the hip.

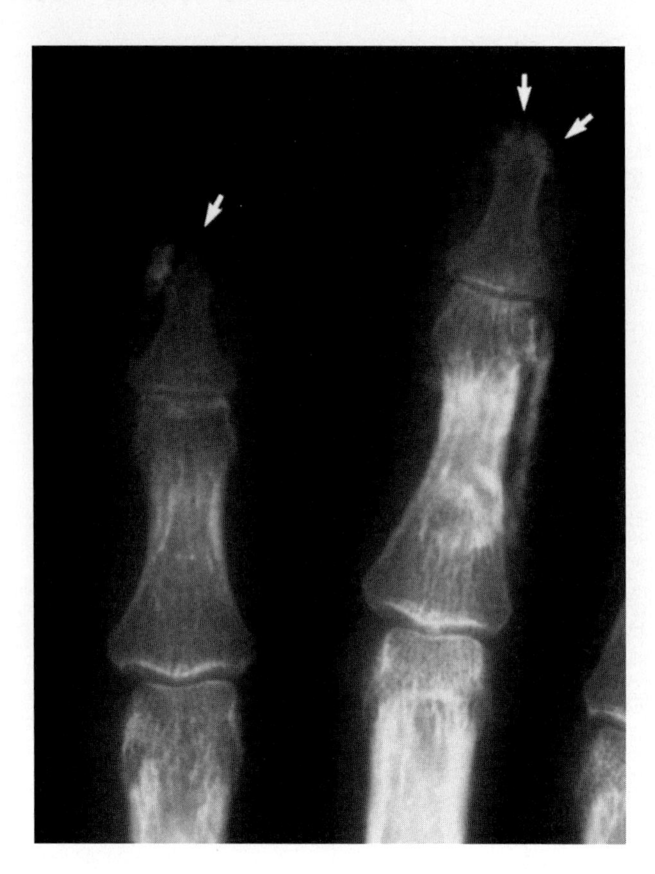

Figure 23–3. Soft tissue atrophy at the distal fingers in scleroderma. This finding is best detected clinically but can also be seen on radiographs *(arrows)*. Note the presence of calcinosis in the fingers, also.

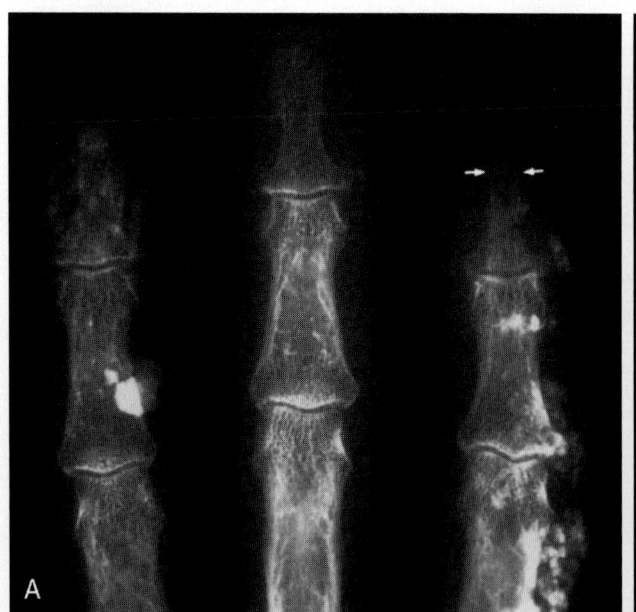

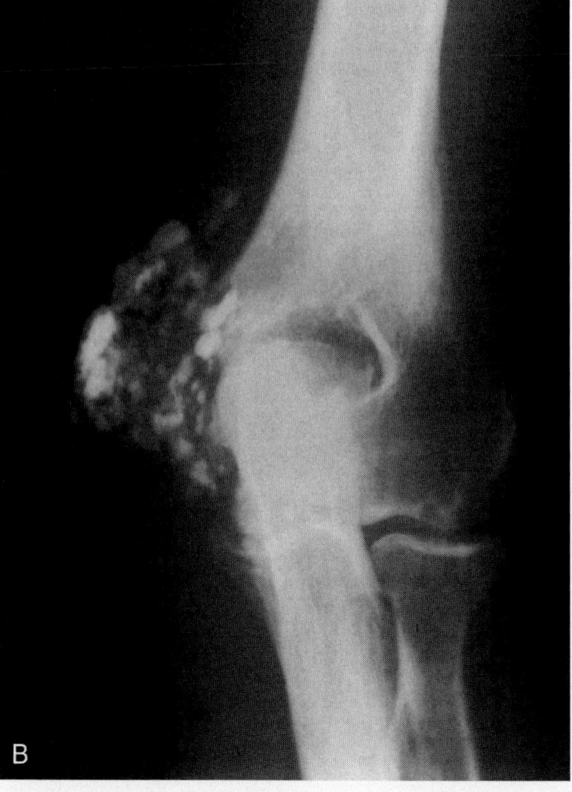

Figure 23–4. Calcific deposits (calcinosis). *A*, Posteroanterior view of the second through the fourth fingers shows calcinosis in the index and fourth finger. There is also soft tissue atrophy at the tip of the index finger as well as bone resorption of the tuft *(arrows)*. *B*, Anteroposterior view of the elbow demonstrates calcinosis around the medial epicondyle in a patient with scleroderma.

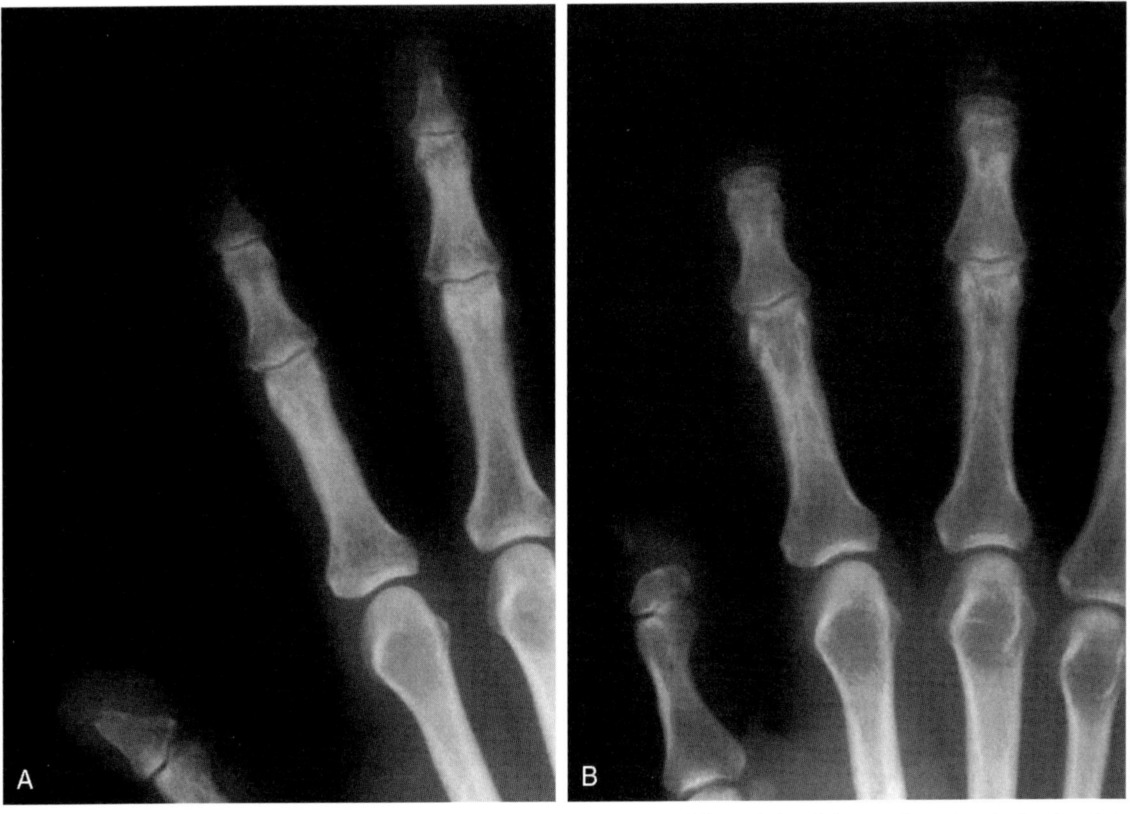

Figure 23–5. Tuft erosions in scleroderma with progression. *A,* Posteroanterior view of the right hand shows tuft erosions in the thumb, index, and long fingers. *B,* Follow-up examination 3 years later shows more extensive erosive changes that affect most of the distal phalanges.

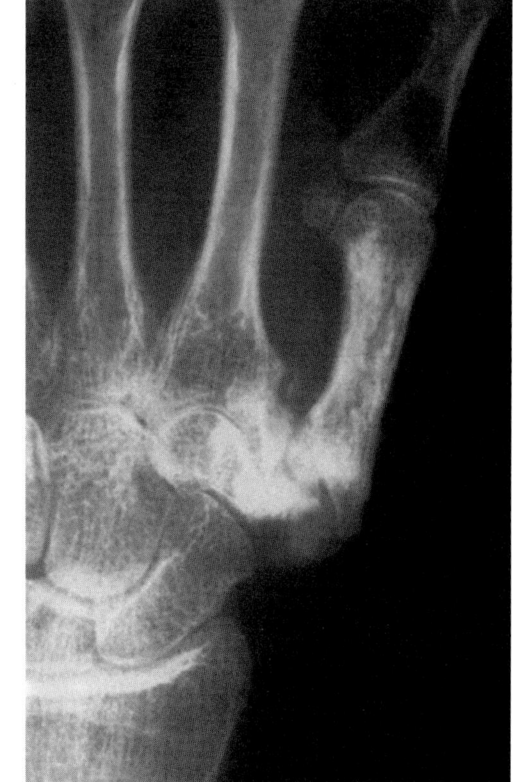

Figure 23–6. Erosive and destructive arthritis at the first carpometacarpal joint is thought to be characteristic of scleroderma.

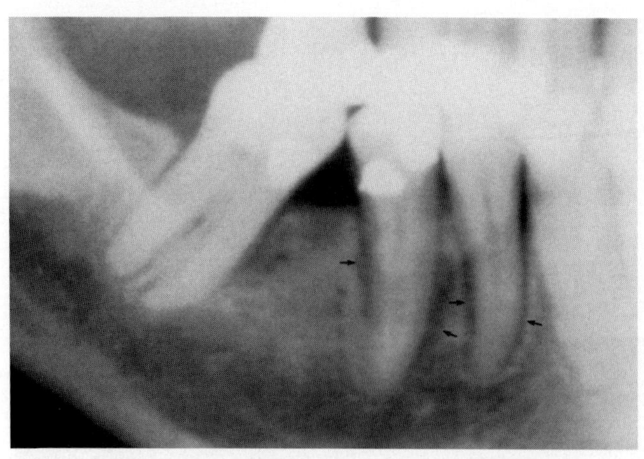

Figure 23–7. Periodontal membrane thickening in scleroderma (*arrows*).

These deposits can also occur in joints (Fig. 23–4*B*), tendons, and intervertebral discs.

4. Bone resorption over the palmar aspect of the tufts of the fingers initially. Later, entire tufts or distal phalanges may resorb (Fig. 23–5). Rarely, resorption of the middle and proximal phalanges can occur. Occasionally, transverse bands of bone resorption in the midshaft of the distal phalanges, simulating acrolysis caused by vinyl chloride toxicity, can be seen with scleroderma.

5. Bone resorption of the posterior superior aspects of the ribs and at the mandibular angles.

6. Joint changes resembling rheumatoid arthritis. These have been described, but are rare. Selective erosive changes at the first carpometacarpal joint have been described in patients with long-standing scleroderma (Fig. 23–6) and are considered characteristic of the disease.

7. Thickening of the periodontal membrane (Fig. 23–7). This is also a characteristic finding of scleroderma.

Recommended Readings

Bassett LW, Blocka KLN, Furst DE, et al: Skeletal findings in progressive systemic sclerosis (scleroderma). AJR Am J Roentgenol 136:1121, 1981.

Brower AC, Resnick D, Karlin C, et al: Unusual articular changes of the hand in scleroderma. Skeletal Radiol 4:119, 1979.

Claman HN: On scleroderma: Mast cells, endothelial cells, and fibroblasts. JAMA 262:1206, 1989.

Resnick D, Scavulli JF, Goergen TG, et al: Intra-articular calcification in scleroderma. Radiology 124:685, 1977.

Resnick D, Greenway G, Vint VC, et al: Selective involvement of the first carpometacarpal joint in scleroderma. AJR Am J Roentgenol 131:283, 1978.

CHAPTER 24

Systemic Lupus Erythematosus

Georges Y. El-Khoury, MD, Mark D. Stanley, MD, and D. Lee Bennett, MD

OVERVIEW

Systemic lupus erythematosus (SLE) is a serious multi-system disease with female predominance of 9:1. The diagnosis is based on the presence of a number of clinical and laboratory findings, including a butterfly rash, Raynaud's phenomenon, alopecia, arthralgias, pleuritis, pericarditis, anemia, and positive LE cell preparation. Articular symptoms are the most common clinical manifestation of SLE as well as the most frequent presentation of the disease.

Radiographic Changes

In SLE, there is typically a relative absence of radiologic changes despite the presence of long-standing arthritic symptoms. Radiographic abnormalities most commonly include periarticular soft tissue swelling and osteopenia (Fig. 24–1). Other abnormalities include alignment abnormalities of the fingers (Fig. 24–2) (swan neck deformities) and subluxation at the interphalangeal joint of the thumb, which are usually reducible, much like Jaccoud's arthropathy. Joint space narrowing and marginal erosions are not

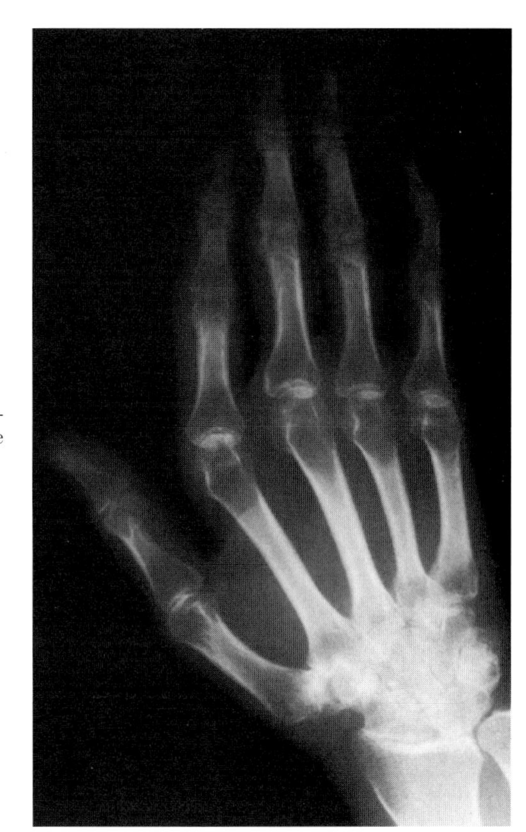

Figure 24–1. Changes in the hand caused by systemic lupus erythematosus. Posteroanterior view of the left hand illustrates severe periarticular osteopenia and soft tissue swelling around the proximal interphalangeal joints of the index and long fingers.

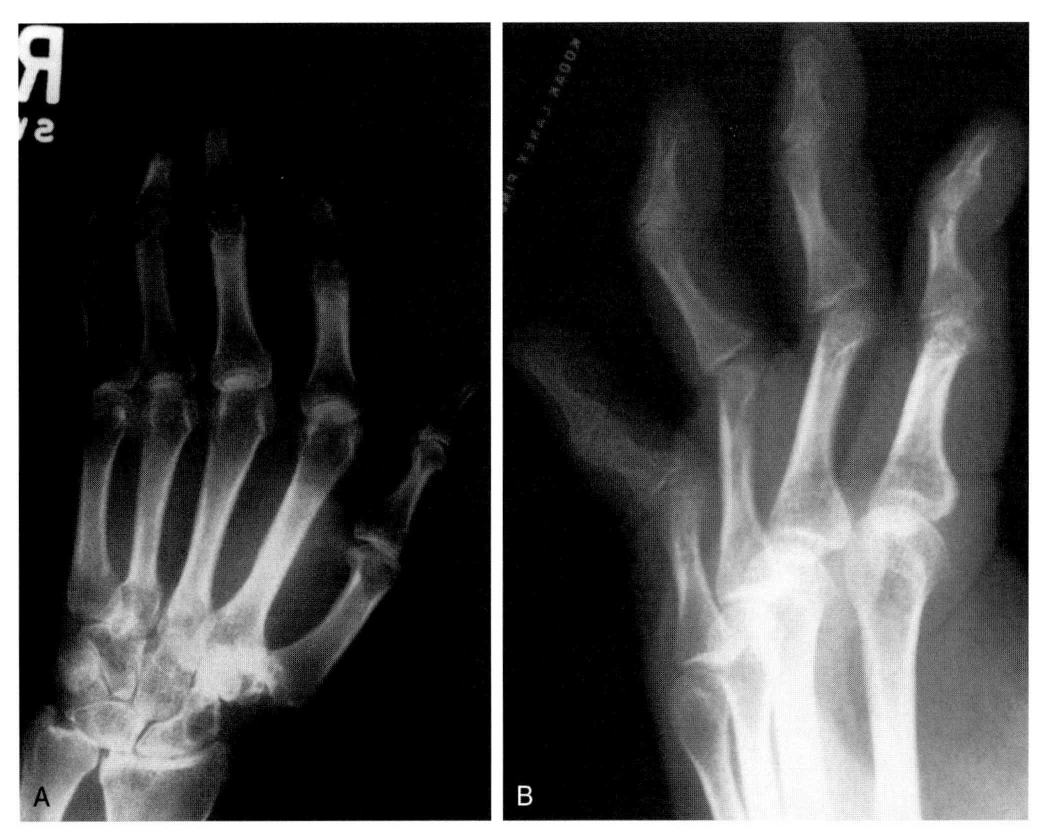

Figure 24–2. Reducible "swan neck" deformities in the finger of the right hand in a patient with systemic lupus erythematosus. *A*, Posteroanterior view of the right hand shows periarticular osteopenia, but the finger deformities are not obvious. *B*, Oblique view of the right hand shows the typical "swan neck" deformities in fingers 2 through 5.

features of SLE. Periarticular calcifications and sclerosis of the terminal phalanges (acral sclerosis) can also be seen with SLE. Avascular necrosis of the femoral heads, femoral condyles, tibial plateaus, and talar domes is fairly common in patients with SLE, but it is not totally clear whether it is due to the SLE or due to the steroid treatment. Both premenopausal and postmenopausal women with SLE treated with steroids are at risk for osteoporosis.

Recommended Readings

Glickstein M, Neustadter L, Dalinka M, et al: Periosteal reaction to systemic lupus erythematosus. Skeletal Radiol 15:610, 1986.
Leskinen RH, Skrifvars BV, Laasonen LS, et al: Bone lesions in systemic lupus erythematosus. Radiology 153:349, 1984.
Russell AS, Percy JS, Rigal WM, et al: Deforming arthropathy in systemic lupus erythematosus. Ann Rheum Dis 33:204, 1974.
Weissman BN, Rappoport AS, Sosman JL, et al: Radiographic findings in the hands in patients with systemic lupus erythematosus. Radiology 126:313, 1978.

CHAPTER 25

Mixed Connective Tissue Disease

Georges Y. El-Khoury, MD, Mark D. Stanley, MD, and D. Lee Bennett, MD

Mixed connective tissue disease (MCTD) is a distinct entity with overlapping clinical features of systemic lupus erythematosus, scleroderma, polymyositis/dermatomyositis, and rheumatoid arthritis. It is characterized by the presence of antibodies against a saline extractable nuclear antigen (ENA). In MCTD, there is lower incidence of renal involvement and a more favorable response to steroid therapy than is seen in any of the involved diseases alone.

Recommended Readings

Silver TM, Farber SJ, Bole GG, et al: Radiological features of mixed connective tissue disease and scleroderma–systemic lupus erythematosus overlap. Radiology 120:269, 1976.
Udoff EJ, Genant HK, Kozin F, et al: Mixed connective tissue disease: The spectrum of radiographic manifestations. Radiology 124:613, 1977.

CHAPTER 26

Polymyositis/Dermatomyositis

Georges Y. El-Khoury, MD, Mark D. Stanley, MD, and D. Lee Bennett, MD

OVERVIEW

Polymyositis and dermatomyositis are inflammatory myopathies characterized by proximal and often symmetrical muscle weakness. Polymyositis is seen after the second decade and is very rare in children, whereas dermatomyositis affects both children and adults with females being affected more than males. In children with dermatomyositis, subcutaneous calcifications can extrude through the skin, causing ulcerations and infections. Joint contractures can also occur in children. In adults with dermatomyositis, the incidence of associated malignancies (usually carcinomas) is believed by some to be increased, although this is controversial. The muscle enzyme, creatine kinase, is usually elevated during the active stage of the disease, but the definitive diagnosis is made by muscle biopsy.

IMAGING FINDINGS

Radiography

Calcinosis (soft tissue calcifications) is a frequent finding seen in up to 74% of patients with juvenile dermatomyositis (Fig. 26–1). Calcinosis can be subcutaneous or it can develop deep in the fascial planes between muscles. Subcutaneous calcinosis is about four times more common than the intramuscular calcinosis (Fig. 26–2). Subcutaneous calcinosis frequently involves the limbs diffusely, but when focal, the elbows, knees, and fingers are more involved (Fig. 26–3). Some investigators have classified the patterns of soft tissue calcifications into four types.

1. Deep linear
2. Deep calcareal (Fig. 26–4)

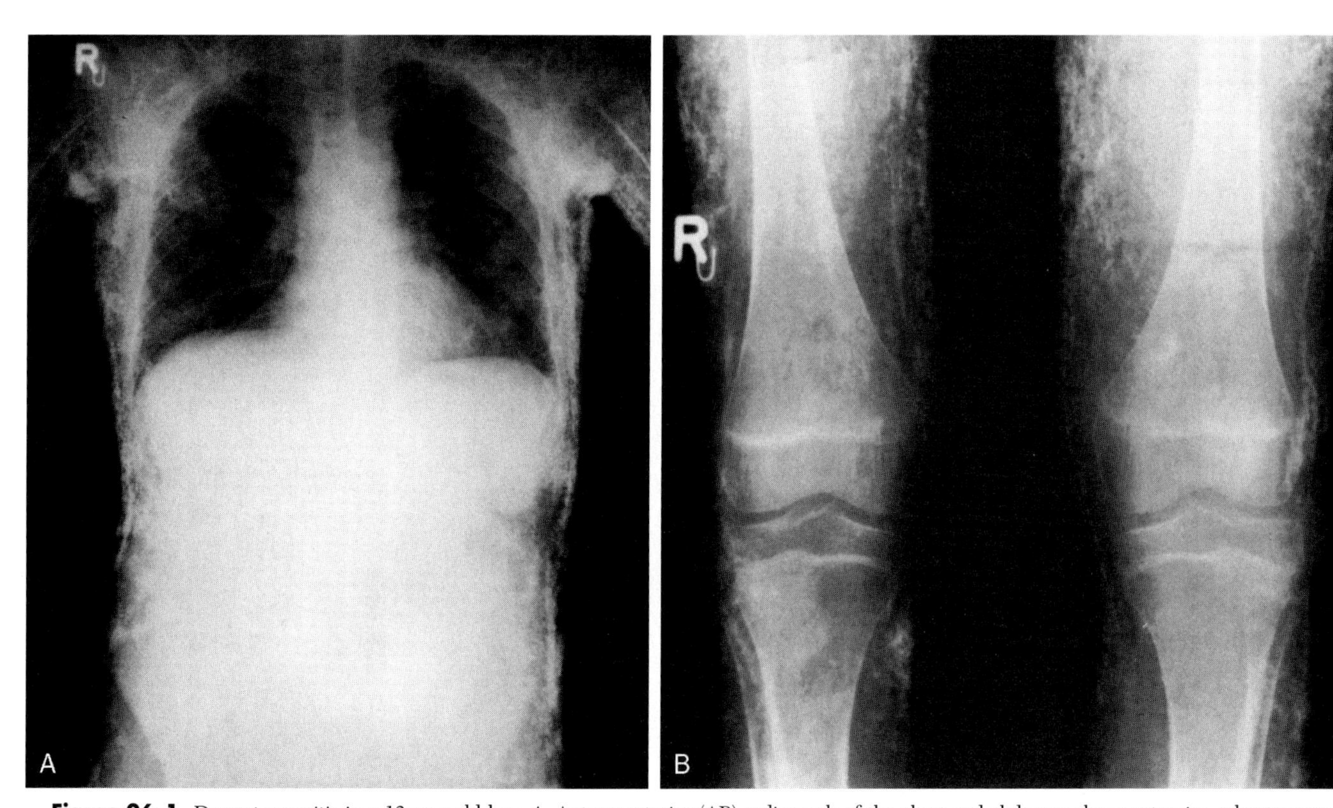

Figure 26–1. Dermatomyositis in a 12-year-old boy. *A,* Anteroposterior (AP) radiograph of the chest and abdomen shows extensive subcutaneous calcifications. *B,* AP view of both knees reveals similar changes.

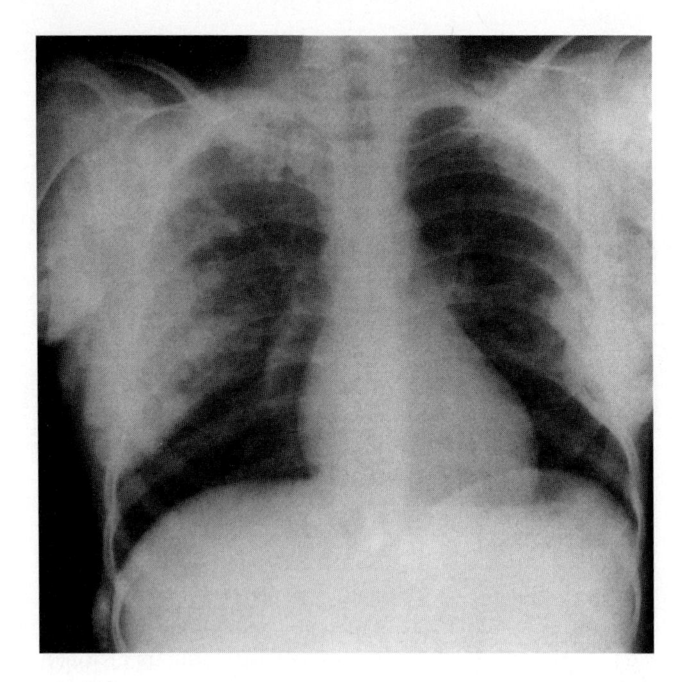

Figure 26–2. Intramuscular calcinosis in an 11-year-old boy. Posteroanterior film of the chest reveals large clumps of calcinosis in the soft tissues of the axilla and chest wall. These are believed to be intermuscular and intramuscular.

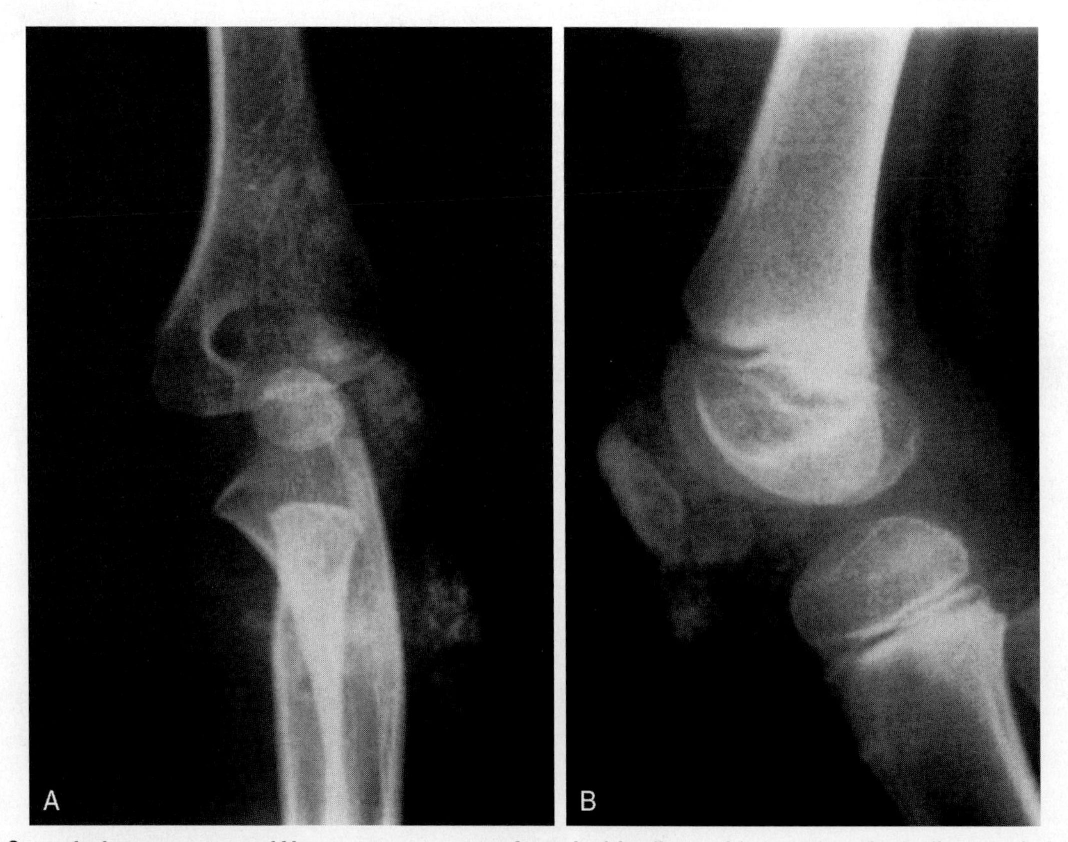

Figure 26–3. Focal calcinosis in a 6-year-old boy. *A,* Anteroposterior radiograph of the elbow and forearm shows focal collections of calcinosis around the elbow. *B,* Lateral view of the knee shows similar changes.

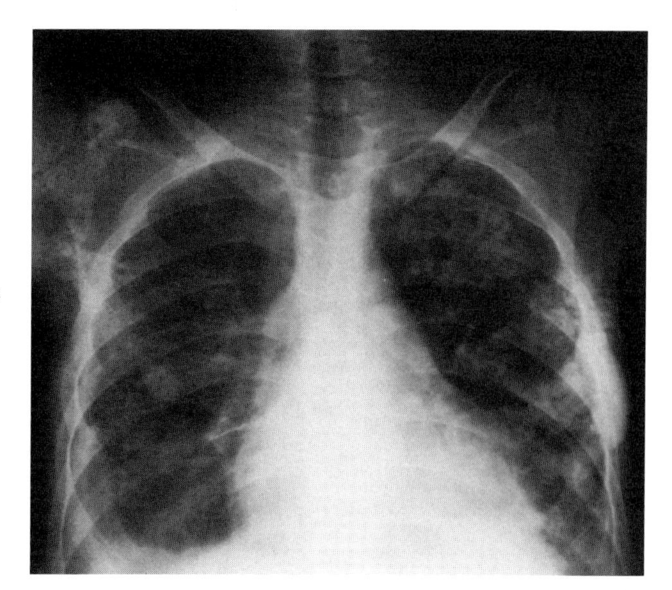

Figure 26–4. Deep calcareal lesions in the chest wall muscles of a 14-year-old patient with dermatomyositis.

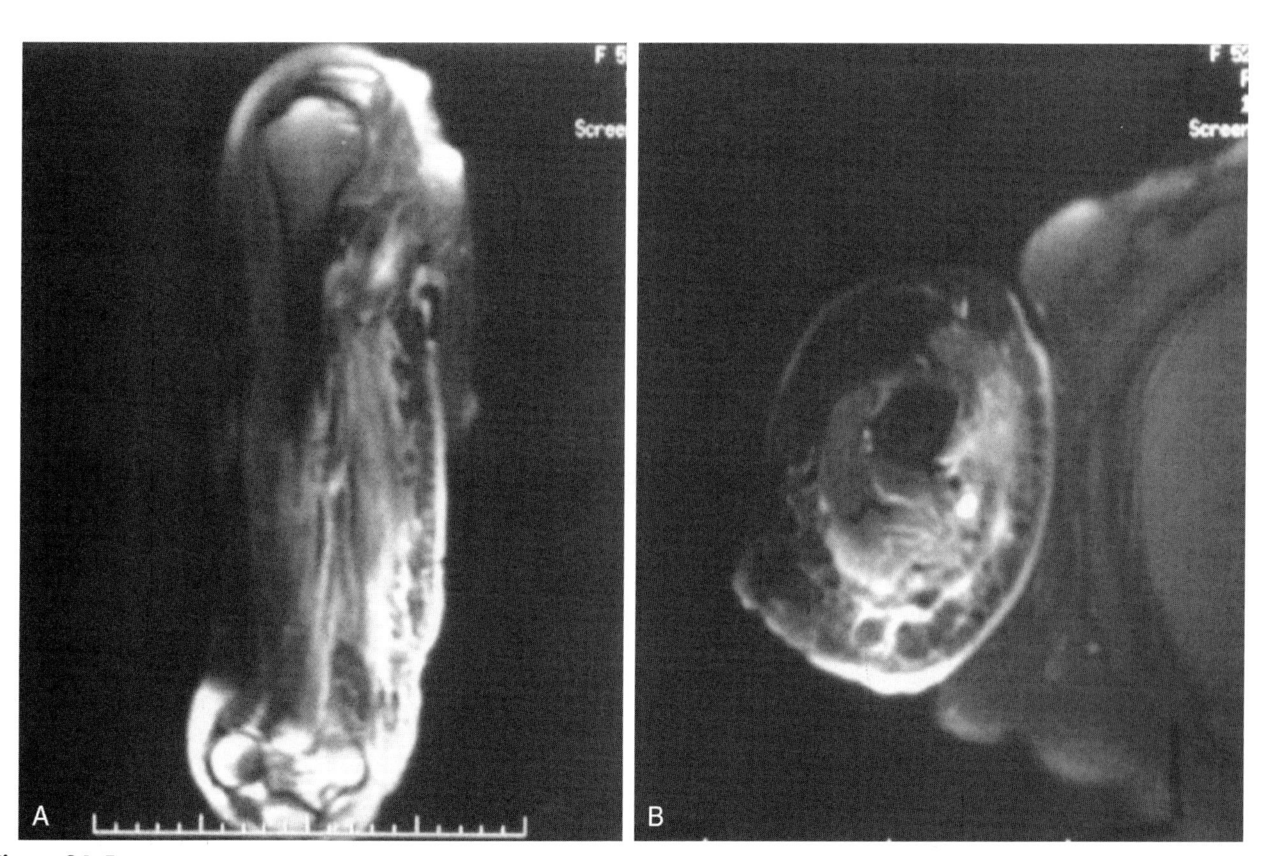

Figure 26–5. Magnetic resonance images show dermatomyositis in a 62-year-old woman. *A* and *B*, Coronal and axial T2-weighted images of the right arm show increased signal intensity in the muscles and subcutaneous tissues, suggesting inflammation and edema.

3. Superficial calcareal (see Fig. 26–3)
4. Lacy or reticular subcutaneous (see Fig. 26–1)

Some children with dermatomyositis develop acrolysis and calcinosis at the finger tips, much like scleroderma. A characteristic abnormality of the thumb in polymyositis consists of subluxation or dislocation at the interphalangeal joint and is known as the "floppy thumb sign."

Computed Tomography

Computed tomography (CT) is useful in demonstrating soft tissue calcifications when radiography fails to show them. CT has also been used to demonstrate "milk of calcium" within the calcareal collections.

Magnetic Resonance Imaging

During the active phase of dermatomyositis, T2-weighted magnetic resonance images (MRIs) show increased signal intensity in the affected muscles and edema in the perimuscular tissue and subcutaneous fat (Fig. 26–5).

Recommended Reading

Blane CE, White SJ, Braunstein EM, et al: Patterns of calcification in childhood dermatomyositis. AJR Am J Roentgenol 142:397, 1984.
Dalakas MC: Polymyositis, dermatomyositis, and inclusion-body myositis. N Engl J Med 325:1487, 1991.
Greenway G, Weisman MH, Resnick D, et al: Deforming arthritis of the hands: An unusual manifestation of polymyositis. AJR Am J Roentgenol 136:611, 1981.
Hernandez RJ, Sullivan DB, Chenevert TL, et al: MR imaging in children with dermatomyositis: Musculoskeletal findings and correlation with clinical and laboratory findings. AJR Am J Roentgenol 161:359, 1993.
Sewell JR, Liyanage B, Ansell BM: Calcinosis in juvenile dermatomyositis. Skeletal Radiol 3:137, 1978.

CHAPTER 27

Septic Arthritis in Adults

Georges Y. El-Khoury, MD, Mark D. Stanley, MD, and D. Lee Bennett, MD

OVERVIEW

Septic arthritis is a disease for which outcome is directly related to the rapidity of diagnosis and early institution of therapy. Delay in diagnosis is fairly common. The most commonly involved joints in order of frequency are the knee, the hip, the shoulder, the wrist, the ankle, and the elbow (Figs. 27–1 to 27–4). The sternoclavicular and sacroiliac joints are typically affected in intravenous drug abusers (Figs. 27–5 and 27–6). In adults, the most common offending organism is *Staphylococcus aureus*. Risk factors predisposing to septic arthritis include advanced age, underlying systemic disease, immunosuppressive therapy, prosthetic joints, joint surgery, arthrocentesis, and intravenous drug abuse. Forty to fifty percent of adults with septic arthritis are older than 60 years of age.

Systemic diseases associated with septic arthritis include rheumatoid arthritis, diabetes mellitus, malignancy, systemic lupus erythematosus, and infections in other sites such as the skin, the lungs, or the urinary tract. Previous joint aspiration or steroid injection has been noted to precede septic arthritis. Septic arthritis following arthroscopy is seen in 0.04% to 4% of patients, especially when intra-articular steroids are used. Most septic arthritis cases are monoarticular (Fig. 27–7); however, 15% to 20% of cases are polyarticular, and the mortality rate is 30% to 40% compared with 4% to 8% for monoarticular cases.

Mode of Infection

Bacteria can reach the synovium via hematogenous spread from a distant focus of infection, local extension

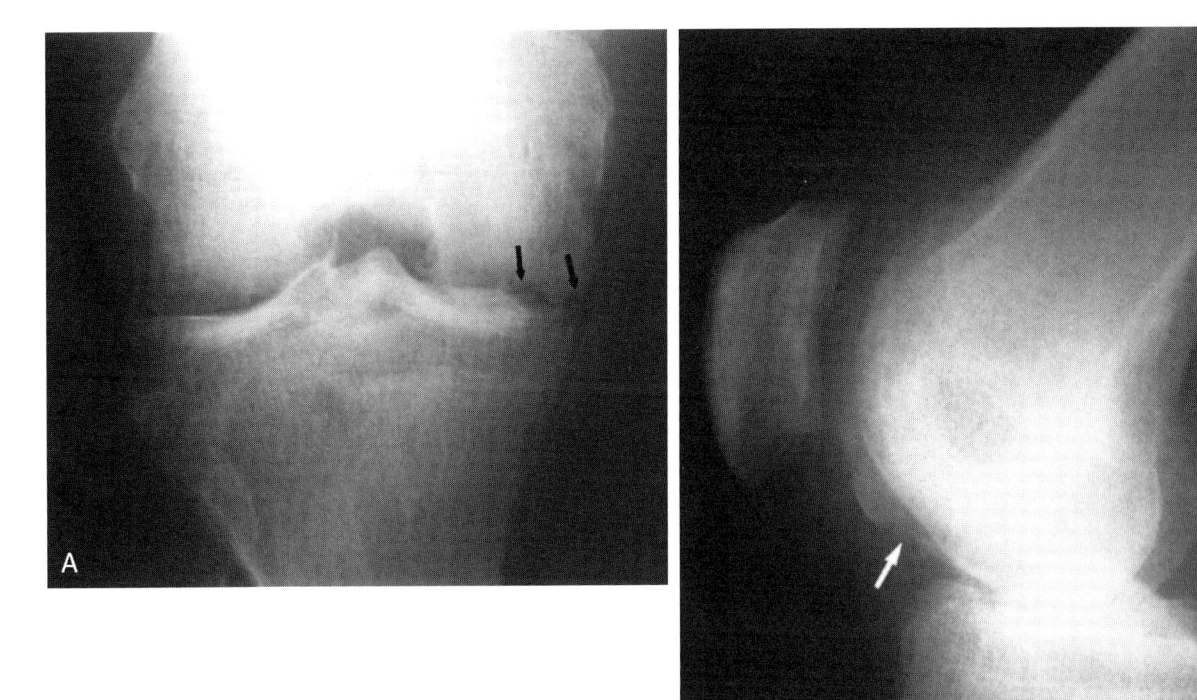

Figure 27–1. Septic arthritis of the knee. *A* and *B*, Anteroposterior and lateral views of the right knee show uniform narrowing of the joint space and subchondral erosions *(arrows)*.

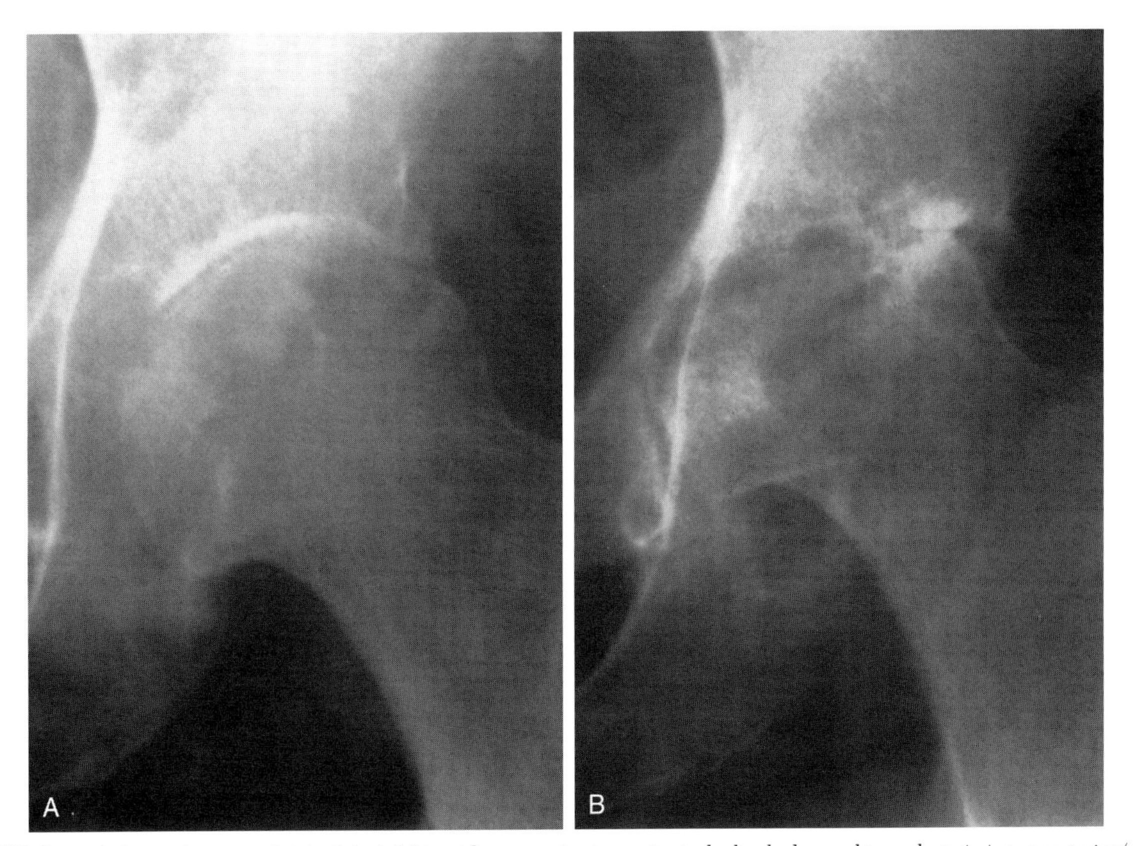

Figure 27–2. Staphyloccocal septic arthritis of the left hip with progression in a patient who has had a renal transplant. *A*, Anteroposterior (AP) view of the left hip shows narrowing of the joint space, osteopenia, and erosions in the weight-bearing portion of the femoral head. *B*, Follow-up AP view of the same hip shows significant destruction of the femoral head and acetabular cavity.

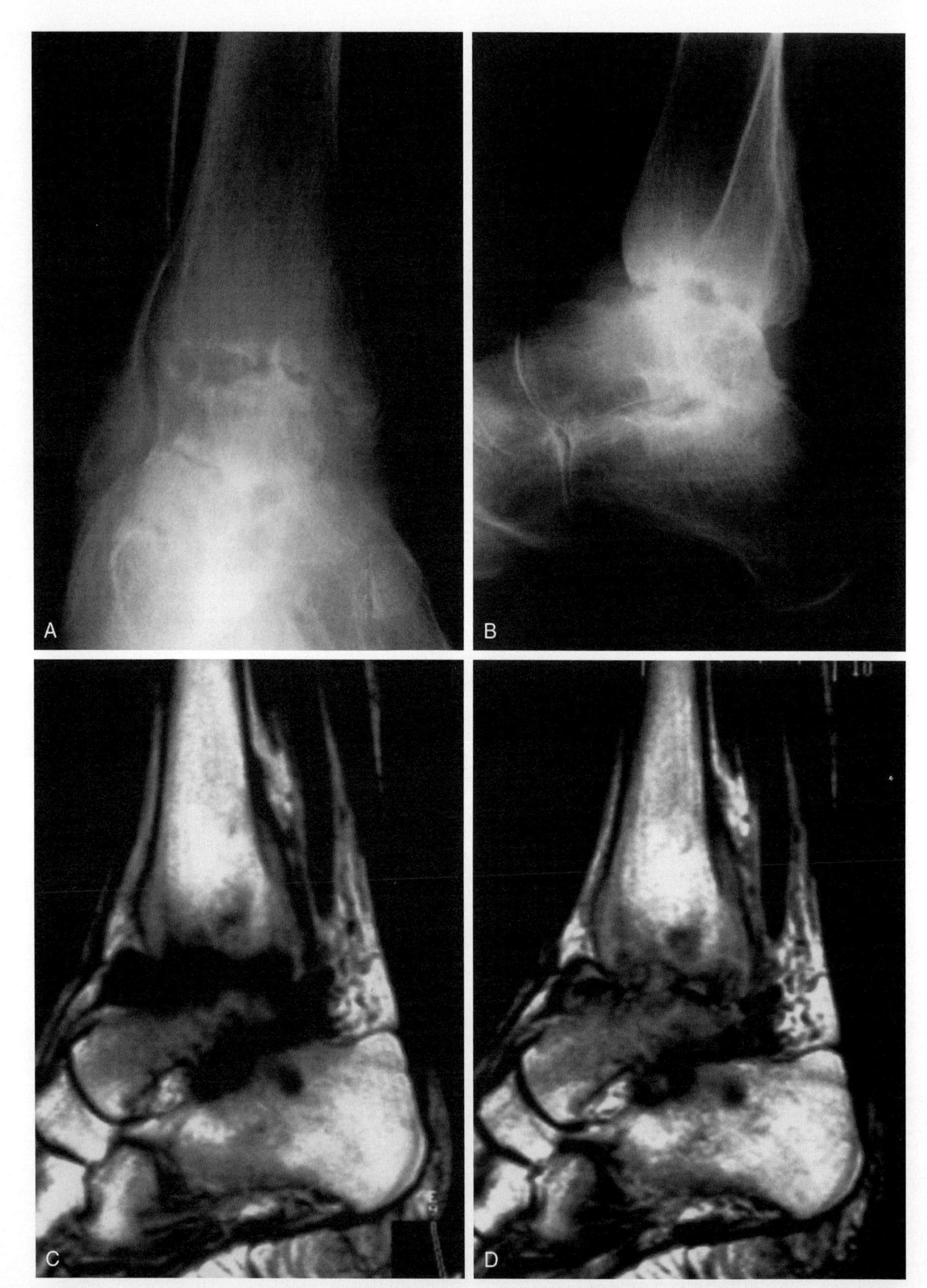

Figure 27–3. A 54-year-old woman with staphylococcal infection of the right ankle and subtalar joint. *A* and *B*, Anteroposterior and lateral views of the ankle show joint space narrowing as well as subchondral erosions and osteopenia. Sagittal T1-weighted MR image *(C)* and T1-weighted image with gadolinium enhancement *(D)* show fluid in the ankle and subtalar joints as well as synovial inflammatory changes.

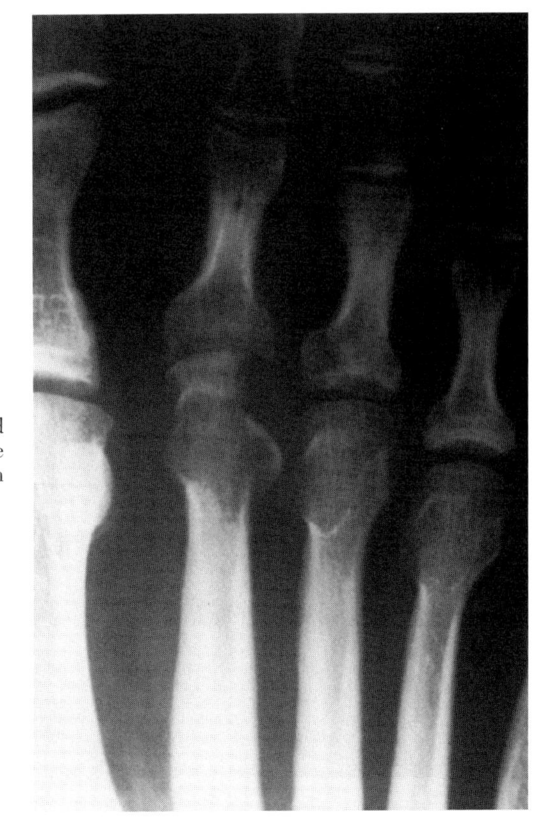

Figure 27–4. A 43-year-old man who stepped on a honey locust thorn and subsequently developed septic arthritis of the second metatarsophalangeal joint. Large erosions are noted in the metatarsal head and base of the proximal phalanx along with joint space narrowing and osteopenia.

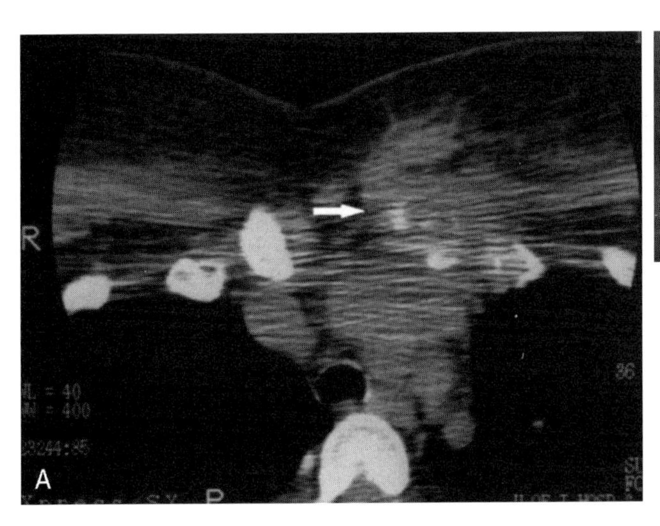

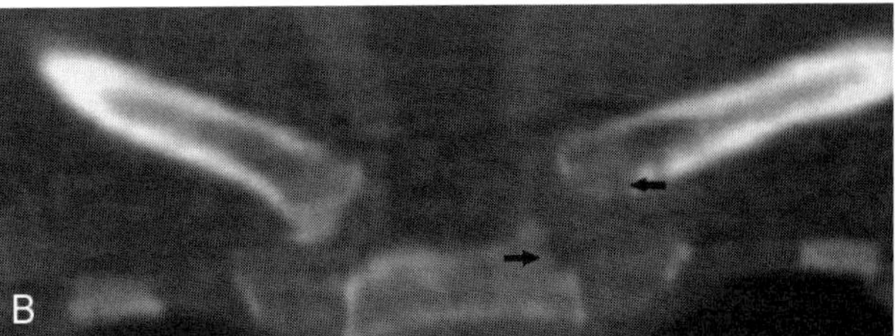

Figure 27–5. Sternoclavicular osteomyelitis. *A*, Axial computed tomography scan with soft tissue window shows bony destruction at the left sternoclavicular joint. This is associated with soft tissue swelling and edema *(arrow)*. *B*, Coronally reconstructed image using bone window reveals bone erosions on both sides of the left sternoclavicular joint *(arrows)*. Needle aspiration of the joints grew *Staphylococcus aureus*.

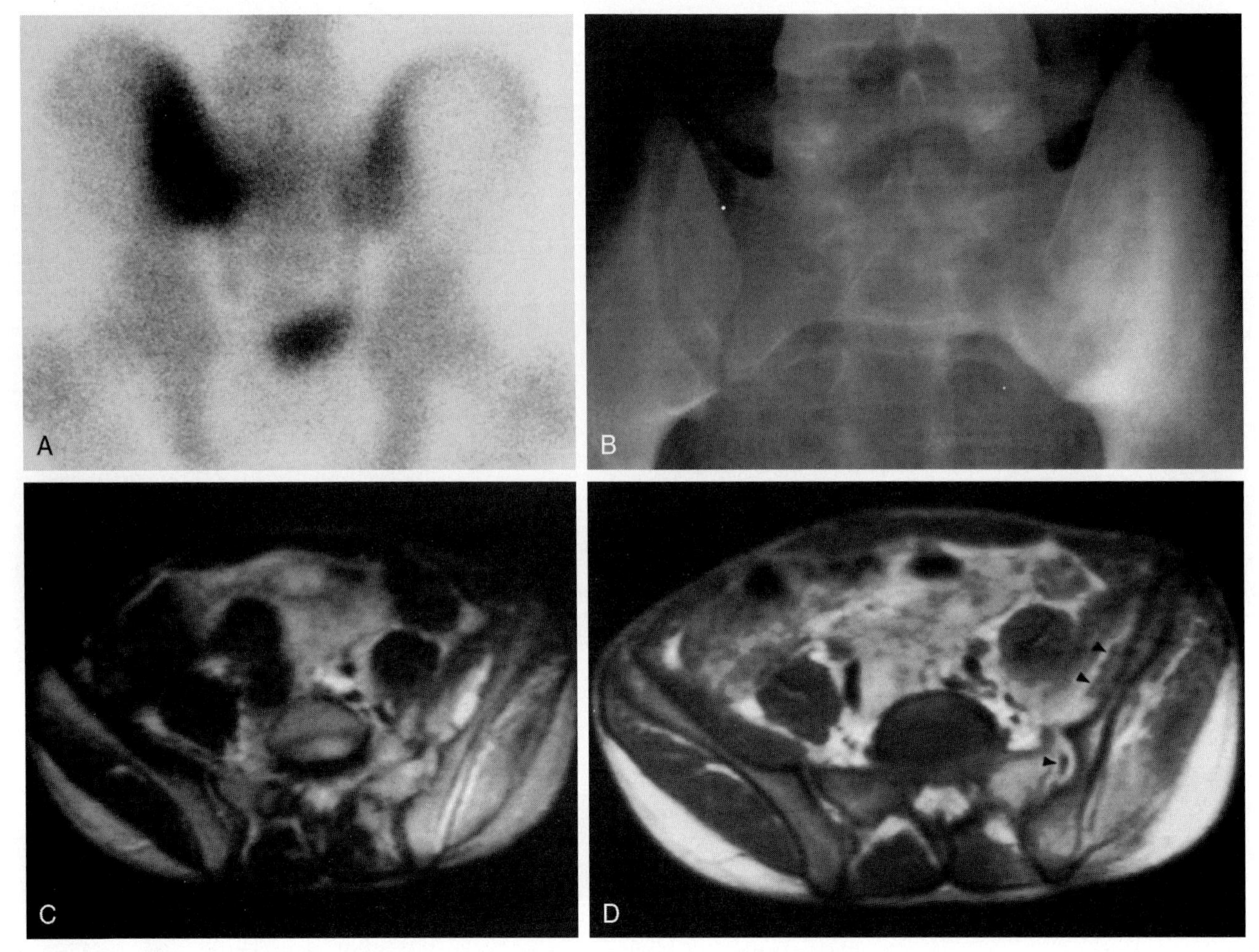

Figure 27–6. Septic arthritis of the left sacroiliac (SI) joint in a 16-year-old drug addict. *A,* Bone scan with posterior view of the pelvis reveals increased uptake in the region of the left SI joint. *B,* Anteroposterior radiograph of the pelvis shows a wide left SI joint owing to erosions associated with a sclerotic reaction on the iliac side. *C,* T2-weighted axial MR image of the pelvis shows increased signal in the marrow and soft tissues around the left SI joint. A fluid collection is suspected between the left iliac wing anteriorly and the iliacus muscle. The psoas muscle on the left is displaced anteriorly by swollen iliacus muscle. *D,* T1-weighted axial MR image following gadolinium enhancement demonstrates the infection in the left SI joint and inflammatory changes in the marrow and soft tissues around the joint. A small amount of nonenhancing fluid is detected in the joint and underneath the iliacus muscle *(arrowheads).*

from a contiguous focus of osteomyelitis, or an adjacent soft tissue abscess, or by direct inoculation from a penetrating wound, surgery, or arthrocentesis.

PROSTHETIC JOINTS (see Chapter 97)

An increasingly occurring joint infection involves prosthetic joints, which radiographically show loosening of the prosthesis (Figs. 27–8 and 27–9). Loosening caused by infection is difficult to differentiate radiographically from mechanical loosening and, in such cases, joint aspiration under fluoroscopic guidance is indicated. Two thirds of prosthetic joint infections occur within the first year after surgery. Such infections can cause morbidity and mortality. Risk of infection is higher in patients with revision surgery owing to a previously infected prosthesis, patients with rheumatoid arthritis, and patients receiving steroid therapy. Also, patients with infection at a distant site are at higher risk for infection in their prosthetic joints.

DIAGNOSIS OF SEPTIC ARTHRITIS

Joint Aspirations

The diagnosis of septic arthritis is often not suspected clinically at initial presentation. However, to preserve joint function, with the slightest clinical suspicion, the joint should be immediately aspirated and the fluid checked microscopically and cultured appropriately. Successful therapy requires prompt and complete drainage of the joint in addition to adequate antibiotic therapy. Percutaneous placement of catheters into large joints allows continuous drainage and irrigation of infected joints.

Radiography

Initially, except for the joint effusion and soft tissue swelling, which are non-specific, plain radiographs may be totally unremarkable. As the disease advances (10–14 days), osteopenia and uniform joint space narrowing can be appreciated. On serial examinations, the osteopenia and joint space narrowing can progress rapidly (see Figs. 27–1 and 27–2). The osteopenia can become profound (see Fig. 27–3). The joint space loss indicates irreversible cartilage damage. Marginal erosions develop concomitant with joint space narrowing (see Figs. 27–1 to 27–4). Cortical erosions in the patients with pyogenic sacroiliitis occur first on the iliac side (see Fig. 27–6). In a deep joint, such as the sacroiliac joint, normal radiographs do not exclude infection. Pyogenic sacroiliitis may mimic intraperitoneal, retroperitoneal, or hip infections (Fig. 27–9).

Bone Scintigraphy

Technetium-99m methylene diphosphate (^{99m}Tc-MDP) bone scintigraphy is a very sensitive method for detecting acute joint infection; however, it is not specific (see Fig. 27–6*A*). Bone scintigraphy is also very helpful when looking for polyarticular disease.

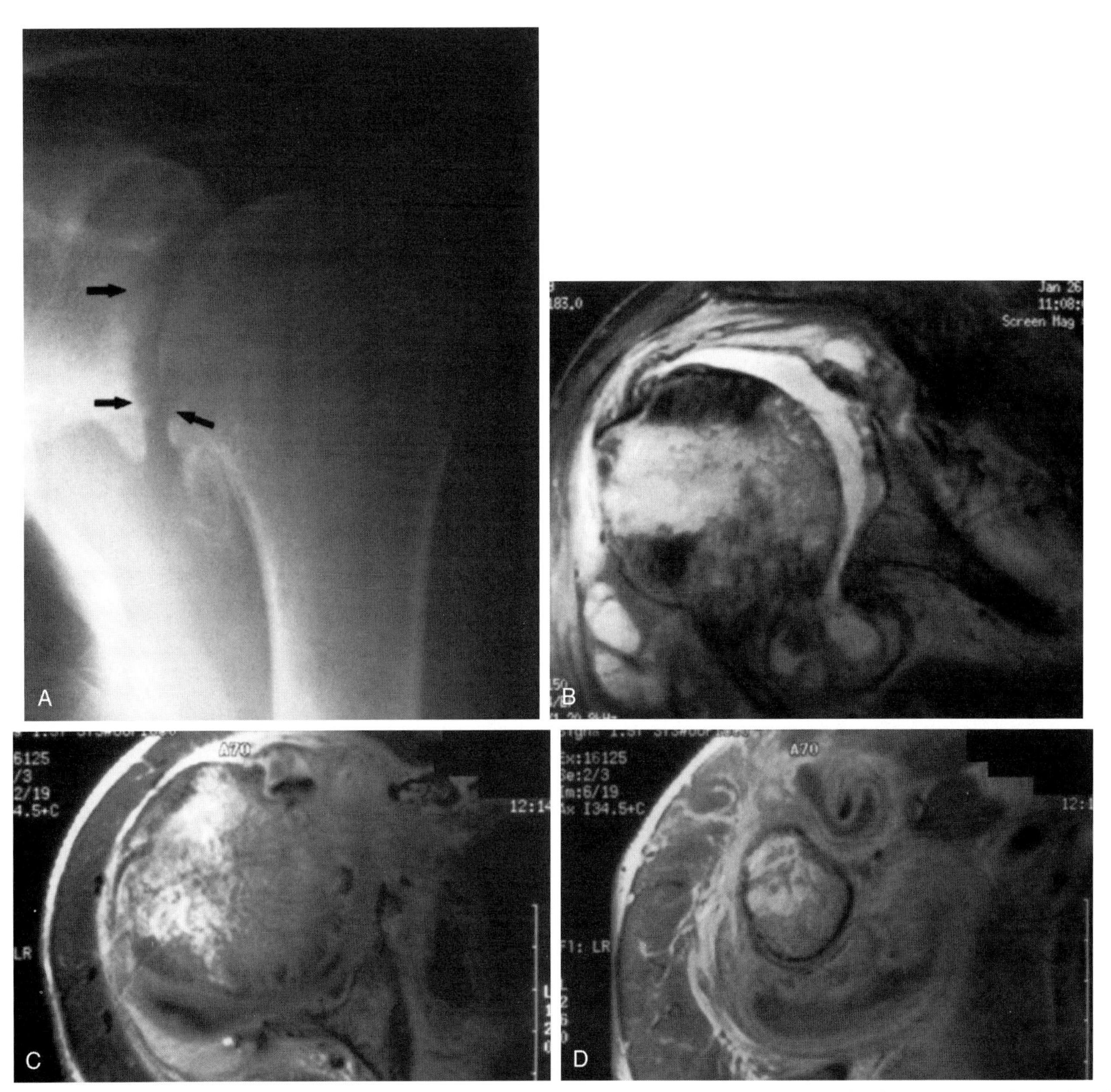

Figure 27–7. Septic arthritis of the left shoulder in a 52-year-old man. *A*, Anteroposterior view of the left shoulder shows erosions on both sides of the joint *(arrows)*. *B*, Coronal T2-weighted MR image of the shoulder shows a large joint effusion and edema in the soft tissues and bone marrow. *C* and *D*, Axial T1-weighted MR images following gadolinium enhancement show an inflamed, thick synovium and subchondral erosions in the humeral head and glenoid. A nonenhancing joint effusion is well demonstrated; there is also fluid in the tendon sleeve around the tendon of the long head of the biceps.

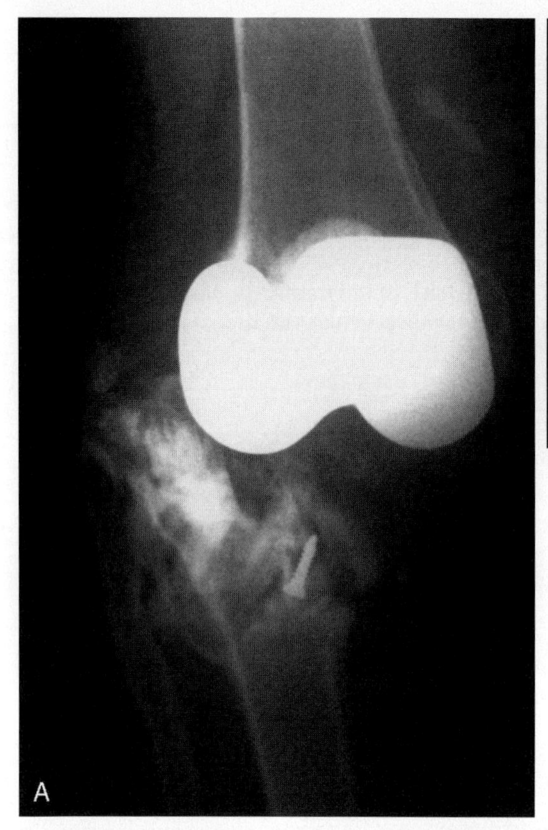

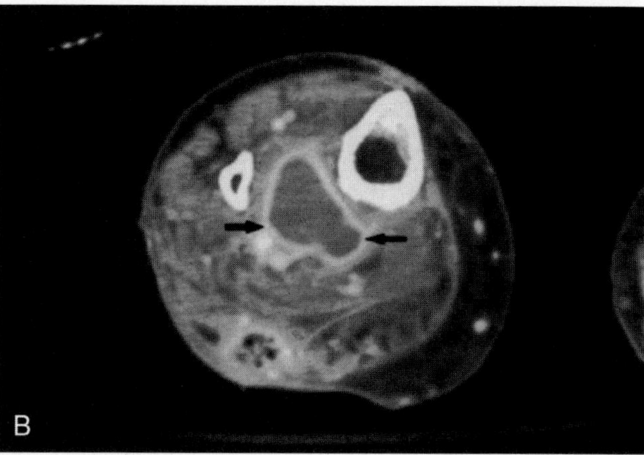

Figure 27–8. Infection involving the tibial component in a total knee replacement. *A*, Anteroposterior view of the right knee shows significant bony erosions involving the proximal tibia medially. The tibia is also subluxed laterally. *B*, Axial computed tomography section of the calf following intravenous contrast injection shows a deep abscess collection in the soft tissues of the calf *(arrows)*.

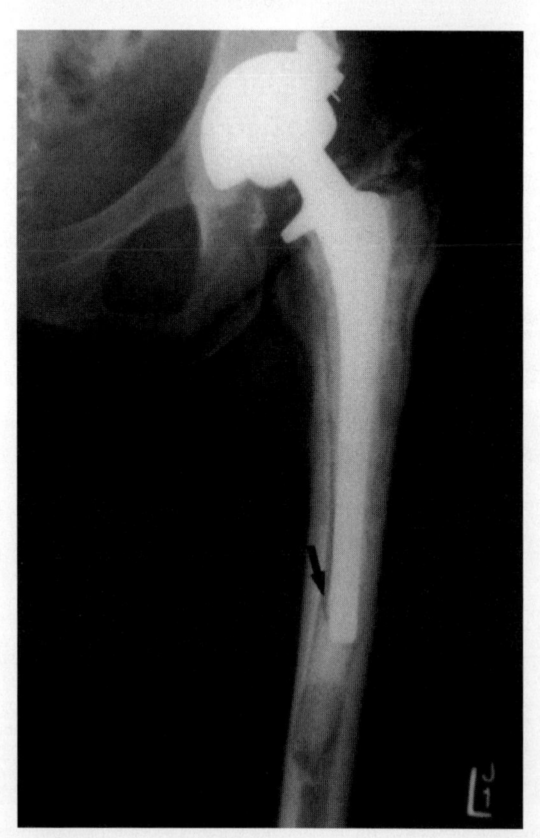

Figure 27–9. Infected total hip prosthesis, proven by needle aspiration under fluoroscopic guidance. Note lucency between the methylmethacrylate mantle and the bone. A fracture is also noted in the methylmethacrylate mantle *(arrows)*.

Computed Tomography

Difficult anatomical areas such as the sternoclavicular joint and SI joint are best evaluated by computed tomography (CT) (see Figs. 27–5 and 27–8). Fluid collections, soft tissue abscess, and early bone destruction can be easily detected by CT (see Fig. 27–8).

Magnetic Resonance Imaging

The findings of early septic arthritis on magnetic resonance imaging include the presence of joint effusion along with intra-articular debris and marrow edema (see Figs. 27–3 and 27–6). Contrast enhancement reveals increased signal intensity of the synovium (see Fig. 27–7) and fluid (abscess) collections (see Fig. 27–6).

Recommended Readings

Goldenberg DL, Cohen AS: Acute infectious arthritis: A review of patients with nongonococcal joint infections (with emphasis on therapy and prognosis). Am J Med 60:369, 1976.

Guyot DR, Manoli A II, Kling GA: Pyogenic sacroiliitis in IV drug abusers. AJR Am J Roentgenol 149:1209, 1987.

Leslie BM, Harris JM III, Driscoll D: Septic arthritis of the shoulder in adults. J Bone Joint Surg Am 71:1516, 1989.

Renner JB, Agee MW: Treatment of suppurative arthritis by percutaneous catheter drainage. AJR Am J Roentgenol 154:135, 1990.

Resnik CS, Ammann AM, Walsh JW: Chronic septic arthritis of the adult hip: Computed tomographic features. Skeletal Radiol 16:513, 1987.

CHAPTER 28

Osteoarticular Tuberculosis

Georges Y. El-Khoury, MD, Mark D. Stanley, MD, and D. Lee Bennett, MD

OVERVIEW

The prevalence of tuberculosis in the United States has been rising since 1981. The reasons cited for this rise include the increase in the number of patients with suppressed immune systems, an aging population, and the development of mycobacterium strains that are resistant to available drug therapy. HIV is currently the leading cause for reactivation of latent tuberculosis. The most common infecting organism is *Mycobacterium tuberculosis.* Osteoarticular tuberculosis occurs in approximately 1% to 5% of patients with pulmonary disease. More than 50% of patients with osteoarticular tuberculosis have no concomitant pulmonary tuberculosis.

TUBERCULOUS SPONDYLITIS

Spinal tuberculosis accounts for approximately 50% to 60% of all skeletal tuberculosis cases. Most cases occur in the thoracic spine followed by the lumbar and then the cervical spine. Spinal involvement is considered to be the most serious form of skeletal tuberculosis because delay in diagnosis can lead to spinal cord compression and irreversible neurologic deficit (see Chapter 46).

TUBERCULOUS ARTHRITIS

Although any joint can be affected, tuberculous arthritis mainly affects the large, weight-bearing joints such as the hips and the knees (Figs. 28–1 and 28–2). The wrist, shoulder, and ankle are less commonly involved (Fig. 28–3). The disease is typically monoarticular and insidious. Delay in diagnosis is common, and synovial biopsy provides the highest diagnostic yield.

Imaging

There are no specific or pathognomonic features of tuberculous arthritis. Typically, there is soft tissue swelling, periarticular osteoporosis, marginal erosions, and joint space narrowing (see Figs. 28–1 and 28–2). Periosteal reaction and bony sclerosis are usually lacking. Similar findings can be seen in pyogenic arthritis, but in pyogenic arthritis the disease progresses faster. Rheumatoid arthritis (RA) can resemble tuberculous arthritis, but RA is typically polyarticular. In children, tuberculous synovitis results in epiphyseal enlargement similar to that seen in juvenile rheumatoid arthritis and hemophilia. The most commonly affected joints are the hip, followed by

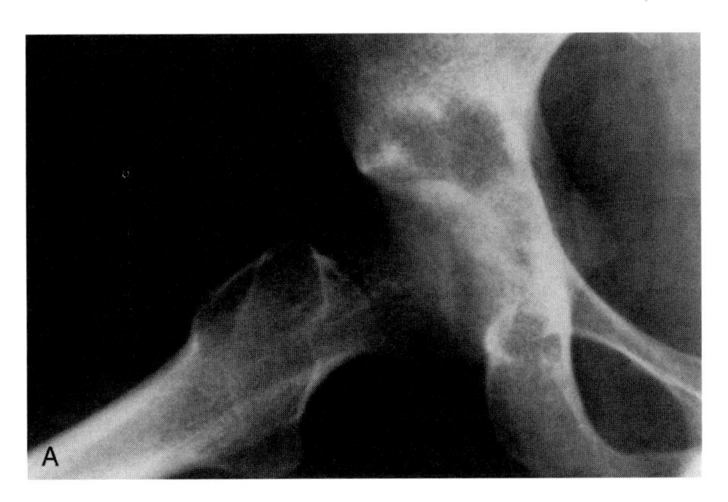

Figure 28–1. Tuberculous arthritis of the right hip in a 50-year-old Vietnamese immigrant. The patient had a 1-year history of right hip pain. Hip aspiration and culture revealed tuberculosis. *A,* A frog-leg anteroposterior view of the right hip shows marked narrowing of joint space with large erosions on the acetabular side.

Illustration continued on following page

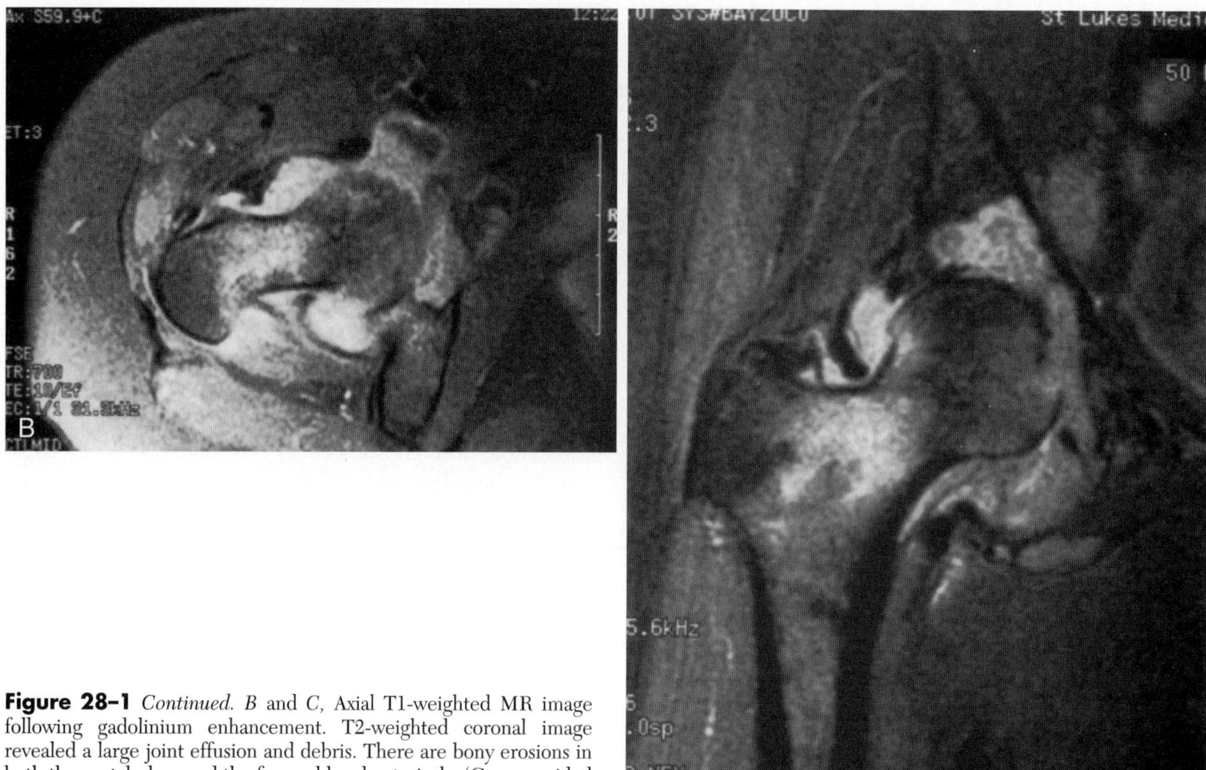

Figure 28–1 *Continued. B* and *C,* Axial T1-weighted MR image following gadolinium enhancement. T2-weighted coronal image revealed a large joint effusion and debris. There are bony erosions in both the acetabulum and the femoral head anteriorly. (Case provided by Dr. Daniel Harrington from Milwaukee, WI.)

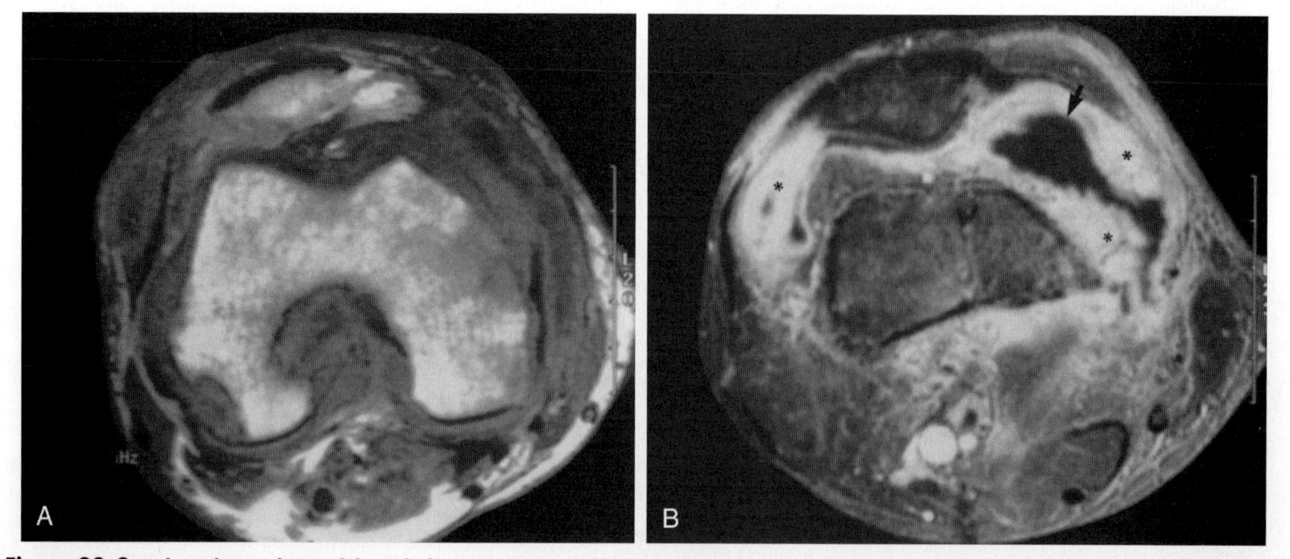

Figure 28–2. Tuberculous arthritis of the right knee in a 54-year-old patient with known chest tuberculosis. *A,* T1-weighted axial MR image shows soft tissue swelling, joint effusion, and bony erosions in the femoral condyles. *B,* T1-weighted axial image with fat suppression following gadolinium enhancement shows marked thickening of the synovium, which brightly enhanced *(asterisks).* There is also a joint effusion *(arrow).* (Case provided by Dr. Kyung Jin Suh from Kyungpook National University Hospital in Korea.)

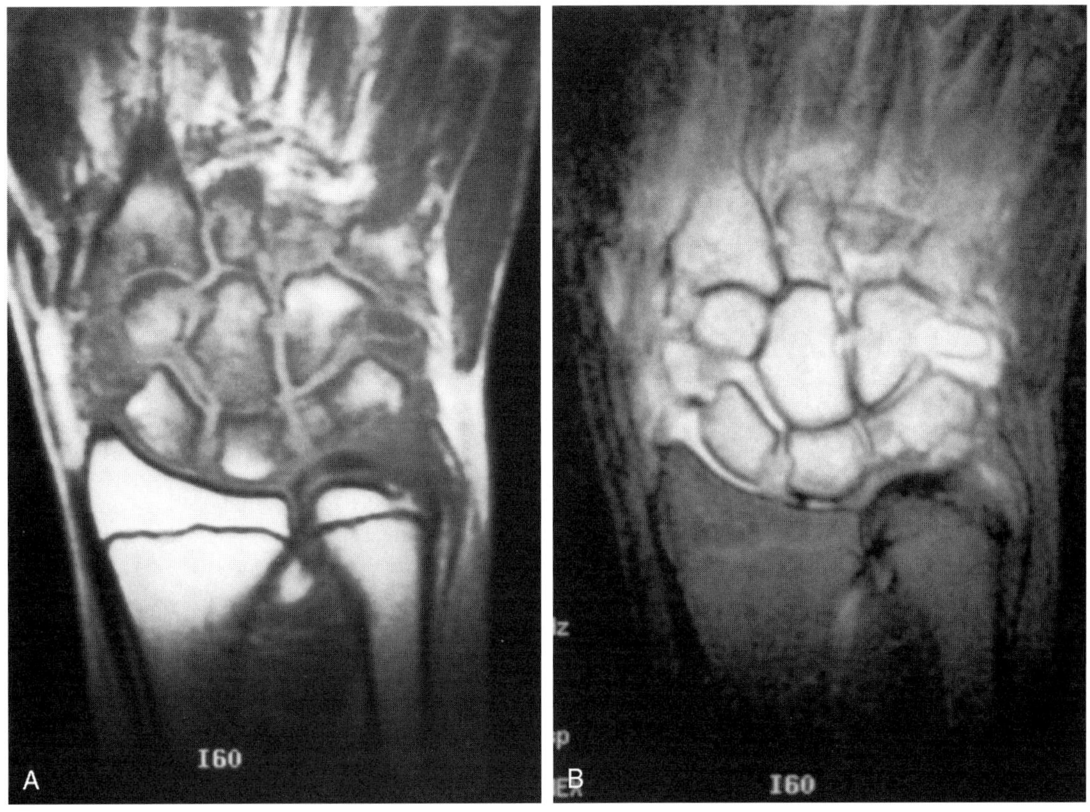

Figure 28–3. Tuberculous arthritis of the left wrist in a 14-year-old girl with a 3-month history of wrist pain. *A*, T1-weighted coronal MR image shows synovial thickening and marrow edema in the carpal bones and proximal metacarpals. *B*, T2-weighted fat suppressed coronal MR image shows increased signal intensity of the synovium and bone marrow in the carpal bones and proximal metacarpals, indicating inflammation. (Case provided by Dr. Kyung Jin Suh from Kyungpook National University Hospital in Korea.)

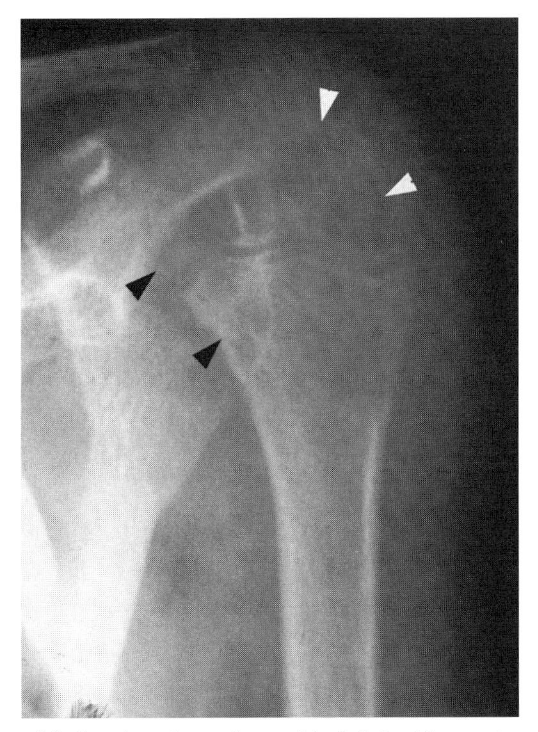

Figure 28–4. Tuberculous arthritis of the left shoulder in a 9-year-old child. There is a lytic lesion destroying the epiphysis of the humeral head and crossing into the metaphysis *(arrowheads)*.

the shoulder and the knee. Tuberculous arthritis can present with a lytic lesion in the epiphysis which crosses the epiphyseal plate in more than one third of cases. Such a finding is characteristic of tuberculous arthritis (Fig. 28–4).

Tuberculous dactylitis (spina ventosa) is the most common form of musculoskeletal tuberculosis in infants. Sickle cell dactylitis is similar radiographically to tuberculous dactylitis, but is characteristically bilateral.

Recommended Readings

Evanchick CC, Davis DE, Harrington TM: Tuberculosis of peripheral joints: An often missed diagnosis. J Rheumatol 13:187, 1986.

Garrido G, Gomez-Reino JJ, Fernandez-Dapica P, et al: A review of peripheral tuberculous arthritis. Semin Arthritis Rheum 18:142, 1988.

Goldblatt M, Cremin BJ: Osteo-articular tuberculosis: Its presentation in coloured races. Clin Radiol 29:669, 1978.

Rathakrishnan V, Mohd TH: Osteo-articular tuberculosis. Skeletal Radiol 18:267, 1989.

Sharif HS, Clark DC, Aabed MY, et al: Granulomatous spinal infections: MR imaging. Radiology 177:101, 1990.

Watts HG, Lifeso RM: Current concepts review: Tuberculosis of bones and joint. J Bone Joint Surg Am 78:288, 1996.

CHAPTER 29

Crystal Arthropathies

Georges Y. El-Khoury, MD, Mark D. Stanley, MD, and D. Lee Bennett, MD

Crystal arthropathies are characterized by the deposition of specific crystals in joints and soft tissues producing a spectrum of clinical and radiographic abnormalities. Crystals implicated in these diseases include monosodium urate, calcium pyrophosphate dihydrate, and calcium hydroxyapatite. The exact role of crystals in producing joint disease is not well understood because crystals can be found in the synovial fluid of asymptomatic patients.

GOUT

Overview

Gout is approximately 20 times more common in men. The peak incidence of onset is in the 40s and 50s. Hyperuricemia is the underlying cause of gout. In more than 90% of patients with gout, the defect is related to diminished excretion of uric acid in the urine. A few patients synthesize and excrete excessive amounts of uric acid. These are typically patients with myeloproliferative or lymphoproliferative disorders. Such patients rarely, if ever, develop radiographically apparent disease. Rarely, some patients develop gout because of enzyme deficiencies affecting purine metabolism resulting in hyperuricemia, as seen in Lesch-Nyhan syndrome. Gouty arthritis is the most frequent rheumatologic disorder in transplant patients receiving cyclosporine.

The diagnosis of gout is made by confirming the presence of monosodium urate crystals in the synovial fluids or connective tissues. Using compensated polarized light microscopy, monosodium urate crystals appear negatively birefringent.

Radiographic Changes

In modern clinical practice, tophaceous gouty arthritis is rare because of the effective use of uricosuric agents and allopurinol. Tophi in the soft tissues of the hand and wrist appear as lumpy, bumpy asymmetrical masses, which occasionally contain cloudy calcifications (Fig. 29–1). Tophi are composed of monosodium urate crystals embedded in a matrix of lipid and glycosaminoglycans. When tophi enlarge, they erode the para-articular bone, producing sharp, punched-out erosions with well-defined cortical margins (Fig. 29–2). The joint space typically is preserved until late in the disease (Figs. 29–2 and 29–3). There is no associated osteoporosis. Bone may grow from the eroded sites over the tophi, producing the *overhanging edges* that are fairly characteristic of gout (see Fig. 29–2). The most common site for gout is the feet, especially the first metatarsophalangeal joint; other common sites are the hands, wrists, and elbows. Bilateral olecranon bursal swelling along with irregular bony erosions of the olecranon are considered virtually diagnostic of gout. Intraosseous calcified tophi occasionally are noted in the hands and wrists (Fig. 29–4). These can simulate enchondromas or bone infarcts. Gout also can present as a single lytic lesion resembling a primary bone tumor. There also are

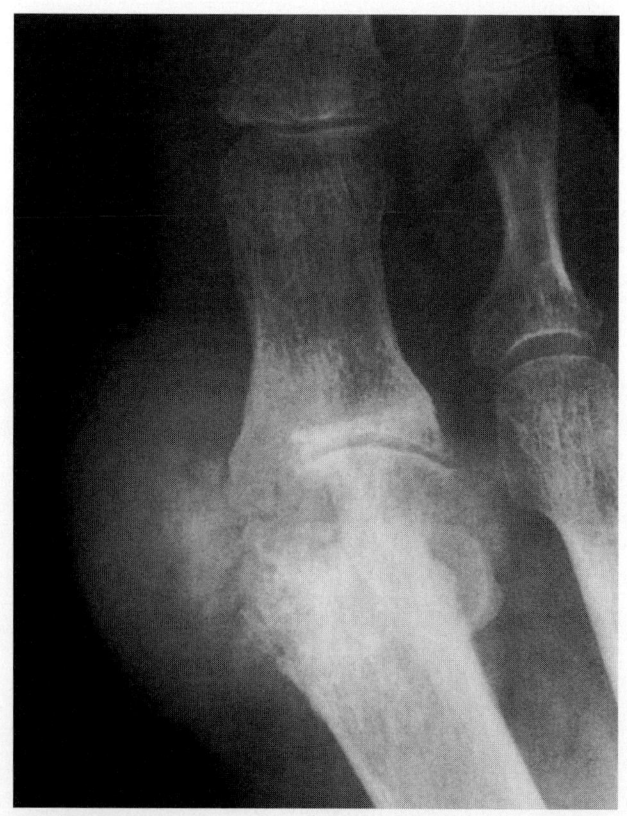

Figure 29–1. Anteroposterior view of the first metatarsophalangeal joint of the left great toe shows a large cloudy mass medial to the joint. The mass contains calcifications and is causing large pressure erosions on both sides of the metatarsal head and base of the proximal phalanx medially.

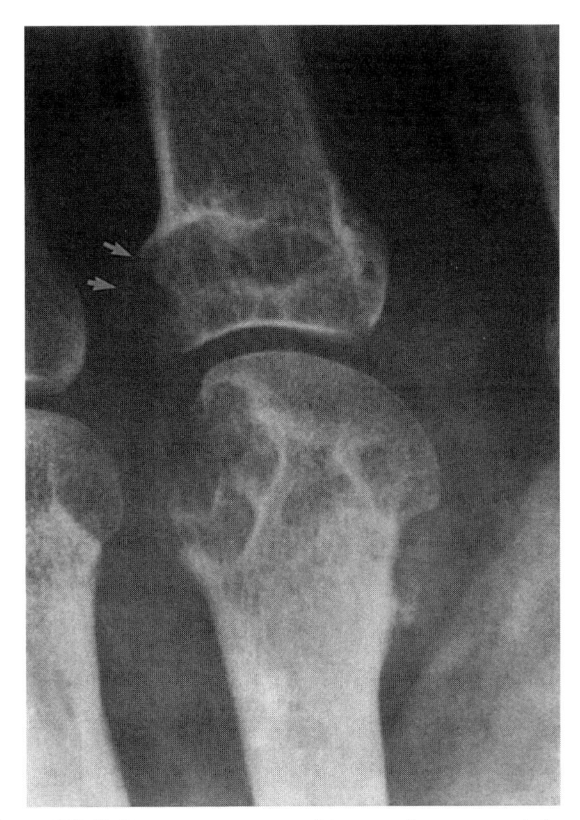

Figure 29–2. Posteroanterior view of the second metacarpophalangeal joint of the right hand shows large punched-out erosions in the head of the second metacarpal and base of the proximal phalanx. An overhanging edge is seen in the proximal phalanx medially *(arrows)*. Note the preserved joint space.

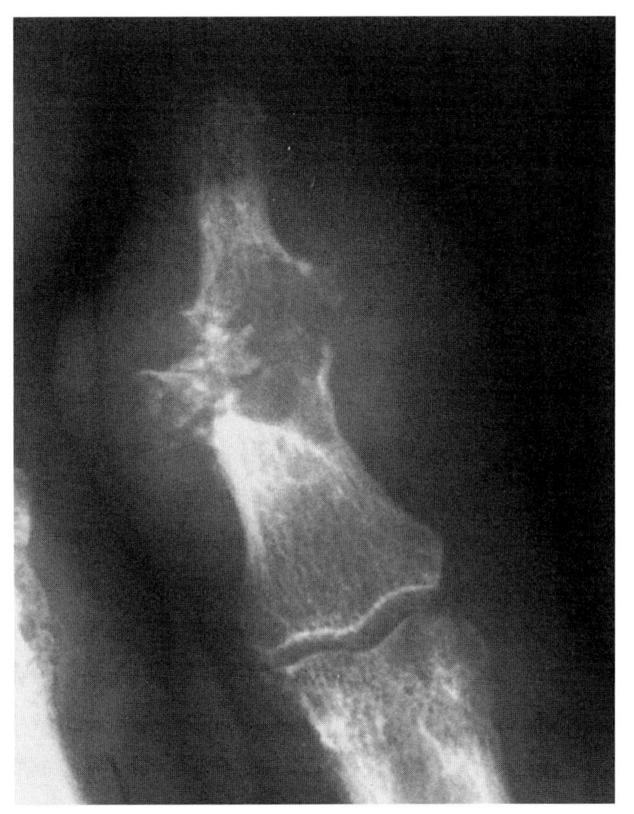

Figure 29–3. Advanced changes of gout with total destruction of the distal interphalangeal joint of the index finger. A large tophus is seen surrounding the joint.

reports of gouty arthritis in the axial skeleton and sacroiliac joints. When the intervertebral disk is involved, it simulates a disk space infection, and a needle biopsy is required to confirm the diagnosis of gout. With effective medical treatment, tophi can dissolve rapidly, resulting in collapse and telescoping of digits, with redundancy of the skin; this clinical picture is known as *main en lorgnette*.

A few reports describe the magnetic resonance (MR) imaging appearance of tophi. On T1-weighted images, tophi usually show a low-to-intermediate signal. On T2-weighted images, tophi have variable signal intensity but are mainly low-to-intermediate signal (Fig. 29–5).

CALCIUM PYROPHOSPHATE DIHYDRATE DEPOSITION DISEASE

Overview

Calcium pyrophosphate dihydrate (CPPD) deposition disease is the most common form of crystal-induced arthropathy. The disease is associated most commonly with aging. More than 25% of individuals older than age 80 years have CPPD crystals in their cartilage. At a younger age, the disease is associated with metabolic

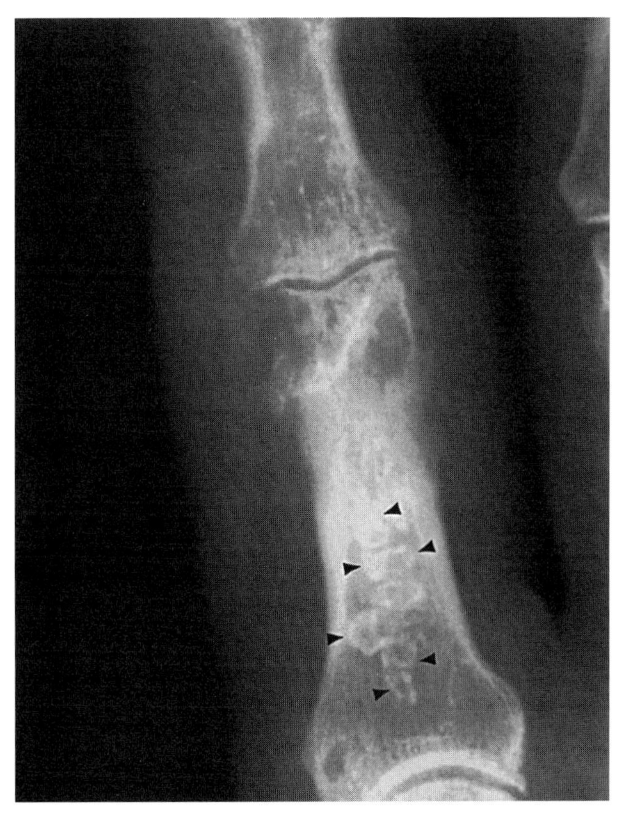

Figure 29–4. Intraosseous crystals of monosodium urate *(arrowheads)* in a patient with advanced tophaceous gout. There also are large erosions in the proximal phalanx at the proximal interphalangeal joint.

abnormalities, such as primary hyperparathyroidism, hemochromatosis, and Wilson's disease. CPPD deposition disease also is associated with hypomagnesemia and hypothyroidism. Rarely the disease is familial. Clinically, CPPD deposition disease can be asymptomatic, or it can simulate rheumatoid arthritis, osteoarthritis, neuropathic arthropathy, or acute gout, hence the term *pseudogout.* The diagnosis is made microscopically by examining the synovial fluid under polarized light looking for weakly birefringent crystals or by radiographically identifying the typical cartilage calcifications in the knee or wrist.

Definitions

Chondrocalcinosis is a generic term that refers to the presence of calcifications in cartilage. The calcifications can be due to CPPD crystals or due to some other type of crystals. *CPPD deposition disease* implies the presence of CPPD crystals in the cartilage or in the soft tissues around a joint. *Pseudogout* is a clinical diagnosis referring to an acute attack of arthritis caused by shedding of CPPD crystals and simulating acute gout.

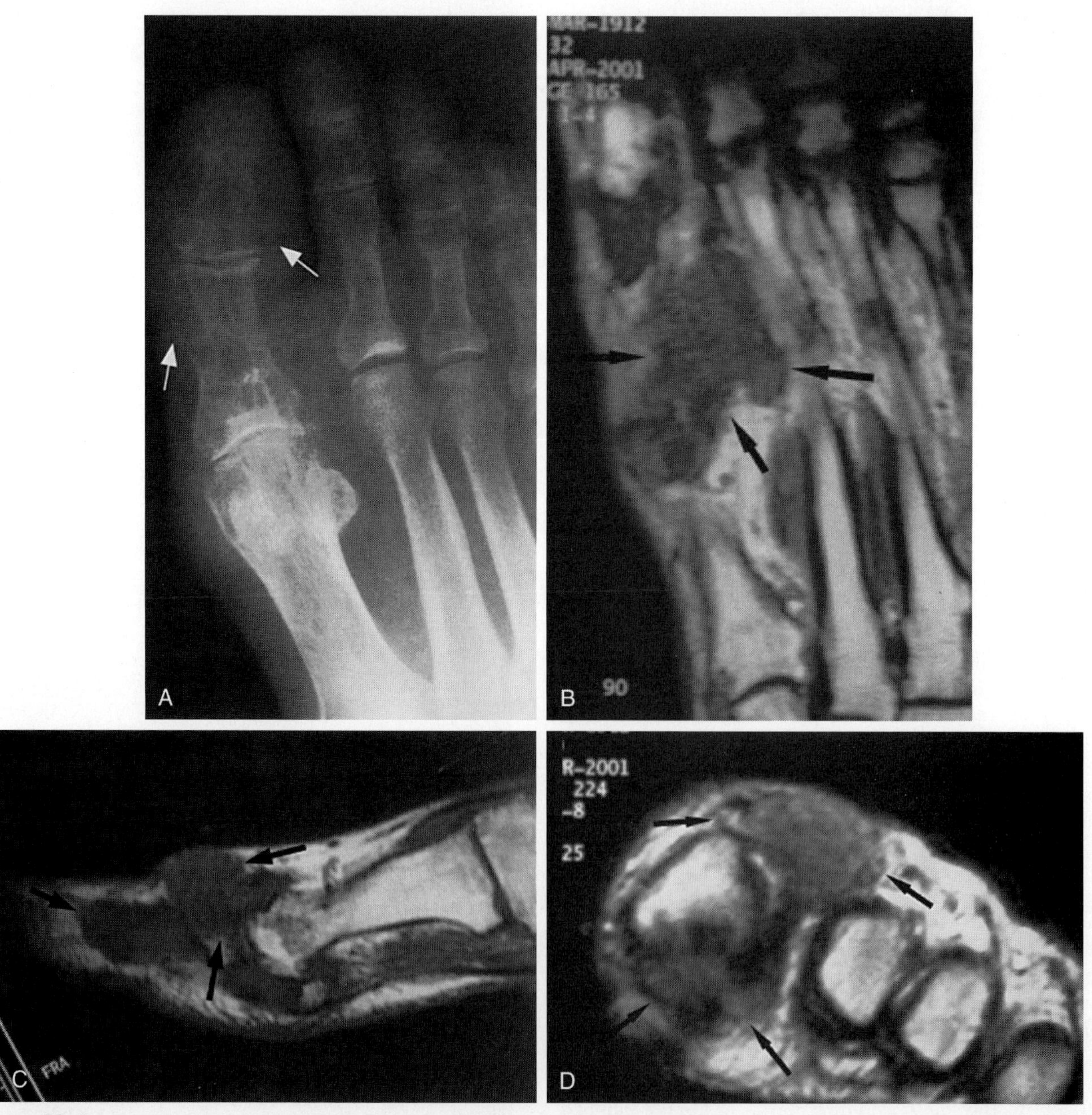

Figure 29–5. MR imaging shows gout in a 79-year-old man presenting with a lobulated mass in the region of the first metatarsophalangeal joint. The patient is otherwise asymptomatic. *A,* Anteroposterior view of the left foot shows lytic lesions in the great toe and first metatarsophalangeal joint *(arrows). B* and *C,* T1-weighted coronal and sagittal images show the lobulated mass, which is almost isointense with muscle *(arrows). D,* On T2-weighted axial image, the lobulated mass shows a moderate increase in signal intensity *(arrows).*

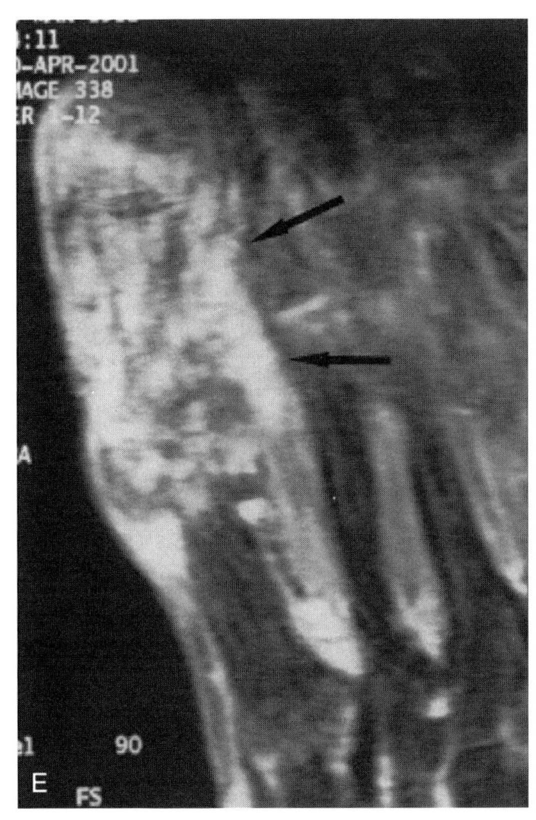

Figure 29–5. *Continued. E,* Fat-suppressed T1-weighted image after gadolinium injection shows significant enhancement of the mass and the affected bone *(arrows).*

Radiographic Findings

The radiographic findings can be classified into two components:

1. Distinctive cartilage and soft tissue calcifications
2. Degenerative joint disease, which superficially resembles osteoarthritis

CPPD initially deposits in the fibrocartilage and hyaline cartilage. Densely calcified menisci in the knee, triangular fibrocartilage in the wrist, and symphysis pubis frequently are present (Fig. 29–6). These three sites sometimes are imaged to screen for CPPD; however, absence of chondrocalcinosis on conventional radiographs does not exclude CPPD. Hyaline cartilage calcifications are fine, linear, and parallel to the subchondral bone (Fig. 29–6*A*). Hyaline cartilage calcifications are seen most commonly in the knee, wrist, elbow, and shoulder. The labra of the glenoid and acetabulum as well as the annulus of the intervertebral disk can also show calcifications (Fig. 29–7). Calcifications can deposit in the synovium, joint capsule,

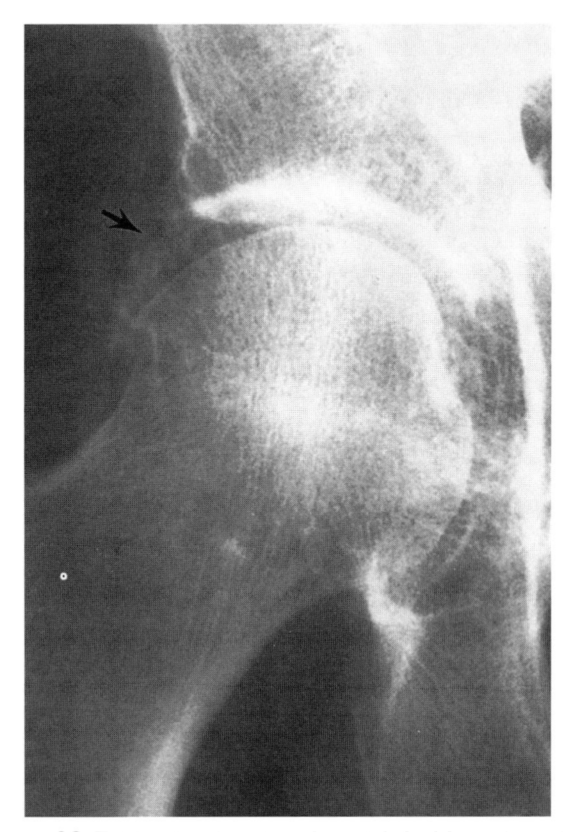

Figure 29–7. Chondrocalcinosis in the acetabular labrum *(arrow).*

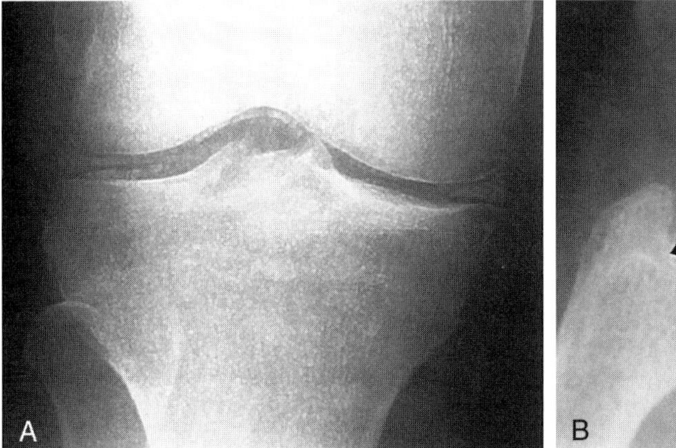

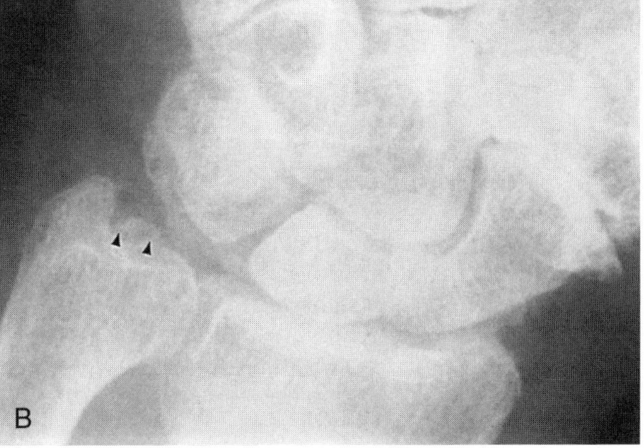

Figure 29–6. Chondrocalcinosis in the knee and wrist. *A,* Anteroposterior view of the knee shows densely calcified menisci and hyaline cartilage. *B,* Posteroanterior view of the right wrist shows faint calcification in the triangular fibrocartilage *(arrowheads)* and linear calcifications in hyaline cartilage of the triquetrum and scaphoid.

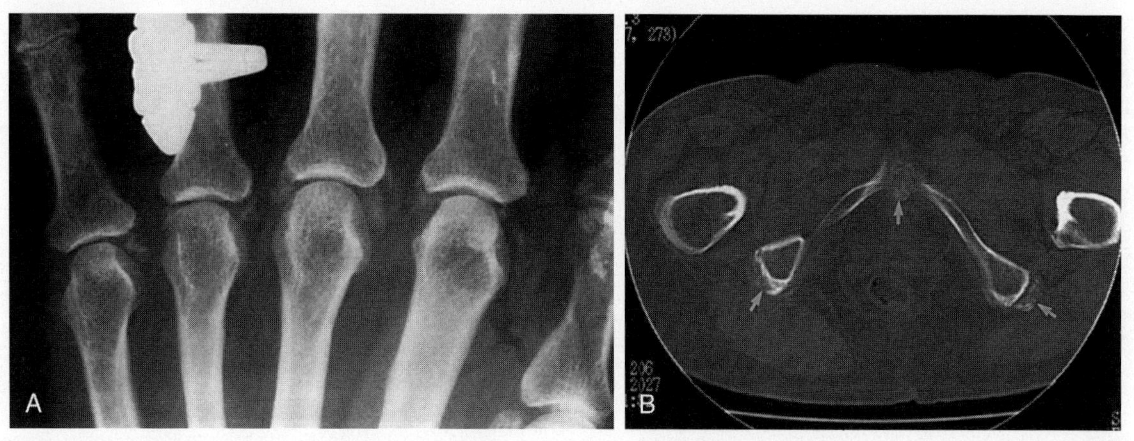

Figure 29–8. CPPD in the joint capsules and tendons. *A*, Posteroanterior view of the right hand shows dense calcifications in the capsules of the metacarpophalangeal joints in a patient known to have calcium pyrophosphate dihydrate deposition disease. *B*, Axial CT section through the lower pelvis shows calcification in the tendons of the hamstrings bilaterally *(lower arrows)*. Dense calcifications are present in the symphysis pubis *(upper arrow)*.

ligaments, and tendons (Fig. 29–8). The supraspinatus, triceps, quadriceps, and Achilles' tendons frequently are involved. Ligamentum flavum can enlarge because of CPPD deposits producing spinal stenosis. Rarely, large masses of CPPD can deposit in the soft tissues around joints. This condition has been termed *tophaceous* or *massive pseudogout* (Fig. 29–9). In our experience, the soft tissues around the dens seems to be a favorite site for tophaceous or massive pseudogout (see Fig. 29–9).

CPPD arthropathy superficially resembles osteoarthritis; at a closer look, however, it definitely has distinctive features. Unusual locations are involved, such as the wrists, elbows, shoulders, metacarpophalangeal joints, and talonavicular joints (Fig. 29–10). The patellofemoral joint shows disproportionate narrowing and degenerative changes compared with the tibiofemoral joint (Fig. 29–11). Other distinctive features include prominent subchondral cysts, collapse of the subchondral bone, loose bodies, and sclerosis resembling neuropathic arthropathy (Fig. 29–12).

HEMOCHROMATOSIS

Overview

Primary (familial, idiopathic) hemochromatosis is a disorder of iron metabolism caused by increased absorption of iron from the gut. Excess iron is deposited as hemosiderin in various tissues, primarily in the liver, pancreas, and heart. As iron concentration increases in the tissues, cellular damage and fibrosis occur. Primary hemochromatosis is 10 times more common in men than in women. Ferric salts promote the formation and deposition of intra-articular calcium pyrophosphate crystals.

Radiographic Findings

Associated arthropathy is seen in about 80% of patients with hemochromatosis, and it typically involves the hands, wrists, hips, and knees. Chondrocalcinosis occurs in about two thirds of patients with hemochromatosis arthropathy. The radiographic features of hemochromatosis arthropathy are almost identical to those of CPPD arthropathy. Some authors have described subtle differences between these two conditions. Findings specific to hemochromatosis include (1) more prevalent degenerative changes in the metacarpophalangeal joints, including the joints of the fourth and fifth fingers; (2) the presence of peculiar hooklike osteophytes on the radial aspect of the metacarpal heads (Fig. 29–13); and (3) less prevalent radiocarpal involvement with scapholunate dissociation in hemochromatosis arthropathy compared with CPPD arthropathy.

CALCIUM HYDROXYAPATITE DEPOSITION DISEASE

Overview

Deposition of calcium hydroxyapatite in tendons is a fairly common condition that is known as *calcific tendinitis* (see Chapter 72). Some authors have detected hydroxyapatite crystals in a high proportion of effusions associated with osteoarthritis. The coexistence of CPPD crystals and hydroxyapatite crystals (mixed crystal deposition disease) has been reported in some patients with destructive joint changes. Of special interest is the *Milwaukee shoulder*, which is a rare destructive arthropathy seen in elderly women and is believed to be caused by hydroxyapatite crystals (Fig. 29–14). The deposition of calcium hydroxyapatite can occur as a primary idiopathic process, or it may be secondary to other underlying diseases, such as end-stage renal disease, collagen vascular disease (scleroderma, CRST [calcinosis cutis, Raynaud's phenomenon, sclerodactyly, and telangiectasia] syndrome), and tumoral calcinosis. Similar to other crystal diseases, hydroxyapatite deposition disease can be asymptomatic.

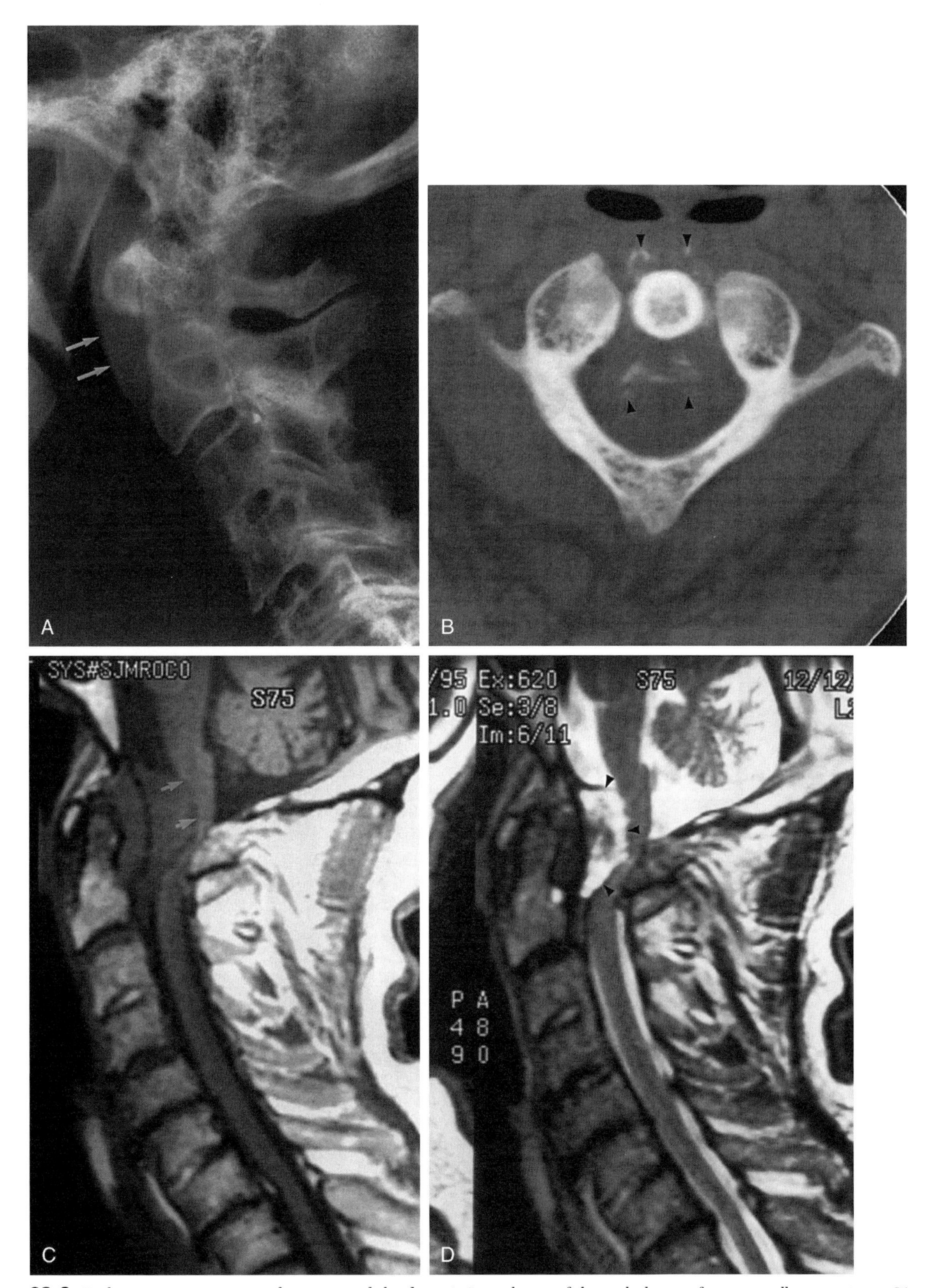

Figure 29–9. Tophaceous or massive pseudogout around the dens. *A*, Lateral view of the neck shows soft tissue swelling anterior to C1 and C2 *(arrows)*. *B*, Axial CT section through C1 shows punctate soft tissue calcifications anterior and posterior to the dens *(arrowheads)*. *C* and *D*, T1- and T2-weighted sagittal MR images show a large mass anterior and posterior to the tectorial membrane, which is displaced posteriorly. The cord and medulla are displaced posteriorly and compressed *(arrows* in *C)*. A small portion of the mass is seen anterior to C2. The mass is isointense with the cord on the T1-weighted image. On the T2-weighted images, the part of the mass in direct contact with the dens is hypodense, but the larger component between the tectorial membrane and the cord is bright with a central hypointense area *(arrowheads* in *D)*.

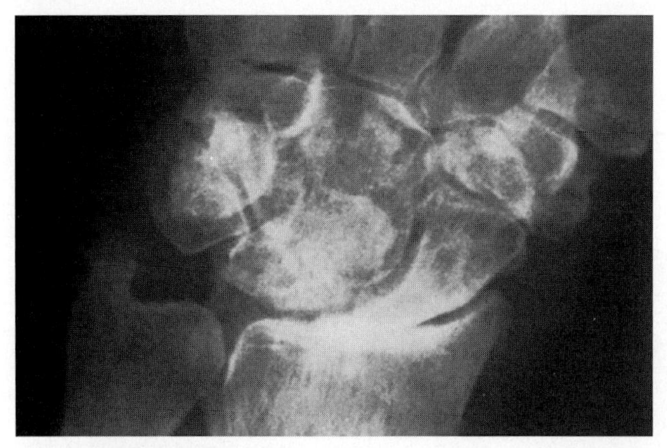

Figure 29–10. Degenerative changes in the wrist resulting from calcium pyrophosphate dihydrate deposition disease. Posteroanterior view of the wrist shows advanced degenerative changes in the radiocarpal and midcarpal joints.

Radiographic Changes

Joint calcifications in patients with intra-articular hydroxyapatite deposition are seen rarely on radiographs. When present, these calcifications are usually periarticular and unrelated to the hyaline cartilage or fibrocartilage. Radiographically the most common finding is an osteoarthritis-like picture or a destructive arthropathy resembling a neuroarthropathy (see Fig. 29–14). The disease is associated with chronic rotator cuff tear and a joint effusion containing hydroxyapatite crystals, collagenases, and proteases. The dominant shoulder typically is involved, although bilateral disease is common.

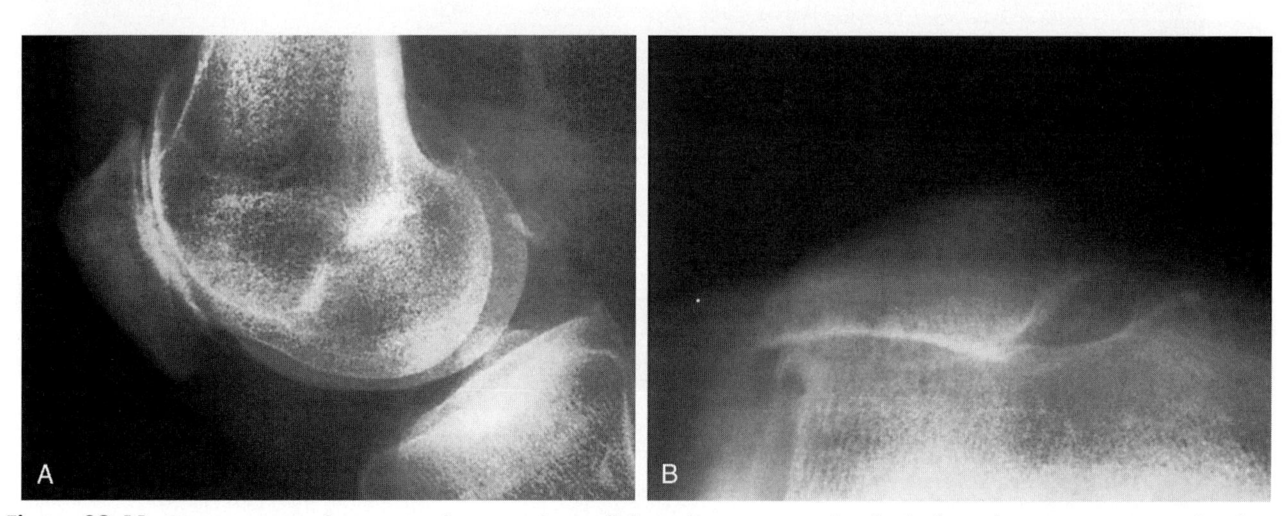

Figure 29–11. Disproportionate degenerative disease in the patellofemoral joint compared with tibiofemoral joint in a patient with calcium pyrophosphate dihydrate deposition disease. *A* and *B*, Lateral and Merchant's views of the knee show marked joint space narrowing and degenerative changes at the patellofemoral joint. The tibiofemoral joint is relatively well preserved.

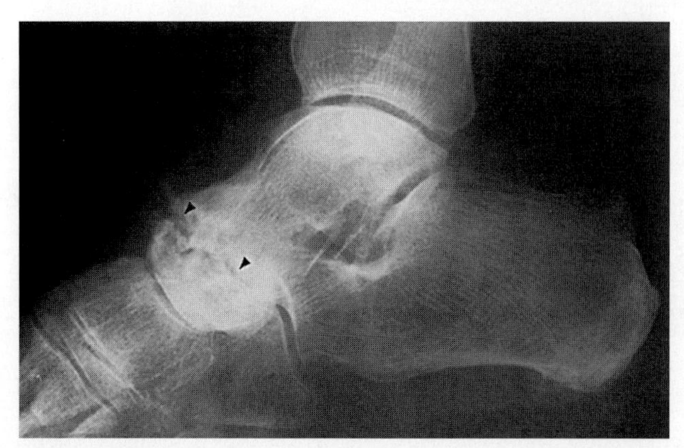

Figure 29–12. Advanced degenerative changes in the talonavicular joint resemble neuropathic arthropathy *(arrowheads)* in a patient with CPPD.

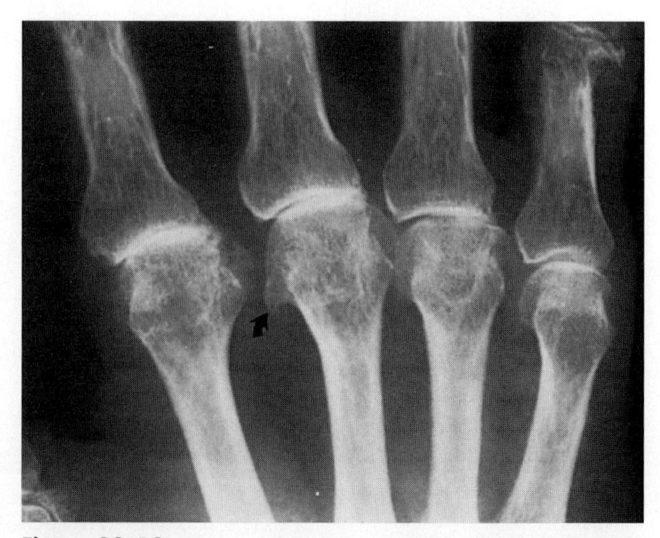

Figure 29–13. Changes of hemochromatosis in the hand. There is advanced degenerative disease in the metacarpophalangeal joints including the fourth and, to a lesser extent, the fifth metacarpophalangeal joints. A hook osteophyte is seen in the third metacarpophalangeal joint *(arrow)*.

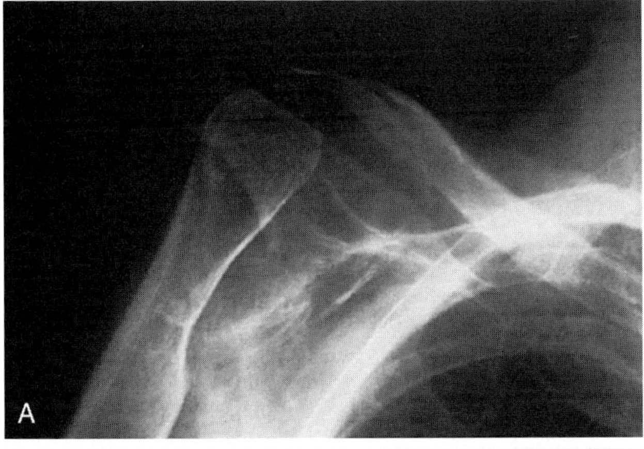

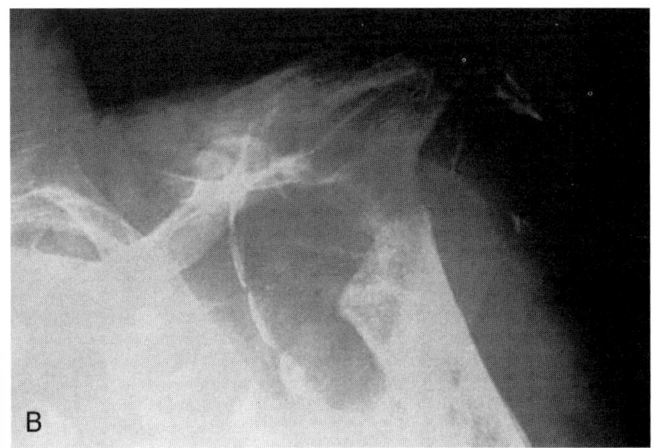

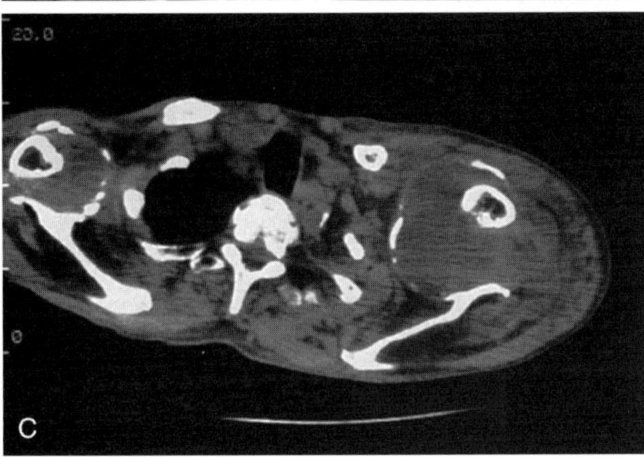

Figure 29–14. Calcium hydroxyapatite deposition disease affecting both shoulders (Milwaukee shoulders). *A* and *B*, Anteroposterior radiographs of the right and left shoulders show a severe destructive arthropathy involving both shoulders. The left shoulder closely resembles a neuroarthropathic joint. *C*, CT scan through the shoulder shows large joint effusions, more pronounced on the left. Capsular calcifications are present bilaterally. (Case provided by Dr. Jung Nguyen, University of Texas Health Sciences at San Antonio, San Antonio, TX.)

Recommended Readings

Adamson III TC, Resnik CS, Guerra Jr J, et al: Hand and wrist arthropathies of hemochromatosis and calcium pyrophosphate deposition disease: Distinct radiographic features. Radiology 147:377, 1983.

Alarcon GS, Reveille JD: Gouty arthritis of the axial skeleton including the sacroiliac joints. Arch Intern Med 147:2018, 1987.

Barthelemy CR, Nakayama DA, Carrera GF, et al: Gouty arthritis: A prospective radiologic evaluation of sixty patients. Skeletal Radiol 11:1, 1984.

Bloch C, Hermann G, Yu T-F: A radiologic reevaluation of gout: A study of 2,000 patients. AJR Am J Roentgenol 134:781, 1980.

Chaoui A, Garcia J, Kurt AM: Gouty tophus simulating soft tissue tumor in a heart transplant recipient. Skeletal Radiol 26:626, 1997.

Chen C, Chandnani VP, Kang HS, et al: Scapholunate advanced collapse: A common wrist abnormality in calcium pyrophosphate dihydrate crystal deposition disease. Radiology 177:459, 1990.

Dieppe PA, Crocker P, Huskisson EC, et al: Apatite deposition disease: A new arthropathy. Lancet 1:266, 1976.

El-Khoury GY, Tozzi JE, Clark CR, et al: Massive calcium pyrophosphate crystal deposition at the craniovertebral junction. AJR Am J Roentgenol 145:777, 1985.

Eustace S, Buff B, McCarthy C, et al: Magnetic resonance imaging of hemochromatosis arthropathy. Skeletal Radiol 23:547, 1994.

Foucar E, Buckwalter J, El-Khoury GY: Gout presenting as a femoral cyst: A case report. J Bone Joint Surg Am 66:294, 1984.

Gottlieb NL, Gray RG: Allopurinol-associated hand and foot deformities in chronic tophaceous gout. JAMA 238:1663, 1977.

Hirsch JH, Killien FC, Troupin RH: The arthropathy of hemochromatosis. Radiology 118:591, 1976.

Hodge JC, Ghelman B, DiCarlo EF, et al: Calcium pyrophosphate deposition within the ligamenta flava at L2, L3, L4, and L5. Skeletal Radiol 24:64, 1995.

Resnick D, Broderick TW: Intraosseous calcifications in tophaceous gout. AJR Am J Roentgenol 137:1157, 1981.

Resnick D, Niwayama G, Goergen TG, et al: Clinical, radiographic and pathologic abnormalities in calcium pyrophosphate dihydrate deposition disease (CPPD): Pseudogout. Radiology 122:1, 1977.

Rubenstein J, Pritzker KPH: Crystal-associated arthropathies. AJR Am J Roentgenol 152:685, 1989.

Steinbach LS, Resnick D: Calcium pyrophosphate dihydrate crystal deposition disease revisited. Radiology 200:1, 1996.

Yu JS, Chung C, Recht M, et al: MR imaging of tophaceous gout. AJR Am J Roentgenol 168:523, 1997.

CHAPTER 30

Neuropathic Arthropathy (Charcot Joint)

Georges Y. El-Khoury, MD, Mark D. Stanley, MD, and D. Lee Bennett, MD

OVERVIEW

Currently the most common cause for neuroarthropathy is diabetes mellitus. Other causes include syringomyelia and rarely a variety of other disorders, such as tabes dorsalis, leprosy, poliomyelitis, multiple sclerosis, myelodysplasia, and congenital insensitivity to pain. Neuroarthropathy is believed to be due to diminished pain sensation and proprioception, which leaves the extremities and spine without protection from repeated microtrauma in the face of continued activity. About one third of patients have pain in the affected joint at the time of presentation with early Charcot changes. One relatively common presentation is an acute form of the neuroarthropathy in which a rapid and severely disorganizing arthritis destroys the joint in a few weeks (Fig. 30–1). The joint typically is swollen, warm, and erythematous. Neuropathic joints occasionally are observed in paralyzed patients who are completely bedridden and are believed to be due to minor trauma from passive exercise and transportation during caring for the patient (Fig. 30–2).

RADIOGRAPHIC FINDINGS

The earliest finding in neuroarthropathy may be a fracture or, much less commonly, a dislocation. The foot and ankle are the most commonly affected sites in diabetes mellitus (Figs. 30–3 and 30–4). In order of frequency, the

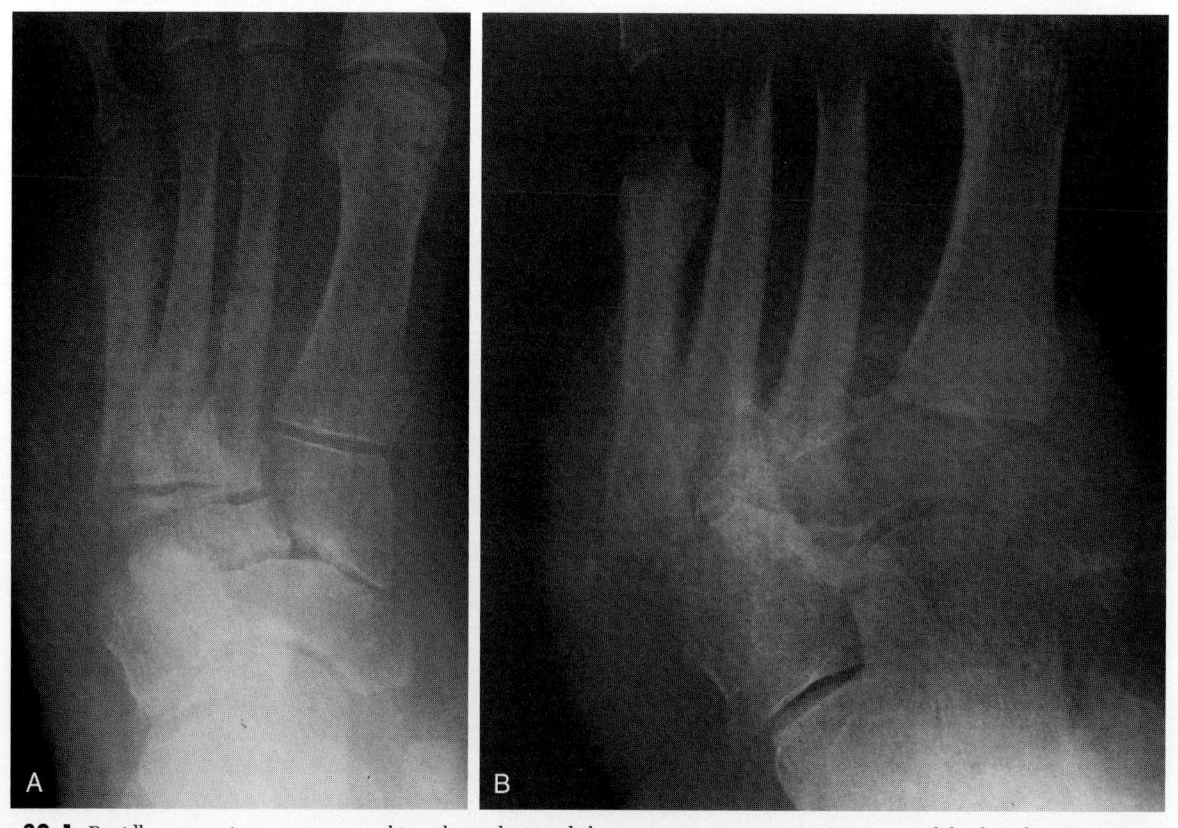

Figure 30–1. Rapidly progressing acute neuropathic arthropathy in a diabetic patient. *A,* Anteroposterior view of the foot shows amputation of the fifth ray and part of the fourth ray as a result of previous infection. *B,* Immediately after the first image was taken *(A),* the foot started to swell, and in 3 months the tarsometatarsal joints became completely disrupted.

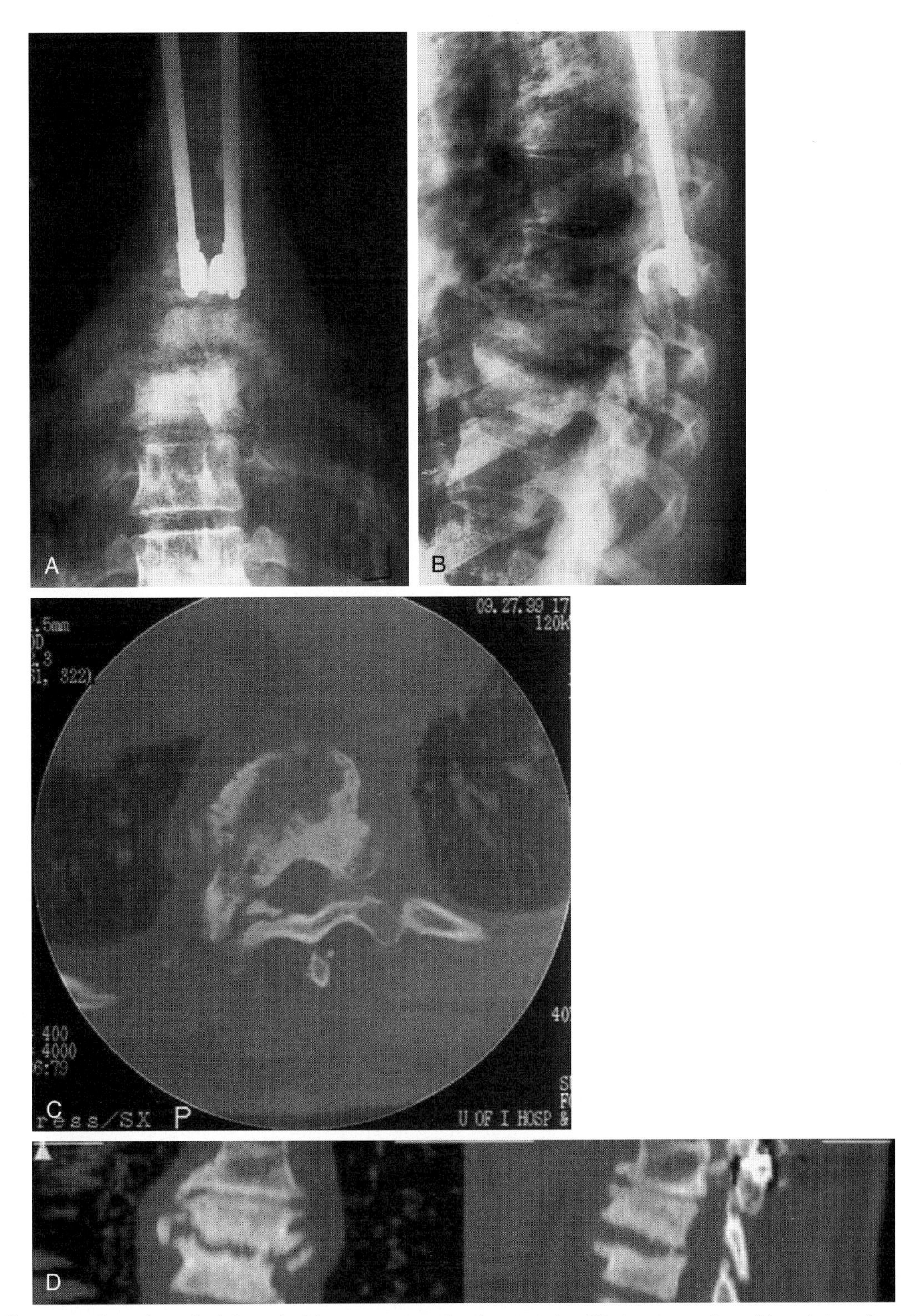

Figure 30–2. Neuropathic arthropathy (Charcot) of the spine in a paralyzed patient. *A* and *B*, Anteroposterior and lateral views show significant bone destruction and sclerosis at the T7-8 level with associated soft tissue swelling. A needle biopsy, performed to rule out infection, was negative. *C*, Axial CT image through T7-8 reveals bone sclerosis and fragmentation surrounded by soft tissue swelling. *D*, Coronal and sagittal CT reconstructions reveal Charcot changes.

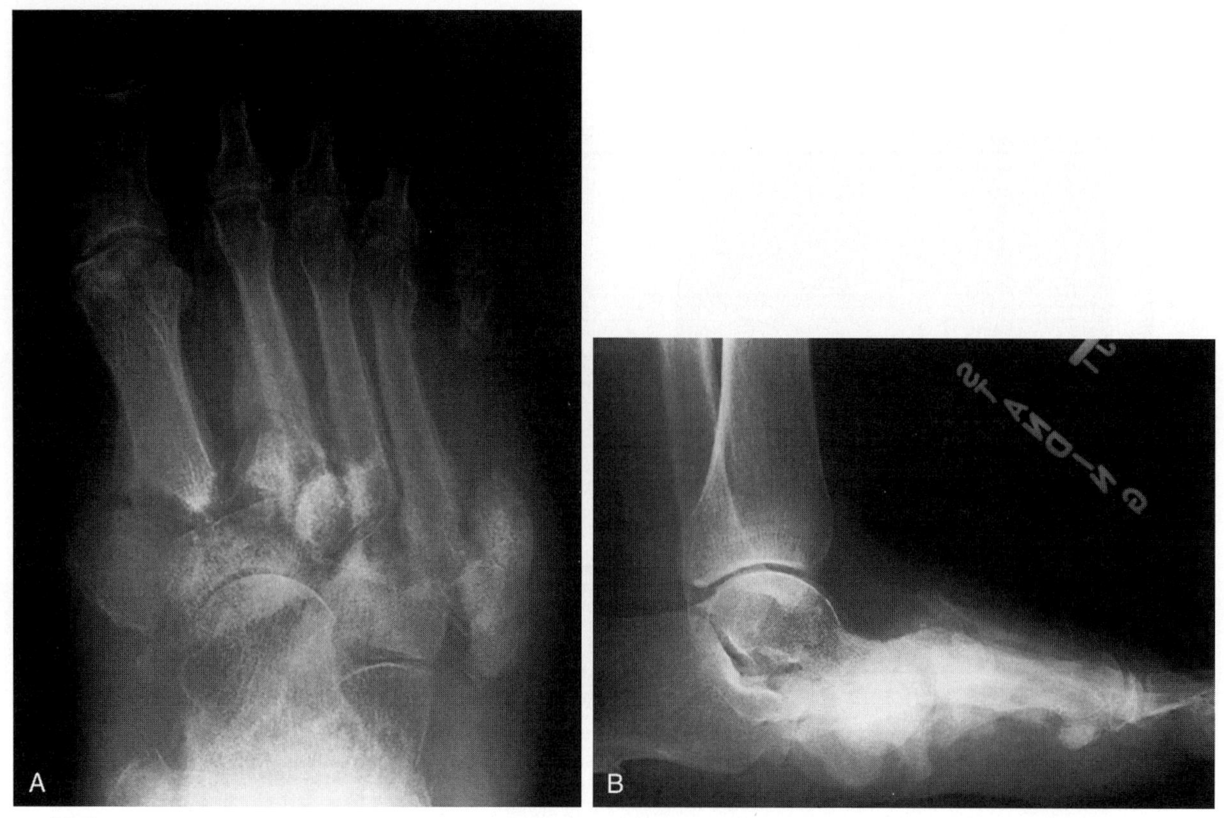

Figure 30–3. Tarsometatarsal Charcot changes. *A* and B, Anteroposterior and lateral views of the left foot in a diabetic patient reveal typical midfoot Charcot changes with bone destruction, fragmentation, and joint disorganization at the tarsometatarsal joints.

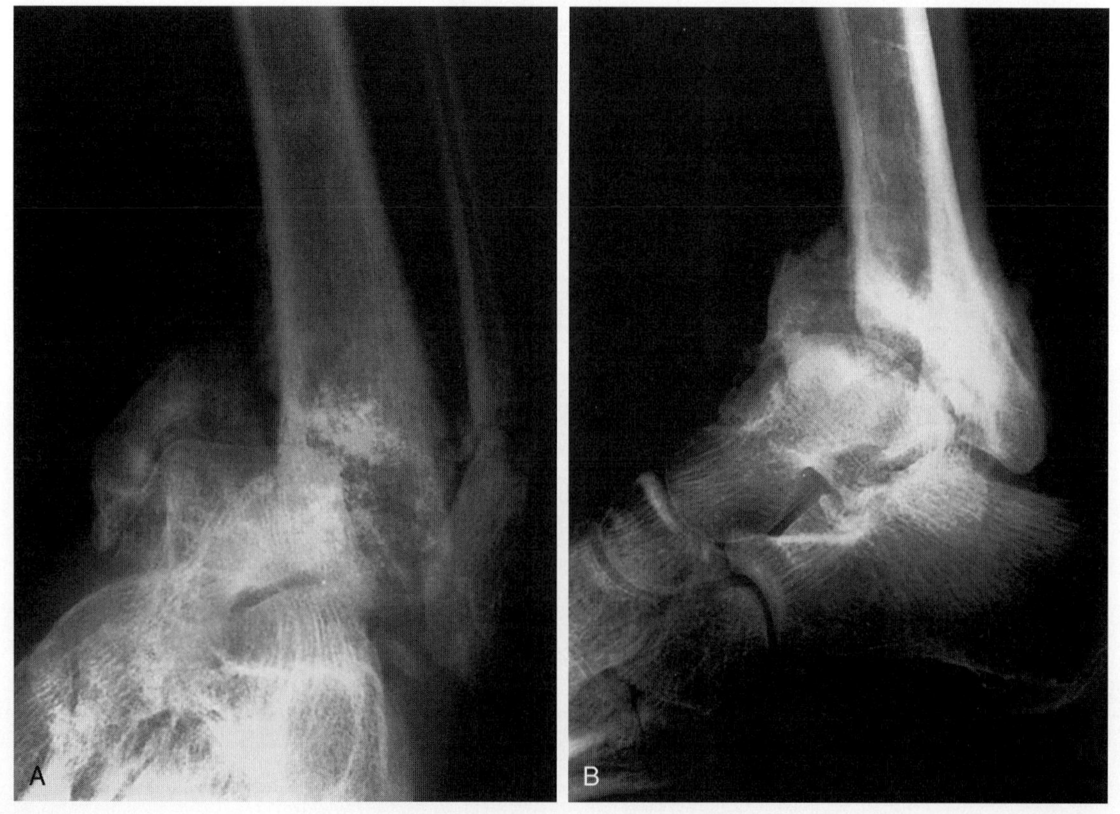

Figure 30–4. Hypertrophic type of neuropathic arthropathy of the ankle in a patient with diabetes. The patient presented with progressive deformity and swelling of the left ankle. There was no definite history of trauma. *A* and *B*, Anteroposterior and lateral views of the ankle show severe Charcot changes with deformity and swelling.

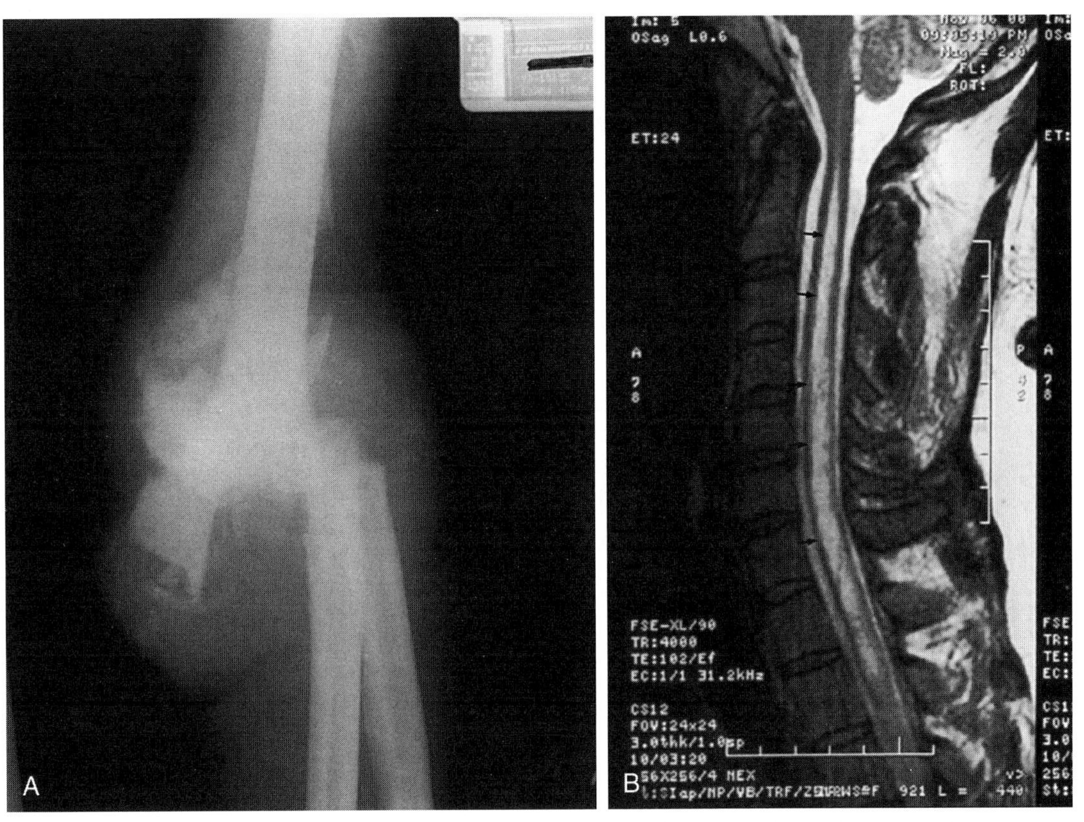

Figure 30–5. Hypertrophic type of neuropathic arthropathy (Charcot) in a patient with a syrinx in the cervical cord. *A,* Anteroposterior view of the elbow shows marked disorganization of the joint, fragmentation, and soft tissue swelling. The lack of osteopenia is remarkable in this elbow. *B,* T2-weighted midsagittal MR image of the cervical spine reveals a large syrinx extending from C2 to T2 *(arrowheads).*

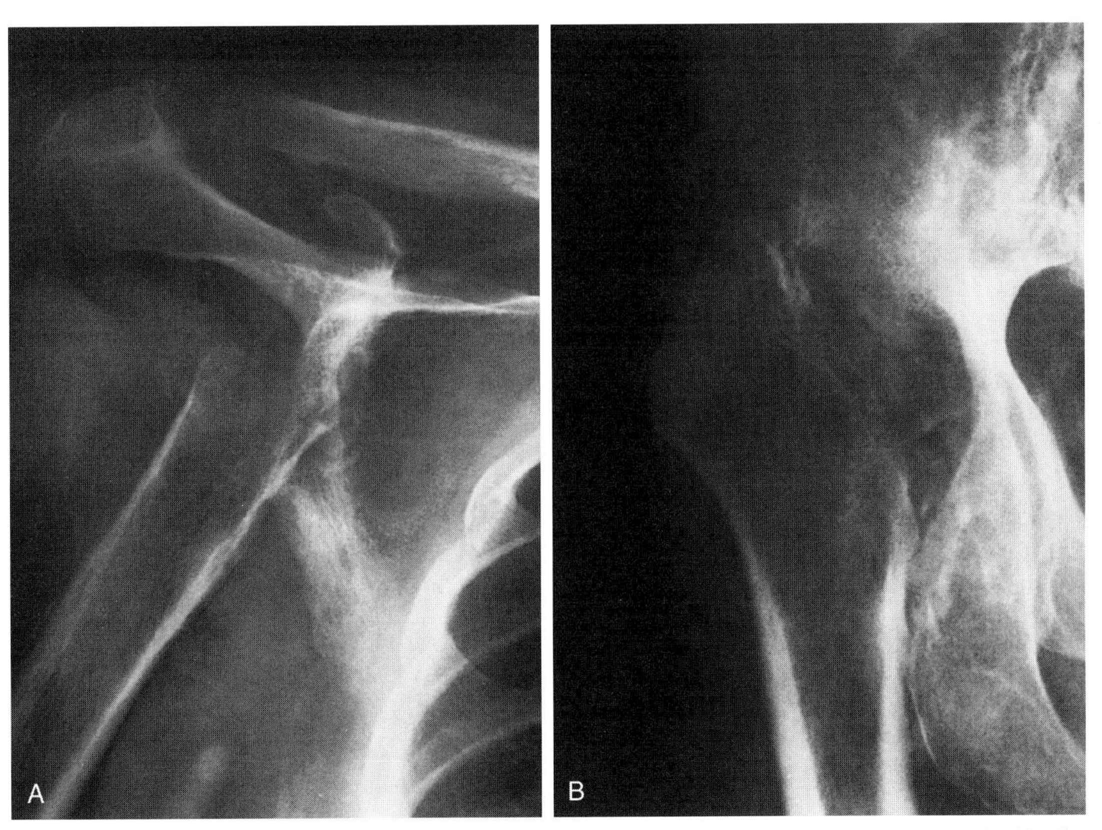

Figure 30–6. Atrophic Charcot joints in a 65-year-old patient with a history of syphilis. *A,* Anteroposterior view of the right shoulder shows complete resorption of the humeral head and neck. The glenoid is resorbed partially. *B,* Anteroposterior view of the right hip in the same patient shows complete resorption of the femoral head and neck. The acetabulum is large and shallow because of bony resorption.

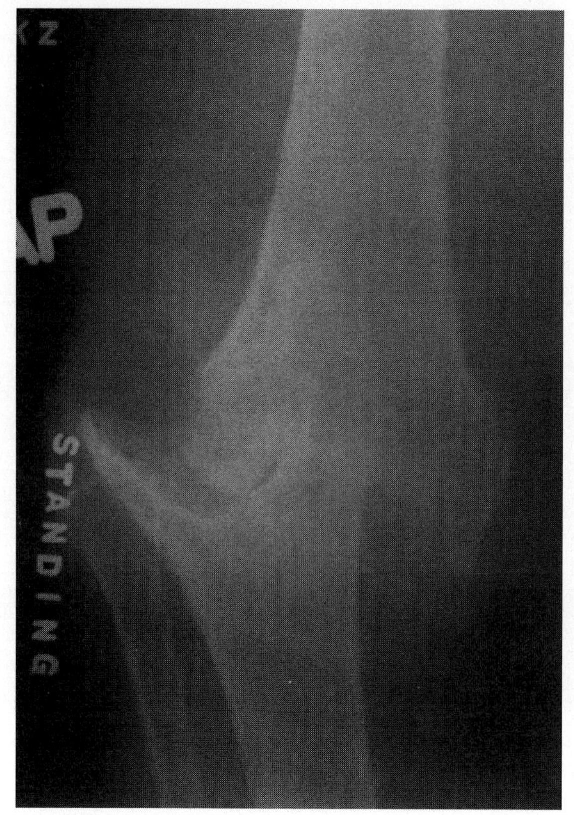

Figure 30–7. Atrophic Charcot joint of the right knee in a patient with long-standing, poorly controlled diabetes.

tarsometatarsal, metatarsophalangeal, and ankle joints are most frequently involved.

Radiographically, there are two types of neuropathic joints: hypertrophic and atrophic. The hypertrophic type is much more common, especially in the weight-bearing joints, in which it has been described to resemble severe osteoarthritis "with a vengeance." The joint typically shows marked disorganization, fragmentation, sclerosis, and huge osteophyte formation (Figs. 30–4 and 30–5). In its advanced stages, hypertrophic neuroarthropathy is easy to diagnose. The atrophic type typically is observed in the shoulder and rarely in the hip (Figs. 30–6 and 30–7). It frequently poses a diagnostic problem because it resembles infection or neoplasm. The proximal humerus often appears as if surgically resected. In the spine, hypertrophic Charcot changes have been reported in paraplegic patients (see Fig. 30–2). These changes start at the disk level, typically in the lumbar spine, and can occur at multiple disk levels below the spinal cord injury.

In our experience, the earliest finding in a Charcot joint

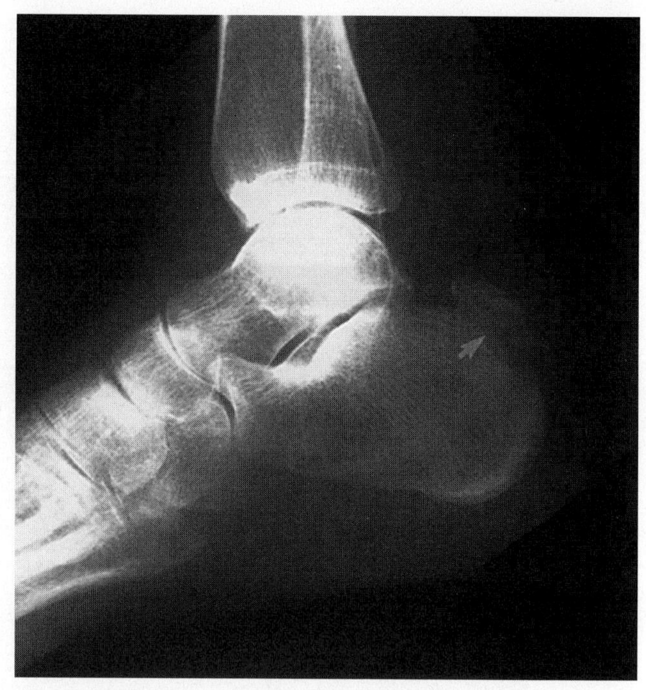

Figure 30–8. Calcaneal insufficiency avulsion fracture *(arrow)*, which is almost pathognomonic of diabetic neuropathy.

is a fracture. At least one fracture appears to be almost pathognomonic of diabetic neuropathy. This fracture occurs at the insertion of the Achilles' tendon to the calcaneal tuberosity, and it is called *calcaneal insufficiency avulsion fracture* (Fig. 30–8).

Recommended Readings

Allman RM, Brower AC, Kotlyarov EB: Neuropathic bone and joint disease. Radiol Clin North Am 26:1373, 1988.

Edelman SV, Kosofsky EM, Paul RA, et al: Neuro-osteoarthropathy (Charcot's joint) in diabetes mellitus following revascularization surgery: Three case reports and a review of the literature. Arch Intern Med 147:1504, 1987.

El-Khoury GY, Kathol MH: Neuropathic fractures in patients with diabetes mellitus. Radiology 134:313, 1980.

Gold RH, Tong DJF, Crim JR, et al: Imaging the diabetic foot. Skeletal Radiol 24:563, 1995.

Hatzis N, Kaar TK, Wirth MA, et al: Neuropathic arthropathy of the shoulder. J Bone Joint Surg Am 80:1314, 1998.

Mitchell ML, Lally JF, Ackerman LV, et al: Case report 697. Skeletal Radiol 20:550, 1991.

Newman JH: Spontaneous dislocation in diabetic neuropathy. J Bone Joint Surg Br 61:484, 1979.

Norman A, Robbins H, Milgram JE: The acute neuropathic arthropathy—a rapid, severely disorganizing form of arthritis. Radiology 90:1159, 1968.

Park Y-H, Taylor JAM, Szollar SM, et al: Imaging findings in spinal neuroarthropathy. Spine 19:1499, 1994.

CHAPTER 31

Amyloidosis

Georges Y. El-Khoury, MD, Mark D. Stanley, MD, and D. Lee Bennett, MD

OVERVIEW

Amyloidosis is a heterogeneous group of disorders characterized by the extracellular deposition of an amorphous material that stains with Congo red. When these deposits replace normal tissue constituents, the function of the affected organ often is compromised.

The currently used classification was proposed at the International Conference of Amyloidosis in 1979, and it is based on the biochemical composition of the amyloid fibrils. The main types are described.

PRIMARY AMYLOIDOSIS

Primary amyloidosis (AL amyloid) is seen in patients with primary or idiopathic amyloidosis and in patients with multiple myeloma. It consists of an immunoglobulin light chain protein (*A* for amyloidosis and *L* for light chain [AL amyloid]). Most patients with primary amyloidosis have a monoclonal protein detectable by immunoelectrophoresis of the serum or urine even in the absence of multiple myeloma. AL amyloid usually deposits in the tongue, heart, and other visceral organs. Rarely, it deposits in the bones and joints, producing imaging findings similar to dialysis-related amyloidosis (DRA).

SECONDARY AMYLOIDOSIS

Secondary amyloidosis (AA amyloid) is associated with chronic infections, such as tuberculosis and osteomyelitis, and with chronic inflammatory conditions, such as rheumatoid arthritis, Crohn's disease, and familial Mediterranean fever. AA amyloid typically deposits in the visceral organs, such as kidneys, spleen, and bowel.

DIALYSIS-RELATED AMYLOIDOSIS

Dialysis-related amyloidosis (DRA) is characterized by increased serum concentration of β_2-microgloblulin and deposition of amyloid fibrils in a variety of tissues, especially bone, tendons, and joints. The prevalence of DRA increases progressively with the duration of dialysis and reaches almost 80% after 15 years of therapy. The reason for preferential accumulation of this amyloid in the musculoskeletal system is not known. One of the most common complications of DRA is carpal tunnel syndrome; it is usually the first symptom of this form of amyloidosis. Amyloid depositing in the carpal tunnel entraps the median nerve and causes carpal tunnel syndrome.

Radiographic Features

Destructive Arthropathy

Destructive arthropathy is a common feature of DRA. It is caused by the deposition of β_2-microglobulin in the bone and soft tissue around joints. This deposition manifests as large erosions and cystic changes affecting the joints symmetrically (Fig. 31–1). The most commonly affected joints are the wrists, hips, shoulders, and elbows. The joint spaces are relatively preserved until late in the disease. The radiographic appearance closely resembles pigmented villonodular synovitis or gout. Intraosseous accumulation of amyloid produces cystic destruction of bone resembling brown tumors, which also are seen in this patient population (Fig. 31–2). These focal lesions are referred to as *intraosseous amyloidomas.* In the vertebral bodies, amyloidomas can cause vertebral collapse and neurologic deficit (Fig. 31–3). Pathologic fractures in the hip have resulted from intraosseous amyloid deposits. Renal transplant has been shown to reduce the joint symptoms but does not seem to reverse the existing bony changes.

Spondyloarthropathy

Spondyloarthropathy closely resembles disk space infection in patients with long-standing hemodialysis. There is disk space narrowing with erosions of the opposing vertebral end plates. Needle biopsy is often necessary to arrive at the diagnosis.

Magnetic Resonance Imaging

Many magnetic resonance (MR) imaging studies have been published evaluating patients on long-term dialysis.

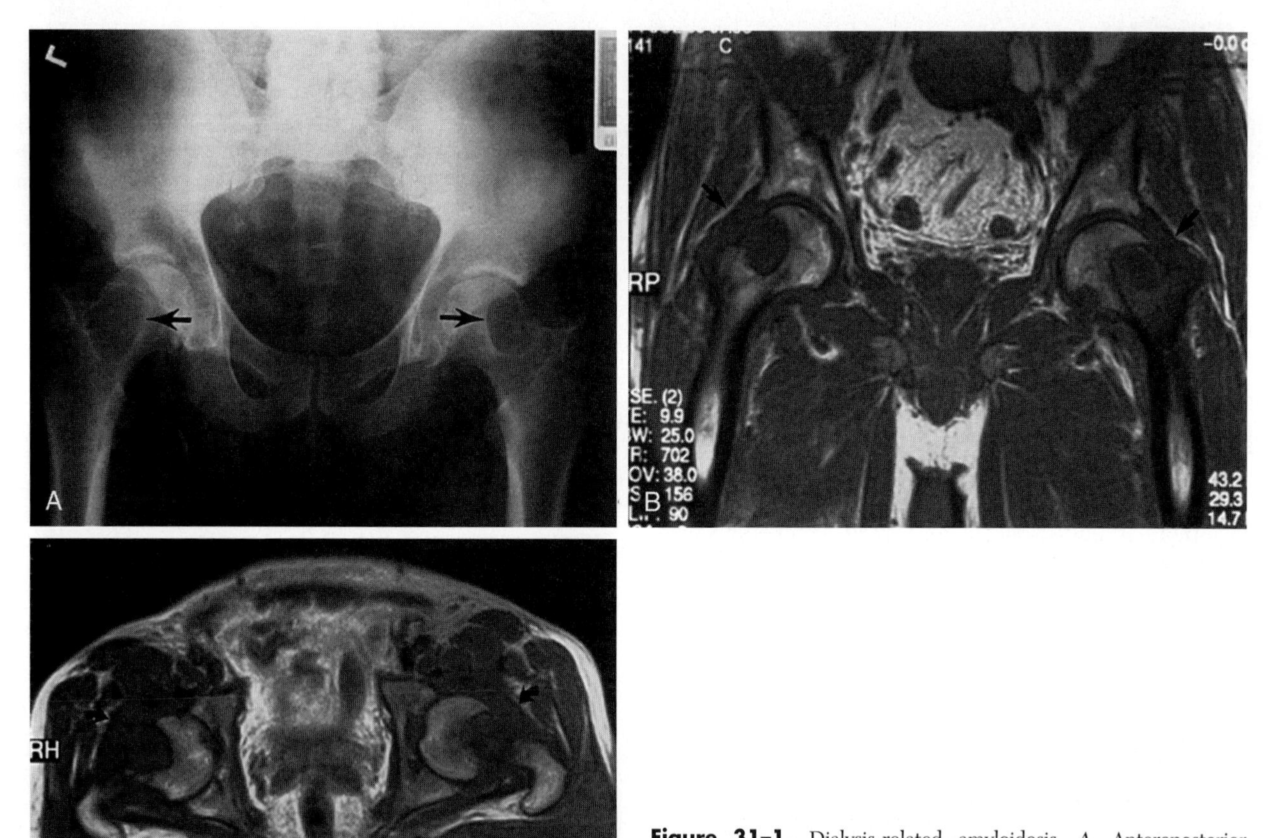

Figure 31-1. Dialysis-related amyloidosis. *A,* Anteroposterior radiograph of the pelvis shows lytic lesions in both femoral necks *(arrows). B,* Coronal T1-weighted MR image through the hips shows the cystic areas. The joint capsule is thickened bilaterally *(arrows). C,* Axial T1-weighted image shows the amyloid deposits anterior to the cystic lesions *(arrows).*

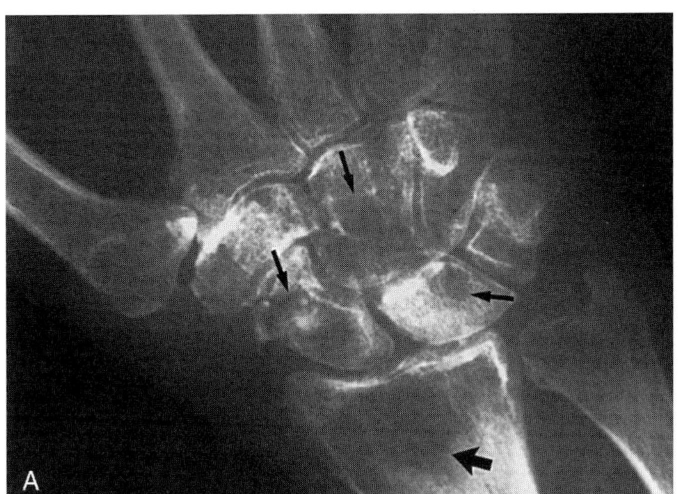

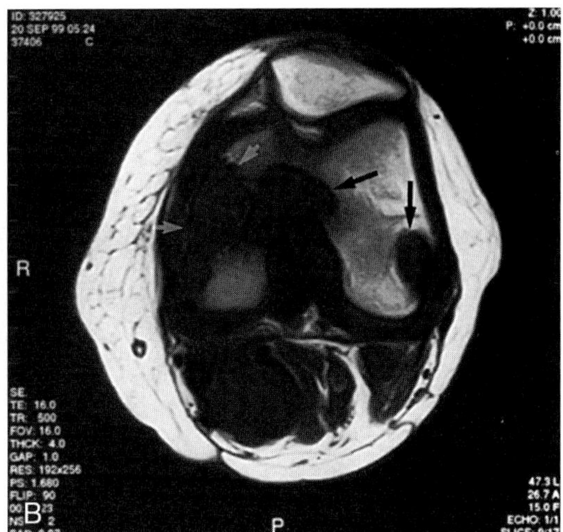

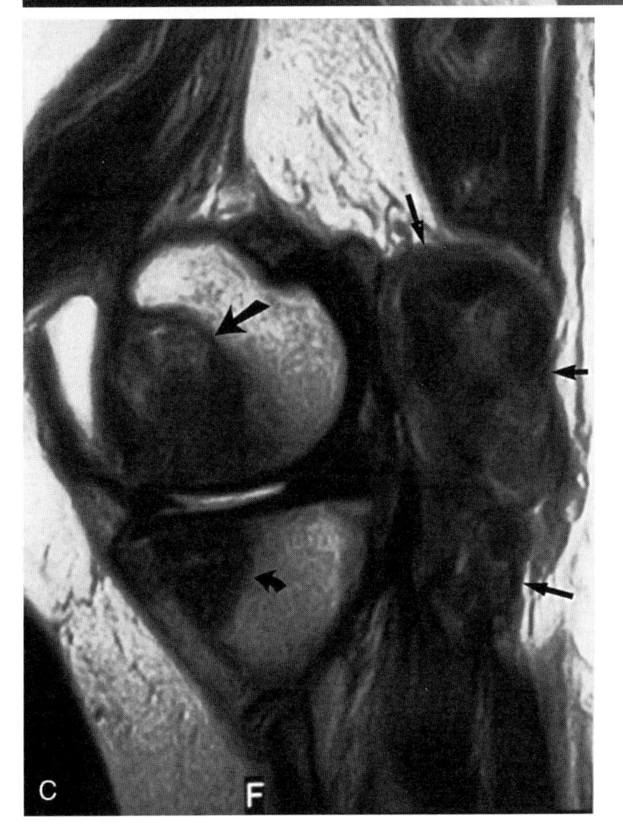

Figure 31–2. Intraosseous deposits in a patient with dialysis-related amyloidosis. *A,* Posteroanterior view of left wrist shows several cystic lucencies in the distal radius, distal ulna, and carpal bones *(arrows).* These cystic changes are believed to be due to intraosseous deposition of amyloid. *B,* T1-weighted axial MR image of the knee shows amyloid deposits in the femoral condyles *(arrows).* Amyloid has low signal intensity on this sequence. *C,* T2-weighted sagittal MR image shows amyloid masses within the bone and in the soft tissues posteriorly *(arrows).* The signal intensity of the amyloid has increased slightly compared with the T1-weighted image, but generally it is still low.

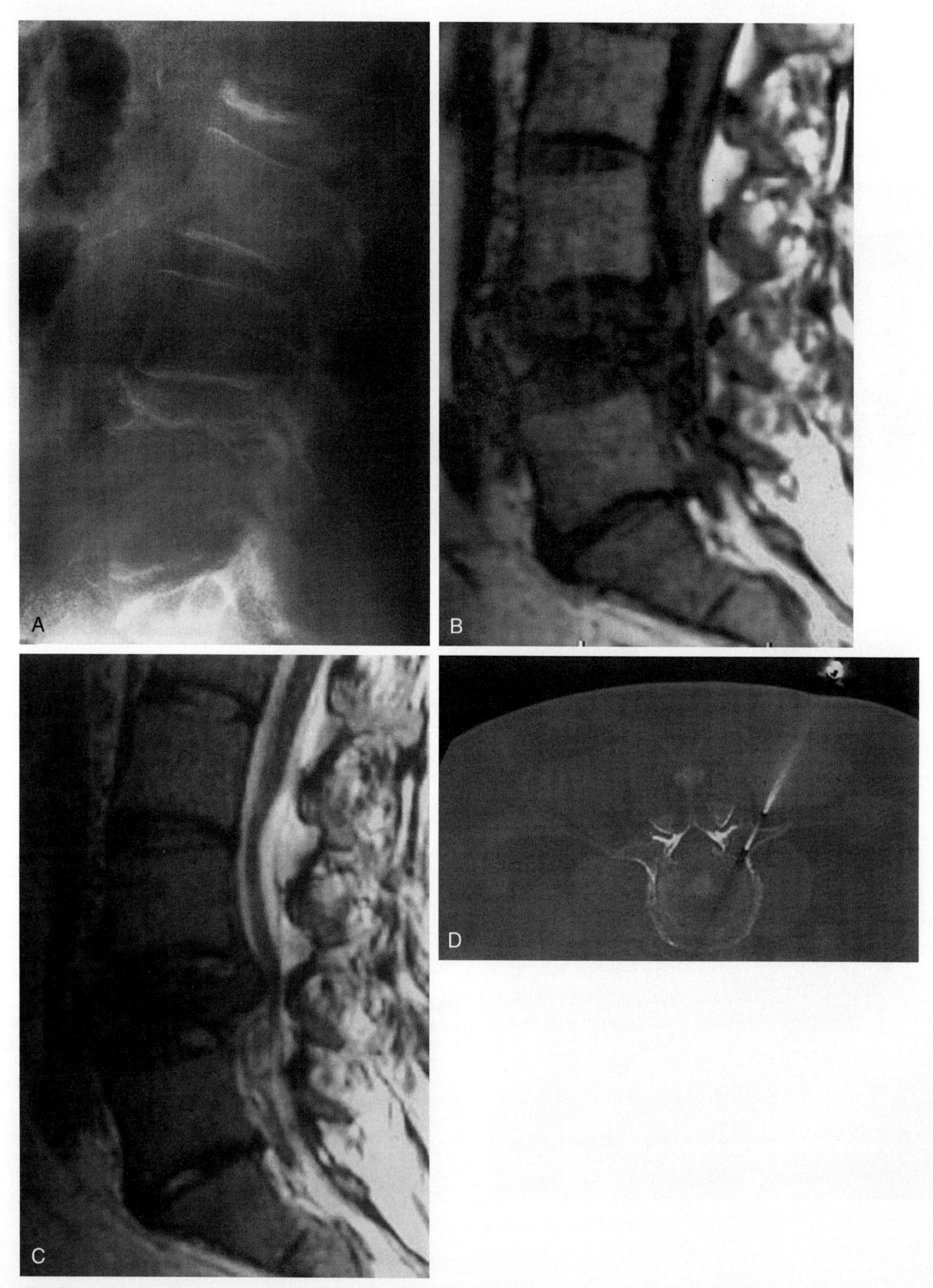

Figure 31–3. Vertebral collapse caused by amyloid deposits (intraosseous amyloidoma) in a patient with dialysis-related amyloidosis. *A,* Lateral view of the lumbar spine shows collapse of the vertebral bodies of L1 and L4. *B* and *C,* Sagittal T1- and T2-weighted MR images of the lumbar spine show infiltration of the body of L4 with low signal intensity material. The collapsed body is retropulsed into the spinal canal and is compressing the thecal sac. *D,* Needle biopsy specimen of L4 revealed amyloid.

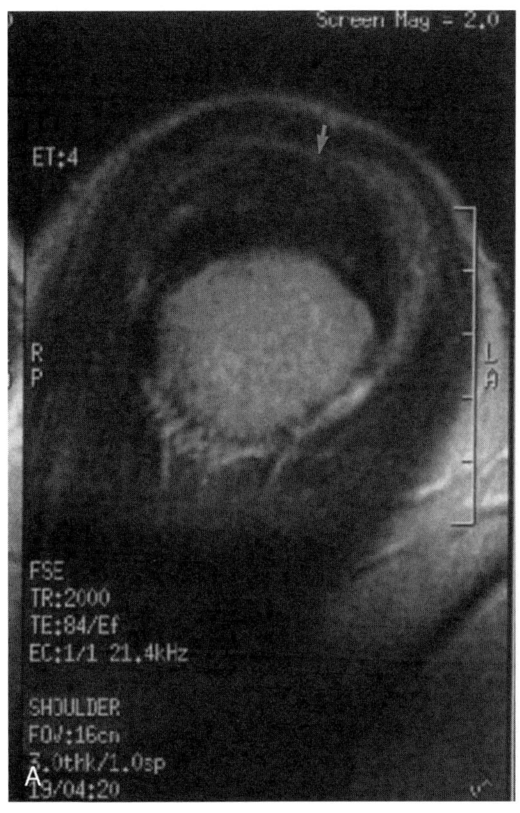

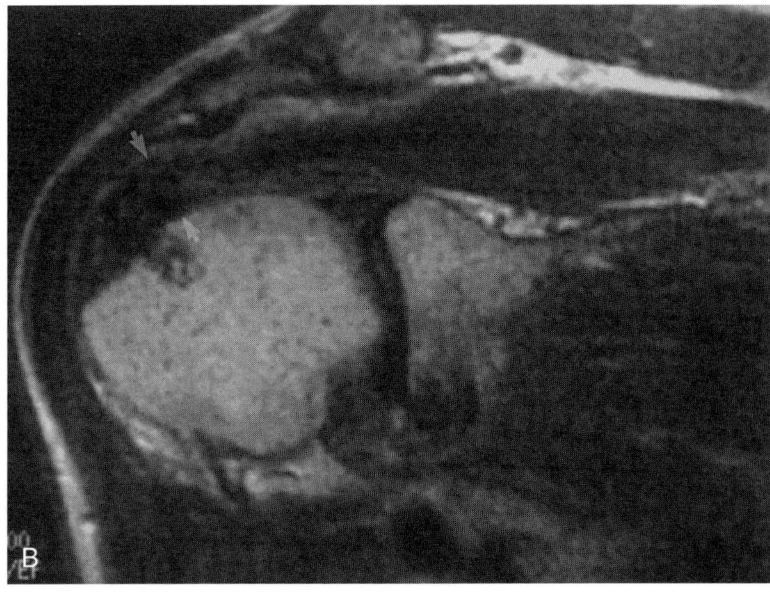

Figure 31–4. Thickening of the supraspinatus tendon in a patient on long-term hemodialysis. *A,* T2-weighted sagittal MR image shows thickening of the supraspinatus tendon believed to be due to amyloidosis *(arrow)*. *B,* T2-weighted oblique coronal image reveals thickening of the supraspinatus tendon *(arrows)* and low signal intensity. There is no rotator cuff tear. A small subchondral cyst is present in the humeral head.

The findings include (1) tendon thickening, especially of the supraspinatus tendon (Fig. 31–4); (2) capsular thickening, best shown in the iliofemoral ligament of the hip (see Fig. 31–1); (3) fluid collection within joints and in bursae around the joints; and (4) focal osseous cystic lesions in joints. Ranging in size from 2 mm to 4 cm, these cystic lesions communicate with the joint in most cases (see Figs. 31–1 and 31–2).

MR signal from amyloid deposits within the soft tissues typically is low on T1- and T2-weighted sequences (see Fig. 31–2). Intraosseous lesions show low signal on T1-weighted images and variable, although usually high, signal on T2-weighted images.

Recommended Readings

Campistol JM, Sole M, Munoz-Gomez J, et al: Pathological fractures in patients who have amyloidosis associated with dialysis. J Bone Joint Surg Am 72:568, 1990.

Escobedo EM, Hunter JC, Zink-Brody GC, et al: Magnetic resonance imaging of dialysis-related amyloidosis of the shoulder and hip. Skeletal Radiol 25:41, 1996.

Fitzpatrick DC, Jebson PJL, Madey SM, et al: Upper extremity musculoskeletal manifestations of dialysis-associated amyloidosis. Iowa Orthop J 16:135, 1996.

Kurer MHJ, Baillod RA, Madgwick JCA: Musculoskeletal manifestations of amyloidosis: A review of 83 patients on haemodialysis for at least 10 years. J Bone Joint Surg Br 73:271, 1991.

Otake S, Tsuruta Y, Yamana D, et al: Amyloid arthropathy of the hip joint: MR demonstration of presumed amyloid lesions in 152 patients with long-term hemodialysis. Eur Radiol 8:1352, 1998.

Ross LV, Ross GJ, Mesgarzadeh M, et al: Hemodialysis-related amyloidomas of bone. Radiology 178:263, 1991.

CHAPTER 32

Hemophilic Arthropathy

Georges Y. El-Khoury, MD, Mark D. Stanley, MD, and D. Lee Bennett, MD

OVERVIEW

Hemophilia is a group of hereditary hemorrhagic disorders characterized by abnormal blood coagulation. It is caused by a deficiency in factor VIII (hemophilia A) or factor IX (hemophilia B or Christmas disease). Degenerative arthropathy and other musculoskeletal abnormalities occur only with severe deficiencies of factor VIII or IX (i.e., when these factors are <5% of the normal level). Hemorrhage seems to start in the synovium, then ruptures into the joint. There is evidence that with recurrent hemarthrosis, large amounts of iron pigment are released into the joint, which are taken up by the synovium and subsequently stimulate the development of a proliferative synovitis (Fig. 32–1). This abnormal synovium is capable of producing proteolytic and collagenolytic enzymes, causing destruction of the articular cartilage. The knee, elbow, and ankle are the most frequently affected joints.

RADIOGRAPHIC FINDINGS

Early in the disease, radiographic findings are limited to soft tissue swelling without skeletal abnormalities. With progressive disease, there is osteopenia, epiphyseal overgrowth, subchondral cyst formation, and subsequent joint space narrowing. In end-stage disease, secondary osteoarthritis develops, which may become severe and crippling.

Patients with progressive hemophilic arthropathy often are followed with joint surveys, and depending on the severity of the disease, each joint is staged. Following is a summary of the Arnold and Hilgartner (1977) staging system:

I. Stage I (Fig. 32–2)
   A. Soft tissue swelling or effusion
   B. No skeletal abnormalities
II. Stage II (Fig. 32–3)
   A. Osteoporosis and epiphyseal overgrowth
   B. No joint space narrowing
III. Stage III (Fig. 32–4)
   A. Subchondral cyst formation
   B. Squaring of patella
   C. Widening of intercondylar notch of knee
   D. Widening of trochlear notch of ulna
   E. Synovial opacification owing to hemosiderin deposits (Fig. 32–5)
   F. No joint space narrowing

All the joint changes in stages I, II, and III are reversible. The changes in stages IV and V are irreversible.

I. Stage IV (Fig. 32–6)
   A. Cartilage destruction with joint space narrowing plus findings from the previous three stages
II. Stage V (Fig. 32–7)
   A. Substantial disorganization of joint structures
   B. Severe loss of joint space and degenerative changes
   C. Extensive enlargement of epiphyses
   D. Fibrous joint contractures

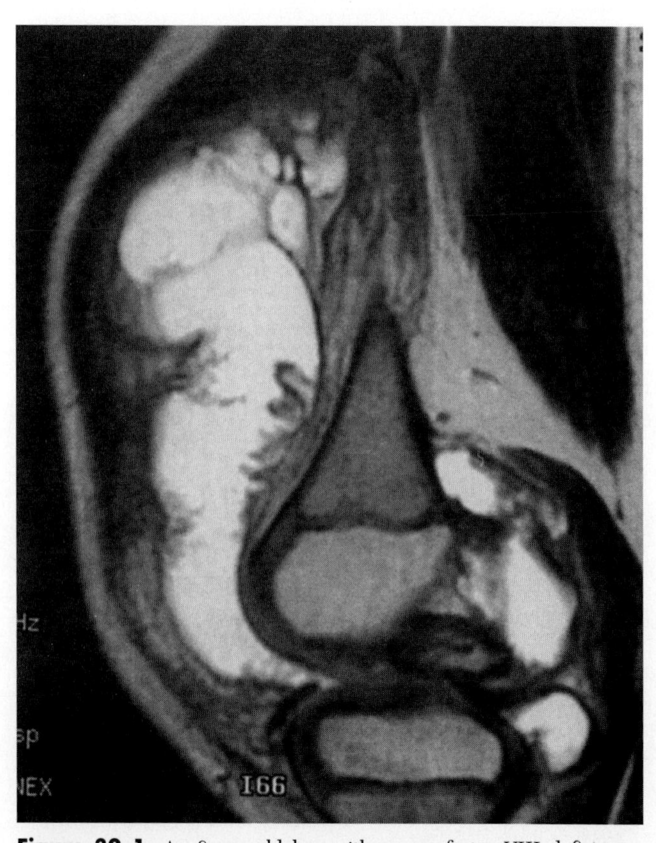

Figure 32–1. An 8-year-old boy with severe factor VIII deficiency complained of recurrent right knee bleeding. Sagittal T2-weighted MR image shows a large joint effusion (fresh blood). The synovium is thickened and shows multiple fronds projecting into the fluid.

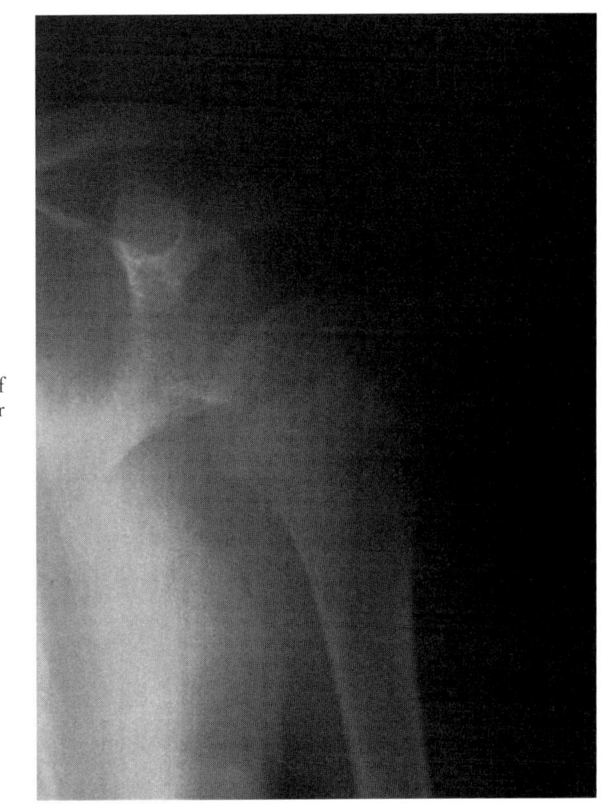

Figure 32–2. Stage I hemophilic arthropathy. There is pseudosubluxation of the left shoulder as a result of an acute intra-articular bleed. There are no other abnormalities.

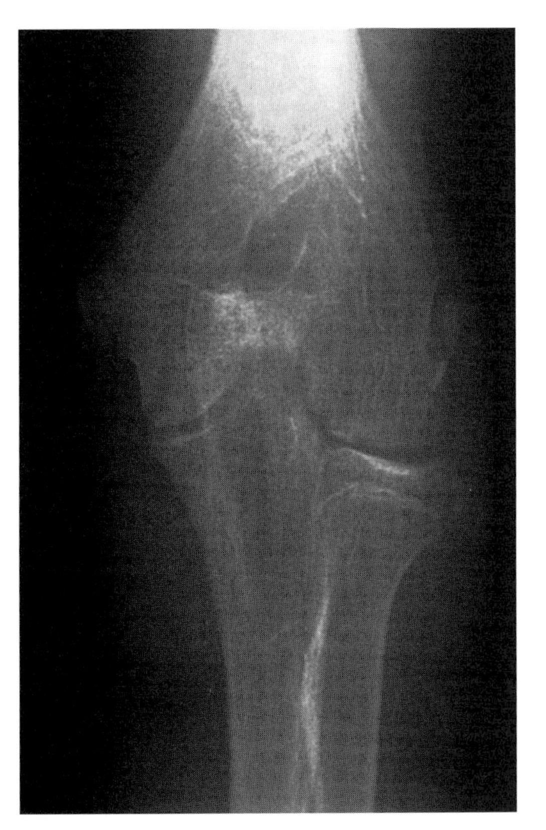

Figure 32–3. Stage II hemophilic arthropathy. Anteroposterior view of the left elbow in a teenage boy with severe factor VIII deficiency shows significant osteopenia and overgrowth of the epiphyses in the elbow, especially the epiphysis of the radial head.

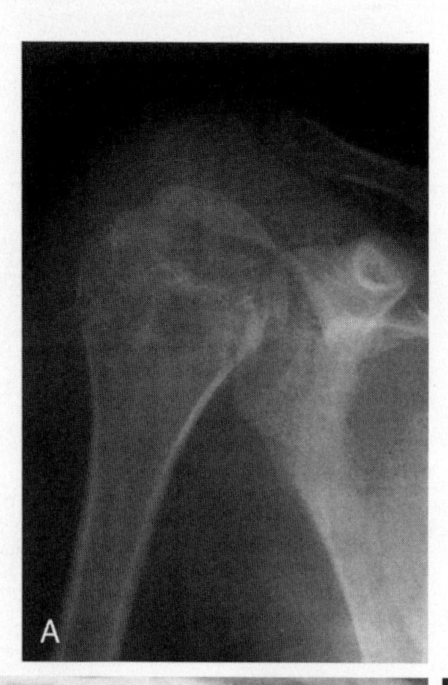

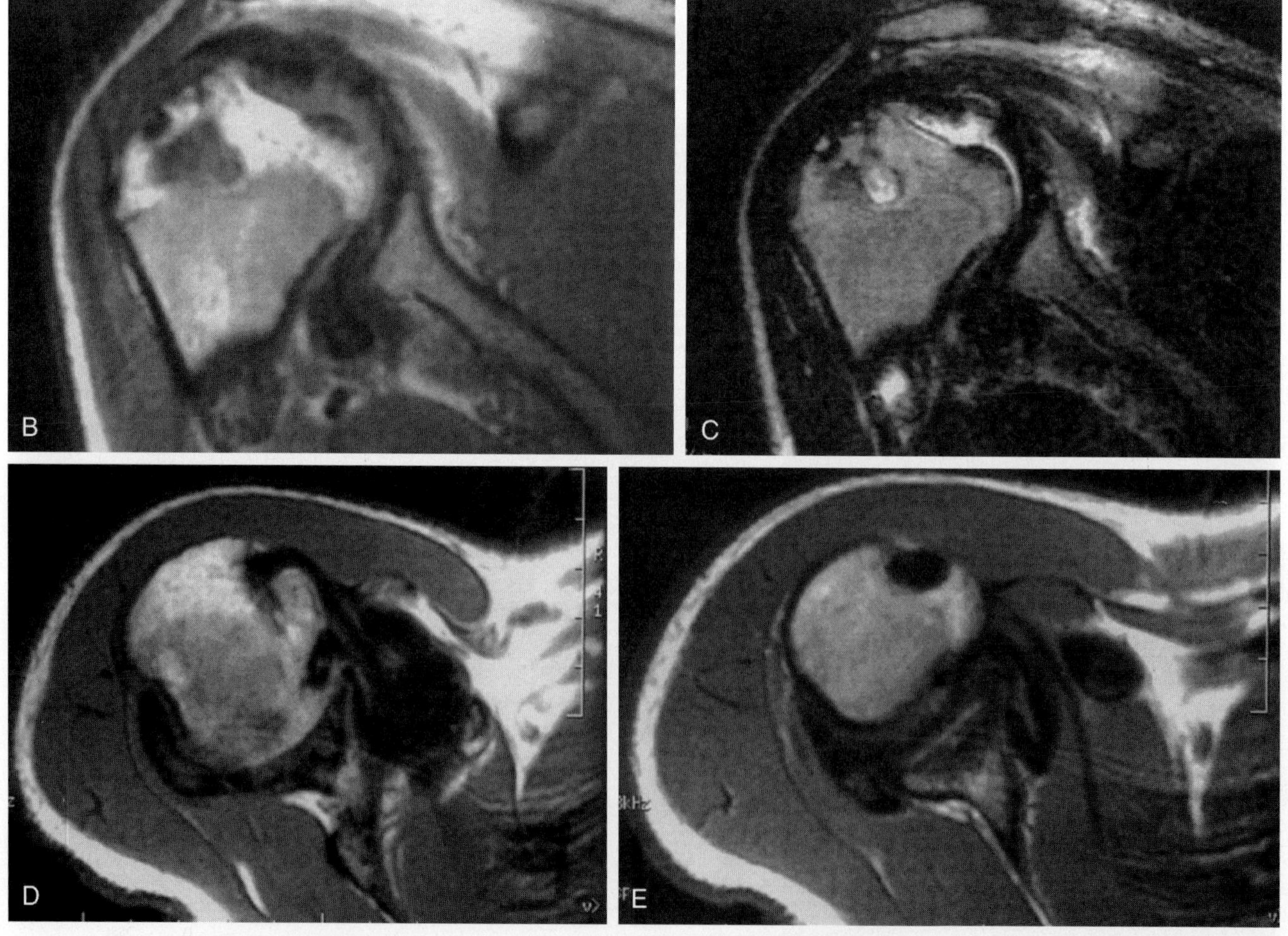

Figure 32–4. Stage III hemophilic arthropathy in a 16-year-old boy. *A*, Anteroposterior view of the shoulder shows deformity of the humeral head and subchondral cysts. There is no evidence of joint space narrowing. *B* and *C*, Oblique coronal T1- and T2-weighted MR images of the right shoulder show the deformity of the humeral head, the subchondral cysts, and low signal intensity in the synovium. *D* and *E*, Axial T1-weighted MR images of the shoulder show thickened and hypertrophied synovium, which appears dark. The appearance of the synovium is attributed to hemosiderin deposition from recurrent bleeding.

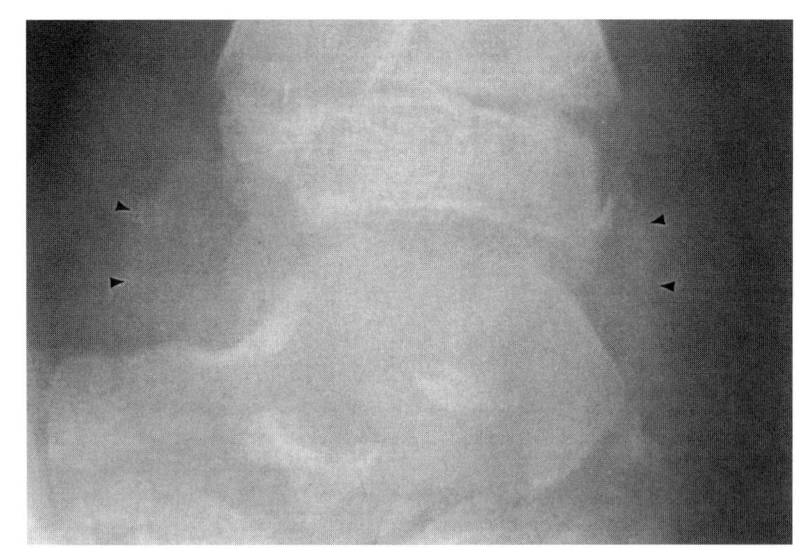

Figure 32–5. Synovial opacification *(arrowheads)* in the ankle of a 6-year-old boy with hemophilia.

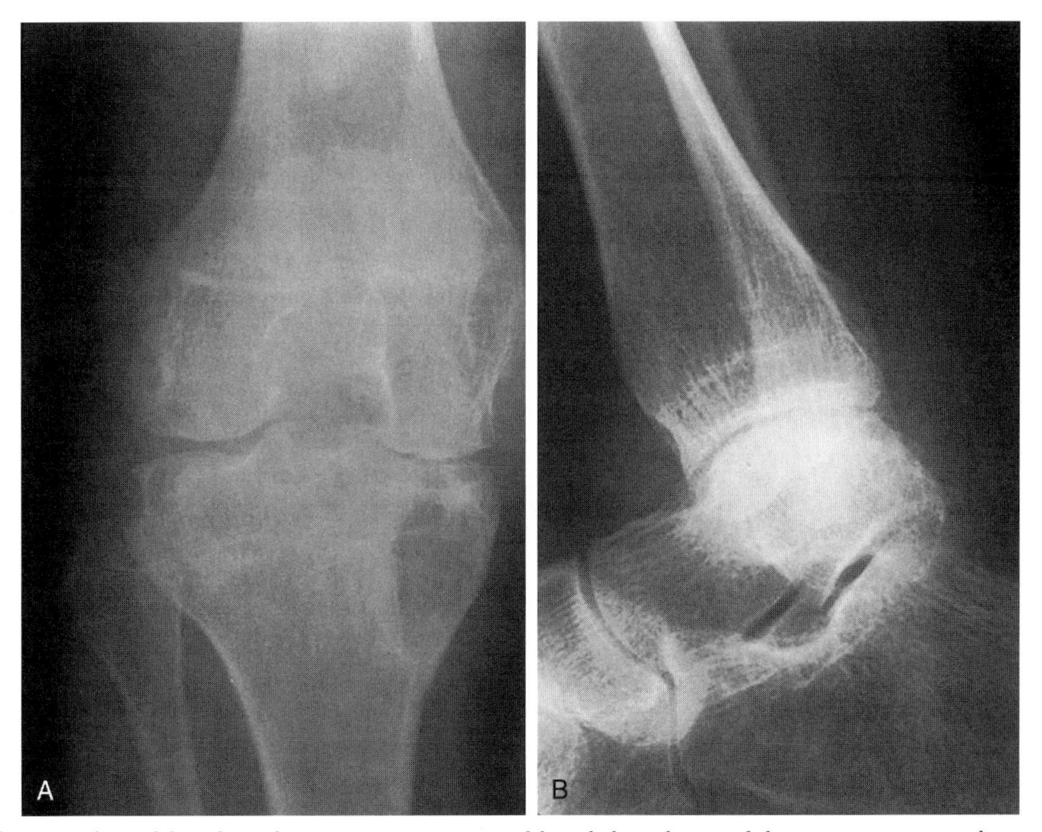

Figure 32–6. Stage IV hemophilic arthropathy. *A*, Anteroposterior view of the right knee shows medial joint space narrowing, a large trochlear notch, and a subchondral cyst in the lateral tibial plateau. The large lucent area in the medial tibial plateau most likely represents an intraosseous bleed. *B*, Lateral view of the ankle from the same patient reveals marked joint space narrowing and degenerative changes.

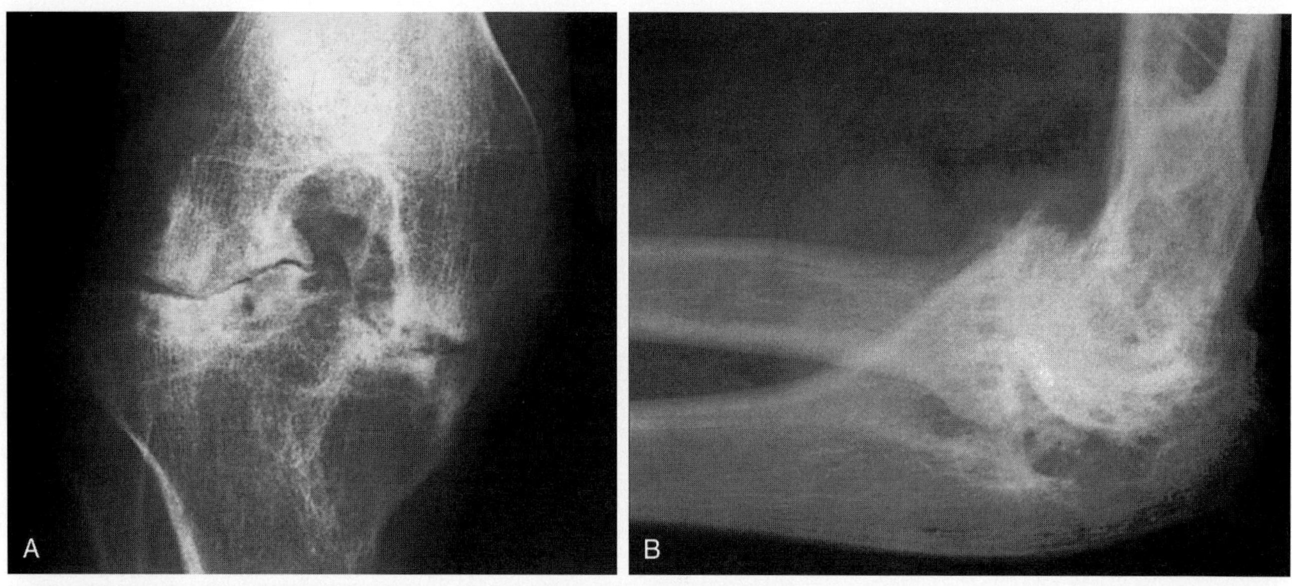

Figure 32–7. Stage V hemophilic arthropathy. *A,* Anteroposterior view of the knee in a 24-year-old man shows advanced secondary degenerative changes with marked narrowing of the joint space. *B,* Lateral view of the elbow from the same patient shows complete loss of the joint space and advanced secondary degenerative changes.

COMPLICATIONS

Hemophilic pseudotumor of bone is a known complication of severe hemophilia that occurs when the clotting factor level is less than 1% of normal. The destructive bone lesions occur because of (1) a muscle hematoma causing underlying bone compression and destruction, (2) subperiosteal hemorrhage, or (3) intraosseous hemorrhage (see Fig. 32–6*A*). The iliopsoas muscle is the most common site for soft tissue bleeding in patients with hemophilia, followed by the abdominal wall. As for intraosseous bleeding, the iliac bone is the most common site for a hemophilic pseudotumor.

References and Recommended Readings

Arnold WD, Hilgartner MW: Hemophilic arthropathy. J Bone Joint Surg Am 59:287, 1977.

Gilchrist GS, Hagedorn AB, Stauffer RN: Severe degenerative joint disease: Mild and moderately severe hemophilia A. JAMA 238:2383, 1977.

Hermann G, Hsu-Chong Y, Gilbert MS: Computed tomography and ultrasonography of the hemophilic pseudotumor and their use in surgical planning. Skeletal Radiol 15:123, 1986.

Mainardi CL, Levine PH, Werb Z, et al: Proliferative synovitis in hemophilia. Arthritis Rheum 21:137, 1978.

Shirkhoda A, Mauro MA, Staab EV, et al: Soft-tissue hemorrhage in hemophiliac patients. Radiology 147:811–814, 1983.

CHAPTER 33

Synovial Osteochondromatosis (Chondromatosis)

Georges Y. El-Khoury, MD, Mark D. Stanley, MD, and D. Lee Bennett, MD

OVERVIEW

Primary synovial osteochondromatosis is a rare arthropathy characterized by the formation of multiple cartilaginous and osteocartilaginous loose bodies within a joint or occasionally within a bursa or tendon sheath (Fig. 33–1). In the literature, several terms have been used to describe this condition, including synovial *chondromatosis, osteochondromatosis, chondrometaplasia,* and *chondrosis.* The exact cause of synovial osteochondromatosis is unknown, although many authors believe that the synovial lining undergoes chondrometaplastic change, producing cartilaginous projections, which later detach and become loose bodies. Adults between the ages of 30 and 50 years typically are affected, and the condition occurs twice as frequently in men as in women. There are some case reports of primary synovial osteochondromatosis in children. Synovial osteochondromatosis most commonly involves the knee (about 50% of cases) (Fig. 33–2), but it also occurs in the elbow, hip, ankle, and shoulder (Figs. 33–3, 33–4, and 33–5). Extra-articular involvement is rare, and it involves the tendon sheaths in the hands and feet or bursae around large joints (see Fig. 33–1). Extra-articular synovial osteochondromatosis usually presents as a soft tissue mass with stippled or lamellar calcific densities (see Fig. 33–1).

At histologic examination, the lesion may contain cartilage only, cartilage and bone, or bone with fatty

Text continued on page 261

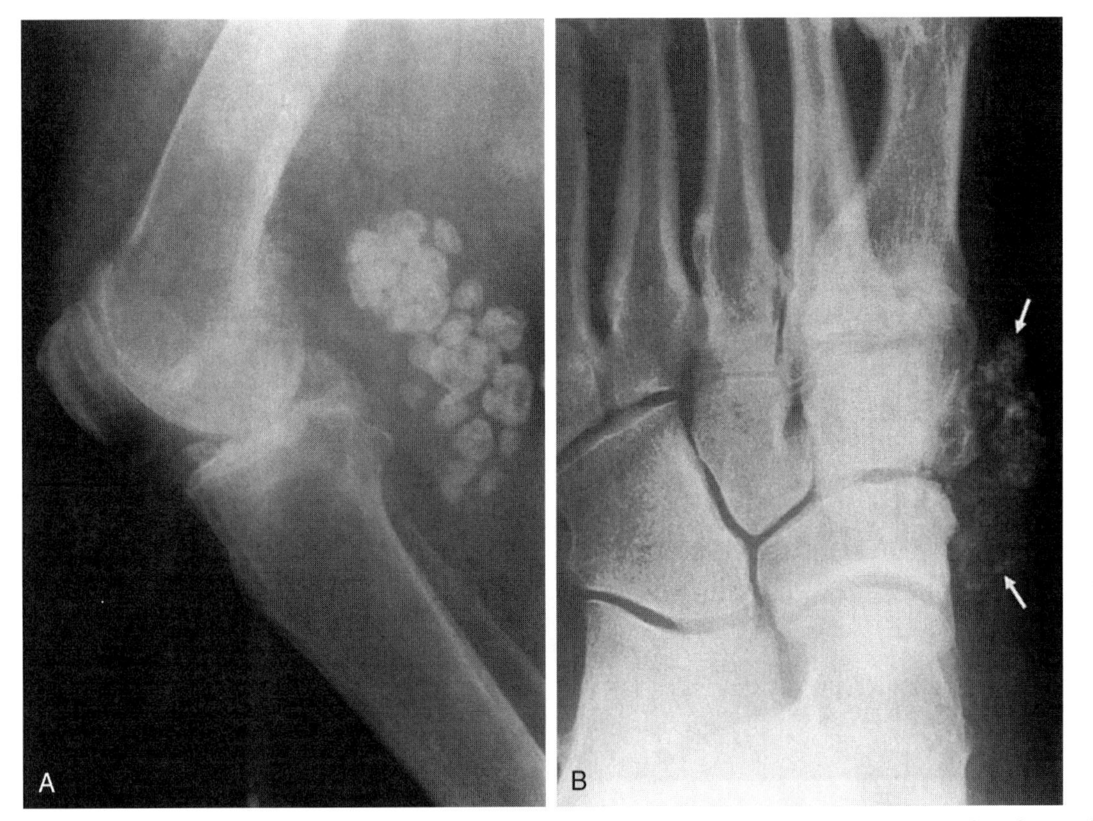

Figure 33–1. Extra-articular synovial osteochondromatosis. *A,* Synovial osteochondromatosis in a Baker's cyst seen on a lateral view of the knee. *B,* Synovial osteochondromatosis in a bursa medial to the tarsal navicular and medial cuneiform *(arrows).*

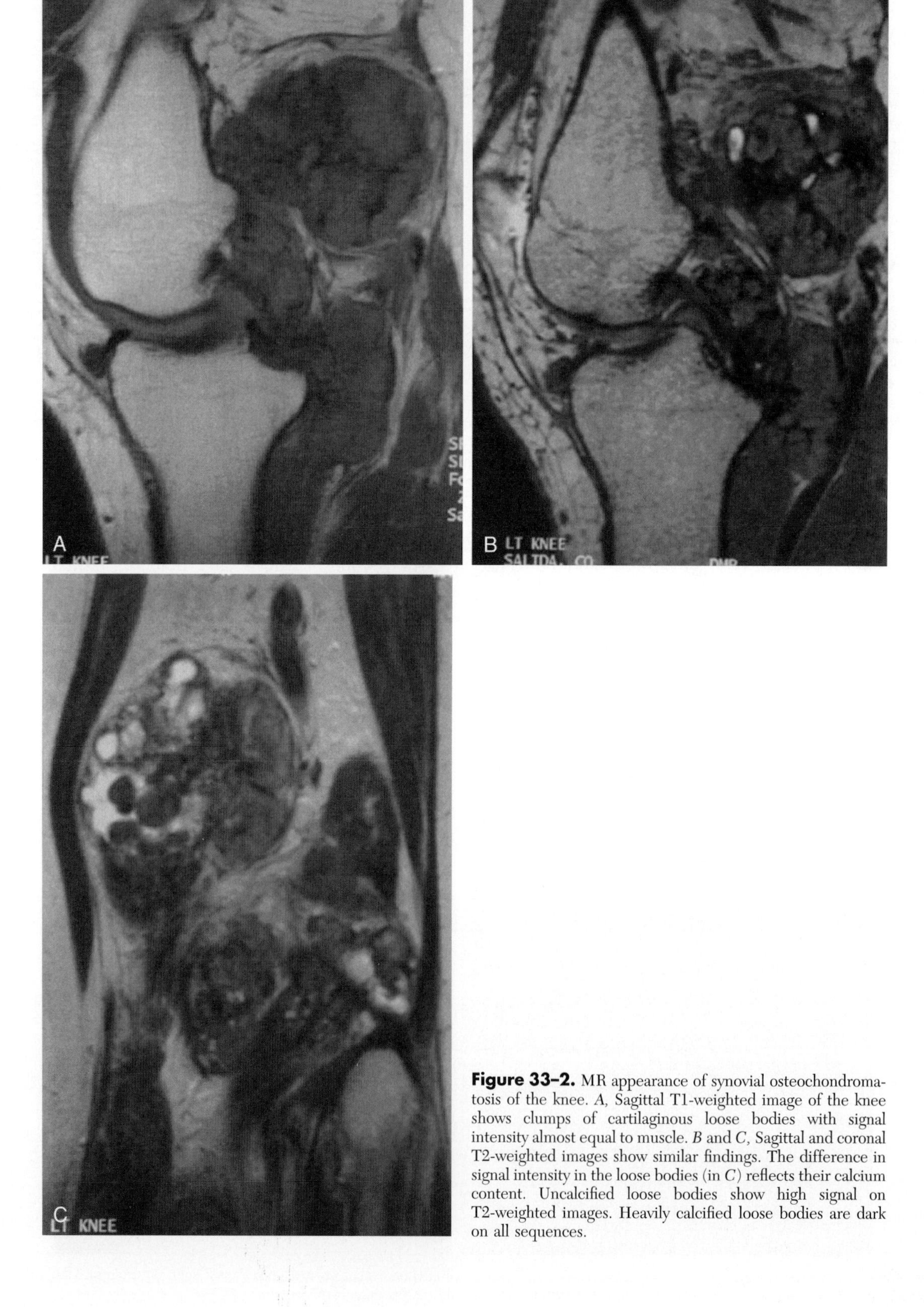

Figure 33–2. MR appearance of synovial osteochondromatosis of the knee. *A*, Sagittal T1-weighted image of the knee shows clumps of cartilaginous loose bodies with signal intensity almost equal to muscle. *B* and *C*, Sagittal and coronal T2-weighted images show similar findings. The difference in signal intensity in the loose bodies (in *C*) reflects their calcium content. Uncalcified loose bodies show high signal on T2-weighted images. Heavily calcified loose bodies are dark on all sequences.

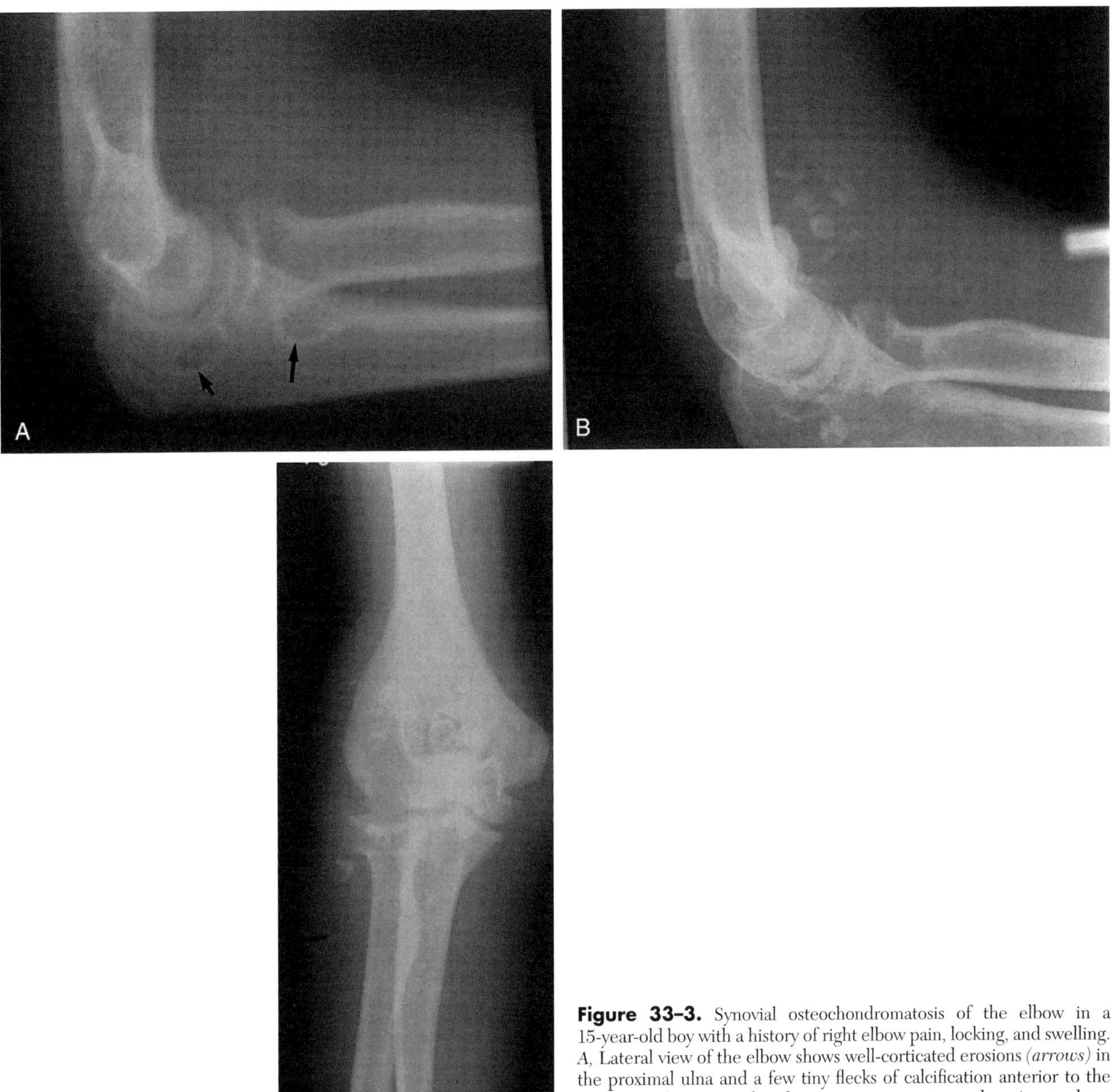

Figure 33–3. Synovial osteochondromatosis of the elbow in a 15-year-old boy with a history of right elbow pain, locking, and swelling. *A*, Lateral view of the elbow shows well-corticated erosions *(arrows)* in the proximal ulna and a few tiny flecks of calcification anterior to the joint. *B* and *C*, Lateral and anteroposterior views taken 4 years later show the number and size of the loose bodies have increased. The diagnosis of primary synovial osteochondromatosis was confirmed at surgery.

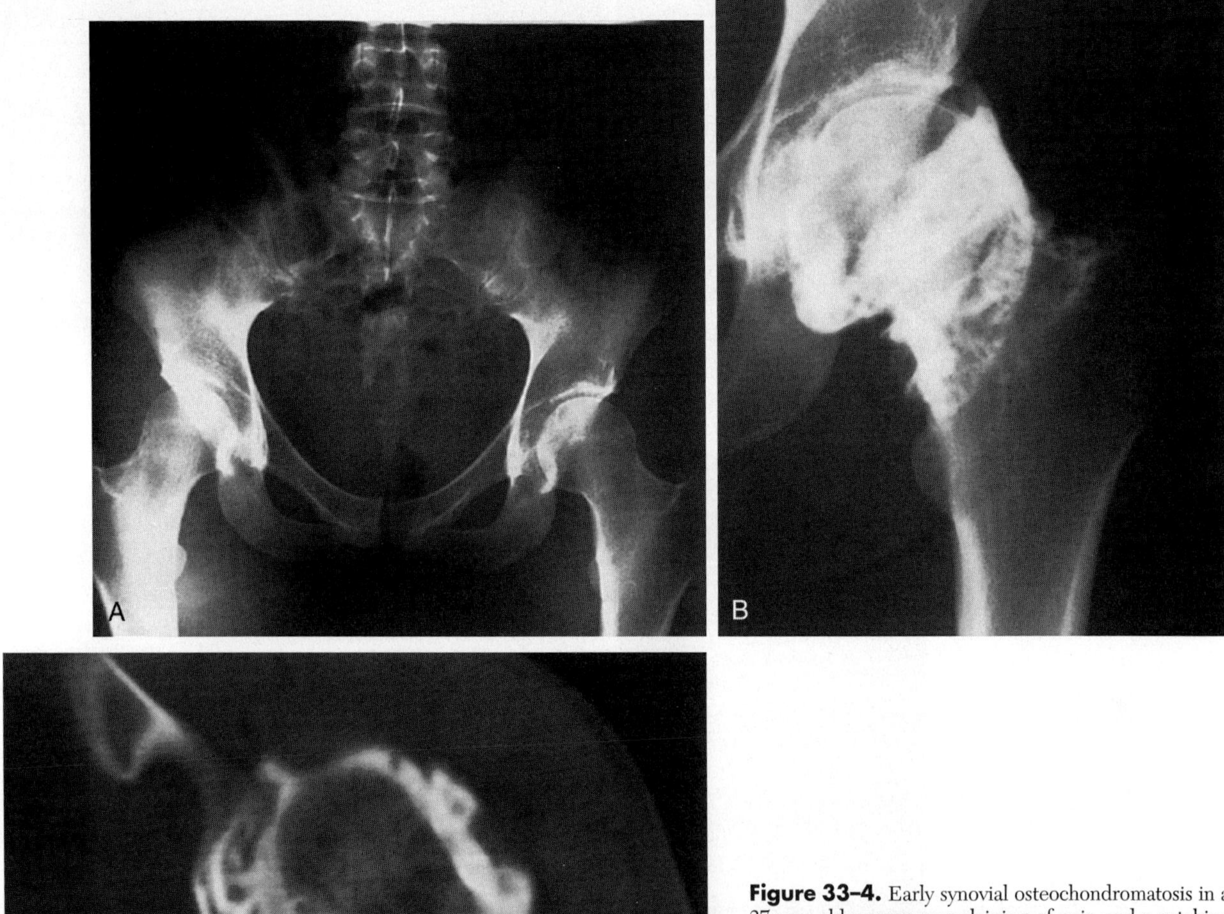

Figure 33–4. Early synovial osteochondromatosis in a 27-year-old woman complaining of pain and a catching sensation in her left hip. *A,* Anteroposterior radiograph of the pelvis shows no abnormalities in either hip. *B,* Anteroposterior radiograph of the left hip after injecting 7 mL of positive contrast material shows multiple small lucent filling defects. *C,* Axial CT section through the hip taken after hip injection confirms the presence of multiple small cartilaginous loose bodies within the joints. The diagnosis of primary synovial osteochondromatosis was confirmed at surgery.

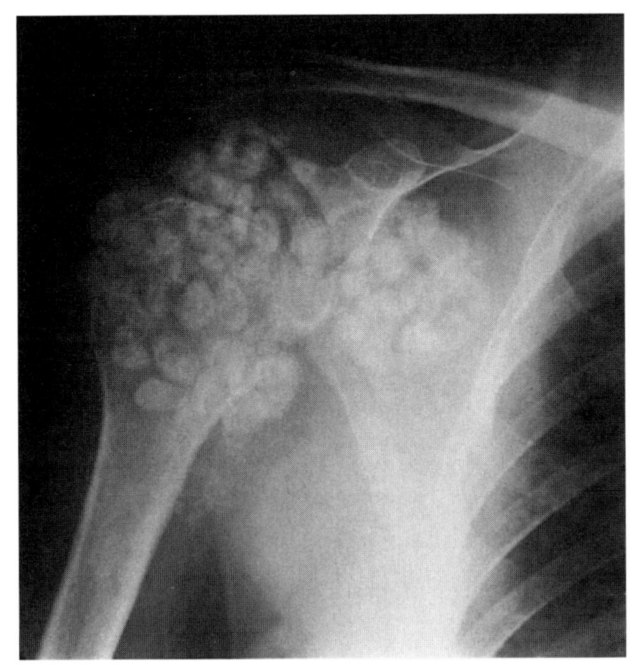

Figure 33–5. Surgically proven primary synovial osteochondromatosis of the right shoulder in a 40-year-old man with a long-standing history of shoulder discomfort and catching sensation.

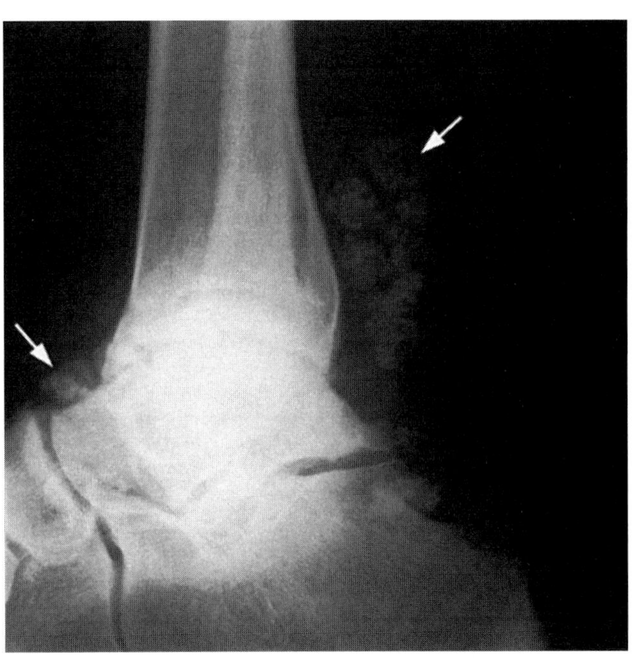

Figure 33–7. Post-traumatic osteoarthritis of the ankle complicated with secondary synovial osteochondromatosis. Lateral view of the ankle shows multiple chondral loose bodies *(arrows)* in the joint.

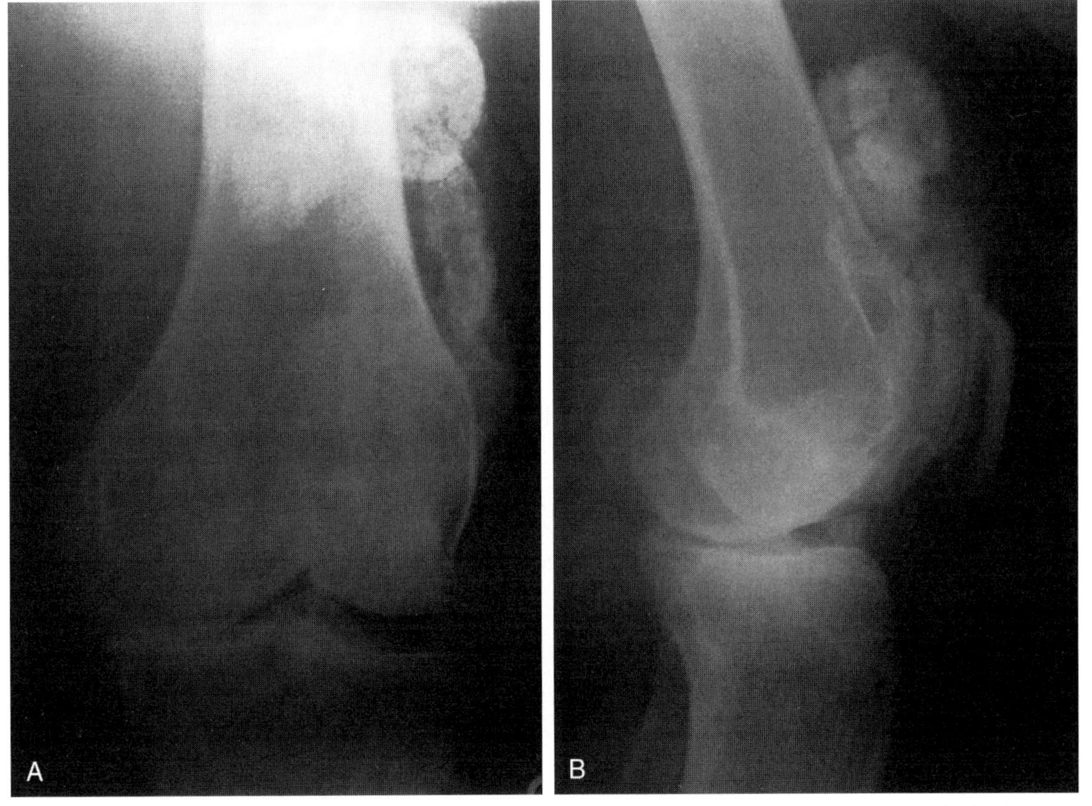

Figure 33–6. *A* and *B*, Secondary synovial osteochondromatosis in a 72-year-old man with advanced osteoarthritis of the left knee. At surgery, the patient had four large chondral loose bodies in the suprapatellar pouch. The synovium was essentially normal.

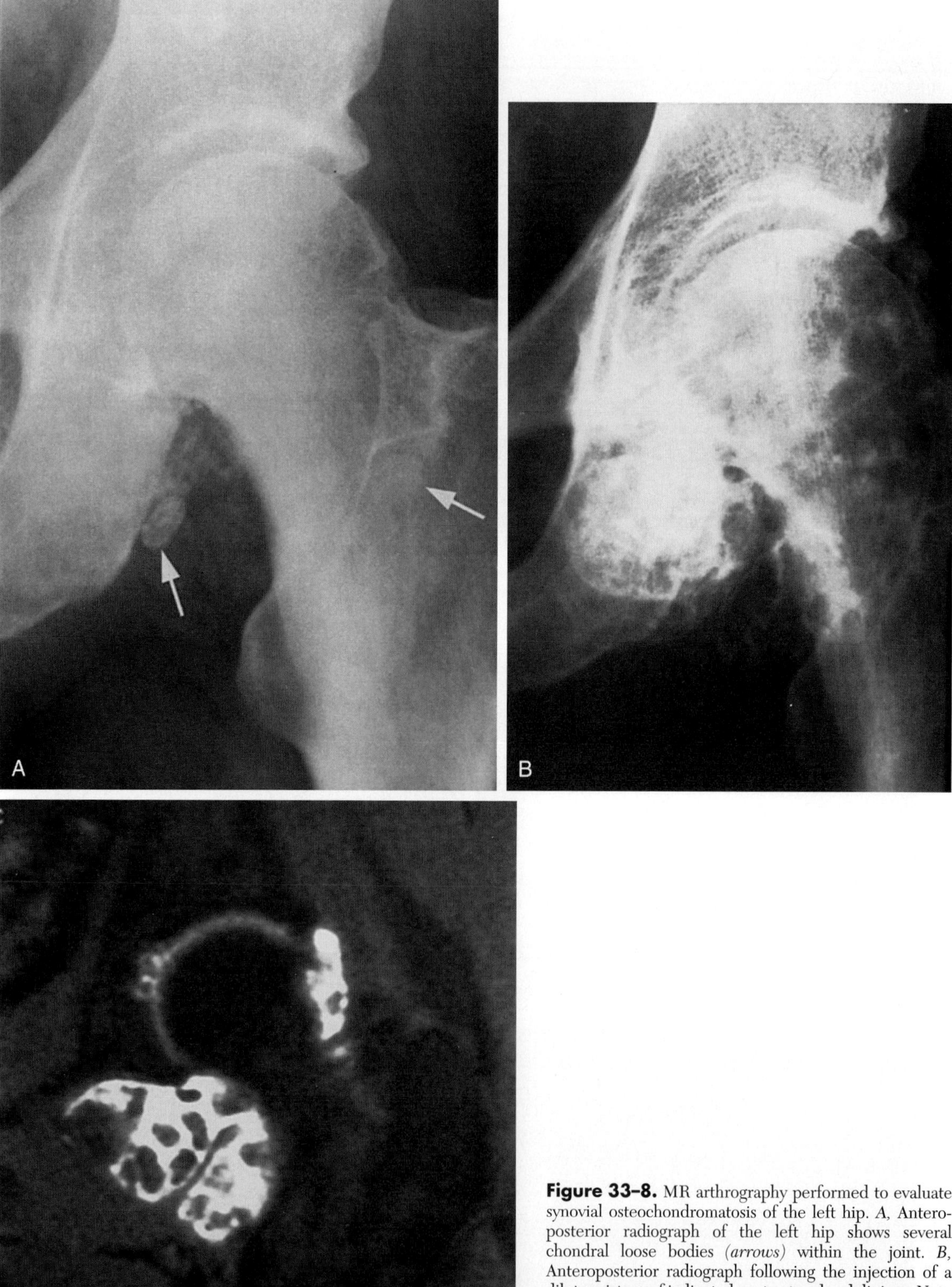

Figure 33–8. MR arthrography performed to evaluate synovial osteochondromatosis of the left hip. *A*, Anteroposterior radiograph of the left hip shows several chondral loose bodies *(arrows)* within the joint. *B*, Anteroposterior radiograph following the injection of a dilute mixture of iodinated contrast and gadolinium. Note that many more loose bodies are seen compared to the previous radiograph *(A)*. *C*, Coronal T1-weighted image with fat suppression shows the loose bodies as dark filling defects within the injected fluid which appears bright. (Case provided by Dr. Hany El-Madbouh from Peterborough, United Kingdom.)

marrow. Malignant transformation into chondrosarcoma has been reported but is rare.

Clinically, patients present with pain, swelling, limitation of motion, locking, crepitus, and palpable loose bodies. Treatment consists of removal of all the free loose bodies and resection of as much of the synovium as possible. Recurrence of the disease is rare.

Secondary synovial osteochondromatosis is more common than the primary type and is seen in association with osteoarthritis, neuropathic joints, osteochondritis dissecans, and osteochondral fractures (Fig. 33–6). The number of loose bodies in the secondary type is small compared with the primary type, and the synovium is normal (Fig. 33–7).

IMAGING

Radiography

When the cartilaginous loose bodies are calcified or ossified, the radiographic diagnosis is fairly simple. The radiographic appearance of primary osteochondromatosis is that of multiple round or ovoid loose bodies, usually of similar size, without underlying diseases that are associated with loose body formation, such as osteoarthritis, Charcot's arthropathy, or osteochondritis dissecans (see Figs. 33–3 through 33–5). In about one third of patients, the loose bodies are not mineralized, and radiographs may be normal or may reveal only soft tissue swelling of the affected joint (see Figs. 33–3*A* and 33–4*A*).

In tight joints, such as the hip, elbow, and ankle and occasionally the shoulder, erosions are common (see Fig. 33–3*A*). When circumferential erosions involve the femoral neck, the hip becomes at risk for pathologic fracture.

Computed Tomography

Air or positive contrast arthrography followed by computed tomography (CT) is an excellent technique to confirm the diagnosis, especially when the calcifications are minimal or absent (see Fig. 33–4). CT without arthrography is more sensitive than radiography in detecting faintly calcified loose bodies.

Magnetic Resonance Imaging

Uncalcified cartilage loose bodies are isointense with muscle on T1-weighted magnetic resonance (MR) images and bright on T2-weighted images (see Fig. 33–2). Densely calcified loose bodies have low signal on all MR imaging sequences. Ossified loose bodies show signal intensity characteristic of marrow fat centrally and cortical bone peripherally. On MR arthrography, the loose bodies are seen as dark filling defects within the bright fluid (Fig. 33–8).

Recommended Readings

Blankestijn J, Panders AK, Vermey A, et al: Synovial chondromatosis of the temporo-mandibular joint: Report of three cases and a review of the literature. Cancer 55:479, 1985.

Coolican MR, Dandy DJ: Arthroscopic management of synovial chondromatosis of the knee. J Bone Joint Surg Br 71:498, 1989.

Goldberg RP, Genant HK: Calcified bodies in popliteal cysts: A characteristic radiographic appearance. AJR Am J Roentgenol 131: 857, 1978.

Karlin CA, DeSmet AA, Neff J, et al: The variable manifestations of extraarticular synovial chondromatosis. AJR Am J Roentgenol 137: 731, 1981.

Kay PR, Freemont AJ, Davies DRA: The aetiology of multiple loose bodies: Snow storm knee. J Bone Joint Surg Br 71:501, 1989.

Norman A, Steiner GC: Bone erosions in synovial chondromatosis. Radiology 161:749, 1986.

Pope Jr TL, Keats TE, deLange EE, et al: Idiopathic synovial chondromatosis in two unusual sites: Inferior radioulnar joint and ischial bursa. Skeletal Radiol 16:205, 1987.

Sim FH, Dahlin DC, Ivins JC: Extra-articular synovial chondromatosis. J Bone Joint Surg Am 59:492, 1977.

CHAPTER 34

Alkaptonuria (Ochronosis)

Georges Y. El-Khoury, MD, Mark D. Stanley, MD, and D. Lee Bennett, MD

OVERVIEW

Alkaptonuria (ochronosis) is a rare inborn error of metabolism of the amino acids phenylalanine and tyrosine. It is an autosomal recessive defect causing absence of the enzyme homogentisic acid oxidase. As a consequence, homogentisic acid accumulates in the body, resulting in pigmentation of the cartilage, sclera, pinna of the ears, tendons, and endocardium. These deposits harden the hyaline cartilage and fibrocartilage and make them fragile. The urine of alkaptonuric patients darkens if left standing probably because of oxidation of homogentisic acid.

RADIOGRAPHIC FINDINGS

Although the disease is genetic and can be detected in childhood, radiographic changes in the joints take until about age 30 to manifest. Ochronotic arthropathy typically involves the vertebral column, where the earliest finding manifests as uniform disk space narrowing throughout the spine. With time, the disks become markedly narrowed and heavily calcified (Fig. 34–1). These findings in the spine are pathognomonic of ochronosis. The large peripheral joints, such as the knees, hips, and shoulders, are involved with osteoarthritis-like changes. These joints

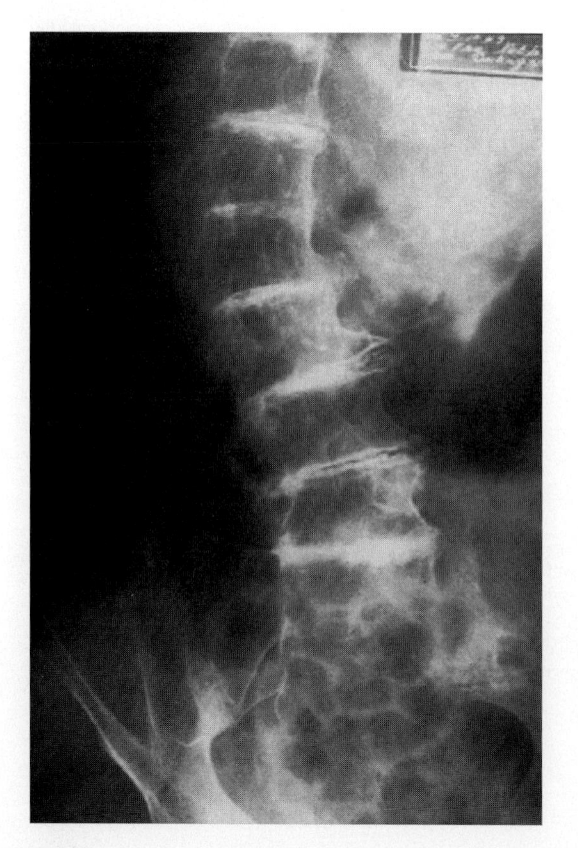

Figure 34–1. Changes of ochronosis in the spine. Note the uniform and severe disk space narrowing throughout the spine. All the disks show calcifications within them.

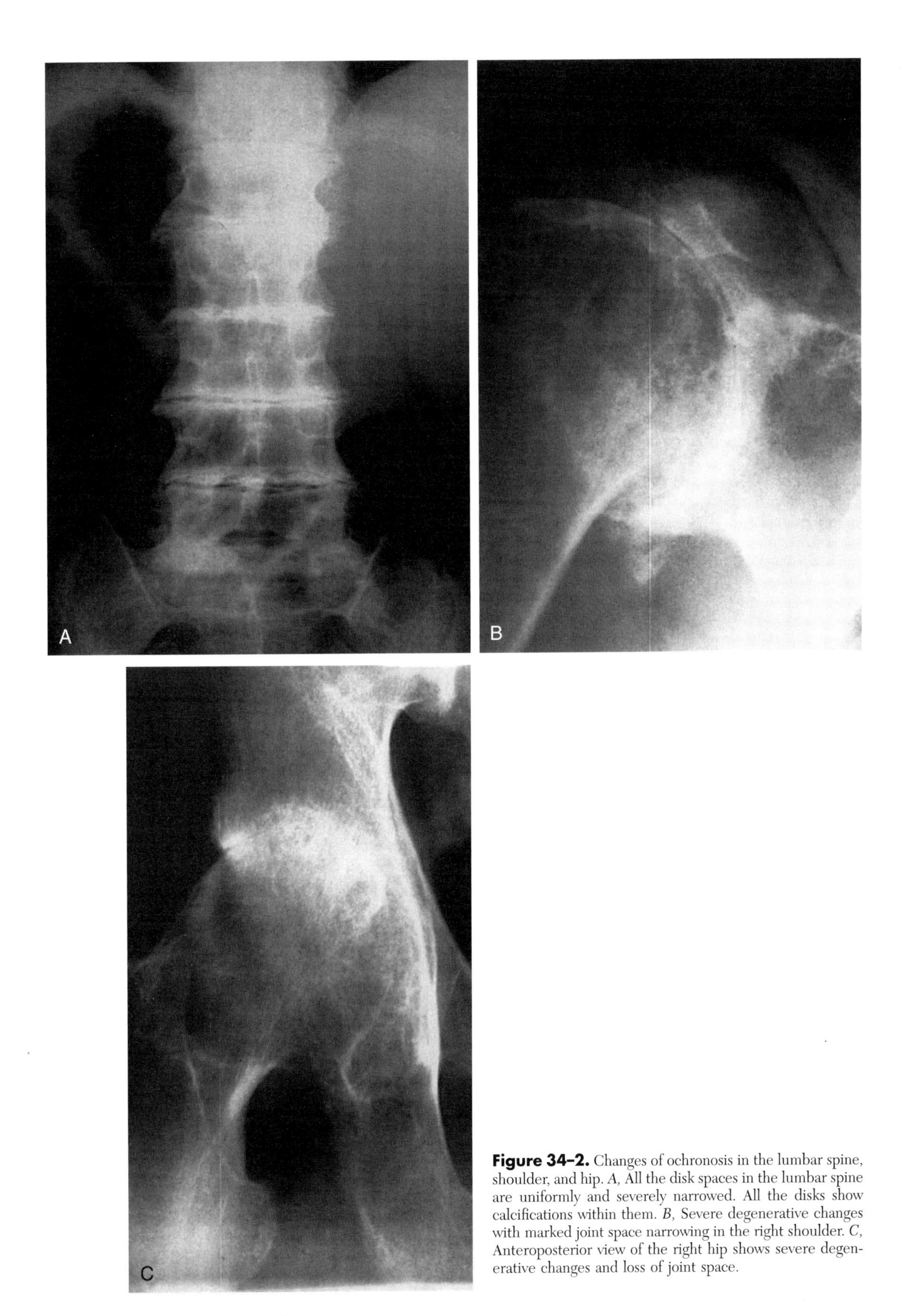

Figure 34–2. Changes of ochronosis in the lumbar spine, shoulder, and hip. *A*, All the disk spaces in the lumbar spine are uniformly and severely narrowed. All the disks show calcifications within them. *B*, Severe degenerative changes with marked joint space narrowing in the right shoulder. *C*, Anteroposterior view of the right hip shows severe degenerative changes and loss of joint space.

become uniformly narrowed, with subchondral sclerosis and osteophyte formation (Fig. 34–2). The menisci in the knees show chondrocalcinosis.

Recommended Readings

Justesen P, Anderson Jr PE: Radiologic manifestations in alcaptonuria. Skeletal Radiol 11:204, 1984.

Lagier R, Steiger U: Hip arthropathy in ochronosis: Anatomical and radiological study. Skeletal Radiol 5:91, 1980.

Schumacher HR, Holdsworth DE: Ochronotic arthropathy: I. Clinicopathologic studies. Semin Arthritis Rheum 6:207, 1977.

CHAPTER 35

Pigmented Villonodular Synovitis

Georges Y. El-Khoury, MD, Mark D. Stanley, MD, and D. Lee Bennett, MD

OVERVIEW

Pigmented villonodular synovitis (PVNS) is a benign proliferative disorder affecting the synovium of diarthrodial joints, tendon sheaths, and bursae. The cause of PVNS is unknown. PVNS of the tendon sheath often is referred to as giant cell tumor of the tendon sheath. This lesion typically affects the flexor tendons of the first three fingers. Intra-articular lesions can present as localized PVNS, which is characterized by a focal nodular or pedunculated mass or as a diffuse process involving the entire synovial lining. The diffuse form is more common, accounting for approximately 75% of all cases. Prognosis with the localized form is good after excision, whereas recurrence is common with the diffuse form. PVNS is typically a monarticular process occurring mainly in young and middle-aged adults (20 to 50 years old) without significant sex predilection. The knee is the most frequently affected joint, accounting for about 66% to 80% of all cases. Sixteen percent of cases are reported in the hip, and the rest occur in ankle, elbow, shoulder, and facet joints, mostly of the lumbar spine. In the knee, the onset is insidious, and the disease manifests with local warmth, swelling, and stiffness and occasionally a palpable mass. Mechanical symptoms suggestive of internal derangement or a loose body occur with the localized form of PVNS. Giant cell tumor of the tendon sheath is one of the most common soft tissue tumors of the hands and feet. Malignant transformation is extremely rare with PVNS.

PATHOLOGY

In the diffuse form, the synovium is yellow-brown in color. There are numerous villous projections and nodules covering the synovial surface. Microscopically the lesion consists of a tumor-like proliferation of mononuclear histiocytes, lymphocytes, plasma cells, and giant cells. The histiocytic elements may contain varying amounts of lipid and hemosiderin. Hemosiderin deposits are seen in intracellular and in extracellular sites. The lesion is often vascular. The localized form of PVNS is similar histologically to the diffuse variety except that it is less vascular, has fewer hemosiderin deposits, and is more fibrotic.

IMAGING

Radiography

Radiographic changes are reported in 79% of cases. The most common abnormality is joint effusion and soft tissue swelling. Lack of calcifications within the joint or in the soft tissues around the joint is an important distinguishing feature of PVNS (Fig. 35–1). There is normal bone density in joints involved with PVNS. Erosive and cystic changes occur in 50% of cases but mainly in joints with tight capsules, such as the hip and elbow (see Fig. 17–17). These cysts and erosions have well-defined sclerotic borders and are typically nonmarginal (Fig. 35–1A). The

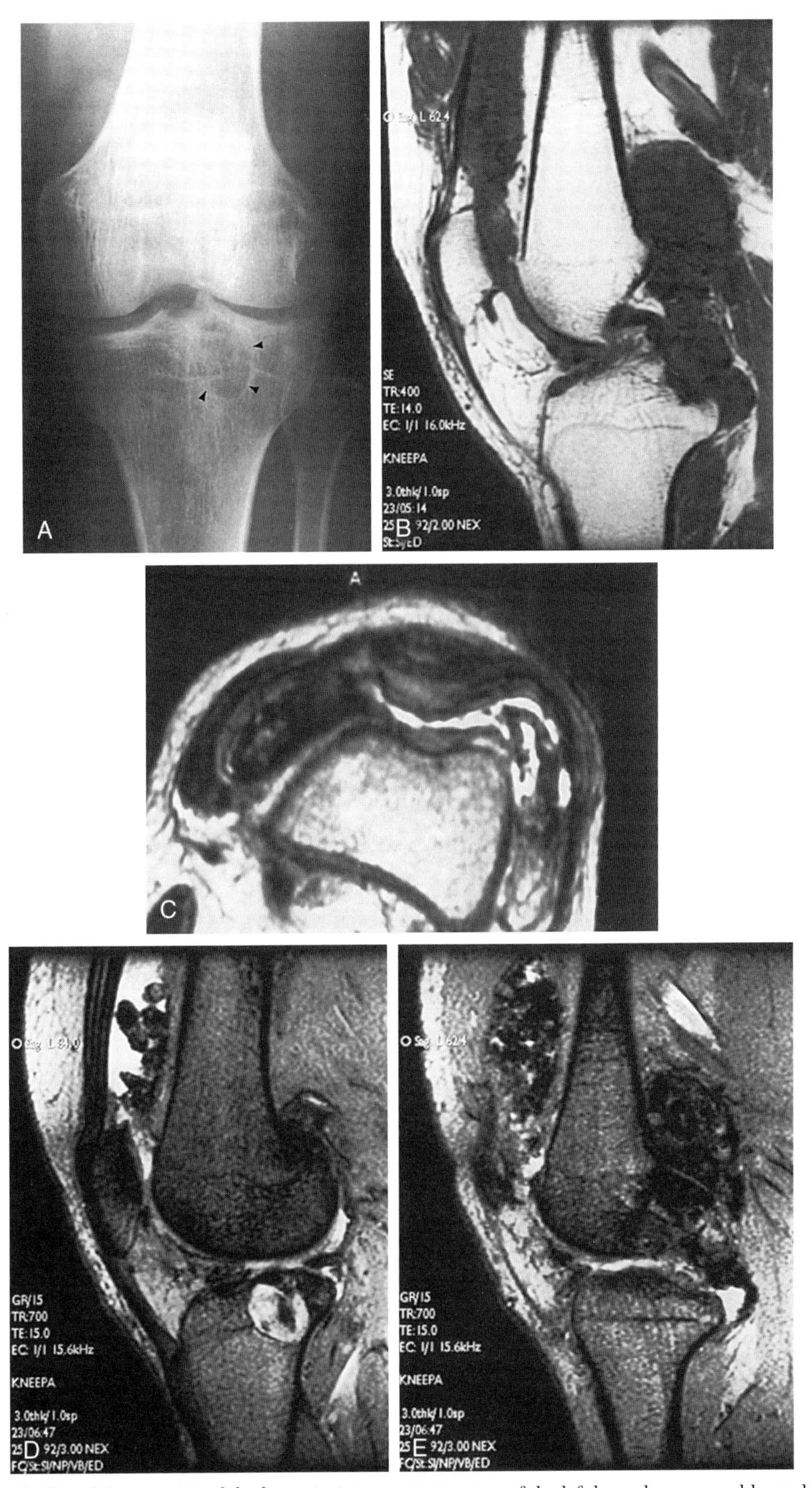

Figure 35–1. Pigmented villonodular synovitis of the knee. *A*, Anteroposterior view of the left knee shows normal bone density and normal joint space. A large lytic lesion is seen in the proximal tibia *(arrowheads)*. *B*, T1-weighted sagittal MR image reveals the presence of fluid and mass effect in the suprapatellar pouch and in the soft tissues posterior to the joint. Low signal specks are noted within the masses. *C*, T2-weighted axial image at the level of the superior pole of the patella shows lobular low signal masses filling the suprapatellar pouch. A joint effusion also is present. *D* and *E*, Sagittal gradient echo images reveal hypodense synovial masses. The decreased signal intensity in these masses is due to the presence of hemosiderin.

joint space is relatively well preserved until late in the disease, especially in the knee, but this is not necessarily true in the hip, where joint space narrowing can occur early. There is a lack of productive osteophytic changes with PVNS, which helps to differentiate it from osteoarthritis.

Computed Tomography

The hypertrophied synovial tissue may show high attenuation because of the presence of hemosiderin. Computed tomography (CT) is useful in showing the erosive and cystic changes in the involved joints.

Magnetic Resonance Imaging

The magnetic resonance (MR) imaging appearance depends on the relative presence of lipid, hemosiderin, fibrous stroma, pannus, and cellular elements. The entire synovium typically is covered with numerous villous projections and tiny nodules (Fig. 35–2). The constant presence of a joint effusion helps in appreciating the abnormal synovial surface (see Fig. 35–2). The involved synovium usually shows intermediate signal intensity about equal to that of muscle on T1- and T2-weighted sequences. There is also tissue enhancement after contrast injection. Areas of low signal intensity on all sequences are often present, which are attributed to the presence of hemosiderin (see Fig. 35–1). Gradient echo sequences are helpful in showing small amounts of hemosiderin because of the susceptibility artifacts produced by the hemosiderin, seen as areas of signal loss (Fig. 35–1*D* and *E*; also see Fig. 17–15). The localized form of PVNS in the knee can occur anywhere in the synovium but is especially common in the infrapatellar fat pad. Other joints can also be involved (Fig. 35–3). PVNS appears as a nodule or a mass with heterogeneous signal intensity on T1- and T2-weighted images (Fig. 35–3). Giant cell tumor of the tendon sheath appears as a lobulated mass attached to the tendon (Figs. 35–4 and 35–5). Signal intensity varies depending on hemosiderin and fibrous tissue content of the lesion.

DIFFERENTIAL DIAGNOSIS

PVNS should be differentiated from synovial sarcoma, which is almost always extracapsular and in about one third of cases contains calcification. Synovial osteochondromatosis may be distinguished from PVNS by the presence of calcifications within the intra-articular loose bodies. Synovial hemangioma usually is associated with phleboliths and large draining vessels. Rheumatoid arthritis, hemophilic arthropathy, and tuberculous arthritis can be distinguished from PVNS on the basis of the clinical and laboratory findings.

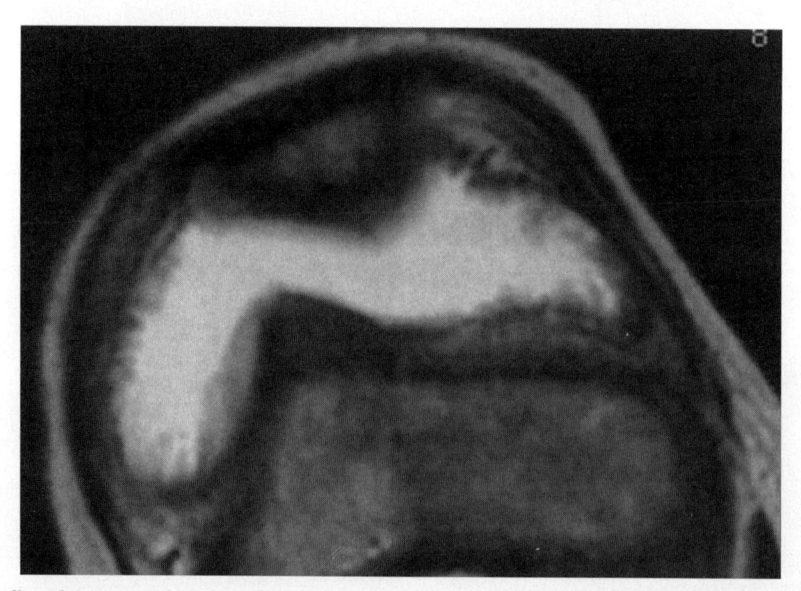

Figure 35–2. Pigmented villonodular synovitis of the right knee. Axial T2-weighted MR image shows a joint effusion with numerous villous synovial projections.

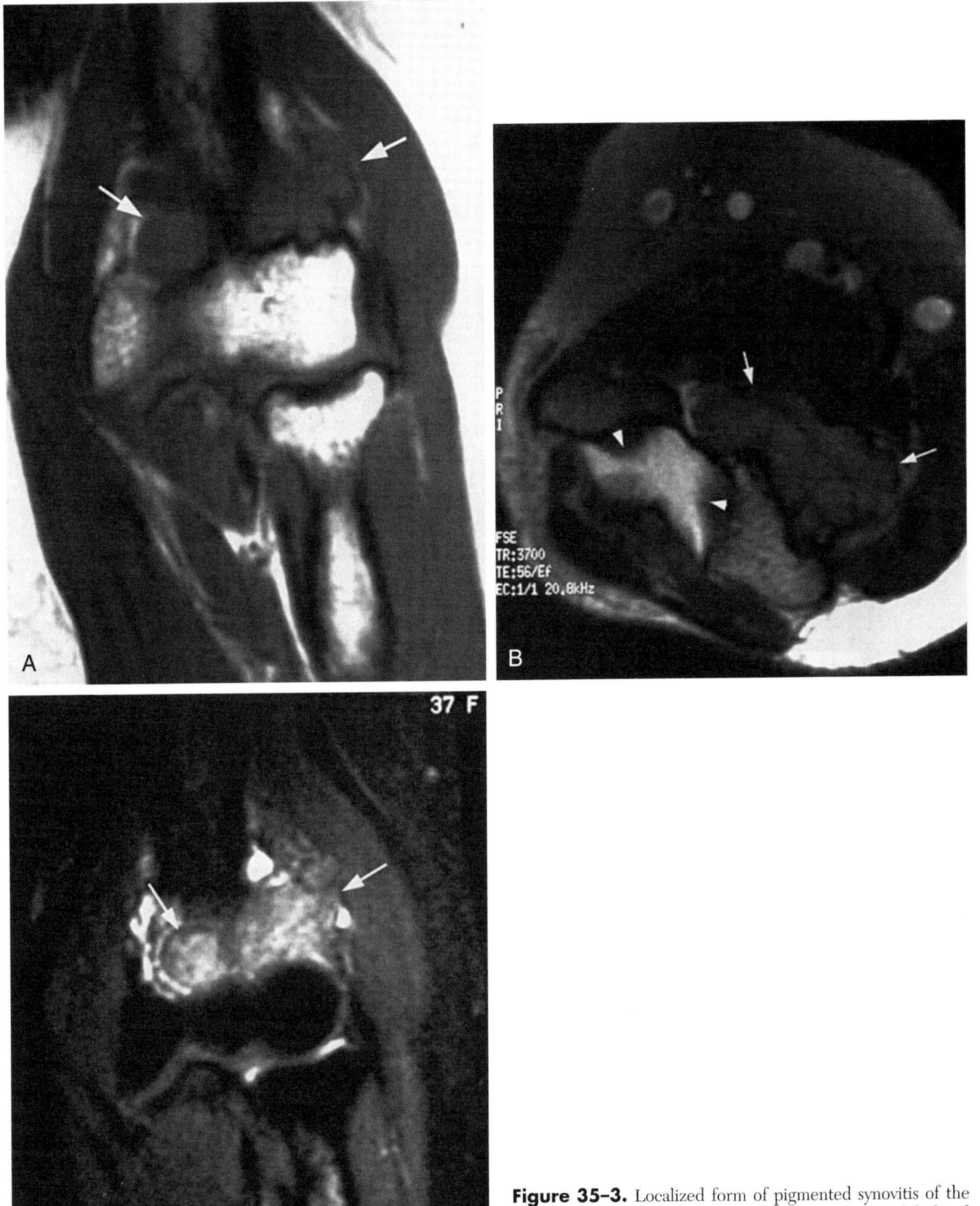

Figure 35–3. Localized form of pigmented synovitis of the elbow joint. *A*, Coronal T1-weighted image shows a lobulated mass *(arrows)* in the anterior joint, which is isointense with muscle. *B*, Axial T2-weighted image with fat saturation shows the mass *(arrows)* and fluid *(arrowheads)* in the olecranon fossa. *C*, Coronal T2-weighted image with fat suppression demonstrates the mass *(arrows)*, which is bright. Note also the presence of a joint effusion.

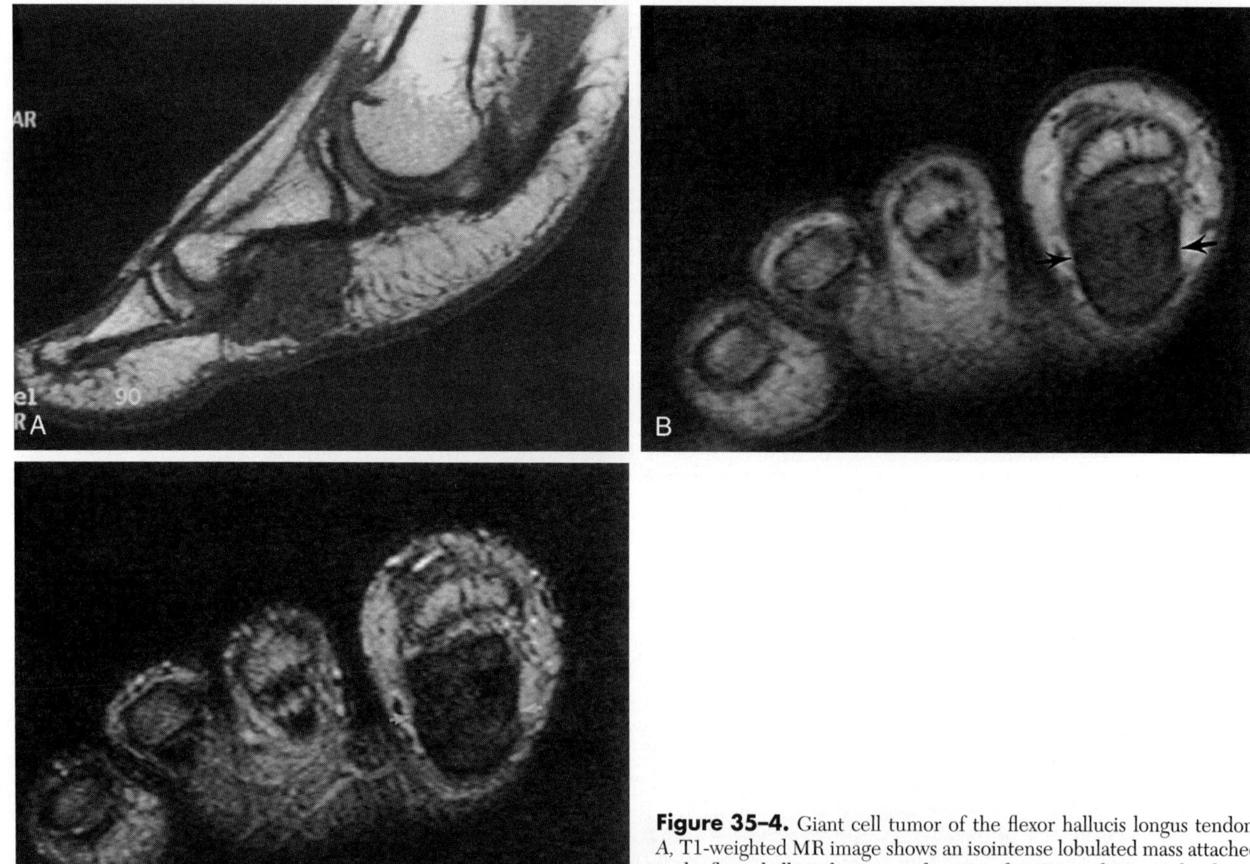

Figure 35–4. Giant cell tumor of the flexor hallucis longus tendon. *A*, T1-weighted MR image shows an isointense lobulated mass attached to the flexor hallucis longus tendon. *B* and *C*, T1- and T2-weighted axial images show the mass *(arrows)*.

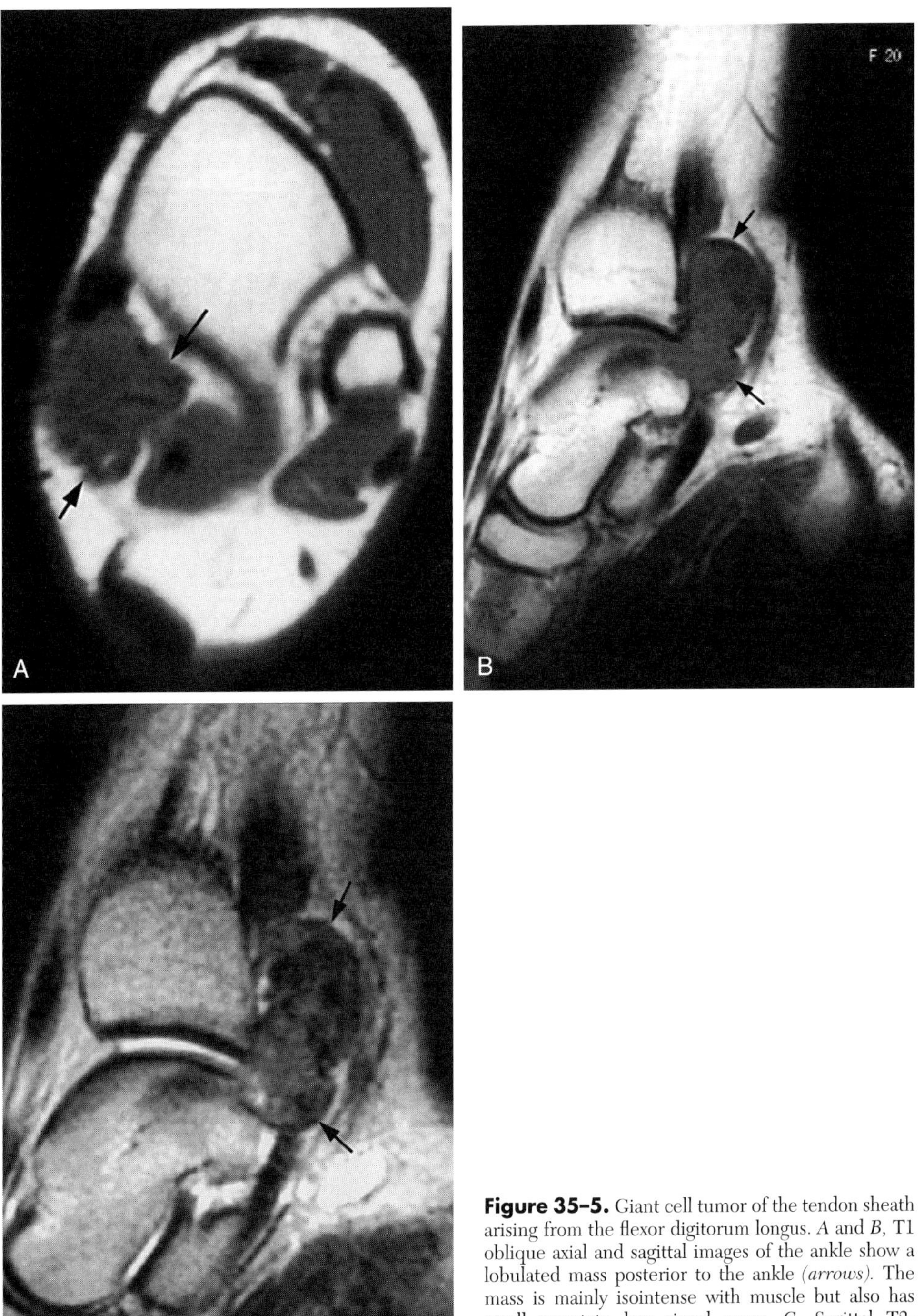

Figure 35–5. Giant cell tumor of the tendon sheath arising from the flexor digitorum longus. *A* and *B*, T1 oblique axial and sagittal images of the ankle show a lobulated mass posterior to the ankle *(arrows)*. The mass is mainly isointense with muscle but also has small punctate low signal areas. *C*, Sagittal T2-weighted image shows the lobulated mass *(arrows)*. The low signal areas represent hemosiderin deposits.

Recommended Readings

Abrahams TG, Pavlov H, Bansal M, et al: Concentric joint space narrowing of the hip associated with hemosiderotic synovitis (HS) including pigmented villonodular synovitis (PVNS). Skeletal Radiol 17:37, 1988.

Cotton A, Flipo R-M, Chastanet P, et al: Pigmented villonodular synovitis of the hip: Review of radiographic features in 58 patients. Skeletal Radiol 24:1, 1995.

Dandy DJ, Rao NS: Benign synovioma causing internal derangement of the knee. A report of nine cases. J Bone Joint Surg Br 72:641, 1990.

Docken WP: Seminars in arthritis and rheumatism. Semin Arthritis Rheum 9:1, 1979.

Dorwart RH, Genant HK, Johnston WH, et al: Pigmented villonodular synovitis of synovial joints: Clinical, pathologic, and radiologic features. AJR Am J Roentgenol 143:877, 1984.

Flandry F, Hughston JC: Current concepts review: Pigmented villonodular synovitis. J Bone Joint Surg Am 69:942, 1987.

Hughes TH, Sartoris DJ, Schweitzer ME, et al: Pigmented villonodular synovitis: MRI characteristics. Skeletal Radiol 24:7, 1995.

Jelinek JS, Kransdorf MJ, Utz JA, et al: Imaging of pigmented villonodular synovitis with emphasis on MR imaging. AJR Am J Roentgenol 152:337, 1989.

Lin J, Jacobson JA, Jamadar DA, et al: Pigmented villonodular synovitis and related lesions: The spectrum of imaging findings. AJR 172:191, 1999.

Lowenstein MB, Smith JRV, Cole S: Infrapatellar pigmented villonodular synovitis: Arthrographic detection. AJR Am J Roentgenol 135:279, 1980.

Spritzer CE, Dalinka MK, Kressel HY: Magnetic resonance imaging of pigmented villonodular synovitis: A report of two cases. Skeletal Radiol 16:316, 1987.

CHAPTER 36

Lipoma Arborescens

Georges Y. El-Khoury, MD, Mark D. Stanley, MD, and D. Lee Bennett, MD

OVERVIEW

Lipoma arborescens is a rare synovial disorder resulting from villous lipomatous proliferation. Fat accumulates in the subsynovial tissues, which produces prominent villous projections. The lesion occurs most commonly in the knee, especially the suprapatellar pouch. Males are more affected than females. Most reported cases have occurred in adults at middle age or beyond. The exact cause is not known, but it is thought to represent a synovial response to trauma and inflammation. Lipoma arborescens can be associated with degenerative joint disease and rheumatoid arthritis. Clinically, patients present with painless joint swelling as a result of synovial thickening and joint effusion. Grossly the lesions appear as finger-like projections of fat covered by synovium. Histologically the lesion is characterized by extensive villous proliferation of the synovial membrane with hyperplasia of the synovial fat (Fig. 36–1).

Matsumoto and colleagues (2001) stressed the importance of differentiating lipoma arborescens from true intra-articular lipoma, which some reports have considered as one entity. True intra-articular lipomas are rare, they occur *de novo* (i.e., without underlying joint disease), and they mostly occur in elderly women. Grossly the lesion appears as a mass consisting of mature adipose tissue enclosed by a thin fibrous capsule (Fig. 36–2).

IMAGING

Radiographic findings in lipoma arborescens are nonspecific; most authors believe, however, that magnetic resonance (MR) imaging is diagnostic. MR imaging shows synovial frondlike masses with signal characteristics of fat on all pulse sequences and an associated joint effusion (see Fig. 36–1; also see Fig. 17–8). Characteristically, there is absence of hemosiderin.

References and Recommended Readings

Feller JF, Rishi M, Hughes EC: Lipoma arborescens of the knee: MR demonstration. AJR Am J Roentgenol 163:162, 1994.

Grieten M, Buckwalter KA, Cardinal E, et al: Case report 873. Skeletal Radiol 23:652, 1994.

Matsumoto K, Okabe H, Ishizawa M, et al: Intra-articular lipoma of the knee joint: A case report. J Bone Joint Surg Am 83:101, 2001.

Ryu KN, Jaovisidha S, Schweitzer M, et al: MR imaging of lipoma arborescens of the knee joint. AJR Am J Roentgenol 167:1229, 1996.

Sola JB, Wright RW: Arthroscopic treatment for lipoma arborescens of the knee: A case report. J Bone Joint Surg Am 80:99, 1998.

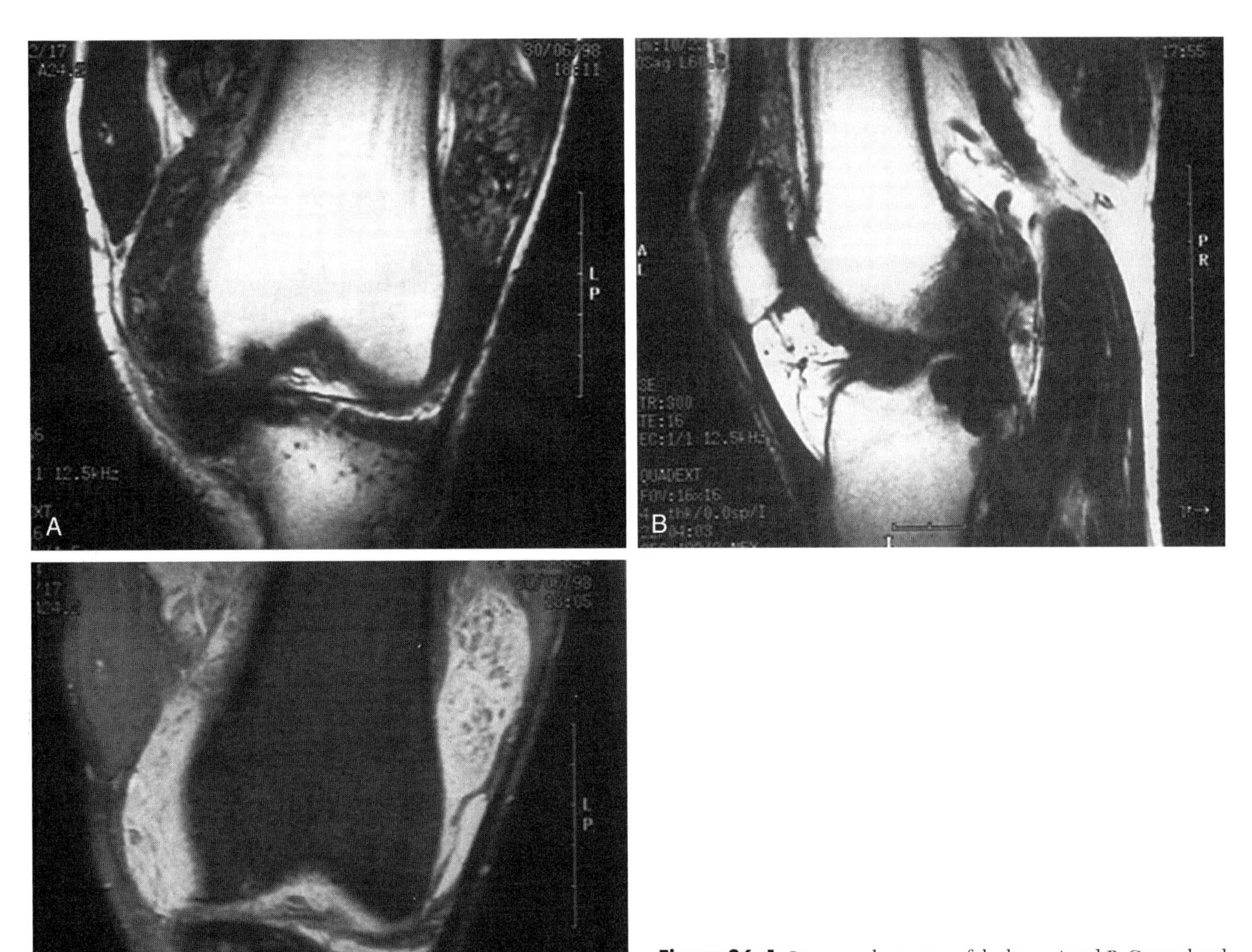

Figure 36–1. Lipoma arborescens of the knee. *A* and *B*, Coronal and sagittal T1-weighted MR images of the knee show numerous villous fronds with signal intensity of fat floating within the synovial fluid. Large erosions are visible in the articular surface of the tibia. *C*, T2-weighted fat-suppressed MR image shows numerous villous fronds floating within the joint effusion. (Case provided by Dr. Luc Des Harnais, Montreal, Canada.)

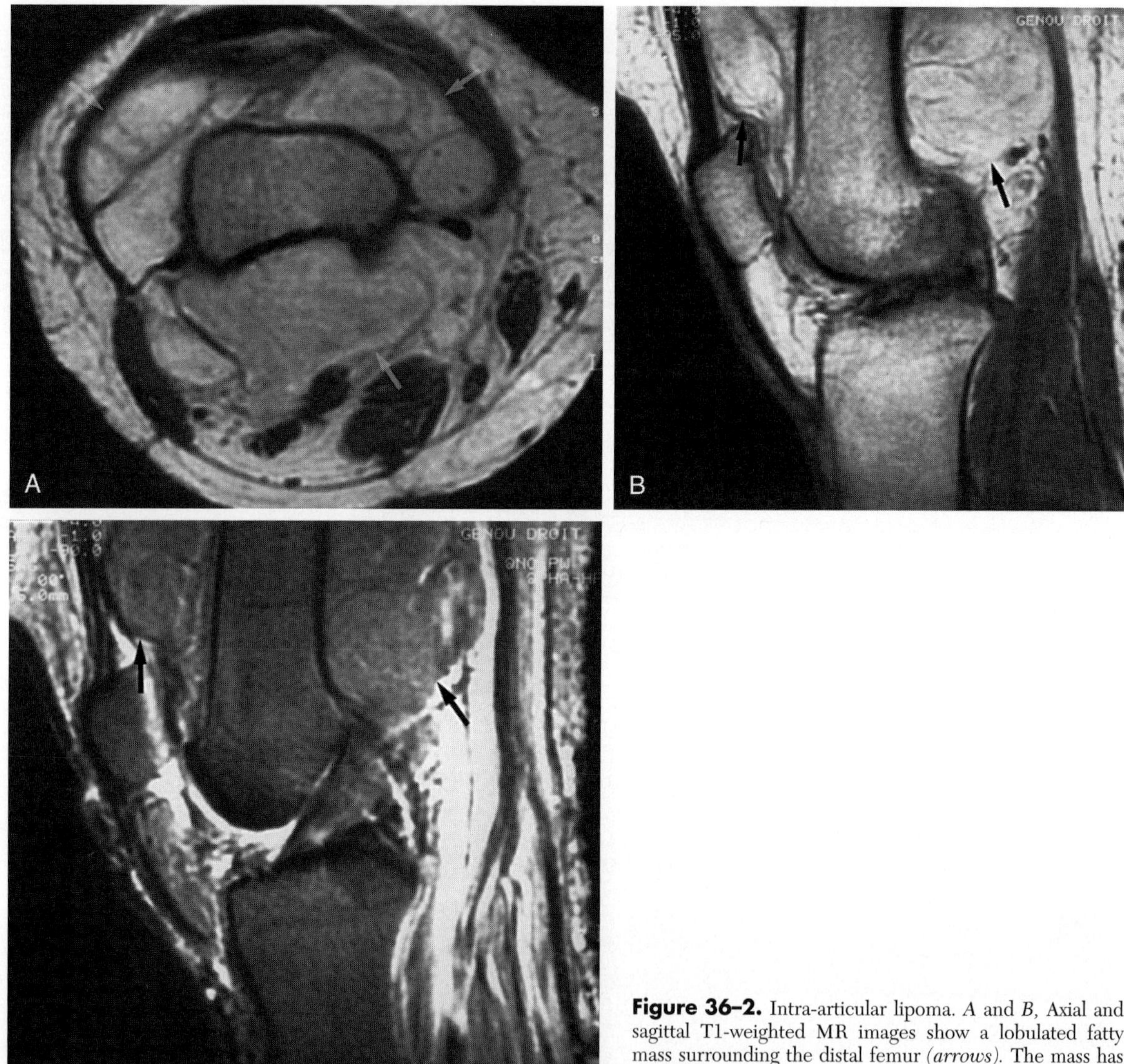

Figure 36–2. Intra-articular lipoma. *A* and *B*, Axial and sagittal T1-weighted MR images show a lobulated fatty mass surrounding the distal femur *(arrows)*. The mass has some fibrous septa. *C*, Sagittal T2-weighted image shows the fatty mass *(arrows)*.

CHAPTER 37

Relapsing Polychondritis

Nabil J. Khoury, MD, and Georges Y. El-Khoury, MD

OVERVIEW

Relapsing polychondritis is a rare inflammatory disease of unknown cause, involving the cartilage of the nose, ears, upper respiratory tract, and joints. It can be primary or associated with another autoimmune disease. Peripheral arthritis occurs in 70% to 80% of cases, whereas costochondral and spinal pain are less frequent. When primary, the associated arthritis is usually transient and nonerosive. Upper respiratory tract involvement affects more than half of patients at some time during the course of the disease; obstruction of the major airways is the most serious component of the disease, accounting for about 50% of the deaths. The diagnosis is made clinically but may be confirmed by a cartilage biopsy.

IMAGING

Radiography

Occasional joint space narrowing in affected joints can be seen. This narrowing typically occurs without erosions, bone destruction, or joint deformity. Mild sacroiliac joint narrowing also has been reported in some patients. In patients with significant erosions and joint destruction, it

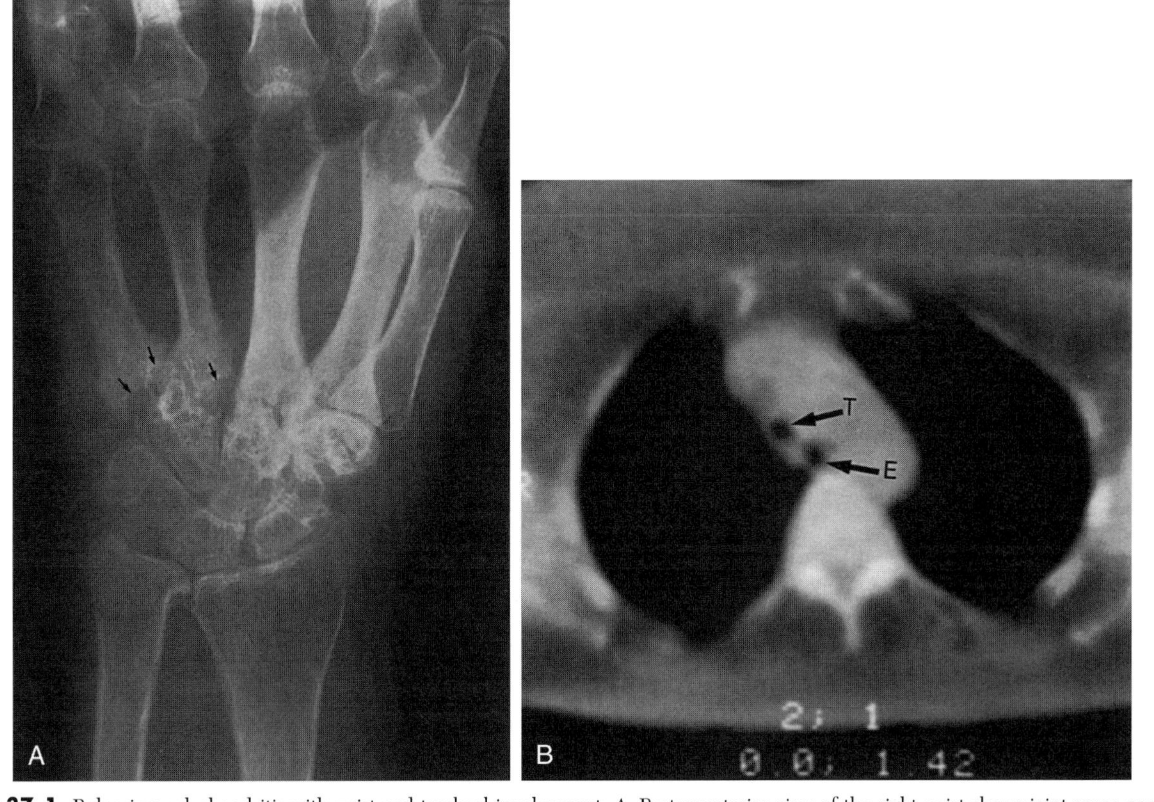

Figure 37–1. Relapsing polychondritis with wrist and tracheal involvement. *A*, Posteroanterior view of the right wrist shows joint space narrowing and erosions. The erosions are most pronounced at the fourth and fifth carpometacarpal joints *(arrows)*. *B*, Dynamic CT study performed on the electron-beam CT unit shows significant narrowing of the trachea (T), especially in expiration. The esophagus (E) also is outlined with air in this section.

has been suggested that an underlying rheumatic disease may have been unrecognized (Fig. 37–1A).

Computed Tomography

Computed tomography (CT) studies are helpful in showing major airway disease in the larynx, trachea, and main stem bronchi, which can be fatal (Fig. 37–1B). These findings include collapse of the major airways as a result of cartilage destruction and narrowing of the lumen secondary to inflammation, edema, and cicatricial changes.

Recommended Readings

Booth A, Dieppe PA, Goddard PL, et al: The radiological manifestations of relapsing polychondritis. Clin Radiol 40:147, 1989.
Braunstein EM, Martel W, Stilwill E, et al: Radiological aspects of the arthropathy of relapsing polychondritis. Clin Radiol 30:441, 1979.
Johnson TH, Mital N, Rodnan GP, et al: Relapsing polychondritis. Radiology 106:313, 1973.
Oddone M, Toma P, Taccone A, et al: Relapsing polychondritis in childhood: A rare observation studied by CT and MRI. Pediatr Radiol 22:537, 1992.
O'Hanlan M, McAdam LP, Bluestone R, et al: The arthropathy of relapsing polychondritis. Arthritis Rheum 19:191, 1976.

CHAPTER 38

Miscellaneous Arthropathies

Georges Y. El-Khoury, MD, Mark D. Stanley, MD, and D. Lee Bennett, MD

SILICONE ARTHRITIS

Overview

Silicone implants have been used since the 1970s as spacers in the wrist and hand and have enjoyed some popularity. Lately the enthusiasm has waned because of reports of complications, particularly related to giant cell synovitis associated with a destructive arthropathy. The synovitis is believed to be due to shedding of small silicone particles, which incite a giant cell reaction. The interval between the surgical procedure and the onset of synovitis varies between 1 and 9 years. The incidence of this complication is not exactly known, but it is believed to be greater than 30%.

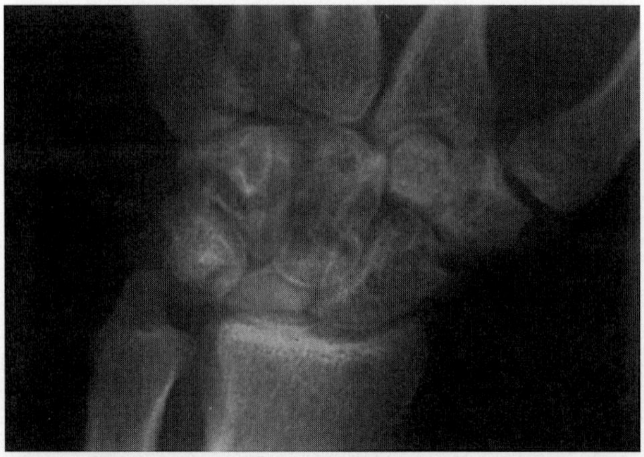

Figure 38–1. Silicone arthritis after using a lunate implant. There are large cystic lesions in the capitate, hamate, and, to a lesser extent, triquetrum. These represent giant cell foreign-body reaction to the silicone particles.

Radiographic Findings

Radiographically, silicone arthritis is characterized by well-defined lytic or cystic areas, sometimes with thin sclerotic margins (Fig. 38–1). The joint spaces are well preserved until late in the disease. Osteopenia is lacking, and the cystic pattern in this disease is strikingly similar to that of pigmented villonodular synovitis. Deformity, shrinkage, and displacement of the implant commonly are associated with this arthritis.

REPLANTATION ARTHROPATHY

Most replantations are for partially or completely amputated digits in the hands. Currently the procedure is successful in about 80% of cases. A peculiar arthropathy in the joints distal to the site of anastomosis has been described. It is characterized by articular irregularity at the subchondral plates, erosions, cystic changes, and secondary osteoarthritis (Fig. 38–2). This arthropathy is believed to be due to ischemia, avascular necrosis, and neuroarthropathy.

ARTHROPATHY ASSOCIATED WITH BURN INJURIES

Overview

Burn injuries are fairly common; however, osteoarticular changes secondary to burns rarely are discussed in the literature. These changes include osteoporosis and heterotopic osssifications. The upper extremities are affected more frequently with these abnormalities than the lower extremities. Heterotopic ossifications typically occur around the elbows, but the shoulder, hip, and knee also can be involved (Fig. 38–3). In the elbow, complete posterior bony bridging with total loss of joint mobility has been reported. Thermal necrosis resulting in joint destruction also can occur, especially in the proximal interphalangeal joints.

Imaging

Radiography

There is no correlation between the extent or degree of the burn and the heterotopic ossification. The radiographic changes do not always occur on the side with the more severe burns. In severe cases, a complete posterior bony bridge develops from the olecranon to the medial epicondyle, resulting in complete loss of mobility at the elbow (see Fig. 38–3). Patients with heterotopic ossification tend to have significant osteoporosis. With deep burns of the hands, joint destruction in proximal interphalangeal joints can be seen, which later progresses into fusion.

Nuclear Medicine

Bone scanning has been used to follow the evolution of heterotopic ossification. Bone scans are thought to be reliable in assessing maturation of heterotopic ossification, which takes about 1 to 1.5 years, at which time the bone scan returns to normal. It has been suggested that surgery should be delayed until maturation occurs to reduce the risk of recurrence. The current thinking stresses, however, that surgical intervention need not be delayed until the bone scan becomes normal. Surgical intervention is indicated without delay for burn patients with marked restriction of range of motion or for progressive nerve compression.

FROSTBITE

Overview

Frostbite is a rare injury that usually is seen when the skin and underlying tissues are exposed directly to bitterly cold temperatures even for a short time. The basic mechanism is believed to be a vascular injury. Superficial frostbite involves the skin and subcutaneous tissues, whereas severe cases involve the muscles, tendons, bone, and cartilage. In the acute stage, clinical assessment of tissue viability is difficult before the development of gangrene. Triple-phase bone scanning has been advocated as a means to assess the extent of frostbite injury and as an indicator of microvascular integrity. Bone and joint changes are observed in the late stages of the disease. The bone and joint deformities are more pronounced in young patients because of the involvement of epiphyseal growth cartilage in children. Articular abnormalities and phalangeal deformities are believed to be due to direct chondrocyte injury after freezing.

Imaging

Radiography

In childhood, typically the epiphyseal growth plates of the distal phalanges and, to a lesser extent, the middle phalanges are involved. There is premature epiphyseal closure resulting in brachydactyly and radial clinodactyly of the fifth finger owing to uneven growth (Fig. 38–4). The thumb is typically normal because it is protected by being clasped in the palm (see Fig. 38–4).

In children and adults, damage to the articular cartilage in the distal interphalangeal and proximal interphalangeal joints leads to arthritis, which in the late stages resembles osteoarthritis (Fig. 38–5). Tuftal resorption and loss of soft tissue at the distal ends of the fingers seem to occur together. Elastic cartilage in the pinna also can become damaged, and it manifests with calcification and ossifications.

Bone Scanning

Triple-phase bone scan is a good indicator of tissue viability as early as 2 days after the cold injury. Absent initial perfusion and no bone uptake on delayed images indicate deep soft tissue and bone infarction requiring amputation (Fig. 38–6).

Recommended Readings

Brunet WG, Munro TG: Destructive joint disease following replantation of digits of the hand. J Can Assoc Radiol 41:210, 1990.

Carrera GF, Kozin F, Flaherty L, et al: Radiographic changes in the hands following childhood frostbite injury. Skeletal Radiol 6:33, 1981.

Gralino BJ, Porter JM, Rosch J: Angiography in the diagnosis and therapy of frostbite. Radiology 119:301, 1976.

Guerra Jr J, Resnick D, Gelberman RH, et al: Replantation of digits or hands followed by destructive joint disease. Radiology 152:591, 1984.

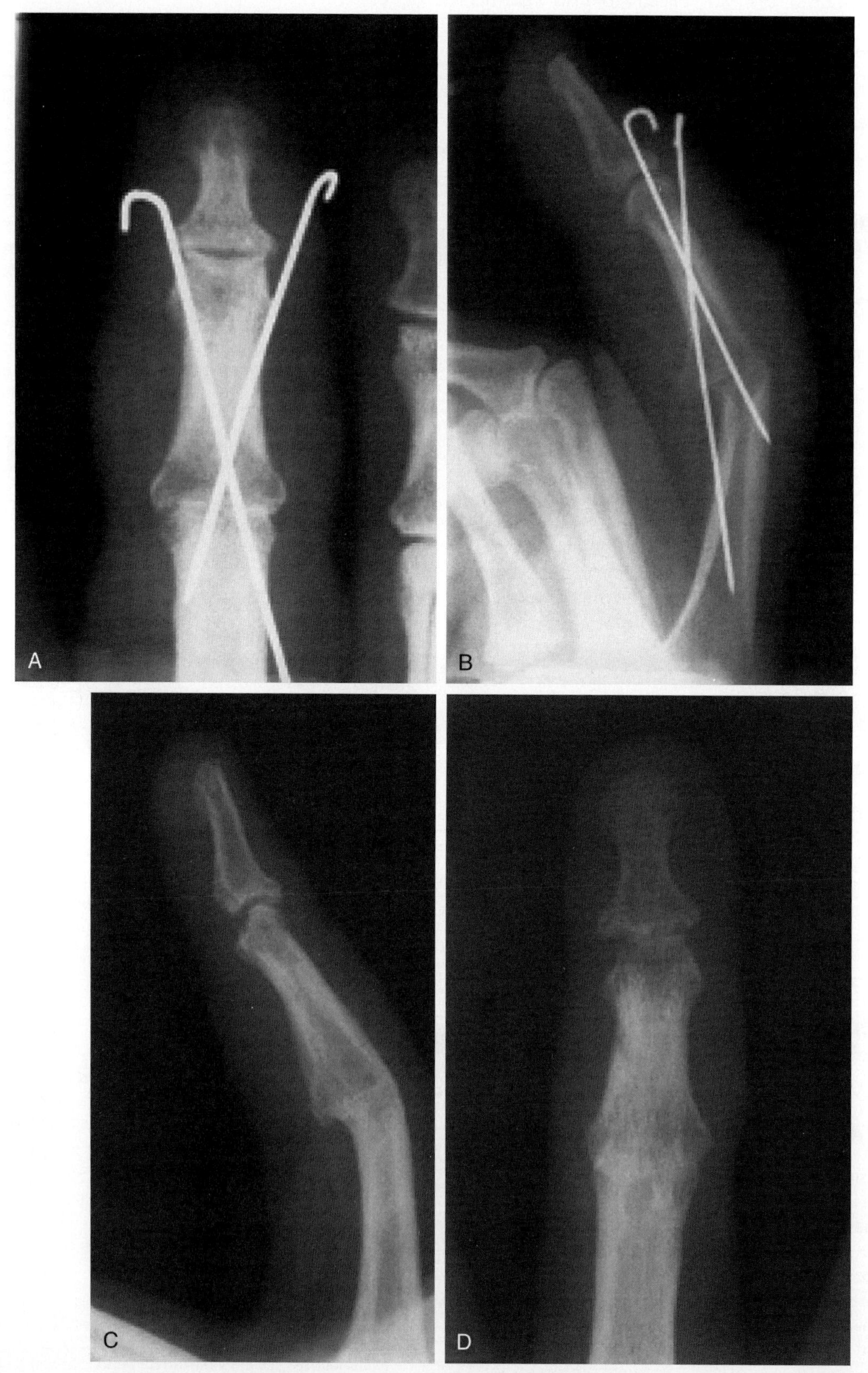

Figure 38–2. Replantation arthropathy of the distal interphalangeal (DIP) joint after the long finger was amputated at the proximal interphalangeal (PIP) joint. *A* and *B*, The amputated stump was replanted, and the PIP joint was fused and transfixed using Kirschner wires. *C* and *D*, The PIP joint fused without complications; however, the DIP joint shows narrowing and irregular subchondral plates on both sides of the joint space.

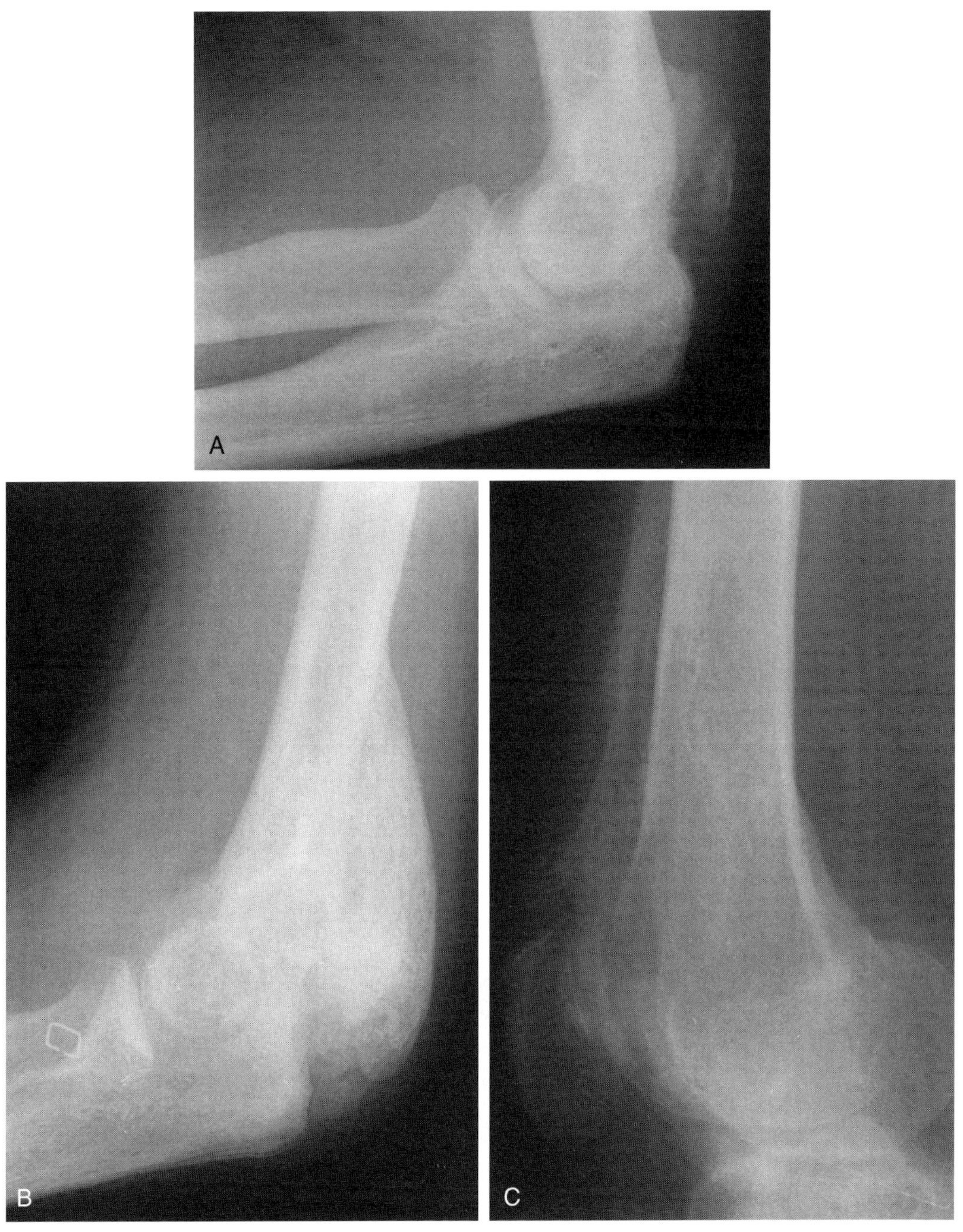

Figure 38–3. Burn arthropathy affecting the elbow. *A*, Moderate involvement with heterotopic ossification on the posteromedial aspect of the elbow. Patient complained of limited extension. *B*, Large heterotopic ossification on the posteromedial aspect of the elbow in another burn patient with severe limitation of elbow extension. *C*, Heterotopic ossification above the knee anteriorly.

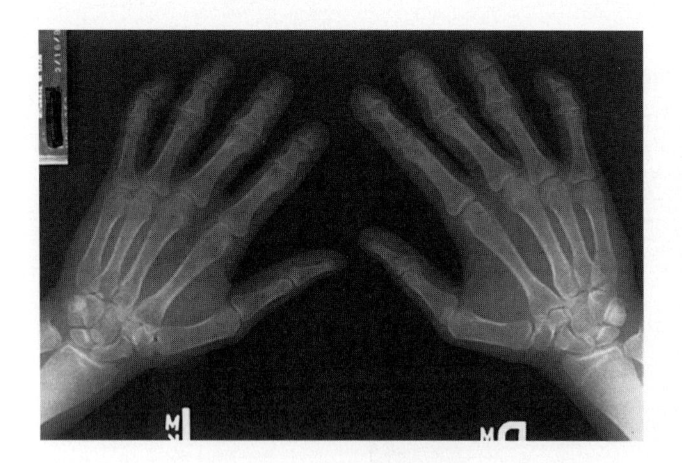

Figure 38–4. Changes of frostbite in the hands of an adolescent who sustained the injury in childhood. The distal and middle phalanges are short because of premature epiphyseal closure. Note also the uneven growth in the middle phalanx of the fourth and fifth fingers bilaterally, resulting in brachydactyly and clinodactyly. The thumbs are normal. (Case provided by Dr. Guilermo Carrera, Medical College of Wisconsin, Milwaukee, WI.)

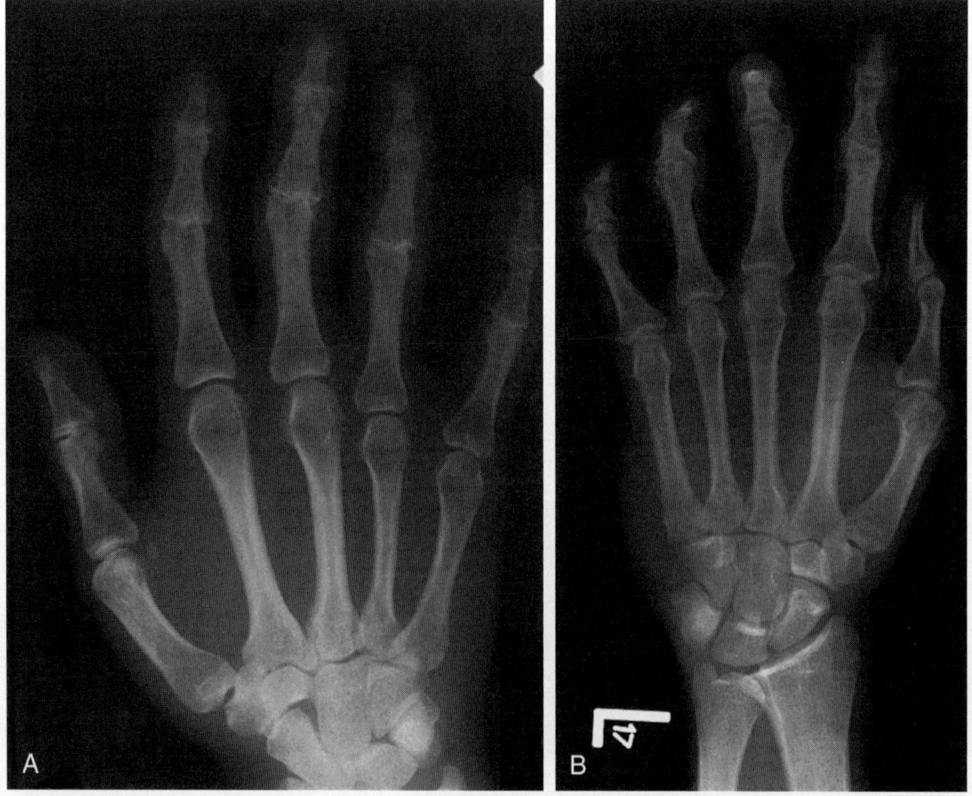

Figure 38–5. *A*, Frostbite in an adult showing arthritis in the distal interphalangeal and proximal interphalangeal joints. The arthritis is similar to osteoarthritis except that the thumb and the first carpometacarpal joint are unaffected. *B*, Changes of frostbite in an adult who was injured in childhood. There was amputation of the distal phalanx of the third finger and deformity of the fourth and fifth fingers.

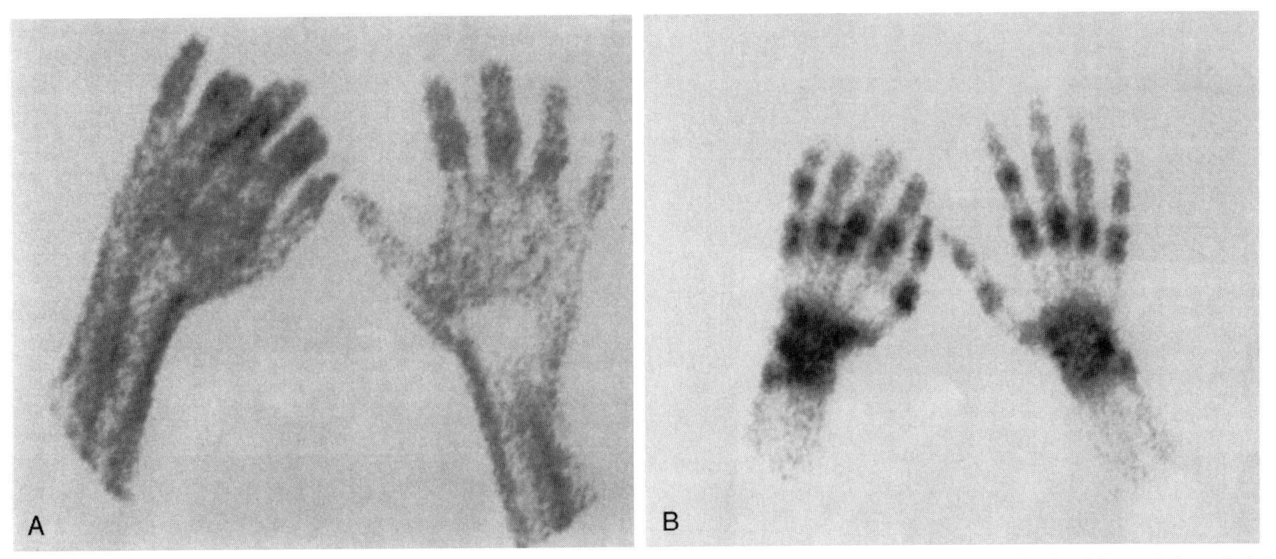

Figure 38–6. Triple-phase bone scan in a patient with severe frostbite. *A*, Early flow images show poor perfusion to the distal fingers bilaterally but more on the right side. *B*, Delayed images show persistence of poor perfusion, indicating infarction of bone and soft tissues in the distal fingers on the right side.

Koch BM, Wu CM, Randolph J, et al: Heterotopic ossification in children with burns: Two case reports. Arch Phys Med Rehabil 73:1104, 1992.

Mehta RC, Wilson MA: Frostbite injury: Prediction of tissue viability with triple-phase bone scanning. Radiology 170:511, 1989.

Peters WJ: Heterotopic ossification: Can early surgery be performed, with a positive bone scan? J Burn Care Rehabil 11:318, 1990.

Rosenthal DI, Rosenberg AE, Schiller AL, et al: Destructive arthritis due to silicone: A foreign-body reaction. Radiology 149:69, 1983.

Schiele HP, Hubbard RB, Bruck HM: Radiographic changes in burns of the upper extremity. Radiology 104:13, 1972.

Schneider HJ, Weiss MA, Stern PJ: Silicone-induced erosive arthritis: Radiologic features in seven cases. AJR Am J Roentgenol 148:923, 1987.

Yeh C-W, Chan KF: Case report 87. Skeletal Radiol 4:49, 1979.

SECTION 

METABOLIC BONE DISEASE

CHAPTER 39

Osteoporosis

Leon Lenchik, MD, and Mitchell Kline, MD

OVERVIEW

Osteoporosis is the most common metabolic disease of bone, affecting approximately 28 million people in the United States. It is a major public health problem mainly because of high morbidity and the economic costs associated with osteoporotic fractures. Each year in the United States, approximately 700,000 vertebral fractures, 250,000 proximal femur fractures, and 240,000 distal radius fractures are attributed to osteoporosis. Having an osteoporotic fracture significantly increases a patient's risk of subsequent fractures; therefore, the goal is to make the diagnosis of osteoporosis prior to the first fracture. Bone densitometry is an essential tool for making this goal a reality.

PATHOPHYSIOLOGY

Osteoporosis is a skeletal disorder characterized by decreased bone mass and increased susceptibility to fractures. It is classified as either primary or secondary. Primary osteoporosis is further subdivided into type 1 (postmenopausal) and type 2 (age-related or senile).

Secondary osteoporosis has an identifiable cause other than menopause and aging. The causes include endocrine conditions (hypogonadism, hyperthyroidism, hyperparathyroidism, pituitary disease, diabetes mellitus, and pregnancy), congenital diseases (osteogenesis imperfecta, Ehlers-Danlos syndrome, homocystinuria, Marfan's syndrome, systemic mastocytosis, and Gaucher's disease), nutritional disorders (alcoholism, malnutrition, calcium deficiency, vitamin D deficiency, and scurvy), and drugs (glucocorticoids, anticonvulsants, and heparin).

Risk Factors

Risk factors for osteoporosis are female gender, increased age, estrogen deficiency, white or Asian ethnicity, family history of osteoporosis, calcium deficiency, vitamin D deficiency, low body weight, history of prior fracture, and smoking.

All of these risk factors are strongly associated with low bone mineral density (BMD). Low BMD is considered the major risk factor for osteoporosis and the strongest predictor of fracture risk. The association between low BMD and risk of fracture is analogous to the relationship between high serum cholesterol and risk of myocardial infarction.

Some risk factors for osteoporotic fractures are unrelated to BMD. They are propensity to fall, low physical function, impaired vision, impaired cognition, and environmental hazards.

CLINICAL FEATURES

Clinical history and physical examination allow diagnosis of osteoporosis in its advanced stages, usually in the setting of a fracture caused by minimal trauma. Other findings are loss of height, kyphosis, respiratory difficulties, gastrointestinal complaints, and depression. Often, however, patients are asymptomatic.

IMAGING FINDINGS

Radiographic Diagnosis

Radiographic diagnosis is possible only in individuals with advanced disease, usually in the setting of osteopenia and fracture. The most common finding is generalized osteopenia, often accompanied by cortical thinning and prominence of weight-bearing trabeculae (Fig. 39–1). Often patients with osteoporosis present with fractures, especially of the vertebrae, proximal femur, distal radius, and proximal humerus. Vertebral fractures and deformities include anterior wedging, posterior wedging, and biconcave end plates (Fig. 39–2). Magnetic resonance (MR) imaging can be useful in distinguishing between compression fractures due to osteoporosis and those due to tumor. Tumor usually completely replaces the normal fatty marrow of the vertebral body, whereas in osteoporotic compression fractures, some normal fat signal remains within the vertebral body on MR images (Fig. 39–3).

In the absence of fractures, conventional radiographs have low sensitivity in detecting osteoporosis. Loss of 30% to 50% of bone mass is required before osteoporosis can be detected on radiographs. Indeed, radiographs are often normal in patients in whom osteoporosis is diagnosed on the basis of bone densitometry (Fig. 39–4).

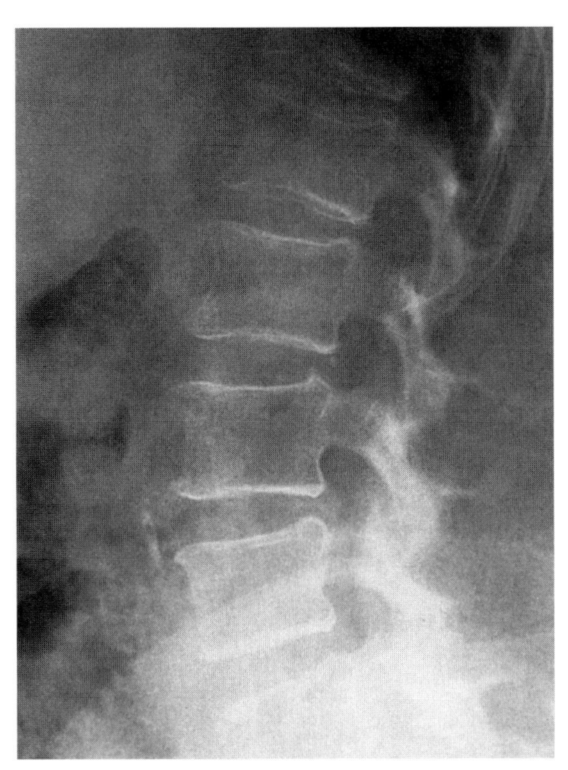

Figure 39–1. Lateral radiograph of the lumbar spine in a woman with osteoporosis shows generalized osteopenia with thinned but well-defined cortical margins.

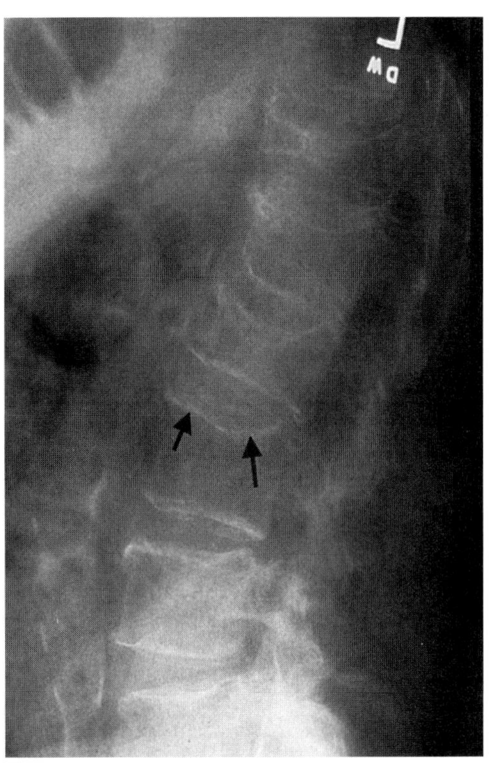

Figure 39–2. Lateral radiograph of the thoracolumbar spine in an older woman with osteoporosis shows generalized osteopenia, compression fractures, and concave vertebral end plates (*arrows*).

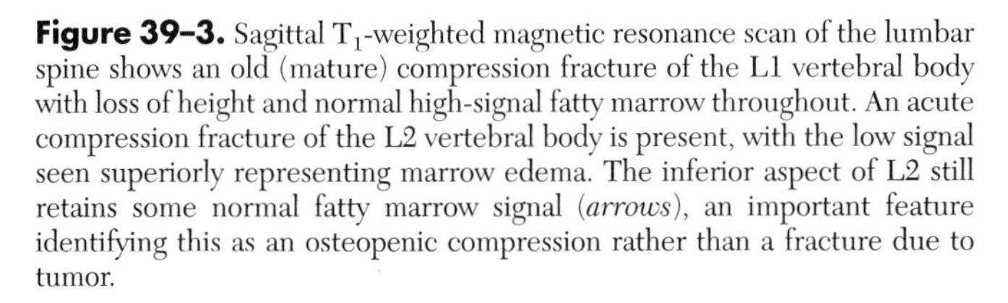

Figure 39–3. Sagittal T_1-weighted magnetic resonance scan of the lumbar spine shows an old (mature) compression fracture of the L1 vertebral body with loss of height and normal high-signal fatty marrow throughout. An acute compression fracture of the L2 vertebral body is present, with the low signal seen superiorly representing marrow edema. The inferior aspect of L2 still retains some normal fatty marrow signal (*arrows*), an important feature identifying this as an osteopenic compression rather than a fracture due to tumor.

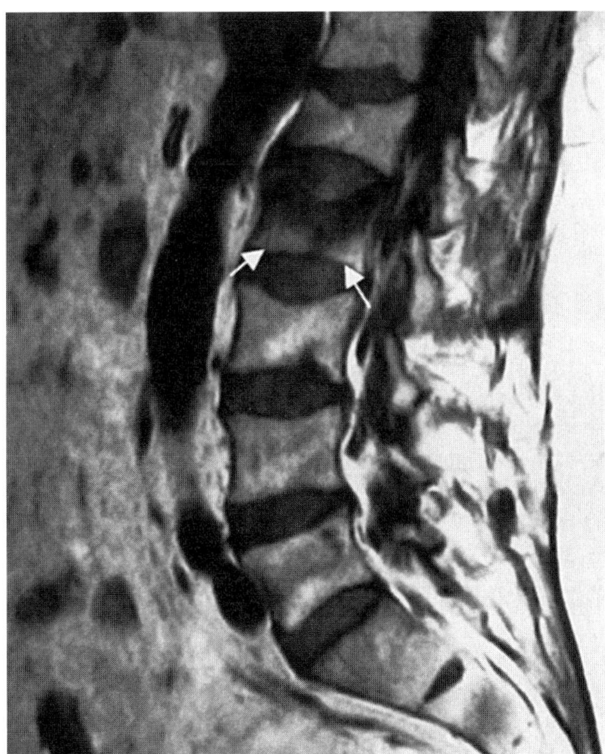

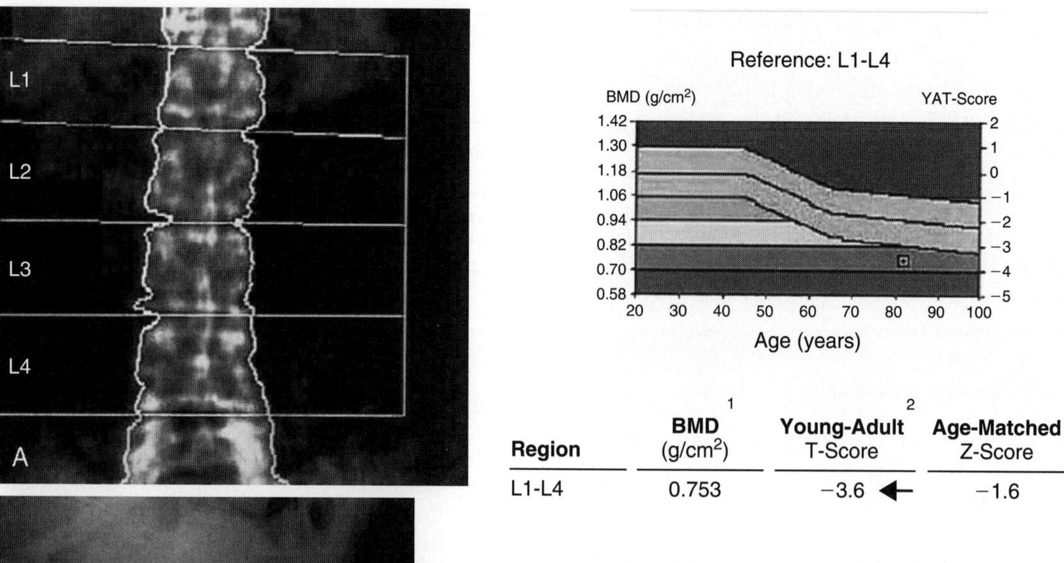

Region	BMD[1] (g/cm²)	Young-Adult[2] T-Score	Age-Matched[3] Z-Score
L1-L4	0.753	−3.6 ←	−1.6

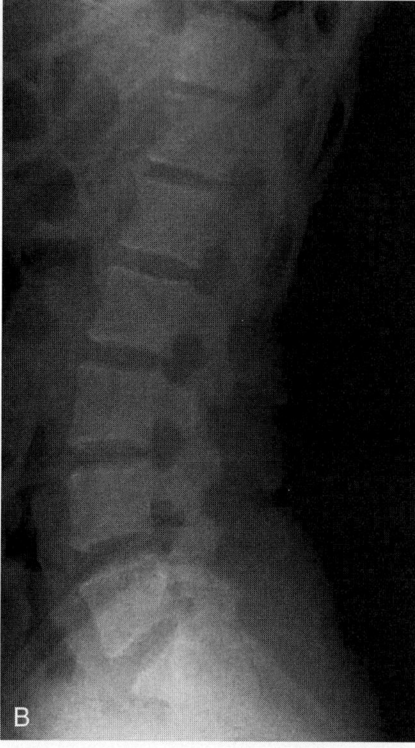

Figure 39–4. *A,* Dual x-ray absorptiometry scan of the lumbar spine in an 81-year-old white woman shows a T-score = −3.6 (*arrow*). Because the T-score is less than −2.5, the woman would be classified as having osteoporosis according to World Health Organization criteria. *B,* Lateral radiograph of the lumbar spine in the same woman shows normal mineralization of the visualized skeleton. (See legend, Fig. 39–6, for explanation of graph in *A.*)

Differential Diagnosis

The differential diagnosis of generalized osteopenia should include osteomalacia, hyperparathyroidism, and multiple myeloma. Glucocorticoid-induced osteoporosis may show exuberant callus and condensation of bone at vertebral end plates as well as complications, including vertebral osteonecrosis and intravertebral vacuum (Fig. 39–5).

BONE DENSITOMETRY

Bone densitometry allows diagnosis of osteoporosis in asymptomatic individuals by providing accurate, reproducible measurement of bone mineral density.

Clinical Indications

Clinical indications for bone densitometry are estrogen deficiency, prolonged glucocorticoid therapy, radiologic osteopenia or fractures, primary hyperparathyroidism, and monitoring of antiresorptive therapy. Currently, the National Osteoporosis Foundation of the United States recommends bone densitometry in all postmenopausal white women younger than 65 years who have at least one risk factor in addition to menopause and in all white women older than 65 years regardless of the presence or absence of other risk factors. The use of bone densitometry for screening of populations at risk is widely debated.

Methods

Densitometric methods are commonly divided into central and peripheral. Central methods allow measurement of the spine, hip, or both; they are dual x-ray absorptiometry (DXA) and quantitative computed tomography (QCT). Peripheral methods allow measurement of the phalanges, forearm, or heel; they are peripheral dual x-ray absorptiometry (pDXA), peripheral quantitative computed tomography (pQCT), and quantitative ultrasonography (QUS).

Dual X-ray Absorptiometry

DXA examinations allow BMD measurement of any portion of the skeleton. Typically, two skeletal sites are measured: the posteroanterior (PA) lumbar spine and proximal femur. In patients with severe degenerative disease, PA spine measurement may overestimate BMD; therefore, measurement of the lateral lumbar spine, distal radius, or total body may be more useful.

Quantitative Computed Tomography

QCT examinations are performed on the lumbar spine. Although BMD measurements are obtained with the use of an imaging CT scanner, a software package and a calibration phantom are required. QCT has the unique advantage of selectively measuring trabecular bone and providing a volumetric density in g/cm^3.

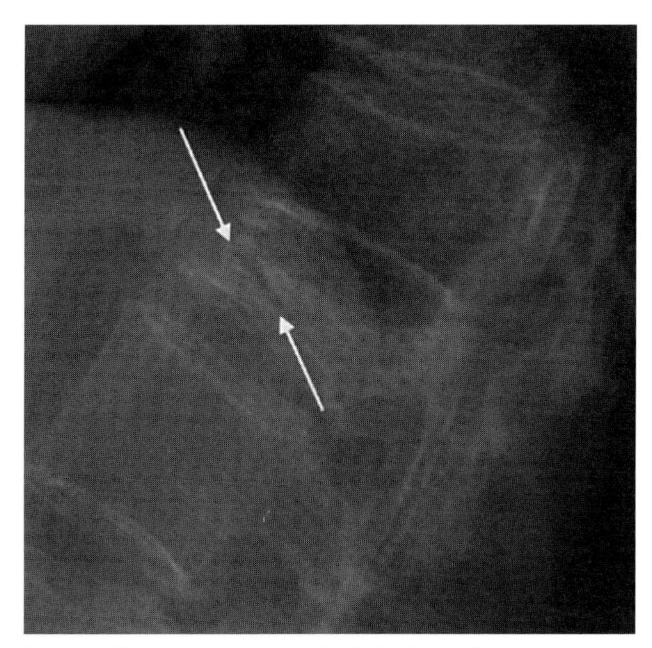

Figure 39–5. Lateral radiograph of the lower thoracic spine in a 65-year-old woman with glucocorticoid-induced osteoporosis shows generalized osteopenia and a compression fracture with intravertebral vacuum (*arrows*). Although the vacuum may be seen with other causes of osteoporosis, its presence is important for exclusion of underlying malignancy.

Table 39–1. World Health Organization Diagnostic Criteria for Postmenopausal Caucasian Women

Diagnostic Categories	T-Score
Normal	≥–1
Osteopenia	<–1 and >–2.5
Osteoporosis	≤–2.5 (without fractures)
Established osteoporosis	≤–2.5 (with fractures)

Peripheral Densitometry

Peripheral methods provide greater portability, greater ease of use, and lower radiation dose at a lower cost than central methods. Peripheral QCT may be used to measure the forearm. Peripheral DXA may be used to measure the forearm, phalanges, or calcaneus. QUS may be used to measure the calcaneus, forearm, or phalanges.

QUS is different from other densitometric methods because it does not measure BMD. Instead, speed of sound and broadband ultrasound attenuation are usually measured. From these measurements, additional parameters, such as stiffness and quantitative ultrasound index, may be calculated. QUS parameters may be used to assess an individual's risk for fracture.

Diagnosis of Osteoporosis, Assessment of Fracture Risk, and Monitoring of Therapy

On printouts of most densitometry devices, BMD measurements are expressed as a T-score (Fig. 39–6). T-scores are calculated by subtracting the mean BMD of young-normal reference population from the patient's BMD and dividing by the standard deviation of the young-normal reference population. Traditionally, the T-score has been considered the clinically most relevant value because it is used for diagnosis of osteoporosis. In postmenopausal white women, a T-score ≤ –2.5 has been widely used to indicate osteoporosis (Table 39–1). This diagnostic threshold is appropriate only for particular densitometric techniques (e.g., DXA) and measurement sites (e.g., PA spine, hip, and forearm). For other techniques (e.g., QCT, QUS) and skeletal sites (e.g., heel, phalanges), diagnostic criteria are currently controversial. Controversy also exists concerning diagnostic criteria in premenopausal women, men, and nonwhite individuals. In fact, the use of T-scores for diagnosis of osteoporosis is currently being debated. In the future, T-scores may be abandoned in favor of diagnostic thresholds based on measured BMD and associated fracture risk.

Assessment of fracture risk using densitometry is possible because studies have shown that bone strength is highly correlated with BMD on biomechanical testing. Further evidence is provided by prospective epidemiologic trials showing that BMD predicts fracture rates in

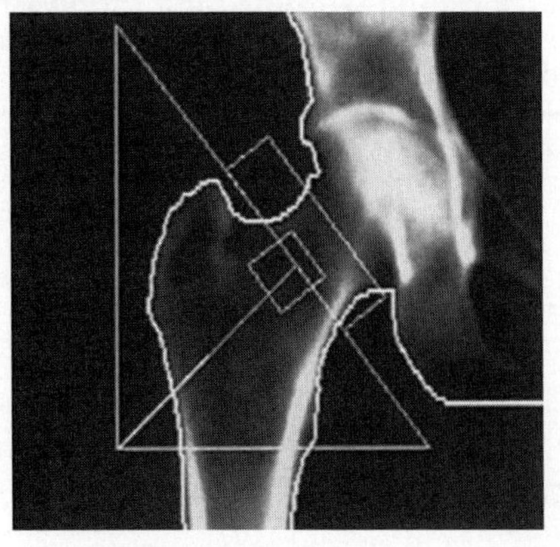

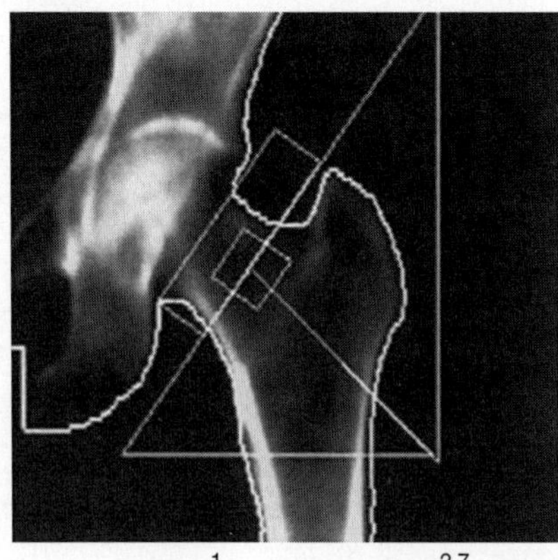

Region Total	BMD [1] (g/cm^2)	Young-Adult [2,7] T-Score	Age-Matched [3] Z-Score
Left	0.680	−2.7 ←	−1.0
Right	0.651	−2.9 ◄	−1.3
Mean	0.665	−2.8	−1.1
Difference	0.029	0.2	0.2

Figure 39–6. Dual x-ray absorptiometry (DXA) scan of both hips in a 73-year-old white woman. She has been receiving hormone replacement therapy and says she has no history of other therapy for osteoporosis, prior fractures, or other risk factors for osteoporosis. The printouts for DXA devices are manufacturer-specific. However, the common features are (1) the image of the skeletal site that was scanned; (2) a graph that plots the patient's age and bone mineral density (BMD); and (3) numerical data, consisting of the BMD value for the region of interest measured, the BMD value expressed as a standard deviation of young-normal reference population (T-score), and the BMD value expressed as a standard deviation of age-matched reference population (Z-score). The most relevant of the numerical information is the T-score. In the patient whose DXA printout is shown here, the left total femur T-score = −2.7 (*arrow*), and the right total femur T-score = −2.9 (*arrowhead*). Because the T-scores are less than −2.5, the patient would be classified as having osteoporosis according to World Health Organization criteria.

populations. In general, each standard deviation decrease in BMD is associated with 1.5-fold to 3.0-fold increase in risk of fracture. Although QUS does not measure BMD, several prospective trials indicate that QUS parameters are predictive of hip fracture rates.

Monitoring of therapy with densitometry is possible provided that the devices used have low rates of precision (i.e., reproducibility) errors and measure skeletal sites that respond well to therapy. Because of excellent precision and greatest increases in BMD with therapy, measurement of the spine with DXA or QCT is preferable to other densitometric methods and measurement sites for monitoring of response to therapy.

PATIENT MANAGEMENT

Medical management usually requires antiresorptive therapy. In the United States, there are many choices of pharmacologic agents, including estrogens, selective estrogen receptor modulators, bisphosphonates, and calcitonin. These medications have been shown to stabilize or increase BMD and thereby to reduce the risk of osteoporotic fractures. Regardless of what medical therapy is chosen, adequate intake of calcium and vitamin D is essential.

After fracture, maximizing physical function is mandatory and often necessitates analgesic therapy. Orthopedic management is usually required after hip, forearm, and humerus fractures. In the elderly, psychosocial support is an important part of patient care.

Recommended Readings

Cummings SR, Black DM, Nevitt MC, et al: Bone density at various sites for prediction of hip fractures. Lancet 341:72, 1993.

Cummings SR, Palermo L, Browner W, et al: Monitoring osteoporosis therapy with bone densitometry: Misleading changes and regression to the mean. JAMA 283:1318–1321, 2000.

Eastell R, Boyle IT, Compston J: Management of male osteoporosis: Report of the UK Consensus Group. Q J Med 91:71–92, 1998.

Eastell R: Treatment of postmenopausal osteoporosis. N Engl J Med 338:736–746, 1998.

Faulkner KG, von Stetten E, Miller P: Discordance in patient classification using T-scores. J Clin Densitom 2:343–350, 1999.

Finsen V, Anda S: Accuracy of visually estimated bone mineralization in routine radiographs of the lower extremity. Skeletal Radiol 17:270, 1988.

Genant HK, Cooper C, Poor G, et al: Interim report and recommendations of the World Health Organization task-force for osteoporosis. J Bone Miner Res 10:259–264, 1999.

Genant HK, Engelke K, Fuerst T: Noninvasive assessment of bone mineral and structure: State of the art. J Bone Miner Res 11:707–730, 1996.

Gluer CC: Monitoring skeletal changes by radiological techniques. J Bone Miner Res 14:1952–1962, 1999.

Gluer CC: Quantitative ultrasound techniques for the assessment of osteoporosis: Expert agreement on current status. The International Quantitative Ultrasound Consensus Group. J Bone Miner Res 12:1280–1288, 1997.

Kanis JA, Melton LJ III, Christiansen C, et al: The diagnosis of osteoporosis. J Bone Miner Res 9:1137, 1994.

Lenchik L, Rochmis P, Sartoris DJ: Optimized interpretation and reporting of dual X-ray absorptiometry (DXA) scans. AJR Am J Roentgenol 171:1509–1520, 1998.

Marshall D, Johnell O, Wedel H: Meta-analysis of how well measures of bone mineral density predict occurrence of osteoporotic fractures. BMJ 312:1254–1259, 1996.

Melton LJ III, Atkinson EJ, O'Fallon WM, et al: Long-term fracture prediction by bone mineral assessed at different skeletal sites. J Bone Miner Res 8:1227, 1993.

Ross PD, Davis JW, Vogel JM, et al: A critical review of bone mass and the risk of fractures in osteoporosis. Calcif Tissue Int 46:149, 1990.

CHAPTER 40

Regional Osteoporosis

Leon Lenchik, MD, and Mitchell Kline, MD

ETIOLOGY

Causes of regional osteoporosis include reflex sympathetic dystrophy syndrome, transient regional osteoporosis, and disuse.

PATHOPHYSIOLOGY

Reflex Sympathetic Dystrophy Syndrome

Reflex sympathetic dystrophy syndrome (RSDS) is mediated by the sympathetic nervous system and is characterized by aggressive osteoporosis, soft tissue swelling progressing to atrophy, and contracture. The cause is usually traumatic, often involving minor trauma, but the disease may also be idiopathic.

Transient Regional Osteoporosis

Transient regional osteoporosis includes transient osteoporosis of the hip and regional migratory osteoporosis. The etiology of these disorders is unknown.

Disuse or Immobilization

Disuse from chronic illness or paralysis, or immobilization due to trauma leads to an increase in bone resorption and a decrease in bone formation, causing osteoporosis.

CLINICAL FEATURES

The hallmark of RSDS is pain in an extremity, often out of proportion to the inciting event. Swelling of the extremity, joint stiffness, and changes in skin color and temperature can occur. These changes do not occur in an anatomic nerve distribution. Men and women are affected equally. The diagnosis is usually made clinically.

Transient osteoporosis of the hip usually affects middle-aged adults, men more often than women. Symptoms appear suddenly and patients often remember the exact time and date when the disease hit them. In men, it can affect either hip or can be bilateral. In women, however, transient osteoporosis preferentially affects the left hip. Transient osteoporosis of the hip was first described in pregnant women, typically in the third trimester. Regional migratory osteoporosis is seen most often in middle-aged men and usually involves one of the lower extremities. Both of these conditions may result in pain and are characteristically self-limiting with good prognosis.

Osteoporosis related to disuse or immobilization is usually asymptomatic. However, affected patients are at risk for insufficiency fractures.

IMAGING FINDINGS

For all of the causes of regional osteoporosis, the most common finding on conventional radiographs is osteopenia, which may be uniform, bandlike, or patchy (Fig. 40–1). The joint spaces are well preserved and there are no marginal or focal subchondral erosions. In transient osteoporosis of the hip, close inspection of the radiographs reveals profound and diffuse osteopenia as well as absence of the subchondral bony plate around the femoral head. Marrow edema is seen on magnetic resonance (MR) imaging—appearing as decreased signal on T1-weighted sequences and increased signal on T2-weighted sequences (Fig. 40–2)—and radiopharmaceutical uptake is increased on bone scanning (Fig. 40–3). Some patients with transient osteoporosis of the hip have a joint effusion. A condition that has MR image findings identical to transient osteoporisis of the hip is termed transient bone marrow edema syndrome. In this syndrome the plain radiographs do not show any osteoporosis. Increased uptake on bone scan can also be seen in RSDS. MR findings are nonspecific in RSDS but include skin thickening, soft tissue edema, soft tissue enhancement, and, rarely, bone marrow edema.

Differential Diagnosis

The lack of joint space narrowing and bony erosions helps distinguish regional osteoporosis from septic arthritis and other inflammatory arthritides. Differentiating transient osteoporosis of the hip from osteonecrosis can be difficult because the radiographic (osteoporosis) and MR (marrow edema) findings can overlap. The marrow edema in transient osteoporosis of the hip, however, is usually more diffuse and tends to extend into the femoral neck (Fig. 40–2*B*). In avascular necrosis the edema is more

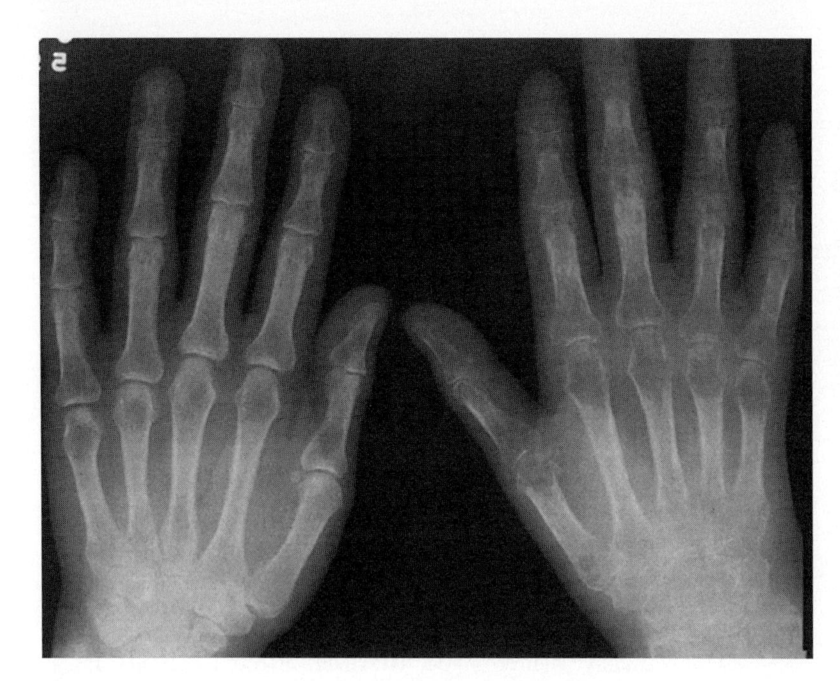

Figure 40–1. Posteroanterior view of both hands shows diffuse osteopenia involving the left hand in a 35-year-old patient with reflex sympathetic dystrophy. The joint spaces are well maintained, and no articular erosions are present.

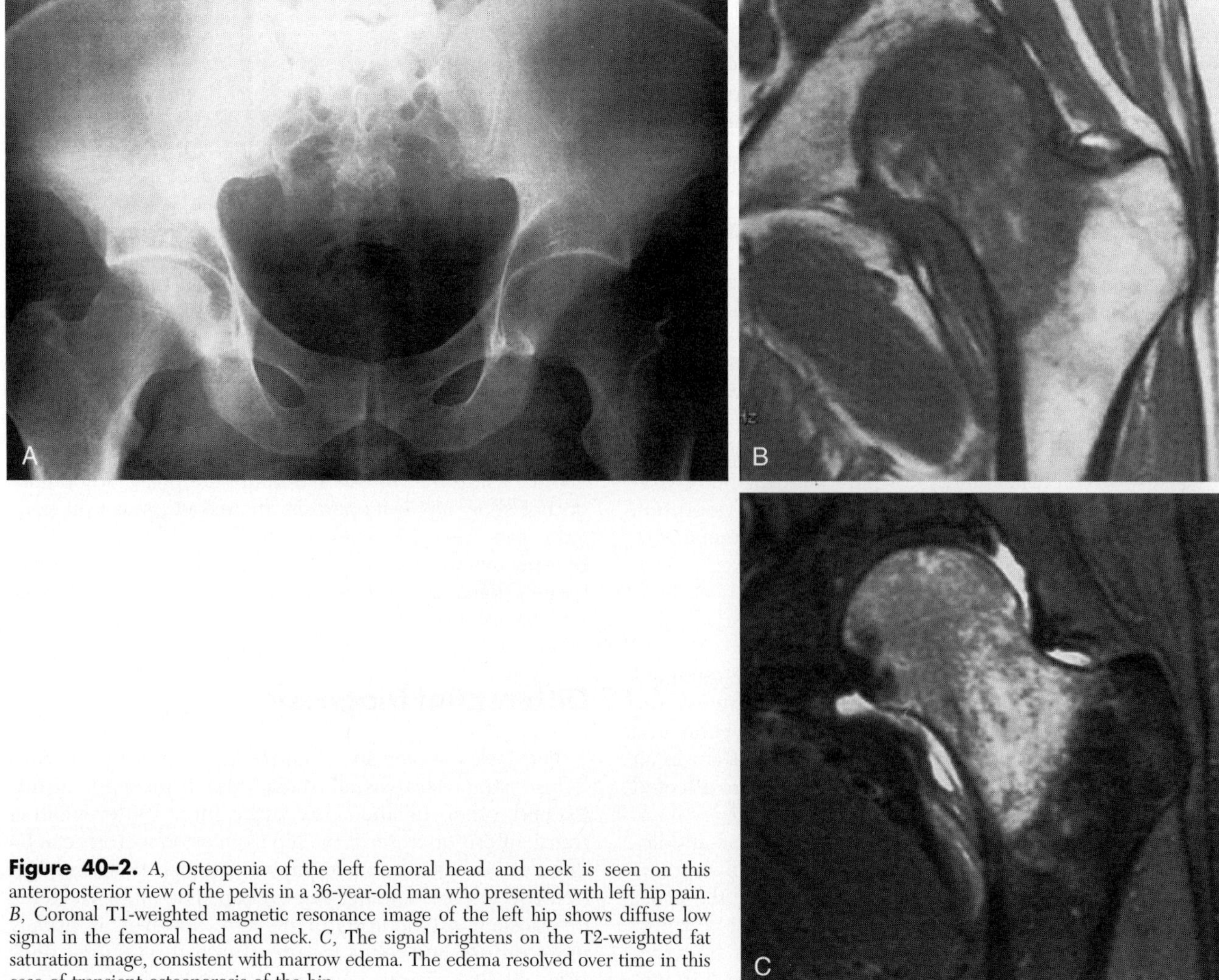

Figure 40–2. *A,* Osteopenia of the left femoral head and neck is seen on this anteroposterior view of the pelvis in a 36-year-old man who presented with left hip pain. *B,* Coronal T1-weighted magnetic resonance image of the left hip shows diffuse low signal in the femoral head and neck. *C,* The signal brightens on the T2-weighted fat saturation image, consistent with marrow edema. The edema resolved over time in this case of transient osteoporosis of the hip.

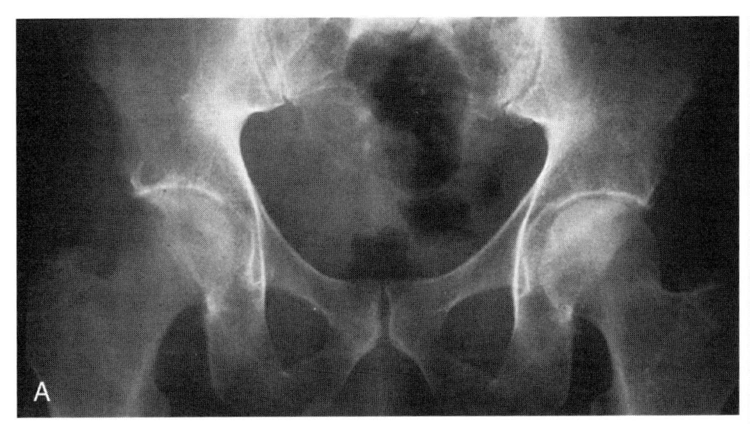

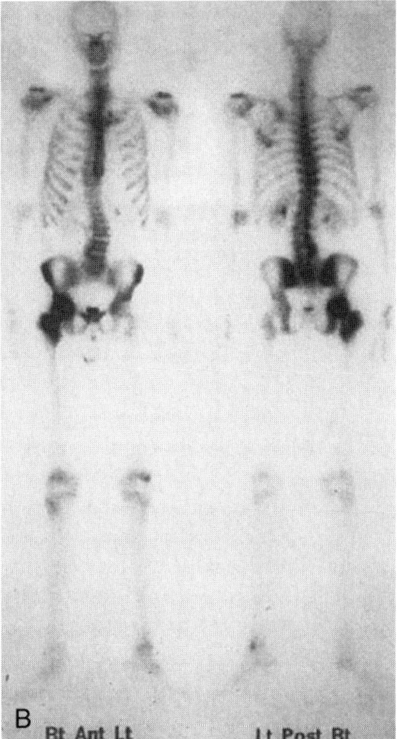

Figure 40–3. *A,* Frontal radiograph of the pelvis in a woman with a 2-month history of right hip pain shows relative osteopenia of the right proximal femur compared with the left proximal femur. *B,* Bone scan in the same patient shows intense radiopharmaceutical uptake in the right proximal femur. Because the patient's symptoms resolved without complications, the findings were attributed to transient regional osteoporosis.

localized to the femoral head, involving mainly its anterior superior quadrant in the early phases of the disease. (See Chapter 44 for more information on avascular necrosis of bone). Also, bone marrow edema on MR images is a nonspecific finding that may be seen in many inflammatory and neoplastic disorders.

Recommended Readings

Arnstein AR: Regional osteoporosis. Orthop Clin North Am 3:585, 1972.

Dunne F, Walters B, Marshall T, et al: Pregnancy associated osteoporosis. Clin Endocrinol 39:487, 1993.

Genant HK, Kozin F, Bekerman C, et al: The reflex sympathetic dystrophy syndrome: A comprehensive analysis using fine-detail radiography, photon absorptiometry and bone and joint scintigraphy. Radiology 117:21, 1975.

Lequesne M, Kerboull M, Bensasson M, et al: Partial transient osteoporosis. Skeletal Radiol 2:1, 1977.

Kozin F, McCarty DJ, Simms J, et al: The reflex sympathetic dystrophy syndrome. I: Clinical and histologic studies: Evidence for bilaterality, response to corticosteroids and articular involvement. Am J Med 60:321, 1976.

McCord WC, Nies KM, Campion DS, et al: Regional migratory osteoporosis: A denervation disease. Arthritis Rheum 21:834, 1978.

Naides SJ, Resnick D, Zvaifler NJ: Idiopathic regional osteoporosis: A clinical spectrum. J Rheumatol 12:763, 1985.

Smith R, Athanasou NA, Ostlere SJ, et al: Pregnancy associated osteoporosis. Q J Med 88:865, 1995.

Smith R, Phillips AJ: Osteoporosis during pregnancy and its management. Scand J Rheumatol Suppl 107:66, 1998.

Vande Berg BE, Malghem JJ, Labaisse MA, et al: MR imaging of avascular necrosis and transient marrow edema of the femoral head. Radiographics 13:501, 1993.

Wilson AJ, Murphy WA, Hardy DC, et al: Transient osteoporosis: Transient bone marrow edema? Radiology 167:757, 1988.

CHAPTER 41

Rickets and Osteomalacia

Leon Lenchik, MD, and Mitchell Kline, MD

DEFINITION

Rickets and *osteomalacia* describe a group of disorders characterized by incomplete mineralization of normal osteoid tissue. Rickets is a disorder of growing bone in which there is hypomineralization of the growth plate. Osteomalacia occurs after the cessation of growth.

PATHOPHYSIOLOGY

Calcium and phosphate are the two main constituents of hydroxyapatite, the mineral part of bone tissue. Normal bone mineralization depends on an adequate supply of these constituents. Vitamin D plays a major role in bone mineralization by maintaining calcium and phosphate homeostasis through its action on bone, gastrointestinal tract, kidneys, and parathyroid glands. The metabolic pathway for vitamin D requires (1) exposure of the skin to ultraviolet light and (2) enzymatic processes in the liver and the kidneys to form the active metabolite. Any interruption in this pathway can lead to vitamin D deficiency and result in rickets or osteomalacia.

Rickets and osteomalacia can also be caused by an inadequate supply of phosphate, usually due to renal tubular loss of phosphate. A variety of disorders cause deficient bone mineralization via this pathway, including vitamin D–resistant rickets (also called X-linked hypophosphatemia) and tumor-related osteomalacia (associated with nonossifying fibromas, hemangiopericytomas, giant cell tumors, and fibrous dysplasia).

Inadequate supply of calcium causing rickets can be seen in premature infants if they are not given enough calcium to meet their tremendous need.

CLINICAL FEATURES

Rickets

The signs and symptoms of rickets are variable and depend on the etiology of the disorder and the age of the patient. Children may present with failure to thrive or with stunted skeletal growth. Tetany, seizures, weakness, or myopathy may occasionally be seen.

Findings on physical examination include delay of fontanelle closure, frontal bossing, softening of the skull, delay in dentition, prominence of the costochondral junctions of the middle ribs (rachitic rosary), kyphoscoliosis, joint enlargement, and bowing of long bones.

Rickets is commonly classified as primarily hypocalcemic or primarily hypophosphatemic. Patients with primarily hypocalcemic rickets have decreases in serum calcium and phosphate levels and urine calcium values and increases in urine phosphate and parathyroid hormone (PTH) levels. Patients with primarily hypophosphatemic rickets have normal serum calcium levels and decreases in serum phosphate levels. Urine calcium may be normal, urine phosphate may be increased or decreased, and PTH may be normal or increased in such patients.

Osteomalacia

Unlike the signs and symptoms of rickets, the signs and symptoms of osteomalacia are subtle. Symptoms include fatigue, malaise, bone pain, muscle pain and muscle weakness. Most patients with osteomalacia have decreased levels of calcium and phosphate in both serum and urine.

IMAGING FINDINGS

Rickets

The most common imaging finding in a patient with rickets is generalized osteopenia. Other findings, apparent at the growth plates, include increased lucency and widening of the physes, and irregularity (fraying) and cupping of the metaphyses (Fig. 41–1). The most involved parts of the skeleton are the distal femur, both ends of the tibia (Fig. 41–2), the distal radius and ulna, the proximal humerus, and the costochondral junctions of the middle ribs.

The differential diagnosis includes hypophosphatasia and metaphyseal chondrodysplasia (Schmid type).

The most common complication of rickets is skeletal deformity. In neonates, posterior flattening and squaring of the skull (i.e., craniotabes) may be seen. In early childhood, bowing deformities of the arms and legs are common (Fig. 41–3). In older children, scoliosis, vertebral end plate deformities, basilar invagination of the skull, triradiate deformity of the pelvis, and slipped capital femoral epiphysis may be seen (Fig. 41–4).

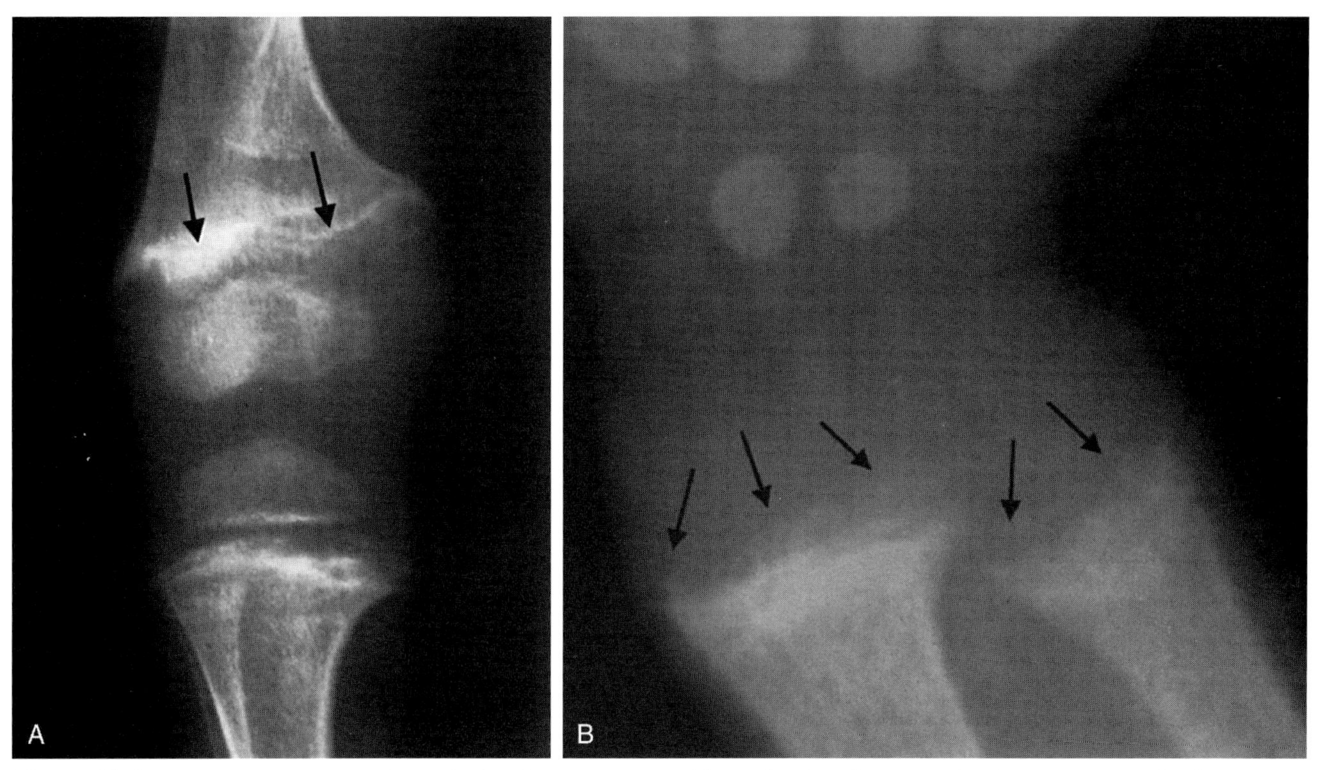

Figure 41–1. *A*, Anteroposterior (AP) radiograph of the knee in a 3-year-old boy with rickets shows generalized osteopenia. Note the characteristic widening of the growth plate and irregularity (*arrows*) of the metaphyses of the distal femur as well as the proximal tibia and fibula. *B*, PA radiograph of the wrist in another boy with rickets shows widening, fraying (*arrows*), and cupping of the metaphyses of the distal radius and ulna.

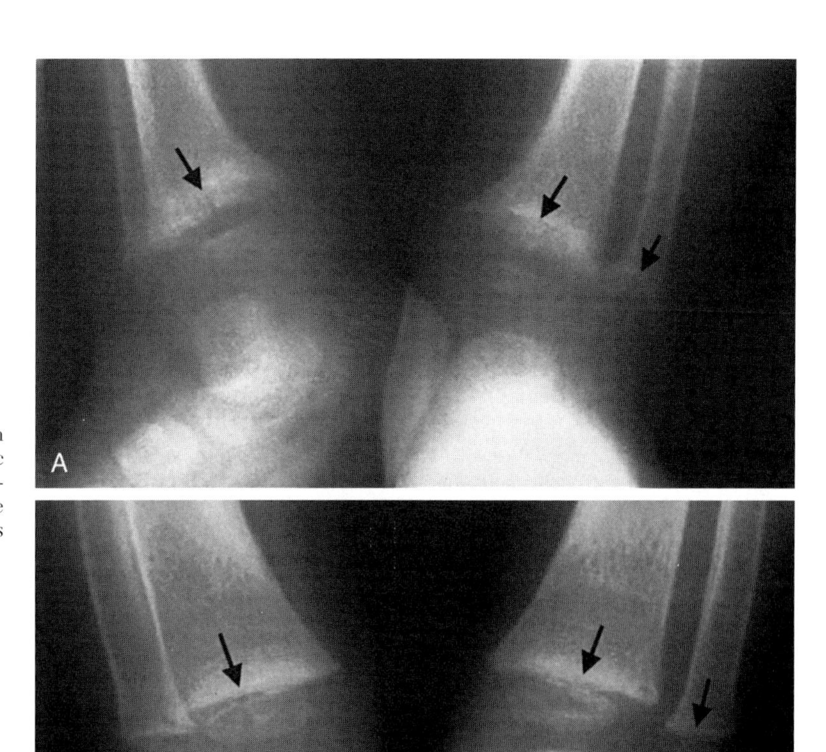

Figure 41–2. *A*, Anteroposterior (AP) view of both ankles in a child with rickets shows characteristic findings—axial growth plate widening, metaphyseal fraying, and metaphyseal cupping (*arrows*). *B*, AP view of the ankles in the same child after treatment for rickets shows normal metaphyses and growth plates (*arrows*).

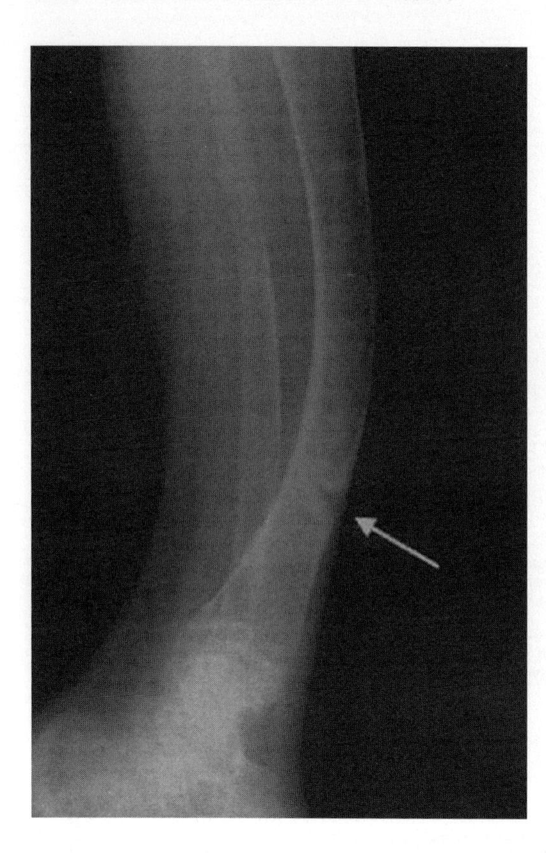

Figure 41–3. Lateral radiograph of the leg in a 6-year-old boy with rickets shows generalized osteopenia, bowing of the tibia, and an insufficiency fracture (*arrow*) in the distal tibia.

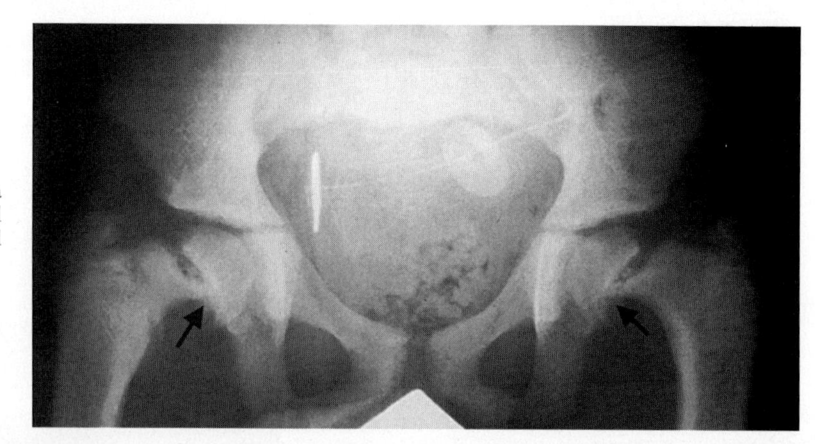

Figure 41–4. Anteroposterior view of the pelvis in a child with rickets due to renal failure shows bilateral slipped capital femoral epiphyses (*arrows*). A peritoneal dialysis catheter overlies the pelvis.

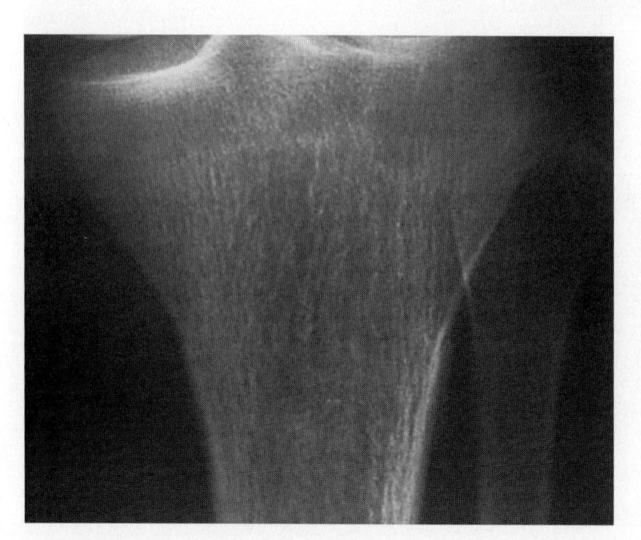

Figure 41–5. Anteroposterior radiograph of the proximal tibial metaphysis in a 55-year-old woman with chronic renal failure shows generalized osteopenia. Note the thick but indistinct bony trabecula characteristic of osteomalacia.

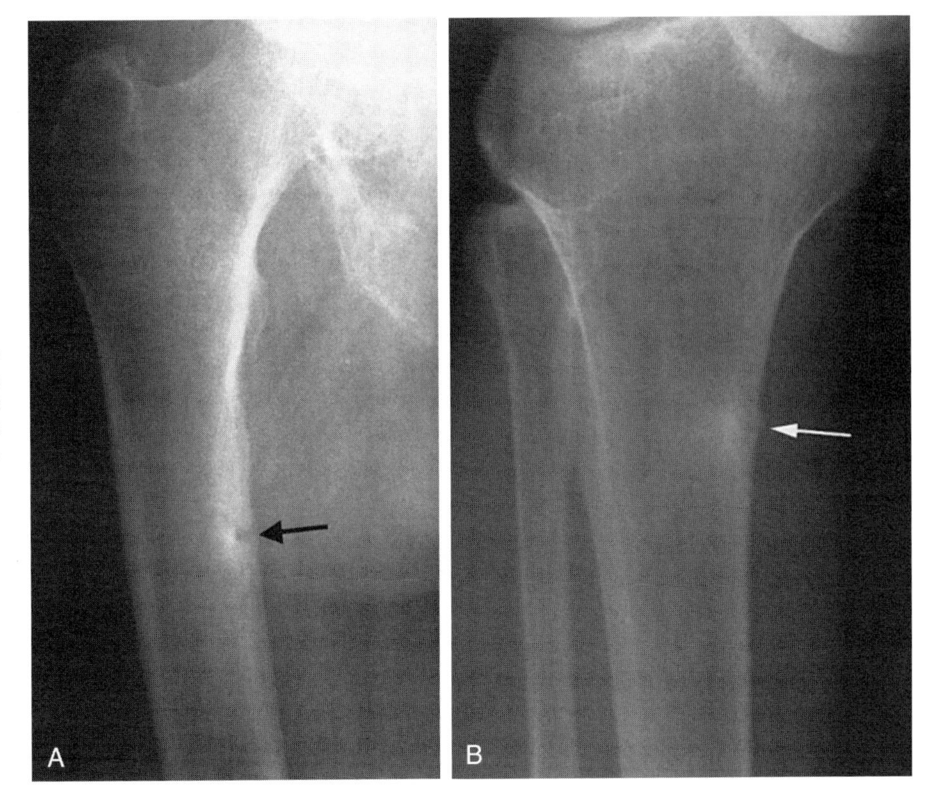

Figure 41–6. *A* and *B*, Anteroposterior radiographs from two separate patients with osteomalacia showing Looser zones (*arrows*) along the medial cortex of the proximal femur (*A*) and proximal tibia (*B*).

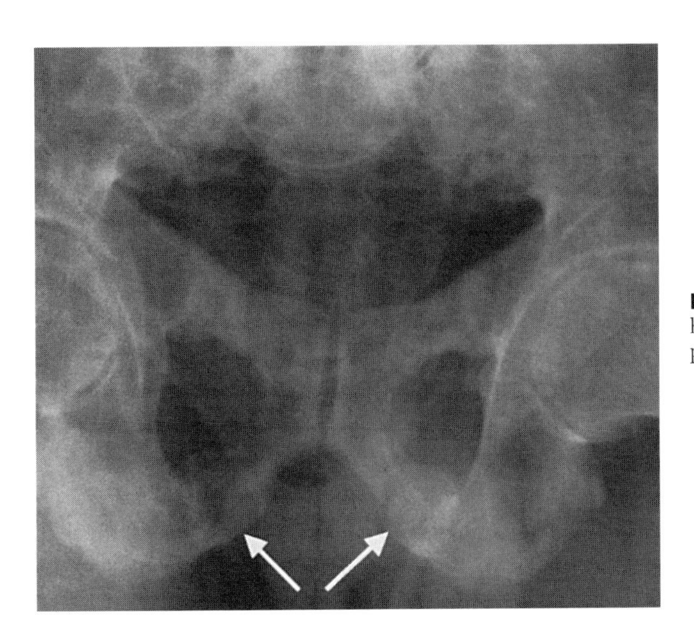

Figure 41–7. Anteroposterior view of the pelvis in a man with X-linked hypophosphatemia shows bilateral Looser zones (*arrows*) involving the pubo-ischial junctions.

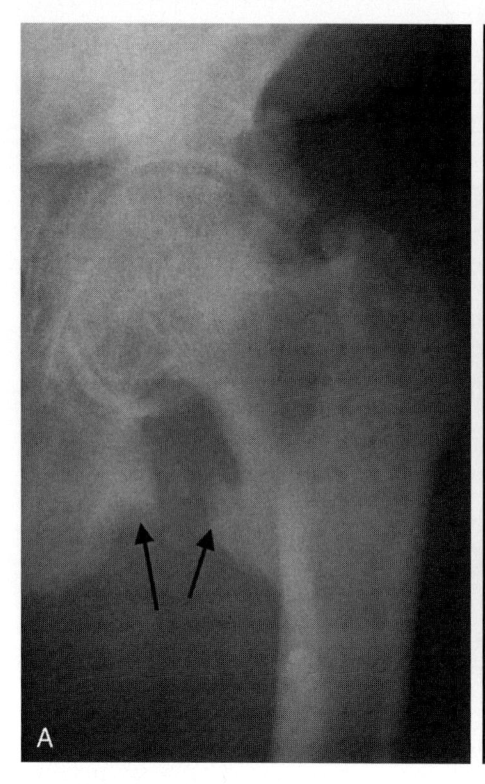

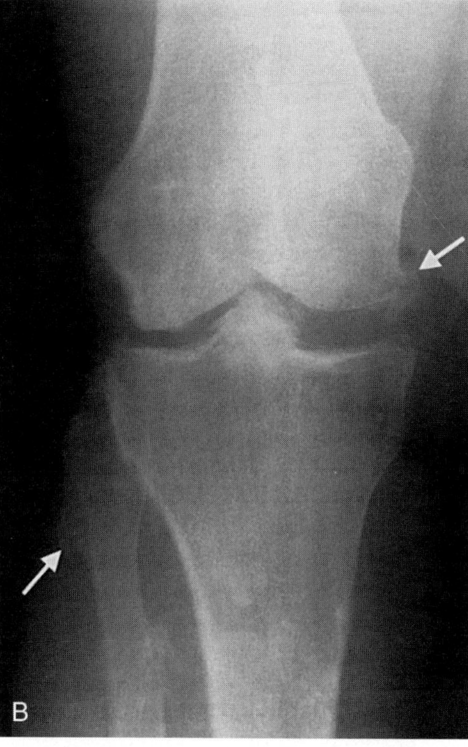

Figure 41–8. *A* and *B*, Radiographs of a man with X-linked hypophosphatemia show large enthesophytes (*arrows*) at multiple ligament and tendon attachment sites about the hip (*A*) and knee (*B*).

Osteomalacia

The most common imaging finding in osteomalacia is generalized osteopenia. The presence of coarsened, indistinct bony trabeculae (Fig. 41–5) and Looser zones suggests the diagnosis. *Looser zones* are insufficiency-type stress fractures that occur in a bilateral and symmetric distribution. The characteristic sites are the medial margins of the proximal femur, the posterior proximal ulna, the axillary margin of the scapula, the pubic rami, and the ribs (Fig. 41–6).

The most common complications of osteomalacia are skeletal deformity (vertebrae and skull) and fractures (vertebral bodies and long bones).

Patients with X-linked hypophosphatemia can have marked enthesopathy at multiple sites and a thick trabecular pattern throughout the skeleton in addition to the usual findings of osteomalacia (Figs. 41–7 and 41–8).

BONE DENSITOMETRY

Patients with rickets and osteomalacia often have low bone mineral density. Bone densitometry may help in monitoring the response to medical therapy.

Recommended Readings

Avila NA, Skarulis M, Rubino DM, et al: Oncogenic osteomalacia: Lesion detection by MR skeletal survey. AJR 167:343–345, 1996.

Ecklund K, Doria AS, Jaramillo D: Rickets on MR images. Pediatr Radiol 29:673–675, 1999.

Finch PJ, Ang L, Eastwood JB, et al: Clinical and histological spectrum of osteomalacia among Asians in South London. Q J Med 83:439–448, 1992.

Gloriex FH: Rickets, the continuing challenge. N Engl J Med 325: 1875–1877, 1991.

Mankin HJ: Rickets, osteomalacia, and renal osteodystrophy: An update. Orthop Clin North Am 21:81, 1990.

Pitt MJ: Rickets and osteomalacia are still around. Radiol Clin North Am 29:97, 1991.

Ryan PJ, Fogelman I: Bone scintigraphy in metabolic bone disease. Semin Nucl Med 3:291–305, 1997.

Reginato AJ, Falasca GF, Pappu R, et al: Musculoskeletal manifestations of osteomalacia: Report of 26 cases and literature review. Semin Arthritis Rheum 28:287–304, 1999.

Steinbach HL, Noetzli M: Roentgen appearance of the skeleton in osteomalacia and rickets. AJR Am J Roentgenol 91:955, 1964.

CHAPTER 42

Hyperparathyroidism and Renal Osteodystrophy

Leon Lenchik, MD, and Mitchell Kline, MD

OVERVIEW

Primary hyperparathyroidism may be caused by a single parathyroid adenoma (80%), hyperplasia (15% to 20%), or, rarely, carcinoma (less than 0.5%). Secondary hyperparathyroidism is usually caused by chronic renal insufficiency. The term renal osteodystrophy includes all of the disorders of bone and mineral metabolism associated with chronic renal insufficiency. It combines features of secondary hyperparathyroidism, rickets, osteomalacia, and osteoporosis. In children with chronic renal failure, the features of rickets dominate, whereas in adults, the features of secondary hyperparathyroidism dominate.

PATHOPHYSIOLOGY

Primary hyperparathyroidism is characterized by incompletely regulated, excessive secretion of parathyroid hormone.

Renal osteodystrophy is a spectrum ranging from primarily high-turnover bone disease (i.e., secondary hyperparathyroidism) to low-turnover bone disease (i.e., osteomalacia).

CLINICAL FEATURES

Primary Hyperparathyroidism

Before the wide use of laboratory screening, patients with hyperparathyroidism often presented with advanced disease, characterized by the triad of renal calculi, diffuse bone pain, and dementia. Today, most patients are asymptomatic. Only 10% to 15% of patients with primary hyperparathyroidism demonstrate skeletal changes. The incidence of primary hyperparathyroidism increases with age and is more common in women.

Primary hyperparathyroidism is characterized by concomitant elevation of serum calcium and parathyroid hormone levels. Elevated serum alkaline phosphatase values, reduced serum phosphate levels, and elevated urinary calcium and phosphate values are common.

Renal Osteodystrophy

The signs and symptoms of renal osteodystrophy are often nonspecific. They include muscle weakness, bone pain, and skeletal deformities. In children, growth retardation is the dominant finding.

Chronic renal failure causes phosphate retention, which results in hypocalcemia and in turn leads to parathyroid hyperplasia and secondary hyperparathyroidism. Laboratory evaluation may reveal the serum level of calcium to be normal, low, or elevated.

IMAGING FINDINGS

Hyperparathyroidism

The most common imaging finding in hyperparathyroidism is generalized osteopenia. Other findings are bone resorption, bone sclerosis, focal lytic (brown tumors) or sclerotic lesions, chondrocalcinosis, soft tissue calcification, and vascular calcification. *Brown tumors* (also called osteoclastomas) are focal regions of giant cells and fibrous tissue found more commonly in primary hyperparathyroidism (Fig. 42–1).

The most characteristic finding is bone resorption. It may be classified as subchondral, trabecular, endosteal, intracortical, subperiosteal, subligamentous, or subtendinous. Subperiosteal resorption, believed to be almost pathognomonic of hyperparathyroidism, is most often seen in the hands and feet (Fig. 42–2). The radial aspects of the middle phalanges of the index and middle fingers and the tufts of distal phalanges (acro-osteolysis) are most commonly affected (Fig. 42–3). Trabecular resorption is most often seen in the diploic space of the skull ("salt-and-pepper" appearance) (Fig. 42–4). Subligamentous and subtendinous resorption is most often seen at the inferior margin of the clavicle, at the insertion of the plantar aponeurosis and Achilles tendon on the calcaneus, at the ischial tuberosities of the pelvis, at the tuberosities of the humerus, and at the femoral trochanters (Fig. 42–5). Subchondral resorption may be seen in the sacroiliac joints, sternoclavicular joints, acromioclavicular joints, symphysis pubis, discovertebral junction, and patella

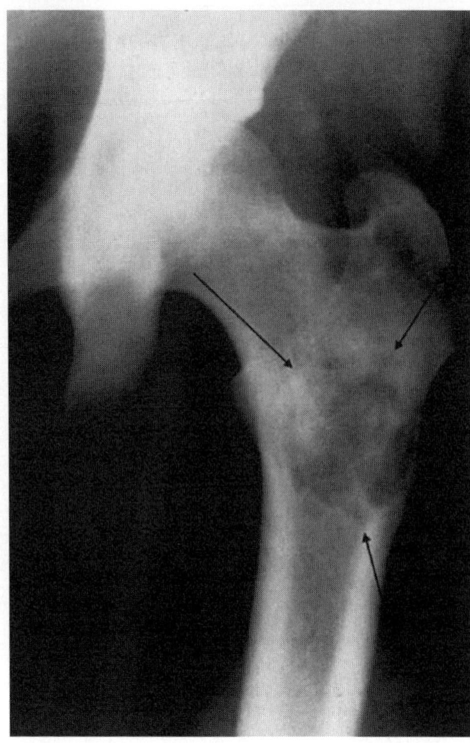

Figure 42–1. Anteroposterior radiograph of the hip in a 45-year-old patient with secondary hyperparathyroidism shows a well-circumscribed lytic lesion (brown tumor) in the proximal femur (*arrows*).

(Figs. 42–6 and 42–7). Erosions may involve the bare areas of the interphalangeal joints, the proximal medial aspects of long bones, and the lamina dura around the roots of the teeth (Fig. 42–8).

Findings that are more common in secondary than in primary hyperparathyroidism are osseous sclerosis and soft tissue calcification (Fig. 42–9). Sclerosis of the spine classically has the appearance of a rugger jersey or soccer shirt—that is, it appears striped (Fig. 42–10). Chondrocalcinosis and brown tumors, however, are more common in primary than in secondary hyperparathyroidism (Fig. 42–11).

Musculoskeletal complications of hyperparathyroidism include rupture or avulsion of tendons and ligaments. Ligamentous laxity often affects the sacroiliac joints, acromioclavicular joints, and spine. Tendon rupture is most common in the quadriceps and patellar tendons (Fig. 42–12).

Renal Osteodystrophy

Imaging findings in renal osteodystrophy are a combination of rickets, osteomalacia, secondary hyperparathyroidism, and osteoporosis.

Patients undergoing dialysis may present with additional findings. Aluminum toxicity may lead to severe osteomalacia, which commonly results in fractures (Fig. 42–13). Carpal and phalangeal cysts may develop, likely as a result of osseous deposits of amyloid (Fig. 42–14). Spondyloarthropathy is also likely due to amyloid deposition and may mimic that of infectious diskitis or neuropathic arthropathy.

Differential Diagnosis

Hyperparathyroidism

Osseous sclerosis in secondary hyperparathyroidism has a broad differential diagnosis, which consists of metastatic disease, radiation-induced bone disease, hypoparathyroidism, myelofibrosis, mastocytosis, sickle cell disease, and Paget's disease. Bony sclerosis causing a rugger jersey

Text continued on page 301

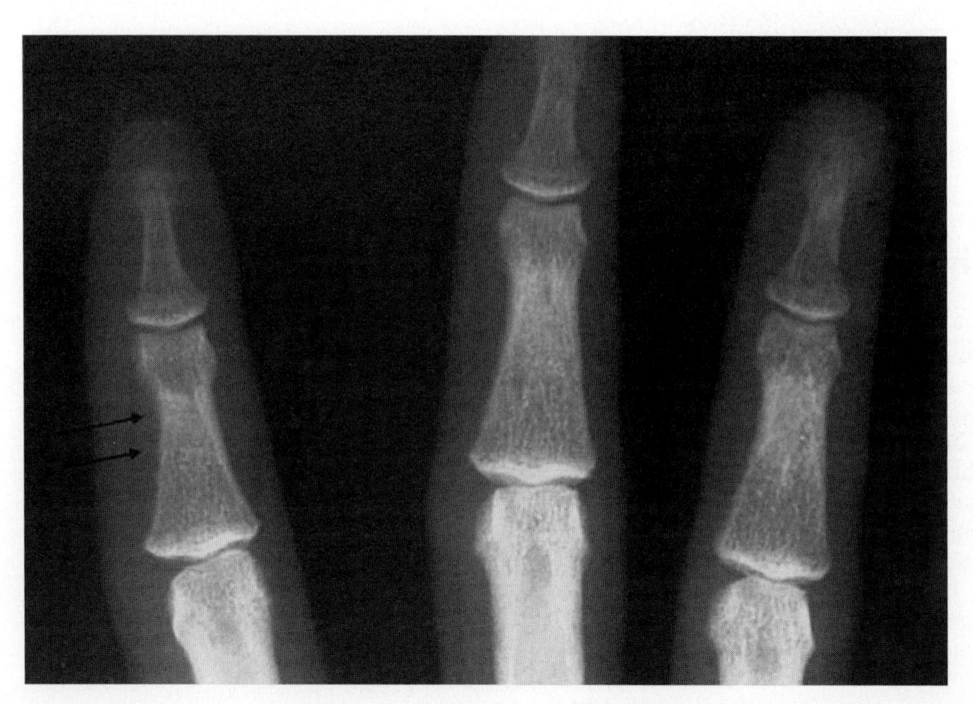

Figure 42–2. Posteroanterior radiograph of the hand in a 56-year-old woman with primary hyperparathyroidism shows subperiosteal bone resorption (*arrows*) along the radial aspect of the middle phalanges, most evident in the index finger. Subperiosteal resorption in this location is characteristic of hyperparathyroidism.

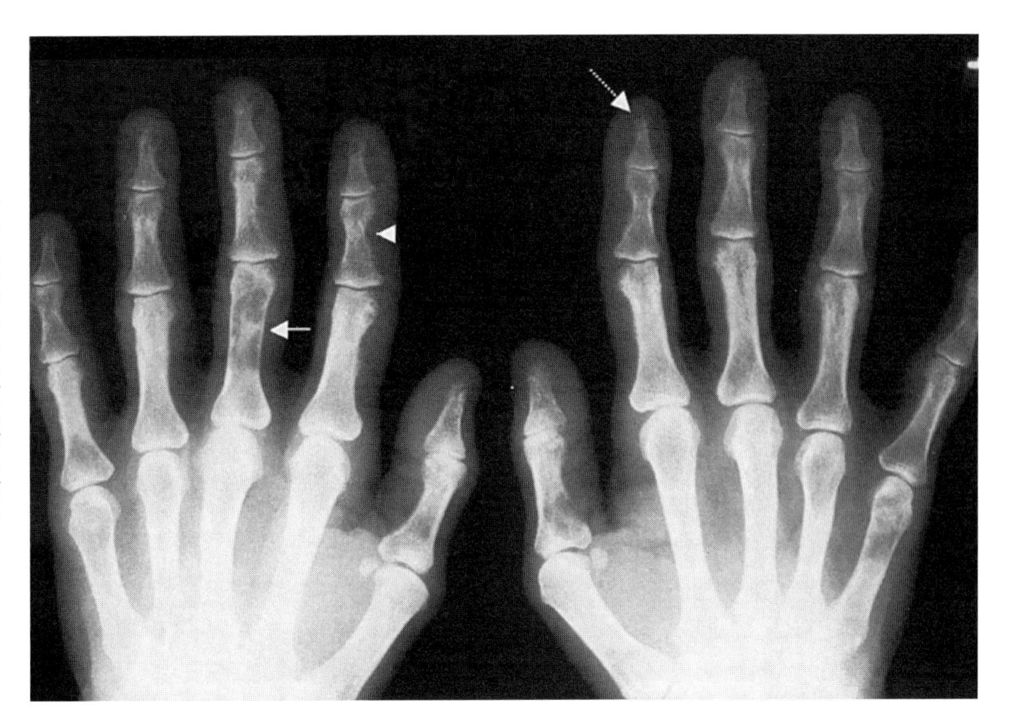

Figure 42–3. Posteroanterior view of both hands in a patient with primary hyperparathyroidism. There are erosive changes of several of the distal phalangeal tufts (*dashed arrow*) as well as subperiosteal bone resorption (*arrowhead*) involving primarily the radial aspects of the middle phalanges of the index and middle fingers. Lytic lesions in the proximal phalanx of the right middle finger (*solid arrow*) and the proximal phalanx of the left thumb and the left fifth metacarpal are brown tumors.

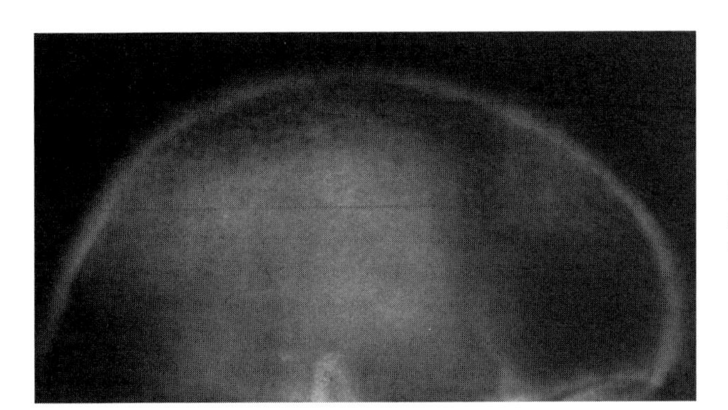

Figure 42–4. Lateral radiograph of the skull in a 53-year-old man with secondary hyperparathyroidism shows trabecular resorption of the diploic space ("salt-and-pepper" appearance).

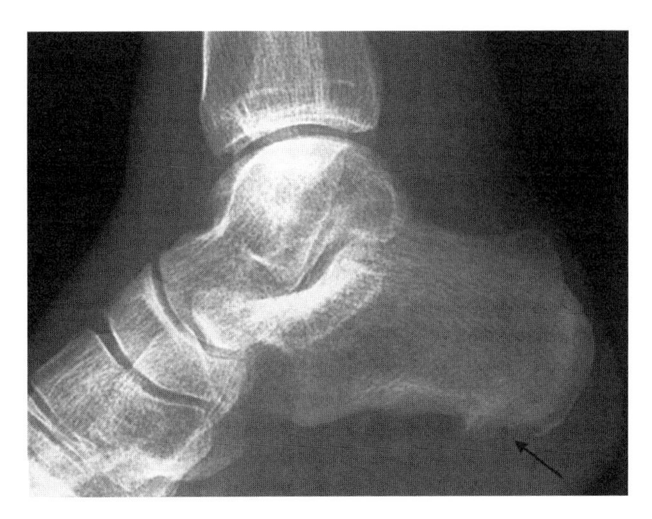

Figure 42–5. Lateral radiograph of the heel in a 67-year-old woman with secondary hyperparathyroidism shows subligamentous resorption involving the origin of the plantar aponeurosis (*arrow*).

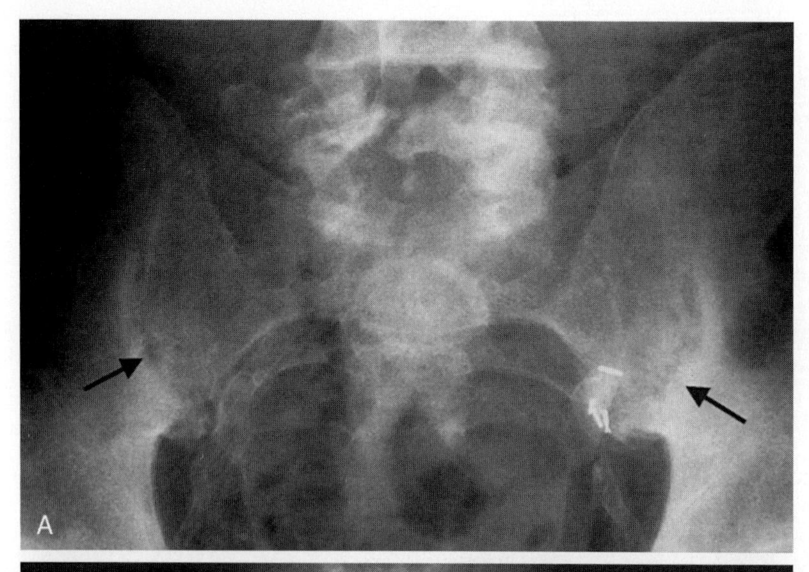

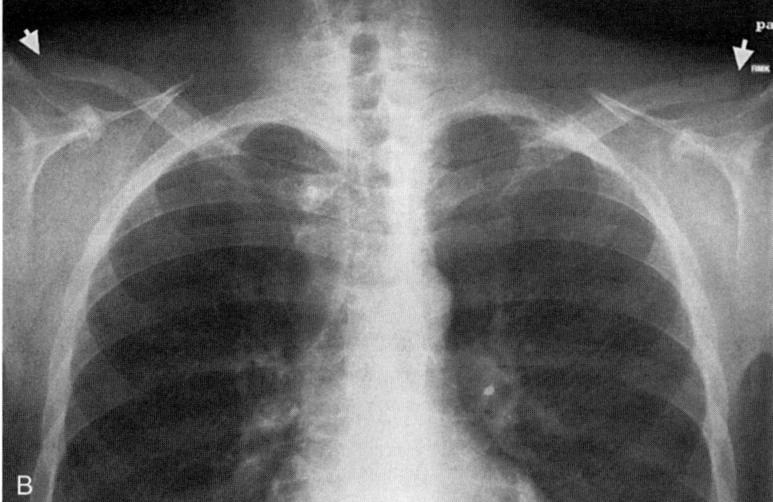

Figure 42–6. Radiographs of a man with secondary hyperparathyroidism. Bilateral resorption with widening of the sacroiliac joints (*arrows*) can be seen on an anteroposterior view of the pelvis (*A*). There is resorption of the distal clavicles (*arrows*) on the posteroanterior view of the chest (*B*).

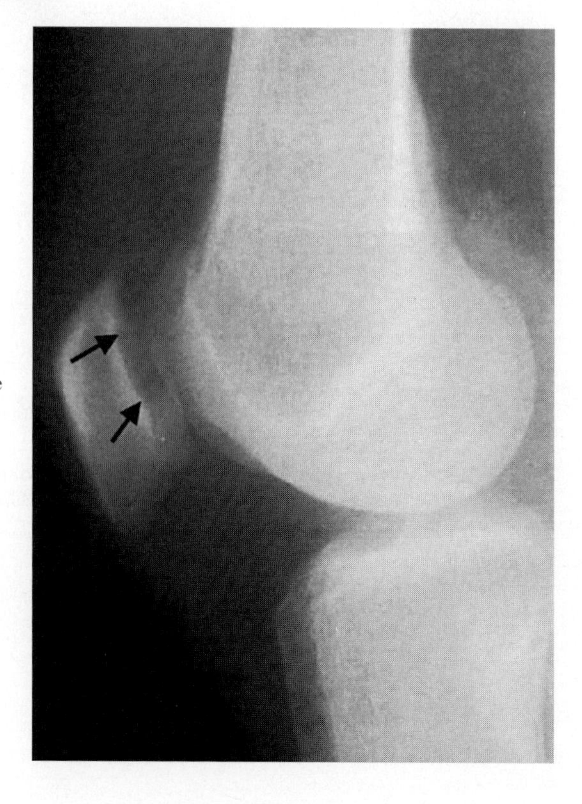

Figure 42–7. Lateral knee view with subchondral resorption (*arrows*) involving the patellar articular surface in a patient with secondary hyperparathyroidism.

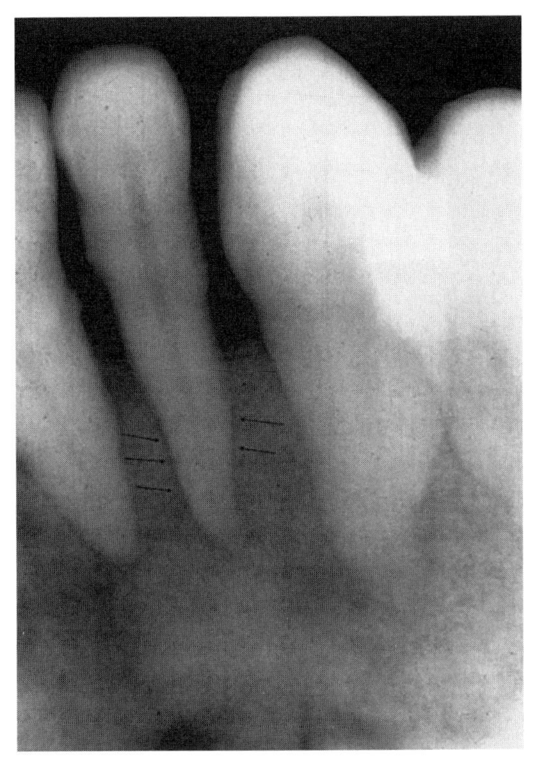

Figure 42–8. Dental radiograph in a patient with secondary hyperparathyroidism shows resorption of the lamina dura of the mandible (*arrows*).

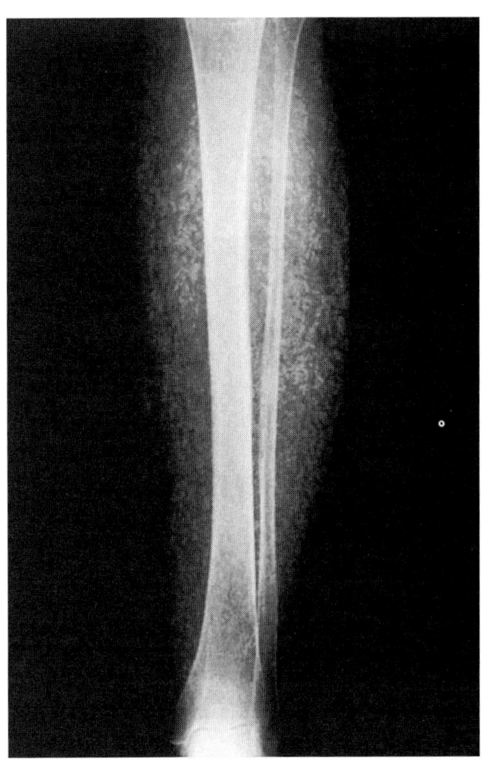

Figure 42–9. Radiograph shows diffuse soft tissue calcification in the calf of a patient with secondary hyperparathyroidism.

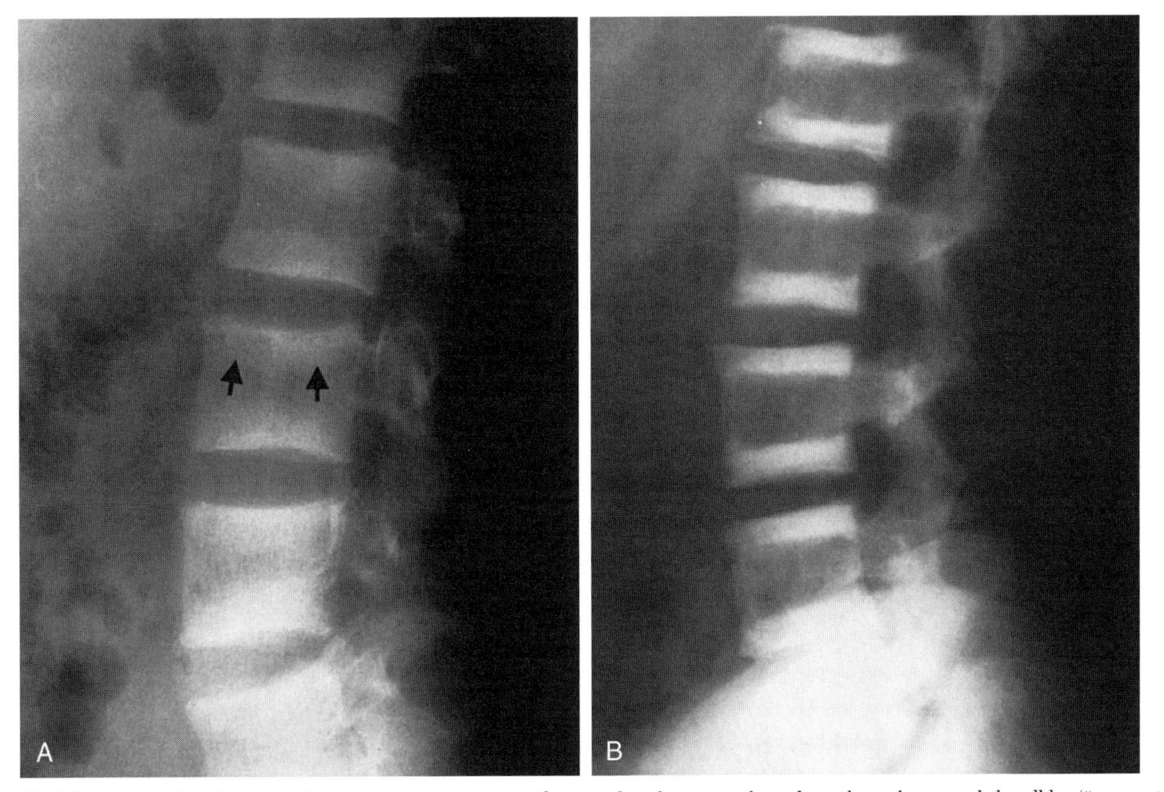

Figure 42–10. *A*, Lateral radiograph of the spine in a person with secondary hyperparathyroidism shows horizontal, bandlike ("rugger jersey") sclerosis of the vertebral bodies (*arrows*). (Courtesy of Dr. Murali Sundaram, Rochester, MN). *B*, For comparison, a lateral radiograph of the spine in a patient with osteopetrosis shows the sclerotic changes along the vertebral end plates to be more dense and better defined than are seen with secondary hyperparathyroidism.

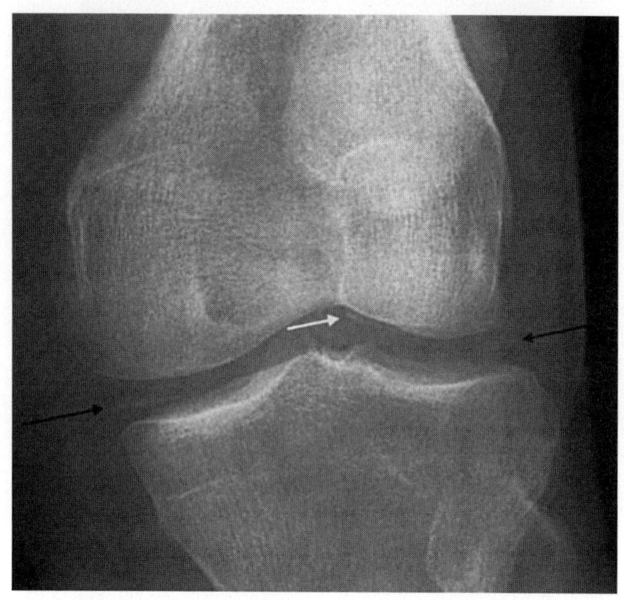

Figure 42–11. Anteroposterior radiograph of the knee in a 63-year-old man with primary hyperparathyroidism shows chondrocalcinosis of the articular hyaline cartilage (*white arrow*) and meniscal fibrocartilage (*black arrows*). Chondrocalcinosis is more common in primary than in secondary forms of hyperparathyroidism.

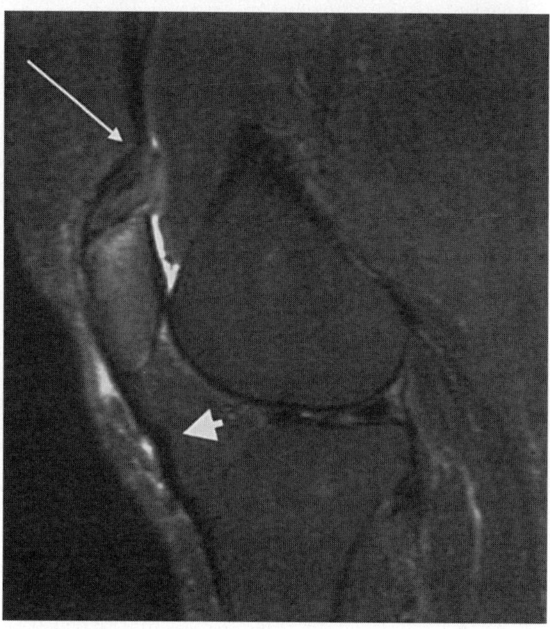

Figure 42–12. Sagittal short tau inversion recovery (STIR) magnetic resonance image of the knee in a woman with secondary hyperparathyroidism associated with chronic renal insufficiency shows a tear of the quadriceps tendon (*arrow*) with associated laxity of the extensor mechanism and waviness of the patellar tendon (*arrowhead*).

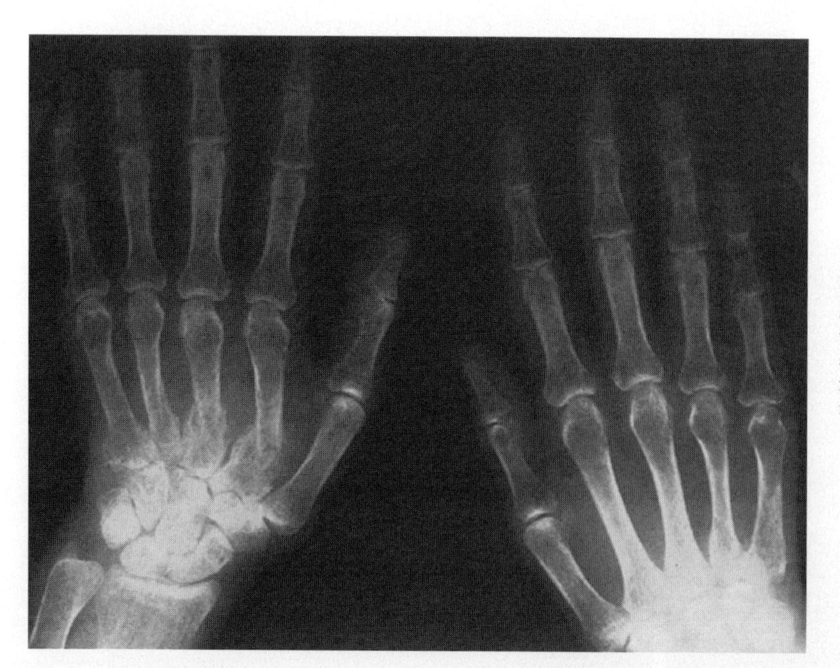

Figure 42–13. Posteroanterior view of both hands in a patient with severe osteomalacia due to aluminum toxicity from chronic renal dialysis. Decreased mineralization can be seen throughout the bones as well as insufficiency fractures of the right second, third, and fourth metacarpals.

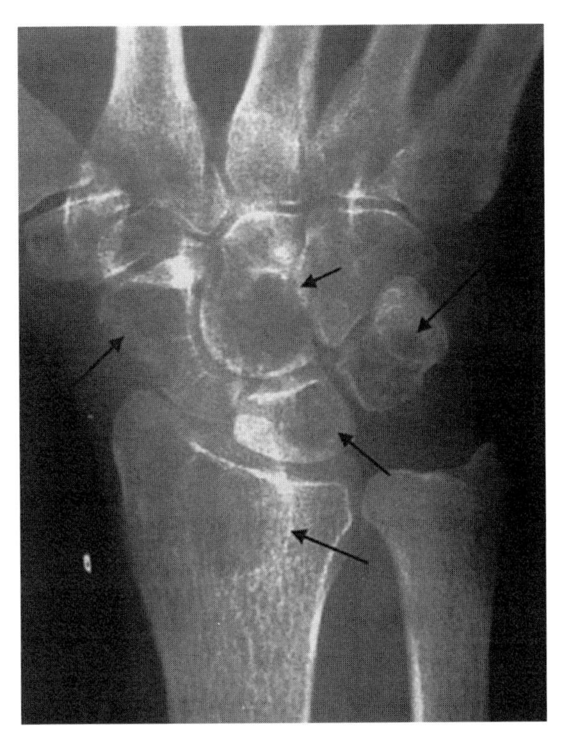

Figure 42–14. Posteroanterior radiograph of the wrist in a 70-year-old man undergoing dialysis shows multiple, well-defined lytic lesions (*arrows*) in the carpal bones, indicating amyloidosis.

(soccer shirt) appearance in the spine can look similar to the changes seen in osteopetrosis (see Fig. 42–10). Chondrocalcinosis may be seen with calcium pyrophosphate dihydrate deposition disease or hemochromatosis. The differential diagnosis for brown tumors includes other lytic lesions, such as giant-cell tumor and fibrous dysplasia.

Renal Osteodystrophy

Findings in secondary hyperparathyroidism associated with renal osteodystrophy may mimic those of primary hyperparathyroidism. Osteomalacia secondary to other causes should be included in the differential diagnosis. Soft tissue calcification has many causes, including collagen-vascular diseases (scleroderma, dermatomyositis), idiopathic tumoral calcinosis, hydroxyapatite crystal deposition disease, and hypervitaminosis D.

Complications

The most common complication of renal osteodystrophy is fracture, which may involve osteomalacic bone, osteoporotic bone, and brown tumors. Tendon rupture is also commonly seen. Complications of long-term hemodialysis include carpal tunnel syndrome and amyloid deposits in the joints. Complications of renal transplantation are osteonecrosis and osteopenia, usually due to steroid therapy.

BONE DENSITOMETRY

Patients with primary hyperparathyroidism often have decreased bone mineral density. In particular, skeletal sites that contain more cortical than trabecular bone (e.g. midradius, femoral neck) are likely to measure low

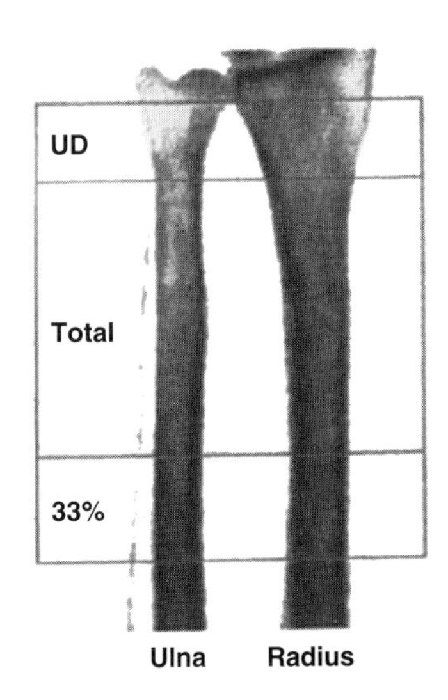

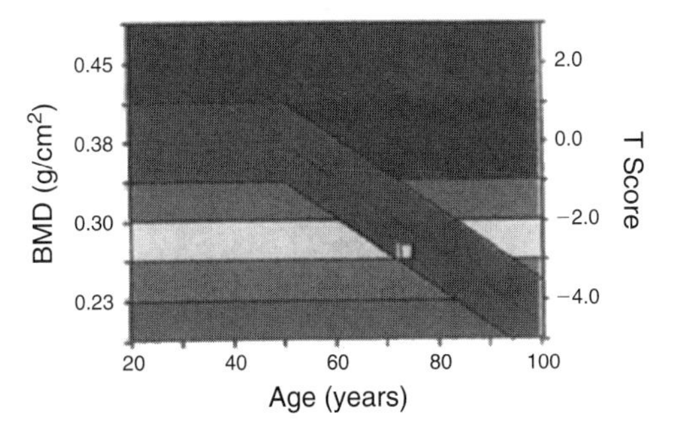

Region		BMD [1,6] (g/cm²)	Young-Adult [2] %	Young-Adult T	Age-Matched [3] %	Age-Matched Z
RADIUS	UD	0.275	73	−2.8 ←	92	−0.6
ULNA	UD	0.231	–	–	–	–
BOTH	UD	0.261	–	–	–	–
RADIUS	33%	0.464	65	−3.5 ◄	83	−1.3
ULNA	33%	0.518	–	–	–	–
BOTH	33%	0.489	–	–	–	–
RADIUS	TOTAL	0.381	69	−3.4	86	−1.2
ULNA	TOTAL	0.363	–	–	–	–
BOTH	TOTAL	0.374	–	–	–	–

Figure 42–15. Dual x-ray absorptiometry scan of the forearm in a 74-year-old woman with primary hyperparathyroidism. The T-score for the diaphysis of the radius ("33% site") (*arrowhead*) is lower than the T-score in the epiphysis ("UD site") (*arrow*). These findings are characteristic because hyperparathyroidism generally affects cortical bone (which is abundant in the diaphysis) prior to trabecular bone (which is abundant in the epiphysis).

(Fig. 42–15). Bone densitometry findings in patients with renal osteodystrophy are less consistent. Occasionally, bone mineral density is increased as a result of the osteosclerosis associated with this disorder.

Recommended Readings

Adams JE: Renal bone disease: Radiological investigation. Kidney Int 56(Suppl 73):S38–S41, 1999.

Consensus Development Conference Panel: Diagnosis and management of asymptomatic primary hyperparathyroidism. Ann Intern Med 114:593–597, 1991.

Genant HK, Heck LL, Lanzl LH, et al: Primary hyperparathyroidism: A comprehensive study of clinical, biochemical, and radiographic manifestations. Radiology 109:513, 1973.

Greenfield GB: Roentgen appearance of bone and soft tissue changes in chronic renal disease. AJR Am J Roentgenol 116:749, 1972.

Johnson DW, McIntyre HD, Brown A, et al: The role of DEXA bone densitometry in evaluating renal osteodystrophy in continuous ambulatory peritoneal dialysis patients. Perit Dial Int 16:34–40, 1996.

Kricun R, Kricun ME, Arangio GA, et al: Patellar tendon rupture with underlying systemic disease. AJR Am J Roentgenol 135:803, 1980.

Olmastroni M, Seracini D, Lavoratti G, et al: Magnetic resonance imaging of renal osteodystrophy in children. Pediatr Radiol 27:865–868, 1997.

Parfitt AM: The hyperparathyroidism of chronic renal failure: A disorder of growth. Kidney Int 52:3–9, 1997.

Resnick D: Abnormalities of bone and soft tissue following renal transplantation. Semin Roentgenol 13:329, 1978.

Rodino MA, Shane E: Osteoporosis after transplantation. Am J Med 104:459–469, 1998.

Shapiro R: Radiologic aspects of renal osteodystrophy. Radiol Clin North Am 10:557, 1972.

Sherrard DJ, Hercz G, Pei Y, et al: The spectrum of bone disease in end-stage renal failure: An evolving disorder. Kidney Int 43:436–442, 1993.

Silverberg SJ, Gartenberg F, Jacobs TP, et al: Longitudinal measurements of bone density and biochemical indices in untreated primary hyperparathyroidism. J Clin Endocrinol Metab 80:723–728, 1995.

Tigges S, Nance EP, Carpenter WA, et al: Renal osteodystrophy: Imaging findings that mimic those of other disorders. AJR Am J Roentgenol 165:143–148, 1995.

Turton DB, Miller DL: Recent advances in parathyroid imaging. Trends Endocrinol Metab 7:163–168, 1996.

CHAPTER 43

Paget's Disease

Leon Lenchik, MD, and Mitchell Kline, MD

OVERVIEW

Paget's disease is a common disorder characterized by excessive and abnormal remodeling of bone. It usually presents in patients older than 45 years, is slightly more common in men than in women, and is more common in whites than in African Americans.

PATHOPHYSIOLOGY

A slow virus may be responsible for the greater size and number of osteoclasts. Genetic factors may also be contributory. The disease progresses slowly and may be monostotic, although more commonly it is polyostotic.

CLINICAL FEATURES

Many patients with Paget's disease are asymptomatic, and their disease is discovered incidentally with radiography. Symptoms and signs depend on the sites of involvement and severity of disease. Patients may have bone pain, progressive bony enlargement, bowing of long bones, or fractures. Deafness may result from cranial nerve compression at the skull base or from involvement of middle ear ossicles. The spinal cord may be compressed as a result of basilar invagination of the skull or enlargement of the vertebrae. A rare but usually fatal complication is the development of a pagetic sarcoma.

IMAGING FINDINGS

Conventional Radiography

Three sequential stages may be seen: lytic, mixed, and sclerotic. Occasionally, different stages coexist. The lytic stage is most common in the skull and long bones. Subarticular involvement with sharply demarcated lysis that advances down the diaphysis is diagnostic (Fig. 43–1). The sclerotic stage often involves the axial skeleton, resulting in bone enlargement, cortical and trabecular thickening, and trabecular distortion (Fig. 43–2).

Characteristic Sites

In the skull, the lytic phase (e.g., osteoporosis circumscripta) may be seen in the frontal or occipital bones

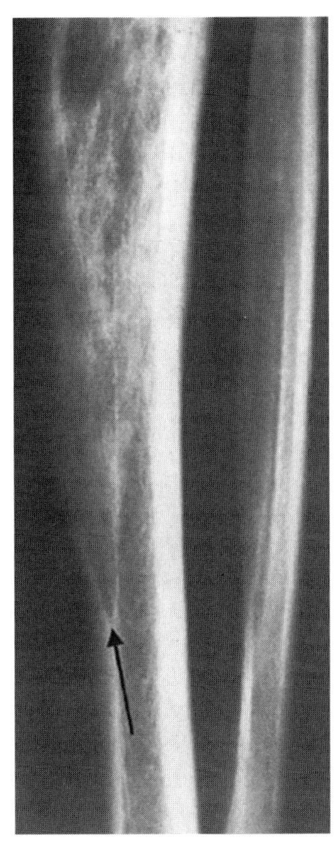

Figure 43–1. Oblique radiograph of the tibia in a 60-year-old man with Paget's disease shows sharply demarcated lysis along the distal anterior cortex (*arrow*) of the tibia ("blade of grass" appearance). The proximal tibia shows bony expansion and has a mixed sclerotic and lytic appearance, indicating more advanced disease.

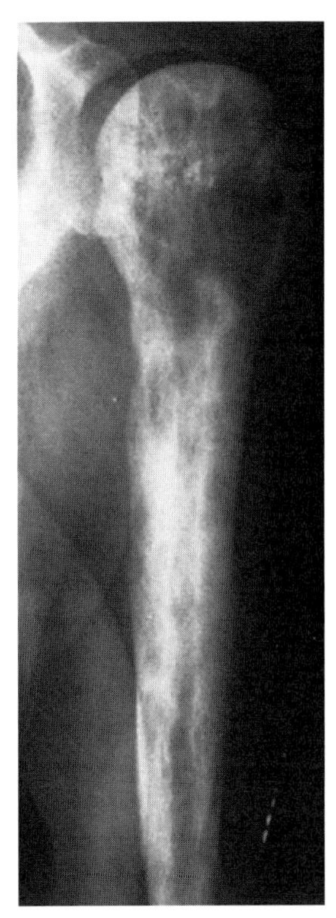

Figure 43–2. Anteroposterior radiograph of the humerus in an 81-year-old woman with Paget's disease shows the characteristic bony sclerosis, enlargement, and deformity.

(Fig. 43–3). More advanced disease in the skull manifests as "cotton wool" appearance (i.e., mixed sclerosis and lysis) (Fig. 43–4). In the spine, "picture-frame" vertebra or ivory vertebra may be seen (Fig. 43–5). In the pelvis, thickening of the iliopectineal line, patchy areas of lucency and sclerosis, and acetabular protrusion may be observed (Fig. 43–6). In the long bones, lysis of subarticular bone is demonstrated initially; it then advances as a wedge of lucency down the diaphysis (e.g., flame shape or "blade of grass" appearance). Coarsened trabecula, bony sclerosis, bony enlargement, and deformity characterize advanced disease (Fig. 43–7). In the tibia, the lytic phase may begin in the diaphysis.

Advanced Imaging

Bone scans are very sensitive and are most commonly used for staging. Findings include intense radiopharmaceutical uptake within the involved bone (Fig. 43–8). Findings on computed tomography are bone enlargement, cortical thickening, and trabecular disorganization (Fig. 43–9). On magnetic resonance imaging, the marrow spaces in Paget's disease of the pelvis and peripheral skeleton have normal signal (Fig. 43–10). In the spine, however, the marrow has variable signal characteristics, possibly related to the stage of the disease. The normal fatty marrow signal seen in Paget's disease can be used to exclude complications such as fracture and neoplasm.

Complications

Complications of Paget's disease include basilar invagination, spinal stenosis, premature osteoarthritis, insufficiency fractures (Fig. 43–11), and neoplasms (Fig. 43–12). Most pagetic sarcomas are osteosarcomas. Rarely, giant cell tumors are seen. They are usually benign and have a predilection for the skull and facial bones. Multiple giant cell tumors in the same person are very suggestive of Paget's disease.

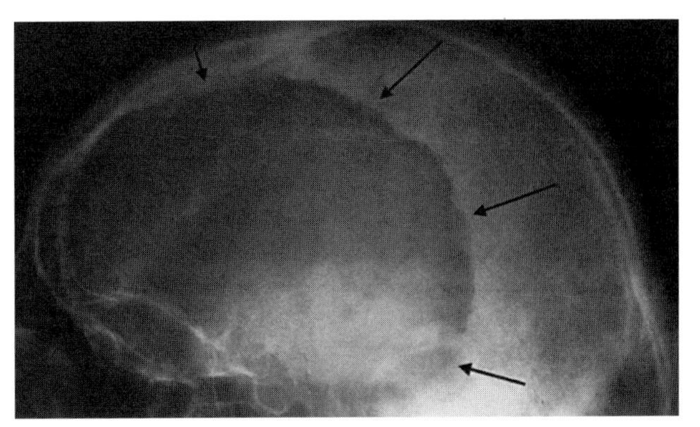

Figure 43–3. Lateral radiograph of the skull in a patient with Paget's disease shows a well-demarcated lytic lesion (*arrows*) representative of osteoporosis circumscripta.

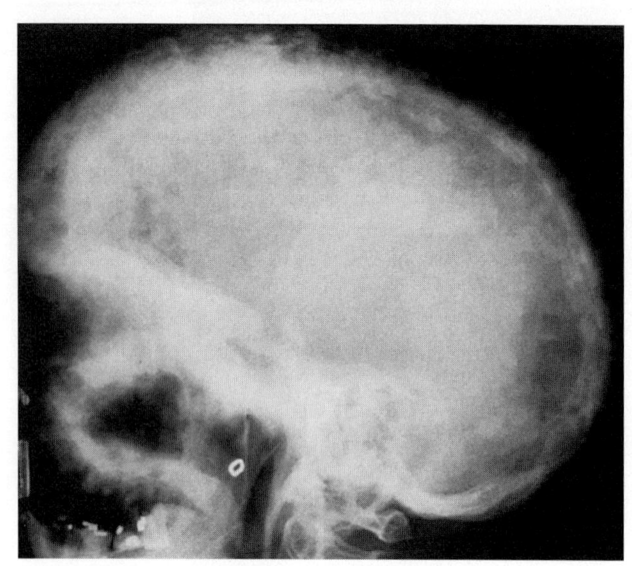

Figure 43–4. Lateral radiograph of the skull showing marked sclerotic changes with a characteristic "cotton wool" appearance seen with Paget's disease.

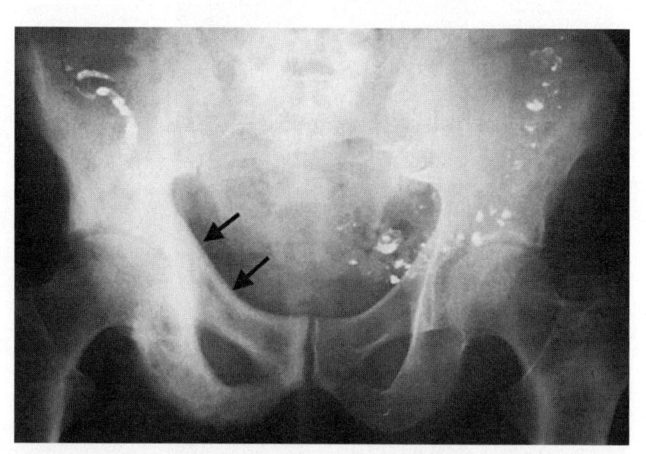

Figure 43–6. Anteroposterior radiograph of the pelvis in a patient with Paget's disease shows thickening of the right iliopectineal line (*arrows*) as well as trabecular coarsening and sclerosis of the right hemipelvis.

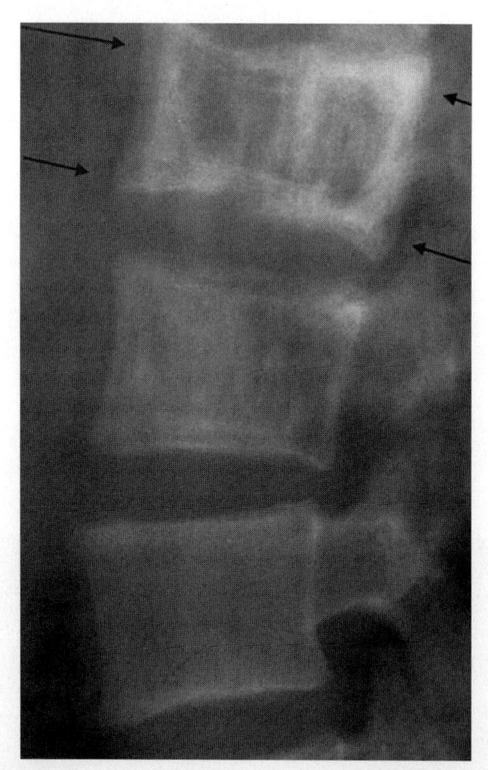

Figure 43–5. Lateral radiograph of the lumbar spine in a 73-year-old woman with Paget's disease shows vertebral body enlargement (*arrows*) and cortical thickening ("picture frame" appearance) as well as trabecular coarsening and distortion.

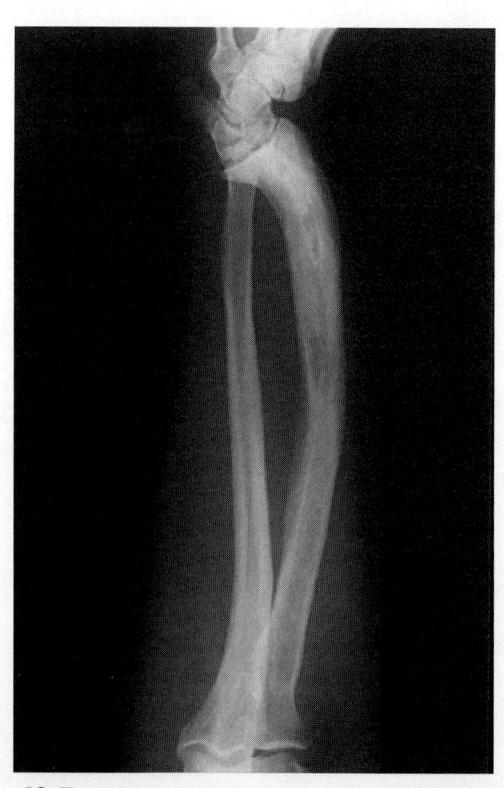

Figure 43–7. Radiograph of the forearm shows cortical expansion, sclerosis, and bowing of the radius characteristic of Paget's disease.

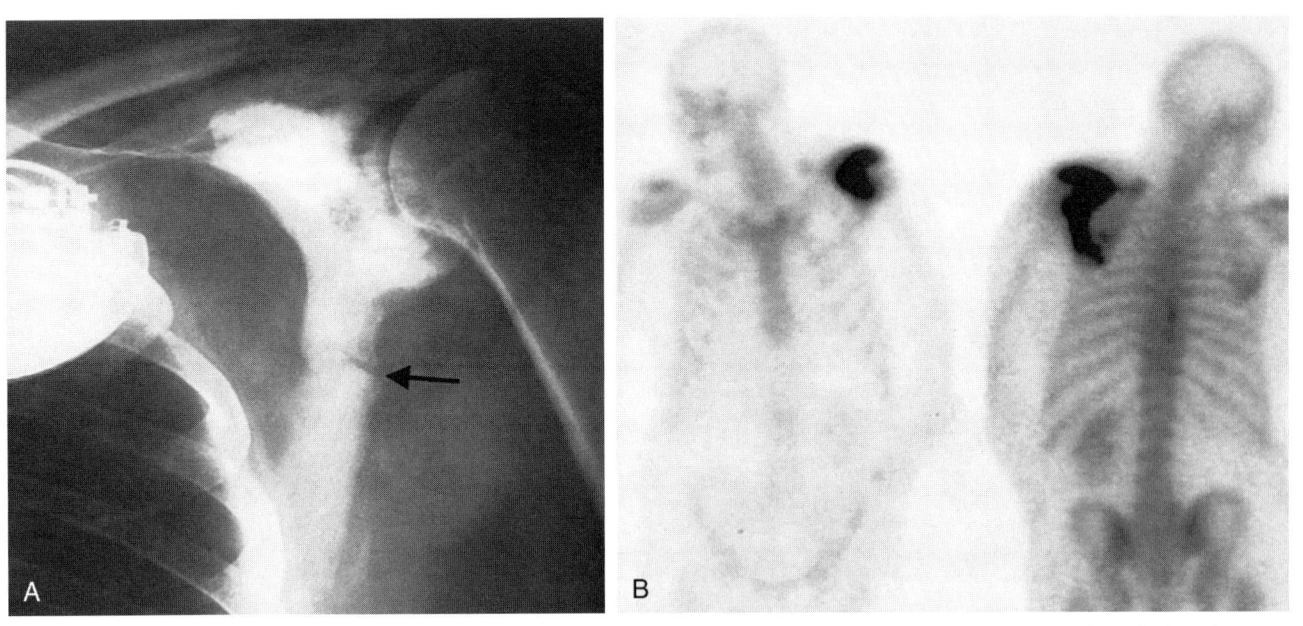

Figure 43–8. *A,* Radiograph shows changes of Paget's disease involving the scapula with an insufficiency fracture (*arrow*) along the lateral margin. *B,* On this nuclear bone scan, there is intense radiopharmaceutical uptake in the region of the Paget's involvement of the lateral scapula.

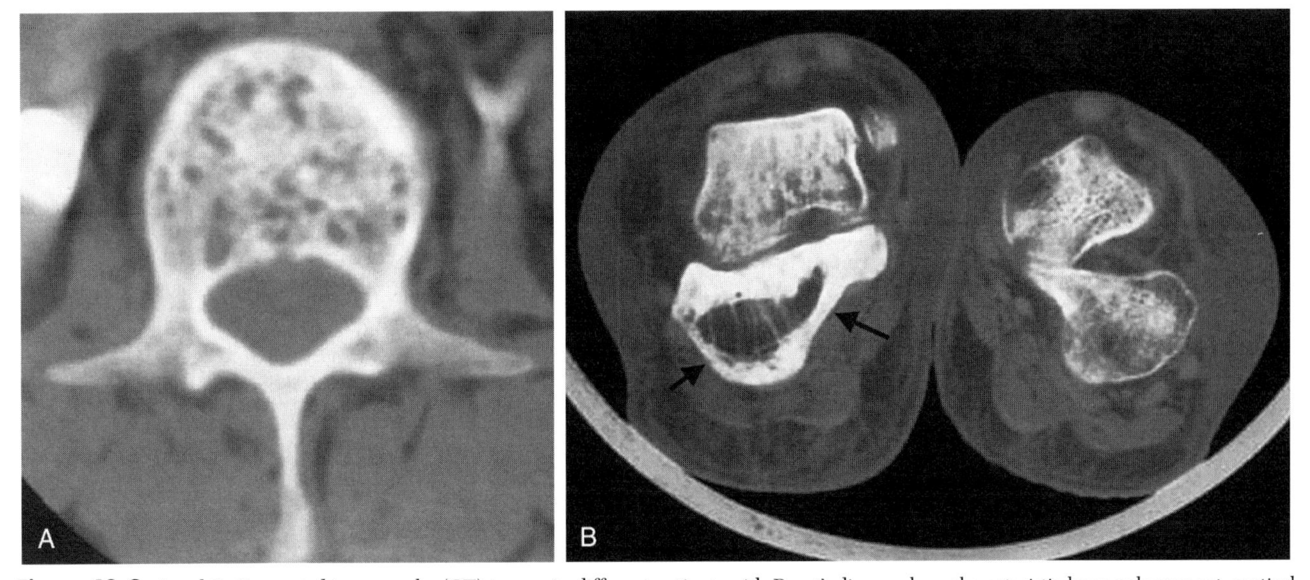

Figure 43–9. *A* and *B,* Computed tomography (CT) images in different patients with Paget's disease show characteristic bone enlargement, cortical thickening, and trabecular coarsening of the vertebral body (*A*) and of the right calcaneus (*arrows*) and talus (*B*).

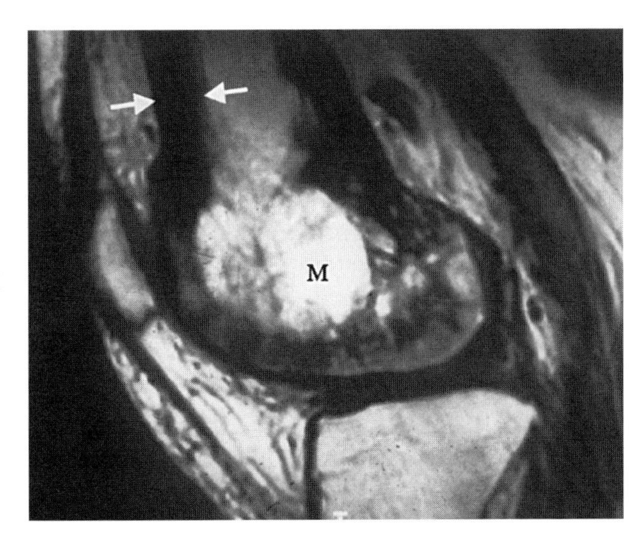

Figure 43–10. Sagittal T1-weighted magnetic resonance image of the knee in a patient with Paget's disease shows characteristic cortical thickening (*arrows*) and trabecular disorganization of the distal femur. The normal fatty marrow signal (M) excludes complications such as fracture and neoplasm.

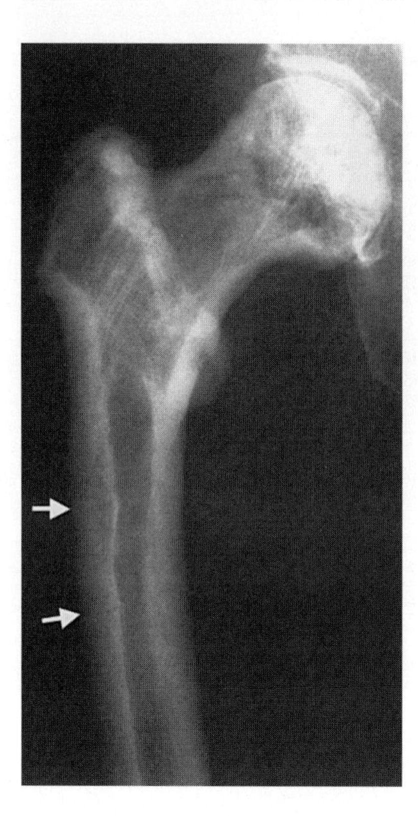

Figure 43–11. Anteroposterior radiograph of the proximal femur shows cortical thickening and trabecular coarsening in this patient with Paget's disease. Insufficiency fractures are present along the convex surface of the bone (*arrows*), a characteristic location.

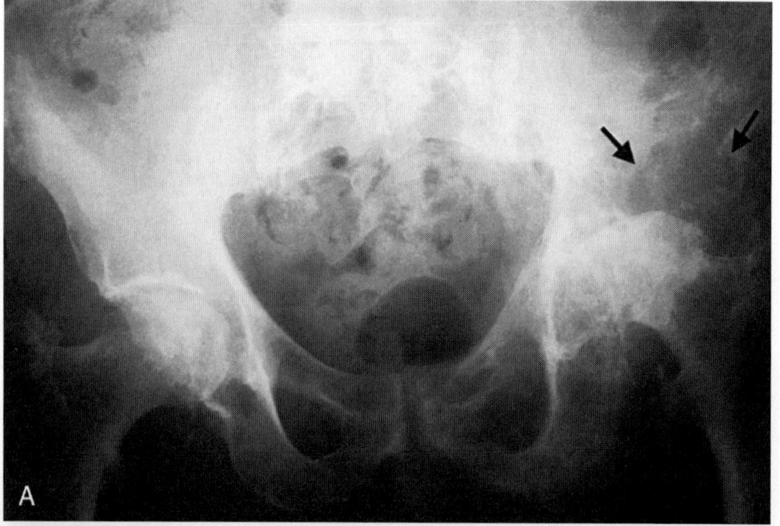

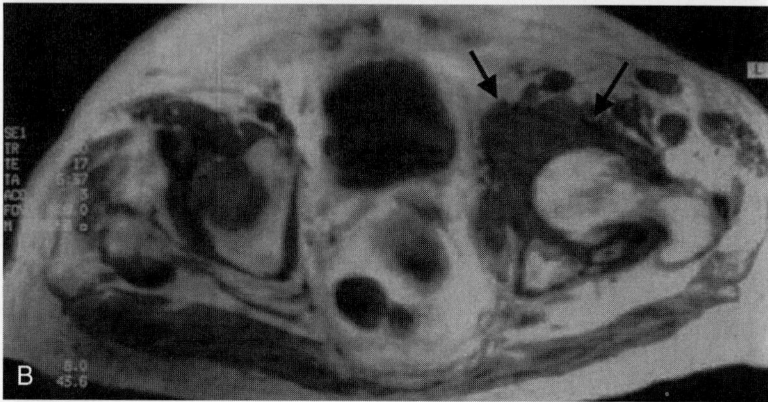

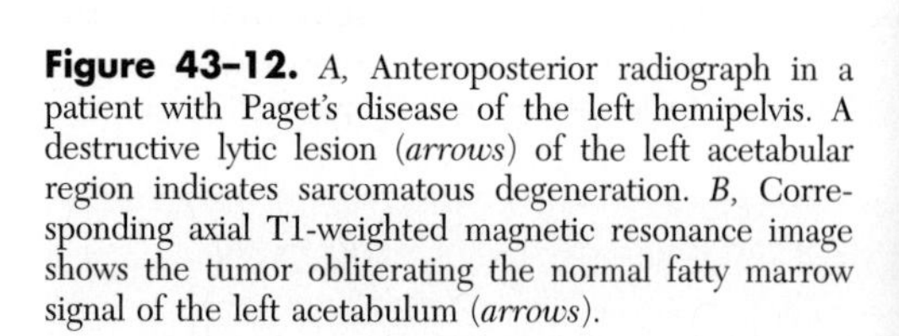

Figure 43–12. *A,* Anteroposterior radiograph in a patient with Paget's disease of the left hemipelvis. A destructive lytic lesion (*arrows*) of the left acetabular region indicates sarcomatous degeneration. *B,* Corresponding axial T1-weighted magnetic resonance image shows the tumor obliterating the normal fatty marrow signal of the left acetabulum (*arrows*).

Differential Diagnosis

The lytic phase of Paget's disease is often pathognomonic. The combination of bony sclerosis and bony enlargement is also quite specific.

The differential diagnosis of bony sclerosis in the absence of bony enlargement includes blastic metastases, myelofibrosis, Hodgkin's disease, mastocytosis, and sclerotic multiple myeloma.

The differential diagnosis of calvarial hyperostosis includes hyperostosis frontalis interna, fibrous dysplasia, and metastatic disease.

Recommended Readings

Altman RD, Collins B: Musculoskeletal manifestations of Paget's disease of bone. Arthritis Rheum 23:1121, 1980.
Boutin RD, Spitz DJ, Newman JS, et al: Complications in Paget disease at MR imaging. Radiology 209:641–651, 1998.
Greditzer HG III, McLeod RA, Unni KK, et al: Bone sarcomas in Paget disease. Radiology 146:327, 1983.
Reid IR, Nicholson GC, Weinstein RS, et al: Biochemical and radiologic improvement in Paget's disease of bone treated with alendronate: A randomized, placebo-controlled trial. Am J Med 171:341–348, 1996.
Resnick D: Paget disease of bone: Current status and a look back to 1943 and earlier. AJR Am J Roentgenol 150:249, 1988.
Singer FR: Update on viral etiology of Paget's disease of bone. J Bone Miner Res 14:29–33, 1999.
Siris ES: Clinical review: Paget's disease of bone. J Bone Miner Res 13:1061–1065, 1998.
Steinbach HL: Some roentgen features of Paget's disease. AJR Am J Roentgenol 86:950, 1961.

CHAPTER 44

Miscellaneous Metabolic Disorders

Leon Lenchik, MD, and Mitchell Kline, MD

HYPOPARATHYROIDISM

Overview

The most common cause of hypoparathyroidism is excision of or damage to the parathyroid glands during thyroidectomy or radical neck dissection. Less common causes are failure of parathyroid gland development, reduced parathyroid gland function due to altered regulation, and impaired parathyroid hormone action.

Patients usually present with neuromuscular irritability precipitated by hypocalcemia. Over the long term, muscle cramps, personality disturbances, dry skin, alopecia, or abnormal dentition may be seen. Decreased serum levels of parathyroid hormone associated with hypocalcemia, hyperphosphatemia, and normal renal function characterize the laboratory findings in hypoparathyroidism.

Imaging Findings

Imaging findings in hypoparathyroidism are varied. They include bony sclerosis, subcutaneous calcification (Fig. 44–1), calvarial thickening, basal ganglia calcification (Fig. 44–2), hypoplastic dentition, premature physeal fusion, and spinal ossifications. The findings in the spine resemble those of diffuse idiopathic skeletal hyperostosis, including calcification and ossification of the anterior longitudinal ligament and spinal osteophytes (Fig. 44–3).

PSEUDOHYPOPARATHYROIDISM

Overview

Pseudohypoparathyroidism (PHP) describes a group of disorders characterized by unresponsiveness of target tissues to parathyroid hormone. Patients with PHP have a characteristic somatotype termed *Albright's hereditary osteodystrophy.* It consists of obesity, round face, short stature, mental retardation, and brachydactyly. Some patients also have dermatologic, ocular, olfactory, and gustatory abnormalities. As in hypoparathyroidism, hypocalcemia and hyperphosphatemia are often present. Serum parathyroid hormone values may be normal or elevated.

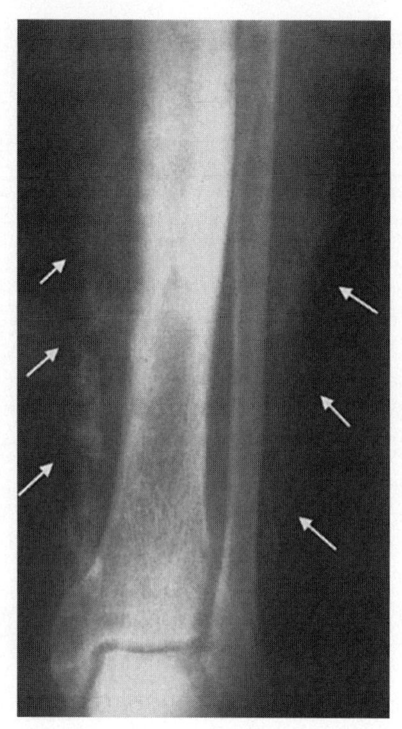

Figure 44–1. Radiograph of the leg in a 41-year-old man with hypoparathyroidism shows extensive subcutaneous calcifications (*arrows*).

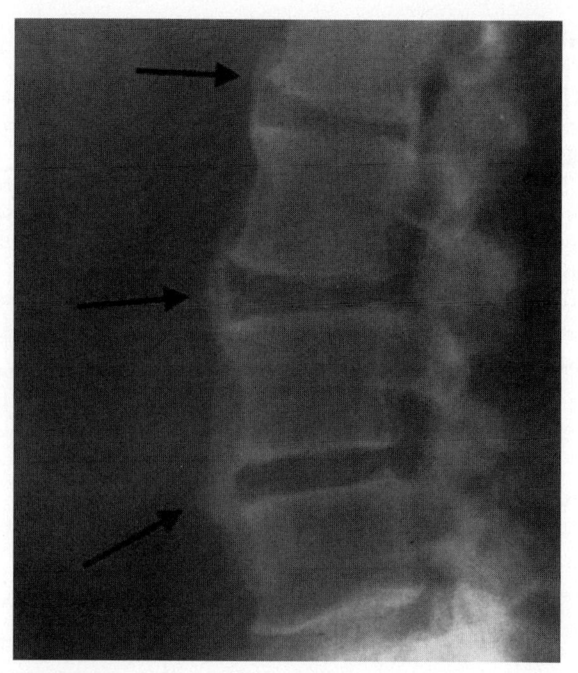

Figure 44–3. Lateral radiograph of the lumbar spine in a man with hypoparathyroidism shows ossification of the anterior longitudinal ligament (*arrows*). This finding can look similar to diffuse idiopathic skeletal hyperostosis (DISH) or chronic retinoid (Vitamin A derivative) therapy (see Chapter 47).

Imaging Findings

Imaging findings in PHP, which are similar to those in hypoparathyroidism, include bony sclerosis, soft tissue calcification, calvarial thickening, and basal ganglia calcification. In addition, short metacarpals, metatarsals, and phalanges, diaphyseal exostoses, and cone-shaped epiphyses may be seen (Fig. 44–4).

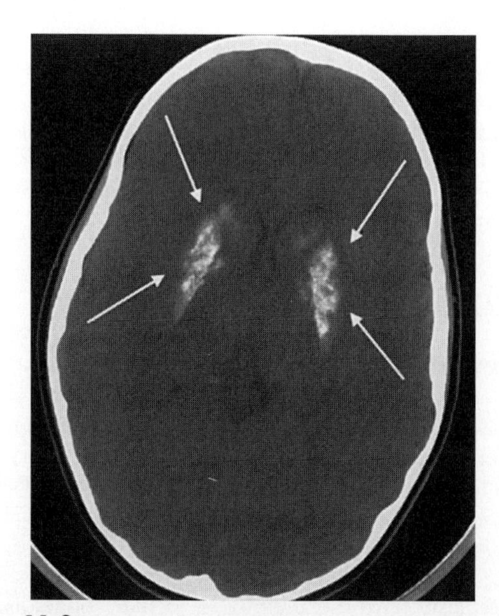

Figure 44–2. Computed tomography image of the brain without intravenous contrast shows calcification of the basal ganglia (*arrows*) characteristic of disorders of calcium-phosphate regulation, such as hypoparathyroidism. (Courtesy of Dr. Thomas Underhill, Winston-Salem, NC.)

Differential diagnosis for short fourth metacarpal includes Turner syndrome and physeal trauma.

HYPERTHYROIDISM

Overview

The most common causes of hyperthyroidism in adults are toxic diffuse goiter and toxic nodular goiter. Excessive thyroid hormone results in stimulation of bone formation and bone resorption; however, bone resorption is dominant.

Clinical Features

In children with hyperthyroidism, accelerated skeletal maturation and advanced bone age may be seen. In adults, generalized osteoporosis leading to vertebral fractures and kyphosis are observed. Other possible clinical findings are weakness, fatigue, myopathy, nervousness, weight loss, tachycardia, palpitations, diarrhea, and hypersensitivity to heat.

Imaging Findings

Generalized osteopenia is the most common imaging finding in hyperthyroidism. Others are accelerated skeletal maturation, kyphosis, insufficiency fractures, and thyroid acropachy. Thyroid acropachy is a rare complication that occurs after treatment of hyperthyroidism. A dense periosteal reaction with a feathery contour, seen in an asymmetric distribution and most prominent along the radial margin of metacarpals and phalanges, is virtually diagnostic (Fig. 44–5).

Differential Diagnosis

Periosteal reaction involving multiple bones may be seen in primary or secondary hypertrophic osteoarthropathy and hypervitaminosis A. In hypertrophic osteoarthropathy, the long bones are considerably more likely to be involved than the hands and feet, and the feathery pattern of periostitis is typically absent (Fig. 44–6).

HYPOTHYROIDISM

Overview

Causes of hypothyroidism include surgery, atrophy, tumors, iodine deficiency, antithyroid medications, and pituitary disorders. In children, hypothyroidism leads to mental retardation, obesity, developmental delay, growth retardation, lethargy, and constipation. In adults with the disorder, dry coarse skin and hair, fatigue, lethargy, paresthesias, constipation, and bradycardia are seen.

Imaging Findings

Imaging findings in hypothyroidism depend on the patient's age. In general, the epiphyses are stippled in infancy, fragmented in childhood (Fig. 44–7), and cone-shaped in adolescence. In infants, absence of the distal femoral and proximal tibial epiphyses should prompt laboratory evaluation to exclude hypothyroidism (Fig. 44–8). In the skull, wormian bones and prolonged separation of sutures may be seen. Slipped capital femoral epiphysis is another possible complication in children. Anteriorly wedged (e.g., bullet) vertebrae may result in kyphosis. In the skull, dentition and pneumatization of the sinuses may be delayed. In the long bones, dense metaphyseal bands may be seen. Unlike the osseous findings in children, imaging findings in adults with hypothyroidism are usually mild. Generalized osteopenia is the most common finding.

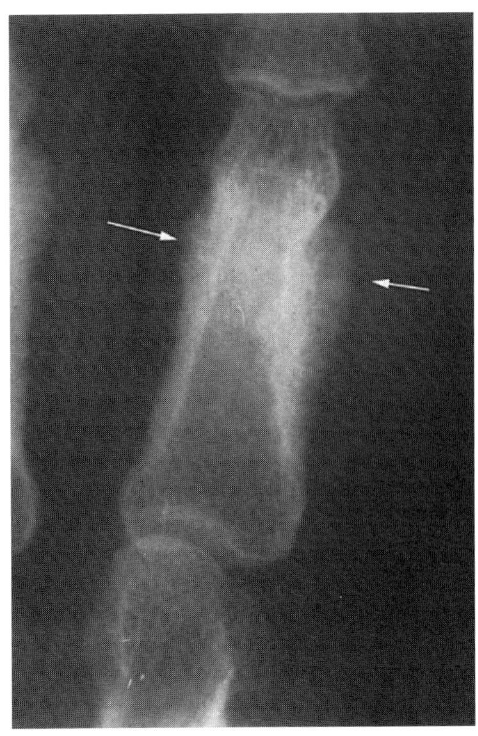

Figure 44–5. Posteroanterior radiograph of the finger in a 45-year-old man with thyroid acropachy. Note the dense solid periosteal reaction with feathery contour along the phalangeal shaft (*arrows*).

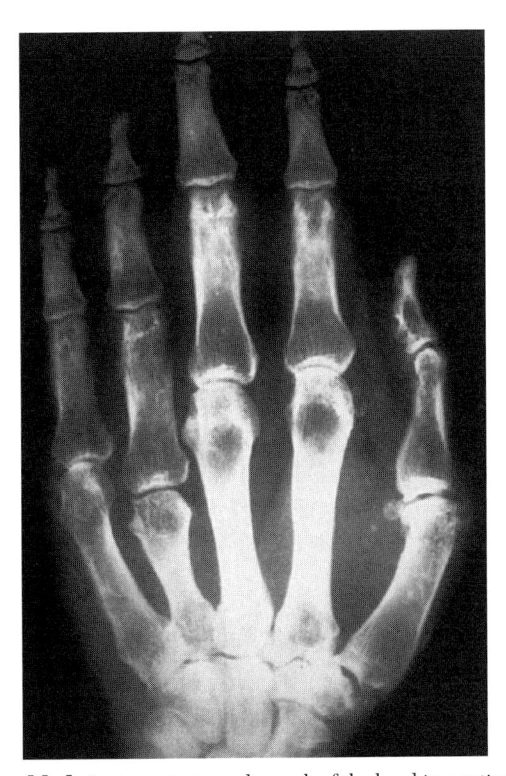

Figure 44–4. Posteroanterior radiograph of the hand in a patient with pseudohypoparathyroidism (PHP) shows a short fourth metacarpal. This is a nonspecific finding that may be seen with other disorders (e.g., Turner syndrome).

ACROMEGALY

Overview

Acromegaly is usually caused by a pituitary adenoma that produces excessive growth hormone, which in turn stimulates bone formation and soft tissue proliferation. In the immature skeleton, the same condition is called *pituitary gigantism* and manifests as delayed skeletal maturation.

Rheumatologic complaints, such as backache and arthropathy of large joints, are common. Other findings are coarse facial features, thick skin, thick calvarium, lantern jaw, prominent tongue, organomegaly, fatigue, lethargy, and painful kyphosis. Carpal tunnel syndrome or

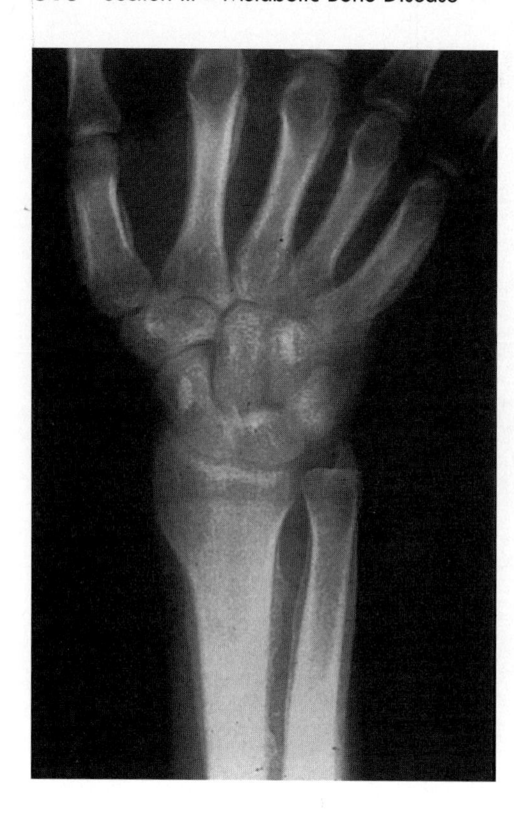

Figure 44–6. Posteroanterior radiograph of the forearm in a patient with hypertrophic osteoarthropathy shows linear periosteal reaction along the diaphyses of the radius, ulna, and metacarpals. Note the absence of the feathery contour seen in Figure 44–5.

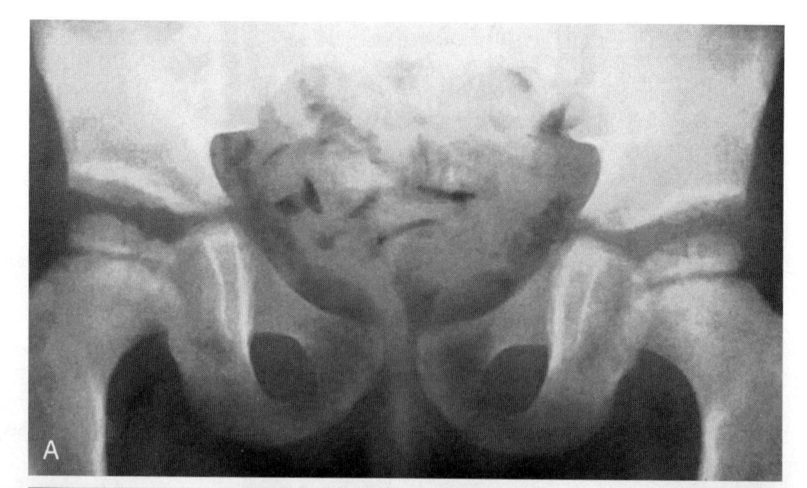

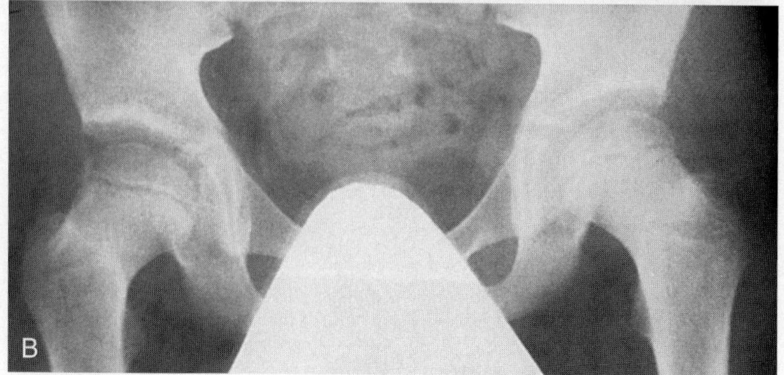

Figure 44–7. Anteroposterior radiograph of the pelvis in a child with hypothyroidism. *A,* The initial study shows irregularity and mild fragmentation of the capital femoral epiphyses, a finding that could mimic Legg-Perthes disease. *B,* Follow-up examination performed 18 months later, after treatment, shows normal-appearing femoral epiphyses.

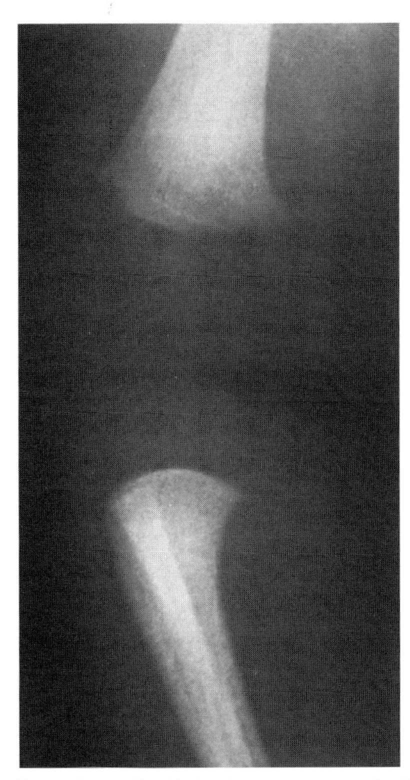

Figure 44–8. Radiograph of the knee in a 7-week-old child with hypothyroidism. Nonossification of the epiphysis of the distal femur and proximal tibia at this age indicates delayed skeletal maturity.

spinal cord compression may result from soft tissue and bony hypertrophy.

Imaging Findings

Imaging findings in acromegaly include soft tissue thickening in the heel pads (Fig. 44–9) and digits. Bony enlargement is most common in the skull (calvarial thickening, prognathic mandible) (Fig. 44–10), vertebrae, and phalangeal tufts (Fig. 44–11). Less common findings

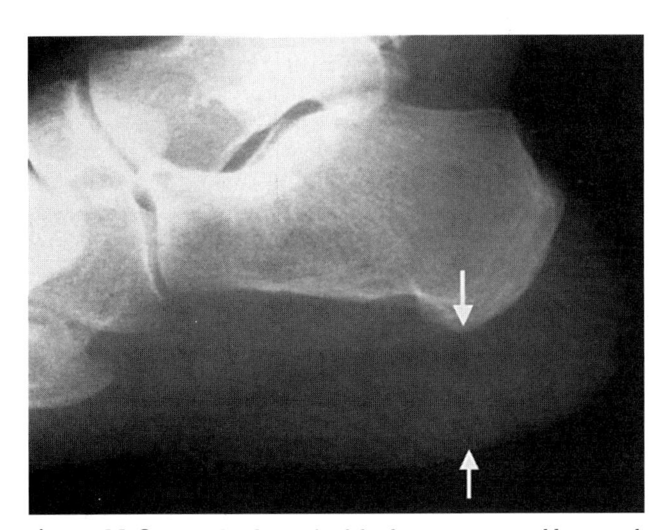

Figure 44–9. Lateral radiograph of the foot in a 30-year-old man with acromegaly shows soft tissue thickening of the heel pad (*arrows*).

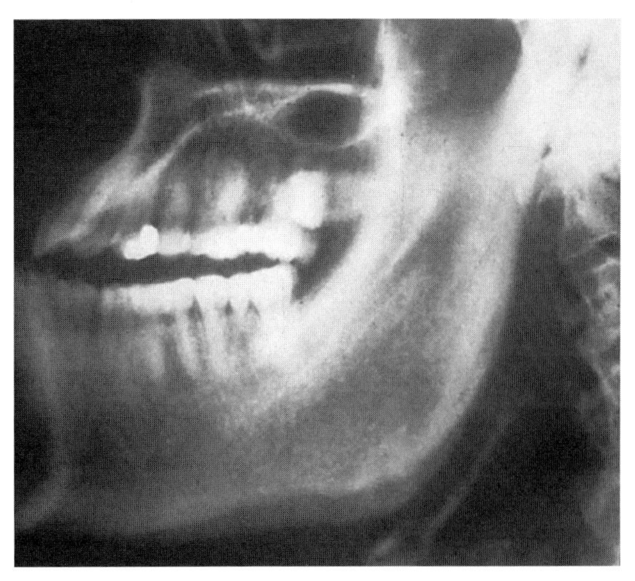

Figure 44–10. Lateral radiograph of the jaw in a patient with acromegaly shows the characteristic prognathic mandible. Note that the mandible is disproportionately larger than the maxilla.

are enlargement of costochondral junctions, the sella turcica, and paranasal sinuses, widening of intervertebral disks, scalloping of posterior vertebrae (Fig. 44–12), spadelike phalangeal tufts, and arthropathy of large joints. Arthropathy is caused by chondrocyte proliferation. In the early stages, joint space widening is seen. Later, as the thickened cartilage is unable to acquire essential nutrients,

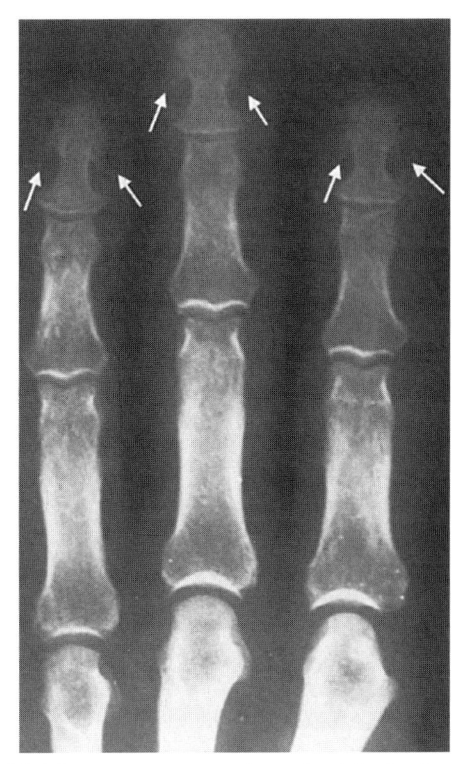

Figure 44–11. Posteroanterior radiograph of the hand in a patient with acromegaly shows spadelike enlargement of the phalangeal tufts (*arrows*) associated with soft tissue overgrowth.

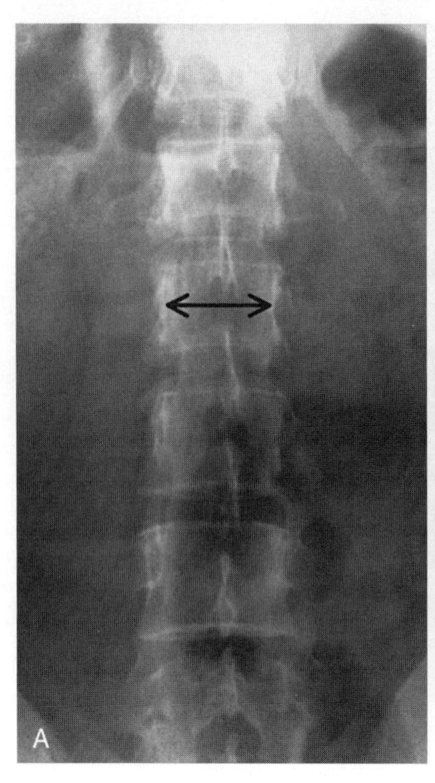

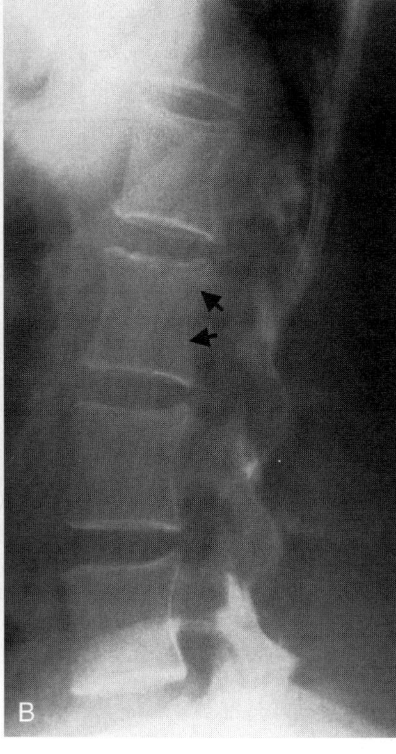

Figure 44–12. Anteroposterior (*A*) and lateral (*B*) radiographs of the lumbar spine in a patient with acromegaly shows widening of the interpedicular distance (*arrows* in *A*) and posterior scalloping of the vertebral bodies (*arrows* in *B*).

it begins to crumble, and radiologic signs of joint space narrowing, subchondral sclerosis, subchondral cysts, and osteophytes become apparent (Fig. 44–13).

Differential Diagnosis

Posterior vertebral scalloping may also be seen in spinal neoplasms, neurofibromatosis, Marfan's syndrome, Ehlers-Danlos syndrome, and achondroplasia.

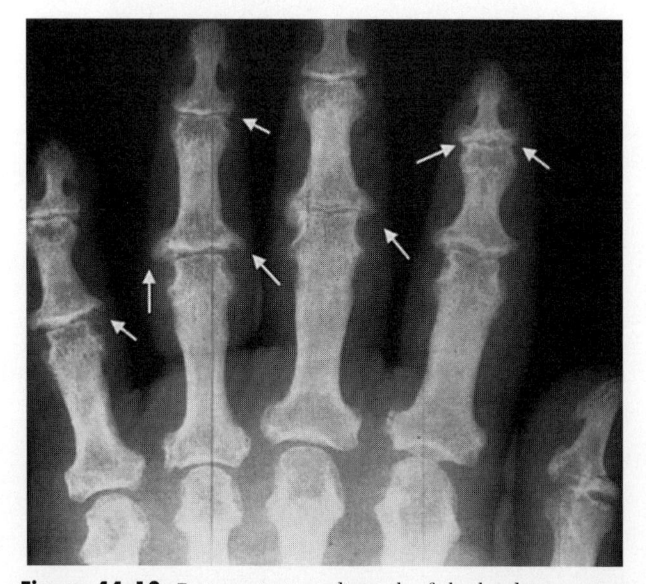

Figure 44–13. Posteroanterior radiograph of the hand in a patient with arthropathy related to acromegaly shows joint space narrowing and osteophytes (*arrows*) involving the interphalangeal joints of the digits.

CUSHING'S DISEASE

Overview

Cushing's disease is due to an excessive amount of corticosteroids within the body. It can be caused by either overproduction of steroids by the adrenal glands (endogenous) or administration of glucocorticoid therapy (exogenous). Adrenal hyperplasia is the most common endogenous cause. Cushing's disease is characterized by weight gain from central fat deposition, muscle weakness, fatigue, and hypertension.

Imaging Findings

Osteoporosis and osteonecrosis are the main radiologic findings in Cushing's disease. Osteoporosis in the spine can show biconcave end plates (fish-type vertebrae) with thickened margins (condensation) (Fig. 44–14). Osteonecrosis is a complication seen more commonly in exogenous Cushing's disease than in endogenous disease. The radiologic findings in osteonecrosis are discussed below. Exuberant callus formation with fracture healing is another characteristic of Cushing's disease.

OSTEONECROSIS

Overview

Osteonecrosis is the end result of the interruption of the blood supply to bone. *Avascular necrosis* often refers to infarction in the subarticular regions of bone. Common sites are the femoral head, humeral head, femoral condyles, talar dome, and scaphoid. *Bone infarct* is used to refer to an ischemic process affecting the metadiaphyseal portions of bone. Osteonecrosis may be caused by the disruption of vessels from fracture or dislocation or by blockage of vessels from various disease processes. An example of the latter is thrombosis of small vessels in sickle cell disease. Other causes of osteonecrosis are glucocorticoid therapy, pancreatitis, alcoholism, collagen-vascular disease, Gaucher's disease, radiation therapy, and Caisson's disease. Spontaneous osteonecrosis is a condition of unknown etiology that affects older individuals. It most often involves the weight-bearing portion of the medial femoral condyle, although it can also involve the femoral head and tarsal navicular.

Both traumatic and atraumatic causes of osteonecrosis are generally associated with pain.

Imaging Findings

In the early stages of avascular necrosis, conventional radiographic findings are normal. With more advanced disease, radiographs show subchondral lucency with underlying bone sclerosis (Fig. 44–15). Eventually, bone fragmentation, articular surface collapse, and secondary osteoarthritis changes dominate. Magnetic resonance (MR) imaging is much more sensitive to the early changes of avascular necrosis. Characteristic findings on T2-weighted MR images consist of bone marrow edema associated with serpentine double lines of high and low signal (Fig. 44–16). The findings on bone scan include initial photopenia representing the lack of vascular supply followed by increased uptake of radiopharmaceutical, which is indicative of the repair process (Fig. 44–17).

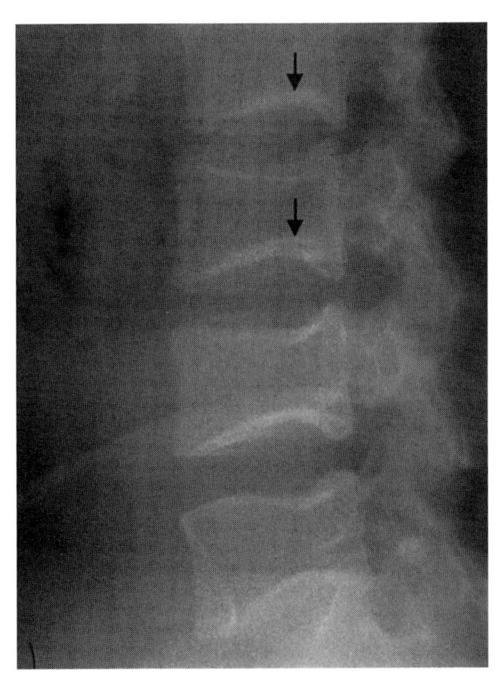

Figure 44–14. Lateral view of the lumbar spine in a man receiving long-term corticosteroid therapy shows the characteristic "fish-type" vertebral bodies with end plate depressions and marginal condensation (*arrows*).

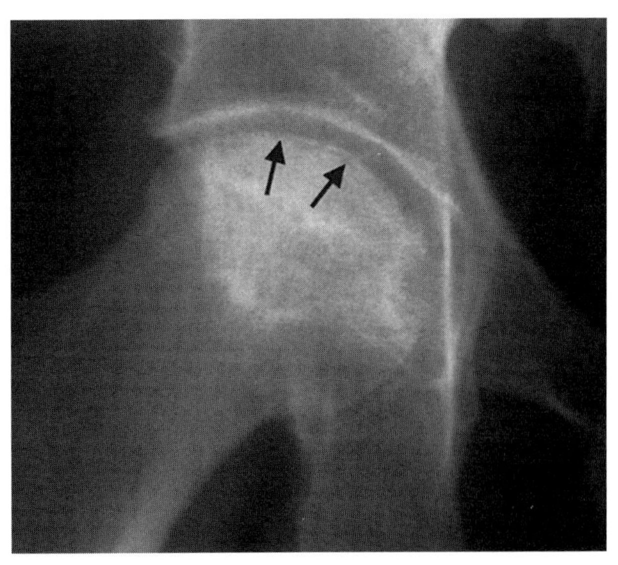

Figure 44–15. Anteroposterior radiograph of the hip in a person receiving corticosteroid therapy shows subtle subchondral lucency (*arrows*) and sclerosis in the femoral head, indicating avascular necrosis.

Bone infarcts in the metadiaphyseal regions show characteristic well-defined sclerotic margins on radiographs (Fig. 44–18). On MR imaging, acute bone infarcts show ill-defined regions, which have low signal on T1-weighted images and high signal on T2-weighted images. As the infarcts mature, a well-defined rim of low signal on both T1- and T2-weighted images surrounds a central region of fat signal (Fig. 44–19).

Spontaneous osteonecrosis predominantly affects the weight-bearing surface of the medial femoral condyle (Fig. 44–20). Imaging findings are similar to those previously described for avascular necrosis of the femoral head, with subchondral lucency, underlying bone sclerosis, and eventual articular collapse.

OSTEOCHONDRITIS DESSICANS

Overview

Osteochondritis dissecans (OCD) is a fracture of subchondral bone with or without a fracture of the overlying articular cartilage. It is believed to most likely be the result of acute or repetitive trauma. Sites that commonly develop OCD are the lateral aspect of the medial femoral condyle, the talar dome, and the capitellum.

In many cases, pain, limited range of motion, joint locking, or soft tissue swelling is present. Some patients with OCD may be asymptomatic. Occurrence of the disorder during adolescence is typical.

Imaging Findings

Fragmentation of the subchondral bone is the most common radiographic finding in OCD. If the overlying cartilage is also fractured, the osteochondral lesion may break off and become an intra-articular loose body

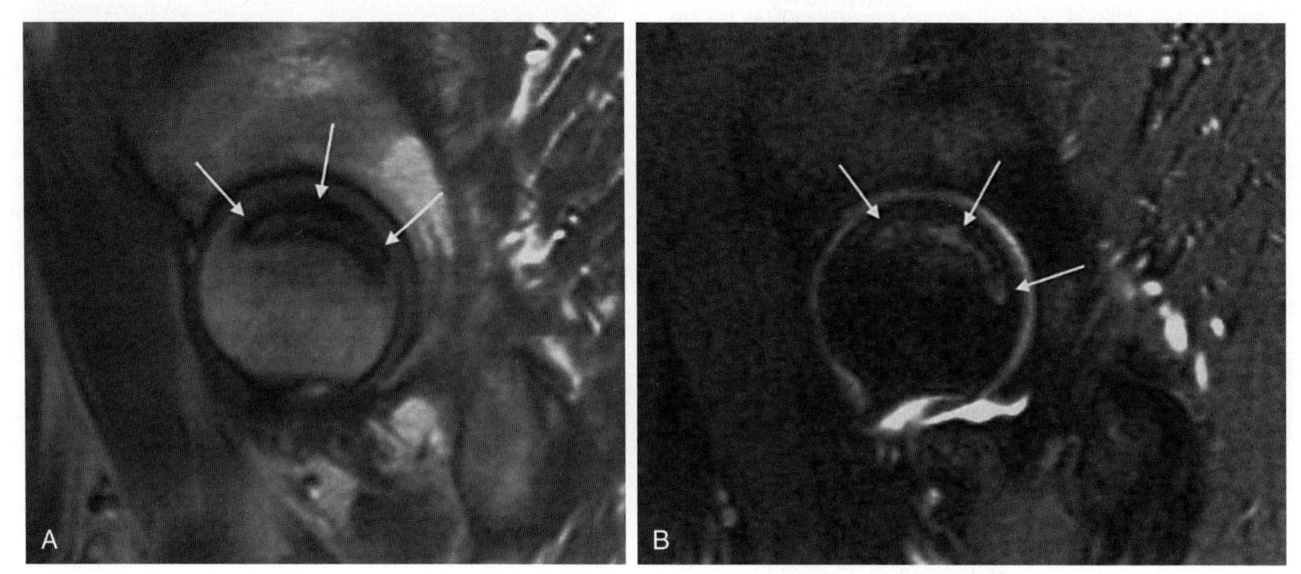

Figure 44–16. *A,* Sagittal T1-weighted magnetic resonance (MR) image of the left hip shows a crescent-shaped black outline in the superior aspect of the femoral head (*arrows*) characteristic of avascular necrosis. *B,* Sagittal short tau inversion recovery (STIR) MR image of the same region (*arrows*) shows the same black outline with central high signal (the double-line sign).

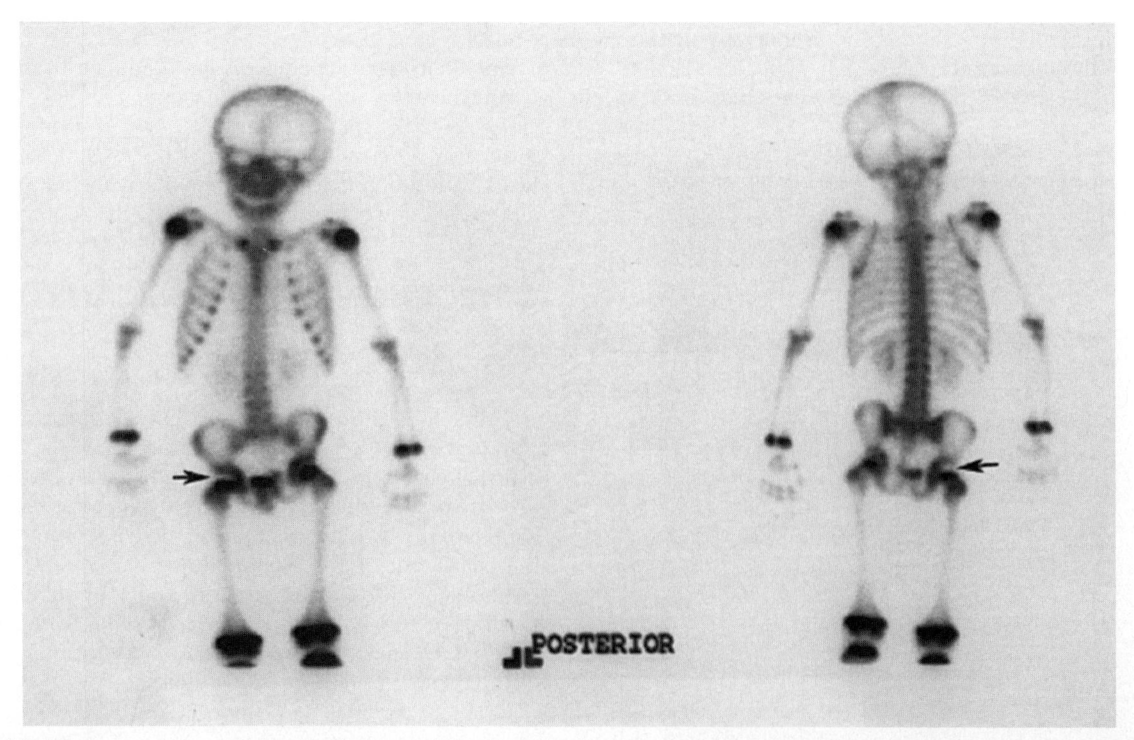

Figure 44–17. Bone scan in a child shows a photopenic defect (*arrow*) in the right femoral head characteristic of early avascular necrosis. (Courtesy of Dr. Nat E. Watson, Jr., Winston-Salem, NC.)

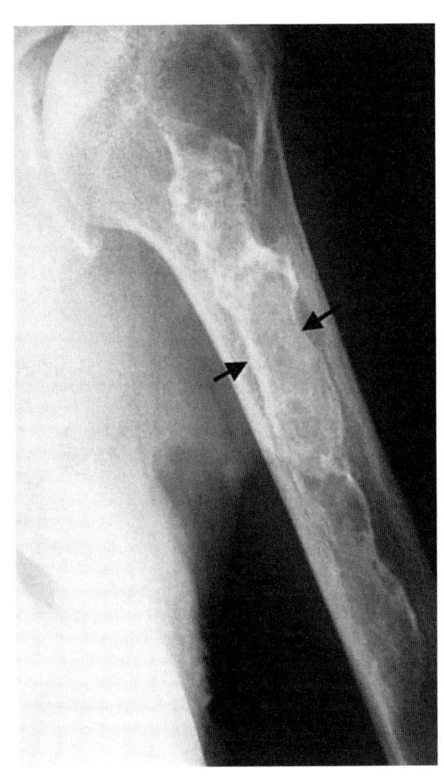

Figure 44–18. Anteroposterior view of the left humerus shows a well-defined sclerotic lesion in the proximal metaphysis/diaphysis characteristic of a bone infarct (*arrows*).

(Fig. 44–21). The loose bodies may grow as the synovial fluid nourishes them. It is therefore very important to differentiate a stable OCD (intact overlying cartilage) from an unstable OCD (fractured cartilage).

MR imaging is useful in the detection and evaluation of OCD. MR arthrography is especially useful when the OCD is suspected to be unstable. The bony margins of stable osteochondral lesions have low signal on both T1- and T2-weighted sequences (Fig. 44–22). On either T2-weighted images or T1-weighted images obtained after injection of a contrast agent, high signal that extends between the lesion and the native bone indicates a break in the overlying cartilage with extension of fluid around the fragment and a greater likelihood that the fragment will become loose (Fig. 44–23).

RADIATION CHANGES OF BONE

Overview

Radiation therapy has long been known to cause changes within bone. Damage to blood vessels as well as injury to osteoblasts and osteoclasts occurs after radiation therapy. The changes are related to the radiation dose and the maturity of the skeletal system. Growth arrest can be seen in the immature skeleton, with doses exceeding 1200 cGy (1200 rad). In mature bone, the dose threshold for radiation changes is 3000 cGy.

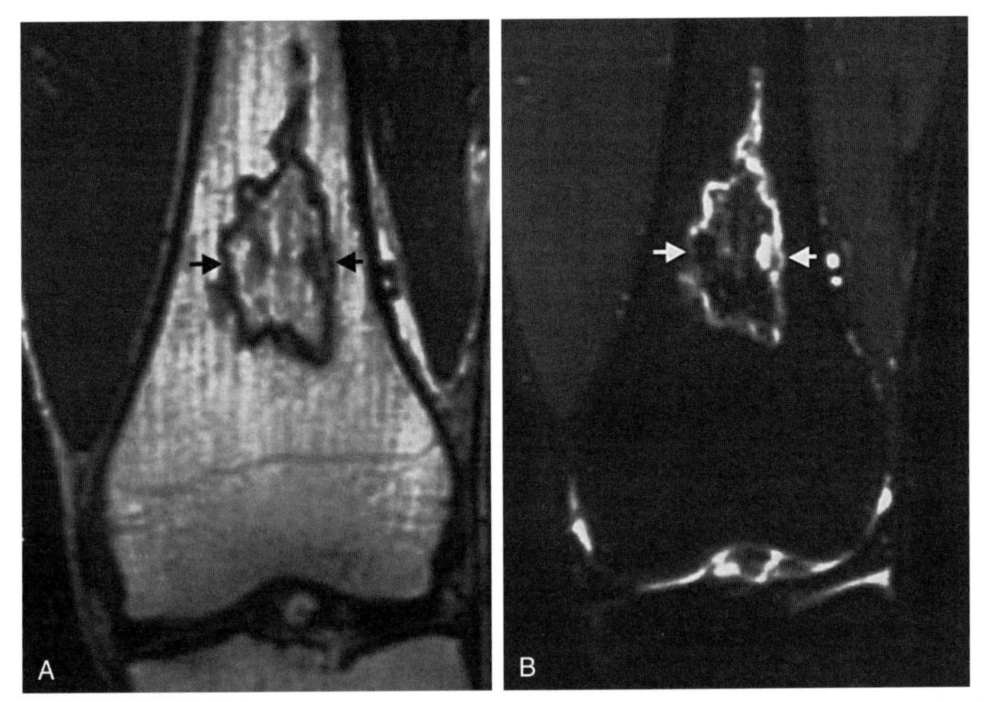

Figure 44–19. *A*, Coronal T1-weighted magnetic resonance image of a mature bone infarct shows a well-defined intramedullary lesion with central fat signal (*arrows*). *B*, The high signal rim on the short tau inversion recovery (STIR) image (*arrows*) is believed to represent granulation tissue at the bone-infarct interface.

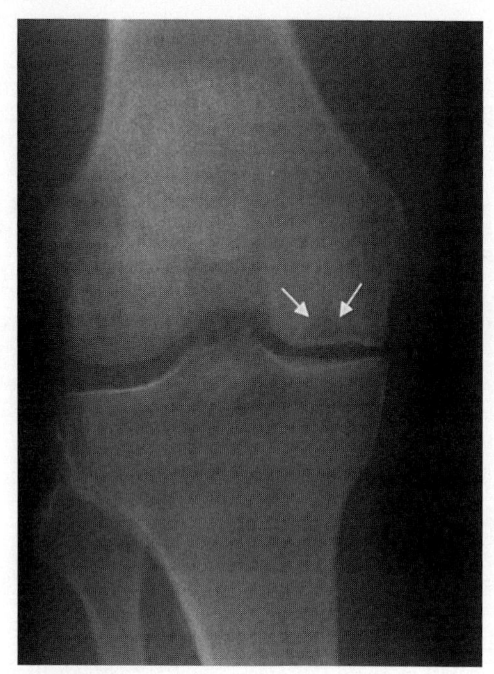

Figure 44–20. Anteroposterior view of the knee in a 64-year-old woman with spontaneous osteonecrosis. The medial femoral condyle is flat and there is a lucent region involving the weight-bearing surface of the medial femoral condyle (*arrows*).

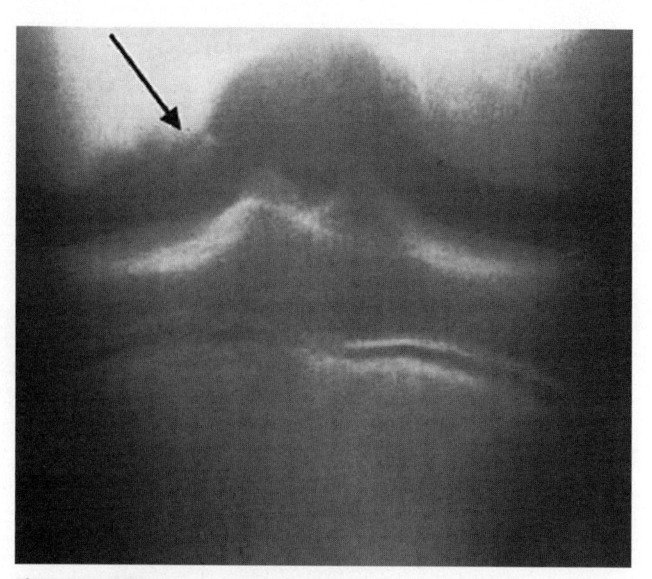

Figure 44–21. Radiograph of the knee in a child who presented with pain and locking of the knee shows an osteochondral lesion involving the medial femoral condyle. Note the associated intra-articular osteochondral body (*arrow*).

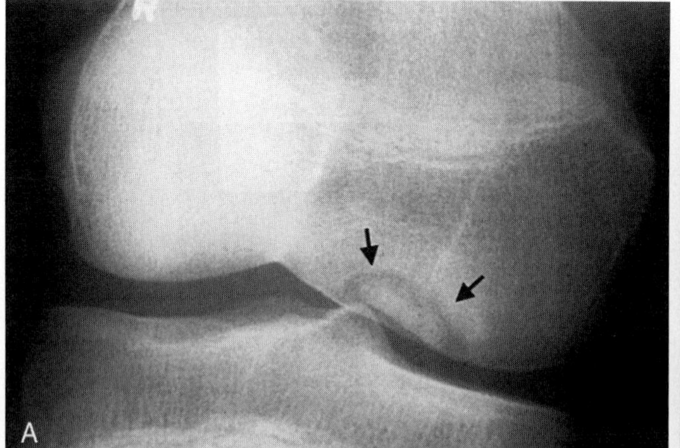

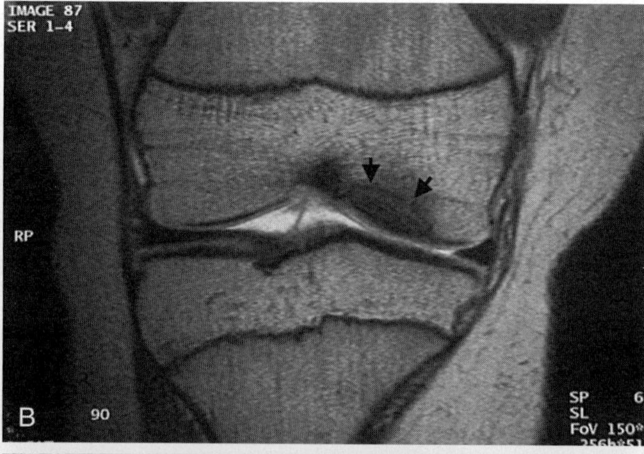

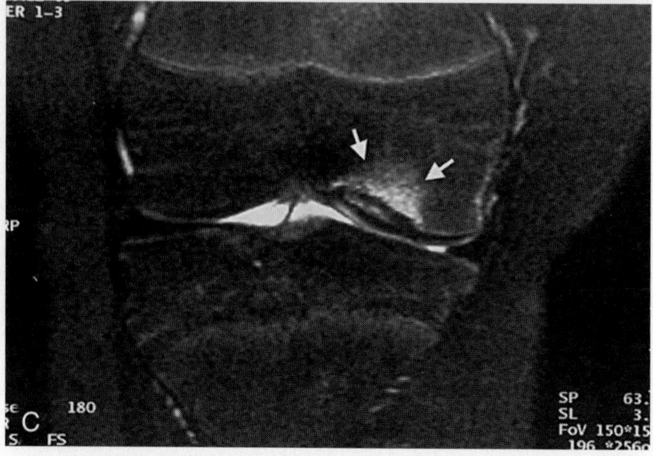

Figure 44–22. *A*, Anteroposterior radiograph of the knee in an adolescent with osteochondritis dissecans showing a lucent lesion (*arrows*) in the lateral aspect of the medial femoral condyle. *B*, On a coronal T1-weighted magnetic resonance (MR) image obtained after intra-articular injection of contrast agent, no fluid extends around the osteochondritis dissecans (*arrows*), indicating a stable fragment. *C*, Coronal T2-weighted MR image with fat saturation shows marrow edema (*arrows*) adjacent to and within the osteochondral fragment but no fluid tracking around the fragment.

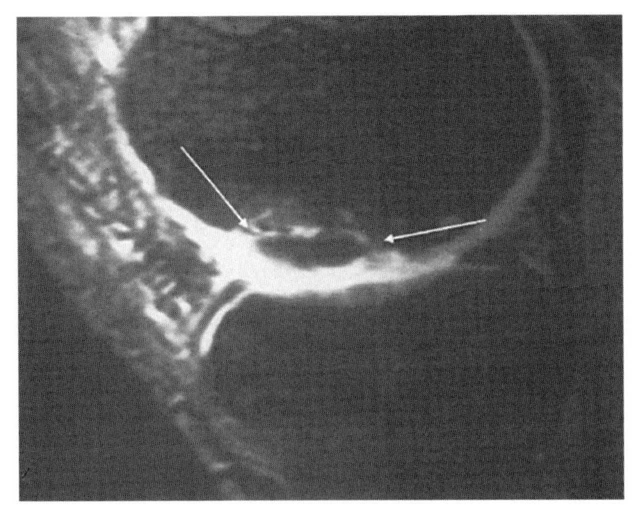

Figure 44–23. Sagittal T2-weighted magnetic resonance image of the knee shows an osteochondral lesion in the medial femoral condyle. Note that the margins of the lesion have high signal (*arrows*), indicating possible instability of the fragment.

The clinical findings associated with radiation changes in bone are due to the complications that arise. They include growth disturbance in the immature skeleton, fractures in children and adults, and benign and malignant neoplasms (see Fig. 4–22).

Imaging Findings

In the immature appendicular skeleton, radiation changes can lead to limb shortening, bowing, and fractures such as slipped capital femoral epiphysis and slipped humeral epiphysis. Metaphyseal changes may resemble those seen in rickets, with fraying, cupping, and growth plate widening. In the axial skeleton, radiation therapy can cause scoliosis or kyphosis, which may not manifest until 5 years after therapy.

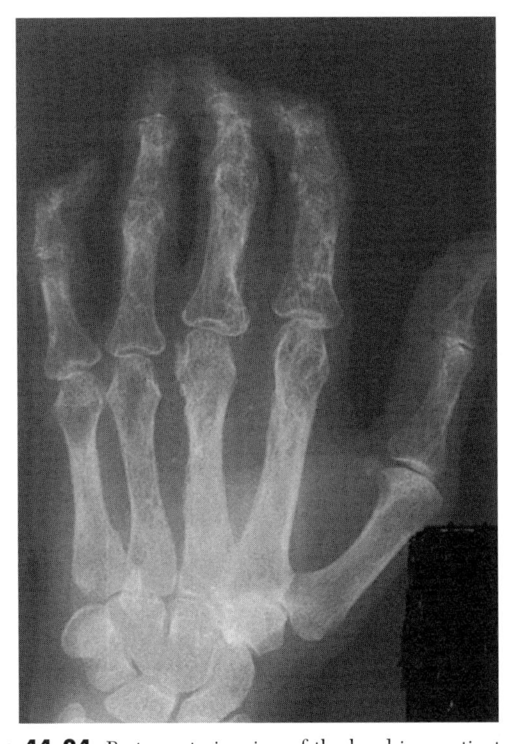

Figure 44–24. Posteroanterior view of the hand in a patient treated with radiation therapy for squamous cell carcinoma of the skin. The radiation has caused patchy areas of demineralization with trabecular thickening and disorganization.

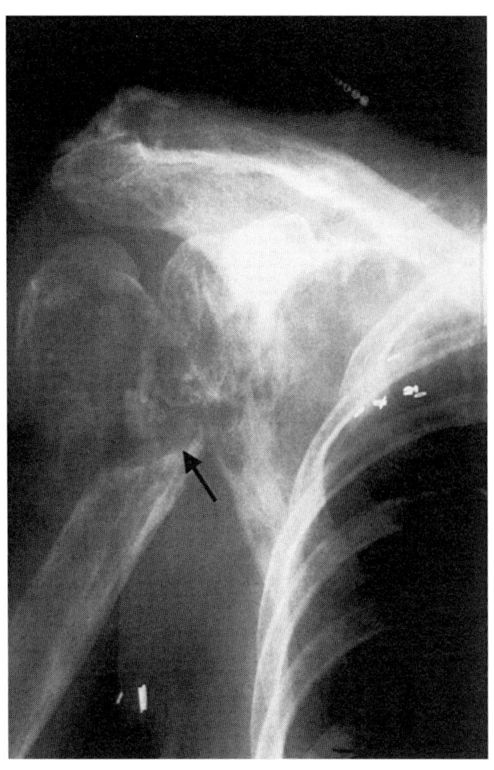

Figure 44–25. Frontal view of the shoulder in a woman treated for breast cancer with radiation to this area. A pathologic fracture is present in the proximal humerus (*arrow*). Regions of cortical lysis and trabecular disorganization are present in the humerus, scapula, and clavicle.

In the mature skeleton, irradiation can cause discrete areas of cortical lysis, disorganization and coarsening of trabeculae, and diffuse demineralization (Fig. 44–24). These changes are usually not seen until 1 year after therapy. Insufficiency fractures may occur but tend to heal without complications.

Osteonecrosis can arise years after radiation therapy; the differential diagnosis of this complication includes recurrent tumor, radiation-induced neoplasm, and infection (Fig. 44–25). Lack of a soft tissue mass and stability of findings over time make osteonecrosis more likely. Radiation-induced neoplasms include osteochondromas, which arise in children irradiated at a young age, and osteosarcoma, the most common malignant neoplasm. A latent period of at least 4 years between irradiation and tumor appearance is necessary to make a diagnosis of radiation-induced tumor.

Bone marrow changes can be identified on MR imaging as early as one week after radiation therapy. Marrow edema is initially detected on short tau inversion recovery (STIR) images as high signal intensity. Fat replacement of irradiated marrow is seen as increased signal on T1-weighted MR images within 2 weeks of therapy. A completely fatty marrow can be present within the radiation field as early as 2 months after therapy (Fig. 44–26).

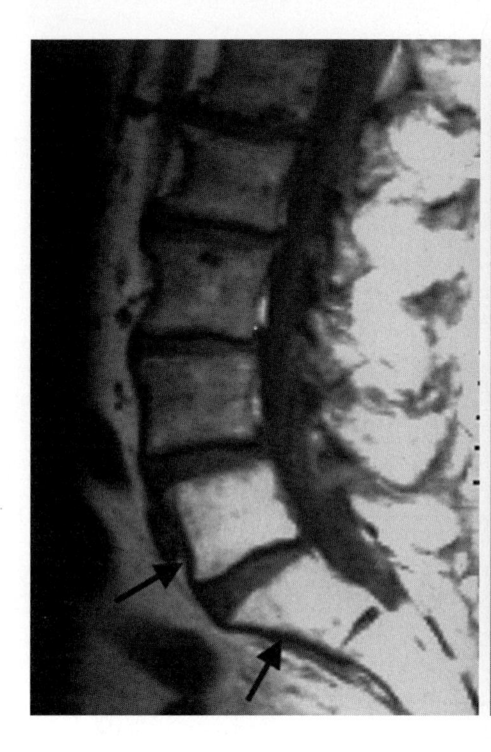

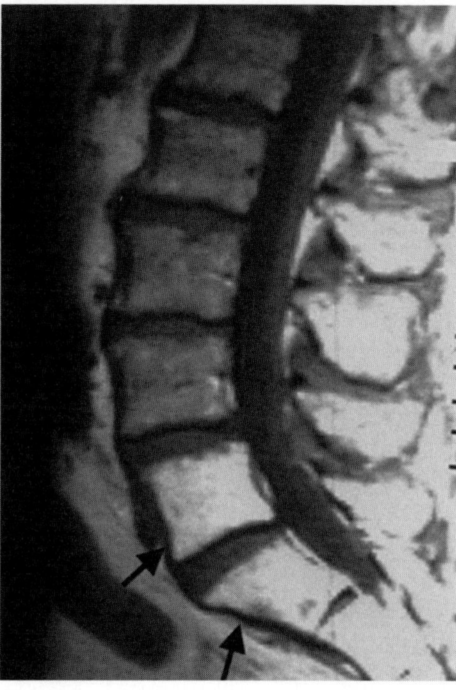

Figure 44–26. Sagittal T1-weighted magnetic resonance images of the lumbar spine in a patient treated with radiation for a pelvic tumor. Homogeneous fatty replacement of the marrow can be seen in the L5 vertebral body and the sacrum (*arrows*), areas that were included in the radiation field.

Recommended Readings

Blomlie V, Rofstad EK, Skjønsberg A, et al: Female pelvic bone marrow: Serial MR imaging before, during, and after radiation therapy. Radiology 194:537, 1995.

Burnstein MI, Kottamasu SR, Pettifor JM, et al: Metabolic bone disease in pseudohypoparathyroidism: Radiologic features. Radiology 155: 351, 1985.

De Smett AA, Fisher DR, Graf BK, et al: Osteochondritis dissecans of the knee: Value of MR imaging in determining lesion stability and the presence of articular cartilage defects. AJR Am J Roentgenol 155:549, 1990.

Kidd GS, Schaaf M, Adler RA, et al: Skeletal responsiveness in pseudohypoparathyroidism: A spectrum of clinical disease. Am J Med 68:772, 1980.

Lang EK, Bessler WT: The roentgenologic features of acromegaly. AJR Am J Roentgenol 86:321, 1961.

Meunier PJ, S-Bianchi GG, Edouard CM, et al: Bony manifestations of thyrotoxicosis. Orthop Clin North Am 3:745, 1972.

Mitchell DG, Rao VM, Dalinka MK, et al: Femoral head avascular necrosis: Correlation of MR imaging, radiographic staging, radionuclide imaging, and clinical findings. Radiology 162:709, 1987.

Mitchell MJ, Logan PM: Radiation induced changes in bone. Radiographics 18:1125, 1998.

Scanlon GT, Clemett AR: Thyroid acropachy. Radiology 83:1039, 1964.

Steinbach HL, Young DA: The roentgen appearance of pseudohypoparathyroidism (PH) and pseudo-pseudohypoparathyroidism (PPH): Differentiation from other syndromes associated with short metacarpals, metatarsals and phalanges. AJR Am J Roentgenol 97:49, 1966.

Steinberg H, Waldron BR: Idiopathic hypoparathyroidism: An analysis of fifty-two cases, including the report of a new case. Medicine 31:133, 1952.

Wietersen FK, Balow RM: The radiologic aspects of thyroid disease. Radiol Clin North Am 5:255, 1967.

SECTION IV

SPINAL DISORDERS

CHAPTER 45

Imaging of Low Back Pain

Georges Y. El-Khoury, MD, and Elias R. Melhem, MD

OVERVIEW

Low back pain (LBP) is one of the most common health problems in our society, being the most common cause of disability for persons younger than 45 years. The lifetime prevalence of LBP is approximately 80%. At any given time, 31 million Americans have LBP; 1 in 20 have associated sciatica. Fifty percent of patients with acute LBP return to normal function within 2 weeks, and only 5% remain disabled after 3 months. Patients with chronic LBP consume 85% to 90% of the direct and indirect costs of managing LBP. The vast majority of cases of acute uncomplicated LBP are benign and self-limiting and do not require any imaging studies.

The causes of LBP can be classified as follows:

1. Degenerative disk diseases:
   a. Herniated nucleus pulposus (HNP).
   b. Spinal stenosis.
   c. Internal disk disruption.
2. Facet syndrome.
3. Spondylolysis and spondylolisthesis.
4. Neoplasm.
5. Disk space infection and epidural abscess (see Chapter 46).
6. HLA-B27 spondyloarthropathies (see Chapter 22).
7. Segmental instability.

IMAGING OF LOW BACK PAIN

Because LBP is a self-limiting disease and its prevalence is so high, imaging of every patient with the complaint would be very costly. LBP does not follow the classic model of disease, in which symptoms closely correlate with imaging abnormalities. A significant proportion of patients with LBP (80%) may have no detectable abnormalities. Asymptomatic individuals, however, may be found to have striking findings on imaging studies. Several government and institutional agencies, including the American College of Radiology (ACR), have developed specific indications for imaging LBP.

Indications for Radiography

According to the Quebec Task Force on Spinal Disorders, the indications for radiography of low back pain are as follows (Spitzer et al, 1987):

1. Neurologic deficit.
2. Age more than 50 or less than 20 years.
3. Fever.
4. Suspected neoplasm.
5. Trauma.

"Red flags" indicating that a patient's LBP may be complicated and, therefore, warrants imaging, according to the ACR's Appropriateness Criteria 1998, Volume 2):

1. Recent significant trauma or milder trauma in patients older than 50 years.
2. Unexplained weight loss.
3. Unexplained fever.
4. Immunosuppression.
5. History of cancer.
6. Intravenous drug use.
7. Prolonged use of corticosteroids, osteoporosis.
8. Age older than 70 years.

Imaging Modalities

Plain Radiography

Any investigation of LBP should start with plain radiography. In certain conditions, such as trauma and osteoporosis, no further investigation is required if plain radiographic findings are normal. In patients with herniated nucleus pulposus, plain radiography is often normal; however, the evaluation is still useful for ruling out other serious conditions, such as a neoplasm and infection, for accurate localization of disk herniation, and for surgical planning, especially in patients with transitional vertebrae.

Magnetic Resonance Imaging

It has become common practice to go directly to magnetic resonance (MR) imaging after radiography. MR imaging is the modality of choice for the evaluation of herniated nucleus pulposus, spinal stenosis, vertebral neoplasm, spondylodiskitis, epidural abscess, and failed back surgery syndrome.

Computed Tomography and Myelography

Computed tomography (CT) is ideal for the evaluation and follow-up of suspected pars defects (spondylolysis) and focal bony lesions such as osteoid osteoma and

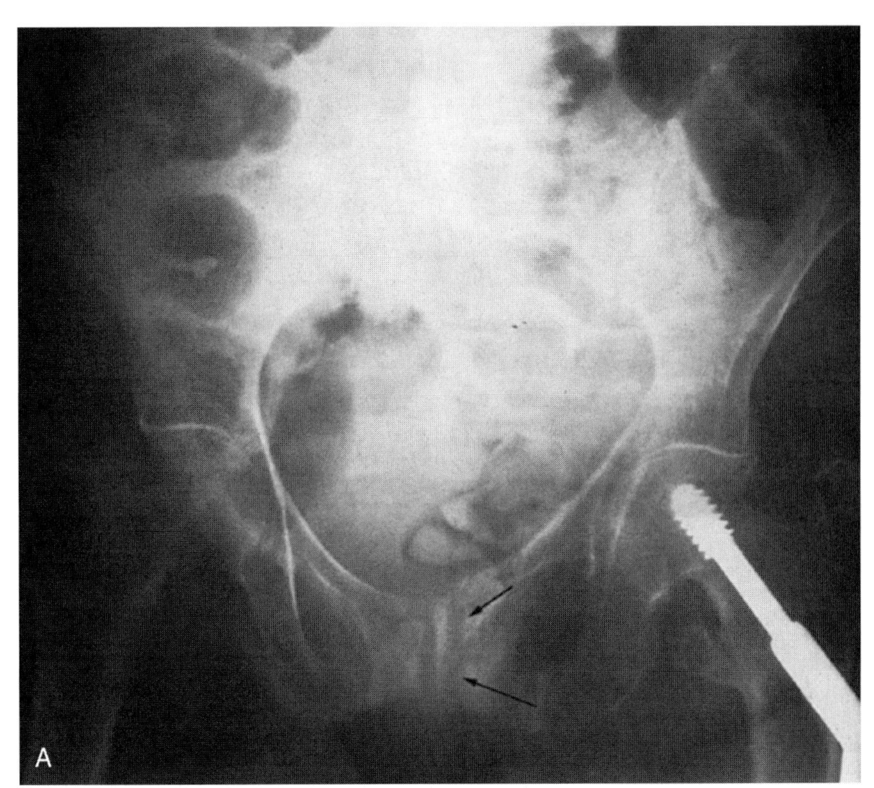

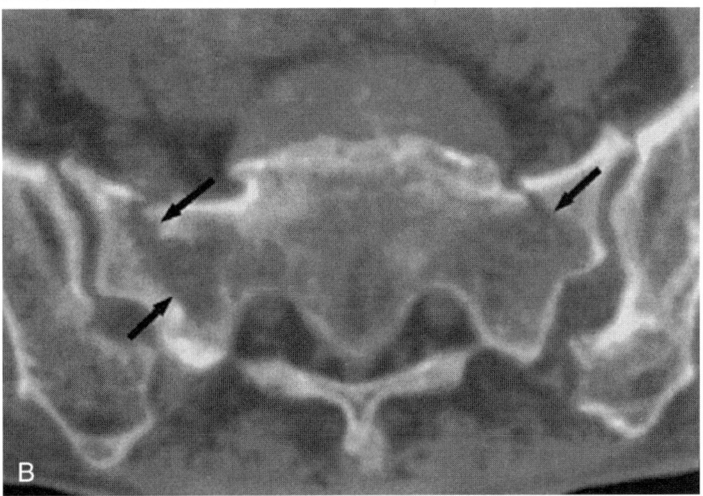

Figure 45–1. Sacral insufficiency fractures in a 64-year-old woman who complained of low back pain and left groin pain brought on by activity. *A,* Anteroposterior radiograph of the pelvis shows fractures in the symphysis pubis on the left side (*arrows*). The visualized bones are osteopenic. *B,* Computed tomography section through the sacrum shows bilateral insufficiency fractures (*arrows*).

osteoblastoma of the vertebral arch. CT is often used to confirm the diagnosis if insufficiency fracture of the sacrum is suspected from plain radiography or bone scan (Fig. 45–1). CT is the best imaging modality for assessing progress in spinal fusion. It is also frequently utilized during spinal interventional procedures, such as vertebral and disk space biopsies.

CT-myelography is not commonly used, but some back surgeons still prefer the technique in the preoperative evaluation of spinal stenosis. CT-myelography is occasionally used instead of MR imaging in patients who have pacemakers or are claustrophobic.

Radionuclide Bone Scanning

The role of radionuclide bone scanning in LBP is limited. The modality is indicated for the detection of acute pars defects (spondylolysis) not visualized on plain radiographs. Frequently, the planar images are supplemented with single photon emission computed tomography (SPECT) images, which improve sensitivity and specificity. Back pain due to suspected osteoid osteoma may be better localized by a bone scan, but the findings are not specific and require confirmation with CT.

Radionuclide bone scanning can detect metastasis or disk space infection, but such lesions are better studied with MR imaging. Positron emission tomography (PET) using fludeoxyglucose F 18 (^{18}FDG) is now widely used to detect primary and metastatic neoplasms throughout the body, including the spine (Fig. 45–2). PET is a powerful diagnostic tool because it goes beyond anatomic imaging—it demonstrates the metabolic activity of the tumor. This property is useful in testing for eradication of tumors treated by irradiation or chemotherapy when

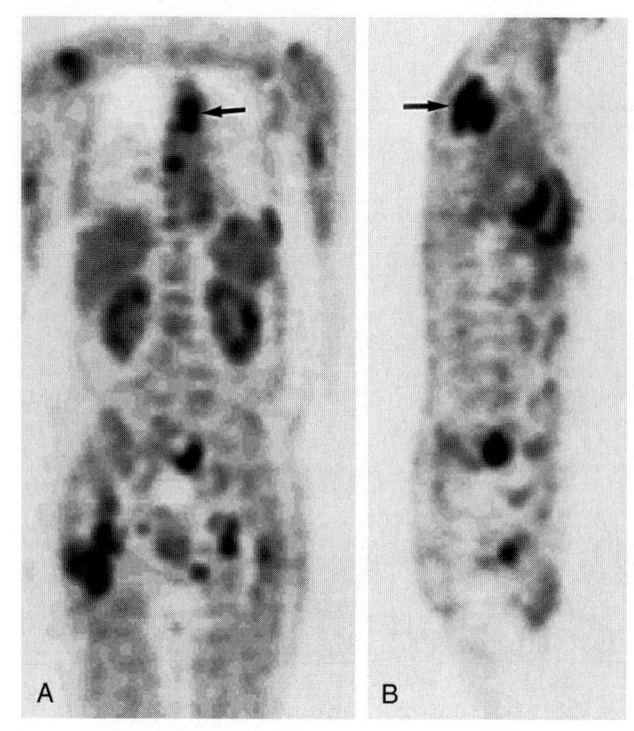

Figure 45–2. Positron emission tomography scans in a 54-year-old man with non-Hodgkin's lymphoma. *A*, Coronal slice through the spine shows multiple foci of abnormal accumulation of ^{18}FDG in proximal femurs, proximal humeri, and thoracic spine (*arrow*). *B*, A sagittal slice through the spine shows abnormal accumulation of ^{18}FDG in the upper thoracic spine (*arrow*).

conventional imaging methods continue to show anatomic abnormalities. Integrated imaging systems using both PET and CT have been introduced into clinical use to improve specificity and increase efficiency.

DEGENERATIVE DISK DISEASE

Overview

The intervertebral disk is the largest avascular structure in the body. It is composed of the nucleus pulposus, the anulus fibrosus, and two cartilaginous end plates (Fig. 45–3). In young individuals, the nucleus pulposus consists of a soft gelatinous material that is eccentrically located, being closer to the posterior disk margin. The anulus fibrosus acts as a the limiting capsule for the nucleus pulposus.

The disk functions in transforming axial loads into tensile forces, which are transmitted via the anular (Sharpey's) fibers to the bone. A healthy disk contains approximately 85% water, the rest being collagen and proteoglycans. The intervertebral disk is essentially an osmotic system sensitive to pressure and concentration of proteoglycans. Sharpey's fibers and the rings of the outer anulus, consisting of type I collagen, are demonstrated well on MR imaging. The inner anulus and the hydrated gel of the nucleus are inseparable on MR imaging. The intranuclear cleft seen in normal adults represents fibrous transformation of the previously gelatinous matrix of the nucleus.

Disk degeneration is a progressive and irreversible process that is universally present in all humans after the age of 30 years. There is no one unifying theory that explains the etiology and mechanism of disk degeneration. The two most commonly discussed theories are the nutritional theory and the loss of confined fluid theory. The disk receives most of its nutrition by diffusion from blood vessels in the marrow of the vertebral bodies and blood vessels on the periphery of the anulus. With age, according to the nutritional theory, the vertebral end plates calcify and become less porous, thus permitting less exchange of nutrients. Therefore, the number of cells (chondrocytes) that support the disk matrix diminish with age.

The loss of confined fluid theory has been advanced and tested by Lipson and Muir (1980), who showed that creating a defect in the anulus leads to rapid degeneration of the disk. In another experiment using cadaveric spines subjected to fatigue loading, the initial injury was found to represent anular disruption, which progressed to radial tears and extrusion of the nucleus.

With degeneration of the disk, biochemical and structural changes take place simultaneously. Progressive water loss is noted, and the water content drops to approximately 70%. Degenerative disk disease continues at middle age and old age, but at a lower rate. Later in life, the spine becomes stiff but pain free. Back pain and sciatica in older patients is the result not of disk disease but of spinal stenosis. The disks become more fibrous and disorganized, with loss of distinction between nucleus and anulus.

Plain Radiography

The initial radiographic finding in disk degeneration is reduction in disk height. The disk height in the lumbar spine normally shows progressive increase in height from the L1-L2 disk to the L4-L5 disk. The L5-S1 disk normally has the least height. Acute disk herniation commonly causes no change in disk height (Fig. 45–4). Early in degenerative disk disease, reduction in disk height is often associated with traction osteophytes (see Fig. 45–16), which later transform to claw osteophytes (Fig. 45–5). Some writers emphasize the importance of traction

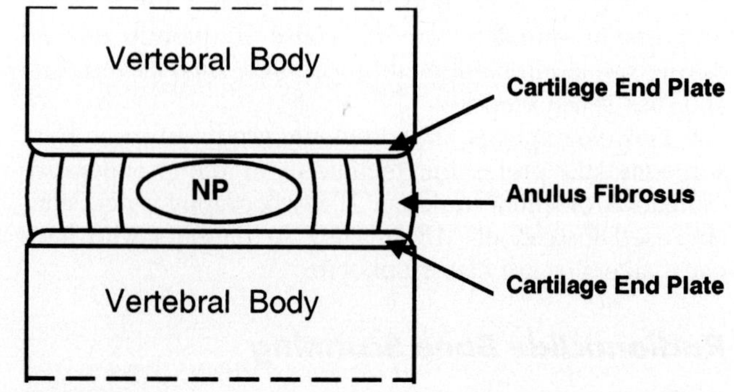

Figure 45–3. Diagram illustrating anatomy of the intervertebral disk. NP, nucleus pulposus.

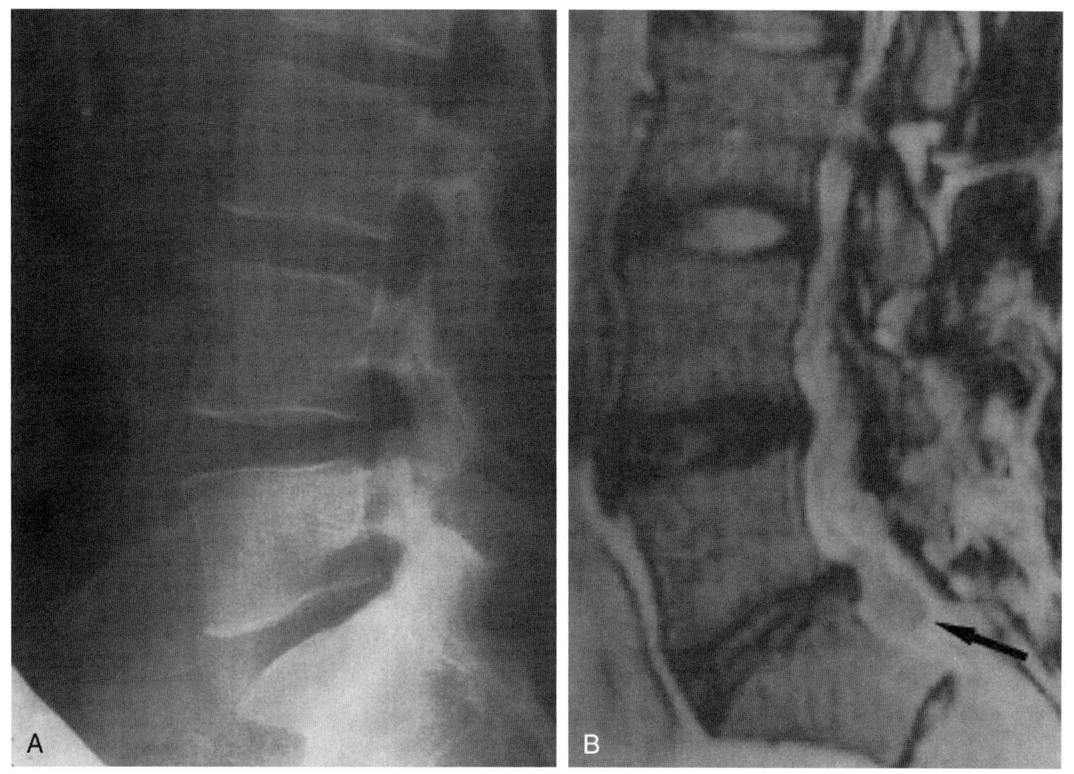

Figure 45–4. Normal lateral lumbar spine. The patient has a surgically documented extruded fragment, presumably arising from L5-S1. *A*, Lateral view of the lumbar spine shows normal disk height at all levels, including L5-S1, which is normally the disk with the least height. *B*, Sagittal T2-weighted magnetic resonance image shows an oval bright mass behind S1 (*arrow*), which at surgery proved to be an extruded disk fragment.

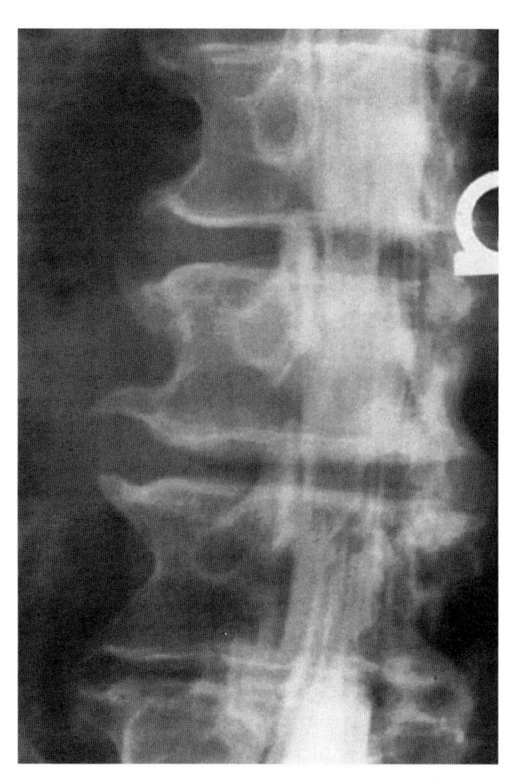

Figure 45–5. Typical claw osteophytes arising from all the visualized lumbar vertebrae.

osteophytes as a sign of segmental instability and a source of LBP. As the disk degeneration progresses, the disk height becomes significantly diminished, osteophytes become prominent, and the motion segments become stable but also less mobile (Fig. 45–6). Vertebral end plates may touch and become sclerotic. Along with the disk degeneration, there is concomitant apophyseal joint osteoarthritis, which on plain radiography manifests as facet hypertrophy due to osteophyte formation and facet sclerosis (see Fig. 45–6).

Magnetic Resonance Imaging

MR imaging has a distinct advantage over other imaging modalities in detecting early degenerative disk disease, the underlying process behind disk herniation. Observing this process led Masaryk and colleagues (1988) to state that "while not all degenerative disks herniate, all herniated disks eventually degenerate." On MR imaging, degenerative disk disease manifests as changes in the disks, vertebral end plate, facet joints, ligamentum flavum, and neural foramina. Early findings reveal decreased signal intensity in the disks on T2-weighted MR images. The earliest disk to show degeneration in this manner is the L5-S1 disk. Decreased signal intensity is followed by diminished disk height and circumferential disk bulging (Fig. 45–7). The facet joints reveal osteoarthritis, facet hypertrophy, subchondral cysts, and occasional

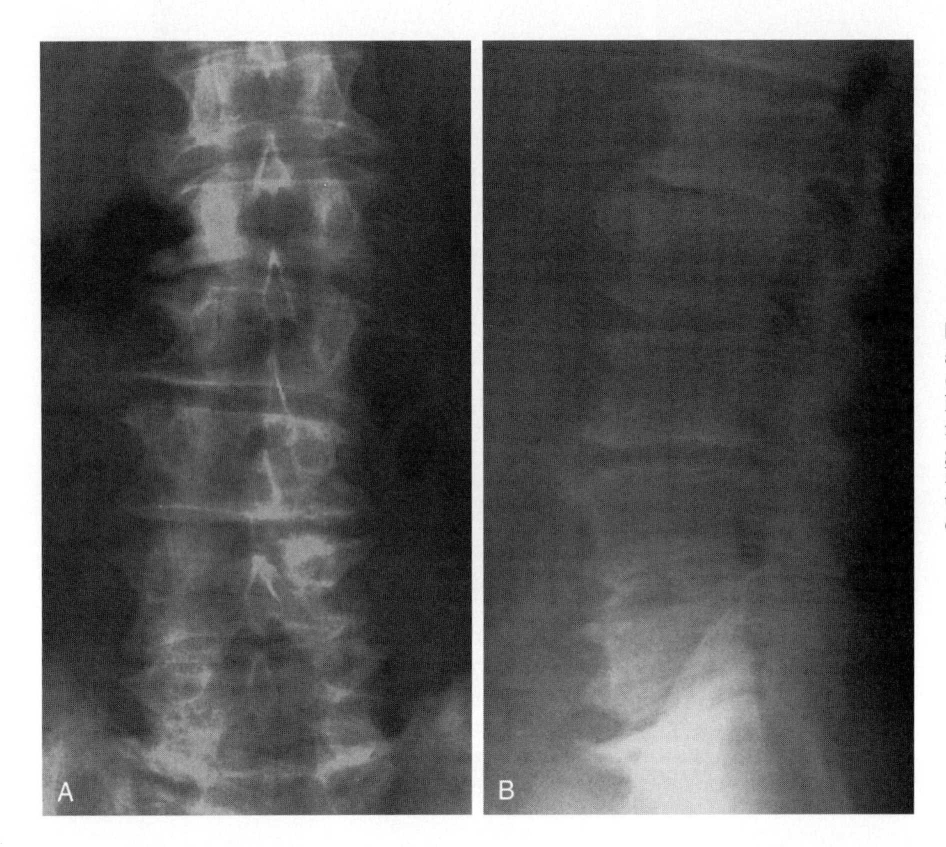

Figure 45–6. Degenerative disk disease and facet osteoarthritis. Anteroposterior (*A*) and lateral (*B*) radiographs of the lumbar spine show all the disk spaces to be narrowed and the presence of large marginal osteophytes. The articular facets in the lower lumbar spine are hypertrophied, and the facet joints show osteoarthritis. Mild degenerative scoliosis is also present.

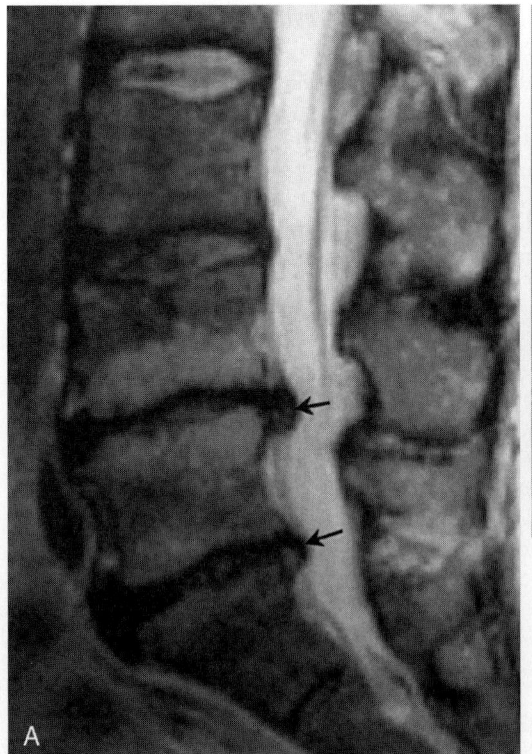

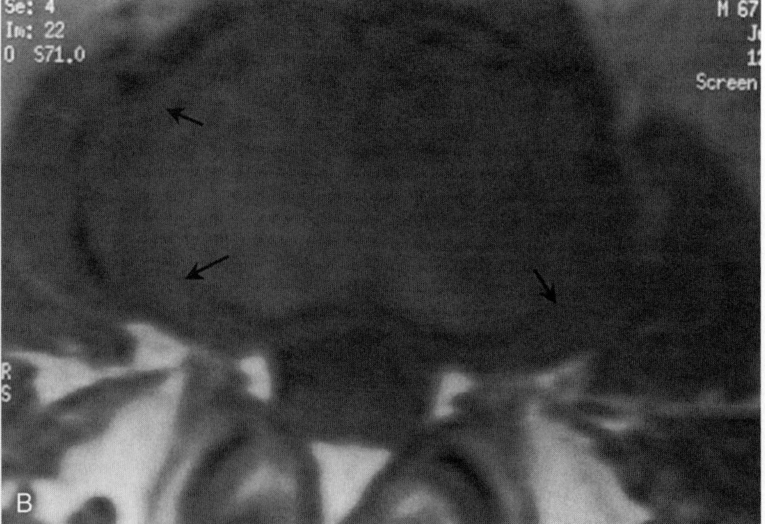

Figure 45–7. Disk degeneration on magnetic resonance imaging. A, Sagittal T2-weighted image of the lumbar spine shows marked decrease in signal intensity and loss of disk height involving L4-L5 and L5-S1 disks. Diffuse disk bulging is also noted at these levels (*arrows*). The L2-L3 disk is bright, indicating that it is still well hydrated. *B*, Axial T_1-weighted image through the L4-L5 disk illustrates diffuse disk bulge beyond the bony margins (*arrows*).

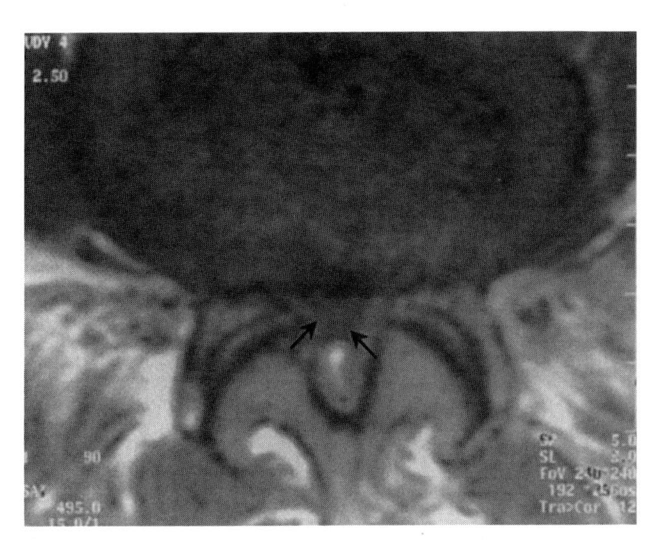

Figure 45–8. Degenerative spinal stenosis due to facet joint osteoarthritis and facet hypertrophy. Axial T_1-weighted magnetic resonance images at the L4-L5 level show a narrow spinal canal and compressed thecal sac (*arrows*).

formation of synovial cysts, which protrude into the spinal canal (Fig. 45–8). The ligamentum flavum thickens and buckles with loss of disk height. As a result of these morphologic changes, the spinal canal and the neural foramina become narrow. The end stage of this process is known as "degenerative spinal stenosis" (see Fig. 45–8). Spinal stenosis is usually most pronounced at L3-L4 and L4-L5 but not at L5-S1, where the spinal canal is fairly wide.

Modic Changes

Modic and associates (1988) described changes in the vertebral end plates and vertebral bodies that are associated with degenerative disk disease as follows:

Type I: Type I disease has decreased signal intensity on T1-weighted MR images and increased signal intensity on T2-weighted images (Fig. 45–9). These changes can be confused with vertebral osteomyelitis or infiltrations of the marrow by metastatic deposits. Histologically, the type I changes represent disruption and fissuring of the end plates along with fibrovascular tissue deposition in the bone marrow. Type I changes tend to enhance with the administration of gadolinium. With time, type I changes have been observed to progress to type II changes.

Type II: In type II degenerative disk disease, signal intensity is increased on T1-weighted MR images, but the changes are isointense to slightly increased in signal intensity on T2-weighted images (Fig. 45–10). Histologically, the type II changes represent fatty replacement of the vertebral marrow adjacent to the degenerated disk. The type II changes tend to be stable and do not progress to type III changes.

Type III: Signal intensity in type III changes is decreased on both T1- and T2-weighted MR images (Fig. 45–11). The Type III changes on MR imaging correlate with bony sclerosis observed on plain radiographs.

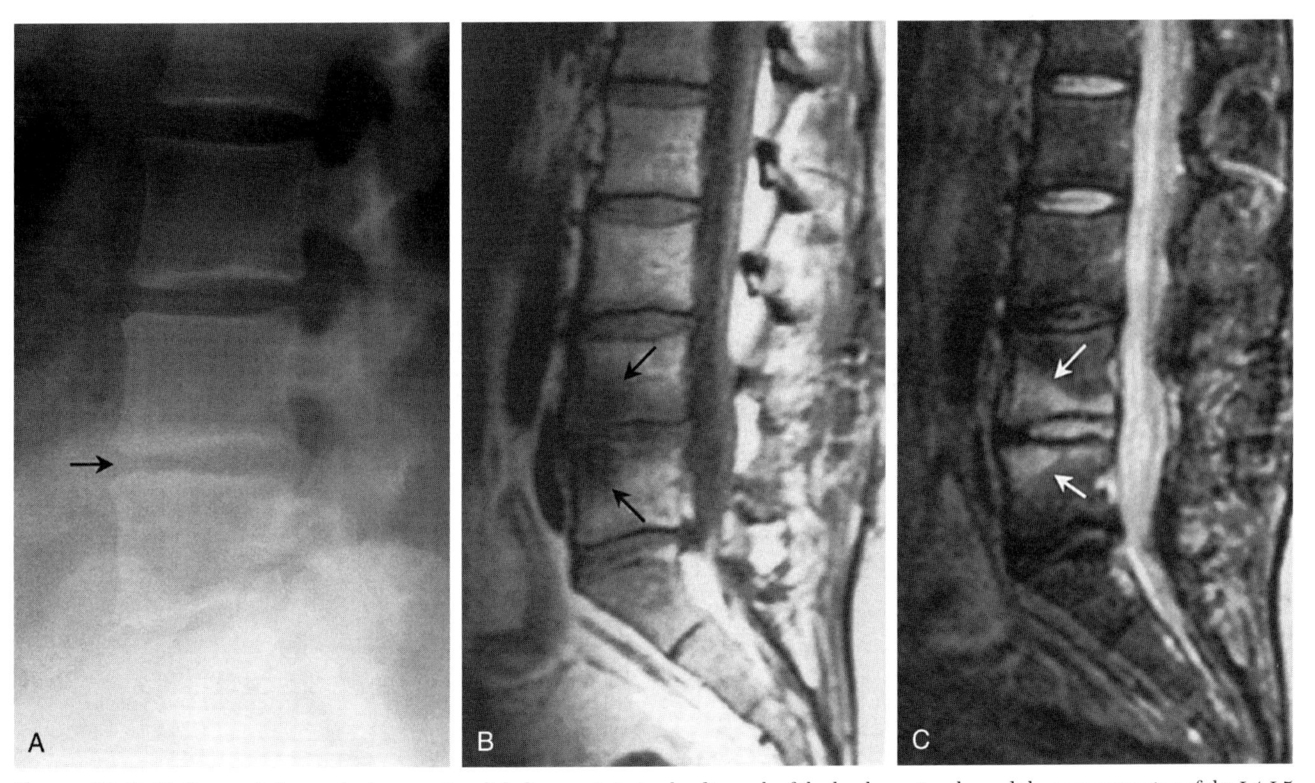

Figure 45–9. Modic type I changes in degenerative disk disease. *A*, Lateral radiograph of the lumbar spine shows disk space narrowing of the L4-L5 disk (*arrow*). *B*, Midline sagittal T1-weighted magnetic resonance (MR) image shows areas of decreased signal intensity in the L4 and L5 vertebrae (*arrows*) across from the L4-L5 disk. C, On a T2-weighted MR image, the same areas are now bright (*arrows*).

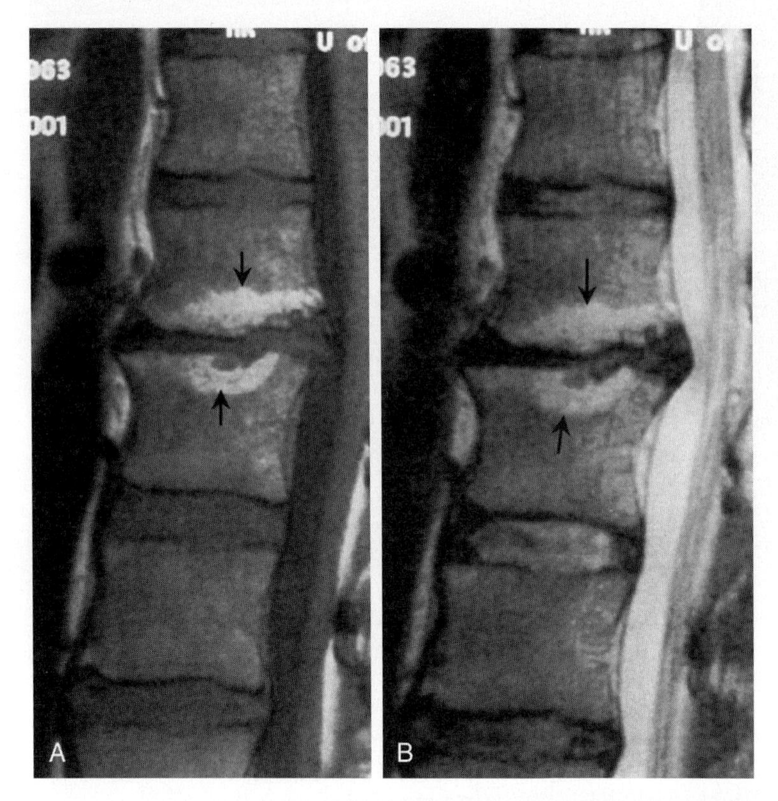

Figure 45–10. Magnetic resonance images showing Modic type II changes at the L1-L2 level in a 37-year-old man with low back pain. *A*, Midsagittal T1-weighted image shows narrowing of the L1-L2 disk and disk bulging. The end plates of L1 and L2 adjacent to the disk are irregular owing to Schmorl's nodes. The marrow is bright on both sides of the disk (*arrows*) because of fatty infiltration (Modic type II changes). *B*, Midsagittal T2-weighted image shows bright signals in the marrow adjacent to the L1-L2 disk (*arrows*).

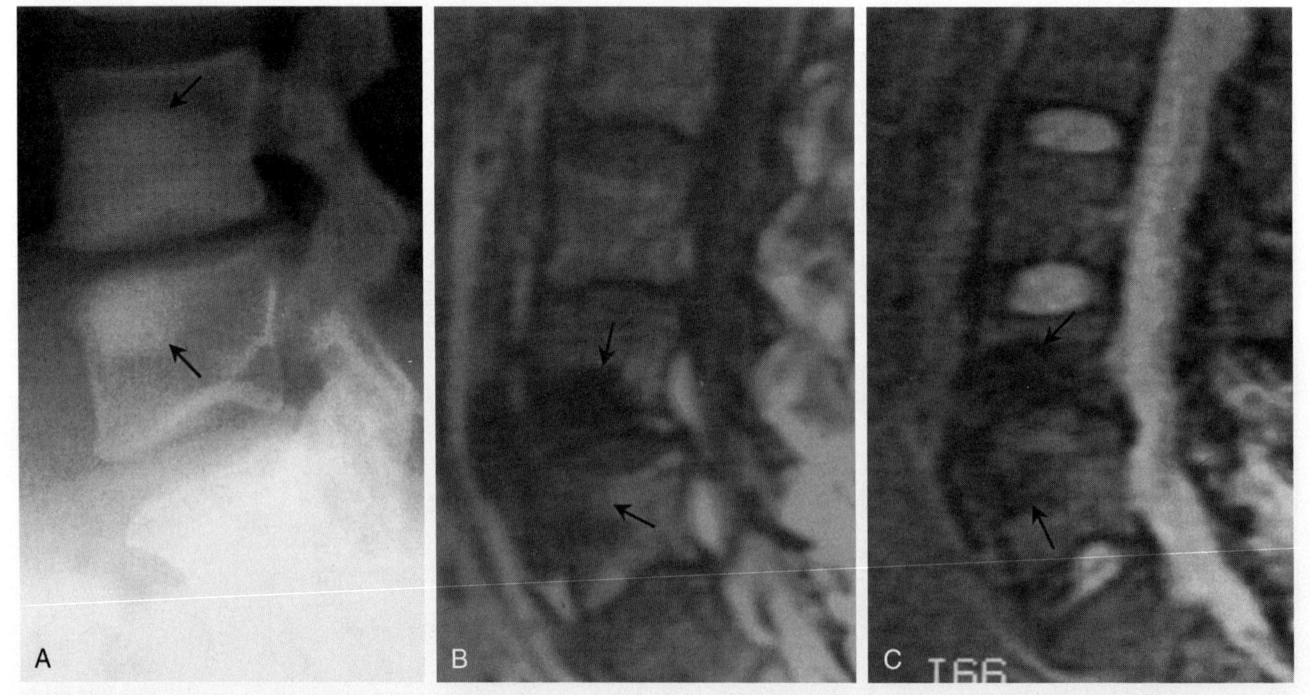

Figure 45–11. Modic type III changes in a 39-year-old woman who was previously treated with chymopapain for disk herniation at the L4-L5 level. *A*, Lateral radiograph shows narrowing of the L4-L5 disk and bony sclerosis in L4 and L5 vertebrae adjacent to the disk (*arrows*). A traction osteophyte is noted arising from the inferior anterior corner of L4. T1-weighted (*B*) and T2-weighted (*C*) sagittal magnetic resonance images show decreased signal intensity in L4 and L5 corresponding to the areas of sclerosis.

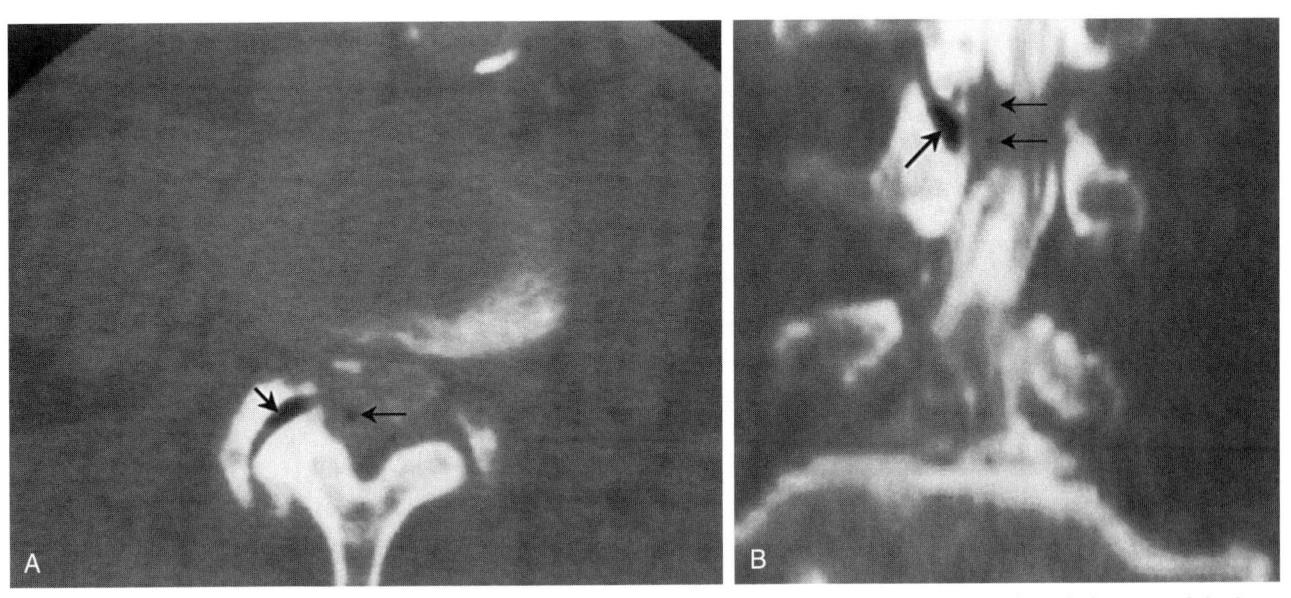

Figure 45–12. Synovial cyst arising from the L3-L4 facet joint on the right. *A*, Computed tomography section through the L3-L4 disk after a myelogram shows an air-containing cyst (*long thin arrow*). The cyst communicates with the right facet joint, which also contains air (*short thick arrow*) and shows osteoarthritis. *B*, Coronal reconstruction demonstrates displacement of the thecal sac by the synovial cyst (*thin arrows*). The facet joint from which the cyst originated also contains air (*thick arrow*).

Synovial Cyst of the Facet Joint

Synovial cyst of the facet joint manifests at middle age or later; affected patients complain of LBP and radicular pain. Symptoms are not likely to improve with conservative therapy, although some writers have reported improvement with injection of steroids into the facet joint.

Imaging Characteristics

Eighty percent of facet joint synovial cysts occur at the L4-L5 level (see Fig. 45–13). A synovial cyst appears as a rounded posterolateral epidural mass adjacent to a degenerated facet joint. The cyst wall occasionally calcifies, and the cyst may contain gas (Fig. 45–12). On MR imaging, the cyst has a hypointense rim, and the rest of the cyst is nearly isointense with cerebrospinal fluid on T1- and T2-weighted images. The rim enhances with administration of gadolinium (Figs. 45–13 and 45–14). The differential diagnosis of synovial cyst of the facet joint includes an extruded disk fragment and cystic neurofibroma.

Degenerative Scoliosis

Degenerative scoliosis results from asymmetric disk space narrowing and vertebral rotation caused by disk degeneration and instability (Fig. 45–15). The risk factors for degenerative scoliosis are female sex, white race, and degenerative disk disease. Patients with degenerative scoliosis have symptoms of spinal stenosis.

Retrolisthesis

Retrolisthesis is a form of segmental instability. Early in degenerative disk disease, the rate of disk degeneration may exceed the rate of osteoarthritis in the facet joints; this process may result in retrolisthesis with subtle posterior slippage of the superior vertebral body on the inferior vertebral body. Retrolisthesis is best evaluated on lateral flexion and extension radiographs (Fig. 45–16). Some writers believe that retrolisthesis is often associated with chronic LBP.

Facet Syndrome

The facet syndrome is a poorly understood entity in which back pain and sciatica are attributed to facet joint degeneration. Expanded synovial recesses and inflammation of the synovial capsule are believed to stimulate the nociceptive nerve endings and compress nerve roots in the spinal canal and neural foramina. Early studies of the injection of facet joint with steroids and local anesthetics were encouraging (Figs. 45–17 and 45–18). Later work has challenged the efficacy of facet joint injection and questioned the entire concept of the facet syndrome.

HERNIATED NUCLEUS PULPOSUS (HNP)

HNP is uncommon in children, in whom disks are healthy and well hydrated. The disease begins to rise in prevalence after age 25 years and peaks around age 40 years.

Plain Radiographs

Acute disk herniation often causes no changes in disk height. Occasionally, a fragment of bone is seen within the spinal canal on a plain radiograph. This represents a disk herniation associated with a vertebral rim avulsion (see

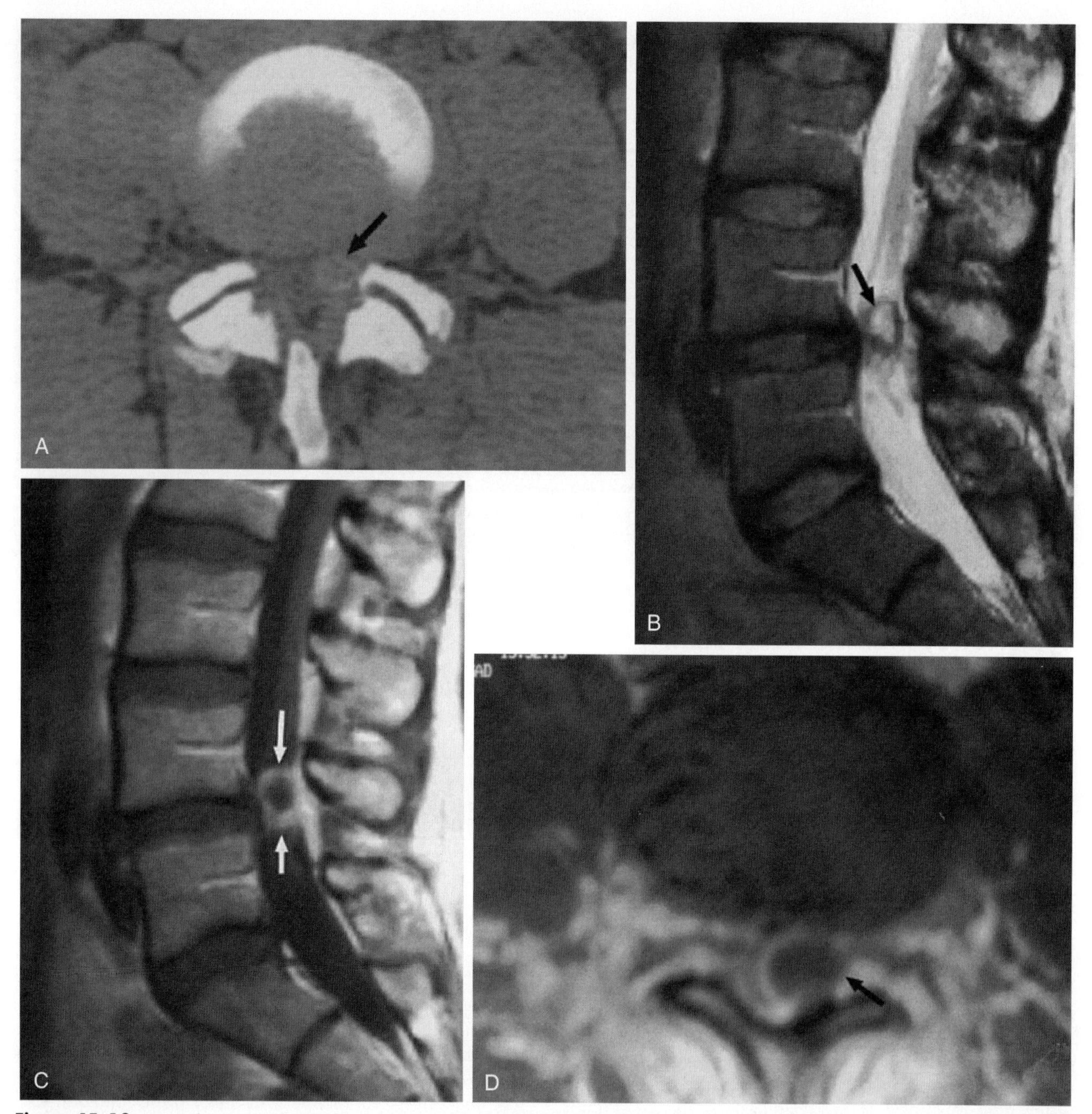

Figure 45–13. Synovial cyst in a 52-year-old woman presenting with low back pain and left radicular symptoms. *A,* Computed tomography section through the L4-L5 disk level shows a 1 × 1-cm cyst with faintly calcified wall (*arrow*). This cyst is situated between the left facet joint and the thecal sac, which is compressed. *B,* Sagittal T2-weighted magnetic resonance (MR) image shows the cyst to contain fluid (*arrow*). Sagittal (*C*) and axial (*D*) T1-weighted MR images obtained after intravenous injection of gadolinium demonstrate enhancement of the cyst wall (*arrows*).

Chapter 48). Vertebral rim avulsion is common in adolescents, who typically present with a history of acute trauma. The condition has been known by different names, including posterior lumbar apophyseal fracture, traumatic displacement of the cartilaginous vertebral rim, and vertebral end plate fracture.

Magnetic Resonance Imaging

Incidental findings on MR imaging that are unrelated to the patient's symptoms are very common. Disk bulge is reported to be present in 52% of asymptomatic volunteers, and disk protrusion in 27%. In the literature, more than 35 terms have been used to describe disk abnormalities. Precise terminology, however, helps separate abnormalities that are significant from those that are not; it also improves communication among physicians. In our practice, we have adopted the following terminology, which was advanced by Jensen and colleagues (1994):

- *Normal:* No disk extension beyond the interspace.
- *Bulge:* Circumferential symmetric extension of the disk beyond the interspace (see Fig. 45–7).
- *Protrusion:* Focal or asymmetric extension of the disk beyond the interspace, with the base against the disk of

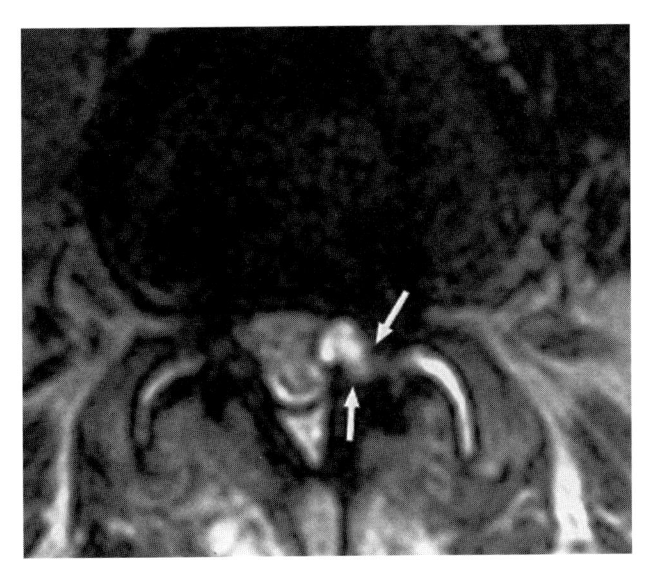

Figure 45–14. T2-weighted magnetic resonance image shows communication between the left facet joint and a synovial cyst (*arrows*). The facet joints are seen to have joint effusions.

origin broader than any other dimensions of the protrusion (Fig. 45–19). Most protrusions are central or paracentral; on rare occasions, the herniated portion can protrude either into the neural foremen or far laterally (Fig. 45–20).

- *Extrusion:* More extreme extension of the disk beyond the interspace, with the base against the disk of origin narrower than the diameter of the extruded material. The vast majority of protrusions and extrusions occur at the L4-L5 and L5-S1 levels. Extruded fragments within the spinal canal can migrate above or below the level of the disk, but usually below (Fig. 45–21).

When disk material is extruded into the epidural space, the nerve roots are mechanically compressed; they are also affected by the inflammatory mediators provoked by the disk material, which acts as a foreign body. The herniated fragment initially takes up fluid and enlarges. Hydration of the extruded fragment can be seen as increased signal intensity on T2-weighted MR images (Fig. 45–22). Sciatica is most pronounced in the first 3 weeks of symptoms, when extruded disk material is largest because of hydration and when the epidural inflammatory response is most intense. To reduce the inflammatory response, patients with acute disk herniation are often treated with epidural steroid injections, which can be easily performed with the use of fluoroscopic guidance. The approach to reach the epidural space is either via the sacral hiatus or through the interlaminar foramen at L2-L3 or L3-L4 (Figs. 45–23 and 45–24).

Extreme Lateral (Far-Out) Disk Protrusion

In extreme lateral disk protrusion, the bulk of the herniated portion protrudes into the neural foramen or beyond (see Fig. 45–20). It is easy to overlook this abnormality on MR imaging, and myelography findings are almost always negative. The incidence of far-out disks is reported to be between 1% and 10% of all disk herniations. Radicular symptoms are more common with extreme lateral disk protrusion than with other types of disk herniations, probably because of the direct contact

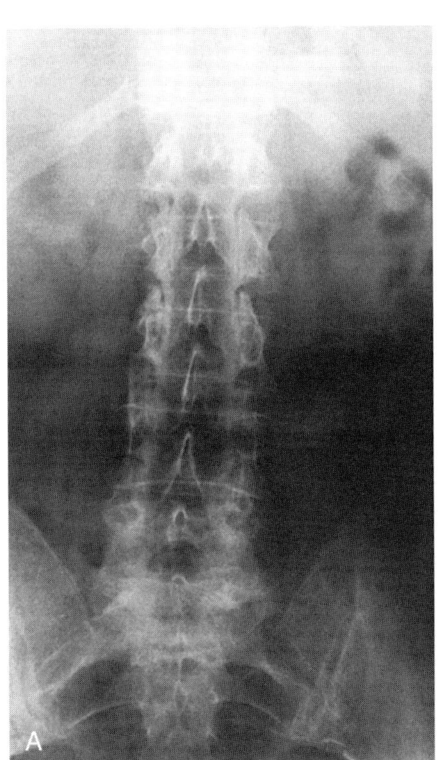

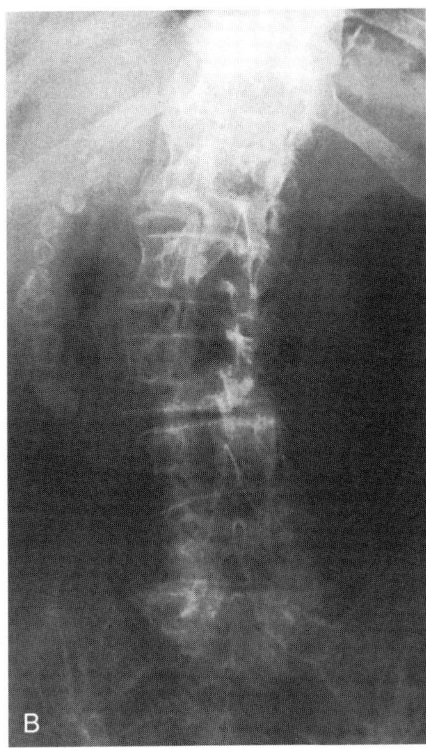

Figure 45–15. Degenerative scoliosis. *A,* Anteroposterior radiograph of the lumbar spine of a 67-year-old patient shows the spine to be fairly straight. *B,* A radiograph of the same patient at age 80 years shows significant degenerative scoliosis convex to the right side. In the interim, the patient developed gallstones.

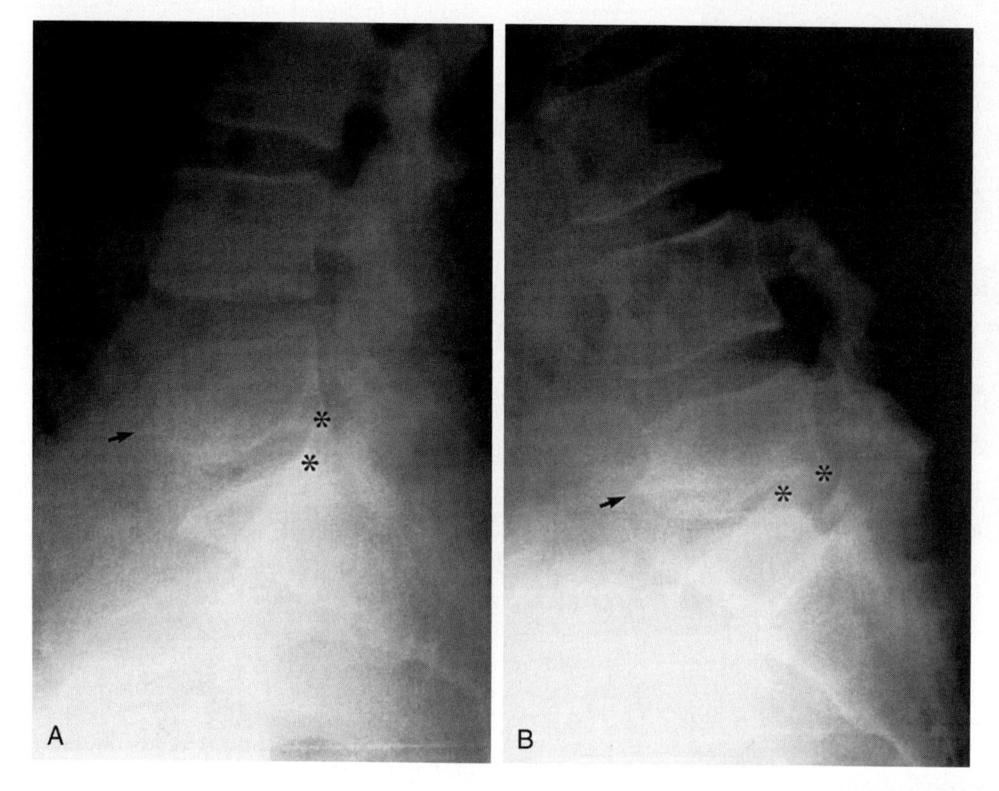

Figure 45–16. Radiographs showing severe retrolisthesis of L5 on S1 and a traction osteophyte arising from L5 (*arrows*) in a 42-year-old man. *Asterisks* indicate the posteroinferior corner of L5 and the posterosuperior corner of S1. *A*, Lateral extension view of the lumbosacral junction reveals minimal retrolisthesis of L5 on S1. *B*, Lateral flexion view shows accentuation of the retrolisthesis. In our experience, retrolisthesis becomes more pronounced in flexion.

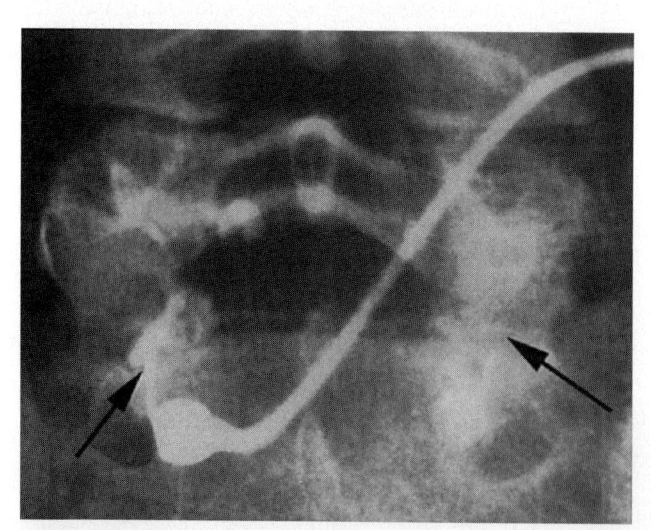

Figure 45–17. Facet joint arthrogram using the inferior recess approach (*arrows*).

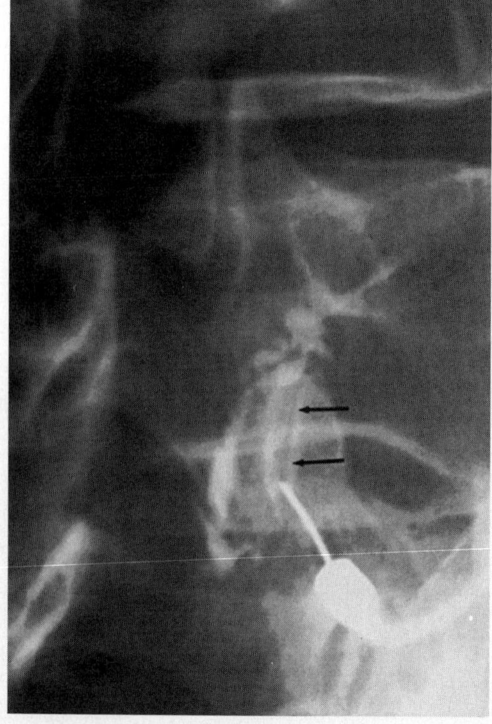

Figure 45–18. Facet joint arthrogram using a direct approach, in which the needle is placed directly into the joint (*arrows*).

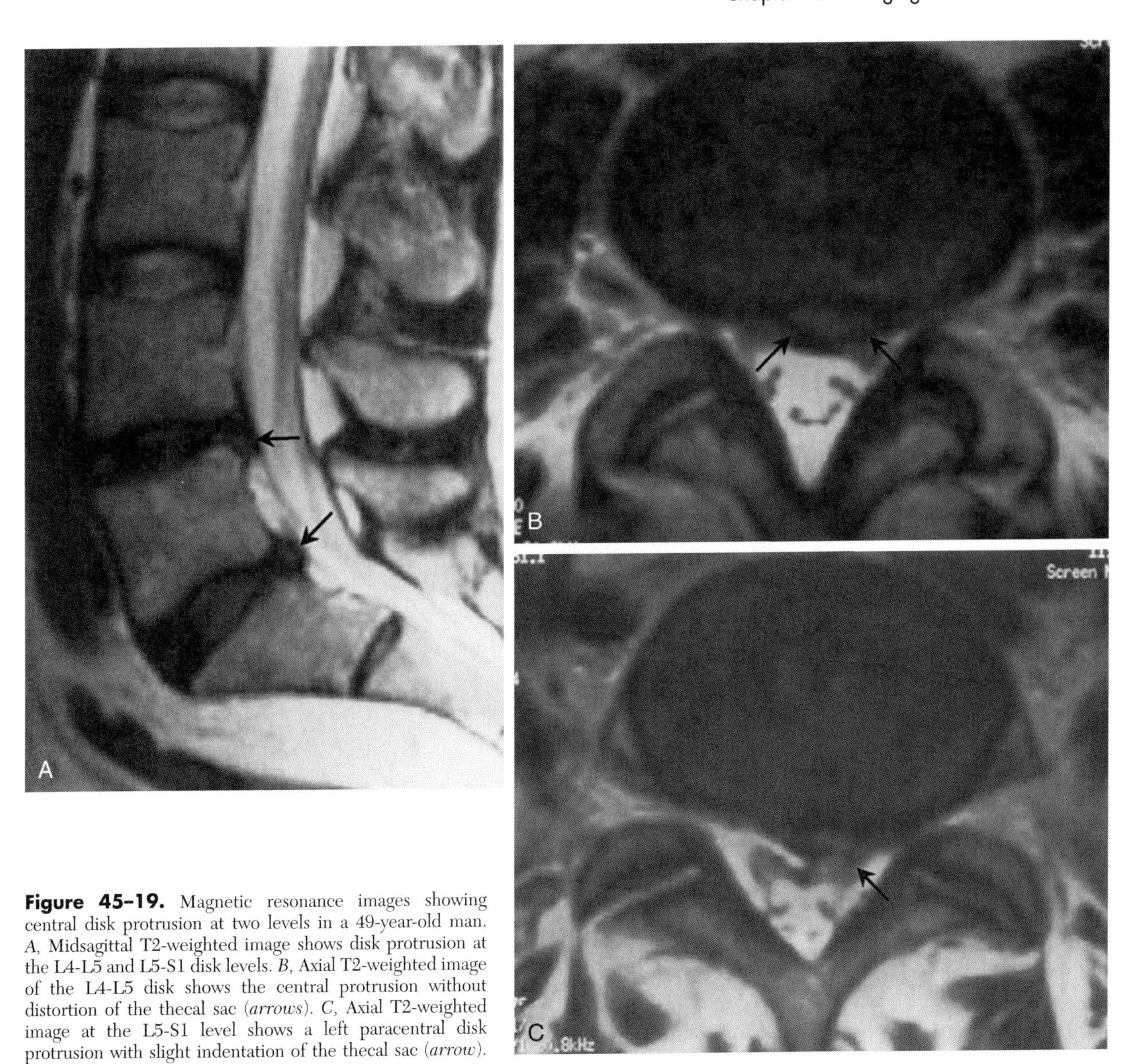

Figure 45–19. Magnetic resonance images showing central disk protrusion at two levels in a 49-year-old man. *A*, Midsagittal T2-weighted image shows disk protrusion at the L4-L5 and L5-S1 disk levels. *B*, Axial T2-weighted image of the L4-L5 disk shows the central protrusion without distortion of the thecal sac (*arrows*). *C*, Axial T2-weighted image at the L5-S1 level shows a left paracentral disk protrusion with slight indentation of the thecal sac (*arrow*).

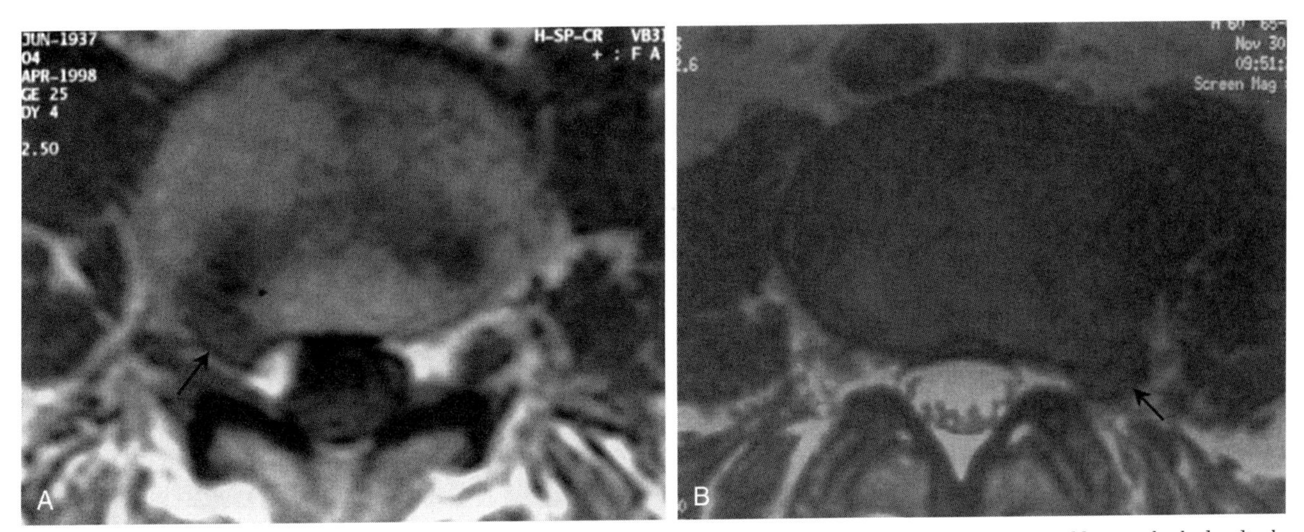

Figure 45–20. Far lateral disk herniation on magnetic resonance imaging. *A*, Far lateral disk protrusion in a 60-year-old man who had radicular symptoms relating to L4 nerve root on the right side. Axial T1-weighted image taken at the L4-L5 disk level shows a far lateral disk protrusion obliterating the neural foramen on the right (*arrow*) and displacing the L4 nerve root. *B*, Far lateral disk protrusion at L4-L5 (*arrow*) in a 32-year-old man presenting with selective L4 nerve root radicular signs and symptoms on the left. This axial T2-weighted image shows focal disk protrusion (*arrow*) just beyond the neural foreman on the left.

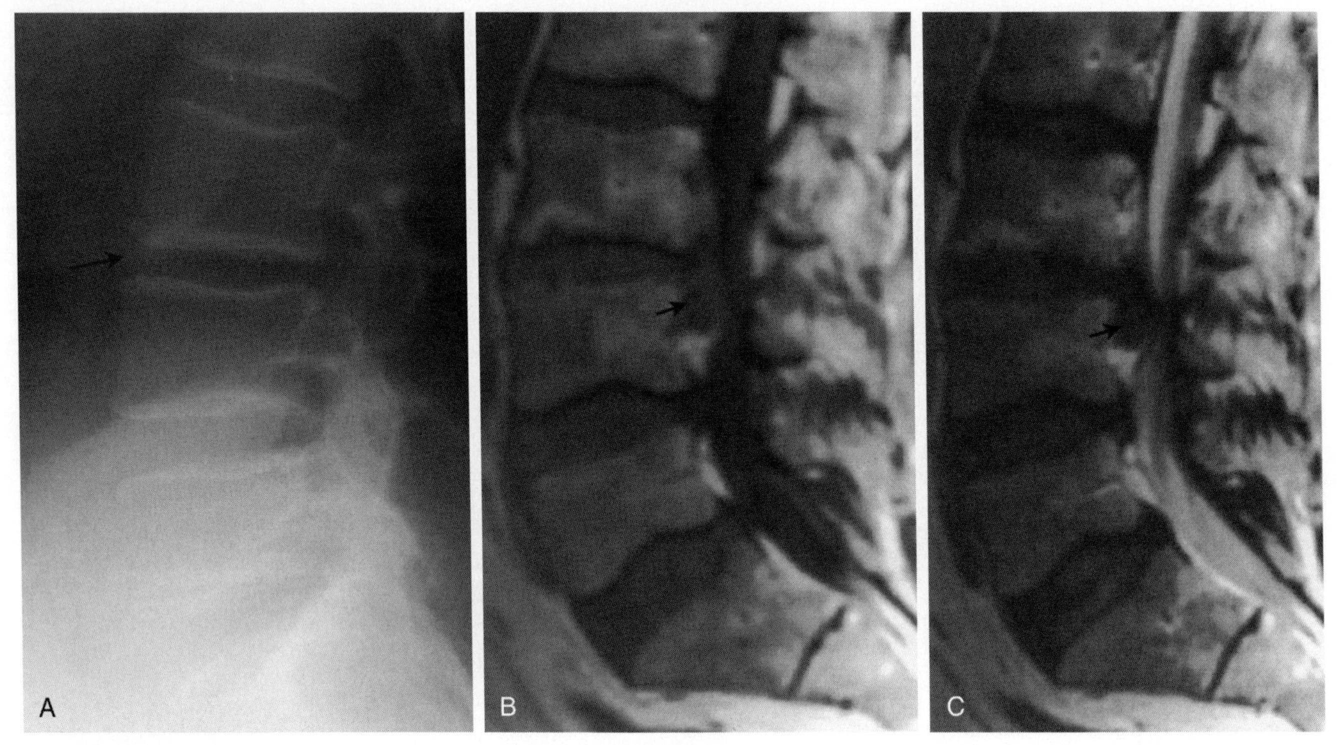

Figure 45–21. Extruded disk fragment in a 56-year-old man presenting with acute low back pain. *A*, Lateral radiograph of the lumbar spine reveals narrowing of the L3-L4 disk (*arrow*). Sagittal T1-weighted (*B*) and T2-weighted (*C*) magnetic resonance images of the lumbar spine show a large extruded disk fragment posterior to the L3-L4 disk and upper L4 body (*arrow*).

with the dorsal root ganglion. Diagnosis of a far-out disk herniation results in modification of the surgical approach.

Cauda Equina Syndrome

The cauda equina consists of peripheral nerves, both motor and sensory, below the level of the conus medullaris and within the spinal canal. About 2% of patients with disk herniation experience severe symptoms, consisting of low back pain, unilateral or bilateral sciatica, motor weakness, urinary or fecal incontinence, and saddle anesthesia. This clinical condition, known as the cauda equina syndrome, is caused by significant compression of the cauda equina, often by a large midline extruded disk, typically at the L4-L5 level. Cauda equina syndrome is probably the only absolute indication for emergent disk excision. Other causes of cauda equina syndrome are spinal trauma and metastatic disease to the spine with epidural invasion.

Anular Tears

Anular tears are frequently seen on MR imaging, but their relationship with LBP is uncertain. Some writers have shown vascular ingrowth and granulation tissue in association with anular tears, suggesting that such tears may play a role in diskogenic pain. These tears are very likely due to mechanical stresses, and they seem to accelerate disk degeneration. T2-weighted MR images and gadolinium-enhanced T1-weighted MR images show increased signal intensity at the site of the anular tear (Fig. 45–25).

Natural History of a Disk Herniation

The majority of herniated disks shrink by more than 75%. The shrinkage occurs in the first month after symptoms start, as a result of hydration followed by quick dehydration (Fig. 45–26). This process explains why some extruded disks appear bright on T2-weighted MR images early in the disease. Only about 8% of disk herniations enlarge (Fig. 45–27). Extrusions and large protrusions have the highest rate of spontaneous shrinkage and symptom improvement.

Internal Disk Disruption

Internal disk disruption is a controversial entity, the proponents of which believe that chronic LBP originates in the disk. The mechanism is thought to be leakage of the nucleus pulposus into the outer anulus or epidural space without frank herniation. The diagnosis is confirmed by abnormal findings on diskography with concordant pain reproduction at the time of the disk injection. Treatment is controversial, but some surgeons advocate spinal fusion.

Imaging Findings versus Clinical Findings

The cause of LBP is often obscure, and in up to 80% of patients, a definitive diagnosis cannot be made. Often, clinical and imaging findings are loosely associated. There

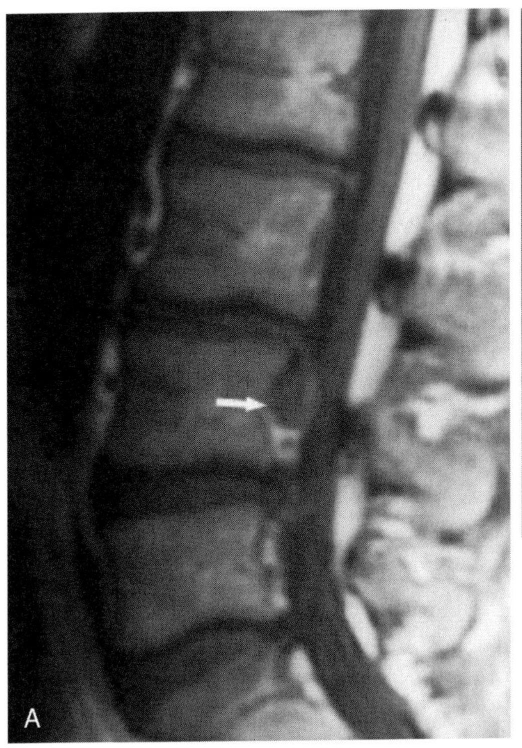

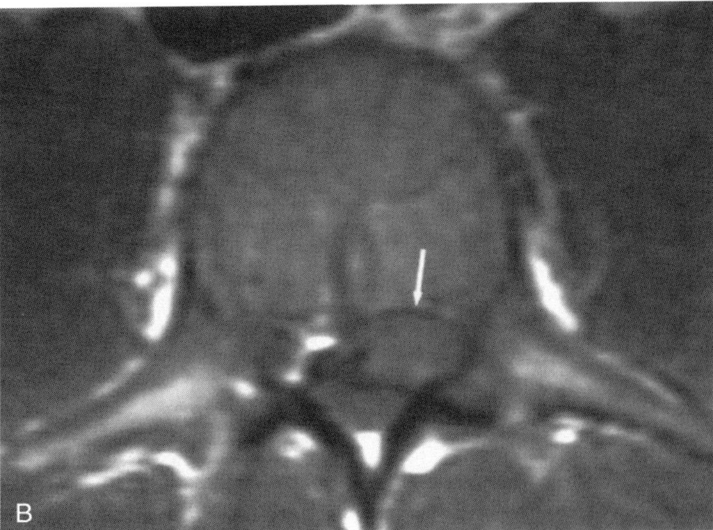

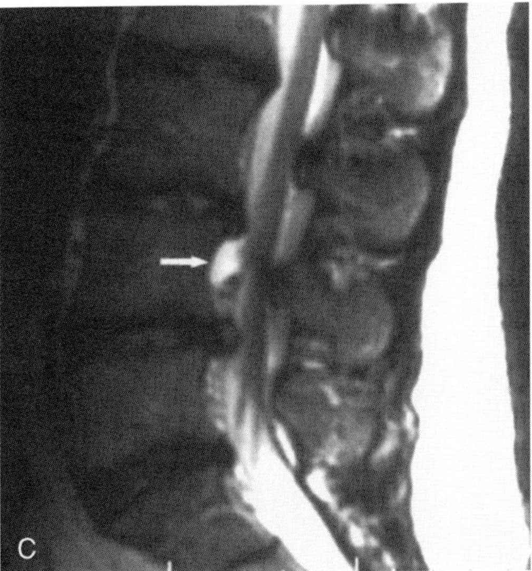

Figure 45–22. Magnetic resonance images showing acute disk extrusion with a hyperbright fragment. Sagittal (*A*) and axial (*B*) T1-weighted images show an extruded disk fragment arising from the L3-L4 disk and lodging behind L4 (*arrow*) on the left. A large extrusion is noted at the L4-L5 level. *C*, Sagittal T2-weighted image demonstrates the extruded fragment to be hyperbright (*arrow*).

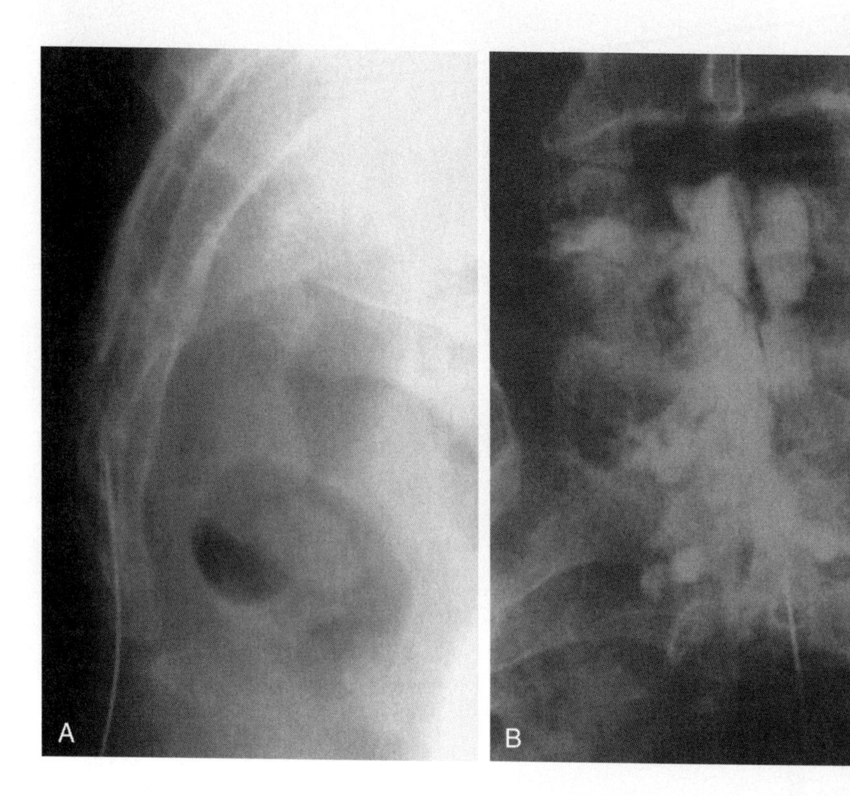

Figure 45–23. Epidural steroid injection performed through the sacral (caudal) hiatus approach. *A*, Lateral radiograph of the sacrum taken after introduction of the needle shows appropriate position of the needle. *B*, Epidurogram performed to document the position of the needle.

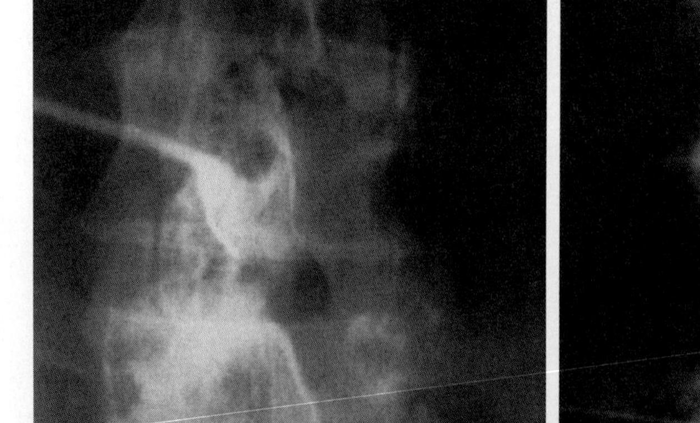

Figure 45–24. *A* and *B*, Radiographs of an epidural steroid injection using the translaminar approach. The lateral view (*B*) confirms that the injection is truly epidural (*arrows*).

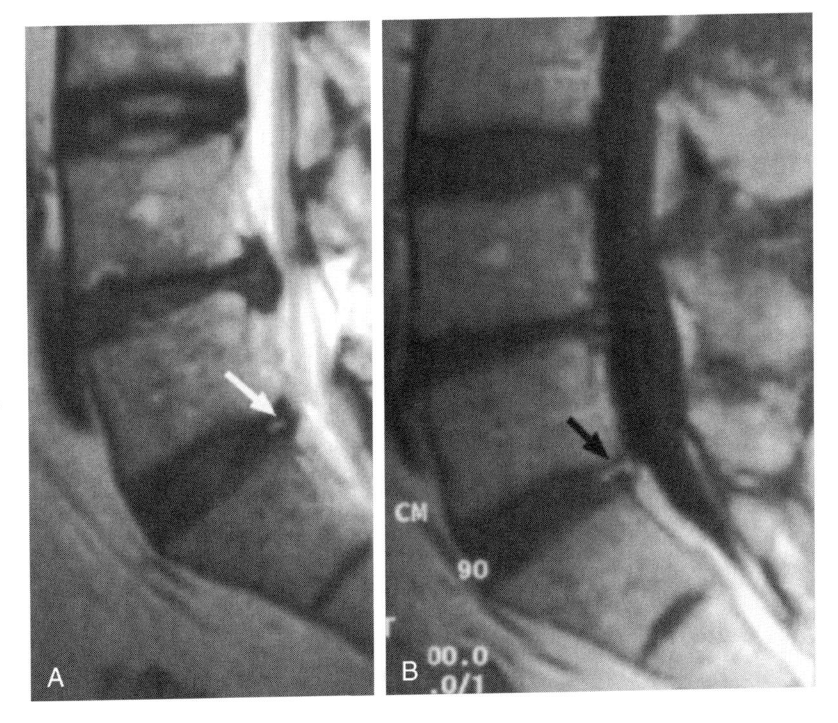

Figure 45–25. Magnetic resonance images of anular tears in a 48-year-old man. *A,* Sagittal T2-weighted image shows an anular tear located posteriorly in the L5-S1 disk (*arrow*). At the L4-L5 level there is central disk protrusion. *B,* T1-weighted image obtained after a gadolinium injection demonstrates enhancement of the annular tear (*arrow*).

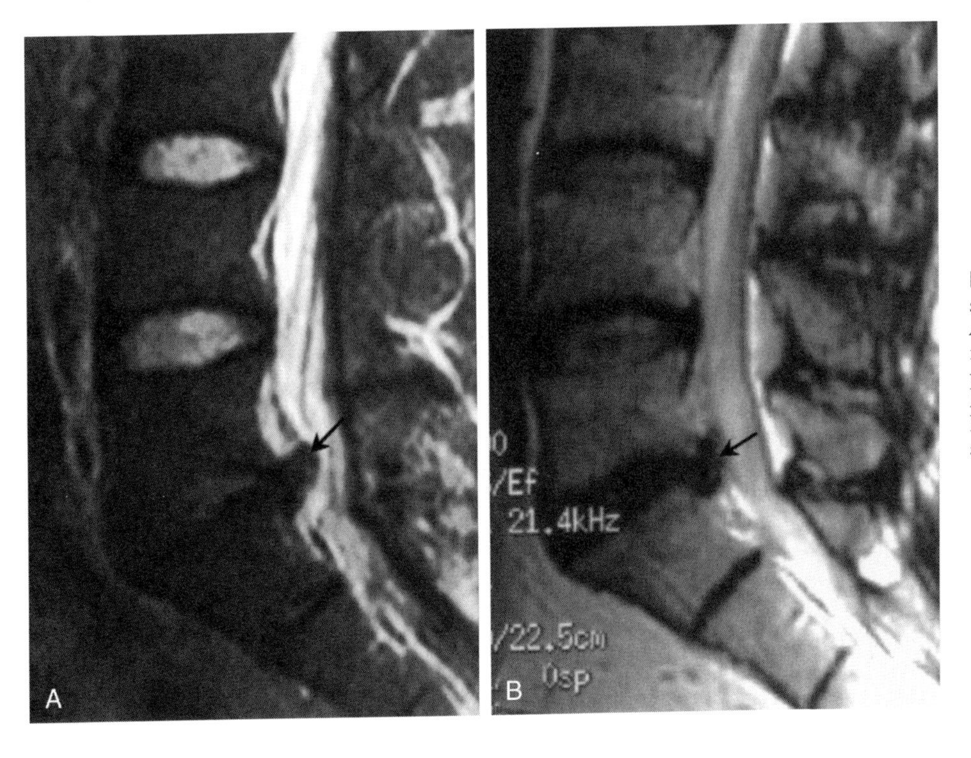

Figure 45–26. Large disk protrusion that is getting smaller with time. *A,* Midsagittal T2-weighted magnetic resonance image shows a large disk protrusion at the L5-S1 level (*arrow*). *B,* A second image obtained 9 months later shows significant reduction in the size of the protrusion (*arrow*).

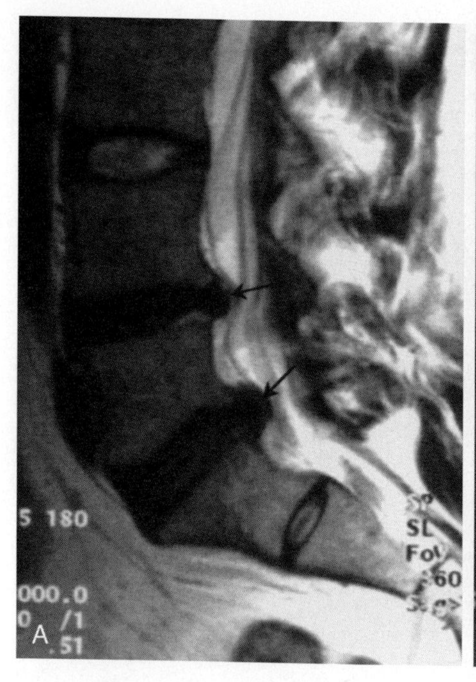

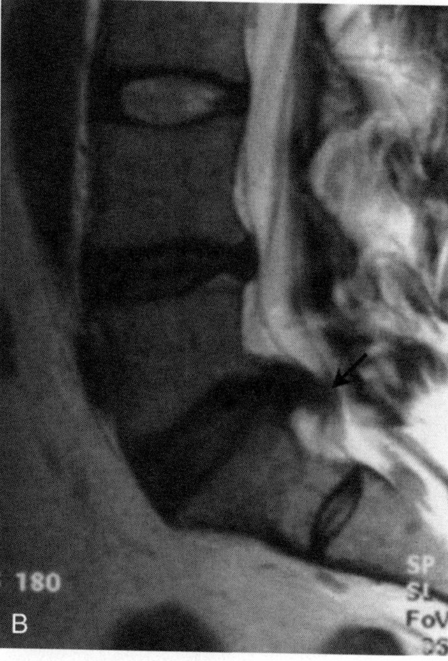

Figure 45–27. Disk herniation in a 39-year-old man that is enlarging with time. A, Midsagittal T2-weighted magnetic resonance image shows disk protrusion at L4-L5 and L5-S1 disk levels (*arrows*). *B*, Six months later, the symptoms became progressively worse, and an MR image obtained at this time shows that the protrusion at L5-S1 had become a large extruded fragment (*arrow*).

is ample evidence from multiple studies that altered morphology does not always produce symptoms. Disk bulging, disk protrusion, and anular tears are common findings in asymptomatic individuals. Nevertheless, large compressive lesions (i.e., disk extrusions) are almost always symptomatic.

DISK SURGERY

Determinants of Successful Surgery

The best surgical results are achieved when (1) a well-defined disorder is demonstrated by imaging studies and (2) the imaging findings are tightly correlated with the clinical history and physical findings. It is important to remember that incidental imaging findings unrelated to the patient's symptoms are very common.

Imaging in the Immediate Postoperative Period

In the immediate postoperative period, MR imaging studies should be avoided, because the findings are confusing and difficult to interpret; they can resemble those in recurrent or residual disk herniation. A waiting period of 2 to 3 months should pass before postoperative MR imaging is attempted.

Failed Back Syndrome

About 300,000 first-time laminectomies are performed every year in the United States. Approximately 15% of the recipients of such procedures continue to have pain and disability. The most common causes of failed back syndrome are recurrent or residual disk herniation and epidural scarring; less common causes are postoperative diskitis (see Chapter 46), arachnoiditis, and instability.

Scar versus Recurrent or Residual Disk

Epidural scarring and adhesions can cause compression and tethering of the nerve roots. Nerve roots and dorsal ganglia are particularly sensitive to mechanical deformation. Nerves encased by a scar suffer from impairment of nutritional support and exoplasmic transport. Surgeons have been using a variety of materials at surgery to inhibit epidural scarring. Although scarring does not always produce pain, there is a high correlation between the extent of scarring and the severity of recurrent radicular pain. Ross and associated (1990) found that patients with extensive scarring are two to three times more likely to experience recurrent radicular pain than those with less extensive scarring. In fact, epidural scarring is responsible for 24% of all failed back operations. Reoperation solely to remove an epidural scar is contraindicated. Surgery in patients with epidural scarring is more demanding, raises the risk of dural tears, and results in more scarring.

Visualization of epidural scars and differentiation between epidural scar and a disk fragment in the epidural space can be highly accurate with gadolinium-enhanced MR imaging. A scar enhances immediately after an intravenous gadolinium injection and typically shows no mass effect (Figs. 45–28 and 45–29). A recurrent or residual disk shows a mass effect, displacing and compressing the thecal sac and nerve roots (Fig. 45–30). It is usually discrete and well marginated, but it can occasionally be fragmented and enmeshed with a scar. Recurrent

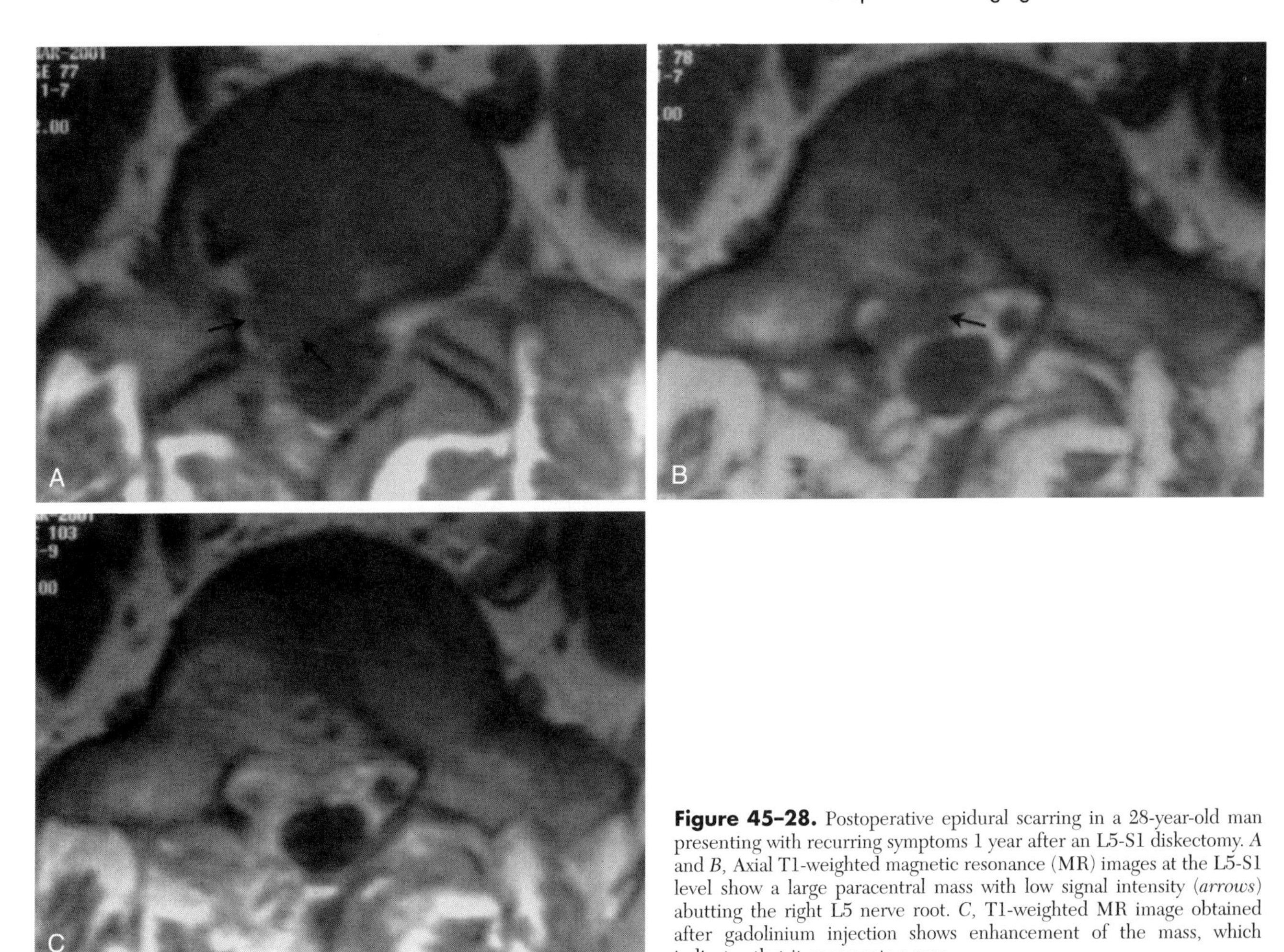

Figure 45–28. Postoperative epidural scarring in a 28-year-old man presenting with recurring symptoms 1 year after an L5-S1 diskectomy. *A* and *B*, Axial T1-weighted magnetic resonance (MR) images at the L5-S1 level show a large paracentral mass with low signal intensity (*arrows*) abutting the right L5 nerve root. *C*, T1-weighted MR image obtained after gadolinium injection shows enhancement of the mass, which indicates that it represents a scar.

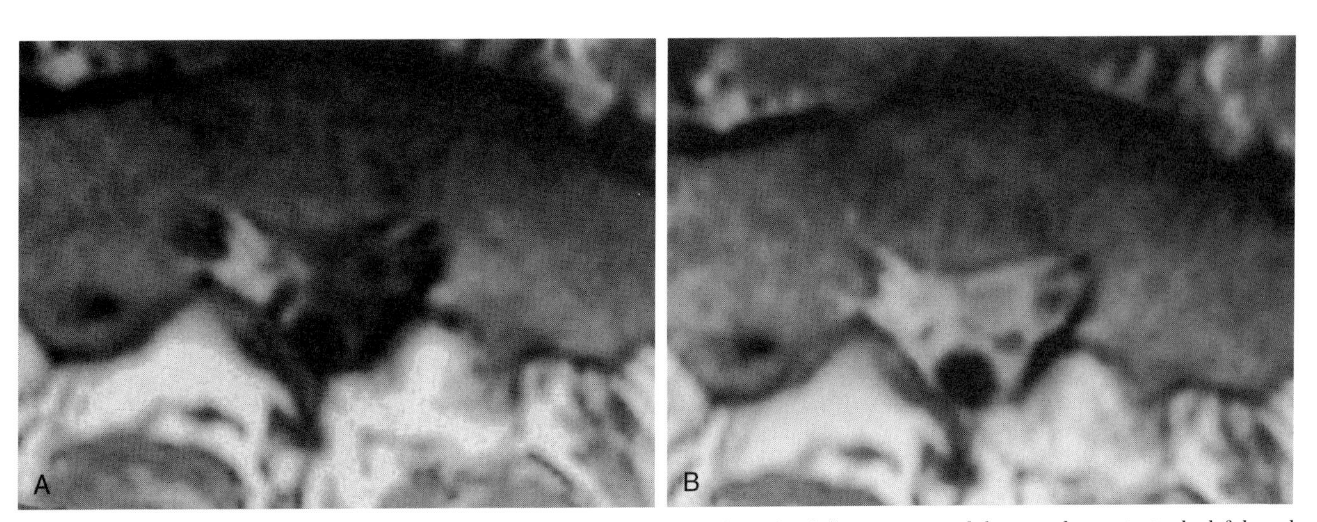

Figure 45–29. Postoperative epidural scarring. *A*, Axial T1-weighted image at the S1 level shows a mass with low signal intensity in the left lateral recess with extension to midline. *B*, A T1-weighted image with gadolinium enhancement taken at the same level as *A* shows appreciable enhancement of the epidural mass. This finding, in the proper clinical setting, makes the diagnosis of postoperative epidural scar almost certain.

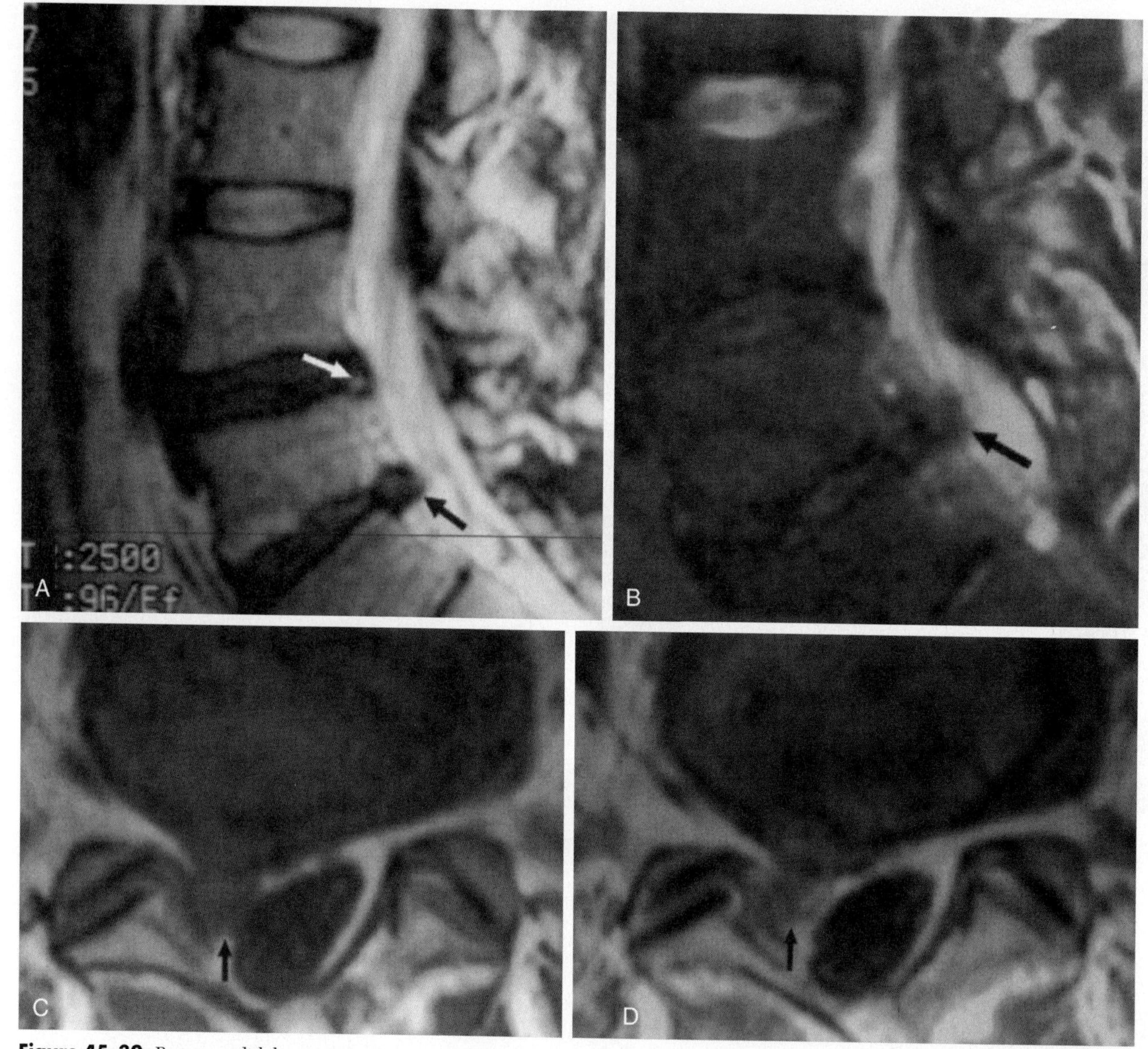

Figure 45–30. Recurrent disk herniation in a 34-year-old man who presented with recurring symptoms 4 months after surgery for disk herniation. *A,* Sagittal T2-weighted magnetic resonance (MR) image obtained at initial presentation shows a large disk protrusion at the L5-S1 level (*black arrow*); this disk was surgically resected. An anular tear is incidentally noted at the L4-L5 level (*white arrow*). *B,* Four months after surgery, the patient's symptoms recurred, and a sagittal T2-weighted MR image shows findings suggestive of recurrent disk at the L5-S1 level (arrow). MR images obtained before (*C*) and after (*D*) gadolinium injection show only mild peripheral enhancement, confirming the presence of a recurrent disk extrusion (*arrows*) at the L5-S1 level.

or residual disk does not enhance initially, although it could be surrounded with an enhancing scar, which would enhance immediately after gadolinium injection (Fig. 45–31).

Postoperative Nerve Root Enhancement with Gadolinium

Enhancement of a nerve root after gadolinium injection is occasionally noted after surgery, but its significance is uncertain. Some surgeons believe that it represents hyperemia due to previous disk herniation and nerve root compression (see Fig. 45–31).

Arachnoiditis

Arachnoiditis is a rare postoperative complication. It is diagnosed on MR imaging or CT-myelography through the demonstration of either clumping of the nerve roots or formation of adhesions between the nerve roots and thecal sac (Fig. 45–32).

SPINAL STENOSIS

Spinal stenosis is defined as narrowing of the spinal canal. The narrowing can involve one or multiple motion segments.

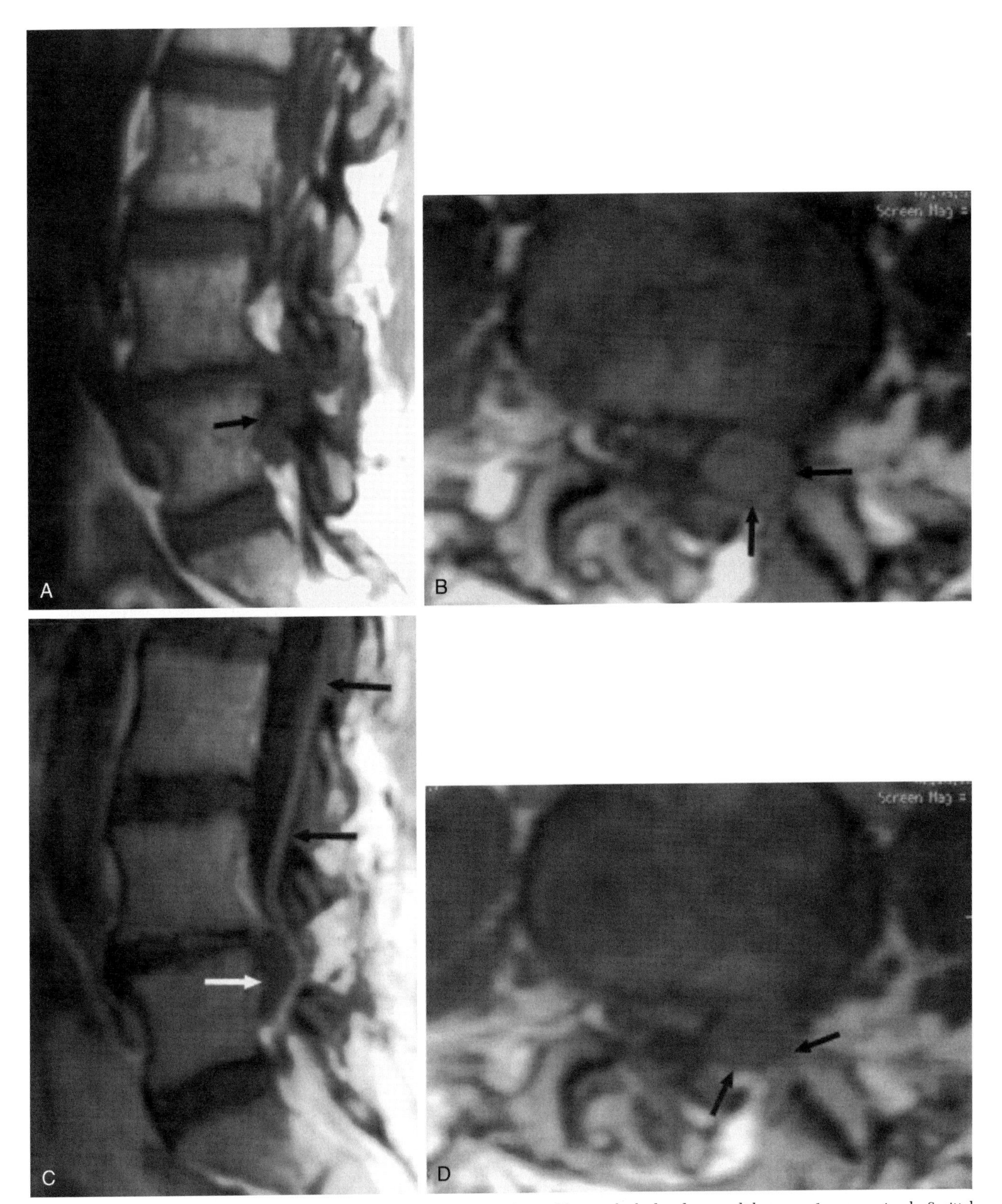

Figure 45–31. Recurrent disk extrusion and enhancing nerve roots in a 45-year-old man who had undergone disk surgery 1 year previously. Sagittal (*A*) and axial (*B*) T1-weighted images taken 1 year after the disk operation show a large mass posterior to the L5 vertebral body (*arrows*) on the left. The L4-L5 disk is narrow, and there is a laminectomy defect on the left side. Sagittal (*C*) and axial (*D*) T1-weighted images obtained after a gadolinium injection show no enhancement of the mass (*white arrow*), a feature confirming that the mass represents a recurrent disk extrusion and not a scar. Note the enhanced spaghetti-like structure on the sagittal image (*C, black arrows*), which represents an enhancing nerve root.

Illustration continued on following page

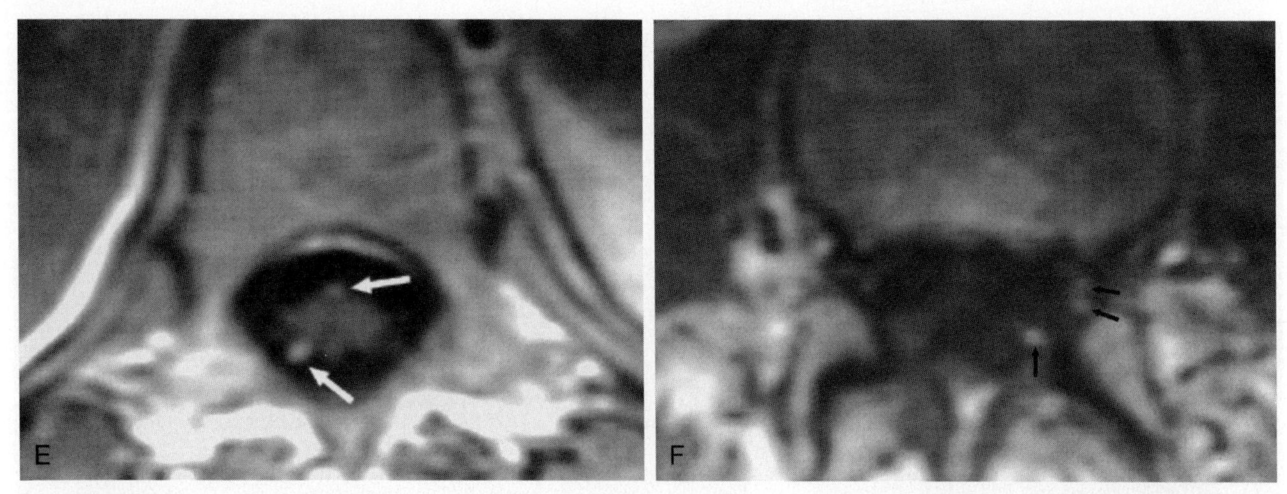

Figure 45–31 *Continued.* Axial T1-weighted images taken just above (*E*) and just below (*F*) the conus after gadolinium injection show enhancing nerve roots (*arrows*).

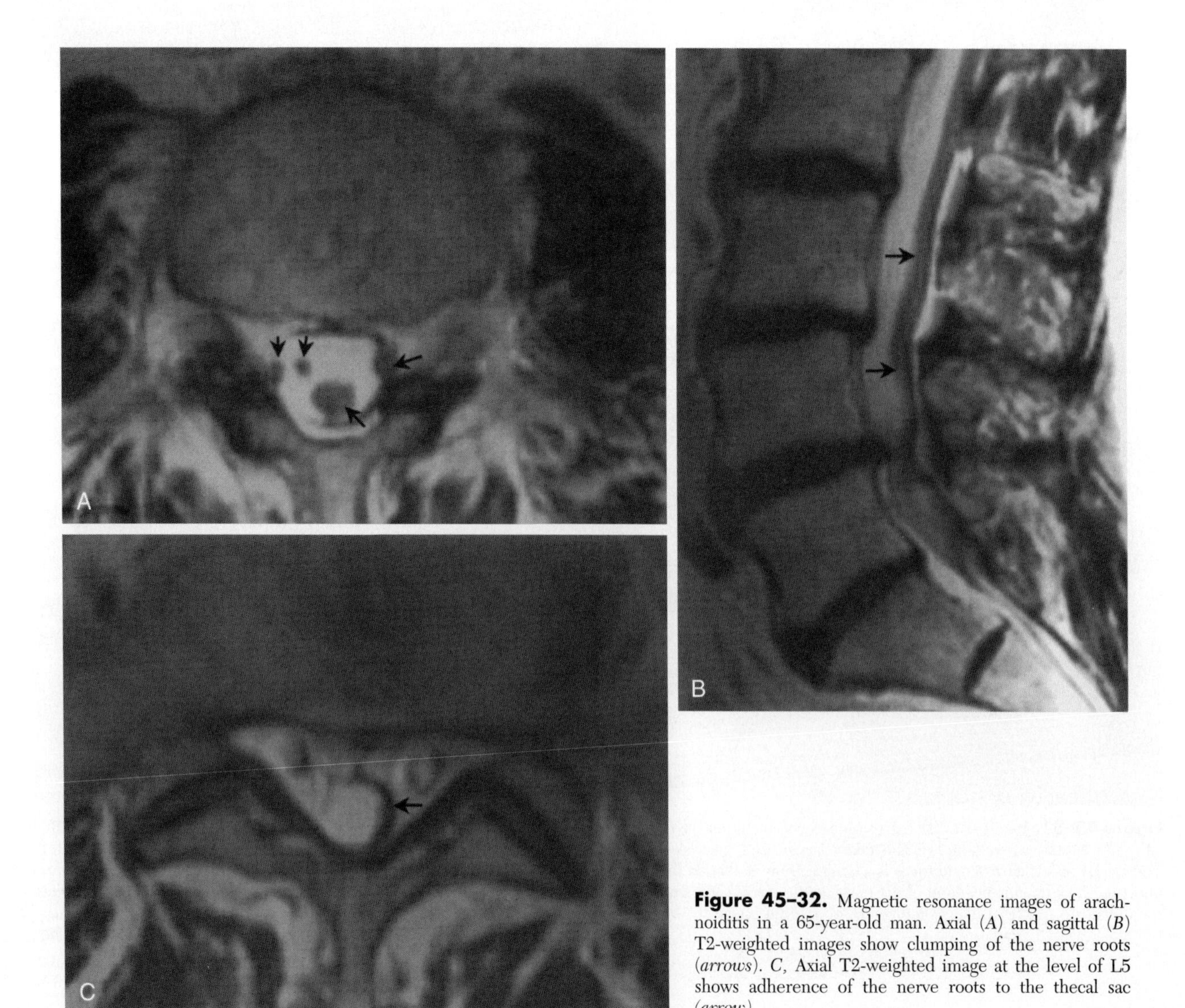

Figure 45–32. Magnetic resonance images of arachnoiditis in a 65-year-old man. Axial (*A*) and sagittal (*B*) T2-weighted images show clumping of the nerve roots (*arrows*). *C,* Axial T2-weighted image at the level of L5 shows adherence of the nerve roots to the thecal sac (*arrow*).

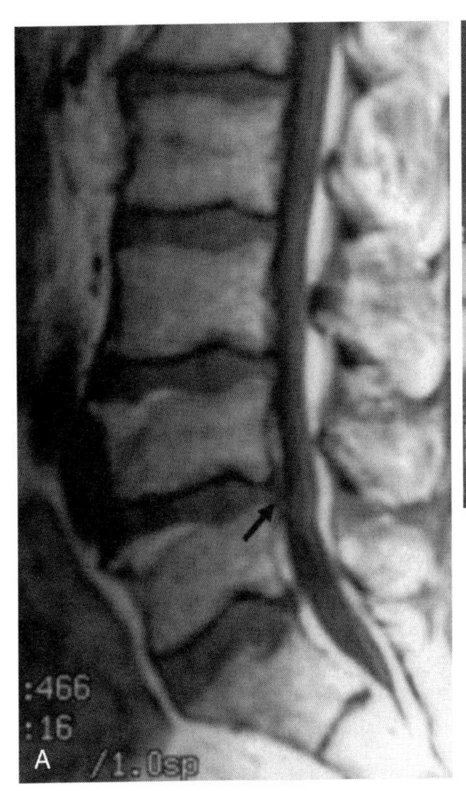

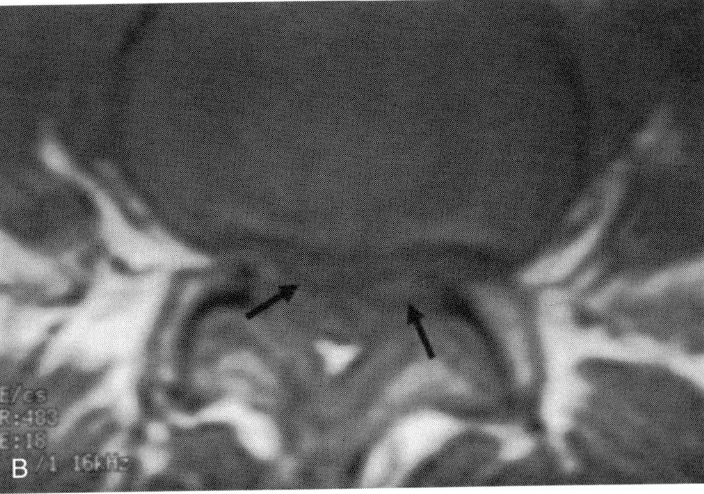

Figure 45–33. Congenital (idiopathic) spinal stenosis. A, Sagittal T1-weighted magnetic resonance (MR) image shows a uniformly narrow spinal canal. The thecal sac is squeezed into a thin ribbon. In addition, there is a central disk protrusion at the L4-L5 level (*arrow*). *B,* Axial T1-weighted MR image shows the disk protrusion at L4-L5 (*arrows*); this makes the spinal stenosis more pronounced.

Classification of Spinal Stenosis

Grabias (1980) has classified spinal stenosis as follows:

1. Congenital or developmental:
   a. Achondroplastic.
   b. Idiopathic (Fig. 45–33).
2. Acquired:
   a. Degenerative (Fig. 45–34).
   b. Combined (degenerative and idiopathic).
   c. Degenerative spondylolisthesis (Fig. 45–35).
   d. Traumatic.
   e. Postfusion.
   f. Miscellaneous (Paget's disease, acromegaly).

The most common types of spinal stenosis are the degenerative type and the spinal stenosis associated with degenerative spondylolisthesis (discussed later in this chapter). Degenerative stenosis is typically most pronounced at the L3-L4 and L4-L5 levels. The narrowing in

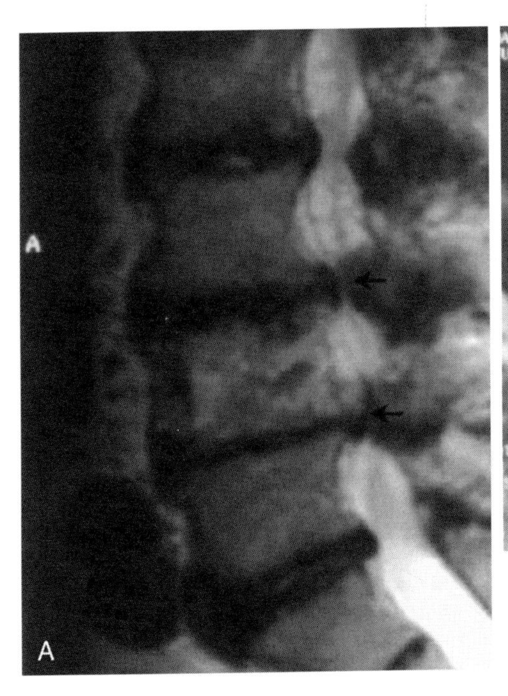

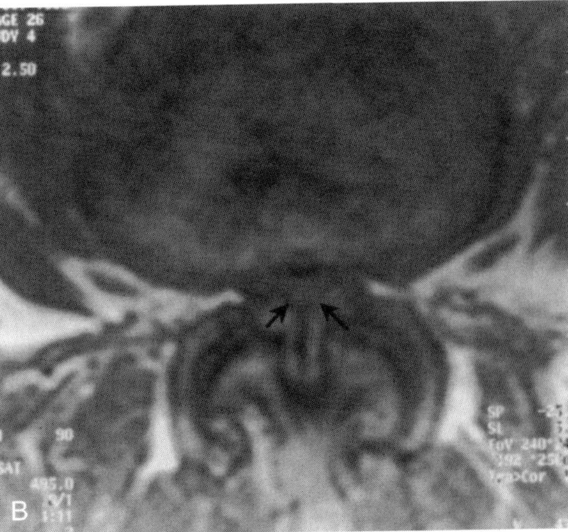

Figure 45–34. Severe degenerative spinal stenosis in a 73-year-old man with symptoms of neurogenic claudication. *A,* Midline sagittal T2-weighted magnetic resonance (MR) image of the lumbar spine reveals severe spinal stenosis at the L3-L4 and L4-L5 levels (*arrows*). *B,* Axial T-weighted MR image at the L4-L5 level shows facet joint osteoarthritis and thickened ligamentum flavum compressing the thecal sac (*arrows*).

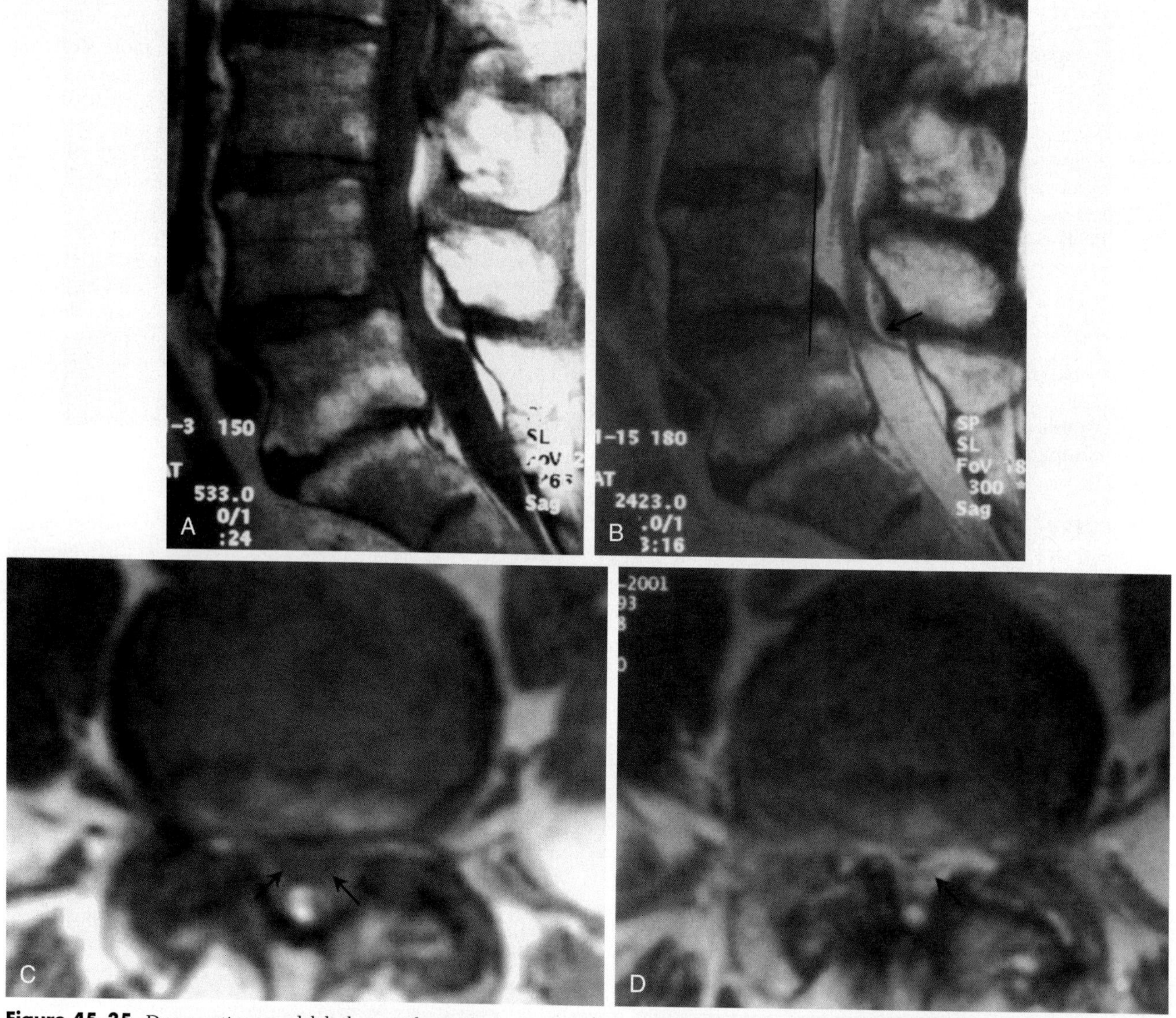

Figure 45–35. Degenerative spondylolisthesis with spinal stenosis in a 66-year-old woman complaining of symptoms of neurogenic claudication. T1-weighted (*A*) and T2-weighted (*B*) magnetic resonance (MR) images through the midsagittal plane. There is about 15% spondylolisthesis of L4 on L5 (a straight line has been drawn on the image to illustrate abnormal alignment) along with significant spinal stenosis. Axial T1-weighted (*C*) and T2-weighted (*D*) MR images taken at the level of the stenosis show the thecal sac compressed (*arrows*) by the hypertrophied, degenerated facets and thick ligamenta flava.

degenerative spinal stenosis can be seen at the spinal canal (central stenosis) and in the lateral recesses or neural foramina, but invariably, the stenoses at these sites progress simultaneously.

Anatomic Changes in Degenerative Spinal Stenosis

Disk degeneration is usually the first step in the degenerative process of the spine. It manifests as disk space narrowing, disk bulging, and marginal osteophyte formation (see Fig. 45–34).

Facet degenerative disease or osteoarthritis is similar to osteoarthritis of other synovial joints. It starts with erosions at the articular cartilage and loss of joint space, osteophyte formation, and facet hypertrophy (see Fig. 45–34).

Buckling or redundancy and thickening of the ligamentum flavum is due to loss of disk height and settling of the facet joints (see Fig. 45–34).

Imaging of Degenerative Spinal Stenosis

Plain radiography is essential in any patient in whom degenerative spinal stenosis is suspected, although by itself, plain radiography does not provide diagnostic information. Degenerative disease of the disks and facets is often observed on plain radiographs.

Magnetic Resonance Imaging

MR imaging is very accurate in showing the level of stenosis. It helps in surgical planning because it can clearly demonstrate the central stenosis, lateral recess, and neural canal stenosis (see Fig. 45–34). The relationships of the compressing components—the bulging disk, the hypertrophied facet, and the ligamentum flavum buckling—with the thecal sac are also well delineated on MR imaging (see Fig. 45–34).

Computed Tomography and Myelography

CT-myelography is not recommended for the routine evaluation of patients with spinal stenosis, although some surgeons still prefer to use CT-myelography for preoperative planning in such patients.

SPONDYLOLYSIS AND SPONDYLOLISTHESIS

Definition

Spondylolysis is also known as a pars defect or a discontinuity in the pars interarticularis. *Spondylolisthesis* is defined as the forward displacement of part or all of one vertebra on another. (The term is derived from Greek words "spondylos," meaning vertebra, and "olisthesis," meaning to slip or slide.)

Classification of Spondylolisthesis

The most widely used classification of spondylolisthesis is the one advanced by Wiltse and coworkers (1981) as follows:

1. Dysplastic or congenital.
2. Isthmic:
   a. Lytic.
   b. Elongated, but intact pars.
   c. Acute fracture of the pars.
3. Degenerative.
4. Post-traumatic.
5. Pathologic.

Isthmic lytic spondylolisthesis and degenerative spondylolisthesis are the most common types.

Dysplastic Spondylolisthesis

In dysplastic spondylolisthesis, there is congenital dysplasia of the upper sacrum or neural arch of L5, resulting in an ineffective locking mechanism between L5 and the sacrum (Fig. 45–36). The body weight produces a forward thrust and gradual slippage of L5 on the sacrum. The pars interarticularis of L5 remains intact, and therefore, slips of the dysplastic type exceeding 35% invariably compress the cauda equina. There is a strong hereditary influence in this type of disease.

Isthmic Lytic Spondylolisthesis

Isthmic lytic spondylolisthesis is the most common variety among young patients. The pars defect is not seen in infants. Many writers now believe that spondylolysis is due to a stress fracture of the pars interarticularis. The pars defect or spondylolysis typically affects the L5 vertebra (Fig. 45–37). It is almost never seen in children before they start to walk; the prevalence then starts to increase until it reaches 6% in the general population.

Symptoms are relatively uncommon with this abnormality, and only a minority of cases progress to severe spondylolisthesis. With unilateral spondylolysis, there is typically no associated spondylolisthesis. Boys are twice as frequently affected as girls, but girls are more prone to severe displacement and tend to have more symptoms than boys. There is a higher incidence of spondylolysis in individuals who perform heavy physical activities or whose activities require them to stand up from a stooping position, such as weight lifters, loggers, gymnasts, and football players, especially linemen. Certain populations, such as Eskimos, have a much higher incidence of spondylolysis than others.

In acute spondylolysis, the edges of the defect are irregular and the gap is narrow. In 20% to 25% of these early cases, the pars defect is on one side, and such defects are more likely to heal (Fig. 45–38). Persistent unilateral pars defect, over time, leads to contralateral pedicle and arch hypertrophy (Fig. 45–39). Slippage or spondylolisthesis progresses more frequently during periods of rapid growth, and progression is seldom seen after age 25 years.

Two forms of slipping occur with spondylolisthesis, tangential slipping (Fig. 45–40) and angular slipping (Fig. 45–41; also see Fig. 45–43*A*, *B*). Tangential slipping merely represents forward translation of L5 on S1 (see Fig. 45–40). Angular slipping involves rotation of L5 with respect to the sacrum (see Fig. 45–41; also see Fig. 45–43 *A*, *B*). This results in localized kyphosis at the lumbosacral junction and occurs only when the tangential slip exceeds 50%. Adoptive morphologic changes occur in the body of L5 and the superior surface of the first sacral segment, which are seen with the more severe grades of spondylolisthesis. Typically, the body of L5 assumes a trapezoidal shape, and the superior surface of S1 becomes rounded in contour (see Fig. 45–36*A*).

Spondylolysis and spondylolisthesis are, in the majority of cases, diagnosable on plain radiographs. Oblique projections are not routinely required. In early cases, spondylolysis maybe suspected clinically but cannot be confirmed radiographically; in such situations, a bone scan with SPECT may be necessary (Fig. 45–42). Positive findings of bone scan or SPECT in patients who have a short clinical history suggests repair activity at the fracture site. Later, bone scans are helpful in distinguishing between patients with long-standing or established nonunion, in whom the bone scan is normal, and those in whom healing is still progressing. On MR imaging, a fibrocartilaginous mass can be identified in the majority of patients at the sites of the spondylolysis. In about 21% of such patients, the mass compresses the thecal sac.

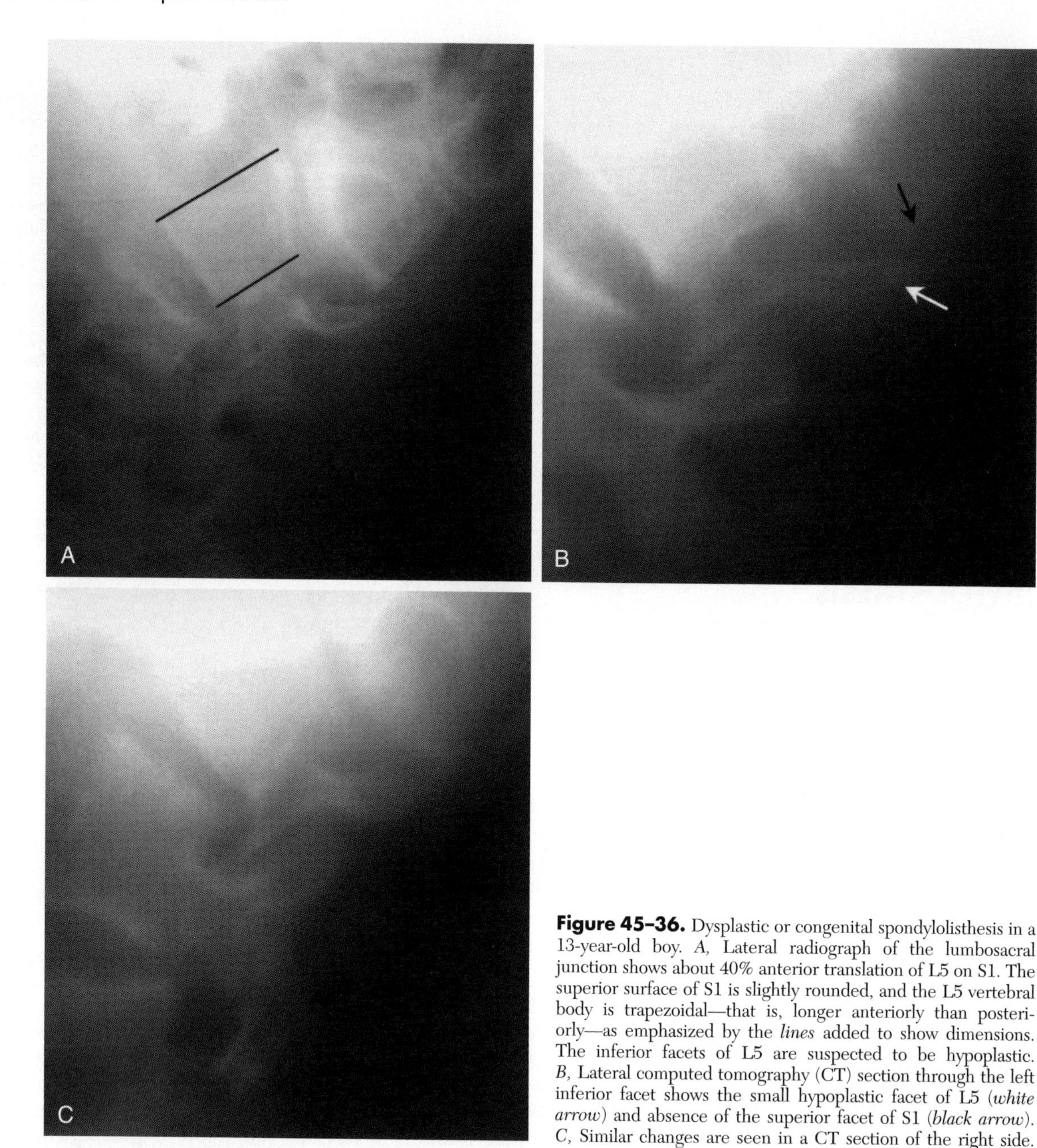

Figure 45–36. Dysplastic or congenital spondylolisthesis in a 13-year-old boy. *A*, Lateral radiograph of the lumbosacral junction shows about 40% anterior translation of L5 on S1. The superior surface of S1 is slightly rounded, and the L5 vertebral body is trapezoidal—that is, longer anteriorly than posteriorly—as emphasized by the *lines* added to show dimensions. The inferior facets of L5 are suspected to be hypoplastic. *B*, Lateral computed tomography (CT) section through the left inferior facet shows the small hypoplastic facet of L5 (*white arrow*) and absence of the superior facet of S1 (*black arrow*). *C*, Similar changes are seen in a CT section of the right side.

Elongated but Intact Pars

Wiltse (1981) believes elongated but intact pars to be fundamentally the same disease as isthmic lytic spondylolisthesis. He speculates that repeated microfractures occur in the pars, which are allowed to heal in an elongated position as the body of L5 slides forward. Cases described in the literature with elongated pars have all been observed at the L5 level (Fig. 45–43).

Acute Pars Fractures

A very rare type of spondylolisthesis, acute pars fracture should not to be confused with traumatic spondylolisthesis. Acute fracture of the pars is caused by severe trauma.

Degenerative Spondylolisthesis

Degenerative spondylolisthesis is probably the most common type of spondylolisthesis. It usually occurs in women older than 50 years and is five to six times more common in women than in men. The slip usually occurs at the L4-L5 level but is also seen at other levels. The slip does not exceed 25%, but because the pars interarticularis is intact, spinal stenosis is associated with the slip (Fig. 45–44; see Fig 45–35). The etiology of degenerative

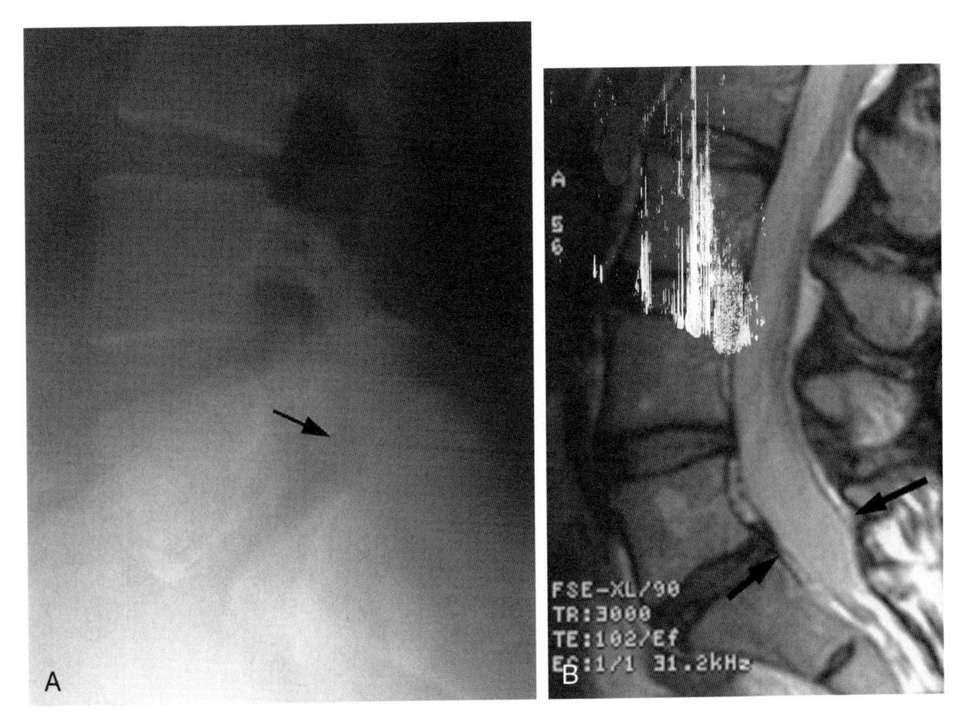

Figure 45–37. Isthmic lytic spondylolisthesis resulting in a wide spinal canal. *A,* Lateral radiograph of the lumbosacral junction shows grade I spondylolisthesis. *Arrow* points to the pars interarticularis defect. *B,* Sagittal T2-weighted magnetic resonance image of the lumbar spine shows a wide spinal canal at the lumbosacral junction (*arrows*).

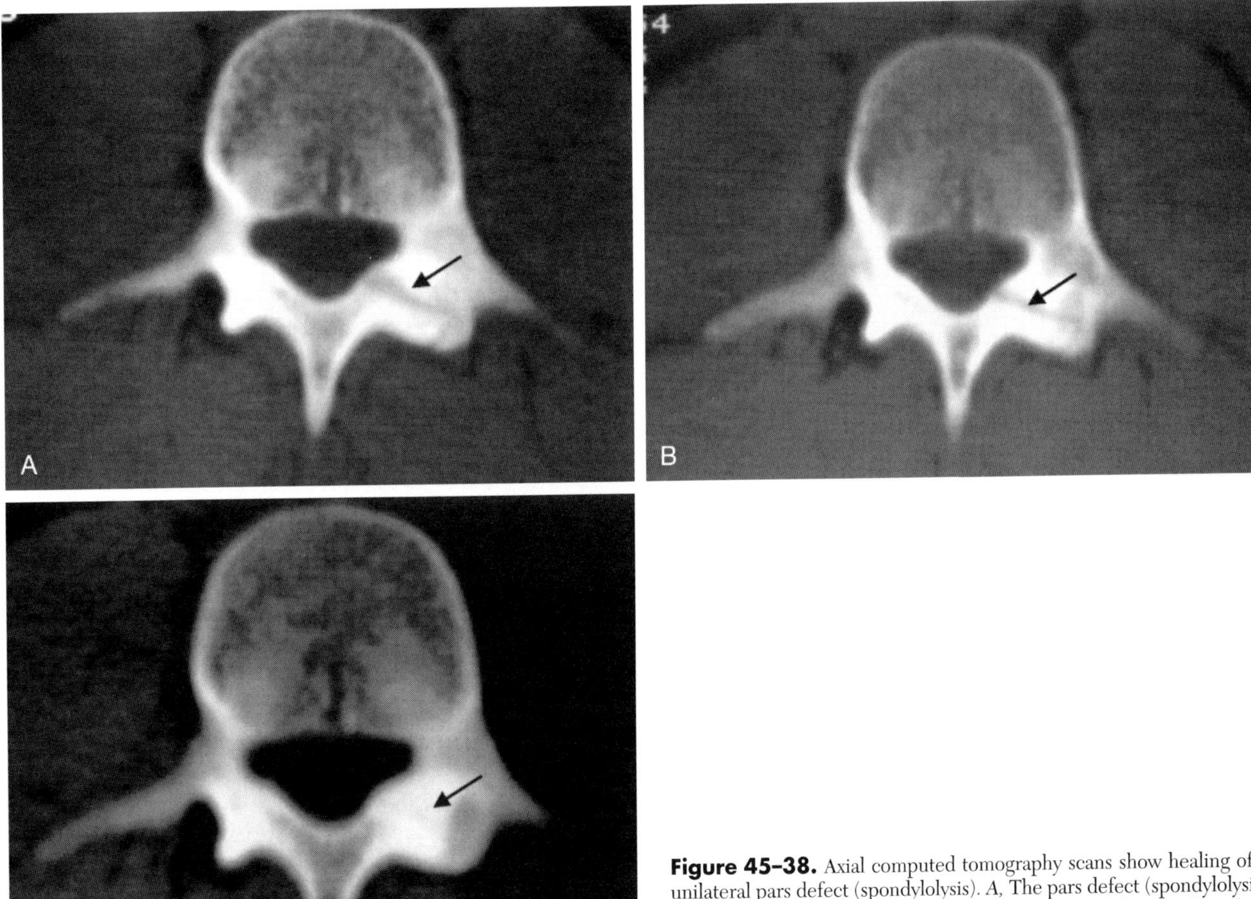

Figure 45–38. Axial computed tomography scans show healing of a unilateral pars defect (spondylolysis). *A,* The pars defect (spondylolysis) is seen on the left (*arrow*). *B,* Five weeks later, the defect is still evident (*arrow*). *C,* Five months later, the pars defect has almost completely healed (*arrow*).

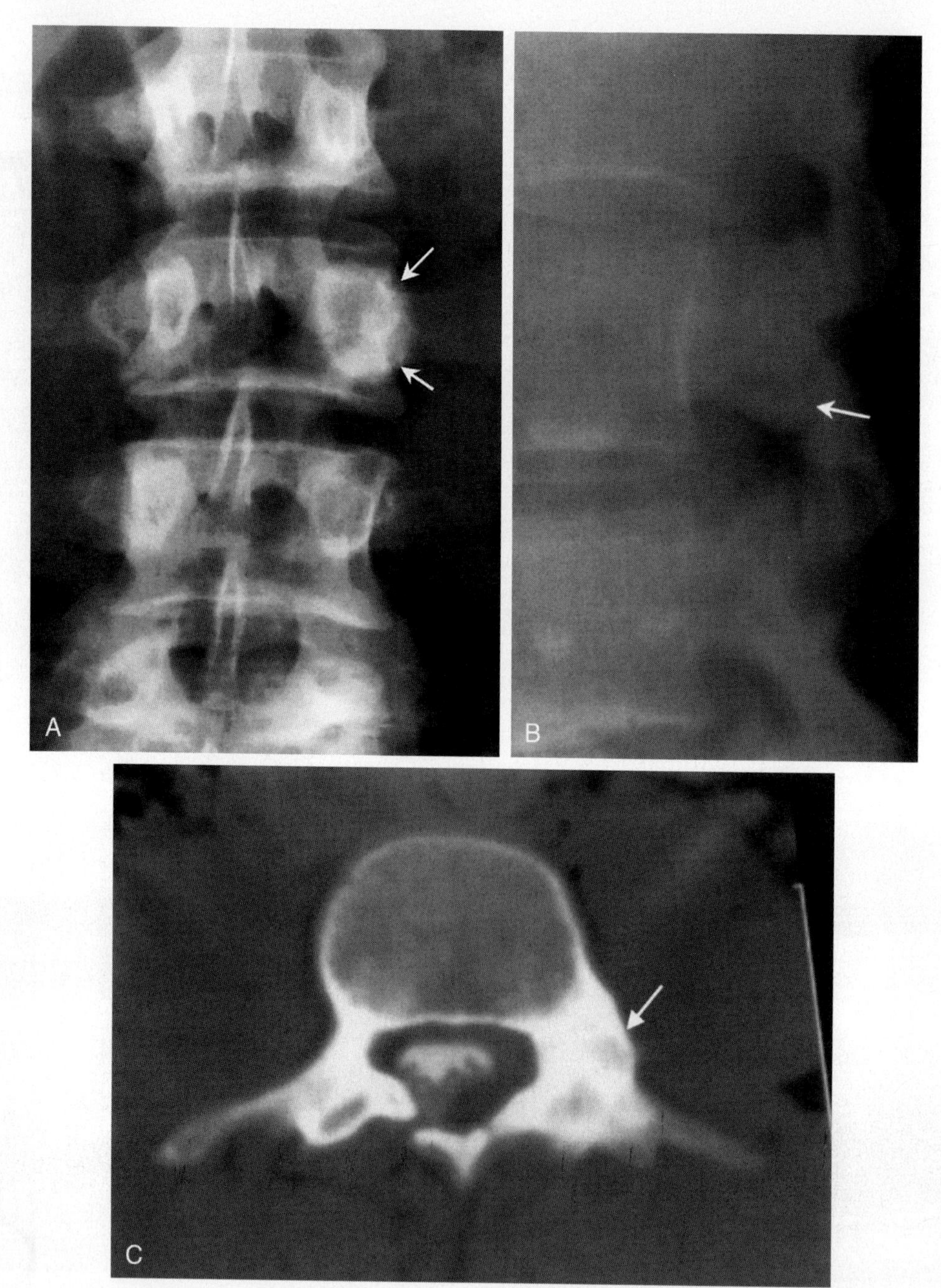

Figure 45–39. Unilateral spondylolysis with contralateral pedicle and arch hypertrophy in a 22-year-old man. *A,* Anteroposterior radiograph of the lumbar spine reveals hypertrophy of the left pedicle in L3 (*arrows*). *B,* Lateral radiograph of the same area shows a pars defect (spondylolysis) in L3 (*arrow*). *C,* Computed tomography section through the arch of L3 demonstrates a pars defect on the right and compensatory hypertrophy of the pedicle and arch on the left (*arrow*).

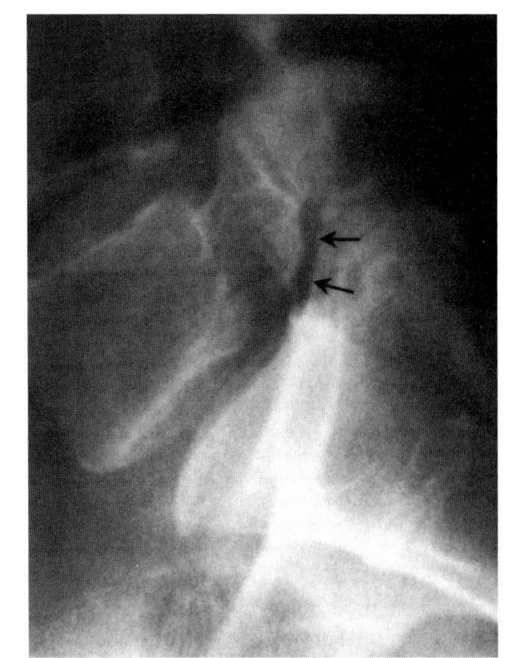

Figure 45–40. Spondylolysis with spondylolisthesis (grade I) of L5. The pars defect is bilateral (*arrows*), and the slippage is tangential.

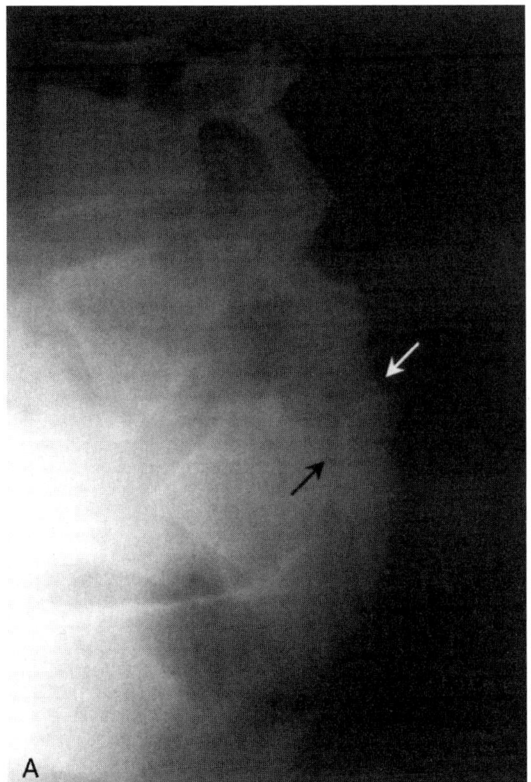

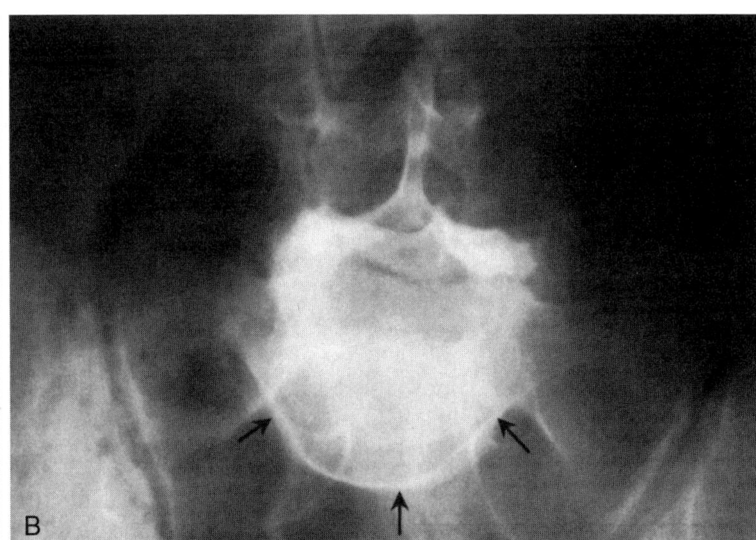

Figure 45–41. Radiographs showing severe spondylolisthesis (grade IV) with both tangential and angular slippage. *A*, Lateral view of the lumbosacral junction shows complete slippage of L5 on S1 (*black arrow*). In addition, there is angular slippage resulting in significant kyphosis. Note the rounded superior surface of S1 (*white arrow*) and the trapezoidal shape of the L5 body. *B*, Anteroposterior view shows the Napoleon's hat deformity (*arrows*) typically seen with severe spondylolisthesis of L5.

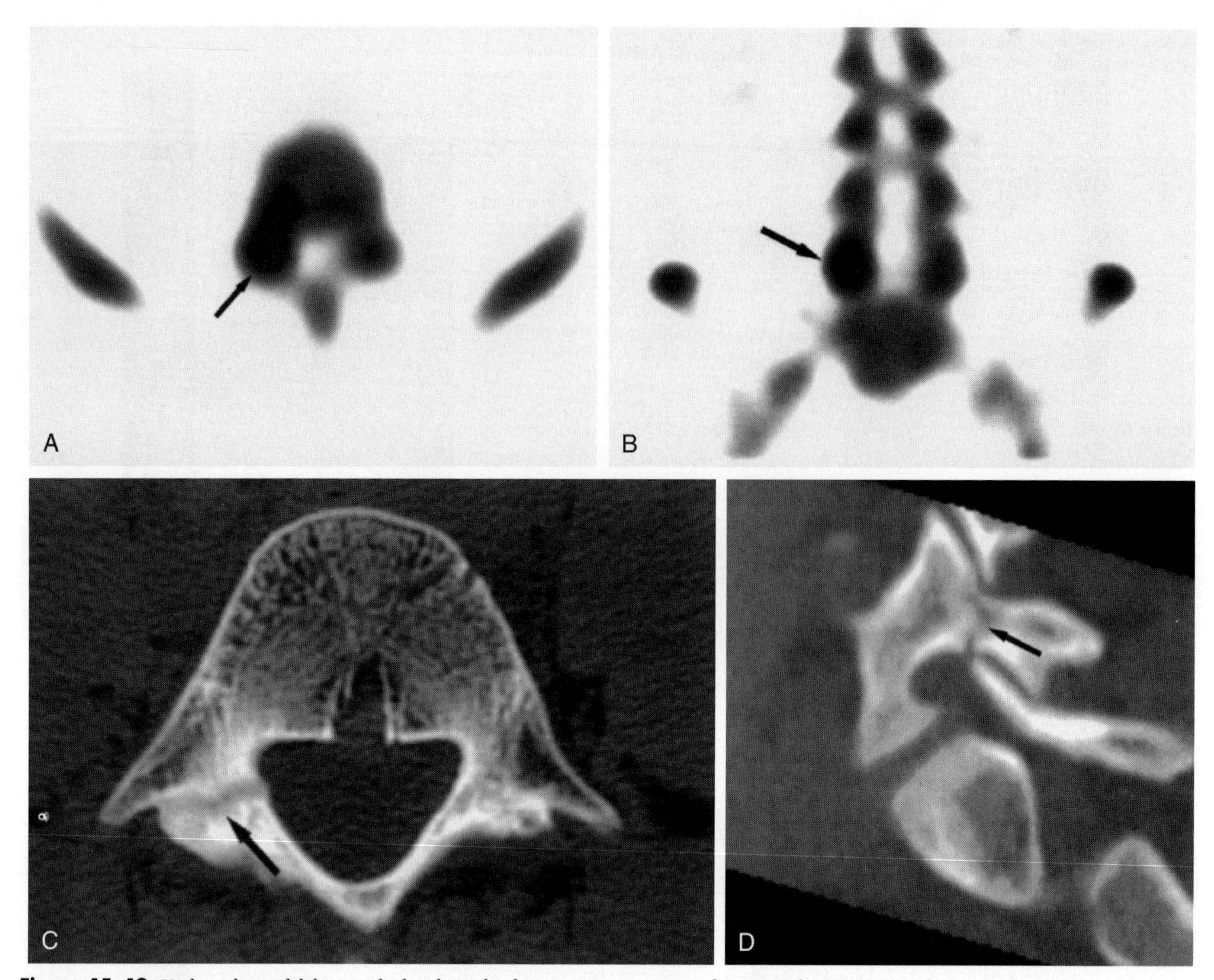

Figure 45–42. Unilateral spondylolysis studied with single photon emission computed tomography (SPECT) and computed tomography (CT). The patient is a 15-year-old, athletically active boy presenting with low back pain mainly on the right side. Axial (*A*) and coronal (*B*) SPECT sections show intensely increased uptake in the region of the pars interarticularis on the right (*arrows*). *C*, CT section through L5 confirms the development of a pars defect on the right (*arrow*). *D*, Sagittal CT reconstruction demonstrates the defect (*arrow*) well.

spondylolisthesis is not fully understood, but some writers believe that orientation of the facet joints in the sagittal plane may be a cause.

Post-traumatic Spondylolisthesis

In post-traumatic spondylolisthesis, a rare type of the disorder, the slip is due to injuries with fractures in the supporting vertebral structures other than the pars. The slip occurs gradually and is not an acute fracture-dislocation.

Pathologic Spondylolisthesis

Pathologic spondylolisthesis is the result of a localized or generalized bone disease affecting the bony hook mechanism between two adjacent vertebrae. The bony hook consists of the pedicle, pars, and superior and inferior facets. When the bony hook fails, the body weight thrusts the vertebra above the hook forward in relation to the vertebra below it.

Quantifying the Amount of Slippage

Two methods are commonly used to quantify the amount of slippage. The first method was introduced by Meyerding in 1932. He classified the severity of the spondylolisthesis into four grades by dividing the superior end plate from anterior to posterior into four equal segments. In grade I, L5 slides on S1 a distance equal to or less than one fourth the anteroposterior diameter of the S1 superior end plate (see Fig. 45–40). In grade IV, L5 has slid completely anteriorly and is barely sitting on the anterosuperior corner of S1 (see Fig. 45–41A).

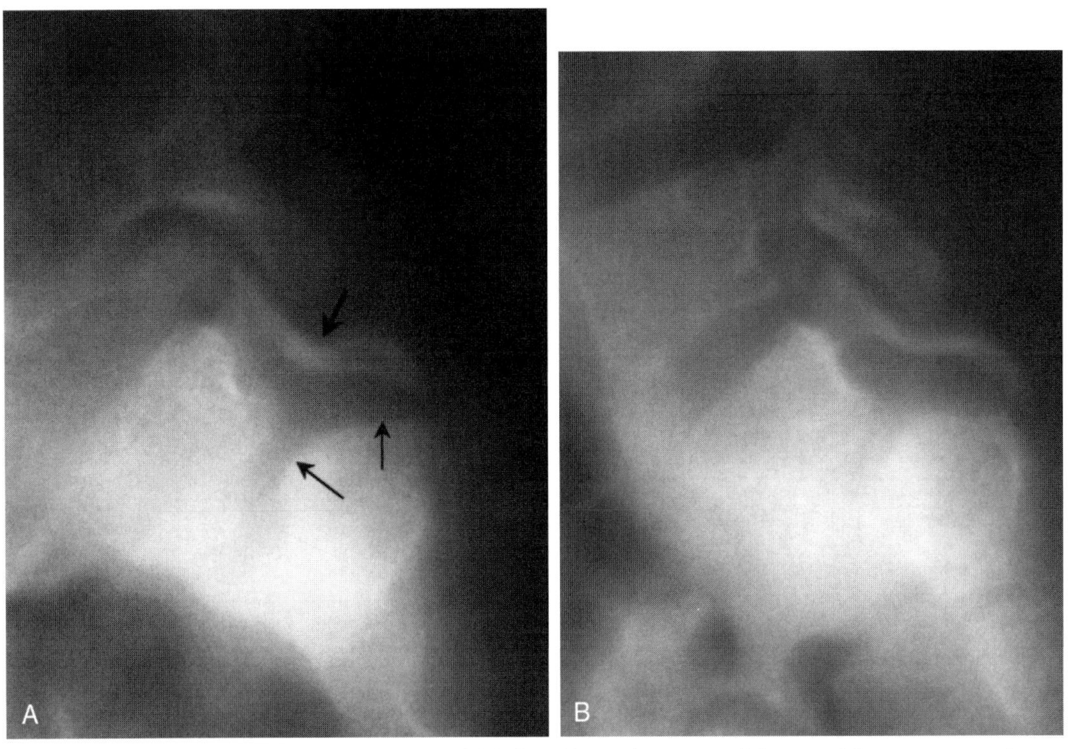

Figure 45–43. Conventional tomography sections showing spondylolisthesis due to elongation and thinning of the pars interarticularis in a 9-year-old girl. *A,* Lateral section through the left pars interarticularis shows elongation and thinning of the pars (*thick arrow*). The superior surface of S1 is rounded (*thin arrows*). The L5 body shows both tangential and angular slippage resulting in kyphosis at the L5-S1 level. *B,* Similar changes are seen on the right side, although the right pars interarticularis is longer and thinner than that on the left.

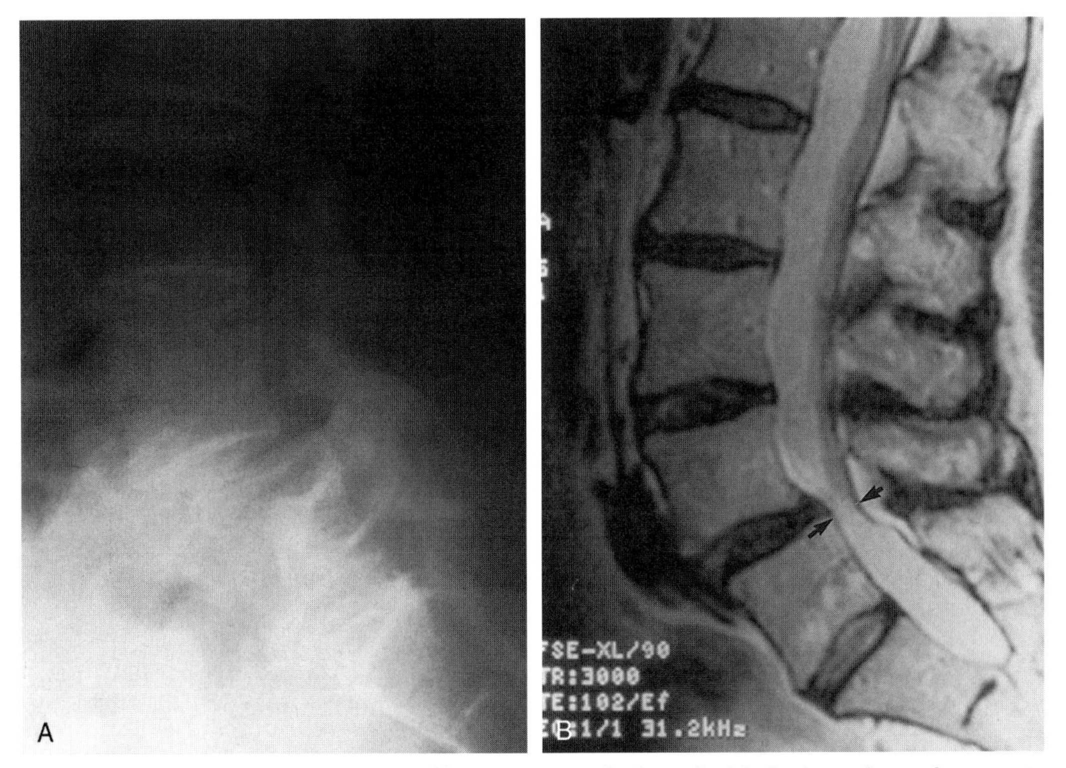

Figure 45–44. Degenerative spondylolisthesis in a 62-year-old woman. *A,* Lateral radiograph of the lumbosacral spine shows anterior translation of L4 on L5 of about 25%. The pars interarticularis is intact. *B,* Midsagittal T2-weighted magnetic resonance (MR) image shows spinal stenosis at L4-L5 level (*arrows*).

Illustration continued on following page

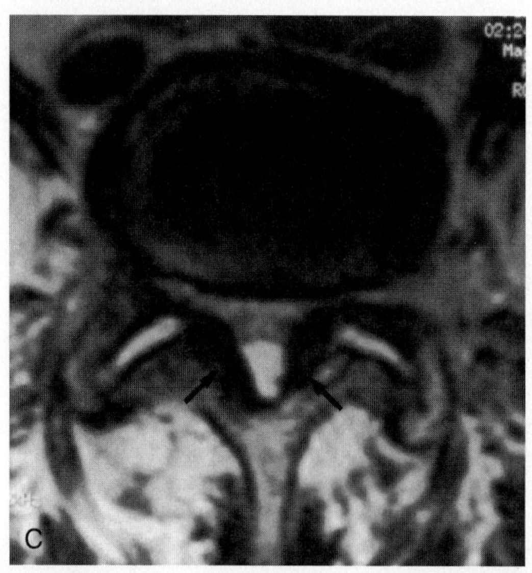

Figure 45–44 *Continued. C,* Axial T2-weighted MR image at the L4-L5 disk level confirms the presence of spinal stenosis. Note the degenerative disease of the facet joints with osteophyte formation, joint effusions, and thickening of the ligamentum flavum on both sides (*arrows*).

The other method of quantifying slippage involves calculating the percentage of anterior displacement of L5 on S1.

Another parameter that is important in patients with severe slippage is the measurement of sagittal rotation or lumbosacral kyphosis. This angle is measured by extending a line from along the anterior cortex of the L5 body until it intersects a second line drawn along the posterior cortex of S1 (Fig. 45–45). In normal individuals, this angle is about zero.

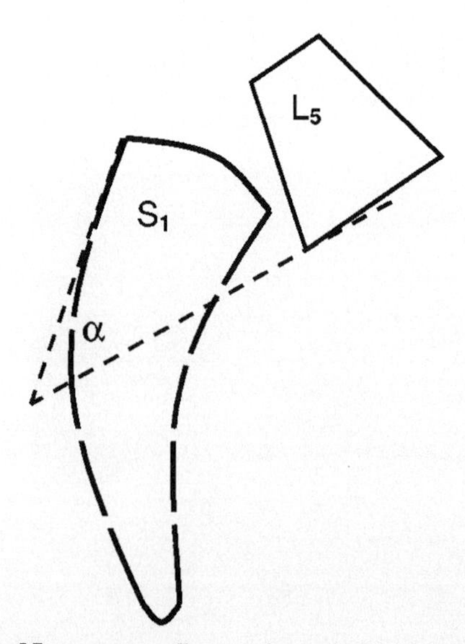

Figure 45–45. Diagram to illustrate the method for measuring angular slippage and resultant kyphosis in spondylolisthesis. One line is drawn along the posterior aspect of S1, and a second line is drawn along the anterior surface of L5. Angle α at the intersection of these two lines determines the degree of angular slippage.

References and Recommended Readings

Ackerman SJ, Steinberg EP, Bryan RN, et al: Persistent low back pain in patients suspected of having herniated nucleus pulposus: Radiologic predictors of functional outcome—implications for treatment selection. Radiology 203:815, 1997.

Adams MA, Hutton WC: Gradual disc prolapse. Spine 10:524, 1985.

Anderson RE, et al: Acute low back pain—radiculopathy. In ACR Task Force on Appropriateness Criteria. *American College of Radiology Appropriateness Criteria, ACR Web Site Edtion* [online text]. Reston, VA: American College of Radiology, 1998. 21 March 2002; http://www.acr.org/departments/appropriatenesscriteria/text.html.

Boden SD: Current concepts review: The use of radiographic imaging studies in the evaluation of patients who have degenerative disorders of the lumbar spine. J Bone Joint Surg [Am] 78-A:114, 1996.

Brant-Zawadzki MN, Dennis SC, Gade GF, et al: What the clinician wants to know: Low back pain. Radiology 217:321, 2000.

Bogduk N, Modic MT: Controversy: Lumbar discography. Spine 21:402, 1996.

Grabias S: Current concepts review: The treatment of spinal stenosis. J Bone Joint Surg [Am] 62-A:308, 1980.

Hilibrand AS, Urquhart AG, Graziano GP, et al: Acute spondylolytic spondylolisthesis. J Bone Joint Surg [Am] 77A:190, 1995.

Jensen MC, Brant-Zawadzki MN, Obuchowski N, et al: Magnetic resonance imaging of the lumbar spine in people without back pain. N Engl J Med 331:69, 1994.

Kerslake RW, Worthington BS: MRI of the spine (editorial). Clin Radiol 43:227, 1991.

Lejeune JP, Hladky JP, Cotton A, et al: Foraminal lumbar disc herniation: Experience with 83 patients. Spine 19:1905, 1994.

Lipson SJ, Muir H: 1980 Volvo Award in Basic Science: Proteoglycans in experimental intervertebral disc degeneration. Spine 6:194, 1981.

Masaryk TJ, Ross JS, Modic MT, et al: High-resolution MR imaging of sequestered lumbar intervertebral disks. AJR Am J Roentgenol 150:1155, 1988.

Meyerding HW: Spondylolisthesis. Surg Gynecol Obstet 54:371, 1932.

Modic MT, Ross JS: Morphology, symptoms, and causality. Radiology 175:619, 1990.

Modic MT, Masaryk TJ, Ross JS, et al: Imaging of degenerative disk disease. Radiology 168:177, 1988.

Modic MT, Ross JS, Obuchowski NA, et al: Contrast-enhanced MR imaging in acute lumbar radiculopathy: A pilot study of the natural history. Radiology 195:429, 1995.

Morgan S: Pictorial Review: MRI of the lumbar intervertebral disc. Clin Radiol 54:703, 1999.

Ogata K, Whiteside LA: 1980 Volvo Award Winner in Basic Science: Nutritional pathways of the intervertebral disc: An experimental study using hydrogen washout technique. Spine 6:211, 1981.

Osti OL, Vernon-Roberts B, Moore R, et al: Annular tears and disc degeneration in the lumbar spine. J Bone Joint Surg [Br] 74-B:678, 1992.

Pritchett JW, Bortel DT: Degenerative symptomatic lumbar scoliosis. Spine 18:700, 1993.

Ross JS, Modic MT, Masaryk TJ: Tears of the anulus fibrosus: Assessment with Gd-DTPA-enhanced MR imaging. AJR Am J Roentgenol 154:159, 1990.

Ross JS, Robertson JT, Frederickson RCA, et al: Association between peridural scar and recurrent radicular pain after lumbar diskectomy: Magnetic resonance evaluation. Neurosurgery 38:855, 1996.

Shapiro S: Medical realities of cauda equina syndrome secondary to lumbar disk herniation. Spine 25:348, 2000.

Spitzer et al: Spine 12(7 Suppl):S1–59, 1987.

Stadnik TW, Lee RR, Coen HL, et al: Annular tears and disk herniation: Prevalence and contrast enhancement on MR images in the absence of low back pain or sciatica. Radiology 206:49, 1998.

Wiltse LL: Classification, terminology and measurements in spondylolisthesis. Iowa Orthop J 1:52, 1981.

CHAPTER 46

Spinal Infections

Elias R. Melhem, MD, and Georges Y. El-Khoury, MD

OVERVIEW

A variety of infections can involve the spine. They include spondylodiskitis, postoperative diskitis, epidural abscess, septic arthritis of the facet joints, and disk space infections (diskitis) in children.

Spinal infections are typically preceded by infections elsewhere in the body, such as the respiratory system, genitourinary system, and the skin. Predisposing factors are diabetes mellitus, intravenous drug abuse, liver disease, kidney failure, and conditions that suppress the immune system.

Patients often present with fever and generalized malaise along with back pain and tenderness over the involved segments of the spine. Some patients have elevations of the leukocyte count, but the majority have elevations in the erythrocyte sedimentation rate, especially those with epidural abscesses. Most studies of spinal infections report delays in diagnosis.

SPONDYLODISKITIS (INFECTIOUS SPONDYLITIS)

Clinical Features

Different terms have been used to describe the same entity, thus adding to the confusion. They include infectious spondylitis, diskitis, disk space infection, and pyogenic vertebral osteomyelitis.

Spondylodiskitis, the most common spinal infection, is usually secondary to hematogenous spread. The disease is most common in the elderly, and the majority of patients are male (3:1). The most frequently isolated organism is *Staphylococcus aureus*, which accounts for 42% to 84% of infections, followed by *Streptococcus* species (20%) and gram-negative bacilli. Multiple organisms are isolated from about 24% of patients. In intravenous drug users, *Pseudomonas* is the most common organism causing spondylodiskitis. In diabetic patients, this infection is sometimes caused by gas-forming organisms such as *Escherichia coli* (Fig. 46–1).

A specific diagnosis in spondylodiskitis is achieved by needle aspiration performed with computed tomography (CT) control, which typically yields positive aspirate cultures in about 75% of patients who have not been treated previously with antibiotics. The most common location for spondylodiskitis is the lumbar spine (Fig. 46–2), followed by the thoracic spine (Fig. 46–3) and the cervical spine (Fig. 46–4). One fourth of the cases occurring in the lumbar spine are complicated by secondary epidural abscesses. The incidence of epidural abscess increases to 33% in patients with spondylodiskitis of the thoracic spine (see Fig. 46–3) and to about 90% in those with spondylodiskitis of the cervical spine (Fig. 46–5). The incidence of serious neurologic deficit is highest in spondylodiskitis complicated by epidural abscess in the thoracic spine, followed by that in the cervical spine. The high incidence is due to the small size of the spinal canal and resultant cord compression in these areas.

Imaging of Spondylodiskitis

Radiography

Plain radiographs may not reveal abnormalities for several weeks. The findings initially are nonspecific until florid disk space destruction becomes evident. Typically, there is disk space narrowing, osteopenia, erosions, and, eventually, destruction of the adjacent vertebral end plate (see Fig. 46–4).

Radionuclide Bone Scans

Radionuclide bone scans are not commonly used in the evaluation of spondylodiskitis. Bone scans are sensitive in the detection of early infection or multiple sites of infection, but they lack spatial resolution and are often nonspecific (Fig. 46–6).

Computed Tomography

The role of computed tomography (CT) is currently limited to being the major modality for guidance of needle biopsy and needle aspiration procedures in patients with spinal infections (Fig. 46–7).

Magnetic Resonance Imaging

Currently, magnetic resonance (MR) imaging is the modality of choice for the imaging of spondylodiskitis. The strength of MR imaging lies not only in its high sensitivity and specificity (>90%) for detecting early disease but also in its ability to demonstrate associated epidural and extraosseous soft tissue extensions of infection (see Figs. 46–3 and 46–5).

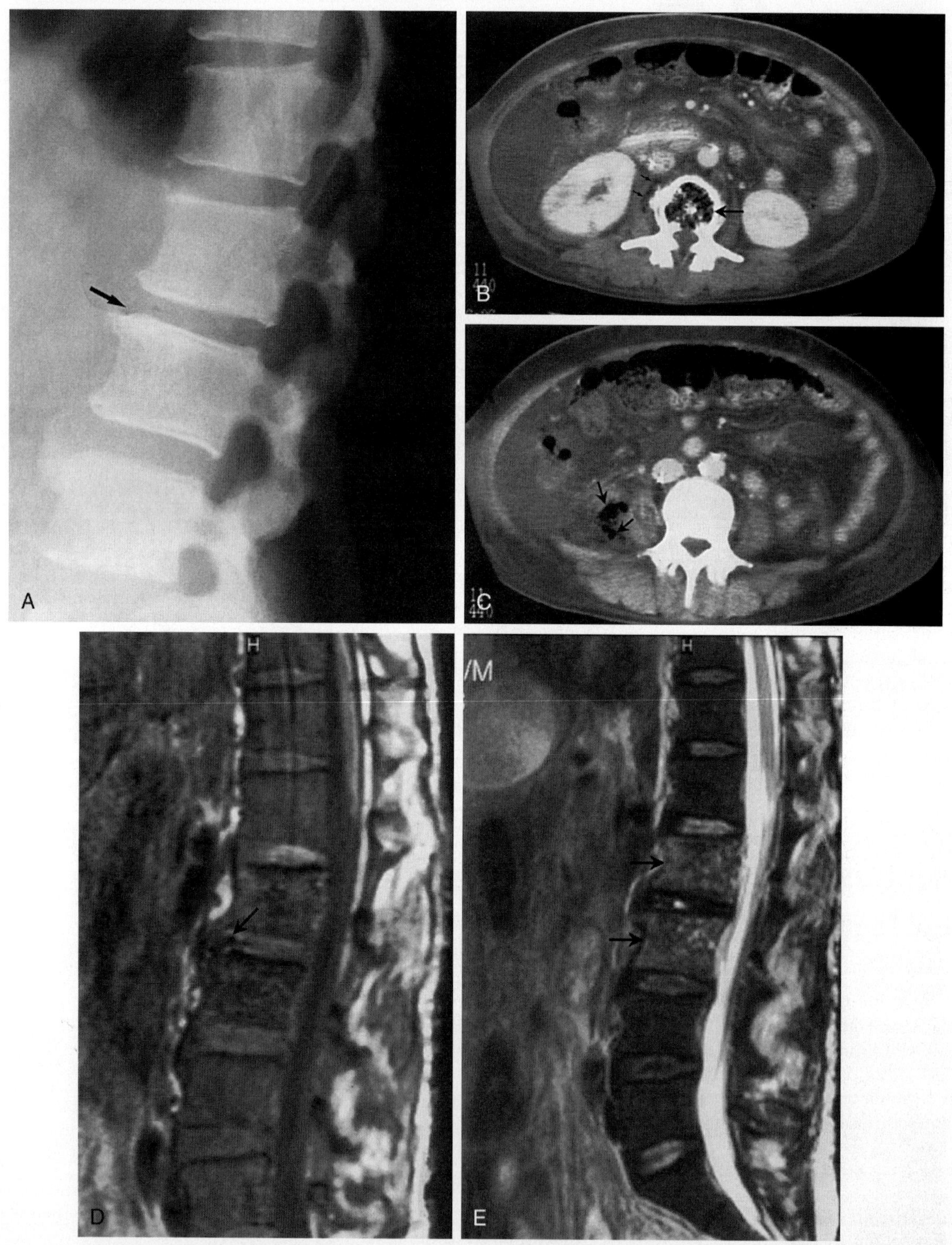

Figure 46–1. Gas-containing spondylodiskitis and psoas abscess caused by *Escherichia coli* in a 54-year-old diabetic woman presenting with back pain and septicemia. *A,* Lateral radiograph of the lumbar spine shows narrowing of L2-L3 disk and gas within the disk (*arrow*). *B,* Computed tomography (CT) section through the L2-L3 disk demonstrates the presence of gas within the disk (*large arrow*) with extension into the right psoas muscle (*small arrows*). *C,* CT section through the L4 vertebra shows a gas collection within a swollen right psoas muscle (*arrows*). *D,* Sagittal T1-weighted magnetic resonance (MR) image of the lumbar spine demonstrates the gas, as dark signals, within the L2-L3 disk (*arrow*). Abnormal marrow signals are also noted in L2 and L3 vertebral bodies. *E,* Sagittal T2-weighted MR image confirms the abnormalities in the L2-L3 disk and adjacent vertebrae (*arrows*). (Case provided by Dr. Kenjirou Ohashi of Iowa City, IA.)

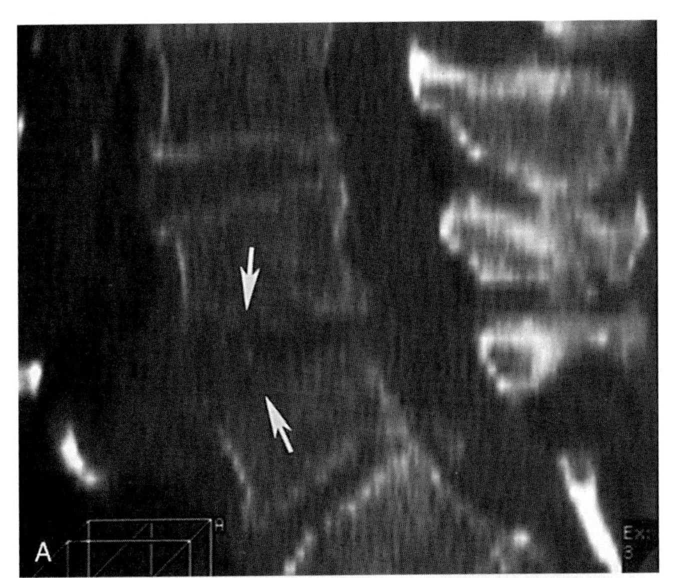

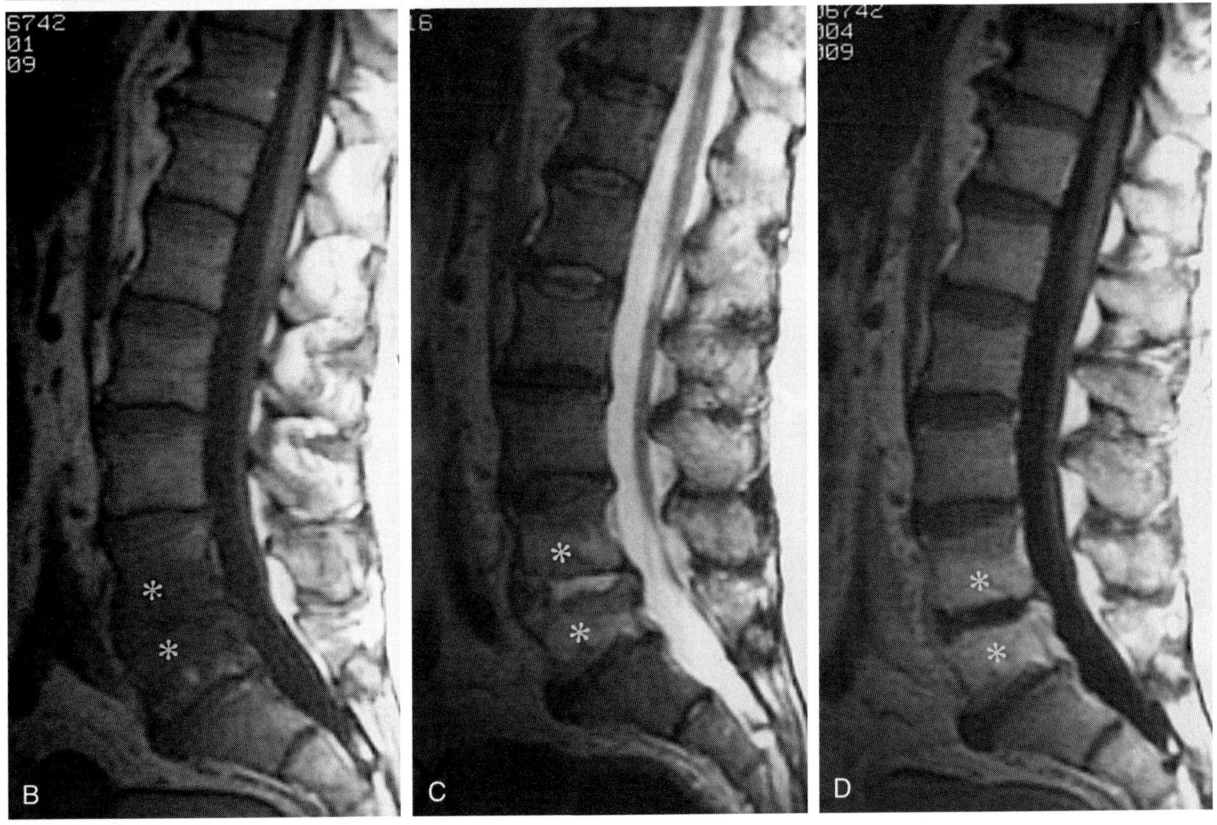

Figure 46–2. Computed tomography (CT) and magnetic resonance (MR) imaging findings in a 76-year-old diabetic man who presented with back pain, malaise, and low-grade fever. *A,* Midline sagittal CT reformation of the lumbar spine demonstrates erosions and destruction of the L4 and L5 vertebral end plates (*arrows*), loss of intervertebral disk height, and mild prevertebral soft tissue swelling, findings compatible with spondylodiskitis. *B,* Midline sagittal T1-weighted MR image shows decreased marrow signal intensity in the L4 and L5 vertebral bodies (*asterisks*) with loss of the vertebral end plates. *C,* Midline sagittal T2-weighted MR image demonstrates increased signal intensity in the L4 and L5 vertebral bodies (*asterisks*) and the corresponding intervertebral disk. The loss of the intranuclear cleft in the L4-L5 disk helps distinguish the infected disk from healthy, well-hydrated ones. *D,* Contrast-enhanced midsagittal T1-weighted MR image demonstrates homogeneous enhancement of the L4 and L5 vertebral bodies (*asterisks*). Lack of disk enhancement suggests liquefaction and abscess formation.

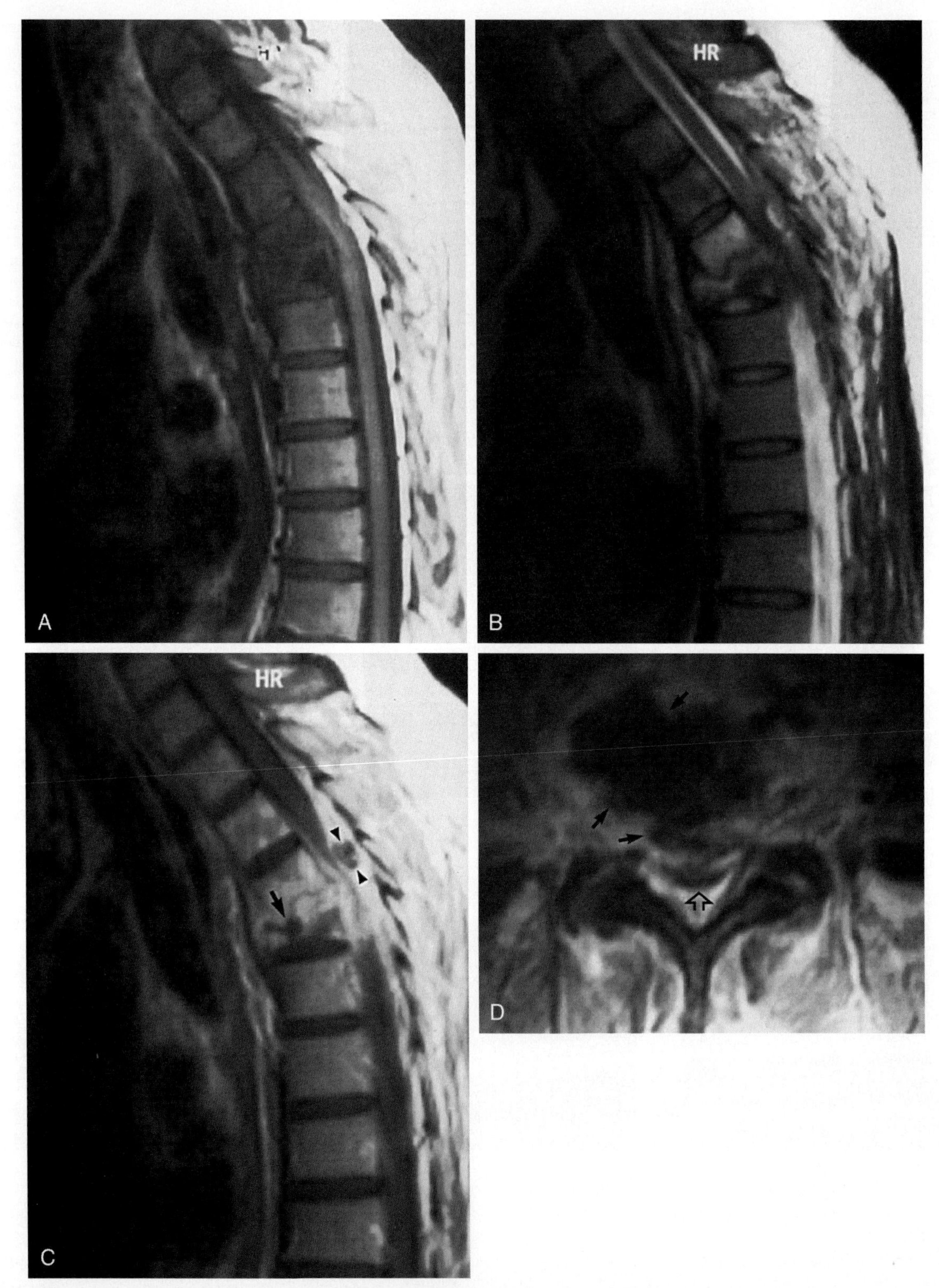

Figure 46–3. Spondylodiskitis complicated by epidural abscess in a 32-year-old man who presented with fever, upper thoracic pain, lower extremity weakness, and sensory loss. Sagittal T1-weighted (*A*) and T2-weighted (*B*) magnetic resonance (MR) images show changes of spondylodiskitis at the T4-T5 level. *C,* Sagittal T1-weighted MR images after gadolinium injection shows an epidural abscess (*arrowheads*) compressing the cord. Pus can also be seen within the disk (*arrow*) as an area of low signal intensity. *D,* Axial T1-weighted MR image after gadolinium enhancement shows fluid (pus) in the disk (*closed arrows*), in the paravertebral soft tissue to the right of the vertebral body, and in the epidural space posterior to the thecal sac (*open arrowhead*).

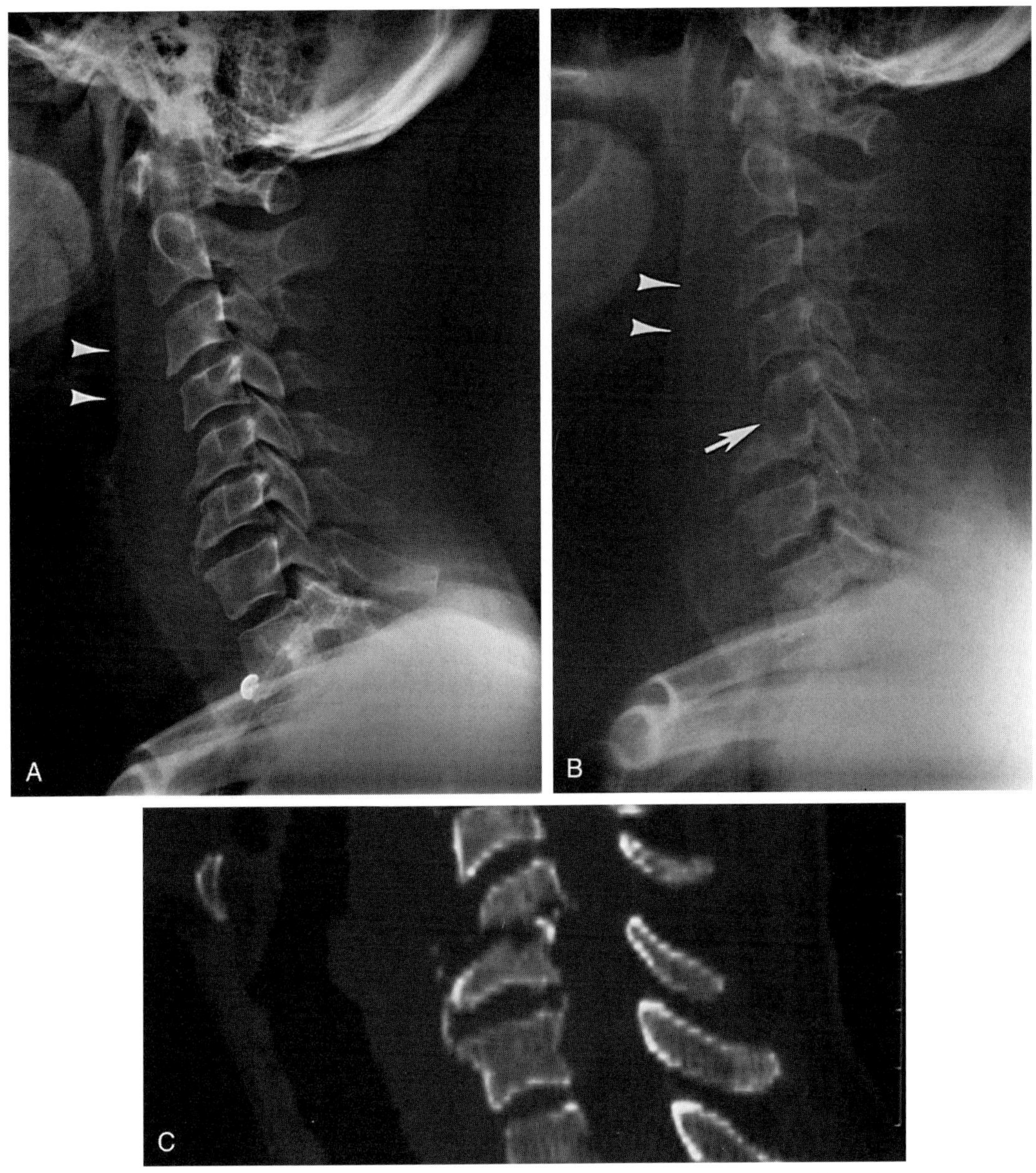

Figure 46–4. Spondylodiskitis of the cervical spine in a 32-year-old woman with history of intravenous drug abuse who presented with neck pain and difficulty swallowing. *A,* Lateral radiograph demonstrates narrowing of the C5-C6 disk and prevertebral soft tissue swelling (*arrowheads*). *B,* Lateral radiograph obtained 1 month later shows erosions and destruction of the C5 and C6 vertebral end plates. The C5-C6 disk is obliterated (*arrow*), and the prevertebral soft tissue swelling is still present (*arrowheads*). *C,* Midline sagittal computed tomography reconstruction demonstrates erosions and destruction of the C5 and C6 vertebral end plates, mild retrolisthesis of C5 vertebra on C6, and prevertebral soft tissue swelling.

The typical findings are decreased signal intensity in the disk and adjacent vertebral bodies on T1-weighted MR images with corresponding increased signal intensity in the disk and adjacent vertebral bodies on T2-weighted images (see Fig. 46–2). The absence of an intranuclear cleft on T2-weighted images differentiates an infected disk from a well-hydrated, healthy disk (see Fig. 46–2). After injection of gadolinium, the disk, adjacent vertebral bodies, and involved paraspinal and epidural soft tissues are all enhanced (see Fig. 46–2). Fluid (pus) collections, whether in the disk, epidural space, or paravertebral soft tissues, remain dark after contrast enhancement (see Figs. 46–2, 46–3, 46–5, and 46–7). It should be noted that not all cases of spondylodiskitis display this typical appearance, and variations have been reported. In some patients, the vertebral bodies adjacent to the infected disk reveal normal or decreased signal on T2-weighted MR images (see Fig. 46–7). The explanation for this finding is not clear, but in some patients, bands of low signal intensity adjacent to the disk correlate with bony sclerosis seen on plain radiographs (Fig. 46–7). Early use of MR imaging is now recommended to avoid delay in diagnosis. Early findings can also be atypical, revealing involvement of one vertebral body (Fig. 46–8) or one vertebral body and one disk. Signal abnormalities can also be atypical early in the disease.

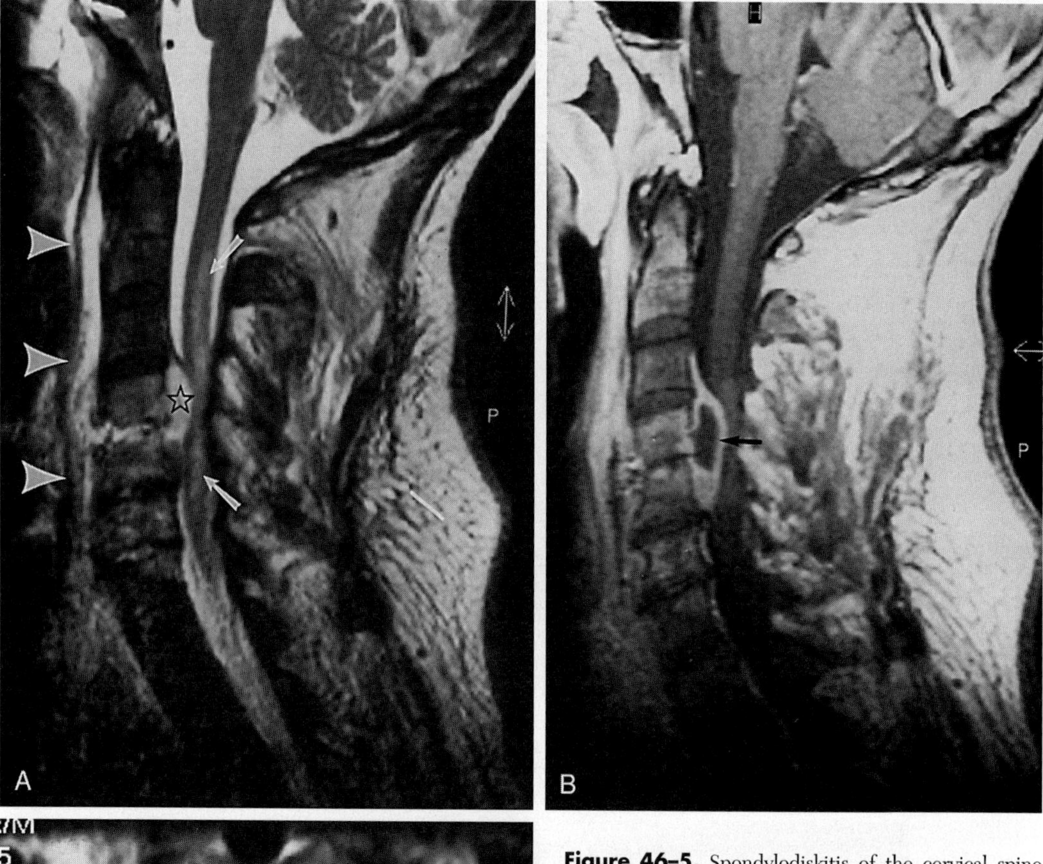

Figure 46–5. Spondylodiskitis of the cervical spine complicated by an epidural and prevertebral abscess in a 44-year-old man who presented with neck pain and progressive arm and leg weakness. *A,* Midline sagittal T2-weighted magnetic resonance (MR) image shows multilevel spondylodiskitis involving the C4, C5, C6, and C7 vertebral bodies and the corresponding intervertebral disks. An epidural abscess collection (*star*) is noted behind the C4 vertebral body compressing the cord. Also, a large prevertebral abscess (*arrowheads*) extends throughout the entire length of the cervical spine. Both abscess collections originated from the C4-C5 disk. Associated cord edema (*arrows*) is seen above and below the level of the epidural abscess. *B,* Contrast-enhanced midline sagittal T1-weighted MR image demonstrates enhancement of the involved vertebral bodies and intervertebral disk. In addition, there is peripheral enhancement of the epidural abscess (*arrow*). *C,* Axial T2-weighted MR image at the C3 level confirms the extension of cord edema (*arrow*) above the level of the epidural abscess.

The role of MR imaging in assessing the response to treatment is not well-defined. This modality may delineate deterioration, while the clinical response to treatment shows improvement. When gadolinium enhancement disappears, active inflammation can be excluded.

One of the earliest signs of a favorable response to treatment manifests as reduction in the soft tissue inflammation, which appears as a decrease in soft tissue enhancement after gadolinium injection.

Long-term follow-up of spondylodiskitis is usually carried out with plain radiographs, which demonstrate sclerosis and bony fusion in 6 months to 2 years (Fig. 46–9). The disk space disappears completely, and the height of the two involved vertebral bodies is reduced to about one vertebral body height.

Differential Diagnosis

Conditions that can mimic spondylodiskitis include Modic type I degenerative disk changes (see Fig. 45–9), ankylosing spondylitis with motion at a single disk level (Anderson lesion) (see Fig. 22–9), and noninfectious

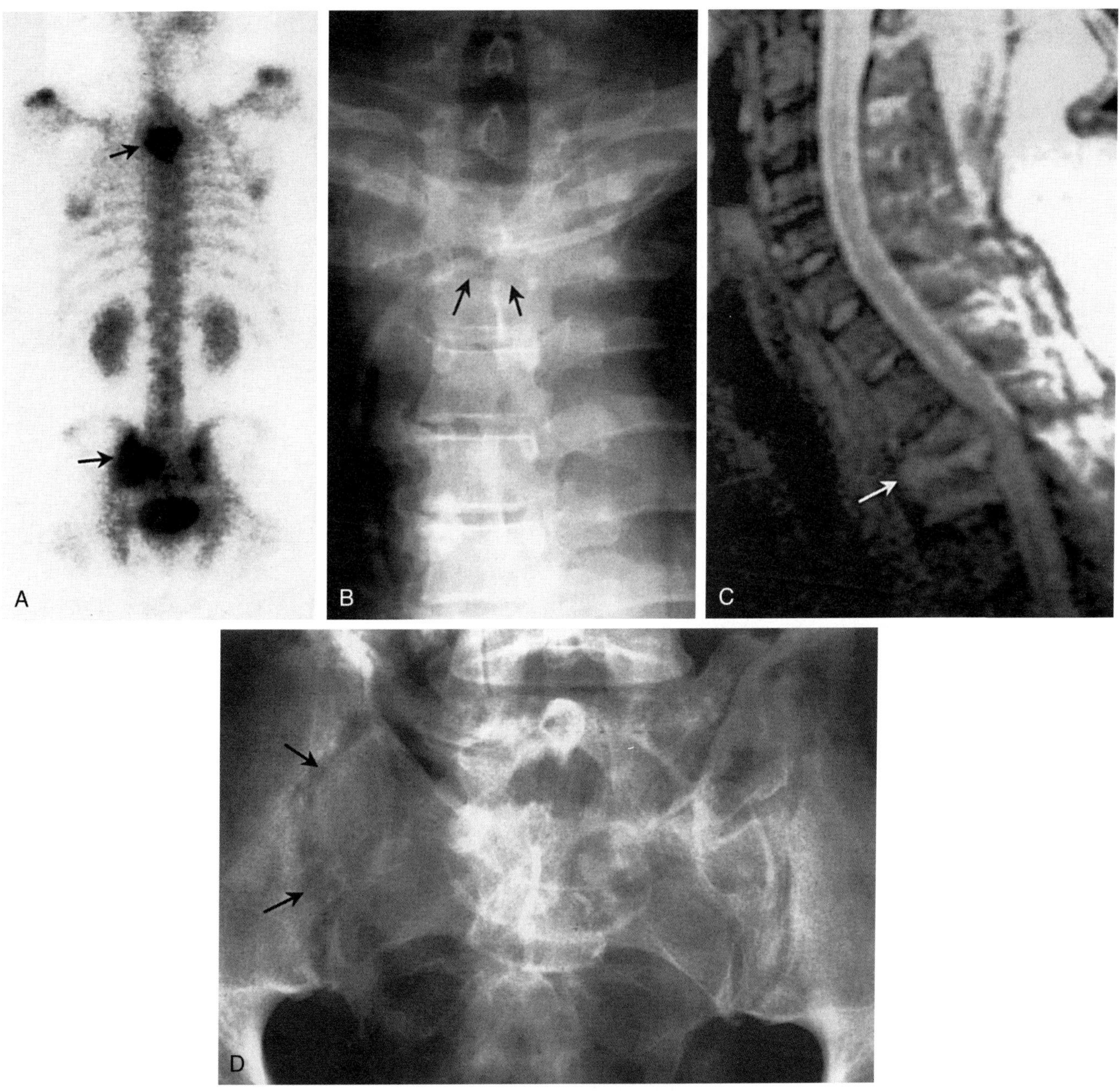

Figure 46–6. Multicentric *Staphylococcus aureus* infections in a 53-year-old diabetic that involve the thoracic spine, right sacroiliac joint, and right foot. *A,* Radionuclide bone scan reveals increased uptake in the upper thoracic spine (*upper arrow*), right sacroiliac joint (*lower arrow*), and right foot (*not shown*). *B,* Anteroposterior radiograph of the upper thoracic spine shows end plate erosions in T2 and T3, a wide T2-T3 disk (*arrows*), and paravertebral soft tissue swelling. *C,* Sagittal T2-weighted magnetic resonance image of the spine demonstrates spondylodiskitis at the T2-T3 level (*arrow*), with significant erosions of the end plates and cord compression. *D,* Anteroposterior radiograph of the sacroiliac joints shows bony erosions on both sides of the right sacroiliac joint but mainly on the iliac side (*arrows*).

spondyloarthropathy, which is observed in patients receiving long-term hemodialysis (Fig. 46–10).

POSTOPERATIVE DISKITIS

Clinical Features

Postoperative diskitis is a recognized complication of disk surgery. Most series report a delay in diagnosis. It is estimated to occur after 0.2% to 3.0% of operations on a lumbar disk. Disk space infections have also been reported after diskography, myelography, and chemonucleolysis.

With postoperative diskitis, the majority of patients become symptomatic in 1 to 4 weeks, after an initial period of pain relief. Patients present with excruciating back pain and paravertebral muscle spasm. Less than half have a fever. *S. aureus* is the most commonly recovered pathogen, although in a significant number of cases, no organism is recovered, suggesting a noninfectious inflammatory process.

Imaging

Plain Radiography

Plain radiography is not sensitive in detecting early disease. Radiographic changes appear about 4 to 6 weeks after the onset of symptoms. The findings include vertebral end plate demineralization, disk space narrow-

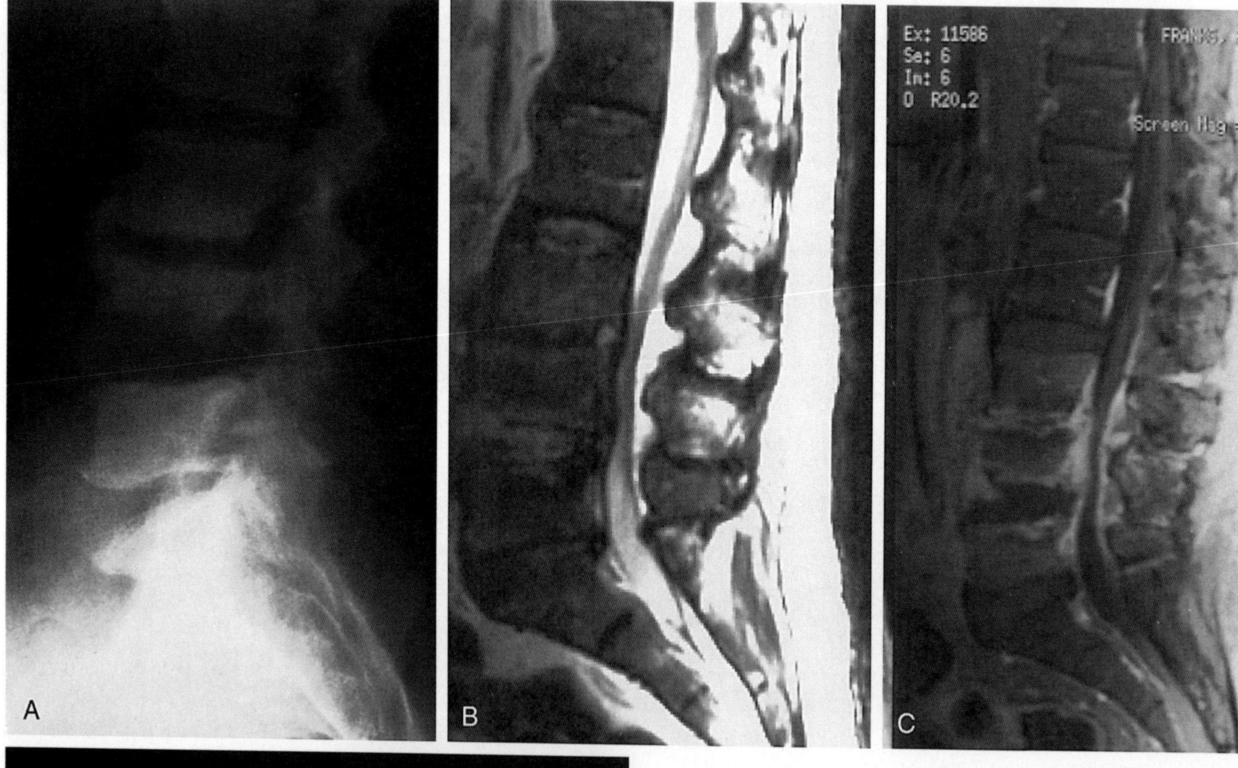

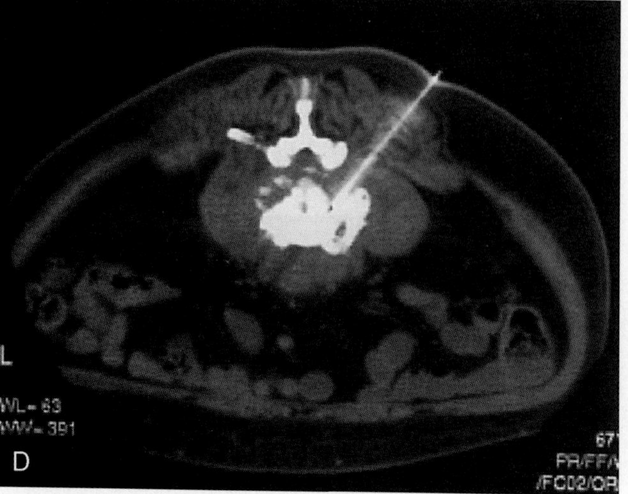

Figure 46–7. Spondylodiskitis involving three vertebrae and two disks. *A,* Lateral radiograph of the lumbar spine shows superior and inferior end plate erosions in L4 and inferior end plate erosions in L3. The L3-L4 disk and L4 vertebra have lost some height. *B,* Sagittal T2-weighted magnetic resonance (MR) image shows changes of spondylodiskitis at L3-L4 and L4-L5 levels. The thecal sac is compressed at these levels. *C,* Sagittal T1-weighted MR image after gadolinium enhancement shows a small amount of fluid in the L3-L4 disk and a large amount in the L4-L5 disk. The thecal sac is encased and compressed by a bright mass, very likely a phlegmon. *D,* The L4-L5 disk was aspirated under CT guidance, which yielded pus. Cultures of the pus specimen grew *Staphylococcus aureus.*

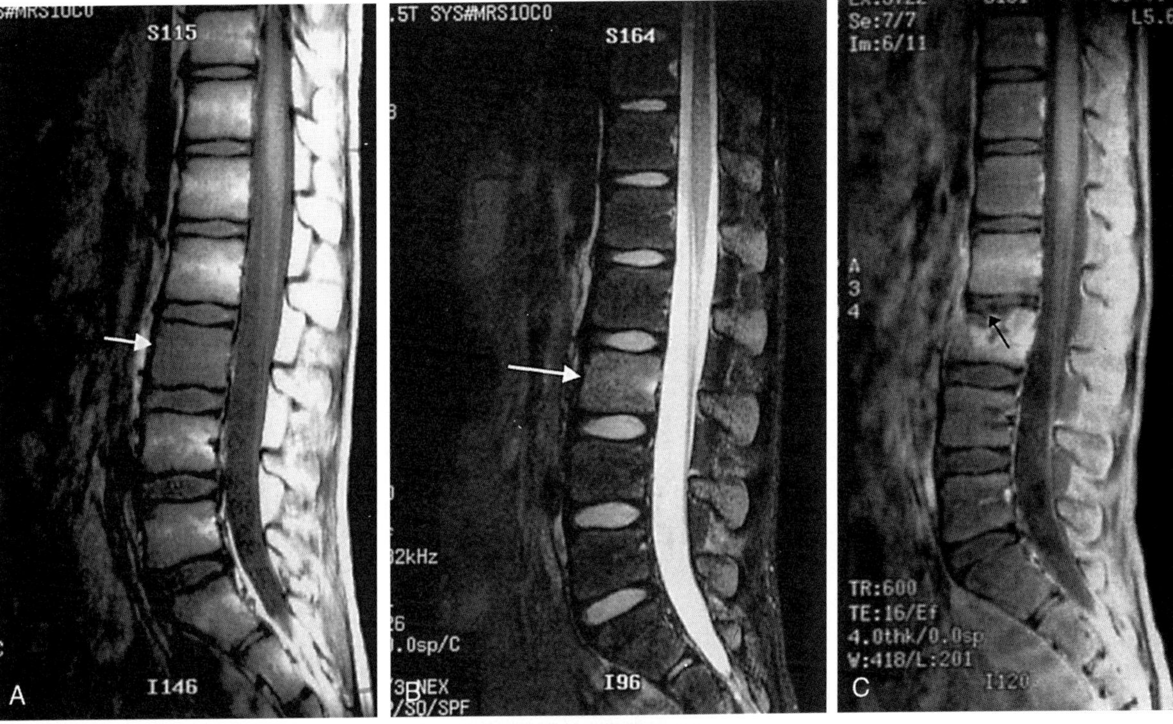

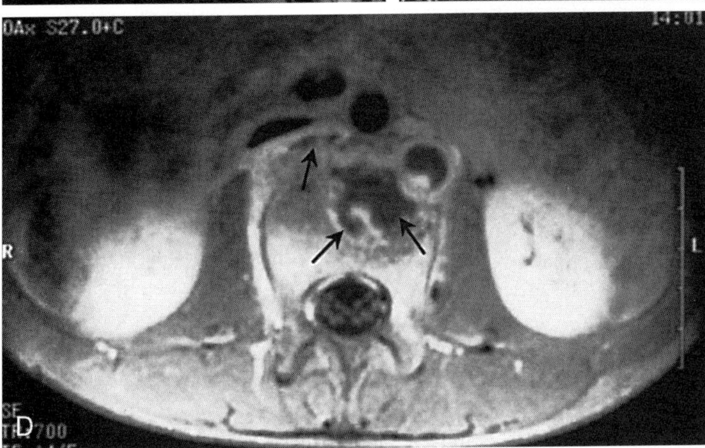

Figure 46–8. Spondylodiskitis that started within a vertebral body and initially did not involve the disk. The patient, a 12-year-old girl, presented with back pain of 1 week's duration. Sagittal T1-weighted (*A*) and T2-weighted (*B*) magnetic resonance (MR) images reveal abnormal marrow signals involving the entire L3 vertebral body (*arrows*). Sagittal (*C*) and axial (*D*) T1-weighted, gadolinium-enhanced MR images obtained 6 weeks later show typical changes of spondylodiskitis, with pus collections within the disk and paravertebral soft tissues (*arrows*).

ing, and irregular vertebral end plates (Fig. 46–11), progressing to disk space obliteration, sclerosis, and bony fusion. It takes about 6 months to 2 years for bony fusion to become solid.

Magnetic Resonance Imaging

It is important to note that in MR imaging of asymptomatic postoperative patients, signal intensity changes within the operated disk and adjacent vertebral bodies are rare, and their presence should not be considered normal postoperative findings; disk enhancement after gadolinium injection is not a normal postoperative finding either.

The MR imaging changes in postoperative diskitis are similar to those in de novo spondylodiskitis. They are characterized by decreased signal intensity within the disk and adjacent vertebrae on T1-weighted images. On T2-weighted images, signal intensity is increased in the disk and the adjacent marrow, with loss of the intranuclear cleft. After gadolinium injection in all patients with diskitis, there is homogeneous enhancement of adjacent bone above and below the disk. This finding is very rare in asymptomatic postoperative patients.

EPIDURAL ABSCESS

Overview

Epidural abscess is a serious infection whose prevalence is on the increase. The incidence of neurologic deficit is fairly high when the epidural abscess involves the thoracic or cervical spine. A true epidural abscess consists of a collection of pus, and it appears to have a worse prognosis than an epidural granulation tissue (phlegmon) (Fig. 46–12). Its is difficult to diagnose an epidural abscess on clinical grounds alone, and if the abscess is left untreated, paraplegia, quadriplegia, or even death may

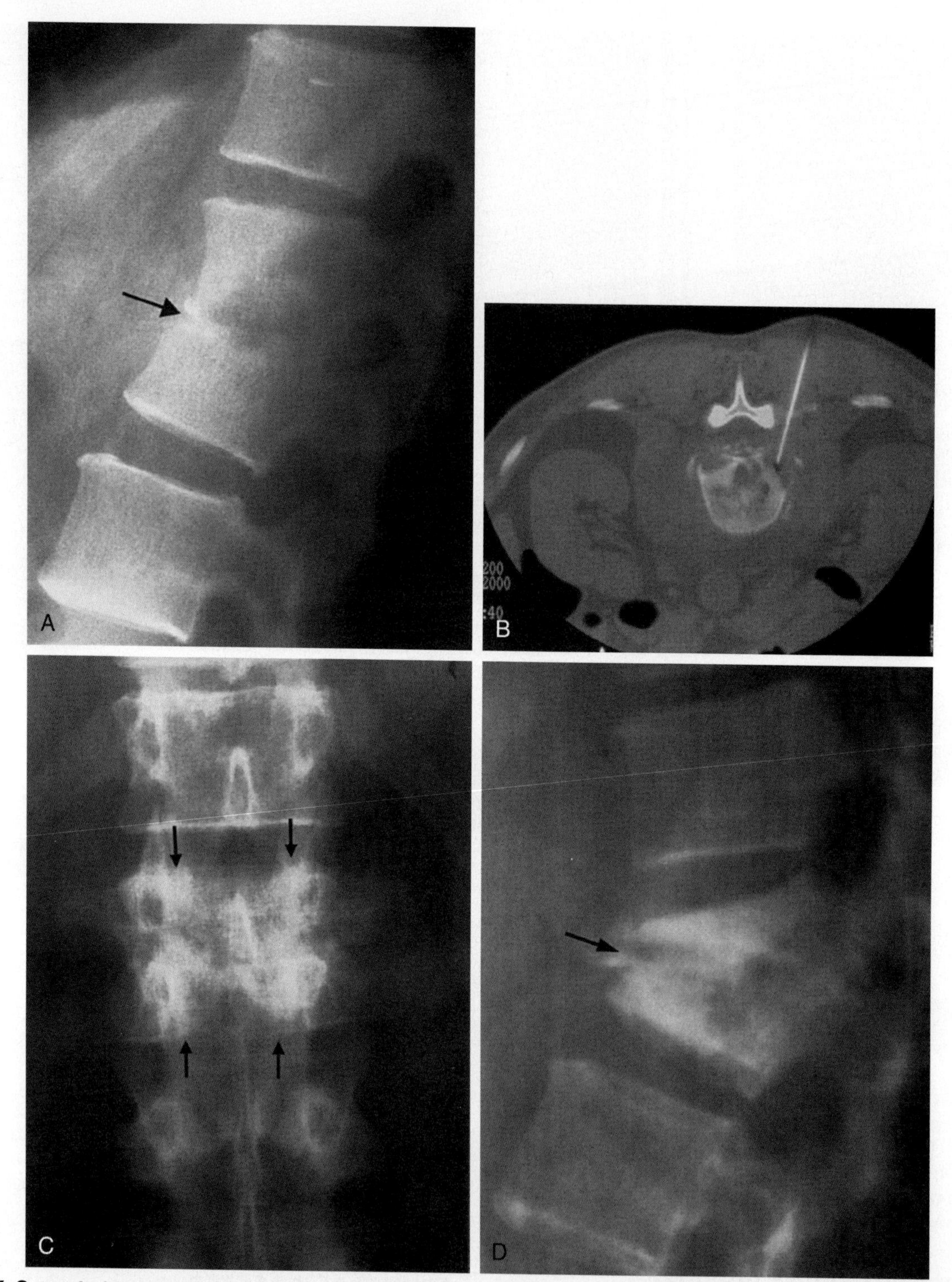

Figure 46–9. Vertebral fusion 4 months after effective treatment of a spondylodiskitis in a 44-year-old woman. *A*, Lateral radiograph of the lumbar spine shows changes of spondylodiskitis at the L1-L2 level, with end plate erosions and disk resorption (*arrow*). *B*, Computed tomography section obtained during a needle biopsy reveals the bone destruction and paravertebral soft tissue swelling associated with the infection. Anteroposterior (*C*) and lateral (*D*) radiographs obtained 4 months later reveal progressive fusion. The disk space is completely obliterated (*arrow* in *D*), and the height of the fusing vertebrae is diminished (*arrows* in *C*).

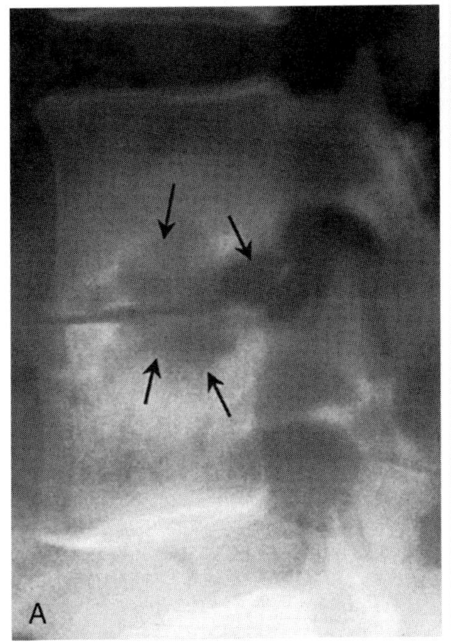

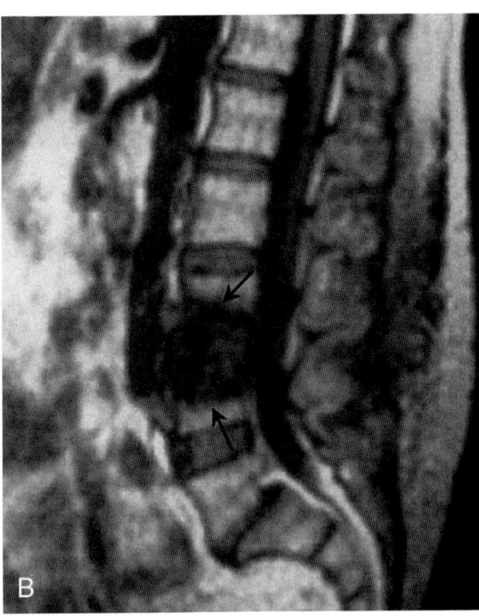

Figure 46–10. Disk changes in chronic uremia resembling spondylodiskitis. The patient, a 62-year-old woman undergoing long-term hemodialysis, presented with low back pain. *A*, Lateral radiograph of the lumbar spine revealed narrowing of the L3-L4 disk space and large erosions at the vertebral end plates (*arrows*). *B*, Sagittal T1-weighted magnetic resonance image of the lumbosacral spine shows areas of low signal intensity (*arrows*) involving the L3-L4 disk and about half of the vertebral bodies on either side of it. There is no associated soft tissue mass or epidural abscess. Cultures of tissue specimens obtained by both computed tomography–guided biopsy and open biopsy did not grow any organisms.

result. Diabetics, drug abusers, and immunosuppressed patients are more likely to experience this disease.

Epidural abscess is usually associated with spondylodiskitis and is then referred to as a *secondary epidural abscess* (Fig. 46–13). Occasionally, an epidural abscess is associated with facet joint infection. When an epidural abscess occurs independent of other spinal infections, it is referred to as a *primary epidural abscess* (Fig. 46–14). Studies using MR imaging have shown that primary epidural abscesses are much less common than secondary abscesses. Primary abscesses constitute about 12% to 29% of all epidural abscesses. Epidural abscesses are most common in the thoracic spine, followed by the lumbar spine and then the cervical spine. An epidural abscess may be focal or may be diffuse, extending to multiple levels and occasionally spreading throughout the entire spinal canal (Fig. 46–15). *S. aureus* is the most common infecting organism.

Imaging

MR imaging with gadolinium enhancement is the best modality for imaging epidural abscesses. They typically demonstrate one of the following three basic imaging patterns: (1) a phlegmon or mass of granulation tissue (see Fig. 46–12) embedded with microabscesses, (2) a localized liquid abscess collection (see Figs. 46–13 and 46–14), and (3) a diffuse abscess extending up and down the epidural space (see Fig. 46–15). Frank abscesses have a worse prognosis than phlegmons. The phlegmon appears as an isointense mass with the cord on T1-weighted MR images; on T2-weighted images and gadolinium-enhanced T1-weighted images, the mass appears uniformly bright (see Fig. 46–12).

The localized liquid abscess collection is essentially a pus collection surrounded by inflammatory tissue. On T1-weighted MR images, this pattern reveals a mass that is isointense with the spinal cord. The mass appears bright on T2-weighted images. It is easy to miss this collection on T2-weighted images because its signal intensity is similar to that of cerebrospinal fluid. Gadolinium-enhanced T1-weighted images are diagnostic because they show a dark central fluid collection (abscess) surrounded by a high-intensity halo (see Figs. 46–13 and 46–14).

The diffuse pattern of epidural abscess is also difficult to see on T1- and T2-weighted images (see Fig. 46–15); however, diffuse collections become more obvious when the interface between the abscess and the dura enhances after injection of gadolinium.

SEPTIC ARTHRITIS OF THE FACET JOINT (PYOGENIC FACET ARTHROPATHY)

Overview

Septic arthritis of the facet joint is a rare entity, with only a few cases described in the literature. The infection occurs by hematogenous spread and is usually caused by *S. aureus.* All the cases reported in the literature so far have involved the lumbar spine (Figs. 46–16 and 46–17). The infection typically occurs at the L4-L5 facet level (see Fig. 46–16). Patients present with severe low back pain and fever. The pain is present at rest and with activity and is typically unilateral. Clinically, it is very difficult to differentiate this condition from spondylodiskitis. Septic arthritis of the facet joint can be associated with an epidural abscess (see Figs. 46–16 and 46–17). Facet joint aspiration performed with CT control followed by culture of the aspirate is the most direct way of making this diagnosis (see Fig. 46–17). When the infection is limited to the facet joint, aspiration may be therapeutic, obviating surgery.

Text continued on page 367

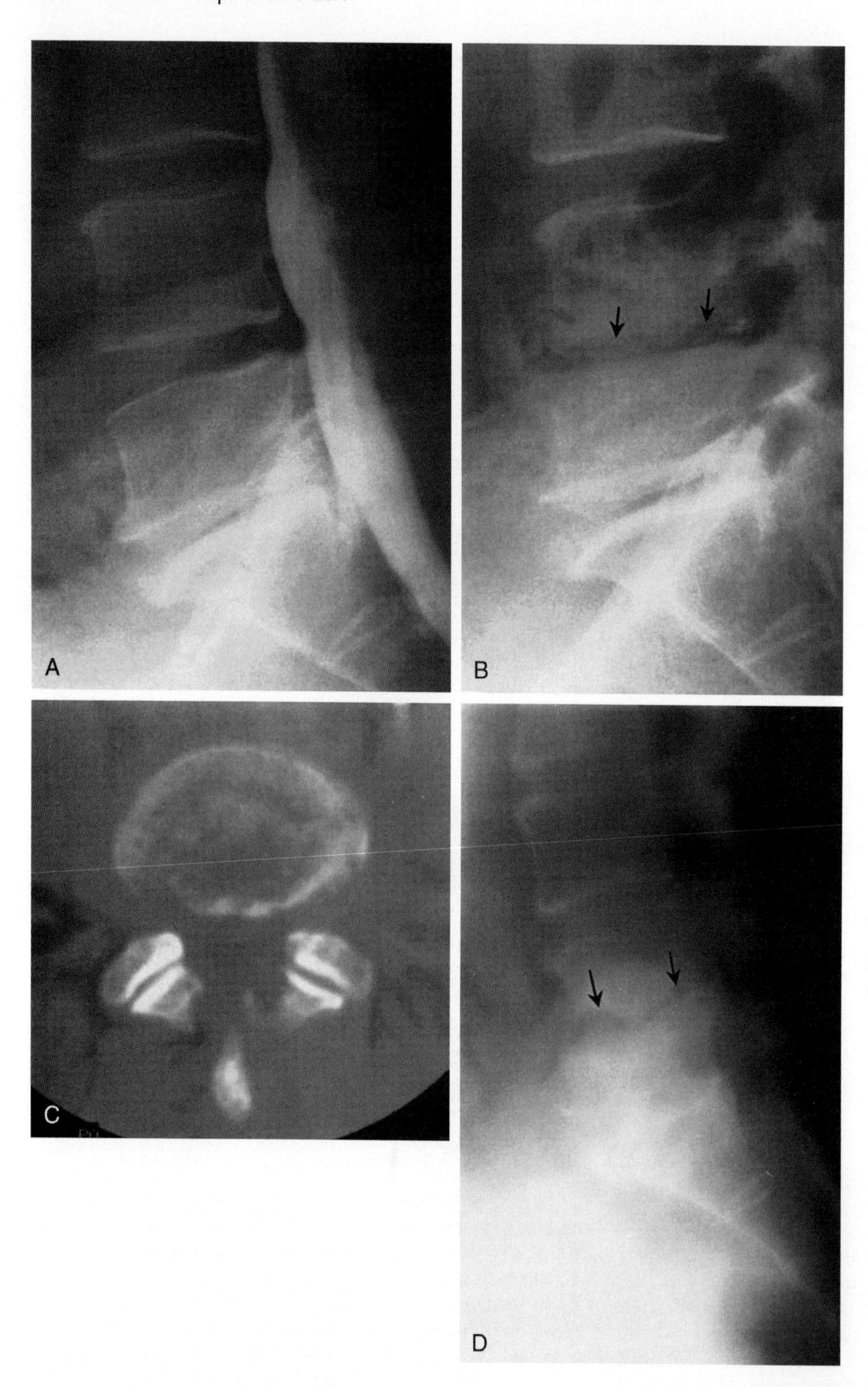

Figure 46–11. Postoperative spondylodiskitis in a 51-year-old man who started to experience symptoms a few weeks after a diskectomy at L4-L5. *A*, Cross-table lateral radiograph obtained during a myelogram before the operation shows a disk protrusion at the L4-L5 level. The protruding fragment was surgically resected, and the immediate postoperative course was uneventful. *B*, Lateral radiograph taken 2 months after the operation and about 5 weeks after the new symptoms of back pain and muscle spasm began. There is significant narrowing of the L4-L5 disk space, and the inferior end plate of L4 appears eroded (*arrows*). *C*, Computed tomography section through L4-L5 disk shows the end plate erosions. Some soft tissue swelling is noted surrounding the L4-L5 disk. *D*, Lateral tomogram obtained about 7 months after the operation shows extensive vertebral erosions (*arrows*) and sclerosis on both sides of the disk. Multiple needle biopsies and cultures of biopsy specimens did not yield an organism.

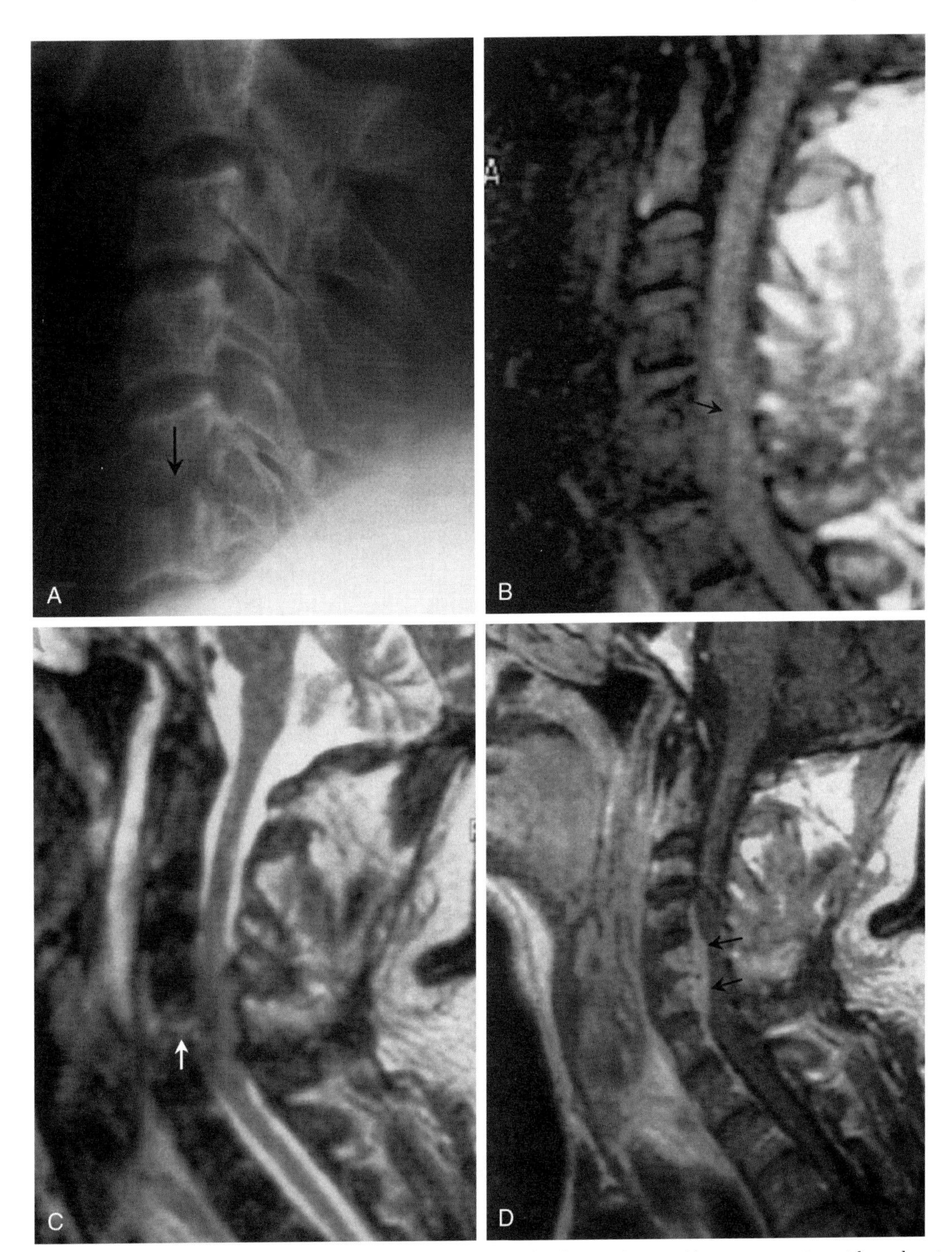

Figure 46–12. Epidural phlegmon secondary to spondylodiskitis at C5-C6 level in a 59-year-old man presenting with neck pain, fever, and paraparesis. *A,* Lateral radiograph of the cervical spine shows changes of spondylodiskitis at C5-C6, with disk narrowing and erosions at the inferior end plate of C5 (*arrow*). *B,* Midline sagittal T1-weighted magnetic resonance (MR) image shows an epidural mass behind C5 and C6 vertebrae (*arrow*). The mass is seen compressing the cord. *C,* T2-weighted MR image shows increased signal in the infected disk and adjacent end plates (*arrow*). The thecal sac is obliterated by the epidural abscess. *D,* T1-weighted MR image obtained after gadolinium injection shows an enhancing mass in the epidural space (*arrows*). This appearance is typical of a phlegmon. The cord is compressed and displaced posteriorly by the phlegmon.

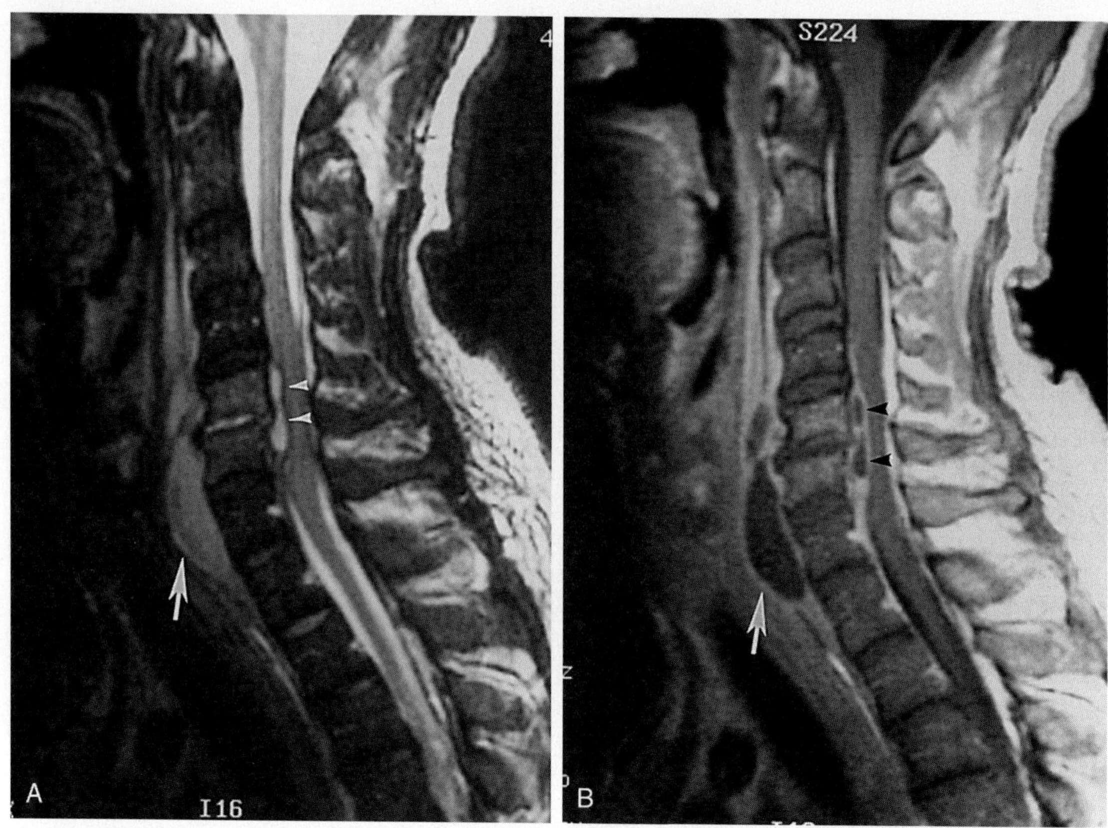

Figure 46–13. Secondary epidural abscess in a 48-year-old man who presented with neck pain and progressive weakness in the arm and leg. *A*, Midline sagittal T2-weighted magnetic resonance (MR) image shows increased signal intensity in the C5 and C6 vertebral bodies and the corresponding intervertebral disk. In addition, there are epidural (*arrowheads*) and prevertebral (*arrow*) abscess collections. The epidural abscess can be seen compressing the spinal cord. *B*, Contrast-enhanced midline sagittal T1-weighted MR image demonstrates homogeneous enhancement of the C5 and C6 vertebral bodies and intervertebral disk. There is also peripheral enhancement of the epidural (*arrowheads*) and prevertebral (*arrow*) abscess collections.

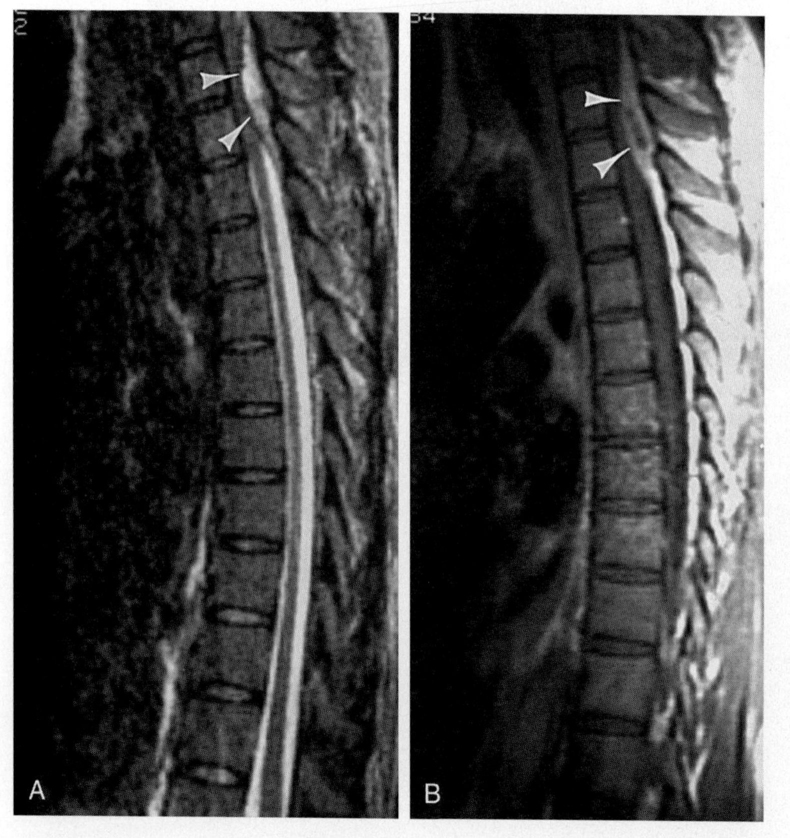

Figure 46–14. Primary epidural abscess of the thoracic spine in a 32-year-old HIV-positive woman who presented with lower extremity hyperreflexia and decreased anal sphincter function. *A*, Midline sagittal T2-weighted magnetic resonance (MR) image demonstrates a small hyperintense posterior epidural abscess (*arrowheads*) in the upper thoracic spine that is causing cord compression. The adjacent vertebrae, disks, and facet joints are all normal. *B*, Contrast-enhanced midline sagittal T1-weighted MR image shows a peripherally enhancing epidural abscess (*arrowheads*) that is compressing the cord. There is no abnormal enhancement in adjacent vertebrae, disks, or facet joints.

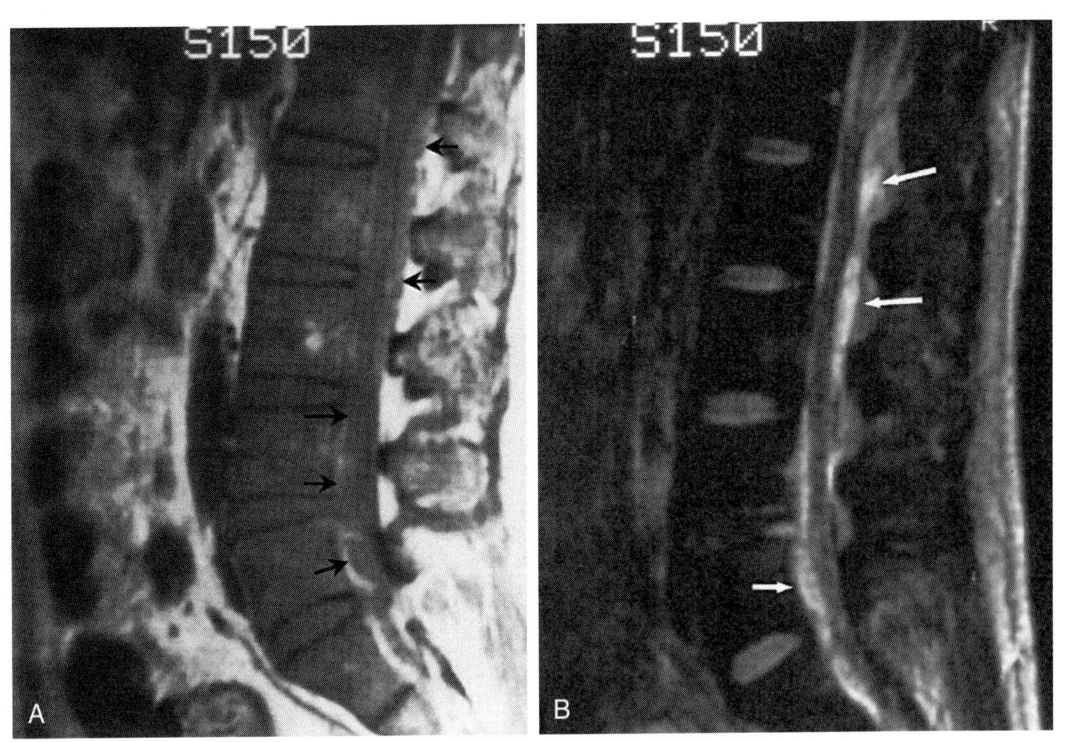

Figure 46–15. Diffuse epidural abscess in a 29-year-old intravenous drug abuser. The patient was referred to our institution for treatment of subacute bacterial endocarditis. At presentation, he had excruciating back pain, *Staphylococcus aureus* bacteremia, and shock. Sagittal T1-weighted (*A*) and T2-weighted (*B*) magnetic resonance images show epidural fluid collections throughout the lumbar and lower thoracic spine (*arrows*). These collections are compressing and distorting the thecal sac. The patient was too sick to allow the examination to continue. At surgery, an extensive epidural abscess was found to extend throughout the lower thoracic and lumbar spine.

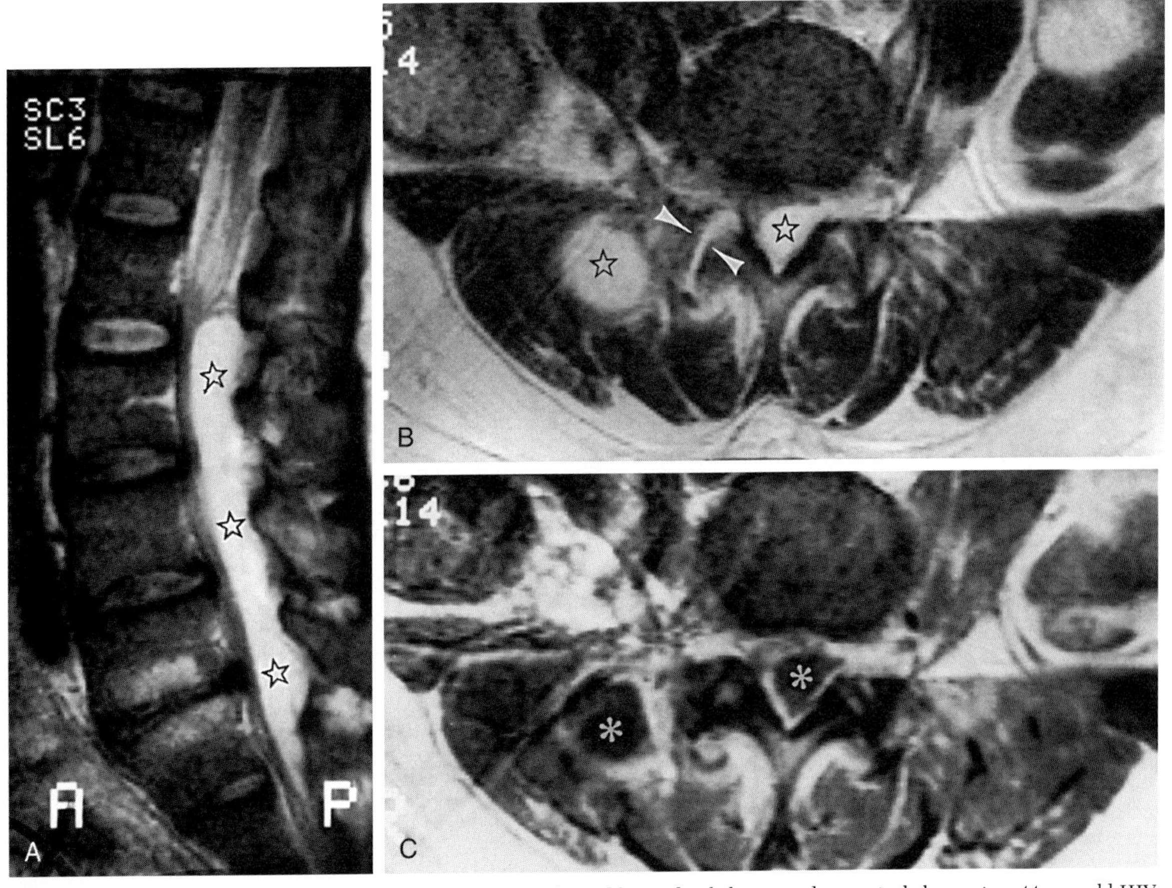

Figure 46–16. Septic arthritis of the right facet joint at L4-L5 complicated by epidural abscess and paraspinal abscess in a 44-year-old HIV-positive man. He presented with lower extremity weakness and urinary incontinence. *A,* Midline sagittal T2-weighted magnetic resonance (MR) image shows a large hyperintense posterior epidural abscess (*stars*) extending from the bottom of L1 to L5 and obliterating the thecal sac. *B,* Axial T2-weighted MR image at the L4-L5 level shows septic arthritis of the right facet (hyperintensity in the facet joint and erosions of the articulating facets) (*arrowheads*) with associated hyperintense posterior epidural and right paraspinal abscess collections (*stars*). *C,* Contrast-enhanced axial T1-weighted MR image at the L4-L5 level shows peripherally enhancing epidural (*asterisk in the spinal canal*) and right paraspinal (*asterisk on the left*) abscesses.

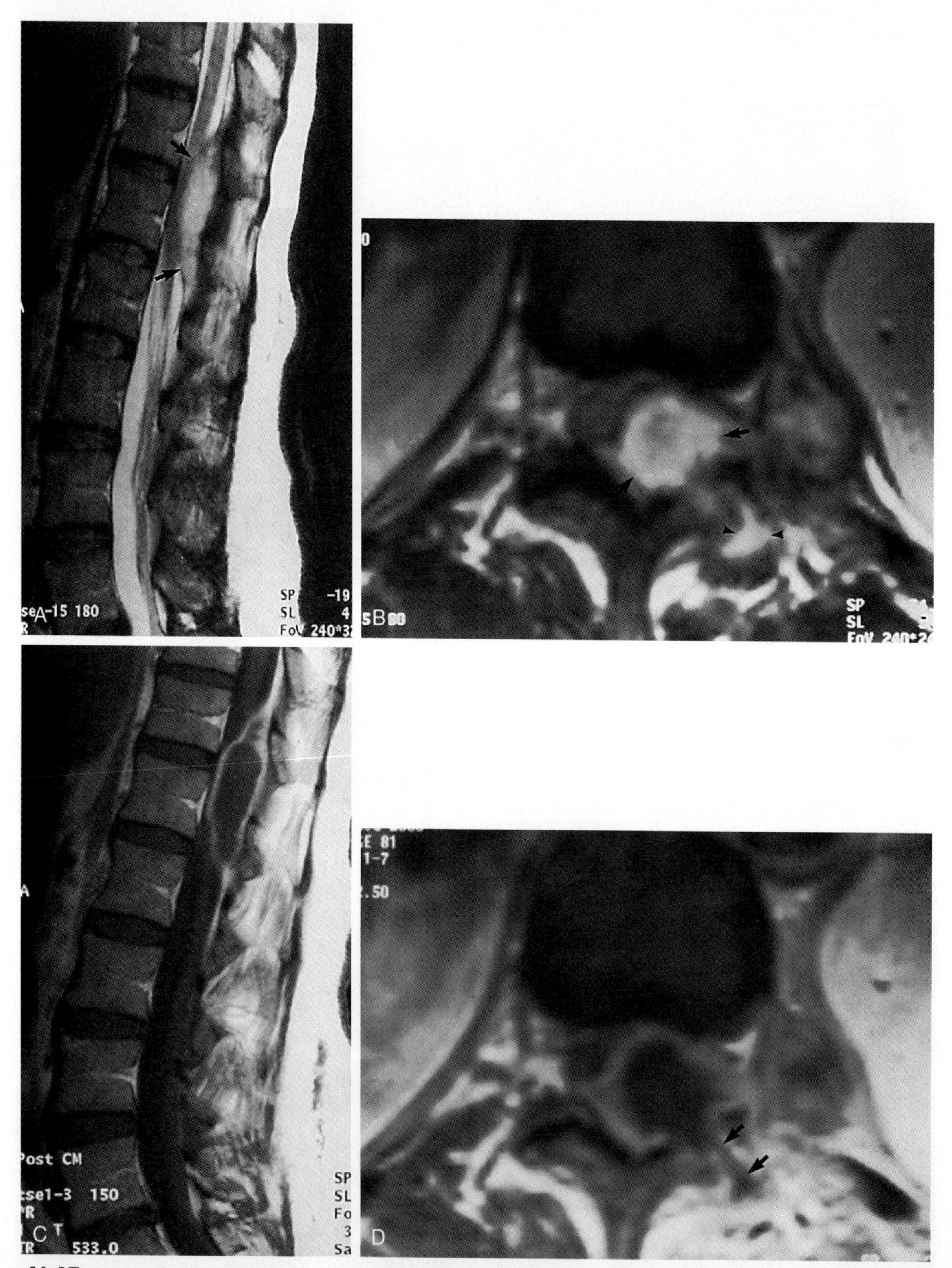

Figure 46–17. Septic arthritis of the left T12-L1 facet joint in a 52-year-old woman with metastatic ovarian cancer who was undergoing chemotherapy. *A*, Midline sagittal T2-weighted magnetic resonance (MR) image of the lumbar spine shows a large fusiform mass compressing the distal cord and conus (*arrows*). *B*, Axial T2-weighted MR image at the T12-L1 disk level shows a hyperintense mass (*arrows*) displacing the cord to the right and compressing it. The left facet joint is distended with fluid (*arrowheads*). Sagittal (*C*) and axial (*D*) gadolinium-enhanced T1-weighted MR images show rim enhancement of the mass. On the axial image (*D*), the mass seems to communicate with facet joint on the left (*arrows*).

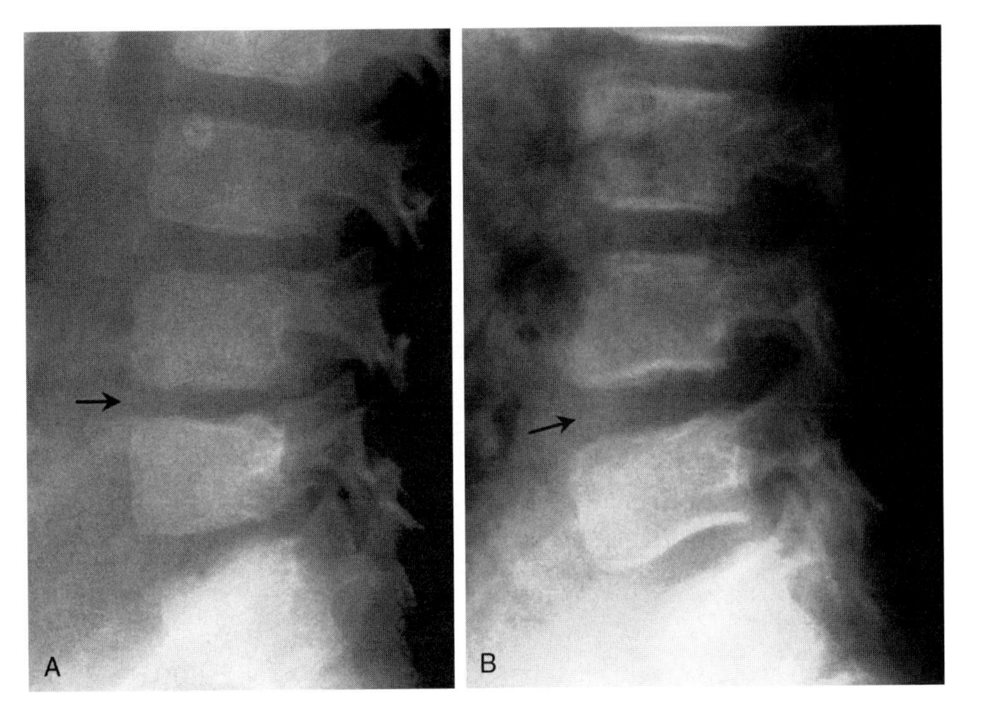

Figure 46–19. Childhood diskitis with complete recovery of the loss in disk height. *A*, Lateral radiograph of the lumbar spine at presentation, when the patient was 5 years old, shows significant narrowing of the L4-L5 disk space (*arrow*). The bony structures are intact. *B*, Lateral radiograph of the lumbar spine obtained when the patient was 10 years old shows that the disk height at L4-L5 has returned to normal (*arrow*).

Imaging

Radiographic Findings

The disease starts either in the anteroinferior part of the vertebral body or within the paravertebral soft tissues. Focal erosions and osseous destruction can be detected in the anteroinferior corner of the vertebral body, a finding characteristic of tuberculous spondylitis (Fig. 46–20). Initially, the process is purely lytic and is almost always confined to the anterior portion and vertebral body and adjacent end plate (see Fig. 46–20). It leads to collapse of the anterior body in the subsequent kyphosis. Less often, the posterior elements of the vertebra can become involved. Multiple vertebral levels may be affected in noncontiguous fashion (skip lesions). Involvement of the adjacent vertebrae occurs by spread of the abscess

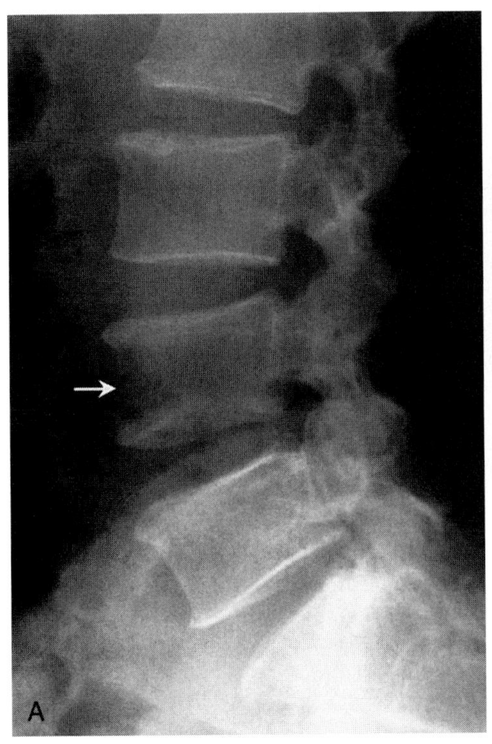

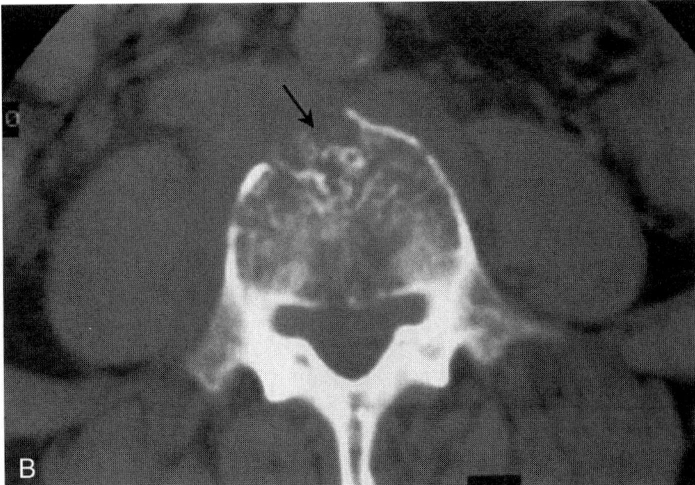

Figure 46–20. Early tuberculous spondylitis manifesting as erosions at the anterior margin of L4. The patient, a 69-year-old Vietnamese man, presented with low back pain unrelated to activity. *A*, Lateral radiograph of the lumbar spine shows a destructive lesion in the anterior cortex of L4 (*arrow*). The remainder of the vertebral body appears osteopenic. *B*, Computed tomography (CT) section shows a large erosion of the anterior cortex of L5 (*arrow*). The erosion is associated with a paraspinal soft tissue mass. The diagnosis was confirmed by CT-guided needle biopsy and culture.

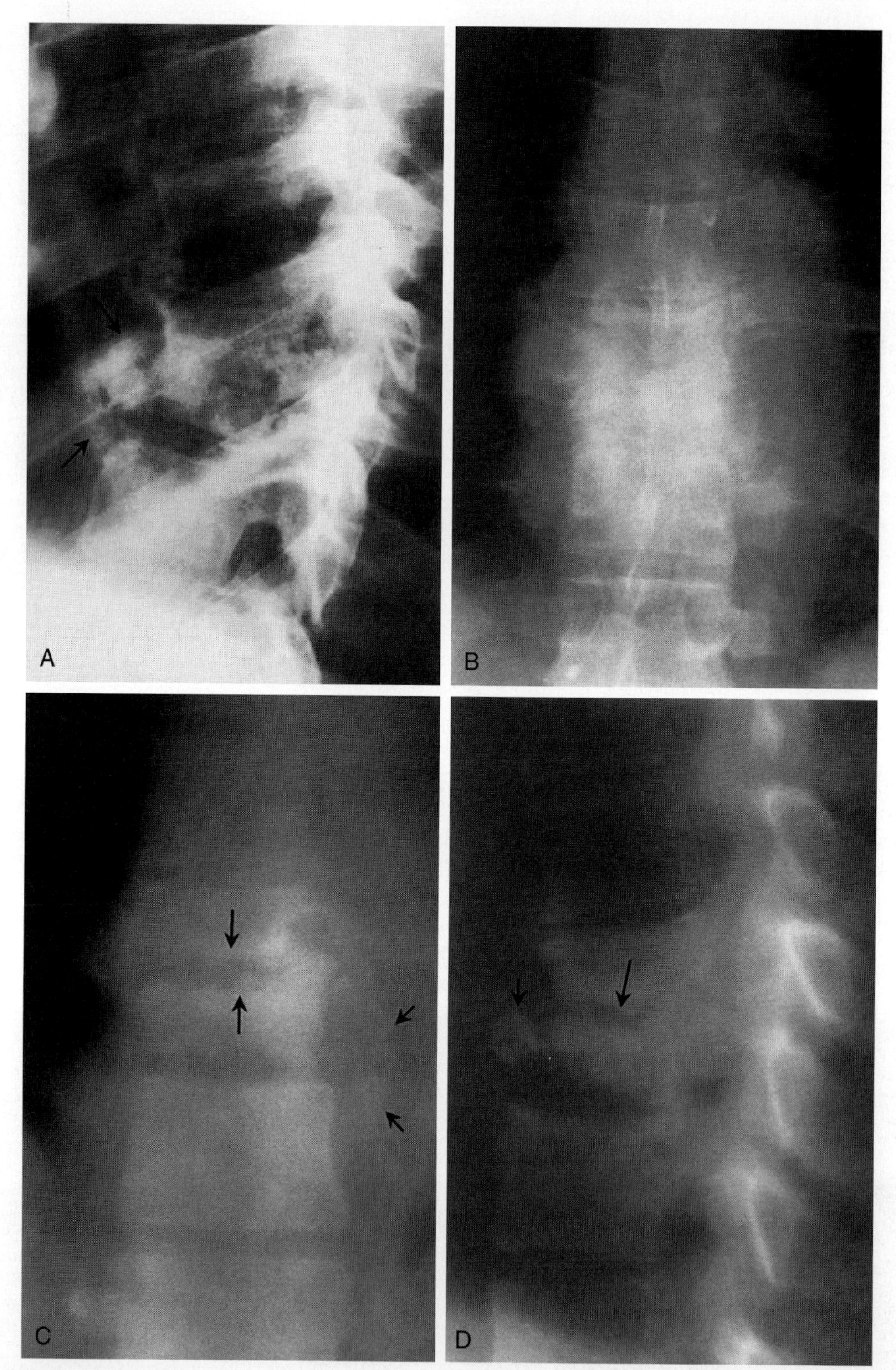

Figure 46–21. Tuberculous spondylitis in a 34-year-old man who presented with a 3-month history of "stiffness in the back." He was neurologically intact. Culture of tissue obtained at open biopsy revealed tuberculosis. Anteroposterior (*A*) and lateral (*B*) radiographs of the thoracic spine show marked disk space narrowing and vertebral erosions at the T8-T9 level. Paravertebral soft tissue swelling and calcifications are also shown (*arrows* in *A*). Anteroposterior (*C*) and lateral (*D*) tomographic views show disk abnormality, end plate erosions at T8-T9 (*long arrows*), and paravertebral calcifications (*short arrows*), findings characteristic of tuberculosis.

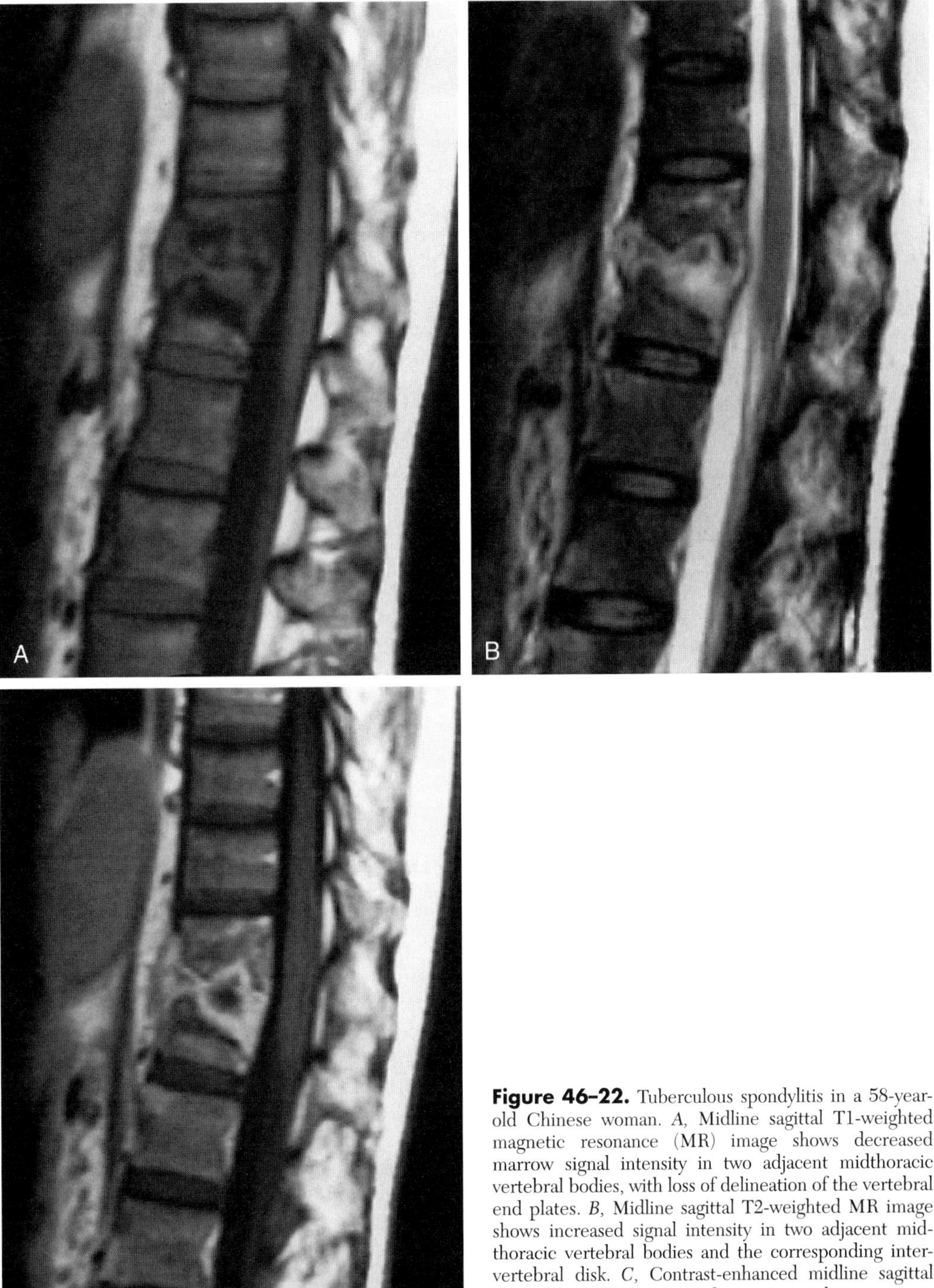

Figure 46–22. Tuberculous spondylitis in a 58-year-old Chinese woman. *A,* Midline sagittal T1-weighted magnetic resonance (MR) image shows decreased marrow signal intensity in two adjacent midthoracic vertebral bodies, with loss of delineation of the vertebral end plates. *B,* Midline sagittal T2-weighted MR image shows increased signal intensity in two adjacent midthoracic vertebral bodies and the corresponding intervertebral disk. *C,* Contrast-enhanced midline sagittal T1-weighted MR image demonstrates homogeneous enhancement of the involved vertebral bodies and ringlike enhancement of the intervertebral disk with central liquefaction.

Illustration continued on following page

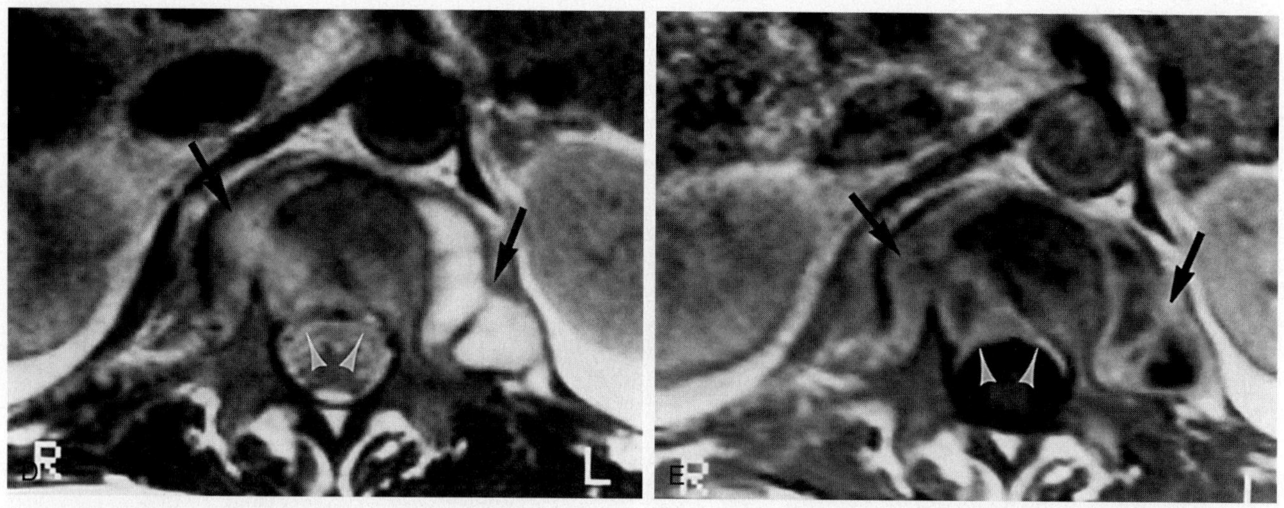

Figure 46–22 *Continued. D,* Axial T2-weighted MR image at the level of the infection demonstrates a collection (abscess) with high signal intensity in the intervertebral disk extending into the paraspinal (*arrows*) and epidural (*arrowheads*) space. *E,* Contrast-enhanced axial T1-weighted MR image at midthoracic level demonstrates a heterogeneously enhancing abscess collection in the intervertebral disk that extends into the paraspinal (*arrows*) and epidural (*arrowheads*) spaces.

beneath the anterior longitudinal ligaments. Early, the disk is not involved by the infection; it tends, however, to protrude into the destroyed vertebra, giving the impression of disk space widening.

Paraspinal abscess formation is common and can be detected on radiographs as areas of fusiform soft tissue swelling around the spine. Calcification within the abscess is virtually pathognomonic of tuberculosis (Fig. 46–21).

Radionuclide Scanning

Findings of both bone scans and gallium scans are often normal in tuberculous spondylitis, despite the presence of active disease.

Computed Tomography

CT is very effective in demonstrating bony and soft tissue changes in spinal tuberculosis (see Fig. 46–21). It also demonstrates the paraspinal and psoas abscesses very well.

Magnetic Resonance Imaging

MR imaging is the modality of choice in patients with neurologic findings. It is the best technique for demonstrating soft tissue abscesses, epidural abscesses, and cord compression. MR imaging has become the main imaging modality for evaluation of patients with tuberculous spondylitis (Fig. 46–22).

Differential Diagnosis

Tuberculous spondylitis can mimic metastatic disease, low-grade pyogenic spondylodiskitis, brucellosis, and fungal infections. Brucellosis is favored by the finding of a small paraspinal abscess or absence of such an abscess. Also with brucellosis, the lower lumbar spine is typically involved and kyphosis is absent.

FUNGAL SPONDYLITIS

Clinical Features

Fungal infections of the spine typically occur as opportunistic infections in immunocompromised patients. Predisposing factors include the use of corticosteroids and immunosuppressive drugs, HIV infection, injudicious use of broad-spectrum antibiotics, and parenteral hyperalimentation. Delay in diagnosis is quite common and often leads to poorer prognosis. The long-term clinical outcome is related not to the specific species of fungus but rather to the time between the onset of symptoms and the initiation of treatment. So that delay in diagnosis may be avoided, fungal cultures should be performed on aspirated material whenever a spinal infection is suspected. *Candida albicans* is the most common infecting organism in some series.

Imaging

The imaging features of fungal infections are similar to those of pyogenic injections. Plain radiographs typically reveal end plate erosions and paravertebral soft tissue swelling. CT demonstrates the bony and soft tissue changes to better advantage (Fig. 46–23). MR imaging is very useful in delineating the epidural extension of the infection as well as the cord compression.

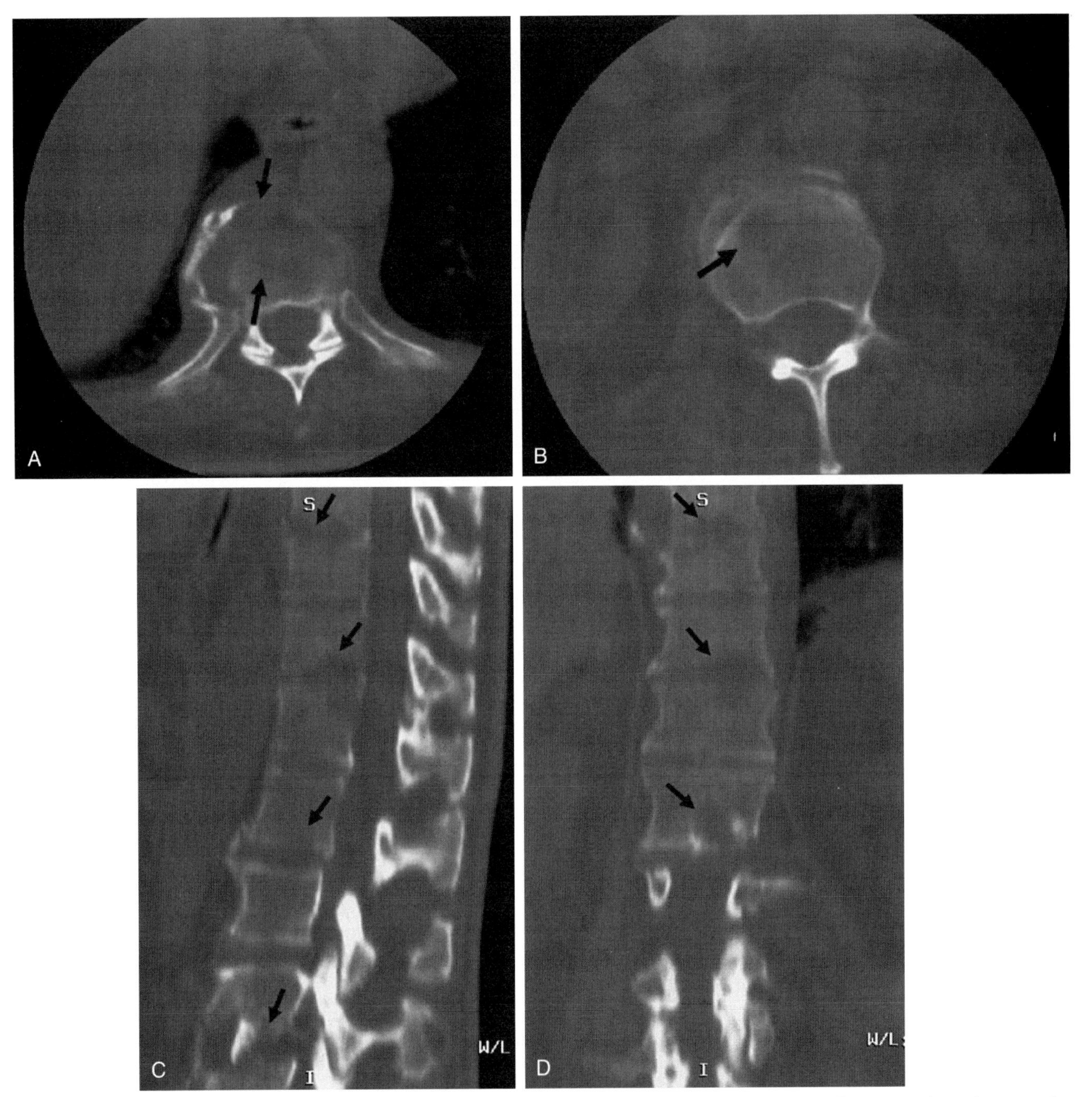

Figure 46–23. Multilevel spondylodiskitis due to *Candida albicans* in a 55-year-old man who is immunologically suppressed. Axial computed tomography (CT) sections from the thoracic (*A*) and lumbar (*B*) spine show areas of bone destruction (*arrows*). Sagittal (*C*) and coronal (*D*) CT reconstructions illustrate multilevel spondylodiskitis (*arrows*). The lower site of involvement at L3-L4 (seen in *C*) has undergone curettage and grafting.

Recommended Readings

Al-Mulhim FA, Ibrahim EM, El-Hassan AY, et al: Magnetic resonance imaging of tuberculous spondylitis. Spine 20:2287, 1995.

Baltz MS, Tate DE, Glaser JA: Lumbar facet joint infection associated with epidural and paraspinal abscess. Clin Orthop 339:109, 1997.

Boden SD, Davis DO, Dina TS, et al: Postoperative diskitis: Distinguishing early MR imaging findings from normal postoperative disk space changes. Radiology 184:765, 1992.

Dagirmanjian A, Schils J, McHenry M, et al: MR imaging of vertebral osteomyelitis revisited. AJR Am J Roentgenol 167:1539, 1996.

Ehara S, Khurana JS, Kattapuram SV: Pyogenic vertebral osteomyelitis of the posterior elements. Skeletal Radiol 18:175, 1989.

Eismont FJ, Bohlman HH, Soni PL, et al: Pyogenic and fungal vertebral osteomyelitis with paralysis. J Bone Joint Surg [Am] 65A:19, 1983.

Frazier DD, Campbell DR, Garvey TA, et al: Fungal infections of the spine. J Bone Joint Surg [Am] 83A:560, 2001.

Friedman DP, Hills JR: Cervical epidural spinal infection: MR imaging characteristics. AJR Am J Roentgenol 163:699, 1994.

Gillams AR, Chaddha B, Carter AP: MR appearances of the temporal evolution and resolution of infectious spondylitis. AJR Am J Roentgenol 166:903, 1996.

Govender S, Rajoo R, Goga IE, et al: Aspergillus osteomyelitis of the spine. Spine 16:746, 1991.
Hadjipavlou AG, Mader JT, Necessary PA-C, et al: Hematogenous pyogenic spinal infections and their surgical management. Spine 25:1668, 2000.
Kramer N, Rosenstein ED: Rheumatologic manifestations of tuberculosis. Bull Rheum Dis 46:5, 1997.
Kricun R, Shoemaker EI, Chovanes GI, et al: Epidural abscess of the cervical spine: MR findings in five cases. AJR Am J Roentgenol 158:1145, 1992.
Kuker W, Mull M, Lothar M, et al: Epidural spinal infection. Spine 22:544, 1997.
Modic MT, Feiglin DH, Piraino DW, et al: Vertebral osteomyelitis: Assessment using MR. Radiology 157:157, 1985.
Numaguchi Y, Rigamonti D, Rothman MI, et al: Spinal epidural abscess: Evaluation with gadolinium-enhanced MR imaging. RadioGraphics 13:545, 1993.
Olson EM, Duberg AC, Herron LD, et al: Coccidioidal spondylitis: MR findings in 15 patients. AJR Am J Roentgenol 171:785, 1998.
Post MJD, Quencer RM, Montalvo BM, et al: Spinal infection: Evaluation with MR imaging and intraoperative US. Radiology 169:765, 1988.
Rawlings CE III, Wilkins RH, Gallis HA, et al: Postoperative intervertebral disc space infection. Neurosurgery 13:371, 1983.
Rombauts PA, Linden PM, Buyse AJ, et al: Septic arthritis of a lumbar facet joint caused by *Staphylococcus aureus*. Spine 25:1736, 2000.
Sandhu FS, Dillion WP: Spinal epidural abscess: Evaluation with contrast-enhanced MR imaging. AJR 158:405, 1992.
Sartoris DJ, Moskowitz PS, Kaufman RA: Childhood diskitis: Computed tomographic findings. Radiology 149:701, 1983.
Sharif HS, Aideyan OA, Clark DC, et al: Brucellar and tuberculous spondylitis: Comparative imaging features. Radiology 171:419, 1989.
Sharif HS, Clark DC, Aabed MY, et al: Granulomatous spinal infections: MR imaging. Radiology 177:101, 1990.
Szalay EA, Green NE, Heller RM, et al: Magnetic resonance imaging in the diagnosis of childhood discitis. J Pediatr Orthop 7:164, 1987.
Weaver P, Lifeso RM: The radiological diagnosis of tuberculosis of the adult spine. Skeletal Radiol 12:178, 1984.
Wong-Chung JK, Naseeb SA, Kaneker SG, et al: Anterior disc protrusion as a cause for abdominal symptoms in childhood discitis: A case report. Spine 24:918, 1999.

CHAPTER 47

Diffuse Idiopathic Skeletal Hyperostosis and Ossification of the Posterior Longitudinal Ligament

Timothy E. Moore, MD

DIFFUSE IDIOPATHIC SKELETAL HYPEROSTOSIS

Overview

Diffuse idiopathic skeletal hyperostosis (DISH) is a common condition characterized by flowing ossifications along the anterolateral vertebral bodies but without significant disk degeneration, facet joint ankylosis, or sacroiliac joint fusion. DISH has been known in the literature by multiple names, but between 1975 and 1978, Resnick and colleagues published several reports describing a large number of patients with spinal and extraspinal manifestations of this condition and coined the term "diffuse idiopathic skeletal hyperostosis," or DISH.

The disease is most common in white men older than 50 years, in whom the prevalence is about 12% to 25%. The etiology of DISH is still uncertain. Patients are typically healthy, and symptoms are rare except for back stiffness and occasional dysphagia (Fig. 47–1). Patients with associated ossification of the posterior longitudinal ligament (OPLL) may have neurologic symptoms as a result of cord compression and myelopathy.

Imaging

Radiography

The radiographic changes in DISH are most common (97%) in the lower thoracic spine centered around T9. By definition, four or more contiguous vertebral bodies show anterolateral flowing osteophytes or ossifications (Fig. 47–2). The disk heights are preserved, and the facet joints and sacroiliac joints remain unfused. Involvement of the cervical and lumbar spine is common, but not without thoracic involvement. In fact, the diagnosis of DISH is difficult to establish without thoracic involvement. The right side of the spine is more involved with these ossifications, probably because aortic pulsations inhibit their formation on the left side (Fig. 47–3). OPLL can

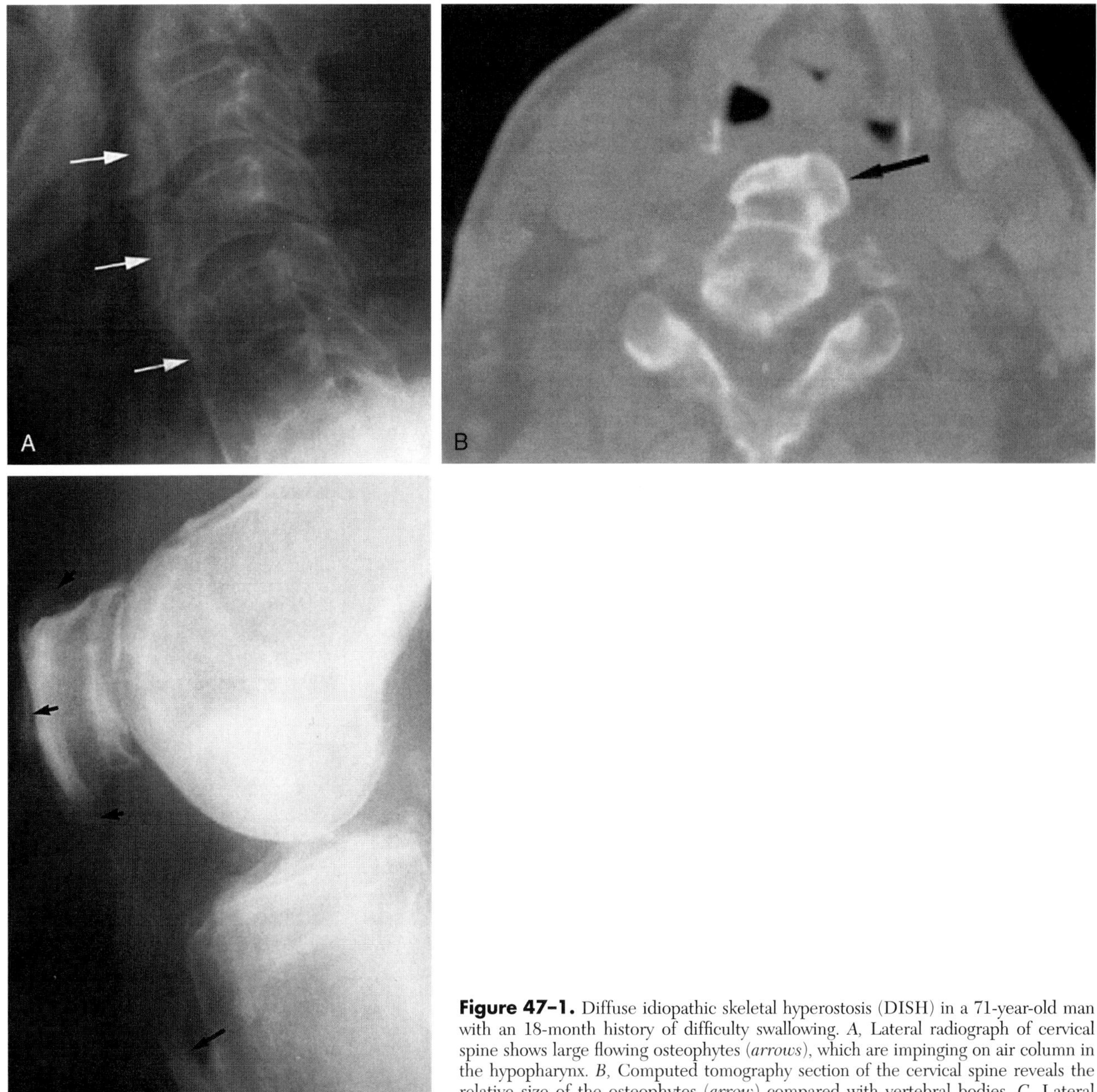

Figure 47–1. Diffuse idiopathic skeletal hyperostosis (DISH) in a 71-year-old man with an 18-month history of difficulty swallowing. *A*, Lateral radiograph of cervical spine shows large flowing osteophytes (*arrows*), which are impinging on air column in the hypopharynx. *B*, Computed tomography section of the cervical spine reveals the relative size of the osteophytes (*arrow*) compared with vertebral bodies. *C*, Lateral radiograph of the knee reveals changes of DISH in the patella and at the tibial tubercle (*arrows*).

occur in association with DISH. This association is reported in 50% of patients with DISH of the cervical spine, where the association can produce cord compression.

Extraspinal DISH

Extraspinal DISH is most commonly seen in the pelvis, where it manifests as bony proliferation at the origins and insertions of ligaments and tendons. These findings are typically encountered at the iliac crest, the ischial tuberosity, and the greater and lesser trochanters (Fig. 47–4; see Figs. 47–1 and 47–3). The calcaneus may also have large posterior spurs, which develop at the Achilles tendon insertion, or inferior spurs at the origin of the plantar aponeurosis.

Differential Diagnosis

DEGENERATIVE DISK DISEASE

Degenerative disk disease is more common in the cervical and lumbar spine, whereas DISH is more common in the thoracic spine. Degenerative disk disease is characterized by disk space narrowing and vacuum phenomenon, which are not seen with DISH. The degenerative osteophytes are also different from the flowing ossifications seen with DISH.

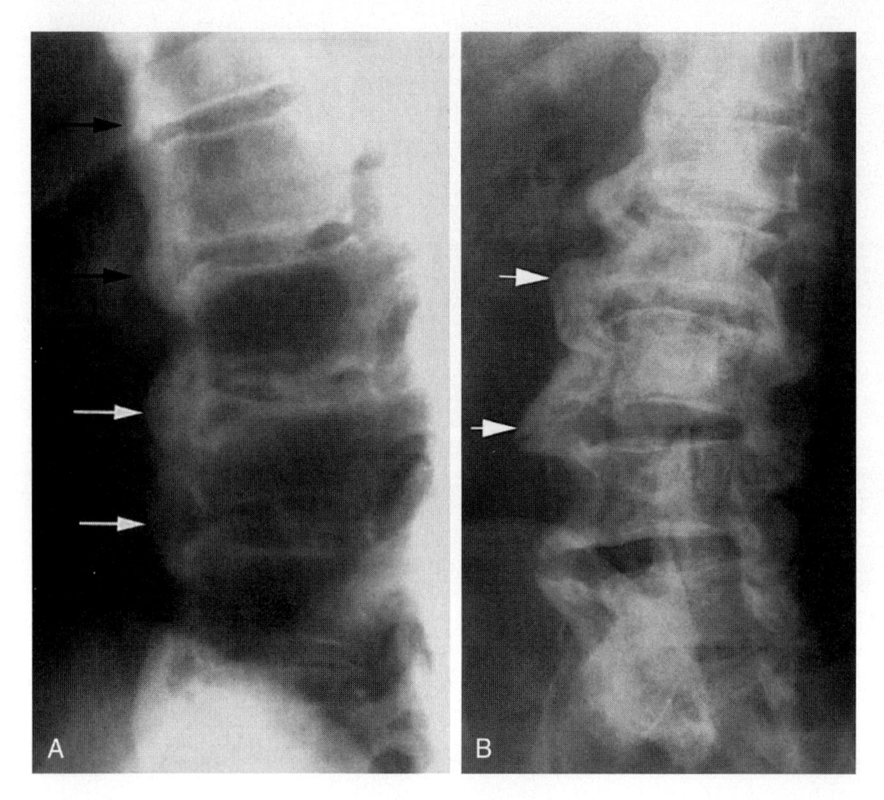

Figure 47–2. Advanced changes of diffuse idiopathic skeletal hyperostosis in the thoracic and lumbar spine. *A,* Lateral radiograph of the thoracic spine shows flowing osteophytes throughout the entire spine (*arrows*). *B,* Oblique radiograph of the lumbar spine shows huge flowing osteophytes (*arrows*). The disk spaces are well preserved.

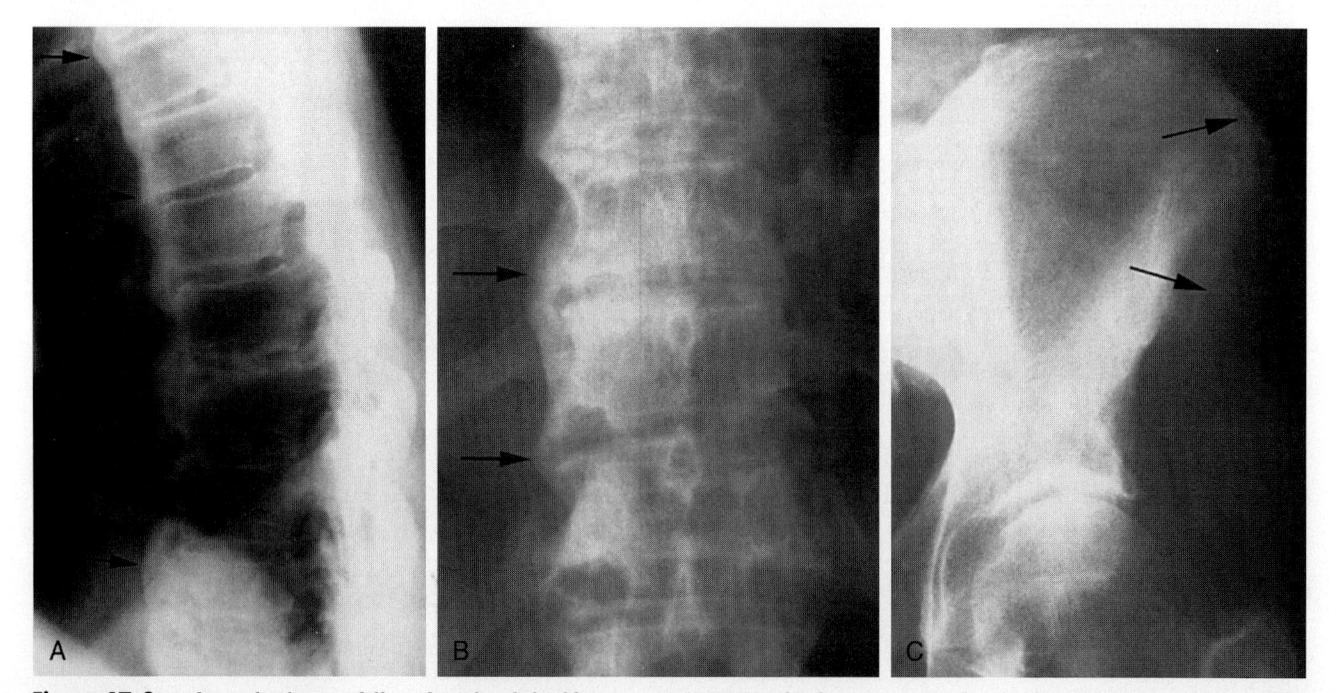

Figure 47–3. Radiographs showing diffuse idiopathic skeletal hyperostosis (DISH) in the thoracic spine and in the iliac crest. *A,* Lateral view of the thoracic spine shows extensive DISH changes with flowing osteophytes throughout the entire thoracic spine (*arrows*). *B,* Anteroposterior view of the thoracic spine reveals that the flowing osteophytes are more pronounced on the right side (*arrows*). *C,* Anteroposterior view of the left hemipelvis shows changes of DISH in the iliac crest (*arrows*).

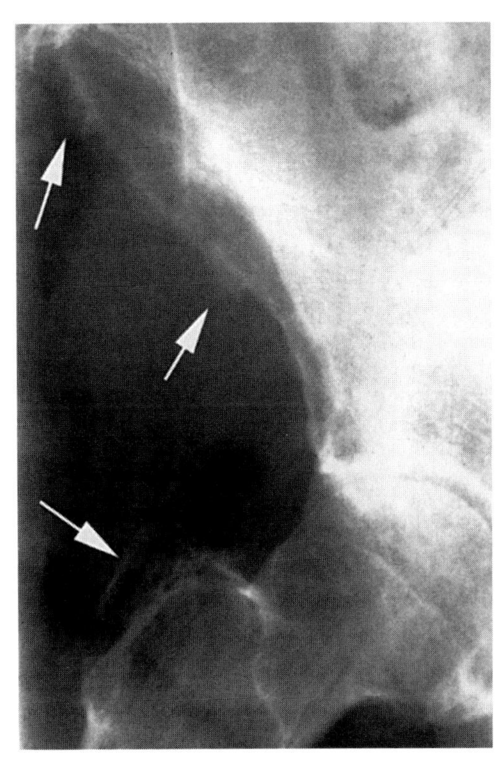

Figure 47–4. Anteroposterior radiograph of the right hip shows extraspinal diffuse idiopathic skeletal hyperostosis affecting the iliac crest and greater trochanter (*arrows*).

ANKYLOSING SPONDYLITIS

Ankylosing spondylitis (AS) is sometimes difficult to distinguish from DISH. The sacroiliac joints and facet joints are involved in ankylosing spondylitis, becoming fused in advanced cases. In DISH, the sacroiliac joints and apophyseal joints remain intact.

In ankylosing spondylitis, syndesmophytes extend between adjacent vertebrae but are initially thin and vertically oriented. There is also osteitis at the corners of the vertebral bodies as well as vertebral squaring. In DISH, the ossifications are thick and irregular, and they occur over the anterolateral aspects of the vertebral bodies.

OSSIFICATION OF THE POSTERIOR LONGITUDINAL LIGAMENT

Overview

An idiopathic condition, OPLL was first recognized as a distinct entity in the 1960s in middle-aged and elderly Japanese men. Subsequently it has been found to occur in other races, in women, and also in association with a variety of disorders of the spine, including diffuse idiopathic skeletal hyperostosis and retinoid-related changes.

OPLL may be asymptomatic but can also cause significant spinal stenosis with cord compression. There is a relationship between the thickness of the ossified ligament and the severity of clinical findings. The ligamentum flavum may also become ossified, further contributing to spinal stenosis.

The midcervical spine is the most commonly involved region and usually C5–C6 is the first level to be involved. The calcification tends to extend in both cranial and caudal directions as the process progresses. However, involvement of the thoracic and lumbar spine is fairly uncommon.

Imaging

Radiography

Lateral views demonstrate the calcified posterior longitudinal ligament as a band of variable thickness lying posterior to the intervertebral disks and vertebral bodies. It may show a segmental distribution, lying posterior to the disks only (Fig. 47–5) or the vertebral bodies only. In other cases, the band may appear as a continuous ossified structure that extends posterior to several vertebral levels (Fig. 47–6). The ossification may be fused to the posterior aspects of the vertebral bodies or may be separated from them by a thin radiolucent line.

Computed Tomography

The inherent tissue contrast in computed tomography (CT) clearly delineates the ossified ligament lying posterior to the vertebral bodies. Sagittal two-dimensional

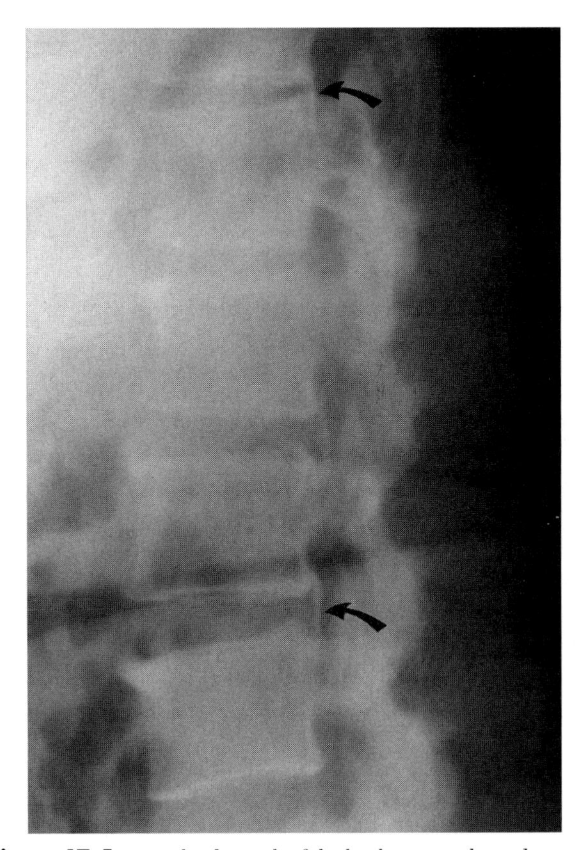

Figure 47–5. Lateral radiograph of the lumbar spine shows discontinuous ossification of the posterior longitudinal ligament (*arrows*) posterior to intervertebral disks.

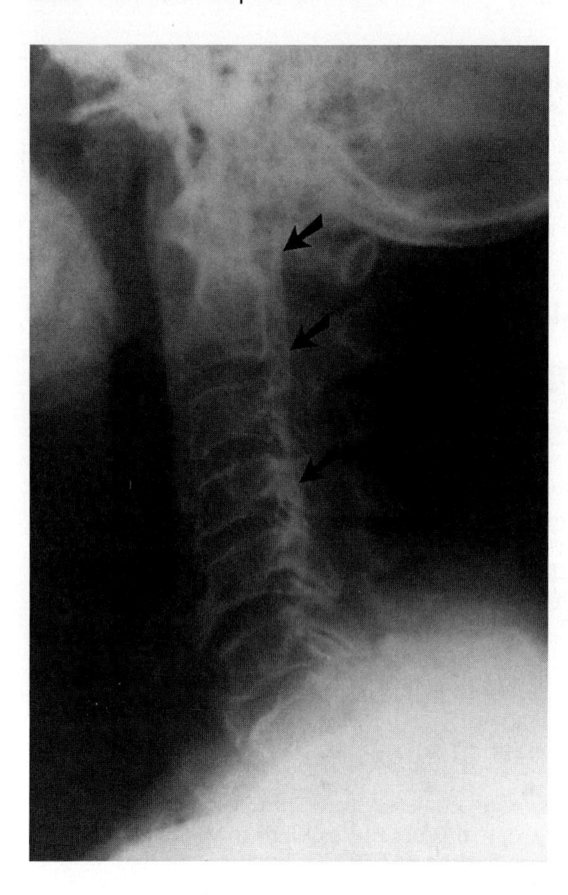

Figure 47–6. Lateral radiograph of the cervical spine shows continuous ossified posterior longitudinal ligament (*arrows*) composed of cancellous bone.

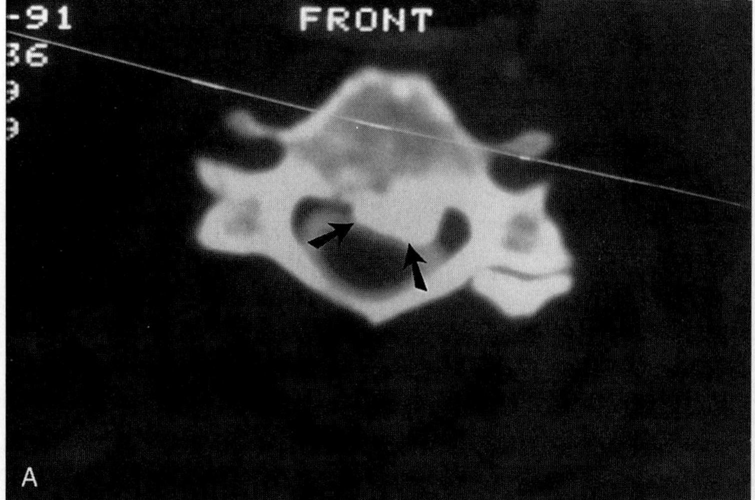

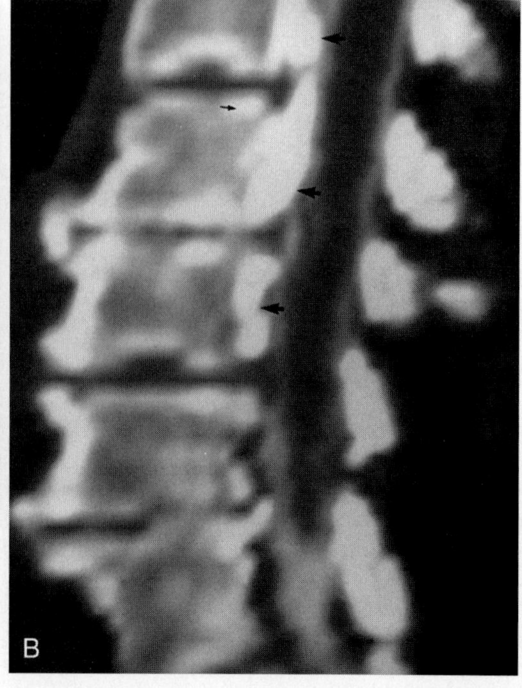

Figure 47–7. Computed tomography myelogram of ossification of the posterior longitudinal ligament. *A*, Axial section in the cervical spine shows thick ossified posterior longitudinal ligament (*arrows*) with spinal stenosis. *B*, Sagittal reconstruction of the lumbar region in same patient shows thick ossification of the posterior longitudinal ligament.

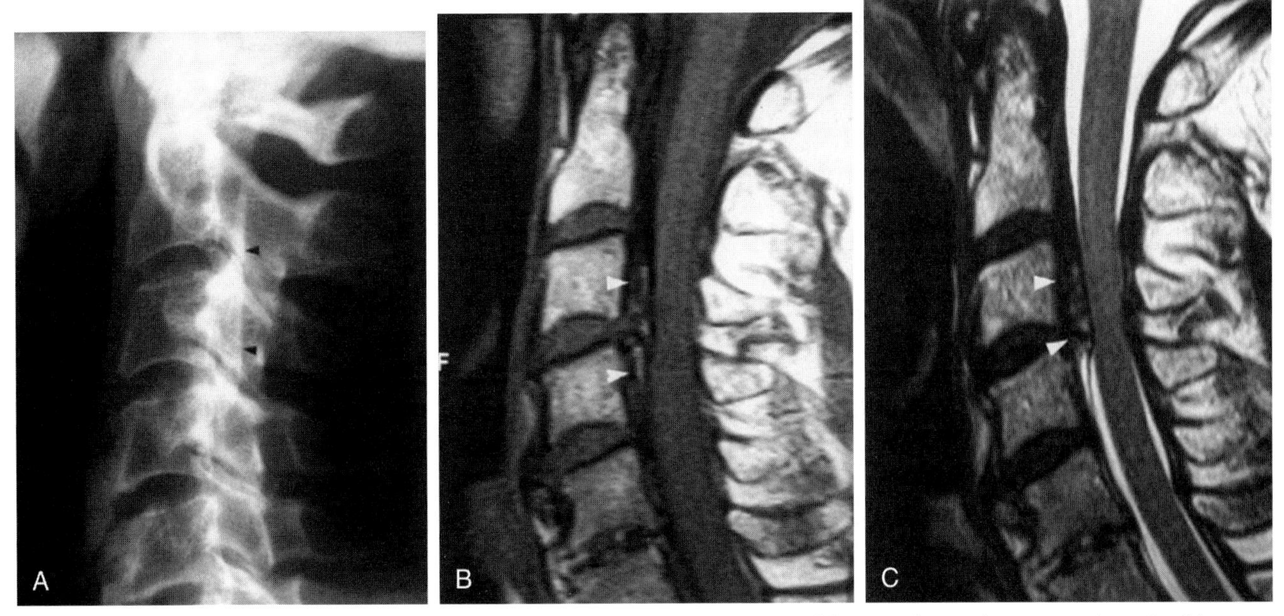

Figure 47–8. MRI image of OPLL in an elderly male with long tract signs. *A*, Lateral view of the cervical spine shows OPLL *(arrowheads)* posterior to the C2, C3, and upper C4 bodies. The patient also has DISH in the lower cervical spine. *B* and *C*, Sagittal T1- and T2-weighted images reveal both marrow and cortical signals within the ossifications *(arrowheads)*. The cord is compressed by the OPLL.

reconstructions may be helpful in appreciating the longitudinal orientation of the process (Fig. 47–7).

Magnetic Resonance Imaging

The main role of magnetic resonance (MR) imaging lies in assessment of spinal stenosis and any associated myelomalacia or syringomyelia. The ossified ligament itself may be composed of either cancellous or cortical bone and may therefore demonstrate signal characteristics of either or both of these structures (Fig. 47–8).

Recommended Readings

Deutsch EC, Schild JA, Mafee MF: Dysphagia and Forestier's disease. Arch Otolaryngol 111:400, 1985.

Griffiths ID, Fitzjohn TP: Cervical myelopathy, ossification of the posterior longitudinal ligament, and diffuse idiopathic skeletal hyperostosis: Problems in investigation. Ann Rheum Dis 46:166, 1987.

McAfee PC, Regan JJ, Bohlman HH: Cervical cord compression from ossification of the posterior longitudinal ligament in non-Orientals. J Bone Joint Surg [Br] 69:569, 1987.

Ono K, Konenobu K, Miyamoto S, Okada K: Pathology of ossification of the posterior longitudinal ligament and ligamentum flavum. Clin Orthop 359:18, 1999.

Resnick D, Guerra J Jr, Robinson CA, et al: Association of diffuse idiopathic skeletal hyperostosis (DISH) and calcification and ossification of the posterior longitudinal ligament. AJR Am J Roentgenol 131:1049, 1978.

Resnick D, Niwayama G: Radiographic and pathologic features of spinal involvement in diffuse idiopathic skeletal hyperostosis. Radiology 119:559, 1976.

Resnick D, Shaul SR, Robins JM: Diffuse idiopathic skeletal hyperostosis (DISH): Forestier's disease with extraspinal manifestations. Radiology 115:513, 1975.

Terayama K, Maruyama S, Miyashita R, et al: Ossification of the posterior longitudinal ligament in the cervical spine. Orthop Surg 15:1083, 1964.

Tsukamoto Y, Onitsuka H, Lee K: Radiographic aspects of diffuse idiopathic skeletal hyperostosis in the spine. AJR Am J Roentgenol 129:913, 1977.

CHAPTER 48

Miscellaneous Spinal Disorders

Timothy E. Moore, MD

SCHMORL'S NODES

Overview

Schmorl's node is usually reserved for larger and more focal forms of intravertebral disk penetration than occurs in Scheuermann's disease and juvenile lumbar osteochondrosis. These foci of disk and end plate cartilage are often referred to as *cartilaginous nodes.* They vary in diameter from a few millimeters to a centimeter. Schmorl's nodes are fairly common and are observed in about 20% of asymptomatic spines. Isolated Schmorl's nodes (as opposed to the multiple small nodes seen in Scheuermann's disease and juvenile lumbar osteochondrosis) are of no clinical significance apart from their occasional confusion with significant pathologic lesions. There is, however, an association with the development of degenerative disease in the adjacent disk.

Schmorl's nodes are usually seen in the thoracic and lumbar spine. They are uncommon in the cervical spine.

Imaging

Radiography

Schmorl's nodes are best seen on lateral thoracic and lumbosacral spine radiographs (Fig. 48–1) but may also be seen on anteroposterior views. They appear as well-defined, rounded, radiolucent lesions immediately adjacent to an end plate. The end plate typically shows a defect, and the adjacent disk may show some loss of height (Fig. 48–1*B*).

Computed Tomography

Like radiography, computed tomography (CT) shows Schmorl's nodes as well-defined, rounded, radiolucent lesions immediately adjacent to an end plate (Fig. 48–2). A multilobulated appearance may be seen, and a small break in the adjacent end plate is sometimes discernible.

Magnetic Resonance Imaging

Sagittal and coronal magnetic resonance (MR) imaging may demonstrate a small interruption in the adjacent end plate through which the disk material has herniated into the vertebral body. The lesions have low to intermediate signal intensity on T1-weighted MR images and variable signal on T2-weighted images, depending on the degree of hydration of the disk (Figs. 48–2 and 48–3). Lesions may occasionally demonstrate some minor peripheral enhancement on T1-weighted images after intravenous injection of gadolinium, but the centers of the lesions do not typically enhance. Sometimes, marrow reaction with edema surrounding the lesion can be detected on T2-weighted MR images (Fig. 48–2*C*).

Differential Diagnosis

Confusion with other more serious lesions can potentially lead to unnecessary tests or even treatment for nonexistent diseases such as metastases. The CT finding of a rounded radiolucent lesion, or the MR imaging signal characteristics of a high water content, can easily be mistaken as metastasis or myeloma. Careful attention must be paid to the association of Schmorl's nodes with the end plate. Lack of enhancement after intravenous injection of contrast media on MR imaging may be helpful, but most Schmorl's nodes show characteristic features, and contrast enhancement is seldom necessary.

SCHEUERMANN'S DISEASE

Overview

Scheuermann's disease, juvenile lumbar osteochondrosis, Schmorl's nodes, and the anterior limbus vertebra are entities that share a common etiology, i.e., intravertebral disk penetration during childhood.

In the growing spine, the cartilaginous end plates have regions of potential weakness. Depending on both the physical activity of the individual and congenital predisposition to more fragile end plates, disk material can penetrate to varying extents through the cartilaginous end plates and into the developing vertebral body, resulting in varying manifestations of skeletal deformity. Variations in severity, number of vertebrae involved, and anatomic region affected contribute to the clinical presentation and diagnostic label applied. All forms of the disease tend to be more common in physically active individuals and are almost unknown in people who have chronic debilitating disease during childhood.

In Scheuermann's disease, the pathologic process is characterized by multiple small penetrations of the disk material into the growing vertebral end plates, leading to an irregular end plate contour with decreased growth, especially anteriorly (Fig. 48–4). By definition, three or

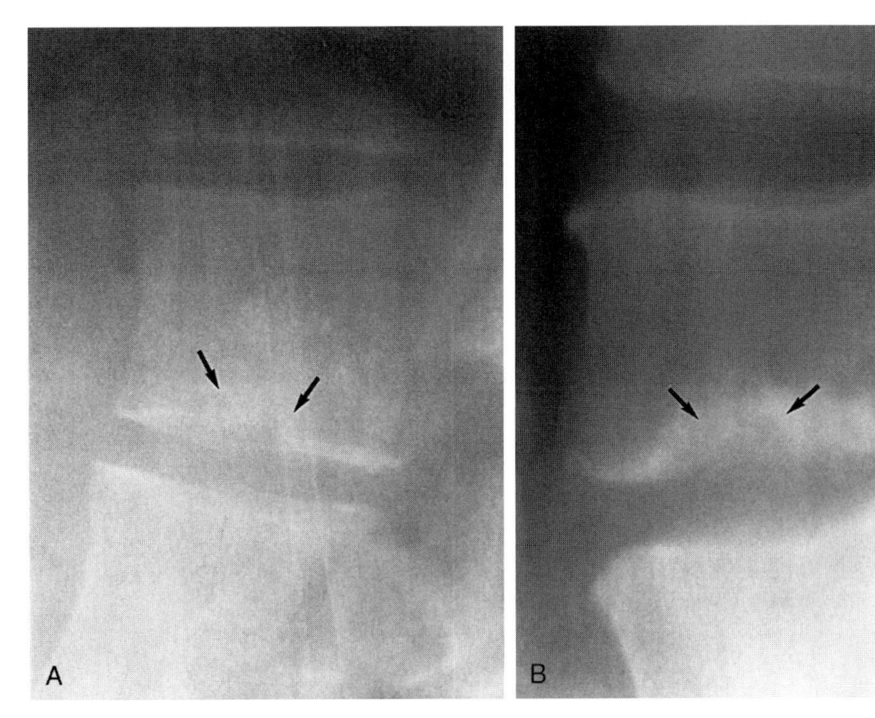

Figure 48–1. *A,* Lateral radiograph of the lumbar spine shows a lobulated Schmorl's node in a 43-year-old woman (*arrows*). The node is seen penetrating the inferior end plate of L1. *B,* A lobulated Schmorl's node in an 83-year-old man. Conventional tomogram shows a defect in the inferior end plate of L4 connected to a lobulated Schmorl's node (*arrows*). The L4-L5 disk is narrow, probably because of herniation of disk material into the vertebral body of L4.

more contiguous, anteriorly wedged thoracic vertebrae should be present for the diagnosis to be made. The growth disturbances cause anterior wedging of vertebrae, with accentuation of thoracic kyphosis and occasionally scoliosis. Intravertebral disk penetration can sometimes cause vertebral end plate irregularity without wedge deformity (Fig. 48–5). Also, solitary, paired, or noncontiguous affected vertebrae are common. They are regarded as nonclassic manifestations of the same disorder.

Scheuermann's disease is commonly of no clinical significance. However, it can cause chronic back pain or kyphosis. In adults, the residual disk and vertebral deformities may contribute to degenerative disk disease, chronic pain, and kyphosis (Fig. 48–6).

The lower thoracic spine and thoracolumbar junction are most commonly affected.

Imaging

Radiography

Lateral thoracic spine radiographs demonstrate irregularity of vertebral end plates, disk narrowing, and anterior vertebral body wedging. A long thoracic kyphosis is often present (see Fig. 48–6). Compensatory lordosis is noted on lateral cervical and lumbar spine radiographs.

Computed Tomography

CT is generally of little value in the diagnosis of Scheuermann's disease; however, CT may be performed for other reasons in patients with either active or "burned out" Scheuermann's disease. It is important to recognize the multiple narrow disks and the irregular end plates that often show multiple small Schmorl's nodes.

Magnetic Resonance Imaging

Sagittal MR imaging findings reflect the morphologic changes seen in lateral radiographs. In addition, the narrowed disks often show low signal intensity in T2-weighted MR images because of premature dehydration. Small foci of disk material may be seen penetrating through the end plates. In adults, associated intervertebral disk changes may be seen, ranging from minor degenerative disk disease and disk bulges to significant extrusions. MR imaging may be helpful in detecting these associated abnormalities, but as with CT, this technique is not necessary for diagnosis.

Juvenile Lumbar Osteochondrosis

Another manifestation of intravertebral disk penetration, juvenile lumbar osteochondrosis is very similar to Scheuermann's disease but occurs in the lumbar spine and can be associated with spinal stenosis and low back pain. Concurrent Scheuermann's disease may be seen in the thoracic spine, whereas Schmorl's nodes and anterior limbus vertebrae are commonly seen alongside end plate irregularity in the lumbar spine.

Imaging

PLAIN RADIOGRAPHS

Standard lateral lumbosacral spine radiographs demonstrate irregularity of vertebral end plates, disk narrowing, and shortened vertebral bodies (see Fig. 48–5). There may be retrolisthesis at one or more levels.

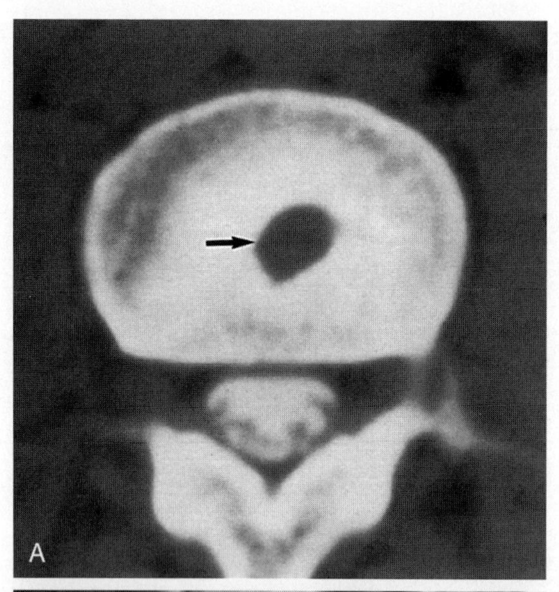

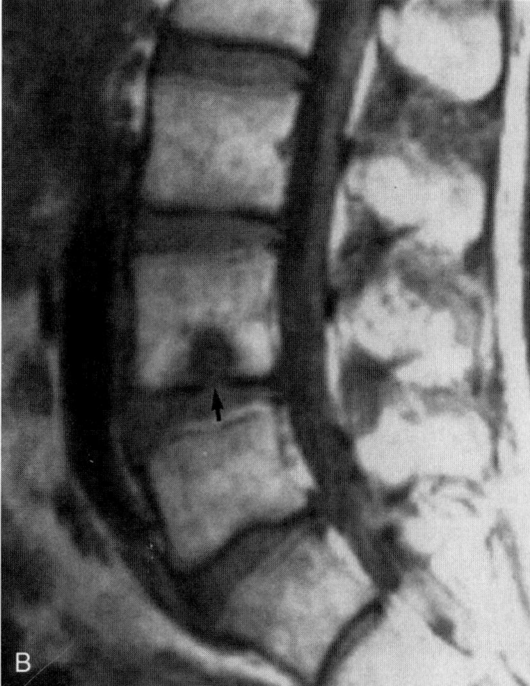

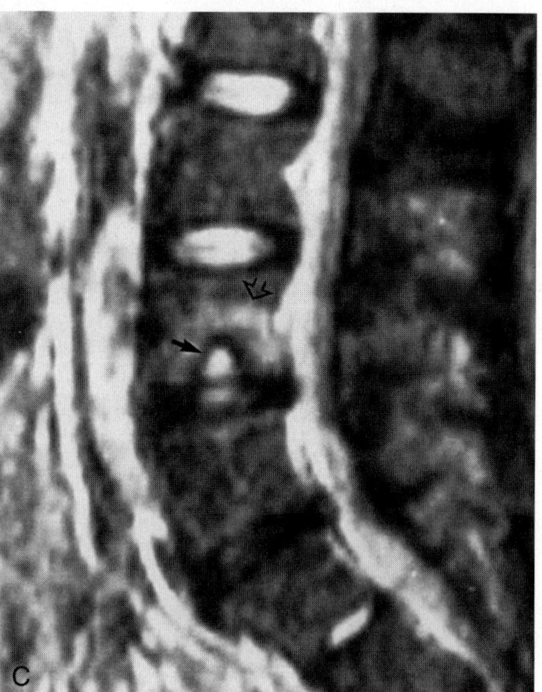

Figure 48–2. Schmorl's node arising from the L4-L5 disk and penetrating the inferior end plate of L4 in a 35-year-old man with low back pain. *A,* Computed tomography section taken just above the inferior end plate of L4 shows a well-defined lucency (*arrow*) surrounded by sclerotic bone. *B,* T1-weighted magnetic resonance (MR) image demonstrates the communication of the Schmorl's node with L4-L5 disk (*arrow*), which is narrow. *C,* Sagittal T2-weighted MR image shows the Schmorl's node to be bright (*solid arrow*). Marrow edema is seen surrounding the Schmorl's node (*open arrowhead*).

COMPUTED TOMOGRAPHY

As with Scheuermann's disease, CT is generally of little value in the diagnosis of juvenile lumbar osteochondrosis, which is better demonstrated by conventional radiography. However, CT is often performed for other reasons or may be ordered as part of a search for a cause of back pain. Multiple narrow disks and irregular end plates are shown on CT sections in addition to multiple small Schmorl's nodes.

MAGNETIC RESONANCE IMAGING

Sagittal MR images show the short, deformed vertebrae seen in lateral radiographs. The narrowed disks often have low signal intensity in T2-weighted MR images because of premature dehydration. Foci of disk material may be seen penetrating through the end plates or between an anterior vertebral rim and the vertebral body if limbus vertebrae are also present.

DIFFERENTIAL DIAGNOSIS

Skeletal dysplasias may be associated with irregular end plates and short vertebral bodies but are also seen with other extraspinal anomalies. Similar deformities could result from old fractures, but a previous history of significant trauma is required to confirm such a diagnosis.

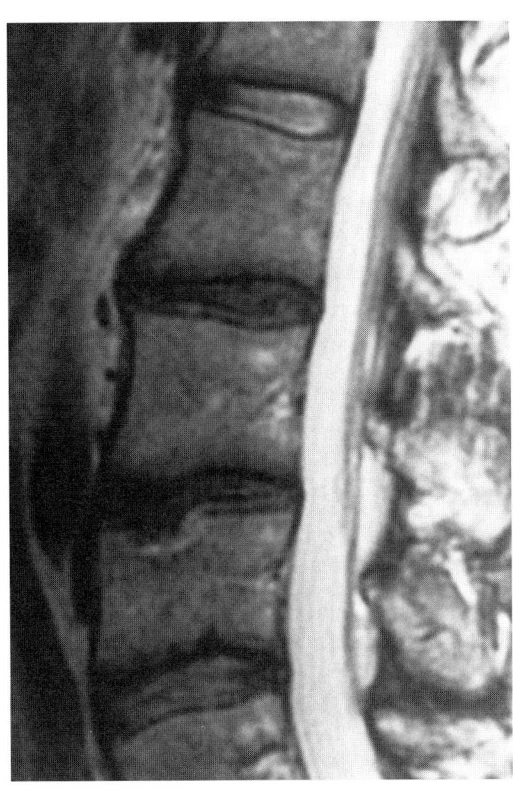

Figure 48–3. Sagittal T2-weighted magnetic resonance image of the lumbar spine shows a Schmorl's node penetrating the superior end plate of L4. The L3-L4 disk and the Schmorl's node have low signal intensity. The L3-L4 disk has lost some height.

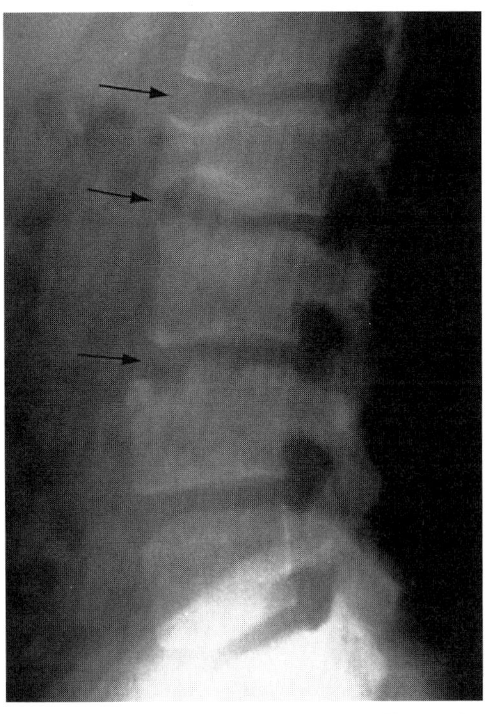

Figure 48–5. Juvenile lumbar osteochondrosis affecting the lumbar spine in an athletically active 14-year-old boy. On this lateral radiograph, the disk spaces are narrow, and the end plates are markedly irregular (*arrows*). The L2 vertebral body is wedged.

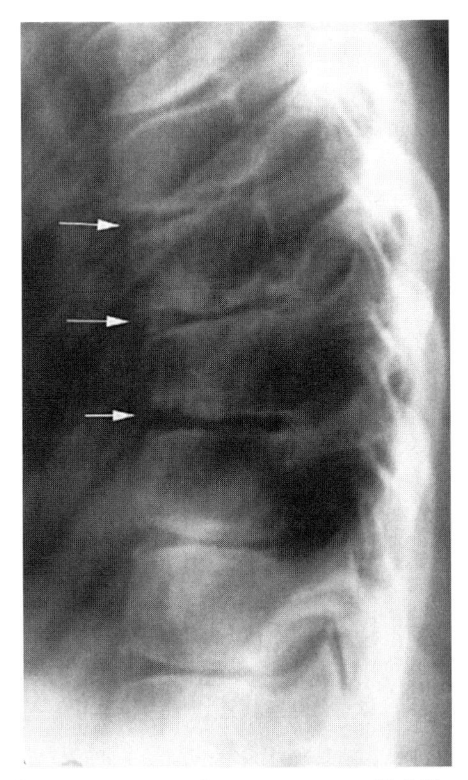

Figure 48–4. Scheuermann's disease in a 14-year-old child. On a lateral radiograph, disks at the involved levels are narrow, and the vertebral end plates are irregular (*arrows*). The vertebral bodies are anteriorly wedged.

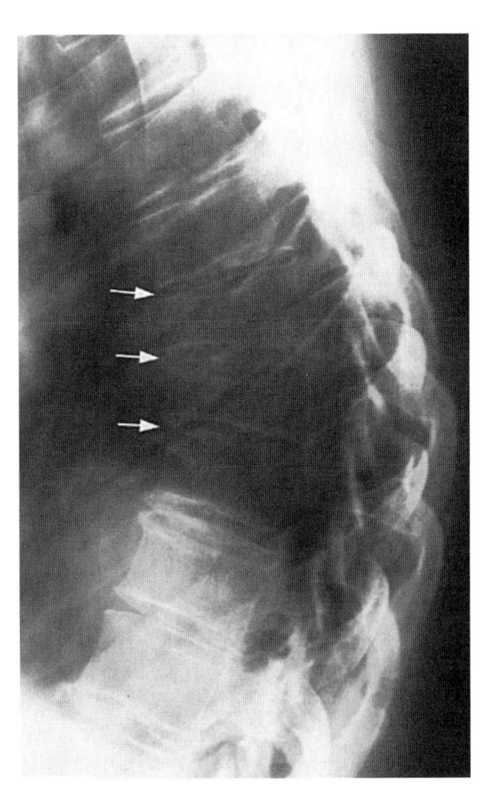

Figure 48–6. Scheuermann's disease affecting the midthoracic spine at three levels in a 35-year-old man. The lateral radiograph reveals moderate kyphosis. Note the disk space narrowing, irregular end plates (*arrows*), and wedging of the vertebral bodies.

LIMBUS VERTEBRA

Overview

Limbus vertebra is the term given to an ossicle that appears triangular when viewed in lateral radiographs and that lies at the anterosuperior or anteroinferior aspect of a vertebral body. The lumbar spine is most commonly affected. This disorder develops in childhood when disk material extrudes between an anterior vertebral rim apophysis and the vertebral body. The apophysis continues to grow but remains separated from the vertebral body by a zone of persistent disk material. This separated apophysis has been shown by diskography to remain continuous with the intervertebral disk. Isolated limbus vertebrae are seldom of any clinical significance, but they may be associated with Scheuermann's disease or juvenile lumbar osteochondrosis.

Imaging

Radiography

Lateral lumbar spine radiographs demonstrate a well-defined triangular ossicle (the limbus vertebra) adjacent to the anterosuperior or anteroinferior corner of a vertebral body (Fig. 48–7). It usually occupies a matching defect in the vertebral body, although often the limbus vertebra is somewhat larger than the defect and may protrude anteriorly.

Computed Tomography

CT shows a well-defined soft tissue plane that is continuous with the disk and separates the vertebral body from the limbus vertebra (Fig. 48–8).

Magnetic Resonance Imaging

Sagittal MR images show the morphological changes seen on lateral radiographs. The ossicle has normal marrow. Nuclear material may be identified between the vertebral body and the limbus fragment.

Differential Diagnosis

Confusion of limbus vertebra with an acute fracture is the most commonly encountered problem. The well-defined and corticated margins should exclude acute fracture as a diagnosis. On MR imaging, the lack of hemorrhage or edema should exclude acute fracture.

HEMISPHERIC SCLEROSIS

Overview

A benign condition, hemispheric sclerosis typically affects the anteroinferior aspect of a vertebral body. The involved part of the vertebra has marrow fibrosis and thickened trabeculae in a hemispheric distribution that extends to the vertebral margins. Frequently, there are degenerative changes in the adjacent intervertebral disk. The most likely etiology appears to be a response to repetitive trauma that is caused by a lack of cushioning effect from the adjacent disk. Hemispheric sclerosis mainly affects middle-aged women and is sometimes a cause of back pain, although it may also be an incidental finding in asymptomatic individuals.

L4 is the most commonly involved vertebra, but hemispheric sclerosis can be seen in other locations and may occasionally occur as a "kissing" lesion with a mirror-image distribution across the disk (Fig. 48–9). When this disorder occurs outside the lumbar spine, it is

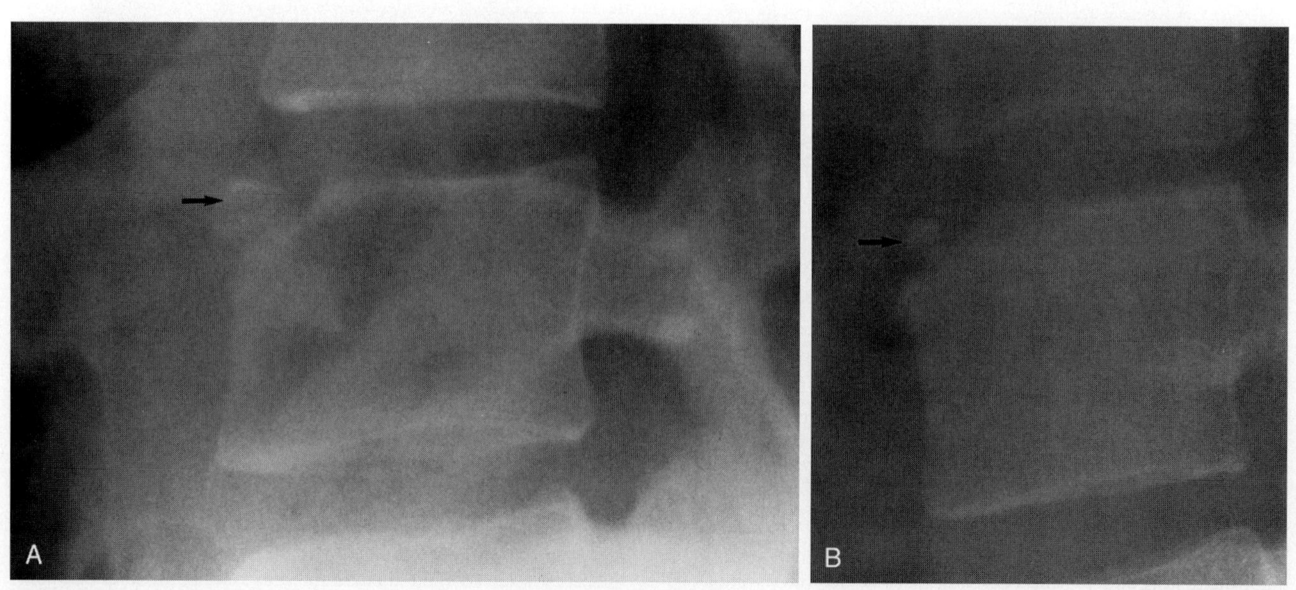

Figure 48–7. *A*, Lateral radiograph of L4 shows anterosuperior limbus vertebra (*arrow*) separated from the vertebral body by a well-defined radiolucent line. *B*, Radiograph of another patient with a limbus vertebra arising from L4 (*arrow*).

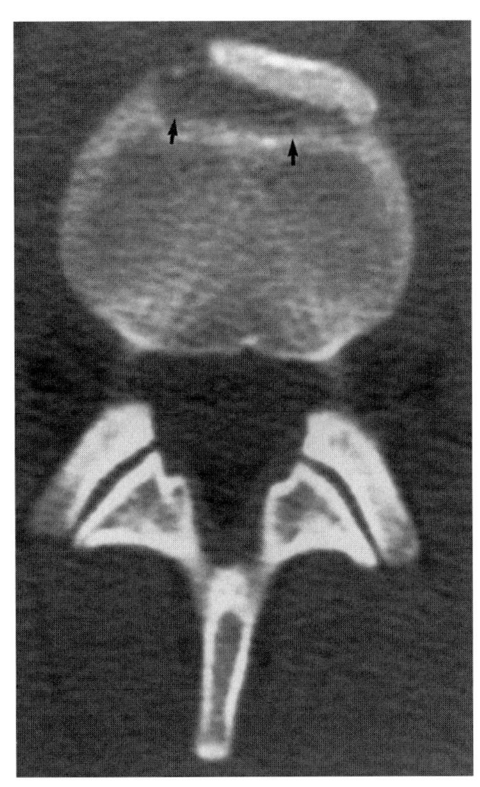

Figure 48–8. Axial computed tomography section through superior end plate of L4 shows anterior separation of the limbus vertebra (*arrows*).

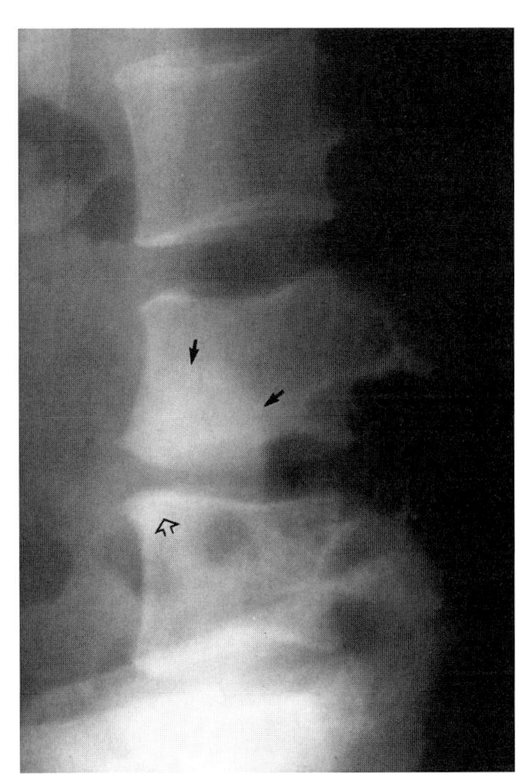

Figure 48–9. Lateral radiograph of the lumbar spine shows hemispheric sclerosis of the anteroinferior aspect of the L4 body (*solid arrows*) along with a smaller "kissing" lesion in the anterosuperior aspect of L5 (*open arrowhead*). The L4-L5 disk space is narrowed.

usually adjacent to a disk that is unstable because of a condition such as rheumatoid arthritis or ligamentous injury or because it is the first unfused vertebra adjacent to a spinal fusion.

Imaging

Radiography

Radiographs usually show a hemispheric region of increased density in the anteroinferior aspect of a vertebral body extending to the vertebral margins (Figs. 48–9 and 48–10). The adjacent disk is frequently narrowed, anterior osteophyte formation may be seen, and there can be minor irregularities in the contour of the affected end plate. Sometimes the sclerosis assumes a broad-based or band-like distribution that abuts the inferior end plate and extends the full anteroposterior dimension of the vertebral body.

Computed Tomography

CT demonstrates increased bone density extending to the end plate.

Magnetic Resonance Imaging

In most cases, MR images show edema within bone marrow in a distribution that is the same as or larger than

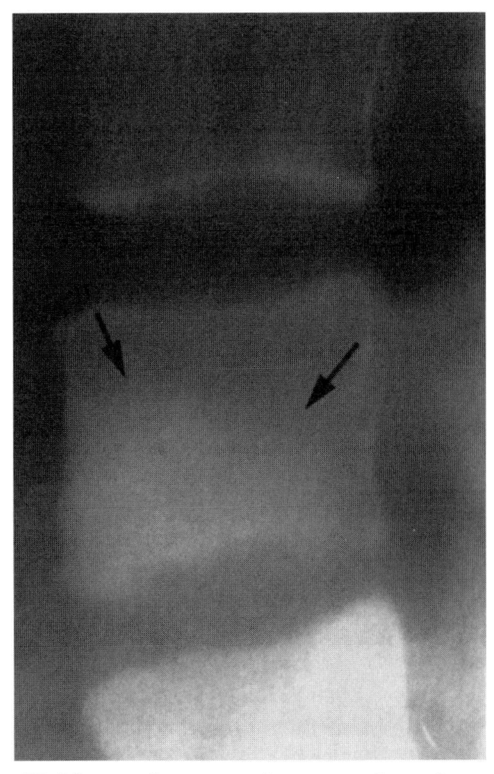

Figure 48–10. Lateral tomogram demonstrates hemispheric sclerosis (*arrows*) in a 45-year-old woman complaining of low back pain.

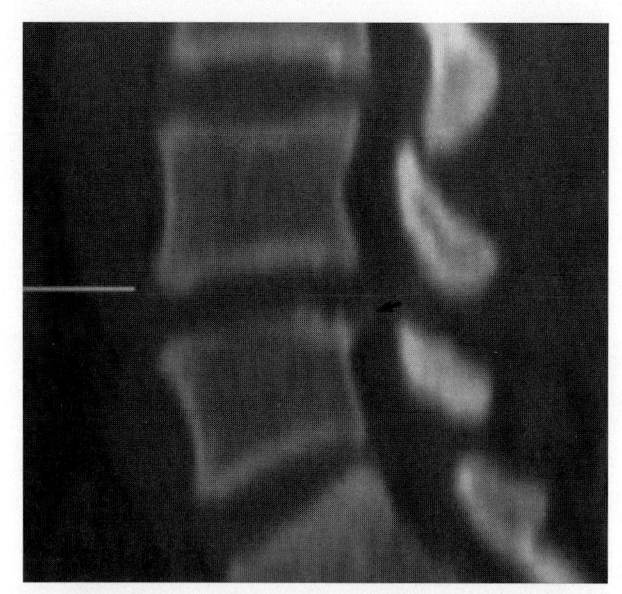

Figure 48–11. Posterior rim avulsion. Reformatted sagittal computed tomography image shows sclerosis and irregularity of posterosuperior rim and a bone fragment in the spinal canal (*arrow*).

that seen in the radiographs. Hemispheric sclerosis may also have low signal in all sequences (see Fig. 45–11); variable amounts of fat and edema can result in a spectrum of MR appearances. The classic hemispheric distribution and adjacent degenerative disk changes are more specific MR imaging findings than the signal characteristics.

Differential Diagnosis

Other conditions resulting in dense vertebral bodies include Hodgkin's lymphoma, blastic metastatic disease, and Paget's disease. However, the characteristic hemispheric distribution should help differentiate this condition, especially if there are ancillary findings such as a "kissing" lesion in the adjacent vertebra or evidence of adjacent degenerative disk disease (see Fig. 48–9). MR imaging findings are generally less specific than the plain film findings.

VERTEBRAL RIM AVULSION

Overview

Vertebral rim avulsion is sometimes referred to as a posterior limbus vertebra. However, the vertebral rim avulsion is the result of a posterior intervertebral disk protrusion, often associated with a specific injury, whereas the anterior limbus vertebra is part of the spectrum of intravertebral disk penetration lesions of childhood (see earlier discussions in this chapter).

Vertebral rim avulsions are usually acute, painful lesions affecting young patients. They occur when excessive tension exerted by Sharpey's fibers avulses a fragment of bone or apophysis from the posterior vertebral rim.

Any part of the spine may be affected, but most lesions occur between L4 and S1.

Imaging

Radiography

The fracture fragment is often difficult to see on radiography. It appears as a small bone density posterior to the superior or inferior vertebral margin. Detecting this finding can represent one of the few occasions in which a disk protrusion may be diagnosed on plain radiographs.

Computed Tomography

CT is the optimum imaging method for detecting these lesions. It is also useful for assessing the associated disk protrusion or extrusion (see Fig. 48–11).

On bone windows, the bone fragments can be clearly seen lying behind the vertebral body (Fig. 48–12). Soft tissue windows demonstrate the disk protrusion (Fig. 48–13).

Magnetic Resonance Imaging

MR imaging is excellent for demonstrating the disk protrusion of vertebral rim avulsion. Also, the extent of marrow edema on T2-weighted MR images may differentiate an acute from a chronic lesion. The bone fragment has low signal intensity on all sequences, so differentiating this lesion from a disk protrusion can be difficult.

Differential Diagnosis

The avulsed rim may resemble an osteophyte or a focus of calcification in the posterior longitudinal ligament.

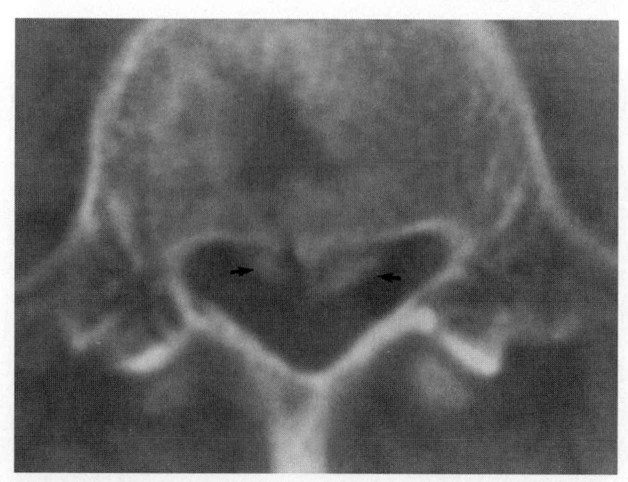

Figure 48–12. Posterior rim avulsion. Axial computed tomography section with bone window demonstrates posteriorly displaced bone fragments (*arrows*) and a fracture of posterosuperior vertebral rim of L5 vertebra.

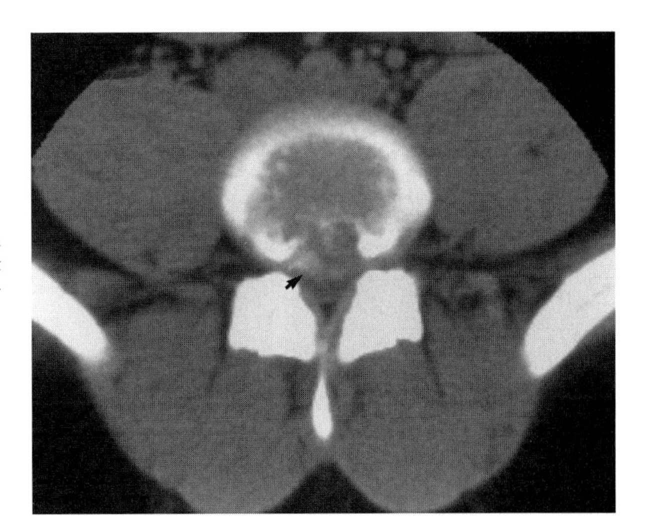

Figure 48–13. Posterior rim avulsion. Axial computed tomography section with soft tissue window shows posterior disk protrusion with associated defect in vertebral body. A bone fragment (*arrow*) is seen avulsed off the posterosuperior corner of the L5 vertebral body.

RETINOID-INDUCED CHANGES IN THE SPINE

Overview

Vitamin A derivatives have been used in the treatment of a number of skin disorders, including acne, severe ichthyosis, Darier's disease, psoriasis, and basal cell nevus syndrome. Retinoid-induced changes of bone do not occur with topical use of vitamin A derivatives nor with limited oral use of these agents for a few months, as in acne treatment. Only when taken by mouth over long periods have vitamin A derivatives been found to promote the growth of bony excrescences. The bony changes after prolonged use of Vitamin A derivatives simulate those seen in diffuse skeletal hyperostosis (DISH) and ossification of the posterior longitudinal ligament (OPLL). Although the spine is the most common site of retinoid-induced changes, extraspinal involvement can also occur. Clinically, these excrescences are associated with diminished spinal mobility and stiffness.

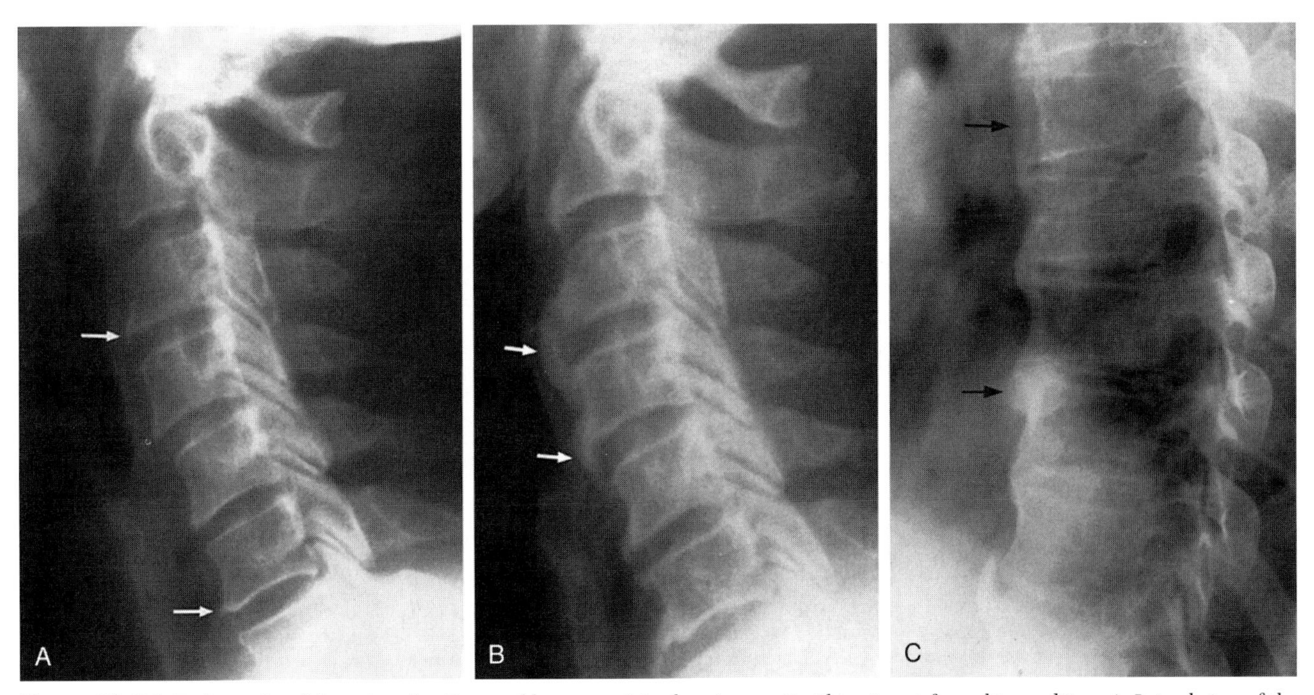

Figure 48–14. Radiographs of the spine of a 40-year-old man receiving long-term retinoid treatment for a skin condition. *A*, Lateral view of the cervical spine taken 4 years after initiation of treatment shows small osteophytes arising from the anteroinferior corners of C3 and C6 (*arrows*). Lateral views of the cervical (*B*) and thoracic (*C*) spine taken 8 years after initiation of treatment demonstrate extensive retinoid-induced changes that closely resemble DISH (*arrows*).

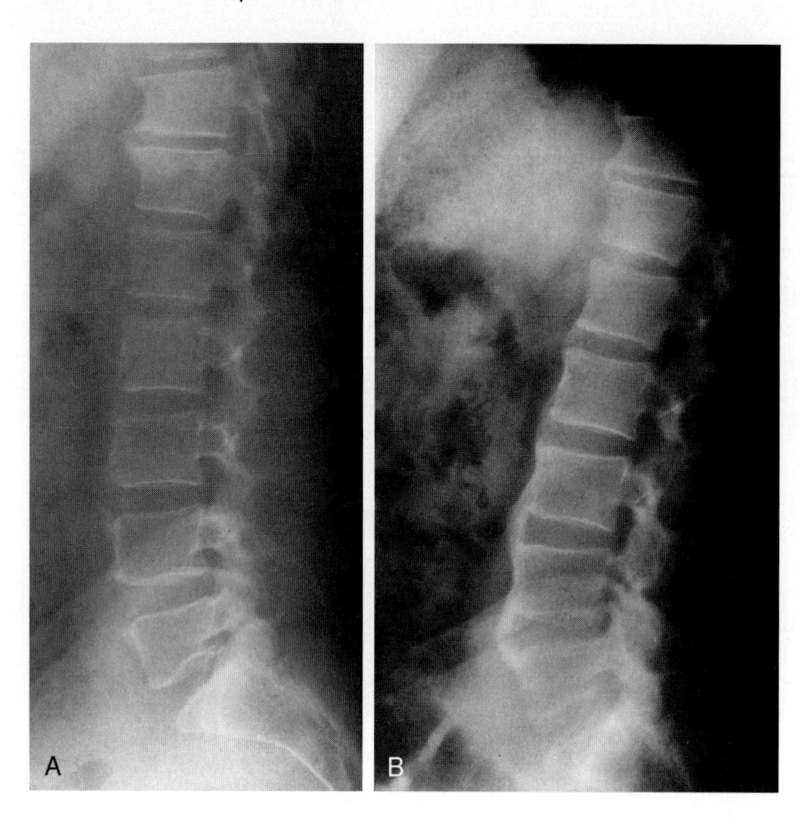

Figure 48–15. Retinoid-induced changes in the spine. *A*, Lateral lumbosacral spine radiograph shows small bone excrescences at the margins of L3-L4 and L4-L5 disks. *B*, A lateral radiograph of the lumbar spine taken 5 years later demonstrates extensive ossification in the anterior longitudinal ligament, a feature that is also typical of diffuse idiopathic skeletal hyperostosis, except that here the changes developed more rapidly and in a much younger patient population.

Imaging

Radiography

The findings of retinoid-induced changes in the spine are essentially the same as those described for DISH, except that the hyperostosis develops much faster and at a younger age than in DISH. The findings include flowing ossifications along the anterior and lateral aspects of the vertebral bodies (Figs. 48–14 and 48–15). Occasionally, patients may demonstrate radiographic findings of OPLL.

Computed Tomography and Magnetic Resonance Imaging

CT and MR imaging are not necessary for the diagnosis of retinoid-induced changes in the spine but may be helpful in the assessment of associated problems, such as spinal stenosis.

Recommended Readings

Banerian KG, Wang A, Samberg LC, et al: Association of vertebral end plate fracture with pediatric lumbar intervertebral disc herniation: Value of CT and MR imaging. Radiology 177:763, 1990.

Dihlmann W: Hemispherical spondylosclerosis: A polyetiologic syndrome. Skeletal Radiol 7:99, 1981.

Dihlmann SW, Eisenchenk A, Mayer HM, et al: The "mirror image" and "two thirds" types of hemispherical spondylosclerosis. Eur Spine J 4:110, 1995.

Epstein NE, Epstein JA: Limbus lumbar vertebral fractures in 27 adolescents and adults. Spine 16:962, 1991.

Ghelman B, Freiberger RH: The limbus vertebra: An anterior disc herniation demonstrated by discography. AJR Am J Roentgenol 127:854, 1976.

Goldfarb MT, Ellis CN, Voorhees JJ: Retinoids in dermatology. Mayo Clin Proc 62:1161, 1987.

Jensen MC, Brant-Zawadzki MN, Obuchowski N, et al: Magnetic resonance imaging of the spine in people without back pain. N Engl J Med 331:69, 1994.

Kilcoyne RF, Cope R, Cunningham W, et al: Minimal spinal hyperostosis with low-dose isotretinoin therapy. Invest Radiol 21:41, 1986.

Kodama T, Matsunaga S, Taketomi E, et al: Retinoid and bone metabolic marker in ossification of the posterior longitudinal ligament. In Vivo 12:339, 1998.

Lawson JP, McGuire J: The spectrum of skeletal changes associated with long-term administration of 13-cis-retinoic acid. Skeletal Radiol 16:91, 1987.

Lowe GL: Scheuermann's disease. Orthop Clin North Am 30:475, 1999.

Modic MT, Steinberg PM, Ross JS, et al: Degenerative disc disease: Assessment of changes in vertebral body marrow with MR imaging. Radiology 166:193, 1988.

Pennes DR, Ellis CN, Madison KC, et al: Early skeletal hyperostoses secondary to 13-cis-retinoic acid. AJR Am J Roentgenol 141:979, 1984.

Pennes DR, Martel W, Ellis CN, et al: Evolution of skeletal hyperostosis caused by 13-cis-retinoic acid therapy. AJR Am J Roentgenol 151:967, 1988.

Swischuk LE, John SD, Allberry SA. Disk degenerative disease in childhood: Scheuermann's disease, Schmorl's nodes, and the limbus vertebra: MRI findings in 12 patients. Pediatr Radiol 28:334, 1998.

Tallroth K, Schlenzka D: Spinal stenosis subsequent to juvenile lumbar osteochondrosis. Skeletal Radiol 19:203, 1990.

Vernon-Roberts B, Schmorl CG: Pioneer of pathology and radiology. Spine 19:2724, 1994.

Yang IK, Bahk YW, Choi KH, et al: Posterior lumbar apophyseal ring fractures: A report of 20 cases. Neuroradiology 36:453, 1994.

SECTION V

PEDIATRIC BONE DISEASES

CHAPTER 49

Developmental Dysplasia of the Hip

Lori L. Barr, MD, and Georges Y. El-Khoury, MD

OVERVIEW

The term *developmental dysplasia of the hip* (DDH) has replaced the term *congenital dislocation of the hip.* DDH is a better term because it includes cases that are not congenital but that developed after birth. It also describes a spectrum of severity, such as subluxation, dislocation, and hip dysplasia.

Normal hip development hinges on a balanced growth between the acetabular triradiate cartilage and a well-centered spherical femoral head. In a normal hip at birth, the femoral head is difficult to subluxate or dislocate using provocative testing (Ortolani's sign) during physical examination. This stable relationship between the femoral head and acetabulum is lost in DDH, and the femoral head can be displaced out, and back into the acetabulum by the examiner. The normal acetabulum is most shallow at birth, but it gradually becomes deeper until after puberty, creating a stable ball-in-socket joint. Screening in newborn nurseries has resulted in the detection of most DDH cases at birth, although a few cases escape early detection or develop DDH after birth. Ultrasonography currently is used widely for DDH screening, but the physical examination still is considered by pediatric orthopedists as the most reliable test for detecting DDH. In more than 90% of cases, the structural changes in DDH are reversible when patients are treated by concentric reduction using abduction braces. When the disease is not detected until after 6 months of age, the chances of successful concentric reduction are diminished.

In children with cerebral palsy and other neuromuscular disorders, the proximal femur and acetabulum are deformed because of unequal muscle forces and spasm. Coxa valga predisposes these children to hip subluxation and dislocation.

The incidence of DDH in children of European descent is about 1.5 per 1000 live births. DDH is rare in other populations. The condition occurs more frequently in females, especially firstborns. With breech presentation, there is a sixfold increase in incidence of DDH over vertex presentation. Oligohydramnios creates an intrauterine environment that predisposes the infants to develop DDH. A family history of DDH is a significant predisposing factor.

An infant screened clinically and found normal at birth needs only routine physical examination until he or she begins to walk. Only a few children who are normal at birth go on to develop DDH. If the patient has predisposing factors, screening with ultrasound usually is employed. A child with Graf I hips by sonography is declared normal and needs no further evaluation (Graf, 1980). The child with Graf IIA hips needs a follow-up sonogram to ensure that the hips have developed normally by the age of 1 month.

IMAGING

Sonography

Hip sonography consists of static coronal and transverse images of both hips and dynamic evaluation using the Barlow maneuver, which includes hip adduction and gentle application of a posterior force. The examiner feels the femoral head displacing out of the acetabulum if it is dislocatable. This technique combines the strengths of Graf's static classification of DDH with Harcke's dynamic evaluation to provide a complete evaluation of the hips. The Graf α angle is measured by drawing a line along the bony acetabulum in the coronal view and a second line delineating the slope of the acetabulum. The more shallow the acetabulum, the smaller the α angle. The normal α angle is greater than 60 degrees and is considered a Graf type I normal hip (Fig. 49–1). In infants less than 1 month old, the α angle can measure 55 to 60 degrees and is considered a Graf IIA or physiologically immature hip. After 1 month of age, any measurement of less than 60 degrees is considered abnormal, and if the angle persists between 55 and 60 degrees, the hip is classified as Graf IIB or mildly dysplastic hip. Such hips are characterized by a shallow acetabulum with a rounded rim. The Graf III hip is subluxed, and the acetabulum is shallower. The Graf IV hip is dislocated, and the acetabulum is flat and has no contact with the femoral head. During the ultrasound examination, an estimation is made of the thickness, echogenicity, and orientation of the cartilaginous labrum.

Radiography

After 4 months of age, anteroposterior and frog-leg views of the pelvis become the diagnostic modality of choice (Fig. 49–2). Radiography also is essential for monitoring the results of treatment in children with DDH (see Fig. 49–2). The acetabular angle is measured by

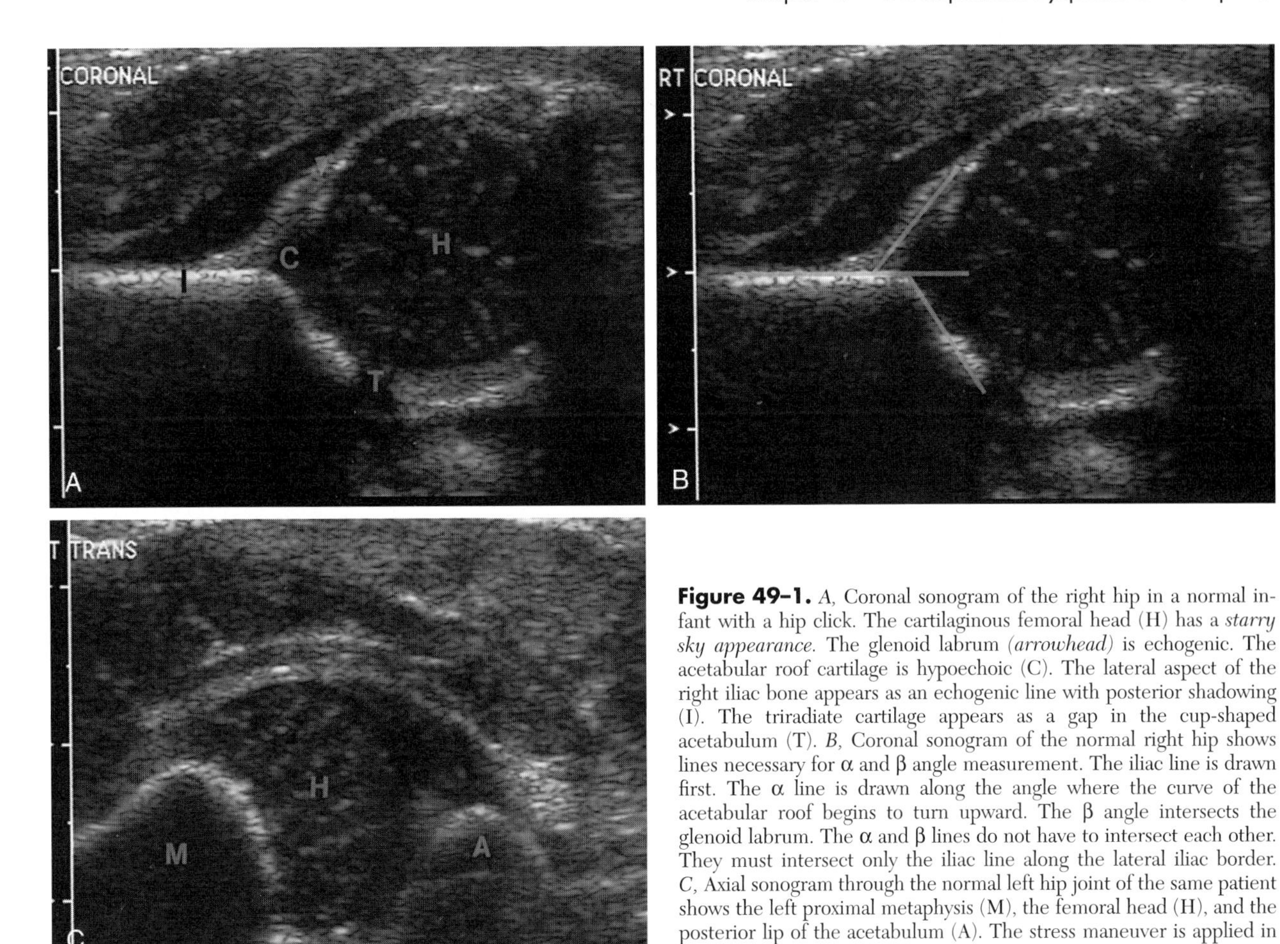

Figure 49–1. *A,* Coronal sonogram of the right hip in a normal infant with a hip click. The cartilaginous femoral head (H) has a *starry sky appearance.* The glenoid labrum *(arrowhead)* is echogenic. The acetabular roof cartilage is hypoechoic (C). The lateral aspect of the right iliac bone appears as an echogenic line with posterior shadowing (I). The triradiate cartilage appears as a gap in the cup-shaped acetabulum (T). *B,* Coronal sonogram of the normal right hip shows lines necessary for α and β angle measurement. The iliac line is drawn first. The α line is drawn along the angle where the curve of the acetabular roof begins to turn upward. The β angle intersects the glenoid labrum. The α and β lines do not have to intersect each other. They must intersect only the iliac line along the lateral iliac border. *C,* Axial sonogram through the normal left hip joint of the same patient shows the left proximal metaphysis (M), the femoral head (H), and the posterior lip of the acetabulum (A). The stress maneuver is applied in this position.

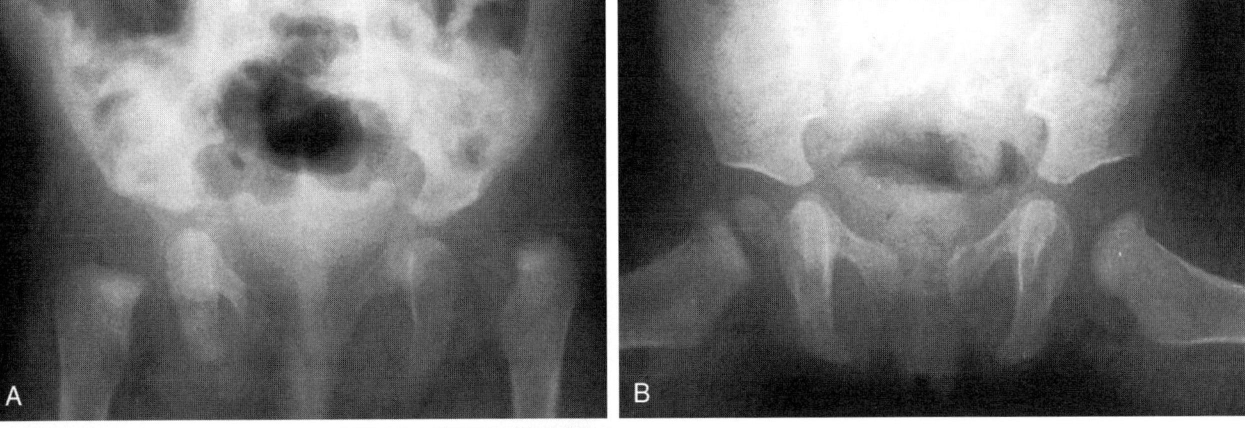

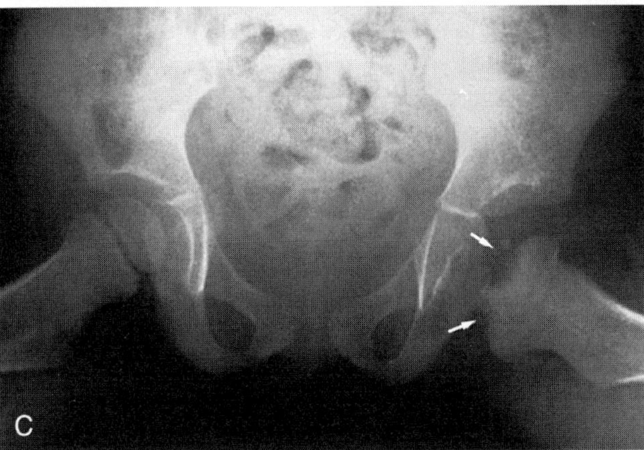

Figure 49–2. A 3-month-old girl who was referred to the pediatric orthopedic clinic with the diagnosis of developmental dysplasia of the hip (DDH) on the left side. *A,* Anteroposterior view of the pelvis shows DDH on left side with a left shallow acetabulum and interrupted Shenton's line. *B,* Frog-leg view taken at age 1 year shows the left femoral head is starting to ossify and is well centered within the acetabulum. The acetabulum is developing normally. *C,* Repeat frog-leg view at age 3 years shows avascular necrosis *(arrows)* of the left femoral head, which is a common complication of treated DDH.

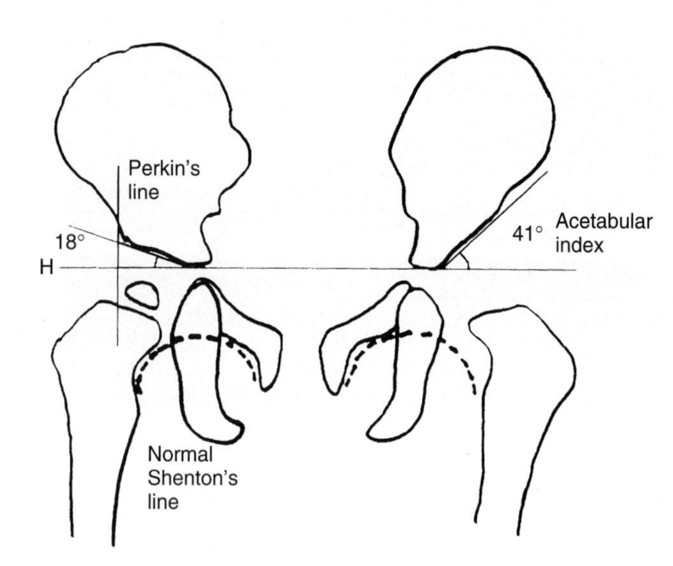

Figure 49–3. Line drawing illustrates the acetabular index, Perkin's line, and Shenton's line. The right hip is normal, whereas the left hip has developmental dysplasia of the hip. (H = line of Hilgenreiner.)

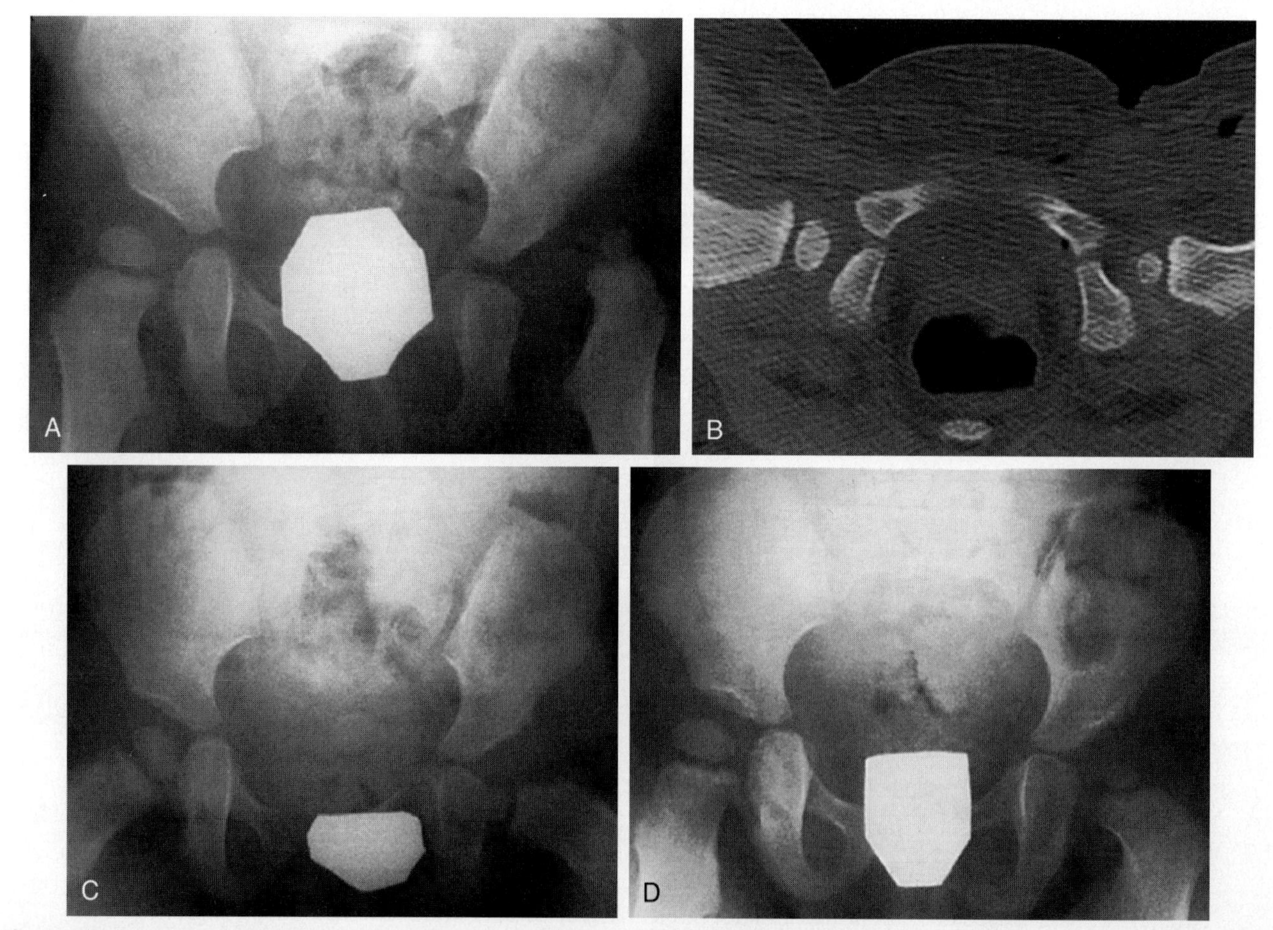

Figure 49–4. Continued hip subluxation in a patient discovered to have developmental dysplasia of the hip on the left side at the age of 16 months. *A*, Anteroposterior view of the pelvis shows the left hip to be completely dislocated. *B*, CT scan performed immediately after surgical reduction reveals adequate reduction of the left hip. *C*, Frog-leg view taken 3 months after the reduction shows the left femoral head to remain well centered within the acetabulum. *D*, Anteroposterior view at age 21 months shows the left femoral head is subluxed laterally, the acetabulum remains shallow, and Shenton's line is grossly interrupted.

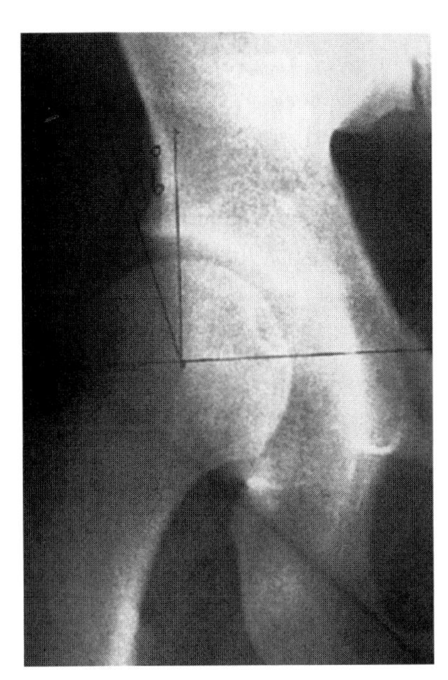

Figure 49–5. Center edge angle of Wiberg (16 degrees) in a 28-year-old woman with right hip dysplasia. The angle is measured by drawing a line from the center of the femoral head to the edge of the acetabulum and another line starting at the center of the femoral head and extending perpendicular to the plane of the pelvis.

drawing the H line of Hilgenreiner through the two acetabula. Then the acetabular line is drawn tangential to the edge of each acetabulum (Fig. 49–3). The angle between the two lines is termed the *acetabular index,* and it measures about 27 degrees at birth and decreases to 22 degrees by the age of 1 year. A measurement greater than 30 degrees is consistent with DDH. Perkin's line is drawn vertically from the lateral aspect of the acetabulum. Normally, most of the femoral head (>80%) should be medial to Perkin's line (see Fig. 49–3). A curved, almost semicircular line can be drawn along the medial border of the femoral neck and the superior border of the obturator foramen, which is referred to as *Shenton's line.* Normally, this line is an uninterrupted arc, but with hip subluxation or dislocation, Shenton's line becomes interrupted (see Fig. 49–3). The frog-leg abduction view allows the determination of whether the subluxed or dislocated hip is reducible. Radiographically the differentiation between hip subluxation and hip dysplasia is usually possible. Subluxated hips show some contact between the femoral head and the acetabulum, but the femoral head is displaced laterally or superolaterally, and Shenton's line is disrupted (Fig. 49–4). Hip dysplasia is defined as inadequate development of the acetabulum. In hip dysplasia, Shenton's line is almost normal, but the center edge angle of Wiberg is less than 20 degrees (Fig. 49–5). Recognizing hip dysplasia at any age is important because acetabular dysplasia can lead to early osteoarthritis, especially in females (Fig. 49–6). Hip dysplasia can go undetected until adulthood, at which time some surgeons may elect to perform osteotomies in all three bones of the acetabulum to rotate the acetabulum and achieve adequate coverage of the femoral head (Fig. 49–7).

Computed Tomography

Computed tomography (CT) is useful after reduction and casting. The child is brought directly to the CT suite after surgery. A limited study (two to three sections) using thick sections (6 to 10 mm) and low mAs through the hip joints is sufficient to determine whether the femoral head is reduced adequately. Immediately after reduction, the reduced femoral head often appears slightly more posterior in position than the normal side (see Fig. 49–4), but this should be considered normal reduction. CT with three-dimensional reconstruction is helpful in the evaluation of adults with hip dysplasia and in follow-up of postoperative osteotomies of the acetabulum (see Fig. 49–7).

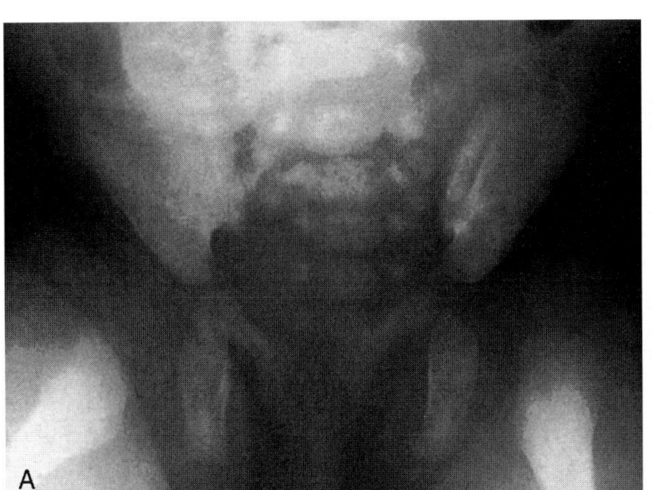

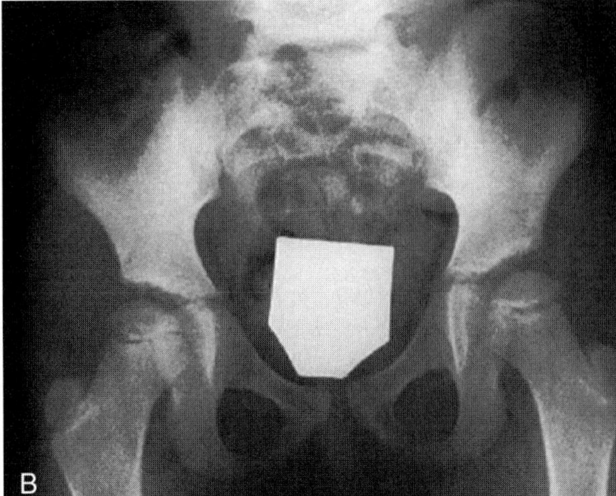

Figure 49–6. Persistent hip dysplasia in a child. *A,* Anteroposterior view of the pelvis in a 1-month-old girl shows bilateral developmental dysplasia of the hip. She was treated with an abduction brace. *B,* Anteroposterior view at age 5 years reveals persistent hip dysplasia on the left side with poor coverage of the left femoral head.

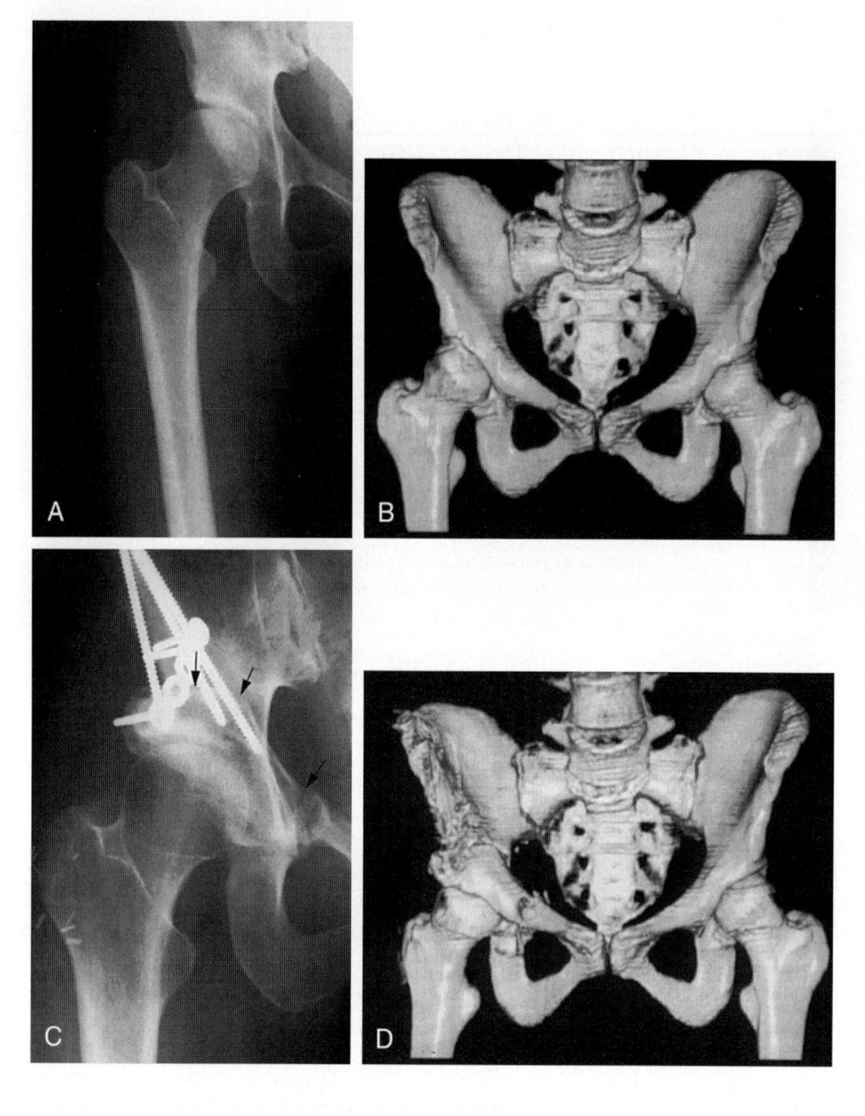

Figure 49–7. Hip dysplasia in a 32-year-old woman who presented to the orthopedic clinic with bilateral hip pain. *A*, Anteroposterior view of the right hip shows incomplete coverage of the femoral head. Shenton's line was within normal range, but the center-edge angle measured 15 degrees. *B*, Three-dimensional CT scan shows the bilateral hip dysplasia with incomplete coverage of the femoral heads on both sides. *C*, The patient underwent an osteotomy *(arrows)* on the right side to improve coverage of the femoral head. *D*, Three-dimensional CT scan shows the improved coverage on the right side compared with the left.

Magnetic Resonance Imaging

Magnetic resonance imaging is beginning to supplant CT for postoperative evaluation of the dysplastic hip. While the child is recovering from the general anesthetic used for general anesthetic used for surgery, a T1-weighted coronal localized is followed by proton density axial images using fat-saturation. Imaging time is less than 5 minutes and the examination involves no radiation.

Recommended Readings

Browning WH, Rosenkrantz H: Computed tomography in congenital hip dislocation. J Bone Joint Surg Am 64:27, 1982.
Graf R: The diagnosis of congenital hip-joint dislocation by the ultrasonic compound treatment. Arch Orthop Trauma Surg 97:117, 1980.
Hansson G, Nachemson A, Palmen K: Screening of children with congenital dislocation of the hip joint on the maternity wards in Sweden. J Pediatr Orthop 3:271, 1983.
Johnson ND, Wood BP, Jackman KV: Complex infantile and congenital hip dislocation: Assessment with MR imaging. Radiology 168:151, 1988.
Laor T, Roy DR, Mehlman CT: Limited magnetic resonance imaging examination after surgical reduction of developmental dysplasia of the hip. J. Pediatr Orthop 20(5):572–574, 2000.
Schlesinger AE, Hernandez RJ: Diseases of the musculoskeletal system in children: Imaging with CT, sonography, and MR. AJR Am J Roentgenol 158:729, 1992.
Simons GW, Flatley TJ, Sty JR, et al: Intra-articular osteocartilaginous obstruction to reduction of congenital dislocation of the hip. J Bone Joint Surg Am 70:760, 1988.
Terjesen T, Bredland T, Berg V: Ultrasound for hip assessment in the newborn. J Bone Joint Surg Br 71:767, 1989.
Weinstein SL: Natural history of congenital hip dislocation (CDH) and hip dysplasia. Clin Orthop 225:63, 1987.

CHAPTER 50

Legg-Calvé-Perthes Disease

Lori L. Barr, MD, and Georges Y. El-Khoury, MD

OVERVIEW

Legg-Calvé-Perthes (LCP) disease, or idiopathic avascular necrosis of the femoral head, usually occurs between the ages of 4 and 8 years, and it tends to affect white boys. Boys are affected four to five times more often than girls, and the disease is bilateral in 10% to 12% of cases. Of children suffering LCP disease, 89% have delayed bone age and are significantly shorter than normal children of the same age. In a small percentage of patients (<3%), the disease is preceded by a transient synovitis. The cause of LCP disease is not known, but the initial insult seems to compromise the capsular branches of the medial circumflex artery, which supplies the femoral head. Initially, there is development of dead bone devoid of osteocytes and surrounding marrow edema. A subchondral fracture and fragmentation of the capital femoral epiphysis follow. The intermediate phase is characterized by resorption of the dead bone and new bone formation. In the late repair phase, normal-looking bone with haversian systems is laid down.

The differential diagnosis includes Meyer's dysplasia (Fig. 50–1); hypothyroidism; and secondary avascular necrosis caused by sickle cell disease, steroid use, or Gaucher's disease. Abduction bracing and osteotomy are performed to prevent lateral subluxation and to keep the femoral head contained within the acetabulum. Overall outcome correlates with the amount of femoral head involvement. Factors associated with poor prognosis include lateral subluxation, older age at diagnosis, metaphyseal cystic areas (Fig. 50–2), and laterally displaced mineralized cartilage during the repair phase.

IMAGING FINDINGS

Radiographs

The earliest findings include arrested growth of the femoral head on the affected side and widening of the joint space medially. The femoral head appears smaller and more sclerotic than the normal side because of diminished blood supply (Fig. 50–3*A*). The widening of the medial joint space is attributed to epiphyseal cartilage hypertrophy (Fig. 50–3*B*). At initial presentation, a bone age often is obtained because many patients also have delayed skeletal maturation.

As the disease progresses, a subchondral fracture and increased density of the capital femoral epiphysis develop (see Fig. 50–3*A*, *B*). Then mottled sclerosis with lucencies develops followed by fragmentation of the epiphysis (see Fig. 50–3*C*). Metaphyseal cystic lucencies are seen in one third of patients (see Fig. 50–2). In later phases, immature bone clumps form beneath a reconstituted epiphyseal outline, and the femoral epiphysis returns to normal, or it may heal with a flattened and irregular appearance (see Fig. 50–3*C*). When the femoral head deformity persists, the acetabulum becomes secondarily deformed. Persistent hip deformities resulting from LCP disease can be seen in adults, including large and flattened femoral head (coxa magna), short and wide femoral neck, premature physeal closure of proximal femoral epiphysis, osteochondritis dissecans of the femoral head, and deformity of the acetabulum.

Scintigraphy

Scintigraphic findings typically precede radiographic findings in LCP disease. Pinhole images in the anteroposterior and frog-leg positions reveal a photopenic defect in the involved femoral head (Fig. 50–4). The size of the defect correlates with outcome, with larger photopenic defects suggesting a worse prognosis. There often is increased activity in the corresponding acetabulum. During later stages of the disease, there are areas of increased activity in the femoral neck; these reflect sites of bone repair and healing.

Magnetic Resonance Imaging

Magnetic resonance (MR) imaging is the most sensitive modality for the early detection of LCP disease but rarely is needed in clinical practice. It shows marrow edema in the affected femoral head, which is the earliest sign of LCP disease.

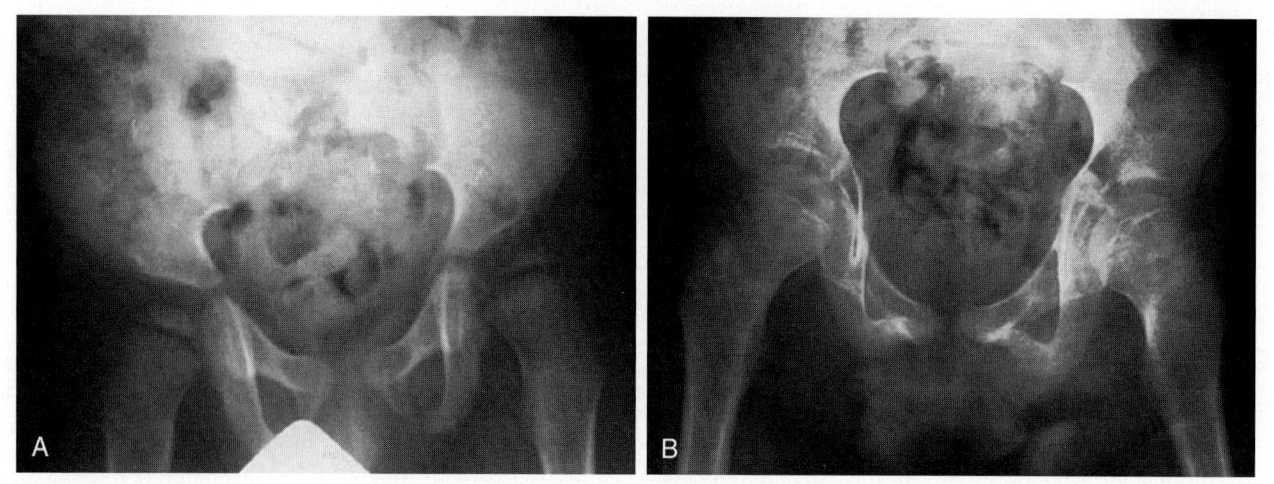

Figure 50–1. Meyer's dysplasia in a 39-month-old child. The diagnosis was made after other diseases, including Legg-Calvé-Perthes disease, were excluded. *A,* Anteroposterior view of the pelvis shows flattening and irregularity of the capital femoral epiphyses. The patient also had mild developmental dysplasia of the hip on the left side. *B,* At age 10 years, both hips appear normal.

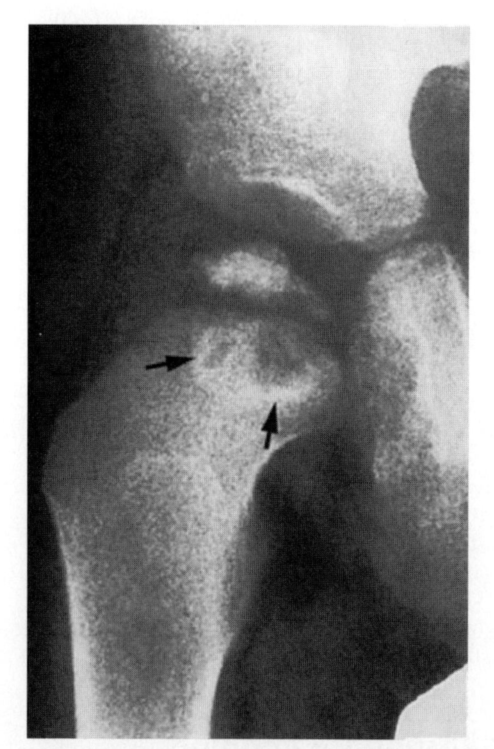

Figure 50–2. Legg-Calvé-Perthes disease in a 3-year-old boy. Anteroposterior view shows collapse and sclerosis of the femoral capital epiphysis. Metaphyseal cystic changes also are visible *(arrows).*

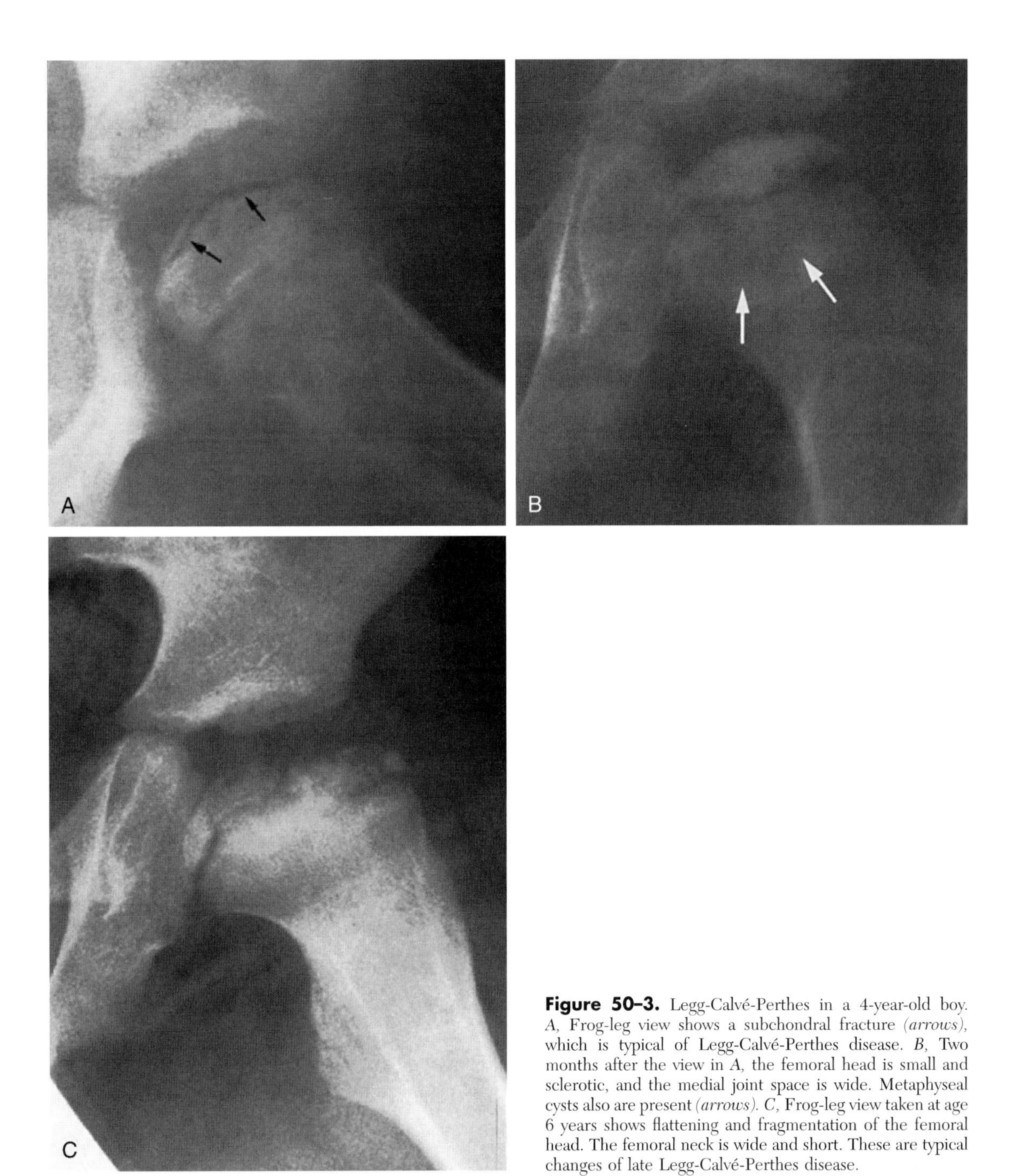

Figure 50–3. Legg-Calvé-Perthes in a 4-year-old boy. *A*, Frog-leg view shows a subchondral fracture *(arrows)*, which is typical of Legg-Calvé-Perthes disease. *B*, Two months after the view in *A*, the femoral head is small and sclerotic, and the medial joint space is wide. Metaphyseal cysts also are present *(arrows)*. *C*, Frog-leg view taken at age 6 years shows flattening and fragmentation of the femoral head. The femoral neck is wide and short. These are typical changes of late Legg-Calvé-Perthes disease.

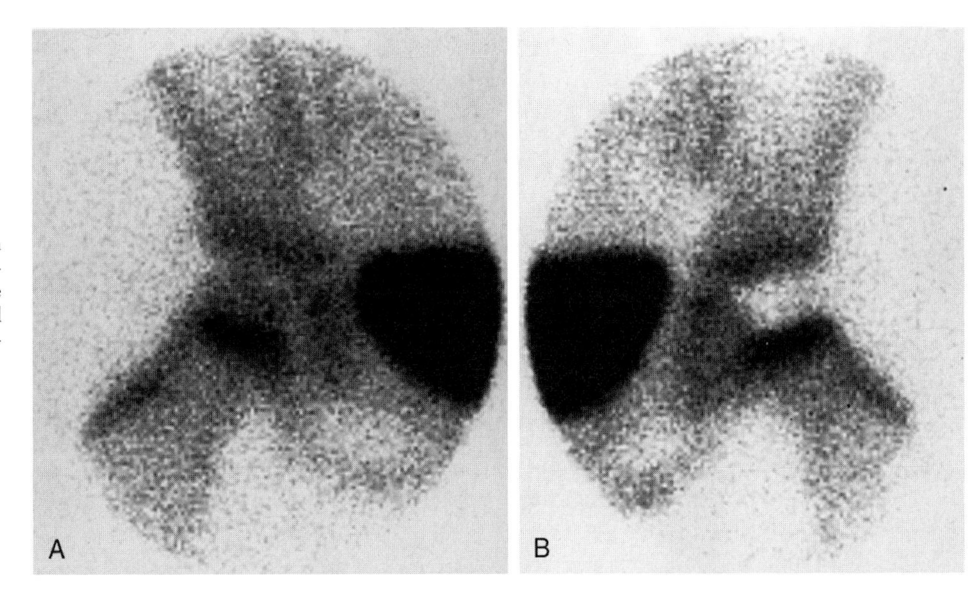

Figure 50–4. Scintigraphy in Legg-Calvé-Perthes disease. *A*, Normal right hip. *B*, Photopenia in the region of the left femoral head caused by Legg-Calvé-Perthes disease.

Recommended Readings

Bos CFA, Bloem JL, Bloem RM: Sequential magnetic resonance imaging in Perthes' disease. J Bone Joint Surg Br 73:219, 1991.

Crossan JF, Wynne-Davies R, Fulford GE: Bilateral failure of the capital femoral epiphysis: Bilateral Perthes disease, multiple epiphyseal dysplasia, pseudoachondroplasia, and spondyloepiphyseal dysplasia congenita and tarda. J Pediatr Orthop 3:297, 1983.

Gershuni DH, Axer A, Hendel D: Arthrographic findings in Legg-Calve-Perthes disease and transient synovitis of the hip. J Bone Joint Surg Am 60:457, 1978.

Goldman AB, Hallel T, Salvati EM, et al: Osteochondritis dissecans complicating Legg-Perthes disease. Radiology 121:561, 1976.

Jaramillo D, Galen TA, Winalski CS, et al: Legg-Calve-Perthes disease: MR imaging evaluation during manual positioning of the hip—comparison with conventional arthrography. Radiology 212:519, 1999.

Jaramillo D, Kasser JR, Villegas-Medina OL, et al: Cartilaginous abnormalities and growth disturbances in Legg-Calve-Perthes disease: Evaluation with MR imaging. Radiology 197:767, 1995.

Salter RB, Thompson GH: Legg-Calve-Perthes disease. J Bone Joint Surg Am 66:479, 1984.

Wenger DR, Ward WT, Herring JA: Current concepts review Legg-Calve-Perthes disease. J Bone Joint Surg Am 73:778, 1991.

CHAPTER 51

Slipped Capital Femoral Epiphysis

Lori L. Barr, MD, and Georges Y. El-Khoury, MD

OVERVIEW

Slipped capital femoral epiphysis (SCFE) is the most common hip disorder affecting adolescents. The disease usually occurs between the ages of 9 and 16 years. When SCFE occurs outside this age range, endocrine and metabolic abnormalities, such as hypogonadism, hypothyroidism, and renal osteodystrophy (Fig. 51–1), should be considered as the underlying cause. SCFE is a Salter-Harris type I fracture of the proximal femoral physis. Most cases are idiopathic and occur in obese and sexually underdeveloped boys. SCFE affects boys twice as frequently as girls and children of African descent more frequently than whites. In boys, the left hip is affected more commonly. Both hips are involved either simultaneously or sequentially in about 60% of cases (Fig. 51–2). The second slip can be asymptomatic in about one third of patients. The separated epiphysis displaces mainly posteriorly, whereas the femoral neck displaces anteriorly, resulting in an apparent varus deformity. Histologically the separation occurs at the hypertrophic zone of cartilage in the physeal plate. Presenting symptoms include hip or groin pain and limp and, less commonly, knee pain. Half of all patients recall a traumatic incident.

Acute unstable SCFE is treated with reduction and pinning in anatomic position. Chronic SCFE is pinned in situ to arrest further slippage. Follow-up radiographs are obtained to look for hardware complications, such as penetration of the screws into the joint, chondrolysis, and avascular necrosis, and to observe the opposite hip for slippage.

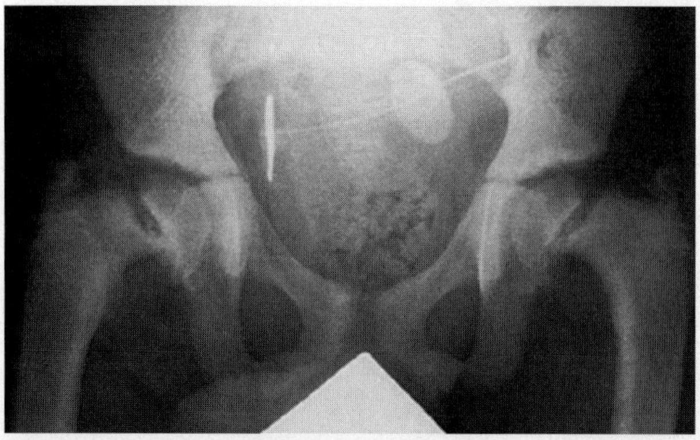

Figure 51–1. Bilateral chronic slipped capital femoral epiphysis in a 15-year-old boy with chronic renal failure treated with peritoneal dialysis. Anteroposterior view of the pelvis shows bilateral slipped capital femoral epiphysis. The bone texture is coarse as a result of renal osteodystrophy.

IMAGING

Radiography

In early SCFE, the epiphysis slips posteriorly on the femoral neck, creating a crescent-shaped area of increased density where the femoral head overlaps the femoral neck. This increased density is referred to as the *metaphyseal blanch sign* (Fig. 51–3). The most common radiographic pattern is what is known as *chronic SCFE.* This pattern is preceded by the preslip stage, in which the radiographic findings include subtle widening, osteopenia, and indistinctness of the physeal plate (Fig. 51–4). During the chronic phase, the femoral epiphysis is displaced posteriorly, and bony remodeling is noted along the medial aspect of the neck (see Fig.

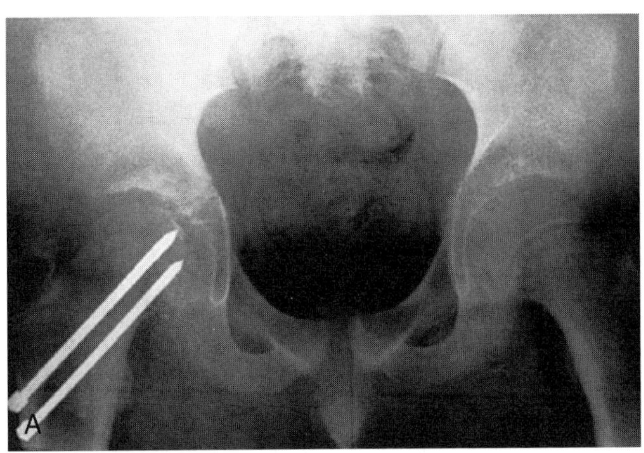

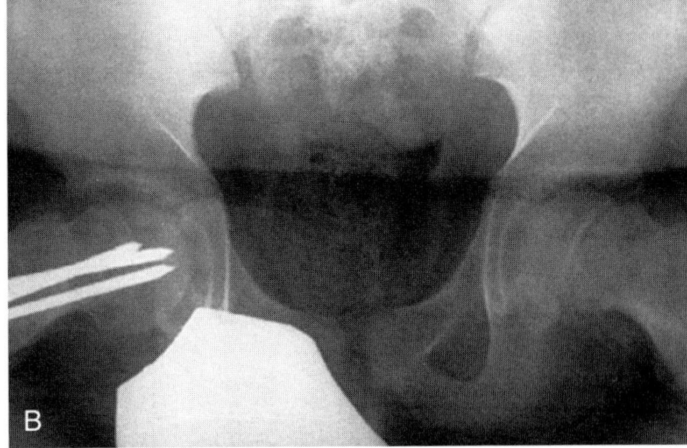

Figure 51–2. Bilateral slipped capital femoral epiphysis, which developed first in the right hip, then 4 months later in the left hip. *A*, Anteroposterior view of the hips shows pinning of the capital femoral epiphysis on the right, which has been reduced to anatomic position. The left hip shows a wide epiphysis but no definite slip. *B*, Frog-leg view confirms the presence of slipped capital femoral epiphysis on the left.

51–3). The physeal plate widening should be appreciated on the anteroposterior view (see Fig. 51–4). A line drawn along the lateral aspect of the femoral neck normally should intersect the lateral portion of the femoral head (Klein's line) (Fig. 51–5). When Klein's line does not intersect the femoral head, SCFE is considered. A true lateral view of the hip shows the extent of posterior displacement of the femoral epiphysis. Some patients present during the *acute-on-chronic phase,* which is characterized by sudden increase in pain and symptoms.

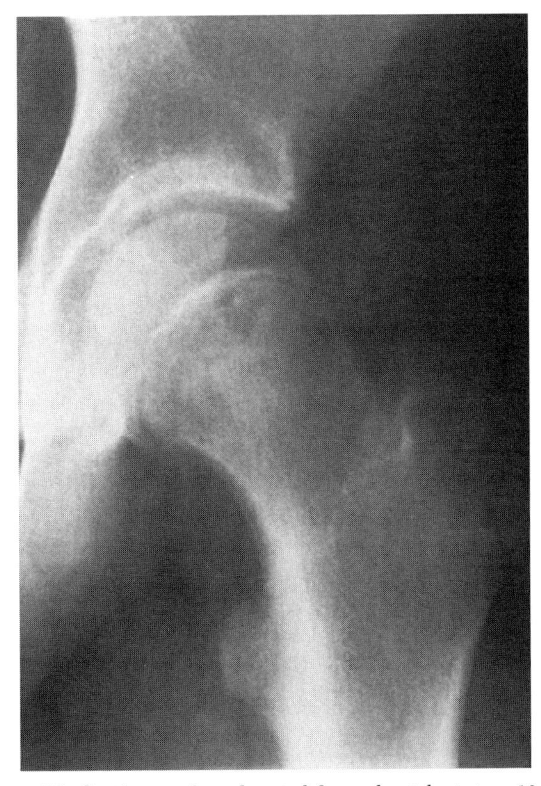

Figure 51–3. Chronic slipped capital femoral epiphysis in a 12-year-old girl. Anteroposterior view of the hip shows increased density in the central portion of the metaphysis, which represents the blanch sign. Note the periosteal new bone formation along the medial aspect of the femoral neck, which is a sign of bone remodeling and chronicity of the slip.

This phase is characterized by acute increase in the severity of the slip. Acute SCFE is suspected when patients present with sudden onset of hip pain (Fig. 51–6). Radiographically the abnormality resembles an acute Salter-Harris type I fracture, with no bony remodeling in the femoral head and no evidence of attempted healing at the site of the slip (see Fig. 51–6).

Magnetic Resonance Imaging

If there is any question concerning radiographic findings, sagittal magnetic resonance (MR) imaging shows the posterior displacement of the femoral epiphysis.

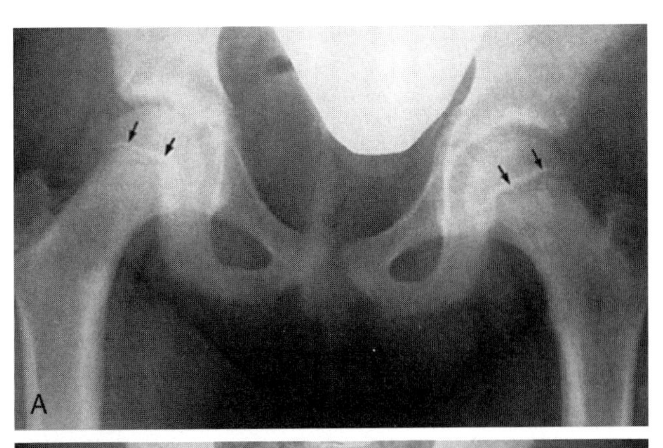

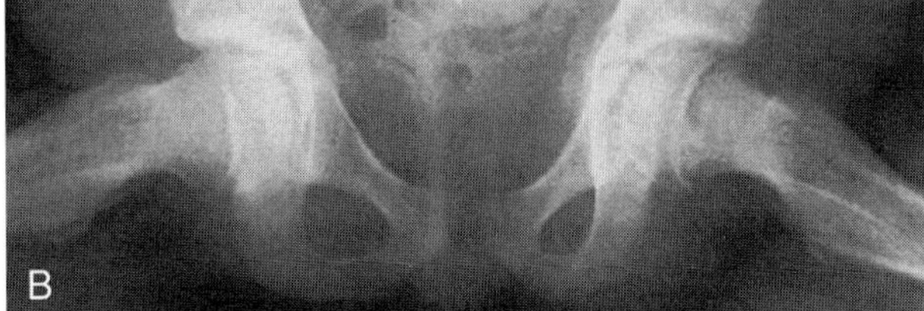

Figure 51–4. Early slipped capital femoral epiphysis or preslip stage on the left in a 10-year-old girl presenting with left hip pain. *A* and *B*, Anteroposterior and frog-leg views of the hips show osteopenia of the left hip and indistinct physeal plate *(arrows)*. The right hip is normal *(arrows)*.

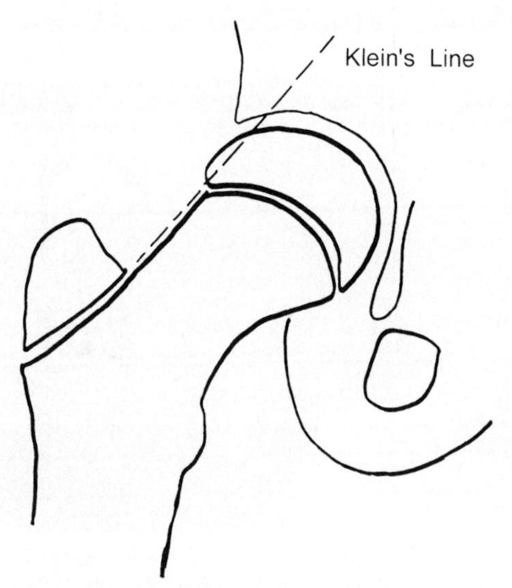

Figure 51–5. Klein's line drawn on a diagram of a normal hip shows the line intersecting the lateral portion of the femoral head.

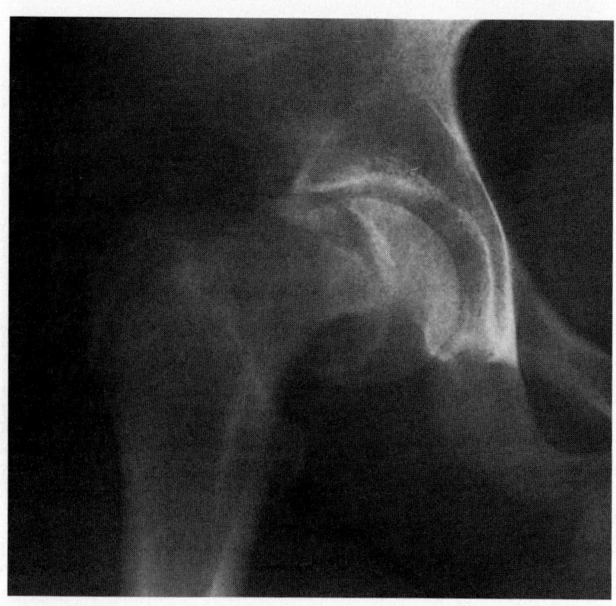

Figure 51–6. Acute unstable slipped capital femoral epiphysis of the right hip in a 16-year-old boy.

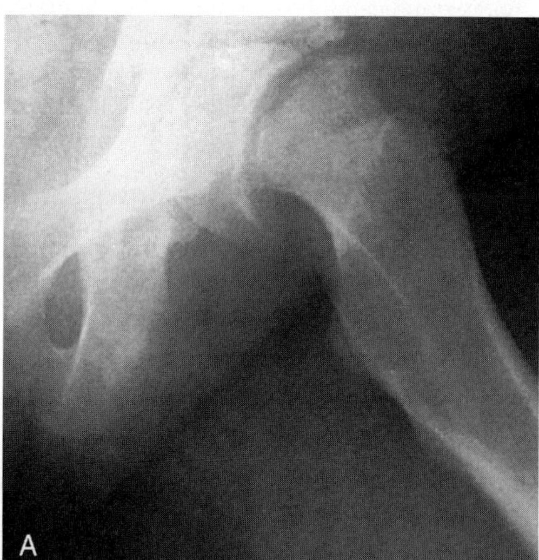

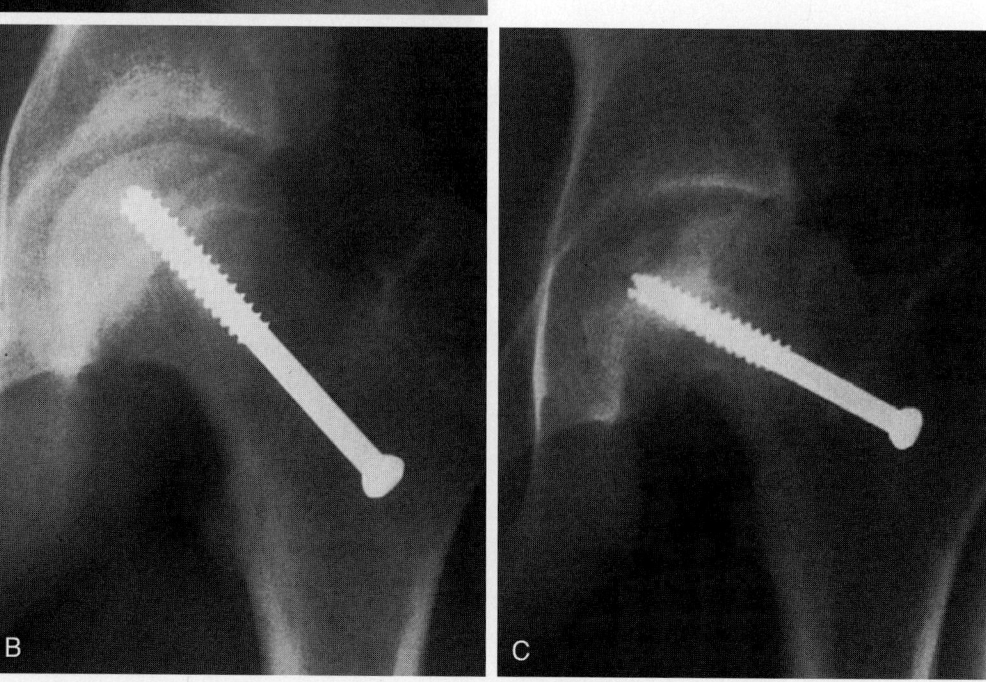

Figure 51–7. Avascular necrosis of the femoral head after reduction and internal fixaton of an acute slipped capital femoral epiphysis. *A*, Frog-leg view of the left hip at initial presentation shows the acute slipped capital femoral epiphysis. *B*, Four months after reduction and internal fixation shows increased density of the femoral head, suggesting avascular necrosis. *C*, Two years after surgery, the femoral head is collapsed and deformed.

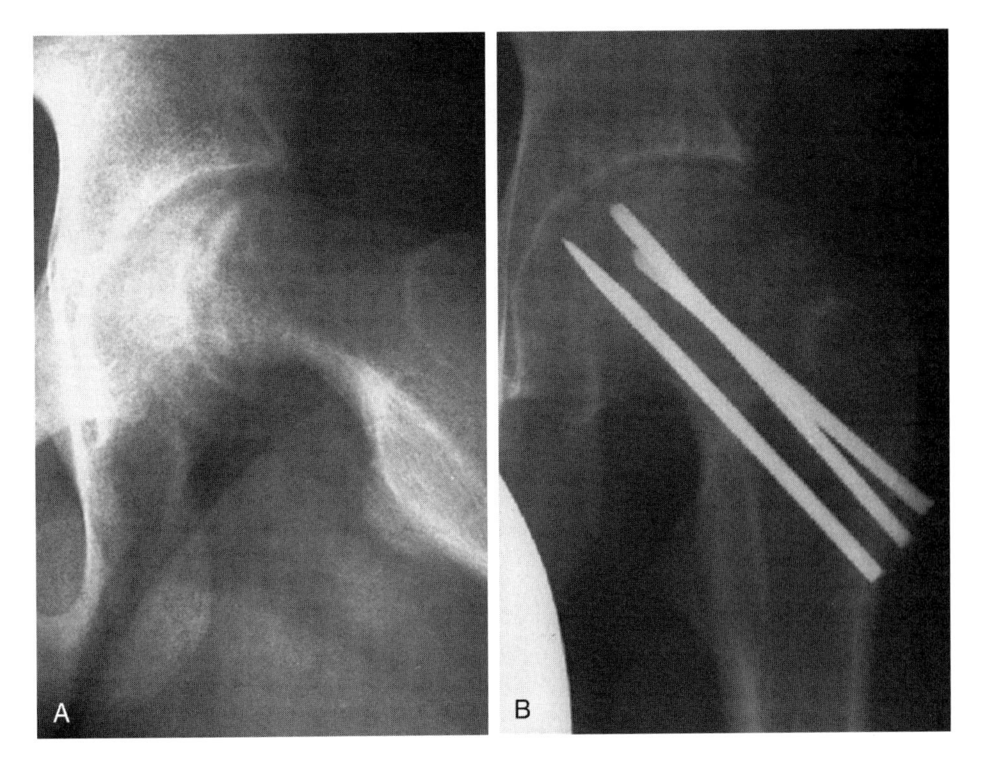

Figure 51–8. Chondrolysis in slipped capital femoral epiphysis affecting the left hip of a 13-year-old boy. *A,* Slipped capital femoral epiphysis of the left hip at presentation. *B,* Sixteen months after internal fixation, anteroposterior view shows some cartilage loss and cystic changes in the acetabulum.

COMPLICATIONS

Two serious complications have been noted to follow SCFE: avascular necrosis of the femoral head (Fig. 51–7) and chondrolysis (Fig. 51–8). Avascular necrosis and chondrolysis can lead to secondary osteoarthritis of the hip at an early age.

Recommended Readings

Boles CA, El-Khoury GY: Slipped capital femoral epiphysis. Radiographics 17:809, 1997.

Carney BT, Weinstein SL, Noble J: Long-term follow-up of slipped capital femoral epiphysis. J Bone Joint Surg Am 73:667, 1991.

Crawford AH: Slipped capital femoral epiphysis. J Bone Joint Surg Am 70:1422, 1988.

El-Khoury GY, Mickelson MR: Chondrolysis following slipped capital femoral epiphysis. Radiology 123:327, 1977.

Goldman AB, Lane JM, Salvati E: Slipped capital femoral epiphyses complicating renal osteodystrophy: A report of three cases. Radiology 126:333, 1978.

Steel HH: The metaphyseal blanch sign of slipped capital femoral epiphysis. J Bone Joint Surg Am 68:920, 1986.

Wolf EL, Berdon WE, Cassady JR, et al: Slipped femoral capital epiphysis as a sequela to childhood irradiation for malignant tumors. Radiology 125:781, 1977.

CHAPTER 52

Bowlegs

Lori L. Barr, MD, and Georges Y. El-Khoury, MD

OVERVIEW

Bowing of the legs is a common orthopedic problem in infants and children. At birth, the knees are typically in about 10 to 15 degrees varus, but this corrects spontaneously by about 14 months of age. The most common cause of bowlegs is physiologic bowing. Other causes include Blount's disease, ricketts, and bone dysplasias.

PHYSIOLOGIC BOWING

In most cases, physiologic bowing has a benign clinical course, and it corrects spontaneously. Differentiating physiologic bowing from Blount's disease (infantile tibia vara) on radiographs can be difficult, especially before age 2 years.

A key finding in physiologic bowing is the presence of bowing in the femora and tibiae (Fig. 52–1), whereas in Blount's disease the bowing is centered at the proximal tibia. Physiologic bowing typically is bilateral, although the severity of the bowing is not necessarily equal. Measurements of the metaphyseal-diaphyseal angle in the proximal tibia can be helpful (Fig. 52–2). There is, however, significant overlap between the two diseases, and it currently is believed that the two conditions are manifestations of different severities of the same disease. Patients with a metaphyseal-diaphyseal angle of less than 10 degrees are considered to have physiologic bowing, and patients with a metaphyseal-diaphyseal angle above 16 degrees have Blount's disease (Fig. 52–3). Patients with measurements between 10 and 16 degrees are followed until the disease declares itself.

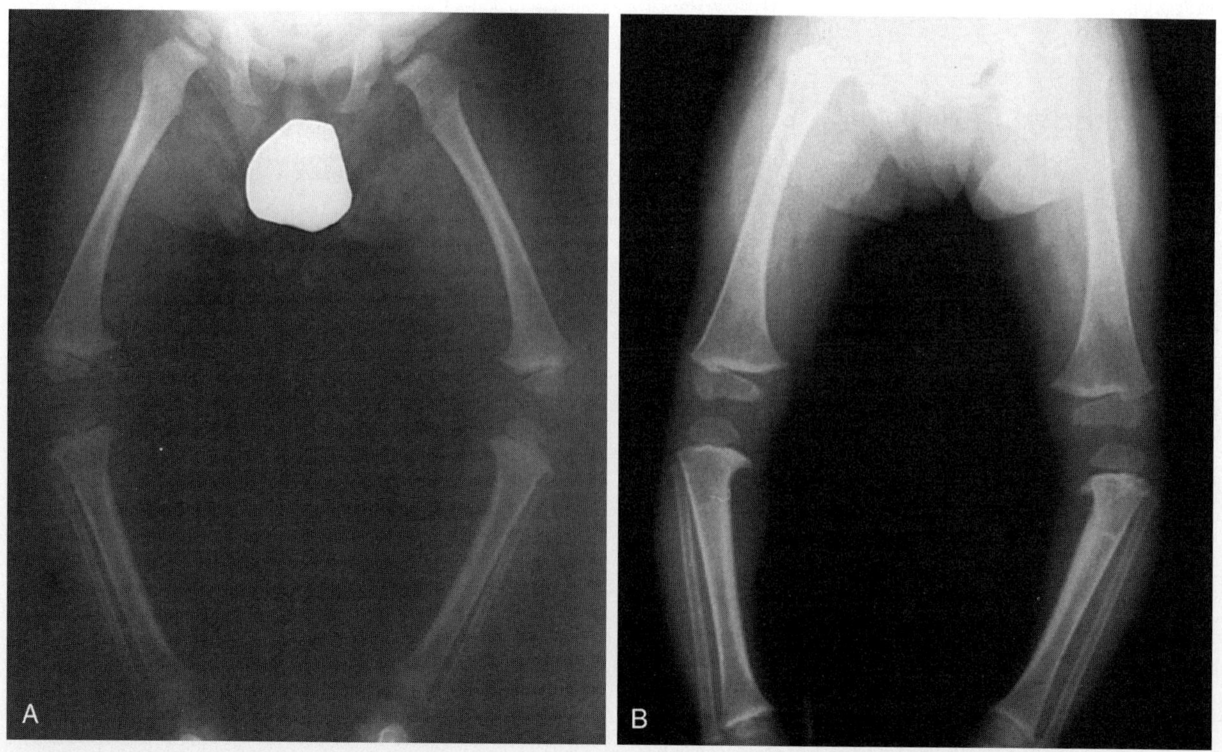

Figure 52–1. Physiologic bowing versus Blount's disease. *A,* Physiologic bowing in an otherwise healthy 2-year-old child. The bowing in this patient involves the femora and the tibiae. *B,* Physiologic bowing on the right side and Blount's disease on the left in a 20-month-old girl.

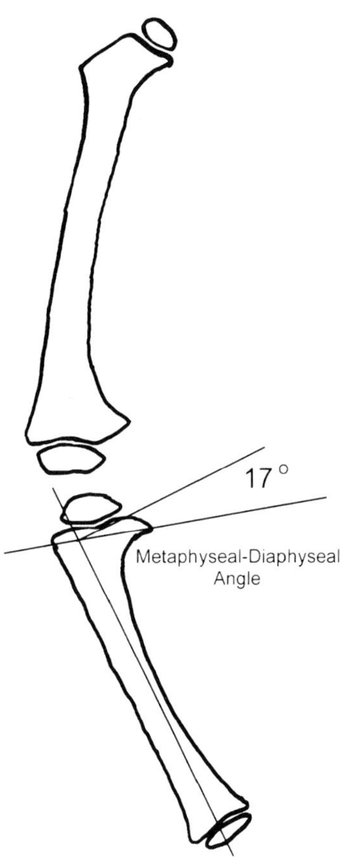

Figure 52–2. Line drawing illustrates the measurement of the metaphyseal-diaphyseal angle.

INFANTILE BLOUNT'S DISEASE

As in physiologic bowing, Blount's disease occurs in children who start walking early. The abnormality is believed to be due to excessive loading of the medial portion of the growth plate and surrounding bone in the proximal tibia. The increased stresses cause the medial epiphyseal and metaphyseal bones to be replaced with cartilage, and the medial portion of the proximal tibial epiphysis and metaphysis develop growth retardation, resulting in tibia vara. The disease is usually bilateral.

The severity of the disease has been graded by Langenskiöld (1989) into six radiographic stages, which have bearing on the treatment. In the earlier stages (I and II), there is mild-to-moderate flattening and irregularity of the medial aspect of the proximal tibial epiphysis and metaphysis (see Fig. 52–3). A bony spur may be seen medially and posteriorly (Fig. 52–4). In more advanced stages, the irregularity and fragmentation of bones become progressively more severe (Fig. 52–5). A bony bar may develop across the medial physis, which further impedes growth. Focal fibrocartilagenous dysplasia is a rare disease of the proximal tibia that has distinctive radiographic features but clinically resembles infantile Blount's disease (Fig. 52–6).

ADOLESCENT BLOUNT'S DISEASE

Blount's disease has a biomodal age distribution. Infantile Blount's disease occurs between the ages of 1 and 6 years. The other type occurs in early adolescence and is

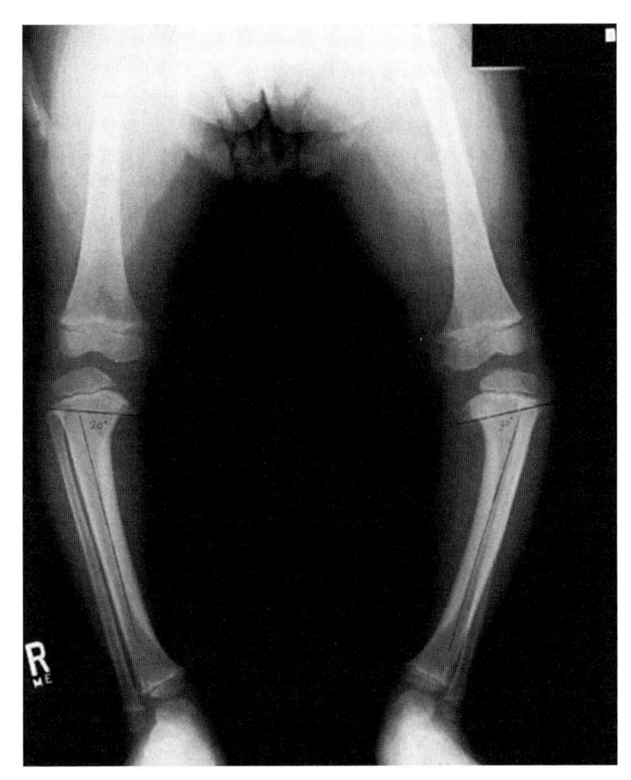

Figure 52–3. Bilateral Blount's disease more severe on the left side. The metaphyseal-diaphyseal angle is 20 degrees on the right and 30 degrees on the left.

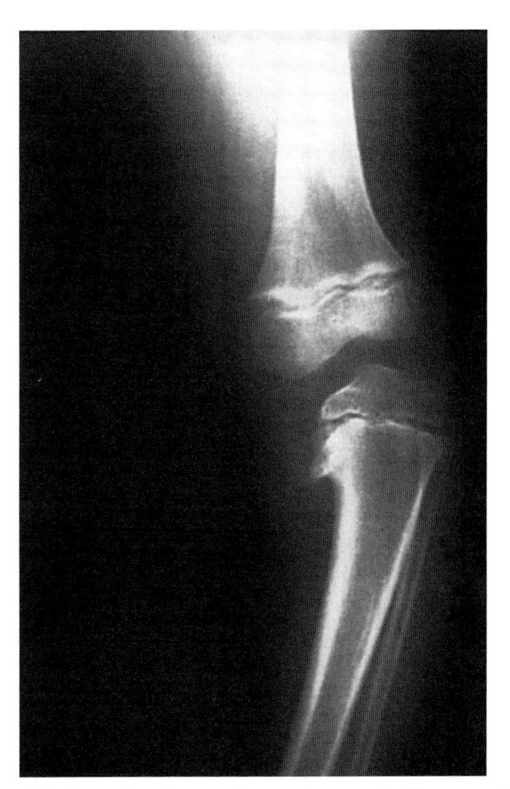

Figure 52–4. Early Blount's disease (Langenskiöld stage II) with medial spur.

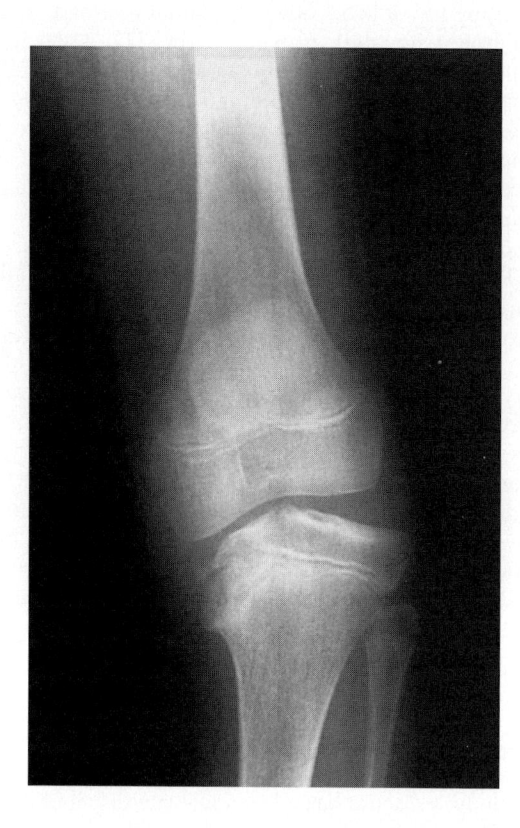

Figure 52–5. Advanced Blount's disease (Langenskiöld stage V).

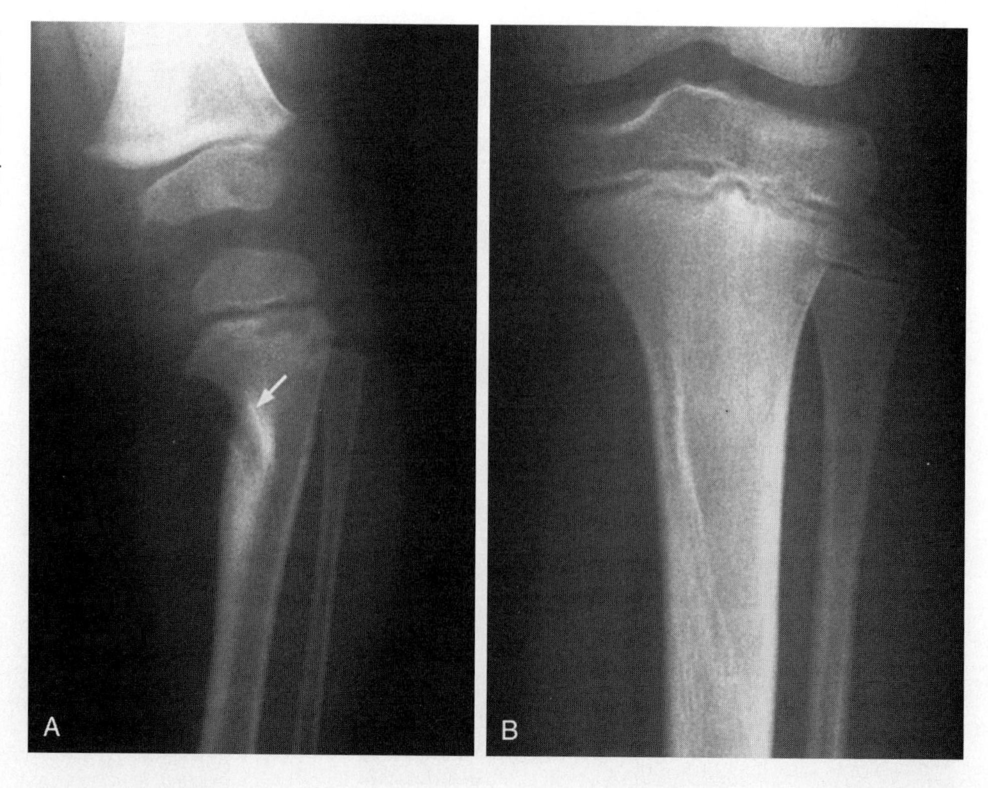

Figure 52–6. Focal fibrocartilaginous dysplasia of the proximal tibia. *A,* Anteroposterior view shows the cartilaginous rest in medial metaphysis of the proximal tibia *(arrow)* in a 17-month-old boy. *B,* Anteroposterior view obtained at the age of 9 years shows complete correction of bowing.

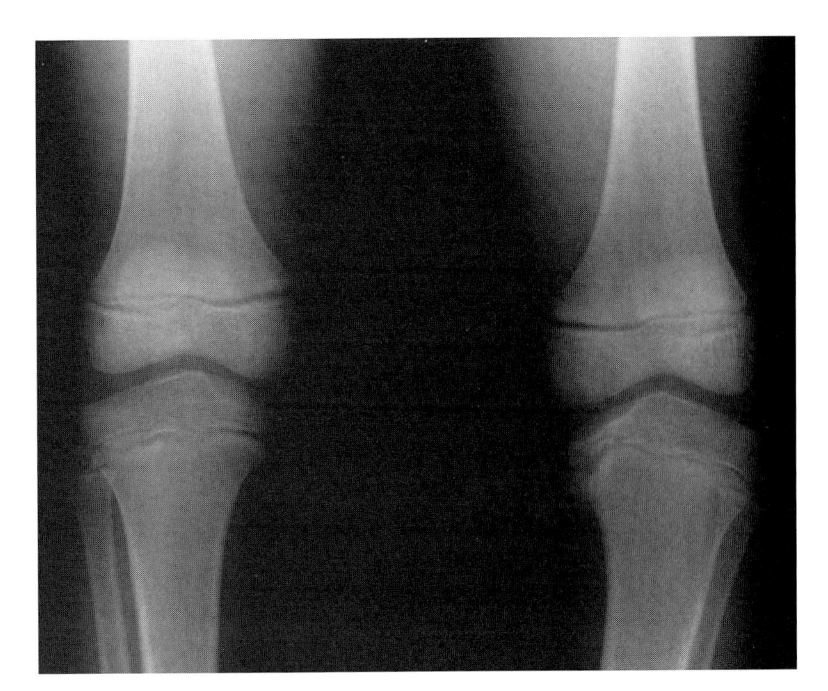

Figure 52–7. Adolescent Blount's disease on the left side in an 11-year-old boy.

less common. In adolescent Blount's disease, the radiographic changes are not as pronounced as in the infantile type because the tibial epiphysis and metaphyses already are well developed (Fig. 52–7). Because of the size of the patient, the deformity should be assessed using long leg films and exposures made in standing position for measurement of the mechanical axis of the lower extremity.

CONGENITAL POSTEROMEDIAL BOWING OF THE TIBIA AND FIBULA

This disease affects the distal third of the tibia and fibula (Fig. 52–8) and is associated with severe calcaneovalgus deformity of the foot. With conservative treatment, the angular deformity in the tibia diminishes, but relative

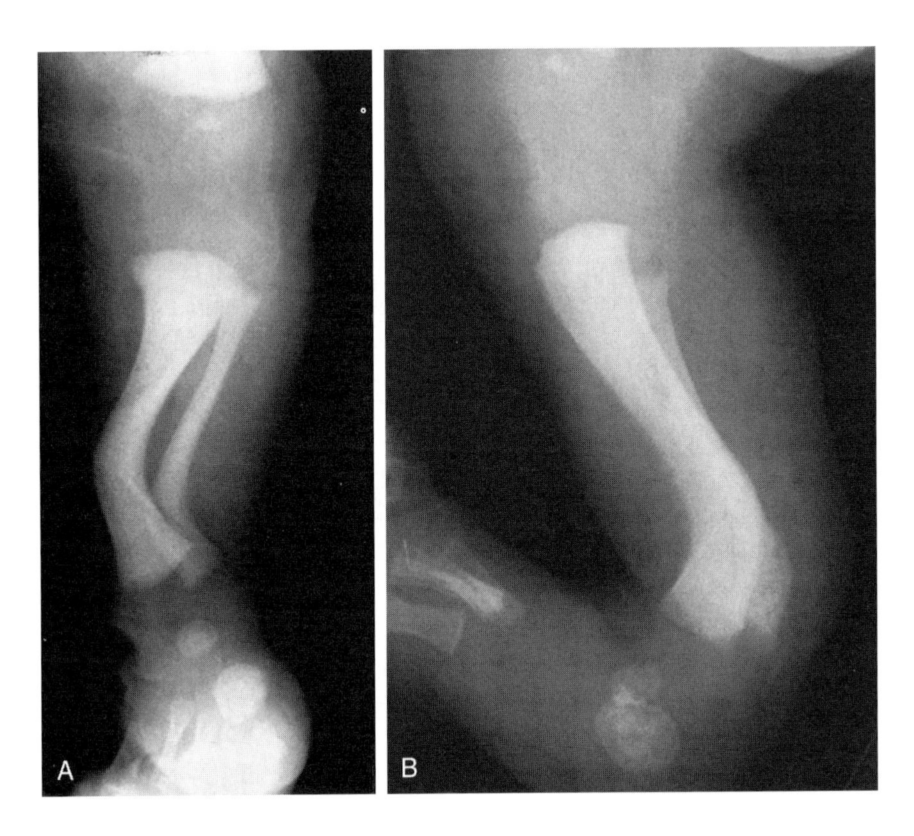

Figure 52–8. *A* and *B*, Anteroposterior and lateral views of the leg in an infant with congenital posteromedial bowing of the tibia and fibula.

shortening and soft tissue abnormalities persist. The cause is uncertain, but different authors believe that the deformity could be due to an intrauterine fracture, abnormal fetal position, or an embryologic defect in the development of the leg.

Recommended Readings

Feldman MD, Schoenecker PL: Use of the metaphyseal-diaphyseal angle in the evaluation of bowed legs. J Bone Joint Surg Am 75:1602, 1993.

Greene WB: Infantile tibia vara. J Bone Joint Surg Am 75:130, 1993.

Langenskiöld A: Tibia vara: A critical review. Clin Orthop 246:195, 1989.

Levine AM, Drennan JC: Physiological bowing and tibia vara: The metaphyseal-diaphyseal angle in the measurement of bowleg deformities. J Bone Joint Surg Am 64:1158, 1982.

Pappas AM: Congenital posteromedial bowing of the tibia and fibula. J Pediatr Orthop 4:525, 1984.

Thompson GH, Carter JR: Late-onset tibia vara (Blount's disease). Clin Orthop 255:24, 1990.

CHAPTER 53

Congenital and Developmental Foot Abnormalities

Lori L. Barr, MD, Georges Y. El-Khoury, MD, and Shigeru Ehara, MD

ANGLE MEASUREMENTS, NORMAL RELATIONSHIPS, AND TERMINOLOGY

In the evaluation of congenital foot abnormalities, it is important to be familiar with the normal relationships between the bones of the foot and to develop a working knowledge of what angles reveal significant information and how to draw these angles.

Anteroposterior View of the Normal Foot

1. A line drawn through the long axis of the talus projects through the long axis of the first metatarsal (Fig. 53–1*A*).
2. A line drawn through the long axis of the calcaneus projects through the long axis of the fourth metatarsal or between the fourth and fifth metatarsals (see Fig. 53–1*A*).
3. The talocalcaneal angle (Kite's angle) is formed by the intersection of the lines described in *1* and *2* (see Fig. 53–1*A*). It ranges from 30 to 50 degrees in young children. It is closer to 50 degrees in infants, but it progressively decreases with age. After age 5 years, the angle ranges from 15 to 35 degrees.

Lateral View of a Normal Foot

1. After age 5 years, a line drawn through the long axis of the talus projects through the long axis of the first metatarsal (Fig. 53–1*B*). In infants and young children, this line passes below the first metatarsal.
2. The talocalcaneal angle on the lateral view ranges from 25 to 45 degrees and is not age dependent (see Fig. 53–1*B*).
3. On a lateral weight-bearing view, the angle formed between a line tangent to the inferior surface of the calcaneus and first metatarsal should not be less than 155 degrees (Hibb's angle) (see later under Cavus Foot [Pes Cavus]).

Terminology

Hindfoot: The talus and the calcaneus.

Midfoot: The tarsal navicular, cuboid, and cuneiforms.

Forefoot: The remainder of the bones of the foot distal to the midfoot.

Equinus: Abnormal elevation of the posterior part of the calcaneus with respect to its anterior aspect; also referred to as abnormal plantar flexion of the hindfoot.

Calcaneus: Abnormal elevation of the anterior aspect of the calcaneus; also referred to as abnormal dorsiflexion of the hindfoot.

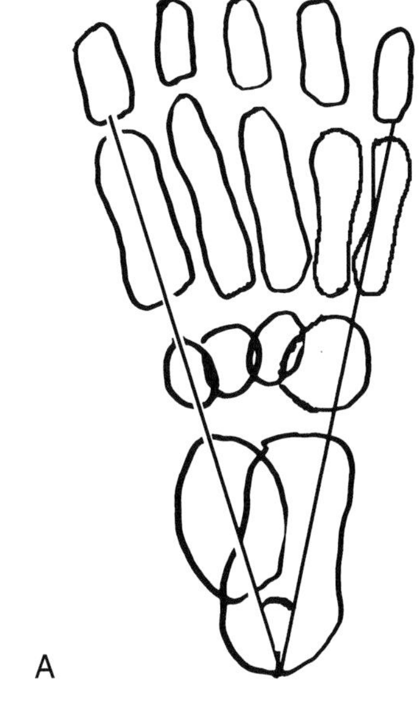

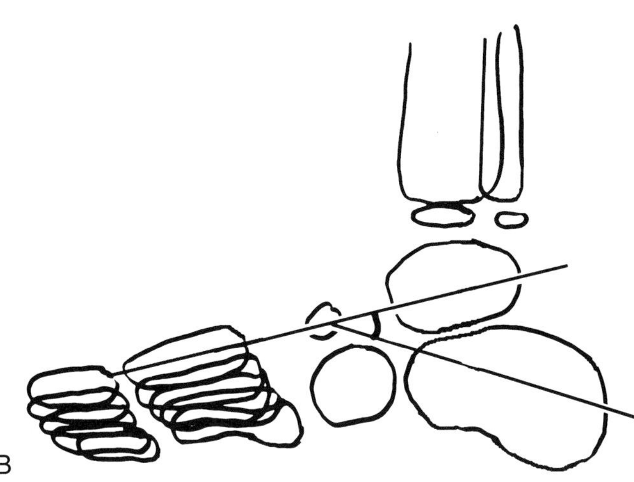

Figure 53–1. Normal foot. *A,* Line drawing of anteroposterior view of the foot depicts normal relationships between the hindfoot and forefoot. The talocalcaneal angle (Kite's angle) is drawn; it normally measures 15 to 35 degrees. *B,* Line drawing of a lateral view of the foot depicts normal relationships between the hindfoot and forefoot. A line drawn through the long axis of the talus should pass through the long axis of the first metatarsal. The talocalcaneal angle also is drawn (range, 25 to 45 degrees).

Valgus: Foot part or bone deviated or turned away from the midline of the body.
Varus: Foot part or bone deviated or turned toward the midline of the body.
Abduction: The foot is drawn away from the midline; sometimes used interchangeably with *valgus.*
Adduction: The foot is drawn toward the midline with no rotation applied; sometimes used interchangeably with *varus.*
Talipes: A deformity of the foot involving the talus; derived from Latin (*tali* = talus, heel, or ankle; *pes* = foot).

METATARSUS ADDUCTUS (METATARSUS VARUS, METATARSUS ADDUCTOVARUS)

Overview

Metatarsus adductus is the most common congenital foot deformity. In most cases, the deformity has a good prognosis, with more than 90% of the affected feet correcting spontaneously. Clinically the hindfoot is in normal position, but the metatarsals are adducted, creating a concave medial border of the foot (Fig. 53–2).

Imaging

Radiographs rarely are requested. The diagnosis is made clinically in infants with this condition, but if the deformity persists, radiographs are performed. They reveal normal alignment of the hindfoot with adductus deformity of the forefoot, which is most severe at the first metatarsal and least severe at the fifth metatarsal (Fig. 53–3).

A variant of metatarsus adductus is the *skewfoot,* in which in addition to forefoot adduction, there is hindfoot valgus. The diagnosis cannot be made at birth, and some believe the skewfoot develops as a result of the treatment of metatarsus adductus. Anteroposterior standing radiographs confirm what has been termed *zigzag* deformity with valgus of the hindfoot and adductus of the forefoot.

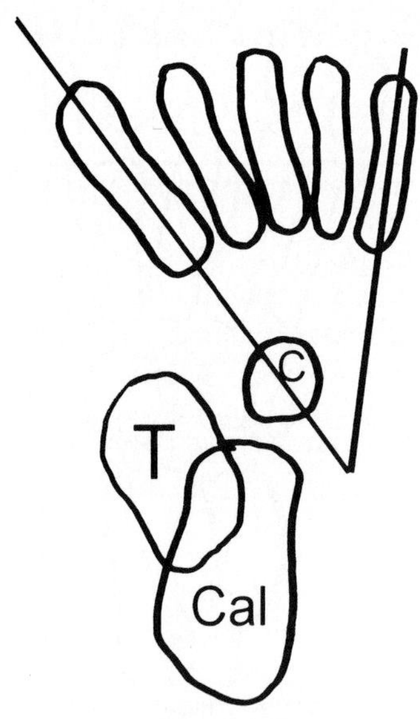

Figure 53–2. Metatarsus adductus. Line drawing depicts the bony relationships in a foot with metatarsus adductus. T, talus; Cal, calcaneus; C, cuboid.

CLUBFOOT DEFORMITY (TALIPES EQUINOVARUS)

Overview

Congenital clubfoot occurs in 1 in 800 births. In about half the cases, the disease is bilateral, and it is twice as common in boys than in girls. There is a genetic tendency, and in families with an affected child, the risk to the next child is about 2% to 5%. Abnormalities associated with clubfoot include atrophy of the calf muscles and absence of dorsalis pedis artery. Clinically the hindfoot is plantar flexed (equinus), the hindfoot and midfoot are inverted, and the midfoot and forefoot are adducted—hence the term *talipes equinovarus.* If untreated, the deformity in clubfoot persists into adulthood, resulting in significant dysfunction.

Imaging

Although the diagnosis of congenital clubfoot is made clinically at birth, radiography commonly is used in follow-up of treatment. The anteroposterior view is obtained with the foot forced against the radiographic cassette to simulate weight bearing. The lateral view is obtained with the foot dorsiflexed and everted.

The anteroposterior view shows the long axis of the talus and calcaneus to be almost parallel, the cuboid is displaced medially, and the metatarsals are adducted markedly. A line drawn along the long axis of the talus passes lateral to the first metatarsal (Figs. 53–4 and 53–5).

On the lateral view, the talus and calcaneus are almost parallel, and the talus is plantar flexed in relation to the tibia. The normal upward pitch of the calcaneus is reversed, and the long axis of the calcaneus goes into equinus. On the lateral view, the metatarsal bones align almost parallel to each other (see Figs. 53–4 and 53–5).

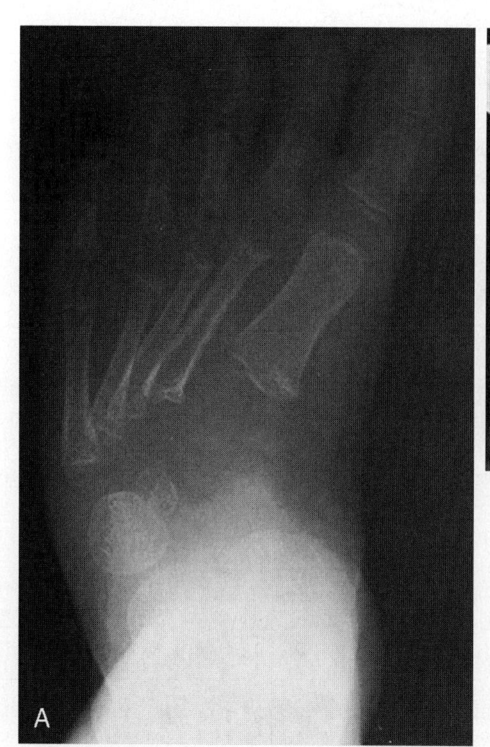

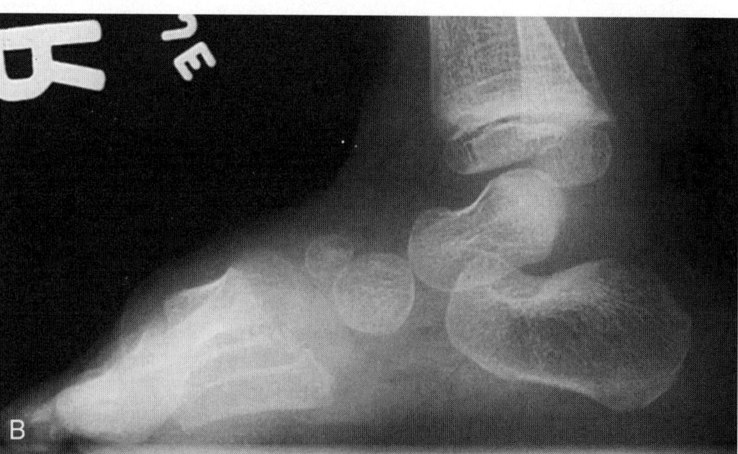

Figure 53–3. Metatarsus adductus in a 3-year-old boy. *A* and *B,* Anteroposterior and lateral views of a foot with persistent metatarsus adductus. The hindfoot is normal on both views; however, the forefoot is severely adducted.

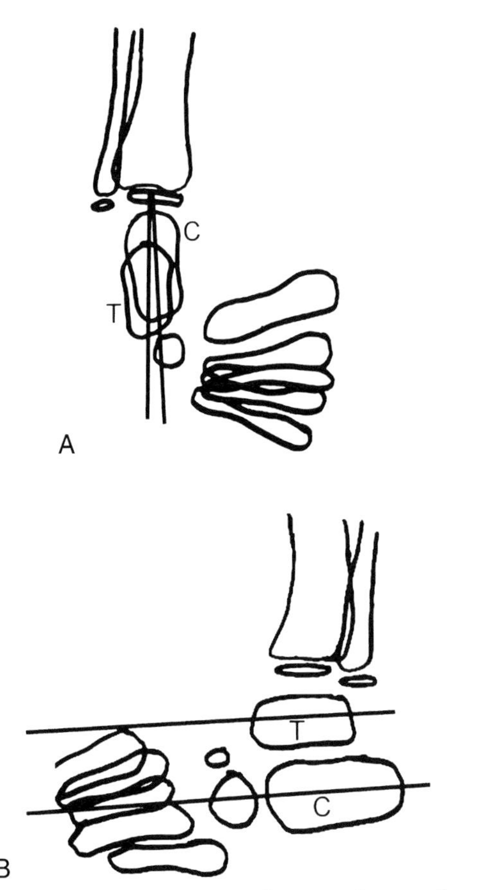

Figure 53–4. Clubfoot deformity. *A* and *B*, Line drawings depict the typical relationships between the bones in clubfoot deformity. The calcaneus is in equinus, and the talus and calcaneus are almost parallel on the anteroposterior and lateral views. T, talus; C, calcaneus.

CONGENITAL VERTICAL TALUS

Overview

Congenital vertical talus is a rare abnormality of the foot that is associated with neuromuscular disease in about 50% of patients. In such patients, the deformity is believed to be due to muscle imbalance. In the rest, the disease is idiopathic. On physical examination, the deformity is rigid, and the arch of the foot is reversed or convex down. The head of the talus can project on the medial and plantar aspect of the foot. The deformity persists if untreated.

Imaging

On the lateral view, the talus is plantar flexed, and the navicular, if ossified, is dislocated dorsally above the head of the talus. The cuboid often is subluxed dorsally in relation to the calcaneus. On the anteroposterior view, the hindfoot is in valgus (Figs. 53–6, 53–7, and 53–8).

FLATFOOT (PES PLANUS)

Overview

Flatfoot is characterized by loss of the longitudinal arch of the foot and hindfoot valgus (Fig. 53–9). It can be divided into two types: (1) flexible flatfoot, in which the joints are mobile, and (2) rigid flatfoot, in which the mobility of the subtalar joint is limited and usually is

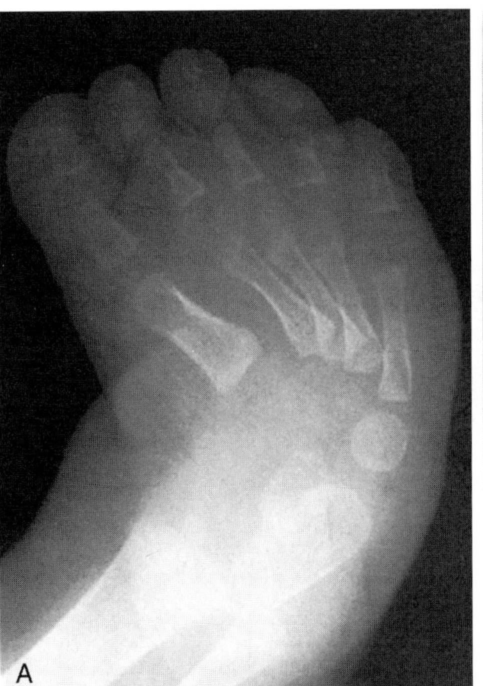

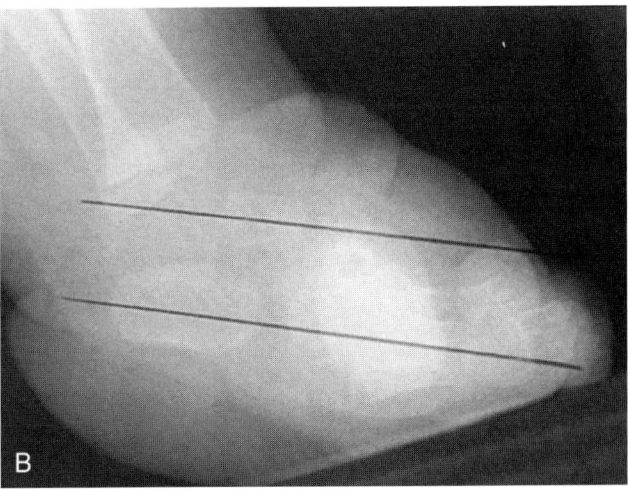

Figure 53–5. Clubfoot deformity in a 6-month-old boy. *A*, Anteroposterior view of the foot reveals severe adduction of the midfoot and forefoot. *B*, Lateral view of the foot taken with forced dorsiflexion shows the calcaneus in equinus. The talus and calcaneus are parallel, and the heel is not touching the floor.

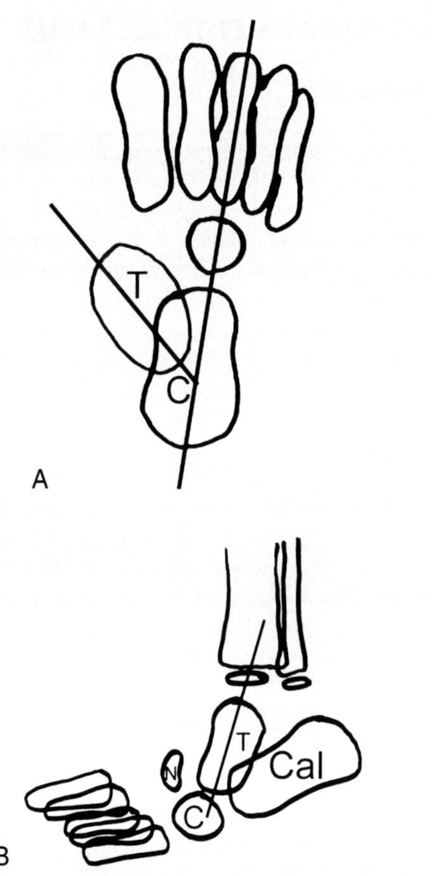

Figure 53–6. Congenital vertical talus. *A,* Line drawing shows the relationships of the hindfoot to the forefoot on anteroposterior projection. *B,* Line drawing shows the relationships of the hindfoot to the forefoot on the lateral projection. T, talus; Cal, calaneus; C, cuboid; N, navicular.

caused by tarsal coalition. Rigid flatfoot typically is associated with foot pain and spasm or contracture of the peroneal tendons (peroneal spastic flatfoot) (see Chapter 54). Flexible flatfoot is more common and rarely is associated with pain or disability. Acquired flatfoot is another related entity that is seen most commonly in middle-aged women and is usually due to posterior tibial tendon dysfunction. Other causes include disruption of the spring ligament or talonavicular ligament. Acquired flatfoot deformity starts as being flexible, but with time it becomes rigid.

Imaging

For flexible flatfoot, imaging rarely is needed, but it often is indicated to rule out a tarsal coalition. The lateral view shows loss of the longitudinal arch of the foot with the calcaneus almost parallel to the floor. The collapse of the longitudinal arch can occur either at the navicular-cuneiform joint (navicular-cuneiform sag) or at the talonavicular joint (talonavicular sag) (Fig. 53–9*A*). On the lateral view, a line drawn through the long axis of the talus passes inferior to the first metatarsal (Fig. 53–9*B*). On the anteroposterior view, the talus projects medially, and the talocalcaneal angle is increased. The navicular shifts laterally. A line drawn through the long axis of the talus passes medial to the first metatarsal. The standing hindfoot alignment view (Buck's view) is ideal for assessing the degree of hindfoot valgus in flatfeet (Fig. 53–9*D*).

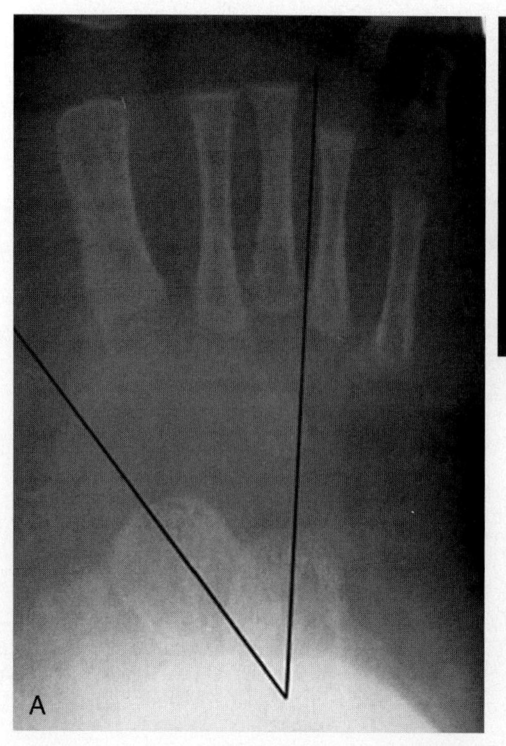

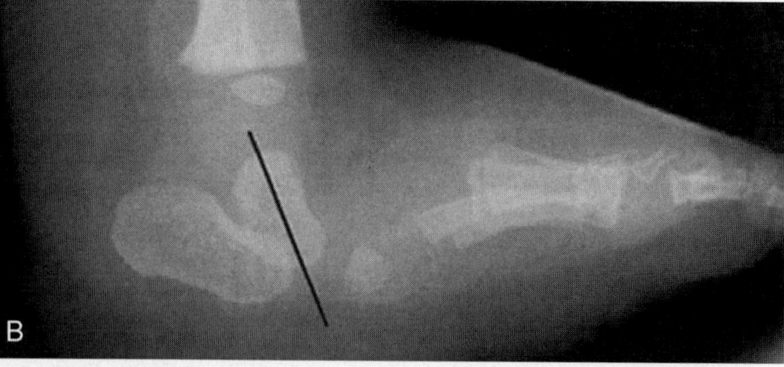

Figure 53–7. Congenital vertical talus deformity in a 13-month-old boy. *A,* Anteroposterior view shows valgus of the hindfoot with the talar head projecting medially. *B,* Lateral view shows the vertical orientation of the talus *(line).* The calcaneus is in equinus, and the arch of the foot is reversed (convex down).

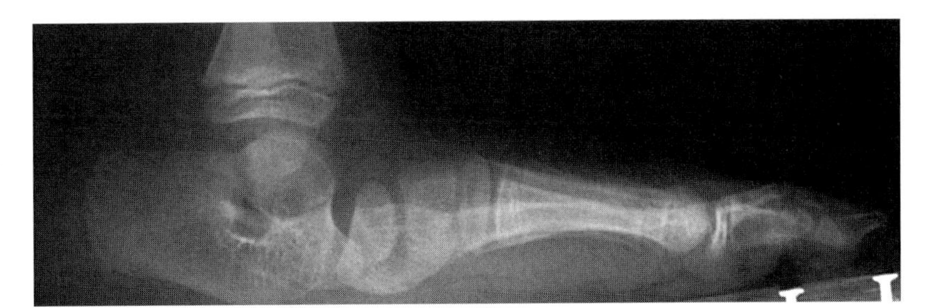

Figure 53–8. Congenital vertical talus in a 7-year-old child. Note the superior dislocation of the navicular bone in relation to the head of the talus.

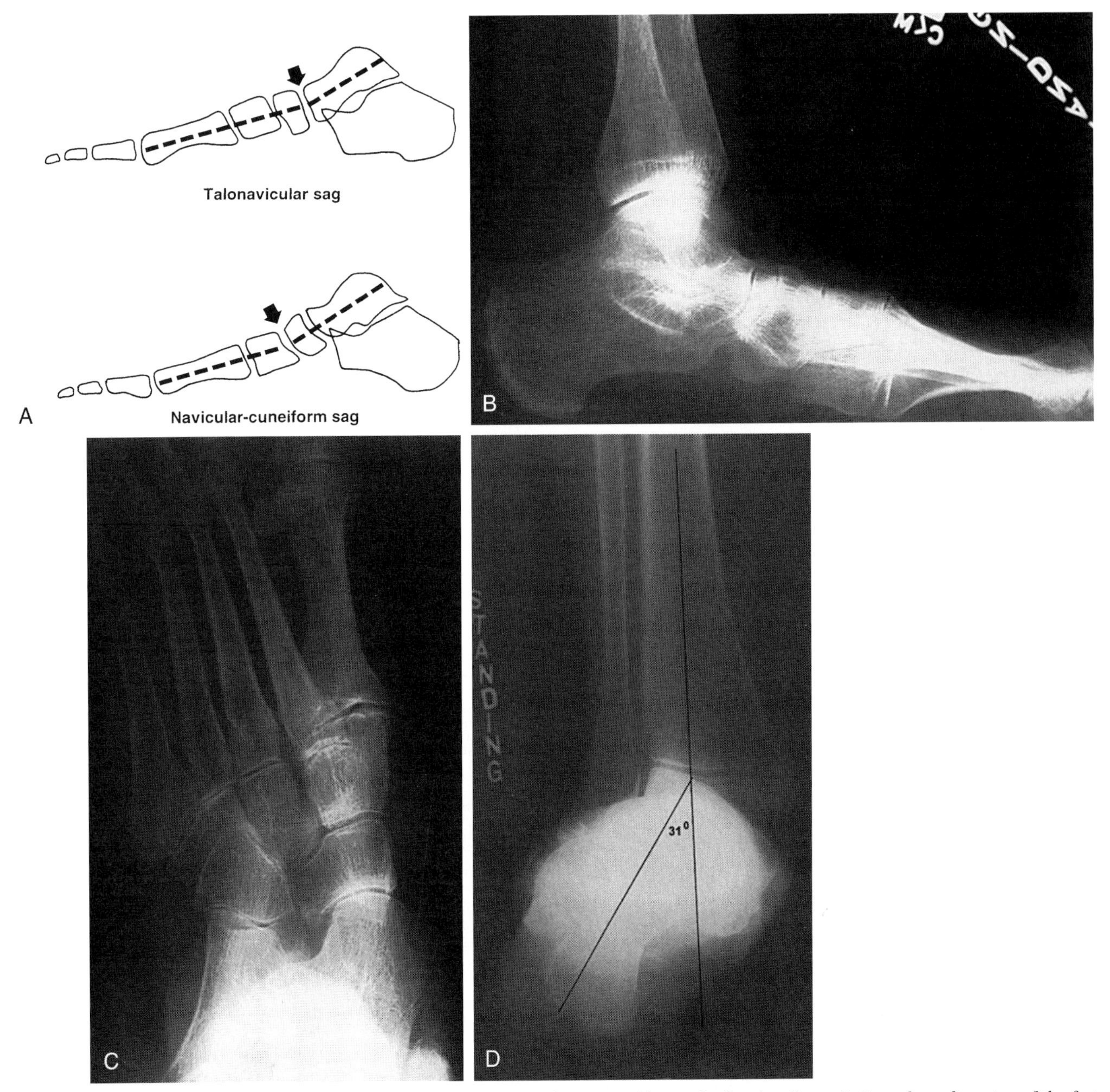

Figure 53–9. Flatfoot (pes planus) deformity. *A*, Line drawing shows the points of longitudinal arch collapse. *B*, Lateral standing view of the foot shows the longitudinal arch to be flat because of a talonavicular sag. *C*, Anteroposterior view shows the talus to be pointing medially, suggesting valgus of the hindfoot. *D*, Standing hindfoot alignment view (Buck's view) reveals significant valgus of 31 degrees. Normally the long axis of the tibia and long axis of the calcaneus are parallel.

CAVUS FOOT (PES CAVUS)

Overview

Cavus foot is characterized by an exaggerated arch of the foot, affecting either the medial aspect of the foot (cavovarus) (Fig. 53–10) or the entire foot (calcaneocavus) (Fig. 53–11). In pes cavovarus, the high arch is due primarily to plantar flexion of the first metatarsal and associated pronation of the forefoot (see Fig. 53–10). In the calcaneocavus foot, the deformity is characterized by dorsiflexion of the calcaneus and plantar flexion of the forefoot. Frequently the deformity is acquired, with an underlying neuromuscular disorder causing muscle atrophy and weakness. Such diseases include Charcot-Marie-Tooth disease, which is the most common cause of cavus foot; syringomyelia; spinal cord tumor; diastematomyelia; cerebral palsy; muscular dystrophy; and meningomyelocele. Patients with Charcot-Marie-Tooth disease typically develop a cavus deformity that affects the medial border of the foot. Acquired cavus foot deformity in a previously healthy child is an indication for thorough investigation of the spine with magnetic resonance (MR) imaging to rule out cord abnormalities. Another common type of cavus foot is congenital cavus foot, in which a family history is present in about one third of patients.

Imaging

Radiographic findings on the standing lateral view vary with the type of cavus deformity. If the entire foot is in cavus, the arch of the foot is high, the calcaneus is pitched up, and the intersection of lines drawn through the long axis of the talus and first metatarsal on the lateral view is angulated dorsally instead of being a straight line (see Fig. 53–11). When only the medial aspect of the foot is involved, the medial arch is high, and the first metatarsal is flexed down; however, the fifth metatarsal lies flat on the floor (see Fig. 53–10). On the standing lateral view of the foot, measurement of Hibb's angle in normal feet should not be less than 155 degrees (see Fig. 53–10). The anteroposterior view of the foot reveals hindfoot varus with almost parallel alignment of the talus and calcaneus. Another method for assessing the presence of a cavus deformity is accomplished by measuring what is known as the calcaneal pitch (also called the calcaneal inclination angle). This angle is constructed by drawing a line parallel to the floor and a second line along the plantar surface of the calcaneus. The angle created by the intersection of these two lines should normally measure 15 to 30 degrees. Angles greater than 30 degrees indicate the presence of a cavus deformity (Fig. 53–10*B*). Angles less than 15 degrees indicate the presence of a flatfoot deformity.

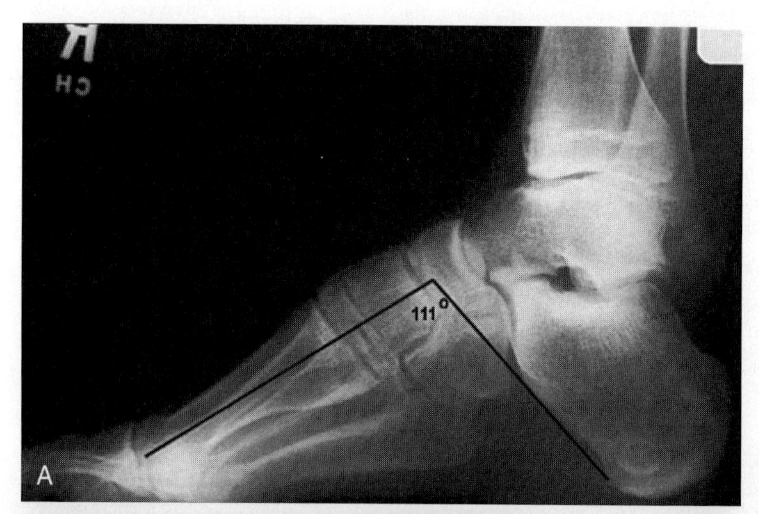

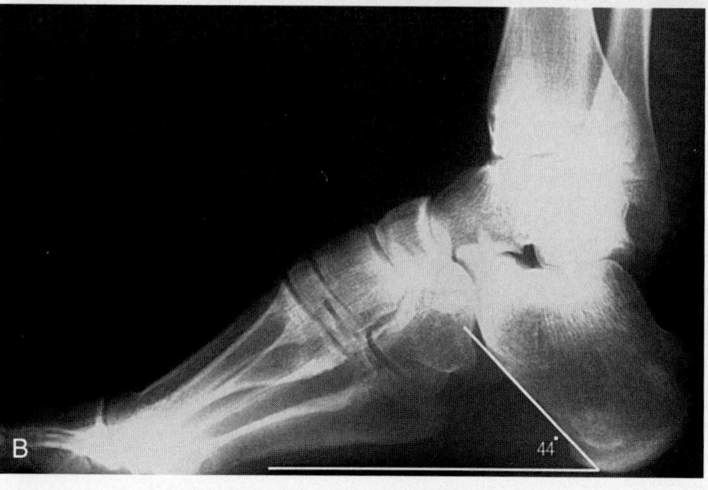

Figure 53–10. Measurement methods for cavus deformity. *A*, Cavus deformity affecting the medial aspect of the foot. Hibb's angle measured 111 degrees, which is much less than normal (155 degrees), confirming the presence of a cavus deformity. *B*, This figure illustrates the method for measuring the calcaneal pitch (normal range is 15 to 30 degrees), which in this case measures 44 degrees.

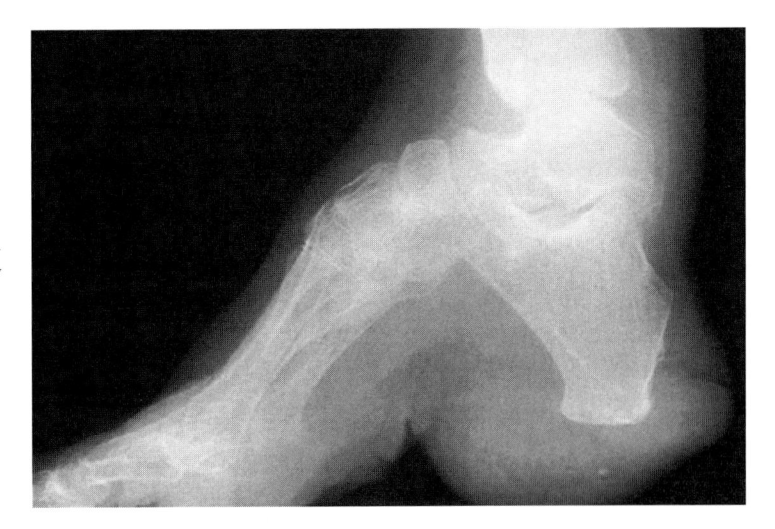

Figure 53–11. Calcaneovarus foot. Cavus deformity affecting the entire foot. Note the pistol grip deformity produced by the severe calcaneus of the hindfoot.

HALLUX VALGUS

Overview

Hallux valgus is defined as the lateral deviation of the great toe. Associated findings include medial displacement of the first metatarsal and painful soft tissue swelling over the medial aspect of the first metatarsophalangeal joint. The condition commonly is called *bunion deformity*.

Hallux valgus is a developmental disorder that progresses with age. Some patients develop a thickened and painful medial capsule (*bunion*) during childhood or adolescence. The age of onset is less than 20 years in 46% and less than 15 years in 31%. Familial occurrence is reported.

In adulthood, the development of hallux valgus is related to wearing ill-fitting shoes. The force applied to the tip of the toe causes lateral shift of the great toe owing to the weakness of the medial capsule of the metatarsophalangeal joint. Other contributing factors to hallux valgus include metatarsus primus varus, muscle imbalance, and pes planus.

Imaging

The radiographic examination includes a weight-bearing anteroposterior view and a weight-bearing lateral view, a non–weight-bearing oblique view, and a sesamoid skyline view (Fig. 53–12). Deformity of the metatarsal head and the first metatarsophalangeal joint and the alignment abnormalities are the main features of hallux valgus.

Deformity of the Metatarsal Head. Forces applied to the great toe produce subluxation at the first metatarsophalangeal joint and medial displacement of the first metatarsal head. The first metatarsal head widens, and the pressure from the phalanx results in a sagittal groove and enlargement of the metatarsal head (Fig. 53–13).

Alignment. Medial deviation of the first metatarsal is associated with pronation, which cannot be evaluated easily on plain radiographs; however, it correlates with the degree of metatarsus primus varus and the intermetatarsal angle. Pronation of the first metatarsal also can be assessed by the degree of subluxation of the sesamoids (Fig. 53–14). The sesamoids are an integral part of the normally functioning first metatarsophalangeal joint. Erosions and chondromalacia of the sesamoids and resultant osteoarthritis can occur at the hallux sesamoid joint (see Fig. 53–14*C*). The change in alignment causes plantar displacement of the abductor hallucis tendon, migration and attenuation of the extensor hallucis longus tendon, and lateral displacement of the flexor hallucis longus tendon.

Angles. Angles used in assessing hallux valgus deformity (Fig. 53–15) include the following:

Hallux valgus angle (the angle made by the long axes of the proximal phalanx and the first metatarsal): normal (<15 degrees), mild hallux valgus (<20 degrees), moderate hallux valgus (20 degrees to 40 degrees), severe hallux valgus (>40 degrees).

Intermetatarsal angle (the angle formed by the long axes of the first and second metatarsals): normally less than 9 degrees. The magnitude of this angle correlates well with the degree of hallux valgus.

Angle of metatarsus primus varus (the angle formed by the long axes of the first metatarsal and the medial cuneiform): normally 10 degrees or less. The metatar-

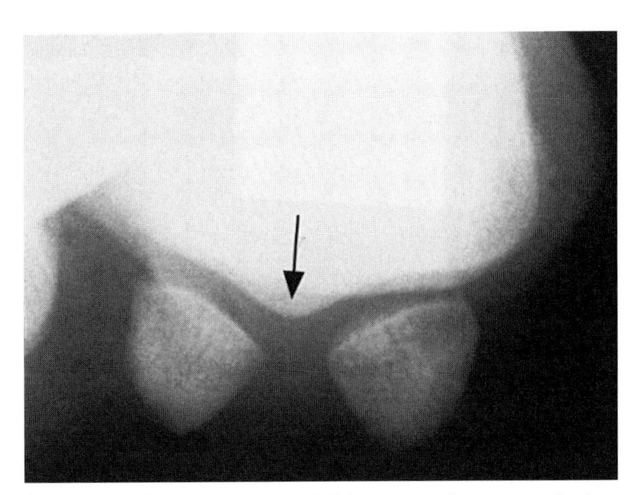

Figure 53–12. Normal sesamoid skyline view. The crista is the bony ridge (*arrow*) separating the two sesamoids.

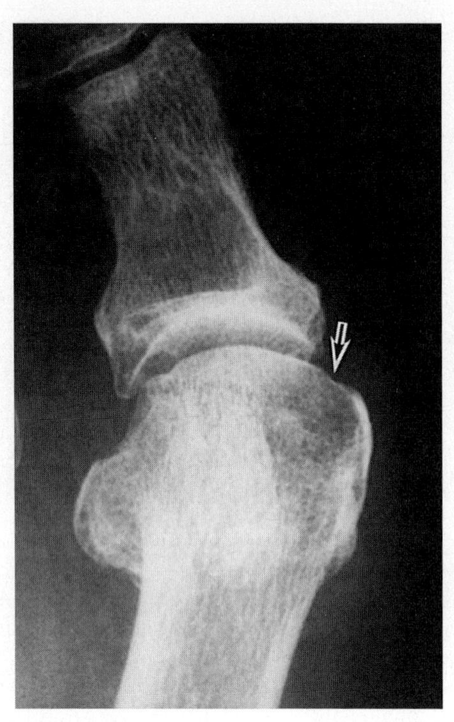

Figure 53–13. Hallux valgus deformity in a 60-year-old woman. Anteroposterior view shows uncovering of the articular surface medially with a pressure margin *(arrow)*.

socuneiform joint is oriented medially, and severe medial obliquity predisposes to metatarsus primus varus.

Postsurgical Concerns. There are many types of surgical procedures, including bunionectomy with medial capsulorrhaphy, arthroplasty, tendon transfer, and a variety of osteotomies. Radiographic evaluation is important to assess postsurgical changes and complications. Stress fractures in the metatarsals have been described after bunion surgery, especially affecting the second metatarsal.

HALLUX RIGIDUS

Overview

Hallux rigidus is a disorder of the first metatarsophalangeal joint characterized clinically by pain and limitation of motion. It is the second most common deformity of the great toe next to hallux valgus. It is usually the result of osteoarthritis involving mainly the dorsal aspect of the first metatarsophalangeal joint. Conditions predisposing to the development of hallux rigidus include anatomic variations (long narrow foot, pronated foot, elevation of the first metatarsal, long first metatarsal), improper shoes, abnormal gaits, and osteochondritis dissecans.

The age at onset ranges from 12 to 57 years. The condition is twice as common in women than in men. Localized swelling and pain over the first metatarsopha-

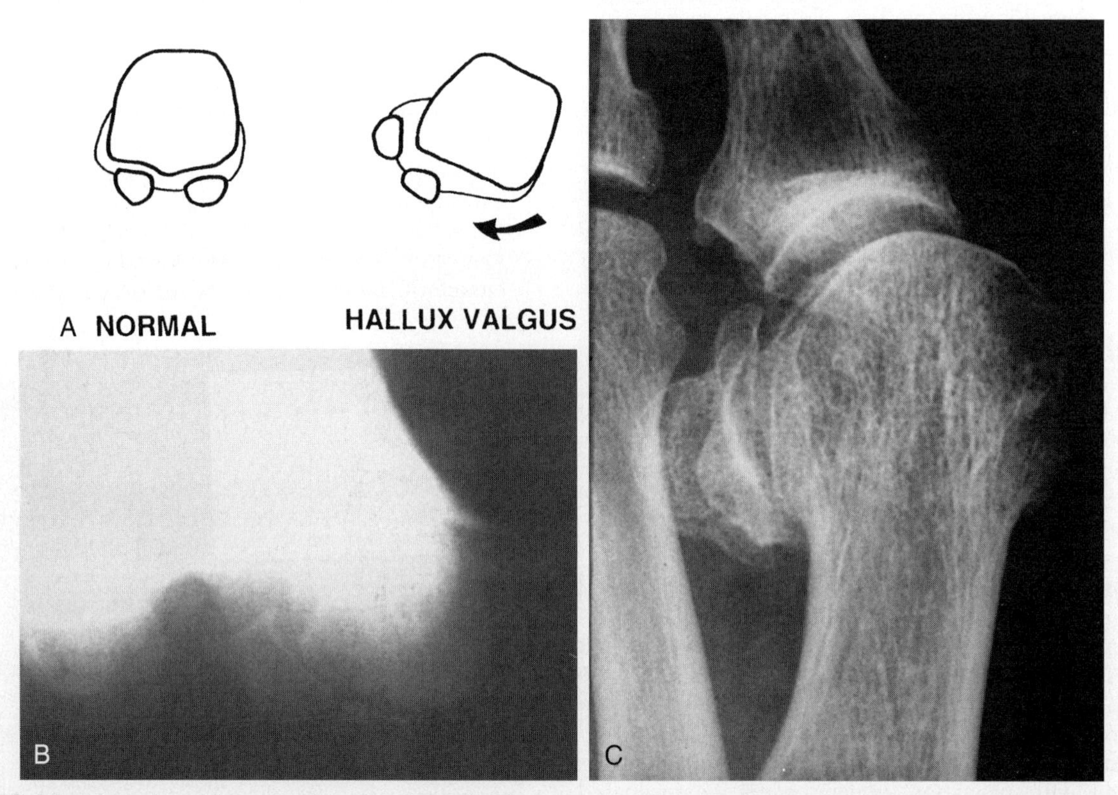

Figure 53–14. Relationship of the sesamoid bones to the head of the first metatarsal. *A,* Line drawing shows the normal and abnormal relationships between the first metatarsal head with the sesamoid bones. *B,* Skyline view shows lateral dislocation of the sesamoids caused by hallux valgus. Note the flattening of the crista. *C,* Oblique view shows advanced osteoarthritis of the hallux-sesamoid joint with osteophyte formation and joint space narrowing.

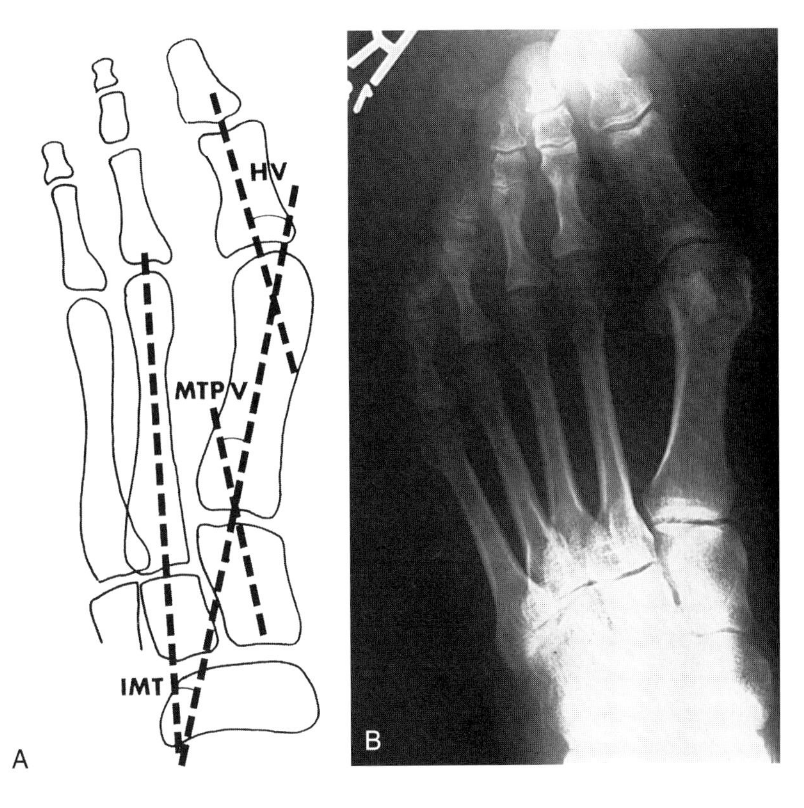

Figure 53–15. Hallux valgus. *A,* Line drawing shows important angle measurements for feet with hallux valgus. HV, hallux valgus angle; IMT, intermetatarsal angle; MTPV, angle of metatarsus primus varus. *B,* Standing anteroposterior view of a foot with hallux valgus.

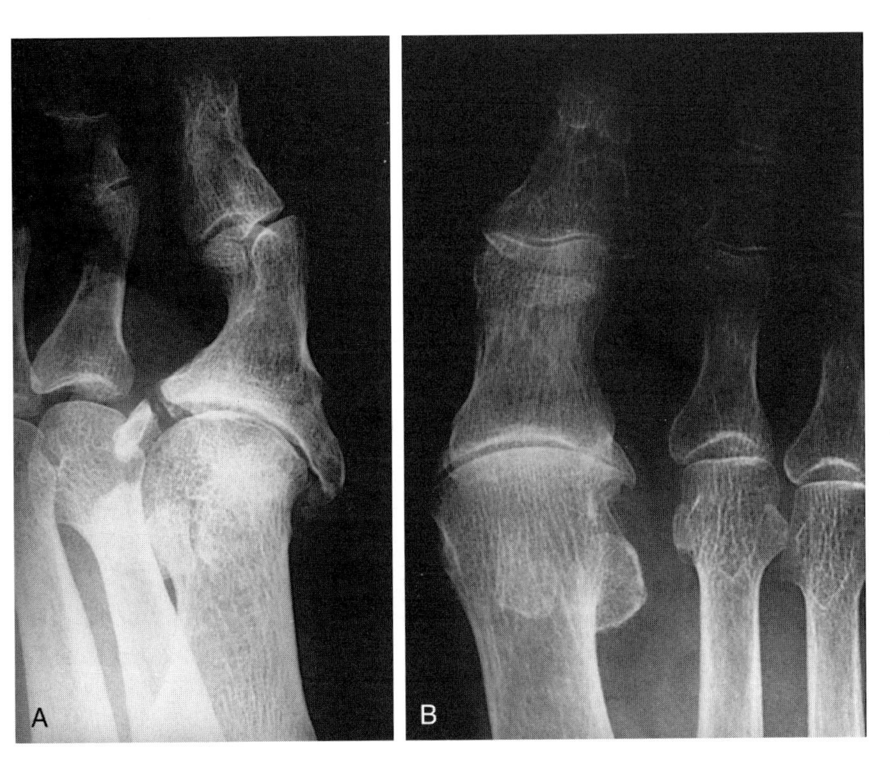

Figure 53–16. Hallus rigidus. *A,* Prominent osteophyte on the dorsal aspect of the proximal phalanx in a 60-year-old man. A detached osteophyte also is seen on the plantar aspect. *B,* Narrowing of the first metatarsophalangeal joint and osteophytes medially and laterally in a 68-year-old man.

langeal joint are common symptoms. Dorsiflexion is restricted, whereas plantar flexion may be unaffected.

Imaging

The radiographic features are those of osteoarthritis. Extensive osteophytes are present at the dorsum of the first metatarsophalangeal joint (Fig. 53–16*A*), but they also extend laterally and medially (Fig. 53–16*B*). The first metatarsophalangeal joint space typically is narrowed. Osteoarthritis of the hallux-sesamoid joint also can be seen.

Recommended Readings

Benacerraf BR, Frigoletto FD: Prenatal ultrasound diagnosis of clubfoot. Radiology 155:211, 1985.

Berg EE: A reappraisal of metatarsus adductus and skewfoot. J Bone Joint Surg Am 68:1185, 1986.

Drennan JC: Congenital vertical talus. J Bone Joint Surg Am 77:1916, 1995.

Gentili A, Mashih S, Yae L, et al: Pictorial review: Foot axes and angles. Br J Radiol 69:968, 1996.

Hardy RH, Clapham JCR: Observation on hallux valgus. J Bone Joint Surg Br 33:376, 1951.

Hattrup SJ, Johnson KA: Subjective results of hallux rigidus following treatment with cheilectomy. Clin Orthop 226:182, 1988.

Heywood AWB: The mechanics of the hind foot in club foot as demonstrated radiographically. J Bone Joint Surg Br 46:102, 1964.

Hubbard AM, Davidson RS, Meyer JS, et al: Magnetic resonance imaging of skewfoot. J Bone Joint Surg Am 78:389, 1996.

Jacobsen ST, Crawford AH: Congenital vertical talus. J Pediatr Orthop 3:306, 1983.

Karasick D, Wapner KL: Hallux rigidus deformity: Radiologic assessment. AJR Am J Roentgenol 157:1029, 1991.

Karasick D, Wapner KL: Hallux valgus deformity: Preoperative radiologic assessment. AJR Am J Roentgenol 155:119, 1994.

Kilcoyne RF, Krych SM, Gloeb H: Radiological measurement of congenital and acquired foot deformities. Appl Radiol 22:35, 1993.

Mosca VS: Flexible flatfoot and skewfoot. J Bone Joint Surg Am 77:1937, 1995.

O'Connor PJ, Bos CFA, Bloem JL: Tarsal navicular relations in club foot: Is there a role for magnetic resonance imaging? Skeletal Radiol 27:440, 1998.

Piggott H: The natural history of hallux valgus in adolescence and early adult life. J Bone Joint Surg Br 42:749, 1960.

Ponsetti IV: Treatment of congenital club foot. J Bone Joint Surg Am 74:448, 1992.

Scott G, Wilson DW, Bentley G: Roentgenographic assessment in hallux valgus. Clin Orthop 267:143, 1991.

Shereff MJ, Baumhauer JF: Hallux rigidus and osteoarthritis of the first metatarsophalangeal joint. J Bone Joint Surg Am 80:898, 1998.

Simons GW: Analytical radiography of club feet. J Bone Joint Surg Br 59:485, 1977.

Vanderwilde R, Staheli LT, Chew DE, et al: Measurements on radiographs of the foot in normal infants and children. J Bone Joint Surg Am 70:407, 1988.

Weseley MS, Barenfeld PA, Shea JM, et al: The congenital cavus foot—a follow-up report. Bull Hosp Jt Dis 42:217, 1982.

CHAPTER 54

Tarsal Coalition

Lori L. Barr, MD, and Georges Y. El-Khoury, MD

OVERVIEW

Tarsal coalition is believed to result from failure of segmentation of the primitive mesenchyme. The condition frequently is associated with peroneal spastic flatfoot, and symptoms typically appear in the teens and 20s. In most patients (>90%), tarsal coalition occurs either at the middle facet of the subtalar joint or between the anterior process of the calcaneus and lateral aspect of the navicular. Coalitions can be fibrous, cartilaginous, or bony. Clinically, calcaneonavicular coalitions are less symptomatic than subtalar coalitions. Feet with calcaneonavicular bars also are less rigid and not as flat as feet with subtalar coalition. The prevalence of these two types of coalitions is almost equal. In about half the cases, the disease is bilateral.

IMAGING

The subtalar joint can be imaged directly using an axial (Harris) view (Fig. 54–1) or an oblique view of the subtalar joint (Broden) (Fig. 54–2) of the hindfoot or by computed tomography (CT) (Fig. 54–3). Bony coalitions show a bony solid bridge connecting the talus to the calcaneus at the sustentaculum tali (Fig. 54–3*A*). Cartilaginous and fibrous coalitions are difficult to differentiate on imaging studies; both show narrowing of the middle subtalar joint with irregularity and sclerosis at the articular surfaces (Fig. 54–3*B*). Indirect radiographic signs of subtalar coalition include the talar beak, on the lateral view, just proximal to the articular surface, at the dorsal aspect of the talar head (Fig. 54–4). The talar beak should not be

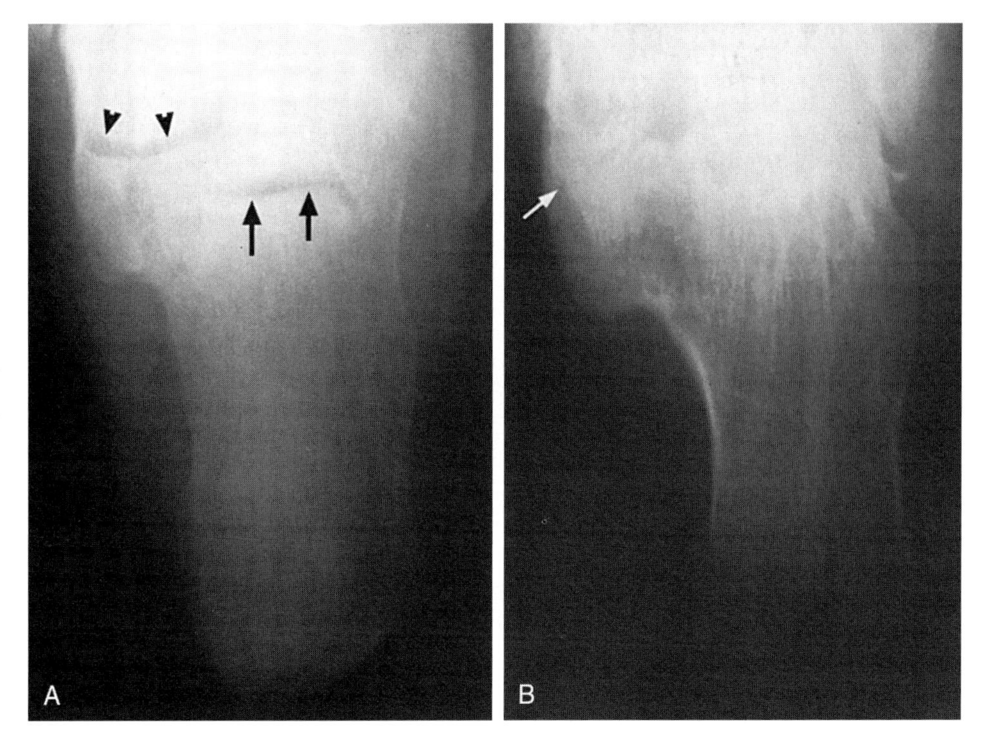

Figure 54–1. Axial view. *A*, Normal axial (Harris) view of the hindfoot. The posterior *(arrows)* and middle subtalar joints *(arrowheads)* are well shown. *B*, Bony coalition *(arrow)* at the middle subtalar joint.

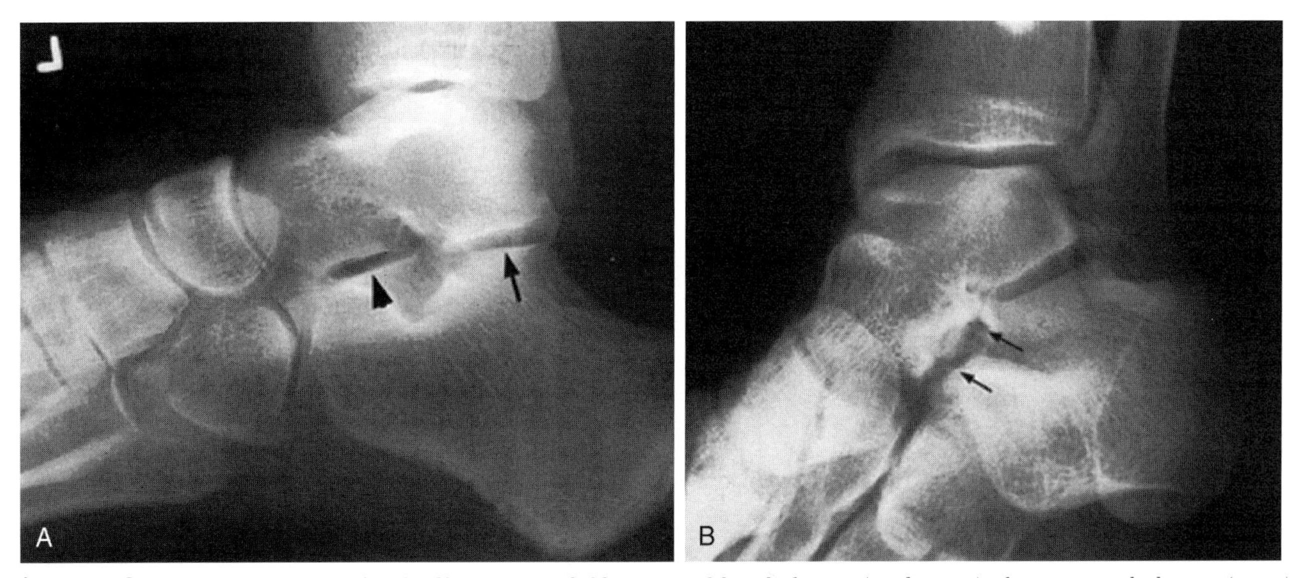

Figure 54–2. Oblique (Broden) view of the hindfoot. *A*, Normal oblique view of the subtalar joint (Broden view). The posterior subtalar joint *(arrow)* and middle subtalar joint are well shown *(arrowhead)*. *B*, Cartilaginous coalition at the middle subtalar joint *(arrows)*.

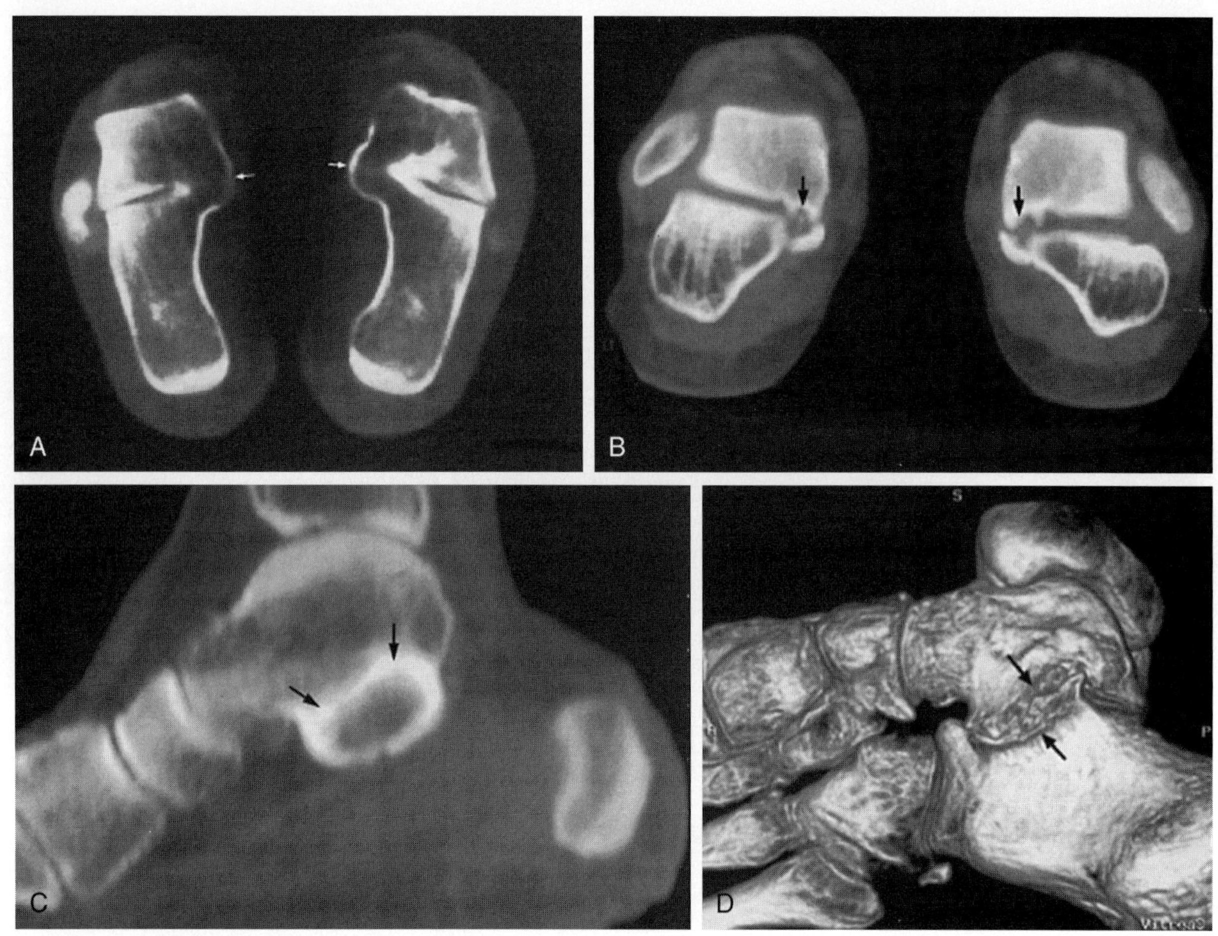

Figure 54–3. CT of the subtalar joint. *A,* Bony coalition of the middle subtalar joints *(arrows)*. *B,* Cartilaginous coalition of the middle subtalar joints *(arrows)*. *C,* Sagittal reconstructed image of the foot shows bony coalition of the middle subtalar joint *(arrows)*. *D,* Three-dimensional image shows a cartilaginous coalition of the middle subtalar joint *(arrows)*.

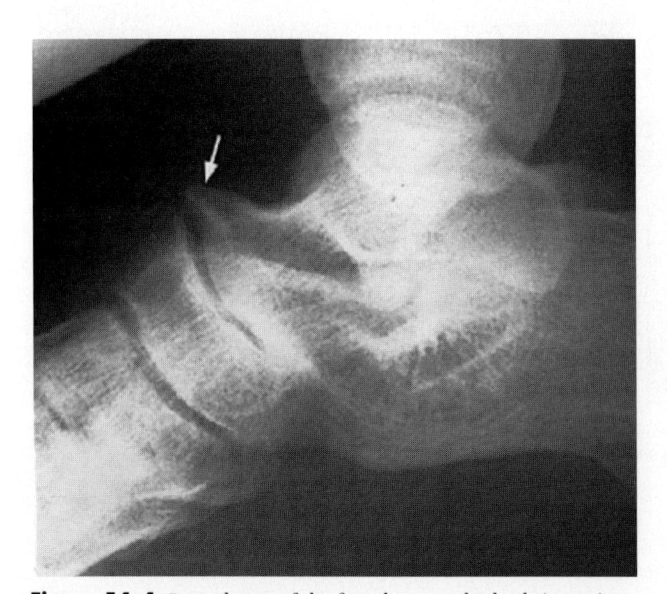

Figure 54–4. Lateral view of the foot shows a talar beak *(arrow)* in a patient with subtalar coalition.

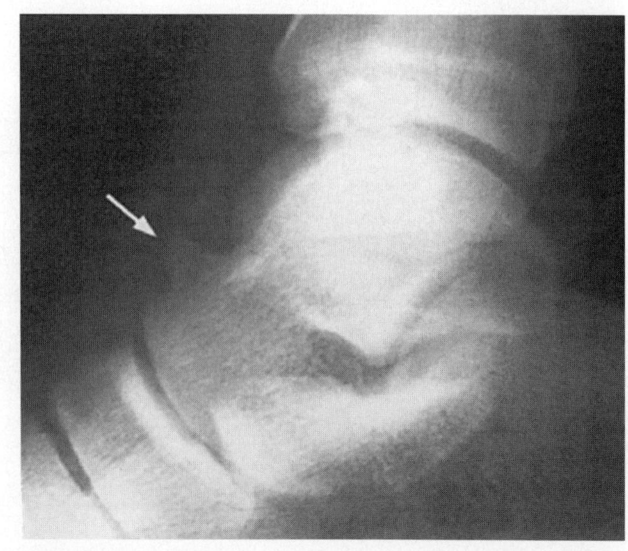

Figure 54–5. Athletic spur at the insertion of the ankle joint capsule *(arrow)*. This spur should not be confused with a talar beak (see Fig. 54–4).

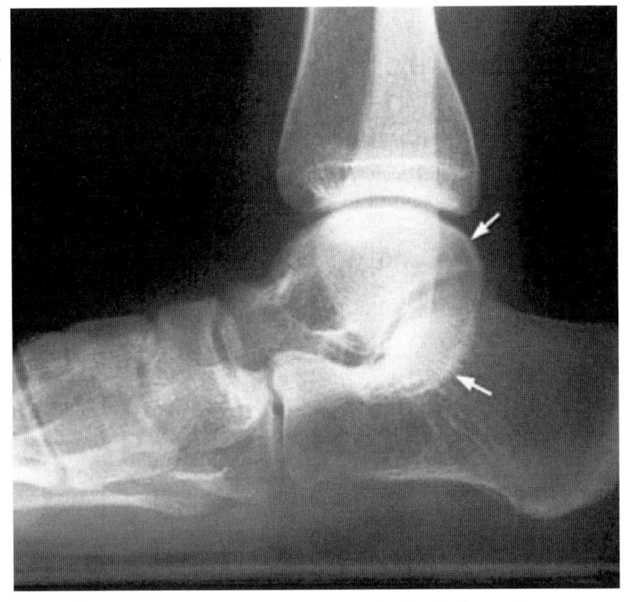

Figure 54–6. The C sign *(arrows)* of subtalar coalition.

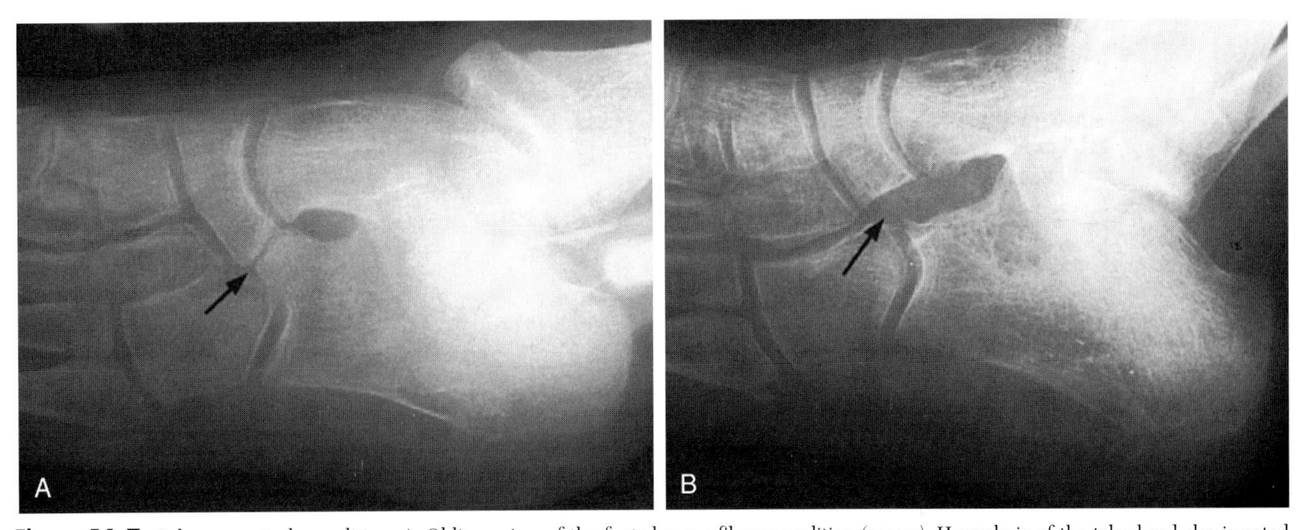

Figure 54–7. Calcaneonavicular coalition. *A*, Oblique view of the foot shows a fibrous coalition *(arrow)*. Hypoplasia of the talar head also is noted. *B*, Oblique view of the foot after resection of the coalition *(arrow)*.

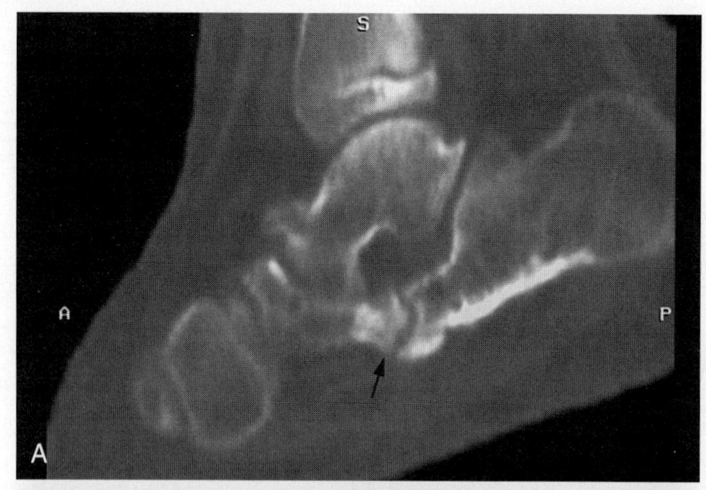

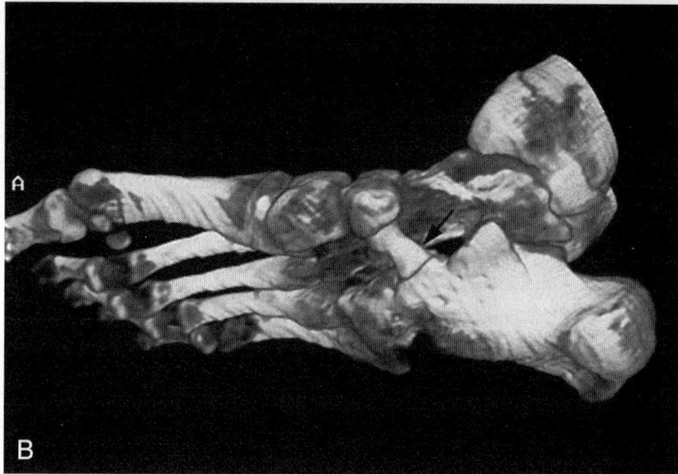

Figure 54–8. Reossification of a calcaneonavicular coalition two years after surgical resection in a 14-year-old boy. *A*, Sagittal reformatted image shows the reossification *(arrow)*. The bony bridging is, however, not complete. *B*, Three-dimensional reconstruction shows the reossification to better advantage *(arrow)*.

confused with a similar lesion, the athletic spur, which develops more proximal on the talar neck at the insertion of the joint capsule of the ankle (Fig. 54–5). The C sign is another indirect sign seen on the lateral view (Fig. 54–6). It is formed by drawing a line along the talar dome and extending it posteriorly and inferiorly to include the inferior margin of the sustentaculum tali.

Calcaneonavicular coalitions can be identified consistently on an oblique view of the foot (Fig. 54–7). The anterior process of the calcaneus typically is elongated as it extends to connect with the lateral aspect of the navicular. A secondary radiographic sign includes hypoplasia of the talar head. Reossification after surgical resection of the coalition is a well-known complication (Fig. 54–8).

Recommended Readings

Deutsch AL, Resnick D, Campbell G: Computer tomography and bone scintigraphy in the evaluation of tarsal coalition. Radiology 144:137, 1982.

Goldman AB, Pavlov H, Schneider R: Radionuclide bone scanning in subtalar coalitions: Differential considerations. AJR Am J Roentgenol 138:427, 1982.

Kumar SJ, Guille JT, Lee MS, et al: Osseous and non-osseous coalition of the middle facet of the talocalcaneal joint. J Bone Joint Surg Am 74:529, 1992.

Lateur LM, Van Hoe LR, Ghillewe KV, et al: Subtalar coalition: Diagnosis with the C sign on lateral radiographs of the ankle. Radiology 193:847, 1994.

Mosier KM, Asher M: Tarsal coalitions and peroneal spastic flat foot. J Bone Joint Surg Am 66:976, 1984.

Wechsler RJ, Karasick D, Schweitzer ME: Computed tomography of talocalcaneal coalition: Imaging techniques. Skeletal Radiol 21:353, 1992.

CHAPTER 55

Anemias

Michel E. Azouz, MD, and Lori L. Barr, MD

SICKLE CELL DISEASE

Overview

Hemoglobin S (HbS) is the most common hemoglobin variant and is prevalent worldwide. Sickle cell disease is hereditary and is seen predominantly in blacks. It is a chronic hemolytic anemia that occurs when HbS is the predominant hemoglobin. HbS-S is associated with the most severe clinical manifestations, including bone abnormalities. Other combinations of hemoglobin variants are possible, such as HbS-C and HbS-thalassemia. In the first few months of life, fetal hemoglobin HbF levels are high and are protective against the development of sickle cell disease. As the levels of HbF fall, HbS rises causing sickling and ischemic episodes. Sickle cell disease is characterized by acute, recurrent, painful ischemic crises, secondary to oxygen deprivation affecting the bone and bone marrow. Bone changes in sickle cell disease are secondary to bone marrow hyperplasia (owing to the chronic hemolysis and resultant anemia) and infarction (owing to sickling, vascular stasis, and capillary obstruction). Red blood cells are deformed and unable to flow through normal capillaries. Osteomyelitis complicating sickle cell disease often is caused by *Staphylococcus aureus*, but the incidence of *Salmonella* bone infections is relatively high (Fig. 55–1).

Imaging

Marrow hyperplasia is seen as marrow spaces expand. There is expansion of the diploic space in the skull, enlargement of the tubular bones, and *squaring* of the metacarpals with thinning of their cortex. With severe anemia, there is reconversion of fatty (yellow) marrow into cellular (red) marrow. In some patients, even the epiphyseal and apophyseal centers contain hematopoietic marrow as verified by magnetic resonance (MR) imaging. In the spine, cortical thinning of the vertebral bodies can lead to compression fractures. Insufficiency fractures also can be seen in long bones.

Ischemia manifests with bone destruction, periosteal reaction, and soft tissue swelling, closely resembling osteomyelitis (Fig. 55–2). Bone infarcts are often metadiaphyseal and may occur in the bones of the hands and feet; this is called the *hand-foot syndrome* and is seen in infants between the ages of 6 months and 2 years (Fig. 55–3). In older children and adolescents, other tubular bones, such as humerus, femur, and tibia, can become infarcted.

Central bone infarcts heal with fibrosis, medullary calcifications that can be seen on plain radiographs (Fig. 55–4). Complications of bone infarcts in growing children are rare but may be serious and include early growth plate fusion, shortening and deformity of tubular bones (Fig. 55–5), infection, and pathologic fracture. Metaphyseal infarcts eventually lead to shortening and deformity because of arrest or asymmetry of growth plate development.

In the spine, the combination of bone softening and infarcts in sickle cell disease may lead to notching involving the central parts of the upper and lower end plates of multiple vertebral bodies. Central cupping gives the vertebral bodies the appearance of an *H* (Fig. 55–6).

Changes of avascular necrosis may be seen in the humeral and femoral heads (Fig. 55–7). On MR imaging, acute infarcts show sharply marginated central zones of low to intermediate signal intensity on T1-weighted images and high signal on T2-weighted images.

On plain radiography, differentiation between acute bone infarct and osteomyelitis is often impossible. Other imaging techniques also are unreliable in distinguishing the two conditions. Acute bone infarcts are, however, far more prevalent than osteomyelitis in patients with sickle cell disease.

THALASSEMIA

Overview

Thalassemic syndromes include a variety of hereditary hemolytic anemias, with the underlying defect being reduced synthesis of one of the globin polypeptide chains. α-Thalassemia is found most often in Asia. β-Thalassemia is also known as *Mediterranean anemia* and can be classified as major, intermediate, or minor. Thalassemia major is the homozygous form, which is characterized by severe anemia, extensive marrow hyperplasia, and iron deposition in the soft tissues.

Imaging

Radiographic changes in the skeleton are the result of marrow hyperplasia secondary to the chronic hemolysis

Text continued on page 428

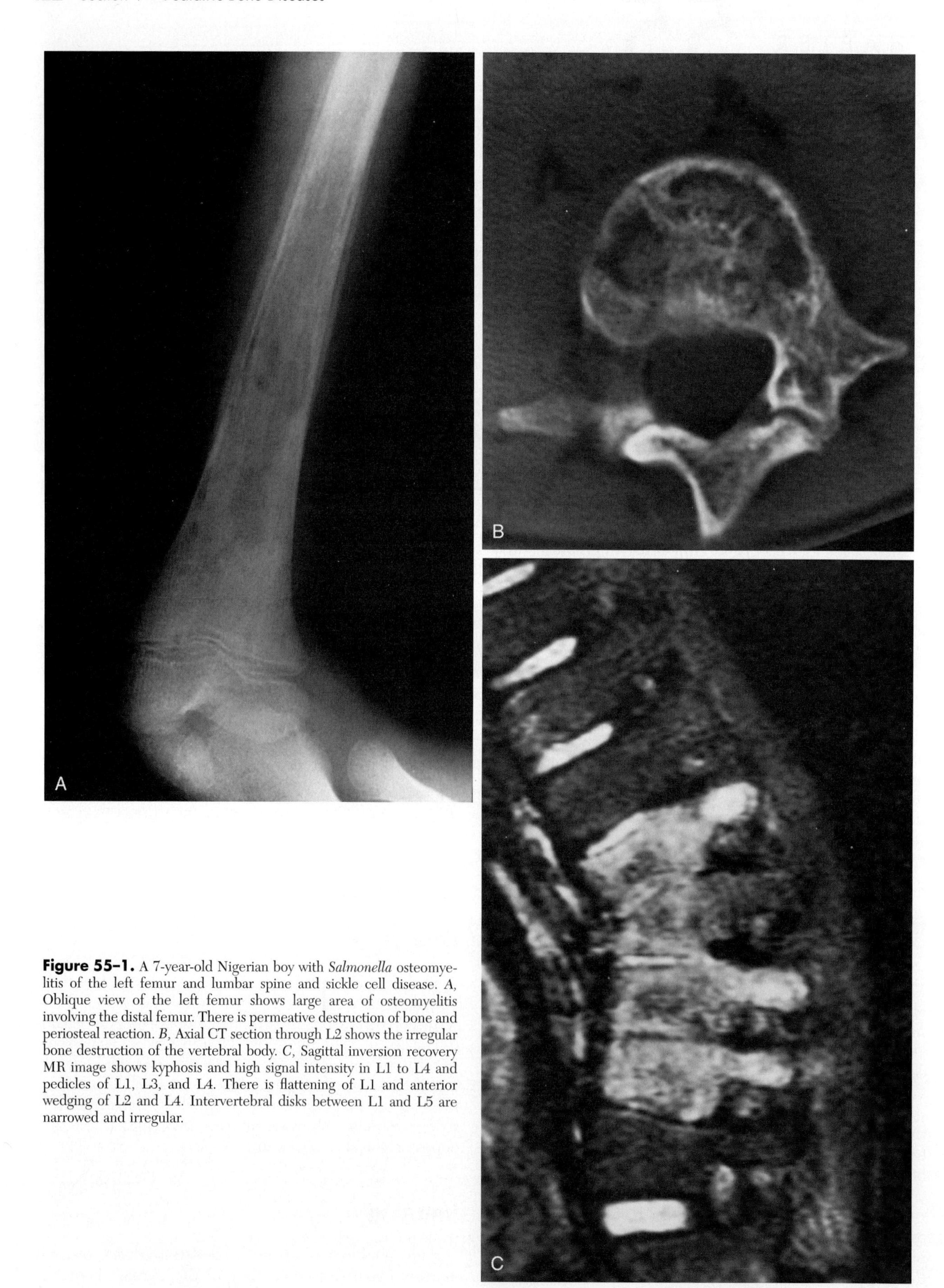

Figure 55–1. A 7-year-old Nigerian boy with *Salmonella* osteomyelitis of the left femur and lumbar spine and sickle cell disease. *A*, Oblique view of the left femur shows large area of osteomyelitis involving the distal femur. There is permeative destruction of bone and periosteal reaction. *B*, Axial CT section through L2 shows the irregular bone destruction of the vertebral body. *C*, Sagittal inversion recovery MR image shows kyphosis and high signal intensity in L1 to L4 and pedicles of L1, L3, and L4. There is flattening of L1 and anterior wedging of L2 and L4. Intervertebral disks between L1 and L5 are narrowed and irregular.

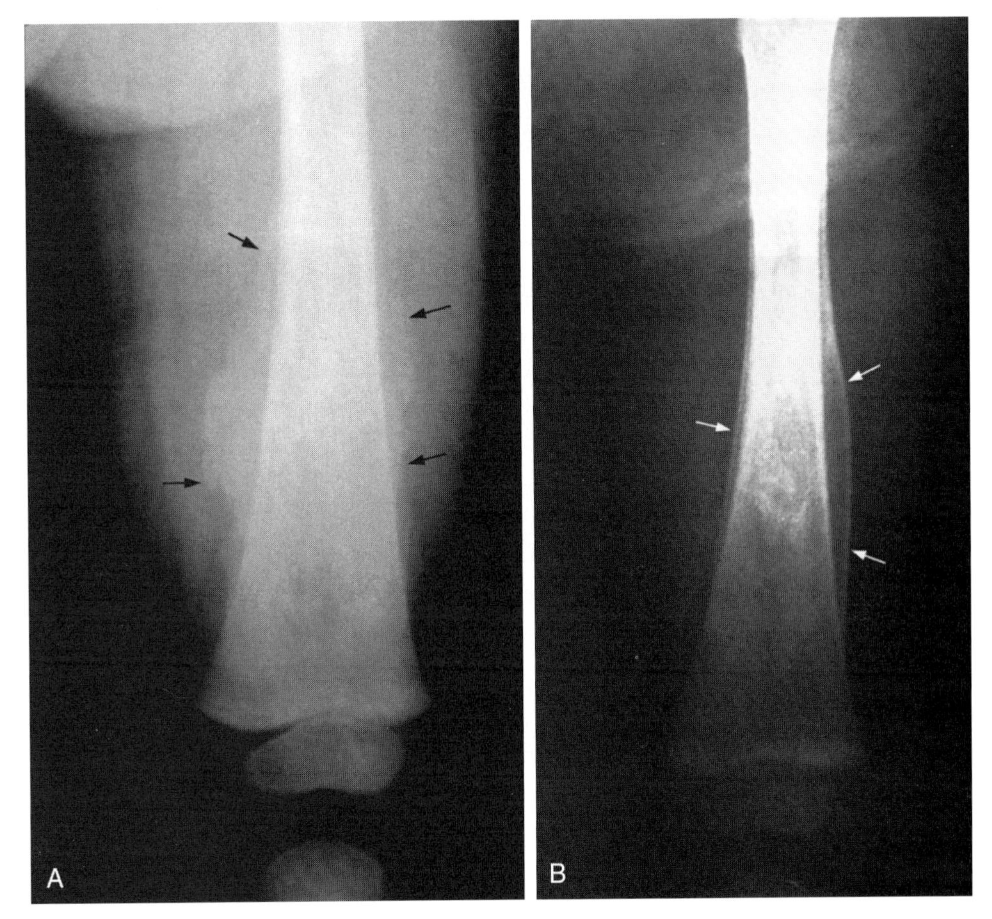

Figure 55–2. Bone infarction resembling ostemyelitis clinically and radiographically in a patient with sickle cell disease. *A*, Anteroposterior view of the left femur shows periosteal reaction *(arrows)* and lytic and blastic lesions within medullary space. *B*, Anteroposterior view of the left femur taken 4 months later shows the infarct in the medullary space. The periosteal reaction has matured *(arrows)*.

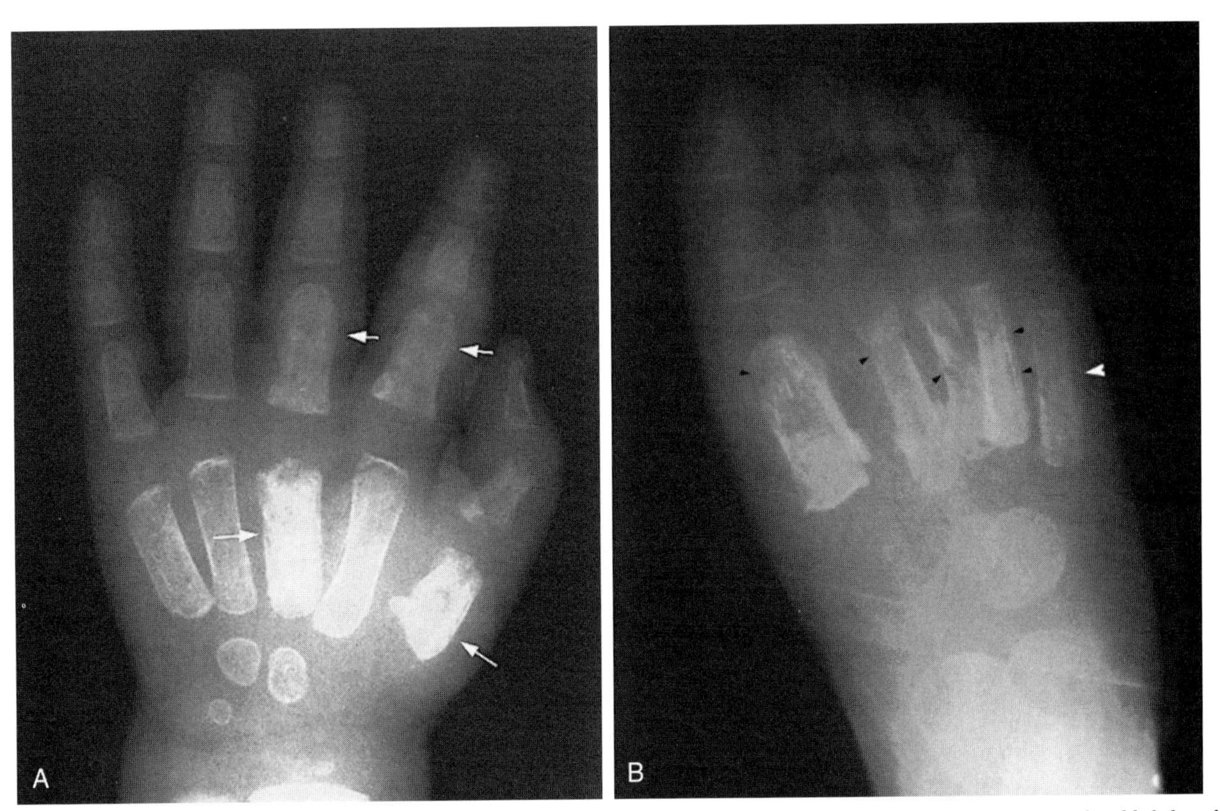

Figure 55–3. Hand-foot syndrome in a 2-year-old child with sickle cell disease. *A* and *B*, Anteroposterior views of the right hand and left foot show the typical changes of hand-foot syndrome with areas of bone destruction and sclerosis *(arrows and arrowheads)*. Note the presence of periosteal reaction and pathologic fractures in the foot.

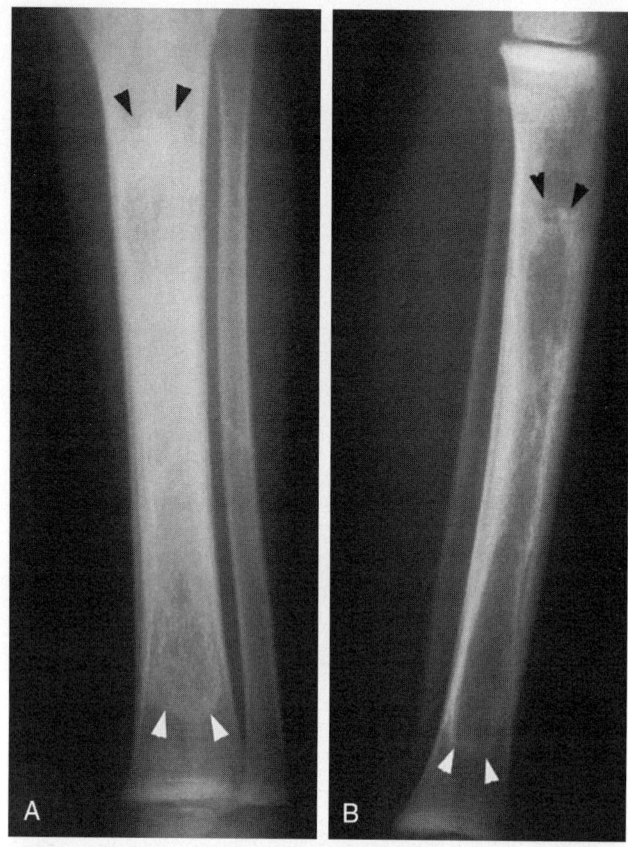

Figure 55–4. *A* and *B,* Anteroposterior and lateral views of the tibia show marrow infarct in sickle cell disease in a 5-year-old child. Anteroposterior view of the tibia shows sclerosis caused by medullary calcifications and fibrosis *(arrowheads).*

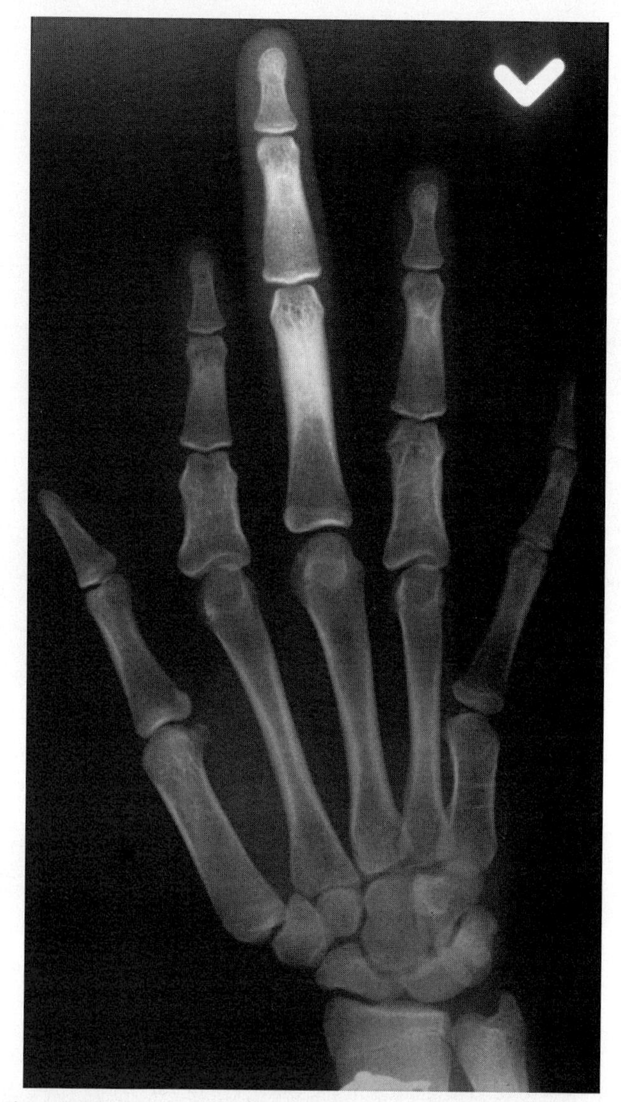

Figure 55–5. A 17-year-old girl with sickle cell disease and bilateral asymmetrical brachydactyly of the hands and feet. Anteroposterior view of the left hand shows significant shortening of the fifth metacarpal and proximal phalanx of the second and fourth fingers. These are likely sequelae of infarcts involving the cartilaginous growth plates and causing early growth plate fusion.

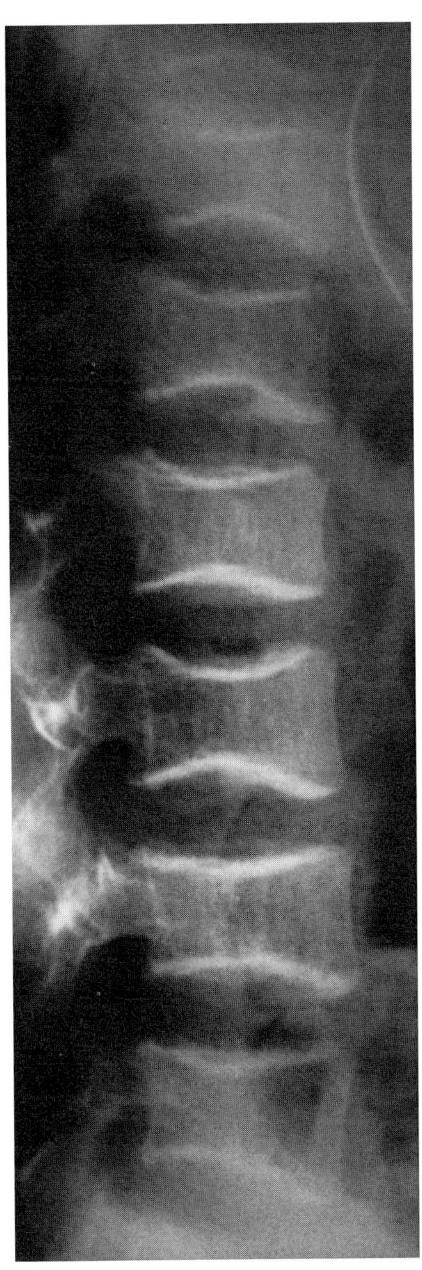

Figure 55–6. A 14-year-old girl with sickle cell disease and H-shaped vertebral bodies seen on a lateral view of the lumbar spine.

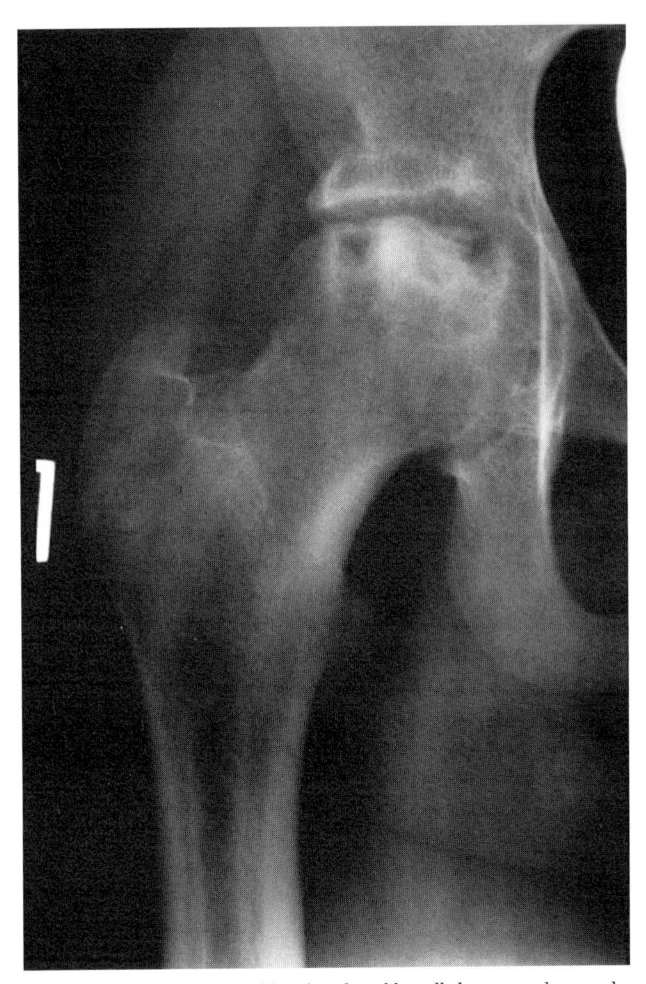

Figure 55–7. A 13-year-old girl with sickle cell disease and avascular necrosis of the humeral head. Long diaphyseal infarcts are also present, where thick sclerotic longitudinal lines are seen paralleling the endosteal surface of the cortex, causing a *bone-within-a-bone* appearance.

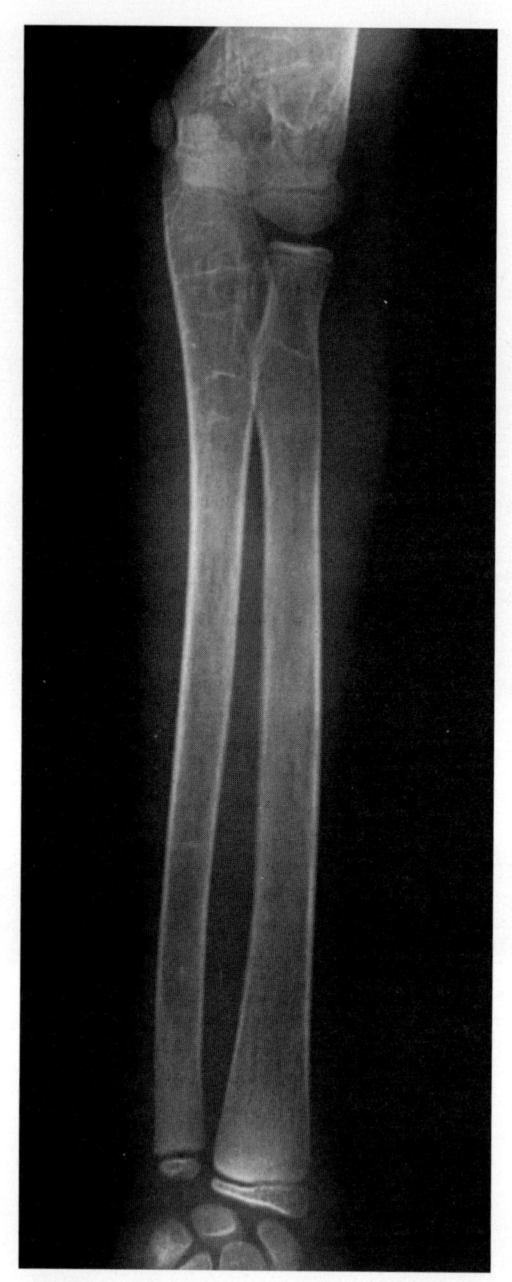

Figure 55–8. A 13-year-old girl, born in Southern Italy, with thalassemia major. Anteroposterior view of the left forearm shows cortical thinning and diaphyseal expansion of the radius and ulna.

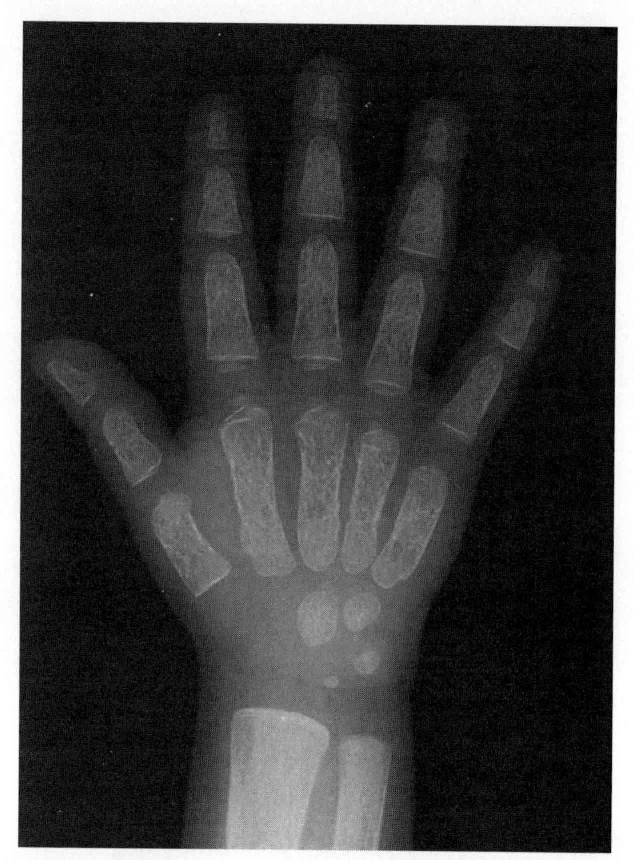

Figure 55–9. Coarse trabeculae through the digits in a 4-year-old child with thalassemia major.

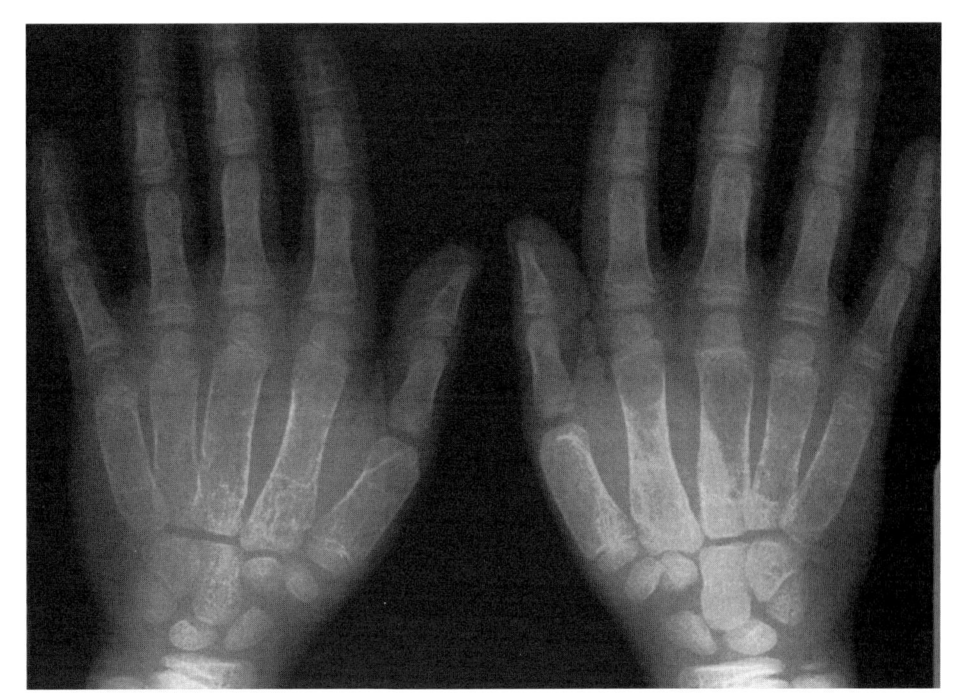

Figure 55–10. Squaring of the metacarpals in a 9-year-old child with thalassemia major.

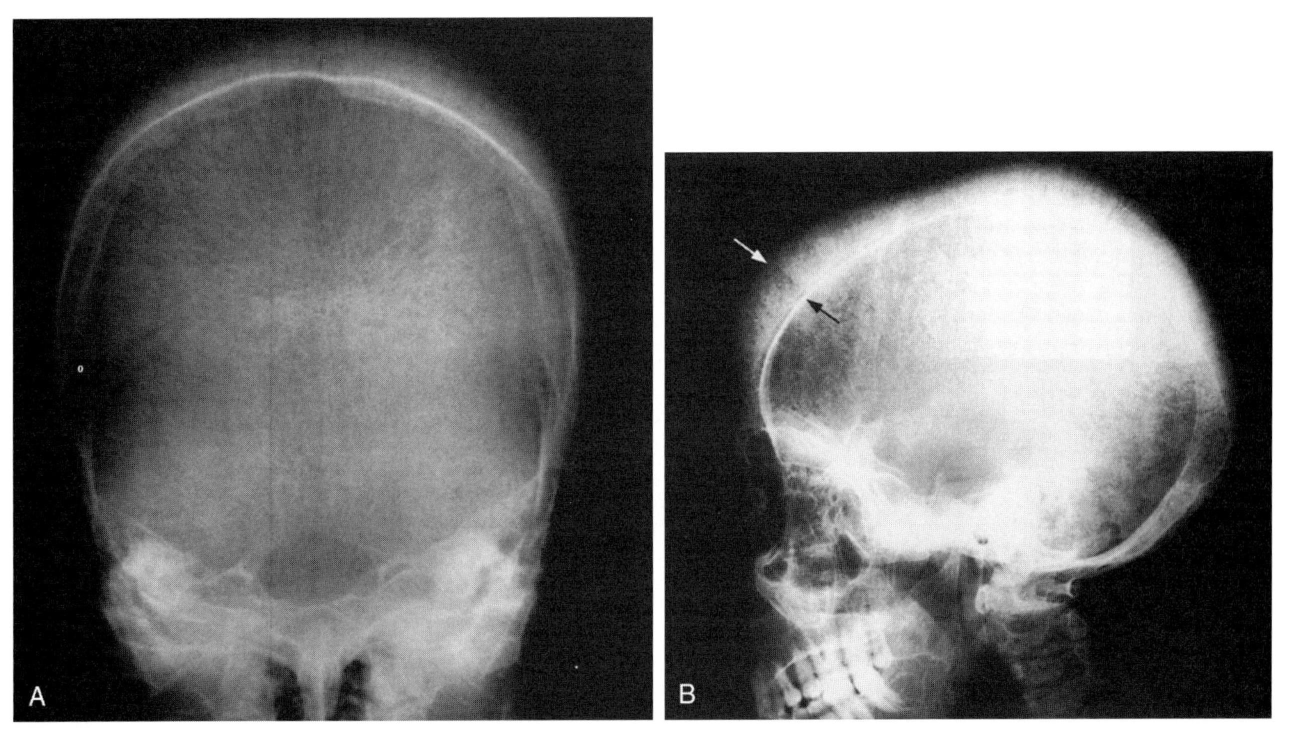

Figure 55–11. *A* and *B*, Towne and lateral views of the skull show significant diploic expansion with "hair-on-end" appearance *(arrows* on *B)*.

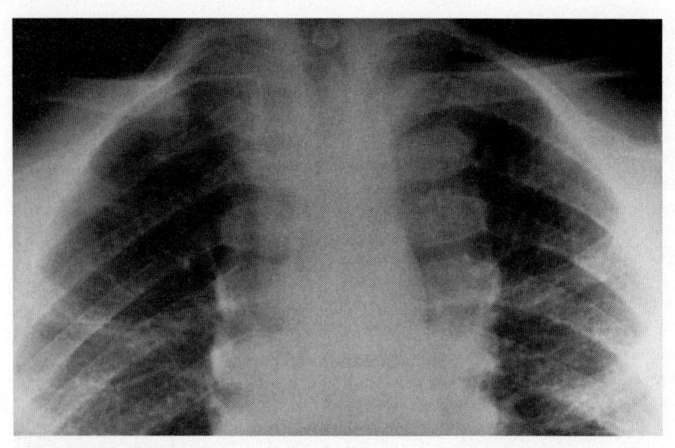

Figure 55–12. An 18-year-old man born in Turkey with thalassemia major. Anteroposterior view of the upper chest shows osteopenia, posterior rib expansion, and bilateral posterior mediastinal masses denoting extramedullary hematopoiesis.

with resultant anemia. Most of the bone changes are seen after the first year of life and include widening of themedullary cavity and thinning of the cortex (Fig. 55–8). Coarsening of the trabecular bone pattern also may be seen (Fig. 55–9). The joints are not altered. In the hands and feet, there is osteopenia and *squaring* of the metacarpals, metatarsals, and proximal phalanges, usually seen after 1 year of age (Fig. 55–10).

In the skull, diploic space widening may be significant, with thinning of the outer table and vertical striations seen in the diploë, giving a "hair-on-end" appearance (Fig. 55–11). Diploic widening usually starts in the frontal bone and progresses posteriorly. The paranasal sinuses and mastoid air cells are relatively small because of adjacent bone marrow overgrowth. There is also relative enlargement of the zygomatic bones, hypertelorism, and hypertrophy of the maxilla causing overbite and protrusion of the upper incisors.

Vertebral bodies show osteopenia and paravertebral soft tissue masses representing extramedullary hematopoiesis (Fig. 55–12). On the frontal chest radiograph, these changes may be seen usually associated with ribbon-like widening of the ribs and thinning of their cortex and a prominent cardiac silhouette, which may be mistaken for primary congenital heart disease. After minor trauma, insufficiency fractures and vertebral compressions are seen as the result of osteopenia.

Avascular necrosis is rare in thalassemia, and premature closure of the growth plates has been reported in the proximal humerus and distal femur.

Deferoxamine is a chelating agent used to treat iron overload in thalassemic patients who have received multiple transfusions. It may cause platyspondyly and metaphyseal growth disturbances, resembling rickets, especially when used in infancy and in high doses.

Recommended Readings

Ben Dridi MF, Oumaya A, Gastli H, et al: Radiological abnormalities of the skeleton in patients with sickle-cell anemia: A study of 222 cases in Tunisia. Pediatr Radiol 17:296–302, 1987.

Bohrer SP: Bone changes in the extremities in sickle cell anemia. Semin Roentgenol 22:176–185, 1987.

Bonnerot V, Sebag G, de Montalembert M, et al: Gadolinium-DOTA enhanced MRI of painful osseous crises in children with sickle cell anemia. Pediatr Radiol 24:92–95, 1994.

Cockshott WP: Dactylitis and growth disorders. Br J Radiol 36:19–26, 1963.

Crowley JJ, Sarnaik S: Imaging of sickle cell disease. Pediatr Radiol 29:646–661, 1999.

Currarino G, Erlandson ME: Premature fusion of epiphyses in Cooley's anemia. Radiology 83:656–664, 1964.

Fernandez M, Slovis TL, Whitten-Shurney W: Maxillary sinus marrow hyperplasia in sickle cell anemia. Pediatr Radiol 25:S209–S211, 1995.

Gumbs RV, Higginbotham-Ford EA, Teal JS, et al: Thoracic extramedullary hematopoiesis in sickle-cell diesase. AJR Am J Roentgenol 149:889–893, 1987.

Kim HC, Alavi A, Russell MO, Schwartz E: Differentiation of bone and bone marrow infarcts from osteomyelitis and sickle cell disorders. Clin Nucl Med 14:249–254, 1989.

Koren (Kurlat) A, Garty I, Katzuni E: Bone infarction in children with sickle cell disease: Early diagnosis and differentiation from osteomyelitis. Eur J Pediatr 142:93–97, 1984.

Lawson JP, Ablow RC, Pearson HA: Premature fusion of the proximal humeral epiphyses in thalasssemia. AJR Am J Roentgenol 140:239–244, 1983.

Levin TL, Sheth S, Berdon WE, et al: Deferoxamine-induced platyspondyly in hypertransfused thalassemic patients. Pediatr Radiol 25:S122–S124, 1995.

Levin TL, Sheth SS, Hurlet A, et al: MR marrow signs of iron overload in transfusion-dependent patients with sickle cell disease. Pediatr Radiol 25:614–619, 1995.

Rao VM, Fishman M, Mitchell DG, et al: Painful sickle cell crisis: Bone marrow patterns observed with MR imaging. Radiology 161:211–215, 1986.

Sebes JI: Diagnostic imaging of bone and joint abnormalities associated with sickle cell hemoglobinopathies. AJR Am J Roentgenol 152:1153–1159, 1989.

Van Zanten TEG, Stratius Van Eps LW, Golding RP, Valk J: Imaging the bone marrow with magnetic resonance during a crisis and in chronic forms of sickle cell disease. Clin Radiol 40:486–489, 1989.

CHAPTER 56

Leukemia

Michel E. Azouz, MD, and Lori L. Barr, MD

OVERVIEW

Leukemia is the most common malignancy of childhood. It accounts for more than one third of all pediatric malignancies. In most cases, pediatric leukemia is of the acute lymphoblastic type.

Patients present with fever and bone and joint pains, which can be mistaken for acute rheumatic fever or juvenile rheumatoid arthritis. Anemia, bruising, and fatigue are other clinical signs and symptoms that may be present. Hepatosplenomegaly and lymphadenopathy are common findings on physical examination. There is no correlation between the extent of the bone lesions and the severity or prognosis of the leukemia. Definitive diagnosis is made on bone marrow aspiration, on which aspirates show replacement of the normal cell population with sheets of leukemia cells (lymphoblasts).

IMAGING

In patients with leukemia, imaging of bony abnormalities is done best with plain radiography. Soft tissue lesions are studied best with magnetic resonance (MR) imaging. Intraosseous granulocytic sarcomas are destructive and can be studied on plain radiographs followed by computed tomography (CT) or MR imaging to delineate their extent and plan for needle biopsy and radiation therapy and in follow-up to assess response to treatment.

Bone and joint changes, when present, are the same for all types of leukemia. These include diffuse osteopenia, metaphyseal lucent bands, osteolytic lesions, periosteal reaction, and rarely sclerotic lesions. Generalized osteopenia is the most common manifestation of childhood leukemia. Insufficiency fractures of long bones and vertebral body compression may be seen. Typically, multiple, partially collapsed vertebrae are present (Fig. 56–1).

Transverse metaphyseal lucent bands may be seen, parallel to the growth plates in the long bones, particularly around the knees, but also in the proximal femora and distal tibiae (Fig. 56–2). They are usually not due to leukemic deposits but rather secondary to the osteopenia that accompanies systemic diseases. These lucent *leukemic* bands also may be seen in flat bones and in the vertebral bodies.

Osteolytic lesions caused by leukemic infiltrates are usually metaphyseal. They present as large, ill-defined focal areas of bone destruction (Fig. 56–3). Another pattern of bone destruction presents as diffuse permeative or moth-eaten lesions (Fig. 56–4). The bone lesions can be seen in the long or flat bones associated with sunburst or lamellar periosteal reaction (Fig. 56–5). With this pattern of bone involvement, metastatic neuroblastoma and Langerhans' cell histiocytosis should be considered in the differential diagnosis. On MR imaging, leukemic infiltrates are of low signal intensity on T1-weighted images, replacing the high signal intensity normal marrow. On T2-weighted images with fat suppression and on short tau inversion recovery (STIR) images, leukemic deposits in the bone marrow stand out as areas of high signal intensity.

Patients with leukemia may have severe thrombocytopenia resulting in subperiosteal hemorrhage. In these cases, periosteal reaction is seen in the diaphyses and can be linear or lamellated. Approximately one third of children with leukemia also have delayed skeletal maturation. Osteosclerotic lesions also have been described but are rare. Metaphyseal growth arrest lines are noted frequently, especially after chemotherapy. Widening of the

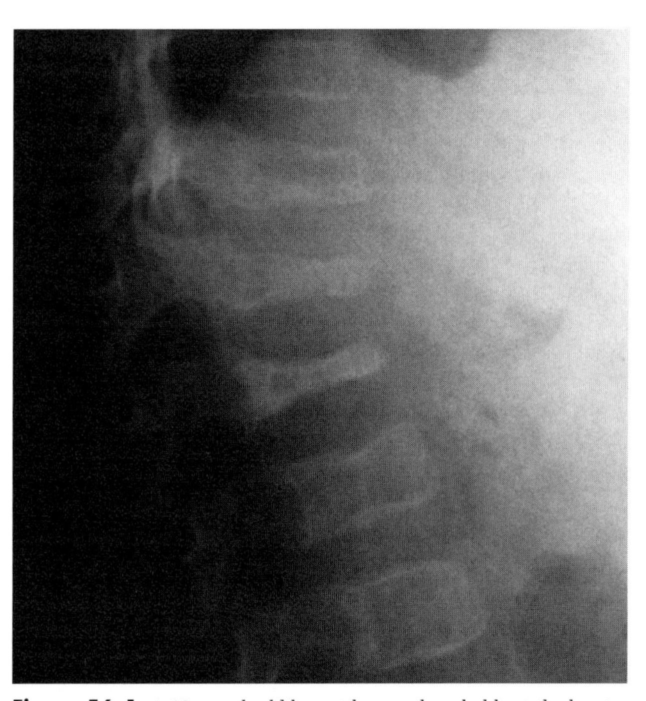

Figure 56–1. A 19-month-old boy with acute lymphoblastic leukemia. Lateral view of the spine at the thoracolumbar junction shows osteopenia and collapse of multiple vertebral bodies.

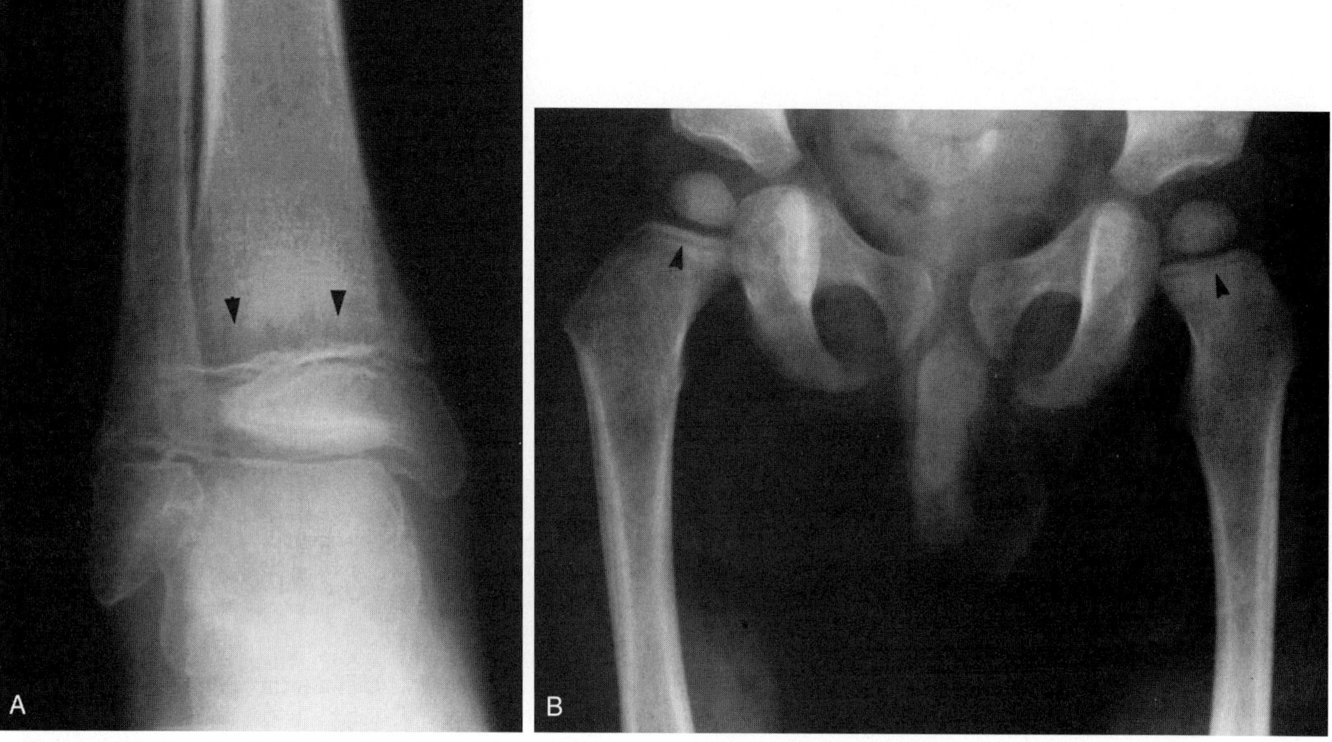

Figure 56–2. Metaphyseal lucent bands in leukemia. *A*, A 2.5-year-old boy with acute lymphoblastic leukemia. Anteroposterior view of the right ankle shows a thick metaphyseal band *(arrowheads)* in the distal tibial metaphysis. *B*, An 18-month-old boy with acute lymphoblastic leukemia. Anteroposterior radiograph of both hips and proximal femora shows transverse lucent metaphyseal bands.

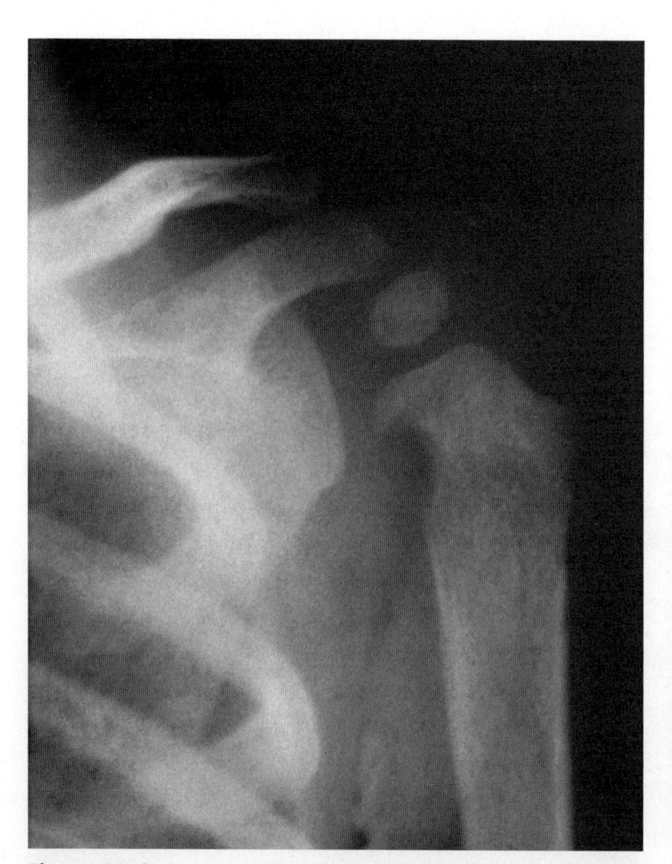

Figure 56–3. A 1-year-old boy with acute lymphoblastic leukemia and widespread, osteolytic destructive lesions in the metaphyses of multiple long bones. Anteroposterior radiograph of the left shoulder shows a destructive lesion in the proximal humeral metaphysis medially. There is no periosteal reaction. This lesion is not a specific finding of leukemia.

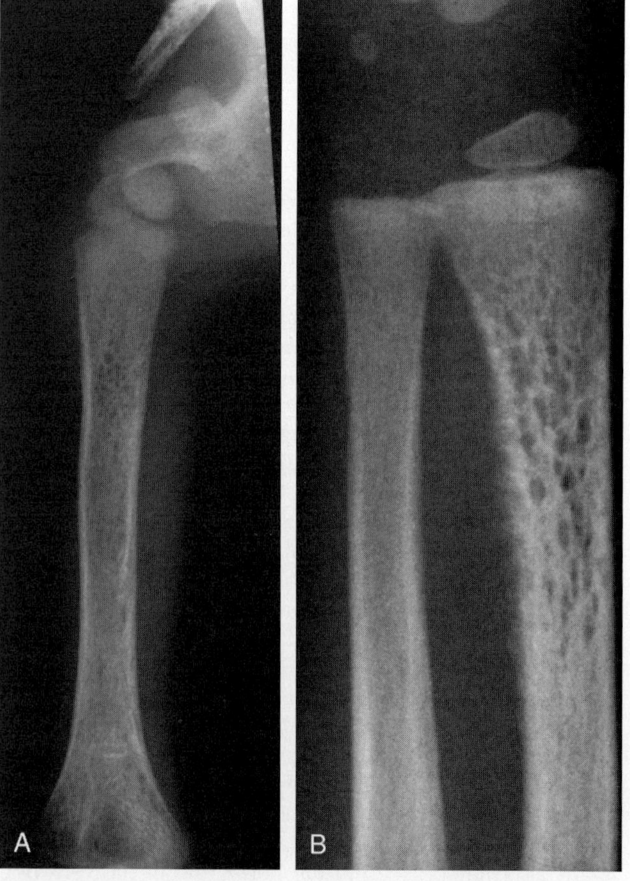

Figure 56–4. A 2.5-year-old girl with diffuse bone pains. A radiographic skeletal survey showed a permeative pattern of destruction of several bones (including the skull, flat bones, and ribs). A diagnosis of leukemia was confirmed by bone marrow biopsy. *A*, Anteroposterior view of the right humerus. *B*, Anteroposterior view of the right distal forearm.

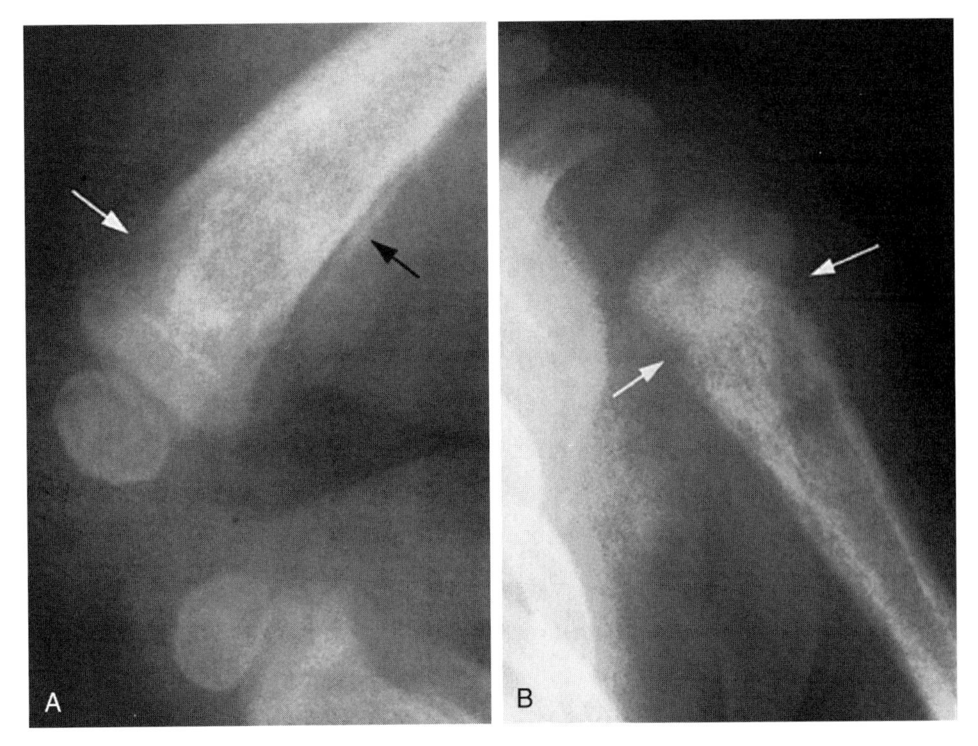

Figure 56–5. A 16-month-old child with leukemia. *A*, Lateral view of the knee reveals metaphyseal erosions in the distal femur *(white arrow)* and periosteal reaction *(black arrow)*. *B*, Anteroposterior view of the shoulder reveals metaphyseal erosions in the proximal humerus *(white arrows)*.

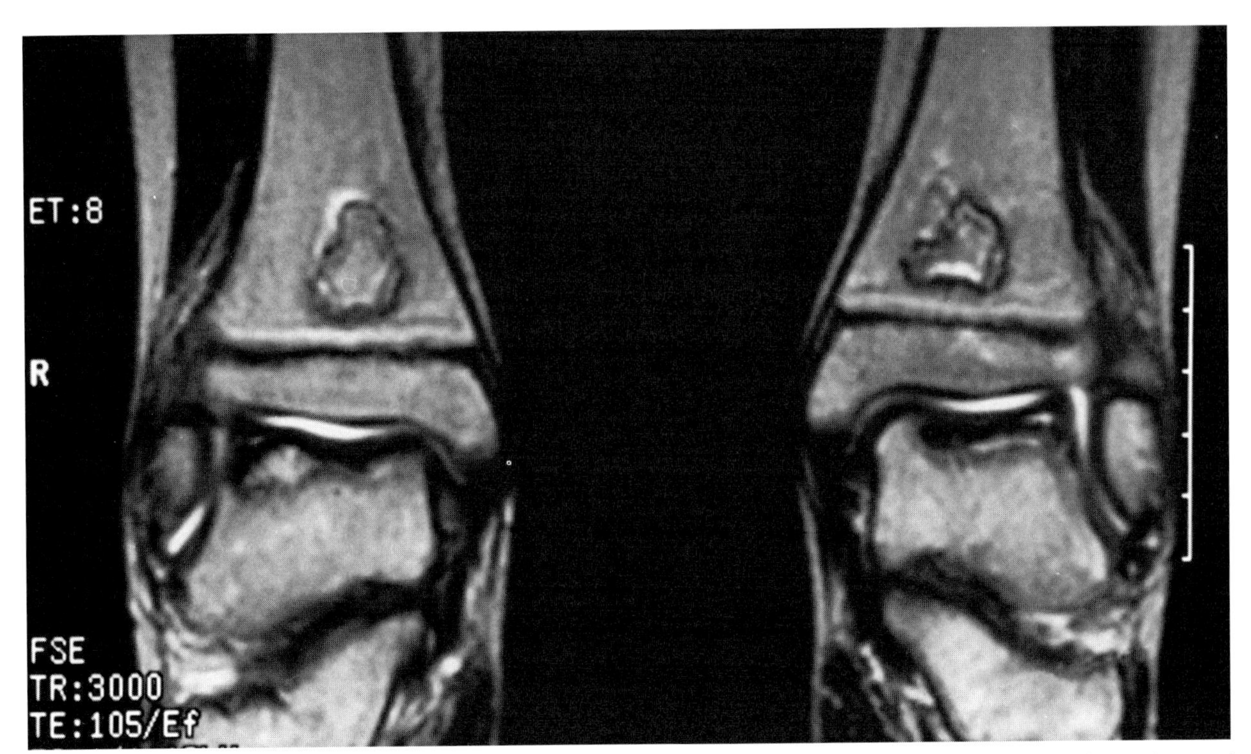

Figure 56–6. An 11-year-old leukemic boy on long-term steroid therapy. Coronal T1-weighted MR image of the ankles shows bone infarcts in distal metaphyses of the tibiae and talar bones.

cranial sutures is seen when there is leukemic infiltration of the meninges or intracranial hemorrhage. Bone infarcts and avascular necrosis, when present, are usually secondary to steroid therapy (Fig. 56–6).

Chloromas (named after their characteristic green color, which is due to the enzyme myeloperoxidase) are solid leukemic tumor deposits, composed of myeloid precursor cells. They are better termed *granulocytic sarcomas* because not all chloromas are green. One or more of these masses may precede the development of myelogenous leukemia by months or years, but generally when a granulocytic sarcoma is discovered, the diagnosis of leukemia is already established. There has been an increase in the number of reported granulocytic sarcomas associated with leukemia as a result of prolonged survival of patients. Granulocytic sarcomas are more common in children than in adults with leukemia, with about 60% of patients younger than age 15 years.

Granulocytic sarcomas tend to occur in the sacrum; facial bones, especially the orbits and paranasal sinuses; sternum; ribs; and soft tissues almost anywhere in the body. These sarcomas are sensitive to radiation or chemotherapy.

Recommended Readings

Benz G, Brandeis WE, Willich E: Radiological aspects of leukaemia in childhood. Pediatr Radiol 4:201–213, 1976.

Gallagher DJ, Phillips DJ, Heinrich SD: Orthopedic manifestations of acute pediatric leukemia. Orthop Clin North Am 27:635–644, 1996.

Guermazi A, Feger C, Rousselot P, et al: Granulocytic sarcoma (chloroma): Imaging findings in adults and children. AJR Am J Roentgenol 178:319–325, 2002.

Nixon GW, Gwinn JL: The roentgen manifestations of leukemia in infancy. Radiology 107:603–609, 1973.

Steiner RM, Mitchell DG, Rao VM, Schweitzer ME: Magnetic resonance imaging of diffuse bone marrow disease. Radiol Clin North Am 31:383–409, 1993.

CHAPTER 57

Lymphoma

Michel E. Azouz, MD, and Lori L. Barr, MD

OVERVIEW

Except for Hodgkin's disease and Burkitt's lymphoma, which have specific histologic features, the classification of lymphomas is difficult and controversial. The most common sites for lymphoma in children are the lymph nodes (in the abdomen, retroperitoneum, mediastinum, and neck) and the thymus. As a group, lymphomas in children are aggressive tumors. In children, non-Hodgkin's lymphoma of bone is rare, and in contrast to primary lymphoma of bone in adults, it is always viewed as a manifestation of a systemic disease. At presentation, lymphoma of bone can be either solitary or multicentric. Primary bone lymphoma represents only about 4% of all malignant bone tumors.

Hodgkin's disease is rare in children younger than age 7 years. The incidence increases after the age of 7 years and becomes almost equal to non-Hodgkin's lymphoma after age 10. In Hodgkin's lymphoma, skeletal involvement typically results from invasion by adjacent diseased lymph nodes. Hodgkin's lymphoma usually involves the sternum, ribs, and vertebral bodies.

African Burkitt's lymphoma is an endemic sarcoma involving the jaw or orbit of children in the tropical parts of Africa. The primary focus usually involves the facial bones and long bones. Non-African Burkitt's lymphoma is sporadic, usually presents as an abdominal or pelvic tumor, but also can involve the face and long bones.

In children with lymphoma, bone marrow involvement is an important indicator of disseminated disease. Rarely, bone lymphoma is associated with post-transplantation lymphoproliferative disorders.

IMAGING

Bone changes in lymphoma are not specific (see Figs. 12–26 through 12–30). They present as permeative or destructive metaphyseal ill-defined lesions resembling the appearance seen in leukemia, metastatic neuroblastoma, and Langerhans' cell histiocytosis. There is typically no associated soft tissue mass. Periosteal reaction and pathologic fracture may be seen. Rarely, sclerotic bone changes are present. The ivory vertebra, a finding in adult Hodgkin's disease, is rare in pediatric patients.

In general, the extent of bone lesions in lymphoma is grossly underestimated on plain radiographs because the edges of individual lesions blend imperceptibly with the adjacent normal bone (Fig. 57–1A). Radionuclide bone scans and computed tomography (CT) help determine the

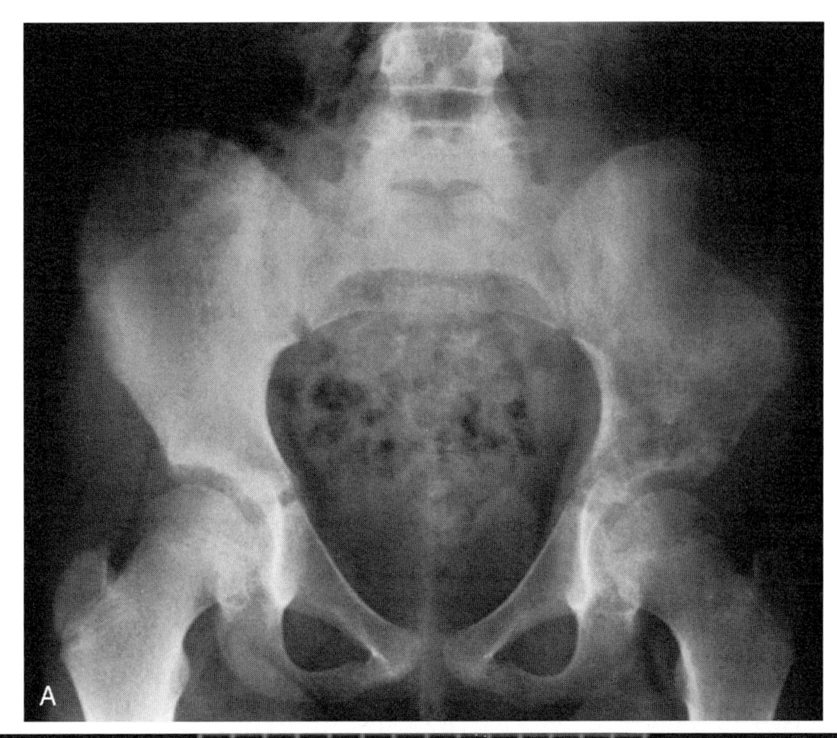

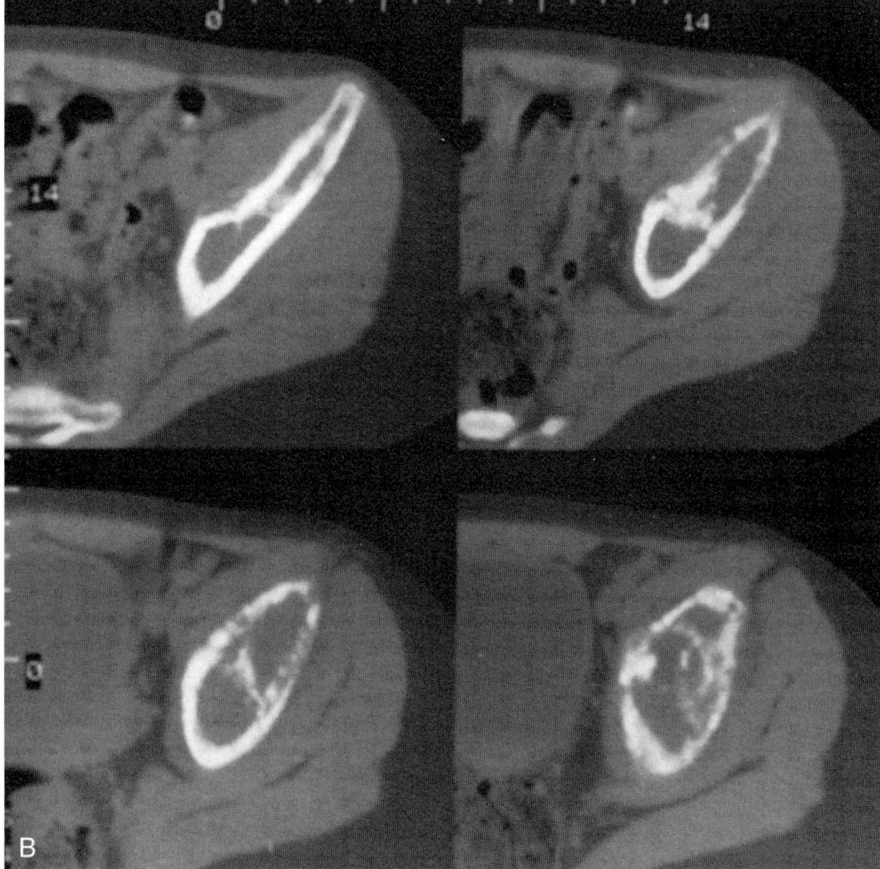

Figure 57–1. An 8-year-old girl with a primary bone lymphoma of the left iliac bone. *A,* Anteroposterior view of the pelvis shows an ill-defined, aggressive-looking lytic lesion in the supra-acetabular part of the ilium. There is a small adjacent (pelvic) soft tissue mass. *B,* CT sections through the lesion show its destructive nature. (Case provided by Dr. Dominique Marton, Montreal, Canada.)

location and degree of bone destruction to better advantage (Fig. 57–1*B*). The combination of a dull-aching bone pain, normal plain radiograph, and abnormal bone scan should suggest the possibility of osseous lymphoma.

In Burkitt's lymphoma and in bone lesions secondary to direct invasion by diseased lymph nodes as seen in Hodgkin's disease, CT best outlines the region of bone destruction, and magnetic resonance (MR) imaging best delineates the abnormal soft tissue masses.

On MR imaging, marrow involvement reveals low signal intensity on T1-weighted images and is hyperintense on T2-weighted and short tau inversion recovery (STIR) sequences. Lymphomatous infiltration of the bone marrow may be patchy or diffuse. MR imaging may be used to identify optimal sites for biopsy or bone marrow aspiration.

Recommended Readings

Gawish HHA: Primary Burkitt's lymphoma of the frontal bone. J Neurosurg 45:712–715, 1976.

Guermazi A, Brice P, de Kerviler E, et al: Extranodal Hodgkin disease: Spectrum of disease. Radiographics 21:161–179, 2001.

Kaushik S, Fulcher AS, Frable WJ, May DA: Posttransplantation lymphoproliferative disorder: Osseous and hepatic involvement. AJR Am J Roentgenol 177:1057–1059, 2001.

Mulligan ME, McRae GA, Murphey MD: Imaging features of primary lymphoma of bone. AJR Am J Roentgenol 173:1691–1697, 1999.

Stiglbauer R, Augustin I, Kramer J, et al: MRI in the diagnosis of primary lymphoma of bone: Correlation with histopathology. J Comput Assist Tomogr 16:248–253, 1992.

Vanel D, Bayle C, Hartmann O, et al: Radiological study of two disseminated malignant non-Hodgkin lymphomas affecting only the bones in children. Skeletal Radiol 9:83–87, 1982.

Whittaker LR: Burkitt's lymphoma. Clin Radiol 24:339–346, 1973.

CHAPTER 58

Osteogenesis Imperfecta

Michel E. Azouz, MD, and Lori L. Barr, MD

OVERVIEW

Osteogenesis imperfecta (OI) represents a group of heritable connective tissue diseases characterized by an abnormal quantity or quality (or both) of type I collagen and an abnormal osteoid matrix. As a result, there is osteoporosis and a variable degree of bone fragility. Fractures may be severe leading to perinatal death, or there may be only mild-to-moderate bone fragility resulting in occasional fractures. Other clinical features include blue sclera, defective dentition (dentinogenesis imperfecta), hearing impairment, and hyperlaxity of ligaments. Based on a variety of clinical findings, family history, mode of inheritance, and radiologic findings, there are at least four types of OI.

OI type II is the perinatal lethal form. OI types I and IV are mild-to-moderate forms. In OI type III, fractures may be present at birth, but they are not numerous. Because it is not possible to classify OI based only on the radiographic findings, radiologists should report their findings based on the old classification (i.e., the rare congenital form of OI, which is usually but not always severe, and the tarda form, which is more common and of variable severity).

OI either is dominantly inherited or occurs sporadically as a new mutation. In general, the tarda form of OI has an autosomal dominant pattern of inheritance. As for the congenital OI type II, almost all cases are new mutations. Recessively inherited forms of OI are rare.

IMAGING

Type II (congenital) OI is detected by prenatal ultrasound usually in the second trimester (Fig. 58–1). OI is one of the most common skeletal dysplasias diagnosed by prenatal ultrasound. Usually there is no family history, and the diagnosis is unexpected. Findings include a delayed and deficient ossification of the skeleton, including the skull (Fig. 58–2), with multiple fractures of tubular bones, ribs, or both. Fetal limbs may be short and deformed (Fig. 58–3).

On prenatal ultrasound, the skull is compressible and shows decreased echogenicity. The intracranial structures are unusually well seen because ossification of the skull is deficient or (rarely) absent. When fetal ultrasound done after 17 weeks' gestation is normal, the diagnosis of OI type II can be excluded confidently. The radiographic diagnosis of congenital OI is usually easy. The long bones (femora and humeri) are described as accordion-like: broad and crumpled (see Fig. 58–3).

Postnatal imaging findings are variable, reflecting the wide spectrum of severity in OI. Radiographic findings include one or more of the following: diminished bone

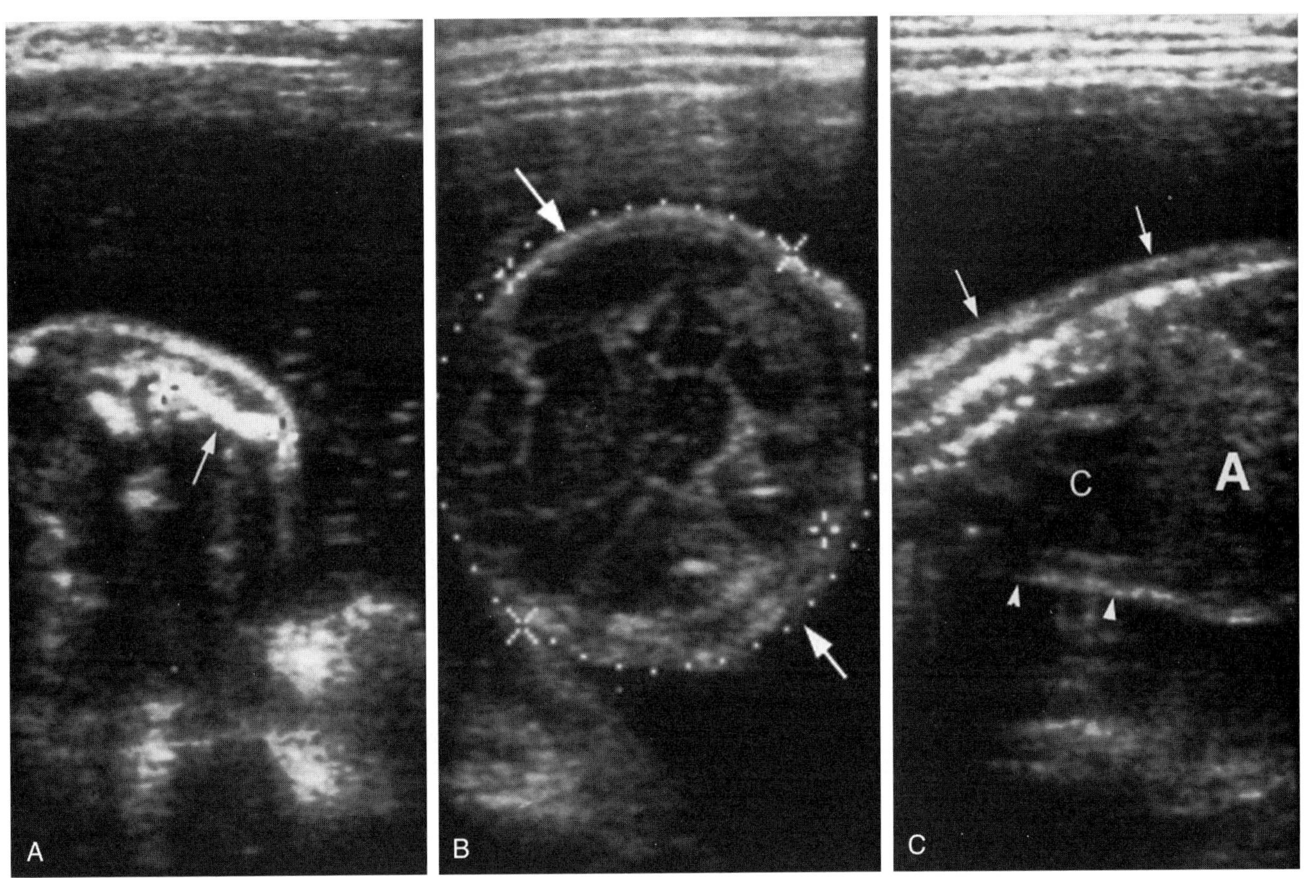

Figure 58–1. Osteogenesis imperfecta diagnosed prenatally using ultrasound. *A*, Ultrasound of the femur at 30 weeks' gestation shows a short, deformed femur *(arrow)*. *B*, Axial view of the fetal head at the level of the thalami shows poor skull ossification with unusual detailed brain visualization. *C*, Sagittal view of fetal torso shows a small chest (C) compared with abdomen (A) (*arrows* point to the spine, and *arrowheads* point to the sternum). Note the polyhydramnios resulting from esophageal compression by the small rib cage. (Case provided by Dr. Monzer Abu Yousef, Iowa City, Iowa.)

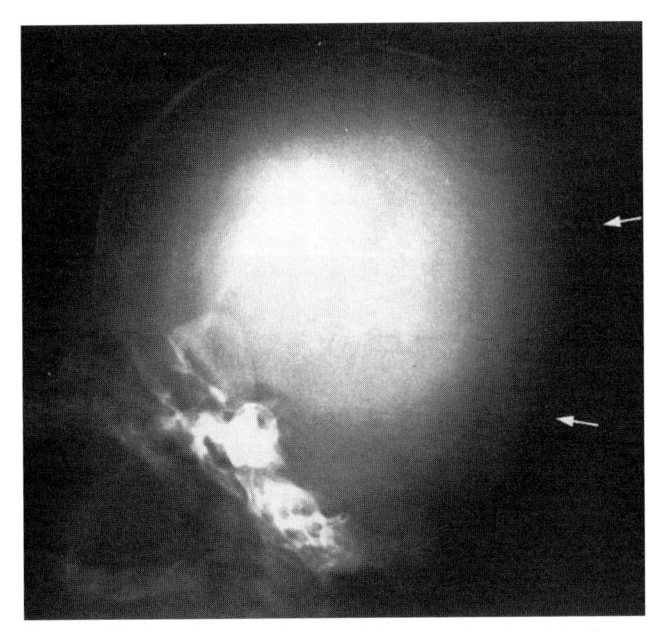

Figure 58–2. Deficient ossification of the parietal and occipital bones *(arrows)* in a newborn with the congenital form of osteogenesis imperfecta.

density, fractures and sequelae of fractures (Fig. 58–4), bowing, shortening, and other deformities of the long bones. The long bones also may appear thin, gracile, and osteopenic. There may be a large head with or without wormian bones (Fig. 58–4*B*, *C*). Bone softening may result in multiple vertebral body wedging or flattening, elongation of the pedicles (Fig. 58–4*D*), spinal deformity (kyphosis, scoliosis, or both), acetabular protrusion (Fig. 58–5), coxa vara, and genu valgum. Multiple contiguous vertebral bodies may show a biconcave appearance, also termed *codfish vertebrae*.

In most cases of OI, bony demineralization is recognized easily on plain radiographs. In some mild cases, osteopenia may require quantitation by bone density measurement. A major neurologic complication is basilar invagination, with brainstem compression. Spinal cord compression has been reported, but this is rare. Migratory transient osteoporosis associated with OI also has been reported. Simple bone cysts are associated rarely with OI. They can occur in the metaphysis of long bones or in flat bones. Popcorn-like calcifications within the metaphyses also have been described in OI. After bisphosphonate treatment of OI, multiple transverse dense (sclerotic)

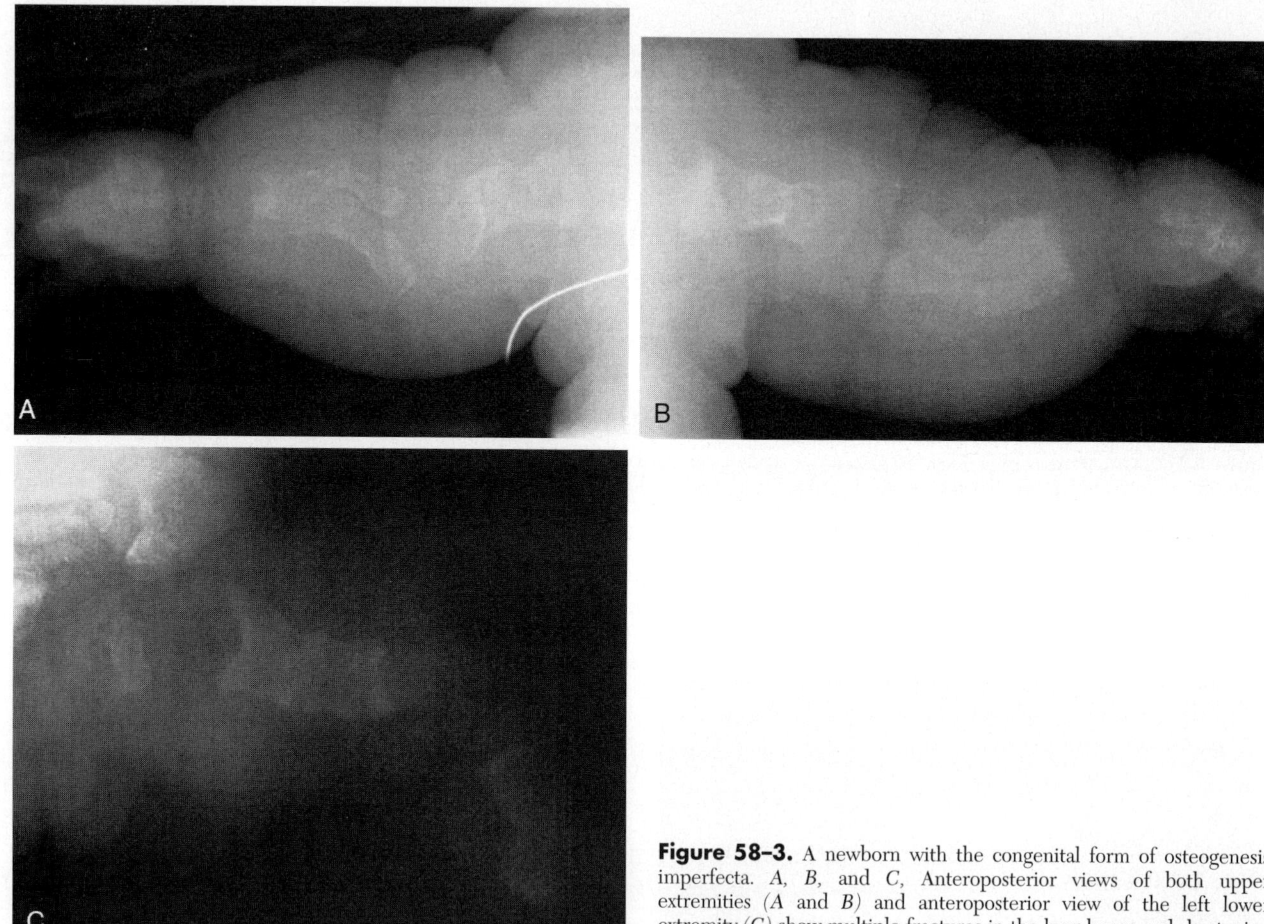

Figure 58-3. A newborn with the congenital form of osteogenesis imperfecta. *A, B,* and *C,* Anteroposterior views of both upper extremities *(A* and *B)* and anteroposterior view of the left lower extremity *(C)* show multiple fractures in the long bones and shortening and deformity of these bones.

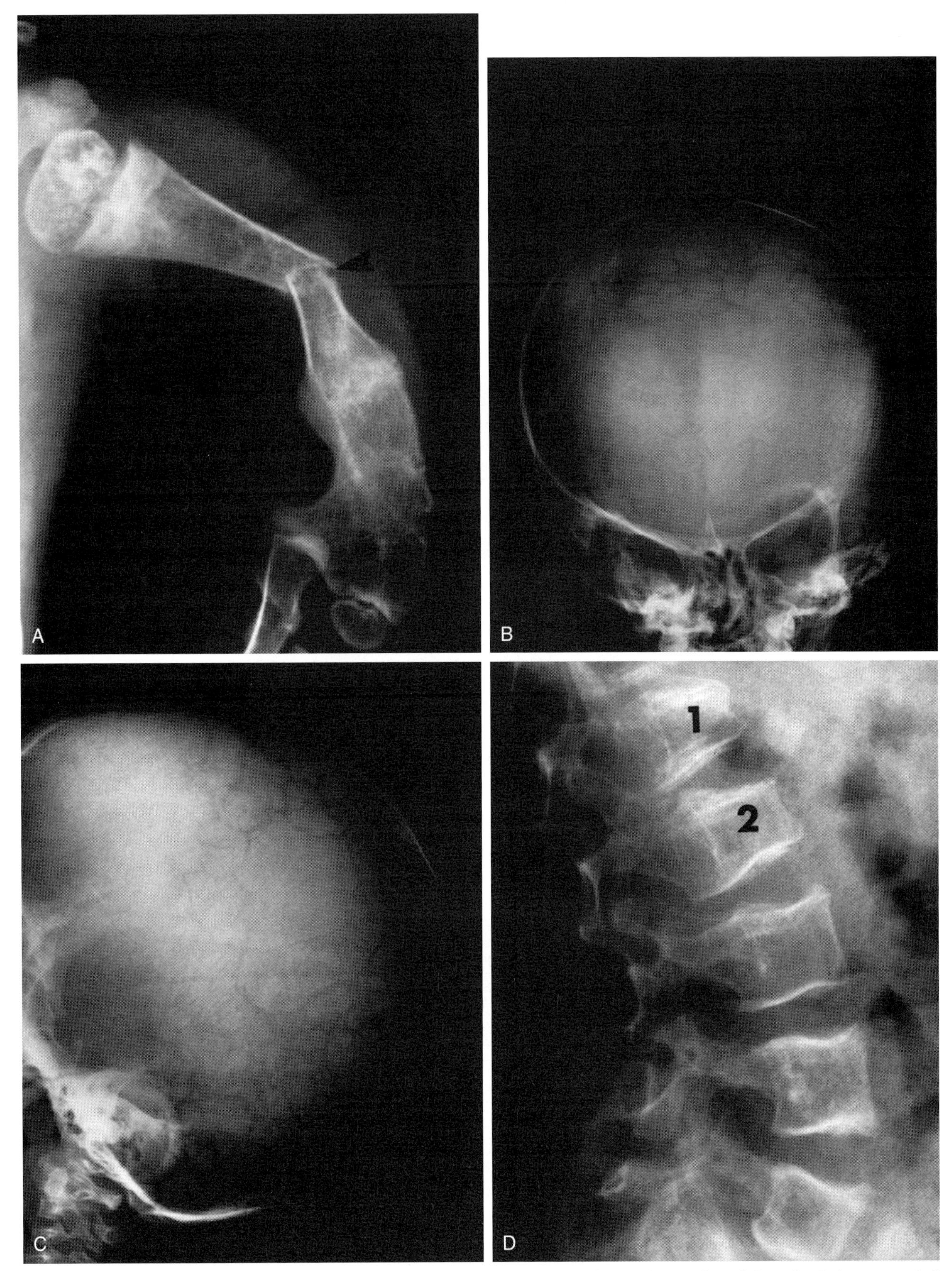

Figure 58–4. A 6-year-old girl with osteogenesis imperfecta. *A*, Anteroposterior view of the arm shows a short, deformed right humerus with thin cortices and a recent mid-diaphyseal fracture *(arrowhead)*. Two other (healing) fractures are seen more distally. *B* and *C*, Anteroposterior and lateral views of the skull from the same patient at age 3 years show a relatively large cranium, osteopenia, and multiple wormian bones. *D*, Lateral view of the lumbar spine shows anterior wedging of L1 (1) and L2 (2) vertebral bodies. Note the elongation of the pedicles caused by softening of the bone.

metaphyseal lines may be seen (Fig. 58–6). Bisphosphonates are potent inhibitors of osteoclastic activity; they may stimulate bone formation and prevent fractures in patients with OI.

Hyperplastic (exuberant) callus is seen occasionally in OI, particularly in the femur and iliac bones (Fig. 58–7). This phenomenon can occur after surgery or with trauma. A fracture may or may not be visible. The callus can appear aggressive on plain radiographs and computed tomography (CT) scan. Biopsy should be avoided, and instead the patient should be followed. On serial radiographs, the callus matures gradually and diminishes in size. Osteosarcoma does occur in association with OI, but it is extremely rare.

DIFFERENTIAL DIAGNOSIS

Child abuse is the main concern in the radiologic differential diagnosis of OI, especially in OI patients with multiple fractures. Wormian bones, osteopenia, and bone deformities favor OI. Metaphyseal corner fractures and bucket-handle fractures in otherwise normal infants indicate nonaccidental trauma. Some mild cases of OI may be confused with child abuse, especially if the teeth and sclera are normal. Skin biopsy for collagen analysis shows normal results in children with multiple fractures of abuse. The presence of OI does not exclude the possibility of child abuse, however. Proven cases of child abuse in children with OI have been reported.

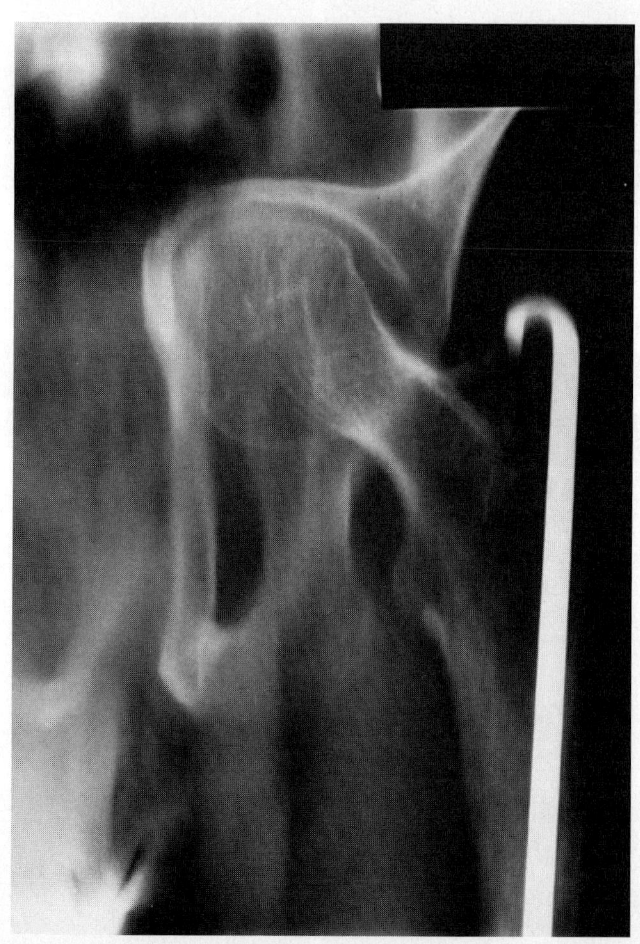

Figure 58–5. A 16-year-old boy with osteogenesis imperfecta and severe left acetabular protrusion seen on anteroposterior conventional tomography of the left hip.

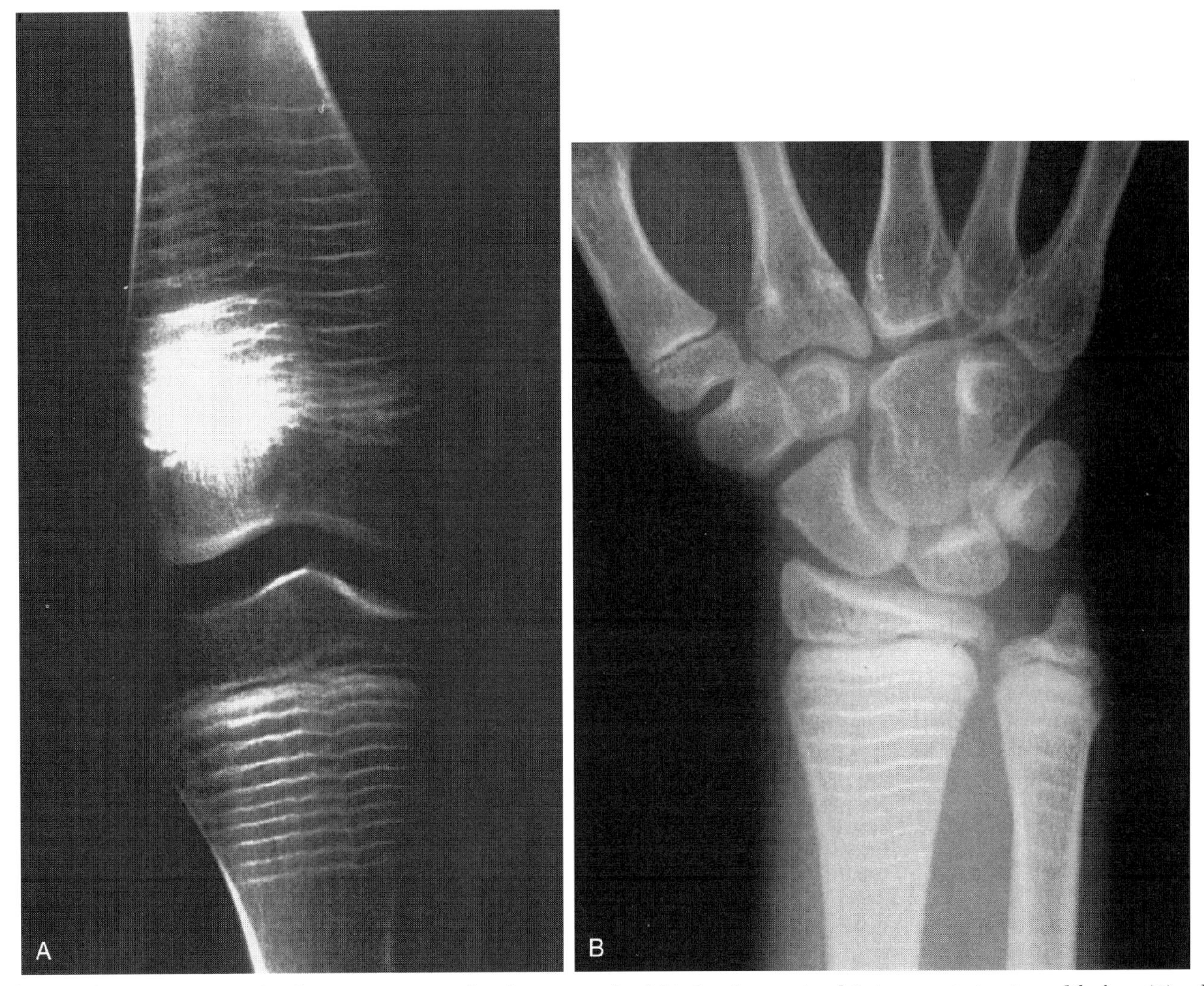

Figure 58–6. A 9-year-old girl with osteogenesis imperfecta being treated with bisphosphonate. *A* and *B*, Anteroposterior views of the knee (*A*) and wrist (*B*) show sclerotic transverse lines, which correspond with the treatment given intermittently every 4 months for the last 3 years. (Images provided by Dr. Karen MacEwan, Corvallis, OR.)

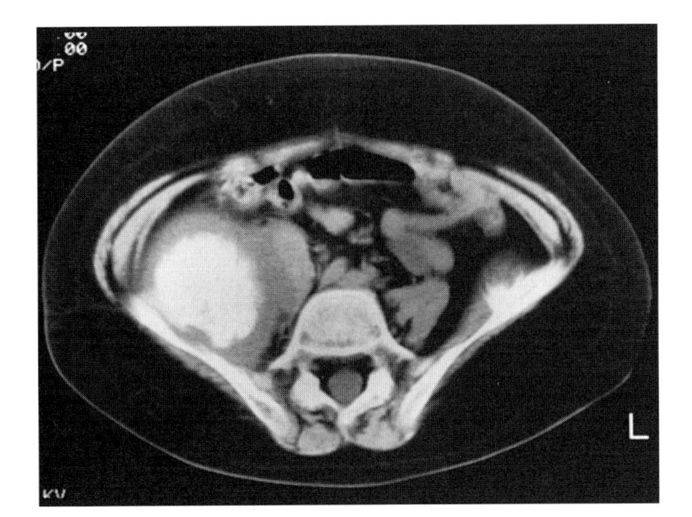

Figure 58–7. A 15-year-old boy with osteogenesis imperfecta and bilateral iliac exuberant callus formation. CT section through the pelvis shows a large, aggressive-looking ossific mass arising on the anterior surface of the right ilium bilaterally. The mass is seen displacing the iliopsoas muscle medially. On the left, the callus is smaller and likely mature. Atrophy of the gluteal muscles is also evident.

Recommended Readings

Ablin DS: Osteogenesis imperfecta: A review. Can Assoc Radiol J 49:110–123, 1998.

Ablin DS, Greenspan A, Reinhart M, Grix A: Differentiation of child abuse from osteogenesis imperfecta. AJR Am J Roentgenol 154:1035–1046, 1990.

Azouz EM, Chen M-F, Khalifé S, et al: New form of bone dysplasia with multiple fractures associated with monosomy X. Am J Med Genet 66:163–168, 1996.

Azouz EM, Fassier F: Hyperplastic callus formation in OI. Skeletal Radiol 26:744–745, 1997.

Gagliardi JA, Evans EM, Chandnani VP, et al: Osteogenesis imperfecta complicated by osteosarcoma. Skeletal Radiol 24:308–310, 1995.

Gahagan S, Rimsza ME: Child abuse or osteogenesis imperfecta: How can we tell? Pediatrics 88:987–992, 1991.

Glorieux FH, Bishop NJ, Plotkin H, et al: Cyclic administration of pamidronate in children with severe osteogenesis imperfecta. N Engl J Med 339:947–952, 1998.

Goldman AB, Davidson D, Pavlov H, Bullough PG: "Popcorn" calcifications: A prognostic sign in osteogenesis imperfecta. Radiology 136:351–358, 1980.

Goodman P, Dominguez R, Wood BP: Radiological cases of the month—expansile bone cyst in osteogenesis imperfecta. Am J Dis Child 144:933–934, 1990.

Karagkevrekis CB, Ainscow DAP: Transient osteoporosis of the hip associated with osteogenesis imperfecta. J Bone Joint Surg Br 80:54–55, 1998

Kozlowski K, Kan A: Intrauterine dwarfism, peculiar facies and thin bones with multiple fractures—a new syndrome. Pediatr Radiol 18:394–398, 1988.

Maroteaux P, Cohen-Solal L, Bonaventure J, et al: Syndromes létaux avec gracilité du squelette. Arch Fr Pediatr 45:477–481, 1988.

Nakamura K, Kurokawa T, Nagano A, et al: Familial occurrence of hyperplastic callus in osteogenesis imperfecta. Arch Orthop Trauma Surg 116:500–503, 1997.

CHAPTER 59

Osteopetrosis and Pyknodysostosis

M. Patricia Harty, MD, and Simon Kao, MD

OSTEOPETROSIS

Overview

Osteopetrosis comprises a spectrum of bone disorders characterized by abnormal osteoclast function and increased bone density. It first was identified in 1904 by the radiologist Albers-Schönberg, who described skeletal sclerosis and multiple fractures in a young man. Osteopetrosis subsequently was classified into autosomal dominant and autosomal recessive types. The autosomal dominant type is divided into autosomal dominant osteopetrosis I (ADO I) and autosomal dominant osteopetrosis II (ADO II), also called *Albers-Schönberg disease.* The autosomal recessive type includes malignant or infantile osteopetrosis and an intermediate type characterized by carbonic anhydrase II isoenzyme deficiency.

Autosomal Dominant Osteopetrosis

Most patients with ADO report a family history of the disease, anemia, and frequent fractures. A few patients have been reported to have short stature, frontal bossing, and exophthalmus, but most patients are normal in appearance. The estimated genetic penetrance of ADO is approximately 75%, although phenotypic expression of the disorder varies. No genetic mutations have been identified in ADO.

Patients with ADO I have more pronounced cranial involvement with a thickened skull and cranial nerve palsies. Patients with ADO II tend to have multiple fractures and a propensity for mandibular osteomyelitis. Bone pain is the most common symptom in patients with ADO, although many patients remain asymptomatic for years. Treatment of ADO is directed toward relief of symptoms.

Radiographically, ADO is characterized by increased density throughout the skeleton with sclerosis progressing with age. Abnormal remodeling of bone is present but to a lesser degree than in autosomal recessive osteopetrosis. Increased sclerosis of the vertebral end plates results in the "rugger-jersey spine," and faulty bone remodeling produces endobones, or the "bone-within-a-bone" appearance, found in the spine and iliac bones (Fig. 59–1). The Erlenmeyer flask deformity that results from abnormal bone remodeling is seen in the distal femoral metaphyses. The principal radiographic finding in ADO I is diffuse sclerosis and thickening of the calvarium and sclerosis of the spine and pelvis. More involvement of the appendicular skeleton is found in ADO II. Fractures are more frequent in this form, and abnormalities in remodeling contribute to delayed healing (Fig. 59–2). The calvarium is less involved in ADO II; however, the skull base is thickened and sclerotic, and cranial nerve compression may occur in this entity.

Magnetic resonance (MR) imaging of the skull in patients with ADO shows decreased signal intensity in the thickened cranial vault in ADO I and in the skull base in ADO II. Cerebellar tonsil herniation, optic nerve sheath dilation, and cephaloceles are seen in ADO I, most likely as a result of increased intracranial pressure from the severely thickened calvarium. Greater involvement of the skull base in ADO II results in optic canal stenosis and optic nerve atrophy. Bone scintigraphy done on patients with ADO typically shows a *superscan*—increased uptake of technetium-99m medronate in the spine and the metaphyses and epiphyses of the long bones with no renal uptake. Bone marrow imaging shows diffusely decreased bone marrow uptake of sulfur colloid. Dual x-ray absorptiometry shows increased bone mineral density in osteopetrosis. Dual x-ray absorptiometry has been used to assess progression of the disease and response to therapy in infantile osteopetrosis.

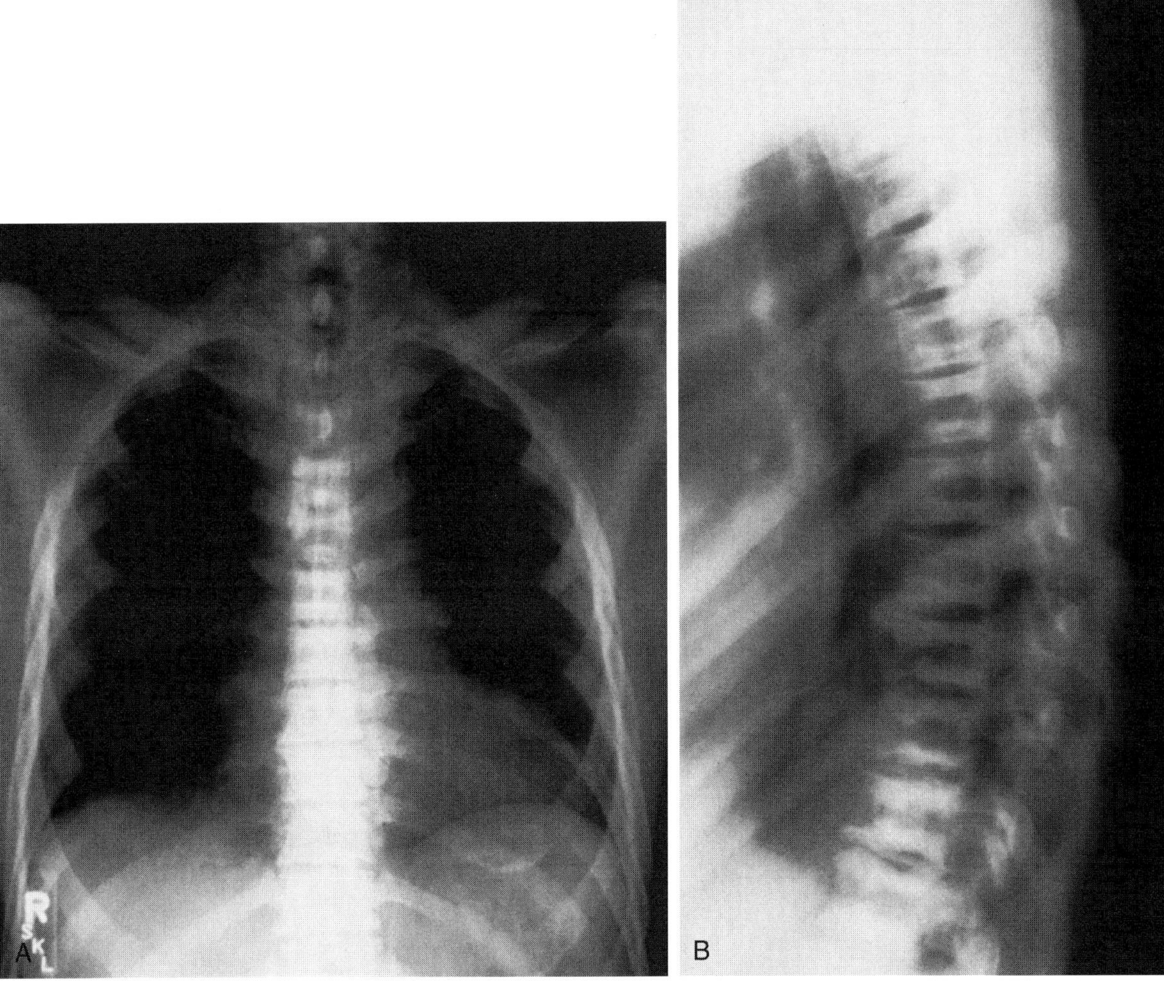

Figure 59–1. Spinal changes in autosomal dominant osteopetrosis. *A* and *B*, Frontal and lateral views of the spine in an adult with autosomal dominant osteopetrosis show increased bands of density in the vertebral end plates, the *rugger-jersey* spine.

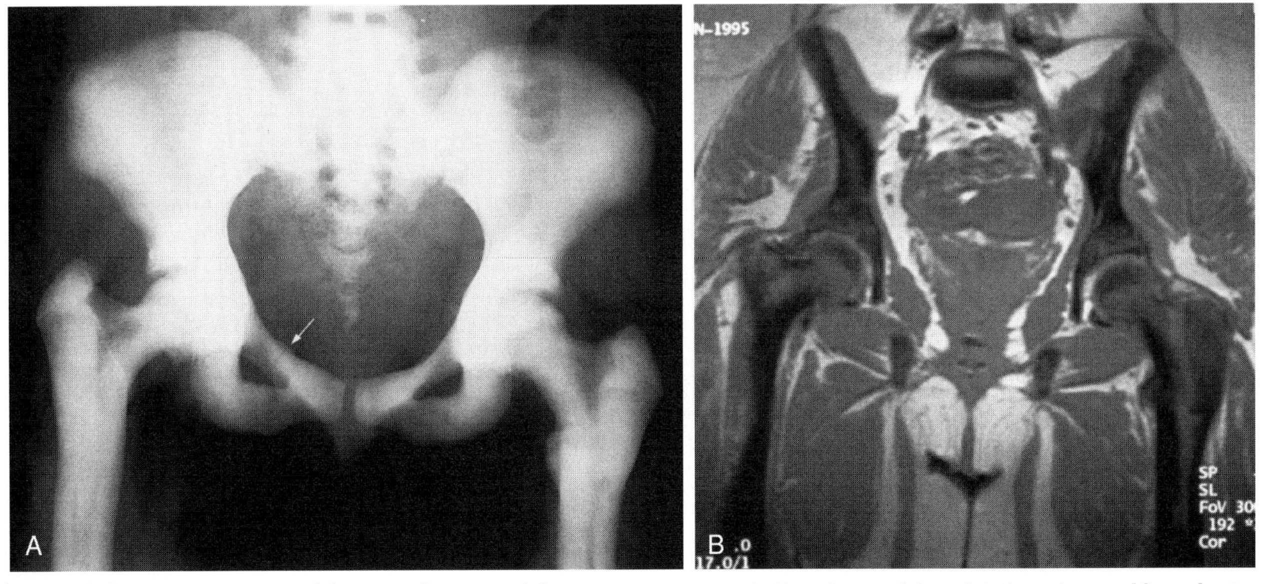

Figure 59–2. MR findings in an adolescent with autosomal dominant osteopetrosis. *A*, Frontal view of the pelvis shows increased bone density, a healing pathologic fracture through the right superior pubic ramus *(arrow)*, and coxa vara on the right. *B*, Coronal T1-weighted MR image through the pelvis shows decreased signal intensity throughout the pelvic and proximal femoral marrow.

Autosomal Recessive Osteopetrosis

The autosomal recessive or infantile malignant form of osteopetrosis typically presents within the first year of life. Initial manifestations vary depending on the severity of the disease in the individual; however, hepatosplenomegaly and anemia are common presenting symptoms, followed by failure to thrive, cranial nerve palsies, infections, fractures, and hypocalcemia. Obliteration of the marrow space by unresorbed bone leads to anemia, extramedullary hematopoiesis, and pancytopenia. Fewer than one third of these patients survive beyond 6 years of age. Interferon-γ has been used to treat autosomal recessive osteopetrosis. Bone marrow transplant from matched sibling donors is the treatment of choice, however, with 70% survival. In successfully treated patients, radiographic abnormalities can return to normal as the donor marrow restores the osteoclast population and normal bone remodeling begins. Radiographic changes become apparent 4 months after bone marrow transplant.

The radiographic features of autosomal recessive osteopetrosis result from decreased or absent osteoclast function with subsequent abnormal bone remodeling and accumulation of calcified osteoid (Fig. 59–3). The bones are dense, and the metaphyses are undertubulated or flared (Fig. 59–4). Corticomedullary differentiation is lost, and transverse lucent bands can be seen in the metaphyses and metadiaphyses. As in ADO, the bone-within-a-bone appearance is seen in the spine and pelvis (Fig. 59–5). The calvarium is thickened, and a "hair-on-end" appearance may be seen. Macrocephaly is typical. Abnormal bone accumulation leads to stenosis of the skull base foramina with resultant cranial nerve dysfunction. In some cases, calcium cannot be released from bones, resulting in hypocalcemia and radiographic evidence of rickets, or osteopetrorickets. Transverse fractures and epiphyseal separations are common.

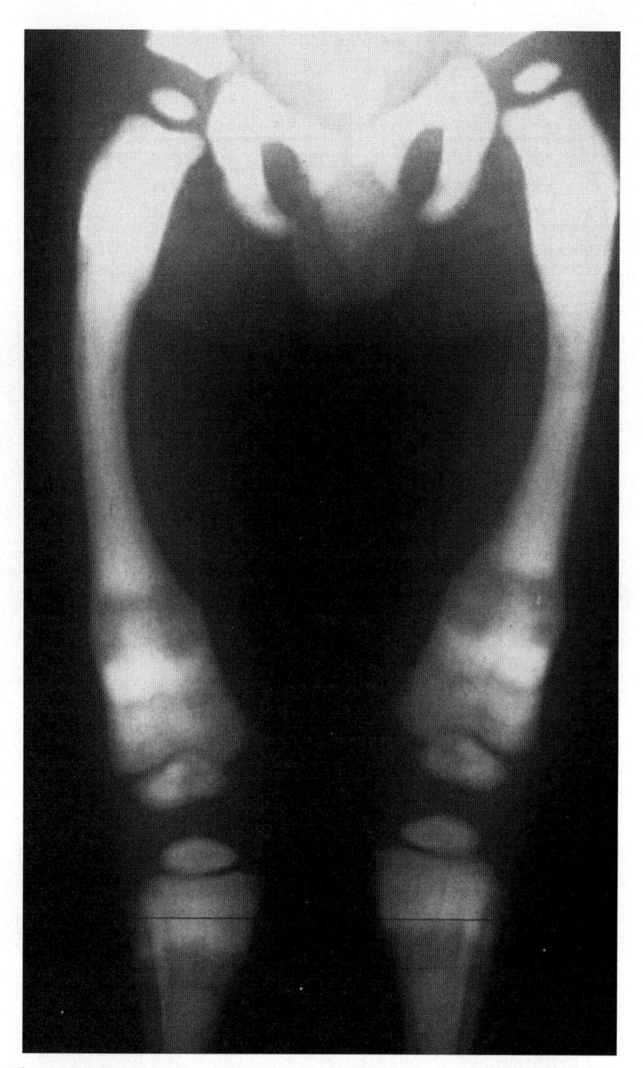

Figure 59–4. Infant with autosomal recessive osteopetrosis. Frontal radiograph of the lower extremities in an infant shows diffusely increased bone density. Undertubulation and transverse lucent bands are seen in metaphyses.

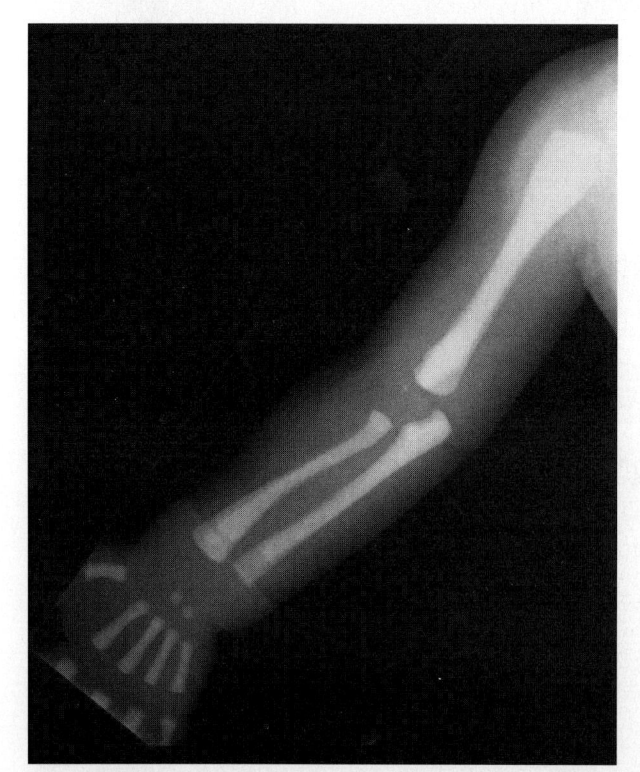

Figure 59–3. Autosomal recessive osteopetrosis. Anteroposterior radiograph of the upper extremity in a 4-month-old infant with autosomal recessive osteopetrosis shows increased bone density with loss of corticomedullary differentiation.

The intermediate form of autosomal recessive osteopetrosis is characterized by dense bones, metaphyseal deformities, renal tubular acidosis, and intracranial calcifications. The bone findings are generally less severe than those in the infantile form, although patients develop coxa vara and sustain pathologic fractures. The genetic disorder responsible for this inborn error of metabolism results in a deficiency of the isoenzyme carbonic anhydrase type II.

PYKNODYSOSTOSIS

Overview

Pyknodysostosis (Greek *pyknos* = dense or thick) is a rare autosomal recessive sclerosing bone dysplasia first described independently by Andren and colleagues and Maroteaux and Lamy in 1962. The French impressionist painter Henri de Toulouse-Lautrec was believed to be afflicted with this condition. The basic defect resides in the cathepsin K gene located in the long arm of chromosome 1 (1q21). Cathepsin K is an enzyme, lysosomal cysteine protease, expressed predominantly in osteoclasts involved in the resorption of bone matrix by cleaving type I and II collagens. This enzyme is important in the dynamic bone remodeling process. Mutation in the cathepsin K gene results in absence of the cathepsin K enzyme and loss of collagenolytic activity in pyknodysostosis.

The condition is characterized clinically by short-limbed dwarfism (with adult height <150 cm or <59 inches), frontal and occipital prominence, persistently open anterior fontanelle, hypoplastic face, bluish sclera, prominent nose, narrow palate, micrognathia, dental abnormalities (persistence of deciduous teeth and delayed eruption of permanent teeth, hypodontia, caries), brittle bones, hypoplasia or aplasia of clavicles, wrinkled skin over dorsa of fingers with grooved or flattened nails, and short fingers. Upper airway obstruction has been reported partly as a result of a relatively long uvula and mandibular deformity. There is defective secretion of growth hormone in some patients, and growth hormone has been used to improve linear growth. In contrast to osteopetrosis, there is no associated anemia and hepatosplenomegaly.

Radiologically, there is generalized osteosclerosis; however, in contrast to osteopetrosis, there is preservation of the medullary canal of the long bone (Fig. 59–6*A*). The bones are brittle despite their density, and pathologic fractures are common (Fig. 59–6*B*). Fractures heal well, however.

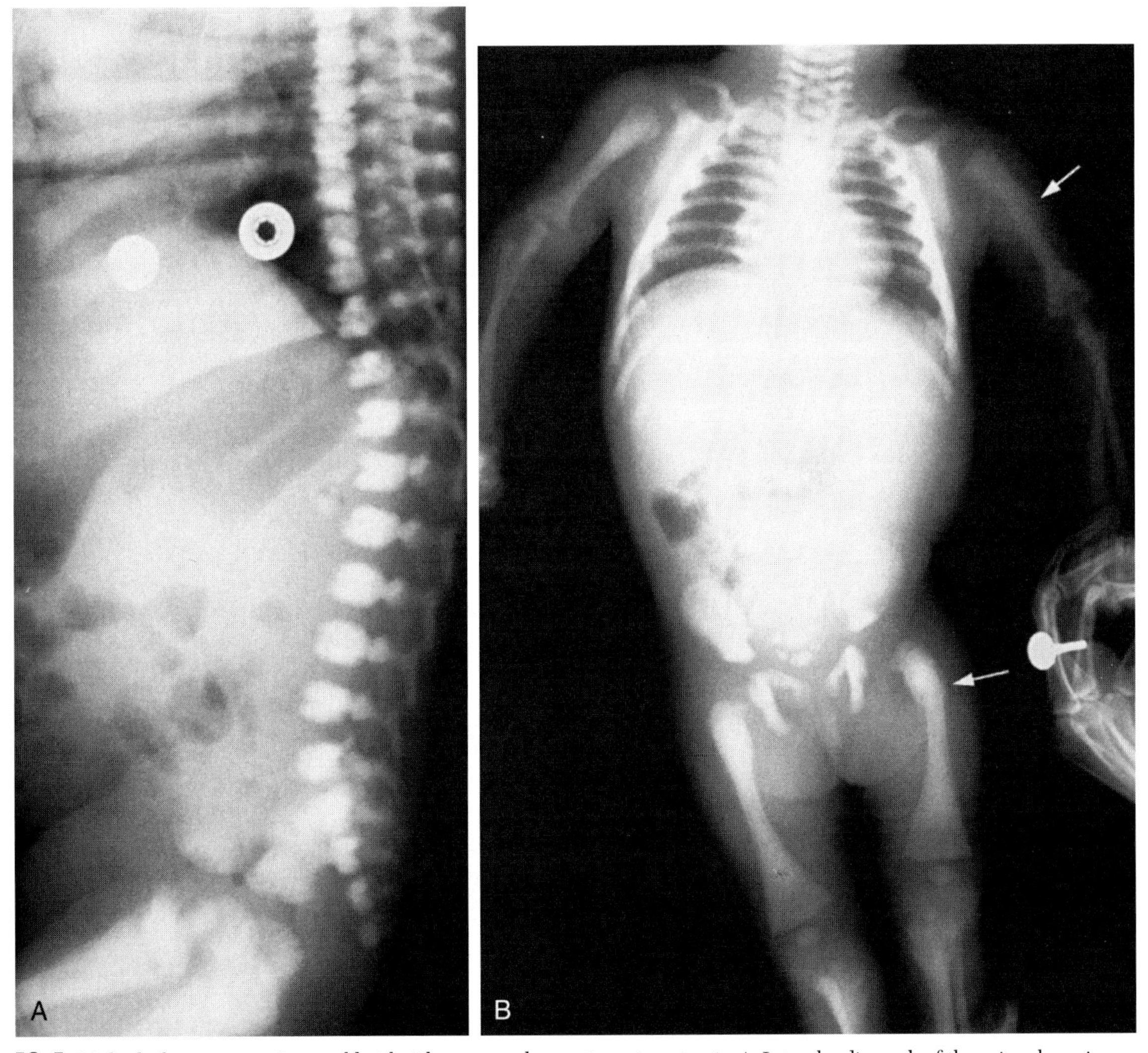

Figure 59–5. Multiple fractures in a 2-year-old girl with autosomal recessive osteopetrosis. *A*, Lateral radiograph of the spine shows increased bony sclerosis and healing pathologic fractures through the proximal femora. *B*, Anteroposterior babygram shows multiple healing pathologic fractures *(arrows)* in the long bones, diffusely increased bone density, and a *bone-within-a-bone* appearance in the iliac bones.

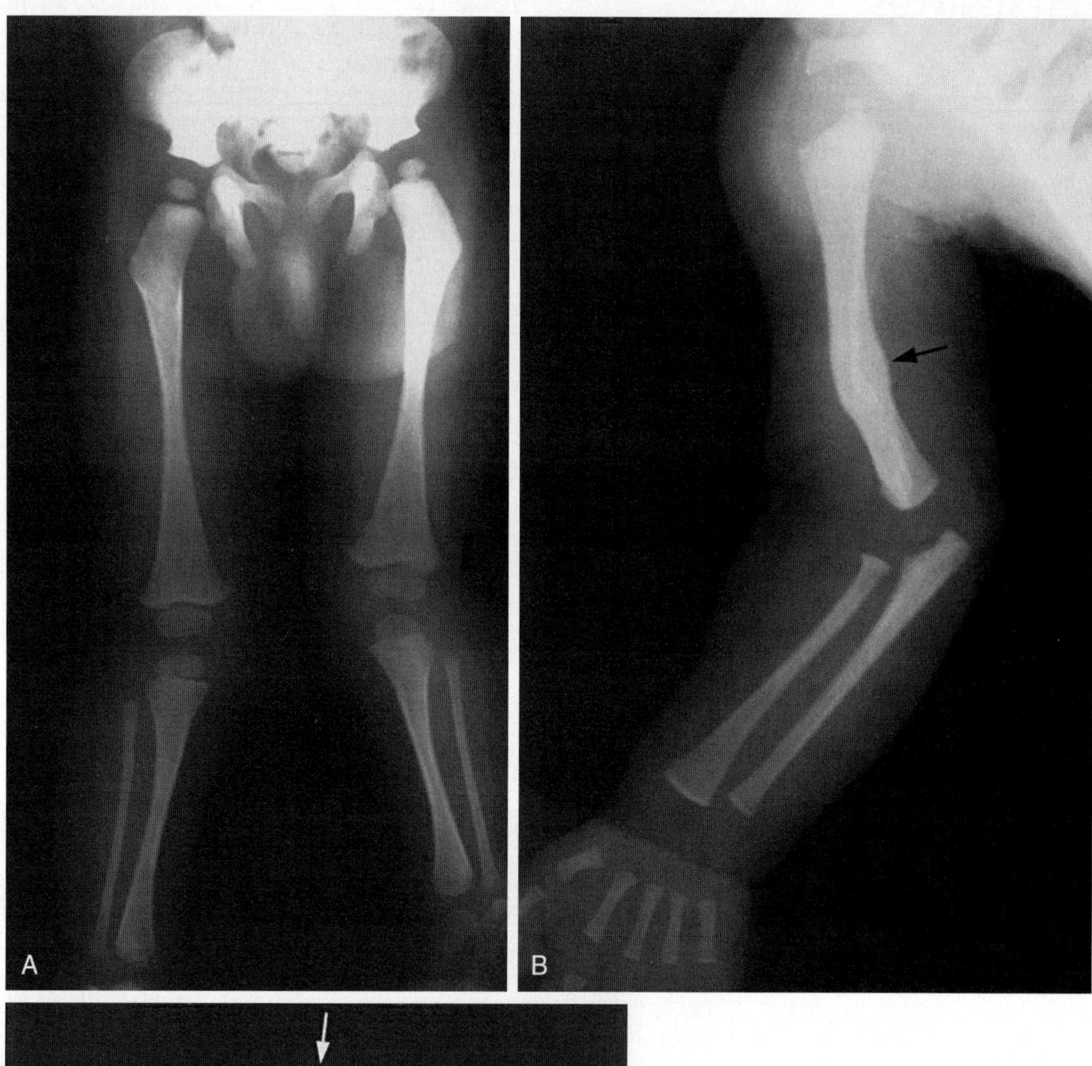

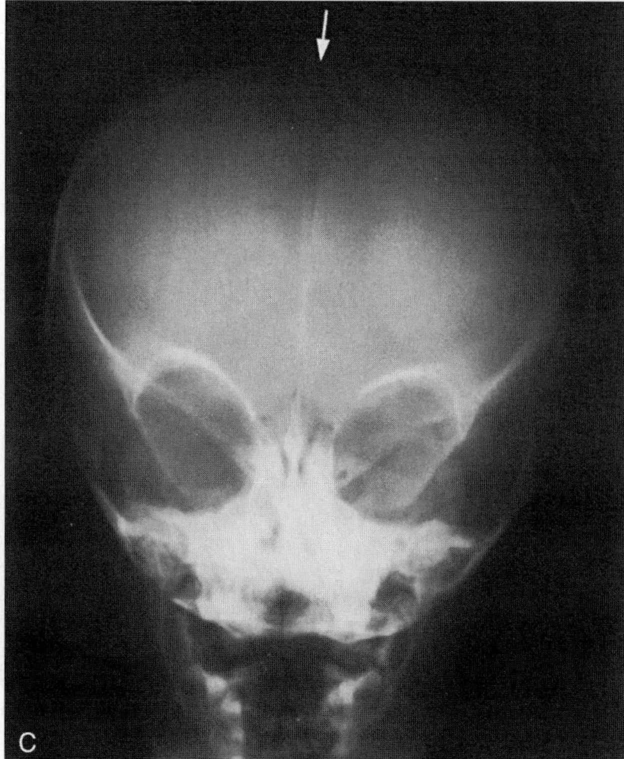

Figure 59–6. Pyknodysostosis. *A,* Frontal radiograph of lower extremities shows generalized osteosclerosis with preservation of medullary canal of long bones. *B,* Frontal radiograph of the right upper extremity shows increased bone density in a healed pathologic fracture *(arrow)* of the right humerus. *C,* Frontal view of skull shows delayed closure of sutures and fontanelles *(arrow)* and osteosclerosis.

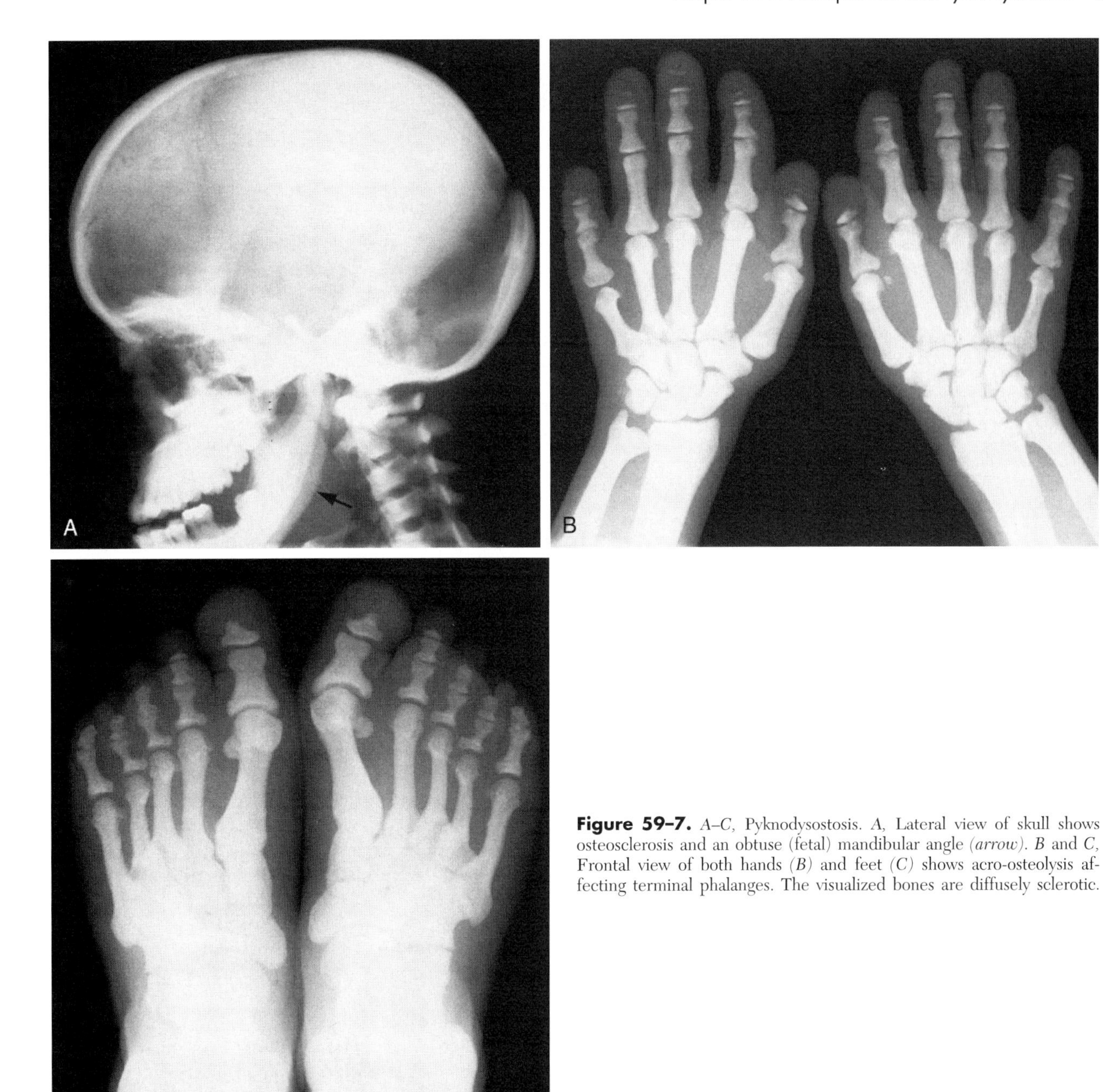

Figure 59–7. *A–C,* Pyknodysostosis. *A,* Lateral view of skull shows osteosclerosis and an obtuse (fetal) mandibular angle *(arrow)*. *B* and *C,* Frontal view of both hands *(B)* and feet *(C)* shows acro-osteolysis affecting terminal phalanges. The visualized bones are diffusely sclerotic.

Craniofacial abnormalities include delayed closure of sutures and fontanelles (Fig. 59–6*C*), wormian bones, cranial hyperostosis (Fig. 59–7*A*), hypoplasia and lack of pneumatization of paranasal sinuses, and obtuse (fetal) mandibular angle (Fig. 59–7*A*). Spinal abnormalities include segmentation anomalies resulting in scoliosis, spondylolysis, and spondylolisthesis. Ilia are narrow. There is partial or total aplasia of the acromial end of clavicle and the distal phalanges of digits (acro-osteolysis) (Fig. 59–7*B*, *C*). Differential diagnoses include cleidocranial dysplasia, osteopetrosis, mandibuloacral dysplasia, and acro-osteolysis syndromes.

Recommended Readings

Adler IN, Stine KC, Kurtzburg J, et al: Dual-energy x-ray absorptiometry in osteopetrosis. South Med J 93:501, 2000.

Albers-Schönberg H: Roentgenbilder Einer Seltenen Knochennerkrankung. Munch Med Wochenschr 51:365, 1904.

Andren L, Dymling K-E, Hogeman K-E, et al: Osteopetrosis acro-osteolytica: A syndrome of osteopetrosis, acro-osteolysis and open sutures of the skull. Acta Chir Scand 124:496, 1962.

Armstrong DG, Newfield JT, Gillespie R: Orthopedic management of osteopetrosis: Results of a survey and review of the literature. J Pediatr Orthop 19:122, 1999.

Bénichou OD, Laredo JD, de Vernejoul MC: Type II autosomal dominant osteopetrosis (Albers-Schönberg disease): Clinical and radiological manifestations in 42 patients. Bone 26:87, 2000.
Cheow HK, Steward CG, Grier DJ: Imaging of infantile osteopetrosis before and after bone marrow transplantation. Pediatr Radiol 81:869, 2001.
Curé JK, Key LL, Goltra DD, et al: Cranial MR imaging of osteopetrosis. AJNR Am J Neuroradiol 21:1110, 2000.
deVernejoul MC, Benichou O: Human osteopetrosis and other sclerosing disorders: Recent genetic developments. Calcif Tissue Int 69:1, 2001.
Gelb BD, Edelson JG, Desnick RJ: Linkage of pycnodysostosis to chromosome 1q21 by homozygosity mapping. Nat Genet 10:235, 1995.
Gelb BD, Shi G, Chapman HA, et al: Pycnodysostosis, a lysosomal disease caused by cathepsin K deficiency. Science 273:1236, 1996.
Greenspan A: Sclerosing bone dysplasias—a target-site approach. Skeletal Radiol 20:561, 1991.
Kim S, Park CH, Byungseok K: Superscan in an autosomal-dominant benign form of osteopetrosis. Clin Nucl Med 26:636, 2001.
Maroteaux P, Lamy M: La pycnosydostose. Presse Med 70:999, 1962.
Soliman AT, Ramadan MA, Sherif A, et al: Pycnodysostosis: Clinical, radiologic, and endocrine evaluation and linear growth after growth hormone therapy. Metab Clin Exp 50:905, 2001.
Yousefzadeh DK, Agha AS, Reinertson J: Radiographic studies of upper airway obstruction with cor pulmonale in a patient with pycnodysostosis. Pediatr Radiol 8:45, 1979.

Infantile Cortical Hyperostosis (Caffey's Disease)

Michel E. Azouz, MD, and Lori L. Barr, MD

OVERVIEW

Since the 1970s, the incidence of infantile cortical hyperostosis (Caffey's disease) has been declining. The disease is characterized clinically by irritability, fever, and areas of painful and tender soft tissue swelling. Infantile cortical hyperostosis typically occurs before the age of 6 months with an average age of onset at about 9 weeks. Incidence in boys and girls is about equal. A swollen jaw is a common presentation, and the mandible is the most frequently affected bone. Leukocytosis, anemia, and an elevated erythrocyte sedimentation rate may be present. Serum calcium and phosphorus levels are normal. The disease is usually self-limited, and the swelling often disappears within a few weeks. It may reappear, however, at the same site or at different sites. A viral cause has been considered for the disease, but no organism has been isolated from blood, bone marrow, or biopsy specimens obtained from established cases of Caffey's disease. Antibiotics do not alter the course of the disease. Familial cases of Caffey's disease have been reported in which the disease is believed to be transmitted as an autosomal dominant trait.

IMAGING

A definitive diagnosis typically is made on plain radiography. A skeletal survey frequently is required. Particular attention should be given to the mandible (Fig. 60–1) and the thoracic cage (Fig. 60–2). The clavicles, scapulae, and long bones also are involved frequently (see Fig. 60–2). Periosteal reaction and diaphyseal cortical thickening are seen in the affected bones (Fig. 60–3). The disease has not been described in the vertebrae or in the phalanges. In familial cases, periosteal new bone formation may be present at birth. Several cases have been recognized in utero on ultrasound studies.

Bilateral involvement is common. Radiographic findings persist for several months, with complete recovery usually reported by the end of the second year of life. Bowing of the long bones may persist longer. Previously affected tubular bones may show expansion of the medullary cavity and relatively thin cortex for months or years (Fig. 60–4). Bone and gallium scans become positive early, before the development of changes seen on plain radiographs. Magnetic resonance (MR) imaging is not necessary. It does reveal, however, marked soft tissue edema and inflammation surrounding the affected bones.

DIFFERENTIAL DIAGNOSIS

Changes in individual bones, such as the scapula, showing enlargement and sclerosis may be mistaken for tumor. In the differential diagnosis, hyperostosis produced by trauma, osteomyelitis, congenital syphilis, hypervitaminosis A, and scurvy should be entertained. Long-term prostaglandin administration in neonates with cyanotic congenital cardiac disease can be associated with symmetrical cortical hyperostosis and periosteal reaction (Fig. 60–5), but the mandible typically is spared.

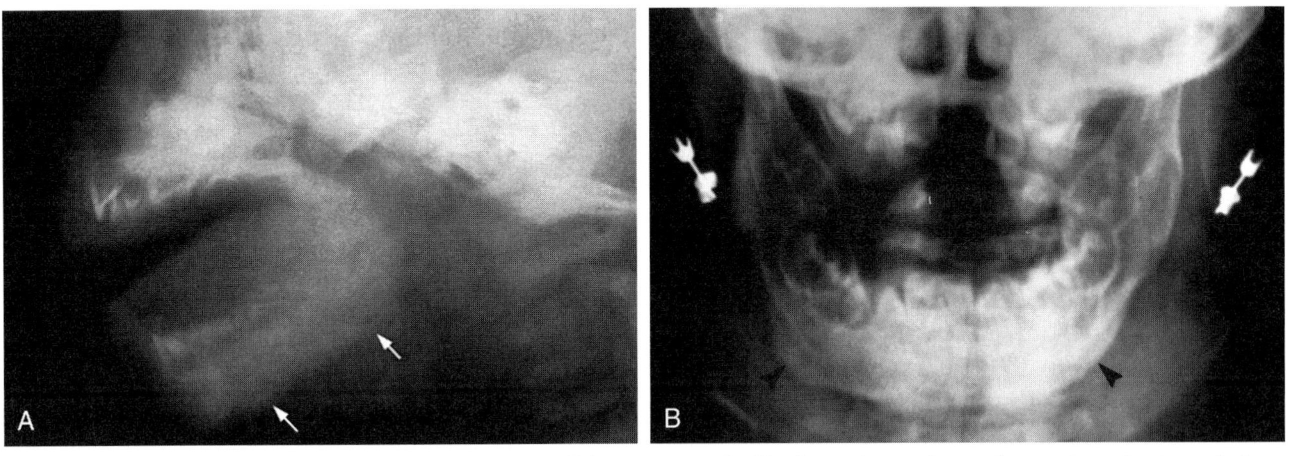

Figure 60–1. Infantile cortical hyperostosis affecting the mandible. *A,* A 2-month-old infant with a swollen tender jaw. Lateral radiograph shows periostitis *(arrows)* and soft tissue swelling. *B,* A 5-month-old girl with infantile cortical hyperostosis showing thick periosteal reaction paralleling the mandible *(arrowheads)*, resulting in overall expansion of the mandible.

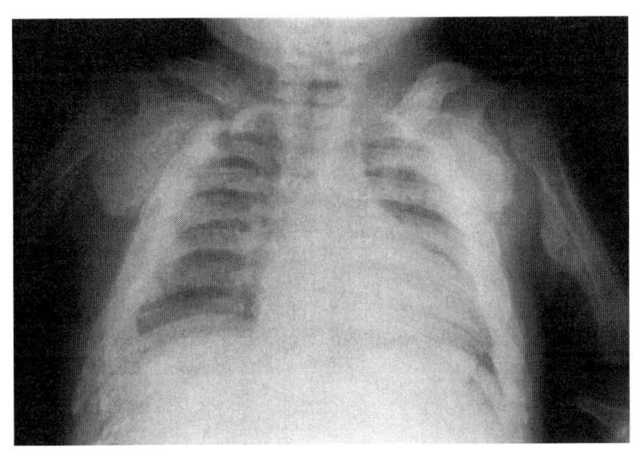

Figure 60–2. Infantile cortical hyperostosis in a 3-month-old infant affecting the rib cage, clavicles, scapulae, and humeri.

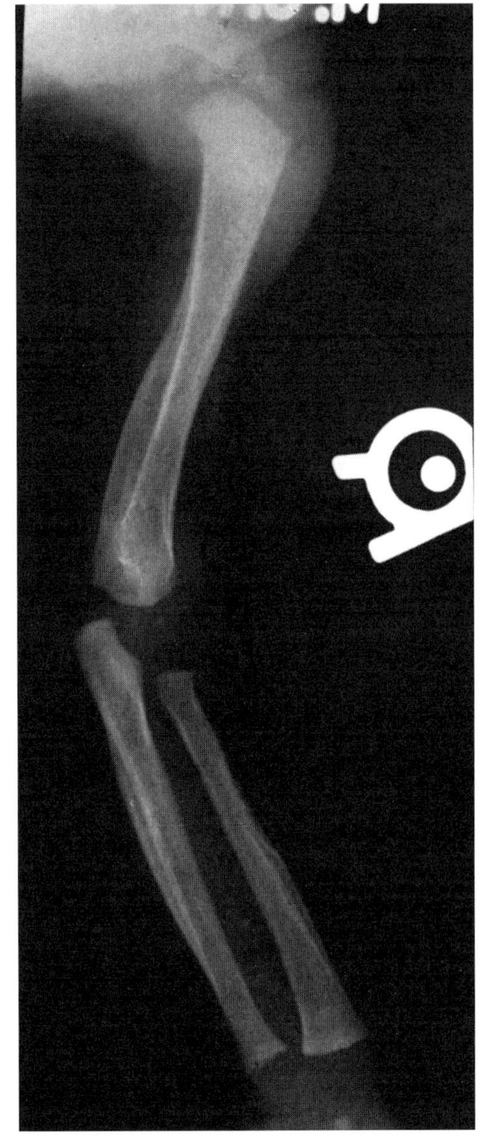

Figure 60–3. A 3-month-old girl with infantile cortical hyperostosis showing thick periosteal new bone formation along the humerus, radius, and ulna.

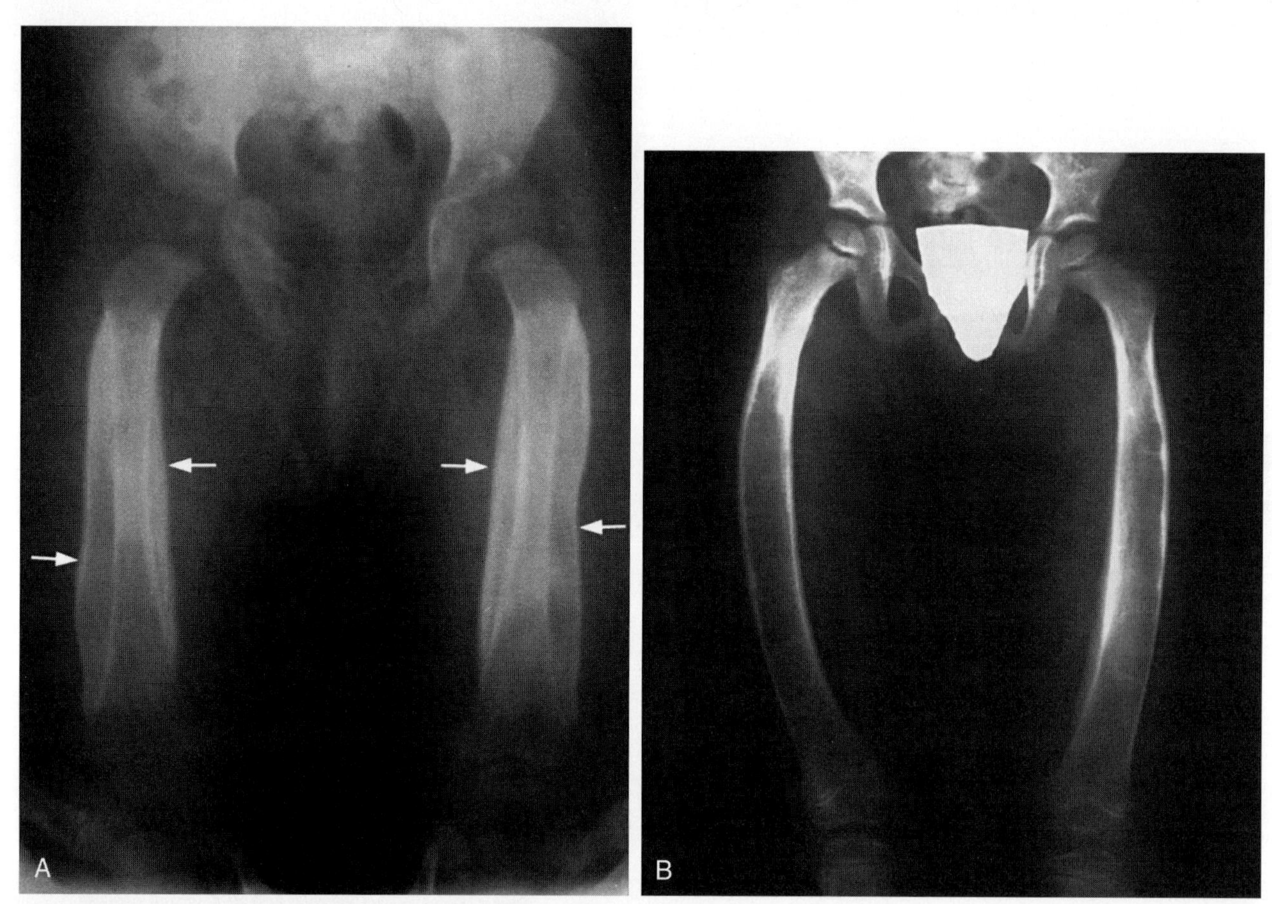

Figure 60–4. Persistent bowing and deformity of the femora in a child who previously had Caffey's disease. *A,* Initial examination at age 5 months shows extensive periosteal thickening *(arrows)* in both femora. *B,* Follow-up examination 2 years later shows bowing and widening of the diaphysis in both femora.

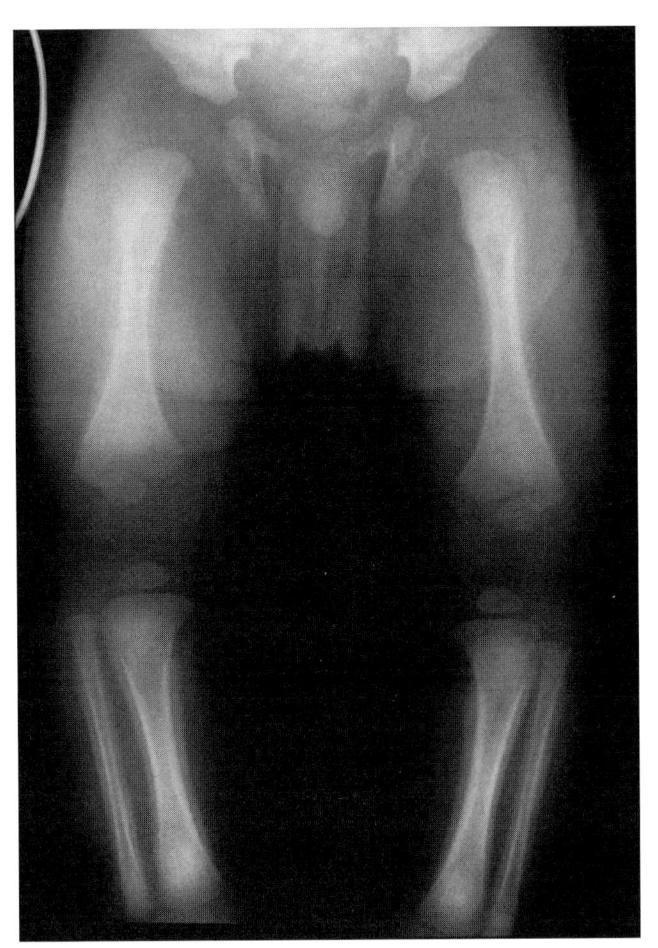

Figure 60–5. A 1-month-old boy on long-term prostaglandin therapy. Anteroposterior view of both lower extremities shows symmetrical, circumferential, and smooth periosteal reaction along the full length of the diaphysis of all long bones. Both upper extremities were similarly involved, but the mandible was normal.

Recommended Readings

Beluffi G, Chirico G, Colombo A, et al: Report of a new case of neonatal cortical hyperostosis: Histological and ultrastructural study. Ann Radiol 27:79–88, 1984.

Blasier RB, Aronson DD: Infantile cortical hyperostosis with osteomyelitis of the humerus. J Pediatr Orthop 5:222–224, 1985.

Caffey J: Infantile cortical hyperostosis. J Pediatr 29:541–559, 1946.

Emmery L, Timmermans J, Christens J, Fryns JP: Familial infantile cortical hyperostosis. Eur J Pediatr 141:58–59, 1983.

Fauré C, Beyssac J-M, Montagne J-P: Predominant or exclusive orbital and facial involvement in infantile cortical hyperostosis (DeToni-Caffey's disease). Pediatr Radiol 6:103–106, 1977.

Katz D, Eller DJ, Bergman G, et al: Caffey's disease of the scapula: CT and MR findings. AJR Am J Roentgenol 168:286–287, 1997.

Maclachlan AK, Gerrard JW, Houston CS, et al: Familial infantile cortical hyperostosis in a large Canadian family. Can Med Assoc J 130:1172–1174, 1984.

Matzinger MA, Briggs VA, Dunlap JH, et al: Plain film and CT observations in prostaglandin-induced bone changes. Pediatr Radiol 22:264–266, 1992.

Saatci I, Brown JJ, McAlister WH: MR Findings in a patient with Caffey's disease. Pediatr Radiol 26:68–70, 1996.

Swerdloff BA, Ozonoff MB, Gyepes M: Late recurrence of infantile cortical hyperostosis (Caffey's disease). AJR Am J Roentgenol 108:461–467, 1970.

Tien R, Barron BJ, Dhekne RD: Caffey's disease: Nuclear medicine and radiologic correlation: A case of mistaken identity. Clin Nucl Med 13:583–585, 1988.

Williams JL: Periosteal hyperostosis resulting from prostaglandin therapy. Eur J Pediatr 6:231–232, 1986.

CHAPTER 61

Osteomyelitis and Septic Arthritis in Infants and Children

Michel E. Azouz and Lori L. Barr

OVERVIEW

Hematogenous osteomyelitis is a fairly common disease in the pediatric age group. A large variety of organisms can cause osteomyelitis, by extension from infected contiguous soft tissues or via the bloodstream from a remote source. The definitive diagnosis is established after identification of the causative organism on blood or tissue culture. The most common infecting agent in children is *Staphylococcus aureus*. Hematogenous osteomyelitis typically involves the highly vascularized metaphyses of fast-growing bones, such as the distal femur, proximal tibia, proximal humerus, and distal radius. In tubular bones, hematogenous pyogenic osteomyelitis starts in the metaphysis, where the blood flow on the venous side of the capillary loops is slow. Organisms are trapped in the terminal loops forming microabscesses in the bone marrow followed by areas of local bone destruction, which gradually coalesce. The intraosseous pressure within the marrow space forces the infection to spread under the periosteum, forming subperiosteal abscesses. Rupture of the subperiosteal abscesses is responsible for extension of the infection into the adjacent soft tissues. Subperiosteal abscesses are more frequent and form more rapidly in infants.

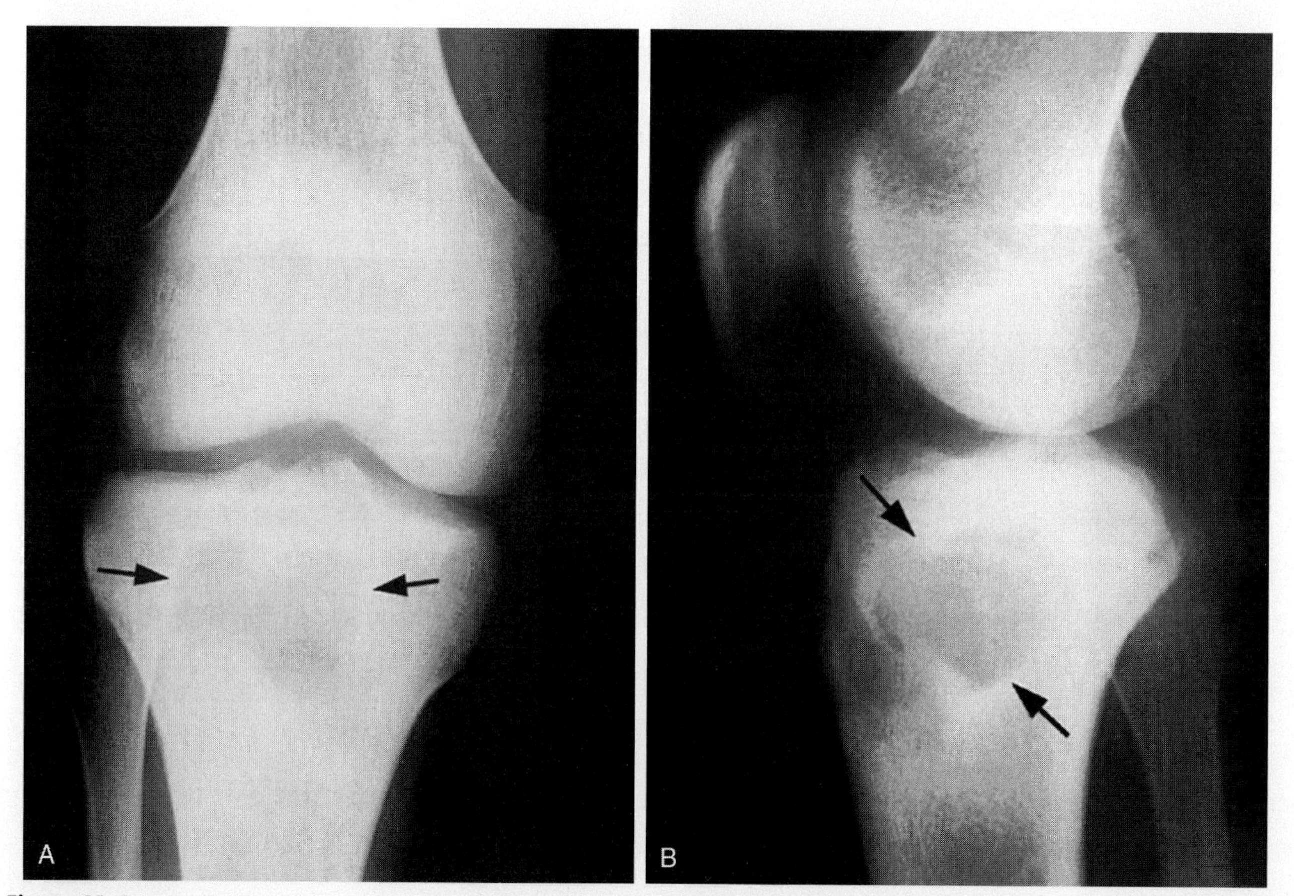

Figure 61–1. Bone (Brodie's) abscess in a 14-year-old girl with knee pain and joint effusion. Her erythrocyte sedimentation rate was high, but she had no fever or chills. *A* and *B*, Anteroposterior and lateral views of the right knee show a focal irregular lucency *(arrows)* in the proximal tibial metaphysis, which yielded pus on needle biopsy.

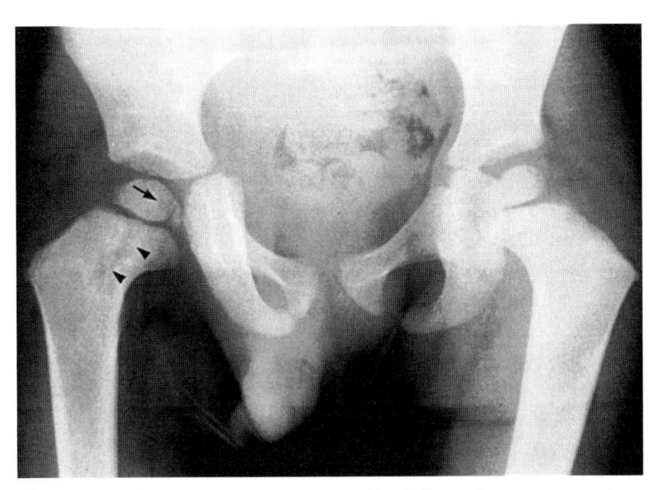

Figure 61–2. Anteroposterior view of both hips shows osteomyelitis spreading from the proximal metaphysis of the right femur *(arrowheads)* to the epiphysis *(arrow)* and right hip joint. Needle aspiration and culture of the right hip joint revealed septic arthritis.

Acute infections can spread throughout the medullary cavity of the affected tubular bone, especially in immunologically compromised children. If the infecting organism is of low virulence and the host resistance is intact, a localized focus of metaphyseal bone destruction persists as a sharply localized bone (Brodie's) abscess (Fig. 61–1). Contiguous spread of a metaphyseal infection into the epiphyses or into the adjacent joint occurs more frequently in infants younger than 16 months of age because normal metaphyseal blood vessels cross the growth plates of tubular bones and supply the epiphyses (Fig. 61–2). In children older than 16 months, there is no communication between the metaphyseal and epiphyseal vessels across the growth plate. Primary epiphyseal osteomyelitis is possible but is rare.

Late epiphysiometaphyseal changes may be seen in children years after a severe meningococcal sepsis because of occlusion of small blood vessels by microthrombi or septic emboli during the acute phase of the disease (Fig. 61–3). Gross deformities with irregular round or triangular-shaped epiphyses are seen, resulting in shortening, bowing, and limb-length discrepancy (Fig. 61–4). These deformities are due to the development of bony physeal bars, which cause growth arrest (see Fig. 61–4).

If an acute bone infection is not treated adequately or the host resistance is low, a subacute and chronic infection may follow the acute episode. At times, subacute bone infections occur de novo with no clinically identifiable acute phase.

In chronic osteomyelitis, abscess cavities, sequestra, and involucra develop (Fig. 61–5). A sequestrum is a devitalized or dead bone fragment surrounded by fluid or granulation tissue (or both) within a bone abscess or cavity. An involucrum is a newly formed sheath of periosteal bone deposited around infarcted bone (see Fig. 61–5). This complication is seen rarely nowadays because antibiotics are universally available.

A special form of chronic osteomyelitis called *chronic recurrent multifocal osteomyelitis* (CRMO) occurs in older children and adolescents. The cause is unknown. CRMO mainly involves metaphyseal sites, and it is destructive and productive (Fig. 61–6). The disease is characterized clinically by episodes of local swelling and pain over the affected bones, particularly in the tibia, fibula, and clavicle. It also may be associated with low-grade fever, leukocytosis, and elevated erythrocyte sedimentation rate. In the spine, CRMO can produce collapse of one or more vertebral bodies resembling vertebra plana, associated with Langerhans' cell histiocytosis. CRMO is a self-limited disease, and only supportive therapy is required for treatment.

Septic arthritis may occur at any age, but it is primarily a disease of infants and children. Infection can spread into the joint from a contiguous osteomyelitis (Fig. 61–7) or hematogenously from an infected lung or skin. Direct implantation of bacteria into a joint or bone also can occur (Fig. 61–8).

The hip is the most frequently affected joint, followed by the knee and shoulder. *S. aureus* is usually the responsible organism. In acute osteomyelitis and septic arthritis, early diagnosis and prompt treatment are crucial to avoid permanent sequelae. When the diagnosis and management of neonatal septic arthritis of the hip are delayed, severe capsular distention because of significant

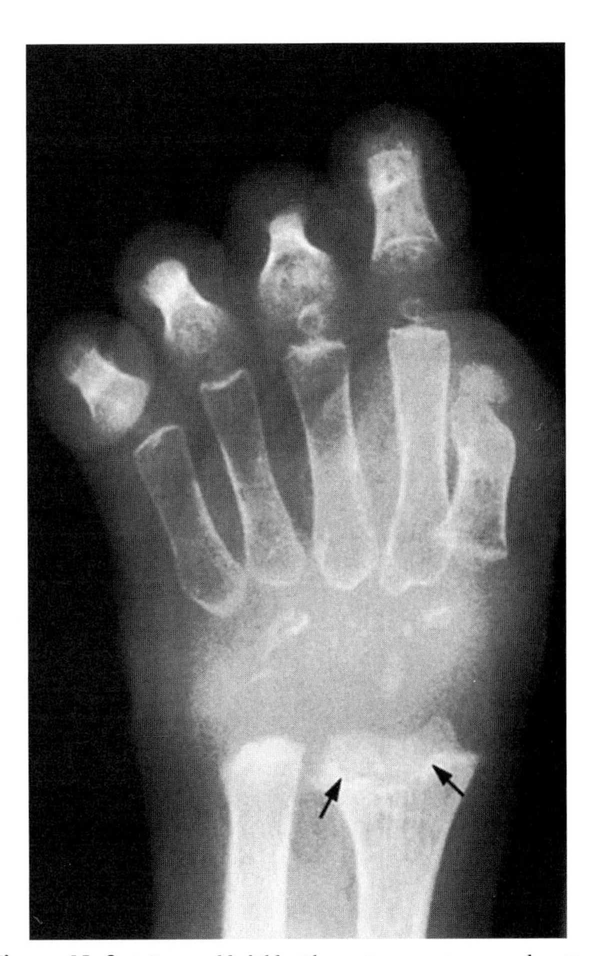

Figure 61–3. A 2-year-old child with previous meningococcal septicemia. Posteroanterior view of the right hand and wrist shows amputation of all the fingers because of vascular compromise. There is also premature closure of the distal radial growth plate *(arrows)*.

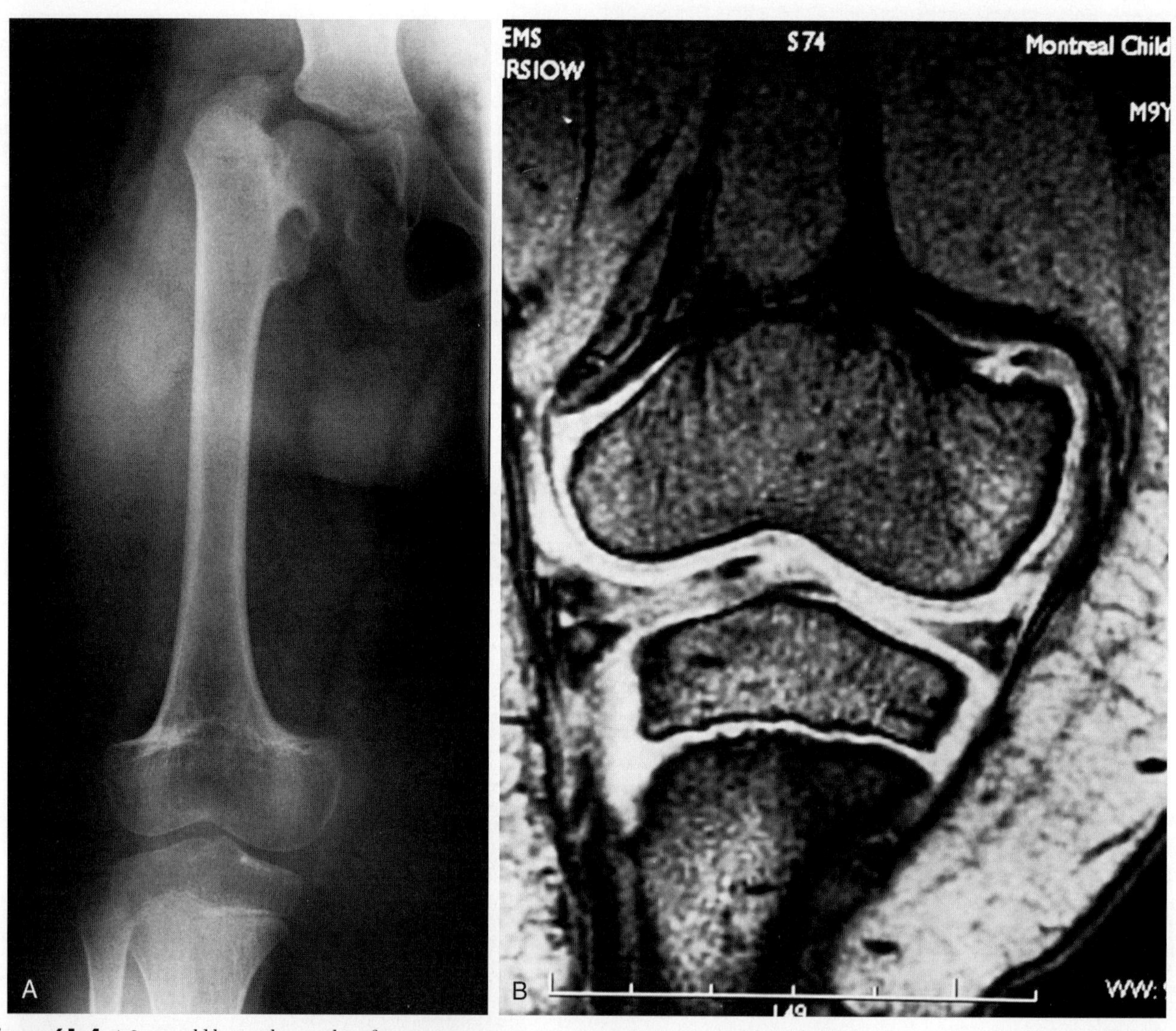

Figure 61–4. A 9-year-old boy with sequelae of meningococcemia. *A*, Anteroposterior view of the right femur shows premature closure of the growth plate in the proximal and distal femur. *B*, Coronal gradient echo sequence of the right knee outlines the large central growth plate fusion and the deformity in the distal femur.

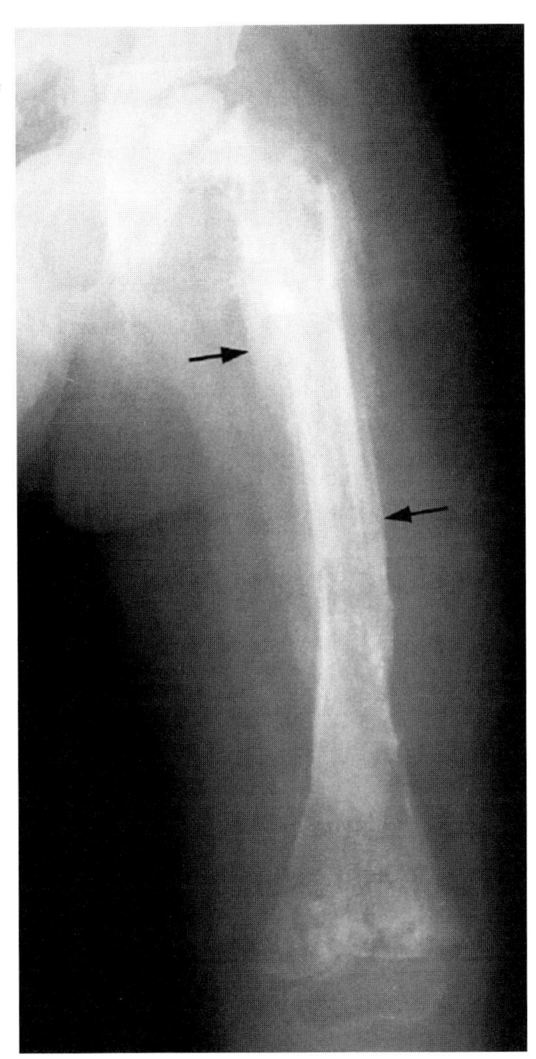

Figure 61–5. Chronic osteomyelitis in a 7-year-old child. Anteroposterior view of the left femur shows involvement of almost the entire femur except for the epiphyses at both ends. Involucrum *(arrows)* is seen surrounding the devitalized shaft (sequestrum).

accumulation of pus under pressure can result in dislocation of the femoral head. Similar changes also can be seen in the shoulder (Fig. 61–9).

IMAGING

Within the first 2 or 3 days after the onset of symptoms, radiographic studies reveal deep soft tissue swelling and edema with obliteration of the intermuscular fascial planes. Later, edema in the subcutaneous fat becomes detectable. Bone destruction and visible periosteal reaction are relatively late findings. They become visible on plain radiographs by 10 days to 2 weeks after the disease starts.

Bone and gallium scintigraphy are more sensitive than plain radiography for early diagnosis of osteomyelitis. The three-phase bone scan is sensitive and is usually positive within 1 or 2 days of the start of symptoms. The focus of bone infection appears as an area of increased tracer activity (Fig. 61–10). In growing children, it is often difficult to detect metaphyseal osteomyelitis in close proximity to the growth plate because both areas show increased tracer activity on the bone scan (see Fig. 61–10). Rarely a photopenic area is seen in the acute phase of osteomyelitis. This photopenic area is believed to be due to fulminant inflammation and rapid increase in the intraosseous pressure causing thrombosis of the vascular channels and acute ischemia.

Computed tomography (CT) is used mainly in chronic osteomyelitis. It is helpful in detecting cortical bone destruction, abscess cavities, and sequestra.

Magnetic resonance (MR) imaging is sensitive for the early diagnosis of osteomyelitis. It detects marrow changes, adjacent soft tissue infection, soft tissue abscesses, and joint effusion (see Fig. 61–10). Subperiosteal abscess formation is shown clearly by MR imaging and by ultrasonography, which is much less costly, is done more

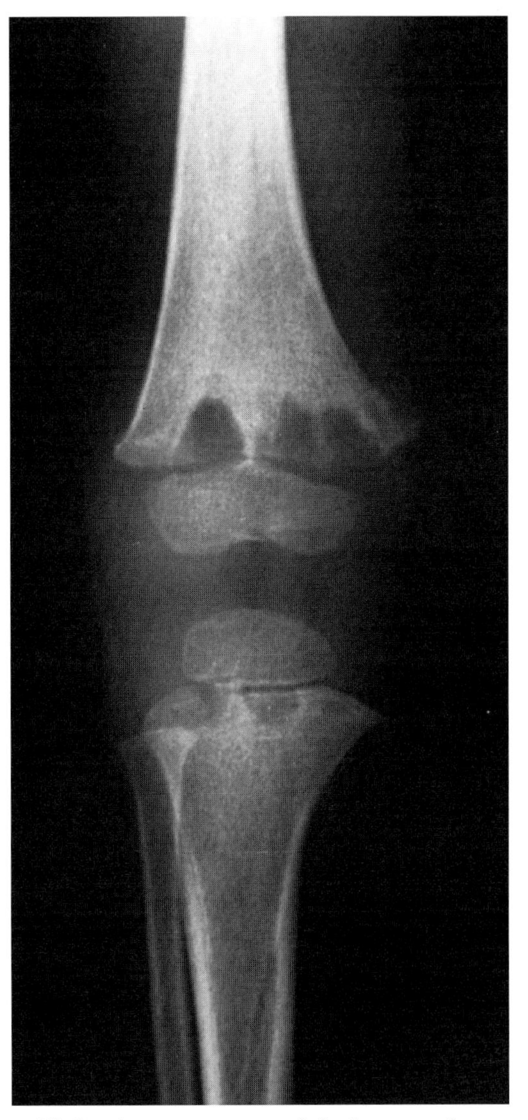

Figure 61–6. Chronic recurrent multifocal osteomyelitis in a 20-month-old boy. Anteroposterior view of the right knee reveals relatively well-defined metaphyseal lucencies. Similar but less extensive changes were seen in the left knee.

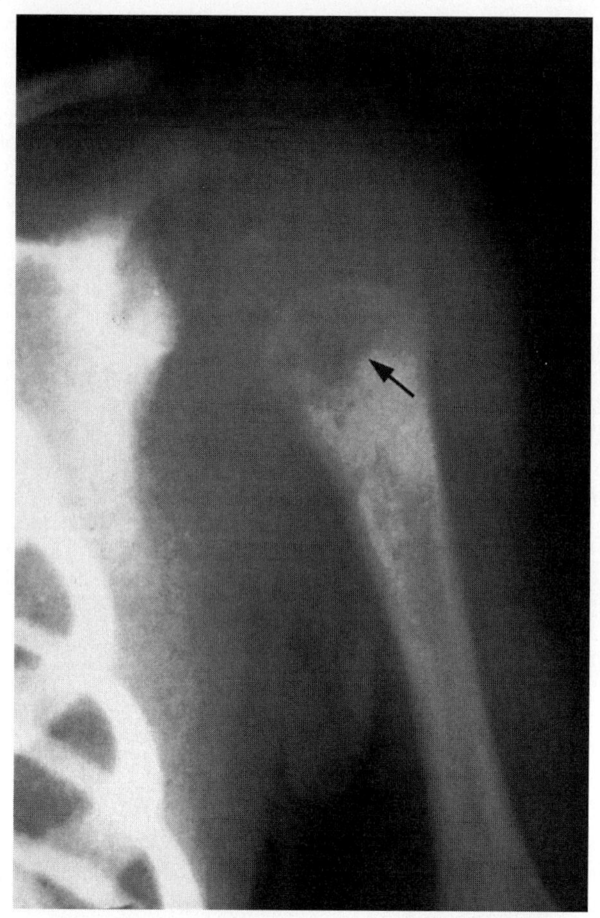

Figure 61–7. Septic arthritis of the left shoulder in a 3-month-old infant, which started as an osteomyelitis in the metaphysis *(arrow)*.

rapidly, and requires no sedation. Needle aspiration of the abscess can be done under ultrasound guidance.

In CRMO, the clavicle, tibia, fibula, and distal femur are affected most frequently (see Fig. 61–6). Affected bones eventually show sclerosis and expansion, and the lesions can be bilateral. Cavities and sequestra are not part of CRMO.

Partial or complete destruction of the growth plate resulting from ischemia or active infection causes growth disturbance with limb shortening and deformity as a result of bone bridging across the growth plate, also known as an *epiphyseal bar* (see Figs. 61–3 and 61–4). This deformity is usually well shown on plain radiographs but may require MR imaging for accurate mapping, especially if surgery is contemplated. Gradient-echo sequences at 15 degrees to 30 degrees flip angle in sagittal and coronal planes show the bony bridge across the cartilaginous plate (see Fig. 61–4*B*).

Brodie's abscess is seen as a well-defined, round or oval metaphyseal bone cavity (see Fig. 61–1), which may cross the growth plate into the epiphysis and which may contain one or two small dense sequestra. Periosteal reaction may be seen, especially in the subacute form or if there is reactivation of the infection (Fig. 61–11). The cavity often is surrounded by an irregular zone of bone sclerosis. Brodie's abscesses are well shown on plain radiographs and CT. MR imaging may be used to assess activity of the infection and extension into the epiphysis and adjacent soft tissues (see Fig. 61–11).

Fluid including pus in a joint is seen easily on ultrasonography. This modality should be used to help guide fluid aspiration for culture. Plain radiography in cases of suppurative arthritis shows deep soft tissue swelling, edema, and displacement of fat planes (see Fig. 61–9). Widening of the joint space may be seen, particularly in the hip and shoulder joint of neonates and infants, where an actual septic hip dislocation may occur (Fig. 61–12). Osteopenia or contiguous bone infection and partial or complete destruction of the femoral head may be seen eventually if treatment is delayed or is not adequate. Alternatively the femoral head may become enlarged as a result of chronic local hyperemia. MR imaging also is sensitive but not specific for the diagnosis of septic arthritis. On MR imaging, fluid is seen as low signal intensity on T1-weighted images and high signal on T2-weighted images and on inversion recovery. Synovial hyperemia is seen after gadolinium administration on

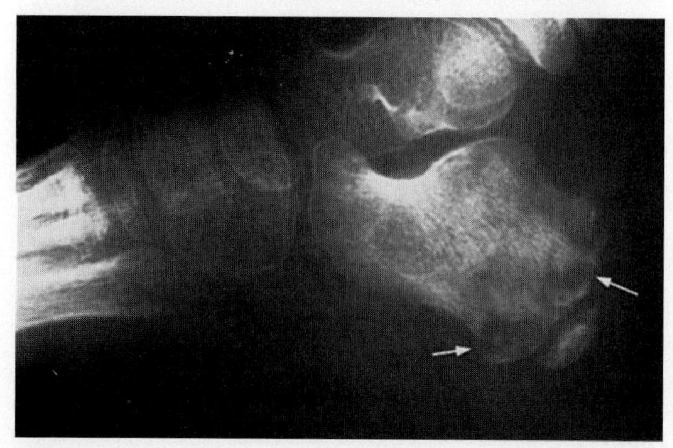

Figure 61–8. Osteomyelitis of the calcaneus in a 5-year-old boy who stepped on a nail. Lateral view of the foot shows areas of bone destruction *(arrows)* 3 weeks after the initial injury.

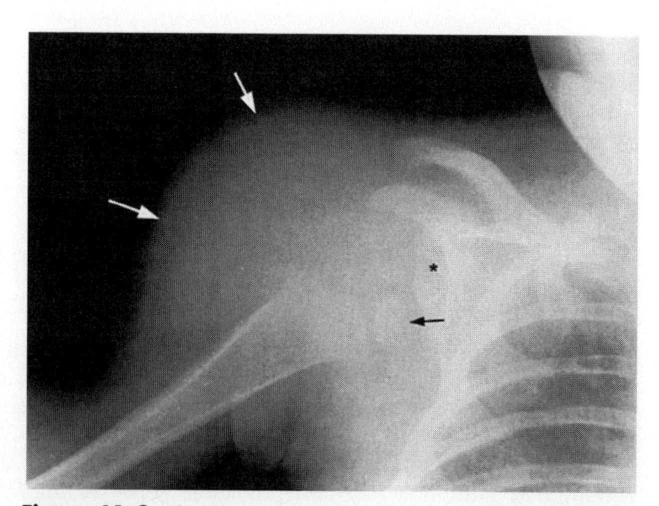

Figure 61–9. Shoulder dislocation in an infant with septic arthritis. Anteroposterior view of the shoulder shows a huge effusion and soft tissue swelling *(white arrows)*. Also noted is the inferior dislocation of the humeral head *(black arrow)*. An asterisk is placed at the center of the glenoid fossa.

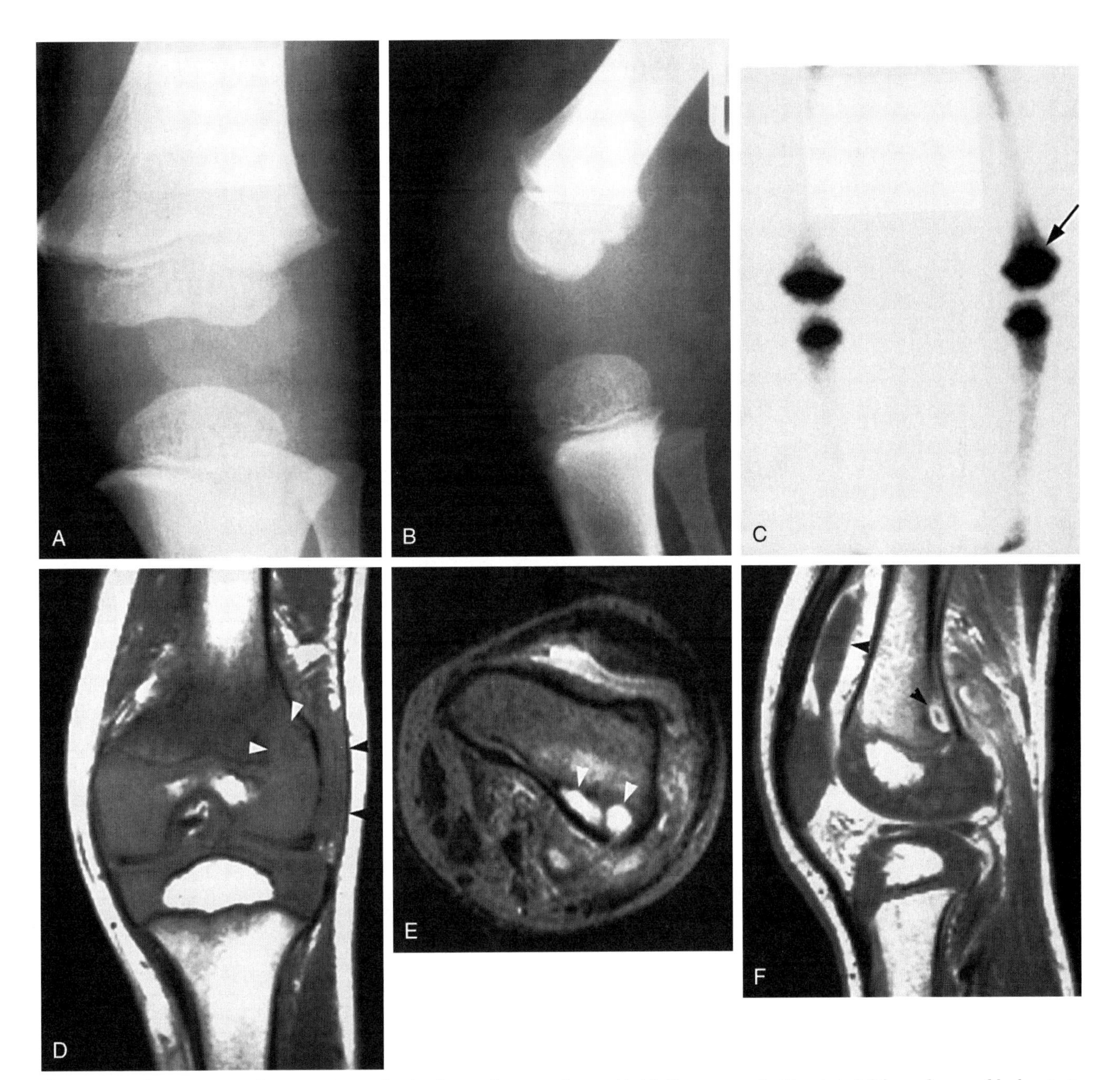

Figure 61–10. Small focus of osteomyelitis in the distal metaphysis in a 2-year-old girl. She presented with pain in left knee, fever, and leukocytosis. *A* and *B*, Anteroposterior and lateral views of the left knee are essentially unremarkable. *C*, Bone scan with technetium 99m medronate shows minimal increase in the radionuclide uptake in the distal metaphysis *(arrow)* of the left femur. *D*, Coronal T1-weighted image shows an abnormal focus of decreased signal intensity on the lateral side of the distal femoral metaphysis *(white arrowheads)*. The soft tissues adjacent to the lesion are edematous *(black arrowheads)*. *E* and *F*, Axial T2-weighted image and sagittal T1-weighted image after gadolinium enhancement clearly show the focus of osteomyelitis *(arrowheads)*. A large reactive joint effusion is also present *(top arrow* in *F)*.

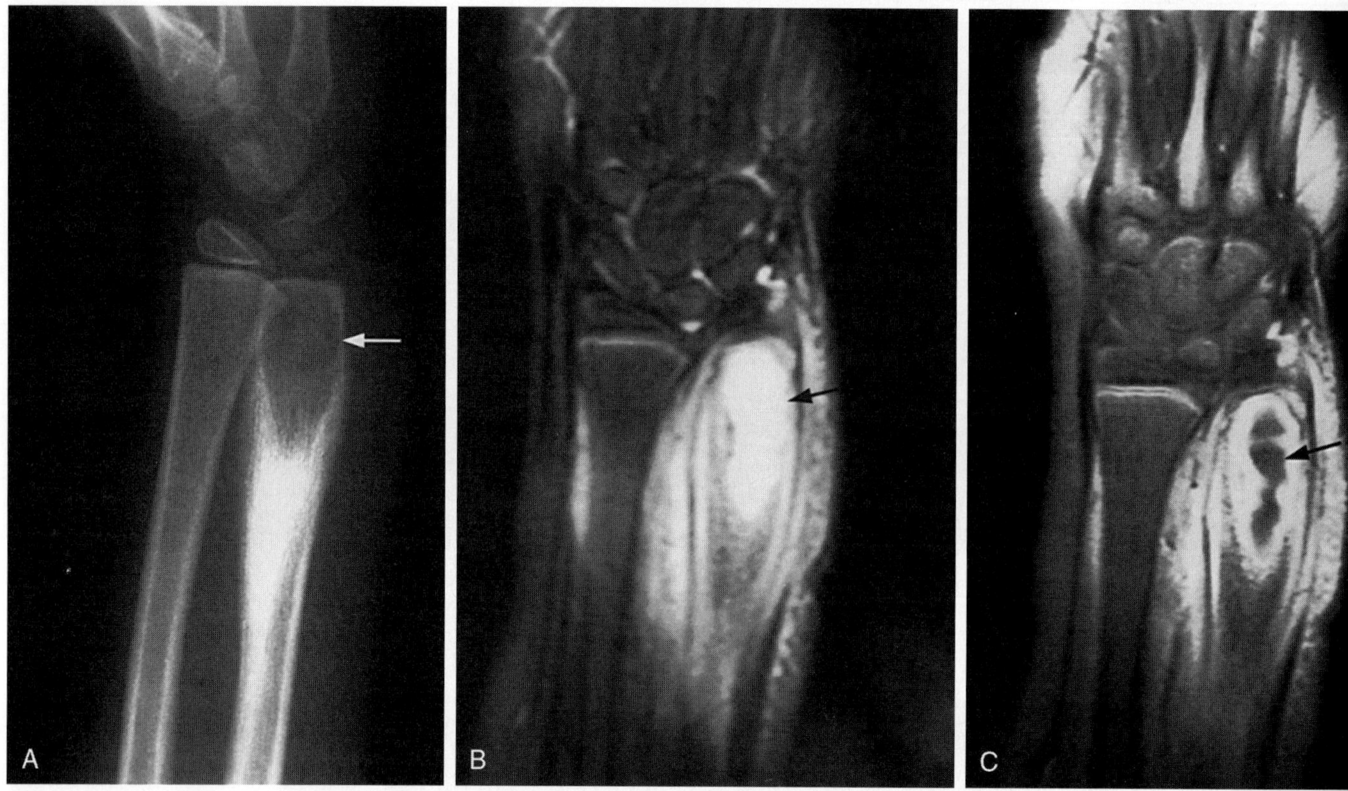

Figure 61–11. Subacute osteomyelitis with abscess formation in the distal ulna. *A*, Oblique view of the wrist shows a lytic expansile lesion in the metaphysis of the distal ulna *(arrow)*. There is thick periosteal reaction and bony sclerosis surrounding the lesion. *B* and *C*, Coronal T2-weighted image and T1-weighted image after gadolinium enhancement (both sequences are fat suppressed) show the abscess collection *(arrow)*, elevation of the periosteum, and soft tissue edema.

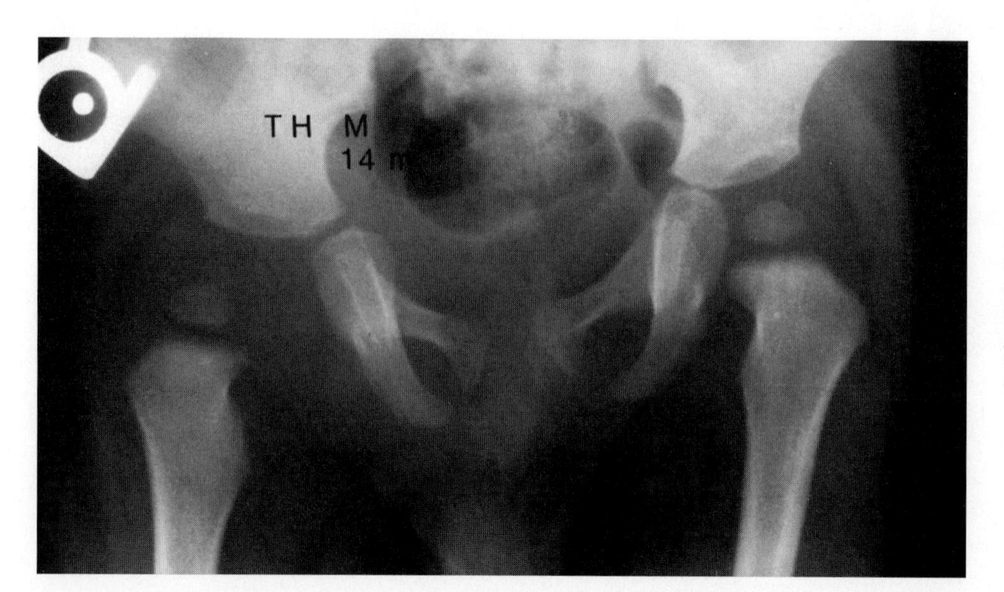

Figure 61–12. Septic arthritis of the right hip in a 14-month-old child. Anteroposterior view of the pelvis shows widening of the hip joint on the right resulting from capsular distention from pus accumulation under pressure.

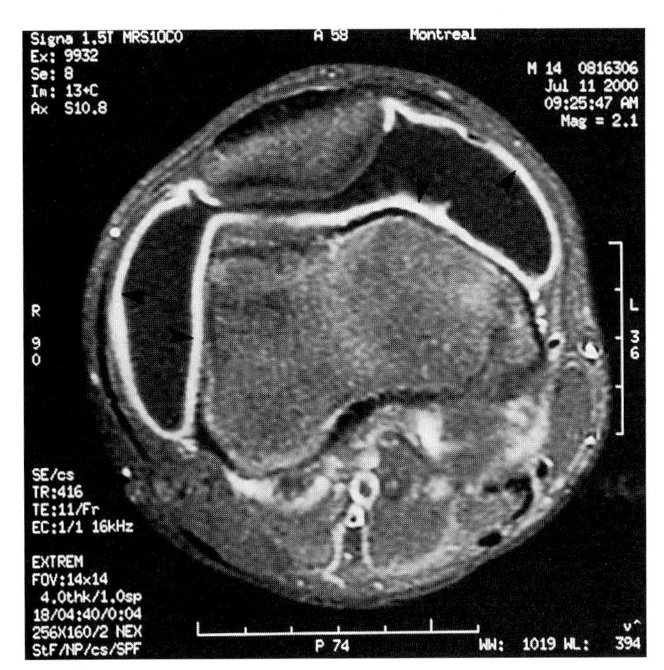

Figure 61–13. Enhancing synovium in a 14-year-old boy with severe septic arthritis. Axial T1-weighted image with fat suppression taken after gadolinium injection shows joint fluid surrounded by thick enhancing (bright) synovium *(arrowheads)*.

T1-weighted images, with fat suppression. The synovium is seen directly as a thin or thick irregular line of increased signal intensity (Fig. 61–13).

Recommended Readings

Azouz EM: Apparent or true neonatal hip dislocation? Radiologic differential diagnosis. Can Med Assoc J 29:595–597, 1983.

Azouz EM, Chhem RK, Lambert R, Oudjhane K: Imaging for bone and joint infection in children and adults. Curr Orthop 8:226–236, 1994.

Azouz EM, Greenspan A, Marton D: CT evaluation of primary epiphyseal bone abscesses. Skeletal Radiol 22:17–23, 1993.

Bureau NJ, Chhem RK, Cardinal E: Musculoskeletal infections: US manifestations. Radiographics 19:1585–1592, 1999.

Fernandez F, Pueyo I, Jimenez JR, et al: Epiphysiometaphyseal changes in children after severe meningococcic sepsis. AJR Am J Roentgenol 136:1236–1238, 1981.

Gylys-Morin V: MR imaging of pediatric musculoskeletal inflammatory and infectious disorders. Magn Reson Imaging Clin N Am 6:537–559, 1998.

Hoffer FA, Emans J: Percutaneous drainage of subperiosteal abscess: A potential treatment for osteomyelitis. Pediatr Radiol 26:879–881, 1996.

Jaramillo D, Treves ST, Kasser JR, et al: Osteomyelitis and septic arthritis in children: Appropriate use of imaging to guide treatment. AJR Am J Roentgenol 165:399–403, 1995.

Oudjhane K, Azouz EM: Imaging of osteomyelitis in children. Radiol Clin North Am 39:251–266, 2001.

Pöyhiä T, Azouz EM: MR Imaging evaluation of subacute and chronic bone abscesses in children. Pediatr Radiol 30:763–768, 2000.

Santos E, Boavida JE, Barroso A, et al: Late osteoarticular lesions following meningococcemia with disseminated intravascular coagulation. Pediatr Radiol 19:199–202, 1989.

Staalman CR, Tania BH: Neonatal osteomyelitis. J Belge Radiol 67:7–11, 1984.

Turpin S, Lambert R: Role of scintigraphy in musculoskeletal and spinal infections. Radiol Clin North Am 39:169–189, 2001.

Zi Yin E, Frush DP, Donnelly LF, Buckley RH: Primary immunodeficiency disorders in pediatric patients: Clinical features and imaging findings. AJR Am J Roentgenol 176:1541–1552, 2001.

CHAPTER 62

Nonaccidental Trauma (Battered Child Syndrome)

Michel E. Azouz, MD, and Lori L. Barr, MD

OVERVIEW

Since the 1970s, there has been increased awareness and concern regarding the ethical, social, and medical problems related to physical abuse as a cause of injury to infants and children. The exact frequency of child abuse and neglect is unknown. Cases that are detected and investigated represent the tip of the iceberg; most victims are unrecognized.

Nonaccidental trauma initially was described by Caffey (1972) and Silverman (1972) and recognized as a major cause of morbidity and mortality in infants and children. Florid cases are relatively easy to recognize, but mild cases may pose significant diagnostic difficulty, resulting in increased frequency of recurrence of the abuse. Of battered children, 5% eventually die of child abuse if they are returned to their original environment. Battered children tend to be younger than 3 years old; of either sex; and of no particular social, economic, or ethnic background.

Soft tissue injuries, such as bruises, hematomas, and burns, may be present on the face, buttocks, and extremities with or without fractures. The presence of old and new injuries in various stages of healing indicates repeated abuse over time.

Other injuries that may be present include trauma to the head and central nervous system, including subdural hematomas (Fig. 62–1) with or without an associated skull fracture, retinal hemorrhage, and papilledema. Trauma to the chest can result in rib fractures, pneumothorax, hemothorax, and subcutaneous emphysema. Abdominal injuries may cause bruising, hematoma, or ruptured viscera, particularly the liver and pancreas. Intracranial and abdominal injuries can be fatal.

The responsibility of the child abuse team includes diagnosing abuse, providing treatment for the physical and psychological injuries sustained by the child, and dealing with the family to avoid recurrence of child abuse. Great care should be taken to avoid false accusations and to establish the correct diagnosis because many clinical

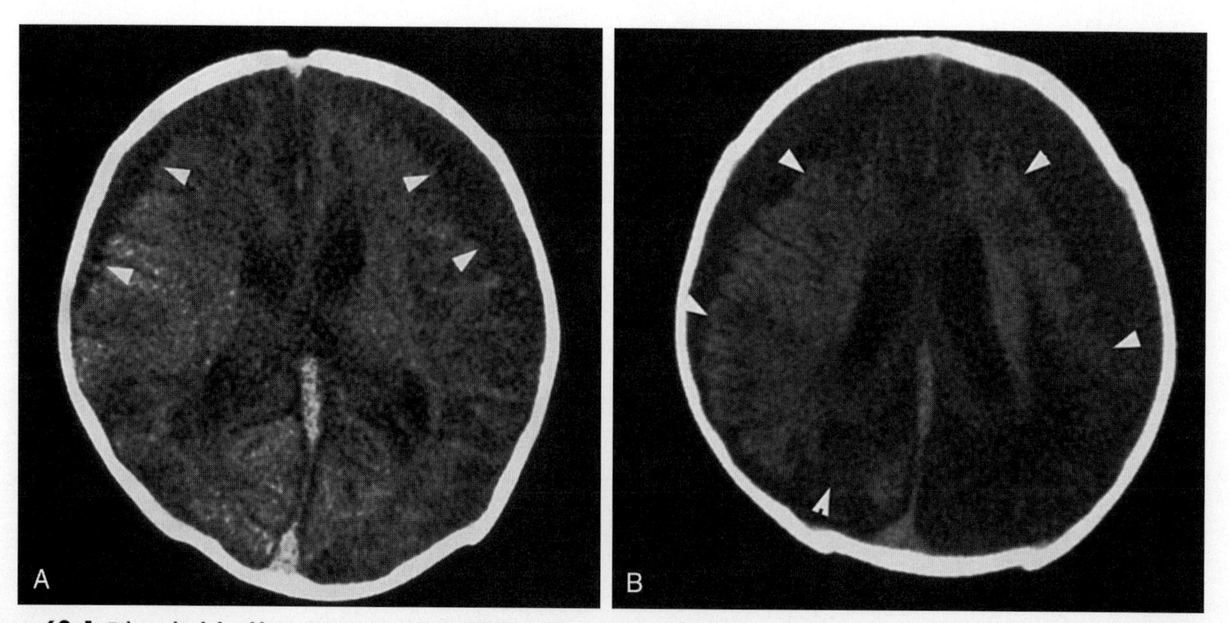

Figure 62–1. Bilateral subdural hemotomas in an abused child. *A,* Head CT scan shows bilateral frontal subdural hematomas *(arrowheads). B,* Repeat CT scan performed 4 weeks later shows brain atrophy with increase in the subdural fluid *(arrowheads).*

conditions may present with fractures or periosteal reaction in infants and children, including birth injuries, myelomeningocele, osteogenesis imperfecta, Caffey's disease, osteomyelitis, congenital syphilis, kinky hair syndrome, and scurvy. An infant or child with any of these entities (e.g., osteogenesis imperfecta) also may be the victim of abuse and neglect.

IMAGING

On plain radiographs, a single or multiple fractures may be seen in different stages of healing with callus and periosteal new bone formation. Growth plate injuries and epiphyseal separations can be seen. The rotatory twisting forces exerted on an extremity are transmitted to ligamentous and periosteal attachments on the metaphysis. This mechanism results in the characteristic metaphyseal corner and bucket-handle fractures of child abuse (Figs. 62–2 and 62–3). Alternatively, such forces can cause spiral diaphyseal fractures of the femur and humerus or may lift the periosteum, resulting in a subperiosteal hematoma, which eventually calcifies (Fig. 62–4). A diaphyseal fracture in an infant who does not walk is suspicious for child abuse. Because of delay in seeking medical help, diaphyseal fractures that are not immobilized can show exuberant callus formation.

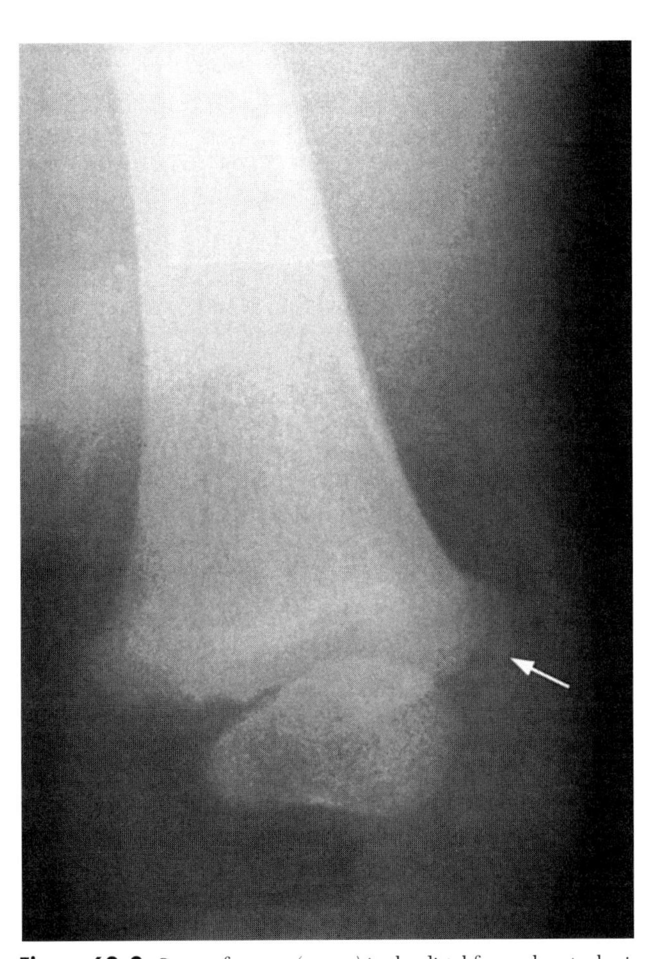

Figure 62–2. Corner fracture *(arrow)* in the distal femoral metaphysis seen on an anteroposterior view of the femur.

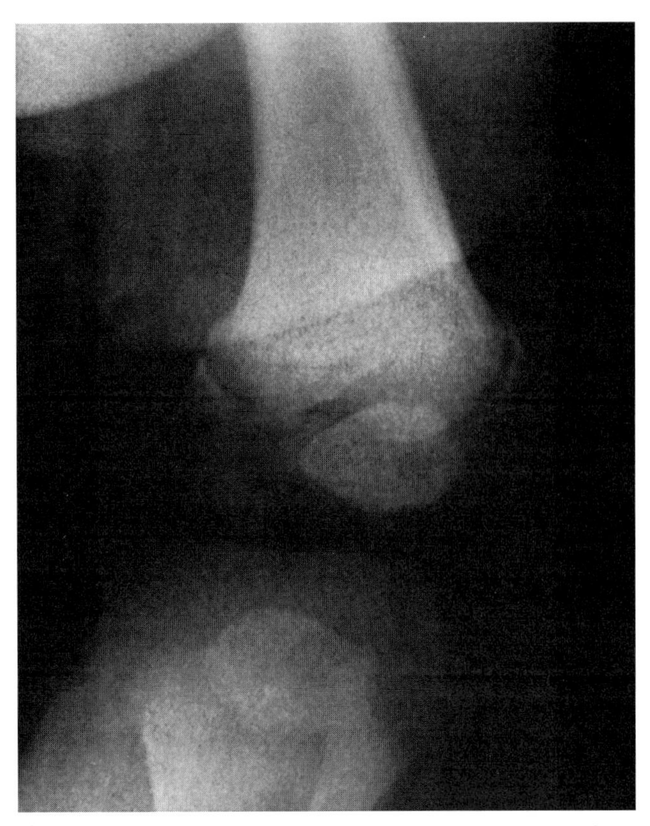

Figure 62–3. Bucket-handle fracture in the distal femoral metaphysis seen on an anteroposterior view of the knee.

In infancy, skull, rib (Fig. 62–5), and metaphyseal fractures predominate. Injuries specific for physical abuse include the classic metaphyseal lesions; posterior rib fractures; and other fractures for which a cause is difficult to explain, such as fractures of the sternum, scapula, and spinous processes. These fractures "speak for the child who cannot speak for himself."

A high-quality skeletal survey is required in suspected cases for screening, diagnosis, and dating of the fractures. A radionuclide bone scan may be used for screening suspected cases of child abuse. A bone scan is more sensitive, and it can detect injuries missed on plain radiographs. In contrast to the skeletal survey, however, a radionuclide bone scan is not readily available, and sedation is required. Fractures near the growth plates may be missed because fractures and growth plates give an increased uptake of the radiotracer in the same region. In general, scintigraphy is used to complement the skeletal survey. Computed tomography (CT) is used frequently to detect subdural hemotomas and subsequent brain atrophy (see Fig. 62–1). Magnetic resonance (MR) imaging may be needed to assess the nature and extent of soft tissue injuries, including subperiosteal hemorrhage injuries complicated by infection and epiphyseal fractures with displacement in infants.

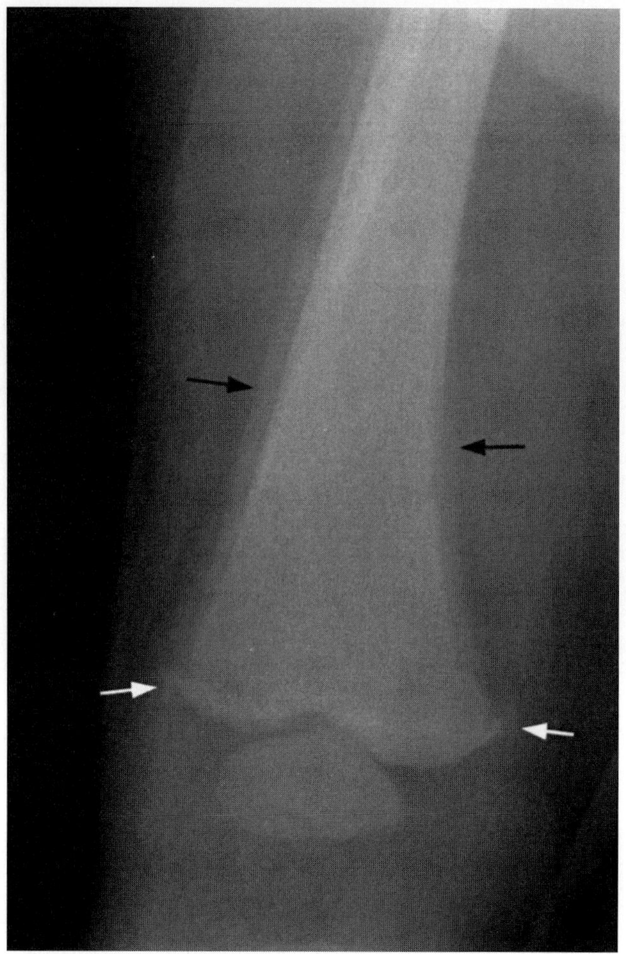

Figure 62–4. Periosteal elevation in an abused child *(black arrows)*. Note the small corner fractures in the distal metaphysis of the femur *(white arrows)*.

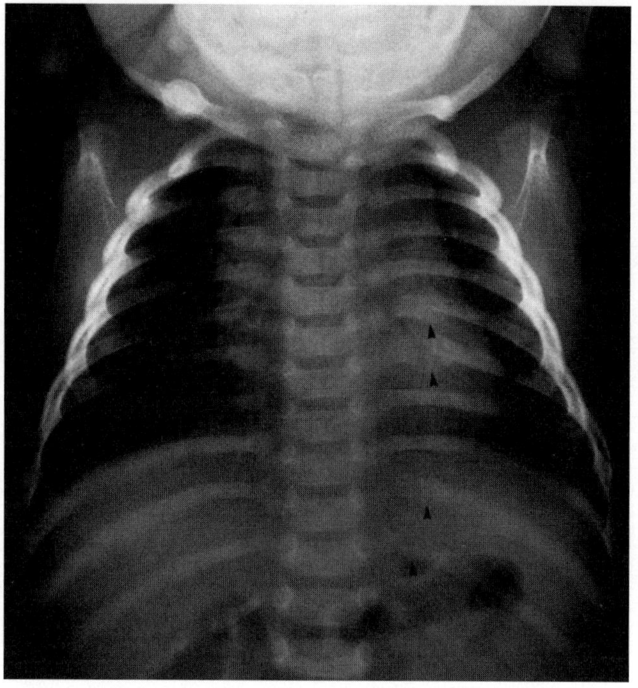

Figure 62–5. Multiple posterior rib fractures *(arrowheads)* in a 2-month-old boy who is known to have been abused.

Recommended Readings

Ablin DS, Greenspan A, Reinhart M, Grix A: Differentiation of child abuse from osteogenesis imperfecta. AJR Am J Roentgenol 154:1035–1046, 1990.

Caffey J: On the theory and practice of shaking infants: Its potential residual effects of permanent brain damage and mental retardation. Am J Dis Child 124:161–169, 1972.

Galleno H, Oppenheim WL: The battered child syndrome revisited. Clin Orthop 162:11–19, 1982.

Kleinman PK: Diagnostic imaging of child abuse. AJR Am J Roentgenol 155:703–712, 1990.

Kleinman PK (ed): Diagnostic Imaging of Child Abuse, 2nd ed. St. Louis, Mosby, 1998.

Merten DF, Radkowski MA, Leonidas JC: The abused child: A radiological reappraisal. Radiology 146:377–381, 1983.

Nimkin K, Kleinman PK: Imaging of child abuse. Radiol Clin North Am 39:843–864, 2001.

Silverman FN: Unrecognized trauma in infants, the battered child syndrome, and the syndrome of ambroise tardieu. Radiology 104:337–353, 1972.

Sty JR, Starshak RJ: The role of bone scintigraphy in the evaluation of the suspected abused child. Radiology 146:369–375, 1983.

CHAPTER 63

Neurofibromatosis and Congenital Pseudarthrosis

Michel E. Azouz, MD, and Lori L. Barr, MD

OVERVIEW

Neurofibromatosis is the most common of the neurocutaneous syndromes. Its clinical expression is highly variable. Bone lesions associated with neurofibromatosis include kyphosis and scoliosis, posterior scalloping of vertebral bodies, enlarged neural foramina, and dysplasia of the sphenoid with partial or complete absence of the greater or lesser wing. Other localized abnormalities include twisted ribbon-like ribs, hypertrophy of an extremity (Fig. 63–1), congenital bowing, pseudarthrosis of the tibia and fibula (or rarely the forearm bones), bone deformity caused by pressure erosions from adjacent plexiform neurofibroma or mesenchymal dysplasia of individual bones, and multiple large nonossifying fibromas. Tubular bone overgrowth rarely may be associated with subperiosteal hemorrhage after minor trauma, owing to loose attachment of the periosteum.

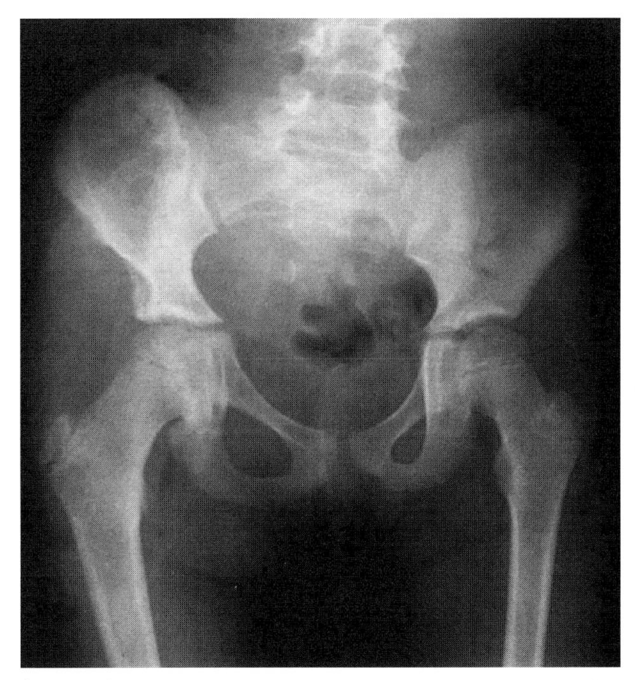

Figure 63–1. Hemihypertrophy in a 7-year-old girl with neurofibromatosis. Anteroposterior view of the pelvis shows hypertrophied right hemipelvis, right hip, and right femur.

In congenital pseudarthrosis, the tibia usually is affected, although the ipsilateral fibula also can be involved (Fig. 63–2), and occasionally it may precede the tibia in pseudarthrosis formation. In the upper extremity, either the radius or the ulna (Fig. 63–3) or rarely both bones of the same forearm may be affected. Pseudarthrosis of the tibia can be idiopathic (e.g., without neurofibromatosis); only about 50% of patients with pseudarthrosis have neurofibromatosis.

Although termed *congenital,* pseudarthrosis is seen rarely at birth. It typically occurs later in infancy or early childhood. A fracture occurs after relatively minor trauma. The fracture does not heal, and nonunion associated with bone dysplasia becomes evident. Surgical correction remains one of the most difficult orthopedic problems. The cause of pseudarthrosis of the tibia is believed to be a mesodermal defect affecting the distal tibia and surrounding soft tissues.

IMAGING

Plain radiography is the modality of choice for the diagnosis and follow-up of congenital pseudarthrosis of the tibia. In the prefracture stage, the affected tibia shows anterior or anterolateral bowing, sclerosis, medullary canal narrowing, and distal diaphyseal constriction. One or more localized cystic lesions may be seen in the distal diaphysis (Fig. 63–4). The prefracture stage eventually leads to fracture and pseudarthrosis. Right and left tibiae are involved equally. Bone sclerosis, resorption and tapering of adjacent fracture fragments, and shortening may occur, especially with repeated recurrences after osteotomy and bone grafting. Cystic lesions may persist (Fig. 63–5). Magnetic resonance (MR) imaging allows assessment of the periosteal and soft tissue changes near the pseudarthrosis.

Recommended Readings

Harlow CL, Kilcoyne RF, Aeling J, Becker MH: Skin and bones: Dermatologic conditions with skeletal abnormalities. Skeletal Radiol 26:201–213, 1997.

Mahnken AH, Staatz G, Hermanns B, et al: Congenital pseudarthrosis of the tibia in pediatric patients: MR imaging. AJR Am J Roentgenol 177:1025–1029, 2001.

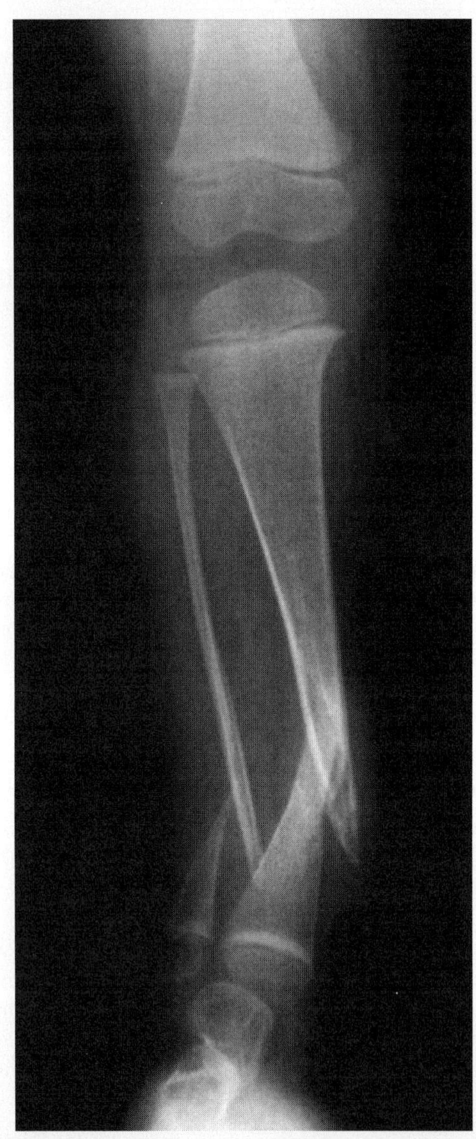

Figure 63–2. A 1-year-old girl with neurofibromatosis and congenital pseudarthrosis of the right tibia and fibula. Anteroposterior view of the right leg shows the pseudarthrosis in the distal diaphysis of the tibia and fibula. There is tapering, overriding, and angulation of the fragments.

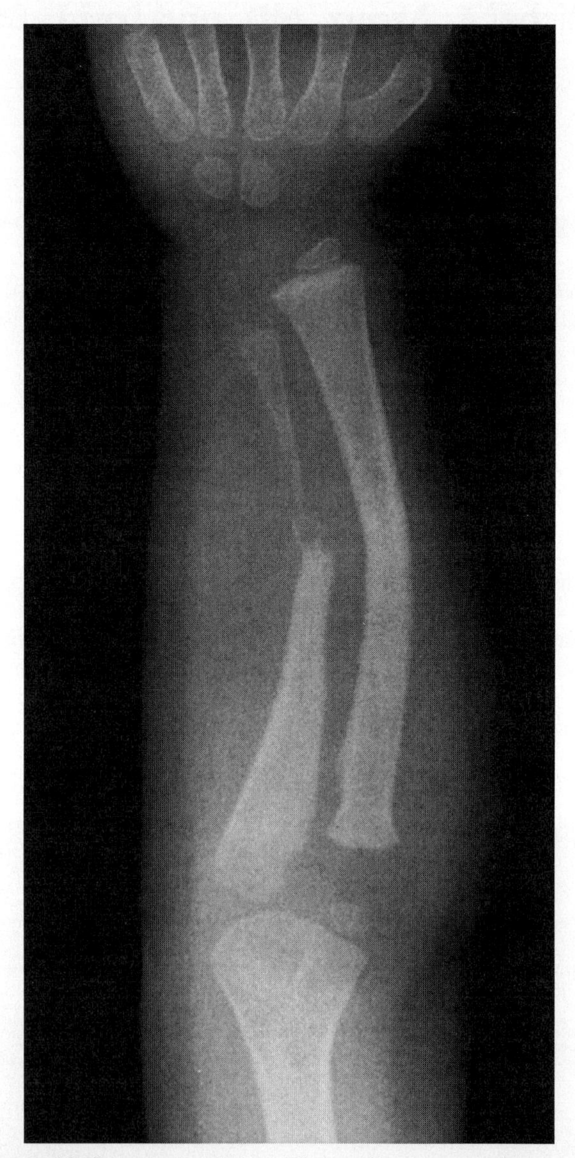

Figure 63–3. A 2.5-year-old boy with neurofibromatosis and congenital pseudarthrosis of the left ulna. Anteroposterior radiograph of the forearm shows interruption of the continuity of the ulna, with significant tapering, and underdevelopment of the distal ulnar fragment. There also is mild lateral bowing of the radius.

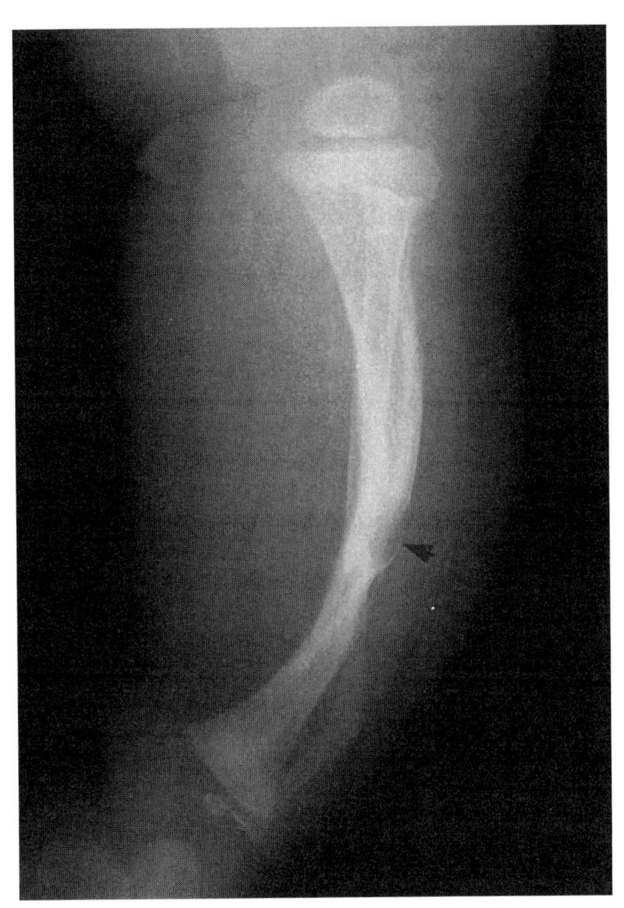

Figure 63–4. A cystic lesion *(arrow)* in the tibia at the prefracture stage. The distal fibula is dysplastic and bowed.

Manske PR: Forearm pseudarthrosis—neurofibromatosis. Clin Orthop 139:125–127, 1979.

Murray HH, Lovell WW: Congenital pseudarthrosis of the tibia: A long-term follow-up study. Clin Orthop 166:14–20, 1982.

Roach JW, Shindell R, Green NE: Late-onset pseudarthrosis of the dysplastic tibia. J Bone Joint Surg Am 75:1593–1601, 1993.

Sprague BL, Brown GA: Congenital pseudarthrosis of the radius. J Bone Joint Surg Am 56:191–194, 1974.

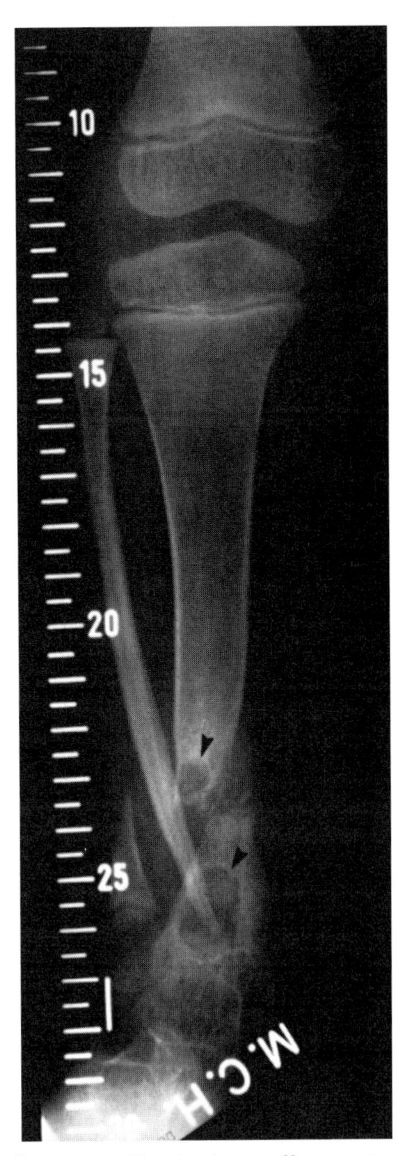

Figure 63–5. A 3-year-old girl with neurofibromatosis and congenital pseudarthrosis of the right tibia and fibula. After multiple surgeries, the tibia became short, and well-defined cystic areas are seen *(arrowheads)* on this anteroposterior view of the lower leg taken with a ruler.

SECTION VI

MR IMAGING STUDIES OF JOINTS, TENDONS, LIGAMENTS, AND MUSCLES

CHAPTER 64

Knee

Carol A. Boles, MD, and Georges Y. El-Khoury, MD

OVERVIEW

The knee is the largest and most complex joint in the body. It is also a commonly injured joint. Larson (1983) characterized the knee, more than any other joint, as the one that requires the normal functioning of all its parts to provide harmony of motion, security of stability, and protection against degenerative changes. Trauma to the knee can result in injury to the bone, articular cartilage, menisci, ligaments, and tendons.

BONE INJURIES

Bone Contusions (Bruises)

Bone contusions represent trabecular microfractures with associated medullary hemorrhage and edema. They are not detected on plain radiography, and only about one third are seen at arthroscopy.

Bone contusions result from a violent collision between two bones or when a bone is struck by a hard object during an accident or a sport activity. Contusions produced by compressive forces are much larger than those produced by traction (Figs. 64–1 and 64–8). Most bone contusions have a benign course and resolve spontaneously within 6 weeks to 3 months. A small percentage of bone contusions occur in the subchondral region, have a well-defined wedge or hemispheric appearance, and are broad based along the articular surface (Fig. 64–1). These are more serious because a high percentage show eventual osteochondral injury (Fig. 64–2). The long-term effects of bone contusions on articular cartilage have not been studied by magnetic resonance (MR) imaging; however, experimental studies showed chondrocyte and cartilage matrix abnormalities at the site of the impact. On MR imaging, most bone contusions have either a geographic or a reticular pattern showing low signal intensity on T1-weighted images and high signal intensity on T2-weighted images (Fig. 64–3) reflecting capillary leakage of fluid and blood into the extracellular marrow spaces. Bone contusions may be difficult to see on T2-weighted fast spin echo sequences because of the high signal of marrow fat obscuring the marrow edema. The addition of fat suppression greatly improves marrow edema detection. With intravenous gadolinium administration, contusions show no or minimal, often delayed enhancement.

Diagnostically, bone contusions are helpful in predicting the mechanism of injury. *Kissing contusions* involving the posterior aspect of the lateral tibial plateau and the subchondral region of the lateral femoral condyle, at about its midportion, are fairly specific for an anterior cruciate ligament (ACL) tear (Fig. 64–4). In children and adolescents, this same pattern of bone contusions can be seen without a tear of the ACL, resulting from the laxity of ligaments at this age. The ACL disrupts when the foot is planted with the tibia fixed, while the lateral femoral condyle rotates posteriorly on the lateral tibial plateau impacting its posterior edge. This mechanism occasionally can be reproduced on physical examination in patients with a deficient ACL. It is referred to as the *pivot shift* sign.

Medial collateral ligament (MCL) disruption produced by valgus stress often is associated with contusions of the lateral femoral condyle and lateral tibial plateau (Fig. 64–5). Transient patellar dislocation also is associated with specific bone contusions, which lead to the diagnosis even when the patella already is reduced.

Bone contusions can direct attention to occult fractures or osteochondral injuries. Fractures can resemble bone contusions on MR imaging except that they are associated with cortical interruption and show fracture lines. Most occult fractures occur in the lateral compartment or involve the medial facet of the patella. These are serious injuries that should be distinguished from bone contusions because the recovery time is longer, and prognosis is poorer.

Bony Avulsions Associated with Internal Knee Derangement

There are key avulsion injuries in the knee that should direct attention to internal knee derangement and instability. A Segond fracture is a cortical avulsion of the lateral tibia at the insertion of the middle section of the lateral capsular ligament (Fig. 64–6). The significance of this fracture is related to its high association with ACL and meniscus tears. Tears of the ACL have been reported in 75% to 100% of patients with a Segond fracture. Medial and lateral meniscus injuries also have been reported in 66% to 70% of patients with this injury.

Segond fractures are easily missed on MR imaging, but they are readily identifiable on anteroposterior or tunnel radiographs of the knee. Small Segond fractures can be detected on computed tomography (CT) (Fig. 64–7). The

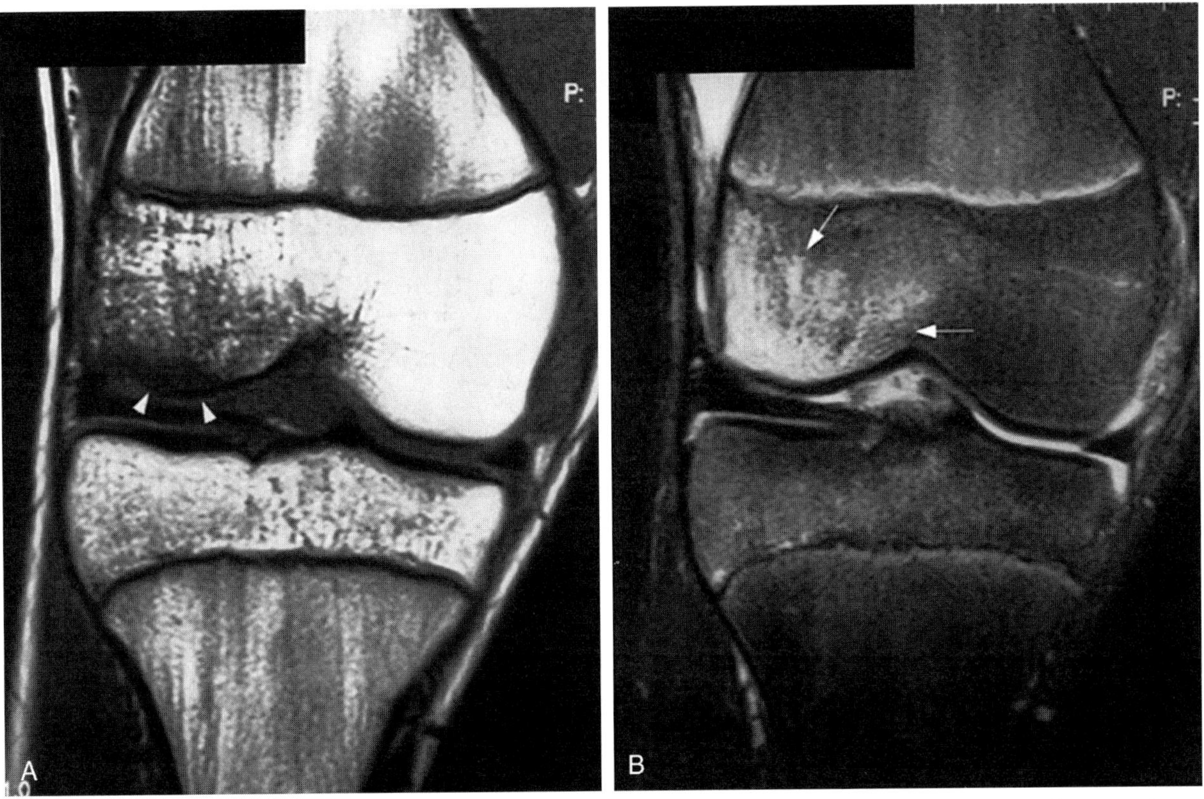

Figure 64–1. Subchondral marrow contusion. *A,* T1-weighted coronal MR image shows an area of markedly decreased signal intensity (geographic pattern) in the subchondral region of the lateral femoral condyle *(arrowheads)* surrounded by a larger area of decreased signal intensity (reticular pattern) representing a marrow contusion. *B,* T2-weighted coronal MR image with fat suppression shows increased signal intensity in the entire region of the contusion *(arrows).*

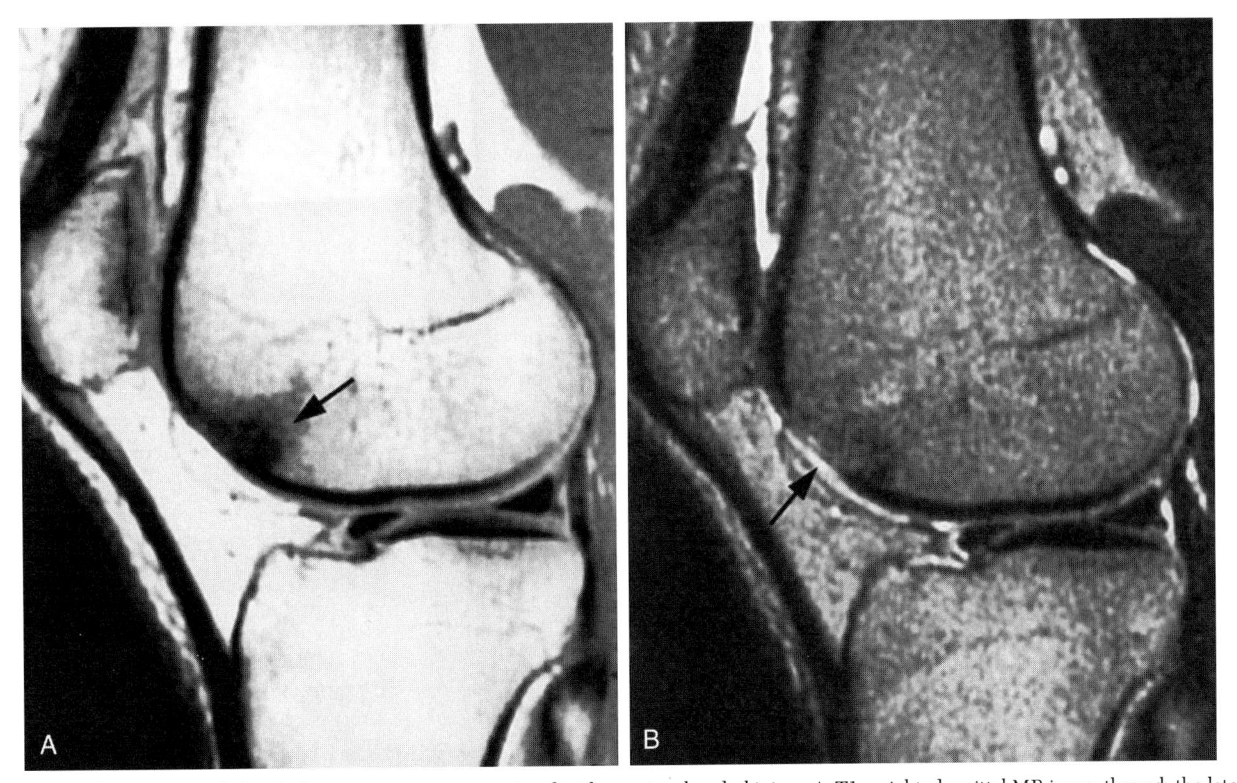

Figure 64–2. Subacute subchondral marrow contusion associated with an osteochondral injury. *A,* T1-weighted sagittal MR image through the lateral femoral condyle shows the low signal region *(arrow). B,* T2-weighted sagittal MR image reveals a large area of osteochondral damage *(arrow).* Part of the contusion remains low signal on T2 weighting, a finding that often correlates with sclerosis on plain radiographs.

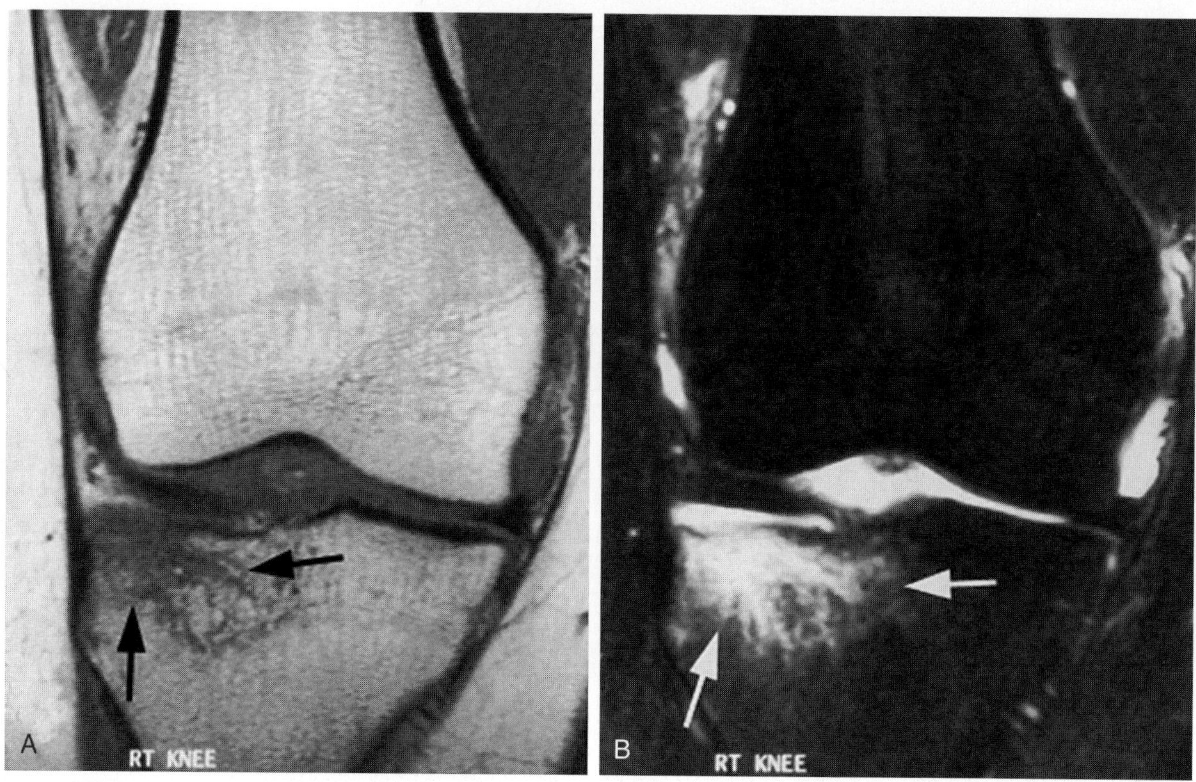

Figure 64–3. Marrow contusion with reticular pattern in the lateral tibial condyle. *A* and *B*, T1-weighted coronal MR image and T2-weighted coronal image with fat suppression show the reticular pattern of the marrow contusion *(arrows)*, best shown on the T1-weighted image.

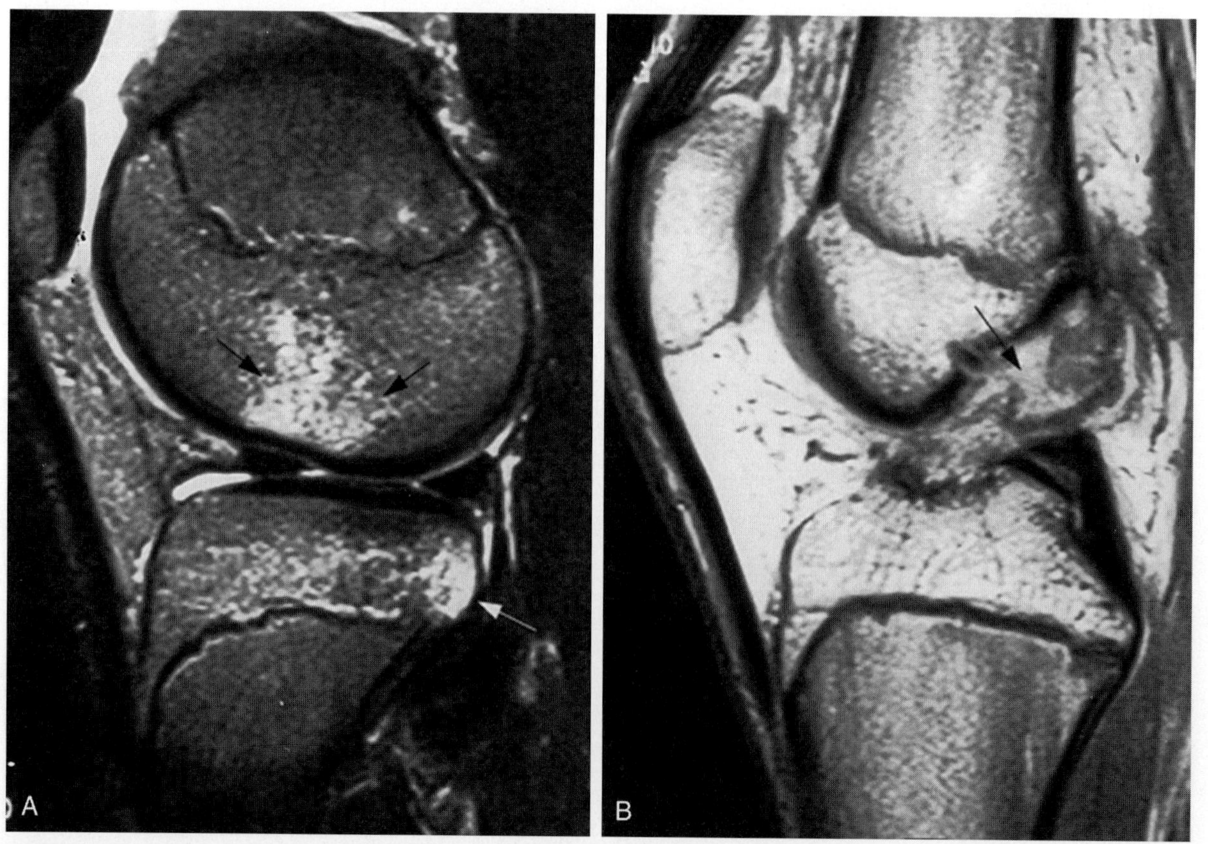

Figure 64–4. *Kissing contusions* suggestive of an anterior cruciate ligament tear. *A*, Sagittal T2-weighted MR image with fat suppression shows increased signal in the lateral femoral condyle *(black arrows)* and lateral tibial plateau posteriorly *(white arrow)*. *B*, Sagittal proton-density MR image shows the disrupted anterior cruciate ligament *(arrow)*.

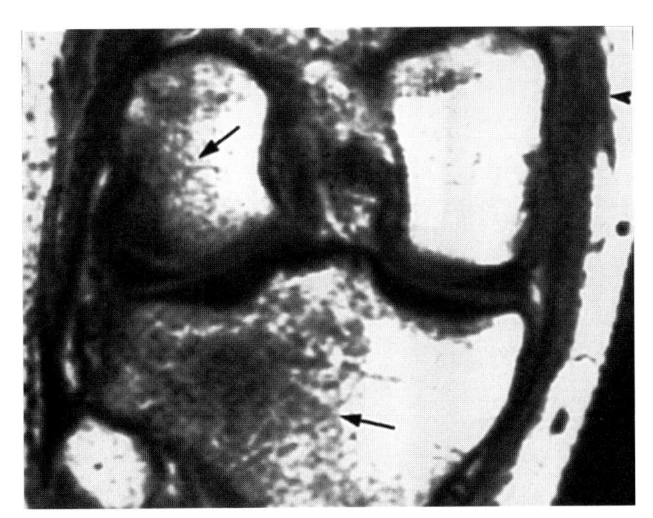

Figure 64–5. T1-weighted coronal MR image of marrow contusion of the lateral femoral condyle and lateral tibial plateau *(arrows)* and medial collateral ligament sprain *(arrowhead)* produced by valgus stress.

avulsed fragment measures 4 to 27 mm, and it originates 2 to 10 mm below the rim of the lateral tibial plateau (see Figs. 64–6 and 64–7).

Avulsion of the intercondylar eminence of the tibia is functionally similar to an ACL tear (Fig. 64–8; see Fig. 64–7). The ACL inserts on the tibia at a wide depressed area just anterior and medial to the intercondyle eminence, but some fibers insert at the base of the tibial eminence. This injury can be seen in adults and children but is more common in children. In children, avulsion of the eminence tends to be an isolated injury, and only about 8% of these avulsions are associated with meniscal or other ligamentous injuries. In adults, 37% of intercondylar eminence avulsions are associated with other soft tissue or bony injuries. Avulsion of the intercondylar eminence sometimes is difficult to visualize on radiographic views, and CT or MR imaging may be required to make a definitive diagnosis (see Figs. 64–7 and 64–8).

Avulsion of the fibular head should direct attention to injury of the posterolateral corner of the knee. The posterolateral corner of the knee is anatomically complex. It is supported by the lateral (fibular) collateral ligament (LCL), arcuate ligament, biceps femoris muscle and tendon, popliteus muscle and tendon, and lateral head of the gastrocnemius muscle. The LCL originates from the lateral femoral condyle just above the groove of the popliteus tendon and inserts distally on the fibular head along with the biceps femoris. Avulsion of the fibular head, also called the *arcuate sign*, indicates LCL disruption and posterolateral corner instability (Fig. 64–9). This injury is produced by force application against the anteromedial aspect of the tibia with the knee in extension. The avulsed fragment can be detected easily on an anteroposterior radiograph of the knee (Fig. 64–9A). Posterolateral instability is a serious injury, and it is commonly associated with posterior cruciate ligament (PCL) and ACL injuries. A focal compression fracture of the anterior surface of the medial tibial plateau is seen commonly with posterolateral corner injuries (Fig. 64–10). The unusual location of this fracture fits the mechanism of the injury, in which an extended knee is hit on its anteromedial aspect. When detected, this fracture is unique enough to warrant the search for a posterolateral corner injury.

Avulsion of the PCL off the posterior tibia is a rare injury, but its presence is specific for PCL disruption (Fig. 64–11). The mechanism of injury is due to hyperextension of the knee or a dashboard injury to the upper tibia with the knee flexed.

Tibial plateau fractures, especially of the lateral tibial plateau, are common injuries in adults. There is a high association of ligamentous injuries (36%) and meniscus injuries (55%) with tibial plateau fractures (Fig. 64–12). Some authors advocate the use of MR imaging instead of CT after the initial radiographs for tibial plateau fractures.

Lipohemarthrosis

Demonstration of a fluid-fluid level on horizontal-beam radiography initially was thought to represent lipohemarthrosis. This finding was attributed to an intra-articular fracture with release of marrow fat and blood into the joint. Until more recently, this sign had not been examined critically, and the presence of a fluid-fluid level on radiography became synonymous with a fat-fluid level of lipohemarthrosis. With the widespread use of MR imaging, it became obvious that a fluid-fluid level does not imply the presence of a lipohemarthrosis but also could represent intra-articular blood separating into serum and cellular elements (Fig. 64–13). A fluid-fluid level can be present in the knee, or in any large joint, without a concomitant intra-articular fracture. To diagnose the presence of lipohemarthrosis within a traumatized joint, three separate fluid signal intensities (double fluid-fluid level) should be present on MR imaging (Fig. 64–14). Single (intra-articular blood) or double (lipohemarthrosis) fluid-fluid levels also can be shown by CT (Fig. 64–15).

Text continued on page 478

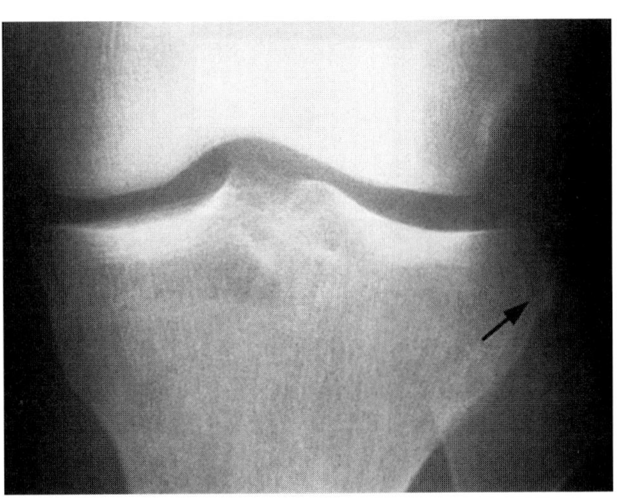

Figure 64–6. Segond fracture *(arrow)* seen on anteroposterior radiograph of the knee. The patient had a disrupted anterior cruciate ligament on MR imaging.

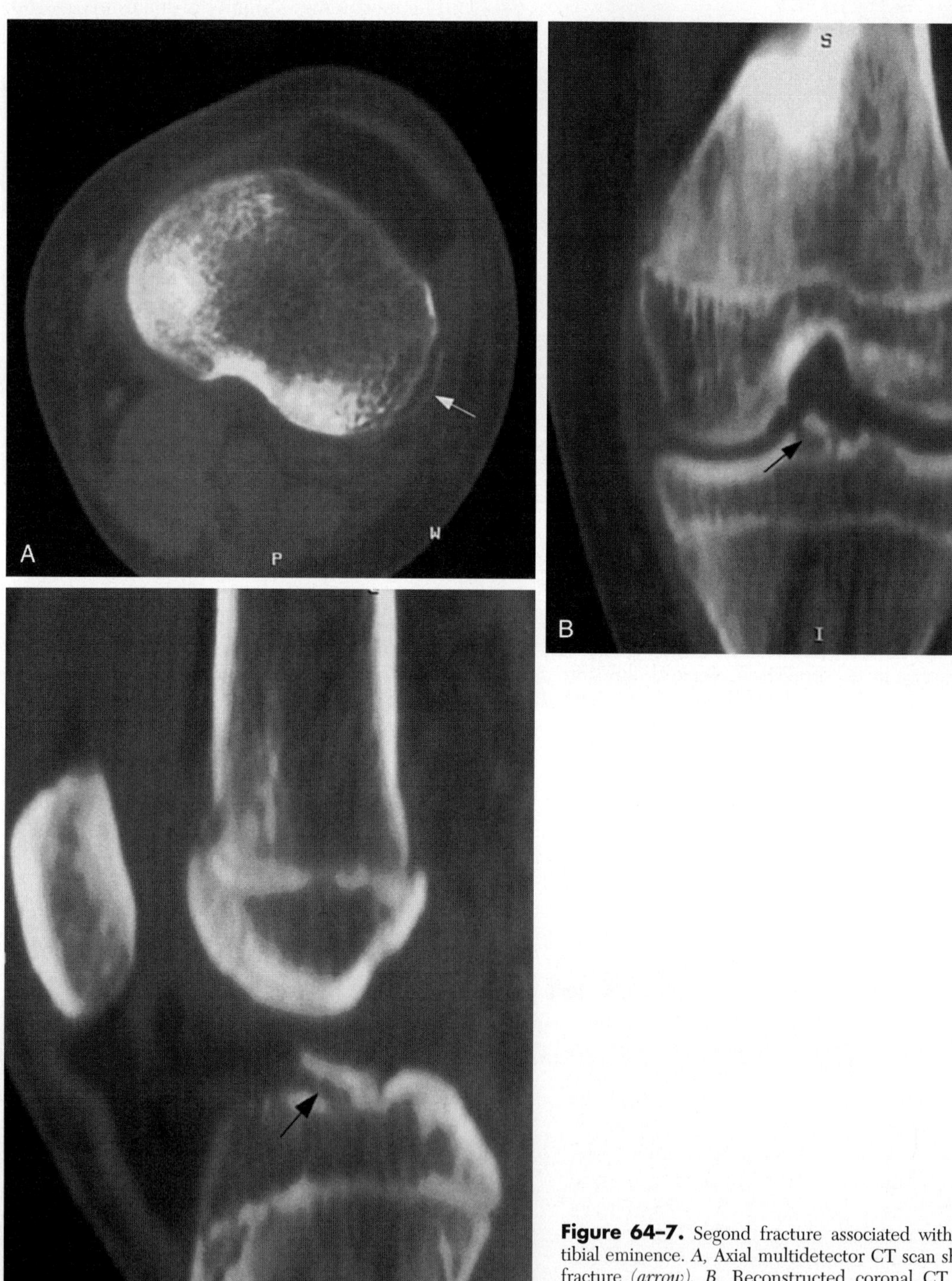

Figure 64–7. Segond fracture associated with avulsion of the tibial eminence. *A,* Axial multidetector CT scan shows the Segond fracture *(arrow). B,* Reconstructed coronal CT scan shows the Segond fracture *(white arrow)* and avulsed tibial eminence *(black arrow). C,* Reconstructed sagittal MR image shows the avulsed tibial eminence *(arrow).*

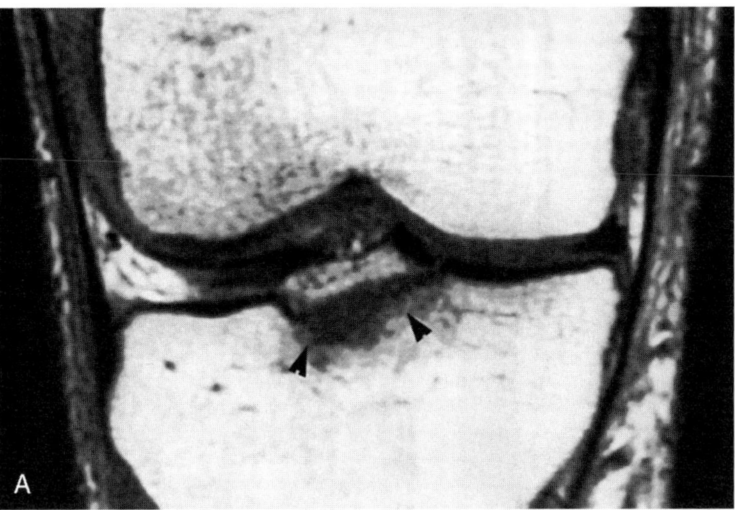

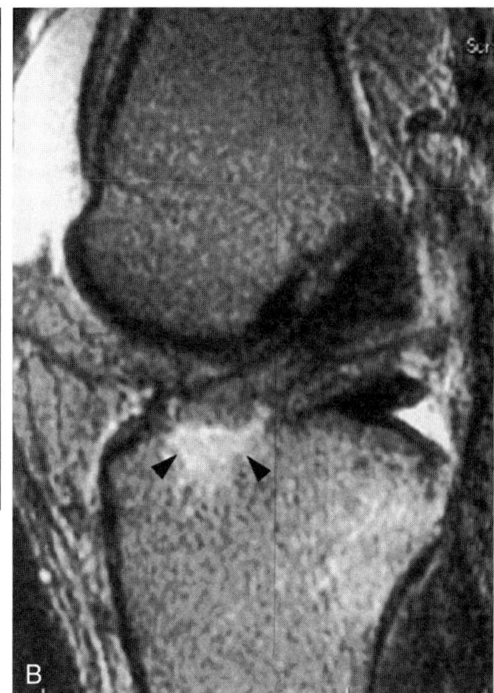

Figure 64–8. Avulsed tibial eminence. *A*, T1-weighted coronal MR image shows a fracture at the base of the tibial eminence *(arrowheads)*. *B*, T2-weighted sagittal MR image shows the fracture surrounded by marrow edema *(arrowheads)*. The contusion is small because it is produced by traction forces compared with contusions produced by compressive forces as in Figure 64–5.

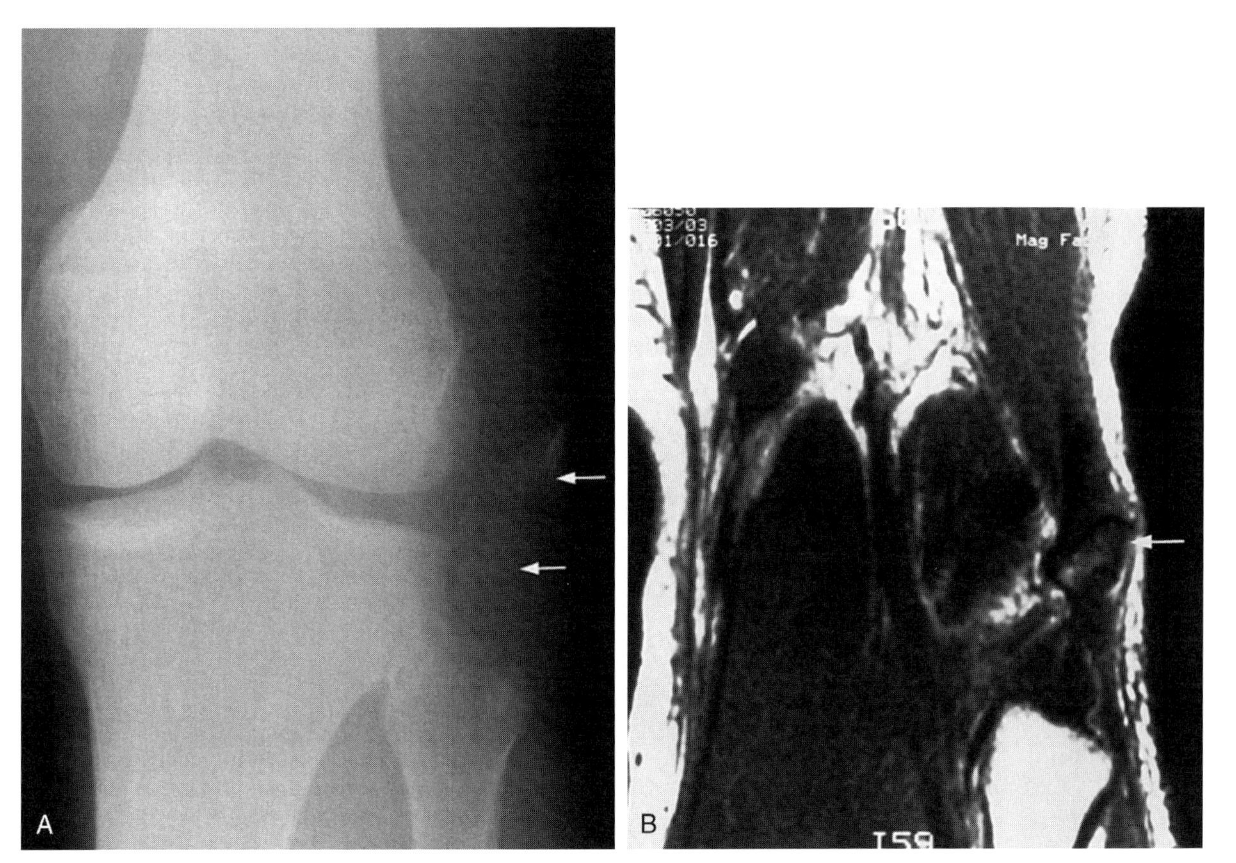

Figure 64–9. Avulsion of the fibular head associated with posterolateral corner injury of the knee. *A*, Anteroposterior radiograph of the left knee shows avulsion of the fibular head with superior displacement of the avulsed fragments *(arrows)*. *B*, Coronal T1-weighted MR image shows a superiorly displaced bone fragment *(arrow)*. The tendon of the biceps femoris is seen inserting on the avulsed fragment.

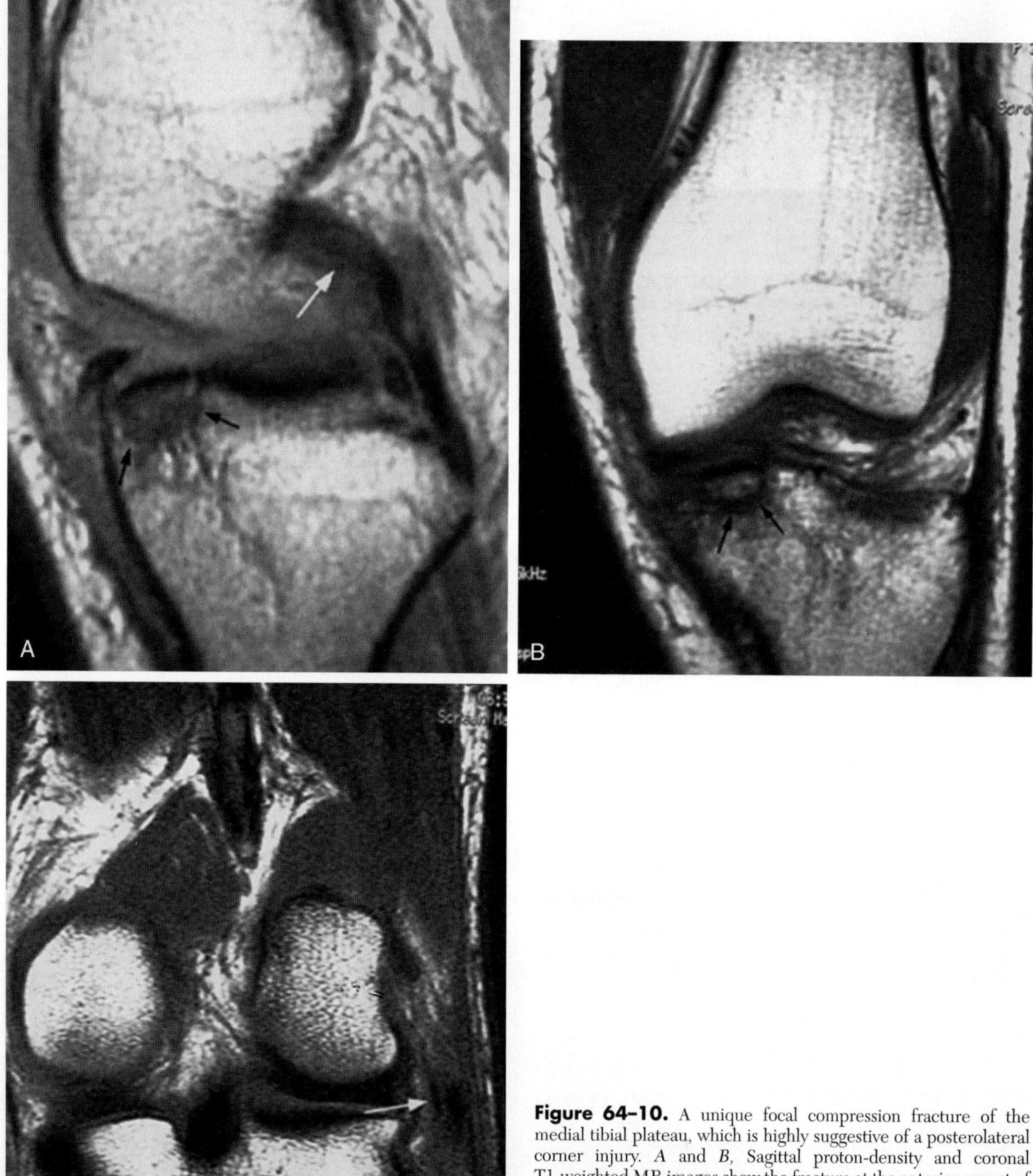

Figure 64–10. A unique focal compression fracture of the medial tibial plateau, which is highly suggestive of a posterolateral corner injury. *A* and *B*, Sagittal proton-density and coronal T1-weighted MR images show the fracture at the anterior aspect of the medial tibial plateau *(black arrows)*. Note the partial tear of the posterior cruciate ligament *(white arrow)*. *C*, Coronal T1-weighted MR image shows the posterolateral corner injury. The lateral collateral ligament is disrupted *(arrow)*.

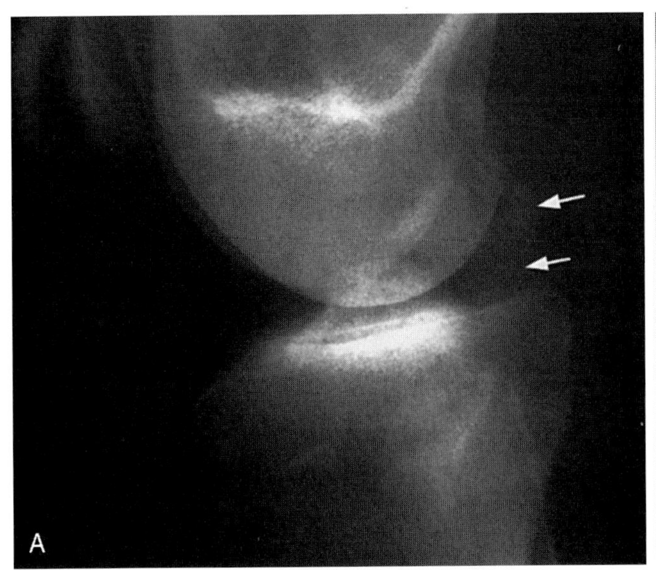

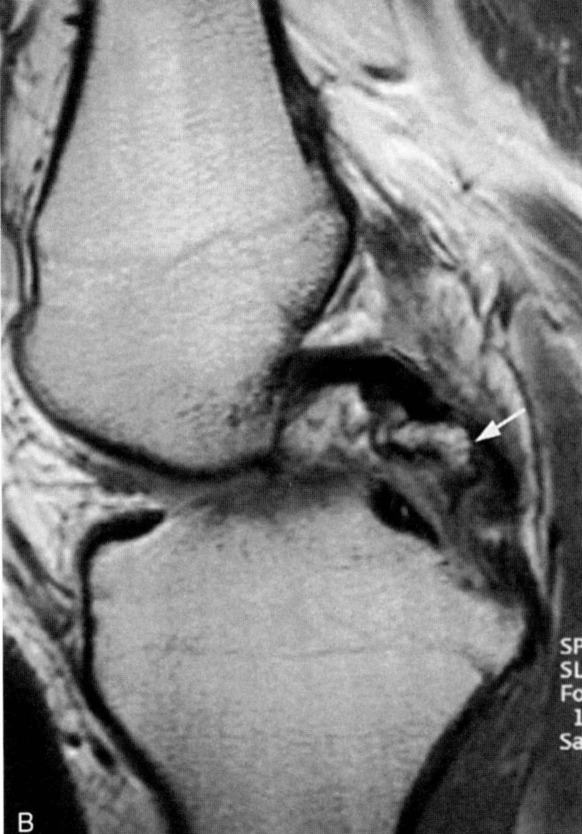

Figure 64–11. Avulsion of the posterior cruciate ligament from its tibial insertion. A, Lateral radiograph of the knee shows fragments of bone projecting over the posterior joint space *(arrows)*. B, Sagittal proton density MR image reveals the avulsed posterior cruciate ligament *(arrow)*.

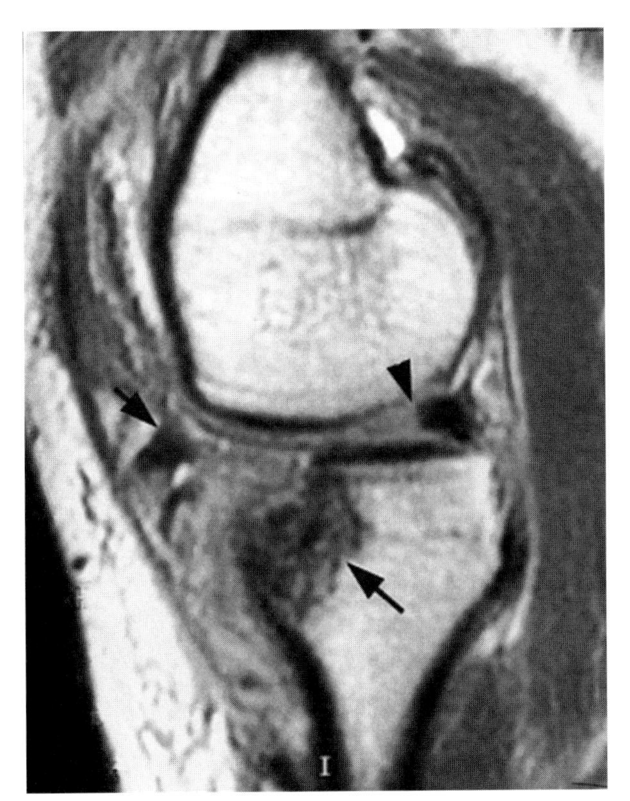

Figure 64–12. Tibial plateau fracture associated with meniscus tears. Sagittal proton-density MR image shows the tibial plateau fracture *(long arrow)*. The posterior horn of the medial meniscus *(arrowhead)* is truncated, and the anterior horn of the medial meniscus *(short arrow)* is displaced and rotated about 180 degrees.

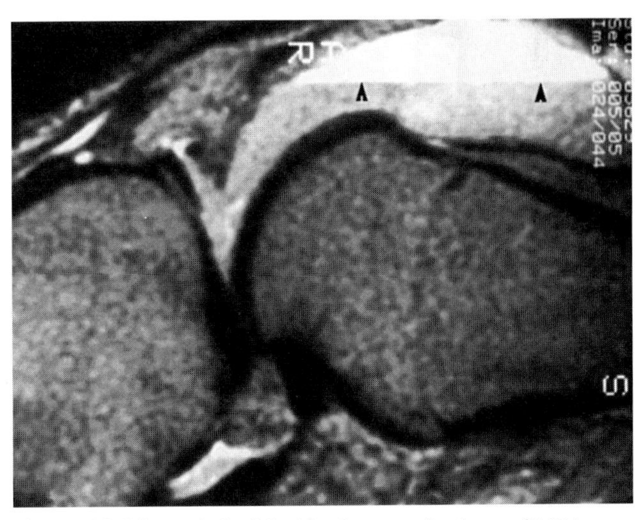

Figure 64–13. Single fluid-fluid level. T2-weighted sagittal MR image with fat suppression shows bright serum superior to the arrowheads and blood cells deep to the arrowheads.

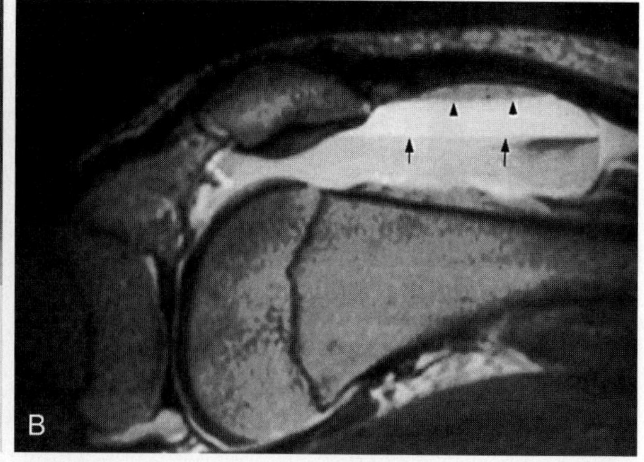

Figure 64–14. Lipohemarthrosis with a double fluid-fluid level in a 14-year-old boy who was hit on the knee by a hard object. *A*, Cross-table lateral radiograph of the knee shows fat-fluid level *(arrowheads)*. The white arrow points to the traumatized site on the medial femoral condyle. *B*, T2-weighted sagittal MR image shows fat floating over the serum *(arrowheads)*. The serum appears bright *(arrows)*, and blood cells form the deep layer.

MENISCUS INJURIES

Overview

The menisci function in load distribution of compressive and torsional forces across the knee joint. About 50% of load bearing is transmitted through the menisci when the knees are in extension and 85% in flexion. The menisci also function as stabilizers against anteroposterior translation. The presence of a tear diminishes the ability of the meniscus to distribute stress. In the 1980s and 1990s, surgeons became interested in preserving as much as possible of the menisci at surgery. After meniscectomy, the extent of osteoarthritis is related directly to the amount of meniscus removed. Currently, partial meniscectomy and meniscus repair are preferred over total meniscectomy.

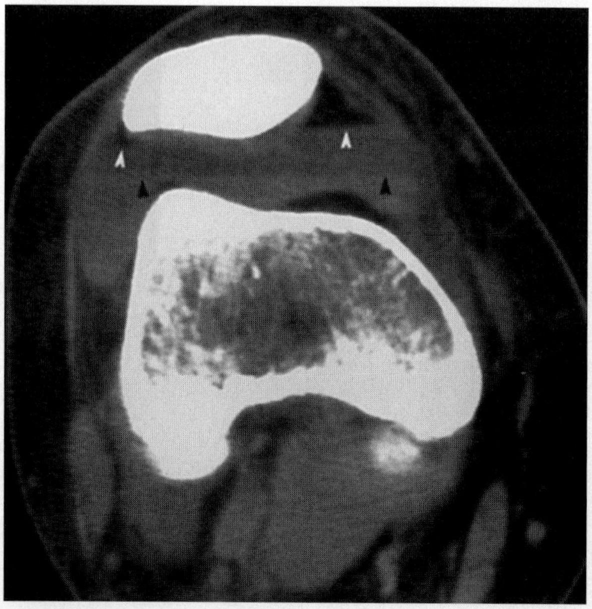

Figure 64–15. Lipohemarthrosis with double fluid-fluid levels *(arrowheads)* seen on axial CT section. The patient had a tibial plateau fracture (not included on this image).

The peripheral third of the meniscus, also called the *red zone*, is well vascularized and is capable of healing by forming scar tissue.

Menisci become torn when they are caught and crushed between the femoral condyle and tibial plateau (Fig. 64–16). The lateral meniscus is torn less frequently than the medial meniscus. The lateral meniscus is less adherent to the joint capsule and more mobile than the medial meniscus (Fig. 64–17).

Meniscus tears can be asymptomatic. The prevalence of meniscus tears in asymptomatic volunteers is 13% in those younger than age 45 years and 33% in those older than age 45.

Imaging

MR imaging currently is the leading modality for noninvasive evaluation of the knee. The diagnostic accuracy for meniscus tears using MR imaging exceeds 90%.

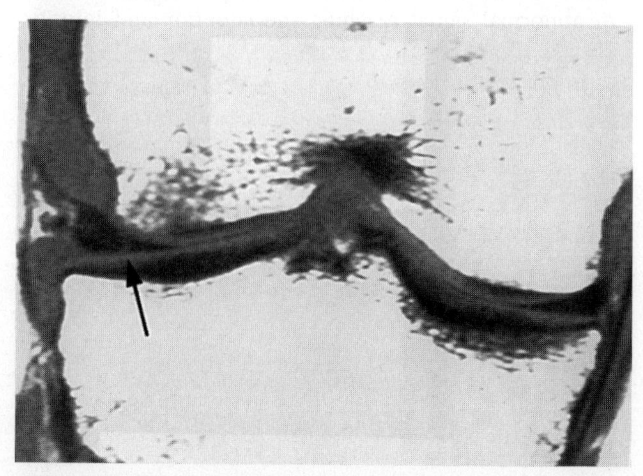

Figure 64–16. Meniscus tear caused by a crush injury between the femoral condyle and tibial plateau. T1-weighted coronal MR image shows the tear at the apex of the lateral meniscus *(arrow)*. The contusion in the lateral femoral condyle resulted from its impact on the tibial plateau.

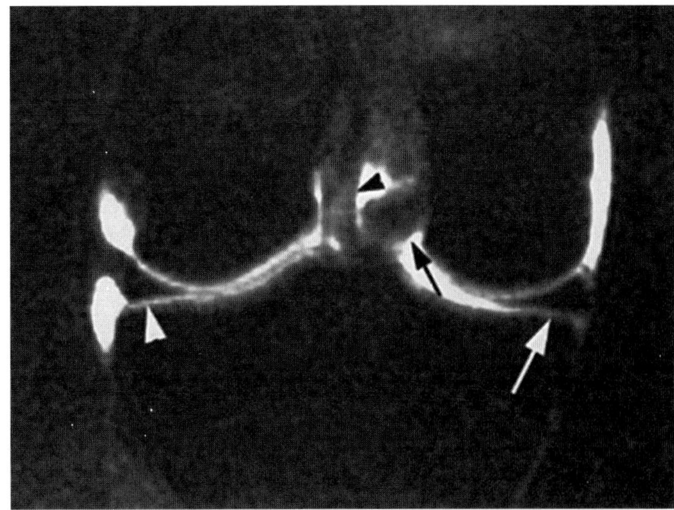

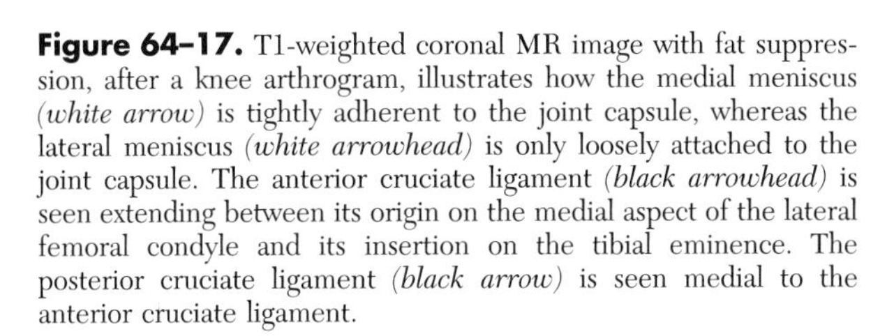

Figure 64–17. T1-weighted coronal MR image with fat suppression, after a knee arthrogram, illustrates how the medial meniscus *(white arrow)* is tightly adherent to the joint capsule, whereas the lateral meniscus *(white arrowhead)* is only loosely attached to the joint capsule. The anterior cruciate ligament *(black arrowhead)* is seen extending between its origin on the medial aspect of the lateral femoral condyle and its insertion on the tibial eminence. The posterior cruciate ligament *(black arrow)* is seen medial to the anterior cruciate ligament.

Normal menisci are seen on MR imaging as wedge-shaped structures, which are low signal intensity on all standard sequences (see Fig. 64–17). There are three imaging signs for the diagnosis of a meniscus tear: (1) increased signal within the meniscus, (2) abnormal shape (or contour) of the meniscus, and (3) abnormal size (or volume) of the meniscus.

Focal or linear regions of high signal that stay within the substance of the meniscus have been shown not to represent meniscus tears. Only high signal that unequivocally reaches the articular surface has been shown on arthroscopy to represent a tear (Fig. 64–18). DeSmet and colleagues (1993) found that the frequency of tears increased with the number of images showing high signal reaching the surface of the meniscus. A tear is less likely if high signal reaching the surface is seen only on one image, and such an abnormality should be diagnosed as a possible tear. High signal contacting the articular surface on a single image has only a 30% to 55% chance of representing a tear at surgery. A meniscus tear should be diagnosed if high signal shows definite contact with meniscal surface on at least two sequential images (see Fig. 64–18). Menisci with abnormal signal reaching the superior and inferior surfaces are significantly more likely to be torn (Fig. 64–19). Often it is difficult to tell with certainty whether a high signal reaches to the articular surface, and such menisci should be interpreted as normal. As for imaging planes that most efficiently reveal these tears, De Smet's study showed that sagittal images are better than coronal images in detecting meniscus tears.

Some normal structures within the knee can mimic meniscus tears. On sagittal images, the transverse meniscal ligament can simulate a tear of the anterior horn of the lateral meniscus (Fig. 64–20). The meniscofemoral ligaments (of Humphrey and Wrisberg) and the popliteus tendon can simulate a tear of the posterior horn of the lateral meniscus (Fig. 64–21). To avoid these pitfalls, one should follow each of these suspicious findings on multiple serial sections, when it becomes obvious that they

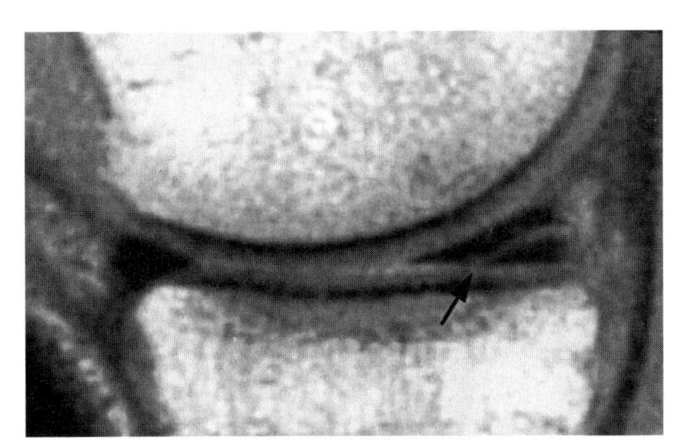

Figure 64–18. Sagittal T1-weighted MR image shows linear high signal in the posterior horn of the medial meniscus reaching the inferior articular surface on sequential images, which was found at arthroscopy to be a tear.

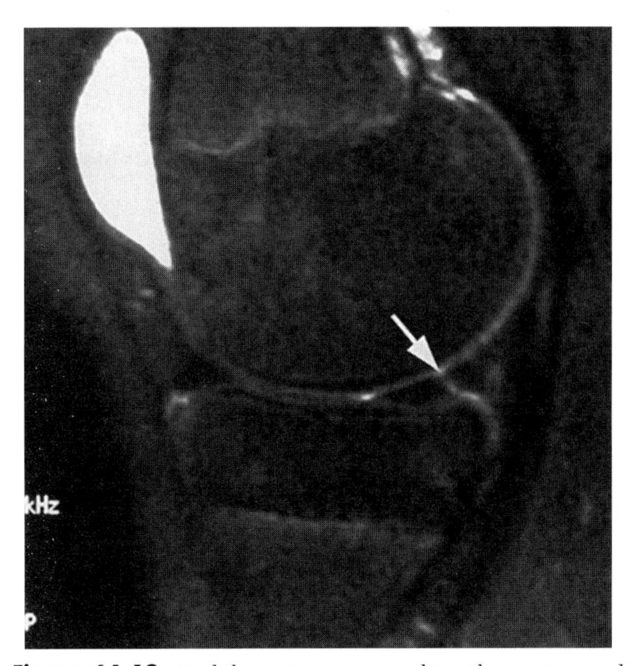

Figure 64–19. Medial meniscus tear reaching the superior and inferior surfaces of the posterior horn *(arrow)* on a sagittal T2-weighted image with fat suppression.

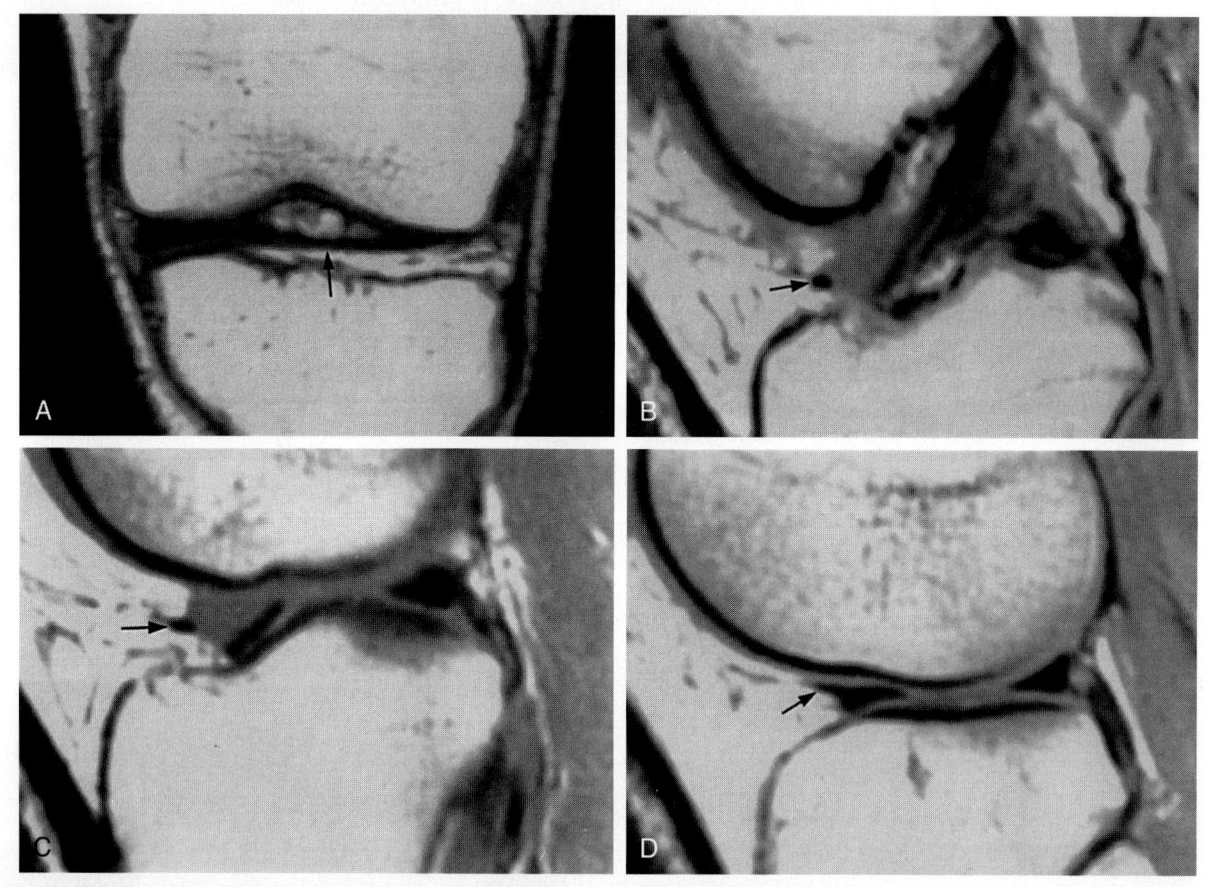

Figure 64–20. Transverse ligament, which can be confused for a tear at its insertion to the anterior horn of the lateral meniscus. *A* and *B*, Coronal T1-weighted MR image and sagittal midline proton-density image show the transverse ligament *(arrow)*. *C* and *D*, Two sagittal proton-density images, which follow the transverse ligament to its insertion on the anterior horn of the lateral meniscus *(arrow)*.

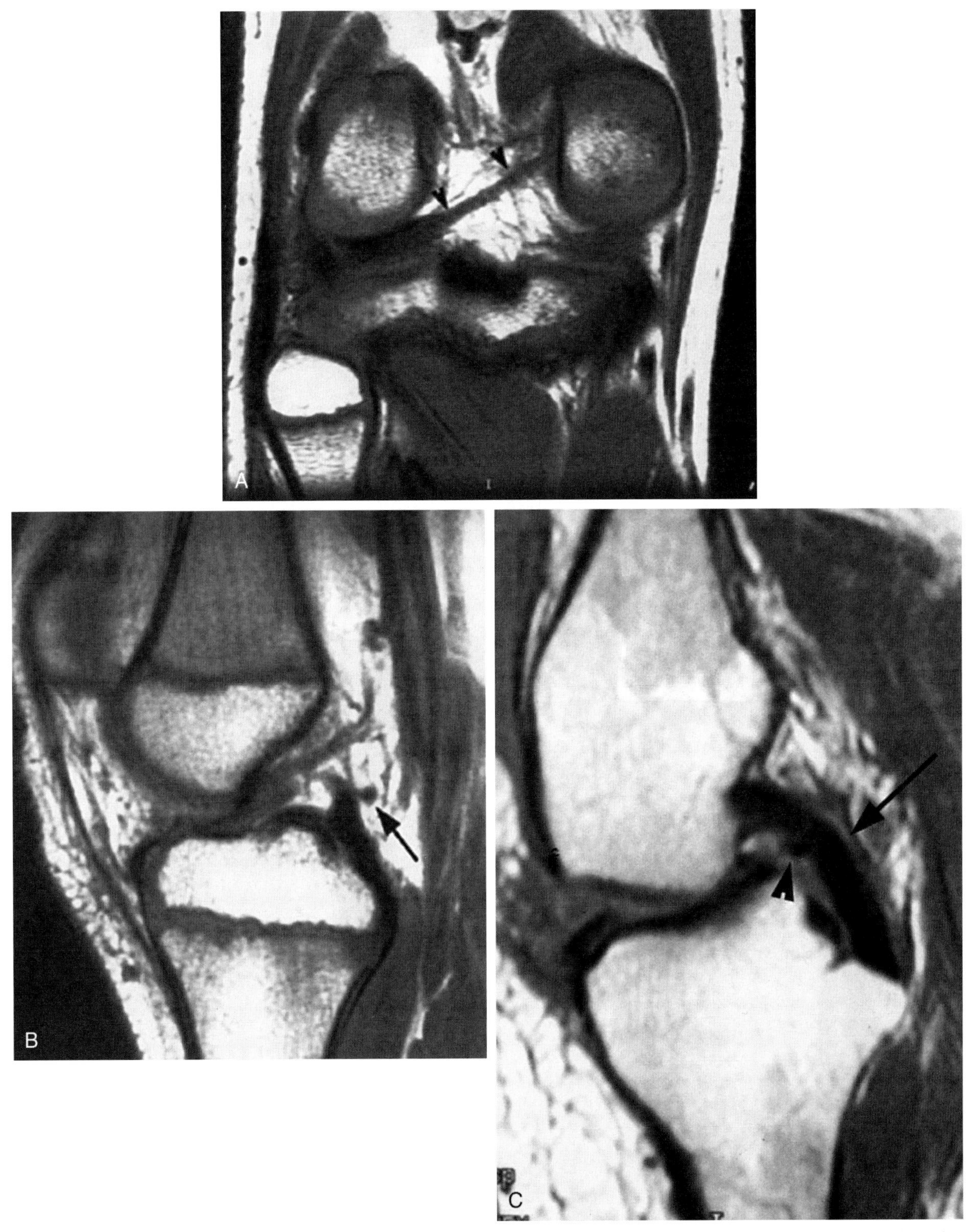

Figure 64–21. The meniscofemoral ligaments. *A* and *B*, Coronal T1-weighted MR image and proton-density sagittal image show Wrisberg's ligament (*arrowheads* in *A* and *arrow* in *B*). *C*, Sagittal proton-density MR image shows Humphrey's ligament (*arrowhead*) anterior to the posterior cruciate ligament (*arrow*).

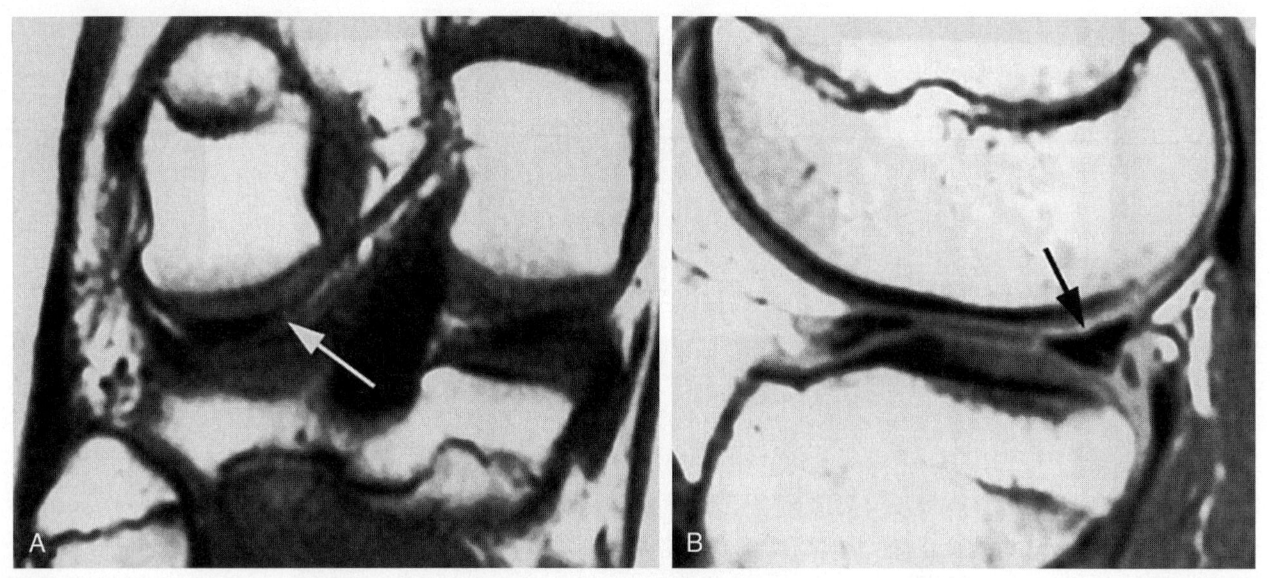

Figure 64–22. Meniscofemoral ligament inserting on the posterior horn of the lateral meniscus simulating a meniscus tear. *A,* Coronal T1-weighted MR image shows Wrisberg's ligament inserting at the posterior horn of the lateral meniscus *(arrow)*. *B,* Sagittal proton-density MR image shows the ligament just before its insertion *(arrow)* simulating a meniscus tear.

represent normal anatomic structures rather than tears (Fig. 64–22; see Fig. 64–20).

Abnormalities of shape and size of the meniscus are important signs of a meniscus tear. On sagittal sections, the anterior and posterior horns of both menisci are triangular in shape. The posterior horn of the medial meniscus is larger than the anterior horn, whereas the posterior horn of the lateral meniscus is usually equal in size to the anterior horn. Anterior horns should not measure more than 6 mm in height. The most frequent pattern of a meniscus tear with displacement is the bucket-handle tear (Fig. 64–23). It often involves the entire meniscus. MR signs of a bucket-handle tear include the double PCL sign (Fig. 64–24), the flipped meniscus sign (Fig. 64–25), and a fragment in the intercondylar notch (Fig. 64–26). With the double PCL sign, a linear low signal intensity structure is shown anterior and inferior to the PCL (see Fig. 64–24). This represents the displaced meniscal fragment, which runs parallel to the PCL. This sign is seen only with bucket-handle tears of the medial meniscus. Bucket-handle tears of the lateral meniscus are prevented from locating underneath the PCL by an intact ACL. The double PCL sign frequently is associated with absence or decreased size of the posterior horn and body of the medial meniscus (see Fig. 64–24).

A bucket-handle tear involving most of the posterior meniscus can flip into the anterior compartment and rest on the anterior horn (see Fig. 64–25) or can flip posteriorly (Fig. 64–27). An anterior flip results in a large anterior horn, greater than 6 mm in height and greater than 6 mm

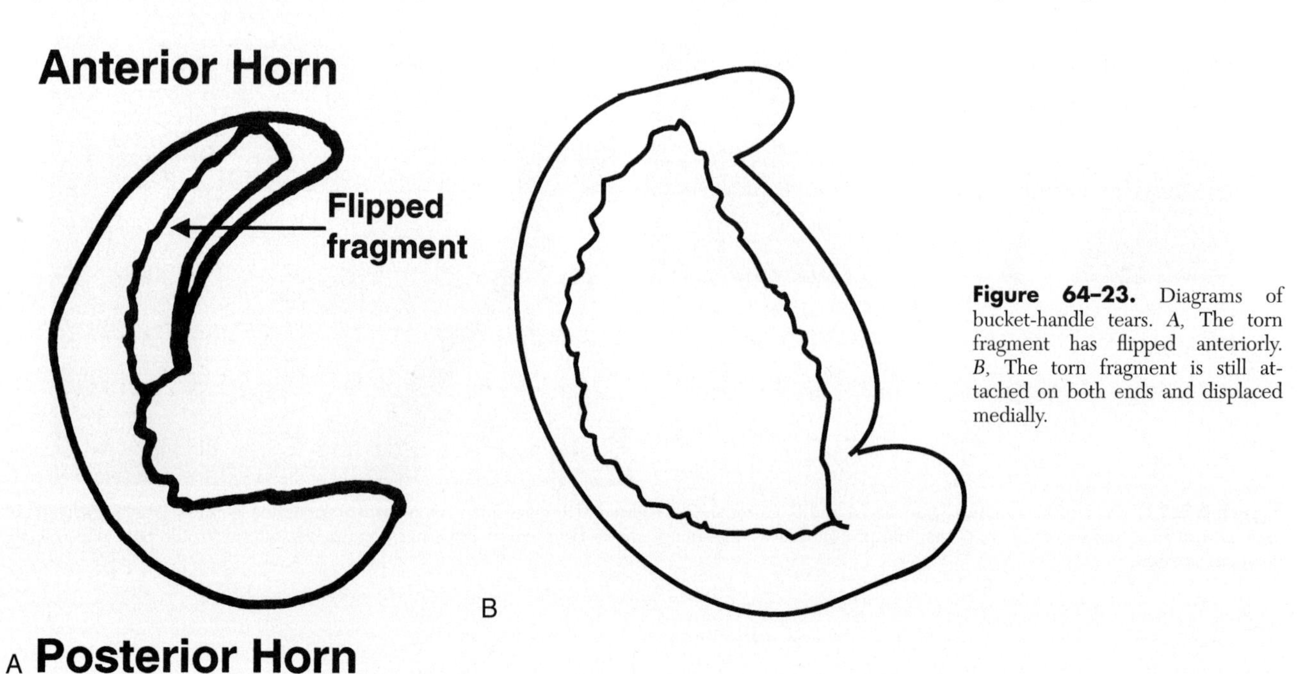

Figure 64–23. Diagrams of bucket-handle tears. *A,* The torn fragment has flipped anteriorly. *B,* The torn fragment is still attached on both ends and displaced medially.

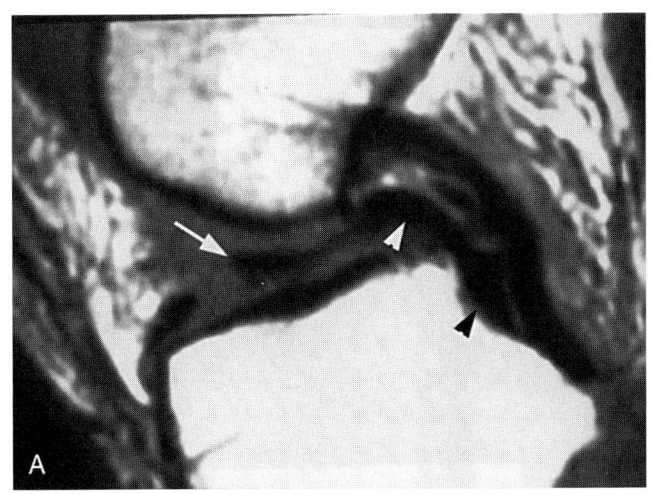

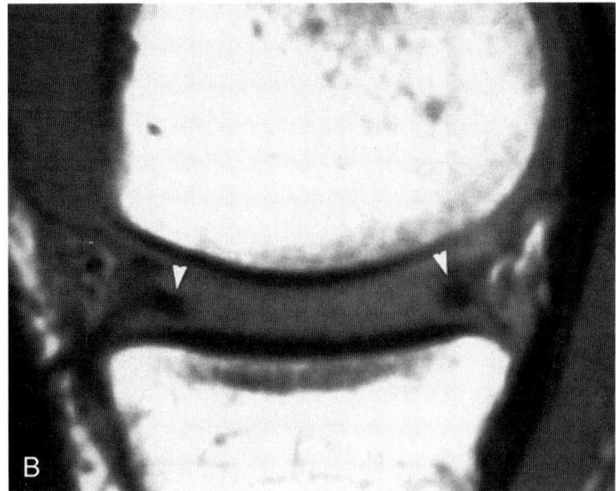

Figure 64–24. The double posterior cruciate ligament sign. *A,* Sagittal proton-density MR image illustrates the double posterior cruciate ligament sign. The arrowheads and white arrow point to the medially displaced bucket-handle fragment. *B,* Sagittal proton-density image taken medial to *A* shows the remnants of the anterior and posterior horns of the medial meniscus *(arrowheads).*

in width in the medial meniscus or 10 mm in the lateral meniscus. Absence or markedly diminished size of the posterior horn is another finding in this type of tear. The anterior horn and the superimposed fragment are often inseparable (Fig. 64–28), but occasionally a line of increased signal intensity is seen between the two structures (see Fig. 64–25). A completely detached or free meniscus fragment can occur on rare occasions. The free fragment acts like a loose body and can displace anywhere in the joint, including subligamentally between the MCL and the medial tibial plateau (Fig. 64–29).

In the acute phase, often the remnant of the bucket-handle tear shows a truncated apex on MR images (Fig. 64–30). Surgeons often try to reshape the deformed truncated remnant to make it look as close to normal as possible. With time, the remnant remodels and becomes triangular, resembling a normal meniscus, although the width of the remodeled remnant is significantly shorter than normal (Fig. 64–31).

Types of Meniscus Tears

Different meniscus tear patterns have been described in the literature, and it is often essential to categorize meniscus tears into specific types because they have different significance. The horizontal cleavage tear is a degenerative tear, which often is seen in elderly patients and can be asymptomatic. It is the most common type of tear seen in patients with meniscus cysts. This tear splits the meniscus at its apex and can extend to the periphery of the meniscus (Fig. 64–32). Longitudinal (vertical) tears follow the long axis of the meniscus and can vary in length (Fig. 64–33). The bucket-handle tear is an extreme form of the vertical tear, in which the torn fragment can displace centrally into the joint or, if one end is detached from the meniscus, the torn fragment could flip anteriorly or posteriorly (see Figs. 64–25 and 64–27). The radial tear

Figure 64–25. Bucket-handle tear with fragment flipped anteriorly *(arrowheads)* on a sagittal proton-density MR image. Note the torn posterior horn *(black arrow)* and the large anterior horn *(arrowheads).*

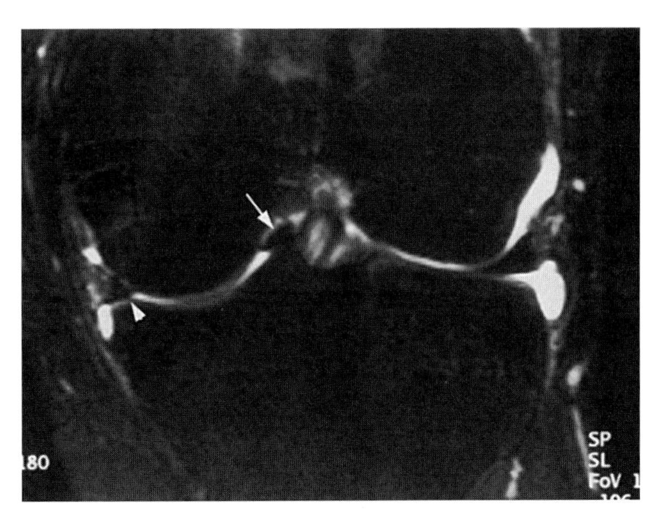

Figure 64–26. Meniscal fragment displaced into the intercondylar notch *(arrow).* T2-weighted coronal MR image with fat suppression shows the meniscal fragment in the intercondylar notch anteriorly. The arrowhead points to the truncated apex of the medial meniscus.

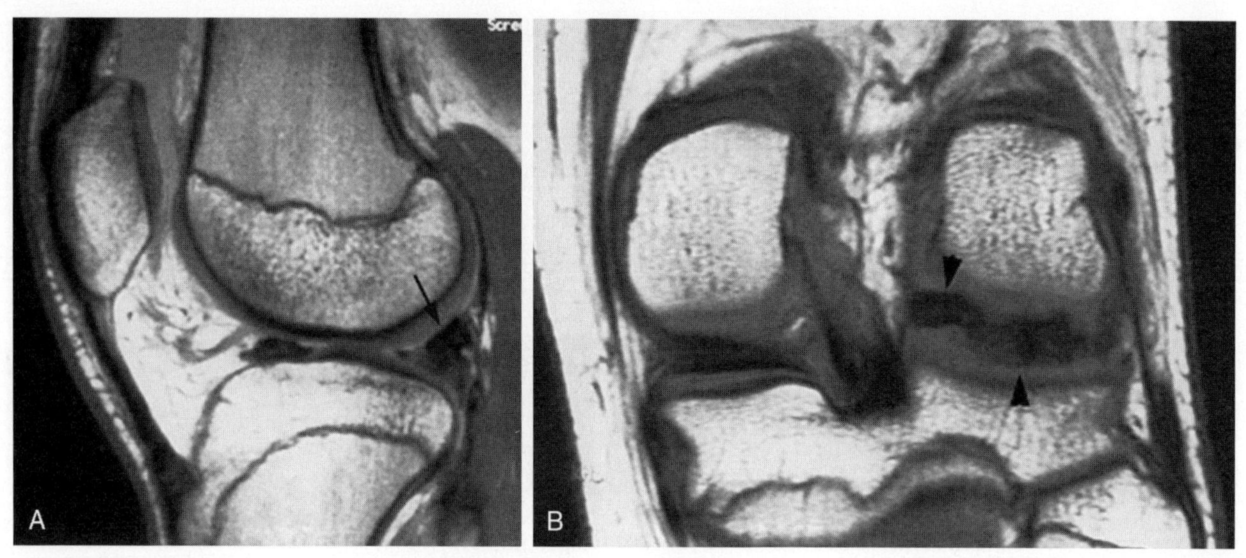

Figure 64–27. Bucket-handle tear where the fragment has flipped posteriorly. *A*, Sagittal T1-weighted MR image shows the large posterior horn *(arrow)*, which actually is two meniscal horns stacked on top of each other. *B*, Coronal T1-weighted image shows the torn fragment has flipped posteriorly *(arrowheads)*.

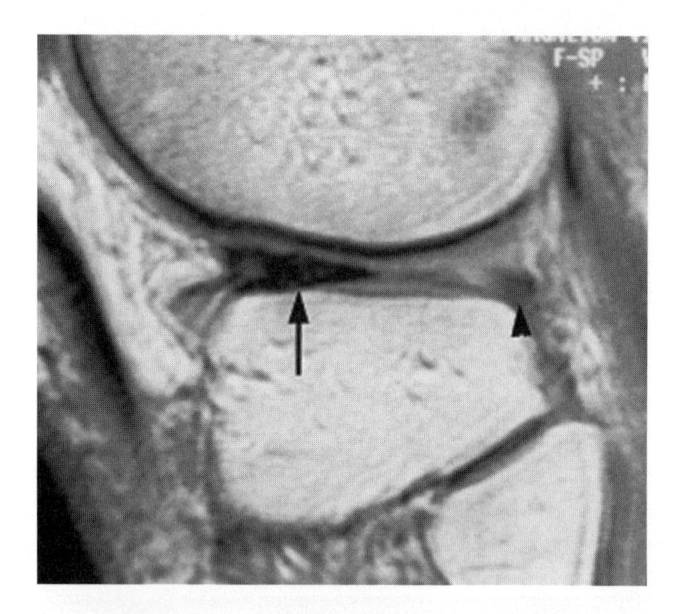

Figure 64–28. Meniscal fragment flipped anteriorly, but the fragments are inseparable on a sagittal proton-density MR sequence. The large anterior horn shows increased width *(arrow)* because it consists of two meniscal horns. The posterior horn is absent *(arrowhead)*. Average width of the lateral meniscus is 10 mm throughout. The average width for the posterior horn of the medial meniscus is 12 mm, and for the anterior horn the average width is 6 mm.

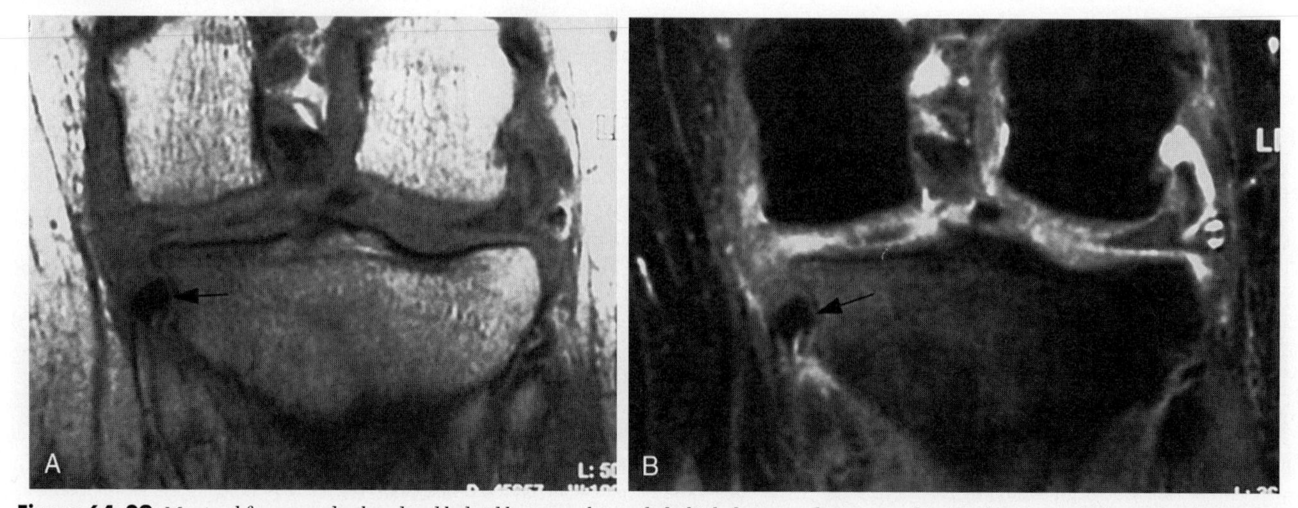

Figure 64–29. Meniscal fragment displaced and lodged between the medial tibial plateau and joint capsule. *A* and *B*, Coronal T1-weighted MR image and T2-weighted image with fat suppression show a fragment of the torn medial meniscus *(arrow)* lodged between the bone and the joint capsule.

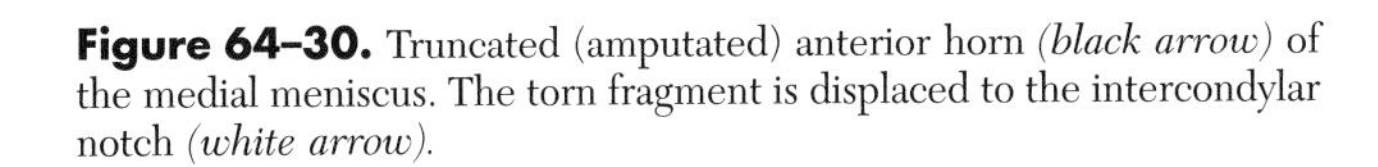

Figure 64–30. Truncated (amputated) anterior horn *(black arrow)* of the medial meniscus. The torn fragment is displaced to the intercondylar notch *(white arrow)*.

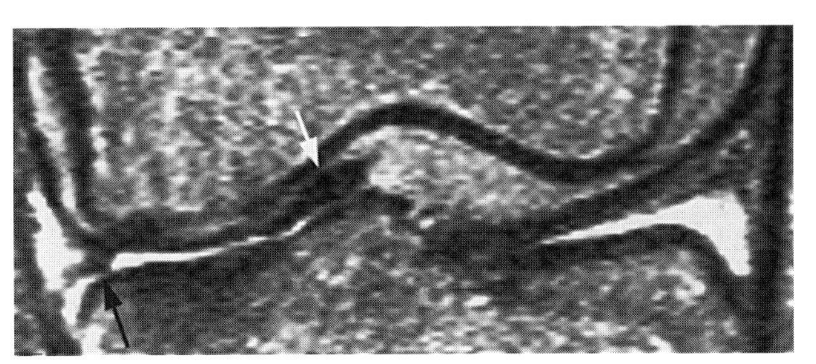

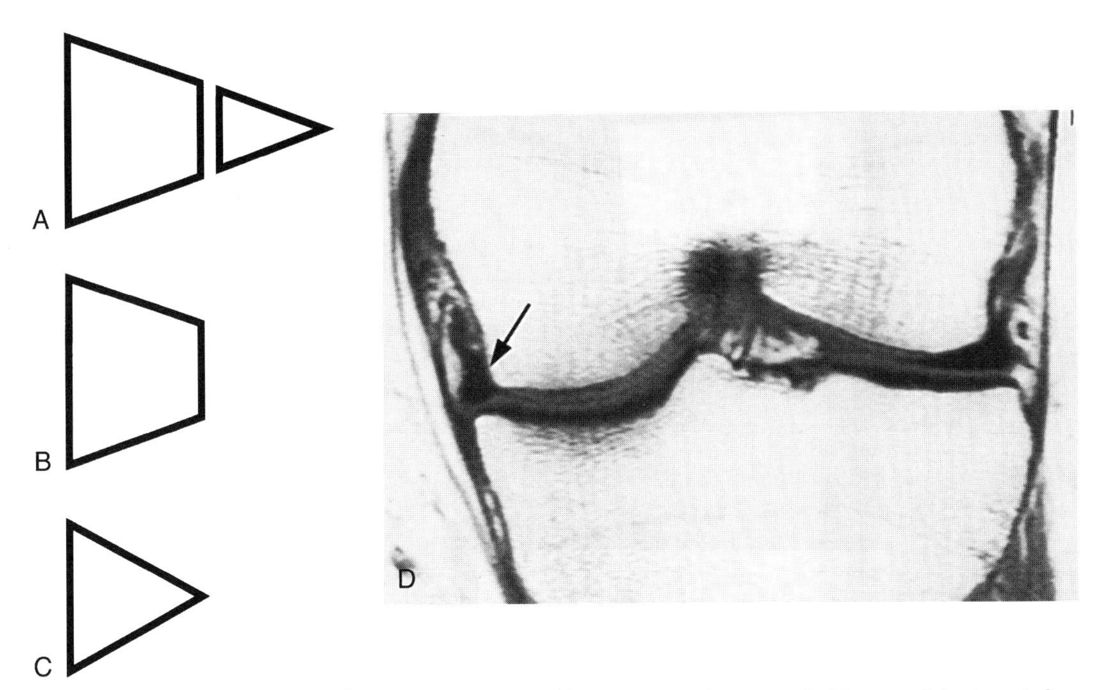

Figure 64–31. Remodeling of the meniscus after a tear. *A*, Diagram illustrating a meniscus tear. *B*, Diagram of the truncated remnant, which surgeons usually attempt to reshape. *C*, Diagram of a remodeled remnant. *D*, Coronal T1-weighted MR image shows a remodeled remnant of a previously torn medial meniscus *(arrow)*.

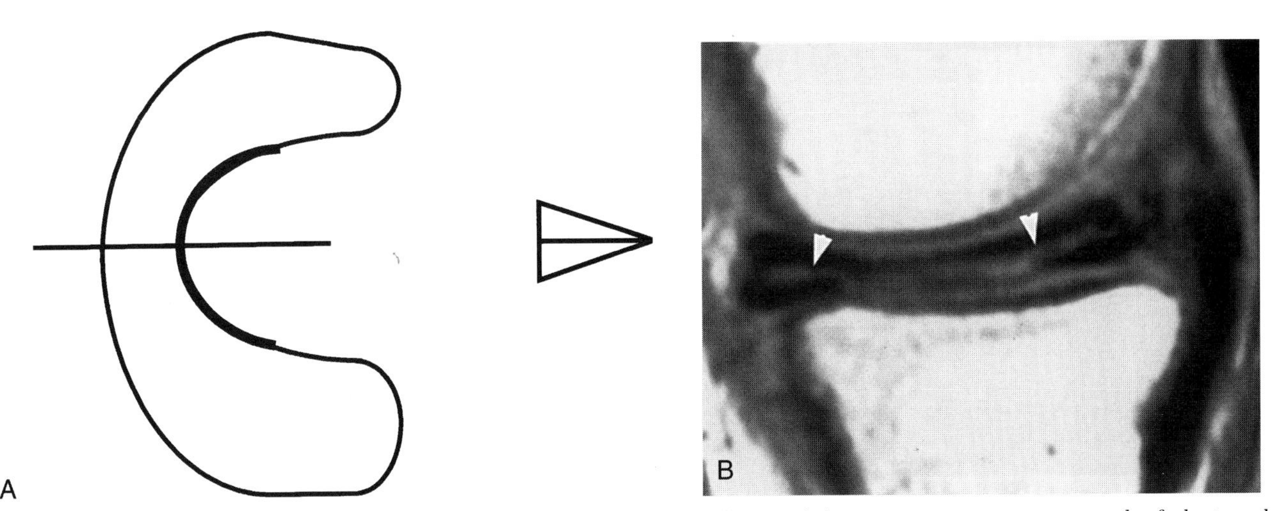

Figure 64–32. Horizontal cleavage tear of the meniscus. *A*, Diagram illustrating a horizontal cleavage tear. *B*, MR imaging example of a horizontal cleavage tear *(arrowheads)* seen on a sagittal proton-density image.

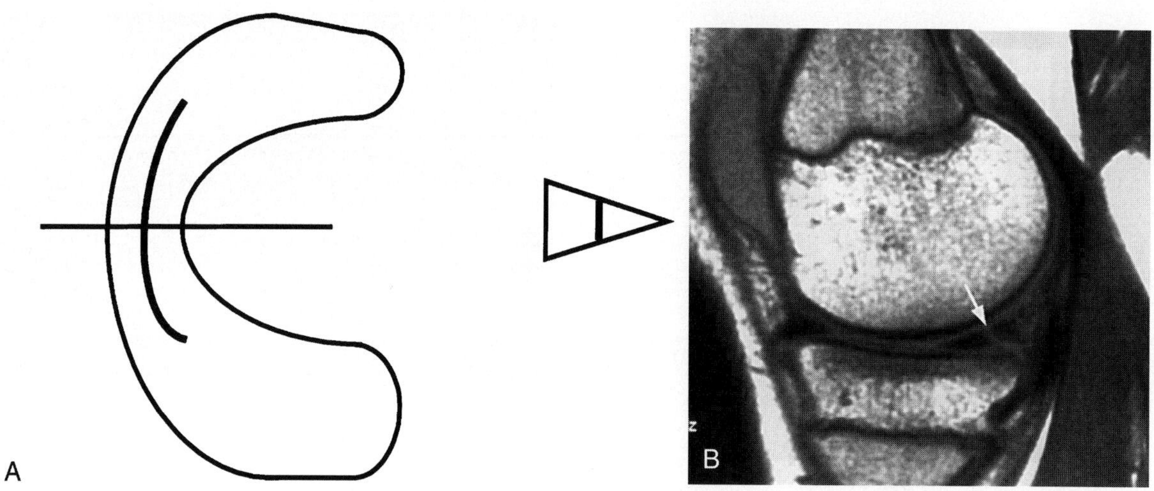

Figure 64–33. Longitudinal tear of the meniscus. *A,* Diagram illustrating a longitudinal tear. *B,* Sagittal proton-density MR image shows a longitudinal tear involving the posterior horn of the medial meniscus *(arrow).*

runs perpendicular to the long axis of the meniscus, and it can be a full-thickness tear, running from the apex to the periphery, or it could be a partial-thickness tear. A parrot beak tear is usually a small curved or obliquely oriented tear with respect to the long axis of the meniscus (Fig. 64–34). Oblique tears are similar to horizontal tears except they run obliquely with respect to the short axis of the meniscus and exit on the superior or inferior surface of the meniscus. Finally, complex tears do not follow a specific pattern and tend to propagate in different planes (Fig. 64–35).

Posterior Horn of the Lateral Meniscus

At the posterolateral corner of the knee, the posterior horn of the lateral meniscus is separated from the joint capsule by the popliteus tendon. At this site, the posterior horn is suspended by delicate bands of fibrous tissue that connect the meniscus to the capsule (Fig. 64–36). These bands are called *fascicles.* The separation of the posterior horn of the lateral meniscus from the joint capsule should not be misinterpreted as a peripheral meniscus tear. Failure to show the superior or inferior fascicle or both on a thorough MR imaging examination indicates the presence of a peripheral meniscus tear (Fig. 64–37).

Imaging of the Postoperative Meniscus

Recurrent symptoms after surgical repair or partial meniscectomy are a fairly common clinical problem. The MR imaging appearance of the postoperative meniscus varies depending on the extent of the original injury and the type of surgery used. Studies by Smith and Totty (1990) and Deutsch and colleagues (1993) showed that in menisci with limited resection, recurrent tears can be diagnosed using the same criteria that are applied in

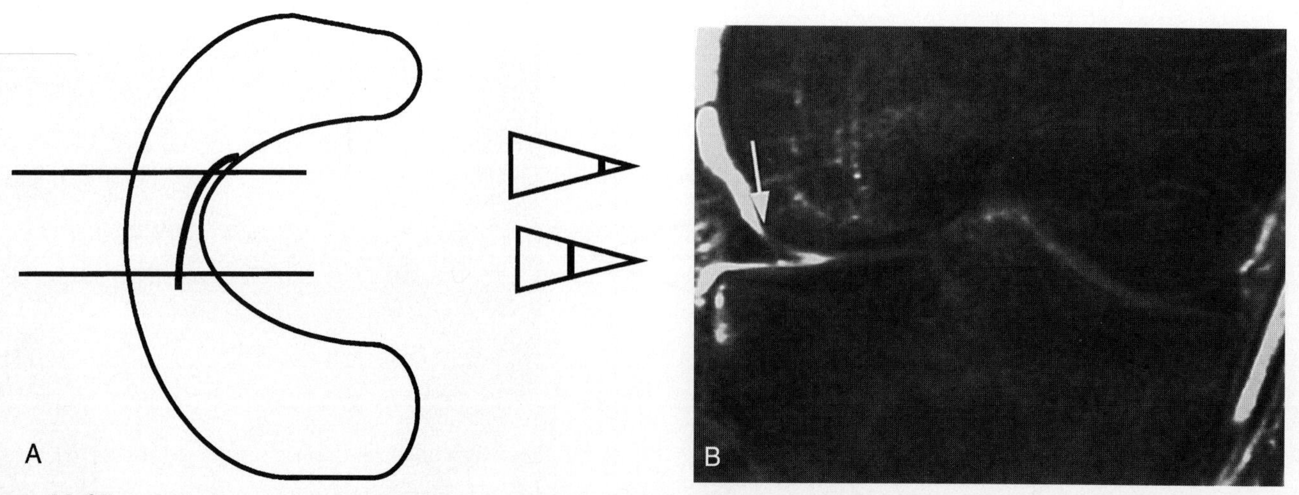

Figure 64–34. Parrot beak tear. *A,* Diagram illustrating a parrot beak tear. *B,* T2-weighted coronal MR image shows a parrot beak tear involving the apex of the anterior horn *(arrow)* of the lateral meniscus.

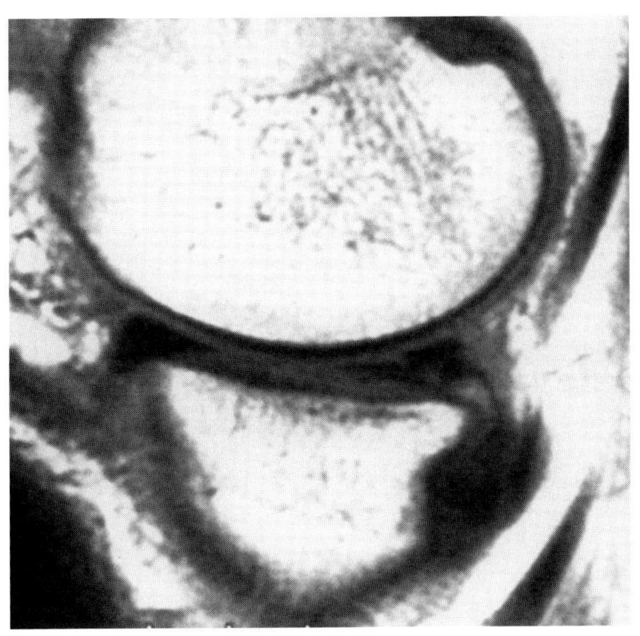

Figure 64–35. Complex meniscus tear involving the posterior horn of the medial meniscus where the tear propagated in different planes. The image is a sagittal proton-density MR image.

nonoperated menisci. Mild contour abnormalities in operated menisci, such as mild blunting of the apex, are considered normal. Conventional diagnostic criteria cannot be used to diagnose recurrent tears in menisci in which extensive partial meniscectomies have been done. In such knees, abrupt contour changes and displaced

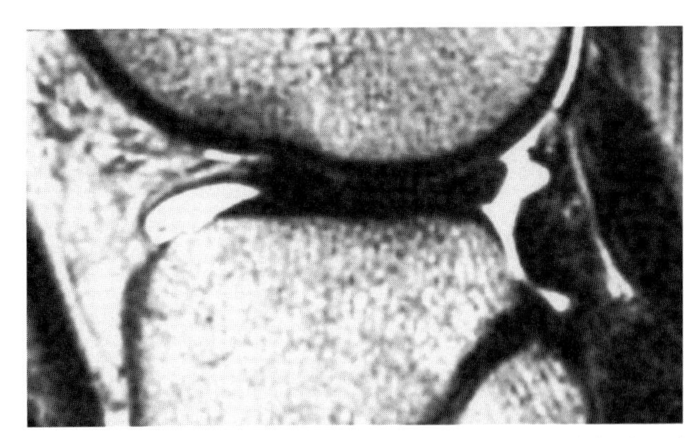

Figure 64–37. Peripheral tear of the posterior horn of the lateral meniscus. Sagittal T2-weighted MR image shows absence of the fascicles suggesting a peripheral tear.

meniscal fragments are more reliable signs of a recurrent tear. Lim and coworkers (1999) showed that finding linear high signal within the meniscus that reaches the articular surface is also a reliable sign of a recurrent tear (Fig. 64–38). Currently, most centers use MR-arthrography in the evaluation of postoperative knees with recurrent symptoms (Fig. 64–39). Distention of the joint with intra-articular contrast material seems to improve the detection of recurrent meniscus tears over MR studies done without contrast. Contrast material extending into the meniscus represents a tear. MR-arthrography also is superior to conventional MR imaging for the detection and evaluation of cartilaginous lesions, osteochondritis dissecans, loose bodies, and plicae.

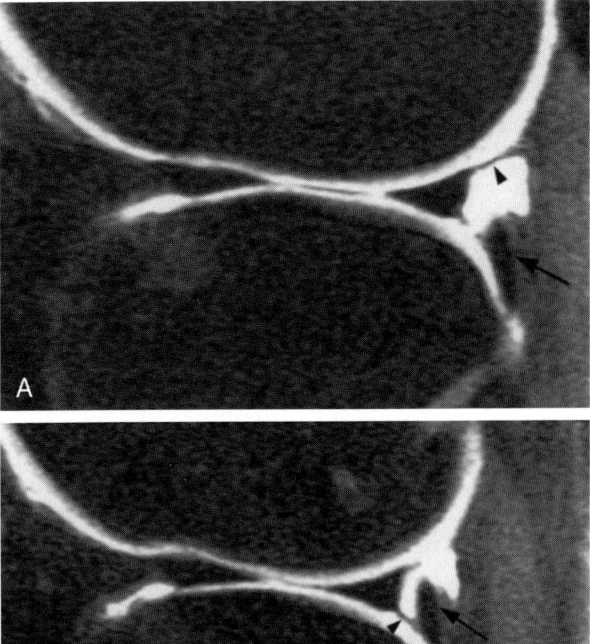

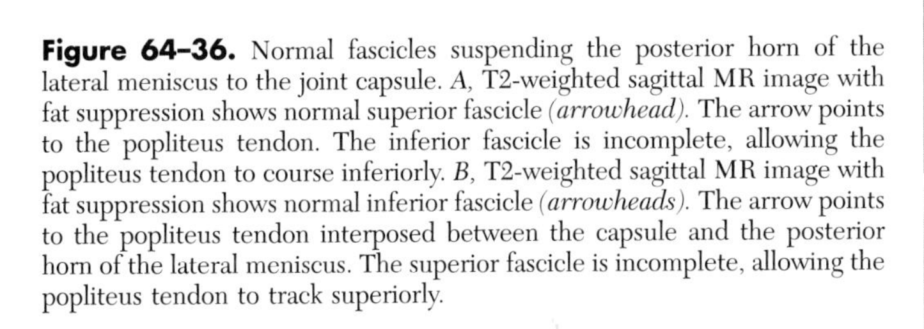

Figure 64–36. Normal fascicles suspending the posterior horn of the lateral meniscus to the joint capsule. *A*, T2-weighted sagittal MR image with fat suppression shows normal superior fascicle *(arrowhead)*. The arrow points to the popliteus tendon. The inferior fascicle is incomplete, allowing the popliteus tendon to course inferiorly. *B*, T2-weighted sagittal MR image with fat suppression shows normal inferior fascicle *(arrowheads)*. The arrow points to the popliteus tendon interposed between the capsule and the posterior horn of the lateral meniscus. The superior fascicle is incomplete, allowing the popliteus tendon to track superiorly.

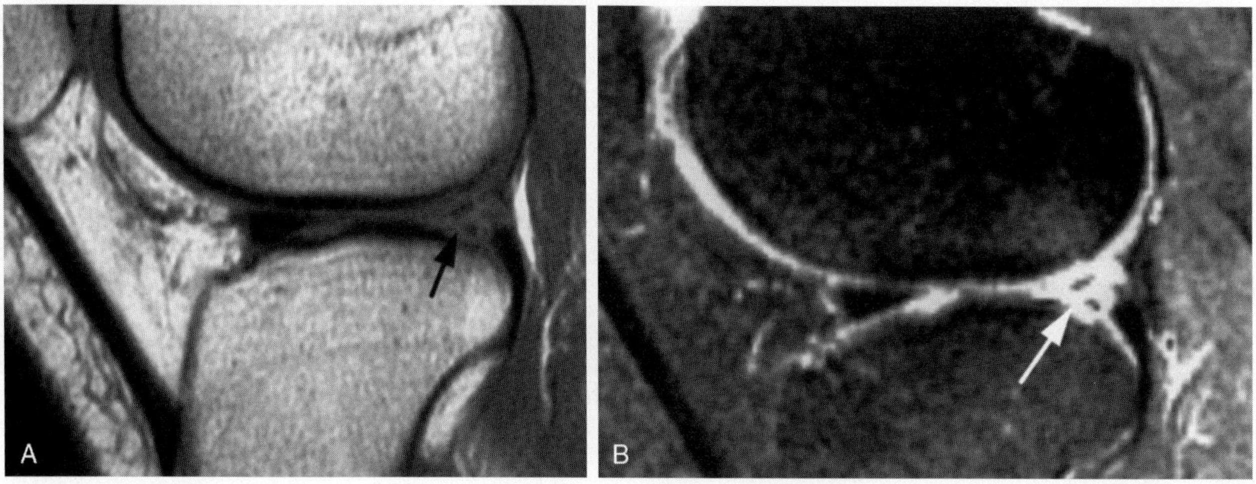

Figure 64–38. Fluid in a torn remnant. *A* and *B*, Sagittal proton-density MR image and T2-weighted sagittal image show a torn remnant of the posterior horn with fluid within the tear *(arrow)*.

KNEE LIGAMENTS

Anterior Cruciate Ligament

The ACL is an intracapsular but extrasynovial structure. It consists of the anteromedial and posterolateral bands. It originates from the posteromedial aspect of the lateral femoral condyle and courses distally and anteriorly to insert on the depressed area just anterior and medial to the intercondylar eminence of the tibia. The ACL tibial attachment is close to the insertion of the anterior horn of the medial meniscus. The ACL runs parallel to Blumensaat's line (a line drawn along the intercondylar roof) (Fig. 64–40). The ACL has functional isometry (i.e., it remains taut throughout the full range of motion), and it acts as the primary restraint to anterior displacement of the tibia.

MR imaging is highly accurate in diagnosing ACL tears (>90%). An intact ACL appears as a straight linear band or bands from origin to insertion. It has low but inhomogeneous signal intensity on T1-weighted, proton-density, and T2-weighted images (see Figs. 64–17 and 64–40).

The ACL is the most frequently injured ligament in the knee, and this injury often is associated with other ligamentous and meniscus tears. The MR imaging signs of an acute ACL tear include discontinuity of the fibers, usually proximally or in the midsubstance, and increased signal intensity within the ligament on T1-weighted, proton-density, and T2-weighted sequences (Fig. 64–41; see Fig. 64–4*B*). Some reports indicate that partial ACL tears constitute 10% to 35% of all ACL injuries; however, there are no tested MR imaging criteria for diagnosing partial tears. It is unlikely that such criteria will be developed because even arthroscopists (arthroscopy is considered the gold standard) find it difficult to confirm the presence or absence of partial tears, and most partial tears are treated nonoperatively.

Secondary MR imaging signs have been used to assist in

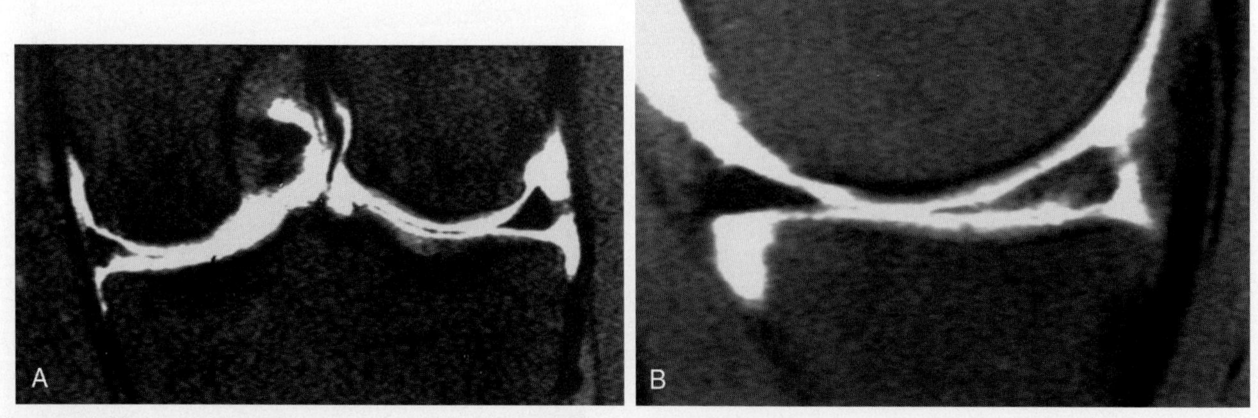

Figure 64–39. MR arthrogram to evaluate a symptomatic knee after a previous partial medial meniscectomy where a small flap was removed and the meniscus was contoured to approximately normal shape. *A* and *B*, Coronal and sagittal T1-weighted MR images with fat suppression show severe deformity of the inferior surface of the posterior horn and body of the medial meniscus, indicating a recurrent tear.

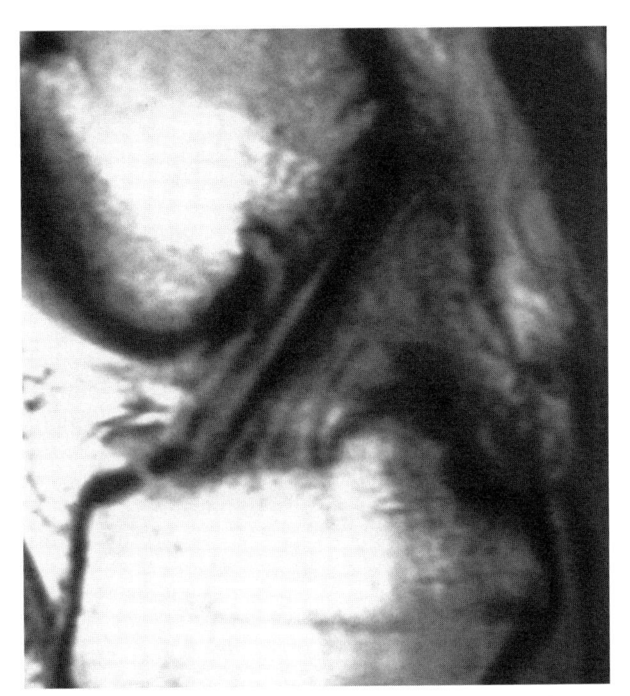

Figure 64–40. Normal anterior cruciate ligament. Sagittal proton-density MR image shows the fibrous bands of a normal anterior cruciate ligament.

the diagnosis of ACL tears. These include kissing bone bruises in the lateral femoral condyle and lateral tibial plateau (see Fig. 64–4*A*); anterior translation of the lateral tibial plateau greater than 5 mm; uncovering of the inferior surface of the posterior horn of the lateral meniscus; hyperbuckled PCL; and deep lateral femoral notch, which can be seen on MR imaging and plain radiographs. Acutely torn ACLs do not heal, and surgical reconstruction is required using an autograft from the central portion of the patellar ligament (Fig. 64–42).

Chronic ACL tears are diagnosed when the ligament appears disrupted or has changed its course or slope. A chronically torn ACL may drop down in position to lie horizontally on the tibial surface (Fig. 64–43). Rarely, it may appear continuous but has focal angulation caused by adhesions with the PCL or lateral femoral condyle. The absence of acute findings, such as hemarthrosis and bone bruises, also helps to differentiate acute from chronic ACL tear.

Radiologists occasionally are requested to evaluate the integrity of a reconstructed ACL on MR imaging. Signal characteristics of the graft could vary initially with time; however, most authors believe that by 12 months postoperatively, the graft should show decreased signal on all sequences. One report indicated that increased graft signal, focal or diffuse, may persist beyond 12 months. Successful reconstruction is guided by two principles, isometry and avoidance of bony impingement (Fig.

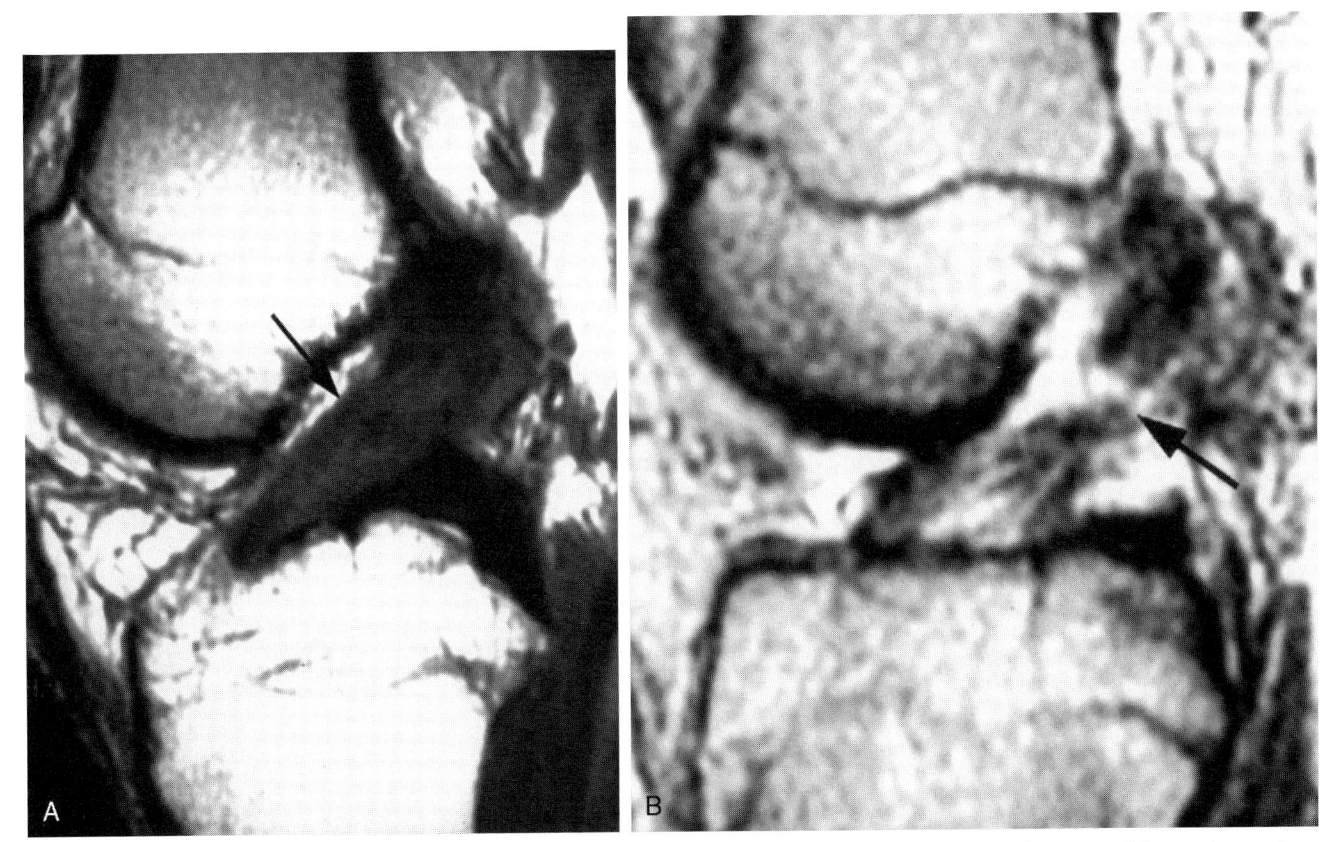

Figure 64–41. Acute anterior cruciate ligament tear. *A* and *B*, Sagittal proton-density and T2-weighted MR images show a tear of the anterior cruciate ligament *(arrow)*. The anterior cruciate ligament is thick and has high signal intensity. The ligament is disrupted in its midportion *(arrow)*. (See Figure 64–4*B* for another example of acute anterior cruciate ligament tear.)

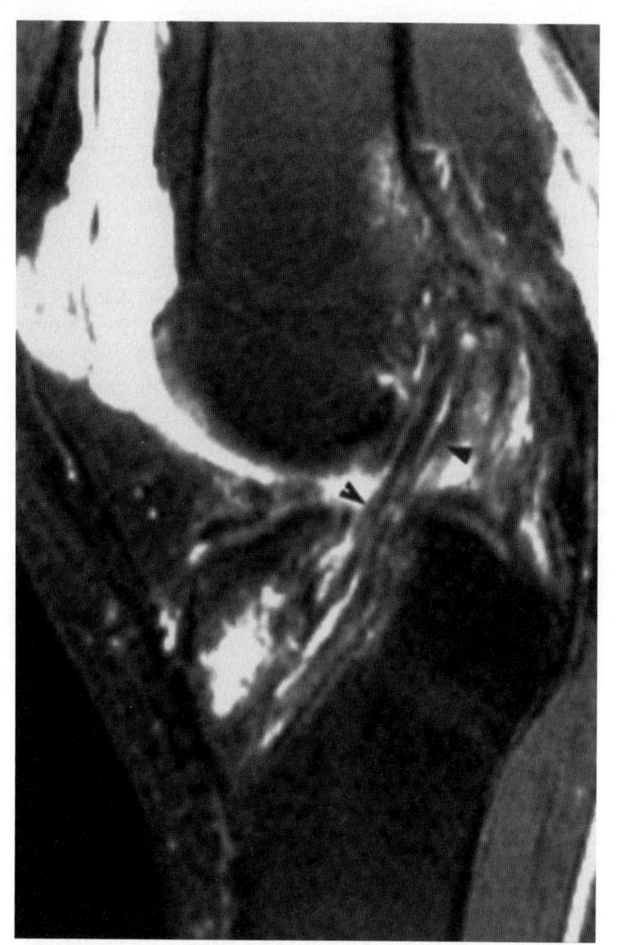

Figure 64–42. Intact reconstructed anterior cruciate ligament. Sagittal T2-weighted image with fat suppression after an arthrogram shows normal position of the tibial tunnel and the fibers of the anterior cruciate ligament to be intact *(arrowheads)*. There is no bony impingement on the graft.

64–44). Isometry implies that the reconstructed ligament remains taut throughout the range of motion. Avoidance of bony impingement on the graft can be accomplished when the tibial tunnel is created parallel to the intercondylar roof (Blumensaat's line) and its proximal opening lies slightly posterior to this line (see Fig. 64–42). A steep tunnel would result in impingement by the distal end of the intercondylar ridge when the knee is in extension.

A shallow tunnel may impinge on the ligament in extension and produces ligament laxity in flexion. Bony impingement eventually results in graft disruption (see Fig. 64–44). Graft disruption is evaluated best in the sagittal plane, where discontinuity of the fibers and bony impingement can be identified (see Fig. 64–44).

Posterior Cruciate Ligament

The PCL is larger than the ACL and twice as strong. The principal function of the PCL is to restrain the tibia from displacing posteriorly. It is evaluated best on sagittal images, where a normal PCL has an arcuate shape and low signal intensity on all MR imaging sequences (see Fig. 64–21*C*). Closely associated with the PCL are the meniscofemoral ligaments (ligaments of Humphrey and Wrisberg) (see Fig. 64–21). The PCL and the meniscofemoral ligaments originate from the lateral surface of the medial femoral condyle. The PCL courses posteriorly and inferiorly to insert on the posterior aspect of the proximal tibia about 1 cm distal to the joint line (see Fig. 64–21). The meniscofemoral ligaments insert on the posterior horn of the lateral meniscus (see Fig. 64–22).

The PCL is torn much less frequently than the ACL. A common mechanism of injury is attributed to a posterior force on the anterior tibia with the knee flexed (dashboard injury). Forced hyperextension of the knee is another mechanism. In contradistinction to the ACL, partial tears and interstitial tears of the PCL are fairly common, and these tears are shown clearly on MR imaging (Fig. 64–45).

The diagnostic criteria of PCL tears on MR imaging are similar to ACL tears. Discontinuity of the ligament with edema and bleeding indicates acute disruption. Most PCL tears occur in the midsubstance (Fig. 64–46). In less than 10% of cases, the PCL avulses off its insertion to tibia along with a piece of bone (see Fig. 64–11). The torn PCL has been shown to heal and reconstitute to almost normal size and shape (Fig. 64–47).

In contrast to the ACL, there are no standard surgical procedures that are widely accepted for repairing torn PCLs. As a result, only a few torn ligaments are treated surgically.

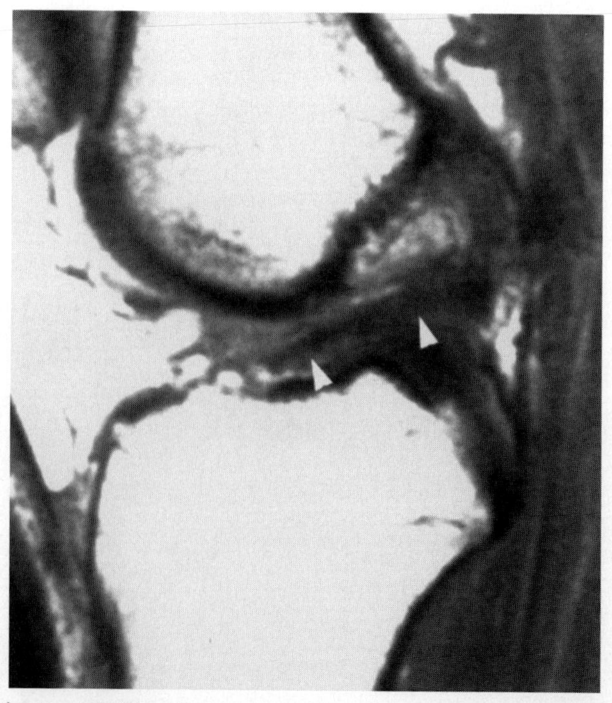

Figure 64–43. Chronic anterior cruciate ligament tear. Sagittal proton-density MR image shows the anterior cruciate ligament to be detached from its origin on the lateral femoral condyle and lying almost horizontally on the tibial surface *(arrowheads)*. There is no edema or hemorrhage in the ligament.

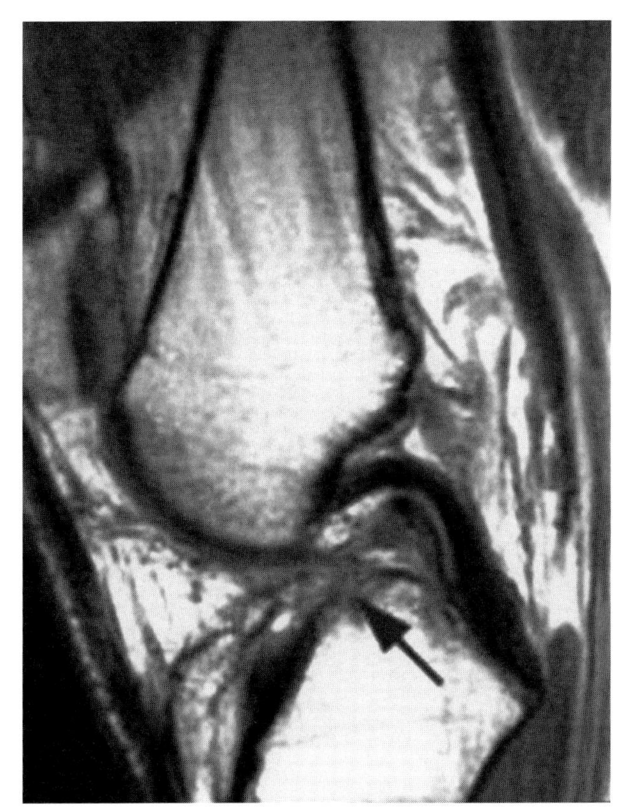

Figure 64–44. Anterior cruciate ligament graft impingement and disruption. Sagittal proton-density MR image shows the tibial tunnel to be anterior to the intercondylar roof, and the graft is disrupted *(arrow)*.

Medial Collateral Ligament

The MCL is a complex structure consisting of superficial and deep fibers, which act as the main restraint for valgus stress. The superficial layer, which is the largest and most important component of the MCL, originates from the superior aspect of the medial femoral condyle and inserts on the proximal shaft of the tibia about 5 to 6 cm distal to the medial tibial plateau (Fig. 64–48). The deep layer originates on the medial meniscus and consists of two parts, the meniscofemoral and meniscotibial ligaments, which insert on the femoral condyle and tibial plateau. The MCL is integrated closely with the joint capsule and medial meniscus, which is connected firmly to the MCL (see Fig. 64–17). This attachment makes the medial meniscus less mobile but more susceptible to injury compared with the lateral meniscus, which is not attached to any ligaments.

MR imaging evaluation of the MCL is done best in the coronal plane, where a normal MCL appears as a dark linear band on all sequences (see Fig. 64–48). Clinically, there are three grades of injury. Grade I represents a sprain, and it shows high signal in the soft tissues around the MCL on T2-weighted images, but the ligament itself appears intact. Grade II represents a partial tear, in which in addition to the edema, some MCL fibers are disrupted. Grade III is a complete disruption of the MCL, which on MR imaging shows discontinuity of the ligament along with significant edema and bleeding (Fig. 64–49). There is a high association of MCL tears with meniscocapsular separation, also called a *peripheral tear* of the medial meniscus. Gravin and coworkers (1993) pointed out that it can be difficult to differentiate partial from complete MCL tears on MR imaging.

Lateral (Fibular) Collateral Ligament

Injuries of the LCL are rare and typically are part of a severe injury involving the posterolateral corner of the knee (discussed earlier). The LCL is a cordlike structure that originates on the lateral femoral condyle; it courses

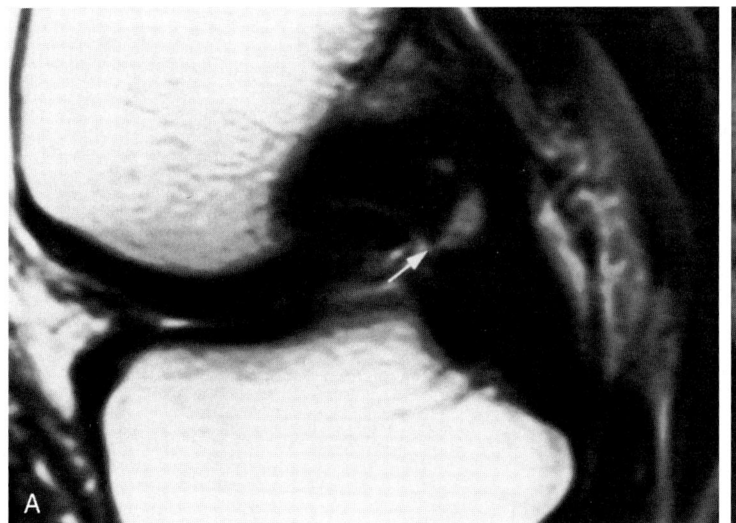

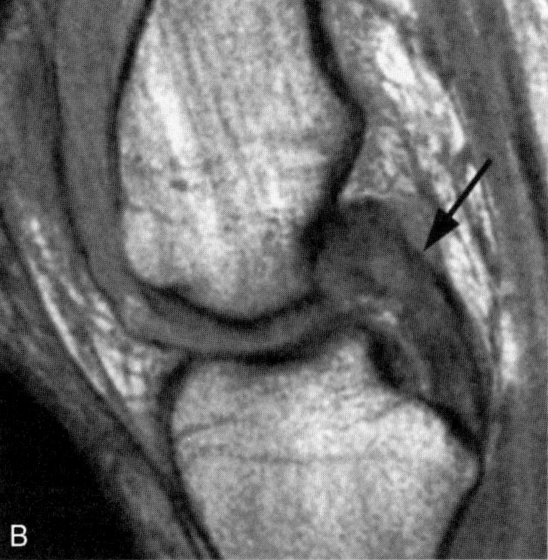

Figure 64–45. Partial and interstitial tears. *A*, Sagittal T1-weighted MR image shows a partial tear *(arrow)* of the posterior cruciate ligament 4 weeks after the initial injury. *B*, Sagittal proton-density image shows an interstitial tear of the posterior cruciate ligament *(arrow)* from another patient. Several of the fibers are still intact, but the ligament is thick and edematous.

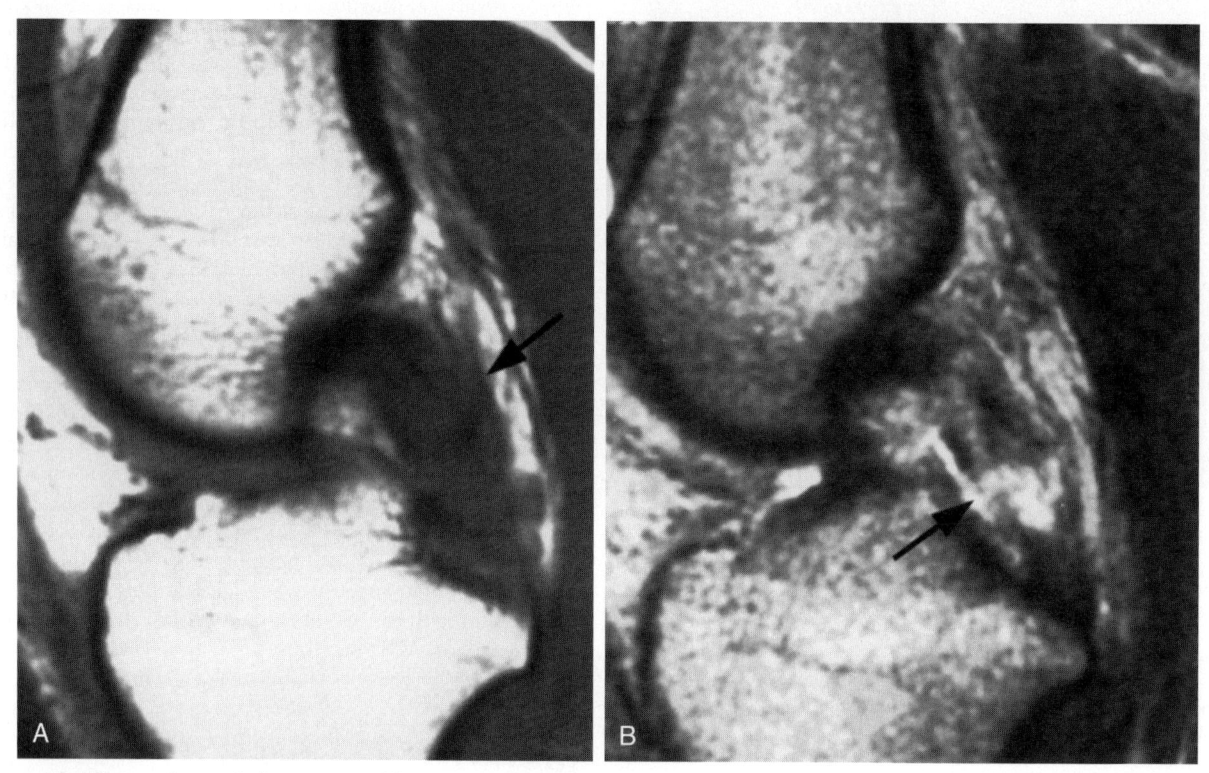

Figure 64–46. Complete midsubstance tear of the posterior cruciate ligament. *A* and *B*, Sagittal proton-density MR image and T2-weighted sagittal image show a full-thickness tear through the midsubstance *(arrow)* of the posterior cruciate ligament.

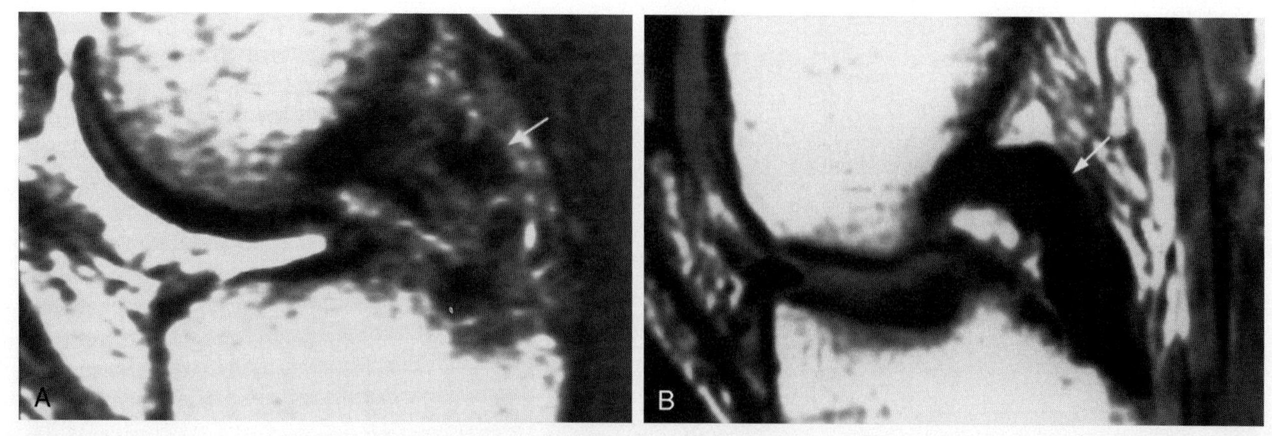

Figure 64–47. Reconstitution of a torn posterior cruciate ligament in a 21-year-old college wrestler. *A*, Sagittal T2-weighted MR image taken after the initial injury shows intrasubstance disruption of the posterior cruciate ligament *(arrow)*. *B*, Sagittal proton-density image taken 2 years later shows complete reconstitution of the posterior cruciate ligament without any surgical intervention *(arrow)*.

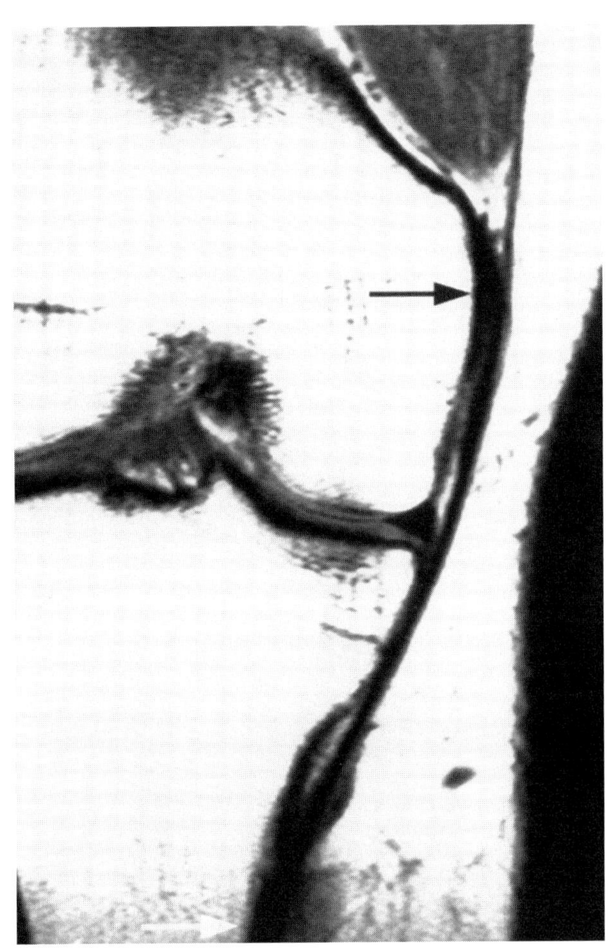

Figure 64–48. Normal medial collateral ligament. Coronal T1-weighted MR image shows the origin *(black arrow)* and insertion *(white arrow)* of the medial collateral ligament.

posteriorly and inferiorly to insert on the head of the fibula along with the tendon of the biceps femoris (Fig. 64–50). It functions as the main restraint to varus stress on the knee.

When the LCL is disrupted, plain radiographs can show an avulsion fracture off the fibular head (see Fig. 64–9). On MR imaging, LCL disruption is shown best on coronal images (see Figs. 64–10 and 64–50).

EXTENSOR MECHANISM

The extensor mechanism consists of the quadriceps femoris muscles, quadriceps tendon, patella, patellar ligament, medial and lateral patellar retinacula, and tibial tubercle.

Quadriceps Tendon

Anatomy

The quadriceps tendon appears multilayered on sagittal MR images; most commonly, three layers are seen, with the superficial layer being the tendon of the rectus femoris, the middle layer being the tendon of the vastus medialis and lateralis, and the posterior layer being the tendon of the vastus intermedius (Fig. 64–51). Typically, quadriceps tendons do not have a uniform number of layers from side to side. Most tendons show more layers on their medial than on their lateral side. The quadriceps tendon is about 8 ± 2 mm thick. Fibers from the rectus femoris tendon tend to bypass the patella anteriorly and become confluent with the patellar tendon.

Quadriceps Tendon Tear

Most patients with quadriceps tendon tear are older than age 40, probably because of age-related collagen degeneration. Biopsy specimens obtained from ruptured tendons almost always show focal tendon degeneration. Men are more affected than women (6:1). Disruptions can be partial or complete; partial tears involve one or two layers, leaving the other layers intact (Fig. 64–52). Partial tears are more difficult to diagnose clinically than complete tears, but even complete tears can be misdiagnosed initially. Systemic disease can be a predisposing factor, especially with bilateral tears. About one third of all patients with bilateral tears have a predisposing factor, such as uremia, hyperparathyroidism, diabetes mellitus, or collagen vascular disease. The mechanism of injury involves violent contraction of the quadriceps muscle against a full body weight. On physical examination, there is loss of active extension of the knee or inability to maintain a passively extended knee against gravity. Surgical repair is recommended for complete tears. Radiographically, there is a soft tissue mass superior to the patella; the patella appears low in position, and its upper pole is tilted anteriorly. On MR imaging, the tendon usually is disrupted at the osteotendinous junction or within 2 cm from the osteotendinous junction (Fig. 64–53). Occasionally, disruption occurs at the musculotendinous junction.

Patella

Anatomy and Function

The patella is the largest sesamoid bone in the body. Most of the quadriceps tendon inserts on the superior pole, whereas the patellar tendon arises from the inferior pole. About three quarters of its posterior surface proximally is covered by hyaline cartilage, which is the thickest hyaline cartilage in the body. On axial sections, the lateral facet is identified by its larger surface area and thin pointed lateral edge, whereas the medial facet is smaller and has thicker bone. In addition to the common traumatic fractures of the patella, rare types of patellar fractures include stress fracture, fracture complicating a surgical procedure on the patella, and osteochondral fracture associated with transient lateral patellar dislocation.

Patellar Stress (Fatigue) Fracture

Spontaneous fractures of the patella in individuals engaged in athletic activities almost always are due to

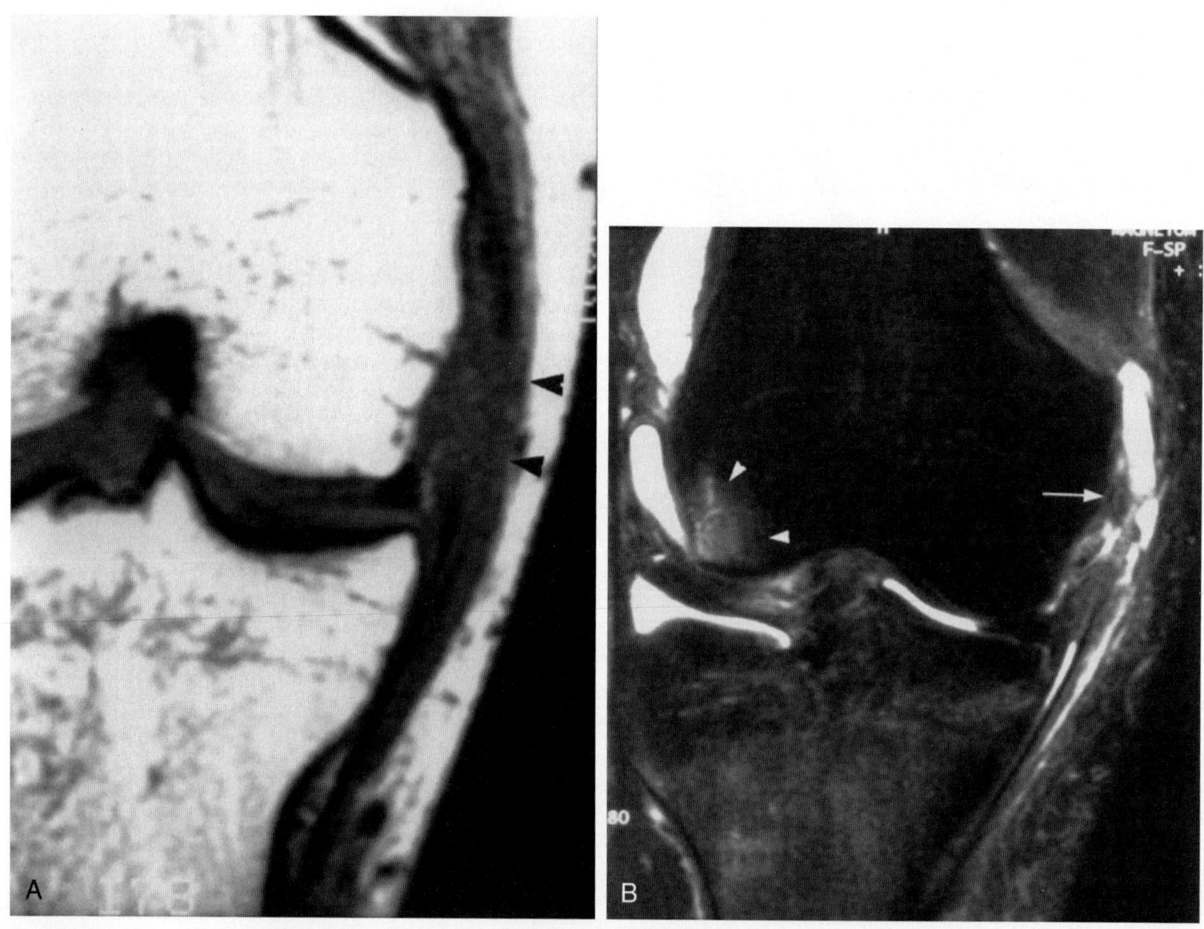

Figure 64–49. Examples of complete disruption of the medial collateral ligament. *A*, Coronal T1-weighted MR image shows complete disruption of the medial collateral ligament *(arrowheads)*. *B*, T2-weighted coronal image (from another patient) shows complete disruption of the medial collateral ligament at its origin from the medial femoral condyle *(arrow)*. The arrowheads point to a marrow contusion from the valgus injury, which caused the medial collateral ligament tear.

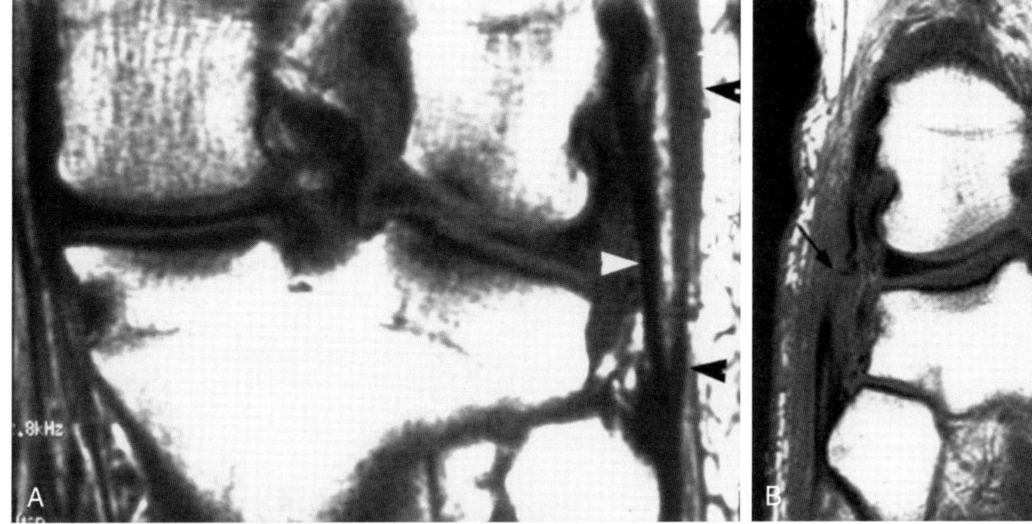

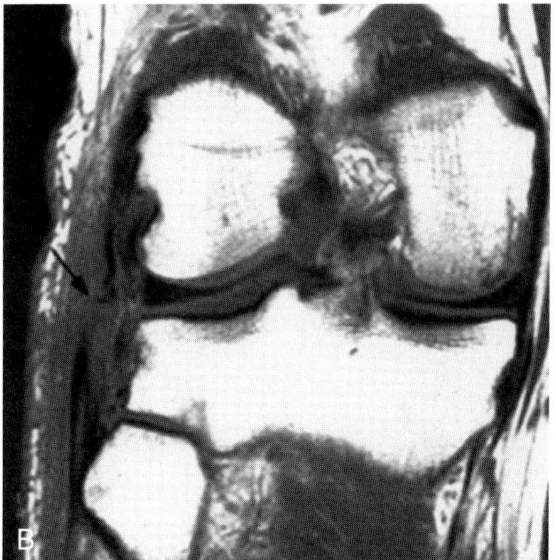

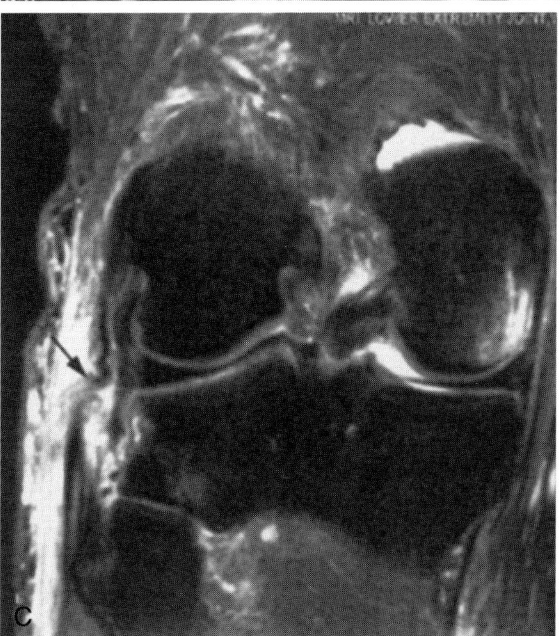

Figure 64–50. Normal and disrupted lateral collateral ligament. *A*, T1-weighted coronal MR image shows a normal lateral collateral ligament *(white arrowhead)* inserting to the lateral aspect of the fibular head along with the biceps femoris tendon *(black arrowheads)*. *B* and *C*, Coronal T1-weighted image and T2-weighted image with fat suppression from another patient show complete disruption of the lateral collateral ligament *(arrow)*.

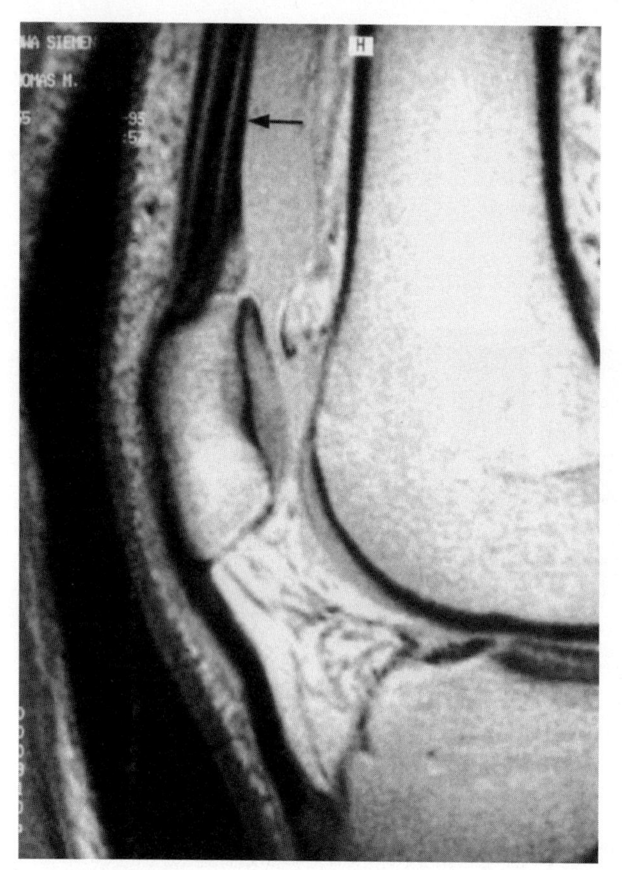

Figure 64–51. Normal quadriceps tendon. Three layers of the tendon *(arrow)* are seen on this sagittal proton-density MR image.

stress fracture. When complete, this fracture is typically a two-part transverse fracture (Fig. 64–54). In the early stages, this fracture can be incomplete, and it always starts on the anterior surface of the patella.

Fracture of the Patella After Anterior Cruciate Ligament Reconstruction

Fracture occurs when the central portion of the patellar tendon is used as a donor site for ACL reconstruction. The patellar tendon graft is harvested with a wedge of bone from the patella, which weakens the patella (Fig. 64–55).

Transient Lateral Patellar Dislocation

Transient lateral patellar dislocation is a commonly missed diagnosis at the time of initial presentation. This injury accounts for about 2% to 3% of all knee injuries and 9% to 16% of acute knee injuries in young athletes who present with hemarthrosis. Patients are usually unaware that they had a lateral dislocation because most dislocations are transient and reduce spontaneously. During the dislocation, the medial patellar facet impacts against the lateral femoral condyle, producing bone contusions, osteochondral fractures, or both. The medial retinaculum usually is disrupted or sprained. All these findings are shown best on axial MR sequences (Fig. 64–56).

Patellar Tendinitis

The normal patellar tendon has uniform low signal intensity on T1-weighted, T2-weighted, and proton-density images. The anteroposterior diameter of the tendon increases slightly from the proximal to the distal end (Fig. 64–57). The anteroposterior diameter of a normal tendon, measured at its patellar insertion, does not exceed 7 mm. Women generally tend to have thicker patellar tendons than men. In patellar tendinitis (also called *jumper's knee*), the condition is characterized by pain at the proximal insertion of the patellar tendon. It is seen often in basketball or volleyball players and is caused by repetitive microtrauma. Jumper's knee occurs during adolescence and early adulthood and is seen more often in boys. The tendon shows increased signal intensity on T1-weighted, T2-weighted, and proton-density images; it also shows increased anteroposterior diameter proximally exceeding 7 mm. The margins of the affected tendons appear indistinct, especially posterior to the thickened segment. In advanced cases of patellar tendinitis, some of the fibers at the osteotendinous junction may be disrupted (Fig. 64–58). If the offending sports activity continues, complete disruption can occur (Fig. 64–59).

Text continued on page 499

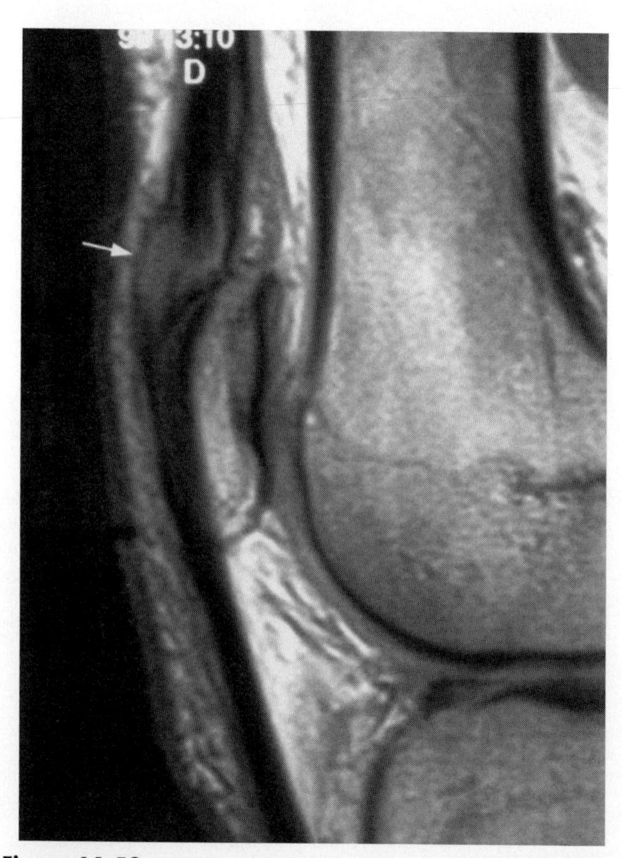

Figure 64–52. Partial tear of the quadriceps tendon in a 54-year-old man. Sagittal proton-density MR image shows thickening of the tendon and increased signal intensity at its insertion *(arrow)*. There is disruption of the anterior two layers of the tendon *(arrow)*.

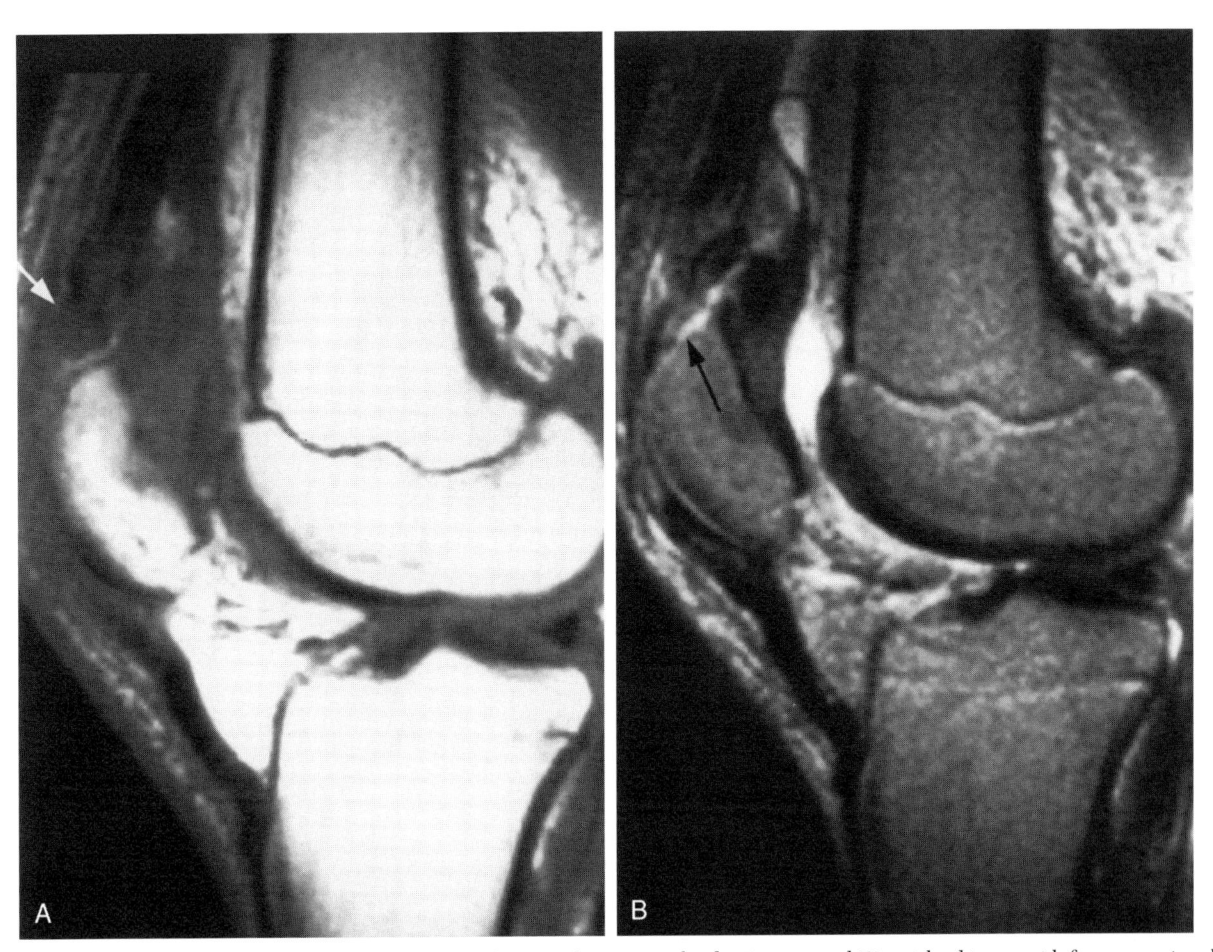

Figure 64–53. Complete disruption of the quadriceps tendon. *A* and *B*, T1-weighted MR image and T2-weighted image with fat suppression show complete disruption of the quadriceps tendon at the osteotendinous junction *(arrow)*. The patella is tilted anteriorly.

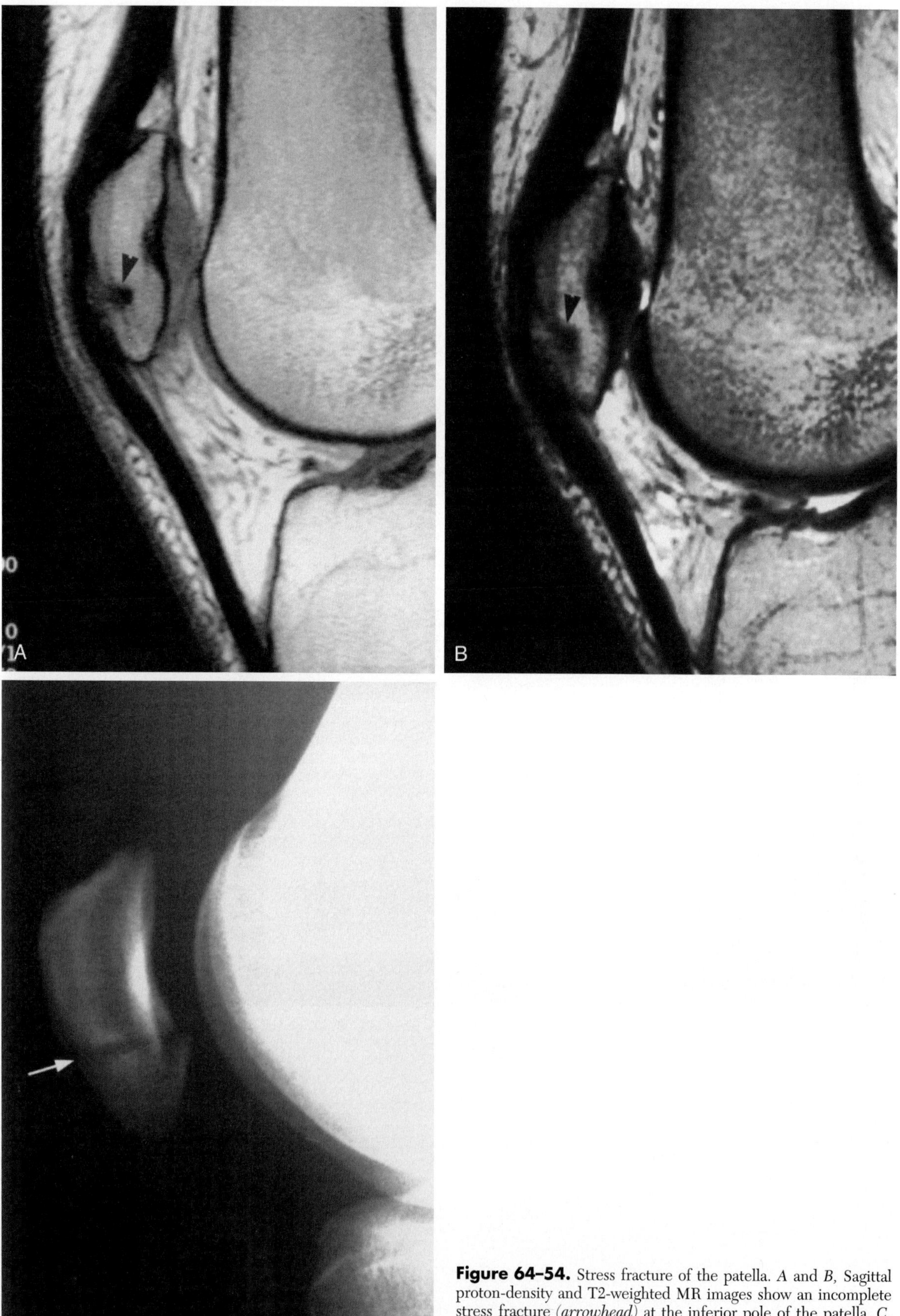

Figure 64–54. Stress fracture of the patella. *A* and *B*, Sagittal proton-density and T2-weighted MR images show an incomplete stress fracture *(arrowhead)* at the inferior pole of the patella. *C*, Lateral radiograph taken 4 weeks later shows the fracture has become complete *(arrow)*.

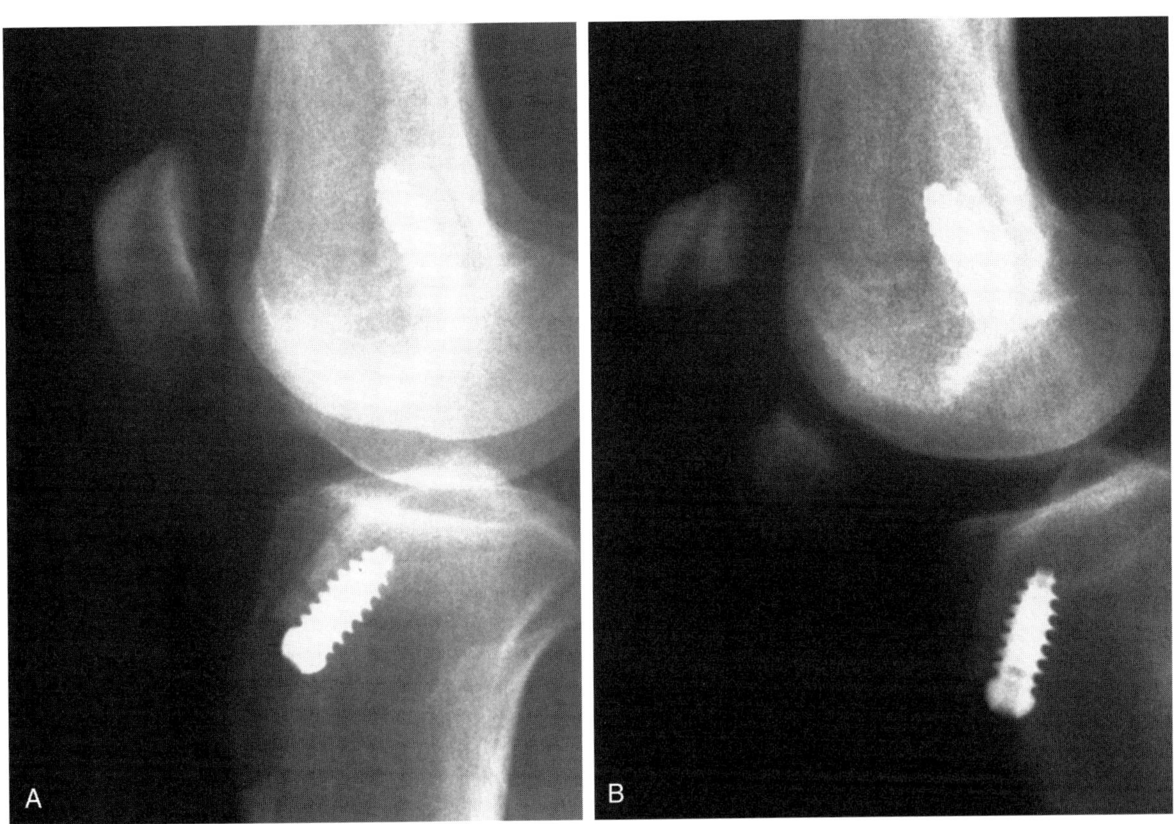

Figure 64–55. Fracture of the patella after anterior cruciate ligament reconstruction. *A*, Lateral radiograph of the patella after anterior cruciate ligament reconstruction shows the anterior cortex of the patella is missing inferiorly, which is a normal finding after harvesting the central portion of the patellar tendon to use as an anterior cruciate ligament graft. *B*, Lateral view taken 10 days later, after a physical therapy session, shows a displaced transverse fracture of the patella.

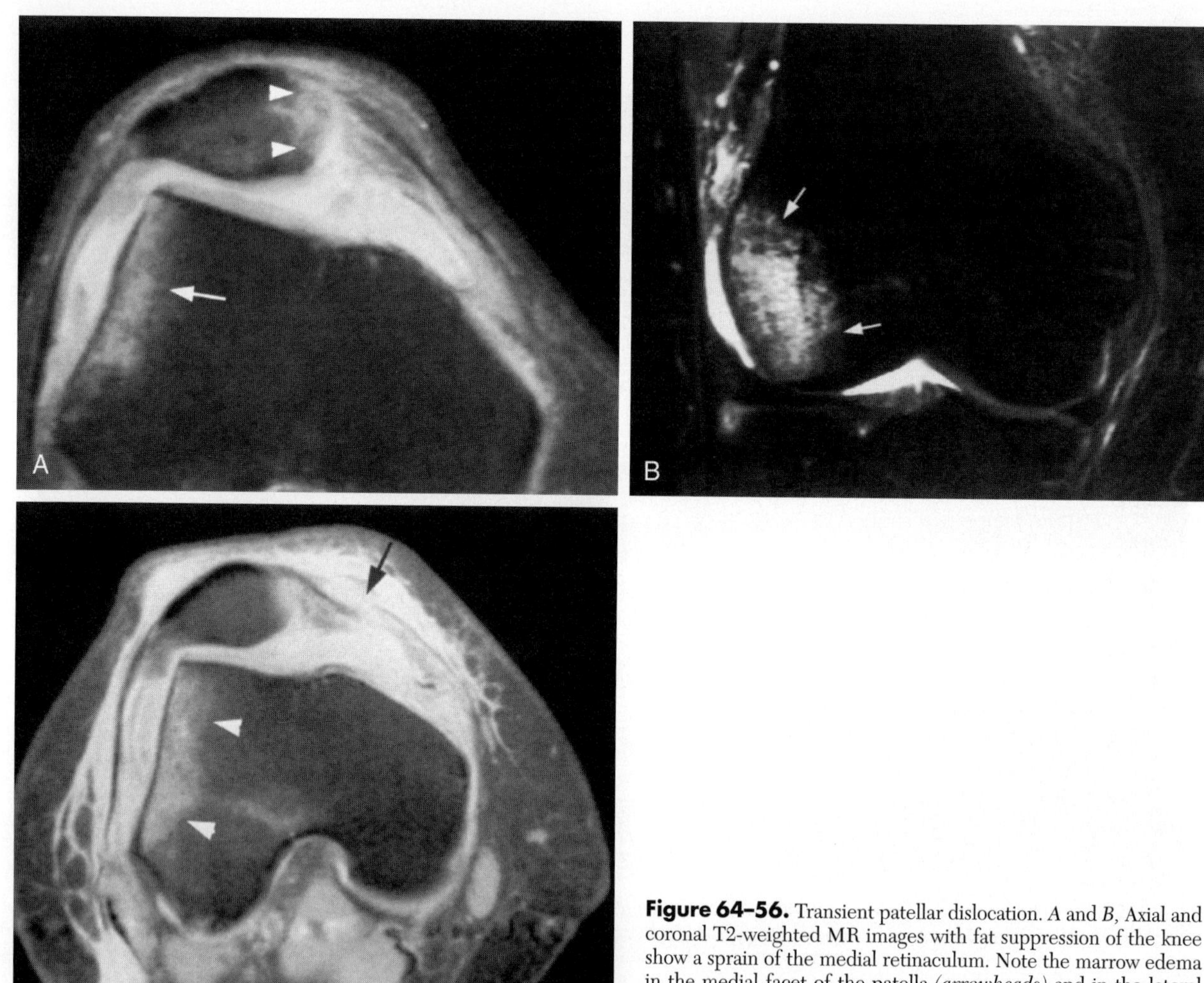

Figure 64–56. Transient patellar dislocation. *A* and *B*, Axial and coronal T2-weighted MR images with fat suppression of the knee show a sprain of the medial retinaculum. Note the marrow edema in the medial facet of the patella *(arrowheads)* and in the lateral femoral condyle *(arrows)*. *C*, Axial image in another patient shows disruption of the medial retinaculum *(arrow)* and edema in the lateral femoral condyle *(arrowheads)*.

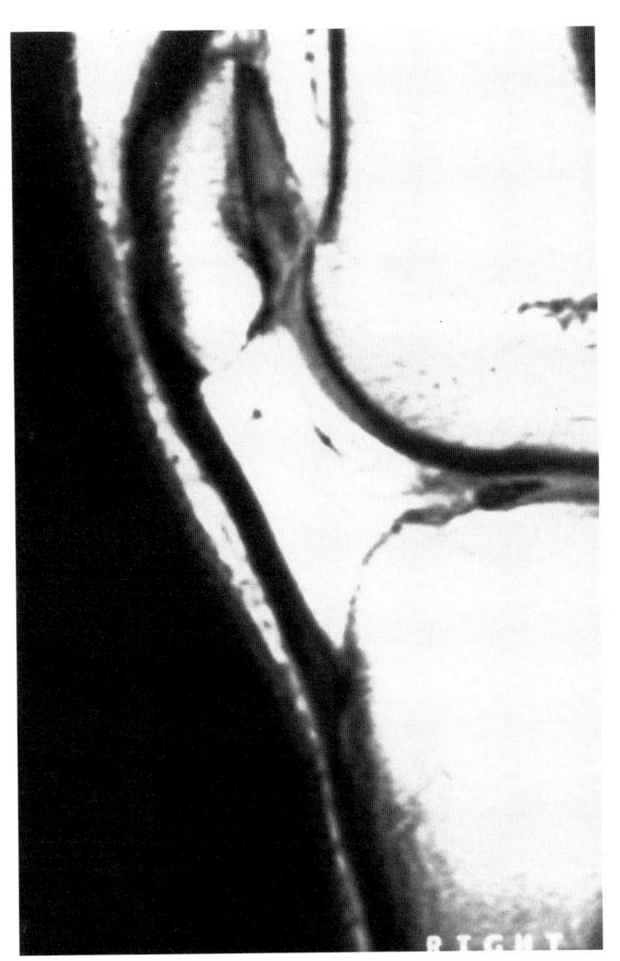

Figure 64–57. Normal patellar ligament. T1-weighted sagittal MR image shows a normal patellar ligament. The ligament is slightly thicker distally.

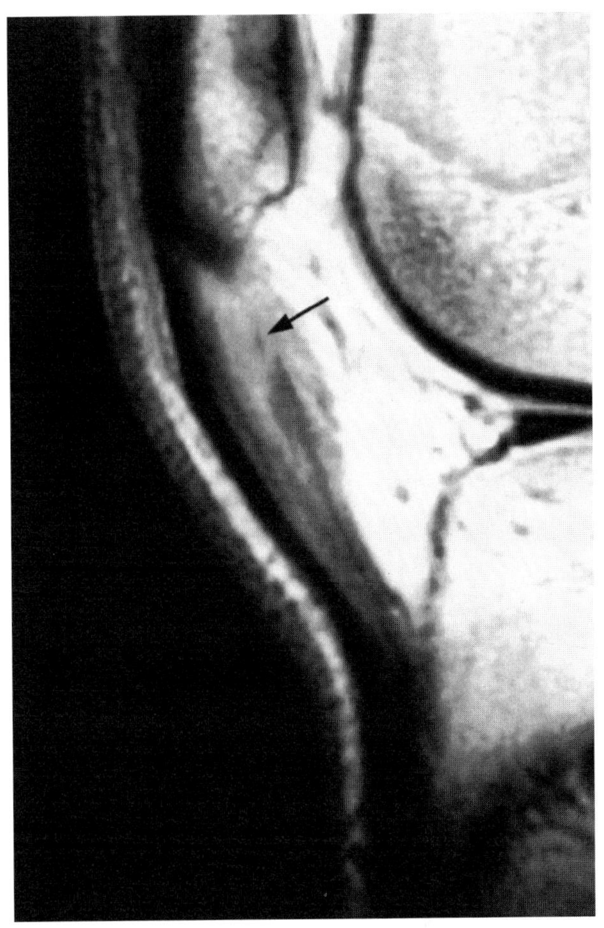

Figure 64–58. Partial tear of the patellar ligament in a 20-year-old gymnast who felt severe pain in the infrapatellar region after landing from a jump. Proton-density sagittal MR image shows thickening, increased signal intensity, and partial tearing of the patellar ligament *(arrow)*.

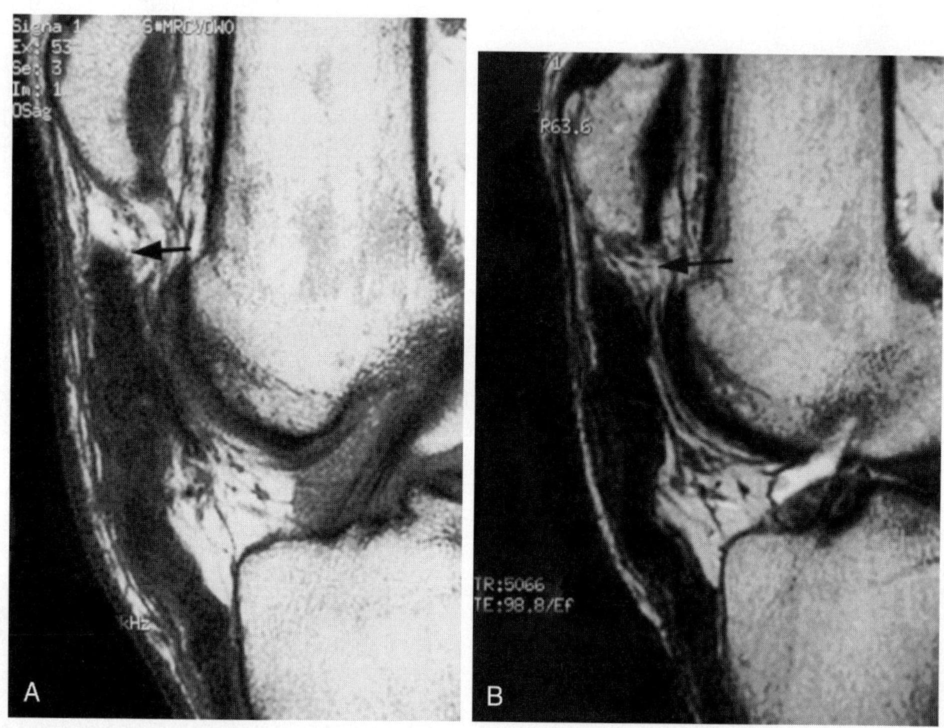

Figure 64–59. Complete tear of the patellar ligament. *A* and *B*, Sagittal proton-density and T2-weighted MR images show complete disruption of the ligament at the inferior pole of the patella *(arrow)*. The tear is old, and no edema or bleeding is present.

MISCELLANEOUS CONDITIONS

Discoid Meniscus

Discoid meniscus is a fairly common problem, occurring in about 1.5% to 3% of the general population. The lateral meniscus typically is involved. The cause of discoid meniscus is not clear, but it is likely to be a developmental problem in which enlargement of the meniscus is due to absence of the posterior attachment, leading to meniscus subluxation, repeated injury, and enlargement. Some studies showed that discoid meniscus is frequently bilateral. The most common symptoms are pain, snapping sound with flexion and extension of the knee, and decreased range of motion. In some patients, the condition is asymptomatic. Both sexes are affected, but most reports show a female predominance.

The classification most orthopedists use is the Watanabe classification (1978); three types of discoid menisci are described: complete, incomplete, and Wrisberg type. Complete and incomplete types can be seen at any age, but the Wrisberg type has been reported only in children younger than age 16. The shape of the meniscus in the Wrisberg type is fairly normal (i.e., not discoid), but the meniscus is reported arthroscopically to show lack of normal attachment to the tibia posteriorly. This lack of normal attachment results in hypermobility of the posterior horn, which subluxes into the joint with flexion and extension, at arthroscopy. The prevalence of the Wrisberg type is unknown, and we do not know of any imaging criteria for diagnosing this abnormality.

Plain radiographs show widening of the lateral joint space and rarely cupping of the lateral tibial plateau. In most cases, the radiographic findings are subtle and easy to miss.

The MR imaging diagnosis of a discoid meniscus is based on showing an abnormally large meniscus, which covers most of the lateral tibial plateau (Fig. 64–60). The complete type, also called the *slab type discoid meniscus,* is easy to diagnose by MR imaging or by arthrography (Fig. 64–61). It covers the entire articular surface of the tibial plateau and extends into the intercondylar notch. An incomplete discoid meniscus can be difficult to differentiate from a large but normal meniscus.

The average width of a normal lateral meniscus is 11.6 mm. On a sagittal MR sequence using 4-mm-thick slices, the anterior and posterior horns should appear connected on about two sections. A discoid meniscus is diagnosed when three or more contiguous slices show continuity between the anterior and posterior horns (Fig. 64–62).

In discoid menisci, abnormal high signal intensity is a frequent finding on T1-weighted, proton-density, and T2-weighted images (see Fig. 64–61). It is, however, often difficult to distinguish diffuse signal abnormalities caused

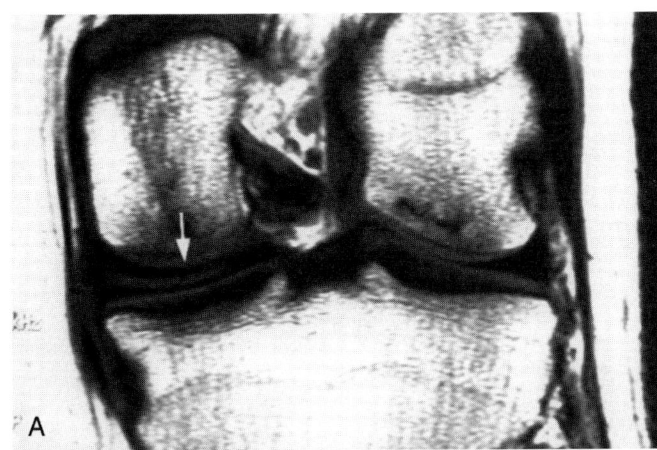

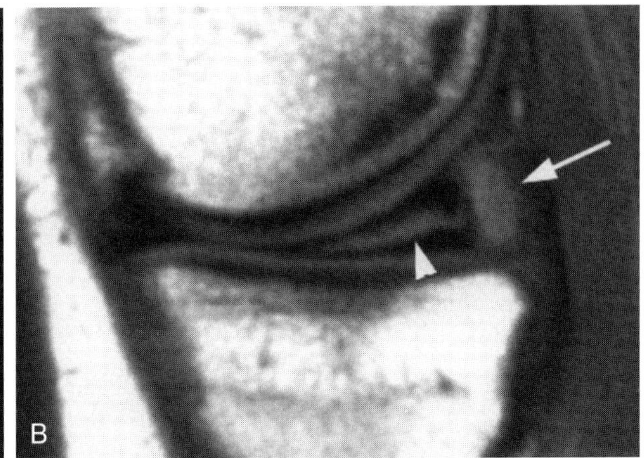

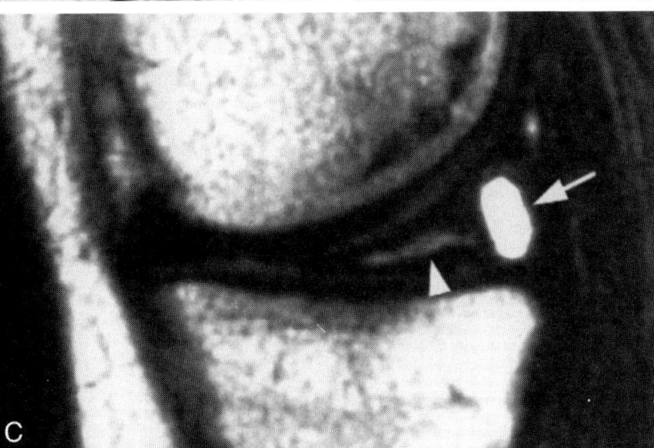

Figure 64–60. Torn incomplete discoid meniscus communicating with a meniscal cyst. *A,* T1-weighted coronal MR image shows a discoid meniscus on the medial side (which is rare) *(arrow).* The meniscus has an extensive tear. *B* and *C,* Sagittal T1-weighted and T2-weighted images show the large meniscus with a tear *(arrowhead)* communicating with a paramenisсal cyst. Incidentally noted are small marrow infarcts in the femoral condyles caused by previous steroid use.

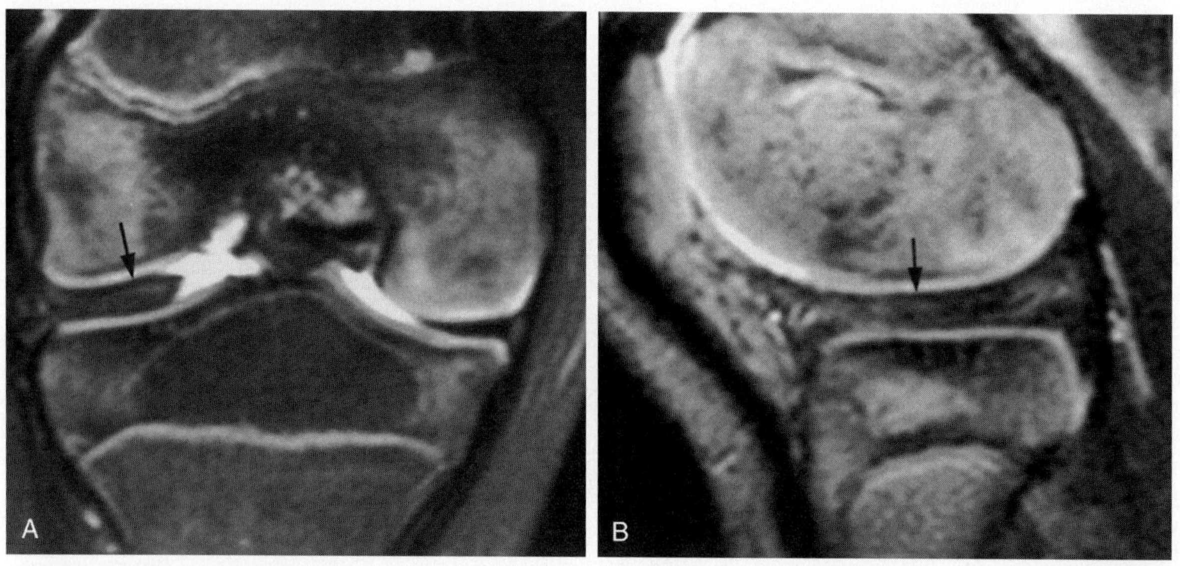

Figure 64–61. Complete (slablike) discoid meniscus in a child with knee pain. *A* and *B*, Coronal and sagittal T2-weighted MR images show a large thick lateral meniscus with increased signal within *(arrow)*.

by myxoid degeneration from those caused by meniscus tears. There also is a higher incidence of meniscus cysts associated with discoid menisci (see Fig. 64–60).

Synovial Plicae and Plicae Syndrome

Plicae are embryologic synovial remnants seen on imaging studies as incidental findings. They present as thin synovial projections arising at specific sites from which they derive their names. These are the mediopatellar (Fig. 64–63*A*), suprapatellar, and infrapatellar (or ligamentum mucosum) (Fig. 64–63*B*, *C*) plicae. The mediopatellar plica can give rise to symptoms owing to chronic irritation and thickening. Over time, an injured or inflamed plica becomes stiff and does not glide normally over the femoral condyle. Symptoms of pain, locking, and clicking can be present. There are no definite MR imaging criteria for this syndrome, but it should be suggested when the mediopatellar plica is thicker than expected or thicker than other normal plicae in the joint (Fig. 64–63*D*).

The suprapatellar plicae normally forms an incomplete septum between the suprapatellar pouch and the joint, allowing for free communication between the two compartments. Rarely the suprapatellar plica is complete, resulting in two separate compartments. In such a situation, a synovial disease can involve one compartment without affecting the other.

Cysts, Ganglia, and Bursae

Meniscus Cysts

Meniscus cysts are seen commonly on MR imaging of the knee. Clinically, patients present with knee pain and possibly a palpable mass. A cyst also can be discovered incidentally on MR imaging in patients coming for evaluation of internal knee derangement. Initially, meniscus cysts were thought to be more common on the lateral side of the knee, where they are easier to palpate, but with the wide use of MR imaging, it is now believed that meniscus cysts are associated more commonly with the posterior horn of the medial meniscus, where they are difficult to palpate. On the medial side, the MCL forces these cysts to migrate away from the site of the meniscus tear.

The underlying cause of a meniscus cyst is a meniscus tear, which allows synovial fluid to be pumped out through the tear and into the soft tissues to accumulate as a cyst (Fig. 64–64). Some authors use the term *parameniscus cyst* to describe these lesions and the term *intrameniscus cyst* to describe fluid collections within the meniscus. Intrameniscus cysts may not show focal enlargement of the meniscus. Intrameniscus cysts are less common than parameniscus cysts.

A long-standing parameniscus cyst can produce a well-corticated pressure erosion in the tibial plateau just

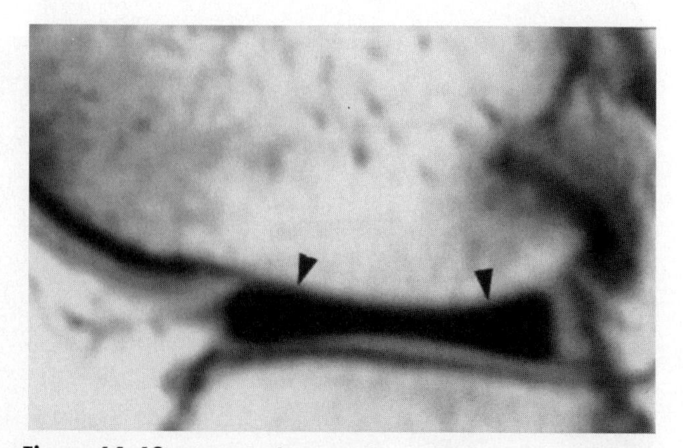

Figure 64–62. Continuity between the anterior and posterior horns as a sign of discoid meniscus, shown here on sagittal T1-weighted MR image *(arrowheads)*.

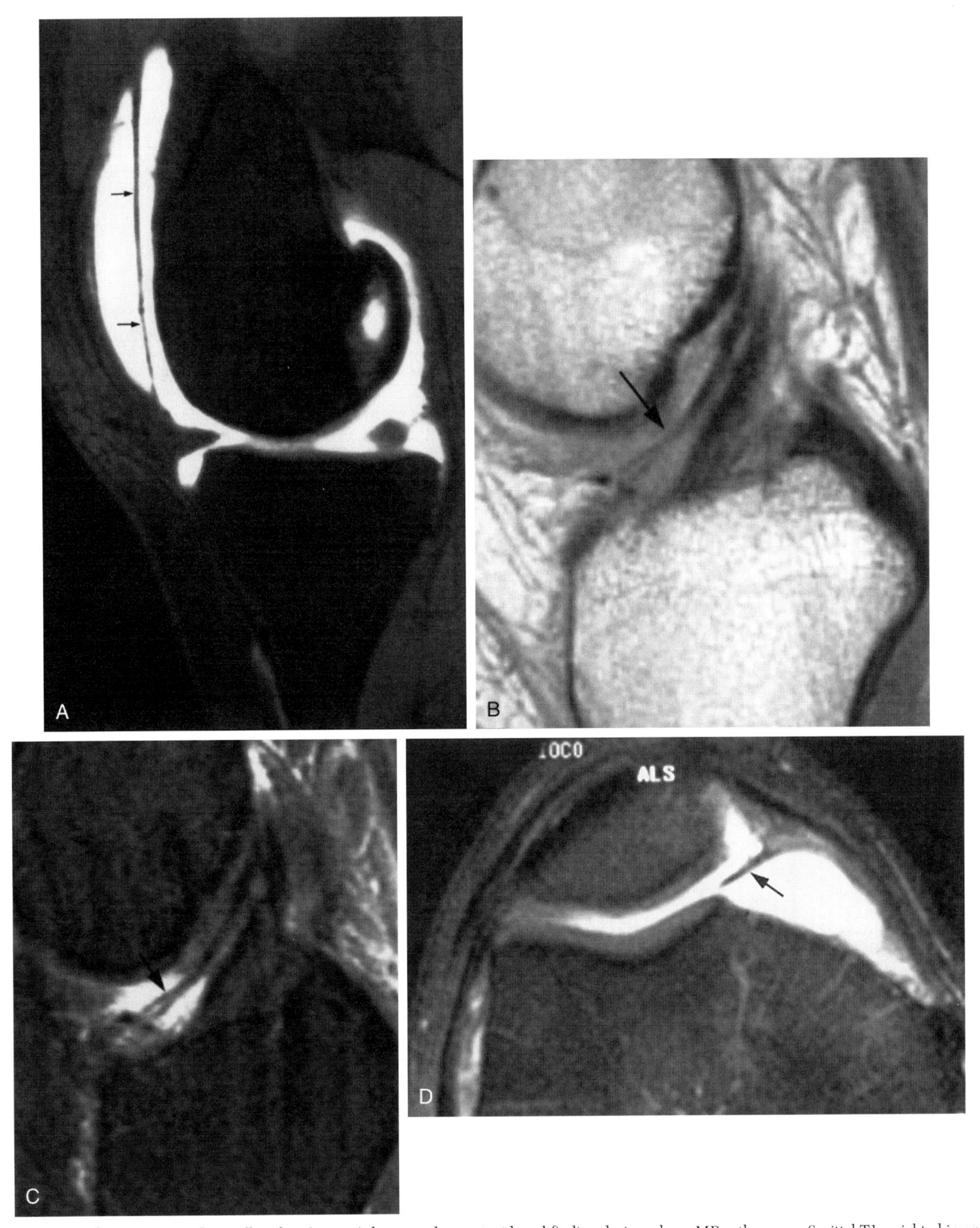

Figure 64–63. Plica. *A*, Mediopatellar plica *(arrows)* discovered as an incidental finding during a knee MR arthrogram. Sagittal T1-weighted image with fat suppression. *B* and *C*, Sagittal proton-density image and T2-weighted image with fat suppression show an infrapatellar plica (ligamentum mucosum) *(arrow)*. It runs parallel to the anterior cruciate ligament, but its insertion on the tibia is anterior to the anterior cruciate ligament. *D*, Mediopatellar plica, which is thick and causing pain. Axial and sagittal T2-weighted MR images with fat suppression show a thickened plica *(arrow)*. A joint effusion is present.

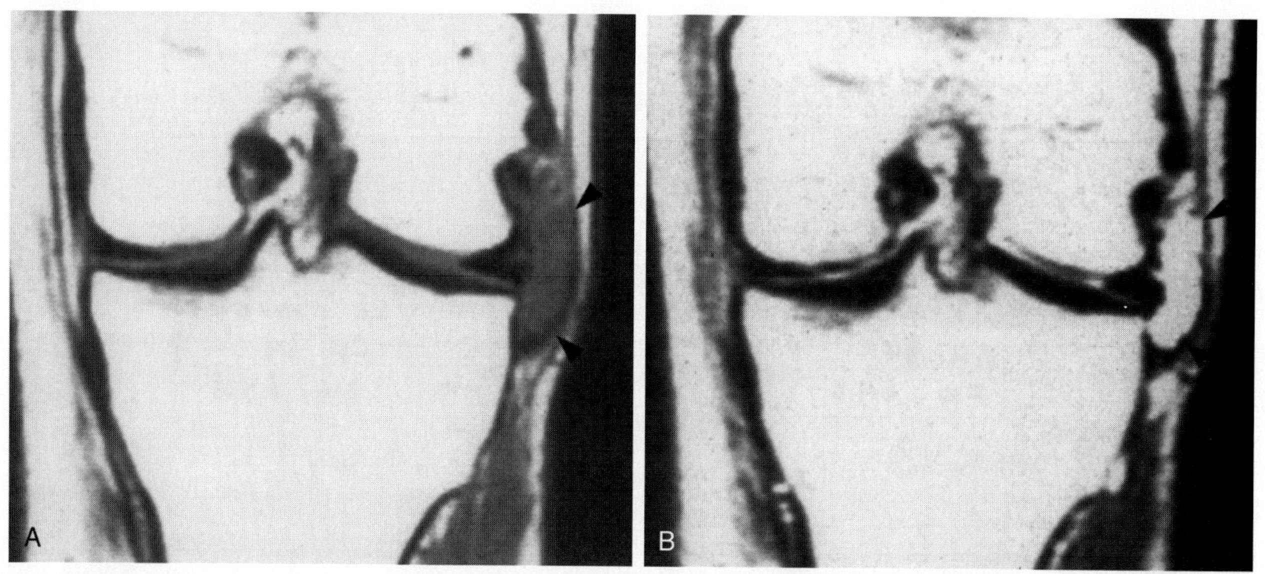

Figure 64–64. *A* and *B*, Parameniscal cyst *(arrowheads)* communicating with a meniscus tear on coronal T1-weighted and T2-weighted MR images.

below the joint (Fig. 64–65). This erosion is fairly characteristic of a parameniscus cyst on plain radiography (Fig. 64–66). The MR imaging appearance of a parameniscus cyst is that of a well-defined fluid collection with low or intermediate signal on T1-weighted images and increased signal on T2-weighted images. Cysts almost always seem to communicate with a horizontal cleavage tear and frequently show septations (Fig. 64–67). It is not known why some meniscus tears produce parameniscus cysts, whereas others do not have this capability.

Baker's (Popliteal) Cyst

Baker's cyst is a commonly encountered cystic mass that occurs at the posteromedial aspect of the knee. It results from an outpouching of the synovium at a weak location in the joint capsule between the medial head of the gastrocnemius and semimembranosus tendons (Fig. 64–68). The cyst can enlarge, or it can rupture into the soft tissues of the calf, where it can be confused with thrombophlebitis. On MR imaging, Baker's cyst has a typical appearance with low signal intensity on T1-weighted images and high signal intensity on T2-weighted images (Fig. 64–69). It can have few septations, and occasionally it can contain thick gelatinous fluid that is difficult to aspirate. Chondral loose bodies are a fairly common finding in these cysts.

Ganglion Cyst

Ganglion cysts do not communicate with joints. They can arise from the tendon sheaths, muscles, and ligaments. Within the knee, they can arise from the cruciate

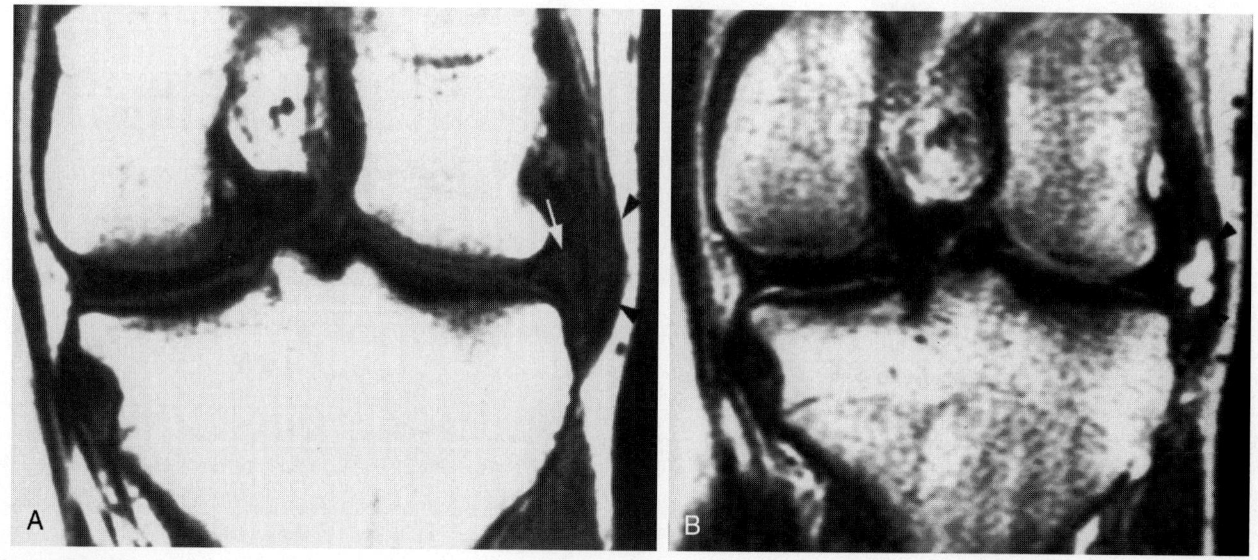

Figure 64–65. *A* and *B*, Parameniscal *(black arrowheads)* and intrameniscal *(white arrow)* cysts on coronal T1-weighted and T2-weighted MR images.

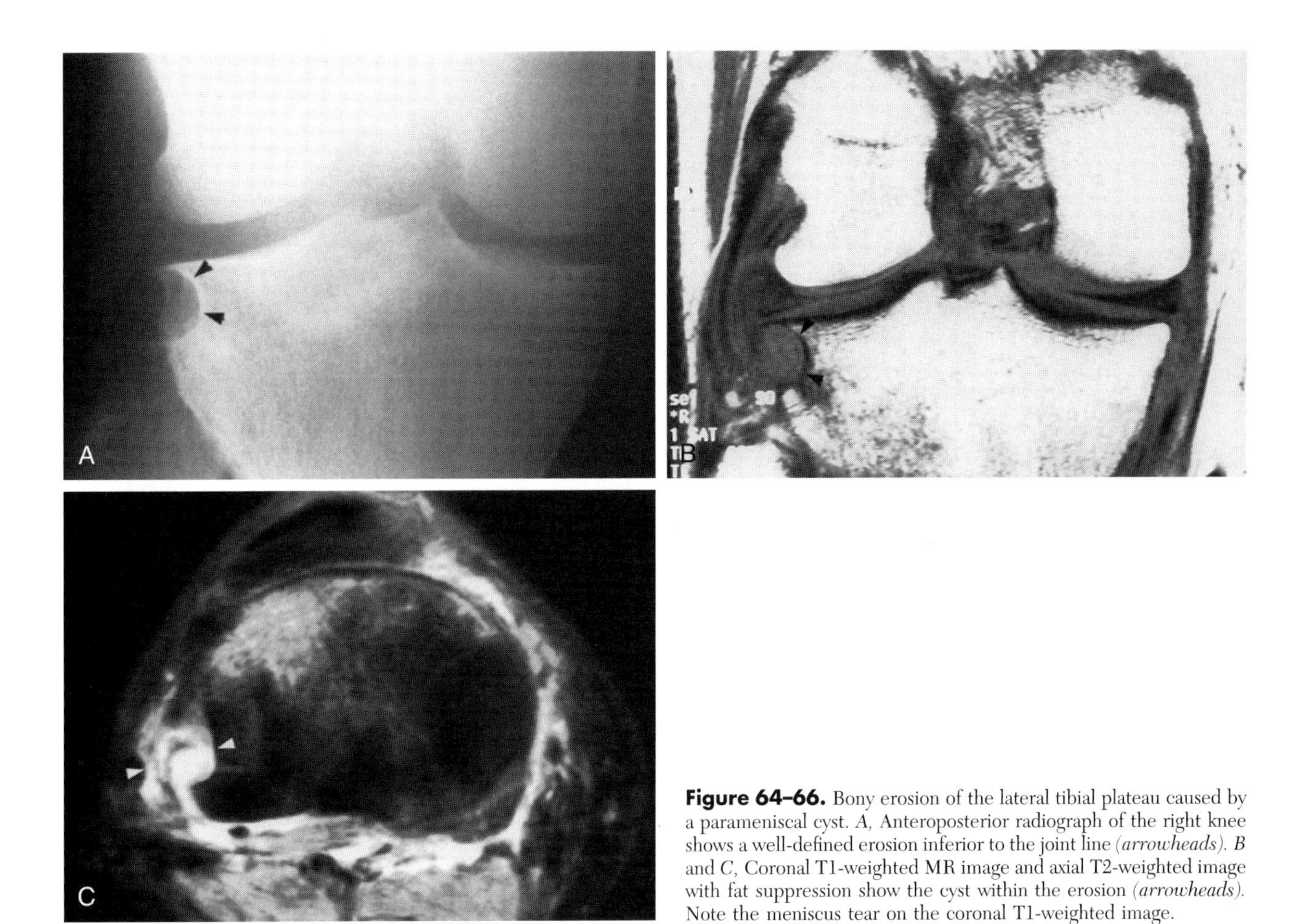

Figure 64–66. Bony erosion of the lateral tibial plateau caused by a parameniscal cyst. *A*, Anteroposterior radiograph of the right knee shows a well-defined erosion inferior to the joint line *(arrowheads)*. *B* and *C*, Coronal T1-weighted MR image and axial T2-weighted image with fat suppression show the cyst within the erosion *(arrowheads)*. Note the meniscus tear on the coronal T1-weighted image.

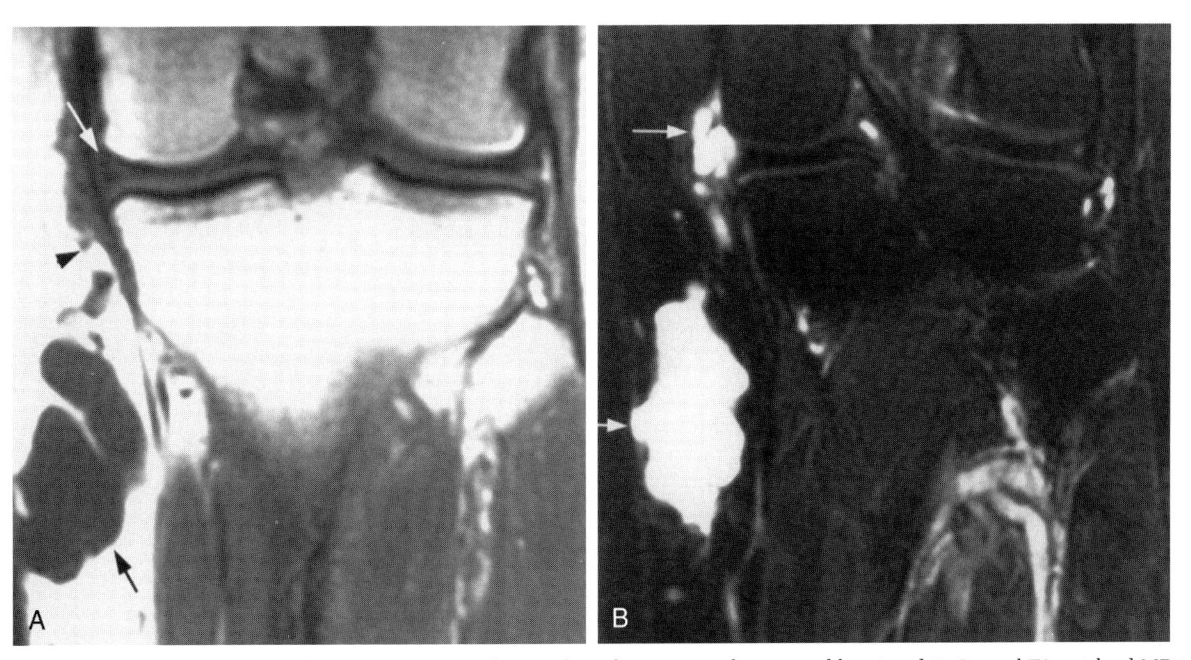

Figure 64–67. Large meniscal cyst with septations. It extends away from the joint into the proximal leg. *A* and *B*, Coronal T1-weighted MR image and T2-weighted image with fat suppression show the large cyst *(arrows* and *arrowheads)* communicating with a torn medial meniscus.

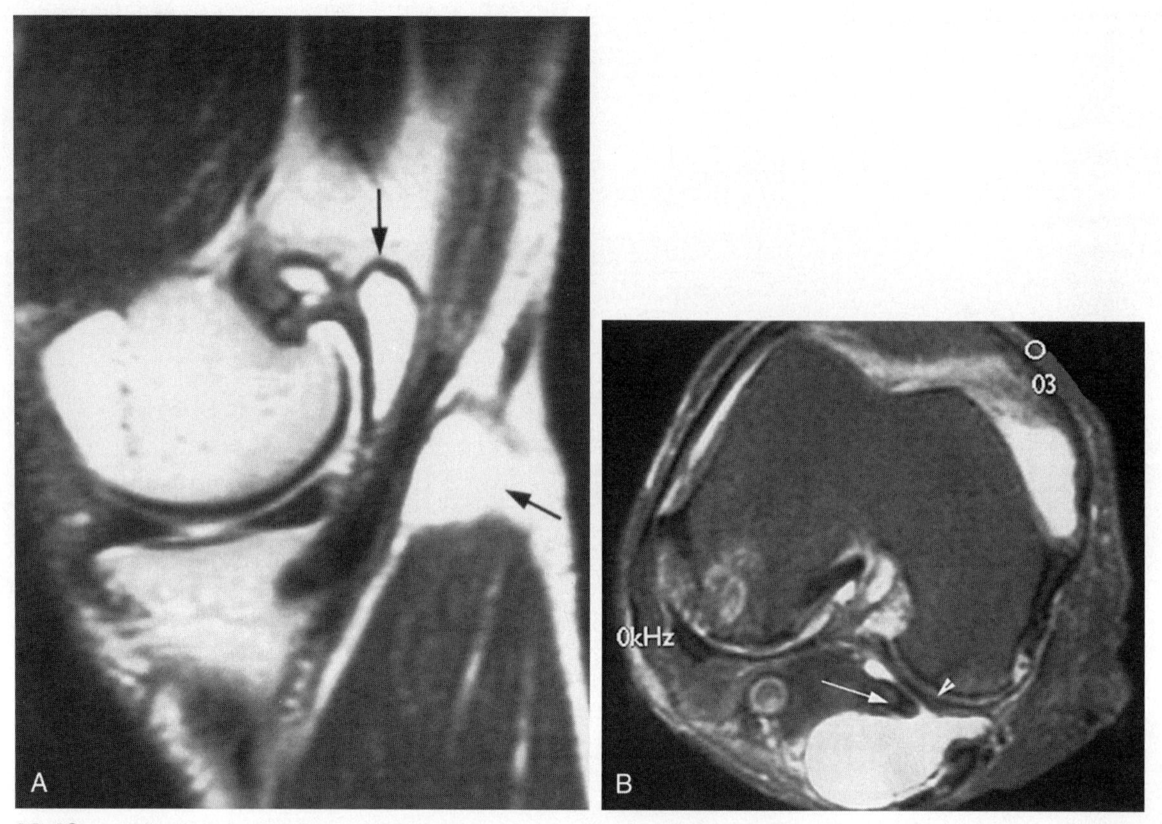

Figure 64–68. A small Baker's (popliteal) cyst developing between the tendons of the medial head of the gastrocnemius and semimembranosus. *A,* T2-weighted sagittal MR image through the semimembranosus tendon shows the Baker's (popliteal) cyst *(black arrows). B,* Axial T2-weighted image of the knee with fat suppression shows a high signal intensity region representing Baker's (popliteal) cyst. The arrow points to the medial head of the gastrocnemius tendon. The arrowhead points to the small communication between the joint and the cyst.

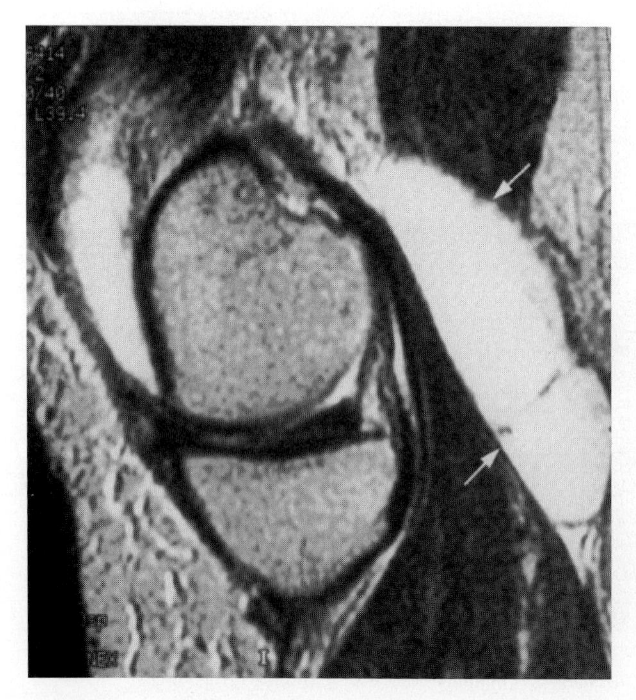

Figure 64–69. Large Baker's (popliteal) cyst *(arrows)* seen on a T2-weighted sagittal MR image.

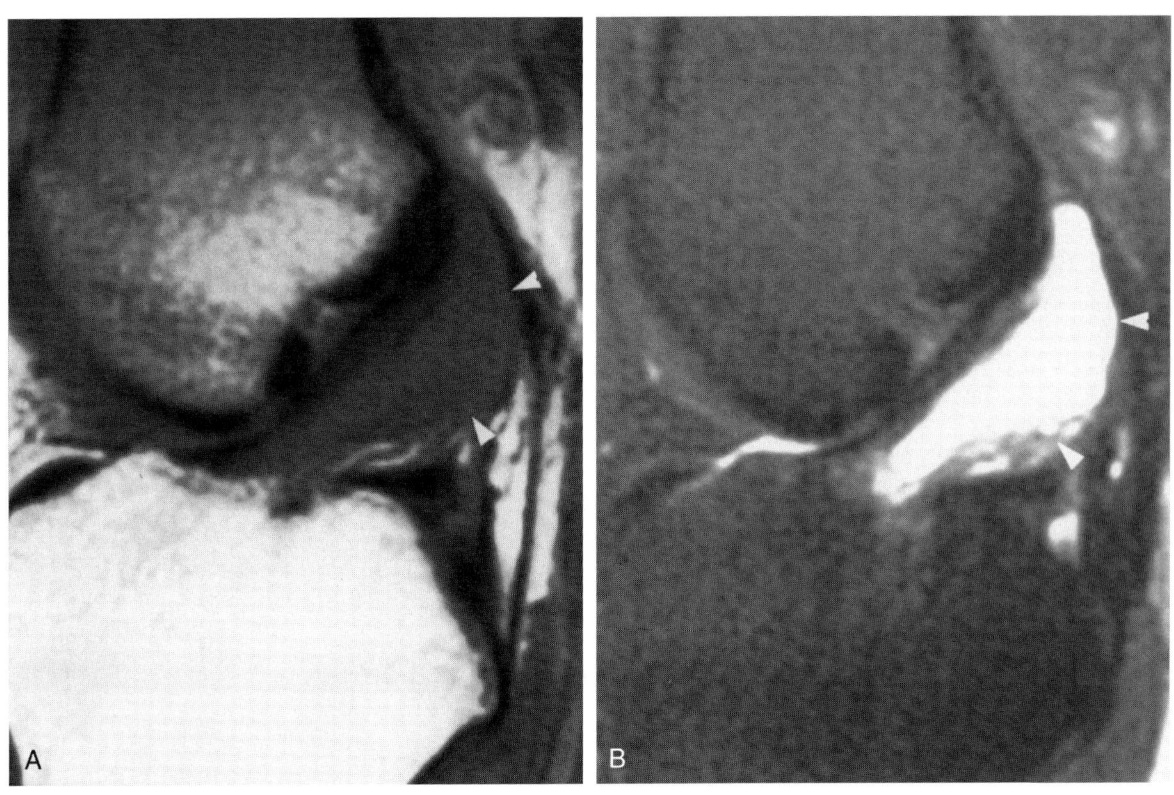

Figure 64–70. Ganglion cyst of the anterior cruciate ligament. *A* and *B*, Proton-density sagittal MR image and T2-weighted sagittal image with fat suppression show a ganglion cyst of the anterior cruciate ligament *(arrowheads)*.

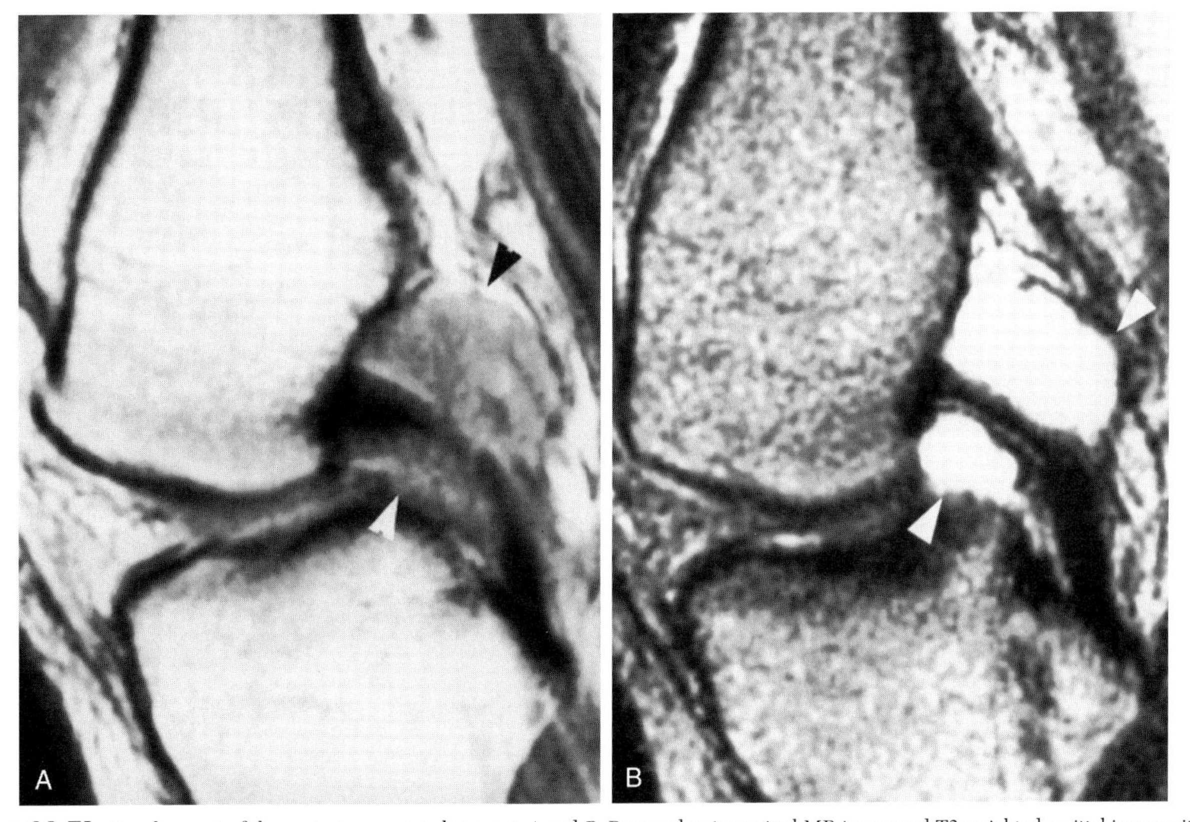

Figure 64–71. Ganglion cyst of the posterior cruciate ligament. *A* and *B*, Proton-density sagittal MR image and T2-weighted sagittal image with fat suppression show a ganglion cyst of the posterior cruciate ligament *(arrowheads)*.

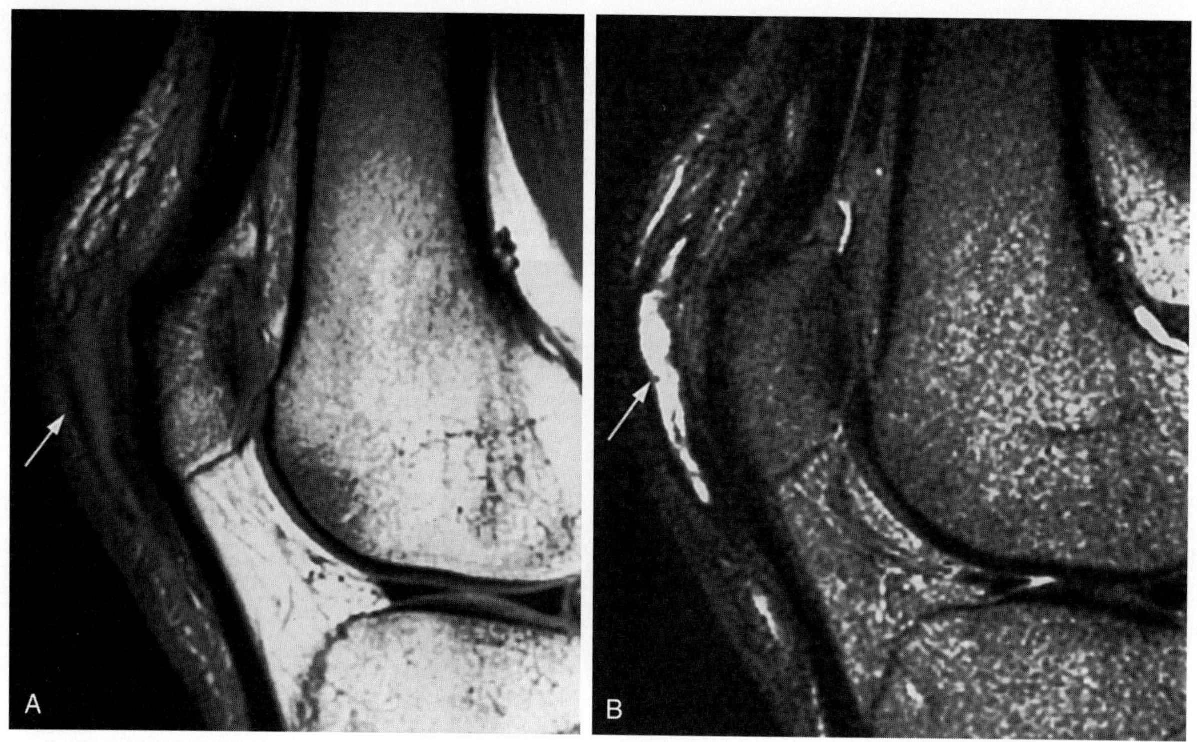

Figure 64–72. Prepatellar bursitis in a young athlete who presented with prepatellar soft tissue swelling and pain. *A* and *B*, Proton-density sagittal MR image and T2-weighted sagittal MR image with fat suppression show a fluid collection anterior to the patella *(arrow)*. There is associated soft tissue edema along with the fluid collection. These findings are typical of prepatellar bursitis.

ligaments, where they can produce vague symptoms. On MR imaging, they appear as well-defined fluid collections closely associated with the cruciate ligaments (Figs. 64–70 and 64–71).

Bursae

There are several bursae around the knee, and these can become inflamed from repeated trauma resulting in traumatic bursitis. The most common form of bursitis in the knee is prepatellar bursitis, which presents as painful swelling anterior to the patella (Fig. 64–72). Imaging is not required for the diagnosis of prepatellar bursitis in most cases, but bursitis sometimes is detected on MR imaging when done for other reasons.

Recommended Readings

Aichroth PM, Patel DV, Marx CL: Congenital discoid lateral meniscus in children: A follow-up study and evolution of management. J Bone Joint Surg Br 73:932, 1991.

Applegate GR, Flannigan BD, Tolin BS, et al: MR diagnosis of recurrent tears in the knee: Value of intraarticular contrast material. AJR Am J Roentgenol 161:821, 1993.

Bates DG, Hresko MT, Jaramillo D: Patellar sleeve fracture: Demonstration with MR imaging. Radiology 193:825, 1994.

Boden SD, Davis DO, Dina TS, et al: A prospective and blinded investigation of magnetic resonance imaging of the knee. Clin Orthop 282:177, 1992.

Brandser EA, Rile MA, Berbaum KS, et al: MR imaging of anterior cruciate ligament injury: Independent value of primary and secondary signs. AJR Am J Roentgenol 167:121, 1996.

Brody GA, Pavlov H, Warren RF, et al: Plica synovialis infrapatellaris: Arthrographic sign of anterior cruciate ligament disruption. AJR Am J Roentgenol 140:767, 1983.

Brophy DP, O'Malley M, Lui D, et al: MR imaging of tibial plateau fractures. Clin Radiol 51:873, 1996.

Burk DL Jr, Dalinka MK, Kanal E, et al: Meniscal and ganglion cysts of the knee: MR evaluation. AJR Am J Roentgenol 150:331, 1988.

Carpenter JE, Kasman R, Matthews LS: Fractures of the patella. J Bone Joint Surg Am 75:1550, 1993.

Christie MJ, Dvonch VM: Tibial tuberosity avulsion fracture in adolescents. J Pediatr Orthrop 1:391, 1981.

Connolly B, Babyn PS, Wright JG, et al: Discoid meniscus in children: Magnetic resonance imaging characteristics. Can Assoc Radiol J 47:347, 1996.

De Smet AA, Norris MA, Yandow DR, et al: MR diagnosis of meniscal tears of the knee: Importance of high signal in the meniscus that extends to the surface. AJR Am J Roentgenol 161:101, 1993.

Deutsch AL, Mink JH, Fox JM, et al: Peripheral meniscal tears: MR findings after conservative treatment or arthroscopic repair. Radiology 176:485, 1990.

Disler DG, Peters TL, Muscoreil SJ, et al: Fat-suppressed spoiled GRASS imaging of knee hyaline cartilage: Technique optimization and comparison with conventional MR imaging. AJR Am J Roentgenol 163:887, 1994.

El-Khoury GY, Wira RL, Berbaum KS, et al: MR imaging of patellar tendinitis. Radiology 184:840, 1992.

Eustace S, Keogh C, Blake M, et al: MR imaging of bone oedema: Mechanisms and interpretation. Clin Radiol 56:4, 2001.

Falchook FS, Tigges S, Carpenter WA, et al: Accuracy of direct signs of tears of the anterior cruciate ligament. Can Assoc Radiol J 47:114, 1996.

Farley TE, Howell SM, Love KF, et al: Meniscal tears: MR and arthrographic findings after arthroscopic repair. Radiology 180:517, 1991.

Fielding JR, Franklin PD, Kustan J: Popliteal cysts: A reassessment using magnetic resonance imaging. Skeletal Radiol 20:433, 1991.

Garvin GJ, Munk PL, Vellet AD: Tears of the medial collateral ligament: magnetic resonance imaging findings and associated injuries. J Can Assoc Radiol 44:199, 1993.

Goldman AB, Pavlov H, Rubenstein D: The Segond fracture of the proximal tibia: A small avulsion that reflects major ligamentous damage. AJR Am J Roentgenol 151:1163, 1988.
Haramati N, Staron RB, Rubin S, et al: The flipped meniscus sign. Skeletal Radiol 22:273, 1993.
Helms CA, Laorr A, Cannon WD Jr: The absent bow tie sign in bucket-handle tears of the menisci in the knee. AJR Am J Roentgenol 170:57, 1998.
Hodler J, Haghighi P, Pathria MN, et al: Meniscal changes in the elderly: Correlation of MR imaging and histologic findings. Radiology 184:221, 1992.
Horton LK, Jacobson JA, Lin J, et al: MR imaging of anterior cruciate ligament reconstruction graft. AJR Am J Roentgenol 175:1091, 2000.
Kaplan PA, Nelson NL, Garvin KL, et al: MR of the knee: The significance of high signal in the meniscus that does not clearly extend to the surface. AJR Am J Roentgenol 156:333, 1991.
Kaplan PA, Walker CW, Kilcoyne RF, et al: Occult fracture patterns of the knee associated with anterior cruciate ligament tears: Assessment with MR imaging. Radiology 183:835, 1992.
Kendall NS, Hsu, SYC, Chan KM: Fracture of the tibial spine in adults and children: A review of 31 cases. J Bone Joint Surg Br 84:848, 1992.
Kirsch MD, Fitzgerald SW, Friedman H, et al: Transient lateral patellar dislocation: Diagnosis with MR imaging. AJR Am J Roentgenol 161:109, 1993.
Kode L, Lieberman JM, Motta AO, et al: Evaluation of tibial plateau fractures: Efficacy of MR imaging compared with CT. AJR Am J Roentgenol 163:141, 1994.
Kosarek FJ, Helms CA: The MR appearance of the infrapatellar plica. AJR Am J Roentgenol 172:481, 1999.
Koval KJ, Helfer DL: Tibial plateau fractures: Evaluation and treatment. J Am Acad Orthop Surg 3:86, 1995.
Larson RL: The knee—the physiological joint. J Bone Joint Surg [Am] 65:143, 1983.
Lecas LK, Helms CA, Kosarek FJ, et al: Inferiorly displaced flap tears of the medial meniscus: MR appearance and clinical significance. AJR Am J Roentgenol 174:161, 2000.
Lektrakul N, Skaf A, Yeh L, et al: Pericruciate meniscal cysts arising from tears of the posterior horn of the medial meniscus: MR imaging features that simulate posterior cruciate ganglion cysts. AJR Am J Roentgenol 172:1575, 1999.
Lim PS, Schweitzer ME, Bhatia M, et al: Repeat tear of postoperative meniscus: Potential MR imaging signs. Radiology 210:183, 1999.
Lugo-Oliverieri CH, Scott WW Jr, Zerhouni EA: Fluid-fluid levels in knees: Do they always represent lipohemarthrosis? Radiology 198: 499, 1996.
Malghem J, Van de berg BC, Lebon C, et al: Ganglion cysts of the knee: Articular communication revealed by delayed radiography and CT after arthrography. AJR Am J Roentgenol 170:1579, 1998.
Mason RW, Moore TE, Walker CW, et al: Patellar fatigue fractures. Skeletal Radiol 25:329, 1996.
Neuschwander DC, Drez D Jr, Finney TP: Lateral meniscal variant with absence of the posterior coronary ligament. J Bone Joint Surg Am 74:1186, 1992.
Palmer WE, Levine SM, Dupuy DE: Knee and shoulder fractures: Association of fracture detection and marrow edema on MR images with mechanism of injury. Radiology 204:395, 1997.
Pavlov H, Goldman AB: The popliteus bursa: An indicator of subtle pathology. AJR Am J Roentgenol 134:313, 1980.
Quinn SF, Brown TR, Demlow TA: MR imaging of patellar retinacular ligament injuries. J Magn Reson Imaging 3:843, 1993.
Recht MP, Applegate G, Kaplan P, et al: The MR appearance of cruciate ganglion cysts: A report of 16 cases. Skeletal Radiol 23:597, 1994.
Recht MP, Resnick D: MR imaging of articular cartilage: Current status and future directions. AJR Am J Roentgenol 163:283, 1994.
Roychowdhury S, Fitzgerald SW, Sonin AH, et al: Using MR imaging to diagnose partial tears of the anterior cruciate ligament: Value of axial images. AJR Am J Roentgenol 168:1487, 1997.
Rubin DA, Britton CA, Towers JD, et al: Are MR imaging signs of meniscocapsular separation valid? Radiology 201:829, 1996.
Ruff C, Weingardt JP, Russ PD, et al: MR imaging patterns of displaced meniscus injuries of the knee. AJR Am J Roentgenol 170:63, 1998.
Schatzker J, McBroom R, Bruce D: The tibial plateau fracture: The Toronto experience. Clin Orthop 138:94, 1979.
Shands PA, McQueen DA: Demonstration of avulsion fracture of the inferior pole of the patella by magnetic resonance imaging. J Bone Joint Surg Am 77:1721, 1995.
Shankman S, Beltran J, Melamed E, et al: Anterior horn of the lateral meniscus: Another potential pitfall in MR imaging of the knee. Radiology 204:181, 1997.
Silverman JM, Mink JH, Deutsch AL: Discoid menisci of the knee: MR imaging appearance. Radiology 173:351, 1989.
Singson RD, Feldman F, Staron R, et al: MR imaging of displaced buckethandle tear of the medial meniscus. AJR Am J Roentgenol 156:121, 1991.
Siwek CW, Rao JP: Ruptures of the extensor mechanism of the knee joint. J Bone Joint Surg [Am] 63:932, 1981.
Smith DK, Totty WG: The knee after partial meniscectomy: MR imaging features. Radiology 176:141, 1990.
Snearly WN, Kaplan PA, Dussault RG: Lateral-compartment bone contusions in adolescents with intact anterior cruciate ligaments. Radiology 198:205, 1996.
Sonin AH, Fitzgerald SW, Bresler ME: MR imaging appearance of the extensor mechanism of the knee: Functional anatomy and injury patterns. Radiographics 15:367, 1995.
Spritzer CE, Courneya DL, Burk DL, et al: Medial retinacular complex injury in acute patellar dislocation: MR findings and surgical implications. AJR Am J Roentgenol 168:117, 1997.
Tasker AD, Ostlere SJ: Relative incidence and morphology of lateral and medial meniscal cysts detected by magnetic resonance imaging. Clin Radiol 50:778, 1995.
Tomczak RJ, Hehl G, Mergo PJ, et al: Tunnel placement in anterior cruciate ligament reconstruction: MRI analysis as an important factor in the radiological report. Skeletal Radiol 26:409, 1997.
Tuckman GA, Miller WJ, Remo JW, et al: Radial tears of the menisci: MR findings. AJR Am J Roentgenol 163:395, 1994.
Vahey TN, Bennett HT, Arrington LE, et al: MR imaging of the knee: Pseudotear of the lateral meniscus caused by the meniscofemoral ligament. AJR Am J Roentgenol 154:1237, 1990.
Watanabe M, Takeda S, Ikeuchi H: Atlas of Arthroscopy, 3rd ed. Tokyo, Iagaku Shoin, 1978, pp 87–91.
Weber WN, Neumann CH, Barakos JA, et al: Lateral tibial rim (Segond) fractures: MR imaging characteristics. Radiology 180:731, 1991.
Yu JS, Petersilge C, Sartoris DJ, et al: MR imaging of injuries of the extensor mechanism of the knee. Radiographics 14:541, 1994.
Zaricznyi B: Avulsion fracture of the tibial eminence: Treatment by open reduction and pinning. J Bone Joint Surg Am 59:1111, 1997.
Zeiss J, Saddemi SR, Ebraheim NA: MR Imaging of the quadriceps tendon: Normal layered configuration and its importance in cases of tendon rupture. AJR Am J Roentgenol 159:1031, 1992.

CHAPTER 65

Shoulder

Craig W. Walker, MD

OVERVIEW

Shoulder pain is a commonly encountered problem in clinical practice, and it can result from a variety of causes. Impingement of the musculotendinous structures of the rotator cuff between the humeral head and the coracoacromial arch is a common cause of shoulder pain in middle-aged and elderly patients and ultimately may result in a rotator cuff tear. Tendon abnormalities also can result from intrinsic degeneration, trauma, and inflammatory arthritis. Anterior shoulder instability is a fairly common condition of young individuals. It is usually the result of injury to the anterior stabilizers of the shoulder. Magnetic resonance (MR) imaging can provide valuable information in the evaluation of patients with suspected rotator cuff or labroligamentous abnormalities.

NORMAL ANATOMY

The osseous structures of the shoulder include the proximal humerus, the scapula, and the distal clavicle. The scapula is composed of a body and three lateral processes: the coracoid, the acromion, and the glenoid. The humeral head articulates with the glenoid fossa of the scapula to form the glenohumeral joint. The distal clavicle articulates with the acromion to form the acromioclavicular joint.

Soft Tissue Structures

The rotator cuff consists of the supraspinatus, infraspinatus, teres minor, and subscapularis tendons (Fig. 65–1). The muscles of the rotator cuff arise from the scapula and insert on the proximal humerus. At their insertion, these tendons are inseparable from each other and from the glenohumeral joint capsule. As a group, these muscles provide dynamic stability to the glenohumeral joint. The supraspinatus and infraspinatus muscles are innervated by the suprascapular nerve, whereas the teres minor muscle is innervated by branches of the axillary nerve. The subscapularis muscle is innervated by the subscapular nerve.

The supraspinatus muscle arises from the suprascapular fossa of the scapula and courses laterally to insert on the superior aspect of the greater tuberosity. Its principal action is to abduct the humerus.

The infraspinatus muscle arises from the infraspinous fossa of the scapula and courses laterally and slightly superiorly to insert on the greater tuberosity just posterior and inferior to the supraspinatus tendon. Its primary function is to externally rotate the humerus.

The teres minor muscle originates from the inferior and larteral border of the scapula to insert on the greater tuberosity inferior to the infraspinatus tendon. It functions as an external rotator of the humerus and helps resist anterior subluxation of the humeral head.

The subscapularis muscle originates from the subscapular fossa of the scapula and courses laterally across the anterior aspect of the glenohumeral joint to insert on the lesser tuberosity. Fibers from the subscapularis tendon continue from the lesser tuberosity across the intertubercular (bicipital) groove as the transverse ligament. The transverse ligament acts in maintaining the long head of the biceps tendon within the bicipital groove and prevents it from dislocating medially. The primary function of the subscapularis muscle is to internally rotate the humerus.

The glenoid labrum and glenohumeral ligaments are important static stabilizers of the glenohumeral joint. Injuries to these structures may result in glenohumeral instability. Three glenohumeral ligaments attach to the

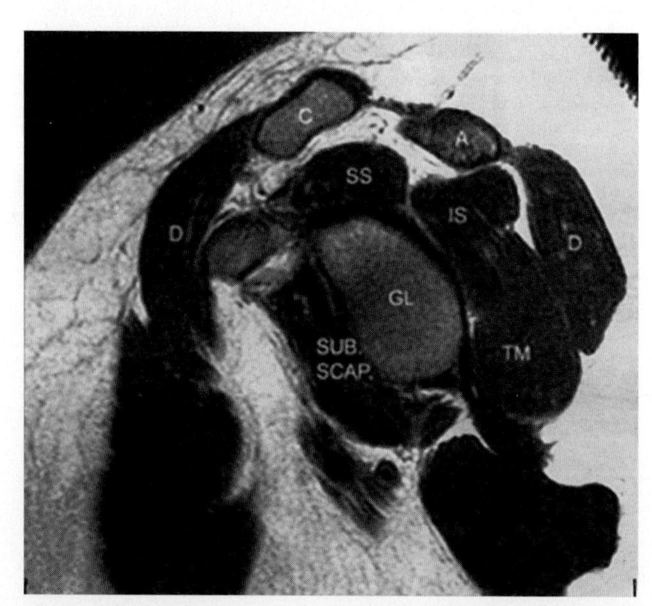

Figure 65–1. Oblique sagittal MR image through the glenoid fossa shows the relationships of the rotator cuff muscles with the glenoid fossa. A, acromion process; C, clavicle; D, deltoid muscle; GL, glenoid fossa; IS, infraspinatus muscle; SS, supraspinatus muscle; SUB SCAP, subscapularis muscle; TM, teres minor muscle.

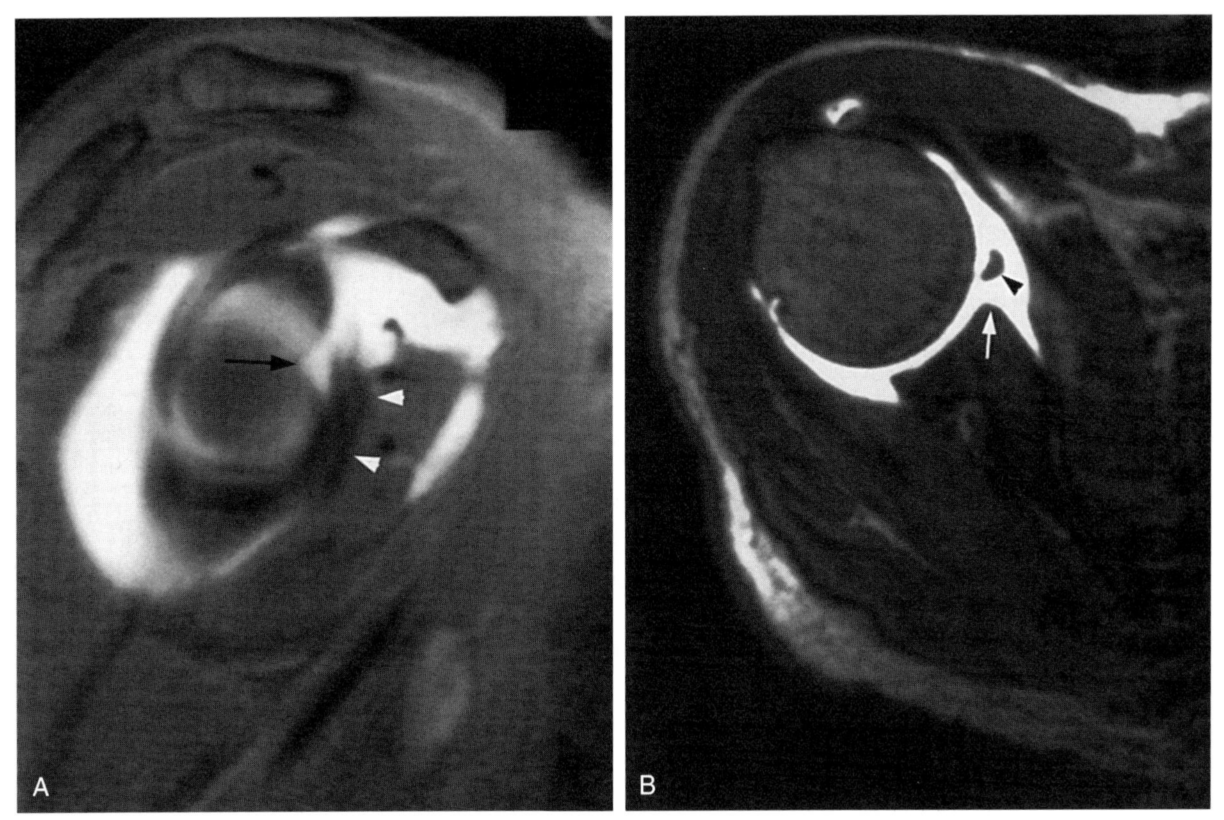

Figure 65–2. Buford complex. *A,* Oblique sagittal fat-suppressed T1-weighted MR image with fat saturation from a shoulder arthrogram illustrates the absence of the labrum *(arrow)* on the anterior and superior aspect of the glenoid. The white arrowheads point to the thick cordlike middle glenohumeral ligament. *B,* Axial image shows the thick middle glenohumeral ligament *(arrowhead)* and the absent anterior labrum *(arrow).*

glenoid labrum and represent thickened bands of the joint capsule.

The glenoid labrum is a fibrocartilaginous triangular structure that attaches to the periphery of the glenoid fossa and deepens it. There are several normal variations to the appearance and attachment of the glenoid labrum to the anterosuperior portion of the glenoid. In 11% to 17% of individuals, the anterosuperior labrum does not attach to the glenoid, creating a foramen that can be misinterpreted as a labral tear.

The superior glenohumeral ligament originates from the superior glenoid, just anterior to the long head of the biceps tendon. It runs parallel to the coracoid process and inserts on the humeral head superior to the lesser tuberosity near the bicipital groove.

The middle glenohumeral ligament originates from the glenoid neck and superior glenoid labrum and inserts near the anatomic neck of the humerus, medial to the lesser tuberosity. The middle glenohumeral ligament limits external rotation in mild degrees of abduction. The middle glenohumeral ligament has the greatest variability of the glenohumeral ligaments and can be absent in approximately 25% of individuals. Its size can vary from thin to cordlike.

The Buford complex is a congenital anomaly affecting the anterosuperior labrum and middle glenohumeral ligament. It occurs in approximately 1% to 2% of individuals and consists of a thick, cordlike middle glenohumeral ligament and an absent or deficient anterosuperior glenoid labrum (Fig. 65–2).

The inferior glenohumeral ligament (IGL) is the largest and most important glenohumeral ligament. The IGL is composed of an anterior band, posterior band, and axillary pouch. It arises from and contributes to the inferior two thirds of the glenoid labrum; it attaches laterally on the humeral neck. The IGL is a major stabilizer when the arm is abducted 90 degrees, limiting anterior and posterior translation of the humeral head.

The coracohumeral ligament extends from the lateral margin of the coracoid process, traveling in the interval between the anterior margin of the supraspinatus tendon and superior margin of the subscapularis tendon to insert on the greater and the lesser tuberosities of the humerus.

ROTATOR CUFF TEAR

The cause of rotator cuff tears is likely multifactorial. Extrinsic factors and intrinsic factors are implicated. The main extrinsic cause for rotator cuff tear is impingement, which causes coracoacromial arch stenosis (Fig. 65–3). Impingement is believed to produce repetitive trauma to the rotator cuff, causing most rotator cuff tears. Intrinsic factors are related to intrasubstance tendon degeneration, which progresses to tendinosis (tendinopathy), partial tear, and eventually full-thickness tear. The underlying causes of tendon degeneration are thought to be advanced age, diminished blood supply to the tendon, and overuse.

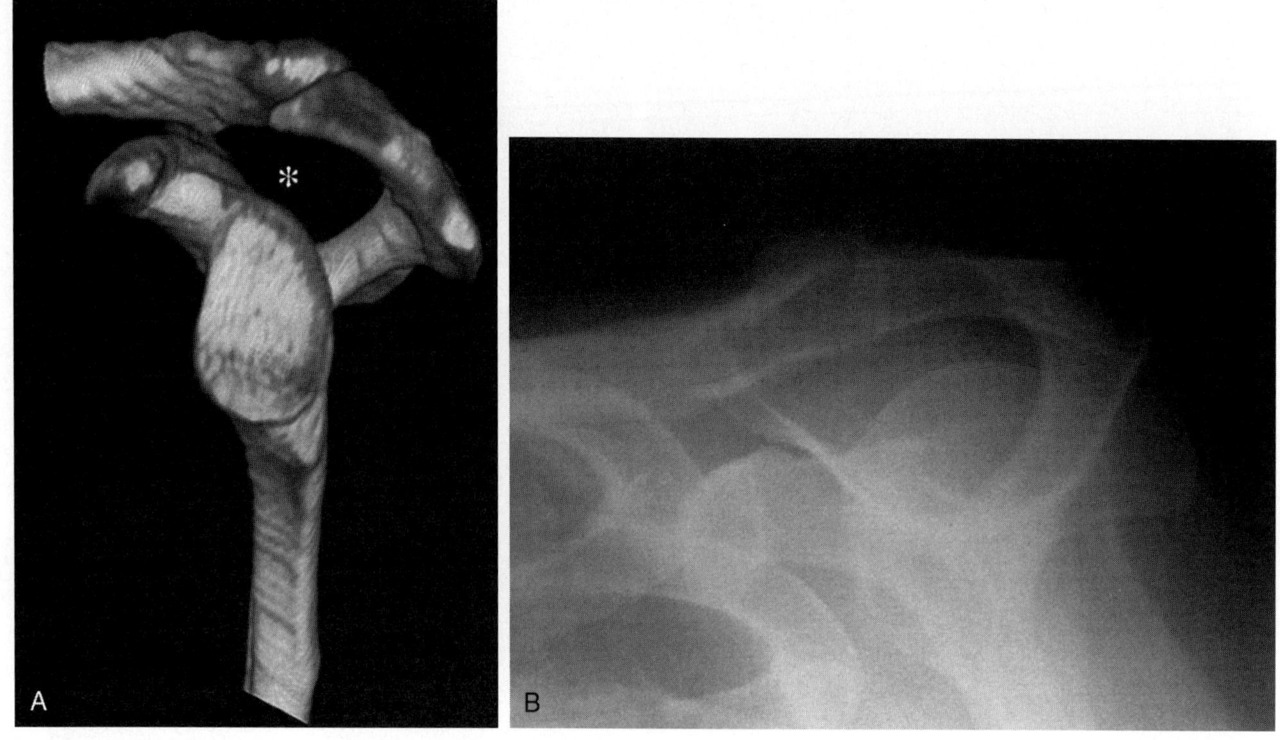

Figure 65–3. Coracoacromial arch. *A,* Three-dimensional CT scan of the coracoacromial arch. The asterisk is in the center of the supraspinatus outlet. *B,* Bigliani view shows a normal coracoacromial arch.

The impingement syndrome is a clinical diagnosis characterized by signs and symptoms that result from compression of the rotator cuff tendons and subacromial bursa between the humeral head and coracoacromial arch. This compression leads to a cycle of tendon injury and repair that ultimately results in decreased tendon tensile strength and tear.

Impingement may result from either primary or secondary extrinsic factors. Primary extrinsic impingement is caused by morphologic variations of the coracoacromial arch that narrow the supraspinatus outlet (see Fig. 65–3). The shape of the acromion process has been implicated in causing rotator cuff tears. Three separate acromial morphologic types (Fig. 65–4*A*) have been described depending on their appearance on conventional radiographs using the supraspinatus outlet (also called *Bigliani view*) (see Fig. 65–3*B*). Type I acromion has a flat inferior margin. Type II acromion has a curved gradual arc along its inferior surface. Type III acromion has an inferiorly hooked, anterior tip (Fig. 65–4*B*). This type of acromion most significantly narrows the supraspinatus outlet and has the highest association with rotator cuff tears. Extrapolating this classification of acromial morphology to MR imaging (Fig. 65–4*C*) has been difficult, resulting in significant interobserver and intraobserver variability.

The orientation of the acromion also is thought to cause rotator cuff abnormalities. Lateral or anterior acromial downsloping also may cause narrowing of the supraspinatus outlet (Figs. 65–5 and 65–6).

Subacromial spurs form along the inferior surface of the acromion or at the coracoacromial ligament insertion and frequently are associated with impingement on the rotator cuff. Spur size seems to correlate with the presence of underlying rotator cuff tears (Fig. 65–7). Os acromiale is an unfused acromial apophysis that can displace caudally during deltoid muscle contraction, resulting in intermittent impingement and supraspinatus outlet stenosis (Fig. 65–8). Acromioclavicular joint osteoarthritis with inferior osteophyte formation projecting from the distal clavicle or acromion may impinge on the supraspinatus myotendinous junction (Fig. 65–9). Coracoacromial ligament thickening has been implicated as a cause and sequela of rotator cuff impingement (Fig. 65–10).

Coracoid process abnormalities are an uncommon cause of impingement that may result from unusually large coracoid processes or a change in orientation after a coracoid fracture. Coracohumeral impingement (Fig. 65–11) results in anterior shoulder pain because of compression on the anterior joint capsule, subscapularis tendon, and subscapularis bursa between the humeral head and the coracoid process.

Secondary extrinsic impingement is a disease of young athletes performing repetitive overhead and throwing activities. The mechanism is believed to be due to fatigue of the dynamic stabilizers of the shoulder (i.e., the rotator cuff muscles and other muscles acting on the shoulder, such as the deltoid, latissimus dorsi, teres major, short and long heads of the biceps, and long head of the triceps). When that happens, the humeral head subluxes superiorly, reducing the subacromial space and causing intermittent impingement. Making this diagnosis can be difficult, and imaging studies fail to show any structural abnormalities. The condition is treated with

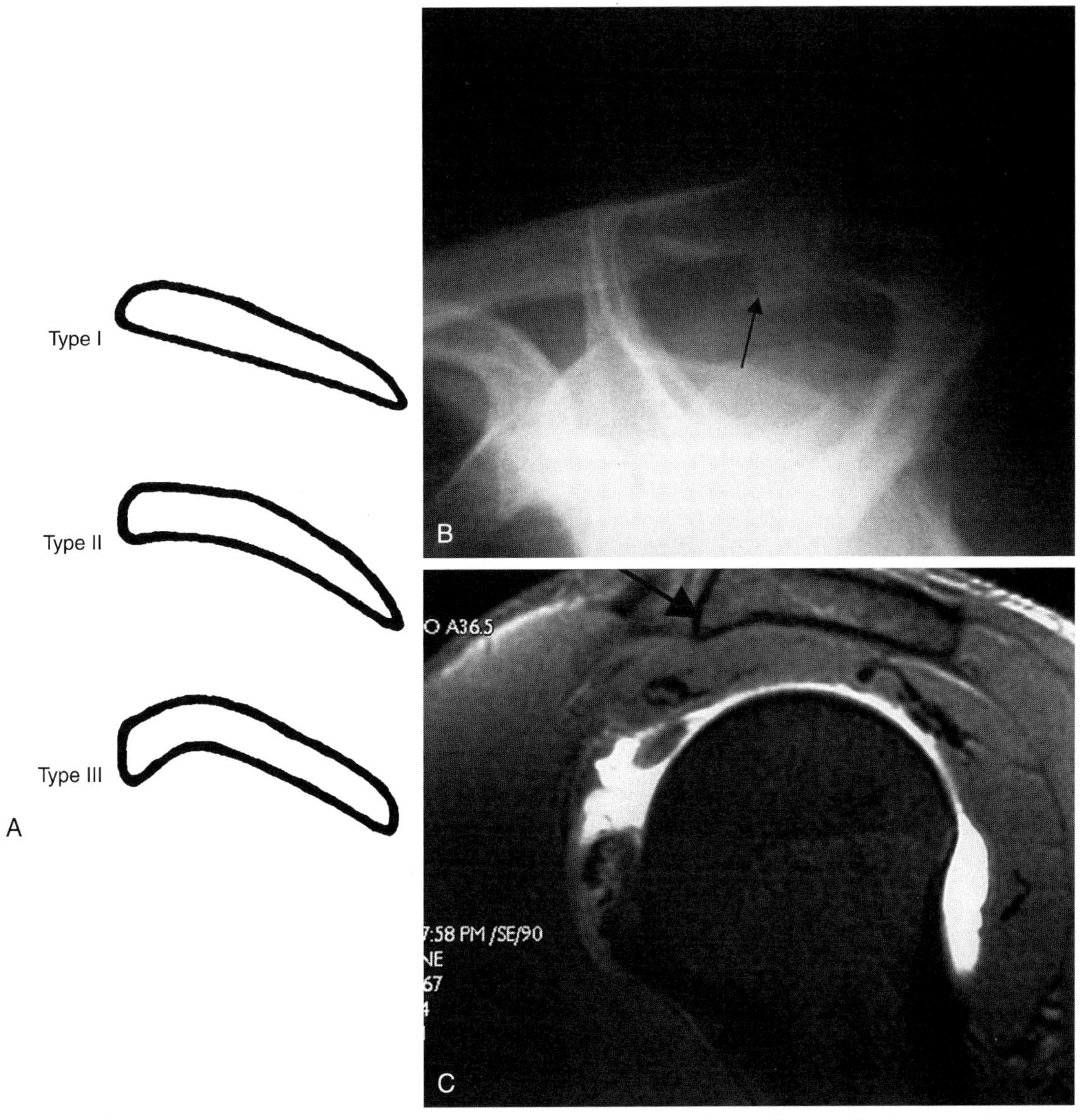

Figure 65–4. Diagram of acromial types and images of hooked (type III) acromion. *A,* Line drawing illustrates the three types of acromia. *B,* Bigliani view shows a hooked acromion (type III) *(arrow)*. *C,* Type III (hooked) acromion *(arrow)* on an oblique sagittal T1-weighted MR image.

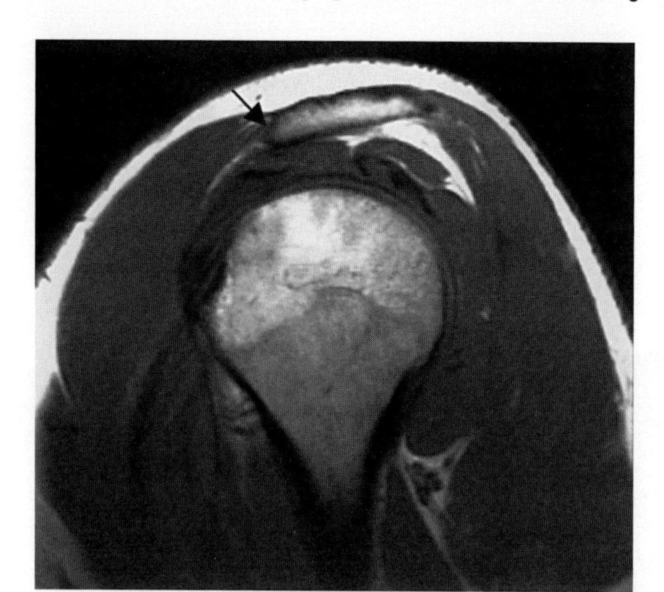

Figure 65–5. Anterior downsloping of the acromion *(arrow)* seen on a T1-weighted oblique sagittal MR image.

physical therapy to strengthen the muscles that stabilize the shoulder joint.

Intrinsic tendon pathology is the alternative theory explaining the cause of rotator cuff tears. In this theory, subacromial spurs and coracohumeral ligament hypertrophy are a result of rotator cuff degeneration rather than the cause. The supraspinatus is a relatively small muscle compared with the long lever arm on which it acts. Repetitive overloading of the supraspinatus can lead to tendon degeneration and tear. This theory is supported by the presence of a hypovascular zone within the supraspinatus tendon, approximately 1 cm from its insertion. Because of the decreased perfusion, this area of the tendon is predisposed to degeneration from repetitive microtrauma, resulting in tendinosis (tendinopathy) and ultimately tear. The vascular supply of the rotator cuff tendon is better on the bursal side, which explains why partial tears are more common on the articular side (80%) than on the bursal side (20%).

The rotator cuff, similar to all other tendons in the body, undergoes gradual degeneration, eventually tearing with age. MR studies done on asymptomatic individuals have shown that rotator cuff abnormalities generally start around age 40 years and become more prevalent with advancing age. Of asymptomatic individuals older than age 60, 50% have either partial-thickness or full-thickness tears. Similar to tendon abnormalities at other sites, rotator cuff degeneration manifests with a spectrum of severity from tendinosis or tendinopathy to full-thickness tear. Often tendinosis is seen in conjunction with partial-thickness or full-thickness tears; healthy tendons do not rupture.

Tendinosis (tendinopathy) is the basic degenerative process underlying all tendon abnormalities. Morphologic changes include tendon thickening and splitting of the collagen bundles. On MR imaging, tendinosis appears as an area of tendon thickening and increased signal intensity on T1-weighted images with or without minimal increase in signal intensity on T2-weighted images (Fig. 65–12). These abnormalities need to be distinguished from signal alterations resulting from the magic angle effect that occurs on short echo time pulse sequences when tendons are aligned at about 55 degrees relative to the main magnetic field. This artifact disappears on sequences with long echo time.

Tendon tears are either partial thickness or full thickness. Large full-thickness tears are associated with muscle retraction, muscle atrophy, and superior migration of the humeral head. Partial-thickness tears may involve the bursal surface or articular surface; occasionally, they may be intrasubstance (or interstitial) (Figs. 65–13, 65–14,

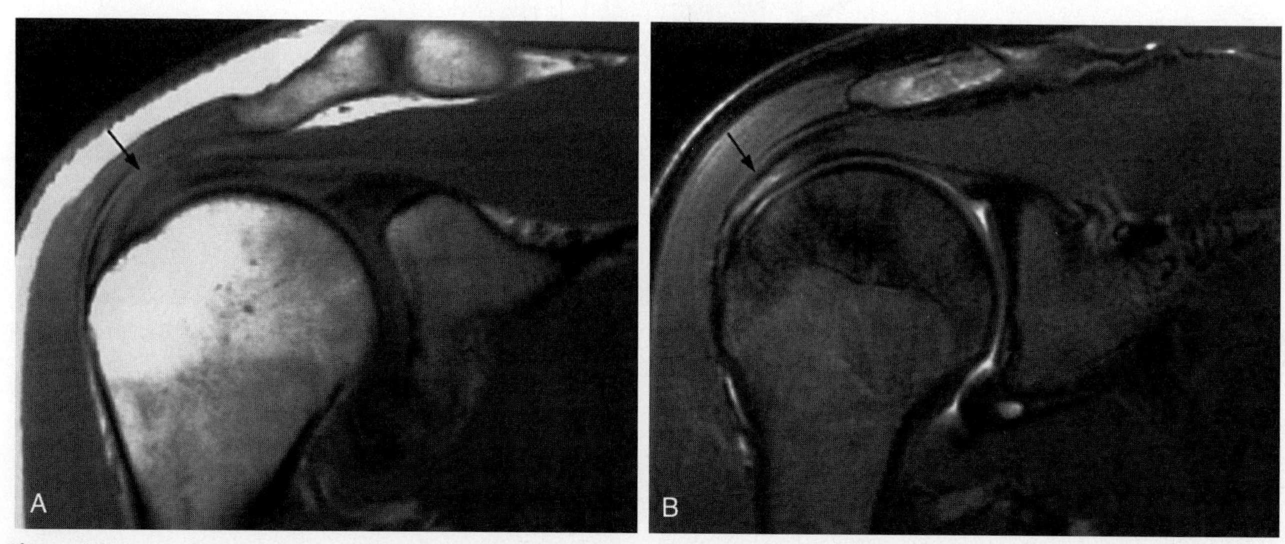

Figure 65–6. Lateral acromial downsloping associated with severe tendinosis or interstitial tear of the supraspinatus tendon. *A*, T1-weighted oblique coronal MR image shows thickening of the tendon and increased signal within it *(arrow)*. Notice the lateral downsloping of the acromion. *B*, T2-weighted fat-suppressed image shows significant increase in signal intensity *(arrow)* within the substance of the tendon suggesting either severe tendinosis or early interstitial tear.

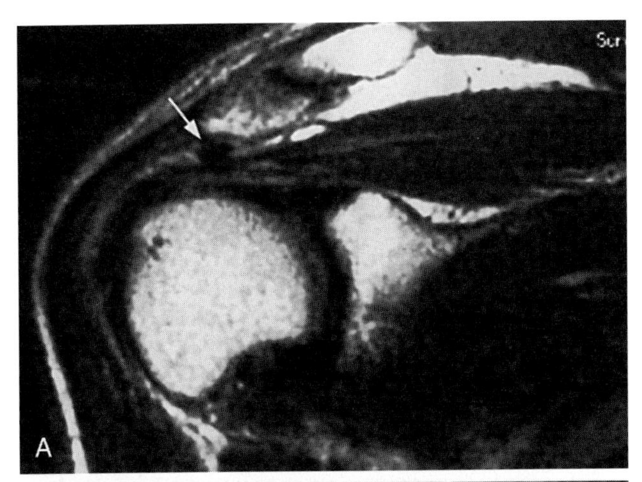

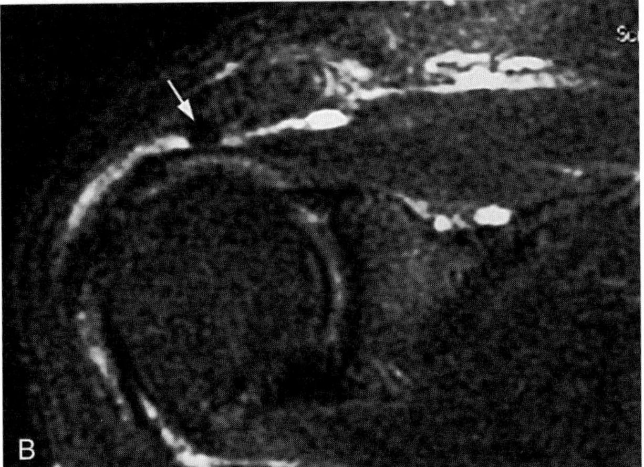

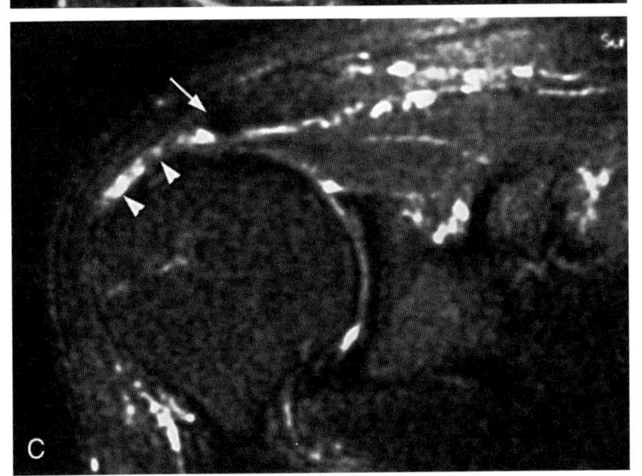

Figure 65–7. Subacromial spur causing impingement and rotator cuff tear. *A,* Oblique coronal T1-weighted MR image shows lateral downsloping of the acromion. At the tip of the acromion, a large osteophyte *(arrow)* is seen impinging on the rotator cuff. *B,* Oblique coronal fat-suppressed T2-weighted MR image with fat saturation shows the osteophyte *(arrow)* and thinning of the tendon at the site of impingement. *C,* Another fat-suppressed T2-weighted oblique coronal MR image, which is anterior to the previous sections (*A* and *B*) shows a large full-thickness tear *(arrowheads)* in the rotator cuff tendon. The arrow points to the impinging osteophyte.

and 65–15) in location. With an interstitial tear, the signal abnormality remains within the tendon substance on T2-weighted images, and the tear does not reach to the surface (see Fig. 65–15). MR arthrography improves the detection of articular surface partial tears. Articular surface partial tears of the rotator cuff are about four times more common than bursal surface partial tears. MR findings of partial-thickness tears are characterized by underlying tendinopathy, stripping of the tendon fibers off the bone (Fig. 65–16), or the development of a focal erosion or an ulcer-like lesion on the articular surface of the tendon (see Fig. 65–13). These changes are seen much less commonly on the bursal side of the tendon. When the partial tear extends over a long distance, the remaining intact fibers of the tendon appear attenuated (see Fig. 65–16). Intrasubstance or interstitial rotator cuff tears are considered a form of partial tear (see Fig. 65–15). On MR imaging, there is tendon thickening along with linear or globular increased signal within the tendon. Intrasubstance or interstitial tears represent splitting of the collagen bundles within the supraspinatus tendon caused by matrix degeneration (see Fig. 65–15).

Full-thickness tears allow communication of the glenohumeral joint with the subacromial-subdeltoid bursa. Fluid usually is identified easily extending from the glenohumeral joint across the tear into the bursa. Tears are characterized by their location and size in the anteroposterior and mediolateral dimensions. On MR imaging, a full-thickness tear is seen as an area of discontinuity within the tendon and increased signal intensity involving the entire thickness of the tendon on T2-weighted images (Figs. 65–17 and 65–18). Multiple tendon involvement and muscle retraction is seen with massive tears. MR imaging is highly sensitive and specific for the diagnosis of full-thickness tears, and there is minimal intraobserver and interobserver variability. With tendinopathy and partial tears, MR imaging is less sensitive and specific; intraobserver and interobserver variability is worse than with full-thickness tears. T2-weighted sequences with fat suppression have been shown to improve the detection and characterization of rotator cuff tears. MR arthrography using diluted gadolinium also improves the visualization of articular surface partial tears.

Tears in Specific Tendons

About 95% of all rotator cuff tears occur in the supraspinatus tendon. Typically, they start in the anterior third of the supraspinatus tendon at or close to its bony insertion (Figs. 65–19 and 65–20). Tears that start in the posterior third of the supraspinatus tendon are much less common. They usually occur in younger patients and are often post-traumatic. Large tears, also called *massive tears,* span the entire anteroposterior width of the supraspinatus tendon and may extend posteriorly and anteriorly to involve the infraspinatus and subscapularis tendons. Massive tears often are associated with rupture of the long head biceps tendon (see Fig. 65–17).

Subscapularis tendon tears typically occur in conjunction with supraspinatus tendon tears or, much less commonly, in isolation. Isolated injuries can be associated with posterior glenohumeral dislocations, typically in young patients. Degeneration and tearing of the subscapu-

Text continued on page 520

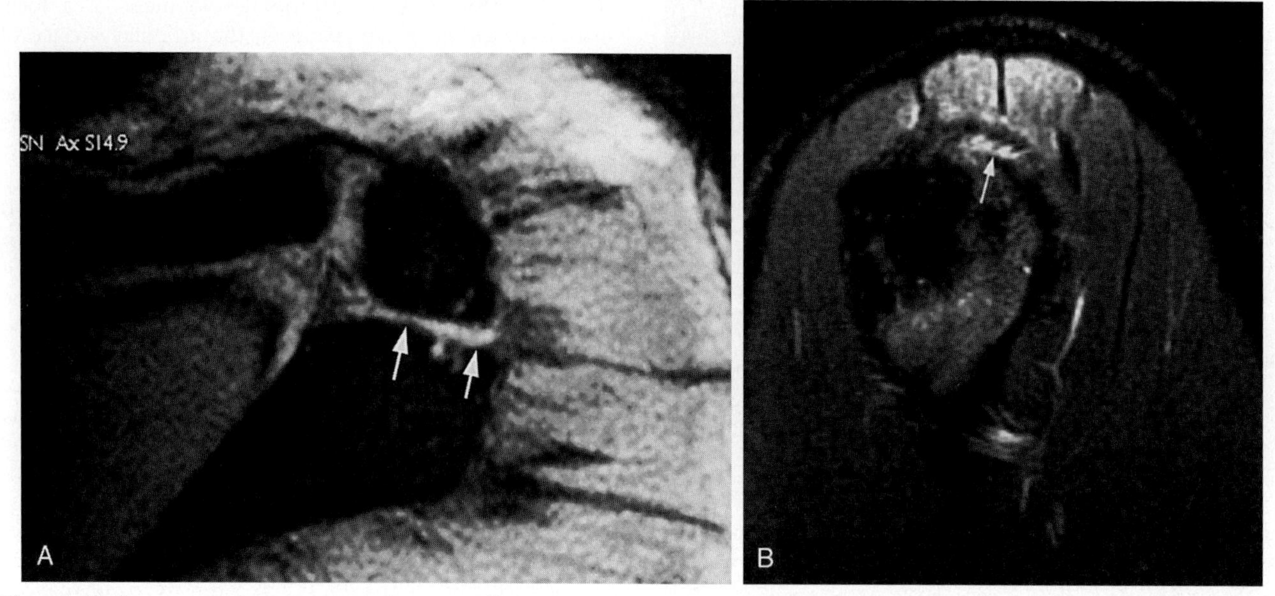

Figure 65–8. Os acromiale with associated rotator cuff tear. *A*, Axial T2-weighted gradient echo MR image shows the os acromiale *(arrows)*. *B*, Oblique sagittal fat-suppressed T2-weighted MR image from the same patient shows an articular surface partial tear *(arrow)* of the supraspinatus tendon.

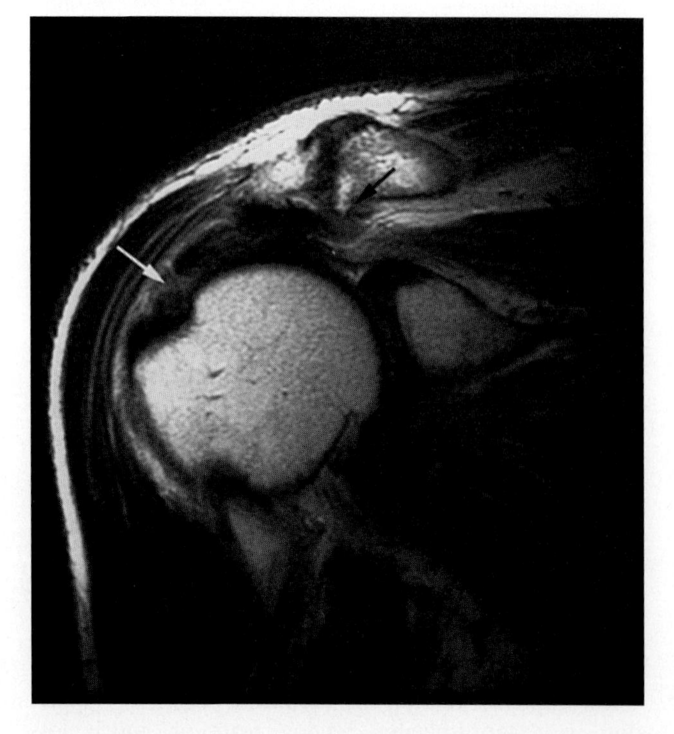

Figure 65–9. Acromioclavicular osteoarthritis causing impingement. Oblique coronal T2-weighted MR image shows an inferior osteophyte arising from the distal clavicle *(black arrow)*. The osteophyte is seen impinging on the myotendinous junction of the supraspinatus. There is a full-thickness tear of the rotator cuff *(white arrow)* at its insertion.

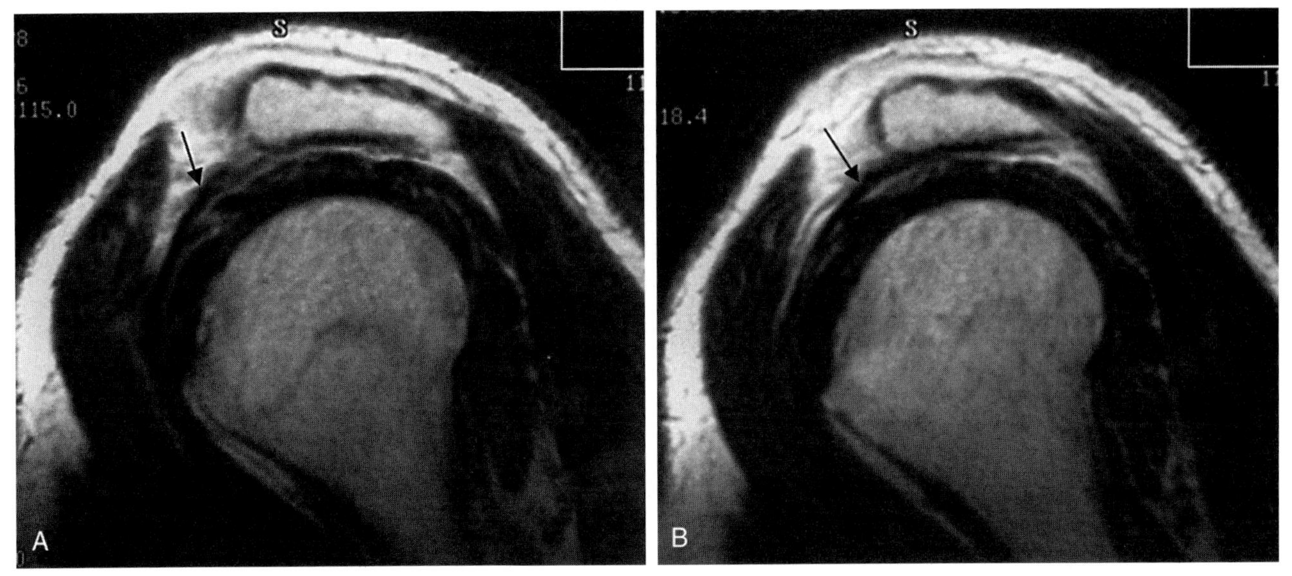

Figure 65–10. Coracoacromial ligament thickening. *A* and *B*, Two consecutive oblique T2-weighted sagittal images show thickening of the coracoacromial ligament *(arrow)*.

Figure 65–11. Coracohumeral impingement resulting in subscapularis tendinosis and tear. Axial T2-weighted MR image shows a deformed hooked coracoid process *(white arrow)* impinging on the subscapularis tendon *(white arrowheads)*, which is thickened and torn off its insertion. A deltoid muscle lipoma is incidentally noted *(black arrow)*.

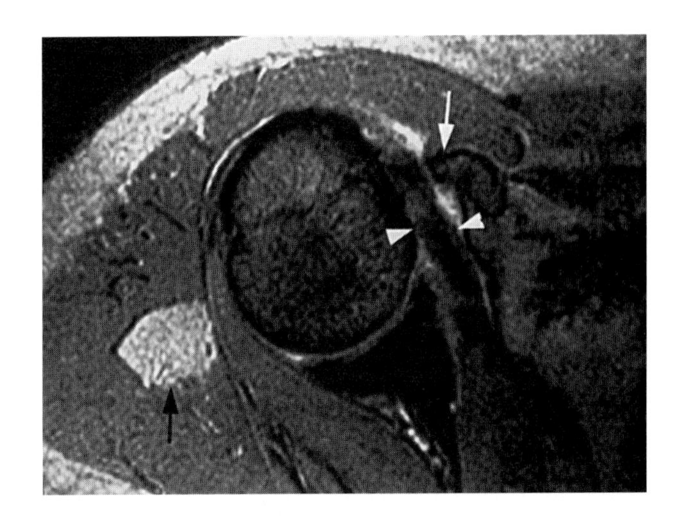

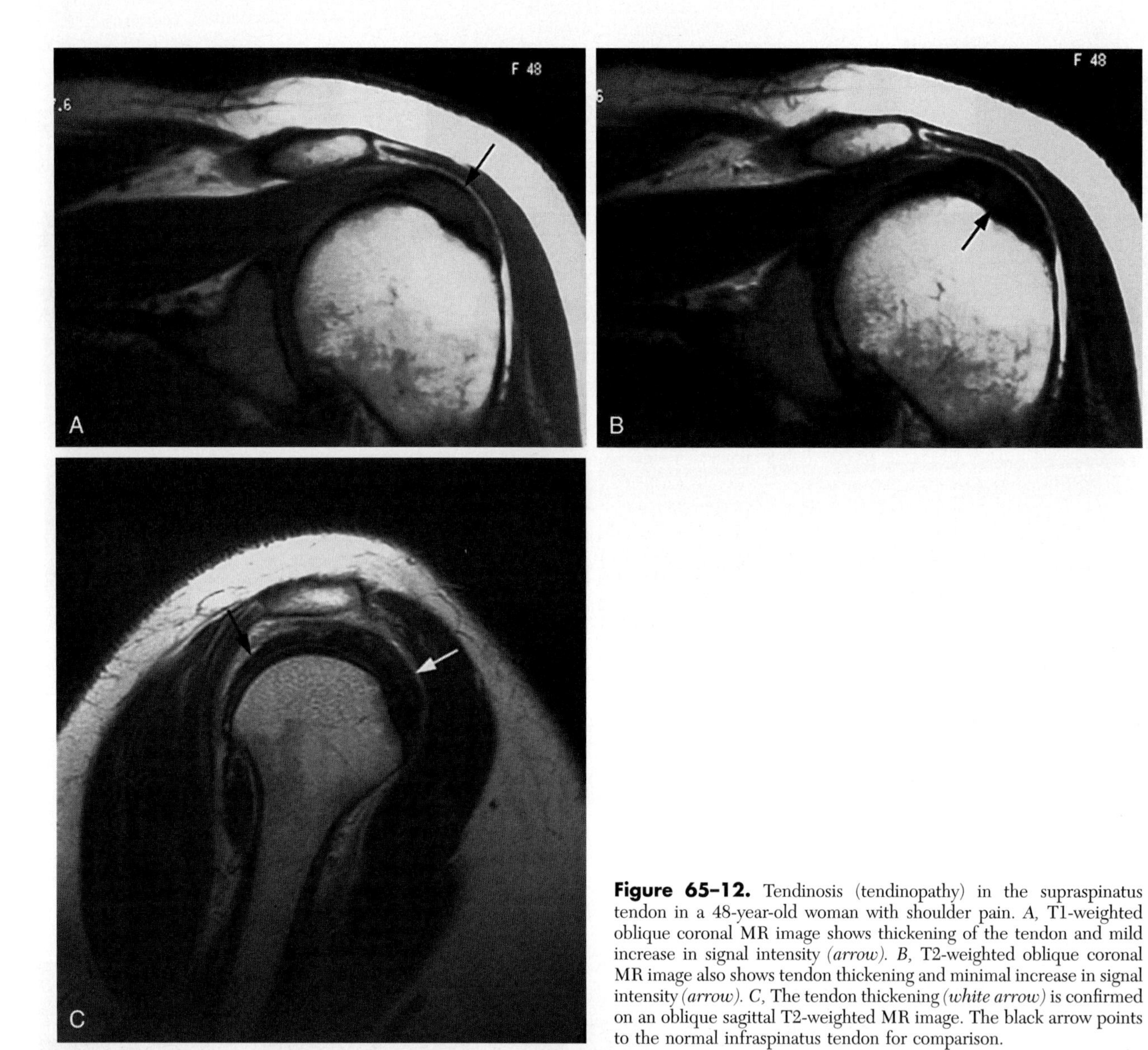

Figure 65–12. Tendinosis (tendinopathy) in the supraspinatus tendon in a 48-year-old woman with shoulder pain. *A,* T1-weighted oblique coronal MR image shows thickening of the tendon and mild increase in signal intensity *(arrow). B,* T2-weighted oblique coronal MR image also shows tendon thickening and minimal increase in signal intensity *(arrow). C,* The tendon thickening *(white arrow)* is confirmed on an oblique sagittal T2-weighted MR image. The black arrow points to the normal infraspinatus tendon for comparison.

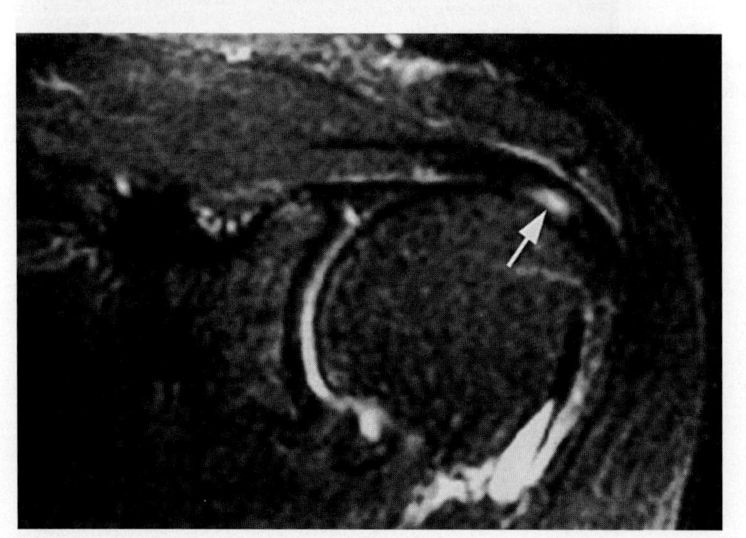

Figure 65–13. Articular surface partial tear *(arrow)* seen as a focal region of high signal on an oblique coronal T2-weighted fat-suppressed MR image. Some fibers are still intact superior to the lesion.

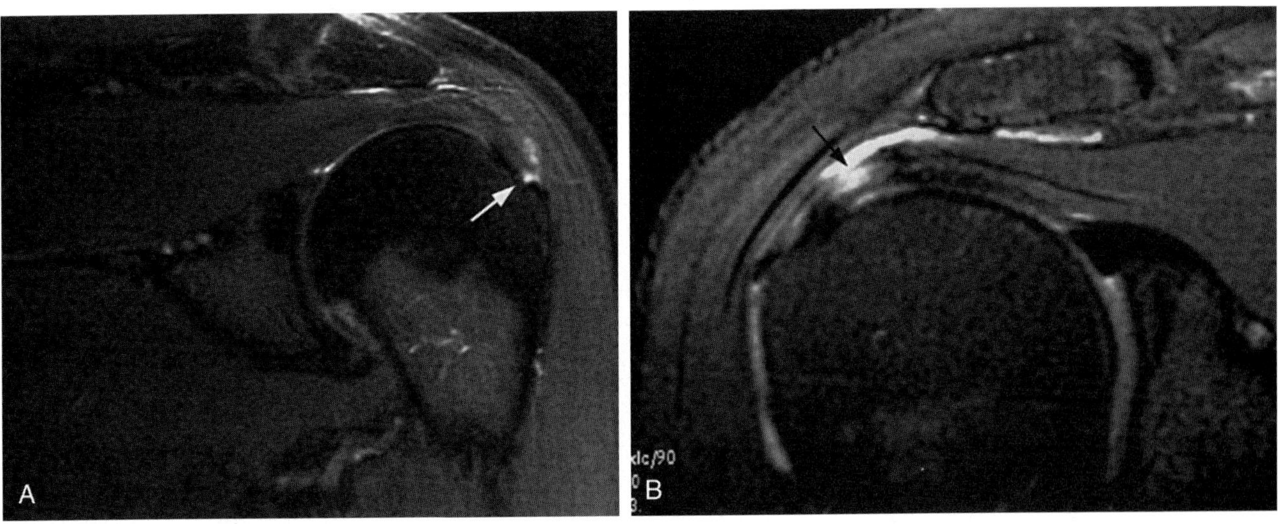

Figure 65–14. Bursal surface partial rotator cuff tears. *A*, Oblique coronal T2-weighted image with fat suppression shows a small bursal surface partial tear *(arrow)* at the tendon insertion. *B*, Another patient with a bursal surface partial tear just proximal to the tendon insertion *(arrow)*. (The image is an oblique coronal T2-weighted image with fat suppression.)

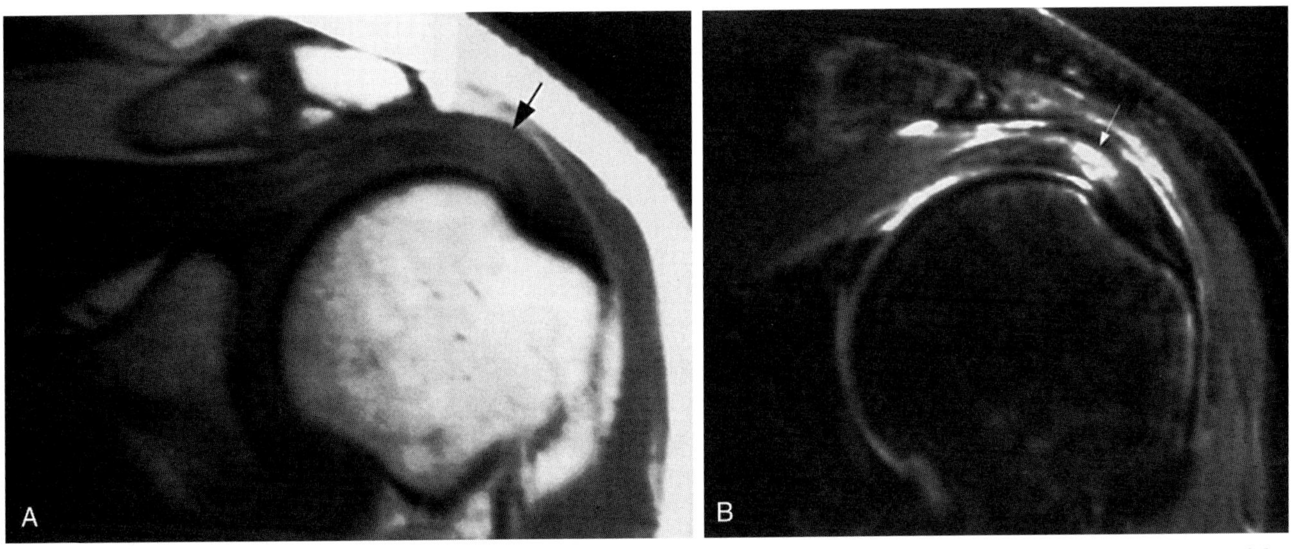

Figure 65–15. Intrasubstance (interstitial) tear of the supraspinatus tendon. *A*, T1-weighted oblique coronal MR image shows thickening of the tendon and increased signal within it *(arrow)*. *B*, T2-weighted fat-suppressed MR image shows significant increase in signal intensity *(arrow)* within the substance of the tendon. There are intact fibers above and below the interstitial tear.

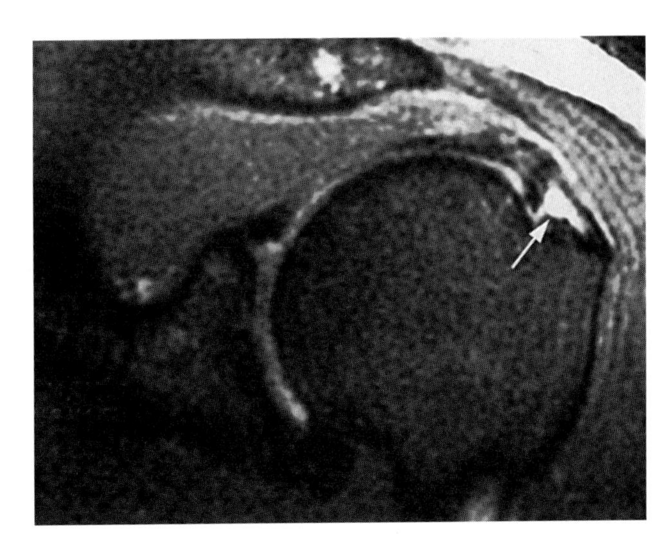

Figure 65–16. Large partial tear *(arrow)* on the articular side of the supraspinatus tendon. Note the intact attenuated fibers remaining superior to the tear. (This image is a fat-suppressed T2-weighted oblique coronal image.)

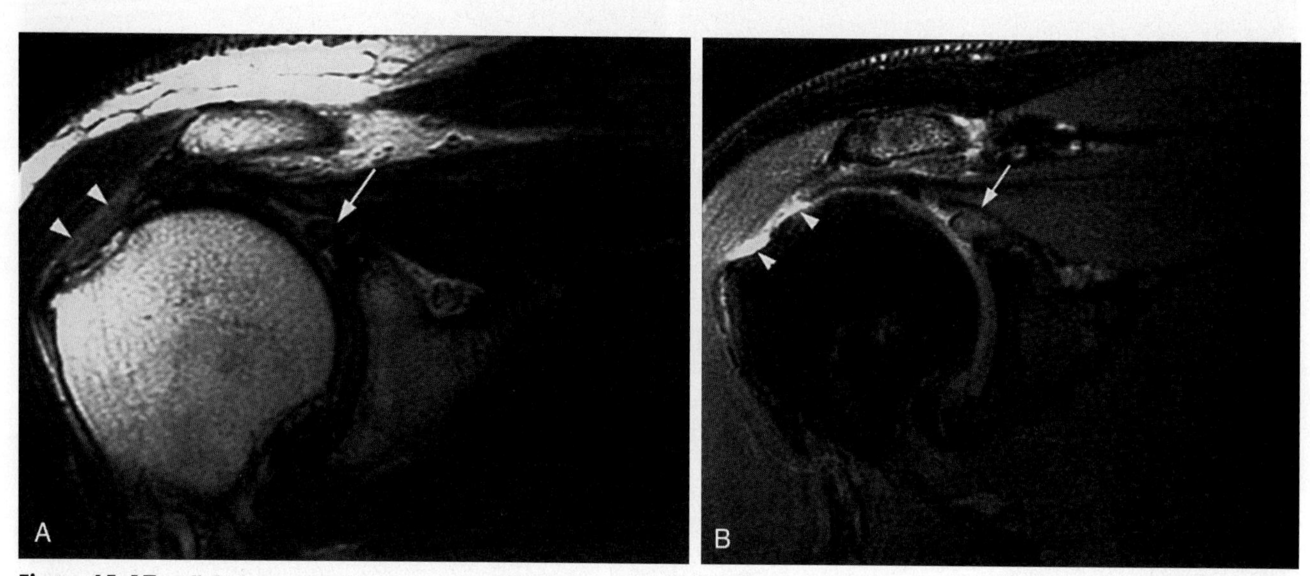

Figure 65–17. Full-thickness tear of the supraspinatus tendon along with disruption of the long head of the biceps tendon. *A* and *B*, Oblique coronal T1-weighted MR image and fat-suppressed T2-weighted image show the large defect in the rotator cuff *(arrowheads)*. The arrow points to a remnant of the torn long head of the biceps tendon.

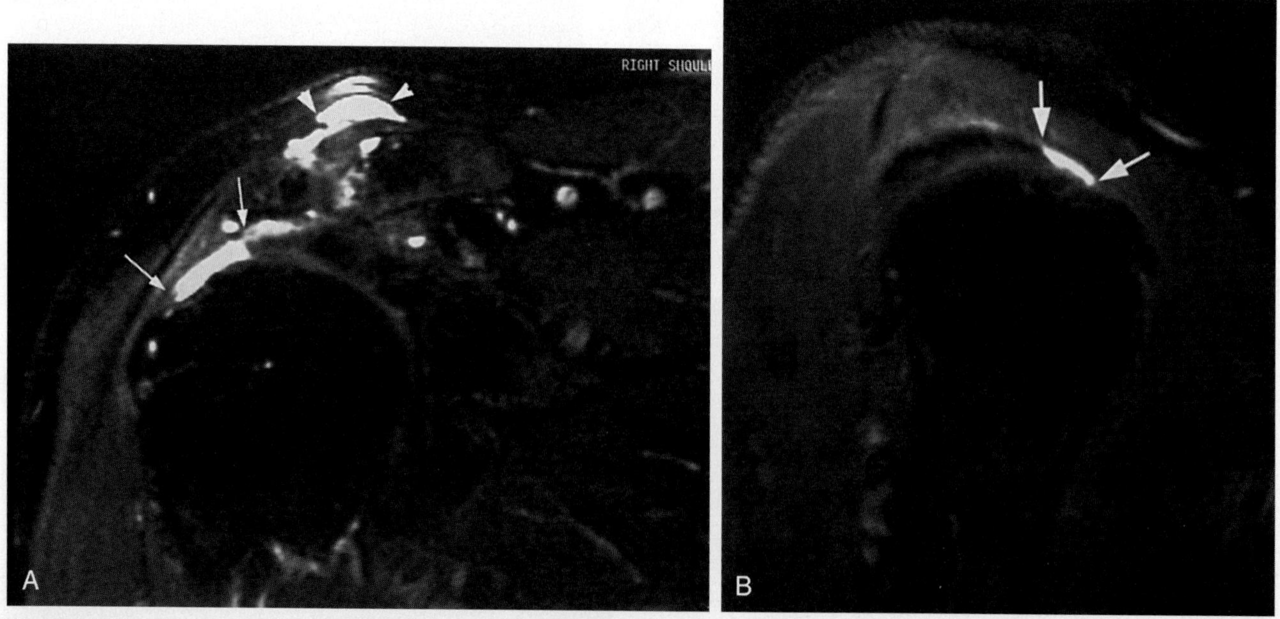

Figure 65–18. Another patient with full-thickness tear of the supraspinatus tendon. *A*, Oblique coronal T2-weighted image with fat suppression shows the tear *(arrows)*. There is also fluid above the acromioclavicular joint *(arrowheads)*. *B*, Oblique sagittal T2-weighted MR image with fat suppression shows full-thickness tear *(arrows)*.

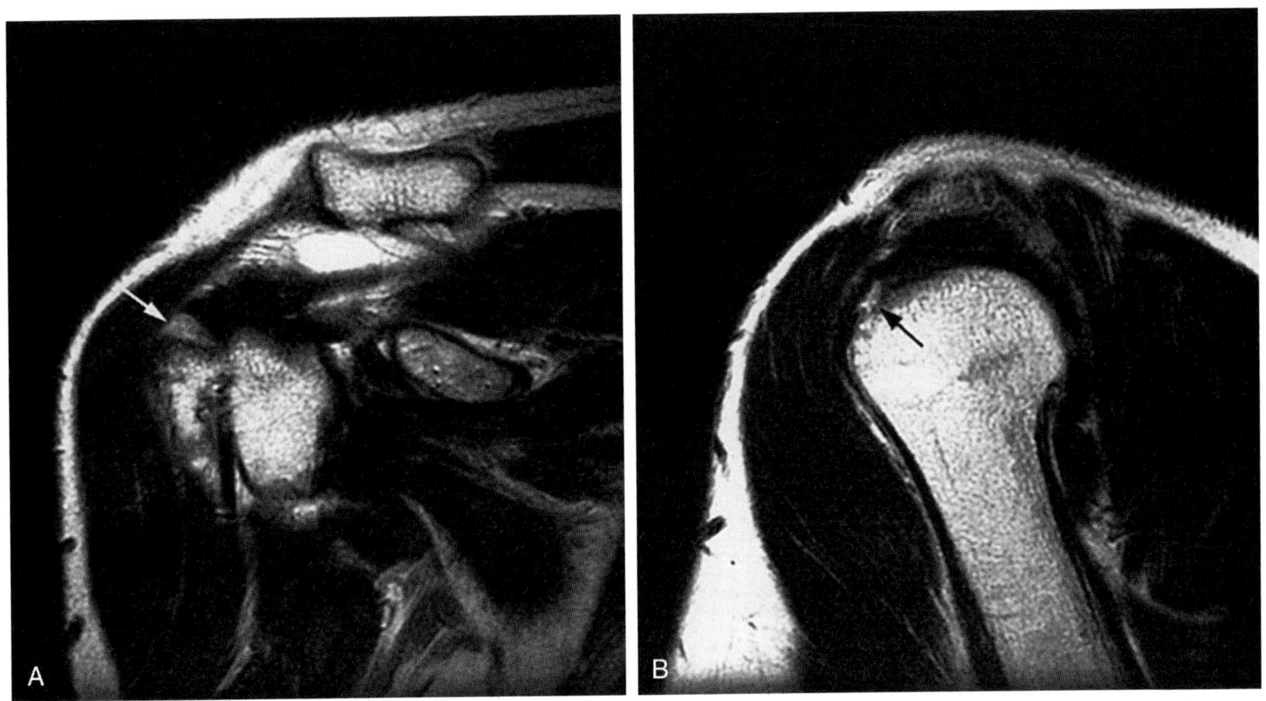

Figure 65–19. Most rotator cuff tears start at the anterior third of the supraspinatus tendon. *A*, T2-weighted oblique coronal MR image shows a tear of the supraspinatus tendon in its anterior third *(arrow)*. *B*, T2-weighted oblique sagittal MR image from another patient shows a similar tear (as in *A*) starting in the anterior third of the supraspinatus tendon *(arrow)*.

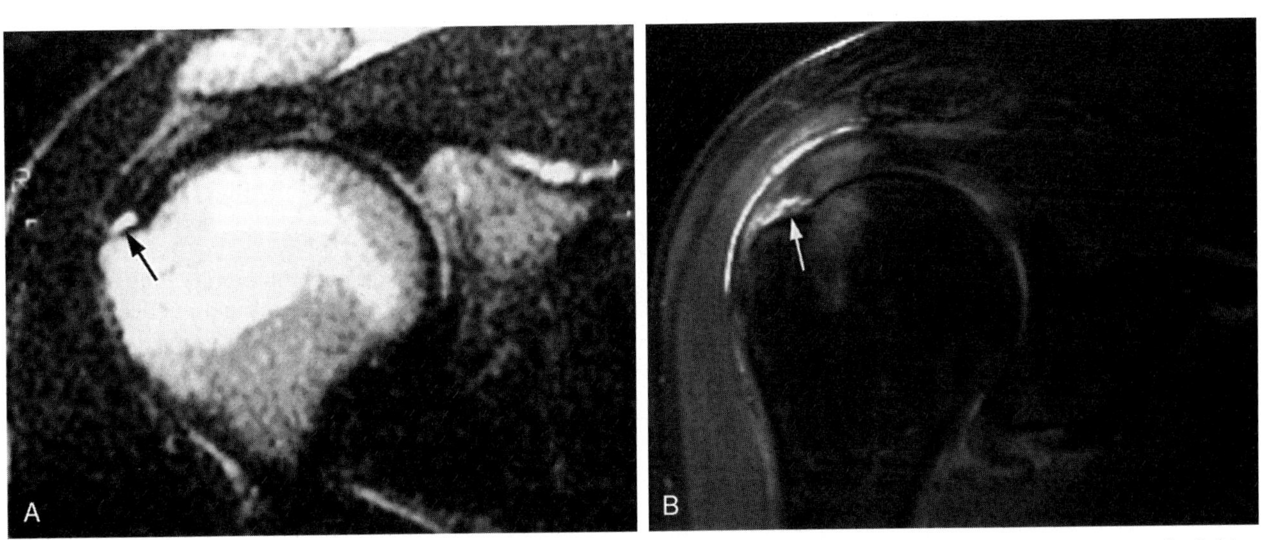

Figure 65–20. Rotator cuff tears starting close to or at the insertion to the greater tuberosity. (*A* and *B* are different patients.) *A*, T2-weighted oblique coronal MR image shows a small tear starting at the insertion of the supraspinatus tendon to the bone *(arrow)*. *B*, Fat-suppressed T2-weighted oblique coronal MR image from another patient shows a partial tear of the supraspinatus tendon at its insertion to the bone *(arrow)*. The supraspinatus tendon also shows tendinosis.

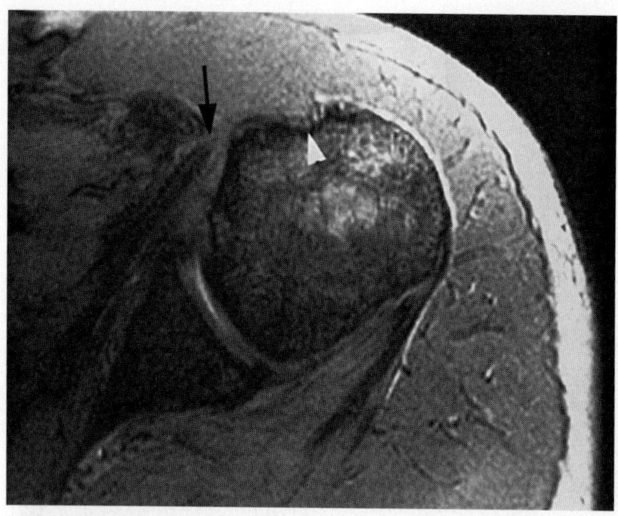

Figure 65–21. Subscapularis tendon disruption *(arrow)* associated with a long head of the biceps tendon tear *(arrowhead)*. Axial T2-weighted MR image shows the discontinuity in the subscapularis tendon *(arrow)*. The bicipital groove is empty *(arrowhead)*, suggestive of a tear of the biceps tendon.

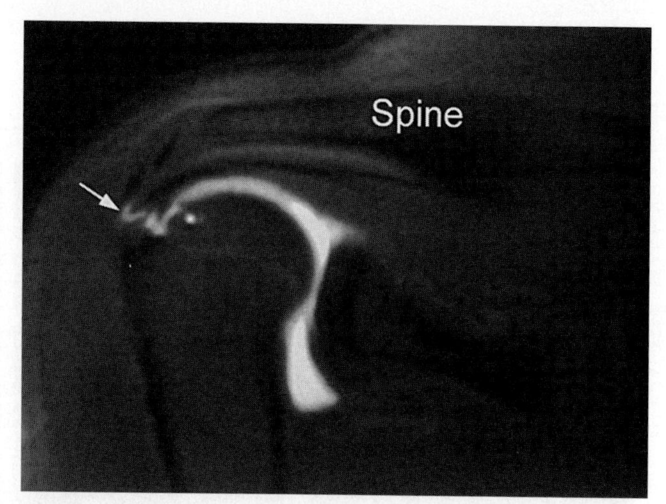

Figure 65–23. Full-thickness tear of the infraspinatus tendon in a 22-year-old woman who is a competitive tennis player. Fat-suppressed T1-weighted oblique coronal MR image after a gadolinium-enhanced arthrogram shows the tear in the infraspinatus tendon *(arrow)*. Spine, scapular spine.

laris tendon can be the result of coracoid impingement (see Fig. 65–11). Subscapularis tendon tears have a strong association with long head of the biceps tendon abnormalities (i.e., disruption or dislocation) (Figs. 65–21 and 65–22). Dislocated long head of the biceps tendons is found either anterior to the shoulder joint or within the joint.

Isolated full-thickness tears of the infraspinatus tendon are rare. They usually are seen in young athletic individuals (Fig. 65–23).

Teres minor tendon tears rarely occur in isolation. They can be associated with posterior glenohumeral dislocations.

Figure 65–22. Dislocation of the biceps tendon. Fat-suppressed T2-weighted axial MR image shows rupture of the subscapularis tendon associated with medial dislocation of the long head of the biceps tendon *(arrow)*.

GLENOHUMERAL INSTABILITY

The shoulder is the joint with greatest range of motion in the body; however, it is the least stable joint. The shoulder is prone to subluxation or dislocation in abduction and external rotation. Structures that promote stability of the shoulder include the glenoid fossa and labrum, which, in the presence of small amounts of synovial fluid, act as a suction cup in maintaining the humeral head in place. Other static stabilizers include the joint capsule and glenohumeral ligaments. The dynamic stabilizers include the rotator cuff muscles, the long head of the biceps muscle and tendon, and the other muscles that act on the shoulder joint.

Glenohumeral instability is typically a disease of young athletic individuals involved in throwing activities or of patients who have had a previous shoulder dislocation with injury to the labrum or anterior glenoid or both. Glenohumeral instability is subdivided into unidirectional (anterior or posterior) and multidirectional categories. Patients with unidirectional instability are evaluated best by MR arthrography. The introduction of diluted gadolinium into the glenohumeral joint provides superior evaluation of the relationship between the glenoid labrum, the glenohumeral ligaments, and the origin of the bicipital tendon. MR arthrography has improved lesion detection and characterization greatly. Imaging parameters are adjusted to take advantage of the intra-articular contrast by using fat-suppressed T1-weighted images in all three planes. Imaging of the shoulder with the arm abducted and externally rotated (ABER position) tightens the anterior band of the IGL and improves detection of abnormalities at the antero-inferior labrum (Fig. 65–24).

Anterior instability is the most commonly encountered type of glenohumeral instability. It results from injuries to the anterior band of the IGL labral complex. Several separate injuries to these structures have been described.

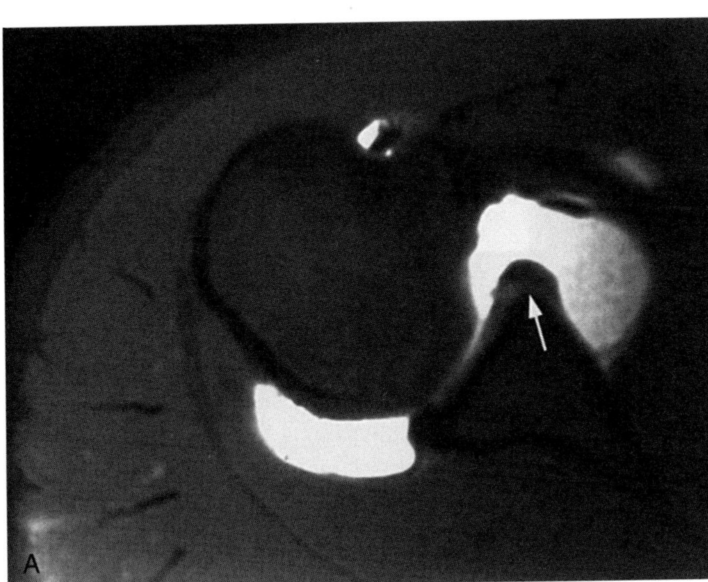

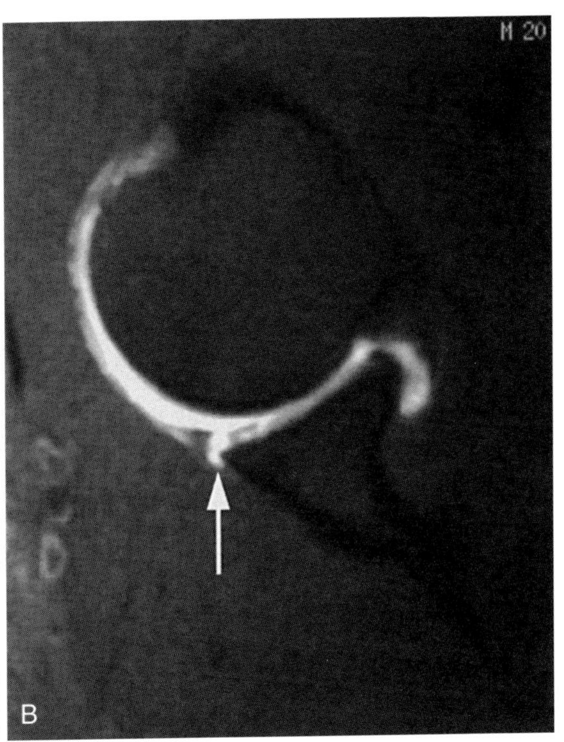

Figure 65–24. Anterior labral tear shown only by the ABER (arm abducted and externally rotated) position in a 20-year-old man who has a clinically unstable shoulder. *A*, Fat-suppressed axial T1-weighted MR arthrogram shows a normal anterior labrum *(arrow)*. *B*, Fat-suppressed T2-weighted MR image with arm in abduction and external rotation (ABER) shows a tear in the anterior labrum *(arrow)*.

A Bankart lesion represents detachment of the anteroinferior glenoid labrum with or without an associated glenoid fracture (Fig. 65–25). On MR images, fluid can be seen separating the anteroinferior labrum from the underlying glenoid. The injury is often the result of anterior shoulder dislocation and is associated frequently with a posterolateral humeral head impaction (Hill-Sachs lesion) (Fig. 65–26).

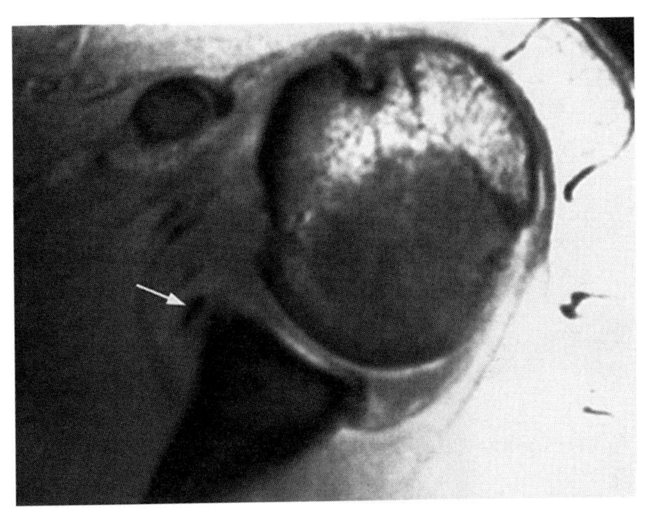

Figure 65–25. Typical Bankart lesion involving the anterior glenoid and anterior labrum. T1-weighted axial MR image shows a small fracture off the anterior glenoid and a torn anterior labrum with displacement *(arrow)*.

Anterior labroligamentous periosteal sleeve avulsion (ALPSA) lesion is similar to the labral Bankart lesion but with an associated intact anterior scapular periosteum. The intact periosteum allows the avulsed glenoid labrum to displace medially and rotate inferiorly.

Glenolabral articular disruption (GLAD) lesion is the result of a forced adduction injury resulting in a partial labral tear associated with a glenoid articular cartilage injury (Fig. 65–27). These patients present with anterior shoulder pain and typically have less instability than patients with more severe glenoid labral injuries.

Humeral avulsion of the glenohumeral ligament (HAGL) is a relatively uncommon cause of anterior shoulder instability. It represents an avulsion of the inferior glenohumeral ligament complex from the humerus.

Posterior instability is uncommon; it accounts for approximately 2% of all cases of shoulder instability. Posterior instability can result from injuries to the posterior band of the IGL. A reverse Bankart lesion represents a posterior labral tear (Fig. 65–28) associated with an anteromedial impaction injury on the superior aspect of the humeral head (reverse Hill-Sachs).

A Bennet lesion represents an area of ossification located in an extra-articular and posterior location. It is most likely the result of traction by the IGL during the deceleration phase of pitching. It is associated with posterior labral tears and posterior rotator cuff injuries.

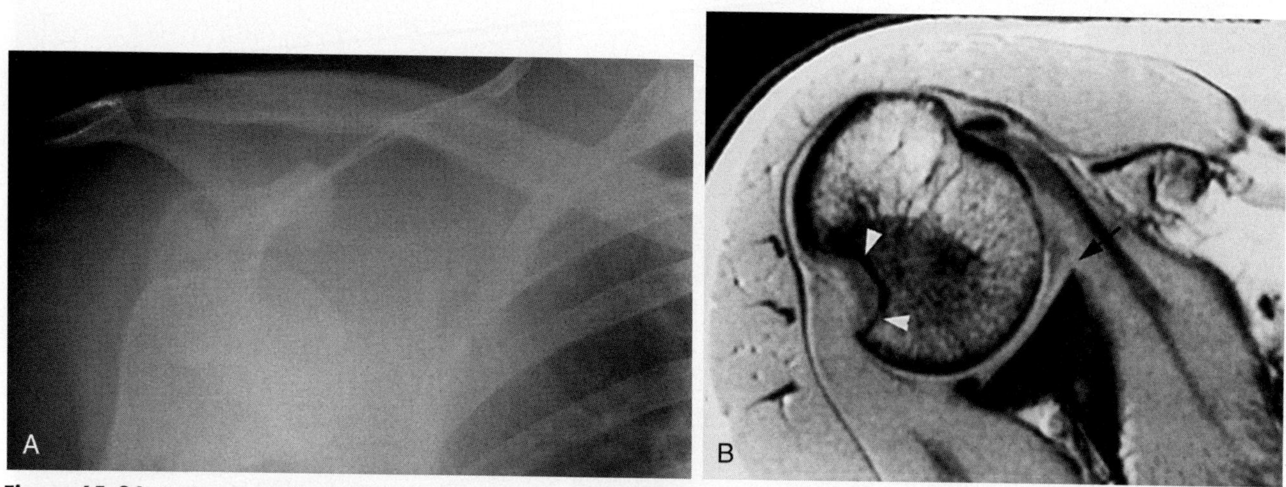

Figure 65–26. Bankart and Hill-Sachs lesions after an acute shoulder dislocation. *A,* Anteroposterior radiograph of the right shoulder shows an anterior dislocation. *B,* Gradient echo MR image shows a large Hill-Sachs lesion in the posterior superior aspect of the humeral head *(arrowheads).* The anterior labrum also is torn *(arrow).*

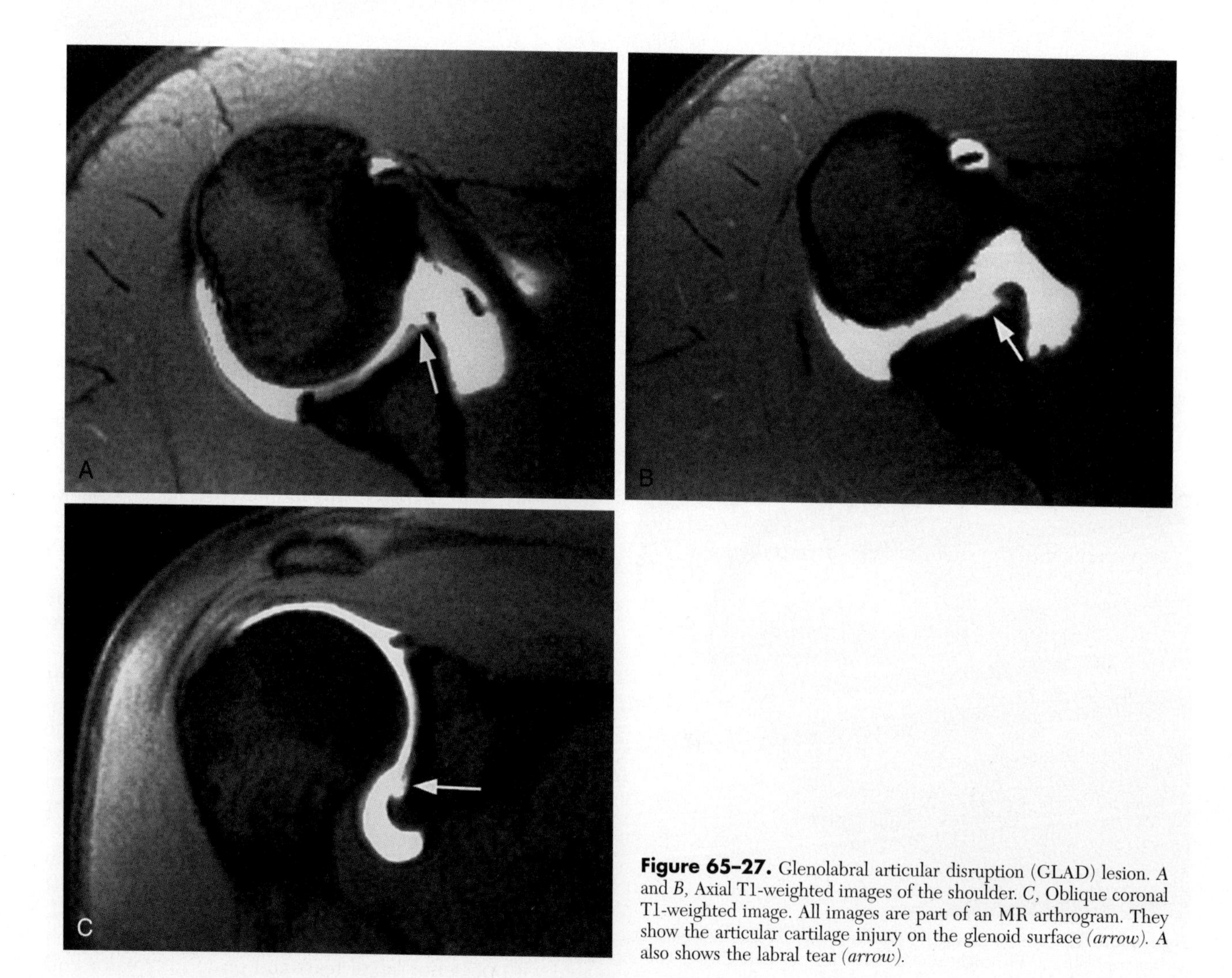

Figure 65–27. Glenolabral articular disruption (GLAD) lesion. *A* and *B,* Axial T1-weighted images of the shoulder. *C,* Oblique coronal T1-weighted image. All images are part of an MR arthrogram. They show the articular cartilage injury on the glenoid surface *(arrow). A* also shows the labral tear *(arrow).*

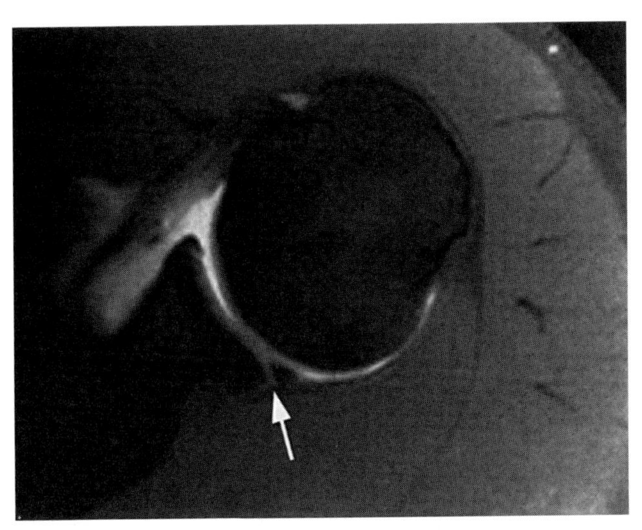

Figure 65–28. Posterior labral tear in a pitcher with symptoms of posterior instability. Axial fat-suppressed T1-weighted MR arthrogram reveals a tear in the posterior labrum *(arrow)*.

SUPERIOR LABRAL ANTERIOR POSTERIOR LESIONS

Superior labral anterior posterior (SLAP) lesions are injuries to the superior labrum at the origin of the long head of the biceps tendon. Clinically, it is difficult to make the diagnosis of a SLAP lesion. The torn labrum can displace into the joint analogous to a bucket-handle meniscal tear producing a painful locking and snapping sensation. SLAP lesions are shown best on MR arthrography. Arthroscopically, SLAP lesions have been classified into four types:

Type I: The biceps tendon is intact, but the superior labrum shows degenerative fraying (Fig. 65–29).

Type II: In addition to labral fraying, there is detachment of the long head of the biceps tendon and superior labrum from the glenoid (Fig. 65–30). The type II SLAP lesion is the most common type (about 40%).

Type III: The long head of the biceps tendon is intact, but the superior labrum shows a bucket-handle tear with displacement of the superior labrum into the joint (Fig. 65–31). Type III lesions constitute about one third of all SLAP lesions.

Type IV: A bucket-handle tear of the labrum is present with extension of the tear into the biceps tendon.

Other types have been described, but they are fairly rare, and it is often difficult to determine on MR imaging which type of SLAP lesion exists. About 40% of patients with SLAP lesions have partial rotator cuff tears. SLAP lesions are evaluated best on the oblique coronal MR sequences and, to a lesser extent, on axial sequences (see Figs. 65–29, 65–30, and 65–31). During scanning, the arm should be placed in external rotation to apply tension on the long head of the biceps. SLAP lesions should be differentiated from a normal variant known as the *sublabral recess* (Fig. 65–32). This recess allows fluid to track in between the superior labrum and the glenoid, in the region of the bicipital-labral complex. Fluid in the recess tends to parallel the surface of the glenoid, whereas a sublabral tear is oriented in the direction of the humeral head (see Figs. 65–30 and 65–31).

MISCELLANEOUS LESIONS

A paralabral cyst (or suprascapular notch ganglion) typically is located posterior and superior to the glenoid and almost always is associated with a labral tear (Fig. 65–33). Cysts are described as being spinoglenoid or suprascapular based on their location and may cause compression of the suprascapular nerve (Fig. 65–34). Suprascapular nerve entrapment may produce a neuropathy with atrophy of the supraspinatus or infraspinatus muscle (see Fig. 65–34). Clinically, patients present with chronic shoulder pain and weakness. Most suprascapular nerve entrapment cases are caused by a paralabral cyst, and only a few are caused by a neoplasm or a hematoma. In some cases, there is neuropathy and muscle atrophy but no identifiable cyst or mass.

The quadrilateral space syndrome is caused by entrapment of the axillary nerve and posterior humeral circumflex artery as they pass through the quadrilateral space. The quadrilateral space is bounded superiorly by the teres minor muscle, inferiorly by the teres major muscle, medially by the long head of triceps muscle, and laterally by the proximal humerus. Entrapment is caused most often by fibrous bands, but ganglia also have been described in this location. Compression of the axillary nerve can lead to teres minor and deltoid muscle atrophy (Fig. 65–35).

Acute brachial neuritis also is known as *Parsonage-Turner syndrome*. This is an uncommon cause of atraumatic shoulder pain and weakness. The shoulder pain experienced by these patients can be severe. Weakness usually occurs as pain is resolving. Patients are treated

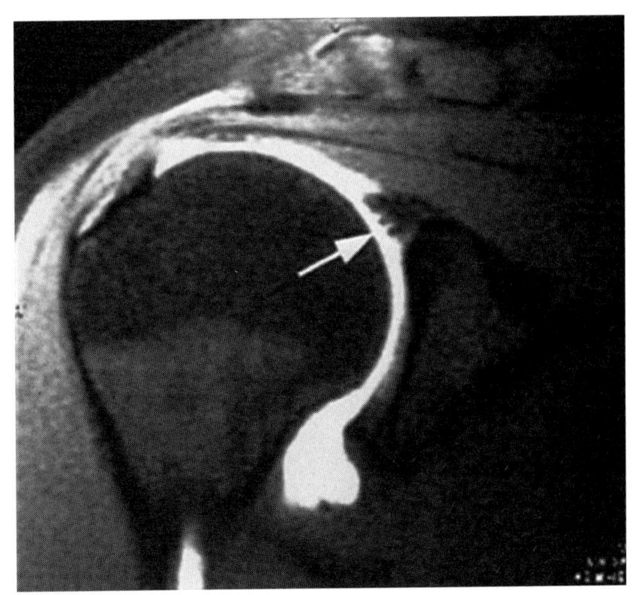

Figure 65–29. Type I superior labral anterior posterior (SLAP) lesion. Oblique coronal T1-weighted image after an MR arthrogram shows fraying and irregularity of the superior labrum typical of type I SLAP lesion *(arrow)*.

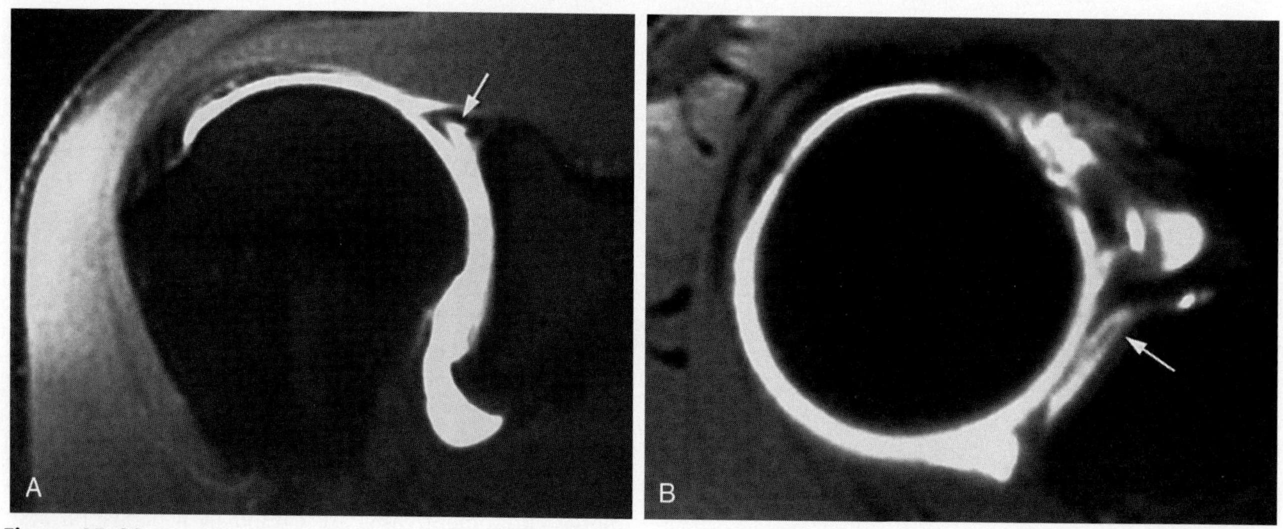

Figure 65–30. Type II SLAP lesion. *A* and *B*, Oblique coronal T1-weighted and axial T1-weighted images with fat suppression from an MR arthrogram show a type II SLAP lesion *(arrow)*.

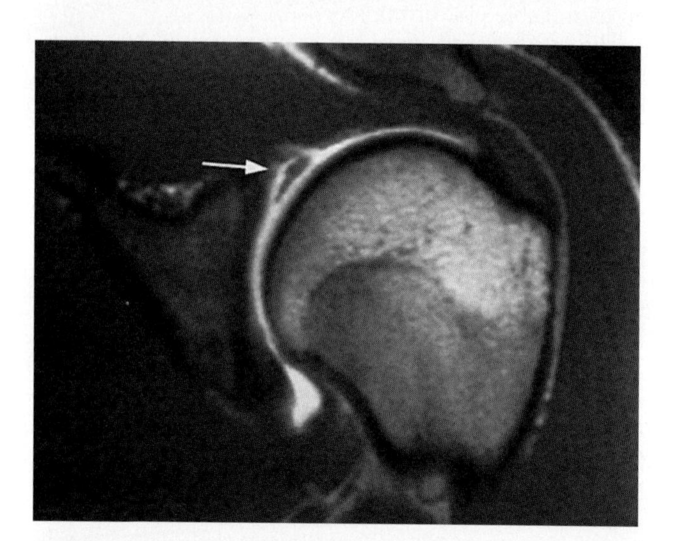

Figure 65–31. Type III SLAP lesion. Fat-suppressed oblique coronal T1-weighted images from an MR arthrogram show a type III SLAP lesion with displacement of the torn labrum into the joint *(arrow)*.

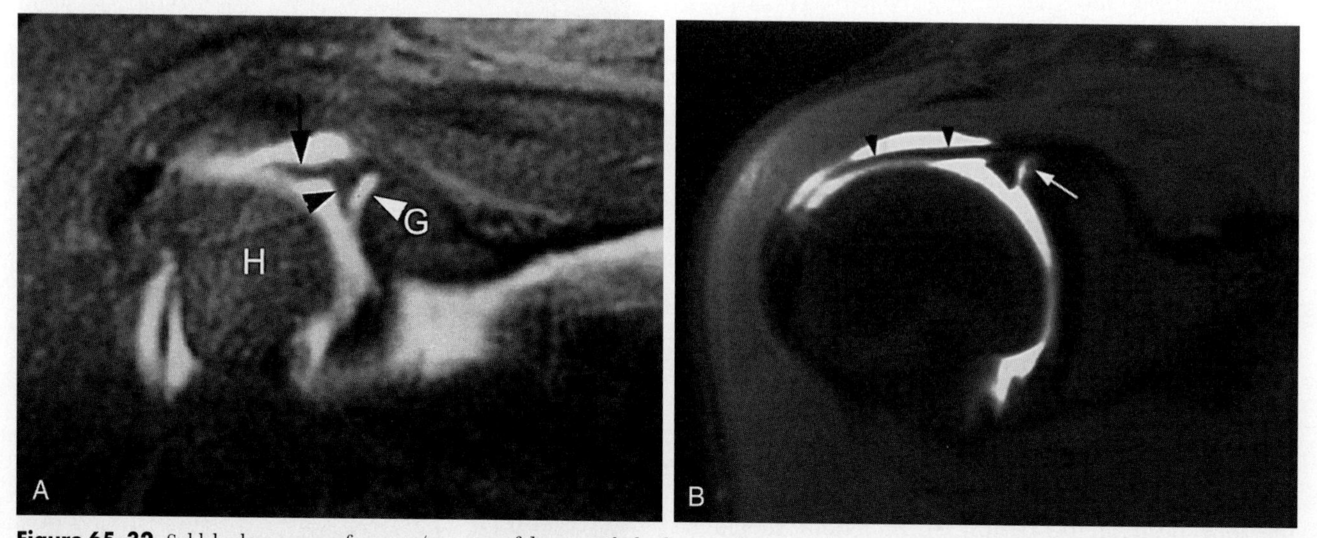

Figure 65–32. Sublabral recess—or foramen (a matter of degree with the foramen being smaller)—a normal variant not to be confused with a SLAP lesion. *A*, Fat-suppressed oblique coronal T1-weighted image from an MR arthrogram. The section is through the anterior glenoid. Note the large sublabral recess *(white arrowhead)* between the labrum *(black arrowhead)* and the glenoid (G). The long head of the biceps tendon *(arrow)* is seen inserting on the superior labrum. H, humeral head. *B*, Similar image from a different patient showing the sublabral recess or foramen *(white arrow)*. The long head of the biceps *(arrowheads)* is seen inserting on the superior labrum and glenoid.

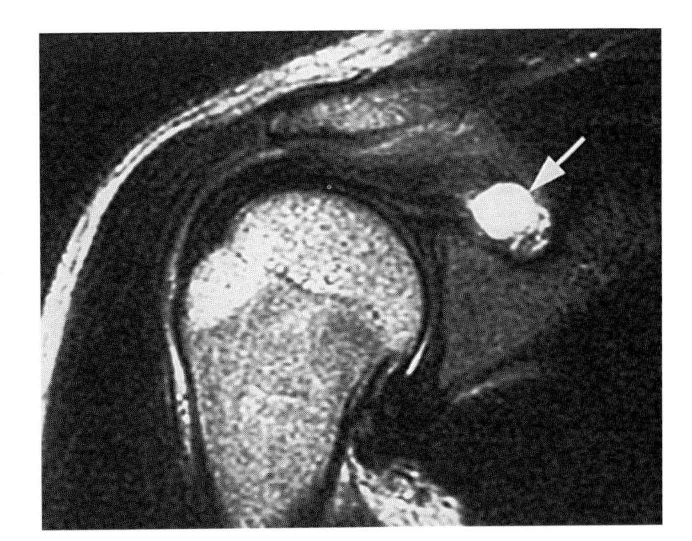

Figure 65–33. Suprascapular ganglion *(arrow)* cyst seen on an oblique coronal T2-weighted MR image.

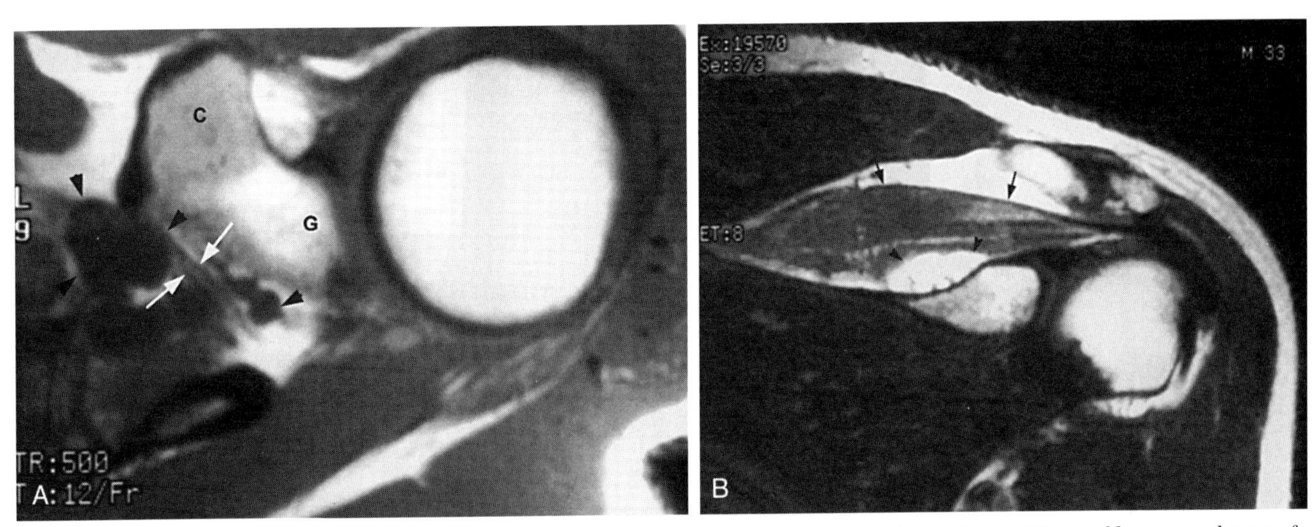

Figure 65–34. Suprascapular ganglia cyst causing suprascapular neuropathy and supraspinatus muscle atrophy in a 33-year-old man complaining of left shoulder pain and weakness. *A,* Axial T1-weighted MR image shows multiple small cysts *(arrowheads)* in the suprascapular notch. The suprascapular nerve and artery *(white arrows)* are compressed by the cysts. C, coracoid process; G, glenoid. *B,* Oblique coronal T2-weighted MR image shows the suprascapular ganglion cysts *(arrowheads).* There is also increased signal within the entire supraspinatus muscle. The size of the supraspinatus muscle *(arrows)* is diminished.

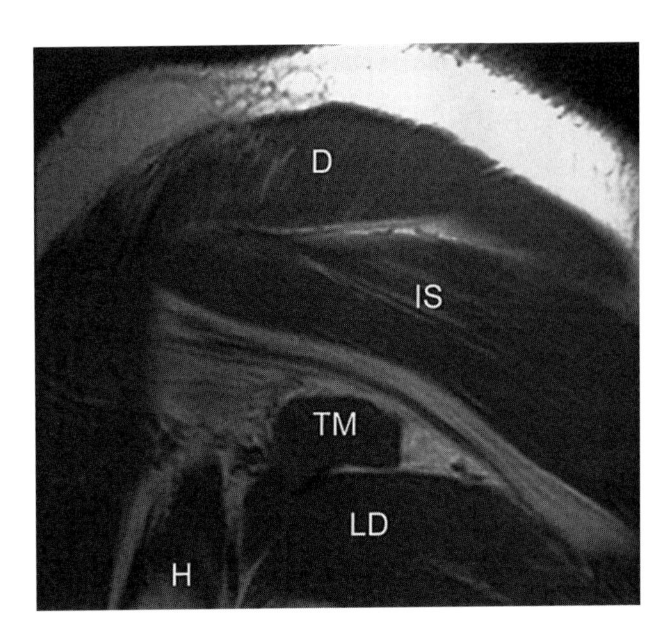

Figure 65–35. Quadrilateral space syndrome resulting in teres minor muscle atrophy. Oblique coronal T1-weighted MR image shows atrophy and fatty replacement of the teres minor muscle. D, deltoid; IS, infraspinatus; TM, teres major; LD, latissimus dorsi; H, humerus.

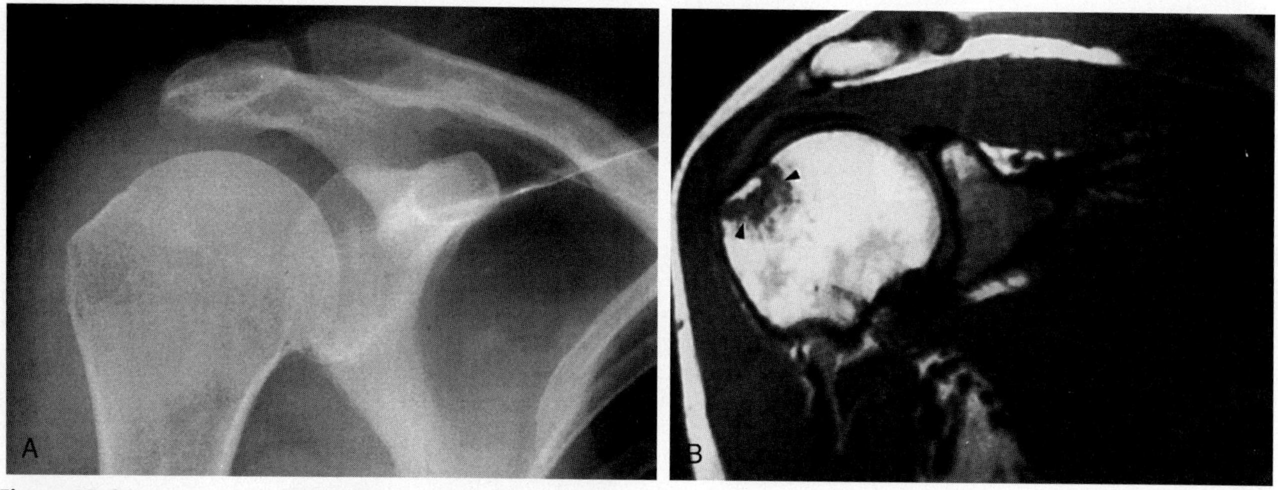

Figure 65–36. Occult fracture of the greater tuberosity in a 42-year-old woman who fell off a horse. She had signs and symptoms of an acute rotator cuff tear. *A*, Anteroposterior radiograph of the right shoulder shows no abnormalities. *B*, Oblique coronal T1-weighted MR image reveals a nondisplaced fracture of the greater tuberosity *(arrowheads)*.

symptomatically, with symptoms resolving spontaneously. MR imaging findings include edema and atrophy of affected muscles.

MR imaging is unparalleled in its ability to reveal bone contusions and occult nondisplaced fractures (Fig. 65–36). Many of these injuries are not visible on conventional radiographs. Bone contusions are seen as poorly defined heterogeneous lesions with low signal intensity on T1-weighted images and high signal intensity on short tau inversion recovery (STIR) or T2-weighted images. Chondral loose bodies also can be detected on MR imaging, especially when the joint contains fluid (Fig. 65–37).

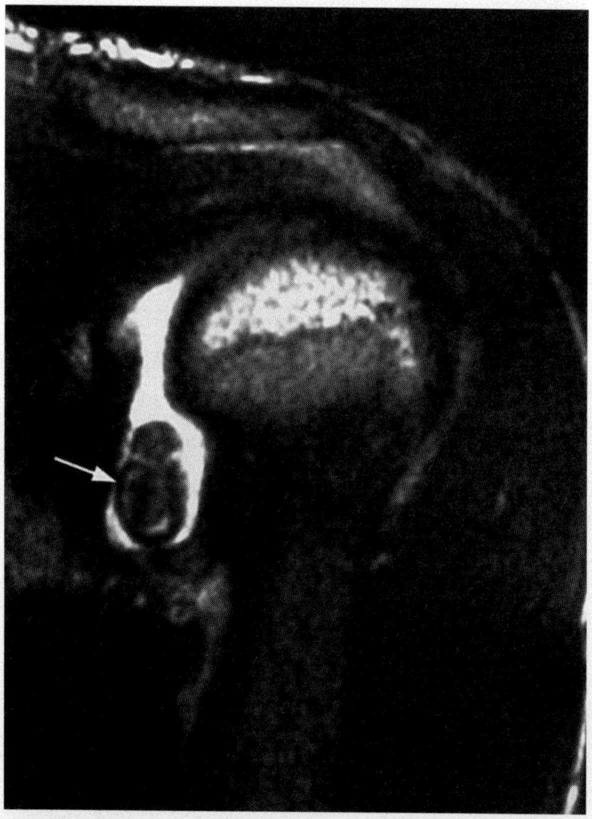

Figure 65–37. Chondral loose body *(arrow)* in a 23-year-old female athelete who complained of pain and a catching sensation in her shoulder. Fat-suppressed oblique coronal T2-weighted MR image shows fluid in the joint and a large loose body *(arrow)* in the axillary recess.

Recommended Readings

Bigliani LU, Levine WN: Subacromial impingement syndrome. J Bone Joint Surg Am 79:1854, 1997.

Bigliani LU, Ticker JB, Flatow EL, et al: The relationship of acromial architecture to rotator cuff disease. Clin Sports Med 10:823, 1991.

Carrino JA, McCauley TR, Katz LD, et al: Rotator cuff: Evaluation with fast spin-echo versus conventional spin echo MR imaging. Radiology 202:533, 1997.

Cartland JP, Crues JV III, Stauffer A, et al: MR imaging in the evaluation of SLAP injuries of the shoulder: Findings in 10 patients. AJR Am J Roentgenol 159:787, 1992.

Chandnani VP, Yeager TD, DeBerardino T, et al: Glenoid labral tears: Prospective evaluation with MR imaging, MR arthrography, and CT arthrography. AJR Am J Roentgenol 161:1229, 1993.

Clark JM, Harryman DT II: Tendons, ligaments, and capsule of the rotator cuff: Gross and microscopic anatomy. J Bone Joint Surg Am 74:713, 1992.

Dines D, Warren RF, Inglis AE, et al: The coracoid impingement syndrome. J Bone Joint Surg Br 72:314, 1990.

Erickson SJ, Fitzgerald SW, Quinn SF, et al: Long head bicipital tendon of the shoulder: Normal anatomy and pathologic findings on MR imaging. AJR Am J Roentgenol 158:1091, 1992.

Ferrari JD, Ferrari DA, Coumas J, et al: Posterior ossification of the shoulder: The Bennet lesion: Etiology, diagnosis, and treatment. Am J Sports Med 22:171, 1994.

Fritz RC, Helms CA, Steinbach LS, et al: Suprascapular nerve entrapment: Evaluation with MR imaging. Radiology 182:437, 1992.

Fritz RC, Stoller DW: Fat-suppression MR arthrography of the shoulder. Radiology 185:614, 1992.

Green MR, Christensen PC: Magnetic resonance imaging of the glenoid labrum in anterior shoulder instability. Am J Sports Med 22:493, 1994.

Gross M, Seeger LL, Smith J, et al: Magnetic resonance imaging of the glenoid labrum. Am J Sports Med 18:229, 1990.

Gusmer PB, Potter HG, Schatz JA, et al: Labral injuries: Accuracy of detection with unenhanced MR imaging of the shoulder. Radiology 200:519, 1996.

Haygood TM, Langlotz CP, Kneeland JB, et al: Categorization of acromial shape: Interobserver variability with MR imaging and conventional radiography. AJR Am J Roentgenol 162:1377, 1994.

Helms CA, Martinez S, Speer KP: Acute brachial neuritis (Parsonage-

Turner syndrome): MR imaging appearance—report of three cases. Radiology 207:255, 1998.
Jee WH, McCauley TR, Katz LD, et al: Superior labral anterior posterior (SLAP) lesions of the glenoid labrum: Reliability and accuracy of MR arthrography for diagnosis. Radiology 218:127, 2001.
Linker CS, Helms CA, Fritz RC: Quadrilateral space syndrome: Findings at MR imaging. Radiology 188:675, 1993.
McCauley TR, Pope CF, Jokl P: Normal and abnormal glenoid labrum: assessment with multiplanar gradient-echo MR imaging. Radiology 183:35, 1992.
Misamore GW, Lehman DE: Parsonage-Turner syndrome (acute brachial neuritis). J Bone Joint Surg Am 78:1405, 1996.
Monu JUV, Pope TL, Chabon SJ, et al: MR diagnosis of superior labral anterior posterior (SLAP) injuries of the glenoid labrum: Value of routine imaging without intraarticular injection of contrast material. AJR Am J Roentgenol 163:1425, 1994.
Morrison DS, Bigliani LU: The clinical significance of variations in acromial morphology. Orthop Trans 11:234, 1987.
Neer CS II: Impingement lesions. Clin Orthop 173:70, 1983.
Neer CS II: Rotator cuff tears associated with os acromiale. J Bone Joint Surg Am 66:1320, 1984.
Neviaser TJ: The anterior labroligamentous periosteal sleeve avulsion lesion: A cause of anterior instability of the shoulder. Arthroscopy 9:17, 1993.
Ohashi K, El-Khoury GY, Albright JP, et al: MRI of complete rupture of the pectoralis muscle. Skeletal Radiol 25:625, 1996.
Pattern RM: Tears of the anterior portion of the rotator cuff (the subscapularis tendon): MR imaging findings. AJR Am J Roentgenol 162:351, 1994.
Quinn SF, Sheley RC, Demlow TA, et al: Rotator cuff tears: Evaluation with fat-suppressed MR imaging with arthroscopic correlation in 100 patients. Radiology 195:497, 1995.
Rafii M, Firooznia H, Sherman O, et al: Rotator cuff lesions: Signal patterns at MR imaging. Radiology 177:817, 1990.
Robinson P, White LM, Lax M, et al: Quadrilateral space syndrome caused by glenoid labral cysts. AJR Am J Roentgenol 175:1103, 2000.
Seeger LL, Gold RH, Bassett LW, et al: Shoulder impingement syndrome: MR findings in 53 shoulders. AJR Am J Roentgenol 150:343, 1988.
Sher JS, Uribe JW, Posada A, et al: Abnormal findings on magnetic resonance images of asymptomatic shoulders. J Bone Joint Surg Am 77:10, 1995.
Smith AM, McCauley TR, Jokl P: SLAP lesions of the glenoid labrum diagnosed with MR imaging. Skeletal Radiol 22:507, 1993.
Snyder SJ, Karzel RP, Del Pizzo W, et al: SLAP lesion of the shoulder. Arthroscopy 6:274, 1990.
Tirman PFJ, Bost FW, Garvin GL, et al: Posterosuperior glenoid impingement of the shoulder: Findings at MR imaging and MR arthrography with arthroscopic correlation. Radiology 193:431, 1994.
Tirman PFJ, Bost FW, Steinbach LS, et al: MR arthrographic depiction of tears of the rotator cuff: Benefit of abduction and external rotation of the arm. Radiology 192:851, 1994.
Tirman PFJ, Feller JF, Janzen DL, et al: Association of glenoid labral cysts with labral tears and glenohumeral instability: Radiologic findings and clinical significance. Radiology 190:653, 1994.
Tuite MJ, Orwin JF: Anterosuperior labral variants of the shoulder: Appearance on gradient-recalled-echo and fast spin-echo MR images. Radiology 199:537, 1996.
Williams MW, Snyder SJ, Buford D: The Buford complex: The "cord-like" middle glenohumeral ligament and absent anterosuperior labrum complex—a normal or anatomic variant. Arthroscopy 10:241, 1994.

The Hip

Carol J. Ashman, MD, Shella Farooki, MD, Jonathan A. Lee, MD, and Joseph S. Yu, MD

ANATOMY

Hyaline cartilage covers the acetabulum and femoral head, whereas the acetabular margin is rimmed by a fibrocartilaginous labrum. The femoral head is devoid of cartilage at the fovea capitus, a depression along the central portion of the femoral head where the ligamentum teres attaches. On the acetabular side, the ligamentum teres inserts on the transverse ligament.

The acetabulum is a cup-shaped structure that is lined by the horseshoe-shaped hyaline cartilage. The cartilage is deficient inferiorly at the acetabular notch, which is bridged by the transverse ligament.

The acetabular labrum is attached to the bony margin of the acetabulum and to the transverse ligament. It is triangular in cross section and is thicker posterosuperiorly and thinner anteriorly. The labrum deepens the acetabulum and provides stability to the hip. At the acetabular notch, there is a normal cleft at the junction of the labrum with the transverse ligament that should not be mistaken for a labral detachment. The location of this cleft differs, however, from that of the typical labral tear.

The joint capsule attaches to the acetabulum at the base of the labrum. Along the superior acetabulum, the capsule attaches 1 to 2 mm above the base of the labrum. Three capsular condensations constitute the iliofemoral, pubofemoral, and ischiofemoral ligaments, which contribute to joint stability. The zona orbicularis is a portion of the capsule that comprises a ring of fibers around the midpoint of the femoral neck.

LABRAL TEARS AND DETACHMENTS

Overview

Labral tears and detachments are the most common form of labral pathology. In patients with hip pain and normal radiographs, there is a high incidence of labral tears. Any portion of the labrum may be involved, but most labral pathology occurs in the anterosuperior aspect of the labrum. Some series report, however, most tears in the posterosuperior labrum.

The acetabular labrum may be torn or detached as a consequence of traumatic injury. Labral tears produced by hip dislocations can prevent concentric reduction of the femoral head when there is entrapment of an intra-articular fragment. Patients with acetabular dysplasia are prone to developing labral tears and detachments because of increased forces on the labrum. Patients with Legg-Calvé-Perthes disease also may develop labral tears. Osteoarthritis can lead to labral tears as a result of increased shearing forces. Patients with labral tears and detachments typically complain of painful clicking, catching, giving way, or locking of the hip.

A labral tear or detachment results in loss of congruity between the femoral head and the acetabulum, leading to increased intra-articular pressure and joint effusion. The increase in intra-articular pressure displaces the synovial fluid through the labral defect into adjacent soft tissues, producing a paralabral cyst. Labral tears and detachments result in the development of a localized stress point on the femoral head and lateral migration of the femoral head because it lacks its peripheral cover. These changes eventually lead to osteoarthritis.

Imaging

Magnetic resonance (MR) arthrography is an accurate and minimally invasive means for evaluating the acetabular labrum. MR arthrography is superior to unenhanced MR imaging and conventional arthrography for the diagnosis of labral tears and detachments (Fig. 66–1). On unenhanced MR imaging, it may not be possible to separate the labrum from the adjacent capsule if there is not enough fluid to distend the perilabral sulcus. Distention of the joint with gadolinium alleviates this problem and allows improved visualization of the labral margins (see Fig. 66–1).

The technique of MR arthrography has been described by Petersilge and colleagues (1996). For optimal signal-to-noise ratio, a surface coil or a phased-array coil should be used. A 14- to 16-cm field of view, a slice thickness of 3 to 5 mm with spin echo images and 1.5 mm for gradient echo sequences, and a matrix of 256 × 256 are recommended. Imaging in all three planes is necessary for adequate evaluation. Coronal oblique images oriented perpendicular to the acetabular opening and axial oblique images oriented parallel to the axis of the femoral neck are essentially perpendicular to the acetabular labrum, optimizing its depiction (Fig. 66–1).

Either spin echo or gradient echo T1-weighted imaging with fat suppression results in greater conspicuity of the labrum. Volumetric gradient recalled echo technique allows thinner slices, mitigating partial volume averaging and greater signal-to-noise ratio than is possible with spin echo techniques.

The normal acetabular labrum is triangular in shape, has smooth margins, and is homogeneously low in signal intensity on all sequences (see Fig. 66–1). In asymptomatic individuals, variations in labral morphology and signal intensity on unenhanced MR images have been reported and are thought to represent normal age-related variations. Morphologic variations include a rounded or irregular shape or absence of labrum. Signal variations include areas of increased signal within the labrum, which may or may not reach the surface. These normal variations are more common in older individuals. Correlation with the clinical presentation is important to determine the significance of these alterations. These findings should not be considered normal in patients with symptoms and signs suggestive of labral degeneration (Fig. 66–2).

Labral detachments are manifested by the presence of linear high signal intensity between the acetabulum and the base of the labrum. The high signal intensity may

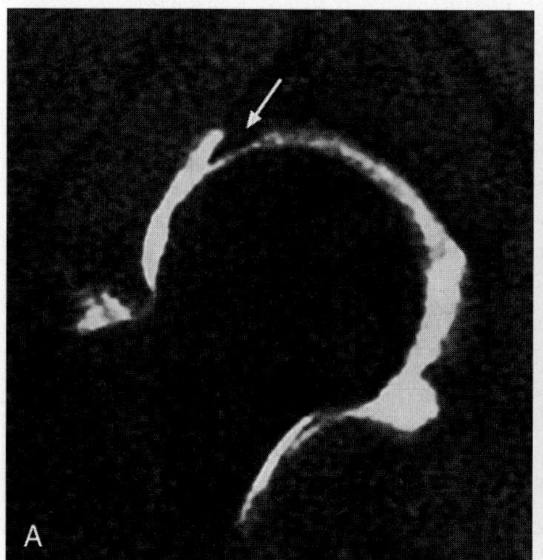

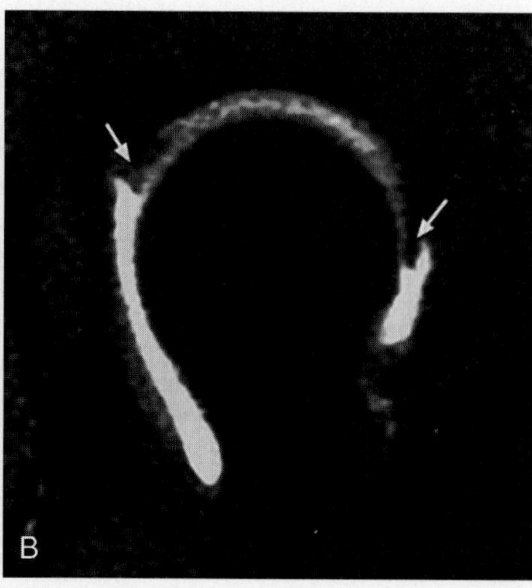

Figure 66–1. Normal labrum. *A,* Oblique coronal T1-weighted MR arthrogram with fat suppression taken after intra-articular injection of gadolinium shows a normal labrum *(arrow). B,* Oblique axial T1-weighted image with fat suppression shows a normal labrum *(arrows).*

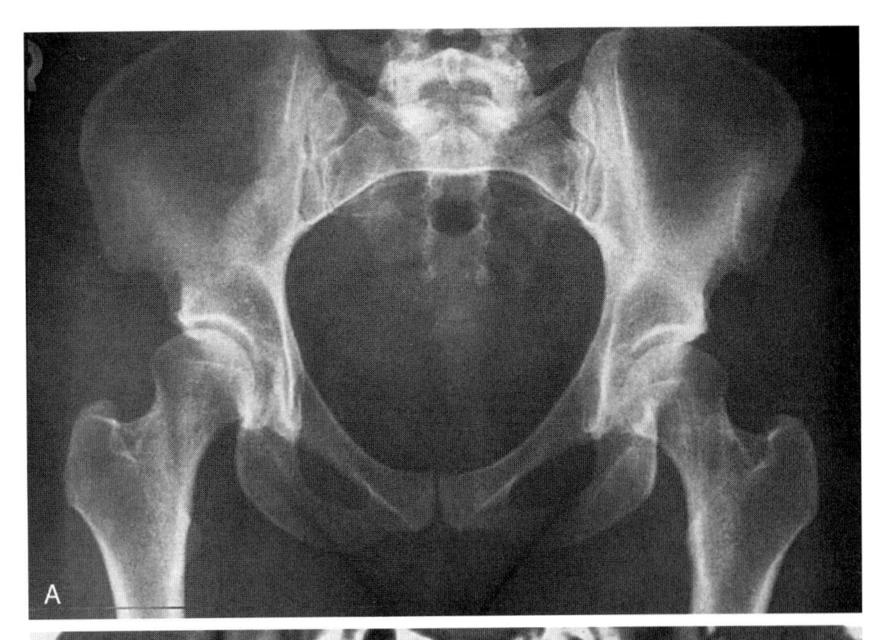

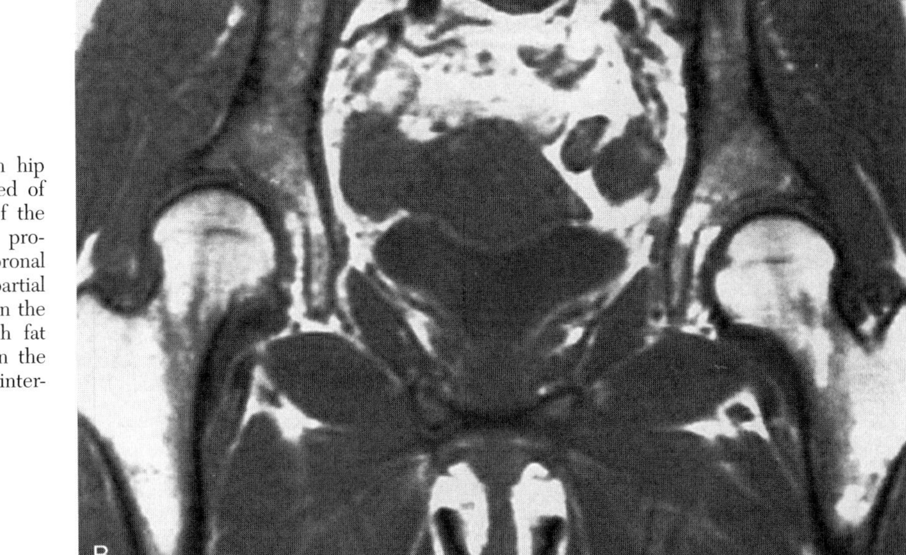

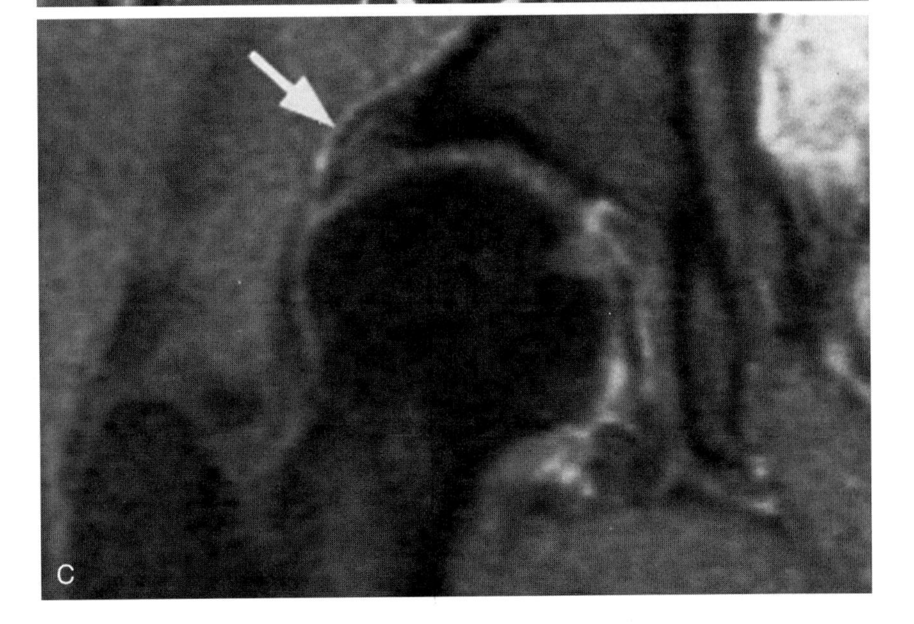

Figure 66–2. Increased labral signal in hip dysplasia. A 39-year-old patient complained of right hip pain. *A*, Anteroposterior view of the pelvis shows bilateral hip dysplasia, more pronounced on the right. *B*, T1-weighted coronal MR image shows small acetabula with partial covering of the femoral heads, especially on the right. *C*, T2-weighted coronal image with fat suppression reveals increased signal within the labrum in the right hip *(arrow)*. This is interpreted to represent labral degeneration.

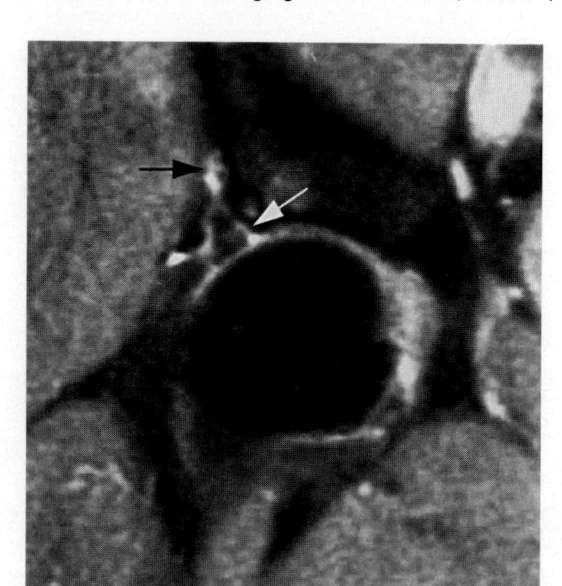

Figure 66–3. Labral detachment in a 52-year-old man who complained of a painful catching sensation in the right hip. Fat-suppressed proton-density coronal MR image reveals a detached labrum *(white arrow)* and an associated paralabral cyst *(black arrow)*.

extend into the substance of the labrum (Fig. 66–3). The presence of high signal intensity traversing the full width of the labrum constitutes a detachment (Figs. 66–3, 66–4, and 66–5); signal abnormality that extends across a portion of the labrum represents a tear. Although most tears, in our experience, involve the anterior superior aspect of the labrum, some occur in the posterior labrum, but this is rare (Fig. 66–5*C* and *D*). Morphologic changes may accompany labral tears and detachments. Irregularity of the labral margin, a rounded contour, and labral enlargement have been described in association with labral tears.

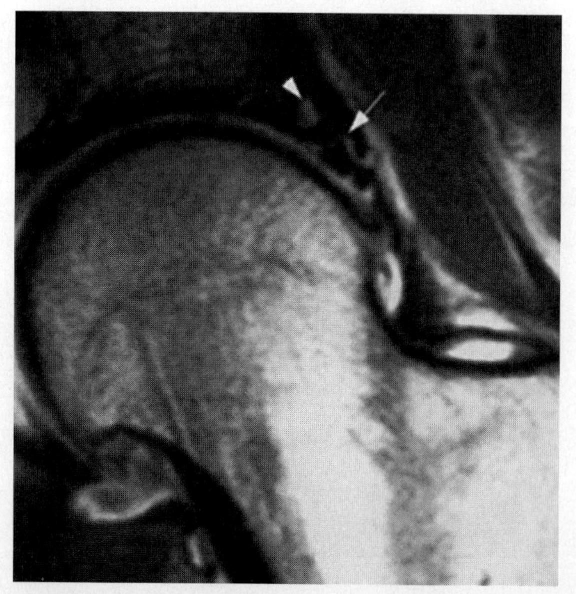

Figure 66–4. Labral detachment seen on an MR arthrogram. Coronal T1-weighted image of the left hip shows the detached labrum *(arrow)*. A small adjacent intraosseous cyst *(arrowhead)* also is shown. (Case provided by Dr. Christian Pfirrman, Zurich, Switzerland.)

There is disagreement regarding the significance of a contrast-filled cleft at the anterosuperior aspect of the labrum. There is also controversy concerning the presence of normal undercutting of the acetabular labrum by hyaline cartilage. Petersilge (2000) reported that undercutting is not of high signal intensity on MR arthrography, in contrast to a labral tear.

Intraosseous or soft tissue paralabral cysts frequently are associated with labral tears and detachments and are considered as secondary signs of labral disruptions (Fig. 66–6; see Figs. 66–3 and 66–4). Such cysts may or may not communicate with the joint or labrum. Cysts that do not communicate with the joint are of low signal intensity on T1-weighted images and high signal intensity on T2-weighted images. Communicating cysts show high signal intensity on T1-weighted images as a result of passage of gadolinium from the joint into the cyst (see Fig. 66–6). Paralabral cysts are characterized by a well-defined, smooth border.

STRESS FRACTURES OF THE HIP

Overview

Stress fractures are produced by repetitive small submaximal injuries rather than a single traumatic event. Stress fractures occur when the trabecular damage overwhelms the normal repair process of bone. In the region of the hip joint, most stress fractures occur in the femoral neck, but fractures also may occur in the femoral head or acetabulum.

There are two types of stress fractures: (1) Fatigue fractures result from abnormal stresses in qualitatively normal bone. These fractures occur in athletes or military recruits. Fatigue fractures are more likely to occur at the inferomedial aspect of the femoral neck. (2) Insufficiency fractures are produced by normal forces acting on qualitatively abnormal bone. Osteoporosis, renal osteodystrophy, and corticosteroid therapy are predisposing conditions. Insufficiency fractures in the elderly are more likely to occur in the subcapital region at the superolateral junction of the head and neck.

The pain of a stress fracture may be gradual in onset. It is produced by the inciting activity and relieved by rest. The patient's pain may be referred to the knee, causing confusion as to the site of the abnormality.

Incomplete stress fractures can progress to become complete (Figs. 66–7 and 66–8). Patients with complete fractures of the femoral neck are at risk for displacement and may require surgical fixation.

Insufficiency fractures of the femoral head are rare. They are confused easily with avascular necrosis on imaging studies (Fig. 66–9). These fractures progress with collapse of the subchondral bone and rapid destruction of the hip joint.

Isolated acetabular insufficiency fractures are uncommon and difficult to diagnose if not suspected. Patients may present with variable symptoms, including incapaci-

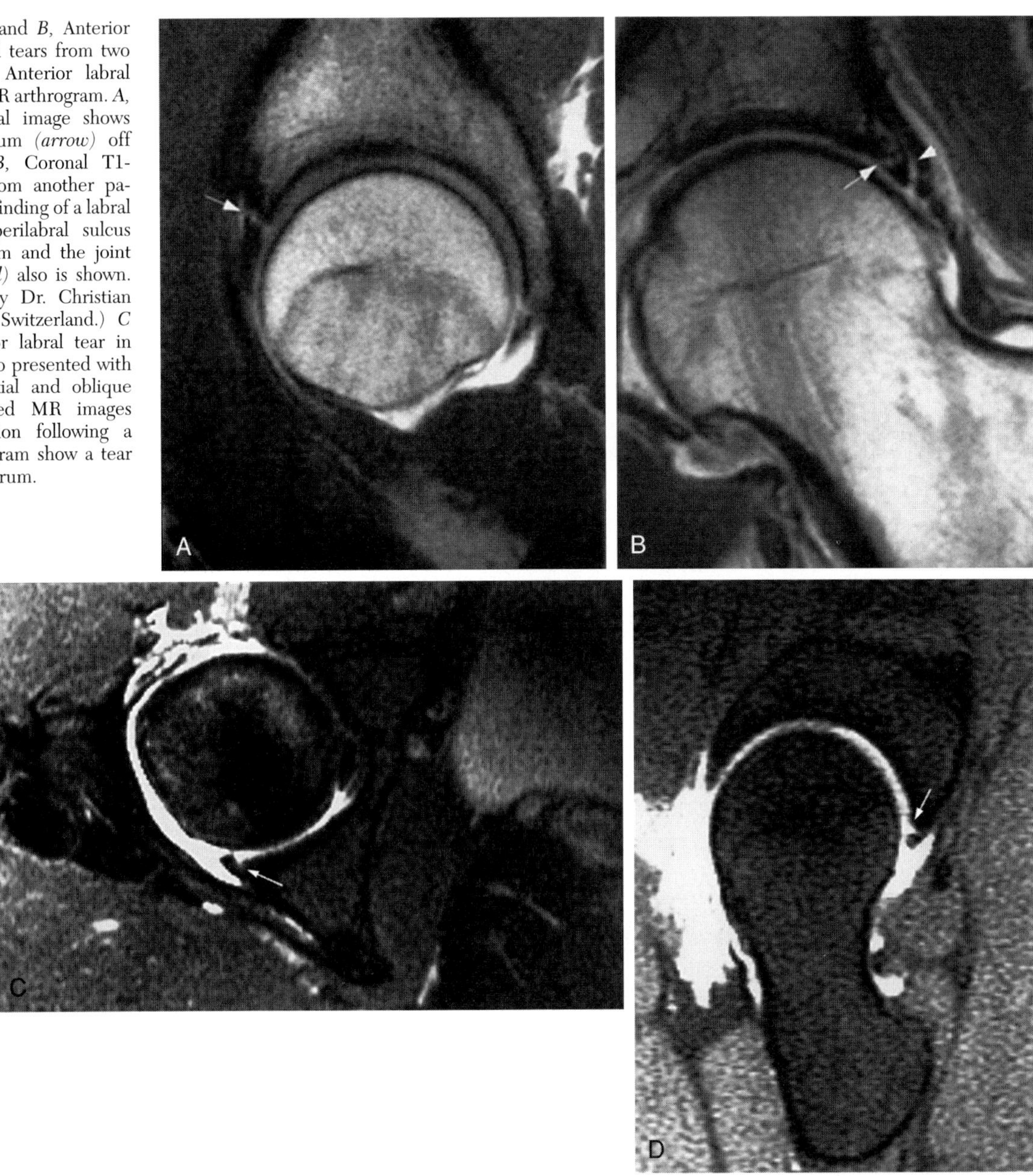

Figure 66–5. *A* and *B*, Anterior and posterior labral tears from two different patients. Anterior labral tear shown on an MR arthrogram. *A*, T1-weighted sagittal image shows the detached labrum *(arrow)* off the acetabulum. *B*, Coronal T1-weighted image from another patient shows similar finding of a labral tear *(arrow)*. A perilabral sulcus between the labrum and the joint capsule *(arrowhead)* also is shown. (Cases provided by Dr. Christian Pfirrman, Zurich, Switzerland.) *C* and *D*, A posterior labral tear in another patient who presented with right hip pain. Axial and oblique sagittal T2-weighted MR images with fat suppression following a gadolinium arthrogram show a tear in the posterior labrum.

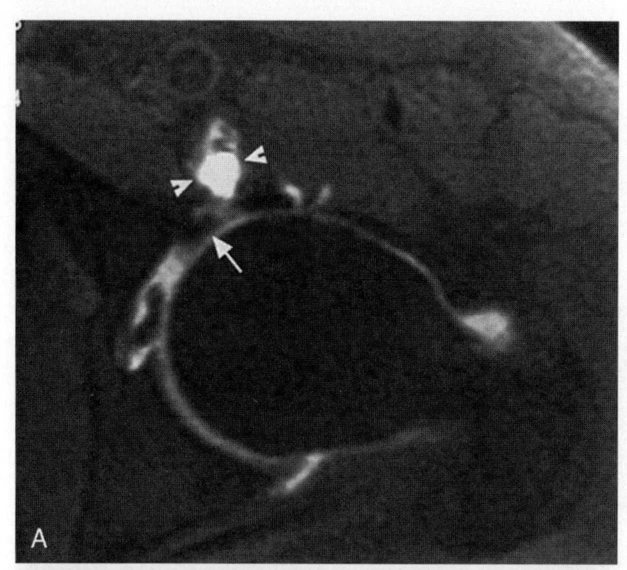

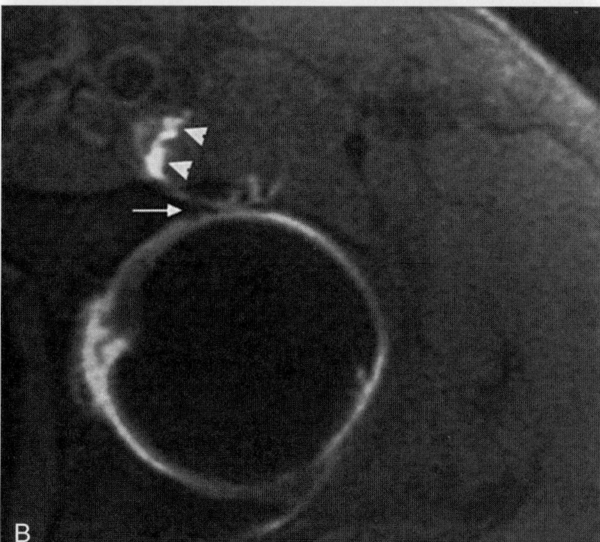

Figure 66–6. Labral tear associated with a paralabral cyst in the soft tissues. *A* and *B*, Axial postgadolinium T1-weighted MR arthrograms with fat suppression reveal a labral tear *(arrow)* and paralabral cyst *(arrowheads)*.

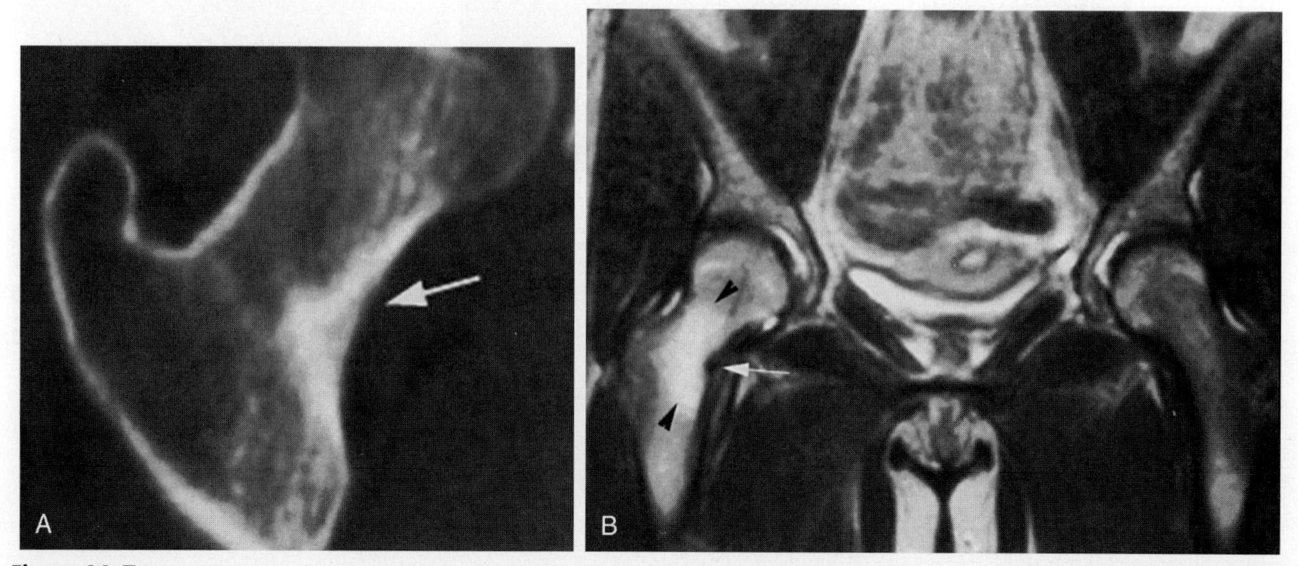

Figure 66–7. Incomplete stress fracture in a 22-year-old female athlete presenting with progressive right hip pain. *A*, Reconstructed coronal CT scan shows a linear lucency in medial cortex of the femoral neck *(arrow)* surrounded by bony sclerosis. *B*, Coronal T2-weighted MR image shows the fracture *(arrow)* surrounded by marrow edema *(arrowheads)*.

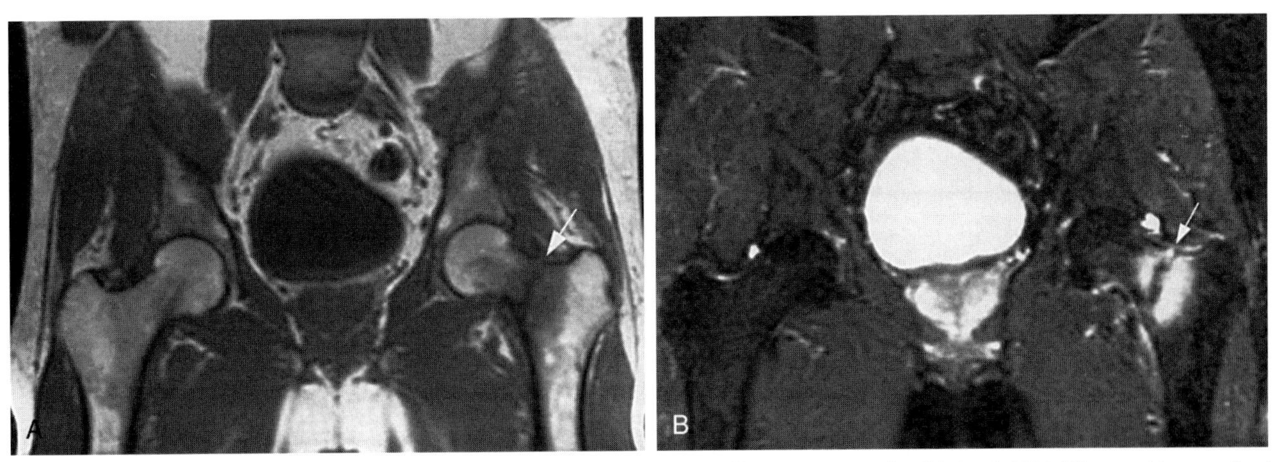

Figure 66–8. Complete stress (fatigue) fracture of the femoral neck in a 42-year-old runner. T1-weighted *(A)* and STIR *(B)* coronal images clearly reveal the fracture traversing the entire femoral neck *(arrow)* surrounded by edema.

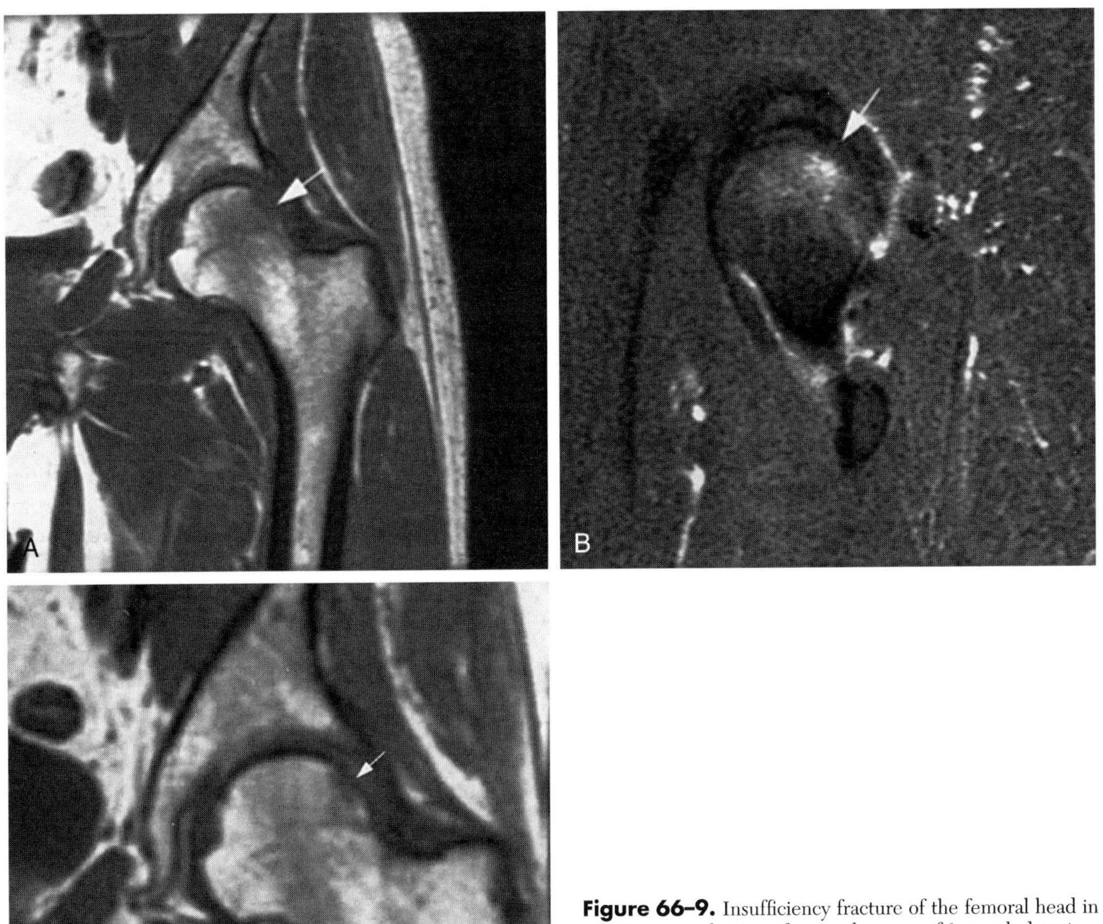

Figure 66–9. Insufficiency fracture of the femoral head in a patient complaining of severe hip pain of 1 month duration. *A,* T1-weighted coronal MR image of the left hip shows a short curvilinear band *(arrow)* of decreased signal intensity in the subchondral region of the femoral head. The bone marrow around this lesion is edematous. *B,* Sagittal T2-weighted fat-suppressed image after gadolinium enhancement shows a corresponding area in the femoral head of increased signal intensity *(arrow)*. *C,* T1-weighted coronal image acquired 3 months later shows significant resolution of the edema, but the presumed fracture is still evident *(arrow)*.

tating hip pain. Early diagnosis is important because most patients respond well to conservative therapy.

Imaging

Early detection of stress fractures in the hip is important. Initial radiographs are insensitive for detecting early stress fractures, which appear normal in 40% of patients. MR imaging has been shown to have greater sensitivity and specificity than bone scintigraphy for detection of radiographically occult fractures. These abnormalities are detected more easily by MR imaging because of the marrow signal changes they produce (see Figs. 66–7 and 66–8). Patients with normal radiographs who are unable to bear weight should undergo MR imaging to exclude the presence of a stress fracture.

Because fractures of the hip joint region may be mimicked by pubic bone fractures, the field of view should extend from the hip to the pubic symphysis (Fig. 66–10).

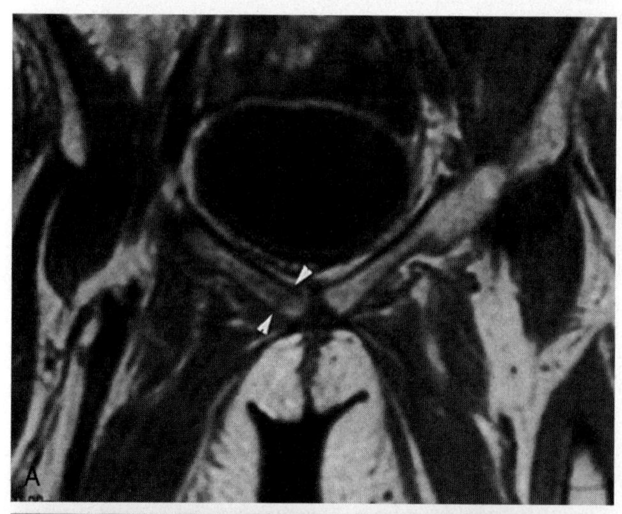

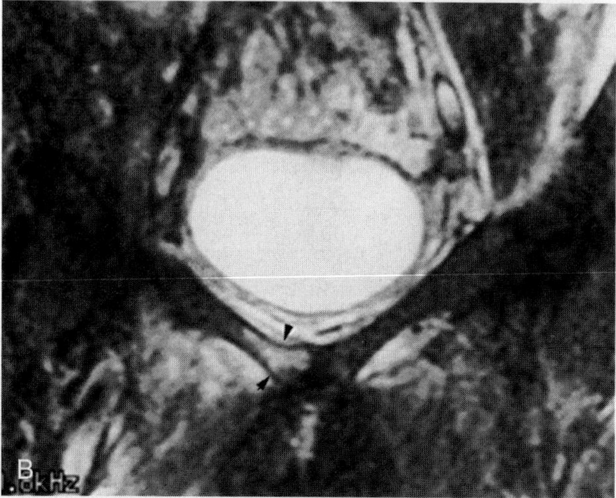

Figure 66–10. Insufficiency fracture of the superior pubic ramus in a 32-year-old woman with anorexia nervosa and profound osteopenia. She presented with right hip and groin pain. *A* and *B*, T1- and T2-weighted coronal MR images show a fracture *(arrowheads)* in the superior pubic ramus on the right.

T1-weighted and T2-weighted images with fat suppression or short tau inversion recovery (STIR) sequences allow optimal evaluation. The coronal plane is the most useful imaging plane for detecting occult femoral fractures. A limited protocol is advocated by some because the examination can be performed relatively quickly and at reduced cost.

The development of an ill-defined area of intramedullary low signal intensity on T1-weighted images and high signal intensity on T2-weighted or STIR images is compatible with bone marrow edema. This finding, referred to as a *stress response,* often precedes the development of a linear low signal intensity abnormality. A stress fracture appears as a linear band of low signal intensity on T1-weighted images, extends to the cortex, and is surrounded by edema (see Figs. 66–7 and 66–8). Edema may be seen in the adjacent soft tissues. Insufficiency fractures of the femoral head are frequently subchondral in location, often paralleling the articular surface, and may be confused with osteonecrosis (see Fig. 66–9). The double line sign is a useful differentiating feature. This finding is characteristic of osteonecrosis but not femoral head insufficiency fractures. Review of the clinical history also is important; absence of risk factors of osteonecrosis makes this diagnosis less likely. Stress fractures of the acetabulum tend to occur above the acetabular roof and are oriented parallel to it (Fig. 66–11).

HIP FRACTURES IN THE ELDERLY

Overview

Hip fractures are a major public health problem. Elderly individuals are predominantly affected, and fractures of the femoral neck and intertrochanteric region are common in this population. By the age of 80 years, 10% of white women and 5% of white men sustain hip fractures. Hip fractures are associated with a high rate of morbidity and mortality, with a 1-year mortality rate of 14% to 36%. Early diagnosis and treatment offer the best means to decrease the likelihood of complication.

Fractures of the femoral neck are almost twice as common as fractures of the intertrochanteric region of the femur. Femoral neck fractures occur predominantly in women; the female-to-male ratio varies from 3:1 to 6:1 in most series. The female-to-male ratio in intertrochanteric fractures is lower, approaching 1:1 in some series. The predilection of hip fractures for the elderly is related to the high incidence of osteoporosis in this group. Hip fractures are most commonly the result of relatively moderate trauma, usually falls, but also may occur spontaneously.

Because of the tenuous blood supply and the absence of periosteal healing of femoral neck fractures, osteonecrosis and nonunion are frequent complications of femoral neck fractures. The incidence of osteonecrosis is almost 24%, and nonunion occurs in about 25% of displaced fractures. Nonunion in intertrochanteric fractures is rare because of the good blood supply.

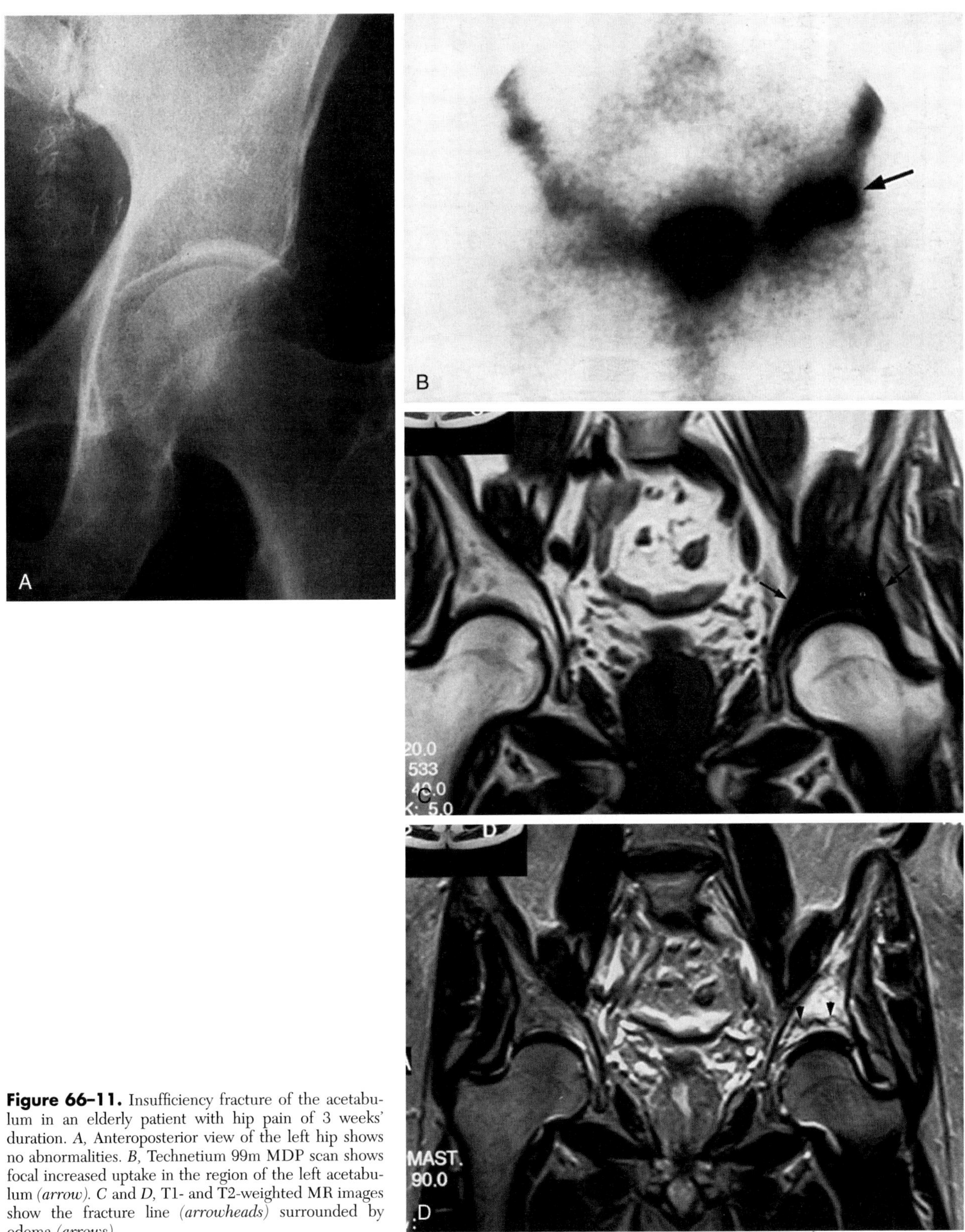

Figure 66–11. Insufficiency fracture of the acetabulum in an elderly patient with hip pain of 3 weeks' duration. *A*, Anteroposterior view of the left hip shows no abnormalities. *B*, Technetium 99m MDP scan shows focal increased uptake in the region of the left acetabulum *(arrow)*. *C* and *D*, T1- and T2-weighted MR images show the fracture line *(arrowheads)* surrounded by edema *(arrows)*.

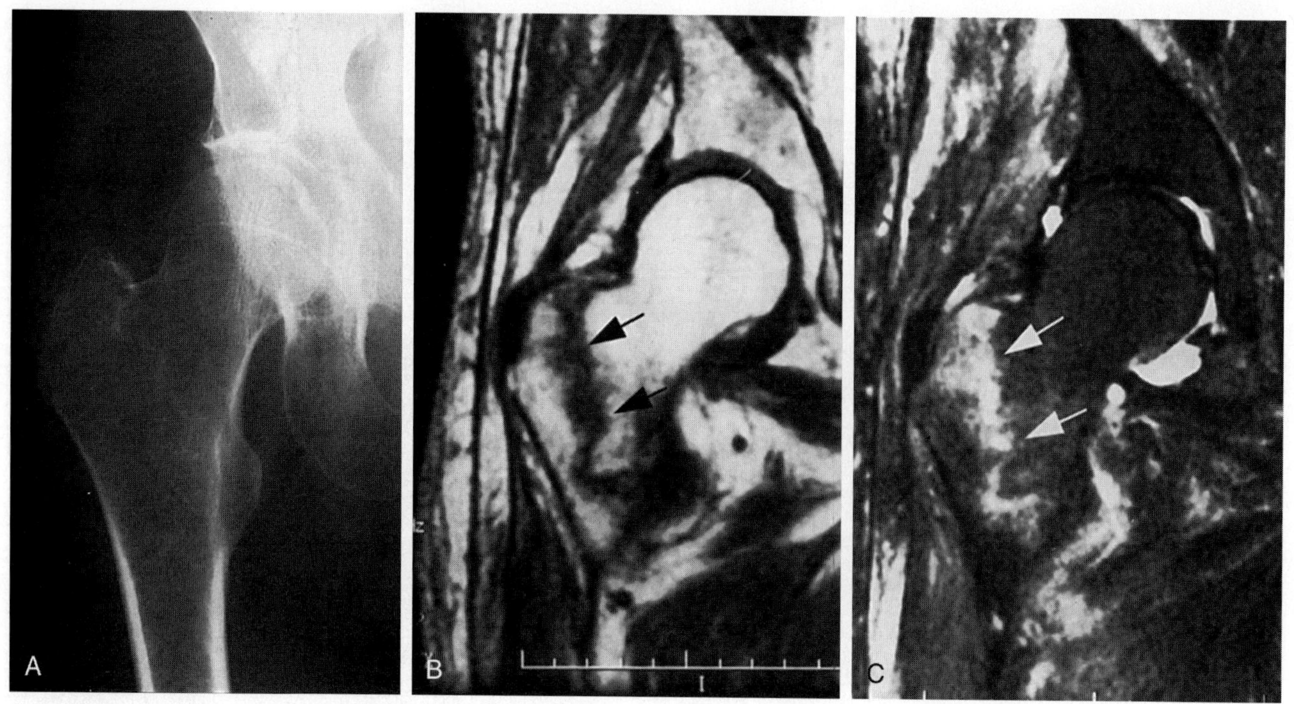

Figure 66–12. Occult intertrochanteric fracture in an elderly woman who missed a step while going down stairs. The injury was minor, but the patient was unable to bear weight on the right hip. *A,* Anteroposterior view of the right hip taken in the emergency department shows osteopenia, but no fractures were detectable. *B* and *C,* T1- and T2-weighted fat suppressed coronal images of the right hip show a complete intertrochanteric fracture *(arrows)* with surrounding edema.

Imaging

Undisplaced hip fractures may not be detectable on plain radiographs in osteopenic patients. MR imaging is highly sensitive, however, for detecting such fractures (Fig. 66–12). Fractures are manifested on MR imaging by signal changes in the bone marrow produced by the edema and hemorrhage caused by cortical and trabecular disruption. An amorphous area of high signal intensity is shown on T2-weighted and STIR images with corresponding low signal intensity on T1-weighted images. A low signal intensity fracture line is visible on T1-weighted images traversing the cancellous bone and extending into the cortex (Fig. 66–12).

Most femoral neck fractures are subcapital, spiraling from the superoposterior junction of the head and neck anteriorly and inferiorly. Intertrochanteric fractures characteristically extend between the two trochanters (see Fig. 66–12). Most intertrochanteric fractures are comminuted, with the lesser trochanter forming a separate fragment. Incomplete intertrochanteric fractures previously were thought to be rare; however, 10% of patients

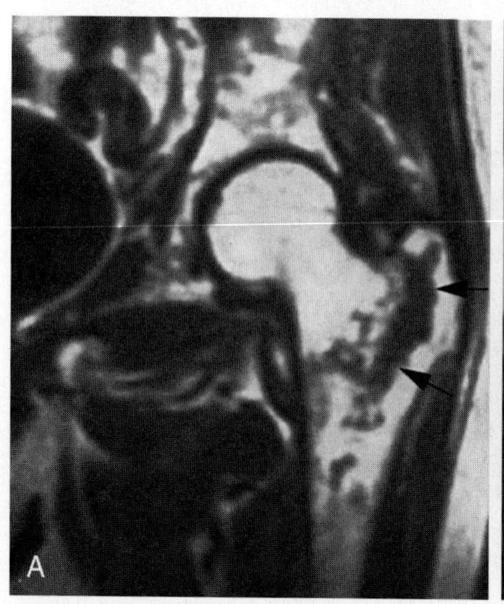

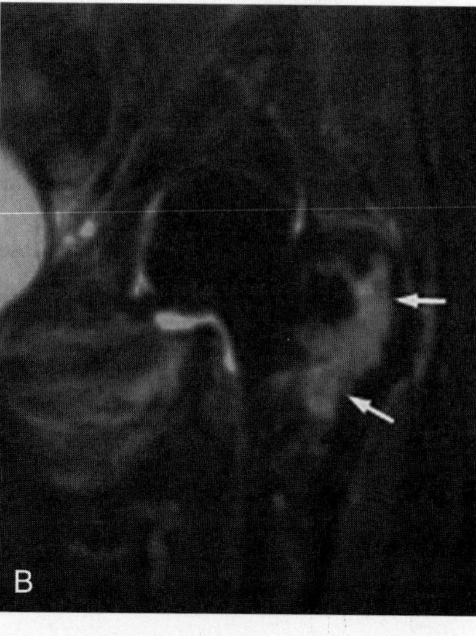

Figure 66–13. Incomplete intertrochanteric fracture in a 95-year-old osteopenic woman who presented with left hip pain. *A* and *B,* T1- and T2-weighted fat suppressed images of the left hip reveal the incomplete intertrochanteric fracture.

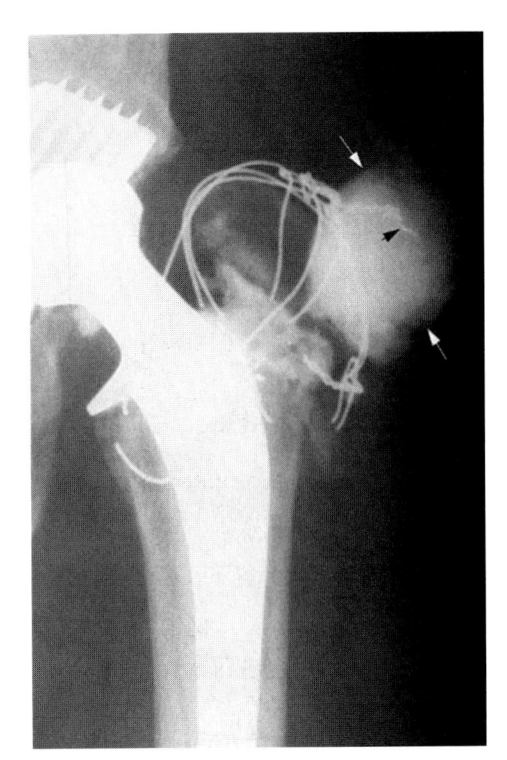

Figure 66–14. Bursitis developing around a broken wire. Anteroposterior view obtained after a hip arthrogram shows a large bursa *(white arrows)* surrounding a broken wire *(black arrowhead)*.

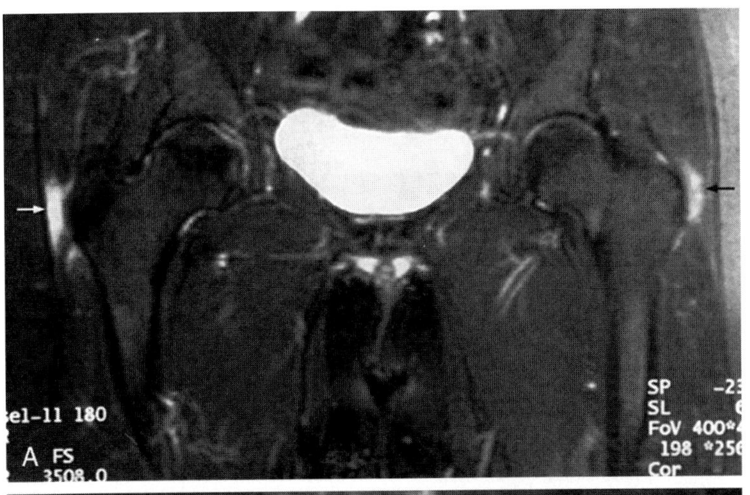

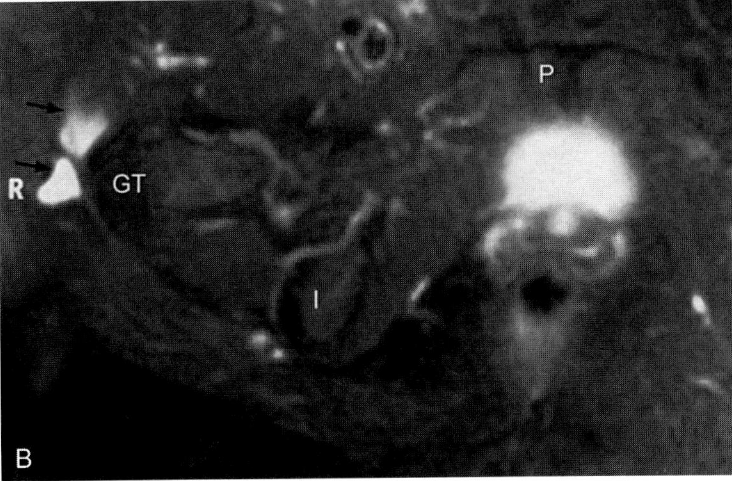

Figure 66–15. A 73-year-old woman with symptoms of bilateral trochanteric bursitis who failed conservative therapy. *A*, T2-weighted coronal MR image with fat suppression shows increased signal intensity in the soft tissues adjacent to the greater trochanter on both sides *(arrows)* but more on the right. *B*, Axial T2-weighted image with fat suppression reveals fluid in the right trochanteric bursa *(arrows)* indicating bursitis. P, pubis; I, ischium; GT, greater trochanter.

with a clinically suspected hip fracture were shown to have an incomplete intertrochanteric fracture on MR imaging (Fig. 66–13). This fracture often is radiographically occult, but MR imaging enables definitive diagnosis.

TROCHANTERIC BURSITIS

Overview

Trochanteric bursitis is a common disease that occurs in middle-aged and elderly patients. Women are more affected than men. Symptoms consist of pain and tenderness, which can be acute or chronic, over the greater trochanteric region. The diagnosis almost always is made clinically, but in difficult cases MR imaging may be helpful in confirming the diagnosis. Bursae also can develop around sharp objects, such as broken wires or sharply pointed hardware (Fig. 66–14).

Imaging

Depending on the stage of the disease and level of inflammation, a trochanteric bursa can show varying amounts of fluid (Fig. 66–15). Edema in the surrounding soft tissue also is seen with this condition. T2-weighted MR images with fat suppression are ideal for showing the intrabursal fluid and surrounding soft tissue inflammation (see Fig. 66–15).

Recommended Readings

Abe I, Harada Y, Oinuma K, et al: Acetabular labrum: Abnormal findings at MR imaging in asymptomatic hips. Radiology 216:576, 2000.

Anderson MW, Greenspan A: Stress fractures: Current concepts. AJR Am J Roentgenol 199:1, 1996.

Berger PE, Ofstein RA, Jackson DW, et al: MRI demonstration of radiographically occult fractures: What have we been missing? Radiographics 9:407, 1989.

Boden BP, Speer KP: Femoral stress fractures. Clin Sports Med 16:307, 1997.

Bosgost GA, Lizerbram EK, Crues JV: MR imaging in evaluation of suspected hip fracture: Frequency of unsuspected bone and soft tissue injury. Radiology 197:263, 1995.

Conway WF, Totty WG, McEnery KW: CT and MR imaging of the hip. Radiology 198:297, 1996.

Cooper KL, Beabout JW, McLeod RA: Supraacetabular insufficiency fractures. Radiology 157:15, 1985.

Czerny C, Hofmann S, Neuhold A, et al: Lesions of the acetabular labrum: Accuracy of MR imaging and MR arthrography in detection and staging. Radiology 200:225, 1996.

Czerny C, Hofmann S, Urban M, et al: MR arthrography of the adult acetabular capsular-labral complex: Correlation with surgery and anatomy. AJR Am J Roentgenol 173:345, 1999.

Daffner RH, Pavlov H: Stress fractures: Current concepts. AJR Am J Roentgenol 159:245, 1992.

Deutsch AL, Coel MN, Mink JH: Imaging of stress injuries to bone: Radiography, scintigraphy and MR imaging. Clin Sports Med 16:275, 1997.

Deutsch AL, Mink JH, Waxman AD: Occult fractures of the proximal femur: MR imaging. Radiology 170:113, 1989.

Evans PD, Wilson C, Lyons K: Comparison of MRI with bone scanning for suspected hip fracture in elderly patients. J Bone Joint Surg Br 76:158, 1994.

Fitzgerald RH: Acetabular labrum tears: Diagnosis and management. Clin Orthop 311:60, 1995.

Grangier C, Garcia J, Howarth NR, et al: Role of MRI in the diagnosis of insufficiency fractures of the sacrum and acetabular roof. Skeletal Radiol 26:517, 1997.

Hochman MG, Murphy S, Hall FM, et al: MR arthrography of the hip: Technique, normal anatomy and intraarticular pathology. Radiology 201:535, 1996.

Hodler J, Yu JS, Goodwin D, et al: MR arthrography of the hip: Improved imaging of the acetabular labrum with histologic correlation in cadavers. AJR Am J Roentgenol 165:887, 1995.

Klaue K, Durin CW, Ganz R: The acetabular rim syndrome. J Bone Joint Surg Br 73:423, 1991.

Lecouvet FE, Vande Berg BC, Malghem J, et al: MR imaging of the acetabular labrum: Variations in 200 asymptomatic hips. AJR Am J Roentgenol 167:1025, 1996.

Leunig M, Werlen S, Ungersbock A, et al: Evaluation of the acetabular labrum by MR arthrography. J Bone Joint Surg Br 79:230, 1997.

Lindholm TS, Osterman K, Vankka E: Osteochondritis dissecans of elbow, ankle and hip: A comparison survey. Clin Orthop 148:245, 1980.

Magee T, Hinson G: Association of paralabral cysts with acetabular disorders. AJR Am J Roentgenol 174:1381, 2000.

May DA, Purins JL, Smith DK: MR imaging of occult traumatic fractures and muscular injuries of the hip and pelvis in elderly patients. AJR Am J Roentgenol 166:1075, 1996.

Otte MT, Helms CA, Fritz RC: MR imaging of supra-acetabular insufficiency fractures. Skeletal Radiol 26:279, 1997.

Nishii T, Nakanishi K, Sugano N, et al: Acetabular labral tears: Contrast-enhanced MR imaging under continuous leg traction. Skeletal Radiol 25:349, 1996.

Petersilge CA: Chronic adult hip pain: MR arthrography of the hip. Radiographics 20:S43, 2000.

Petersilge CA, Haque M, Petersilge WJ, et al: Acetabular labral tears: Evaluation with MR arthrography. Radiology 200:231, 1996.

Plotz GMJ, Brossmann J, Schunke M, et al: Magnetic resonance arthrography of the acetabular labrum. J Bone Joint Surg Br 82:426, 2000.

Potter HG, Montgomery KD, Heise CW, Helfet DL: MR imaging of acetabular fractures: Value in detecting femoral head injury, intraarticular fragments and sciatic nerve injury. AJR Am J Roentgenol 163:881, 1993.

Rafii M, Mitnick H, Klug J, Firooznia H: Insufficiency fracture of the femoral head: MR imaging in three patients. AJR Am J Roentgenol 168:159, 1997.

Schnarkowski P, Steinbach LS, Tirman PF, et al: Magnetic resonance imaging of labral cysts of the hip. Skeletal Radiol 25:733, 1996.

Schultz E, Miller TT, Boruchov SD, et al: Incomplete intertrochanteric fractures: Imaging features and clinical management. Radiology 211:237, 1999.

Sterling JC, Edelstein DW, Calvo RD, Webb II R: Stress fractures in the athlete: Diagnosis and management. Sports Med 14:336, 1992.

Tarr RW, Kaye JJ, Nance Jr EP: Insufficiency fractures of the femoral neck in association with chronic renal failure. South Med J 81:863, 1988.

Tehranzadeh J, Vanarthos W, Pais MJ: Osteochondral impaction of the femoral head associated with hip dislocation: CT study in 35 patients. AJR Am J Roentgenol 155:1049, 1990.

Yamamoto T, Schneider R, Bullough PG: Subchondral insufficiency fracture of the femoral head: Histopathologic correlation with MRI. Skeletal Radiol 30:247, 2001.

CHAPTER 67

The Elbow Joint

Joseph G. Craig, MBChB, and John Ly, MBBS

ANATOMY

The elbow joint comprises three joints with a common capsule. The radius and ulna articulate with the condyles of the distal humerus, the capitellum, and the trochlea in a hinge joint, allowing flexion and extension. The head of the radius articulates with the radial notch of the ulna forming a pivot joint. The radial head is surrounded by the annular ligament, which is attached to the anterior and posterior margins of the radial notch.

The elbow joint is strengthened medially and laterally by the ulnar collateral ligament (UCL) and radial collateral ligament (RCL). The UCL is composed of anterior and posterior bundles and a transverse segment. The clinically important anterior bundle arises from the medial epicondyle and inserts on a tubercle on the medial aspect of the coronoid process. On magnetic resonance (MR) imaging, the anterior bundle is seen best on coronal images with the elbow in extension (Fig. 67–1). The lateral collateral ligamentous complex is made up of the RCL, the annular ligament, and the lateral UCL. The RCL arises from the lateral epicondyle and inserts into the annular ligament (Fig. 67–2). The lateral UCL arises from the lateral epicondyle and runs behind the posterior aspect of the proximal radius to insert on a ridge on the lateral aspect of the proximal ulna (Fig. 67–3; see also Fig. 67–12).

The muscles surrounding the elbow can be divided into anterior, posterior, medial, and lateral compartments. The anterior muscles consist of the biceps and brachialis. The biceps runs superficial to the brachialis and inserts via the distal tendon on the radial tuberosity (Fig. 67–4). The brachialis arises from the distal two thirds of the anteromedial and anterolateral surfaces of the humerus and inserts into the capsule of the elbow joint, the roughened anterior surface of the coronoid process, and the tuberosity of the ulna (Fig. 67–5).

The posterior muscles comprise the triceps and the anconeus. The triceps covers the posterior aspect of the humerus and inserts into the posteroinferior surface of the olecranon (see Fig. 67–5). The anconeus is a small

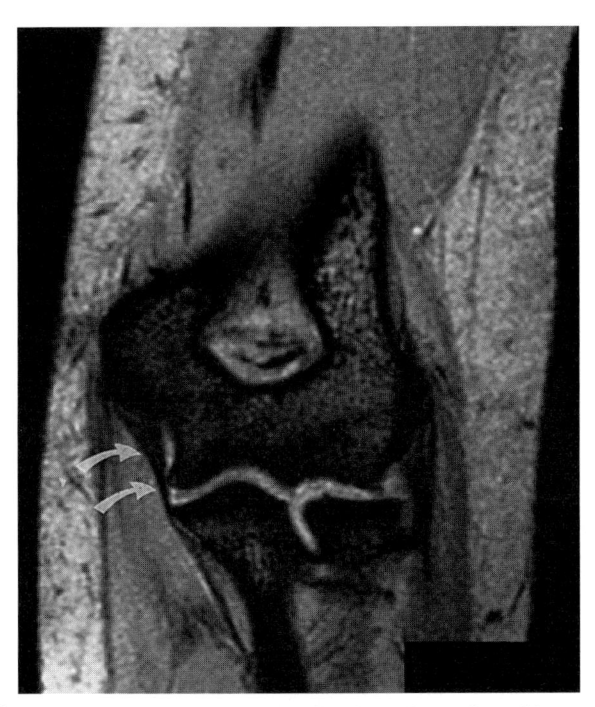

Figure 67–1. Normal anterior bundle of the ulnar collateral ligament *(arrows)*. Coronal gradient echo MR image shows the ulnar collateral ligament arising from the medial epicondyle and inserting on the medial aspect of the coronoid process.

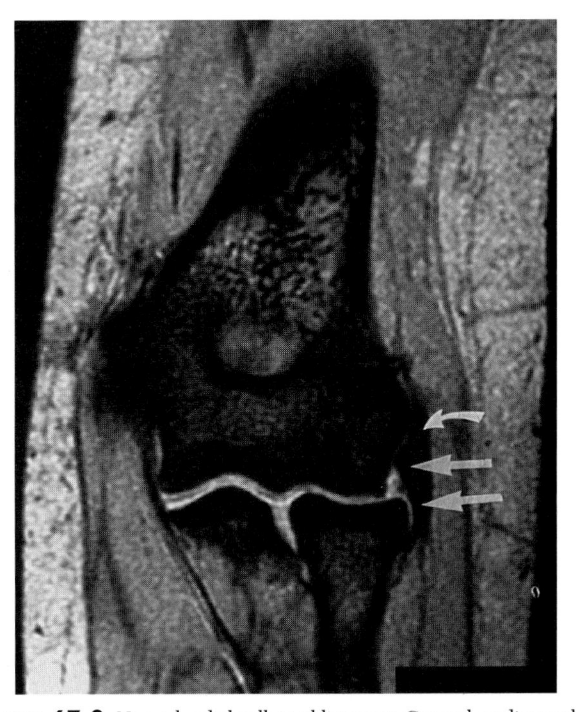

Figure 67–2. Normal radial collateral ligament. Coronal gradient echo MR image shows the normal radial collateral ligament *(straight arrows)* arising from the lateral epicondyle and inserting on the annular ligament. The curved arrow points to the common extensor tendon.

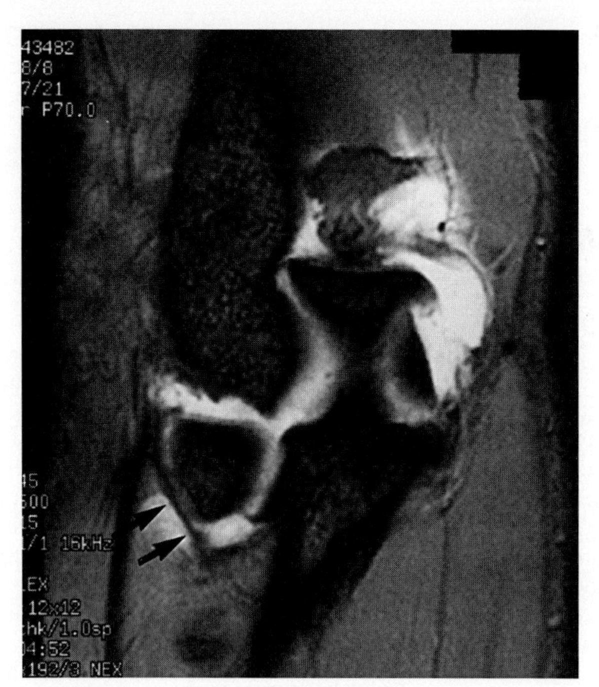

Figure 67–3. Normal ulnar collateral band of the radial collateral ligament *(arrows)* seen on a coronal gradient echo MR image, running posterior to the neck of the radius to insert on the supinator crest of the ulna.

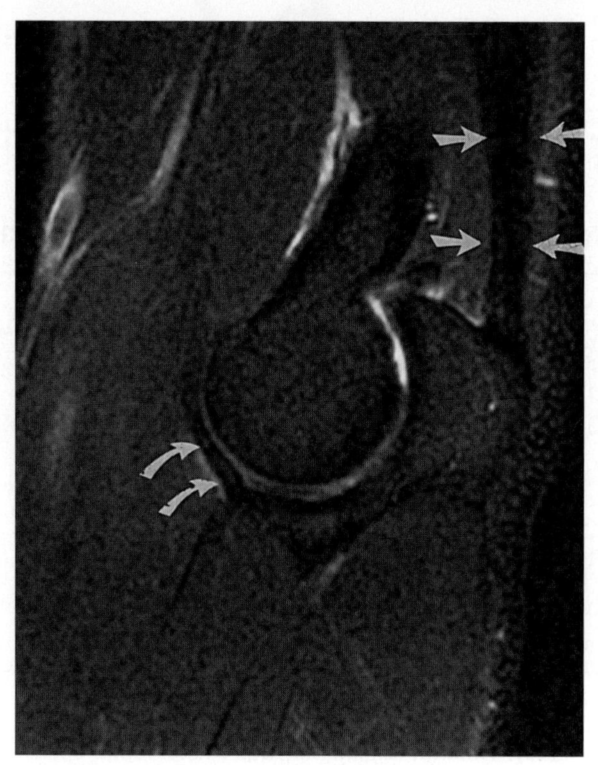

Figure 67–5. Normal triceps tendon *(straight arrows)* and brachialis insertion *(curved arrows)* on a T2-weighted sagittal MR image.

muscle that arises from the posterior aspect of the lateral epicondyle of the humerus and inserts into the lateral surface of the olecranon and posterior surface of the proximal ulna.

The medial compartment of the elbow comprises the flexors of the hand and wrist, the pronator teres, and the palmaris longus. The pronator teres arises from the medial supracondylar ridge and the medial epicondyle. A second deeper head originates from the coronoid process. It inserts on the roughened area in the middle of the lateral surface of the radius. The other flexor muscles, the flexor carpi radialis, the palmaris longus, the flexor carpi ulnaris, and the flexor digitorum superficialis, arise from the common flexor tendon at the medial epicondyle.

The lateral compartment comprises the supinator, the

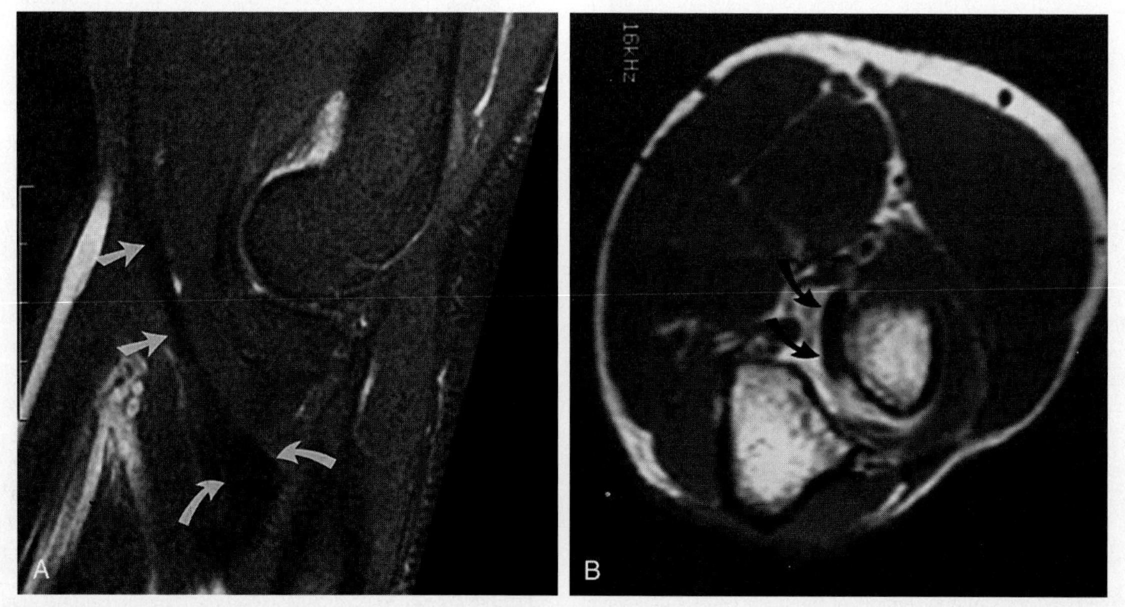

Figure 67–4. Normal biceps tendon. *A,* The normal distal biceps tendon *(straight arrows)* seen on a sagittal T2-weighted MR image. The tendon widens just before its insertion *(curved arrows)*. *B,* Axial T1-weighted image shows the distal biceps tendon inserting on the radial tuberosity *(arrows)*.

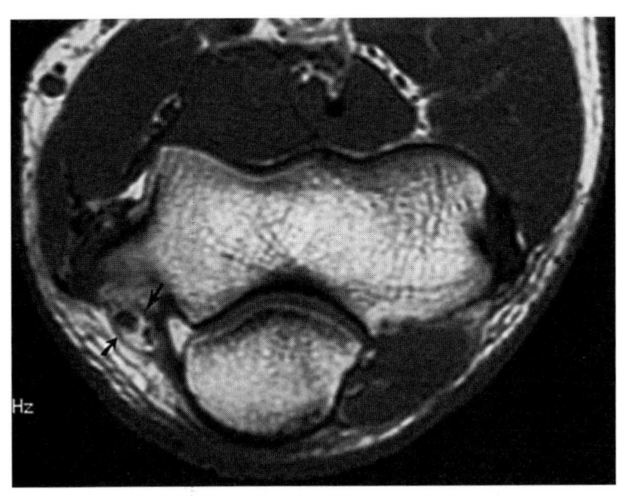

Figure 67–6. Normal ulnar nerve within the cubital tunnel on a T1-weighted axial image. The floor *(straight arrow)* and roof *(curved arrow)* of the cubital tunnel are shown.

extensor muscles of the digits and wrist, and the brachioradialis. The supinator arises from the lateral epicondyle of the humerus. The brachioradialis arises from the upper part of the lateral supracondylar ridge of the humerus, and the extensor carpi radialis longus arises from the lower part of the supracondylar ridge; the extensor carpi radialis brevis, the extensor digitorum, the extensor digiti minimi, and the extensor carpi ulnaris all arise by a common tendon from the lateral epicondyle.

The nerves running across the elbow are the ulnar, radial, and median nerves. The ulnar nerve travels posterior to the medial epicondyle and through the cubital tunnel (Fig. 67–6). It then enters the forearm between the two heads of the flexor carpi ulnaris. The median nerve in the distal arm courses along the medial side of the brachial artery. In the cubital fossa, it lies behind the median cubital vein and enters the forearm between the two heads of the pronator teres. The radial nerve in the distal arm sits deep in the groove between the brachialis medially and the brachioradialis laterally. At or near the level of the lateral epicondyle, it divides into the superficial and deep branches.

TECHNIQUE

We use a flexible coil wrapped around the elbow. The scan can be obtained with the patient supine or prone, but patients are more comfortable in the supine position. The field of view is 10 to 12 cm. Coronal T1-weighted or gradient echo sequences allow optimal visualization of the medial and lateral collateral ligaments. The axial plane allows optimal visualization of the muscles, nerves, proximal radioulnar joint, and annular ligament. We obtain coronal thin-section gradient echo, sagittal, and axial T1-weighted sequences followed by coronal and sagittal fast spin echo fat-suppressed T2-weighted sequences.

With MR imaging of the elbow, it is possible to overlook subtle abnormalities, but MR arthrography can improve the accuracy of diagnosis. Indications for MR arthrography include assessment of osteochrondritis dissecans, loose bodies, osteochondral fractures, and collateral ligament tears.

ULNAR (MEDIAL) COLLATERAL LIGAMENT

Overview

Degeneration and tearing of the UCL is seen in throwing athletes, particularly javelin throwers and baseball pitchers. Although injury may be acute, it more commonly is secondary to chronic microtrauma from repetitive valgus stresses. The UCL comprises anterior and posterior bundles and a transverse section. The anterior bundle is the primary medial constraint of the elbow. Eighty percent of medial elbow stability is provided by the anterior bundle.

Imaging

The anterior bundle of the UCL is seen best on coronal sections, running from the medial epicondyle to the medial aspect of the coronoid process (see Fig. 67–1). The ligament is wider at its attachment to the medial epicondyle and narrows distally. On MR imaging, full-thickness tears are seen as an area of discontinuity in the anterior bundle (Figs. 67–7 and 67–8).

On T2-weighted or short tau inversion recovery (STIR) images, increased signal is seen in the region of the tear. On MR arthrography, fluid crosses the tear and extravasates into the soft tissues. The UCL commonly ruptures in its midsubstance (see Fig. 67–7*B*). The diagnosis of partial tear of the UCL is more difficult but is important because these patients may require surgical repair. On MR imaging, partial-thickness tear is seen as an area of undercutting at the coronoid attachment of the UCL (see Fig. 67–8). On MR arthrography, contrast material leaks around the detachment of the UCL but is contained within the intact superficial layer of the UCL and capsule; this has been called the *T sign* and first was described on computed tomography (CT) arthrography. In chronic degeneration of the UCL, the ligament is diffusely thickened and weakened (Fig. 67–9).

Secondary changes may be seen in the elbow from chronic valgus instability in throwing athletes, including medial epicondylitis, ulnar neuropathy, posteromedial olecranon impingement, and radiocapitellar overload. Impingement of the olecranon process on the medial aspect of the trochlea is associated with marrow edema, posteromedial osteophytes, and loose body formation.

Reconstruction of the anterior bundle of the UCL involves the use of an autologous tendon graft, such as the palmaris longus tendon. The ulnar nerve usually is transposed to an anterior submuscular position at the same time.

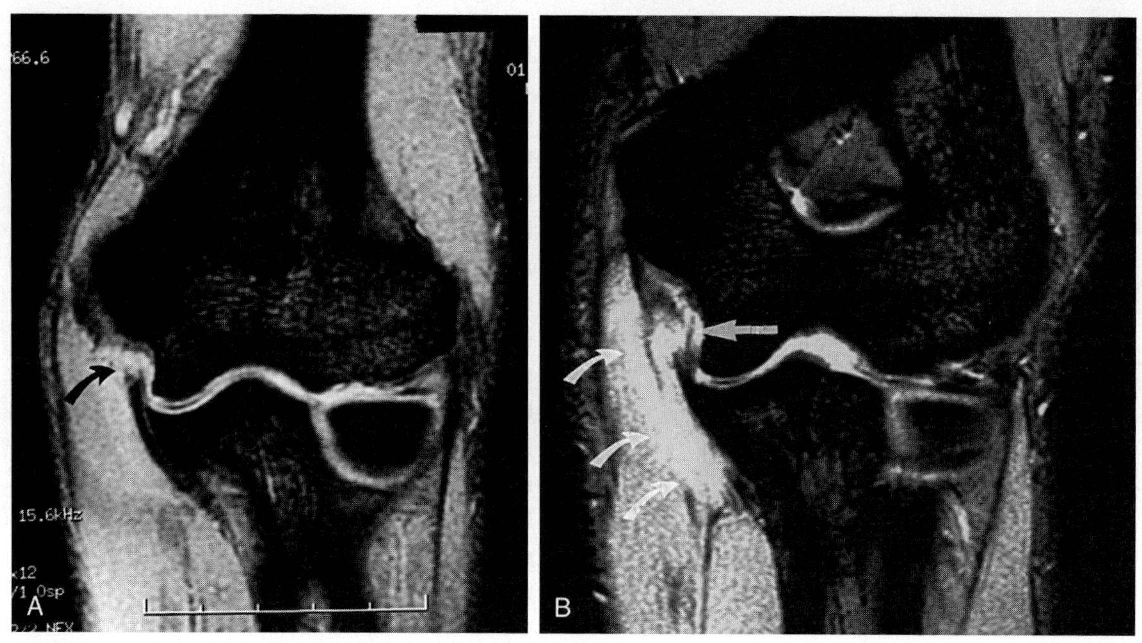

Figure 67–7. Complete tear of the ulnar collateral ligament. *A*, Coronal gradient echo MR image shows complete tear of the ulnar collateral ligament at its proximal attachment *(arrow)*. The patient is a 17-year-old pitcher who complained of medial elbow pain. *B*, Coronal gradient echo image shows complete midsubstance tear *(straight arrow)* in a 19-year-old pitcher who experienced sudden onset of medial elbow pain while pitching. Note the fluid and soft tissue edema adjacent to the tear *(curved arrows)*.

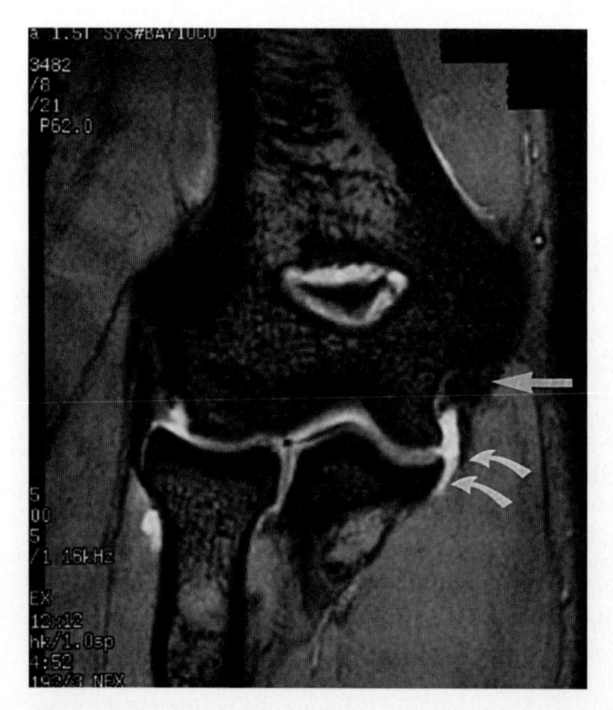

Figure 67–8. Partial tear of the ulnar collateral ligament in a 22-year-old pitcher who experienced increasing medial elbow pain. Coronal gradient echo MR image shows thickening of the proximal portion of the ulnar collateral ligament *(straight arrow)*. There is undercutting of the ulnar collateral ligament at its attachment to the coronoid process *(curved arrows)* consistent with a partial tear.

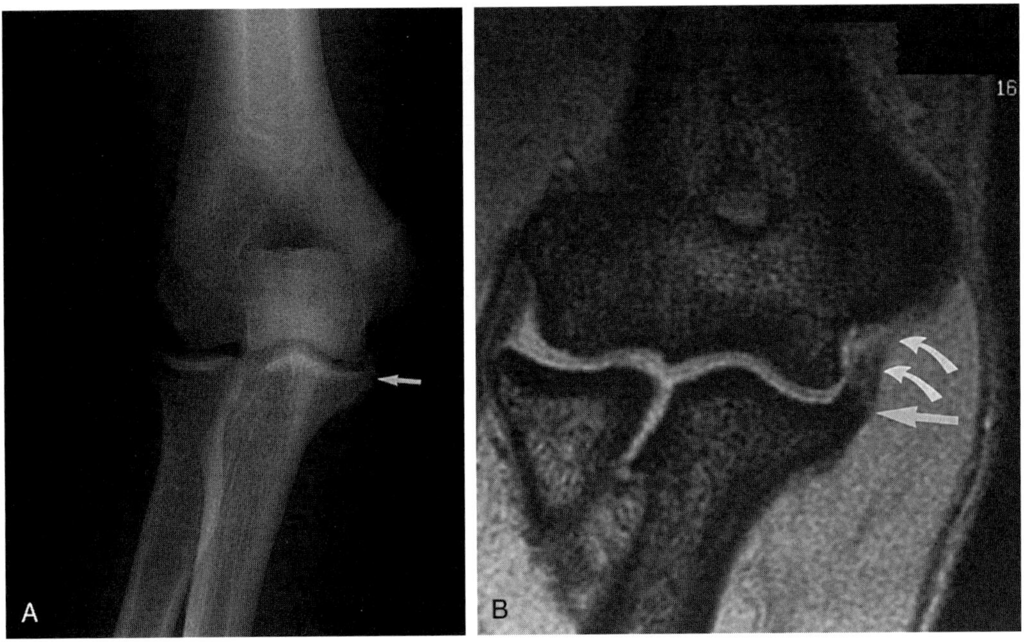

Figure 67–9. Chronic degeneration of the ulnar collateral ligament in a 24-year-old professional baseball player who presented with increasing right elbow pain. *A,* Anteroposterior view of the elbow shows a traction spur at the insertion of the ulnar collateral ligament *(arrow). B,* Coronal gradient echo MR image shows marked thickening of the anterior bundle of the ulnar collateral ligament *(curved arrows)* consistent with chronic degeneration. The spur at the olecranon is visible *(straight arrow).*

RADIAL (LATERAL) COLLATERAL LIGAMENT

The ulnar band of the RCL is an important stabilizer of the elbow. Posterolateral rotatory instability occurs as a result of injury to this ligament, usually from traumatic subluxation or dislocation of the elbow. Patients experience painful clicking, snapping, and a feeling the joint is going to dislocate on extension and supination of the elbow. The ulnar band of the RCL can be visualized on coronal images (see Fig. 67–3). Tears of this band are most common in the proximal portion of the ligament and are seen as an area of disruption in the ligament (Fig. 67–10).

LATERAL EPICONDYLITIS

Overview

Lateral epicondylitis, often referred to as *tennis elbow,* is caused by degenerative change in the origin of the common extensor tendon. The origin of the extensor carpi radialis brevis tendon is particularly affected. The condition is associated with chronic overuse and microtearing of the common extensor tendon fibers. It is a common injury among racquet sport athletes and occurs in 50% of tennis players at some time during their careers. The onset is usually around 40 years of age. Patients present with lateral elbow pain and have focal tenderness over the lateral epicondyle. The lateral elbow pain is aggravated by activities that require resisted wrist extension. Most cases respond to conservative treatment. In patients needing surgery, MR imaging can assess the severity of the tendon abnormality and aid in surgical planning. Invasion of the tendon by fibroblasts and vascular granulation tissue has been described on histologic evaluation.

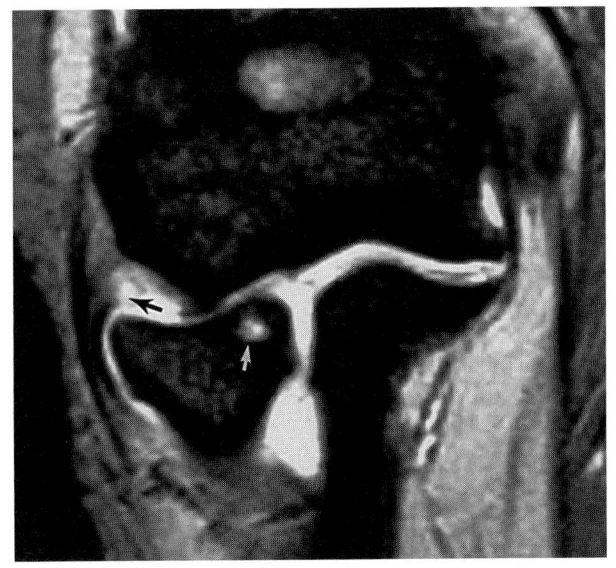

Figure 67–10. Tear of the ulnar band of the radial collateral ligament in a 41-year-old man who sustained a previous radial head fracture. The patient complained of persistent elbow pain. Coronal gradient echo MR image shows a tear of the proximal ulnar band of the radial collateral ligament *(black arrow).* A small subchondral cyst is noted in the radial head *(white arrow).*

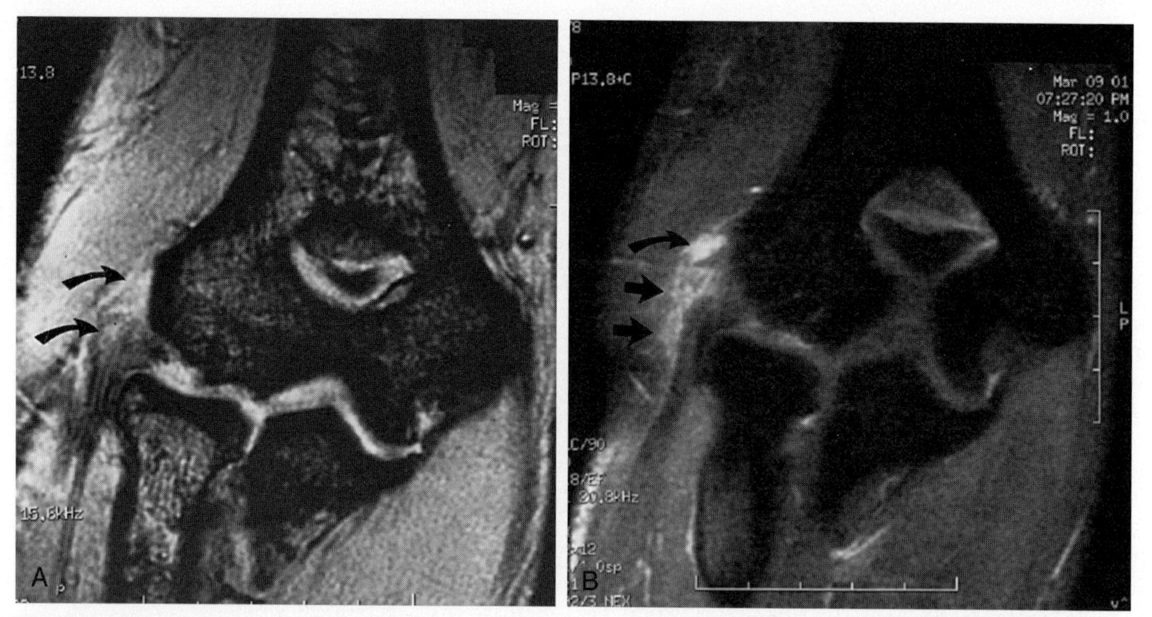

Figure 67–11. Lateral epicondylitis in a 36-year-old woman who complained of lateral elbow pain. *A,* Coronal gradient echo MR image shows diffusely increased signal within the extensor carpi radialis brevis tendon *(arrows). B,* Fat-suppressed T1-weighted image after gadolinium enhancement shows increased signal intensity *(arrows)* in the extensor carpi radialis brevis tendon.

Imaging

The coronal plane is the key to MR imaging, although useful information also can be obtained from axial images. T1-weighted, T2-weighted, STIR, and gradient echo sequences all provide diagnostic information (Fig. 67–11). Changes seen on MR imaging have been classified as tendinosis, partial tear, or complete tear. Tendinosis is associated with increased signal on T1- and T2-weighted images at the origin of the common extensor tendon from the lateral epicondyle. Partial tears are characterized by thinning of the tendon with disruption of some of its fibers. Complete tears are characterized by disruption of the tendon from its bony attachment. Unsuspected tears of the lateral ligamentous complex may occur in association with tears of the common extensor tendon.

MEDIAL EPICONDYLITIS

Overview

Medial epicondylitis is much less common than lateral epicondylitis. The condition also has been referred to as *medial tennis elbow, golfer's elbow,* and *pitcher's elbow.* It is characterized by degeneration of the common flexor tendon, particularly the pronator teres and flexor carpi radialis tendons. The same spectrum of changes is seen on histologic evaluation as in lateral epicondylitis.

Clinically, patients experience pain and tenderness over the medial epicondyle. Resisted wrist flexion and pronation often reproduces the symptoms. Commonly associated signs and symptoms of ulnar nerve neurapraxia may be present. Most patients respond to conservative treatment, although difficult cases may require surgery.

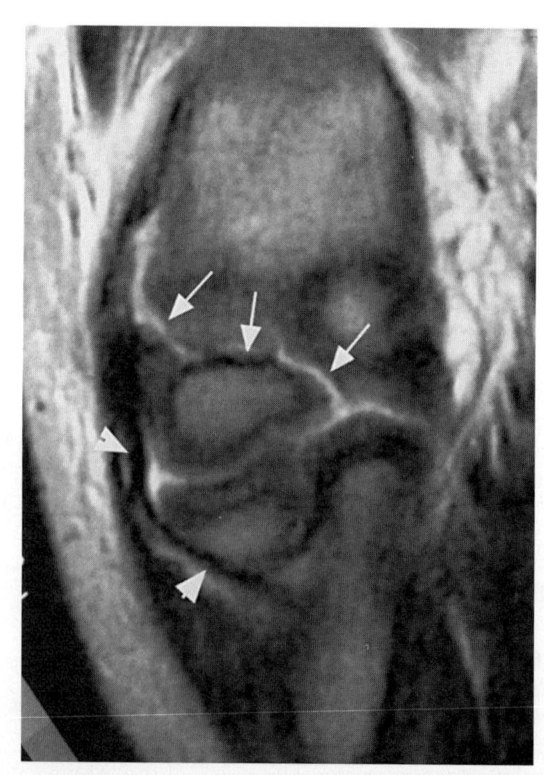

Figure 67–12. Salter-Harris type IV fracture of the lateral condyle *(arrows)* in a 7-year-old girl. T2-weighted coronal MR image shows the fracture. The ulnar collateral band of the radial collateral ligament *(arrowheads)* is intact.

Imaging

MR imaging is useful in patients resistant to conservative therapy and in whom surgery is being considered. Coronal images are key to the diagnosis. The MR imaging findings of tendinosis and partial-thickness and full-thickness tear of the common flexor tendon are the same as those of the common extensor tendon. MR imaging is useful in evaluating associated conditions, in particular abnormalities of the UCL and the ulnar nerve.

TRAUMA

Most fractures around the elbow can be diagnosed by radiography. In a small subset of patients, MR imaging can be useful after trauma. Lateral condylar fractures in children are the most common Salter-Harris type IV fracture (Fig. 67–12). The fracture frequently is misdiagnosed as Salter-Harris type II injury because the intra-articular portion of the fracture is through unossified cartilage. MR imaging is able to categorize this fracture clearly and assess the amount of displacement (see Fig. 67–12).

Stress fractures of the upper extremity are uncommon. They may be either fatigue or insufficiency fractures (Fig. 67–13). Stress fractures of the olecranon occur in adolescent and adult athletes, particularly baseball pitchers (see Fig. 67–13).

RUPTURE OF DISTAL BICEPS TENDON

Overview

Rupture of the distal biceps tendon is an uncommon injury with most cases occurring in middle-aged men. The mechanism of injury is forced hyperextension of the elbow against a flexed forearm. The injury typically occurs in weightlifters (Fig. 67–14).

As with other tendons, tendinosis increases the risk of a tear (Fig. 67–15). Clinically, patients present with pain and bruising in the antecubital fossa. There may be a palpable lump in the arm because of the retracted biceps muscle. Subacute or chronic tears can be more difficult to diagnose. The bicipital bursa normally is present adjacent to the distal biceps tendon at its insertion on the radial tuberosity. Chronic inflammation of the bicipital bursa has also been implicated as a possible cause of distal biceps tendon rupture (see Fig. 67–15). Partial tears of the distal biceps tendon are less common and are more difficult to diagnose.

Imaging

On MR imaging, the axial plane is most useful in determining if the tear is partial or complete. The sagittal plane is most useful for determining the amount of retraction of the tendon. MR imaging features in full-thickness tears include absence of the distal tendon, a fluid-filled gap, edema within the biceps muscle, and a mass in the antecubital fossa (see Fig. 67–14). In chronic tears, atrophy of the biceps muscle is noted. With partial tear, thinning or thickening of the distal tendon can be present (see Fig. 67–15). Prompt diagnosis of an acute injury is important because the tendon usually is repaired surgically.

DISTAL TRICEPS TENDON TEAR

Overview

Spontaneous rupture of the distal triceps tendon can occur in association with systemic diseases, such as chronic renal failure with secondary hyperparathyroidism, osteo-

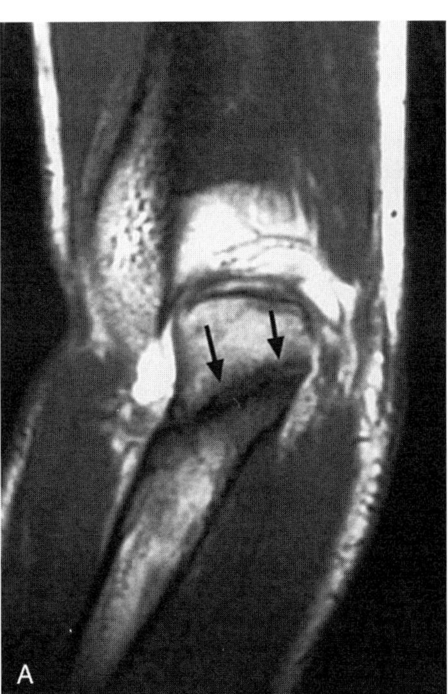

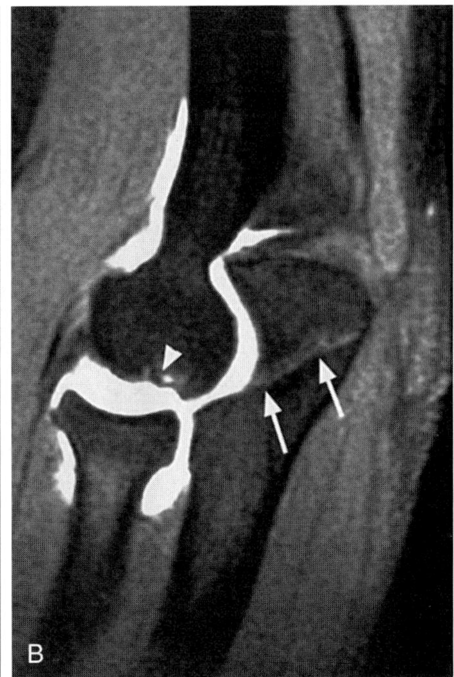

Figure 67–13. Stress fracture of the olecranon in a 21-year-old pitcher. *A*, Coronal T1-weighted MR image shows an oblique fracture line *(arrows)* in the olecranon process. *B*, T2-weighted fat suppressed sagittal image shows the fracture line *(arrows)*. Note the osteochondral fracture of the capitellum *(arrowhead)*. The fluid in the joint is from an MR arthrogram.

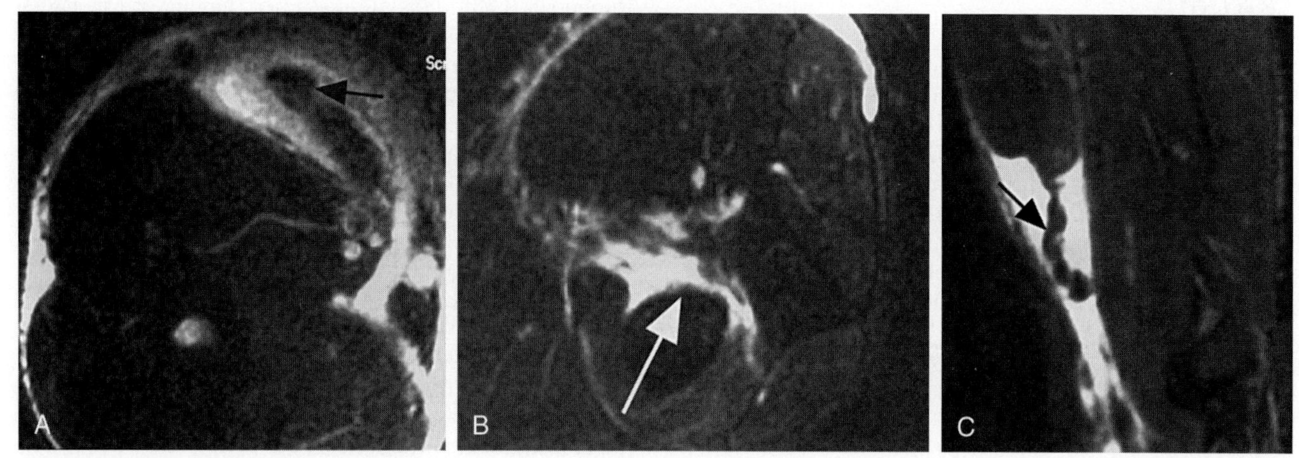

Figure 67–14. Complete rupture of the distal biceps tendon in a 43-year-old man. He presented with swelling of the distal right arm and elbow after an acute extension injury performing weightlifting. *A,* Axial T2-weighted fat-suppressed MR image of the distal arm shows significant thickening of the distal biceps tendon *(arrow).* Fluid and edema surround the tendon. *B,* Axial T2-weighted fat-suppressed MR image through the radial tuberosity reveals absence of the inserting biceps tendon *(arrow). C,* Sagittal T2-weighted fat-suppressed MR image through the distal arm and elbow reveals complete disruption of the distal biceps tendon, along with retraction of the tendon *(arrow)* and biceps muscle.

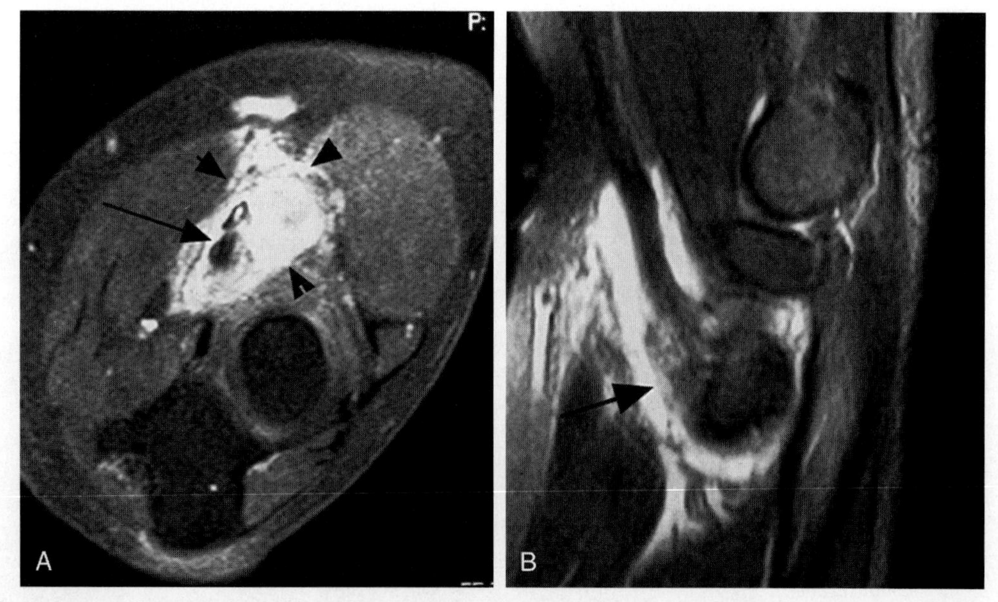

Figure 67–15. Tendinosis and partial tear of the distal biceps tendon. The patient is a 58-year-old sheet metal worker who performs heavy lifting at work. *A* and *B,* Axial and sagittal T2-weighted fat-suppressed MR images show thickening of the distal biceps tendon *(arrow).* A large amount of fluid surrounds the tendon *(arrowheads).*

genesis imperfecta, systemic lupus erythematosus, rheumatoid arthritis, and steroid therapy. Traumatic ruptures are the result of forced elbow flexion against a contracted triceps muscle. Ruptures typically are seen at the tendon insertion to the olecranon but also may occur at the myotendinous junction. Partial tears are usually at the olecranon.

Imaging

Acute tears at the olecranon process can be associated with an avulsion fracture on radiographs. On MR imaging, a full-thickness tear is seen as a discontinuity in the triceps tendon with adjacent edema and fluid. Partial tear is seen as focal disruption of part of the tendon (Fig. 67–16).

OSTEOCHONDRITIS DISSECANS

Overview

Osteochondritis dissecans refers to an osteochondral injury on the anterior aspect of the capitellum that is seen in adolescents, typically baseball pitchers and gymnasts. It is attributed to chronic impaction of the radial head against the capitellum.

Imaging

On conventional radiographs, an area of subchondral lucency in the capitellum may be seen. Prognosis and management depend on the status of the articular cartilage and the stability of the fragment. Accuracy of diagnosis is improved with MR arthrography (Fig. 67–17). The unstable variety of osteochondritis dissecans shows fluid tracking around the osteochondral fragment (Fig. 67–17*A*), and this lesion is capable of shedding loose bodies into the joint (Fig. 67–17*B*). The use of a dedicated cartilage sequence, such as the three-dimensional fat-suppressed spoiled GRASS (SPGR) sequence, can improve assessment of the osteochondral fragment, particularly the integrity of the overlying cartilage (Fig. 67–18).

Osteochondritis dissecans should be distinguished from Panner's disease, which is classified as an osteochondrosis and occurs in a younger age group, typically children between 5 and 10 years old. This condition probably is due to osteonecrosis of the capitellum secondary to a traumatic insult. Regeneration of the capitellum usually occurs, and in most cases no residual deformity is seen (Fig. 67–19).

Osteochondritis dissecans and Panner's disease should be differentiated from the pseudodefect that occurs on the posterolateral aspect of the capitellum (Fig. 67–20) and is considered a normal variant. The pseudodefect can be seen on sagittal and coronal images, and it is more prominent when there is fluid within the joint. The pseudodefect is seen posteriorly (see Fig. 67–20); Panner's disease and osteochondritis dissecans occur in the anterior capitellum (see Fig. 67–17*A*).

NERVE INJURY

Overview

The ulnar nerve lies within the cubital tunnel posterior to the medial epicondyle (see Fig. 67–6). The ulnar nerve is the most easily identified nerve at the elbow, and it is the most frequently injured. The nerve may be injured from

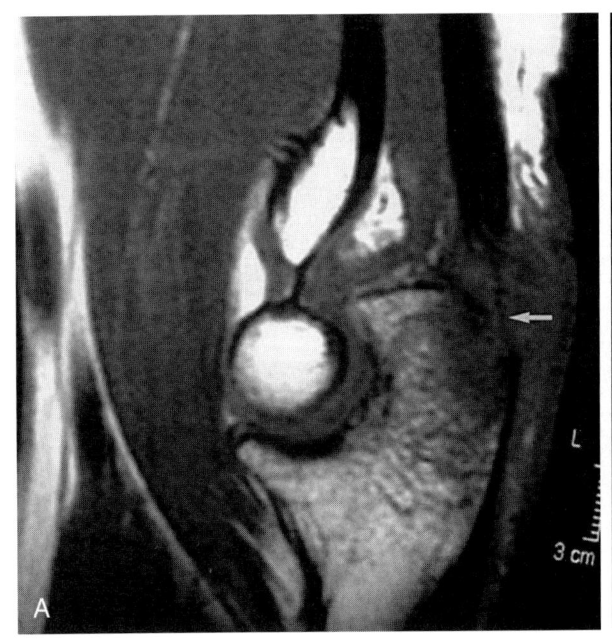

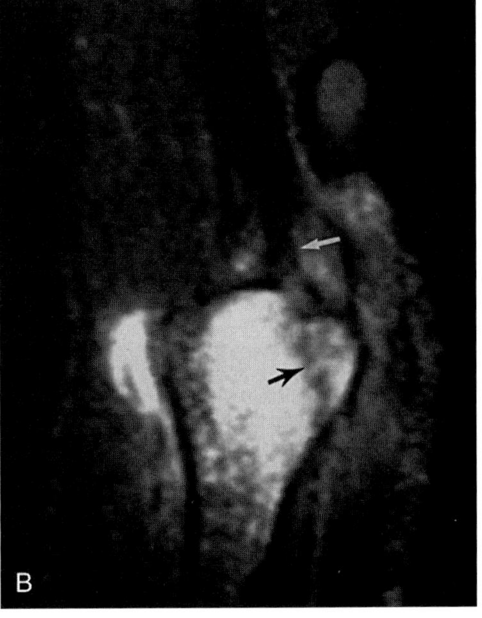

Figure 67–16. Partial and complete distal triceps tendon tears. Partial tear of the distal triceps tendon in a 44-year-old man who was assaulted 3 weeks earlier. *A*, T1-weighted sagittal MR image shows partial disruption of the triceps tendon at its insertion *(arrow)*. Soft tissue and bone marrow edema are present. *B*, Coronal STIR image reveals the partial tendon tear *(white arrow)*, which is surrounded by soft tissue and bone marrow edema. A fracture of the olecranon process *(black arrow)* also is present.

Illustration continued on following page

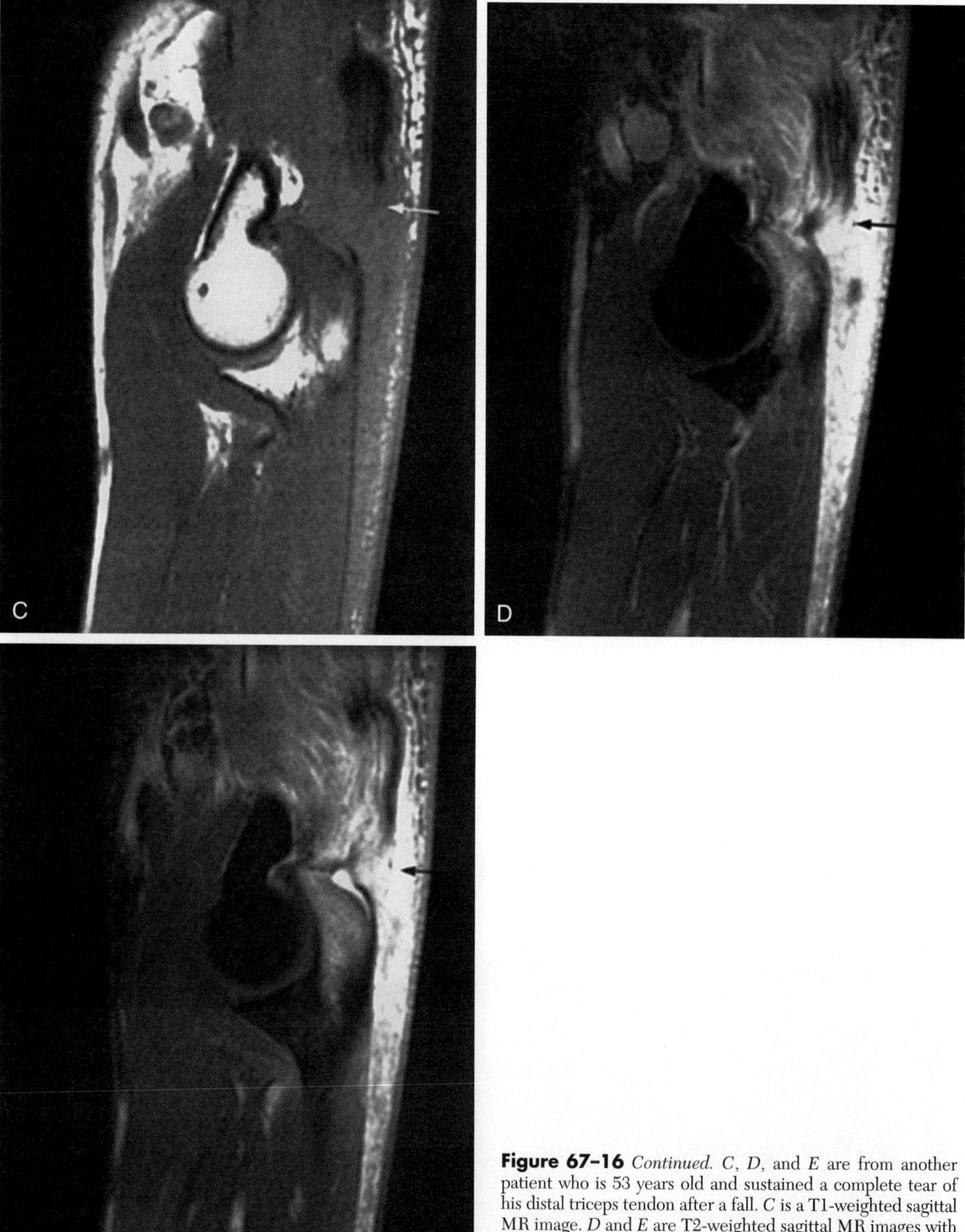

Figure 67–16 *Continued. C, D,* and *E* are from another patient who is 53 years old and sustained a complete tear of his distal triceps tendon after a fall. *C* is a T1-weighted sagittal MR image. *D* and *E* are T2-weighted sagittal MR images with fat suppression. All three images show complete disruption of the tendon at its insertion to the olecranon *(arrow)*.

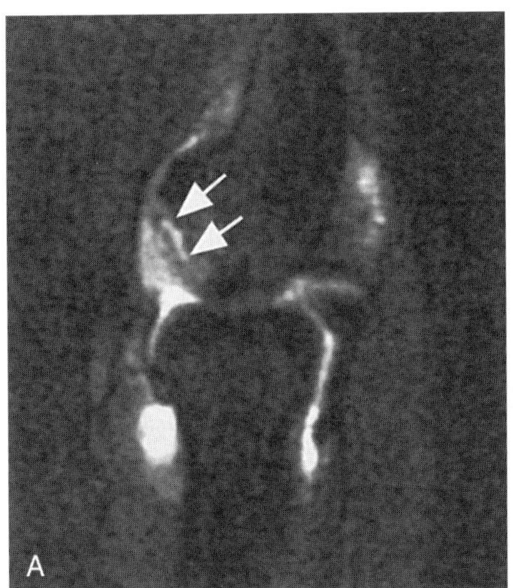

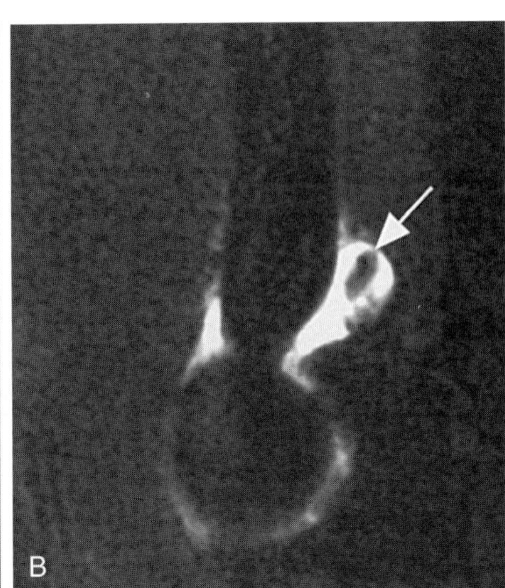

Figure 67–17. Unstable osteochondritis dissecans in a 15-year-old pitcher who complained of pain and locking of the right elbow. *A*, Sagittal T1-weighted with fat suppression image from an MR arthrogram shows tracking of contrast material around the osteochondral defect *(arrows)* in the capitellum. *B*, More medial image from the same study shows a loose body within the joint *(arrow)*, confirming the fact that the osteochondritis is unstable and shedding loose bodies.

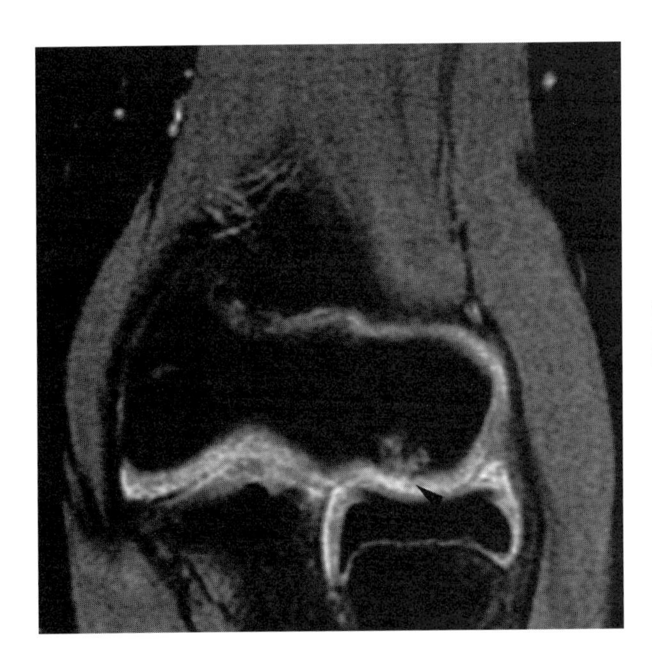

Figure 67–18. Stable osteochondritis dissecans in a 13-year-old boy. Coronal three-dimensional fat-suppressed spoiled GRSS (SPGR) image shows intact hyaline cartilage covering the osteochondral defect in the capitellum *(arrow)*.

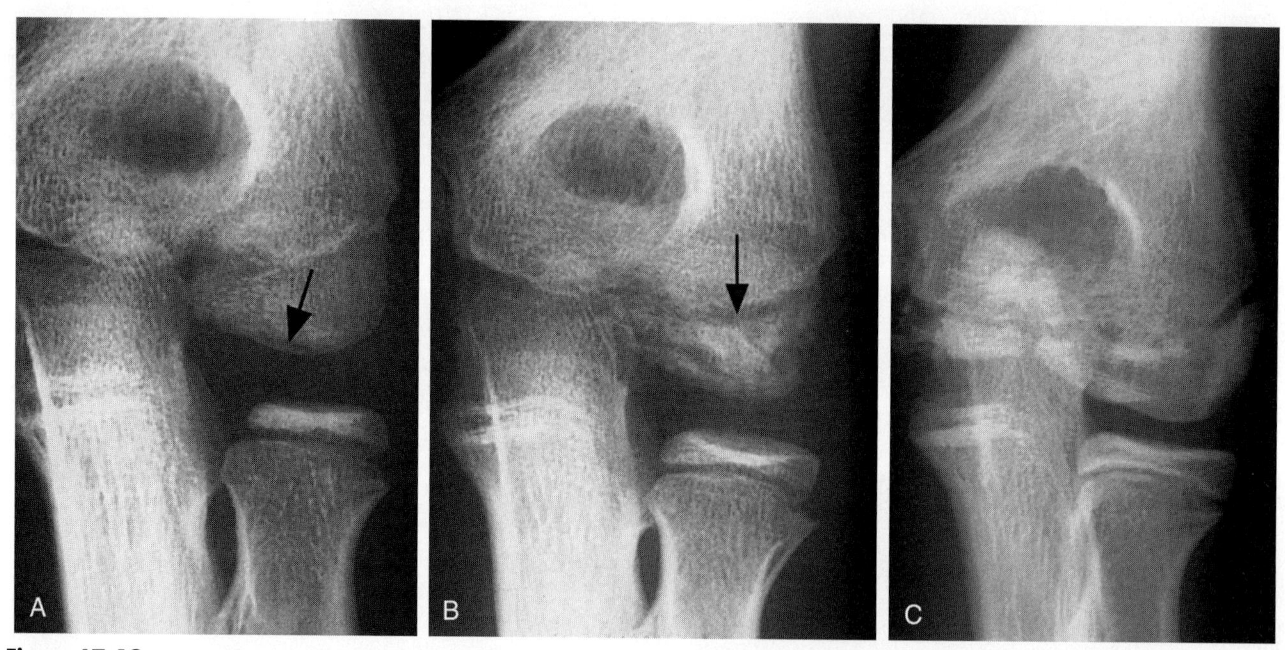

Figure 67–19. Panner's disease with complete regeneration of the capitellum in an 11-year-old boy. *A*, Anteroposterior view of the elbow taken after initial symptoms shows a linear lucency in the subchondral bone of the capitellum *(arrow)*. *B*, Repeat anteroposterior view 10 months later shows complete collapse and fragmentation of the capitellum *(arrow)*. *C*, Anteroposterior view taken 2.5 years after the initial examination reveals complete regeneration of the capitellum. The trabecular pattern of the capitellum is entirely normal. The irregular ossification of the trochlea is a normal variant.

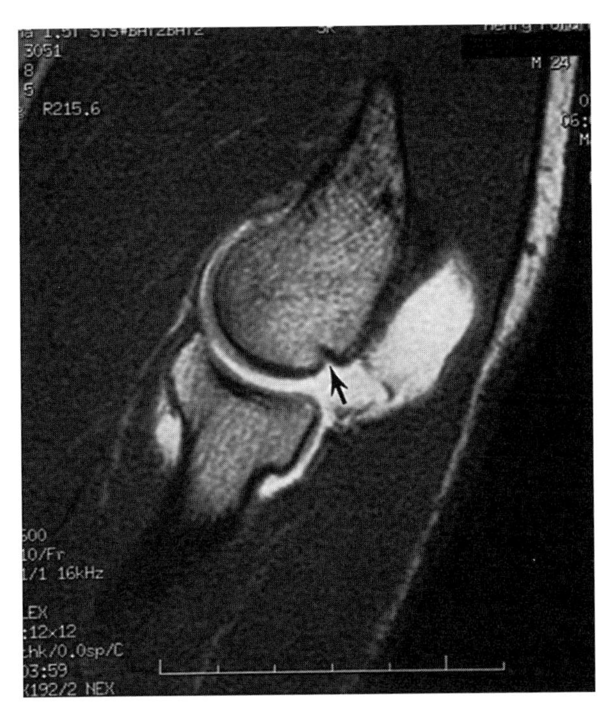

Figure 67–20. Pseudodefect of the capitellum *(arrow)* seen on a sagittal T1-weighted MR image after an intra-articular gadolinium injection. This normal variant should not be confused with an osteochondral abnormality.

direct trauma, particularly in athletes, in whom the superficial location of the nerve makes it vulnerable to a direct blow. Overhead throwing athletes also are susceptible to ulnar nerve injury because of traction on the nerve.

Imaging

Nerves around the elbow display intermediate-to-low signal intensity on all pulse sequences. On T1-weighted images, they are similar to muscle in signal intensity, whereas on T2-weighted images, the signal may increase slightly when compared with muscle, but it is always lower than fat.

The cubital tunnel retinaculum is lax in extension but stretches during flexion, resulting in diminished cubital tunnel volume (see Fig. 67–6). The cubital tunnel can be compromised by a thickened cubital tunnel retinaculum, anomalous anconeus muscle (anconeus epitrochlearis), masses within the cubital tunnel, osteophytes, and previous fractures of the medial epicondyle. Thickening of the cubital tunnel retinaculum is referred to as *Osborn's lesion.* During flexion and extension of the elbow, the ulnar nerve moves and slides within the cubital tunnel. Any condition that narrows the cubital tunnel interferes with normal ulnar nerve motion. The retinaculum of the cubital tunnel may be absent, allowing the ulnar nerve to sublux or dislocate over the medial epicondyle causing secondary friction neuritis. Signs and symptoms of ulnar neuritis include muscle weakness in the flexor carpi ulnaris and the flexor digitorum profundus to the ring and little fingers, and intrinsic muscle weakness. On MR imaging, ulnar nerve entrapment shows increased signal intensity on T2-weighted images, fusiform or focal enlargement, and deviation of the nerve (Figs. 67–21 and 67–22). Muscle denervation may result in diffuse high signal within the respective muscles on T2-weighted or STIR images. Surgical procedures designed to decompress ulnar nerve entrapment include medial epicondylectomy; decompression of the nerve; and subcutaneous, intramuscular, or submuscular nerve transfer (see Fig. 67–22).

RADIAL AND MEDIAN NERVE ENTRAPMENT

The deep branch of the radial nerve, the posterior interosseous nerve, can be compressed as it passes between the superficial and deep parts of the supinator muscle, resulting in the supinator syndrome. Superficial radial nerve injury can occur from blunt trauma to the lateral aspect of the elbow.

At the elbow, the medial nerve can be compressed from the distal humerus proximally to the flexor digitorum superficialis distally. A bony spur, the supracondylar process, can arise from the anteromedial surface of the distal humerus in 3% of the population. A band of fibrous tissue, the ligament of Struthers, may connect the supracondylar process to the medial epicondyle and compress the medial nerve. Muscle variations and soft tissue masses also may compress the nerve.

LOOSE BODIES

Overview

Loose bodies are common in the elbow joint and often are secondary to prior trauma. They may vary from a few millimeters in size to more than 1 cm. Large joint bodies usually can be identified on conventional radiographs. Smaller joint bodies are more difficult to diagnose. Clinically, patients present with symptoms of pain and locking of the elbow joint.

Imaging

On MR imaging, visualization of joint bodies is improved by the presence of a joint effusion or by performing an MR arthrogram (see Fig. 67–17*B*). Osteophytes and synovial folds may mimic small loose bodies. Loose bodies are most common in the anterior elbow except in throwing athletes, in whom they are found more commonly in the posterior compartment.

Recommended Readings

Beltran J, Rosenberg ZS, Kawelblum M, et al: Pediatric elbow fractures: MRI evaluation. Skeletal Radiol 23:227, 1994.

Coel M, Yamada CY, Ko J: MR imaging of patients with lateral epicondylitis of the elbow (tennis elbow): Importance of increased signal of the anconeus muscle. AJR Am J Roentgenol 161:1019, 1993.

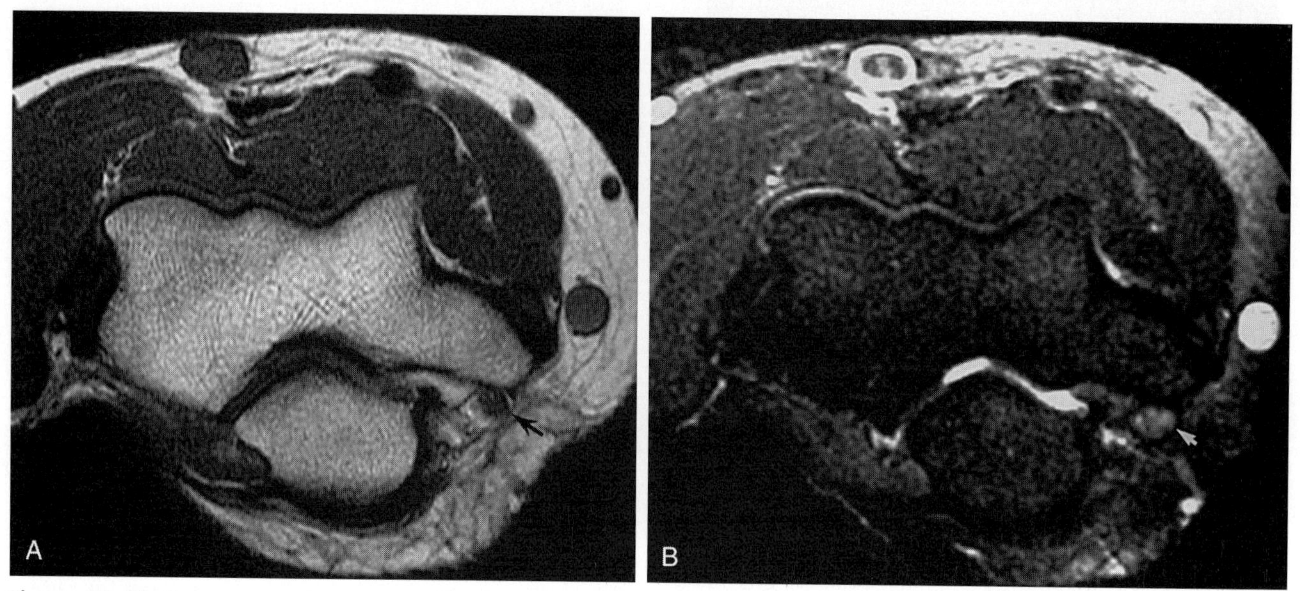

Figure 67–21. Post-traumatic ulnar nerve neuritis in a 35-year-old man who received a direct blow to the elbow and experienced ulnar nerve symptoms. *A,* Axial T1-weighted image shows moderate swelling of the ulnar nerve *(arrow)*. *B,* Axial fat-suppressed T2-weighted image shows swelling and increased signal within the ulnar nerve *(arrow)* consistent with traumatic neuritis.

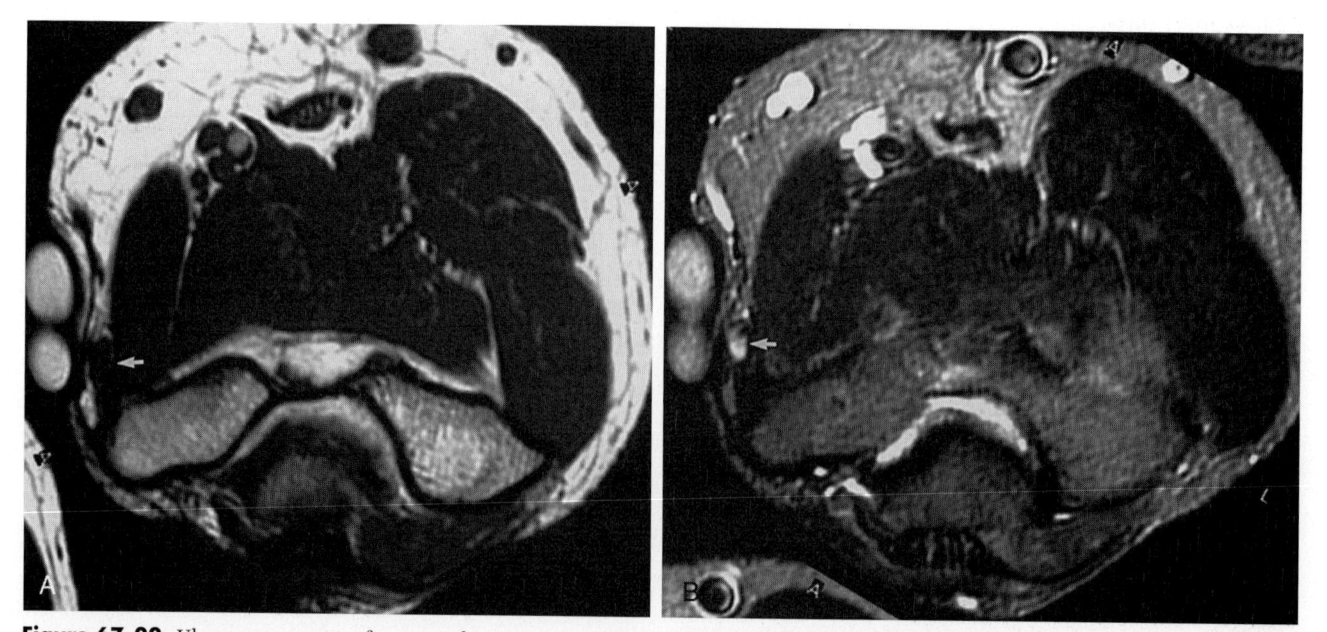

Figure 67–22. Ulnar nerve neuritis after surgical transposition of the nerve. A 37-year-old man who had transposition of the ulnar nerve 5 years earlier experienced further ulnar nerve symptoms. *A,* Axial T1-weighted image shows a swollen ulnar nerve sitting in the subcutaneous tissue of the medial right forearm *(arrow)*. *B,* Axial fast STIR image shows the transposed swollen nerve with increased signal *(arrow)*, consistent with neuritis.

Conway JE, Jobe FW, Glousman RE, et al: Medial instability of the elbow in throwing athletes. J Bone Joint Surg Am 74:67, 1992.
Cotten A, Jacobson J, Brossmann J, et al: Collateral ligaments of the elbow: Conventional MR imaging and MR arthrography with coronal oblique plane and elbow flexion. Radiology 204:806, 1997.
Cotten A, Jacobson J, Brossmann J, et al: MR arthrography of the elbow: Normal anatomy and diagnostic pitfalls. J Comput Assist Tomogr 21:516, 1997.
Field LD, Altchek DW: Elbow injuries. Clin Sports Med 14:59, 1995.
Friedman ACGL, White PG: Pictorial review: Magnetic resonance imaging of the paediatric elbow. Clin Radiol 52:582, 1997.
Fritz RC: Magnetic resonance imaging of the elbow. Semin Roentgenol 30:241, 1995.
Fritz RC, Steinbach LS: Magnetic resonance imaging of the musculoskeletal system: Part 3. The elbow. Clin Orthop 324:321, 1996.
Fritz RC, Steinbach LS, Tirman PFJ, et al: MR imaging of the elbow: An update. Radiol Clin North Am 35:117, 1997.
Gaary EA, Potter HG, Altchek DW: Medial elbow pain in the throwing athlete: MR imaging evaluation. AJR Am J Roentgenol 168:795, 1997.
Gardner E, Gray DJ, O'Rahilly R: Anatomy: A Regional Study Of Human Structure, 4th ed. Philadelphia, WB Saunders, 1975.
Griffith JF, Roebuck DJ, Cheng JCY, et al: Acute elbow trauma in children: Spectrum of injury revealed by MR imaging not apparent on radiographs. AJR Am J Roentgenol 176:53, 2001.
Herzog RJ: Efficacy of magnetic resonance imaging of the elbow. Med Sci Sports Exerc 26:1193, 1994.
Ho CP: Sports and occupational injuries of the elbow: MR imaging findings. AJR Am J Roentgenol 164:1465, 1995.
Kim YS, Yeh LR, Trudell D, et al: MR imaging of the major nerves about the elbow cadaveric study examining the effect of flexion and extension of the elbow and pronation and supination of the forearm. Skeletal Radiol 27:419, 1998.
Kramer J, Stiglbauer R, Engel A, et al: MR contrast arthrography (MRA) in osteochondrosis dissecans. J Comput Assist Tomogr 16:254, 1992.
Le TB, Mont MA, Jones LC, et al: Atraumatic osteonecrosis of the adult elbow. Clin Orthop 373:141, 2000.
Miller TT: Imaging of elbow disorders. Orthop Clin North Am 30:21, 1999.
Mirowitz SA, London SL: Ulnar collateral ligament injury in baseball pitchers: MR imaging evaluation. Radiology 185:573, 1992.
Morrey BF, An KN: Functional anatomy of the ligaments of the elbow. Clin Orthop 201:84, 1985.
Nakanishi K, Masatomi T, Ochi T, et al: MR arthrography of elbow: Evaluation of the ulnar collateral ligament of elbow. Skeletal Radiol 25:629, 1996.
Nuber GW, Diment MT: Olecranon stress fractures in throwers: A report of two cases and a review of the literature. Clin Orthop 278:58, 1992.
O'Driscoll SW, Bell DF, Morrey BF: Posterolateral rotatory instability of the elbow. J Bone Joint Surg Am 73:440, 1991.
O'Driscoll SW, Horii E, Carmichael SW, et al: The cubital tunnel and ulnar neuropathy. J Bone Joint Surg Br 73:613, 1991.
Patel VV, Heidenreich FP, Bindra RR, et al: Morphologic changes in the ulnar nerve at the elbow with flexion and extension: A magnetic resonance imaging study with 3-dimensional reconstruction. J Shoulder Elbow Surg 7:368, 1998.
Patten RM: Overuse syndromes and injuries involving the elbow: MR imaging findings. AJR Am J Roentgenol 164:1205, 1995.
Peiss J, Adam G, Casser R, et al: Gadopentate-dimeglumine-enhanced MR imaging of osteonecrosis and osteochondritis dissecans of the elbow: Initial experience. Skeletal Radiol 24:17, 1995.
Potter HG: Imaging of posttraumatic and soft tissue dysfunction of the elbow. Clin Orthop 370:9, 2000.
Potter HG, Hannafin JA, Morwessel RM, et al: Lateral epicondylitis: Correlation of MR imaging, surgical, and histopathologic findings. Radiology 196:43, 1995.
Potter HG, Weiland AJ, Schatz JA, et al: Posterolateral rotatory instability of the elbow: Usefulness of MR imaging in diagnosis. Radiology 204:185, 1997.
Resnick D, Kang HS: Internal Derangement of Joints. Philadelphia, WB Saunders, 1997.
Rosenberg ZS, Beltran J, Cheung YY, et al: The elbow: MR features of nerve disorders. Radiology 188:235, 1993.
Rosenberg ZS, Beltran J, Cheung YY: Pseudodefect of the capitellum: Potential MR imaging pitfall. Radiology 191:821, 1994.
Schenk M, Dalinka MK: Imaging of the elbow: An update. Orthop Clin North Am 28:517, 1997.
Schwartz ML, Al-Zahrani S, Morwessel RM, et al: Ulnar collateral ligament injury in the throwing athlete: Evaluation with saline-enhanced MR arthrography. Radiology 197:297, 1995.
Seiler JG, Parker LM, Chamberland C, et al: The distal biceps tendon. J Shoulder Elbow Surg 4:149, 1995.
Sonin AH, Fitzgerald SW: MR imaging of sports injuries in the adult elbow: A tailored approach. AJR Am J Roentgenol 167:325, 1996.
Steinbach LS, Schwartz M: Elbow arthrography. Radiol Clin North Am 36:635, 1998.
Steinborn M, Heuck A, Jessel C, et al: Magnetic resonance imaging of lateral epicondylitis of the elbow with a 0.2-T dedicated system. Eur Radiol 9:1376, 1999.
Sugimoto H, Ohsawa T: Ulnar collateral ligament in the growing elbow: MR imaging of normal development and throwing injuries. Radiology 192:417, 1994.
Timmerman LA, Andrews JR: Undersurface tear of the ulnar collateral ligament in baseball players: A newly recognized lesion. Am J Sports Med 22:33, 1994.
Timmerman LA, Schwartz ML, Andrews JR: Preoperative evaluation of the ulnar collateral ligament by magnetic resonance imaging and computed tomography arthrography: Evaluation in 25 baseball players with surgical confirmation. Am J Sports Med 22:26, 1994.
Wilson FD, Andrews JR, Blackburn TA, et al: Valgus extension overload in the pitching elbow. Am J Sports Med 11:83, 1983.

CHAPTER 68

Wrist

Shella Farooki, MD, Carol J. Ashman, MD, and Joseph S. Yu, MD

ANATOMY

The wrist is composed of eight carpal bones that are aligned in two groups of four bones comprising the proximal and distal carpal rows. The proximal carpal row consists of the scaphoid, lunate, triquetrum, and pisiform, whereas the distal row consists of the trapezium, trapezoid, capitate, and hamate. The pisiform is a sesamoid bone located within the distal end of the flexor carpi ulnaris tendon. The carpus is concave on its palmar surface, forming the roof of the carpal tunnel, and, with the flexor retinaculum on the volar surface, confines the flexor tendons and median nerve. On the dorsum of the wrist, the extensor tendons, surrounded by synovial sheaths, are separated into six compartments. The joints in the wrist are separated into compartments as follows: the distal radioulnar, radiocarpal, pisiform-triquetral, midcarpal, common carpometacarpal, first carpometacarpal, and intermetacarpal compartments.

INDICATIONS FOR IMAGING

The indications for magnetic resonance (MR) imaging of the wrist include suspected triangular fibrocartilage complex (TFCC) tear, ligamentous disruption, carpal tunnel syndrome, infection, and osteonecrosis. MR imaging is also performed to evaluate nonspecific pain following trauma and palpable masses in the wrist.

TRIANGULAR FIBROCARTILAGE COMPLEX

Overview

The TFCC consists of the triangular fibrocartilage (TFC), dorsal and volar radioulnar ligaments, sheath of the extensor carpi ulnaris tendon, ulnocarpal ligaments, ulnar collateral ligament, and ulnomeniscal homologue (Fig. 68–1). The TFC separates the radiocarpal and inferior radioulnar compartments. The TFC is thicker peripherally than centrally (see Fig. 68–1*B*), and it may show a central fenestration. The TFCC is an important soft tissue structure that serves as a cushion between the carpal bones and distal ulna and contributes to the stability of the inferior radioulnar joint. TFCC tears can result in ulnar-sided pain, weakness, or crepitus.

TFC tears have been classified by Palmer (1989) into two types—traumatic (class I) and degenerative (class II). The tears are subdivided further by location into radial and ulnar tears. Class IA tears involve the radial side of the disk and usually are slitlike. Class IB tears are avulsions of the TFC from the ulna. An avulsion of the ulnar attachment with injury of the ulnocarpal ligaments is classified as a class IC lesion. Class ID tears are avulsions of the radial attachment. Degenerative tears are classified on the basis of severity, with class IIA lesions being the least severe, with thinning of the TFC, and class IIE lesions being most severe, with TFC perforations, osteoarthritis, and ligamentous disruptions.

Imaging

Arthrography

Tricompartmental arthrography traditionally has been considered to be the gold standard for the detection of TFC abnormalities. At arthrography, a TFC tear is diagnosed when contrast material communicates between the inferior radioulnar joint and radiocarpal joint. A perforation of the TFC is a common injury in young symptomatic patients and accounts for about 25% of patients with ulnar-sided wrist pain. Positive arthrograms can occur in asymptomatic older individuals with degenerative perforations. One third of individuals with painful wrists have been shown to have pathologic communications between the radiocarpal and midcarpal compartments, indicating abnormalities of one or both proximal intercarpal ligaments. The likelihood of a tear of the TFC or lunotriquetral ligament increases if the patient complains only of ulnar-sided pain.

Magnetic Resonance Imaging

Patients are imaged either prone with the wrist held over the head or supine with the wrist at the patient's side. The latter position is more comfortable for the patient. Using dedicated extremity coils, small field of view (8 to 12 cm), and thin slice thickness, MR imaging can show the TFCC and other soft tissues and osseous structures. Three-dimensional gradient echo and two-dimensional spin echo sequences are effective for imaging the TFCC.

The TFCC is visualized best on coronal images (see Fig. 68–1). The TFC component is normally a low signal

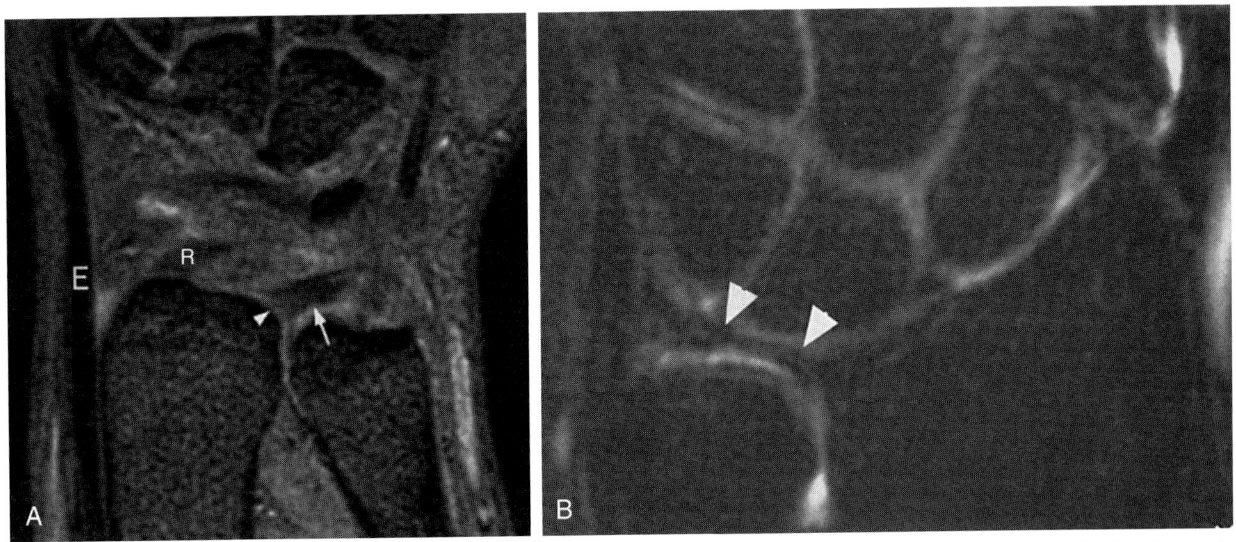

Figure 68–1. Normal triangular fibrocartilage (TFC) MR anatomy. *A*, Coronal three-dimensional gradient-echo MR image of the dorsal wrist shows the extensor carpi radialis tendon (E), dorsal radiotriquetral ligament (R), and the normally low signal TFC *(arrow)*. High signal *(arrowhead)* at the insertion to the radius is a normal finding. *B*, Coronal T2-weighted image of the wrist with fat suppression shows a normal TFC. The TFC is thicker peripherally than centrally *(arrowheads)*.

intensity, disklike structure, but it may show intermediate signal foci on T1-weighted images because of myxoid degeneration. The radial attachment to the hyaline cartilage of the distal radius usually is shown clearly, but the ulnar attachment sometimes can be difficult to visualize because of the surrounding loose connective tissue. The dorsal and volar radioulnar ligaments are seen best on axial images coursing from the distal radius to the distal ulna.

Complete tears or perforations of the TFC appear as regions of high signal on T2-weighted sequences that extend from the radiocarpal joint to the distal radioulnar joint. Traumatic tears are more common at the radial attachment (Fig. 68–2). Partial tears are seen as focal regions of high signal on T2-weighting that affect only the radiocarpal or the distal radioulnar surface of the disk. The TFC insertion to the hyaline cartilage of the distal radius exhibits high signal on T2-weighted sequences. Care must be taken not to mistake this normal attachment of the TFC for a tear (see Figs. 68–1A and 68–2). Fluid within the distal radioulnar joint is a nonspecific finding, although it can represent a secondary sign of a TFC tear.

Degeneration of the TFC is characterized by intermediate signal abnormality on proton-density sequences that remains intermediate on T2-weighted or gradient echo sequences (Fig. 68–3). Degeneration of the TFC has been shown to be age dependent, with 55% of individuals in their 30s and 100% of individuals in their 50s showing degenerative changes. Degenerative tears are more com-

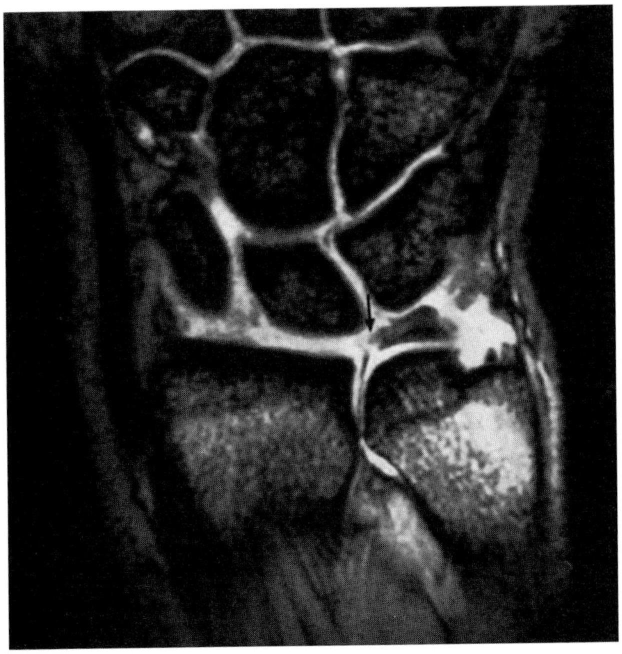

Figure 68–2. Triangular fibrocartilage traumatic tear. Typical traumatic tear at the radial insertion of the triangular fibrocartilage *(arrow)* seen on a T2-weighted, fat-suppressed MR image.

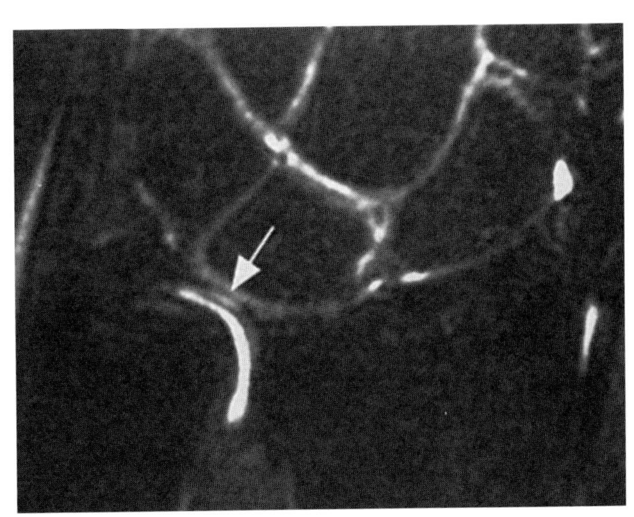

Figure 68–3. MR imaging of triangular fibrocartilage (TFC) degeneration. T2-weighted coronal MR image with fat suppression in a 31-year-old woman shows increased signal intensity within the TFC. This finding is believed to represent TFC degeneration.

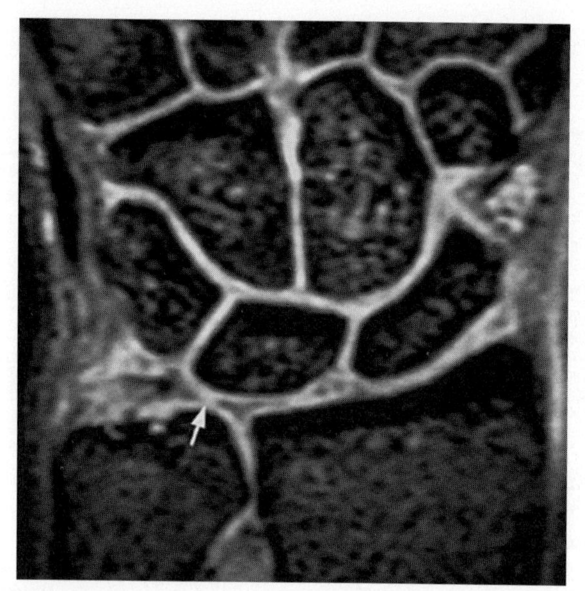

Figure 68–4. Central degenerative tear *(arrow)* of the triangular fibrocartilage seen on a coronal gradient echo sequence.

mon in the central portion of the TFC, where it is the thinnest (Fig. 68–4), whereas traumatic tears tend to occur on the radial side of the TFC (see Fig. 68–2).

WRIST TENDONS

Anatomy

Tendons of the wrist consist of the flexor and extensor groups (Fig. 68–5). The extensor tendons course along the dorsum of the wrist within six separate compartments, which are numbered from the radial to ulnar direction. The first compartment is located lateral to the radial styloid process and contains the abductor pollicis longus and extensor pollicis brevis tendons. Lister's tubercle, an osseous protuberance on the dorsal surface of the distal radius, separates the second and third compartments. The second compartment is lateral to Lister's tubercle and contains the extensor carpi radialis brevis and longus tendons. The third compartment contains the extensor pollicis longus tendon and is medial to Lister's tubercle.

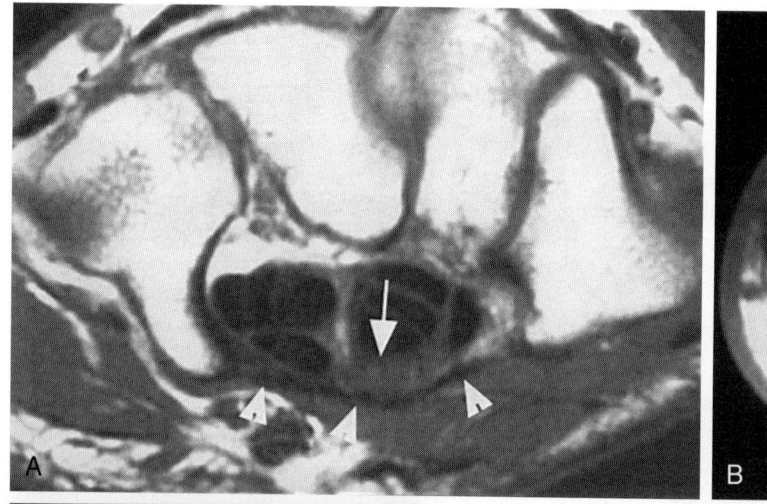

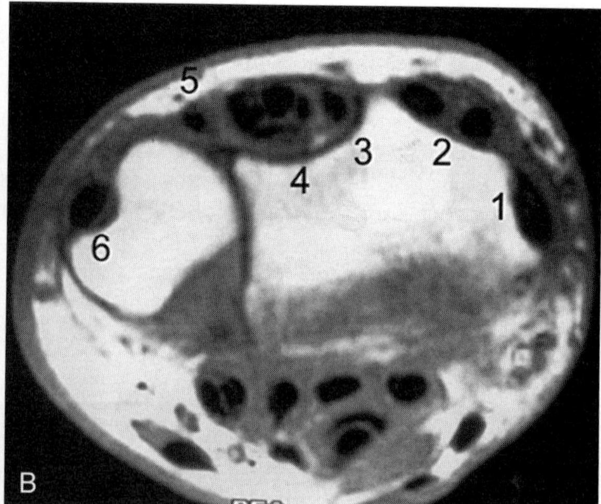

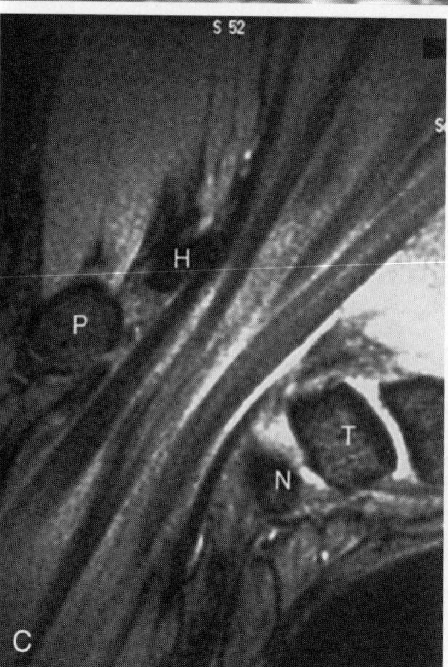

Figure 68–5. Normal tendons of the wrist. *A,* T1-weighted axial image (with palm of hand facing up) shows the flexor tendons within the carpal tunnel. The median nerve *(arrow)* is seen within the carpal tunnel just beneath the flexor retinaculum *(arrowheads)*. *B,* T1-weighted axial image (with palm of hand facing down) shows the six extensor compartments. *C,* Coronal gradient echo MR image shows the flexor tendons within the carpal tunnel. H, hook of the hamate; P, pisiform; T, trapezium; N, scaphoid tuberosity.

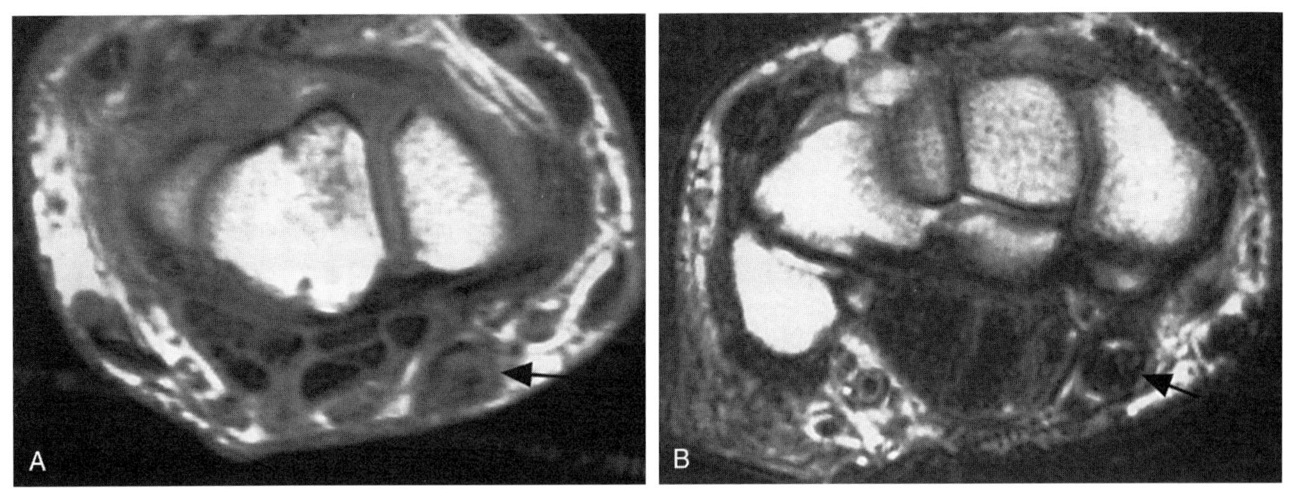

Figure 68–6. Partial tear and tendinopathy (tendinosis) of the flexor carpi radialis tendon. *A* and *B*, Axial T1- and T2-weighted MR images through the distal part of flexor carpi radialis tendon *(arrow)* show thickening, increased signal, and splitting (partial tearing) of the tendon.

The fourth compartment contains the extensor indicis proprius and extensor digitorum communis tendons, and the fifth compartment contains the extensor digiti minimi. The sixth compartment contains the extensor carpi ulnaris and is located medial to the ulnar styloid (see Fig. 68–5*B*). The flexor tendons are located on the volar aspect of the wrist within or adjacent to the carpal tunnel (see Fig. 68–5*A*, *C*).

Imaging

Tendon abnormalities may present as a complete rupture or as partial tears. On MR imaging, complete tears show discontinuity of the tendon with free edges. Partial tears appear as linear high intrasubstance signal on T2-weighted images, irregular thickening, or attenuation of the tendon (Fig. 68–6). Increased signal within the tendon substance and thickening of the tendon indicate degenerative changes (tendonopathy or tendinosis) (see Fig. 68–6). *Magic angle phenomenon* is a common problem in tendon imaging; it presents as increased signal intensity within the tendon on short TE sequences, which should disappear on long TE sequences.

Tenosynovitis can be post-traumatic, infectious, or due to an inflammatory process, such as rheumatoid arthritis. Tenosynovitis involving the flexor compartment can result in carpal tunnel syndrome (Fig. 68–7). MR imaging shows fluid within the tendon sheaths, and the contents of the carpal tunnel appear widely separated (see Fig. 68–7).

de Quervain's disease refers to tenosynovitis involving the sheaths of the abductor pollicis longus and extensor pollicis brevis (the first extensor compartment) at the level of the radial styloid and may result in chronic radial-sided wrist pain (Fig. 68–8). Typically, patients have a history of inflammatory arthritis such as rheumatoid arthritis or overuse injury with resultant synovial hypertrophy and fibrosis.

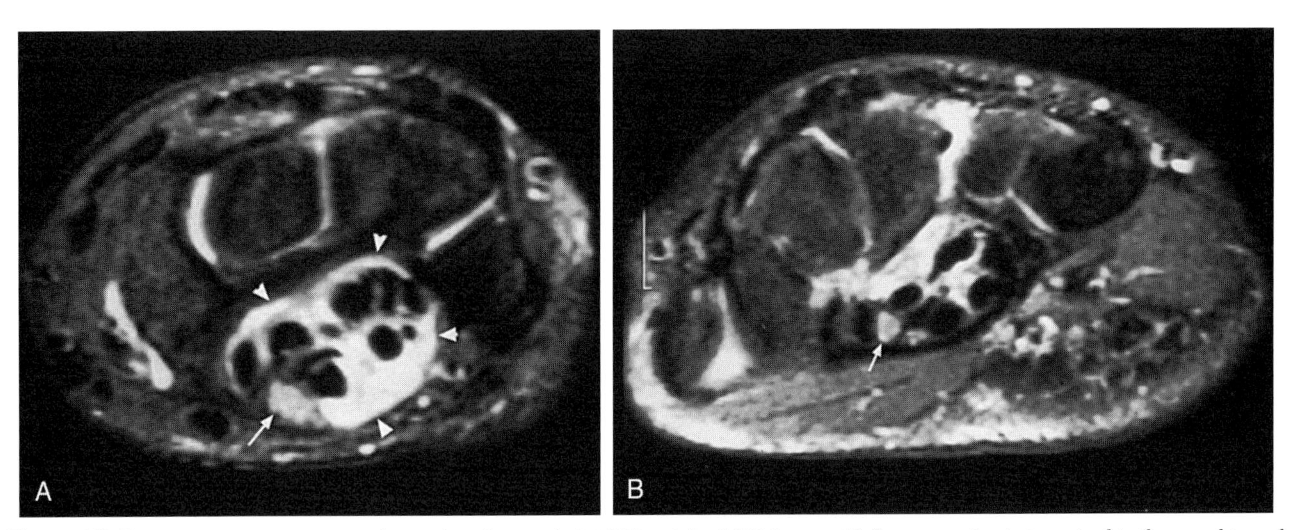

Figure 68–7. Tenosynovitis causing carpal tunnel syndrome. *A*, Axial T2-weighted MR image with fat suppression just proximal to the carpal tunnel shows distention of the tendon sheath with fluid *(arrowheads)* and separation of the tendons. The median nerve *(arrow)* is swollen. *B*, Axial T2-weighted MR image with fat suppression from the same patient through the palm of the hand shows the tenosynovitis to be less pronounced, and the median nerve is not swollen *(arrow)*.

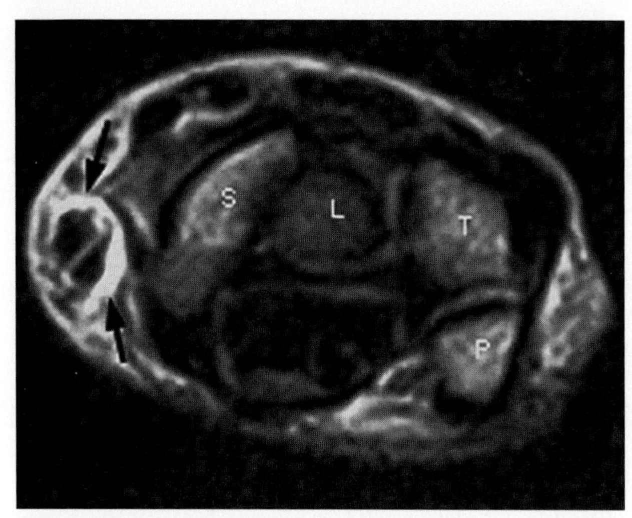

Figure 68–8. de Quervain's disease. Axial T2-weighted MR image shows fluid in the first extensor compartment consistent with de Quervain's tenosynovitis *(arrows)*. The abductor pollicis longus and extensor pollicis brevis tendons appear to be within normal limits. P, pisiform; T, triquetrum; L, lunate; S, scaphoid.

CARPAL TUNNEL SYNDROME

Overview

Carpal tunnel syndrome is the most common form of entrapment neuropathy. It is more common in women and is bilateral in about 50% of cases. Most patients are in their 20s to 40s. The dominant hand usually is affected, and occupational or overuse injuries are common causes. Other predisposing factors include amyloidosis, pregnancy, diabetes mellitus, acromegaly, and systemic lupus erythematosus. Masses within the carpal tunnel, such as a ganglion cyst, lipoma, hemangioma, or nerve tumor, also can result in carpal tunnel syndrome. Clinically, patients present with weakness and paresthesias of the first through the fourth fingers in the palmar aspect of the hand exacerbated by repeated flexion.

Anatomy

The volar boundary of the carpal tunnel is the flexor retinaculum, which is a broad low signal intensity structure, extending from the hook of the hamate and pisiform medially to the scaphoid and trapezial tuberosities laterally (see Figs. 68–5*A* and 68–6). The dorsal boundary is formed by the carpal bones. Within the carpal tunnel are the flexor digitorum profundus and superficialis tendons and the flexor pollicis longus tendon. The flexor pollicis longus tendon is located radial to the rest of the tendons but within the carpal tunnel. The median nerve is located in the superficial radial portion of the carpal tunnel just deep to the flexor retinaculum (see Fig. 68–5*A*); it is round or flat and has intermediate signal intensity on T1-weighted images. Axial images with T1- and T2-weighted sequences are best for evaluating the contents of the carpal tunnel. Disease processes that increase the volume of the carpal tunnel contents or decrease the size of the tunnel predispose to carpal tunnel syndrome (see Fig. 68–7).

Imaging

In most patients with carpal tunnel syndrome, imaging studies are not needed. MR imaging has been used when an underlying mass or tendon pathology is suspected. Other indications for MR imaging include discrepant electromyogram and clinical findings, symptoms after carpal tunnel release, and preoperative assessment for endoscopic surgery.

MR imaging findings of carpal tunnel syndrome include palmar bowing of the flexor retinaculum and enlargement or increased signal intensity of the median nerve on T2-weighted images (edema) at the level of the pisiform (see Fig. 68–7). When tenosynovitis occurs, there is resultant crowding of the contents of the carpal tunnel, which causes entrapment neuropathy of the median nerve (see Fig. 68–7). In chronic cases, the nerve may show low signal intensity because of fibrosis, and associated thenar muscle atrophy may be present.

In postoperative cases, MR imaging is used in patients with persistent or recurrent symptoms to assess the status of the flexor retinaculum. After surgical release, MR imaging shows discontinuity in the flexor retinaculum and volar displacement of the carpal tunnel contents (Fig. 68–9).

GUYON'S CANAL

Guyon's canal is an approximately 4-cm-long fibro-osseous tunnel in the anteromedial wrist that contains the ulnar nerve and accompanying artery and vein. The ulnar nerve can become entrapped in Guyon's canal by masses or post-traumatic osseous deformities. Other common causes of ulnar nerve entrapment include ganglion cysts (Fig. 68–10), anomalous or accessory muscles, and pseudoaneurysm of the ulnar artery.

TRAUMA

Intrinsic Ligaments

The intrinsic carpal ligaments originate and insert on the carpal bones and act to restrict motion between these bones. The two most important interosseous ligaments are the scapholunate and lunotriquetral ligaments, which connect the bones in the proximal row, separating the radiocarpal from midcarpal compartment (Fig. 68–11). These ligaments tend to be C-shaped and are thicker dorsally and ventrally. The central portion is membranous. The intrinsic carpal ligaments are not shown consistently on MR imaging. Disruption of the scapholunate or lunotriquetral ligaments often leads to wrist instability (Fig. 68–12).

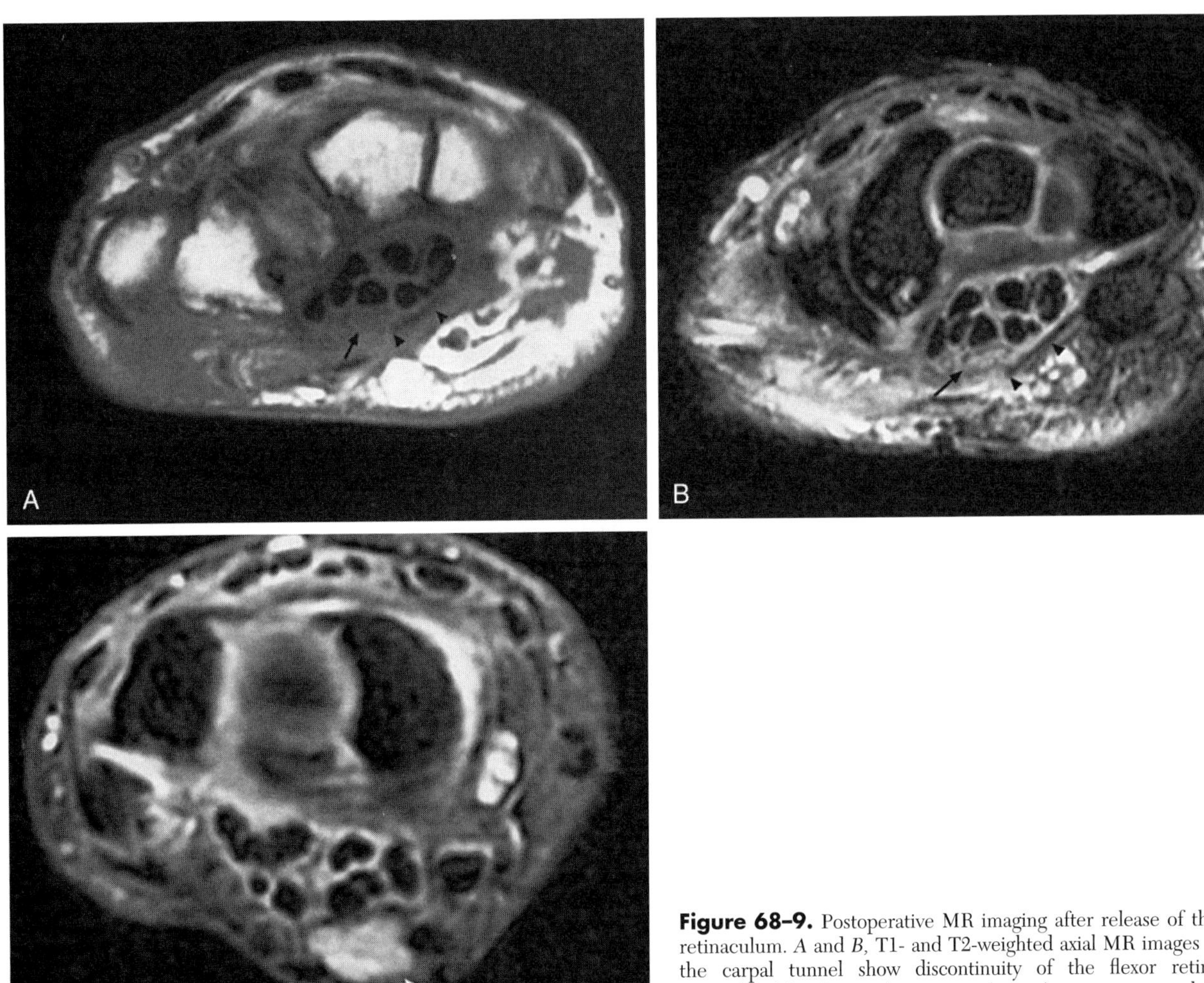

Figure 68–9. Postoperative MR imaging after release of the flexor retinaculum. *A* and *B*, T1- and T2-weighted axial MR images through the carpal tunnel show discontinuity of the flexor retinaculum *(arrowheads)*. The median nerve *(arrow)* appears normal. *C*, T2-weighted MR axial image from another patient with median nerve neuritis *(arrow)* after carpal tunnel release.

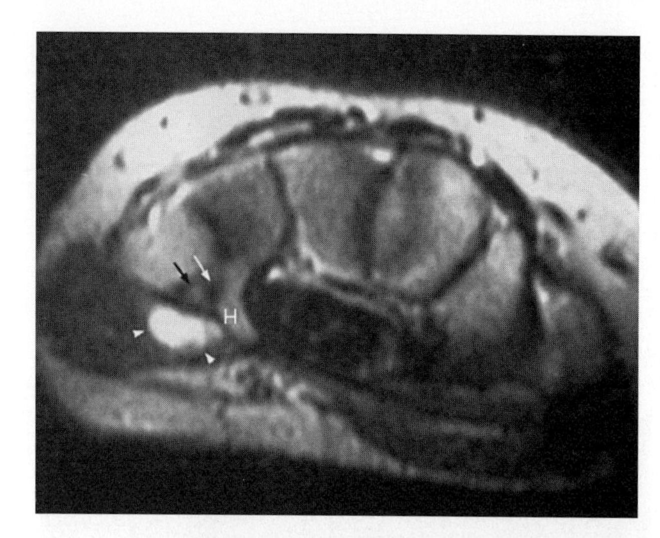

Figure 68–10. Axial T2-weighted MR image shows entrapment of the ulnar nerve within Guyon's canal by a ganglion cyst *(arrowheads).* The *black arrow* shows the ulnar nerve, and the *white arrow* shows the ulnar artery. H, hook of the hamate.

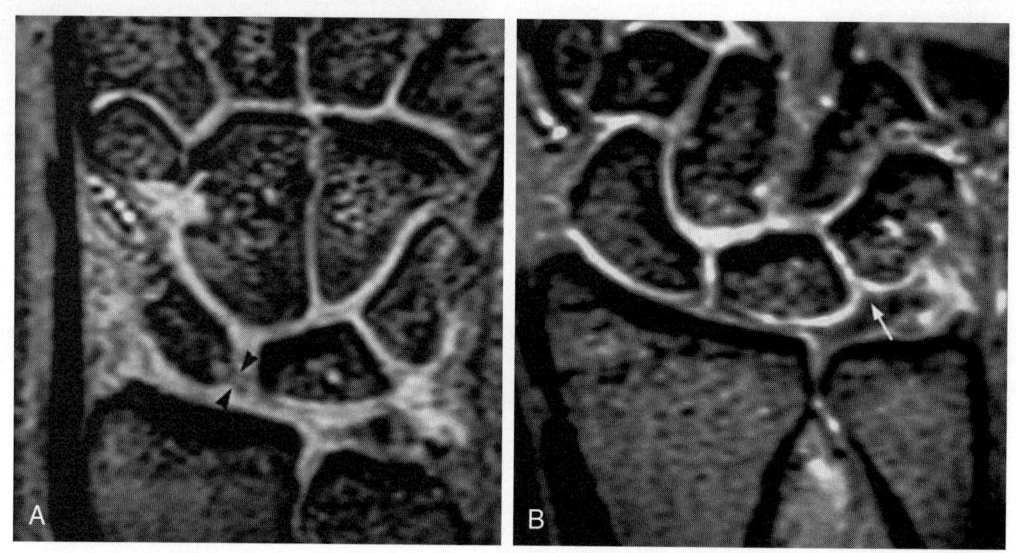

Figure 68–11. Normal intercarpal (or intrinsic carpal) ligaments. *A,* Coronal gradient echo image shows the scapholunate ligament *(arrowheads). B,* Coronal gradient echo image shows the lunotriquetral ligament *(arrow).*

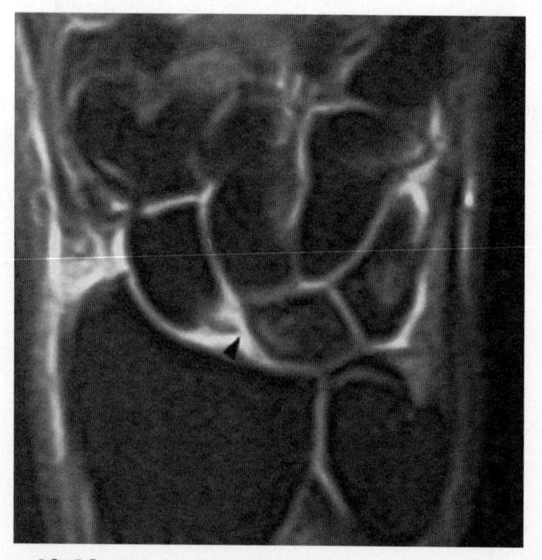

Figure 68–12. Disrupted scapholunate ligament *(arrowhead)* resulting in scapholunate dissociation. The abnormality is demonstrated on a coronal T1 fat-suppressed image post Gd arthrogram.

Carpal Fractures

The most common carpal fracture is a scaphoid fracture, constituting about 60% to 70% of all fractures involving the carpus. Nearly 70% of scaphoid fractures occur at the waist and are nondisplaced. MR imaging has been advocated as a cost-effective method for evaluating patients with persistent symptoms and normal radiographs (Fig. 68–13). Fractures are seen as low signal intensity linear abnormalities on T1-weighted images and high signal on T2-weighted images (see Fig. 68–13). There is usually surrounding bone marrow edema around the fracture line.

Stress Fractures

Stress fractures of the wrist are rare. Stress fractures of the hook of the hamate are the most common (Fig. 68–14). Golfers, tennis players, and baseball players are at risk owing to the forces created by the racquet or

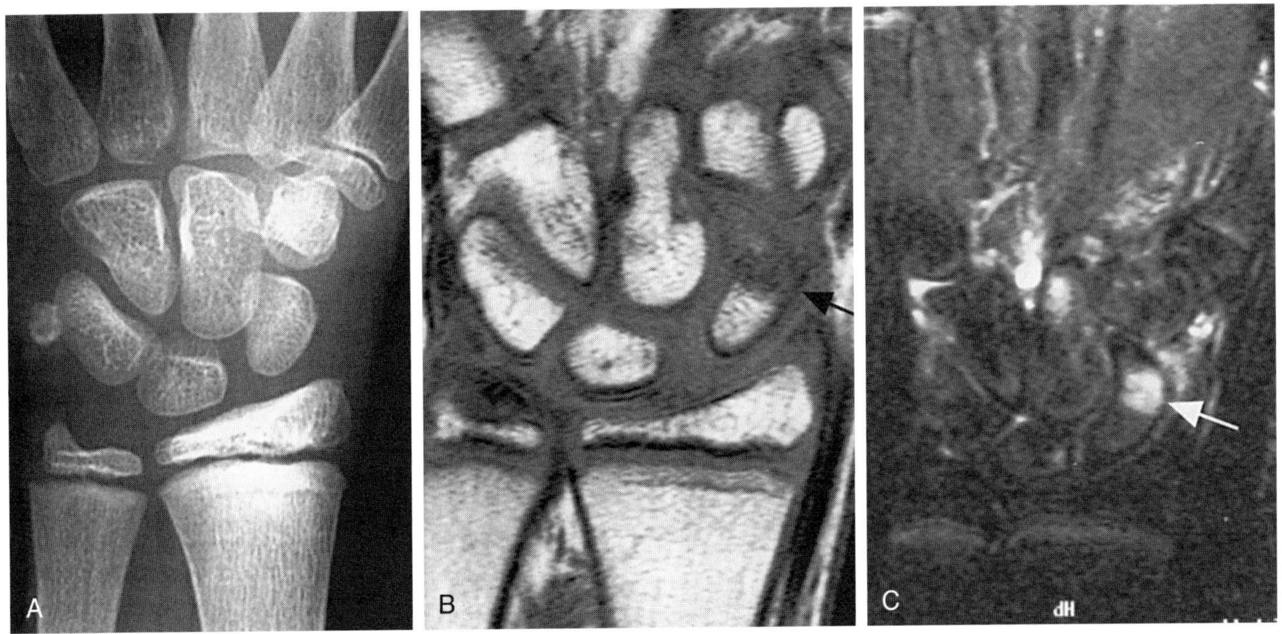

Figure 68–13. Schaphoid fracture in an 11-year-old boy that was not evident on radiography but is easily shown by MR imaging. *A,* Plain posteroanterior radiograph is normal. *B* and *C,* Coronal T1-weighted MR image and T2-weighted MR image with fat suppression show the fracture *(arrow)* and surrounding edema.

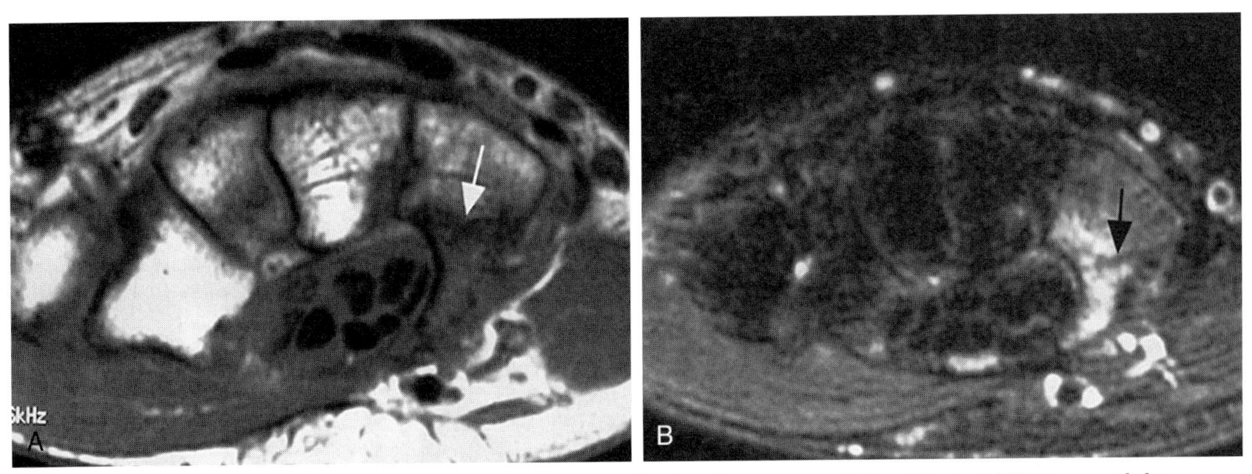

Figure 68–14. Stress fracture of the hook of the hamate. *A* and *B,* T1-weighted axial MR image and T2-weighted axial MR image with fat suppression show the fracture line and surrounding edema in a golfer *(arrow).* The fracture was not seen on plain radiographs.

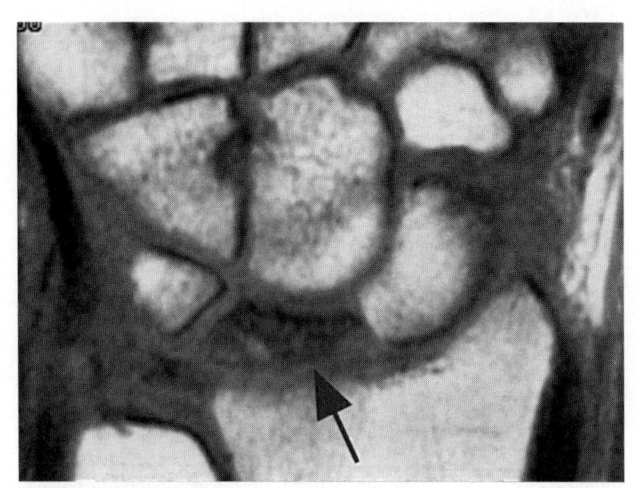

Figure 68–15. Kienböck's disease. T1-weighted coronal MR image shows collapse and decreased signal intensity within the lunate *(arrow)*.

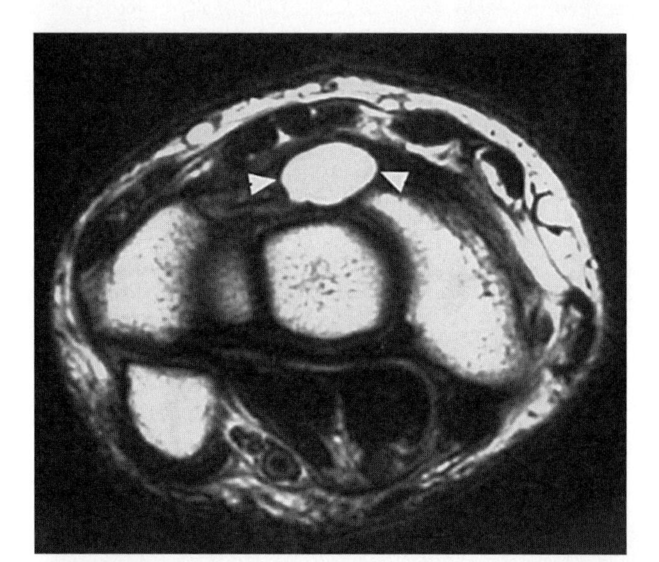

Figure 68–16. Ganglion cyst deep to the extensor tendons *(arrowheads)* shown on a T2-weighted axial image.

club when they strike the hook of the hamate. When this injury is suspected, a carpal tunnel view and slightly supinated lateral view of the wrist are the projections of choice, although computed tomography (CT) and MR imaging (see Fig. 68–14) have gained acceptance as initial diagnostic modalities.

KIENBÖCK'S DISEASE

Overview

The lunate bone is susceptible to avascular necrosis. Loss of blood supply to the lunate has been attributed to fracture, repetitive trauma causing microfractures, and traumatic injury to the ligaments that carry blood supply to the lunate. Kienböck's disease most commonly affects males in the teens through the 30s.

Imaging

Initially the lunate may appear normal on radiography, but with time it shows sclerosis, loss of height, fragmentation, and eventual collapse. An association between Kienböck's disease and ulna minus variance has been described. MR imaging can diagnose the disease earlier than radiography. The MR findings in Kienböck's disease include marrow edema (low signal on T1-weighting, high signal on T2-weighting) and fibrosis (low signal on both T1 and T2) (Fig. 68–15). In contradistinction to ulnolunate impaction or lunate fractures, the marrow signal abnormality in Kienböck's disease involves almost all of the lunate.

WRIST MASSES

The most commonly encountered mass in the wrist is a ganglion cyst. Ganglion cysts are easy to diagnose clinically and are imaged only when the presentation is atypical or

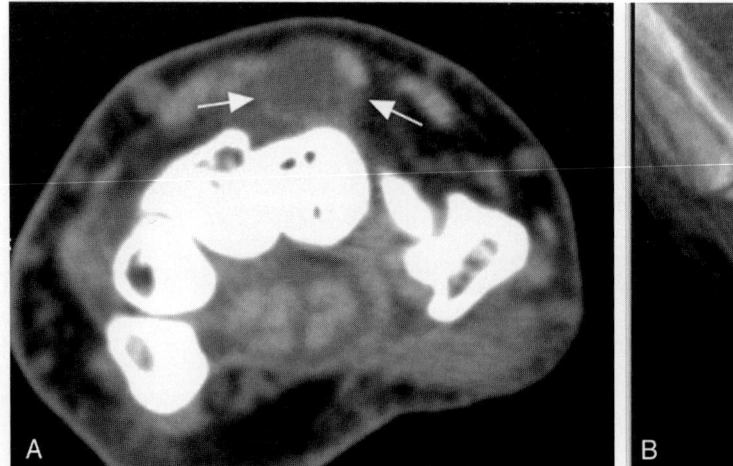

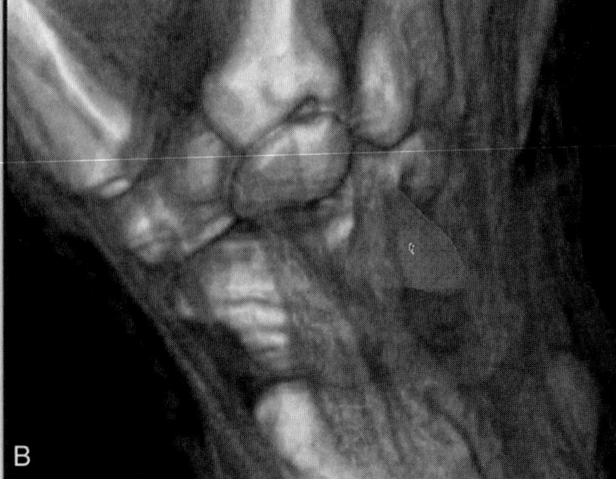

Figure 68–17. Ganglion cyst depicted on three-dimensional CT (multislice). *A,* Axial CT image shows a lesion on the dorsum of the wrist *(arrows)*. The lesion has a lucent center suggesting a fluid collection. *B,* Three-dimensional CT scan of the wrist with the ganglion cyst protruding between the extensor tendons.

the diagnosis is uncertain. MR imaging (Fig. 68–16), and ultrasound are effective in making the diagnosis of a ganglion cyst. Multislice CT has also been tried in evaluating soft tissue masses (Fig. 68–17).

References and Recommended Readings

Breitenseher MJ, Metz VM, Gilula LA, et al: Radiographically occult scaphoid fractures: Value of MR imaging in detection. Radiology 203:245, 1997.

Brown RR, Fliszar E, Cotten A, et al: Extrinsic and intrinsic ligaments of the wrist: Normal and pathologic anatomy at MR arthrography with three-compartment enhancement. Radiographics 18:667, 1998.

Glajchen N, Schweitzer M: MRI features of de Quervain's tenosynoviits of the wrist. Skeletal Radiol 25:63, 1996.

Hunter JC, Escobedo EM, Wilson AJ, et al: MR imaging of clinically suspected scaphoid fractures. AJR Am J Roentgenol 168:1287, 1997.

Imaeda T, Nakamura R, Shionoya K, et al: Ulnar impaction syndrome: MR imaging findings. Radiology 201:495, 1996.

Maurer J, Bleschkowski A, Tempka A, et al: High resolution MR imaging of the carpal tunnel and the wrist. Acta Radiol 41:78, 2000.

Metz VM, Schratter M, Dock WI, et al: Age-associated changes of the triangular fibrocartilage of the wrist: Evaluation of the diagnostic performance of MR imaging. Radiology 184:217, 1992.

Monagle K, Dai G, Chu A, et al: Quantitative MR imaging of carpal tunnel syndrome. AJR Am J Roentgenol 172:1581, 1999.

Munk PL, Lee MJ, Logan PM, et al: Scaphoid bone waist fractures, acute and chronic: Imaging with different techniques. AJR Am J Roentgenol 168:779, 1997.

Netscher D, Cohen V: Ulnar nerve compression at the wrist secondary to anomalous muscles: A patient with a variant of abductor digiti minimi. Ann Plast Surg 39:647, 1997.

Oneson SR, Scales LM, Timins ME, et al: MR imaging interpretation of the Palmer classification of triangular fibrocartilage complex lesions. Radiographics 16:97, 1996.

Oneson SR, Timins ME, Scales LM, et al: MR imaging diagnosis of triangular fibrocartilage pathology with arthroscopic correlation. AJR Am J Roentgenol 168:1513, 1997.

Palmer AK: Triangular fibrocartilage complex lesions: A classification. J Hand Surg Am 14:594, 1989.

Pretorius ES, Epstein RE, Dalinka MK: MR imaging of the wrist. Radiol Clin North Am 35:145, 1997.

Rominger MB, Bernreuter WK, Kenney PJ, et al: MR imaging of anatomy and tears of wrist ligaments. Radiographics 13:1233, 1993.

Ruocco MJ, Walsh JJ, Jackson JP: MR imaging of ulnar nerve entrapment secondary to an anomalous wrist muscle. Skeletal Radiol 27:218, 1998.

Scott JR, Cobby M, Taggart I: Magnetic resonance imaging of acute tendon injury in the finger. J Hand Surg Br 20:286, 1995.

Smith DK: Scapholunate interosseous ligament of the wrist: MR appearances in asymptomatic volunteers and arthrographically normal wrists. Radiology 192:217, 1994.

Smith DK, Snearly WN: Lunotriquetral interosseous ligament of the wrist: MR appearances in asymptomatic volunteers and arthrographically normal wrists. Radiology 191:199, 1994.

Sowa DT, Holder LE, Patt PG, et al: Applications of magnetic resonance imaging to ischemic necrosis of the lunate. J Hand Surg Am 14:1008, 1989.

Subin GD, Mallon WJ, Urbaniak JR: Diagnosis of ganglion in Guyon's canal by magnetic resonance imaging. J Hand Surg Am 14:640, 1989.

Sugimoto H, Miyahi N, Ohsawa T: Carpal tunnel syndrome: Evaluation of median nerve circulation with dynamic contrast-enhanced MR imaging. Radiology 190:459, 1994.

Sugimoto H, Shinozaki T, Ohsawa T: Triangular fibrocartilage in asymptomatic subjects: Investigation of abnormal MR intensity. Radiology 191:193, 1994.

Totterman SMS, Miller RJ: Triangular fibrocartilage complex: Normal appearance on coronal three-dimensional gradient-recalled-echo MR images. Radiology 195:521, 1995.

Totterman SMS, Miller RJ: Scapholunate ligament: Normal MR appearance on three-dimensional gradient-recalled-echo images. Radiology 200:237, 1996.

Totterman SMS, Miller RJ, McCance SE, et al: Lesions of the triangular fibrocartilage complex: MR findings with a three-dimensional gradient-recalled-echo sequence. Radiology 199:227, 1996.

Zeiss J, Jakab E, Khimji T, et al: The ulnar tunnel at the wrist (Guyon's canal): Normal MR anatomy and variants. AJR Am J Roentgenol 158:1081, 1992.

Zeiss J, Skie M, Ebraheim N, et al: Anatomic relations between the median nerve and flexor tendons in the carpal tunnel: MR evaluation in normal volunteers. AJR Am J Roentgenol 153:533, 1989.

CHAPTER 69

The Foot

Shigeru Ehara, MD

TARSAL TUNNEL SYNDROME

Anatomy

The tarsal tunnel is a fibro-osseous structure extending from the medial malleolus to the navicular bone. It is divided into the upper tibiotalar tunnel and the lower talocalcaneal tunnel of the hindfoot. The floor of the tarsal tunnel consists of the medial surface of the talus, the sustentaculum tali, and the calcaneus; the roof is formed by the flexor retinaculum (Fig. 69–1). The tunnel is most narrow at the distal portion, where it is compartmentalized by several deep fibrous septa. The posterior tibial nerve courses distally between the flexor hallucis longus tendon sheath and flexor digitorum longus tendon sheath in the tarsal tunnel. The posterior tibial nerve typically divides, within the tarsal tunnel, into three branches (see Fig. 69–1): (1) the medial calcaneal nerve, which can be

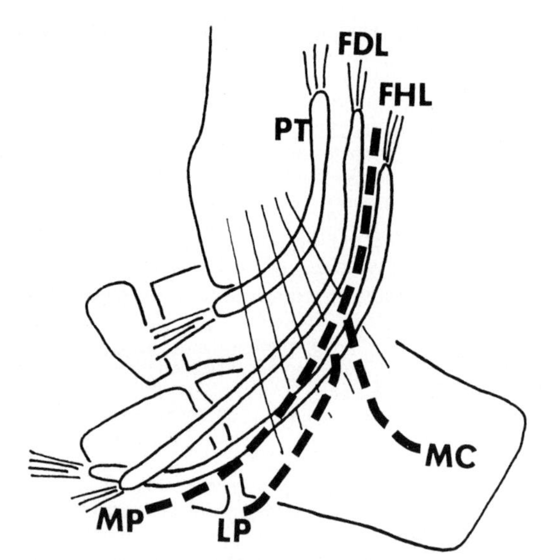

Figure 69–1. Diagram of the tarsal tunnel. Posterior tibial nerve *(dashed line)* runs underneath the flexor retinaculum *(thin oblique lines)* between the flexor hallucis longus (FHL) and flexor digitorum longus (FDL) tendons, and it branches within the tarsal tunnel into the medial calcaneal nerve (MC), lateral plantar nerve (LP), and medial plantar nerve (MP). PT, posterior tibial tendon.

the first branch of the posterior tibial nerve or can arise as a branch of the lateral plantar nerve; (2) the lateral plantar nerve, which branches beneath the flexor retinaculum; and (3) the medial plantar nerve, which is the largest branch of the posterior tibial nerve.

Overview

The tarsal tunnel syndrome is an entrapment neuropathy of the posterior tibial nerve or one of its branches within the tarsal tunnel. Symptoms are variable depending on the level of compression and chronicity. The proximal syndrome occurs in the upper tibiotalar tunnel and is caused by compression of the entire posterior tibial nerve before it divides (Fig. 69–2). Compression in the lower talocalcaneal tunnel causes dysfunction in one of the plantar nerves (Fig. 69–3). Patients present with burning pain on the plantar aspect of the foot aggravated by activity and relieved by rest. On physical examination, Tinel's sign is positive along the posterior tibial nerve and its branches.

Intrinsic compression of the nerve can be due to space-occupying lesions, including a ganglion cyst (see Figs. 69–2 and 69–3), varicosities, soft tissue tumors, exostosis, tarsal coalition with prominent medial talocalcaneal bar, tenosynovitis of the flexor tendons, fibrosis after fracture, and synovial hypertrophy from rheumatoid arthritis. Extrinsic causes include trauma, hypertrophy of the abductor hallucis muscle, and biomechanical alteration resulting in increased pressure within the tarsal tunnel (talocalcaneal coalition, pronated flatfoot, excessive pronation while running). Trauma can cause hemorrhage within the tunnel, resulting in adhesion. Accessory muscles can increase the pressure within the tarsal tunnel. These accessory muscles include the peroneocalcaneus internus, tibiocalcaneus internus, accessory soleus, and the flexor digitorum accessorius longus.

Imaging

Magnetic resonance (MR) imaging is ideally suitable for detecting lesions in the tarsal tunnel along the course of the posterior tibial nerve and its branches (see Figs. 69–2 and 69–3). In one series, MR imaging revealed the cause in 17 of 19 cases with clinical symptoms of tarsal tunnel syndrome.

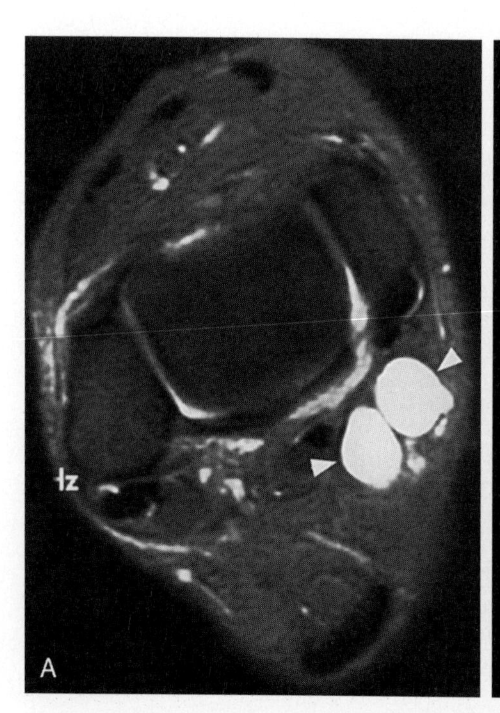

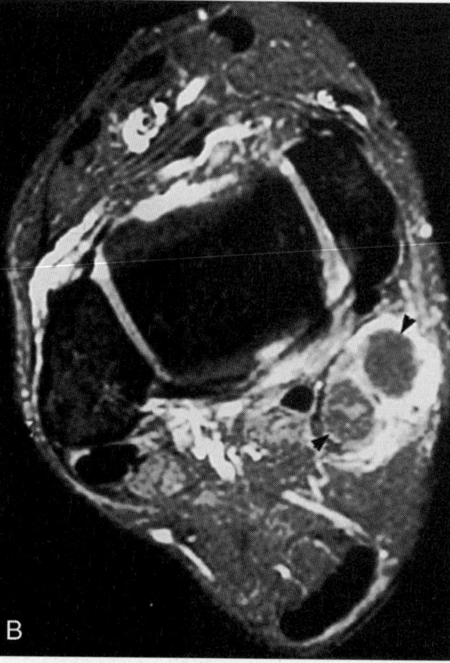

Figure 69–2. Proximal tarsal tunnel syndrome caused by a ganglion in the upper tibiotalar tunnel. *A* and *B*, T2- and T1-weighted fat-suppressed MR images with gadolinium enhancement show a septated ganglion cyst *(arrowheads)* within the tarsal tunnel.

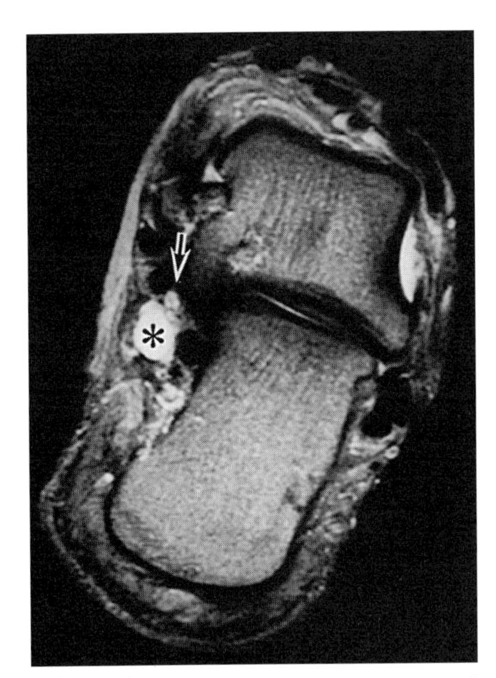

Figure 69–3. Tarsal tunnel syndrome resulting from a ganglion cyst in the lower talocalcaneal tunnel. T2-weighted MR image shows a ganglion (*) in the tarsal tunnel, adjacent to the posterior tibial nerve *(arrow)*.

ANTERIOR TARSAL TUNNEL SYNDROME

The anterior tarsal tunnel is a flattened space between the Y-shaped inferior extensor retinaculum and the fascia overlying the talus and the navicular (Fig. 69–4). The deep peroneal nerve courses distally within this fibro-osseous tunnel. The anterior tarsal tunnel syndrome is an entrapment neuropathy of the deep peroneal nerve or its branches at the anterior tarsal tunnel. Symptoms include paresthesia with pain and numbness radiating from the dorsum of the foot to the great and second toes. The causes include mechanical stress; fractures; and inappropriate footwear, such as high boots and tight shoestrings. Imaging principles applied in other entrapment neuropathies also can be used to diagnose anterior tarsal tunnel syndrome.

SINUS TARSI SYNDROME

Anatomy

The sinus tarsi is a cone-shaped space between the talus and calcaneus situated just anterior to the posterior subtalar joint. It opens anterolaterally and posteromedially. The sinus tarsi is filled with fat and contains extracapsular ligaments, which include the talocalcaneal interosseous ligament, the inferior extensor retinaculum (with lateral, intermediate, and medial roots), and the cervical ligament. These ligaments support the lateral aspect of the ankle (Fig. 69–5). The inferior extensor retinaculum is Y-shaped, with the upper limb extending from the medial malleolus and the lower limb from the medial cuneiform and sole of the foot. Blood vessels within the sinus tarsi include branches of the posterior tibial artery.

Overview

The diagnostic criteria are subjective and include lateral ankle pain, feeling of instability, and pain relief after injecting local anesthetic into the sinus tarsi. Patients typically have no pain at rest or walking on flat ground but complain of pain after walking on irregular surfaces or walking up and down stairs and jumping. Swelling over the orifices of the sinus tarsi also may be seen.

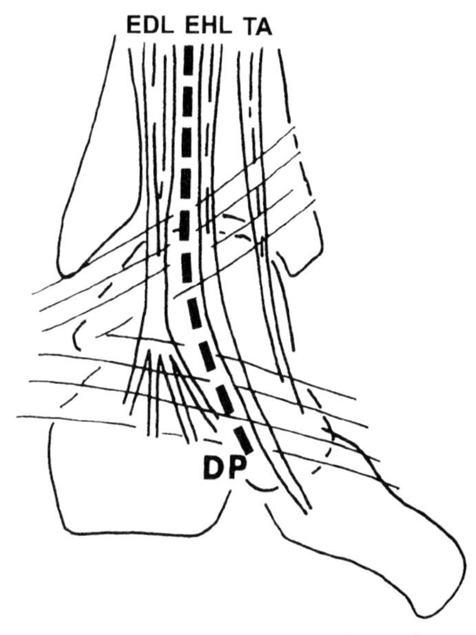

Figure 69–4. Diagram of the anterior tarsal tunnel. Deep peroneal nerve (DP) *(dashed line)* runs underneath the inferior extensor retinaculum *(thin oblique lines)*. TA, tibialis anterior tendon; EHL, extensor hallucis longus tendon; EDL, extensor digitorum longus tendon.

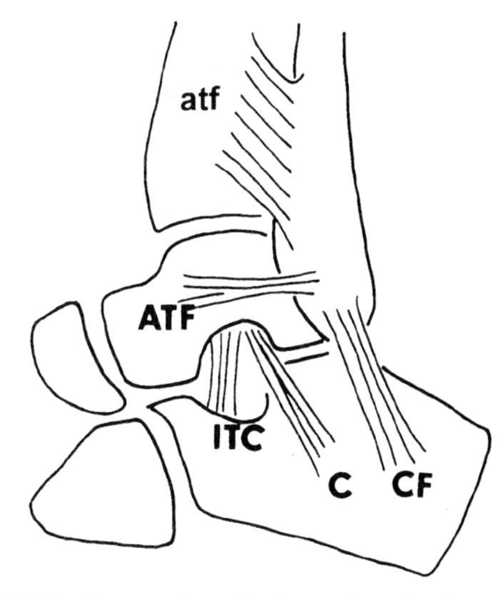

Figure 69–5. Diagram shows the ligamentous structures supporting the lateral aspect of the subtalar joint. In addition to these five ligaments (atf, anterior tibiofibular ligament; ATF, anterior talofibular ligament; CF, calcaneofibular ligament; C, cervical ligament; ITC, interosseous talocalcaneal ligament), the inferior extensor retinaculum also acts as a lateral supporting structure.

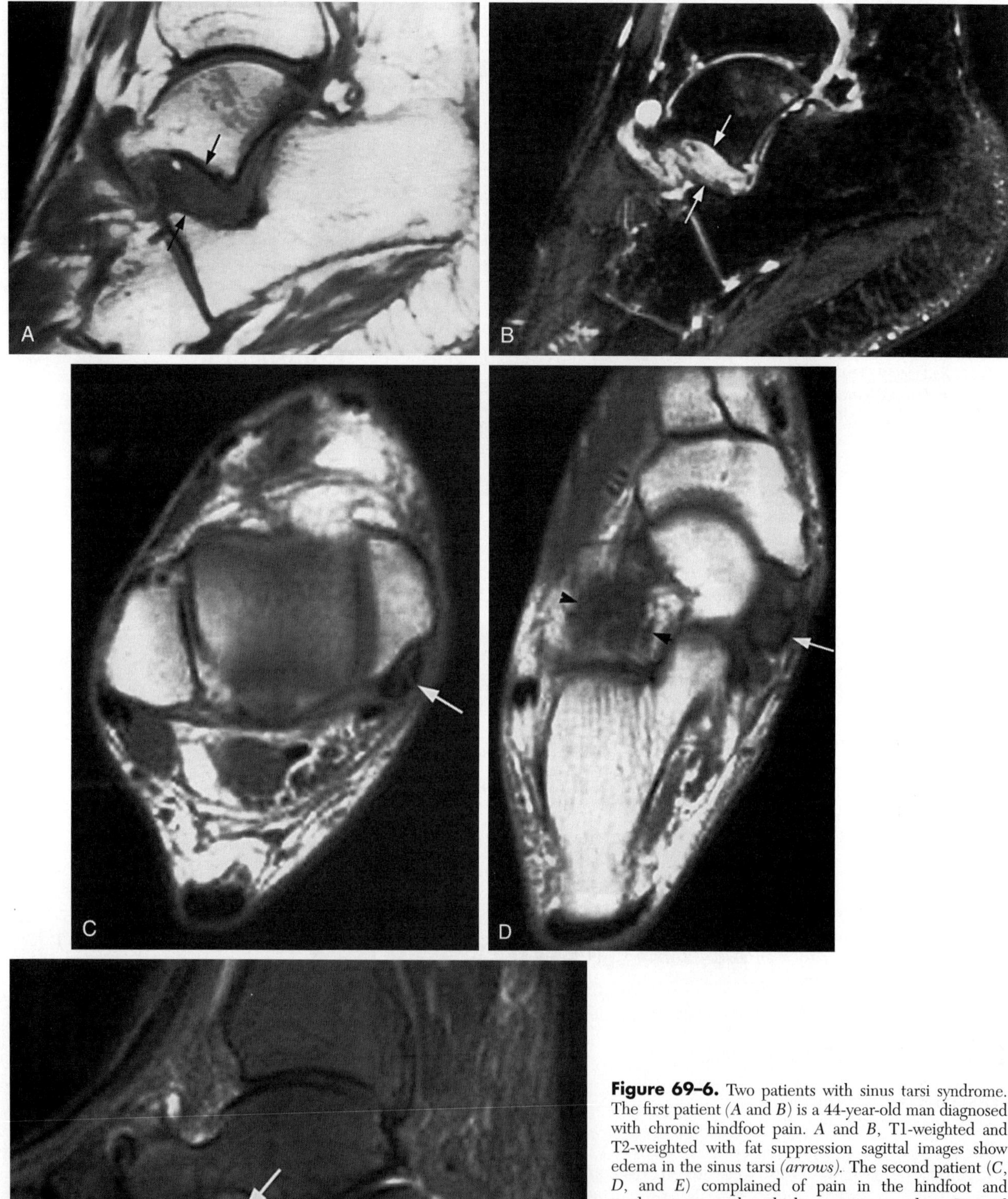

Figure 69–6. Two patients with sinus tarsi syndrome. The first patient (*A* and *B*) is a 44-year-old man diagnosed with chronic hindfoot pain. *A* and *B*, T1-weighted and T2-weighted with fat suppression sagittal images show edema in the sinus tarsi *(arrows)*. The second patient (*C*, *D*, and *E*) complained of pain in the hindfoot and tenderness over the tibialis posterior tendon. *C*, Axial T1-weighted image at the level of the ankle joint shows tendinosis *(arrow)* involving the tibialis posterior tendon. *D*, Axial T1-weighted image through the hindfoot shows tendinosis in the tibialis posterior tendon *(arrow)* and edema in the sinus tarsi *(arrowheads)*. *E*, Sagittal T2-weighted image with fat suppression reveals edema in the sinus tarsi *(arrow)* surrounding the interosseous talocalcaneal ligament *(arrowheads)*. (Case provided by Dr. Robert H. Choplin, Indianapolis, Indiana.)

The most common causes of sinus tarsi syndrome are inversion injury of the ankle and less commonly synovial inflammation, such as rheumatoid arthritis and gout. Subtalar sprain is the most common cause for rupture of the interosseous and cervical ligaments. Tear of the lateral collateral ligaments of the ankle is seen frequently in patients with abnormal sinus tarsi on MR imaging.

Imaging

Typically, no radiographic findings are noted. MR imaging reveals changes related to fibrosis, chronic synovitis, or fluid in the sinus tarsi (Fig. 69–6). MR imaging also can show associated tears of the lateral ankle ligaments, most often the calcaneofibular ligament, which is reported in about 79% of patients with sinus tarsi syndrome. Of patients with lateral ligamentous injury, 39% have sinus tarsi syndrome. Tear of the tibialis posterior tendon also is associated with sinus tarsi syndrome (Fig. 69–6 *C*, *D*, and *E*).

DIABETIC FOOT

Overview

Diabetic foot is a broad term applied to a variety of foot problems in patients with diabetes mellitus, related to vasculopathy, neuropathy, and infections. Vascular disorders of the feet include large and medium vessel atherosclerosis and microangiopathy of the skin and muscles. Arterial occlusion occurs commonly at multiple levels below the popliteal region. Metatarsal arteries are occluded in 60% of patients with diabetes.

Neuropathy in diabetic patients includes metabolic neuropathy, vascular neuropathy, and nerve compression syndromes (see also Chapter 30). Chronic symmetrical distal polyneuropathy is the most common of metabolic neuropathies. Deformities of the toes are more frequent in diabetic patients and are related to the neuropathy; they include hammertoe, claw toe, and angular deformities. Demineralization and osteolysis at the ends of phalanges and distal metatarsals also occur. Neuropathic (Charcot's) arthropathy involves the forefoot (metatarsophalangeal and interphalangeal joint deformities) (Fig. 69–7), midfoot (Lisfranc's joint) (Fig. 69–7*B*), and hindfoot (subtalar and ankle joints). Fracture-dislocation of Lisfranc's joint is the most common feature of neuropathic arthropathy in the diabetic foot (see Fig. 69–7*B*). Avulsion of the posterior calcaneal tuberosity at the insertion of Achilles' tendon is fairly common and is almost pathognomonic of diabetes mellitus (Fig. 69–8).

Generalized and local factors contribute to foot infections in patients with diabetes mellitus, including poor blood supply, abnormal sensation in the foot, and minor trauma. Soft tissue infections most often start in the toes. Dorsal cellulitis and abscess formation commonly are the result of surface infection (Fig. 69–9). Necrotizing cellulitis or fasciitis is a rare but serious soft tissue infection.

Staphylococcus aureus is the most common pathogen, and aerobic gram-positive cocci are the next most common. Anaerobic infections are rare but serious. The presence of gas on radiography most likely is due to gas-forming gram-negative bacteria. Clostridial infections (*gas gangrene*) occur much less commonly in diabetic patients but are considered to be serious.

Osteomyelitis is typically a chronic infection contiguous with a skin ulcer (Fig. 69–10). Osteomyelitis is more likely, if the associated ulcer is large and deep; however, osteomyelitis may occur without ulceration (Fig. 69–11).

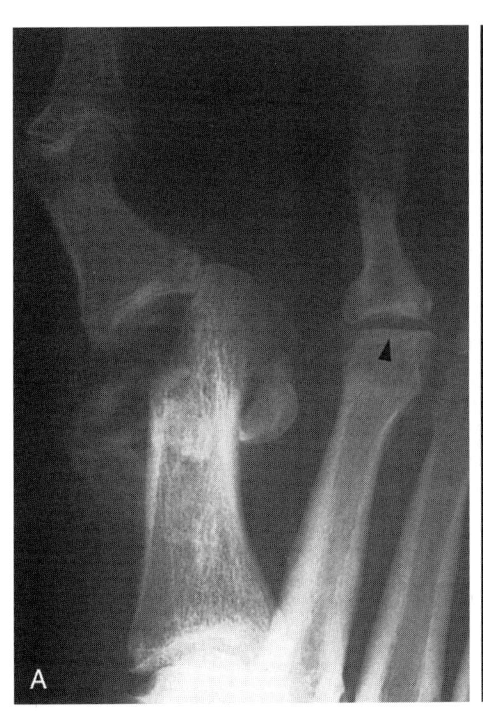

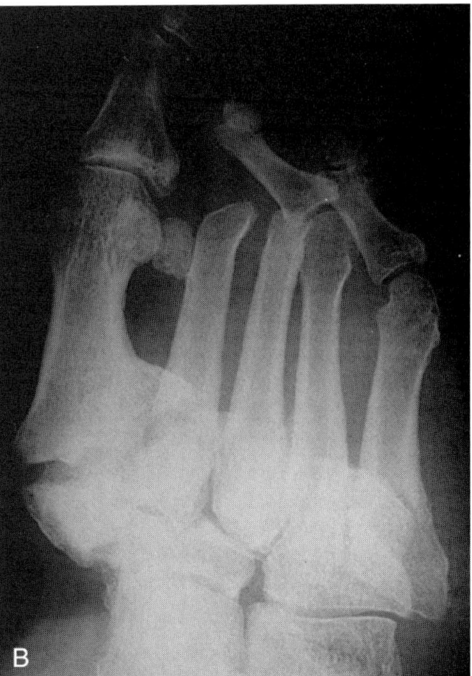

Figure 69–7. Neuropathic arthropathy of the foot in diabetic patients. *A*, Dislocation and disorganization of the first metatarsophalangeal joint in a 41-year-old diabetic patient with peripheral neuropathy. Note the avascular necrosis of the second metatarsal head *(arrowhead)* (Freiberg's disease), which is a common finding in patients with diabetes mellitus. *B*, A 53-year-old diabetic man with dislocation of the second through the fifth tarsometatarsal joints. The second and third toes were amputated because of gangrene.

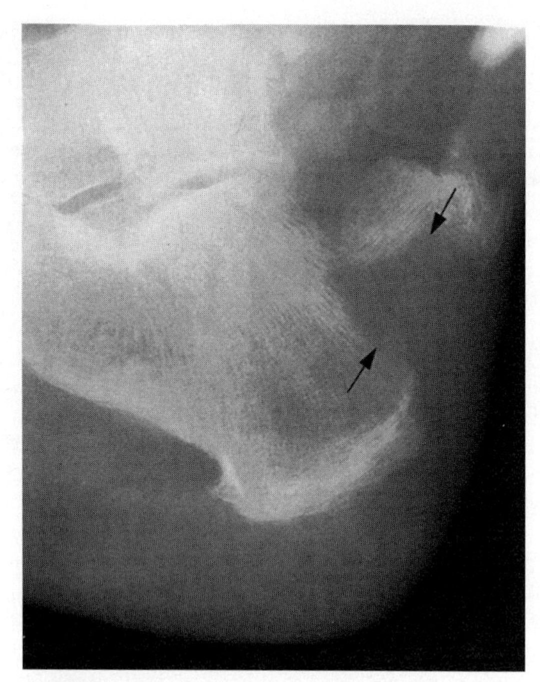

Figure 69–8. Avulsion of the calcaneal tuberosity *(arrows)* in a 26-year-old diabetic woman with peripheral neuropathy.

Ulceration is most common in areas with high pressure or friction, particularly over bony prominences (Fig. 69–12). It often is modified by associated vasculopathy. The infection starts at the cortex and probably is better termed *osteitis* because most patients with a variety of forms of osteomyelitis lack generalized symptoms.

Abscess formation is common in the soft tissue of the foot (see Fig. 69–9) and in the toes, metatarsals, and calcaneus. Bone fragmentation and sequestrum formation are common findings. Small sinuses with large intraosseous abscesses may be characteristic.

Interphalangeal and metatarsophalangeal joints adjacent to ulcers often are affected (see Fig. 69–10). Infection spreads to the adjacent bones, resulting in osteomyelitis (see Fig. 69–10). Radiologic findings include local osteoporosis, subchondral bone resorption, and collapse of the articular surface (see Fig. 69–10).

Imaging

Radiography still is used in the initial examination. It is usually adequate to detect neuropathic (Charcot's) arthropathy and deformities (see Chapter 30) but not sensitive enough to detect early stages of infection. Bone scintigraphy with technetium-99m diphosphonate reflects osteoblastic activity and blood supply. A three-phase bone scan is useful for differentiating soft tissue infection and osteomyelitis, if radiography is negative. If other bone changes are present, however, the specificity of the three-phase bone scan decreases. An indium-111-labeled white blood cell scan is useful in detecting active osteomyelitis, particularly when there is no radiographic sign of neuroarthropathy. An indium-111-labeled white blood cell scan is probably better than MR imaging for the detection of osteomyelitis. MR imaging is most suitable to detect soft tissue abscesses and bone marrow involvement (see Figs. 69–9 and 69–11). Acute neuropathic changes in the midfoot could resemble osteomyelitis closely on MR imaging. The acute fractures associated with Charcot's arthropathy produce marrow edema, which often is misinterpreted as representing osteomyelitis. Percutaneous core bone biopsy is used to differentiate the two conditions and to isolate pathogens for specific antibiotic therapy.

MORTON'S NEUROMA

Overview

Morton's neuroma (Morton's metatarsalgia, interdigital neuroma) is a neural degeneration with perineural fibrosis of the interdigital nerve resulting from mechanical stresses, causing burning and toe pain. Morton's neuroma is not considered to be a neoplasm.

Pathologically, there is a swollen nerve with atrophic nerve fibers surrounded by hypertrophic connective tissue

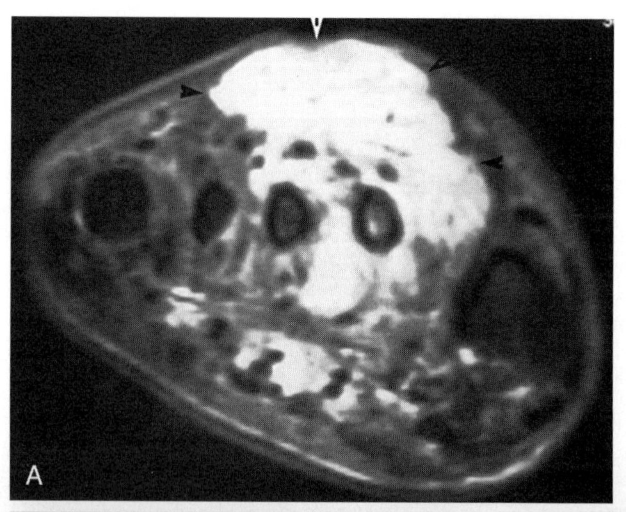

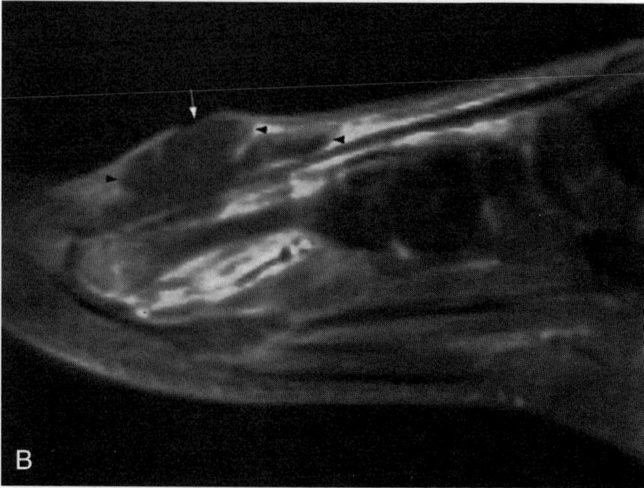

Figure 69–9. Abscess on the dorsum of the foot associated with a skin ulcer. *A,* Axial T2-weighted MR image shows a large area of increased signal intensity on the dorsum of the foot *(black arrowheads).* Note the skin ulcer *(white arrowhead). B,* Gadolinium-enhanced sagittal T1-weighted image with fat suppression shows the abscess collection *(black arrowheads)* and the skin ulcer *(white arrow).*

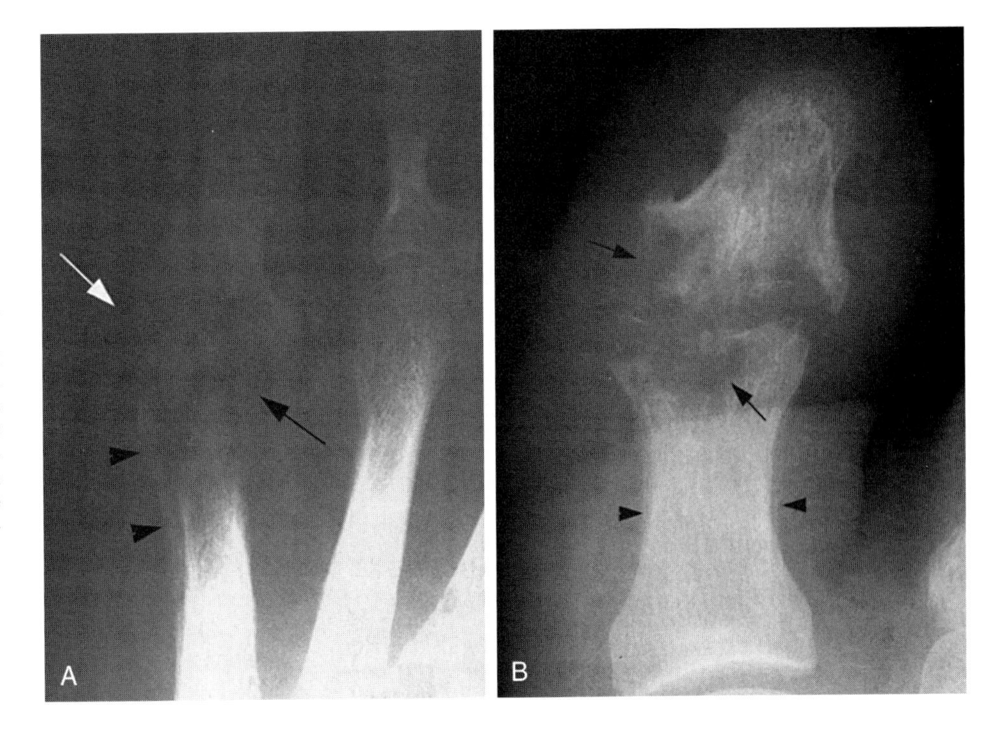

Figure 69–10. Diabetic osteomyelitis with septic arthritis. *A*, Anteroposterior radiograph of the little toe shows a skin ulcer *(white arrow)* with adjacent septic arthritis *(black arrow)* and osteomyelitis with periosteal reaction *(black arrowheads)*. The fifth metatarsophalangeal joint is dislocated. *B*, Anteroposterior radiograph of the great toe from another patient shows erosions *(black arrows)* with destruction of the interphalangeal joint. The infection has extended into the shaft of the proximal phalanx producing periosteal reaction *(arrowheads)*.

elements. Renaut's bodies are a common finding, suggesting compressive damage to the nerve.

Burning pain localized to the plantar aspect of the foot typically is aggravated by activity and by wearing tight-

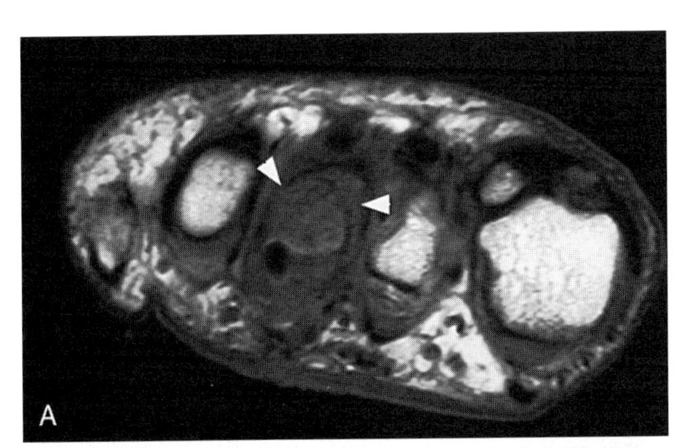

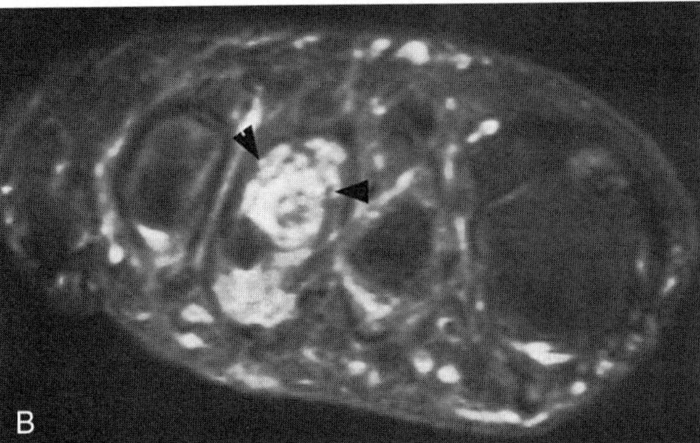

Figure 69–11. Osteomyelitis of the third metatarsal in a 42-year-old diabetic patient. *A*, Axial T1-weighted MR image shows low signal intensity in the third metatarsal *(arrowheads)*. The soft tissues around the bone also show low signal intensity. *B*, Axial T1-weighted MR image after gadolinium enhancement and fat suppression shows the osteomyelitis *(arrowheads)* and surrounding soft tissue reaction.

fitting high-heeled shoes. Some investigators have used MR imaging to show that asymptomatic Morton's neuromas are fairly common. The lesion is seen more frequently in women than in men (10:1), usually in their 40s and 50s. Morton's neuroma is most common between the third and fourth metatarsal heads (55%) and between the second and third metatarsal heads (45%).

Imaging

MR imaging is the study of choice. Coronal and axial images are most useful (Fig. 69–13; also see Fig. 17–14).

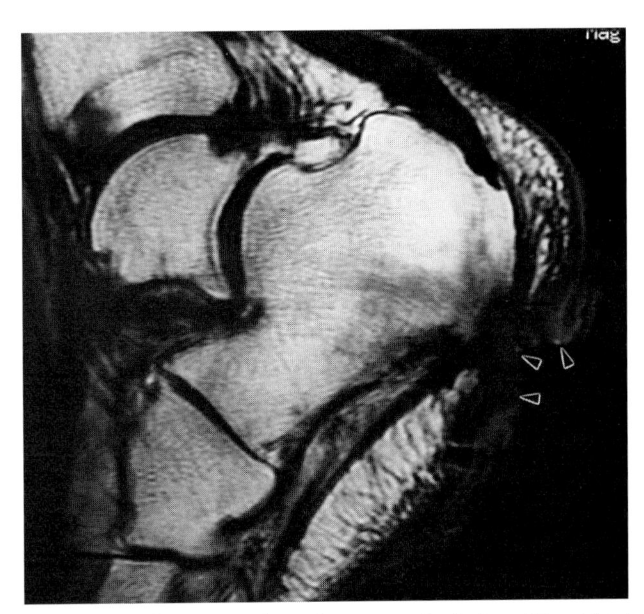

Figure 69–12. Osteomyelitis adjacent to a heel ulcer. T1-weighted sagittal MR image shows a deep ulcer at the heel *(arrowheads)* and decreased signal intensity in the bone marrow adjacent to the ulcer, which is highly suggestive of osteomyelitis.

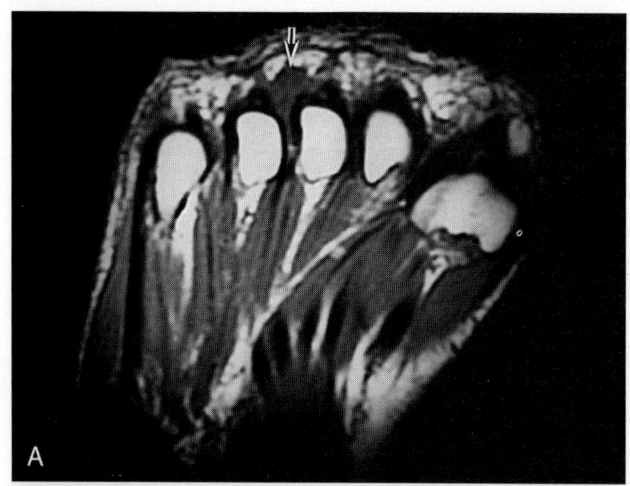

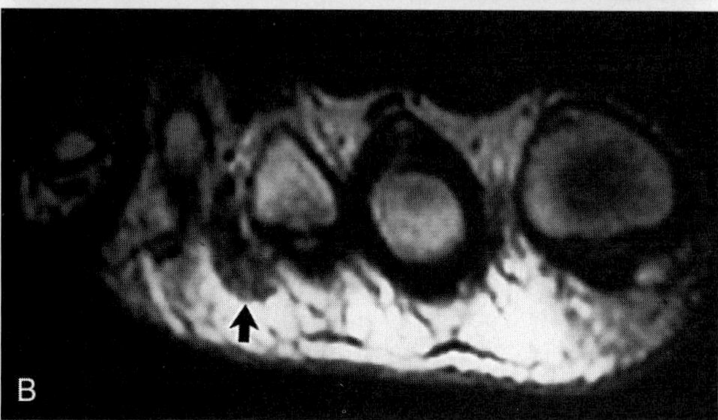

Figure 69–13. Morton's neuroma in the third web space. *A*, Coronal T1-weighted MR image shows a low signal intensity. A soft tissue mass between the third and fourth metatarsal heads is visible *(arrow)*. *B*, Axial T2-weighted image shows a low signal intensity mass between the third and fourth metatarsal heads *(arrow)*.

The lesion can be difficult to visualize on MR imaging, and it may have low signal intensity on T1- and T2-weighted images. Typically, Morton's neuromas show moderate-to-marked signal enhancement after gadolinium administration.

PLANTAR FASCIITIS

Overview

The plantar fascia is multilayered, consisting of medial, central, and lateral components. It arises from the medial calcaneal tuberosity, and its dominant central portion fans out to insert at the plantar plates of the metatarsophalangeal joints. The thickness of the plantar fascia is normally about 3 mm.

In the acute phase, patients present with pain localized to the deep region of the heel at the origin of the plantar fascia, and in the chronic phase, the pain may extend more distally. The disease is common in middle-aged women and young athletic men. Obesity is a frequent finding in women with plantar fasciitis. It is usually unilateral but can be bilateral in 15% of patients. Risk factors include rapid weight gain, obesity, sudden increase in athletic activity, a tight Achilles' tendon, shoes with poor cushioning, and prolonged weight bearing. Achilles' tendon tightness predisposes to plantar fasciitis because of limited dorsiflexion of the foot.

Microtearing of the plantar fascia resulting from mechanical stress and fascial or perifascial inflammation are believed to be the underlying abnormalities. Biopsy specimens have revealed necrosis of the collagen fibers and repair. Inflammatory arthritis, such as Reiter's syndrome, ankylosing spondylitis, and psoriatic arthritis rarely can cause plantar fasciitis.

Imaging

Radiography is performed to rule out stress fractures, erosions caused by inflammatory bursitis, enthesopathy, and other bony lesions. Calcaneal spur may be present; it is not more common in patients with plantar fasciitis than in the normal population, and its relationship to the symptoms is controversial.

MR imaging rarely is indicated, but it may reveal thickening of the plantar fascia and surrounding inflammatory reaction (Fig. 69–14). The thickness of the plantar fascia is increased significantly; it is normally 2 to 4 mm but becomes 6 to 10 mm in patients with plantar fasciitis.

STRESS FRACTURES

Stress fractures in the foot occasionally can be difficult to diagnose clinically or by conventional imaging modalities. In high-power athletes, MR imaging is helpful to make an early diagnosis of stress fractures in the tarsal and metatarsal bones (Figs. 69–15 and 69–16).

Recommended Readings

Beltran J, Munchow AM, Khabiri H, et al: Ligaments of the lateral aspect of the ankle and sinus tarsi. Radiology 177:455, 1990.

Berkowitz JF, Kier R, Rudicel S: Plantar fasciitis: MR finding. Radiology 179:665, 1991.

Bertran J: Sinus tarsi syndrome. Magn Reson Imaging Clin North Am 2:59, 1994.

Bridges Jr RM, Deitch EA: Diabetic foot infections: Pathophysiology and treatment. Surg Clin North Am 74:537, 1994.

Erikson SJ, Canale PB, Carrera GF, et al: Interdigital (Morton) neuroma: High-resolution MR imaging with a solenoid coil. Radiology 181:833, 1991.

Erikson SJ, Quinn SF, Kneeland JB, et al: MR imaging of the tarsal tunnel and related spaces: Normal and abnormal findings with anatomic correlation. AJR Am J Roentgenol 155:323, 1990.

Gold RH, Tong DJF, Crim JR, et al: Imaging of the diabetic foot. Skeletal Radiol 24:563, 1995.

Guiloff RJ, Scadding JW, Klenerman L: Morton's metatarsalgia: Clinical, electrophysiological and histological observations. J Bone Joint Surg Br 66:586, 1984.

Kerr R, Frey C: MR imaging of tarsal tunnel syndrome. J Comput Assist Tomogr 15:280, 1991.

Kier R: Magnetic resonance imaging of plantar fasciitis and other causes of heel pain. Magn Reson Imaging Clin North Am 2:97, 1994.

Klein MA, Spreitzer AM: MR imaging of the tarsal sinus and canal: Normal anatomy, pathologic findings, and features of the sinus tarsi syndrome. Radiology 186:233, 1993.

Lipsky BA: Osteomyelitis of the foot in diabetic patients. Clin Infect Dis 25:1318, 1997.

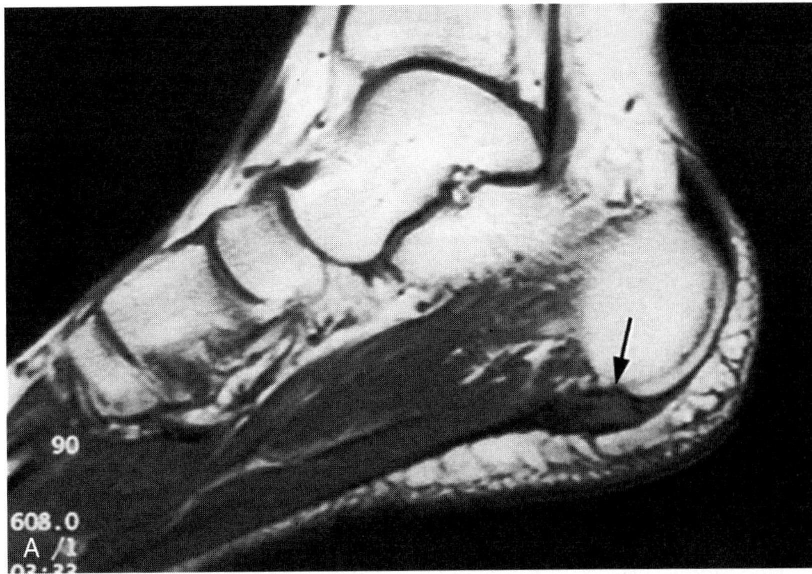

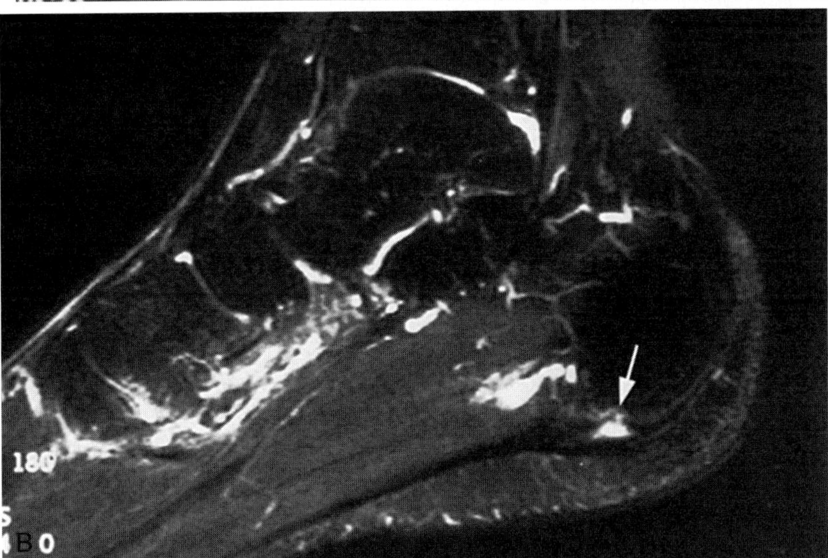

Figure 69–14. Plantar fasciitis in a 52-year-old woman. T1-weighted sagittal and T2-weighted sagittal images with fat suppression show increased signal, thickening, and tearing of some of the fibers in the plantar fascia *(arrows)*.

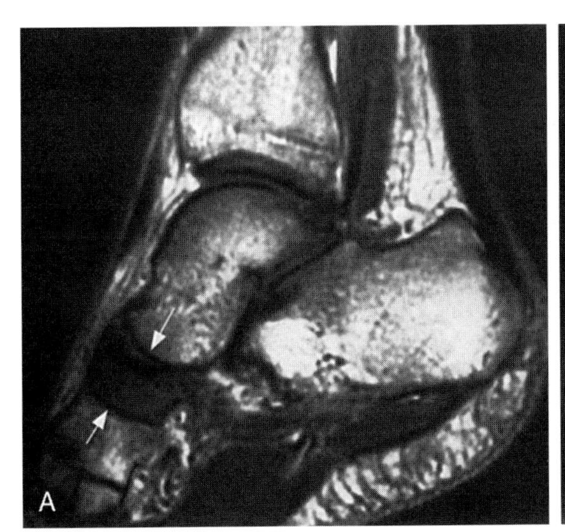

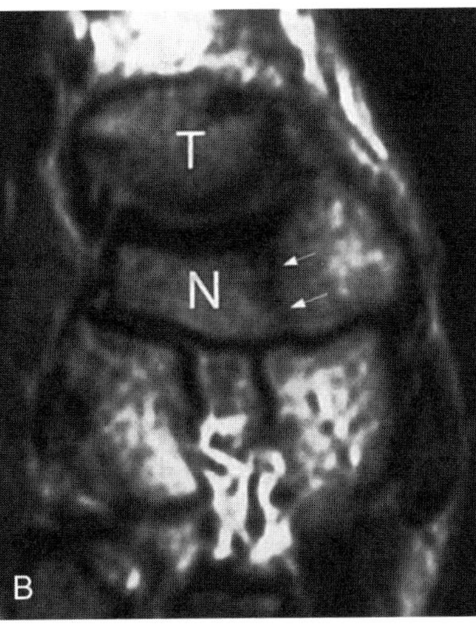

Figure 69–15. Stress fracture of the tarsal navicular in a 16-year-old athlete. *A,* Sagittal T1-weighted MR image shows decreased signal in the navicular *(arrows)*. *B,* Coronal proton-density image shows the fracture line *(arrows)*. T, talar head; N, tarsal navicular.

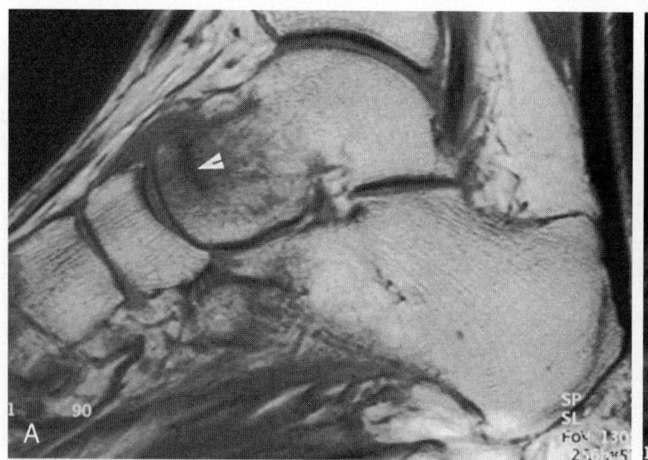

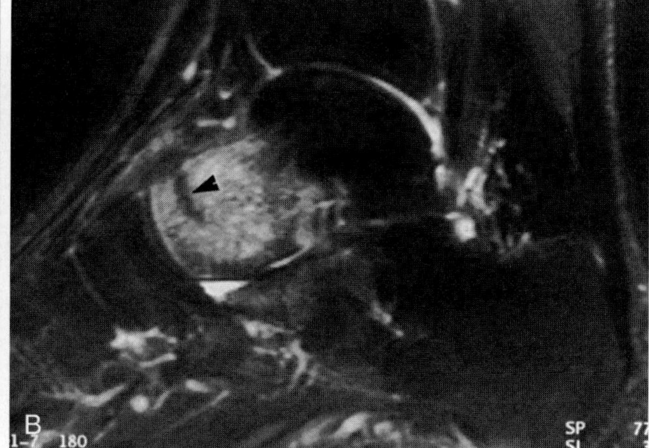

Figure 69–16. Stress fracture of the talar head. *A* and *B*, Sagittal T1- and T2-weighted MR images with fat suppression show the fracture line *(arrowhead)* and surrounding edema.

Meyer JM, Garcia J, Hoffmyer P, et al: The subtalar sprain: A roentgenographic study. Clin Orthop 226:169, 1988.

Shereff MJ, Grande DA: Electron microscopic analysis of the interdigital neuroma. Clin Orthop 271:296, 1992.

Singh D, Angel J, Bentley G, et al: Plantar fasciitis. BMJ 315:172, 1997.

Taillard W, Meyer J-M, Garcia J, et al: The sinus tarsi syndrome. Int Orthop 5:117, 1981.

Zanetti M, Lederman T, Zollinger H, et al: Efficacy of MR imaging in patients suspected of having Morton's neuroma. AJR Am J Roentgenol 168:529, 1997.

Zanetti M, Strehle JK, Zollinger H, et al: Morton neuroma and fluid in the intermetatarsal bursae on MR images of 70 asymptomatic volunteers. Radiology 203:516, 1997.

Zongzhao L, Jiansheng Z, Li Z: Anterior tarsal tunnel syndrome. J Bone Joint Surg Br 73:470, 1991.

CHAPTER 70

Internal Derangements of the Temporomandibular Joint

Timothy E. Moore, MD

OVERVIEW

Although the temporomandibular joint (TMJ) is subject to all the disorders common to any synovial joint, there are specific internal derangements that only affect the TMJ and that are fairly common. Internal derangement of the TMJ may be defined as an abnormal relationship of the articular disk relative to the mandibular condyle, the articular eminence, or the glenoid fossa. TMJ internal derangement occurs in 28% of adults and can be a major cause of pain and dysfunction. It is particularly common in relatively young women, with some series reporting a female-to-male ratio greater than 10:1.

The underlying abnormality is typically a dislocation of the articular disk. An understanding of the normal meniscocondylar movements is essential for the interpretation of imaging studies. Normally the biconcave disk is interposed between the articular surfaces of the mandibular condyle and the temporal bone, with the thin central portion, also called the *intermediate zone,* occupying the narrowest part of the joint between the articular eminence and the condyle. The thicker anterior and posterior bands lie in front of and behind the intermediate zone. On opening of the mouth, the condyle translates forward over the eminence and carries the disk with it. To maintain normal articular contours, the disk also moves posteriorly relative to the condyle. With the mouth fully opened, the intermediate zone still occupies the narrowest point between the eminence and the condyle. Maintaining the correct disk position is determined by the shape of the disk and by a balance between the anterior pull of the superior head of lateral pterygoid muscle and the posterior pull of the bilaminar zone that connects the posterior band of the disk to the tympanic plate and the mandibular neck (Fig. 70–1).

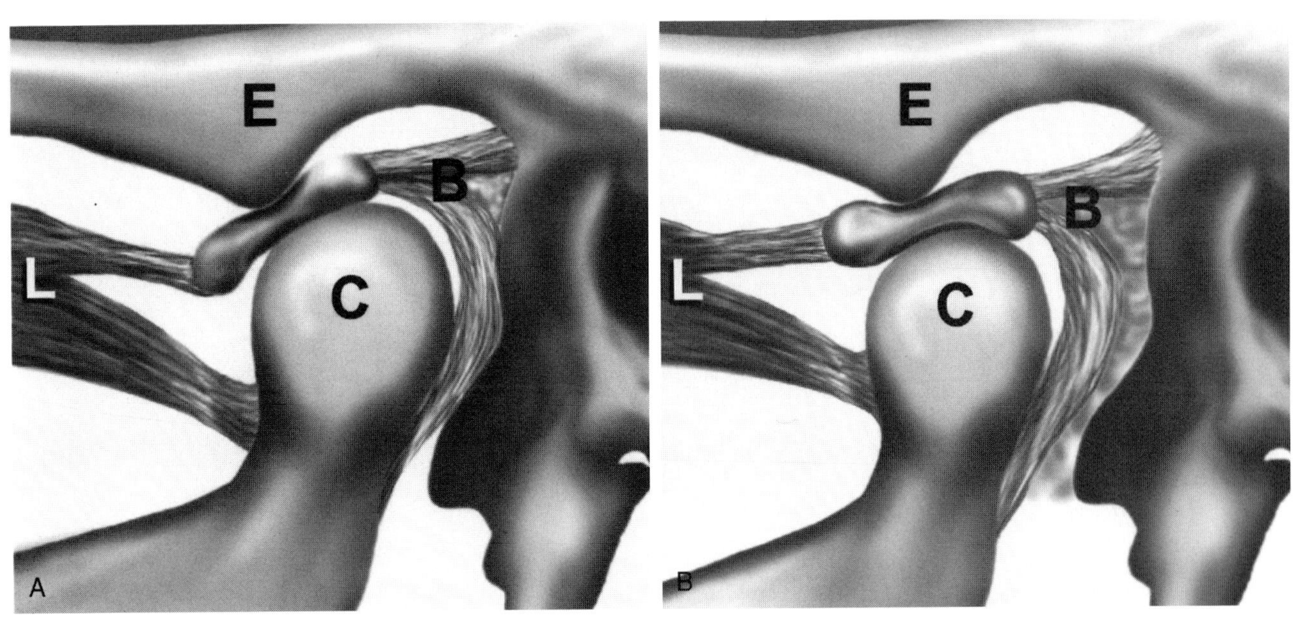

Figure 70–1. Diagram of normal temporomandibular joint movements. *A*, Closed mouth. The condyle is in the glenoid fossa, and the disk is interposed between the articular surfaces with the anterior band in front of the condyle and the posterior band behind. *B*, Open mouth. The condyle has translated anteriorly. The disk remains interposed between the articular surfaces. E, articular eminence; C, mandibular condyle; B, bilaminar zone; L, lateral pterygoid muscle.

INTERNAL DERANGEMENT

With internal derangement, the disk usually is displaced anteriorly or anteromedially, and two main disk displacement patterns can occur.

Anterior Disk Displacement with Reduction. Anterior disk displacement with reduction (Fig. 70–2) is the less advanced pattern. In its mildest form, there is anterior disk displacement with early reduction to the normal position on opening of the mouth. With the mouth closed, the posterior band is dislocated anterior to the condyle. On opening of the mouth, the condyle passes forward over the posterior band so as to reduce the dislocation and articulate normally with the intermediate zone. More advanced stages are associated with reduction of the disk at progressively later stages of opening. On closing of the mouth, the disk returns to the abnormal anterior position. Patients usually experience an audible click from the affected joint on opening of the mouth caused by the sudden return of the posterior band to its normal posterior position. Occasionally a patient experiences a second click on closing the mouth, caused by the disk dislocating anteriorly. This is known as the *reciprocal click*.

Anterior Displacement Without Reduction. With repeated anterior disk dislocations, the bilaminar zone becomes overstretched, and it loses its elasticity (Fig. 70–3). As a result, the bilaminar zone no longer can hold the disk in a posterior enough position to permit the condyle to pass over the posterior band and reduce the dislocation. The disk remains dislocated anteriorly with the posterior band in front of instead of behind the condyle. This dislocation causes mechanical obstruction to the normal anterior translation of the condyle, which in turn prevents full opening of the mouth. With further overstretching, the bilaminar zone may become perforated or detached from the posterior band, and degenerative changes eventually ensue. As patients progress from the anterior displacement with reduction to the anterior displacement without reduction stages of this disorder, they experience a change from a clicking jaw to a jaw that no longer clicks but that will not open fully. This is known as a *closed lock*.

IMAGING

Conventional Imaging Techniques and Computed Tomography

Basic imaging principles, as discussed elsewhere, may be applied to imaging of arthritis, infections, fractures, and dislocations of the TMJ. Tomography, especially orthopantomography, computed tomography (CT), and conventional radiography all play valuable parts in such imaging. Internal derangements are not shown with these imaging methods, however. TMJ arthrography became popular in the late 1970s and early 1980s and was the first imaging method to show effectively the dynamic relationship between the articular disk, the mandibular condyle, and the temporal bone. Arthrography now is essentially obsolete, however, because MR imaging can detect the clinically relevant findings without the discomfort and inherent morbidity of an invasive procedure.

Magnetic Resonance Imaging

MR imaging reliably can show the articular disk as a low signal biconcave structure interposed between the mandibular condyle and the temporal bone. Its relationship to these structures is shown readily by imaging the joint in the mouth open and mouth closed positions.

Thin slice sagittal images are obtained simultaneously

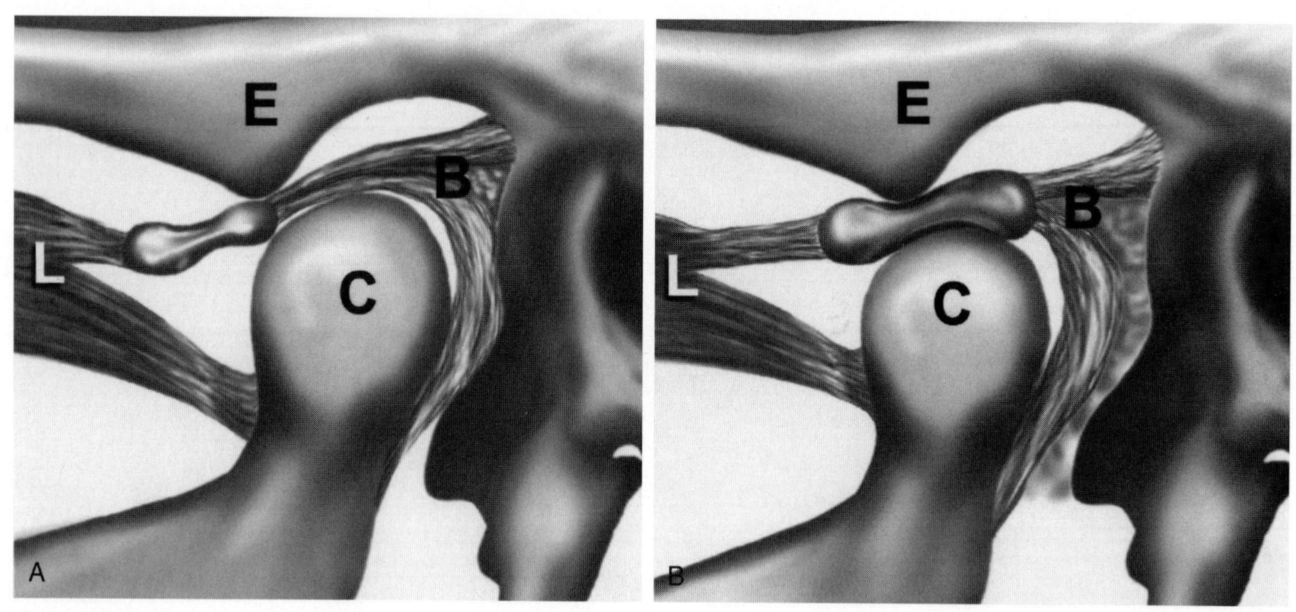

Figure 70–2. Diagram of anterior disk displacement with reduction. *A*, Closed mouth. The condyle is in the glenoid fossa, but the disk is dislocated anteriorly with the posterior band anterior to the condyle. *B*, Open mouth. The condyle has translated anteriorly. The disk has reduced and is seen in the normal position with the posterior band behind the condyle. E, articular eminence; C, mandibular condyle; B, bilaminar zone; L, lateral pterygoid muscle.

on both sides with the mouth closed and with the mouth open. The two sets of sagittal images in each series may be parallel to one another, which has the advantage of following the anterior translation of the condyles on opening of the mouth. Alternatively, if the equipment permits, oblique sagittal images may be obtained parallel to the angle of the mandible (Fig. 70–4); this has the advantage of displaying the anatomy so that the lateral pterygoid muscle, articular disk, and bilaminar zone line up more advantageously. A suggested protocol is as follows: One series of T1-weighted sagittal images with the mouth closed is followed by a series with the mouth opened as far as is reasonably comfortable. An adjustable bite-block is used. The bite-block keeps the mouth open and records the number of millimeters between the incisors. Oral surgeons often wish to know this measurement, and it can be annotated on the images. These two series are sufficient for most imaging purposes, but an

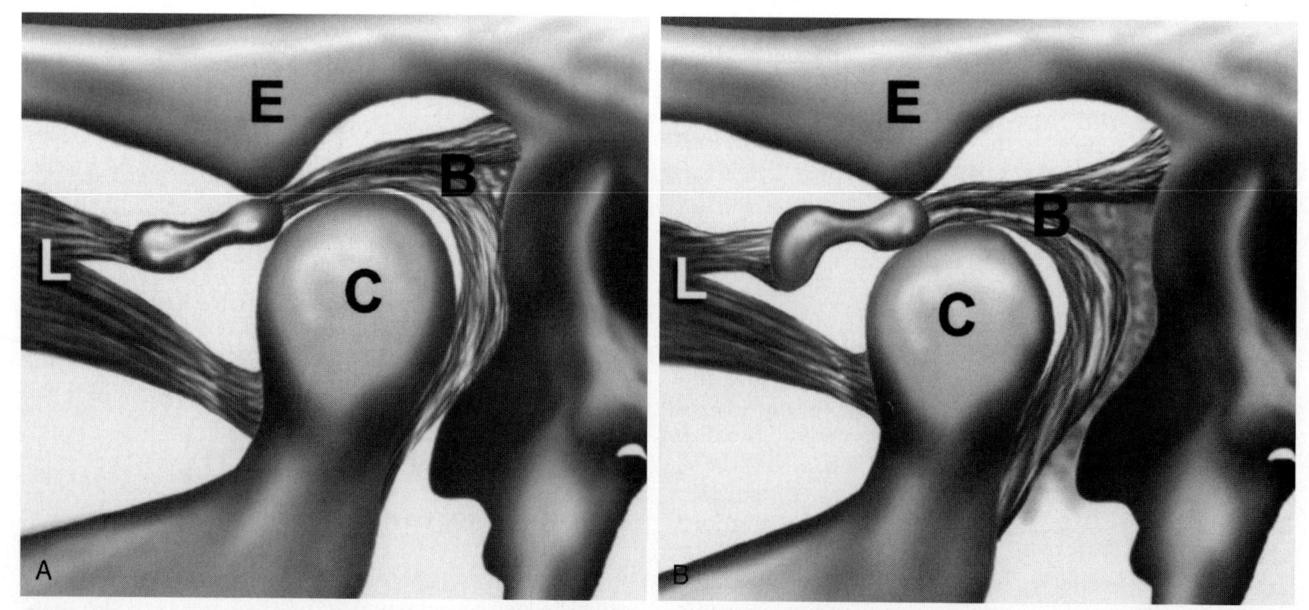

Figure 70–3. Diagram of anterior disk displacement without reduction. *A*, Closed mouth. The condyle is in the glenoid fossa. The disk is dislocated anteriorly with the posterior band anterior to the condyle. *B*, Open mouth. The condyle has translated a limited amount anteriorly. Further translation is blocked by the disk, which remains forward with the posterior band anterior to the condyle. E, articular eminence; C, mandibular condyle; B, bilaminar zone; L, lateral pterygoid muscle.

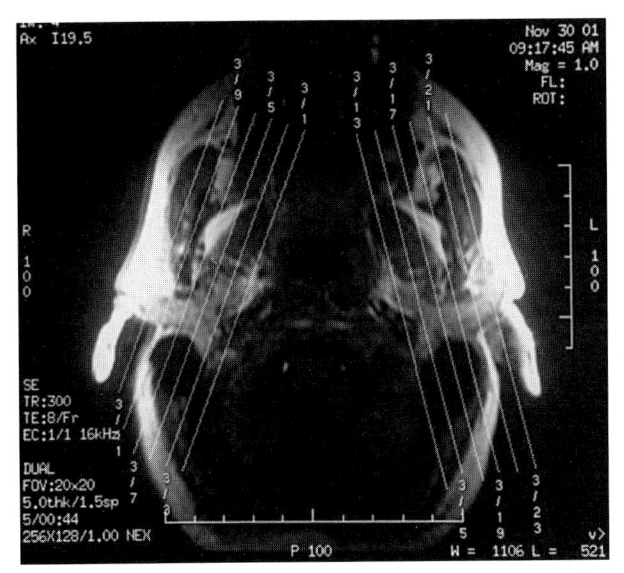

Figure 70–4. Axial scout MR imaging shows orientation of oblique sagittal images. Parameters may be either parallel to the body of the mandible or at 90 degrees to the medial-lateral axis of the mandibular condyle.

additional 30-degree flip gradient echo closed mouth series has the advantage of detecting other abnormalities, such as joint effusion or edema. This series also serves as another chance to identify the disk, which sometimes is difficult to see on T1-weighted closed mouth series. Occasionally a coronal series may be useful in identifying a disk that has dislocated medially or laterally.

An advantage to arthrography over MR imaging of the TMJ is the ability of arthrography to show real-time movements of the condyle and disk, especially reduction and dislocation of the disk. For this reason, some radiologists employ either MR dynamic or pseudodynamic imaging methods. Echo-planar imaging has been used for dynamic imaging but still is not widely available, and the disk can be difficult to identify. A more common method is pseudodynamic imaging, in which a series of single rapid-sequence images is obtained at varying stages of opening of the mouth using the adjustable bite-block. Images can be viewed sequentially on a monitor and *scrolled* through to create the pseudodynamic effect of the mouth opening and closing (Fig. 70–5).

Magnetic Resonance Imaging Appearance of the Normal Disk

Closed mouth sagittal images (Fig. 70–6*A*) show the mandibular condyle located centrally within the glenoid fossa. The low signal fibrocartilaginous disk shows a bow tie shape and is seen interposed between the condyle and the eminence. The junction of the posterior band of the disk and the bilaminar zone should be at about 12 o'clock.

Open mouth images (Fig. 70–6*B*) show the condyle in a more anterior position and located close to the articular eminence. The disk also has moved forward, but the intermediate zone maintains its location between the condyle and the eminence. The disk does not move as far forward as the condyle.

Anterior Disk Displacement with Reduction

Closed mouth images (Fig. 70–7*A*) show the condyle located normally in the glenoid fossa. The disk is displaced forward with the posterior band anterior to the condyle. The bilaminar zone is interposed between the condyle and the eminence. Open mouth images (Fig. 70–7*B*) are normal because the disk has reduced to its normal position.

Anterior Disk Displacement Without Reduction

Closed mouth images (Fig. 70–8*A*) reveal findings identical to those seen in anterior disk displacement with reduction (i.e., the disk appears dislocated anteriorly). On open mouth images (Fig. 70–8*B*), the disk remains dislocated anteriorly. The condyle usually does not translate as far forward on the articular eminence, and the disk may deform as a result of the presence of the condyle impacting on the posterior band.

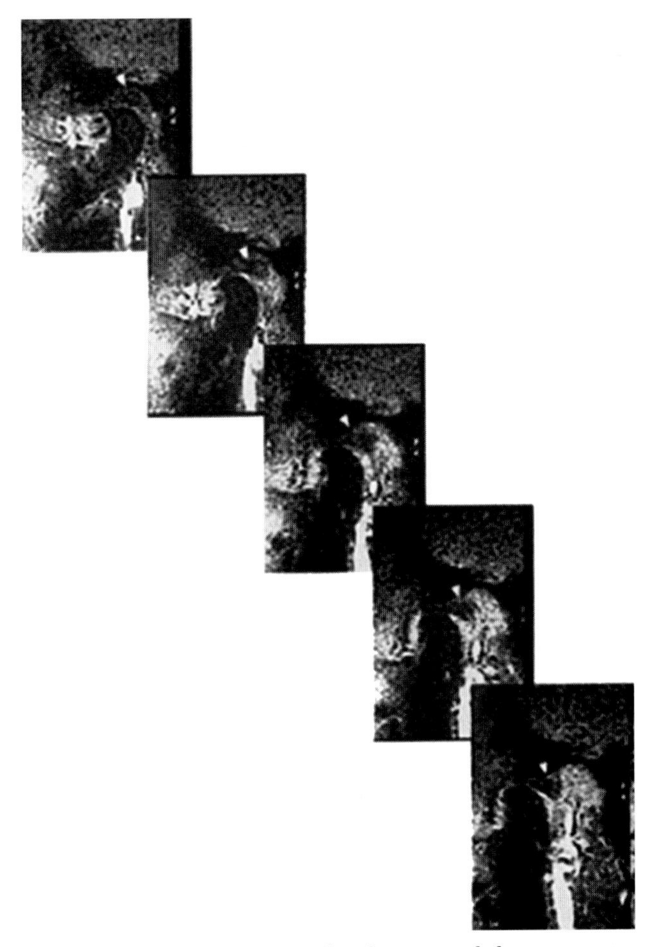

Figure 70–5. Selected images taken from a pseudodynamic imaging series of a normal temporomandibular joint. Top left image shows the mouth closed, and in the bottom right the mouth is opened. Note how the disk *(white arrrowhead)* retains its normal relationship to the condyle.

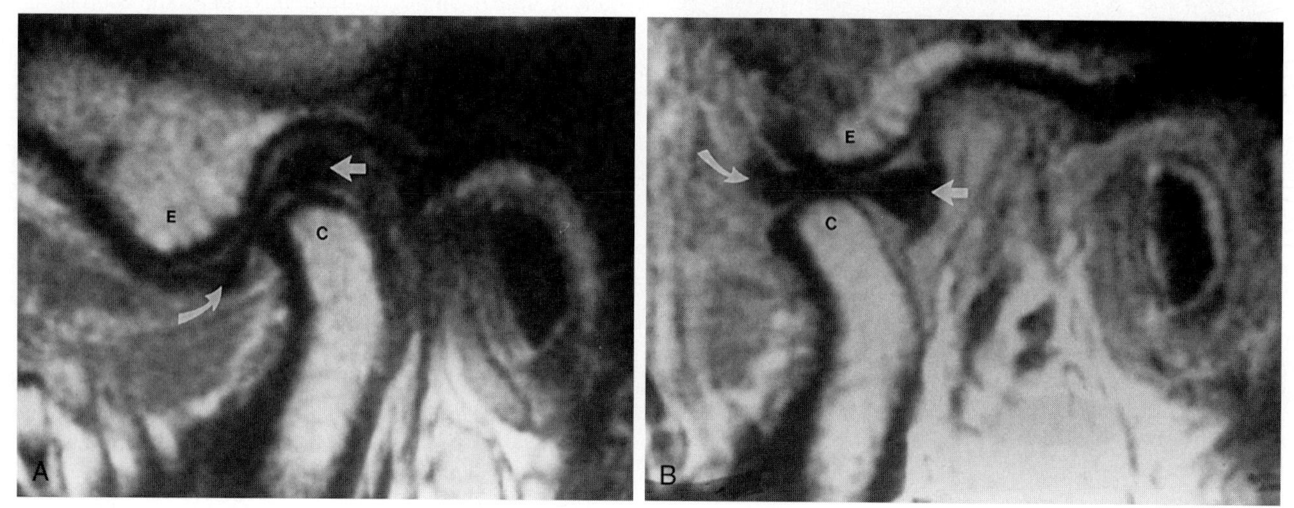

Figure 70–6. Normal sagittal T1-weighted MR images of the temporomandibular joint. *A*, Closed mouth. The condyle lies centrally within the glenoid fossa. The disk lies anterosuperior to the condyle. *B*, Open mouth. The condyle has moved forward on the articular eminence, and the disk has remained interposed between convex articular surfaces of the mandibular condyle and articular eminence. E, articular eminence; C, mandibular condyle; straight arrow, posterior band of disk; curved arrow, anterior band.

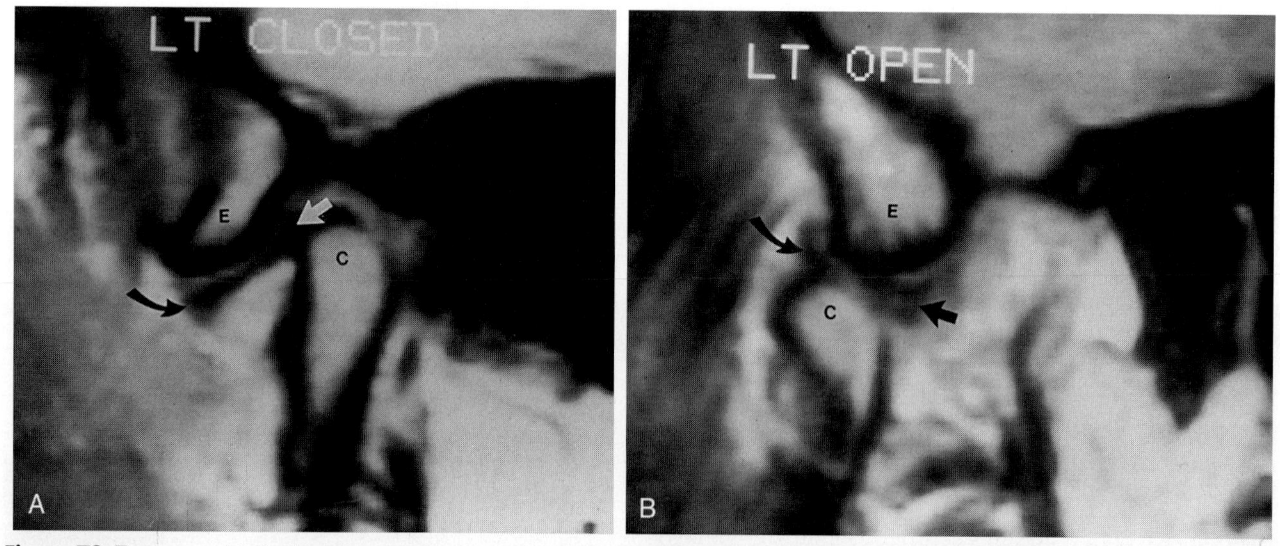

Figure 70–7. Anterior disk displacement or dislocation with reduction. *A*, Closed mouth. The disk is displaced anteriorly so that the posterior band is anterior to the condyle and the concave inferior surface of the disk no longer articulates with the condyle. *B*, Open mouth. The disk has returned to a normal position between the mandibular condyle and articular eminence. E, articular eminence; C, mandibular condyle; straight arrow, posterior band of disk; curved arrow, anterior band.

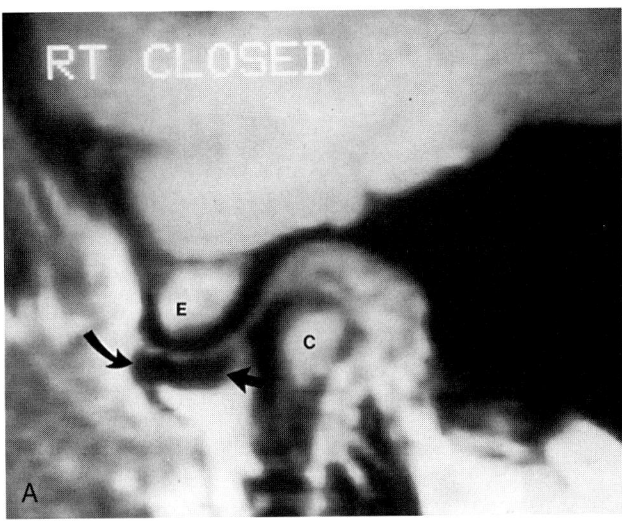

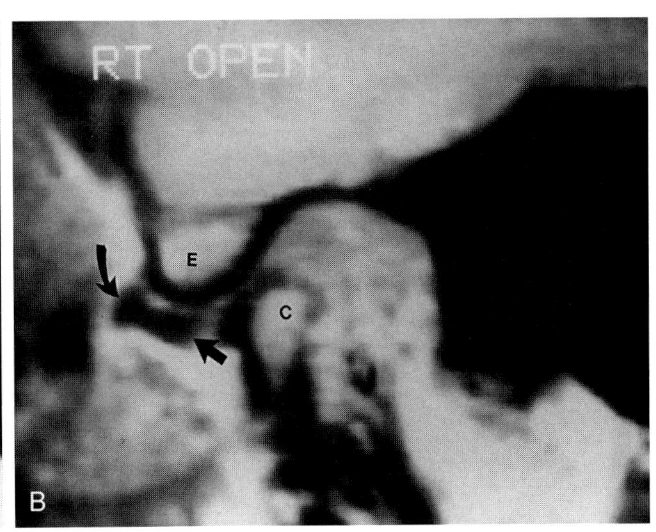

Figure 70–8. Anterior disk displacement or dislocation without reduction. *A,* Closed mouth. The disk is dislocated anterior to the mandibular condyle *(arrows). B,* Open mouth. The disk has remained anterior to the condyle and shows some foreshortening as a result of compression by the condyle. Mechanical obstruction by the dislocated disk has limited anterior translation of the condyle markedly. E, articular eminence; C, mandibular condyle; straight arrow, posterior band of disk; curved arrow, anterior band.

Recommended Readings

Kaplan PA, Helms CA: Current status of temporomandibular joint imaging for the diagnosis of internal derangements. AJR Am J Roentgenol 152:697, 1989.

Larheim TA, Westesson P, Sano T: Temporomandibular joint disk displacement: Comparison in asymptomatic volunteers and patients. Radiology 218:428, 2001.

Masui T, Isoda H, Mochizuki T, et al: Pseudodynamic imaging of the temporomandibular joint: SE versus GE sequences. J Comput Assist Tomogr 20:448, 1996.

Milano V, Desiate A, Bellino R, et al: Magnetic resonance imaging of temporomandibular disorders: Classification, prevalence and interpretation of disc displacement and deformation. Dentomaxillofac Radiol 29:352, 2000.

Solberg WK, Woo MW, Houston JB: Prevalence of mandibular dysfunction in young adults. J Am Dent Assoc 98:25, 1979.

CHAPTER 71

Tendon Disorders

Georges Y. El-Khoury, MD, Nabil J. Khoury, MD, and Bernard Roger, MD

TENDON ANATOMY

The tendon is a densely packed connective tissue structure consisting of type I collagen fibrils embedded in a matrix of proteoglycans. Tendons are relatively hypocellular, containing mature fibroblasts (or tenocytes), which are the predominant cell type. The tenocytes are responsible for producing and maintaining a healthy matrix. The biochemical constituents of the tendon include collagen (30%), elastin (2%), and water (68%). Collagen fibrils are arranged in closely packed bundles to form fascicles. These fascicles are bound together by loose connective tissue that permits longitudinal movement of collagen fascicles and supports blood vessels, lymphatics, and nerves. Tendons that move in a straight line, such as Achilles' tendon, are surrounded by loose areolar connective tissue called the *paratenon.* The paratenon is usually not visible on imaging studies unless it is surrounded by fluid on both sides (Fig. 71–1). Tendons that bend sharply around corners, such as the long flexors of the foot, are subjected to compressive forces; they are enclosed by tendon sheaths, which act as pulleys to direct the path of

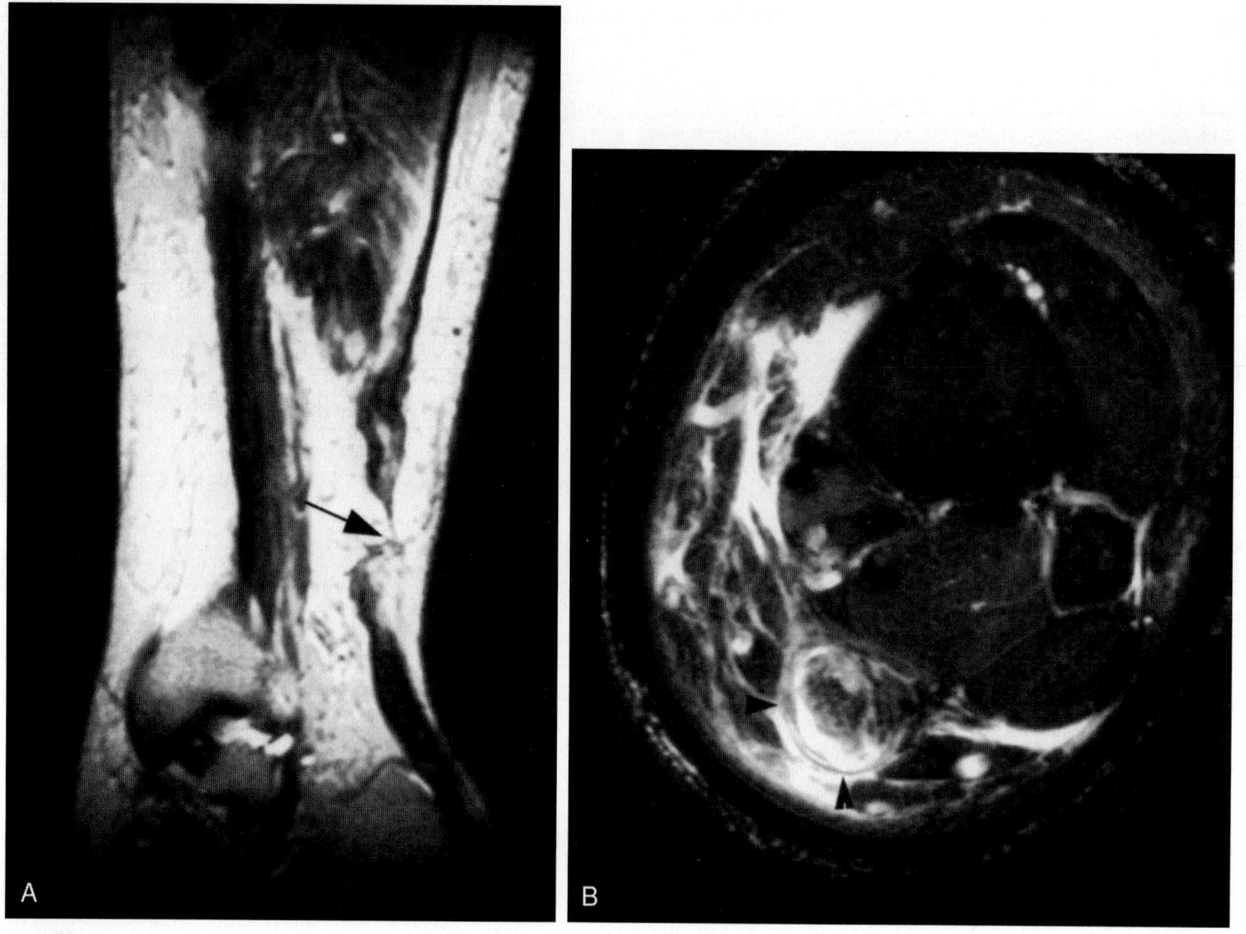

Figure 71–1. Paratenon of the Achilles tendon in a 37-year-old man with acute rupture. *A,* Sagittal T2-weighted MR image shows the disruption of the Achilles tendon *(arrow)*. *B,* Axial T2-weighted image with fat suppression shows the paratenon *(arrowheads)* because it is surrounded by fluid on both sides.

the tendons (Fig. 71–2). To reduce friction, tendon sheaths are lubricated by synovial fluid, which is secreted by synovial cells lining the tendon sheaths. Tendons are relatively hypovascular; they receive their blood supply from vessels in the perimysium, at the periosteal insertion, and from surrounding soft tissues.

PHYSIOLOGY AND BIOMECHANICS

Tendons transmit forces from muscles to bone across joints. Tendons can withstand large tensile forces; they have the highest tensile strength of any soft tissue in the body. Exercise exerts positive, long-term effects on the structural and mechanical properties of the tendon by stimulating the synthesis of collagen fibrils. Maximum load capacity of a tendon and resistance to tearing peaks in the 20s, then decreases with age.

A tendon elongates in response to stress, and if forces are within the physiologic range, the tendon returns to normal length after the forces are released. During relaxation to normal length, 90% to 95% of the absorbed energy is returned to the system as kinetic energy; the rest is dissipated within the tendon as heat. With repetitive loading, the core temperature of the tendon can rise to

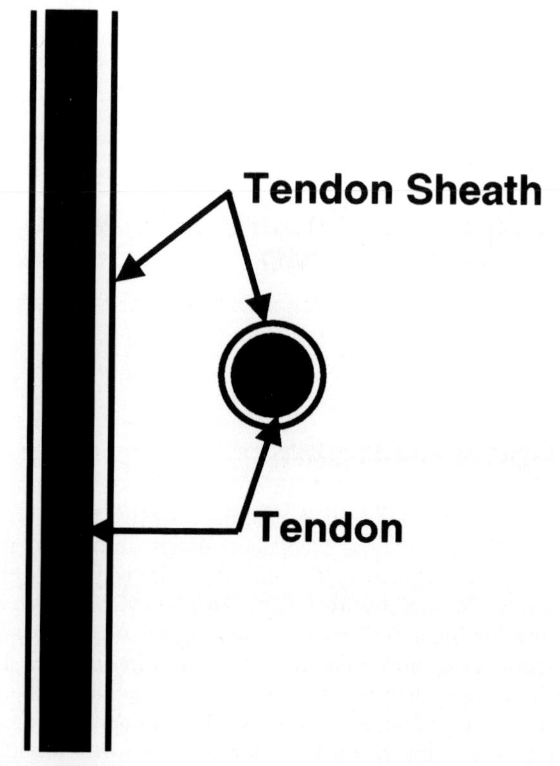

Figure 71–2. Diagram of a normal tendon with its tendon sheath.

nearly 45°C. The heat causes tenocyte injury and promotes tendon degeneration.

Not all tendons have the same elastic properties. The digital flexors and extensors of the hand and wrist are stiff and show little change in length with muscle contraction. In contrast, tendons involved with locomotion, such as Achilles' tendon, are more elastic. Tendons that transmit large loads under eccentric or elastic conditions are more subject to injury. Stretching a tendon more than 8% of its total length usually results in an acute tear.

TENDON RESPONSE TO INJURY

Novacheck (1998) described tendons as responding to stress either physiologically (by regenerative response) or pathologically (by degenerative response). Tissue produced during a regenerative response is structurally and functionally identical to normal tissue, whereas tissue produced during a degenerative response is of lower structural and functional quality. It is not known why collagen formed during the degenerative response fails to mature. In the degenerative response, the matrix cells are overwhelmed by the injury and are unable to repair the matrix. The inadequate matrix synthesis leads to tendon degeneration or tendinosis. If the abusive activity continues, gross structural abnormalities develop in the form of partial and complete tendon tears. The term *tendinosis* currently is preferred over tendinitis because the histologic changes are those of a degenerative process rather than inflammation. With tendinosis, the matrix becomes disorganized, losing its tightly arranged collagen bundles. The fibroblasts become more numerous, and the vascularity is increased. It is now believed that tendinosis is the result of increased activity, hypoxia, and aging.

Autopsy studies have revealed that tendinosis can be present in asymptomatic individuals, especially with increasing age. Tendon degeneration may represent part of the normal aging process.

CLASSIFICATION AND IMAGING CHARACTERISTICS OF TENDON ABNORMALITIES

There is significant confusion in the literature regarding the classification and naming of tendon abnormalities. This confusion stems from the fact that tendon anatomy varies from one tendon to another, and disease manifestations in the different tendons are not uniform. The classification presented here is derived from the orthopedic literature; we find this classification easy to use, and it facilitates our communication with our orthopedic colleagues.

Paratenonitis and Tenosynovitis

Paratenonitis is inflammation limited to the paratenon resulting from friction in tendons that have a straight course, such as Achilles' tendon. The condition is seen frequently in medium-distance and long-distance runners. Clinically, there is pain, swelling, tenderness, and crepitus. On magnetic resonance (MR) imaging, there is edema in the areolar tissue around the tendon (Fig. 71–3). In tendons that have a tendon sheath, tenosynovitis presents with excessive fluid within the tendon sheath (Fig. 71–4). With chronic tenosynovitis, as in rheumatoid arthritis, there also may be synovial proliferation and hypertrophy (Fig. 71–5). Small amounts of fluid often are seen within tendon sheaths around the ankle in asymptomatic individuals. This finding is particularly common with the flexor hallucis longus and should not be considered to represent tenosynovitis.

Tendinosis (Tendinopathy)

Tendinosis is intratendinous degeneration (fatty, mucoid, or hyaline) that can be painful or asymptomatic; a nodule or swelling can be palpated. On MR imaging, there is tendon thickening with or without change in signal intensity (Fig. 71–6).

Partial Tear

In a partial tear, there is underlying tendinosis plus splitting of the collagen bundles as seen in tendons around the ankle (Fig. 71–7). In the patellar tendon and supraspinatus tendon, actual disruption of some fibers can be detected at the tendo-osseous junction (Fig. 71–8). Tendon attenuation is sometimes present. Clinically the tendon can feel thick, rubbery, and tender to palpation. On MR imaging, there is thickening of the tendon with linear increased signals along the long axis of the tendon secondary to splitting of the collagen bundles (Fig. 71–8). Affected tendons that have tendon sheaths also show fluid in the tendon sheath.

Complete Tear

Healthy tendons do not disrupt. There is always underlying tendinosis, which predisposes tendons to rupture. Clinically, there is loss of function related to the muscle attached to the disrupted tendon. On MR imaging, the abnormality depends on the duration of the tear. Acutely the tendon shows discontinuity, bleeding, and edema within the surrounding soft tissues. Chronic disruption manifests with retraction and thickening of the disrupted ends of the tendon. Torn tendons that have a tendon sheath reveal an empty tendon sheath at the site of the tear (Fig. 71–9).

SPECIFIC TENDONS

Achilles' Tendon

Anatomy

Achilles' tendon is the largest tendon in the body; it is about 4 to 6 mm thick (Fig. 71–10). Injection studies have

Text continued on page 586

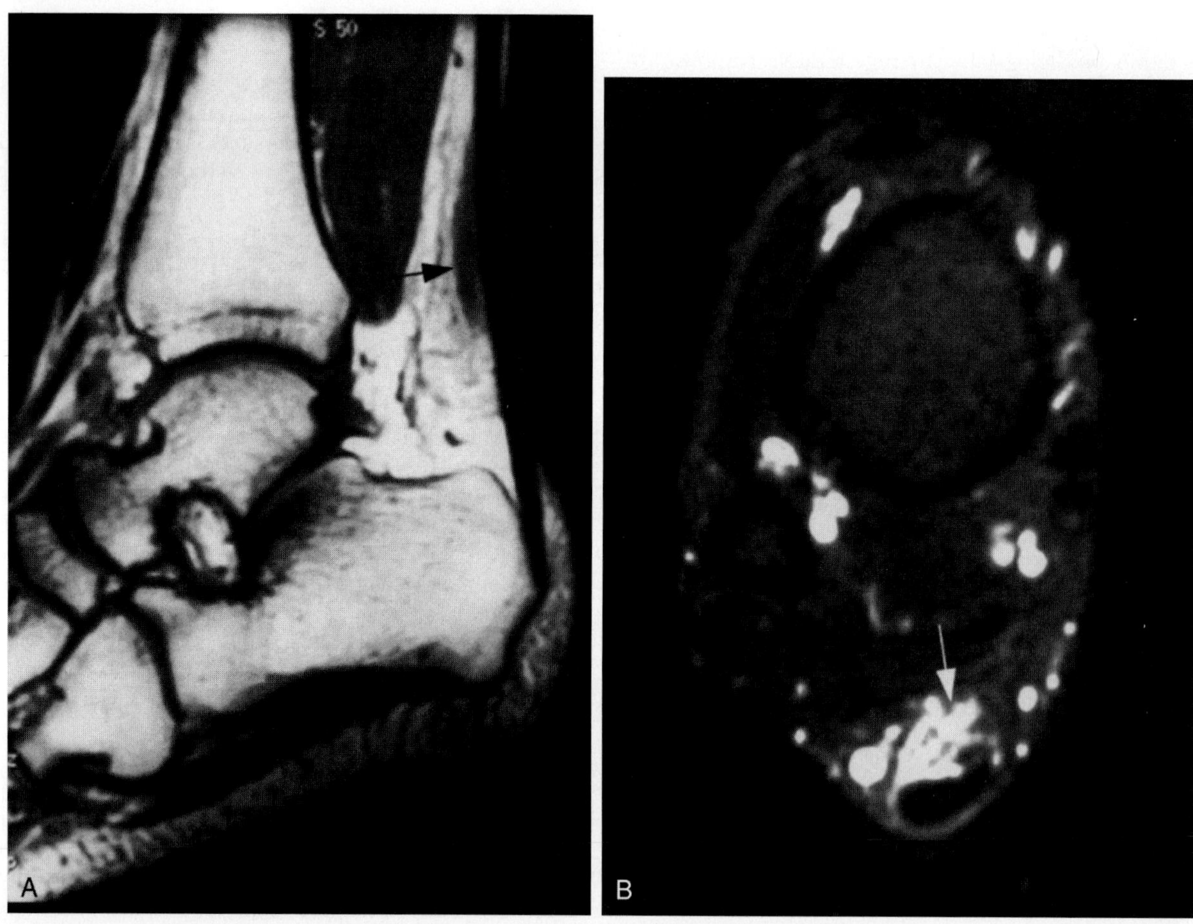

Figure 71–3. Paratenonitis in a 19-year-old college athlete. *A*, T1-weighted sagittal MR image shows an area of decreased signal intensity (edema) anterior to Achilles' tendon *(arrow)*. *B*, Axial T2-weighted MR image with fat suppression shows increased signal intensity *(arrow)* anterior to Achilles' tendon. These findings are characteristic of paratenonitis.

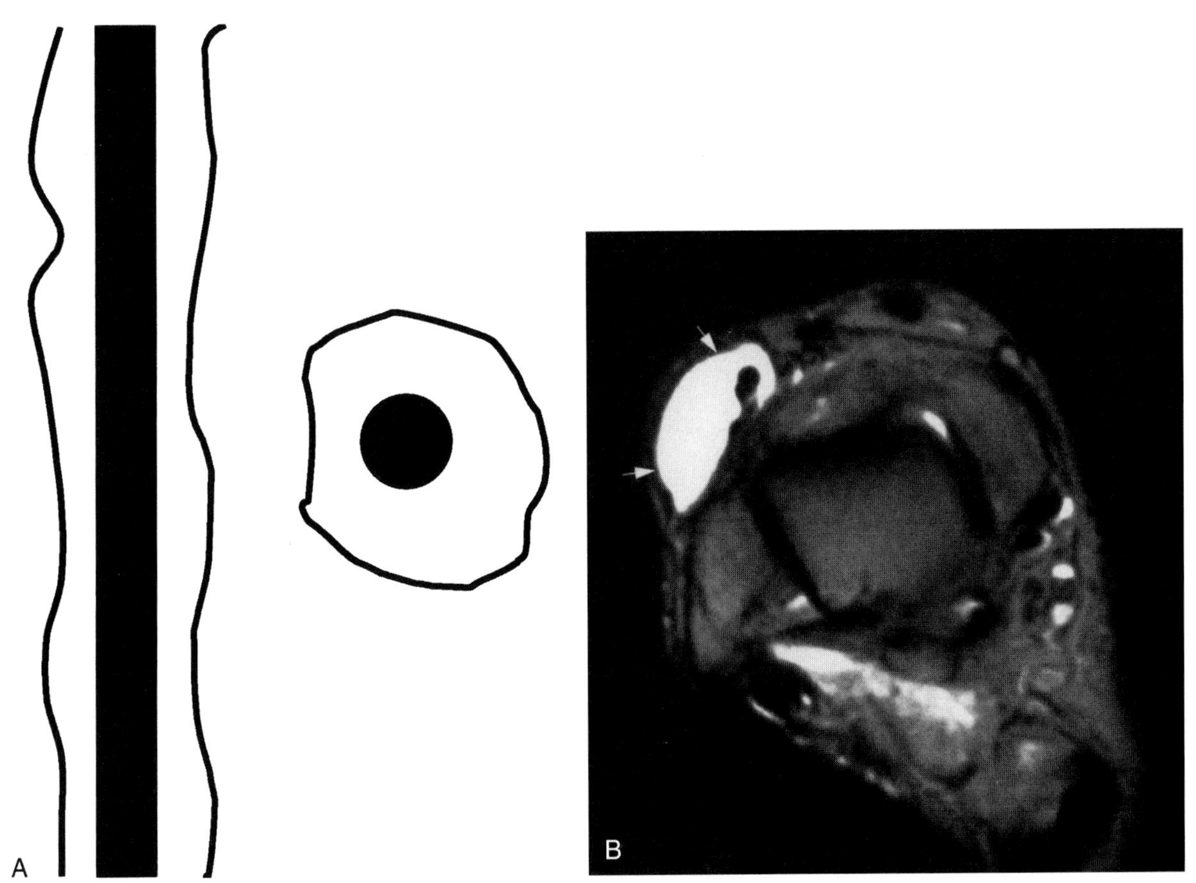

Figure 71–4. Tenosynovitis. *A,* Diagram depicting tenosynovitis. The tendon is normal, but the tendon sheath is distended with fluid. *B,* Acute tenosynovitis of the extensor digitorum longus in a young athlete wearing poorly fitting shoes. T2-weighted oblique axial MR image shows marked distention of the tendon sheath with fluid *(arrows).*

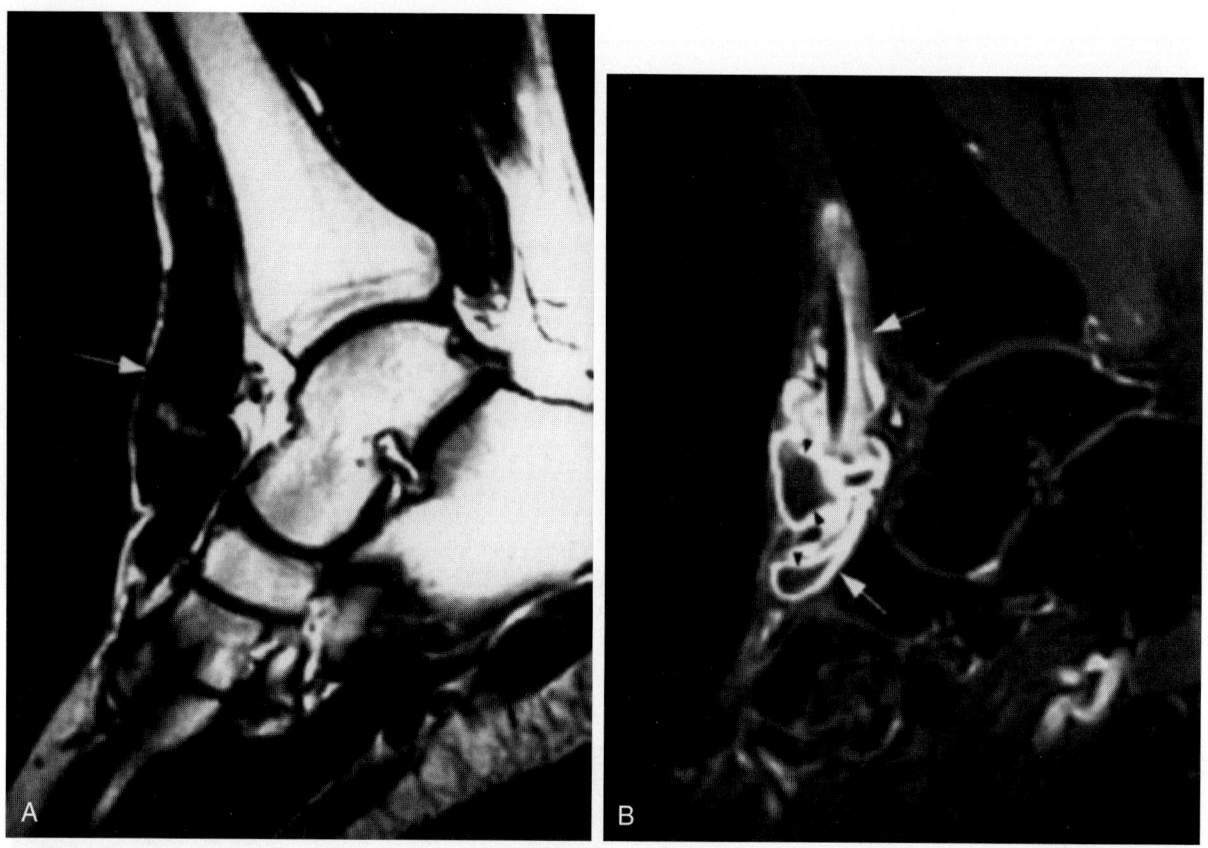

Figure 71–5. Chronic tenosynovitis (pathologically proven) of the tibialis anterior tendon sheath. *A*, Sagittal T1-weighted MR image shows a large fusiform mass *(arrow)* anterior to the ankle joint. The mass has low signal intensity. *B*, Sagittal T1-weighted MR image with fat suppression taken after intravenous injection of gadolinium shows marked thickening and enhancement of the synovial lining of the tendon sheath *(arrows)*. Some fluid also is noted within the tendon sheath *(arrowheads)*.

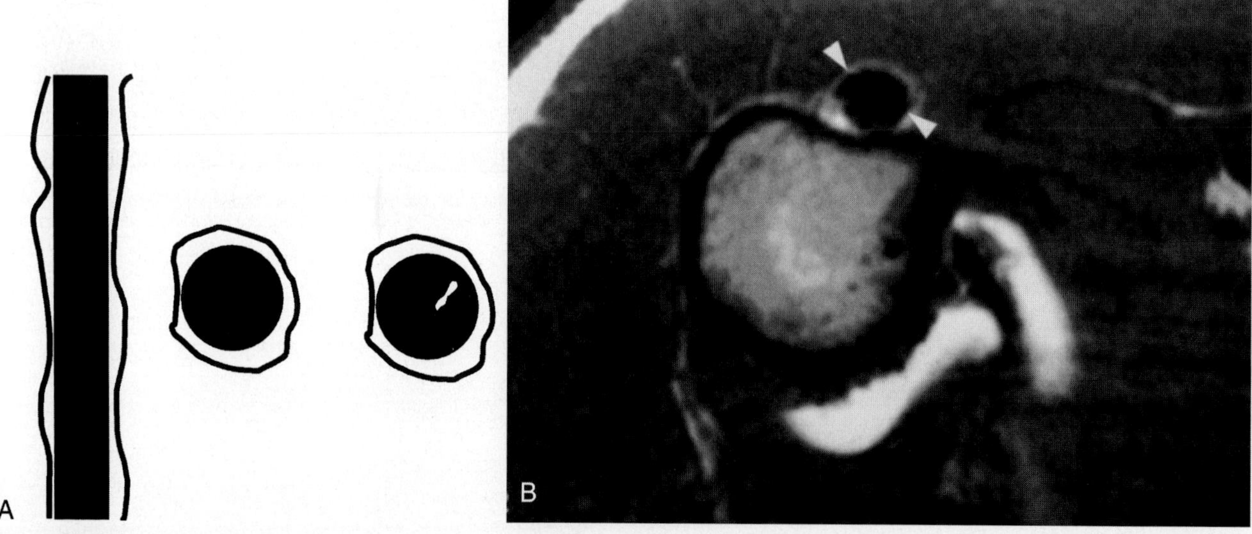

Figure 71–6. Tendinosis (tendinopathy). *A*, Diagram depicting the thickening of the tendon when tendinosis is present. In more advanced cases, there also is some increased signal intensity within the tendon. *B*, Tendinosis affecting the bicipital tendon (long head of the biceps tendon). Note the marked thickening of the tendon *(arrowheads)* and the small areas of increased signal intensity within the tendon.

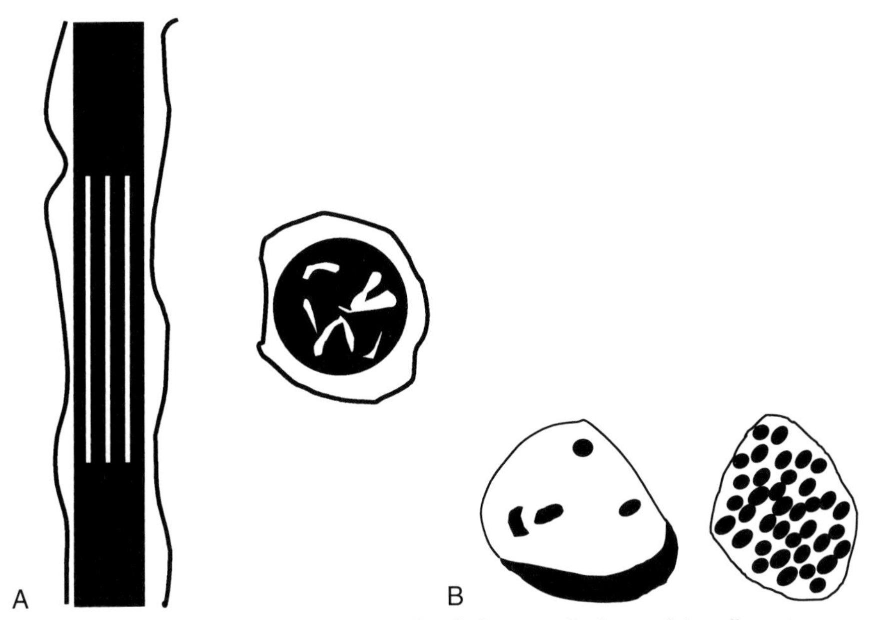

Figure 71-7. Partial tear of the tendon. *A*, Diagram depicting the tendon thickening and splitting of the collagen into separate bundles. *B*, Diagram depicting severe splitting of the tendon and adhesions between some bundles with the tendon sheath as viewed on axial images.

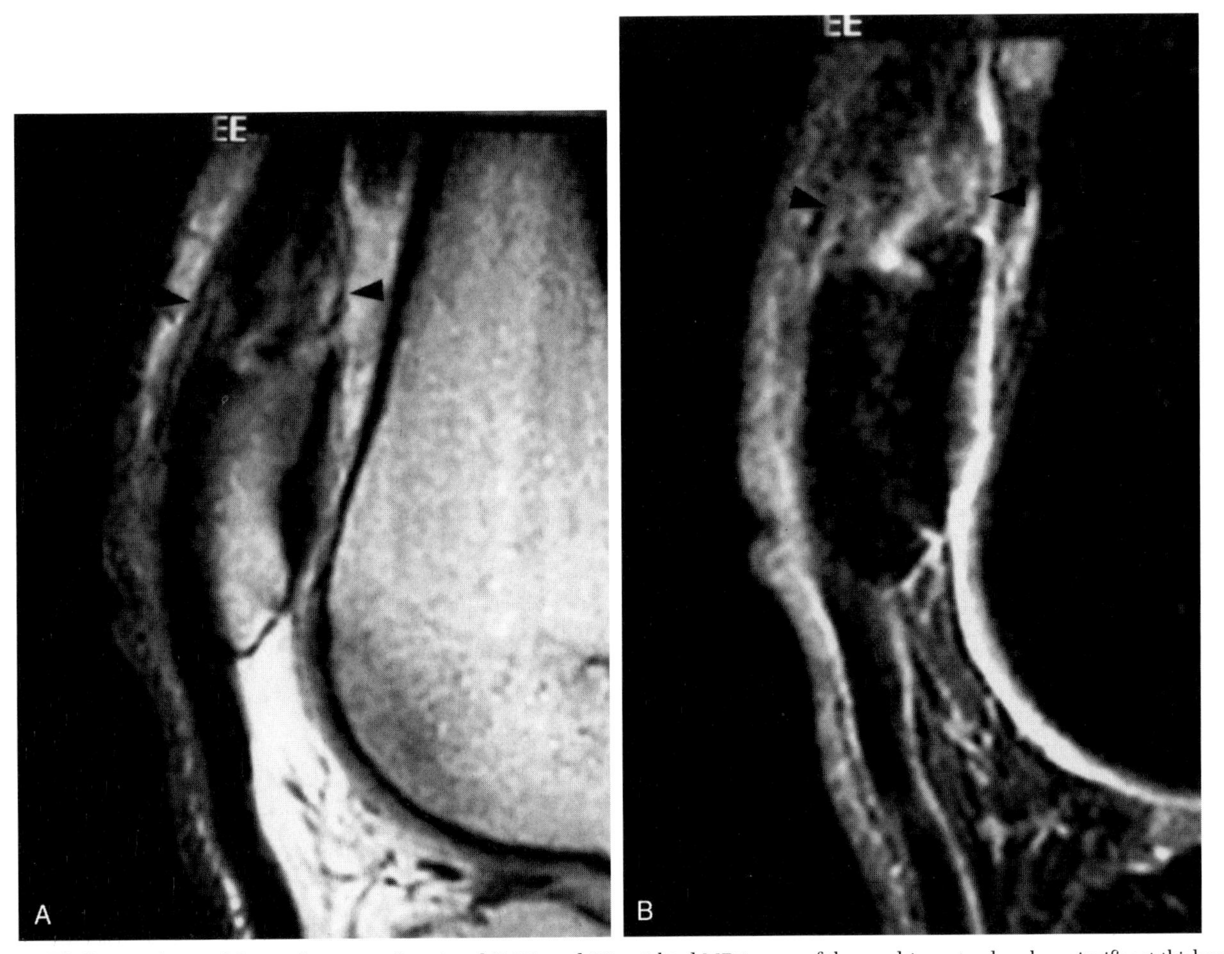

Figure 71-8. Partial tear of the quadriceps tendon. *A* and *B*, T1- and T2-weighted MR images of the quadriceps tendon show significant thickening and splitting of the tendon *(arrowheads)*. The tendon also shows increased signal intensity between its split bundles.

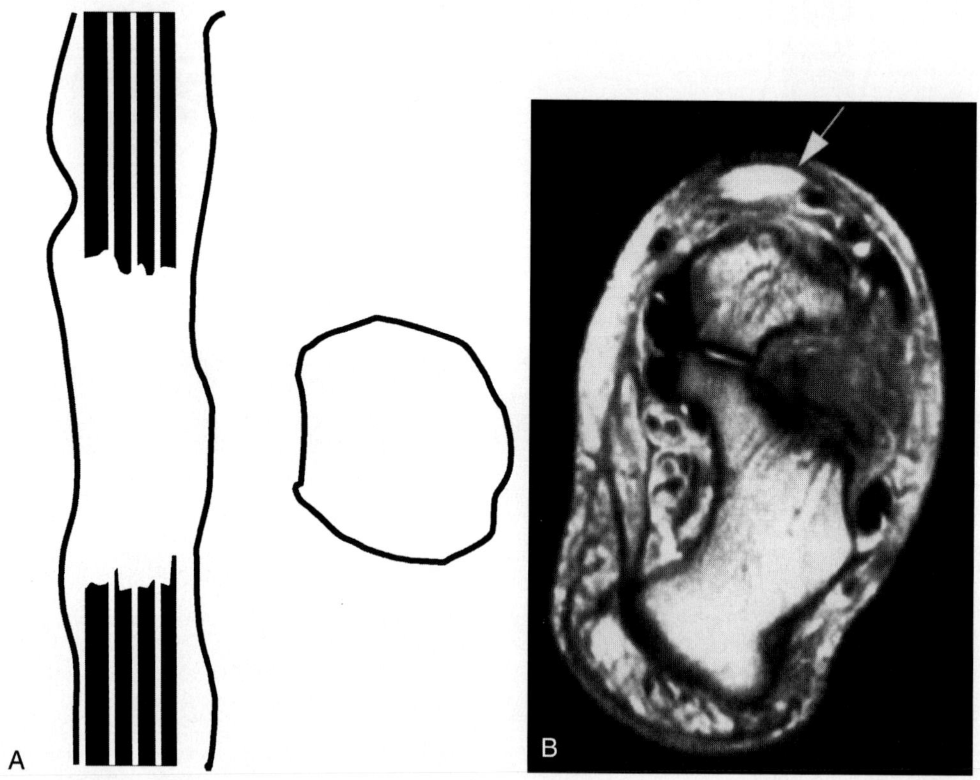

Figure 71–9. Complete tendon tear. *A,* Diagram depicting a completely torn tendon shows thickening and splitting of the stumps. The tendon sheath, on axial views, appears empty and distended with fluid. *B,* T2-weighted oblique axial MR image shows an empty, fluid-distended, tibialis anterior tendon sheath *(arrow).*

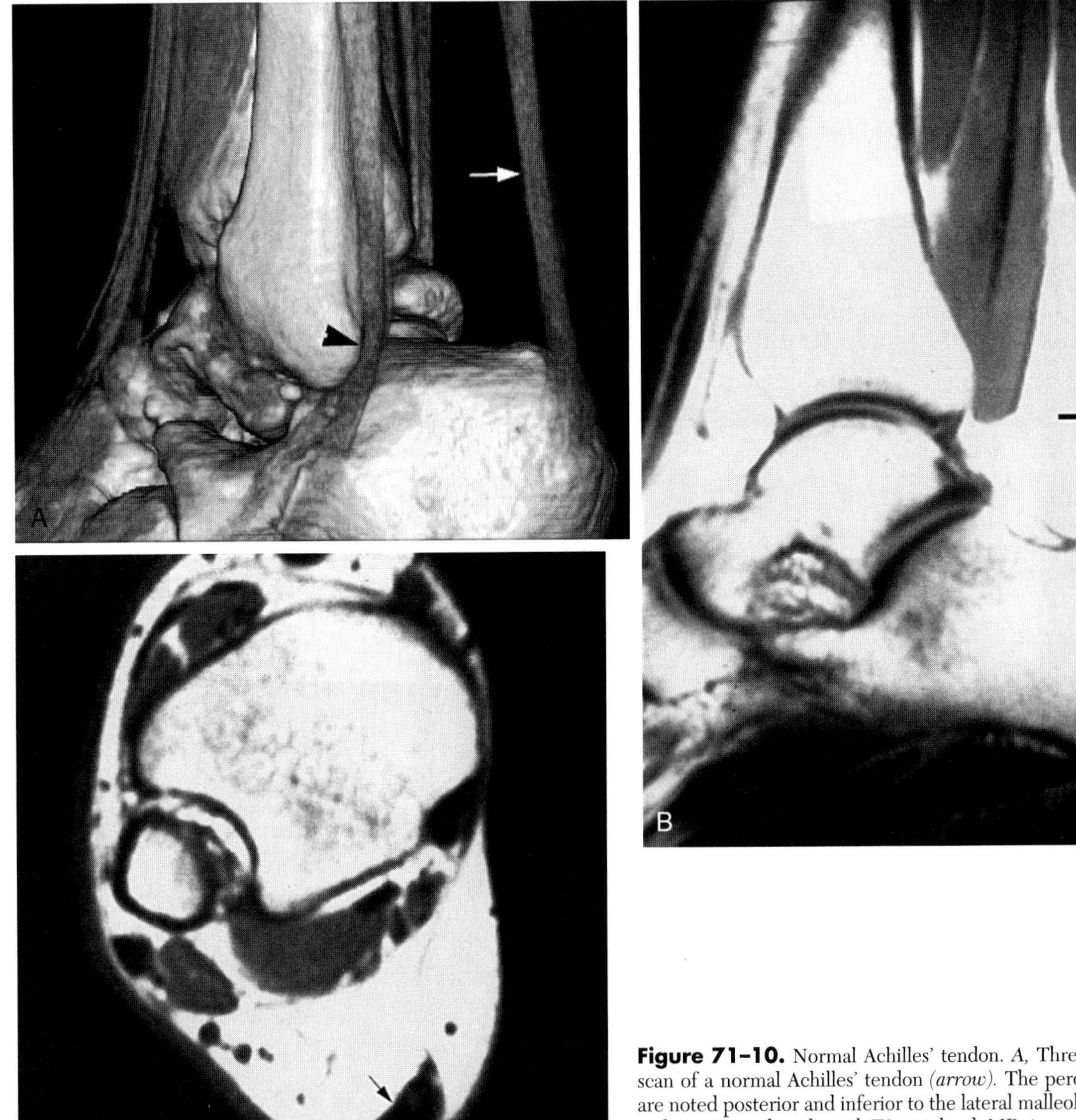

Figure 71–10. Normal Achilles' tendon. *A,* Three-dimensional CT scan of a normal Achilles' tendon *(arrow).* The peroneal tendons also are noted posterior and inferior to the lateral malleolus *(arrowhead). B* and *C,* Sagittal and axial T1-weighted MR images show a normal Achilles' tendon *(arrow).* Note the concave anterior surface *(black arrow)* on the axial image *(C).*

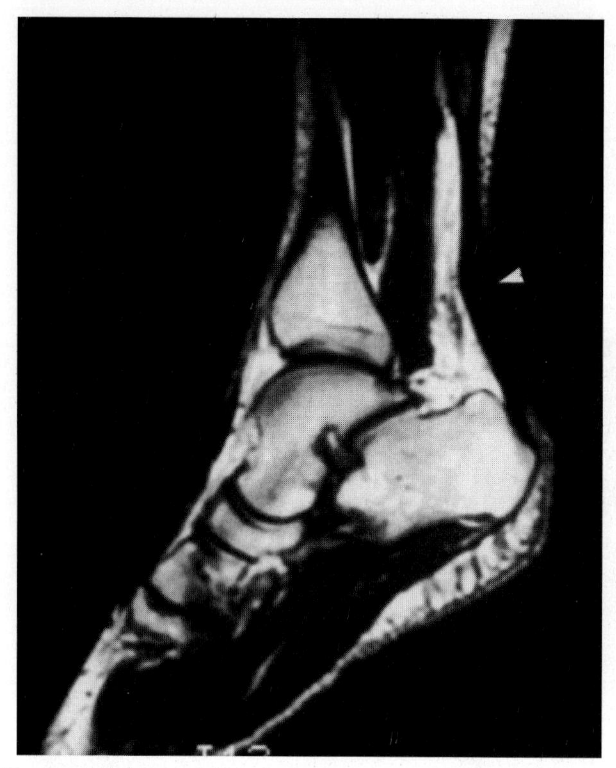

Figure 71–11. Achilles' tendon tendinosis (tendinopathy). T1-weighted sagittal MR image shows fusiform thickening of the tendon a few centimeters proximal to its insertion *(arrowhead)*. There are no signal abnormalities noted within the tendon.

shown diminished blood supply 2 to 6 cm above its insertion, where tears typically occur. On axial sections, Achilles' tendon is curved gently with concavity facing anteriorly (Fig. 71–10*C*).

Paratenonitis

Paratenonitis of Achilles' tendon is particularly common in runners; patients typically complain of pain and crepitus. On MR imaging, there is edema around the tendon but no change in size and no signal abnormalities within the tendon (see Fig. 71–3).

Tendinosis

Achilles' tendon tendinosis is the result of cyclical loading or overuse. Some medications are known to be toxic to the tenocytes, and long-term use can result in tendinosis and tendon rupture. The Food and Drug Administration warns health providers of the potential for Achilles' tendon tendinosis and rupture in patients using fluoroquinolones, such as ciprofloxacin (Cipro). On MR imaging, there is fusiform thickening of the tendon centered 2 to 6 cm above the tendon insertion without significant signal abnormalities (Fig. 71–11). A variant of this abnormality is known as *insertional tendinopathy* (or tendinosis), in which tendon thickening occurs at the tendon insertion, and the lesion often is associated with a calcaneal spur (Fig. 71–12).

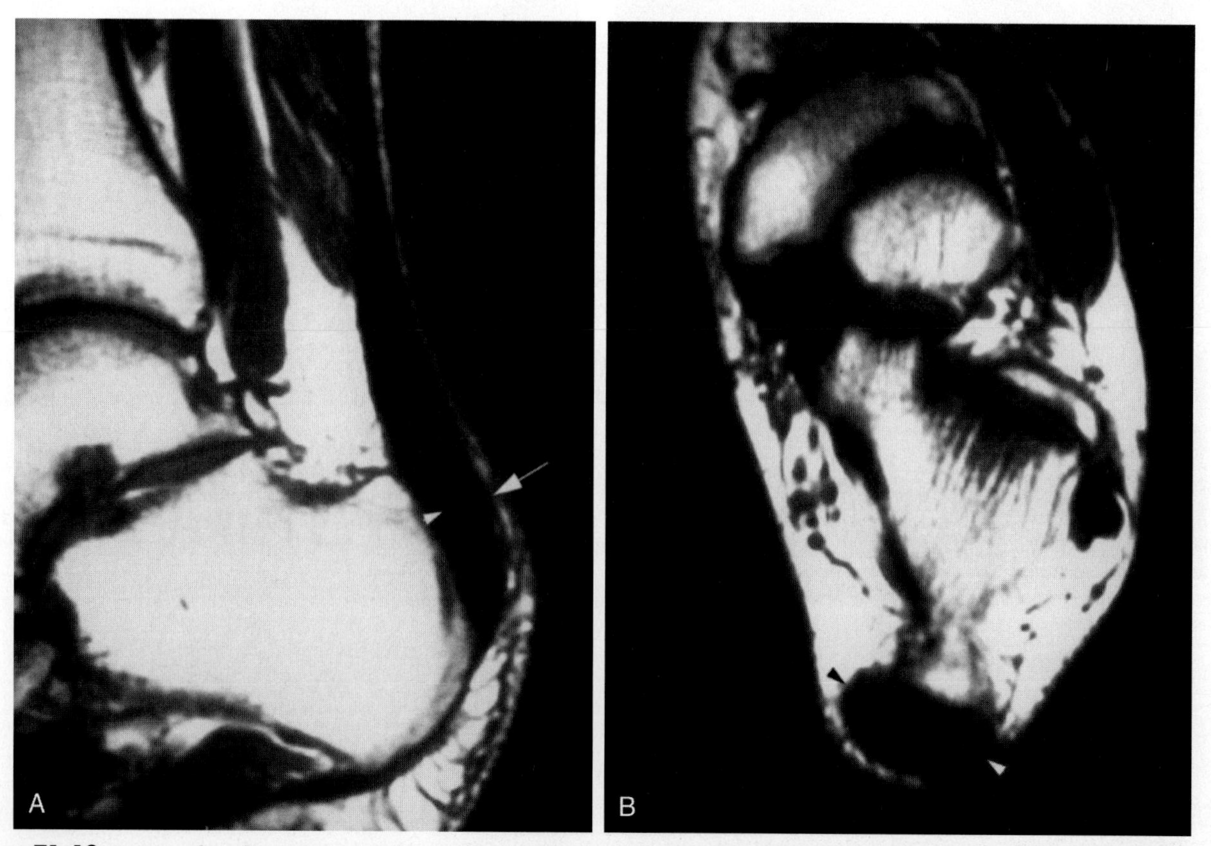

Figure 71–12. Insertional tendinopathy of Achilles' tendon. *A*, Sagittal T2-weighted MR image shows thickening of Achilles' tendon at its insertion *(arrow)*. Note the linear increased signal intensity within the tendon at its insertion *(arrowhead)*. *B*, Axial T1-weighted MR image taken at the Achilles' tendon insertion reveals significant thickening of the tendon *(arrowheads)*.

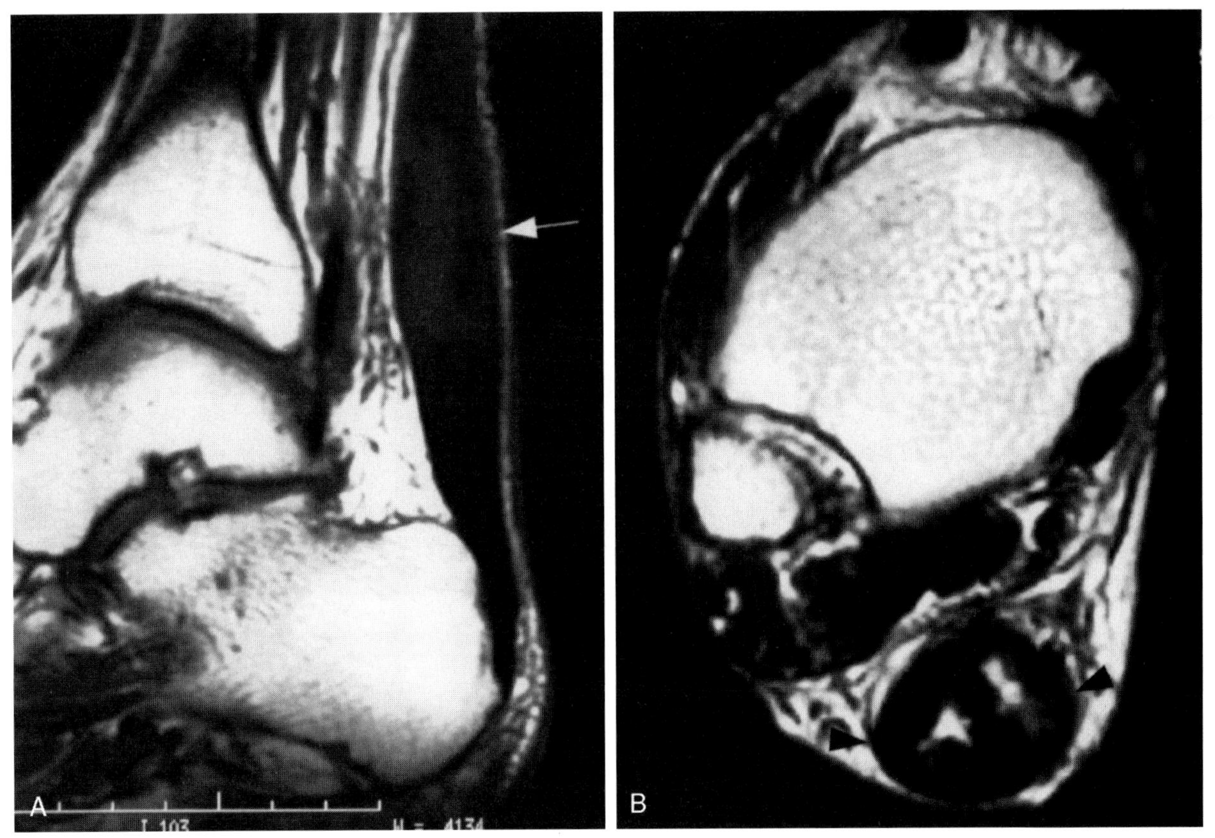

Figure 71–13. Partial tear of Achilles' tendon. *A,* Sagittal T1-weighted MR image of Achilles' tendon shows marked thickening of the tendon *(arrow)* and mild increase in signal intensity. *B,* Axial T2-weighted MR image shows increased signal intensity within the thickened tendon *(arrowheads)*.

Partial Tear

Partial tear is a more severe manifestation of the spectrum of tendon degeneration. There is always underlying tendinosis with splitting of the collagen bundles. On MR imaging, the tendon is thickened, and it shows increased signal intensity within its substance, especially on T2-weighted images and to a lesser extent on T1-weighted images (Fig. 71–13).

Complete Tear

Complete tear is a disabling condition that can be acute or chronic. Acute tears typically occur in sedentary middle-aged men engaged in episodic overactivity. They result from forceful dorsiflexion of the foot. The rate of misdiagnosis by clinical examination is 20% to 30%. On MR imaging, an acute tear shows discontinuity in the tendon, edema, and hemorrhage (Fig. 71–14). With chronic tears, there is discontinuity of the tendon but no associated edema or hemorrhage. History of significant trauma may be lacking.

Posterior Tibial Tendon

Anatomy and Function

The posterior tibial tendon (PTT) passes behind the medial malleolus and inserts primarily on the navicular tuberosity (Fig. 71–15). The PTT is the main inverter of the subtalar joint; it also elevates the medial longitudinal arch of the foot. When the hindfoot is in valgus or the forefoot is in varus, excessive forces are placed on the PTT.

Dysfunction

PTT dysfunction is a constellation of signs and symptoms consisting of ankle pain, instability, and foot deformity. It is associated with either partial or complete PTT tear; as a result, the longitudinal arch of the foot collapses, the subtalar joint everts, the heel assumes a valgus position, and the foot abducts producing what is known as *acquired flatfoot.* Stretching or disruption of the spring, deltoid, and talocalcaneal interosseous ligaments also is implicated in causing acquired flatfoot. PTT dysfunction usually is encountered in middle-aged obese women, and it often is associated with systemic disease, such as obesity, diabetes mellitus, seronegative spondyloarthropathies, and steroid intake.

Posterior Tibial Tendon Tears

PTT tears typically start behind the medial malleolus as a result of the shear forces generated as the tendon sharply changes direction. On MR imaging, tendinosis and partial tears show increased tendon girth, splitting, and increased signal intensity (Fig. 71–16). With complete tear, the tendon sheath fills with fluid, but no tendon is identified

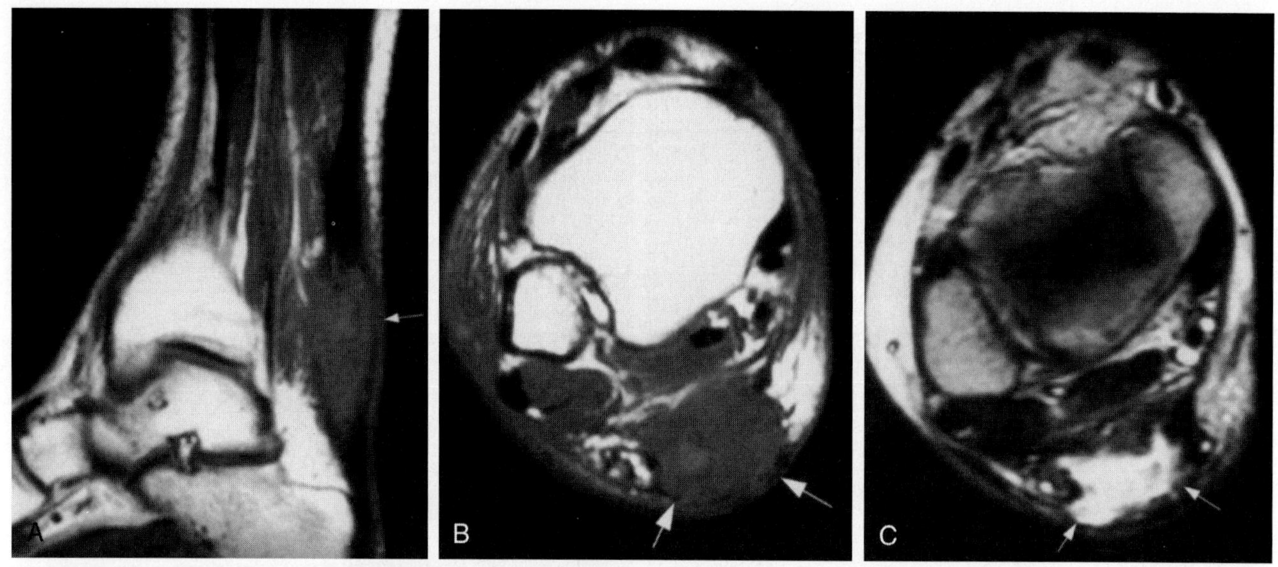

Figure 71–14. Acute Achilles' tendon disruption in a 50-year-old football coach weighing 98 kg. *A*, Sagittal T1-weighted image shows complete disruption of Achilles' tendon *(arrow)*. The increased signal is due to blood at the site of the rupture. *B* and *C*, Axial T1- and T2-weighted MR images show the acutely disrupted tendon *(arrows)*.

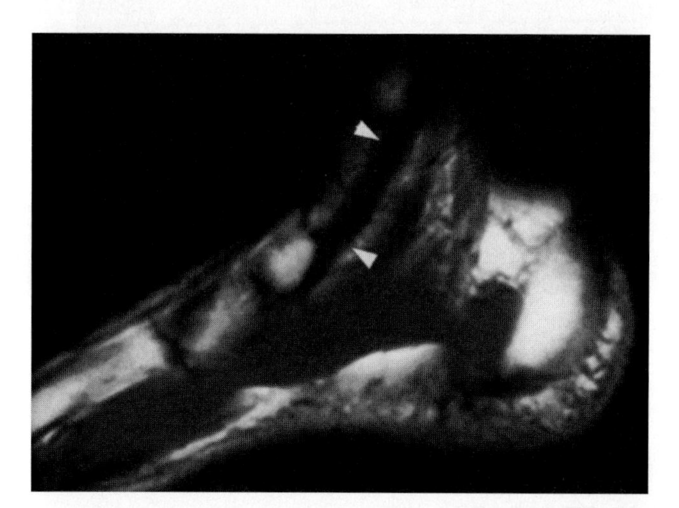

Figure 71–15. Normal posterior tibial tendon *(arrowheads)* on T1-weighted sagittal MR image. The tendon is seen passing posterior to the medial malleolus and inserting on the tuberosity of the navicular.

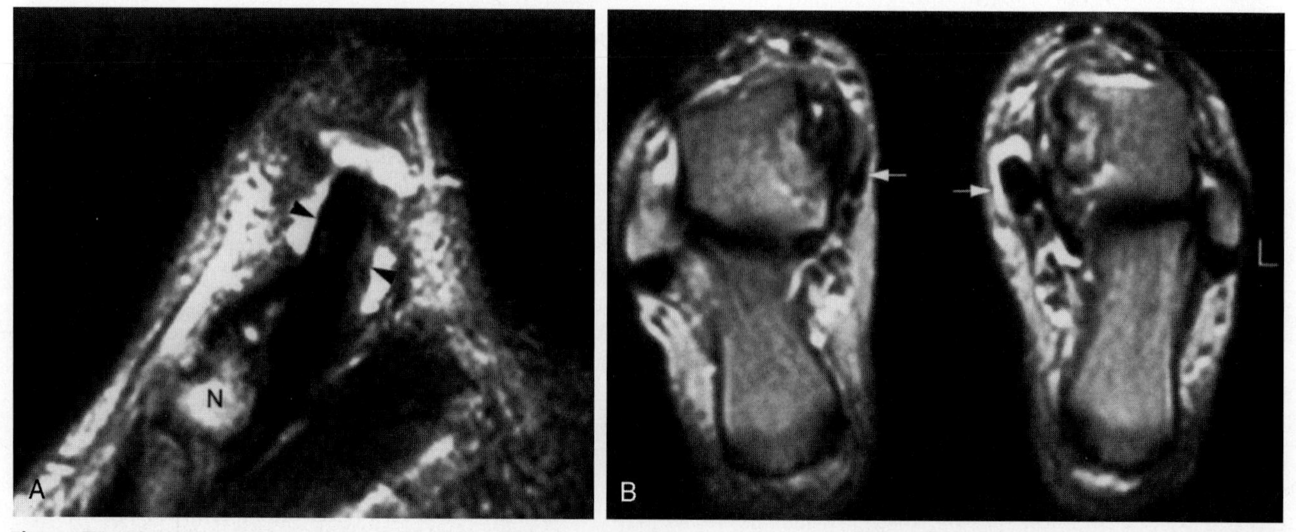

Figure 71–16. Severe posterior tibial tendon (PTT) tendinosis (tendinopathy). *A*, Sagittal T2-weighted MR image shows thickening of the left PTT *(arrowheads)* and fluid in the tendon sheath. N, navicular tuberosity. *B*, Oblique axial T2-weighted MR image of both ankles with arrows pointing to the PTT on both sides. The normal tendon on the right side is about one third the size of the abnormal left tendon *(arrows)*. The PTT on the left is surrounded by fluid.

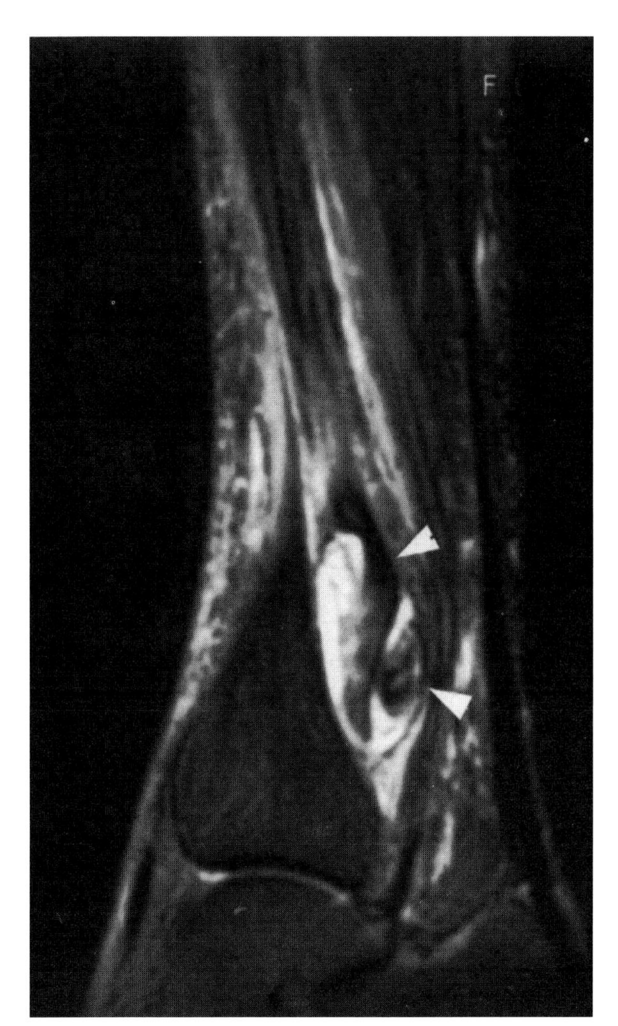

Figure 71–17. Complete tear of the posterior tibial tendon *(arrowheads)* seen on sagittal T2-weighted MR image. The disrupted tendon is curled *(arrowheads)* above the medial malleolus.

within the tendon sheath (Fig. 71–17). Secondary signs of PTT tears include osteophytes at the posteromedial aspect of the medial malleolus (Fig. 71–18), hypertrophy of the navicular tubercle, and presence of os tibiale externum.

Peroneal Tendons (Peroneus Brevis and Longus)

Anatomy and Function

The peroneal tendons course in the fibular groove, with the peroneus brevis tendon being sandwiched between the lateral malleolus and the peroneus longus tendon (Fig. 71–19; see also Fig. 71–10). Mild flattening or crescentic appearance of the peroneus brevis tendon at the distal fibula is considered normal. The peroneal tendons have a common synovial sheath, which separates at the level of the midcalcaneus. The peroneal tendons function as lateral stabilizers of the ankle; they also pronate and abduct the foot.

Tendinosis and Tears of the Peroneal Tendons

Complete tears are uncommon; typically, there is tendinosis (Fig. 71–20), and partial tears with longitudinal splits are seen (Fig. 71–21). Occasionally, tendon attenuation is present. Peroneus brevis is involved more commonly with tears, but often both tendons show abnormalities at the same time. The MR imaging changes of tendinosis and partial tears of the peroneal tendons are similar to those seen in the PTT (see Fig. 71–21).

Acute peroneus longus tendon disruption at the level of the os peroneus can occur during vigorous athletic activity, but this injury is extremely rare. This injury is diagnosed when a fracture of the os peroneus is detected radiographically with distraction of the fracture fragments (Fig. 71–22).

Tendon Subluxation

Subluxation of the peroneal tendons is seen occasionally in skiers. The condition is believed to be due to a flat or convex fibular groove. The diagnosis usually is made clinically but can be confirmed by ultrasound, computed tomography (CT), or MR imaging. Ultrasound is the

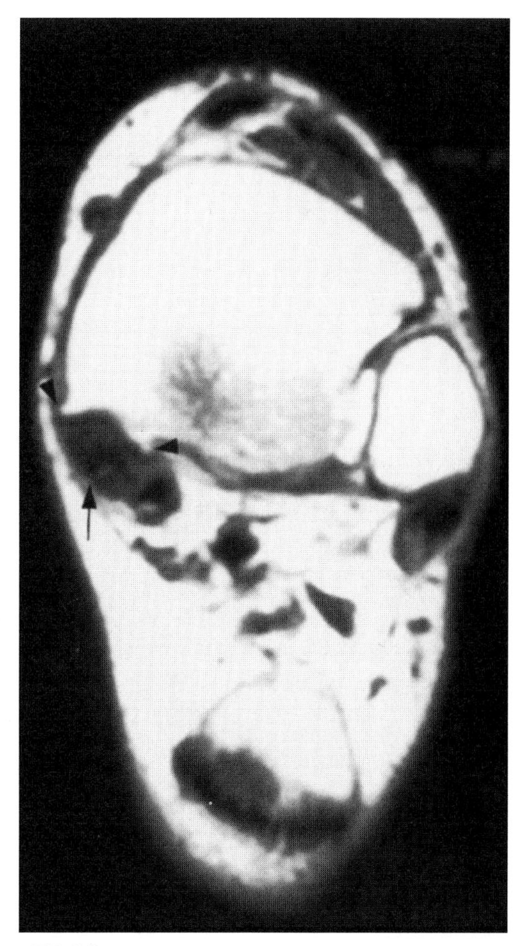

Figure 71–18. Oblique axial T1-weighted MR image shows osteophytes *(arrowheads)* arising from the medial malleolus. Note the severe tendinosis of the posterior tibial tendon *(arrow)*.

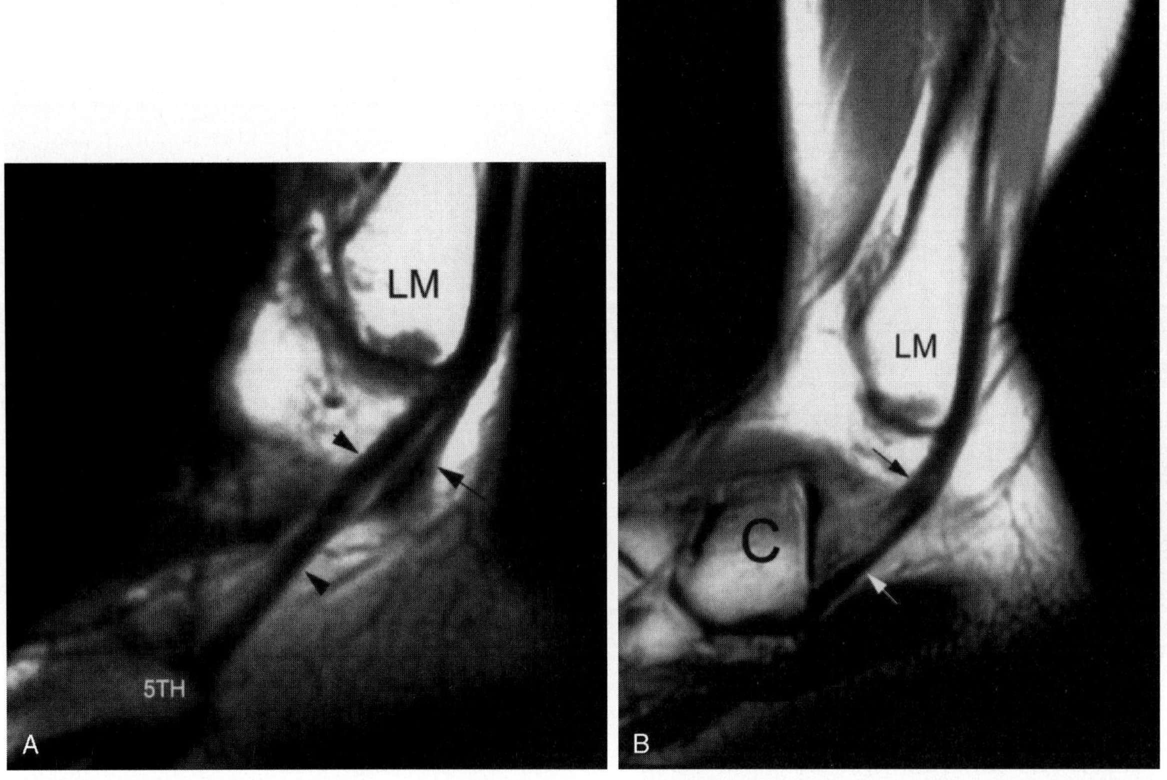

Figure 71–19. Normal peroneal tendon shown on sagittal T1-weighted MR images of the ankle. *A,* The peroneus brevis tendon *(arrowheads)* inserting at the tuberosity of the fifth metatarsal bone. The peroneus longus *(arrow)* courses behind the peroneus brevis. *B,* The peroneus longus *(white arrow)* passes in the peroneal groove of the cuboid (C) to insert on the medial aspect of the midfoot. The black arrow points to the peroneus brevis. LM, lateral malleolus.

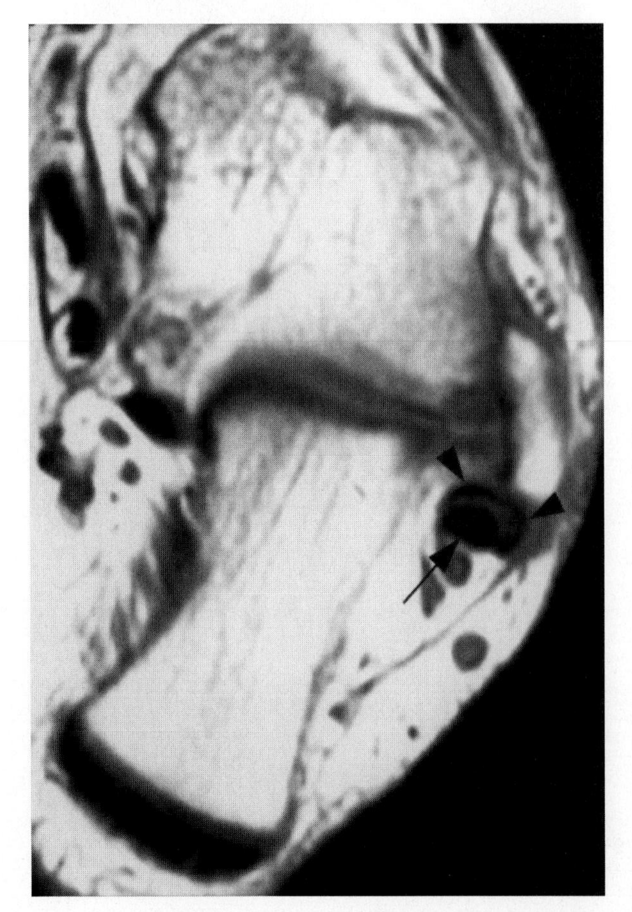

Figure 71–20. Tendinosis of the peroneal tendons shown on a T1-weighted oblique axial MR image. The peroneus brevis *(arrowheads)* is flattened and curved and has a boomerang shape. The peroneus longus *(arrow)* is thickened.

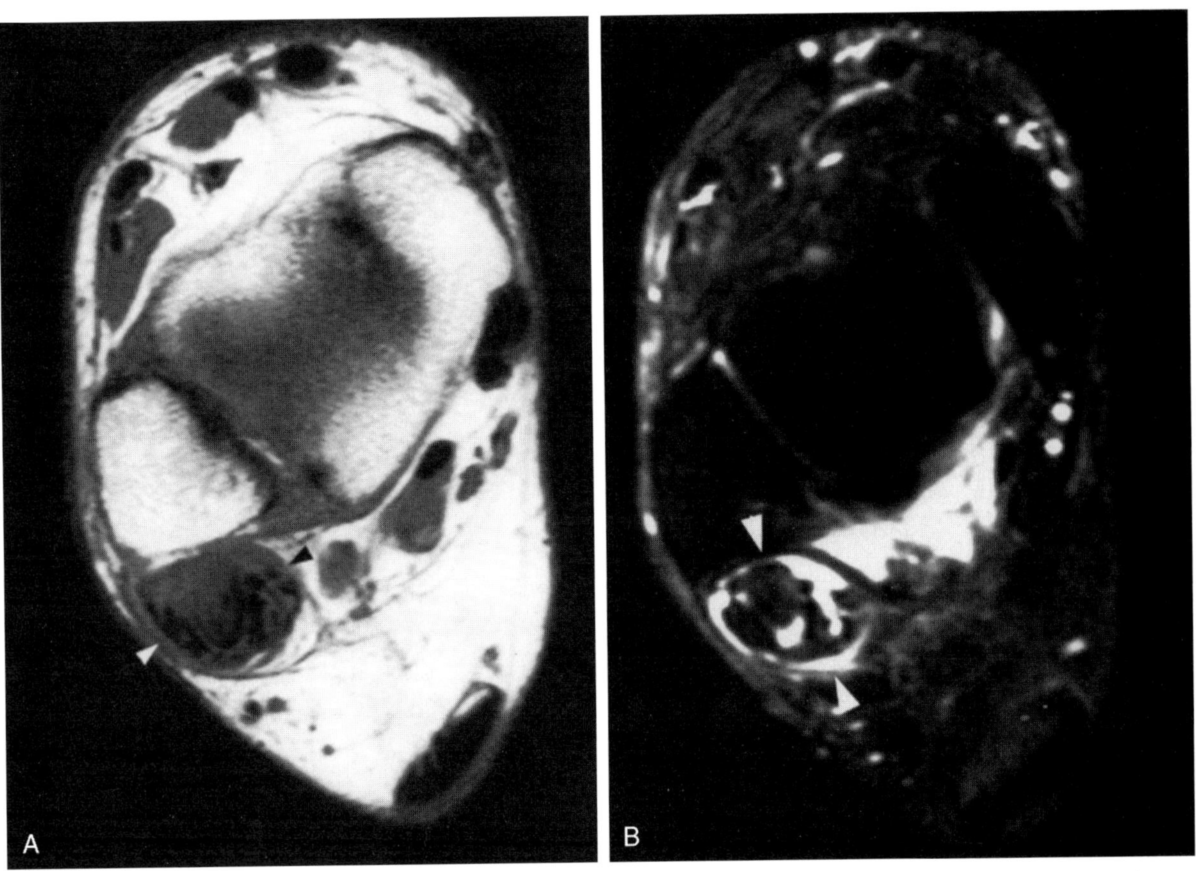

Figure 71–21. Partial tears of both peroneal tendons. *A* and *B*, Oblique axial T1- and T2-weighted MR images of the right ankle show marked thickening and splitting of both peroneal tendons. The tendon sheath is distended with fluid *(arrowheads)*.

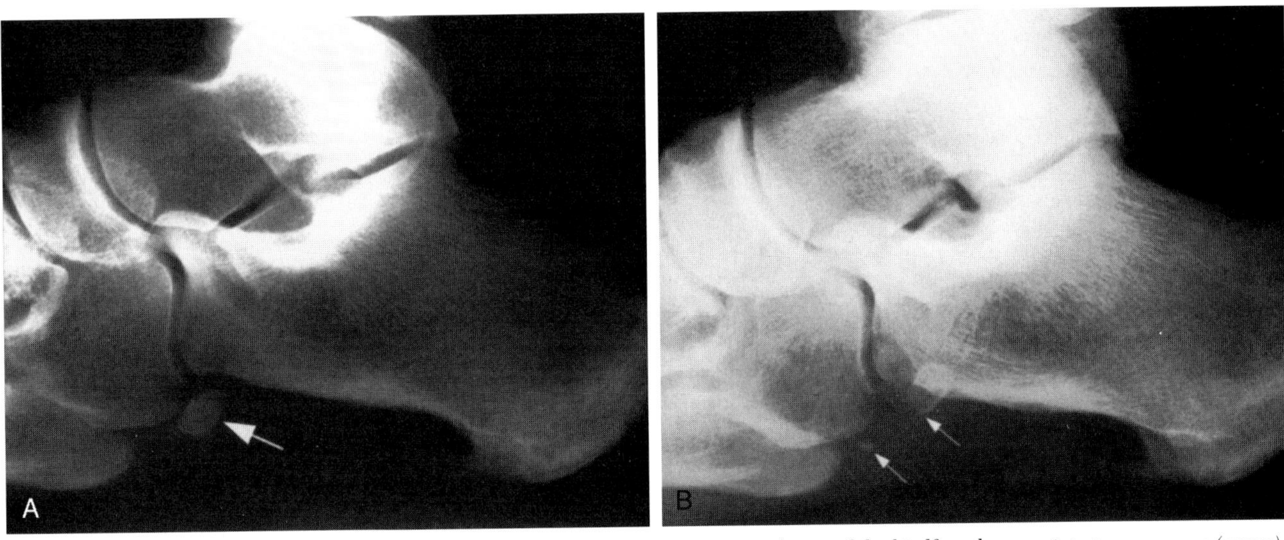

Figure 71–22. Acute peroneus longus tendon disruption at the os peroneus. *A*, Lateral view of the hindfoot shows an intact os peroneus *(arrow)*. *B*, Repeat view after the injury reveals a fracture *(arrows)* through the os peroneus with retraction of the proximal fragment.

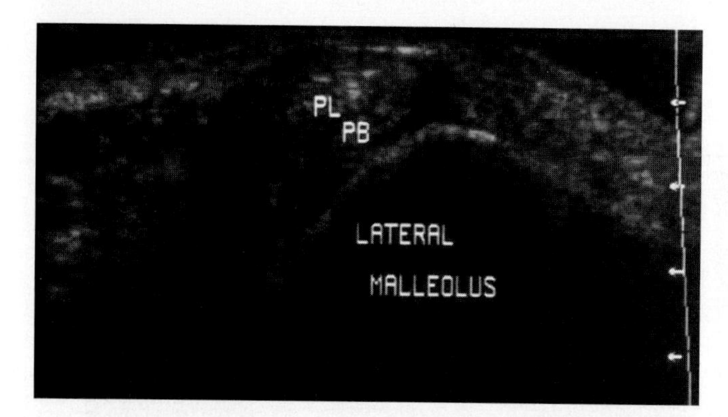

Figure 71–23. Subluxation of the peroneal tendons. Transverse ultrasound image of the lateral malleolar region shows the peroneal tendons laterally subluxed from their normal posterior location. The echogenic peroneus longus (PL) and peroneus brevis (PB) tendons are surrounded by anechoic fluid in the tendon sheath. (Case provided by Dr. Joseph G. Craig, Detroit, MI.)

cheapest and most available modality. The tendons slip out of their normal position behind the lateral malleolus to a more anterior position, alongside the lateral malleolus (Fig. 71–23).

Stenosing Tenosynovitis

Stenosing tenosynovitis is a rare condition that has been described to occur in the peroneal tendons. The tendon sheath becomes thickened, and adhesions develop between the visceral and parietal layers of the tendon sheath. These abnormalities are shown best by tenography; the presence of adhesions and sacculations are key findings (Fig. 71–24).

Anterior Tibial Tendon

Anatomy and Function

The anterior tibial tendon inserts on the plantar and medial aspects of the first metatarsal and medial cuneiform. It functions in producing dorsiflexion and inversion of the foot. Eccentric contraction of the tibialis anterior muscle controls descent of the foot to the floor at heel strike (i.e., prevents foot slap or footdrop). Concentric contraction of the tibialis anterior muscle allows the foot to clear the floor in the swing phase.

Anterior Tibial Tendon Tear

Closed tendon tears are rare, and most anterior tibial tendon ruptures are caused by lacerations. Closed ruptures usually occur at the inferior edge of the extensor retinaculum. Clinically, patients present with footdrop, which is confused with a neurologic problem. On MR imaging, there is discontinuity of the tendon with retraction of the proximal segment (Fig. 71–25).

Recommended Readings

Khoury NJ, El-Khoury GY, Saltzman CL, et al: Rupture of the anterior tibial tendon: Diagnosis by MR imaging. AJR Am J Roentgenol 167:351, 1996.

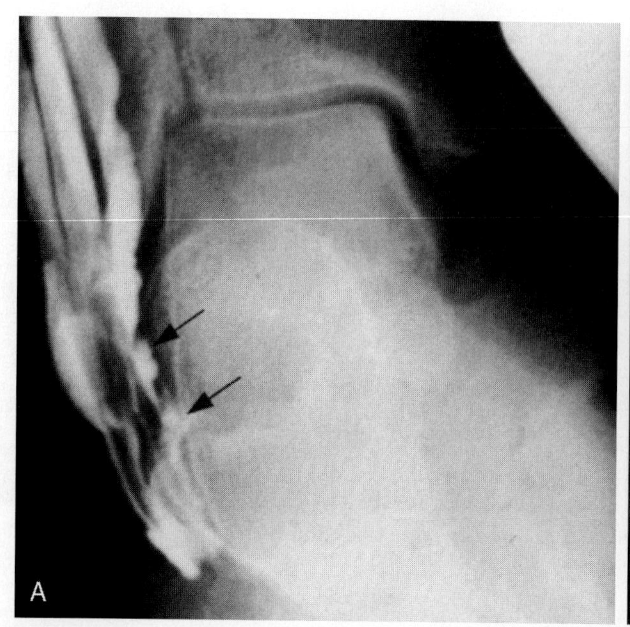

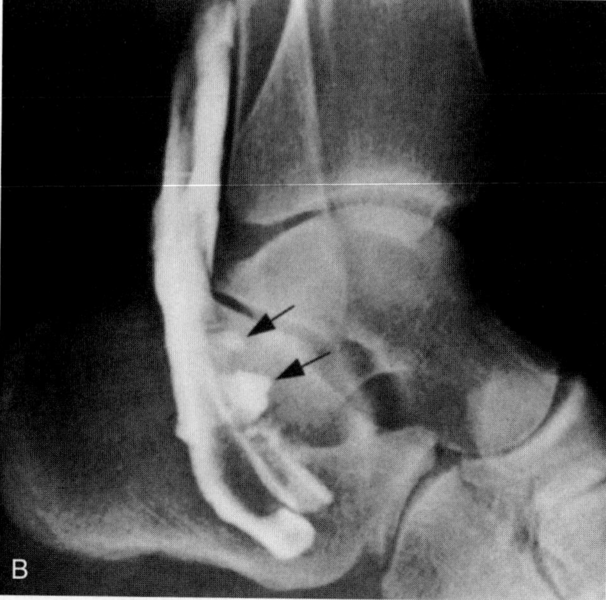

Figure 71–24. Stenosing tenosynovitis. *A* and *B*, Anteroposterior and lateral views show adhesions and sacculations *(arrows)* between the parietal and visceral synovium of the tendon sheaths.

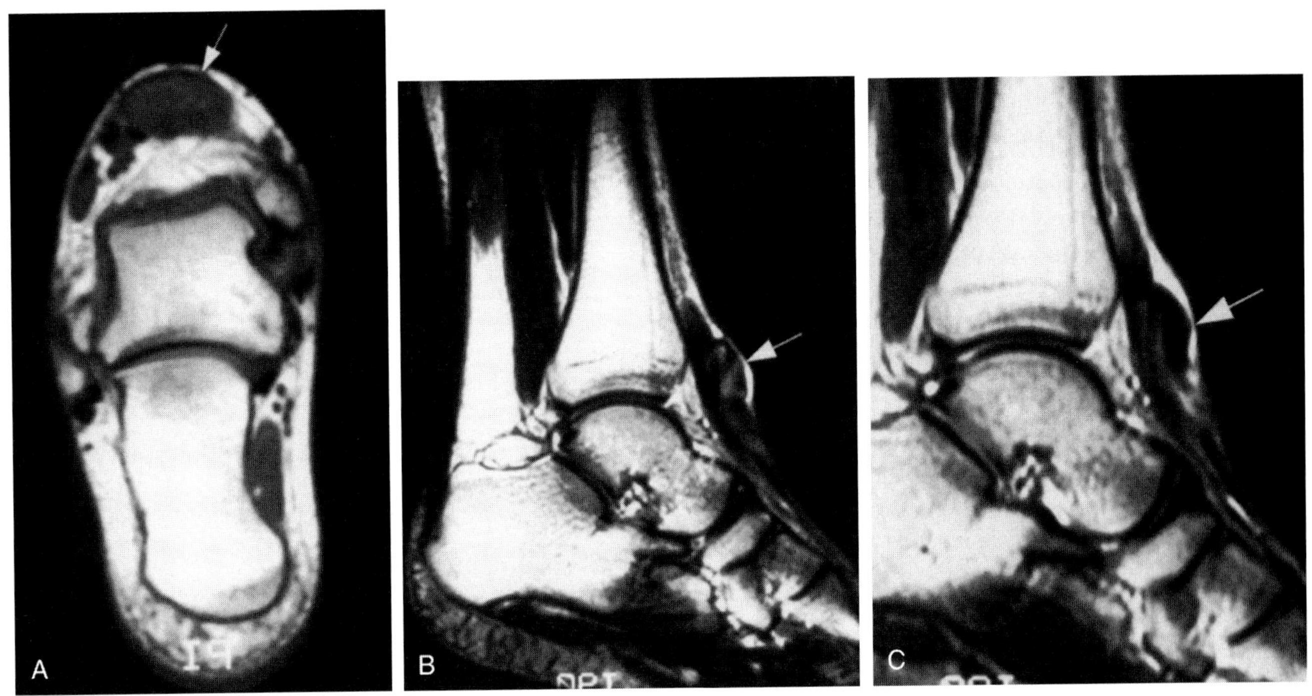

Figure 71–25. Anterior tibial tendon tear. *A*, Oblique axial T1-weighted MR image shows a large distended tendon sheath *(arrow);* there is no identifiable tendon within the tendon sheath. *B* and *C*, Sagittal T1- and T2-weighted MR images reveal the tendon interruption *(arrows)* and the distention of the tendon sheath.

Khoury NJ, El-Khoury GY, Saltzman CL, et al: MR imaging of posterior tibial tendon dysfunction. AJR Am J Roentgenol 167:675, 1996.

Khoury NJ, El-Khoury GY, Saltzman CL, et al: Peroneus longus and brevis tendon tears: MR imaging evaluation. Radiology 200:833, 1996.

Lutter LD, Mizel MS, Pfeffer GB: Tendon problems of the foot and ankle. In Lutter LD, Mizel MS, Pfeffer GB (eds): Foot and Ankle. Rosemont, IL, American Academy of Orthopaedic Surgeons, 1994, pp 269–282.

Maffulli N: Current concepts review: Rupture of the Achilles tendon. J Bone Joint Surg Am 81:1019, 1999.

Mota J, Rosenberg ZS: Magnetic resonance imaging of the peroneal tendons. Topics in Magnetic Resonance Imaging 9:273, 1998.

Myerson MS, McGarvey W: Disorders of the insertion of the Achilles tendon and Achilles tendinitis. J Bone Joint Surg Am 80:1814, 1998.

Novacheck TF: Running injuries: A biomechanical approach. J Bone Joint Surg Am 80:1220, 1998.

Pomeroy GC, Pike RH, Beals TC, et al: Acquired flatfoot in adults due to dysfunction of the posterior tibial tendon. J Bone Joint Surg Am 81:1173, 1999.

Rosenberg ZS, Beltran J, Cheung YY, et al: MR features of longitudinal tears of the peroneus brevis tendon. AJR Am J Roentgenol 168:141, 1997.

Sammarco GJ: Peroneus longus tendon tears: Acute and chronic. Foot Ankle Int 16:245, 1995.

Teitz CC, Garrett WE Jr, Miniaci A, et al: Tendon problems in athletic individuals. J Bone Joint Surg Am 79:138, 1997.

CHAPTER 72

Calcific Tendinitis

Georges Y. El-Khoury, MD, Mark D. Stanley, MD, and Bernard Roger, MD

OVERVIEW

Calcific tendinitis is a self-limiting condition that most commonly affects the supraspinatus tendon and other tendons of the rotator cuff (Figs. 72–1 through 72–4). Other less common locations for calcific tendinitis include tendons of the gluteus maximus (Figs. 72–5 and 72–6), rectus femoris, vastus lateralis, quadriceps, pectoralis major, deltoid, and adductor magnus, and tendons of the wrist, hand, neck, and ankle. In many of these atypical locations, especially when associated with cortical erosion, it can be a diagnostic challenge to differentiate calcific tendinitis from infection or malignant disease.

Calcific tendinitis has a peak incidence during the 30s through 50s. In patients with calcific tendinitis of the supraspinatus tendon, most investigations have found a female predominance and that sedentary individuals have an increased incidence over manual laborers.

Presentations vary from asymptomatic incidental findings on radiographs to acute painful crises. Calcific tendinitis in the area of the pelvis and lower extremity may simulate radicular pain from herniated intervertebral disk. Clinical findings in some cases include functional disability, local swelling, and mild pyrexia. Serum calcium, phosphate, and electrolyte abnormalities have been consistently absent. Patients experiencing acute onset of pain often describe the pain as a sudden heavy sensation. Gradually the pain resolves as a hardening soft tissue mass develops. In most cases, the intensity of symptoms parallels the local inflammation and laboratory abnormalities.

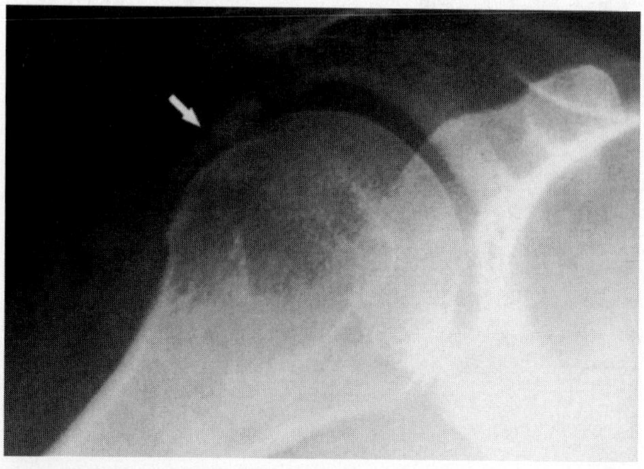

Figure 72–1. Calcific tendinitis of the supraspinatus tendon *(arrow)*.

PATHOGENESIS

Calcific tendinitis historically has been thought to be a degenerative process. The clinical history, the self-limiting nature of the disease, and the decreased incidence in individuals after their 50s weigh against a degenerative process being the underlying mechanism. Calcific tendinitis of the supraspinatus tendon evolves through three distinct stages: precalcific, calcific, and postcalcific.

During the precalcific stage, the site of future calcification undergoes chondrocyte-mediated fibrocartilage transformation. A segment of the supraspinatous tendon 1.5 cm proximal to the insertion is prone to fibrocartilage transformation. This segment, which also is known as the *critical zone*, is vulnerable to hypoxia and mechanical stresses. It is believed that the hypoxia and mechanical stresses are the inciting events for the fibrocartilage transformation. There are no known imaging findings during the precalcific stage and patients are typically asymptomatic during this phase.

The second stage, termed the *calcific stage*, is divided into two phases: a formative phase and resorptive phase.

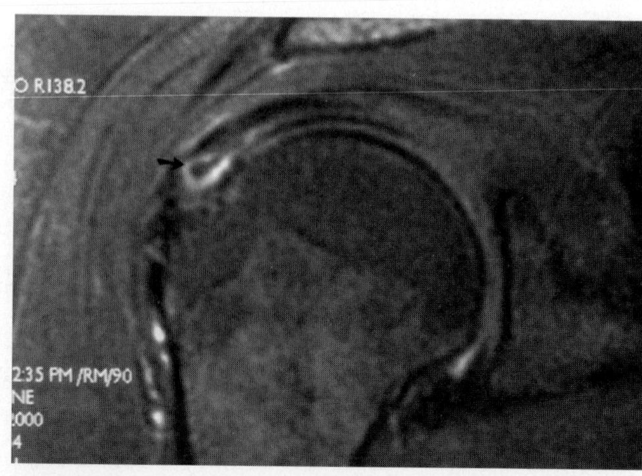

Figure 72–2. MR imaging appearance of calcific tendinitis of the supraspinatus tendon in a 45-year-old woman with acute shoulder pain. T2-weighted coronal image shows a small low signal intensity lesion surrounded by edema *(arrow)* within the supraspinatus tendon.

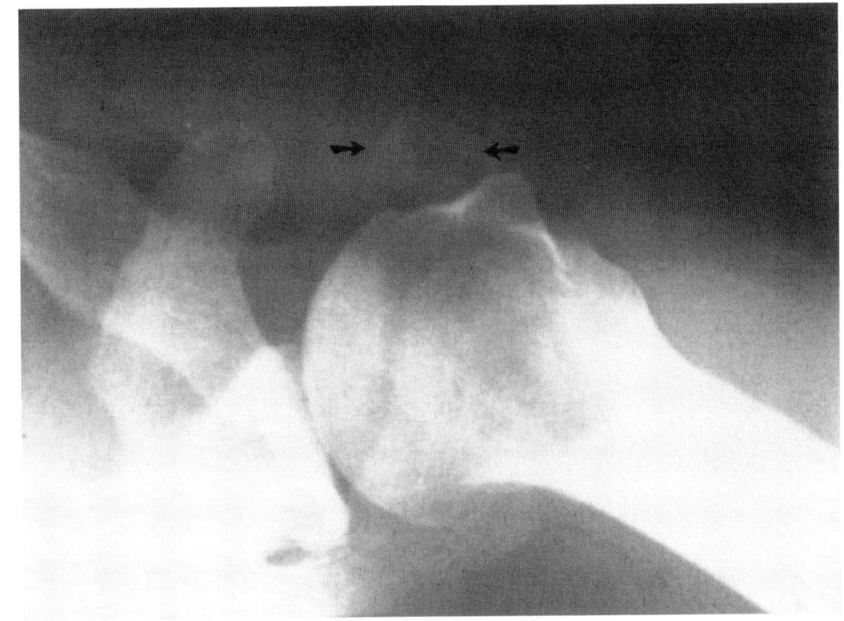

Figure 72–3. Axillary view of the left shoulder shows a large calcific density in the subscapularis tendon *(arrows)*.

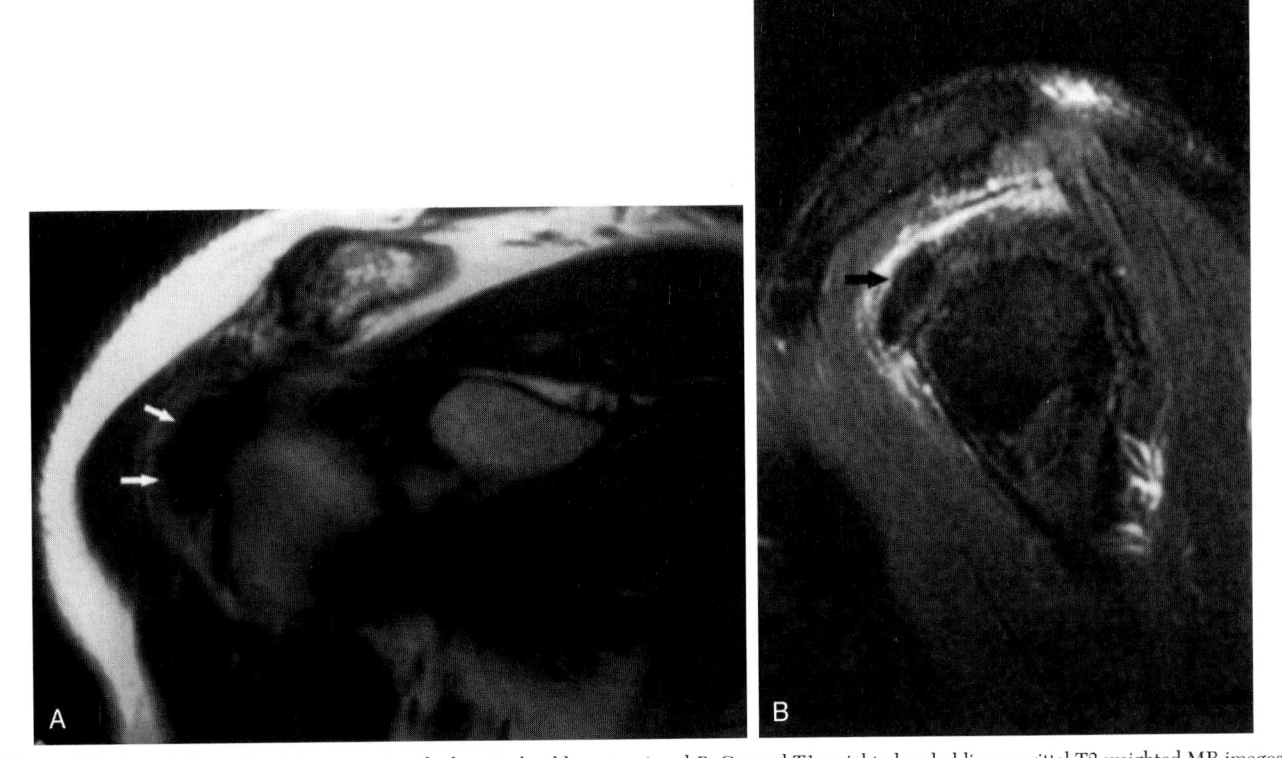

Figure 72–4. Calcific tendinitis in a patient with chronic shoulder pain. *A* and *B*, Coronal T1-weighted and oblique sagittal T2-weighted MR images show a large area of decreased signal intensity *(arrows)*, representing the calcium deposits in the supraspinatus tendon.

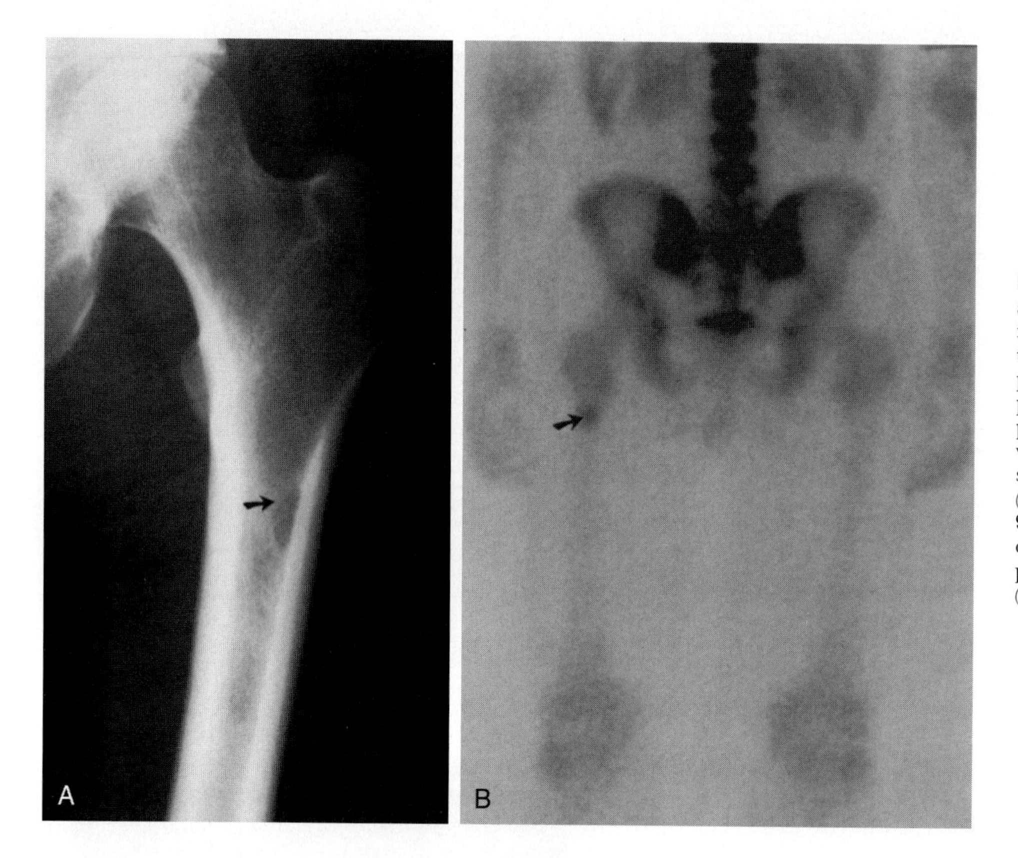

Figure 72–5. Bone scan appearance of calcific tendinitis at the insertion of the gluteus maximus tendon in a middle-aged man. The patient presented with excruciating pain in the posterior aspect of the proximal thigh. *A,* Anteroposterior view of the proximal femur shows a small lytic lesion in the femoral shaft *(arrow)*. *B,* Bone scan (technetium 99m medronate) reveals focal increased radionuclide uptake in the proximal shaft of the left femur *(arrow)*.

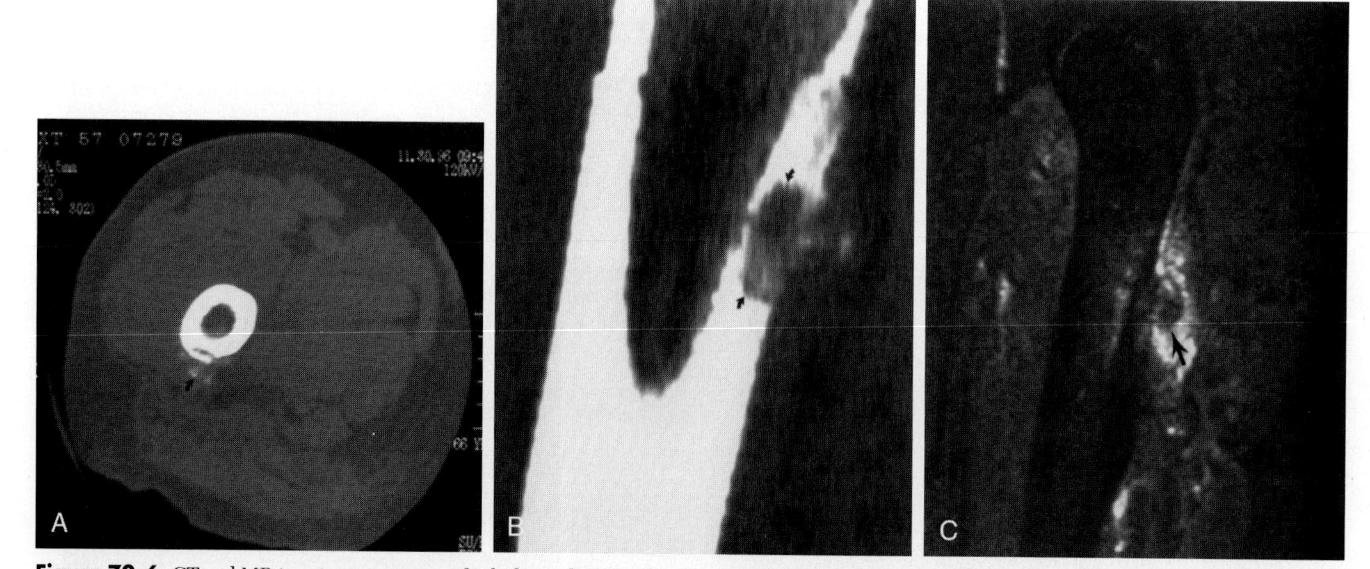

Figure 72–6. CT and MR imaging appearance of calcific tendinitis at the insertion of the gluteus maximus in a middle-aged man who presented with intense pain in the proximal thigh. He initially was thought to have a metastatic lesion in the proximal femur. *A,* CT scan of the lesion shows focal destruction of the cortex with associated calcific mass *(arrow)*. *B,* Reconstructed sagittal CT scan of the right femur shows a deep cortical erosion *(arrows)* with adjacent soft tissue calcification. *C,* Sagittal T2-weighted MR image with fat suppression reveals soft tissue calcification *(arrow)* and surrounding soft tissue edema. There is some marrow edema present.

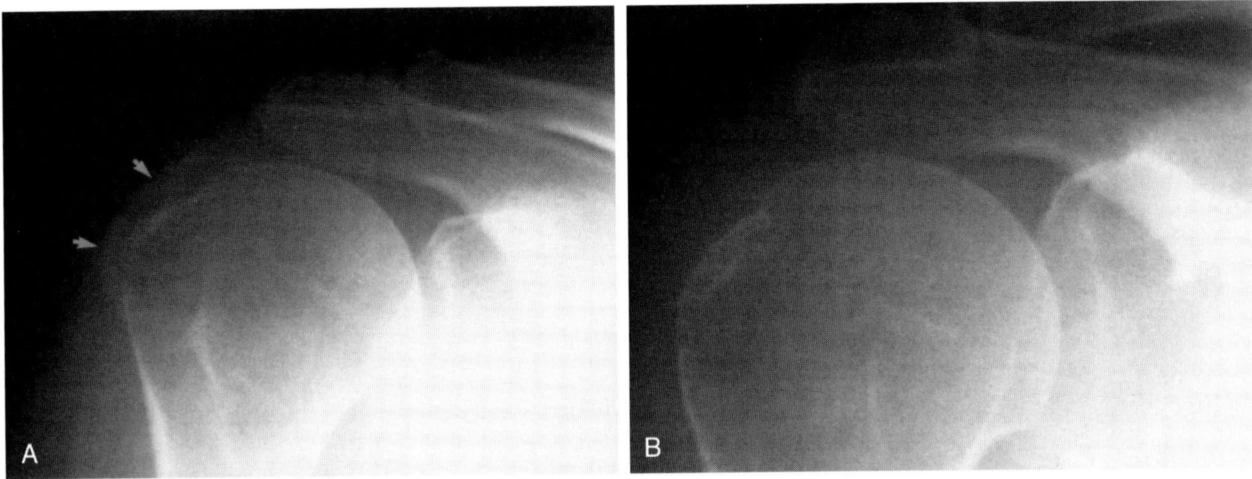

Figure 72–7. Disappearance of the tendon calcification 2 months after the acute onset of pain. *A,* Initial radiograph taken during the acute crisis shows a clump of ill-defined calcification *(arrows)* at the insertion of the supraspinatus tendon. *B,* Radiograph obtained 2 months later when only minimal symptoms were present shows complete disappearance of calcific deposits.

During the formative phase, the previously formed fibrocartilage is replaced by calcific deposits. Radiographs obtained during the formative phase may reveal a homogeneous, well-defined deposit. During the resorptive phase, macrophages and giant cells phagocytose the calcific deposits. The formative and the resorptive phases may exist concurrently at different foci within the tendon. The resorptive phase is often painful, and symptoms may be incapacitating. At surgery, a toothpaste-like substance extrudes from the tendon after it has been incised. The acute pain may be due to increased intratendinous pressure. Radiographs obtained during the resorptive phase may reveal an ill-defined, amorphous deposit, or this phase may lack findings altogether. The amorphous, ill-defined appearance clinically correlates with acute attacks of pain.

The last phase is the postcalcific stage, in which the calcific deposits become phagocytosed and replaced by granulation tissue. The granulation tissue eventually matures into a scar. Chronic low-intensity pain may persist for months during this stage.

IMAGING

On plain radiographs, soft tissue calcifications have a *comet tail* appearance (Fig. 72–7; see also Figs. 72–1 and 72–2). These calcifications can be difficult to show without multiple views, including internal and external rotation or oblique views. Because the peak of symptoms occurs during the resorptive phase of the calcific stage, symptoms often do not correlate with radiographic findings. Completely asymptomatic cases of calcific tendinitis are discovered incidentally during radiographic evaluation for other conditions. Fewer than 10% of patients with supraspinatus calcifications ever develop symptoms of acute calcific tendinitis. If calcific deposits are seen on plain radiographs during an acute painful episode, radiologic resolution may be expected to occur 6 to 10 weeks after the acute presentation (see Fig. 72–7).

Cortical erosion and periosteal reaction have been described in association with calcific tendinitis in a variety of locations, most commonly the insertion of the gluteus maximus (see Figs. 72–5 and 72–6) and pectoralis major. Cortical erosions associated with calcific tendinitis can be confused with a neoplastic process (see Figs. 72–5 and 72–6). Cortical metastases, although less common than metastases to the bone marrow, are observed with different primary sources, the most common being bronchogenic carcinoma. Cortical metastasis typically is not accompanied, however, by mineral deposition in adjacent soft tissues (see Fig. 72–6).

Computed tomography (CT) is the imaging modality of choice for evaluating the cortical destruction and tendinous calcifications (see Fig. 72–6). The CT appearance of calcific tendinitis has been described as *flame shaped* following the course of the involved tendon. CT is recommended when plain radiography fails to establish a definitive diagnosis.

MR imaging may be confusing because the soft tissue and marrow edema may be misinterpreted as infection or a neoplastic process (see Fig. 72–6*C*). MR imaging also is suboptimal in assessing soft tissue calcifications and subtle cortical erosions.

Recommended Readings

Chow HY, Recht MP, Schils J, et al: Acute calcific tendinitis of the hip. Arthritis Rheum 40:974, 1997.

Hayes CW, Rosenthal DI, Plata MJ, et al: Calcific tendinitis in unusual sites associated with cortical bone erosion. AJR Am J Roentgenol 147:967, 1987.

Mizutani H, Ohba S, Mizutani M, et al: Calcific tendinitis of the gluteus maximus tendon with cortical bone erosion: CT findings. J Comput Assist Tomogr 18:310, 1994.

Seeger LL, Butler DL, Eckardt JJ, et al: Tumoral calcinosis-like lesion of the proximal linea aspera. Skeletal Radiol 19:579, 1990.

Speed CA, Hazleman BL: Calcific tendinitis of the shoulder. N Engl J Med 340:1582, 1999.

Uthoff HK: Calcifying tendinitis. Ann Chir Gynaecol 85:111, 1996.

CHAPTER 73

Muscle Injuries

Georges Y. El-Khoury, MD, and Bernard Roger, MD

OVERVIEW

Skeletal muscle constitutes the largest tissue in the body, making up nearly 50% of the total body weight. Muscle injuries are best studied as part of the muscle-tendon unit, which consists of the muscle fibers and tendons, and their bony attachments. In children and adolescents, muscles and tendons originate and insert on apophyses. The apophyseal growth plates are the weakest link in the muscle-tendon unit, whereas tendons are the strongest component.

MORPHOLOGY AND PHYSIOLOGY

Muscle fibers have no end-to-end connections; a muscle fiber in the sartorius muscle can be 60 cm long. The myotendinous junction is a highly specialized connection; most muscle injuries occur at or near the myotendinous junction (Fig. 73–1). The myotendinous junction is a folded membrane at the muscle-tendon interface that increases the contact area, reducing stress between the muscle and tendon. Muscles generally cross one joint, but many muscles cross two joints. Muscles function in producing and modulating joint motion. Histochemical stains can be used to classify muscle fibers according to their structural, biochemical, and physiologic characteristics.

Type I (slow-twitch) fibers have an aerobic system of enzymes. They are rich in blood supply and mitochondria. They are slow to contract and slow to relax but are resistant to fatigue.

Type II (fast-twitch) fibers are divided into subtypes IIA and IIB. Type II fibers have an anaerobic system of enzymes and few mitochondria. They have fast contraction and fast relaxation times, but they fatigue faster than type I fibers.

The performance of a muscle is related to its content of the different fiber types. Muscle biopsy specimens from long-distance runners show predominance of type I fibers, whereas biopsy specimens from successful sprinters show predominance of type II fibers. It is not determined if muscle composition is genetic or related to training.

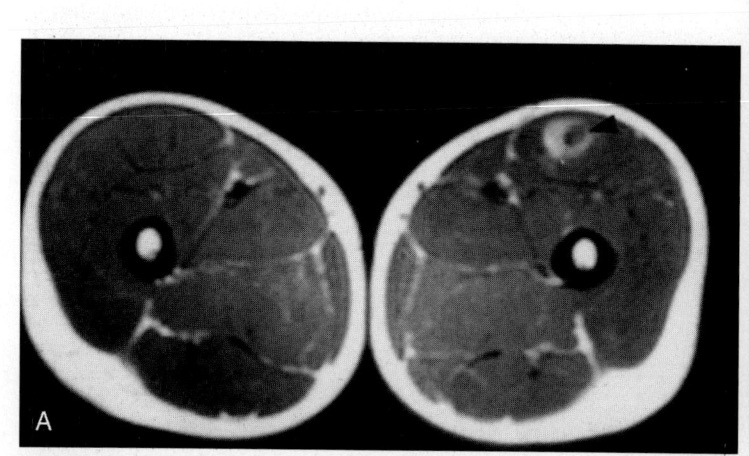

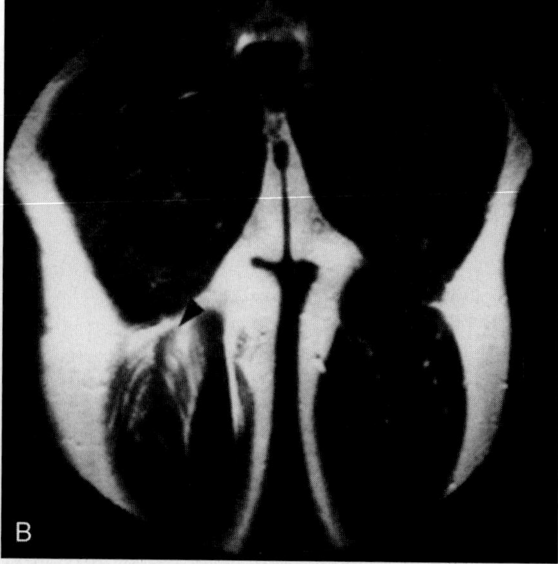

Figure 73–1. Muscle strains at the myotendinous junction from two different patients. Both injuries occurred during athletic activity. *A*, T1-weighted MR image from a soccer player who developed acute pain in the anterior thigh on the left. The tendon of the rectus femoris is seen *(arrowhead)* surrounded by high signals, which represent blood. *B*, Coronal T2-weighted MR image of the thigh in a football player who developed excruciating pain in the posterior thigh while running. The image shows the injury at the myotendinous junction *(arrowhead)* where the muscle fibers are seen attaching to the tendon.

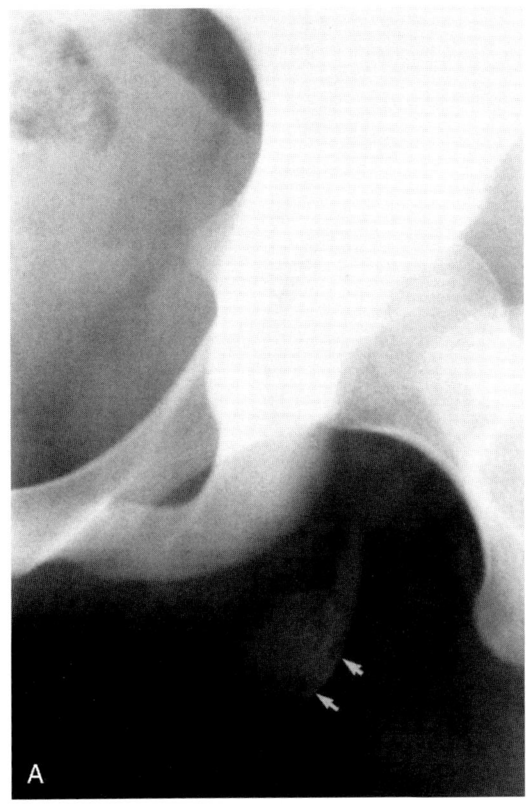

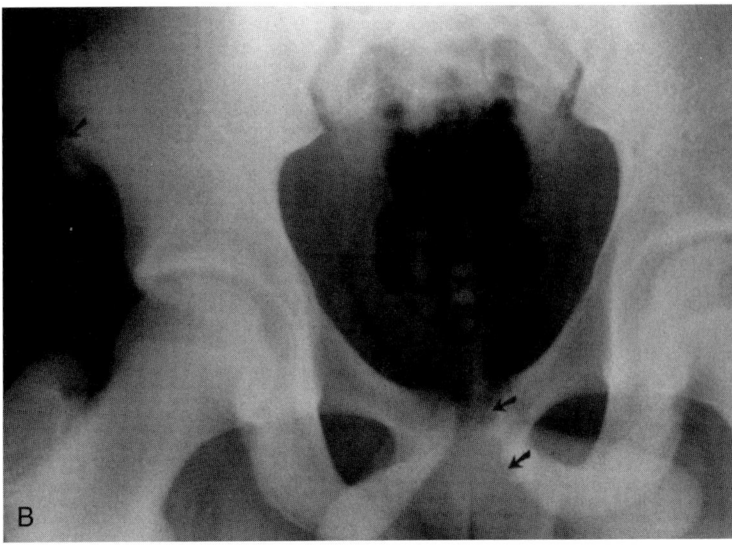

Figure 73–2. Apophyseal avulsion injuries in two different teenage athletes. *A,* Anteroposterior view of the left hip from a 14-year-old cheerleader who was performing a split. She developed severe pain in the lower buttock and could not stand up. The image shows a large apophyseal avulsion off the left ischium. *B,* A 13-year-old star soccer player with chronic pain in the left groin. He recently developed a new pain above the right hip. The image shows chronic avulsive changes in the left pubic bone *(arrows)* and acute avulsion of the anterior superior iliac spine *(arrow)*.

TYPES OF MUSCLE CONTRACTION

Isometric Contraction

Isometric muscle contraction is contraction without change in muscle length (e.g., lifting a weight with the elbow maintained in a partially flexed position).

Concentric Contraction

Concentric muscle contraction is contraction where the resisting load is less than the force or tension generated by contracting muscle; muscle shortens as it contracts (e.g., performing curls with dumbbells). Isometric contraction can be converted to concentric contraction by recruiting more muscle fibers until the muscle starts to shorten.

Eccentric Contraction

Eccentric muscle contraction is contraction where the resisting load is greater than the force generated by the muscle so the muscle lengthens as it contracts. Excessive tension is generated within muscle during eccentric contraction, rendering it more susceptible to rupture or tearing (e.g., contraction experienced in the quadriceps muscle after a jump from a height).

SITES OF INJURY IN MUSCLE-TENDON UNIT

The site of injury is age related. In children and adolescents, the apophyseal growth plates are common sites of injury (Fig. 73–2). Just after adolescence (age 16 to 21 years), relative weakness persists at previous apophyseal plates. Multiple avulsion sites can be seen in one third of young patients (Fig. 73–2*B*). When the muscle-tendon unit is stretched to failure experimentally, it typically fails at the myotendinous junction (see Fig. 73–1).

MUSCLE INJURIES

Muscle Laceration and Contusion

Imaging is rarely if ever indicated for muscle lacerations or contusions. These injuries are noted incidentally when imaging is performed for other reasons, such as femoral neck or pelvic fractures.

Muscle Strain (Tear or Rupture)

Weak or fatigued muscles are more likely to rupture because such muscles absorb less energy. Previously

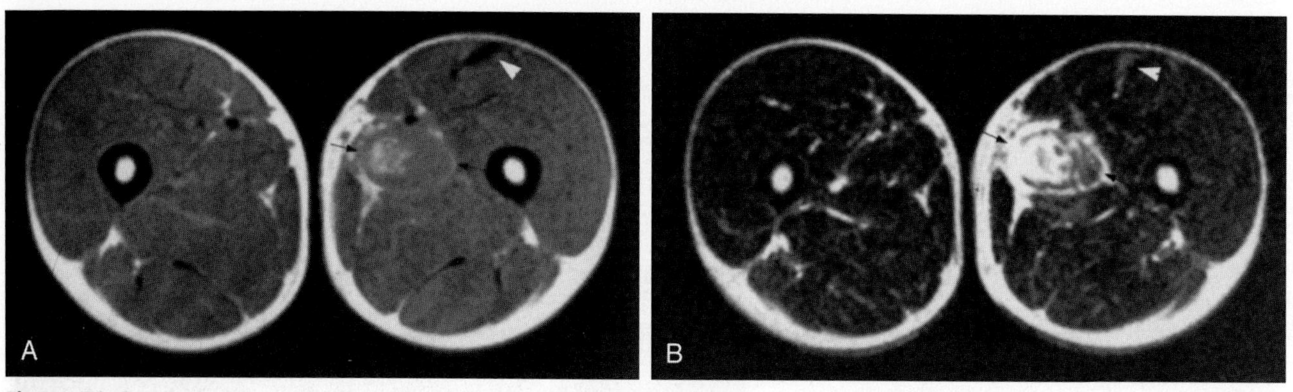

Figure 73–3. Muscle tear of the adductor longus 4 weeks after injury to the rectus femoris in a 21-year-old football player. *A* and *B*, T1- and T2-weighted axial MR images show an acute strain of the adductor longus muscle *(arrows)* and an old strain of rectus femoris *(arrowhead)* in the left thigh.

injured muscles that did not heal completely put other muscles at risk for injury when the athlete prematurely resumes his or her sports activities (Fig. 73–3).

The mechanism of injury is sudden and forceful eccentric muscle contraction. Muscle groups most commonly affected with muscle strains are as follows:

Muscles that function eccentrically
Muscles that cross more than one joint
Muscles with high content of type II fibers

Assessment of Severity of Muscle Strains

Partial tears (Fig. 73–4) are more common than complete tears, but they are less severe. Partially torn muscles show some loss of function and diminished strength.

Complete muscle disruption typically occurs at the myotendinous junction and frequently is associated with muscle retraction and complete loss of function (Fig. 73–5). Complete muscle tears do not always represent a true disruption within the muscle substance but rather a tendon avulsion off the bone or a complete tendon tear producing what is known as the *mop-end* appearance on magnetic resonance (MR) imaging (Fig. 73–6).

Magnetic Resonance Imaging Findings in Muscle Strain

Muscle Deformity

In the acute phase, there is a focal mass resulting from bleeding and edema at the sites of muscle disruption. In the chronic phase, a defect is seen at the site of the tear, and a mass is produced by the retracted muscle (Fig. 73–7).

Signal Changes

In the acute phase, there is increased signal on T1-weighted MR images as a result of fresh bleeding (see Fig. 73–1). In the chronic phase, a torn retracted muscle

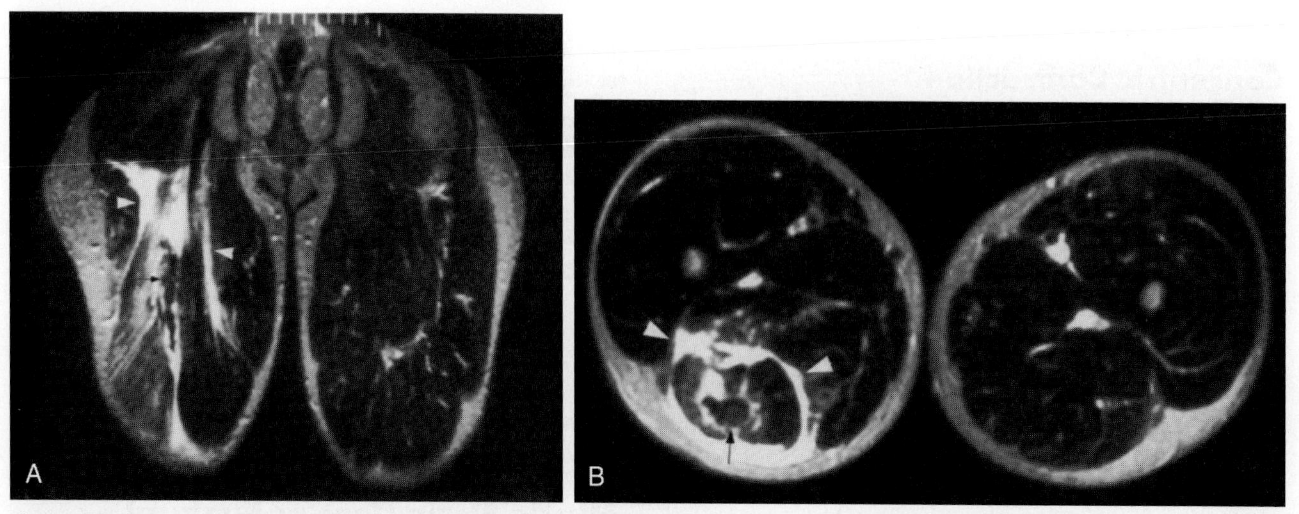

Figure 73–4. Partial subacute tear of the hamstring in a baseball player who developed severe incapacitating pain while dashing toward second base. *A* and *B*, Coronal and axial T2-weighted MR images reveal blood and edema around *(arrowheads)* and within *(arrow)* the right hamstring muscle. A low signal intensity blood clot is seen within the hamstring *(arrow)*. The low density is due to the presence of hemosiderin.

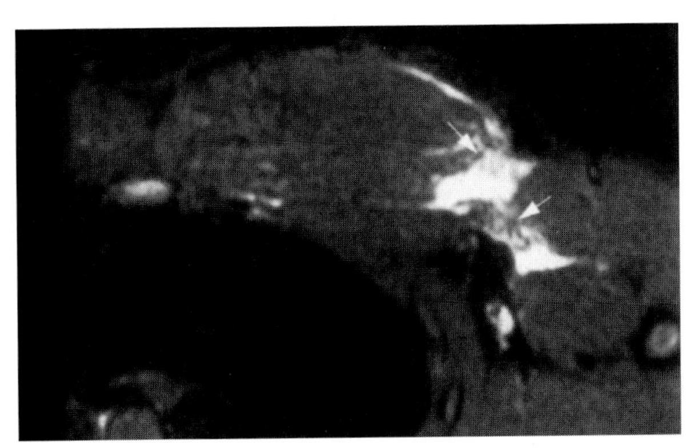

Figure 73–5. Complete tear of the pectoralis major muscle in a weightlifter. Axial T2-weighted MR image shows disruption of the pectoralis major muscle at the myotendinous junction *(arrows)*. The muscle is retracted medially.

may show atrophy and fatty replacement. Hemosiderin deposition is seen often in subacute and chronic muscle tears, and it appears dark on all sequences (see Fig. 73–4).

Indications for Magnetic Resonance Imaging

Most muscle injuries are minor and do not require any imaging. The indications for MR imaging include the following:

1. Delineate extent of injury in high-performance athletes (see Figs. 73–5 and 73–6)
2. Patients presenting with soft tissue mass but no clear history of trauma (see Fig. 73–7)
3. When a specific diagnosis is crucial for initiation of therapy (see Fig. 73–6)
4. Severe and disabling delayed-onset muscle soreness (DOMS) (Fig. 73–8).

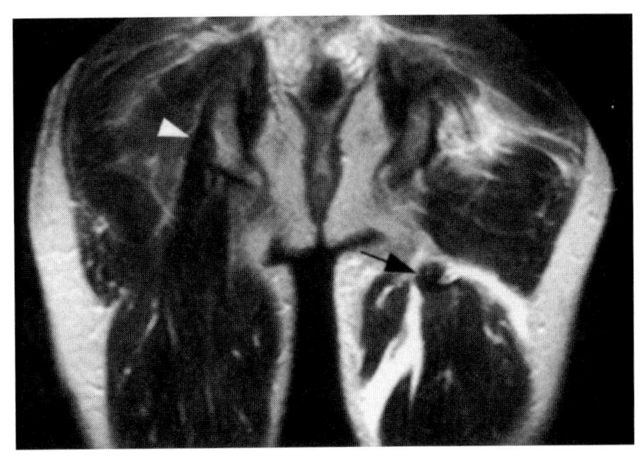

Figure 73–6. Complete disruption of the common hamstring tendon in a 20-year-old college athlete. Coronal T2-weighted MR image shows a complete tear *(arrow)* of the common hamstring tendon on the left. The muscle has retracted distally, giving the mop-end appearance. Note the normal common hamstring tendon on the right side *(arrowhead)*. The treatment for this injury in a high-performance athlete is surgical reattachment of the tendon.

DELAYED-ONSET MUSCLE SORENESS

In DOMS, muscle pain starts 24 to 72 hours after exercise; patients present with muscle pain, swelling, and joint stiffness. DOMS typically follows excessive eccentric muscle activity, such as running downhill. Severity of symptoms depends on the duration and intensity of the exercise. Muscle biopsy specimens from patients with DOMS reveal tissue damage with ultrastructural abnormalities in the Z band and elevation of serum creatine kinase enzyme level. These findings led some investigators to believe that DOMS may represent a mild form of rhabdomyolysis. On MR imaging, the affected muscles appear swollen and edematous (i.e., bright on T2-weighted images) (see Fig. 73–8).

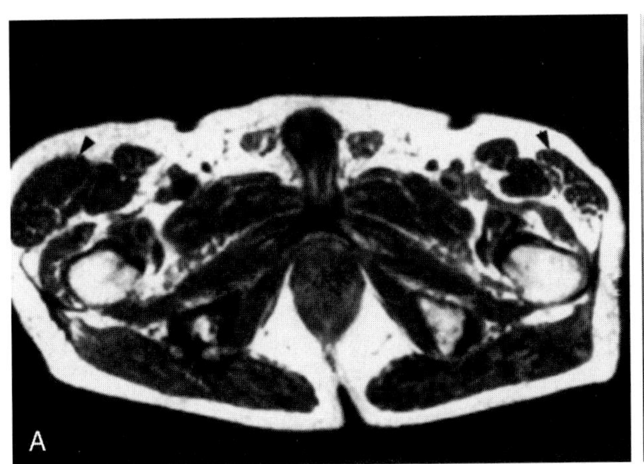

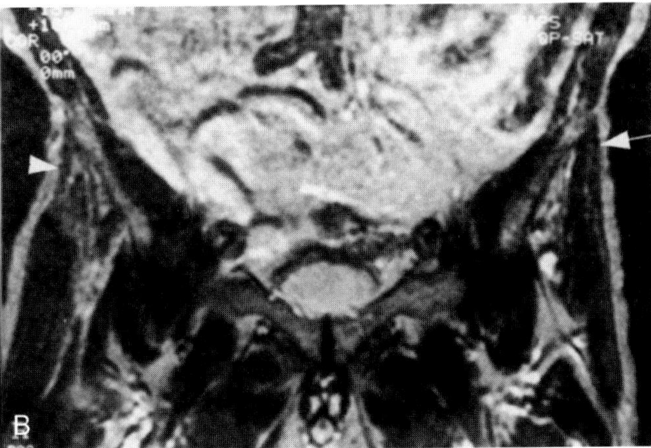

Figure 73–7. Mass produced by a ruptured, retracted tensor fascia lata in a 72-year-old patient who fell several months ago. *A*, Axial T1-weighted MR image at the level of the lesser trochanter shows tensor fascia lata muscle on both sides *(arrowheads)*, but on the right the muscle is much larger because of retraction. *B*, T2-weighted coronal MR image shows the muscle tear *(arrowhead)* off the anterior superior iliac spine. Note the normal muscle origin on the left side *(arrow)*.

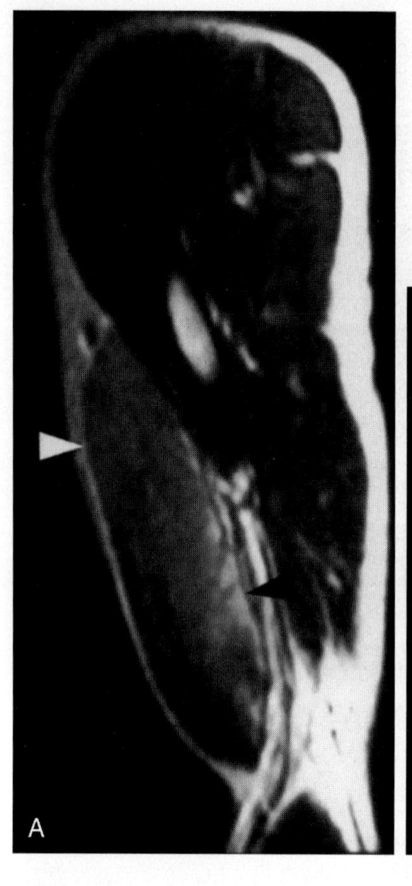
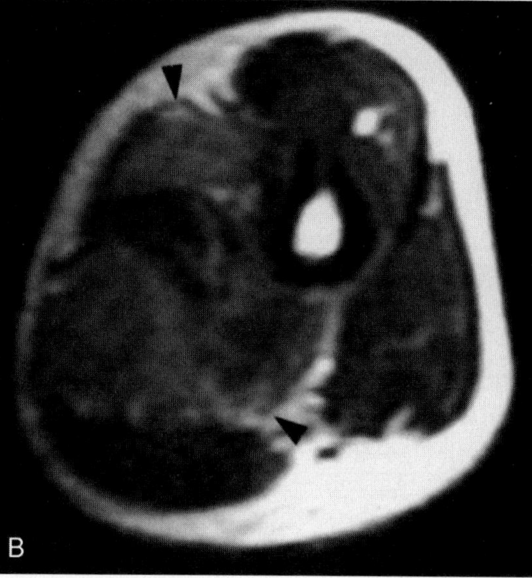

Figure 73–8. Delayed-onset muscle soreness in a 21-year-old bodybuilder who was attempting to increase his muscle mass rapidly using eccentric exercises. *A* and *B*, Sagittal and axial T2-weighted MR images of the left arm show swelling and edema *(arrowheads)* affecting segments of the biceps brachii and brachialis muscles. The affected areas show increased signal intensity *(arrowheads)*.

MUSCLE HEMATOMA AND FASCIAL BLEEDS

Muscle hematoma and fascial bleeds are fairly common, and they have specific imaging characteristics. Acute bleeding usually appears bright on T1-weighted MR images. Hemosiderin from old hematomas appears dark on all sequences. Acute and chronic bleeding often coexist (Fig. 73–9), and blood collections of unknown duration can present as masses that are easily confused with neoplasm clinically and on imaging studies.

MUSCLE HERNIATION

Muscle herniation is a rare condition affecting mainly the anterior tibial compartment, although other sites can be involved. Muscle herniation usually is caused by blunt

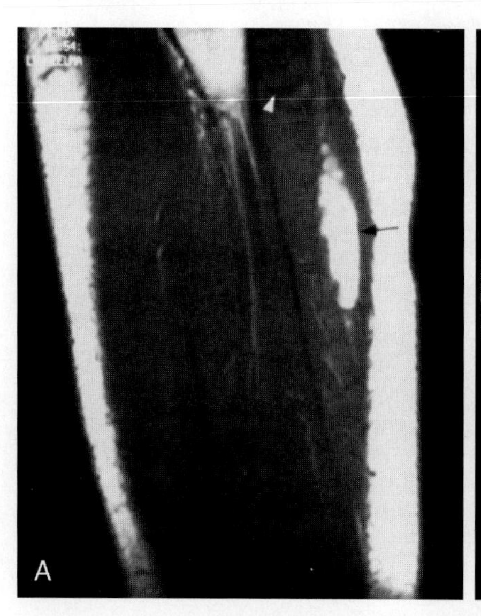
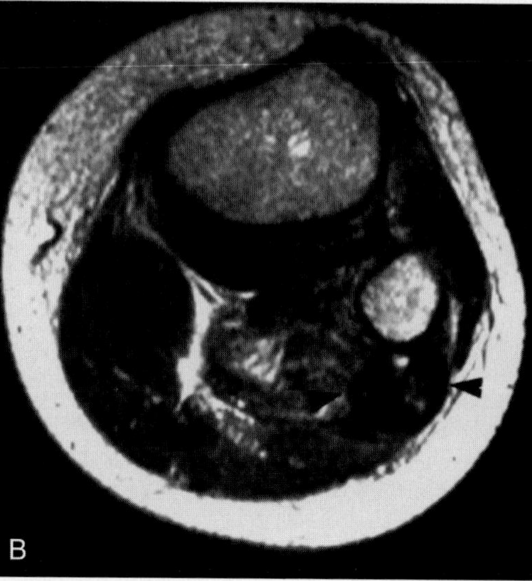

Figure 73–9. Chronic and acute hematomas in calf muscles of a 13-year-old athlete. *A*, T1-weighted sagittal MR image shows an old (dark) hematoma *(arrowhead)* and acute bright bleed *(arrow)* in the lateral head of the gastrocnemius. *B*, T2-weighted axial MR image shows the old hematoma *(arrowheads)* as a dark area.

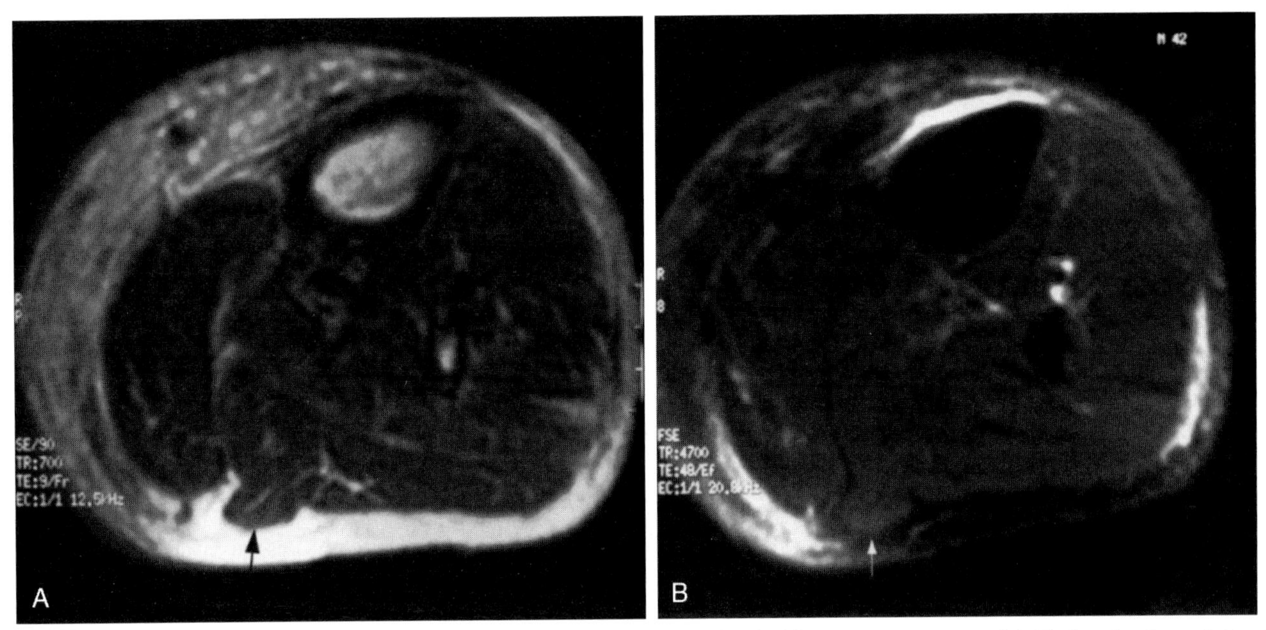

Figure 73–10. Muscle herniation, through a rent in the deep fascia, of the lateral head of the gastrocnemius. *A* and *B*, The herniated muscle is well shown on T1- and T2-weighted axial MR images *(arrow)*. The signal characteristics of the herniated portion are identical to normal muscle on both MR sequences.

trauma. Clinical examination is diagnostic, but when in doubt MR imaging provides a definitive diagnosis. The herniated portion of the muscle shows change in size between contraction (larger) and relaxation (smaller). Signal characteristics of herniated muscle are identical to normal muscle on all MR sequences (Fig. 73–10).

Recommended Readings

DeSmet AA: Magnetic resonance findings in skeletal muscle tears. Skeletal Radiol 22:479, 1993.

El-Khoury GY, Brandser EA, Kathol MH, et al: Imaging of muscle injuries. Skeletal Radiol 25:3, 1996.

El-Khoury GY, Daniel WW, Kathol MH: Acute and chronic avulsive injuries. Radiol Clin North Am 35:747, 1997.

Fernbach SK, Wilkinson RH: Avulsion injuries of the pelvis and proximal femur. AJR Am J Roentgenol 137:581, 1981.

Fleckenstein JL, Weatherall PT, Parkey PW, et al: Sports-related muscle injuries: Evaluation with MR imaging. Radiology 172:793, 1989.

Kathol MH, El-Khoury GY, Moore TE, et al: Calcaneal insufficiency avulsion fractures in patients with diabetes mellitus. Radiology 180:725, 1991.

McMaster PE: Tendon and muscle ruptures. J Bone Joint Surg 15:702, 1933.

Metzmaker JN, Pappas AM: Avulsion fractures of the pelvis. Am J Sports Med 13:349, 1985.

Micheli LJ, Fehlandt AF Jr: Overuse injuries to tendons and apophyses in children and adolescents. Clin Sports Med 11:713, 1992.

Shellock FG, Fukunaga T, Mink JH, et al: Exertional muscle injury: Evaluation of concentric versus eccentric actions with serial MR imaging. Radiology 179:659, 1991.

Sundar M, Carty H: Avulsion fractures of the pelvis in children: A report of 32 fractures and their outcome. Skeletal Radiol 23:85, 1994.

Zarins B, Ciullo JV: Acute muscle and tendon injuries in athletes. Clin Sports Med 2:167, 1983.

Zeiss J, Ebraheim NA, Woldenberg LS: Magnetic resonance imaging in the diagnosis of anterior tibialis muscle herniation. Clin Orthop 244:249, 1989.

CHAPTER 74

Muscle Disorders

Georges Y. El-Khoury, MD, D. Lee Bennett, MD, and Robert H. Choplin, MD

MUSCLE INFECTIONS

Overview

Muscle infections are confusing because multiple terms are used to describe them and the different organisms involved produce a variety of imaging manifestations. Terms such as *pyomyositis, tropical pyomyositis, bacterial or infectious myositis, myonecrosis, gas gangrene, fasciitis,* and *necrotizing fasciitis* all have been used.

Pyomyositis

Pyomyositis, tropical pyomyositis, bacterial myositis, and myonecrosis are used interchangeably in the literature; however, *pyomyositis* is now the accepted term to describe bacterial muscle infections with abscess formation. Pyomyositis often is associated with fasciitis and cellulitis. Pyomyositis typically occurs in the large muscles of the buttocks and lower extremities. It also occurs in the paraspinal muscles, deltoid, and triceps. *Staphylococcus aureus* is the causative agent in 90% of cases, and group A streptococcus causes most of the remaining infections. The condition initially was called *tropical pyomyositis* because it used to be rare outside the tropics. Recently more cases have been observed in temperate climates, especially in patients with diabetes mellitus, HIV infection, chronic steroid usage, and connective tissue disorders.

Imaging

Magnetic resonance (MR) imaging with gadolinium enhancement is effective in making the diagnosis. Early in the disease, MR imaging shows an intramuscular mass, but the signal intensity changes are nonspecific, and the lesion can resemble a sarcoma or a variety of benign lesions. When an abscess starts to form, rim enhancement becomes obvious, especially on fat-suppressed T1-weighted images with gadolinium enhancement (Fig. 74–1). Liquefaction of the infected tissue and pus collections are hyperintense on T2-weighted MR images (Figs. 74–1 and 74–2). Associated fasciitis and cellulitis frequently are present. Computed tomography (CT) with intravenous contrast also shows a characteristic low-attenuation mass with rim enhancement when a muscle abscess is present (Fig. 74–2A). In diabetic patients, pyomyositis is difficult to differentiate from diabetic muscle infarction (DMI) clinically and on MR imaging.

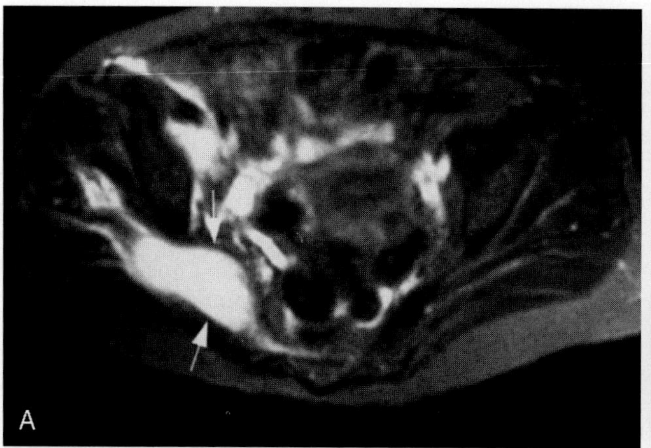

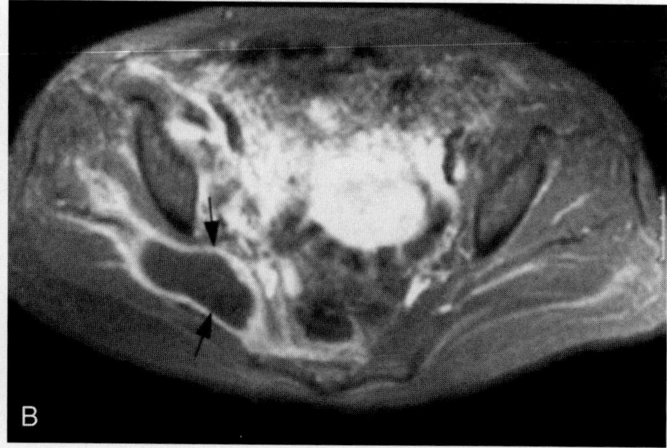

Figure 74–1. Muscle abscess in a 50-year-old woman with fever and leukocytosis. *A,* T2-weighted axial MR image with fat suppression of the pelvis shows a localized area of increased signal intensity in the right buttock *(arrows)* and a diffuse area of increased signal intensity within the right pelvic area. *B,* T1-weighted axial MR image with fat suppression and gadolinium enhancement shows the buttock abscess *(arrows)* as an area of decreased signal with an enhancing rim. The pelvic inflammatory changes also are seen as an area of increased signal.

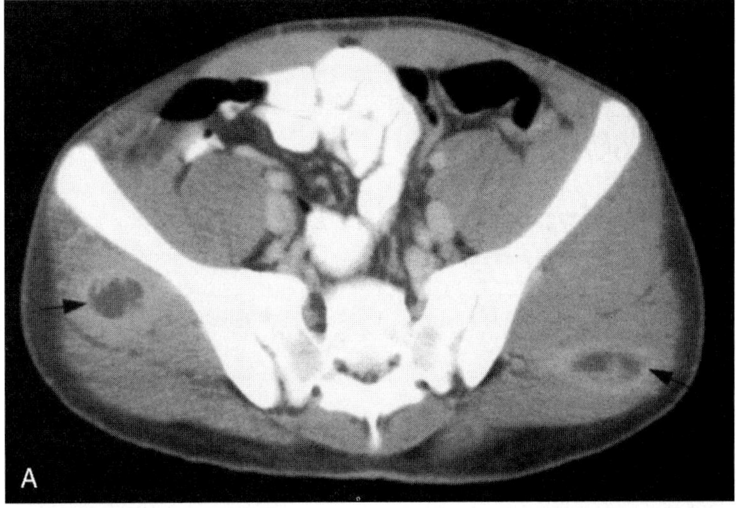

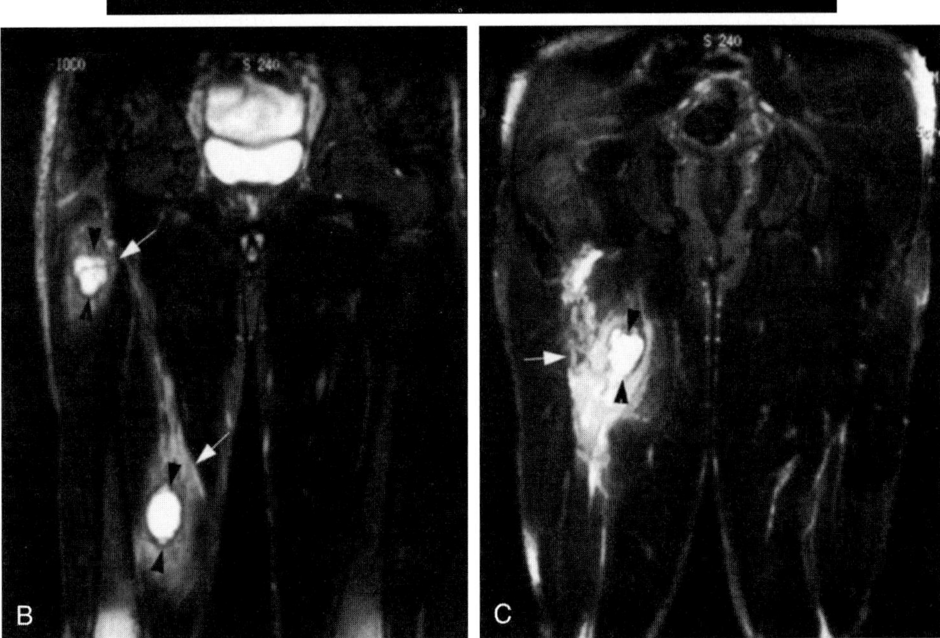

Figure 74–2. Multiple muscle abscesses in a 31-year-old patient with AIDS. *A,* Contrast-enhanced CT scan shows areas of low attenuation in the gluteus medius on the right and gluteus maximus on the left *(arrows).* The low-attenuation areas (abscesses) are surrounded by a faint rim of contrast enhancement. *B* and *C,* Coronal T2-weighted MR images with fat suppression show three abscess collections in the thigh muscles *(arrowheads)* surrounded by inflamed muscle *(arrows).*

Necrotizing Fasciitis

Necrotizing fasciitis is a rare soft tissue infection characterized by rapidly progressive necrosis of the subcutaneous tissue and deep fascia. It usually is accompanied by severe systemic toxicity. In the early stages, underlying muscle often is spared, but as the disease progresses, muscle involvement can be extensive. Without early surgical intervention, the mortality rate with necrotizing fasciitis is high; early recognition and aggressive surgical débridement are crucial for successful treatment. Most patients have an underlying debilitating illness, such as diabetes mellitus. About 55% of patients with necrotizing fasciitis have soft tissue gas visible on radiography or CT (Fig. 74–3). The gas dissects between the fascial planes in necrotizing fasciitis in contradistinction to clostridial and nonclostridial myonecrosis (gas gangrene), in which gas is seen between the muscle fibers (Fig. 74–4). The gas in necrotizing fasciitis typically is produced by *Clostridium perfringens* and by anaerobic bacteria. Necrotizing fasciitis received a lot of publicity recently because of a cluster of cases in England, caused by a group A streptococcus bacterium, which resulted in a 70% mortality rate.

Gas Gangrene

Gas gangrene should be suspected when the pathognomonic feather-like or herringbone gas pattern is noted in the soft tissues radiographically (see Fig. 74–4). This appearance is due to gas dissecting between muscle fibers and is considered to be a sign of myonecrosis. The disease is potentially lethal, and early diagnosis and treatment are essential to reduce morbidity and mortality. Infective agents include *Clostridium,* group A streptococcus, and gram-negative organisms. Identification of the infecting agent is essential because *Clostridium* responds well to extensive surgical débridement or amputation, specific antimicrobial agents, and hyperbaric oxygen therapy. Hyperbaric oxygen therapy may be ineffective for gram-negative infections.

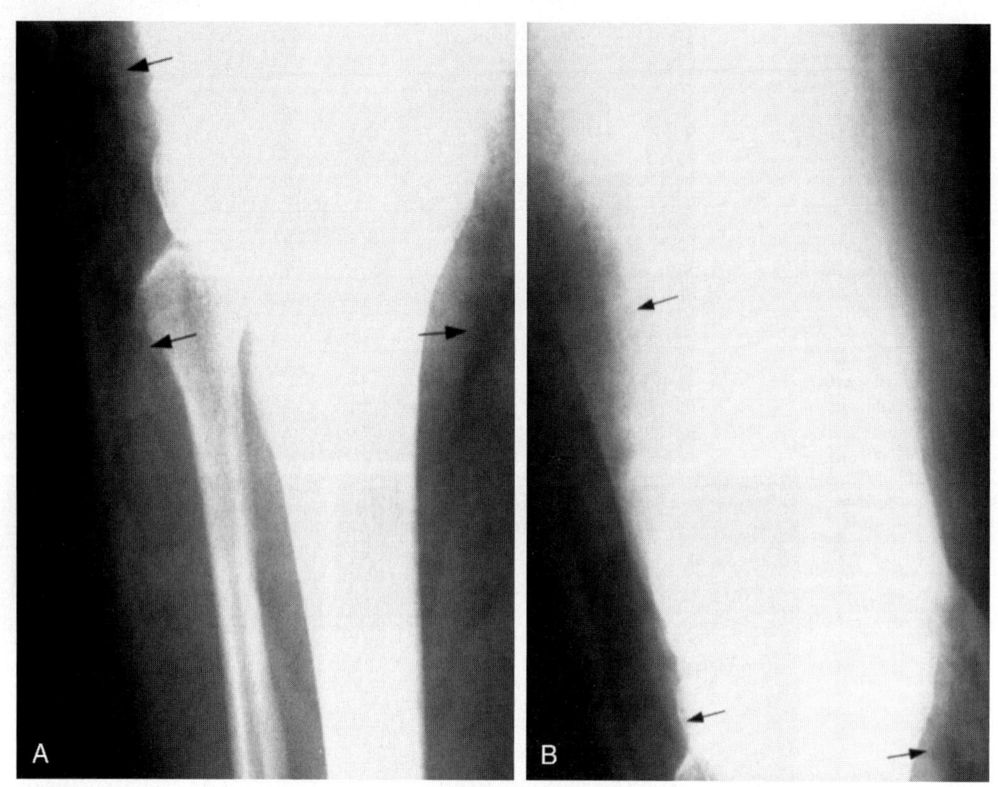

Figure 74–3. Necrotizing fasciitis of the right lower extremity in a 56-year-old diabetic patient with fulminant infection of the right foot, which extended to involve the entire lower extremity. *A* and *B*, Anteroposterior views of the knee and distal thigh show air dissecting between the fascial planes (*arrows*).

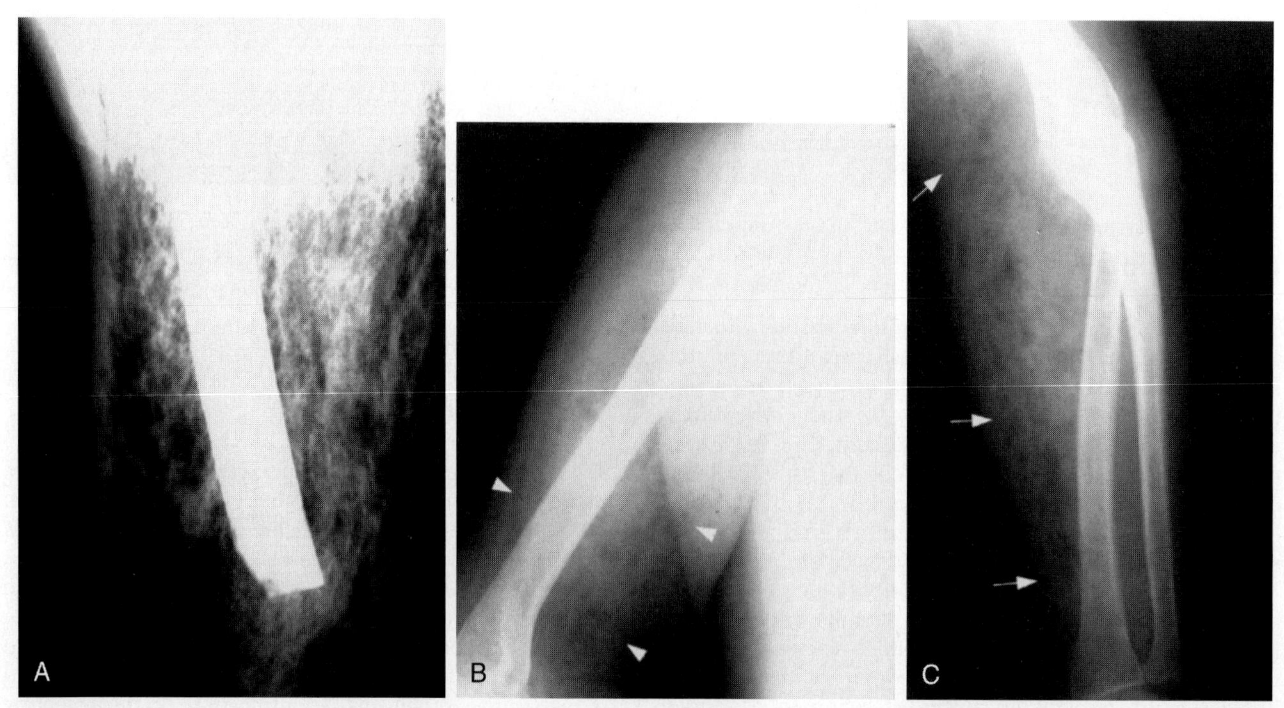

Figure 74–4. Gas gangrene in two diabetic patients. *A*, *Clostridium* infection of the lower extremity, which did not respond to above-knee amputation. Anteroposterior view of the right thigh shows air between the muscle fibers of the thigh, which is characteristic of a *Clostridium* infection. *B* and *C*, Mixed gram-negative and gram-positive infection of the right upper extremity. The gas within the muscles has a bubbly appearance (*arrowheads* on *B* and *arrows* on *C*). At surgery, all the muscles of the right upper extremity were necrotic.

MUSCLE DENERVATION

Overview

Denervation of skeletal muscle is the loss of nerve stimulation derived from lower motor neurons. This loss of stimulation can be caused by lesions at the anterior horn of the spinal cord, nerve root, plexus, or peripheral nerve. The diagnosis usually is made clinically and by electromyography.

Imaging

Imaging studies have contributed little to the diagnosis of muscle denervation except in chronic cases, in which denervated muscle shows atrophy and fatty replacement (Fig. 74–5). In acute denervation, MR imaging is unreliable in depicting the injury. With subacute denervation, there is increased signal intensity on T2-weighted and short tau inversion recovery (STIR) images (Figs. 74–5 and 74–6).

There are a few reports of hypertrophy and pseudohypertrophy of muscle in response to denervation. There is no clear explanation for true muscle hypertrophy. However, it may be related to increased workload by the remaining muscle fibers. Pseudohypertrophy reflects the accumulation of excess fat and connective tissue within the muscle.

RHABDOMYOLYSIS

Overview

Rhabdomyolysis is an acute disorder resulting from injury to skeletal muscle followed by release of cell contents into the extracellular fluid. Causes include excessive physical activity, prolonged muscle compression, crush injuries, seizures, sepsis, hyperthermia, extreme cold, coma, and drug or alcohol overdose.

Laboratory findings include elevation of serum creatine phosphokinase, hypocalcemia, hyperkalemia, and myoglobinuria. Rhabdomyolysis can lead to acute renal failure, and so the use of iodinated contrast material should be avoided in patients suspected of having rhabdomyolysis.

Imaging

Radionuclide Scanning

During the acute phase of rhabdomyolysis, radionuclide scanning shows marked uptake of technetium 99m pyrophosphate in the affected muscles. This technique is especially helpful for showing the extent of rhabdomyolysis. The mechanism of technetium 99m pyrophosphate localization in injured muscle is not clear, but it may be due to calcium deposition.

Computed Tomography

Affected muscles show areas of decreased attenuation (hypodensity) on CT compared with normal muscles (Fig. 74–7).

Magnetic Resonance Imaging

The signal intensity of affected muscles on T1-weighted MR images is variable. On T2-weighted MR images, involved muscles show increased signal intensity.

DIABETIC MUSCLE INFARCTION (DMI)

Overview

DMI is a relatively rare complication of diabetes mellitus. It tends to occur in poorly controlled type 1 insulin-dependent diabetics, with most suffering from one or more diabetic complications, such as retinopathy, nephropathy, or neuropathy. Most reported patients had diabetes for at least 5 years. DMI is seen in men and

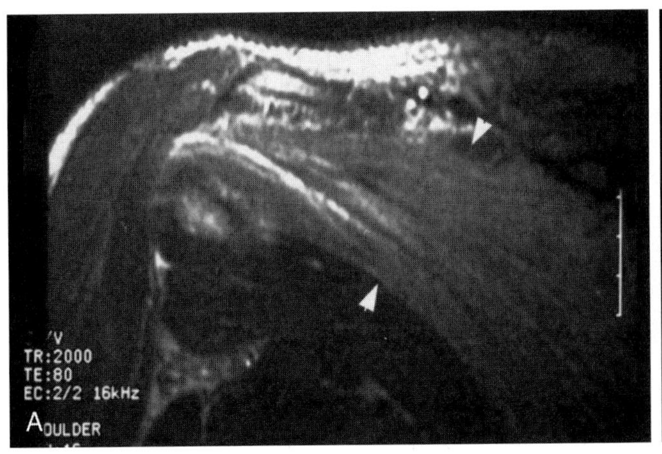

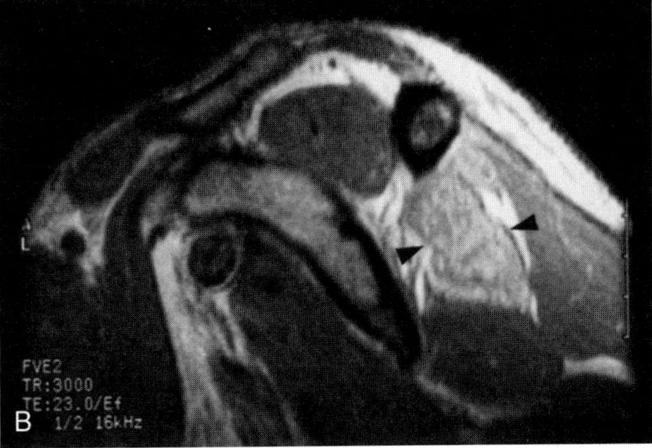

Figure 74–5. Infraspinatus muscle atrophy resulting from entrapment neuropathy of the suprascapular nerve. *A* and *B*, T2-weighted oblique coronal and proton-density oblique sagittal MR images of the right shoulder reveal increased signal intensity suggesting denervation and atrophy of the infraspinatus muscle *(arrowheads)*.

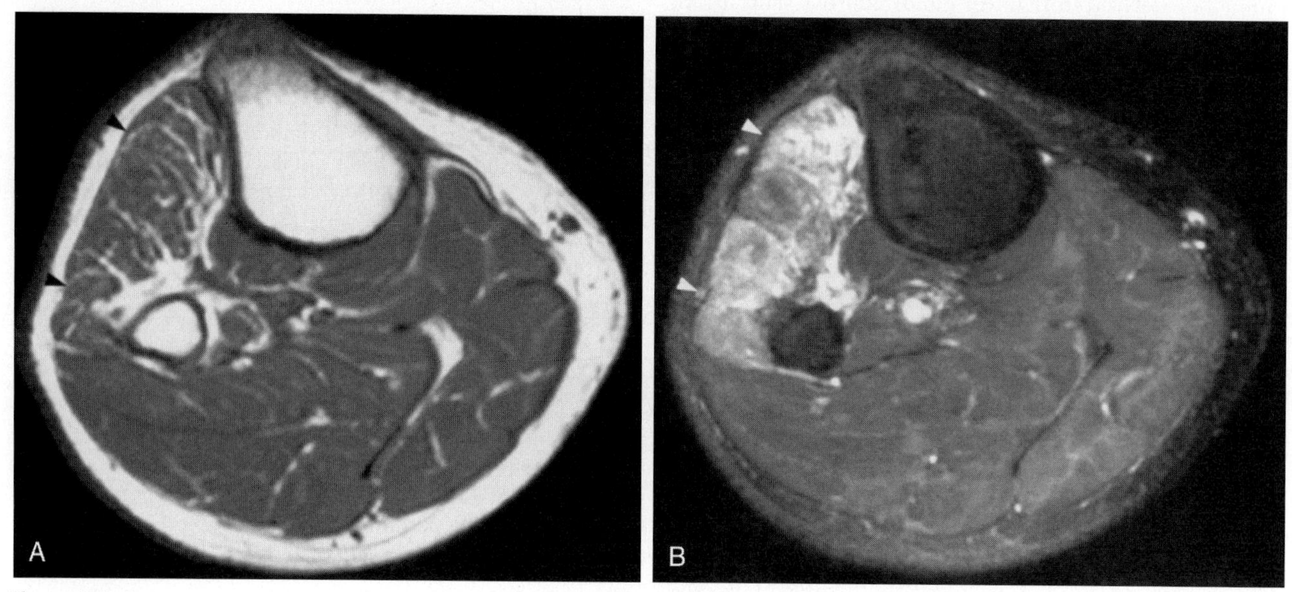

Figure 74–6. Denervation of the anterior compartment of the leg resulting from deep peroneal nerve injury. *A* and *B*, Axial T1- and fat suppressed T2-weighted MR images show increased signal intensity in the anterior compartment *(arrowheads)*. Muscle atrophy and fatty replacement are noted on the T1-weighted image *(A)*.

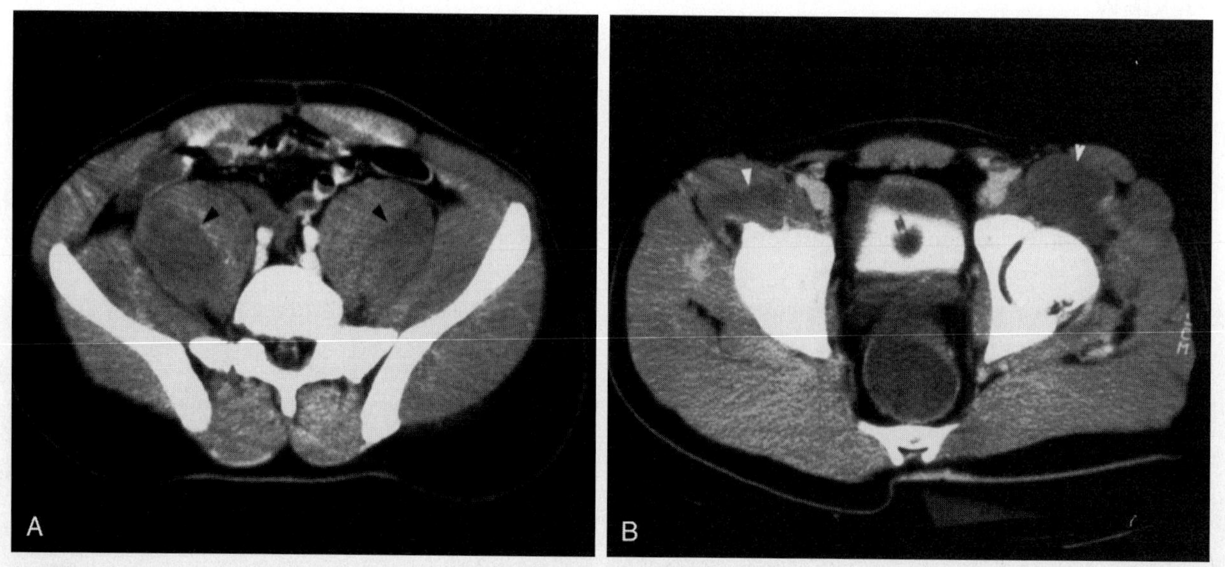

Figure 74–7. Rhabdomyolysis in a 35-year-old man who was in prolonged shock after abdominal trauma. The patient died of renal failure, which probably was related to myoglobinuria. *A* and *B*, CT scans through the pelvis show areas of decreased attenuation in the psoas and iliopsoas muscles *(arrowheads)* caused by rhabdomyolysis. (Case provided by Dr. Stuart Mirvis, Baltimore, Maryland.)

women, with slight female preponderance. Muscles of the thigh, buttock, and calf are involved most commonly, and in about one third of patients, the disease is bilateral. Clinical findings include sudden onset of pain, tenderness, and swelling, with or without a focal mass. These findings resolve gradually without antibiotics. DMI should be differentiated from deep venous thrombosis, early pyomyositis, and neoplasm. Laboratory findings reveal elevated erythrocyte sedimentation rate; however, the white blood cell count is typically normal, which helps in differentiating DMI from pyomyositis. The serum muscle enzyme creatine kinase level can be normal or elevated, and this may be related to the time of testing for creatine kinase with respect to the onset of the muscle infarct. The presumed cause of DMI is ischemia resulting from extensive thrombosis of medium and small arteries.

Imaging

MR imaging is the only useful modality in diagnosing DMI. It shows enlargement of the affected muscle or muscle group. T1-weighted MR images show loss of the normal fatty intramuscular septa, and affected muscles appear isointense with unaffected muscles. Tiny foci of increased signal intensity can be seen in some patients; this is believed to be due to focal hemorrhage. T2-weighted MR images show subcutaneous edema and subfascial fluid in most patients. T2-weighted images and gadolinium-enhanced T1-weighted images show similar hyperintensity within the affected muscles (Fig. 74–8). About two thirds of patients show areas of rim enhancement on gadolinium-enhanced images, with central dark areas. These areas are believed to represent foci of necrosis in ischemic muscle. It often is difficult to differentiate these areas from muscle abscesses. Core needle biopsy can be done to differentiate the two conditions.

CALCIFIC MYONECROSIS OF THE LEG

Overview

Calcific myonecrosis is a rare lesion of muscle usually resulting from a post-traumatic compartment syndrome and ischemia. The lesion presents as a slowly expanding mass. In one study with 19 cases, 8 presented within 1 year of the initial injury. Three of 19 cases presented more than 10 years after injury. The muscles of the anterior compartment of the leg are the most commonly involved, although the entity also has been described with the tibialis posterior and tensor fascia lata. This disorder has been associated with femoral or tibial fractures with ischemia, fibular fracture with peroneal nerve injury, gunshot wounds, and blunt trauma with anterior compartment syndrome.

Typically, calcific myonecrosis is a fusiform cystic mass. It has a fibrous wall and a variable amount of necrotic muscle, bone, fascia, nerve, and tendon tissue. There is blood and fibrin within the cyst. If the cyst becomes infected, it may present as an abscess.

These cysts are treated with wide débridement. If they are not completely débrided, superinfection and persistent drainage may result.

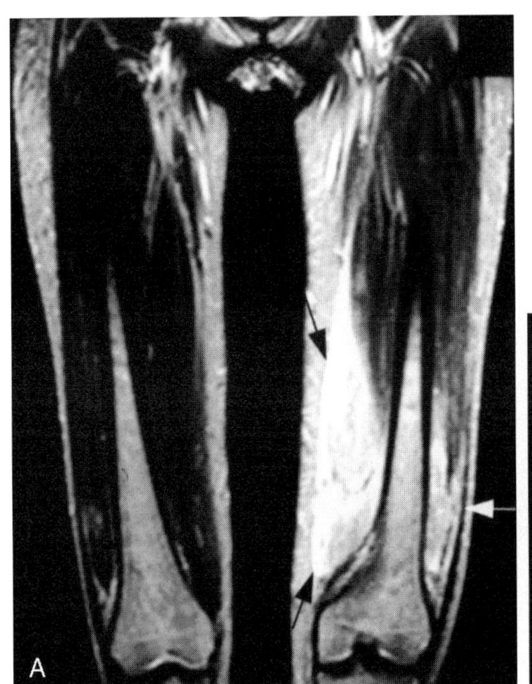

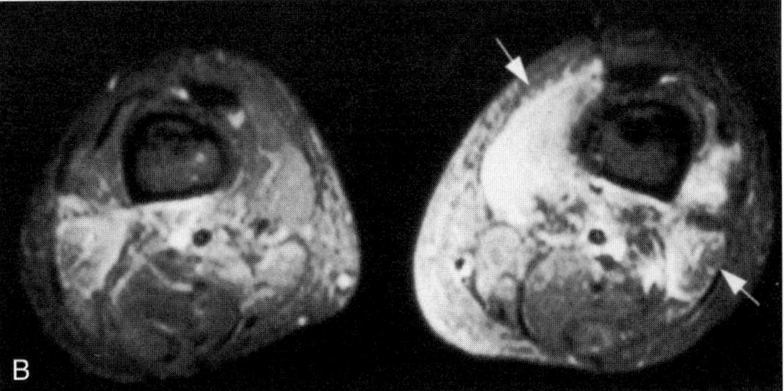

Figure 74–8. Diabetic muscle infarction of the left thigh muscles in a patient with poorly controlled diabetes mellitus. *A* and *B*, T2-weighted coronal and fat suppressed T2-weighted axial MR images reveal increased signal in the left thigh muscles *(arrows)* and to a lesser extent in the right, which on follow-up proved to be diabetic muscle infarction.

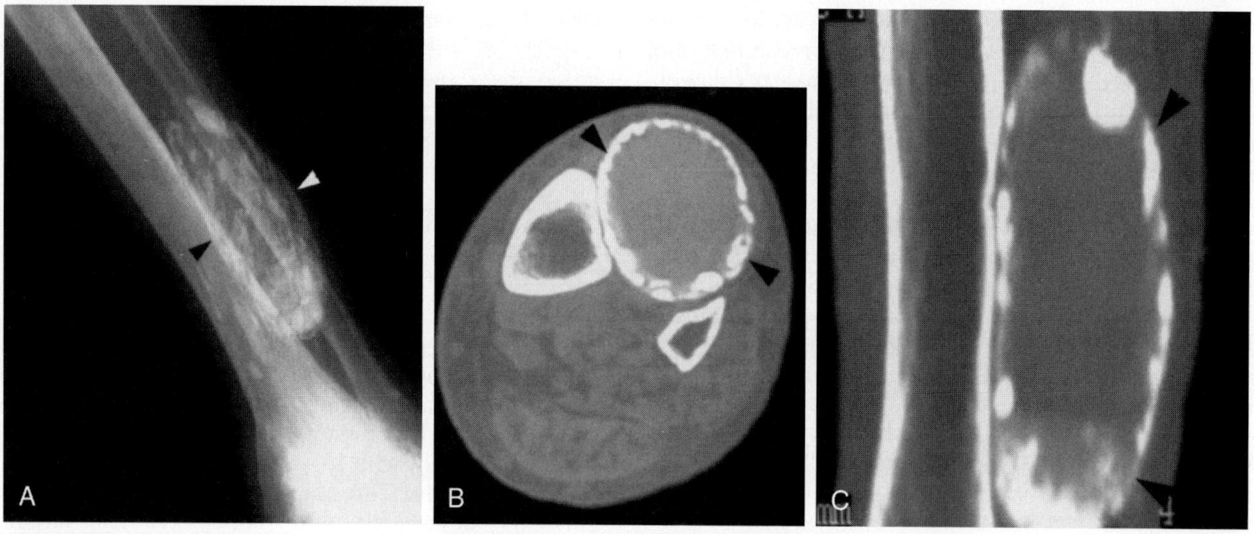

Figure 74–9. Calcific myonecrosis in a 77-year-old man with a remote history of trauma to the leg now presenting with an enlarging anterior leg mass. *A,* Anteroposterior view of the left leg shows a fusiform calcified mass in the leg *(arrowheads)*. *B* and *C,* Axial and coronal reconstructed CT scans show a cystic lesion with a calcified wall *(arrowheads)* occupying the anterior compartment of the leg. Bone remodeling is noted adjacent to the cyst.

Imaging

Radiographs show a fusiform mass most commonly in the anterior compartment of the leg (Fig. 74–9). The mass may have peripheral calcification, or there may be layering calcifications within the mass. Signs of previous trauma, such as fractures, gunshot wound, or surgery, are seen in more than half of the patients. Bone remodeling resulting from pressure by the cyst may be present in cases with a long history (Fig. 74–9*B*). Thick periosteal reaction is seen in some patients. CT or MR imaging shows a cystic mass frequently with a calcified wall (Fig. 74–9*B, C*). Multiple muscles in the anterior compartment may be involved. Differential diagnosis includes synovial sarcoma, epithelioid sarcoma, soft tissue osteosarcoma, myositis ossificans, dystrophic calcification associated with a muscle abscess, polymyositis, and dermatomyositis.

SKELETAL MUSCLE METASTASES

Overview

Skeletal muscle is the largest tissue in the body, forming nearly 50% of the body weight; however, metastases to skeletal muscles are rare. Metastases to skeletal muscle tend to occur in advanced stages of the neoplasm, when most patients (87%) have evidence of widespread metastatic disease (Fig. 74–10). Most muscle metastases are discovered incidentally because these lesions are neither painful nor palpable. It is not clear why metastatic tumors to skeletal muscles are rare, but some believe that factors related to blood flow and high tissue pressure may be the reason. Carcinomas from lung, pancreas, and kidney account for most of these metastases. Lymphoma, leukemia, and malignant melanoma (see Fig. 74–10) also have

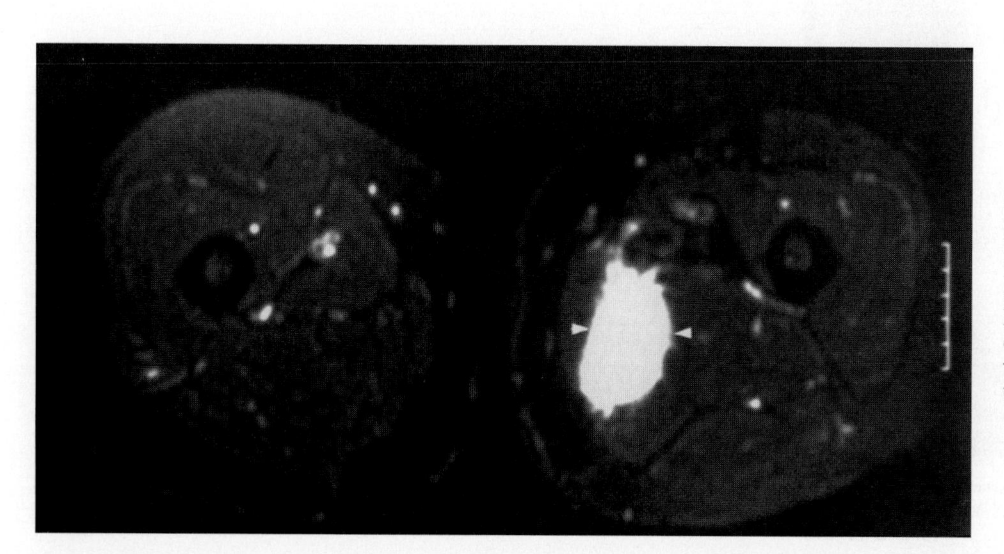

Figure 74–10. Skeletal muscle metastasis from a malignant melanoma of the skin. T2-weighted axial MR image with fat suppression shows a large high signal intensity lesion in the adductor magnus muscle on the left side *(arrowheads)*, which on needle aspiration proved to be a metastatic melanoma.

been reported. The most frequently involved muscles are the large, more central muscles, such as the iliopsoas, erector spinae, gluteals, thigh muscles, and diaphragm.

Imaging

The findings of skeletal muscle metastases on MR imaging are not pathognomonic; they can resemble primary muscle sarcoma, hematoma, and abscess (see Fig. 74–10). On contrast-enhanced CT, skeletal muscle metastasis presents as a rim-enhancing mass with central hypoattenuation.

Recommended Readings

Barloon TJ, Zachar CK, Harkens KL, et al: Rhabdomyolysis: Computed tomography findings. J Comput Assist Tomogr 12:193, 1988.

Chason DP, Fleckenstein JL, Burns DK, et al: Diabetic muscle infarction: Radiologic evaluation. Skeletal Radiol 25:127, 1996.

DeBeucketeer L, Vanhoenacker F, DeSchepper A Jr, et al: Hypertrophy and pseudohypertrophy of the lower leg following chronic radiculopathy and neuropathy: Imaging findings in two patients. Skeletal Radiol 28:229, 1999.

Fleckenstein JL, Watumull D, Conner KE, et al: Denervated human skeletal muscle: MR imaging evaluation. Radiology 187:213, 1993.

Janzen DL, Connell DG, Vaisler BJ: Calcific myonecrosis of the calf manifesting as an enlarging soft tissue mass: Imaging features. AJR Am J Roentgenol 160:1072, 1993.

Jelinek JS, Murphey MD, Aboulafia AJ, et al: Muscle infarction in patients with diabetes mellitus: MR imaging findings. Radiology 211:241, 1999.

Khoury NJ, El-Khoury GY, Kathol MH: MRI diagnosis of diabetic muscle infarction: Report of two cases. Skeletal Radiol 26:122, 1997.

Patel R, Mishkin FS: Technetium-99m pyrophosphate imaging in acute renal failure associated with nontraumatic rhabdomyolysis. AJR Am J Roentgenol 147:815, 1986.

Petersilge CA, Pathria MN, Gentili A, et al: Denervation hypertrophy of muscle: MR features. J Comput Assist Tomogr 19:596, 1995.

Pretorius ES, Fishman EK: Helical CT of skeletal muscle metastases from primary carcinomas. AJR Am J Roentgenol 174:401, 2000.

Ramirez H Jr, Brown JD, Evans HW Jr: Case report 225. Skeletal Radiol 9:223, 1983.

Ryu KN, Bae DK, Park YK, et al: Calcific tenosynovitis associated with calcific myonecrosis of the leg: Imaging features. Skeletal Radiol 25:273, 1996.

Sledz KM, Granke DS, Chendrasekhar A, et al: Necrotizing soft tissue infections of the body wall: Computed tomographic evaluation. Emergency Radiology 3:261, 1996.

Sridhar KS, Rao RK, Kunhardt B: Skeletal muscle metastases from lung cancer. Cancer 59:1530, 1987.

Stock KW, Helwig A: MR imaging of acute exertional rhabdomyolysis in the paraspinal compartment. J Comput Assist Tomogr 20:834, 1996.

Uetani M, Hayashi K, Matsunaga N, et al: Denervated skeletal muscle: MR imaging. Radiology 189:511, 1993.

Vukanovic S, Hauser H, Wettstein P: CT localization of myonecrosis for surgical decompression. AJR Am J Roentgenol 135:1298, 1980.

Williams JB, Youngberg RA, Bui-Mansfield LT, et al: MR imaging of skeletal muscle metastases. AJR Am J Roentgenol 168:555, 1997.

Wysoki MG, Santora TA, Shah RM, et al: Necrotizing fasciitis: CT characteristics. Radiology 203:859, 1997.

SECTION VII

MISCELLANEOUS MUSCULOSKELETAL DISORDERS

CHAPTER 75

Myositis Ossificans

Nabil J. Khoury, MD, and Georges Y. El-Khoury, MD

OVERVIEW

Myositis ossificans is heterotopic formation of bone in soft tissues, mainly muscles. Most patients have had a previous traumatic injury. Other predisposing factors include paraplegia, burns, and tetanus. Myositis ossificans presents clinically as a tender soft tissue mass that becomes gradually hard. Histology is similar to normal bone with a proliferative mesenchymal zone that undergoes differentiation into mature bone from the periphery to the center. During the active proliferative phase, myositis ossificans resembles osteosarcoma histologically; recognizing this entity in the early stages by imaging helps avoid a biopsy that could be misleading.

LOCATION

Any muscular area can be involved by myositis ossificans. In post-traumatic cases, the large muscles of the thigh, especially the quadriceps (Fig. 75–1), and the arm, especially the brachialis muscle, most commonly are involved.

IMAGING

Plain Radiographs

In the early phase, there is soft tissue swelling of the involved area. After about 1 month, ill-defined faint

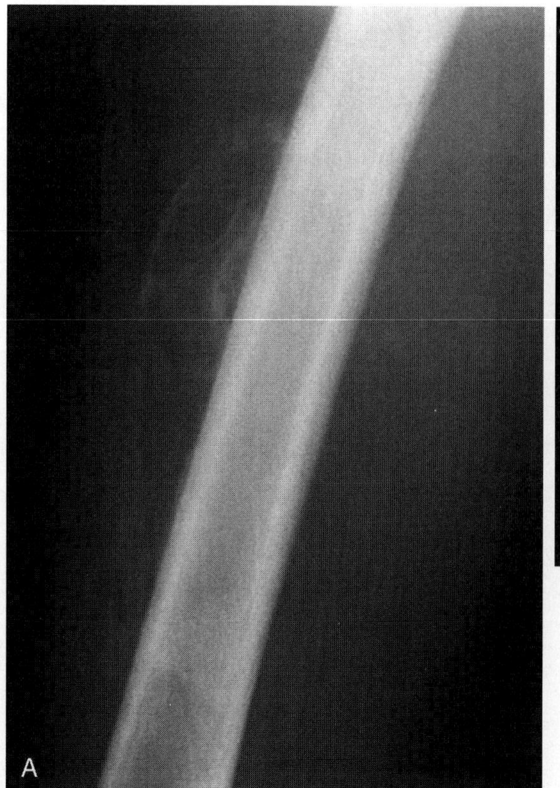

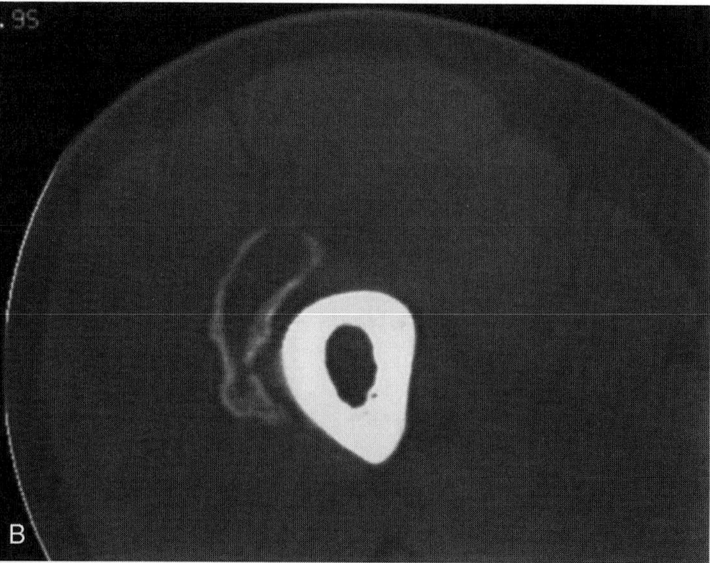

Figure 75–1. Typical appearance of myositis ossificans. *A*, Radiograph of the right thigh shows a large area of soft tissue calcification in the vastus intermedius muscle. *B*, CT scan of the same area shows the calcification starting in the periphery of the lesions (zoning phenomenon).

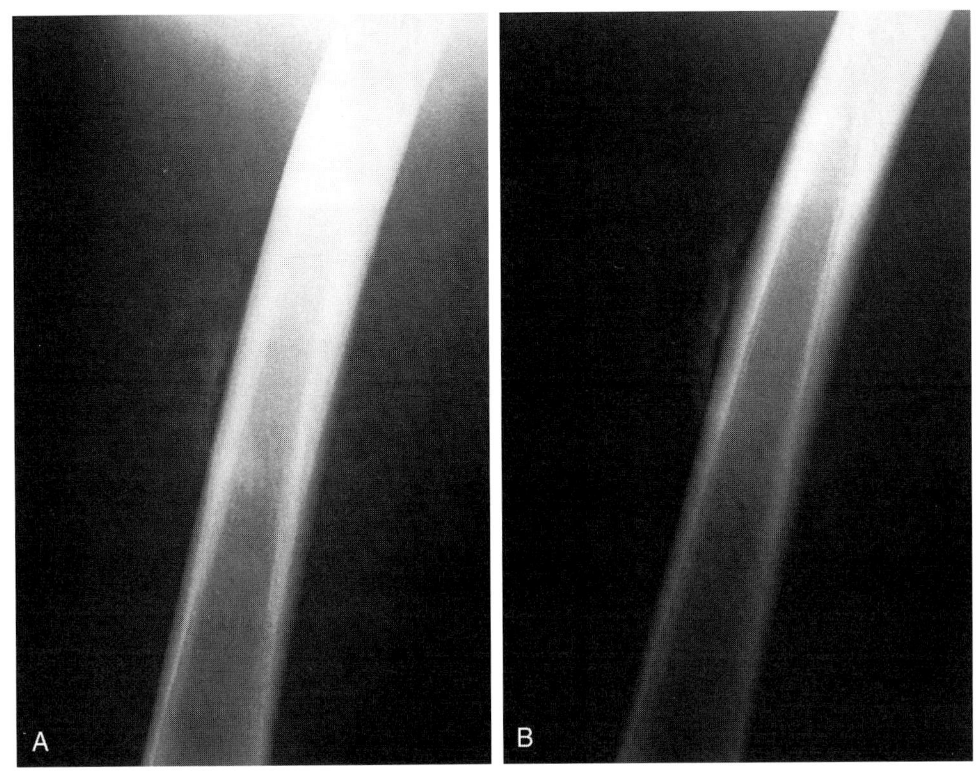

Figure 75–2. Myositis ossificans shrinking in size. *A,* Myositis ossificans in the anterior thigh about 1 month after the injury. *B,* Two months after the injury, the mass is smaller, and bone within the mass appears more mature.

calcifications start to appear. One to 4 months later there is gradual loss of volume of the mass (Fig. 75–2) and formation of a well-defined calcified rim and a central lucency, which is typical of myositis ossificans (zoning phenomenon) (see Fig. 75–1*B*). The lesion is separated from the adjacent bone by a lucent line, and the underlying cortex is always intact (see Fig. 75–2). With time, myositis ossificans becomes adherent to the cortex. A periosteal reaction can develop in association with myositis ossificans (Fig. 75–3). The periosteal reaction can appear lamellated but typically uninterrupted and benign looking.

Computed Tomography Scan

Computed tomography (CT) is the best modality to show the zoning phenomenon and the lucent soft tissue plane between the lesion and the adjacent bone (see Fig. 75–1*B*). Identifying the zoning phenomenon is key in differentiating myositis ossificans from surface osteosarcoma.

Magnetic Resonance Imaging

Magnetic resonance (MR) imaging should be avoided, especially early in the disease, when myositis ossificans can resemble a neoplasm. The affected muscle shows extensive edema (increased signal on T2-weighted images) involving almost the entire muscle (Fig. 75–4). A rounded mass with low signal intensity develops within the edematous muscle (Fig. 75–4*B*). As the ossific mass matures and shrinks in size, the high signal intensity in the soft tissue around the mass diminishes. Rarely, there is bone marrow edema in the adjacent bone.

DIFFERENTIAL DIAGNOSIS

On radiography and CT scan, myositis ossificans should be differentiated from parosteal osteosarcoma and extraosseous osteosarcoma. These tumors ossify in the center first, whereas myositis ossificans starts ossifying in the periphery (see Fig. 75–1*B*). The appearance of early myositis ossificans on MR imaging can be confusing; it can mimic a soft tissue neoplasm or an inflammatory mass.

Recommended Readings

Amendola MA, Glazer GM, Agha FP, et al: Myositis ossificans circumscripta: Computed tomographic diagnosis. Radiology 149:775, 1983.

De Smet AA, Norris MA, Fisher DR: Magnetic resonance imaging of myositis ossificans: Analysis of seven cases. Skeletal Radiol 21:503, 1992.

Drane WE: Myositis ossificans and the three-phase bone scan. AJR Am J Roentgenol 142:179, 1984.

Goldman AB: Myositis ossificans circumscripta: A benign lesion with a malignant differential diagnosis. AJR Am J Roentgenol 126:32, 1976.

Kransdorf MJ, Meis JM, Jelinek JS: Myositis ossificans: MR appearance with radiologic-pathologic correlation. AJR Am J Roentgenol 157: 1243, 1991.

Norman A, Dorfman HD: Juxtacortical circumscribed myositis ossificans: Evolution and radiographic features. Radiology 96:301, 1970.

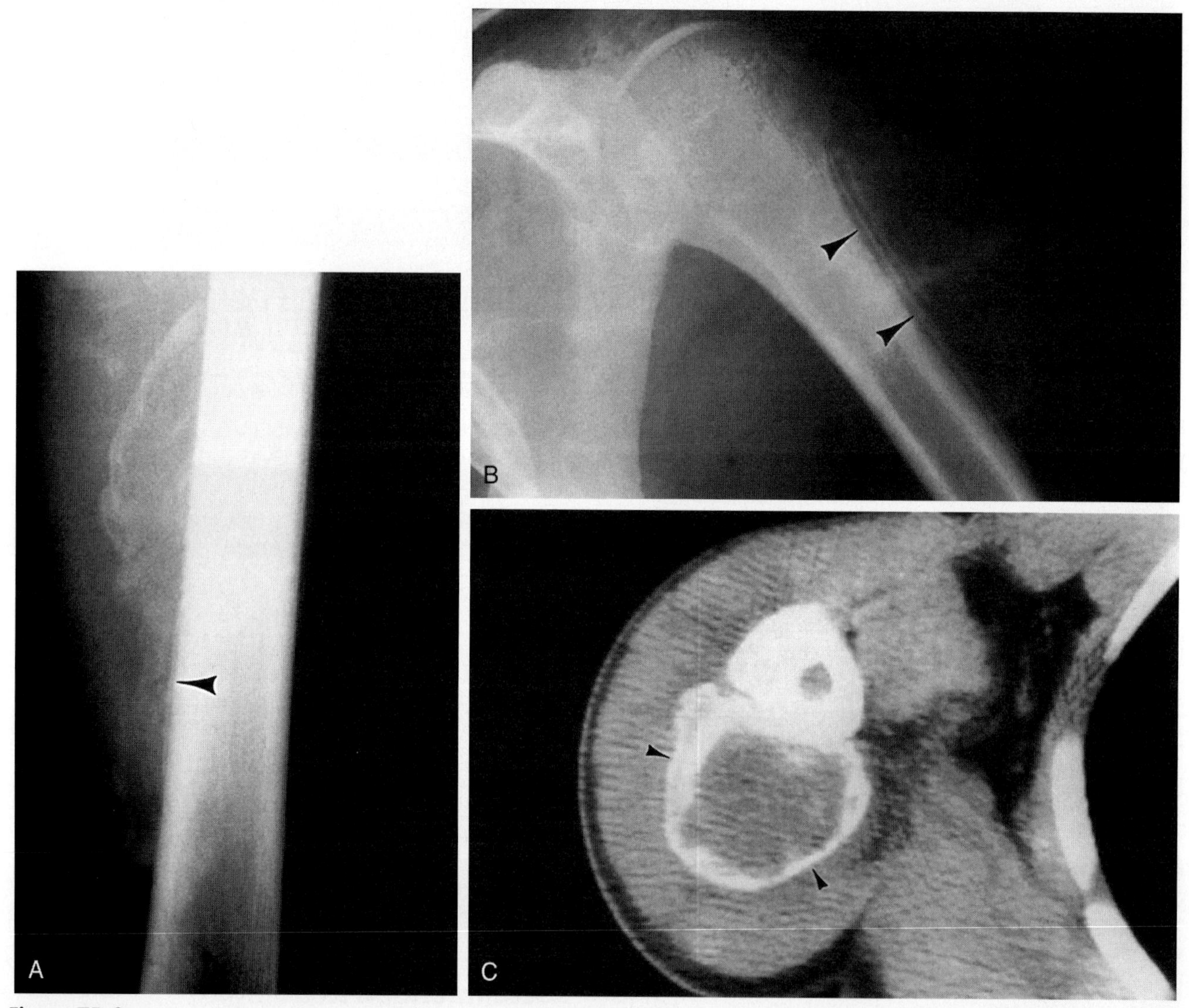

Figure 75–3. Periosteal reaction associated with myositis ossificans. *A,* Note the thin continuous periosteal reaction *(arrowhead). B,* Lamellated periosteal reaction in myositis ossificans *(arrowheads).* This is a rare finding in myositis ossificans. *C,* CT scan of the arm in the same patient as in *B* shows typical findings of myositis ossificans with zoning phenomenon *(arrowheads).* (Case provided by Dr. Hirokazu Mizutani from Nagoya, Japan.)

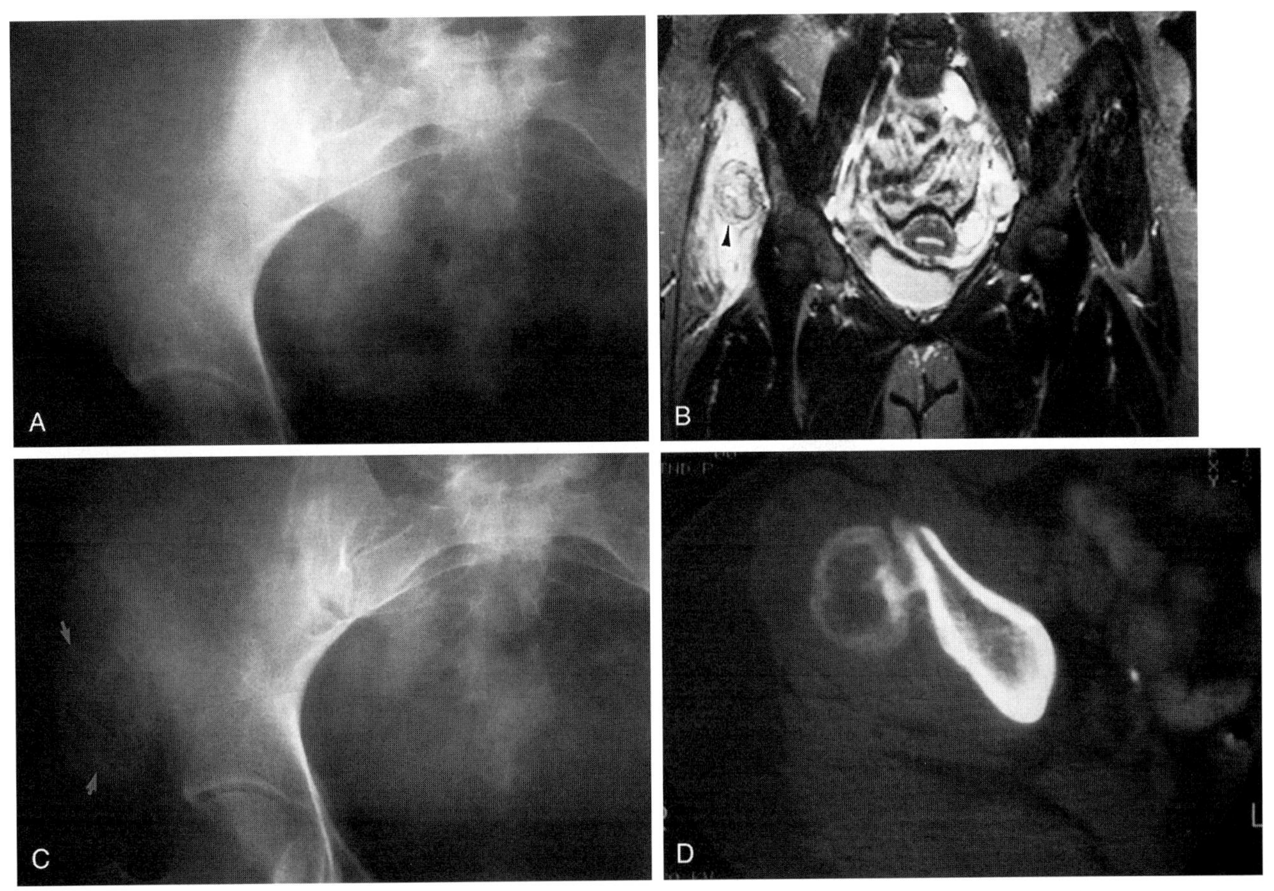

Figure 75–4. Early myositis ossificans misinterpreted clinically and on MR imaging for a soft tissue tumor. *A*, Anteroposterior radiograph obtained 3 weeks after an injury on the football field shows no abnormalities. *B*, T2-weighted coronal MR image of the pelvis performed at the same time reveals extensive edema within the gluteus medius and minimus muscles. There is a rounded mass in the center of the edematous muscles, with low signal intensity in its rim *(arrowhead)*. *C*, Anteroposterior radiograph was repeated 3 weeks later. It shows a calcific mass attached to the ilium *(arrows)*. *D*, CT scan confirms the diagnosis of myositis ossificans based on the zoning phenomenon.

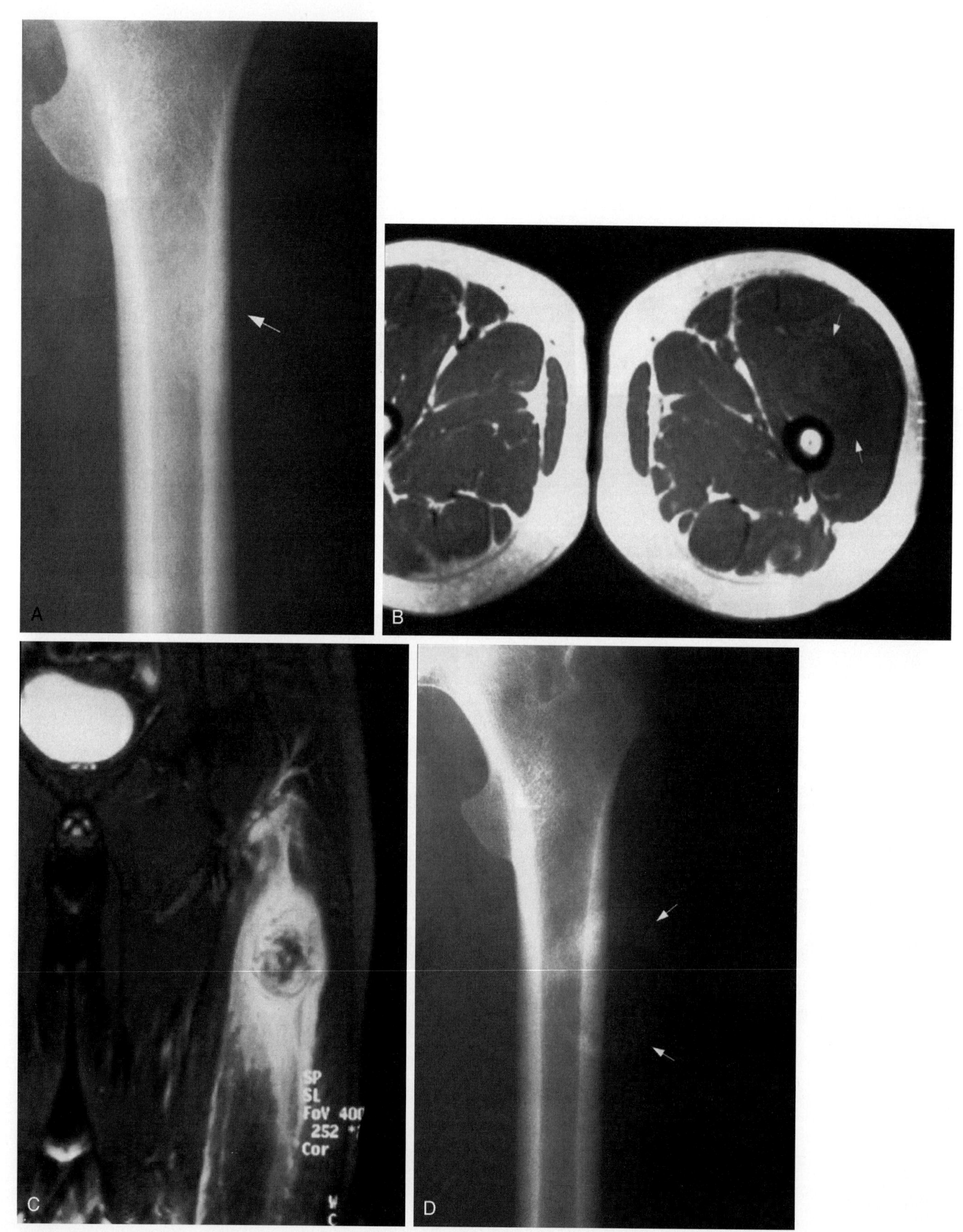

Figure 75–5. Typical MR imaging appearance of early myositis ossificans in a 22-year-old athelete presenting with a painful mass in the anterior thigh. *A,* Initial anteroposterior radiograph of the left thigh shows a small faint calcification in the soft tissues of the anterior thigh *(arrow). B* and *C,* T1-weighted axial and T2-weighted coronal MR images performed 10 days after the initial radiograph show a large rounded mass with low signal intensity rim *(arrows)* surrounded by edema. *D,* Repeat anteroposterior radiograph 1 month later shows the typical appearance of myositis ossificans with peripheral calcifications, or possibly ossification *(arrows),* also known as the zoning phenomenon.

Nuovo MA, Norman A, Chumas J, et al: Myositis ossificans with atypical clinical, radiographic, or pathologic findings: A review of 23 cases. Skeletal Radiol 21:87, 1992.
Ogilvie-Harris DJ, Fornasier VL: Pseudomalignant myositis ossificans: Heterotopic new-bone formation without a history of trauma. J Bone Joint Surg Am 62:1274, 1980.
Peck RJ, Metreweli C: Early myositis ossificans: A new echographic sign. Clin Radiol 39:586, 1988.
Shirkhoda A, Armin AR, Bis KG, et al: MR imaging of myositis ossificans: Variable patterns at different stages. J Magn Reson Imaging 5:287, 1995.

CHAPTER 76

Fibrodysplasia Ossificans Progressiva

Nabil J. Khoury, MD, and Georges Y. El-Khoury, MD

OVERVIEW

Fibrodysplasia ossificans progressiva (FOP) is a rare congenital mesodermal disorder. The disease is characterized by progressive ossification of striated muscles, tendons, ligaments, and fasciae resulting in diminished mobility and progressive disability. The pattern of inheritance is unknown but may be autosomal dominant with a wide range of expressivity. Males and females are equally affected. Onset is usually between birth and 5 years of age. Histologically, there are inflammatory changes, followed by mesenchymal proliferation and membranous and lamellar bone formation. Biopsy should be avoided because it can stimulate additional bone formation.

CLINICAL PRESENTATION

Any area of the body can be involved in FOP, with the most common being the shoulder, thoracic cage, and paraspinal regions. The facial muscles, tongue, diaphragm, visceral smooth muscles, and abdominal wall muscles are all spared. Early clinical symptoms include heat, edema, and painful soft tissue mass. With progression, the mass ossifies, and contractures develop, leading to restricted mobility. The most frequent presenting symptom is torticollis. Congenital skeletal abnormalities of the thumb and great toe are seen in 90% of patients.

IMAGING

On plain radiographs, heterotopic ossifications in FOP appear as ossified bony bridges between different bones (Fig. 76–1). Secondary joint subluxations and fractures of the ossified bridges may occur followed by the formation of pseudarthroses. Congenital bony abnormalities appear as a malformed first metatarsal with hallux valgus and short great toes (Fig. 76–2). A short first metacarpal also is described (Fig. 76–3). These malformations are characteristic and can be detected at birth.

Computed tomography (CT) shows the early changes of FOP as areas of soft tissue swelling (Fig. 76–3A) followed by rounded or linear foci of ossification noted within the areas of swelling. In adults, the ossifications can become extensive and dense (Fig. 76–4). Radionuclide bone scan is more sensitive than plain radiographs in showing early soft tissue ossifications.

Magnetic resonance (MR) imaging may be helpful in detecting early soft tissue changes before the development of ossifications. The MR abnormalities are nonspecific, however; they consist of swelling of the involved muscles and connective tissues simulating a mass lesion (see Fig. 76–2A). The signal intensity is isointense to muscles on T1-weighted images and increased on T2-weighted images. With progression of the disease, there is decrease in the degree of swelling and in the signal intensity on T2-weighted images denoting ossification and fibrosis.

Recommended Readings

Carter SR, Davies AM, Evans N, et al: Value of bone scanning and computed tomography in fibrodysplasia ossificans progressiva. Br J Radiol 62:269, 1989.
Cohen RB, Hahn GV, Tabas JA, et al: The natural history of heterotopic ossification in patients who have fibrodysplasia ossificans progressiva: A study of forty-four patients. J Bone Joint Surg Am 75:215, 1993.
Connor JM, Evans DAP: Fibrodysplasia ossificans progressiva: The clinical features and natural history of 34 patients. J Bone Joint Surg Br 64:76, 1982.
Kaplan FS, Strear CM, Zasloff MA: Radiographic and scintigraphic features of modeling and remodeling in the heterotopic skeleton of

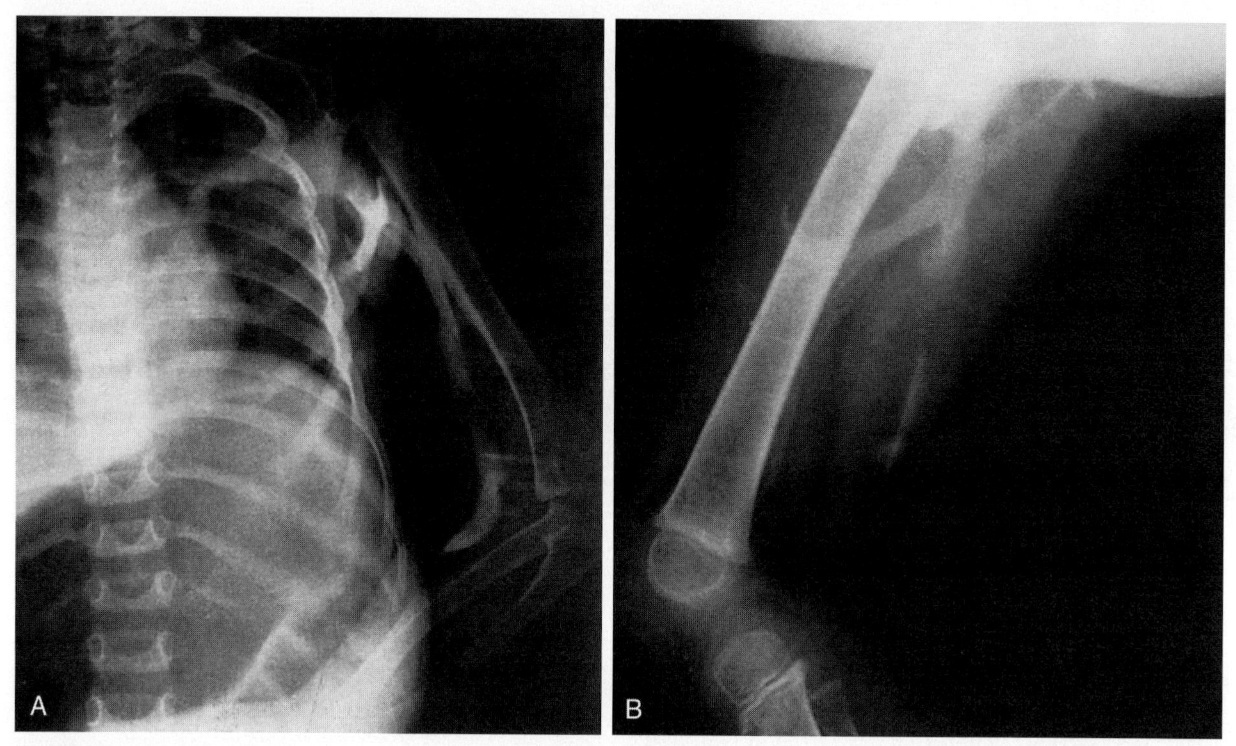

Figure 76–1. Radiographic changes of fibrodysplasia ossificans progressiva in a child. *A,* The left chest wall, axilla, arm, and elbow have ridges of mature bone. *B,* The muscles of the posterior thigh are involved with the same process.

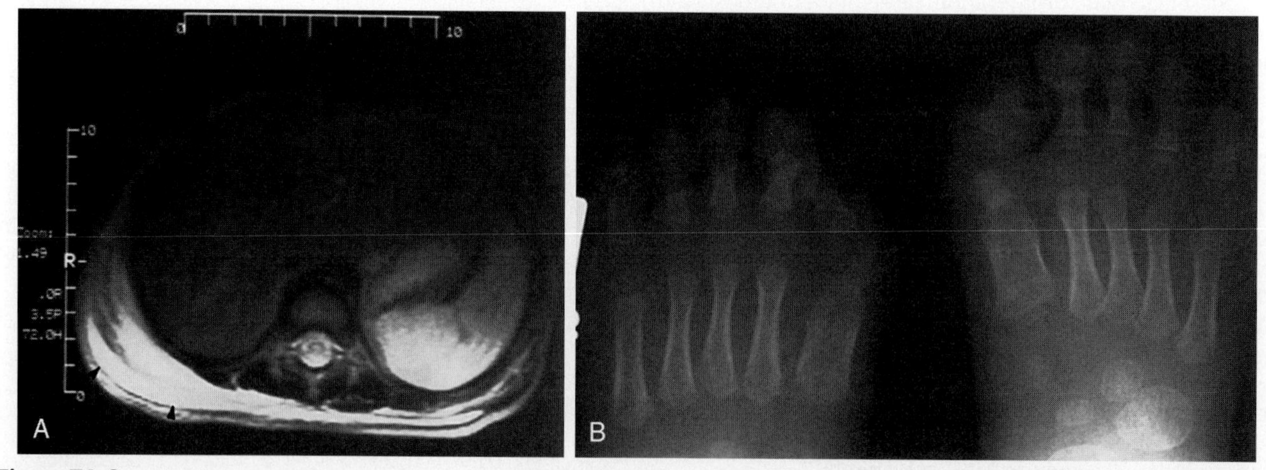

Figure 76–2. Early inflammatory changes of fibrodysplasia ossificans progressiva. *A,* T2-weighted MR image of the abdomen shows edema in the back muscles *(arrowheads)* on the right side. *B,* Posteroanterior radiograph of the feet shows short deformed great toes bilaterally, which are diagnostic of fibrodysplasia ossificans progressiva.

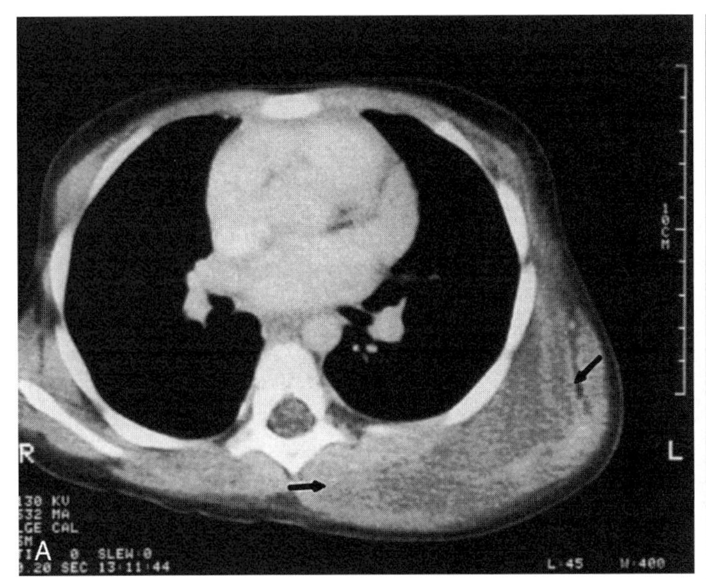

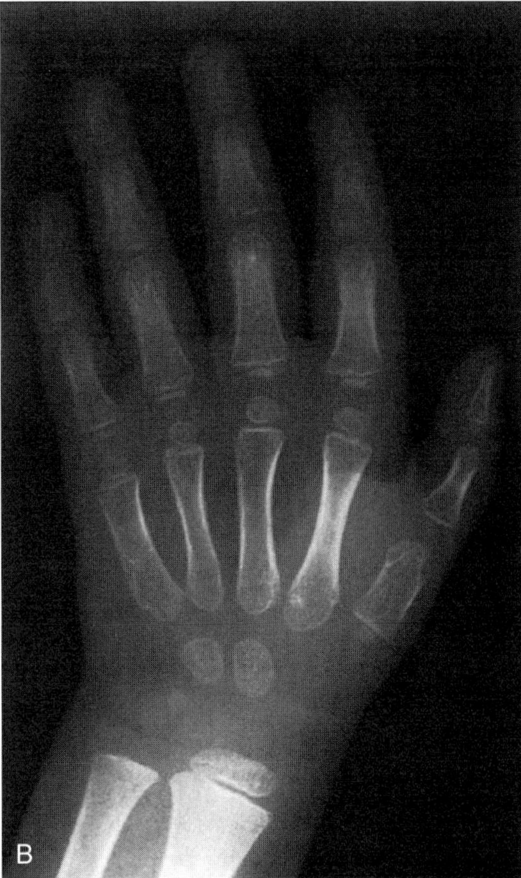

Figure 76–3. Early inflammatory changes of fibrodysplasia ossificans progressiva detected on CT scan. *A*, CT scan of the chest wall in a child with a large soft tissue mass below the left scapula shows significant soft tissue swelling compared with the right *(arrows)*. No calcifications are identified at this early stage. *B*, Radiograph of the right hand shows short first metacarpal.

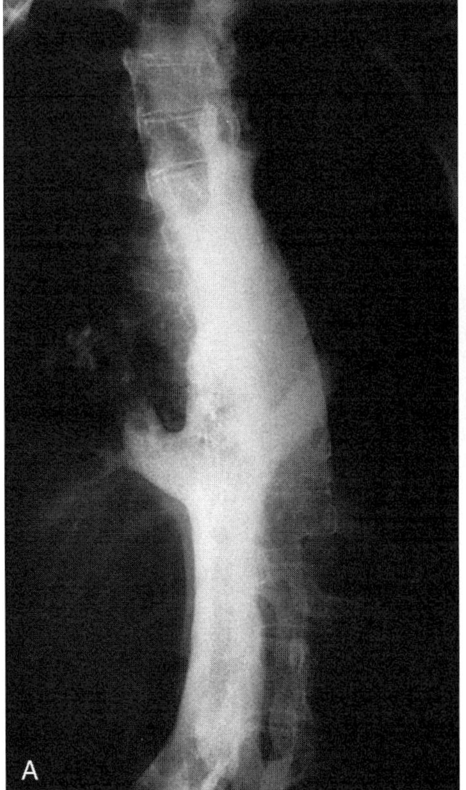

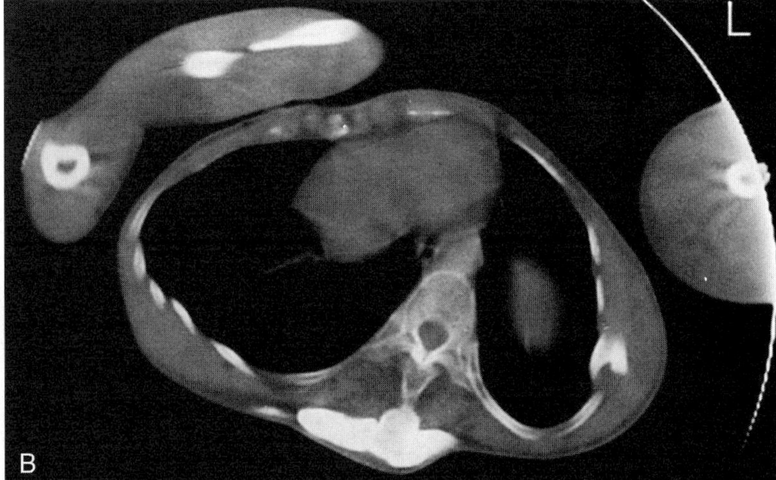

Figure 76–4. Mature soft tissue ossifications in an adult with fibrodysplasia ossificans progressiva. *A*, Anteroposterior radiograph of the spine reveals extensive ossifications in the soft tissues posterior to the spine. *B*, CT scan shows soft tissue ossifications posterior to the spine and in the left chest wall. The rib cage is deformed by fibrodysplasia ossificans progressiva.

patients who have fibrodysplasia ossificans progressiva. Clin Orthop 304:238, 1994.
Lindhout D, Golding RP, Taets van Amerongen AHM: Fibrodysplasia ossificans progressiva: Current concepts and the role of CT in acute changes. Pediatr Radiol 15:211, 1985.
Reinig JW, Hill SC, Fang M, et al: Fibrodysplasia ossificans progressiva: CT appearance. Radiology 159:153, 1986.
Smith R: Fibrodysplasia (myositis) ossificans progressiva: Clinical lessons from a rare disease. Clin Orthop 346:7, 1998.
Thickman D, Bonakdar-pour A, Clancy M, et al: Fibrodysplasia ossificans progressiva. AJR Am J Roentgenol 139:935, 1982.

CHAPTER 77

Condensing Bone Diseases

Nabil J. Khoury, MD, and Georges Y. El-Khoury, MD

Condensing bone diseases, or sclerosing bone dysplasias, are a group of developmental bone abnormalities of unknown cause, characterized by enchondral or intramembranous bone formation. These dysplasias probably have common defects of bone formation and resorption; this is best illustrated by what is known as *mixed sclerosing bone dystrophy,* which is an *overlap syndrome* in which two or more of these dysplasias occur in the same patient (Fig. 77–1). The most frequently encountered dysplasias are osteopoikilosis, melorheostosis, and osteopathia striata.

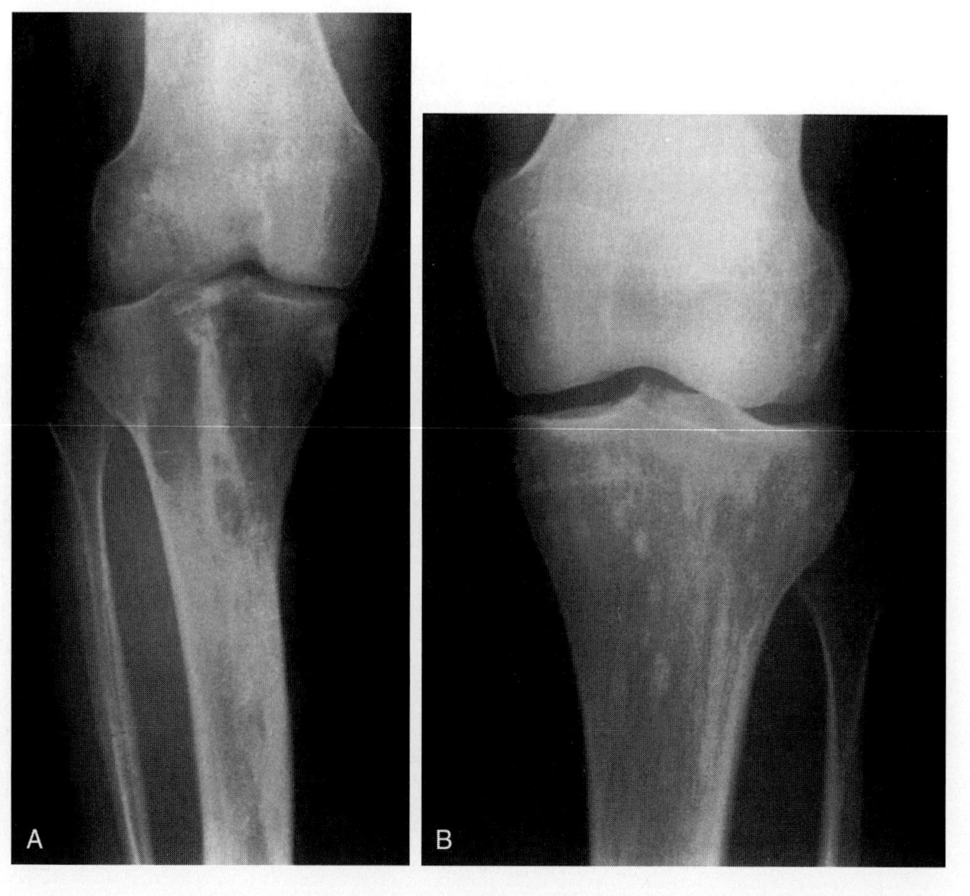

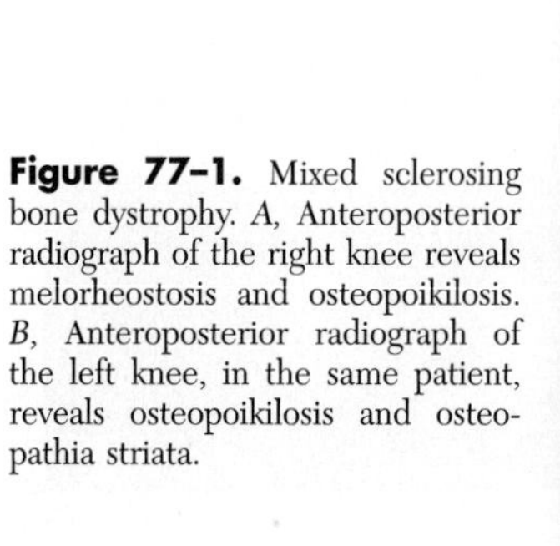

Figure 77–1. Mixed sclerosing bone dystrophy. *A,* Anteroposterior radiograph of the right knee reveals melorheostosis and osteopoikilosis. *B,* Anteroposterior radiograph of the left knee, in the same patient, reveals osteopoikilosis and osteopathia striata.

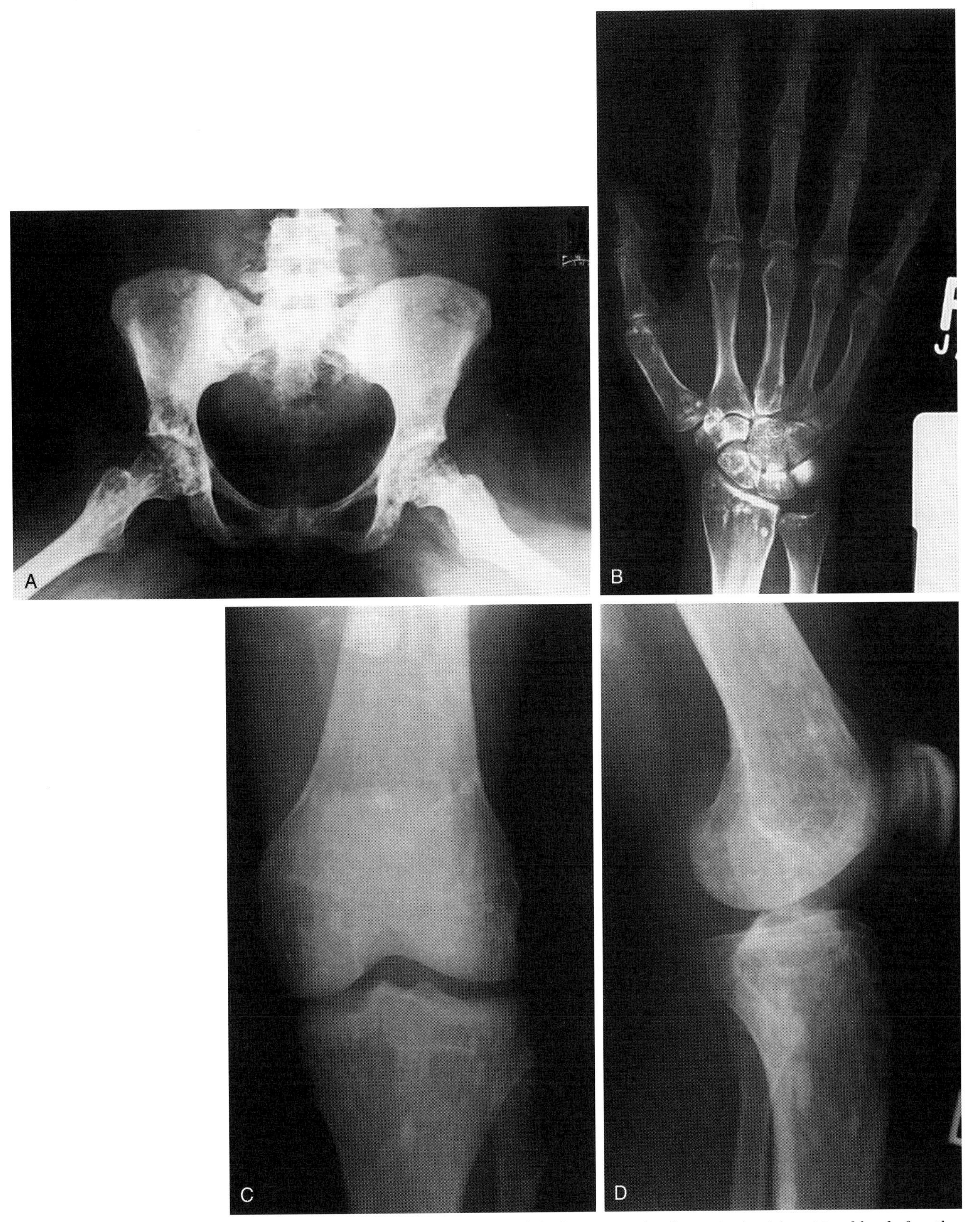

Figure 77–2. Osteopoikilosis. *A*, The small bone islands are centered around the hips. *B*, Similar changes in the right wrist and hand of another patient. *C* and *D*, Anteroposterior and lateral views of the knee in another asymptomatic patient in whom osteopoikilosis was discovered accidentally.

Illustration continued on following page

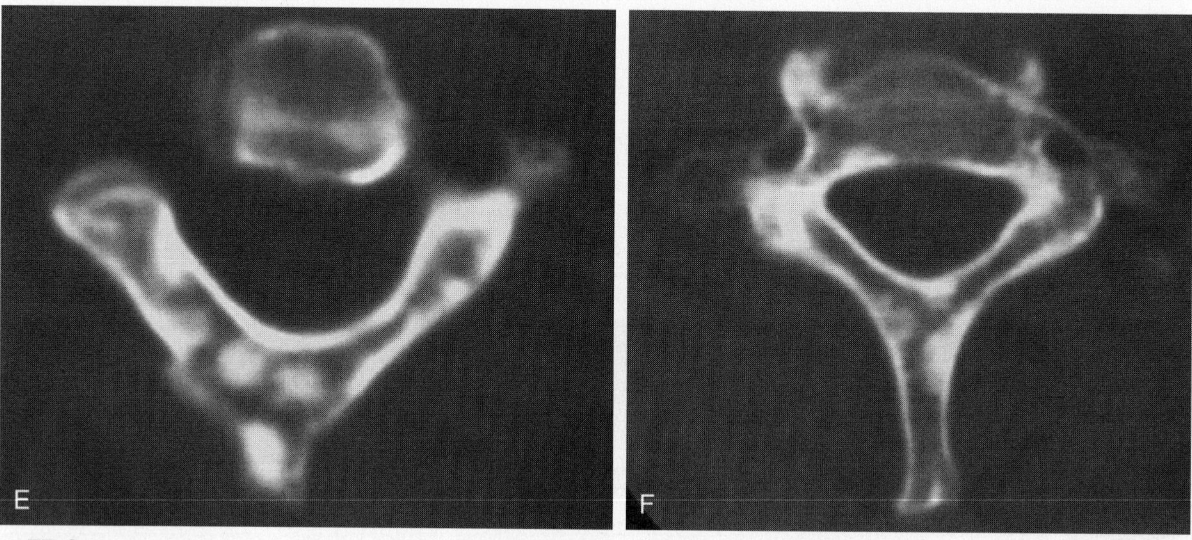

Figure 77–2 *Continued. E* and *F,* CT of the cervical spine shows osteopoikilosis involving the arches of C2 and C7. This patient is a 36-year-old female who was involved in a car accident.

OSTEOPOIKILOSIS

Overview

Osteopoikilosis is a rare asymptomatic condensing bone disease of unknown cause. Osteopoikilosis may be inherited as an autosomal dominant disease. It is characterized by the presence of multiple small sclerotic bone lesions, which resemble bone islands, and the disease usually is discovered incidentally. The involvement is almost always bilateral but not equal. Histologically the lesions reveal compact lamellar bones.

Skeletal Involvement

Almost all bones can be involved by osteopoikilosis, but the spine and the skull are generally exempt, although the authors have seen one case where the spine was involved (Fig. 77–2*E* and *F*). The long bones and carpal and tarsal bones are most commonly affected. In long bones, the lesions occur around joints, and the lesions are located within the medullary space.

Imaging

Radiographs are diagnostic, showing multiple, round or oval sclerotic lesions (Fig. 77–2). These lesions are of variable size but usually small. Symmetrical distribution is frequent. The lesions may change in size and number and may disappear. They can progress in young individuals and regress in older individuals. Bone scans are typically normal.

Differential Diagnosis

Osteopoikilosis may be confused with blastic metastasis in patients with a known primary malignancy (Fig. 77–3). The distribution of the lesions around joints, their small size, and the negative bone scan should help in differentiating osteopoikilosis from blastic bone metastasis.

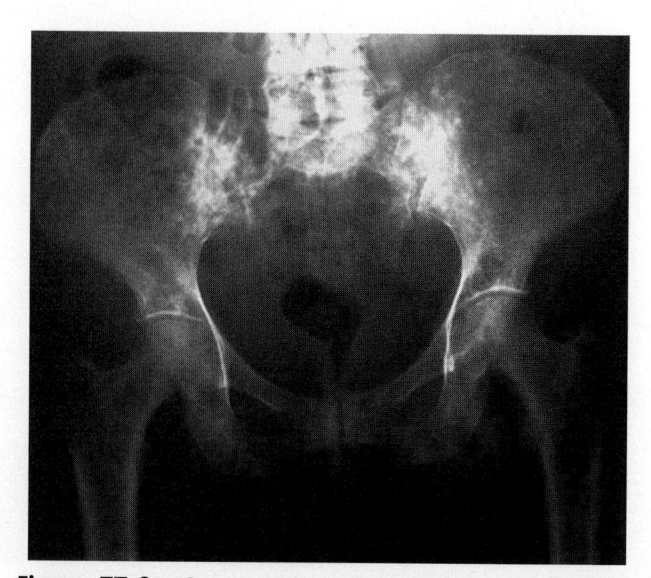

Figure 77–3. Blastic metastasis from carcinoma of the breast resembling osteopoikilosis. There are small blastic lesions throughout the pelvis and to a lesser extent in the hips. The distribution, size, and density of the lesions are different from osteopoikilosis.

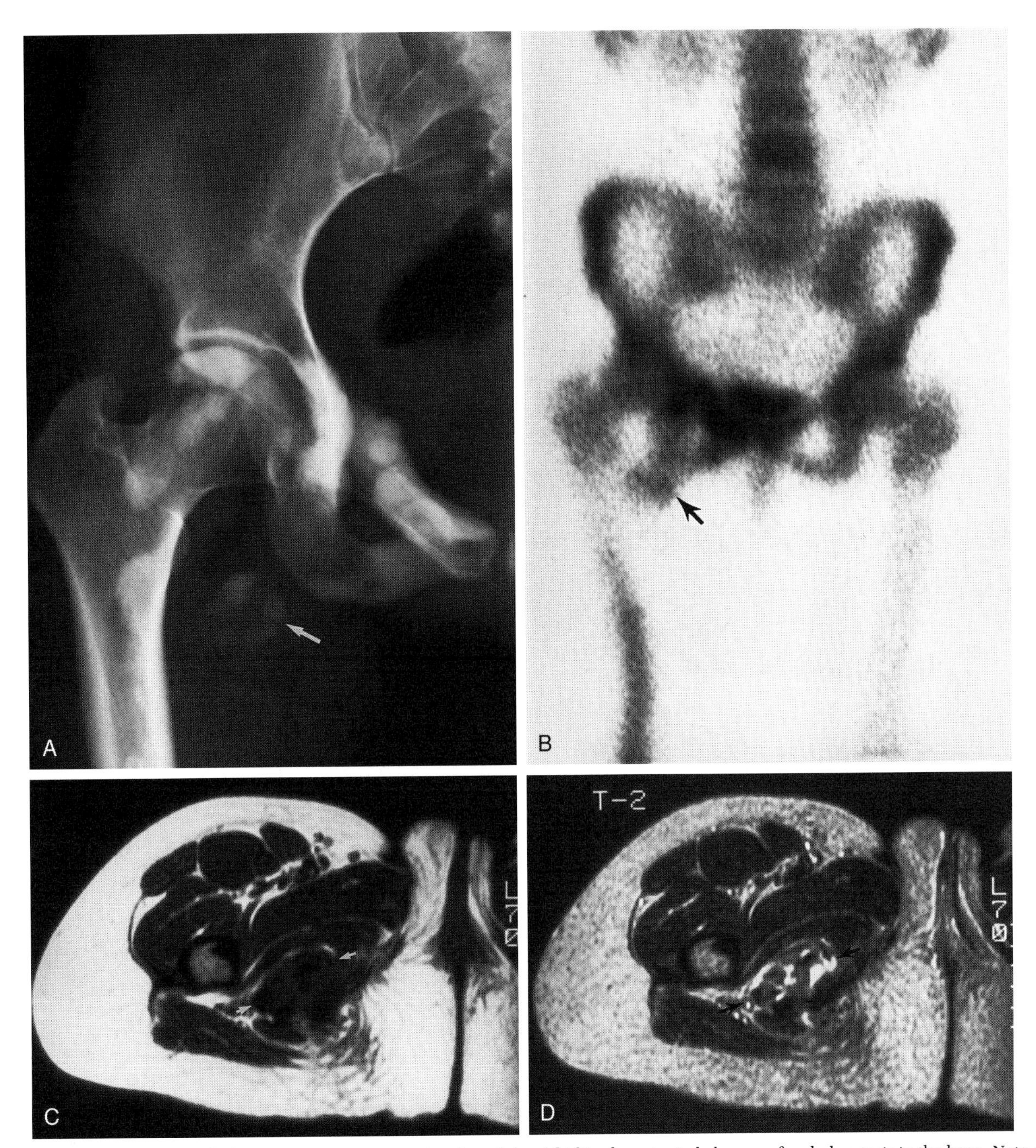

Figure 77–4. Soft tissue melorheostosis. *A*, Anteroposterior view of the right hip shows typical changes of melorheostosis in the bone. Note also the soft tissue ossifications *(arrow)*, which represents soft tissue melorheostosis. *B*, Radionuclide (Tc 99m MDP) bone scan shows moderately increased radiotracer uptake in the involved bones and soft tissues *(arrow)*. *C* and *D*, T1- and T2-weighted axial MR images of the proximal thigh demonstrate the soft tissue melorheostosis *(arrows)*. The areas of markedly increased signal intensity on the T2-weighted image are difficult to explain but could represent either fluid or slow flow in surrounding veins (Images provided by Dr. Mark Murphy, Washington DC.)

MELORHEOSTOSIS

Overview

Melorheostosis is a benign nonheritable mesodermal disorder of unknown cause characterized by dense bone proliferation along the cortical margin of the bones. The diagnosis is made based on the characteristic radiographic findings. The histologic features are nonspecific. The disease can be seen at any age but is detected mostly in young adults. It progresses rapidly in children and progresses slowly or remains stable in adults. Males and females are equally affected. Melorheostosis can be associated with other mesenchymal abnormalities, including soft tissue and skin fibrosis, linear scleroderma, and para-articular soft tissue ossification. Clinically, patients either are asymptomatic or complain of some pain, deformity, and limited range of motion.

Skeletal Involvement

Melorheostosis usually involves the bones in the extremities. Any part of the bone can be involved, including the epiphysis. Usually the disease is unilateral and only rarely bilateral. In the leg and forearm, typically one bone is involved. Bony involvement frequently follows a sclerotomal distribution, which is a zone innervated by a single sensory nerve. Changes in the spine have been described, but they are unusual. Soft tissue ossifications also can occur, and these are more common than generally appreciated. These soft tissue masses are nearly always para-articular or adjacent to areas of affected bone (Fig. 77–4). They are reported most commonly medial to the hip and posterior to the knee. These masses should not be confused with myositis ossificans or soft tissue osteosarcoma.

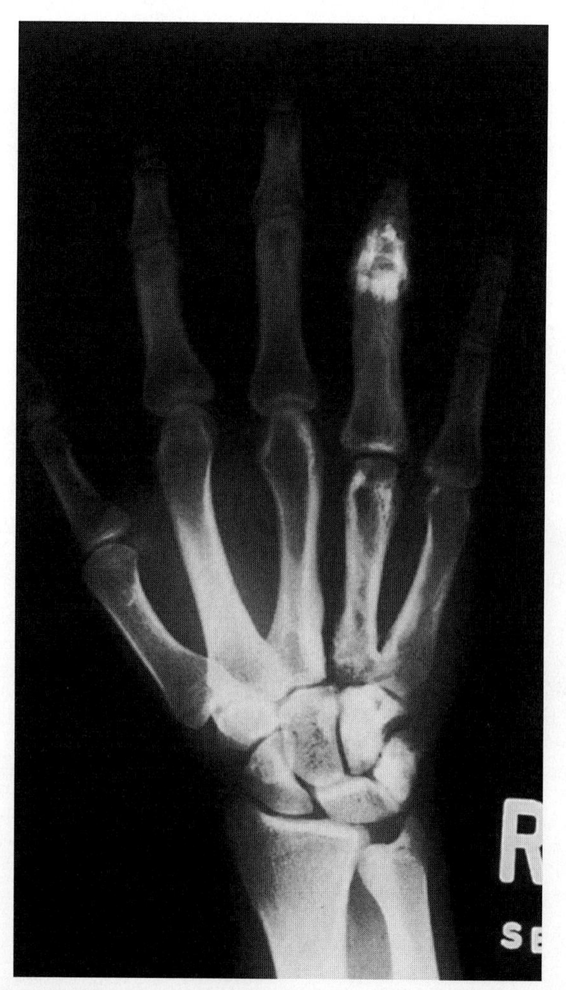

Figure 77–5. Melorheostosis affecting the triquetrum, the hamate, and the entire fourth ray.

Imaging

Melorheostosis has characteristic radiographic features. It shows cortical thickening with irregular, wavy, dense hyperostosis flowing along the long axis of the cortical bone (Figs. 77–4 and 77–5). There is a tendency for one side of the bone to be involved, and this has the typical appearance of "dripping candle wax." Endosteal sclerosis also is noted within the cancellous bone, appearing as streaky or patchy sclerosis.

Computed tomography (CT) scan and magnetic resonance (MR) imaging are not needed for establishing the diagnosis. When CT is done, it clearly shows the dense cortical thickening and the endosteal sclerosis (Fig. 77–6). Thick dense bone has low signal intensity on all MR sequences. On bone scan, melorheostosis shows a moderate increase in radiotracer uptake (see Fig. 77–4*B*).

OSTEOPATHIA STRIATA (VOORHOEVE'S DISEASE)

Overview

Osteopathia striata is a rare bone dysplasia characterized by the presence of linear striated bone densities oriented along the long axis of long bones. Osteopathia striata is an autosomal dominant disease. Affected patients are asymptomatic and have no characteristic laboratory findings. The diagnosis is made as an incidental finding on radiographic studies. Osteopathia striata can be associated with focal dermal hypoplasia; this association is known as Goltz syndrome.

Skeletal Involvement

Osteopathia striata has been described in all bones except the skull and the clavicles. The most common location is the metaphysis of long tubular bones, most frequently those of the lower extremities. The pelvic bones also are involved frequently. Osteopathia striata is almost always bilateral.

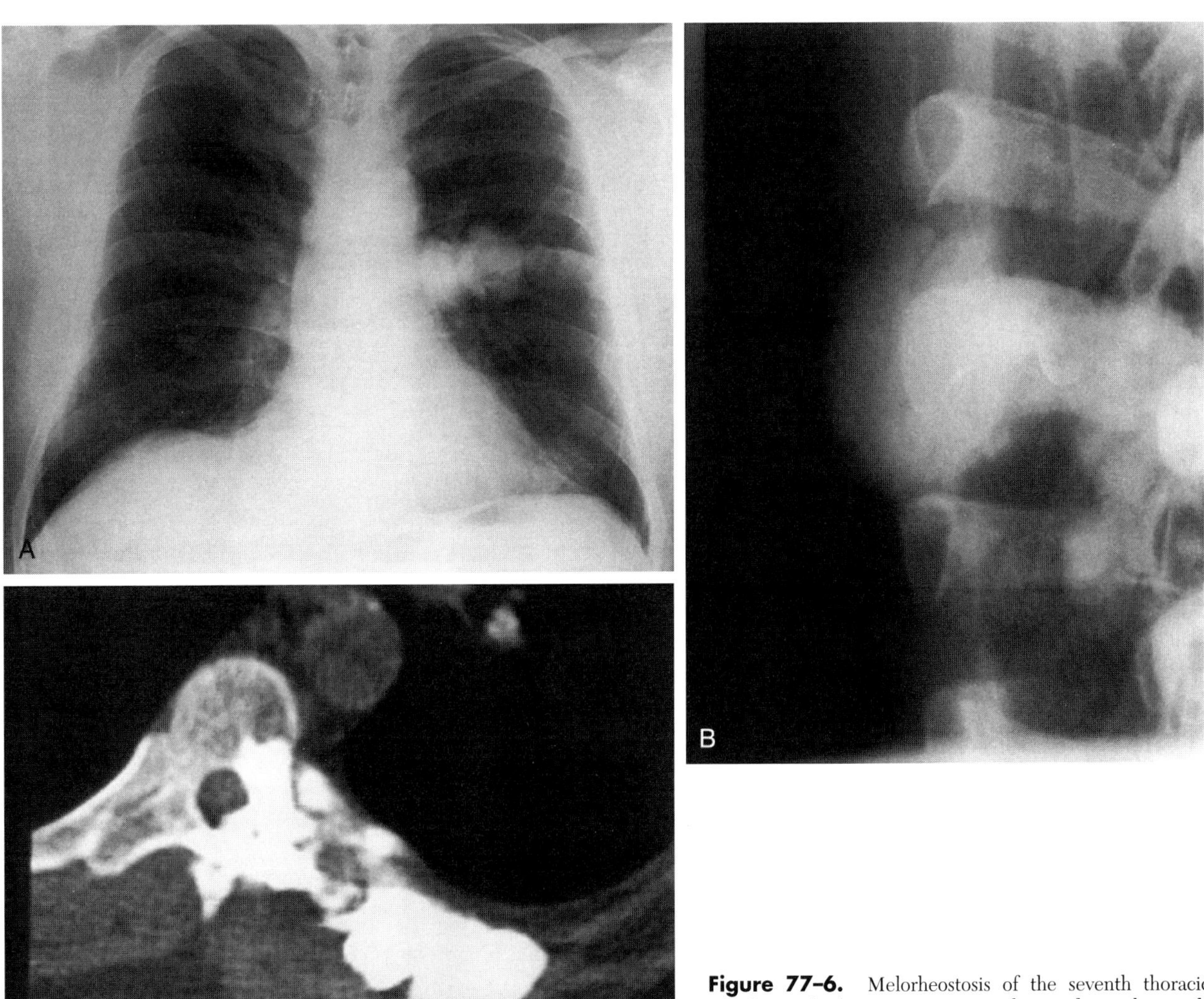

Figure 77–6. Melorheostosis of the seventh thoracic vertebra and rib. *A*, Preoperative chest radiograph reveals what was thought initially to be a hilar lesion. *B*, Oblique radiograph shows rib involvement with a sclerotic lesion. *C*, CT scan reveals involvement of the seventh thoracic vertebra and left seventh rib with melorheostosis.

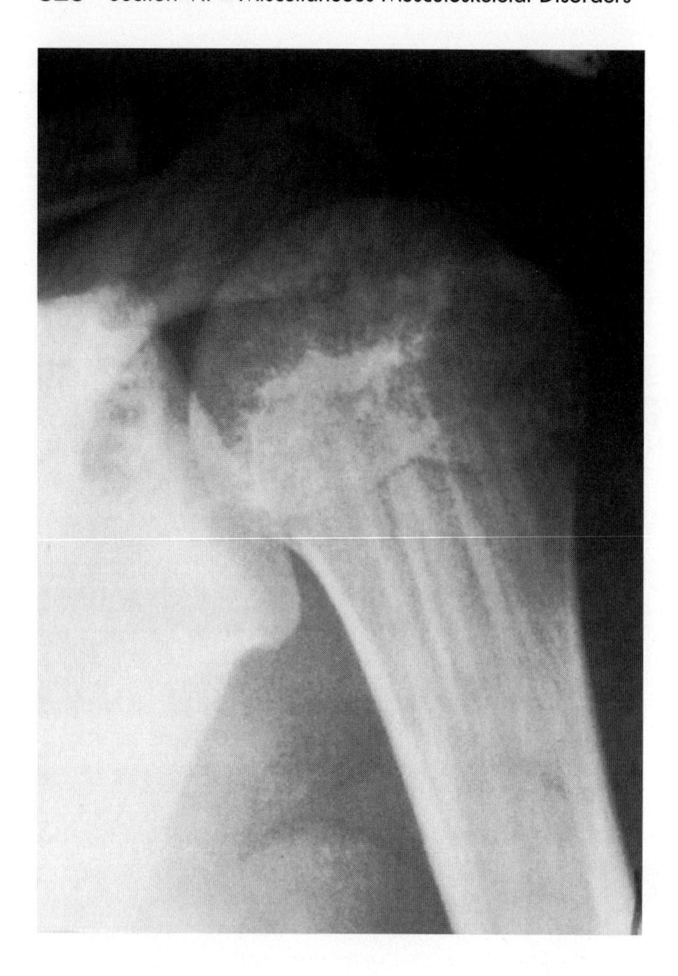

Figure 77–7. Osteopathia striata discovered accidentally in a 14-year-old child. Linear bone densities are seen in the proximal metaphysis of the humerus.

Imaging

Radiographically, osteopathia striata is characterized by fine, linear, striated medullary densities that are oriented along the long axis of the long bones (Fig. 77–7). In the iliac bones, they have a fan-shaped appearance. The length of the striations varies in proportion to the rate of bone growth. The longest striations are in the distal femur. The involved bones are normal in contour, density, and cortical thickness (see Fig. 77–7). Bone scan is normal.

Recommended Readings

Belzunegui J, Plazaola I, Uriarte E, et al: Mixed sclerosing bone dystrophy: Report of a case and review of the literature. Clin Rheumatol 15:378, 1996.

Benli IT, Akalin S, Boysan E, et al: Epidemiological, clinical and radiological aspects of osteopoikilosis. J Bone Joint Surg Br 74:504, 1992.

Bueno AL, Ramos FJ, Bueno O, et al: Severe malformations in males from families with osteopathia striata with cranial sclerosis. Clin Genet 54:400, 1998.

Campbell CJ, Papademetriou T, Bonfiglio M: Melorheostosis. J Bone J Surg Am 50:1281, 1968.

Carlson DH: Osteopathia striata revisited. J Can Assoc Radiol 28:190, 1977.

Chigira M, Kato K, Mashio K, et al: Symmetry of bone lesions in osteopoikilosis: Report of 4 cases. Acta Orthop Scand 62:495, 1991.

Gay Jr BB, Elsas LJ, Wyly JB, et al: Osteopathia striata with cranial sclerosis. Pediatr Radiol 24:56, 1994.

Gehweiler JA, Bland WR, Carden Jr TS, et al: Osteopathia striata—Voorhoeve's disease: Review of the roentgen manifestations. Am J Roentgenol Radium Ther Nucl Med 118:450, 1973.

Greenspan A: Sclerosing bone dysplasias—a target-site approach. Skeletal Radiol 20:561, 1991.

Greenspan A, Azouz EM: Bone dysplasia series: Melorheostosis: Review and update. Can Assoc Radiol J 50:324, 1999.

Lagier R, Mbakop A, Bigler A: Osteopoikilosis: A radiological and pathological study. Skeletal Radiol 11:161, 1984.

Mungovan JA, Tung GA, Lambiase RE, et al: Tc-99m MDP uptake in osteopoikilosis. Clin Nucl Med 19:6, 1994.

Murray RO, McCredie J: Melorheostosis and the sclerotomes: A radiological correlation. Skeletal Radiol 4:57, 1979.

Spieth ME, Greenspan A, Forrester DM, et al: Radionuclide imaging in forme fruste of melorheostosis. Clin Nucl Med 19:512, 1994.

Whyte MP, Murphy WA, Fallon MD, et al: Mixed sclerosing bone dystrophy: Report of a case and review of the literature. Skeletal Radiol 6:95, 1981.

Whyte MP, Murphy WA, Siegel BA: 99m-Tc-pyrophosphate bone imaging in osteopoikilosis, osteopathia striata, and melorheostosis. Radiology 127:439, 1978.

CHAPTER 78

Hypertrophic (Pulmonary) Osteoarthropathy

Nabil J. Khoury, MD, and Georges Y. El-Khoury, MD

Hypertrophic osteoarthropathy is a clinicoradiologic syndrome characterized by the presence of periostitis, digital clubbing, and painful swollen joints. Two forms are described: the primary form, also called *pachydermoperiostitis,* and the secondary form, which is far more common and is addressed first.

SECONDARY HYPERTROPHIC OSTEOARTHROPATHY

Overview

Most cases of secondary hypertrophic osteoarthropathy occur in patients with a variety of pulmonary and pleural diseases, including primary and metastatic lung tumors, chronic infections (e.g., tuberculosis), and cystic fibrosis. Nonpulmonary causes of hypertrophic osteoarthropathy also are encountered but are less common. These include diseases of the digestive tract (inflammatory bowel disease, lymphoma, polyposis, carcinoma), cardiovascular system (cyanotic congenital heart disease), hepatobiliary system (post–liver transplantation, biliary atresia, cirrhosis), and nasopharyngeal carcinoma. Hypertrophic osteoarthropathy also has been observed distal to infected arterial grafts (e.g., aortic, aortobifemoral, axillary). Secondary hypertrophic osteoarthropathy was termed *hypertrophic pulmonary osteoarthropathy* in the past because the underlying disease in adults is most commonly a lung carcinoma (80% of cases) usually of the squamous cell type. The incidence of hypertrophic osteoarthropathy in patients with bronchogenic carcinoma varies between 0.7% and 12%. There are a few reports of hypertrophic osteoarthropathy in children, but only about 12% of children with hypertrophic osteoarthropathy have neoplastic disease, whereas about 92% of adults have a neoplasm.

Clinically, hypertrophic osteoarthropathy can be confused with the common arthritides because of the joint pain and swelling. The only distinguishing feature of hypertrophic osteoarthropathy is the clubbing of the fingers. Effective therapy depends on resection of the primary tumor. In patients in whom tumor regression is achieved with chemotherapy or surgery, clinical symptoms of hypertrophic osteoarthropathy recede promptly. Intrathoracic vagotomy also is reported to produce relief of symptoms. Following therapy of the primary tumor, radiographic changes in bones may persist for several months after symptoms clear.

Pathologically, there is edema of the periosteum, round cell infiltration, and subperiosteal new bone formation. The synovium and articular capsules adjacent to involved bones are edematous and inflamed. The exact mechanism for hypertrophic osteoarthropathy is not clear. An increase in peripheral blood flow is thought to cause the periostitis either by release of hormones or by reflex vasodilation transmitted by the vagus nerve.

Skeletal Involvement

Hypertrophic osteoarthropathy involves mainly the long bones and the short tubular bones (Fig. 78–1). Any part of the long bones can be affected but mainly the distal ends. Other bones can show changes of hypertrophic osteoarthropathy, including the skull, maxillae, mandibles, scapulae, patellae, clavicles, ribs, and pelvis. The disease is usually symmetrical. Asymmetry sometimes is observed with arterial graft sepsis. There are no reported cases of spine or sacrum involvement with hypertrophic osteoarthropathy.

Imaging

Radiography

The appearance of periostitis depends on the duration of the process. Early in the disease the periosteal reaction is thin and delicate, but later on in the disease the periosteal reaction becomes thicker and dense (Fig. 78–2). The periostitis appears usually as a smooth periosteal reaction of variable thickness with single or multiple layers of proliferative new bone formation. It also may be wavy or undulating (Figs. 78–3 and 78–4). At the finger tufts, bone hypertrophy and soft tissue swelling are seen in older patients, whereas acro-osteolysis more commonly is present in younger patients. All these changes usually decrease with effective treatment of the primary disease.

Radionuclide Bone Scanning

Bone scans are highly sensitive in detecting hypertrophic osteoarthropathy even before periosteal new bone formation can be detected on plain radiographs. Typically, there is increased linear uptake along the cortex of the involved bones (Fig. 78–5*A*). This uptake may appear patchy, however, resembling metastasis (Fig. 78–5*B*). When synovitis is present, increased uptake in and around the joints also is observed.

Magnetic Resonance Imaging

Experience with magnetic resonance (MR) imaging is limited. Periostitis is seen as increased signal intensity on T2-weighted images around the cortical margins, whereas the dense periosteum appears as a hypointense rim. Joint effusions also are encountered sometimes.

Differential Diagnosis

A few conditions may mimic the changes of periostitis in hypertrophic osteoarthropathy, including thyroid acropachy, leukemia, and hypervitaminosis A.

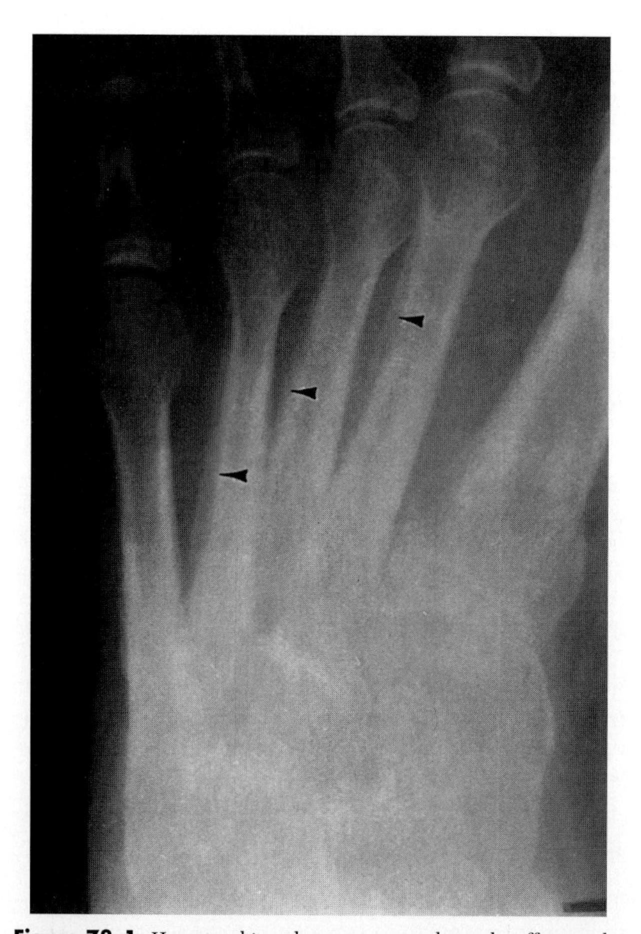

Figure 78–1. Hypertrophic pulmonary osteoarthropathy affecting the feet in a patient with sarcoidosis. Thin, delicate periosteal reaction involves the metatarsals *(arrowheads)*.

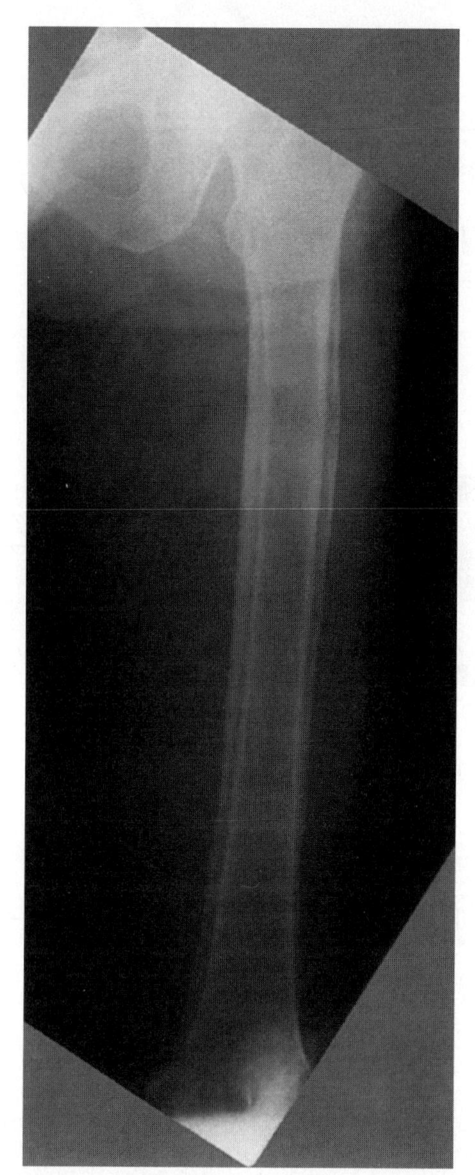

Figure 78–2. Hypertrophic pulmonary osteoarthropathy in a 69-year-old woman with carcinoma of the lung. Thick, dense periosteal reaction involves the femoral diaphysis.

PRIMARY OR IDIOPATHIC HYPERTROPHIC OSTEOARTHROPATHY

Overview

Primary or idiopathic hypertrophic osteoarthropathy is not associated with other disease processes. The disease is rare, accounting for less than 5% of all cases of hypertrophic osteoarthropathy. The disease usually starts in adolescence and progresses for about 10 years before arresting spontaneously. Associated skin changes are often, but not always, present, including pachyderma, cutis vertices gyrata, thickening of the skin in the face and forehead, and excessive sweating. When hypertrophic osteoarthropathy is associated with skin changes, the

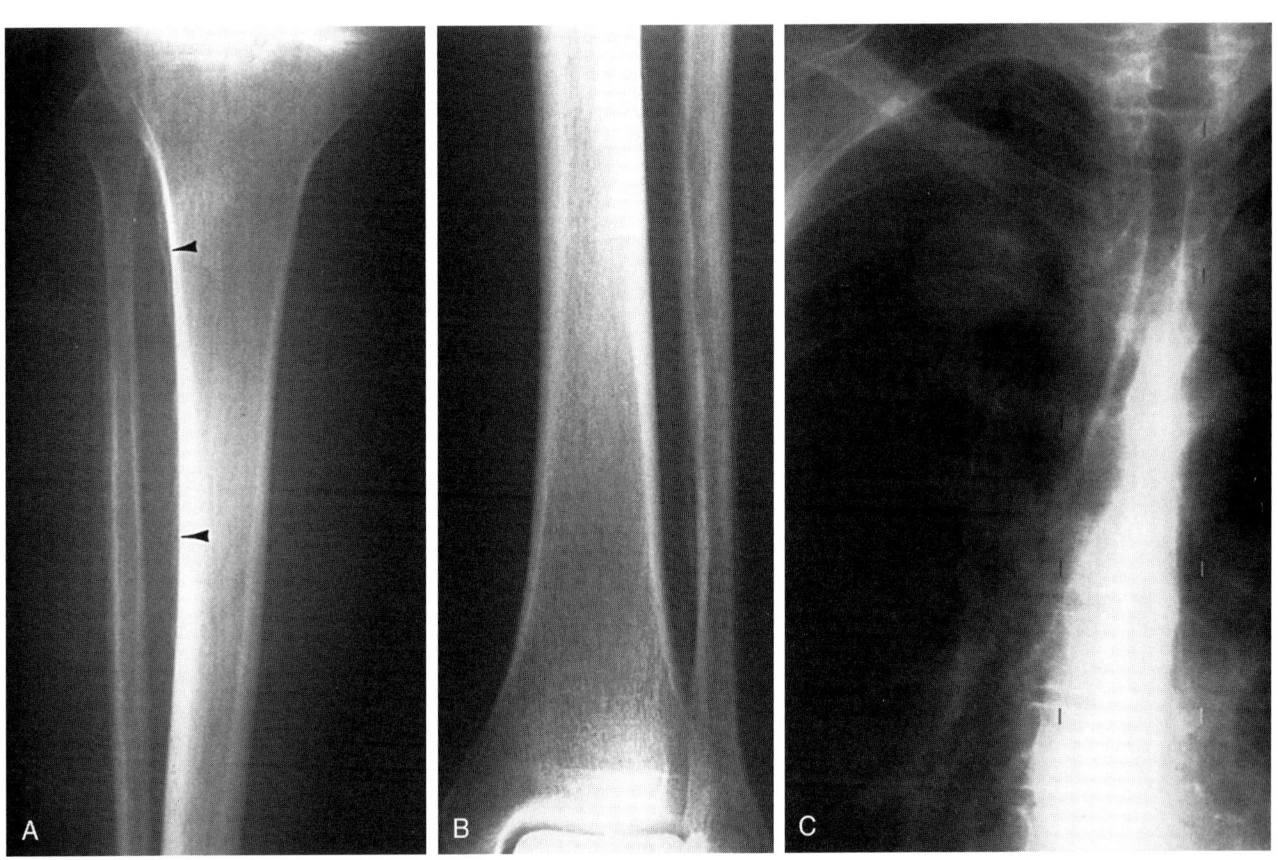

Figure 78–3. Typical changes of hypertrophic pulmonary osteoarthropathy in a patient with carcinoma of the lung. *A,* Wavy periosteal reaction involves the right tibia *(arrowheads)* and fibula. *B,* Periosteal reaction is noted in the left fibula. *C,* Chest radiograph reveals a 4-cm nodule in the right upper lobe, which later was found to be a carcinoma.

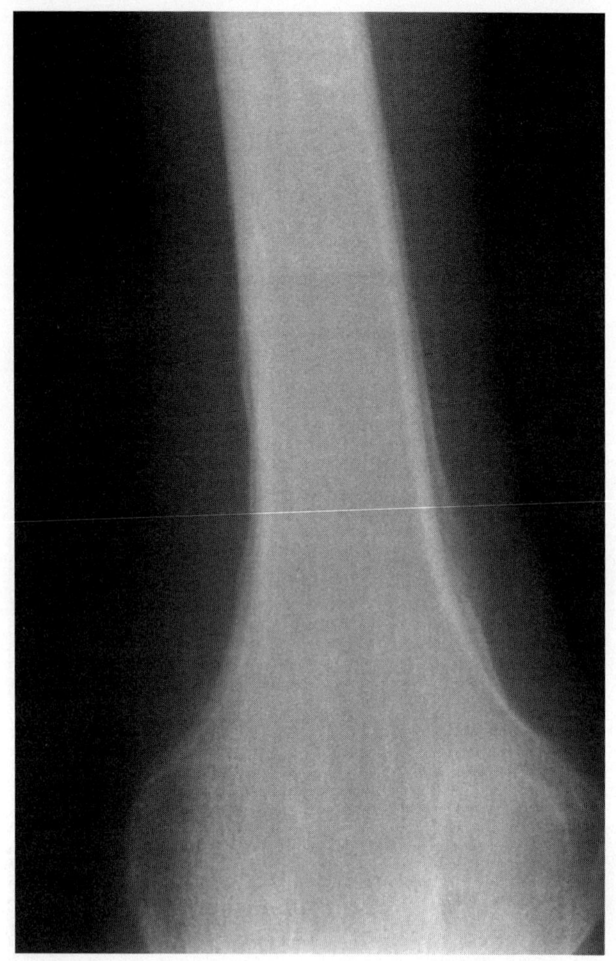

Figure 78–4. Hypertrophic pulmonary osteoarthropathy in the distal femur resulting from carcinoma of the lung. There is thick periosteal reaction around the distal femur.

disease is referred to as *pachydermoperiostitis*. Pachydermoperiostitis can be familial or sporadic, and it occurs primarily in males. From 40% to 50% of patients have a family history of this condition.

Imaging

Radiography is the modality of choice for studying periosteal new bone formation involving the distal third of the long bones in the legs and forearms (Fig. 78–6). The bone ends, such as the malleoli and epicondyles, also are involved. This involvement differentiates pachydermoperiostitis from secondary hypertrophic osteoarthropathy, in which periosteal reactions typically involve the diaphyseal portions of long bones. Acro-osteolysis also has been reported in patients with pachydermoperiostitis (see Fig. 78–6*B*).

Differential Diagnosis

Pachydermoperiostitis in adults should be differentiated from the secondary type of hypertrophic osteoarthropathy, acromegaly, and thyroid acropachy. In children, pachydermoperiostitis should be differentiated from congenital syphilis, infantile cortical hyperostosis (Caffey's disease), hypervitaminosis A, and Engelman's disease.

Recommended Readings

Ali A, Tetalman MR, Fordham EW, et al: Distribution of hypertrophic pulmonary osteoarthropathy. AJR Am J Roentgenol 134:771, 1980.

Bhate DV, Pizarro AJ, Greenfield GB: Idiopathic hypertrophic osteoarthropathy without pachyderma. Radiology 129:379, 1978.

Capelastegui A, Astigarraga E, Garcia-Iturraspe C: MR findings in pulmonary hypertrophic osteoarthropathy. Clin Radiol 55:72, 2000.

Davies RA, Darby M, Richards MA: Hypertrophic pulmonary osteoarthropathy in pulmonary metastatic disease: A case report and review of the literature. Clin Radiol 43:268, 1991.

DeVries N, Datz FL, Manaster BJ: Case report 399. Skeletal Radiol 15:658, 1986.

Freeman MH, Tonkin AK: Manifestations of hypertrophic pulmonary osteoarthropathy in patients with carcinoma of the lung: Demonstration by 99mTc-pyrophosphate bone scans. Radiology 120:363, 1976.

Guyer PB, Brunton FJ, Wren MWG: Pachydermoperiostosis with acro-osteolysis. J Bone Joint Surg Br 60:219, 1978.

Martinez-Lavin M, Pineda C, Valdez T, et al: Primary hypertrophic osteoarthropathy. Semin Arthritis Rheum 17:156, 1988.

Morgan B, Coakley F, Finlay DB, et al: Hypertrophic osteoarthropathy in staging skeletal scintigraphy for lung cancer. Clin Radiol 51:694, 1996.

Pineda CJ, Martinez-Lavin M, Goobar JE, et al: Periostitis in hypertrophic osteoarthropathy: Relationship to disease duration. AJR Am J Roentgenol 148:773, 1987.

Reginato AJ, Schiapachasse V, Guerrero R: Familial idiopathic hypertrophic osteoarthropathy and cranial suture defects in children. Skeletal Radiol 8:105, 1982.

Spruijt S, Krijgsman AA, Van den Broek JAC, et al: Hypertrophic osteoarthropathy of one leg—a sign of aortic graft infection. Skeletal Radiol 28:224, 1999.

Stevens M, Helms C, El-Khoury G, Chow S: Unilateral hypertrophic osteoarthropathy associated with aortobifemoral graft infection. AJR Am J Roentgenol 170:1584, 1998.

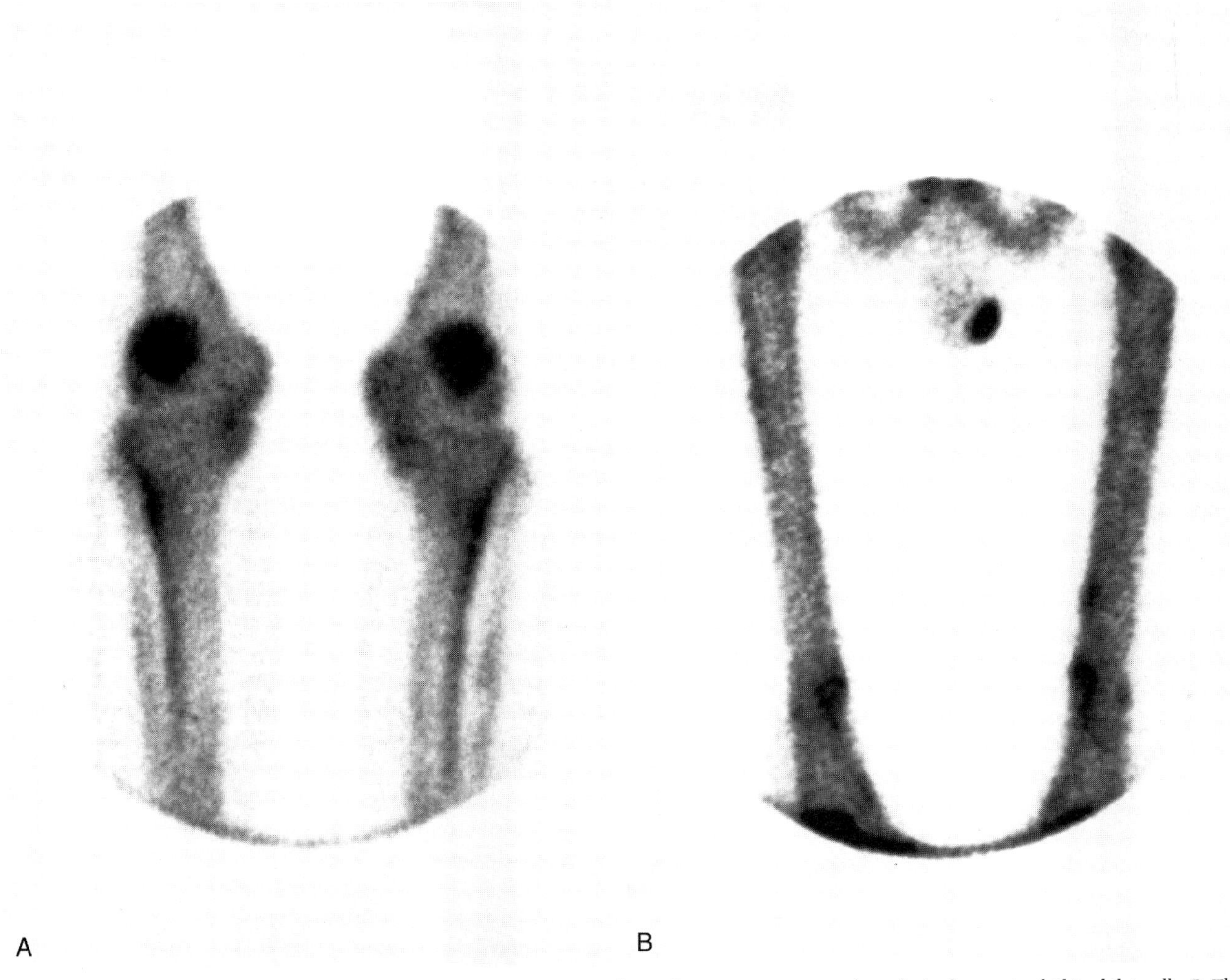

Figure 78–5. Bone scan appearance of hypertrophic pulmonary osteoarthropathy. *A*, Linear increased uptake in the proximal tibiae bilaterally. *B*, The uptake in the femora is more spotty, resembling metastasis.

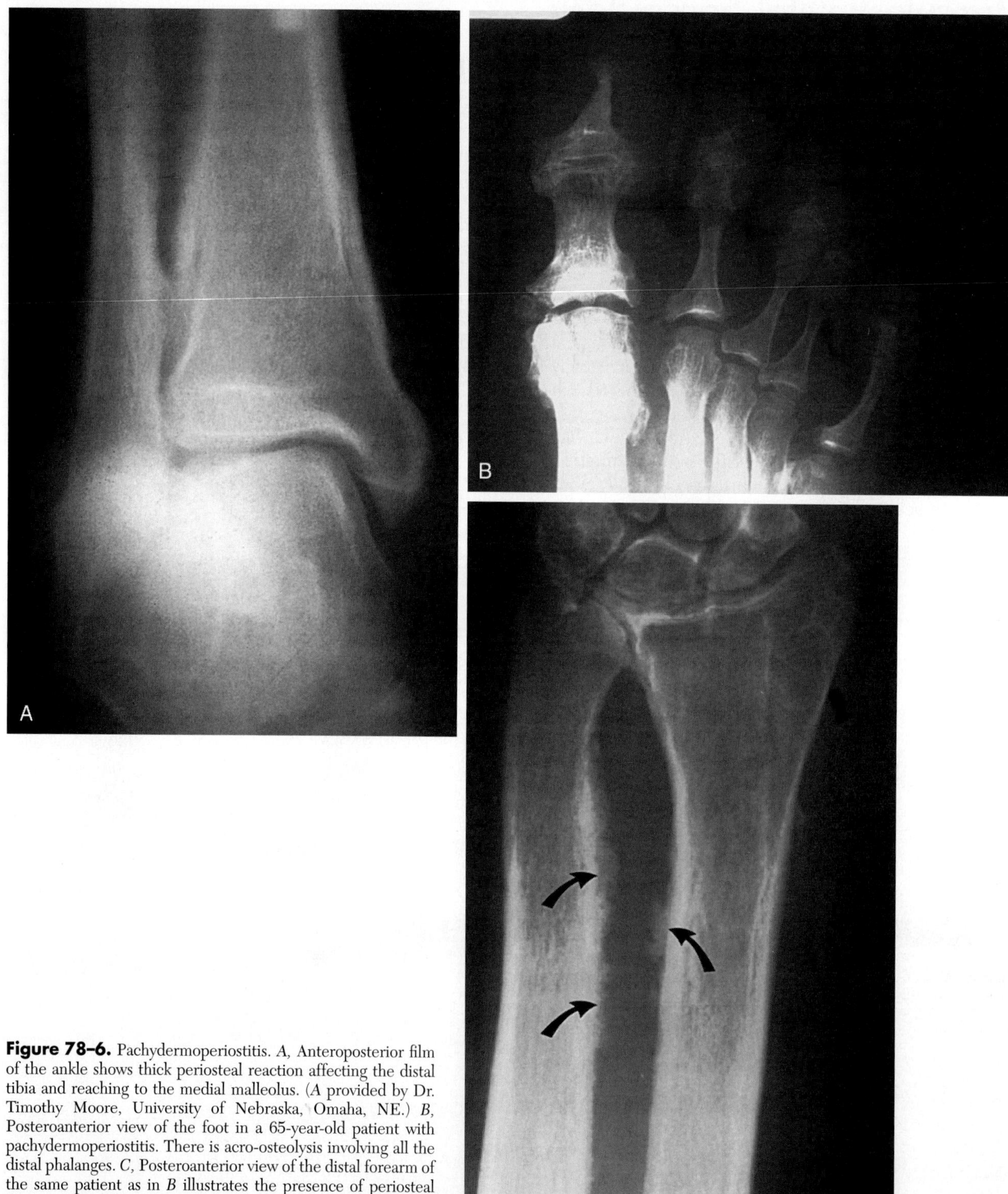

Figure 78–6. Pachydermoperiostitis. *A,* Anteroposterior film of the ankle shows thick periosteal reaction affecting the distal tibia and reaching to the medial malleolus. (*A* provided by Dr. Timothy Moore, University of Nebraska, Omaha, NE.) *B,* Posteroanterior view of the foot in a 65-year-old patient with pachydermoperiostitis. There is acro-osteolysis involving all the distal phalanges. *C,* Posteroanterior view of the distal forearm of the same patient as in *B* illustrates the presence of periosteal hypertrophic changes at the distal radius and ulna *(arrows).* These changes persisted unchanged for many years.

CHAPTER 79

Musculoskeletal Sarcoidosis

Nabil J. Khoury, MD, and Georges Y. El-Khoury, MD

OVERVIEW

Sarcoidosis is a chronic multisystem inflammatory disease characterized by the presence of noncaseating granulomas. Both sexes are affected equally, but the disease is more prevalent in African Americans and patients of Northern European origin. Onset is usually in early adulthood and occasionally during childhood. Clinically, 80% to 90% of cases with bone involvement have associated lung, mediastinal, and supraclavicular nodal disease. Skin (30%) and eye (20%) involvement also commonly occur. Erythema nodosum commonly is associated with acute sarcoid arthritis. In the musculoskeletal system, sarcoidosis can involve joints, muscles, and bones.

JOINTS

Two forms of arthritis are associated with sarcoidosis—acute and chronic. Acute arthritis is usually self-limited, lasting a few weeks to 3 months. Knees and ankles are the most commonly affected joints. The arthritis is nonerosive and nondeforming. Chronic arthritis is uncommon and usually is associated with long-standing sarcoidosis. The shoulders, hands, wrists, ankles, and knees are affected, and the arthritis can be deforming; it resembles psoriatic arthritis or Jaccoud's arthropathy.

MUSCLES

Muscular sarcoidosis is rare and usually asymptomatic. Symptomatic forms can be either nodular or myopathic. Patients with the nodular form have a single or multiple nodules within the skeletal muscles. Patients with the myopathic form have myalgia, proximal muscle weakness, and muscle atrophy similar to patients with polymyositis. The nodular type has a characteristic appearance on magnetic resonance (MR) imaging (Fig. 79–1). The nodules reveal a star-shaped central area of low signal

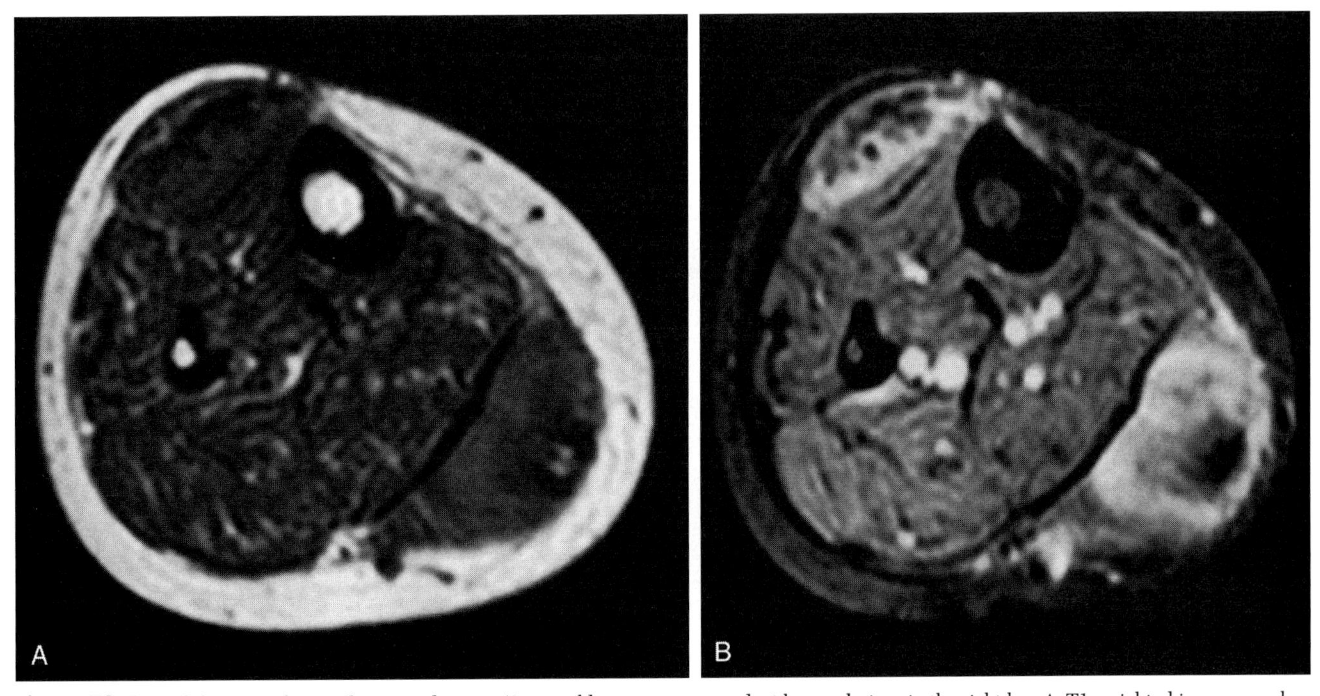

Figure 79–1. Nodular type of muscular sarcoidosis. A 56-year-old woman presented with mass lesions in the right leg. *A*, T1-weighted image reveals the isointense mass lesions in the tibialis anterior and medial head of the gastrocnemius muscles. The masses contain dark star-shaped structures within them. *B*, T2-weighted image shows increased signal intensity in the mass lesions. The central areas remain dark. This appearance is thought to be characteristic of sarcoid granulomas. (Images provided by Dr. Shoichiro Otake, Toki Municipal General Hospital, Japan.)

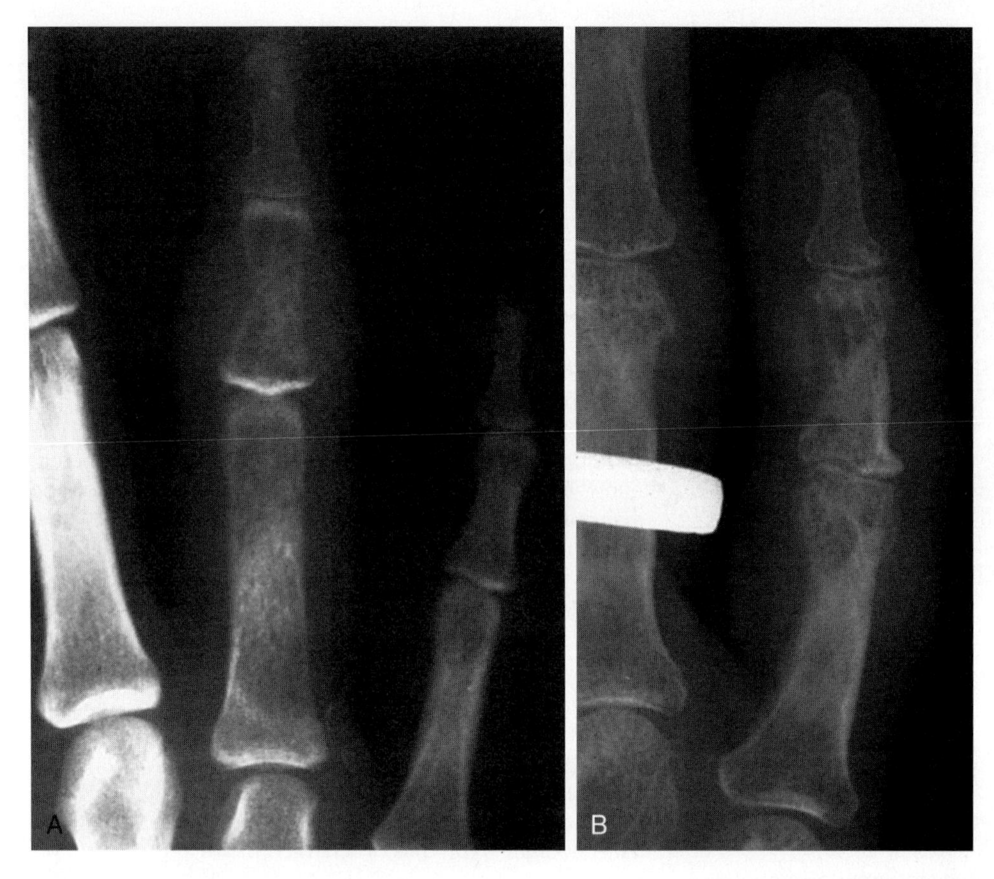

Figure 79–2. Sarcoidosis of bone. *A*, Posteroanterior view of the hand shows permeative destruction of the middle phalanx of the ring finger resulting in what is known as the *lacelike* appearance. Note the soft tissue swelling resulting from the presence of a granuloma. *B*, Radiograph of another patient with sarcoidosis of the little finger shows punched-out lesions and lacy trabecular pattern.

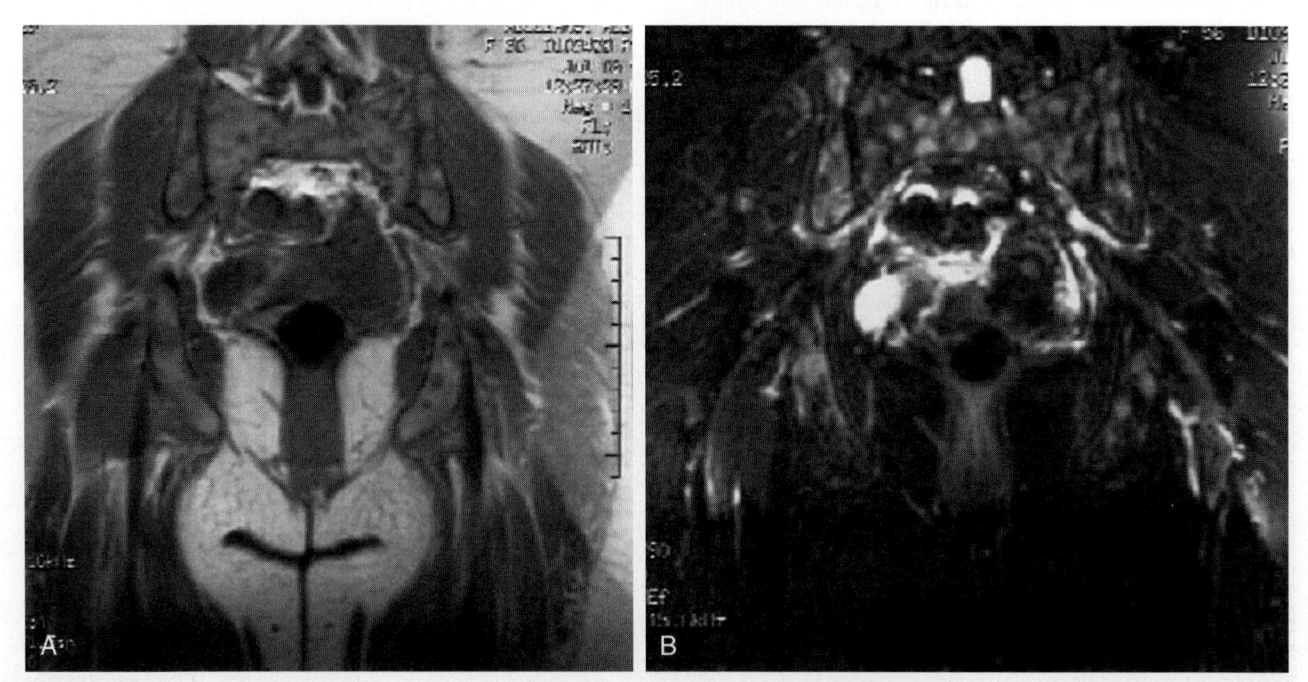

Figure 79–3. Sarcoidosis of the bone marrow. *A*, T1-weighted image shows marrow infiltration with small nodules. *B*, T2-weighted image with fat suppression reveals increased signal intensity in the sarcoid granulomas within the marrow. (Images provided by Dr. Sandra Moore, Mount Sinai Hospital, New York, NY.)

intensity surrounded by an area of high signal intensity. Histologically the central area represents fibrosis surrounded by inflammatory tissue. The myopathic type is best studied with gallium-67 scintigraphy, which shows diffuse uptake in the involved muscles. The MR imaging changes in the myopathic type are nonspecific, and in chronic cases the findings can be confused with steroid myopathy.

BONE

Bone involvement has been reported in 1% to 13% of sarcoidosis patients; it typically affects the bones of the hands and feet and much less commonly the pelvis, spine, ribs, long bones, skull, and nose. Involvement of the hands and feet is usually asymptomatic. Different radiographic patterns have been described, including diffuse marrow replacement, permeative pattern, lytic lesions, and sclerotic pattern. The permeative pattern is characteristic of sarcoidosis, showing a reticular lacelike appearance (Fig. 79–2). Localized areas of infiltration show discrete areas of lytic destruction. Sclerotic lesions are rare, and they typically involve the axial skeleton (skull, spine, ribs, pelvis, and proximal femur). In contrast to the lesions in hands and feet, sclerotic axial lesions are almost always symptomatic. Sclerotic sarcoid lesions should be differentiated from osteoblastic metastasis, lymphoma, and systemic mastocytosis.

MR imaging studies of the marrow in patients with sarcoid showed marrow infiltration with sarcoid granulomas that resemble metastases; the lesions are, however, smaller than metastatic lesions and almost equal in size, giving a "starry sky" appearance (Fig. 79–3). They appear dark on T1-weighted images and bright on T2-weighted images, and they enhance with gadolinium injection.

Recommended Readings

Abdelwahab IF, Norman A: Osteosclerotic sarcoidosis. AJR Am J Roentgenol 150:161, 1988.

Golzarian J, Matos C, Golstein M, et al: Case report: Osteosclerotic sarcoidosis of spine and pelvis: Plain film and magnetic resonance imaging findings. Br J Radiol 67:401, 1994.

Hall FM, Shmerling RH, Aronson M, et al: Case report 705. Skeletal Radiol 21:182, 1992.

Otake S: Sarcoidosis involving skeletal muscle: Imaging findings and relative value of imaging procedures. AJR Am J Roentgenol 162:369, 1994.

Resnik CS, Young JWR, Aisner SC, et al: Case report 594. Skeletal Radiol 19:79, 1990.

Yaghmai I: Radiographic, angiographic and radionuclide manifestations of osseous sarcoidosis. Radiographics 3:375, 1983.

CHAPTER 80

Idiopathic Tumoral Calcinosis and Secondary Tumoral Calcinosis

Nabil J. Khoury, MD, and Georges Y. El-Khoury, MD

OVERVIEW

Idiopathic tumoral calcinosis is a rare disease of unknown cause occurring in otherwise healthy individuals. The disease tends to be familial in about one third of cases, and it is more common in African Americans. Idiopathic tumoral calcinosis is characterized by the presence of large calcified masses mostly around the hips, elbows, shoulders, and gluteal and thigh regions. Involvement of the hands and knees is notably uncommon (Fig. 80–1). The disease usually is detected in individuals in their teens and 20s, although patients with tumoral calcinosis have ranged in age from 10 months to 77 years.

Idiopathic tumoral calcinosis is believed to be due to an error in phosphorus metabolism causing extracellular deposition of calcium hydroxyapatite crystals. Biochemical analysis typically reveals hyperphosphatemia. The kidney function, serum calcium, parathyroid hormone, and alkaline phosphatase levels all are normal.

The masses consist of cystic structures separated by thick septae. Calcium-suspended hydroxyapatite crystals are suspended in these fluid-filled cysts. The lesions grow slowly and are painless, unless they ulcerate, draining chalky material with the consistency of toothpaste. Large masses around joints can produce limitation of motion; however, the joints themselves are not involved with the disease.

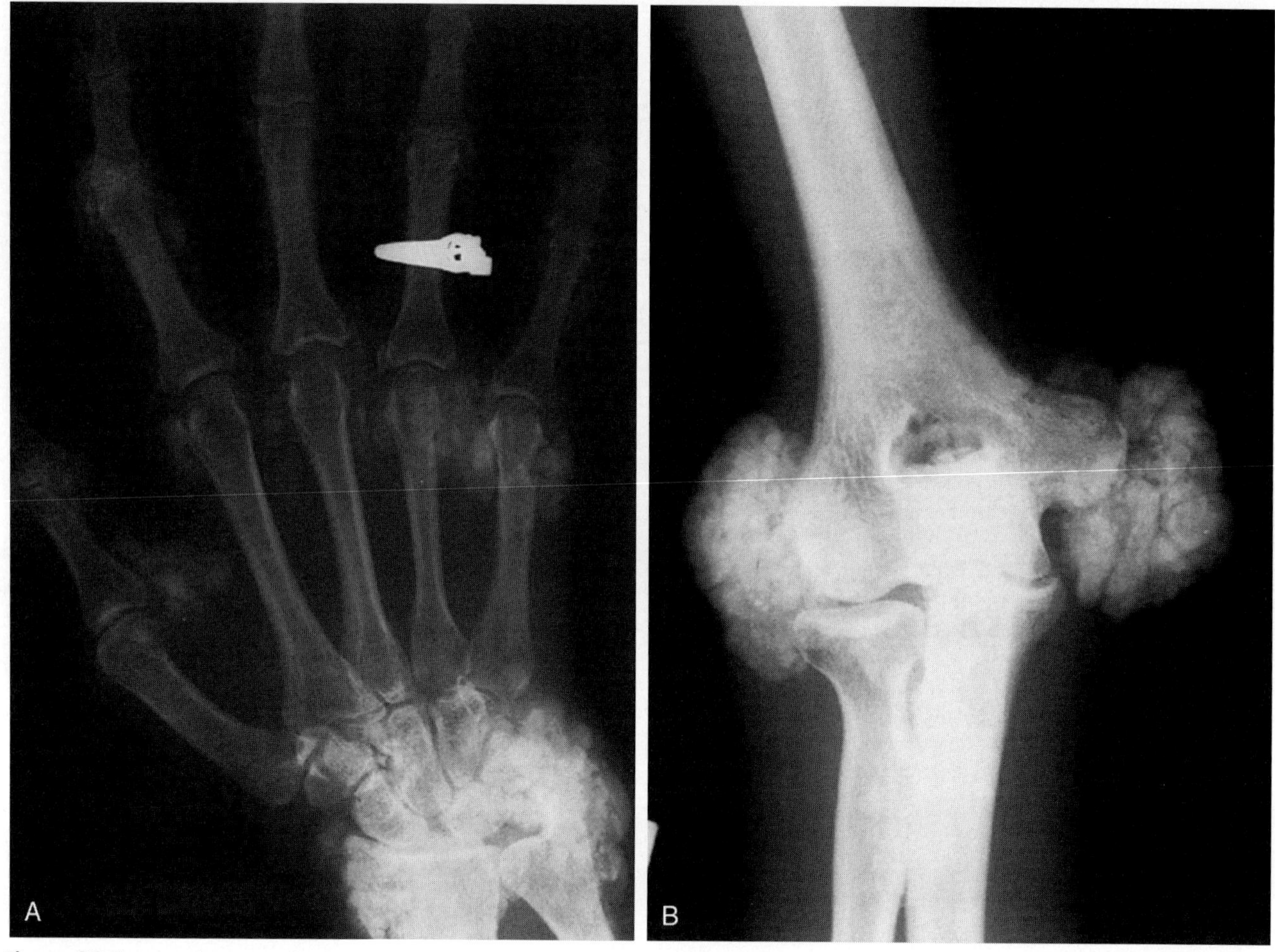

Figure 80–1. Secondary tumoral calcinosis affecting the hand and elbow in a patient with chronic renal failure. Idiopathic tumoral calcinosis does not usually involve the hand. *A*, Posteroanterior view of the left hand and wrist shows multiple calcific collections. *B*, Anteroposterior view of the left elbow shows two large calcific collections in the soft tissue on either side of the elbow.

IMAGING

Radiography and Computed Tomography

Radiography and computed tomography (CT) show well-defined multinodular collections of calcific material in the soft tissues adjacent to joints (Fig. 80–2). Radiolucent septae separate the individual cystic collections. The calcium hydroxyapatite crystals tend to precipitate to the dependent portions of the cysts producing fluid-calcium levels, also known as the *sedimentation sign.* This sign is seen best on CT but also can be detected on radiographs obtained with a horizontal beam. Bilateral symmetrical periarticular involvement can be seen radiographically, but this may be clinically silent. Usually idiopathic tumoral calcinosis does not erode the adjacent bone.

Magnetic Resonance Imaging

On magnetic resonance (MR) imaging, the nodular cystic collections show low signal intensity on T1-weighted images and variable signal intensity on T2-weighted images (Fig. 80–3). The fluid in the cysts may appear hyperintense, but the sediment in the dependent portion appears hypointense. The septae usually enhance with gadolinium injection.

Radionuclide Bone Scan

Bone scans reveal increased uptake in the affected areas.

DIFFERENTIAL DIAGNOSIS

Radiographically, idiopathic tumoral calcinosis is indistinguishable from metastatic soft tissue calcifications often seen in patients with chronic renal failure complicated with secondary hyperparathyroidism. In such patients, the calcium-phosphorus product is typically greater than 75. This condition is referred to as *secondary tumoral calcinosis,* and it is more common than idiopathic tumoral calcinosis (see Fig. 80–1). Calcinosis circumscripta in patients with scleroderma or CRST syndrome (calcinosis

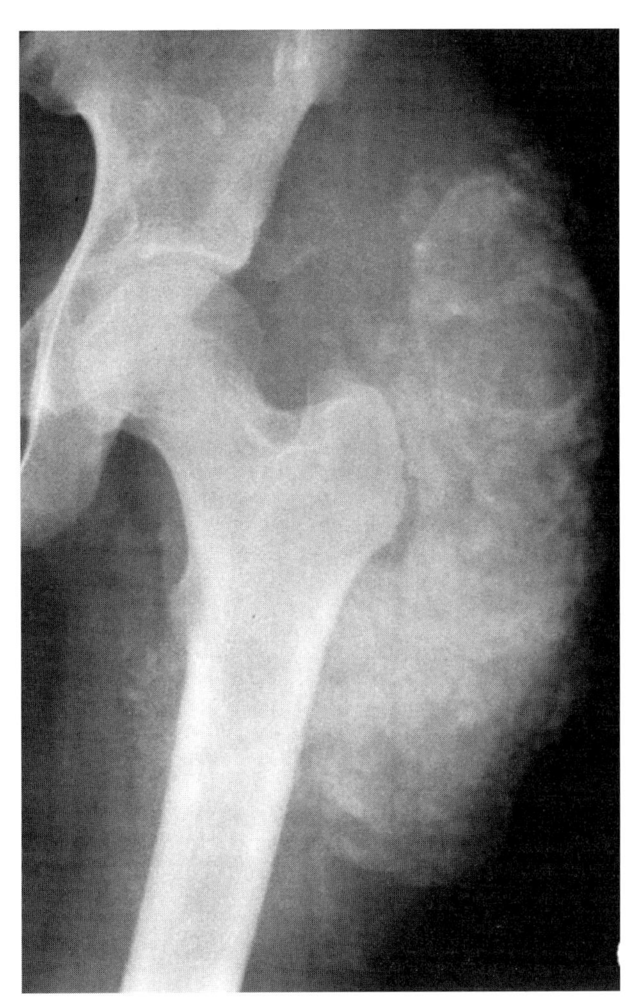

Figure 80–2. Idiopathic tumoral calcinosis of the left hip. There is a large multilocular collection of calcific material in the soft tissues lateral to the hip.

cutis, Raynaud's phenomenon, sclerodactyly, and telangiectasia) can resemble tumoral calcinosis. Occasionally, soft tissue chondromas and chondrosarcomas can resemble tumoral calcinosis.

Recommended Readings

Bishop AF, Destouet JM, Murphy WA: Tumoral calcinosis: Case report and review. Skeletal Radiol 8:269, 1982.

Chew FS, Crenshaw WB: Idiopathic tumoral calcinosis. AJR Am J Roentgenol 158:330, 1992.

Geirnaerdt MJ, Kroon HM, Van der Heul RO, et al: Tumoral calcinosis. Skeletal Radiol 24:148, 1995.

Martinez S, Vogler III JB, Harrelson JM, et al: Imaging of tumoral calcinosis: New observations. Radiology 174:215, 1990.

Resnick CS: Tumoral calcinosis. Arthritis Rheum 32:1484, 1989.

Steinbach LS, Johnston JO, Tepper EF, et al: Tumoral calcinosis: Radiologic-pathologic correlation. Skeletal Radiol 24:573, 1995.

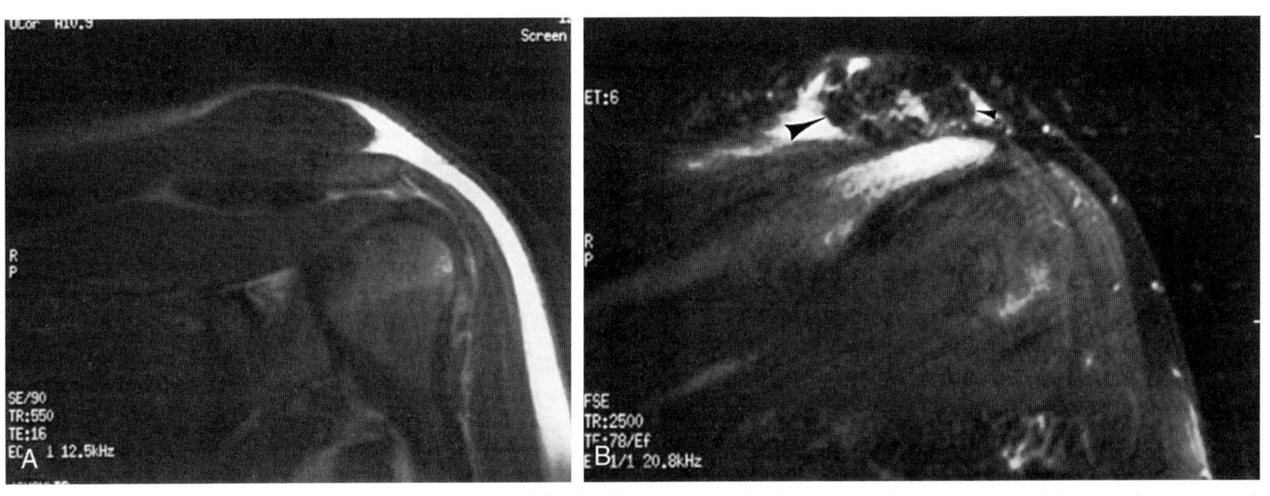

Figure 80–3. MR imaging of secondary tumoral calcinosis of the shoulder. *A*, T1-weighted image reveals a large mass with low signal intensity in the soft tissue superior to the shoulder joint. *B*, T2-weighted image shows the calcific density has low signal intensity on this sequence also *(arrowheads)*.

SECTION VIII

ULTRASOUND OF THE MUSCULOSKELETAL SYSTEM

CHAPTER 81

General Considerations

Nathalie J. Bureau, MD, and Étienne Cardinal, MD

OVERVIEW

Along with radiographs, fluoroscopy, computed tomography (CT), and magnetic resonance (MR) imaging, ultrasound is an integral part of the imaging arsenal available to radiologists and can be used judiciously to diagnose and treat a wide range of disorders of the musculoskeletal system. This chapter reviews the main indications of ultrasound in the investigation and treatment of disorders of the musculoskeletal system.

ADVANTAGES

Ultrasound is a low-cost, widely available imaging technique that provides a unique contact between the radiologist and the patient during the ultrasound examination. Ultrasound is superior to MR imaging in depicting soft tissue calcification. In contrast to CT and MR imaging, ultrasound performance is not degraded by metallic artifact. Ultrasound also provides real-time imaging that can be used for dynamic evaluation of anatomic structures and to guide interventional procedures. This technique is highly accurate in determining the solid or cystic nature of a soft tissue lesion.

LIMITATIONS

Linear array transducers, normally used in the evaluation of the musculoskeletal system, yield an image field of view as large as their width, which is typically about 4 to 6 cm. Ultrasound is limited in its capacity to image large anatomic segments or lesions when compared with other cross-sectional imaging modalities such as CT and MR imaging. Ultrasound extended-field-of-view imaging technology enables depiction of larger anatomic areas. Because of the absence of ultrasound transmission through bone, ultrasound cannot characterize intraosseous lesions and is limited in its ability to define the relationships between a lesion and adjacent bone or joint.

TECHNIQUE

Because of their wide field of view and their high resolution in the near field, linear array transducers are the most appropriate probes for the evaluation of musculoskeletal disorders. Higher frequency probes (7 to 13 MHz) are indicated for the evaluation of superficial structures or subcutaneous masses, whereas deeper lesions are examined best with 5- to 7-MHz and sometimes 3.5-MHz transducers. When evaluating superficial structures or lesions, it is helpful to use a stand-off gel pad or a thick coat of gel. Care should be taken to adjust the focal zone of the transducer according to the structure that is being examined to ensure the highest resolution. An ultrasound examination may include a static and a dynamic evaluation and evaluation of the contralateral side for comparison.

In most cases, it is recommended to obtain a radiograph of the anatomic area being evaluated by ultrasound because both imaging modalities provide complementary information. Radiographs may disclose soft tissue calcification or ossification diagnostic of a soft tissue hemangioma or myositis ossificans. Radiographs may reveal the presence of an osseous lesion or show osseous involvement, such as pressure erosions, periosteal reaction, or overt osseous destruction, difficult to appreciate with ultrasound.

Recommended Readings

Bureau NJ, Cardinal E, Chhem RK: Ultrasound of soft tissue masses. Semin Musculoskelet Radiol 2:283, 1998.

Fornage BD: Soft-tissue masses. Clin Diagn Ultrasound 30:21, 1995.

Weng L, Tirumalai AP, Lowery CM, et al: US extended-field-of-view imaging technology. Radiology 203:877, 1997.

CHAPTER 82

Tendons

Nathalie J. Bureau, MD, and Étienne Cardinal, MD

OVERVIEW

With its high spatial resolution and its dynamic capabilities, ultrasound is particularly well suited for the evaluation of tendon pathology. (Further information on tendon imaging is found in Chapters 71 and 72.) Tendons are structures composed of dense bundles of collagen, elastin, and reticulin oriented in a parallel fashion. Normal tendons are hyperechoic and show a fibrillar appearance at ultrasound, which represents the parallel fibrils of the tendon (Fig. 82–1).

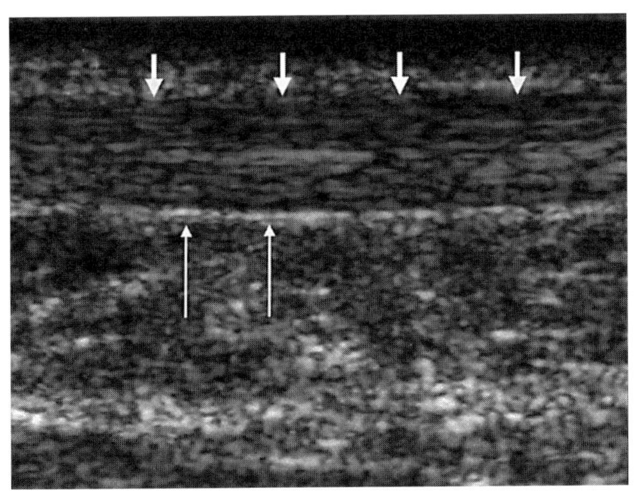

Figure 82–1. Normal Achilles' tendon. Longitudinal ultrasound of Achilles' tendon shows the fibrillar hyperechoic pattern of a normal tendon *(short arrows)*. The paratenon appears as a fine hyperechoic line that surrounds the tendon *(long arrows)*.

Tendons usually are evaluated in two orthogonal planes (Fig. 82–2), and abnormalities seen should be confirmed in both planes. Evaluation of the rotator cuff tendons in the longitudinal and transverse planes allows for accurate determination of the dimensions and location of a tear. Except for Achilles' tendon, which is evaluated best in the longitudinal plane, the ankle tendons are evaluated best in the transverse plane. This plane provides optimal definition of the tendon contours and yields accurate assessment of the amount of fluid in the synovial sheath.

One must be aware of the artifact of anisotropy, which may cause the tendon to appear hypoechoic when the ultrasound beam is oblique relative to the long axis of the tendon fibers (Fig. 82–3). To avoid this artifact, care must be taken to keep the transducer surface perpendicular to the tendon when doing the ultrasound examination. Conversely, this artifact sometimes may be useful to identify a tendon masked by a background of hyperechoic soft tissues. While imaging the tendon in a transverse

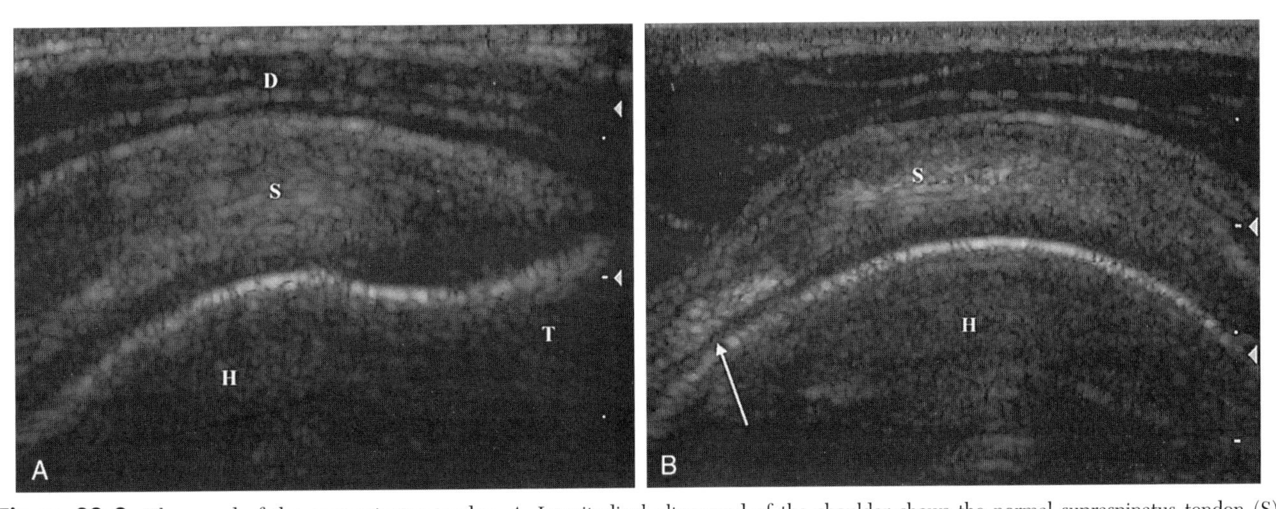

Figure 82–2. Ultrasound of the supraspinatus tendon. *A*, Longitudinal ultrasound of the shoulder shows the normal supraspinatus tendon (S) overlying the humeral head (H). The tendon appears as an echogenic band of tissue, which tapers as it inserts on the greater tuberosity (T) giving the normal beak appearance. D, deltoid. *B*, Transverse ultrasound of the shoulder shows the supraspinatus tendon (S) posterior to the intra-articular portion of the long head of the biceps tendon (LHB) *(arrow)*, which appears as a hyperechoic oval-shaped structure. The first 1.5 cm of rotator cuff located behind the LHB tendon represents the supraspinatus tendon, followed immediately by the infraspinatus located more posteriorly. H, humeral head.

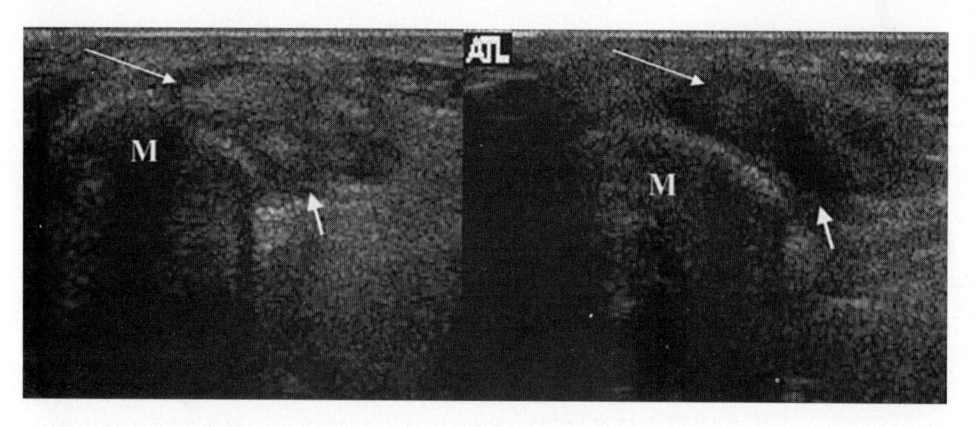

Figure 82–3. Peroneal tendons anisotropy. Dual transverse ultrasound of the peroneal tendons at the level of the lateral malleolus (M), obtained with slightly different angulation of the transducer, shows the peroneus longus *(long arrow)* and peroneus brevis *(short arrow)* tendons as oval-shaped, hyperechoic-appearing structures on the left and as hypoechoic-appearing structures on the right.

plane, gently rocking the probe makes the tendon appear alternately hypoechoic and hyperechoic.

ROTATOR CUFF

Anatomy

The rotator cuff is composed of four tendons. The supraspinatus tendon inserts on the greater tuberosity, the infraspinatus and teres minor tendons insert on the posterolateral aspect of the humeral head, and the subscapularis tendon inserts on the lesser tuberosity. At ultrasound, on a transverse view of the rotator cuff, the first 1.5 cm of the cuff located immediately behind the long head of the biceps (LHB) tendon represents the supraspinatus tendon (see Fig. 82–2*B*). The infraspinatus tendon is located immediately posterior to the supraspinatus tendon.

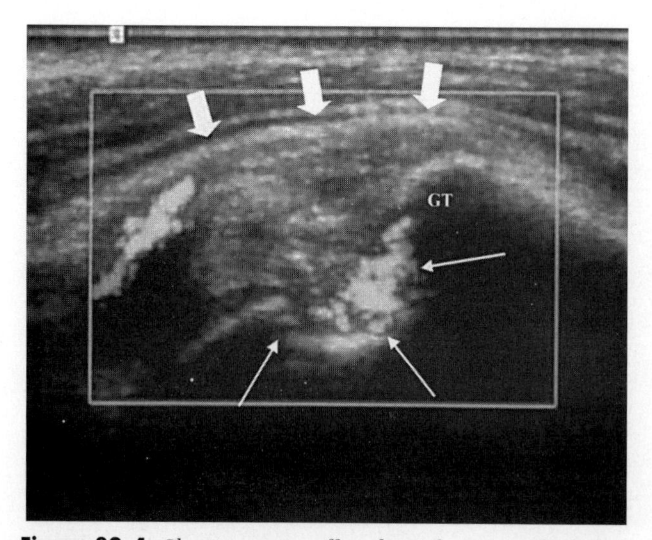

Figure 82–4. Chronic rotator cuff tendinopathy. Longitudinal ultrasound of the rotator cuff with color Doppler shows a supraspinatus tendon *(thick arrows)* with mixed echogenicity and hypervascularity. Note the large cortical erosion of the humeral head *(thin arrows)* adjacent to the greater tuberosity (GT).

Tendinopathy

Chronic degenerative changes of the rotator cuff tendons may exist without evidence of a tear. In chronic tendinopathy or tendinosis, the rotator cuff tendons show diffuse hypogenicity or mixed echogenicity with normal or increased tendon thickness (Fig. 82–4). Cortical irregularities of the humeral head at the sites of insertion of the rotator cuff tendons may be associated with chronic tendinopathy.

Tears

Most rotator cuff tears occur in the supraspinatus tendon, at or within 2 to 3 mm from its insertion on the greater tuberosity. In the hands of a properly trained sonographer, ultrasound yields a sensitivity greater than 90% for the diagnosis of full-thickness rotator cuff tears. A full-thickness tear is characterized by nonvisualization of the tendon or focal discontinuity of tendon fibers appearing as a hypoechoic area between the two edges of the tendon (Fig. 82–5). This appearance should be confirmed in two orthogonal (longitudinal and transverse) planes. The subacromial-subdeltoid bursa and deltoid muscle usually protrude into the tendon defect. Secondary signs include focal thinning of the tendon (Fig. 82–6), loss of the convex superior margin of the tendon, and enhancement of the echogenicity of the articular cartilage (Fig. 82–7).

A partial-thickness tear appears as a mixed hyperechoic and hypoechoic region (Fig. 82–8) or a hypoechoic discontinuity (Fig. 82–9) in the rotator cuff tendons. The lesion involves either the bursal or the articular surface of the tendon. Intrasubstance partial tear appears as a hypoechoic lesion not extending to either the humeral or the bursal surface of the tendon.

Calcifications

The presence of calcific deposits in the rotator cuff tendons is a frequent finding that may or may not be associated with symptomatic tendinopathy. These deposits appear as linear, curvilinear, or globular echogenic areas usually showing posterior acoustic shadowing and

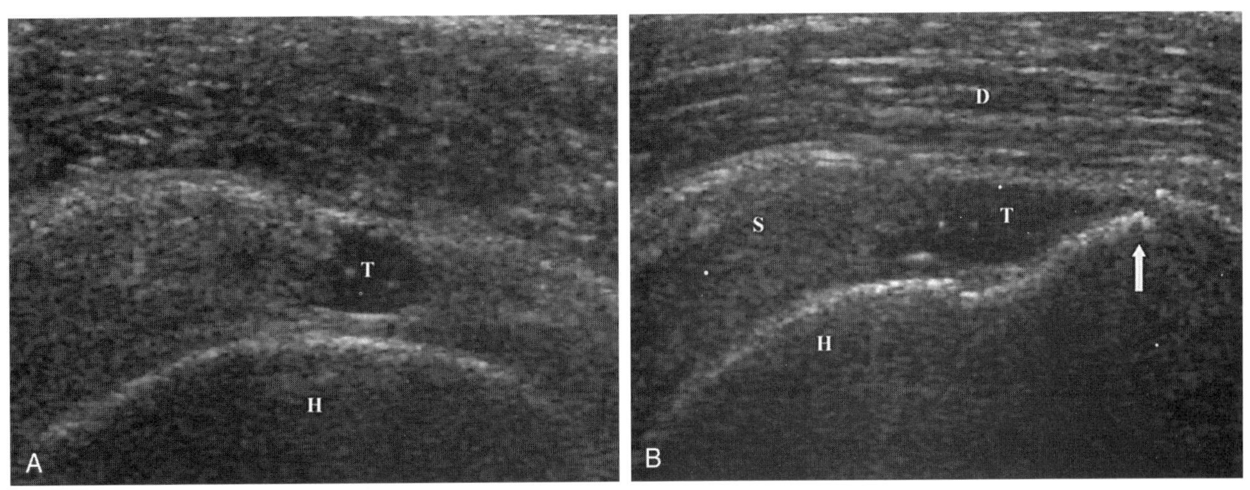

Figure 82–5. Full-thickness tear of the supraspinatus tendon. *A*, Transverse ultrasound of the rotator cuff shows a hypoechoic fluid-filled defect (T) between the torn ends of the supraspinatus tendon. H, humeral head. *B*, Longitudinal ultrasound shows the location of the tear (T) over the greater tuberosity *(arrow)*, while the supraspinatus tendon (S) is retracted over the humeral head (H). The size of the tear and its location can be assessed in both planes. D, deltoid muscle.

sometimes causing focal enlargement of the tendons (Fig. 82–10).

ANKLE

Anatomy

The anterior ankle tendons from medial to lateral are the tibialis anterior tendon, the extensor hallucis longus tendon, and the extensor digitorum longus tendon. They are the primary extensors of the ankle. The tibialis anterior tendon is the largest of the anterior tendons and inserts on the medial cuneiform. The extensor hallucis longus tendon inserts on the first toe. The extensor digitorum longus tendon is seen proximally as a single tendon and distally divides into four independent tendons that insert on the second through fifth toes.

The lateral ankle tendons are the peroneus brevis and the peroneus longus tendons. They are the primary abductors and everters of the foot. The peroneus brevis tendon is located anterior and medial relative to the peroneus longus tendon, immediately behind the lateral malleolus (Fig. 82–11). The peroneus brevis tendon inserts on the base of the fifth metatarsal. The peroneus longus tendon inserts on the plantar aspect of the foot, the medial cuneiform, and base of the first metatarsal.

The medial ankle tendons are the posterior tibial (PT) tendon, the flexor digitorum longus (FDL) tendon, and the flexor hallucis longus (FHL) tendon (Fig. 82–12). They are the primary inverters of the ankle. The PT tendon is the largest of the three medial ankle tendons and

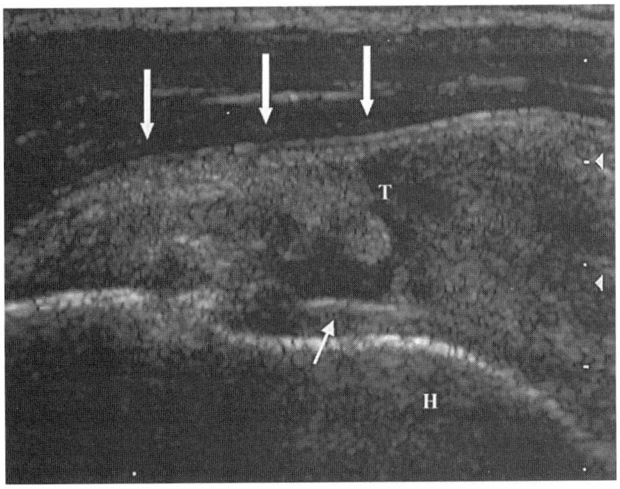

Figure 82–6. Full-thickness tear of the rotator cuff with focal thinning of the tendon. Longitudinal ultrasound shows hypoechoic full-thickness tear (T) of the supraspinatus tendon with loss of the normal convexity and thinning of distal end of the tendon *(long arrows)*. Applying pressure over the rotator cuff with the transducer may reveal abnormalities of the tendon contours and increase the conspicuousness of more subtle rotator cuff tears. Note the uncovered cartilage sign *(short arrow)*. H, humeral head.

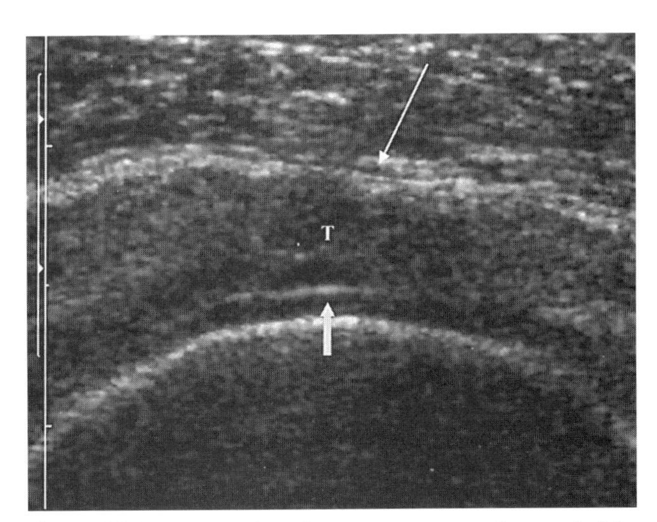

Figure 82–7. Uncovered cartilage sign. Transverse ultrasound of the rotator cuff shows a small fluid-filled, full-thickness tear (T). There is increased sound transmission through the tear, and the underlying cartilage appears as a bright hyperechoic line *(short arrow)*. Note the focal defect in the convexity of the surface of the tendon *(long arrow)*.

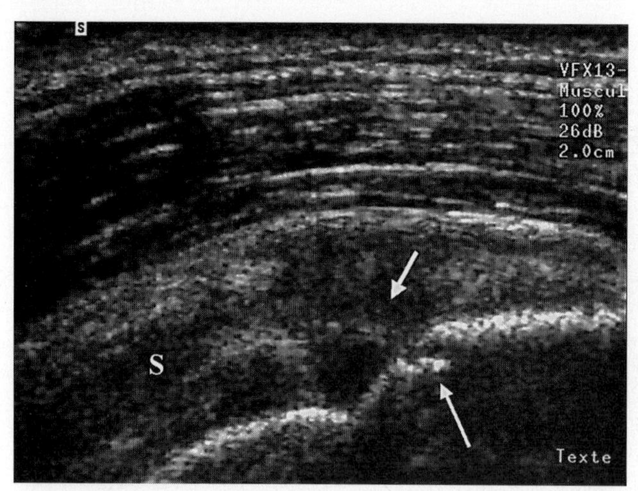

Figure 82–8. Partial-thickness rotator cuff tear. Longitudinal ultrasound of the rotator cuff shows a mixed hyperechoic and hypoechoic defect *(short arrow)* involving the articular surface of the supraspinatus tendon (S) corresponding to a partial-thickness tear. Note the small focal irregularity in the cortex of the humeral head underlying the tear *(long arrow)*.

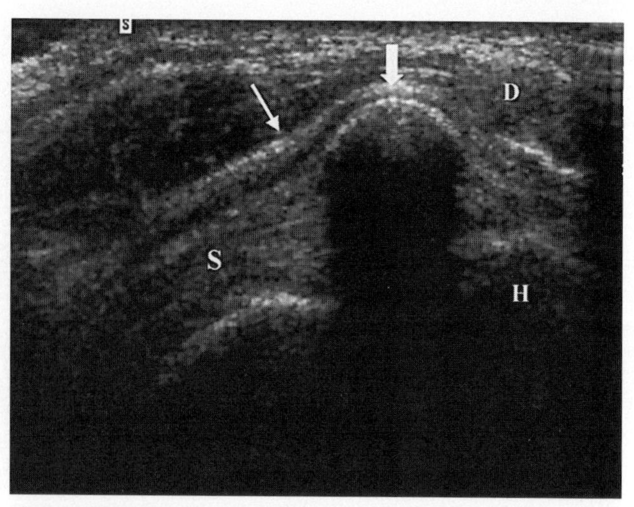

Figure 82–10. Calcific tendinitis. Longitudinal ultrasound of the supraspinatus tendon (S) shows a curvilinear hyperechoic structure *(thick arrow)* with strong posterior acoustic shadowing, within the distal aspect of the tendon. The calcification protrudes on the superior aspect of the tendon, causing displacement of the overlying subacromial-subdeltoid bursa *(thin arrow)* and deltoid muscle (D). H, humeral head.

is the strongest ankle tendon after Achilles' tendon. The PT tendon is located immediately behind the medial malleolus, then, moving posterolaterally, the FDL tendon, the neurovascular bundle with the tibial nerve, and the FHL tendon are successively encountered. The PT tendon inserts on the medial aspect of the navicular. The FHL tendon inserts on the plantar aspect of the distal phalanx of the first toe, and the FDL tendon divides into four separate tendons that insert on the plantar aspects of the second to fifth toes.

Achilles' tendon inserts on the posterior aspect of the calcaneal tuberosity and acts to plantar flex the foot (Fig. 82–13). It is formed by the confluence of the gastrocnemius and soleus tendons. It measures approximately 10 to 15 cm long in adults. Between Achilles' tendon and the calcaneus lies the deep retrocalcaneal bursa. There is a second bursa located superficially to the distal Achilles' tendon. The normal maximal anteroposterior diameter of Achilles' tendon is less than 7.1 mm. Achilles' tendon has no synovial sheath but is enveloped by a fibrous tissue called the *paratenon*.

Tendinopathy

In tendinopathy, the tendon shows focal or diffuse enlargement and appears less echogenic than normal. There is increased hypoechoic space between the tendon fibrils (Fig. 82–14). Color Doppler or power Doppler imaging may show regions of hyperemia within the tendon (Fig. 82–15). In cases of chronic tendinopathy, especially involving Achilles' tendon, intratendinous calcifications may be present (Fig. 82–16).

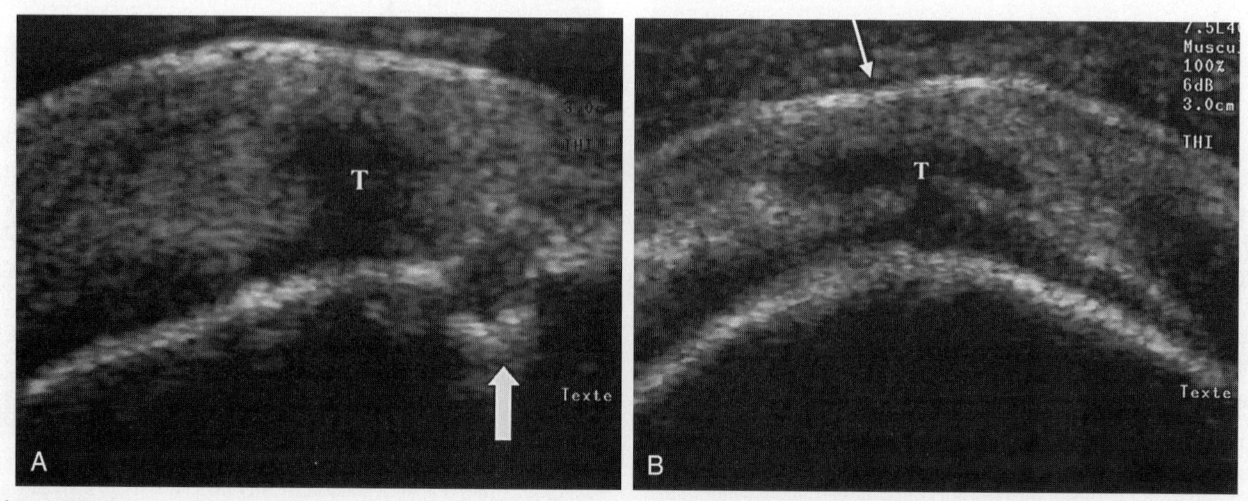

Figure 82–9. Partial-thickness articular surface tear of the rotator cuff. Longitudinal ultrasound *(A)* and transverse ultrasound *(B)* show a hypoechoic defect (T) between the torn ends of the tendon. The tear involves the articular surface of the tendon and does not extend through the entire thickness of the tendon. Notice the cortical erosion underlying the partial tear *(arrow on A)* and the subtle focal defect of the tendon convexity *(arrow on B)*.

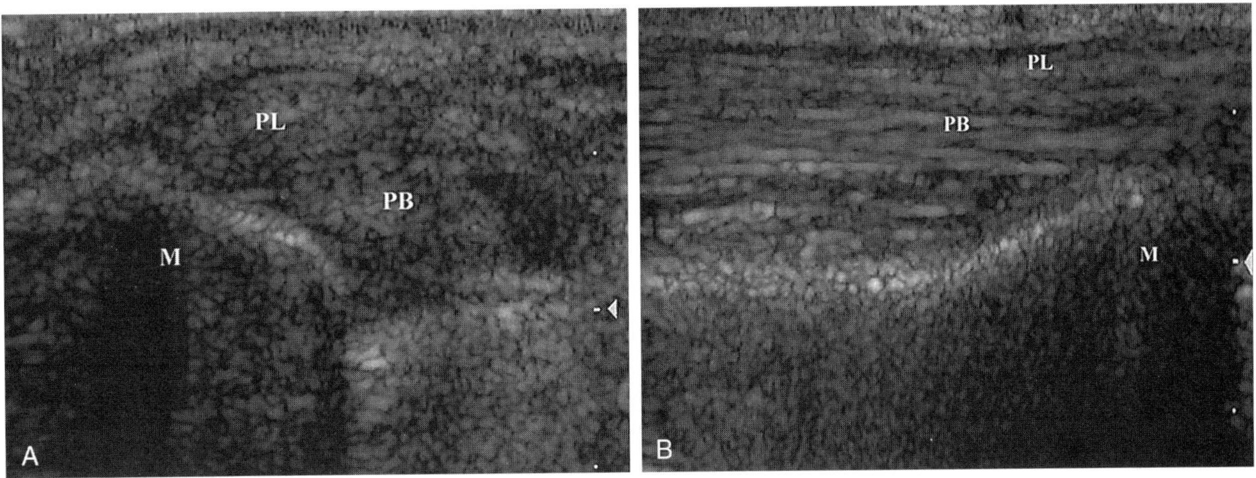

Figure 82–11. Normal peroneus longus and peroneus brevis tendons. *A*, Transverse ultrasound of the lateral aspect of the ankle shows the normal location of the peroneus brevis (PB) and peroneus longus (PL) tendons behind the lateral malleolus (M). *B*, Longitudinal ultrasound of the lateral aspect of the ankle shows the peroneus brevis (PB) tendon interposed between the lateral malleolus (M) and the peroneus longus (PL) tendon.

Tenosynovitis and Paratenonitis

Acute inflammation of the tendon sheath (or the paratenon in the case of Achilles' tendon) is characterized by the accumulation of anechoic fluid within the tendon sheath. It is frequently associated with signs of tendinopathy.

Chronic tenosynovitis is characterized by hypoechoic thickening of the tendon sheath, which is noncompressible when pressure is applied with the ultrasound transducer. A small amount of anechoic fluid is often present (Fig. 82–17).

Up to 4 mm of anechoic fluid can be found, usually posteriorly, in the synovial sheath of the normal PT tendon over its segment distal to the medial malleolus. Up to 3 mm of anechoic fluid can be found in the common peroneal tendon sheath of normal individuals, just distal to the lateral malleolus. The tibiotalar joint and the tendon sheath of the FHL may communicate in normal individuals. In the presence of a tibiotalar joint effusion, anechoic fluid in the FHL tendon sheath is not diagnostic of a tenosynovitis.

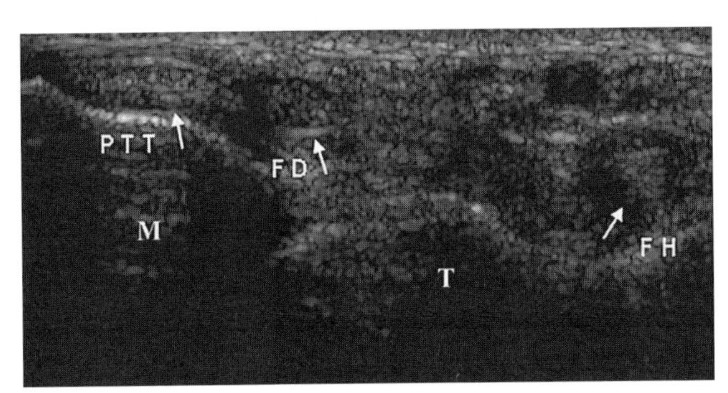

Figure 82–12. Normal medial ankle tendons. Transverse ultrasound of the medial aspect of the ankle shows the posterior tibialis tendon (PTT) immediately adjacent to the cortex of the medial malleolus (M). Moving posterolaterally, the flexor digitorum longus (FD) tendon and flexor hallucis longus (FH) tendon are seen at the level of the talus (T).

Tears

A complete tear manifests by a total discontinuity in the tendon. In the acute setting, this gap may be filled with a hypoechoic hematoma, whereas in chronic cases the gap

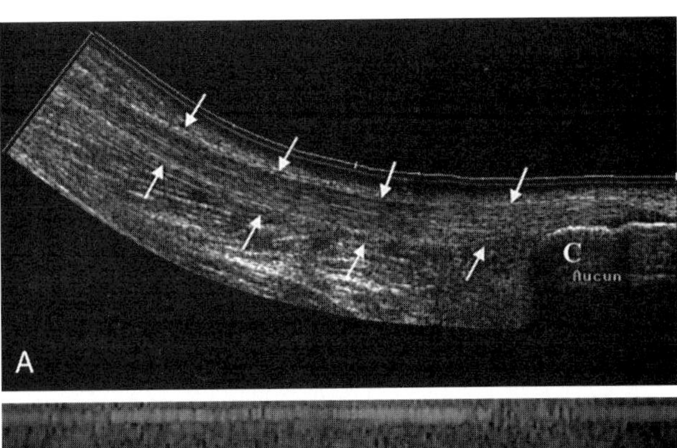

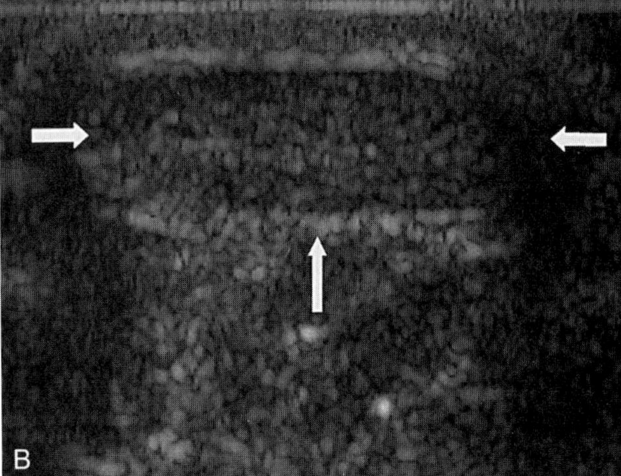

Figure 82–13. Normal Achilles' tendon. *A*, Extended-field-of-view longitudinal ultrasound of the posterior ankle shows the entire length of the Achilles' tendon *(arrows)* down to its insertion on the calcaneal tuberosity (C). *B*, Transverse ultrasound depicts the punctate hyperechoic pattern of the normal tendon *(arrows)*. The anteroposterior diameter of the tendon is evaluated best in the transverse plane.

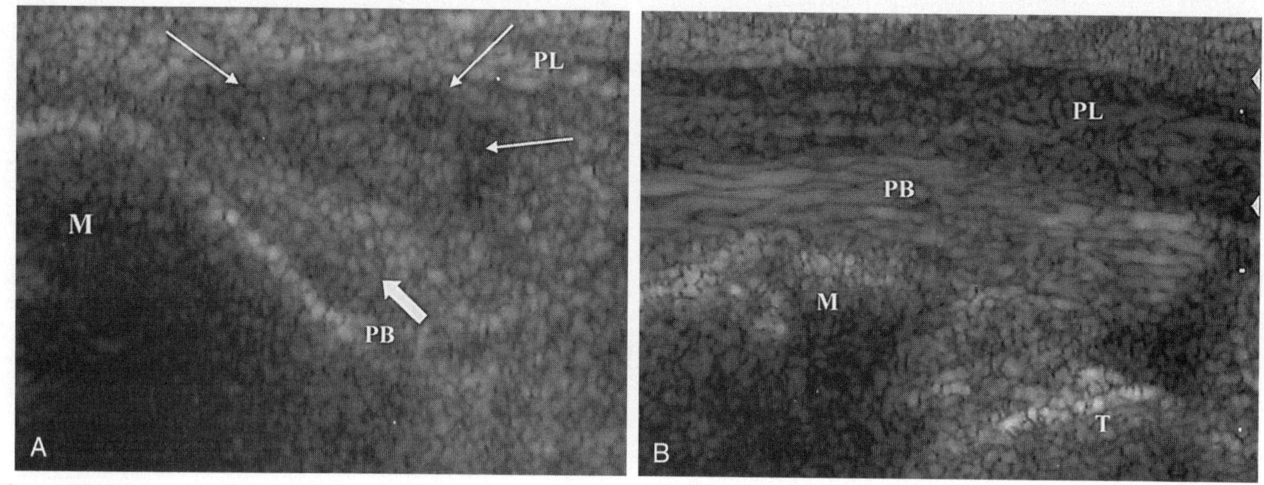

Figure 82–14. Peroneus longus tendinopathy. Transverse ultrasound *(A)* and longitudinal ultrasound *(B)* at the level of the lateral malleolus (M) show a diffusely hypoechoic and swollen peroneus longus tendon (PL) against the echogenic peroneus brevis (PB) tendon, which shows a normal fibrillar appearance.

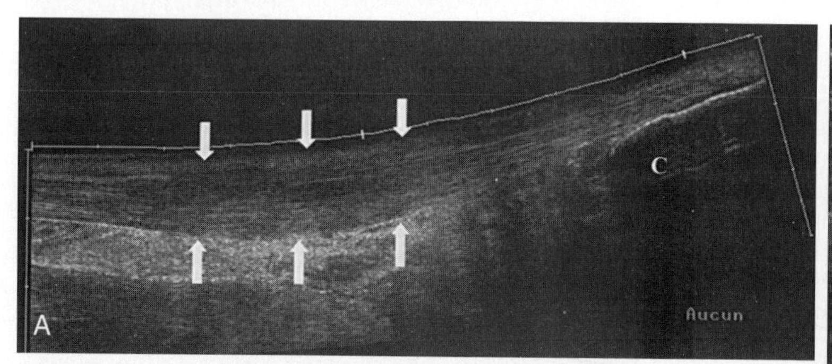

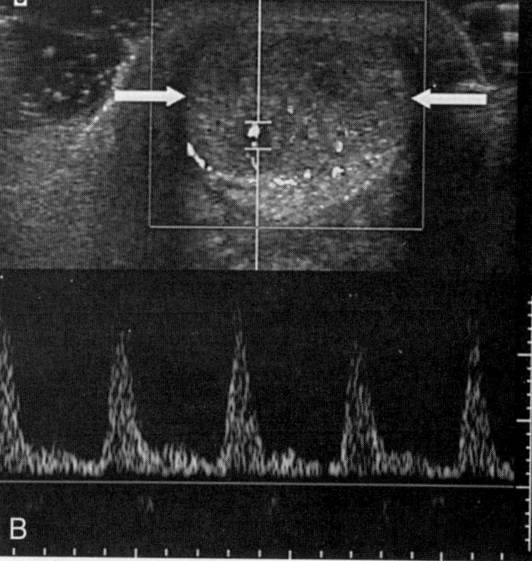

Figure 82–15. Tendinopathy of Achilles' tendon. *A*, Extended-field-of-view longitudinal ultrasound of Achilles' tendon shows enlargement and abnormal hypoechogenicity of a segment of the tendon *(between arrows)* proximal to its insertion on the calcaneal tuberosity (C). *B*, Transverse ultrasound with color and pulsed Doppler studies documents hyperemia within the tendon.

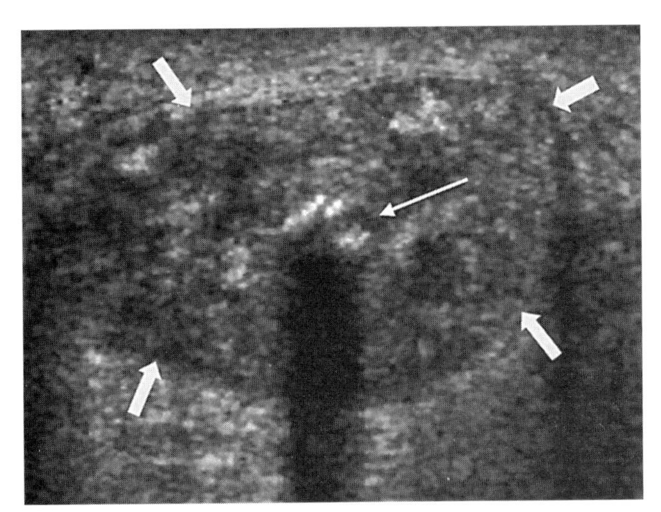

Figure 82–16. Chronic calcifying Achilles' tendinopathy. Transverse ultrasound of the posterior ankle shows a diffusely enlarged Achilles' tendon *(short arrows)* with mixed echogenicity and internal hyperechoic foci *(long arrow)* with acoustic shadowing representing intratendinous calcifications.

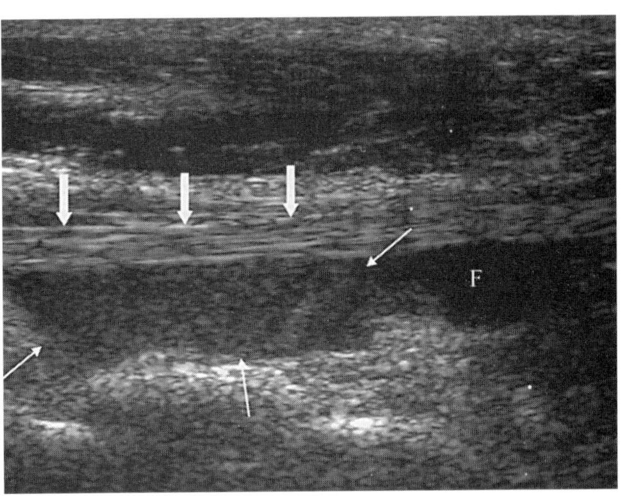

Figure 82–17. Chronic flexor hallucis longus tenosynovitis. Longitudinal ultrasound of the posterior aspect of the ankle shows synovial thickening of the flexor hallucis longus tendon sheath *(thin arrows)* and a small amount of fluid (F) surrounding a normal tendon *(large arrows)*.

may be filled with scar tissue, which is usually hyperechoic and noticeable because of focal distortion and absence of the normal fibrillar appearance of the tendon. Retraction of the proximal and distal ends of the tendon may be present. A partial tendon tear is seen as a focal hypoechoic area of interruption of the tendon fibrils, which may be intratendinous or may extend to the tendon surface (Fig. 82–18).

Tears of the peroneus longus and peroneus brevis tendons usually present as longitudinal splits. Longitudinal tears of the peroneus brevis tendon frequently originate within the fibular groove, at the level of the lateral malleolus, and the peroneus longus tendon can insinuate into the tendon fissure of the peroneus brevis.

Tears of the PT tendon usually occur at the level of the medial malleolus. The spectrum of abnormalities involving the PT tendon range from tenosynovitis, tendinopathy, and partial tears to complete tears. Rupture of the PT tendon typically develops as a chronic process in middle-aged or elderly women, who present with pain and swelling over the medial aspect of the ankle with acquired flatfoot deformity. Complete tears of the PT tendon are usually transverse and appear as a hypoechoic gap between the two ends of the tendon.

Achilles' tendon is the most frequently injured tendon of the ankle. "Weekend athletes" are particularly at risk for Achilles' tendon tears. Injuries usually occur approximately 2 to 6 cm proximal to the calcaneal insertion, corresponding to an area of relative avascularity of the tendon. At that level, the tendon fibers also form a spiral twist resulting in an area of decreased resistance to tensile forces. Ultrasound has shown its value in the diagnosis of Achilles' tendon tendinopathy, paratenonitis, partial tears, and complete tears (Fig. 82–19).

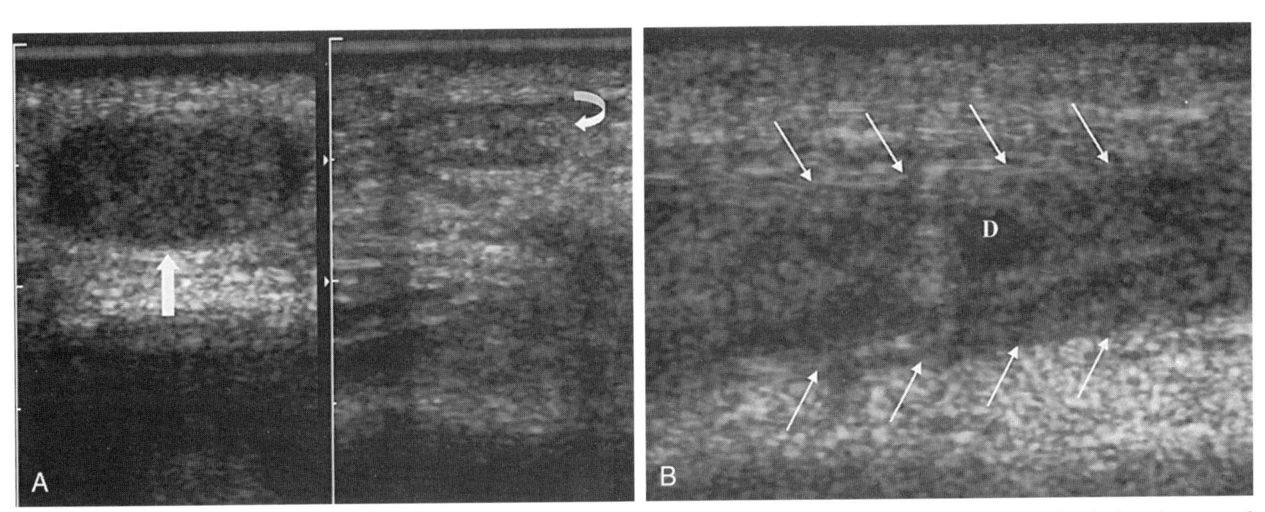

Figure 82–18. Partial intrasubstance tear of tibialis anterior tendon. *A*, Dual transverse ultrasound. The symptomatic right tibialis anterior tendon *(arrow)* appears hypoechoic and is enlarged almost to three times the normal left side *(curved arrow)*. *B*, On longitudinal ultrasound, a focal intrasubstance hypoechoic defect (D) is identified within the diffusely hypoechoic swollen tendon *(arrows)*.

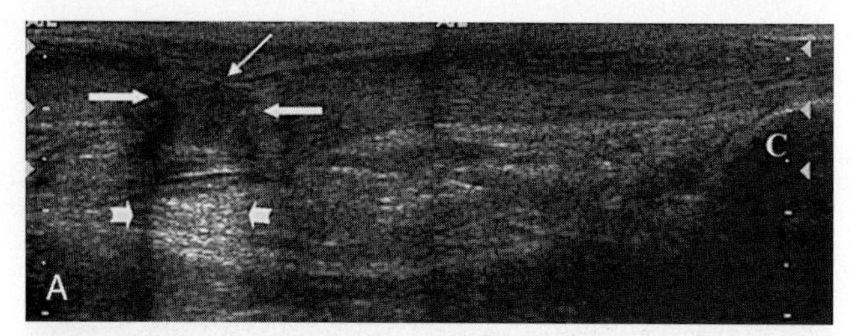

Figure 82–19. Acute rupture of Achilles' tendon. Dual longitudinal ultrasound of the posterior aspect of the ankle shows a complete tear of Achilles' tendon. The gap between the retracted ends of the tendon *(large arrows)* is filled with relatively hypoechoic material *(thin arrows)* representing a fresh hematoma. Note the posterior acoustic enhancement *(notched arrows)* behind the hematoma. C, calcaneal tuberosity.

Subluxation and Dislocation

The peroneal tendons are the ankle tendons most frequently involved in subluxation or dislocation. These injuries commonly occur in sports-related activities, such as skiing, ice skating, basketball, and soccer, and are seen with rupture of the peroneal retinaculum. In cases of dislocation, one or both tendons are seen lateral or anterior to the lateral malleolus. Dynamic ultrasound examination during dorsiflexion and eversion of the ankle can reveal transient dislocation or subluxation of the peroneal tendons.

LONG HEAD OF THE BICEPS (LHB) TENDON

Anatomy

The LHB tendon is intra-articular but extrasynovial. It takes its origin from the superior glenoid tubercle and travels through the rotator cuff interval between the supraspinatus tendon and the subscapularis tendon before entering the bicipital groove. The transverse ligament, the coracohumeral ligament, and, more distally, the falciform ligament (a tendinous expansion from the humeral insertion of the pectoralis major) all participate in maintaining the LHB tendon within the bicipital groove.

Evaluation of the LHB tendon is an integral part of the ultrasound examination of the shoulder. At ultrasound, the segment of the LHB tendon located in the bicipital groove is evaluated easily, whereas the segment proximal to the bicipital groove is not entirely visible at ultrasound. When imaged with the ultrasound probe perpendicular to the bicipital groove of the proximal humerus, the LHB tendon appears as an echogenic oval structure in the groove and often is surrounded by a small amount of hypoechoic fluid (Fig. 82–20). The maximal anteroposterior diameter of the LHB tendon is 3.3 to 4.7 mm. When imaged in the longitudinal plane, the tendon shows a hyperechoic fibrillar appearance (Fig. 82–21).

Tenosynovitis and Tendinopathy

The presence of fluid within the biceps tendon sheath is a nonspecific finding and is associated in 90% of cases with pathology elsewhere in the glenohumeral joint. The diagnosis of bicipital tenosynovitis should not be considered if the presence of fluid within the biceps tendon sheath is an isolated finding and the LHB tendon appears normal.

In cases of biceps tendinopathy, the presence of fluid in the tendon sheath is associated with abnormalities of the tendon itself. The tendon may be enlarged, may be hypoechoic, and may show irregular margins (Fig. 82–22). Focal tenderness may be elicited by palpation with the ultrasound transducer.

Tears

A longitudinal intrasubstance tear of the LHB tendon may appear as a fusiform cystic cavity within the tendon (Fig. 82–23). A partial-thickness tear appears as a focal

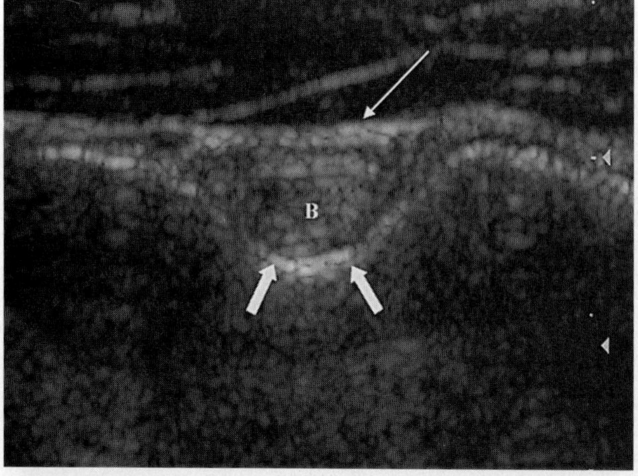

Figure 82–20. Normal long head of the biceps tendon in the transverse plane. Transverse ultrasound at the level of the bicipital groove *(large arrows)* shows the biceps tendon (B) as an oval-shaped echogenic structure within the tendon groove. The transverse ligament appears as a fine hyperechoic line *(long arrow)* covering the biceps tendon groove.

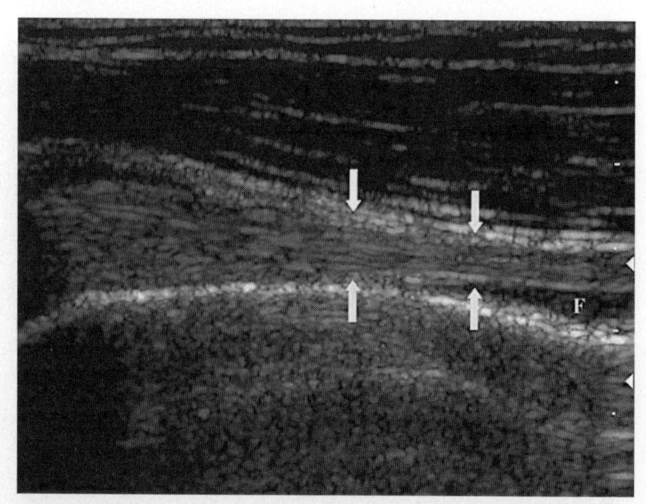

Figure 82–21. Normal long head of the biceps tendon in the longitudinal plane. Longitudinal ultrasound at the level of the bicipital groove. The biceps tendon is seen as an hyperechoic fibrillar structure *(arrows)*. Note the small amount of fluid (F) seen in the most dependent portion of the tendon groove, which may be seen in normal subjects.

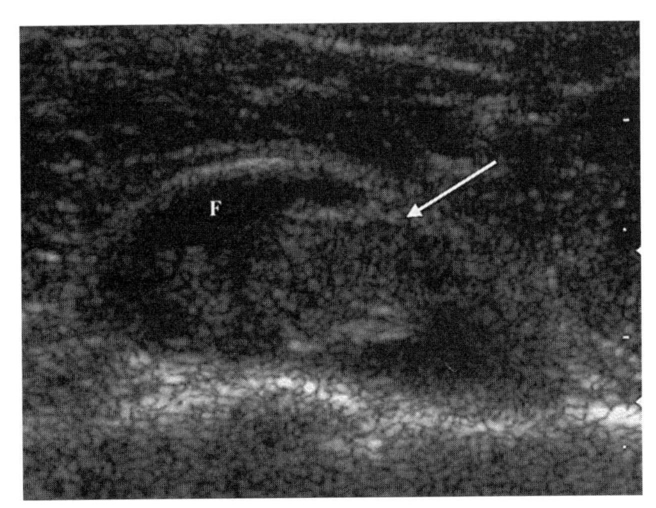

Figure 82–22. Long head of the biceps tendinitis. Transverse ultrasound slightly distal to the bicipital groove shows a swollen, hypoechoic tendon *(arrow)* with irregular margins. The tendon sheath is distended with fluid (F).

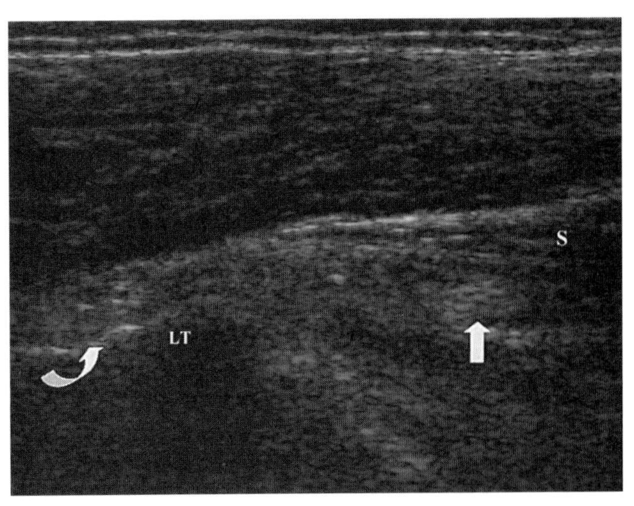

Figure 82–24. Long head of the biceps tendon dislocation. Transverse ultrasound at the level of the midbicipital groove *(curved arrow)* shows the long head of the biceps tendon *(straight arrow)* dislocated medial to the lesser tuberosity (LT) and located deep to the subscapularis tendon.

hypoechoic area corresponding to focal discontinuity of tendon fibrils.

The diagnosis of a complete rupture of the LHB tendon usually is made clinically. When the patient flexes the elbow, the biceps muscle belly retracts and forms a globular mass on the anterior aspect of the distal humerus. An abrupt concavity also may be palpated on the proximal aspect of the muscle. Ultrasound sometimes may be needed to differentiate a complete rupture from a partial tear. In cases of complete rupture, the bicipital groove appears empty. On longitudinal scan, the retracted distal end of the tendon may be identified.

Dislocation and Subluxation

Dislocation of the LHB tendon frequently is associated with a tear of the subscapularis tendon. The LHB tendon dislocates medially and is identified on transverse ultrasound scan, medial to the empty bicipital groove (Fig. 82–24). Occasionally the LHB tendon is displaced only partially out of the bicipital groove (Fig. 82–25). Dynamic ultrasound examination with external rotation of the shoulder facilitates the detection of the abnormal displacement of the LHB tendon.

QUADRICEPS AND PATELLAR TENDONS

Further information on the extensor mechanism of the knee is found in Chapter 64.

Anatomy

The quadriceps tendon inserts on the superior pole of the patella (Fig. 82–26). The suprapatellar bursa, which

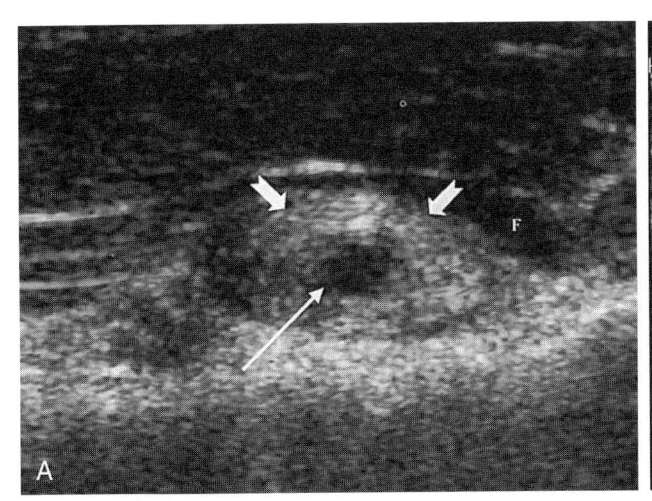

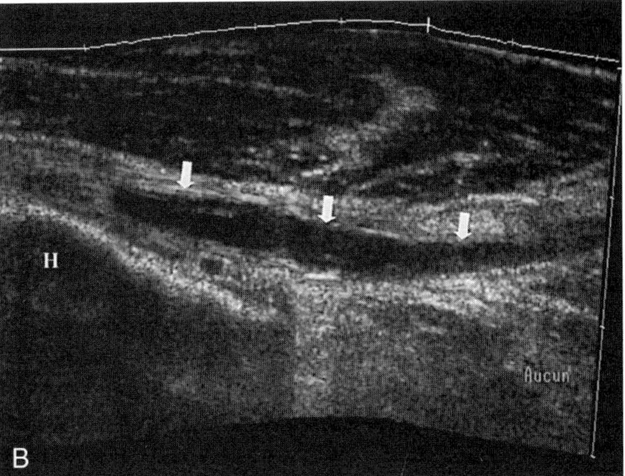

Figure 82–23. Intrasubstance tear of the long head of the biceps tendon. *A*, Transverse ultrasound at the level of the bicipital groove shows a central, hypoechoic, cavity-like partial tear of the tendon. There is a small amount of fluid within the tendon sheath (F). *B*, Longitudinal extended-field-of-view ultrasound shows the long cystic cavity *(arrows)*, which extends distally to the myotendinous junction of the biceps. H, humerus.

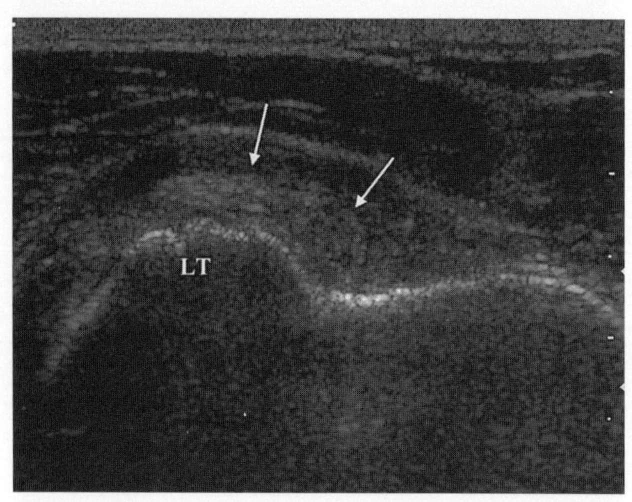

Figure 82–25. Long head of the biceps tendon subluxation. Transverse ultrasound at the level of the proximal bicipital groove shows the long head of the biceps *(arrows)* sitting on top of the lesser tuberosity (LT), partially in and partially out of the bicipital groove.

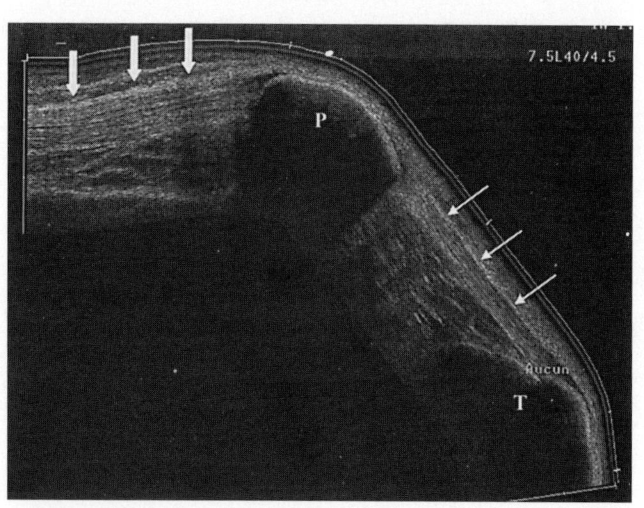

Figure 82–26. Normal quadriceps and patellar tendons. Extended-field-of-view longitudinal ultrasound of the knee in slight flexion shows the normal fibrillar appearance of the quadriceps tendon *(large arrows)*, which attaches on the superior pole of the patella (P). The normal patellar tendon *(thin arrows)* extends from the inferior pole of the patella to the tibial tuberosity (T).

freely communicates with the knee joint, is interposed between the quadriceps tendon and the anterior surface of the distal femur.

The patellar tendon is the continuation of the central portion of the quadriceps tendon, which extends from the patella to the tibial tuberosity. It is a strong, flat band, approximately 8 cm long, which courses obliquely downward to its tibial insertion. The deep infrapatellar bursa is interposed between the distal tendon and underlying tibial tuberosity.

Tears

Partial or complete tears of the quadriceps tendon are uncommon injuries and usually occur in individuals older than 40 years. They may be post-traumatic, but most occur in a tendon weakened by tendinopathy. Recognized risk factors that weaken the quadriceps tendon are diabetes, chronic renal insufficiency in patients on dialysis, systemic or local injection of steroids, rheumatoid arthritis, gout, and systemic lupus erythematosus.

In most cases, a complete rupture of the quadriceps tendon is diagnosed accurately clinically on the basis of loss of extensor function and the presence of a palpable mass or a defect above the patella. Ultrasound may be useful to confirm the diagnosis and help plan surgical repair by showing the location, extension of the tear, and quality of the tendon remnants (Fig. 82–27). In cases of complete rupture of the quadriceps tendon, ultrasound shows complete interruption of the tendon fibers with retraction of the free ends. In the acute phase, the gap usually is filled with hypoechoic or anechoic material representing a hematoma. Gentle distal traction on the patella during ultrasound examination increases the gap between the ends of the tendon and may help confirm the presence of a complete tear.

Partial quadriceps tendon tears are more difficult to evaluate on physical examination. They occur near the

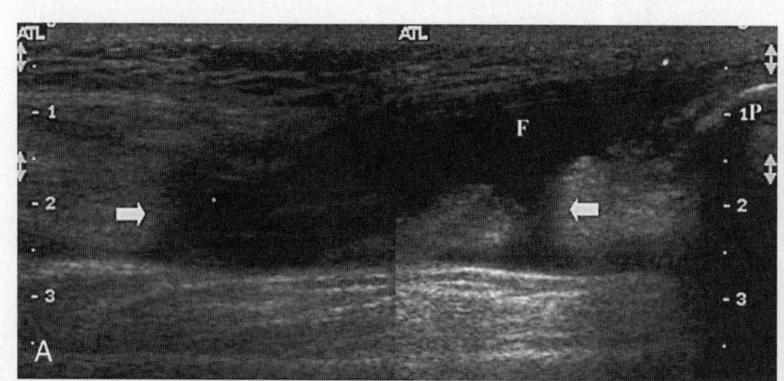

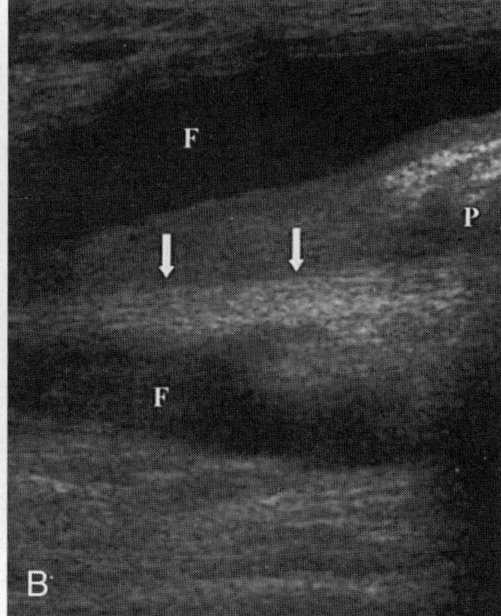

Figure 82–27. Extensive partial tear of the quadriceps tendon. *A*, Dual longitudinal ultrasound of the distal anterior thigh, proximal to the superior pole of the patella (P), documents a tear of the quadriceps tendon and a large fluid-filled (F) gap between the retracted ends *(arrows)* of the tendon. *B*, Longitudinal ultrasound medial to *A* shows a thin hyperechoic band of tendon remaining intact *(arrows)* and surrounded by hypoechoic fluid (F).

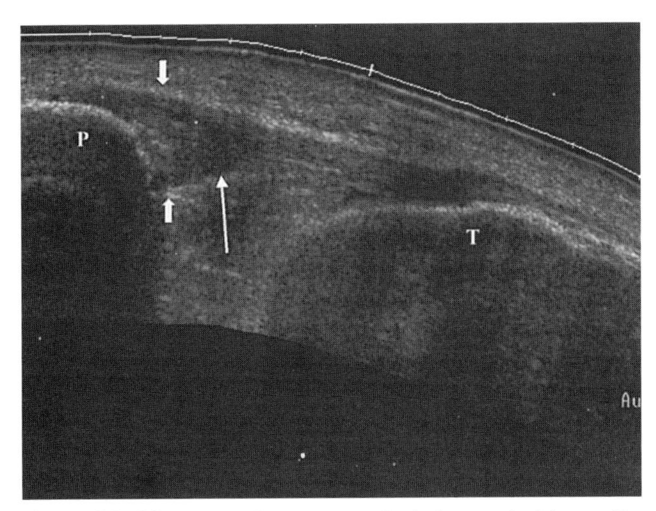

Figure 82–28. Jumper's knee. Longitudinal ultrasound of the patellar tendon shows a focal hypoechoic area in the deep portion of the proximal patellar tendon *(long arrow)*. The proximal tendon also is diffusely swollen and hypoechoic *(short arrows)*. P, patella. T, tibial tuberosity.

superior pole of the patella or at the myotendinous junction of the rectus femoris muscle. At ultrasound, a focal hypoechoic defect in the tendon, not involving the full thickness of the tendon and with at least one tendinous lamina remaining intact, is seen. If the tear occurs proximally, retraction of the muscle belly may be shown.

Patellar Tendinosis and Partial Tears

Patellar tendinosis (referred to as *jumper's knee*) and partial tears are seen in young individuals involved in activities in which the extensor mechanism of the knee is put under significant stress by repetitive actions of acceleration, deceleration, jumping, and landing. Patients present with anterior knee pain associated with tenderness of the patellar tendon near its insertion on the lower pole of the patella.

Ultrasound shows a hypoechoic area in the proximal portion of the patellar tendon with a variable amount of swelling of the surrounding tendon (Fig. 82–28). Hyperechoic foci representing calcification or dystrophic ossification may be present.

SNAPPING HIP SYNDROME

Overview

The snapping hip syndrome is defined as a painful audible snap occurring in the hip on motion. A snapping hip may be asymptomatic. Causes of snapping hips may be intra-articular (loose bodies, labral tear) or extra-articular. Extra-articular causes are divided further into medial and lateral varieties, resulting from snapping iliopsoas tendon over the iliopectineal eminence and iliotibial band or gluteus maximus over the greater trochanter. Ultrasound is well suited for the evaluation of extra-articular causes of snapping hip because the hip may be examined dynamically with ultrasound during joint motion.

Medial Extra-Articular Causes of Snapping Hip

The normal iliopsoas tendon is located anterior to the hip joint and appears as an oval-shaped hyperechoic structure when examined on a transverse plane (Fig. 82–29*A*). It also can be examined in a longitudinal plane as it curves over the hip joint (Fig. 82–29*B*) and appears hyperechoic and fibrillary. Normally, when examined in a transverse plane, the tendon glides smoothly back and forth, medial to lateral during hip rotation. A variety of different hip motions can generate a hip snap and depend on individual patients. Most commonly, extending the hip from an abducted, flexed, and externally rotated (frog-leg) position generates the snap, and the sonographer is able to correlate the clicking sound with an abrupt movement of

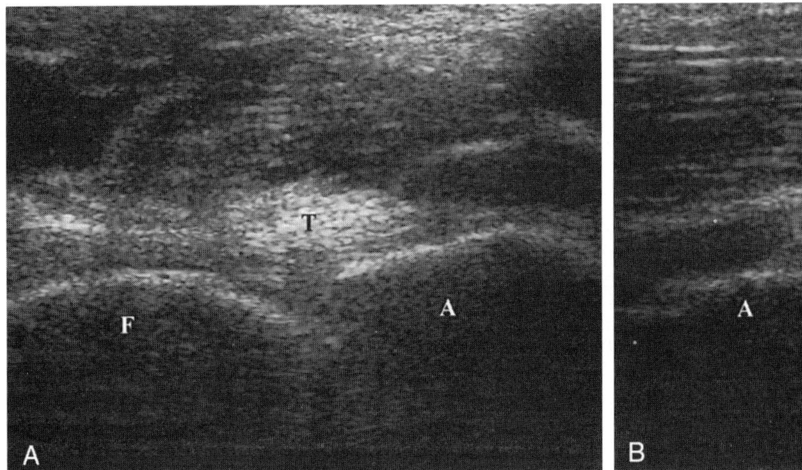

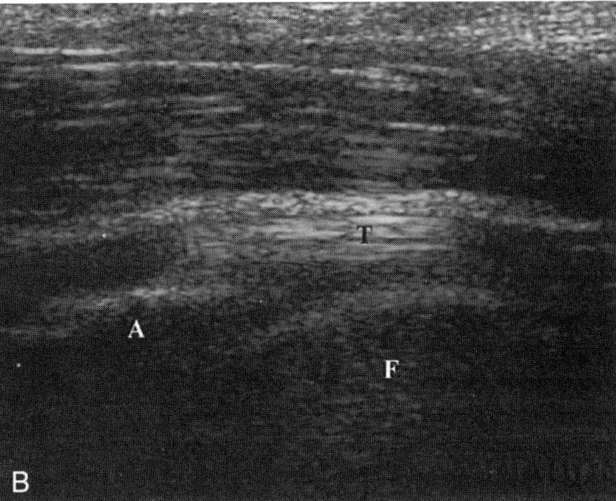

Figure 82–29. Normal iliopsoas tendon. *A*, Transverse ultrasound of the right hip shows normal iliopsoas tendon as a hyperechoic oval-shaped (T) structure anterior to the hip joint. *B*, Longitudinal ultrasound of the right hip shows iliopsoas tendon as a fibrillary hyperechoic band (T) curving anterior to the hip joint. The tendon at the extremity of the image is hypoechoic because of anisotropy. A, acetabulum; F, femoral head.

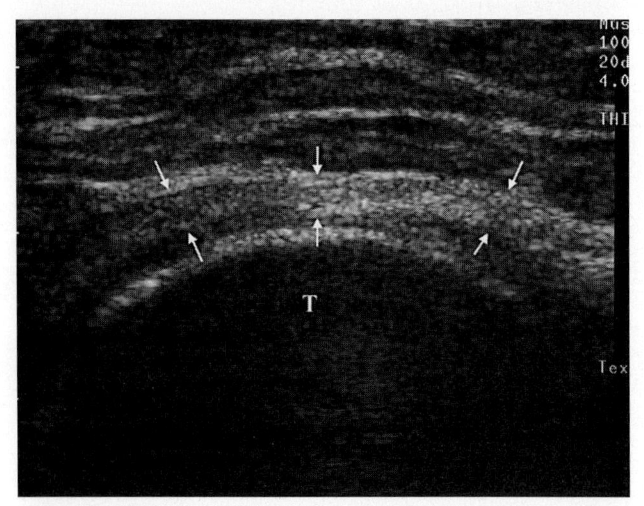

Figure 82–30. Normal iliotibial band. Transverse ultrasound of the lateral aspect of the left thigh shows the iliotibial band as a stripe of hyperechoic tissue *(arrows)* of homogeneous thickness covering the greater trochanter (T).

the tendon from lateral to medial. When the snapping hip is painful, the sonographer is able to correlate the abnormal snap with the elicitation of pain.

Lateral Extra-Articular Causes of Snapping Hip

For lateral causes of extra-articular snapping hip, the transducer is positioned in a transverse plane over the greater trochanter of the snapping hip as the patient is lying in lateral decubitus on the contralateral hip. The iliotibial band is identified as a hyperechoic stripe covering the greater trochanter (Fig. 82–30). As the patient flexes and extends the hip, the iliotibial band and the most anterior portion of gluteus maximus are seen gliding smoothly anteriorly (during flexion) and posteriorly (during extension). In cases of snapping hip, the thickened iliotibial band or gluteus maximus abruptly snaps over the greater trochanter correlating with the clicking sound. Occasionally the hip has to be examined while the patient is standing up because contraction of the gluteus muscle is necessary to generate the snap.

Recommended Readings

Bencardino J, Rosenberg ZS, Delfaut E: MR imaging in sports injuries of the foot and ankle. Magn Reson Imaging Clin N Am 7:131, 1999.

Bianchi S, Zwass A, Abdelwahab IF, et al: Diagnosis of tears of the quadriceps tendon of the knee: Value of sonography. AJR Am J Roentgenol 162:1137, 1994.

Binnie JF: Snapping hip (hanche à ressort; schnellende hufte). Ann Surg 58:59, 1913.

Bureau NJ, Roederer G: Sonography of the Achilles tendon xanthomas in patients with heterozygous familial hypercholesterolemia. AJR Am J Roentgenol 171:745, 1998.

Cardinal É, Buckwalter KA, Capello WN, et al: US of the snapping iliopsoas tendon. Radiology 198:521, 1996.

Colosimo AJ, Bassett FH: Jumper's knee diagnosis and treatment. Orthop Rev 19:139, 1990.

Farin P, Jaroma H: Sonographic detection of tears of the anterior portion of the rotator cuff (subscapularis tendon tears). J Ultrasound Med 16:221, 1996.

Fessell DP, Vanderschueren GM, Jacobson JA, et al: US of the ankle: Technique, anatomy, and diagnosis of pathologic conditions. Radiographics 18:325, 1998.

Fornage BD: Achilles tendon: US examination. Radiology 159:759, 1986.

Fornage BD, Rifkin MD: Ultrasound examination of tendons. Radiol Clin North Am 26:87, 1988.

Khan KM, Bonar F, Desmond P, et al: Patellar tendinosis (jumper's knee): Findings at histopathologic examination, US, and MR imaging. Radiology 200:821, 1996.

Khoury NJ, El-Khoury GY, Saltzman CL, et al: MR imaging of posterior tibial tendon dysfunction. AJR Am J Roentgenol 167:675, 1996.

Khoury NJ, El-Khoury GY, Saltzman CL, et al: Peroneus longus and brevis tendon tears: MR imaging evaluation. Radiology 200:833, 1996.

Mack LA, Gannon MK, Kilcoyne RF, et al: Sonographic evaluation of the rotator cuff: Accuracy in patients without prior surgery. Clin Orthop 234:21, 1988.

Mack LA, Nyberg DA, Matsen III FA: Sonograhic evaluation of the rotator cuff. Radiol Clin North Am 26:161, 1988.

Middleton WD, Reinus WR, Totty WG, et al: Ultrasonographic evaluation of the rotator cuff and biceps tendon. J Bone Joint Surg Am 68:440, 1986.

Nazarian LN, Rawool NM, Martin CE, et al: Synovial fluid in the hindfoot and ankle: Detection of amount and distribution with US. Radiology 197:275, 1995.

Nunziata A, Blumenfeld I: Cadera a resorte: A proposito de una variedad. Prensa Med Argent 32:1997, 1951.

Pelsser V, Cardinal E, Hobden R, et al: Extraarticular snapping hip: Sonographic findings. AJR Am J Roentgenol 176:67, 2001.

Ptasznik R, Hennessy O: Abnormalities of the biceps tendon of the shoulder: Sonographic findings. AJR Am J Roentgenol 164:409, 1995.

Reinherz RP, Zawada SJ, Sheldon DP: Recognizing unusual tendon pathology at the ankle. J Foot Surg 25:278, 1986.

Steinmetz A, Schmitt W, Schuler P, et al: Ultrasonography of Achilles tendons in primary hypercholesterolemia: Comparison with computed tomography. Atherosclerosis 74:231, 1988.

Teefey SA, Middleton WD, Yamaguchi K: Shoulder sonography: State of the art. Radiol Clin North Am 37:767, 1999.

van Holsbeeck M, Introcasso JH: Appendix: Table of normal values. In van Holsbeeck M, Introcasso JH (eds): Musculoskeletal Ultrasound. St. Louis, Mosby–Year Book, 1991, p 316.

van Holsbeeck MT, Kolowich PA, Eyler WR, et al: US depiction of partial-thickness tear of the rotator cuff. Radiology 197:443, 1995.

Wiener SN, Seithz WHJ: Sonography of the shoulder in patients with tears of the rotator cuff: Accuracy and value for selecting surgical options. AJR Am J Roentgenol 160:103, 1992.

CHAPTER 83

Ligaments and Fascia

Étienne Cardinal, MD, and Nathalie J. Bureau, MD

LIGAMENTS

Anatomy

Ligaments are bands of fibrous tissue made of parallel fibers tightly packed together that connect bone to bone. At ultrasound, ligaments generally present as thin hyperechoic stripes with a fibrillary appearance when imaged in a longitudinal plane parallel to the surface of the transducer (Fig. 83–1). Similar to tendons, ligaments also may display anisotropy and appear hypoechoic when imaged obliquely relative to the transducer's surface. Ligaments are difficult to image with ultrasound because of their small size and because they often are surrounded by hyperechoic fat. With higher resolution transducers (7 to 13 MHz), some ligaments, such as the anterior talofibular and anterior tibiofibular ligaments of the ankle and the collateral ligaments of the knee, can be evaluated with ultrasound. The medial collateral ligament attaches proximally on the medial femoral epicondyle and distally on the medial aspect of the proximal tibial metaphysis. It has a trilaminar appearance, with its superficial and deep hyperechoic components being separated by a thin stripe of less echogenic tissue (Fig. 83–2). The lateral collateral ligament attaches proximally on the lateral femoral epicondyle and distally on the fibular head with fan-shaped extremities.

Tears

If there is a partial tear, the ligament appears hypoechoic and thickened with disruption of its fibrillary pattern and has poorly defined contours (Fig. 83–3). When fiber discontinuity is documented and is associated with a hematoma that separates the fibers, a diagnosis of acute complete ligament tear can be established. Hyperechoic foci may occur secondary to bone avulsion or chronic ligamentous calcification.

PLANTAR FASCIA

Anatomy

The plantar fascia consists of three components: a central, a lateral, and a medial portion. The central portion is the thickest and attaches on the inner tubercle of the calcaneus. It is thickest proximally and becomes broader and thinner distally, where it divides into five processes,

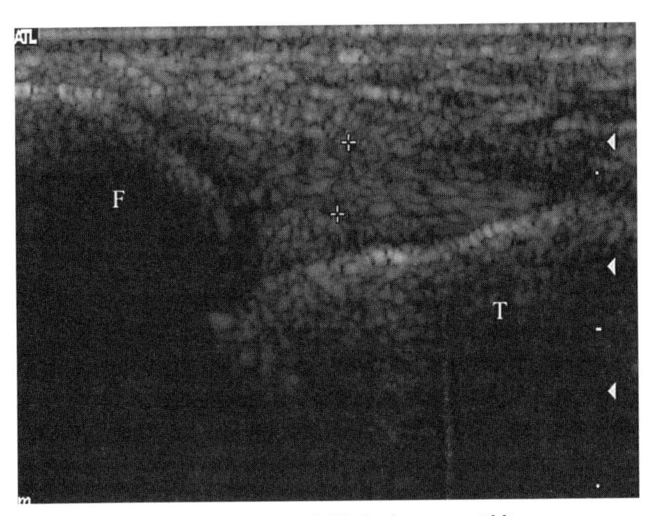

Figure 83–1. Normal anterior talofibular ligament. Oblique transverse ultrasound of left ankle shows a normal anterior talofibular ligament *(between cursors)* presenting as a hyperechoic fibrillary band attaching the distal fibula (F) to the talus (T).

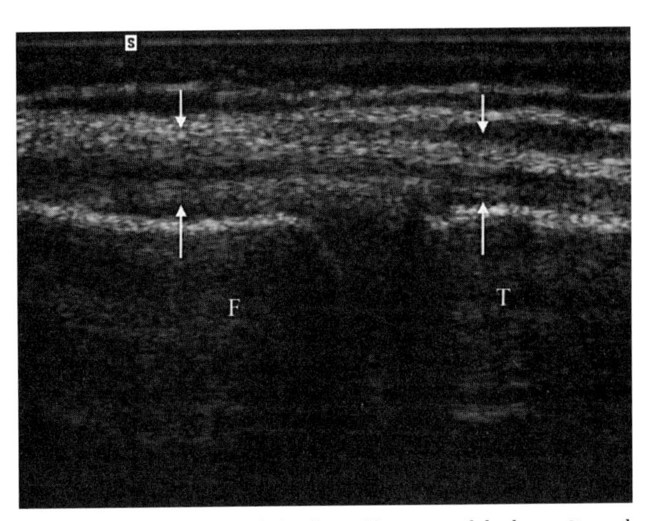

Figure 83–2. Normal medial collateral ligament of the knee. Coronal ultrasound scan of the knee shows the trilaminar appearance of the medial collateral ligament *(between arrows)* with its hyperechoic deep and superficial component separated by a hypoechoic line. F, medial femoral condyle; T, proximal tibia.

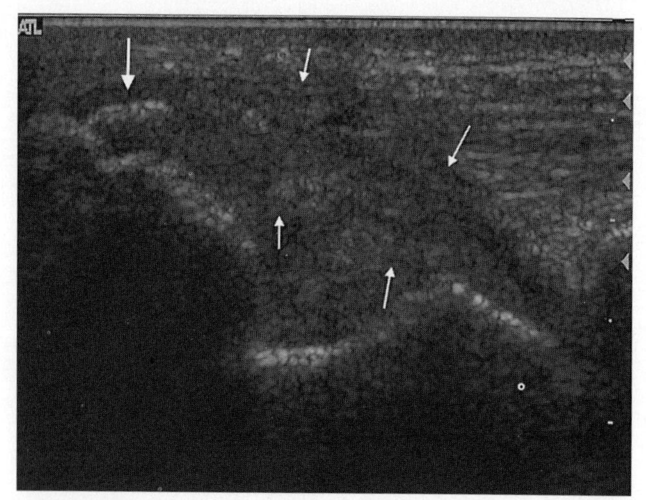

Figure 83–3. Partial tear of ankle deltoid ligament. Oblique transverse ultrasound of the ankle shows severe partial tear of the deltoid ligament with increase in size and hypoechogenicity of the ligament, loss of its fibrillary pattern, and poor definition of its contours *(small arrows)*. The tear is associated with a small bone avulsion of the tibia at the proximal attachment site of the ligament presenting as a hyperechoic focus *(large arrow)*.

one for each toe. The lateral portion attaches proximal on the calcaneus and distally at the base of the fifth metatarsal. The medial component is thin.

To examine the plantar fascia, the patient is positioned prone on the examination table, with the sole of the foot facing up or with the foot hanging at the end of the table. At ultrasound, the plantar fascia presents normally as a hyperechoic fibrillary stripe thickest proximally (<4 mm) (Fig. 83–4) and getting progressively thinner distally, measuring about 1 mm. It is subject to the anisotropy artifact and has to be examined with the transducer parallel to its axis. The normal fascia appears hyperechoic and is surrounded by fatty tissue, which is also hyperechoic. The fibrillary appearance of the fascia allows its localization. The most proximal fibers of the plantar fascia are convex on their plantar aspect as they insert on the plantar aspect of the calcaneus and may appear normally hypoechoic because of anisotropy. To enhance visualization of the plantar fascia, the great toe may be extended dorsally, to stretch the plantar fascia.

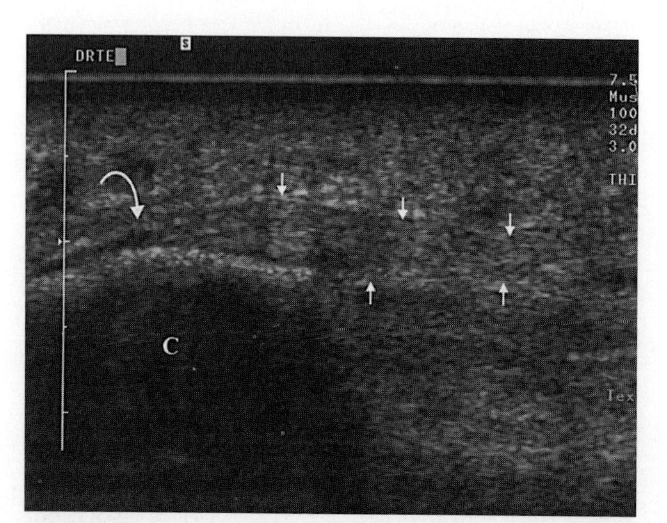

Figure 83–4. Normal plantar fascia. Longitudinal ultrasound of the plantar aspect of the heel shows normal hyperechoic fibrillary plantar fascia *(small arrows)* attaching proximally on the plantar aspect of the calcaneus (C). The most proximal fibers of the plantar fascia are hypoechoic because of anisotropy *(curved arrow)*.

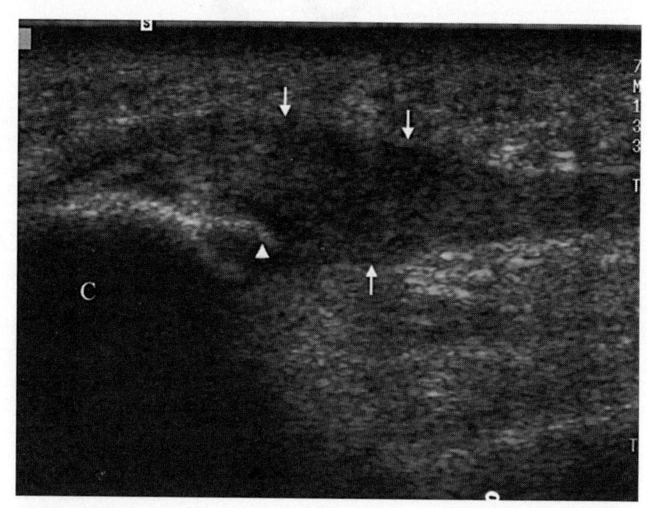

Figure 83–5. Plantar fasciitis. Longitudinal ultrasound of the plantar aspect of the heel shows abnormal thickened hypoechoic fascia from plantar fasciitis *(small arrows)*. Associated plantar spur *(arrowhead)* of the calcaneus (C) can be seen.

Plantar Fasciitis

Plantar fasciitis is a common cause of heel pain. Sonographic signs of plantar fasciitis are hypoechogenicity of the fascia and thickening of the fascia greater than 4 mm (Fig. 83–5). Plantar fasciitis may be associated with calcifications or surrounding fluid.

Recommended Readings

Cardinal E, Chhem RK, Beauregard CG, et al: Plantar fasciitis: Sonographic evaluation. Radiology 201:257, 1996.

Gray H: The articulations. In Pick TP HR (ed): Anatomy, Descriptive and Surgical. New York, Bounty Books, 1977, p 217.

Vanderschueren G, Fessel DP, van Holsbeeck MT: Ankle. In Chhem RK, Cardinal E (eds) Guidelines and Gamuts in Musculoskeletal Ultrasound. New York, Wiley-Liss, 1999, p 213.

van Holsbeeck MT, Introcasso JH: Sonography of Ligaments. In van Holsbeeck MT, Introcasso JH (eds): Musculoskeletal Ultrasound. St. Louis, Mosby–Year Book, 1991, p 124.

CHAPTER 84

Muscles

Étienne Cardinal, MD, and Nathalie J. Bureau, MD

OVERVIEW

Skeletal muscles represent 40% to 45% of the total body mass. Normally the ultrasound image of muscle in the longitudinal plane is that of a pennate structure with hypoechoic muscular bundles being separated by bright linear hyperechoic septa of the perimysium (Fig. 84–1). On transverse scan, these septa present as bright hyperechoic dots in a more hypoechoic background of muscular fibers, which gives the appearance of a "starry sky"(Fig. 84–2). The muscle is surrounded by the epimysium or muscle fascia, which is thin and hyperechoic.

TEARS

Skeletal muscles frequently are involved in trauma and are subject to tear and hematoma formation. Although it might not be needed in minor trauma, ultrasound can be used to detect a muscle tear and to evaluate its extent. Ultrasound also can be used for follow-up and evaluation of healing. The appearance of the muscle tear varies according to its severity.

Small contusions might not be detectable with ultrasound and are clinically benign. Small partial tears appear as a small focal hypoechoic area with muscle fiber discontinuity and focal loss of the normal pennate architecture of the muscle. A small hematoma may also be present (generally <1 cm in diameter).

Moderate-size muscle tears present with a larger area of muscle fiber discontinuity separated by a hematoma that may be focal or dissect within muscle fibers or along fascial planes (Fig. 84–3). The tear involves less than a third of the muscle area on a transverse scan.

Severe partial tears involve more than a third of the area of the muscle on transverse plane. Complete muscle tears are associated with large hematomas separating muscle fibers. The distal end of the retracted muscle floating in the hematoma has the appearance of a "bell clapper" (Fig. 84–4).

HEMATOMA

Muscle hematomas may result from trauma, although they can occur spontaneously in patients with bleeding disorders or receiving anticoagulant therapy. The ultrasound presentation of a hematoma varies according to age. Acute hematoma may be hyperechoic if imaging is done hours after injury. The hematoma quickly becomes hypoechoic with posterior acoustic enhancement. Fluid-fluid level may be seen if the limb is left immobile long

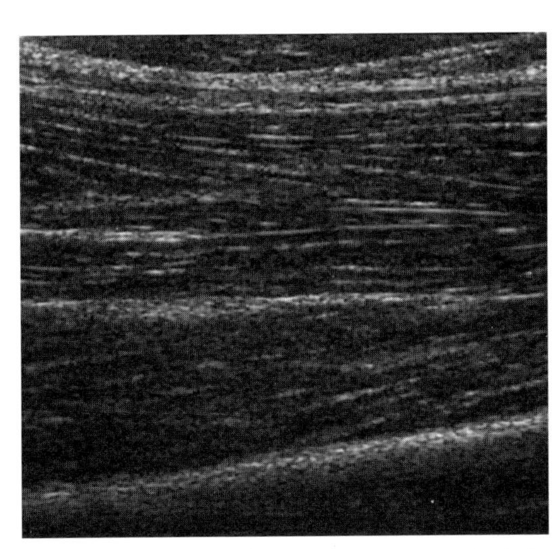

Figure 84–1. Normal rectus femoris muscle. Longitudinal ultrasound shows a penniform appearance of a normal rectus femoris muscle with hypoechoic muscle bundles separated by hyperechoic lines and septa.

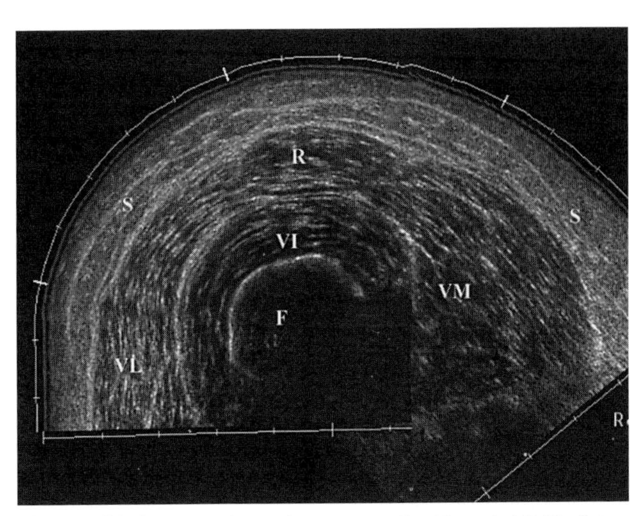

Figure 84–2. Normal quadriceps muscle. Extended-field-of-view transverse ultrasound shows the components of the quadriceps muscle (R, rectus femoris muscle; VI, vastus intermedius muscle; VM, vastus medius muscle; VL, vastus lateralis muscle) as hypoechoic structures with disseminated hyperechoic dots reminiscent of a *starry sky*. F, femur; S, subcutaneous fat.

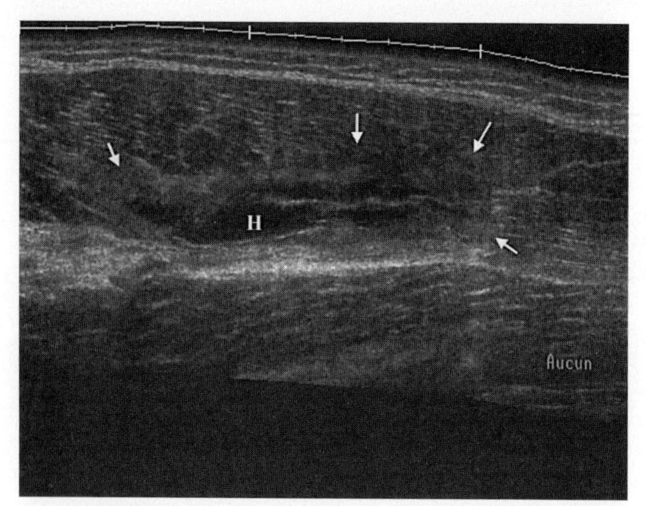

Figure 84–3. Partial tear of rectus femoris muscle. Extended-field-of-view longitudinal ultrasound shows loss of penniform pattern of the muscle *(arrows)* with disruption of fibers and formation of a hypoechoic hematoma containing a few hyperechoic linear septations (H).

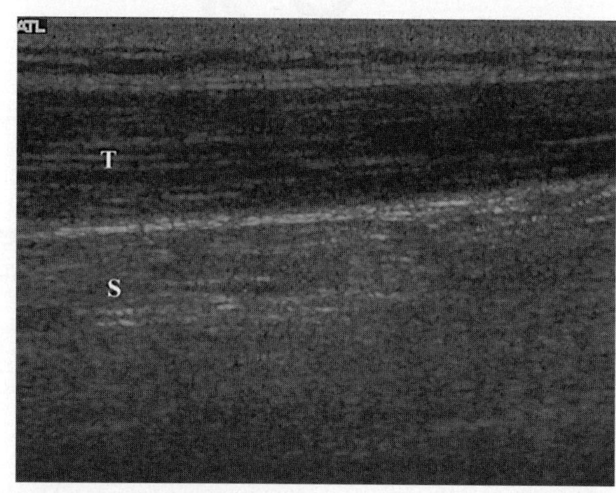

Figure 84–5. Muscular atrophy. Coronal ultrasound of the supraspinatus fossa shows diffuse atrophy of the supraspinatus muscle (S) from chronic rotator cuff disease. Note the diffuse hyperechogenicity of the muscle with loss of the normal penniform pattern compared with normal overlying trapezius muscle (T).

enough for settling of cellular debris in the dependent portion of the collection. Septations may be seen (see Fig. 84–3). Compression of the collection with the probe mobilizes the content. With healing, the hematoma decreases in size, organizes, and becomes heterogeneous and noncompressible with the ultrasound probe. It eventually regresses with muscle reorganization.

MUSCLE HERNIATION

A fascial tear may occur with or without underlying muscle tear and appears as thinning of the fascia or as a defect. Interruption of the fascia may cause the muscle to herniate through the defect into the subcutaneous tissue. The anterolateral aspect of the leg is the most common site of muscle hernia. Muscle hernia may simulate a soft tissue mass clinically. With real-time ultrasound examination, muscle herniation through the fascial defect can be shown during active muscle contraction.

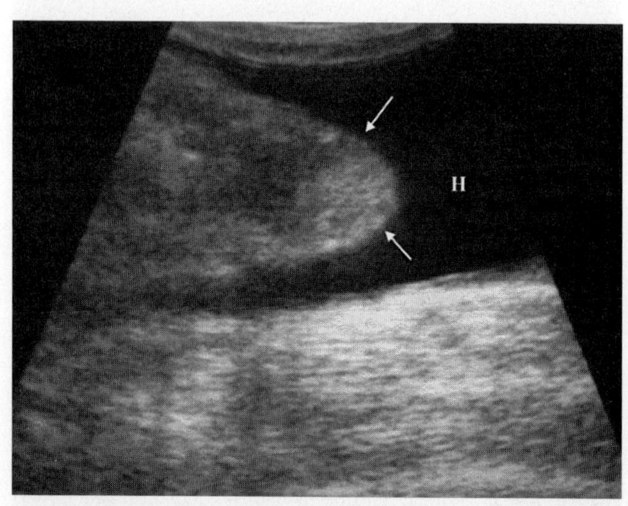

Figure 84–4. Complete muscle tear. Longitudinal ultrasound of the anterior thigh shows the distal end of the torn and retracted proximal rectus femoris muscle *(arrows)* floating in a large hypoechoic hematoma (H). This appearance is reminiscent of a *bell clapper.* (Courtesy of Germain C. Beauregard, MD, Montreal, Canada.)

MYOPATHIES

The muscle appears hyperechoic in cases of congenital or acquired myopathy, such as Duchenne's dystrophy, polymyositis, or atrophy secondary to disuse (Fig. 84–5). The increased echogenicity of the muscle is secondary to the increased content of collagen or to fatty infiltration of the muscle. This hyperechoic appearance of the muscle is not specific. Ultrasound can be used to guide needle biopsy for a more specific diagnosis.

Recommended Readings

Heckmatt JZ, Dubowitz V: Ultrasound imaging and directed needle biopsy in the diagnosis of selective involvement in muscle disease. J Child Neurol 2:205, 1987.

Lefebvre E, Pourcelot L: La pathologie musculaire. In Lefebvre E, Pourcelot L (eds): Echographie Musculo-Tendineuse. Paris, Masson, 1991, p 64.

Kransdorf MJ, Jelinek JS, Moser RP: Imaging of soft tissue tumors. Radiol Clin North Am 31:359, 1993.

CHAPTER 85

Soft Tissue Masses

Nathalie J. Bureau, MD, and Étienne Cardinal, MD

OVERVIEW

The value of ultrasound evaluation of soft tissue masses goes beyond the differentiation between a cystic and a solid mass. Although ultrasound cannot reliably distinguish between benign and malignant soft tissue masses in all cases, in many cases, a specific diagnosis can be established.

The purpose of imaging soft tissue masses, either with ultrasound or other modalities, is threefold: (1) to identify and characterize the lesion, (2) to establish a final or differential diagnosis, and (3) to stage the lesion. The imaging investigation of a soft tissue tumor should start with radiographs.

Ultrasound may suggest a specific diagnosis in cases of pseudotumors, such as hematomas, muscle tears, muscle hernias, tendon ruptures, asymmetrical amounts of subcutaneous fat (Fig. 85–1), or anatomic variants such as accessory soleus muscle, all of which can present as soft tissue tumors. In cases of certain benign soft tissue tumors, such as subcutaneous lipomas, hemangiomas, peripheral neurogenic tumors, or fibromatosis, a specific diagnosis usually can be established by ultrasound.

If a lesion is well visualized by ultrasound but a definite diagnosis cannot be established on the basis of its features, an ultrasound-guided fine-needle aspiration or a biopsy can be done without the need of other diagnostic tests. If it is not possible to delineate a lesion by ultrasound, other imaging studies may be required before doing the biopsy.

CYSTIC LESIONS

Synovial Cysts

Synovial cysts are synovium-lined para-articular fluid collections that may or may not communicate with the joint. They usually are secondary to an intra-articular pathology, which causes an increase in intra-articular pressure. Frequently associated arthropathies include rheumatoid arthritis, osteoarthrosis, and crystal deposition diseases.

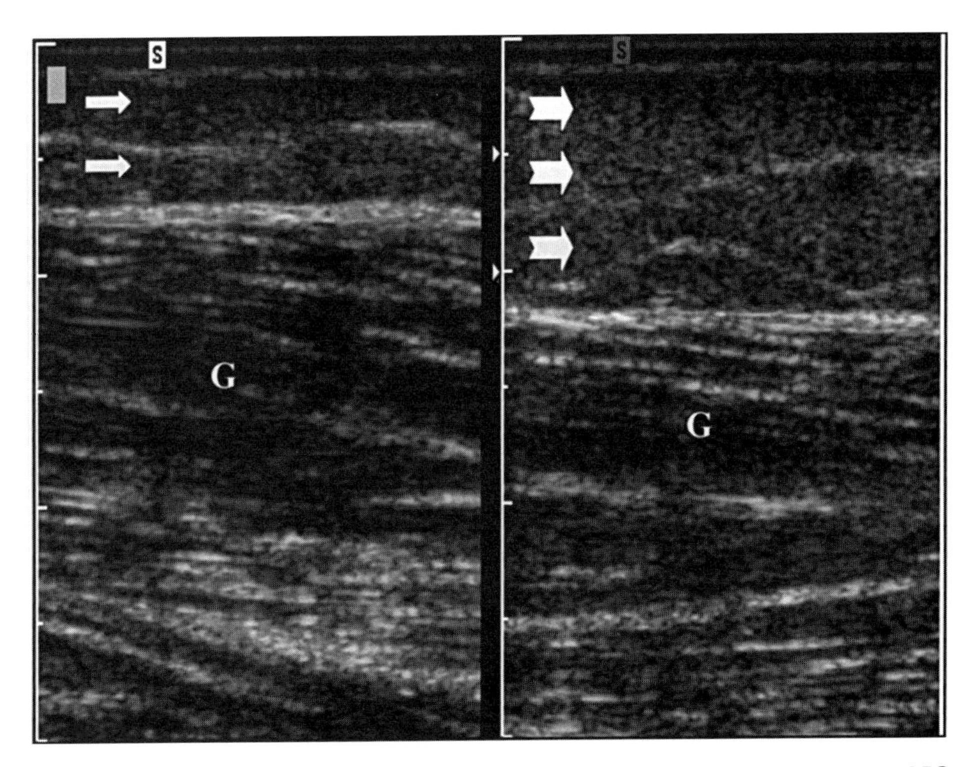

Figure 85–1. Asymmetrical amount of subcutaneous fat. Dual transverse ultrasound of the right and left buttocks in a female patient complaining of a palpable mass. Ultrasound reveals an asymmetrical distribution of subcutaneous fat with thicker subcutaneous adipose tissue on the left side *(notched arrows)* compared with the right side *(arrows)*. G, gluteus maximus.

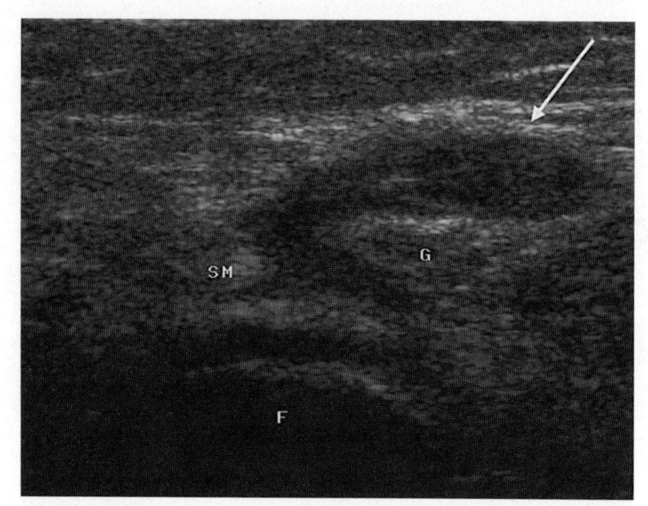

Figure 85–2. Popliteal cyst. Transverse ultrasound of the posteromedial aspect of the knee at the level of the femoral condyle (F) shows the origin of the hypoechoic popliteal cyst *(arrow)* between the medial gastrocnemius muscle and tendon (G) and the semimembranosus tendon (SM).

Synovial cysts are most frequent about the knee. The popliteal (Baker's) cyst results from a valvelike mechanism that allows communication between the knee joint and the gastrocnemius-semimembranosus bursa in response to increased intra-articular pressure.

The diagnosis of a popliteal cyst is made readily at ultrasound. Identification of the origin of the mass between the medial gastrocnemius muscle and tendon and the semimembranosus tendon is diagnostic of a popliteal cyst (Fig. 85–2). These cysts may rupture and produce symptoms that simulate thrombophlebitis clinically. Ultrasound excludes the diagnosis of venous occlusion and in the acute setting shows fluid dissecting into the subcutaneous tissues and between the muscles of the calf. The distal aspect of the cyst may show a pointed configuration as opposed to the normal rounded appearance, and gentle pressure applied with the transducer may express a small amount of fluid through the defect, which can be seen with real-time sonography.

Ganglion Cysts

Ganglion cysts are believed to be the result of myxomatous degeneration of periarticular connective tissue and usually are attached to a tendon sheath or joint capsule, particularly in the hands, wrists, and feet. The lesions have fibrous walls with no discernible internal lining cell type and contain mucoid material. They appear as well-defined hypoechoic structures at ultrasound, sometimes showing internal septations (Fig. 85–3). Ultrasound has shown its value in the identification of occult ganglia in the wrist. In the wrist, ganglia are most frequent on the dorsal aspect of the scapholunocapitate junction (Fig. 85–4). They frequently measure only 2 to 3 mm in diameter and may be symptomatic. Ultrasound-guided aspiration and steroid injection of ganglia may provide an effective alternative treatment to surgery.

Paraglenoid Labral Cysts

There is an association between labral tears of the shoulder and the presence of paralabral cysts. Cysts may be located in the suprascapular notch or spinoglenoid notch (Fig. 85–5), and if they are large, they may compress the suprascapular nerve or axillary nerve and cause denervation of the external rotator muscles, mainly the infraspinatus muscle. These cysts are believed to result from the extension of joint fluid through a capsulolabral tear into the paraglenoid soft tissues. Paraglenoid labral cysts can be diagnosed at ultrasound and should be sought for systematically during ultrasound evaluation of the shoulder. Because ultrasound is not reliable to detect labral tears, the investigation should be pursued with

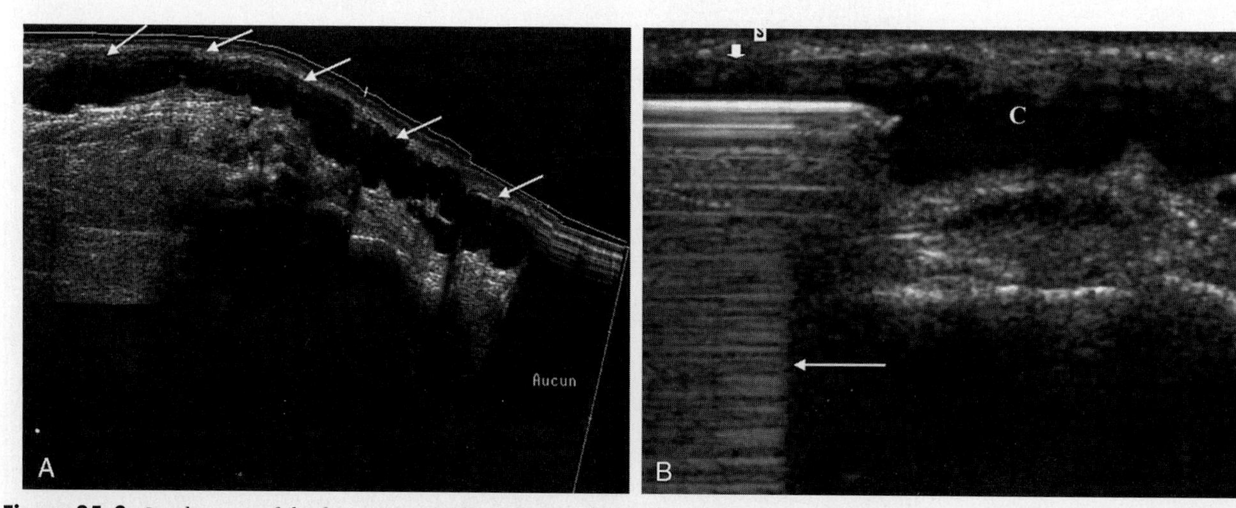

Figure 85–3. Ganglion cyst of the foot. *A,* Longitudinal ultrasound of the lateral aspect of the dorsum of the foot shows an 8-cm-long hypoechoic tubular structure *(arrows)* extending from the fourth interdigital space to the lateral malleolus. The lesion originated from the tendon sheath of the extensor tendon of the fifth digit. *B,* Under ultrasound guidance, the cyst (C) was punctured and thick gelatinous material was removed, which is consistent with the diagnosis of a ganglion cyst. The needle *(thick arrow)* appears as a linear hyperechoic structure with posterior ring down artifacts *(thin arrow).*

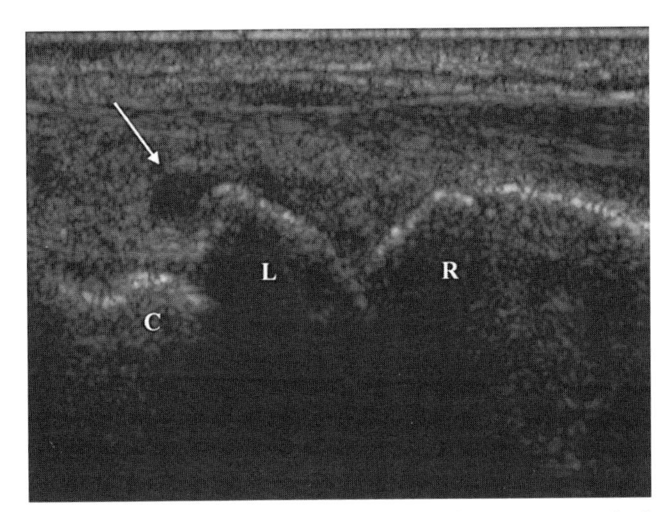

Figure 85–4. Occult dorsal ganglion cyst of the wrist. Longitudinal ultrasound of the dorsal aspect of the wrist at the level of the lunocapitate joint in a patient with wrist pain reveals a small hypoechoic lesion *(arrow)* representing an occult ganglion cyst. R, radius; L, lunate; C, capitate.

magnetic resonance (MR) arthrography to identify the labral tear.

Meniscal Cysts

Meniscal cysts are associated with a horizontal meniscal tear and occur most frequently in relation to the anterior horn of the lateral meniscus. As in the case of paraglenoid cysts of the shoulder, joint fluid is forced through the tear and extends in the parameniscal soft tissues.

Although MR imaging is superior to ultrasound in the evaluation of meniscal lesions, patients may present clinically with a firm or fluctuant, usually painful soft tissue mass at the level of the joint line, which does not suggest a meniscal tear. In this particular setting, ultrasound may be useful to show the fluid collection, which is often multiloculated and extends to the adjacent meniscus.

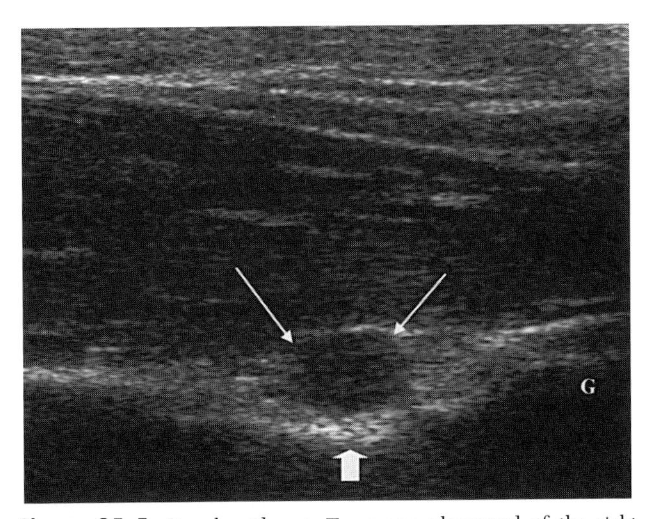

Figure 85–5. Paraglenoid cyst. Transverse ultrasound of the right shoulder immediately below the spinous process of the scapula shows a well-defined hypoechoic lesion *(long arrows)* sitting in the spinoglenoid notch *(short arrow)* adjacent to the glenoid (G). When such a lesion is found during an ultrasound shoulder examination, the investigation should be pursued with CT-arthrogram or MR-arthrogram of the shoulder to exclude a labral tear.

SOLID LESIONS

Lipoma

Lipoma is the most common soft tissue tumor of adulthood. Lipomas are classified according to location as superficial (subcutaneous) or deep. Subcutaneous lipomas are usually oval-shaped and show a homogeneous echotexture at ultrasound (Fig. 85–6). They present with variable echogenicity on ultrasound. Most superficial lipomas are isoechoic or hyperechoic compared with subcutaneous fat, but they also may appear hypoechoic or show mixed echogenicity. When they are encapsulated, a fine echogenic line can be seen at ultrasound. Subcutaneous lipomas appear avascular at Doppler studies.

Hemangioma

Hemangiomas are classified pathologically as capillary, cavernous, arteriovenous, or venous according to the predominant type of vascular channel seen at histologic examination. They account for 7% of all benign tumors. Most hemangiomas evaluated radiographically are intramuscular. Patients may present with a painful palpable soft tissue mass or ill-defined muscular pain sometimes increased after exercise. Radiographs occasionally may reveal the presence of phleboliths.

At ultrasound, intramuscular hemangiomas may appear hypoechoic or hyperechoic and show a variable morphology, ranging from well-defined oval-shaped lesions to multiloculated lesions with irregular margins. Real-time ultrasound may show filling of vascular channels and enlargement of the lesion when the limb is placed in a dependent position (Fig. 85–7). Variable patterns of blood flow may be shown on color Doppler. Applying compres-

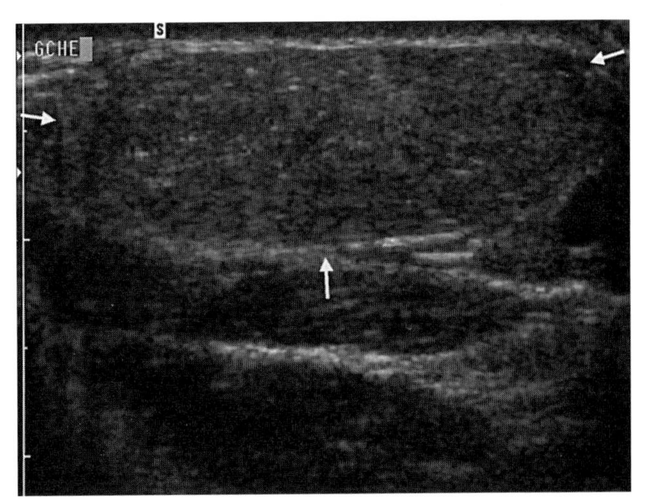

Figure 85–6. Subcutaneous lipoma. Longitudinal ultrasound of the left proximal forearm shows an oval, isoechoic, homogeneous subcutaneous mass *(arrows)* superficial to the musculature.

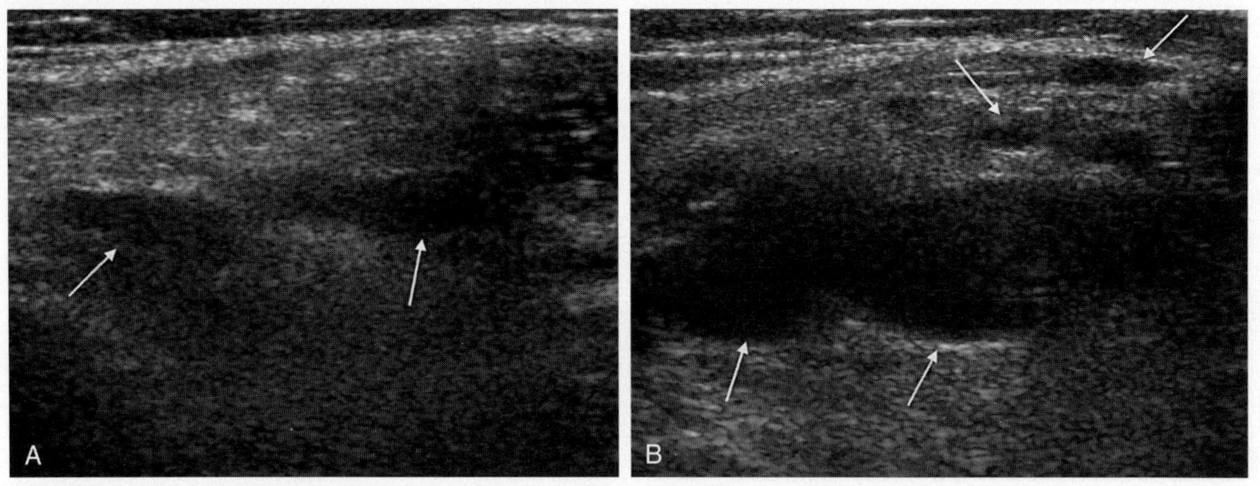

Figure 85–7. Intramuscular hemangioma. Transverse ultrasound studies of the anterior thigh done with the patient supine *(A)* and done with the patient standing up *(B)* show hypoechoic areas *(arrows)* within a predominantly hyperechoic ill-defined mass. The hypoechoic areas represent vascular channels that enlarge significantly when the patient is standing up.

sion to the lesion with the ultrasound probe often helps show the presence of flow within the blood channels.

Morton's Neuroma

Interdigital neuromas are not true neoplasms. They are believed to be the result of a fibrotic degeneration of a digital branch of the medial or lateral plantar nerve. They typically are located in the third intermetatarsal space at the level of the metatarsal head, just distal to the intermetatarsal ligament. Women are affected more frequently than men.

Ultrasound examination is done from the plantar aspect of the foot, and each interspace is scanned in the coronal and sagittal plane over an area between the distal third of the metatarsals to the base of the proximal phalanges. The hand of the ultrasonographer not holding the ultrasound probe assists in visualizing the intermetatarsal space by applying pressure on the dorsal aspect of each interspace to widen the space between the metatarsal heads. Morton's neuromas usually appear as small hypoechoic lesions of variable shape, although sometimes they appear anechoic or with mixed echogenicity (Fig. 85–8). Continuity between the mass and the interdigital nerve sometimes can be shown and improves diagnostic confidence. Ultrasound has shown its value in the diagnosis of Morton's neuroma with reported sensitivities between 85% and 98%.

Peripheral Nerve Sheath Tumors

At ultrasound, peripheral nerves appear as hyperechoic tubular structures with internal linear echoes and are shown best with high-frequency linear array transducers. Peripheral nerve sheath tumors are uncommon. Benign tumors are most frequent, and pathologically they can be divided into two different histologic types: Neurilemomas or schwannomas originate from the nerve sheath cells, whereas neurofibromas arise from Schwann cells in the endoneurial space. Neurofibromas and schwannomas appear as hypoechoic or mixed echogenicity superficial soft tissue masses with well-defined margins, in the expected distribution of a nerve (Fig. 85–9*A*). Posterior acoustic enhancement may be seen in schwannomas because of the uniform cellular pattern in these tumors. Demonstration of continuity between the nerve and the lesion increases the level of confidence in the diagnosis (Fig. 85–9*B*).

Identification of a hypoechoic, homogeneous mass with posterior acoustic enhancement at the distal end of an amputation stump is fairly pathognomonic of a postamputation neuroma, especially if continuity with the nerve can be shown. These lesions are generally painful, and symptoms are reproduced by pressure with the transducer.

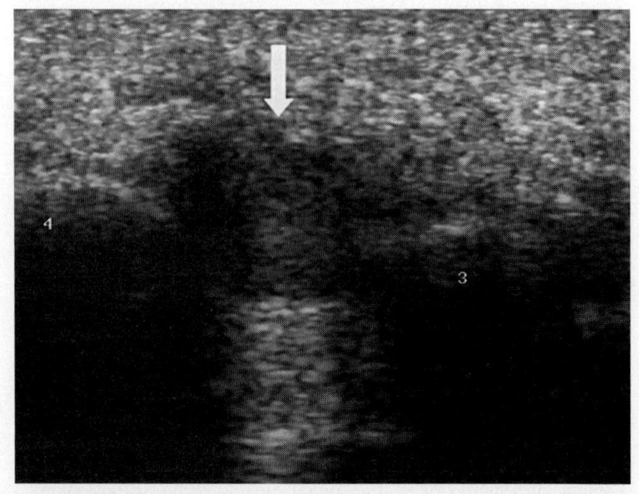

Figure 85–8. Morton's neuroma. Transverse ultrasound at the level of the third interdigital space of the left foot shows a small homogeneous hypoechoic lesion *(arrows)* between the third and fourth metatarsal heads. Note the strong posterior acoustic enhancement.

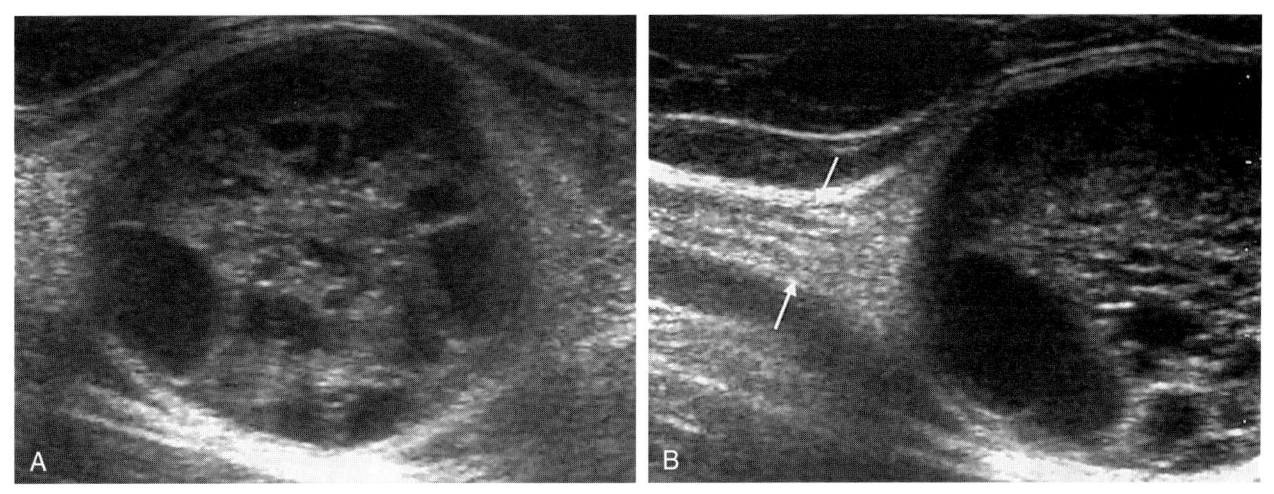

Figure 85–9. Median nerve schwannoma. *A*, Longitudinal ultrasound of the arm of a 71-year-old woman shows a well-circumscribed solid lesion with mixed echogenicity. *B*, The continuity between the nerve *(arrows)* and the lesion is depicted, which increased the level of confidence in the diagnosis. Palpation of the lesion produced pain in the distribution of the median nerve.

Superficial Plantar Fibromatosis (Ledderhose Disease)

Ledderhose disease is the result of the proliferation of fibrous tissue within the plantar fascia or aponeurosis. It affects mostly adults in their 20s to 40s, and the involvement is often bilateral. It usually presents clinically as a palpable nodule over the sole of the foot. Ultrasound is particularly well suited to investigate such a lesion and shows single or multiple oval-shaped, relatively homogeneous hypoechoic solid lesions in relation to the plantar aponeurosis and subcutaneous tissue (Fig. 85–10).

Myositis Ossificans

Myositis ossificans is encountered more often in adolescents and young adults. It involves more frequently the thigh, buttock, and elbow and less frequently the shoulder and calf. Most patients have a history of trauma, although 40% do not recall any trauma. On computed tomography (CT), the peripheral rim or calcification and lucent center (zoning phenomenon) is fairly characteristic.

At ultrasound, a characteristic zonal pattern also is identified that varies depending on the age of the trauma. In the immature phase, the zonal pattern shows a hypoechoic center surrounded by an undulating echogenic ring, again surrounded by an outer hypoechoic ring (Fig. 85–11). At this early stage, radiographs are normal. The central hypoechoic region corresponds to hemorrhage and necrosis, and the echogenic rim corresponds to calcified osteoid matrix. In the mature phase of myositis ossificans, the echogenic ring becomes more hyperechoic, and calcifications become visible on radiographs. The zonal pattern is less apparent. In the consolidated phase, the mass is completely hyperechoic because of mature ossification.

Elastofibroma Dorsi

Elastofibroma dorsi is a rare pseudotumor of connective tissue typically located in the subscapular region. It is found most commonly in the elderly with a prevalence of 2%.

The ultrasound evaluation of this mass is made with the patient's arm elevated, which raises the scapula and reveals the full size of the lesion. At ultrasound, elastofibroma dorsi presents as a mass with linear or streaky hypoechoic strands against an echogenic background. This pattern reflects the alternating pattern of fibroelastic deposit and fat. The sonographic appearance of elastofibroma dorsi is characteristic and, in the proper clinical setting, obviates the need for biopsy or other imaging test.

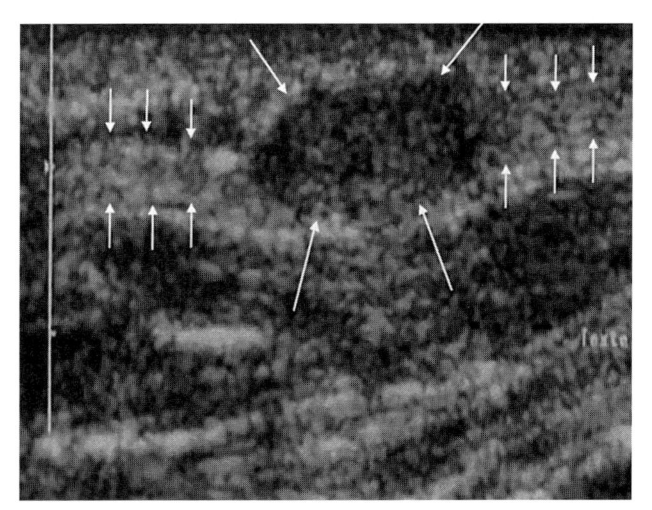

Figure 85–10. Plantar fibromatosis. Longitudinal ultrasound of the plantar aspect of the foot shows an oval-shaped, well-defined, hypoechoic solid lesion *(long arrows)* in continuity with the hyperechoic, fibrillary plantar fascia *(short arrows)*.

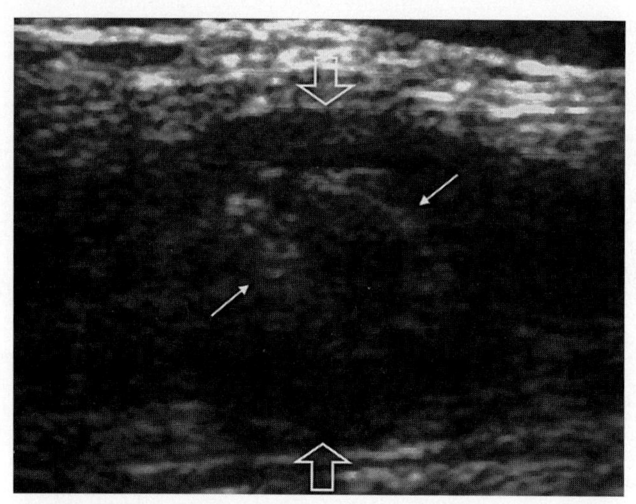

Figure 85–11. Myositis ossificans. Longitudinal ultrasound of the rectus femoris muscle shows a mass with a hypoechoic center surrounded by an echogenic ring *(arrows)*, again surrounded by an outer hypoechoic ring *(open arrows)*. This ultrasound appearance represents the characteristic *zonal pattern* seen in myositis ossificans. (Courtesy of Germain C. Beauregard, MD, Montreal, Canada.)

Tumor Recurrence

Ultrasound is useful for the follow-up of patients after tumor surgery. The value of ultrasound is especially apparent when hardware is present because ultrasound of soft tissue is not degraded by metallic artifact as much as MR imaging or CT. Color Doppler and power Doppler imaging are useful to assess malignancy of a tumor, tumor response to chemotherapy, and postoperative recurrence.

Recommended Readings

Alexander AA, Nazarian LN, Feld RI: Superficial soft-tissue masses suggestive of recurrent malignancy: Sonographic localization and biopsy. AJR Am J Roentgenol 169:1449, 1997.

Baudrez V, Malghem J, Van de Berg B, et al: Ultrasonography of dorsal elastofibroma: Apropos of 6 cases. J Radiol 79:549, 1998.

Bergman GA: Synovial lesions of the hand and wrist. Radiol Clin North Am 3:265, 1995.

Bianchi S, Martinoli C, Abdelwahab IF, et al: Elastofibroma dorsi: Sonographic findings. AJR Am J Roentgenol 169:1113, 1997.

Brandser EA, Goree JC, El-Khoury GY: Elastofibroma dorsi: Prevalence in an elderly patient population as revealed by CT. AJR Am J Roentgenol 171:977, 1998.

Breidahl WH, Adler RS: Ultrasound-guided injection of ganglia with corticosteroids. Skeletal Radiol 25:635, 1996.

Bureau NJ, Cardinal E, Chhem RK: Ultrasound of soft tissue masses. Semin Musculoskeletal Radiol 2:283, 1998.

Bureau NJ, Kaplan PA, Dussault RG: MRI of the knee: A simplified approach. Curr Probl Diagn Radiol 24:1, 1995.

Cardinal E, Buckwalter KA, Braunstein EM, et al: Occult dorsal carpal ganglion: Comparison of US and MR imaging. Radiology 193:259, 1994.

Choi H, Varma DGK, Fornage BD, et al: Soft tissue sarcoma: MR imaging vs. sonography for detection of local recurrence after surgery. AJR Am J Roentgenol 157:353, 1991.

Derchi LE, Balcon G, De Flaviis L, et al: Sonographic appearances of hemangiomas of skeletal muscle. J Ultrasound Med 8:263, 1989.

Fehrman D, Orwin J, Jennings R: Suprascapular nerve entrapment by ganglion cysts: A report of six cases with arthroscopic findings and review of the literature. Arthroscopy 11:727, 1995.

Fornage BD: Peripheral nerves of the extremities: Imaging with US. Radiology 167:179, 1988.

Fornage BD, Eftekhari F: Sonographic diagnosis of myositis ossificans. J Ultrasound Med 8:463, 1989.

Fornage BD, Lorigan JG: Sonographic detection and fine-needle aspiration biopsy of nonpalpable recurrent or metastatic melanoma in subcutaneous tissues. J Ultrasound Med 8:421, 1989.

Fornage BD, Tassin GB: Sonographic appearances of superficial soft tissue lipomas. J Clin Ultrasound 19:215, 1991.

Garant M, Sarazin L, Cho KH, et al: Soft-tissue recurrence of osteosarcoma: Ultrasound findings. Can Assoc Radiol J 46:305, 1995.

Kaplan PA, Matamoros JA, Anderson JC: Sonography of the musculoskeletal system. AJR Am J Roentgenol 155:237, 1990.

Kransdorf MJ, Jelinek JS, Moser RP: Imaging of soft tissue tumors. Radiol Clin North Am 31:359, 1993.

Kransdorf MJ, Murphey MD: Imaging of Soft Tissue Tumors. Philadelphia, WB Saunders, 1997.

Moore T, Fritts H, Quick D, et al: Suprascapular nerve entrapment caused by supraglenoid cyst compression. J Shoulder Elbow Surg 6:455, 1997.

Murphey MD, Fairbairn KJ, Parman LM, et al: Musculoskeletal angiomatous lesions: Radiologic-pathologic correlation. Radiographics 15:893, 1995.

Peetrons P, Allaer D, Jeanmart L: Cysts of the semilunar cartilages of the knee: A new approach by ultrasound imaging: a study of six cases and review of the literature. J Ultrasound Med 9:333, 1988.

Provost N, Bonaldi VM, Sarazin L, et al: Amputation stump neuroma: Ultrasound features. J Clin Ultrasound 25:85, 1997.

Quinn TJ, Jacobson JA, Craig JG, et al: Sonography of Morton's neuromas. AJR Am J Roentgenol 174:1723, 2000.

Resnick D, Niwayama G: Soft tissues. In Resnick D (ed): Diagnosis of Bone and Joint Disorders, vol 6, 3rd ed. Philadelphia, WB Saunders, 1995, p 4567.

Rosenberg AE: Skeletal system and soft tissue tumors. In Cotran RS, Kumar V, Robbins SL (eds): Pathologic Basis of Disease, 5th ed. Philadelphia, WB Saunders, 1994, p 1213.

Shapiro PP, Shapiro SL: Sonographic evaluation of interdigital neuromas. Foot Ankle 16:604, 1995.

Silvestri E, Martinoli C, Derchi LE, et al: Echotexture of peripheral nerves: Correlation between US and histologic findings and criteria to differentiate tendons. Radiology 197:291, 1995.

Thomas EA, Cassar-Pullicino VN, McCall IW: The role of ultrasound in the early diagnosis and management of heterotopic bone formation. Clin Radiol 43:190, 1991.

Tirman P, Feller J, Janzen D, et al: Association of glenoid labral cysts with labral tears and glenohumeral instability: Radiologic findings and clinical significance. Radiology 190:653, 1994.

Tung GA, Entzian D, Stern JB, et al: MR imaging and MR arthrography of paraglenoid labral cysts. AJR Am J Roentgenol 174:1707, 2000.

Van der Woude H-J, Vandershueren G: Ultrasound in musculoskeletal tumors with emphasis on its role in tumor follow-up. Radiol Clin North Am 37:753, 1999.

van Holsbeeck M, Introcaso JH: Musculoskeletal ultrasonography. Radiol Clin North Am 30:907, 1992.

van Holsbeeck M, Powell A: Ankle and foot. In Fornage BD (ed): Musculoskeletal Ultrasound. New York, Churchill Livingstone, 1995, p 221.

CHAPTER 86

Infections

Nathalie J. Bureau, MD, and Étienne Cardinal, MD

OVERVIEW

Delay in establishing the diagnosis is an important prognostic factor in musculoskeletal infections, especially in the case of septic arthritis. Ultrasound is sensitive in the detection of fluid collections and allows real-time guidance for immediate needle aspiration. Because it is innocuous, ultrasound can be repeated easily for follow-up of infections.

SOFT TISSUE ABSCESSES

Superficial abscesses are a common complication of cellulitis. Ultrasound is the modality of choice to detect drainable soft tissue collections and to guide fine-needle aspiration. Abscesses may have different ultrasound appearances. The collection may appear as an anechoic or diffusely hypoechoic mass with increased posterior acoustic enhancement. Conversely the lesion may be hyperechoic (Fig. 86–1) or isoechoic relative to surrounding tissues. Sometimes the lesion does not create any mass effect. Dynamic evaluation of the soft tissue area by palpation or gentle compression with the ultrasound probe is particularly important in this setting to reveal motion of the purulent echogenic material. The margins of the lesion may be well circumscribed or blend in with the surrounding tissues. Septa may be present and internal echoes, which represent debris or gas. Color Doppler imaging may be used to show hyperemia at the periphery of the mass and absence of flow in the center.

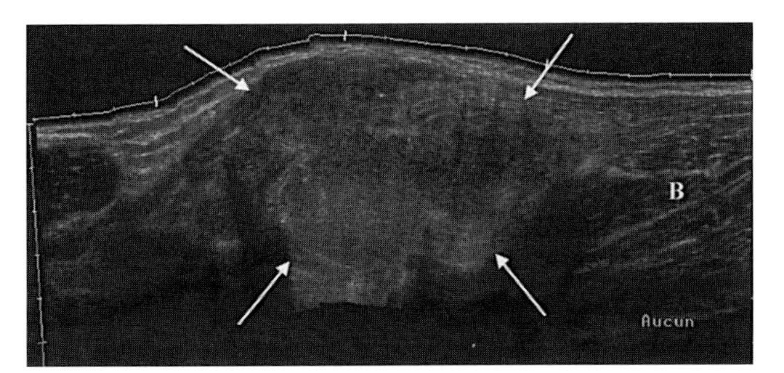

Figure 86–1. Large abscess of the biceps muscle. Extended-field-of-view longitudinal ultrasound of the arm in an intravenous drug abuser shows a large mass *(arrows)* slightly hyperechoic relative to the biceps muscle (B). The abscess was removed under ultrasound guidance using a pigtail catheter for drainage.

SEPTIC ARTHRITIS

Bacterial arthritis is the most rapidly destructive joint disease and results in irreversible loss of joint function in 25% to 50% of patients. *Staphylococcus aureus* is the most frequently involved pathogen. The infection is usually the result of hematogenous seeding of a joint during bacteremia.

Ultrasound is particularly sensitive in detection of a joint effusion and may be helpful in the hip, wrist, or shoulder, where physical examination is less reliable and radiographs are often noncontributory in the acute setting (Fig. 86–2). Ultrasound does not allow differentiation of a noninfected joint effusion from septic arthritis, but it may be used to guide aspiration of sometimes minute amounts of joint fluid. Ultrasound-guided arthrocentesis may reduce the risk of contamination of other anatomic compartments, particularly in the hand, wrist, or foot.

INFECTIOUS BURSITIS

Bursae are small pouches lined by synovium, and they normally contain a small amount of synovial fluid. Their purpose is to reduce friction by creating a space between two closely apposed structures that move relative to one another. Septic bursitis most frequently involves the olecranon (Fig. 86–3) and prepatellar bursae. These bursae are located in the subcutaneous tissue and are more prone to trauma. The diagnosis of septic bursitis normally is suggested by the clinical presentation, but in the presence of extensive cellulitis, physical examination may be difficult. Ultrasound may help in the diagnosis by showing a fluid collection in the vicinity of a joint, while the joint itself is free of an effusion, ruling out the diagnosis of a septic arthritis.

The ultrasound appearance of septic bursitis is not specific. A bursal effusion may be secondary to trauma, infection, and inflammatory or crystal-induced arthropathies. Color Doppler may show hyperemia in the walls of the collection. The diagnosis is confirmed by aspiration of fluid and microbiologic examination. Ultrasound facilitates fluid aspiration.

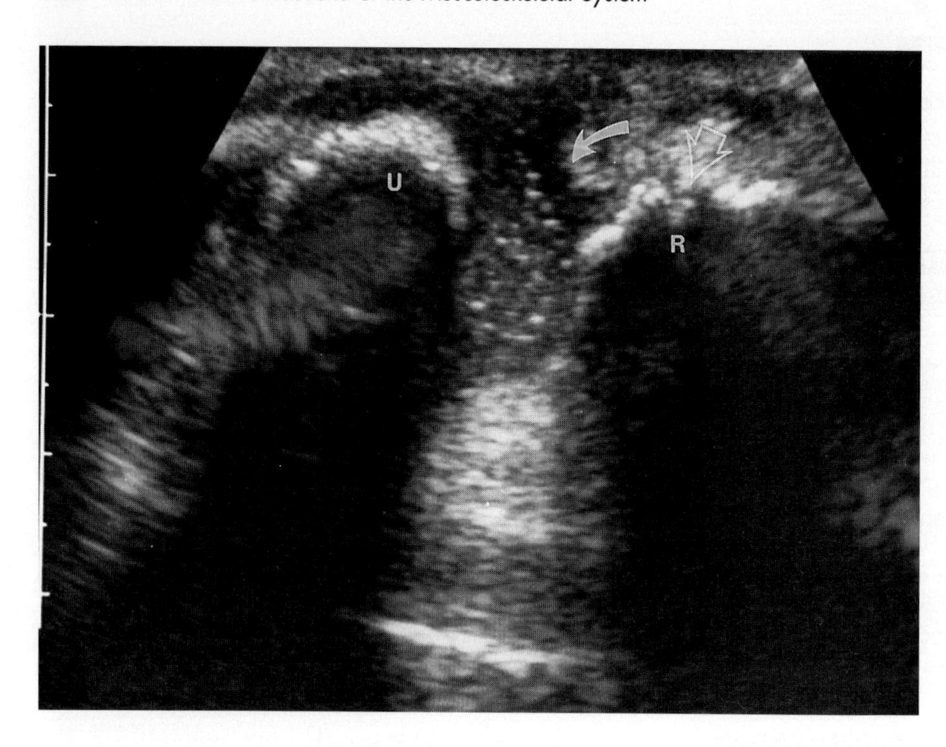

Figure 86–2. Septic arthritis of the wrist. A patient with dermatomyositis, on immunosuppressive therapy and who had multiple surgical tenosynovectomies. This patient presented with a swollen, red distal forearm and wrist. Transverse sonogram at the level of the distal radioulnar joint shows the joint capsule distended with pus *(curved arrow)*. There are cortical erosions of the distal radius *(open arrow)*. *Staphylococcus aureus* was cultured from the aspirated thick material. R, radius; U, ulna. (From Bureau NJ, Ali SS, Chhem RK, Cardinal E: Ultrasound of musculoskeletal infections. Semin Musculoskeletal Radiol 2:299–306, 1998.)

INFECTIONS AROUND METALLIC HARDWARE

In contrast to magnetic resonance (MR) imaging and computed tomography, ultrasound is not degraded by metallic artifact and may be useful in cases of infection complicating a joint prosthesis or metallic fixation. Differentiating mechanical loosening of a prosthesis from septic loosening is a common clinical dilemma, which is difficult to resolve on the basis of the clinical presentation and radiographic findings alone. Fluoroscopically guided aspiration of the prosthetic joint is often necessary to establish a definite diagnosis of infection.

Ultrasound is sensitive in detecting the presence of joint fluid but cannot differentiate between a joint effusion resulting from mechanical loosening, septic arthritis, or other causes. In a painful prosthetic hip, demonstration of a large joint effusion (mean bone-to-capsule distance, 10.2 mm) associated with extra-articular fluid collections at ultrasound has been shown to be highly suggestive of an infection.

Ultrasound may be of value in cases of osteomyelitis complicating metallic fixation in an extremity. In this setting, ultrasound may show loosening of metallic hardware and fluid collections or sinus tracts in the soft tissues of the involved extremity (Fig. 86–4). If a fluid collection is detected, it should be aspirated under ultrasound guidance and the specimen sent for microbiologic examination.

OSTEOMYELITIS

MR imaging is the imaging modality of choice in suspected cases of osteomyelitis because it allows evaluation of soft tissues and osseous structures. The role of ultrasound in the diagnosis of osteomyelitis is limited. In the appropriate clinical setting, ultrasound may show a fluid collection adjacent to the affected bone in adults or a subperiosteal fluid collection in children and younger patients. Identification of such fluid collections is not diagnostic of osteomyelitis, but the advantage of ultrasound is to allow rapid, precise aspiration of the fluid collection to obtain a specimen that is sent for microbiologic examination. In cases of chronic osteomyelitis, ultrasound may show the presence of sinus tracts or abscess formations, which are characteristics of active osteomyelitis (Fig. 86–5). MR imaging is the modality of choice to evaluate the extent of the soft tissue and osseous involvement.

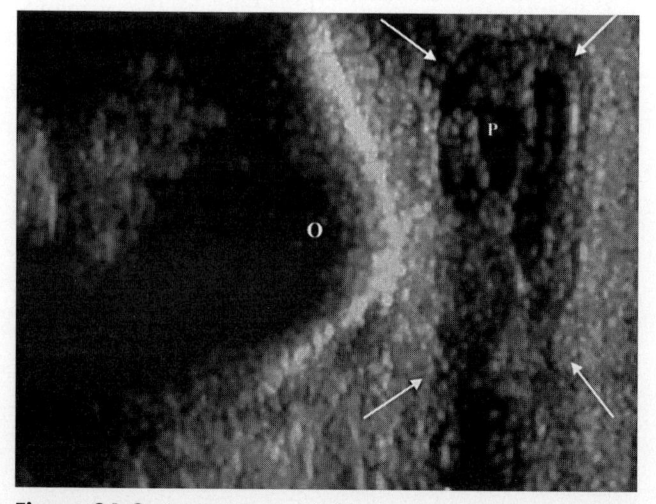

Figure 86–3. Septic bursitis of the olecranon bursa. Longitudinal ultrasound of the posterior aspect of the elbow shows a distended olecranon bursa (O) with synovial thickening *(arrows)* and hypoechoic pus (P). *Staphylococcus aureus* was cultured from the fluid obtained with ultrasound-guided aspiration.

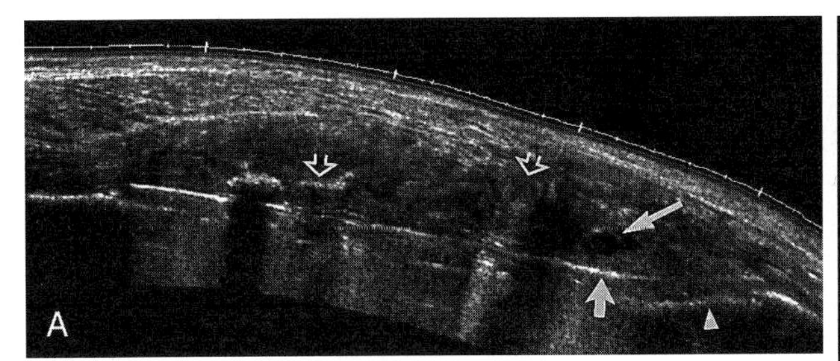

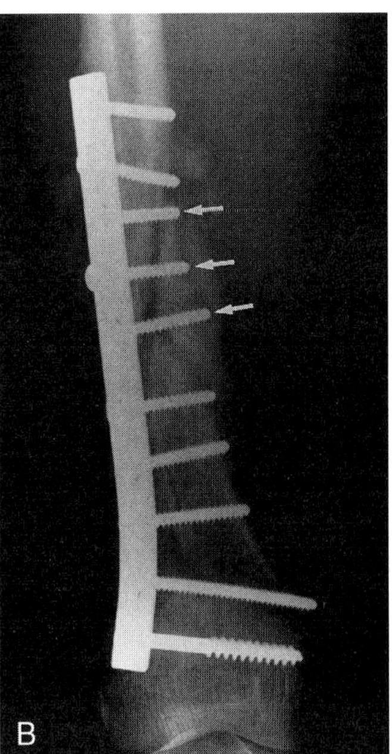

Figure 86–4. Osteomyelitis complicating metallic fixation of the right femur in a 35-year-old man. *A*, Extended-field-of-view longitudinal ultrasound of the lateral aspect of the femur shows a hypoechoic fluid collection *(long solid arrow)* adjacent to a metallic plate *(short solid arrow)* and the underlying cortex of the distal femoral diaphysis *(arrowhead)*. Loosened screws also are identified *(open arrows)*. *B*, Radiograph obtained within 5 days of the ultrasound examination shows metallic fixation of a comminuted fracture of the distal femoral diaphysis. Lucent areas surrounding some of the screws *(arrows)* are suggestive of loosening. *Staphylococcus aureus* was cultured from fluid obtained with ultrasound-guided aspiration of the hypoechoic collection, and osteomyelitis of the femur with loosening of the metallic hardware was confirmed at surgery. (From Bureau NJ, Chhem RK, Cardinal E. Musculoskeletal Infections: US Manifestations. Radiographics 19:1585–1592, 1999.)

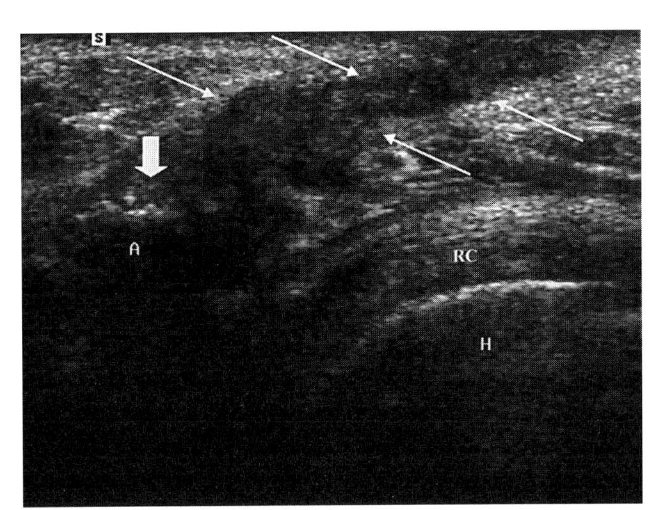

Figure 86–5. Chronic osteomyelitis of the acromion. A 56-year-old woman presented with increasing pain in the left shoulder. She underwent acromioplasty 9 months before. Coronal ultrasound of the left shoulder shows a hypoechoic heterogeneous sinus tract *(thin arrows)* extending from the margins of the acromion (A) to the skin. Note the irregularities *(thick arrow)* of the cortex of the acromion compared with the smooth hyperechoic cortical surface of the humeral head (H). Osteomyelitis of the acromion was confirmed at surgery. RC, rotator cuff.

Recommended Readings

Applegate GR, Cohen AJ: Pyomyositis: Early detection utilizing multiple imaging modalities. Magn Reson Imaging 9:187, 1991.

Bureau NJ, Ali SS, Chhem RK, et al: Ultrasound of musculoskeletal infections. Semin Musculoskeletal Radiol 2:299, 1998.

Bureau NJ, Chhem RK, Cardinal E: Musculoskeletal infections: US manifestations. Radiographics 19:1585, 1999.

Bureau NJ, Dussault RG, Keats T: Imaging of bursae around the shoulder joint. Skeletal Radiol 25:513, 1996.

Canoso JJ, Barza M: Soft tissue infections. Rheum Dis Clin North Am 19:293, 1993.

Fessell DP, Jacobson JA, Craig J, et al: Using sonography to reveal and aspirate joint effusions. AJR Am J Roentgenol 174:1353, 2000.

Goldenberg DL: Septic arthritis. Lancet 351:197, 1998.

Loyer EM, DuBrow RA, David CL: Imaging of superficial soft-tissue infections: Sonographic findings in cases of cellulitis and abscess. AJR Am J Roentgenol 166:149, 1996.

Loyer EM, Kaur H, David CL: Importance of dynamic assessment of the soft tissues in the sonographic diagnosis of echogenic superficial abscesses. J Ultrasound Med 14:669, 1995.

Mikhail IS, Alarcon GS: Nongonococcal bacterial arthritis. Rheum Dis Clin North Am 19:311, 1993.

van Holsbeeck MT, Eyler WR, Sherman LS, et al: Detection of infection in loosened hip prostheses: Efficacy of sonography. AJR Am J Roentgenol 163:381, 1994.

CHAPTER 87

Interventional Ultrasound

Nathalie J. Bureau, MD, and Étienne Cardinal, MD

Ultrasound is recognized as a useful imaging technique in interventional radiology. Because ultrasound is a readily available, nonionizing imaging modality with real-time imaging capabilities, it offers a significant advantage over fluoroscopy, computed tomography scan, and magnetic resonance imaging to guide interventions in the musculoskeletal system. Ultrasound localizes the lesion or the anatomic structure and allows visualization of the needle at all times during the procedure.

Ultrasound may be used to evaluate the efficacy of the procedure and for follow-up imaging. Several procedures may be done under ultrasound guidance, including aspiration of cysts (see Fig. 85–3), fluid collections, and abscesses; arthrocentesis; installation of drainage catheter; biopsy; steroid injection; treatment of calcific tendinitis (Fig. 87–1); and foreign body retrieval.

Recommended Readings

Aima R, Cardinal E, Bureau NJ, et al: Calcific Shoulder Tendinitis: Treatment with modified US-guided fine-needle technique. Radiology 221:455–461, 2001.

Cardinal E, Beauregard CG, Chhem RK: Interventional musculoskeletal ultrasound. Semin Musculoskeletal Radiol 1:311, 1997.

Cardinal E, Chhem RK, Beauregard CG: Ultrasound-guided interventional procedures in the musculoskeletal system. Radiol Clin North Am 36:597, 1998.

Dodd III GD, Esola CC, Memel DS, et al: Sonography: The undiscovered jewel of interventional radiology. Radiographics 16: 1271, 1996.

Farin PU, Jaroma H, Soimakallio S: Rotator cuff calcifications: Treatment with US-guided technique. Radiology 195:841, 1995.

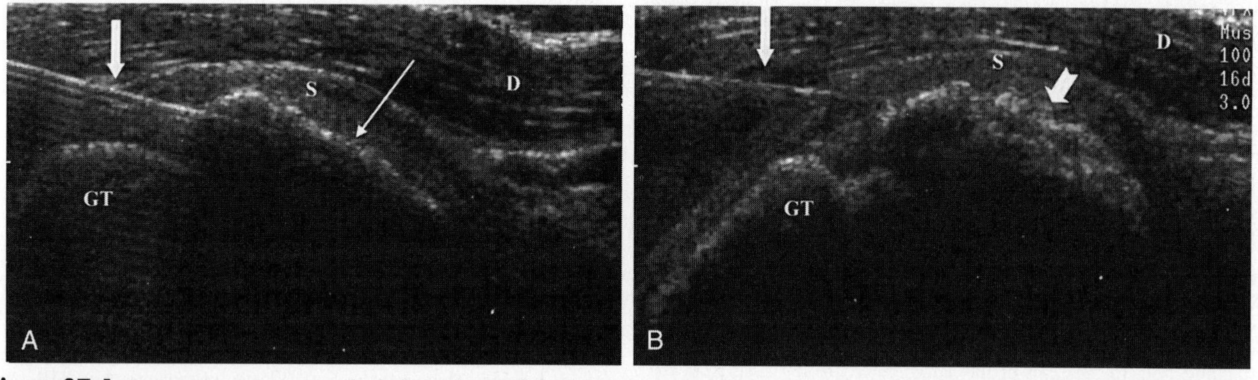

Figure 87–1. Percutaneous treatment of calcific tendinitis of the supraspinatus tendon. *A,* Longitudinal ultrasound of the right shoulder shows the extremity of the needle *(thick arrow)* within the calcification *(thin arrow)* of the supraspinatus tendon (S). *B,* Under real-time imaging, calcium is being removed by aspiration. Irregularities appear on the superior margin of the calcification *(notched arrow)* as it is being fragmented. D, deltoid; GT, greater tuberosity.

SECTION IX

INTERVENTIONAL PROCEDURES IN THE MUSCULOSKELETAL SYSTEM

CHAPTER 88

Epidural Steroid Injection

Donald L. Renfrew, MD

OVERVIEW AND PREOPERATIVE PLANNING

Rationale

Multiple studies have shown the efficacy of epidural steroid injection in the treatment of patients with back and leg pain and neck and arm pain. Injection should be performed with fluoroscopic guidance and contrast material injection. Sedation of the patient should be avoided whenever possible, particularly in cervical injections, because severe complications may ensue if the patient is not able to complain of pain on injection.

Equipment

C-arm fluoroscopic equipment with a targeting laser is recommended. The laser allows precise localization of the central ray of the x-ray beam, minimizing parallax. To use the laser, arrange the c-arm and patient so that the crosshairs of the targeting device are aimed at the appropriate target. Place the needle tip on the skin surface at the location of the laser, then angle the needle until the laser intersects the central portion of the needle hub.

Informed Consent

Inform the patient of the risks and benefits of the procedure and possible alternative therapies. Risks include contrast reaction, infection, neurologic damage, and a wet tap, which in some patients results in a severe positional headache and may require an epidural blood patch for treatment. Side effects of the steroids include insomnia, fever, hot flashes, agitation, and appetite and personality changes. Injected local anesthetic may cause numbness in the buttocks or legs, and occasionally patients have temporary weakness of one or both lower extremities lasting 1 to 3 hours. Patients may achieve excellent pain relief with the local anesthetic, only to have the pain return after the local anesthetic wears off, and it may be 2 to 3 days before the steroids are effective in relieving back and leg pain. Provide patients with instructions regarding who to call if pain becomes severe or if they experience any unusual symptoms, including loss of bowel or bladder control, loss of lower extremity sensation or motor control, fever, or neck rigidity.

CAUDAL EPIDURAL STEROID INJECTION

Patient Selection

Caudal epidural steroid injection is indicated for patients with low back pain, particularly coccydynia.

Procedure Description

Review magnetic resonance (MR) images if available and ensure that the thecal sac ends in the upper sacral spinal canal. Place the patient in the prone position. Palpate the sacral hiatus, which can be felt as two hard, bony prominences at the top of the natal cleft. Prepare and drape the skin in sterile fashion, and obtain local anesthesia with 1% lidocaine. Perform lateral fluoroscopy, and judge the appropriate angle of approach to enter the caudal epidural space. This angle varies with the degree of sacral lordosis. The needle should parallel the plane of the sacral canal. Project backward onto the skin surface and penetrate the skin with a 22-gauge, 3.5-inch (10-cm) spinal needle. Advance the needle until it is anchored in the skin and subcutaneous tissue and retains a fixed angle after release. Repeat fluoroscopy, make appropriate adjustments to needle course, and advance the needle until it is in the sacral epidural space at approximately the S3 level. Switch the fluoroscopy plane to frontal, and confirm a midline position of the needle tip.

Inject 1 to 5 mL of contrast material to confirm placement in the epidural space. Contrast material injected during all the injections described in this section should be nonionic and compatible with intrathecal administration. Iohexol (Omnipaque 180) is one such contrast material. Always read the package insert for any injected materials before administration. Contrast material should flow away from the needle tip and into the central spinal canal and the sacral nerve circumneural sheaths, forming an inverted *Christmas tree* pattern (Fig. 88–1). Record an image. Inject steroid and dilute local anesthetic. Injected steroid should be compatible with epidural injection. Betamethasone (Celestone) works well but may be difficult to obtain. Methylprednisolone (Depo-Medrol) may be used, but the solution must not contain preservatives. Injected local anesthetic should be compatible with epidural injection. Options include lidocaine and bupivacaine (Marcaine). Record a

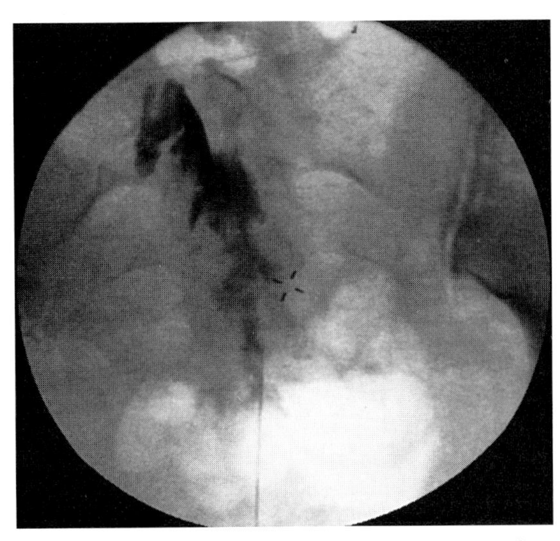

Figure 88–1. Frontal view of the sacrum after injection of contrast material via a caudal approach. Note the inverted *Christmas tree* appearance to the contrast material as it fills the caudal epidural space and circumneural sheaths of the sacral nerves.

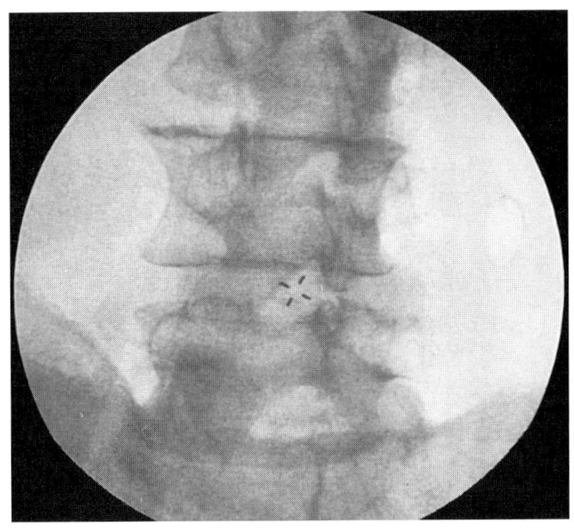

Figure 88–2. Oblique anteroposterior view of the lumbar spine, with the crosshairs of the laser-aiming device centered halfway between the inferior margin of the L4 lamina and the superior margin of the L5 lamina.

second image. The total volume of injected material should be at least 10 mL to ensure distribution through the sacral and lower lumbar epidural spaces. Monitor the patient, and record the degree of pain relief 15 to 30 minutes after the procedure.

LUMBAR INTERLAMINAR EPIDURAL STEROID INJECTION

Patient Selection

Lumbar interlaminar epidural steroid injection is indicated for patients with low back pain with or without leg pain.

Procedure Description

Review the patient's history and lumbar MR images to identify the possible pain generator and to choose a level where there is no severe central canal stenosis. Place the patient in the prone position. Identify the appropriate level of injection by counting up from the L5-S1 level or down from the last rib. Place the c-arm in an oblique position so that the inferior aspect of the upper lamina and the superior aspect of the lower lamina at the level of injection are seen easily (Fig. 88–2). Prepare and drape the skin in a sterile fashion, and advance a 22-gauge, 3.5-inch (10-cm) spinal needle aiming midway between the superior and inferior laminar margins. In larger patients, a 5-inch (15-cm) or 7-inch (22-cm) needle may be necessary.

Anchor the needle in approximately 2 cm of subcutaneous tissue so that it does not move when released, and confirm needle position with fluoroscopy. Adjust the needle course if necessary so that it is on a line to avoid the lamina and enter the spinal canal at approximately the midline (Fig. 88–3). Switch to lateral fluoroscopy and advance the needle until it is approximately 2 to 3 mm posterior to the spinal canal. Remove the stylus from the needle, and hook up to a syringe of nonionic contrast material via a connecting tube. Inject 0.1 to 0.2 mL of contrast material, and confirm that the needle is in the posterior parafascial space (the contrast material should pool around the needle tip). Unhook the connecting tube, and withdraw the plunger of the syringe slightly so that 2 to 3 cm of air is in the distal tube. Reattach the tube to the needle. Put gentle pressure on the syringe plunger, and confirm that the plunger *bounces back* when the pressure is removed (confirming that the needle tip is within soft tissue). Advance the needle tip in 1-mm intervals, carefully watching the interface between the air and contrast material in the connecting tube. When the epidural space is entered, this interface moves slightly toward the patient.

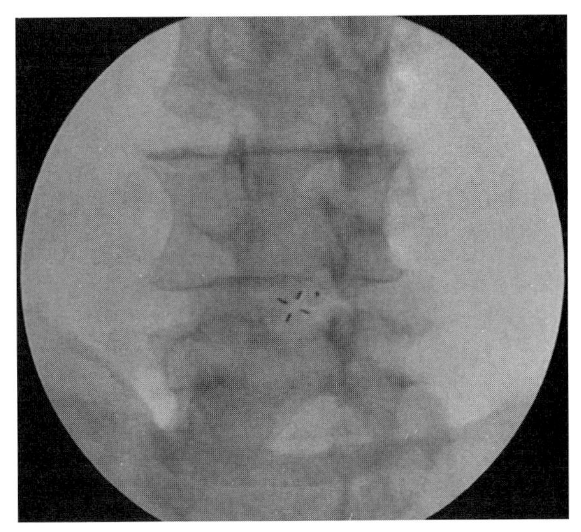

Figure 88–3. Oblique view of the lumbar spine, with the injecting needle advanced along an appropriate course. The needle is seen on-end slightly superior and medial to the crosshairs. The needle is in the appropriate position; the c-arm was moved slightly to show the needle better.

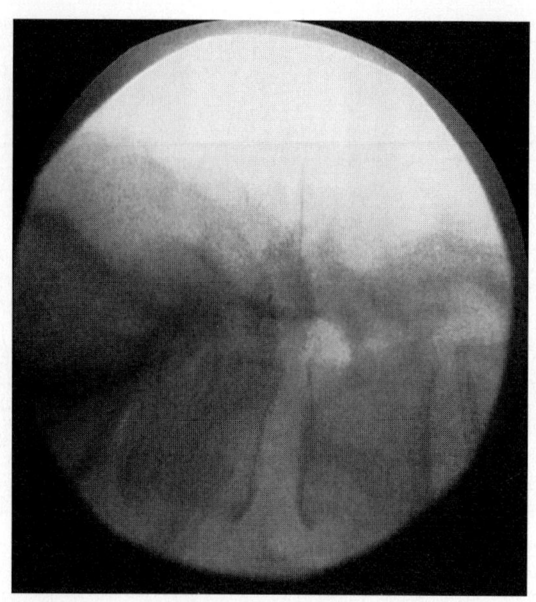

Figure 88–4. Lateral view of the lumbar spine, with the injecting needle within the epidural space and contrast material spreading along the posterior aspect of the epidural space.

Apply gentle pressure on the syringe plunger with each advance of the needle tip to confirm continued resistance to injection. Perform intermittent lateral fluoroscopy to confirm that the needle tip is in the posterior aspect of the spinal canal. If the needle tip is advanced past the posterior one third of the canal and it is not possible to inject easily, switch fluoroscopy planes and ensure that the needle is at or near the midline. On loss of resistance to injection (which should occur when the interface between the air and contrast material in the connecting tube moves), inject 0.5 to 1.0 mL of contrast material to confirm that the needle tip is in the epidural space (Fig. 88–4). Inject 3.0 mL of contrast material, and record lateral and frontal images (Figs. 88–5 and 88–6). Inject steroid and anesthetic, and repeat frontal and lateral images to document the pattern of distribution of injected materials (Figs. 88–7 and 88–8). The patient may have reproduction of typical symptoms or new and different back and leg pain during injection. If this is severe, decrease the rate of injection. Monitor the patient after the procedure, and record response to local anesthetic 15 to 30 minutes after injection.

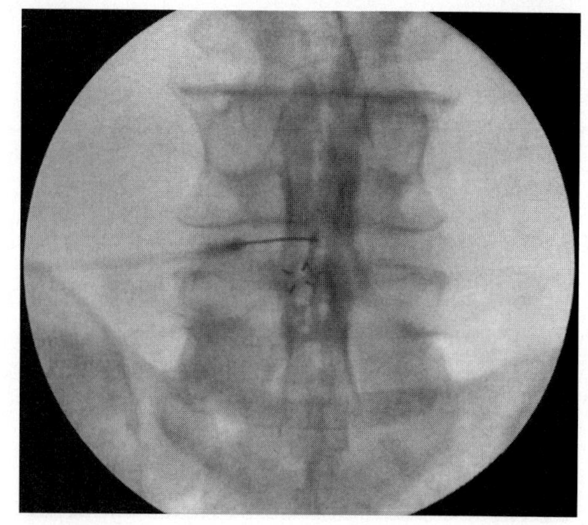

Figure 88–6. Frontal view of the lumbar spine, with 3.0 mL of nonionic contrast material injected showing the epidural space.

LUMBAR TRANSFORAMINAL EPIDURAL STEROID INJECTION

Patient Selection

Lumbar transforaminal epidural steroid injection is indicated for patients with predominantly single-level

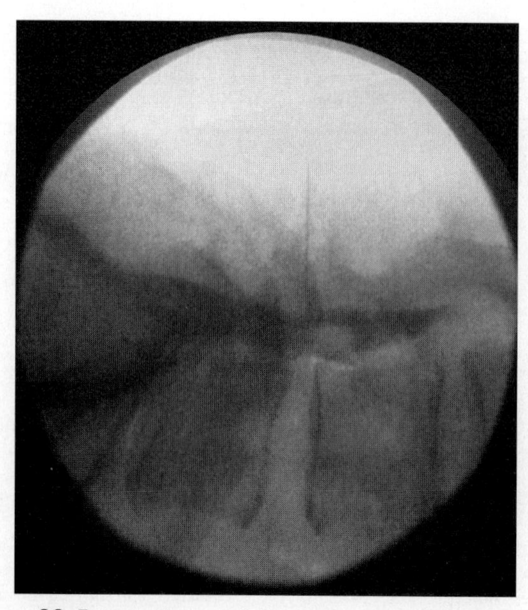

Figure 88–5. Lateral view of the lumbar spine, with 3.0 mL of nonionic contrast material injected showing the epidural space.

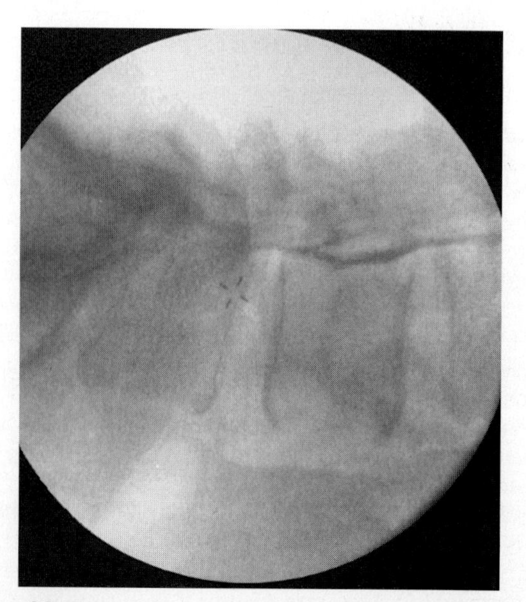

Figure 88–7. Lateral view of the lumbar spine after injection of contrast material, steroids, and anesthetic. Note further distribution of injected contrast material.

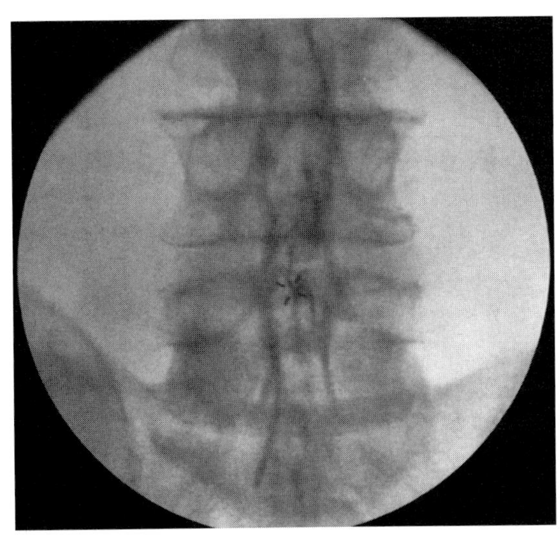

Figure 88–8. Frontal view of the lumbar spine after injection of contrast material, steroids, and anesthetic. Note further distribution of injected contrast material.

radicular leg pain. Transforaminal injections tend to be more painful but may provide better relief of symptoms in patients with single-level radiculopathy.

Procedure Description

Review the patient's history and lumbar MR images to identify the possible pain generator and to choose the appropriate level of injection. Place the patient in the prone position. Identify the appropriate level of injection by counting up from the L5-S1 level or down from the last rib. Place the c-arm in an oblique position so that the pedicle of the appropriate level is seen on end as the *eye* of the *Scotty dog*. Prepare and drape the skin in a sterile fashion, and advance a 22-gauge, 5-inch (15-cm) spinal needle aiming directly below the midpoint of the pedicle. Almost all patients require at least a 5-inch (15-cm) needle, and some large patients require a 7-inch (22-cm) needle. Anchor the needle in approximately 2 cm of subcutaneous tissue so that it does not move when released, and confirm needle position with fluoroscopy. Adjust the needle course as necessary to maintain a position just below the pedicle at the appropriate level. Continue to advance the needle until either the bone of the lateral vertebral body is contacted or the patient experiences severe leg pain.

If the patient experiences severe leg pain, ascertain if this is the usual location of the patient's pain, inject 0.5 mL of lidocaine, and wait 1 minute by the clock for the anesthetic to go into effect. If the patient does not experience leg pain but the needle tip encounters firm, bony resistance, switch to a frontal fluoroscopy plane, and confirm that the needle tip is beneath the pedicle and alongside the vertebral body. Inject 0.5 to 1.0 mL of contrast material, and record the image (Fig. 88–9). Contrast material should flow along the circumneural sheath of the exiting spinal nerve and may reflux into the epidural space centered at the level of injection (Fig. 88–10). If contrast material does not flow along the

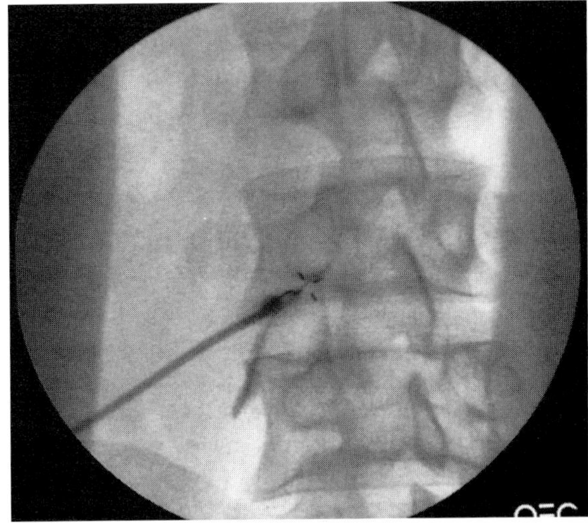

Figure 88–9. Oblique view of the lumbar spine, with contrast enhancement with left L3 circumneural sheath. The needle tip is beneath the *eye* (pedicle) of the *Scotty dog*, and contrast material tracks proximally and distally.

circumneural sheath, adjust the needle tip until it is in appropriate position. When contrast material flows along the circumneural sheath, inject 1.5 to 2.0 mL of local anesthetic and 2.0 to 3.0 mL of steroid-containing solution. Record another image. Monitor the patient after the procedure, and record response to local anesthetic 15 to 30 minutes after injection.

THORACIC INTERLAMINAR EPIDURAL STEROID INJECTION

Patient Selection

Thoracic interlaminar epidural steroid injection is indicated in patients with midback pain.

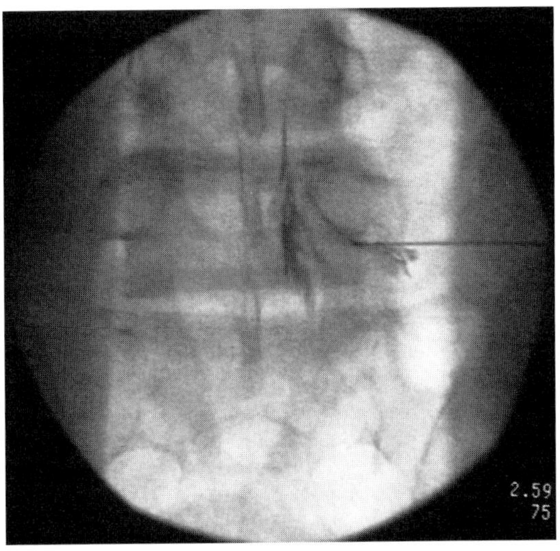

Figure 88–10. Frontal view of the lumbar spine, with the needle tip beneath the right L5 pedicle and contrast material tracking distally along the circumneural sheath and refluxing into the epidural space. The needle tip is along the lateral aspect of the L5 vertebral body.

Procedure Description

This procedure is virtually identical to lumbar interlaminar epidural steroid injection. The lamina may be more difficult to define at this level, and the posterior margin of the spinal canal may be difficult to visualize on the lateral examination because of overlapping ribs. Special care should be taken to prevent the needle tip from entering the spinal canal because of the risk of spinal cord damage.

CERVICAL INTERLAMINAR EPIDURAL STEROID INJECTION

Patient Selection

Cervical interlaminar epidural steroid injection is indicated for patients with neck or arm pain.

Procedure Description

Review the MR images to see if there is a tumor or severe spinal canal stenosis at the C7-T1 level. Place the patient prone on the fluoroscopy table. Perform fluoroscopy in the anteroposterior plane, and tilt the beam so that the central ray is parallel to the spinous processes and target the C7-T1 level. Prepare and drape the skin in sterile fashion, and insert a 25-gauge, 3.5-inch (10-cm) spinal needle into the posterior soft tissues at the C7-T1 level, midway between the spinous processes at these levels (Fig. 88–11). Anchor the needle in approximately 1.5 to 2.0 cm of tissue so that it does not move after no longer being held, and switch to a lateral fluoroscopy plane. The needle tip should project between the C7 and T1 spinous processes (Fig. 88–12). Advance the needle so that the tip is 2 to 3 mm posterior to the spinolaminar line.

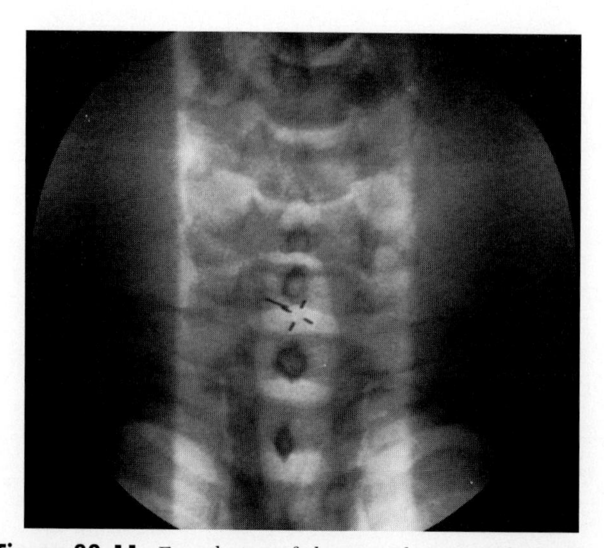

Figure 88–11. Frontal view of the cervical spine, with the laser crosshairs at the C7-T1 level and the injecting needle passing from lateral to medial in the central portion of the crosshairs, appropriately positioned to pass just beneath the C7 spinous process.

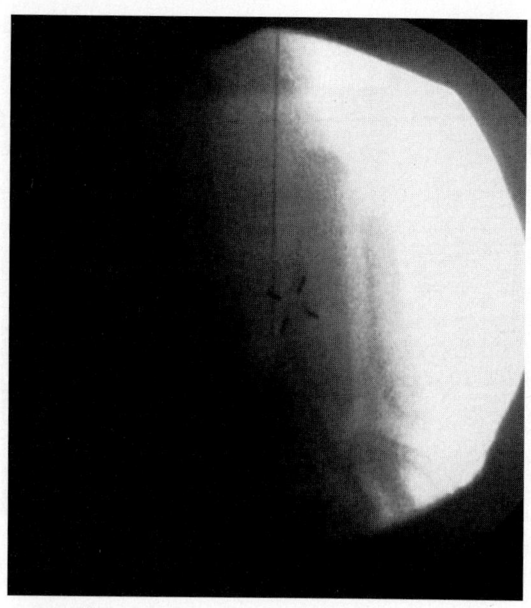

Figure 88–12. Lateral view of the cervical spine, with the needle between the C7 and T1 spinous processes. The needle tip is 1 to 2 mm posterior to the spinal canal.

Remove the needle stylus, and hook up to a connecting tube and 5-mL syringe filled with nonionic contrast material. Inject 0.2 to 0.3 mL, and confirm that the contrast material pools in the posterior parafascial tissues. Advance the needle tip in 1-mm increments, injecting 0.1 mL of contrast material with each advance, until the contrast material flows readily away from the needle tip and into the epidural space. In contrast to the situation in the lumbar spine, where the epidural space is larger and a

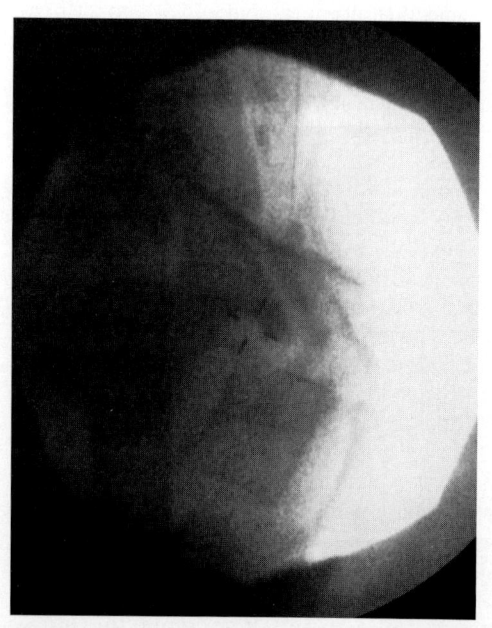

Figure 88–13. Lateral view of the cervical spine, with the needle in the posterior epidural space at the C7-T1 level. There is contrast material posteriorly between the C7 and T1 spinous processes from test injections and in the posterior epidural space (compare with Fig. 88–12).

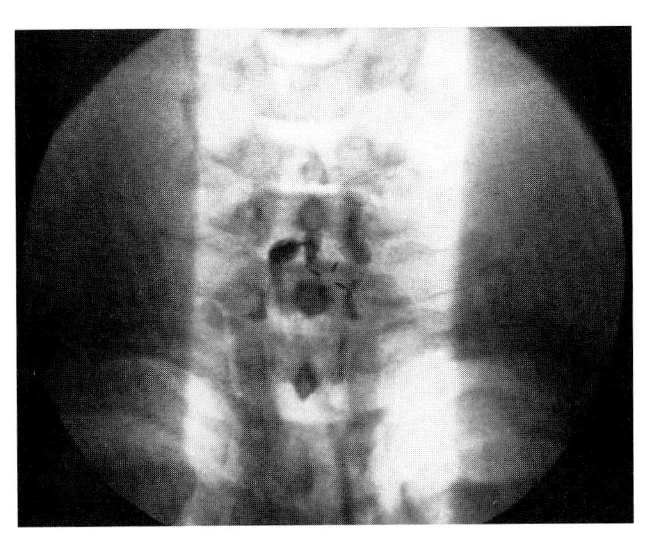

Figure 88–14. Frontal view of the cervical spine with the needle in position and contrast material flowing in the epidural space, filling some of the circumneural sheaths.

22-gauge spinal needle is used, spontaneous movement of an air–contrast material interface introduced into the connecting tube cannot be relied on. The contrast material should flow superiorly and inferiorly from the needle tip in a dense black line; if it is difficult to visualize the injected contrast material, a subarachnoid position of the needle tip should be suspected. Inject 1.0 to 3.0 mL of contrast material, and record lateral and anteroposterior images (Figs. 88–13 and 88–14). Inject 2.0 to 3.0 mL of steroid-containing solution into the epidural space, and record lateral and anteroposterior images to document distribution of injected materials in the epidural space. *Do not inject* local anesthetic into the cervical epidural space. Monitor the patient for any adverse consequences for approximately 30 minutes before discharge.

CERVICAL TRANSFORAMINAL EPIDURAL STEROID INJECTION

Cervical transforaminal epidural steroid injection also may be performed. The level of difficulty and possible complications are significantly higher than with lumbar transforaminal and cervical interlaminar epidural steroid injections. This procedure should be performed only after extensive experience with other injection procedures.

Recommended Readings

Bush K, Hillier S: A controlled study of caudal epidural injections of triamcinolone plus procaine for management of intractable sciatica. Spine 16:572, 1991.

Dilke TFW, Burry HC, Grahame R: Extradural corticosteroid injection in management of lumbar nerve root compression. BMJ 2:635, 1973.

Hodges SD, Castleberg RL, Miller T, et al: Cervical epidural steroid injection with intrinsic spinal cord damage: Two case reports. Spine 23:2137, 1998.

Link SC, El-Khoury GY, Guilford WB: Percutaneous epidural and nerve root block and percutaneous lumbar sympatholysis. Radiol Clin North Am 36:509, 1998.

Lutz GE, Vad VB, Wisneski RJ: Fluoroscopic transforaminal lumbar epidural steroids: An outcome study. Arch Phys Med Rehabil 79:1363, 1998.

Renfrew DL, Moore TE, Kathol MH, et al: Correct placement of epidural steroid injections: Fluoroscopic guidance and contrast administration. AJNR Am J Neuroradiol 12:1003, 1991.

Shulman M: Treatment of neck pain with cervical epidural steroid injection. Reg Anesth 11:92, 1986.

Warfield CA, Biber MP, Crews DA, et al: Epidural steroid injection as a treatment for cervical radiculitis. Clin J Pain 4:201, 1988.

White AH, Derby R, Whynne G: Epidural injections for the diagnosis and treatment of low-back pain. Spine 5:78, 1980.

CHAPTER 89

Nerve Root Blocks

Donald L. Renfrew, MD

RATIONALE

Nerve root blocks are performed to diagnose (or confirm the diagnosis of) single-level radiculopathy from nerve root impingement or irritation (van Akkerveeken, 1993). Nerve root blocks also provide pain relief.

PATIENT SELECTION

Nerve root blocks are indicated for patients with single-level radiculopathy.

PROCEDURE DESCRIPTION

The procedure is identical to transforaminal epidural steroid injection (see Chapter 88) except that a slightly more lateral position of the needle tip is favored and slightly lower volumes of agents are injected (0.5 to 1.0 mL of contrast material, local anesthetic, and steroid). More concentrated local anesthetic and no steroid may be used for strictly diagnostic (nontherapeutic) blocks. Consider performing a second *control* block at an adjacent level if the results of the block are to be used to direct surgical treatment.

Reference

van Akkerveeken PF: The diagnostic value of nerve root sheath infiltration. Acta Orthop Scand Suppl 251:64, 1993.

CHAPTER 90

Sacroiliac Joint Injection

Donald L. Renfrew, MD

RATIONALE

Sacroiliac joint injection is performed to diagnose and treat pain from the sacroiliac joint.

PATIENT SELECTION

Sacroiliac joint injection is indicated for patients with low back paraspinal pain or referred hip pain or groin pain.

PROCEDURE DESCRIPTION

Place the patient prone on the fluoroscopic table. Perform fluoroscopy of the sacroiliac joints, maneuvering the plane of fluoroscopy from one oblique to the other or from craniad to caudad until the inferior aspect of the sacroiliac joint is visualized optimally as a well-seen lucency between the articular margins of the sacrum and pelvis. Prepare and drape the skin in a sterile manner, and advance a 22-gauge, 3.5-inch (10-cm) spinal needle until it is anchored in approximately 2 cm of soft tissue and does

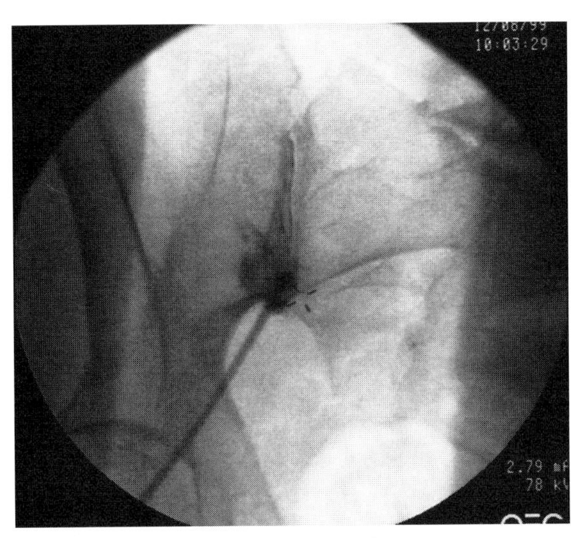

Figure 90–1. Frontal view of the left sacroiliac joint. Contrast material is pooled around the inferior aspect of the joint from an extra-articular injection done before needle adjustment. An arthrogram was obtained from injection after needle adjustment.

not move after being released. Aim for the inferiormost aspect of the sacroiliac joint. When firm resistance is encountered, confirm the needle tip position with fluoroscopy. A distinct difference in the sensation may be transmitted through the needle when the joint is entered or when articular cartilage is pierced. Inject a small amount of contrast material. If contrast material pools around the needle tip, adjust the needle position so that it lies within the sacroiliac joint and reinject to confirm an intra-articular position of the needle tip with an arthrogram (Fig. 90–1). Inject 1.0 to 3.0 mL of local anesthetic and 1.0 to 3.0 mL of steroid. If the injection elicits pain, record whether this pain is the patient's typical pain or different from the patient's typical pain. Monitor the patient for 30 minutes after the procedure, and document the extent of pain relief.

Recommended Reading

Schwarzer AC, Aprill CN, Bogduk N: The sacroiliac joint in chronic low back pain. Spine 20:31, 1995.

CHAPTER 91

Facet Joint Procedures

Donald L. Renfrew, MD

OVERVIEW

The facet joints may be a source of low back or leg pain in approximately 15% to 20% of patients with low back pain. No specific constellation of physical examination findings is predictive of whether the facet joints are the source of pain in a given patient, and imaging findings of degeneration are nonspecific and insensitive. Response to provocative and therapeutic injection procedures of the facet joints may provide the best method of evaluating whether these joints are the cause of a patient's pain.

LUMBAR MEDIAL BRANCH BLOCKS

Patient Selection

Lumbar medial branch blocks are indicated for patients with predominantly central low back pain with some radiation to the hips or buttocks but little leg pain. Medial branch blocks generally are performed to evaluate patients for possible rhizotomy (see later).

Procedure Description

Choose the level of injection based on the pain pattern or imaging studies or both. Generally the L4-L5 and L5-S1 levels may be chosen, with inclusion of higher levels in patients with severe facet arthropathy. If the pain is central or bilateral, inject both sides; for unilateral pain, inject the symptomatic side. Place the patient prone on the fluoroscopic table, and identify the general region of the injection levels with anteroposterior fluoroscopy. Prepare and drape the skin in sterile fashion. Identify the location of the most superior medial branch to be blocked, which often is L4, at the junction of the L5 transverse and superior articular processes. Place a 22-gauge, 3.5-inch

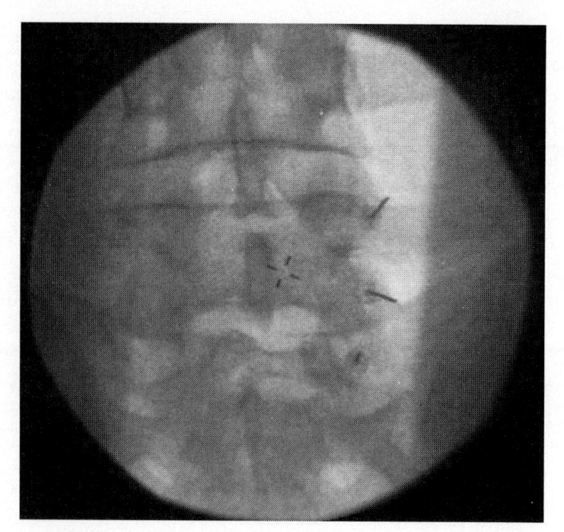

Figure 91–1. Frontal view of the lumbosacral spine, with needles appropriately placed at the right L4, L5, and S1 positions.

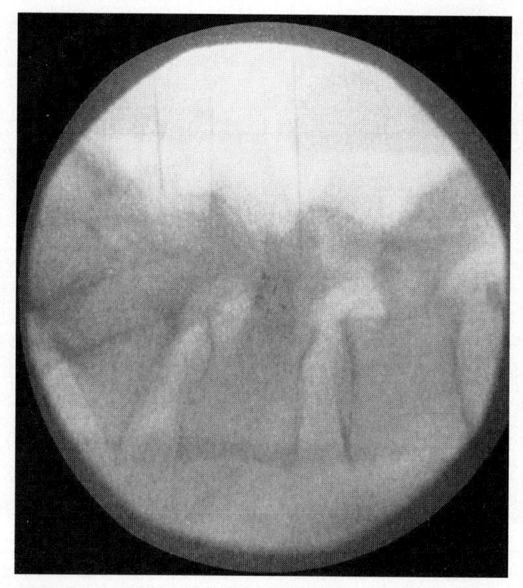

Figure 91–2. Lateral view of the lumbosacral spine, with needles appropriately placed at the L4, L5, and S1 positions. The needle tips should not project anterior to the transverse process level; if they do, anesthetization of the ventral ramus of the spinal nerve may result.

(10-cm) needle along the plane of the central beam toward the target. Occasionally, 5-inch (15-cm) needles are required in larger patients.

Without changing to the lateral plane, change the central beam position to the next target, and anchor along this target line. Continue until needles are aimed at all medial branches to be blocked on one side (Fig. 91–1). The position for blocking the S1 contributing branches is on the dorsal aspect of the sacrum, and an appropriately positioned needle would be against the dorsal sacrum and encounter firm, bony resistance. Switch to a lateral fluoroscopy plane and adjust needle tips (if necessary) so that they are at the level of the transverse processes (the S1 needle is along the dorsal sacral margin) (Fig. 91–2). Return to the anteroposterior plane, and confirm that the needle tips remain appropriately positioned. Inject 0.2 mL of nonionic contrast material to confirm a nonvascular position of the needle tips (Fig. 91–3). Inject 0.75 to 1.0 mL of local anesthetic and (optional) 0.2 mL of steroid at each level. Repeat the procedure for the contralateral side if indicated. Monitor the patient for 30 minutes after the procedure, and document the degree of improvement of low back pain.

CERVICAL MEDIAL BRANCH BLOCKS

Patient Selection

Cervical medial branch blocks are indicated for patients with predominantly neck pain with referred head, shoulder, or upper back pain. Medial branch blocks generally are performed to evaluate patients for possible rhizotomy (see later).

Procedure Description

Choose the level of injection based on the pain pattern or imaging studies or both. In contrast to the situation in the lumbar spine, in which most of the degenerative facet changes occur at the lowest two levels and are bilateral, symptomatic cervical facet joints more frequently involve upper levels of the cervical spine and more often are unilateral. Place the patient prone on the fluoroscopy table, and perform fluoroscopy in the anteroposterior plane. Angle the c-arm caudally so that the mild indentations along the midportion of the lateral masses may be seen. Identify the general target area, and prepare and drape the skin in a sterile fashion. Identify the specific target point along the waist of the articular process above and below the facet joint to be blocked. Place a 25-gauge, 3.5-inch (10-cm) spinal needle along the central ray aimed at this target. Try to graze the lateralmost aspect of the articular pillar with the needle tip (Fig. 91–4). When the

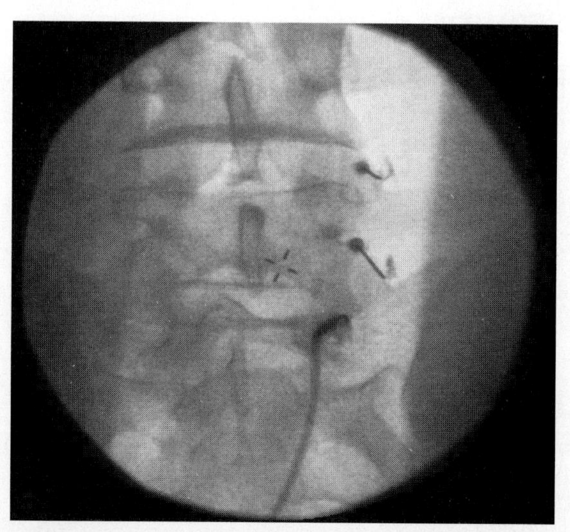

Figure 91–3. Frontal view of the lumbosacral spine, with needles appropriately placed at the right L4, L5, and S1 positions after injection of 0.2 mL of contrast material. The contrast material pools around the needle tips, confirming a nonvascular location.

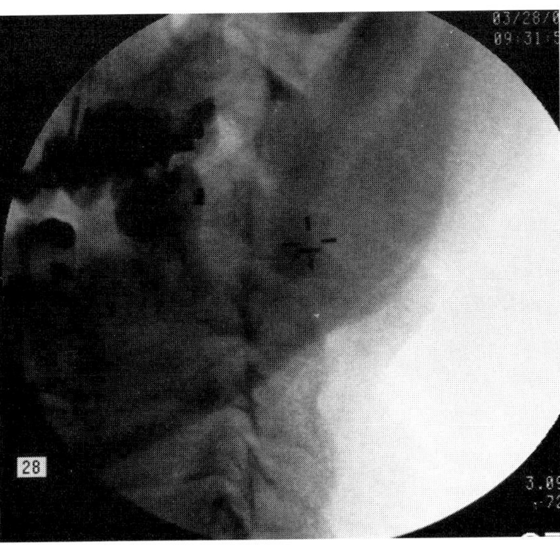

Figure 91–4. Frontal view of the cervical spine shows the crosshairs of the laser-aiming device centered at the C3 lateral mass and a needle in place so that the tip is at the waist of the lateral mass.

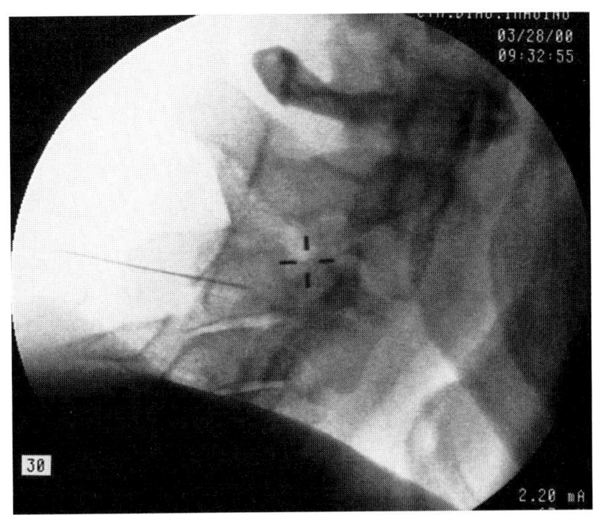

Figure 91–5. Lateral view of the cervical spine shows that the needle tip seen in Figure 91–4 projects at the waist of the C3 lateral mass, in the location of the C3 medial branch of the C3-4 facet.

needle is anchored in the soft tissues, switch to a lateral fluoroscopy plane. Confirm that the needle tip is at the correct location (Fig. 91–5). If the needle tip is not at the level of the lateral mass, advance until the needle tip is along the lateral aspect of the articular pillar or until the needle tip encounters bone. When the needle tip is positioned appropriately, place additional needles as needed. Record frontal and lateral images on final needle position, and inject 0.2 mL of nonionic contrast material through each needle to confirm a nonvascular location of the needle tips. Inject 0.5 to 0.8 mL of local anesthetic and (optional) 0.2 mL of steroid at each level. Monitor the patient, and record pain response at 30 minutes.

LUMBAR INTRA-ARTICULAR FACET JOINT INJECTION

Patient Selection

Lumbar intra-articular facet joint injection is indicated for patients with predominantly central low back pain with some radiation to the hips or buttocks but usually little leg pain.

Procedure Description

Choose the level of injection based on the pain pattern or imaging studies or both. If the pain is central or bilateral, inject both sides; for unilateral pain, inject the symptomatic side. Place the patient prone on the fluoroscopic table, and identify the general region of the injection levels with anteroposterior fluoroscopy. Prepare and drape the skin in sterile fashion. Perform oblique fluoroscopy to visualize the joint space optimally. Aim for the inferiormost aspect of the joint (there is a relatively redundant capsule at the inferior recess of the joint, which makes it easier to obtain an intra-articular injection). Advance a 22-gauge, 5-inch (15-cm) needle until it is within the joint. A change in tactile sensation on entry of the joint may be appreciated.

Inject 0.25 to 0.5 mL of nonionic contrast material, and record an image (Fig. 91–6). If the procedure is being done for diagnostic purposes, the total volume of injectate should be kept less than 1.5 mL so that all material remains intracapsular. In this situation, inject appropriate volumes of local anesthetic and steroid. If it is not necessary to keep all injected materials intracapsular, 1.0 to 1.5 mL of local anesthetic and 1.0 to 1.5 mL of steroid may be injected at each level, keeping the total volume of steroid injected for all levels less than 3.0 mL. Injection of greater than 1.5 mL usually results in capsular rupture and spread of materials in the epidural space. In cases in which there is marked degenerative change of the facet joint or an oblique approach is unsuccessful, use a direct posterior

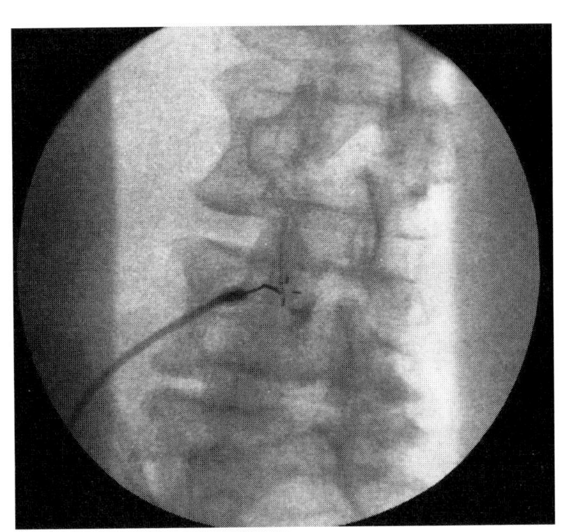

Figure 91–6. Oblique view of the lumbar spine, with a needle in the left L3-4 facet joint. A thin, sharp line of contrast material is seen between the inferior articular process of L3 and the superior articular process of L4; contrast material also pools in the inferior recess of the joint.

approach to the joint, with the target point being the inferior aspect of the inferior articular facet from the vertebra above.

CERVICAL INTRA-ARTICULAR FACET JOINT INJECTION

Rationale

Cervical intra-articular facet joint injection is performed to diagnose and treat pain from the cervical facet joints.

Patient Selection

Cervical intra-articular facet joint injection is indicated for patients with predominantly neck pain with referred head, shoulder, or upper back pain.

Procedure Description

Choose the level of injection based on the pain pattern or imaging studies or both. As noted previously (see under "Cervical Medial Branch Blocks"), symptomatic cervical facet joints often involve upper levels of the cervical spine and more often are unilateral than in the lumbar spine. Place the patient prone on the fluoroscopy table, and perform fluoroscopy in the anteroposterior plane. Angle the c-arm caudally so that the facet joints are well seen as parallel lucent bands separated by the lateral masses. Identify the general target area, and prepare and drape the skin in a sterile fashion. Identify the specific facet joint to be blocked. Place a 25-gauge, 3.5-inch (10-cm) spinal needle along the central ray aimed at the joint. When the needle is anchored in the soft tissues, switch to a lateral fluoroscopy plane. Confirm that the needle tip is at the correct level. Advance the needle tip so that it is inside the joint. Inject 0.25 to 0.50 mL of contrast material, and record anteroposterior and lateral (Fig. 91–7) images showing the arthrogram. If the procedure is being done for diagnostic purposes, the total volume of injectate should be kept less than 1.25 mL so that all material remains intracapsular. In this situation, inject appropriate volumes of local anesthetic and steroid. If it is not necessary to keep all injected materials intracapsular, 0.75 to 1.0 mL of local anesthetic and 0.75 to 1.0 mL of steroid may be injected at each level, keeping the total volume of steroid injected for all levels at less than 3.0 mL. Monitor the patient, and record pain response at 30 minutes.

SYNOVIAL CYST INJECTION

Patient Selection

Synovial cyst injection is indicated for patients with radiculopathy secondary to synovial cysts.

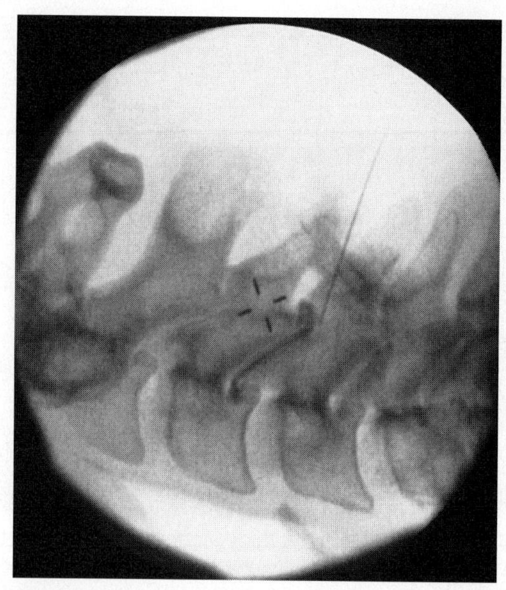

Figure 91–7. Lateral view of the cervical spine, with a needle in the posterior aspect of the C3-4 facet joint and contrast material within the joint.

Procedure Description

The procedure is identical to intra-articular facet joint injection except that a high volume of injectate is used purposefully to attempt to rupture the symptomatic synovial cysts. Even in patients whose cysts remain unruptured, the injection may provide long-term or permanent pain relief, apparently because injected materials cause the synovial cyst to collapse or resorb. Rupture of the cyst may be felt as the sudden *giving way* of resistance during injection. These injections frequently are quite painful and reproduce the patient's symptoms. The initial response to injection (whether typical pain is elicited or not), whether the cyst apparently has undergone rupture, and whether the patient experiences pain relief approximately 30 minutes after the procedure all should be recorded.

SPONDYLOLYSIS DEFECT INJECTION

Rationale

Spondylolysis defect injection is performed to diagnose and treat patients with spondylolysis.

Patient Selection

Spondylolysis defect injection is indicated for patients with low back pain secondary to spondylolysis.

Procedure Description

The procedure is essentially the same as an intra-articular facet joint injection except that the target is the

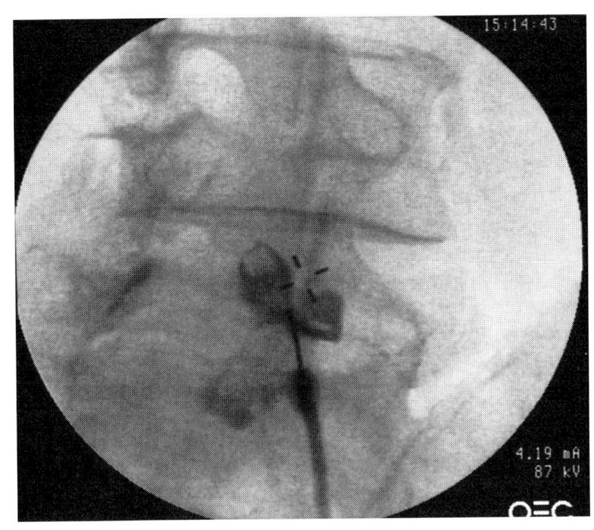

Figure 91–8. Oblique view of the lumbar spine shows a needle in place at the inferior aspect of the L4-5 facet joint and contrast material within an L5 pars defect.

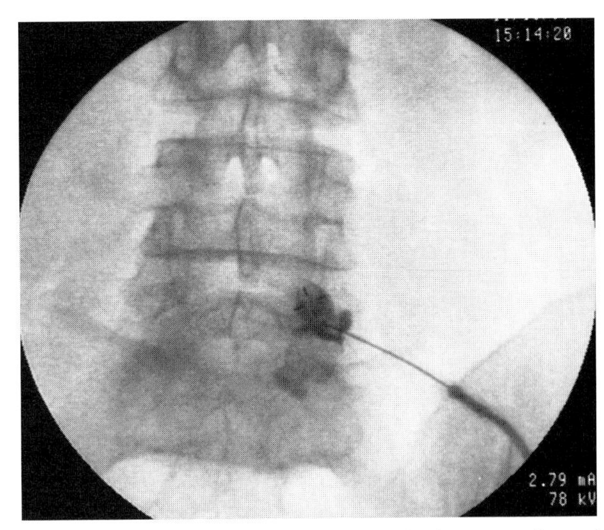

Figure 91–9. Anteroposterior view of the lumbar spine shows the needle at the level of a right L5 pars defect and contrast material in the defect.

spondylolysis defect rather than the joint space. Injection into the facet joint above or below the defect may result in material tracking into the defect as well. Contrast material shows a localized collection in the defect (Figs. 91–8 and 91–9). Occasionally, contrast material tracks across the midline to a contralateral defect at the same level.

RADIOFREQUENCY DENERVATION (RHIZOTOMY)

Patient Selection

Radiofrequency denervation is indicated for patients with low back pain secondary to facet arthropathy. Patients should have responded favorably to facet joint blocks (see under "Lumbar Medial Branch Blocks" and "Cervical Medial Branch Blocks"), preferably on two different occasions with short-acting and long-acting anesthetic, with pain relief appropriate in timing to drug duration of action.

Procedure Description

The procedure is technically similar to medial branch blocks, and it is preferable to have the same operator using the same technique for needle placement perform the blocks and the rhizotomy to ensure consistent needle tip placement. After needle tip placement is confirmed via c-arm fluoroscopy in two planes, pass a stimulating current through the needle tip. If this current elicits radiculopathy, reassess needle tip position, and choose a more posterior position. If elicited sensation is confined to the central low back or buttocks (or neck, in the case of cervical rhizotomy), proceed by giving 1.0 mL of local anesthetic, and perform radiofrequency lesioning. Increase the temperature of the tissues in the vicinity (2- to 3-mm radius) of the needle tip to approximately 90°C for 90 seconds. Create two colinear lesions (only one lesion is created if the needle tip is resting on bone, as is always the case at the S1 level and as may be the case at other levels). The patient should be advised that although pain relief may follow immediately, many patients undergoing rhizotomy experience a *flare* of persistent or increased low back pain for several weeks after the procedure.

Recommended Readings

Dwyer AB, Aprill C, Bogduk N: Cervical zygapophyseal joint pain patterns: I. A study in normal volunteers. Spine 15:453, 1990.
Kuslich SD, Ulstrom CL, Michael CJ: The tissue of origin of low back pain and sciatica: A report of pain response to tissue stimulation during operations on the lumbar spine using local anesthesia. Orthop Clin North Am 22:181, 1991.
Mooney V, Robertson J: The facet syndrome. Clin Orthop 115:149, 1976.
Parlier-Cuau C, Wybier M, Nizard R, et al: Symptomatic lumbar facet joint synovial cysts: Clinical assessment of facet joint steroid injection after 1 and 6 months and long-term follow-up in 30 patients. Radiology 210:509, 1999.
Sarazin L, Chevrot A, Pessis E, et al: Lumbar facet joint arthrography with the posterior approach. Radiographics 19:92, 1999.
Schwarzer AC, Aprill CN, Derby R, et al: Clinical features of patients with pain stemming from the lumbar zygapophyseal joints: Is the lumbar facet syndrome a clinical entity? Spine 19:1132, 1994.
van Kleef M, Barendse GAM, Kessels A, et al: Randomized trial of radiofrequency lumbar facet denervation for chronic low back pain. Spine 24:1937, 1999.

CHAPTER 92

Diskography

Donald L. Renfrew, MD

OVERVIEW

Spine surgeons have used diskography to diagnose internal disruption of the disk for more than 50 years. In 1968, Holt questioned the validity of diskography, claiming in particular that the examination produced many false-positive results. More recently, Walsh and colleagues (1990) published results indicating that when correctly performed by an experienced diskographer, false-positive results are not a problem. The North American Spine Society published a position statement that summarizes many of the issues involved in diskography (Guyer and Ohnmeiss, 1995).

LUMBAR DISKOGRAPHY

Patient Selection

Lumbar diskography is indicated for patients with predominantly axial low back pain suspected to be secondary to internal disruption of the disk. Diskography is performed most frequently either to confirm that a disk seen to be abnormal at magnetic resonance (MR) imaging is the source of the patient's symptoms (with an adjacent nonpainful level serving as a control) or to decide which of many morphologically abnormal disks is the symptom-producing lesion.

Procedure Description

Review the patient's history and lumbar MR images to identify abnormal disks and to choose the appropriate level of injection. Confer with the referring physician regarding levels to be injected and ground rules for how many levels need to be injected (see later). Place the patient in the prone position. Identify the appropriate level of injection by counting up from the L5-S1 level or down from the last rib and correlating with the MR images. Place the c-arm in the oblique position so that the superior articular process of the inferior vertebra at the injection level is at approximately the midpoint of the disk. Angle the tube so that the end plates of the disk are well defined and the disk is seen as a transverse lucency. The target point is just anterior to the superior articular process, entering the disk from inferior to superior. Prepare and drape the skin in a sterile fashion. Puncture the skin with a 22-gauge, 5-inch (15-cm) or 7-inch (22-cm) needle (the longer needles may be needed for larger patients) along the target line.

Advance the needle 0.5 cm at a time, confirming the correct course with oblique fluoroscopy (Fig. 92–1). When the margin of the disk is encountered, a difference in resistance should be felt. At this point, switch to anteroposterior fluoroscopy, and advance the needle so that it is at the midsagittal level (Fig. 92–2). Switch to lateral fluoroscopy and confirm that the needle tip is between the posterior and anterior thirds of the disk. Instruct the patient to forget about all pain caused by needle placement and to concentrate on any new pain during injection. Remove the needle stylus and attach a connecting tube and syringe filled with a mixture of contrast material and antibiotic solution. Begin injecting, and note pain behavior (Fig. 92–3). If no pain is elicited, inject until either a firm end point is reached or 3.0 to 3.5 mL of the contrast material and antibiotic mixture is injected, and record anteroposterior and lateral examinations. If pain is elicited, note the amount of contrast material injected at the time of pain production and the ease of contrast material injection (no resistance, some resistance, or marked

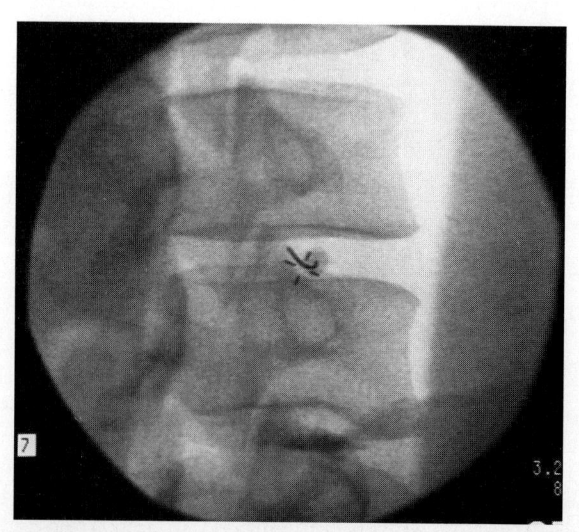

Figure 92–1. Oblique view of the lumbar spine for diskography at L3-4. The angle is such that the superior articular process is in the middle of the intervertebral disk, the vertebral end plates are parallel, and the disk is seen as a lucency in the middle of the field. The laser pointer is just anterior to the base of the superior articular process, and the needle curves from anteroinferior to superoposterior, which is the best course for avoiding the ventral ramus of the segmental nerve at the level of the injection.

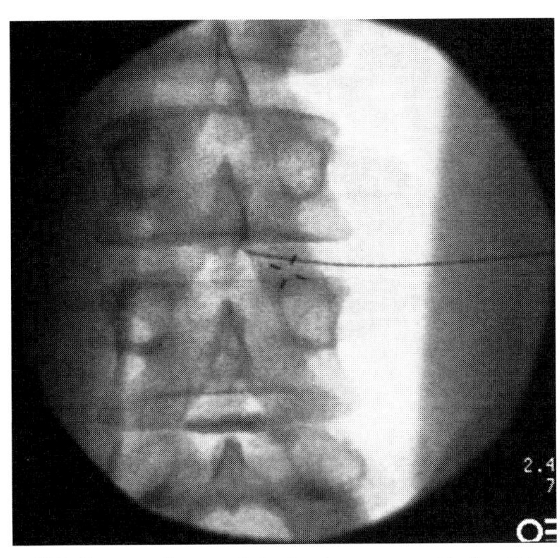

Figure 92–2. Frontal view of the lumbar spine, with the diskography needle in the middle of the L3-4 disk.

resistance to injection). Ask the patient to rate the pain on a scale of 0 to 10, with 0 being no pain at all and 10 being excruciating pain. Ask the patient if the pain is familiar and concordant with his or her usual pain or different from his or her usual pain. If pain is elicited during injection of contrast material, inject 0.5 to 1.5 mL of local anesthetic, and record if the patient has worsening of pain beyond the numerical rating offered so far. Remove the needle. Record anteroposterior and lateral spot films.

Inject at least two levels, one of which is symptomatic and one of which elicits no symptoms or a minimal (1 or 2) pain rating. If both of the first two injections produce 5 or greater pain on the 0-to-10 scale, perform a third injection.

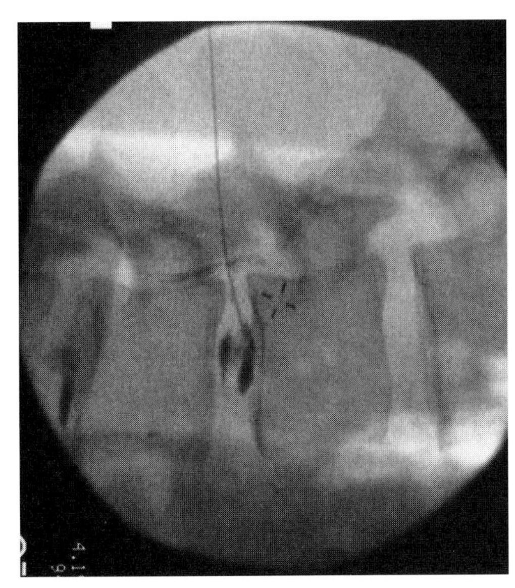

Figure 92–3. Lateral view of the lumbar spine with the diskography needle in the posterior third of the disk (the nucleus is usually posterior to the central portion of the intervertebral disk). There is contrast material in the nucleus and along the posterior margin of the disk and adjacent vertebral bodies. There also is contrast material at the L4-5 level from diskography performed at this level.

If this injection also is symptomatic at a greater than 5 level, consider another injection. If there are four positive levels, it usually is unnecessary to look further for a control level because most orthopedic surgeons would not fuse more than three levels. Establish such ground rules with the referring physician before embarking on disk injection. Report each level of the diskogram separately, with description of each of the following: (1) the amount of material injected, (2) whether the injection was intranuclear, (3) the pain rating of the injection on a scale of 0 to 10 (as indicated previously), (4) whether the pain was concordant or discordant with the patient's usual symptoms, (5) the location of the pain, (6) whether pain was elicited early/with minimal pressure or late/with greater pressure, (7) the degree of resistance to injection, and (8) the morphology of the disk. Disks that are painful early in injection or with minimal injection pressure may respond better to anteroposterior fusion with diskectomy, whereas disks painful only late in injection with higher degrees of pressure may do well with only posterior fusion.

After the final injection, obtain a computed tomography (CT) scan of the injected levels, and report (at each level) whether the injection is intranuclear, the location and severity of annular fissures, whether contrast material reached the exterior of the disk, and the location of any such contrast material (Fig. 92–4). CT in such cases serves at least three purposes: (1) it documents an intranuclear injection; (2) it shows the location of annular fissures; and (3) it provides morphologic information, which may have changed since the previously obtained MR images or which is not evident on the previously obtained MR images (because of a poor-quality MR examination).

CERVICAL DISKOGRAPHY

Cervical diskography also may be performed. As with cervical transforaminal epidural steroid injections and nerve blocks of the cervical spine, the level of difficulty and possible complications are significantly higher than

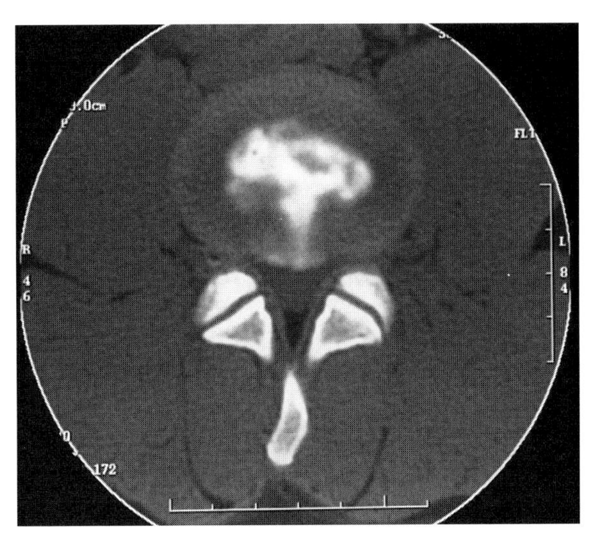

Figure 92–4. CT scan of the disk injected in Figures 92–1 through 92–3. There is a full-thickness posterior annular fissure extending to the disk margin.

with similar lumbar procedures. Cervical diskography should be performed only after extensive experience with other injection procedures.

References and Recommended Readings

Derby R, Howard HW, Grant JM, et al: Predictive pressure-controlled discography. Spine 24:364, 1999.

Guyer RD, Ohnmeiss DD: Contemporary concepts in spine care lumbar discography: Position statement from the North American Spine Society Diagnostic and Therapeutic Committee. Spine 20:2048, 1995.

Holt EP: The question of discography. J Bone Joint Surg Am 50:720, 1968.

Walsh TR, Weinstein JN, Spratt KF, et al: Lumbar discography in normal subjects. J Bone Joint Surg Am 72:1081, 1990.

CHAPTER 93

Percutaneous Needle Biopsy

Donald L. Renfrew, MD

OVERVIEW

Radiographically guided percutaneous needle biopsy of musculoskeletal lesions has a long history. Percutaneous needle biopsy has been shown to be diagnostically efficacious and cost-effective. Percutaneous intervertebral disk biopsy virtually always is performed to exclude disk infection and is described separately. The differential diagnosis of most other musculoskeletal lesions includes neoplasm and infection, and recommendations here are similar to those offered by Logan and colleagues (1996).

LUMBAR DISK

Patient Selection

Lumbar disk percutaneous needle biopsy is indicated for patients with a high suspicion of diskitis, either because of a strong clinical history (e.g., intravenous drug abuse, immunocompromise, severe unremitting nonmechanical back pain) and magnetic resonance (MR) imaging findings or because of highly suspicious MR imaging findings, back pain, and elevated erythrocyte sedimentation rate or C-reactive protein or both.

Procedure Description

The procedure is similar to lumbar diskography, with the following differences:

1. Before beginning the procedure, the patient may be given sedative or analgesic agents (because evaluation of pain response is not necessary).
2. Instead of a 22-gauge spinal needle, a larger coaxial needle set is used.
3. Before placement of the coaxial trocar, local anesthetic is placed along the path of the biopsy needle using a 5-inch, 25-gauge spinal needle.
4. The trocar of the biopsy kit is advanced into the margin of the disk, and aspiration is performed. If gross pus is aspirated, the aspirated material is sent for culture and sensitivity, and the disk should be injected with antibiotic and contrast material.
5. If no pus is aspirated through the trocar, the coaxial biopsy needle is placed through the trocar, and three to five biopsy specimens are obtained. The retrieved tissue is sent for culture and sensitivity and histopathologic analysis for degenerative versus inflammatory changes.
6. At the end of the procedure, a contrast material and antibiotic mixture is injected through the trocar (which remains at the margin of the disk throughout the procedure), and anteroposterior and lateral examinations are recorded.

SOFT TISSUE MASS

Patient Selection

Soft tissue mass percutaneous needle biopsy is indicated for patients with soft tissue masses. These masses may be discovered on physical examination but not be amenable to palpation-guided biopsy because of proximity to vital structures. Alternately, such lesions may be discovered on imaging procedures performed to evaluate for metastatic deposits or the cause of symptoms such as pain.

Procedure Description

Most of these procedures should be performed using computed tomography (CT) guidance. If several lesions are present (e.g., in a patient with suspected metastatic disease wherein one lesion must be confirmed histologically before therapy), choose the most easily and safely biopsied lesion. Position the patient, taking into account patient comfort and the mechanical necessities of needle placement within the scanning gantry. Place the patient so that scanning may be performed with the needle en route to the lesion. Provide patient sedation as necessary. With the patient in an appropriate position on the CT scanner table and the lesion and course of the needle established, mark the skin using the positioning laser light, and measure from an easily palpated or visualized landmark (e.g., the spinous process) or use a localizing grid placed on the patient's skin. After establishing the correct skin entrance site, use a 1.5- to 3.0-cm, 25-gauge straight needle to anesthetize the skin and subcutaneous soft tissues, and repeat the scan to confirm that the correct skin puncture site has been chosen (Fig. 93–1). Measure from the skin surface to the margin of the lesion. Place a needle stop at the appropriate position on the trocar of a coaxial biopsy device. Insert the trocar in 2 to 3 cm of soft tissue so that it remains in place when released, and repeat the CT scan to confirm that the trocar is on the correct course to the lesion. Advance the trocar in 1-cm intervals, with scanning on each advance, until the lesion margin is reached (Fig. 93–2). Remove the trocar stylus, and place the biopsy needle to the edge of the lesion. Fire the biopsy device, and perform a CT scan to document biopsy specimen location (Fig. 93–3). Culture the obtained specimens if infection is a clinical consideration. Place the specimens in appropriate containers for pathologic analysis. Obtain at least three specimens of tissue from the mass. After obtaining all specimens and removing the trocar, scan from several centimeters above to several centimeters below the lesion to document the absence of hematoma and other complications. If there is even a remote chance of penetrating the pleural space, obtain an expiratory midchest CT slice to confirm the absence of pneumothorax. Follow up with the referring physician in 3 to 5 days to discuss the histology of the sample obtained.

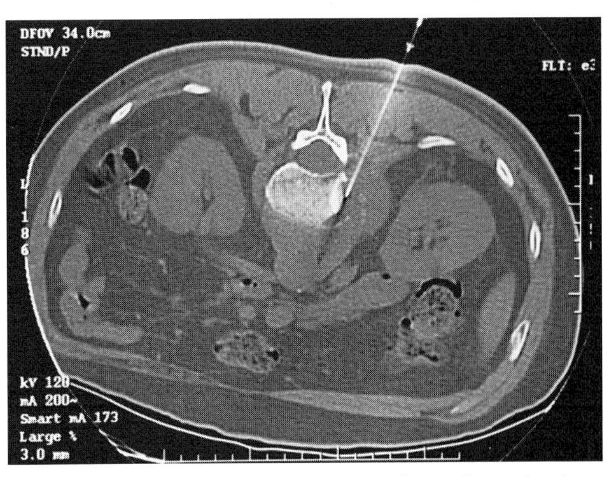

Figure 93–2. The trocar of the biopsy kit has been advanced and its tip is at the margin of the vertebral body.

BONE LESION

Patient Selection

Indications for bone lesion percutaneous needle biopsy are similar to those for soft tissue lesions.

Procedure Description

Use the technique described previously for bone lesions that consist of soft tissue with no overlying barrier of bone (between the chosen skin surface site and the lesion). For lesions that have a bony barrier between the lesion and

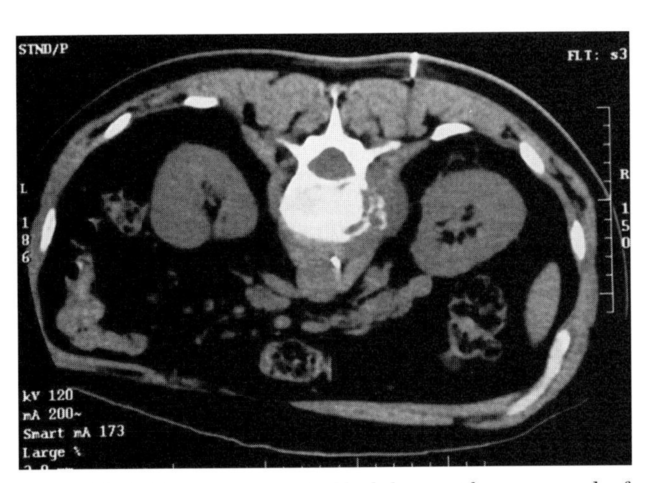

Figure 93–1. CT scan of a vertebral body lesion with an associated soft tissue mass. A 25-gauge needle used for local anesthesia is in place. This scan shows that the needle course should be adjusted slightly medially to intersect the vertebral body margin.

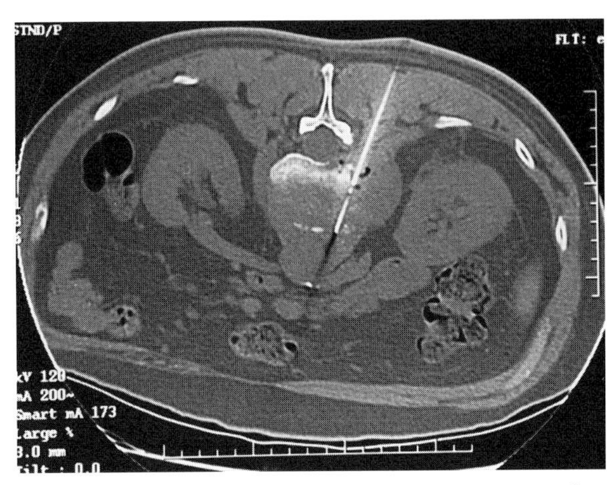

Figure 93–3. CT scan of the same case as in Figures 93–1 and 93–2 after firing the spring-loaded device. The needle has obtained a biopsy specimen from alongside the vertebral body and the peripheral aspect of the disk.

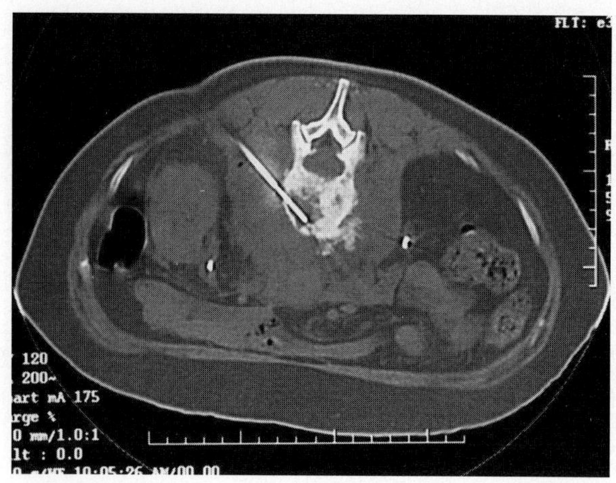

Figure 93–4. CT scan of a bone biopsy needle in place through the cortex of the vertebral body. The larger size of the needle results in greater artifact. The junction of the outer trocar and inner cutting needle is seen adjacent to the vertebral body margin.

chosen skin puncture site or that consist exclusively of bone, alter the biopsy technique as follows:

1. Use a special needle kit designed for bone biopsy.
2. Advance the trocar of the needle kit to the bony barrier, separating the lesion from the skin site (or the abnormality to be biopsied if the lesion consists entirely of bone).
3. Drive the biopsy needle through the lesion, which usually requires a twisting action on a handle. As with the use of spring-loaded biopsy devices, document needle location during biopsy with CT (Fig. 93–4). Samples should be handled per instructions of the pathologist. Save aspirated blood for analysis as well because such aspirated blood may contain diagnostic tissue. The tissue obtained with a bone biopsy device usually needs to be extruded forcibly from the device; the biopsy kits contain metallic plungers for this purpose.
4. When a hole in the bony barrier has been created, spring-loaded biopsy guns may be introduced coaxially through the trocar, and the biopsy procedure may proceed as with a soft tissue mass biopsy.

References and Recommended Readings

Fraser-Hill MA, Renfrew DL: Percutaneous needle biopsy of musculoskeletal lesions: 1. Effective accuracy and diagnostic utility. AJR Am J Roentgenol 158:809, 1992.

Fraser-Hill MA, Renfrew DL, Hilsenrath PE: Percutaneous needle biopsy of musculoskeletal lesions: 2. Cost-effectiveness. AJR Am J Roentgenol 158:813, 1992.

Hewes RC, Vigorita VJ, Freiberger RH: Percutaneous bone biopsy: The importance of aspirated osseous blood. Radiology 148:69, 1983.

Hopper DK: Percutaneous, radiographically guided biopsy: A history. Radiology 196:329, 1995.

Logan PM, Connell DG, O'Connell JX, et al: Image-guided percutaneous biopsy of musculoskeletal tumors: An algorithm for selection of specific biopsy techniques. AJR Am J Roentgenol 166:137, 1996.

White LM, Schweitzer ME, Deely DM: Coaxial percutaneous needle biopsy of osteolytic lesions with intact cortical bone. AJR Am J Roentgenol 166:143, 1996.

CHAPTER 94

Percutaneous Vertebroplasty

John B. Weigele, MD, PhD

RATIONALE

Percutaneous vertebroplasty is a promising new, minimally invasive therapy for painful vertebrae resulting from osteoporotic compression fractures and destructive benign and malignant tumors. Invented in France in 1984 by Gailbert and Deramond, the procedure virtually was ignored in the United States until more recently, when interest in vertebroplasty exploded in this country. The procedure involves injecting freshly mixed polymethyl methacrylate (PMMA) as a semisolid paste into a painful vertebral body through a percutaneously placed large-gauge needle. The PMMA solidifies as it polymerizes, strengthening and stabilizing the weakened vertebra. The mechanical effects are thought to provide pain relief and to prevent further vertebral collapse. Chemical and thermal effects from the PMMA, the solvent, and the exothermic polymerization reaction also may play roles in pain relief by destroying nociceptive receptors and nerves.

INDICATIONS

Percutaneous vertebroplasty initially was used in Europe to treat benign (painful hemangioma) and malignant vertebral body tumors (multiple myeloma, metastases). Avascular necrosis (Kümmell's disease), lymphoma, eosinophilic granuloma, and osteoporotic compression fracture also have been treated. To date in the United States, most vertebroplasties have been performed to treat osteoporotic compression fractures.

PATIENT SELECTION

The underlying pathologic process influences the selection criteria for vertebroplasty. Patients with a malignant tumor causing severe focal back pain are candidates if there is no epidural disease. Partial osteolysis of the posterior vertebral body wall is not a contraindication if there is no cord compression or epidural tumor. Neither previous nor intended radiation therapy is a contraindication. Large osteolytic malignant lesions occasionally are treated prophylactically to prevent vertebral collapse.

Patients with painful hemangiomas are candidates for vertebroplasty; however, nonaggressive, asymptomatic lesions do not require treatment. Aggressive hemangiomas may require alcohol ablation or surgery or both in addition to vertebroplasty. Vertebroplasty may be used for structural reinforcement of an involved vertebra before surgery.

Vertebroplasty initially was offered to patients with painful osteoporotic compression fractures after the failure of conservative medical therapy. The pain associated with an acute compression fracture typically resolves after 2 weeks to 3 months. Most of these patients can be managed successfully with initial immobilization, bracing, and analgesics, followed by graduated physical therapy and exercise. Some patients experience persistent, severe debilitating pain, however, that may last indefinitely and are candidates for vertebroplasty. There has been a growing trend to offer vertebroplasty earlier, in part because of the greater acceptance of the procedure, the low procedural morbidity, and the high probability of a positive therapeutic response and also in an effort to minimize the complications of conservative medical care related to immobilization, such as deep venous thrombosis, pulmonary embolism, and pneumonia. Some patients with acute osteoporotic compression fractures experience such severe pain that maximal narcotic therapy is ineffective. These patients may benefit greatly from an immediate vertebroplasty.

TECHNIQUE

Percutaneous vertebroplasty currently is an evolving technique, and many procedural variations are in use. The author's present protocol is as follows. Patients referred to the interventional neuroradiology service for evaluation are seen in an outpatient clinic. A directed history and physical examination are obtained, and relevant imaging studies are reviewed. Pain history, relevant medical and surgical conditions, and analgesic use are recorded.

The indications for vertebroplasty are assessed critically. The pain associated with a symptomatic osteoporotic compression fracture typically is localized at the level of the involved vertebra and described as deep and intense. Standing and bending usually exacerbate the pain. Radicular pain is atypical and should prompt a thorough search for another cause. On examination, pain elicited by palpation of the spinous process of the involved vertebra is a strongly supportive sign.

Serial plain films are reviewed to identify the age of the compression fracture. Magnetic resonance (MR) images are evaluated to exclude other causes of back pain, such as a herniated disk, degenerative disk disease, and foraminal or spinal stenosis. Bone edema helps identify a recent, symptomatic compression fracture. Metabolic activity on bone scan is a helpful marker of the compression fracture causing pain and highly predictive (93%) of a positive therapeutic response. Computed tomography (CT) is useful to evaluate retropulsion of bone fragments and cortical destruction. Potential contraindications to vertebroplasty include coagulopathies, greater than 20% canal compromise by retropulsed bone fragments, vertebra plana, systemic infection, and long-standing back pain (>1 year).

The author performs vertebroplasty on an outpatient basis. The patient is directed to fast after midnight. Informed consent is obtained, including the risks of infection, hemorrhage, fracture, and cement leakage possibly causing spinal cord or nerve root compression requiring urgent surgical decompression, pain, paralysis, pulmonary embolism, and death. Conscious sedation is administered with intravenous fentanyl and midazolam (Versed). Physiologic monitoring is performed with cardiac, pulse oximeter, and blood pressure recording. Nasal cannula oxygen is provided as required.

The procedure is performed under strictly sterile conditions. The patient is placed prone, and the overlying skin is prepared and draped in a sterile fashion. Intravenous cefazolin (Ancef), 1 g, is given. The symptomatic vertebra is identified fluoroscopically, and 1% lidocaine is infiltrated from the skin down to the periosteum over the pedicle. For a lumbar vertebroplasty, an 11- or 13-gauge Osteo-Site needle (Cook, Bloomington, IN) is advanced percutaneously through the pedicle into the vertebral body with biplane fluoroscopic guidance. An oblique course in the posteroanterior plane is used to direct the needle medially as it traverses the vertebra (Fig. 94–1*A*, *B*). The tip of the needle is positioned in the anterior third of the vertebral body near the midline (Fig. 94–1*C*, *D*). This positioning optimizes the chance of obtaining adequate filling through a single needle. For a thoracic vertebroplasty, the initial access is performed in an identical fashion with a 15-gauge Ostycut needle (C. R. Bard Inc, Covington, GA) and exchanged over a stainless steel guidewire for an Osteo-Site needle advanced coaxially over a Geremia needle (Cook, Bloomington, IN).

A vertebrogram is performed with 3 mL of iohexol (Omnipaque 300). If significant communication with the basivertebral plexus or paralumbar veins is visualized, the venous connection is blocked by instillation of 1 to 2 mL of autologous blood clot formed in vitro with a mixture of autologous blood, contrast material, and collagen fibers (Avitene, Davol Inc, Cranston, RI).

The cement is prepared immediately before use. Methyl methacrylate powder (Surgical Simplex P, Howmedica, Rutherford, NJ) is mixed in a 3:1 ratio with barium sulfate powder (Lafayette Pharmaceuticals, Lafayette, IN) in a plastic bowl. Chilled liquid methyl methacrylate (Surgical Simplex, Howmedica, Rutherford, NJ) is stirred until a semisolid paste is obtained with the consistency of pancake batter. The paste is loaded into the barrel of an injection device (Osteo-Force High Pressure Injector, Cook, Bloomington, IN) with 1-mL syringes. The device has a screw-type plunger and high-pressure tubing for connection to the needle. The PMMA is injected slowly and carefully into the vertebral body under continuous fluoroscopic observation (see Fig. 94–1*C, D*). The instillation is continued until the cement approaches the posterior wall of the vertebra (Fig. 94–1*E, F*). This usually takes 1 to 3 minutes. The injection is stopped if there is marked resistance, extravasation, or venous leakage (basivertebral, epidural, paralumbar veins). In some circumstances, the cement injection may be reattempted cautiously after allowing the cement in the vertebra to harden or after withdrawing the needle slightly. If the vertebral body is not filled sufficiently, the same procedure is repeated with a second needle placed through the other pedicle (Fig. 94–2).

After the procedure, the patient is kept on bed rest for 4 hours, then discharged home in the care of a responsible adult. Patients are encouraged to attempt to decrease analgesic use and to substitute nonsteroidal anti-inflammatory drugs for narcotic analgesics as tolerated. They return to the clinic in 3 weeks for a follow-up visit.

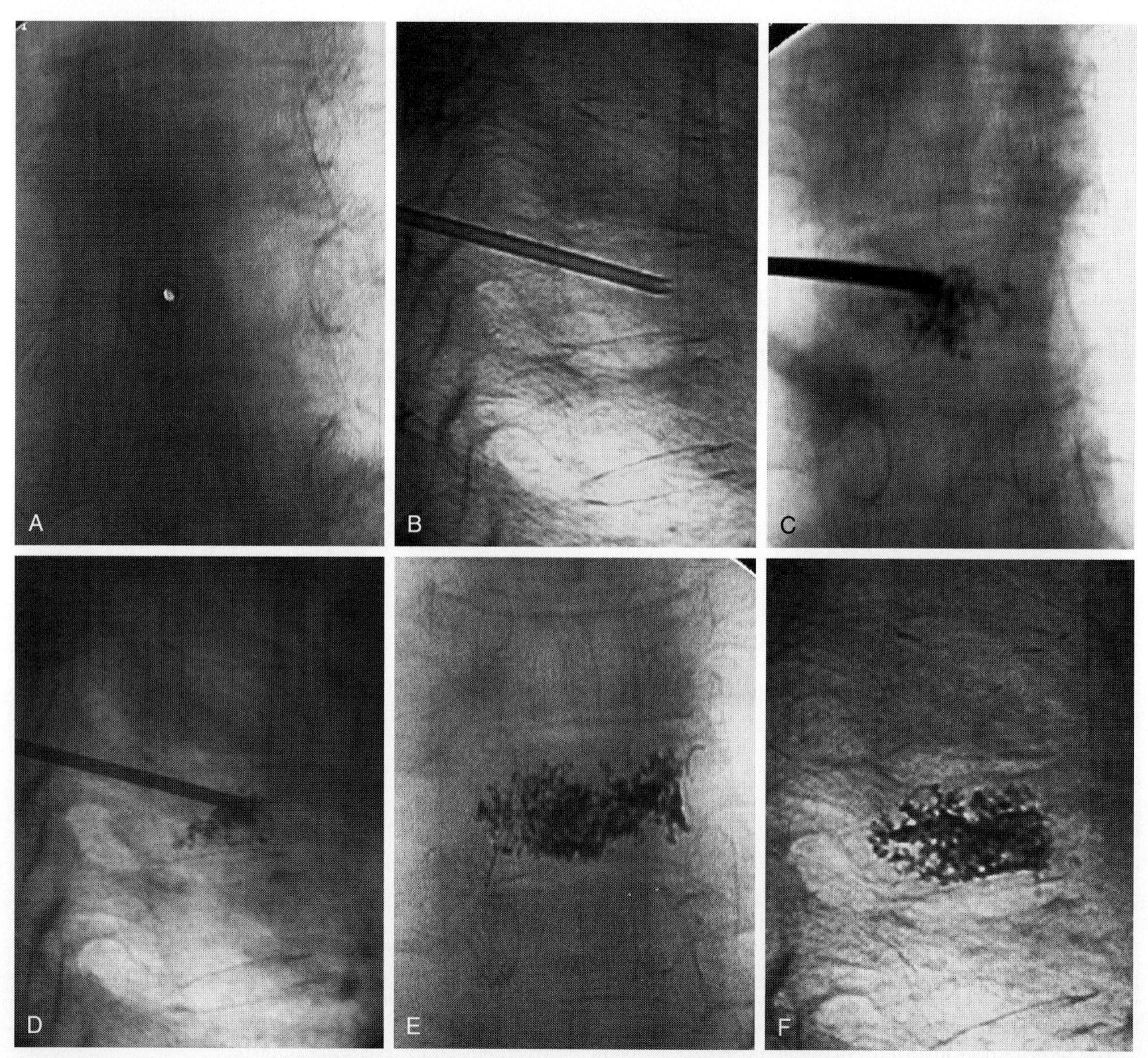

Figure 94–1. Vertebroplasty for a painful T6 osteoporotic compression fracture. *A* and *B*, Right anterior oblique and lateral radiographs show an oblique needle course through the pedicle into the anterior aspect of the vertebral body. *C* and *D*, Anteroposterior and lateral images during polymethyl methacrylate (PMMA) injection show the needle tip in the anteromedial aspect of the vertebra, optimal for filling the vertebra through a single needle. Note the cement crossing the midline of the vertebral body *(D)*. *E* and *F*, Final anteroposterior and lateral radiographs show most of the vertebral body homogeneously filled with PMMA.

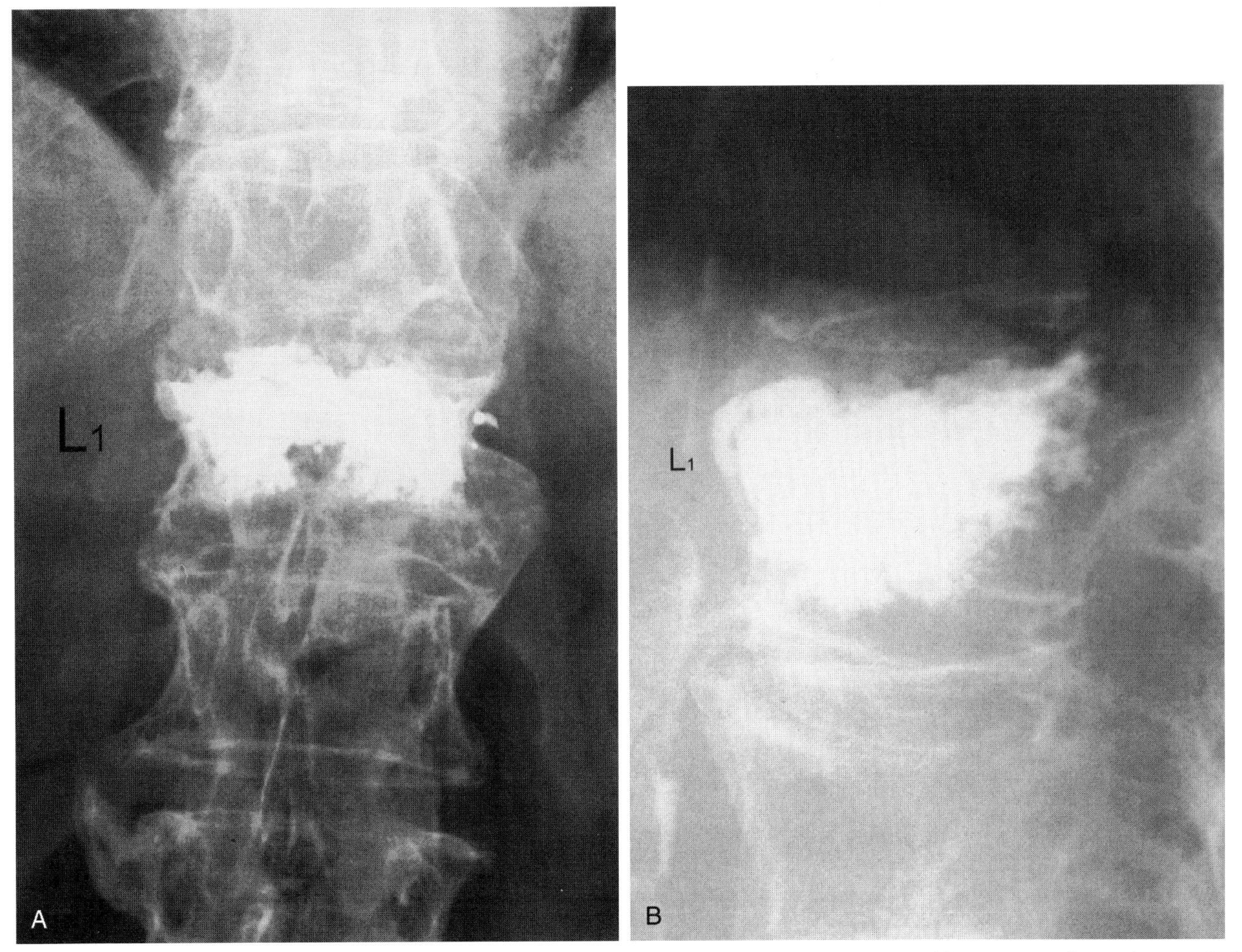

Figure 94–2. PMMA injected through both pedicles (two needle injections) in a 69-year-old woman with a painful osteopenic collapse of L1 vertebral body. AP and lateral views show good filling of the vertebral body with PMMA.

RESULTS

The most important potential benefit of vertebroplasty is the relief of pain. Most published clinical results are uncontrolled retrospective case studies. Prospective, randomized clinical trials are planned (Jensen and Dion, 2000). In large retrospective series, greater than 80% to 90% of patients with metastases and multiple myeloma (Cotten et al, 1998; Murphy and Deramond, 2000) and greater than 90% of patients with hemangiomas and osteoporotic compression fractures experienced significant or complete pain relief (Cotten et al, 1998; Deramond et al, 1998; Jensen and Dion, 2000). The therapeutic benefit usually occurred within a few days of treatment. Reported complication rates were 1.3% for treatment of osteoporotic compression fractures, 2.5% for treatment of hemangiomas, and 10% for treatment of metastases (Murphy and Deramond, 2000). Known complications have included nerve root and spinal cord compression, pulmonary embolism, fractures (rib, transverse process, and pedicle), paravertebral hematoma, epidural abscess, and sedation-related ventilatory problems.

For patients with osteoporotic compression fractures or myeloma, the therapeutic effect was not proportional to the amount of PMMA filling. There was marked pain relief even when the vertebra was not filled completely with cement. Epidural, intradiskal, and venous leaks in those patients were common but usually not clinically significant. Foraminal leaks occasionally caused nerve root compression requiring decompressive surgery. Most (75%) patients treated for painful metastases and myeloma reported continued pain relief at 6-month follow-up.

SUMMARY

Percutaneous vertebroplasty is a minimally invasive, safe, and effective technique to treat painful vertebrae caused by osteoporotic compression fractures and destructive benign and malignant tumors.

Recommended Readings

Cortet B, Cotten A, Boutry N, et al: Percutaneous vertebroplasty in patients with osteolytic metastases or multiple myeloma. Rev Rhum Eng Ed 66:177, 1997.

Cotten A, Boutry N, Cortet B, et al: Percutaneous vertebroplasty: State of the art. Radiographics 18:311, 1998.
Cotten A, Dewatre F, Cortet B, et al: Percutaneous vertebroplasty for osteolytic metastases and myeloma: Effects of the percentage of lesion filling and the leakage of methyl methacrylate at clinical follow-up. Radiology 200:525, 1996.
Deramond H, Debussche C, Pruvo JP, et al: La vertébroplastie. Feuillets de Radiol 30:262, 1990.
Deramond H, Depriester C, Gailbert P, et al: Percutaneous vertebroplasty with polymethylmethacrylate: Technique, indications and results. Radiol Clin North Am 36:533, 1998.
Gailbert P, Deramond H, Rosat P, et al: Preliminary note on the treatment of vertebral angioma by percutaneous acrylic vertebroplasty. Neurochirurgie 33:166, 1987.
Jensen M, Dion J: Percutaneous vertebroplasty in the treatment of osteoporotic compression fractures. Neuroimaging Clin N Am 10:547, 2000.
Murphy K, Deramond H: Percutaneous vertebroplasty in benign and malignant disease. Neuroimaging Clin N Am 10:535, 2000.

CHAPTER 95

Needle Procedures in the Peripheral Musculoskeletal System

Thomas D. Berg, MD, and Georges Y. El-Khoury, MD

OVERVIEW

Before any joint injection procedure, it is essential to place the patient in a comfortable position so that the patient can tolerate remaining still for the duration of the procedure. The joint or structure to be injected is localized fluoroscopically, and the entry site is marked on the skin. The skin is prepared with a povidone-iodine (Betadine) solution, which is allowed to dry. Some radiologists prefer to add alcohol to cleanse the area further. The skin and underlying soft tissues are infiltrated with local anesthesia.

HIP ASPIRATION AND INJECTION PROCEDURES

Hip aspiration to exclude infection is a commonly requested procedure for patients with native hips and those with total hip arthroplasty. Diagnosis of a native hip infection is considered an emergent procedure because intra-articular infection can lead to rapid lysis and destruction of cartilage, with underlying subchondral bone loss and resultant permanent disability. Early diagnosis is also essential in an infected total hip arthroplasty to help avoid additional bone destruction, which further complicates and delays a revision procedure.

To perform a native hip aspiration, the area is prepared and draped, and the skin is infiltrated with local anesthetic. A sandbag is used to place the foot in internal rotation, and the femoral pulses are palpated before needle placement. The needle position is guided under fluoroscopy (Fig. 95–1). A 20-gauge needle is advanced until it encounters the bone. The needle is rotated 180 degrees, and aspiration is attempted with a 10-mL syringe. If joint fluid is aspirated, anaerobic and aerobic cultures and cell count samples are obtained and sent immediately to the laboratory for evaluation. If no fluid is obtained, the

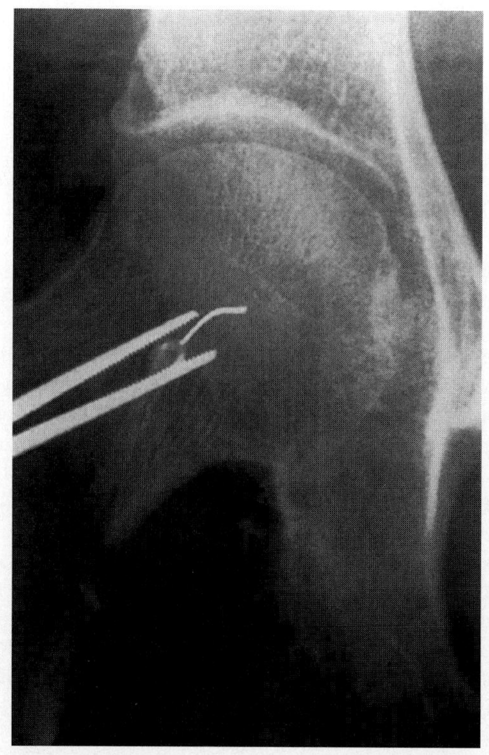

Figure 95–1. Anteroposterior view of the right hip shows correct needle position for native hip joint injection or aspirations.

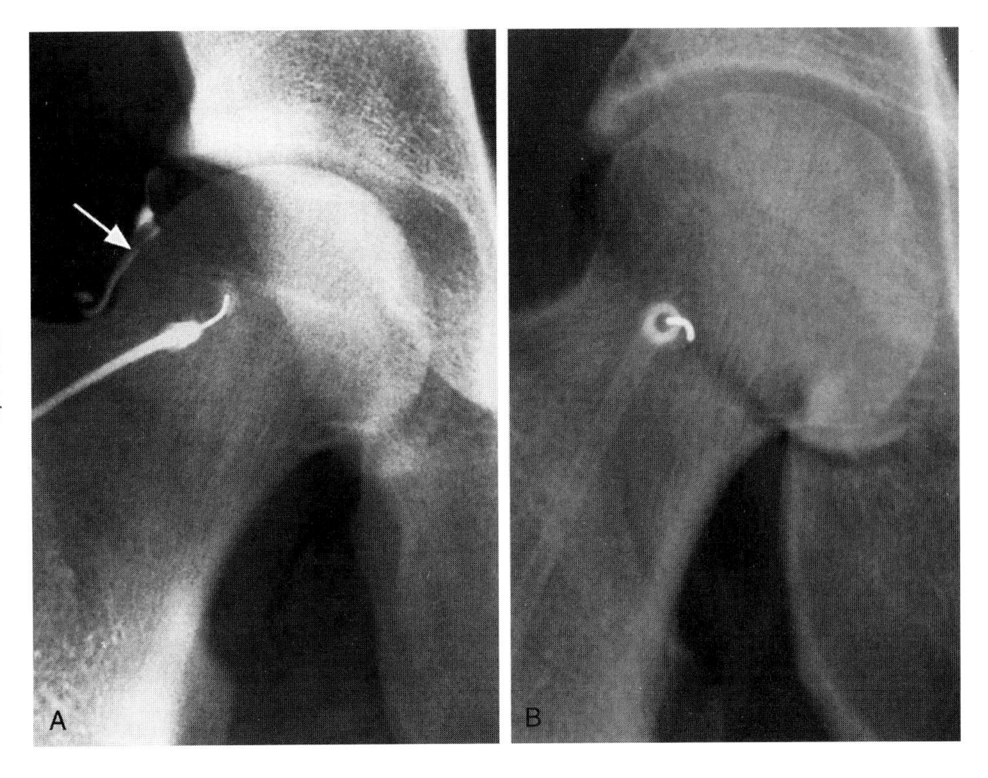

Figure 95–2. *A,* Contrast material is injected to ensure proper position of the needle *(arrow)*. *B,* Air arthrogram confirms the proper position of the needle.

needle is repositioned, and aspiration is reattempted. To confirm the intra-articular position of the needle, 3 to 4 mL of nonionic contrast material or air can be injected (Fig. 95–2). Preliminary data have shown no effect on bacterial growth or viability with usage of dilute iohexol (Omnipaque 180) on common gram-positive and gram-negative bacteria.

Diagnostic hip injection before total hip arthroplasty is an important diagnostic tool. The procedure is especially helpful when there is coexisting spine pathology, an atypical clinical presentation, or other diagnostic uncertainty regarding the origin of the pain. A 22-gauge needle is advanced under fluoroscopic guidance, and the femoral neck is targeted (see Fig. 95–1). Intra-articular position of the needle is confirmed by injecting 3 to 4 mL of contrast material or air (see Fig. 95–2). Approximately 7 to 10 mL of local anesthetic and steroid (0.25% bupivacaine [Marcaine] and betamethasone [Celestone]) is injected. Some advocate bupivacaine alone; however, the authors typically add 1 mL of betamethasone to 9 mL of bupivacaine for longer term relief for the patient during the period before surgery. The amount of steroid injected should not exceed 1 mL (6 mg) of betamethasone or roughly 25 to 30 mg of methylprednisolone (Depo-Medrol). The patient should be asked to record his or her symptoms on a scale of 1 to 10 before and immediately after the procedure. If the source of the pain is from the hip, the patient typically reports immediate pain relief. The pain intensity does not change after the injection if the source of the pain is the lumbar spine. The patient's response should be documented in the radiology procedure note, and the referring physician ideally should examine the patient within 1 hour of the injection.

Total hip arthroplasty aspiration is performed with a 20-gauge needle positioned as shown in Figure 95–3. The needle is advanced to the hub, with the tip advanced beyond (or posterior to) the hip prosthesis. Aspiration is attempted as the needle is withdrawn slowly. It is important to advance the needle fully to the hub because the fluid generally is located within the most dependent

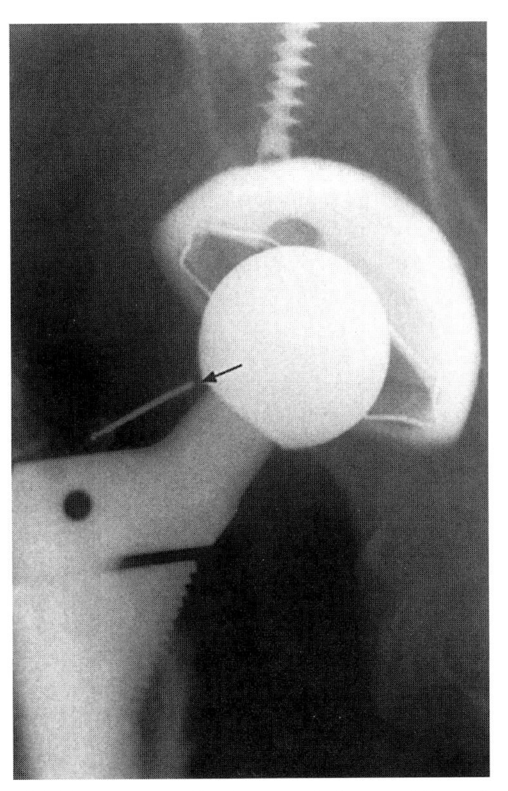

Figure 95–3. Correct needle placement before total hip arthroplasty aspiration. The *arrow* points to the proper position for the needle.

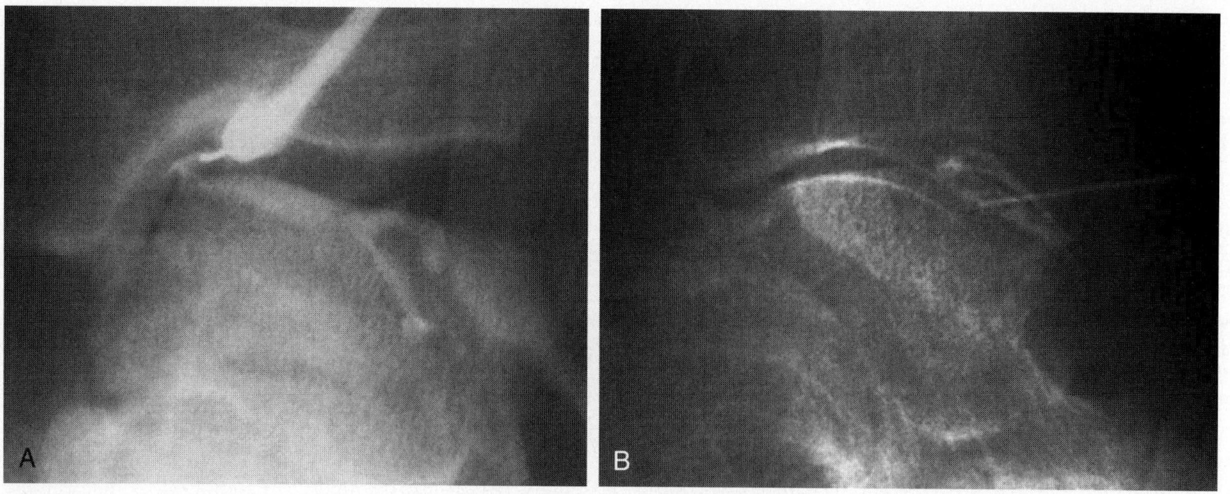

Figure 95–4. Ankle joint injection. *A*, Anteroposterior view. *B*, Lateral view, which is most helpful in accessing the position of the needle within the joint.

portion of the pseudocapsule. When fluid is obtained, it should be sent immediately for anaerobic and aerobic cultures and cell count.

INJECTIONS OF THE ANKLE AND FOOT

Chronic ankle and foot pain is a difficult clinical problem, with multifactorial causes ranging from post-traumatic degenerative joint disease to inflammatory or infectious causes. Accurate assessment of the source of symptoms can be difficult because of the close proximity of the different joints. Plain radiography and computed tomography (CT) do not always identify the source of pain. A normal-appearing joint at radiography does not preclude that joint as a source of pain. Diagnostic injections of the foot and ankle frequently are requested to assist in confirmation of clinically suspected symptoms related to a specific joint before surgical fusion.

The joint is localized with fluoroscopic guidance, the skin is marked, and the entry site is infiltrated with local anesthetic. The joint is entered with a 22-gauge needle, and the needle position is confirmed with 0.5 to 1 mL of contrast material. Bupivacaine 0.25% or lidocaine 1% is injected until resistance is encountered or the patient feels joint fullness. Occasionally, steroid is added to the anesthetic to treat an inflammatory joint condition.

Commonly requested sites for injection include the ankle; talonavicular, subtalar, and calcaneocuboid joints; os trigonum synchondrosis; and metatarsophalangeal joint. Occasionally a tenogram or a tendon sheath injection is requested.

The ankle joint (Fig. 95–4) is injected from an anterior approach. The patient is placed supine, with the plantar surface of the foot on the fluoroscopic table and ankle joint slightly flexed. To ensure easy access to the joint, the ankle is turned to the lateral position, and the needle is advanced under fluoroscopic control (Fig. 95–4*B*). Care is taken to avoid the extensor tendons and dorsalis pedis artery by palpating the injection site.

The posterior subtalar joint can be entered from a posterior lateral or lateral approach (Fig. 95–5). The lateral approach is probably the easier of the two and is shown in Figure 95–5*B*.

If the posterior approach is undertaken, the needle entry site is just lateral to Achilles' tendon (see Fig. 95–5*A*). Approximately 10% to 15% of subtalar joints communicate with the ankle joint. If this occurs, it should be documented, and pain response to the injection should be recorded.

The medial facet of the subtalar joint is entered from a medial approach by localizing the sustentaculum tali and placing the needle just superior to the bony protuberance. Occasionally a middle or anterior subtalar joint communicates with the talonavicular joint. If this occurs, it should be recorded, and the patient's response should be documented.

Before injecting the talonavicular joint, the extensor tendons and dorsalis pedis artery are palpated, and the skin is marked away from these structures. The talonavicular joint is localized best in the lateral view, and the 22-gauge needle is advanced under intermittent fluoroscopic visualization. A small amount of contrast material is injected to confirm the position (Fig. 95–6), and the joint is injected with 2 to 3 mL of a diluted solution containing anesthetic and steroid. Purely diagnostic procedures are performed exclusively with local anesthetic. The calcaneocuboid (Fig. 95–6*C*), navicular–first cuneiform (Fig. 95–6*D*), and tarsometatarsal joints (Fig. 95–6*E*) are approached in a similar fashion and under direct fluoroscopic control.

Metatarsophalangeal joint injections can be performed as diagnostic and therapeutic injections. The injections are performed with a 25-gauge needle, which is directed at the joint space from a dorsal approach assisted by fluoroscopy in a steep oblique or lateral projection. The proximal approach is important, especially in degenerative joints, because the overlying edge of the proximal phalanx impedes entrance to the joint. Approximately 1 mL of contrast material is used to confirm the intra-articular position of the needle. Then 1 to 2 mL of a 50% solution containing local anesthetic and steroid (0.25% bupivacaine and betamethasone) is injected. Occasionally, a metatarsophalangeal arthrogram is performed to exclude a capsular (plantar plate) rupture.

The os trigonum can be another source of posterior ankle pain, which worsens in dorsiflexion. The needle is directed fluoroscopically into the synchondrosis between the posterior aspect of the talus and the os trigonum. Studies have shown that patients with good response to the injection often have full symptomatic relief after resection of the os trigonum.

The peroneus longus and brevis tendon sheath injections are performed to diagnose stenosing tenosynovitis and confirm or treat symptomatically peroneal tendon disorders. The tendon sheaths are localized from the lateral side by placing the medial malleolus on the table with the lateral malleolus facing the radiologist. The distal fibula is palpated, and a 25-gauge needle is placed posterior and inferior to the fibular tip. Figure 95–7 shows the normal common tendon sheath and tendons as lucent filling defects. Stenosing tenosynovitis would appear as poor distention with multiple adhesions between the visceral and parietal surfaces of the tendon sheath (Fig. 95–8).

MISCELLANEOUS INJECTIONS

Some less commonly requested injection sites include the biceps tendon sheath, conjoint hamstring tendon, iliopsoas tendon sheath, and symphysis pubis, especially in athletes. The biceps tendon sheath is located within the intertubercular groove. Magnetic resonance (MR) imaging findings or clinical examination may localize the symptoms to the biceps tendon sheath. A diagnostic and therapeutic injection can help confirm patient symptoms and treat the inflammatory process. Figure 95–9 shows contrast material within the biceps tendon sheath.

Injecting the conjoint hamstring (Fig. 95–10) tendon should be performed under CT guidance, with caution to avoid the sciatic nerve. A 50% solution of 0.25% bupivacaine and betamethasone, 1 to 2 mL, is used for the injection.

The iliopsoas tendon injection may be performed for symptomatic relief of snapping iliopsoas syndrome. The iliopsoas tendon sheath is localized by placing a 22-gauge needle just proximal to the tendon insertion on the lesser trochanter. A 50% solution containing 0.25% bupivacaine and betamethasone, 2 to 3 mL, is injected after localization with 1 to 2 mL of nonionic contrast material. Figure 95–11 shows correct needle placement.

Symphysis pubis injections are performed targeting the center of the joint. The needle is advanced 3 to 5 mm after the tip hits the fibrocartilage, which feels firm. The injection contains 3 mL of 0.25 bupivacaine, followed by 1 mL (6 mg) of betamethasone or 30 mg of methylprednisolone (Fig. 95–12).

SINOGRAPHY

Sinograms are useful in assessing the extent of a sinus and whether it communicates with an abscess cavity or a joint. A stiff catheter typically is selected and is pushed until it meets resistance. Diluted contrast material (1:1) with saline is used for the injection. The opening of the sinus should be blocked or squeezed to prevent backflow of the contrast material. The injection is followed fluoroscopically, and spot filming is performed as needed (Fig. 95–13).

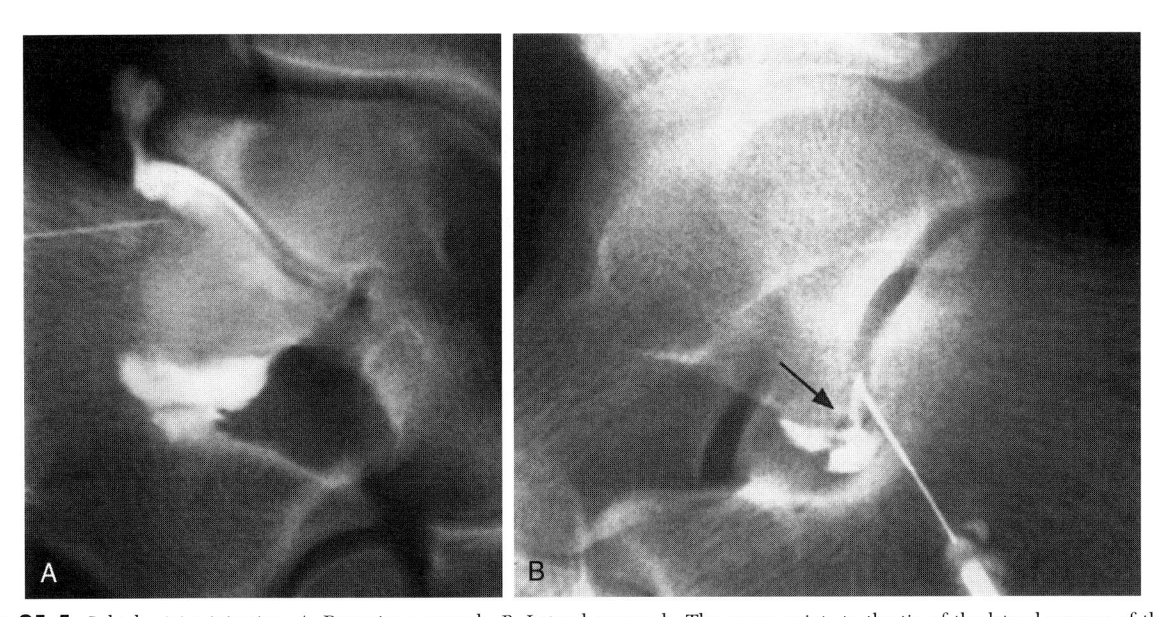

Figure 95–5. Subtalar joint injection. *A*, Posterior approach. *B*, Lateral approach. The arrow points to the tip of the lateral process of the talus.

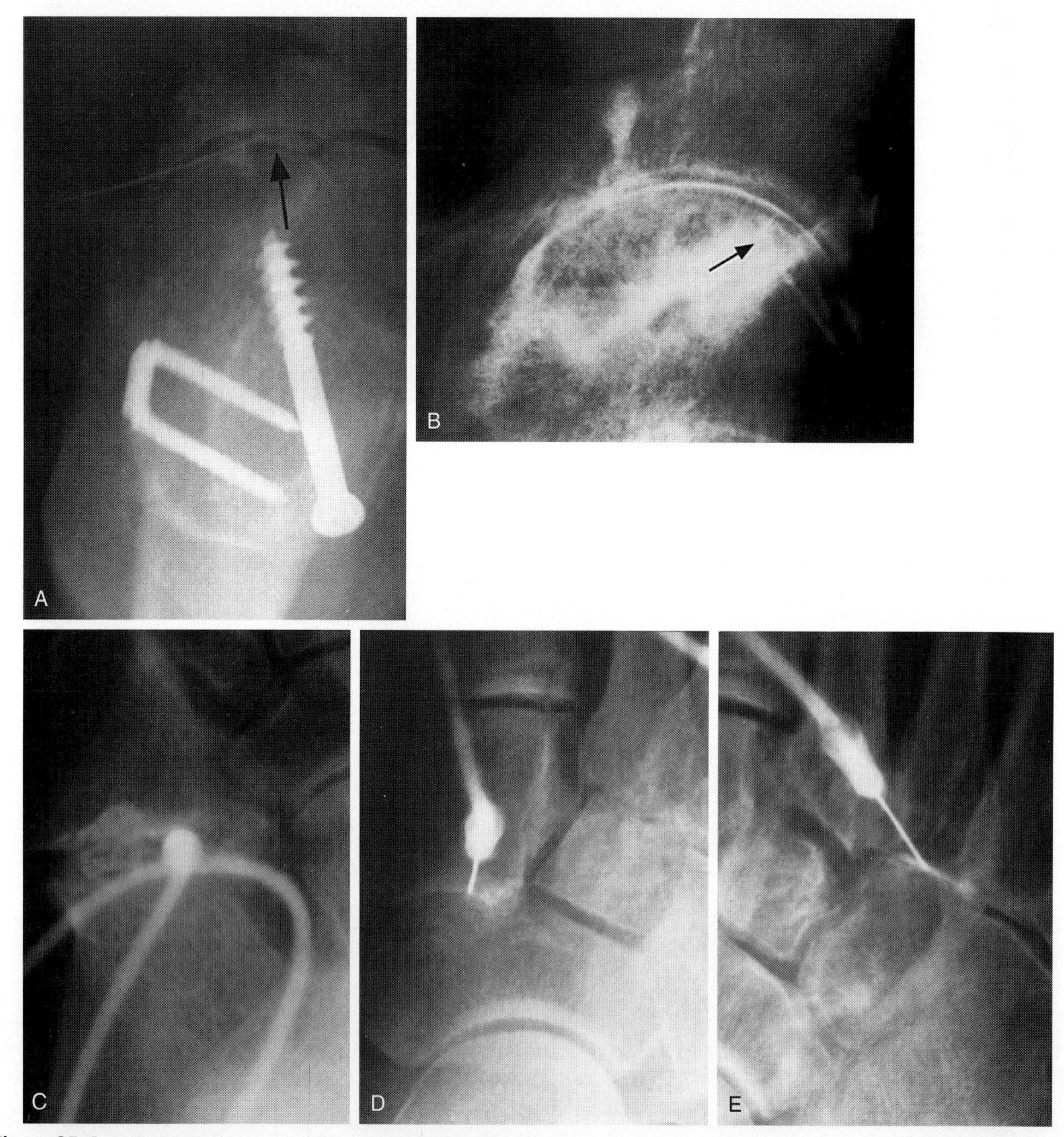

Figure 95–6. Midtarsal injections. *A,* Anteroposterior view shows the needle position within the talonavicular joint *(arrow)*. *B,* Steep oblique view of the talonavicular joint. The arrow is pointing to the contrast media flowing freely within the joint. *C,* Calcaneocuboid injection. *D,* Navicular first cuneiform injection. *E,* Lateral cuneiform third metatarsal joint injection.

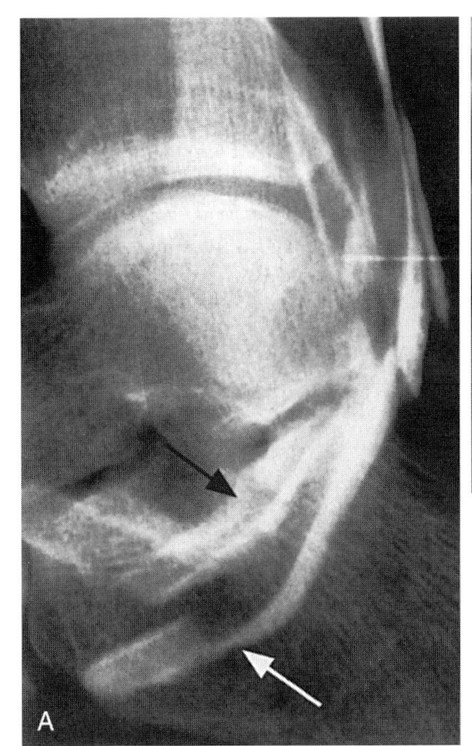

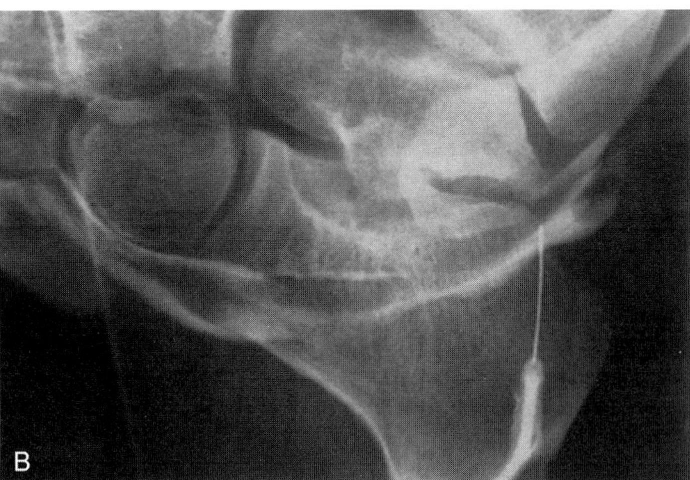

Figure 95–7. Tenograms of the peroneal tendons. *A*, Lateral view shows the needle in the peroneus brevis tendon sheath. Both tendon sheaths are filling: peroneus brevis *(black arrow)* and peroneus longus *(white arrow)*. *B*, Normal tenogram from another patient.

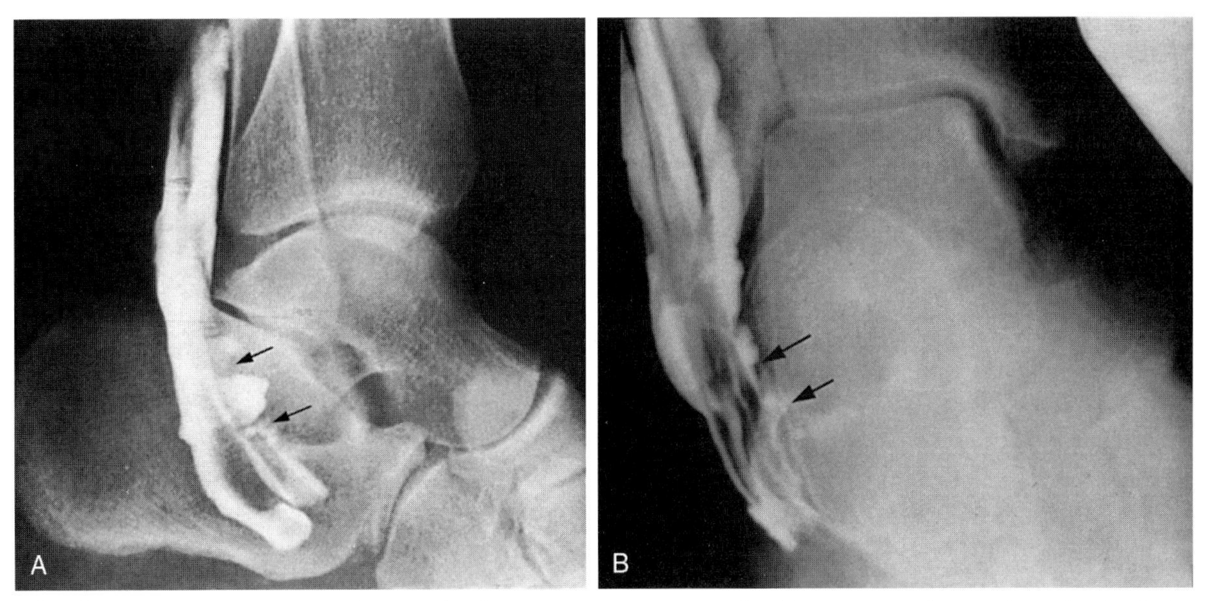

Figure 95–8. Stenosing tenosynovitis. *A* and *B*, Lateral and anteroposterior views. Note the adhesion and sacculation in the tendon sheaths *(arrows)*.

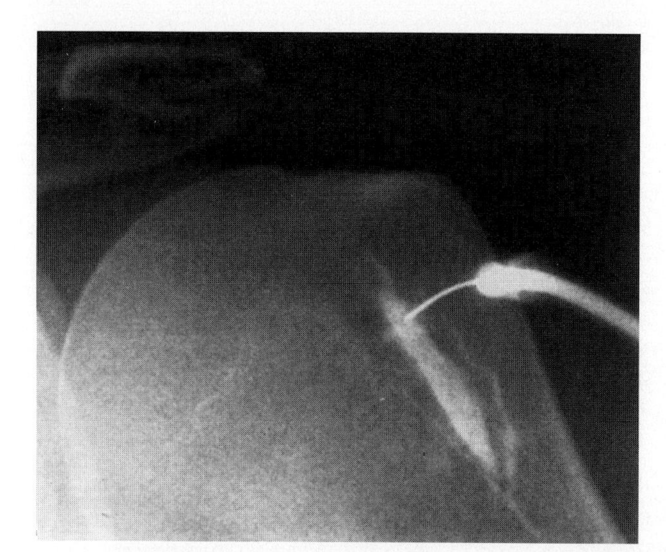

Figure 95–9. Bicipital tendon (longhead of biceps) sheath injection.

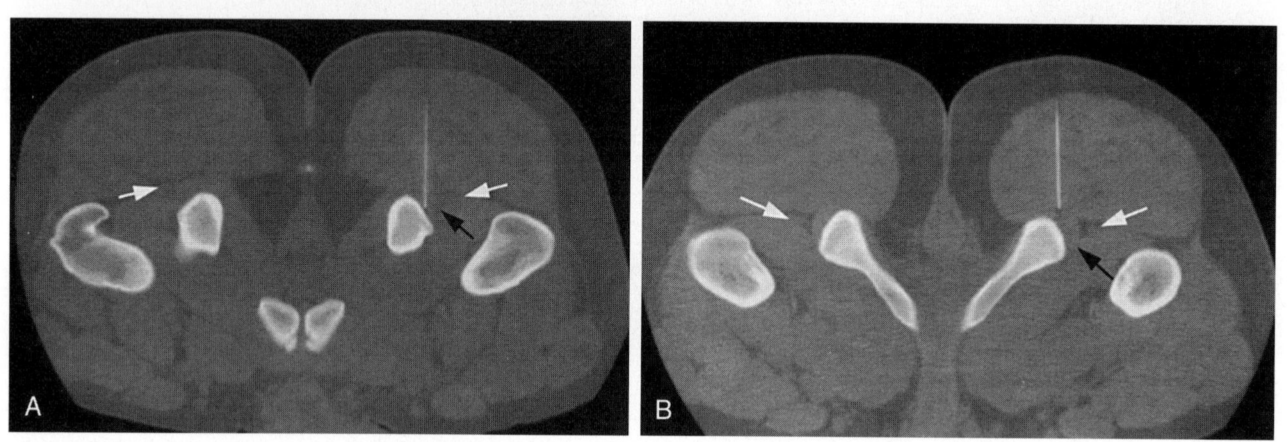

Figure 95–10. Injections into the origin of the hamstring. *A*, Injection into the proximal tendon. *B*, Injection into the more distal origin of the tendon. The *black arrows* point to the hamstring tendon, and the *white arrows* point to sciatic nerve.

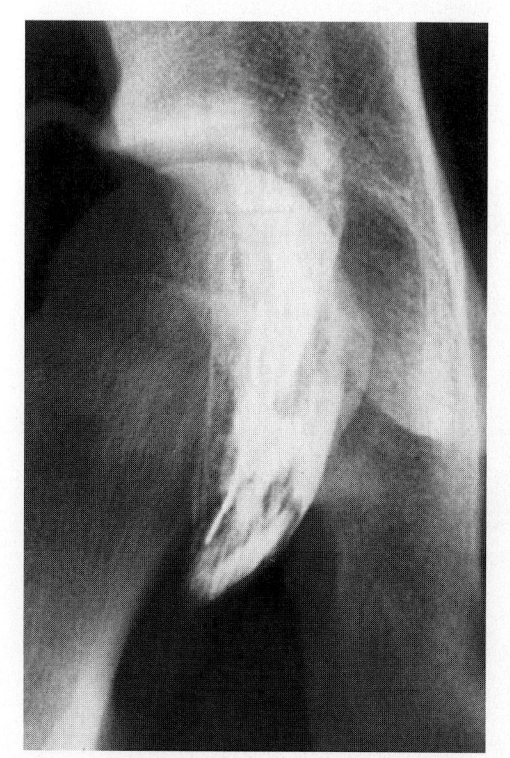

Figure 95–11. Contrast material injection illustrating opacification of the iliopsoas myotendinous junction.

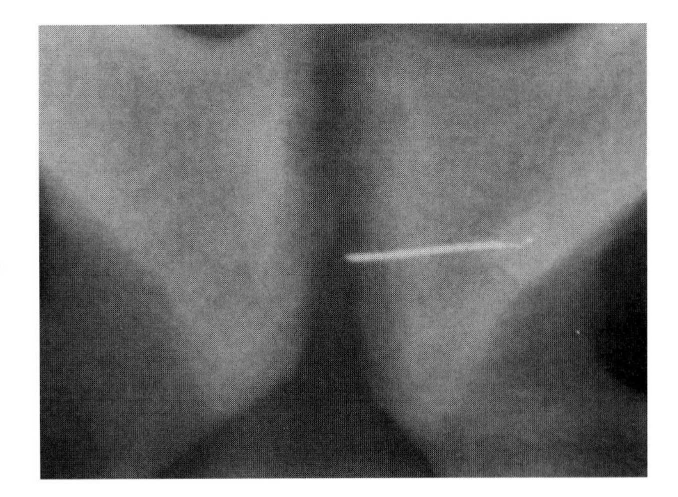

Figure 95–12. A 20-year-old college football player with pain in the symphysis pubis. Plain radiographs show irregularity and erosions at the symphysis. The symphysis was injected, and the patient had symptomatic relief.

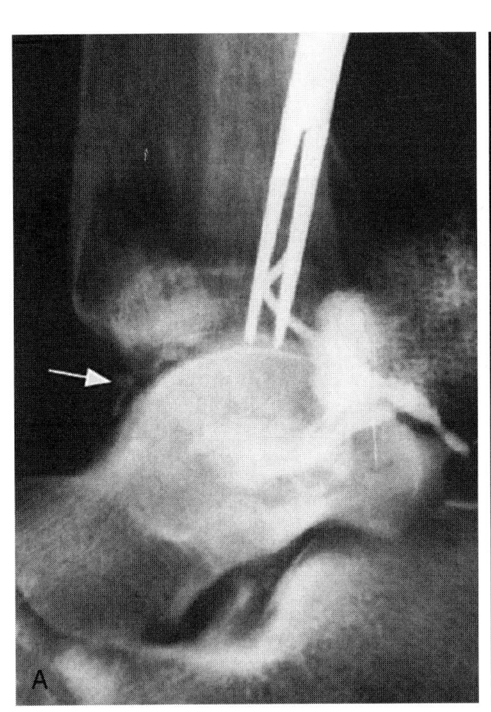

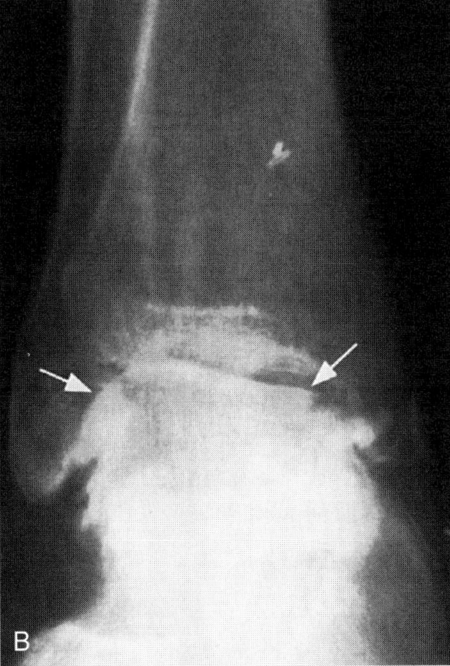

Figure 95–13. Sinography. A sinus was injected over the lateral aspect of the ankle. The sinus developed after a crush injury to the foot 1 year previously. *A,* Lateral view taken during the injection shows the contrast material within the ankle joint *(arrow). B,* Anteroposterior view taken after the injection was completed. The entire joint is outlined with contrast material *(arrows).*

Recommended Readings

Chow S, Brandser E: Diagnostic and therapeutic foot and ankle injections. Semin Musculoskeletal Radiol 2:421, 1998.

Crawford RW, Ellis AM, Gie GA, et al: Intraarticular local anesthesia for pain after hip arthroplasty. J Bone Joint Surg Br 79:796, 1997.

DeMichaelis B, Burrows L, Berg T, et al: Intraarticular injections of the ankle and foot. In Davies A, Jenkins J, et al (eds): Medical Radiology. New York, Springer-Verlag, 2000, p 8.

Gilula LA, Oloff LM, Caputi R, et al: Ankle tenography: A key to unexplained symptomatology: Part II. Diagnosis of chronic tendon disabilities. Radiology 151:581, 1984.

Hugo PC, Newberg AH, Newman JS, et al: Complications of arthrography. Semin Musculoskeletal Radiol 2:345, 1998.

Khoury NJ, El-Khoury GY, Saltzman CL, et al: Intraarticular foot and ankle injection to identify source of pain before arthrodesis. AJR Am J Roentgenol 167:669, 1996.

Kleiner JB, Thorne RP, Curd JG: The value of bupivacaine hip injection in differentiation of coxarthrosis from lower extremity neuropathy. J Rheumaztol 18:422, 1991.

Lucas PE, Hurwitz SR, Caplan PA, et al: Fluoroscopically guided injections into the foot and ankle: Localization of the source of pain as a guide to treatment—prospective study. Radiology 304:411, 1997.

Manusco CA, Ranawat CS, Esdaile JM, et al: Indications for total hip and total knee arthroplasties. J Arthroplasty 11:34, 1996.

Mitchell MJ, Bielecki D, Bergman AG, et al: Localization of specific joint causing hind foot pain: Value of injecting local anesthetics into individual joints during arthrography. AJR Am J Roentgenol 164:1473, 1995.

Nerr CS, Welsh RP: The shoulder in sports. Orthop Clin North Am 8:583, 1977.

Newberg AH: Anesthetic and corticosteroid joint injections: A primer. Semin Musculoskeletal Radiol 2:415, 1998.

Pavlov H: Talo-calcaneonavicular arthrography. In Freiberger RH, Aaye JJ, Spiller J (eds): Arthrography. New York, Appleton-Century-Croft, 1979, p 257.

Ruhoy MK, Newberg AH, Wodlowski ML, et al: Subtalar joint arthrography. Semin Musculoskeletal Radiol 2:433, 1998.

Teng MMH, Destouet JM, Gilula LA, et al: Ankle tenography: A key to unexplained symptomatology: Part I. Normal tenographic anatomy. Radiology 151:575, 1984.

CHAPTER 96

Percutaneous Radiofrequency Coagulation of Osteoid Osteoma

Thomas J. Gilbert, Jr., MD

OVERVIEW

Osteoid osteoma typically presents with chronic localized pain in patients 15 to 25 years old. The pain typically is worse at night and is relieved by salicylates. The imaging features of osteoid osteoma are typical, and in most cases biopsy is not required for diagnosis. A small (<2 cm) osteolytic lesion is seen, often with a small central radiodense nidus, and can be seen in medullary bone, in the cortex, or in a subperiosteal location. Variable amounts of sclerosis and periosteal new bone formation are seen. Cortical lesions within the diaphysis of a long tubular bone typically show marked new bone formation, whereas intramedullary lesions within an epiphysis may show little if any sclerosis or new bone formation. On a radionuclide bone scan, the lesion shows marked focal increased uptake, with a peripheral zone of more moderate uptake. On magnetic resonance (MR) imaging, the nidus can be difficult to identify and may have a variable appearance. Abnormal increased signal intensity invariably is seen on T2-weighted images (using spin-echo, fast spin-echo, or STIR sequences) within adjacent medullary bone, overlying periosteum, and adjacent soft tissues. Epiphyseal lesions may be associated with fluid or synovial hyperplasia within the adjacent joint.

Long-term medical treatment for osteoid osteoma has been described, but it is not tolerated by most patients. Surgical removal can be complicated by difficulty with intraoperative localization of the nidus or may require an en bloc resection. Percutaneous radiofrequency ablation has been described, and its success has been shown to be equivalent to surgery.

PATIENT SELECTION

Before the procedure, the patient's clinical history and imaging examinations should be reviewed. Many patients who are referred for ablation have an alternative diagnosis, such as cortical osteomyelitis or Brodie's abscess. The radiolucent component of these lesions typically is less well circumscribed and may be serpiginous. These lesions also may show a central photopenic area on radionuclide bone scan. Benign cortical irregularities and benign cystic lesions can be confused with osteoid osteoma in patients with a suggestive pain pattern. These lesions should not show an increase in uptake on radionuclide bone scan and should show normal signal intensity in adjacent bone or soft tissues on MR imaging. In patients with atypical or indeterminate radiographic appearance, radiofrequency ablation still might be considered if the patient's pain pattern is classic. Patients with atypical clinical and radiographic features should be referred for biopsy or excisional resection.

PROCEDURE

Radiofrequency ablation can be performed with local anesthesia if the lesion is peripheral. General anesthesia is required for more central lesions. The procedure is performed with computed tomography (CT) guidance (Fig. 96–1). Core biopsy initially is performed using an Ackermann bone biopsy needle and coaxial technique (Fig. 96–2). The biopsy needle is removed, and a radiofrequency needle with a 5-mm exposed tip is placed in the needle tract through the trocar. The needle position is confirmed with CT, and radiofrequency ablation is performed at 90°C for 6 minutes (Fig. 96–3). Care should be taken that the entire lesion be included in the 6-mm burn radius, and if the lesion is larger than 12 to 13 mm in

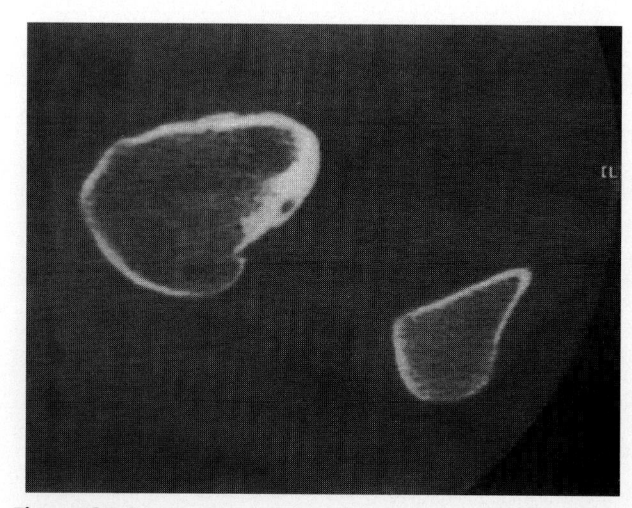

Figure 96–1. CT section through the femoral neck shows a medial cortical osteoid osteoma within the posterior aspect of the right femoral neck.

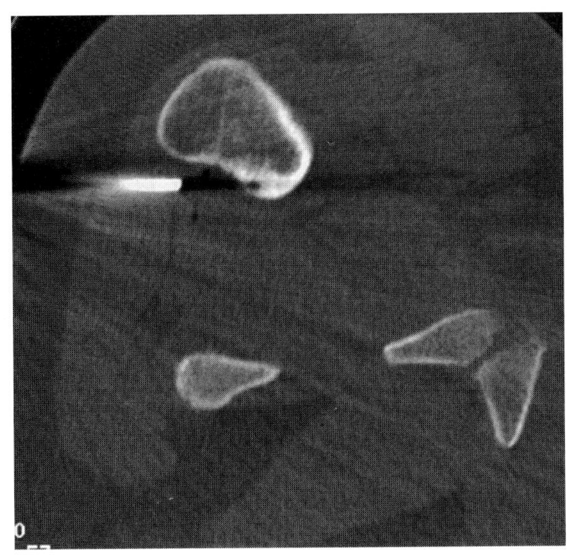

Figure 96–2. CT section with the trocar in place to assess the trajectory for the biopsy.

diameter, two separate burns should be made. Bupivacaine 0.25%, 2 mL, is injected into the biopsy site, and the needle is removed.

POSTOPERATIVE INSTRUCTIONS AND FOLLOW-UP

Patients are given an outpatient prescription for immediate postoperative pain control because some may experience significant discomfort on the same evening of the procedure. Most patients experience significant improvement in symptoms by 12 to 24 hours, although 30% experience residual pain or discomfort that is presumably related to the biopsy. These symptoms are typically less severe than their preoperative pain and does not show the same response to salicylates. These symptoms usually resolve over a period of weeks. Patients are instructed to avoid high-impact activities to minimize the risk of a stress fracture at the site of ablation. The recurrence rate is 9%.

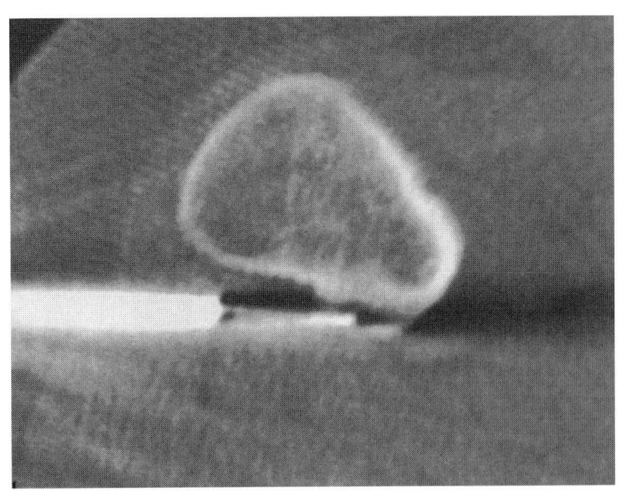

Figure 96–3. Radiofrequency needle positioned within the lesion immediately before ablation.

Recommended Readings

Kneisl JS, Simon MA: Medical management compared with operative treatment for osteoid-osteoma. J Bone Joint Surg Am 74:179, 1992.

Rosenthal DI, Hornicek FJ, Wolfe MW, et al: Percutaneous radiofrequency coagulation of osteoid osteoma compared with operative treatment. J Bone Joint Surg Am 80:815, 1998.

SECTION X

IMAGING OF MAJOR PROSTHESES

CHAPTER 97

Imaging of Total Hip Replacement from the Surgeon's Perspective

John J. Callaghan, MD

OVERVIEW

The total hip arthroplasty operation was developed in Europe in the late 1950s but was not widely accepted in the United States until after the mid-1970s. Today, more than 150,000 hip replacements are done each year in the United States alone. Radiologists frequently are asked to image and evaluate total hip arthroplasties. Radiographic criteria to assess proper placement of the prosthetic components are well established (Fig. 97–1). The hip prosthesis is a durable construct, and a well-done replacement is expected to last more than 15 years. Complications are suspected clinically and confirmed on imaging studies and joint aspiration. Major complications include mechanical loosening, wear of the polyethylene liner, histiocytic reaction (also called *osteolysis, particle disease,* or *foreign body reaction*), and infection.

Historically the early hip replacements all were fixed with lucent cement. Polyethylene was used as the surfacing material on the acetabular side, and a femoral stem was placed in the medullary canal of the femur. In the early 1970s, barium was added to the cement, making

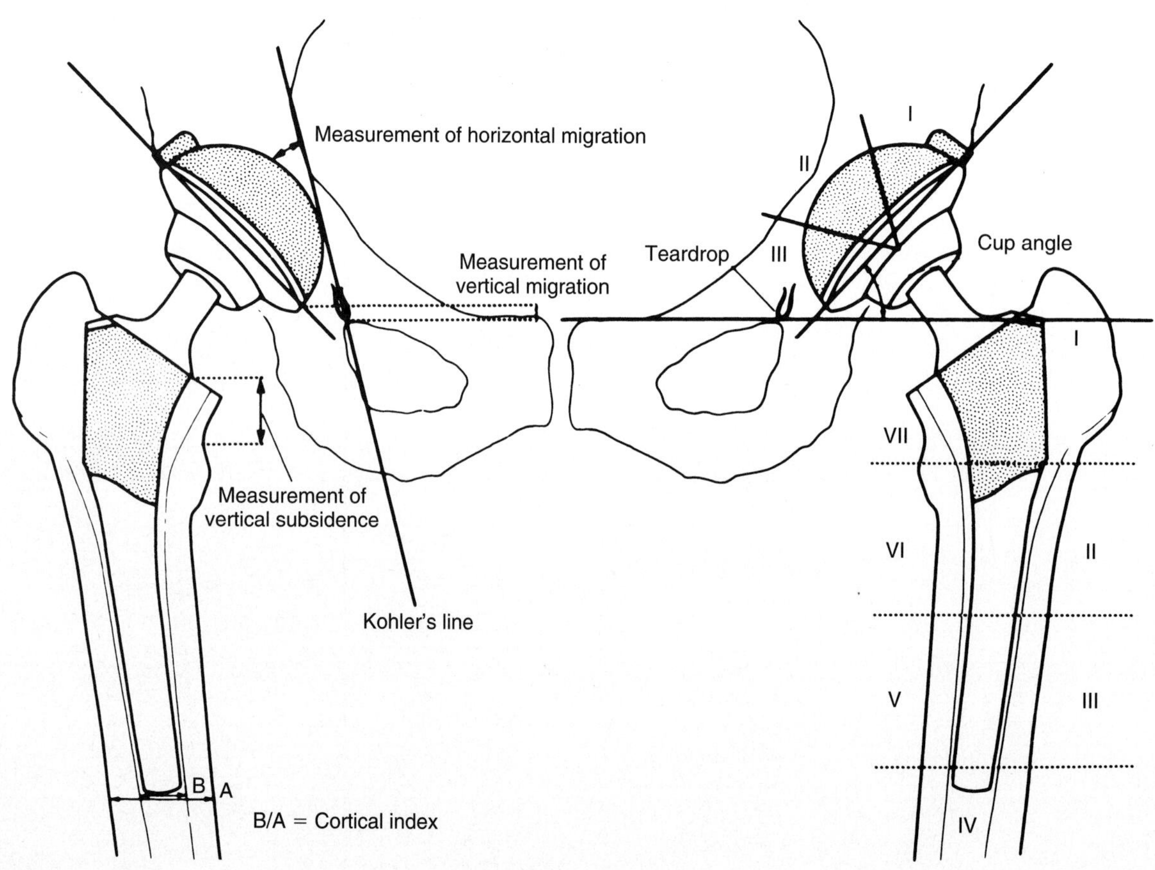

Figure 97–1. Landmarks and measurements used to evaluate total hip replacement. On the acetabular side, the relationship of the acetabular component to the teardrops is used. On the femoral side, the relationship of the proximal medial femoral component to the lesser trochanter is used. Radiolucencies and osteolysis are documented in three zones around the acetabular component and seven zones around the femoral component. The acetabular component and proximal third of the stem are depicted as being porous.

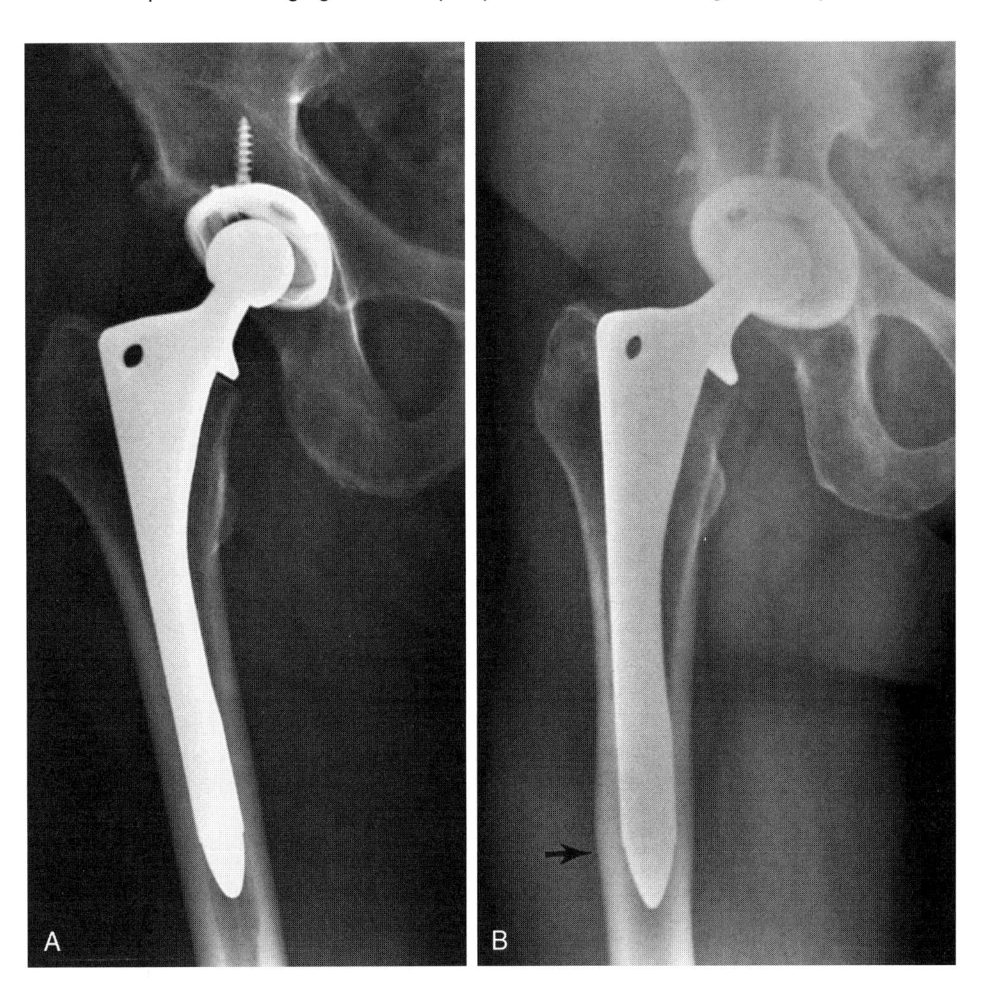

Figure 97–2. Secure cementless components. Postoperative *(A)* and 6-year follow-up *(B)* anteroposterior radiographs of the right hip show secure components with no radiolucent lines around the acetabular or femoral components. Note the distal cortical hypertrophy around the tip of the stem *(arrow)*.

it radiopaque and much easier to evaluate. Because of cement failure, cementless components were developed later whereby porous surfaces were created on the components so that bone could grow into these surfaces. On the acetabular side, screws and lugs were added to secure the acetabular component to the bone. The porous or ingrowth surfaces were made by attaching metal beads and meshes to the stems. Some stems had beaded surfaces throughout their length, whereas others had beaded surfaces or mesh only in the upper third of the stem (see Fig. 97–1). Some stems had porous surfaces circumferentially around the stem, whereas others had only patches of ingrowth surfaces.

The metallurgy used in the original cemented devices on the femoral side included stainless steel and chrome-cobalt alloys. In the mid-1980s, titanium became popular in total hip arthroplasty. Titanium devices are half as stiff as the stainless steel and chrome-cobalt devices, producing less bone loss resulting from stress shielding. Stress shielding is related to the fact that material used in the stem is 10 to 20 times stiffer than bone, and loads are transferred from the bone to the metal leading to bone atrophy. In the mid-1980s, modularity became available whereby femoral heads are manufactured independent of the stem. This development enabled the surgeon the versatility of selecting femoral heads of different sizes and metallurgy. For this reason, the density of the femoral head may be different than that of the stem. Currently the most common combination is a less radiodense titanium stem and a more radiodense chrome-cobalt femoral head.

In the late 1970s, metal-backed acetabular components with a polyethylene liner became available. Initially, prosthetic acetabula were cemented and nonmodular, but later they became cementless and modular whereby the plastic liner is captured by the metal shell intraoperatively (Fig. 97–2). Some of the capturing mechanisms have a wire, which can be detected radiographically. An understanding of these various metallurgic and design features can help the radiologist evaluate the radiographs of total hip prostheses better.

IMAGING OF THE TOTAL HIP PROSTHESIS IN THE POSTOPERATIVE PERIOD

In the immediate postoperative period (e.g., in the recovery room), a low anteroposterior view of the pelvis and a cross-table lateral view are taken (both views should include the acetabular and femoral components down to the tip of the stem). These views are obtained to rule out intraoperative complications, particularly fractures, dislocations, and perforation of the femoral cortex, all of which require immediate attention.

At 6 weeks postoperatively, anteroposterior and cross-table lateral views are obtained as a baseline examination for long-term follow-up. The ideal position of the acetabular cup on the anteroposterior view is 35 degrees to 50 degrees slant of the lateral opening in relation to the pelvis (see Fig. 97–1). On the cross-table lateral view, a mild amount of anteversion is desirable (10 degrees to 20 degrees). There is, however, no universally accepted method on assessing acetabular anteversion. For a cementless prosthesis, the stem of the femoral component ideally fills the medullary space. It should be placed in a neutral position with respect to the long axis of the femur (Fig. 97–2*A*). If the prosthesis is cemented, the stem should be placed in a neutral position and surrounded by a uniform 2 to 3 mm of cement.

LONG-TERM FOLLOW-UP OF CEMENTED PROSTHESES

All devices implanted in the United States initially were cemented on the acetabular and the femoral sides. The most important radiographic sign in determining if a component is fixed or loose hinges on whether it moved over time. To evaluate a total hip arthroplasty optimally, the reviewer should have serial radiographs dating back to the immediate postoperative examination. Optimally, comparable radiographic views should be available throughout the follow-up interval. Investigators have shown that changes in the position of the components of 5 mm or less should be attributed to differences in the radiographic technique.

On the acetabular side, the teardrop of the acetabulum (see Fig. 97–1) or, if the teardrop is not present, the superior rim of the obturator foramen can be used as a reference point to determine superior migration of the acetabular component. To assess medial migration of the acetabular component, a useful landmark is a line drawn perpendicular to the line connecting the teardrops on either side. Movement of the center of the acetabular shell medially or superiorly indicates migration and loosening of the implant (Fig. 97–3).

The center of the prosthetic femoral head should be identified, and measurements are made to the acetabular shell. Migration of the head superiorly over time indicates wear of the polyethylene liner (Fig. 97–4).

Another important evaluation in the follow-up of the cemented acetabular component is whether or not lucencies between the cement and bone have developed and progressed with time. It is rare to see a radiolucency between the acetabular component and cement; however, even a minor lucency here suggests component loosening. Cracks in the cement mantle are a sign of loosening of the prosthesis. In reporting these complications, the radiologist should specify the location according to the three zones described by DeLee and Charnley (1976). The lateral third of the sphere is zone 1, the middle is zone 2, and the medial third is zone 3 (see Fig. 97–1). Radiolucencies are evaluated for their thickness and whether or not they progress. Progressive radiolucencies greater than 1 to 2 mm, especially when circumferential, are thought to represent loosening of the implant. In one study, 95% of the acetabular implants that had a circumferential radiolucency at the bone-cement interface were loose at the time of surgery.

On the femoral side of the construct, there is no definite landmark to evaluate for migration of the component. The tops of the lesser trochanters have been used for this purpose by measuring the distance from the proximal medial aspect of the femoral component, which may have a collar, to the top of the lesser trochanter. If a metal wire has been implanted in the lateral aspect of the femur or greater trochanter, it also can be used as a reference point to determine movement of the prosthesis (Fig. 97–5). In general, a reduction in the distance between the medial aspect of the implant and the lesser trochanter over time

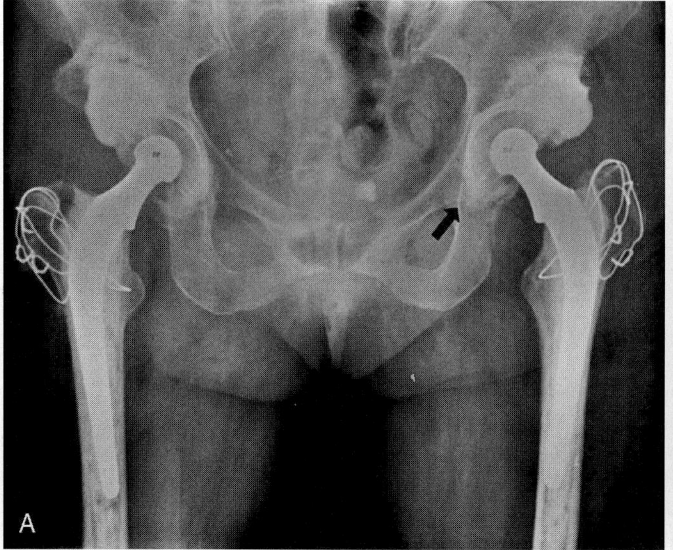

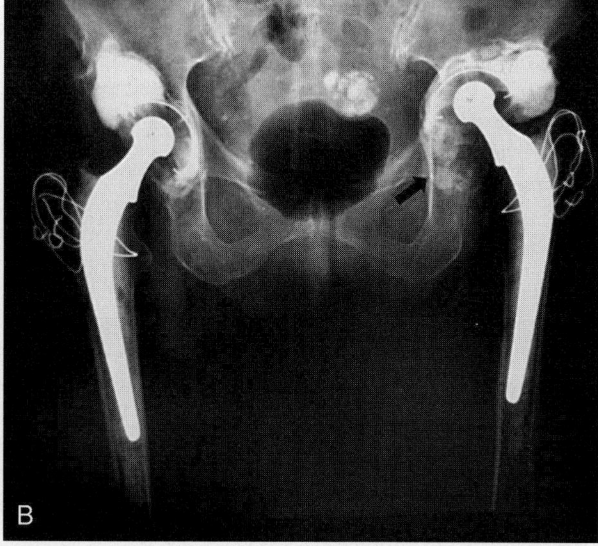

Figure 97–3. Medial and superior migration of the acetabular component. Postoperative (*A*) and 25-year follow-up (*B*) anteroposterior radiographs of the pelvis in a patient with left acetabular component migration. Note the change in the relationship of the teardrop (*arrow*) to the inferior border of the component. The acetabular component has migrated superiorly and medially.

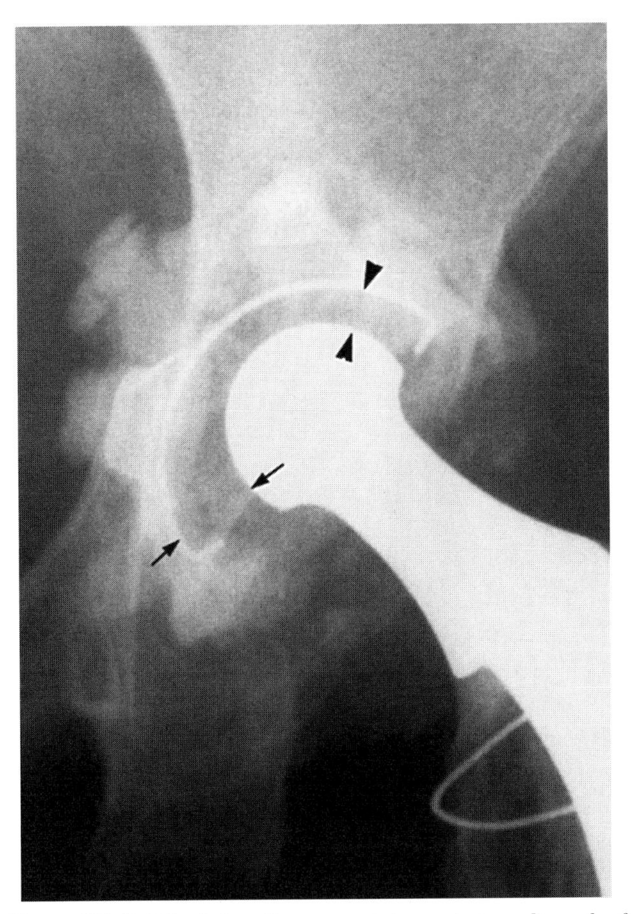

Figure 97–4. Polyethylene liner wear. Anteroposterior radiograph of the left hip shows thinning of the polyethelene liner superiorly *(arrowheads)*. The femoral head normally should be equidistant from the wire marker in the acetabulum *(arrows)*.

is a reliable measure for subsidence (or settling) of the femoral component (see Fig. 97–5).

Cement failure, which manifests radiographically as cracks in the cement mantle, is a common mode for loosening of a femoral component (Fig. 97–6). Other radiographic criteria for loosening include a radiolucent line between the stem and the cement, which is usually seen superolaterally (zone 1) (see Fig. 97–6). As with the acetabular component, determinations of progressive bone-cement radiolucencies also are evaluated. The location of these lucencies should be documented as being in one or more of the seven zones around the femoral component (see Fig. 97–1). Zone 1 is the area of the greater trochanter, zone 2 refers to the lateral midstem, zone 3 refers to the distal lateral stem, zone 4 is around the tip of the prosthesis, zone 5 is around the distal medial aspect of the prosthesis, zone 6 is at the midstem medially, and zone 7 refers to the proximal medial aspect of the stem. The presence of a circumferential bone-cement lucency around the femoral component, especially when the lucency measures more than 1 to 2 mm thick or shows progression over time, is highly suggestive of loosening (see Fig. 97–6).

Remodeling of bone related to the use of a stiff metal implant is difficult to detect on the acetabular side; however, on the femoral side it is fairly easy to detect. Because the metal implant is much stiffer than bone, it is common to see a decrease in the bone density in the proximal femur over time. This decrease is especially apparent in the medial femoral cortex and is known as *stress shielding*. Stress shielding is a positive sign for stability in cemented (Fig. 97–7) and cementless implants because it shows that the load has been transferred from the bone around the proximal implant distally (see Fig. 97–7). Another radiographic proof of secure fixation is the presence of cortical hypertrophy around the tip of the stem, which is further evidence of stress transfer (see Fig. 97–7).

An important finding in the radiographic evaluation of the femoral component relates to the fact that with advancing age, the cortex of the femur loses endosteal bone, and the medullary canal expands (Fig. 97–8; see 97–3). As the medullary canal widens, an apparent lucency can be seen between the cement and bone, which should not be misinterpreted as loosening of the stem (see Figs. 97–3 and 97–8). Components stay secure during this endosteal canal widening because a neocortex forms at the bone-cement interface. This phenomenon has been documented on autopsy specimens.

LONG-TERM FOLLOW-UP OF CEMENTLESS PROSTHESES

Cementless acetabular and femoral components have been widely used only since the mid-1980s. Radiographic criteria for the evaluation of stable or secure fixation of cementless components have been developed. For the femoral component, the principal criterion for stable fixation relates to absence of movement in relationship to the lesser trochanter during the follow-up period. Another sign of stable fixation is the absence of radiodense lines around the porous-coated portion of the implant, although these lines are common around the smooth portion. A third sign relates to bone formation at the junction of the porous coating and the smooth portion of the implant. This sign is called *spot weld phenomenon* (Fig. 97–9). A fourth sign of stable fixation of the femoral component is the presence of stress shielding (or loss of bone density) in the proximal femur, which is a positive sign that the loads have been transferred from bone to the implant, which is much stiffer (see Fig. 97–9). When this happens, the cortex around the distal stem hypertrophies, producing a *spot weld* phenomenon (see Figs. 97–2*B* and 97–9).

For cementless femoral components, an intermediate form of fixation, also called *stable fibrous fixation*, occasionally can be seen (Fig. 97–10). Radiographs of components with stable fibrous fixation show radiolucent lines parallel to the porous-coated portion of the stem, which do not progress over long-term follow-up (see Fig. 97–10). They are usually less than 1 to 2 mm thick. In stable fibrous fixation, stress shielding is not observed, and the implant does not migrate in relationship to the lesser trochanter with time.

A loose cementless femoral component is characterized by lucencies that progress over time (Fig. 97–11). Loads do not transmit from the bone to the implant in a loose

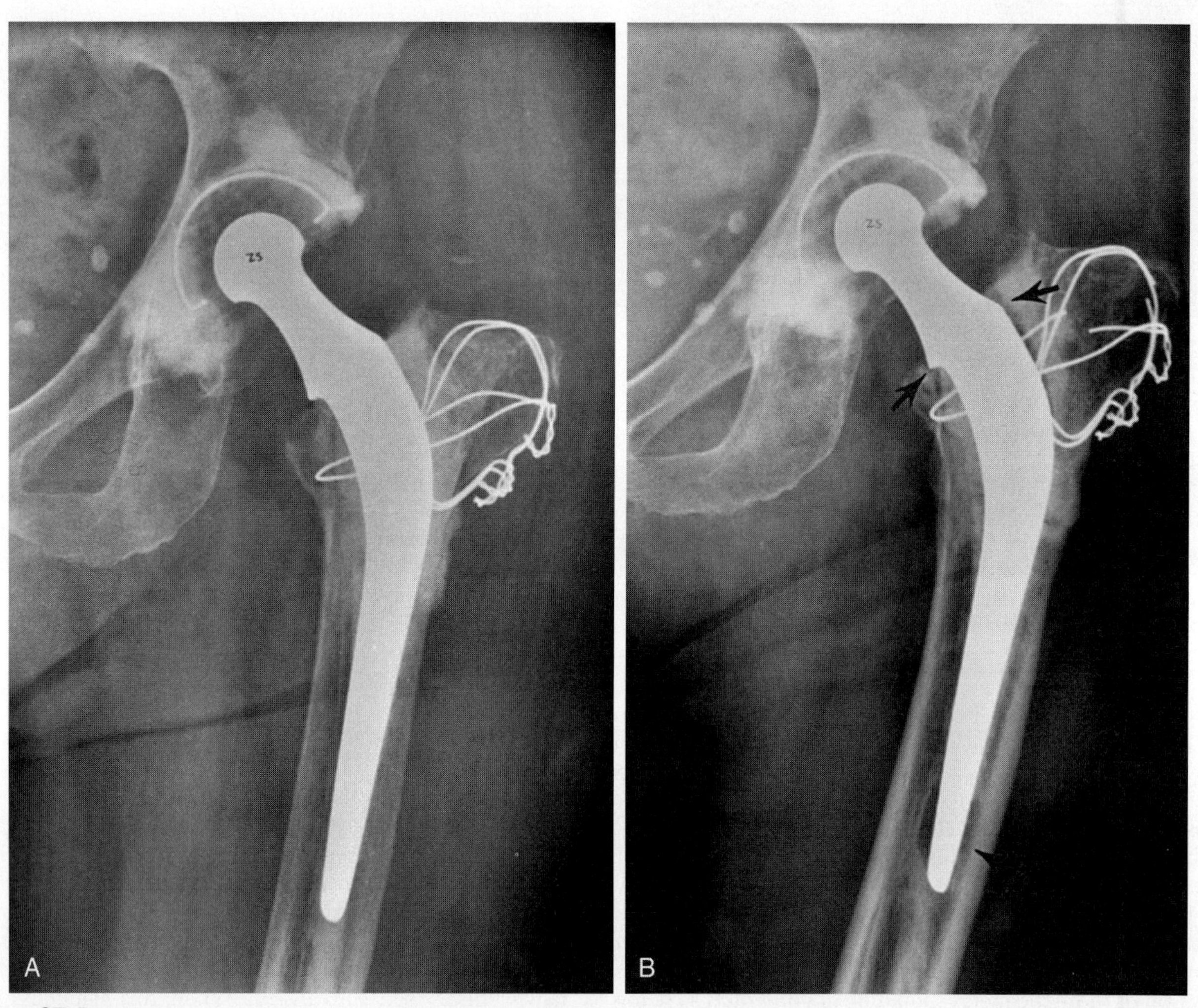

Figure 97–5. Femoral component loosening. Postoperative *(A)* and 25-year follow-up *(B)* radiographs show femoral component loosening *(arrows)*. Note the medial decrease in distance between the femoral component and medial wire, the superolateral prosthesis cement radiolucency, and the osteolysis distally.

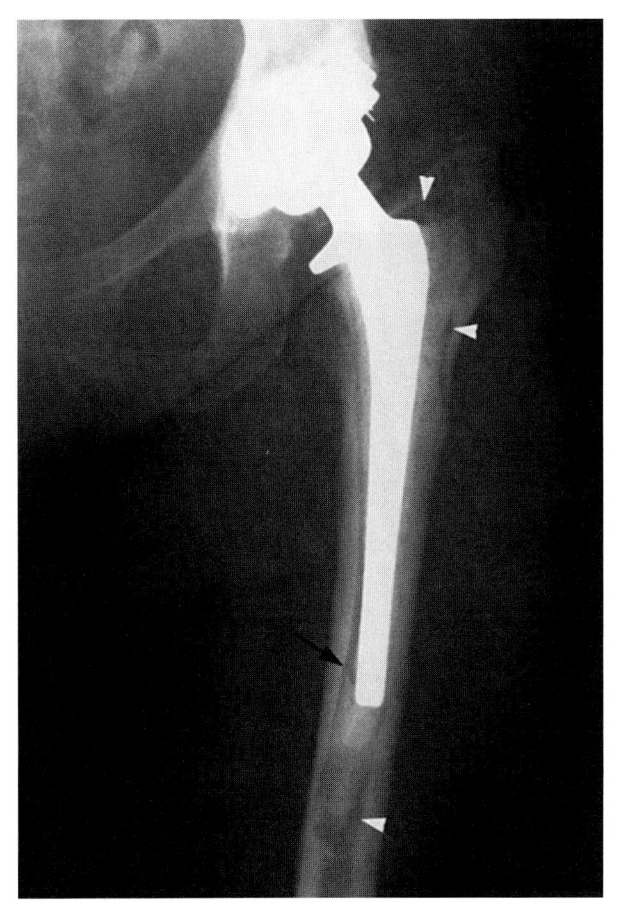

Figure 97–6. Crack in the cement and loosening of the stem. Anteroposterior view of the left hip shows lucency between the cement and bone *(arrowheads)*, which in some areas measures more than 2 mm. Note the break in the cement around the tip of the stem *(arrow)*.

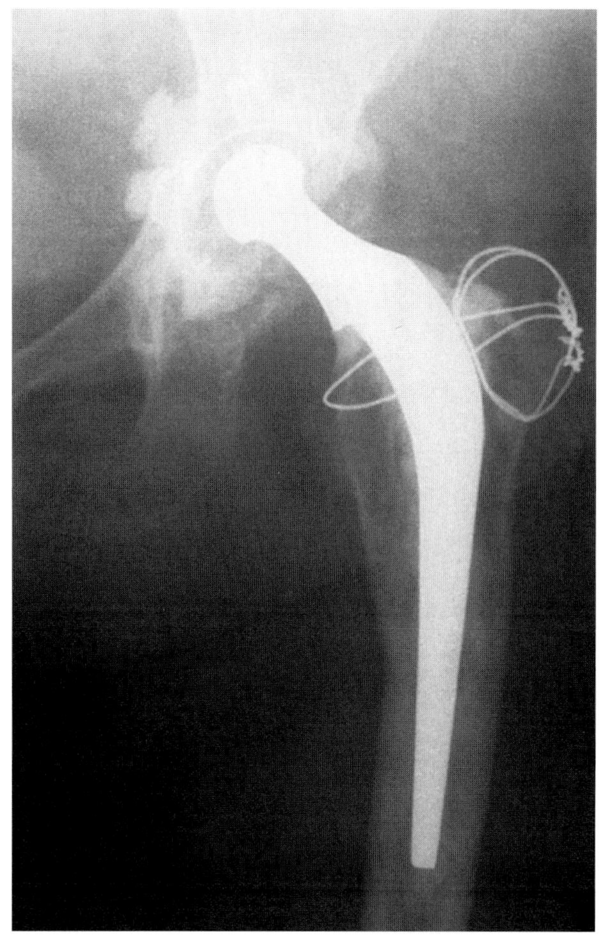

Figure 97–7. Stress shielding around a cemented stem. Anteroposterior radiograph of the left hip shows thick cortex in zones III, IV, and V. The cortex is thinned in zones I, II, VI, and VII because of stress shielding.

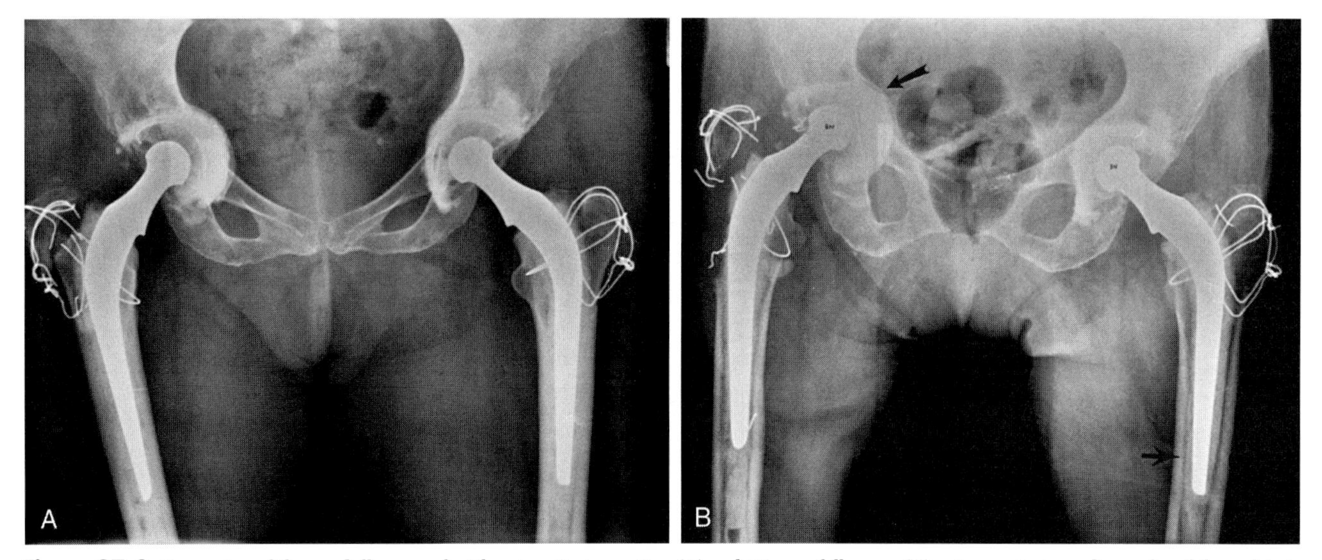

Figure 97–8. Expansion of the medullary canal with aging. Postoperative *(A)* and 25-year follow-up *(B)* anteroposterior radiographs of the pelvis in a patient with a stable long-term radiographic fixation of her implants. At follow-up, the right acetabular component has an incomplete bone-cement radiolucent line *(arrow)*, and the right greater trochanter has gone on to nonunion with breakage of the wires. In both femora, there is thinning of the cortical bone *(arrow)*, and the medullary canals expanded.

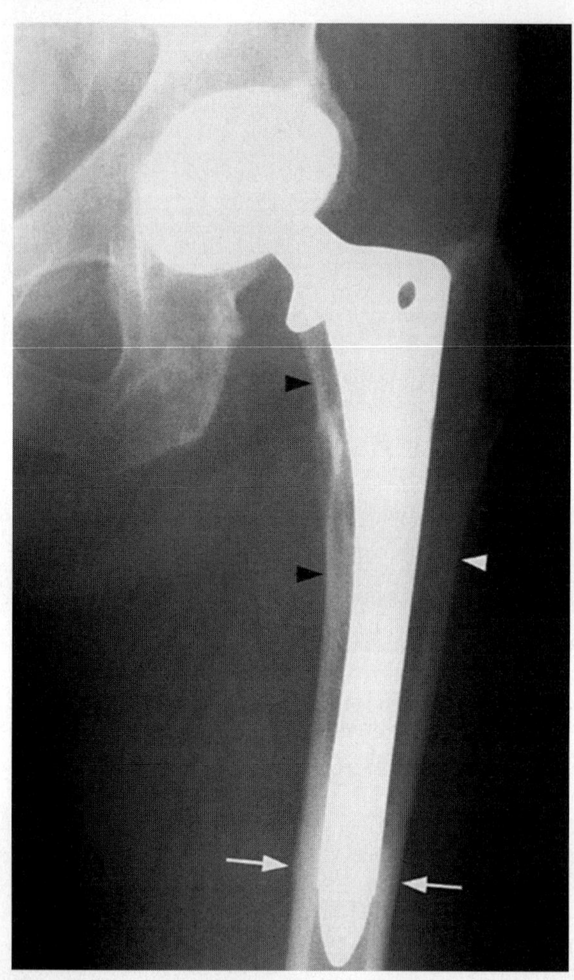

Figure 97–9. Anteroposterior view of the left hip shows stable (secure) fixation of an uncemented prosthesis. *Spot weld* phenomenon (thickening of the cortex) is seen around the distal stem *(white arrows)* and stress shielding (bone atrophy) around the proximal stem *(arrowheads)*.

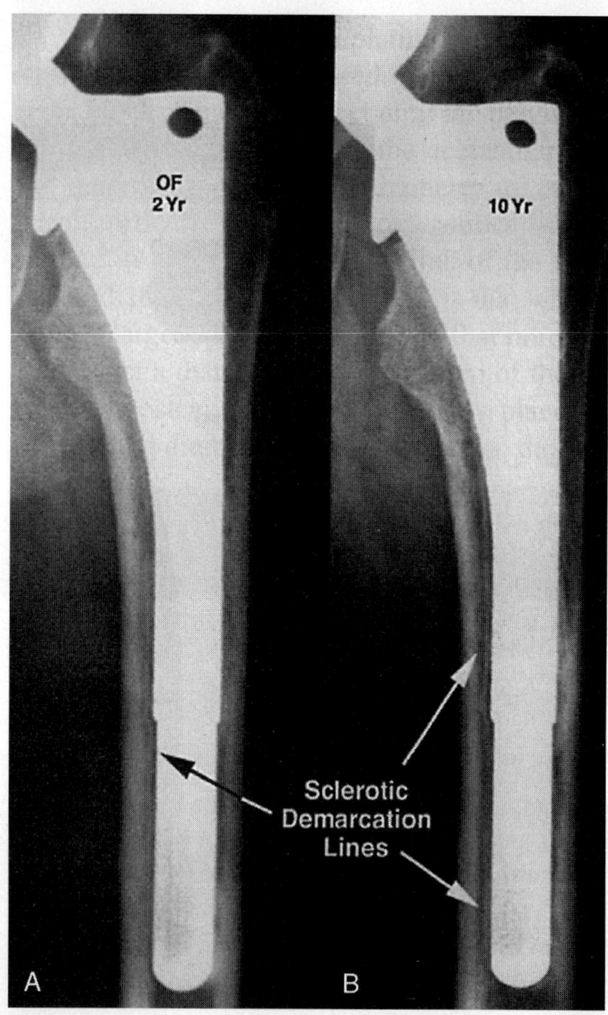

Figure 97–10. Stable intermediate or fibrous fixation of a cementless prosthesis. *A,* Anteroposterior view of the left hip taken 2 years postoperatively shows no complications. *B,* At 10 years' follow-up, a stable sclerotic line is seen surrounding the stem, indicating intermediate or fibrous fixation.

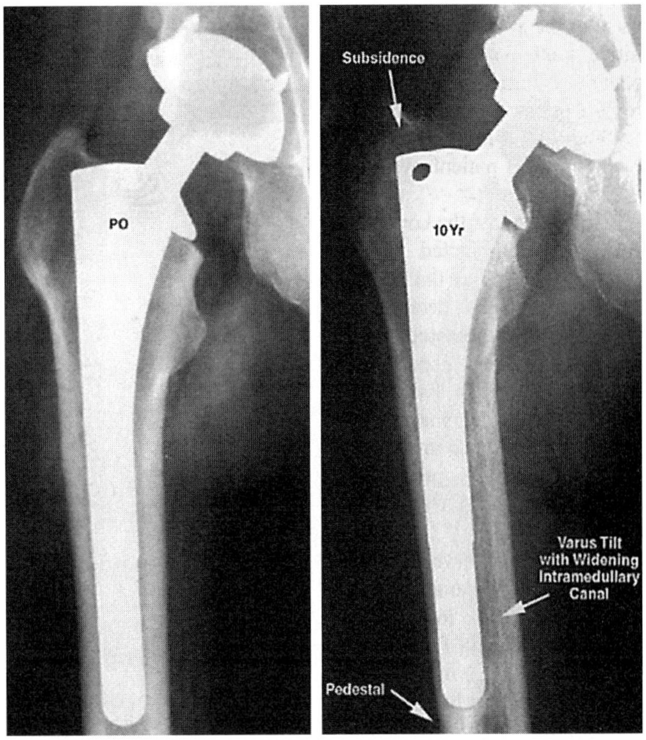

Figure 97–11. Loosening of a cementless stem over 10 years. *A*, Anteroposterior view of the right hip taken postoperatively shows no complication. *B*, Examination repeated 10 years later shows the stem has drifted into varus and settled deeper into the femur (subsidence).

prosthesis, and stress shielding does not develop. A loose stem can rotate within the femur, and cortical thickening can develop at the tip of the stem; this phenomenon is called the *bone pedestal* (Fig. 97–11). The shedding of metallic beads from the surface of the stem is another sign that the stem is moving and likely loose (Fig. 97–12).

The principal sign of loosening of a cementless acetabular component is the presence of migration or a change in the relationship of the component to the teardrops in the vertical or horizontal direction. A second sign relates to the presence of circumferential lucencies between the bone and the prosthesis, especially lucencies that measure 1 to 2 mm wide. Small noncircumferential lucencies can be seen between a cementless acetabular prosthesis and bone, and these should not be interpreted as representing loosening. A final sign of loosening in the acetabular component includes the presence of screw breakage and shedding of beads or mesh coatings off of the prosthesis (see Fig. 97–12). Bead or mesh shedding also can occur with loose femoral components.

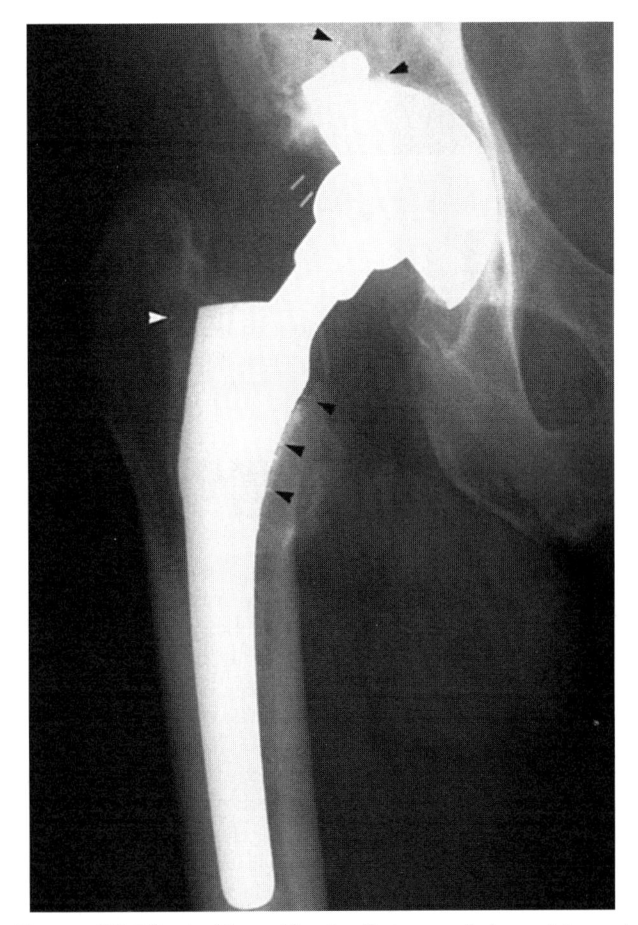

Figure 97–12. Shedding of beads off the acetabular and femoral components as a sign of early loosening of these components. Anteroposterior view of the right hip shows small beads adjacent to the acetabular and femoral components *(arrowheads)*.

OTHER CAUSES OF IMPLANT FAILURE

The radiologist should be aware of other failure modes of total hip implants. In the cementless stem, in which the metal is weakened by sintering of the beads or fiber mesh, a fatigue fracture can develop (Fig. 97–13). Early on, these metal fatigue fractures can be subtle or impossible to show without fluoroscopy. With modular cementless components, the polyethylene liner can detach from the acetabular shell, and the locking mechanisms between the polyethylene liner and metal cup can fracture. Detached fragments from the locking mechanism can be detected

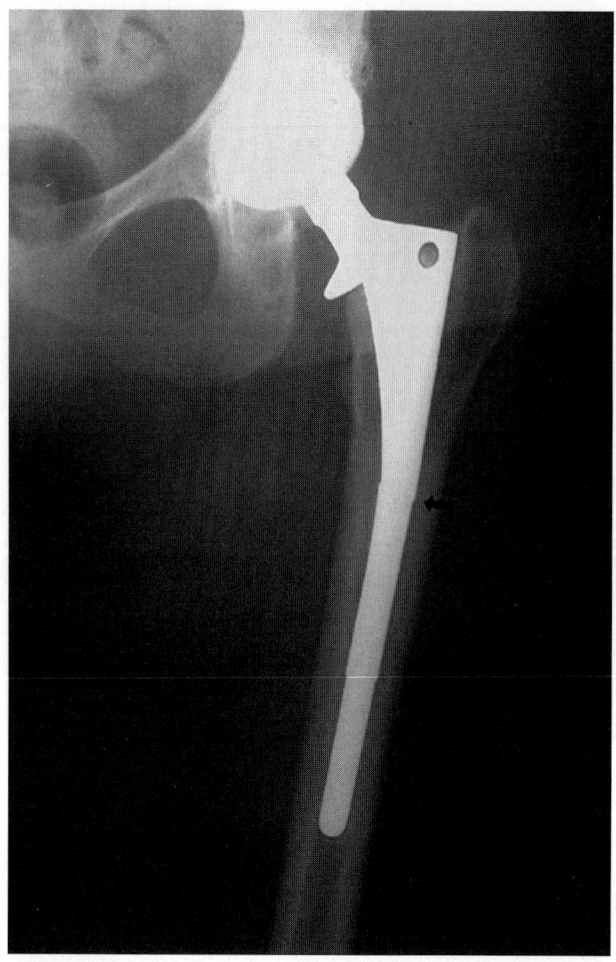

Figure 97–13. Fatigue fracture of the stem 9 years postoperatively. Anteroposterior view of the left hip shows a subtle fracture of the stem *(arrow)*.

radiographically (Fig. 97–14). The femoral head also can dissociate from the femoral stem trunion, or the femoral neck trunion can break at the junction with the femoral head.

WEAR OF THE POLYETHYLENE LINER AND ITS SEQUELAE

The importance of wear in the failure of total hip arthroplasty became more widely recognized in the 1990s. Wear at the bearing surface between the femoral head and the acetabular polyethylene liner is recognized as a major contributor of failure in cemented and cementless devices. Wear that occurs at the polyethylene liner generates micrometer- and nanometer-sized polyethylene particles, which migrate to the bone-prosthesis or bone-cement interface and stimulate a histiocytic response (also called *osteolysis* or *particle disease*), causing either linear bone loss or expansile lytic bone lesions (Fig. 97–15). The joint space begins to expand until it encompasses the entire prosthesis. This is especially true on the acetabular side. Currently, there is increased awareness of wear, especially because patients are living 20 to 30 years after implantation of the device. To evaluate wear, the joint should be assessed for eccentric migration of the femoral head superiorly or medially in relationship to the metallic shell over time (see Fig. 97–4). This migration is relatively easy to detect in cementless devices if a chrome-cobalt femoral head and a titanium acetabular shell are used (the most common construct). Quantifying wear can be done by comparing the distance between the femoral head and the acetabular shell inferiorly with the distance between the femoral head and the shell superiorly (see Fig. 97–4). Immediately postoperatively the head is centered perfectly within the shell. If several years later wear takes place, the head migrates superiorly, and the superior measurement is less than the inferior measurement (see Fig. 97–4). Edge detection techniques have been developed to measure wear more accurately by precisely delineating the edge of the femoral head and the edge of the acetabular component.

The most common site for osteolysis caused by a histiocytic reaction is in the proximal medial aspect of the femur. It is difficult sometimes, however, to distinguish

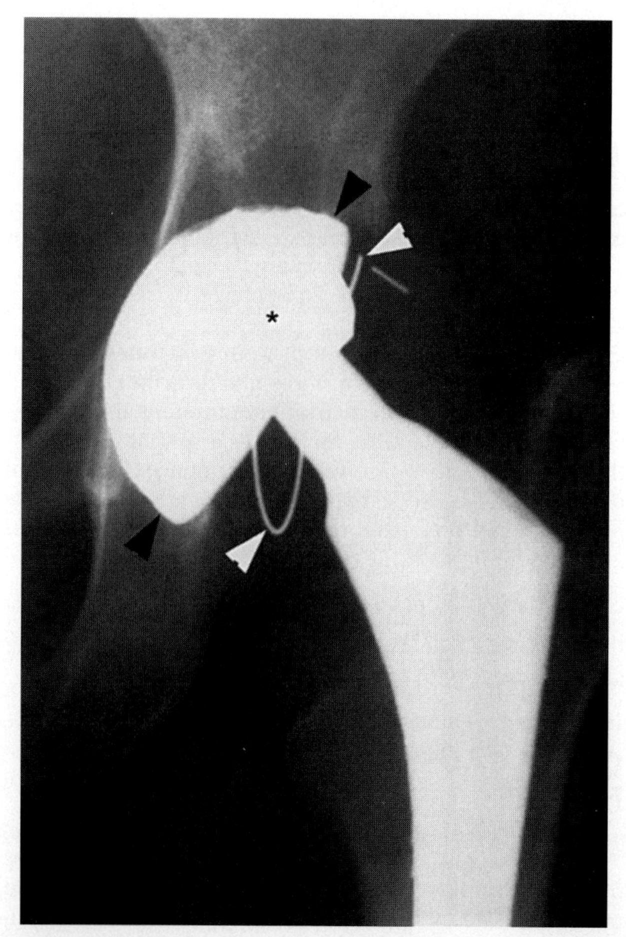

Figure 97–14. Fracture in the locking mechanism between the acetabular shell and the polyethylene liner. Anteroposterior view of the hip shows the break in the wire for the locking mechanism *(white arrowheads)*. The liner is lucent, but it is presumed to be displaced because of the eccentric positioning of the femoral head. An asterisk is at the center of the head, and the black arrowheads show that the head has shifted superiorly.

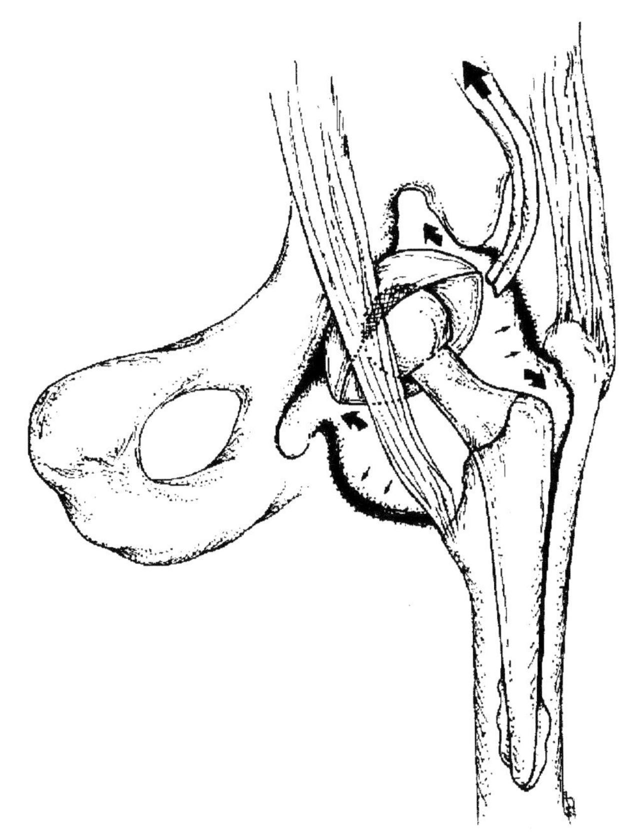

Figure 97–15. Expansion of the effective joint space in particle disease. Diagram shows wear and the release of wear particles into the effective joint space. Capsular tissue has some capacity to transport particles through the lymphatic system *(large straight arrow)*. In the effective joint space, joint fluid and wear particles follow the path of least resistance. The pseudocapsule becomes thickened *(small straight arrows)* because of phagocytosis of wear particles and the development of foreign-body granulomas. The effective joint space can extend along interfacial planes, expand into bone, or assume a variety of combinations *(curved arrows)*.

whether bone loss proximally is related to stress shielding or to histiocytic reaction. With histiocytic bone resorption, there is usually scalloping of the cortex along with a sclerotic margin similar to an expansile benign bone tumor. Large lesions in the greater trochanter can lead to a fracture. Osteolysis (or histiocytic reaction) around the stem is characterized by endosteal scalloping (Fig. 97–16). This scalloping is seen most commonly with cementless femoral components in which patch porous coating rather than circumferential porous coating is used. It has been shown that wear particles are pumped around the stem beside the smooth surface of the implant.

On the acetabular side, osteolysis between bone and cement manifests as a lucency with a sclerotic margin, although expansile lesions also can occur (Fig. 97–17). Expansile lesions are more common with cementless devices, and many times occur centrally rather than peripherally because of screw holes, and screws provide access for the polyethylene debris into the bone. To see adequately the extent of osteolysis on the acetabular side of the prosthesis, oblique views or computed tomography (CT) may be necessary (Fig. 97–18). Extensive acetabular osteolysis also can cause fractures around the acetabulum, especially of the posterior column.

INFECTION OF TOTAL HIP PROSTHESIS

Infection is one of the most dreaded complications of total hip replacement, mainly because it is difficult to treat. In the immediate postoperative period (first 6 weeks), problems with the wound, such as persistent drainage or wound infection, should raise the suspicion of an infected prosthesis. Joint aspiration under fluoroscopic guidance should be done to confirm or rule out infection (see Fig. 95–3).

After a total hip replacement, patients should become pain-free on weight bearing by about 3 months postoperatively. If hip pain persists after the third month postoperatively, however, infection should be considered. Laboratory studies, such as serial erythrocyte sedimentation rate and C-reactive protein determinations, can be helpful in drawing attention to the possibility of infection. Normally the erythrocyte sedimentation rate should not exceed 30 mm/h, and the C-reactive protein should not exceed 1.5. Aspirated fluid should be cultured for aerobic and anaerobic bacteria. A white blood cell count should be done on the aspirated fluid; the normal count should remain less than 10,000 white blood cells/mL. In the first 3 months, the most common infecting agent is *Staphylococcus aureus;* after 3 months, *Staphylococcus epidermidis* is most common.

DIFFERENTIATING ASEPTIC LOOSENING FROM INFECTION

In patients followed for a long time, it may be difficult to differentiate aseptic or mechanical loosening from septic loosening of a total hip arthroplasty. Generally, in a well-done total hip replacement, aseptic loosening on the femoral side starts at the prosthesis-cement interface rather than at the bone-cement interface. Early and progressive circumferential bone-cement lucencies suggest infection. Infection should be considered in a cementless prosthesis developing early progressive lucencies; however, these radiolucencies may be a sign that the cementless component never achieved ingrowing bone on the prosthesis.

Periosteal reaction on the femoral side should raise the possibility of infection. Periosteal reaction should be distinguished from cortical hypertrophy around the distal stem (see Figs. 97–2 and 97–9). On the acetabular side, aseptic loosening is usually a long-term rather than a short-term process. It occurs at the cement-bone interface as a biologic response to wear debris. This is in contradistinction to the prosthesis-cement interface failure in the femoral component, which occurs because of the mechanical failure and fractures in the cement (Fig. 97–19). Progressive circumferential lucency at the bone-cement interface of a cemented acetabular component in

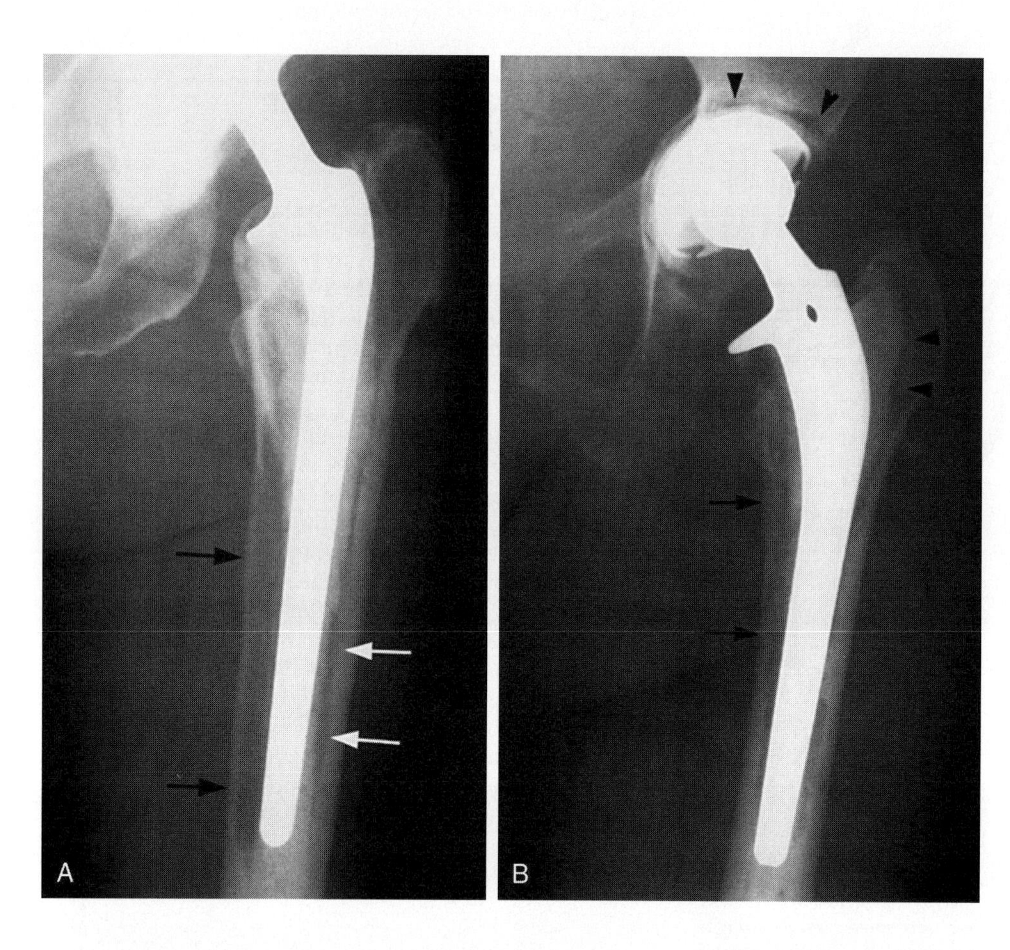

Figure 97–16. Osteolysis (particle disease) in two different patients with cemented prostheses. *A,* Extensive bone resorption *(black arrows)* medially and endosteal scalloping *(white arrows)* laterally. *B,* Changes of osteolysis around the acetabular and femoral components *(arrows* and *arrowheads).*

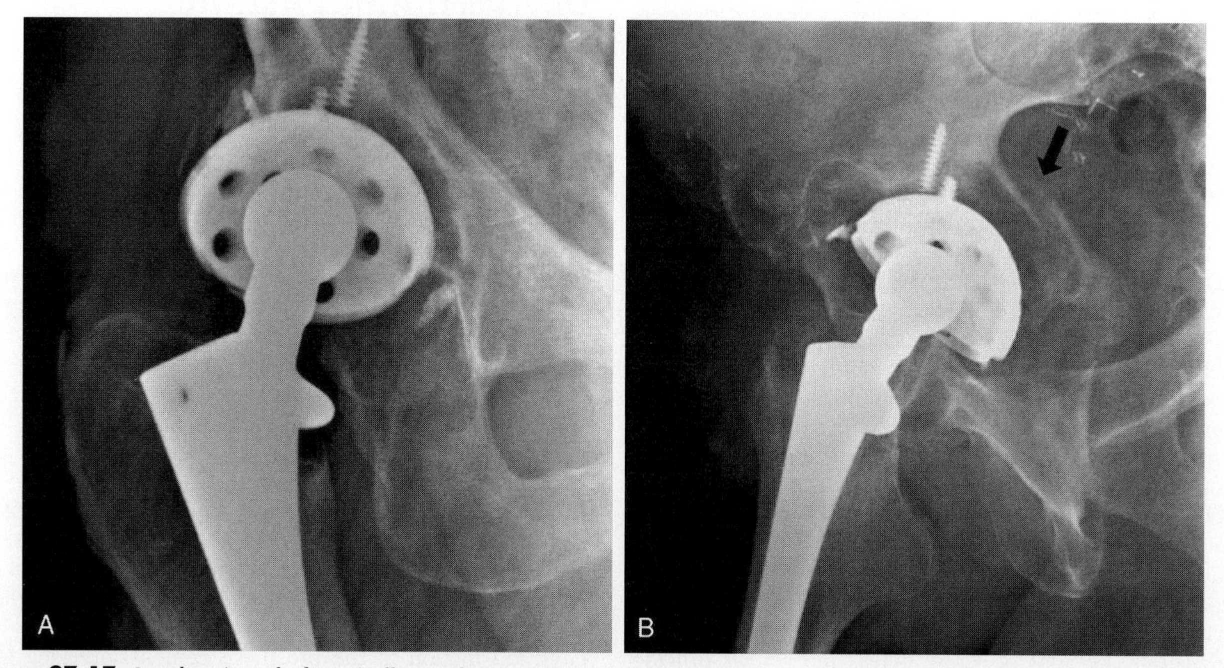

Figure 97–17. Osteolysis (particle disease) affecting the acetabular component. *A,* Anteroposterior view of the hip shows acetabular component loosening and acetabular osteolysis demonstrated by breakage of a screw and expansile bone loss. *B,* Oblique view of the right hip shows a fracture in the posterior acetabular column *(arrow).*

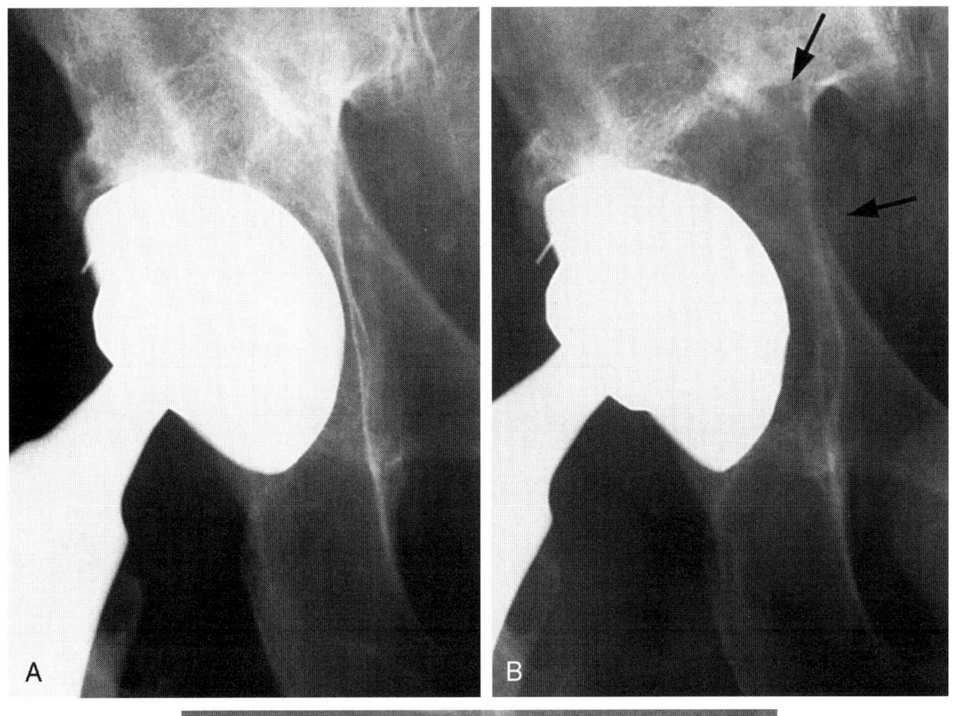

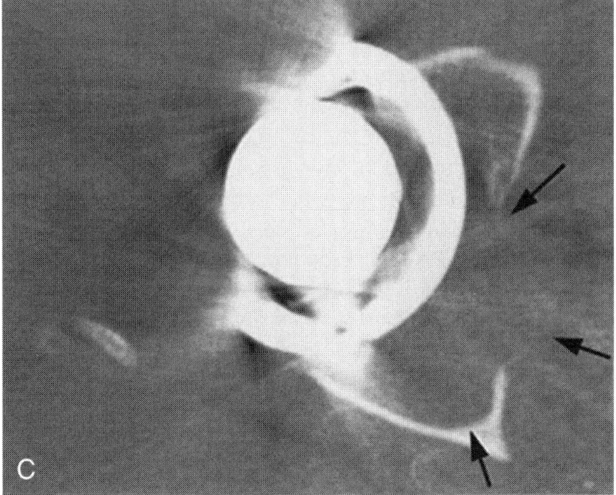

Figure 97–18. Osteolysis resembling a malignant bone tumor. *A,* Postoperative anteroposterior view of the right hip shows a cementless acetabular component with no evidence of complications. *B,* Follow-up anteroposterior view 3 years later shows a focal destructive lesion in the medial acetabulum *(arrows).* Although osteolysis was the primary diagnosis, metastasis also was considered. *C,* CT scan shows the lytic lesion and the soft tissue component *(arrows).* At surgery, the lesion proved to be osteolysis caused by particle disease.

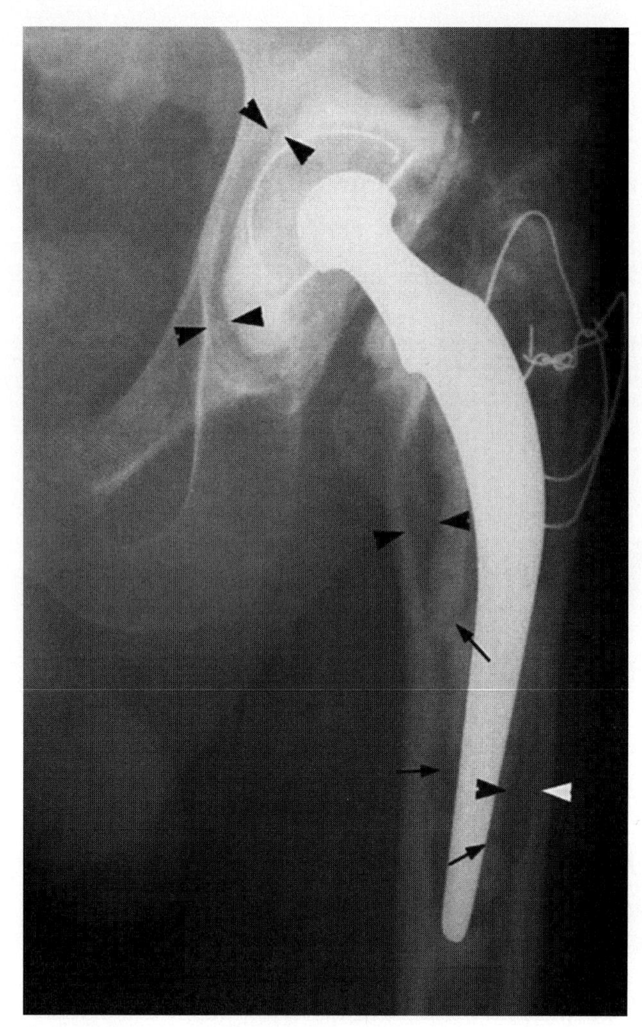

Figure 97–19. Infected prosthesis. Anteroposterior view of the left hip shows significant lucencies between the cement and the bone (>3 mm) in both acetabular and femoral components *(arrowheads)*. The prosthesis is obviously loose, and the cement is broken *(arrows)*.

the first year is suspicious for infection (see Fig. 97–19). The acetabular cementless component rarely shows a circumferential 1- to 2-mm lucency in the first year or two. Cementless acetabular components that do not develop bone ingrowth and show lucencies of 1 to 2 mm at the bone-metal interface in the first 2 years always should be suspected to be infected until proved otherwise.

HETEROTOPIC BONE FORMATION

Heterotopic bone formation around a prosthetic hip, enough to limit joint motion, is a rare problem. Patients who are at risk for this complication include patients with history of heterotopic bone formation after a previous surgery, patients with diffuse idiopathic skeletal hyperostosis, and patients with hypertrophic osteoarthritis. The diagnosis radiographically is straightforward and is based on the development of clumps of mature bone around the neck of the prosthesis (Fig. 97–20). Patients at risk are treated prophylactically with 10 days of indomethacin or a single dose of radiation, 700 to 800 rad, with shielding of the ingrowth surfaces. The treatment is given in the first few days postoperatively.

CONCLUSION

In general, implant stability is evaluated best by judging whether the component has moved within the bone over time. On the acetabular side, the teardrop is the most commonly used reference point for assessing migration. On the femoral side, the lesser trochanter is the most commonly used reference point. Other long-term concerns include wear at the weight-bearing surface shown by superior or superomedial migration of the femoral head within the acetabular shell. Associated with wear is the histiocytic reaction (or particle disease), which is most common in the proximal medial calcar area of the femur and in the greater trochanter. Bone loss around the acetabular component manifests with central migration of the prosthesis. Multiple radiographic views may be necessary to define accurately the extent of the acetabular osteolysis. Finally, infection is a serious complication in which imaging and joint aspiration play a significant role in early diagnosis.

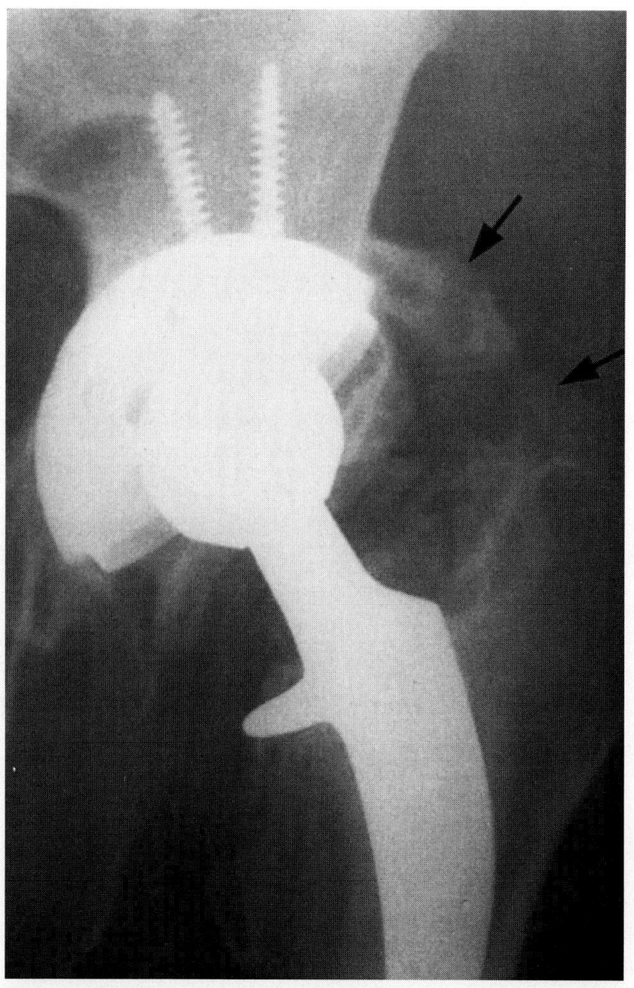

Figure 97–20. Heterotopic ossifications causing restricted motion of the left hip. Anteroposterior view of the left hip shows mature heterotopic ossifications *(arrows)*, especially over the lateral aspect of the joint.

Recommended Readings

Berry DJ, Lewallen DG, Hanssen AD, et al: Pelvic discontinuity in revision total hip arthroplasty. J Bone Joint Surg Am 81:1692, 1999.
Bloebaum RD, Mihalopoulus NL, Jensen JW, et al: Postmortem analysis of bone growth into porous-coated acetabular components. J Bone Joint Surg Am 79:1013, 1997.
Brand RA, Yoder SA, Pedersen DR: Interobserver variability in interpreting radiographic lucencies about total hip reconstructions. Clin Orthop 192, 1985.
Brooker AF, Bowerman JW, Robinson RA, et al: Ectopic ossification following total hip replacement: Incidence and a method of classification. J Bone Joint Surg Am 55:1629, 1973.
Bugbee WD, Culpepper WJ 2nd, Engh CA Jr, et al: Long-term clinical consequences of stress-shielding after total hip arthroplasty without cement. J Bone Joint Surg Am 79:1007, 1997.
Callaghan JJ, Albright J, Goetz D, et al: Charnley total hip arthroplasty with cement: Minimum twenty-five-year follow-up. J Bone Joint Surg Am 82:487, 2000.
Callaghan JJ, Dysart SH, Savory CG: The uncemented porous-coated anatomic total hip prosthesis: Two-year results of a prospective consecutive series. J Bone Joint Surg Am 70:337, 1988.
Capello W, D'Antonio J, Feinberg J, et al: Hydroxyapatite-coated total hip femoral components in patients less than fifty years old. J Bone Joint Surg Am 79:1023, 1997.
Comadoll JL, Shermen RE, Gustilo RB, et al: Radiographic changes in bone dimensions in asymptomatic cemented total hip arthroplasties, results of nine to thirteen year follow up. J Bone Joint Surg Am 70:433, 1988.
DeLee JG, Charnley J: Radiological demarcation of cemented sockets in total hip replacement. Clin Orthop 121:20, 1976.
Engh CA, Bobyn JD: The influence of stem size and extent of porous coating on femoral bone resorption after primary cementless hip arthroplasty. Clin Orthop 231:7, 1988.
Engh CA, Hooten JP, Zettl-Schaffer KF, et al: Evaluation of bone ingrowth in proximally and extensively porous-coated anatomic medullary locking prostheses retrieved at autopsy. J Bone Joint Surg Am 77:903, 1995.
Engh CA, Massin P, Suthers KE: Roentgenographic assessment of the biologic fixation of porous-surfaced femoral components. Clin Orthop 257:107, 1990.
Engh CA, McGovern TF, Bobyn JD, et al: A quantitative evaluation of periprosthetic bone-remodeling after cementless total hip arthroplasty. J Bone Joint Surg Am 74:1009, 1992.
Engh CA, Zettl-Schaffer KF, Kukita Y, et al: Histological and radiographic assessment of well functioning porous-coated acetabular components: A human postmortem retrieval study. J Bone Joint Surg Am 75:814, 1993.
Gruen TA, McNeice GM, Amstutz HC: "Modes of failure" of cemented stem-type femoral components: A radiographic analysis of loosening. Clin Orthop 141:17, 1979.
Harris WH, McCarthy JC, O'Neill DA: Femoral component loosening using contemporary techniques of femoral cement fixation. J Bone Joint Surg Am 64:1063, 1982.
Harris WH, Schiller AL, Scholler JM, et al: Extensive localized bone resorption in the femur following total hip replacement. J Bone Joint Surg Am 58:612, 1976.
Heekin RD, Engh CA, Herzwurm RJ: Fractures through cystic lesions of the greater trochanter: A cause of late pain after cementless total hip arthroplasty. J Arthroplasty 11:757–760, 1996.
Hodgkinson JP, Shelley P, Wroblewski BM: The correlation between the roentgenographic appearance and operative findings at the bone-cement junction of the socket in Charnley low friction arthroplasties. Clin Orthop 228:105, 1988.
Jasty M, Maloney WJ, Bragdon CR, et al: Histomorphological studies of the long-term skeletal responses to well fixed cemented femoral components. J Bone Joint Surg Am 72:1220, 1990.
Jasty MJ, Floyd WE, Schiller AL, et al: Localized osteolysis in stable nonseptic total hip replacement. J Bone Joint Surg Am 68:912, 1986.
Johnston RC, Fitzgerald RH Jr, Harris WH, et al: Clinical and radiographic evaluation of total hip replacement: A standard system of terminology for reporting results. J Bone Joint Surg Am 72:161, 1990.
Kwong LM, Jasty M, Mulroy RD, et al: The histology of the radiolucent line. J Bone Joint Surg Br 74:68, 1992.
Livermore J, Ilstrup D, Morrey BF: Effect of femoral head size on wear of the polyethylene acetabular component. J Bone Joint Surg Am 72:518, 1990.
Loudon JR, Charnley J: Subsidence of the femoral prosthesis in total hip replacement in relation to the design of the stem. J Bone Joint Surg Br 62:450, 1980.
Maloney WJ, Jasty M, Harris WH, et al: Endosteal erosion in association with stable uncemented femoral components. J Bone Joint Surg Am 72:1025, 1990.
Maloney WJ, Jasty M, Rosenberg A, et al: Bone lysis in well-fixed cemented femoral components. J Bone Joint Surg Br 72:969, 1990.
Maloney WJ, Peters P, Engh CA, et al: Severe osteolysis of the pelvis in association with acetabular replacement without cement. J Bone Joint Surg Am 75:1627, 1993.
Maloney WJ, Sychterz C, Bragdon C, et al: The Otto Aufranc Award: Skeletal response to well fixed femoral components inserted with and without cement. Clin Orthop 333:15, 1996.
Massin P, Schmidt L, Engh CA: Evaluation of cementless acetabular component migration: An experimental study. J Arthroplasty 4:245, 1989.
McCaskie AW, Brown AR, Thompson JR, et al: Radiological evaluation of the interfaces after cemented total hip replacement: Interobserver and intraobserver agreement. J Bone Joint Surg Br 78:191, 1996.
Pidhorz LE, Urban RM, Jacobs JJ, et al: Quantitative study of bone and soft tissues in cementless porous-coated acetabular components retrieved at autopsy. J Arthroplasty 8:213, 1993.
Ritter MA, Zhou H, Keating CM, et al: Radiological factors influencing femoral and acetabular failure in cemented Charnley total hip arthroplasties. J Bone Joint Surg Br 81:982, 1999.
Schmalzried TP, Callaghan JJ: Current concepts review: Wear in total hip and knee replacements. J Bone Joint Surg Am 81:115, 1999.
Schmalzried TP, Jasty M, Harris WH: Periprosthetic bone loss in total hip arthroplasty. J Bone Joint Surg Am 81:849, 1992.
Schmalzried TP, Kwong LM, Jasty M, et al: The mechanism of loosening of cemented acetabular components in total hip arthroplasty. Clin Orthop 274:60, 1992.
Shaver SM, Brown TD, Hillis SL, et al: Digital edge-detection measurement of polyethylene wear after total hip arthroplasty. J Bone Joint Surg Am 79:690, 1997.
Udomkiat P, Wan Z, Dorr LD: Comparison of preoperative radiographs and intraoperative findings of fixation of hemispheric porous-coated sockets. J Bone Joint Surg Am 83:1865, 2001.
Urban RM, Jacobs JJ, Sumner DR, et al: The bone-implant interface of femoral stems with non-circumferential porous coating. J Bone Joint Surg Am 78:1068, 1996.
Zicat B, Engh CA, Gokcen E: Patterns of osteolysis around total hip components inserted with and without cement. J Bone Joint Surg Am 77:433, 1995.

Index

Note: Page numbers followed by the letter f refer to figures; those followed by the letter t refer to tables.